INTERSECTIONS BETWEEN PARTICLE AND NUCLEAR PHYSICS

3rd CONFERENCE on the INTERSECTIONS
between
PARTICLE and NUCLEAR PHYSICS
May 14-19, 1988

ROCKPORT
MAINE

AIP
CONFERENCE
PROCEEDINGS 176

RITA G. LERNER
SERIES EDITOR

INTERSECTIONS BETWEEN PARTICLE AND NUCLEAR PHYSICS

ROCKPORT, ME 1988

EDITOR:

GERRY M. BUNCE
BROOKHAVEN NATIONAL
LABORATORY

AMERICAN INSTITUTE OF PHYSICS NEW YORK 1988

L.C. Catalog Card No. 88-62535
ISBN 0-88318-376-5
DOE CONF-8805142

Printed in the United States of America.

FOREWORD

This conference was distinguished by many excellent presentations, and I hope that this proceedings will be useful as a general source of information in this overlap region of physics. We were particularly fortunate to be holding this conference at this time when many new and striking results were being debated, including observations of the supernova, the spin information carried by quarks in the proton, the first results from searches for exotic K decays, and early intriguing results from the CERN and AGS heavy ion programs.

These meetings are designed around a number of afternoon mini-workshops which are organized by two coordinators for each topic, one from the nuclear physics community and one from particle physics. The coordinators, listed on a following page, and the conference chairman, Erich Vogt, did most of the work in developing the scientific program and deserve credit for creating this very successful conference. Vogt, in the following introduction, credits the excellent work done by the secretarial staff assembled for this conference. I would also like to thank Ms. Joan Depken, the secretary for the proceedings whose considerable effort has led to what is, I believe, a successful book.

Gerry Bunce
July 1988

Introduction

This third international Conference on the Intersections Between Particle and Nuclear Physics continues the tradition of the first two which combined a noble purpose with a formula for success.

The noble purpose is to bring together the particle physics community and the nuclear physics community to discuss their new common interests. The advent of the Standard Model of quarks, leptons and unified forces has given enormous stimulus to the field of subatomic physics. There are important questions of what lies beyond the Standard Model which are addressed, especially, by experiments proposed for the new generation of high-energy colliders. In addition, there are a range of questions pertaining to possible improvements of the Standard Model and to applying it to the description of systems of strongly interacting particles. It is this range of issues which now bring together particle and nuclear physicists in common pursuits which primarily involve the energy range of 1–100 GeV. This third Intersections Conference appears to have uncovered a great deal of exciting physics in this energy domain, as can be discovered from the various plenary and invited papers of these proceedings and from the very eloquent summary of Robert Jaffe.

The interests of the Intersections Conference expanded at this meeting to embrace a number of common issues in astrophysics which are shared by nuclear and particle physics. This triple intersection worked very well and promised to be fruitful in the future. Highlighting it, there was a special evening session devoted to the joint work of Hans Bethe and Gerry Brown in which Bethe spoke about stellar collapse and Brown about equations of state for nuclear matter at high densities and pressures. Although not included in these proceedings, these talks added a great deal to the conference.

The formula for success used for this meeting is largely the invention of Alan Krisch, who was the chairman of the first two Intersections Conferences. It begins, of course, with noble purpose and exciting physics, but its essence then is to choose a beautiful and isolated site and to create an interesting mixture of plenary papers, invited papers and participants.

The first two Intersections Conferences were held in mountain sites: Steamboat Springs in Colorado (1984) and Lake Louise in Alberta (1986). Although different, the Maine Coast is endowed with great natural beauty. The Samoset Resort is excellent and the conference enjoyed most splendid service from all the staff. The conference is indebted to Stanley Kowalski who pointed us in the direction of Samoset.

A very special part of the Krisch formula is to arrange to have all the many administrative and clerical matters of the conference handled with great skill and fine style by professional staff seconded from the various major laboratories which sponsor the conference. The conference secretary, Terry Murphy, from my own laboratory was extraordinarily resourceful and able in managing so much of the conference preparation and its operation. At Rockport she was assisted by a very fine complement of staff: Sheila Dodson from Bates and Lorraine King from TRIUMF, who also helped greatly in advance of the conference, Debbie Burton from Michigan and Joan Depken from Brookhaven, Roberta Marinuzzi and Karen Poelakker from LAMPF, and Karen's sister, Edna Baron, who volunteered to spend her holiday from Minnesota in assisting the staff. All the participants benefited from their competent and cheerful effort.

E.W.Vogt

CONTENTS

PLENARY SESSIONS

SUMMARIES

ACCELERATOR PHYSICS

ANTIPROTON PHYSICS

Electron and Muon Physics

Hadron Scattering

HADRON SPECTROSCOPY

MESON AND LEPTON DECAYS

Neutrino Physics

Nonaccelerator and Astrophysics

Relativistic Heavy-Ion Physics

Spin Physics

Session Coordinators

Accelerator Physics	A. Chao C. Leemann
Antiproton Physics	P. Barnes R. Cester
Electron and Muon Physics	H. Montgomery W. Donnelly
Hadron Scattering	S. Heppelmann O. Häusser
Hadron Spectroscopy	D. Geesaman D. Hitlin
Meson and Lepton Decays	M. Zeller N. Reay
Neutrino Physics	W. Haxton D. White
Non-Accelerator and Astrophysics	E. Beier S. Kahana
Relativistic Heavy Ion Physics	J. Harris T. Matsui
Spin Physics	T. Roser J. Moss

Organizing Committee

E. W. Vogt (Chairman)
V. W. Hughes (Co-Chairman)
G. Bunce
P. A. Carruthers
R. H. Dalitz
G. R. Farrar
G. Fidecaro
D. F. Geesaman
R. Hofstadter
N. Isgur
S. E. Koonin
A. D. Krisch
T. D. Lee
T. W. Ludlam
M. H. MacFarlane
J. S. McCarthy
F. E. Mills
R. E. Mischke
E. J. Moniz
T. A. O'Halloran
L. S. Schroeder
L. C. Teng
W. T. H. van Oers
B. Winstein

Sponsors

Brookhaven National Laboratory
CEBAF
Fermilab
Los Alamos National Laboratory
University of Michigan
MIT Bates Linear Accelerator Center
TRIUMF
Yale University
U.S. National Science Foundation

CONFERENCE SUMMARY*

R. L. Jaffe
Center for Theoretical Physics
Laboratory for Nuclear Science
and Department of Physics
Massachusetts Institute of Technology
Cambridge, Massachusetts 02139 U.S.A.

INTRODUCTION

This has been a very full and very interesting conference. It's more like the *union* of particle and nuclear physics than the intersection, with a good dose of astrophysics thrown in, too. The excitement testifies to the vitality of the field, but the challenge to summarize it is daunting. Confronting the menu of topics I might mention, I have been forced to choose only a few and leave many interesting and important subjects out completely. Much like a fine restaurant, one must eventually make a choice — leave some wonderful dish for next time — and dig in. I suppose one could attempt a sort of intellectual *dim sum*, including a taste of everything, leading most likely to intellectual indigestion, so I will be more conservative, begin with some brief hors d'doeuvres, add a few first courses which will be more substantial but still leave room for the entreé, which (considering the location of this conference) might be described as crustacean: hard, spiney and ill-tempered. Finally, since astrophysics seems to aid the digestion, I will close with a couple of exciting developments in nuclear and particle astrophysics.

— *MENU* —

I. Hors d'oeuvres:
 - I.1 CP Violation in $K_L \to \pi\pi$.
 - I.2 Missing τ-Decays.
 - I.3 Cabbibo, Kobayashi, Maskawa Parameters in B-Decay.
 - I.4 Hanbury–Brown–Twiss Correlations in High Energy Heavy Ion Collisions.

II. First Courses:
 - II.1 Application of Perturbative QCD to Exclusive Processes.
 - II.2 Positron Peaks in Heavy Ion Collisions.
 - II.3 Mesons in the 1 – 2 GeV Region.

---

* This work is supported in part by funds provided by the U. S. Department of Energy (D.O.E.) under contract #DE-AC02-76ER03069.

II.4 Structure Functions.

III. Entreé:

Cracks in the Foundations of the Quark Model.

IV. Dessert:

IV.1 Inhomogeneous Nucleosynthesis.

IV.2 Astronomy with Ultraviolet Energy Cosmic Rays.

I.1 CP Violation in $K_L \to \pi\pi$

To begin, here is evidence that our field does indeed progress. Only a few years ago we had no idea how small ϵ' might be. Now, as reported by D. Marlow,[1] two experiments appear to have converged on a non-zero value for ϵ'. Figure 1, taken from his talk, shows the history of attempts to measure ϵ'/ϵ. The figure suggests that theory (labeled "standard model") has sat idly by as experiment slowly conformed. Quite the contrary, predictions of ϵ'/ϵ have been forced to adapt to the increasingly restrictive limits set by experiment. More data on ϵ'/ϵ will be forthcoming both from NA31 at CERN and from E731 at FNAL.

There are still important issues to be resolved in $K_L \to 2\pi$. In particular, CPT requires the phases of η_{+-} and η_{00} to be the same within a degree.[2] At present $|\phi_{+-} - \phi_{00}| = 9.8° \pm 5.4°$.[3] An approved experiment at FNAL will provide information on this phase in the near future. Other rare K-decay searches were described here and much more lies ahead for the next Intersections Conference.

I.2 Missing τ-Decays

Marty Perl gave a disquieting report on τ-decay.[4] The topological one prong branching ratio for τ-decay is now very well-measured

$$B_\tau(1 \text{ prong}) = 86.5 \pm 0.3\% \quad .$$

The sum over all identified one prong modes combined with bounds on unobserved modes with multiple neutral mesons accounts for at most $80.8 \pm 1.3\%$ of τ-decays. This leaves about 5% of one-prong τ-decays unaccounted for. Historically, such puzzles have been very fruitful: The neutrino resolved a missing energy problem. The "energy crisis" in the SPEAR data of the mid-1970's was due to the neutrino rich decays of the c-quark and τ-lepton. On the other hand, for a while there was a similar problem with missing J/ψ decays. That did not lead to anything new and exciting; instead a collection of small shifts in branching ratios removed the discrepancy. This is certainly a subject to watch for the future.

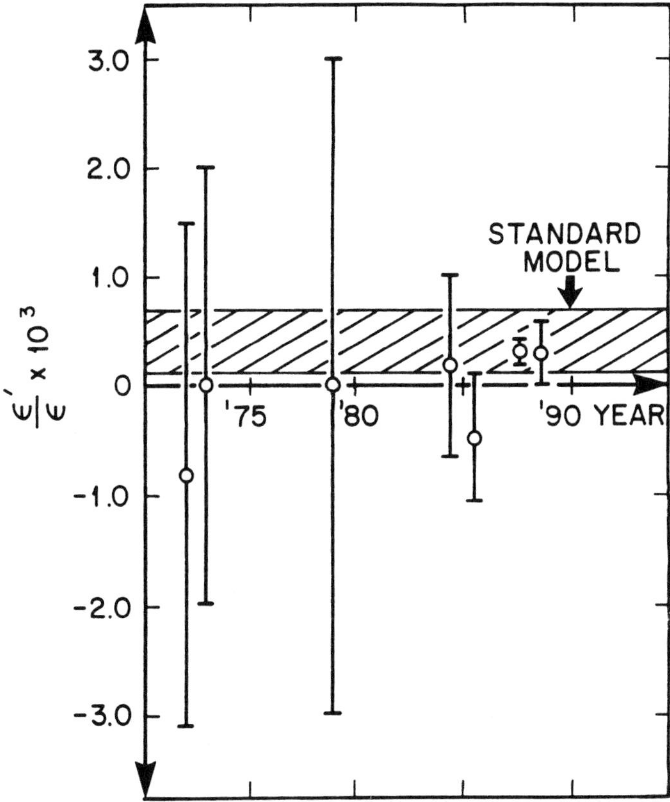

Fig. 1: $\epsilon'\epsilon$ as a function of time.[1]

I.3 CKM Parameters in B-Decay

The future of B-decay physics promises to be quite exciting. We have heard much about $B^0\bar{B}^0$ mixing, B-factories, CP-violation and tests of the standard model in B-decay. I cite one example: Schubert[5] reported recent results from ARGUS which have begun to pin down a very important element of the Cabibbo, Kobayashi–Maskawa matrix. Two exclusive, non-charm decay modes have been observed:

$$B(B^+ \to p\bar{p}\pi^+) = (5.2 \pm 1.4 \pm 1.9) \times 10^{-4}$$
$$B(B^0 \to p\bar{p}\pi^+\pi^-) = (6.0 \pm 2.0 \pm 2.2) \times 10^{-4}$$

indicating $V_{ub} \neq 0$. On the basis of these, Schubert quoted $V_{ub} = 0.0055 \pm 0.0025$, which, I believe, should be taken with caution. It is very difficult to extract V_{ub} from measurement of B-decay into non-charm final states. The method most amenable to theoretical analysis is to look at semi-leptonic B-decay, $B \to Xe\nu$, for $M_X < M_D$. The major problem here is to model the hadronic matrix element $\langle X | J_\mu^{\mathrm{weak}} | B \rangle$. Two approaches bound the available alternatives. One, advocated by Altarelli and collaborators,[6] calculates the decay at the quark level ignoring hadronization. The other, espoused by Grinstein, Isgur and Wise,[7] saturates the final state with a tower of meson resonances. The two approaches give rather different results at the present time. One can see the complementary strengths and weaknesses in both approaches. I'd feel more confident if they converged on a single answer. These cautionary remarks can only be amplified for such a complicated exclusive decay as $B^+ \to p\bar{p}\pi^+$. Until we know better how to handle the hadronic physics, the numerical value of V_{ub} is uncertain.

I.4 Hanbury–Brown–Twiss in High Energy Heavy Ion Collisions

Time does not permit me much mention of heavy ion physics, but I would not want to pass without noting briefly the first reported measurements of Hanbury–Brown–Twiss correlations between negative pions in heavy ion collisions at CERN.[8] This seems to be a promising way to get at the coordinate space structure of heavy ion collisions, measuring the shape and space-time development of the hadronic matter, even on an event-by-event basis. Figure 2 shows $\pi^-\pi^-$ correlations measured by NA35 for several different rapidity bites. The data are fit to a model parametrized by transverse and longitudinal size, R_T and R_L, and a coherence parameter λ. There are peculiarities in this data: R_T is surprisingly large ($\sim$ 8 fm) in mid-rapidity, λ is small ($\lambda \approx 1$ is expected). Even more spectacular is the HBT correlation for a single event (Fig. 2d) demonstrating that shape and space-time development information will be available for each event (and with considerable precision at the huge multiplicities expected at RHIC).

II.1 Application of Perturbative QCD to Exclusive Processes

The conference has been enlivened by yet another installment in the continuing debate over the application of perturbative QCD (pQCD) to exclusive processes. In his talk here, Isgur[9] once again challenged this application of pQCD at ordinary energies ($Q^2 \sim 10$ GeV^2). His criticisms were phrased in a language all of us can understand; let us hope pQCD advocates can reply with equal clarity.

It is important to remember that some of the scaling laws of pQCD have had considerable experimental success. Predictions like $F_\pi(Q^2) \to \mathcal{O}(1/Q^2)$ and $G_M^N(Q^2) \to \mathcal{O}(1/Q^4)$ work, and the quark model derivation is given as an elementary problem in graduate particle physics courses. At this conference we saw the prediction that $\gamma d \to np$ goes like s^{-11} confirmed by experiment.

Isgur and Llewellyn–Smith[10] point out that most of the contribution to these predictions comes from regions of momentum space in which the invariant mass of virtual gluons is very small. To measure these, they define the "legal contribution" to a form factor, or other exclusive quantity, as the part which

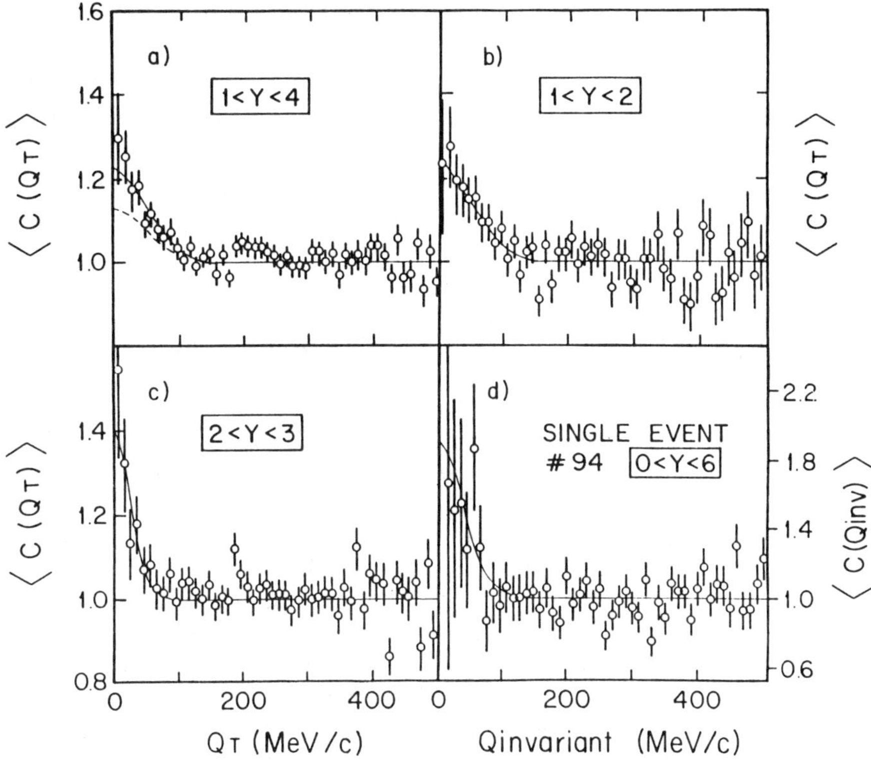

Fig. 2: $\pi^-\pi^-$ correlations in heavy ion collions.[8]

remains when they remove from loop integrations those regions in which any gluon invariant mass becomes less than some parameter m_0^2. For example

$$F_\pi^{\text{Legal}}\left(Q^2, m_0^2\right) \equiv \int_0^1 dx \int_0^1 dx'\, \theta(xx'Q^2 - m_0^2) \frac{\phi_\pi^*(x)\phi_\pi(x')}{xx'Q^2}$$

$F_\pi^{\text{Legal}}(Q^2, 0) \equiv F_\pi(Q^2)$ is the standard pQCD calculation. $F_\pi^{\text{Legal}}(Q^2, m_0^2)$ approaches the usual pQCD result as $Q^2 \to \infty$ for fixed m_0^2. But for state of the art wavefunctions, and m_0^2 as low as 0.25 GeV^2, Isgur and Llewellyn-Smith find $F_\pi^{\text{Legal}} << F_\pi$ at $Q^2 \sim 10 - 25\,\text{GeV}^2$. Things are even worse for the nucleon where $\lesssim 1\%$ of G_M^N is "legal" at $Q^2 = 25\,\text{GeV}^2$ and $m_0^2 = 0.25\,\text{GeV}^2$.

One may worry that a sharp cut-off at m_0^2 is rather extreme. [Perhaps one should give the gluon a mass m_0^2, thereby suppressing the infrared domain.] One can also worry whether the whole procedure has a quantitative gauge invariant meaning. Nevertheless, the *qualitative* point remains that these integrals are dominated by regions where perturbative QCD is suspect. This criticism does not apply to the traditional successes of pQCD in inclusive processes like e^+e^- annihilation, deep inelastic lepton scattering and μ-pair production.

II.2 Positron Peaks in Low Energy Heavy Ion Collisions

Sharp positron lines emitted in low energy collisions of heavy ions have been reported for years. They do not seem likely to go away. Data from the EPOS,[11] and ORANGE[12] spectrometer collaborations both at GSI were reported here. Both see several peaks for a variety of nuclear species, some in coincidence with electrons. As reported by Koenig,[12] some peaks reported by one group are confirmed by the other, but there is not yet any consensus on the experimental phenomena. The peaks are definitely target and projectile species dependent, and I gather anecdotally that the excitation function for some peaks is known to depend sensitively on the projectile energy.

The most naive interpretation of these e^+e^- peaks is that they signal the production and subsequent decay of some new neutral boson, X^0. Although this doesn't provide an adequate explanation of the observations (several lines are seen, the X^0 would have to be produced nearly at rest, *etc.*), nevertheless searches have been undertaken for evidence of the X^0 in other processes. Beam dump searches at FNAL[13] and SLAC[14] combine with limits on the $X^0 e\bar{e}$ coupling from $(g-2)_e$ to exclude an X^0 of mass $\sim 1.8\,\text{MeV}$ by many standard deviations. To avoid these limits one must invoke a large absorption cross-section (so the X^0 doesn't make it out of the dump); or a form factor at the $X^0 e\bar{e}$ vertex which varies several orders of magnitude between $Q^2 = M_X^2$ (at GSI) and $Q^2 \approx 0$ (at beam dumps). The former requires a new long-range force, the latter a new low mass interaction scale.

Koenig reported very preliminary results from Bhabha scattering experiments ($e^+e^- \to e^+e^-$) mounted specifically to look for an X^0 progenitor of the GSI peaks. One group sees a peak at an energy suggested by the EPOS results, but the other does not. Much more data will be available soon. E. Danzmann[15] reported a search for peaks in the $\gamma\gamma$ spectrum in $\sim 6\,\text{MeV/A}$ uranium on thorium collisions at the Berkeley Super Hilac. Initial data showed a peak corresponding to an X^0 mass of $1062 \pm 2\,\text{KeV}$, an energy unknown in the GSI experiments, but a new run failed to increase the significance of the signal.

I concur with Chodos'[11] prejudice that the progenitor of the GSI e^+ phenomena is not a new particle. There are too many different lines, too much variation with beam and target species and with bombarding energy, and the beam dump limits are too hard to circumvent. Chodos suggested that the e^+e^- peaks may be relics of a QED phase transition driven by the intense fields created in heavy ion collisions. I am skeptical of this because I can't see why the natural scale of this exotic phase should be m_e (the positron line energies in the ion center of mass frame range from $\sim$ 250 keV to $\sim$ 425 keV) which is, after all, a parameter intrinsic to the usual phase of QED.

If the peaks are not associated with a new particle, they are probably highly specific to heavy ion collisions. It is therefore most important to map out the dependence of the phenomenon on the nature of the target and projectile and especially on the beam energy, since this might give us a clue what configuration of charged matter is responsible for the effect.

II.3 Meson Spectroscopy in the 1–2 GeV Region

This subject has been around for a long time, long before I began doing particle physics. The reason I have chosen to mention it here is that new experiments at new accelerators have (and will) accumulated a wealth of precise data sensitive to complementary aspects of the 1–2 GeV meson spectrum. Given a few more years of running, it should be possible to make major progress on this thorny problem. The complementary processes to which I refer are:

1. Resonance production in J/ψ radiative decays ($J/\psi \to \gamma X$) — likely to be a source of mesons rich in gluons.
2. Resonance formation in photon-photon scattering ($\gamma\gamma \to X$) — a likely source of C-even quark states excluding $J = 1$.
3. Resonance formation in photon-virtual photon scattering ($\gamma\gamma^* \to X$). Again a source of quark states, with $J = 1$ allowed. Especially valuable in comparison with (2).
4. Resonance production in $p\bar{p}$-annihilation at rest ($p\bar{p} \to \pi X$ etc.). A potentially powerful tool, waiting to be exploited in the ACOL era at LEAR.
5. Resonance production in meson-nucleon scattering ($mN \to XN$, etc.). The old workhorse, still yielding new information especially from the LASS program at SLAC and the MPS program at BNL.

There were many important reports to this conference. Here is a brief selection

- From the GAMS collaboration at Serpukhov: Good photon resolution allows study of multi-η and $\pi^0\eta$ final states. Alde[16] reported long-standing evidence for the $G(1590)$ decaying to $\eta\eta$ and $\eta\eta'$, most likely $J^{PC} = 0^{++}$. He also showed indications, which generated some controversy, for a p-wave $\eta\pi^0$ state ($J^{PC} = 1^{-+}$) at 1405 MeV. This state is particularly interesting because it's forbidden to a non-relativistic $q\bar{q}$-pair. The $\eta\pi^0$ channel is available at LEAR via $p\bar{p} \to \eta\pi^0\pi^0$, a natural for the Crystal Barrel.[17].
- From LASS at SLAC: Good kaon identification, high statistics and lots of experience in PWA. LASS has been particularly effective in identifying the strange mesons which pin down new $SU(3)$ multiplets. Here Dunwoodie[18] reported radial excitations of the $K(890)(1^{--})$, $K(1430)(2^{++})$ and $\kappa(1410)(0^{++})$. He also shows evidence for a 5^{--} K^*-state at 2380 MeV.

  LASS has made a study of non-strange mesons produced in $K^-p \to K\bar{K}\Lambda$, likely to be $\bar{s}s$-states. Dunwoodie reported two resonances near 1530 MeV, an $f_0(1530)0^{++}$ and an $f_1(1530)(1^{++})$. The former has the right mass and quantum numbers to be the $\bar{s}s$-member of a nonet containing the $\epsilon(1300)$ and $\kappa(1410)$. This leaves the $S^*(975)$ out of the usual quark model classification scheme and well-suited to interpretation as a $q^2\bar{q}^2$-state (or $K\bar{K}$ "molecule"). The latter fits nicely as the $\bar{s}s$-state in a 1^{++} nonet with the A_1, Q_A and D-mesons, but this leaves the old $E(1420)$ unclassified, perhaps to be reassigned as a $\bar{K}K^*$ "molecule" (or $q^2\bar{q}^2$-state) at $\bar{K}K^*$ threshold in close analogy to the S^*.

- From MARK III at SLAC: More on radiative J/ψ decays. Wisniewski[19] reported further attempts to understand the shape of the $\iota(1440)$ in $K\bar{K}\pi$. In particular, the shape of the $K\bar{K}\pi$ mass spectrum could be understood as a single resonance provided the influence of the $K\bar{K}$ threshold on the shape of the $\delta(980)$ was properly included. This has the effect of sharpening a continuing mystery: The ι appears to decay into $K\bar{K}\pi$ entirely through $i \to \delta\pi$, $\delta \to K\bar{K}$, but the decay $i \to \delta\pi$, $\delta \to \eta\pi$ has not been seen although $\eta\pi$ is a known, significant mode of decay of the δ.

- MPS at BNL: High resolution, good particle identification and high statistics. Blessing[20] showed evidence for activity in both 0^{-+} and 1^{++} partial waves near 1285 MeV. The 1^{++} is presumably the familiar $D(1285)$. The 0^{-+} is further evidence for a radial excitation of the η. This leaves the 0^{-+} $\iota(1440)$ out of the usual $q\bar{q}$-classification and opens the door for a glueball assignment.

  Blessing also reported activity in both the 1^{++} and 0^{-+} partial waves near 1420 MeV, possibly consistent with the MARK III observations of the $\iota(1440)$ and the TPC-2γ evidence for a $J = 1$ resonance at the $E(1420)$.

- TPC-2γ at SLAC: Two photon physics. A natural way to make quark states but not a strong glueball channel. Barker[21] reported studies of $\gamma\gamma$ and $\gamma\gamma^*$ resonance formation. $\gamma\gamma$ shows no activity near 1420 MeV, consistent with a $J = 1$ assignment for the $E(1420)$ and a glueball assignment for the $\iota(1440)$. $\gamma\gamma^*$ shows evidence for both the $D(1285)$ and the $E(1420)$.

I am not going to try to make a unified presentation of this data. But over the next few years I expect one will be possible. A summary of the implications of all this for glueballs has been given here, with many caveats, by Meshkov.[22]

II.4 Deep Inelastic Structure Functions

The first measurements of the structure functions of the proton in the deep inelastic domain were reported by an MIT–SLAC collaboration twenty years ago.[23] It is impossible to underestimate the importance of these and subsequent measurements of nucleon and nuclear structure functions with electron, muon and neutrino beams. It is, however, easy (especially for theorists) to forget how difficult these experiments are: the cross-sections are small, systematic errors are difficult to control and the relative normalization among different experiments has proved difficult to determine. Although accuracy has improved over the years, it is wise not to take very seriously the unconfirmed reports of a single experiment.

a) The EMC Effect — There is now considerable agreement on the differences between nuclear and nucleon structure functions:

- Shadowing is seen at low-x, and it appears to scale.[24]
- The large enhancement at $x \lesssim 0.2$ first reported by EMC[25] appears to have been overestimated. Now all groups agree on a smaller, $\lesssim 5\%$ enhancement at low-x. A new analysis of radiative corrections to the SLAC data agrees with this picture.
- No A-dependence of $R = \sigma_L/\sigma_T$ was seen in a dedicated SLAC experiment.

Theory has also progressed: With the disappearance of the low-x enhancement, pion models are in retreat. Calculations of nuclear binding effects have been improved, modified for relativistic $\mathcal{O}\left(p^2/M^2\right)$ effects[27] and appear to account for some, but not all of the depletion in the nuclear quark distribution at large-x. No single mechanism seems to explain the EMC effect at all x-values. Instead a recipé which includes Fermi motion (to produce an enhancement) at $x \gtrsim 0.7$, shadowing (to give depletion) at $x \lesssim 0.1$, and partial deconfinement, or softening of the quark momentum distribution, to produce a depletion $(0.3 \lesssim x \lesssim 0.7)$ and corresponding enhancement $(0.1 \lesssim x \lesssim 0.3)$, seems to be required. The microscopic origin of deconfinement is still hotly debated, but probably cannot be settled on the basis of deep inelastic scattering experiments alone.

b) Structure Functions and QCD — Many high statistics, high energy measurements of the structure functions of many targets have become available over the past few years. The result is confusion. Rith[28] reported disagreement among different data sets on almost every target studied in the recent past. I will not attempt to repeat his summary, for which you should see his excellent talk, but here are some of the main points:

1. $F_2^p(x,Q^2)$ — BCDMS and EMC disagree. EMC is low with respect to BCDMS at low-x and high at high-x. The BCDMS data fit nicely straight lines in $\log Q^2$ but these lines do not fit the old SLAC data when extrapolated to low-Q^2. The EMC data match the SLAC data better.
2. $F_2^n(x,Q^2)$ — EMC at $x = 0.45$ and 0.58 show $F_2^n(x,Q^2)$ rising with $\ln Q^2$ not falling as expected from QCD. Nevertheless, the errors are such that the data are not inconsistent with QCD.
3. $xF_3^{Fe}(x,Q^2)$ — CDHS data have been fit to QCD at large W^2 ($W^2 \geq 11\,\mathrm{GeV}$). The fits are good, but do not pass through lower W^2 (*i.e.* lower Q^2) data when extrapolated back. The discrepancy is most noticeable at low-x $(0.08 \leq x \leq 0.225)$ and in the highest-x bin $(x = 0.650)$.
4. Finally, the EMC have studied the Q^2-dependence of the ratio of iron and deuterium structure functions. This is particularly interesting since it is a direct test of the QCD prediction that anomalous dimensions

(which determine Q^2-dependence) are independent of target. The plot of $d\left(\ln F_2^{Fe}/F_2^D\right)/d\ln Q^2$ shows much more rapid decrease with x than predicted by QCD. Target mass corrections improve the agreement slightly. Higher twist, if included, improves it more. Rith cautions, however, that the discrepancy is not too great if systematic errors are taken into account.

Some of these discrepancies are doubtless experimental in origin. Other may be evidence for higher-twist effects. Higher twist has a bad name. The term has become a catch-all for corrections to asymptotic QCD predictions in all processes. It is used frequently by theorists to explain away descrepancies between these asymptotic predictions and data at heroic but still finite energies and momentum transfers. This is unfortunate because in deep inelastic lepton scattering, "higher twist" has quite a precise meaning and there is much to learn if it could be measured. "Twist" is a quantum number (twist $\equiv$ scale dimension – spin) which determines the power of Q^2 at which a given operator contributes to deep inelastic scattering. An operator of twist-T goes like $\left(Q^2\right)^{2-T}$, modulo logarithms, at large Q^2. Twist-two operators determine the structure functions F_2, F_3, etc. $\mathcal{O}(1/Q^2)$ corrections to scaling are associated with operators of twist-4. [Twist-3 operators are suppressed by a factor of quark mass in spin independent deep inelastic scattering.] The complete set of twist-4 operators were determined years ago,[29] and a "parton-like" interpretation of twist-4 effects was derived.[30,31] Roughly speaking, $\mathcal{O}(1/Q^2)$ effects are associated with correlated quark-gluon distributions within the target[30] or, alternatively (by use of QCD equations of motion) with quark and gluon transverse momentum distributions.[31] Perhaps the discrepancy between the BCDMS and SLAC data and the anomalous Q^2-dependence seen by CDHS are evidence for significant twist-4 contributions. If so, one should be quite optimistic that twist-4 can be studied in detail in future experiments. Of course, not even higher twist effects can explain inconsistencies between experiments at the same Q^2 and x.

I am well-aware that nearly every summary talk on structure functions contains pious statements about the significance of higher twist. At the present time, no data exist with sufficient precision and a sufficient Q^2-range (say $0.1 < Q^2 < 10\,\mathrm{GeV}^2$) to make a serious assault on higher twist. The possibility of such a program should figure prominently in the menu of any new lepton facility.

III. CRACKS IN THE FOUNDATIONS OF THE QUARK MODEL

The quark model predates QCD and provides of some of the simplest and most persuasive explanations of hadron properties. Perhaps, someday, it will be derived from QCD, perhaps by lattice calculations on supercomputers. The quark model is based on the apparent fact that nearly unrenormalized quark degrees of freedom persist down to the confinement scale. Nucleons behave as bound states of three quarks, mesons (the pion, perhaps excepted) as quark-antiquark. The list of hadron properties explained this way is very long: μ_p/μ_n, $2M_p\mu_p$, M_ρ/M_p, g_A/g_V, $(F/D)_{g_A/g_V}$, $(F/D)_\mu$, $(F/D)_M$, $SU(6)\times O(3)$ spectroscopy, ... Watching QCD devolve from the short-distance, perturbative domain one would hardly expect such simplicity. As $1/Q^2$ increases $\alpha_s(Q^2)$ grows until at 1 GeV one might expect some complex quasi-particle picture to emerge. If it does, the quasi-particles are quarks (perhaps chiral quarks in the fashion described by Georgi and Manohar[32]). The simplicity of the quark model is perhaps the most persuasive reason to seek some deeper physical picture of confinement in QCD and not leave it to the CRAY's.

Recently, the simplicity of the quark model has been challenged by two measurements. One, the sigma-term in πN scattering, has been around (in some form) for nearly a decade. The other, the integrated spin asymmetry in deep inelastic electron scattering, was reported by the EMC last year. Both are measurements of nucleon forward matrix elements of quark operators. Both have been interpreted to require anomalously large strange quark components in the nucleon. Both are subject to considerable experimental and theoretical uncertainty.

TABLE 1: Nucleon Matrix Elements of bilinear quark operators.

COMMON NAME	OPERATOR	SCALE	NON-STRANGE QUARKS	STRANGE QUARKS	COMMENTS
Vector Charge	$\bar{q}\gamma_\mu q$	—†	3	0	Pedagogical
"Momentum" Sum Rule	$\bar{q}\gamma_\mu D_\nu q$	$\sim 10\,\mathrm{GeV}^2$	.438 ±.002	.026 ±.006	Higher spin operators, too. Note small errors. From dimuons.
Axial Charge	$\bar{q}\gamma_\mu\gamma_5 q$	$\sim 10\,\mathrm{GeV}^2$	.21 ±.19	−.17 ±.10	Deep inelastic spin asymmetry. Extraction problems. Q^2-dependence. $\nu p \to \nu p$ corroboration.
Sigma Term	$m\bar{q}q$	—†	55 ± 8 MeV	~ 325 MeV	Extraction problems? Problems with chiral dynamics?
Magnetic Moment	$\frac{1}{2}\vec{r} \times \bar{q}\vec{\gamma}q$	—†	2.88 $+F_2^{(0)}$	0.12 $+F_2^{(0)}$	Measure $F_2^{(0)}$ in $\nu p \to \nu p$.[34]

†Operator is renormalization scale invariant — no scale need be set.

To place these measurements in context, I would like to look systematically at what is known about the nucleon forward matrix elements of quark operators, especially the dependence on quark flavor, strange versus non-strange. Everything is summarized in Table I. The rows label different operators. The

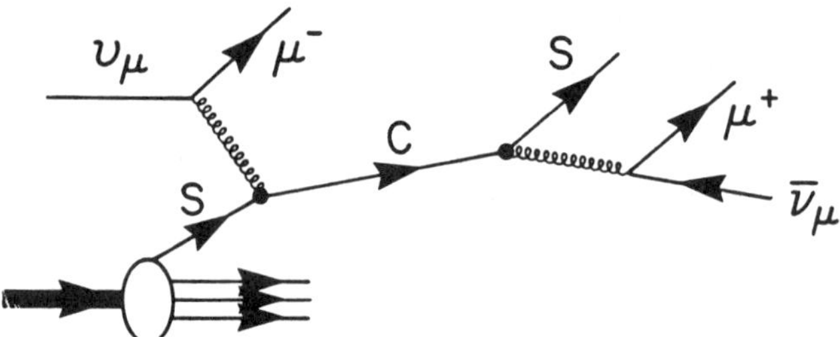

Fig. 3: Dimuon production in inelastic neutrino scattering.

first two columns define the operator. If the operator matrix element is scale-dependent, column three gives the scale of the measurement. In columns four and five the experimental numbers are given for non-strange and strange quarks, respectively. Some comments, expanded upon in the text, are given in the last column.

III.1 Vector Charge

This operator is included for pedagogical purposes only. The matrix element $\langle N|\bar{s}\gamma_\mu s|N\rangle$ measures the nucleon's strangeness, which is zero:

$$\langle N|\bar{s}\gamma_\mu s|N\rangle = 2p_\mu S = 0 \ . \tag{III.1}$$

The non-strange version, $\bar{u}\gamma_\mu u + \bar{d}\gamma_\mu d$, can be related to three times the nucleon's baryon number by means of (III.1)

$$\langle N|\bar{u}\gamma_\mu u + \bar{d}\gamma_\mu d|N\rangle = \langle N|\bar{u}\gamma_\mu u + \bar{d}\gamma_\mu d + \bar{s}\gamma_\mu s|N\rangle = 3(2p_\mu) \ .$$

This little trick will be repeated often below. Here, there is no question that $\bar{s}\gamma_\mu s$ can be added to $\bar{u}\gamma_\mu d + \bar{d}\gamma_\mu d$ because $\langle N|\bar{s}\gamma_\mu s|N\rangle$ is strictly zero. Later, the same manipulation will require the assumption that there are few s-quark pairs in the nucleon or that they do not affect the operator in question.

III.2 "Momentum" Sum Rule

This operator matrix element is measured in deep-inelastic scattering, where it corresponds, in parton language, to the fraction of the target's momentum carried by quarks in an infinite momentum frame. It is determined by an integral over $F_2(x)$ which converges quickly at low-x. [The higher moments of the quark's momentum distribution are measured by higher moments of $F_2(x)$ and determine matrix elements of higher spin twist-two operators. We ignore them because they are poorly measured for s-quarks.] Strange quarks can be separated from up and down quarks by measuring charm production in neutrino scattering. The idea, illustrated in Fig. 3, is that neutrino scattering from an s-quark produces ($\cos\theta_c \approx 1$) a leading charm quark which decays $\sim 20\%$ of the time to $\mu^+\bar{\nu}_\mu X$ producing a leading μ^+ along $\vec{q}$. [In fact, the excess of opposite sign dimuons in ν-scattering was a puzzle resolved by the discovery of charm.] In this way CDHS[33] has measured that $2.6 \pm 0.6\%$ of the nucleon's momentum is carried by s and $\bar{s}$ quarks.

$\bar{u}$ and $\bar{d}$ quarks together carry twice as much momentum. All together, $u, d, \bar{u}$, and $\bar{d}$ quarks carry $43.8 \pm 0.2\%$ of the nucleon's momentum. These numbers have been confirmed by the CCFRR collaboration.[33]

These measurements are quite accurate (by the standards of this field) and support a picture of a nucleon whose "intrinsic" wave function is devoid of s (and $\bar{s}$) quarks, and which develops a small $s\bar{s}$-component by means of the perturbative evolution from the intrinsic scale ($\lesssim 1$ GeV) to the Q^2 at which the experiments are performed. Indeed, the measurements indicate that the strange "sea" in the nucleon is suppressed by about a factor two compared to the non-strange pair sea:

$$\frac{\langle s + \bar{s} \rangle}{\langle \bar{u} + \bar{d} \rangle} = 0.52 \pm 0.09 \quad .$$

In summary, this operator is well-studied and fulfills quark model expectations. It will be a point of reference for other, more problematic operators.

III.3 The Axial Charge

Various operators belonging to the $SU(3)_f$-multiplet

$$A^a_\mu = \bar{q}\gamma_\mu\gamma_5\frac{\lambda^a}{2}q$$

($a = 1, 2, \ldots 8$, $\mathrm{Tr}\,\lambda^a\lambda^b = 2\delta^{ab}$) have been measured in the Gamow–Teller part of hyperon and nucleon β-decay. These are consistent with $SU(3)$ symmetry (within experimental errors) with F and D invariant matrix elements

$$F = 0.45 \pm 0.02$$

$$D = 0.80 \pm 0.02 \quad .$$

[I have used the F and D values extracted by Kaplan and Manohar[34] for reasons explained by them.] Using $SU(3)$ symmetry it is possible to extract the nucleon matrix elements of two flavor combinations:

$$G^{(3)} s_\mu \equiv \frac{1}{2} \langle Ns | \bar{u}\gamma_\mu\gamma_5 u - \bar{d}\gamma_\mu\gamma_5 d | Ns \rangle$$

$$G^{(8)} s_\mu \equiv \frac{1}{2\sqrt{3}} \langle Ns | \bar{u}\gamma_\mu\gamma_5 u + \bar{d}\gamma_\mu\gamma_5 d - 2\bar{s}\gamma_\mu\gamma_5 s | Ns \rangle$$

with the result

$$\begin{aligned} G^{(3)} &= \frac{1}{2}(F + D) = 0.625 \pm 0.020 \\ G^{(8)} &= \frac{1}{2\sqrt{3}}(3F - D) = 0.159 \pm 0.018 \end{aligned} \tag{III.2}$$

To obtain the strange and non-strange operator matrix elements separately more information is necessary. The spin asymmetry in deep inelastic lepton scattering ($g_1^{ep}(x)$) has long been known to be a source of the needed information.[35] Recently the EMC reported a measurement of $g_1^{ep}(x)$ which combined with Eqs. (III.2) indicates a surprisingly large nucleon matrix element of $\bar{s}\gamma_\mu\gamma_5 s$.[36] The quantity measured is schematically

$$A_1 = \frac{\mu^\uparrow p^\uparrow - \mu^\uparrow p^\downarrow}{\mu^\uparrow p^\uparrow + \mu^\uparrow p^\downarrow}$$

where $\uparrow$ and $\downarrow$ denote polarization along the beam momentum. A_1 scales in the deep inelastic limit and is related to the structure functions g_1 and F_1 by

$$g_1(x) \simeq A_1 F_1(x)/(1+R)$$

which is approximate because the higher twist structure function $g_2(x)$ (expected to be small) has been ignored.

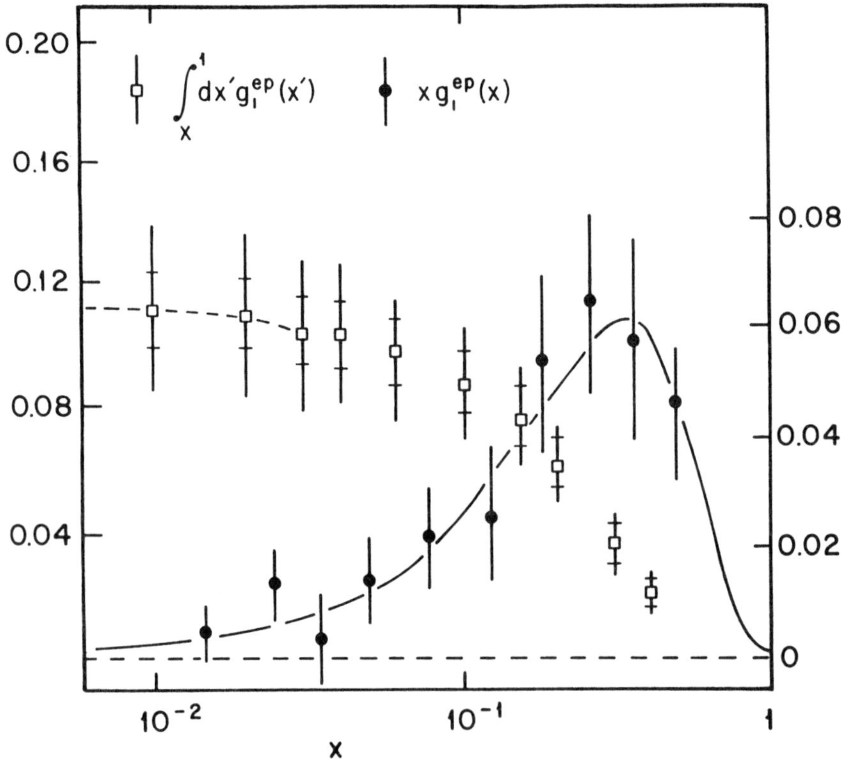

Fig. 4: $g_1^{ep}(x)$ and $\int_x^1 g_1^{ep}$ as determined by EMC.[36]

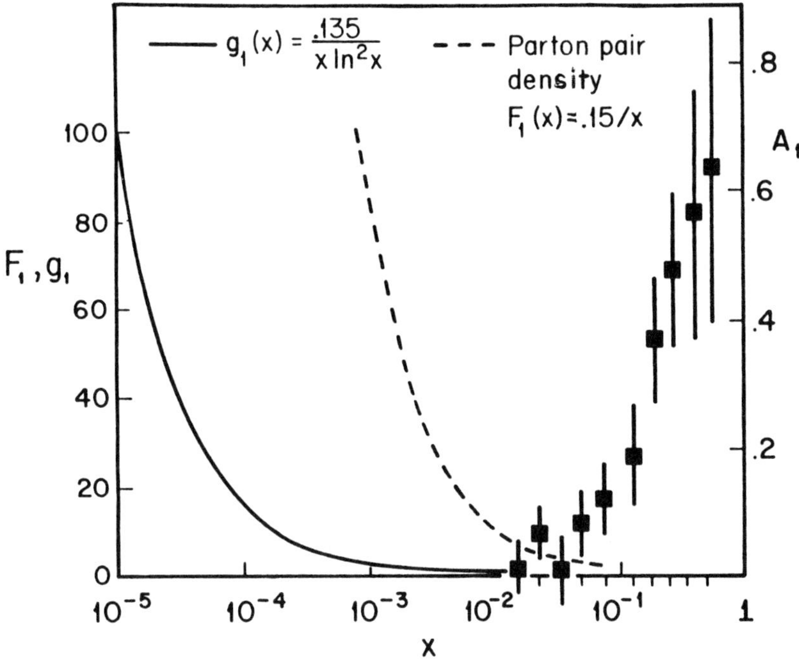

Fig. 5: $A_1(x)$, $g_1(x)$ and $F_1(x)$ illustrating the relation $g_1 \simeq A_1 F_1$ and the associated extrapolation uncertainties.

The experimenters construct the sum rule[36]

$$\int_0^1 dx\, g_1^{ep}(x) = 0.114 \pm 0.012 \pm 0.026 \quad .$$

Before discussing the significance of this result, let's look at it's reliability. The data extend down to $x \approx 0.02$ at high Q^2. The data $(xg_1(x))$ and the integral from x to 1 are shown in Fig. 4. To complete the sum rule it is necessary to extrapolate to $x = 0$. Close and Roberts have emphasized the danger of this extrapolation.[36] It can be illustrated by Fig. 5. The experiment measures A_1, which is observed to decrease with decreasing x. To obtain $g_1(x)$ it is necessary to multiply A_1 by $F_1(x)$ which appears (experimentally) to diverge at small x like $1/x$. $F_1(x) = 0.15/x$, which fits the EMC data[39] is shown in Fig. 5. Clearly the value of $g_1(x)$ obtained by multiplying A_1 by F_1 in the domain $x < 0.02$ depends very sensitively upon the form assumed for the extrapolation. The EMC assumed $xg_1(x) \sim x$. Close and Roberts made two fits, (I) like EMC, assumed $xg_1(x) \propto x$ but with a slightly different fit to the data for $x > 0.02$ and with the result shown in the second row of Table II; (II) they took the low-x behavior of $g_1(x)$ suggested by one interpretation of Regge theory (see Fig. 6),[39] with the result shown in the third row of Table II. The low-x behavior of $g_1(x)$ is not known on fundamental grounds, so one must not take any particular fit very seriously. I think one is forced to conclude that the true uncertainty in $\int_0^1 dx\, g_1^{ep}(x)$ is considerably larger than the quoted errors.

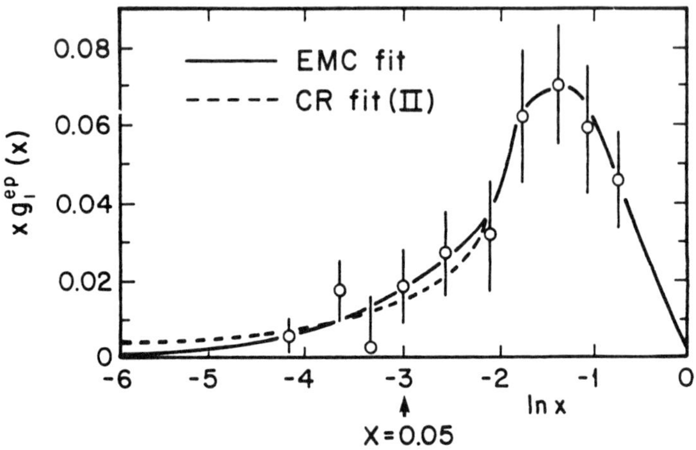

Fig. 6: Two fits to $xg_1^{ep}(x)$.[37]

TABLE II: Different extractions of the strange and non-strange components of the nucleon's axial vector charge.

FIT TO EMC DATA	$G^{(0)}\left(10\,\mathrm{GeV}^2\right)^{\dagger}$	$G^{s}\left(10\,\mathrm{GeV}^2\right)^{\dagger}$	$G^{\ell}\left(10\,\mathrm{GeV}^2\right)^{\dagger}$
(I) EMC	$.014 \pm .097$	$-.170 \pm .101$	$.211 \pm .192$
(II) Close & Roberts[36] $xg_1 \sim x$	$.098 \pm .114$	$-.086 \pm .118$	$.379 \pm .228$
(III) Close & Roberts[36] $xg_1(x) \sim 1/\ln^2 x$	$.172 \pm .135$	$-.012 \pm .138$	$.527 \pm .270$
(IV) EMC + F/D from np & Λp only	$.025 \pm .097$	$-105 \pm .115$	$.181 \pm .194$
Kaplan & Manohar[34] analysis of $\nu p \to \nu p$ *elastic* scattering	$.03 \pm .08$	$-.15 \pm .08$	$.24 \pm .06$

†Note scale dependence.

The significance of this integral is a sum rule one can derive from the operator produce expansion in QCD:[35]

$$\int_0^1 dx\, g_1^{ep}(x,Q^2) = \left\{\frac{1}{6}G^{(3)} + \frac{1}{6\sqrt{3}}G^{(8)} + \frac{1}{3}G^{(0)}\right\}\left(1 - \frac{\alpha_s(Q^2)}{\pi} + \mathcal{O}\left(\alpha_s^2\right)\right) \tag{III.3}$$

where $G^{(3)}$ and $G^{(8)}$ have been defined and $G^{(0)}$ is the analogously defined matrix element of the flavor singlet axial charge

$$G^{(0)} S_\mu = \frac{1}{3}\left\langle Ns|\bar{u}\gamma_\mu\gamma_5 u + \bar{d}\gamma_\mu\gamma_5 d + \bar{s}\gamma_\mu\gamma_5 s|Ns\right\rangle \quad .$$

Since $G^{(3)}$ and $G^{(8)}$ are known from β-decay, the sum rule can be considered a measurement of $G^{(0)}$. In the parton model, the structure function $g_1(x)$ measures the probability distribution (weighted by 1/2 the squared charge) for quarks with helicity parallel minus those antiparallel to the target's helicity in an infinite momentum frame:

$$\int dx\, g_1^{ep}(x) = \frac{4}{18}\Delta u + \frac{1}{18}\Delta d + \frac{1}{18}\Delta s$$

where

$$\Delta q = \int_0^1 dx\left[q_\uparrow(x) - q_\downarrow(x) + \bar{q}_\uparrow(x) - \bar{q}_\downarrow(x)\right]$$

and

$$\Delta q s_\mu = \langle ps|\bar{q}\gamma_\mu\gamma_5 q|ps\rangle \quad .$$

Some subtleties associated with the sum rule (III.3) are discussed with other theoretical issues below.

The "canonical" quark model expectation would be that the nucleon contains few, if any, strange quarks. Even if it contains s-quarks, so long as they are unpolarized, $\Delta s = 0$. So one expects $\Delta s \approx 0$, which implies

$$G^{(0)} \simeq 2\sqrt{3}\, G^{(8)} = 0.55 \pm 0.06 \quad . \tag{III.4}$$

Instead, from the EMC result one obtains

$$G^{(0)} = -0.02 \pm 0.09$$

or

$$\Delta s = -0.20 \pm 0.09$$
$$\Delta u + \Delta d = 0.15 \pm 0.17 \quad .$$

The values of Δs and $\Delta u + \Delta d$ for different evaluations of $\int g_1^{ep}(x)$ and for a different choice of F/D are shown in Table II.† Depending on the fit to $\int g_1^{ep}(x)$

† I have not attempted a proper error analysis, so the errors quoted in the table must be considered illustrative, not definitive.

one sees a one to two standard deviation departure of Δs from zero. For the EMC fit $|\Delta s| \approx |\Delta u + \Delta d|$ and strikingly, $G^{(0)} \approx 0$, which has spawned the catch-phrase that the EMC results show that the quark's spin accounts for none of the nucleon's spin! Perhaps we should remember that the canonical quark model result is that the quark's spin gives only about half the nucleon's spin (Eq. III.4), the rest being in orbital angular momentum or in the spin of gluons.

An interesting corroboration of the EMC result has been pointed out by Kaplan and Manohar.[34] They note that the matrix element of $G^{(0)}$ also enters elastic νp-scattering because the Z^0 couples to the axial current $\bar{u}\gamma_\mu\gamma_5 u - \bar{d}\gamma_\mu\gamma_5 d - \bar{s}\gamma_\mu\gamma_5 s$ which is not traceless in $SU(3)_f$. After studying corrections from c-, b- and t-quarks (the Z^0 axial current *is* traceless in each generation) they extract $G^{(0)} = 0.03 \pm 0.08$ from the data of Ahrens *et al.*[40] I have added a row to Table II accordingly. This source of information on $G^{(0)}$ does not suffer from the extrapolation uncertainties of $\int g_1^{ep}(x)$, though it has its own problems concerning the parametrization of form factors and the paucity of data, discussed in Ref. [34].

All of this discussion is summarized by the third row of Table I.

III.4 The Sigma Term

This problem has been around for years and has been reviewed recently,[41] so I will be brief. $\Sigma_{\pi N}$, defined by

$$\Sigma_{\pi N} = \frac{1}{3}\sum_{a=1}^{3} \langle N \,|[Q_5^a, [Q_5^a, H]]|\, N\rangle \quad ,$$

[Q_5^a are the isospin axial charges.], is "measurable" by extrapolating low energy πN scattering data to an unphysical point. The answer is generally agreed to be about 60 MeV, although there are dissenters: Ericson,[42], for example, has argued recently that a more stable extrapolation method yields $\Sigma_{\pi N} \simeq 45$ MeV.

In QCD,

$$\Sigma_{\pi N} = \bar{m}\left\langle N|\bar{u}u + \bar{d}d|N\right\rangle \quad ,$$

$\left(\bar{m} = \frac{1}{2}\left(m_u + m_d\right)\right)$. It measures the nucleon mass shift due to $SU(2)$ chiral symmetry breaking to lowest order in $\bar{m}$. Closely related to $\Sigma_{\pi N}$ are the octet of scalar densities $\bar{q}\frac{\lambda^a}{2}q$ for $a = 1, 2, \ldots 8$, and the term in the QCD Hamiltonian that breaks $SU(3)_f$-symmetry

$$H_{\rm SB}\left(SU(3)\right) = \frac{\bar{m} - m_s}{\sqrt{3}}\left(\bar{u}u + \bar{d}d + \bar{s}s\right) = \frac{\bar{m} - m_s}{\sqrt{3}}\bar{q}\lambda^8 q \qquad \text{(III.5)}$$

To lowest order in $H_{\rm SB}\left(SU(3)\right)$, $\bar{q}\lambda^8 q$ can be related to baryon mass differences

$$\left(\bar{m} - m_s\right)\left\langle N\left|\bar{q}\lambda^8 q\right| N\right\rangle = \sqrt{3}\left(M_\Lambda - M_\Xi\right) \quad .$$

Combined with the value of $\Sigma_{\pi N}$ and the usual PCAC estimate of quark mass ratios, $m_s/\bar{m} = 2\mu_K^2/\mu_\pi^2$, we obtain

$$m_s \langle N | \bar{s}s | N \rangle \simeq 325\,\mathrm{MeV}$$

or, equivalently

$$\langle N | \bar{s}s | N \rangle \simeq \frac{1}{2} \left\langle N \left| \frac{\bar{u}u + \bar{d}d}{2} \right| N \right\rangle \ .$$

These results are surprisingly large: The nucleon mass increases 325 MeV when m_s is increased from zero to its physical value; the scalar density of s-quarks is *half* that of the u and d quarks, which get valence contributions.

It is important to stress that the extraction of $\Sigma_{\pi N}$ is difficult, and furthermore the use of lowest order perturbation theory to calculate $\langle N | \bar{q}\lambda^8 q | N \rangle$ may be unreliable. In several models,[43] when a nucleon is constructed with a large $\bar{s}s$ content at $m_s = 0$, the s-quarks are quenched as m_s increases, as shown in Fig. 6. The reason is simple quantum mechanics: as m_s increases, the nucleon wave function sees a repulsive potential in the strange direction in flavor space, which it avoids by reducing the $\bar{s}s$ content of its wave function. This effect cannot be seen in lowest order perturbation theory where the nucleon wavefunction is unchanged from the symmetry limit. Of course, anyone who invokes non-linearities in $H_{\mathrm{SB}}\,(SU(3))$ is obliged to account for the successes of this approximation, such as the Gell–Mann–Okubo formula.

If it turns out that $m_s \langle N | \bar{s}s | N \rangle$ is this large, the consequences for QCD are enormous: As I mentioned, the nucleon mass shift due to $SU(3)$ symmetry breaking is $\gtrsim$ 300 MeV. Lattice calculations using the quenched approximation (in which $\langle N | \bar{s}s | N \rangle = 0$) will fail to reproduce the physical nucleon mass. The $SU(2)$ Skyrme model is unreliable. Kaons may condense in nuclear matter at rather low density.[44] Unfortunately, no one as yet has found a direct way to confirm the value of $m_s \langle N | \bar{s}s | N \rangle$ obtained from the sigma term analysis.

III.5 Magnetic Moments

Another well-studied bilinear quark operator is the flavor $SU(3)$ octet of magnetic moments

$$\vec{\mu}^a = \frac{1}{2}\bar{q}\vec{r} \times \vec{\gamma}\frac{\lambda^a}{2}q \qquad a = 1, 2, \ldots 8 \ .$$

In addition, we can define the "baryomagnetic" moment

$$\vec{\mu}^0 = \frac{1}{6}\bar{q}\vec{r} \times \vec{\gamma} q \ .$$

Interestingly, Kaplan and Manohar have pointed out that nothing is known about $\left\langle N | \vec{\mu}^0 | N \right\rangle$. The ordinary isoscalar and isovector magnetic moments of electromagnetism are related to $\vec{\mu}^8$ and $\vec{\mu}^3$, respectively, because

$$Q = \mathrm{diag}\,(2/3, -1/3, -1/3) = T^3 + \frac{1}{\sqrt{3}} T^8 \ .$$

Kaplan and Manohar point out that the Z_0 vector current has an $SU(3)$ flavor singlet piece, so elastic neutrino scattering is, in principle, sensitive to $\vec{\mu}^0$. No analysis of $\nu p \to \nu p$ has paid attention to this distinction, so there is nothing yet known about $\langle N|\vec{\mu}^0|N\rangle$. Nucleon model enthusiasts (quark, Skyrme, bag, hybrid bag, flux tube, ...) should hurry to get their predictions for $\langle N|\vec{\mu}^0|N\rangle$ in before the experimental value is known!

III.6 Theory

So far there has been no entirely satisfying response from defenders of quark models nor, on the other hand, have supporters of alternative models of hadron phenomenology given a convincing explanation of $\Sigma_{\pi N}$ and $\int g_1^{ep}$. I think it is important to view the set of bilinear quark operators as a whole: it is not good enough merely to admix s and $\bar{s}$ quarks into the nucleon wavefunction as required to explain $\Sigma_{\pi N}$ and $\int g_1^{ep}$; it is also necessary to explain, in the same model, why the s-quarks do not show up in the nucleon's (null-plane) momentum distribution, and why other quark model results like the F/D ratio for magnetic moments and axial vector charges work so well. Here I will only comment on a few theoretical issues.

- Q^2-Dependence of $G^{(0)}$

There is a sublety concerning $\int g_1^{ep}$ not evident in the sum rule (Eq. (III.3).[45] The octet axial charges $G^{(a)}$ are constants independent of Q^2. This is a special case of the general result that conserved currents have vanishing anomalous dimension. The axial currents, A^a_μ, would be conserved only if the quarks were massless, but that's good enough. $G^{(0)}$ is different. A^0_μ is not conserved, even for massless quarks, because of the triangle anomaly:[46]

$$\partial_\mu A^{\mu 0} = \sqrt{\frac{2}{3}}\,\frac{3g^2}{16\pi^2}\epsilon_{\lambda\sigma\alpha\beta}\,\mathrm{Tr}\,G^{\lambda\sigma}G^{\alpha\beta}$$

and consequently $G^{(0)}$ may depend on Q^2. The Q^2-dependence has been calculated to lowest order and it is weak.[47] But the Q^2-dependence of $G^{(0)}$ may be strong at small Q^2 where perturbative QCD is irrelevant. One could *imagine* a scenario in which $G^{(0)} \approx 1/2$ (the canonical quark model value) at low-Q^2, and $G^{(0)} \approx 0$ (the EMC result) at $Q^2 \approx 10\,\mathrm{GeV}^2$. This is illustrated in Fig. 7, and can be neither proved nor disproved with the present state of theory. If this were the case one would say that the quarks lose spin to the gluons via the triangle anomaly as Q^2 evolves. The analogous statement about momentum: that quarks lose momentum to gluons via the trace anomaly as Q^2 evolves, is one way of summarizing the standard Alterelli–Parisi evolution of quark distributions.

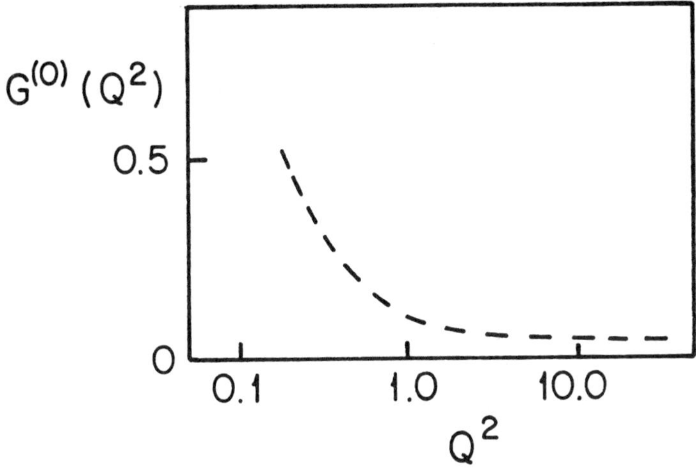

Fig. 7: Possible Q^2-evolution of $G^{(0)}(Q^2)$.

• Current verses Constituent Quarks

It is an ancient idea[48] that the quarks which couple in a point-like fashion to local currents are not the same objects which are bound by three's into baryons. An obvious source of the difference is the chiral symmetry breaking phase transition which redefines the quark vacuum at a mass scale $\Lambda_\chi \sim 1$ GeV. Above Λ_χ the elementary fermions are "current" quarks which couple simply to leptons; below Λ_χ, the "constituent" quarks are the economical effective fields. The two are related by a generalized Bogoliubov transformation: a constituent quark, Q, looks like a current quark, q, accompanied by a cloud of $q\bar{q}$-pairs. "Current" $\bar{s}s$-pairs might mix into the constituent U quark. Donoghue and Nappi[49] and more recently, Kaplan and Manohar[34] have invoked this sort of flavor mixing as a possible solution to the $\Sigma_{\pi N}$ and $\int g_1^{ep}$ puzzles.

There are two pitfalls of this approach. First, it's not at all clear why $\bar{s}s$ and $\bar{s}\gamma^\mu\gamma^5 s$ should have large nucleon matrix elements but $\bar{s}\gamma^\mu D^\nu s$ remains small. Second, when chiral symmetry breaks, flavors may mix in many ways. Recently, Bernard, Meißner and I studied the problem in the Nambu–Jona-Lasinio model[43] and found that a likely scenario is that the $\bar{q}q$ cloud in a constituent U quark is dominantly $\bar{u}u$. In the unusual case that there is a large $\bar{s}s$ component in the U-quark it quenches quickly as m_s increases from zero to its physical value.

• Spin and Flavor in the Skyrme Model

I have a few comments on attempts to understand $\Sigma_{\pi N}$ and $\int g_1^{ep}$ in the Skyrme model. Donoghue and Nappi[49] estimated $\Sigma_{\pi N}$ and in a three flavor Skyrme model and found $\sim$ 60 MeV. Brodsky, Ellis and Karliner[50] claim

$G^{(0)} \approx 0$ is natural in a three flavor Skyrme model. First, one must remember that these results call into question the many successes[53] of the usual two-flavor Skyrme model which, of course, ignores any strangeness content of the nucleon. Second, no one has yet looked at $\langle N|\bar{s}\gamma^{\mu}D^{\nu}s|N\rangle$ in the Skyrme model. This matrix element can be estimated using methods developed by Goldstone and Wilczek[54] and it is crucial for the model that $\langle N|\bar{s}\gamma^{\mu}D^{\nu}s|N\rangle$ be small even though $m_s\langle N|\bar{s}s|N\rangle$ and $\langle Ns|\bar{s}\gamma_{\mu}\gamma_5 s|Ns\rangle$ are large. Finally, these matrix elements probe the flavor singlet sector of the Skyrme model which has always been one of its weak points. For example, the result of Ref. [50] is that $G^{(0)} \to 0$ as $N_c \to \infty$, but so should $M_{\eta'}/M_N \to 0$ as $M_c \to \infty$, and yet $M_{\eta'}/M_N > 1$ in the real world. It would be very interesting to know the first non-vanishing contribution to $G^{(0)}$ in (generalized) Skyrme models.

IV.1 Inhomogenous Nucleosynthesis

In a sleepy parallel session, towards the end of the conference, Applegate[51] presented some lively speculations concerning the effect of a first order confining QCD phase transition on nucleosynthesis in the early universe. The story begins at a temperature just above the phase transition temperature, T_c. Quarks and gluons are unconfined and distributed more or less homogeneously. At or just below T_c, bubbles of new, confining phase nucleate and grow. As Witten argued some years ago, baryon number is less soluble in the new phase than the old: in the confining phase each net unit of baryon number costs a free energy $f \sim M_N$ which is much greater than the free energy of a baryon in the deconfined phase $f_0 \sim T_c$. Baryons, then, tend to accumulate outside the bubbles of new vacuum. Eventually, when the phase transition is complete, protons and neutrons are distributed in lumps. The average size of these lumps depends sensitively on the amount of supercooling and other details of the phase transition,[52] but it is likely to be macroscopic even though the underlying dynamics is on the fermi scale.

Later, as the universe cools, neutrons diffuse but protons, coupled electromagnetically to the plasma, do not. As a result, at the time of nucleosynthesis, there are both relatively neutron rich and neutron poor regions. Nucleosynthesis proceeds rather differently in neutron rich regions. More deuterium and ^{7}Li and less ^{4}He are formed than in a homogeneous model with the same Ω_B. The most interesting outgrowth of this line of speculation is its implication for dark matter candidates. Standard, homogeneous nucleosynthesis can account for primordial deuterium and ^{4}He abundance only if $\Omega_B \sim 0.1$. This suggests that much of the mass of the universe may be in the form of some exotic, dark matter. Inhomogenous, neutron-rich nucleosynthesis may be able to account for deuterium and ^{4}He abundances with $\Omega_B \sim 1$. However, the scenario may have trouble with so-called "r-process" elements, heavy elements formed in neutron rich regions. Inhomogeneities large enough to give $\Omega_B \sim 1$ may give too large a primordial abundance of r-process elements.

This idea is in its infancy. If the QCD confining phase transition led to significant inhomogeneities at the time of primordial nucleosynthesis, then many aspects of nuclear cosmology will have to be considered.

IV.2 Point Sources of Ultra-High Energy Cosmic Rays

In the same parallel session, E. Loh[53] reported on the observation of cosmic ray air showers with energies in excess of 5×10^{17} eV, with the "Fly's Eye" detector. For the first time (to my knowledge) it appears possible to identify specific sources of these ultra-high energy particles. Using timing and the geometry of the detector array, it is possible for the Fly's Eye to determine the direction of the primary with a resolution of order a degree. The data were converted into a sky map by plotting each event on the celestial sphere as an elliptical Gaussian density function. The data were then compared with randomized "background" data. The resulting distribution shows a significant excess in the galactic plane. Within the plane, there is an apparent excess in the direction of Cygnus X-3! The data show other significant variations and certainly require further analysis. Loh also showed a contour map of region around Cygnus X-3 which appeared to show a significant excess. There is some indication that this signal is correlated with the Cygnus X-3 4.8 hour orbital period peaking at .3 and .6 in the phase.

Only a neutral particle could be responsible for these events. Despite the extremely high energy, any charged particle would curve so much in the galactic magnetic field that it would no longer be correlated with the source direction. Photons are one candidate. Neutrons are another: at this energy the survival rate for neutrons travelling $\gtrsim$ 30,000 light years from Cygnus X-3 (with a $\sim$ 1,000 sec lifetime) is greater than 10%. The Fly's Eye will soon install muon chambers to help distinguish hadron and photon initiated showers. This report provided a glimpse of future Intersection Conferences when astrophysics in the extreme reaches of the spectrum will be a significant topic.

REFERENCES

1. D. Marlow, these proceedings.
2. V. V. Varmin *et al.*, *Nucl. Phys.* **B247**, 293 (1984).
3. Particle Data Group, *Rev. Mod. Phys.* **56**, 59 (1984).
4. M. Pert, these proceedings.
5. K. Schubert, these proceedings.
6. A. Altarelli, M. Cabibbo, G. Corbo and L. Maiani, *Nucl. Phys.* **B207**, 3651 (1982);
7. B. Grinstein, N. Isgur and M. Wise, *Phys. Rev. Lett.* **56**, 258 (1986).
8. T. Humanic, these proceedings.
9. N. Isgur, these proceedings.
10. N. Isgur and C. H. Llewellyn-Smith, to be published.
11. A. Chodos, these proceedings.

12. W. Koenig, these proceedings.

13. FERMILAB Experiment E605.

14. SLAC Experiment E141.

15. E. Danzmann, these proceedings.

16. D. Alde, these proceedings.

17. H. Koch, these proceedings.

18. W. Dunwoodie, these proceedings.

19. W. Wisniewski, these proceedings.

20. S. Blessing, these proceedings.

21. A. Barker, these proceedings.

22. S. Meshkov, these proceedings.

23. W. H. H. Panofsky, in *Proceedings of the 14th International Confrence on High Energy Physics* (Vienna, 1968).

24. P. Renton, these proceedings.

25. J. J. Aubert *et al.*, *Phys. Lett.* **123B**, 275 (1983).

26. R. Boested, these proceedings.

27. H. Jung and G. A. Miller, *Phys. Lett. B* **200**, 251 (1988); M. I. Strickman and L. L. Frankfurt, *Phys. Let.* **B183**, 253 (1987); R. L. Jaffe, in *Relativistic Dynamics and Quark Nuclear Physics*, M. B. Johnson and A. Picklesheimer, eds. (Wesley, New York, 1986), p. 537.

28. K. Rith, these proceedings.

29. R. L. Jaffe and M. Soldate, in *Perturbative Quantum Chromodynamics*, D. W. Duke and J. F. Owens, eds. (Tallahassee, 1981); *Phys. Lett.* **105B**, 467 (1981); *Phys. Rev.* **D26**, 49 (1982); S. P. Luttrell, S, Wada and B. R. Webber, *Nucl. Phys.* **B188**, 219 (1981).

30. R. L. Jaffe, *Nucl. Phys.* **B229**, 205 (1983); *Phys. Lett.* **116B**, 437 (1982).

31. R. K. Ellis, W. Furmanski and R. Petronzio, *Nucl. Phys.* **B207**, 1 (1982); **B212**, 29 (1983); J. Qiu, Argonne preprint ANL-HEP-PR-88-10.

32. H. Georgi and A. Manohar, *Nucl. Phys.* **B234**, 189 (1984).

33. H. Abramowicz, *et al.* (CDHS), *Z. f. Physik* **C15**, 19 (1982); K. Lang *et al.* (CCFRR), paper submitted to the 1985 International Symposium on Lepson and Photon Interactions at High Energies (Kyoto, 1985). For a review see F. Sciulli, in *Proceedings of the 1985 International Symposium on Lepton and Photon Intersections at High Energies*, M. Konuma and K. Kakahashi, eds. (Kyoto, February, 1986).

34. D. Kaplan and A. Manohar, Harvard University preprint HUTP-88/A024.

35. J. Ellis and R. L. Jaffe, *Phys. Rev.* **D9**, 1447 (1974).

36. J. Ashman *et al.* (EMC collaboration), *Phys. Lett.* **206B**, 364 (1988).

37. F. E. Close and R. G. Roberts, *Phys. Rev. Lett.* **60**, 1471 (1988).

38. I thank P. Renton for this fit.

39. A. Kuti, MIT preprint CTP #234 (1971), unpublished; L. Galfi, P. Gnadig, J. Kuti, F. Niedermayer and A. Patkos, *Acta Phys. Acad. Sci.* (Hung.) **31**, 85 (1971).

40. L. A. Ahrens *et al.*, *Phys. Rev.* **D35**, 785 (1987).

41. C. L. Korpa and R. L. Jaffe, *Comm. Nucl. Part. Phys.* **17**, 163 (1987); G. Gasser, talk presented at the $\pi N(2)$ Workshop at Los Alamos, Bern preprint BUTP-87/13.

42. T. E. O. Ericson, *Phys. Lett.* **195B**, 116 (1987)

43. V. Bernard, R. L. Jaffe and U.–G. Meißner, *Phys. Lett.* **198B**, 92 (1987), MIT preprint MIT-CTP-1547, to be published in *Nuc. Phys. B*.

44. D. B. Kaplan and A. E. Nelson, *Phys. Lett.* **175B**, 57 (1986), **179B** (E), 409 (1986).

45. R. L. Jaffe, *Phys. Lett.* **193B**, 101 (1987).

46. J. Schwinger, *Phys. Rev.* **82**, 664 (1951).

47. J. Kodaira, *Nucl. Phys.* **B165**, 129 (1979).

48. H. J. Melosh, *Phys. Rev.* **D9**, 1095 (1974).

49. J. Donoghue and C. Nappi, *Phys. Lett.* **168B**, 105 (1986).

50. S. J. Brodsky, J. Ellis and M. Karliner, *Phys. Lett.* **206B**, 309 (1988).

51. J. Appelgate, these proceedings, J. Appelgate, C. Hogan and R. Scherrer, *Phys. Rev.* **D35**, 1151 (1987).

52. G. M. Fuller, G. J. Mathews and C. R. Alcock, *Phys. Rev.* **D37**, 1350 (1988).

53. E. Loh, these proceedings.

Spin-Isospin Nuclear Response

K.H. Hicks
TRIUMF, 4004 Wesbrook Mall, Vancouver, B.C., V6T 2A3

ABSTRACT

The scattering of protons off individual nucleons in the nucleus, known as 'quasi-free' scattering, gives information on the modification of the nuclear force inside the nuclear medium. In order to extract this information, we need to know the nucleon-nucleon (N-N) interaction in free space, and the propagation of the proton through the nucleus (from elastic scattering). Direct comparison of the analyzing powers and polarizations for quasi-free scattering as opposed to N-N scattering show large (40%) differences. Conventional (non-relativistic) calculations are unable to reproduce this effect. Relativistic calculations, which use large, opposing vector and scalar potentials reproduce this effect for both isoscalar and isovector inclusive reactions and for exclusive $(p, 2p)$ reactions.

INTRODUCTION

A fundamental question in nuclear and particle physics is how the nuclear medium modifies the strong (and weak) force compared with what is found in free space. An example of so-called 'medium modifications' is the well-known EMC effect,[1] where the structure function for scattering of muons from ^{54}Fe is very different from scattering of muons from deuterium. In the present study, we investigate medium modifications for proton-nucleus scattering at intermediate energies (200-500 MeV). Rather than looking for quark effects, our goal is to learn about the force on a nucleon in nuclear matter, and to test existing models of proton-nucleus scattering.

We will concentrate on 'quasi-free' (or quasielastic) scattering. This region of excitation energy in the (p, p') spectrum lies above the giant resonances, as shown in Fig. 1a. The quasi-free continuum results from knock-out of a nucleon from the nucleus by the projectile, as shown in Fig. 1b. This is essentially doing nucleon-nucleon (N-N) scattering in the nuclear medium, which can be compared to N-N data in free space. This gives a direct measure of the modification of the N-N interaction by the nuclear medium.

More explicit information can be obtained by measuring the $(p, 2p)$ reaction, where the original orbital of the knocked-out proton can be reconstructed from the measured energies and angles of both protons. The largest contribution to inclusive (p, p') quasielastic scattering will be from surface orbitals, whereas the exclusive $(p, 2p)$ reaction will separate contributions from surface orbitals and orbitals deeper inside the nucleus.

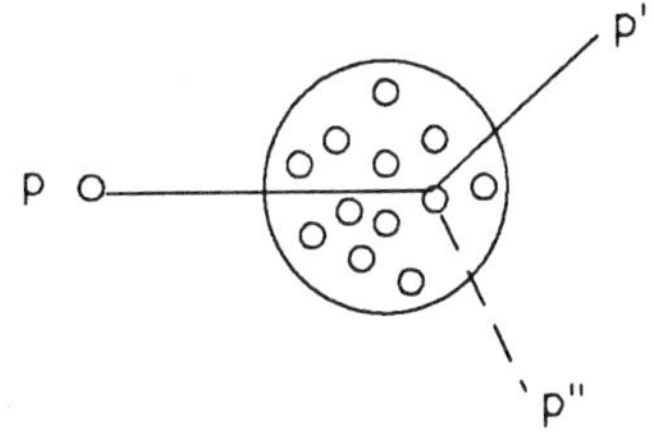

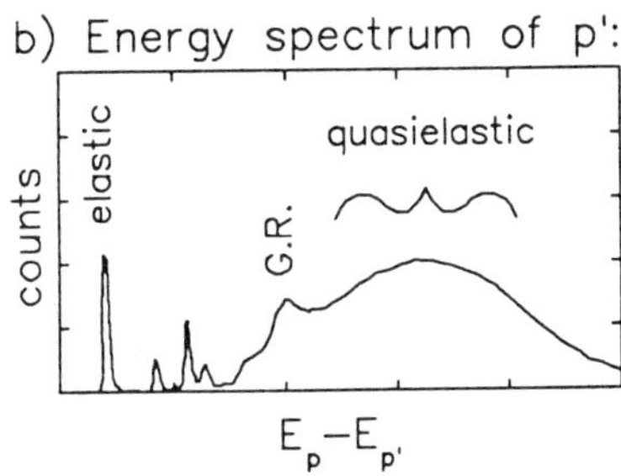

Fig. 1 a) Schematic representation of quasielastic scattering. b) Energy spectrum for the inclusive (p, p') reaction.

In order to extract information about nuclear matter (and *how* the N-N force gets modified), it is useful to examine the success or failure of various models of the proton-nucleus force. Also, we should examine models of free-space N-N scattering, for comparison with quasi-free scattering. To build a model of the proton-nucleus force, we need a description of nuclear matter, and a description to propagate the proton through nuclear matter. The former can be tested independently using static properties of the nucleus, and the latter can be tested with elastic proton-nucleus scattering data. Existing theories of nuclear matt[illegible] reviewed, where we look for a consistent description of both quasi-free scattering and nuclear saturation properties.

Nucleon-Nucleon Scattering

In principle, all we need are the N-N data for comparison with quasielastic data to observe medium modifications. If we are to understand *why* the nuclear medium modifies the N-N force, we need to understand something about how the N-N interaction works.

A proper description of the N-N force is one of the oldest problems in nuclear physics. The traditional approach[2] is to build a model with an exchange of mesons, and use the Feynman rules to calculate a cross section. This approach works reasonably well for the long-range part of the N-N force (where single pion-exchange dominates), but fails for the short-range part of the force where the hard "core" cannot be reproduced. Our present understanding that the nucleon is made up of 3 quarks gives some insight into this failure, and dooms the pure meson-exchange model unless quark effects are somehow included. Furthermore, the proton is not a point particle, which is implicitly assumed in the Feynman rules. Both of these problems can be approximately fixed by introducing[2] form factors (or "cut-off" parameters) at the vertex of the Feynman diagrams. With this *caveat*, a very successful picture of the N-N interaction at energies below

300 MeV has been built,[2] known as the Bonn potential.

The Bonn potential is not the only model of the N-N interaction, but it is perhaps the most satisfying one at energies below 300 MeV. This is because it uses only real (experimental) meson masses, along with the well-established Feynman field theory formalism, with only a dozen or so fitted parameters. Also, the set of diagrams provided by a field theoretic model forms a sound basis for a consistent generalization to many-body forces[2] which lead to medium modifications (this is not true for a dispersion-relation model such as the Paris potential). It is important to point out that the Bonn potential uses of the *relativistic* Blankenbecler-Sugar equation in order to solve the ambiguity of getting a potential from 2-pion exchange. This attractive scalar potential is large and opposes the repulsive vector potential from the ω-meson. These same features will arise again later in this paper. The 2-pion exchange is often replaced, with good approximation, in a one-boson exchange potential (OBEP) by a fictitious σ-meson to simplify the calculations. In principle, one could always use the full potential if questions of this approximation arise.

The fits to the N-N phase shifts (below 300 MeV) by the Bonn potential are excellent, as shown in Ref. 2. It should be noted that another good description of N-N scattering is given by the recent model of van Faassen and Tjon,[3] which may be valid at energies as high as 1 GeV. This demonstrates the usefulness of applying a meson-exchange theory for the intermediate-energy N-N interaction. Hence, the N-N interaction below 300 MeV is sufficiently well understood for our purposes here to compare with quasi-free scattering. We can now examine how the nuclear medium modifies this force in a semi-microscopic manner.

Nuclear Matter

An even older problem than finding a proper theoretical model for N-N scattering is that of a proper description of nuclear matter. The static properties of nuclear matter have been known for some time; they are: 1) the binding energy per nucleon $E/N \simeq 16$ MeV, 2) the density in the center of heavy nuclei $\rho \simeq 0.17\, fm^{-3}$, and 3) the compression modulus $K \simeq 210$ MeV. The density is often given in terms of its equivalent fermi momentum of $k_F \simeq 1.35\, fm^{-1}$.

The conventional approach[2] to this problem is to start with a realistic N-N potential, and apply Brueckner theory for the many-body description. Recent calculations of this type have been done by Day,[4] and are also given in Ref. 2. The results are summarized in Fig. 2, where the dashed lines show the conventional non-relativistic results, and the open symbols show the saturation minima for different N-N potentials. A so-called Coester band is formed, for which no N-N potential predicts nuclear matter saturation correctly. Extensions such as isobaric degrees of freedom or many-body forces do not help this situation.[2]

Relativistic calculations of nuclear matter, which use the Dirac equation for the single-particle motion, can be applied to the Brueckner formalism. Calculations of this type are also shown in Fig. 2 (from Ref. 2) by the solid lines, which correctly reproduce the saturation properties of nuclear matter. These calculations use a OBEP which produce essentially the same N-N scattering (and deuteron properties) as the full Bonn potential; similar results are obtained by replacing the σ-meson with the more realistic 2-pion exchange.[2]

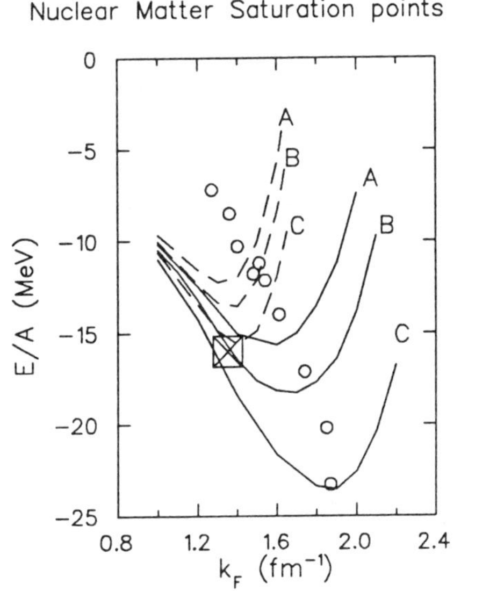

Fig. 2 Conventional (solid curves and points) and relativistic (dashed) Brueckner calculations[2] for different N-N potential parameters.

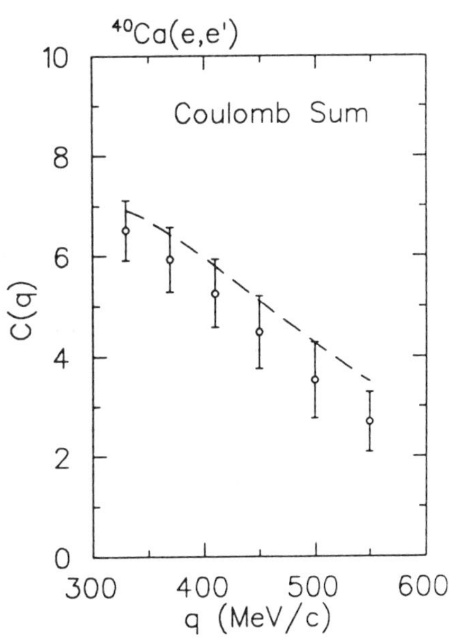

Fig. 3 Integrated longitudinal response from ^{40}Ca(e, e') with the calculation of Horowitz.[5]

It is useful to note that a relativistic approach led to the success of both N-N scattering and nuclear matter models. The Dirac equation is given by

$$[\not{p} - V_v - (M + V_s)]u_s(p) = 0 \tag{1}$$

where

$$u_s(p) = \sqrt{\frac{E^* + M^*}{2M^*}} \begin{pmatrix} 1 \\ \frac{\vec{\sigma}\cdot\vec{p}}{E^*+M^*} \end{pmatrix} \chi_s \tag{2}$$

with $M^* = M + V_s$ and $E^* = \sqrt{p^2 + (M^*)^2}$. The success of the Dirac-Brueckner approach comes from the extra binding due to the decrease of M^* with increasing density ($V_s = 0$ in free space). The large scalar potential V_s is essential to the success of nuclear matter saturation, and is related to the attractive scalar potential from 2-pion exchange in N-N scattering. It is interesting to look for other signatures of the large and opposing scalar and vector potentials, which are relativistic effects. Specifically, we are interested in medium modifications due to these relativistic effects.

The integrated longitudinal response function for quasielastic electron scattering is known to be substantially less (by about 50%) than the Coulomb sum rule value of Ze^2 (the number of protons times the square of the electron charge). Recent calculations[5] taking into account the change in the vacuum polarization response of the Dirac sea because of the decrease in the relativistic effective mass M^* are shown in Fig. 3 by the solid curve compared to data for ^{40}Ca. These calculations are in much better agreement with the data than previous non-relativistic calculations that give essentially the Coulomb sum rule value. We observe that the same feature that gives proper saturation of nuclear matter also modifies the response of nuclear matter to virtual photons in agreement with experiment. We will look for similar features in quasielastic proton scattering.

Proton-Nucleus Elastic Scattering

Most of the research on intermediate energy proton-nucleus scattering in the past decade has concentrated on elastic scattering, and excitation of the low-lying discrete states and giant resonances.[6] The most dramatic success of models of proton-nucleus scattering has been for elastic scattering.[7,8] Models based on both the Schrödinger equation and the Dirac equation give calculations that predict all observables (cross section $d\sigma/d\Omega$, analyzing power A_y, and spin-rotation parameter Q) of elastic proton-nucleus data, usually within 10% (or better). These calculations work for a range of nuclei at incident proton energies between 200-800 MeV.

Elastic scattering calculations are usually performed in the impulse approximation (IA), which is essentially a single-scattering approximation. This approximation is valid only above 100 MeV, where the N-N force approaches a minimum. From the point of view of quasielastic scattering, where we want to see protons scattered from a single nucleon in the nucleus, the success of the IA for elastic scattering in the intermediate energy range is important.

The ingredients for a model of elastic scattering in the IA are: 1) a description of the N-N interaction, and 2) a description of the nuclear density. Although more sophisticated models using Hartree calculations of the nuclear density (using results similar to the previous section on infinite nuclear matter, but for finite nuclei) have been done,[7,9] it is usually sufficient to use the nuclear charge density as determined by electron scattering.[10] It is particularly useful to use the charge density for non-closed shell nuclei where Hartree calculations often give poor results. Similarly, models using the full Bonn potential are available,[11] but it is sufficient to use a pseudo-potential.[9,10]

With these simple inputs, an excellent prediction of elastic scattering data is obtained, as shown in Fig. 4. The success of these calculations demonstrates the validity of the single-scattering approximation, and shows that we understand how to propagate intermediate-energy protons through the nucleus (*i.e.*, we know the 'distortions' of the proton wave-function as it enters the nuclear environment).

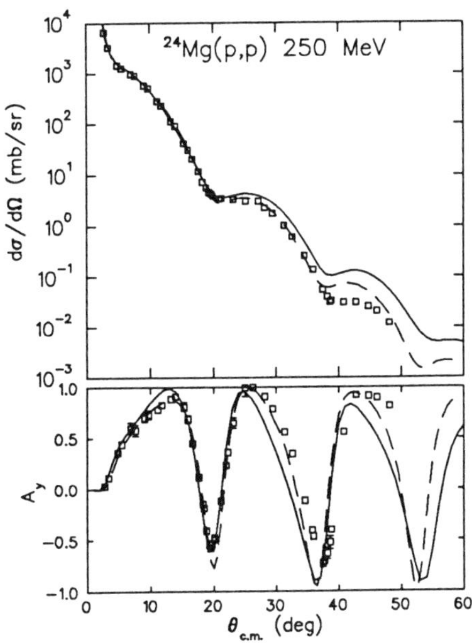

Fig. 4 Sample of elastic proton scattering data with nonrelativistic (solid) and relativistic (dashed) calculations.

Both relativistic and non-relativistic models of elastic scattering are compared to the data in Fig. 4. Both models do quite well, and no dramatic relativistic effects are observed (except for better agreement with the spin-observables at forward angles). This was *not* the case for earlier calculations that did not include knock-on exchange and density dependent effects (the relativistic model was favored above 400 MeV for these earlier calculations). Elastic scattering is sensitive mainly to the central force, which is derived (in the relativistic model) from a cancellation between the large scalar and vector potentials (from the σ and ω mesons). No large effect is seen from the effective mass M^* of a nucleon *in the nucleus* from a proton which is passing by (we note that elastic scattering occurs at the surface, and is not sensitive to the nuclear interior, due to the strong N-N interaction causing absorption from the elastic channel).

Quasielastic Scattering

Some of the earlier (p, p') data for the quasielastic region were taken at LAMPF, notably the S_{nn} measurements by Glashausser[12] and the complete set of spin-observables by Carey.[13] Measurements of S_{nn} , in principle, determine the spin-response of the nucleus (*i.e.*, how the nuclear environment changes the amount of $\Delta S = 1$ spin-flip of the incident proton compared with the amount of spin-flip from free N-N scattering). The results of Ref. 12 suggest that the amount of non-spin-flip $\Delta S = 0$ transitions are heavily suppressed, compared with the amount of $\Delta S = 0$ seen for N-N scattering.[12] However, these results are model dependent, and appear only at certain angles. Similar results, but to a lesser extent, are seen for TRIUMF data.[14] A conventional explanation for the suppression of $\Delta S = 0$ is due to the RPA response of the nucleus (see next section). The RPA response helps, but does not fully reproduce the data.[12,14]

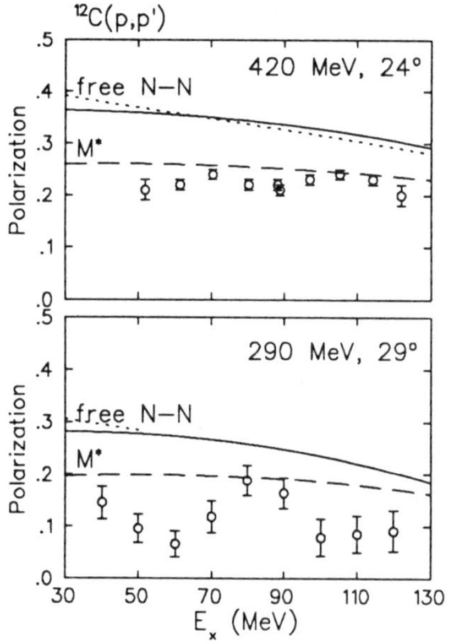

Fig. 5 Data from Ref. 15 along with free N-N (solid), RPA (dotted) and relativistic model[16] (dashed) curves.

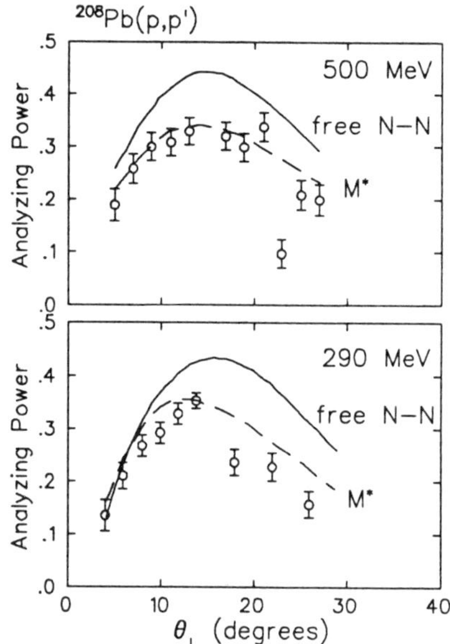

Fig. 6 Data from Ref. 17 as a function of proton lab angle for the quasielastic peak. See text for curves.

The complete set of spin-observable measurements[13] showed little deviation from the free N-N values, except for the polarization, which is substantially suppressed. Since the polarization is one of the simplest observables, it is worth examining this effect in more detail. The spin-observables for quasielastic scattering on ^{12}C has been measured at TRIUMF by the University of Toronto group,[15] and a similar suppression of P is seen. These data are presented in Fig. 5 as a function of the excitation energy of the nucleus, where the solid line represents the isospin-averaged polarization for free N-N scattering. This rather surprising suppression is evidence for strong medium modifications of the N-N force. These results are true for both ^{208}Pb at 500 MeV[13] and for ^{12}C at both

420 and 290 MeV. Also shown by the solid lines in Fig. 5 are the predictions from a relativistic model[16] which uses an effective mass of M^* for a nucleon in the nucleus, similar to the description of nuclear matter given above. This model will be discussed in the next section.

Further evidence for this effect is seen in measurements of the analyzing power A_y for quasielastic scattering on ^{208}Pb. These data, measured at TRIUMF by the Oregon group,[17] are shown in Fig. 6 for incident proton energies of 290 and 500 MeV. In this figure, the A_y is plotted as a function of proton lab angle θ_L for a bin of excitation energy near the peak of the quasielastic cross section. Similar to the polarization measurements, these data show significantly reduced values of A_y . Also shown in Fig. 6 are calculations from the relativistic model[16] given by the dashed curves. These calculations predict the correct magnitude (and shape) for the angular distribution. No realistic nonrelativistic calculation to date has been able to reproduce these data.

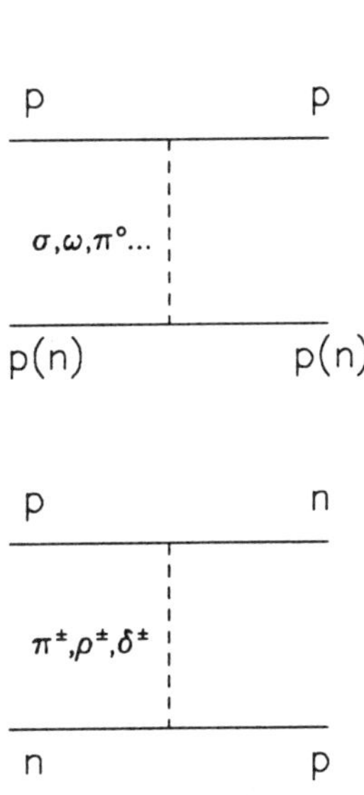

Fig. 7 Schematic representation of the direct meson-exchange part to the isoscalar (top) and isovector (bottom) channels.

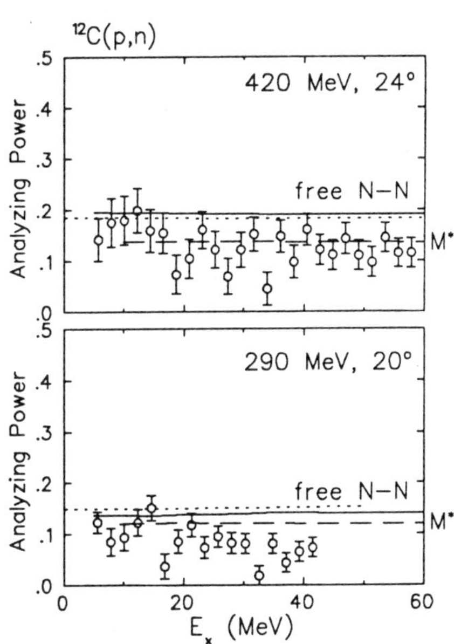

Fig. 8 Data from Ref. 18 for the isovector charge-exchange reaction. See Fig. 5 for explanation of curves.

The (p, p') data in Fig. 5 and 6 are sensitive mainly to the isoscalar part of the N-N force from exchange of σ and ω mesons. The isovector part of the N-N force from exchange of charged mesons can be investigated with the (p, n) reaction (at higher energies where knock-on exchange becomes less important); this is shown schematically in Fig. 7. Recent data[18] for A_y in the quasielastic region for the (p, n) reaction are shown in Fig. 8. The target, angles, and energies were chosen to match the (p, p') data shown in Fig. 5. The free N-N values of A_y , given by the solid lines, are smaller for the charge-exchange channel than for the isoscalar channel. The medium modifications, given by the difference between the data and the solid lines, are correspondingly smaller. Again, the point to be made is that the data show significant medium modification effects, independent of any model. If we look at the model predictions, we see the relativistic calculations[16] (dashed lines) are in reasonable agreement with the data, whereas the nonrelativistic model with an RPA response[19] (dotted lines) are not very different from the free N-N values.

Models of Quasielastic Scattering

The nonrelativistic surface response model of Esbensen and Bertsch[20] has been successful in predicting the quasielastic cross sections for nucleon-nucleus scattering. This model has been extended by Smith[19] and compared with (p, n) cross sections with good results. This same model has been compared with a complete set of spin-observables from $^{54}Fe(p, p')$ quasielastic data,[14] and does not predict the absolute magnitude of the D_{ij} but give slopes (as a function of excitation energy) in good agreement with the data. This is particularly true for both the polarization and the analyzing power of the ^{54}Fe data, where the data are suppressed by about 50% compared to the calculations. The slopes of these calculations result from intermediate particle-hole excitations of the nucleus, as shown in Fig. 9. These are modifications of the N-N force due to the nuclear medium, and are calculated with the random phase approximation. This is the so-called RPA response of the nucleus.

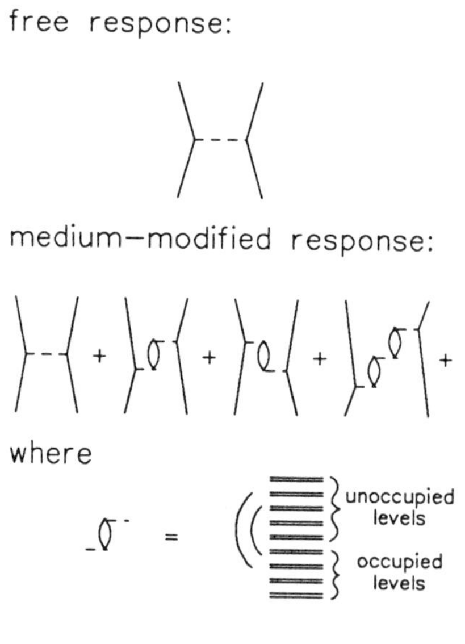

Fig. 9 Schematic representation of the RPA response of the nucleus.

The relativistic model of Horowitz[16] does not include an RPA response, but rather concentrates on relativistic effects from the large, opposing vector and scalar potentials in the nucleus. It is found that these relativistic effects change the absolute magnitude of the spin-observables in the direction required by the data. Thus, the calculated polarization (and also the analyzing power) are suppressed by 30-40%. The reason for this is evident from examining the form for A_y at small angles,

$$A_y \simeq \frac{p^2 \mathrm{Im}[t_s^* t_v] \sin\theta_L}{|M^* t_s + E^* t_v|^2} \tag{3}$$

where M^* and E^* are the same as in equation (2) and t_s and t_v are the scalar and vector amplitudes from σ and ω exchange in the N-N interaction. As M^* decreases, the sensitive cancellation between t_s and t_v is changed, thereby decreasing A_y.

The large effects of the RPA response for heavier nuclei suggest that the RPA should be put into the relativistic model. Work on a proper relativistic RPA is in progress,[21] and results for quasielastic calculations are expected soon. Other improvements, such as 2-step (2-nucleon knock-out) contributions, which are included in nonrelativistic calculations, should also be investigated in a relativistic model.

The Exclusive $(p, 2p)$ Reaction

The effects seen in quasielastic scattering should also be evident in the exclusive $(p, 2p)$ reaction where both the scattered and the knocked-out proton energies and angles are measured. Peaks corresponding to knock-out from specific orbitals in the nucleus are found in the summed-energy spectrum of both detected protons. This allows an examination of the density dependence of medium modifications.

Data[22] for the analyzing power for $(p, 2p)$ from a ^{40}Ca target at 300 MeV are shown in Fig. 10 for knock-out from the $d_{5/2}$ and $d_{3/2}$ orbitals. These data are compared to the zero-range nonrelativistic and relativistic calculations of Kudo.[23] Also shown are preliminary exact finite-range calculations that are being developed by Cooper and Maxwell.[24] Both relativistic calculations show a large difference from the nonrelativistic calculations, and are in better agreement with the data. The difference is attributed[24] to distortions from the large, opposing vector and scalar potentials in the nucleus. Also available are new $(p, 2p)$ data[25] for an ^{16}O target at 500 MeV, which are shown in Fig. 11 for knock-out from the $s_{1/2}$ and $p_{1/2}$ orbitals. The calculations shown were done by Miller and Dymarz,[25] and show effects similar to the ^{40}Ca data. Finite-range calculations for these data are expected shortly.

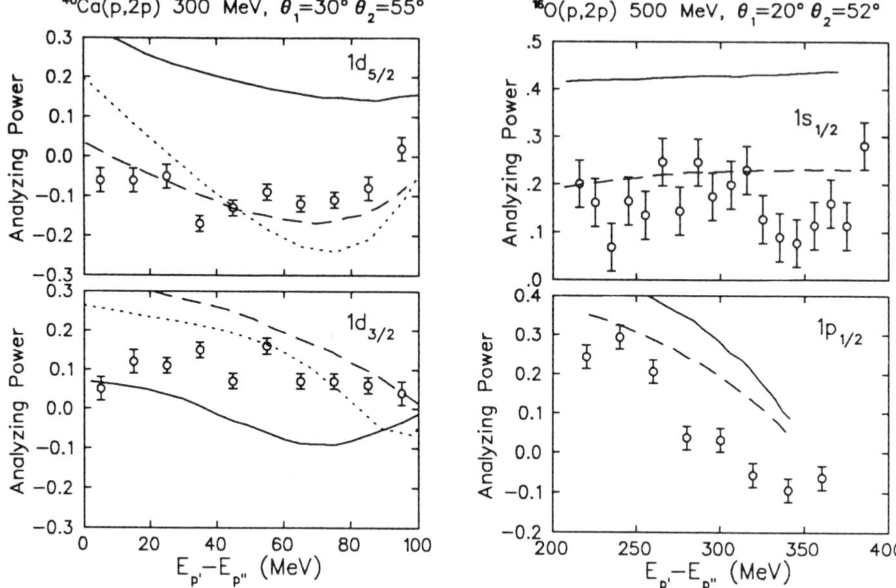

Fig. 10 Data from Ref. 22 along with nonrelativistic (solid) and relativistic (dashed and dotted) calculations.

Fig. 11 Data from Ref. 25 along with nonrelativistic and relativistic calculations in a different model.

Summary and Conclusions

The spin-response of the nucleus shows modifications to the N-N force inside the nuclear medium. The largest effects are seen for the analyzing power and the polarization for quasielastic scattering as compared to the A_y for free N-N scattering. The suppression of A_y is seen for both the isoscalar (p,p') and for the isovector (p,n) isospin channels. These data are in good agreement with predictions from relativistic models of proton-nucleus scattering, as opposed to nonrelativistic models which are close to the free N-N values of A_y . The success of the relativistic model is due to the large, opposing vector and scalar potentials that are calculated with the impulse approximation. These potentials are related to the strong repulsive force from ω-exchange and to the strong attractive force from 2π-exchange in meson-exchange models of N-N scattering. These potentials are consistent with those needed to give the proper saturation properties of nuclear matter for relativistic Brueckner calculations. Realistic nonrelativistic Brueckner calculations cannot reproduce these results.

Strong relativistic effects are also seen for $(p,2p)$ calculations of A_y above

300 MeV where the data are quite different from expectations based on free N-N scattering. More $(p, 2p)$ data above 300 MeV (perhaps with other spin-observables) would be helpful in examining in detail the spin-isospin response of the nucleus.

References

[1]J.J. Aubert, et al., Phys. Lett. **123B**, 275 (1983); K. Rith, contribution to this conference.

[2]R. Machleidt, Adv. Nucl. Phys. **19**, J.W. Negele and E. Vogt, eds., Plenum Press, New York, 1988.

[3]E.E. van Faassen and J.A. Tjon, Phys. Rev. **C30**, 285 (1984).

[4]B.D. Day, Phys. Rev. **C24**, 1203 (1981).

[5]C.J. Horowitz, submitted to Phys. Lett. B.

[6]G.R. Satchler, *Direct Nuclear Reactions*, OUP, Oxford, 1983.

[7]J.R. Shepard, J.A. McNeil, and S.J. Wallace, Phys. Rev. Lett. **50**, 1443 (1983);

[8]L. Ray and G.W. Hoffmann, Phys. Rev. **C31**, 538 (1985).

[9]D.P. Murdock and C.J. Horowitz, Phys. Rev. **C35**, 1442 (1987).

[10]K.H. Hicks, et al., accepted in Phys. Rev. C.

[11]K. Nakayama and W.G. Love, submitted to Phys. Rev. C.

[12]C. Glashausser, et al., Phys. Rev. Lett. **58**, 2404 (1987).

[13]T.A. Carey, et al., Phys. Rev. Lett. **53**, 144 (1984).

[14]M. Vetterli, contribution to this conference.

[15]C. Chan, et al., submitted to Phys. Lett. B.

[16]C.J. Horowitz and D.P. Murdock, Phys. Rev. **C35**, 1442 (1987); C.J. Horowitz and M.J. Iqbal, Phys. Rev. **C33**, 2059 (1986).

[17]X.Y. Chen, et al., Phys. Lett. **205B**, 436 (1988), and private communication.

[18]K.H. Hicks, et al., to be published.

[19]R.D. Smith and J. Wambach, Phys. Rev. **C36**, 2704 (1987).

[20]H. Esbensen and G.F. Bertsch, Phys. Rev. **C34**, 1419 (1985).

[21]Conference on Spin and Isospin Excitations of the Nucleus, Telluride, 1988.

[22]P. Kitching, et al., in *Advances in Nuclear Physics*, J.W. Negele and E. Vogt, eds. (Plenum, New York, 1985), Vol. 15, p. 43.

[23]Y. Kudo, to be published; see also Y. Kudo and K. Miyazaki, Phys. Rev. **C34**, 1192 (1986).

[24]E.D. Cooper and O.V. Maxwell, in Ref. 21.

[25]C.A. Miller, et al., to be published.

ASSOCIATED PRODUCTION OF HYPERNUCLEI WITH (π^+, K^+) REACTION

J. C. Peng
Los Alamos National Laboratory
Los Alamos, New Mexico 87545
USA

ABSTRACT

Status of associated production of hypernuclei is discussed. Recent (π^+, K^+) results from AGS to study the spectroscopy of Λ single particle states in heavy systems are presented.

I. INTRODUCTION

The status of hypernuclear physics was reviewed by Barnes[1] at the Steamboat Spring meeting and more recently by Gal[2] at the Lake Louise meeting. In this talk, I will not attempt to give a review on the whole field. Instead, I will focus on the subject of associated production of hypernuclei in which significant progress has been made recently. Other progress in hypernuclear physics have been reported in several recent conferences.[3,4]

Let us start by giving a simple definition of hypernucleus. A hypernucleus is a nucleus consisting not only of ordinary nucleons (protons and neutrons) but also of hyperons $(\Lambda, \Sigma, \Xi, \Lambda\Lambda, \ldots)$ or heavy-flavored baryons $(\Lambda_c, \Lambda_b, \ldots)$. At this time the existence of Λ-hypernuclei is established beyond any doubt. Somewhat controversial evidences on the existence of Σ-[5-7] and $\Lambda\Lambda$-hypernuclei[8] have also been reported. The exciting possibilities of studying heavy-flavored hypernuclei have been discussed,[9] but experimental evidence of such hypernuclei remains to be established.[10]

To appreciate the physics motivation for hypernuclear studies, it is instructive to make the analogy between hypernucleus and muonic atom. In the case of muonic atom, one implants a different type of leptons into an ordinary atom. The absence of Pauli-blocking allows muon to probe the interior of the electron orbitals. Similarly, a hypernucleus is formed by sticking a different type of baryon inside an ordinary nucleus. The hyperon can therefore dive deep inside the nucleus without being inhibited by the Pauli exclusion principle. This offers the opportunity to probe the nuclear interior, a region usually inaccessible by hadronic beams. Hypernuclear studies therefore allow us to address many interesting issues such as the modification of hyperon properties (lifetime, weak decay modes, magnetic moment, ...) inside a nucleus, and the alteration of nuclear properties by the presence of hyperon.

The history of hypernuclear physics can be roughly divided into three periods according to the experimental techniques used to produce hypernuclei. The first clear evidence of hypernucleus was reported[11] in an emulsion experiment in 1953. A ${}^{7}_{\Lambda}Li$ hyperfragment produced by K^- in flight was observed to travel some distance before decaying into ${}^4He + {}^3He + \pi^-$. For the next twenty years or so, all experimental data on hypernuclei were exclusively obtained from emulsion and bubble chamber measurements. These data provided valuable information on mass, spin-parity, lifetime, and decay modes of hypernuclei. Unfortunately, this technique is restricted to the study of ground state properties of light hypernuclei ($A < 16$). To remedy this situation, one entered the second period of hypernuclear physics.

In early 1970, the strangeness-exchange reaction (K^-, π) was successfully measured at CERN.[12] The selective excitation of the so-called substitutional states led to the determination[13] of small spin-orbit splitting in p-shell Λ-hypernuclei. The first evidence of Σ-hypernuclear states was later obtained in the (K^-, π) experiment.[5] Difficult coincidence experiments such as $(K, \pi\gamma)$[14] and $(K, \pi\ \mathrm{p})$[15] have also been performed to study electroweak decays of hypernuclei. Despite these interesting findings, the progress in hypernuclear physics suffered severely from the meagre kaon flux at existing facilities. Indeed, one of the major motivations[16] for constructing a high intensity accelerator is to obtain a much higher kaon flux for hypernuclear studies.

Before a kaon factory is built, it is important to explore new experimental techniques for hypernuclear studies. Various reactions including (p, K^+),[17] (π^+, k^+),[18] (γ, K^+),[19] $(\bar{p}, \bar{\Lambda})$[20] have been discussed as alternative ways to produce hypernuclei. These reactions all fall into the category of associated production of strangeness, in which a pair of $s\bar{s}$ quarks are produced with a non-strange beam. The (γ, K^+) reaction will be pursued at CEBAF and the (p, K^+) reaction is being studied at SATURNE II. The focus of this talk is on the (π^+, K^+) experiment, recently performed at the AGS accelerator. The success of the (π^+, K^+)

experiment marked the arrival of the latest period of hypernuclear physics, characterized by using non-strange beam to produce hypernuclei with associated production mechanism.

II. THE (π^+, K^+) REACTION

Theoretical aspects of the (π, K) reaction were first studied by Dover, Ludeking, and Walker.[18] In many ways, the (π^+, K^+) complements the (K^-, π) reactions. Table I shows the comparison between these two reactions.[21] Probably the most important difference between these two reactions is in their kinematics. Characterizing all associated production reactions, the (π, K) reaction imparts large momentum to the produced hyperon. Figure 1 shows the momenta transfer in the (π^+, K^+) and (K^-, π^-) reactions as a function of beam energy. The momentum transfer in the $n(\pi^+, K^+)\Lambda$ reaction is in general much greater than in the $n(K^-, \pi^-)\Lambda$ reaction. This implies that high-spin hypernuclear states are more favorably excited with the (π^+, K^+) reaction, in striking contrast to the selective excitation of low spin states by the (K^-, π^-) reaction at forward angles. The (π^+, K^+) reaction is also expected to probe the nuclear interior somewhat better than the (K^-, π^-) reaction, since the K^+ has a relatively long mean-free-path in nucleus. The peak $n(\pi^+, K^+)\Lambda$ cross section of 0.9 mb/sr, occurring around 1050 MeV/c, is roughly five times smaller than the peak cross section of the $n(K^-, \pi^-)\Lambda$ reaction. Nevertheless, Table I shows that the much higher pion flux more than compensates this difference and makes the (π^+, K^+) reaction more efficient than the (K^-, π^-) reaction for producing high-spin hypernuclear states. The good purity of the pion beam and the absence of kaon decay background also make the (π^+, K^+) experiment rather clean.

Feasibility of the (π^+, K^+) reaction for producing Λ-hypernuclei was first demonstrated in an experiment[22] performed at AGS. Figure 2 shows that the ${}^{12}_{\Lambda}C$ ground state and the excited state at $E_x = 11$ MeV are clearly observed. DWIA calculations[21,23] using realistic optical potentials and spectroscopic factors describe the observed cross sections well. The success of this pioneering experiment suggests that the (π^+, K^+) reaction is ideal for studying heavy hypernuclei, since the states in heavy hypernuclei are usually of high-spin nature and should be favorably excited by the (π^+, K^+) reaction. Of particular interest are the deeply-bound states in heavy hypernuclei. The possibility that partial-deconfinement of Λ, caused by the u and d quarks being Pauli-blocked, could

TABLE I. Comparison between the $n(k^-, \pi^-)\Lambda$ and the $n(\pi^+, K^+)\Lambda$ Reactions

	$n(k^-, \pi^-)\Lambda$	$n(\pi^+, K^+)\Lambda$
momentum transfer	<100 MeV/c	≈300 MeV/c
selectivity	low spin states	high spin states
mean free path	$\pi^- = 1.9$ fm $k^- = 2.0$ fm	$\pi^+ = 1.6$ fm $k^+ = 4.6$ fm
cross section	$\sigma_{max}(0^0) = 5$ mb/sr	$\sigma_{max}(0^0) = 0.9$ mb/sr
beam intensity (purity)	$< 10^5$/beam spill (k/all ≃ 7%)	$\approx 10^7$/beam spill (π/all ≃ 80%)

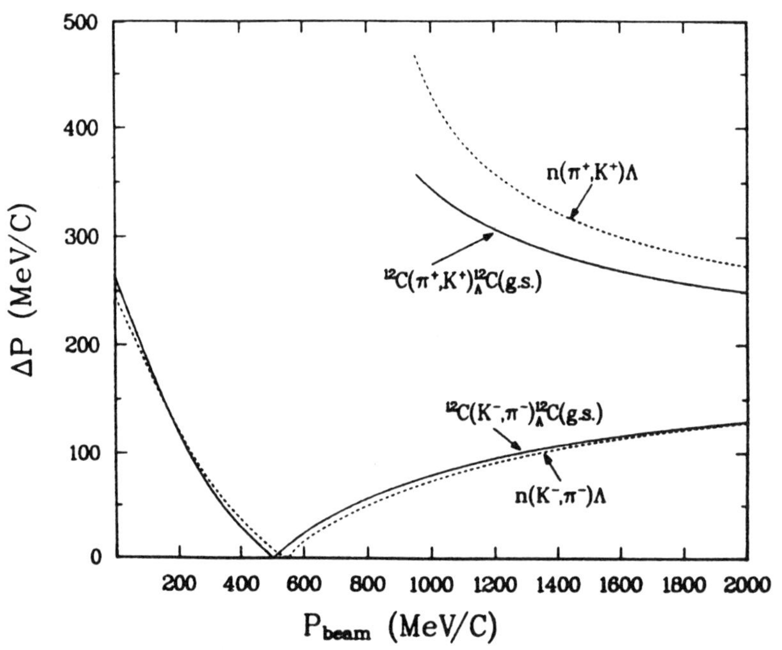

Fig. 1. Momentum transfer for various reactions at $\theta_{lab} = 0^0$.

occur for deeply bound hypernuclear states has been discussed.[24,25] One possible manifestation of such partial-deconfinement effects could be an anomalously small binding energy such as observed for ${}^{5}_{\Lambda}He$.[24] The existing (K^-, π^-) data on heavy nuclei[26] provide almost no information on the bound states. In fact, the ground state binding energy of Λ-hypernuclei is only known accurately up to ${}^{15}_{\Lambda}N$. Two (π^+, K^+) experiments were proposed at KEK[27] and at AGS[28] to extend the study of (π^+, K^+) reaction to heavy nuclei. In the following, we will discuss the (π^+, K^+) experiment recently performed at AGS.[28]

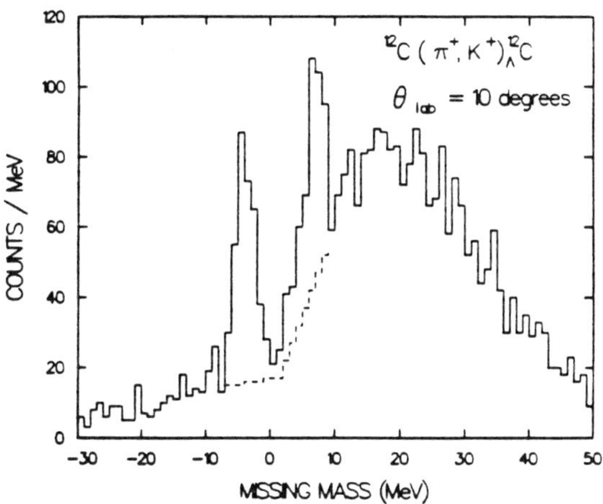

Fig. 2. Spectra of ${}^{12}C(\pi^+, K^+)$ at $\theta_{lab} = 10^0$ from the first AGS (π^+, K^+) experiment.

III. (π^+, K^+) EXPERIMENT AT AGS

In early 1987, the AGS experiment E798 received about six weeks of beam time. This allowed us to measure the (π^+, K^+) reaction over a number of target nuclei as shown in Table II. The measurement was made using the AGS Low Energy Separated Beam (LESB-I) and the hypernuclear spectrometer "Moby-Dick." The beam energy was around 1035 MeV/c and the kaon spectrometer was placed at 10° throughout the measurements. Up to 10^7/sec beam intensity was used. High rate drift chambers[29] placed at the mass slit and around the target handled this relatively high-rate environment properly. The mass separator and the TOF (time-of-flight) provided good π^+ separation from K^+ and proton. A thin lead-plate degrader placed near the mass slit significantly reduced the e^+

TABLE II. Relevant Parameters for the (π^+, K^+) Measurements

Target	Thickness (gm/cm^2)	Run time (hrs)	π^+ intensity (1/sec)
^9Be	2.35	27	2×10^6
^{12}C	2 – 4	13.5	2×10^6, 1×10^7
^{13}C	2.0	23	1×10^7
^{16}O	3.0	434	2×10^6
^{28}Si	4.03	73.3	2×10^6
^{40}Ca	4.03	109.5	2×10^6, 1×10^7
^{51}V	2.89	23	1×10^7
^{89}Y	4.05	88	1×10^7

contamination in the beam. Identification of kaons was achieved by using liquid nitrogen Cherenkov counters vetoing fast pions and by the TOF information. For a fraction of the high-rate measurements, a lucite Cherenkov detector was also placed downstream of the target to veto events accompanied by fast pions.

Figure 3 shows the "tagged" $^{12}C(\pi^+, K^+)$ spectrum obtained in this experiment. The tagging was made by requiring a large energy deposition in the active scintillation target. This tagging technique effectively reduces the Λ quasi-free background which usually leaves little energy in the target. The Λ-hypernuclear bound states, on the other hand, are usually accompanied by large energy release through its weak-decay. A comparison between Fig. 3 and Fig. 2 demonstrates the usefulness of this technique.

The (π^+, K^+) spectra obtained on ^{28}Si and ^{89}Y targets are shown in Fig. 4. The dashed lines show that experimental backgrounds are effectively reduced in the (π^+, K^+) experiment, allowing a sensitive measurement of deeply-bound hypernuclear states. A series of peaks, marked by arrows in Fig. 4, are clearly observed. These peaks are interpreted as Λ-particle neutron-hole states with Λ occupying various single particle orbitals together with a neutron hole in the va-

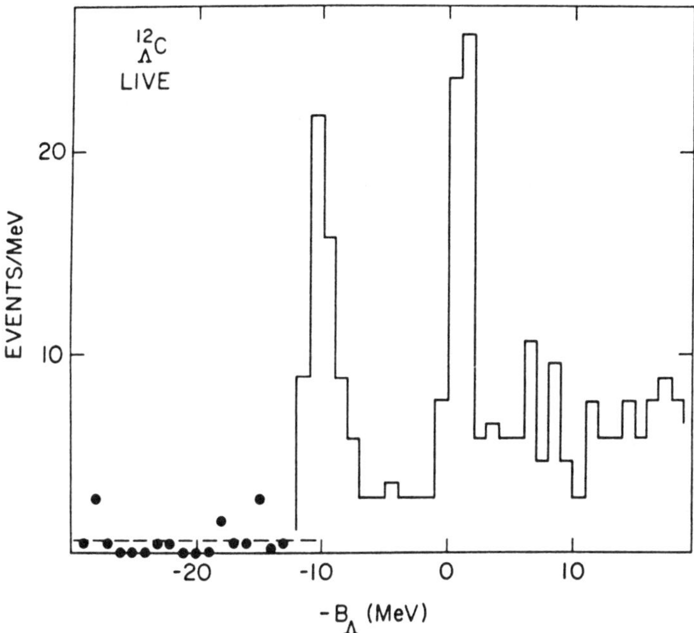

Fig. 3. Spectra of $^{12}C(\pi^+, K^+)$ from the recent AGS experiment. Large energy deposition in the scintillator target is required.

lence shell. Such particle-hole states are expected to be excited by the (π^+, K^+) reaction which turns a neutron into a Λ particle. The particle-hole configurations assigned for these various peaks are supported by the DWIA calculations[30] which generate theoretical spectra shown in Fig. 5. Neutron pick-up strength determined from the $^{89}Y(p,d)$ and $^{28}Si(p,d)$ reactions is used in these calculations. Comparison between Fig. 4 and Fig. 5 shows that both the position and the strength of these hypernuclear states are well predicted by calculations. A similar calculation by Motoba[31] is also capable of describing the data.

The (π^+, K^+) spectra measured on the closed shell nuclei ^{16}O and ^{40}Ca are shown in Fig. 6. Distinct peaks are observed in the $^{16}O(\pi^+, K^+)$ spectrum. In contrast, no clear peaks stand out in the $^{40}Ca(\pi^+, K^+)$ spectrum. The relatively featureless spectrum of $^{40}_{\Lambda}Ca$ is attributed to the fact that single particle states built on either the $d_{5/2}^{-1}$ or the $d_{3/2}^{-1}$ neutron hole are excited in the reaction with comparable strength. Spacing between the $d_{3/2}^{-1}$, $d_{5/2}^{-1}$ hole states is roughly half of the shell spacing of Λ-orbitals. The valley between the single particle states

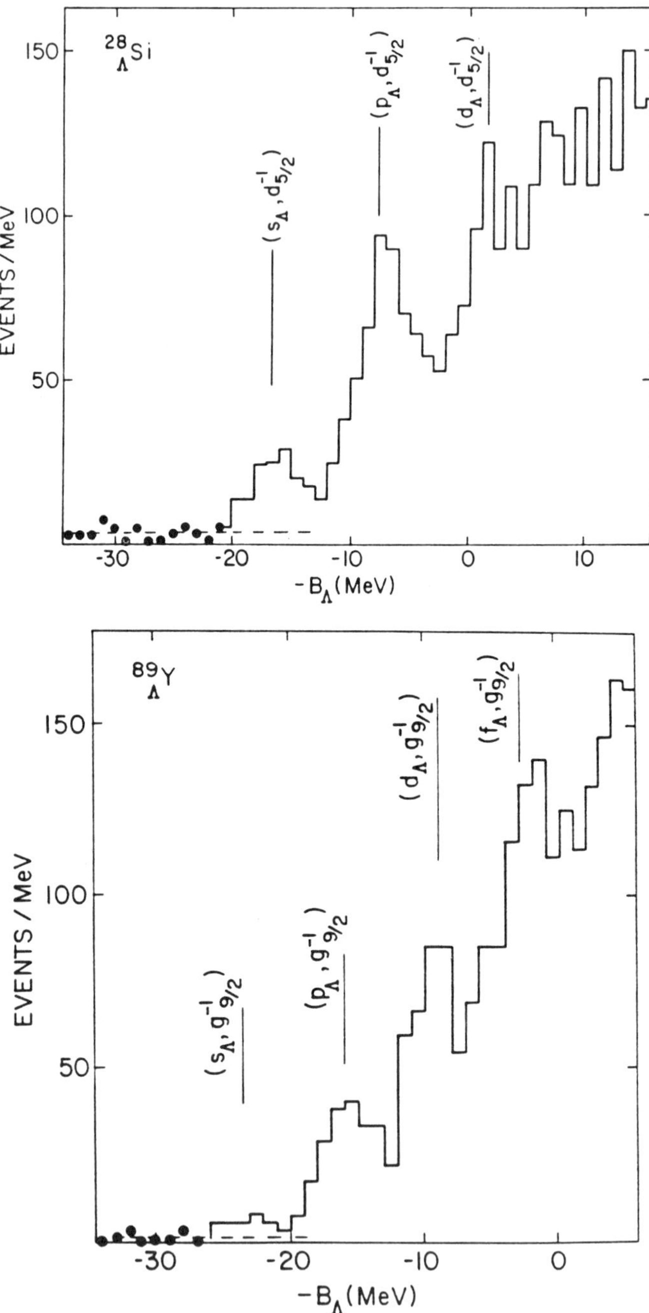

Fig. 4. (π^+, K^+) spectra on ^{28}Si and ^{89}Y measured at 1035 MeV/c and $\theta_{lab} = 10^0$.

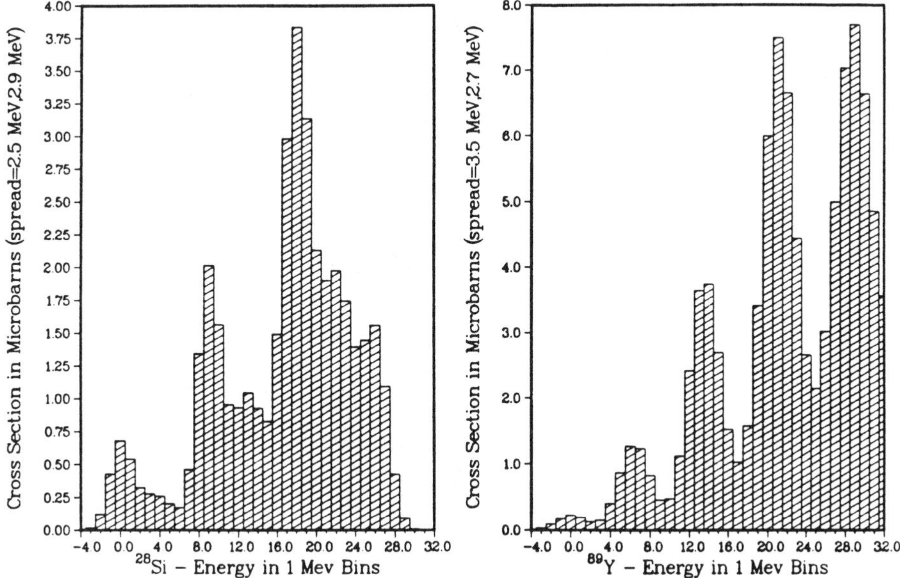

Fig. 5. Predicted (π^+, K^+) spectra by Millener.[30]

based on $d_{3/2}^{-1}$ hole is therefore filled by the states based on $d_{5/2}^{-1}$, resulting in a featureless spectrum. The calculation by Motoba[31] on the $^{16}O(\pi^+, K^+)$ and $^{40}Ca(\pi^+, K^+)$ spectra describe the data well.

The systematics of the Λ single particle binding energies is recently examined by Millener, Dover, and Gal.[32] Figure 7 collects all experimental data on B_Λ for mass greater than 9. In a square-well potential, the binding energy of the hypernuclear ground state is given as

$$B_\Lambda = D_\Lambda - \pi^2\hbar^2/2MR^2 = D_\Lambda - (\pi^2\hbar^2/2Mr_0^2)A^{-\frac{2}{3}} \tag{1}$$

where D_Λ is the depth of the potential and $R = r_0 A^{1/3}$ is assumed. Figure 8 shows clearly that the mass dependence of B_Λ deviates from $A^{-2/3}$. Using a Woods-Saxon potential with a depth of 29.34 MeV, Dover[33] showed that all the binding energy data can be well reproduced. It is remarkable that a potential depth independent of the mass and the angular momentum can fit all the B_Λ data. A more detailed analysis using a density-dependent non-local potential has been performed. The results are shown as the solid curves in Fig. 7. An ℓ-independent value of 28 MeV is obtained for D_Λ in this analysis.

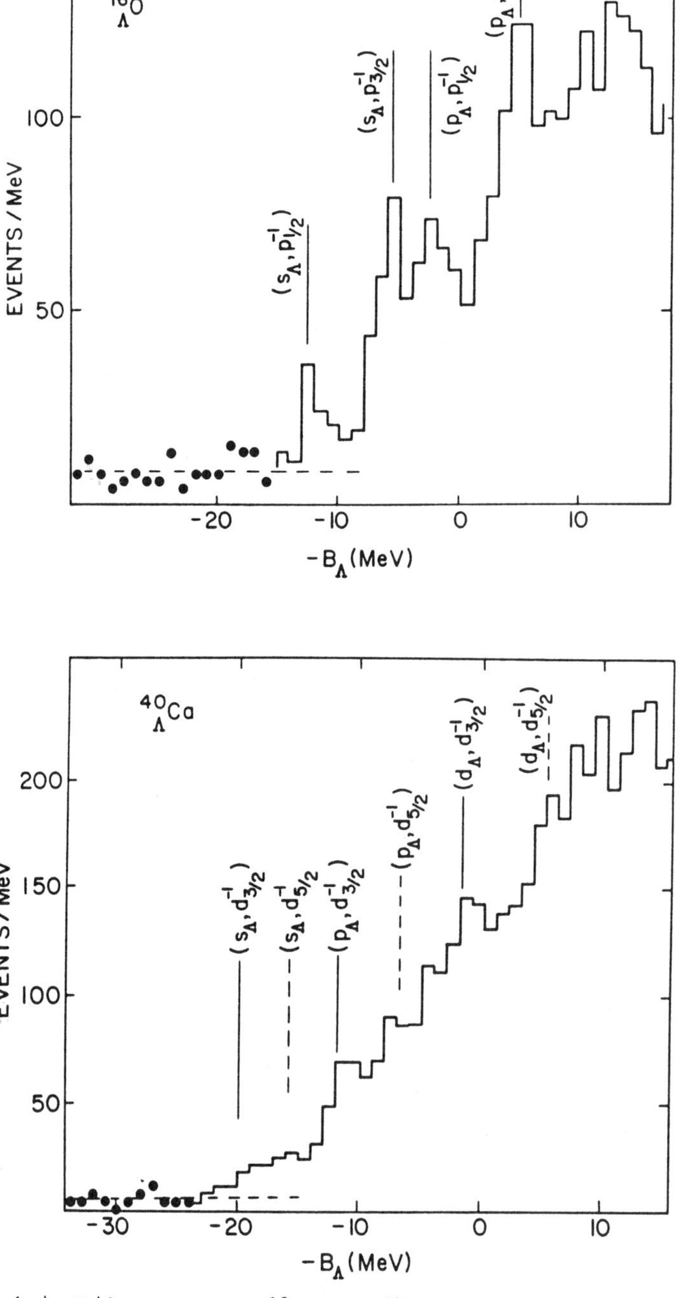

Fig. 6. (π^+, K^+) spectra on ^{16}O and ^{40}Ca measured at 1035 MeV/c and $\theta_{lab} = 10^0$.

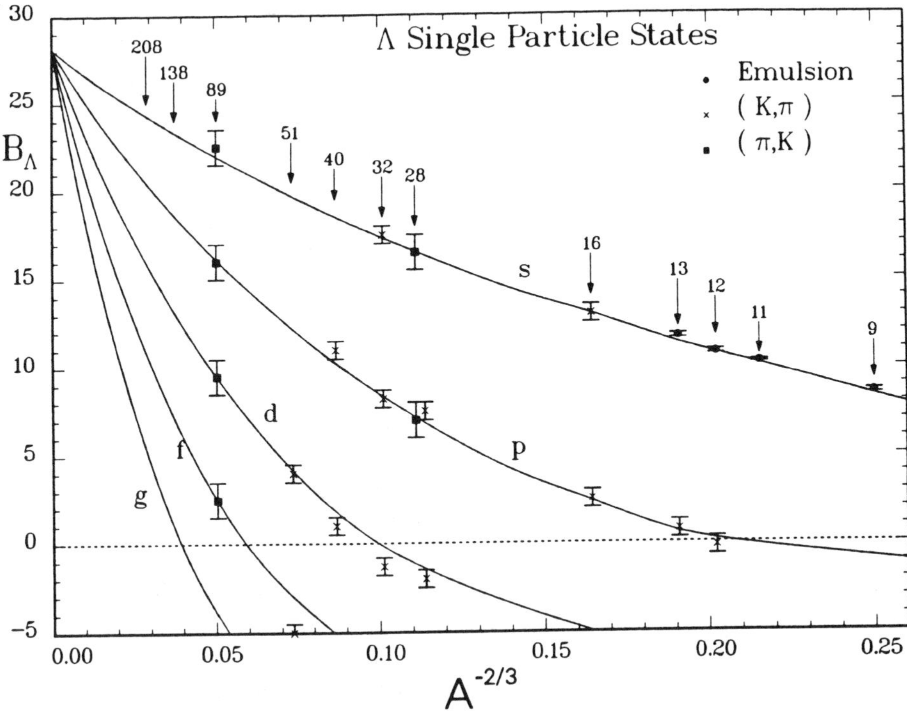

Fig. 7. Binding energies (B_Λ) data as a function of $A^{-2/3}$. The curves correspond to the solution obtained for the density dependent and non-local potential in Ref. 32.

The present (π^+, K^+) data show the first evidence of shell structures for deeply bound states in heavy nuclei. Similar attempts to locate the positions of nucleon single particle states have been made with the $(e, e'p)$ and $(p, 2p)$ reactions.[34] Figure 8 shows the compilation of proton separation energies determined from these experiments. Due to the large intrinsic width of these deep hole states, it becomes very difficult to determine their location for nuclei heavier than Nickel. The situation is more favorable for probing the nuclear interior with hypernuclei formation. The widths of these deeply bound states are narrow and the systematics of their binding energies can be mapped out by (π^+, K^+) or (γ, K^+) reactions all the way to the heaviest stable nuclei.

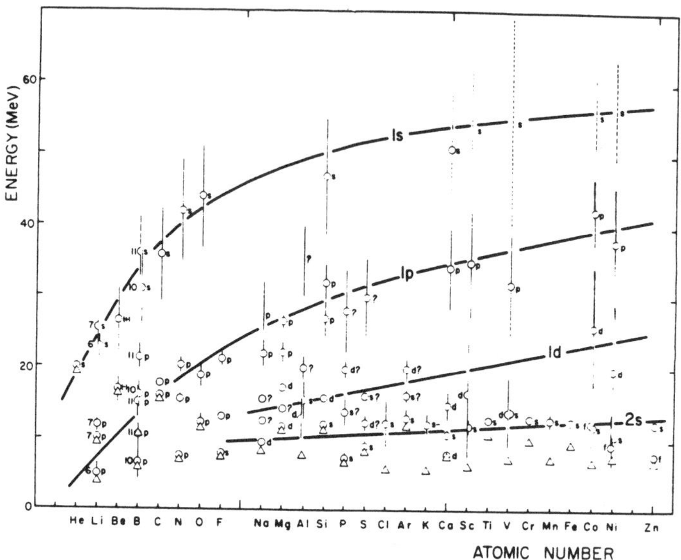

Fig. 8. Separation energies as a function of A. The data are from quasi-free scattering reactions (p, 2p) and $(e, e'p)$.[34]

The success of the shell model to describe the Λ binding energies in heavy nuclei tends to support the view that Λ particle maintains its identity deeply inside the nuclei. However, it is probably unreasonable to expect large effects caused by possible partial-deconfinement of Λ, and it is important to further improve the accuracy (such as energy resolution) for a more sensitive search of such effects.

IV. FUTURE PROSPECT AND CONCLUSION

At KEK, an ambitious project[27] is underway to construct a dedicated beam line and spectrometer to study the (π^+, K^+) reaction. The proposed program[35] will cover the production of Λ-hypernuclei and Σ-hypernuclei. In addition, coincidence measurements to detect the lifetime and weak decay modes of hypernuclei are planned. The possibility of measuring the magnetic moment of ${}^{13}_{\Lambda}C$ has also been discussed.[36] Recent studies suggest that hypernuclear states produced in the (π^+, K^+) reaction have much larger polarization than those produced in the (K^-, π^-) reaction.[37] If the magnetic moments of Λ-hypernuclear states can be successfully measured, they would provide extremely valuable information on the medium modification of Λ inside nucleus. A first (π^+, K^+) experiment has already been performed at KEK on ^{12}C and ^{56}Fe targets.[38] It is expected that the next generation of (π^+, K^+) data will mostly come from KEK.

A new kaon beam line[39] is being constructed at AGS for searching the H-particle. This beam line will also supply abundant pions up to 1.8 GeV/c. This beam line offers good opportunity to extend the (π^+, K^+) measurements to higher beam energies and heavier targets. In addition, the (π^-, K^+) reaction is suitable for studying the production of Σ^--hypernuclei.

It is also interesting to consider the possibility of producing charm-hypernuclei with associated production. A possible reaction is $(\pi^+, \bar{D}^\circ)$ which converts a neutron into a Λ_c. The momentum transfer in this reaction is $\approx$ 1.5 GeV/c, making it difficult to form the charm-hypernuclei directly. Nevertheless, the $(\pi^+, \bar{D}^\circ)$ reaction would be very useful to tag on the production of charm-hypernuclei.

In conclusion, associated production of hypernuclei is anticipated to be a major experimental technique in the future for hypernuclear studies. The feasibility of the (π^+, K^+) reaction is well established by two recent experiments at AGS. The usefulness of the (π^+, K^+) is demonstrated by the first observation of deeply-bound Λ hypernuclear states in heavy system. Other methods of associated production of hypernuclei such as (γ, K^+), (p, K^+), and $(\pi^+, \bar{D}^\circ)$ should be pursued in the future.

V. ACKNOWLEDGEMENTS

I would like to thank Drs. R. E. Chrien, P. H. Pile and D. J. Millener for helpful discussions. The collaboration of the AGS E798 experiment consists of P. H. Pile, R. E. Chrien, S. Bart, R. J. Sutter, N. Tsoupas from BNL; J. C. Peng, C. S. Mishra from LANL; T. Kishimoto, L. G. Tang, Z. Xu, W. Von Witsch, E. V. Hungerford from Houston; K. Maeda from Tohoku; D. Gill from TRIUMF; R. L. Stearns from Vasser; B. Quinn, R. McCrady, J. Seydoux from CMU; H. Plendl from Florida State and J. Reidy from Mississippi.

REFERENCES

1. P. Barnes, *AIP Conference Proceedings*, **No. 123** (1984) 265.

2. A. Gal, *AIP Conference Proceedings*, **No. 150** (1986) 127.

3. Proceedings of the 1986 INS International Symposium on Hypernuclear Physics, INS, University of Tokyo (1986).

4. Proceedings of the International Symposium on Strangeness in Hadronic Matter, Nucl. Phys. A479 (1988) 1C.

5. R. Bertini *et al.*, *Phys. Lett.* **90B** (1980) 375.

6. H. Piekarz *et al.*, *Phys. Lett.* **110B** (1982) 428.

7. T. Yamazaki *et al.*, *Phys. Rev. Lett.* **54** (1985) 102.

8. M. Danysz *et al.*, *Nucl. Phys.* **49** (1963) 121.

9. N. I. Starkov and V. A. Tsarev, *Nucl. Phys.* **A450** (1986) 507C.

10. Yu. A. Batusov *et al.*, *JETP Lett.* **33** (1981) 56.

11. M. Danysz and J. Pniewski, *Phil. Mag.* **44** (1953) 348.

12. M. Faessler *et al.*, *Phys. Lett.* **46B** (1973) 468.

13. W. Brückner *et al.*, *Phys. Lett.* **79B** (1978) 157.

14. M. May *et al.*, *Phys. Rev. Lett.* **51** (1983) 2085.

15. R. Grace *et al.*, *Phys. Rev. Lett.* **55** (1985) 1055.

16. The Physics and a Plan for a 45 GeV Facility, LA-10720-MS (1986).

17. S. Shinmura, *Nucl. Phys.* **A450** (1986) 147C.

18. C. B. Dover, L. Ludeking, and G. E. Walker, *Phys. Rev.* **C22** (1980) 2073.

19. S. R. Cotanch and S. S. Hsiao, *Nucl. Phys.* **A450** (1986) 419C.

20. J. C. Peng, Proc. Third LAMPF II Workshop, Los Alamos, July 1983, LA-9933-C, p.531.

21 . J. C. Peng, *Nucl. Phys.* **A450** (1986) 129C.

22. C. Milner *et al.*, *Phys. Rev. Lett.* **54** (1985) 1237.

23. D. Halderson, Y. Mo, and P. Ning, *Phys. Rev. Lett.*, **57** (1986) 1177.

24. E. V. Hungerford and L. C. Biedenharn, *Phys. Lett.*, **142B** (1984) 232.

25. T. Goldman, AIP Conference Proceedings, No. 123 (1984) 799.

26. R. Bertini *et al.*, *Nucl. Phys.* **A360** (1981) 315.

27. KEK proposal, Spokesmen: O. Hashimoto and T. Shibata (1985).

28. AGS proposal 798, Spokesmen: J. C. Peng and P. Pile (1984).

29. P. Pile, Proc. Third LAMPF II Workshop, Los Alamos, July 1983, LA-9933-C, p. 875.

30. D. J. Millener, private communication 1987.

31. T. Motoba, *Nucl. Phys.*, **A479** (1988) 227C.

32. D. J. Millener, C. B. Dover, and A. Gal, uppublished 1988.

33. C. B. Dover, in Proceedings of the International Symposium on Medium Energy Physics, Beijing, June 1987.

34. S. Frullani and J. Mongey, Advances in Nuclear Physics, 14 (1984) 1.
G. Jacob and T. A. J. Maris, *Rev. Mod. Phys.* **45** (1973) 6.

35. O. Hashimoto, in Proceedings of 1986 INS International Symposium on Hypernuclear Physics, p. 196.

36. H. Ejiri *et al.*, in Proceedings of 1986 INS International Symposium on Hypernuclear Physics, p. 223.

37. H. Ejiri *et al.*, *Phys. Rev.* **C36** (1987) 1435.

38. H. A. Thiessen, private communication (1988).

Kaon Decay Physics

Daniel R. Marlow
Department of Physics, Princeton University, Princeton, NJ 08544

Introduction

The study of kaon decays has long played a central role in the development of particle physics. Indeed, early experimental observations of kaons and hyperons in cosmic ray data helped establish the field as one distinct from nuclear physics. Since that time, kaons have led us to the discoveries of strangeness, parity non-conservation, *CP* violation, and have provided the first hints of extra quark generations. It is remarkable that three decades after the first accelerator-based experiments, the study of kaon decays still promises the possibility of unmasking new physical phenomena.

Much of the kaon's utility stems from the low mass of its active ingredient, the strange quark. Unlike particles containing the much heavier charm and bottom quarks, kaons can be copiously produced with (relatively) low energy proton beams and live long enough to be formed into beams that can be momentum analyzed and/or purified. The mass of the kaon is also ideally situated, in that it is heavy enough that a wide variety of final states are possible, but light enough that none of them are very complicated. This helps to simplify the design of experiments and avoids difficulties with combinatorial backgrounds.

The modern experiments that will be the subject of this talk are, however, not simple in an absolute sense. In fact one of the reasons that they are being performed only now—the BNL AGS, after all, has been in business for almost thirty years—has do with progress in detector technology. In particular, the introduction of the wire chamber and its variants, coupled with the reduction of the size and cost of detector electronics and steady improvements in computing and data acquisition systems, give us the ability to screen vast numbers of kaons in search of extremely subtle effects. A second, probably more significant, reason for the current interest is the success enjoyed by the Standard Model of electroweak interactions. It is amusing to note the lengths to which experimentalists will go in hope of moving beyond this theory. Perhaps this is merely desperation. History indicates, however, that kaon decays may be just the right place to look.

CP Violation—Status of ϵ'/ϵ

I would be remiss in not including the important topic of *CP* violation—long a centerpiece of K physics—in a talk on kaon decays. After a quick rehashing of a few basics, which in light of the existence of a number excellent reviews of the subject[1,2] will be fairly schematic, I will say a few words about two recent precision measurements of ϵ'/ϵ.

Neutral kaons for *CP* violation experiments are typically produced via strong interaction (associated) production mechanisms, which create states of definite strangeness—i.e., the K^0 and its antiparticle, the $\overline{K^0}$. Since kaons have zero spin, it follows that

$CP|K^0\rangle = |\overline{K^0}\rangle$ and vice versa. The CP +1 and CP −1 states are

$$|K_1\rangle = \frac{1}{\sqrt{2}}(|K^0\rangle + |\overline{K^0}\rangle) \qquad \text{and} \qquad |K_2\rangle = \frac{1}{\sqrt{2}}(|K^0\rangle - |\overline{K^0}\rangle) \tag{1}$$

If left to propagate in vacuum, the K^0's meet their demise via weak decay. The weak interaction changes strangeness but is (almost) CP conserving. Both two and three pion final states are energetically possible, the former being CP +1 and the latter CP −1. Since phase space favors the two pion mode by a about a factor of 1700, the decay process effectively filters out the K_1 component leaving a beam of mostly slowly decaying K_2's. A careful look at the long lived component, however, reveals that well after the K_1 should have decayed away, a small number of $K^0 \to \pi\pi$ decays are observed. This effect was first seen in 1964 in the celebrated experiment of Christenson, Cronin, Fitch, and Turlay[3], which provided the first evidence for CP non-conservation.

There are two ways to arrive at the observed effect. The first is "mixing", whereby the long and short lived components are actually mixtures of K_1 and K_2, that is

$$|K_L\rangle = \frac{|K_2\rangle + \varepsilon|K_1\rangle}{\sqrt{1+\varepsilon^2}} \qquad \text{and} \qquad |K_S\rangle = \frac{|K_1\rangle + \varepsilon|K_2\rangle}{\sqrt{1+\varepsilon^2}} \tag{2}$$

A prime example of mixing is the "superweak" model[4] of Wolfenstein, wherein the K_1 admixture in the K_L arises from $|\Delta S| = 2$ transitions between K^0 and $\overline{K^0}$. Although all—with the possible exception of the most recent[5,6,7] (see below)—experimental data is consistent with mixing alone, "direct" CP violation, whereby the K_2 decays directly into two pions is also a possibility. Among other places, direct CP violation arises in the widely employed Kobayashi-Maskawa (KM) six-quark mixing scheme[8] and has been the focus of a number of experimental CP violation studies.

The standard litmus test for direct CP violation is a difference in the ratios

$$|\eta_{+-}|^2 = \frac{\Gamma(K_L \to \pi^+\pi^-)}{\Gamma(K_S \to \pi^+\pi^-)} \qquad \text{and} \qquad |\eta_{00}|^2 = \frac{\Gamma(K_L \to \pi^0\pi^0)}{\Gamma(K_S \to \pi^0\pi^0)} \tag{3}$$

(If mixing is all that is happening, we expect $\eta_{+-} = \eta_{00}$ since all two pion decays are just decays of a K_1.) These measurements are effectively measurements of ϵ'/ϵ , where ϵ and ϵ' are the complex amplitude ratios

$$\epsilon = \frac{\langle I=0|T|K_L\rangle}{\langle I=0|T|K_S\rangle} \qquad \text{and} \qquad \epsilon' = \frac{1}{\sqrt{2}}\frac{\langle I=2|T|K_L\rangle}{\langle I=0|T|K_S\rangle} \tag{4}$$

By decomposing the $|\pi^+\pi^-\rangle$ and $|\pi^0\pi^0\rangle$ final states into their $I=0$ and $I=2$ components ($I=1$ is forbidden by Bose statistics), invoking the $|\Delta I| = 1/2$ rule, and doing a little algebra, one can show that

$$\text{Re}\,\frac{\epsilon'}{\epsilon} \simeq \frac{1}{6}\left[1 - \left|\frac{\eta_{00}}{\eta_{+-}}\right|^2\right] \tag{5}$$

Note that a, say, 3% measurement of the double ratio $|\eta_{00}/\eta_{+-}|^2$ determines ϵ'/ϵ to .5%.

Calculations of ϵ'/ϵ in the KM model are subject to some uncertainty, and yield values the range[1] $.001 < |\epsilon'/\epsilon| < .007$ (versus zero for superweak models), thus a $\sim .1\%$ measurement may be needed to really pin things down. Even with Nature's factor-of-six, this is obviously a tough experimental challenge. There has been steady improvement in measurements of ϵ'/ϵ over the years, as is shown in figure 1. The two most recent points on the plot represent CERN experiment[5] NA31 and experiment[6] E731 at the Fermilab Tevatron. The two experiments share the common feature of using high momentum ($<p_K>\sim 100$ GeV/c) K^0 beams, but in other respects are quite different (see reference 7 for further details).

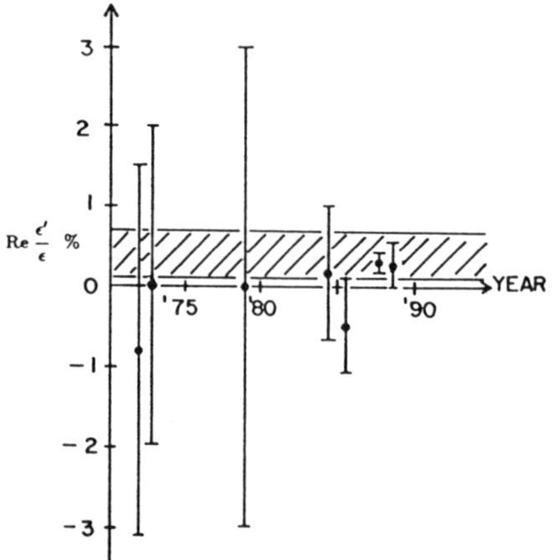

Figure 1. Measurements of ϵ'/ϵ versus year. The points from left to right are from references 9,10,11,12,13,5 and 6 respectively. Central values were determined using equation (5); the error bars are combined systematic and statistical. The shaded area shows the range of theoretical values in the six-quark KM model[1].

E731, shown in figure 2, uses a magnetic spectrometer followed by a lead glass array and muon filter to simultaneously detect decays from two parallel neutral beams, one of which contains a regenerator to produce K_S's. The use of this clever dual beam technique results in the cancellation of systematic errors arising from time dependent efficiencies that might otherwise distort the ratio being measured. As an added precaution, the regenerator is periodically switched from beam to beam. Neutral mode decays are studied in separate runs, during which a thin lead sheet is placed at the end of the decay region to convert photons from π^0 decay. As is typical of this type of experiment, the statistical accuracy is limited by the number of reconstructed $K_L \to 2\pi^0$ decays, which in the case of E731 is 6747. The relative paucity of these events arises in part from the requirement that at least one photon convert in the thin lead sheet. Possible systematic worries include the different decay distributions for K_L and K_S and background from diffractive and inelastic effects in the regenerator. E731's published[6] value is $\epsilon'/\epsilon = 0.0032 \pm 0.0028 \pm 0.0012$ where the first error is statistical and the second systematic. E731 now has a much larger data sample on tape and should be able to significantly reduce their statistical error in the near future.

CERN SPS experiment NA31 alternates between collinear K_L and K_S beams and uses a calorimeter to simultaneously detect charged and neutral decay modes (see figure 3). K_S are produced in separate runs by directing a beam of about 10^7 protons per

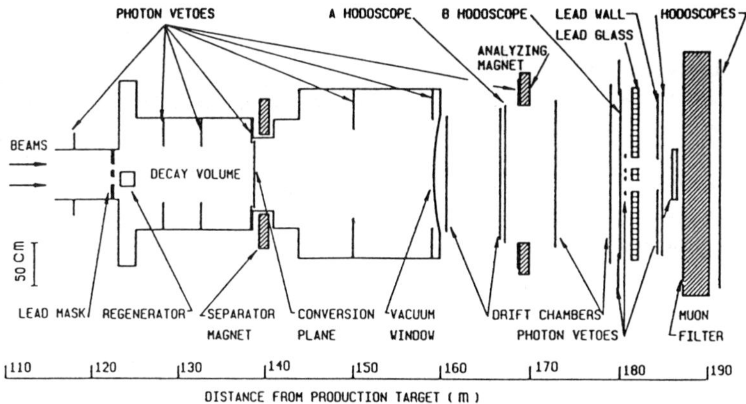

Figure 2. Elevation view of Tevatron Experiment E731. (From reference 6.)

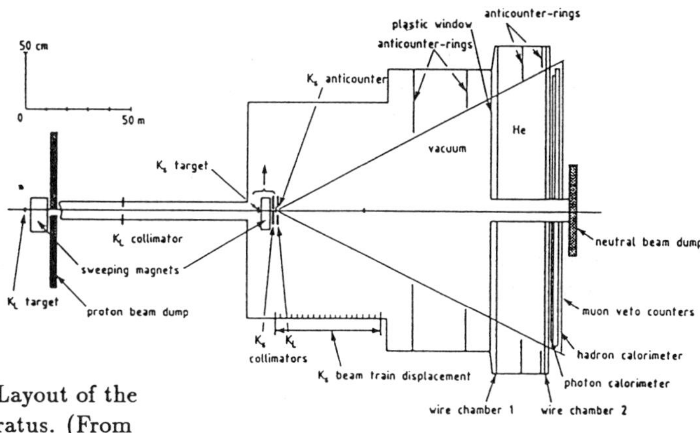

Figure 3. Layout of the NA31 apparatus. (From reference 5).

pulse to a target mounted on a trolley that moves throughout the decay volume in such a way as to mock up the K_L decay distribution. The calorimetric technique results in an increased acceptance for $K_L \to 2\pi^0$ decays—NA31 has 109,000 such events— at the cost of worse mass resolution ($\sigma_M = 22$ MeV/c^2 for NA31[14] vs. $\sigma_M = 3$ MeV/c^2 for E731) for charged mode decays. NA31's result of $\epsilon'/\epsilon = .0033 \pm .0007 \pm .0008$ represents a three standard deviation difference from zero, which is the first statistically significant experimental evidence for direct *CP* violation. In view of the importance of the result, not to mention the difficulty of the measurement, further confirmation would be useful. It is indeed fortunate that two independent and complementary experiments have been undertaken.

Searches for Lepton Flavor Violation in K Decays

An observation of lepton flavor non-conservation, assumed to be strictly forbidden in the Standard Model, would be conclusive evidence for new physics. Examples include $\mu \to e\gamma$, $\mu \to eee$, $K_L \to \mu e$ and $K^+ \to \pi^+ \mu e$. These reactions can be mediated in a variety of ways, and are predicted to occur by most models that attempts to go beyond the standard model. Some theories incorporate horizontal gauge bosons (HGB's) as intermediaries, while others call for the exchange of lepto-quarks or constituents. Details can be found in a number of places[15,16,17].

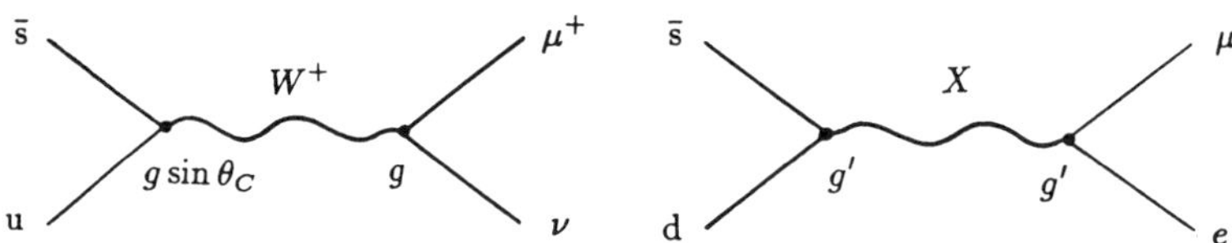

Figure 4. Tree level graphs for $K^+ \to \mu^+\nu$ and $K_L \to \mu e$.

A frequently made (and important) remark is that by comparing, say, the reactions $K^+ \to \mu^+\nu$ and $K_L \to \mu e$ (see figure 4) —the first mediated by W^+ exchange and the second by a hypothetical HGB, X—it is evident that the ratio of rates for the two reactions will be roughly

$$\frac{\Gamma(K_L \to \mu e)}{\Gamma(K^+ \to \mu^+\nu)} \simeq \left(\frac{g'}{g}\right)^4 \frac{1}{\sin^2\theta_c}\left(\frac{M_W}{M_X}\right)^4 \tag{6}$$

Putting in the measured values for $\Gamma(K^+ \to \mu^+\nu)$, τ_{K^0}, and θ_c, and making the not unreasonable *assumption* that the couplings g and g' are equal, we deduce X masses given by

$$M_X \simeq 2M_W \left(\frac{1}{B(K_L \to \mu e)}\right)^{\frac{1}{4}} \simeq \begin{cases} 17\ \text{TeV}, & B = 10^{-8} \\ 170\ \text{TeV}, & B = 10^{-12} \end{cases} \tag{7}$$

where the two values for B represent the state of things before the current round of lepton flavor violation experiments[18-21] got underway, and where things should be once they are completed. While not an alternative to the SSC, these experiments are capable of unveiling physics at SSC and higher mass scales at a fraction of the cost.

$K_L \to \mu e$ experiments are also sensitive to $K_L \to \mu^+\mu^-$ and $K_L \to e^+e^-$, the former being a famous example of a suppressed but measurable[22] $B \simeq 10^{-8}$ flavor changing neutral current, and the latter suffering an additional helicity suppression to the few 10^{-12} level.

Three $K_L \to \mu e$ experiments are now completed or underway, two at Brookhaven[18,19] and a third at KEK[20]. The first AGS experiment, E780, was done using a modified version of an earlier *CP* violation apparatus (see figure 5). In this experiment, the decay products of a $< P_{K_L} > \sim 9$ GeV/c beam passed through an analyzing magnet and were tracked by four 3 mm cell mini-drift chamber arrays. Electrons were identified using an H_2 Čerenkov

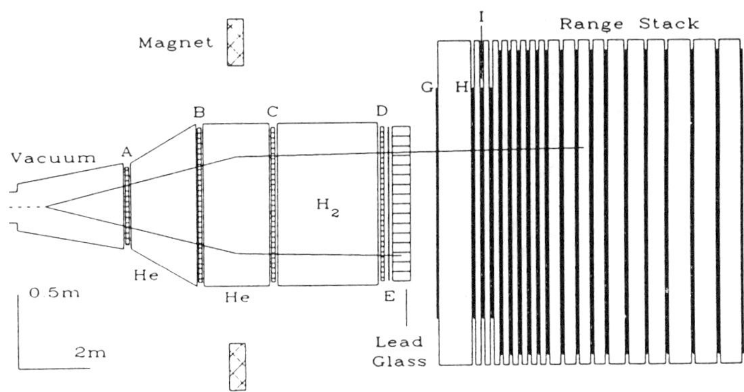

Figure 5. Plan view of the E780 detector. (From reference 18.)

detector and a lead glass array, while muons were tagged by a steel filter followed by a steel/scintillator range stack.

The leading background to $K_L \to \mu e$ is $K_L \to \pi e \nu$ where the pion is mistaken for a muon, either because of in-flight decay or failure of the particle ID system. This puts a premium on mass resolution, since mistaking a heavier particle for a lighter one (i.e., a π for a μ) will cause the reconstructed K_L mass to come out low—in this case by at least 8 MeV. An important exception is the unfortunate case where the π decays in the bend plane of the analyzing magnet in such away as to increase its apparent momentum. Luckily, this will generally throw off the non-bend plane and the vertex reconstructions, but attempts to get below the $\sim 10^{-10}$ level must take this background into account.

One can deduce the direction of the incident K_L by using the reconstructed decay vertex and the known position of the K_L production target. For genuine $K_L \to \mu e$ events, this direction should be the same as that of the total momentum vector of the reconstructed μe final state, so that experiments typically present their data in the form of a plot of the square of the angle between the two directions versus the μe invariant mass.

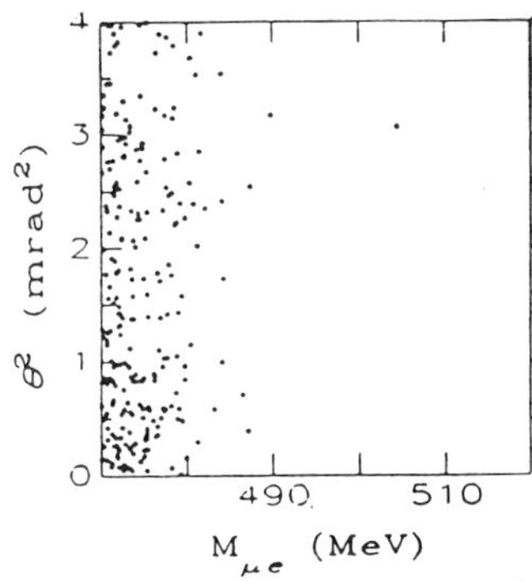

Figure 6. θ^2 vs $M_{\mu e}$ (ref. 18).

Figure 6 shows such a plot for the data from E780. The acceptance region (not drawn) has been determined by an analysis of $K_L \to \pi^+\pi^-$ decays. The absence of events near M_K translates into an upper limit of $B(K_L \to \mu e) < 6.7 \times 10^{-9}$ (90% confidence level). E780 also sets a limit of $B(K_L \to e^+e^-) < 4.5 \times 10^{-9}$ and observes two $K_L \to \mu^+\mu^-$ events, consistent with the known branching ratio. In a recent run, E780 acquired 5-10 times more data, which is currently being analyzed.

Brookhaven experiment E791[19] is a next generation $K_L \to \mu e$ search. This experiment is similar in spirit to E780, but uses an additional bending magnet to guard against the bend region π decay background by making an extra momentum measurement and comparing it against the first. The extra measurement also improves the momentum resolution by about the factor of $\sqrt{2}$ that one would naively expect (E791 boasts an impressive 1.4 MeV/c^2 mass resolution). The ultimate goal of E791 is a $B(K_L \to \mu e)$ sensitivity of $\sim 10^{-12}$. A 1987 test run yielded $B(K_L \to \mu e) < 1.1 \times 10^{-8}$ (90% C.L.) and they have already presented a beautiful mass plot showing 26 $K_L \to \mu^+\mu^-$ events (this doubles the world's supply) from a partial analysis of their 1988 data. There is another search[20] for $K_L \to \mu e$ being conducted at KEK in Japan, from which results should soon be forthcoming. This experiment also makes a double momentum measurement, but with a different configuration of magnets.

In addition to $K_L \to \mu e$, one can search for the related reaction $K^+ \to \pi^+\mu^+e^-$. The shorter lifetime of the K^+ and the smaller phase space for the three body final state reduce the branching ratio for the latter reaction by about a factor of 50 relative to $K_L \to \mu e$, making it a less favorable place to look. However, if the relevant reaction is purely vector in nature, then $K^+ \to \pi^+\mu^+e^-$ will go, while $K_L \to \mu e$ won't. In view of the lack (the surplus?) of theoretical guidance, it seems wise to pursue all avenues. Finally, $K^+ \to \pi^+\mu^+e^-$ experiments bring a *different* set of systematics; the two reactions complement one another nicely.

The only dedicated $K^+ \to \pi^+\mu^+e^-$ experiment to date is BNL experiment E777[21]. The apparatus (figure 7) consists of a pair of dipole magnets, the first of which separates decay products from the incident beam of 5.8 GeV/c (unseparated) K^+ and deflects the π^+ and μ^+ to the right and the e^- to the left. The second dipole, along with four MWPC packages, is used for momentum analysis of the daughter particles. Background reduction counts heavily on particle identification and the builders of E777 have emphasized this in their design.

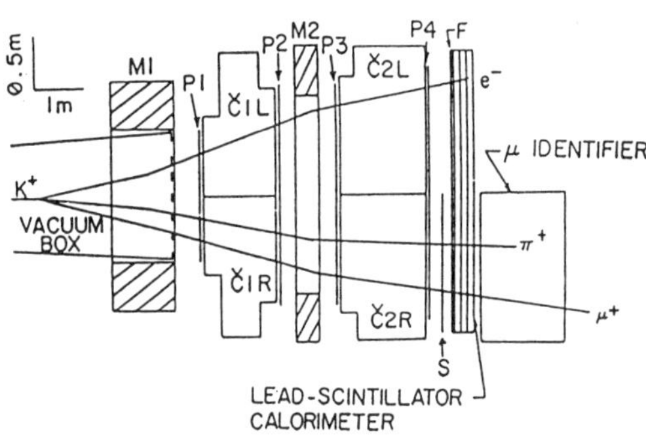

Figure 7. Plan view of the E777 apparatus (ref. 23).

There are two Čerenkov counters on each side of the apparatus. The Čerenkov counters on the right (π^+ & μ^+) side of the apparatus are filled with CO_2 so as to efficiently veto electrons, whereas the counters on the left (e^-) side are filled with H_2 to avoid accidental firings by μ's or π's. Further particle identification is achieved by a Pb-Scintillator sandwich array, placed after the last MWPC package and a steel/proportional-tube muon identifier on the right hand side of the apparatus.

Figure 8 shows the final sample of $K^+ \to \pi^+\mu^+e^-$ candidates obtained from E777. The value of $S \equiv \sqrt{\sum S_i^2}$, where S_i is the distance between the i'th track and the best common vertex, versus the reconstructed $\pi^+\mu^+e^-$ mass is plotted for each event.

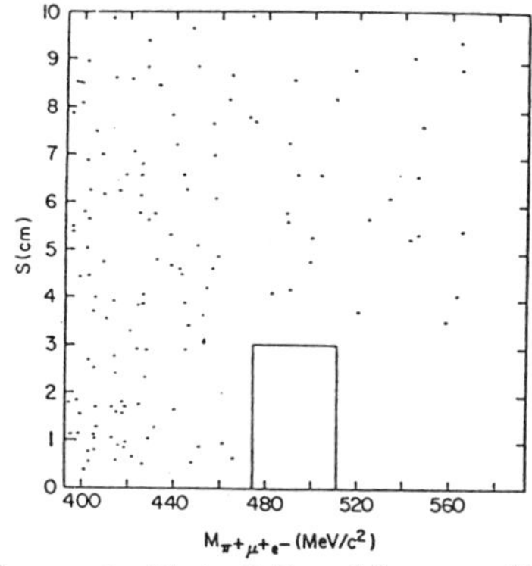

Figure 8. Plot of S vs $M_{\pi^+\mu^+e^-}$ (from ref. 24).

The signal box is determined from an analysis of $K^+ \to \pi^+\pi^+\pi^-$ events, suitably corrected by Monte Carlo. The absence of candidates in the box yields a 90% C.L. upper limit[24] of $B(K^+ \to \pi^+\mu^+e^-) < 1.1 \times 10^{-9}$. The same data are also used to deduce an upper limit of $B(\pi^0 \to \mu^+e^-) < 7.8 \times 10^{-8}$. E777 has acquired more data in a recent run and hopes to improve these limits by a factor of eight or so.

As a by-product, E777 also obtains a sample of $K^+ \to \pi^+e^+e^-$ events, which has a measured[22] branching ratio of $B = (2.7 \pm .5) \times 10^{-7}$. This mode is particularly topical in light of observations of correlated narrow electron-positron peaks in heavy ion collisions[25], which suggest the existence of a $M = 1.8$ MeV/c^2 state. Some interpretations of the observed peaks call for short-lived axion variants[26,27] (the original axion , first introduced to solve the strong CP problem[28,29,30] has long been ruled out by studies of radiative quarkonium decays [31,32,33]). The short-lived axion variants should show up in kaon decay[34] as $K^+ \to \pi^+A^0 \to \pi^+e^+e^-$

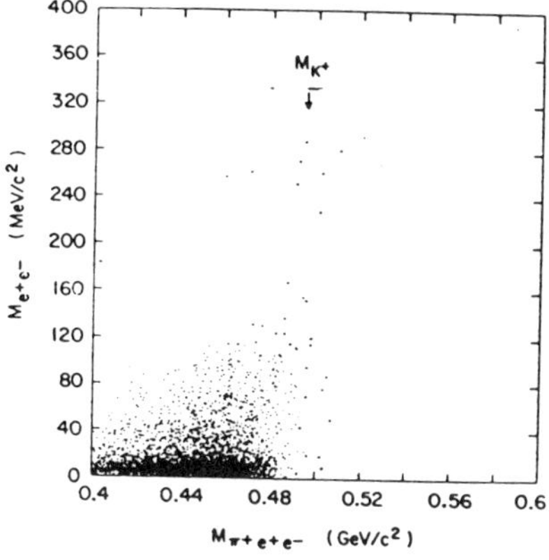

Figure 9. Plot of $M_{e^+e^-}$ vs $M_{\pi^+e^+e^-}$ (from reference 23.)

Figure 9 shows a plot of the reconstructed e^+e^- mass versus the mass of the $\pi^+e^+e^-$ system. The large number of events at low M_{ee} and $M_{\pi ee} < M_K$ arise from Dalitz decays ($K^+ \to \pi^+\pi^0 \to \pi^+\gamma e^+e^-$) where the photon goes undetected. Based on an analysis[23] of the plot in the region $M_{\pi ee} \simeq M_K$ and $M_{ee} \sim 0$, E777 sets a limit on the branching ratio to short-lived ($\tau_{A^0} \leq 10^{-13}$ s) A^0 particles of $B(K^+ \to \pi^+A^0) < 4.7 \times 10^{-7}$ (90% C.L.). This data, taken together with limits[37] on $\pi^+ \to A^0e^+\nu$ rules out the proposed axion model variants.

Although no signal in the $K \to \pi^+\pi^0 \to \pi^+e^+e^-$ chain is evident in the plot of figure 9, with the improved statistics of their 1988 run, E777 should have enough events to see a signal. This would allow E777 to pin down a value for $B(\pi^0 \to e^+e^-)$. Past experimental attempts[35,36], which suffer from either limited statistics or large background subtractions average[22] to $B(\pi^0 \to e^+e^-) = (1.8 \pm .7) \times 10^{-7}$, a value that is two or more times higher than predicted by most theories[35]. At the central value, E777 should get ~ 100 events from their recent run. It is gratifying that an experiment aimed primarily at "long shot" physics, can also produce results on less speculative topics.

Search for the Reaction $K^+ \to \pi^+ +$ nothing

Unlike $K_L \to \mu e$ and $K^+ \to \pi^+\mu^+e^-$, the reaction $K^+ \to \pi^+ +$nothing, where "nothing" means undetected neutrals, is not forbidden in the Standard Model. The Standard Model version of this process ($K^+ \to \pi^+\nu\overline{\nu}$) is forbidden at the tree level, but allowed in higher order. The first order cancellation is produced by the GIM mechanism[38], which is most easily understood in the context of two quark doublets.

$$\begin{pmatrix} u \\ d_c \end{pmatrix} \begin{pmatrix} c \\ s_c \end{pmatrix} \quad \text{where} \quad \begin{pmatrix} d_c \\ s_c \end{pmatrix} = \begin{pmatrix} \cos\theta_c & \sin\theta_c \\ -\sin\theta_c & \cos\theta_c \end{pmatrix} \begin{pmatrix} d \\ s \end{pmatrix} \tag{8}$$

where d_c and s_c are the "Cabbibo rotated" (weak eigenstate) quarks and $\theta_c \simeq 13°$ is the Cabbibo angle, which parametrizes their mixing with the d and s mass eigenstate quarks. Using the above expressions (and leaving out the $V - A$ operators etc.) it is easy to see that the charged weak current, which goes like

$$J^+ \sim \overline{u}d_c + \overline{c}s_c = (\overline{u}d + \overline{c}s)\cos\theta_c + (\overline{u}s - \overline{c}d)\sin\theta_c \tag{9}$$

can change strangeness, whereas the weak neutral current, which goes like

$$J^0 \sim \overline{u}u + \overline{d}_c d_c + \overline{c}c + \overline{s}_c s_c = \overline{u}u + \overline{c}c + \overline{d}d + \overline{s}s \tag{10}$$

contains no flavor changing parts. This cancellation, which follows from the unitarity of the mixing matrix, also applies to the six-quark KM mixing scheme.

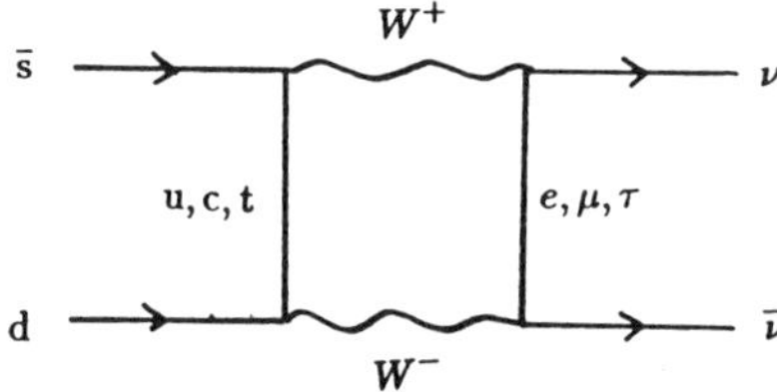

Figure 10. Higher order box graph contribution to $K^+ \to \pi^+\nu\overline{\nu}$.

For higher order processes, such as the one shown in figure 10, the cancellation is spoiled by differences in the quark masses, so that $K^+ \to \pi^+\nu\overline{\nu}$ can proceed, albeit at a greatly reduced rate.

Theoretical calculations [39] yield

$$
\begin{aligned}
B(K^+ \to \pi^+ \nu_i \bar{\nu}_i) &= \frac{.61 \times 10^{-6}}{|V_{us}|^2} \left| \sum_{j=c,t} V_{js}^* V_{jd} D(x_j) \right|^2 \\
&\simeq .61 \times 10^{-6} \left| D(x_c) + s_2 \left(s_2 + s_3 e^{i\delta}\right) D(x_t) \right|^2
\end{aligned}
\tag{11}
$$

where the V_{ij} are the Kobayashi-Maskawa mixing matrix elements, and the $s_i \equiv \sin\theta_i$ are the sines of the corresponding angles that parametrize the matrix. The function $D(x_j) = D(m_j^2/m_W^2)$ is given by

$$
D(x) = \frac{1}{8}\left(1 + \frac{3}{(1-x)^2} - \frac{(4-x)^2}{(1-x)^2}\right) x \log x + \frac{x}{4} - \frac{3}{4}\frac{x}{1-x} \tag{12}
$$

and gives the dependence of $B(K^+ \to \pi^+ \nu\bar{\nu})$ on the mass of the contributing quarks.

Using measurements, of the b-quark lifetime, recent ARGUS B^0 mixing data[52], and the values of $B(K_L \to \mu^+\mu^-)$ and the $K_L - K_S$ mass difference to constrain the KM mixing matrix elements, calculations[40,41,42] yield a range of allowed values of $B(K^+ \to \pi^+ \nu\bar{\nu})$ of $(1 - 7) \times 10^{-10}$ for three neutrino generations. The precise value depends on m_t and the KM mixing angles and is subject to small QCD corrections. The range of allowed values is shown in figure 11, where $B(K^+ \to \pi^+ \nu\bar{\nu})$ is plotted versus the top quark mass. Despite the uncertainties, which should be reduced as the KM matrix elements are better determined by future experiments, $K^+ \to \pi^+ \nu\bar{\nu}$ is probably the cleanest low energy test of higher order weak processes.

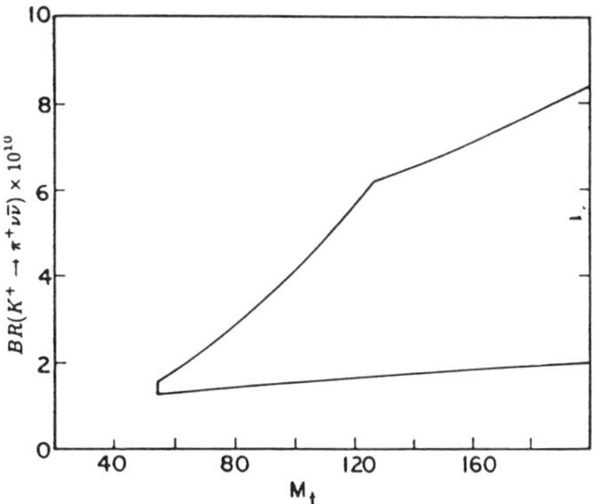

Figure 11. Range of allowed values for $B(K^+ \to \pi^+ \nu\bar{\nu})$ as a function of the top quark mass. The allowed limits change slightly, depending assumptions about values of parameters in the $B\bar{B}$ system (from ref. 41).

The current experimental limit[47] of $B(K^+ \to \pi^+ \nu\bar{\nu}) < 1.4 \times 10^{-7}$ 90% C.L. on this process lies well above the Standard Model range, leaving a window open for the emergence of "new physics". Perhaps the *least* exotic possibility that might lead to higher than expected values for $B(K^+ \to \pi^+ \nu\bar{\nu})$ is the existence of a fourth generation of fermions[43,44]. Although the increase over the three-quark Standard Model value is not necessarily large, certain scenarios, where the mixing between the third and fourth generations is large, can produce a several fold increase.

In addition, the HGB's and lepto-quarks discussed above in connection with $K_L \rightarrow \mu e$ and $K^+ \rightarrow \pi^+\mu^+e^-$ might also mediate $K^+ \rightarrow \pi^+\nu\bar{\nu}$, although such contributions tend to be constrained by the $K_L - K_S$ mass difference.

It is also possible that the unobserved neutrals may be something other than neutrinos. In supersymmetry, for example, the reaction $K^+ \rightarrow \pi^+\tilde{\gamma}\tilde{\gamma}$ may produce a considerable enhancements in the $K^+ \rightarrow \pi^+ +$ nothing rate[45], provided the photinos are light enough and the squark masses are smaller than 70 GeV/c^2. UA1 data[46] largely exclude such light $m_{\tilde{q}}$ values, but leave some room for values above 60 GeV/c^2.

Two body final state processes, of the form $K^+ \rightarrow \pi^+X^0$ where X^0 goes undetected, comprise another class of $K^+ \rightarrow \pi^+ +$ nothing reactions. The original Peccei-Quinn-Wilczek-Weinberg axion would fit the bill nicely, although as discussed above, this has already been ruled out by a variety of experiments (searches[47] for $K^+ \rightarrow \pi^+ +$ nothing, among them). However, the familon[48]—a neutral, weakly interacting Nambu-Goldstone boson arising from the spontaneous breakdown of family symmetries—proposed by Wilczek would have the same experimental signature. Since the familon is essentially massless, it would make itself known by the appearance of a sharp spike in π^+ energy at the endpoint of the $K^+ \rightarrow \pi^+\nu\bar{\nu}$ spectrum. Given this clean signature, and a predicted $K^+ \rightarrow \pi^+f$ branching ratio in the range of 2.6×10^{-8} to 2.6×10^{-11} (the current experimental upper limit[47] is $B(K^+ \rightarrow \pi^+f) < 4.5 \times 10^{-8}$), the familon should either be discovered or ruled out by the current $K^+ \rightarrow \pi^+ +$ nothing search (see below).

Thus far, two dedicated $K^+ \rightarrow \pi^+ +$ nothing searches[47,49,50] have been conducted. Most recently, Asano et. al., have set a 90% C.L. upper limit of $B(K^+ \rightarrow \pi^+\nu\bar{\nu}) < 1.4 \times 10^{-7}$.

In early 1988, an AGS experiment[51] started taking data using a large magnetic solenoid detector. The detector bears a superficial resemblance to the typical detector used at e^+e^- colliders, but has been optimized for the detection of π^+ mesons in the 200 MeV/c momentum range. A sketch of the apparatus is shown in figure 12. A beam of 800 MeV/c K^+ enter the detector from the left and pass through a BeO degrader before being brought to rest in a stopping target, formed from optically clad scintillating fibers. The fibers are grouped into 379 bundles-of- six and aligned parallel to the beam.

Decay π^+'s leave the stopping target and pass through a five layer drift chamber, where their curvature in the 10 kG axial field is measured to determine their momentum. The drift chamber is of the "jet chamber" variety, with the wires of each layer being grouped into a number of six sense-wire cells. The 2'nd and 4'th layers are stereo layers, which allows a measurement of a track's trajectory in z.

Once through the drift chamber, the π^+'s are brought to rest in a 15 layer plastic scintillator range stack. The stack is segmented 24 ways in azimuth and each counter is read from both ends by two inch phototubes. One novel feature of the apparatus is the readout scheme for the phototubes, wherein the anode current of each range stack phototube is continuosly recorded by a transient recorder consisting of an 8 bit, 500 Mhz flash

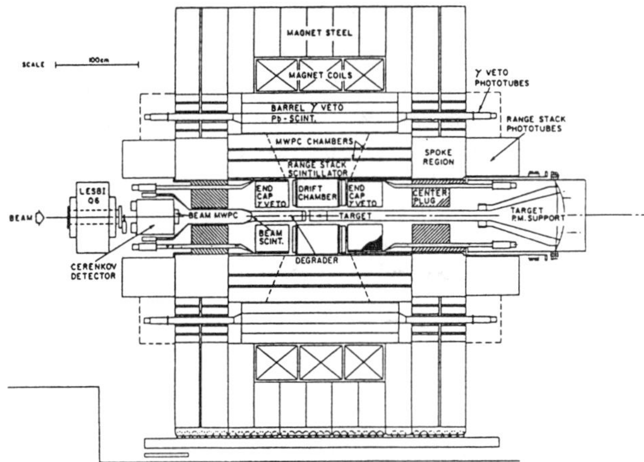

Figure 12: Side view of the E787 detector.

ADC driving a high speed ECL memory. This creates a detailed history of each event that is written to tape for subsequent analysis. Among other things, this information is useful for detecting the $\pi^+ \rightarrow \mu^+ \rightarrow e^+$ decay sequence that takes place in the π^+ stopping layer. This tag is essential for the rejection of the topologically identical $K^+ \rightarrow \mu^+\nu$ background, which occurs at rate that is 10 orders-of-magnitude above the anticipated sensitivity (this is not an experiment for the faint-hearted !).

$K^+ \rightarrow \mu^+\nu$ events are also rejected by measurements of range, energy, and momentum. Furthermore, correlations between any two of these variables, momentum/range being the most powerful, can be used to separate π^+'s from μ^+'s, even at the same momentum.

The range stack is surrounded by an array of Pb/scintillator shower counters (the "Barrel Veto") that is used to detect photons from the second major background, $K^{\pm} \rightarrow \pi^+\pi^0 \rightarrow \pi^+\gamma\gamma$. The photon detection coverage is completed by a of pair Pb/scintillator endcaps. Because shower fluctuations and photo-nuclear absorption limit the π^0 detection inefficiency to the 10^{-5} to 10^{-6} range, only the part of the $K^+ \rightarrow \pi^+\nu\bar{\nu}$ spectrum above the $K^+ \rightarrow \pi^+\pi^0$ peak will be useful, since nuclear interactions of the stopping π^+'s can reduce their energy in an undetectable way. This "spectrum cut" reduces the $K^+ \rightarrow \pi^+\nu\bar{\nu}$ detection efficiency by a factor-of-five, but does not affect the $K^+ \rightarrow \pi^+ f$ efficiency.

E787 acquired its first useful physics data in an early 1988 shakedown run. The detector worked as planned, and the group collected a sample of triggers that should allow them to improve on the Asano limit by roughly an order-of-magnitude. The ultimate goal of the experiment is a sensitivity of $B(K^+ \rightarrow \pi^+\nu\bar{\nu}) < 2 \times 10^{-10}$. The general purpose nature of the detector will allow the group to search for or detect a number of other rare K decays; $K^+ \rightarrow \pi^+\mu^+\mu^-$ and $K^+ \rightarrow \pi^+\gamma\gamma$, being two examples, are already included in the experimental trigger.

Summary and Conclusion

The field of kaon decay physics continues to thrive. Recent *CP* violation experiments have for the first time produced evidence for a non-vanishing ϵ'/ϵ value. A number of dedicated rare kaon decay experiments are now reporting results or are well underway. These measurements are a sensitive probe of physics at high mass scales and serve to constrain models of elementary particles.

Acknowledgements

The author gratefully acknowledges useful discussions with R. Cousins, G. Gollin, D. Lazarus, P. Meyers, W. Morse, J. Okamitsu, A.J.S. Smith, R. Whyley, and M. Zeller.

References

1) L. Wolfenstein, Ann. Rev. Nucl. Sci. 36, 137 (1986).
2) J. W. Cronin, Rev. Mod. Phys. 53, 373 (1981).
3) J. H. Christenson, J. W. Cronin, V. L. Fitch, and R. Turlay, Phys. Rev. Lett. 13, 138 (1964).
4) L. Wolfenstein, Phys. Rev. Lett. 13, 569 (1964).
5) H. Burkhardt et. al., NA31 collaboration, CERN preprint EP/88-47 /break (submitted to Phys. Lett. B) (1988).
6) M. Woods et. al., Phys. Rev. Lett. 60, 1695 (1988).
7) Talk by George Gollin this conference.
8) M. Kobayashi and T. Maskawa, Prog. Theo. Phys. 49, 652 (1973).
9) M. Banner et. al., Phys. Rev. Lett. 28, 1597 (1972).
10) M. Holder et. al., Phys. Lett. 40B, 141 (1972).
11) J. H. Christenson et. al., Phys. Rev. Lett. 43, 1209 (1979).
12) J. K. Black et. al., Phys. Rev. Lett. 54, 1628 (1985).
13) R. H. Bernstein et. al., Phys. Rev Lett. 54, 1631 (1985).
14) M. Kasemann, PhD. Thesis, (1987).
15) R. N. Cahn and H. Harari, Nucl. Phys. B176, 135 (1980¿
16) Physics with LAMPF II, Proposal LA-9798-P, (1983).
17) T. Goldman pg. 895, Proc. of Conference on Intersections Between Particle and Nuclear Physics, Steamboat Springs (1984).
18) H. B. Greenlee et. al., Phys. Rev. Lett. 60, 893 (1988). Talk by W. Morse, this conference.
19) AGS E791 (UCLA, LASL, Penn, Stanford, Temple, Wm.& Mary collaboration), Talk by R. Whyley, this conference.
20) KEK proposal for $K_L \to \mu e$, T. Inagaki et. al.
21) AGS E777 (BNL, PSI, Washington, Yale collaboration), talk by D. Lazarus, this conference.

22) M. Aguilar-Benitez et. al., Particle Data Group, Phys. Lett. 170B, (1986).
23) N. J. Baker et. al., Phys. Rev. Lett. 59, 2832 (1987).
24) C. Campagnari et. al., submitted to Phys. Rev. Lett. (1988).
25) T. Cowan et. al., Phys. Rev. Lett. 56, 444 (1986).
26) R.D. Peccei, T.T. Wu, and T. Yanagida, Phys. Lett. 172B, 435 (1986).
27) L. M. Krauss and F. Wilczek, Phys. Lett. 173B, 189 (1986).
28) R. D. Peccei and H. R. Quinn, Phys. Rev. Lett. 38, 1440 (1977); Phys. Rev. D16, 1791 (1977).
29) S. Weinberg, Phys. Rev. Lett. 40, 223 (1978).
30) F. Wilczek, Phys. Rev. Lett. 40, 279 (1978).
31) C. Edwards et. al., Phys. Rev. Lett. 48, 903 (1982).
32) M. Sivertz et. al., Phys. Rev. D26, 717 (1982).
33) F. Porter and K. Königsmann, Phys. Rev. D25, 1993 (1982), D26, 716E (1982).
34) L. M. Krauss and M. Zeller, Phys. Rev. D34, 3385 (1986).
35) J. S. Frank et. al., Phys. Rev. D28, 423, (1983).
36) J. Fischer et. al., Phys. Lett. 73B, 364 (1978).
37) R. Eichler et. al. Phys. Lett. 175B, 101 (1986).
38) S. L. Glashow, J. Illiopoulos, and L. Maiani, Phys. Rev. D2, 1285 (1970).
39) T. Inami and C.S. Lim, Prog. Theo. Phys. 65, 297 (1981).
40) F.J. Gilman and J.S. Hagelin, Phys. Lett. 133B, 433 (1983).
41) J. Ellis, J.S. Hagelin and S. Rudaz, Phys. Lett. 192B , 201 (1987).
42) J. Ellis, J.S. Hagelin, S. Rudaz, and D.-D. Wu, CERN preprint TH.4816/87.(submitted to Nucl. Phys. B).
43) U. Türke, Phys. Lett. 168B, 296 (1986).
44) W. J. Marciano and Z. Parsa, Ann. Rev. Nucl. Part. Sci. 36, 171 (1986).
45) S. Bertolino and A. Masiero, Phys. Lett. B174, 343 (1986).
46) C. Albajar et. al., Phys. Lett. B198, 261 (1987).
47) Y. Asano et. al., Phys. Lett. 107B, 159 (1981).
48) F. Wilczek, Phys. Rev. Lett. 49, 1549 (1982).
49) J. H. Klems et. al., Phys. Rev. D4, 66 (1971).
50) G. D. Cable et. al., Phys. Rev. D8, 3807 (1973).
51) AGS E787, (BNL, Princeton, TRIUMF, collaboration). Talk by R. Strand, this conference.
52) H. Albrecht et. al., Phys. Lett. 192B, 245 (1987).

CHIRAL SYMMETRY IN QCD

John F. Donoghue
University of Massachusetts, Amherst, MA 01002

ABSTRACT

The subject of chiral symmetry and the techniques for utilizing it are overviewed.

It is often said that there is no rigorous techniques for dealing with low energy processes in QCD, and hence one must resort to models, such as the quark model or pole models. However, this is not true. Chiral symmetry techniques follow rigorously from QCD and form a rich and subtle way to handle very low energy physics. To a large extent, however, this beautiful theory is very poorly understood by large portions of the community. In many ways it is better known to theorists doing superstrings in 11 dimensions then it is by the physicists who study QCD processes like π N scattering. In the talk, I hope to provide a brief introduction to chiral perturbation theory, along with a couple of examples from π N physics.

In case that one needs to be convinced that QCD is a highly nontrivial theory, one need only look at the approximate symmetries of the QCD Lagrangian. Here I will pretend that there are only three quarks u,d,s. Their masses are small enough that we can consider the limit when $m_i \to 0$. In this case, the action of QCD,

$$S = \int d^4x \; L_{QCD}$$

$$L_{QCD} = -\frac{1}{4} F^A_{\mu\nu} F^{A\mu\nu} + \bar{\Psi}\, i \not{D} \Psi$$

$$\Psi = \begin{pmatrix} u \\ d \\ s \end{pmatrix} \tag{1}$$

has the following invariances:

1) Scale invariance. There are no dimensionful parameters in the action, and hence it is invariant under a change of scale.

$$\Psi(x) \to \Psi(x) + \lambda^{3/2}\, \Psi(\lambda x)$$

$$A^A_\mu(x) \to A^A_\mu(X) = \lambda\, A^A_\mu(\lambda x) \tag{2}$$

Since this is a continuous invariance, Noether's theorem tells us that there is a classically conserved current

$$J_D^\mu = T^{\mu\nu} x_\nu$$

$$\partial_\mu J_D^\mu = T^\mu_\mu = 0 \tag{3}$$

where $T^{\mu\nu}$ is the energy momentum tensor. In fact, the energy momentum tensor is easily seen to be classically traceless. The consequence is that single particle states must be massless, as kinematically the energy momentum tensors trace is the particle mass

$$< P | T^\mu_\mu | P > = m = 0 \tag{4}$$

2) $SU(3)_L \times SU(3)_R$ chiral symmetry. If one splits the fermion fields into then left- and right-handed components

$$\Psi_{\substack{L \\ R}} = \frac{1}{2} (1 \pm \gamma_5) \Psi \tag{5}$$

the two sets of fields decouple, since

$$\bar{\Psi} \not{D} \Psi = \bar{\Psi}_L \not{D} \Psi_L + \bar{\Psi}_R \not{D} \Psi_R \tag{6}$$

There are separate SU(3) invariances for each set

$$\Psi_L \rightarrow e^{i\alpha \cdot \lambda} \Psi_L$$

$$\Psi_R \rightarrow e^{i\beta \cdot \lambda} \Psi_R \tag{7}$$

where λ^A, A = 1, 2 ... 8, are the SU(3) Gell Mann matrices acting on the flavor indices (u,d,s), and α^A, β^A are arbitrary constants. This is the SU(3) chiral symmetry.

3) U_A (1) axial symmetry. There is an extra invariance beyond the above SU(3) symmetry. If one rotates all fields by a common γ_5 transformation

$$\Psi \rightarrow e^{ia\gamma} 5 \, \Psi \tag{8}$$

the Lagrangian is unchanged. Here the symmetry current is

$$A_\mu^0 = \bar{u} \gamma_\mu \gamma_5 u + \bar{d} \gamma_\mu \gamma_5 d + \bar{s} \gamma_\mu \gamma_5 s \ , \tag{9}$$

which is classically conserved.

These symmetries of the Lagrangian are not visible in the particle spectrum, and any calculational scheme which uses QCD must deal with their fates. Two of them, the scale and axial $U_A(1)$ invariances, turn out not to be even approximate symmetries of the full quantum (1) theory. Noether's theorem has mislead us and even though the action is invariant the full path integral defining the theory is not. As operator relations, the current divergences become[1]

$$\partial_\mu J^\mu_D = \theta^\mu{}_\mu = \beta\, F^A_{\mu\nu} F^{A\mu\nu} + m_u \bar{u} u + m_d \bar{d} d + m_s \bar{s} s$$

$$\partial_\mu A^{0\mu} = \frac{3\alpha_s}{4\pi} F^A_{\mu\nu} \tilde{F}^{A\mu\nu} + m_u \bar{u} \gamma_5 u + m_d \bar{d} \gamma_5 d + m_s \bar{s} \gamma_5 s \qquad (10)$$

where β is related to the QCD β function

$$\beta = -\frac{(33 - 2N_f)}{24\pi} \alpha_s \qquad (11)$$

and the effects of quark masses has been restored. The gluonic portions of the current divergences are often referred to as the trace anomaly and axial anomaly respectively. They are very important for the overall structure of QCD.

The chiral SU(3) symmetry has a different fate. It remains an approximate symmetry, but in a mode with "dynamical symmetry breaking". This is not a "breaking" of the symmetry as would be produced by large quark masses, but rather a hiding of the symmetry.[2] This occurs when the vacuum state is not invariant under the symmetry and (pseudo-) Goldstone bosons are produced. The pions, kaons, and eta would be the massless Goldstone particles if the quark masses vanished. Even in the symmetry limit, the symmetry is not visible in the spectrum. Conventional symmetries, such as isospin, have the property that a symmetry transformation relates a single particle state to another single particle state in the same multiplet. In this case, particles must fall in degenerate multiplets, such as (P, N) or (π^+, π^0, π^-) for isospin symmetry. With a dynamically hidden symmetry, the symmetry transformation is realized by relating states with differing numbers of the Goldstone particles. Thus in the symmetry limit a transformation of the vacuum produces a state with a pion in it, degenerate in energy with the vacuum because the pion is massless. Likewise the transformation of the proton state relates it to states with extra pions. There need not be single particle multiplets, and hence, one cannot see the symmetry in the spectrum, aside from the existence of the Goldstone bosons. However there is still a rich body of predictions from the symmetry.

The "old fashioned" technique to obtain predictions from chiral symmetry involves the soft pion theorem,[3] involving a pion of vanishing four-momentum

$$\lim_{q_u \to 0} < \beta\, \pi^i(q) \,|\, O \,|\, \alpha > = -\frac{i}{F_\pi} < \beta \,|\, [Q^i_5, O\,] \,|\, \alpha > \qquad (12)$$

where $F_\pi = 93$ MeV is the pion decay constant. One typically expands matrix elements in powers of the moments and then applies this formula in order to relate matrix elements. In some cases this is easy, but others, especially involving more than one pion, can be quite complicated.

There is a more convenient and ultimately more powerful technique which is now heavily used. It involves the use of nonlinear phenomenological Lagrangians.[4,5,6] The basic idea is that if a prediction follows from symmetry consideration alone then any Lagrangian which has the same symmetry will yield the same predictions. The procedure then involves writing out the most general possible Lagrangian consistent with chiral symmetry. One then calculates all possible diagrams using this Lagrangian, renormalizes the parameters, and the result is the most general structure compatible with the symmetry. In practice this daunting procedure is made much easier at low energy by dimensional considerations.

We want a compact way to write a Lagrangian, involving only the Goldstone bosons, which has the desired symmetry. Because all numbers of pions are involved, the result will always be nonlinear. Most authors use the exponential representation

$$U = \exp\left(i \frac{\lambda^A \phi^A}{F} \right) \tag{13}$$

where ϕ^A $A = 1, 2 \ldots 8$ are the octet of pseudoscalar mesons and λ^A are the SU(3) matrices. F is a constant with dimensions of mass. In order to see that this form is appropriate, imagine a coupling to the chiral sources, quarks,

$$\bar{\Psi}_L U \Psi_R + \text{h.c.} \tag{14}$$

The conditions which we want is that U contain the 8 meson fields and be invariant under a transformation

$$\Psi_L \to \Psi'_L = L \Psi_L \tag{15a}$$

$$\Psi_R \to \Psi'_R = R \Psi_R \tag{15b}$$

$$U \to U' = L U R^+ \tag{15c}$$

where L, R are separate SU(3) matrices. Since U is also an SU(3) matrix, the latter transformation can indeed be form invariant. (Other representations are possible, but are not widely used.)

Using this SU(3) matrix for the fields, one wants the most general possible Lagrangian consistent with the global symmetry Eq. 15c. There are, in fact, an infinite number of terms. However, in the limit of exact symmetry there are no terms which just involve the matrix U without any derivatives. For such a term to be invariant it would have to be of the form Tr $(U U^\dagger U U^\dagger \ldots U U^\dagger)$ which is just a pure number, since $U U^\dagger = 1$. There is one possible form with two derivatives, Tr $(\partial_\mu U \partial^\mu U^\dagger)$, and increasing numbers with more derivatives. The general Lagrangian then has the form.

$$L_0 = \frac{F_0^2}{4} \mathrm{Tr}\,(\partial_\mu U\, \partial^\mu U^\dagger) + \frac{\alpha_1}{4} \left[\mathrm{Tr}\,(\partial_\mu U\, \partial^\mu U^\dagger)\right]^2$$

$$+ \frac{\alpha_2}{4} \mathrm{Tr}\,(\partial_\mu U\, \partial_\nu U^\dagger)\, \mathrm{Tr}\,(\partial^\mu U\, \partial^\nu U^\dagger) + \ldots \tag{16}$$

where F_0, α_1, α_2 are constants. In order that this parameterization describe correctly normalized pion fields when we expand it using Eq. 13,

$$L = \frac{F_0^2}{2F^2} \partial_\mu \phi^i \partial^\mu \phi^i + \ldots \tag{17}$$

We choose $F_0 = F$ and later consideration of the decay $\pi \to e\,\nu$ identifies $F = F_\pi = 94$ MeV at this order. [These identifications need to be done for the renormalized quantities when carrying out loop calculations]. The quark masses break the chiral symmetry, with a particular pattern

$$\bar{\Psi}\, m\, \Psi = \bar{\Psi}_L\, m\, \Psi_R + \bar{\psi}_R\, m\, \Psi_L$$

$$\bar{\Psi}_L\, m\, \Psi_R \to \bar{\Psi}_L\, L\, m\, R\, \Psi_R \tag{18}$$

This mass matrix needs to be incorporated into the Lagrangian in a power series,

$$L = L_0 + L_m$$

$$L_m = \frac{F_\pi^2 B_0}{4} \mathrm{Tr}\,(mU + U^\dagger m) + B_0^2 \beta_1 \left[\mathrm{Tr}\,(mU + U^\dagger m)\right]^2 + \ldots \tag{19}$$

where B_0 is a constant.

The key ingredient which makes this framework manageable is the idea of the expansion in powers of the energy. In matrix elements, derivatives become factors of the four momenta, q_μ, and each matrix element will become an expansion in the momenta, schematically

$$M_{if} = F_0^2 q^2 + \alpha_1 q^4 + \alpha_2 q^4 + O\,(q^6, q^8) \ldots \tag{20}$$

[A similar expansion for the mass terms exists.] In principle there is always an energy below which the effect of $q^4, q^6, \ldots$ terms are small compared to q^2 terms, and hence the values of all but one of the infinite number of coefficients do not need to be known. The lowest order terms in the energy expansion

$$L_{l.o.} = \frac{F_\pi^2}{4} \mathrm{Tr}\,(\partial_\mu U\, \partial^\mu U^\dagger) + \frac{F_\pi^2 B_0}{2} \mathrm{Tr}\,(m\,U + U^\dagger\, m) \tag{21}$$

yield exactly the low energy predictions of chiral symmetry in terms of F_π, m_π and m_K, and reproduce the results of the soft pion techniques. In practice, the expansion in the real world seems to be in powers of $\frac{q^2}{1\ \mathrm{GeV}^2}$ or perhaps $\frac{q^2}{m_\rho^2}$, so that "low energy" refers to the energy being much less than 1 GeV. At moderate energies, corrections to the lowest order predictions are given by the terms of order q^4 and require new coefficients in the effective Lagrangian (α_1, α_2, ...). However, the number of terms is small, and these coefficients can be determined from the data.

At this point one can already learn about the quark masses. Expansion of Eq. 21 tells us that

$$m_\pi^2 = (m_u + m_d)\, B_0 \equiv 2\hat{m}\, B_0$$

$$m_K^2 = (m_u + m_s)\, B_0$$

$$m_{\eta_8}^2 = \frac{4}{3}\, m_K^2 - \frac{1}{3}\, m_\pi^2 \qquad \text{(the Gell Mann Okubo relation)} \tag{22}$$

While the absolute values of the quark masses are never obtained from chiral symmetry, their ratio is

$$\frac{\hat{m}}{m_s} = \frac{m_\pi^2}{2m_K^2} \approx \frac{1}{25} \tag{23}$$

in lowest order. This result is not changed much at higher order.[5] This is the best source of information about quark masses.

Note that the effective Lagrangian of QCD is not the same as that of the linear sigma model, another renormalizable theory with chiral symmetry. It is not even the same as that obtained with the sigma field "integrated out", i.e. summing overall effects of sigma exchange and loops, leaving only pion fields in the Lagrangian. It is true that both theories have the same lowest order form, Eq. 21, as do <u>all</u> theories with the same patterns of chiral SU(3) symmetry. However, the coefficients of the terms of order q^4 depend on the underlying theory, and will then differ between QCD and the linear sigma model. The results for the linear sigma model have been worked out and disagree strongly with those found in nature.[5] [The coefficients of QCD have not yet been obtained, but my presumption here is that they are equal to what we see in the data.]

We now must deal with the question of loop diagrams involving pions. The effective Lagrangian is a nonlinear, nonrenormalizable theory and one might anticipate difficulties when attempting to renormalize such a theory. However, loops are necessary in order to generate the full set of intermediate states required by unitarity. The procedure is in fact rather simple. Using the lowest order Lagrangian in a loop diagram generates effects at order E^4. The general proof is given in a very nice paper by Weinberg,[4] and this will be seen explicitly in the example of $\pi\pi$ scattering given later. Divergences present can be absorbed into the coefficients at order E^4, with renormalized coefficients being extracted from experiment. Higher loops will renormalize yet higher terms in the energy expansion. It is in this sense that the theory is nonrenormalizable, in that to completely specify the full theory one would need to determine an infinite number of coefficients. However, the low energy expansion comes to our rescue. At low energy only a finite (small) number of coefficients need to be identified.

In practice, then, the procedure goes as follows. At very low energies one needs only consider the lowest order Lagrangian, Eq. 21, treated at tree level. To determine the first order corrections to these results, one includes one loop diagrams and the E^4 terms in the Lagrangian. A set of renormalized coefficients is determined from the data, and then further predictions can be made. The general Lagrangian at order E^4 has been determined by Gasser and Leutwyler,[5] and they have performed the renormalization procedure. This method is often called chiral perturbation theory, and it appears as rigorous as any perturbative expansion can be.

Those processes which utilize only chiral SU(2) can be studied with greater precision than those using chiral SU(3) only because they can occur at lower energy. SU(3) predictions always involve the K or the η and then occur at energies of $\frac{1}{2}$ GeV. We can test the reliability of the energy expansion at this energy by looking at processes such as $\pi\pi$ scattering. At $\frac{1}{2}$ GeV the E^4 terms become visible, but the energy expansion has not broken down. One can expect 30% corrections (generically, but with individual variations) to lowest order relations. Note that chiral SU(3) predictions at $\sqrt{s} = \frac{1}{2}$ GeV are no worse then SU(2) predictions at the same energy. $K \to 3\pi$ and $\pi\pi \to \pi\pi$ at $\sqrt{s} = \frac{1}{2}$ GeV should and do both work reasonably well but $K\bar{K} \to \pi\pi$ and $\pi\pi \to \pi\pi$ at $\sqrt{s} = \frac{1}{2}$ GeV will both likely fail.

It is worth mentioning some techniques which are not chiral perturbation theory. One is Skyrmions,[7] solitons which can appear in an effective chiral Lagrangian. These are not part of the low energy expansion. In forming a soliton, the destabilizing pressure of $L_{l.o.}$ is balanced by the stabilizing pressure of higher order terms at order E^4 by finding a configuration where they are both equal. However, this means that yet higher order terms (six derivatives and more) are also then of comparable magnitude. However, these are unknown. Solitons lie at high energy, where the energy expansion has broken down. Although fascinating, they do not share the rigor of the low energy expansion. Another technique which is in common use is to treat loops keeping only the leading nonanalytic effects (such as $m^2 \ln m^2$) but dropping constants and the E^4 coefficients. While not senseless, this has been seen to be in practice incorrect, as it almost never yields a phenomenologically acceptable result. The E^4 coefficients are important.

As a first example of chiral perturbation theory, consider $\pi\pi$ scattering, where the renormalization has been carried out by Gasser and Leutwyler,[4] and the phenomenology has been discussed recently in Ref. 8. Here I will give the formulas in the massless limit for the sake of compactness, although the results quoted are for the full theory. Because of the symmetry of the scattering process for $\pi^\alpha + \pi^\beta \to \pi^\gamma + \pi^\delta$, the amplitude is given in terms of a single function

$$M(\pi^\alpha + \pi^\beta \to \pi^\gamma + \pi^\delta) = A(s, t, u)\, \delta_{\alpha\beta}\, \delta_{\gamma\delta} + A(t, s, u)\, \delta_{\alpha\gamma}\, \delta_{\beta\delta} + A(u, t, s)\, \delta_{\alpha\delta}\, \delta_{\beta\gamma}$$

$$s = (p_\alpha + p_\beta)^2$$

$$t = (p_\alpha - p_\gamma)^2$$

$$u = (p_\alpha - p_\delta)^2 \tag{24}$$

This has the following form

$$A(s, t, u) = \frac{s - m_\pi^2}{F^2} + B(s, t, u) + C(s, t, u) \tag{25}$$

where the first term is due to the lowest order Lagrangian and

$$B(s, t, u) = \frac{1}{6F^4}\left\{ 3s^2\, K(s) + t(t-u)\, K(t) + u(u-t)\, K(u) - \frac{1}{96\pi^2}\left(21s^2 + 5(t-u)\right)\right\}$$

$$K(s) = \frac{-1}{16\pi^2}\left[\ln\left(\frac{-s}{\mu^2}\right) - 2\right]$$

$$C(s, t, u) = \frac{1}{2F^4}\left[4\alpha_1^{ren}\, s^2 + \alpha_2^{ren}\left(s^2 + (t-u)^2\right)\right] \tag{26}$$

where the loops have been calculated using dimensional regularization. In dimensional regularizations the renormalized couplings are related to the bare ones by an infinite renormalization of order $\frac{1}{d\text{-}4}$. Note that the renormalized coefficients extracted from experiment will depend on the choice of the scale μ present in B(s, t, U)'. B and C have a factor of $\frac{1}{F_\pi^4}$ and hence must be of order E^4 strictly for dimensional reasons.

Data on ππ scattering comes from two basic sources. At low energy the ππ interaction can be extracted from $K_L \to \pi\pi\, e\nu$, with the analysis being under good theoretical control. At somewhat higher energies, $\pi N \to \pi\pi N$ can provide information by extrapolation of the amplitude to the pion pole in the t channel, as in Fig. 1. Here caution is advised as often theoretical assumptions are made in the extrapolation in order to increase the statistical power of the experiment. However, since all too often experiments disagree with each other well outside of their quoted errors, it becomes clear that these assumptions are not above suspicion. However, only the general trends of the data will be important for our purposes.

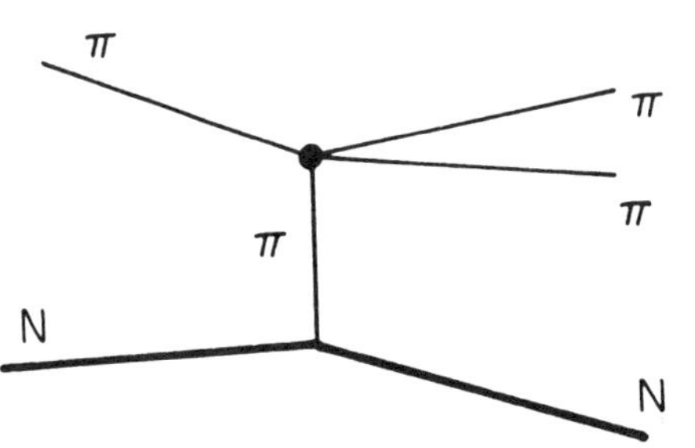

The data is given in Fig. 2. Here the dotted line is the lowest order chiral symmetry prediction. While, under closer inspection than is possible in these figures, it is found to do well near threshold, it clearly becomes worse at moderate energies violating unitarity at 700 MeV. The solid line is the full prediction at order E^4, using

$$\alpha_1^{\text{ren}}\left(\frac{1}{2}\text{ GeV}\right) = -.010$$

$$\alpha_2^{\text{ren}}\left(\frac{1}{2}\text{ GeV}\right) = +.010 \qquad (27)$$

As expected, it merges with the lowest order result near threshold, but becomes important at moderate energies. These effects are clearly present in the data. By its very nature, the low energy expansion must fail as it passes the ρ resonance (the bump in T:) as an infinite number of terms must be present in order to reproduce a Breit-Wigner shape. However, the low energy tail of the ρ is correctly present in the coefficients of the E^4 Lagrangian.

A second example also comes from pion nucleon scattering, i.e. the sigma term in $\pi N \to \pi N$. Here, in the symmetry limit, if both pions were at zero four momentions the amplitude would vanish. In the presence of quark masses, the

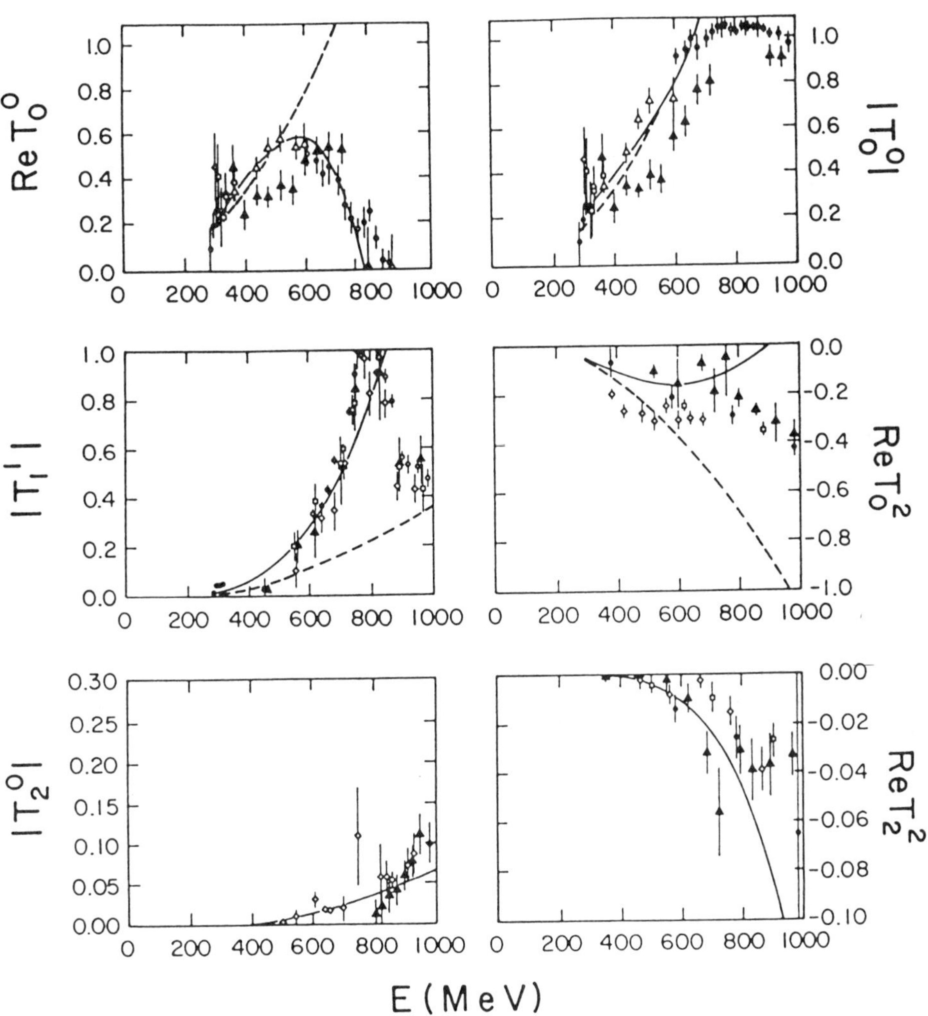
Re T$_0^0$
$|T_0^0|$
$|T_1^1|$
Re T$_0^2$
$|T_2^0|$
Re T$_2^2$
E (MeV)

amplitude extracted to the Cheng Dashen point, $\nu = \frac{s-t}{4m_p} = 0$ and $t = 2m_\pi^2$, is a measure of the symmetry breaking, i.e. the sigma term.

$$\sigma = \langle P | m_u \bar{u} u + m_d \bar{d} d | P \rangle \tag{28}$$

There are several issues here. One is the theoretical relation between the "experimental" amplitude at the Cheng Dashen point $\Sigma\left(2m_\pi^2\right)$ and the σ matrix element above. There has been recent progress here as Gasser, Sainio and Švarc[9] have calculated this correction in chiral perturbation theory, with the result

$$\sigma(0) = \Sigma(2m_\pi^2) - (4.5 \pm 1.2)\ \text{MeV} \tag{29}$$

Another issue is the extraction of the Σ matrix element from experiment. This is a subtle procedure but results seem to be stable around 60 MeV,

$$\Sigma = 64 \pm 8\ \text{NeV} \qquad \text{(Ref. 10)}$$

$$= 61\ \text{MeV} \qquad \text{(Ref. 11)}$$

or

$$\sigma \approx 55 \pm 8\ \text{MeV} \tag{30}$$

In baryons, the energy expansion seems to be in terms of variables such as $(p + q)^2 - p^2 \approx 2m_N E_\pi$ rather than $E_\pi{}^2$ as in mesons. This parameter gets large quickly, such that pion threshold $(= 2m_N m_\pi)$ is comparable to $s = (500\ \text{MeV})^2$ in the meson regime. Corrections become more important in baryons. There is room for caution in the extrapolation, but at present there is no known flaw.

The third issue in this connection is the theoretical expectation for the σ matrix element.[12] Use of the quark mass ratio, Eq. 23, plus hyperon mass splitting leads to

$$\delta \equiv \langle P | \hat{m}\left(\bar{u} u + \bar{d} d - 2 \bar{s} s\right) | P \rangle = \frac{3m_\pi^2}{2m_k^2}(m_\Xi - m_\Lambda) \approx 25\ \text{MeV} \tag{31}$$

to first order in SU(3). A direct comparison of σ and δ reveals that unless $\langle P | \bar{s} s | P \rangle \neq 0$ there is no way to reconcile these values. This issue is receiving considerable scrutiny at present, with attempts to scrutinize the extraction of σ, the validity of the SU(3) analysis and the possibility of generating the proton's $\bar{s} s$ matrix element. As long as the data does not change, it is clear that some aspects of our standard intuition will have to change. To a large extent the small value of δ comes from just the additive quark model in baryons, plus the quark mass ratio. For example, hyperon mass splittings are expected to be roughly linear in the strange quark mass,

$$m_\Lambda - m_p \approx \langle \Lambda | m_s \bar{s} s | \Lambda \rangle - \langle P | m_s \bar{s} s | \Lambda \rangle$$

$$\approx \langle \Lambda | m_s \bar{s} s | \Lambda \rangle \tag{32}$$

so that

$$\sigma = \langle P | \hat{m}(\bar{u} u + \bar{d} d) | P \rangle \approx \frac{3\hat{m}}{m_s} \langle \Lambda | m_s \bar{s} s | P \rangle \approx 25 \text{ MeV} \tag{33}$$

Some portion of this heuristic derivation must be wrong if the data is correct. The use of first order QCD would suggest that the flaw is in the assumption that the proton's $\bar{s}s$ matrix element is small.

The "contents" of the nucleon is a subtle concept, although in principle matrix elements of various operators are well defined and can give us some insight into the structure of the proton. Use of the trace anomaly relation allows a division

$$m_N = m_0 + \sigma$$

$$m_0 = \langle P | \ \beta F^A_{\mu\nu} F^{A\mu\nu} + m_s \bar{s} s \ | P \rangle = 880 \text{ MeV}$$

$$\sigma = \langle P | \ m_u \bar{u} u + m_d \bar{d} d \ | P \rangle = 55 \text{ MeV} \tag{34}$$

Use of the same input which lead to Eq. 31 plus the experimental value of σ leads to a further division

$$\langle P | \beta F^A_{\mu\nu} F^{A\mu\nu} \ | P \rangle = 510 \pm 100 \text{ MeV}$$

$$\langle P | m_s \bar{s} s \ | P \rangle = 370 \pm 100 \text{ MeV} \tag{35}$$

where the error bars come from the quoted range of σ. In this split, most of the nucleon mass "comes from" non valence components. The interpretation of this requires some caution. Consider, for example, how the analysis would go if heavy quarks were included. The corresponding changes in the trace relation would be the heavy quark mass terms plus the change of the β function to that of six quarks.

However, the effect of the heavy quark can be calculated in QCD perturbation theory, where the results just shifts the coefficient in the β function, i.e.

$$<P|\beta_6 F^2 + m_c \bar{c} c + m_t \bar{t} t + m_b \bar{b} b + ... |P> = <P|\beta_3 F^2 + ... |P> + O\left(\frac{1}{m_H}\right) \tag{36}$$

or, to lowest order using $\beta \alpha$ (33 - 2 $N_{flavors}$),

$$<P|m_c \bar{c} c + m_t \bar{t} t + m_b \bar{b} b|P> \approx \frac{6}{27} <P|\beta_3 F^2 |P> \approx 100 - 200 \text{ MeV} \tag{37}$$

Is this contribution to the proton mass "due to" heavy quarks, as indicated by the left hand side, or gluons, as on the right side? What this tells us is in many ways as expected: In an interacting field theory all of the fields are present at all times. A similar example is set by the vacuum matrix elements such as

$$<0|m_s \bar{s} s |0> = -F_\pi^2 m_K^2 = -(220 \text{ MeV})^4 \tag{38}$$

in lowest order chiral symmetry. In reality, the puzzle is not why these non valence matrix elements can be large, but why the naive quark model does so well. The answer must be that the effective "valence" quarks of the quark model are composites of all quarks and gluons with properties that approximate a massive fermion. The quark model has then really parameterized the end result, but has not helped us understand the real connection to QCD.

In this brief talk, I have not been able to present much of the richness of the theory of chiral symmetry. The subject is undergoing a theoretical revival as it is clear that it is an essential, rigorous and useful aspect of QCD. The areas of study include trying to understand the effective Lagrangians to $O(E^4)$ in order to quantify low energy QCD, to understand the connection with the underlying theory, and to use chiral symmetry as a tool in the search for new physics (such as rare K decays or CP violation). It is a somewhat subtle theory and theorists are still learning how to apply it. Experiment has not yet refocused its attention on some of these issues. However, the interest in intense hadron facilities (π, K, $\bar{p}$ ... machines) and rare decays may yet add more information. Essentially every process involving π, K, η or photons at low energy bears on the study of chiral symmetry. I am convinced that the language of chiral symmetry is now a permanent feature of the field and it should be learned by any serious student of the standard model.

REFERENCES

1. S. Adler, Phys. Rev. 177, 2426 (1969).
   J. Bell and R. Jackiev, Nuovo Cimento A60, 47 (1969).
   K. Fujikawa, Phys. Rev. D21, 2848 (1980); Phys. Rev. D23, 2262 (1981).
   R.J. Crewther, Phys. Rev. Lett. 28, 1421 (1972).
   M.S. Chanowitz and J. Ellis, Phys. Rev. D7, 2490 (1973).
2. S. Coleman, in "Laws of Hadronic Matter" ed by N. Zichichi (Academic London, 1975), reprinted in S. Coleman, "Aspects of Symmetry" (Cambridge, 1985).
3. S.L. Adler and R. Dashen, "Current Algebras and Applications to Particle Physics", (Benjamin, NY, 1968).
4. S. Weinberg, Physica 96A, 327 (1979).
5. J. Gasser and H. Leutwyler, Am. Phys. (N.Y.) 158, 142 (1984) Nucl. Phys. B250, 465 (1985).
6. H. Georgi, "Weak Interactions and Modern Particle Theory" (Benjamin, Menlo Park, 1984).
   C.G. Callen, S. Coleman, J. Wess and B.Zumino, Phys. Rev. 177, 2247 (1969).
   S. Coleman, J. Wess and B. Zumino, Phys. Rev. 177, 2239 (1969).
7. T.H.R. Skyrme, Proc. Roy. Soc. Lon. A260, 127 (1961); Nucl. Phys. 31, 556 (1962).
   G. Adkins, C. Nappi and E. Witten, Nucl. Phys. B228, 552 (1983).
8. J.F. Donoghue, C. Ramirez and G. Valencia, University of Massachusetts preprint UMHEP-296.
9. J. Gasser, M.E. Sainio and A. Švarc, University of Bern preprint, BUTP-87/17.
10. G. Höhler, "Pion Nucleon Scattering" in Landolt-Börnstein V of ± / 9b ed. H. Schopper, Springer Heidelberg (1983).
    R. Koch, Z. Physik C15, 161 (1982).
11. J. Gasser, Proceedings of the Second International Workshop on πN Physics, ed. by W. Gibbs and B. Nefkins (Los Alamos, LA - 11/84 - C, 1987), p. 266.
12. T.P. Cheng, Phys. Rev. D13, 2161 (1976).
    R.L. Jaffe, (Phys. Rev. D21, 3215 (1980).
    C.A. Dominguez and P. Langacker, Phys. Rev. D24, 1905 (1981).
    J.F. Donoghue and C.R. Nappi, Phys. Lett. 168B, 105 (1986).

PERTURBATIVE QCD IN EXCLUSIVE PROCESSES

Nathan Isgur*
CERN, 1211 Geneva 23
Switzerland

C.H. Llewellyn Smith
Department of Theoretical Physics
1 Keble Road
Oxford, OX1 3NP,
U.K.

(presented by Nathan Isgur)

Abstract

We have re-examined our arguments against the dominance of perturbative QCD over soft, non-perturbative effects in exclusive processes at currently available Q^2. We find that claims that our earlier objections can be overcome are based on the illegitimate use of perturbative QCD, and that the perturbative contributions are much smaller than soft contributions (which are capable of explaining the data) in the simple test cases of the pion and nucleon electromagnetic form factors. We see no reason why our results should not generalize, and conclude that there is no justification for the continued application of perturbative QCD to exclusive processes.

Since we first sounded a warning [1] against the use of perturbative QCD for exclusive processes [2-5] there have been many developments in both theory and experiment. We have therefore re-examined the issues involved. Here I summarise our results : a detailed discussion is given elsewhere [6].

The factors which prompted this re-examination include:

1) Reanalyses [7-10] of the perturbative calculation, which reveal a sign error that reverses, for a given nucleon wave function, the predicted sign of the nucleon magnetic form factors. This implies that only asymmetric and very sub-asymptotic wave functions have any hope of describing the data for G_M^p and G_M^n.
2) Sum rule calculations [9], supported by lattice Monte Carlo calculations [11, 12], which find distribution amplitudes of the type apparently needed to explain both the sign

* permanent address: Department of Physics, University of Toronto, Toronto, Canada

and magnitude of the nucleon magnetic form factors, and the magnitude of the pion electric form factor F_π, in terms of perturbative QCD. On the other hand, direct sum rule calculations [13-16] of the pion and nucleon form factors find, in agreement with our earlier conclusion [1], that these form factors are dominated by non-perturbative effects at accessible values of Q^2.

3) New data for exclusive hadronic [17,18] and gamma-gamma [19] reactions, which generally disagree with the predictions of perturbative QCD.

We concentrate on form factors, since they constitute the one case in which advocates of perturbative QCD claim unmitigated success for their calculations. Our main point is that, as anticipated in [1], these calculations cannot be believed since they involve an illegitimate use of perturbation theory. On the other hand, we find [1,6] that non-perturbative effects can easily explain the data.

As before [1], we start by observing that according to perturbative QCD

$$Q^4 G_M^p \sim \alpha_S^2 f \tag{1}$$

where f, which has dimensions m^4, varies logarithmically with Q^2 and would be expected to be of the order of some typical hadronic mass scale such as $< p_T^2 >^2$. This guess is borne out by simple models [1] which lead to values of G_M^p that are about two orders of magnitude less than the value of about 1 GeV^4 observed for Q^2 between 3 and 31 GeV^2. Simple models combined with perturbative QCD also underestimate the pion form factor, by a factor of about three.

We now explain how asymmetric wave functions apparently generate perturbative contributions to form factors much greater than the order of magnitude estimate above and why this enhancement is an artefact due to the illegitimate use of perturbation theory. We consider first the pion form factor which is given asymptotically by

$$F_\pi(Q^2) = \frac{4\alpha_s(Q^2)}{3\pi^2} \int_0^1 dx' \int_0^1 dx \frac{\phi_\pi^*(x')\phi_\pi(x)}{x' x Q^2} \tag{2}$$

where the distribution amplitude ϕ_π is related to the $S = 0$, $L_z = 0$ component of the (infinite momentum or light-cone) pion q-$\bar{q}$ wave function ψ_π^0 by

$$\phi_\pi(x) = \int d^2 p_T \, \psi_\pi^0(x, \vec{p}_T) \tag{3}$$

where x is the quark momentum fraction in the infinite momentum frame, ψ_π^0 satisfies

$$\int dx d^2 p_T \mid \psi_\pi^0(x, \vec{p}_T) \mid^2 = P_{q\bar{q}}^0, \tag{4}$$

$P_{q\bar{q}}^0$ being the associated q-$\bar{q}$ probability, and we have used the symmetry of the wave function for $x = x_1 \longleftrightarrow x_2 = 1 - x$ in writing (2). In ref. [1] we found that with the asymptotic form $\phi_\pi^{as} = Ax(1-x)$, or other forms suggested by simple models with probability peaked at $x = 1/2$, eqn.(2) cannot explain the data unless A corresponds to an impossibly large value of $< p_T^2 >$. However, sum rules suggest [9] a distribution $\phi_\pi^{CZ} = Bx(1-x)(1-2x)^2$ which concentrates probability near $x = 0$ and $x = 1$ and generates a much bigger form factor than ϕ_π^{as} for a given $< p_T^2 >$ because of the factor $1/(x'x)$ in (2).

Eqn.(2) is obtained by convoluting the wave functions for the incoming and outgoing pion with a hard scattering amplitude T_H which, to leading order in $1/Q^2$, can be calculated perturbatively as a power series in α_S for $Q^2 \to \infty$ with x and x' fixed. To leading order in α_S, T_H is given by the one gluon exchange contributions shown in fig.1. It is intuitively clear that the momentum transfer flowing through the gluons, which is given by $x_2' x_2 Q^2$ for the first two diagrams and $x_1' x_1 Q^2$ for the second two, must be large in order for perturbation theory to be applicable. In particular, if Q^2 is not sufficiently large, there will be important higher twist contributions to T_H. An illustrative model of such contributions is given by including an effective gluon mass m_g and using $1/(x'xQ^2 + m_g^2)$ for the propagators in fig.1; in such a model the higher twist contributions are proportional to $m_g^2/(x'xQ^2)$. The message is that $xx'Q^2$ must be large compared to the typical hadronic mass scale in order for perturbation theory to be applicable. This condition is not satisfied in calculations based on ϕ_π^{CZ} at available Q^2.

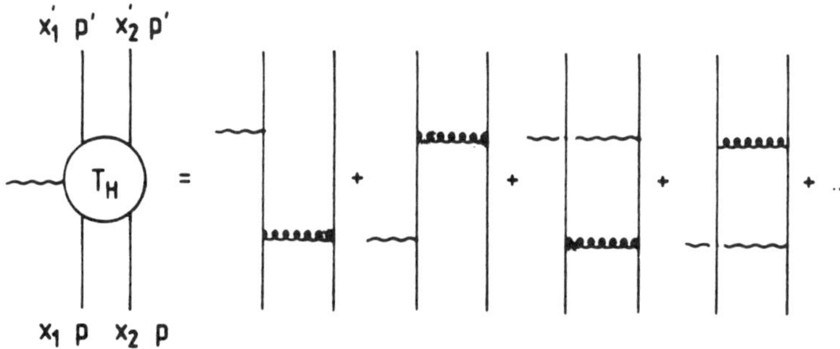

Fig. 1 : The hard scattering amplitude which controls the asymptotic behaviour of the pion form factor; $p(p')$ is the four-momentum of the initial (final) pion and $x_1 + x_2 = x_1' + x_2' = 1$.

To show just how misleading perturbation theory is when combined with this asymmetric wave function, we follow two procedures -

a) we completely exclude the regions where perturbation theory is a priori unreliable by including factors $\theta(x'xQ^2 - m_{min}^2)$ in the calculation;
b) we include a gluon mass m_g when calculating the amplitudes corresponding to fig.1.

In both cases the difference between the results obtained with these procedures and the asymptotic result is a higher twist effect. The results are given in Table 1 for several plausible values* of m_{min}^2 and m_g^2 for Q^2 in the range where perturbative QCD has been claimed [5] to be applicable to the existing data. These results suggest that when ϕ_π^{CZ} is

* It is presumably the scale of glueball masses that controls higher twist effects in the t-channel, so that it is probably safe to set m_{min}^2 or $m_g^2 \gtrsim 1GeV^2$. However, we would not consider it ridiculous to assume a value as small as $0.25GeV^2$ given that perturbative QCD is not totally misleading in deep inelastic scattering for $Q^2 \sim 1GeV^2$.

used, the asymptotic formula (2) overestimates the perturbative contribution by about an order of magnitude. The perturbative contribution exceeds the higher twist effects only for $\epsilon = m^2_{min}/Q^2$ or m^2_g/Q^2 less than about 0.01 (see Ref. 6).

procedure :	(a)		(b)	
m^2_{min} or m^2_g : (GeV^2)	0.25	1.0	0.25	1.0
$Q^2(GeV^2) = 1$	2	0	10	0
2	8	1	16	6
4	16	2	23	10
8	27	8	32	16

Table 1. The percentage of the full contribution of ϕ^{CZ}_π to Eqn.(2) which *remains* after regulating higher twist effects.

Exactly analogous considerations apply in the case of nucleon form factors. If perturbation theory is used uncritically, amplitudes which concentrate probability at small x lead to large form factors because the hard scattering amplitude contains factors $1/x$. However, it is not legitimate to use perturbation theory unless $x_i x'_j Q^2$ is large, and this condition is not satisfied unless Q^2 is enormous. Once again we quantify the degree to which perturbation theory is misleading a) by introducing theta functions which exclude the regions where perturbation theory is a priori unreliable (for details see [6]), and b) including a gluon mass in the calculation of the hard scattering amplitude. The suppressions that result for G^p_m (the effects on G^n_m are equally devastating) are given in Table 2. We see that in this case the asymptotic formula for G^p_m, when used with the distribution amplitude ϕ^{CZ}_p of Ref. [9], overestimates the perturbative contribution by about two orders of magnitude. The perturbative contribution exceeds the higher twist effects only for ϵ less than about 3x10^{-4} (see Ref.6); this value of ϵ corresponds to an m^2_{min} or m^2_g of about 0.003 GeV^2 at $Q^2 = 10\ GeV^2$ or to a Q^2 of about 1000 GeV^2 for m^2_{min} or m^2_g equal to 0.25 GeV^2 !

Having argued that perturbative QCD cannot explain the observed form factors, we must ask what mechanism is responsible. The answer given by us earlier [1], and discussed in great detail in [6] for both "intuitive" and asymmetric (CZ) wave functions, is that overlap of the "soft" components of the wave functions for the initial and final particles generates a higher twist contribution which, although very model dependent, has the same magnitude as the data in all cases.

Finally we must ask whether our conclusions preclude the use of perturbation theory at available Q^2 in other processes. As far as inclusive processes such as deep inelastic

procedure :	(a)		(b)	
m^2_{min} or m^2_g : (GeV^2)	0.25	1.0	0.25	1.0
$Q^2(GeV^2) =$ 5	1	0	3	1
10	1	0	4	2
25	1	1	7	3

Table 2: The percentage of the full contribution of ϕ_p^{CZ} to the asymptotic formula for G_m^p which *remains* after regulating higher twist effects.

scattering or the Drell Yan production of massive mu-pairs is concerned, the answer is no[6]. In these cases the expansion parameter is Q^2 (except for $x \to 1$) and the leading order terms, which are of order 1, dominate the higher twist contributions, which are of order m^2/Q^2, for Q^2 above a few GeV^2. In the case of form factors, the leading order terms are of order α_S^p, with $p = 1$ for the pion and $p = 2$ for the nucleon, which is to be compared to higher twist terms of order $m^2/\hat{Q}^2$ where $1/\hat{Q}^2$ is greater than $1/Q^2$ by factors such as $< (1/x) >$; it follows immediately that we would expect much higher Q^2 to be needed to test form factors. In the case of other exclusive processes, more powers of α_S are involved in general and the situation is even worse.

We conclude that, very unfortunately, the many elegant tests of perturbative QCD in exclusive processes that have been proposed are irrelevant at experimentally accessible values of momentum transfer.

Acknowledgements

This work was begun while N.I. was a visitor at the Department of Theoretical Physics, Oxford University. He is grateful to the Department and to the Science and Engineering Research Council for making this visit possible. He would also like to acknowledge the hospitality of the Theory Division at CERN, where it was completed.

This research was funded in part by grants from the Science and Engineering Council of the UK and the Natural Sciences and Engineering Research Council of Canada.

References

[1] N. Isgur and C.H. Llewellyn Smith, *Phys. Rev. Lett.* **52** (1984) 1080.

[2] G.R. Farrar and D.R. Jackson, *Phys. Rev. Lett.* **43** (1979) 246.
A.V. Efremov and A.V. Radyushkin, *Phys. Lett.* **94B** (1980) 245.
A. Duncan and A.H. Mueller, *Phys. Lett.* **90B** (1980) 159 and *Phys. Rev. D.* **21** (1980) 1626.

[3] G.P. Lepage and S.J. Brodsky, *Phys. Rev.D.* **22** (1980) 2157.

[4] G.P. Lepage, S.J. Brodsky, T. Huang and P.B. MacKenzie in *Particles and Fields 2*, eds. A.Z. Capri and A.N. Kamal (Plenum Press, Inc., New York, 1983), p.83.
S.J. Brodsky, T. Huang, and G.P. Lepage, *ibid*, p.143.
S.J. Brodsky and G.P. Lepage, *Phys. Rev.* **D24** (1981) 1808.
G.R. Farrar, E. Maina and F. Neri, *Nucl. Phys.* **B259** (1985) 702.

[5] S.J. Brodsky in *Proc. Int. Symp. on Medium-Energy Physics, Beijing, China, 1987*; and *SLAC-PUB*-4387 (1987).

[6] N. Isgur and C.H. Llewellyn Smith, *CERN preprint TH- 5013/88, 1988.*

[7] I.G. Aznauryan, S.V. Esaybegyan and N.L. Ter-Isaakyan, *Phys. Lett.* **B90** (1980) 151 and **B92** (1980) 371(E).

[8] V.M. Belyaev and B.L. Ioffe, *Sov. Phys.-JETP* **56** (1982) 493.

[9] V.L. Chernyak and A.R. Zhitnitsky, *Phys. Rep.* **112** (1984) 174; *Nucl. Phys.* **B246** (1984) 52.

[10] N. Isgur and C.H. Llewellyn Smith, *unpublished* (1983).

[11] A.S. Kronfeld and D.M. Photiadis, *Phys. Rev.* **D31** (1985) 2939.

[12] S. Gottlieb and A.S. Kronfeld, *Phys. Rev. Lett.* **55** (1985) 2531; *Phys. Rev.* **D33** (1986) 227.

[13] B.L. Ioffe and A.V. Smilga, *Phys. Lett.* **114B** (1982) 353; *Nucl. Phys.* **B216** (1983) 373.

[14] V.A. Nesterenko and A.V. Radyushkin, *Phys. Lett.* **115B** (1982) 410; *Sov. Phys.-JETP Lett.* **35** (1982) 488.

[15] V.A. Nesterenko and A.V. Radyushkin, *Phys. Lett.* **128B** (1983) 439; *Yad. Fiz.* **39** (1984) 1287.

[16] A.V. Radyushkin, *Acta Phys. Pol.* **B15** (1984) 403.

[17] P.R. Cameron et al., *Phys. Rev.* **D32** (1985) 3070.

[18] S. Heppelmann et al., *Phys. Rev. Lett.* **55** (1985) 1824.

[19] W.G.J. Langeveld, *Proc. 23rd Int. Conf. on High Energy Physics, Berkeley, 1986*, ed. S. Loken (World Scientific, Singapore, 1987), p.1207.
M. Althoff et al., *Phys. Lett.* **130B** (1983) 449.
H. Aihara et al., *Phys. Re. Lett.* **57** (1986) 404.

NEW RESULTS FROM THE BROOKHAVEN AND CERN HEAVY ION PROGRAMS

William A. Zajc
Columbia University, New York, NY 10027

Abstract

Recent results from the Brookhaven and CERN heavy ion programs are presented. The motivation for using nuclear beams is briefly discussed, followed by an analysis of some of these early data.

1 Introduction

The recent successes of the BNL and CERN nuclear programs have provided, for the first time, large amounts of statistically significant, high quality data describing the reactions of truly relativistic heavy-ion collisions. While most results are preliminary in nature, important glimpses of this exciting world of high-density/large-volume physics may still be obtained from them. This article is *not* intended as a comprehensive review of these experiments. Instead, it is designed to provide a selective introduction to the field, emphasizing both various aspects of the basic reaction mechanisms and new, not yet understood, phenomena.

1.1 Motivation

In the last decade, quantum-chromodynamics has advanced from "leading candidate" status to that of "uncontended champion". That this is true in spite of the near-absence of analytic solutions is tribute to the internal beauty and self-consistency of the theory. Various theorists, seeking to turn a vice into a virtue, have turned to numerical lattice QCD calculations as a tool for exploring properties of bulk hadronic matter. Nearly every such calculation has indicated the presence of a "deconfining phase transition" occurring at $T_C \sim 200\,\mathrm{MeV}$ and energy density $\mathcal{E} \sim$ 2-4 $\mathrm{GeV/fm^3}$. (See Karsch [1] and references therein for a recent review relevant to high temperature QCD.)

Under these conditions of high temperature and energy density, the individual hadrons lose their identity, and the quarks and gluons within them may directly interact with those from other hadrons, rather than by exchange of virtual hadrons. It is important to note that the quarks and gluons are not "free"– they remain strongly interacting, but now over a volume larger than that of a single hadron. The most common name for this deconfined state is the "quark-gluon plasma", or QGP.

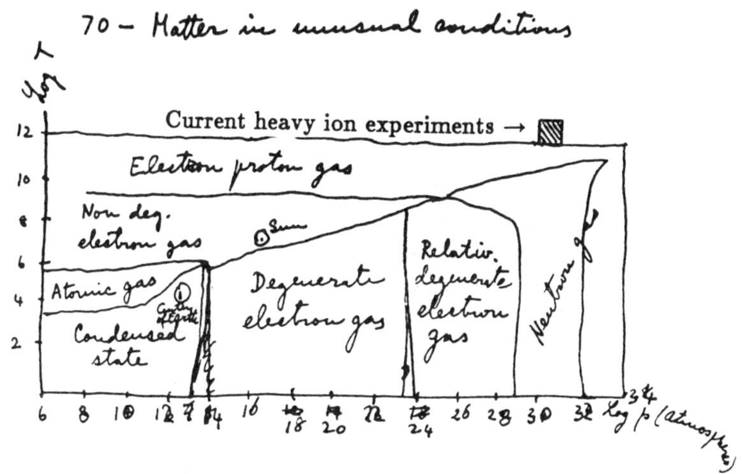

Figure 1: The phase diagram of "Matter under unusual conditions", taken from *Notes on Thermodynamics and Statistics,* by Enrico Fermi (University of Chicago Press, 1966). Superimposed on this diagram is the region explored by current heavy-ion experiments.

The most common method for attempting to create this state is via the collision of two heavy nuclei at high energy. This, then, is a major goal of most current high energy heavy ion experiments: creation and detection of the quark-gluon plasma.

However, it is important to note that relativistic heavy ion physics is an intrinsically interesting endeavor, independent of any theoretical underpinnings obtained from lattice QCD theories. For instance, early studies were motivated by the predictions of novel states of *nuclear* matter by Lee and Wick[2]. Alternatively, an attempt to study nuclear matter in the regime where the inter-nucleon spacing becomes comparable to the hard-core in the nucleon-nucleon interaction (0.4-0.5 fm) would lead naturally to the requirement for energy densities of order 10 $\mathrm{GeV/fm^3}$. Finally, any new opportunity to study matter under previously unobtainable conditions represents an important intellectual challenge. Such studies were anticipated nearly four decades age by Fermi, who mapped out the entire territory shown in Fig. 1. I have taken the liberty of modifying this diagram to show the domain obtainable by the BNL and CERN heavy-ion experiments. It is quite striking that these machines already offer the ability to study energy and density regimes that are truly *terra incognita.*

At Brookhaven, 14.5 $A \cdot \mathrm{GeV}$ beams of ^{16}O and ^{28}Si are provided by the AGS, using the Tandem accelerator as an injector. (A test run of the E814 experiment has also used a 10.0 $A \cdot \mathrm{GeV}$ ^{28}Si beam.) At CERN, beams of 60 and 200 $A \cdot \mathrm{GeV}$ ^{16}O and ^{32}S have been delivered to several experiments. In this report, I will concentrate on a few selected results from each program that illustrate the breadth

Table 1

Observables for Heavy Ion Collisions

Variable	Advantages	Disadvantages
Multiplicity	Well-defined pp data is well-known	Required segmentation is costly
E_T	Straightforward Predicted by many models	Energy scale systematics Predicted by many models
Single-particle momenta spectra	Precise Co-requisite to PID	Costly Tracking/scanning efficiencies??
Like-particle interferometry	Provides source geometry Sensitive to dynamics	Pair tracking difficulties Interpretation difficult
Flavor ratios	Well-defined	Requires (costly) PID
Dileptons	Clean, early signal Well-understood interaction	Low rates

and depth already obtained by the first-generation experiments. Before doing so, it is worth examining the menu of physics observables accessible to measurement.

1.2 Observables

Suppose that a novel state of matter were formed in a high energy nuclear collision. How would we detect it? This is by no means a trivial question, since the new state (call it a plasma) is expected to live only ~ 10 fm/c before re-hadronizing into "ordinary" hadrons (barring the formation of truly exotic particles, such as multiply-strange droplets). Detection strategies may be roughly categorized as those that 1.) seek to completely characterize the event in some variable, such as transverse energy, multiplicity, the single-particle momentum spectrum, etc.; or 2.) focus on some particular signal thought to be both sensitive to the early stages of the event (i.e., when the plasma was formed) *and* to be unmodified by later-stage dynamics. Strategy 1 is clearly the path to understanding the backgrounds associated with any collision, exotic or not. It suffers in that one must look for deviations from the "expected" result, which is seldom calculable or model-able in any rigid fashion. The difficulty with Strategy 2 of course, is that it depends on some model for the very existence of its signal. Table 1 attempts to classify many of the common observables according to these criteria.

2 Basic Measurements

This section discusses results from measurements of multiplicity transverse energy. The word "basic" in the title of this section is intended not in the sense of "trivial", but as in "essential".

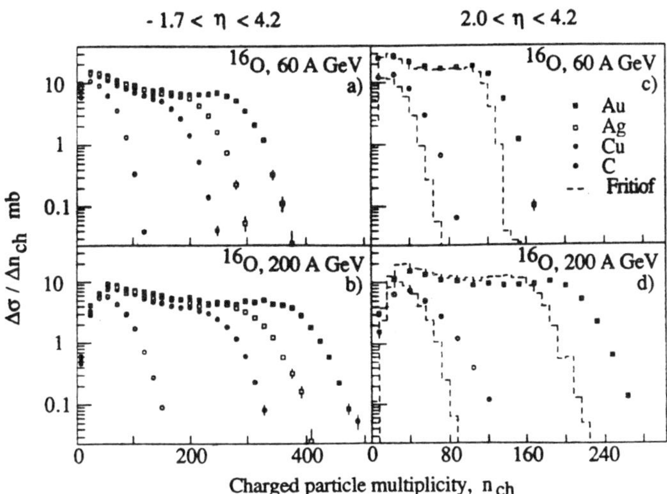

Figure 2: Multiplicity distributions as measured by the WA80 collaboration in two different rapidity intervals.

2.1 Multiplicity Distributions

The multiplicity distribution of charged particles emerging from a reaction is one of the most fundamental parameters characterizing an event. In p-p physics, where such distributions are loosely called "KNO distributions", an analysis of the fluctuations in multiplicity has provided insight into the number of statistically independent processes (clusters). In heavy ion physics, we may use the existing p-p data as a calibration to (perhaps) determine the number of nucleon-nucleon collisions contributing to a given nucleus-nucleus collision.

The WA80 collaboration has measured the charged particle multiplicity distributions over $\sim 4\pi$ by combining results from the Plastic Ball (target fragmentation region) and an array of Iarocci-type streamer tubes with pad readout (segmentation 22,000). Their results[3] for 60 and 200 $\mathrm{A \cdot GeV}$ ^{16}O on a variety of targets are shown in Fig. 2. One is immediately struck not only by the large multiplicities involved, but also by the "plateau": The differential cross section to produce a given multiplicity n_{ch} is roughly constant from $n_{ch} = 100$ to 400. (This for 200 $\mathrm{A \cdot GeV}$ ^{16}O on a Au target.)

The shape of the multiplicity distribution, while quite different from the typical negative binomial observed in p-p collisions, is easily understood in terms of the geometry of colliding nuclei. Since this point will reoccur for the case of E_T distributions, it is worth examining in some detail. Assume that ν independent samples of some (p-p?) multiplicity distribution P_n with average $\langle n \rangle$ are combined to give a

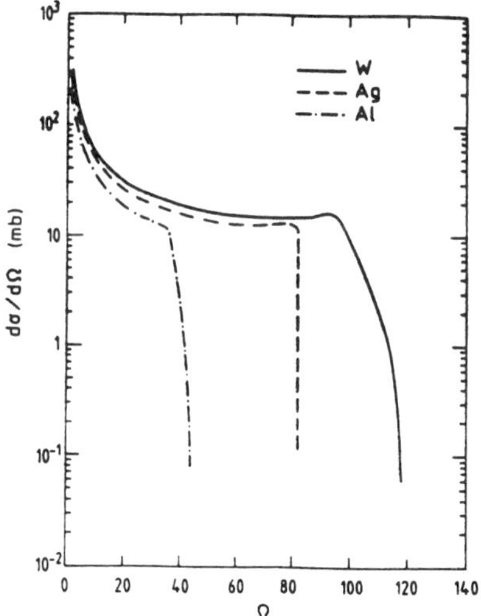

Figure 3: The distribution in the number of collisions N/σ_{pp} for ^{16}O incident on various targets, as calculated by NA34.

total multiplicity N. We then have

$$\langle N \rangle = \langle \nu \rangle \langle n \rangle \tag{1}$$

$$\langle N^2 \rangle - \langle N \rangle^2 \equiv D^2 = \langle \nu \rangle d_n^2 + d_\nu^2 \langle n \rangle \quad , \tag{2}$$

where d_n and d_ν are the dispersions in P_n and the number of samples, respectively. If we regard ν as labeling the number of independent nucleon-nucleon collisions, this quantity has a large relative dispersion, induced by the average over the impact parameter. A typical example of this is shown in Fig. 3, where the broad plateau structure is seen in the number of collisions. Thus, the dominant term in determining the width of the convoluted multiplicity distribution comes from averaging over this geometric dispersion, not from the intrinsic width governed by P_n.

One may learn about the underlying dynamics only by attempting to *fix* the geometry. The NA35 collaboration has done this by using various central event triggers[5], which are thought to correspond to only the most central impact parameters. Their data for the dispersion versus the mean are shown in Fig. 4, both for minimum bias and centrally-triggered data. Also shown is a prediction for the dispersion based on the p-A data:

$$\frac{D_-}{n_-} = \frac{1}{\sqrt{16}} \cdot \frac{D_{p-A}}{\langle n_{p-A} \rangle} \quad , \tag{3}$$

where it has been assumed that 1.) the geometry has been effectively fixed at $\nu_p = 16$, and 2.) that the relevant independent distribution is the *p-A* multiplicity distribution. From this, we conclude that no new, coherent phenomena are observed at this level; the nucleus-nucleus result looks rather like that from 16 independent p-A collisions. (Although the reduced dispersion from p-A collisions is not greatly different

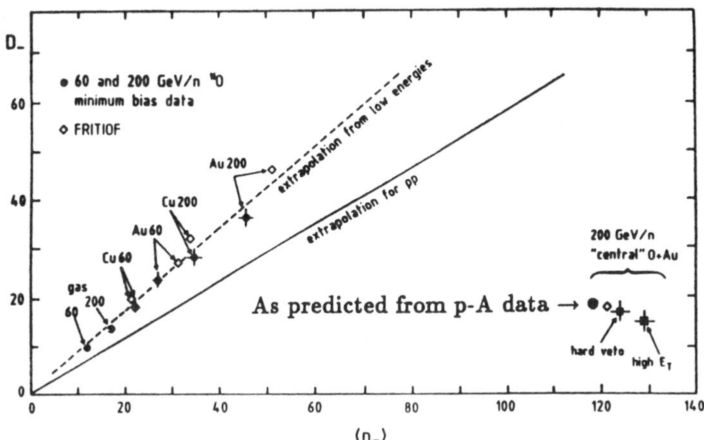

Figure 4: The dispersion versus the mean of negative tracks as measured by NA35. The minimum bias results are consistent with those from lower energies (dashed line). The dispersions for two different central triggers are also shown, along with predictions from p-A collisions, assuming all 16 oxygen nucleons interact.

than that for p-p, it is important that one can refer back only to the p-A results. Attempting a first-principles prediction of the dispersion based on the assumed number of *nucleon-nucleon* collisions would underpredict the observed dispersion by a factor of five.)

2.2 Transverse Energy

Transverse energy is defined as

$$E_T \equiv \sum_i E_i \cdot \sin\theta_i \quad , \tag{4}$$

where the sum runs over all particles in some section of phase space. It is typically measured via calorimetry (although it is possible to explicitly construct the sum for charged particles in a spectrometer), where the sum is then performed module-by-module, so that the association with a given particle is lost. In most cases, the transverse energy remains an approximate Lorentz invariant,

$$E_T \approx \sum_i \sqrt{m_i^2 + (p_T)_i^2} \quad , \tag{5}$$

but the validity of this approximation is sensitively dependent on the details of the particle spectrum (in energy and species) and the calorimeter response (hence the entry regarding energy scale systematics in Table 1).

Assuming these difficulties may be overcome (typically after Monte Carlo correction), the transverse energy then provides a measure of the extent to which the

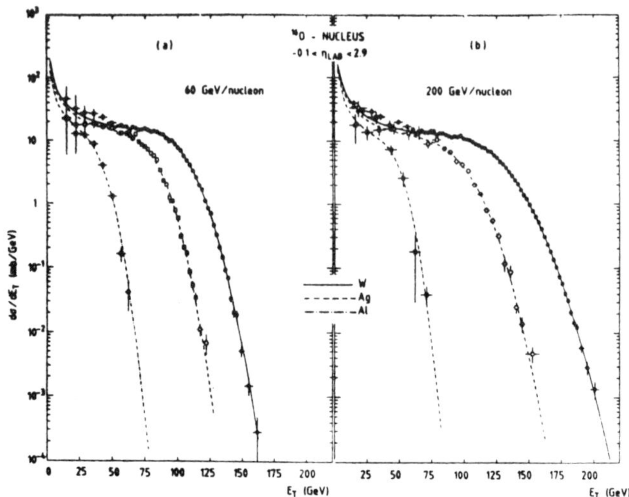

Figure 5: The transverse energy distribution measured by NA34 in the backward direction for various target nuclei.

beam energy has been deflected from its original direction. In the case of heavy ion collisions, it is safe to further assume that this deflection is a random process– no collimated flows (jets) have yet been observed in these reactions. Finally, one often considers this *randomization* of the initial beam energy as a *thermalization.* This should perhaps be regarded as a convenient mnemonic, rather than an accurate prediction of the momentum and flavor distributions expected for true chemical equilibrium.

If the particle emission is a random, uncorrelated process, and if the the mean transverse momentum $\langle p_T \rangle$ is independent of the energy of the event (this second assumption has been approximately verified by WA80[4]), then the $\mathrm{E_T}$ distribution can be thought of as a convolution of the (neutral+charged) multiplicity distribution with the single-particle transverse energy distribution. The considerations of the previous section would then lead us to again expect a shape for the transverse energy distribution dominated by geometry. The characteristic shapes of the $\mathrm{E_T}$ distributions, presented in Figures 5-7, suggest strongly that this is the case. (These data are taken from references[6,4,7].)

At the CERN energies, one sees a steady increase in the maximum value of $\mathrm{E_T}$ as the target mass is increased (see especially Fig. 6). This is not seen at the AGS (Brookhaven energies) by either E802 ([8], Figure 8) for 14.5 $\mathrm{A \cdot GeV}$ ^{16}O beams, or by E814 ([9], Figure 9) for ^{28}Si beams at 10.0 $\mathrm{A \cdot GeV}$. This saturation in the yield has been interpreted[8] as evidence for stopping of the projectile by the target, leading one to expect a central region containing high baryon density. These authors also noted that the ratio of ≈ 6 in absolute cross section observed between the Au and Cu targets (for $E_T > 50$ GeV) is in good agreement with the factor of ≈ 5 estimated from the ratio of nuclear areas that can stop the ^{16}O beam for the

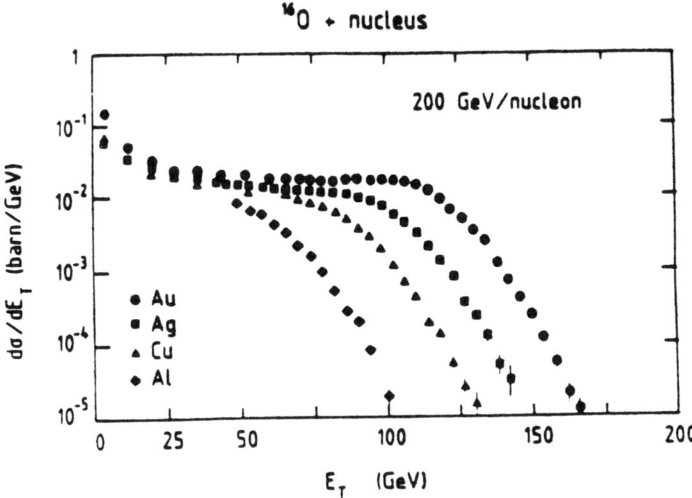

Figure 6: The transverse energy distribution measured by NA35 in the central region $(2.2 < \eta < 3.6)$ for various target nuclei.

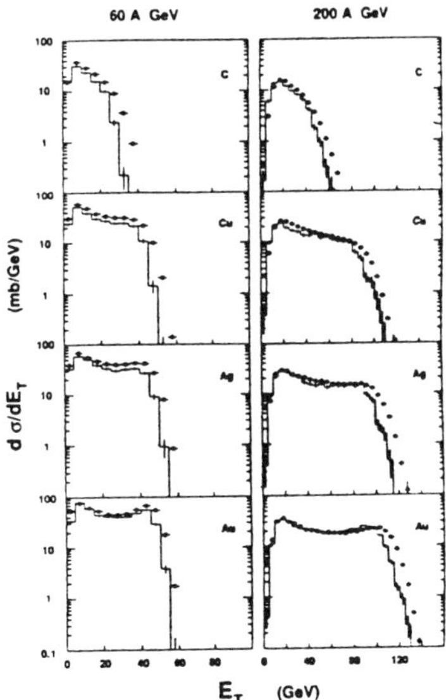

Figure 7: The transverse energy distribution measured by WA80 in the forward hemisphere $(2.4 < \eta < 5.5)$.

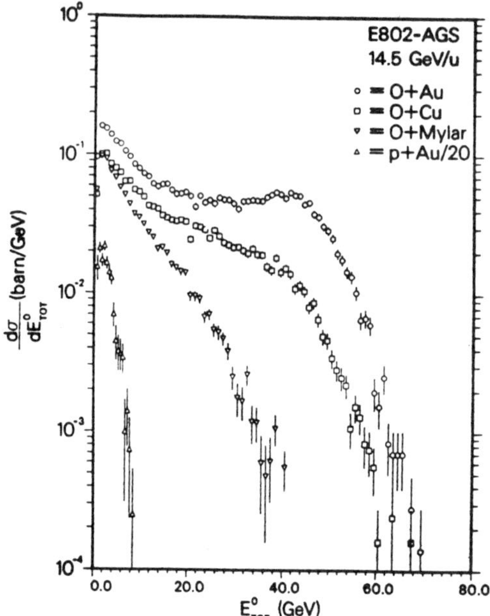

Figure 8: The neutral energy distribution as measured by E802 in the region $1.2 < \eta < 2.4$.

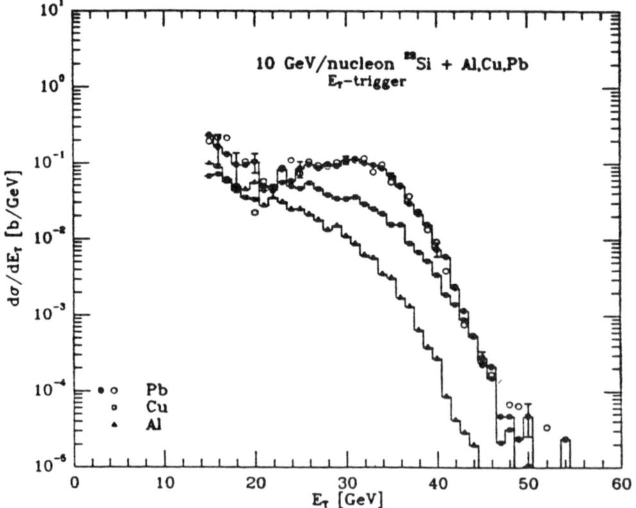

Figure 9: The transverse energy distribution as measured by E814 in the region $1.5 < \eta < 2.5$.

two targets:

$$\text{ratio} = \frac{(R_{Au} - R_O)^2}{(R_{Cu} - R_O)^2} \quad . \tag{6}$$

2.3 Energy Densities

It has become customary to describe the energies measured by a given experiment in terms of a result due to Bjorken[10]:

$$\mathcal{E} = \frac{1}{\pi R^2 \tau} \cdot \frac{dE_T}{dy} \quad , \tag{7}$$

where R is the nuclear radius and τ is an assumed time for hadronization ~ 1 fm/c. Derived as it is in terms of a boost-invariant dynamics, such a formula, when applied to CERN or BNL energies, must be regarded as only schematic in nature. Furthermore, since it directly depends on the assumed hadronization time (unknown to a factor of 2) and on the assumed transverse nuclear region (uncertain, but perhaps larger than R[11]), the systematic uncertainties associated with this result multiply. Nonetheless, having issued the appropriate caveats, it is of interest to compare the energy densities obtained at CERN and BNL. One finds that at CERN, $\mathcal{E} \sim$ 2-4 GeV/fm^3 , whereas at BNL $\mathcal{E} \sim 1$ GeV/fm^3 . These numbers, while uncertain, are at least in the vicinity of the expected phase transition (see Section 1.1).

3 Non-Basic Measurements

This section discusses three new results that either look in exotic channels, or are not explainable by extrapolations of known p-A physics, or both.

3.1 K/π Ratios

Whether or not chemical equilibrium is established, the ratio of production rates for strange versus non-strange mesons is sensitive to the reaction dynamics (see Koch[12] and references cited therein for details). A measurement of these ratios has been provided by the E802 collaboration at Brookhaven, using a sophisticated multiparticle spectrometer.

Particle identification is achieved by combining time-of-flight measurements with magnetic tracking. The time-of-flight system consists of in-beam start counters and a 160 element scintillator hodoscope, providing timing resolution ≤ 100 ps. When combined with the momentum information from four drift chambers, this system permits π-K separation at the 5σ level out to $P_{lab} \simeq 2.0$ GeV/c[13]. Clear bands for protons and deuterons and protons are also seen, as shown in Fig. 10.

It is apparent from Fig. 10 that the K^+ 's are much more numerous than the K^- 's. For collisions of 14.5 A $\cdot$ GeV ^{28}Si on a Au target, the ratio for tracks found in the region $14^o < \theta_{lab} < 28^o$ is $K^+/\pi^+ = 19 \pm 5\%$ versus $K^-/\pi^- = 6 \pm 5\%$ for minimum bias events, and $24 \pm 5\%$ versus $4^{+4}_{-2}\%$, respectively, for central events.

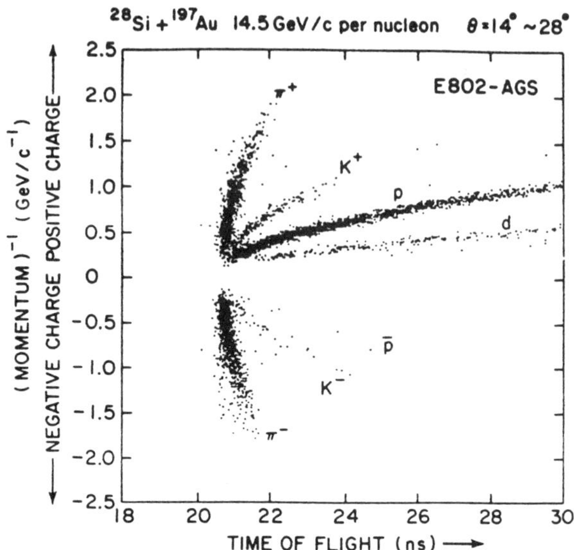

Figure 10: Scatter plot of TOF versus $1/p$ for tracks found in the E802 spectrometer.

Similar results have been obtained from an analysis of a later data set, and when cuts were imposed to obtain a region of uniform acceptance in rapidity and transverse momentum for both the pions and kaons[14].

Since no directly comparable proton-nucleus data set (yet) exists, it is difficult to estimate what value should be expected for these ratios. A rough extrapolation from existing data would suggest $(K^+/\pi^+)_{p-A} \leq 10\%$, and $(K^-/\pi^-)_{p-A} \simeq 4\%$. Until p-A data for direct comparison is available, it is not possible to determine if the large K^+/π^+ ratio results in a straightforward way from associated production in the nuclear environment (plus perhaps rescattering effects), or if it is a direct signal of the high baryon densities inferred for these collisions. Further comparison runs (including p-A), Monte Carlo studies and acceptance corrections are now being vigorously pursued to understand this effect.

3.2 J/Ψ Production

Another interesting flavor channel to probe is the production of charm. This is particularly true since the high mass of the charmed quark makes $c\bar{c}$ pairs an effective short-range probe of the quark-gluon plasma. The disappearance of the J/Ψ, resulting from the loss of confinement and color screening in the plasma, was predicted by Matsui and Satz in 1986[15]. At CERN, a dedicated experiment, NA38, was mounted to look for this signal (or rather absence of a signal) in 200 A · GeV ^{16}O on uranium interactions by detecting dimuons from the decay $J/\Psi \rightarrow \mu^+\mu^-$.

The apparatus consists of the old NA10 dimuon spectrometer, augmented with a compact lead-scintillating fiber calorimeter covering $2.8 < \eta < 4.2$ (for event characterization) and an active target. Dimuon pairs of mass $M_{\mu^+\mu^-} > 0.5$ GeV/c are detected by the apparatus. The signal for opposite-sign pairs is obtained from

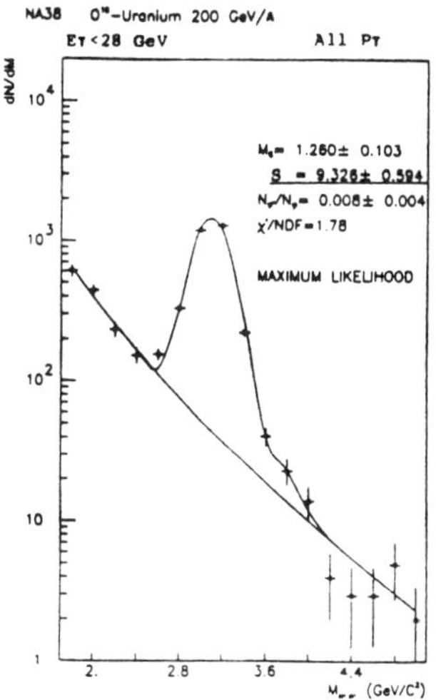

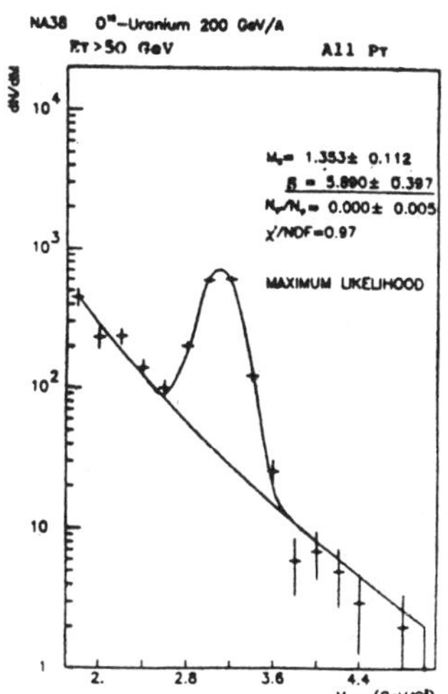

Figure 11: The dimuon invariant mass spectrum, after background subtraction obtained by NA38. The spectrum on the left is for low E_T events, that on the right for events with $E_T > 50$ GeV.

a subtraction designed to remove combinatoric backgrounds resulting from meson decay:

$$N_S \equiv N_{\mu^+\mu^-} - 2\sqrt{N_{\mu^+\mu^+} \cdot N_{\mu^-\mu^-}} \quad . \tag{8}$$

The resulting μ-pair mass spectrum (Fig. 11) is then fit in the region of the J/Ψ peak to extract the *ratio* of the J/Ψ yield to continuum background. If one calls this ratio S,

$$S \equiv \frac{N_{J/\Psi}}{N_{cont}} \quad ,$$

it is found[16] that S is strongly dependent upon E_T :

$$\frac{S(E_T > 50\,\mathrm{GeV})}{S(E_T < 28\,\mathrm{GeV})} = 0.64 \pm 0.06 \quad , \tag{9}$$

where E_T is the transverse energy in the events leading to J/Ψ formation. Thus, the J/Ψ is suppressed in collisions with high energy density, just as predicted by Matsui and Satz! Furthermore, the effect is not seen in p-U collisions while it persists in ^{32}S +U interactions[17], and it exhibits a p_T dependence[16] consistent with that calculated by Karsch and Petronzio for J/Ψ 's moving through a quark-gluon plasma[18].

Has the quark-gluon plasma been found? Unfortunately, it is (much) too early to say. A key piece of required data is the absolute production cross-sections for J/Ψ 's as a function of E_T , thereby removing the potentially dangerous normalization to the continuum. Closely related to this is an ongoing effort to understand the contributions to the continuum in the region of the J/Ψ via Monte Carlo simulation.

A key piece of required theory is a thorough study of effects that could mimic this signal. The observed p_T dependence is certainly reminiscent of various Cronin effects known to exist in p-A production. Furthermore, as emphasized by S. Frankel at this meeting, the J/Ψ production cross-section is strongly energy-dependent in this region ($\sqrt{s_{pp}} = 20$ GeV), so that details of the energy-degradation of the incoming projectile must be taken into account. Obviously, interesting work lies ahead for both theorists and experimentalists before this effect is understood.

3.3 Two-pion Interferometry

In Section 2. it was apparent that multiplicity and transverse energy spectra contained an implicit reflection of the geometry of nuclei. An *explicit* determination of geometry (of the pion-emitting source at least) is obtainable via two-pion interferometry (a.k.a. Bose-Einstein correlations). This technique exploits the tendency for identical bosons to "clump" in phase space[19]. Specifically, the correlation function $C(\vec{q})$, proportional to the probability to find two pions at relative momentum $\vec{q}$, is given by the square of the Fourier transform of the pions source $\rho(\vec{r})$,

$$C(\vec{q} \equiv \vec{p}_2 - \vec{p}_1) \equiv \frac{N(\vec{p}_2, \vec{p}_1)}{N(\vec{p}_1)N(\vec{p}_2)} = 1 + \lambda|\tilde{\rho}(\vec{q})|^2 \quad , \tag{10}$$

where λ is a factor that would be unity in the absence of other correlations *not* attributable to Bose statistics (of a random source). More sophisticated versions of Eq. 10 provide the ability to measure longitudinal and transverse values of the source radius, and/or lifetimes[20]

Using a streamer chamber, the NA35 collaboration has measured approximately all negative tracks in a set of photographs of 200 A · GeV ^{16}O + Au collisions. By assuming that all such tracks are π^- 's, they are able to form the correlation function as defined above, thereby extracting the source radii and lifetimes[21]. These data are shown as a function of the pion rapidity interval in Fig. 12; the results to the fits are summarized in Table 2.

Table 2

NA35 Interferometry Results

Rapidity Interval	$R_\perp$ (fm)	$R_\parallel$ (fm)	λ
$1 < y < 4$	4.1 ± 0.5	$3.1^{+0.7}_{-0.4}$	$0.31^{+.07}_{-.03}$
$1 < y < 2$	4.3 ± 0.6	2.6 ± 0.6	$0.34^{+.09}_{-.06}$
$2 < y < 3$	8.1 ± 1.6	$5.6^{+1.2}_{-0.8}$	0.77 ± 0.19
$3 < y < 4.5$	$4.3^{+1.2}_{-0.8}$	5.8 ± 2.2	0.55 ± 0.20

Upon examining Table 2, one is first struck by the large radii obtained in the central region. Typically, in nuclear collisions one measures a source size on the order of the smaller of the two nuclei involved. This is the case here for the target and

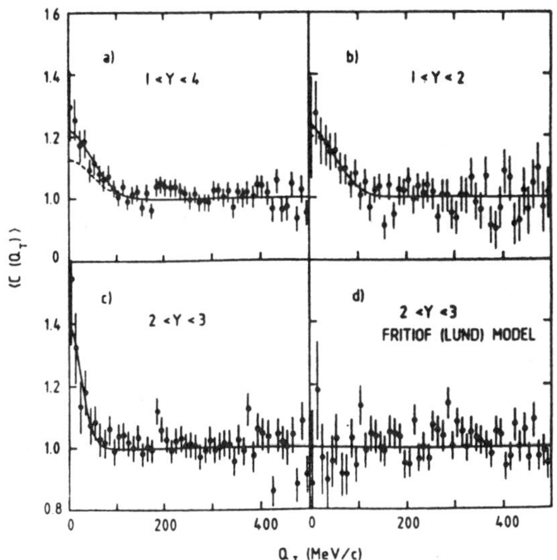

Figure 12: The NA35 data pion interferometry data, as a function of the pion rapidity.

perhaps the projectile region; only the mid-rapidity data show the large enhancement. Complicating the interpretation of these results is the systematic variation of λ with extracted radius. While a reduced value of λ does not necessarily corrupt the geometric content of a Bose-Einstein analysis, some care must be used in understanding the biases when the value of λ and extracted radius appear to be correlated. What is clear is that something interesting is going on– only further work will tell. Particularly intriguing is the ability of this experiment to measure the radius of a single event. Additional results along these lines are eagerly awaited.

4 Conclusions

In this article, I have attempted to demonstrate the vigor of the Brookhaven and CERN programs by focusing on a few selected results. Interesting intellectual challenges now present themselves in analyzing both the "mundane" and exotic(?) physics. Especially appealing in this field is the renewal of close interactions between theorists and experimentalists in understanding these data. This must be seen as a direct consequence of the availability of so much data in so short a time.

5 Acknowledgments

It is pleasure to acknowledge many discussions with my colleagues on Brookhaven Experiment E802. These conversations have considerably expanded my horizons in the field of heavy ion physics. This manuscript has benefitted tremendously from a careful reading by A. Therrien; any remaining errors are my own. This work was supported by Department of Energy Contract DE-FG02-87ER40281.

References

[1] F. Karsch, Z. Phys. **C38**, 15 (1988).

[2] T.D. Lee and G.C. Wick, Phys. Rev. **D9**, 2291 (1974).

[3] "Charged Particle Distributions in ^{16}O Induced Nuclear Reactions at 60 and 200 A · GeV ", R. Albrecht *et al.*, GSI preprint GSI-87-81, December 1987 (submitted to Phys. Lett. B).

[4] R. Albrecht *et al.*, Phys. Lett. **B199**, 297 (1987).

[5] "Charged Particle Multiplicities and Inelastic Cross Sections in High Energy Nuclear Collisions", A. Bamberger *et al.*, NA35 preprint (1987).

[6] F. Corriveau *et al.*, Z. Phys. **C38**, 15 (1988). See also CERN preprint CERN-EP/87-16 (1987).

[7] W. Heck *et al.*, Z. Phys. **C38**, 19 (1988).

[8] T. Abbott *et al.*, Phys. Lett. **B197**, 285 (1987).

[9] W. Heck *et al.*, Z. Phys. **C38**, 19 (1988).

[10] J.D. Bjorken, Phys. Rev. **D27**, 140 (1983).

[11] L. McLerran, private communication.

[12] P. Koch, Z. Phys. **C38**, 269 (1988).

[13] Y.Miake *et al.*, Z. Phys. **C38**, 135 (1988).

[14] See contribution by R. Ledoux in these proceedings.

[15] T. Matsui and H. Satz, Phys. Lett. **B178**, 416 (1986).

[16] A. Bussiere *et al.*, Z. Phys. **C38**, 117 (1988).

[17] G. Landaud, presented at this meeting.

[18] F. Karsch and R. Petronzio, Phys. Lett. **B193**, 105 (1987).

[19] For a recent review, see "Bose-Einstein Correlations– From Statistics to Dynamics", in *Multi-particle Production*, P. Carruthers, ed., World Scientific Press (1988).

[20] K. Kohlehmainen and M. Gyulassy, Phys. Lett. **B180**, 203 (1986).

[21] T.J. Humanic *et al.*, Z. Phys. **C38**, 79 (1988).

NUCLEAR QUASIELASTIC RESPONSE

ROBERT W. LOURIE
Department of Physics and Laboratory for Nuclear Science
Massachusetts Institute of Technology
Cambridge, MA 02139 USA

ABSTRACT

At intermediate energies, all nuclei exhibit a broad peak in their inelastic electron scattering response in which incoherent scatterings from individual nucleons play a major role. Description of this response relies on both an understanding of the nuclear many-body problem with its consequent multiparticle correlation physics as well as knowledge of the appropriate elementary-particle degrees of freedom from which to construct the nucleus. A review will be given of the progress achieved in these areas by means of both inclusive and coincidence electron scattering experiments. Some current theoretical approaches will also be described.

INTRODUCTION

The topic of this talk is the nuclear quasielastic response as revealed by electron scattering experiments. It will focus on two, not entirely independent, aspects of this question. The first is the nature of the nuclear constituents themselves. In particular, are nucleons modified by the nuclear medium and do they, along with nucleon resonances and mesons, comprise the appropriate degrees of freedom? The second aspect concerns the particular combination of these constituents to form a nucleus. We have the unique situation of a strongly-interacting many-body problem which also embodies the strong-coupling confinement aspects of QCD.

After a brief discussion of the relevant features of (e,e′), we will review what has been learned about nucleons in nuclei from quasielastic electron scattering. Then we will examine the question of the reaction mechanism in the quasielastic region, i.e. a study of the processes by which the nucleus absorbs the energy and momentum transferred by the electron. Here one expects many-body correlations to play an important role.

Electron-nucleus scattering is assumed to proceed via the exchange of a single virtual photon. The electron transfers energy ω and momentum $\vec{q}$ to the nuclear system. In inclusive (e,e′) only the scattered electron is detected; one integrates over all allowed final states of the unobserved hadrons. Assuming Lorentz covariance, parity invariance, and electromagnetic current conservation, the (e,e′) cross section depends on only two independent nuclear structure (response) functions: the longitudinal, R_L, which results mainly from interaction with charge and the transverse, R_T, which results from interaction with convection currents and magnetization. The cross section can be written as:

$$\frac{d\sigma}{d\Omega_e d\omega} = \sigma_M \left[v_L R_L(\vec{q},\omega) + v_T R_T(\vec{q},\omega)\right] \quad , \tag{1}$$

where the the kinematic factors $v_L = (q_\mu/|\vec{q}|)^4$ and $v_T = \frac{1}{2}(q_\mu/|\vec{q}|)^2 + \tan^2\theta_e/2$. The four-momentum transfer $q_\mu^2 = |\vec{q}|^2 - \omega^2, \sigma_M$ is the Mott cross section for electron scattering from a point charge and θ_e is the electron scattering angle. The response functions are determined by matrix elements of the nuclear electromagnetic current operator $\hat{J}_\mu = (\hat{\vec{J}}, \hat{\rho})$:

$$R_L(\vec{q}, \omega) = \sum_f |\langle \Psi_f | \hat{\rho} | \Psi_i \rangle|^2 \delta(\omega - E_f + E_i)$$
$$R_T(\vec{q}, \omega) = \sum_f |\langle \Psi_f | \hat{\vec{J}_T} | \Psi_i \rangle|^2 \delta(\omega - E_f + E_i) \tag{2}$$

where $\hat{\rho}$ is the charge operator and $\hat{\vec{J}_T}$ is the transverse (to $\vec{q}$) current operator. Insofar as the nucleus consists of independent nucleons, the nuclear charge and current operators will just be sums of the individual nucleonic ones. A central question, then, is what spinors and form factors should one use to construct these nucleon currents. For example, meson-baryon field theories [1] predict that the lower components of the nucleon spinor will be sizeably enhanced over their free-space values. Additionally, twelve, instead of the usual two, structure functions are present in the photon-nucleon vertex function if the nucleon is off-shell both before and after the electron scattering process. [2] Clearly, we must understand the nature of this off-shell vertex before the phrase "modified nucleons" can be meaningfully interpreted.

QUASIELASTIC ELECTRON SCATTERING

The nuclear response at intermediate energies exhibits a broad bump, the quasielastic peak, as can be seen in figure 1. It occurs at an energy transfer $\omega \approx q_\mu^2/2M_N$ where M_N is the nucleon mass. This is the kinematic condition for elastic scattering from a nucleon at rest, thus incoherent scatterings from constituent nucleons is a natural interpretation of this structure. The width of the peak is then attributed to Fermi motion of the nucleons.

Early attempts [3] to understand the quasielastic peak focussed on the the Fermi gas model. In this model one has non-interacting on-shell nucleons, one of which absorbs all the energy and momentum transferred by the electron and is ejected from the nucleus, i.e. a 1-particle–1-hole (1p-1h) state is created in the target. The response function for such a system may be calculated exactly. [4] Experimental results for the total, i.e. unseparated, (e,e′) cross section [5] on a variety of nuclei are shown in figure 1. The Fermi gas model predictions are indicated by the solid curves. The calculation seems, at first glance, to work very well though a closer look at the figure hints at possible shortcomings of this simple model. In the region of higher energy loss (the so-called "dip" region) there is an excess of cross section. Unless the processes responsible for the excess go to zero very quickly at lower ω (which is unlikely) they will extend underneath the quasielastic peak and spoil the agreement between the Fermi gas model and the data.

A significant experimental advance was made in the early 1980's when the quasielastic (e,e′) cross section on several nuclei was separated into the individual longitudinal and transverse response functions. [6] The longitudinal response was expected to provide a good characterization of the nuclear one-body properties whereas the transverse response should be more sensitive to the nuclear meson fields since mesonic currents constitute a first order relativistic correction to the current operator while they are of second order in the charge operator. [7] The separated response functions showed that the earlier agreement with the Fermi gas model was indeed fortuitous; the longitudinal response was overestimated by theory while the transverse was underestimated. Significant discrepancies still remained after the Fermi gas model was augmented with pion exchange currents. [8] Questions arose regarding the influence of correlations

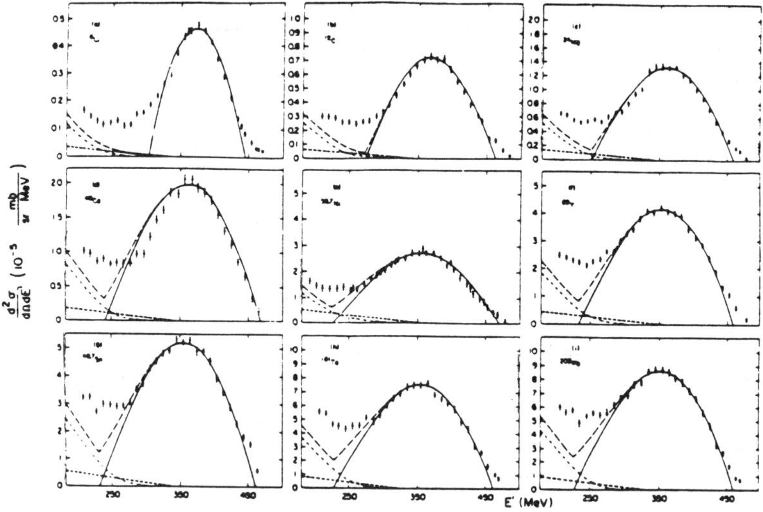

Figure 1. *Quasielastic electron scattering on a variety of nuclei from ^6Li to ^{208}Pb.*[5] *The solid line shows the Fermi gas model fit.*

(multiparticle-multihole components in the reaction mechanism) and the possibility of modified nucleon structure.

NUCLEON STRUCTURE IN NUCLEI

In the quasifree picture the longitudinal nuclear response is determined by the charge form factor of the proton, $F_N(q_\mu^2)$, multiplied by a quantity only dependent on nuclear structure. For a system of non-relativistic nucleons the response function has the form:[9]

$$R_L(\vec{q},\omega) = |F_N(q_\mu^2)|^2 \sum_f |\langle f| \sum_{j=1}^{Z} e^{i\vec{q}\cdot\vec{r}_j} |i\rangle|^2 \delta(\omega - E_f + E_i) \quad . \tag{3}$$

From the longitudinal response function one can form the so-called Coulomb Sum Rule (CSR).[10] The inelastic CSR is obtained by integrating the longitudinal response over energy transfer after dividing out the nucleon form factor:

$$C(\vec{q}) = \int_{0+}^{\infty} \frac{R_L(\vec{q},\omega)}{|F_N(q_\mu^2)|^2} d\omega \quad , \tag{4}$$

where the limit 0^+ implies elastic scattering from the nucleus is excluded. Using closure over the complete set of final states ($\sum_f |f\rangle\langle f| = 1 - |i\rangle\langle i|$) one obtains:

$$\begin{aligned} C(\vec{q}) &= \langle i| \sum_{j,k=1}^{Z} e^{i\vec{q}\cdot(\vec{r}_j - \vec{r}_k)} |i\rangle - |\langle i| \sum_{j=1}^{Z} e^{i\vec{q}\cdot\vec{r}_j} |i\rangle|^2 \\ &= Z\left(1 - F_{el}^2(|\vec{q}|)\right) + \langle i| \sum_{j\neq k}^{Z} e^{i\vec{q}\cdot(\vec{r}_j - \vec{r}_k)} |i\rangle \\ &= Z\left(1 - F_{el}^2(|\vec{q}|)\right) + \rho_2(-\vec{q},\vec{q}) \end{aligned} \tag{5}$$

with ρ_2 the Fourier transform of the two-proton distribution function and F_{el} the elastic form factor of the nucleus (negligible at high $\vec{q}$). The CSR measures the total charge of the nucleus, modified by correlation effects, and depends explicitly on the model for the nucleon form factor. Some experimental results[6a,6b][11] [12] on the Coulomb sum rule are shown in figure 2. The three-body systems ^{3}H and ^{3}He satisfy the sum rule and 'exact' calculations [13] describe the data well. All nuclei heavier then ^{3}He exhibit a considerable suppression (~30–40%) compared to the sum rule value of Z. Correlation effects alone seem insufficient to account for the discrepancy. [14] A recent calculation [15] ascribes the suppression to vacuum polarization effects in a meson-baryon field theory. It was this longitudinal suppression that first lead Noble, [16] and later Shakin, [17] to suggest that the nucleon form factor might be modified in nuclei so that one made an error in constructing the experimental sum rule using the free space form factor. An increase of the charge radius by $\sim$ 15% could account for the suppression.[14] The discovery of the EMC effect [18] lent further support to the concept of medium-modified nucleons.

Several points regarding the CSR must, however, be kept in mind. It was derived for non-relativistic nucleons, formulation of an analogous sum rule for a relativistic system [19] has proven more difficult. At higher momentum transfers magnetic effects are present in the *longitudinal* response whereas the sum rule refers to the *coulomb* (charge) response so that the nucleonic contribution F_N is not given simply by G_E. Also, the sum rule requires an integration to $\omega = \infty$ while 1) the electron scattering process is restricted to $\omega \leq |\vec{q}|$ and 2) the experiments sampled an even smaller range of ω up to $\approx |\vec{q}|/2$. Nevertheless, the observed suppression of the CSR has stimulated a great deal of experimental and theoretical investigation of both nuclear response functions and the nuclear constituents.

The separated (e,e′) data may have more to tell us. The impulse approximation leads to the concept of y-scaling [20] in which the cross section no longer depends on $\vec{q}$ and ω separately but instead is a function only of the single scaling variable $y(\vec{q},\omega)$. In the simplest treatment, one finds that the (e,e′) cross section can be factored into an electron-nucleon piece and a term containing only nuclear structure information. The nuclear response functions are then determined by the nucleonic ones:

$$\begin{aligned}\frac{d^2\sigma}{d\Omega_e d\omega} &= K\,(Z\sigma_{ep} + N\sigma_{en})\,2\pi \int_{|y|}^{\infty} n(k)\,k\,dk \\ &= K\,(Z\sigma_{ep} + N\sigma_{en})\,f(y) \quad ,\end{aligned} \tag{6}$$

where $n(k)$ is the nucleon momentum distribution, K is a kinematic factor and y is the minimum value of the nucleon's initial momentum that satisfies energy conservation. It is defined by

$$\omega + M_A = \sqrt{M_p^2 + (y + |\vec{q}|)^2} + \sqrt{M_{A-1}^2 + y^2} \quad . \tag{7}$$

(More details, including scale-breaking effects due to nuclear binding, may be found in ref. [21]). To determine the scaling function $f(y)$ from experimental data requires dividing out the electron-nucleon cross section and thus should be sensitive to nucleon modifications. Indeed, Sick [22] has used y-scaling to argue that the nucleon radius can increase by no more than 6% in ^{3}He if the scaling of the data is to be maintained. However, by varying the parameter in the dipole form of the nucleon form factor he not only changed the radius (a low momentum transfer property) but the behavior at all momentum transfers. Ciofi degli Atti [23] has examined y-scaling using a chiral bag model [24] for the nucleon form factor where the low-q_μ behavior is determined by the

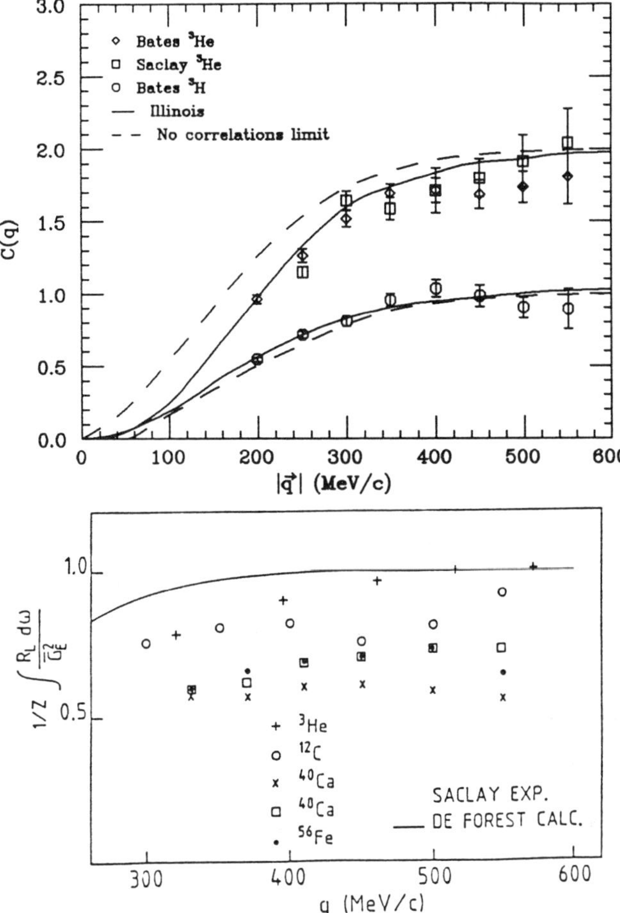

Figure 2. *A) The Coulomb Sum Rule in the three-body systems from Bates*[11] *and Saclay*[12]*. The effect of proton-proton correlations are evident in the ^{3}He data though both nuclei attain the sum rule value of Z. B) The CSR on a variety of heavier nuclei*[6a,6b] *which exhibit a 30–40% suppression. The calculation by DeForest*[9] *is for ^{12}C.*

pion cloud and the high momentum part by the bag radius. The proton electric form factor is parameterized as:

$$G_E(q_\mu^2) = \left(1 - C^2 q_\mu^2\, e^{-q_\mu^2 B^{-2}/6}\right) / \left(1 + R^2 q_\mu^2/12\right)^2 \quad .$$

At high momentum this form 'heals' to the standard dipole form while the polarization of the pion cloud increases the radius to $\langle r^2 \rangle^{1/2} = \sqrt{R^2 + C^2}$. The effect of modifying the nucleon form factor on y-scaling can be seen in figure 3. This plot shows $f(y)$ at $y = 0$ vs. q^2; scaling implies this should be q^2-independent. One sees this is the case when the free form factor is used whereas scaling is destroyed by a 20% increased radius in the standard dipole form factor. The bag model form factor can, however, tolerate a 20% radius increase and still be compatible with scaling.

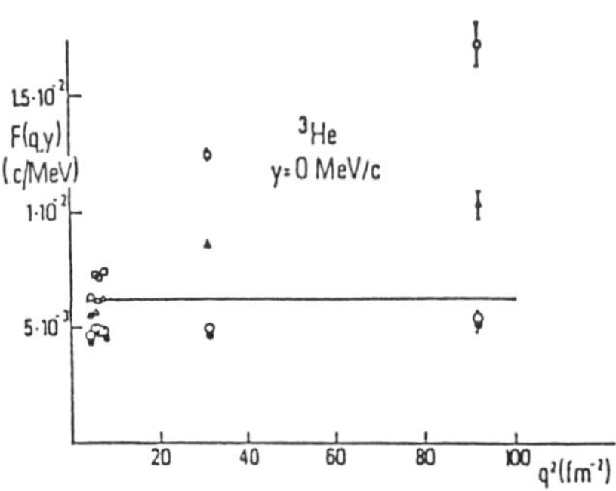

Figure 3. *The scaling function in ^3He at $y = 0$ vs. q^2. Open circles: free nucleon form factor, open squares: radius increased by 20% in dipole form, triangles: radius increased by 20% and static magnetic moment increased, solid circles: radius increased by 20% according to bag model*[24] *parameterization.*

The $f(y)$ discussed above was obtained from the unseparated total cross section data. From eq. 6 ones sees that the two response functions should be given by:

$$R_L^{nuclear} = K\left(ZR_L^{proton} + NR_L^{neutron}\right) f_L(y)$$
$$R_T^{nuclear} = K\left(ZR_T^{proton} + NR_T^{neutron}\right) f_T(y) \quad , \tag{8}$$

and that in this approximation $f_L(y) = f_T(y)$. Figure 4 shows $f_L(y)$ and $f_T(y)$ extracted [25] from the ^{12}C(e,e′) data[6a] from Saclay. Note that f_L and f_T are indeed each functions of the single scaling variable y but that, contrary to the simple impulse approximation, $f_T \neq f_L$. Again, one might be tempted to invoke modified nucleon structure as the explanation. However, a crucial point in this approximation occurs when the (momentum-dependent) nucleon response functions were taken outside the integral over nucleon momenta. A recent study [26] of the scaling properties of the relativistic Fermi gas suggests that this is not a good approximation and that R_L and R_T should in fact approach different scaling functions though the expected difference is smaller than is observed.

In addition, specifically transverse processes such as meson exchange currents are also present in this kinematical region. Such processes can lead to two or more nucleon knockout though the empirical observation of scaling places some constraints on the behavior of any additional reaction components. The simplicity of inclusive (e,e′), obtained by integrating over all final states, also has the liability of mixing the issue of target structure with that of reaction mechanism. One is thus led to perform more exclusive experiments where a hadron (typically a proton) is detected in coincidence with the scattered electron and a *specific* nuclear final state is selected. A program of such experiments has been carried out at the electron scattering facilities at MIT-Bates, NIKHEF-K and Saclay.

For coincidence measurements two additional response functions appear in the general (e,e′X) cross section:

$$\frac{d^4\sigma}{d\Omega_e d\Omega_x d\omega dE_x} = \sigma_M\left[v_L R_L + v_T R_T + v_{LT} R_{LT}\cos\phi_x + v_{TT} R_{TT}\cos 2\phi_x\right] \quad , \tag{9}$$

where $R_{LT}(R_{TT})$ results from an interference between the longitudinal and transverse (two transverse) components of the electromagnetic current. The angle ϕ_x is formed by the (e,e′) plane and the plane containing $\vec{q}$ and the outgoing hadron. Each response

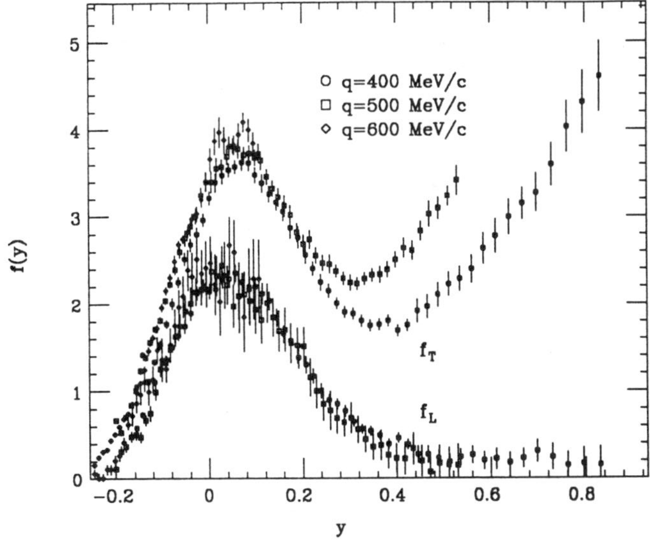

Figure 4. *Longitudinal and transverse scaling functions for* $^{12}C(e,e')$.

function is a function of $\vec{q}, \omega, \theta_x$ (the angle between $\vec{q}$ and the momentum vector of the detected hadron) and E_x (the hadron energy). In the special case of "parallel kinematics" where hadrons are detected in the direction of $\vec{q}$ only R_L and R_T contribute to the cross section.

In the Plane Wave Impulse Approximation (PWIA) the electron is assumed to interact with a quasifree hadron which is then ejected from the nucleus. The wavefunction of the knocked out particle is treated as a plane wave. Considering now the specific case of (e,e′p), (a complete discussion of which can be found in the review article by Frullani and Mougey [27]), the cross section factors into an elementary (e,p) cross section, σ_{ep}, times a quantity depending only on nuclear structure:

$$\frac{d^4\sigma}{d\Omega_e d\Omega_p d\omega dE_p} = K\sigma_{ep} S(\vec{p}, \epsilon_m) \quad . \tag{10}$$

The (e,p) cross section must take into account the fact that the proton is moving with momentum $\vec{p}$ and is off-shell in the initial state. The nuclear spectral function, $S(\vec{p}, \epsilon_m)$ represents the joint probability of finding a proton with momentum $\vec{p}$ and separation energy ϵ_m in the nucleus. In PWIA, $\vec{p}$ and ϵ_m, the so-called missing (or initial) momentum and missing energy, are determined from the experimental quantities as: $\vec{p} = \vec{p}_f - \vec{q}$ and $\epsilon_m = \omega - T_p - T_R$ where $\vec{p}_f$ (T_p) is the momentum (kinetic energy) of the detected proton and T_R is the recoil energy of the A−1 system. Proton knockout leading to bound states of the residual system is characterized by a series of narrow peaks in a missing energy spectrum. By performing experiments at fixed $\vec{p}$ and ϵ_m one can keep the nuclear structure content of eq. 10 fixed and attempt to measure the electron-proton interaction via σ_{ep}. Of course the strongly-interacting proton does not really emerge as a plane wave; in practice it is treated as a distorted wave generated in a complex proton-nucleus optical potential, this gives rise to the Distorted Wave Impulse Approximation (DWIA). The statements above remain approximately true when this final state interaction (FSI) between the proton and residual nucleus is taken into account.

In the last few years, two types of (e,e′p) experiments have been performed to study the electron-proton interaction inside nuclei: longitudinal/transverse (L/T) separations of the (e,e′p) cross section and measurements of its momentum transfer dependence. Both types of measurements have been performed for bound states below the two-particle emission threshold (where, presumably, the reaction mechanism is simpler) and for the continuum above this threshold. In the latter case one uses the (e,e′p) reaction to study the reaction mechanism itself; such experiments will be discussed later on.

The results of L/T separations in (e,e′p) are conveniently expressed as the ratio

$$R_G = \sqrt{\frac{2M_p^2\vec{q}^{\,2}}{q_\mu^4}\frac{R_T}{R_L}} \tag{11}$$

since this ratio has the value $G_M/G_E = \mu_p = 2.79$ for scattering from a free proton (at $q_\mu^2 = 0$). Experimental results for R_G for $1p_{3/2}$ proton knockout from ^{12}C [28] and the $1p_{3/2}$ and $1s_{1/2}$ states in ^{6}Li [29] are shown in figure 5. Some deviation from the free space value of 2.79 is observed. It would, however, be premature to label this as evidence for nucleon modification. Several authors [30] find just such a deviation is to be expected from their DWIA calculations. Suzuki[30b] showed that the spin-orbit component of the proton-nucleus interaction produces a differential effect between the longitudinal and transverse responses. Cohen *et al.*[30a] found that $R_G \neq \mu_p$ occurs in both the Dirac and Schrödinger approaches to nucleon-nucleus interactions. Their results are shown in figure 6. Mesonic currents[30c] may also affect R_G. In particular, a recent relativistic calculation [31] of *1p-1h* MECs found the transverse response to be enhanced by $\sim$ 40%. More precise data over a considerably broader kinematic range are needed to discriminate amongst the various possibilities.

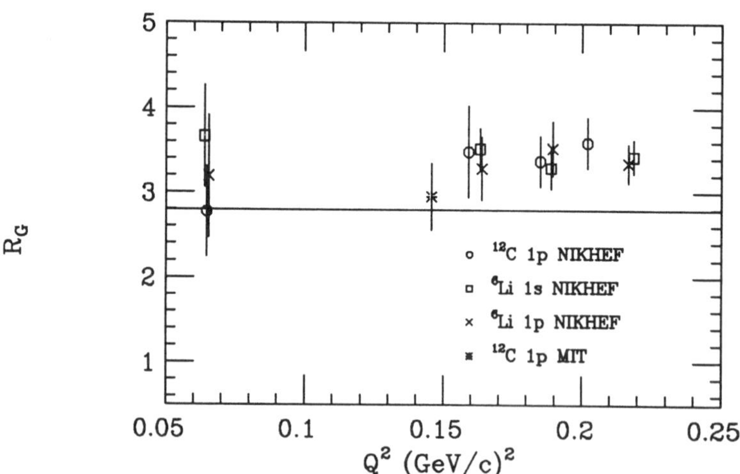

Figure 5. *Experimental results*[28,29] *of longitudinal/transverse separations in (e,e′p) on* ^{12}C *and* ^6Li *expressed as the ratio* R_G *(see text). The solid line indicates the free nucleon value of 2.79.*

A somewhat greater dynamic range can be covered in the other experimental approach, that of measuring the momentum transfer dependence of σ(e,e′p). Measurements have been made at MIT-Bates for $1p_{3/2}$ and $1s_{1/2}$ knockout from ^{12}C at momentum transfers

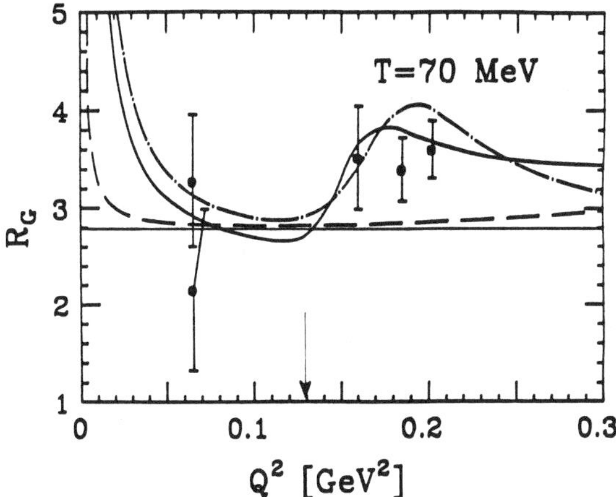

Figure 6. *The ratio R_G for ^{16}O(e,e'p) calculated in PWIA (dashed), Dirac DWIA (solid) and Schrödinger DWIA (dot-dashed) from ref. 30a using the free nucleon current. The data are for ^{12}C(e,e'p) from ref. 28b.*

of 400 MeV/c [32][28a], 600 and 800 MeV/c [33] and 1000 MeV/c. [34] All but one of the measurements were close to the maximum of the quasielastic peak (the Bjorken scaling variable $x_B \equiv q_\mu^2/2M_p\omega$ was in the range $0.64 \leq x_B \leq 0.88$ except for one experiment in the dip region[32] where $x_B = 0.32$). Comparison with theory is in the form of a ratio between the experimental cross section and a DWIA prediction [35] employing an off-shell (e,p) cross section from DeForest. [36] If the dependence on momentum transfer is that of σ_{ep} then a constant ratio should result. This is indeed the case as figure 7 shows. Note that the (e,p) cross section has fallen by a factor of 100 over this range. A similar result has been obtained at Saclay [37] on ^{40}Ca(e,e'p) over the momentum transfer range $330 \leq |\vec{q}| \leq 825$ MeV/c.

REACTION MECHANISMS

We have thus far assumed that the reaction process in the quasielastic region is single-nucleon knockout. The question of how the nucleus absorbs the energy and momentum of the virtual photon will now be addressed. Of course, the mere existence of a quasielastic peak in (e,e′) and discrete shell structure in (e,e′p) indicates that the quasifree knockout process plays an important role. However, the nucleons do not move entirely in independent particle orbits; there are residual two-body interactions (whether due to static nucleon-nucleon potentials, mesonic currents or even quark exchange/antisymmeterization effects) that give rise to dynamical correlations between nucleons. What role do correlated pairs (or more complex clusters) play in the quasielastic peak?

Inelastic electron scattering on the low-ω side of the quasielastic peak in ^{3}He provided early evidence [38] that additional reaction components were required. Calculations that included only coupling to quasifree nucleons significantly underestimated the experimental cross section (see figure 8). Much better agreement was achieved by including the interaction with correlated pairs. Laget [39] obtained his result in a traditional nuclear

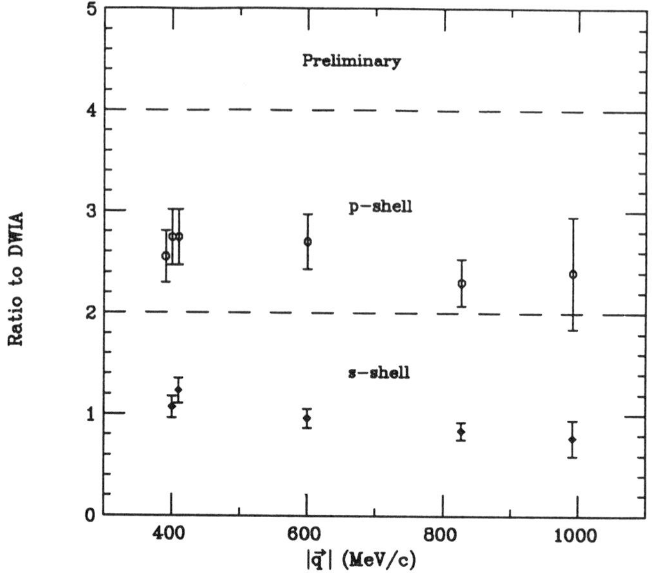

Figure 7. *Ratio to DWIA calculations of the (e,e′p) cross sections for* $1p_{3/2}$ *and* $1s_{1/2}$ *proton knockout from* ^{12}C*. The ratio being constant indicates that the momentum transfer dependence of the data is correctly given by the (e,p) cross section employing free nucleon form factors.*

physics context in which photon absorption on correlated n-p pairs leads to deuteron knockout. In a quark model, Pirner and Vary [40] found that the presence of 6- and 9-quark bags in ^{3}He, in addition to the nucleonic 3-q bag, could also explain the excess low-ω strength. The results of both approaches are shown in figure 8.

Deeper insight into the reaction mechanism can be gained from coincidence (e,e′X) reactions. One approach is to directly observe the knockout of correlated clusters such as deuterons or alpha particles. Another is to examine the missing energy spectra in (e,e′p) reactions since in a pure independent particle model only discrete peaks can occur in a ϵ_m spectrum whereas multinucleon removal via interaction with correlated clusters and their subsequent breakup will populate a continuum of missing energies. This is illustrated in figure 9 where the ^{3}He(e,e′p) spectral function obtained in a Faddeev calculation by Meier-Hajduk *et al.* [41] is shown. The nucleon-nucleon correlations produce a spectral response extending beyond 100 MeV in missing energy.

Evidence for the direct coupling of the virtual photon to correlated proton-neutron pairs has been obtained in the ^{3}He(e,e′d) reaction measured at NIKHEF. [42] They chose $\omega \sim |\vec{q}|^2/4M_N\,(x_B \sim 2)$ in order to emphasize the role of this process. The data at a fixed $|\vec{q}|$=380 MeV/c are shown as a function of initial momentum in figure 10a. A calculation [43] that ignores the coupling to n-p pairs underestimates the low-initial-momentum data by a factor of 50. On the other hand, allowing absorption of the photon on both T=0 and T=1 pairs that emerge as deuterons produces much better agreement with the data. Figure 10b shows that the interaction is with a two-nucleon pair that has the basically the same electromagnetic structure as a free deuteron. Here, the initial momentum was fixed and the momentum transfer dependence of the cross

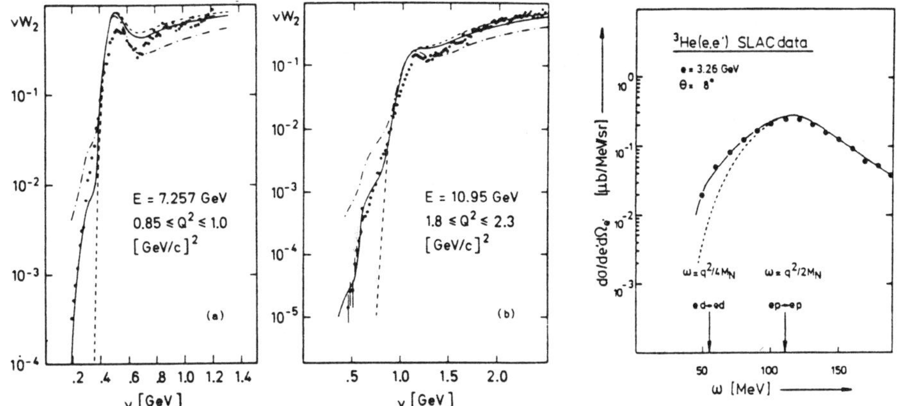

Figure 8. *Calculations of ^{3}He(e,e′) at high momentum transfer. The left two panels show the six- and nine-quark cluster calculations of Pirner and Vary*[40] *where the dashed curve corresponds to a bag radius of 0 fm (nucleons only), the solid curve to a radius of 0.45 fm and the dot-dashed one to a 0.9 fm radius. The rightmost panel shows a more traditional nuclear physics calculation by Laget.*[39] *Quasifree proton knockout alone results in the dashed curve while the solid one also includes the coupling to correlated proton-neutron pairs resulting in deuteron knockout.*

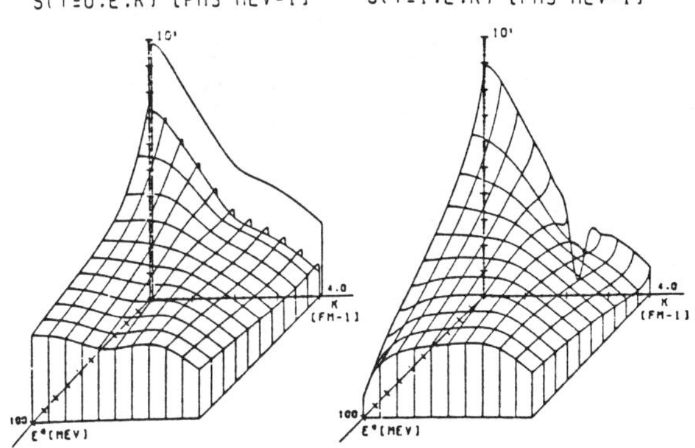

Figure 9. *Spectral function for ^{3}He from ref. 41. The left (right) panel is for T=0 (T=1). The solid line at low ϵ_m in the T=0 channel is from the two-body p+d breakup.*

section measured. One sees that it follows the free (e,d) cross section fairly well while the (e,p) cross section (dotted line) behaves quite differently.

The (e,e′d) experiment probed the two-nucleon correlations at relatively low internal momenta since the kinematics were near $x_B = 2$. The region of high initial momenta is another kinematic regime where correlation effects are expected to be dominant. In this case one typically breaks up the pair and produces continuum states in the residual system. The basic relationship between high-momentum components and continuum strength is indicated by the calculations of Ciofi degli Atti *et al.*[44] displayed in figure 11. The nucleon momentum distribution,

$$n(|\vec{k}\,|, \epsilon_{max}) = \int_0^{\epsilon_{max}} S(\vec{k}, \epsilon_m)\, d\epsilon_m \quad , \tag{12}$$

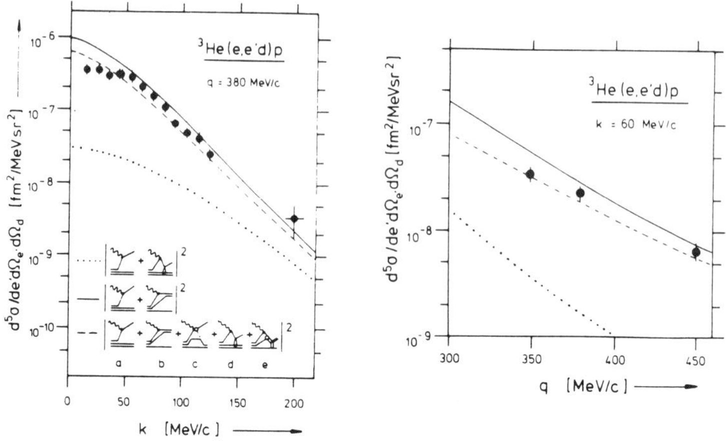

Figure 10. *The ^{3}He(e,e'd) cross section[42] vs. a) initial momentum at fixed momentum transfer and b) vs. momentum transfer at fixed initial momentum. The dotted curve shows the (e,p)+FSI result, the solid curve includes both the (e,p) and (e,pn) processes in PWIA while the full calculation[43] including FSI is shown by the dashed curve.*

in ^{3}He is shown for $\epsilon_{max} = \infty$ (the full distribution), 300, 50 and 12 MeV by the solid, dotted, dot-dashed and short-dashed curves respectively. The contribution from the two-body p+d breakup channel is shown as the long-dashed curve. One sees that 1) the high-momentum strength is completely dominated by the correlations and 2) that the strength is spread over a large continuum in missing energy.

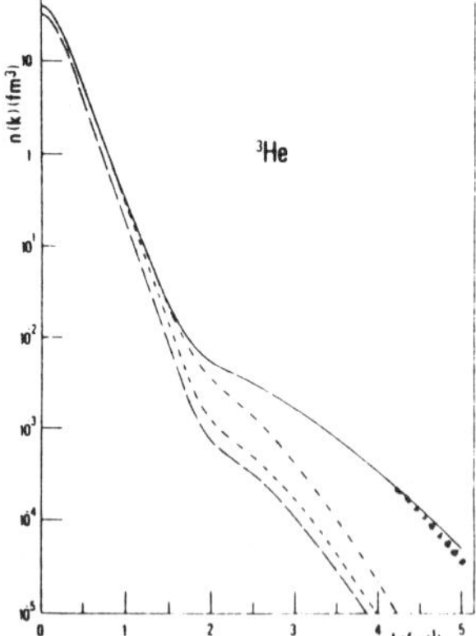

Figure 11. *The proton momentum distribution in ^{3}He.[44] The long-dashed curve is the contribution of the two-body p+d breakup channel while the short-dashed, dot-dashed and dotted result from integrating the continuum strength up tp 12, 50 and 300 MeV respectively. The full $n(|\vec{k}|)$ obtained by integrating to infinity is indicated by the solid curve.*

Just such effects were observed in a recent measurement of ^{3}He(e,e'p) at Saclay.[45] Missing energy spectra for initial momenta of 316, 401 and 458 MeV/c are shown in figure 12. Their kinematics are beyond the quasielastic peak, in the dip region. At low

ϵ_m the narrow peak resulting from $1s_{1/2}$ proton knockout is present but the spectra are dominated by the broad peak populating the continuum out to $\epsilon_m \sim 90$ MeV. The broad structure has the kinematic signature expected for interaction with a two-nucleon pair; it moves to higher ϵ_m as the initial momentum is increased. The arrows in the figure indicate the position expected from disintegration of a pair at rest. The width then reflects the center-of-mass motion of the pair. A calculation [46] of this process, augmented by exchange currents, is in reasonable accord with the data.

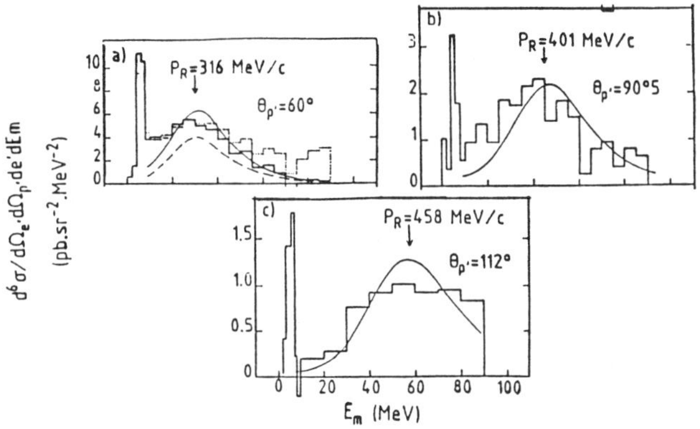

Figure 12. *Missing energy spectra*[45] *for ^{3}He(e,e'p) at high initial momenta. The narrow peak is from $1s_{1/2}$ proton knockout. The solid curves show calculations*[46] *for disintegration of n-p pairs in ^{3}He including meson exchange currents. The arrow indicates the position expected from pairs at rest.*

Of course two-body correlation effects are, if anything, expected to be more important in heavier nuclei. As an example, we show in figure 13 calculations of Alberico *et al.* [47] for the transverse response function in ^{56}Fe(e,e′). These authors use an RPA-correlated Fermi gas model. Note that the 1p-1h component only accounts for roughly 60% of the total transverse response and that considerable 2p-2h strength is predicted throughout the quasielastic peak. Most of the 2p-2h strength results from nucleon dynamical correlations, meson exchange currents, though important, provide only a modest fraction of the total 2p-2h strength. Mulders, [48] in a more schematic calculation that treats nucleon-nucleon correlations as six-quark bags, also finds a substantial, predominantly transverse, two-nucleon contribution over the entire quasielastic peak. Whether it's nucleons and mesons or six-quark bags it is clear that all components of the (e,e′) reaction must be understood before we can confidently interpret deviations from our impulse approximation ideas.

Direct evidence of an additional transverse component to the quasielastic peak has recently been obtained in a longitudinal/transverse separation of the ^{12}C(e,e′p) reaction at MIT-Bates.[28a] The kinematics of $|\vec{q}| = 400$ MeV/c and $\omega = 120$ MeV correspond to a point nearly at the maximum of the quasielastic peak. The separated response functions R_T and R_L are shown in the top two panels of figure 14. One observes the narrow peak from $1p_{3/2}$ knockout and the broad bump usually attributed to $1s_{1/2}$ knockout. The presence of an additional process is clear from the figure; R_L is consistent with zero for $\epsilon_m > 45$ MeV while R_T remains non-zero out to the highest measured missing energies. It is important to realize that final state interactions of the knocked-out proton are unable to produce such a large differential effect between the two response

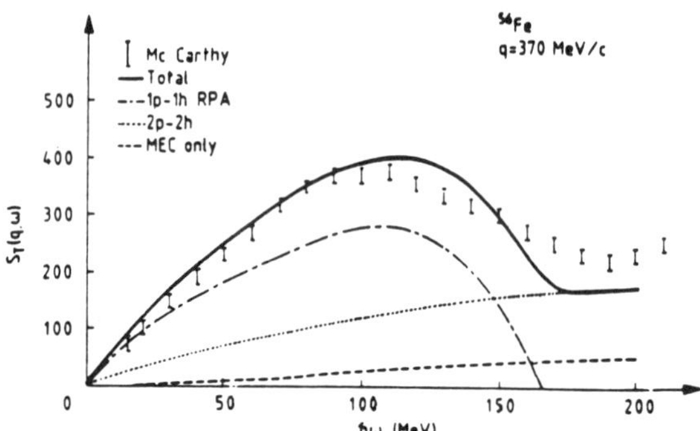

Figure 13. *The transverse response function in iron calculated by Alberico et al.*[47]. *The dash-dot curve is the 1p-1h RPA result, the dotted curve is the total 2p-2h response while the dashed curve shows the MEC contribution. The solid curve indicates the total response function.*

functions. The third panel of figure 14 shows the difference of the two spectral functions S_T and S_L extracted from the two response functions (the (e,e′p) analogs of f_T and f_L). This difference starts increasing above $\epsilon_m = 28$ MeV, the threshold for two-nucleon emission. A recent measurement of ^{12}C(e,e′p) at NIKHEF [49] confirms the increased transversality and its missing energy dependence.

These results are not what one would expect in a one-body model with modified nucleon form factors (either magnetic or coulombic). In such a model the spectral distributions would remain the same. A new transverse current, most likely of two-body character, is present with significant strength even at the maximum of the quasielastic peak and appears to be responsible for the large L/T anomaly observed in (e,e′).

The transverse continuum strength becomes even more pronounced in the higher energy loss dip region. (Referring back to figure 4 one sees that there is little longitudinal strength in the region of $y \sim 0.4$). The missing energy spectrum for ^{12}C(e,e′p) in the dip region[32] ($|\vec{q}| = 400$ MeV/c, $\omega = 200$ MeV) is shown in figure 15. As in the ^{3}He(e,e′p) experiment[45] under similar kinematics, the continuum strength is the dominant contribution to the spectrum. On the other hand, the yield in ^{12}C extends nearly uniformly to large (150 MeV) missing energy.

Absorption of the photon on a correlated pair of nucleons accounted for all of the continuum in ^{3}He since there only the two-nucleon knockout process is available to populate the continuum. In a complex nucleus, reaction channels that involve more nucleons are possible, at least in principle. Naturally, the first multinucleon processes in ^{12}C(e,e′p) to receive theoretical attention have been the two-nucleon ones. Initial-state pair correlations have been estimated by Laget [50] in a quasideuteron model. His result is shown by the solid curve in figure 15. This model provides no strength above 100 MeV. The effect of nucleon-nucleon scattering in the final state has been calculated by T. Takaki. [51] The process whereby only two nucleons emerge was treated fully quantum-mechanically and with a realistic interaction. He finds that this process is also confined to missing energies below 100 MeV and it is a small ($< 20\%$) contribution at the lower missing energies. His results are shown in figure 16. He has also given a qualitative characterization of what one might expect from three-nucleon processes. Employing a zero-range (δ-function) interaction between the three nucleons Takaki obtained the dashed curve in figure 16. Such a process is able to naturally populate the deep ϵ_m

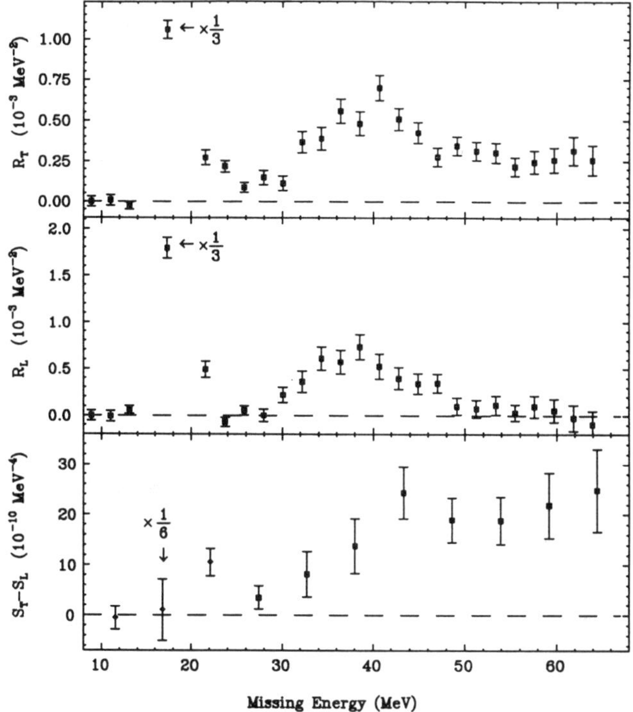

Figure 14. *Separated $^{12}C(e,e'p)$ response functions and their difference.*[28a] *Transverse (a) and longitudinal (b) response functions and the difference in the spectral functions (c) vs. missing energy.*

region. It now appears that reaction mechanisms involving only two nucleons, either as initial-state correlations or final-state scatterings, are inadequate to account for the yield observed at large ϵ_m.

SUMMARY AND OUTLOOK

The quasielastic peak has been found to contain more than simply the incoherent scatterings from quasifree nucleons. Experiments have provided unambiguous signatures of a new transverse current and of two-nucleon correlation effects. Furthermore, reaction processes where several nucleons participate in absorbing the energy of the photon seem to be required to account for the (e,e'p) yield in the dip region on complex nuclei.

Though a very intriguing topic, modification of nucleons in nuclei remains somewhat ambiguous. None of the present experimental results demand modified nucleons. It is not even entirely clear that it is a well-posed question. Is there a unique signature to look for, either in existing or future data, or might many-body effects in the current

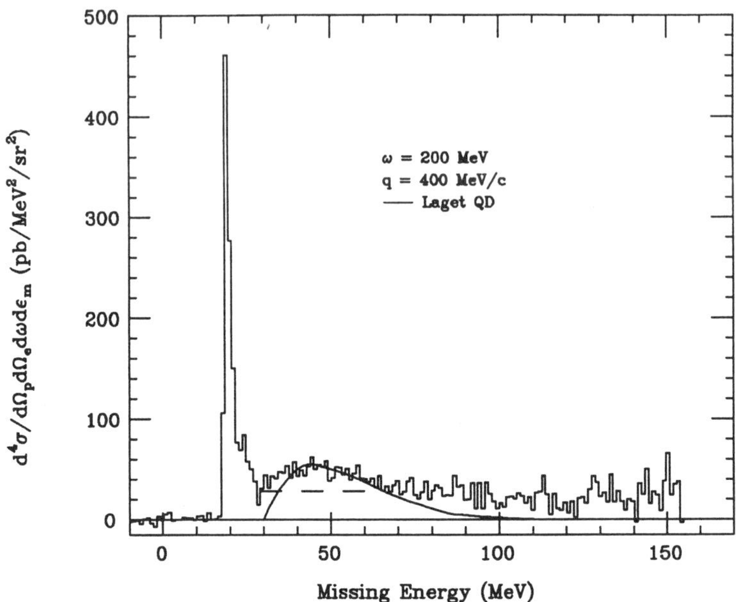

Figure 15. *Missing energy spectrum for $^{12}C(e,e'p)$ in the dip region.*[32] *The area above the dashed line is attributed to $1s_{1/2}$ proton knockout while the solid curve shows the quasideuteron model result of Laget.*[50]

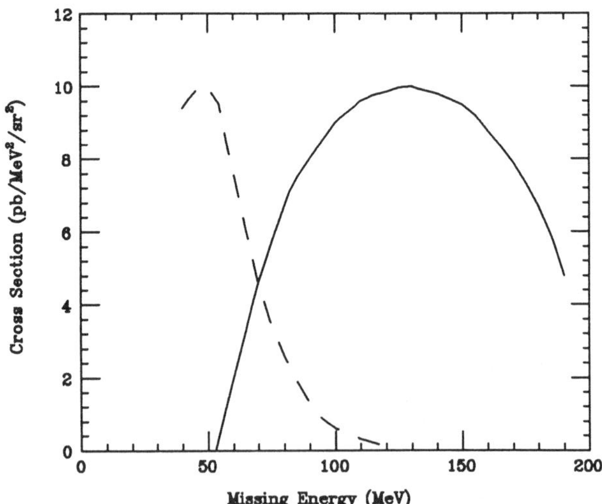

Figure 16. *Preliminary results of T. Takaki for the $^{12}C(e,e'p)$ missing energy spectrum in the dip region. The dashed and solid curves show the shapes of the two-nucleon and three-nucleon knockout contributions respectively. The left-hand scale only applies to the two-nucleon curve; the normalization of the three-nucleon result is arbitrary.*

operator and/or the nuclear wavefunction account for any deviations we observe? Given that we only measure matrix elements, might not the problem be reorganized so that what was a modified current mimics the free current in a different basis? Progress will require greater understanding of the off-shell electron-nucleon vertex function along with more complete knowledge of the nuclear pion fields. On the other hand, some structural change of the nucleon is to be expected on quite general grounds. Oka and Amado [52] showed that any composite quantum system in an attractive potential will develop an increased rms radius, i.e. 'swell'.

Differentiation amongst competing explanations will require experiments that achieve more extreme and broader kinematic conditions than are possible today. We hope to obtain deeper insight into all aspects of the nuclear quasielastic response at the new generation of 100% duty-factor electron machines. They will provide greatly increased sensitivity and dynamic range. The power of the electromagnetic interaction lies in the ability to describe any process by a limited number of invariant response functions. Polarization observables and out-of-plane spectrometry will allow a much more complete decomposition of the response surface than had previously been possible. Experiments that detect two or more hadrons in the final state will provide detailed information on nuclear correlation functions and reaction mechanisms.

ACKNOWLEDGEMENTS

I would like to thank the organizers of the conference for the opportunity to discuss these topics and am pleased to acknowledge many informative discussions with Profs. W. Bertozzi, J.M. Finn, J.W. Van Orden, E.J. Moniz, Drs. T.W. Donnelly, P.E. Ulmer, W. Boeglin and G. van der Steenhoven. I thank L. Weinstein and J. Morrison for allowing their results to be presented prior to publication. This work is supported in part by the U.S. Department of Energy under contract DE-AC02-76ER03069.

REFERENCES

[1] B.D. Serot and J.D. Walecka, *The Relativistic Nuclear Many-Body Problem*, Adv. Nucl. Phys. Vol. 16, Eds. J.W. Negele and E. Vogt, Plenum Press, (1986)

[2] H.W.L. Naus and J.H. Koch, Phys. Rev. **36**, 2459 (1987).

[3] E.J. Moniz *et al.*, Phys. Rev. Lett. **26**, 445 (1971).

[4] E.J. Moniz, Phys. Rev. **184**, 1154 (1969).

[5] R.R. Whitney *et al.*, Phys. Rev. **C9**. 2230 (1974).

[6] a) P. Barreau *et al.*, Nucl. Phys. **A402**, 515 (1983); b) Z.E. Meziani *et al.*, Phys. Rev. Lett. **52**, 2130 and **54** 1233 (1984); c) M. Deady *et al.*, Phys. Rev. **C33**, 1897 (1986); d) R. Altemus *et al.*, Phys. Rev. Lett. **44**, 965 (1980).

[7] J.L. Friar, *Electron and Pion Interactions with Nuclei at Intermediate Energies*, Eds. W. Bertozzi, S. Costa and C. Schaerf, Harwood (1980).

[8] J.W. Van Orden and T.W. Donnelly, Ann. Phys. (N.Y.), **131**, 145 (1980).

[9] T. DeForest Jr., Nucl. Phys. **A414**, 347 (1984).

[10] K.W. McVoy and L. van Hove, Phys. Rev. **125**, 1034 (1966).

[11] K. Dow *et al.*, To be published.

[12] C. Marchand *et al.*, Phys. Lett. **153B**, 29 (1985).

[13] R. Schiavilla *et al.*, Nucl. Phys. **A473**, 267 (1997).

[14] W.M. Alberico, P. Czerski, M. Ericson and A. Molinari, Nucl. Phys. **A462**, 269 (1987).

[15] C.J. Horowitz, Indiana Univ. preprint IU/NTC 88-4 (1988).

[16] J.V. Noble, Phys. Rev. Lett. **46**, 412 (1981).

[17] C.M. Shakin, Nucl. Phys. **A446**, 323c (1985).

[18] J.J. Aubert *et al.* (EMC), Phys. Lett. **123B**, 275 (1983).

[19] G. Do Dang, M. L'Huillier, N. Van Giai and J.W. Van Orden, Phys. Rev. **C35**, 1637 (1987).

[20] G.B. West, Phys. Rep. **18C**, 269 (1975); Y. Kawazoe, G. Takeda and H. Matsuzaki, Prog. Theor. Phys. **54**, 1394 (1975).

[21] C. Ciofi degli Atti, Il Nouvo Cimento, **76A**, 330 (1983).

[22] I. Sick, Phys. Lett. **157B**, 13 (1985).

[23] C. Ciofi degli Atti, *Proceedings of the European Workshop on Few-Body Physics*, Eds. C. Ciofi degli Atti *et al.*, Springer-Verlag (1986).

[24] E. Oset, R. Tegen and W. Weise, Nucl. Phys. **A426**, 456 (1984).

[25] J.M. Finn, R.W. Lourie and B.H. Cottman, Phys. Rev. **C29**, 2230 (1984).

[26] W.M. Alberico, A. Molinari, T.W. Donnelly, E.L. Kronenberg and J.W. Van Order, To appear in Phys. Rev. C.

[27] S. Frullani and J. Mougey, *Single-Particle Properties of Nuclei through (e,e′p) Reactions*, Adv. Nucl. Phys. Vol. 14, Eds. J.W. Negele and E. Vogt, Plenum (1984).

[28] a) P.E. Ulmer *et al.*, Phys. Rev. Lett. **59**, 2259 (1987); b) G. van der Steenhoven *et al.*, Phys. Rev. Lett. **57**, 182 (1986).

[29] G. van der Steenhoven *et al.*, Phys. Rev. Lett. **58**, 1727 (1987).

[30] a) T.D. Cohen, J.W. Van Orden and A. Picklesimer, Phys. Rev. Lett. **59**, 1267 (1987); b) T. Suzuki, Phys. Rev. **C37**, 549 (1988); c) M. Kohno, Universität Mainz preprint.

[31] P.G. Blunden and M.N. Butler, contribution to the *International Conference on Spin Observables of Nuclear Probes,* Telluride, CO. (1988).

[32] R.W. Lourie *et al.*, Phys. Rev. Lett. **56**, 2364 (1986).

[33] L.B. Weinstein *et al.*, to be published.

[34] J. Morrison *et al.*, to be published.

[35] S. Boffi, C. Giusti and F.D. Pacati, Nucl. Phys. **A336**, 437 (1986).

[36] T. DeForest Jr., Nucl. Phys. **A392**, 232 (1983).

[37] D. Reffay-Pikeroen *et al.*, Phys. Rev. Lett. **60**, 776 (1988).

[38] D. Day *et al.*, Phys. Rev. Lett. **43**, 1143 (1979).

[39] J.M. Laget, Can. J. Phys. **62**, 1046 (1984).

[40] H.J. Pirner and J.P. Vary, Phys. Rev. Lett. **46**, 1376 (1981).

[41] H. Meier-Hajduk, CH. Hajduk, P.U. Sauer and W. Theis, Nucl. Phys. **A395**, 332 (1983).

[42] P.H.M. Keizer *et al.*, Phys. Lett. **157B**, 255 (1985).

[43] J.M. Laget, Phys. Lett. **151B**, 325 (1985).

[44] C. Ciofi degli Atti, E. Pace and G. Salmè, Phys. Lett. **141B**, 14 (1984).

[45] C. Marchand *et al.*, Phys. Rev. Lett. **60**, 1703 (1988).

[46] J.M. Laget, *New Vistas in Electronuclear Physics*, Eds. E. Tomusiak *et al.*, Plenum (1986).

[47] W.M. Alberico, M. Ericson and A. Molinari, Ann. Phys. **154**, 356 (1984).

[48] P.J. Mulders, Nucl. Phys. **A459**, 525 (1986).

[49] G. van der Steenhoven *et al.*, to appear in Nucl. Phys. (1988).

[50] J.M. Laget, private communication.

[51] T. Takaki, private communication.

[52] M. Oka and R.D. Amado, Phys. Rev. **C35**, 1586 (1987).

B–Meson Decays

K.R. Schubert
Insitut für Experimentelle Kernphysik
Universität Karlsruhe

The third generation of quarks, anticipated in 1973 by Kobayashi and Maskawa [1] as one of the possibilities to explain CP violation of the K^o–meson system within the framework of the Standard Model and in 1975 by the discovery of the third generation of leptons [2], was found at FNAL in 1977. It showed up in form of the $\Upsilon(1S)$ resonance [3], the lowest–energy 1^{--} state of the b–quark $\bar{b}$–antiquark system. Experiments at DESY in 1978 [4] showed that the b–quark has charge 1/3. These and further experiments showed that the $b\bar{b}$ system follows the same spectroscopic and strong–interaction rules as the $c\bar{c}$ system. The physics of the $b\bar{b}$ system has been studied extensively at the DESY e^+e^- storage ring DORIS using the detectors DASP-II, PLUTO, LENA, Crystal Ball, and ARGUS, and from 1980 on at the Cornell e^+e^- ring CESR using CLEO and CUSB. Today we know twelve states of the $b\bar{b}$ system in contrast to seven of the $c\bar{c}$ system and only four in the e^+e^- bound state system positronium.

The nine lightest known $b\bar{b}$ states decay into final states without b–flavour, e.g.

$$\Upsilon(1S) = 1^3S_1 b\bar{b} \rightarrow ggg \rightarrow \text{ light hadrons,}$$

$$\Upsilon(2S) = 2^3S_1 b\bar{b} \rightarrow \Upsilon(1S)\pi^+\pi^-$$

$$\Upsilon(2S) \rightarrow \gamma + \chi_{bJ}(1P);\ \chi_{bJ}(1P) = 1^3P_J,\ J = 0,1,2.$$

The three heaviest states, $\Upsilon(4S), \Upsilon(5S)$, and $\Upsilon(6S)$, decay into final states with b–flavour, i.e. ground–state or excited B–mesons. The main source for experiments with B–mesons up to now is the $\Upsilon(4S)$ state, $4^3S_1 b\bar{b}$, with a mass of $m = (10580 \pm 3)\ MeV$ and a width of $\Gamma = (24 \pm 2)\ MeV$. This state has a negligeable decay fraction into three gluons (some 10^{-3}) and decays predominantly into a B^+B^- or a $B^o\bar{B}^o$ pair without additional particles. It is produced in e^+e^- annihilation with a cross section of about 1 nb, i.e. the CESR ring with an average luminosity of $\mathcal{L}_{av} = 500/pb/y$ produces $500000 B\bar{B}$ pairs per year and DORIS with $\mathcal{L}_{av} = 100/pb/y$ produces 100000. There is no $B_S\bar{B}_S$ production on the $\Upsilon(4S)$ since this state is below the $B_S\bar{B}_S$ threshold.

The six B–mesons mentioned above, $B^+ = 1^1S_o\bar{b}u$, $B^- = 1^1S_ob\bar{u}$, $B^o = 1^1S_o\bar{b}d$, $\bar{B}^o = 1^1S_ob\bar{d}$, $B_S = 1^1S_o\bar{b}s$, and $\bar{B}_S = 1^1S_ob\bar{s}$, decay only by weak interaction. Their average life time, $< \tau(B^+, B^o, B_S) >$, is about $10^{-12}\ s$.

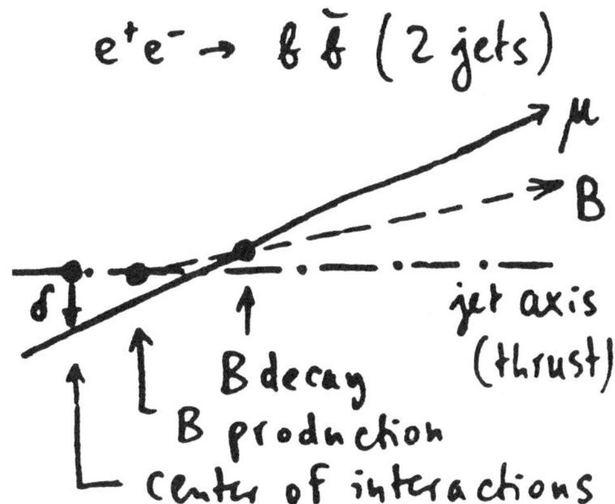

Figure 1: Definition of the impact parameter δ for a muon from a B–meson decay in a $\bar{b}$ jet.

Whereas most of the B–meson properties have been determined with experiments at CESR and DORIS, life time results come from PEP and PETRA. All these results use essentially the same indirect method, the impact parameter method as sketched in fig. 1. At PEP and PETRA, using e^+e^-annihilation at a center–of–mass energy around 30 GeV, B–mesons are produced in jets. Their production points are distributed in a certain volume around the nominal interaction center. Their direction of flight is close to the jet axis which is reconstructable with good precision by the "thrust" method. Decays of a B–meson are selected by requiring an electron or muon with large transverse momentum with respect to the jet axis. An "impact parameter" δ can be determined for each such e or μ following the definition in fig. 1. The parameter is positive if the e/μ direction intersects the jet axis as drawn in fig. 1, and negative if it intersects the jet axis behind the center of interactions. The statistical average $< \delta >$ of all impact parameters would be zero without finite B–meson lifetime, and a positive $< \delta >$ is proportional to the lifetime. First non–zero results were obtained in 1983 at PEP by the collaboration MAC [5] and MARK–II [6]. The results have been updated with more statistics, and all present results are summarized in table 1. The average, taking into account correlations in the systematic error, is

$$< \tau(B^+, B^o, B_S) >= \tau_b = (1.15 \pm 0.15) \cdot 10^{-13} s.$$

A recent δ distribution is shown in fig. 2 [11].

It has to be pointed out that this result is a weighted average of three different mean lives or even more since weakly–decaying b–flavoured baryons may also be present in the selected events. The difference of $\tau(B^+)$ and $\tau(B^o)$ is an interesting topic for the future; it must be recalled that

$$\tau(K^+)/\tau(K_S^o) = 139,$$

$$\tau(D^+)/\tau(D^o) = 2.4,$$

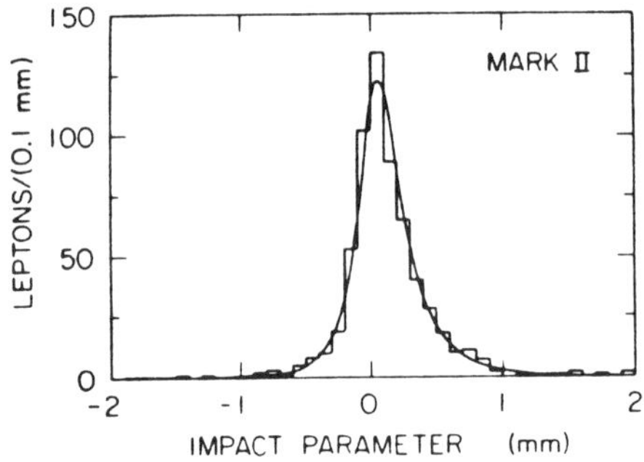

Figure 2: Distribution of impact parameters in a recent B–meson lifetime experiment [11].

Experiment	$\tau(b)$ in 10^{-13} s
JADE 1986 [7]	$18.0^{+5.0}_{-4.0} \pm 4.0$
MAC 1987 [8]	$12.9 \pm 2.0 \pm 2.1$
HRS 1987 [9]	$10.2^{+4.2}_{-3.9}$
TASSO 1987 [10]	$13.6 \pm 1.4 \pm 3.0$
MARK–II 1987 [11]	$9.8 \pm 1.2 \pm 1.3$
DELCO 1988 [12]	$11.7^{+2.7}_{-2.2}{}^{+1.7}_{-1.6}$

Table 1: b–quark mean life results

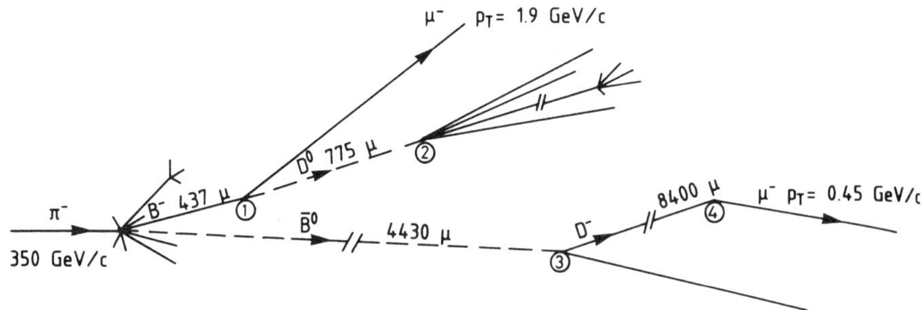

Figure 3: The only known event [13] with reconstructed B–meson production and decay vertices.

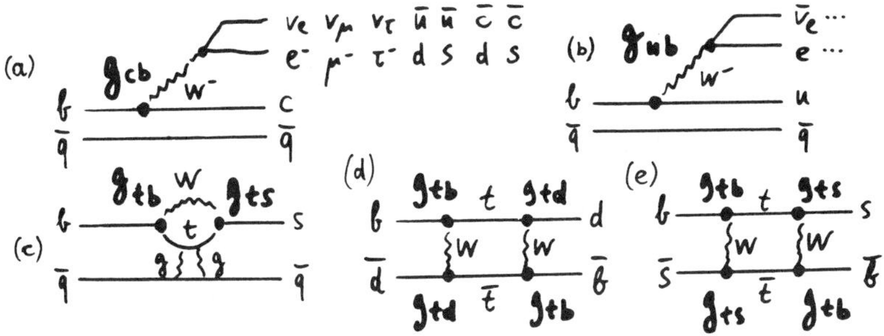

Figure 4: B—meson decay and oscillation diagrams, (a) spectator graph with $b \to c$, (b) spectator graph with $b \to u$, (c) penguin graph, (d) box graph for $\bar{B}^o \to B^o$, (e) box graph for $\bar{B}_S \to B_S$.

and that these differences are due to the strong interaction and are not fully understood until now. The only known two B–mesons with separately reconstructed production and decay vertices have been found in an emulsion experiment at CERN [13]. They are both in the same event, a $\bar{B}^o$ with an individual life time of $5 \cdot 10^{-13}$ s and a B^- with $0.7 \cdot 10^{-13} s$; the event is shown in fig. 3.

Decays of B–mesons are mediated by the standard weak interaction, the relevant diagrams are shown in fig. 4a, b, c. The neutral B–mesons B^o and B_S may show the effect of flavour oscillations before decaying, a B^o may transform itself into a $\bar{B}^o$ (fig. 4d) and a B_S into a $\bar{B}_S$ (fig. 4e) through second–order standard weak interaction. These tree and loop diagrams are sensitive to 5 of the 9 coupling constants g_{ij} of the charged weak quark current. In the Standard Model we have

$$\begin{pmatrix} g_{ud} & g_{us} & g_{ub} \\ g_{cd} & g_{cs} & g_{cb} \\ g_{td} & g_{ts} & g_{tb} \end{pmatrix} = g \cdot \begin{pmatrix} V_{ud} & V_{us} & V_{ub} \\ V_{cd} & V_{cs} & V_{cb} \\ V_{td} & V_{ts} & V_{tb} \end{pmatrix}$$

with the universal coupling constant g as precisely determined from the mean life of the muon and related to the Fermi coupling constant G_F by

$$g^2/M_W^2 = 4\sqrt{2}\,G_F,$$

and with a unitary matrix V_{ij}. This matrix has been introduced by Kobayashi and Maskawa [1] following the gauge principle and ideas of Cabibbo [14], Bjorken–Glashow [15] and Glashow–Iliopoulos–Maiani [16]. It is often called the Cabibbo–Kobayashi–Maskawa (CKM–)matrix. Being unitary it is fully described by only four observable parameters, conventionally chosen as three angles and one phase. In my preferred parametrization [17] these are ϑ_{12}, ϑ_{13}, ϑ_{23}, and φ, where ϑ_{ij} couples the i'th generation to the j'th generation of quarks

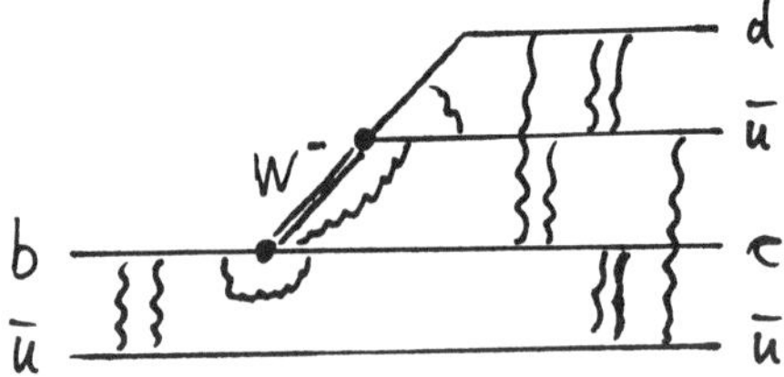

Figure 5: Illustrating the influence of the strong interaction on a weak decay.

and ϑ_{12} is the Cabibbo angle [14]:

$$V_{ij} = \begin{pmatrix} c_{12}c_{13} & s_{12}c_{13} & s_{13}e^{-i\varphi} \\ -s_{12}c_{23} - c_{12}s_{13}e^{i\varphi}s_{23} & c_{12}c_{23} - s_{12}s_{13}e^{i\varphi}s_{23} & c_{13}s_{23} \\ s_{12}s_{23} - c_{12}s_{13}e^{i\varphi}c_{23} & -c_{12}s_{23} - s_{12}s_{13}e^{i\varphi}c_{23} & c_{13}c_{23} \end{pmatrix}$$

In this notation $c_{12} = cos\vartheta_{12}$, $s_{13} = sin\vartheta_{13}$ etc.

The parameters of the CKM–matrix are four out of twenty fundamental parameters of the Standard Model, unpredictable by the Model as the coupling constants g^2, α, α_S or the quark and lepton masses.

There are essentially three reasons to study B–meson decays experimentally:

- to determine the values of the three fundamental parameters ϑ_{13}, ϑ_{23} and φ,
- to further develop our understanding of confinement and nonperturbative QCD including reliable calculation methods,
- to search for new physics, i.e. new interactions which are weaker than the standard V–A weak interaction.

I do not want to elaborate the third point here. No new interaction has been found so far, some recent pertinent results are [18]:

$$\begin{aligned} B(B^o \to e^+e^-) &< 8 \cdot 10^{-5}, \\ B(B^o \to \mu^+\mu^-) &< 5 \cdot 10^{-5}, \\ B(B^o \to l^+l^-X) &< 1.2 \cdot 10^{-4}, \\ B(B^o \to e^\pm\mu^\mp) &< 5 \cdot 10^{-5}, \end{aligned}$$

where B is the decay fraction (branching ratio).

I also want to be very short on the second point here. Weak nonleptonic meson decays, especially into final states with two mesons only, are the simplest laboratory for studying QCD [19] since these processes involve only two valence quarks in the initial and four in the final state. As sketched in fig. 5, the weak process $b\bar{u} \to c\bar{u}d\bar{u}$ is not only governed by the emission of a W^- but also by the exchange of hard and soft gluons. The renormalization of W emission

	CLEO	ARGUS
$B^o \to \bar{D}^o\pi^=\pi^-$	< 4.0 (90%)	
$D^-\pi^+$	$0.6 \pm 0.3 \pm 0.14$	
$D^{*-}\pi^+$	$0.32^{+0.17}_{-0.13}{}^{+0.11}_{-0.07}$	$0.35 \pm 0.18 \pm 0.13$
$D^{*-}\pi^+\pi^+\pi^-$	< 4.8 (90%)	$4.3 \pm 1.2 \pm 2.0$
$D^{*-}\pi^+\pi^o$		$2.0 \pm 1.0 \pm 1.0$
$J/\psi K^{o*}$	0.41 ± 0.19	0.33 ± 0.18
$B^+ \to \bar{D}^o\pi^+$	$0.48^{+0.16}_{-0.13}{}^{+0.11}_{-0.07}$	
$D^-\pi^+\pi^+$	$0.25^{+0.41}_{-0.23}{}^{+0.24}_{-0.07}$	
$D^{*-}\pi^+\pi^+$	$0.21^{+0.15}_{-0.13}{}^{+0.08}_{-0.05}$	$0.6 \pm 0.3 \pm 0.4$
$D^{*-}\pi^+\pi^+\pi^o$		$5.6 \pm 1.7 \pm 3.4$
$J/\psi K^+$	0.09 ± 0.06	0.07 ± 0.04
$J/\psi K^+\pi^+\pi^-$		0.11 ± 0.07
$\psi' K^+$		0.22 ± 0.17

Table 2: Decay fractions of nonleptonic B–meson decays in percent.

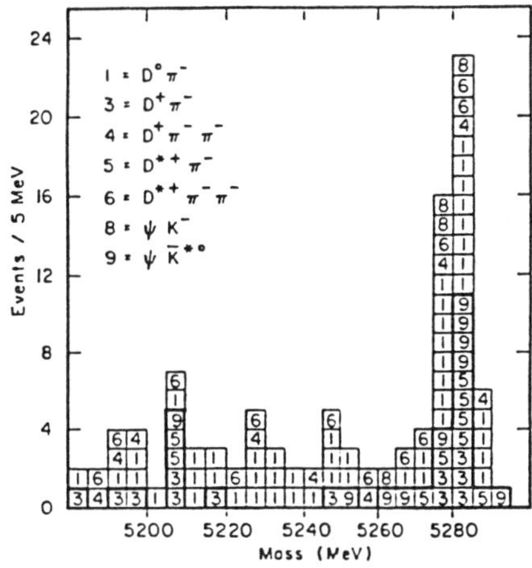

Figure 6: Mass spectrum of fully reconsructed B–meson decay candidates from CLEO [21].

by hard gluons is well treated by perturbative QCD whereas the various soft gluon exchanges can only be handled in a model–dependent way. Exclusive few–body decays of B–mesons have been observed by ARGUS [20] at DORIS and CLEO [21] at CESR. Fig. 6 shows the mass spectrum of decay candidates in seven few–body channels as observed by CLEO [21]. There is a clear peak at $m = 5280\ MeV$, and the obtained decay fractions are in good agreement with those obtained by ARGUS as shown in table 2. The experimental rates

$$\Gamma_i = B_i/\tau_b$$

agree within errors with predictions of a factorizable relativistic formfactor model [22] and $|V_{cb}| \approx 0.05$. Taking the decay $B^+ \to \bar{D}^o\pi^+$ as an example, this model absorbs soft gluon effects in a formfactor $f_+(0)$ for the description of $B \to \bar{D}W^+$ and the pion decay constant f_π for $W^+ \to \pi^+$. The value of $f_+(0)$ is calculated as overlap integral between the b–quark wave function in the B–meson and the c–quark wave function in the D meson, and f_π is the experimental value from $\pi^+ \to \mu^+\nu_\mu$.

The rest of this talk covers the first point, the determination of Standard Model parameters. From weak interactions between u, d, c, and s quarks we know that [18]

$$\begin{aligned} V_{ud} &= 0.9747 \pm 0.010, \\ V_{us} &= 0.220 \pm 0.002, \\ |V_{cd}| &= 0.21 \pm 0.03, \\ |V_{cs}| &= 1.03 \pm 0.16. \end{aligned}$$

If ϑ_{13} and ϑ_{23} are small, these four observables fix only one of the four CKM–matrix parameters, the Cabibbo angle ϑ_{12}. The two other angles are determined from B–meson decays, ϑ_{13} by the rate of $b \to u$ decays, proportional to $|V_{ub}|^2$, and ϑ_{23} by the rate of $b \to c$ decays, proportional to V_{cb}^2. Using unitarity, the cosine of the phase φ can in princple be determined from $|V_{ub}|$ and $|V_{td}|$ as shown at the end of this talk. A crucial test of the unitarity of the CKM–matrix, as required by the Standard Model with three generations of quarks, will in future be possible with the comparison of $B^o\bar{B}^o$ and $B_S\bar{B}_S$ oscillations.

The only present observables of $B\bar{B}$ oscillation are the time–integrated oscillation parameters χ_o for $B^o\bar{B}^o$ and χ_S for $B_S\bar{B}_S$. They are defined as

$$\chi_o = \chi(B^o) = \Gamma(B^o \to \bar{B}^o \to \ell^- X)/\Gamma(B^o \to \ell^\pm X),$$

and χ_S accordingly. χ_o is the fraction of all produced B^o mesons which decay as a $\bar{B}^o$ meson between $t = 0$ and $t = \infty$. It is equal to

$$\chi_o = \frac{(\Delta M)^2 + (\Delta\Gamma/2)^2}{2\Gamma^2 + (\Delta M)^2},$$

where ΔM is the mass difference between the two mass eigenstates $B_1 \approx (B^o + \bar{B}^o)/\sqrt{2}$ and $B_2 \approx (B^o + \bar{B}^o)/\sqrt{2}$, Γ is the average total decay rate of these two states, $\approx 1/\tau(b)$, and $\Delta\Gamma$ is the total decay rate difference between the two states. In the Standard Model with three generations of quarks, ΔM is given by [23]

$$\Delta M = \frac{G_F^2}{6\pi^2} \cdot m_t^2 \cdot \left\{1 + o\left(\frac{m_t}{m_W}\right)\right\} \cdot V_{tb}^2 \left|V_{td}\right|^2 m_B f_B^2 \eta_B B_B,$$

where f_B is the B–decay constant describing $B \to \ell\nu$, η_B is the hard QCD correction, and B_B is a soft QCD correction. Since m_t and V_{td} are unknown, no real prediction for χ_o has been possible.

In e^+e^- annihilation as well as in high energy hadronic reactions like $\bar{p}p$, all b–flavoured hadrons originate from a $b\bar{b}$ pair. Both b quarks have to decay weakly, and at their moments of decay they are either $b\bar{b}$, bb, or $\bar{b}\bar{b}$. The experimental signature for $B\bar{B}$ oscillations is a non–zero value of

$$R_2 = N(bb + \bar{b}\bar{b})/N(b\bar{b} + bb + \bar{b}\bar{b}),$$

where N is the observed number of flavour–identified decays from $t = 0$ to $t = \infty$ between production and decay.

$R_2 = 2\chi(1 - \chi)$, if $b\bar{b}$ is produced incoherently at high energies,
$R_2 = \chi$, if coherent $B^o\bar{B}^o$ pairs from $\Upsilon(4S)$ decays are observed,

$\chi = \chi_o$, if only B^o and $\bar{B}^o$ decays are observed (no $B^\pm$ etc.),
$\chi = f_o\chi_o$, if all B–decays in $\Upsilon(4S)$ are observed (i.e. B^o and $B^\pm$),
$\chi = f_o\chi_o + f_S\chi_S$, if all B–decays in high energy jets are observed.

On the $\Upsilon(4S)$, f_o is the fraction $N(B_o)/N(B)$ which is about 0.45. At high energy, f_o and f_S are defined in the same way; fragmentation studies suggest $f_o \approx 0.38$ and $f_S \approx 0.15$. These f values have still to be modified by branching ratios if not all decay modes are observed.

In March 87, the ARGUS group [24] presented the observation of a surprisingly large $B^o\bar{B}^o$ oscillation rate. With 88000 $\Upsilon(4S)$ decays, the group uses three strategies for selecting flavour–tagged B–decay–pairs, events with two leptons and events with one or two fully reconstructed B–mesons. The "inclusive" di–lepton sample, selected with $p_e > 1.4\ GeV/c$, $p_\mu > 1.4\ GeV/c$, and with hadron misidentifications of 0.5% for e and 2% for μ, consists of $N_1 = 295.1$ lepton pairs with $N_1^{\pm\pm} = 24.8 \pm 8.0$ like–sign lepton pairs after all corrections (J/ψ, charm decays, misidentified hadrons). The second strategy looks for events where one neutral B–meson is fully reconstructed in its $D^{*\pm}n\pi$ decay mode with $n = 1, 2, 3$ or in its $D^{*\pm}\ell^\mp\nu$ mode, and the other B–decay is flavour–tagged by a charged lepton only. The use of $D^{*\pm}\ell^\mp\nu$ decays in this strategy is only possible because B–mesons from the $\Upsilon(4S)$ resonance are nearly at rest. The "semi–inclusive" sample with one reconstructed B and one lepton from the other B contains $N_2 = 24.9$ events with $N_2^{\pm\pm} = 4.1 \pm 2.2$ of the like–sign type. There is an overlap of $N_{12} = 11.1$ and $N_{12}^{\pm\pm} = 1.7$ of events found in both samples. The third strategy looks for events with two neutral B–mesons fully reconstructed. One such event, a B^oB^o decay pair and hence a clear evidence for the existence

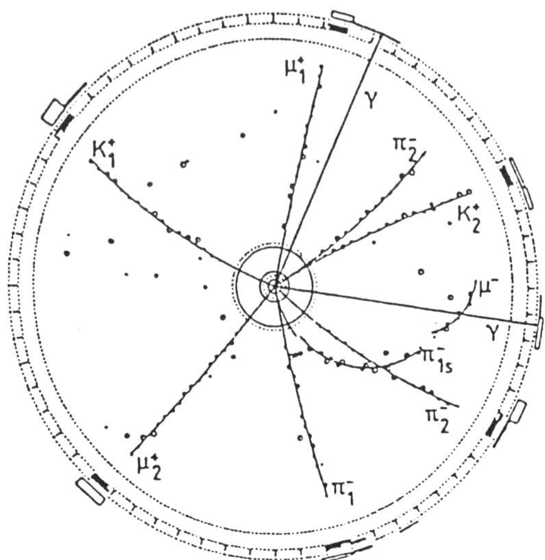

Figure 7: The ARGUS event [24] $e^+e^-(10.58\ GeV) \rightarrow \Upsilon(4S) \rightarrow B^o\bar{B}^o$, $B^o \rightarrow D_1^{*-}\mu_1^+\nu$, $\bar{B}^o \rightarrow B^o \rightarrow D_2^{*-}\mu_2^+\nu$.

of $B^o\bar{B}^o$ oscillations, has been found and is shown in fig. 7. The probability that the interpretation of this event is wrong has been estimated to be $< 10^{-4}$. The final result of the experiment may be expressed as

$$\chi_o = \frac{N_1^{\pm\pm} + N_2^{\pm\pm} - N_{12}^{\pm\pm}}{(N_1 - N_{12})(1 + f_+B_+^2/f_oB_o^2)^{-1} + N_2}.$$

Since the lepton pairs in strategy 1 come from $B^o/\bar{B}^o$ and from $B^\pm$ decays, the correction factor $f_+B_+^2/f_oB_o^2$ is required, where $f_+ = B(\Upsilon(4S) \rightarrow B^+B^-)$, $B_+ = B(B^+ \rightarrow \ell\nu X)$, and f_o, B_o correspondingly. The result and its dependence on $f_+B_+^2/f_oB_o^2$ is displayed in fig 8. Assuming $f_+B_+^2/f_oB_o^2 = 1.2 \pm 0.2$, the value is $\chi_o = 0.18 \pm 0.06$. The statistical significance of the effect corresponds to more than four standard deviations.

$B\bar{B}$ oscillations have also been searched by UA1 [25], CLEO [26], MARK–II [27], JADE [28] and MAC [29]. CLEO, working on the $\Upsilon(4S)$ resonance, is sensitive to only $B^o\bar{B}^o$ like ARGUS, the other four groups to a combination of $B^o\bar{B}^o$ and $B_S\bar{B}_S$ oscillations. UA1 sees an effect of nearly three standard deviations, their result may be expressed as

$$\chi \approx 0.38\chi_o + 0.15\chi_S = 0.121 \pm 0.047.$$

None of the other four groups finds a positive effect, but the limits obtained are compatible with the results of ARGUS and UA1 as shown in fig. 9. The average of all six observations is obtained by multiplying their likelihood functions in the (χ_o, χ_S) plane. This is done, in a Gaussian approximation, in the last part of fig. 9. From this we conclude

$$\chi(B^o) = 0.16 \pm 0.04,$$

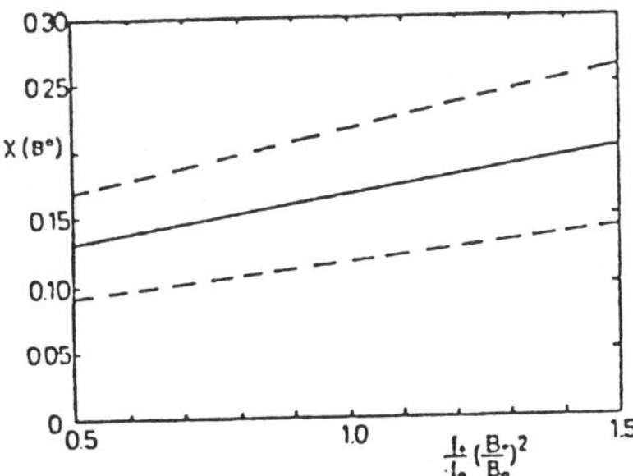

Figure 8: The ARGUS χ_o result [24] as fuction of $f_+B_+^2/f_oB_o^2$.

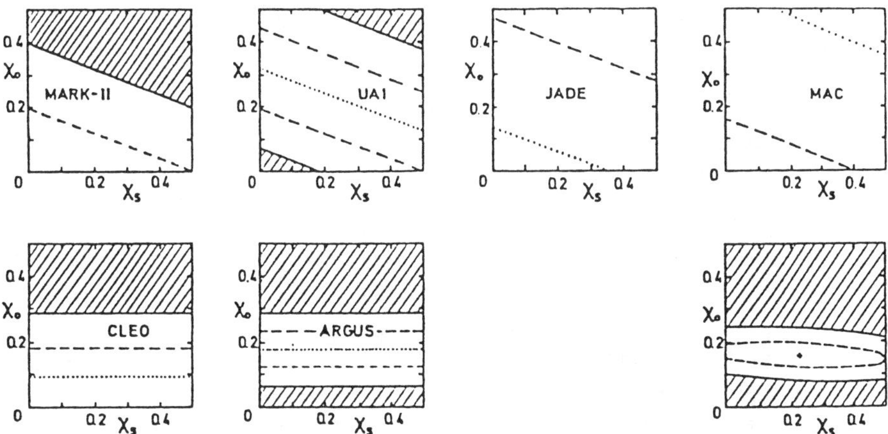

Figure 9: Summary of $B\bar{B}$ oscillation results. The dotted curves are central values, the dashed curves correspond to one and the solid curves to two standard deviations. The last part of the figure is the Gaussian average of the six experiments. Hatched areas are excluded with 95% confidence.

$$\chi(B_S) = \text{ any value within one st. dev.}$$

This average is compatible with the three quark generations Standard Model. Assuming box graph dominance and taking

$$f_B = (130 \pm 20)\ MeV,\ B_B = 0.8 \pm 0.2,\ \eta_B = 0.85,$$

leads to

$$m_t V_{tb}|V_{td}| = (1.6 \pm 0.4)\ GeV.$$

From the unitarity of the CKM–matrix, which is experimentally well fulfilled as shown below, we may use $V_{tb} = 1$ and $|V_{td}| = 0.010 \pm 0.008$ to conclude

$$m_t > 50\ GeV\ (95\% CL).$$

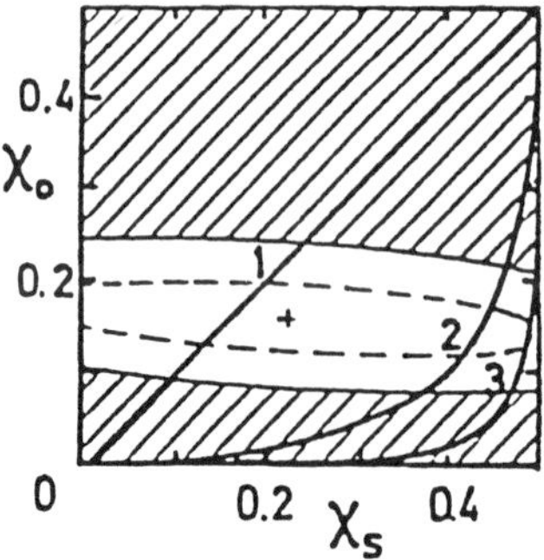

Figure 10: The predicted relation between $\chi(B^o)$ and $\chi(B_S)$ with $|V_{ts}/V_{td}| =$ 1, 2 and 3. The experimental confidence contours are the same as in the last part of fig. 9.

The relative amount of $B_S\bar{B}_S$ and $B^o\bar{B}^o$ oscillations is to large extend independent of m_t, f_B, η_B, and B_B. Applying the box graph evaluation for the B_S system leads to

$$\Delta M(B_S) \approx |V_{ts}/V_{td}|^2 \cdot \Delta M(B^o),$$
$$\frac{\chi_S}{1-2\chi_S} \approx |V_{ts}/V_{td}|^4 \cdot \frac{\chi_o}{1-2\chi_o}.$$

This relation is shown in fig. 10 for $|V_{ts}/V_{td}| = 1$, 2 and 3 together with the experimental bounds from the last part of fig. 9. Even with a ratio of 2, which is the present lower bound from the CKM–matrix unitarity in the Standard Model with the three quark generations, χ_S is constrained to be nearly maximal. It is therefore very important to measure χ_S with sufficient precision. This measurement is one of the most crucial tests of the Standard Model at present.

The matrix elements V_{cb} and $|V_{ub}|$ are best determined from semileptonic B–meson decays. Up to now, only one exclusive channel has been observed, $B^o \to D^{*-}\ell^+\nu$. The ARGUS collaboration [30] finds a decay fraction of

$$B(B^o \to D^{*-}\ell^+\nu) = (7.0 \pm 1.2 \pm 1.9)\%.$$

This result has been obtained from about 90000 $\Upsilon(4S)$ decays by requiring a reconsructed D^{*-} meson in the decay chain $D^{*-} \to \bar{D}^o\pi^-$, $\bar{D}^o \to K^+\pi^-$ with $|\vec{p}(\bar{D}^o\pi^-)| < 2.5\ GeV/c$ and a positive lepton (e or μ) with $|\vec{p}(\ell)| > 1.0\ GeV/c$. From the measured momenta one can calculate a missing mass MM from

$$MM^2 = (E_B - E_{D^*} - E_\ell)^2 - (\vec{p}_B - \vec{p}_{D^*} - \vec{p}_\ell)^2$$

which is equal to the neutrino mass, i.e. zero, for proper $B \to D^*\ell\nu$ decays. The energy of the B–meson is $E_B = m(\Upsilon(4S)/2$, and $\vec{p}_B$ is set to zero which is a good approximation since B–mesons from $\Upsilon(4S)$ mesons formed in a symmetric e^+e^- storage ring are nearly at rest. The obtained MM^2 spectrum is shown in fig. 11. There are 47 ± 8 proper $B^o \to D^{*\pm}\ell^\mp\nu$ events, the width of the

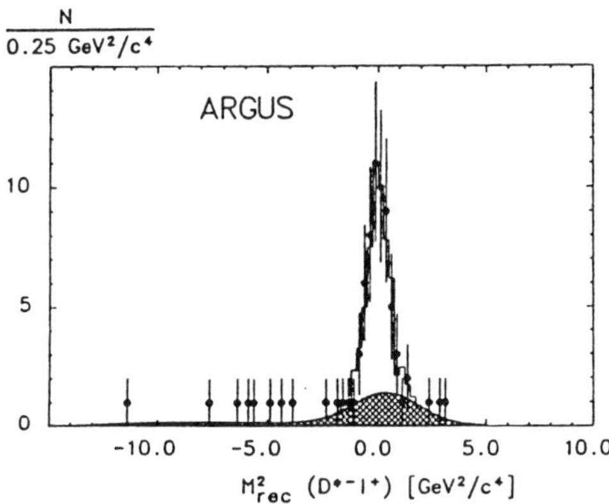

Figure 11: Missing mass spectrum of $D^*\ell$ combinations in ref. [30].

observed peak around zero corresponds to the Monte Carlo expectation. With the obtained decay fraction, as given above, and assuming that the B^o mean life is equal to the average B–meson life time of (1.15 ± 0.15) ps, we obtain

$$\Gamma(B^o \to D^{*-}\ell^+\nu) = (6.1 \pm 2.1) \cdot 10^{10}/s$$

Formfactor models [31] [32] [33] predict

$$\Gamma(B^o \to D^{*-}\ell^+\nu) = V_{cb}^2 \cdot (3 \pm 1) \cdot 10^{13}/s.$$

This leads to

$$V_{cb} = 0.045 \pm 0.011.$$

Inclusive semileptonic decays, $B \to \ell\nu X$, give a slightly higher precision as will be shown later.

For the $b \to u$ decay analogous to the $b \to c$ decay $B^o \to D^{*-}\ell^+\nu$, $B^+ \to \rho^o\ell^+\nu$, form factor models [31] [32] predict

$$\Gamma(B^+ \to \rho^o\ell^+\nu) \approx |V_{ub}|^2 \cdot 1.0 \cdot 10^{13}/s$$

Both CLEO and ARGUS have searched for this decay mode. The CLEO momentum spectrum [34] of all leptons (e and μ) produced at the storage ring energy of the $\Upsilon(4S)$ in coincidence with a $\pi^+\pi^-$ pair near the ρ^o mass is shown by the points in fig. 12. The momentum region between 2.4 and 2.6 GeV/c is free of leptons from $b \to c$ decays and contains 11% of all $B \to \rho\ell\nu$ decays. The histogram in fig. 12 is the lepton spectrum obtained at a storage ring energy below the $\Upsilon(4S)$ mass and represents the non–B–decay leptons which are present when taking data at the $\Upsilon(4S)$ energy. There is no lepton excess on the $\Upsilon(4S)$ above 2.4 GeV/c, and from this absence, CLEO concludes an upper limit on the decay fraction of

$$B(B^+ \to \rho^o\ell^+\nu) < 0.25\% \, (90\% \, C.L.).$$

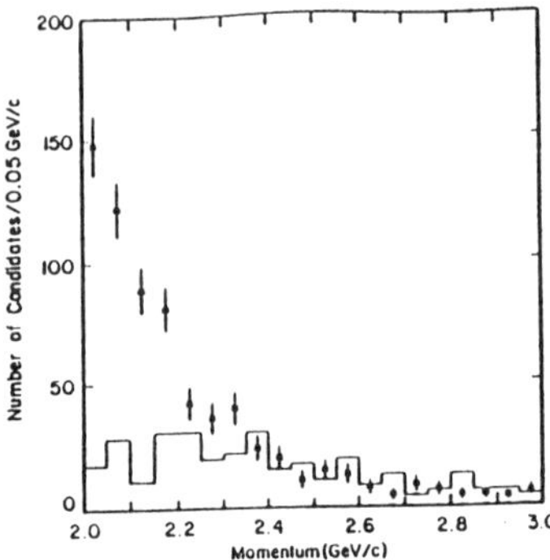

Figure 12: Lepton spectrum from events with $B^+ \to \rho^o \ell^+ \nu$ candidates [34]. The points are from data on the $\Upsilon(4S)$ resonance, the histogram from off–resonance data.

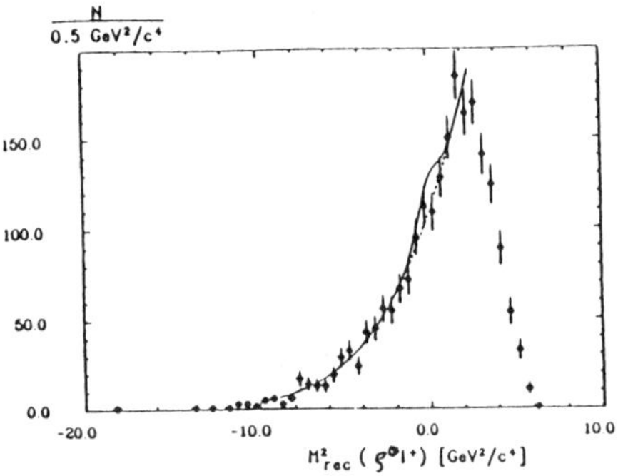

Figure 13: Missing mass spectrum of $\rho^o \ell$ combinations in ref. [35].

ARGUS [35] uses the missing mass technique in their search for this decay channel in complete analogy to the $B^o \to D^{*-}\ell^+\nu$ analysis described above. Fig. 13 shows MM^2 for all $\pi^+\pi^-\ell^\pm$ coincidences observed in about 360000 events taken at the storage ring energy of the $\Upsilon(4S)$, 90000 $\Upsilon(4S)$ decays and 270000 continuum, fulfilling the following selection criteria: $m(\pi^+\pi^-) = m(\rho^o) \pm \Gamma(\rho^o)$, $1.0 \leq |\vec{p}(\pi^+\pi^-)| \leq 2.6\ GeV/c$, $1.5 \leq |\vec{p}(\ell)| \leq 2.6\ GeV/c$, $|cos\alpha| < 0.5$. The angle α is defined between the lepton direction and the thrust axis of axis of all particles in the event except π^+, π^-, and ℓ; the cut on $cos\alpha$ keeps 50% of all $B \to \rho\ell\nu$ events but rejects a large fraction of non–B–decay events from the e^+e^- continuum. The solid curve in fig. 13 shows the Monte Carlo expectation with a $B \to \rho\ell\nu$ signal around $MM^2 = 0$. The data show no indication of such a signal, and ARGUS concludes the following upper limit

$$B(B^+ \to \rho^o \ell^+ \nu) < 0.22\%\,(90\%\,C.L.).$$

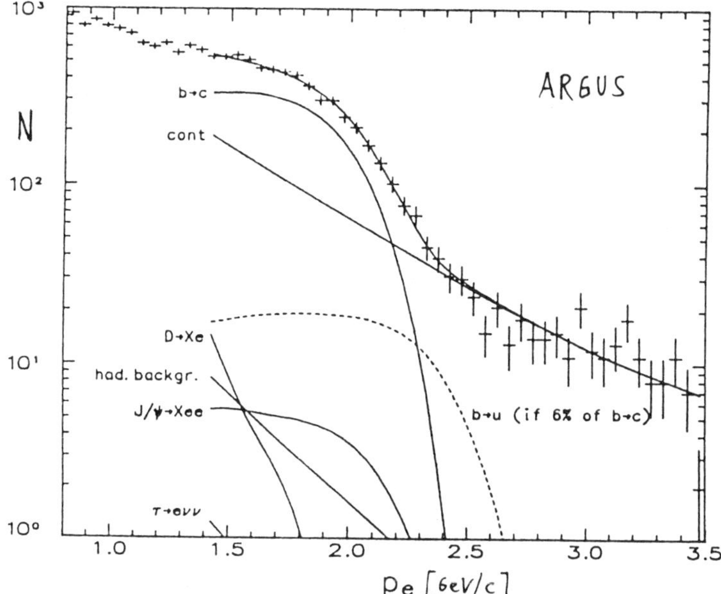

Figure 14: Final ARGUS results on the inclusive electron spectrum from $\Upsilon(4S)$ decays. The curves are the various expected contributions to this spectrum as explained in the text.

The consequence of this result is

$$|V_{ub}| < 0.014,$$
$$|V_{ub}/V_{cb}| < 0.3.$$

Inclusive semileptonic B–meson decays have been studied by CLEO [36], CUSB [37], Crystal Ball [38], and ARGUS [39]. In the following I want to present the final ARGUS analysis from an integrated luminosity of $\mathcal{L} \approx 100/pb$, 90000 $\Upsilon(4S)$ decays, with close–to–final results. When summing over all semileptonic final states, one observes with very good approximation the decays of a free b–quark:

$$\Gamma(\bar{B} \to \ell^- \bar{\nu} X) \approx \Gamma(b \to u\ell^- \bar{\nu}) + \Gamma(b \to c\ell^- \bar{\nu}).$$

The observed electron spectrum from

$$B^o,\, B^{\pm} \to \ell^{\pm} X \text{ at } \sqrt{s} = m(\Upsilon 4S)$$

is shown in fig. 14. The solid curves show the different expected contributions to this spectrum: "$b \to c$" from the decays $b \to c\ell\nu$, "$b \to u$" from $b \to u\ell\nu$, here drawn for the hypothesis $\Gamma(b \to u\ell\nu)/\Gamma(b \to c\ell\nu) = 0.06$, "cont" from non–$B$–decay events present at the $\Upsilon(4S)$ energy including hadrons in these events misidentified as electrons, "had. bg" =misidentified hadrons from B–decay events, "$D \to X\ell$" = electrons from the cascade decay $B \to D + X$, $D \to \mu + X$, "$J/\psi \to ee$" = electrons from $J\psi$ decays, and "$\tau \to e$" electrons from $\tau \to e\nu\bar{\nu}$. Figs. 15a, b show the electron and muon spectrum after subtraction of non–B–decay leptons and misidentified hadrons and after efficiency correction. Electron and muon identification is done with the help of all detector components in a combined likelihood function. In the momentum region above 1.5 GeV/c hadrons are misidentified with about 0.5% probability as electrons and 1.4% as muons. Both the decay fraction $B(B \to \ell^+\nu X)$ and the $b \to u/b \to c$ ratio $\Gamma(b \to u\ell^-\nu)/\Gamma(b \to c\ell^-\nu)$ are obtained by fitting various model expectations to the spectra in figs. 15a, b. The solid curves in the figures show the GIW expectation [32] with zero $b \to u$.

The fits of the GIW model to the ARGUS data give

$$\begin{aligned} B(B \to \mu\nu X) &= (10.7 \pm 0.3 \pm 1.0)\%, \\ B(B \to e\nu X) &= (9.9 \pm 0.3 \pm 0.7)\%, \end{aligned}$$

where the first error is statistical including the extrapolation to $0 \leq p_\ell \leq 1.5$ GeV/c and the second error is the experimental systematics, mainly from acceptance and misidentification uncertainties. The $\Gamma(b \to u\ell\nu)/\Gamma(b \to c\ell\nu)$ ratios are compatible with zero, the GIW fits give

$$\begin{aligned} \Gamma(b \to u\ell\nu)/\Gamma(b \to c\ell\nu) &= (+1.8 \pm 2.0)\% \quad \text{for } \mu^\pm X \\ &= (-0.2 \pm 1.5)\% \quad \text{for } e^\pm X \\ &= (+0.9 \pm 1.3)\% \quad e + \mu \text{ combined.} \end{aligned}$$

The last result corresponds to $\Gamma(b \to u\ell\nu)/\Gamma(b \to c\ell\nu) < 2.5\%$ with 90% confidence, and including experimental systematics $< 3.0\%$ (90% $C.L.$) The theoretical uncertainty is quoted to be a factor of two in ref. [32], a safe upper limit from the ARGUS $e + \mu$ results is therefore

$$\Gamma(b \to u\ell\nu)/\Gamma(b \to c\ell\nu) < 6.0\% \ (90\% \ C.L.).$$

Fits with the Altarelli model [40] give very similar decay fraction results and a similar upper limit on $b \to u/b \to c$ if one leaves two of the three parameters of the model free, $m(c)$ and the quark Fermi momentum in the meson, and restricts the third parameter, $m(u)$, to be smaller than 0.4 GeV/c^2.

Combining the above decay fraction results with those from CLEO, Crystall Ball and CUSB and averaging also electrons and muons leads to

$$B(B \to \ell\nu X) = (11.0 \pm 0.5)\%,$$

and together with the average B–meson lifetime

$$\Gamma(B \to \ell\nu X) = (9.6 \pm 1.3) \cdot 10^{10}/s.$$

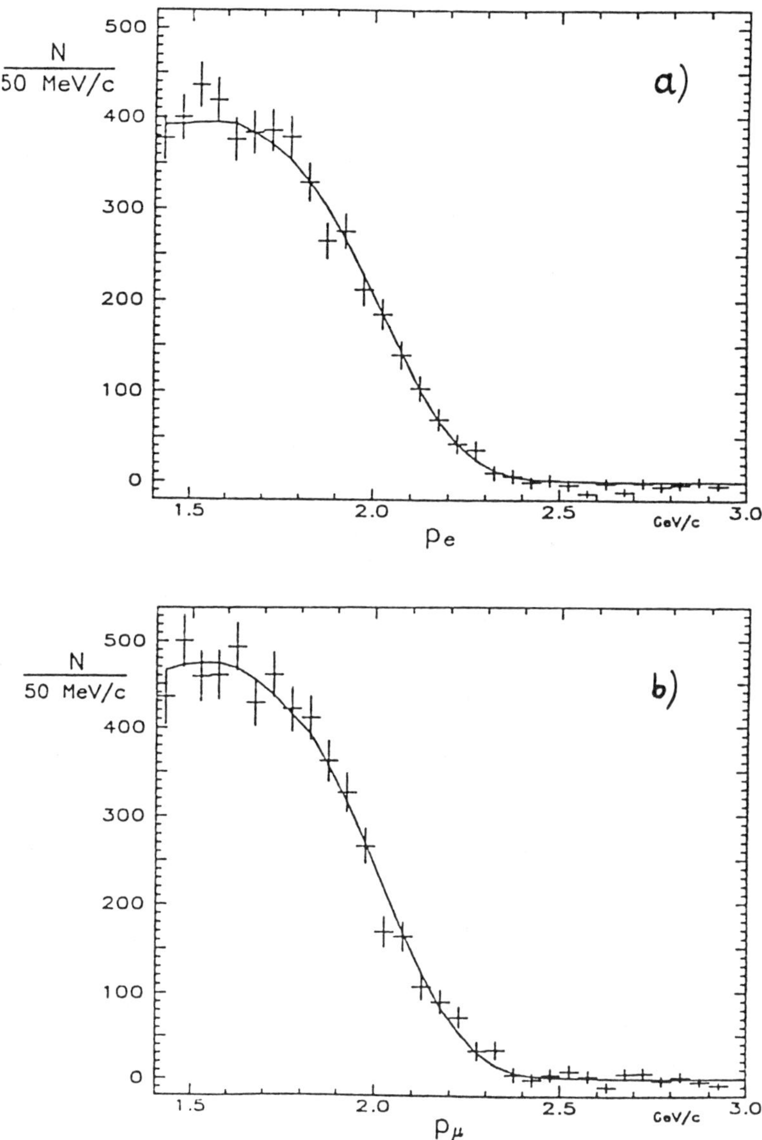

Figure 15: Final ARGUS results on the (a) electron and (b) muon spectrum of inclusive semileptonic B decays in the $\Upsilon(4S)$ rest frame. The data points are after continuum and hadron–background subtraction and after efficiency correction. The curves are the $b \to c\ell\nu$ expectation [32] with zero $b \to u\ell\nu$.

The free quark decay model predicts

$$\Gamma(b \to c\ell\nu) = \frac{G_F^2 m_b^5}{192\pi^3} \cdot \phi_c \cdot V_{cb}^2,$$
$$\Gamma(b \to u\ell\nu) = \frac{G_F^2 m_b^5}{192\pi^3} \cdot \phi_u \cdot |V_{ub}|^2,$$

where the QCD + phase space coefficients ϕ_c and ϕ_u [41] depend on the quark masses m_b, m_c, and m_u. With $m_b = (4.94 \pm 0.05)$ and $m_c = (1.58 \pm 0.06)\ GeV/c^2$ from the Altarelli fit above we obtain $\phi_c = 0.41 \pm 0.03$. The variation of ϕ_u with m_u is weak; small u–quark massses give $\phi_u = 0.80$. Setting $\Gamma(B \to \ell\nu X) = \Gamma(b \to u\ell\nu) + \Gamma(b \to c\ell\nu)$ and using $m_b = (4.94 \pm 0.15)\ GeV/c^2$ leads to

$$V_{cb} = 0.048 \pm 0.005,$$
$$|V_{ub}/V_{cb}| < 0.18\ (90\%\ C.L.),$$
$$|V_{ub}| < 0.009.$$

The first evidence for $V_{ub} \neq 0$ was presented by ARGUS in July 87 [42], the final results have been published recently [43]:

$$B(B^+ \to p\bar{p}\pi^+) = (5.2 \pm 1.4 \pm 1.9) \cdot 10^{-4},$$
$$B(B^o \to p\bar{p}\pi^+\pi^-) = (6.0 \pm 2.0 \pm 2.2) \cdot 10^{-4}.$$

The results have been obtained from 90000 $\Upsilon(4S)$ decays, and $p\bar{p}\pi(\pi)$ candidates have been selected as follows:

- The charged pions are uniquely identified by time–of–flight and specific ionisation measurements.
- Two oppositely charged high–momentum particles ($1.8 < |\vec{p}| < 2.6\ GeV/c$) are assumed to be p and $\bar{p}$; unique identification of individual particles is not possible for this momentum range in ARGUS.
- The rest of the event besides $p\bar{p}\pi(\pi)$ has to have 3 charged particles or more.
- The topological quantity H_2 (second Fox–Wolfram moment) of the rest has to be less than 0.3.
- $|cos\alpha| < 0.8$, where α is the angle between the thrust axis of the $p\bar{p}\pi(\pi)$ combination and the thrust axis of the rest.
- The angle between p and $\bar{p}$ has to be larger than 168°.
- The observed energy of the candidate, $E_{obs} = \sum_{i=1}^{3(4)} \sqrt{\vec{p}_i^2 + m_i^2}$ with the assumed masses m_i, has to be the beam energy E_B within $\pm 2\sigma_E$, where the measurement error σ_E is typically 30 MeV.
- The energy measurement error σ_E has to be smaller than 60 MeV.

All remaining candidates are then submitted to a beam–energy constraint fit, i.e. the measured momenta $\vec{p}_i$ are modified $p_i \to \tilde{p}_i$, until $\tilde{E}_{obs} = E_B$. This fit results

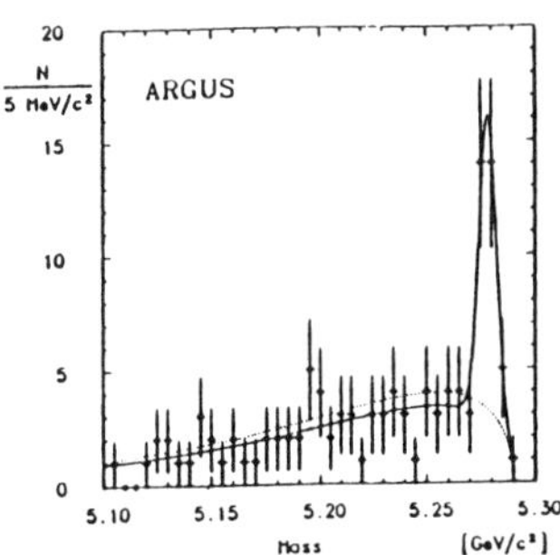

Figure 16: Distribution of beam–energy constraint mass values for all selected $B^{\pm} \to p\bar{p}\pi^{\pm}$ and $B^{o} \to p\bar{p}\pi^{+}\pi^{-}$ candidates in ref. [43].

in a mass value M_{fit} for the candidate, $M_{fit}^2 = E_B^2 - (\sum \vec{p_i})^2$. The distribution of these fitted mass values for all selected $p\bar{p}\pi^{\pm}$ and $p\bar{p}\pi^{+}\pi^{-}$ candidates is shown in fig. 16. There are 34 candidates at the right B–meson mass in an interval where the expected background is 9 with an uncertainty of ± 3. The signal has a statistical significance of more than 4 standard deviations. It could still be faked by either B–decays which are not $p\bar{p}\pi(\pi)$ or by other types of events which have the kinematic peculiarity to produce M_{fit} values near M_B when selected with the criteria above. A large number of fake possibilities has been excluded by ARGUS:

1. The signal is not produced by non–B–decay events from the e^+e^- continuum, since there is no signal in data taken at off–$\Upsilon(4S)$–energies.

2. The σ_E cut prevents that the selected $p\bar{p}\pi(\pi)$ combination is an uncomplete B–decay and one particle is missing.

3. The σ_E cut prevents also the opposite, i.e. that one of the 3(4) selected particles is an extra particle added to a B–decay.

4. The σ_E cut excludes also that the p or $\bar{p}$ candidate is a pion or kaon from a true 3(4)–body B–decay.

5. A combination of possibilities 2 and 4 could give fakes because of opposite changes in the energy balance. This is excluded by performing energy–constraint fits with $E \neq E_B$; these fits do not produce a mass peak.

6. The possibility that the $p\pi(\pi)$ or $\bar{p}\pi(\pi)$ combination in the $p\bar{p}\pi(\pi)$ system is a charmed particle has been ruled out by investigating explicitely all combinations.

Remains the possibility that some of these fake reasons (and additional ones) work together with statistics in a conspiring way. This has been tested by a full $b \to c$ Monte Carlo simulation with 11 times more events than in the real data.

Selecting $p\bar{p}\pi(\pi)$ candidates with the criteria of the real data does not produce a narrow peak in the mass spectrum. Additional information may be found in ref. [44].

Having not found any fake reason, the observed signal is interpreted to originate from genuine $B \to p\bar{p}\pi(\pi)$ decays. These can be mediated by W emission with $V_{ub} \neq 0$ as shown in fig. 4b or by the penguin mechanism as shown in fig. 4c. The penguin mechanism should produce 20 times more $p\bar{\Lambda}\pi(\pi)$ events than $p\bar{p}\pi(\pi)$ events. This is excluded by the following limits observed by ARGUS [43]:

$$B(B^+ \to p\bar{\Lambda}\pi^+\pi^-) < 1.8 \cdot 10^{-4} \ (90\% \, C.L.),$$
$$B(B^o \to p\bar{\Lambda}\pi^-) < 2 \cdot 10^{-4} \ (90\% \, C.L.);$$

therefore the observed signal is the first evidence for a weak decay with the coupling strength of V_{ub}. There are no reliable predictions for a relation between $\Gamma[B \to p\bar{p}\pi(\pi)]$ and $|V_{ub}|^2$. A very rough estimate [42] leads to

$$|V_{ub}/V_{cb}| \geq 0.07,$$
$$|V_{ub}| \geq 0.003.$$

A reliable way to determine $|V_{ub}|$ in the future is to measure the rates of exclusive semileptonic $b \to u$ decays like $B^o \to \pi^-\ell^+\nu$ or $B^+ \to \rho^o\ell^+\nu$. To get these events without or with reasonably small background, one has to use a tagging method. The easiest way to tag is production of $B(\bar{B})$ mesons in $\Upsilon(4S)$ decays where the rest of the event is a fully reconstructed $\bar{B}(B)$ meson. This method is not possible with presently available storage ring luminosities; an estimate with $|V_{ub}/V_{cb}| = 0.10$ leads to the requirement of 10^7 $\Upsilon(4S)$ events. A future "B–Meson–Factory" [45] may reach this aim in a data taking time of about one year.

Our present knowledge of the CKM–matrix can be summarized as follows:

$$|V_{ij}| = \begin{pmatrix} 0.9747 \pm 0.0010 & 0.220 \pm 0.002 & 0.006 \pm 0.003 \\ 0.21 \pm 0.03 & 1.03 \pm 0.16 & 0.048 \pm 0.005 \\ (1.6 \pm 0.4) \ GeV/m_t & & \end{pmatrix}.$$

Unitarity is well fulfilled; the squares in the first and second row add up perfectly to one and $\sum_j V^*_{u_j} V_{c_j}$ can be made zero with an appropriate choice of signs. Assuming unitarity leads to $V_{tb} \approx 1$, $V_{ts} \approx -V_{cb}$, and $V_{cd} \approx -V_{us}$. (The above entry $|V_{td}| = (1.6 \pm 0.4) \ GeV/m_t$ uses already $V_{tb} \approx 1$). All V_{ij} are essentially real with the exception of V_{ub} and V_{td}. The unitarity condition

$$\sum_j V^*_{uj} V_{tj} = 0,$$

with present precision expressed as

$$1 \cdot V_{td} - 0.010 + V^*_{ub} \cdot 1 = 0,$$

can be drawn in the complex plane in the form of "Bjorken's triangle" [46], see fig. 17. Present experiments restrict V^*_{ub} to the dashed area, the outer bound by

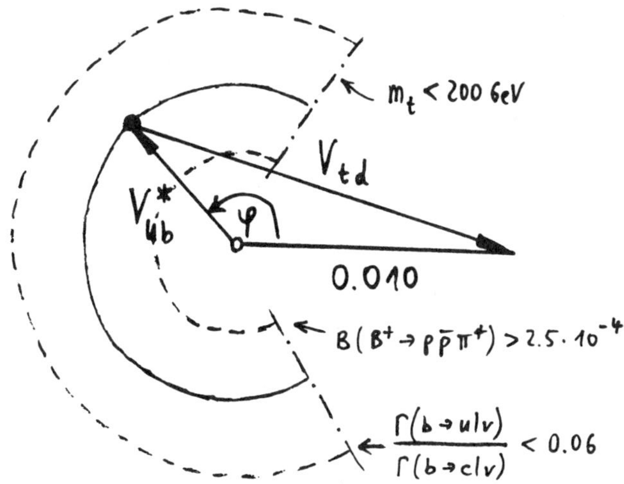

Figure 17: The "Bjorken Triangle" illustrating that the cosine of the phase φ in the CKM–matrix can in principle be determined from rate measurements only.

the inclusive semileptonic spectrum results, the inner bound by the $B \to p\bar{p}\pi(\pi)$ result. The right part of the area is excluded by $M_t < 200\ GeV$ which can be concluded from a calculation of the t–quark contribution to the radiative corrections of $sin^2\vartheta_W$ [47]. More precision on $|V_{ub}|$ and on the oscillation rate $\chi(B^o)$ will pin down the triangle's size and shape in the future, especially if the t–quark mass will be known. It will then be possible to determine $cos\varphi$ and to test if this phase determination coincides with results obtained from CP–violating experiments in the K^o–meson sector.

I would like to thank E. Vogt for the invitation to this very interseting conference.

References

[1] M. Kobayashi and T. Maskawa, Progr. Theor. Phys. 49(1973)652

[2] M.L. Perl et al (MARK–I), Phys. Rev. Letters 35(1975)1489

[3] S.W. Herb et al, Phys. Rev. Letters 39(1977)252

[4] C.W Darden et al (DASP–2), Phys. Letters 76B(1978)246
Ch. Berger et al (PLUTO), Phys. Letters 76B(1978)243
C.W.Darden et al (DASP–2), Phys.Letters 78B(1978)364
J.K. Bienlein et al (LENA), Phys.Letters 78B(1978)360

[5] E. Fernandez et al (MAC), Phys. Rev. Letters 51(1983)1022

[6] N.S. Lockyer et al (MARK–II), Phys. Rev. Letters 51(1983)1316

[7] W. Bartel et al (JADE), Z. Physik C31(1986)349

[8] W.W. Ash et al (MAC), Phys. Rev. Letters 58(1987)640

[9] J.M. Brom et al (HRS), Phys. Letters B195(1987)301

[10] K.U. Pösnecker (TASSO), Proc. Int. Europhysics Conf. on High Energy Physics, Uppsala 1987, ed. by O. Botner

[11] R.A. Ong (MARK–II) SLAC–PUB–4550 (1988)

[12] D.E. Klem et al (DELCO), Phys. Review D37(1988)41

[13] J.P. Albanese et al (WA75), Phys. letters 158B(1985)186

[14] N. Cabibbo, Phys. Rev. Letters 10(1963)531

[15] J.D. Bjorken and S.L. Glashow, Phys. Letters 11(1964)255

[16] S.L. Glashow, J. Iliopoulos, and L. Maiani, Phys. Review D2(1970)1285

[17] L.-L. Chau and W.-Y. Keung, Phys. Rev. Letters 53(1984)1802
and many others

[18] for references see K.R. Schubert, Proc. Int. Europhysics Conf. on High Energy Physics, Uppsala 1987, ed. by O. Botner, p.791

[19] for a recent review see B. Stech, "What we learnt from weak decays of heavy quarks", Heidelberg preprint HD–THEP–88–11(1988)

[20] H. Albrecht et al (ARGUS), Phys. Letters B185(1987)218
H. Albrecht et al (ARGUS), Phys. Letters B199(1987)451

[21] C. Bebek et al (CLEO), Phys. Review D36(1987)1289

[22] M. Bauer, B. Stech, and M. Wirbel, Z. Physik C34(1987)103

[23] as for the K^o in M.K. Gaillard and B.W. Lee, Phys. Rev. D10(1974)897

[24] H. Albrecht et al (ARGUS), Phys. Letters 129B(1987)245

[25] C. Albajar et al (UA1), Phys. Letters 186B(1987)247

[26] A. Bean et al (CLEO), Phys. Rev. Letters 58(1987)183

[27] T. Schaad et al (MARK–II), Phys. Letters 160B(1985)188

[28] W. Bartel et al (JADE), Phys. Letters 146B(1984)437

[29] H.R. Band et al (MAC), Phys. Letters B200(1988)221

[30] H. Albrecht et al (ARGUS), Phys. Letters B197(1987)452

[31] M. Wirbel, B. Stech, and M. Bauer, Z.Physik C29(1985)637

[32] B. Grinstein, N.Isgur, and M.Wise, Phys. Rev. Letters 56(1986)298

[33] T. Altomari and L. Wolfenstein, Phys. Rev. Letters 58(1987)1583

[34] A. Bean et al (CLEO), contributed paper to the Int. Symp. Lepton Photon Interactions, Hamburg 1987

[35] H. Schröder (ARGUS), Proc. 23rd Rencontre de Mariond (1988)

[36] A. Chen et al (CLEO), Phys. Rev. Letters 52(1984)1084
S. Behrends et al (CLEO), Phys. Rev. Letters 59(1987)407

[37] C. Klopfenstein et al (CUSB), Phys. Letters 130B(1983)444

[38] K. Wachs (Crystal Ball), contributed paper to the Int. Europhysics Conf. on High Energy Physics, Uppsala 1987

[39] preliminary results in K.R. Schubert, Proc. XXIIIrd Int. Conf. on High Energy Physics, Berkely 1986, ed. by S. Loken, p.781

[40] G. Altarelli et al, Nucl. Physics B208(1982)365

[41] A. Ali and E. Pietarinen, Nucl. Physics B154 (1979)519

[42] W. Schmidt–Parzefall (ARGUS), Proc. Int. Symp. Lepton Photon Interactions, Hamburg 1987

[43] H. Albrecht et al (ARGUS), DESY88–056(1988) and ref. 44

[44] D. MacFarlane, these Proceedings

[45] R. Eichler et al, SIN–PR–86–13(1986)

[46] J.D. Bjorken, presented e.g. at the SSC Detector Workshop, Berkely, July 1987.

[47] U. Amaldi et al, Phys. Rev. D36(1987)1385

AN OVERVIEW OF NON-ACCELERATOR PHYSICS

M. L. Marshak
School of Physics and Astronomy, University of Minnesota,
Minneapolis MN 55455

Abstract

The study of very high energy (VHE) and ultrahigh (UHE) energy quanta from specific astrophysical sources may yield otherwise unobtainable information about processes in the macroscopic particle physics laboratory environment of a neutron star. Considerable existing data indicate that such quanta have already been observed. However, some data suggest that primaries from at least two neutron stars behave according to unknown physics mechanisms.

Introduction

The title of this talk is at best presumptuous. It would be difficult, if not impossible, to give a reasonable overview of all of non-accelerator physics in the allotted time, even if I limited the review to the subject matter of this conference--nuclear and elementary particle physics.

At the last Rochester conference (held at Berkeley), Maurice Goldhaber[1] presented the list shown in Table 1. The notable distinction between then and now is, of course, the actual observation of neutrinos in close time coincidence with optical observations of Supernova 1987a. This topic was already reviewed, so I shall not comment on it further. There has been progress, but not in my opinion any major breakthroughs, in some of the other areas. A few such new results have been reported at this meeting.

VHE and UHE Physics

I intend in this talk to focus on a contemporary view of very and ultra high energy physics (VHE and UHE physics) and the relationship between these subjects and accelerator-based nuclear and elementary particle physics. The usual definition is that "very high energies" are the order of 1 TeV or above, that is $>10^{12}$ electron volts, while "ultra

Table 1: M. Goldhaber's Listing of "Some Non-Accelerator Experiments"

Proton decay	a. Supernovae
neutron/anti-neutron oscillations	b. H. E. ν's
Cygnus X-3	New particles (KGF)
ν masses	Dark matter
ββ decay (with and without 2 ν's)	Gravitational waves
ν oscillations	Neutral current effects in atoms
Axions	Fifth force
WIMPS	Electric dipole moment (n, e, atoms)
Atmospheric ν's	Free quarks
Solar ν's	Magnetic monopoles
Extra solar ν's	

high energies" are 1 PeV (that is, 10^{15} electron volts) or higher. VHE and UHE physics are subsets of a area which is sometimes described as astroparticle physics. The goal of this discipline, if indeed it exists, is to use the universe itself as a particle physics laboratory whose properties cannot be duplicated in terrestrial laboratories. The elementary particle physics of the "big bang" is, for example, another subset of this field. As we shall see, VHE and UHE astrophysical physics is primarily concerned with the macroscopic elementary particle physics environment created in non-thermal star systems, particularly isolated or binary neutron stars.

A good place to start this review is with a quick summary of conventional cosmic ray physics and a discussion of how that subject fits into this same context. The outstanding questions of cosmic ray physics for more than 50 years have been

1. What is the nature of the cosmic rays? (flux, energy spectrum, composition, anisotropies)
2. What is the nature of the interactions between cosmic rays and the earth's atmosphere? (search for new particles and processes, air shower studies)
3. What is the source of the cosmic rays? (particular sources, acceleration mechanisms, propagation phenomena)

Much progress has been made in answering question 1, although there are still many uncertainties, particularly at the highest energies—above $\approx 10^{15}$ eV. The cosmic ray community has traditionally devoted considerable energy to question 2. I believe the advent of the Fermilab collider and the very real prospect of a relativistic heavy ion collider suggest that the time has mostly passed for studying cosmic rays for this purpose. For example, the Fermilab collider provides 2 TeV in the center-of-mass for pp collisions, corresponding to 2 PeV cosmic ray-air interactions. Cosmic rays do enable experiments at higher energies, but with precious little flux and considerable systematic difficulties.

The Neutron Star as a High Energy Laboratory

The elementary particle physicists who have strayed into this field in recent years have mostly avoided air showers for these reasons. Instead, they have focussed on question 3—the mechanisms for cosmic ray production. These studies are not generic, as has been traditional in cosmic ray physics. Rather, there is a recent, determined effort to extend observations of specific astrophysical objects to higher energy quanta. The rationale is that a non-thermal star provides a unique high energy laboratory, one whose properties will not be duplicated on earth in the foreseeable future. The study of high energy quanta, those directly or almost directly emitted by the central processes in such an object, should yield information which is not provided by the existing measurements of low energy quanta. The information content of the low energy quanta are presumably diminished by the equilibration processes which result in their production. It is further presumed that maximal information yield comes from the study of sources one by one, by a correlation of the information provided by flux quanta over about 20 orders of magnitude in energy.

It is useful to consider some order of magnitude estimates of the nature of these non-thermal star laboratories[2]. The sun, for example, is about 1/2 degree or 10 mr in diameter at a distance of about 1.5 x 10^8 km. The sun's diameter is thus about 1.5 x 10^6 km. The magnetic field of a star such as the sun is several orders of magnitude larger than the earth's field, say ≈100 gauss or 0.01 Tesla. In the collapse of a main sequence star such as our sun to a neutron star, the stellar diameter is reduced to about 10 to 20 km. The surface area decreases by the square of the diameter change or about 10 orders of magnitude. Conservation of magnetic flux requires that the magnetic field increases proportionately, that is to ≈10^{12} G or ≈10^8 T. (Measured magnetic fields are 5 x 10^{12} G for the Crab pulsar, 4-6 x 10^{12} G for Hercules X-1 and 2 x 10^{12} G for 4U 0115+63.) Simultaneously, conservation of angular momentum increases the angular rotational velocity by ≈10 orders of magnitude. Typical neutron star rotational periods range from 10^{-3} to 10 s.

Large, rotating magnetic fields are an obvious application for Faraday's law. A Faraday loop around the surface of a neutron star encloses an area of ≈π x 10^8 m^2. The magnetic flux through this loop is ≈π x 10^{16} webers. If the flux reverses direction several times per second due to rotation of the neutron star, the potential generated around such a loop is ≈10^{17} volts. Of course, this handwaving argument ignores many effects which must be considered by a detailed model of particle acceleration by neutron stars. These effects include synchrotron radiation from accelerated electrons and absorption of high energy photons by $\gamma\gamma$ scattering off the neutron star magnetic field. This simple estimate does, however, show the possibility that a very high energy acceleration mechanism could exist. It also illustrates some features of the extreme environment of temperature, pressure, velocity and electromagnetic field which we seek to study.

Many but not all neutron stars exist in a binary system, in orbit with a ordinary, thermal companion. An isolated neutron star must spin down as its rotational kinetic energy is converted into radiation. A binary system has an additional source of energy—the accretion of mass from the companion into the gravitational potential well of the neutron star. This process may be facilitated by mass ejection from the companion, resulting from its bombardment by radiation from the neutron star. The accretion process in detail is not simple. Angular momentum conservation forbids a direct mass transfer from the companion to the neutron star. The infalling mass may temporarily form an accretion disk around the binary, with the accretion process completed only after this mass loses its angular momentum as a result of non-conservative forces.

Again, an order of magnitude estimate yields a better understanding of the enormity of this process. A neutron star is typically ≈1.4 solar masses or 2.8 x 10^{30} kg and has the earlier stated radius of ≈10 km. The gravitational potential energy yielded by the accreted mass is then 2 x 10^{16} joules/kg or about 20 percent of the mass-energy of the accreted matter. A typical energy output for such a system is 10^{30} joules-s^{-1} (or 10^{37} ergs-s^{-1} in the cgs units that astronomers like to use). The most energetic known binary SMC X-1 has an x-ray luminosity of ≈5 x 10^{31} joules-s^{-1}. The typical luminosity of 10^{30} joules-s^{-1} requires the accretion of 10^{-9} solar masses per year. The number of emitted quanta depends on the energy spectrum. For example, if one percent of a luminosity of 10^{30} joules-s^{-1} appeared as quanta with an energy of ≈1 Tev, the total number of emitted quanta near that energy would equal 6 x 10^{34} s^{-1}. At a typical galactic distance of 10 kpc, the corresponding measured flux for an isotropic emitter would be 5 x 10^{-8} quanta-m^{-2}-s^{-1}. Such a flux is near but above threshold for several existing detectors. Higher and more easily measurable fluxes might occur for short-term bursts.

The question of whether high energy (>100 MeV) quanta from specific astrophysical objects have actually been observed is controversial. There is, however, general agreement on the properties of such quanta. (1) They must be neutral or their trajectories would not point back to their source because of galactic magnetic fields (for momenta $<10^{18}$ GeV/c). (2) They must be stable, that is $\tau < 10^{12}$ s/γ. (3) They must be detectable with reasonable efficiency; otherwise, measurable fluxes imply excessive source luminosities. In the context of known particles, criterion (1) eliminates all particles other than neutrons, neutrinos and photons. Criterion (2) eliminates neutrons, at least for energies $<10^{18}$ eV. Criterion (3) eliminates neutrinos. The high energy quanta are by process of elimination considered as photons, although there is no direct experimental evidence for this identification. Indeed, as we shall see, there is some evidence to the contrary.

The Detection of High Energy Quanta

Whatever their nature, high energy quanta do not penetrate through the atmosphere to the surface of the earth without interaction. Below 100 MeV, fluxes are sufficiently large to permit direct primary detection by satellite-based instruments. At higher energies, small fluxes necessitate ground-based detectors. However, the earth's atmosphere down to sea level represents ≈11 nuclear interaction lengths and ≈28 radiation lengths of material. In almost every case, an incident high energy particle will produce an air shower with an initial interaction at an altitude of 25-50 km above the earth's surface. As indicated in Fig. 1, these air showers differ depending on the incident particle. Photon-induced showers generally have a higher altitude initial interaction than hadron-

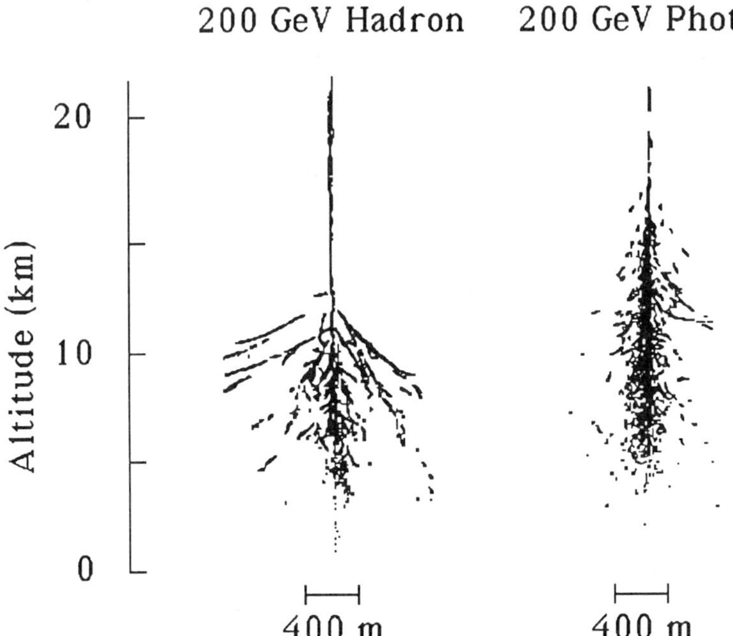

Figure 1—Monte Carlo calculation by A. M. Hillas of charged particle tracks in 200 GeV hadron-induced and photon-induced air showers.

induced showers because the radiation length of air is almost 3 times shorter than the nuclear interaction length. At a similar point in their development, photon-induced showers are more tightly collimated in the lateral dimension than hadron-induced showers. The large lateral spread of hadronic showers primarily results from the average transverse momentum of ≈300 MeV/c imparted imparted in the production of each successive hadronic generation. Another difference is that photon-induced showers contain few muons, while hadronic showers contain muons resulting mostly from pion decay. The few muons found in photon-induced showers are predominantly from the decay of hadroproduced pions[3]. Detailed Monte Carlo calculations suggest muon flux differences of order 10^3 between photon-induced and hadron-induced showers of similar energies.

Despite these differences, the experimental distinction between high energy photon-induced and hadron-induced air showers is difficult and has never been conclusively demonstrated by any detector. The primary difficulty is fluctuations which dominate both types of showers. Fluctuations are generally more significant in hadronic showers because the pion mass is much larger than the electron mass, resulting in a smaller number of particles in the hadronic case. However, regardless of the primary, a few large or many small fluctuations can make one type of shower look like the other. For example, the transfer of a large amount of energy to a π^0 (which decays to two photons) early in the development of a hadron-induced shower will cause the shower to appear photon-induced. The ratio of hadron-induced to photon-induced showers is so large that this process can easily swamp a true photon-induced shower signal.

There is a myth in the VHE/UHE literature that showers can be distinguished by fitting the shower shape to an analytic form and extracting a number called the "age parameter", which is related to the lateral spread of the shower[4]. The general form of the myth is that large "age parameters" (that is large lateral spreads at the detector) are associated with photon-induced showers. The reasoning comes from the literal interpretation of the concept of "age parameter." Photon-induced showers originate higher in the atmosphere and thus are "older" when they impact the earth. As indicated above, hadron-induced showers, although "younger", also have large lateral spread when they impact the earth's surface. Detailed Monte Carlo calculations by Hillas[5] show that photon-induced showers which are actually detected have a smaller "age" than hadron-induced showers of similar size, although this distinction is subject to considerable fluctuation. The "age parameter" thus appears an unreliable discriminant despite its use in many analyses[6].

Two types of detectors have been used to detect high energy air showers, including those possibly associated with quanta from specific astrophysical sources. The *air shower array* consists of a lattice of particle detectors (generally scintillators) which directly sample the charged particles of air shower which reach the surface of the earth. A typical array will consist of ≈1 m^2 detectors arranged with a lattice spacing of ≈10 m. Optimal spacings depend on the detector altitude and the primary energy threshold sought for the detector. Directionality in such an array comes from the determination of the time differences for detection of the air shower particles across the array. Angular resolutions of ≈1° are possible with a sufficiently large array and reasonably small, thick scintillation counters as the detector elements. Properties of typical air shower arrays include an energy threshold of ≈100 TeV for the primary, an active area limited to the size of the populated detector lattice, the possibility of 100 percent active time, no sensitivity to either ambient light or optical photons from the source which one is trying to study and the ability to record data from multiple sources simultaneously. Examples of such arrays include the CYGNUS experiment experiment at Los Alamos, the Baksan array in the Caucasus Mountains and the Akeno array in Japan.

The *atmospheric Cerenkov detector* is sensitive to the highly collimated Cerenkov light produced by the charged air shower particles as they pass through the atmosphere. Because the index of refraction of air is so close to unity, this light disperses in angle by only $\approx 1^{\circ}$ from the primary direction. The Cerenkov light is mostly from the region of the shower maximum and reaches the surface of the earth shortly after the particle remnants of the shower itself. Because these detectors are sensitive to the particle density throughout the shower (and not just the tails that reach the earth's surface), they have a far lower energy threshold ($E_{min} > 0.1$ TeV) than air shower arrays. On the other hand, atmospheric Cerenkov detectors, of course, are very sensitive to ambient light and generally operate from after twilight in the evening to before twilight in the morning, when the moon is sufficiently dim and when there are no clouds. They are also sensitive to star light, either from the observed star or others. The location of many interesting objects in the galactic plane (the "Milky Way") emphasizes the problem of ambient light. Atmospheric Cerenkov detectors also use a focussed optical system whose angular aperture is similar to the angular dispersion of light from a shower. The angular resolution of these devices is determined by these optical properties. These detectors thus view only showers from one source at any particular time. The Whipple-Mt. Hopkins detector, the Haleakala detector and the Crimean Observatory detector (which first observed high energy flux from Cygnus X-3) are all examples of atmospheric Cerenkov detectors.

The simplest strategy for detecting radiation from a particular source with either of these types of detectors is by an integrated flux analysis. A time-integrated, excess flux

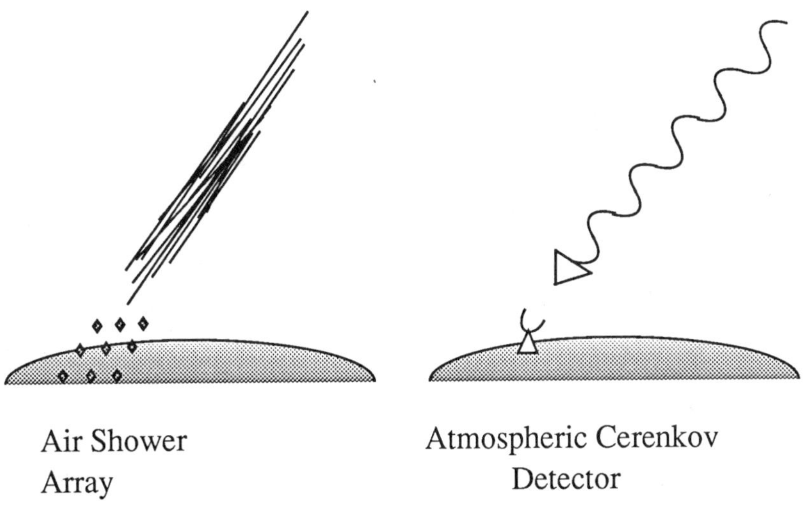

Figure 2—Schematic of charged air shower particles approaching an air shower array (left) and Cerenkov light from an air shower approaching an atmospheric Cerenkov detector (right).

from the direction of a particular astrophysical object as compared to flux from systematically similar but non-source directions would confirm the hypothesis of source-related flux. Systematically similar directions are generally those with equal exposure time, equal ambient light and equal angle above the horizon. Some but few long-term, integrated-flux observations have yielded positive results. For most such claimed detections, the statistical and systematic significance is not considered overwhelming. One possible difficulty is source variability, which is common for observations of non-thermal stars using lower energy quanta. (For example, radio flux from Cygnus X-3 is known to vary by 3 orders of magnitude[7].) For these reasons, researchers in this area have sought to improve their signal-to-background ratios by either selectively reducing their detection of hadronic (and presumably background) showers or by time-locking (either periodic or episodic) to specific sources.

There are several methods (other than the dubious "age parameter") to reject hadron-induced air showers. For an air-shower array, the most straight-forward method is a muon detector array co-located with the air shower array. The CYGNUS apparatus[8] uses two LAMPF detectors located under the air shower array to track muons. Muons are identified by their ability to penetrate the $\approx$3000 gm-cm^{-1} of material which shields the LAMPF experimental area. The Chicago-Michigan detector, currently under installation in Utah, uses a similar approach. A second method for rejecting hadron-induced showers is applicable to atmospheric Cerenkov detectors. Monte Carlo calculations show a different pattern of Cerenkov light impacting the earth's surface for photon-induced and hadron-induced showers. The differing patterns result from the differing shower development discussed above. If the Cerenkov light is imaged and analyzed as a "picture" pixel-by-pixel, a different pattern is expected in the two cases. The Whipple-Mt. Hopkins detector has been instrumented to permit such an analysis. The first attempt used 37 pixels; new data will be taken with more than 100 pixels. The success of this approach is not yet clear.

The temporal analysis of flux is a complex strategy which has considerable power to improve signal-to-background ratios but which can also easily result in "false positive"

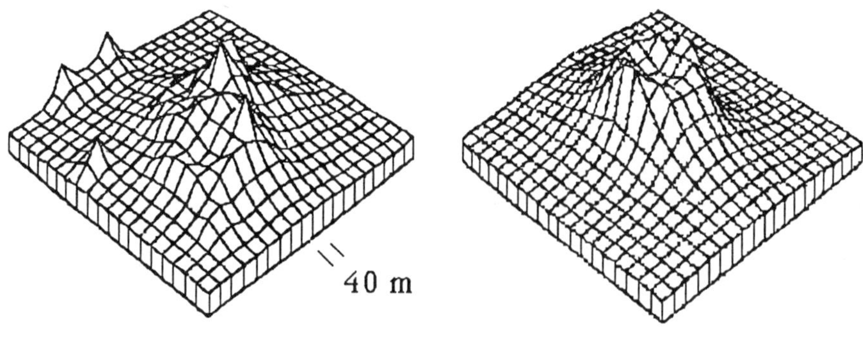

Figure 3—Monte Carlo calculation of the intensity of atmospheric Cerenkov light at impact on the earth's surface for a typical hadron-induced and a typical photon-induced air shower (by D. J. Macomb).

observations. Generally, high energy fluxes are searched for periodicity at periods measured for x-ray flux from the same object. The goal is to discriminate between the hypothesis H_0: *the flux is random; there is no source* and H_1: *there is a source*. The methods are statistical and thus can never "prove" H_1. They are limited to demonstrating that H_0 is extremely unlikely followed by the reasonable inference that H_1 must be true. Episodic analyses have also been performed, both in coincidence with specific events and as a general search procedure. An example of the former is the report by several groups of high energy flux from Cygnus X-3 coincident with the October 1985 radio flare of that object[9].

The High Energy Sky

The outstanding problems in VHE and UHE astroparticle physics today are

- Have we observed VHE and/or UHE quanta from specific astrophysical sources?
- If so, why do they not behave like photons?

Before discussing VHE and UHE flux from specific objects, it is interesting to survey the sky at lower energies. Fig. 4 shows the keV x-ray sky with its significant clustering of strong sources in the galactic plane, particularly towards the region of the galactic center. For photons with $E > 100$ MeV (see Fig. 5), the flux distribution is similar, with ≈5 times more flux from the galactic center region than from other areas of the galactic plane. This plot displays point sources observed by the COS-B satellite[10], which did not note any significant point source flux from the direction of Cygnus X-3. The SAS-2 detector, which had similar sensitivity in energy and flux, recorded data at a different time[11]. It reported Cygnus X-3 as one of the strongest sources in the sky in this same energy range. The apparent Cygnus X-3 flux variability between observing epochs of these two detectors illustrates one of the most severe problems in making systematic observations of high energy sources.

The VHE and UHE sky consists of about a dozen sources observed with various degrees of confidence. Ref. 12 provides an extensive review of the observational data, along with exhaustive references. I will discuss two of the best-studied sources—Cygnus X-3 (Cyg X-3) and Hercules X-1 (Her X-1). Cyg X-3 is a compact binary with a neutron star orbital radius $\approx 10^9$ m and an orbital period (today) of 4.792493 h. The neutron star rotation period (pulsar period) has been tentatively identified as 12.59 ms, although this observation has never been confirmed. The distance from the earth to Cyg X-3 is measured as >10 kpc from absorption of the radio signal on galactic dust clouds which lie at known distances from our solar system. (These same dust clouds prevent optical observations of the companion star in Cyg X-3.) Cyg X-3 is known as a variable flux emitter. The radio flux is characterized by giant flares about 2 weeks in duration every few years and large numbers of lesser, shorter and more frequent flares[7]. The radio flux of the largest flares exceeds the steady-state flux by >1000. The x-ray flux also shows variation of an order of magnitude, not necessarily in coincidence with the radio flares.

There exist more claimed observations of high energy flux from Cyg X-3 than any other source. The high energy data have appeared inconsistent, perhaps because of source flux variation. The source is sometimes observed and sometimes not, with no

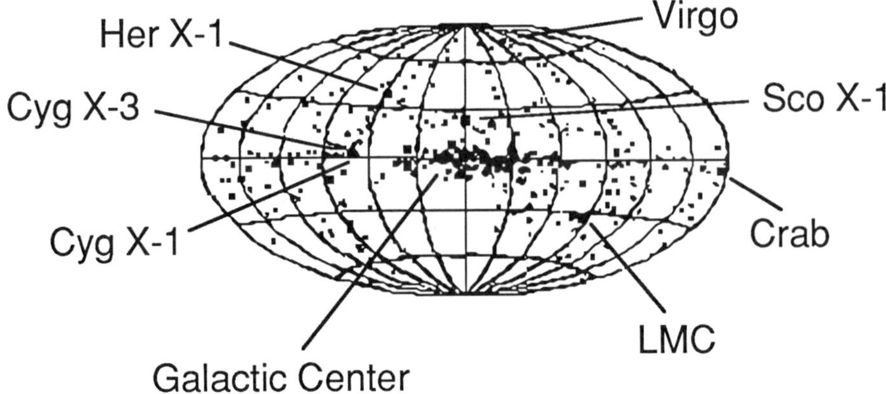

Figure 4—Some x-ray sources shown by galactic latitude and longitude. Longitude in this projection increases to the left along the equitorial plane, starting from the galactic center.

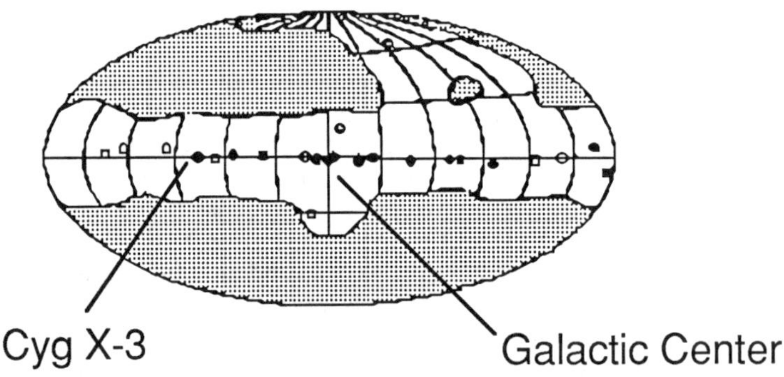

Figure 5—The sources reported by the COS-B detector shown in a galactic latitude and longitude projection similar to the one in Fig. 4. The shaded areas are directions not studied in the COS-B mission. Open symbols indicate fluxes $< 1.3 \times 10^{-6}$ cm^{-2} s^{-1}; closed symbols represent larger fluxes.

apparent pattern which permits an *a priori* determination of the probability of observation. The consensus of positive observations of Cyg X-3 is that the flux is ≈10^{-7} quanta-s^{-1} at energies >1 GeV and that the energy spectrum is dn/dE = $AE^{-2.1}$. In recent years, there is also an emerging consensus on the phase modulation of VHE and UHE flux from Cyg X-3. Fig. 6 is a review of UHE results[9]. The majority of recent observations, both at these energies and VHE energies, show a flux peak at phase ≈0.7 (using the ephemeris of Mason[13]).

Cyg X-3 is the best studied source for a combination of reasons—some significant and some accidental. It is apparently one of the most luminous sources in our galaxy, particularly at the higher energies because of a very flat energy spectrum. For example, the luminosity[12] at >1 PeV is ≈6 x 10^{29} joules-s^{-1}. Indeed, one can speculate that Cyg X-3 alone could account for most of the cosmic rays observed in the galaxy. A less significant reason for the multiple observations is that the declination of Cyg X-3 is 41° North, so it passes directly overhead in the most populated and wealthy areas of the earth.

An interesting and potentially important aspect of Cyg X-3 is the number of anomalous observations associated with this source. The first such observation[14] was made in the late 1970's using the Kiel detector—an air shower array with muon detectors locat-

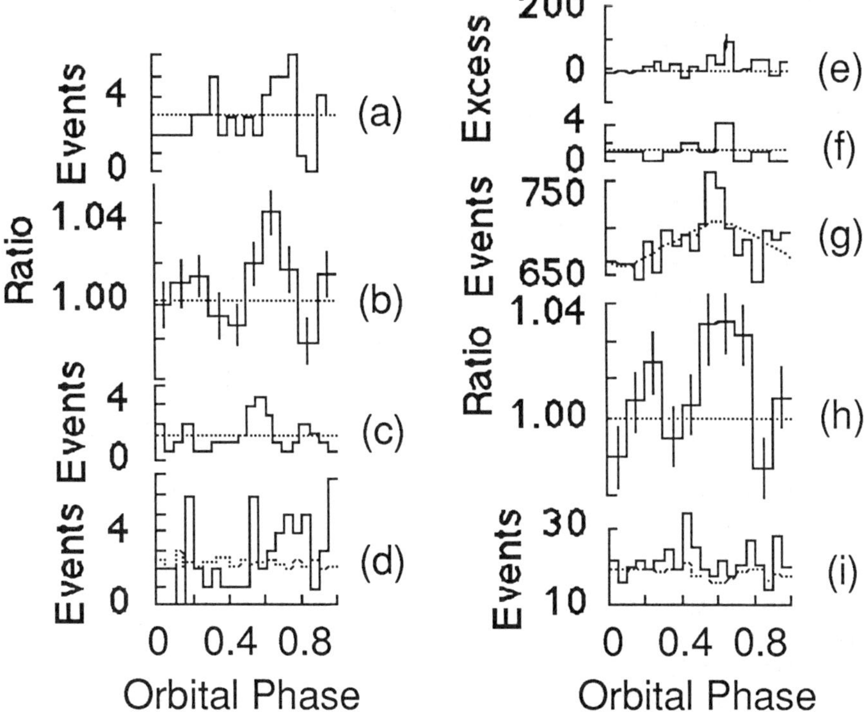

Figure 6—Phase plots reported at the 20th International Cosmic Ray Conference for UHE air shower observations of Cyg X-3. Further details are presented in Ref. 9. The detectors are (a) Ooty, (b) Baksan, (c) Akeno, (d) Kolar Gold Field, (e) Fly's Eye, (f) Ohya, (g) Akeno, (h) Baksan and (i) Haverah Park.

ed in a World War II bunker on the shores of the Baltic. This detector reported a long-term signal from Cyg X-3 in the time-averaged flux at PeV energies. The signal was enhanced by a phase-lock analysis using the Cyg X-3 x-ray period. The anomalous aspect of these data was the observation that the Cyg X-3 signal had ≈80 percent of the muon content of background, hadron-induced showers. The expected muon content for such a detector is <1 percent. Possible problems with the Kiel detector include "punch-through" of electromagnetic showers into the muon detectors and a questionable ephemeris for the phase-locked analysis. However, no other similar detectors were operating simultaneously so there is no possibility for direct confirmation or denial of these results.

In 1985, two deep underground proton decays[15,16] (Soudan 1 and NUSEX) also reported an anomalous flux of muons apparently associated with Cyg X-3. In both cases, there was an excess flux of muons from the Cyg X-3 direction and the signal-to-background ratio was enhanced by a phase-lock analysis. For these two detectors, the problem of shower "punch-through" was not relevant. Both experimental groups also used the generally-accepted Cyg X-3 x-ray ephemeris for their phase-lock analysis. These reports have been criticized because of the statistical analyses which they contain and because both detectors used angular cuts considerably larger than their reported angular resolutions. If the Kiel, Soudan 1 and NUSEX observations are correct, however, they imply some physical mechanisms of which we are not currently aware.

Of course, if such mechanisms exist, they should affect quanta from sources other than Cyg X-3. Recent reports of observations of Her X-1 suggest similar mechanisms. Her X-1 is perhaps the best observed accreting binary in the entire sky. The thermal companion is the optical variable star HZ Hercules. The Her X-1 systems exhibits at least three periodicities—a neutron star spin period of 1.23778 s, an orbital period of 1.700 d and a superperiod, possibly due to

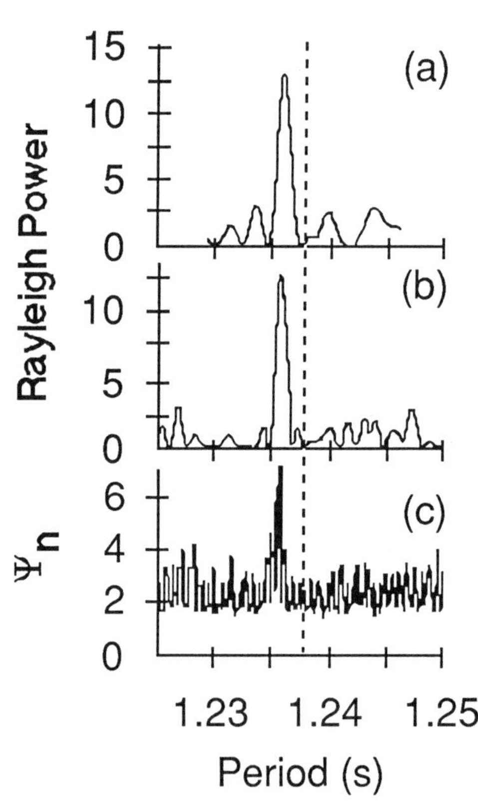

Figure 7—Periodograms (similar to Fourier transforms) for events from the direction of Her X-1 recorded during 1986 bursts by the (a) Haleakala, (b) Whipple-Mt. Hopkins and (c) Los Alamos-CYGNUS detectors. The power at each period is measured with the Rayleigh procedure in (a) and (b) and with the Protheroe procedure in (c).

precession of the orbital plane, of ≈34.9 d. The almost circular neutron star orbit has a radius of ≈14 light-s. In 1986, high energy quanta from Her X-1 were apparently observed by three detectors (Haleakala[17], Whipple-Mt. Hopkins[18], and Los Alamos-CYGNUS[19]), all on different days in May, June and July. In all, four flux "bursts" were observed, each one from 15 to 30 m in duration. A remarkable aspect of these observations was that each found the same neutron star spin period. Fig. 7 shows periodograms (similar to power distribution function obtained by squaring the Fourier transform). The Haleakala and Whipple-Mt. Hopkins both determine power in the periodogram by use of the Rayleigh statistic. The Los Alamos-CYGNUS group uses the Protheroe procedure, which is more sensitive to narrow phase enhancements. In each case, the spin period was 0.017 percent smaller than the x-ray period. Such a shift corresponds to a velocity of 5 x 10^5 km-s^{-1} and suggests a different source location in the binary system for high energy and x-ray quanta.

Two of these three detectors have the capability to distinguish between photon-induced and hadron-induced showers. The Whipple-Mt. Hopkins detector finds that the atmospheric Cerenkov light images of these apparently Her X-1-related events are similar to those of other, presumably hadron-induced showers and different from their expectations for photon-induced showers[20]. The CYGNUS-Los Alamos detector shows a muon multiplicity in their apparently Her X-1-related events which is unambiguously different from their expectation for photon-induced events[19]. Indeed, with a probability of 99 percent, these data suggest that Her X-1-related showers have a higher muon content than showers associated with average (presumably hadron-induced) events. An analysis also shows a 98 percent probability that these events have a higher energy than would be expected for background events. These Her X-1 data, if the experimenter's interpretations are correct, support the earlier Cyg X-3 data in suggesting the existence of either a new particle or a new interaction for photons or neutrinos.

Conclusions

• VHE and UHE physics is an important element of astroparticle physics with apparent direct access to information about unique macroscopic particle physics laboratories.

• There is strong (but not indisputable) evidence that VHE and UHE quanta have been observed.

• There are multiple reports for two sources that such quanta do not interact in the earth's atmosphere in the manner expected for conventional photons.

This work has been supported by the U. S. Department of Energy.

References

[1] M. Goldhaber, Proceedings of the XXIII Intl. Conf. on High Energy Phys., S. C. Loken, ed., World Scientific, 1987, p. 248.
[2] S. L. Shapiro, S. A. Teukolsky, Black Holes, White Dwarfs and Neutron Stars, John Wiley and Sons, New York, 1983.
[3] F. Halzen, Ann. N.Y. Acad. Sciences **490**, 237 (1987).
[4] K. Greisen, Prog. Cosmic Ray Physics, v. 3, J. G. Wilson, ed., Amsterdam, 1956.

[5] A. M. Hillas, private communication.
[6] T. Kifune, *et al.*, Astrophys. J. **301**, 230 (1986).
[7] K. Johnston, *et al.* Astrophys. J. **309**, 707 (1986).
[8] B. L. Dingus, *et al.*, Phys. Rev. Lett. **60**, 1785 (1988).
[9] R. Protheroe, 20th Intl. Cosmic Ray Conf. **8**, 21 (1988).
[10] B. N. Swanenburg *et al.*, Astrophys. J. **243**, L69 (1981).
[11] R. C. Hartman, *et al.*, Astrophys. J. **230**, 597 (1979).
[12] T. C. Weekes, Phys. Reports **160**, 1 (1988)
[13] K. Mason, private communication ($p_0 = 0.1996851$ d; $dp/dt = 7.84 \times 10^{-10}$; t_0 = JD 2442946.739).
[14] M. Samorski and W. Stamm, Astrophys. J. **268**, L17 (1983).
[15] M. L. Marshak, *et al.*, Phys. Rev. Lett. **54**, 2079 (1985) and **55**, 1965 (1985).
[16] G. Battistoni *et al.*, Phys. Lett. **155B**, 465 (1985).
[17] L. K. Resvanis, *et al.*, Astrophys. J. **328**, L9 (1988).
[18] R. C. Lamb, *et al.*, Astrophys. J. **328**, L13 (1988).
[19] G. B. Yodh, *et al.*, Proc. Intl. Conf. on Quark-Gluon Plasma, Bombay, 1988 (UCI Neut. 88-13).

ASPECTS OF HARD INTERACTIONS IN NUCLEI

John P. Rutherfoord*
University of Washington, Seattle, WA 98195

ABSTRACT

High energy interactions involving hard processes with small cross section from *nucleon* targets are generally interpreted in terms of the quark-parton model. The hard process cross section is modified if the target nucleon is embedded in a *nucleus*. When an attempt is made to interpret both published and new, unpublished data in terms of a model describing partons traversing nuclear matter to and/or from a hard interaction, inconsistencies in the picture arise.

INTRODUCTION

One of the early experiments at the Fermilab accelerator was a measurement of the inclusive cross section for high p_T hadrons using *nuclear* targets.[1] Theoretical predictions for this hard interaction suggested that the cross section should depend on target atomic number, A, as A^α where $\alpha \approx 1.0$, i.e. that the nucleus would not differ significantly from a collection of free nucleons. The surprising result, sometimes referred to as the "Cronin effect", indicated large deviations from this naive expectation and is generally interpreted in terms of multiple scattering of the active parton(s) in nuclear matter.[2] Following a qualitative discussion of the expected nuclear effects, data from four hard interaction processes, some of which are presented here for the first time, will be examined to see if a consistent picture of the "Cronin effect" is possible. Stopping short of a detailed phenomenological calculation, this appears unlikely.

EXPECTATIONS

For this discussion I've chosen to examine four hard interaction processes. The choice is dictated by the availability of data and the likelihood of a straight forward interpretation. In Fig. 1 are very schematic representations of the four hard interactions, *i*) lepto-production of hadrons, *ii*) "Drell-Yan" lepton pair production, *iii*) high p_T inclusive hadrons, and *iv*) high p_T hadron pairs. In each case a fat dot indicates the

---

* Permanent address after 1 September, 1988: Department of Physics, University of Arizona, Tucson, AZ 85721

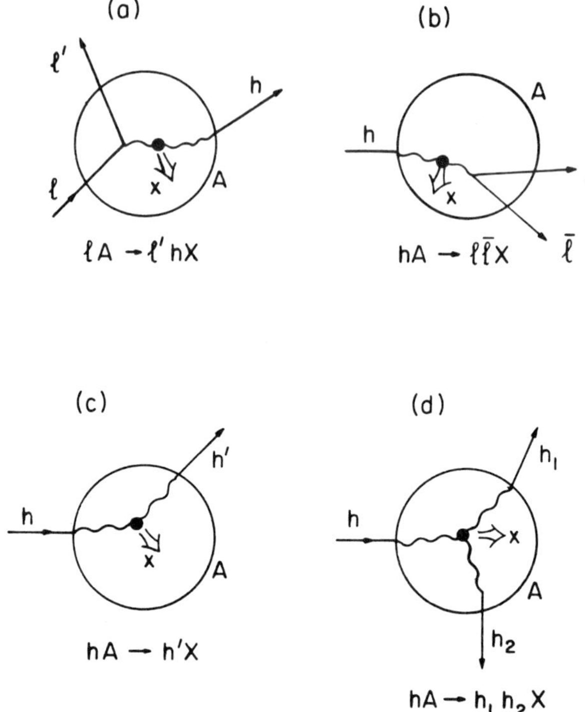

Fig. 1. Hard interaction processes depicted in the nucleus rest frame. a) Lepto-production of hadrons, b) "Drell-Yan" lepton pair production, c) High p_T inclusive hadrons, and d) high p_T hadron pairs.

location of the hard interaction and an arrow indicates other, unobserved debris. And each picture is shown in a rather unconventional frame, the nucleus rest frame. We will attempt to choose data in which hadrons have high momenta, say $p \gtrsim 100GeV/c$ in this frame and, of course, where the interaction is quite hard, i.e. there are large momentum transfers. In some sense the cross sections for the four hard processes of Fig. 1 are "understood" within the quark-parton model and, for some, with perturbative QCD corrections when the target is a single hadron. This means that using parameterizations of parton distribution functions and fragmentation functions, the phenomenological predictions agree, within reason, with the data in a self-consistent way.

Is it possible to understand how these hard processes are modified when the target nucleon is confined within a nucleus? There are many phenomenological papers addressing various aspects of the problem but none attempts to address all the relevant data. In the following I will outline how such a study might be conducted.

It's generally accepted that when a hadron beam particle impinges on a nucleus, the beam hadron "breaks up" at the nuclear surface as indicated in Fig. 2. Again the picture is in the nucleus rest frame. The valence partons in the beam hadron are shown Lorentz contracted but the sea partons are not. The invariant mass of the valence partons is excited by soft interactions with the partons in the nucleus to excited states larger than the beam hadron mass. But the stiff valence partons propagate through the nucleus with little change in momentum. The lower part of Fig. 2 attempts to suggest the valence partons moving on with no change in momentum but the sea partons experiencing collisions.[3]

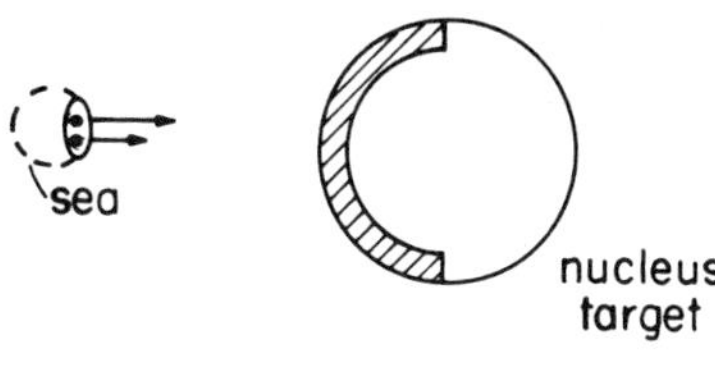

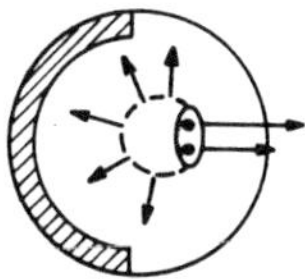

Fig. 2. The beam hadron "breaks up" at the nuclear surface but only the sea partons suffer significant interactions.

It's also generally accepted that the stiff partons "dress" themselves with stiff hadrons well outside the nucleus. The argument goes as follows: a parton leaving a hard interaction "dresses" itself on a time scale $\tau = 1/m$ in the hadron center-of-mass where m is a typical hadron mass, $m \approx 0.5 GeV$. In the nucleus rest frame the time is $\gamma\tau = p/m^2$ where p is the hadron momentum in the nucleus rest frame. So the distance the parton travels from the hard interaction within the nucleus before it "dresses" itself is $\gamma c\tau \approx p$ fermi when p is expressed in GeV/c. When $p = 100 GeV/c$, then $\gamma c\tau \approx 100$ fermi, well outside the largest nucleus as suggested in Fig. 3.

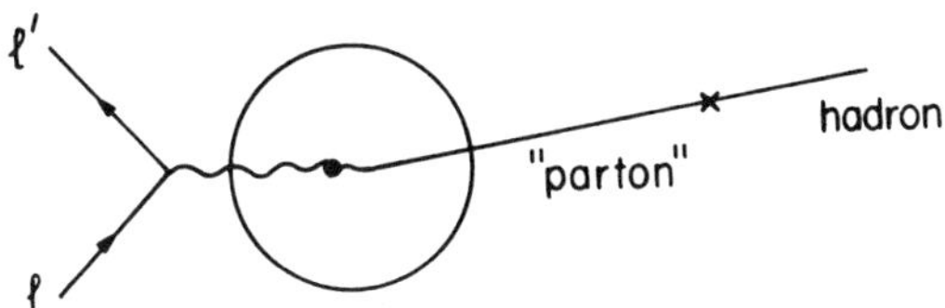

Fig. 3. At sufficiently large lab momentum, a final state parton will "dress" itself outside of the largest nucleus.

Let me now better define the question I will address here. It is: How does a parton propagate through nuclear matter to/from a hard interaction? But let me first address a different but related question in order to make more clear why we will restrict our attention to hard interactions. How does a parton propagate through nuclear matter without hard interactions? Is it the same as with a hard interaction? The answer is "probably not". Fig. 4 depicts a hadron incident on a nuclear target, again in the nucleus rest frame. To focus attention on the valence partons only, I've

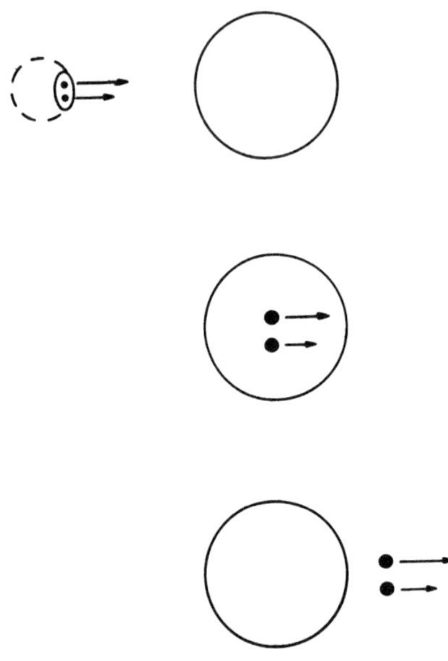

Fig. 4. A beam hadron "breaks up" at the nuclear surface with the stiff valence partons suffering little change in momentum.

not indicated the sea partons except in the first part of the figure. The figure suggests that the partons traverse the nuclear matter without interaction. Now let us review the relevant data. Fig. 5 shows early data from a series of experiments by Busza *et al.*[4]. The hadrons at leading rapidity (which presumably come from stiff partons in the beam) seem to depend little on the size of the nucleus. The data was interpreted to say that, other than the break-up of the beam hadron on the nuclear surface, the rest of the nuclear volume had little impact on the leading hadrons. The term "nuclear transparency" was coined to describe this conclusion. As it turned out "the conclusion about nuclear transparency if not completely wrong is certainly misleading, and the qualitative character of the data is anything but surprising. In a nutshell, the data were not sufficiently focussed to draw the conclusion about transparency ...".[5] Busza goes on to explain that the binning was inadequate and that pseudorapidity is not a good choice of variable for such studies. Later data[6] shown in Fig. 6 suggests a significant suppression of the leading hadrons (at large y_{lab}) as A increases. One interpretation is that the valence partons in the beam hadron lose energy as they penetrate the nucleus. As we will see, this is likely quite different from the hard interactions we will study here.

So back to the main question: how does a parton propagate through nuclear matter to/from a hard interaction? Many phenomenological papers approach this question from a strict quark-parton model point of view represented in Fig. 7. The nucleus is made up of partons within nucleons whose distribution functions are measured. Parton fragmentation functions are also measured. And all the lowest order QCD parton-parton scattering cross sections have been calculated.[7] So one need only turn the crank and get out results. But there are several problems with this approach which are useful to discuss. First, the QCD sub-process cross sections have poles in $\hat{s}$ and $\hat{t}$. These poles are within the region of integration so they must be regulated in some reasonable way. Related to the pole in $\hat{s}$ is the divergence at small x_{Bj} encountered when integrating over the parton momentum fraction in the target. Various papers use different techniques to deal with these infinities giving potentially quite different results were it not for the fact that they tune their procedures to give reasonable agreement with the data they choose to consider. It is unfortunate that the calculations are sensitive to details of the regulation schemes used and the regulation

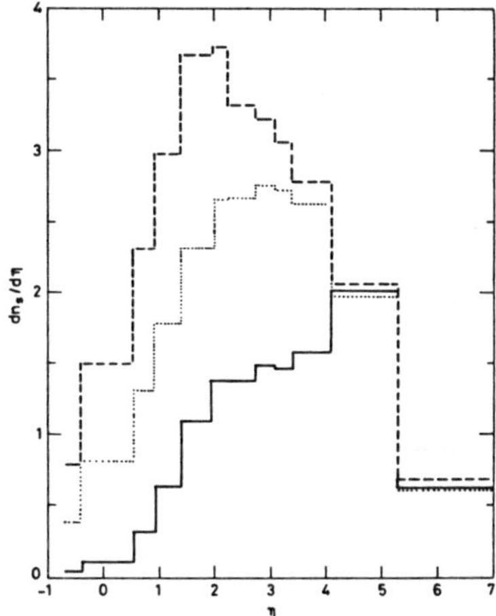

Fig. 5. Charged particle multiplicity in pseudorapidity bins from Ref. 4. A 200 GeV π^- is incident on proton (solid line), Cu (dots), and Pb (dashes) targets respectively. This data was thought to demonstrate "nuclear transparency" in the highest η bin.

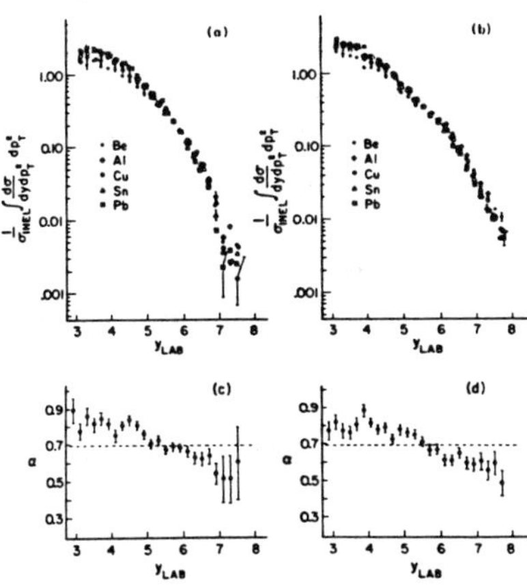

Fig. 6. Charged particle multiplicity as a function of rapidity for a variety of nuclei in a $\sim$ 200 GeV neutron beam from Ref. 6. a) and c) are for negative hadrons while b) and d) are for positive hadrons in the final state.

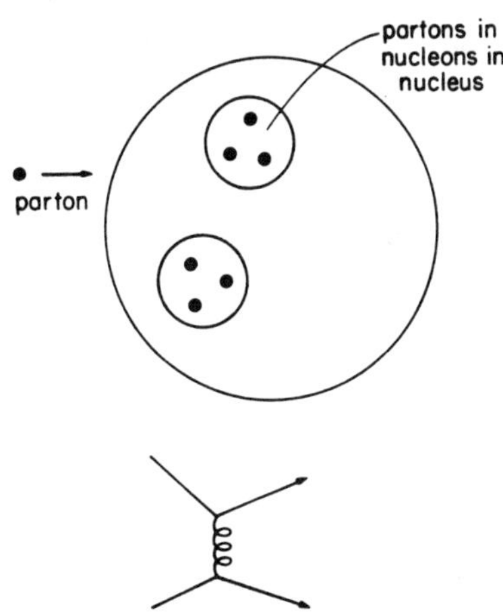

Fig. 7. A high momentum parton incident on a nucleus made up of a collection of partons, each with an associated distribution function.

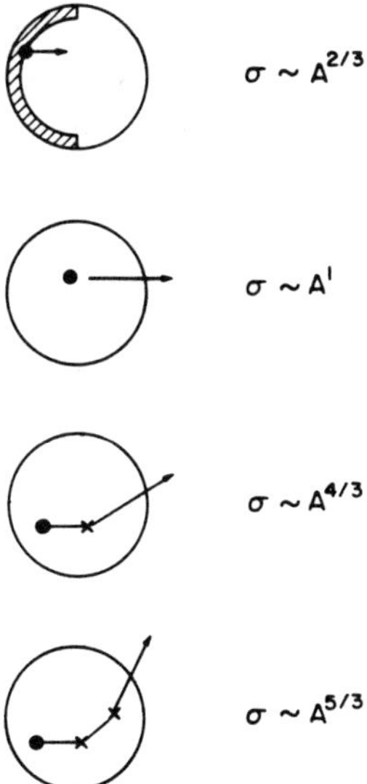

Fig. 8. The nuclear dependence of the cross section can be represented as the sum of terms with different powers of A.

schemes have only qualitative theoretical motivation.

Part of the problem comes from the naive parton model itself. We never actually have a beam parton isolated as suggested as in Fig. 7. It is accompanied by other partons which will provide some color screening which cuts off the long range interactions at low $\hat{s}$ and $\hat{t}$. And final state partons such as those in Fig. 1a will be accompanied by a color string (in the Lund model point of view) connected to a parton in the target fragmentation region which will lead to color screening for this parton as well. Color screening probably has a transverse length cut-off of order 1 fermi.

But probably more important is the fact that the naive parton model does not always properly account for quantum mechanics. Here is an example where coherent effects are important and are not apparent in the semi-classical naive parton model point of view. Bodwin, Brodsky, and Lepage have shown[8] that, in the Drell-Yan process, initial state energy loss of the incident parton is suppressed when

$$Q^2 \gtrsim (0.25GeV^2)A^{2/3}$$

and they suggest analogies to electrodynamic situations. We will assume that the same arguments can be extended to the other hard processes in Fig. 1. Note that for large nuclei, one requires $Q^2 \gtrsim 10GeV^2$ before initial state energy loss can be neglected. This means that dilepton invariant masses in the ψ region may be too low for this approximation to hold.

In summary in hard interactions we expect that 1) the hard process occurs anywhere within the nuclear volume with equal probability because the hard process cross section is so small and there is no attenuation, 2) the parton traversing nuclear matter going to or leaving the hard interaction point will receive small transverse momentum kicks of order $300MeV/c$ per nucleon encountered, i.e.

$$< k_T^2 > \approx (300MeV/c)^2 A^{1/3} f$$

where f is the fraction of nucleons in A encountered and 3) the parton will suffer almost negligible longitudinal momentum loss

$$\Delta p_L \approx (300MeV/c)A^{1/3} f$$

DEFINITIONS

Here we will define the parameters used to represent the experimental data and comment on them. The most commonly used parameters are R_A and α. We will use A to represent both a nucleus of atomic number A and the number of nucleons within that nucleus, i.e. its atomic number. Further, N will represent a single nucleon. Then the data can be parameterized by

$$\frac{\sigma(target A)}{\sigma(target N)} = AR_A = A^{\alpha}$$

or more generally by

$$\frac{\sigma(target A_1)}{\sigma(target A_2)} = \frac{A_1}{A_2} R(A_1, A_2) = (\frac{A_1}{A_2})^{\alpha}$$

For soft processes (with large cross sections) that occur on the nuclear surface as depicted at the top of Fig. 8, we expect $\alpha \approx 2/3$. For hard processes (with small cross sections) that occur anywhere within the nuclear volume with equal probability as shown in the second part of Fig. 8, we expect $\alpha \approx 1$. For hard processes with one other interaction with the nucleus as shown in the third part of Fig. 8, we expect $\alpha \approx 4/3$, and for hard processes with two other interactions with the nucleus as shown in the last part of Fig. 8, we expect $\alpha \approx 5/3$, etc.[9] So in general

$$\sigma \approx c_1 A^{2/3} + c_2 A^1 + c_3 A^{4/3} + c_4 A^{5/3} + \ldots$$

where $c_i = c_i(Q^2, \ldots)$. In principle c_3 can be negative which might happen when energy loss effects overtake multiple scattering effects in high p_T single hadron reactions for instance.

A QUALITATIVE LOOK AT THE DATA

We are going to see that within the picture outlined above, even a qualitative look at the data suggests some inconsistencies. Let us start with lepto-production of hadrons. The first part of Fig. 9 reproduces Fig. 1a suggesting that the momentum of the final state parton is approximately ν and that the parton fragments outside the nucleus into a hadron of momentum fraction z. The fact that the parton does not travel a straight line within the nuclear volume is meant to indicate that it might suffer interactions as it exits the nucleus. The second part of Fig. 9 is a parton model diagram of the same process showing the final state interaction of the parton with a spectator parton within the nucleus. Such a diagram is of "higher twist".

Let us first consider transverse momentum of the observed hadron relative to the virtual photon direction. From a nucleon target several important mechanisms likely contribute. For the present discussion all that is relevant is that $< p_T^2 > \approx 0.3 GeV^2$. The full distribution from two nuclear targets is shown in Fig. 10 taken from Ref. 10. Two features of this data should be kept in mind. First that $< p_T^2 >$ is small (relative to the Drell-Yan process) so that the distribution is quite sensitive to multiple scattering of the final state parton, and second, that there is no discernable difference in the two distributions for the two target nuclei. The situation is summarized in Fig. 11 where $< p_T^2 >$ is plotted versus kinematic variables. We're interested in the highest ν values where $< p_T^2 >$ seems independent of A. This suggests that multiple scattering is too small to see at this level of precision.

Now let us turn to energy loss. Early data[11] shown in Fig. 12 suggested that energy loss was significant. But the values of ν were quite low. More recent data from the EMC collaboration[12] shown in Fig. 13 and 14 imply that energy loss in the

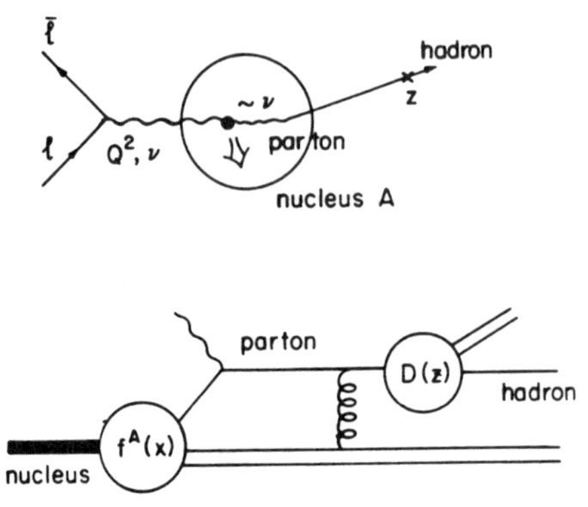

Fig. 9. Lepto-production of hadrons with final state interactions of the struck parton.

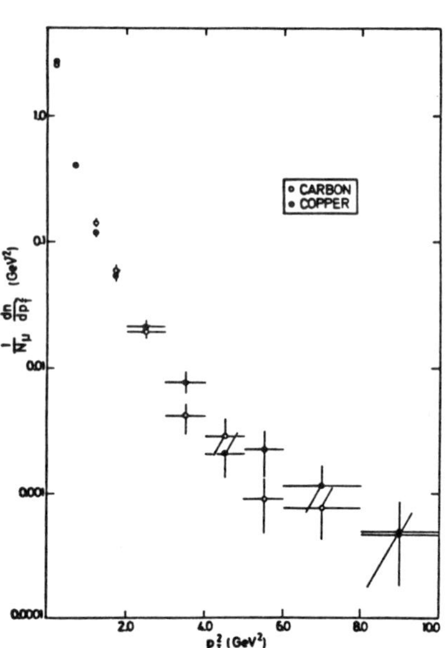

Fig. 10. Charged hadron multiplicity in lepto-production from Ref. 10 versus p_T^2 relative to the virtual photon direction.

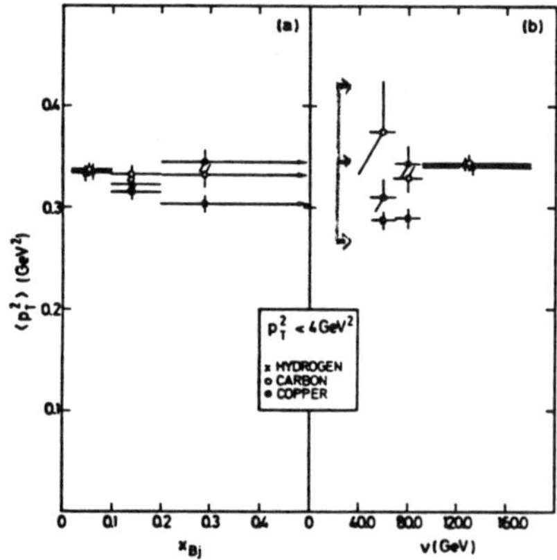

Fig. 11. Average p_T^2 relative to the virtual photon direction versus kinematics from Ref. 10. The vertical bar with arrows indicates the size of the effect seen in Ref. 15.

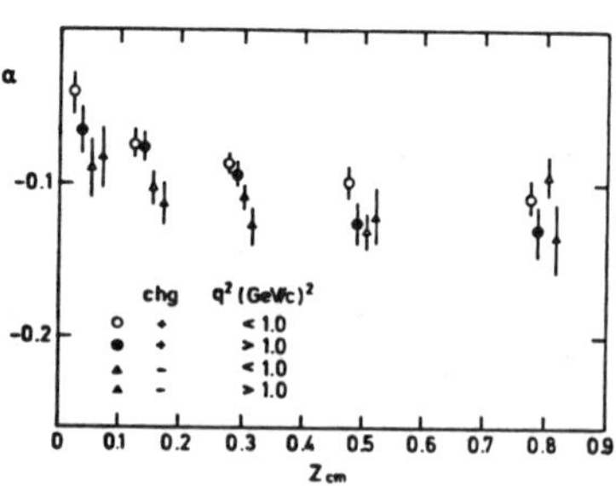

Fig. 12. Nuclear dependence of charged hadron multiplicity in lepto-production from Ref. 11.

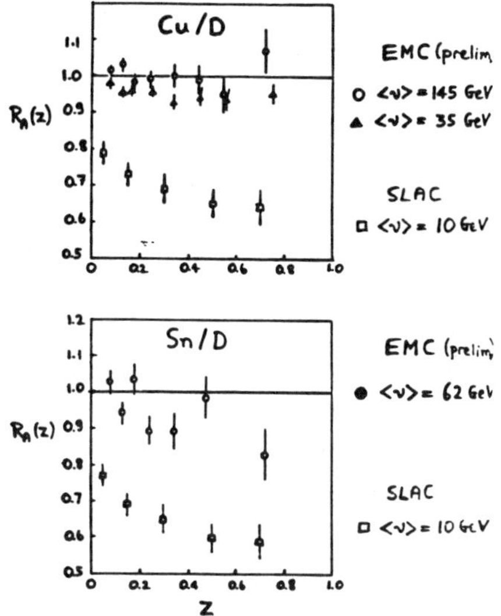

Fig. 13. Nuclear dependence of charged hadron multiplicity in lepto-production from Ref. 12.

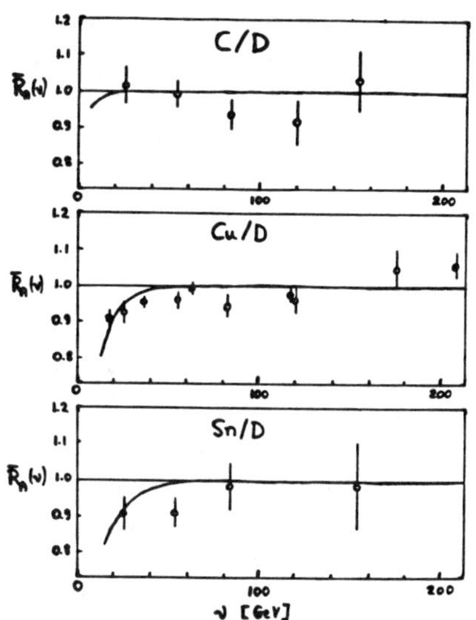

Fig. 14. Average nuclear dependence of charged hadron multiplicity in lepto-production from Ref. 12.

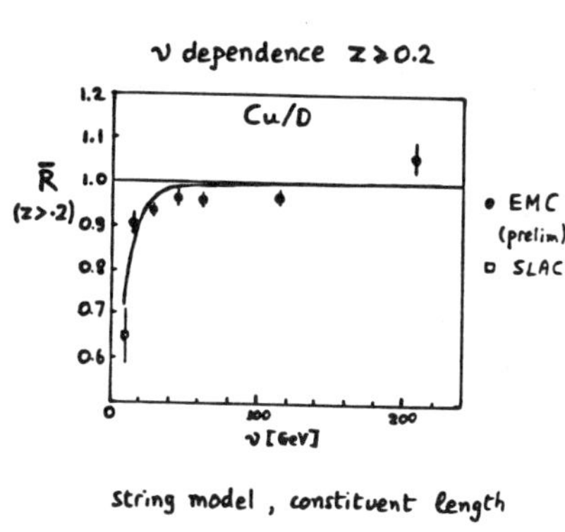

Fig. 15. Average nuclear dependence of charged hadron multiplicity in lepto-production from Refs. 11 and 12 with a string model prediction.

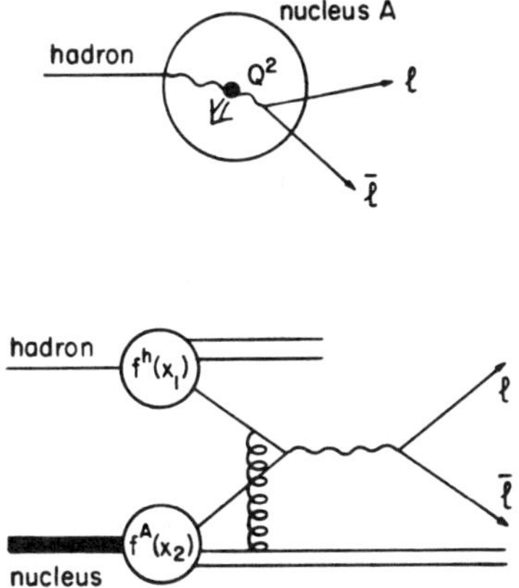

Fig. 16. Drell-Yan lepton pair production with initial state interactions of the annihilating beam parton with spectator partons in the nucleus.

kinematic range of interest to us is quite small, consistent with zero. Fig. 15 shows both the older SLAC data of Ref. 11 with the new EMC data and suggests that the energy loss turns off sharply as ν increases above about $40GeV$.

Now let us turn to a reaction which is closely analogous to deep inelastic scattering, the Drell-Yan process depicted in Fig. 16. The upper part of the figure is the same as in Fig. 1b where the incoming parton follows a wavering path indicating possible initial state interactions. The lower part of Fig. 16 shows a parton model diagram of the Drell-Yan process with a spectator initial state interaction, also a higher twist diagram.

Nuclear effects have been examined in the Drell-Yan continuum for a long time but not at the precision possible in deep inelastic scattering. Fig. 17 shows early proton-nucleus data[13] suggesting that no multiple scattering of the initial parton is seen in this process. The same point is made by the π-nucleus data[14] in Fig. 18. However new high precision data from the NA10 collaboration[15] shown in Fig. 19 clearly exhibit a nuclear effect consistent with multiple scattering of the incident parton in a Tungsten nucleus of order $< k_T^2 > \approx 0.16GeV^2$ as shown in Fig. 20.

Fermilab experiment E772 was designed specifically to look for nuclear effects in the Drell-Yan process with high precision. They have just completed their data run and have some preliminary results[16] which are being shown here for the first time. Their data in Fig. 21 suggests that there is negligible multiple scattering of the incident parton within the precision of the measurement. While the NA10 experiment uses a low energy π^- beam which contains valence anti-quarks and E772 uses an 800 GeV proton beam which contains sea anti-quarks, it is hard to see why these two experiments should disagree so much within the point of view adopted here. For purposes of further comparison I have indicated on Figs. 11, 17, and 18 the difference between Hydrogen (or Deuterium) and the heavy target that should have been seen if the NA10 result can be interpreted as initial state parton multiple scattering. It is clear that there is some inconsistency either in the picture suggested here and/or in the data.

The high precision Drell-Yan data from nuclear targets shown by P. McGaughey at this conference[16] puts strong constraints on any energy loss that might be suffered by the incident parton. The fact that this data is able to address the EMC effect in the Drell-Yan process must mean that energy loss is completely negligible. Otherwise extracting structure functions from such data would be meaningless.

Now I turn to hadronic production of high p_T hadrons, the data in which the "Cronin effect" was first seen. The first part of Fig. 22 reproduces Fig. 1c while the second part is a parton model diagram showing both initial and final state interactions. The theoretical discussion is complicated by the fact that a final state hadron is observed rather than a parton. Since high p_T jet data from nuclear targets,[17] as shown in Fig. 23, exist why not consider this as well? The reason is suggested by the labels in the figure. It is not clear at these energies that the experiments are measuring properties of jets. Fig. 24 shows that the nuclear effect changes dramatically as the "jet" topology changes. Further discussion of the very interesting physics in such

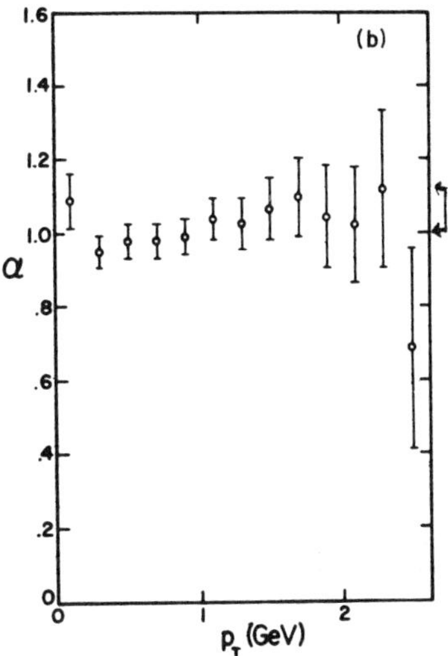

Fig. 17. Nuclear dependence of Drell-Yan continuum events versus p_T from Ref. 13. The vertical bar with arrows indicates the size of the effect seen in Ref. 15.

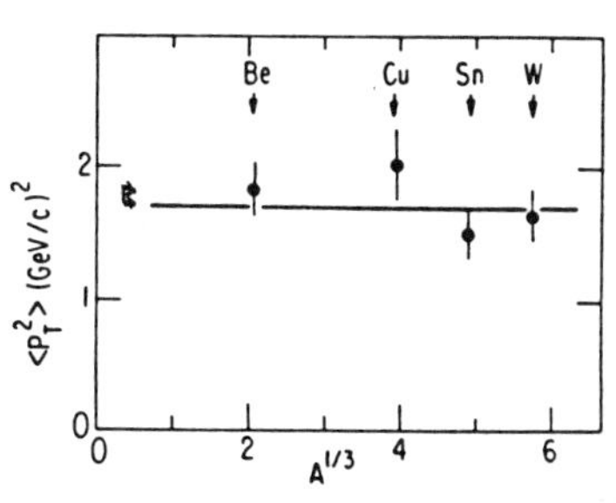

Fig. 18. $< p_T^2 >$ in Drell-Yan continuum events versus target from Ref. 14. The vertical bar with arrows indicates the size of the effect seen in Ref. 15.

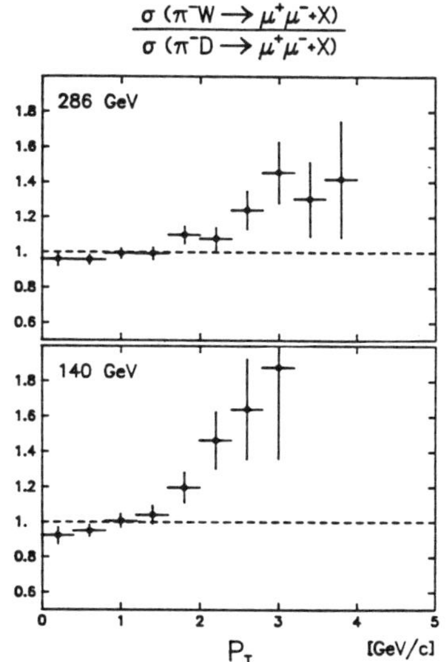

Fig. 19. Nuclear dependence of Drell-Yan continuum events versus p_T from Ref. 15.

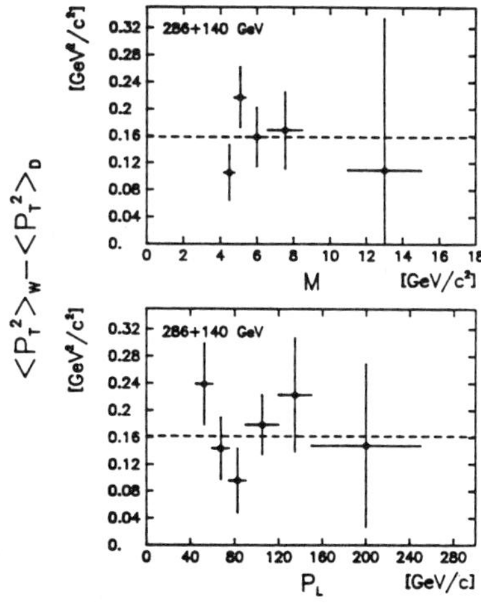

Fig. 20. Difference in $< p_T^2 >$ from W and D targets for Drell-Yan continuum events from Ref. 15.

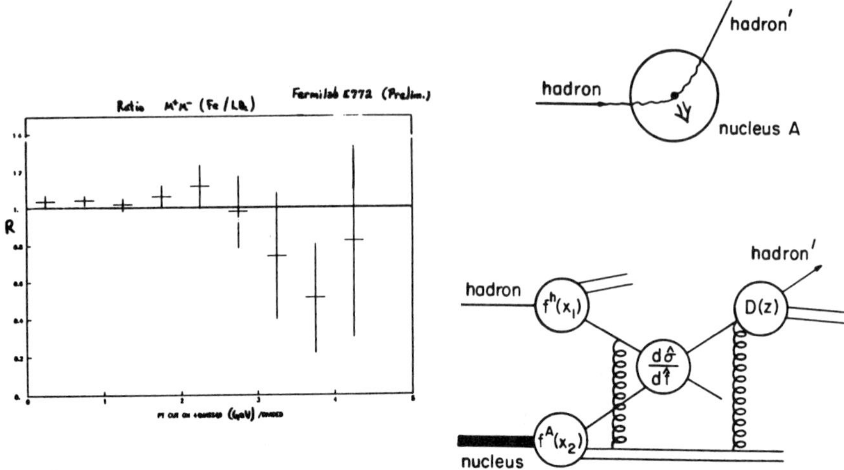

Fig. 21. Very preliminary ratio of Drell-Yan continuum cross sections from Fe and D targets versus p_T from Ref. 16.

Fig. 22. High p_T hadron production with initial and final state interactions of the active partons with spectator partons in the nucleus.

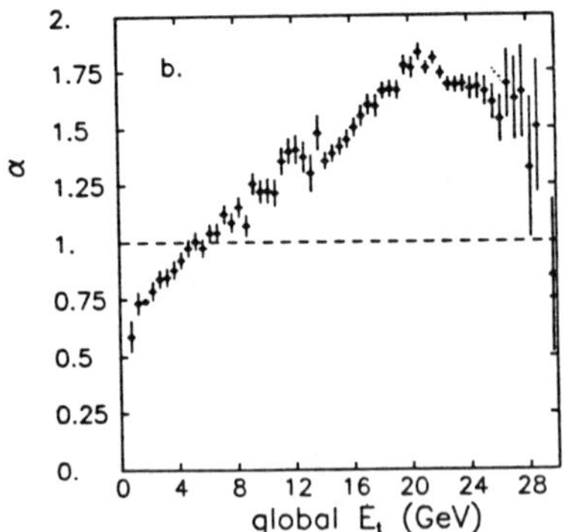

Fig. 23. Nuclear dependence of "jet" cross sections from Ref. 17.

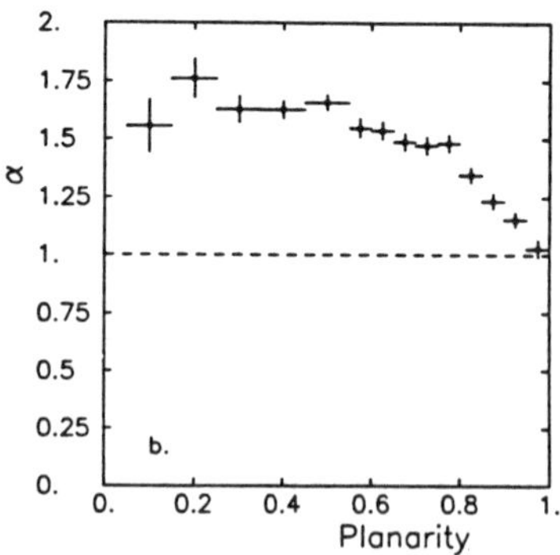

Fig. 24. Nuclear dependence of the "jet" topology from Ref. 17.

data can be found elsewhere in these proceedings.[18] Fermilab experiment E557/672 identifies high p_T single hadrons as well as jets. This data,[19] shown in Figs. 25, is consistent with other data demonstrating a nuclear dependence to the cross section. Recent, unpublished data[20] is shown in Fig. 26, 27, and 28 along with older published data.[1,21] Several features in this data are worth noting. The region where the cross section is large, and the interactions are soft, exhibits a nuclear effect with $\alpha < 1$ as expected. The "Cronin effect" starts at around $p_T = 2GeV$ where $\alpha > 1$ and can be explained by some sort of parton multiple scattering mechanism.[22]. At larger values of p_T the nuclear effect seems to go away. While this is expected, it is a bit surprising that it dies away so rapidly. It is likely that some energy loss mechanism will be required to fit this quantitatively. Fig. 28 suggests that the data at different energies also behaves as expected and that the apparent discrepancies in Figs. 26 and 27 are actually due to the fact that the data were taken at different energies. The reason the values of α are higher for protons, anti-protons, and K^- in the final state is that these cross sections fall more steeply than for the pions and K^+. The same amount of multiple scattering folded with these cross sections would be expected to modify the more steeply falling cross sections by a larger amount.

Hadron pair production by a hadron beam can provide additional constraints to any phenomenological model. The process is depicted in Fig. 1d and in Fig. 29. The data exist in a kinematic regime where the two final state hadrons come off approximately back to back in ϕ with roughly equal values of $|p_T|$. When p_L is about the same for each hadron, then the invariant mass of the pair is given approximately by $m \approx |p_{T1}| + |p_{T2}|$. Some references use this approximation. Fig. 30 shows a compendium of data on this process.[20,21,23,24,25] If we ignore the data of Ref. 24 then we are tempted to conclude that any multiple scattering must be in the initial state and not in the final state since the invariant mass can only be modified by final state interactions. The data at the highest invariant masses from Ref. 20 that go below $\alpha = 1$ can be explained by an energy loss mechanism.

More information is available in this data. Fig. 31 shows the center-of-mass kinematics for the reaction and defines the variable p_{out} which is almost one component of p_T of the pair. As such it might be compared with the p_T distributions in the Drell-Yan process. Fig. 32 shows a clear nuclear dependence in this data suggesting a multiple scattering mechanism.

CONCLUSIONS

It was hoped that all the data shown here would bear on the same problem and would show a certain consistency. Multiple scattering of the same order of magnitude and no energy loss was expected for each of the four hard processes examined. That this hope failed may be due to the imperfect picture used here in an attempt to make qualitative predictions or, possibly, to inaccurate data, some of which is very preliminary. The following table is a score card for the four hard processes, comparing predictions and observations for multiple scattering and energy loss. Question marks

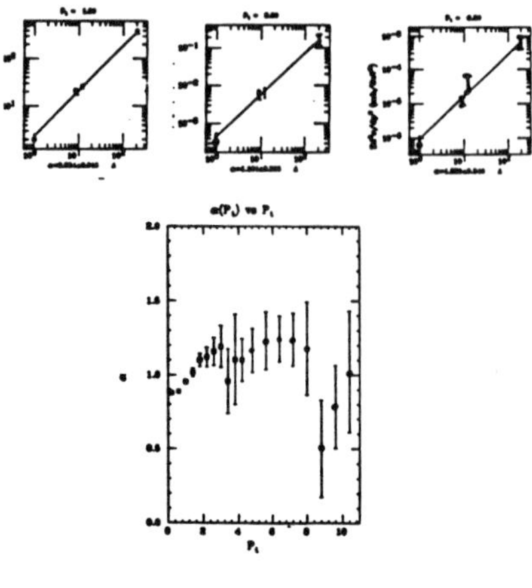

Fig. 25. Nuclear dependence of high p_T single hadrons from Ref. 19.

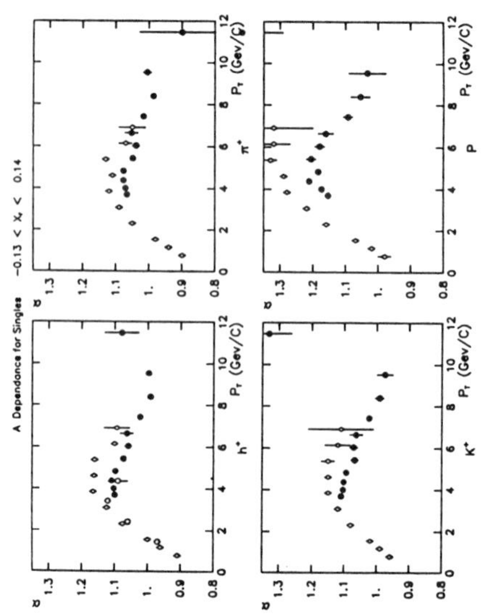

Fig. 26. Nuclear dependence of high p_T positive hadrons. Open diamonds are from Ref. 1, open circles from Ref. 21, and solid circles from Ref. 20.

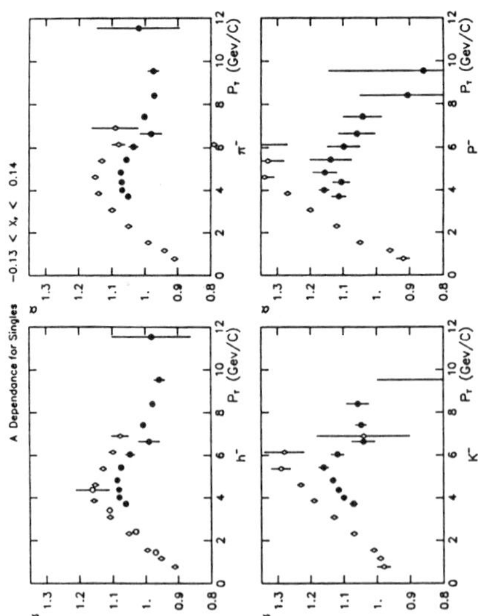

Fig. 27. Same as Fig. 26 for negative hadrons.

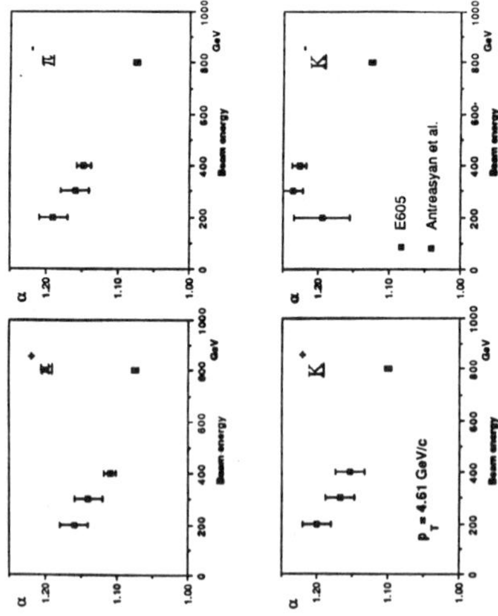

Fig. 28. Energy dependence of the nuclear dependence of high p_T hadrons. The data at 200, 300, and 400 GeV are from Ref. 1 while the 800 GeV data are from Ref. 20.

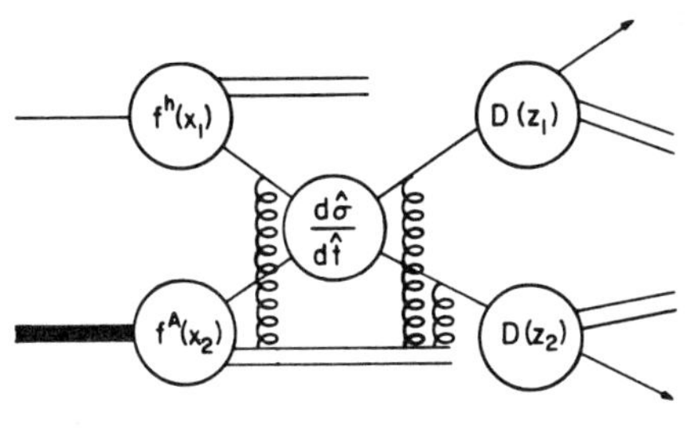

Fig. 29. Production of pairs of hadrons at high p_T with initial and final state interactions of the active partons with spectator partons in the nucleus.

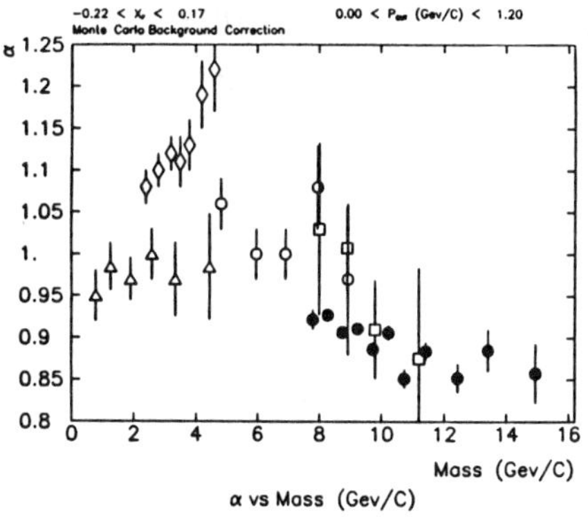

Fig. 30. Nuclear dependence of hadron pair production versus the invariant mass of the pair. The triangles are from Ref. 25, the diamonds from Ref. 24, the open circles from Ref. 21, the squares from Ref. 23, and the solid circles from Ref. 20.

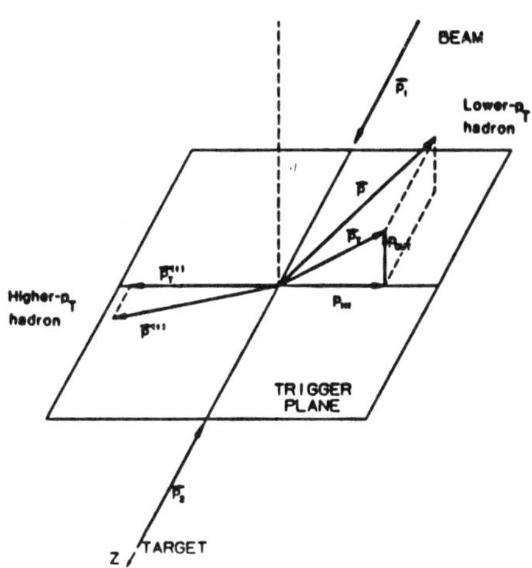

Fig. 31. Definition of the vector p_{out}.

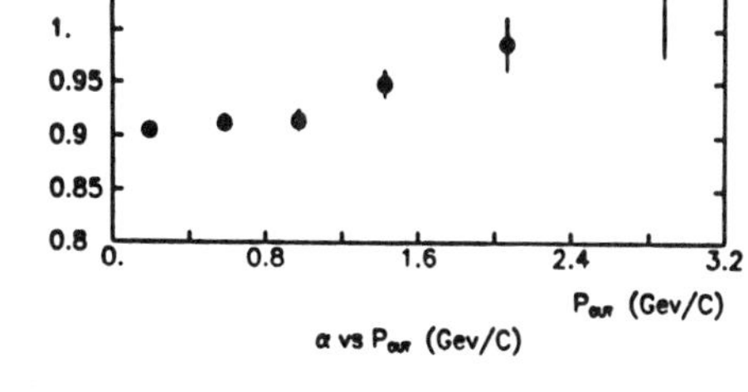

Fig. 32. Nuclear dependence of hadron pair production versus p_{out} from Ref. 20.

indicate either that the data doesn't clearly show the sought effect or that I lack confidence in the prediction. The one entry "Yes/No" means that two experiments are in apparent disagreement.

	Mult. Scat.		Energy Loss	
	Pred.	Obs.	Pred.	Obs
Lepto-prod. of hadrons	Yes	No	No?	No
Drell-Yan	Yes	Yes/No	No	No
High p_T singles	Yes	Yes	No??	Maybe
High p_T pairs	Yes	Yes	No??	Maybe

ACKNOWLEDGMENTS

I'd like to thank Frank Close, Steve Ellis, Sherman Frankel, Gerry Miller, and Al Mueller, for helpful discussions.

REFERENCES

[1]J. W. Cronin *et al.*, Phys. Rev. **D11**, 3105 (1975); L. Kluberg *et al.*, Phys. Rev. Lett. **38**, 670 (1977); D. Antreasyan *et al.*, Phys. Rev. **D19**, 764 (1979).

[2]A more subtle aspect dubbed the "EMC effect" is likely due to a modification of the nucleon's parton distribution functions when the nucleon is bound in a nucleus and/or to Fermi motion. This important area of nuclear physics is explored in other papers at this conference. Because it is a rather small effect compared to those considered here, it will be ignored. See for instance F. E. Close, R. G. Roberts, and G. G. Ross, Z. Phys. **C26**, 515 (1985).

[3]J. D. Bjorken in "Current Induced Reactions" International Summer Institute on Theoretical Particle Physics in Hamburg, 1975, Ed. J. G. Körner, G. Kramer, and D. Schildknecht, Springer-Verlag, Berlin, Heidelberg, New York, 1976.

[4]J. E. Elias *et al.*, Phys. Rev. **D22**, 13 (1980).

[5]Wit Busza, Nucl. Phys. **A418**, 635c (1984).

[6]D. Cheney *et al.*, Phys. Rev. Lett. **40**, 71 (1978).

[7]B. L. Combridge, J. Kripfganz, and J. Ranft, Phys. Lett. **70B**, 234 (1977). John Babcock, Evelyn Monsay, and Dennis Sivers, Phys. Rev. Lett. **40**, 1161 (1978).

[8]Geoffrey T. Bodwin, Stanley J. Brodsky, and G. Peter Lepage, Phys. Rev. Lett. **47**, 1799 (1981). Note that some conclusions drawn in this paper require additional qualifications.

[9]See for example J. H. Kühn, Phys. Rev. **D13**, 2948 (1976).

[10]A. Arvidson *et al.*, EMC Group, Nucl. Phys. **B246**, 381 (1984).

[11]L. S. Osborne *et al.*, Phys. Rev. Lett. **40**, 1624 (1978).

[12]Peter Renton, EMC collaboration, in these proceedings.
[13]A. S. Ito *et al.*, Phys. Rev. **D23**, 604 (1981).
[14]M. L. Swartz *et al.*, Phys Rev. Lett. **53**, 32 (1984).
[15]P. Bordalo *et al.*, Phys. Lett. **193B**, 373 (1987).
[16]P. McGaughey, Fermilab E772 collaboration, private communication and in these proceedings.
[17]R. Gomez *et al.*, Phys. Rev. **D35**, 2736 (1987).
[18]A. Zieminski, Fermilab E557/672 collaboration, in these proceedings.
[19]A. Sambamurti, Fermilab E557/672 collaboration, in these proceedings.
[20]P. B. Straub, Fermilab E605 collaboration, private communication; J. A. Crittenden, Proc. 22nd Recontre de Moriond, Vol. 2, 529 (1987).
[21]H. Jöstlein *et al.*, Phys. Rev. **D20**, 53 (1979); R. McCarthy *et al.*, Phys. Rev. Lett. *40*, 213 (1978).
[22]A more recent phenomenological calculation can be found in M. Lev and B. Petersson, Z. Phys. **C21**, 155 (1983). See also references therein.
[23]Y. B. Hsiung *et al.*, Phys. Rev. Lett. **55**, 457 (1985).
[24]D. Finley *et al.*, Phys. Rev. Lett. **42**, 1031 (1979).
[25]V. Abramov *et al.*, JETP Lett. **38**, 352 (1983).

HIGH-ENERGY HADRON SCATTERING

Giuseppe Fidecaro
European Organization for Nuclear Research, Geneva, Switzerland

ABSTRACT

Open questions in pp and p$\bar{p}$ elastic scattering at high energy, and in pp spin physics are discussed, following a brief review of the two subjects.

INTRODUCTION

It would be hard to summarize here the wealth of information which is by now available on high-energy hadron scattering. I shall rather concentrate on the recent measurement of the real part of the p$\bar{p}$ elastic scattering amplitude at the CERN p$\bar{p}$ Collider. I shall also review what has recently been achieved at the AGS, in a rather different kinematical region, when tagging the spin of the protons.

While results on spin effects have been discussed in the parallel sessions of this conference, the first topic was one of the highlights of the Conference on Elastic and Diffractive Scattering, held last October at the Rockefeller University[1]. There Giorgio Matthiae[2] reported on the measurements performed at $\sqrt{s} = 546$ GeV by the UA4 Collaboration[3] to determine the ratio of the real to imaginary part of the forward p$\bar{p}$ elastic scattering amplitude. The value obtained, much larger than expected, raised a wave of interest because of the possible implications should the UA4 figure be confirmed by other experiments. I shall try to summarize here these implications. Before dealing with the main subject I shall also recall a few points about pp and p$\bar{p}$ elastic collisions at high energy. I shall mention the trend of the data as energy increases, the experimental methods specific to colliding-beam machines, and the bounds set by scattering theory. More details can be found in the review paper by Castaldi and Sanguinetti[4] and in the references quoted there.

A SHORT OVERVIEW OF HIGH-ENERGY pp and p$\bar{p}$ SCATTERING

At very low momentum transfer $|t|$ the differential elastic cross-section $d\sigma/dt$ is dominated by the Coulomb interaction. At large values of $|t|$ it is the hadronic amplitude which prevails. The maximum interference between the two amplitudes takes place at $|t| = 0.0017$ GeV^2 for the centre-of-mass energy $\sqrt{s} = 24.3$ GeV and at $|t| = 0.0011$ GeV^2 for $\sqrt{s} = 546$ GeV. The measurement of the real part of the forward hadronic scattering amplitude concerns this region. I shall come back to this point when discussing the UA4 experiment.

Figure 1a shows the behaviour of $d\sigma/dt$ as a function of $-t$ in the region where the hadronic interaction is dominant. The differential cross-section first decreases rapidly (diffraction peak) following approximately an exponential law $d\sigma/dt \sim \exp(Bt)$ through several orders of magnitude. To be more precise, the local (logarithmic) slope

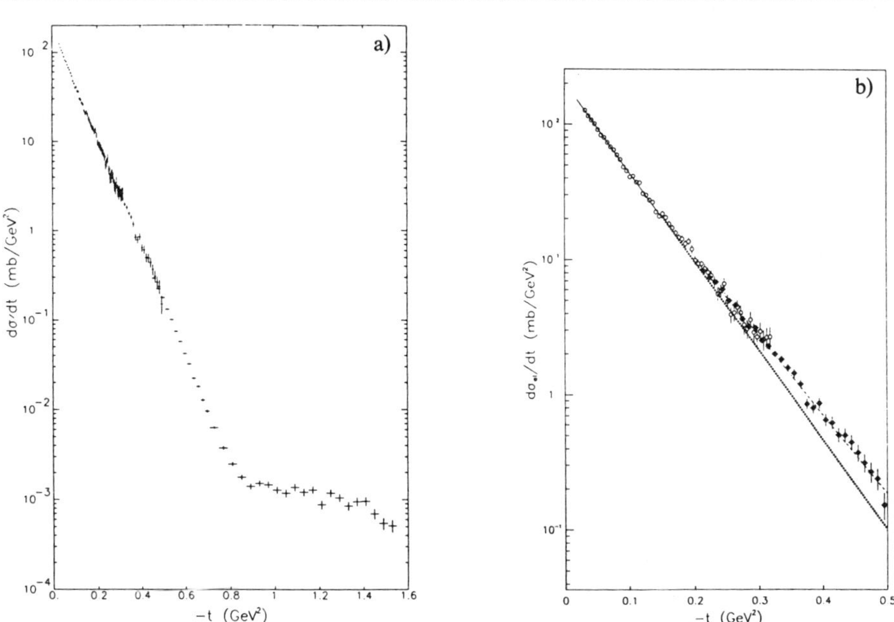

Figure 1: dσ(p$\bar{p}$)/dt at $\sqrt{s}$ = 546 GeV: a)[5] for |t|≤1.55 GeV^2; b)[6] for |t| < 0.5 GeV^2.

$B(t) = (d/dt)\ln(d\sigma/dt)$ varies somewhat: leaving aside the lower energies, where the presence of a small continuous variation of the slope may require a t^2 correction term in the exponent, a sudden change of slope is observed at the CERN ISR and p$\bar{p}$ Collider energies, respectively around $|t| = 0.1$ GeV^2 and $|t| = 0.14$ GeV^2. The existence of such an abrupt change is visible in Fig. 1b and in Fig. 2. Figure 3 shows a comparison of slopes for two different values of t, for pp and p$\bar{p}$ elastic interactions, as a function of the beam momentum. The different behaviour of protons and antiprotons is well seen: the pp diffraction peak keeps 'shrinking' while the energy increases; the p$\bar{p}$ peak, after the initial phase of 'anti-shrinking', shows the same behaviour as the pp peak, and at ISR energies the pp and p$\bar{p}$ experimental points fall indeed very close together. A new measurement of the slope at Fermilab[8] has given, at $\sqrt{s}$ = 1.8 TeV, $B = 17.2 \pm 1.3$ GeV^{-2} in the interval $0.025 < |t| < 0.13$ GeV^2. This measurement confirms that shrinking continues at higher energy in the case of p$\bar{p}$ collisions.

For large values of $|t|$ the elastic pp differential cross-section is characterized by the development of a dip when going from 24 to 200 GeV/c, as shown in Fig. 4a. At 200 GeV/c the dip is well visible, at $|t| \approx 1.5$ GeV^2. At ISR energies the dip goes on shrinking with increasing energy, while becoming less pronounced, as shown in Fig. 4b. In the case of p$\bar{p}$ collisions the dip is developed already at 50 GeV/c, as shown in Fig. 5a. However, it fills progressively going

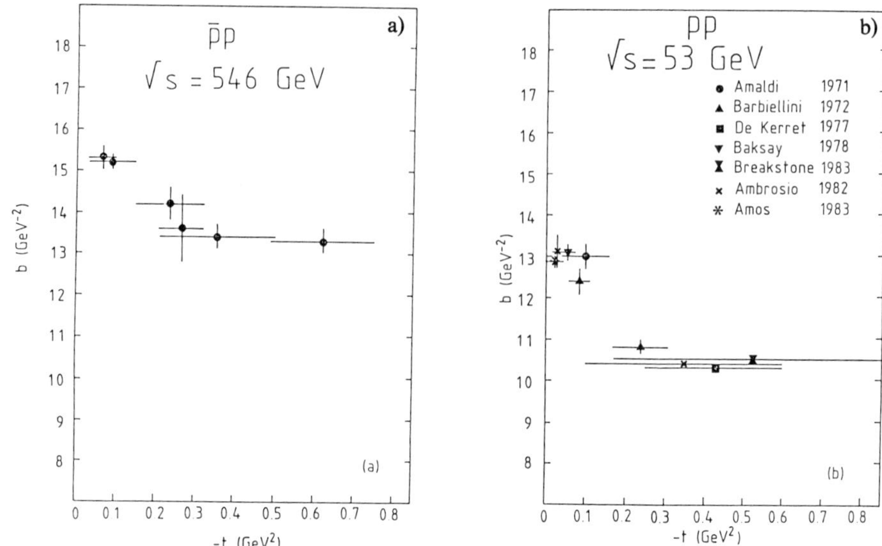

Figure 2: The slope parameter B as a function of the momentum transfer[7]. The horizontal bars indicate the t-interval in which the fit was performed.

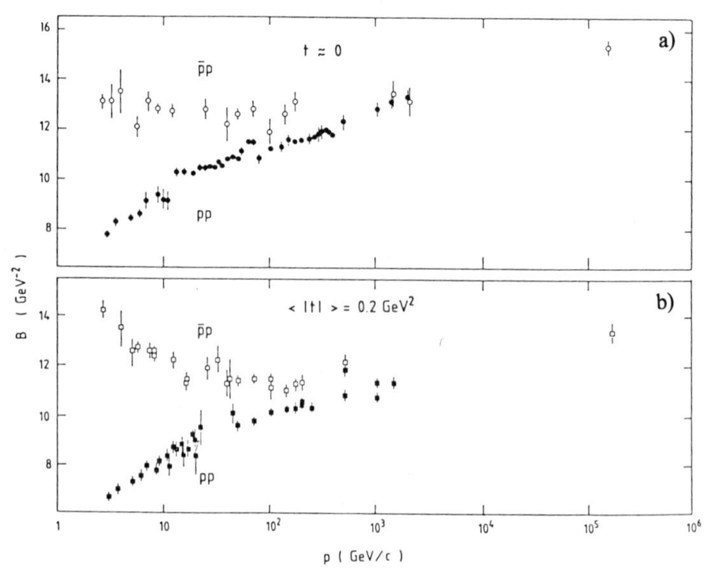

Figure 3: The slope parameter B as a function of beam momentum[4].

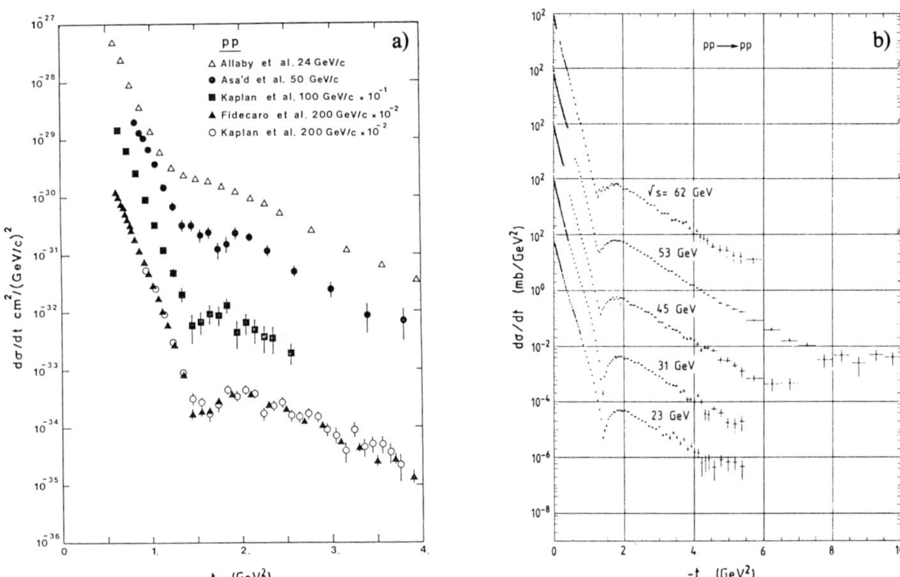

Figure 4: Development of the dip with energy in pp scattering: a) fixed-target[9]; b) ISR[4].

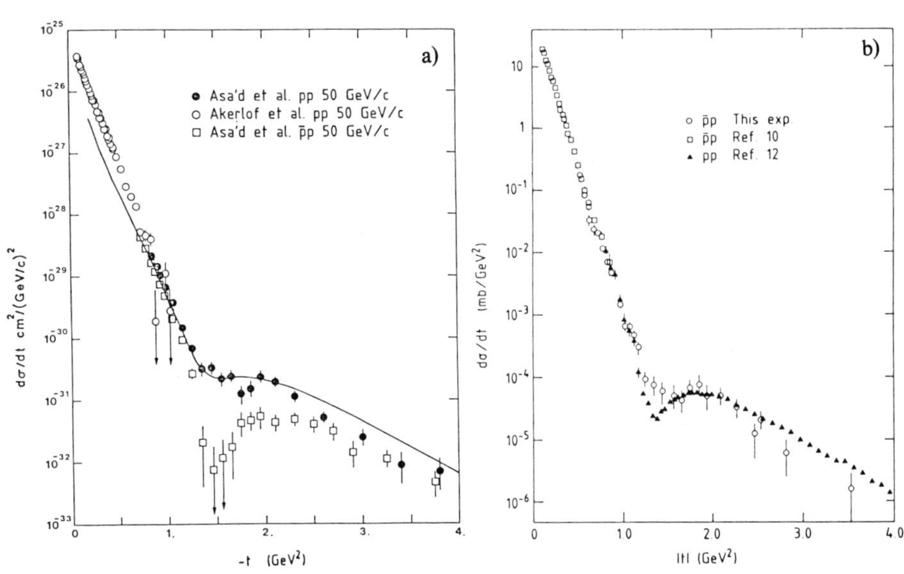

Figure 5: Comparison of the pp and p$\bar{p}$ dip: a) fixed-target[9]; b) ISR[10].

to the ISR and Collider energies, as shown in Fig. 5b and Fig. 1a respectively. Fig. 5b also shows that the shoulders for pp and p$\bar{p}$ are comparable at ISR energies (nearly equal total cross-sections). A comparison with Fig. 1 indicates that at the Collider the p$\bar{p}$ shoulder is much higher, possibly because of the larger total cross-section.

Beyond the dip region the cross-section continues its exponential fall with a much smaller parameter, roughly equal to 2 GeV^{-2}. No secondary minimum in the cross-section has been observed up to $|t| = 10$ GeV^2 at the ISR[11] ($\sqrt{s} = 53$ GeV), and up to $|t| = 14$ GeV^2 at Fermilab[12] (400 GeV/c).

The ISR data had been interpreted in the framework of the geometrical scaling hypothesis, which implies that $(1/\sigma_{tot})d\sigma/d\tau$ should be a universal function of the variable $\tau = t\sigma_{tot}$. This is quantitatively true, except in the region of the dip, when comparing pp cross-sections at $\sqrt{s} = 62$ GeV and $\sqrt{s} = 23$ GeV, as shown in Fig. 6. The filling of the dip should be related to the ratio ρ of the real to the imaginary part of the scattering amplitude. However, the scaling seems to fail when extended to the p$\bar{p}$ Collider data: the ratio σ_{el}/σ_{tot} (Fig. 7a) increases at the Collider, indicating that the opacity is no longer constant. Clearly the asymptotic region has not yet been reached; this is also proved by the fact that the cross-sections σ_{el} (Fig. 7b) and σ_{tot} (Fig. 8) keep increasing well beyond the ISR energies.

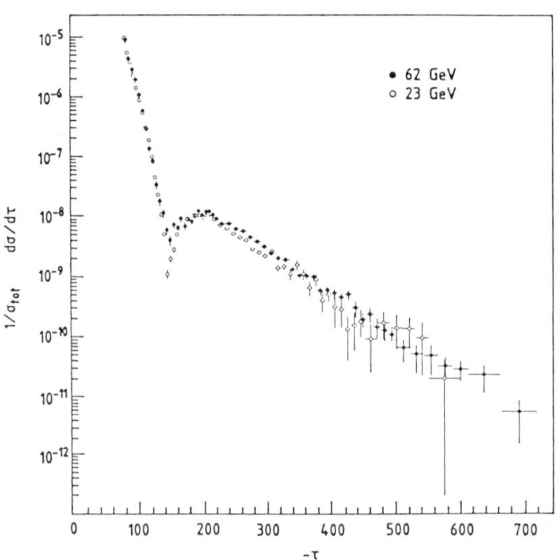

Figure 6: Geometrical scaling at ISR energies for pp elastic scattering[11].

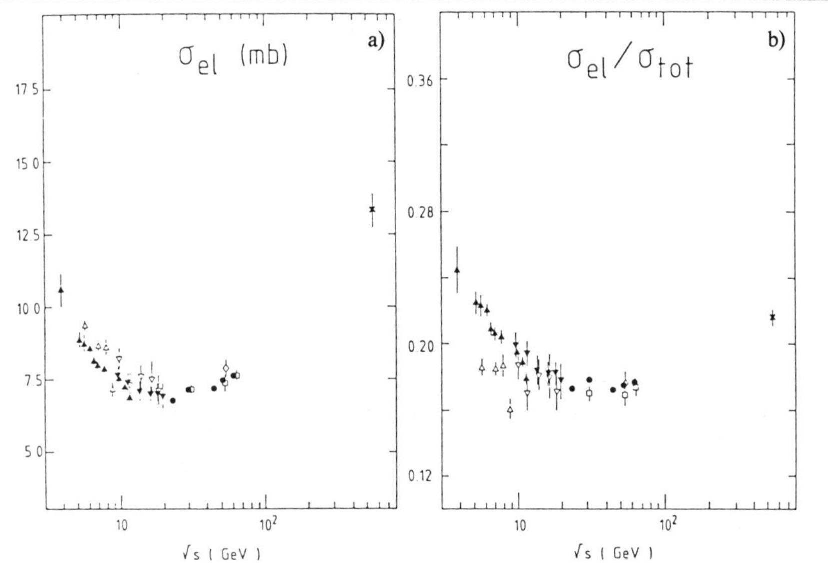

Figure 7: Energy dependence of (a) σ_{el} and (b) σ_{el}/σ_{tot} for pp and $p\bar{p}$ collisions[7].

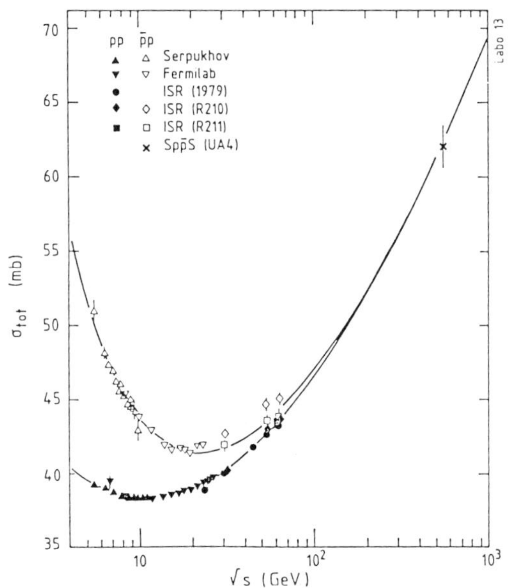

Figure 8: Energy dependence of σ_{tot} for pp and $p\bar{p}$ collisions[4].

EXPERIMENTAL METHODS AT THE COLLIDERS

The methods used to measure differential and total cross-sections at colliding-beam machines are clearly different from those used in fixed-target experiments; it required great skill and ingenuity to develop these methods. Colliding beams of equal energy only are considered here.

In the case of $p\bar{p}$ colliders the crossing beams are collinear, and the elastic events are selected by requiring collinearity of the scattered particles on the two sides of the crossing region (in the case of the ISR the colliding beams meet at an angle, and the angular correlation required for the scatterd particles is not so straightforward). For very small scattering angles the telescopes are normally placed several metres away from the crossing region, and mounted inside the vacuum chamber very close to the circulating beams. The trajectories of the colliding particles have to be controlled, the more accurately the smaller the scattering angle, and this requires the use of so-called high-β insertions in the crossing region. If the scattered particles traverse lattice elements of the machine and are deviated by the magnetic fields, corrections have to be made to their trajectories. All this will be discussed in more detail when the UA4 determination of the real part of the forward scattering amplitude is described.

For the total cross-sections, I shall not consider methods requiring absolute luminosity measurements, and shall only describe the method used by the UA4 Collaboration[13] (previously developed at the ISR[14]). According to the optical theorem, the total cross-section can be written, in terms of the forward differential elastic rate:

$$\sigma_{tot}^2 = \frac{1}{L}\,\frac{16\pi(\hbar c)^2}{1+\rho^2}\cdot\frac{dN_{el}}{dt}\bigg|_{t=0} \tag{1}$$

where ρ is the ratio of the real to the imaginary part of the forward elastic scattering amplitude and L the luminosity. Also,

$$\sigma_{tot} = (N_{el}+N_{inel})/L \tag{2}$$

where N_{el} and N_{inel} are the elastic and inelastic rates respectively. Combining Eqs. (1) and (2) one eliminates the luminosity L, and

$$\sigma_{tot} = \frac{16\pi(\hbar c)^2}{1+\rho^2}\,\frac{dN_{el}/dt\big|_{t=0}}{N_{el}+N_{inel}}, \tag{3}$$

The term ρ^2 represents a small correction and can be obtained from a fit simultaneously performed on total cross-section and ρ data at smaller energies using dispersion relations.

The ratio ρ of the real to the imaginary part is obtained by measuring $d\sigma/dt$ in the region where the known Coulomb scattering amplitude interferes with the hadronic amplitude, as is shown below.

The differential cross-section of elastic scattering is:

$$d\sigma/dt = \pi\left|f_c + f_h\right|^2. \tag{4}$$

Neglecting spin effects, and assuming that the real and the imaginary parts have the same exponential $|t|$ dependence, the hadronic amplitude f_h at low momentum transfer can be parametrized as

$$f_h = (\sigma_{tot}/4\pi)\,(\rho + \mathrm{i})\exp(-B\,|t|/2), \tag{5}$$

where the imaginary part of $f_h(t=0)$ has been replaced by $\sigma_{tot}/4\pi$ (optical theorem). The Coulomb amplitude can be written as

$$f_c = \pm 2\alpha G^2(t)\exp(\mp \mathrm{i}\alpha\phi)/|t|, \tag{6}$$

where α is the fine structure constant, and $G(t)$ the proton electromagnetic form factor. The Coulomb phase is $\phi = \ln(0.08/|t|) - 0.577$. The upper and lower signs in Eq. (6) refer to $p\bar{p}$ and pp respectively.

The ratio ρ is then determined by fitting the experimental data to Eq. (4).

ASYMPTOTIC CONSTRAINTS

The optical theorem relates the total cross-section σ_{tot} to the imaginary part of the forward elastic scattering amplitude. This theorem, first used to study very low energy nuclear reactions, is also valid at very high energy, in spite of the large number of new channels which open up. I should like to stress here that the application of the general principles of scattering theory (unitarity, analyticity, crossing symmetry) can give interesting information on the behaviour of the total cross-sections for $s \to \infty$. Before the ISR were commissioned, it was believed that the total pp and $p\bar{p}$ cross-sections had reached their asymptotic limit. It is now evident that this was not the case, and the asymptotic theorems are still very useful to help in understanding how far the high-energy limit could be.

The Froissart-Martin upper bound, derived from axiomatic field theory, states that $\sigma_{tot} \geq (\pi/m_\pi^2)\ln^2 s$. The available data from the ISR and the $p\bar{p}$ Collider indicate an increase of σ_{tot} compatible with $\ln^2 s$, but with a smaller coefficient. However, σ_{el}/σ_{tot} still increases beyond ISR energies, and the present $\ln^2 s$ behaviour could turn out to be just meaningless.

According to the present-day reformulation of the Pomeranchuk theorem (originally based on dispersion relations for the forward scattering amplitude) $\sigma_{tot}(p\bar{p})/\sigma_{tot}(pp) \to 1$, if either $\sigma_{tot}(p\bar{p}) \to \infty$ or $\sigma_{tot}(pp) \to \infty$. This does not imply that $\Delta\sigma = \sigma_{tot}(p\bar{p}) - \sigma_{tot}(pp)$ should go to zero. However, if the real and the imaginary part of the odd-signature amplitude $F^- = F^{(\bar{p}p)} - F^{(pp)}$ have asymptotically the same sign, as seems to be the case at ISR energies, then $\Delta\sigma$ should go to zero (Fischer theorem).

Other constraints have been worked out, under various hypotheses (Cornille-Martin, MacDowell-Martin, Auberson-Kinoshita-Martin). It is not the purpose of this talk either to review all the work done in the field of asymptotic theorems, or to discuss the various models (asymptotic scaling, eikonal models, etc.). For this I refer to the review paper[4] already mentioned.

THE REAL PART OF THE $p\bar{p}$ SCATTERING AMPLITUDE AT THE CERN COLLIDER[3]

The interest of measuring the real part of the forward scattering amplitude at a given energy resides in the fact that by fitting the experimental values to dispersion relations, one can get predictions of the behaviour of the total cross-section at much larger energy. For example, the total cross-section determined at the CERN $p\bar{p}$ Collider ($\sqrt{s} = 546$ GeV) had been predicted fairly accurately by measurements of the real part of the forward scattering amplitude done ten years earlier at a centre-of-mass energy an order of magnitude smaller (ISR energies); see Fig. 8. Thus it is believed that a measurement of the real part of the forward scattering amplitude at $\sqrt{s} = 546$ GeV can give information on the behaviour of the total cross-sections at LHC and SSC energies. For the measurement of ρ the UA4 Collaboration used the method sketched above.

At the energy of the CERN Collider the hadronic and the Coulomb amplitudes become equal when the scattering angle $\theta \approx 120$ µrad ($|t| \approx 1.1\times10^{-3}$ GeV^2). This angle is about five times smaller than the smallest angle previously reached by UA4 in their measurements of $d\sigma/dt$, and this is an indication of the difficulty of the experiment. The minimum observable angle θ_{min} depends on the betatron amplitude function at the crossing point, as is shown below.

The transverse coordinate y_1 of a particle scattered by an angle θ, at the location of the detector, is:

$$y_1 = \theta \sqrt{\beta^* \beta_1} \sin \Delta\mu, \tag{7}$$

where β^* and β_1 are the betatron amplitude functions in the measuring plane at the crossing point and at the detector respectively. $\Delta\mu$ is the betatron phase shift between these points. Except for the case $\Delta\mu = 90°$, the dependence of y_1 on the transverse coordinate of the particle at the crossing point is negligible. The minimum distance from the beam y_{min}, at which one can place the so-called Roman pots containing the detectors, defines θ_{min}. This distance y_{min} is proportional to the r.m.s. size of the beam σ_1 at the detector:

$$y_{min} = K\sigma_1 = \frac{K}{2} \sqrt{\varepsilon\beta_1}, \tag{8}$$

where ε is the beam emittance, and the factor K was always found to be about 14 at the CERN Collider, no matter the optics used and the detector position. From Eqs. (7) and (8) one gets

$$\theta_{\min} = \frac{K\sqrt{\varepsilon}}{2\sqrt{\beta^*}\sin\Delta\mu}, \tag{9}$$

which does not depend on β_1. Eq. (9) shows that the smaller $\theta_{\min}$, the larger the β-function has to be at the crossing point. The new very high β insertion gave a β-function ten times larger than that previously used by UA4, with $\beta_h = 1086$ m and $\beta_v = 22$ m in the horizontal and the vertical plane, respectively.

The experimental set-up consisted of a system of telescopes placed in the Collider plane symmetrically on both sides of the vacuum pipe, close to the nearest minimum of β_h. On the incoming proton side each telescope consisted of two Roman pots 60 cm apart, 43 m away from the crossing region. On the incoming antiproton side the two telescopea consisted of a single Roman pot, 94 m away from the crossing, because of lack of space. Since the scattered particles traversed quadrupoles and a bending magnet (the latter on the antiproton side) their trajectories had to be calculated using standard transfer matrix formalism. The effective distances of the detectors from the crossing point in the two planes turned out to be: $L_h = 58$ m and $L_v = 27.5$ m (proton side); $L_h = 127$ m and $L_v = 15.7$ m (antiproton side).

Each Roman pot contained a trigger counter, a hodoscope with eight horizontal scintillation counters, and a chamber with wires in the vertical direction. The chamber had four drift planes to measure the horizontal coordinate (130 µm resolution), and a proportional plane to measure the vertical coordinate by charge division (0.4 mm resolution). With respect to the beam axis the inner edges of the Roman pots were at ~ 8 mm (p-side) and at ~ 12 mm ($\bar{p}$-side), corresponding to an angle of 140 µrad ($|t| = 1.5\times10^{-3}$ GeV^2).

The experiment lasted a few days. Position and emittance of the beams were regularly monitored. Typical r.m.s. values of the transverse dimensions of the proton bunches were 4.6 mm and 0.7 mm in the horizontal and the vertical plane, respectively, corresponding to an angular spread of 4.3 µrad and 30 µrad. The figures for the antiproton bunches were about 20 % smaller. The r.m.s. value of the length of the crossing region was 18 cm. The average luminosity was 2×10^{26} cm^{-2} s^{-1}, resulting in a trigger rate of 5–10 Hz. About 8.5×10^5 triggers were collected.

In order to select events which originated in the interaction region a cut was made on the horizontal coordinate $x_0(\bar{p})$ of the antiproton trajectory at the centre of the crossing region. The lever arm of the scattered antiproton telescopes allowed a determination of this coordinate with an accuracy of about 1 cm. A similar cut could not be made for the proton trajectory as the scattered proton telescope consisted of only one Roman pot.

Two more cuts were made on the collinearity distribution in the vertical and in the horizontal plane. The collinearity was defined by the quantities $\Delta\theta_v = y(\bar{p})/L_v(\bar{p}) - y(p)/L_v(p)$, and $\Delta\theta_h = x(\bar{p})/L_h(\bar{p}) - x(p)/L_h(p)$, computed separately for each set of telescopes. The r.m.s. values of the $\Delta\theta_v$ and $\Delta\theta_h$ distributions are about 55 µrad and 8 µrad, respectively. The distributions are not centred around zero because of the transverse displacement of the beam. This was determined to be 6 mm horizontally and 1.6 mm vertically from the distribution of the intercept at the crossing point of the reconstructed line of the collinearly scattered particles. The position along the beam of the crossing region was determined from the time-of-flight difference of the two scattered particles, with an accuracy of about 3 cm.

After these cuts were made, at 3.5 standard deviations, about 10^5 events were left. The results of a Monte Carlo study were then used to obtain the t-distribution, corrected for geometrical acceptance, detector efficiency, and experimental t-resolution. About 7×10^4 events survived in the interval between $|t|_{min} = 0.002$ GeV^2 ($\theta = 165$ µrad) and $|t|_{max} = 0.035$ GeV^2. The t-resolution, mainly determined by the angular spread of the beam, increases from about 5×10^{-4} GeV^2 at $|t|_{min}$ to about 4×10^{-3} GeV^2 at $|t|_{max}$.

The angle of the beam with respect to the nominal machine axis, required to obtain the scattering angle, was determined from the asymmetry of the collinear track distribution in the horizontal and the vertical plane, to be $\alpha_h = 11\pm 8$ µrad and $\alpha_v = 90\pm 10$ µrad, respectively. The correction for the displacement of the vertical direction was not important. The background, obtained from the tail of the $\Delta\theta_h$ distributions, resulted in an average correction of about 0.5%, rising to 1.5% for very low values of t.

The experimental data dN/dt were fitted to the theoretical form of the differential cross-section using the relation $dN/dt = \mathscr{L}\, d\sigma/dt$. $\mathscr{L}$ is the unknown luminosity integrated over the data-taking period; $d\sigma/dt$ is given by Eq. (4). The value $\sigma_{tot}(1+\rho^2) = 63.3\pm 1.5$ mb, previously measured by UA4[13], was always used for the fits.

The quantities B, ρ, and $\mathscr{L}$ were first left as free parameters, and the results $B = 15.5\pm0.8$ GeV^{-2} and $\rho = 0.23\pm0.04$ were obtained from the fit to the whole t-range $0.002<|t|<0.035$ GeV^2. A second fit in the reduced interval $0.006<|t|<0.035\ GeV^{-2}$ (leaving out the lower-t data the correlation between B and ρ is strongly reduced) gave $B = 15.5\pm0.7$ GeV^{-2}. As the value for B obtained from these two fits agreed very well with the value previously measured by UA4 in the interval $0.03<|t|<0.10$ GeV^2, the latter - more accurate - value $B = 15.5\pm0.3$ GeV^{-2} was taken, and used for the final fit with ρ and $\mathscr{L}$ as free parameters. The result was $\rho = 0.238\pm0.024$, the error being statistical. The measured differential cross-sections are shown in Fig. 9a together with the contribution of the Coulomb and of the hadronic terms. The parameter $\mathscr{L}$ determined by the fit fixes the absolute normalization. The ratio $R(t) = [\,(d\sigma/dt)_{meas}/(d\sigma/dt)_{\rho=0} - 1\,]$, showing the effect of the interference term, is plotted in Fig. 9b.

Various systematic uncertainties affect the measurement of ρ, mainly the correlation between ρ, σ_{tot}, and B, and the error on the relative position of the pots (100 µm). The former, given by $\partial\rho/\partial\sigma_{tot} = 0.009$ mb^{-1}, and by $\partial\rho/\partial B = -0.04$ GeV^2, contributes an error $\Delta\rho \simeq 0.02$ from the measurements of σ_{tot} and B. The latter contributes $\Delta\rho \simeq 0.015$. The quadratic combination of these errors gives $\Delta\rho_{syst} \simeq 0.025$. Thus the final result becomes $\rho = 0.24\pm0.04$, plotted in Fig. 9c together with a compilation of results from pp and p$\bar{\text{p}}$ exper-

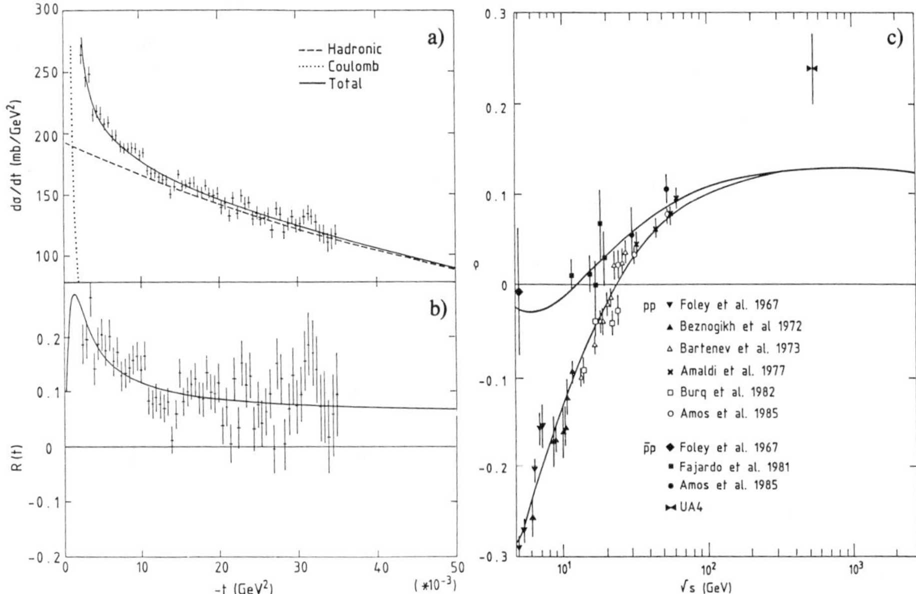

Figure 9: UA4 results[2]. a) the measured dσ/dt. The solid line represents the best fit to the data. The Coulomb and hadronic terms are also shown; b) the ratio R as defined in the text; c) the parameter ρ measured by UA4, shown together with lower energy data for pp and $p\bar{p}$ scattering. The solid line represents the dispersion relation fit not including the UA4 point.

iments at lower energy. With this new value for ρ the total cross-section previously measured by UA4 has to be corrected to become $\sigma_{tot} = 60 \pm 2$ mb.

The present result was derived assuming a constant slope B and a constant ρ in the t-interval considered. Models exist predicting non-negligible changes of the logarithmic slope with t, and the real part might decrease with t faster than the imaginary part. UA4 tested the sensitivity of their result to these effect. If the real and the imaginary part have the same t-dependence, and the slope changes taking the values 15.3, 15.7, 17.2 GeV^{-2} for $|t| = 0.06, 0.04, 0$ GeV^2 respectively, then the value of ρ should be decreased by 0.02. If the slope of the real part is twice as large as that of the imaginary part, then ρ should be decreased by 0.03.

It should be noted that if the total cross-section is parametrized with a leading term proportional to $\ln^2 s$, then the predicted value should be $\rho \simeq 0.15$. If the total cross-section flattens up in the early TeV region, then the predicted value for ρ should be close to 0.1. The UA4 value, being definitely higher than these predictions, by two and three standard deviations respectively, seems therefore to give, if confirmed, a fairly strong indication of an appreciable increase of the cross-section in the energy region beyond the CERN Collider. However, one can note, by an inspection of Fig. 9c, that if all predictions are ignored the UA4 result might look satisfactory to an eye-ball fit. Unfortunately, the UA4 experiment has been dismantled and we have now to wait for the CDF and E-710 experiments at Fermilab ($\sqrt{s} = 1.8$ TeV), to have a confirmation of the CERN result.

Before closing this section I should like to mention that measurements of ρ at the CERN Collider using an internal hydrogen jet target were also reported at the Rockefeller Conference by Breedon[16] (UA6 Collaboration). The jet target was not located in an intersection region, so that the p and p̄ bunches passed at different times: the p̄ bunch arrived 690 ns after the p bunch. The differential cross-section was measured by a set of solid-state detectors, which could be swung from one side of the target to the other side in such a way as to detect pp or pp̄ collisions. The measurements were done in the t-interval $0.001 \leq |t| \leq 0.05$ GeV^2. The preliminary results at $\sqrt{s} = 24.3$ GeV were $\rho_{pp} = 0.014 \pm 0.007 \pm 0.007$ and $\rho_{p\bar{p}} = 0.045 \pm 0.007 \pm 0.007$, in good agreement with previous results. The first error is statistical, the second reflects uncertainties on σ_{tot} and B.

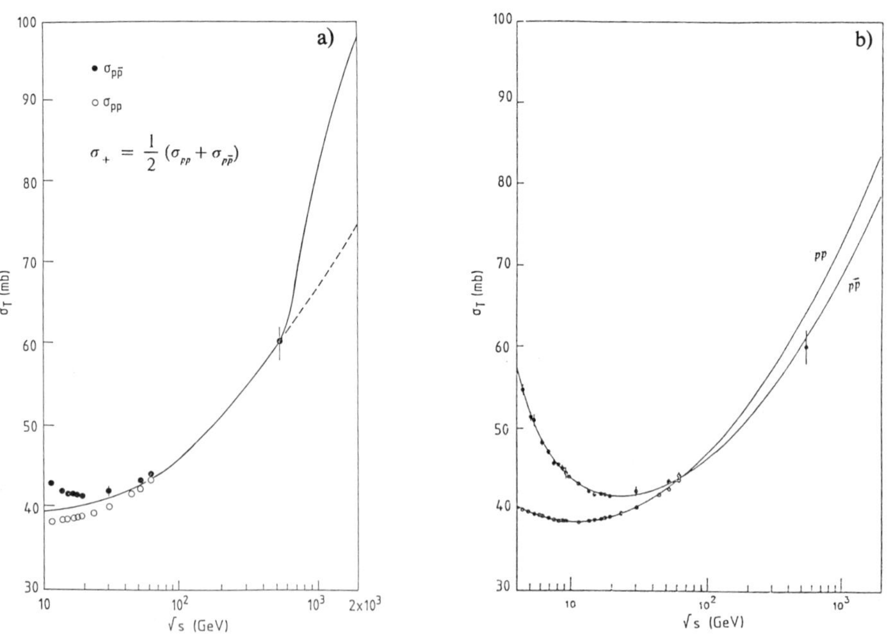

Figure 10: pp and pp̄ total cross-sections predicted by: a) Martin[17]; b) Nicolescu[18].

BEYOND THE ENERGY OF THE CERN COLLIDER

Assuming that the UA4 result is correct, then various theoretical explanations have been considered, also discussed at the Rockefeller Conference[17-19].

i) The total cross-section might rise much faster than $(\ln s)^2$ in the few TeV region, with an unchanged asymptotical behaviour of the pp and p$\bar{\text{p}}$ cross-sections. This could indicate the opening of a new channel in the total cross-section[17]. See Fig. 10a and Fig. 11a. The reasons why such a new threshold should appear are not clear.

ii) The pp and p$\bar{\text{p}}$ total cross-sections could be asymptotically different[18] (odderon hypothesis), crossing each other above the energy region of the ISR. See Fig. 10b and Fig. 11b.

iii) The parametrization of the differential cross-sections could turn out not to be correct. In particular, the slope of the differential cross-section might increase for $t \to 0$. Moreover the slope of the real part might be larger than the slope of the imaginary part.

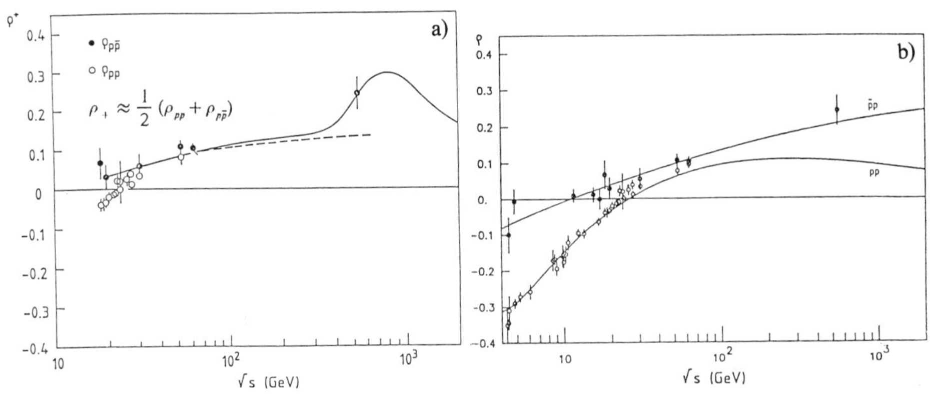

Figure 11: The parameter ρ for pp and p$\bar{\text{p}}$ as predicted by: a) Martin[17]; b) Nicolescu[18].

A discussion of the first two hypotheses was presented by Leader[19] in the Summary talk of the Rockefeller Conference. According to Leader, a measurement of $\sigma_{tot}(p\bar{p})$ at $\sqrt{s} = 1.8$ TeV might help to clarify the situation. He predicts a p$\bar{\text{p}}$ total cross-section of 75 mb, or ≥ 85 mb in the case where the odderon picture, or the new threshold picture, respectively, is correct. A measurement of ρ should help one to choose between the odderon ($\rho \sim 0.24$) and the standard picture ($\rho \leq 0.1$).

More experimental work is now needed: we might be in a situation similar to that existing at the time the ISR operation started, when no surprise was expected.

SPIN EFFECTS

When reviewing the behaviour of high-energy pp and p$\bar{p}$ differential and total cross-sections, it is in general possible to ignore spin interactions, as spin effects tend to be washed out, unless at least one of the interacting particles is in a known spin state. However, experiments immediately become much more difficult, and require more sophisticated equipment.

If a polarized target and a polarized beam are used, the scattering cross-section $\sigma(P_B, P_T)$ can be written[21] as

$$\sigma(P_B, P_T) = \sigma_0 [1 + P_B A + P_T A + P_B P_T A_{nn}], \tag{10}$$

where P_B and P_T are the beam and the target polarization, respectively. Measurements of the analysing power A, using a polarized target and an unpolarized beam, or a polarized beam and an unpolarized target

$$A = \frac{1}{P_T} \frac{\sigma(0\uparrow) - \sigma(0\downarrow)}{\sigma(0\uparrow) + \sigma(0\downarrow)}, \tag{11a}$$

$$A = \frac{1}{P_B} \frac{\sigma(\uparrow 0) - \sigma(\downarrow 0)}{\sigma(\uparrow 0) + \sigma(\downarrow 0)}, \tag{11b}$$

and measurements of the spin-spin correlation parameter A_{nn}

$$A_{nn} = \frac{1}{P_B P_T} \frac{\sigma(\uparrow\uparrow) + \sigma(\downarrow\downarrow) - \sigma(\uparrow\downarrow) - \sigma(\downarrow\uparrow)}{\sigma(\uparrow\uparrow) + \sigma(\downarrow\downarrow) + \sigma(\uparrow\downarrow) + \sigma(\downarrow\uparrow)} \tag{12}$$

have been performed at various laboratories (Argonne, CERN, RAL, Serpukhov), up to about 40 GeV/c. The analysing power A has also been measured in the few-hundred-GeV region at CERN and Fermilab[22-25].

Polarized protons have been accelerated up to 22 GeV/c with 42% polarization at the AGS[26], overcoming as many as 45 strong depolarizing resonances by means of pulsed quadrupoles installed in each of the 12 AGS superperiods. The recent experience with the acceleration of polarized protons at the AGS has been reported in the parallel sessions. At 18.5 GeV/c the maximum reached polarization was 52%, with a peak intensity of 2.5×10^{10} protons per pulse. However, because of the increasing number of depolarizing resonances, the AGS method of tuning out resonances becomes unpractical at much higher energy. The use of Siberian snakes seems much more promising in the multi-hundred-GeV and multi-TeV region, and some tests now in progress at Indiana[27] have been reported at this conference. The status of the 200 GeV/c polarized proton beam (from Λ decay) at Fermilab has also been reported. Here I should like to mention that the setting-up of a polarized proton beam from Λ decay has been envisaged at the future 3 TeV Serpukhov accelerator[28] (UNK) in the momentum interval 500-2600 GeV/c, with an average polarization of 40 %. The expected intensity is 2×10^9 p↑ for 10^{14} incoming protons[29]. Plans also exist for using a polarized proton jet target[30]. Related questions were discussed by Preparata[31].

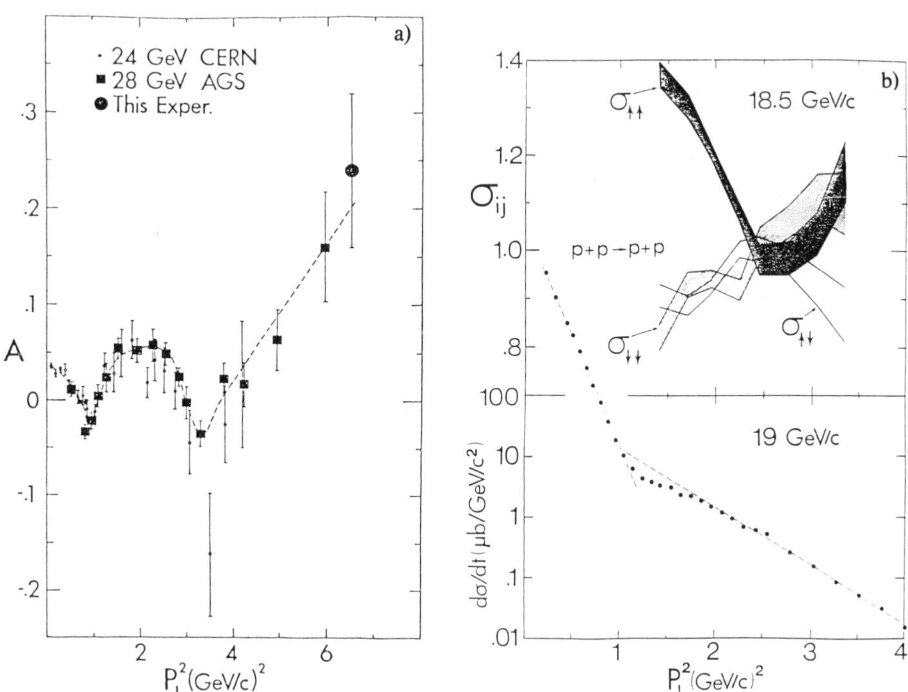

Figure 12: Spin effects as a function of the square momentum transfer: a) the analysing power[32] A; b) the ratio σ_{ij} of the cross-sections $\sigma(\uparrow\uparrow)$, $\sigma(\downarrow\downarrow)$, $\sigma(\uparrow\downarrow)$ to the spin averaged cross-section. The errors are given by the width of each band. The spin averaged cross-section is also shown[20].

New results from the AGS have been reported in the parallel sessions, for which one should refer to the Spin Physics Summary given by Roser and to the specific papers. I shall give, here, a short overview of the work in this energy range[21 26].

Figure 12a shows measurements of A at 28 GeV/c, as a function of the square transverse momentum $P_\perp^2$, and Fig. 12b shows the effect of the spin on the pp cross-section when beam and target are both polarized.

After the minimum at about 3.5 $(GeV/c)^2$ is reached (Fig. 12a), the analysing power A keeps increasing up to the maximum measured transverse momentum. This does not seem to be in agreement with Perturbative Quantum Chromo Dynamics. In fact, if this theory is applicable, then in the Quark Interchange Model the chiral symmetry of massless QCD implies helicity conservation. Consequently, the helicity amplitude terms corresponding to spin flips leading to change of helicity from the initial to the final states must be zero. It is not easy under these conditions to get non-zero values for the analysing power A. Some authors tried to solve this difficulty using the Massive Quark Model, where the constituents are massive and they interact via an infinite sequence of meson exchanges. However, this model leads to other difficulties as it predicts $A_{nn}(90°) = 97\%$, in contradiction with the experiments. Since there is no clear indication about the energy and $P_\perp^2$ above which the Quark Interchange Model is applicable, a way out may perhaps be to perform new measurements, at larger $P_\perp^2$ and/or larger energy. Plans exist to extend the AGS measurements up to 8 $(GeV/c)^2$, and to extend the previous CERN and Fermilab measurements[22-25] up to this transverse momentum, at 800 or 900 GeV/c beam momentum[33].

Concerning A_{nn}, if $\sigma_{\uparrow\uparrow} = \sigma(\uparrow\uparrow)+\sigma(\downarrow\downarrow)$ and $\sigma_{\uparrow\downarrow} = \sigma(\uparrow\downarrow)+\sigma(\downarrow\uparrow)$, the ratio $\sigma_{\uparrow\uparrow}/\sigma_{\downarrow\downarrow}$ for $P_T=P_B=1$ can also be used in place of A_{nn} to represent spin-spin correlation. This ratio is given by:

$$\frac{\sigma_{\uparrow\uparrow}}{\sigma_{\uparrow\downarrow}} = \frac{1+A_{nn}}{1-A_{nn}} \tag{13}$$

Figure 13 shows the existence of a rather rich structure with a maximum value for A_{nn} of 60% ($\sigma_{\uparrow\uparrow}/\sigma_{\uparrow\downarrow} = 4$). Independent measurements of A_{nn} as a function of $P_\perp^2$, either at 11.75 GeV/c fixed momentum or at 90° fixed angle in the centre-of-mass, have given the same result above $P_\perp^2 = 1.5$ $(GeV/c)^2$, showing the existence of a dependence on the transverse momentum, but leaving open the question of an enhancement related to particle-identity effect at 90° c.m. pointed out by Bethe and Weisskopf. The existence of a structure for A_{nn} is not in agreement with the predictions of the Quark Interchange Model that A_{nn} should be 1/3, independently of the angle and the beam momentum. Oscillations of A_{nn} at 90° c.m., and also at angles different from 90°, however, have been predicted by Hendry, and Troshin and Tyurin, as discussed in Ref. 26. Khiari et al.[26] now plan to extend their AGS measurements in order to determine the behaviour of A_{nn} at 22 GeV/c and at large values of $P_\perp^2$, and to measure spin-spin effects at 90° c.m. up to the highest possible energy.

I should like to conclude that despite its apparent simplicity pp and p$\bar{\text{p}}$ elastic scattering is still challenging to experimental and theoretical physicists. Much effort is at present going on to

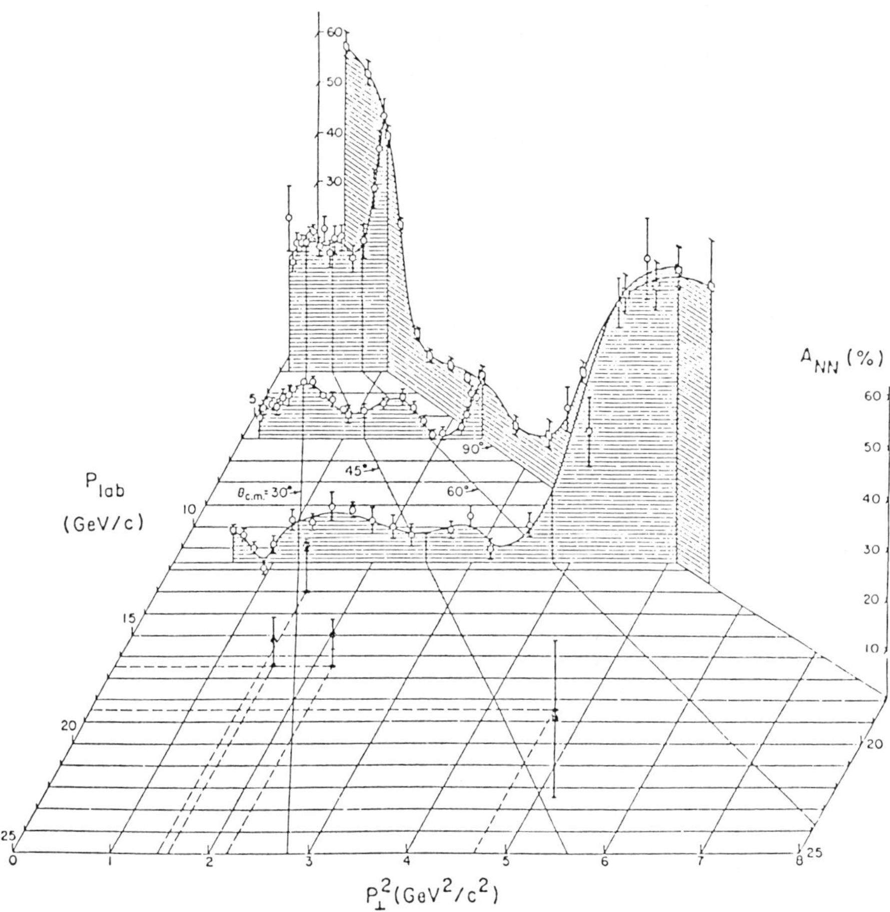

Figure 13: A three-dimensional plot[26] of $\sigma_{\uparrow\uparrow}/\sigma_{\uparrow\downarrow}$ as a function of the beam momentum and of the square transverse momentum.

develop polarized beams and targets: surprises may be around the corner. I am grateful to my friends Alan Krisch, André Martin, and Giorgio Matthiae, who put me up to date on these matters.

REFERENCES

1. Proc. Int. Conf. on Elastic and Diffractive Scattering, Rockefeller University, New York, October 1987, to be published.
2. G. Matthiae, in Ref. 1.
3. D. Bernard et al. (UA4), Phys. Lett. **198B** 583 (1987).
4. R. Castaldi and G. Sanguinetti, Ann. Rev. Nucl. Part. Sci. 35 351 (1985).

5. M. Bozzo et al. (UA4), Phys. Lett. **155B** 197 (1985).
6. M. Bozzo et al. (UA4), Phys. Lett. **147B** 385 (1984).
7. G. Matthiae, Proc. 15th Symposium on Multiparticle Dynamics, Lund, 1984, eds. G. Gustafson and C.Peterson (World Scientific, Singapore, 1984), p. 127.
8. N.A. Amos et al. (E-710), Fermilab-Pub-88/38-E, March 1988.
9. Z. Asa'd et al. (WA7), Nucl. Phys. **B255** 273 (1985).
10. A. Breakstone et al., Phys. Rev. Lett. **54** 2180 (1985).
11. E. Nagy et al., Nucl. Phys. **B150** 221 (1979).
12. W. Faissler et al., Phys. Rev. **D23** 33 (1981).
13. M. Bozzo et al. (UA4), Phys. Lett. **147B** 392 (1984).
14. CERN-Pisa-Rome-Stony Brook Collaboration, Phys. Lett. **62B** 460 (1976).
15. P.E. Faugeras, CERN SPS/84-7 (ARF), March 1984.
16. R. Breedon (UA6), CLMR Memo 221, February 1988, and Ref. 1.
17. A. Martin, CERN-TH.4852/87, September 1987, and Ref. 1.
18. B. Nicolescu, IPNO/TH 88-02, January 1988, and Ref. 1.
19. E. Leader, in Ref. 1.
20. D.G. Crabb et al., UM HE 88-4, March 1988.
21. T. Roser, in Ref. 1.
22. R.V. Kline et al., Phys. Rev. **D22** 553 (1980).
23. M. Corcoran et al., Phys. Rev. **D22** 2624 (1980).
24. G. Fidecaro et al. (WA6), Nucl. Phys. **B173** 513 (1980).
25. G. Fidecaro et al. (WA6), Phys. Lett. **105B** 309 (1981).
26. F.Z. Khiari et al., UM-HE-87-36, April 1988.
27. A. Krisch et al., IUCF proposal, Indiana University, April 1987.
28. Proc. Workshop on the Experimental Programme at UNK, Protvino, September 1987.
29. V.L. Solovianov, in Ref. 28, p. 191.
30. A.N. Vasilyev, in Ref. 28, p. 279.
31. G. Preparata, in Ref. 28, p. 107.
32. P.R. Cameron et al., Phys. Rev. **D32** 3070 (1985).
33. G.R. Court et al., Fermilab P-768, March 1986.

NEUTRINO MASSES, MIXING, MOMENTS, AND MATTER*

William J. Marciano
Brookhaven National Laboratory
Upton, New York 11973

ABSTRACT

The present status of neutrino masses, mixing, and electromagnetic moments is surveyed. Potential enhancements of neutrino oscillations, decay, and spin-flavor precession due to their interactions with matter are described.

1. NEUTRINO MASSES

In the standard $SU(2)_L \times U(1)$ model, the photon and all three species of neutrinos have zero mass. For the photon, masslessness is a natural consequence of exact electromagnetic gauge invariance; its validity being well verified experimentally by the present bound[1] $m_\gamma < 3 \times 10^{-27}\ eV$. However, the masslessness of neutrinos is not on such firm theoretical or experimental footing. Theoretically, $m_\nu = 0$ because only the left-handed component ν_L of each neutrino specie is employed (the right-handed component ν_R is assumed not to exist) and lepton number conservation is required. Relaxing either of these constraints can lead to $m_\nu \neq 0$. Indeed, the present experimental bounds[2,3,4]

$$m_{\nu_e} < 18 \text{ eV}, \tag{1a}$$

$$m_{\nu_\mu} < 0.25 \text{ MeV}, \tag{1b}$$

$$m_{\nu_\tau} < 35 \text{ MeV}, \tag{1c}$$

leave considerable room for speculation that neutrinos actually do possess mass. I might add that the experimental bounds in (1) have improved by factors of $2 \sim 5$ since I last reviewed[5] this subject in 1981. Progress is slow, but we can expect further improvement by factors of ~ 3 during the next few years. A worthy goal would be a further factor of 5 to 10 reduction.

In the standard model charged leptons and quarks acquire masses by coupling their left- and right-handed components through a Higgs scalar isodoublet which obtains a vacuum expectation value. (Mass terms change chirality, so one must couple left-to-right in order to generate mass.) The fermion masses generated by the Higgs mechanism are totally arbitrary, their values are chosen to agree with experiment. If right-handed components exist, then neutrinos can also be given arbitrary masses by the standard Higgs mechanism. The right-handed components will be sterile under $SU(2)_L \times U(1)$ gauge interactions; so,

---

* This manuscript has been authored under contract number DE-AC02-76CH00016 with the U.S. Department of Energy. Accordingly, the U.S. Government retains a non-exclusive, royalty-free license to publish or reproduce the published form of this contribution, or allow others to do so, for U.S. Government purposes.

they do not upset phenomenology. Of course, in such a scenario one must address the question: Why are neutrino masses so much smaller than the masses of other fermions in their generation?

Even without introducing right-handed neutrinos, the neutrino can be given a Majorana mass by coupling ν_L to its (right-handed) charge conjugation ν_L^c (which I subsequently call $\overline{\nu}$, the antineutrino). The mass term would appear as

$$\frac{1}{2} m_\nu \left(\overline{\nu}_L^c \, \nu_L + h.c.\right). \tag{2}$$

(It requires anticommuting fields.) A Majorana mass is not possible for other fermions because it would violate electric charge conservation. In the case of the neutrino, lepton number is violated $\Delta L = 2$ by Eq. (2).

The simplest way of introducing a Majorana mass is to couple the neutrino-charged lepton $SU(2)_L$ doublet to a Higgs triplet.[6] The implications of such a scenario are the occurrence of neutrinoless double beta decay (so far not observed) and the existence of additional physical scalars associated with the Higgs triplet. In particular, a light scalar and pseudoscalar (Majoran) can exist in such a scenario. This mechanism can accommodate a small neutrino mass, but also fails to naturally explain small masses.

In my opinion, the most compelling rationale for small neutrino masses is found by combining the two mass generating mechanisms mentioned above in the see-saw[7] mechanism. The basic idea is motivated by grand unified theories (GUTS) which often contain very heavy right-handed Majorana neutrinos, ν_R. Since ν_R is a singlet under $SU(2)_L \times U(1)$, it is expected that large Majorana masses M_R will arise for such fields. On the other hand, the ordinary Higgs mechanism should lead to a Dirac mass, m_f, which couples ν_R to ν_L. Of course, m_f is expected to be a typical fermion mass scale, i.e., of order m_e, m_u or m_d for the first generation. Therefore, one expects a mass matrix of the form

$$\left(\overline{\nu}_L^c \quad \overline{\nu}_R\right) \begin{pmatrix} 0 & m_f \\ m_f & M_R \end{pmatrix} \begin{pmatrix} \nu_L \\ \nu_R^c \end{pmatrix} \tag{3}$$

Diagonalization leads to a heavy mass $O(M_R)$ Majorana state which is primarily ν_R and a light Majorana state with mass m_f^2/M_R which is the ordinary neutrino. Since $M_R >> m_f$, the induced mass is naturally very small. In such a scenario, one expects the three neutrinos to have a hierarchical mass relationship

$$m_3^2 >> m_2^2 >> m_1^2, \tag{4}$$

since

$$m_1^2 : m_2^2 : m_3^2 :: m_u^4 : m_c^4 : m_t^4, \tag{5}$$

(or $m_e^4 : m_\mu^4 : m_\tau^4$) in many scenarios. If that is the case, then one might (approximately) anticipate

$$m_1^2 : m_2^2 : m_3^2 :: 1 : 10^{10} : 10^{16}. \tag{6}$$

That hierarchy could have important consequences for neutrino oscillations as we shall see. Of course, at issue is what value of M_R sets the neutrino mass scale? Is it the GUT unification scale $\sim 10^{15} \sim 10^{18}$ GeV or something smaller? If the GUT scale enters, then neutrino masses are likely to be very small and extremely hard to determine.

2. MIXING AND OSCILLATIONS

If neutrinos have mass, then one expects them to mix. In analogy with quark mixing, the weak interaction states ν_e, ν_μ and ν_τ will be related to the mass eigenstates ν_1, ν_2 and ν_3 by a unitary matrix. It is convenient to parameterize the mixing by[8]

$$\begin{pmatrix} \nu_e \\ \nu_\mu \\ \nu_\tau \end{pmatrix} = \begin{pmatrix} C_1C_3 & S_1C_3 & S_3e^{-i\delta} \\ -S_1C_2 - C_1S_2S_3e^{i\delta} & C_1C_2 - S_1S_2S_3e^{i\delta} & S_2C_3 \\ S_1S_2 - C_1C_2S_3e^{i\delta} & -C_1S_2 - S_1C_2S_3e^{i\delta} & C_2C_3 \end{pmatrix} \begin{pmatrix} \nu_1 \\ \nu_2 \\ \nu_3 \end{pmatrix}$$

$$C_i \equiv \cos\theta_i, \quad S_i \equiv \sin\theta_i, \quad i = 1,\ 2,\ 3. \tag{7}$$

Although it is probably best to keep an open mind about neutrino mixing, if the quark analogy holds, one expects small θ_i.

Neutrino masses and mixing can lead to many interesting phenomena in laboratory experiments, astrophysics, and cosmology. For example, neutrino decay (e.g., $\nu_\mu \to \nu_e + \gamma$) may occur or neutrinos may have small electromagnetic moments. A particularly popular possibility is neutrino oscillation from one specie to another. I will briefly survey that broad subject.

If neutrinos have masses and mix, then neutrino oscillations will occur. A neutrino ν_ℓ produced at time 0 may at a later time be observed as a $\nu_{\ell'}$. To illustrate the basic features, first consider the simplified case of only two species ν_ℓ and $\nu_{\ell'}$ such that

$$\begin{pmatrix} \nu_\ell \\ \nu_{\ell'} \end{pmatrix} = \begin{pmatrix} \cos\theta & \sin\theta \\ -\sin\theta & \cos\theta \end{pmatrix} \begin{pmatrix} \nu_1 \\ \nu_2 \end{pmatrix} \tag{8}$$

with ν_1 and ν_2 mass eigenstates with masses m_1 and m_2 respectively. If at time $t = 0$, a $|\nu_\ell> = |\nu_1> \cos\theta + |\nu_2> \sin\theta$ is produced with momentum $\vec{p}$, then at a latter time t, it will evolve to

$$i\frac{d}{dt} |\nu_i> = E_i |\nu_i>, i = 1,\ 2$$

$$|\nu(t)> = e^{-iE_1t} |\nu_1> \cos\theta + e^{-iE_2t} |\nu_2> \sin\theta \tag{9}$$

The probability of detecting a ν_ℓ at distance $R \simeq t$ (for $\beta \simeq 1$) is therefore given by

$$P(R)_{\nu_\ell \to \nu_\ell} = |<\nu_\ell | \nu(t)>|^2 = 1 - \sin^2 2\theta \sin^2\left[\frac{\Delta m_{21}^2 R}{4|\vec{p}|}\right] \tag{10a}$$

while the probability of observing a ν'_ℓ is

$$P(R)_{\nu_\ell \to \nu'_\ell} = \mid < \nu'_\ell \mid \nu(t) > \mid^2 = \sin^2 2\theta \sin^2 \left[\frac{\Delta m^2_{21} R}{4 \mid \vec{p} \mid}\right] \tag{10b}$$

where $\Delta m^2_{21} \equiv m^2_2 - m^2_1$. (I have used $E_i \simeq \mid \vec{p} \mid + \frac{1}{2} m^2_i / \mid \vec{p} \mid$.) It oscillates between ν_ℓ and ν'_ℓ with a characteristic oscillation length L_{21}

$$L_{21} = \frac{4\pi \mid \vec{p} \mid}{\Delta m^2_{21}} \tag{11}$$

So, low energy neutrinos are more likely to oscillate. If one believes that Δm^2_{ij} are all small as in the see-saw mechanism, then one must indeed employ neutrinos with an extremely large R/E_ν to have any hope of observing oscillations. At present, laboratory reactor and accelerator experiments looking for $\nu_e \leftrightarrow \nu_\mu$ oscillations have ruled out[9] $\Delta m^2_{21} \gtrsim 10^{-2}$ eV^2 for $\sin^2 2\theta \gtrsim 0.1$ and $\Delta m^2_{21} \gtrsim 1$ eV^2 for $\sin^2 2\theta \gtrsim 0.01$. Bounds on $\nu_\mu \to \nu_\tau$ are not as good (see fig. 1). To explore much smaller Δm^2_{21} requires atmospheric or astrophysical neutrino experiments in which R can be very large.

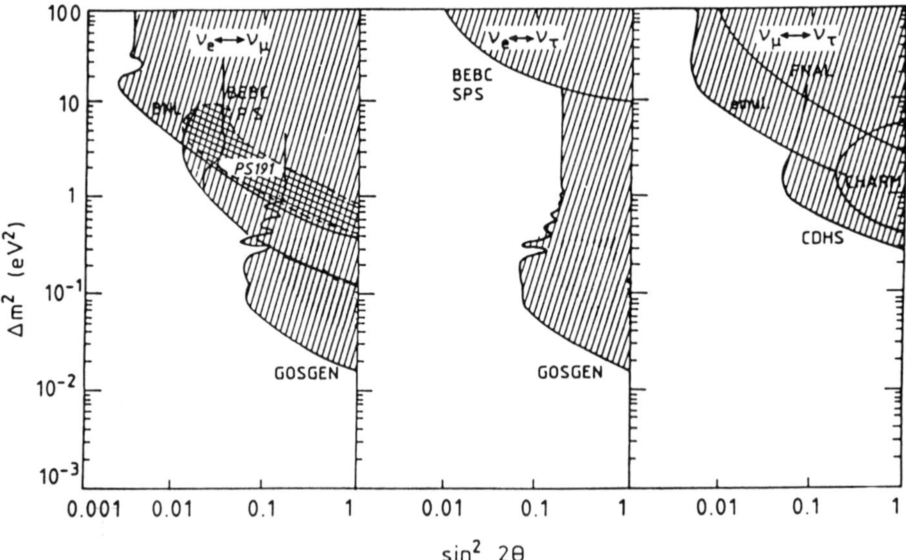

Fig. 1: Bounds on neutrino oscillations.[9] These constraints were summarized by F. Vannucci in 1987. The only update that I have included is deletion of an oscillation signal at the Bugey reactor. Recent bounds from BNL and LAMPF on $\nu_\mu \leftrightarrow \nu_e$ oscillations are similar to the BEBC PS results.

In the real world, there are (at least) three neutrinos, and their mixing can be quite complicated. A full description of neutrino oscillations in the 3 generation scenario will depend on 6 parameters, θ_1, θ_2, θ_3, δ, Δm_{31}^2 and Δm_{21}^2 (see (7)). However, things are considerably simplified in the hierarchy case $m_3^2 >> m_2^2$, m_1^2. Then oscillations effectively become a 2x2 problem with "effective" mixing parameters. The relevant Δm^2 will depend on R/E_ν. To illustrate how decoupling occurs, consider the case $R/E_\nu \sim 0\left(1/\Delta m_{31}^2\right) << 1/\Delta m_{21}^2$ which may be most relevant for terrestrial experiments. In that case, the oscillation probabilities are simply governed by[8]

$$P\left(R\right)_{\nu_\mu \leftrightarrow \nu_e} \simeq \sin^2\theta_2 \sin^2 2\theta_3 \sin^2\left[\frac{\Delta m_{31}^2 R}{4E_\nu}\right] \tag{12a}$$

$$P\left(R\right)_{\nu_\mu \leftrightarrow \nu_\tau} \simeq \sin^2 2\theta_2 \cos^4\theta_3 \sin^2\left[\frac{\Delta m_{31}^2 R}{4E_\nu}\right] \tag{12b}$$

$$P\left(R\right)_{\nu_e \leftrightarrow \nu_e} \simeq 1 - \sin^2 2\theta_3 \sin^2\left[\frac{\Delta m_{31}^2 R}{4E_\nu}\right] \tag{12c}$$

The oscillation length L_{21} is too long to effect the evolution. For small mixing angles, one finds

$$P\left(R\right)_{\nu_\mu \leftrightarrow \nu_\tau} \simeq P\left(R\right)_{\nu_\mu \leftrightarrow \nu_e} / \sin^2\theta_3 \tag{13}$$

This suggests that accelerators which produce ν_μ beams may be much better suited to search for $\nu_\mu \to \nu_\tau$ oscillations rather than $\nu_\mu \to \nu_e$. So far, most experiments have concentrated on $\nu_\mu \to \nu_e$.

For the other extreme $R/E_\nu \sim 0\left(1/\Delta m_{21}^2\right) >> 1/\Delta m_{31}^2$, (very long distances or small energies) oscillations are governed by Δm_{21}^2 in an effective 2 neutrino system. So, for example one finds

$$P\left(R\right)_{\nu_e \to \nu_e} = \sin^4\theta_3 + \cos^4\theta_3 \left(1 - \sin^2 2\theta_1 \sin^2\frac{\Delta m_{21}^2 R}{4E_\nu}\right) \tag{14}$$

It is interesting to compare this formula with the observed flux of solar neutrinos where a long standing problem exists. One expects neutrinos produced in the solar core to give rise to $\nu_e +^{37} Cl \to e^- +^{37} Ar$ at the rate of[10]

$$7.9 \pm 2.5 SNU \qquad \left(1SNU = 10^{-36}\text{captures/atom} - \text{s}\right) \tag{15}$$

However, Davis and collaborators have observed for the last 17 years an average flux of[11]

$$2.2 \pm 0.3\ SNU \tag{16}$$

The discrepancy constitutes the solar neutrino problem. What happened to the missing flux? A simple solution is that it was never present, i.e., the theoretical prediction in (15) is wrong. More exotic is the possibility that the ν_e flux was depleted by oscillations. That could occur if Δm_{21}^2 is fine-tuned to the earth-sun distance. More naturally for $\Delta m_{21}^2 >> R/E_\nu$, one finds from (15) a survival probability average

$$P(R)_{\nu_e \to \nu_e} = \sin^4\theta_3 + \cos^4\theta_3\left(\cos^4\theta_1 + \sin^4\theta_1\right) \tag{17}$$

which can be quite small for particular mixing angles. The smallest value of (17) is 1/3 for $\sin^2\theta_1 = 1/2$, $\sin^2\theta_3 = 1/3$. Such a depletion would solve the solar neutrino problem; but it still appears somewhat contrived and requires relatively large mixing which runs counter to our experience with quark mixing and theoretical prejudices.

A truly ingenious solution to the solar neutrino puzzle has been proposed by Mikheyev and Smirnov.[12] Employing an analysis of matter effects on neutrino oscillations by Wolfenstein,[13] they showed that for a large range of neutrino masses and mixing parameters, neutrino oscillations between ν_e and ν_μ or ν_τ in the sun's interior could be significantly enhanced. In fact, for some neutrino energies one can get a nearly complete transformation of ν_e into another neutrino flavor. That scenario (referred to as the MSW effect) provides an elegant natural solution to the solar neutrino puzzle and hence has become very popular. It enters into the planning for future solar neutrino experiments as well as their interpretation. I describe in section 4 some of its salient features and potential implications. I might note that for the last year Ray Davis[11] has been averaging 5 SNU. One does not know whether that represents a statistical fluctuation or an important hint regarding resolution of the solar neutrino puzzle.

3. ELECTROMAGNETIC MOMENTS

Although neutrinos are electrically neutral, they can have electromagnetic form factors. In the case of 4 component Dirac neutrino[14]

$$\begin{aligned}<\nu \mid J_\mu^{em} \mid \nu> = \overline{u}_\nu[(F(q^2) + \gamma_5\, G(q^2))\gamma^\alpha(g_{\alpha\mu} - \frac{q_\alpha q_\mu}{q^2}) \\ + (M(q^2) + \gamma_5 E(q^2)) i\sigma_{\mu\alpha}q^\alpha]u_\nu\end{aligned} \tag{18}$$

The F and G form factors are charge radii while $M(0)$ is the magnetic dipole moment and $E(0)$ is the electric dipole moment of the neutrino. In the case of Majorana neutrinos (which are self-conjugate), $F = M = E = 0$, i.e., only $G\left(q^2\right)$ can be non-vanishing.

Both Dirac and Majorana neutrinos can also have transition moments that are off diagonal, i.e., connect mass eigenstates ν_2 and ν_1. Magnetic and electric transition moments can give rise to neutrino decay or flavor precession as we shall see.

I will say a few words about the neutrino electromagnetic and transition moments. First consider a transition moment $\kappa_{21}\, e/2m_e$ between ν_2 and ν_1 such that the decay amplitude for $\nu_2 \to \nu_1 + \gamma$ is given by[15]

$$\mathcal{M} = \kappa_{21}\frac{e}{2m_e}\mathcal{E}^\mu q^\nu \overline{\nu}_1\sigma_{\mu\nu}(1-\gamma_5)\,\nu_2 \tag{19}$$

In that case, the decay rate for $\nu_{2_L} \to \nu_{1_R} + \gamma$ is

$$\Gamma\left(\nu_{2_L} \to \nu_{1_R} + \gamma\right) = \frac{\alpha}{2m_e^2}\left(\frac{m_2^2 - m_1^2}{m_2}\right)^3 \kappa_{21}^2 \tag{20}$$

Of course, for κ_{21} and $m_2 - m_1$, very small, as expected, the predicted decay rate will be tiny. Indeed, it can easily exceed the age of the universe.

If the neutrino is 4 component (Dirac), it can have a (diagonal) magnetic and/or electric dipole moment. Of course, the electric dipole moment would violate T and would most likely be the smaller. In the standard $SU(2)_L \times U(1)$ with massive Dirac neutrinos, one finds that one loop induced magnetic dipole moment $\mu = \kappa e/2m_e$ is[15,16]

$$|\mu| = \frac{3eG_\mu m_\nu}{8\sqrt{2}\pi^2} \leq 3 \times 10^{-19}\left(m_\nu/1\text{eV}\right) e/2m_e \tag{21}$$

which is very small. (Transition moments are generally even smaller since they involve flavor mixing.) One, therefore, expects $\kappa \lesssim 10^{-19}$ for very light neutrinos, unless some new chiral changing physics enters.

Magnetic, electric, and transition moments can be bounded by laboratory experiments and/or astrophysics experiments. In the case of $\nu - e$ scattering, the existence of any such moment would increase the cross- section by (neglecting neutrino mass)[17]

$$\Delta\frac{d\sigma(\nu e)}{dy} = \frac{\pi\alpha^2\kappa^2}{m_e^2}\left(\frac{1}{y} - 1\right) \tag{22}$$

where $y = \left(E_e' - m_e\right)/E_\nu$. Note that it is rather insensitive to the initial ν energy. Therefore, low energy experiments where the usual cross-section $\sim E_\nu$ are most sensitive to κ. From existing experimental data, one finds[18,19]

$$|\kappa_{\nu_\mu}| < 10^{-9} \tag{23}$$

$$|\kappa_{\nu_e}| < 4 \times 10^{-10} \tag{24}$$

(Those bounds also apply to electric and transition moments.) It will be difficult to push those direct experimental bounds much farther, since the effect in (22) goes like κ^2. Better bounds,[20]

$$|\kappa_\nu| < 8.5 \times 10^{-11} \tag{25}$$

are obtainable from stellar evolution arguments. Recently, Fukugita and Yazaki[21] have argued that such bounds can be extended to $\kappa \lesssim 10^{-11}$ using observed data on Red Giants. A value of $\kappa \gtrsim 10^{-11}$ would have important implications for solar neutrinos, as we shall see in the next section. Therefore, laboratory experiments should also aim to achieve that level of sensitivity. I might note that rather conservative analyses[22] of supernova dynamics combined with the observation of expected neutrino flux from 1987a give the bound $\kappa < 10^{-12}$. That bound does not apply to Majorana transition moments.

4. MATTER EFFECTS

To understand the effect of matter on neutrino propagation, consider the amplitude for low energy neutrino, ν_ℓ, scattering off a fermion f.

$$\mathcal{M}\left(\nu_\ell f \to \nu_\ell f\right) = -i\frac{G_\mu}{\sqrt{2}}\, \overline{\nu}_\ell\, \gamma^\alpha\left(1-\gamma_5\right)\nu_\ell \overline{f}\gamma_\alpha\left(C^V_{\nu_\ell f} + C^A_{\nu_\ell f}\gamma_5\right) f \tag{26}$$

where C^V and C^A are constants and $G_\mu = 1.16636 \times 10^{-5}\ GeV^{-2}$. In the case of an unpolarized medium, the coherent forward scattering of neutrinos with momentum p_ν can be described by an index of refraction n_{ν_ℓ} given by[23]

$$p_\nu\left(n_{\nu_\ell} - 1\right) = -\sqrt{2}\, G_\mu \sum_{f=e,u,d} C^V_{\nu_\ell f} N_f \tag{27}$$

where the N_f is the fermion number density and

$$C^V_{\nu_\ell f} = T_{3f} - 2Q_f \sin^2\theta_W \qquad f \neq \ell \tag{28a}$$

$$C^V_{\nu_\ell \ell} = 1 + T_{3\ell} - 2Q_\ell \sin^2\theta_W \tag{28b}$$

with $T_{3f} = \pm\frac{1}{2}$ the weak isospin. (Quantum loop corrections to (28) are given in ref. 24). Note that $\nu_e e$ scattering differs from $\nu_\mu e$ and $\nu_\tau e$ scattering because of an additional charged current component in the amplitude. That effect leads to an important difference in indices of refraction

$$-p_\nu\left(n_{\nu_e} - n_{\nu_\mu}\right) = -p_\nu\left(n_{\nu_e} - n_{\nu_\tau}\right) = \sqrt{2}\, G_\mu N_e \tag{29}$$

which can significantly modify oscillations. An alternative description of the difference in (29) can be given by writing the amplitude in (26) as an effective Lagrangian

$$\mathcal{L} = -\sqrt{2}\, G_\mu\, \overline{\nu}_\ell\, \gamma^\alpha\, \frac{1-\gamma_5}{2}\, \nu_\ell\, \overline{f}\, \gamma_\alpha\left(C^V_{\nu_\ell f} + C^A_{\nu_\ell f}\gamma_5\right) f \tag{30}$$

We can average the Lagrangian over the background matter medium using

$$< C^V_{\nu_\ell f}\overline{f}\gamma_o f >= C^V_{\nu_\ell f} N_f \tag{31}$$

whereas all other terms average to zero. One can, therefore, interpret the medium as providing an external potential

$$V = \sqrt{2}\, G_\mu \sum_f C^V_{\nu_\ell f}\, N_f \tag{32}$$

felt by neutrinos. With such a potential, neutrino propagation is described by replacing

$$i\frac{d}{dt} \to i\frac{d}{dt} - V \tag{33}$$

in the equation of motion. (For antineutrinos $V_{\overline{\nu}} = -V_\nu$.)

Althought the magnitude of the potential is small

$$\sim 4 \times 10^{-14}\ \text{eV}\left(N_f/6 \times 10^{23} cm^{-3}\right),$$

it can have truly remarkable consequences when it interferes with an equally small vacuum energy difference $\Delta m^2/2E_\nu$. Several examples will be illustrated below.

4.1 MSW Oscillations

Only the difference of potentials affects neutrino oscillations in matter. In ordinary matter that difference is due to an additional repulsive interaction for the ν_e with electrons in the medium coming from charged current interactions. In that way, the coupled equations of motion in matter become (neglecting quantum corrections)[24] up to common diagonal terms

$$i\frac{d}{dt}\begin{pmatrix}\nu_e\\ \nu_\mu\\ \nu_\tau\end{pmatrix} = \left[U\begin{pmatrix}0 & 0 & 0\\ 0 & \Delta m_{21}^2/2p_\nu & 0\\ 0 & 0 & \Delta m_{31}^2/2p_\nu\end{pmatrix}U^{-1} + \begin{pmatrix}\sqrt{2}G_\mu N_e & 0 & 0\\ 0 & 0 & 0\\ 0 & 0 & 0\end{pmatrix}\right]\begin{pmatrix}\nu_e\\ \nu_\mu\\ \nu_\tau\end{pmatrix} \tag{34}$$

where U is the unitary matrix in (7).

To illustrate the MSW effect, consider the two generation limit, i.e., $\nu_e - \nu_\mu$ oscillations with $S_2 = S_3 = 0$ in (7). For that case (34) becomes

$$i\frac{d}{dt}\begin{pmatrix}\nu_e\\ \nu_\mu\end{pmatrix} = \begin{pmatrix}\sqrt{2}\ G_\mu\ N_e & \frac{-\Delta m_{21}^2}{4p_\nu}\sin 2\theta\\ -\frac{\Delta m_{21}^2}{4p_\nu}\ \sin 2\theta & \frac{\Delta m_{21}^2}{2p_\nu}\cos 2\theta\end{pmatrix}\begin{pmatrix}\nu_e\\ \nu_\mu\end{pmatrix} \tag{35}$$

As we saw before, in a vacuum with $N_e = 0$, $\nu_e \to \nu_\mu$ oscillations are governed by (see (10))

$$P\left(R\right)_{\nu_e\to\nu_\mu} = \sin^2 2\theta \sin^2\left[\pi R/L_{12}\right] \tag{36a}$$

with

$$L_{21} = 4\pi p_\nu/\Delta m_{21}^2 \tag{36b}$$

the vacuum oscillation length. Solving (35) for constant N_e, one finds the modified matter oscillation result

$$P\left(R\right)_{\nu_e\to\nu_\mu} = \sin^2 2\theta_m\ \sin^2\left(\frac{\pi R}{L_m}\right) \tag{37a}$$

where

$$\sin^2 2\theta_m = \frac{\sin^2 2\theta}{1 - 2\frac{L_{21}}{L_0}\cos 2\theta + \left(\frac{L_{21}}{L_0}\right)^2} \tag{37b}$$

$$L_m = \frac{L_{21}}{\left[1 - 2\frac{L_{21}}{L_o}\cos 2\theta + \left(\frac{L_{21}}{L_o}\right)^2\right]^{1/2}} \tag{37c}$$

$$L_o = \frac{2\pi}{\sqrt{2}\, G_\mu\, N_e} \tag{37d}$$

so, we see that the matter mixing angle θ_m can be large even if the vacuum angle θ is small. In fact, if the condition

$$L_{21}/L_o = \cos 2\theta \quad \text{(Resonance Condition)} \tag{38}$$

which corresponds to an electron density

$$N_e = \frac{\Delta m_{21}^2 \cos 2\theta}{2\sqrt{2}\, G_\mu\, p_\nu} \tag{39}$$

is satisfied, one finds $\theta_m = 45^o$, maximal mixing. Of course, with the full 3 generation mixing matrix, the situation is more complicated but the physics is basically the same. The potential in matter can cause energy level crossings and thereby enhance oscillations. (It can also, of course, quench oscillation if the splitting is increased by the matter potential.)

The MSW effect actually comes about from a changing electron density profile in the sun. Neutrinos produced in the solar core start out at a high electron density, but experience a decreasing density as they propagate outward (i.e., the potential changes). If the resonance condition in (39) is satisfied during transit, ν_e can resonate completely into ν_μ (or ν_τ). Assuming that the vacuum angle θ is small, they will not convert back to ν_e as they propagate to the earth.

The MSW solution to the solar neutrino problem is very elegant and naturally attractive. It can resolve the longstanding low neutrino flux observations for Δm^2 in the range $10^{-7} \sim 10^{-4}$ eV^2 and for a large range of mixing angles $0.001 < \sin^2\theta \leq 0.4$ (see fig. 2). It has also inspired many other applications of the effect of matter on neutrinos Two applications, neutrino decay and spin-flavor precession will be briefly surveyed.

4.2 Neutrino Decay

If neutrinos have mass and mix, then the heavier neutrino species should decay into lighter ones. Some vacuum decays such as $\nu_{2_L} \to \nu_{1_R} + \gamma$ (see (20)) will, however, be significantly altered in dense matter because ν_{2_L} and ν_{1_R} experience different potentials. Even more important, in a hot dense plasma, photons acquire an effective mass, the plasma frequency. That mass will inhibit and can quite easily eliminate the possibility of radiative neutrino decay in matter.

A more interesting scenario considered by Berezhiani and Vsotsky[26] involves the decays $\nu_e \to \overline{\nu}_e + M$ (or $\overline{\nu}_e \to \nu_e + M$ in neutron-rich matter)

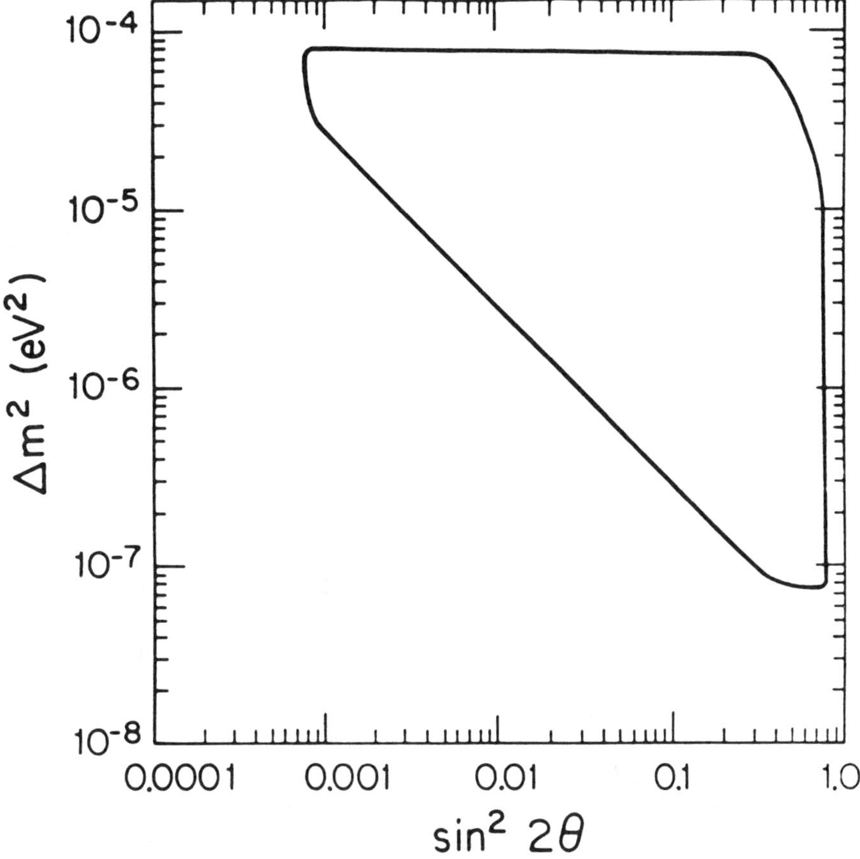

Fig. 2: 30% survival contour for solar ν_e neutrinos undergoing MSW oscillations. For mixing and mass differences inside the contour more than 70% of the ν_e flux is lost.

and $\overline{\nu}_\mu \rightarrow \nu_\mu + M$ in matter with M the hypothetical spin-0 Majoran previously mentioned. Such decays are forbidden in the vacuum because CPT requires ν and $\overline{\nu}$ to have the same mass. However, ordinary neutral matter with $N_e > N_n$ provides a repulsive potential for ν_e and $\overline{\nu}_\mu$ and attractive potential for $\overline{\nu}_e$ and ν_μ (see (28)). Those potentials give rise to effective mass differences that govern the decay rate. The rates for such decays in the earth, sun and supernova were estimated by Berezhiani and Vysotsky using experimental bounds on the Majoran couplings to neutrinos. The most interesting possibility seems to be in a supernova. There the neutron-rich core region can lead to $\overline{\nu}_e \rightarrow \nu_e + M$, $\overline{\nu}_\mu \rightarrow \nu_\mu + M$, $\overline{\nu}_\tau \rightarrow \nu_\tau + M$ and thus skew the flux toward a higher neutrino and depleted antineutrino flux rate. Such an effect would be energy dependent and could be affected by flavor mixing.

4.3 Neutrino Spin-Flavor Precession

One speculative solution to the solar neutrino flux problem was suggested long ago (1971) by Cisneros.[27] He argued that if neutrinos had a magnetic moment $\kappa \sim 10^{-14}$ and large transverse magnetic fields $\sim 10^6 G$ existed in the sun, then neutrinos could undergo spin precession $\nu_{e_L} \to \nu_{e_R}$ as they traversed the solar radius. Since ν_{e_R} is sterile, one would observe an effective depletion of neutrino flux by $\sim 1/2$ in accord (approximately) with the Davis experiment. That scenario has been revived and improved by Okun, Voloshin and Vysotsky (OVV).[28] Their motivation came from the observation that there appears to be an anticorrelation between neutrino flux and variations in sunspot activity, i.e., large magnetic field disturbances.[28] Indeed, there is mounting evidence that solar neutrino flux may follow the 11-year solar cycles.

The precession scenario for solar neutrinos has been extensively studied by OVV. They noted that magnetic or electric dipole moments could give $\nu_{e_L} \to \nu_{e_R}$ precession, while flavor transition moments could result in the combined spin-flavor precession $\nu_{e_L} \to \nu_{\mu_R}$ or ν_{τ_R} (for Dirac neutrinos) or $\nu_e \to \overline{\nu}_\mu$ or $\overline{\nu}_\tau$ (for Majorana neutrinos). In all cases, however, they concluded that a relatively large moment $\kappa \sim 10^{-10} \sim 10^{-11}$ was necessary for realistic solar magnetic fields $10^3 \sim 10^4 G$. Such a scenario renders $\kappa \lesssim 10^{-11}$ an interesting domain for direct experimental searches.

I would like to conclude by describing work done in collaboration with C-S. Lim[29] on the effect of matter on neutrino spin-flavor precision (similar observations have been made by E. Akhmedov).[30] We noted that in a vacuum, neutrino mass differences can quench spin flavor precession. The different potentials felt by distinct flavor neutrinos in matter is also capable of quenching precession. However, when both a mass difference and potential difference exist, they can cancel one another. In that case, a resonance situation can arise and $\nu_L \to \nu'_R$ can proceed unimpeded (provided magnetic fields are present). For a varying density, one can even have an MSW effect in which essentially complete conversion can occur. Let me illustrate how that could come about for Majorana neutrinos (the more realistic case) by considering $\nu_e \to \overline{\nu}_\mu$ spin-flavor precession. In that case, (ignoring other species and components), the precession evolution in a strong magnetic field B is governed by[29]

$$i\frac{d}{dt}\begin{pmatrix}\nu_e \\ \overline{\nu}_\mu\end{pmatrix} = \begin{pmatrix} V_{\nu_e} & \mu^* B \\ \mu B & \frac{\Delta m_{21}^2}{2E_\nu}\cos 2\theta - V_{\nu_\mu}\end{pmatrix}\begin{pmatrix}\nu_e \\ \overline{\nu}_\mu\end{pmatrix} \tag{40}$$

where θ is the mixing angle, and V_{ν_e}, V_{ν_μ} are matter potentials and μ is a transition moment.

For a neutral medium $N_e = N_p$, one finds (N_n = neutron density)

$$V_{\nu_e} = \frac{G_\mu}{\sqrt{2}}\,(2N_e - N_n)$$

$$V_{\nu_\mu} = \frac{G_\mu}{\sqrt{2}}(-N_n) \tag{41}$$

Assuming that B and the number densities are static, one finds that the transition probability for spin-flavor precession is given by

$$P(t)_{\nu_e \to \overline{\nu}_\mu} = \frac{(2\mu B)^2}{\Delta^2 + (2\mu B)^2} \sin^2\left(\sqrt{\Delta^2 + 4\mu^2 B^2}\, t/2\right) \tag{42}$$

$$\Delta = \sqrt{2} G_\mu (N_e - N_n) - \frac{\Delta m_{21}^2}{2E_\nu} \cos 2\theta \tag{43}$$

For the special condition $\Delta = 0$, spin-flavor precession is unimpeded and (42) becomes

$$P(t)_{\nu_e \to \overline{\nu}_\mu} = \sin^2(\mu B t) \tag{44}$$

i.e., maximal spin-flavor precession.

That condition is achieved when the density $N_e - N_n$ just allows a cancellation in (43). Of course, for an adiabatically changing density profile, it is possible to start with $\Delta > 0$ and encounter a resonance region, $\Delta \simeq 0$, where essentially all ν_e of a given energy are converted to $\overline{\nu}_\mu$ and then reach a region where either $\Delta < 0$ or B is no longer large enough to give a significant precession. In that case essentially total conversion of $\nu_e \to \overline{\nu}_\mu$ can occur.

For the above resonant spin-flavor precession scenario to proceed, one needs non-zero flavor transition moments, large magnetic fields and dense matter. In the sun, where $< B > \simeq 10^3 \sim 10^4 G$ is expected, the transition moment κ must be relatively large $\sim 10^{-10} - 10^{-11}$ for large spin-flavor precession of neutrinos with $E_\nu \sim 10$ MeV to occur. Nevertheless, it will be interesting to see whether the hints of a time variation in the neutrino flux correlated with magnetic field fluctuations is confirmed by subsequent experiments. A more likely candidate to look for spin-flavor precession is a supernova. In that case, very large $B \sim 10^{12} \sim 10^{15} G$ can occur. Hence, one is sensitive to $\kappa \sim 10^{-19} \sim 10^{-23}$ a realistic range of transition moments. The signature of such a phenomenon will be an interchange of ν_e and $\overline{\nu}_\mu$ (or $\overline{\nu}_\tau$) supernova spectra. In addition, for $N_n \gtrsim N_e$ one may have $\overline{\nu}_e \to \nu_\mu$ or ν_τ near the core. Observing such an effect would be extraordinary. Will we be so lucky to have one occur close enough to observe and will our detectors be ready if one occurs? I hope so.

5. CONCLUSION

Neutrino physics research has entered an exciting new era. The potential effects of matter on oscillations, decay and spin-flavor precession have inspired many interesting theoretical conjectures with testable experimental consequences. In most cases high densities over large distances are required for such effects to be manifested. That means that we must use the sun, earth, or some other heavenly body (supernova) as a catalyst.

The coming of age of neutrino astronomy nicely complements speculations on neutrino exotica. Following up the patient pioneering work of Ray Davis with new solar neutrino detectors will tell us more about the internal dynamics of our sun and provide a unique laboratory for probing neutrino properties.

Furthermore, if a supernova in our galaxy should occur, a wealth of data will become available.

At the terrestrial level, new experiments are now starting to be mounted at accelerator facilities. Those experiments will probe the standard model parameters and search for new physics in scattering processes such as $\overset{(-)}{\nu}_\mu + e \to \overset{(-)}{\nu}_\mu + e$. In addition, they provide an alternative more controlled means of searching for oscillations and electro- magnetic moments. In my opinion, the next generation of such experiments should focus on $\nu_\mu - \nu_\tau$ oscillations. An appearance experiment in which one observed ν_τ by τ production would be particularly welcome. Many theoretical and experimental ideas have been put forward; but I would guess that new clever possibilities will be found as experiments provoke creative thinking.

Neutrino physics is very hard, but the discovery potential is high and well worth the effort.

REFERENCES

1. Particle Data Table(1986).
2. M. Fritschi *et al.*, Phys. Lett. B173, 485 (1986). For a review see P. Langacker, "Massive Neutrinos" DESY preprint 88-023 (1988).
3. R. Abela *et al.*, Phys. Lett. B146, 431 (1984).
4. ARGUS Collaboration.
5. W.J. Marciano, Comments on Nucl. & Part. Phys. IX, No. 5, 169 (1981).
6. G. Gelinini and M. Roncadelli, Phys. Lett. B99, 411 (1981);
   H. Georgi, S. Glashow, and S. Nussinov, Phys. Lett. B193, 297 (1983).
7. M. Gell-Mann, P. Ramond, and R. Slansky, in Supergravity, eds. F. van Nieuwenhuizen and D. Freedman, (North Holland, Amsterdam, 1979), p. 315; T. Yanagida, Prog. Th. Phys. B135, 66 (1978).
8. W. Marciano and C-S. Lim in BNL Neutriono Workshop proceedings (1987); L. Maiani, Proc. Int. Symposium on Lepton and Photon Interactions, p. 877, Hamburg (1977).
9. P. Vannucci, BNL Neutrino Workshop proceedings (1987). Figure 1 is taken from that talk.
10. J. Bahcall and R. Ulrich, Rev. Mod. Phys. 60, 297 (1988). For a review see A. Baltz and J. Weneser, BNL preprint (1988)submitted to Comments on Nuclear and Particle Physics.
11. R. Davis, private communication.
12. P. Mikheyev and A. Yu. Smirnov, Nuovo Cimento C9, 17 (1986).
13. L. Wolfenstein, Phys. Rev. D17, 2369 (1978).
14. B. Kayser, Phys. Rev. D26, 1662 (1982).
15. W. Marciano and A. Sanda, Phys. Lett. 67B, 303 (1977);
    B.W. Lee and R. Shrock, Phys. Rev. D16, 1444 (1977); S.T. Petcov, Yad. Fiz. 25, 641 (1977)[Sov. J. Nucl. Phys. 25, 340 (1977)].

16. M.A.B. Beg, W. Marciano and M. Ruderman, Phys. Rev. D17, 1395 (1978); K. Fujikawa and R. Shrock, Phys. Rev. Lett. 55, 963 (1980).
17. W.J. Marciano and Z. Parsa, Annual Rev. Nucl. Part. Sci. 36, 171 (1986); A.V. Kyuldjiev, Nucl. Phys. B243, 387 (1984).
18. K. Abe *et al.*, Phys. Rev. Lett. 58, 636 (1987).
19. F. Reines, H.S. Gurr and H.W. Sobel, Phys. Rev. Lett. 37, 315 (1976).
20. P. Sutherland *et al.*, Phys. Rev. D13, 2700 (1976).
21. M. Fukugita and S. Yazaki, Phys. Rev. D36, 3817 (1987).
22. J. Lattimer and J. Cooperstein, Stony Brook preprint (1987); D. Nötzold, Max-Planck-Inst. preprint (1988).
23. F. Langacker, J.P. Leveille and J. Sheiman, Phys. Rev. D27, 1228 (1983).
24. F. Botella, C-S. Lim and W. Marciano, Phys. Rev. D35, 896 (1987).
25. E. Beier *et al.*, Proceedings of the 1986 Snowmass Summer Study.
26. Z. Berezhiani and M.Vysotsky, Phys. Lett. B199, 281 (1987).
27. A. Cisneros, Astrophys. Space Sci. 10, 87 (1971).
28. L.B. Okun, M.B. Voloshin and M. Vysotsky, Sov. J. Nucl. Phys. 44, 440 (1986); Soviet Phys. JETP 64, 446 (1986).
29. C-S. Lim and W.J. Marciano, Phys. Rev. D37, 1368 (1988).
30. E. Akhmedov, Kurchatov Inst. preprint (1988).

SPIN DYNAMICS IN ACCELERATORS AND STORAGE RINGS

L. C. Teng
Fermi National Accelerator Laboratory, Batavia, IL 60510

INTRODUCTION

The understanding of charged particle orbital dynamics in an electromagnetic field is necessary for the acceleration and the storage of particles. The understanding of spin (or magnetic moment) dynamics is essential for the acceleration and the storage of polarized particle beams. In this paper we review the analysis of the dynamics of the spin and the various schemes and devices proposed up-to-date to manipulate the spin to conserve the polarization while accelerating and storing a polarized beam. We shall concentrate on hadrons such as protons, antiprotons and heavier ions although most of the discussions apply equally well to leptons. Electron and positron beams are self-polarizing through spin-flip synchrotron radiation[1] at high energies and hence the acceleration of polarized $e^{\pm}$ beams is less relevant.

RELATIVISTIC SPIN EQUATION AND DEPOLARIZING RESONANCES

In covariant form the relativistic spin equation is[2, 3]

$$\frac{ds^{\alpha}}{d\tau} = \frac{e}{mc}\left[(1+G)F^{\alpha\beta}s_{\beta} + \frac{G}{c^2}\, u^{\alpha}s_{\lambda}F^{\lambda\mu}u_{\mu}\right] \tag{1}$$

where

τ = proper time = time in rest-frame,
s^{α} = spin 4-vector,
u^{α} = velocity 4-vector,
$F^{\alpha\beta}$ = electromagnetic field 4-tensor,
e and m = charge and mass of particle,
c = speed of light,

$$G = \frac{g-2}{2} = \text{anomalous gyromagnetic ratio} = \begin{cases} 1.793 & \text{for proton} \\ 0.00116 & \text{for electron.} \end{cases}$$

The 3 spatial components can be written as

$$\frac{d\vec{s}}{dt} = \frac{e}{\gamma mc}\,\vec{s}\times\left[(1+\gamma G)\vec{B}_\perp + (1+G)\vec{B}_\parallel + \left(\frac{\gamma}{\gamma+1}+\gamma G\right)\frac{\vec{E}\times\vec{v}}{c}\right] \tag{2}$$

where

t = time,

$\vec{s}$ = proper spin = spin in rest-frame,

γ = energy in units of mc^2,

$\vec{v}$ = velocity,

$\vec{B}_\perp$ and $\vec{B}_\parallel$ = magnetic field perpendicular and parallel to $\vec{v}$,

$\vec{E}$ = electric field,

all quantities except $\vec{s}$ are expressed in the lab frame. In accelerators and storage rings the electric term is generally negligible. We will also transform the independent variable t to the turning angle θ around the ring by the relations

$$v dt = (1+\frac{x}{\rho})ds \quad \text{and} \quad \frac{ds}{\rho} = d\theta$$

where

ρ = radius of curvature of the planar closed orbit,

s = distance along the closed orbit,

and we used the right-handed coordinates

x = normal in the horizontal plane of the closed orbit,

y = tangent to the closed orbit,

z = vertical.

The transformed equation is

$$\frac{d\vec{s}}{d\theta} = \vec{s}\times\vec{\Omega} \tag{3}$$

where the precession vector is

$$\vec{\Omega} = \frac{\rho}{(B\rho)}(1+\frac{x}{\rho})\left[(1+\gamma G)\vec{B}_\perp + (1+G)\vec{B}_\parallel\right]_{\text{particle}}$$

$$\equiv \zeta_R\hat{x} - \zeta_I\hat{y} - \kappa\hat{z} = \frac{1}{2}(\zeta\hat{X} + \zeta^*\hat{X}^*) - \kappa\hat{z}$$

and, as defined

$$\zeta = \zeta_R + i\zeta_I \qquad \text{and} \qquad \hat{X} = \hat{x}+i\hat{y}.$$

To calculate the components ζ and κ of $\vec{\Omega}$ one must first solve the orbit equation

$$\frac{d\vec{v}}{dt} = \frac{e}{\gamma mc}\,\vec{v}\times\vec{B} \tag{4}$$

with given $\vec{B}(x,z,s)$ for the orbital motion $x = x(s)$, $z = z(s)$, then obtain $(\vec{B}_{\perp})_{particle}$ and $(\vec{B}_{\parallel})_{particle}$ for use in the spin equation (3). Or one could express $\vec{B}_{\perp}$ and $\vec{B}_{\parallel}$ directly in terms of the particle coordinates. The result[4] is, to first order in x and z.

$$\begin{cases} \kappa = \gamma G = \text{vertical precession angular velocity} \\ \zeta = -\ (1+\gamma G)\rho z'' - i[(1+\gamma G)z' - \rho(1+G)(\frac{z}{\rho})'] \equiv \Sigma\ \epsilon e^{-i\kappa_o\theta} \\ \qquad (\epsilon \equiv |\epsilon|e^{i\lambda}, \quad \text{and} \quad \text{prime} \equiv \frac{d}{ds}) \end{cases} \tag{5}$$

The (complex) horizontal precession angular velocity ζ is zero for particles traveling on the ideal planar closed orbit on which the magnetic field is everywhere vertical. Thus, ζ is the coupling between up and down spins responsible for depolarizing the beam. As indicated, it is multiperiodic with frequencies

$$\kappa_o = \text{integer} = k$$

arising from vertical closed orbit distortions caused by magnet imperfections and

$$\begin{aligned} \kappa_o &= \text{superperiodicity} \pm \text{vertical oscillation tune} \\ &= kS \pm \nu_z \end{aligned}$$

arising from the vertical oscillation of the particle in a magnet lattice with S superperiods. There may also be frequencies involving ν_x and ν_y entering through higher order terms but they are generally small. The amplitude $|\epsilon|$ of ζ has the energy dependence.

$$|\epsilon| \propto \gamma \qquad \text{for "imperfection" terms } \kappa_o = k$$

and

$$|\epsilon| \propto \sqrt{\gamma/\beta} \qquad \text{for "intrinsic" terms } \kappa_o = kS\pm\nu_z.$$

Resonance occurs when the vertical precession frequency equals one of the frequencies contained in ζ, or

$$\kappa \equiv \gamma G = \begin{cases} k & \text{imperfection resonance} \\ kS \pm \nu_z & \text{intrinsic resonance} \end{cases} \tag{6}$$

On a resonance the horizontal precession may accumulate to large values and depolarization may result.

Computer programs are available to evaluate ϵ for specific ring lattices using Eq. (5). The computed resonance strengths for several synchrotrons and storage rings are plotted in Fig. 1. The γ dependencies of the imperfection and the intrinsic resonances are evident. The values of $|\epsilon|$ range from 10^{-2} at the low energy end of the AGS booster to over 10 at the high energy end of the SSC.

SPINOR FORMULATION AND RESONANCE CROSSING

To proceed further it is more convenient to use the spinor formulation. For this we write

$$\vec{s} = \psi^+ \vec{\sigma} \psi$$

where

$$\begin{cases} \psi = \psi(\theta) = \text{2-component spinor,} \\ \vec{\sigma} = \text{Pauli spin matrices,} \\ \qquad \sigma_x = \begin{pmatrix} 0 & 1 \\ 1 & 0 \end{pmatrix}, \quad \sigma_y = \begin{pmatrix} 0 & -i \\ i & 0 \end{pmatrix}, \quad \sigma_z = \begin{pmatrix} 1 & 0 \\ 0 & -1 \end{pmatrix}, \end{cases}$$

and where + denotes hermitian conjugation. The spin equation (3) then becomes the spinor equation

$$\frac{d\psi}{d\theta} = \frac{i}{2} H\psi \tag{7}$$

where the "hamiltonian" is

$$H = \vec{\Omega}\cdot\vec{\sigma} = \zeta_R\sigma_x - \zeta_I\sigma_y - \kappa\sigma_z = \begin{pmatrix} -\kappa & \zeta \\ \zeta^* & \kappa \end{pmatrix}$$

This is equivalent to going from the Heisenberg picture to the Schrödinger picture in quantum mechanics. ψ is the state vector. In this formulation vertical polarization is defined as

$$P_z \equiv \langle\psi^+\sigma_z\psi\rangle$$

For a single resonance $\zeta = \epsilon e^{-i\kappa_o \theta}$, we transform to the resonance frame by

$$\psi = e^{-\frac{i}{2}\kappa_o \theta \sigma_z} \phi$$

and get

$$\frac{d\phi}{d\theta} = \frac{i}{2} K\phi \tag{8}$$

where

$$K = \vec{\omega}\cdot\vec{\sigma} = \epsilon_R \sigma_x - \epsilon_I \sigma_y - \delta\sigma_z = \begin{pmatrix} -\delta & \epsilon \\ \epsilon^* & \delta \end{pmatrix},$$
$$(\delta \equiv \kappa - \kappa_o).$$

If δ and ϵ are constants (or adiabatic invariants) the formal solution is

$$\phi(\theta) = e^{\frac{i}{2}\theta\vec{\omega}\cdot\vec{\sigma}} \phi(0) \equiv M\phi(0).$$

The exponential operator defined as

$$M \equiv e^{\frac{i}{2}\theta\vec{\omega}\cdot\vec{\sigma}} = e^{\frac{i}{2}(\omega\theta)(\hat{\omega}\cdot\vec{\sigma})}$$
$$\equiv \cos\frac{\omega\theta}{2} + i(\hat{\omega}\cdot\vec{\sigma})\sin\frac{\omega\theta}{2}$$

is a rigid rotation of angle $\omega\theta$ about the axis $\hat{\omega}$. In general for a succession of rotations

$$M = M_k M_{k-1} \cdots\cdots M_2 M_1 \equiv e^{\frac{i}{2}\Theta(\hat{n}\cdot\vec{\sigma})}$$

The equivalent single rotation angle Θ and axis $\hat{n}$ are given by

$$\begin{cases} \cos\frac{\Theta}{2} = \frac{1}{2}\,\mathrm{Tr}(M) \\ \hat{n} = \dfrac{1}{2i\sin\frac{\Theta}{2}}\,\mathrm{Tr}(\vec{\sigma}M) \end{cases} \tag{9}$$

The steady state solutions of Eq. (8) (eigenstates of K) give at $\delta = 0$ the polarizations

$$P_z(\delta) = \pm\frac{\delta}{\sqrt{\delta^2 + |\epsilon|^2}}$$

Instantaneous transitions of these states to $-\delta$ give

$$\frac{P_z(-\delta)}{P_z(+\delta)} = \frac{\delta^2 - |\epsilon|^2}{\delta^2 + |\epsilon|^2}$$

Thus the strength $|\epsilon|$ is also the "width" of the resonance.

For a linear crossing of the resonance we write $\delta \equiv \kappa - \kappa_0 = \alpha\theta$ and writing

$$\phi \equiv \begin{pmatrix} f e^{-i\alpha\theta^2/4} \\ g e^{i\alpha\theta^2/4} \end{pmatrix}$$

we obtain from Eq. (8).

$$\begin{cases} \dfrac{df}{d\theta} = \dfrac{i}{2}\,\epsilon g e^{\frac{i}{2}\alpha\theta^2} \\[2ex] \dfrac{dg}{d\theta} = \dfrac{i}{2}\,\epsilon^* f e^{-\frac{i}{2}\alpha\theta^2} \end{cases} \tag{10}$$

The solutions of Eq. (10) are expressible in terms of confluent hypergeometric functions and give the Froissart-Stora[5] asymptotic relation for polarization

$$\frac{P_z(+\infty)}{P_z(-\infty)} = 2\exp(-\pi|\epsilon|^2/2\alpha) - 1 \tag{11}$$

To cross the resonance without depolarization we can do either of two things:

1. Make $|\epsilon|^2/\alpha \to 0$ and hence $P(+\infty) = P(-\infty)$. For intrinsic resonances $\gamma G = kS \pm \nu_z$ we can use pulsed quadrupoles to shift ν_z rapidly to obtain a fast crossing (large α). For imperfection resonances $\gamma G = k$ one uses trim dipoles to correct the vertical closed orbit distortion to reduce $|\epsilon|$. One can reduce $|\epsilon|$ also for intrinsic resonances using trim quadrupoles[6] but there are more conditions to be satisfied.

2. Make $|\epsilon|^2/\alpha \to \infty$ and hence $P(+\infty) = -P(-\infty)$. This is the obvious complementary procedure of 1. The slow adiabatic passage of a strong resonance results in a total polarization flip.

Both procedures have been applied to obtain the polarized beams at ZGS[7], AGS[8], Saturne[9] and KEK-PS[10].

ELIMINATING RESONANCES BY SPIN ROTATORS (SIBERIAN SNAKES)

A. Principles and descriptions

The possibility of eliminating the resonances altogether for the 3-dimensional spin precessional motion was first pointed out by Derbenev and Kondratenko[11]. The principle of the method is simply illustrated for a 1-dimensional resonance in Fig. 2. The upper curve shows an oscillatory motion driven resonantly by the oscillatory force shown as the lower curve. If the phase of the motion could be shifted by 180° every k oscillations, the amplitude would then be alternately excited and damped and the resonance becomes ineffective.

In practice, the phase shifting is not possible for the 1-dimensional oscillation, but quite straightforward for the 3-dimensional spin motion. It is equivalent to simply precessing the spin by 180° about an axis in the horizontal plane. With the 180° spin rotation applied regularly (every revolution, say) the horizontal depolarizing precession will grow for one revolution being excited by the resonance and damp for the next revolution. The series of dipoles placed in a straight section of the ring lattice to produce the 180° spin rotation is known as the Siberian snake. We will refer to it here simply as the Spin Rotator, a more meaningful name. The spin rotator must, of course, leave the particle orbit exterior to the rotator unaffected. Inside the rotator which may be more than 10 m long there will be transverse excursions of the particle trajectory away from the original closed orbit. These excursions are larger at lower energies and make greater demands on the dipole apertures.

A pair of evenly spaced rotators with orthogonal rotating axes form a Double Rotator and is most useful. Such a pair could comprise of a longitudinal rotator with axis along $\hat{y}$ (known as Type 1 Siberian snake invented by K. Steffen and shown in Fig. 3) and a transverse rotator with axis along $\hat{x}$ (Type 2 Siberian snake invented by A. Turrin and shown in Fig. 4). Although a rotator with an arbitrary axis can be designed, this pair of longitudinal and transverse rotators is adequate for all practical purposes. Note that a solenoid acts as a longitudinal rotator. But the field of the solenoid must be proportional to γ and the non-zero focal action of the solenoid on the beam must be compensated.

All rotators shown here employ only horizontal and vertical dipoles. Clearly, dipoles with arbitrary roll angles can also be applied. In fact, helical rotators using dipoles with continuously advancing roll-angles have been proposed by Ya. S. Derbenev and by E. D. Courant. So far, no systematic procedure for the design of rotators has been found. Rotators are still invented by clever cutting and fitting.

The amount of horizontal precession induced by ϵ between rotators (the analog of the amplitude growth in the 1-dimensional illustration) should not get too large. This implies that for strong resonances more than one rotator is required. For weak resonances the illustration of Fig. 2 indicates that if the total horizontal precession in k passages of the rotator is tolerable the rotation per passage needs be only $> \pi/k$. Such a partially excited rotator first suggested by T. Roser, has the advantage of having smaller orbit excursions in the dipoles thereby reducing the aperture requirements at lower energies just when the resonances are weaker.

B. Actions of rotators

We now examine the actions of the rotators quantitatively. The rotation matrix of a full transverse rotator is

$$M_o = e^{\frac{i}{2}\pi\sigma_x} = \cos\frac{\pi}{2} + i\sigma_x \sin\frac{\pi}{2} = i\sigma_x \tag{12}$$

With one such rotator the rotation matrix for one revolution is

$$\begin{aligned} M &= M_{\text{half-turn}} M_o M_{\text{half-turn}} \\ &= e^{\frac{i}{2}\pi(\vec{\omega}\cdot\vec{\sigma})}(i\sigma_x)e^{\frac{i}{2}\pi(\vec{\omega}\cdot\vec{\sigma})} \\ &= i\sigma_x - 2\,\frac{\epsilon_R}{\omega}\sin\frac{\pi\omega}{2}\left(\cos\frac{\pi\omega}{2} + i\sigma_\omega \sin\frac{\pi\omega}{2}\right) \end{aligned} \tag{13}$$

where, as was already defined before

$$\vec{\omega} \equiv \omega\hat{\omega} = \epsilon_R\hat{x} - \epsilon_I\hat{y} - \delta\hat{z}$$

$$\omega \equiv |\vec{\omega}| = \sqrt{\delta^2 + |\epsilon|^2}, \qquad \sigma_\omega \equiv \hat{\omega}\cdot\vec{\sigma}$$

The equivalent one-turn precession angle θ is given by

$$\cos\frac{\theta}{2} = \frac{1}{2}\,\mathrm{Tr}(M) = -\frac{\epsilon_R}{\omega}\sin\pi\omega = -\cos\lambda\,\sin\pi|\epsilon| \qquad (14)$$

where λ is the phase angle of $\epsilon \equiv |\epsilon|e^{i\lambda}$ and where the last expression is the maximum value at $\delta=0$. To reach resonance $\theta=0$ and the right-hand expression must reach unity. Even for $\lambda = \pi$ or $-\cos\lambda = 1$ no resonance can be reached if

$$|\epsilon| < \frac{1}{2} \qquad (15)$$

i.e. all resonances with $|\epsilon| < 1/2$ are eliminated. The same is true with a single longitudinal rotator with matrix $M_\pi = i\sigma_y$. (See article C. below for an explanation of the subscript on M.)

For a double rotator

$$\begin{aligned} M &= M_\pi\, M_{\text{half-turn}}\, M_o\, M_{\text{half-turn}} \\ &= (i\sigma_y)\, e^{\frac{i}{2}\pi(\vec{\omega}\cdot\vec{\sigma})}\,(i\sigma_x)\, e^{\frac{i}{2}\pi(\vec{\omega}\cdot\vec{\sigma})} \\ &= i\sigma_z - 2\,\frac{\epsilon_R}{\omega}\sin\frac{\pi\omega}{2}\left(i\sigma_y\cos\frac{\pi\omega}{2} - \sigma_y\sigma_\omega\sin\frac{\pi\omega}{2}\right) \end{aligned} \qquad (16)$$

and

$$\begin{aligned} \cos\frac{\theta}{2} &= -2\,\frac{\epsilon_R}{\omega}\,\frac{\epsilon_I}{\omega}\,\sin^2\frac{\pi\omega}{2} \\ &= -\sin 2\lambda\,\sin^2\frac{\pi|\epsilon|}{2} \qquad (\text{max. at } \delta=0) \end{aligned} \qquad (17)$$

The condition for not reaching resonance even at $-\sin 2\lambda = 1$ is then

$$|\epsilon| < 1 \qquad (18)$$

As expected the double rotator is "twice as good" as a single rotator. Another advantage of the double rotator is that its "closed spin-trajectory" (analog of the closed orbit for the orbital motion) is mostly vertical. Hence the spins of a vertically polarized beam will stay near the closed spin-trajectory to form a "paraxial" dynamics.

C. <u>Multiple rotators for strong resonances</u>

For strong resonances with $|\epsilon|>1$ we need more than one pair of double rotators. If we have N pairs the rotation matrix for each pair is

$$M_{pair} = (i\sigma_y)\, e^{\frac{i}{2}\frac{\pi}{N}(\vec{\omega}\cdot\vec{\sigma})}\, (i\sigma_x)\, e^{\frac{i}{2}\frac{\pi}{N}(\vec{\omega}\cdot\vec{\sigma})} \tag{19}$$

and we have

$$\cos\frac{\theta}{2N} = -\sin 2\lambda \sin^2\frac{\pi|\epsilon|}{2N} \qquad (\text{max. at } \delta = 0) \tag{20}$$

The condition for not reaching resonance ($\theta=0$) at $-\sin2\lambda = 1$ is modified to

$$|\epsilon| < N \tag{21}$$

from Eq. (18) for a single pair. Thus to eliminate all resonances with $|\epsilon|$ up to 10 in SSC one would need more than 10 pairs of double rotators evenly distributed around the circumference.

A look at the general arrangement[12] of multiple rotators of arbitrary types is illuminating. A rotator with an arbitrary axis in the horizontal plane may be considered as an α-rotation about $\hat{z}$ followed by a π-rotation about $\hat{x}$. The matrix of such a rotator is

$$M_\alpha = e^{\frac{i}{2}\pi\sigma_x}\, e^{\frac{i}{2}\alpha\sigma_z} = i(\sigma_x\cos\frac{\alpha}{2} + \sigma_y\sin\frac{\alpha}{2}) \tag{22}$$

namely a π-rotation about an axis oriented at angle $\alpha/2$ from the transverse axis $\hat{x}$. A general rotator is shown in Fig. 5.

In the most general arrangement of 2N rotators the no-coupling ($\epsilon=0$) vertical precession angle in one turn is

$$\theta = \Sigma\, (-1)^k\, (\alpha_k + \psi_k)$$

where ψ_k is the vertical precession angle due to the ring lattice preceding the kth rotator and is energy dependent. The alternating signs account for the π-rotation about $\hat{x}$ at each rotator. To make θ independent of energy we must have

$$\Sigma\, (-1)^k\, \psi_k = 0. \tag{23}$$

then we have

$$\theta = \Sigma\, (-1)^k\, \alpha_k \tag{24}$$

The arrangement of N evenly distributed pairs of transverse (α=0) and longitudinal ($\alpha=\pi$) rotators corresponds to the case where all ψ_k are equal. Eq. (23) is then automatically satisfied and Eq. (24) gives $\theta = N\pi$.

D. <u>Partial rotators for weak resonances</u>

The one-revolution rotation matrix for a partial (precession angle = πa with a<1) transverse rotator is

$$M = e^{\frac{i}{2}\pi w \sigma_w} e^{\frac{i}{2}\pi a \sigma_x} e^{\frac{i}{2}\pi w \sigma_w}$$

and we have at $\delta = 0$

$$\cos\frac{\theta}{2} = \frac{1}{2}\,\mathrm{Tr}(M) = \cos\frac{\pi a}{2}\cos\pi|\epsilon| - \cos\lambda\,\sin\frac{\pi a}{2}\sin\pi|\epsilon| \quad . \qquad (25)$$

At $\lambda=\pi$ which maximizes this expression we have

$$\cos\frac{\theta}{2} = \cos\pi(\frac{a}{2}-|\epsilon|) . \qquad (26)$$

the condition for not reaching resonance is then

$$|\epsilon| < \frac{a}{2} \qquad (27)$$

agreeing with Eq. (15) at a = 1. This simply states that the rotator must be just strong enough to keep the "operating point" outside (twice) the width of the resonance. At low energies when the resonances are weak, the rotators can be turned on at low "strength" so that the orbit excursion which is expected to scale as a will not be excessively large. For example in the AGS the second lowest intrinsic resonance at $\gamma G = \nu_z = 8.75$ or $\gamma = 4.88$ is relatively strong with $|\epsilon| = 0.015$. A pair of partial double rotator with a = 0.03 will be more than adequate. The orbit excursion in a 1.8 tesla field transverse rotator is 2.6 m/$\beta\gamma$ = 55 cm unacceptable large, but at a = 0.03 the excursion is only 55 cm $\times$ 0.03 = 1.65 cm.

Let us examine the general scaling of the orbit excursion with energy. If we set a $\propto$ $|\epsilon|$ then the orbit excursion in a partial rotator will scale as

$$\text{excursion} \propto |\epsilon|/\beta\gamma \propto \begin{cases} \beta^{-1} & \text{imperfection res.} \\ (\beta^3\gamma)^{-1/2} & \text{intrinsic res.} \end{cases} \qquad (28)$$

At $\gamma = 1.5$ we have $\beta^{-1} = 1.34$ and $(\beta^3\gamma)^{-1/2} = 1.27$, both acceptable values. Thus, the rotator resonance elimination scheme is practicable for all energies above $\gamma = 1.5$. If more than one pair of rotators are needed at the upper end of the energy range, each rotator can be operated at even lower "strength" at the low energy end. Thus, one can reach down to even lower γ values without entailing excessively large orbit excursions.

Most of the analyses of the actions of rotators described throughout this Section have been double-checked by numerical tracking.

It is, by now, more than ten years since the spin rotator was first proposed, yet no experimental test of the scheme has been made. An experiment to test a solenoid longitudinal rotator is in progress[13, 14] on the cooling ring of the Indiana University Cyclotron Facility and will soon be conducted. We expect that it will provide the first experimental confirmation of the analyses.

The above discussions show that the spin rotator (Siberian snake) resonance elimination scheme is practical and effective from γ values below 1.5 up to arbitrarily high values limited only by the number of rotators used. For very strong and especially, overlapping resonances this is the only viable scheme to avoid depolarization. In designing future accelerators and storage rings it is advisable to have space provided in the magnet lattice to accommodate the rotators.

In addition to rotators, research and development efforts on polarized sources and polarimeters are also in great need. The intensity of existing polarized ion sources is some three orders of magnitude below that of unpolarized sources. High intensity and high degree of polarization sources, and fast and non-destructive polarization measurements are very desirable.

REFERENCES

1. A. A. Sokolov and I. M. Ternov, Sov. Phys. Doklady. 8, 1203 (1964)
2. L. H. Thomas, Phil. Mag. 3 (1927)
3. V. Bargmann, L. Michel, and V. L. Telegdi, Phys. Rev. Lett. 2, 439 (1959)
4. E. D. Courant and R. D. Ruth, Brookhaven Nat. Lab. report BNL-51270 (1980)
5. M. Froissart and R. Stora, Nucl. Instr. and Meth. 7, 297 (1960)
6. A. W. Chao, AIP Conf. Proc. No. 95, p.458 (1983)
7. T. Khoe et al., Particle Accelerators 6, 213 (1975)
8. L. G. Ratner, Proc. 7th Int. Symp. on High Energy Spin Phys., Protvino 2, 206 (1987)
9. E. Grorud et al., AIP Conf. Proc. No. 95 p.407 (1983)
10. S. Hiramatsu et al., Proc. 7th Int. Symp. on High Energy Spin Phys., Protvino 2, 119 (1987)
11. Ya. S. Derbenev and A. M. Kondratenko, Proc. 10th Int. Conf. on High Energy Accel., Protvino 2, 70 (1977)
12. K. Steffen, AIP Conf. Proc. No. 145, p.154 (1985)
13. A. D. Krisch et al., Proposal of IUCF Experiment No. CE-05 (Sept. 1986)
14. T. Roser, Proc. Int. Workshop on Hadron Fac. Tech., Los Alamos Nat. Lab. report LA-11130-C, p.317 (1987)

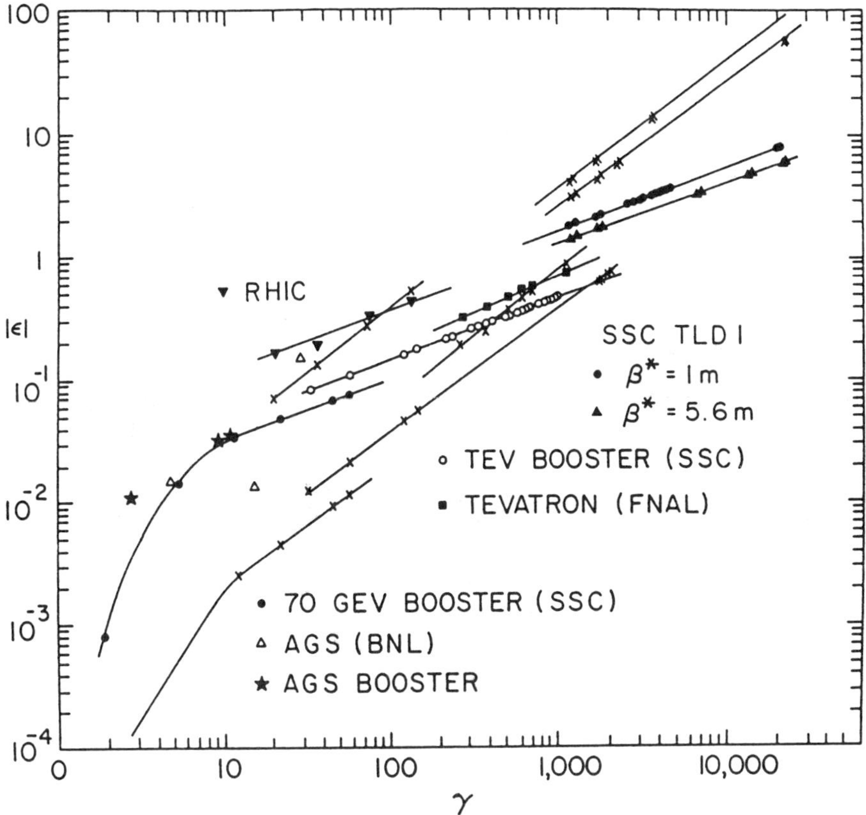

Figure 1. Strengths or widths $|\epsilon|$ of depolarizing resonances in several synchrotrons and storage rings plotted against energy γ. For intrinsic resonances (various symbols as denoted) the normalized invariant of the vertical oscillation is assumed to be 10π mm-mrad. For imperfection resonances (shown as crosses) the maximum alignment and gradient errors are assumed to be $\Delta x = \Delta z = 0.1$ mm and $\Delta B'/B' = 10^{-3}$.

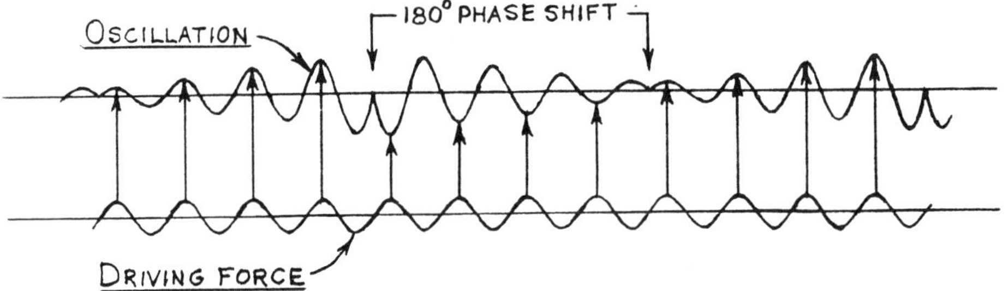

Figure 2. A harmonic oscillation driven by a force in resonance. The motion shifts phase 180° every k (4 as shown) oscillations. The force alternately excites and damps the oscillation between phase shifts.

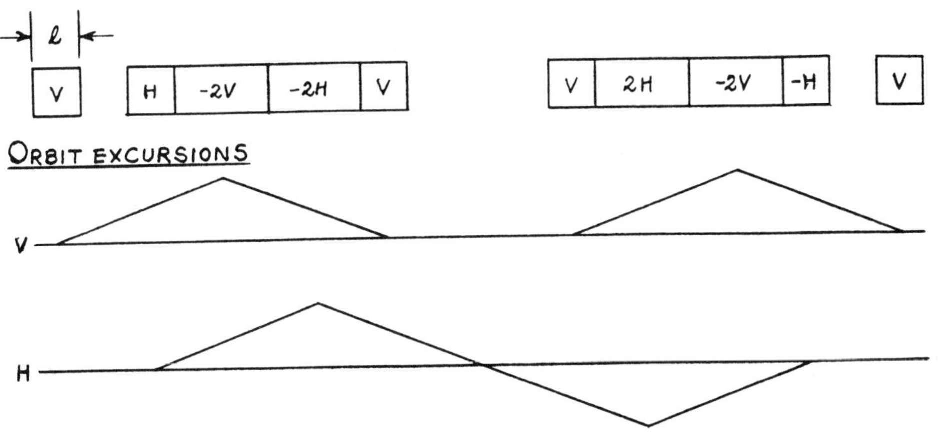

	Conventional dipole	Superconducting dipole
Field B	1.83 T	5.48 T
Unit length ℓ	0.75 m	0.25 m
Total length 19 ℓ	14.25 m	4.75 m
Orbit excursion $\Delta x = \Delta z$	1.15 $m/\beta\gamma$	0.38 $m/\beta\gamma$

Figure 3. Longitudinal rotator (Type 1 Siberian snake) precesses the spin 180° about the longitudinal ($\hat{y}$) axis. For the proton each unit has $B\ell$ = 1.37 Tm and precesses the spin 45°. H and V denote horizontal and vertical orbital deflections.

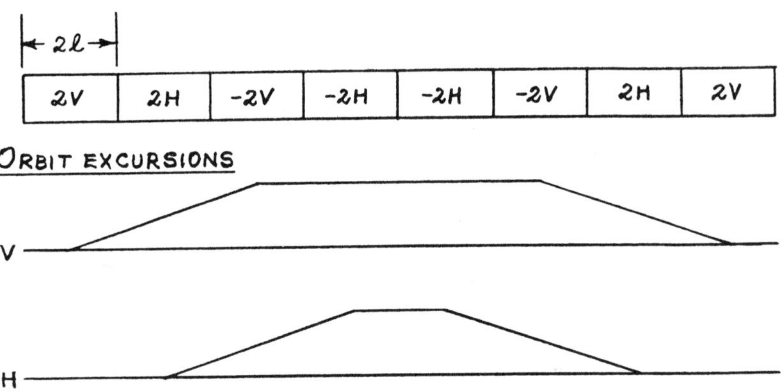

	Conventional dipole	Superconducting dipole
Field B	1.83 T	5.48 T
Unit length ℓ	0.75 m	0.25 m
Total length 16 ℓ	12 m	4 m
Orbit excursion $\Delta x = \Delta z$	2.63 m/$\beta\gamma$	0.88 m/$\beta\gamma$

Figure 4. Transverse rotator (Type 2 Siberian snake) precesses the spin 180° about the transverse ($\hat{x}$) axis. For the proton each unit has $B\ell = 1.37$ Tm and precesses the spin 45°. H and V denote horizontal and vertical orbit deflections.

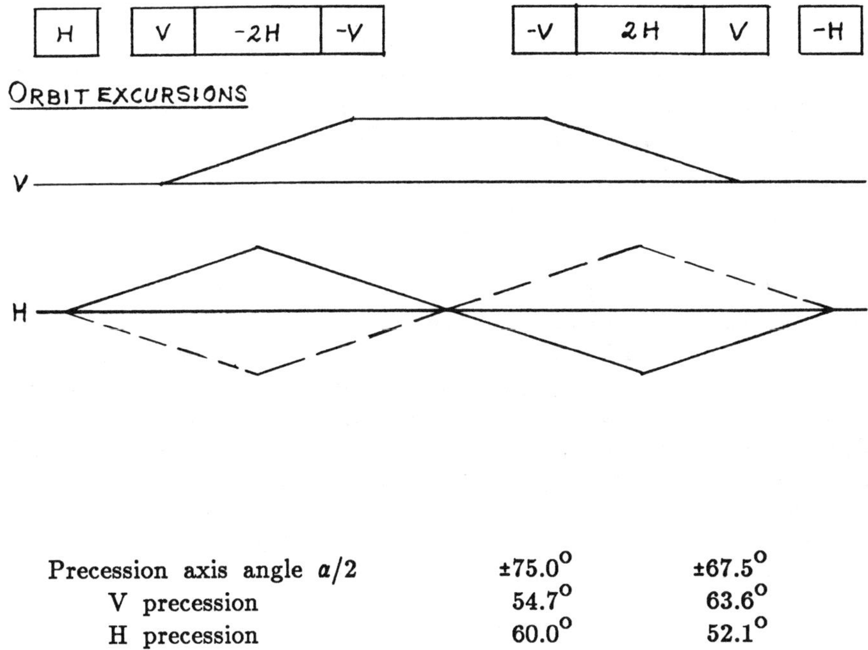

Precession axis angle $\alpha/2$	$\pm75.0^{\circ}$	$\pm67.5^{\circ}$
V precession	54.7°	63.6°
H precession	60.0°	52.1°

Figure 5. Rotators with intermediate precession axes, neither transverse ($\alpha/2=0$) nor longitudinal ($\alpha/2=90^{\circ}$). Rotators with other axial directions can also be designed.

ELECTRON AND MUON PHYSICS SESSIONS: SUMMARY

H.E. Montgomery
Fermi National Accelerator Laboratory
P.O.Box 500, Batavia, Il 60510, U.S.A.

ABSTRACT

A short summary is given of the Electron and Muon Physics parallel sessions at the Conference on the Intersections of Nuclear and Particle Physics, Samoset, Maine in May 1988. The sessions occupied approximately 200 minutes, the summary occupied 15 minutes.

INTRODUCTION

The electromagnetic interaction needs no introduction as a probe of the structure of systems on many scales. The continued use of this technique dominated the sessions on Electron and Muon Physics at the Samoset Meeting. The experimental results continue to stimulate large numbers of theorists and the results on polarized deep inelastic muon scattering and their various interpretations permeated beyond these sessions. The breadth of physics attacked with electrons and muons makes a summary, such as this, rather peculiar. As one of my nuclear physics friends (I think) commented after my summary, "it was interesting to see Nuclear Physics from a long distance with the telescope inverted". The comment may well be applied to this written version of the summary talk.

WHAT GIVES THE NUCLEON IT'S SPIN?

This year is the 20th anniversary of the presentation[1] of the data from SLAC on the scaling of the inelastic scattering cross-section of electrons from nucleons. One of the most fundamental discoveries which followed in the years that followed was that while all the scattering was from quarks, not all the momentum of the nucleon could be accounted for in the scattering. This of course led to the Gluon being introduced. It is therefore interesting to note that one of the possible conclusions from the recent results on polarized muons scattering from polarized protons is that a small fraction of the spin of the nucleon results from the quarks. The data themselves[2] are a measurement of a polarization asymmetry, which when folded with structure function measurements yield a polarized structure function. There are predictions for the integral, over all X_{Bj} of this structure function. It is the construction and interpretation of this integral, which goes beyond the data themselves, which is causing the heat and controversy[2,3,4].

I mentioned above one of the candidate conclusions from these data because it is a favorite observation of mine, that the data on several important structure functions is lacking at low X_{Bj} and moderate momentum transfer. This places the evaluation of several quark sum rules[5] on shaky experimental ground.

STRUCTURE FUNCTIONS, FORM FACTORS; INTERACTIONS IMPORTANT

A full picture of the understanding of structure functions in deep inelastic scattering was presented in the Plenary Sessions. In the parallel session the new results from EMC[7] at CERN and from SLAC-E140[8] now present a rather consistent picture of the ratio between the structure functions of nucleons as measured in moderately heavy nuclei, iron and copper, and those measured on deuterium. One of the components which contributes to the confidence in the result is a determination of the difference between the longitudinal contribution to the scattering from the different targets[8]. The result is consistent with zero and incidentally is a good absolute measurement of the longitudinal to transverse ratio which probably forces a reevaluation of all the structure function data in the SLAC kinematic range. One notes that in all the discussions it is not sufficient to consider the constituents, the partons, but their interactions expressed in terms of QCD corrections or otherwise, are important to describe the data. This is seen also in the new data[9] on the electromagnetic form factor of the deuteron. The electric form factor is comparatively easily described by the models using the basic deuteron constituents however the magnetic form factor requires[10] a full range of exchange currents.

NUCLEONS IN THE NUCLEUS; y SCALING OR NOT?

Changing scales, in momentum transfer and energy, drops us into a region where the target is a nucleus and the constituents are nucleons but lots of analogies with high energies are possible. Experimentally, separation of the longitudinal and transverse components of the scattering is normal practice. In general the picture for light nuclei[10,11] is that the quasi elastic scattering data are moderately well described by models and the Coulomb Sum Rule looks reasonable although nucleon correlation effects are evident at low momentum transfer. y scaling is analogous to Bjorken scaling in that the resulting distributions give a representation of the constituent, nucleon, momentum distribution. There are significant differences[12], of perhaps 40%, between heavy and light nuclei and there appears to be inconsistency between longitudinal and transverse responses[13]. However among the heavy nuclei there is little variation. Model calculations[14] in this field suggest that there is no reason to expect major manifestations of quark structure as distinct from a conventional description although it is clear that y-scaling is broken by the data.

When a final state particle, in addition to the lepton, is observed there are opportunities to further separate the components of the interaction. Such techniques with the wide angle bremmstrahlung as the final state particle now appear to be feasible[15]. The data[16] that exist with a final state proton, detected in the direction of the virtual photon, have many of the problems with the longitudinal transverse ratio shown by the single arm data. However understanding of the data in terms of the nucleon anomalous magnetic moment can be achieved at the 10% level with some effort.

REAL PHOTONS; SUPPORT FOR C.I.M.

Real photons are of course merely a particular subset of virtual photons so they yield very similar types of information. There are now data[17] which compare interactions with an ^{16}O target which yield a final state neutron with those yielding a proton. There are similarities between the data but there is a clear deviation from model expectations at 90 degrees when the photon energy is above about 150 MeV. A class of models used to describe pion production from nuclei is the Δ-hole model. It appears that rather good agreement can be achieved with the data on π^0 production but charged pion production is by no means as well described[18].

It is interesting to note that some data[19] on deuteron photo-disintegration can be well described by models based on constituent interchange exchange. These models are traditionally associated with high energies and high transverse momenta and these data provide a puzzling twist to the discussion[20] of the validity of extensions of these ideas and QCD to exclusive processes in general.

WEAK INTERACTION EFFECTS

There have been several important experiments on weak interactions using electron or muon beams, it was gratifying to see that the Parity violation experiment[21] at Bates using polarized electrons scattering from a Carbon target is approaching the level of sensitivity at which it should see the weak-electromagnetic interference.

CONCLUSIONS

The interest shown in the electron, muon sessions, the number of contributed papers fully justify the important role which this broad filed of physics continues to play. The discussion was lively and open and I for one learned a lot. I acknowledge the efforts of Bill Donnelly who was the joint organizer of this session and who helped me with aspects of the presentation of this summary.

REFERENCES

[1] W.K.H. Panofsky, 14th Int. Conf. on High Energy Physics, Vienna 1968, Ed.

J. Prentki, J. Steinberger, CERN 1968. E.Bloom et al, Paper #363 submitted to that Conference.

[2] V. Papavassilliou, talk presented at this conference.

[3] M. Karliner, talk presented at this conference.

[4] F. Close, talk presented at this conference.

[5] See eg. F. Sciulli, Proc. Int. Symp. on Lepton and Photon Interactions at High Energies, Kyoto 1985. Ed. M.Konuma, K.Takahashi, NISHA,Kyoto,1986.

[6] K. Rith, Plenary Talk presented at this conference.

[7] P. Renton, talk presented at this conference.

[8] P. Bosted, talk presented at this conference.

[9] P. Bosted, (second)talk presented at this conference.

[10] R. Dymarz, talk presented at this conference.

[11] K. Dow, talk presented at this conference.

[12] R. McKeown, talk presented at this conference.

[13] R. Lowry, Plenary Talk presented at this conference.

[14] S. Kumano, talk presented at this conference.

[15] C. Papanicolas, talk presented at this conference.

[16] G. van der Steenhoven, talk presented at this conference.

[17] E. Beise, talk presented at this conference.

[18] J. Koch, talk presented at this conference.

[19] J. Napolitano, talk presented at this conference.

[20] N. Isgur, Plenary Talk presented at this conference.

[21] P. Souder, talk presented at this conference.

SUMMARY OF NEUTRINO PRESENTATIONS

D. Hywel White
Los Alamos National Laboratory

ABSTRACT

This summary is divided into two sections. First, we concentrate on conventional neutrino physics interpreted in the context of standard electroweak theory. Second, we discuss double beta decay where gross violations of the predictions of the theory might appear, and also we discuss specific searches for consequences of finite neutrino mass.

SUMMARY

I. Two measurements were reported at this meeting of interaction cross sections of neutrinos on nuclei ^{12}C. The first, reported by X-Q Lu used the 800 MeV proton beam at LAMPF to produce an intense source of neutrinos through π^+ production and decay and also through subsequent μ^+ decay in the beam stop. The experiment was designed to measure neutrino-electron scattering and verify that the interference between charged and neutral-current contributions is destructive as predicted by the standard model. In addition the background to this measurement is due to charged-current scattering on carbon in the detector and so a measurement of this cross section was made. An additional trick allowed isolation of the scattering to the ^{12}N ground state; ^{12}N beta decays with a 16-ms lifetime and an end-point energy of 15 MeV, so that the decay electron is readily observable in the detector. In Fig. 1 is shown the time distribution of pulses following an electron event with beam on and

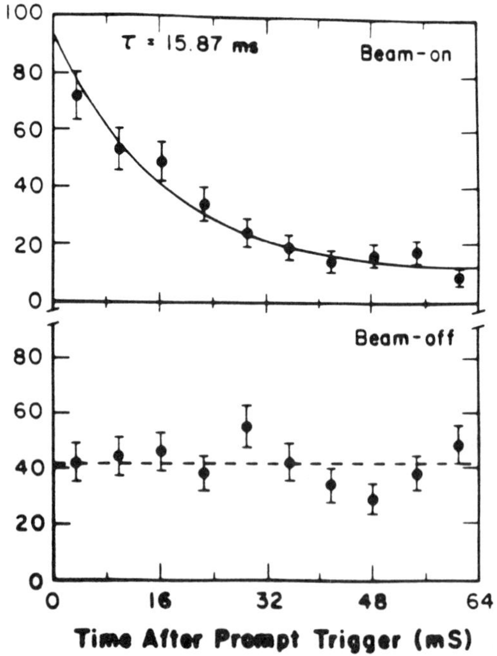

Fig. 1

beam off. The presence of the 16 ms component of beam on data is apparent. The neutrino flux is known from calculation, and so an absolute cross section is determined. There has existed a calculation of these cross sections in the literature by T. W. Donnelly for some time and the measured rates are compared with the calculation in Table I. The agreement is satisfactory although, now that data are available, perhaps further refinement of the calculation may be expected.

TABLE I

Measured Cross Sections

$$\sigma_{gs} = (1.25 \pm 0.11 \pm 0.14) \times 10^{-41} cm^2$$

$$\sigma_{inc} = (1.85 \pm 0.22 \pm 0.23) \times 10^{-41} cm^2$$

T. W. Donnelly Calculation

$$\sigma_{gs} = 0.94 \times 10^{-41} cm^2$$

$$\sigma_{inc} = 1.31 \times 10^{-41} cm^2$$

Robert Manweiler reported on a second experiment that is similar in some ways except that the neutrinos are derived from decay of pions in flight,also at LAMPF. The neutrinos are higher in energy, typically 150 MeV. Events are detected by observing the muon decay after coming to rest. The range of the final-state muon is small, and the decay electron gives a muon lifetime shown in Fig. 2 for both exclusive events, i.e., to the ground state of ^{12}N as above, and to all final nuclear states. The inclusive events have a small background, but the reactions leading to the ground state seem particularly clean. The muon kinetic-energy distributions shown in Fig. 3 are similar for both reaction classes. Recall that the muon kinetic-energy distribution is a little less than 105 MeV below the incident-neutrino energy. Again the neutrino-energy distribution and absolute flux are calculated, and then a cross section is derived. In Table II is shown a comparison of cross section with calculation.

TABLE II

$$\sigma(\nu_\mu + {}^{12}C \rightarrow \mu^- + X) = (10.9 \pm 1.8) \times 10^{-39} cm^2$$

Process	Events	σ_{exp}	σ_{theory}	
Inclusive	63	10.9 ± 2.8	8.2	(DW & O'Connell)
12N(g.s.)	19	3.4 ± 1.6	0.08	(Donnelly)

The inclusive cross section seems reasonable, but the Donnelly calculation predicts that the experiment would not have seen any exclusive events and 3.4 ± 1.6 are seen. This is a somewhat troubling situation, although the experimental statistical precision is limited.

At Brookhaven, a program to measure ν_μ - e scattering to determine the electroweak mixing angle is complete. Milind Diwan reported on the results of the final

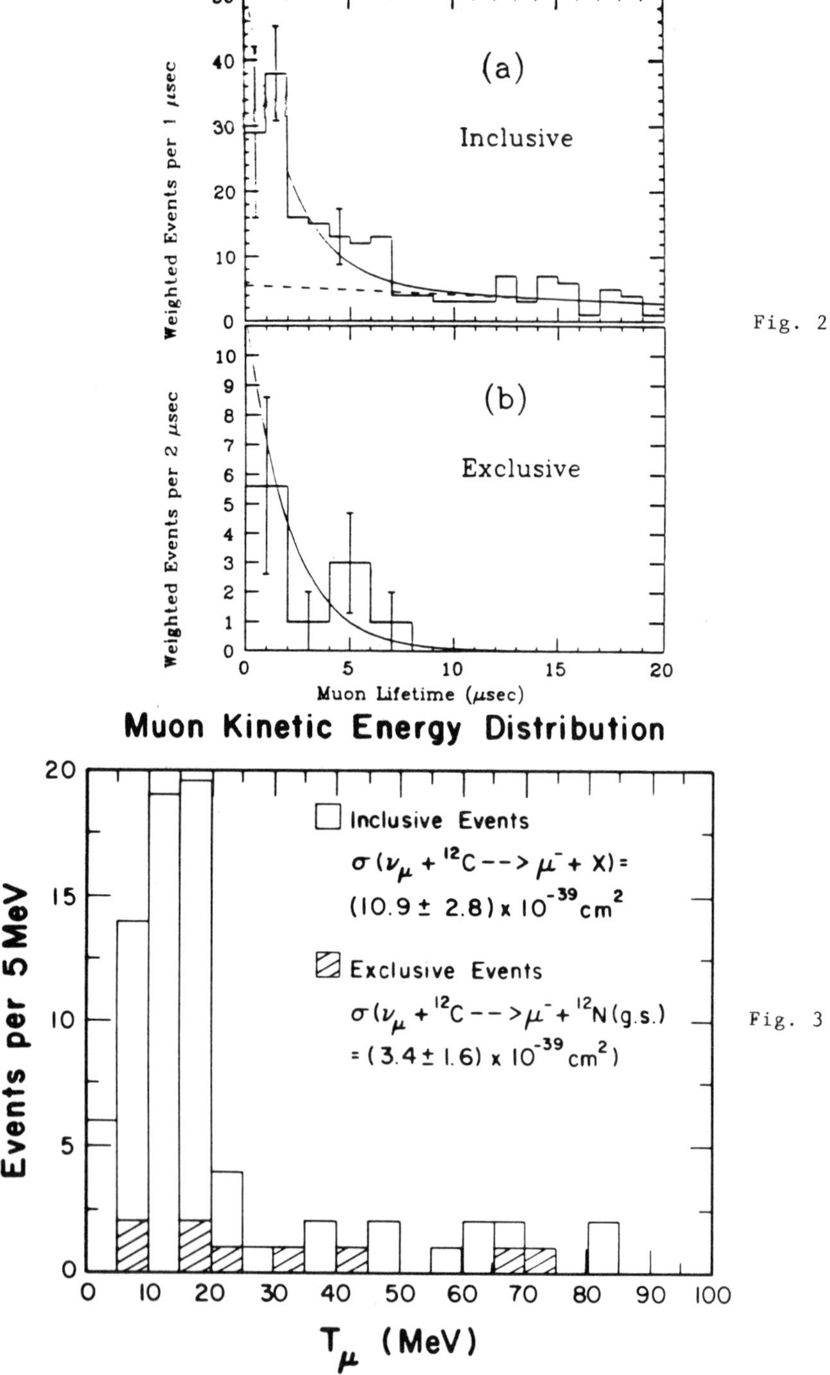
(a)
Inclusive
Weighted Events per 1 μsec
(b)
Exclusive
Weighted Events per 2 μsec
Muon Lifetime (μsec)
Muon Kinetic Energy Distribution
Inclusive Events
σ(ν_μ + ¹²C --> μ⁻ + X) =
(10.9 ± 2.8) x 10⁻³⁹ cm²
Exclusive Events
σ(ν_μ + ¹²C --> μ⁻ + ¹²N(g.s.)
= (3.4 ± 1.6) x 10⁻³⁹ cm²)
Events per 5 MeV
T_μ (MeV)

Fig. 2

Fig. 3

analysis. The beam is from a decay-in-flight source, and events are detected in a highly segmented detector with capability to isolate recoil electrons and measure angle and energy. Neutrino-electron-scattering events are identified by the smallness of the angle of the recoil electron and the neutrino direction; this distribution is shown in Fig. 4. The signal peak at small angles is clear and the cross-section ratio for neutrino and antineutrino scattering is made with precision. This group now has as many events as all previously published data, although more data are expected from CHARM II shortly. In order to normalize the individual neutrino and antineutrino cross sections quasi-elastic scattering

$$\nu_\mu + n \rightarrow \mu^- + p$$
$$\bar{\nu}_\mu + p \rightarrow \mu^+ + n$$

was used, for events predominantly at low Q^2. The event sample is shown in Table III together with the beam energies for both neutrinos and antineutrinos.

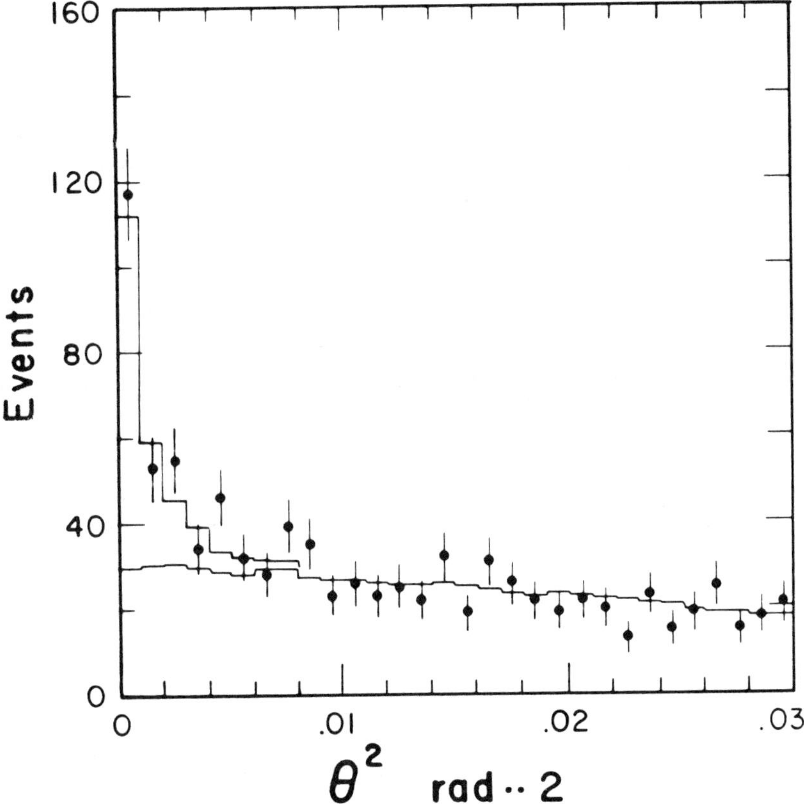

Fig. 4

TABLE III

ELECTRON	ν_μ	$\bar{\nu}_\mu$
observed	159.5 ± 17.3 ± 7.3	96.7 ± 13.2 ± 8.8
efficiency corrected	322.6 ± 35.5 ± 24.5	163.3 ± 22.6 ± 16.8
MUON		
observed	79082	59241
efficiency corrected	10.1×10^5 (12.3%)	4.26×10^5(8.6%)
BEAM		
$< E_n >$	1.274 ± 0.022 GeV	1.227 ± 0.022
$< \sigma(QE) >$	0.919×10^{-38}	0.381×10^{-38}

The value of $\sin^2\theta_W$ is

$$0.197 \quad {}^{+\,0.020}_{-\,0.021} \pm 0.017 ,$$

which is to be compared with 0.228 ± 0.007 ± 0.002 from the masses of W and Z and 0.233 ± 0.003 ± 0.005 from deep-inelastic neutrino scattering. These numbers come from the compilation of Amaldi et al.[1] It is probably too early to worry about a possible discrepancy, but the universality of $\sin^2\theta_W$ after radiative corrections is indeed a cornerstone of the electroweak theory.

II. It is assumed in the standard electroweak model that the mass of the neutrino is zero. This is primarily to assure that neutrinos are left handed and the right-handed sector is absent. Boris Kayser presented a challenge to this point of view and pointed out that there are good reasons to expect neutrino masses to be small, but not zero. The seesaw model gives reason for neutrino masses to be very tiny compared to present measurement capability. The conventional Dirac neutrino also came under scrutiny with a strong possibility that neutrinos may be Majorana; their own antiparticle. This array of physical possibilities makes the neutrino a very interesting particle, although classically difficult to study. Information on neutrino mass can come from the following topics

1. β decay
2. ν oscillations
3. ββ decay
4. Solar neutrinos
5. Supernovae.

There were presentations on Topics 1, 2, 3, and 4 in this neutrino session.

John Wilkerson discussed the state of experiments seeking to observe evidence of finite neutrino mass in tritium beta decay

$$^3H \rightarrow {}^3He + e^- + \nu_e$$

There are a large number of efforts devoted to this problem; they are summarized in Table IV.

TABLE IV

TRITIUM EXPERIMENTS

Investigator	Location	Spectrometer	Source
+Lyubimov	Moscow	Toroidal Magnet	Valine-T
+Kundig	Zurich	Toroidal Magnet	C-T
+TJB,RGHR,JFW	Los Alamos	Toroidal Magnet	T,T2 gas
+Ohshima	Tokyo	p2 Mag	"Acid"-T
Fackler	Livermore	Retarding E-S	T2 solid
Clark, Frisch	IBM	Retarding E-S	?
Jelley	Oxford	Cylindrical Mirror	Ca-Palmitate-T
+Sun	Beijing	Iron Core Magnet	LiT,PdT,organic
Lobashov	Moscow	Retarding E-S	T,T2 gas
Stoeffl	Livermore	Toroidal magnet	T,T2 gas
Daniel	Munchen	Iron Core Magnet	Hf-T
Boyd	Ohio State	Retarding E-S	?
Wellenstein	Brandeis	Cylindrical Mirror	T2 Gas
Kalbfleisch	Oklahoma	?	?
Otten	Mainz	Retarding E-S	T Gas
Derbin-Popeka	Leningrad	Si	Si-T
*Simpson	Guelph	Si	Si-T

* Experiment Completed
+ Have Results

The reason that all this activity is occurring is that there has been a positive result from the Lyubimov group for some time; the data are shown in Fig. 5.

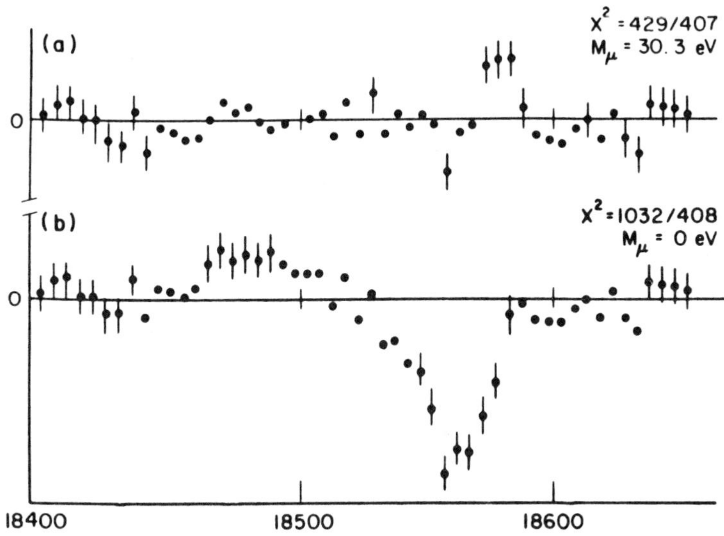

Fig. 5

Statistically, the finite-mass effect is clear, but of course all the problems of these experiments are systematic. The statistical precision is critically dependent on a background-free experiment. It is unfortunate that the groups that have opted to use electrostatic spectrometers have suffered with more background than the magnetic-spectrometer experiments. As an example of the differences between experiments, the resolution functions of a number of experiments are shown in Fig. 6, with low-energy tails in these functions from energy loss in the target depending on the final molecular states and, hence, the structure of the target material. John Wilkerson emphasized the advantage of a gaseous source of T_2 and we look forward to more data from the Los Alamos group. The consensus seems to be that the electron-neutrino mass is certainly below 35 eV, but a mass in the eV range is still open.

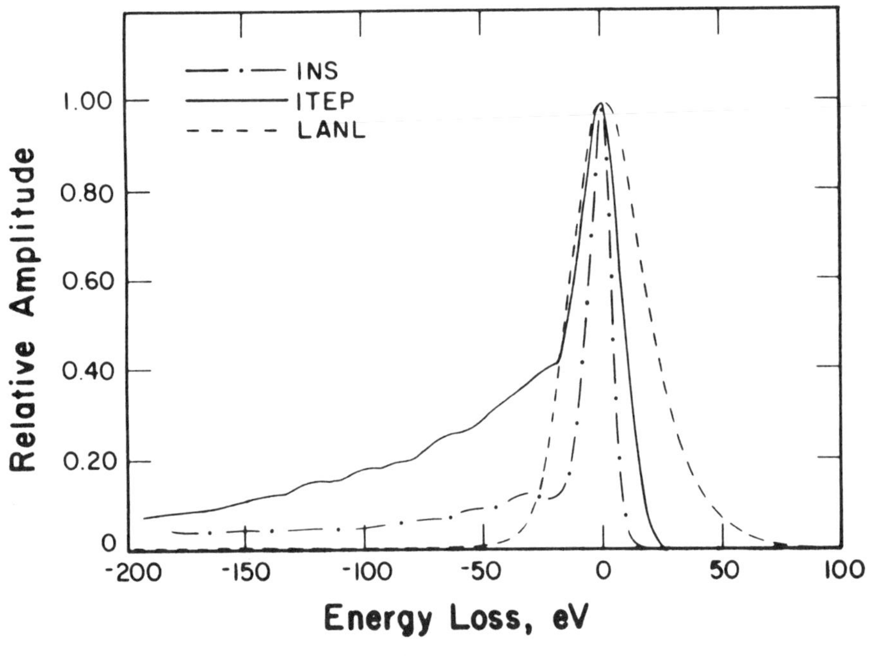

Fig. 6

Two experiments were presented searching for neutrino oscillations. Richard Seto presented data from a Brookhaven narrow-band beam in which oscillations of the type $\nu_\mu \rightarrow \nu_e$ would appear as anomalous charged-current events with electrons in the final state. The energy spectrum of the identified electron events is shown in Fig. 7a with the Monte Carlo estimate in Fig. 7b. The experimenters conclude there is no evidence for anomalous events leading to the limits shown in Fig. 8. Stan Durkin showed data from a Los Alamos measurement, which looked for $\bar{\nu}_\mu \rightarrow \bar{\nu}_e$ oscillations. At the beam stop at LAMPF there are no $\bar{\nu}_\mu$ because π^- are all absorbed before decay. Moreover, ν_e charged-current interactions on ^{12}C are suppressed relative to ν_e on free protons and cut off at a lower energy. In Fig. 9 is shown a distribution of identified electron energies in the detector with a Monte Carlo estimate of the expected rate assuming no $\bar{\nu}_\mu \rightarrow \bar{\nu}_e$ oscillations. The limit implied by this measurement is similar to that in Fig. 8. Further running by this experiment might improve the limit by about a factor of two.

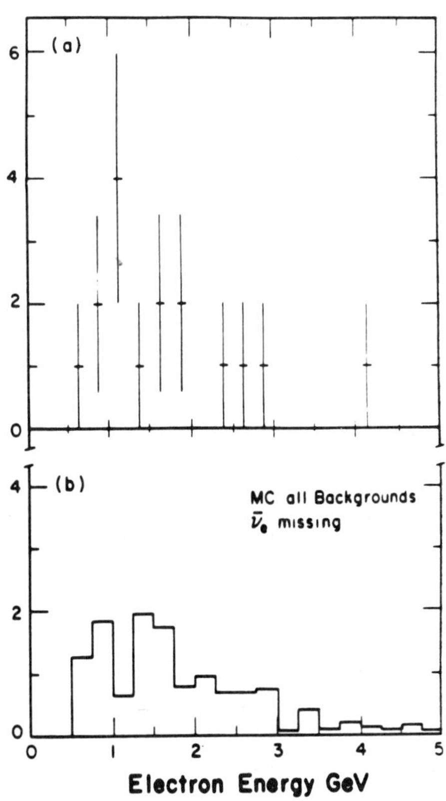

Fig.7(a)

Fig.7(b)

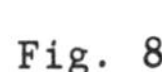

Fig. 8

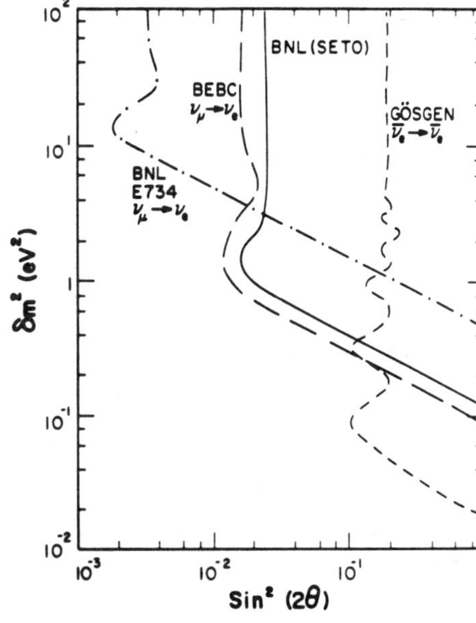

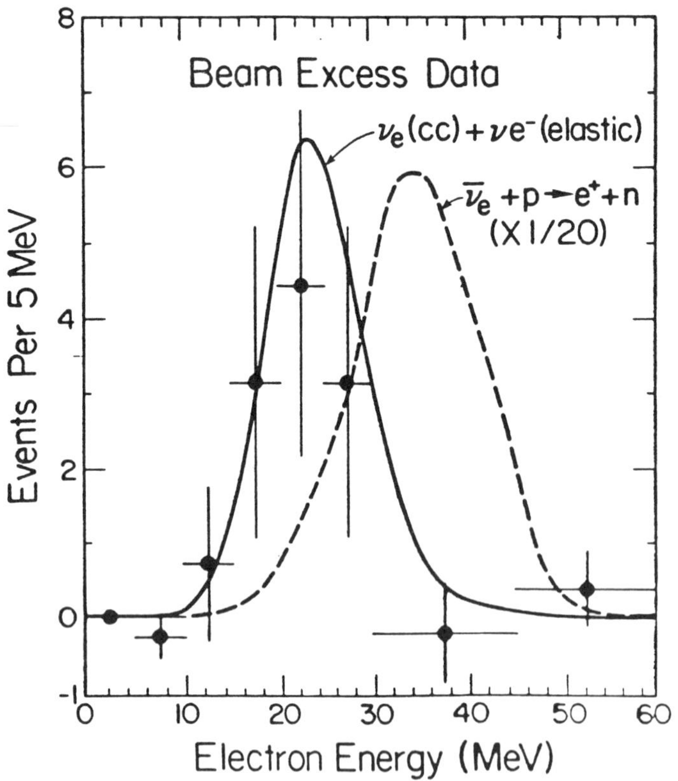

Fig. 9

Double beta decay with two neutrinos in the final state is a process that is expected to occur conventionally in the standard electroweak theory. In the case that single beta decay is forbidden by energy conservation then double beta decay is observable when two nucleons participate in the transition. It has been assumed that these two transitions are independent and that the rate is given by the nuclear matrix element with a second-order weak transition. A problem with this view is that geochemical studies in tellurium have indicated that the decay rate is substantially weaker than expected from a calculation by Haxton and Stevenson. Williams presented a calculation done with Haxton in which the axial-charge term was included and noted to interfere destructively with the principal transition terms. However, this term is too small to account for the Te discrepancy, although it is a step in the right direction. Stevenson made some general remarks that in calculating nuclear matrix elements for double beta decay the correlations between nucleons should be taken into account at least, because double charge exchange has shown the correlations to be significant. In the context of this problem the Irvine experiment on ^{82}Se presented by Alan Hahn was very interesting. This experiment uses a Time-Projection Chamber to reconstruct the tracks from the two electrons in the double-beta-decay transition. The sum of the two electron energies is shown in Fig. 10 for two versions

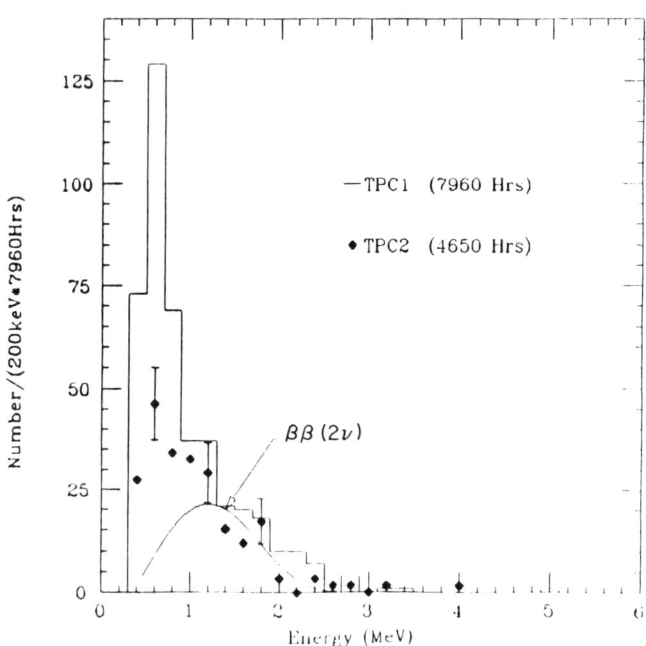
Sum Spectrum Comparison
—TPC1 (7960 Hrs)
◆TPC2 (4650 Hrs)
ββ(2ν)
Number/(200keV*7960Hrs)
Energy (MeV)
0
25
50
75
100
125
0
1
2
3
4
5
6

Fig. 10

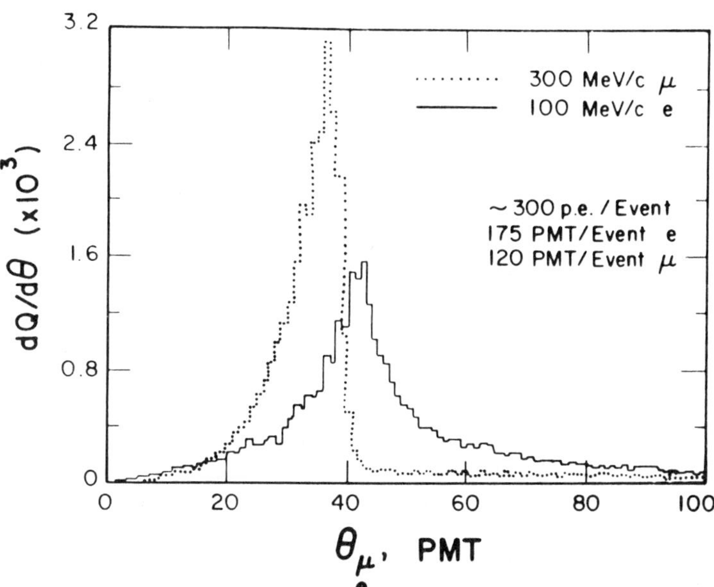
300 MeV/c μ
100 MeV/c e
~300 p.e. / Event
175 PMT/Event e
120 PMT/Event μ
dQ/dθ (x10³)
θμ,e PMT
0
0.8
1.6
2.4
3.2
0
20
40
60
80
100

Fig. 11

of the apparatus; TPC 2 is a lower-background experiment and a fit to the spectrum gives the curve $\beta\beta(2\nu)$ for the decay. The lifetime from this fit is

$$\tau(2\nu) = (1.4 \; ^{+0.6}_{-0.3}) \times 10^{20} \text{ yrs} \; .$$

This value is in approximate agreement with theoretical estimates of the rate, and the Te puzzle remains. Frank Avignone gave a summary of the double-beta-decay work in progress, mentioning in particular the ^{76}Ge experiments looking for neutrinoless decay, notably the UC Santa Barbara - LBL collaboration, which has the highest limit at the moment of 7 x 10^{23} yr. For neutrinoless double beta decay to occur, the neutrino mass must be non-zero and right-hand currents must exist. Major mischief to the standard model would result. The flurry of interest in a possible observation of double beta decay with a majoron has subsided.

A puzzling observation by the Kamioka-II collaboration was reported by William Frati. Fully contained neutrino events are observed in the Kamioka detector that are produced by cosmic-ray interactions in the atmosphere. In simple terms, since the π μ decay chain produces two ν_μ and one ν_e, then it is expected that twice as many muon events over electron events would be seen. Detailed effects reduce this number to somewhat close to unity, according to the standard calculation of Gaisser et al. Muons in the detector produce a different ring pattern to electrons largely because multiple scattering is different at the same deposited energy as shown in Fig. 11. Separation of the two classes of events in this energy range is straightforward. In Table V are shown the data for an exposure of 2.87 kiloton-years with $p_e > 100$ MeV/c and $p_\mu > 200$ MeV/c.

TABLE V

	DATA		MC	
	Total	μ decay	Total	μ decay
Single Ring	178	60	232.5	110.3
m-like	85	52	144.0	103.8
e-like	93	8	8.5	6.5
Multi Ring	87	34	86.2	37.1
Total	265	94	318.7	147.4

The ratio for muon-to-electron events for data divided by the same ratio for Monte Carlo is 0.56 ± 0.08. If the cut on momentum is made equal for electrons and muons at 200 MeV/c then this ratio changes to 0.53 ± 0.08. Beier pointed out in the question period that before too much is made of this result, a concern he communicated about the calculation of cosmic ray neutrino flux should be alleviated.

Finally, Balentekin described an analytical method for calculating the solar neutrino flux in the case that the MSW effect is important. The calculation is straightforward and offers significant saving of time for a wide range of mass-difference squared and mixing parameters.

SUMMARY OF THE RELATIVISTIC HEAVY ION SESSIONS

John W. Harris
Nuclear Science Division, Lawrence Berkeley Laboratory,
University of California, Berkeley, CA 94720, USA

The topics covered in the Relativistic Heavy Ion Sessions span 4 orders of magnitude in energy in the laboratory and a few more in theory. In the two years since the last Intersections conference, experiments in the field of very high energy heavy ion research have begun at CERN and Brookhaven. The prime motivation for these experiments is the possibility of forming *quark matter*. The formation of a quark-gluon plasma in very high energy collisions of nuclei is predicted to occur when the energy density in a collision increases above a critical value where the quark constituents of the incident nucleons, bound in nuclei, form an extended volume of freely interacting quarks, antiquarks and gluons. Recent lattice quantum chromodynamics calculations[1−3] predict a phase transition from a hadronic gas to a quark-gluon plasma at an energy density $\epsilon^{critical} \approx 2$ GeV/fm^3, and temperature $T^{critical} \approx$ 200-250 MeV. At lower energies progress continues towards understanding the behavior of *nuclear matter* in terms of the nuclear equation of state from the low baryon densities and temperatures of a possible liquid-gas phase transition through the densities and temperatures of stable and excited nuclei, neutron star formation and supernova explosions.

From among the various signatures predicted for the quark-gluon plasma, *suppression of J/ψ production* has received most of the recent attention. Matsui and Satz[4] predicted that screening of the quark color charge will prevent $c\bar{c}$ binding into a J/ψ in the deconfinement region. In hadron-hadron collisions $c\bar{c}$ pairs can be produced in either gg or $q\bar{q}$ interactions. However, if a plasma is formed in nucleus-nucleus collisions the $c\bar{c}$ will appear in a deconfining environment. In analogy to Debye screening in electrodynamics, if the Debye radius is less than the radius of the J/ψ at temperature T, then formation of the J/ψ is prohibited assuming the plasma lifetime is longer than the formation time of the J/ψ from $c\bar{c}$. Thus observation of J/ψ suppression in nucleus-nucleus interactions might signify formation of the plasma. The NA38 Collaboration[5] has reported preliminary finding of J/ψ suppression in high transverse energy ($E_\perp$) events, which are associated with small impact parameter collisions. Using a muon pair spectrometer with associated calorimetry the group observes a reduction in the J/ψ peak-to-continuum ratio of approximately 50 percent for high $E_\perp$ events. This is observed in 200 GeV/nucleon ^{16}O + U and ^{32}S + S collisions but not for p + U collisions. The group also sees no $E_\perp$ dependence of the continuum below the J/ψ. The J/ψ suppression is reported to occur for low transverse momentum pairs of $\mu^+\mu^-$ as predicted[6] for plasma formation. These results are very tantalizing, but the NA38 group states that the continuum below the J/ψ as measured by the experiment must still be understood before solid conclusions can be drawn. Furthermore, recent results[7] from Fermilab in 125 GeV/c $\bar{p}$ and π^- interactions on nuclear targets exhibit a similar suppression of the J/ψ. Much theoretical work is underway to quantitatively understand the J/ψ signal from the NA38 measurements. Recent theoretical efforts have concentrated on understanding the detailed evolution of the $c\bar{c}$ in a plasma[8] and hadron gas[9] as well as the role of nuclear absorption in the J/ψ signal.

Two pion interferometry measurements, which lend insight into the underlying collision dynamics have been performed by the NA35 Collaboration[10] at CERN. This interferometry technique was first proposed by Hanbury Brown and Twiss[11] to determine the size of distant stars by measuring the interference pattern of two photons. It is now used to measure the size of the pion-emitting source in high energy collisions of particles and nuclei. One approach[10] involves a Gaussian parameterisation of the pion-emitting source where R_T is the transverse radius, R_L the longitudinal radius and Λ the chaoticity parameter. When $\Lambda = 1$ the emission is totally chaotic. Values of $\Lambda < 1$ correspond to decreasing chaoticity, with $\Lambda = 0$ total coherence. Pions near midrapidity in the effective ^{16}O + Au center-of-mass are observed to originate from a large ($R_T = 8.1 \pm 1.6$ fm), almost spherical($R_L \approx R_T$), and chaotic (almost thermal) source ($\Lambda \approx 0.8$). Pions over the entire rapidity range of the collision reflect a much smaller source ($R_T = 4.1 \pm 0.4$fm), with much less chaoticity($\Lambda \approx 0.3$). This picture suggests the formation of a thermalised fireball at midrapidity. Away from midrapidity the transverse size is near that of the incident projectile, the longitudinal size is small reflecting a correlation length[12] of one unit of rapidity, and the chaoticity parameter is low, near that of e^+e^- and hadron-hadron collisions. Techniques have recently been suggested[13,14] to determine the effect of the lifetime of the source on the measured source sizes. Such measurements may even be sensitive to the expected increase in lifetime if a quark-gluon plasma were formed. The spacetime evolution of these collisions is of course quite complicated and will be the source of much interest and debate for some time to come.

Transverse energy ($E_\perp$) distributions reflect the degree to which the incident energy of the nuclei thermalises in the collision process, i.e. the "stopping power". Results from CERN ion experiments NA34,[15] NA35[16] and WA80[17] and from Brookhaven ion experiment E814[18] are consistent with a description of the central nucleus-nucleus $E_\perp$ distributions by an A_p-fold convolution of the proton-nucleus $E_\perp$ distributions, where A_p is the number of nucleons in the projectile. Most models which reproduce the proton-nucleus $E_\perp$ distributions and are extended to include the nucleus-nucleus collision geometry usually fit the measured nucleus-nucleus $E_\perp$ distributions. However, the models must also be able to predict the behavior of other observables in the collision process. Werner[19] has shown that string models are able to fit the $E_\perp$ and multiplicity distributions. Using a model based upon the formation and breakup of hadronic fireballs, Stachel[18] concludes that the $E_\perp$ distributions exhibit complete stopping of the ^{28}Si projectile incident on targets from Al to Pb at Brookhaven energies of $\approx$10 GeV/nucleon, whereas at the higher CERN energy of 200 GeV/nucleon the ^{16}O + Au data show only partial stopping of approximately 60 percent of the incident energy. The latter is consistent with previous estimates from various CERN experiments.[2] The incident energy where it is possible to convert the most energy of relative motion into energy density in the center-of-mass will be of most interest for studying the question of the existence of the quark-gluon plasma. Furthermore, this should be studied for the heaviest nuclear systems. If the $E_\perp$ distributions from current CERN experiments are translated into energy densities[20] using the Bjorken formula.[21], fairly large energy densities are reached with values at the extrema near those necessary for plasma formation.

Particle abundances and spectra are important to determining the flavor content and mean transverse momenta (temperatures) of the various constituents of the collisions. This information as a function of rapidity would provide a wealth of information on the dynamics of the collision process. Many of the results on particle abundances and spectra are preliminary due to the tremendous difficulty in tracking and particle identification at such high multiplicities and energies. Results from the NA34,[15] NA35,[22] and WA80[17] Collaborations show that the pion spectra in 200 GeV/n ^{16}O and ^{32}S -nucleus collisions are identical to those of proton-nucleus collisions at the same incident energy. This is not surprising since the pion spectra basically reflect conditions after hadronisation and freezeout. Direct photon or lepton spectra might provide information on the temperature at earlier stages of the collision process. However, preliminary results from NA34[15] on photon spectra show that there is little, if any, direct photon signal observed since the measured photon spectra are reproduced by emission from hadronic sources.

One of the earliest predictions for a signature of the deconfinement transition is an enhancement of s and $\bar{s}$ quarks in a quark-gluon plasma in thermal and chemical equilibrium.[23] The strangeness enhancement is a result of suppression of $u\bar{u}$ and $d\bar{d}$ pair production in favor of $s\bar{s}$ pairs in the initial u and d-rich environment remaining from the incident nuclei. Furthermore, the $\bar{u}$ and $\bar{d}$ quarks annihilate with u and d quarks, while the $\bar{s}s$ annihilation occurs less frequently until saturation of the s and $\bar{s}$ abundances. Most calculations predict a significant enhancement in the observed $\bar{s}$ yield as a signature of plasma formation while s quark yields, although enhanced, differ only slightly in a plasma compared to a hadron gas. The actual observation of $\bar{s}$ enhancement after plasma formation is severely complicated by the spacetime evolution of the collision process,[24] making it difficult to discriminate between the hadronisation products of a quark-gluon plasma and those of a chemically-equilibrated hadron gas. However, it is highly unlikely that hadronic processes in the nonplasma phase are best represented by an equilibrium hadron gas. Dynamical approaches have recently predicted[25,26] a separation of strangeness during the hadronisation process in a baryon-rich region. During the mixed phase, consisting of plasma and hadron gas and following the plasma phase, the hadron gas becomes $\bar{s}$-rich due to radiation-cooling from K^+ and K^0 emission. This leaves behind an s-enriched plasma and may manifest itself in the slopes of the observed K^+ and K^- spectra[27] as well as the possible production of strange droplets of matter.[28] Results on *strange particle production* are at a preliminary stage of analysis. The E802 group has reported[29] K^+/π^+ and K^-/π^- results for ^{28}Si + Au at 14.5 GeV/c per nucleon in a limited solid angle range. They find $K^+/\pi^+ = 19$ and 23 percent in peripheral and central interactions, respectively, with typical errors of $\pm\, 5$ percent. These ratios are measured to be in the 5 to 10 percent range for proton-nucleus collisions in the same phase space region measured by the E802 experiment. The K^-/π^- ratios from E802 are found to be 5 percent for both types of interactions. The actual cross sections for strangeness production over a larger solid angle is anxiously being awaited. The NA35 Collaboration has reported[30] Λ, $\bar{\Lambda}$ and K^0 ratios for 60 and 200 GeV/n ^{16}O + Au and p + Au reactions. The ratios are similar for the the nucleus-nucleus and proton-nucleus cases, although both exhibit somewhat higher Λ production and slightly lower $\bar{\Lambda}$ production than observed in proton-proton studies.

The work at the Bevalac accelerator in Berkeley has centered around determining the *equation of state of nuclear matter* at high densities and temperatures although somewhat lower than those where the hadron gas to quark matter phase transition is predicted to occur. Progress in this field has been made in determining systematics of the observed collective sidewards flow[31,32] of matter in collisions of heavier systems ($A \geq 40$) at incident energies of 100 MeV/n to 2.1 GeV/n. The flow is observed to increase with the mass and energy of the incident system as well as the centrality of the collision. The data provide strong evidence for compression effects in these collisions and are best described by a stiff nuclear equation of state. This finding is also supported by recent measurements[33] of the entropy production and the transverse energy distributions in these collisions. Progress has been made on the theoretical side with the inclusion of the momentum dependence of the nuclear potential which has a large effect on these observables. It is imperative that microscopic calculations which include momentum dependent potentials be performed and compared to the large body of observables available in the data in order to better determine the equation of state. Experimental and theoretical work on dilepton production at Bevalac energies is also underway. It has been predicted[34] that dilepton production at these energies is very sensitive to the initial stage of the collisions as is the case at higher energies. The leptons escape the interacting nuclear matter with very little absorption or interaction in the surrounding matter. Thus the dilepton spectrum should be very sensitive to the temperature and density of the system at the time of their creation. This could provide an experimental handle on the actual densities and temperatures that are reached in the compressional stage of the collision process.[34]

The *narrow positron lines* first observed at the GSI in Darmstadt in collisions of very heavy ions just above the Coulomb barrier remain a focal point of much cross-disciplinary research. Some of the positron peaks observed in singles[35] have now been observed as back-to-back e^+e^- coincidences.[36] The data suggest that the lines are a product of a two-body decay near the center-of-mass of the colliding system but at a time when the decaying object is far away from the nuclei. There is in fact evidence for more than one peak, perhaps three have been observed so far. Recently, experimentalists have measured Bhabha scattering of positrons from electrons in attempts to observe narrow resonances at the energies of the positron lines produced in the heavy ion collisions.[36] The results from two different groups are still inconclusive at this time. A third experiment which has the capability to measure with much-increased statistics is presently underway. Another group has measured the production of coincident photons[37] in order to investigate a possible photon decay mode of the object producing the narrow positron lines. These experiments find no coincident photon peaks at the energies of the observed positron peaks. However, a line for emission of back-to-back photons at much lower energy (1062 KeV), slightly above the positronium energy (1022 KeV) was observed in one of the three nuclear systems studied. Theoretically there is no consistent description for the observed systematics of the measurements, especially the connection to the nuclear properties of the system.[38] Furthermore, results from the e^+e^- experiments and further analysis of the photon experiments may provide additional information. Most explanations involve the decay of a neutral particle with internal structure. The particles properties and how it was formed are the subject of more exotic theoretical inquiries.

This work was supported by the Director, Office of Energy Research, Division of Nuclear Physics of the Office of High Energy and Nuclear Physics of the U.S. Department of Energy under contract DE-AC03-76SF00098.

References

1. J. Cleymans, R.V. Gavai and E. Suhonen, Phys. Rep. 130 217(1986).
2. see Proceedings of the Sixth International Conference on Ultra- Relativistic Nucleus-Nucleus Collisions - Quark Matter 1987, Nordkirchen, FRG, 24-28 August 1987, Z. Phys. C38 (1988).
3. J. Kogut, H. Matsuoka, M. Stone, H.W. Wyld, S.Shenker, J. Shigemitsu and D.K. Sinclair, Phys. Rev. Lett. 51 869(1983).
4. T. Matsui and H. Satz, Phys. Lett. B178 416(1986).
5. G. Landaud, Proceedings of this Conference; A. Bussiere et al., Z. Phys. C38 117(1988).
6. F. Karsch and R. Petronzio, Phys. Lett. B193 105(1987).
7. S. Katsanevas et al., FERMILAB-Pub-87/57-E (1987) to be published in Phys. Rev. Lett.
8. M. Chu, Proceedings of this Conference; M. Chu and T. Matsui to be published in Phys. Rev. D; see also J.P. Blaizot and J.Y. Ollitrault, Phys. Lett. B199 499 (1987).
9. S. Gavin and M. Gyulassy, private communication.
10. T. Humanic, Proceedings of this Conference; A. Bamberger et al., Phys. Lett. B203 320(1988).
11. R. Hanbury Brown and R. Twiss, Nature 177 27(1956) and Nature 178 1046(1956).
12. B. Andersson and W. Hofmann, Phys. Lett. B169 364(1986).
13. S. Pratt, Proceedings of this Conference; S. Pratt, Phys Rev. D33 72(1986).

14. G. Bertsch, M. Gong and M. Tohyama, MSU Preprint (1987) to be published in Brief Reports of Phys. Rev. C (1988).
15. B. Jacak, Proceedings of this Conference.
16. A. Bamberger et al., Phys. Lett. B184 271(1987).
17. A. Franz, Proceedings of this Conference.
18. J. Stachel, Proceedings of this Conference.
19. K. Werner, Proceedings of this Conference.
20. R. Albrecht et al., Phys Lett. B199 297(1987).
21. J.D. Bjorken, Phys. Rev. D27 (1983) 140.
22. H. Stroebele et al., Z. Phys. C38 89(1988).
23. R. Hagedorn and J. Rafelski, Phys. Lett. 97B 180(1980).
24. K.S. Lee, M.J. Rhoades-Brown and U. Heinz, Stony Brook Preprint (1987).
25. U. Heinz, K.S. Lee and M.J. Rhoades-Brown, Mod. Phys. Lett. A2 153(1987).
26. C. Greiner, P. Koch and H. Stöcker, U. Frankfurt Preprint UFTP 189 (1986).
27. U. Heinz, K.S. Lee and M.J. Rhoades-Brown, Phys. Rev. Lett. 58 2292(1987).
28. E. Witten, Phys. Rev. D30 272(1984).
29. R. Ledoux, Proceedings of this Conference.
30. G. Vezstergombi et al., Z. Phys. C38 129(1988).
31. H.G. Ritter, Proceedings of this Conference; see also Proceedings of the 8th High Energy Heavy Ion Study, 16-20 November 1987, Berkeley, California, LBL report LBL-24580 (1988).
32. K.G.R.Doss et al., Phys. Rev. Lett. 59 2720(1987).
33. H.R. Schmidt et al., Proceedings of the 8th High Energy Heavy Ion Study, 16-20 November 1987, Berkeley, California, LBL report LBL-24580 142(1988); K.H. Kampert et al., Proceedings of the 8th High Energy Heavy Ion Study, 16-20 November 1987, Berkeley, California, LBL report LBL-24580 154(1988).
34. C. Gale, Proceedings of this Conference.
35. J. Greenberg, Intersections between Particle and Nuclear Physics, Lake Louise, Canada 1986, published by the AIP New York 1986, p. 112.
36. W. Koenig, Proceedings of this Conference.
37. K. Danzmann, Proceedings of this Conference.
38. A. Chodos, Proceedings of this Conference.

HIGHLIGHTS FROM THE ANTIPROTON PHYSICS SESSION

R. Cester
Istituto di Fisica - Universita' di Torino
10125 Torino, Italy

The status of low-medium energy antiproton physics has been presented in five busy sessions.

I will not attempt to summarize all that has been said, instead I will review only a few papers, warning the reader that the definition of highlight is based only on personal taste.

Of the twenty-four papers presented, only few reported on new data. This crisis is due mostly to the fact that LEAR, which has been a rich source of antiproton data in recent years has been off the air for many months during the upgrade of the antiproton source at CERN. In contrast we have heard about many new beautiful experiments which are at present in an advanced state of preparation and that will be in data taking during 1989, again most at LEAR. They should make the Antiproton session in the next IPNP conference an extremely live one.

Given this situation I will subdivide this talk into three parts:

1. recent experimental results

2. theoretical ideas to interpret some of the main features of the data from the first generation of LEAR experiments.

3. experiments in preparation and new experimental techniques.

Results from experiment PS-185 at LEAR have been presented on the reaction: $\bar{P}+P \rightarrow \bar{\Lambda}^{\circ}+\Lambda^{\circ}$ with precise measurements of differential cross sections and Λ° polarizations (as a function of angle) at 6 energy values. It is found that the Λ° and the $\bar{\Lambda}^{\circ}$ are produced with polarizations which show a large dependance on the center of mass energy and angle of emission and reach values in excess of 50%.

The Nijmegen group has succeeded in fitting all the data with a coupled channel potential model which assumes the exchange of four kinds of particles (K,κ(1000), K*(892) and K*(1430)) and requires adjustment of 14 parameters.

In the same reaction CP violation effects have been searched for. The relevant observable are: the transition rate: $\Gamma \sim |S|^2 + |P|^2$, the differential cross section:

$$\frac{d\sigma}{d\Omega} = (1+a\ P \cos\theta) \text{ with } \alpha = \frac{2\mathrm{Re}(S^*P)}{|S|^2+|P|^2} \text{ and P the } \Lambda^{\circ} \text{ polarization}$$

and the asymmetry parameter for the proton from the Λ decay:

$$\beta = \frac{2\ \mathrm{Im}(S^*P)}{|S|^2+|P|^2} .$$

The CP testing ratios are:

$$\Delta = \frac{\Gamma - \Gamma}{\Gamma + \Gamma} = A = \frac{\alpha + \alpha}{\alpha - \alpha} \text{ and } B = \frac{\beta + \beta}{\beta - \beta} \text{ or } B' = \frac{\beta + \beta}{\alpha - \alpha}$$

which should all be zero if CP is a good symmetry.

Donoghue in a very interesting talk on CP violation, predicts a value of A = (.3 to 1.) 10^{-4} based on the assumption that CP violation arises in the decay mixing matrix and taking for ϵ'/ϵ the central value from the NA31 experiment (ϵ' and ϵ parametrize CP violation in the K sector).

The experimental result, from PS-185 is $\langle A \rangle = -.07 \pm .09$, still at a sensitivity level which is too low to be interesting. The experimenters of PS-185 and other groups plan to continue this kind of very challenging experiments possibly also looking in the $\bar{P} + P \to \bar{\Xi} + \Xi$ channel.

Finally PS-185 has measured the cross section for $\bar{P}+P \to K_S+K_S$ at a center of mass energy $\sqrt{s}$ ~2230 MeV close to the mass of the ξ particle. They measure a cross section of 2 μb to be compared with 50 μb for the $K^+ + K^-$ final state.

Quinn has summarized the status of the search for the ξ (2230) which was first seen by MARK3 at SPEAR in the radiative decay of the J/Ψ. In a similar experiment at Orsay the DM2 group as failed to confirm the evidence for such a state.

This state has been searched for, in formation experiments from a P,P initial state, both at BNL and LEAR, looking at the $K^+ + K^-$ final state. It has not been found so far, and limits have been set on the product of branching ratios:

$$B(\xi \to \bar{P} + P) \times B(\xi \to K^+ + K^-) \leq 2\ 10^{-3}.$$

PS-185 will continue its search for this state in the $K_S + K_S$ decay channel and claims that it will reach a sensitivity of ~2 10^{-4} for the product of the branching ratios, if the total width is 3 MeV < Γ_ξ < 60 MeV.

The ξ, if it exists, is interpreted as a Glueball or a high angular momentum $s\bar{s}$ state. In the second case its formation from $\bar{P}P$ should be suppressed.

The results of an experiment carried out at KEK, which studies strange particle production in the annihilation of 4 GeV/c $\bar{P}$ in the Ta nucleus have been analyzed by Rafelsky, who boldly sets up to demonstrate that there is evidence for cold quark matter formation in these data.

At 4 GeV/c, because of the decrease in the antiproton cross-section, the incoming particle is expected to interact well inside the nucleus (1 to 2 fermi) where, it is conjectured, the deposition of the ~2 GeV annihilation energy could cause anomalous matter formation.

The data from the KEK experiment show a large production rate of Λ°, which exceeds by a factor ~2 the estimates obtained from the knowledge of the elementary $\bar{P}$ + Nucleon → Λ° +... cross sections. From the rapidity distribution of the Λ° produced, one can derive the mass of the object which takes part in the reaction: $A_{target} = 13 \pm 1$. The average excitation energy E_x and entropy, S, per baryon can then be calculated $E_x/b=380 \pm 20$, S/b ~4.5. The temperature of a plasma, at which a minimum appears at the calculated E_x/b value, is T = 65 MeV. The corresponding baryon density is 50% above normal. If from the calculated properties of this super cooled phase of quark matter one attempts to estimate the fraction of strange quarks, rates of strange particle production in agreement with the experimental observations are found.

Some questions on this interpretation come from the results of PS-179 for Λ° - production in $\bar{P}$ + Ne at 600 MeV/c. In this low statistics experiment (43 events detected - efficiency corrected to 149 ±34) the same trend of high rate for strange particle production is observed. However in this case the annihilation occurs mostly on the surface of a light nucleus and the hypothesis of quagma formation is harder to accept.

Some interesting papers have been presented on the dynamics of $\bar{N}N$ annihilations. In particular Carl Dover gave a good review of the present understanding of the two body annihilation channels from PP at rest.

A state-of-the-art statistical model, (which incorporates spin and statistic factors and weighs the strange modes to account for the kinematical suppression due to the strange quark mass) allows the computation of branching ratios for two body final states which, with some exception, are in good agreement with the observed values. The statistical model fails badly in the prediction of the relative rate for $(K^{\pm}K^{*+})$ vs (K^+K^-), for (π_{ch}, M_{ch}) vs (π°, M°) (where $M \to \rho, A_2$, B) and for (M^+M^-) vs (M°, M°). Attempts are made to explain these discrepancies by introducing plausible dynamical selection rules.

As Genz pointed out the simplest and more reliable type of analysis is based on quark flavor flux diagrams which avoids the intricacy of gluon exchange mechanisms. The game being played is that of comparing the predictions based on different type of graphs. Specifically one assumes that either annihilation (A) or rearrangement (R) graphs dominate, and computes the branching ratios predicted for two-body (or three-body) final states in the given hypothesis. Comparison with data then allows to decide if R-type or A-type graphs dominate.

The method is sound, however sometimes the data are not; as a consequence two analysis based on different sets of data lead to opposite conclusion, with Genz favoring A2 and Klempt favoring R2.

The request from theorists for better and more complete experiments, extending to antineutron-nucleon interactions and using also polarized targets and/or beams, is likely to be met in the next generation of experiments.

Two new detectors, the Crystal Barrel (PS197) and OBELIX (PS201) are being assembled at LEAR with a research program which emphasizes the study of q qbar spectroscopy and searches for exotic states. A third experiment by the JETSET collaboration which addresses essentially the same problematics, uses an H2 gas jet target intersecting the antiproton beam circulating inside the LEAR ring, and features an ambitious detector design which includes a fast RICH counter for particle identification.

The internal jet target technique yields high luminosities and a sharp definition of the energy of the initial state and is therefore well suited to the study of narrow resonance formation. John Peoples reported on an experiment being installed in the tunnel of the antiproton source at Fermilab where a high luminosity (10^{31}) $\bar{P}P$ source is obtained by intersecting with a molecular H jet the beam circulating in the Accumulator Ring. The first goal of the E760 experiment is that of completing the study of the charmonium spectrum, improving on the accuracy of mass and width measurements of known states and searching for the missing ones (^1P, ^1D, ^3D).

Finally a beautiful experiment is in preparation at LEAR to measure the CP violation parameters in the K sector. The experiment will study the time evolution of K^o, $\bar{K}^o$, originating from $\bar{P}+P=K^o+K^{\pm}+\pi^{\pm}$. The identification of the charged K provides strangeness tagging for the neutral partners. The experiment will extend the measurement of CP violating parameters also to the 3π, K decay channel.

Heaberli gave a status report on polarized hydrogen targets to be used internally at LEAR and other storage rings and discussed the progress made in the construction of storage cells where one hopes to obtain target thickness in excess of 10^{14} atoms/cm^2.

An interesting idea which is being pursued is that of polarizing a circulating antiproton beam by letting it interact repeatedly with a polarized target. It is expected that the two spin components of beam particles will have different interaction cross sections, so that they will be differently attenuated and a net beam polarization will result.

I cannot conclude this review without mentioning the experimental activity with antiproton beams of thermal energy. In a very nice talk Hynes described the progress made in the construction of the thermal source and reviewed the experiments which will be accessible once the source becomes operational. They include a precise measurement of the antiproton gravitational mass (PS200) and have a wide range of applications in condensed matter and atomic physics.

Hadron Scattering Summary

Steven Heppelmann
Penn State University
University Park, PA 16802

The papers presented in the five hadron scattering parallel sessions cover a wide range of topics and energy scales. However, I believe that many of the talks in these sessions connect together into a unified framework. The goal of this summary will be to subjectively discuss some of these connections.

The scattering of hadrons within nuclei provides an opportunity to study sequential interactions between hadron states at small separation. Such studies reveal unique information about free hadron interaction mechanisms and about new aspects of the nuclear force and nuclear structure. The goal is to find aspects of the strong force which must be described by quark and gluon degrees of freedom.

In nuclear physics, there is great interest in probing nuclei with new techniques to reveal structure beyond that predicted by nonrelativistic shell models. In particle physics, the possibility of preparing hadronic state and studying them on very short time and distance scales is extremely interesting. For example, short lived resonances can be produced in nuclear matter and their subsequent interactions in that same nucleus can be observed. Other short lived quark and gluon configurations are produced in very high p_t interactions between hadrons. A short time after such a collision, the quarks and gluons will reform into hadrons and propagate as ordinary hadronic matter. The formation time is $c\tau \simeq \gamma \times [1fm]$ (with γ the relativistic time dilation factor) and can be greater than or equal to the nuclear distance scale. During this formation time the system propagating through the nucleus is a collection of quark and gluon matter which can have a very different nature than ordinary hadrons. Secondary interactions in the nucleus tell about quarks and gluon interactions in a context other than that characterized by ordinary hadron states.

Finally, as these experiments are increasingly interpreted as probes of quark and gluon degrees of freedom, we also must work to isolate and understand the predictions of the older models like the shell model and pre-QCD hadronic scattering models. We must understand the predictions of these theories before interpreting the data in terms of quark and gluon degrees of freedom or more exotic models.

Papers were presented in this session which touched on each of these areas of hadronic nuclear interactions. A question about the presence of exotic quark structures in nuclei has perhaps been resolved by the pion double charge exchange measurements on the Ca isotopes reported by Baer. Previous observations of forward peaking in these angular distributions were thought to require short distance nucleon correlations. It had been speculated that this was evidence for six quark states. It now seems clear that the correlations predicted by the shell model are just the ones needed to explain the data. This result points out the heightened importance in understanding the implications of the non-quark theories before interpreting evidence for new effects.

I believe that the presentation of DeSwart which challenges the assumed charge invariance in the pion-nucleon coupling and the discussion by Matthaey on experimental status and inconsistencies in the extraction of the sigma term (the theory for which was discussed in a plenary talk by Donoghue) can be viewed in the spirit of more carefully specifying the nature of the traditional low energy nuclear force. The intricate spin measurements described by Vetterli on N-nucleus or by Igo on the p-d system are pushing these standard nuclear models (perhaps with relativistic corrections) to find the extent of their predictive power and ultimately their limitations.

The search for differences between the predictions of models with traditional nuclear force acting between meson and baryon degrees of freedom and QCD interaction between quark degrees of freedom has focused on a particular state, the $\Lambda(1405)$. The mass of this state is just below the kp thresholds, tempting nuclear theorists to treat it as a kp state bound with conventional nuclear force. On the other hand, if it is removed from the list of 3 quark baryons it leaves a missing state in the spectrum. Predictions were made for the ratio of decay widths kp $\rightarrow \Lambda\gamma$: kp $\rightarrow \Sigma\gamma$ in models reflecting these two views as discussed by Fearing. Data were presented by McIntyre, with that ratio shown to be of order unity. The predictions of the two views were originally quite different but have recently converged to near the observed value.

The observation within nuclei of the total interaction cross section of short lived hadrons was discussed by Nakai along with presentation of new data on the cross sections for ρ^0 and f^0 mesons. The picture emerging is one in which the absorption cross section is simply related to a geometrical size of the system. Most of these cross sections are within a factor of two of the nominal $\pi \times (1\ fm)^2$. The exception is of course the J/Ψ which is expected to be a system much smaller than 1 fm and is observed to have a very small cross section. It is observed that the f^0 meson, the only orbital angular momentum 1 state studied, has absorption larger than the π^0, ρ^0 or ω^0. Like the total cross sections which can be observed directly with more stable hadrons, the energy dependence of these cross sections is small for energies above a few GeV.

In a similar spirit, the secondary interactions after violent or high p_t scattering of hadrons have been studies by fixed target experiments at Fermi Lab. The plenary talk by Rutherfoord gives a review of this field. The very higt p_t hadronic interactions produce jets, or single partons which penetrate the nucleus while only slightly perturbed by their passage through the nuclear matter. The probability of stopping these fast quarks is predicted to become vanishingly small. The most naive expectation is that the cross section for producing high p_t jets or hadrons should scale as A^1 reflecting the linear increase with A in the number of scattering centers but with no reduction in the cross section for secondary absorption. Observations of A^α with α greater that 1.5 has been the usual result. The apparent explanation is that while these jets are not absorbed, they do scatter by small angles. The cross sections fall so rapidly with p_t that these secondary scatters tend to move events from low p_t to high p_t. There is more scattering in larger nuclei which leads to $\alpha > 1$.

Zieminski has discussed data from Fermilab in which very coplaner back to back jets were studied. They had found that for the most coplaner events, presumably those events for which secondary scattering was least evident, the value of a seemed to return to 1. He points out however, that the exact nature of this effect is still in question. The evolution to quark-hadronic systems through nuclear matter in the formation zone is a very active area of study (see papers by Sambamurti, Toothacker, and Frankel) and has implications reaching into many areas including the search for quark-gluon plasma.

In contrast to single quark or jet production, on rare occasions the very violent collisions between protons in nuclei result in the scatter of both protons without break up. There is great interest in extracting pp elastic scattering in nuclei at large q^2. When these rare processes occur it is the expectation from QCD that the protons which participate have a very small size, inversely correlated with the q^2 scale. There should again be a formation time associated with both shrinking down to that size and expanding back to the original size. As q^2 increases (or for fixed CM angle scattering, as the energy increases) QCD predicts vanishing nuclear absorption. This effect, related to the self shielding nature of the small color neutral system, has been referred to as "Color Transparency". The expectation is that at low energy the nucleon-nucleon absorption cross sections will be the nominal value of $\simeq \pi \times (1\ fm)^2$ but at higher energy this absorption will vanish.

The theoretical discussion was presented by Farrar and the experimental results were summarized by Carroll. Indeed, the absorption seems to decrease as the incident energy is raised from 6 to 9.5 GeV. However, beyond 9.5 GeV the absorption increases again. DeTeramond discussed models which mix the contributions from hard QCD amplitudes with dibaryon resonance contributions (for which there should be no color transparency effects) to describe both this energy dependence and the energy dependence of the A_{nn} spin correlations observed in large and pp scattering.

The validity of the perturbative QCD approach for calculating these large angle exclusive cross sections has been much debated, most recently in the plenary talk by Isgur. The mixture of successes and failures of the QCD approach should be viewed as very provocative. While much has been said about this formalism, only recently have actual cross sections been calculated and they were presented by Zhang.

I have not been able to discuss each result presented in these sessions and for that I apologize. Instead I hope I have emphasized the richness intrinsic to the observation of hadronic interactions in nuclei, illuminating the very current and fundamental issues in both particle and nuclear physics.

REVIEW OF THE MESON AND LEPTON DECAY SESSION

M. E. ZELLER
Physics Department, Yale University, New Haven, CT 06511

ABSTRACT

The session on meson and lepton decays is reviewed.

INTRODUCTION

The parallel sessions on meson and lepton decays covered a broad spectrum of phenomena, but the underlying theme was that of testing the standard model. Since the last conference in this series a large amount of data has been collected by experiments which were then only beginning to come on line. The bulk of the contributions were thus devoted to presenting and discussing these new results. As an overall summary of the session we display some of these results in Table I, as well as published measurements since the last meeting, Lake Louise 1986, and the reviewer's educated guess of what we will find at the next meeting in 1991. While this table does not do justice to the total content of the session, it does give a sense of the recent accomplishments in this field, and of the wealth of new information on mesons and leptons which is accumulating.

TESTING THE STANDARD MODEL

In the early 1980's theoretical models, constructed to cure the gauge hierarchy problem caused by a scalar Higgs particle, made suggestions that lepton flavor might not be conserved. These models allowed decays such as $K^0 \rightarrow \mu e$, $K^+ \rightarrow \pi^+ \mu^+ e^-$, and $\mu \rightarrow eee$ to occur at rates which could be observed with present technology. Thus is was at the first meeting on the intersections of particle and nuclear physics, Steamboat Springs 1984, that plans were discussed for several experiments to test these predictions. This year recent results of many of those efforts were presented.

The session began with a review by R. Shrock of the current theoretical picture of rare π, μ, and K decay physics. While unresolved issues of the standard model are still present,(*e.g.* the origin of fermion masses, numbers of generations, properties of Higgs particles,*etc.*), refinements in the early models have reduced the level of expected lepton flavor violation. He pointed out, however, that the lack of theoretical guidence to physics beyond the Standard Model adds to the importance of continuing the searches.

A review of the activities at the Paul Sherrer Instutute, PSI formerly SIN, was given by R. Engfer which was devoted to π and μ decays. The $\mu \rightarrow eee$ results from the SINDRUM detector have been refined so that a branching ratio limit of 1×10^{-12} (90% CL) could be reported. In an extended technicolor model of Dimopoulos and Ellis this result raises the mass of the propagator above $\sim 93\,\mathrm{TeV/c^2}$.

Other results from SINDRUM were measurements of the branching ratio for $\pi^+ \rightarrow e^+\nu e^+e^-$ and of the decay parameters in the $\mu^+ \rightarrow e^+\nu_e\bar{\nu}_\mu e^+e^-$. Also, an experiment to measure the $\pi^0 \rightarrow e^+e^-$ was completed, resulting in an upper limit which brings experiment into better agreement with theory.

From TRIUMF the recent experiment searching for μ to e conversion was discussed by D. Bryman. Results quoted in Table I are for μ^- Ti $\rightarrow e^-$ Ti, but results for conversion on Pb at $< 4.9 \times 10^{-10}$ were also presented. Included in the talk by Engfer were plans at PSI for a SINDRUM II detector to search for μ to e conversion to the 10^{-13} level of sensitivity. Much of Bryman's talk was concerned with his experiences and cautionary views on the difficulty in going further.

Also from TRIUMF were results presented by D. Britton of a new measurement of the branching ratio for $\pi^+ \rightarrow e^+\nu$. Along with improving the accuracy of this ratio, and showing good agreement with radiatively corrected theory, this experiment set new limits on the mixing of ν_e's with more massive neutrinos in the mass range from 50 to 150 $\mathrm{MeV/c^2}$.

Results on rare K decay physics from BNL were presented in talks by W. Morse (BNL 780: $K^0 \rightarrow \mu e$), R. Whyley (BNL 791: $K^0 \rightarrow \mu e$), and D. Lazarus (BNL 777: $K^+ \rightarrow \pi^+\mu^+e^-$). These experiments are now pushing the limits to theoretically interesting levels. Peripheral results were also discussed: BNL 780 now places an upper limit on the CP violating decay $K^0{}_L \rightarrow \pi^0 e^+e^-$; BNL 777 has about 800 events from the decay $K^+ \rightarrow \pi^+e^+e^-$ and has begun to study the mechanisms for that decay; and BNL 791 has equalled the world sample of $K^0{}_L \rightarrow \mu\mu$ events with approximately 1/4 of their data analyzed.

Another lepton flavor violating process, conversion of muonium to antimuonium, $\mathrm{M\bar{M}}$, has been explored by two groups. Limits on the strength of the coupling of this process, $\mathrm{G_{M\bar{M}}}$, as measured by a group at TRIUMF, presented by G. Marshall, are now at the level of the standard weak interaction coupling, $\mathrm{G_F}$. This group expects to lower this limit by a factor of ten in their next run.

R. Schaefer reported results from a LAMPF measurement of $\mathrm{G_{M\bar{M}}} < 7.5\,\mathrm{G_F}$, and told of plans to reduce the limit to $10^{-2}\,\mathrm{G_F}$ in an upcoming experiment. The new effort is a departure from previous experiments in that it involves a spectrometer system to measure the momenta of both the positron from muonium disassociation and the electron from the muon decay.

CP VIOLATION

Advances in measurements of ϵ'/ϵ in K^0 decays have been great in recent years. G. Gollin reviewed the progress of both his Fermilab experiment, E731, and the CERN effort, NA31. The CERN group, which accumulated

$\sim$ 100K $K^0{}_L \rightarrow \pi^0\pi^0$ events, has published $\epsilon'/\epsilon = .0033 \pm .0007 \pm .0008$, and the Fermilab group, using a test run with 7000 $2\pi^0$ events, has published a value of $.0032 \pm .0028 \pm .0012$. The full sample of $2\pi^0$'s in E731 is $\sim$ 250K events, and analysis is proceeding. Gollin emphasized the sensitivity of ϵ'/ϵ to systematic effects, and discussed the relative merits of the two experiments in this respect.

A. Sanda then reviewed the theoretical status of CP violation. While great progress has been made with the K^0 system, he pointed out the difficulty of observation of "standard" CP violation with light particles other than neutral K's. It is in the B sector that he feels the next advances will occur, and exhorted the experimentalists to come up with "the good idea".

HEAVY MESON DECAYS

At this year's meeting, this session was expanded to include decays of mesons containing c and b quarks. C. Sliwa gave new results from a Fermilab experiment, E691, on semileptonic D^0 and D^+ decays, non- strange $D^+{}_S$ and D^+ decays, and lepton flavor violating D decays. This experiment has gathered a sufficient number of charmed meson decays to allow complete analysis of the exclusive semileptonic decays $D^0 \rightarrow Ke\nu$, $D^+ \rightarrow K^*e\nu$, and $D^+ \rightarrow K\pi e\nu$. With another experiment planned to increase the sample by a factor of ten, we see the D's taking on the role of K decays of the previous decade.

From the ARGUS detector, D. MacFarlane described their efforts to determine V_{ub} in the KM matrix. This study involved understanding the lepton spectrum in $B^0 \rightarrow D^{*-}l^+\nu$ decays and observation of $B^+ \rightarrow \rho^0 l^+\nu$ and $B \rightarrow p\bar{p}\pi^+(\pi^-)$ events. The conclusion of this analysis is: $0.07 \leq |V_{ub}/V_{cb}| \leq 0.20$.

THE FUTURE

While this meeting was rich with recent results, it previewed future experiments, and showed that yet more progress can be expected. From LAMPF, the MEGA project, as described by M. Cooper, will begin to gather data next year in a more sensitive search for $\mu \rightarrow e\gamma$; R. Strand described BNL E787 which just completed its first run and will push the limit for observation of $K^+ \rightarrow \pi^+\nu\bar{\nu}$ below 10^{-9} in the near future; also at BNL, E791 will improve the limit on $K^0 \rightarrow \mu e$; a new experiment, E845, will make a dedicated search for $K^0 \rightarrow \pi^0 ee$; and E777 will significantly increase the sample of K^+ decays with an e^- in the final state. In addition, at PSI the SINDRUM II program is beginning to take shape. Add to this the enhanced future for heavy meson experiments, and one sees that the meson lepton decay session at the next intersections meeting will be as exciting as this year's.

Table I Experimental Results

Measurement[1]	Last Meeting	This Meeting	Next Meeting[2]
$\mu \to e\gamma$	$< 4.9 \times 10^{-11}$		$< 10^{-12}$
$\mu \to eee$	$< 2.4 \times 10^{-12}$	$< 1. \times 10^{-12}$	
μ Ti $\to e$ Ti	$< 1.6 \times 10^{-11}$	$< 4.6 \times 10^{-12}$	$< 10^{-12}$
$\mu^- e^+ \to \mu^+ e^-$	$G_{M\bar{M}} < 42\ G_F$	$G_{M\bar{M}} < 1.1\ G_F$	$G_{M\bar{M}} < 10^{-2}\ G_F$
$\pi^0 \to ee$	$(1.8 \pm 0.7) \times 10^{-7}$	$< 1.2 \times 10^{-7}$	$\sim$ 100 events
$\pi^+ \to e^+\nu$	$(1.228 \pm .022) \times 10^{-4}$	$(1.218 \pm .014) \times 10^{-4}$	
$\pi^+ \to e^+\nu e^+ e^-$	$< 5 \times 10^{-9}$	$(3.3 \pm 0.6) \times 10^{-9}$	
$(M_{\pi^-} - M_{\pi^0})^3$	4.6043 (37)	4.59362 (35)	
$K^0 \to e\mu$	$< 6 \times 10^{-6}$	$< 6.7 \times 10^{-9}$	$< 10^{-10}$
$K^0 \to ee$	$< 2 \times 10^{-7}$	$< 4.5 \times 10^{-9}$	$< 10^{-10}$
$K^0 \to \mu\mu$	$(9.1 \pm 1.9) \times 10^{-9}$	$\sim$100 events	$\sim$500 events
$K^0{}_L \to \pi^0 e^+ e^-$	$< 2.3 \times 10^{-6}$	$< 4 \times 10^{-7}$	$< 10^{-9}$
$K^+ \to \pi^+ \mu^+ e^-$	$< 5 \times 10^{-9}$	$< 1.1 \times 10^{-9}$	$< 1.7 \times 10^{-10}$
$K^+ \to \pi^+ e^+ e^-$	$(2.7 \pm 0.5) \times 10^{-7}$	$\sim$800 events	$\sim$3000 events
$D^0 \to \mu e$		$< 5 \times 10^{-5}$	$< 5 \times 10^{-6}$
$D^0 \to ee$		$< 5 \times 10^{-5}$	$< 5 \times 10^{-6}$

1. Branching ratio unless otherwise noted; quoted for 90% CL.
2. Estimated by the reviewer.
3. MeV/c^2

SUMMARY FOR ASTROPHYSICS AND NON-ACCELERATOR PHYSICS

Sidney H. Kahana
Brookhaven National Laboratory, Upton, NY 11973

INTRODUCTION

Summarising, in real time, a parallel session in a conference of this size and scope serves an obvious purpose. Committing such a summary to paper would appear of less value, since the several contributors to the session have themselves presented their offerings in writing. With this in mind, I will not overly detail the separate talks in this year's Non–Accelerator and Astrophysics section, but will instead reiterate the conclusions (or absence of conclusions) arrived at by these speakers and comment on these in light of their relevance to the theme of the Intersections Conference and to the scientific world at large. I will include within the scope of my comments the plenary talks on closely related topics.

SUPERNOVA 1987a

One group of speakers (Bethe, Brown, Kirshner, Baron, Masaike, and Pennypacker) concentrated on the observation of, modelling of, or search for supernovae. SN1987a, of course, played a highly visible role in all of this, stimulating much activity in astrophysics, in the nuclear physics associated with gravitational collapse and in neutrino physics.

Kirshner displayed in graphic and entertaining fashion the output of a wide variety of detectors brought to bear on the electromagnetic radiation, optical, and otherwise, emitted by SN1987a. We have, uniquely, photographs and spectra from the progenitor [Sanduleak 69–202] of a type II (or any other type) supernova; a surprising progenitor at that, being a superblue rather than a superred giant. As Kirshner noted and Beier later expanded upon, this supernova was the first to be seen as a neutrino emitter[1,2]. The detection of electron antineutrinos from 1987a, a most fortunate happening tied to the perspicacity of the nucleon decay watchers, opens a new and glorious phase in the evolution of neutrino astronomy. My former colleagues at Brookhaven, Ray Davis and collaborators[3], were responsible for the first phase in this discipline. Paradoxically, their measurements have created puzzles, while 1987a has generated interesting limits on the neutrino mass, lifetime, magnetic moment, and other characteristics. At the very least 1987a has dramatically confirmed theoretical ideas about gravitational collapse.

Hans Bethe developed the physics of the extant theoretical mechanisms for producing supernovae subsequent to gravitational collapse (type II). Gerry Brown considered the role in the supernova, and the derivation, of the equation of state for hadronic matter of supranormal density. Baron presented the nature and status of the latest simulations of the prompt mechanism by a Brookhaven – Stony Brook collaboration[4]. Although the present evidence[5,6] from the analysis of the light curve of SN1987a is not conclusive, it appears a vigorous explosion occurred, with total energy near or above 1.5×10^{51} ergs. A prompt mechanism, i.e., direct explosion by the hydrodynamic shock generated during collapse, seems favoured.

Despite the many "moving parts" of these or any other similar calculations, our understanding of the input physics is such that the results are sensitive to a limited number of parameters. Indeed, one can say that if the initial "iron" core mass (i.e., at the instant before collapse) were known and the equation of state for matter up to 3 or 4 times normal saturation were specified, then the fate of the shock could be determined. Aside from these few physics inputs, the most uncertain part of the simulation is the technical treatment of neutrino transport. One can, presently, map out a region of parameter space for an equation of state for which the prompt mechanism will successfully generate supernova explosion. Improvements in the handling of neutrino transport will somewhat alter the boundaries of this region of viability.

The downward evolution of iron core masses in the work of Hashimoto and Nomoto[7] or of Woosley and Weaver[8], makes certain the existence of progenitors with main sequence mass (M_{ms}) for which type II supernova will go by the prompt mechanism. In particular, the current Woosley–Weaver model for $M_{ms} = 18 M_{\odot}$ possesses a core mass $\approx 1.36\ M_{\odot}$, and such cores were found to "explode" in past calculations[4], while the Nomoto–Hashimoto 13 $M_{\odot}$ model for which $M_{core} = 1.18\ M_{\odot}$ will certainly explode.

Bethe and Baron very clearly outlined the steps taken to place the simulations on a good footing. The more difficult problem of deriving a high density description of nuclear matter is still with us. Brown pointed out how high the stakes are and showed how one could reconcile[9] the situation in stellar interiors with conclusions drawn from relativistic heavy ion collisions performed at LBL[10,11]. Knowledge of the state of high density matter is also relevant to the expected transition [in the case of high density and low temperature probably a chiral–restoration] to quark–gluon matter. These two fields of study can usefully feed on each other.

NEUTRINOS FROM SN1987a

Probably the most unusual aspect of 1987a has been the actual observation of neutrinos. This remarkable eventuality was accomplished with detectors designed for other purposes and only fortuitously in place at the right instant. Nevertheless, the rewards have been many. E. Beier went over the setups and results at Kamioka and IMB and reminded us of the startling confirmation of the theories of collapse provided. He discussed the planned upgrade of KII and the possible heavy water detector at Sudbury in Canada (SNO)[12]. Supernova neutrino observatories must likely find other things to do to justify their existence, and, for example, SNO's primary purpose is to examine solar neutrinos; perhaps disentangling the neutral and charged currents.

It is interesting to contrast the limits on neutrino mass discussed in other sessions of this meeting [see B. Kayser and J. Wilkerson in these Proceedings] with the limit deduceable from SN1987a. The laboratory experiments suggest a neutrino mass presently no higher than 18 – 25 ev, while the limit obtained from the supernova is conservatively quoted as between 10 and 15 ev. No one in his right mind would discourage further terrestrial experiments on such a basis, but the advantages of a long flight parth are evident.

Further limits established by the IMB and KII neutrinos include, quixotically, an equality to about a part in 10^9 between the "speed of light" for neutrino and that for photons, a limit on masses for neutrinos into which electron neutrinos can oscillate if any supernova mechanism is to be possible, and a whole host of others described elsewhere[13].

Among all the fanciful imaginings produced by theorists examining the 19 neutrinos seen, there remains a hint of something unusual, tied to the final three late–time sightings by Kamiokande II. These neutrinos, labelled #10, 11, 12 are assigned relatively high energies and consequently are highly unlikely to have all been background (10^{-4} probability). The time sequence in KII alone puts a gap of seven seconds between an initial 1.92 second pulse and these last events. This gap is reduced to five seconds if the IMP and KII detections are joined together in the most straightforward fashion, but still leaves one with a bit of a puzzle. This is especially so when one combines these data with the theoretical difficulties in extending the neutrino–visible cooling period beyond several seconds. It is tempting to conclude the late time neutrinos are the result of an appreciable reheating in the proto–neutron star core. Again, the theoretical time scale for material falling back into the core is on the order of a few seconds. A phase transition from hadronic to quark–gluon matter could result in a change to higher core density, smaller radius and additional gravitational binding energy. The transition, if it were to, say, three–flavoured quark matter, might indeed require a time of the order of many seconds to be spread over the entire, dense core — a tantalizing possibility.

X-RAYS AND γ-RAYS FROM SN1987a

As I have pointed out, observations from SN1987a have by no means ceased. Kirshner mentioned the x–rays detected by GINGA[14], and Masaike displayed very recent observations by this Japanese x–ray satellite. The initial x–ray observations in September showed two components — the harder part of the spectrum above 10 kev can be understood[15] as down–scattering of the two cobalt γ–rays at 0.88 MeV and 1.23 MeV, while the softer x–rays might have arisen from collision between the exploding stellar atmosphere and previous circumstellar matter. Theorists have had a hard struggle with the data, suggesting first in September observations would not be made for several months and then having no easy explanation for the sudden rise (or fall) in intensity seen by GINGA in January.

Masaike made a claim for observation of extremely high energy γ–rays from SN1987a on two days in January also, by a combined Japanese – New Zealand team working with scintillation arrays and Cerenkov mirrors at a site in New Zealand. With respect to the x-ray seen by GINGA, Kirshner warned of the existence of a nearby, perhaps unresolved, source, Cygnus X–1. This variable source might be a problem for the γ–ray experiment as well. In any case, the reader can judge for himself by examining Masaike's figures showing the x–ray flux seen by JANZOS during a 42–hour period in January and the excess over background they extract, with apparent maxima on Januaruy 14 and 15. Surely, these courageous experiments must eventually give fruit, but the ability of the neutron star to produce such high energy particles relies on the existence of a pulsar or at least of a strong magnetic field. Just when this happens we are not sure. Kirshner did tell us that the optical light curve is dropping away from the exponential decay driven by Cobalt–56 decay and that there is little enough time till at least the Cobalt γ–rays are directly visible. The direct observation of these γ–rays and subsequently of, say, iron x–rays from the surface of the neutron star are eagerly anticipated. Accurate measurement of red shifts of Fe x–ray lines would contain a wealth of information about neutron star sizes and hence about the hadronic equation of state at high density.

AUTOMATED, OPTICAL SUPERNOVA SEARCHES

The optical supernova searches by the Berkeley group, reported on here by Carl Pennypacker, are part of another effort to be sustained over a longish time, with admirable goals and several presently interesting results. The differential observation of large numbers of galaxies, automated with a system including a PC to control the telescope and principal collector, and an eventual analysis by a microvax has already produced eight supernova discoveries, six not seen by other observers. Eventually the group expects to see more than fifty supernovae per year. The two present best discoveries, 1986i and 1987k, speak to the recent sightings of more than just 1987a. A most interesting result is the preliminary evaluation of the Hubble constant at 70 km/sec/MPC by using the observed supernovae as distance measures.

ULTRA HIGH ENERGY COSMIC RAYS; FLY'S EYE

From Masaike on γ–rays from pulsars to Loh on high energy γ–rays from anywhere in the galaxy is only a small step. The FLY'S EYE detector in Utah is a massive undertaking, directed towards a fundamentally interesting question, the sources of very high and ultra high energy gamma rays. The detector itself and the analysis used by this group of scientists are both unusual and can be followed in detail in Loh's contribution to these sessions. Loh's presentation of data at this meeting spoke to a momentous observation if sustained, the detection of γ–rays (or other neutrals) at energies $\sim 5 \times 10^{17}$ ev and apparently from the direction of Cygnus X–3. At such an evergy even the short–lived neutrons might arrive relatively unscathed over the 40,000 or so light years from

Cygnus X-3, in fact 14% of neutrons at this energy survive. Once again the data is more tantalizing than conclusive, but with time must acquire considerable significance.

MONOPOLES, PRIMORDIAL NUCLEOSYNTHESIS, FIFTH FORCES and ELECTRIC DIPOLE MOMENT

In this final group of talks, we certainly move away from the subject of supernova or of cosmic rays from remnants of supernovae. Magnetic monopoles, quantized by Dirac, seen by some and not by others, are an inevitable result of GUT theories but perhaps in very small unobservable numbers. Groom, who is himself involved with one group of monopole gazers, presented us with an entertaining and succinct picture of such searches. His most important message was that limits set on monopole fluxes are strongly linked to assumed properties, i.e., aside from the obvious magnetic charge. Only induction detectors are free from bias. The best hope for the future seems to focus on the MACRO detector of Gran Sasso to be started up in part this summer. Again, this detector will give good flux limits for monopoles possibly possessing electric charge but which do not catalyse nucleon decay.

Applegate's indication that a first order QCD phase transition in the early universe, at 10^{-4} seconds in the evolution, might be observable as a relic in the footprint of r–process elements is certainly worthy of careful consideration. An important feature of his collaborative work with Hogan and Scherrer is the possibility of closing the universe mainly with baryons while still achieving the observed deuterium to helium-4 ratio. The criticism of this work by some on its prediction for primordial ^{7}Li is weakened somewhat by the fragility of this element in almost any environment. A preliminary assay of r–process elements in galactic–halo stars has not as yet uncovered any evidence of the phase transition.

I come now to the eminently clear talk by Boynton on a subject which many of us wish (unscientifically) had been left dormant. Fifth, sixth, and other fundamental forces may one day emerge from the primordial slime, and surely our experimental science must be able to withstand these tests, but the present situation is more muddled than ever, surely than before elevators and Eötvös. Boynton described in detail essentially two experiments, one by Eckhardt et al on little g measured up television towers, and the other INDEX I (and II) by his own group at a site near Seattle. Eckhardt has anomalous results for g at various altitudes above the ground, disagreeing to parts in 10^7 with values predicted by numerical continuation of measurements on a surface grid around the tower. Boynton, in INDEX I, using a torsion balance near a 130 meter cliff, obtains $\xi\lambda = -0.023 \pm 0.007$ meters for the product of non–gravitational coupling and range. The sign of this result is inconsistent with the new force being proportional in strength to baryon number alone, and requires a dependence on isotopic spin as well. The positive results of INDEX I and of

Thieberger are seemingly close to being contradicted by the conclusions of Fitch and collaborators. Neither positive result here provides any simple explanation for Eckhardt.

Finally, Bruce McKellar outlined calculations of non–vanishing neutron electric dipole moment from several models. I refer you to his talk for a description of the consequences to both standard and other models from different scales of edm's.

On behalf of Gene Beier and myself I would at the end like to thank all of the speakers who contributed so vigorously and creatively to our sessions. For a more complete list of references on these subjects I refer you to the specific talks in these Proceedings.

REFERENCES

1. K.Hirata et al., Phys. Rev. Lett. 58, 1490 (1987).
2. R. M. Bionta et al., Phys. Rev. Lett. 58, 1494 (1987).
3. J. K. Rowley, B. T. Cleveland, R. Davis Jr., American Inst. Phys. Conf. Proc. 126, 1 (1985).
4. E. Baron, J. Cooperstein, and S. H. Kahana, Phys. Rev. Lett. 55, 126 (1985);
   E. Baron, H. A. Bethe, G. E. Brown, J. Cooperstein, and S. H. Kahana, Phys. Rev. Lett. 59, 736 (1987).
5. S. E. Woosley, P. A. Pinto, and L. Ensman, Astrophys. J. (in press).
6. T. Shigeyama, K. Nomoto, M. Hashimoto, and D. Sugimoto, Nature 328, 320 (1987);
   T. Shigeyama, K. Nomoto, and M. Hashimoto, Astronomy and Astrophys. (in press, Nov. 1987).
7. K. Nomoto and M. Hashimoto, private communication.
8. S. E. Woosley, P. A. Pinto, P. G. Morton, and T. A. Weaver (in press);
   S. E. Woosley, private communication.
9. T. L. Ainsworth, E. Baron, G. E. Brown, J. Cooperstein, and M. Prakash, Nucl. Phys. A464, 740 (1987).
10. D. W. Harris et al., Phys. Lett. 153 (1982);
    H. A. Gustafsson et al., Phys. Rev. Lett.52, 1590 (1984).
11. J. T. Molitoris and H. Stöcker, Phys. Rev. C32, 346 (1985);
    J. Aichelin, A. Rosenhauer, G. Peilert, H.Stöcker, and W. Greiner, Phys. Rev. Lett. 58, 1926 (1987).
12. G. Aardsma et al., (Sudbury Neutrino Observatory Collaboration)Phys. Lett. 194B, 321 (1987).
13. J. M. Lattimer and J. Cooperstein, Astrophys. J. (in press);
    A. Burrows and J. Lattimer, " Neutrinos from SN1987a", University of Arizona preprint (April 1987);
    S. H. Kahana, J. Cooperstein, and E. Baron, Phys. Lett. B196, 254 (1987);
    M. J. Turner, Phys. Rev. Lett., 1797 (1988).
14. T. Dotani et al., Nature 330, 6145 (1987).
15. M. Ioh, S. Kumagi, T. Shigeyama, K. Nomoto, and J. Nishimura, (preprint).

SUMMARY OF SPIN PHYSICS SESSIONS

Thomas Roser
The University of Michigan, Ann Arbor, MI 48109-1120

INTRODUCTION

The list of topics included in the many talks given during the Spin Physics sessions is nearly as long as the one of this conference:

P and T Violation
NN Interaction
πp and πd Elastic Scattering
Nuclear Matter Spin Effects
High Energy Spin Effects
Muon (g - 2)
Polarized Proton Beams
Polarized Gas Targets

This points to the almost trivial fact that spin is fundamental to our understanding of nuclear and particle physics. I will discuss in some detail only four of these topics. Needless to say this choice is very much personally biased and I apologize to all the speakers whose excellent contributions I did not include.

P AND T VIOLATION

According to the 'standard' picture parity non-conserving (PNC) NN interaction is dominated at low energy by meson exchange with one weak meson nucleon vertex. Each of the six possible vertices are characterized by a weak meason nucleon coupling constant:

$$f_\pi^1, h_\rho^0, h_\rho^1, h_\rho^2, h_\omega^0, h_\omega^1$$

In a by now almost classical paper Desplanques, Donoghue, and Holstein (DDH)[1] calculated these coupling constants from QCD renormalized Weinberg-Salam operators using SU(6) wave functions.

M. Simonius described in detail how to analyze the PNC observable A_z of pp scattering in terms of these coupling constants. Using the two combinations:

$$h_\rho^{pp} = h_\rho^0 + h_\rho^1 + h_\rho^2/\sqrt{6}; h_\omega^{pp} = h_\omega^0 + h_\omega^1$$

he finds:

$$A_z(45MeV) = \alpha_1(h_\rho^{pp} + h_\omega^{pp}) = (-1.50 \pm .22) \times 10^{-7} \approx -2 \times 10^{-7}(DDH)$$

$$A_z(230MeV) = \alpha_2 h_\rho^{pp}$$

The proportionality constant in front of the coupling constants depends on strong scattering phase shifts and the geometry of the experiments.

Clearly a measurement at 230 MeV would allow the determination of h_ρ^{pp} and h_ω^{pp}. J. Birchall presented a proposal to measure A_z (230 MeV) at TRIUMF to an accuracy of 0.2 x 10^{-7}. It will be very interesting to hear their result.

Although the agreement between the measurement of A_z (45 MeV) and the prediction of DDH is very good, M. Simonius argued that there is an inconsistency at short range between the PNC potential and the strong interaction models used for the second meson nucleon vertex and for the scattering wave functions.

V. Zeps presented the result of a recent measurement of A_z at the 0^+ resonance in $^{13}C(\vec{p}, p)$ where mixing with the nearby 0^- level of ^{14}N creates considerable enhancement of the PNC effect. The measured value (86 ± 59) x 10^{-7} does not agree with the DDH prediction (-224 x 10^{-7}). However, rather than pointing to a problem in the DDH calculation, this measurement presents a test for the shell model wave functions used in the analysis.

Even larger enhancement factors were found in p-wave resonance scattering of thermal neutrons from ^{139}La[2]. Although this will not add much to the understanding of the PNC NN interaction, it was suggested that this is a unique place to look for T violation in weak NN interaction. D. Bowman and

A. Masaike reported on such efforts underway at Los Alamos and KEK, respectively. In preparation for this experiment A. Masaike measured A_z to be $10.5 \pm 0.6\%$, somewhat bigger than the value of Reference 2.

POLARIZED GAS TARGETS

Recently increased effort has gone into the development of polarized gas targets as a new tool to study spin effects. Unlike solid polarized targets, gas targets are clean in the sense that they consist only of the polarized target atoms. Used as an internal target luminosities greater than 10^{31} cm $^{-2}$ s^{-1} seem possible.

At very low temperatures electron spin polarized atomic hydrogen can be trapped in a high magnetic field and accumulated to high densities. R.S. Raymond described efforts to form an ultra-cold jet of polarized atomic hydrogen by using microwave to extract the atoms from the trap. A target thickness of 10^{14} cm^{-2} could be reached. This would allow to study one and two spin effects in high $p_\perp$ processes.

An alternative scheme that was proposed by W. Haeberli[3] involves a storage cell to accumulate the polarized atoms from a standard atomic beam source. The circulating beam would pass through low conductance openings of the cell. R. Millner presented a proposal to use such a target, as well as a cell filled with optically pumped metastable ^{3}He to measure neutron and proton spin dependent structure functions at the HERA electron storage ring. He anticipates target thicknesses between 10^{14} and 10^{15} cm^{-2}.

A thickness greater than 10^{20} cm^{-2} was achieved by T. Chupp with his high pressure ^{3}He gas target. He polarized the ^{3}He by spin exchange with optically pumped rubidium. The rather long polarizing time precludes it from being used in an open, Haeberli-type cell. However, T. Chupp is proposing to use it as an external target at Bates to measure G_e^n. Another interesting application would be to study spin effects in proton neutron scattering[4].

HIGH ENERGY SPIN EFFECTS

Large spin effects found several years ago in high energy processes are still not understood. We had two reports of experiments using the polarized proton beam at Brookhaven, that continued a systematic study of two areas of unexplained spin effects. With no theoretical explanation at hand, such new data is urgently needed to provide clues for a future understanding.

D.G. Crabb reported on the first systematic measurement of A and A_{NN} in pp-elastic scattering at 18.5 GeV/c covering a $p_{\perp}^2$-range from 1.4 to 3.2 $(GeV/c)^2$. Both A and A_{NN} show pronounced dips in this kinematic range that are not reflected in the spin averaged cross-section. It is planned to extend these measurements to higher momentum transfer where, at lower energy, unexplained large values for A_{NN} were previously observed at Argonne.

The large polarizations measured in inclusive hyperon production is still not explained. However, in a simple fragmentation picture all the polarization data can be parameterized successfully. This model predicts that no polarization should be transferred from the proton to a lambda hyperon, but $\sim$ 70% should be transferred in the case of a sigma hyperon. F. Nessi-Tedaldi described a measurement of polarization transfer in hyperon production. The prediction for the lambda was confirmed, but the result for the sigma (38 $\pm$ 18%) is in disagreement with the fragmentation picture.

THE FERMILAB POLARIZED PROTON BEAM

We heard from D. Underwood of the recent success at Fermilab to form a polarized proton beam from the weak decay of lambdas. 10^7 polarized protons were produced every 20 second spill with an average momentum of 185 GeV/c. The polarization of the protons depends on the angle between their direction and the direction of the lambdas, which results in a polarization distribution that extends from $^-$60% to $^+$60%. A Primakoff polarimeter with an analyzing power of $\sim$ 57% was used and good agreement with the prediction was found. This new facility will soon allow us a first look at two spin effects at very high energies.

This work was supported by the U.S. Department of Energy.

REFERENCES

1. B. Desplanques, D.F. Donoghue, and B.R. Holstein, Ann. Phys. 124 , 449 (1980).
2. V.P. Alfimenkov, et al., Pis'ma Zh. Eksp. Teor. Fiz. 35 , 42 (1982) [JETP Lett. 35 , 51 (1982)].
3. W. Haeberli, 3rd Conference on the Intersections between Particle and Nuclear Physics, Rockport, Maine (May 14-19, 1988).
4. O. Häusser, private communication.

POLARIZED ANTIPROTONS WITH THE SPIN SPLITTER

Robert Rossmanith
CEBAF, Newport News, VA 23606
representing the Spinsplitter Collaboration[1]

ABSTRACT

In this paper a method for polarizing protons, antiprotons and ions using the Stern-Gerlach effect will be discussed. A test of this effect, foreseen for the low energy antiproton storage ring LEAR at CERN, is described. In this test particles with different spin directions are separated by a combination of a solenoid together with several skew quadrupoles: this device is called a spin splitter.

INTRODUCTION

Until now only electrons and positrons could be polarized during storage in storage rings. Electrons and positrons are polarized via the emission of photons according to the Sokolov-Ternov effect[2]. Polarized protons can only be stored when the beam is produced by a polarized source[3] and accelerated to the final energy. In general the beam has to cross many depolarizing resonances[4]. In order to keep polarization relatively high, precautions for reducing the strength of the depolarizing resonances have to be developed[5].

In this paper a selfpolarization mechanism for proton and ion storage rings is described. This is especially required for stored polarized antiprotons. For polarized antiprotons only very weak sources exist[6]. These sources utilize the fact that Lambdas decay into polarized antiprotons via the weak interaction. Several years ago at a workshop in Bodega Bay, Cal., methods for storing polarized antiproton beams were discussed[7]. At that meeting the idea of producing polarized antiprotons in a storage ring using the Stern-Gerlach effect was introduced[8].

THE STERN-GERLACH EFFECT IN A STORAGE RING

The cell of a storage ring is very similar to the magnet assembly used in the Stern-

Gerlach-Rabi effect (fig. 1). In this effect a gradient field is used to kick particles with different spin orientations into different directions. Afterwards an homogenous field rotates the spin by 180 degrees and an additional inhomogenous field adds an additional kick. In the storage ring the elements with the inhomogenous fields are quadrupoles and the spin rotating fields are bending magnets.

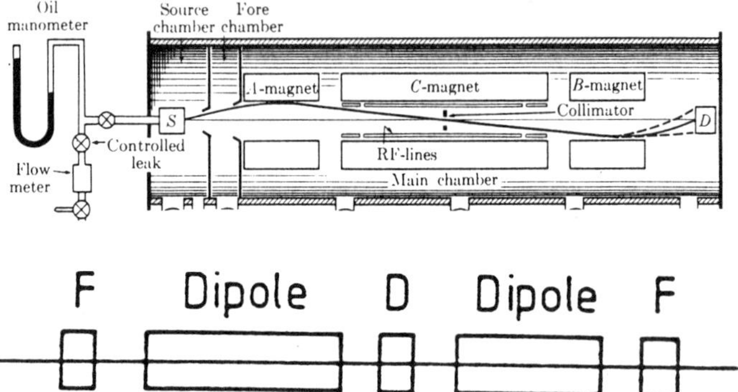

Fig. 1 Comparison of a magnet assembly in a Stern-Gerlach-Rabi experiment (upper drawing) with a cell in the storage ring. The magnets A and B in the Stern-Gerlach Rabi experiment have gradient fields, C is a dipole magnet. In the storage ring lattice F and D are quadrupoles.

Nevertheless, despite these similarities the two assemblies differ in the following two points:

-whereas the Stern-Gerlach-Rabi experiment is performed with two gradient magnets, the storage ring consists of many cells and therefore of many quadrupoles and

-whereas the Stern-Gerlach-Rabi experiment is performed with neutral particles, the particles in the storage ring are charged.

The first statement has the consequence that the kicks can only add when a resonance condition is fulfilled.

$$G\gamma = n \pm Q$$

where G is anomalous magnetic moment (1.793 for protons), γ is the Lorentz factor, n is any integer and Q is the Q-value of the machine.

The above mentioned equation describes not only a condition for adding many kicks over many revolutions but also decribes a condition for a depolarizing resonance. In order to avoid depolarization on the depolarizing resonance the Stern-Gerlach kicks can only be added over a particular range of the machine. An example of this idea will be discussed later.

The second point is a more fundamental one. Since the particles in a storage ring are charged and the beam has a transverse extension each particle is deflected in a different way by the gradient (quadrupole) field. This individually different deflection adds to the Stern-Gerlach kick and smears the deflection out. Fortunately the Stern-Gerlach kicks can add and the kicks according to the field cancel over many revolutions. A more general quantum mechanical treatment of this statement can be found in[9]. For this experiment J. Bell discussed this assumption in more detail[10]. Recently the validity of this argument was proved in a very elegant experiment[11] with free electrons.

The idea is to inject an unpolarized beam into a storage ring. After a certain time particles with different spin directions perform betatron oscillations. The phase of the betatron oscillation is mainly influenced by the spin via the Stern-Gerlach effect. One spin direction is dumped and the remaining beam is polarized.

THE SPIN SPLITTER

Two years ago investigations were started to apply this idea for producing polarized antiprotons in LEAR. LEAR is a low energy antiproton storage ring. The principle layout is shown in fig. 2. In order to separate the two spin directions in one of the straight sections a device called a spin splitter can be installed[12].

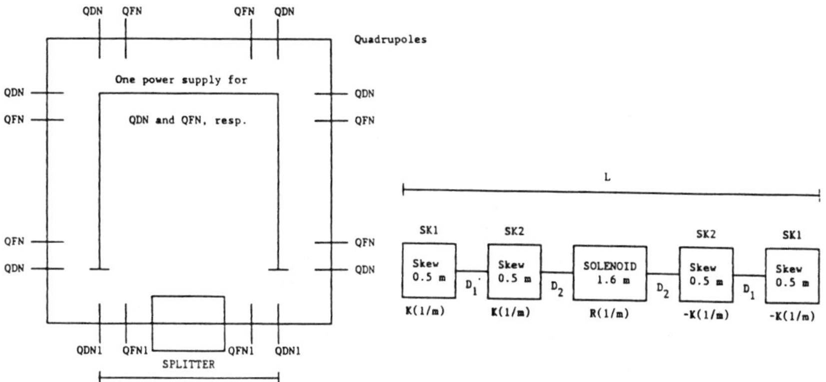

Fig. 2 The position of the spin splitter in LEAR (left) and the arrangements of the magnets in the spin splitter (right)

The spin splitter consists of two skew quadrupoles on each side of a solenoid. The whole device is optically transparent. The particles are kicked in the skew quadrupoles in a direction which depends on the spin orientation of the particle. The spin is rotated in the solenoid by 180 degree so that both the quadrupoles in front and after the solenoid can contribute to the Stern-Gerlach effect.

The strength of the kick $\delta x'$ is

$$\delta x' = \frac{b\mu}{p\beta}$$

where b is the gradient in the quadrupole, μ is $1,41.10^{-23}$ erg/Gauss for protons, p is the momentum of the particle and β is v/c. The separation speed of the spins is therefore $\approx$ 2.5 mm/hour at a beam energy of 200 MeV/c with the spin splitter.

The optics of LEAR with the integrated spin splitter is shown in fig. 3.

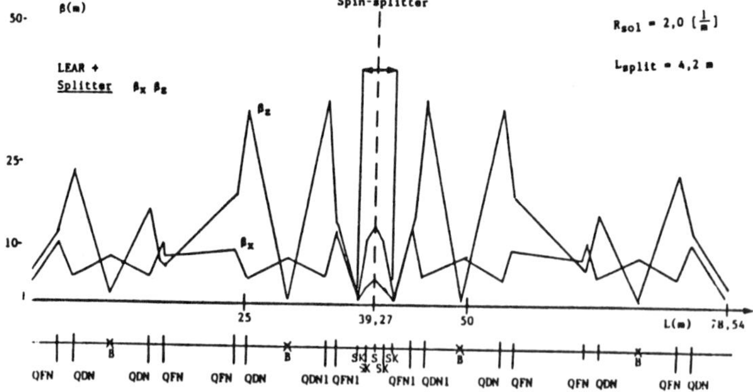

Fig. 3 The LEAR optics with the integrated spin splitter

SPIN DYNAMICS IN A MACHINE WITH A SPIN SPLITTER

The spin dynamics of the particles in a machine with a spin splitter is complicated. The spin splitter acts similarly to a Siberian snake[13]. The closed solution for the spin ($\vec{n}$-axis) lies in the horizontal plane. As shown above the kicks have to coincide with the betatron frequency. Since a storage ring does not operate at an integer Q-value particles with spins aiming in the direction of the $\vec{n}$-axis cannot be excited by the Stern-Gerlach effect. For this effect the spins have to oscillate around the n-axis.

In a ring with a snake the oscillations around the n-axis are no longer free (when only horizontal spin motions are considered): after two revolutions the spin comes back to its original position[14].

As a result the resonce condition mentioned at the beginning has to be modified to

$$m \pm 1/2 = n \pm Q$$

independent of the spin tune $G\gamma$ in the machine. Nevertheless, the strength of the excitation is related to the magnitude $G\gamma$:

$$\delta x' \approx \frac{|1 - cos(G\gamma)|}{2}$$

According to this condition the Q-value of the machine must be 1/2. Since a storage will not work with a spin tune of 1/2 the following trick can be used. When the spin is somewhere vertically deflected in the storage ring, the snake only compensates this deflection after two revolutions when the perturbation is opposite to the snake[15]. Thus, a small vertical spin deflection somewhere in the ring (e.g. a closed orbit bump) can rotate the spin slightly around the momentum axis and change the resonance condition:

$$m \pm 1/2 - \Delta = n \pm Q$$

During the last months the influence of the Q-spread on the degree of the polarization was studied. If particles have a Q-value deviating from the Q-value in the resonance condition, the degree of separation remains small. In order to investigate this effect the nonlinear spin tracking program SITROS[16], originally developed for electrons, was modified for proton beams. The modified program was baptized PROTEUS. With PROTEUS a set of particles and their spins were tracked over many revolutions. PROTEUS takes into account second order optical effects and higher order spin effects. The second order magnetic field components produce the well known Q-spread. According to this program each particle has a different Q-value from revolution to revolution. The distribution of the Q-value is shown in fig. 4. It can be clearly seen that the actual Q-value oscillates around the Q-value of the linear machine.

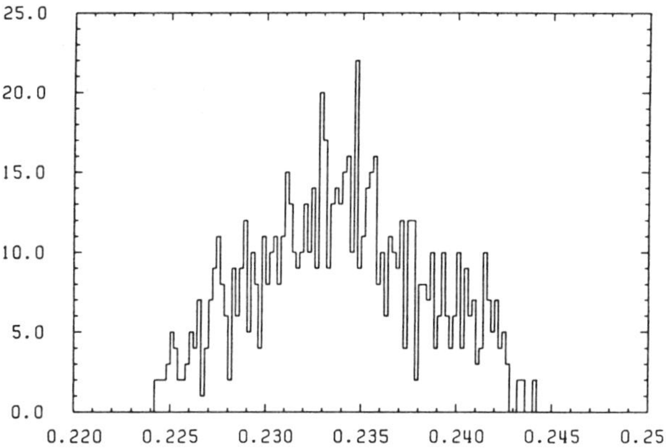

Fig. 4 The distribution of the Q-value of a single particle. The particle is tracked over many revolutions. The horizontal axis shows the Q-value

Oscillations of this type do not affect the build up of the polarization. The reason is the following. The Stern-Gerlach kicks are not distributed over the whole machine but are concentrated at one point. Therefore the resonance has a certain bandwidth. If the Q-value variation is also within this bandwidth and the build-up time of the separation is long compared with these oscillations the resonance condition is fulfilled. In that case the Q-spread does not influence the separation.

When taking higher order fields into account, e.g. octupoles, this statement will not be valid under certain circumstances. The remaining effects however are so small that it is possible to compensate them experimentally by varying the amplitude of a beam bump in an octupole.

In summary it may be said that the production of polarized protons or antiprotons by the Stern-Gerlach effect seems to be possible. With the spin splitter in LEAR the separation time of different spin directions is in the order of half an hour. The Q-spread does not seem to influence the separation time. One problem which still needs investigations is the diagnostics of the beam with separated spins.

The author wishes to thank the many people who made the investigations possible or contributed to this work: J. Bell from CERN, T. Walcher from Mainz, E. Steffens from Heidelberg, U. Strohbusch from Hamburg, T. Limberg from Cern and C. Leeman from CEBAF. We have also have to thank A. Skrinski and J. Shatunotov from Novosibirsk, A. Chao from Berkeley, A. Krisch from Michigan, E. Courant and L. Ratner from BNL for many discussions.

[1] The Spin Splitter Group: D. Cline, M. Conte, M. Gazzaly, R. Giacomich, M. Giorgi, J. Hall, J. Kewisch, H. Kreiser, K. Kuroda, T. Niinikoski, Y. Onel, G. Pauletta, A. Penzo, A. Pisent, M. Pusterla, P. Riley, R. Rossmanith, N.M. Stewart and X. J. Wang

[2] A. A. Sokolov and I. M. Ternov, Sov. Phys. Dokl., 8, 1203 (1964)

[3] Workshop on Polarized Sources and Targets, Helvetica Physica Acta, vol. 59, No. 4

[4] E. D. Courant, S. Y. Lee and S. Tepikian, see ref. 7

[5] A. D. Krisch, Proc. VII Internat. Symp. on High Energy Spin Physics, Protvino 1986

[6] A. Yokosawa, see ref. 7

[7] A. D. Krisch, A. Ling and O. Chamberlain (ed.), AIP conference Proc. No. 145 Particles and Fields Series 34

[8] T. O. Niinikoski and R. Rossmanith, see ref. 7

[9] N. Ramsey, IEEE Trans. IM 36, 155 (1987)

[10] J. Bell, CERN TH-group, private communication

[11] R. van Dyke, P.Schwinberg and H. Dehmelt, Phys. Rev. D, 34(1986)722

[12] A. Penzo, Proc. Physics at LEAR with low energy antiprotons, Nucl. Sc. Res. Conf. Series, Vol. 14, Harwood Academic Publishers

[13] Y. Derbenev and A. Kondratenko, Part. Acc. 8,115

[14] Y. Onel et al., see ref. 5

[15] A. D. Krisch, S. R. Mane and T. Roser, Michigan preprint UM HE 87-04

[16] J. Kewisch, DESY report 83-032(1985)

TOWARDS AN ADVANCED HADRON FACILITY AT LOS ALAMOS

Henry A. Thiessen
Los Alamos National Laboratory, Los Alamos, NM 87545

ABSTRACT

In the 1987 AHF Workshop, it was pointed out that activation of the accelerator is a serious problem. At this workshop, it was suggested that a new type of slow extraction system is needed to reduce the activation. We report on the response to this need. The Los Alamos plan is reviewed including as elements the long lead-time R&D in preparation for a 1993 construction start, a menu of accelerator designs, improved losses at injection and extraction time, active participation in the development of PSR, an accelerated hardware R&D program, and close collaboration with TRIUMF. We review progress on magnets and power supplies, on ceramic vacuum chambers, and on ferrite-tuned rf systems. We report on the plan for a joint TRIUMF-Los Alamos main-ring cavity to be tested in PSR in 1989. The problem of beam losses is discussed in detail and a recommendation for a design procedure for the injection system is made. This recommendation includes taking account of single Coulomb scattering, a painting scheme for minimizing foil hits, and a collimator and dump system for containing the expected spills. The slow extraction problem is reviewed and progress on an improved design is discussed. The problem of designing the accelerators for minimum operation and maintenance cost is briefly discussed. The question of the specifications for an advanced hadron facility is raised and it is suggested that the Los Alamos Proposal of a dual energy machine - 1.6 GeV and 60 GeV - is a better match to the needs of the science program than the single-energy proposals made elsewhere. It is suggested that design changes need be made in all of the world's hadron facility proposals to prepare for high-intensity operation.

REVIEW OF 1987 AHF WORKSHOP

I think that those of you who attended the 1987 AHF Workshop will remember Pete Miller's enthusiastic introduction and his invitation to attend the groundbreaking of AHF in the Fall of 1992. There were two other important outcomes of this meeting. Baconnier reminded us of the experience of present-day machines, which are operating near the maximum level of activation which is tolerable for "hands-on" maintenance. Baconnier's paper clearly documented the problems that the designers of the new high-intensity machine must solve. Teng, in his conference summary, pointed out that the traditional methods of slow extraction are not adequate for the high-efficiency extraction required in a hadron facility and that an invention is needed to improve the extraction efficiency. Both of these speakers presented a challenge to the audience. We discuss some of the responses to their challenge below.

There are a number of problems discussed at the 1987 AHF Workshop which are still with us. At the PSR, there is a 1.5% beam loss which has bothered us for the past year. This problem has been explained by Colton in his talk at this conference. There is also the unexplained transverse instability of the PSR, the need to design an efficient collimation system for PSR and the hadron facilities, and the problem of choosing an architecture which minimizes the lifetime cost of a hadron facility. Solutions to these problems are discussed below.

THE LOS ALAMOS PLAN

Our plan is based on preparing for a construction start in FY-1993 after CEBAF is complete and RHIC is well underway. We are redesigning LAMPF II to include a spallation source, to improve the losses at injection and extraction, and we plan to take advantage of all that has been learned from the PSR commissioning. Our near-term goal is to fill in a menu of accelerator designs with preliminary technical designs and cost estimates so that there will be some options for use during the political and scientific discussions surrounding the funding of an AHF. We also are increasing the pace of hardware development. You will see examples of the work on ferrite-tuned rf, on ceramic vacuum chambers, and on magnets and resonant power supplies at this workshop. We will collaborate with all of the other hadron facilities for this development work. We plan to work particularly closely with the TRIUMF group during the next few years, as is discussed below.

1. Collaboration with TRIUMF

In the Fall of 1987, an agreement was made that the TRIUMF and Los Alamos groups will work together on the R&D for a next generation hadron facility. In particular, it was agreed that a single main-ring cavity will be developed that will meet the tuning range and voltage requirements of both the TRIUMF and Los Alamos main-rings. The Los Alamos group will build the cavity and provide the basic rf-power system. The TRIUMF group will provide the driver amplifier and the feedback control system. This cavity will be tested in the PSR starting with the 1989 running period. In addition to testing the performance of the hardware, the test in PSR will provide information about beam loading, 50-MHz bunching, longitudinal and transverse painting, coupled bunch instabilities, and synchrotron-betatron oscillations. The information obtained will be valuable for all of the proposed hadron facilities.

As a part of this agreement, the present Los Alamos booster cavity and 4 of the 6 existing ferrite toroids will be sent to TRIUMF for further research and development. The TRIUMF group was invited to participate in PSR development program and provided some staffing for all of the development runs of PSR in 1987. TRIUMF has also agreed to build a new beam-position-monitor control chassis for PSR that will process information in parallel with the existing controller. We are looking forward to a productive collaboration no matter which direction the funding of hadron facilities goes in the near future.

2. Menu of Accelerator Options

We propose to fill in a menu of accelerator options so that there will be flexibility when the time comes for a funding decision on AHF. For the high-energy machine, there are basically 4 options, namely:

a) Main Ring Only
b) Full-size Booster and Main-Ring;
c) Half-Size Booster, Half-Size Collector and Main-Ring;
d) Fourth-Size Booster, Full-Size Collector and Main-Ring.

For each of the four options, we must decide whether a stretcher is needed. We must also choose the proper injection energy for each design. A preliminary parameter list for each of these four options is presented in Table I.

The compressor ring for the spallation source and neutrino source shares the LAMPF linac and any afterburner linac. For the purposes of this workshop, we have generated a preliminary compressor design based on 1.6-GeV H^- injection. We chose a 50-MHz rf system for this compressor ring in order to share rf technology and a front-end rfq with the higher-energy machines.

Table I. Menu of Accelerator Options (May '88)

Scenario	Main Only	Full-size Boost	Full-size Main	1/2-size Boost	1/2-size Main	1/4-size Boost	1/4-size Main
Rep Rate (Hz)	6	6	6	12	6	24	6
Inj E (GeV)	2.2	1.6		1.6		1.6	
Energy (GeV)	60	10	60	10	60	6	60
nux	13.2	13.3	12.4	14.2	12.4	7.2	12.4
nuy	12.7	12.7	14.2	12.2	14.2	6.2	14.
Gamma T	22i	22i	8	∞	8	∞	22i
Lattice	yes	yes	yes	yes	yes	no	yes
ε_n (π mm-mrad)	10.5	18	18	9	9	2.7	2.7
long (eV-s)	14	10	10	10	10	10	10
Volts/turn(MV)	7.7	2.8	6.9	1.3	6.9	0.7	7.1
rf power (MW)	6.9	1.3	7.9	0.5	6.0	.26	6.4
df/f (%)	4.8	7.2	0.4	7.2	0.4	6.6	0.9
Sync Tune(max)	0.06	0.07	0.009	0.04	.009	0.02	0.01

nux is horizontal tune
nuy is vertical tune
Gamma T is transition gamma
Lattice = yes means that a preliminary lattice is available
ε_n is normalized space charge emittance for a Laslett tune shift of 0.17 maximum
df/f is tuning range required for rf system
Synch Tune is synchrotron tune (maximum)

TENTATIVE LINAC FRONT-END DESIGN

The linac must deliver H^- with twice the peak current presently available at LAMPF. This current should be available with 201.25 MHz microstructure for the compressor ring and at 50.3 MHz for the high energy rings. We are investigating a new "Volume Type" ion source with cesium converter and a 201.25 MHz RFQ to replace the first tank of the LAMPF drift-tube LINAC. This RFQ will be followed by a 50.3-MHz chopper to obtain the necessary bunch-to-bucket transfer for the high energy rings.

EXPANDED HARDWARE R&D PROGRAM

We are attempting to increase the pace of our hardware R&D effort. This expanded effort is described in the sections which follow:

1. <u>Magnets and Power Supplies</u>

The proposed advanced hadron facility is based on rapid-cycling (6-60-Hz) magnets The main-ring magnets raise the most problems as they must provide a 2.1-2.2-

Tesla field in order to fit the 60-GeV accelerator into the LAMPF site. Calculations are being presented in the hardware working group which show an initial design of a suitable high-field dipole and quadrupole. In order to eliminate eddy-current losses in the coils, a stranded, indirectly cooled conductor should be used in these magnets. This conductor is similar to that discussed by Prof Sasaki at the 1987 AHF Workshop and was discussed by the hardware working group at the 1988 Accelerator Design Workshop.

The Los Alamos group has done the pioneering R&D on the dual-frequency magnet power supply suggested by Praeg. The proof-of-principle power supply has now been thoroughly tested. We have demonstrated that it is possible to build a dual-frequency supply with adjustable length flattop and flatbottom. The final report of this work was presented at the 1988 Accelerator Design Workshop in Los Alamos. In the next year, we plan to work on the question of precision control and regulation of a multi-unit power supply. This work will be done for us by Prof. George Karady of ASU. For this study, we will work with a 1/10 size scaled version of the necessary power supplies.

2. Ceramic Vacuum Chamber

Ceramic vacuum chambers are needed for all of the rapid cycling magnets of AHF and of the TRIUMF Kaon factory. In order that the coupling impedance be minimized without interference with the ac guide field, these chambers must have conducting stripes to carry the image of the beam current (0.1 MHz-10 GHz). A capacitor at one end of each stripe assures a low impedance for these high frequencies while simultaneously providing a high impedance for the guide field (6-60 Hz). Our measurements of the coupling impedance of ceramic chambers show that the stripes must be on the inside of the vacuum chamber. In the next year, we plan to continue the development of these chambers in four steps. First, we will make a one-meter prototype with fixed flanges and internal stripes. This chamber will be fully vacuum tested but there will be no capacitors on the stripes. The second model will be a one-meter-long test of distributed capacitors. The third prototype will be a one-meter unit with capacitors and flanges. One of the two flanges will be demountable so that it will be possible to insert the vacuum chamber in a magnet without splitting the magnet in half. The fourth model will be a three-meter-long model which has all the features of the third unit. In addition, this long model will be curved to follow the beam without increasing the vacuum chamber dimensions to take care of the sagitta of curved magnets. Dr. Michael Featherby of SAIC is doing the ceramic vacuum chamber development under a contract with Los Alamos.

3. Ferrite-Tuned rf

The Los Alamos group has been working on a ferrite-tuned booster cavity for several years. Just after the 1987 AHF Workshop, we achieved 140 kV on a single gap with 15% duty factor, 20% tuning range, and R/Q of 35 Ohms. This cavity was tested for the voltage and tune program required for the TRIUMF booster. The TRIUMF rf group determined that its performance is adequate for their application. After this test, we increased the duty factor to 50% at the same voltage. Unfortunately, we experienced a mechanical failure of two of the ferrite toroids during the higher-duty-factor test. It was possible to repair the broken section (1/6) of each toroid, but because there were no spares available, we were not able to run the cavity again until September. We are now working with the complete cavity and are making temperature measurements of the ferrite during high power operation.

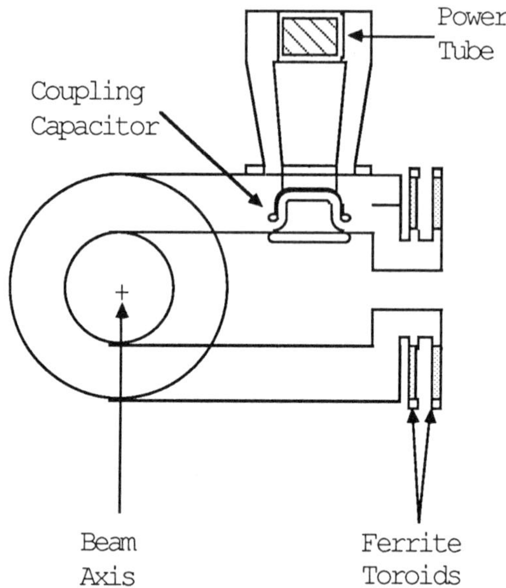

Figure 1. Sketch of proposed joint TRIUMF / Los Alamos main-ring cavity. The beam axis is normal to the plane of the paper. The ferrite is located in a separate tuner as shown. The power tube is located on the tuner rather than

on the beam line so that it will be possible to mount several accelerators one above the other (as in the TRIUMF design) without mechanical interference with the power tube. This cavity is being prepared for a test in the PSR in 1989 as discussed in Section 5.1.

In October 1987, we undertook the design and construction of a second-generation cavity in collaboration with TRIUMF. We chose a main-ring cavity because the smaller tuning range is more compatible with a planned test in PSR. A sketch of this cavity is presented in Figure 1. George Swain of Los Alamos has done the electrical design of the new cavity.

4. Kickers

Fast kickers play an essential role in all of the hadron facility designs. In addition to their use for ring-to-ring beam transfer, they are needed for fast abort systems. At Los Alamos, we have taken on the task of designing a prototype kicker magnet for a possible upgrade of the proton storage ring. We have chosen a module with 10.6-cm high by 20-cm wide gap (the field crosses the 10.6 cm gap). A 60-cm module with 400-Gauss field looks like a reasonable design. It should be pointed out that for this large-aperture kicker, our estimated 5-95% rise time is 100 nanoseconds. Even to achieve this rise time, it was necessary to drive the kicker with two pulsers. The 50-nanosecond gap proposed for most of the hadron facilities should be reexamined with the goal of determining kicker feasibility.

5. Other Necessary Hardware R&D

There are many other items requiring R&D in advance of construction. Foremost among these are the high-field magnets. We expect to start work on these magnets when we are confident of the required apertures. We should be ready to start work on the magnets in six months' or one year's time. Among the other accelerator components requiring R&D are beam diagnostics, controls, injection, and extraction hardware. We must wait to work on these items because of funding and manpower limitations.

A large amount of work must be done on experimental-area equipment. The target cells and remote-handled equipment needs many years of advance work. There must be a serious effort devoted to the cold-neutron source and to the neutrino source. RF separators will be required for the beam lines. These should be superconducting devices, which require a large amount of R&D. None of this work has as yet been started because of funding and manpower limitations. Because no new work has been done recently, we have not included experimental areas in this talk.

H^- INJECTION

Our work on H^- injection is based on the PSR commissioning effort. Because this work is included in the paper of E. Colton in these proceedings, only the conclusions are summarized here. Colton pointed out that the present 1.5% losses (0.6 μA) are the maximum which can be tolerated for hands-on maintenance of the machine components. He also concludes that that the slow beam losses of PSR can be explained by single and multiple scattering in the stripper foil. The injection system of PSR (with an H^o stripper magnet and resulting mismatch) nearly fills the aperture at the moment of injection. Two-thirds of the beam loss is due to multiple Coulomb scattering of beam that is near the edge of the acceptance. The remaining one-third of the losses are explained by single Coulomb scattering of beam that is injected near the center of the aperture. Inclusion of energy loss (dE/dx) is unimportant for the PSR losses for normal extraction, although it must be included to explain the losses seen if an attempt is made to store the beam for many milliseconds.

1. Losses due to Interactions with the Stripper Foil

The important losses from a new accelerator will be those resulting from single Coulomb scattering as long as there is a sufficient aperture to contain the small-angle multiple Coulomb scattered beam. The large-angle scatters can be computed from the formalism in many textbooks. Our favorite reference is Jackson. The integrated projected angle distribution is given by

$$2\int_{\theta}^{\infty} P_s(\theta')d\theta' = 2\pi Nt\left(\frac{zZe^2}{pv}\right)^2\frac{1}{\theta^2}$$

This equation is taken from Jackson (equation 13.114), where $P_S(q)$ is the projected scattering angle distribution, Nt is the number of atoms per unit area, z and Z are the charge of the projectile and atom, respectively, p is the momentum of the projectile, and v is the velocity of the projectile. The factor of two in this equation takes into account both positive and negative scattering angles.

There is another phenomenon which must be considered for the design of the next-generation machines. This is nuclear scattering. The nuclear scattering can be represented by a total cross section because the angle of scattering is large compared with Coulomb scattering. The angle of scattering is so large that it is unreasonable to contain the nuclear-scattered beam in the acceptance of an accelerator or storage ring. The stripping cross section is nearly independent of energy for energies of 800-MeV and higher (note that stripping is a process like dE/dx, which has a minimum value near velocities near 0.9c). A stripper foil of 250 µgram/square centimeter will have a 95% efficiency for conversion of H^- into H^+ at 800 MeV. Because the total cross section is also approximately independent of energy, we note that the fraction of beam which interacts with the stripper foil is roughly $6x10^{-5}$ per hit. The single Coulomb and nuclear interaction probabilities are shown in Figure 2.

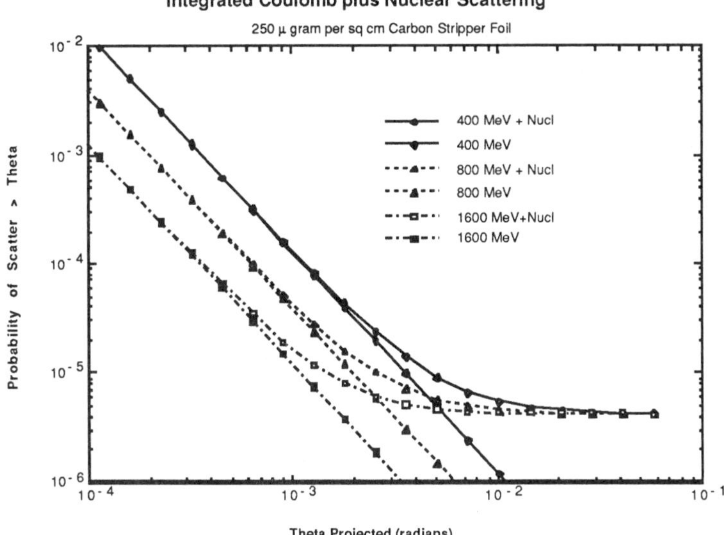

Figure 2. Integrated scattering probability for single Coulomb scattering as a function of angle. Also shown is the sum of single Coulomb scattering and nuclear scattering.

In previous design studies, insufficient attention was given to the interaction of the beam and the stripper foil. At PSR in development runs, there are more than 1000 foil hits per injected particle and a 3% loss. This is easily explained with a $1x10^{-5}$ loss probability (and a factor of 3 enhancement for multiple scattering) or an acceptance of about 2 milliradians.

The lesson is clear. We must keep the losses below 0.6 μA at 800 MeV. The stripper foil must be located at a low-beta double waist. Then the acceptance must be sufficient to accept scattering out to the angle at which Coulomb and nuclear scattering probabilities are approximately equal (2 mrad for 250 μg per square cm of carbon at 800 MeV). Even this may not be sufficient if a large number of turns of injection are required (as at PSR, the new AHF compressor, or the TRIUMF accumulator ring). In these cases, it is necessary to find a painting scheme that minimizes the number of foil hits. If the foil hits cannot be reduced sufficiently, the only alternative is to put a collimator in the downstream portion of the injection straight section to contain the spill. The region between the stripper foil and the collimator must be shielded, remote handling must be provided, and no active components should be located in this portion of the ring.

In designing the injection system, we must be careful not to forget the fraction of the beam which is not fully stripped or which misses the foil. For a 250 μgram per square centimeter foil, at least 5% of the beam will remain as H^o. At PSR, a 200 μgram foil is used. Our experience at PSR is that 10% of the beam is neutral and that up to 5% additional beam misses the foil and remains as H^-. It is absolutely essential to provide a dump for both the H^o and H^- beams which leave the foil. These dumps are not difficult to design since the phase space which must be contained is that of the injected beam, not the beam stored in the ring.

It is the opinion of the Los Alamos group that meeting all of the injection constraints simultaneously requires an accelerator designed to fit around the injection system and that the injection system cannot be simply inserted into a pre-existing straight section. For this reason, all of the new Los Alamos designs have a long injection-straight-section consisting of several cells. For the AHF Booster, we favor a racetrack-shaped lattice.

2. Injection Painting

At Los Alamos, Eugene Colton has made a preliminary study of the fraction of time that a particle hits the foil for several painting schemes. The results are tabulated in Table 2. In addition to meeting the criterion of minimizing foil hits, the painting scheme must result in a beam matched to the accelerator with the desired phase space and G factor. Meeting all of the constraints simultaneously may result in a larger number of foil hits.

Table 2. Foil hits per particle per turn for several painting schemes. (Warning, results are very sensitive to the distance of the beam from the edge of the foil!).

Injection Scheme	Foil Hits per Turn
Fixed Brush (2D)	25%
X-Y Offset and Skew Quad	10%
X-Y Offset and Bumps	10%
TRIUMF Painting Scheme	8%
Combine 3). & 4). ?	6%?

SLOW EXTRACTION: AN INVENTION NEEDED

At the 1987 AHF Workshop, Lee Teng emphasized the need for an invention to improve the efficiency of the slow-extraction system. Present day slow-extraction systems spill on the order of 1% of the beam. The problem is 75 times worse than the injection-loss problem because of the higher beam energy at extraction. Thus we must limit extraction losses to 0.008 μA, or $0.8\text{x}10^{-4}$ for a 100 μA machine Both the Los Alamos and TRIUMF groups are working on improved efficiency schemes. The Los Alamos proposal consists of locating a massless magnetic septum 90° upstream of the electrostatic septum to reduce losses on the electrostatic septum. The TRIUMF alternative is a special thin (10-micron) and short (1-meter) electrostatic septum in place of the massless magnetic septum. It is expected that a reduction of the losses by an order of magnitude to $1\text{x}10^{-3}$ is possible by either of these techniques. Since this is insufficient to limit the extraction losses to the required level, it will be necessary to include a collimator in the extraction system to reduce the losses in the remainder of the ring by an additional factor of 12.

CONSIDERATION OF LIFETIME COST

In the present set of designs for hadron facilities, only the LAMPF II design took account of the operation and maintenance of the accelerators. The design should account for the lifetime cost of the machine including R&D, design, construction, installation, operation and maintenance, and power. In a 30-year projected lifetime of an accelerator, the dominant cost is the operation and maintenance cost. This subject is discussed in the following section.

1. Operation and Maintenance Cost

The operation and maintenance cost of a machine is difficult to predict in advance. Among the factors which influence the result are the complexity of the design, the degree of standardization, the quality of the design and execution, and the desired beam availability. It is unlikely that we can make a quantitative estimate of all these factors. I asked Roy Billinge to use the experience of the PS division at CERN and make a subjective estimate of the expected manpower requirements. I then used the experience of Los Alamos ($145k/full time equivalent) to convert the manpower estimate into cost. The results are presented in Table 3.

Table 3. Estimated Manpower Cost for Several Designs

Design (TRIUMF Nomenclature)	People	Annual Cost
- D (Main-ring Only	60	$ 8.7 Million
- D+E	100	$14.5 Million
- B+D (LAMPF II)	130	$18.9 Million
- B+D+E	170	$24.7 Million
- B+C+D+E (EHF)	210	$30.5 Million
- A+B+C+D+E (TRIUMF)	260	$37.7 Million

In designing the accelerator, the operation and maintenance cost should not be the dominant consideration. Indeed, first priority should go to a conservative, reliable design which meets all the requirements of the physics program. Nevertheless, some weight should be given to simpler designs which will be cheaper to operate. In the 30-year lifetime of a typical machine, the difference in operating cost between the simplest and most complicated design in Table 3 is $870 million.

THE LOS ALAMOS VIEW OF THE SPECIFICATIONS FOR AN ADVANCED HADRON FACILITY

All of the existing proposals for hadron facilities discuss a beam power near 3 MW (100 μA at 30 GeV). We believe that 60 GeV is a better choice for beam used for hadron spectroscopy. We also point out that 1.6-GeV is a better energy for neutrino physics. The 1.6-GeV beam energy is well matched to the needs of the neutron scattering community. Our proposal of 60 GeV at 25 μA with two additional 600 μA beams at 1.6 GeV results in a higher total power (3.5 MW) than the other proposed facilities and a better match to the needs of the users.

SUMMARY AND CONCLUSIONS

The Los Alamos plan consists of performing the long-lead-time R&D required to prepare for a construction start in 1993. To this end, we are preparing a menu of accelerator options, are actively engaged in the development of the PSR, and are accelerating the pace of hardware R&D. This work is being performed in close collaboration with TRIUMF, especially the development of a joint main-ring cavity that is being prepared for testing in PSR in 1989.

The PSR commissioning work has led to an understanding of the source of the observed beam losses. Single Coulomb scattering must be considered in the design of all of the next generation machines. A painting scheme which minimizes foil hits is also required. In order to keep the efficiency of the slow-extraction scheme high enough, it is necessary to use a pre-septum and to add a collimator to contain the residual losses. It is suggested that more consideration be given to an accelerator architecture that minimizes operation and maintenance cost. For these reasons, we believe that all of the proposed hadron facilities should be redesigned for high-intensity operation.

The MIT-Bates South Hall Ring

J. B. Flanz and MIT-SHR Staff
MIT-Bates Accelerator Center

3rd. Conference on the Intersections between Particle and Nuclear Physics

INTRODUCTION

The South Hall Ring (SHR) under construction at the MIT-Bates Linear Accelerator Center will provide high duty factor electron beams. There will capabilities for internal targets and external beams. This is the natural culmination of the Bates developments enabling coincidence experiments in the medium energy range up through 1 GeV. Those recently or nearly completed developments, including the recirculator, the polarized source, and the existing set of large spectrometers, position Bates for taking advantage of unique opportunities to address fundamental issues in nuclear science.

The SHR is being constructed to intersect the existing South experimental hall as shown in Figure 1. With the linear accelerator/recirculator system as the injector, it will produce extracted beams of up to 50 microamps with a duty factor of ~85%. Also under construction is an energy compression system which reduces the beam energy spread to ~0.04%. The capability of providing longitudinally polarized beams both externally and for internal target work is included. With two turn injection the ring will store 80mA for internal target physics. Table 1 summarizes the present Bates parameters and the expected beam parameters to come from the SHR.

BEAM PARAMETERS

	Linac Beam SP,SIM,HT	Internal Target Ring Beam	Extracted Ring Beam
Energy (MeV)	50-1000	300 - 1000	300 - 1000
Peak I (mA) MAX.	40, 5, 40	80	0.05
Average I (uA) MAX.	50	80000	50
Pulse Width (uS) MAX.	15, 15, 1.3	-	-
Rep Rate (PPS) MAX.	600,600,1000	1000	1000
Duty (%)	1, 1, .13	99	85
dE/E (%)	.3, (.04 ECS)	(.04 ECS)	(.04 ECS)

Emittance at 500MeV

Horiz π(mm-mr)	.01	.01 -1.0	.01
Vert π(mm-mr)	.01	.01	.01

SHR LATTICE DESIGN

The ring circumference is dictated both by the constraint that a suitable internal target area exist in the South Hall, and the pulse length available for injection. The accelerator single pass beam is variable in pulse length up to 20 usec. However in order to maximize the peak current to be injected into the ring it is necessary to recirculate the beam in head-to-tail mode where the pulse length will be given by the accelerator/recirculator path length of 1.3 usec. The 190 m circumference will be suitable for two turn injection. With an accelerated peak current of 40mA at 1Khz, this provides a circulating current of 80mA and an average extracted current of 50uA.

The SHR lattice functions are shown in Figure 2. The ring is composed of four basic regions. These include:

1. The achromatic bend region is a four cell section comprising a second order achromatic.

2. The short straight section connects the two bend regions to form a symmetry corrected second order achromatic 180 degree bend. The result is that it is possible to adjust the ring chromaticity over a wide range without disturbing the symmetry corrected second order centroid shifting aberrations at the extraction and internal target locations.

3. The injection/Internal Target straight section incorporates a flat beta region for injection and a small beta region for internal target operation. The former is chosen to maximize the injected beam displacement from the injection septum and allow for reasonable strength injection kickers while keeping the septum far enough out of the path of the extracted beam. The latter region is important for maintaining a reasonable lifetime for a stored beam with an internal target. The low beta reduces the angular growth of the beam caused by small angle and multiple scattering in the target.

4. The Extraction straight section prepares for a high beta region to obtain a large displacement of the extracted beam from the closed orbit in order to extract the beam easily. Furthermore the position tolerances and heating effects on the extraction electrostatic septum are reduced.

EXTRACTED BEAM

The SHR is designed to provide an extracted beam with a high duty factor, small emittance, small energy spread and high average current. It is important to obtain this with high throughput efficiency and reasonable tolerance. The extraction mechanism employed is that of half-integer resonant extraction. During extraction the horizontal tune of the SHR is adjusted towards a half integer tune with ramping air-core quadrupoles. An octupole operating at a constant strength, separates the particle phase space into a stable and unstable region. Initially the beam is injected into the stable region. During extraction the phase space area of the stable region is reduced and the beam enters the unstable region. Its spacial amplitude eventually increases to pass in the field region of an electrostatic septum when it is directed outside the ring.

Figure 3 shows an example of the extraction process. The intersecting parabolas indicate the theoretical boundaries between the stable and unstable regions of phase space. Each plot represents a successive time interval during the extraction process. The beam inside the stable area is show as a snapshot during that time period, the beam outside is accumulated in the figure. After a correcting for a correlation in the extracted beam angles as a function of time, the resulting beam emittance is on the order of 0.01 mm-mr. The time dependence of the intensity of the extracted beam is shown in figure 4. The resulting duty factor is 85%.

INJECTION

The recirculator accelerator pulse (1.3usec) is twice the length of the ring (.65 usec) and therefore up to two turns may be injected into the ring. This can be repeated as often as 1 millisecond. For extraction, the beam is injected with a 3mm displacement from the septum (and closed orbit). It is possible to inject closer to the closed orbit. Another approach for single turn injection has the closed orbit displaced from the septum and the beam injected on the closed orbit axis. These injection schemes produce a beam at the internal target region as shown by the phase space plot in figure 5. The smallest phase space represents the one-turn-on-axis injection. While the large annulus is the phase space of the beam occupied for extraction. The intermediate beam area is defined by the minimum space required for the injected beam to cleanly pass by the injection septum during two-turn injection. Note that the beam is always injected on axis in the vertical phase space and is thus a minimum size in the ring.

INTERNAL TARGET OPERATION

The above discussion indicated the possible stored beam sizes at the internal target location for various methods of injection. The presence of an internal target will affect the nominal beam size. One criteria for the applicability of the ring for use with internal targets is the beam quality and the time that an adequate beam quality is maintained with an internal target in the ring. It is useful to consider some of the mechanisms which affect the beam quality and ultimately determine the experiments which are feasible. This will also yield information on the possible modes of operation of the ring. The mechanisms include the effects of the target on the beam, the effect of the ring components on the beam and the effects of the beam on the target.

Consideration of the above mechanisms can be used to establish the ring operational limits. For any internal target experiment, the background should be manageable. This will be determined by the beam size, halo and ring apertures. Also, the emittance of the beam must be consistent with the physics requirement for tract reconstruction or resolution. For example, assuming the use of a non-dispersion matching spectrometer with a focal plane dispersion of 10 cm%, the beam emittance must be less than 4 mm-mrad to enable a resolution consistent with the 0.04% energy spread of the beam. (This will be called condition 2 in Figure 6). In order to minimize backgrounds, the beam emittance must not grow so large as to intercept any apertures which will produce beam spray at the internal target. With the extraction septum, the emittance should be less than 4 mm-mrad. Without the extraction septum, the emittance should be less than 4 mm-mrad. In either case, it may be possible to absorb beam spray from these sources. However, if the beam hits the internal target conductance limiting apertures, or a target from in the case of a self supporting target, it will be difficult to deal with that background. Any thin aperture will produce an effective target thickness of well over 10^5 that of the target. Clearly, the emittance must be limited to a growth such that the boundary of the beam containing 1.0×10^{-5} of the beam does not intersect the aperture. This upper limit requires an emittance growth to remain below 0.5 mm-mrad. (This will be called condition 1 in Figure 6). Finally, if one requires that the emittance not grow by more than 10%, extraction could be carried out while an internal target is in place. (This will be called condition 0 in Figure 6).

Calculations were made including the effects of scattering and wakefields produced by target collimator apertures together with the ring damping. The emittance growth was determined as a function of internal target density and plotted as a function of time. Figure 6 shows results for 880 MeV. With a beta equal to one meter at the internal target region, all reasonable operating conditions listed above limit the emittance growth of the beam to under 1.E-6 M-Rad.

A fraction of the beam loses energy via bremsstrahlung. That part of the beam which loses more than 20% of the energy is lost near the target. Beam that has lost energy greater than a percent will be lost in the ring. Calculation of this effect gives a contribution to the beam lifetime. If 1/2 the beam is lost, the upper line on Figure 7 indicates the relevant lifetime. The beam lost due to this effect in one second would be equivalent to an average current loss of 0.05 microamps.

The horizontal lines on Figure 6 represent the limits imposed by the constraints discussed above for the different modes of operation. By noting the time it takes to reach those lines, the plot of storage times in Figure 7. can be formed. The indications are that a wide range of operational modes is possible in the ring. Given the fast refill time possible, targets as thick as 10^{19} cm^{-2} are feasible, and targets as thick as 10^{18} are possible in a parasitic mode. It is also possible to store beam for many seconds and possibly longer (depending upon ring behavior) with thinner targets.

A powerful and unique tool for nuclear physics is under construction. The South Hall Ring will be used for internal target physics, under which conditions it offers:

*Large Admittance

*Flexibility (beta Functions)

*High space/element ratio
*Excellent beam quality

The external beam that will be produced will be highly competitive even with beams from CW machines such as microtrons with its *low emittance and *small energy spread. Perhaps just as important is that now, prospective users can provide input to help shape the experimental capabilities, color scheme, and target development.

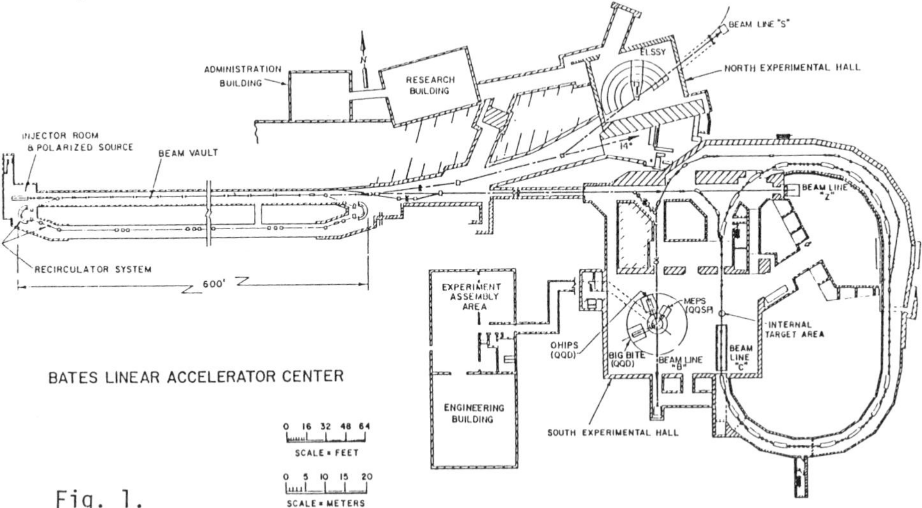

Fig. 1.

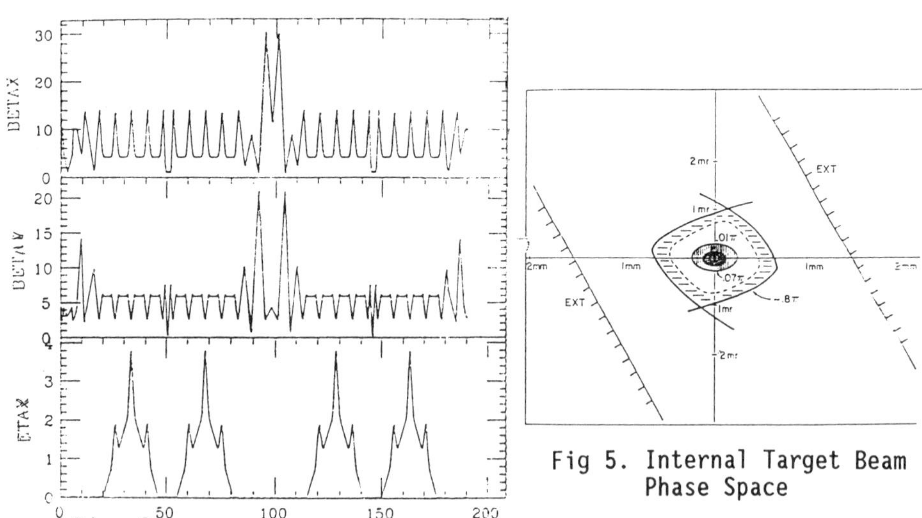

Fig. 2.

Fig 5. Internal Target Beam Phase Space

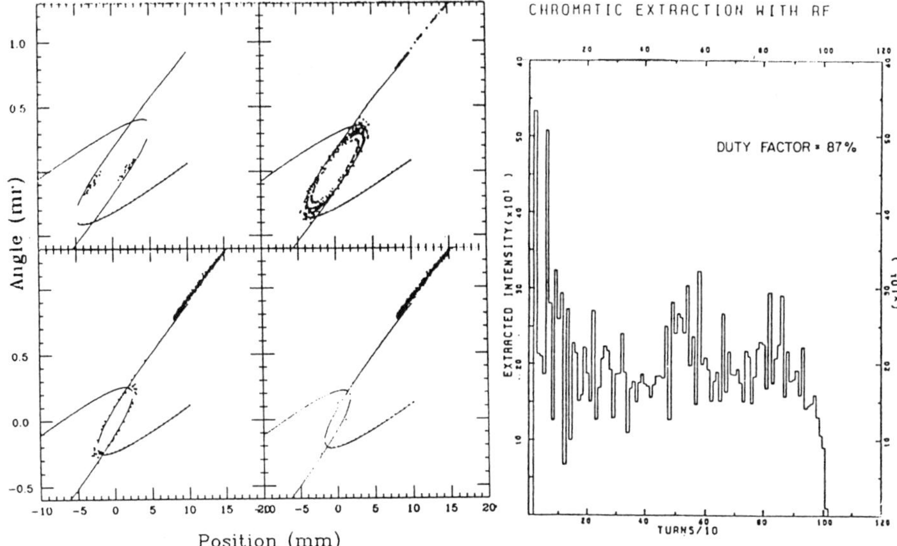

Fig 3. Extraction Phase Space

Fig 4. Extraction Intensity

Fig 6. Internal Target Emittance Growth

Fig 7. Beam Lifetime with Internal Target

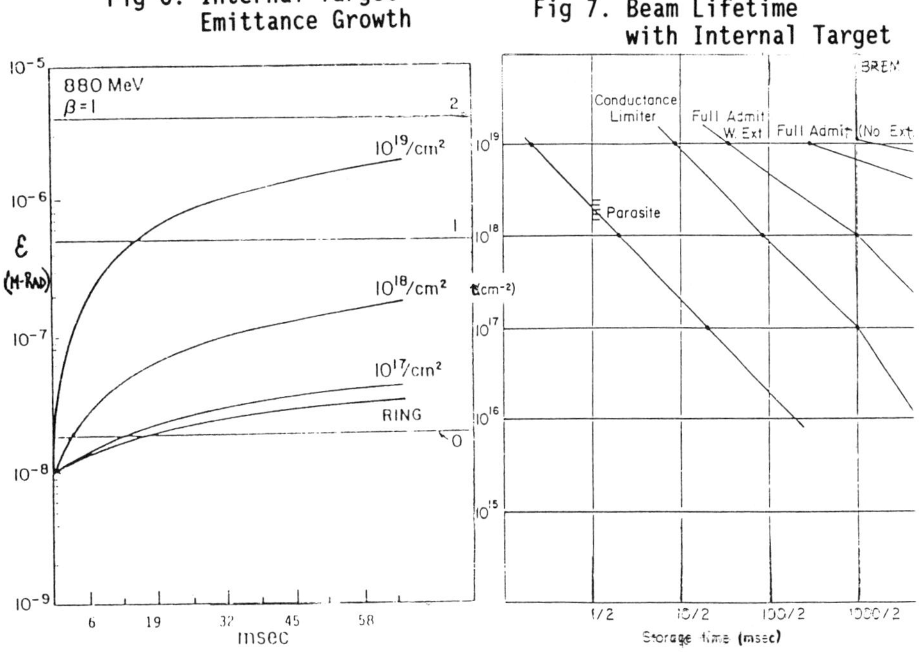

Potential Low Energy $\bar{p}$ Source at BNL

D. C. Peaslee
Department of Physics and Astronomy
University of Maryland
College Park, Maryland 20742

Abstract

The AGS Booster, under construction at Brookhaven National Laboratory, is discussed as a source of low energy antiprotons with momenta from 0.65 to 5.2 GeV/c. Under ideal conditions a production rate of 10^{14}-10^{15} $\bar{p}$/yr. can be envisioned. The premier particle physics study enabled by this beam is CP violation in cascade decays resulting from $\bar{p}p \to \bar{\Xi}\Xi$. For this purpose the multiparticle spectrometer at BNL appears an excellent detector to enhance the data acquisition rate. Considerable exploration of ordinary hadron spectroscopy can simultaneously be accomplished in the poorly known region $\sqrt{s}$ = 2-3 GeV.

1. Introduction

The question of low energy $\bar{p}$ sources in continental North America has attracted recent attention, beginning with a workshop at Fermilab about 2 years ago.[1] The most feasible research at Fermilab turned out to be a one-off experiment on production of charmonium by $\bar{p}p$ collisions on a hydrogen gas jet in the $\bar{p}$ accumulator. A full-fledged $\bar{p}$ program at low energies appeared to be somewhat outside Fermilab's charter as a single-purpose high energy laboratory - and also rather expensive, evoking thumbnail cost estimates of order $50 million.

Four months later a Summer Workshop on Antiproton Beams was held at Brookhaven.[2] There the emphasis was on long beam lines for decontaminating the $\bar{p}$ by decay of the π^- and K^- cohort. It was also noted, however, that the advent of the Booster preaccelerator to the AGS provided an unusual opportunity with regard to antiprotons. The Booster is just 1/4 the size of the AGS and during normal operation with protons is idle most of the cycle. This leaves a period of about 2 seconds in which some protons can be extracted from the AGS and targeted to produce $\bar{p}$'s, which are then run backward through the Booster. The result is a highly purified $\bar{p}$ beam with excellent resolution over a wide momentum range. This production proceeds parasitically when the AGS runs on protons. The principal limitations are that the $\bar{p}$ must be cleared from the Booster each AGS cycle to allow for reloading of the AGS, and that intensity is severely reduced at low extracted momenta because of emittance blowup in the beam.

In 1987 a two-part conference was hosted by the RAND Corporation to explore the options for a substantial source (10^{14}-10^{15}/yr) of very low energy ($\lesssim$ 50 KeV) antiprotons for a number of ultimate applications. At that time it appeared that the potential yield of such $\vec{p}$'s from both BNL and FNAL would be comparable: 10^{13}-10^{14}/yr. The Booster option at BNL might have appeared more immediate in time and cost, but it was recognized that this goal would require stochastic cooling in the Booster of the transverse momentum components in order to avoid beam loss during deceleration. A scheme was detailed for this with extraction into a dedicated RFQ[3]; and a pro forma cost estimate of order $15 million was suggested[4].

This summarizes recent history up to early 1988. The following section will review the concept of the BNL Booster as a $\vec{p}$ source and will discuss some of its uses for elementary particle research.

2. Antiprotons from the Booster

Figure 1 shows the general concept. Each AGS cycle begins by filling the Booster with 3 bunches of 0.5 x 10^{13} protons and ejecting them into the AGS, repeating the process 4 times. The Booster is then idle for about 2 seconds till the start of the next cycle: plenty of time to extract 3 bunches of 28 GeV protons from the AGS to a target station at I 10 and use them to produce antiprotons. The remaining 4.5 x 10^{13} protons per pulse are available for other purposes. A lithium lens is to collect the $\vec{p}$ and a beam line is to transport them back to the Booster, where they are injected through the proton extraction channel. The Booster rf system can be applied to vary the $\vec{p}$ momentum over a range 0.65-5.2 GeV/c. The antiprotons are then extracted in one straight section with a moderately thick septum tangent to the AGS and transported to an experimental area. The extraction must occur every AGS cycle, since the Booster cannot function in any way as a $\vec{p}$ accumulator.

The target area indicated is the old 80-inch bubble chamber building. This has already been chosen as the experimental area for a muon g-2 experiment; but an extension to the existing building could provide $\vec{p}$ working space at low cost by use of existing services. Table I shows some properties of the $\vec{p}$ beam in the Booster. The momentum spread of ± 1% delivered from the $\vec{p}$ production target can be reduced to < 10^{-3} by debunching, and probably further by phase displacement acceleration during extraction. This process compresses the Δp of the total accepted $\vec{p}$ flux without loss, and a double advantage results from the tunability of Δp: wide Δp for search and scan, narrow Δp for study of a resonance already located.

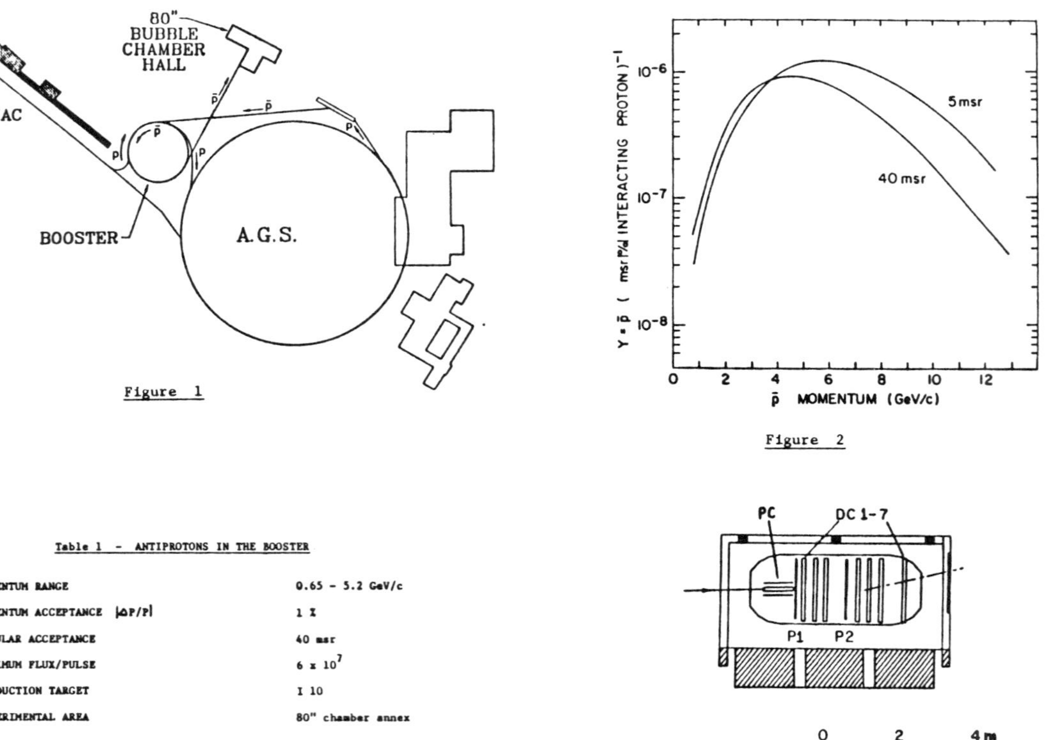

Figure 1

Figure 2

Figure 3

Table 1 - ANTIPROTONS IN THE BOOSTER

MOMENTUM RANGE	0.65 - 5.2 GeV/c
MOMENTUM ACCEPTANCE $\vert\Delta P/P\vert$	1 %
ANGULAR ACCEPTANCE	40 msr
MAXIMUM FLUX/PULSE	6×10^7
PRODUCTION TARGET	I 10
EXPERIMENTAL AREA	80" chamber annex

The yield of $\bar{p}$ in Table 1 is estimated as follows. The production rates of $\bar{p}$ by 28 GeV p on Be are estimated from Sanford-Wang formulas[5], which agree with production rates observed at CERN. These are shown in Fig. 2 for two different angular acceptances. We assume the generally credited factor of about 1/3 for the fraction of beam protons that interact in the target, and a further reduction factor of 1/3 for the capture efficiency of the Li lens. For transport to the Booster around 3 GeV/c,

$$Y \approx 0.9 \times 10^{-6} \times 1/3 \times 1/3 \times 1 \times 40 \times 1.5 \times 10^{13} \qquad (1)$$

$$\approx 6 \times 10^{7}\ \bar{p}/\text{pulse}$$

If this yield is obtained parasitically during 15 weeks/yr. of SEB running at the AGS, the source intensity is

$$S = 6 \times 10^{7}\ \bar{p}/\text{pulse} \times 10^{3}\ \text{pulse/hr} \times 10^{2}\ \text{hr/wk} \times 15\ \text{wk/yr} \qquad (2)$$

$$\approx 10^{14}\ \bar{p}/\text{yr}$$

If the AGS operated an additional 10 weeks/year in a dedicated $\bar{p}$ mode with doubled pulse rate, the source strength in Eq. (2) would be increased by one order of magnitude:

$$S' \sim 10^{15}\ \bar{p}/\text{yr} \qquad (3)$$

The extra running costs would be of order $1 million.

Although these are ideal estimates and hence liable to downward revision in practice, they suffice to indicate a potential source of appreciable integrated intensity.

3. Particle physics research

Practical applications, particularly with regard to antimatter, have been discussed in recent conferences at RAND and Karlsruhe. Here we consider how the Booster $\bar{p}$ source may contribute to particle physics research.

The most striking possibility is the study of CP violation in hyperon decay. This was early recognized as a channel to study T(= CP) violation in weak decays, but estimates of its small size were quite discouraging[6]. Observation of the effect among neutral kaons[7] depended on their otherwise perfect resolution into K_S and K_L, which is unique to bosons in their ground states. The hyperon option was resurrected by the observation[8] that hyperon-antihyperon pairs produced by $\bar{p}p$ annihilation had a high degree of CP symmetry, as with neutral bosons; advantage was also taken of the self-analyzing features of hyperon decay.

There are 3 different measures of CP violation in such reactions: $\Delta\Gamma$, the difference in hyperon and antihyperon decay rates; $\Delta\alpha$, the difference in polarization observed in their decays; and $\Delta\beta$, the difference in spin correlations to be found in decay cascades like $\Xi \to \pi + \Lambda$, $\Lambda \to \pi + p$ and the corresponding anticascades. It turns out that the sensitivity to CP violation increases by about 1 order of magnitude per step in the above sequence. The compromise so far adopted is to work with $\bar{p}p \to \bar{\Lambda}\Lambda$ some 50 MeV above threshold, at a laboratory momentum of $p_L = 1.55$ GeV/c. A first run[9] with some 10^{10} antiprotons attained an accuracy of about 0.1 units in a parameter estimated to be of order 10^{-3}. This led to an estimate of 10^{14} $\bar{p}$ as the basic unit of flux needed to approach a positive measurement.

The Booster scheme imposes a penalty in intensity if the $\bar{p}$ are slowed down without cooling. Thus, if the $\bar{p}$ are collected at 3 GeV/c, there is a loss of intensity by a factor 4 in reaching ~ 1.5 GeV/c for $\bar{\Lambda}\Lambda$ production. On the other hand, slowing down to ~ 2.7 GeV/c for $\bar{\Xi}\Xi$ production just above threshold would cut the intensity only about 20%. The corresponding production cross section is poorly known but appears to be about 1 order of magnitude below the $\bar{\Lambda}\Lambda$ cross section just above threshold[10]. If these figures can be trusted along with the theoretical estimates that $\Delta\beta/\Delta\alpha \sim 10$, then compared with $\bar{p}p \to \bar{\Lambda}\Lambda$ the Booster is about 10–100 times more effective for $\bar{p}p \to \bar{\Xi}\Xi$, making it the experiment of choice. The total flux requirement for a CP violating signal to appear is thus reduced to 10^{12}–10^{13} $\bar{p}$; using Eqs. (2), (3) suggests that in a very favorable case this much data could be acquired in a week.

Although such a conclusion may be overly optimistic, it indicates that accurate measurement of CP violation would be possible on a time scale commensurate with other weak decay experiments. Indeed, the suggestion remains that it might take only a few years instead of many.

In addition to the Booster source the Brookhaven locale could also supply the non-negligible advantage of an excellent detector for this purpose, shown schematically in Fig. 3: the multiparticle spectrometer (MPS). It provides a large volume of magnetic field up to 1 Tesla, well supplied with drift chamber and proportional wire detectors plus readouts and algorithms for handling 6 or more charged particles. The target location relative to the magnet is entirely free, and special counters to surround the target are available or can be fabricated. With this detector one can imagine operating somewhat above the $\bar{\Xi}\Xi$ threshold, obtaining an order of magnitude increase in cross section - by analogy with $\bar{\Lambda}\Lambda$. Also, the use of a relatively long liquid hydrogen target might be feasible, with the promise of a rate increase of up to 10^2 over that on which the 10^{14} $\bar{p}$ requirement was predicated. These factors should compensate for a counting rate loss oforder 10^2 because the wire chambers can handle only about 5×10^5 events/sec.

4. Prospect of SU_3 hadron spectroscopy

Production of $\bar{\Xi}\Xi$ will represent only some 10^{-6} of the total $\bar{p}p$ cross section. Most annihilations will yield cascades of ordinary π, η, K ... mesons. The MPS is an ideal apparatus for collecting and analyzing these cascades and thus studying SU_3 hadron spectroscopy. Especially if it is found that the MPS encompasses $\bar{\Xi}\Xi$ production well above threshold, a sensitive scan for hadron production vs. $\bar{p}p$ energy will be possible. Suitably detailed analysis should reveal several prominent mesons resonances accessed by $\bar{p}p$. It is generally true that projectiles with threshold mass M are most effective in producing resonances in the near region above threshold. For $\bar{p}p$ this implies major effectiveness in the center of mass range 2-3 GeV, corresponding to incident $\bar{p}$ momenta of 0.75-3.8 GeV/c, exactly in the Booster range.

References

1. Proc. 1st. Workshop on Antimatter Physics at Low Energy, Fermilab (April, 1986).
2. Proc. 1986 Summer Workshop on Antiproton Beams, BNL 52082 (May, 1987).
3. Y. Y. Lee and D. I. Lowenstein, RAND Conference, Santa Monica (October, 1987).
4. D. C. Peaslee, Antimatter '87 Conference, Karlsruhe (December, 1987).
5. J. R. Sanford and C. L. Wang, BNL (11479 (May, 1967).
6. T. D. Lee, Preludes in Theoretical Physics (ed. de Shalit, Feshbach and Van Hove) North-Holland (Amsterdam, 1966).
7. J. H. Christenson et al., Phys. Rev. Lett. 13, 138 (1964).
8. J. F. Donoghue and S. Pakvasa, Phys. Rev. Lett. 55, 162 (1985).
9. P. D. Barnes et al., Phys. Lett. 199B, 147 (1987).
10. V. Flaminio et al., Compilation of Cross-Sections III; p and $\bar{p}$ Induced Reactions, CERN-HERA 84-01 (April, 1984).

STATUS REPORT ON THE LOS ALAMOS PROTON STORAGE RING

Eugene Colton, D. Neuffer, H. A. Thiessen, H. Butler, G. R. Swain, A. Lombardi*, D. Fitzgerald, F. Mariam, M. Plum, R. Ryder, R. Macek, D. Clark, J. Hurd, R. Hutson, J. McGill, T. Hardek, R. Shafer, G. Lawrence, A. Jason, B. Blind, R. Hardekopf
Los Alamos National Laboratory, Los Alamos, NM 87545

ABSTRACT

The proton storage ring currently operates at an average current of 30 μA corresponding to 1.25 x 10^{13} protons per pulse (ppp) at a repetition rate of 15 Hz. The design operating current for the machine is 100 μA. We are limited to running at the reduced yield because of beam losses during the accumulation period. These losses are understood and arise mainly from emittance growths during the injection and multiple scattering in the stripping foil during the storage. During beam studies we have succeeded in accumulating in excess of 3.7 x 10^{13} ppp. We have also observed a coherent transverse instability at high charge levels. The signature for the instability is rapid coherent growth of the transverse beam size followed by a loss of beam in the machine. The threshold for the instability depends most strongly upon rf voltage and beam size.

INTRODUCTION

Construction of the Proton Storage Ring (PSR) was completed in Spring, 1985. The ring provides up to a factor of 4000 time compression for production of thermal neutrons. These neutrons are useful for condensed matter physics, chemical physics, materials science, crystallography, and structural biology. The proton beam also serves to create a copious neutrino flux. A thorough review of the PSR operation through the beginning of 1987 has been presented previously.[1] In addition, we have prepared a paper on the transverse instability seen in PSR.[2] During 1987 we performed further experiments and developed further insight into the mechanisms responsible for beam lòss during accumulation. These losses currently limit operation to an average current of 30 μA. In this work we briefly review the principles of operation, then discuss some of the measurements. Next we discuss the mechanisms responsible for the accumulation (slow) losses. Following this, we present empirical results for the coherent instability. Finally we discuss plans for a proposed upgrade which should reduce the accumulation losses to a tolerable level, which we have defined as 100 nA; with this implemented we should be able to raise the delivered current to 100 μA.

---

*Collaborator, CERN, Geneva, Switzerland

PRINCIPLES OF OPERATION

Figure 1 shows a plan view of the PSR complete with the local injection and extraction beam transport systems. Table I lists the relevant parameters of the machine. With the exception of the chromaticities listed, all of the given values are design parameters. In fact, the design chromaticities were -0.8 and -1.3, respectively, in disagreement with the measured values.

Table I. PSR Parameters

Proton Kinetic Energy	797 MeV
Circumference	90.2 m
Focusing Structure	FODO, 10 cell
Betatron Tunes Q_x, Q_y	3.23, 2.22
Acceptance A_x,A_y	124, 167 mm-mrad
Transition Gamma	3.1
Average Beam Pipe Radius	0.05 m
Measured Chromaticities ξ_x, ξ_y	-1.3, -1.0
Buncher Harmonic, Frequency	1, 2.795 MHz
Peak RF Voltage	15 kV
Maximum Synchrotron Tune Q_s	0.0006
Bunch Length	62 m
Stored Beam	5.2×10^{13}
Repetition Rate	12 Hz

Referring to Fig. 1, H^- are transported from the LAMPF Linac and arrive at the PSR through the injection beamline. The H^- are stripped to H^0 in the fringe field of the stripper magnet; the horizontal emittance is increased by a factor of three during this process. The H^0 then drift to a 200 $\mu g/cm^2$ carbon foil where 90% of them are stripped to protons. The surviving H^0 continue downstream to the H^0 beam dump. These processes continue through the injection period, meanwhile the stored protons orbit repeatedly (up to 2800 turns for a 1 msec injection). The circulating current has reached 16.5 amperes while the 1 msec LAMPF macropulse has been compressed to 250 nsec (a reduction of 4000). At the end of injection the two injection kickers are synchronously fired and the intense proton bunch is directed into the extraction beamline. This line is merged into the line D and the protons sent to the experimental targets. The circulating beam in the PSR has a gap of about 28 m in order to facilitate lossless extraction.

MEASUREMENTS

The PSR uses a host of diagnostics to perform basic measurements necessary to maintain operation. The system is sufficiently automated so that we have online representations for proton flux, accumulation losses,

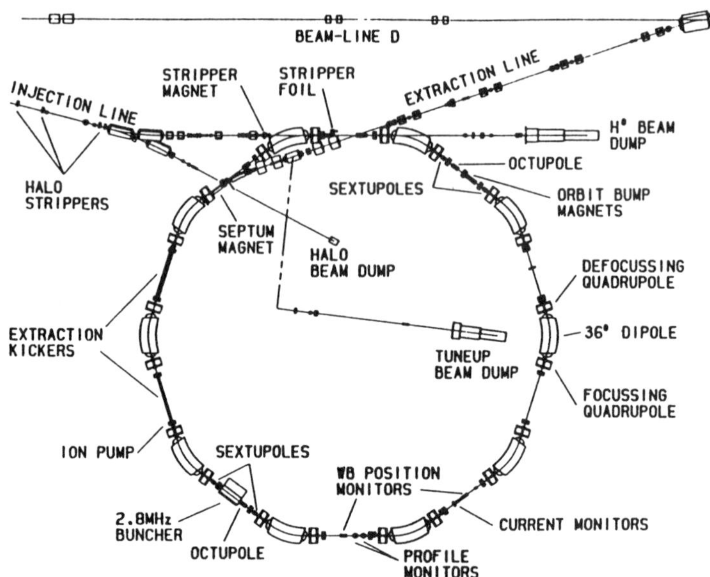

Fig. 1. Plan view of PSR.

longitudinal bunch shape, and closed-orbit locations. During machine studies we have measured the synchrotron tune vs. applied rf voltage, horizontal beta functions at several locations, dispersions, space-charge tune shifts, losses vs. quadrupole settings, leading to a betatron resonance map, betatron tunes vs. relative momentum spread, and many parameters related to the coherent instabilities. We empirically found a best fit to the data for the betatron tunes vs. dp/p over the range $|dp/p| < 0.005$. We found

$$Q_x(dp/p) = Q_x(0) - 1.25\ Q_x(0)\ dp/p + 111\ (dp/p)^2$$

and

$$Q_y(dp/p) = Q_y(0) - 0.96\ Q_y(0)\ dp/p - 84\ (dp/p)^2$$

It appears that there is a zeroth-harmonic contribution of both sextupole and octupolar fields. We have also observed integral harmonic contributions indirectly through the excitation of betatron resonances. Evidence was obtained for the sum resonances $Q_x + 2Q_y = 8$, $2Q_x + Q_y = 9$, and $2Q_x + 2Q_y = 11$. The strong integer $Q_x = 3$ and $Q_y = 2$ resonances are also observed. These resonances do not affect operation at our nominal tunes.

SLOW LOSSES

During production we inject beam for 375 microsec which is about 1050 turns. During this period we lose about 1.6%. These losses are quantified as being 0.2% first turn, 0.15% from nuclear and large angle coulomb scattering in the foil, 0.5% from action of the rf cavity, and 0.85% from emittance growth from foil scattering in the absence of rf. The losses take place at several horizontal aperture restrictions, namely the foil holder and extraction septa. The two stage injection into PSR involves a horizontal emittance growth at the stripper magnet as well as a horizontal plane mismatch of the beam to PSR in both dispersion and phase space at the injection stripper foil. The net result of all this is a first-turn loss, and a beam that is too wide and has non-Gaussian tails. The rf cavity just makes the beam wider, and narrower, as time proceeds due to synchrotron motion, i.e., it is just a dp/p effect. When the beam is widened losses occur at the horizontal restrictions. The rms emittance growth in the foil is simply given by the expression

$$\varepsilon = \varepsilon_0 + \beta_f < \theta^2 > fN_t/2$$

where ε_0 is the initial rms emittance, β_f is the beta function at the foil, $< \theta^2 >$ is the mean squared scattering angle for a single foil traversal, f is the probability of a foil traversal on a turn, and N_t is the number of turns. The emittance grows linearly with time, with the result that $<x^2>$ the mean square value of the beam width increases linearly with time. We have, in fact, observed such a growth in beam width. Presently, the foil traversal probability is high, greater than 50% even when we steer the injected beam so as to reduce the foil traversals. We have observed that the losses increase quadratically with time - a doubling of the injection period results in 4 times the losses. Even with a time averaged current of 30 μA, we are seeing losses of about 0.5 μA which are 5 times the desired maximum.

COHERENT INSTABILITY

A sudden loss of beam current is observed when we exceed a certain current threshold with typically half of the beam vanishing in about 50 microsec. A rapid coherent transverse beam growth is observed just before the loss. Figure 2 are oscilloscope traces showing the time dependences of the beam current for injected fluxes below and above the threshold. These pictures were taken for a coasting beam, i.e. with the rf cavity off; the threshold varies from 0.4-1.1 x 10^{13} protons depending upon the injected vertical beam size. The instability threshold can be pushed up by energizing the rf cavity - the higher the voltage, the higher the threshold. In fact, additional measures, such as inclusion of a second-harmonic component to the cavity voltage, and the use of sextupole and octupole magnets, allowed us to reach 3.7 x 10^{13} ppp; we are source limited at this value because we dispense with 30% of the beam during injection in order to maintain the beam gap for the kicker rise time.

The instability threshold can be represented by the standard coasting beam formula[3] which is expressed as an inequality

$$|Z_\perp| \, I_{thresh} > 4 \, m_0 c^2 \beta\gamma \frac{Q}{R} [(n-Q)\,\eta + \xi\, Q] \frac{\delta p}{p}$$

where $Z_\perp$ is the transverse coupling impedance, Q is the betatron tune, R is the machine radius, n is an integer > Q, $\eta = \gamma_t^{-2} - \gamma^{-2}$, and $\delta p/p$ is the relative momentum spread (fwhn) of the beam. The major impedance of the ring is the space charge impedance, given by

$$Z_\perp = \frac{iRZ_0}{\beta^2\gamma^2}\left(\frac{1}{a^2} - \frac{1}{b^2}\right)$$

where Z_0 is the impedance of free space (approximately 377 ohms), and a and b are the beam radius and beam pipe radius, respectively. The magnitude of $Z_\perp$ is in the neighborhood of 5-10 Mohms/meter. Increasing a reduces $Z_\perp$ and tends to stabilize the beam. Increasing the rf voltage raises the relative momentum spread of the beam, thus also stabilizing the beam; an increase in betatron Q also helps too. The instability growth rate is quite large, of the order of 2×10^4 sec-1; it is directly proportional to the real part of the transverse coupling impedance which has to be of the order of 0.5-1.0 Mohm/meter for such a large growth rate. We have searched for potential contributors to $Re(Z_\perp)$ with little success. We have eliminated the orbit bumper magnets, rf cavity, and the housing containing the stripper foil as active agents. Furthermore, it does not appear that there is a single resonator which is responsible for the short growth time.

A PROPOSED UPGRADE

We are currently planning a staged modification to the PSR. The first is a direct H^- injection system to replace the present two-stage injection system. The use of direct H^- injection has some distinct advantages over the present scheme. The obvious gains are: We can totally match the linac beam to the PSR at the injection point; the emittance growth during the injection will be reduced. The smaller injected emittances can be used to reduce the foil traversals in a number of ways which include linear coupling, collapsing orbit bumps, etc. There is more flexibility in filling of phase space so we can, e.g., balance between space charge effects and accumulation losses. Figure 3 shows a possible rework of section 0, the injection straight, so as to permit direct H^- injection. The lattice bends are each reduced by 3 degrees at the opposite ends of the straight section and a weak combining dipole with 6 degree bend merges the circulating protons and injected H^- at the midpoint; a H^- to proton stripping foil follows thereafter. The detailed design is under study-so far it looks promising. Another important change will be an increased horizontal aperture in the machine - this requires a new extraction system. The increased aperture in the machine can be obtained by a redesign of the fast extraction area; we are planning to replace the stripline extraction kickers with a fast ferrite kicker supplying 18 mrad of kick. After this, the extraction septa can be moved so that the aperture of the machine will be 50 mm radius throughout.

CONCLUSIONS

For the immediate future, we hope to increase the average current via techniques involving reduction of the foil traversals as well as through optimization of the betatron tunes. We will continue to research the coherent instability. We hope to have completed the above improvements by FY 90. After this we look forward to reaching the design current of 100 μA.

REFERENCES

1. G. P. Lawrence, IEEE Particle Accelerator Conference, 87CH2387-9, 825 (1987).

2. D. Neuffer, et al, "Transverse Collective Instability in the PSR," to be published in Particle Accelerators (1988).

3. See e.g., G. Guignard, "Selection of Formulae Concerning Proton Storage Rings," CERN 77-10 (1977).

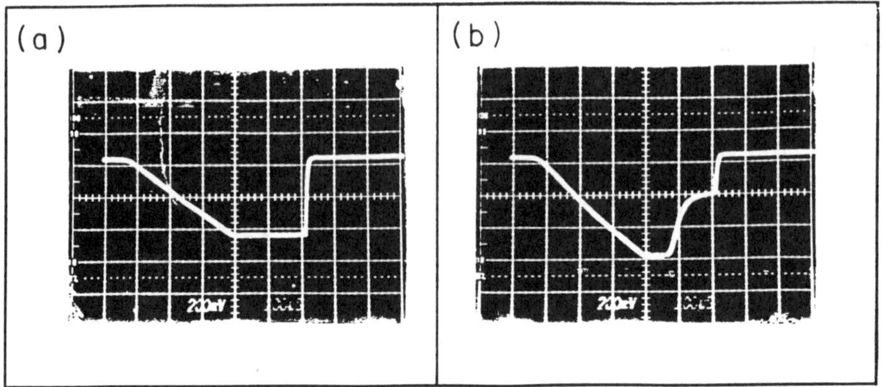

Fig. 2. Oscilloscope tracer indicating circulating current in the PSR for a coasting beam. Injection period is 300 μsec followed by a 200 μsec coast and fast extraction.

(a) Below instability threshold.
(b) Above instability threshold.

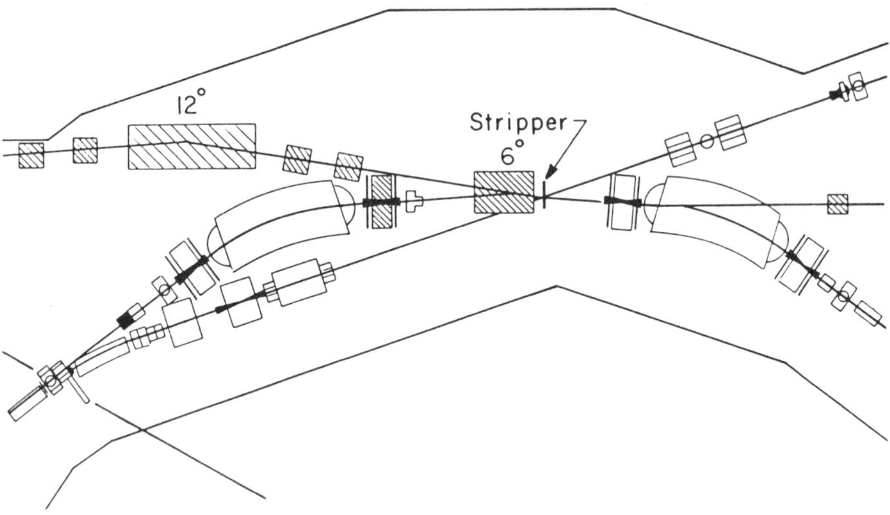

Fig. 3. Suggested rework of the injection section to accommodate direct $H^- \rightarrow H^+$ injection.

BEAM BREAKUP IN SUPERCONDUCTING RECIRCULATING LINACS

Joseph J. Bisognano
Continuous Electron Beam Accelerator Facility
12000 Jefferson Avenue, Newport News, Virginia 23606

INTRODUCTION

The performance and operational flexibility of superconducting recirculating linacs can be limited by a variety of collective phenomena which are grouped under the name beam breakup. These coherent effects include both multipass beam breakup, which is driven predominantly by the high-Q superconducting cavities and limits the average current, and single-bunch beam breakup, which is driven by the short-time (Q-independent) wakefields and limits the total bunch charge. The superconducting radio frequency (SRF) environment has a number of features which distinguish it from room temperature configurations with regard to these beam-environment interactions. Typically the unloaded Qs of the lower higher order modes (HOM) are at the 10^9 level and require significant damping through couplers. High gradient CW operation, which is a principal advantage of SRF, allows for better control of beam quality, which for its preservation requires added care with respect to collective phenomena. Gradients are significantly higher than those attainable with copper in CW operation but remain significantly lower than those obtainable with pulsed copper cavities. A CW superconducting linac is recirculated for several passes to optimize for cost. The focusing in later passes is limited by the requirement of stable transport on the first pass, and this constraint exacerbates the effects of short-time wakefields both through the weakened focusing strength and the inflexibility inherent in having multiple beams in a single focusing channel. In this note the various beam breakup phenomena found in recirculating SRF linacs are described and appraised relative to beam performance.

Q Dependence of Collective Effects

The Q dependence of the various beam-cavity interactions provides a fundamental differentiation which is in many ways more basic in the underlying accelerator physics than whether the phenomena occur in a linear or circular accelerator. At one extreme are coupled-bunch effects (cumulative and multipass BBU and storage ring coupled-bunch instabilities) which are made more severe by high-Q modes. For bunch-to-bunch coupling, wakefields must linger for at least the duration of the bunch spacing, and, moreover, a long or CW train of bunches can coherently excite a mode for a time of order Q/ω. The mode Z/Q is essentially a geometric property of a cavity and is insensitive to wall material or de-Q-ing efforts through couplers. It describes the single-pass excitation of a cavity mode by a beam bunch. On the other hand, $Z = (Z/Q)Q$ describes the coherent summation of multiple excitations over the filling time of the mode. Therefore, it is the impedance $Z = (Z/Q)Q$, not Z/Q, which determines the average current limits for tolerable emittance degradation or the onset of exponential growth, and the coupled-bunch phenomena are extremely sensitive to Q.

At the other extreme are single-bunch, single-pass effects which are driven by the short-time behavior of wakefields. For short bunches (nanoseconds to picoseconds) the inherently short sampling time results in lost frequency resolution (smearing of hundreds of megahertz to hundreds of gigahertz) which effectively averages frequency dependent, possibly high-Q impedances over the mode spectrum. In addition, for picosecond linac bunches, the wakefield at short times is dominated by the impedances at frequencies well above cutoff where modes can be expected to overlap. Thus, it is the impedance Z/Q, not Z, which determines the peak current limits for tolerable emittance degradation, and these limits are insensitive to Q.

Regenerative Beam Breakup

A train of bunches passing offset through an accelerating structure can excite transverse deflecting modes through a longitudinal electric field which increases linearly off axis for such modes. For a low energy beam, the excitation can be strong enough to deflect the beam significantly during transit through a single structure into regions of higher longitudinal field and increased coupling. Therefore, there can be closure of a feedback loop within a single structure: an excited cavity mode deflects a beam which in turn further excites the mode. Stability comes from the damping of the fields through wall losses and couplers. This effect is known as regenerative beam breakup, and for a standing wave structure has a threshold current given by[1]

$$I_{\text{th}} = \frac{\pi^3 E k}{2Z'' L^2} \tag{1}$$

where Z'' is the transverse impedance per unit length, L is the length of the RF structure, E is the beam energy and k is the wavenumber of the HOM. Above threshold, there is exponential growth of the cavity excitation; any offset will ultimately grow to the point of destruction of the beam. Since SRF structures usually are only a few cells long to prevent the appearance of trapped modes, the dependence of the threshold current on the square of the length lessens the importance of this current limit in practice. At a few MeV, however, an undamped ($Q \sim 10^9$), gigahertz SRF cavity would be limited to currents typically of the order of a milliampere. Note that the threshold current depends on Z'', not Z''/Q, and is therefore inversely proportional to the damped Q.

Multipass Multibunch Phenomena

When a linac beam is recirculated through an accelerator structure, the feedback system formed by the cavities and beam is closed and instability can result at sufficiently high currents. Suppose that the first cavity has an excited HOM which deflects the beam. Recirculation brings the beam back through that first cavity for one or more passes. Since the lattice can translate the deflection into a transverse offset, the excited mode can now be enhanced as was the case in regenerative BBU. In addition, many cavities can participate in this instability if HOM frequencies overlap. Figure 1 shows the bunch displacement of succeeding bunches at the exit of a recirculating linear accelerator with bunch number, at a current above the threshold for this *multipass beam breakup.*[2]

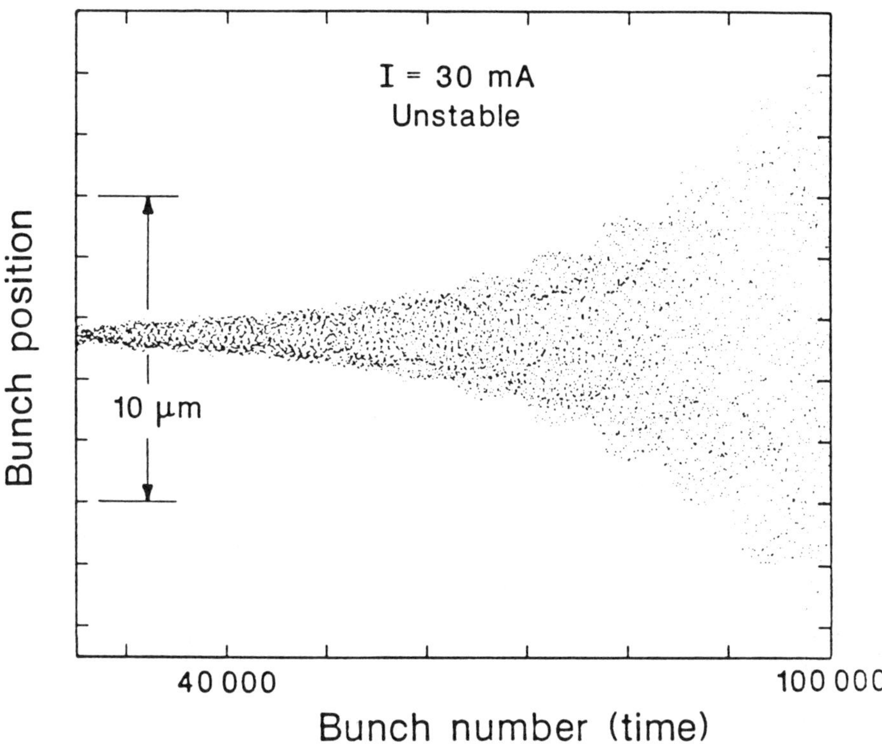

Figure 1 Exponential growth of beam excitation above threshold of multipass beam breakup

Multipass beam breakup in linacs and coupled-bunch instabilities in storage rings are both caused by the interaction of the beam bunches with high-Q higher order modes. The basic difference is that multipass beam breakup involves an infinity of bunches each of which undergoes a finite number of passes, whereas the coupled-bunch instability involves a finite number of bunches each of which undergoes essentially an infinity of passes. Also, given the isochronicity of typical recirculating linac lattices, transverse deflecting modes are of primary importance in linacs, whereas both longitudinal and transverse modes can drive instabilities in storage rings.

The growth rates and threshold current for multipass beam breakup are given by solution of an eigenvalue equation[3]

$$\begin{aligned} D_i = I \sum_{p=2}^{n_p} \sum_{r<p} \sum_{\ell=1}^{n_0} \left(T_{i,\ell}^{pr} \right)_{12} e^{M_0 \Omega \tau (p-r)} Z_\ell h_\ell(\Omega) D_\ell \\ + I \sum_{p=1}^{n_p} \sum_{\ell=1}^{i-1} \left(T_{i,\ell}^{pp} \right)_{12} Z_\ell h_\ell(\Omega) D_\ell \end{aligned} \tag{2a}$$

$$\text{where} \qquad h_n(\Omega) = \frac{H_n(\Omega)\sin(\omega_n \tau)}{1 + H_n(\Omega)^2 - 2\,H_n(\Omega)\cos(\omega_n \tau)} \tag{2b}$$

$$H_n(\Omega) = e^{-\frac{\omega_n \tau}{2Q}} e^{-i\Omega\tau}$$

and

$T^{pq}_{i,\ell}$	is the transfer matrix for (x, p_x) from pass ℓ to pass i			
Z_n	$= \frac{Z''_n \ell e}{2Q f_b}$	ℓ	$=$	cavity length
n_0	$=$ number of cavity sites	n_p	$=$	passes
f_b	$= \frac{1}{\tau} =$ bunching frequency	I	$=$	average beam current
$s_k(\omega_n \tau)$	$= e^{\frac{k\omega_n \tau}{2Q}} \sin(k\omega_n \tau)$			
Z''_n	$=$ transverse impedance of cavity/unit length			
M_0	$=$ number of bunches in one recirculation			

The inverse of the threshold current appears as the eigenvalue. To find thresholds for instability, a plot of complex current eigenvalues has been found useful. First the coherent frequency Ω is swept in real frequency with an arbitrarily small imaginary part corresponding to growth. For n_0 cavities there are n_0 families of eigenvalues, with the actual threshold current corresponding to the smallest possible real value obtained. Figure 2 illustrates the eigenvalue structure for a multicavity, multipass linac. The lowest positive current crossing gives the threshold current. Note that there is a finite value of current below which there is stability.

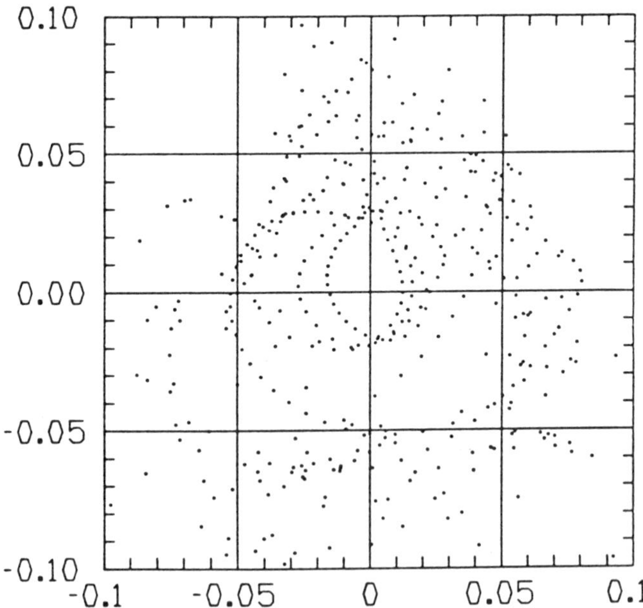

Figure 2 Complex current plot (amperes) for the determination of threshold current of multipass beam breakup driven by 400 cavities in a four-pass configuration

For an HOM of a single cavity, the threshold current is inversely proportional to the transverse impedance Z, and since (Z/Q) is constant, inversely proportional to the damped Q. For a many cavity machine, however, HOM mode frequencies can be expected to overlap. Then, as Q increases, fewer modes overlap and the threshold current does not drop as rapidly with $1/Q$. Also, for a few passes, the threshold current is found to be insensitive to pass numbers for fixed final energy. Below the threshold of multipass beam breakup there exists a steady-state cavity excitation which displace bunches and in some cases can lead to an effective emittance degradation. Since this steady state is a forced oscillation of the cavity, the cavity excitation is at a harmonic of the bunching frequency for equally spaced, equally charged bunches. Thus, all bunches receive the same kick, the phase of which depends on the relationship between the bunch harmonic frequency and the HOM frequency. This kick in principle can be compensated with DC magnetic elements. Clearly, HOMs which are near harmonics of the bunch frequency are of most concern. For user flexibility, bunches can be differentially loaded to provide, for example, different end stations of a linac with different bunch current.[2] For example, a 1500 MHz bunching could be differentially loaded at 500 MHz to form three classes of bunches (1,4,7,10,...), (2,5,8,11,...), and (3,6,9,12,...). Since the basic periodicity of the system is at 500 MHz, HOMs neighboring on harmonics of 500 MHz are now of particular importance. For 500 MHz harmonics which are not also harmonics of 1500 MHz, each of the three classes of bunches receives a different kick. The resulting offset cannot be corrected by DC elements, and thus there is an effective emittance increase during transport through the linac which requires an RF device (e.g., RF separators) to sort out. For average currents at the hundred microampere level, this effect is negligible.

The extreme case of differential bunch loading occurs when short trains (possibly one) of very highly charged bunches are used. The strong wakefields generated can influence both the transverse and longitudinal phase space of succeeding members of the high-charged train or of low-charged fellow-traveling bunches. Control of the Q of both transverse and longitudinal HOMs is required to relieve this possible source of phase space degradation.

Single-Bunch, Single-Pass Effects

In addition to the coupled-bunch phenomena which have been described so far, there is a large class of single-bunch, single-pass effects which limit the peak current (more precisely, bunch charge and length) handling capabilities. In a single pass through an RF structure, there is insufficient time for the bunch to experience the long-term ringing associated with the high Qs characteristic of superconducting cavities. Thus, these current limits are not particularly sensitive to the success or failure of damping the HOM Qs. However, there do remain some features of the collective phenomena which distinguish superconducting cavities from room temperature ones.

Because of the efficiency of the cavities (maximization of shunt impedance is not as crucial as with room temperature cavities) and the requirement of coupling through the beam pipe to avoid sites for multipactoring and breakdown, a typical SRF cavity has a large aperture. This increase in opening reduces the coupling impedance of higher order modes; for transverse modes this coupling can be an order of magnitude below that which would be expected for an optimized room temperature cavity. Secondly, although the gradient achieved in SRF cavities is larger than that which can be achieved in CW room temperature cavities, it remains significantly less than that achieved in pulsed room temperature cavities. Thus, for CW operation, SRF machines have a clear advantage from an impedance point of view over room temperature machines, since the SRF impedance per megavolt is lower. However, in comparison to pulsed, room temperature linacs, the situation is not as clearcut where the SRF cavities offer less impedance per unit length, but depending on gradient, may in fact offer more impedance per megavolt.

Energy spread is induced by the variation of the longitudinal wakefield across the bunch. At 5 MV/m gradient, the induced energy spread is found to be less than 1% for a 1 mm long bunch at peak currents of 500 A in a typical 1500 MHz cavity. With adjustment of the RF phase to remove the linear part of the variation, residual spread at the 0.1% level can be expected. Improvement of achieved gradient directly improves the residual energy spread.

Recirculating SRF linacs also are susceptible to single-bunch transverse beam breakup which is analogous to that which has been extensively simulated and analyzed for single pass linear colliders. There are, however, several issues which are special to the SRF devices. First, given the smooth, large aperture design of SRF cavities, the contribution of bellows, valves, and other discontinuities gain relatively more importance than for small aperture room temperature cavities, and care must be taken to minimize the strength of such impedance sources if the full potential of the superconducting environment is to be realized. Secondly, although SRF cavities, because of their open structure and typically low frequencies, offer reduced transverse impedance, the weak focusing on higher passes inherent in a multipass device can exacerbate this head-tail beam breakup. Here it is found that constant phase advance lattices (in the first pass) with their added focusing strength offer considerably better performance at low injection energies than constant magnetic gradient configurations. Finally, it appears difficult, if not impossible to arrange for the pseudo-Landau-damping which is thought to be effective in limiting emittance degradation in single-pass linacs because of the differences in focusing experienced by the overlapping beams of different energy. For the most part, however, the considerably lower transverse impedance offered by low frequency SRF cavities outweighs these limitations.

ACKNOWLEDGMENT

This work was supported by the U.S. Department of Energy under contract DE-AC05-84ER40150.

REFERENCES

1. P.B. Wilson, "High Energy Electron Linacs," in Physics of High Energy Accelerators, AIP Conference Proceedings No. 87, 1982.
2. G.A. Krafft and J.J. Bisognano, "Two Dimensional Simulations of Multipass Beam Breakup," CEBAF-PR-87-008.
3. J.J. Bisognano and R.L. Gluckstern, "Multipass Beam Breakup in Recirculating Linacs," CEBAF-PR-87-008.

Issues, Options, and Performance Expectations for Storage Ring Based B$\bar{B}$ Factories

David Rice
Cornell University, Ithaca, NY 14853

ABSTRACT

The recent widespread interest in B Factories with very high (>10^{33} cm^{-2}-sec^{-1}) luminosity has encouraged proposals from organizations in several continents. Storage ring based machines have the advantage of many machine-decades of experience upon which to base designs. Peak luminosities of 10^{32} cm^{-2}-sec^{-1} have been achieved at a storage ring operating at energies appropriate for B meson production.

The high currents necessary for these luminosities present the most difficult technical problems in operating such a machine. Large higher mode power, synchrotron radiation power, and fundamental RF power coupled to the beam push technology well beyond that achieved in present machines. In addition, careful control of optics in double rings is required to achieve maximum luminosity.

Recent developments in accelerator technology permit construction of a 2 x 10^{33} cm^{-2}-sec^{-1} luminosity B factory with a high level of confidence in achieving design goals.

PHYSICS RATIONALE

The heavy mass of the B$\bar{B}$ system and the fact that the b quark is presently the heaviest accessible quark have made b physics extremely productive during the past 8 years. Many parameters of the Standard Model have been measured in studies of decays of upsilon particles[1]. At the same time, the much lower production cross section (1/40) of the B's compared to the psi has prevented investigating B decays to a similar depth as psi's.

While more luminosity is better for measurement of B decays in general, the possibility of observation of c-p violation in the b system has stimulated several less vague estimates for integrated luminosity requirements[1-3]. These figures vary by a factor of 10 either way from 50 fb^{-1}. Present capability of e+e- storage rings is about 0.1 fb^{-1} /month (a very good month!) so that on the order of 40 years of steady running would be required to see c-p violation using the estimates above. An order of magnitude improvement in luminosity is clearly needed.

PLANS FOR B FACTORIES

During the last year several plans have been discussed for construction of high luminosity colliders covering the 9 to 11 GeV/c cm range. These are listed in Table 1 below. If the luminosity averaged over time is 50% of peak luminosity, 1 x 10^{33} peak luminosity will yield 1,000 pb^{-1} per month.

Most of these B factories use separate rings for electrons and positrons. The PSI (formerly SIN) proposal[4] is at the most advanced stage. A feasibility study has been completed and a technical review is scheduled for June, 1988. The other plans have been widely discussed; however, I am not aware of any formal proposals.

The outline of the proposed PSI machine is shown in Figure 1. Injection is accomplished with 200 MeV electron and positron linacs, an accumulator ring for both, and a full energy booster synchrotron. RF cavities are placed symmetrically about the 2 interaction points, and horizontal dispersion is zero at the interaction points and RF cavities. Maximum design energy is 7 GeV/beam.

Institution	Type	E_{beam}	Circum	I_{beam}	Luminosity
PSI	race	5 x 5	648	485	1×10^{33}
Cornell	circ	5 x 5	760	1340	2×10^{33} *
KEK	circ	5 x 5	377	50	8×10^{31}**
DESY	circ	2 x 14	115,2300	180,60	1×10^{32}
SLAC	circ	2 x 12	155,2200	540,270	5×10^{32}
"	"	5 x 5	2200	354	1×10^{33}
Novosibirsk	race	5 x 5	500	600	$2\text{-}5 \times 10^{33}$
(Amaldi)	race†	5 x 5	1630	0.1/.03	1.4×10^{33}
"	"	7.5x7.5	"	0.1/.06	1×10^{34}
UCLA	linear	5 x 5	∞	0.02	1×10^{34}
		(GeV)	(meters)	(mA)	(cm^{-2}-sec^{-1})

* with future second ring; present planned upgrade to 5×10^{32} luminosity
** per bunch, total # bunches not specified
† racetrack linear accelerator

Table 1 – B Factory Concepts

CIRCULAR VS. LINEAR

While discussion will in general be limited to circular colliders, a brief comparison with linear schemes is appropriate.

Circular B factories are limited by beam-beam tune shift, synchrotron radiation, and RF higher mode power dissipation and single beam instabilities. Linear colliders are limited by cost of RF power, accelerating structures, and real estate, which drive designers toward very small beam sizes and very high gradients. As a result, major developments must be made in higher frequency RF components, particularly power sources and accelerating structure design.

Limits to the circular collider technologies are well known. Performance may be accurately predicted relative to a linear collider. A linear collider is a much more complex machine requiring control of many rf systems and multiple feedback systems to effectively collide micron size beams. On the other hand, one of the well known limits of circular colliders is the maximum energy, determined primarily by synchrotron radiation losses which scale as $P \propto E^4/\rho^2$. Thus a linear collider may serve as a stepping stone to higher energy machines.

CIRCULAR B FACTORY DESIGN

The luminosity of a storage ring is given by:

$$L = \frac{f^* N_b^2}{4\pi \sigma_x \sigma_y} \quad (1)$$

where:

f^* = number of collisions/second/ir

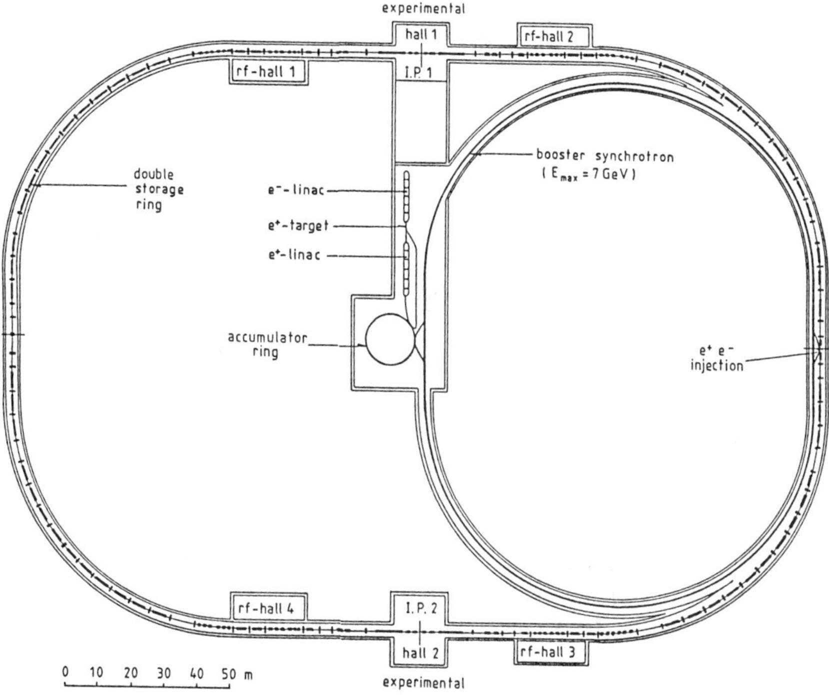

Figure 1 – Proposed P.S.I. B Factory Layout

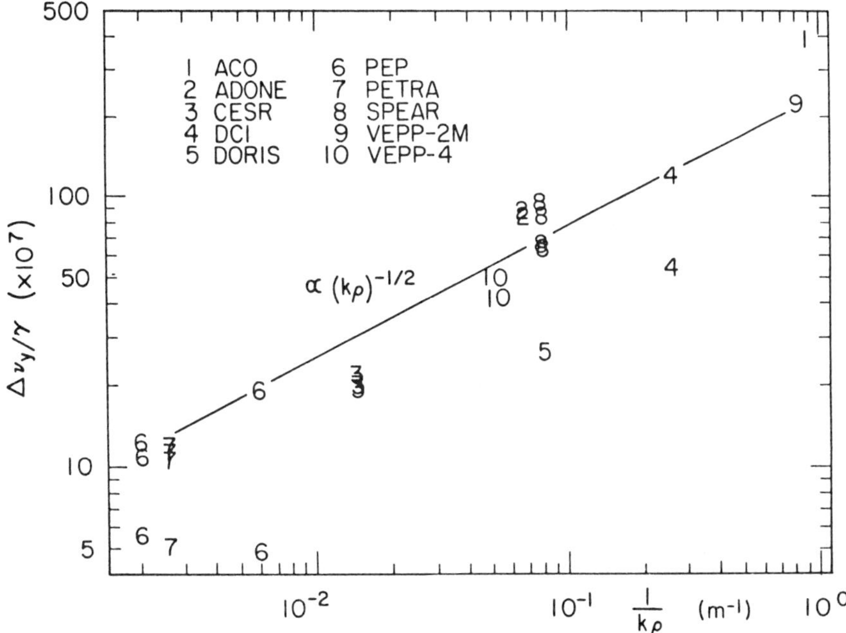

Figure 2 – Scaling of Linear Tune Shift with Number of IR's and Bend Radius

N_b = number of particles per bunch

σ_x, σ_y = horizontal, vertical beam standard deviation at interaction point

The luminosity is maximized by increasing f^* and N_b, or reducing $\sigma_x\sigma_y$. Accelerator physics considerations limit optimization of these parameters in three different categories.

1. $N_b / \sigma_x\sigma_y$ is limited by interaction of the electromagnetic field of one beam on particles in the opposing beam. This is the beam-beam limit.
2. f^* x N_b is limited by single beam instabilities, synchrotron radiation, and higher mode power.
3. f^* is limited by ability to separate the electron beam from the positron beam near the interaction points; i.e., we must prevent unwanted interaction between bunches away from the primary ip's. Other limitations on f^* are imposed by multibunch instabilities and control of these by feedback.

BEAM-BEAM LIMIT

As the charge density of a bunch at the interaction point approaches a critical density, the vertical size of the opposing bunch begins to increase in response to the perturbation caused by the first bunch. If symmetry is maintained between the electron and positron beam, both will increase in vertical size to keep the charge density constant at the limiting value. (This picture assumes flat beams starting with small vertical emittance so the critical charge gradient is reached first in the vertical plane.)

The effect of the electro-magnetic field of one beam on the other is described by the beam-beam tune shift parameter ξ, where:

$$\xi_i = \frac{N_b \, r_e \, \beta_i^*}{2\pi\gamma \, (\sigma_x+\sigma_y)\sigma_i} \tag{2}$$

The subscript i refers to either vertical or horizontal as appropiate, β^* is the beam envelope function at the interaction point, and r_e= 2.8x10^{-15} meters. If $\sigma_y << \sigma_x$, equation (2) may be used to eliminate the beam size in equation (1), with the result:

$$L = \frac{f^* \, \gamma \, N_b \, \xi_y}{2 \, r_e \, \beta_y^*} \tag{3}$$

This equation gives us the simple but discouraging picture that once β_y^* is known (there are rather hard limits on how small β_y^* can be made with respect to beam bunch length), and maximum achievable ξ_y is found, the luminosity is proportional to total current only. There is possibly a factor of 2 improvement to be gained in ξ_y/β_y^* beyond that achieved in present storage rings. All other luminosity gains require proportionately more beam current.

As current is increased beyond the beam-beam tune shift limit described above, the vertical emittance of the beams increases. If the beam-beam effect limits the maximum current of the storage ring, this will happen either when the vertical aperture of the storage ring is reached (usually sooner than a Gaussian particle density distribution would suggest) or when the horizontal tune shift reaches a limit, causing an increase in horizontal beam size. The horizontal tune shift parameter may be approximated by:

$$\xi_x = \frac{N_b\, r_e}{2\pi\, \gamma\, \Gamma\, \varepsilon_x} \tag{4}$$

where Γ = (total σ_x)^2/(σ_x due to emittance only)^2 and is equal to unity unless dispersion is used to enlarge the horizontal beam size. ε_x is the horizontal emittance.

Accurately predicting beam-beam tune shift limits has historically been difficult. Perhaps the best guide is provided by parameterization of data from operating storage rings. Seeman[5] has found good correlation between the linear tune shift between collision points (see reference 5 for definition) and the reciprocal of the number of interactions per turn times the damping time. Data for many currently operating storage rings are shown in Figure 2.[6]

SINGLE BEAM CURRENT LIMITS

Since most of the increase in luminosity over existing machines requires proportionalely higher beam currents, single beam effects are a serious consideration in B factory design. Shortage of space prohibits doing justice to these phenomena. The references cited provide a starting point for further reading.

Both synchrotron radiation ($P_{sr} \propto f^* N_b E^4/r^2$) and higher mode losses ($P_{hm} \propto f^* N_b^2 G_{hm}$) require additional power to be provided by the RF system. Furthermore, this power must then be removed from the dissipating components. Standard electrostatic separator structures may dissipate 10 kilowatts or more at currents in planned B factories. Bellows and steps in vacuum chamber profile can result in damage due to local heating. RF cavities represent the largest fraction of the higher mode impedance. Compromises may be necessary in accelerating mode shunt impedance to reduce these parasitic losses. Figure 3 shows calculated scaling of the accelerating mode and the higher order longitudinal and transverse shunt impedance in the PEP cavities as the beam pipe iris size is varied[7]. Accelerating mode shunt impedance falls slower than transverse and longitudinal higher mode impedances as the beam hole radius is increased.

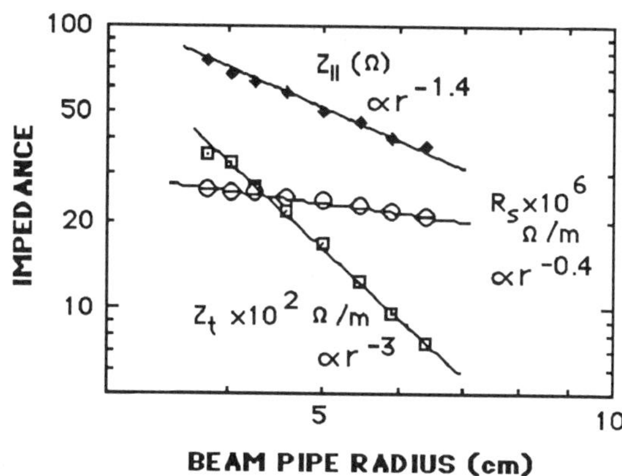

Figure 3 – Impedances in PEP cavity as function of beam hole radius - data from reference 7.

Synchrotron radiation power densities can approach 10 kW/meter in some of the B factory concepts listed above. Photons striking the chamber wall produce photoelectrons which, in turn, cause desorption of gas molecules from the surface, creating a serious vacuum problem[8]. Using localized absorbers with high capacity vacuum pumps as proposed in the ESRF light source may be more effective than conventional distributed ion pumps.

Ion trapping in the potential well of the electrons must be considered. The clearing electrodes used in present synchrotron radiation sources would be destroyed by higher mode heating. If no other mechanism to control ion trapping is found, low loss electrodes must be engineered.

Single beam instabilities become more severe in proportion to the transverse or longitudinal wall impedances[9,10], and in the transverse case, also scale with the local β function value. Accurate estimates of threshold currents, particularly when considering coupled bunch instabilities, must be done by a detailed analysis of wakefields or higher order resonant frequencies[11].

SEPARATION OF E^+ and E^- ORBITS / INJECTION

The maximum collision frequency, f^*, is determined in part by the minimum distance in which the electron and positron orbits can be sufficiently separated to avoid beam-beam interaction at the nearest crossing point to the interaction point. This distance is small if the bend angle of the separator is large, the separator is close to the i.p., and advantage is taken of the optics to separate the orbits faster. These separators must be run very conservatively, however, since they are an integral part of the guide field. Any sparks will likely result in lost beams. RF voltages induced by beams will add to the d.c. voltages applied to the separators.

An additional consideration when electrons and positrons have separate guide field elements is decreased tolerance to regulation errors in magnet power supplies. In multi-bunch CESR we have found that a vertical orbit difference of ± 0.3 mm between electrons and positrons can reduce performance by 20-30%. All elements become critical to reduce orbit shifts and tune differences between the two beams.

Integrated luminosity, not peak, is the final objective. An efficient injector is required to maintain average luminosity close to the peak. This is particularly difficult with large stored currents. A full energy injector capable of topping off the bunches evenly must be provided. Since beam lifetimes are likely to be on the order of 1-2 hours, a top-off will be needed every hour or less. It must be done in well under 10 minutes.

AN OPERATIONAL BENCHMARK

Where do operating colliders stand with respect to the parameters outlined in Table 1? I will use CESR (Cornell Electron Storage Ring) as an example since it most closely approaches those goals and I am familiar with its operation.

CESR has operated with 7 bunches per beam since February, 1987. A single vacuum chamber is shared by the two beams and unwanted collisions are avoided by 3λ horizontal closed orbit perturbations created by electrostatic separators. Table 2 below lists some operational parameters. Performance of CESR under these conditions is summarized in Table 3. Most of the data are for E_{beam}= 5.43 GeV (Y(5S) resonance).

Single beam experience in CESR suggests that factors of 2 improvement in control of impedances may allow currents similar to those planned in B factories. 170 mA has been stored in 14 bunches using positrons. No instabilities limited the current. Occasional coupled bunch quadrupole mode instabilities have been observed both in transverse and longitudinal directions. However, these phenomena are not reproduceable. An anomalous horizontal instability has been observed at low currents only. This is believed to be caused by a non-linear effect from the distributed vacion pumps.

Circumference	768 m	bunch length σ_l	0.017 m
Particles/bunch N_b	1.6×10^{11}	β_v^*	0.015 m
bunches/beam k	7	β_h^*	1.0 m
RF frequency	500 MHz	dispersion at ip η_h^*	0.55 m
RF Cavities	2 x 14 cells	ξ_v (max)	0.021*
Synch Rad Pwr P_{sr}	140 kW	ξ_h (max)	0.026
Higher Mode Pwr P_{hm}	50 kW	typical filling time	20 minutes
Copper Losses P_{diss}	360 kW	energy spread σ_E/E	0.6×10^{-3}

* corrected for "hourglass" effect of beams at interaction region

Table 2 – Operational Parameters of CESR

Peak Luminosity		1.02×10^{32} cm^{-2}-sec^{-1}
Best Integrated Luminosity:	/day	4.7 pb^{-1}
	/week	27 pb^{-1}
	/month	90 pb^{-1}

Table 3 – Best Performance of CESR

CONCLUSION

Storage ring based B factories producing 50-100 pb^{-1}/day are technically achievable. Development must be directed toward 1) improving higher mode properties of RF cavities and separator structures, 2) developing better absorbers for synchrotron radiation and desorbed gases, 3) understanding ion effects, and 4) improving the beam-beam tune shift limit.

REFERENCES

1. K. Berkelman,"CESR as a B Factory," les Rencontres de Physique de la Vallee d'Aoste, La Thuile, Italy, March, 1988
2. R. Eichler et. al., "Motivation and Design Study for a B-Meson factory with High Luminosity," SIN PR-86-13, November, 1986
3. S. Stone, "Search for CP Violation via $e^+ e^- \rightarrow B^o \bar{B}^o \gamma$," CLNS 87/110, Nov. 1987
4. K. Wille, "Feasibility Study of a B Meson Factory," PSI PR-88-01, 1988
5. J. Seeman, "Observations of the Beam-Beam Interaction," Sardinia, 1985, Lecture Notes in Physics, Vol. 247, Springer-Verlag
6. J. Seeman, private communication, May, 1988.
7. R. Siemann, "Cavity Geometry Effects,", unpublished, Oct., 1987
8. N. Mistry, "Ultrahigh Vacuum Systems for Storage Rings and Accelerators," SLAC Summer School on Particle Accelerators, 1985, p. 1443
9. F.Sacherer, "A Longitudinal Stability Criterion for Bunched Beams," IEEE Trans. Nucl. Sci., NS-20 (1973) p. 825
10. J.L.Laclare, 11th Conf. on High Energy Accelerators, CERN (1980) p. 526
11. R. Siemann, "Computer Models of Instabilities in Electron Storage RIngs," 3rd Summer School on High Energy Particle Accelerators, Stony Brook (1984)

CHARMONIUM SPECTROSCOPY WITH $p\bar{p}$ ANNIHILATIONS

J. Peoples
SSC Central Design Group*
c/o Lawrence Berkeley Laboratory
University of California
Berkeley, CA 94720

June 1988

Charmonium, the set of bound states of a charmed quark and its antiquark, have provided considerable insight into the strong forces between quarks and antiquarks.[1] The spacing of the energy levels and the number of bound states can be obtained from a QCD inspired phenomenological potential function. Unfortunately the determination of α_s (at this mass scale) from the measured transition rates of radiative decays of charmonium is roughly 0.3. As a result, the second order corrections to first order processes are large and the theoretical dependence of these rates on α_s is not reliably known. Nevertheless if these nonperturbative calculations can be improved one can expect to extract a more accurate value of α_s because all of the bound states of charmonium should be accessible to experiments.

In spite of the preceding caveats, the $Q\overline{Q}$ potential inspired by QCD given in (1) describes the level structure of charmonium, as well as bottomonium, very effectively.[1]

$$V(Q^2) = \frac{16\pi}{(33-2n_f)Q^2 \ln\left[1+Q^2/\Lambda^2\right]}, \tag{1}$$

n_f = number of open flavors below Q^2 and Λ = scale parameter.

The widths of the charmonium states below the threshold for charmed meson production are very narrow compared to states of comparable mass made of light quarks. For example the two lightest states, the η_c and the ψ, ${}^1S_0$ and ${}^3S_1$ respectively, have experimentally measured widths of 11 MeV and 60 keV.[2] Because the charge conjugation and parity of the η_c is even, it decays primarily into two gluons and its decay rate is proportional to α_s^2. Because the ψ, the ${}^3S_1$ state, has odd charge conjugation and parity, it decays into three gluons and thus its decay rate is proportional to α_s^3. In addition to an extra power of α_s, the three gluon decay amplitude contains the factor of (π^2-9) which reduces the three gluon decay widths still further. Perturbative QCD qualitatively explains these very different widths very successfully. The amplitude for a decay into a gluon and a light $q\bar{q}$ pair is also of order α_s^3. States which can decay into this channel are expected to have widths which are of the order of 1 MeV. The level diagram for charmonia is shown in Fig. 1.

---

*Operated by the Universities Research Association for the Department of Energy.

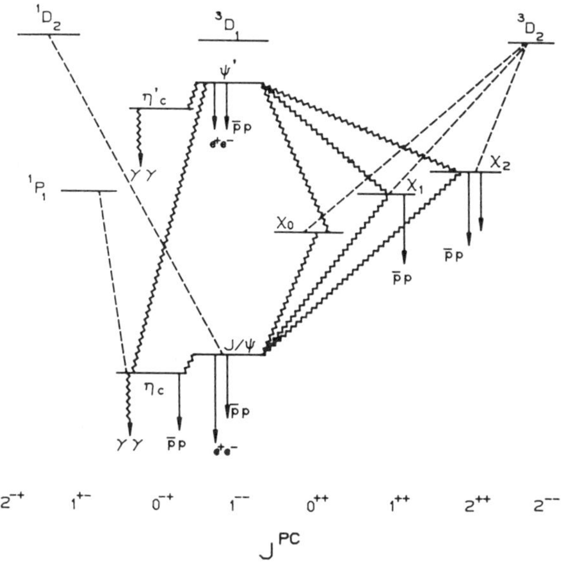

Fig. 1. Energy level diagram for the bound states of charmonium.

Until recently it was possible to explore only states which were either produced directly in e^+e^- annihilation or which were formed from the decay of a state formed in e^+e^- annihilation. The former are limited to the n^3S_1 states since they have the same spin, parity, and charge conjugation as the photon, 1^{--}. The ^3P_J, χ_J states, were first observed in the radiative decay of the ψ'. Since the ψ', the 2 $^3S_1$ state, cannot decay into the $^1P_1$ state by a radiative transition, this state has not been observed in the debris of e^+e^- collisions. The widths of the electromagnetic transitions are of interest because they are directly related to the $Q\overline{Q}$ potential and the wave functions of the initial and final states. The branching ratios for the electromagnetic transitions from the 2 $^3S_1$ to the ^3P_J states and from the ^3P_J states to the $^3S_1$ with the exception of the $^3P_0$ to $^3S_1$ transition are large and accurately known. Nevertheless the partial widths for the decay of ^3P_J to the $^3S_1$ are not known because the widths of the ^3P_J are very poorly known. We expect to remedy this situation with E760 by measuring them to a fraction of an MeV. A summary of the experimental values for masses and widths is given in Table I.

Since the D, the lightest pseudoscalar meson containing a charmed quark and a light quark, has a mass of 1865 MeV/c^2, any $C\overline{C}$ state heavier than 3730 MeV/c^2 with quantum numbers 1^{--}, will decay preferentially into a $D\overline{D}$ pair unless the decay violates parity or charge conjugation. Since the ψ (3770) has quantum numbers 1^{--} and a mass 40 MeV/c^2 greater than the threshold for decaying into a $D\overline{D}$ pair, it decays into $D\overline{D}$. Its width

Table I
Masses and Widths of the Experimentally Observed Bound States of Charmonium[2]

Spectroscopic Notation*	J^{PC}	Mass (MeV)	Total Width	Order of α_s in Hadronic Decay Amplitude
$^1S_0$ (η_c)	0^{-+}	2984±2	11±4 MeV	α_s^2
$^3S_1$ (J/ψ)	1^{--}	3096±.1	63±9 keV	$\alpha_s^3 \cdot (\pi^2 - 9)$
$^3P_0$ (χ_0)	0^{++}	3414.9±1.1	17±4 MeV	α_s^2
$^3P_1$ (χ_1)	1^{++}	3510.7±.5	< 1.3 MeV	α_s^3
$^3P_2$ (χ_2)	2^{++}	3556.3±.4	2.8±2 MeV	α_s^2
$2\,^3S_1$ (ψ')	1^{--}	3686.1±.1	215±40 keV	$\alpha_s^3 \cdot (\pi^2 - 9)$
$^3D_1$ (ψ'')	1^{--}	3769.9±2.4	25±2 MeV	–

*The symbols in parenthesis are the familiar particle data book names

is 25 MeV/c^2 in spite of the small phase space for the decay. Given its broad width the fraction of the $\psi(3770)$ mesons which decay into bound $C\overline{C}$ states is very small. The $^3D_3$, $^3D_2$, and $^1D_2$ cannot be formed in e^+e^- annihilations since they would have to come from the radiative decay of the higher lying $n\,^3S_1$, $n > 2$ states. Since these states like the $\psi(3770)$ decay almost exclusively into open charm, the probability of observing them in e^+e^- annihilations is vanishingly small. Because the $^3D_2$ and $^1D_2$ states have spin, parity, and charge conjugation of 2^{-+} and 2^{--} respectively they cannot decay strongly into charmed pseudoscalar mesons without violating parity. If, as is expected, the $^3D_2$ and $^1D_2$ have masses between 3800 MeV/c^2 and 3850 MeV/c^2, they should have a significant branching ratio for decaying into lighter $C\overline{C}$ states either by a radiative transition to a P-state or by multiple pion emission to the J/ψ or η_c.[1]

In waning days of the ISR another experimental technique for measuring the properties of charmonium was pioneered by R-704.[3] It relied on the fact that it is possible to form all of the states of charmonium in $\bar{p}p$ collisions directly. The experiment relied on the fact the inclusive decay of a resonance R into a J/ψ which subsequently decayed into a e^+e^- could be easily distinguished from the background. The basic reaction is given in (2)

$$\bar{p} + p \to R \to J/\psi + X\,. \tag{2}$$

R704 observed $^3P_1$ and $^3P_2$, and in so doing provided the first measurement of the $^3P_2$ width and a smaller lower bound for the $^3P_1$ width than had been previously observed.[3] E760, the experiment that will be described in this paper is the successor to that experiment. The objectives of E760 include the observation of the unobserved states: $^1P_1$, $2\,^1S_0$, $^3D_2$, and $^1D_2$ and the measurement of masses and widths of the known bound states to a precision of a few tenths of an MeV.

THE EXPERIMENTAL APPROACH USED IN E760

The charmonium state of interest is formed when a a beam of antiprotons ($\bar{p}$'s) collides with a beam of protons (p's) and the center of mass energy of the colliding $\bar{p}p$ system is equal to the energy of the state of interest. The state is observed through its decay either into an inclusive J/ψ with a subsequent decay into e^+e^- or an exclusive two γ decay. The former decay channel is effective when searching for the formation and subsequent decay of states such as the ^3P_J states, while the latter is useful for detecting the formation and subsequent decay of the even spin states into two γ's. Estimates of the signal to background levels indicate that these decay modes can be used to detect the states of interest. Other decay modes may be detectable depending on the backgrounds. A search will be made for the decay of $^1P_1$ into $\gamma + \eta_c$, followed by the η_c decaying into two γ-rays. Other decays to which E760 detector is sensitive are the decays $\eta\eta$ and $\eta\eta'$. There may be some surprises in these modes since they could couple to gluonium.

Since many of the details of using the Fermilab Accumulator to bring a circulating beam of $\bar{p}$'s into collision with a proton beam of a gas jet have been described previously,[4] only a brief summary will be presented here. It is worth noting that all cross sections measured in E760 contains the product of the partial width for decay into $\bar{p}p$ times the branching ratio into the final state of interest. In order to determine a particular matrix element from a cross section either the $\bar{p}p$ branching ratio or the final state branching ratio must be known. E760 cannot disentangle the product of the two branching ratios as can be done in e^+e^-, because the background in the elastic $\bar{p}p$ channel is so large. A circulating beam of 15 mA of antiprotons in the Accumulator Ring of the Fermilab Antiproton Source passing through a jet of hydrogen with a density of 10^{14} atoms/cm^2 will have a luminosity of 10^{33} cm^{-2} sec^{-1}.. The beam and the H_2 jet are both expected to have an RMS radius of 1.5 mm where the two beams intersect.

If nothing were done to counteract the multiple scattering of the $\bar{p}$ beam as it passes through the gas jet, the beam emittance would grow rapidly, ultimately causing the beam to be lost. The multiple scattering will be counteracted by the transverse cooling systems. The beam lifetime of 50 hours will be dominated by $p\bar{p}$ collisions in the gas jet.

The momentum spread of the $\bar{p}$ beam will be kept to 1 MeV or less by stochastic cooling. It tends to grow because of intrabeam scattering, and fluctuations in the energy loss in $\bar{p}\,e^-$ collisions. Under these conditions the mass resolution will be about ± 300 keV/c^2. This should be sufficient to measure the widths of the ^3P_J, $^1P_1$, $^1S_0$ and 2 $^1S_0$ states.

The final states of interest are detected by the apparatus shown in Fig. 2. Since the detector has been described previously only a very brief description will be given here.[5] It consists of two sets of trigger counter hodoscopes, H_1 and H_2, two sets of track chambers, and two calorimeters. Between the inner and outer track chamber there is a sixteen cell Cerenkov counter. Electrons

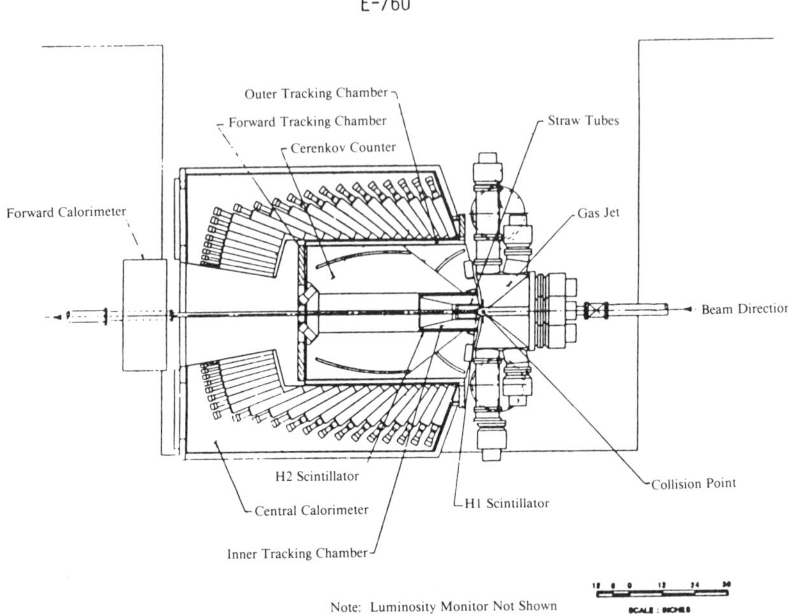

Fig. 2. E760 Detector.

or positrons with azimuthal angles between 16° and 75° will be detected, whereas any pion within these angles between will be below the threshold for counting throughout the $\bar{p}$ energy range of interest. Since the Cerenkov counter cannot distinguish low mass e^+e^- pairs from either a single electron or positron, a radial projection chamber, one of the inner tracking chambers, will be used to make an ionization measurement of each track with sufficient accuracy to distinguish these cases.

The energy of γ-rays, e^+, and e^- will be measured by two electromagnetic calorimeters; the barrel calorimeter and the forward calorimeter. The barrel calorimeter consists of lead glass blocks which are arranged in 64 wedges and each wedge contains 20 blocks subtending azimuthal angles for 15° to 72°. The forward direction from 15° to 2° is covered by a 144 element calorimeter, arranged in a 12×12 array. Each element consists of a sandwich of scintillator and lead. The luminosity is determined from the measurement of $\bar{p}p$ elastic scattering. A silicon wafer detector will be used to make this measurement.

STATUS

The Accumulator was modified to allow the beam to be decelerated from 8.89 GeV/c to 3.5 GeV/c. In addition a special set of core cooling pickups has been installed and commissioned to permit cooling of the beam on the central orbit. These modifications have been thoroughly tested with protons. It is worth noting that the $^3P_0$, $\psi(3094)$, and $\eta_c(2984)$ lie below transition in the Accumulator, or at least the normal transition energy of the Accumulator. In order to reach these states transition must be crossed. Because the guide field of the Accumulator cannot change rapidly it is not possible to cross transition by decreasing the beam energy with the rf system. It is crossed by changing the transition energy, a property of the accumulator lattice, while keeping the beam energy constant. Low intensity beams have crossed transition in this manner with 100% efficiency. To date the total efficiency of the deceleration cycle from 8834 to 4934 including the γ_T jump has reached 83%. Further improvements can be expected.

The gas jet has been operated at a density of 8×10^{13} atoms/cm^2 with a proton beam current of 17 mA. Under these conditions a luminosity 10^{31} cm^{-2} sec^{-1} was achieved. The beam lifetime at 8834 GeV/c was 100 hours as predicted.

Except for the barrel calorimeter the entire detector has been built and installed. Commissioning of these detectors is in progress. The entire detector should be completed during the 1989 fixed target run. At that time it should be possible to fulfill the promise of this technique.

ACKNOWLEDGEMENT

I thank Valerie Kelly for helping me fit these remarks into the procrustean six pages.

REFERENCES

1. W. Kuong, J. Rosen, and C. Quigg, Ann. Rev. Nucl. Part. Sci. 37: 325–82 (1987).
2. M. Aguilar-Benitez, (Particle Data Group). Phys. Lett. 170B: 1–350 (1986).
3. C. Baglin, et al., Phys. Lett. 171B:135–41 (1986); 172B: 455–60 (1986); 187B: 191–97 (1987).
4. J. Peoples, in Low Energy Antimatter, World Scientific Publishing Co., (1986): 144.
5. J. Peoples, in Physics at LEAR with Low-Energy Antiproton Detectors, Harwood Academic Publisher (1987).

STATUS OF THE SEARCH FOR $\xi(2230)$

Presented by Brian Quinn
PS185 collaboration: P.D.Barnes[A], P.Birien[D], B.E.Bonner[G], W.H.Breunlich[I], G.Diebold[A], W.Dutty[D], R.A.Eisenstein[E], G.Ericsson[H], W.Eyrich[C], R.v.Frankenberg[C], G.Franklin[A], J.Franz[D], N.Hamann[D], G.Harig[D], D.Hertzog[A], A.Hofmann[C], T.Johansson[H], K.Kilian[B,F], C.Maher[A], M.Meyerhofer[C], N.Naegele[I], W.Oelert[F], S.Ohlsson[H], H.W.Ortner[C], P.Pawlek[I], B.Quinn[A], E.Rossle[D], H.Schledermann[D], H.Schmitt[D], G.Sehl[F], J.Seydoux[A], F.Stinzing[C], J.Szymanski[A], X.Yi[A], T.Zeludziewicz[F]
A)Carnegie-Mellon Univ. B)CERN C)Univ. Erlangen D)Univ. Freiburg E)Univ. of Illinois F)Juelich KFA G)Los Alamos Nat. Lab. H)Univ. of Uppsala I)IMPE Osterr. Akad. Wissensch

Abstract
The present experimental status of the $\xi(2230)$ is summarized and several possible interpretations are discussed.

The first evidence of the $\xi(2230)$ was observed in 1983 by the Mark III group[1] as a resonance in the $K\bar{K}$ mass distribution in radiative decays of the J/Ψ (Fig. 1) in both the $J/\Psi \rightarrow \gamma K^+K^-$ and $J/\Psi \rightarrow \gamma K_sK_s$ channels. With the additional statistics of a later data sample the parameters describing the $\xi(2230)$ are:

K^+K^- : $M = 2230 \pm 6 \pm 14$ MeV
$\Gamma = 26 ^{+20}_{-16} \pm 17$ MeV
$B(J/\Psi \rightarrow \gamma\, \xi) B(\xi \rightarrow K^+K^-) = (4.2 ^{+1.7}_{-1.4} \pm .8) \times 10^{-5}$

K_sK_s : $M = 2232 \pm 7 \pm 7$ MeV
$\Gamma = 18 ^{+23}_{-15} \pm 10$ MeV
$B(J/\Psi \rightarrow \gamma\, \xi) B(\xi \rightarrow K_sK_s) = (3.1 ^{+1.6}_{-1.3} \pm .7) \times 10^{-5}$

Where the first error is statistical and the second is systematic. The assigned quantum numbers are I=1, J^{PC}= even^{++}, with spin 2^{++} or 4^{++} preferred over 0^{++}. The Mark III group saw no evidence of structure in other channels.

The appearance of such a narrow resonance with a high mass suggested the possibility of an unusual structure for the ξ. A number of possible explanations for a narrow width were advanced. Before considering these it will be useful to examine some experimental limits which can be set on individual branching ratios.

Assuming the ratio K^+K^-: K_sK_s : K_LK_L is given by the expected isoscalar ratio of 2:1:1, a fit to the observed branching ratios gives:

$$B(J/\Psi \rightarrow \gamma\, \xi) B(\xi \rightarrow K^+K^-, K_sK_s, K_LK_L) \approx (9.6 \pm 3) \times 10^{-5}$$

Since the total branching ratio of ξ to kaons can not exceed unity, this sets a lower limit on $B(J/\Psi \rightarrow \gamma\xi)$ of 10^{-4}. From the non-observation of the ξ in inclusive J/Ψ photon spectra (Fig. 2) by the Crystal Ball collaboration[2], an approximate upper limit on B(J/Ψ→γξ) can be set, by inspection of the raw inclusive cross section, at about one tenth of B(J/Ψ→γη'), or B(J/Ψ→γξ)<4×10^{-4}. This sets a lower limit of 0.25 on the total branching ratio of ξ to kaons. So we have:

$$1\times10^{-4} < B(J/\Psi\to\gamma\xi) < 4\times10^{-4}$$

$$B(\xi\to K^+K^-, K_sK_s, K_LK_L) > 0.25$$

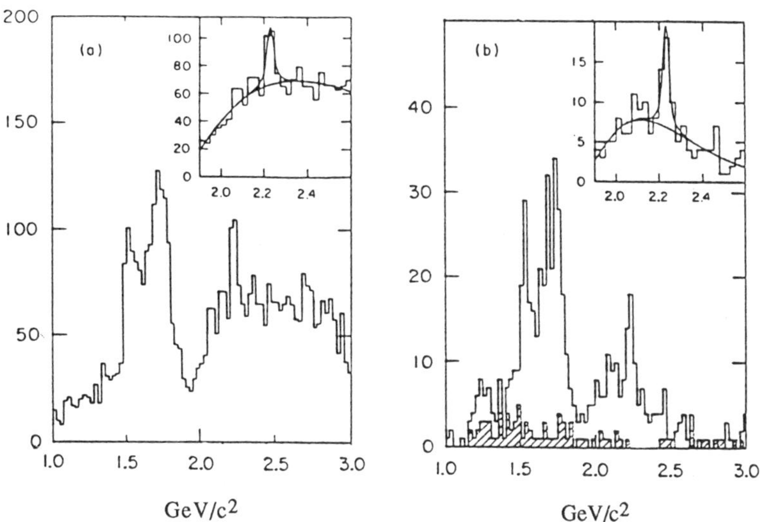

Fig. 1. Mark III spectra vs. invariant mass of the kaons for a)K^+K^- b)K_sK_s

Because the ξ was discovered in the radiative decay of the J/Ψ, which proceeds through two gluons, the possibility that it was a bound state of two gluons, a glueball, was suggested[3]. The search for such particles had been one of the prime motivations for the study of these decays. It has been argued[4] that a glueball can be expected to have a width of as little as 10 to 30 MeV because its decay is O.Z.I. suppressed. This interpretation of the O.Z.I. rule is controversial[5], however. Also, the mass of the ξ is somewhat greater than the 1 to 1.6 GeV predicted for glueballs by M.I.T. bag and lattice gauge calculations but is probably not excluded since there is no phenomenological input to constrain calculations for such states. One property of the ξ which may appear to be at odds with the glueball interpretation is the fact that the Mark III group did not observe decay to two pions, setting an upper limit of about one fifth of the observed rate to two kaons. Since the glue should be insensitive to flavor, one might expect these two branching ratios to be roughly equal. Ward[3], however has done a calculation within the framework of the M.I.T. bag model which shows that remarkable sensitivity to the difference in mass of the quark flavors is possible. Due to kinematic effects he predicts a factor of 500 suppression of the two-pion channel relative to the two-kaon channel for decay of a 0^+ glueball. The suppression is negligible for a 2^+ glueball, however.

A more conventional nature of the ξ, a high spin $s\bar{s}$ meson, has been advocated by Godfrey et al.[6] who find that two such states predicted by their relativized quark model are in reasonable agreement with the properties of the ξ. A $^3F_{2^+}$ state and a $^3F_{4^+}$ state of the $s\bar{s}$ system, with predicted masses of 2270 and 2210 MeV respectively, each have a predicted width of about 50 MeV. The narrowness of the state might be explained by a high orbital angular momentum which inhibits decay. Because of the strange quark content, each is expected to decay preferentially into kaons, in agreement with the observation that $B(\xi \to K\bar{K}) > .25$ found above. The suppression of pion channels also follows naturally. A test of the model is possible since they predict the other F state and an ω-like partner of the ξ with a mass 200 MeV lower. The authors do not quote a value for the expected rate of decay of the J/Ψ into the ξ. As shown above, existing measurements determine this partial width to be 2×10^{-4} of the 64 keV/c^2 width of the J/Ψ, within a factor of two. This may be a significant test of the model as one might expect that the same angular momentum arguments which suppress the decay of the ξ would also suppress its production. This suppression of the width is controversial, Ono[7] predicts partial widths of 370 and 290 MeV for the $^3F_2$ and $^3F_4$ $s\bar{s}$ mesons from his phenomenological model.

A variety of other possibilities have also been suggested for the nature of the ξ. A light Higgs scalar was considered[8], assuming the existence of more than one Higgs doublet. This now appears to be ruled out by the evidence that the ξ has spin 2 or 4. Such exotic structures as $s\bar{s}s\bar{s}$ or $q\bar{q}g$ have also been suggested[9,10]. These models are difficult to test as there is little background for predicting the properties of such systems. A bound state of a Λ and a $\bar{\Lambda}$ has also been suggested[7], although it is difficult to see how such a system could be narrow, even in a high spin state, because of the long range of the annihilation potential expected.

Experimental interest was also sparked by the Mark III results. While Mark III was taking the second set of data which seems to confirm their preliminary findings, the DM2 group at Orsay[11] took similar data which seems to contradict them. The DM2 results, on a somewhat larger sample of J/Ψ's, but with poorer resolution, show no evidence of the ξ in either of the two-kaon channels. Their upper limit on branching ratio grows with increasing width of the J/Ψ, however, and the two results might be in reasonable accord if the ξ is roughly 50 MeV wide. It is interesting to note that DM2 observes structure at 2220 MeV in the $\phi\phi$ mass spectrum[12] in J/Ψ

radiative decays. They find that this is fit better as a 0^- state than as a 2^+, however. Mark III also sees evidence[13] of a 0^- resonance in $\phi\phi$.

At least two other experiments may have seen evidence of the ξ. The LASS spectrometer group[14] observed a resonance at about 2230 MeV in the K_sK_s invariant mass spectrum for the reaction $K^-p\rightarrow\Lambda K_sK_s$. Although the statistics are quite limited, the observed mass spectrum is similar to that obtained by Mark III in the same channel (Fig. 3). No peak is seen in the K^+K^- channel, but phase shift analysis[15] suggests the existence of a 4^{++} resonance in this mass range. Observation of the ξ in such a production channel with a strange quark in the initial state would strongly favor the interpretation of the ξ as an $s\bar{s}$ meson.

On the other hand, the GAMS collaboration[16] also sees evidence of a narrow resonance at about 2220 MeV in the $\eta\eta'$ invariant mass spectrum in the reaction $\pi^-p\rightarrow n\eta\eta'$. Some evidence of structure is seen in data at 100 GeV beam energy, and a very clear signal, though with a cross section of only 50±17 nb, is seen for a 38 GeV beam (Fig. 4). While the $\eta\eta'$ final state could be reached through simple quark-rearrangement diagrams, production through an $s\bar{s}$ meson resonance would be O.Z.I. suppressed. A reaction proceeding through a glueball intermediate state would also be O.Z.I. suppressed, but it seems possible that the suppression factor might be smaller for a glueball intermediate state than for an $s\bar{s}$ meson.

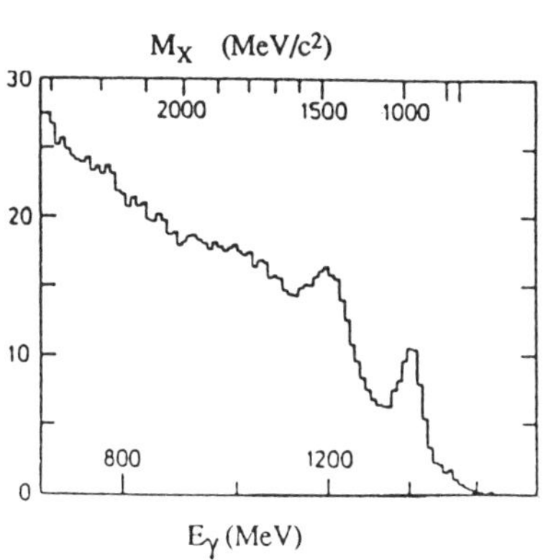

Fig. 2. Inclusive γ spectrum[2] from J/Ψ decay.

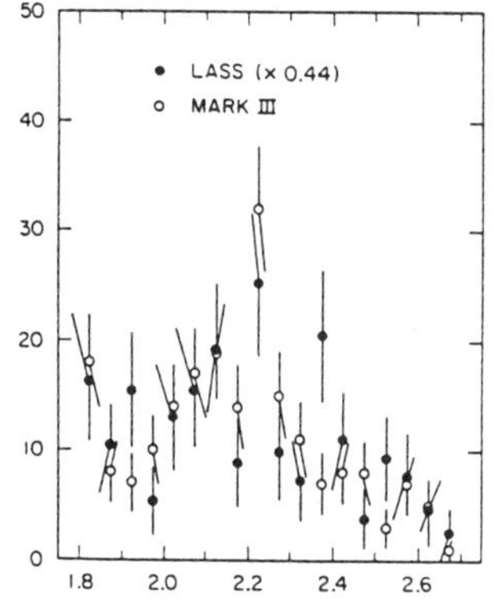

Fig. 3. Comparison of K_sK_s invariant mass spectra from LASS[14] and Mark III[1].

Several experiments have been performed to search for the ξ as an intermediate state in the reaction $\bar{p}p\rightarrow K\bar{K}$. The cross section for such a reaction can be written as:

$$\sigma_{peak}(\bar{p}p\rightarrow\xi\rightarrow K\bar{K})=\frac{4\pi\hbar c}{s-4m_p}(2J+1)\ B(\xi\rightarrow\bar{p}p)\times B(\xi\rightarrow K\bar{K})$$

where $B(\xi \rightarrow K\bar{K})$ is the branching ratio to the final state of interest. $B(\xi \rightarrow K^+K^-)$ will be twice $B(\xi \rightarrow K_sK_s)$ assuming the ratio is determined by the isospin decomposition of an isoscalar state. Because of the factor (2J+1) in the above formula, all branching ratio products will be quoted for an assumed J=0 resonance. Scaling by (2J+1) will give the approximate product for $J \neq 0$, neglecting variation in acceptance and efficiency with angle.

Two experiments have reported the non-observation of the ξ in $\bar{p}p \rightarrow K^+K^-$. The PS170 collaboration[17] at LEAR reports upper limits of 2.7×10^{-3}, 1.8×10^{-3} and 0.48×10^{-3} on the branching ratio product for assumed widths of 1, 3 and 30 MeV respectively. A measurement performed at BNL[18] quotes limits of 1.2×10^{-3} and 0.56×10^{-3} for assumed widths of 7 and 35 MeV respectively. In all cases the quoted values are 3σ limits on the product of branching ratios, $B(\xi \rightarrow \bar{p}p) \times B(\xi \rightarrow K^+K^-)$.

The PS185 collaboration at LEAR has taken data to search for the ξ in the reaction $\bar{p}p \rightarrow K_sK_s$. Two disadvantages of this channel, compared to K^+K^- are the factor of two lower signal expected and the much greater technical difficulties in detecting and reconstructing the neutral particles. These are offset by the much smaller background expected. The non-resonant cross section for $\bar{p}p \rightarrow K^+K^-$ is $50\mu b$[19] while that for the K_sK_s reaction was known to be only a few μb. Preliminary analysis of the PS185 data indicates that this background is only about $2\mu b$. Since the PS185 collaboration already had apparatus designed for reconstructing neutral final states of $\bar{p}p \rightarrow \bar{Y}Y$ by kinematic reconstruction of the V^0's, only relatively minor modifications were required to permit the measurement of the K_sK_s final state. This consisted mainly of adding a set of three concentric layers of streamer tubes surrounding the target to extend the acceptance to large laboratory angles, inaccessible to $\bar{Y}Y$ systems (Fig. 5).

A Monte Carlo study of the expected sensitivity, based upon the actual distribution of beam energies and running time, indicates that the ξ should show up as at least a 3σ effect if $B(\xi \rightarrow \bar{p}p) \times B(\xi \rightarrow K_sK_s) > 9 \times 10^{-5}$ for a width in the range of 6 to 30 MeV. The sensitivity falls off gradually for widths outside this range. Even with $B(\xi \rightarrow K_sK_s) = .5 \times B(\xi \rightarrow K^+K^-)$, this represents a factor of three to five improvement over the existing limits on the coupling of ξ to $\bar{p}p$.

For comparison, one may crudely estimate the expected branching ratio of the ξ to $\bar{p}p$ if its structure is not based upon strange quarks. For the known decays of $c\bar{c}$ mesons into $\bar{p}p$, discounting the $c\bar{c} \rightarrow c\bar{c}$ decays, the fraction of O.Z.I. suppressed transitions which leads to $\bar{p}p$ final states is in the range 0.1% to 1.0% for η_c, J/Ψ, χ_2, and $\Psi(3685)$. Combining a resulting estimate of $B(\xi \rightarrow \bar{p}p) > 0.1\%$ with the lower limit on $B(\xi \rightarrow K_sK_s)$ found above by consideration of the Crystal Ball data, gives

$B(\xi \to \bar{p}p) \times B(\xi \to K_s K_s) > 5 \times 10^{-5}$. So, PS185's sensitivity covers most of the expected range of branching ratios, if the ξ does not have a strange-quark structure.

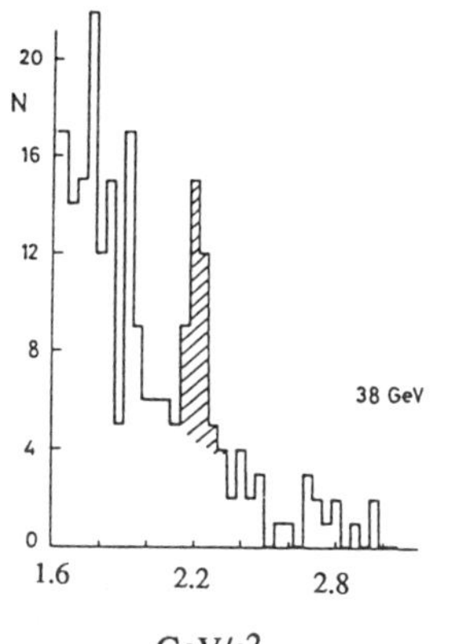

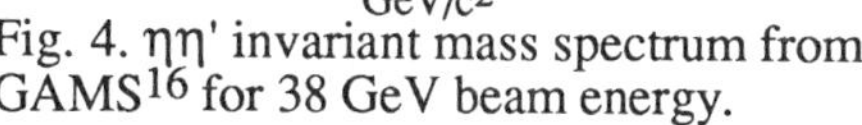

Fig. 4. $\eta\eta'$ invariant mass spectrum from GAMS[16] for 38 GeV beam energy.

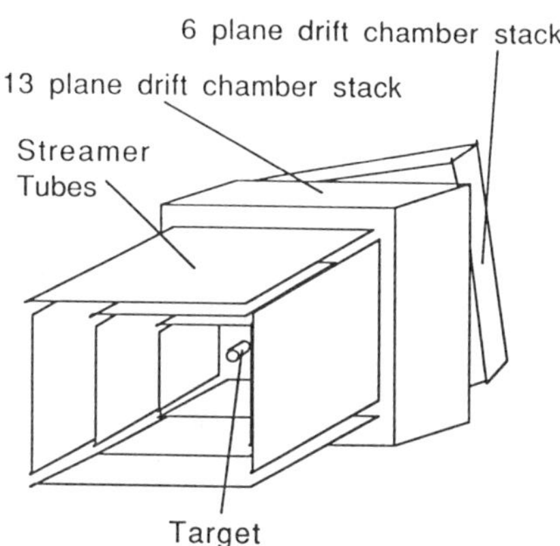

Fig. 5. Layout of main PS185 detectors. Streamer tubes were added for $K_s K_s$.

The classification of $\xi(2230)$ remains unclear. If it is observed in the data of PS185 the $s\bar{s}$ meson interpretation will largely be ruled out. Furthermore, if it is seen, much more precise information on its mass and width will be available because of the high resolution made possible by the LEAR beam. Analysis of that data is on-going, and results are expected later in the year.

REFERENCES

1)R.M. Baltrusaitis et al., Phys. Rev. Lett. **56**, 107 (1986)
2)E.D. Bloom et al., Journal de Physique, **C3**, 407(1983)
3)B.F.L. Ward, Phys. Rev. **D31**, 2849 (1985)
4)C.E. Carlson et al., Phys. Lett.**B99**, 353(1981)
5)A.W. Hendry and D.B. Lichtenberg, Fortschr. Phys. **33**, 5(1985)
6)S. Godfrey, R. Kokoski, and N. Isgur, Phys. Lett. **141B**, 439 (1984)
7)S. Ono, Phys. Rev. **D35**, 944 (1987)
8)R.S. Willey, Phys. Rev. Lett. **52**, 585 (1984)
9)S. Pakvasa, M. Suzuki, and S.F. Tuan, Phys. Lett. **B145**, 135(1984)
10)M.S. Chanowitz and S.R. Sharpe, Phys. Lett. **B132**, 413(1983)
11)J.E. Augustin et al., Phys. Rev. Lett. **60**, 2238(1988)
12)D. Bisello et al., Phys. Lett. **B179**, 294(1986)
13)W. Wisniewski, this conference
14)D. Aston et al., SLAC preprint DPNU86-46 (1986)
15)W. Dunwoodie, this conference
16)D. Alde et al., Yad. Fiz. **45**, 405 (1987)
17)G. Bardin et al., Phys. Lett. **B195**, 292 (1987)
18)J. Sculli et al., XXIII Int. Conf. on High Energy Physics, Berkeley; S.C. Loken, Editor (1986)
19)E. Eisenhandler et al., Nucl. Phys. **B96**, 109 (1975)

SEARCH FOR EXOTIC STATES WITH JETSET AT LEAR

David W. Hertzog
University of Illinois at Urbana-Champaign

representing the JETSET Collaboration:

T. Anderl[6], D. Bassi[3], P. Birien[2], R.K. Bock[1], A. Buzzo[3], E. Chesi[1], M. Dameri[3], P.T. Debevec[4], R. Dobinson[1], R.A. Eisenstein[4], T. Fearnley[7], R. von Frankenberg[6], J. Franz[2], N. Hamann[2], R. Harfield[1], D. Hertzog[4], S. Hughes[4], D. Jeffery[1], A. Johansson[8], T. Johansson[8], K. Kilian[6], K. Kirsebom[3], A. Klett[2], F. Levrero[3], A. Lundby[1], M. Macri[3], M. Marinelli[3], P. Martinengo[3], L. Mattera[3], B. Mouëllic[1], G. Muratori[1], W. Oelert[6], S. Ohlsson[8], Y. Onel[5], B. Osculati[3], M.G. Pia[3], F. Piuz[1], M. Price[1], P. Reimer[4], E. Rössle[2], A. Santroni[3], H. Schmitt[2], B. Stugu[7], R. Tayloe[4], S. Terreni[3]

[1] CERN, CH-1211 Geneva 23
[2] University of Freiburg, D-7800 Freiburg
[3] Genoa University and INFN, I-16132 Genoa
[4] University of Illinois at Urbana-Champaign, Urbana, IL 61801
[5] University of Iowa, Iowa City, IA 52242
[6] Inst. Kernphysik KFA Jülich , D-5170 Jülich 1
[7] University of Oslo, Blindern N-1000 Oslo 3
[8] University of Uppsala, S-75121 Uppsala 1

ABSTRACT

Several experiments at LEAR concern themselves with meson spectroscopy and searches for exotics in the annihilation of antiprotons with protons at rest. In JETSET we will exclusively study in-flight annihilations in a search for gluonic states and exotics in the mass region from 1.96 to 2.43 GeV. The experiment uses a molecular hydrogen gas jet target and an advanced general purpose detector, which surrounds the interaction point. One of the unique features of the detector is its excellent kaon identification at the trigger level. This will enable us to achieve our first physics goal of searching for resonances in the excitation function of the OZI-forbidden, "gluon rich" reaction $\bar{p}p \rightarrow \phi\phi$. The physics interest in the $\phi\phi$ channel and additional goals of the collaboration will be outlined.

INTRODUCTION

At this conference and in the last few years we have seen a wealth of new data pertaining to searches for possible gluonic resonances (gg and $g\bar{q}q$) and to "ordinary" quarkonium ($\bar{q}q$) states. Many of these data suggest interesting new physics at masses above 2 GeV. We have begun a dedicated experimental program at LEAR (PS202) measuring $\bar{p}p \rightarrow M_1 M_2$ and we will begin taking data in 1990. The initial (Phase I) focus of this program is a detailed study of the in-flight production of $K^+K^-K^+K^-$, including ϕK^+K^- and most importantly $\phi\phi$. In Phase II, the detector will include additional sophisticated components which will open up a wide variety of physics channels, some of which are discussed below.

JETSET AT A GLANCE

The JETSET experiment at LEAR will be unique among the new generation post-ACOL efforts in that it will be built around one of the straight sections of the antiproton facility itself. A molecular hydrogen gas jet target will cross the ring, providing a "massless" pure proton target for in-flight interactions. With a gas jet density of 8×10^{13} atoms/cm^2, a LEAR $\bar{p}$ revolution frequency of $3.2\times10^6 s^{-1}$ and an assumed injection of 4×10^{10} $\bar{p}$ into the storage ring, a peak luminosity of $10^{31}\text{cm}^{-2}\text{s}^{-1}$ will be achieved. This luminosity exceeds the most optimistic external-beam experiments by about a factor of 20 or more and opens up the possibility of doing high statistics measurements in rare channels.

Due to continuous stochastic cooling, the momentum uncertainty of the LEAR beam is quite small, $\Delta p/p \approx 10^{-3}$. With this small momentum spread, even very narrow width states can be probed effectively by varying the beam momentum, the practical range is from 600 to 2000 MeV/c, while measuring the production cross section. The experimental mass resolution with this method is about 0.5 MeV. With high statistics and a full solid angle detector, a partial-wave analysis becomes possible, allowing a determination of the quantum numbers of any new state.

From an experimental point of view, the environment is particularly hostile. At LEAR energies, $\sigma_{total}(\bar{p}p)$ is approximately 100 mb, implying an interaction rate of about 10^6 Hz. In order to realize our first goal of collecting $\bar{p}p \rightarrow K^+K^-K^+K^-$ events ($\sigma \approx 4\mu b$) and $\bar{\text{p}}\text{p} \rightarrow \phi\phi$ events (which have never been observed!), a very compact, fast, and modern detector has been designed. With such a device and an assumed cross section for $\bar{\text{p}}\text{p} \rightarrow \phi\phi$ of about $0.4\mu b$, we would collect a few times 10^4 events per day, far exceeding the world total for any final-state "$\phi\phi$" events.

The detector will surround the beam pipe and the gas jet, and will be divided into a "barrel" and a "forward" region of essentially identical components. The Phase I detector, which is shown in the bottom half of Figure 1, will be ready by early 1990. It will have no magnet and will include

- an innovative set of interwoven, multilayered plastic scintillation counters used as a high granularity charged multiplicity trigger. Unique (θ, ϕ) coordinates are determined in less than 100 ns and will be used with a fast kinematic filter in order to select only those events with a correct $4K^{\pm}$ topology.

- a fast, segmented threshold Čerenkov counter in the barrel region in order to reject the dominate charged pion background. Additionally, there will be a ring-imaging Čerenkov counter (RICH) in the forward region. The RICH will provide needed redundancy on the momentum assignment of measured tracks.

- a precision wire tracker made from self-supporting "straw tube" technology. There will be about 1600 such cylindrical drift tubes in the barrel region. They will be made from rolled aluminized mylar and are packed and glued tightly together around the beam pipe. The total structure will have excellent mechanical rigidity and will not require massive endplates characteristic of ordinary cylindrical wire chamber designs. Each straw tube will house a single resistive wire and will be equipped with appropriate electronics to determine the transverse coordinate $\sigma_t \leq 100\mu$m (by drift time) and the longitudinal coordinate $\sigma_l \leq 0.5$ cm (by charge division).

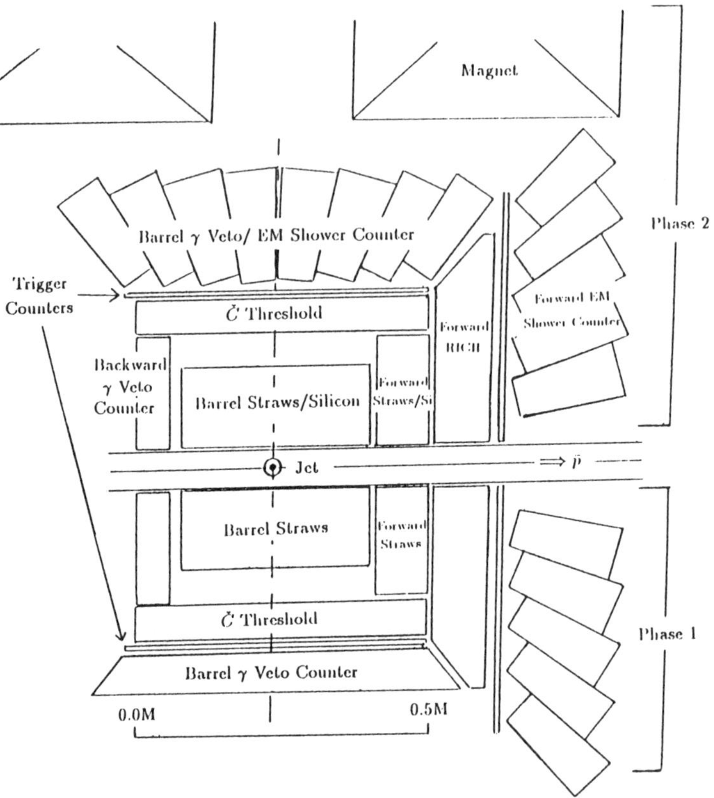

Figure 1: JETSET detector in Phase I (bottom) and Phase II (top).

- a fast γ veto detector with a low threshold surrounding the barrel in order to identify events with π^0s. In the forward region, there will be a fast, dense electromagnetic calorimeter having excellent spatial resolution and good energy resolution ($\sigma_{xy} \approx 5\text{mm}, \sigma(E)/E \approx 9\%/\sqrt{E}; E$ in GeV). This calorimeter will be based on the new technology of plastic scintillating fibers embedded within specially formed Pb foils in a close-packed geometry. The detectors will be oriented such that the fibers point toward the interaction region, allowing photomultiplier tube readout on the outside of the detector, while also eliminating dead spaces. The phase I detector will have a complete γ veto counter in the forward region, which will also aid in particle identification by a crude energy measurement for charged particles.
- a fast, modular data acquisition system based on multiple, VME-based, CERN Valet-Plus [1] systems interconnected with a transputer network. Event data will flow from the respective detector components to individual data acquisition systems and will be gathered by a final master Valet-Plus system where they will be logged to tape.

We are already developing several important components in order to enhance the overall performance of the basic detector and to lead to a broader physics program. The complete general purpose device for Phase II is shown schematically in the top half of Figure 1 and includes additionally

- a warm, solenoidal magnet which will surround the entire detector. The LEAR team is modeling the effect of such an element on the circulating beam (which couples the horizontal and vertical oscillations). With the use of a pair of opposite-polarity compensating solenoids, Baird [2] has demonstrated that a magnet with an $\int B \cdot dl \leq 1.5$ T·m appears to be acceptable.

- a high-precision Si μstrip tracker capable of a transverse resolution of better than 50μm. This will be used in conjunction with the magnet and the straws in order to make an accurate measurement of $p_\perp$ for charged tracks.

- the electromagnetic calorimeter extended to the barrel region. This will allow a much wider physics program by including identifiable neutrals in the final state. With the resolution as projected, neutrals-only final states such as $\pi^0\pi^0$, $\pi^0\eta$, and $\eta\eta$ will become realistic objectives.

LIFE ABOVE $\sqrt{s} = 2m_p$

The $\bar{p}p \rightarrow M_1M_2$ "in-flight" reaction at LEAR can explore the mass range from 1.96 to 2.43 GeV in direct formation experiments. While this range represents a rather narrow window of opportunity (for example, compared to the natural width of strongly decaying high-mass states), it is actually relatively unexplored and potentially rich in new physics. The reasons for the experimental neglect may be loosely tied to the fact that much of the older data coming from $\bar{p}p$ and the newer LEAR data are based on annihilations at rest. Spectroscopy coming from radiative J/Ψ decays, while capable of seeing structure all the way up to the η_c, tends to be rather low in statistics in the high-mass region. Also, many of the past experiments have lacked the ability to look into channels involving neutrals.

A few tantalizing and promising signs of life above $2m_p$ have recently appeared from both the experimental and theoretical side. A partial list of unexplained, observed resonances in this region at this writing includes

- three broad, $J^{PC} = 2^{++}$ states labeled $g_T(2011), g_{T'}(2297)$, and $g_{T''}(2339)$ and seen as $\phi\phi$ resonances at Brookhaven [3] in the OZI-forbidden reaction $\pi^-p \rightarrow \phi\phi n$. These states were found through a sophisticated partial-wave analysis and are interpreted by the authors as potential glueball states due, in part, to their copious production compared to other OZI-allowed background reactions.

- structure in the $\phi\phi$ invariant mass spectrum from an inclusive experiment at the CERN Ω Spectrometer which studied the reaction $\pi^-Be \rightarrow \phi\phi X$ at 85 GeV (the WA67 [4] collaboration). The structure is compatible with the Brookhaven measurement at least for the higher-mass states.

- a single 0^- state at about 2.2 GeV which was seen by the MARK III collaboration [5] and (possibly?) the DM2 collaboration [6] in radiative J/Ψ decays to $\phi\phi$ final states. These data are interesting both for the structure which is seen and, moreover, for the non-observation of the prominent, "glueball candidate" g_T's.

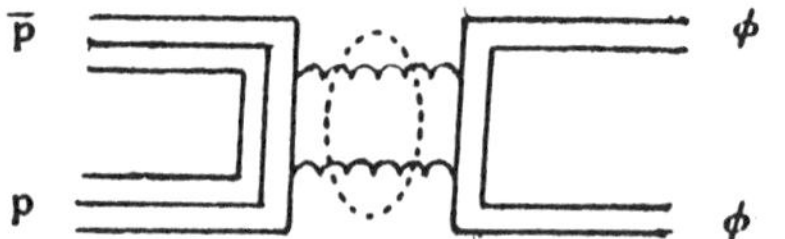

Figure 2: Possible diagram for $\bar{p}p \to \phi\phi$.

- the narrow $\xi(2230)$ state first seen by MARK III [8] in $J/\Psi \to \gamma \bar{K}K$, including $K_s K_s$ and K^+K^-, which remains a mystery [7]. However, there now appears to be at least some independent confirmation from the GAMS [9] collaboration in $\pi^- p \to \eta\eta' n$ and from the LASS [10] collaboration in the reaction $K^- p \to \bar{K}K\Lambda$. The $\eta\eta'$ invariant mass plot from GAMS shows a clear narrow peak, and in the partial-wave analysis of the LASS K^+K^- data, a narrow structure stands out with a J^{PC} of 4^{++}. Are these all the $\xi(2230)$? If these are, then the $^3F_4$ $s\bar{s}$ meson interpretation of the ξ by Godfrey and Isgur [11] is the likely explanation. We would expect approximately a 10% branching ratio to a $\phi\phi$ final state and a family ($^3F_2$, $^3F_3$) nearby in mass which would also decay to $\phi\phi$.

- possible (??) peaks around 2.1-2.3 GeV in the MARK III data shown by Wisniewski [5] at this conference in the reactions $J/\Psi \to \gamma\eta\eta$ and $J/\Psi \to \gamma\bar{K}^*K^*$. These spectra were quite preliminary and the bumps in the data were not identified by the collaboration as an indication of new states.

In summary, this "unexplored" region between $2m_p$ and the η_c is already showing signs of lots of life. Certainly, more dedicated studies would be welcome.

WHY $\phi\phi$?

The reaction $\bar{p}p \to \phi\phi$ provides a very special filter to the world of hidden strangeness and gluonic intermediate states. The reaction can proceed through several competing processes. First, an $s\bar{s}$ pair can be scattered from the quark sea (the strange quark content of the proton). Second, an allowed reaction can proceed via the non-strange component of the ϕ ($\approx$ 1%) which then decays through the dominate $s\bar{s}$ component by $\omega - \phi$ mixing. Finally, as in Figure 2, coherent annihilation of the incoming non-strange quarks and antiquarks is followed by production of a $\phi\phi$ pair through a 2-gluon intermediate state. It is in this reaction that one hopes for production of glueballs. But as mentioned above, ordinary $s\bar{s}$ quarkonium states which decay to $\phi\phi$ are also expected in our mass region.

If one sees a resonance in a gluon-rich channel, is it a glueball? Not necessarily, but it immediately becomes a candidate on which a number of traditional tests can be made. These traditional tests include: flavor independent decays ($gg \to \phi\phi, \rho\rho, \omega\omega$ approximately equally), narrow widths ($\sqrt{OZI}$ rule implies $\Gamma \approx$ 10-30 MeV), and predicted radiative decay ratios ($gg \to \gamma\rho : \gamma\omega : \gamma\phi =$ 9:1:2), however the first two tests are now questioned and the last is nearly impossible experimentally! Following the suggestion of

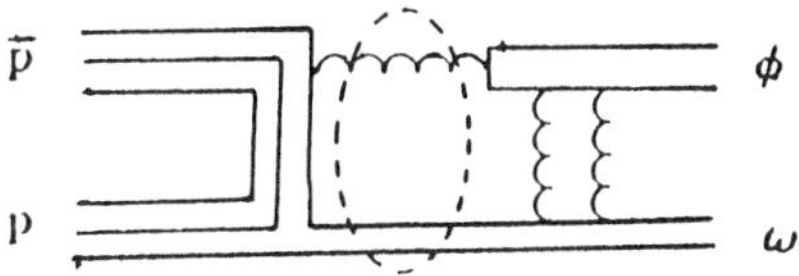

Figure 3: Diagram for $\bar{p}p \to \phi\omega$.

Close [12], we can make tests which might exclude rather than confirm a glueball candidate. In the JETSET context, such anti-tests might include observation of the same resonance in both $K\bar{K}$ and in $K^*\bar{K}$, or into both $\phi\phi$ and $\phi\eta$, or $\phi\pi^0$ final states. We will be sensitive to K_sK_s events from the beginning. Other tests have been suggested and can be found in excellent reviews on the subject [13]. The overall message is that it will take a multi-channeled, multi-experimental approach in order to confirm fully a state as a pure gluonic hadron, and that more experimental inputs are needed in this field.

WHAT ELSE CAN WE DO...BEYOND $\phi\phi$

There are several attractive approaches to finding exotics, which we are beginning to study within the capabilities of our complete general-purpose detector. The reaction $\bar{p}p \to \phi\omega$ is an OZI-forbidden decay from any ordinary $q\bar{q}$ state and therefore promises to be a hunting ground for interesting states of matter. Close [14] predicts that the natural "pecking order" for decays to $\phi\omega$ would be $qq\bar{q}\bar{q}, gq\bar{q}, gg$, then $q\bar{q}$. Chanowitz and Sharpe [15] have suggested that the $\xi(2230)$ might be a hybrid state which would have an appreciable branching fraction to $\phi\omega$. Under the assumption that the color-rescattering graph shown in Figure 3 has a non-negligible probability, then production of hybrid states as described in the bag model would be a possibility within the JETSET direct production approach.

The clearest signature for an exotic is one which has a J^{PC} combination forbidden to any $q\bar{q}$ state $(0^{+-}, 1^{-+}, 2^{+-}, ...)$. Indeed, at this conference the GAMS [16] collaboration showed (controversial!) data indicating a possible "CP-exotic" 1^{-+} state at 1405 MeV. For JETSET, no such state can be directly formed ($\bar{p}p$ is like $\bar{q}q$). However, with LEAR operating at 2 GeV/c, data taken in association with an outgoing π^0 such as $\bar{p}p \to (\phi\pi^0)\pi^0$, or $(\phi\eta)\pi^0$ or $(\bar{K}K^*(1420))\pi^0$ could lead to sensitive and interesting tests. Isgur *et al.* [17] predict that the latter reaction would lead to a hybrid meson with exotic quantum numbers at a mass of 2.1 GeV and a width of about 0.35 GeV.

SUMMARY

In summary, the JETSET collaboration is actively building a state-of-the-art detector capable of probing rare, exclusive reactions from $\bar{p}p$ annihilations in flight. We plan to exploit fully our unique experimental situation and to expand our Phase I detector capabilities in order to meet the challenge of this very promising, yet largely unexplored, physics arena.

We are particularly grateful for the stimulating dialog which has begun with F. Close and N. Isgur at our most recent JETSET collaboration meeting.

REFERENCES

[1] *Valet-Plus User's Guide* CERN DD Division OC Group, (September, 1987) and references therein.

[2] S. Baird *et al.*, Proceedings of the IV LEAR Workshop, Villars-sur-Ollon, Sept. 1987, Edited by C. Amsler, G. Backenstoss, R. Klapisch, C. Leluc, D. Simon, and L. Tauscher.

[3] S. Lindenbaum, this conference, and R.S. Longacre *et al.*, Proceedings of the Second International Conference on Hadron Spectroscopy, April 1987, Edited by Y. Oyanagi, K. Takamatsu, and T. Tsuru.

[4] P.S.L. Booth *et al.*, Nucl. Phys. **B273** (1986) 677.

[5] W.J. Wisniewski this conference and in the Proceedings of the Second International Conference on Hadron Spectroscopy, April 1987, Edited by Y. Oyanagi, K. Takamatsu, and T. Tsuru.

[6] D. Bisello *et al.*, Phys. Lett. **179B** (1986) 294.

[7] See the review at this conference by B. Quinn.

[8] R.M. Baltrusaitis *et al.*, Phys. Rev. Lett., **56** (1986) 107.

[9] D. Alde *et al.*, Phys. Lett. **177B** (1987) 120.

[10] W. Dunwoodie, this conference, and D. Aston *et al.*, Proceedings of the Second International Conference on Hadron Spectroscopy, April 1987, Edited by Y. Oyanagi, K. Takamatsu, and T. Tsuru.

[11] S. Godfrey and N. Isgur, Phys. Lett., **141B** (1984) 439.

[12] F.E. Close, *Gluonic Hadrons*, RAL 87-072.

[13] See, for example, several review articles in the Proceedings of the Second International Conference on Hadron Spectroscopy, April 1987, Edited by Y. Oyanagi, K. Takamatsu, and T. Tsuru.

[14] F.E. Close, Private Communication.

[15] M.S. Chanowitz and S.R. Sharpe, Phys. Lett. **132B** (1983) 413.

[16] D. Alde, this conference.

[17] N. Isgur, R. Kokoski, and J. Paton, Phys. Rev. Lett. **54** (1985) 869.

CP VIOLATION IN KAON, HYPERON AND B-MESON SYSTEMS

John F. Donoghue

Department of Physics and Astronomy, Univ. of Mass., Amherst, MA 01003

ABSTRACT

The effect of recent developments on CP violation is discussed and some possible future directions for experimental progress are compared.

The motivation for studying violations of CP symmetry remains strong. It is one of the few major unsolved problems in particle physics and is potentially sensitive to physics at the TeV scale associated with the symmetry breaking mechanism. However, the relevant experiments are of the high precision, low energy category, which nicely compliments the push to higher energies with the SSC. I refer the reader to reviews for the basic phenomenology and standard notation.[1]

The main problem in the field at present is to find new signals of CP violation which will distinguish among the competing theories (or perhaps rule them all out). All theories have a free parameter which they adjust to fit the $K_L \to 2\pi$ value of ε. In principle, theories then make differing predictions for other observables. The study of these then selects the correct theory. In practice there are two difficulties with this program. One is experimental in that no other nonzero observable has been measured. The experiments are difficult. Hopefully we are nearing the end of this and we may soon have a firm measurement of $\frac{\varepsilon'}{\varepsilon}$. The theoretical difficulty is that the predictions of the models are presently not precise. The calculations involve hadronic matrix elements, which we presently have no reliable method of calculating. However, a very important point to keep in mind is that this will ultimately change. Lattice methods have the potential of reliably calculating these matrix elements and significant progress is presently being made.[2] My personal estimate is that reasonably firm results will be available within three years for kaon nonleptonic matrix elements, with baryons perhaps a few years later. This is about the same time scale as the next generation of experiments. The predictions will be firm at about the time experimental results (hopefully nonzero) are available. This possibility of reliable theory can influence the direction of experiment.

There are two main recent developments. 1) The hint of $\frac{\varepsilon'}{\varepsilon}$ from the CERN group[3]

$$\frac{\varepsilon'}{\varepsilon} = (3.3 \pm 1.1) \times 10^{-3} \tag{1}$$

This is tantalizing and the FNAL group should be able to confirm or refute it at this level. 2) $B^o\ \overline{B}^o$ mixing appears to have been seen by ARGUS,[4] with

$$\frac{\Delta m_B}{\Gamma_B} = 0.73 \pm 0.18 \tag{2}$$

This also may be confirmed or refuted soon, by the CLEO collaboration. $B^o\ \overline{B}^o$ mixing is not CP violating but gives information on the box diagram and KM angles. Although we should reserve final judgement on these observations, in this talk I will assume that both are true in order to examine the consequences. From $\overline{B}B$ mixing we can extract[5] the t quark contribution to ε, up to some KM angles

$$\varepsilon \approx \frac{0.05}{\sqrt{2}}\ s_3\, s_\delta \sqrt{1 - \left(\frac{s_3\, s_\delta}{0.05}\right)^2}\ \left[\frac{f_K^2\, B_K}{f_B^2\, B_B}\right]\ \frac{\Delta m_B}{\Delta m_K} \tag{3}$$

The factor in square brackets is a ratio of hadronic matrix elements. Even allowing the full range of these matrix elements, one obtains an upper bound on the KM angles[5]

$$s_3\, s_\delta < 2 \times 10^{-2}$$

$$s_1\, s_2\, s_3\, s_\delta < 3 \times 10^{-4} \tag{4}$$

The measurement of ε' in turn implies a lower bound on the angles

$$\frac{\varepsilon'}{\varepsilon} = 0.2\ \ s_2\, s_3\, s_\delta\ \ P\ \ H(m_\tau) \tag{5}$$

where P is a matrix element and $H(m_\tau)$ is the top quark mass dependence.[6] Using $\frac{\varepsilon'}{\varepsilon}$ $= 0.0033$ we get

$$s_1\, s_2\, s_3\, s_\delta > 1.5 \times 10^{-4} \tag{6}$$

for the full range of calculations of P. This is a narrow range of allowed angles which in turn implies better predictions of CP observables than previously possible. [If either the value Δm_B or ε' decreases, the allowed range becomes wider]. It is also worth pointing out that once hadronic matrix elements are under control and we know the mass of the top quark, we can obtain $s_1\, s_2\, s_3\, s_\delta$ directly from ε.

The firming of the measurement of ε' is of course the main priority in the field at present. If the present value is confirmed this will immediately remove several potential theories, and will ultimately be a solid test of the KM model. After ε', what could come next? I will mention three classes of experiments which are presently being explored: 1) rare K decays such as $K_L \to \pi^o e^+ e^-$, 2) Hyperon decays and 3) B-meson decays.

In kaon decays, the process $K_L \to \pi^o e^+ e^-$ has attracted considerable attention because of the possibility of a large non-ε signal.[7,8,9] The theory on this decay is presently in flux, although I expect that we should be in better shape in a few months. The CP violating part has been evaluated using the results of Gilman and Wise[10] on the QCD corrections, with the result[7]

$$B(K_L \to \pi^o e^+ e^-)\,\big|_{\not{CP}} = \begin{cases} 5 \times 10^{-13} & \text{Superweak} \\ 4 \times 10^{-12} & \text{KM} \end{cases} \tag{7}$$

However the Gilman Wise technique of treating the charm threshold has been criticized[11] in other contexts and the QCD corrections need to be rechecked for this calculation. The main new action has been in the CP conserving amplitude which proceeds through a two photon state. The amplitude has the form

$$A(K_L \to \pi^o e^+ e^-) = D m_e \bar{u} v + E(p_+ \cdot k - p_{-o} k)\, \bar{u} \not{k} v \tag{8}$$

The D term always has a factor of m_e and is hence suppressed.[7] The E term is hard to generate, but early estimates indicated that it is not large,[8] so that the CO conserving rate was tiny. Recently Sehgal and Morozumi and Iwasaki[9] have proposed a diagram which yields a branching ratio of 10^{-12} to 10^{-11}. [My collaborators and I have found some flaws in their numbers, but the general conclusion seems to hold.] This is presently being studied by several theory groups, and we will hopefully agree soon. Overall this is good news in that the charge asymmetry, a direct measure of CP violation, will be large. The next generation of $K_L \to \pi^o e^+ e^-$ experiments[12] will push the limit down significantly, and probably come close to the predicted rate. It will remain to be seen whether further improvements make the observation possible.

In baryons one can compare the decay parameters of hyperon and antihyperons in order to test CP. This is best done in $p\,\bar{p}$ reactions where the symmetry of the initial state makes it possible to measure CP odd effects by final state asymmetries.[13] The predictions here come in at the 10^{-4} level. The calculations[14] involve quark model matrix elements of the nonleptonic Hamiltonian. These work in magnitude to within a factor of 2-3 of experiment, although they miss the SU(3) structure (yielding $\frac{D}{F} = -1$ instead of the physical $\frac{D}{F} = -\frac{1}{2}$). These calculations are then presently just estimates. As an example of the model dependence, I have chosen two schemes to illustrate some of the variations of model dependence. Using the quoted range of $s_1\, s_2\, s_3\, s_\delta$ and 50 GeV < mt < 200 GeV, I find

$$(\alpha + \bar{\alpha})_{\Lambda \to p\pi^-} = \begin{cases} -(0.3 - 1.0) & \text{Bag model (Ref. 15)} \\ -(1.2 \to 4.0) & \text{Non relativistic (Ref. 16)} \end{cases}$$

$$(\alpha + \bar{\alpha})_{\Xi^- \to \Lambda\pi^-} = \begin{cases} -(0.4 \to 1.2) & \text{Bag model (Ref. 15)} \\ -(1.6 \to 4.8) & \text{Non relativistic (Ref. 16)} \end{cases} \qquad (9)$$

in units of 10^{-4}.

In B meson decays there are two categories of tests. One[17] uses B $\bar{B}$ mixing to interfere $B^o \to f$ with $\bar{B}^o \to f$, leading to partial rate differences. This class already has reasonably firm predictions in the KM model. Hadronic matrix elements cancel and the M_t dependence appears only hidden in the measured value of Δm, leaving only the KM angles. The rate asymmetries are sizeable but only in very rare processes. The other category[18] involves partial rate asymmetries without mixing. These are more problematic as they require two different diagrams of comparable strength and large final state interactions. One might be lucky and see one of these modes but there is no guarantees. Either category requires large numbers ($\sim 10^8$) of useful B mesons.

If we try to balance these different prospects, there are advantages to each. On $K_L \to \pi^o e^+ e^-$ the positive aspects are that it involves known techniques, the non ε may be enhanced and the next rounds of experiments may get close. A negative aspect is that we may need to disentangle ε' from ε (by interference or measurement of $K_L \to \Pi^o e^+ e^-$) unless theory improves. In hyperons there is no ε effect, the reactions are well known and the lattice methods may give good matrix elements. However, the experiments to measure a 10^{-4} effect are not easy and require precise control of systematics. In B mesons there is an advantage that this is a new area of study and hence will probe models in different ways. Some predictions are reasonably reliable and several potential decay modes are available. The disadvantage is again experimental in that the ability to reconstruct the required number of B mesons is beyond known capabilities. All of these reactions are hard to study, and there is no design for any of them (that I know of) that can yet claim to be able to observe the predicted effects. However, in each case experimental groups are now proposing ways to push down the limits on these reactions and are trying to see if actual measurements can be made. These may easily call for more than one generation of experiment. The experience with $\frac{\varepsilon'}{\varepsilon}$ is instructive, as heroic efforts over many years appear now to becoming successful. The further studies of CP signals are valuable and fundamental experiments which don't have to be done at the SSC in kilo-author collaborations. For the health of our field, we should all hope that one of these future directions is also successful.

REFERENCES

1. J.F. Donoghue, B.R. Holstein and G. Valencia, Int. J. Mod. Phys. A2, 319 (1987).
   L. Wolfenstein, Ann. Rev. Nucl. Part. Sci. 36, 137 (1986).
2. S. Sharpe, A. Soni, and G. Martinelli, in Proc. Workshop on Hadronic Matrix Elements and Weak Decays, Ringberg 1988 (to be published).
3. H. Burkhardt et. al., CERN preprint EP/88-47.
4. H. Albrecht et. al., Phys. Lett. 192B, 245 (1987).
5. J.F. Donoghue, T. Nakada, E.A. Paschos and D. Wyler, Phys. Lett. 195, 285 (1987).
6. J. F. Donoghue, E. Golowich, and B.R. Holstein, Physics Reports 131, 319 (1986).
7. J.F. Donoghue, B.R. Holstein, and G. Valencia, Phys. Rev. 35, 2769 (1987).
8. E. de Rafael, G. Ecker, and A. Rich, to be published in Nucl. Phys.
9. L. Sehgal, Aachen preprint.
   T. Morozumi and H. Iwasaki, KEK preprint.
10. F. Gilman and M. Wise, Phys. Rev. D21, 3150 (1980).
11. W. Bardeen, A.J. Buras, and J.M. Gerard, Phys. Lett. 180B, 133 (1986).
12. There are proposals at BNL and KEK.
13. J.F. Donoghue, B.R. Holstein, and G. Valencia, Phys. Lett. 178B, 319 (1986).
14. J.F. Donoghue, X.G. He, and S. Pakvasa, Phys. Rev. D34, 833 (1986).
15. J.F. Donoghue, E. Golowich, B.R. Holstein, and W. Ponce, Phys. Rev.
16. A.L. Yaouanc et. al., Nucl. Phys. B149, 321 (1979).
17. A. Sanda and A. Carter, Phys. Rev. D23, 1567 (1981).
    I. Bigi and A. Sanda, Nucl. Phys. B193, 86 (1981).
18. L.L. Chau and H.Y. Cheng, Phys. Rev. Lett. 53 1037 (1984).
    M. Bander, D. Silverman, and A. Soni, Phys. Rev. Lett. 43, 197 (1979).

ANTIPROTON SPIN PHYSICS WITH INTERNAL POLARIZED TARGET

W. Haeberli
University of Wisconsin-Madison, Madison, WI 53706

ABSTRACT

A project to insert a polarized internal hydrogen or deuterium target in LEAR is discussed. To reach the desired target thickness of 10^{14} polarized atoms/cm^2 requires the use of a storage cell. Successful tests of storage cells that permit 400 wall collisions without significant depolarization are reported. The proposed spin filter experiment to partially polarize the circulating $\bar{p}$ beam by spin-selective attenuation of $\bar{p}$ in the polarized hydrogen target is reviewed and examples of specific planned experiments are presented.

INTRODUCTION

Even at low energies, where few partial waves contribute, the $\bar{p}p$ scattering amplitudes can not be determined from cross section measurements alone, so that the study of spin observables is essential. Recent experiments[1] at LEAR, where the external $\bar{p}$ beam was scattered from a polarized hydrogen target, have yielded much improved measurements of the analyzing power $A_y(\theta)$ in $\bar{p}p$(pol) scattering in the momentum range 500 to 1500 MeV/c. Below, we consider the insertion of a polarized hydrogen gas target in the circulating $\bar{p}$ beam as a means to extend the range of possible antiproton spin experiments. It may become possible, in this way, to study the $\bar{p}p$ spin dependence in the momentum range below 300 MeV/c, where measurements of the ρ-parameter (ratio of real to imaginary part of the forward elastic $\bar{p}p$ amplitudes) show an intriguing, unidentified structure[2]. The proposal by the FILTEX (filter experiment) collaboration[3], to partially polarize the circulating $\bar{p}$ beam by spin selective attenuation of $\bar{p}$ in an internal polarized hydrogen target, will offer other interesting opportunities.

EXTERNAL VS. INTERNAL POLARIZED TARGET

Considerable development work will be needed to address the various technical problems associated with the use of internal targets in storage rings. Even though it is not yet clear in what situations internal targets have compelling advantages over traditional techniques, the following rough comparison based on post-ACOL performance figures of LEAR may be of interest. A polarized butanol target of thickness 1 g/cm^2 represents a target thickness $t = 10^{23}$ $protons/cm^2$. For an extracted beam of 10^7 $\bar{p}/s$ the corresponding luminosity is $L = 10^{30}$ $cm^{-2}s^{-1}$. By comparison, for an internal target of 10^{14} polarized $atoms/cm^2$, and 4×10^{10} stored $\bar{p}$ (revolution frequency $3 \times 10^6 s^{-1}$), the luminosity is $L = 10^{31}$ $cm^{-2}s^{-1}$. Thus the luminosity of the internal target is competitive even if the target thickness were to fall short of the design goal by an order of magnitude. Maybe more important are other advantages of an internal polarized target: chemical and isotopic purity, high energy resolution, large vector and tensor polarization for deuterons, rapid reversal of the target polarization (~10 ms); weak guide field (~5 Gauss), and free choice of polarization axis.

POLARIZED JET TARGET VS. STORAGE CELL

In principle, the beam from a conventional hydrogen atomic beam apparatus can be used as a polarized hydrogen gas target. Typical observed atomic beam intensities are 3×10^{16} atoms/s with an average velocity of 10^5 m/s. For an atomic beam diameter of 1 cm this corresponds to a target thickness $t = 4 \times 10^{11}$ $atoms/cm^2$. At Novosibirsk[4], a jet target was recently used as an internal target in an electron storage ring, using a modern atomic beam apparatus with superconducting six-pole magnets. The observed target thickness was 1×10^{11} $atoms/cm^2$.

While some gain in the output of atomic beam sources may still be possible, experience suggests that other means be sought to overcome the missing factor of 10^2 to 10^3 in target thickness. For this reason we have proposed[5] to inject the atomic beam into a thin-walled storage cell, which has openings for entry of the atoms, and for entry and exit of the circulating beam. The density $\rho = N/C$ of atoms in the cell depends on the number N of atoms injected per second, and on the combined gas conductance C of the three openings. A cell of suitable dimensions to accept the circulating beam at LEAR (after electron cooling) has a conductance of about $C = 10\ \ell/s$ provided the cell is cooled to 80 K in order to reduce the conductance[3]. Thus $N = 10^{17}$ atoms/s is expected to yield $\rho = 10^{13}$ $atoms/cm^3$, or a target thickness $t = 10^{14}$ $atoms/cm^2$ if the target length is chosen as 10 cm.

The use of a storage cell has a number of negative aspects. The cell must be opened for loading of the beam (clam-shell cell). The geometry must be carefully chosen such that the cell does not interfere with the circulating beam, yet maintains as low a gas conductance as possible. The cell must have sufficiently thin walls to permit passage of reaction products through the cell walls. Wall materials must be found that permit several thousand wall collisions without significant depolarization. In addition it must be demonstrated that an ionizing beam passing though the cell has no adverse effect either on the polarization of the gas directly (excitation, ionization) or on the wall properties.

TESTS OF STORAGE CELLS

In order to test storage cells under reasonably realistic conditions, without tying up a large accelerator, the following method was developed. A 50 keV D^+ beam of about 0.5 mA intensity is passed through the storage cell. Pickup of polarized electrons by the D^+ ions leads to fast D^0 atoms, which are polarized in electron spin. Through the hyperfine interaction the electron polarization leads to a partial alignment of deuterons. The alignment P_{zz} (or T_{20}) of the deuterons is subsequently detected as an anisotropy of neutrons from the $^3H(d,n)^4He$ reaction, when the D^0 beam strikes a tritiated titanium target. The measured ratio R of neutron yields at 0° and 90° with respect to the magnetic field at the ^3H target is proportional to the target polarization. While this method tests only the electron polarization of the target atoms, any depolarization will certainly proceed by depolarization of the electrons, rather than by direct interaction with the (small) magnetic moment of the proton in the target atoms.

Best results were obtained with cell walls coated with a thin layer of fluorocarbon (Teflon or Viton), for which the polarization in the cell was found to be essentially the same for cells with 25, 120 or 400 wall collisions[6]. Comparison with measurements on a free atomic beam (no wall depolarization) suggests that the target polarization is above 80%. The target thickness was deduced from the neutron count rate, compared to a known density of hydrogen gas in the vacuum chamber. The observed gain in target thickness with use of the storage cell compared to the same atomic beam without the cell is a factor 250.

A new system is under construction to test cells under clean vacuum conditions, and to extend the temperature range of the cell walls to 20 K to effect further increase in target density.

SPIN FILTER

The basic idea was proposed twenty years ago by Csonka[7]: if the circulating beam in a storage ring passes through an internal polarized target, the stored beam will become successively polarized, provided the total removal cross section is different when the spins of target and beam are parallel than when the spins are antiparallel. Application of the idea to the $\bar{p}$ beam in LEAR was discussed by Kilian and Möhl[8]. It was concluded that the method required a polarized target thickness a factor of 100 larger than those obtained with polarized jet targets. Later, Steffens[3] pointed out that the storage cell method[5] might provide the necessary target thickness, and a proposal for installation of a spin filter at LEAR was presented by the FILTEX collaboration in 1985. The proposal was given conditional approval, the condition being that feasibility of the method be demonstrated with protons in another ring. It is planned to carry out this test in the TSR (test storage ring) at the MPI-Heidelberg.

In order to discuss the operation of the spin filter, we need to consider the total removal cross sections σ^+ and σ^-, for the two different relative spin orientations of the protons in the target and the antiprotons in the ring:

parallel spin: $\sigma_+ = \sigma_0 + \sigma_1$

antiparallel spin: $\sigma_- = \sigma_0 - \sigma_1$

Obviously, a large ratio σ_1/σ_0 is favorable for the spin filter method. Unfortunately this ratio is not known, and the determination of σ_1/σ_0 will be an important goal of the proposed experiment. The calculations below are based on the value $\sigma_1/\sigma_0 = 0.06$, predicted by Dover[9] from the G-parity transformation of the Paris potential. It is clear that as the polarization of the circulating beam builds up, the intensity of the beam (number of stored particles) keeps decreasing. For best statistical accuracy of experiments one wants to optimize $P^2N(t)$. For $\sigma_1/\sigma_0 \ll 1$ the optimum is reached after 2τ, where τ is the attenuation time

$$\tau = \frac{1}{nf\sigma_0}.$$

The polarization at this time is $P = 2\sigma_1/\sigma_0$. We conclude that after a time 2τ, $1/e^2$ or roughly 10% of the circulating beam will have acquired a polarization of about $P = 0.1$.

Operation of the spin filter places considerable demand on the beam life time of the ring and on the density of the internal polarized target. For the target thickness mentioned above, 10^{14} atoms/cm^2, the attenuation time for a beam momentum of 340 MeV/c ($E = 60$ MeV) is 4×10^4 s or 11 hours. We have added 0.05 b to the strong interaction total cross section (0.2 b) to account for Coulomb losses, and assumed a revolution frequency of 10^6 s^{-1}.

It would be incorrect to assume that a polarization of P = 0.1 is insufficient to permit accurate measurements e.g. of $\bar{p}p$ spin correlation coefficients. In reality, the accuracy is dictated by the event rate (luminosity) and by the magnitude of systematic errors. Since the target polarization can be reversed rapidly and without effect on the target density or the properties of the circulating beam, systematic errors will be much easier to control than in conventional experiments. Some examples of typical experiments are discussed briefly in the following section.

PROPOSED FIRST EXPERIMENTS

Measurement of target polarization: The first question to be answered is whether the large target polarization observed in the present bench tests is maintained in use in a storage ring. This question will be studied extensively as part of the planned feasibility measurements using the Heidelberg test storage ring. In addition, the condition of the target cell will be monitored by a Balmer-line polarimeter[10].

Analyzing power in $\bar{p}$ + p(pol) elastic scattering: Measurement of the left-right intensity ratio yields the proton analyzing power A_p. Detector assymmetries cancel to very high accuracy by reversal of the target polarization, which requires $\lesssim$10 ms. Reversal can be accomplished by radiofrequency transitions acting on the atomic beam before it enters the storage cell, without making any other changes in the system.

Initially, measurements of A_p will be made at beam momenta where high quality results have already been obtained with conventional techniques. This will serve as a valuable check of the new method and of the target polarization determinations. The rates are very favorable. For a detector solid angle of $\Delta\Omega$ = 0.1 sr, a target of 10^{14} atoms/cm^2 and 10^{10} $\bar{p}$ stored in the ring, the rate in each detector is 10^3 s^{-1} for a laboratory cross section of 10 mb/sr. For a one minute measurement, the statistical error in A_p is about $\pm 5 \times 10^{-3}$, assuming ~80% target polarization. Once A_p is confirmed to high accuracy, it can be used as a continuous monitor of the target polarization.

Measure buildup of the $\bar{p}$ polarization: It is proposed to set up large solid-angle detectors (~1 sr each) to detect elastic $\bar{p}p$ scattering to the left, the right, as well as up and down. The total yield in the four detectors is monitored during the first hour of stored beam operation, after which the target polarization is reversed for four minutes. Subsequently, the target polarization is returned to the normal orientation to continue polarization buildup in the $\bar{p}$ beam. Comparing the yield during those four minutes with the yield during normal operation should show a relative difference of $(1\pm 0.3)\times 10^{-3}$, resulting from the accumulated $\bar{p}$-polarization during the first hour, $P_{\bar{p}}(1hr) = 5\times 10^{-3}$. The difference in elastic cross section for parallel and antiparallel spins, which was assumed in the above calculation, is based on the values for the total elastic cross section calculated by Dover[9]. Unfortunately the angular dependence has not been reported, but presumably for suitably chosen angles the effect will be larger than for the total cross section. Calculations of the angular dependence would be helpful as a guide to the choice of detector angle. Once measurable buildup of $\bar{p}$ polarization has been established, an absolute measurement of the $\bar{p}$ polarization is planned by measuring the left-right asymmetry with an unpolarized target. This measurement requires that the analyzing power $A_{\bar{p}}$ is known. This is indeed the case, since CP invariance assures that $A_{\bar{p}}$ (polarized $\bar{p}$ on unpolarized p) is equal to the known analyzing power A_p (unpolarized $\bar{p}$ on polarized p).

Spin correlation parameters: If we assume that significant $\bar{p}$ polarization is obtained, the spin correlation parameters are measured by reversing the target polarization every few seconds. Measurements can be made with transverse as well as longitudinal target polarization. Particularly interesting will be the observation of the charge exchange channel, for which very large effects are predicted.

Variation in $\bar{p}$ energy: Measurements are not limited to the particular energy for which the buildup of $\bar{p}$ polarization is most favorable. Rather, following polarization buildup, we would propose to ramp the ring. During ramping, the gas target is in a retracted position, to be inserted again when the new beam energy is reached and the beam is cooled.

CONCLUSIONS

The development of polarized gas targets for storage rings most likely will involve the use of storage cells in order to achieve the required target thickness. Initial tests of storage cells have resulted in an increase of target thickness, compared to a polarized jet, of a factor 250, with a target polarization of at least 80%. The proposed use of this type of target in LEAR still faces many technical problems, some of which will be addressed by inserting the target in the Heidelberg test storage ring.

It should be noted that the target under development will provide the highest luminosity polarized internal target so far proposed, and in fact is competitive with the planned unpolarized internal gas jet target in LEAR. Thus, even if it turned out that the measured cross section ratio σ_1/σ_0 for parallel and antiparallel spins is too small to produce useful polarization in the circulation antiproton beam, the polarized internal target can form the basis of a challenging research program. Studies of interest include the measurement of analyzing powers at momenta lower than can be studied with conventional techniques, as well as studies of polarization transfer from the target proton to the outgoing reaction products. It should be attractive, for instance to demonstrate the dramatic polarization transfer predicted[9,11] for the charge exchange reaction on a polarized proton target.

Finally, it should be mentioned that another method has been proposed[12] to polarize the circulating beam in LEAR by the accumulation of small, spin-dependent deflections in quadrupole magnets in the ring. Recent papers by Penzo[12] and by Rosmanith[13] discuss some of the conceptional problems of this proposal, including the well-known argument that the uncertainity principle prevents Stern-Gerlach spin separation of charged particles. Nevertheless, the discussion presented in the present paper shows that alternate proposals are of interest even if they were to achieve a $\bar{p}$ polarization of only a few percent.

ACKNOWLEDGMENTS:

I should like to acknowledge discussions with the spokesperson of the FILTEX collaboration, Dr. E. Steffens. The work on storage cells was carried out in collaboration with T. Wise, A. Converse and S. Price at the University of Wisconsin-Madison. The work was supported by the U. S. Department of Energy and the Graduate School of the University of Wisconsin-Madison.

REFERENCES

1. C.I. Beard et al.,in Proc. IV LEAR Workshop 1987, edited by C. Amslev et al., Nuclear Science Research Conference Series, Vol. 14 (Harwood, Chur, 1988), p.305.
2. P. Kroll and W. Schweiger, in Proc. IV LEAR Workshop 1987, Nuclear Science Research Conference Series, Vol. 14, (Harwood, Chur, 1988) p.239.
3. E. Steffens, et al., in Physics with Antiprotons at LEAR in the ACOL Era, edited by U. Gastaldi, et al., (Editions Frontieres, Gif-sur-Yvette, 1986) p.245.
4. V.T. Dmitriev et al., Phys. Lett. 157B, 143 (1985).
5. W. Haeberli, in Nuclear Physics with Stored, Cooled Beams,

edited by P. Schwandt and H.O. Meyer, (American Institute of Physics, New York, 1985), p.251.

6. W. Haeberli, in Proc. IV LEAR Workshop 1987, edited by C. Amslev et al., Nuclear Science Research Conference Series, Vol. 14, (Harwood, Chur, 1988) p.217.
7. P.L. Csonka, Nucl. Instrum. Meth., 63, 247 (1968).
8. K. Kilian and D. Möhl, in Physics at LEAR with Low-Energy Cooled Antiprotons, edited by U. Gastaldi and R. Klapisch, (Plenum Press, New York 1984) p.701.
9. C.B. Dover, in Low Energy Antimatter, edited by D.B. Cline, (World Scientific Publishing Co., Singapore, 1986) p.1.
10. G. Graw, et al., in Proc. IV LEAR Workshop 1987, edited by C. Amslev et al., Nuclear Science Research Conference Series, Vol. 14, (Harwood, Chur, 1988) p.221.
11. C.B. Dover and J.M. Richard, Phys. Rev. C25, 1952 (1982)
12. A. Penzo, in Proc. IV LEAR Workshop 1987, edited by C. Amslev et al., Nuclear Science Research Conference Series, Vol. 14, (Harwood, Chur, 1988) p.183.
13. R. Rosmanith, this workshop.

$\Lambda\bar{\Lambda}$ PRODUCTION; EXPERIMENT AND THEORY

Presented by G B. Franklin

PS185 collaboration: P.D.Barnes[A], P.Birien[D], B.E.Bonner[G], W.H.Breunlich[I], G.Diebold[A], W.Dutty[D], R.A.Eisenstein[E], G.Ericsson[H], W.Eyrich[C], R.v.Frankenberg[C], G.Franklin[A], J.Franz[D], N.Hamann[D], G.Harig[D], D.Hertzog[A], A.Hofmann[C], T.Johansson[H], K.Kilian[B,F], C.Maher[A], M.Meyerhofer[C], N.Naegele[I], W.Oelert[F], S.Ohlsson[H], H.W.Ortner[C], P.Pawlek[I], B.Quinn[A], E.Rossle[D], H.Schledermann[D], H.Schmitt[D], G.Sehl[F], J.Seydoux[A], F.Stinzing[C], J.Szymanski[A], X.Yi[A], T.Zeludziewicz[F]

A)Carnegie-Mellon Univ. B)CERN C)Univ. Erlangen D)Univ. Freiburg E)Univ. of Illinois F)Juelich KFA G)Los Alamos Nat. Lab. H)Univ. of Uppsala I)IMPE Osterr. Akad. Wissensch

ABSTRACT

The associated strangeness production reaction $\bar{p} + p \rightarrow \bar{\Lambda} + \Lambda$ is discussed. Recent differential cross sections and polarization data from LEAR experiment PS185 are presented and the status of the theoretical models is reviewed.

ASSOCIATED STRANGENESS PRODUCTION

The $\bar{p} + p \rightarrow \bar{\Lambda} + \Lambda$ reaction has much in common with other "associated strangeness production" reactions such as $\pi^+ + n \rightarrow K^+ + \Lambda$ and $\gamma + p \rightarrow K^+ + \Lambda$. The production of an s quark with an associated $\bar{s}$ typically involves large momentum transfers even at threshold and is therefore sensitive to short range effects. This large momentum transfer is in contrast with the "strangeness exchange" reaction $K^- + n \rightarrow \pi^- + \Lambda$ in which the Λ can be produced at very low q.

The physics probed by these is illustrated by the ingredients of the calculations. They are generally performed in the framework of a DWBA or coupled channels model. The transition potentials can be based on one-kaon exchanges with g_{KNL} being fitted to KN data. However, the relative importance of boson exchanges of higher mass, initial and final state corrections, and intermediate resonances depends on the reaction. The $\gamma + p \rightarrow K^+ + \Lambda$ reaction has no significant initial state distortions but is believed to be sensitive to nucleon resonances. In contrast, the $\bar{p} + p \rightarrow \bar{\Lambda} + \Lambda$ reaction has been found to be extremely sensitive to initial and final state distortions. These reactions, therefore, complement each other and together form a thorough testing grounds for our understanding of short range hadronic interactions.

$\bar{\Lambda}\Lambda$ PRODUCTION MEASUREMENTS

Although data have been available for some time, new results from the PS185 collaboration[1] at LEAR have stimulated a recent increase of theoretical studies of the $\bar{p} + p \rightarrow \bar{\Lambda} + \Lambda$ reaction.. The intensity and quality of the LEAR beam allow high statistics measurements not previously available. In addition, the measurements can be made near threshold, limiting the influence of high partial waves.

The $\bar{\Lambda}\Lambda$ production experiments make use of the self-analyzing property of the Λ; the proton in the decay $\Lambda \rightarrow p + \pi^-$ is correlated with the Λ spin direction. The spin physics can be obtained simply by observing the charged decay products of the Λ and $\bar{\Lambda}$. The observables are the total cross section σ, the differential cross section $d\sigma/d\Omega(\theta_{\bar{\Lambda}})$, the polarization perpendicular to the scattering plane $P_Y(\theta_{\bar{\Lambda}})$, and the spin correlation coefficients C_{ij}, all as a function of incident $\bar{p}$ momentum. It is the availability of these detailed data that allows one to test the various models and parameterizations.

The PS185 experiment is shown in Fig. 1. The incident $\bar{p}$ is tagged with scintillator S1 as it enters the five cell target. The scintillators $S2_i$ and $S3_i$ are used to identify events in which a charged $\bar{p}$ enters a cell but no charged particles leave the cell. Four of the cells contain CH_2, which serves as a proton target, and the fifth cell contains a carbon target for background monitoring. The hodoscope detects at least two of the decay products from $\bar{\Lambda}\Lambda \rightarrow \bar{p}\pi^+ \, p\pi^-$ and is used in the on-line trigger along with the target cell scintillators.

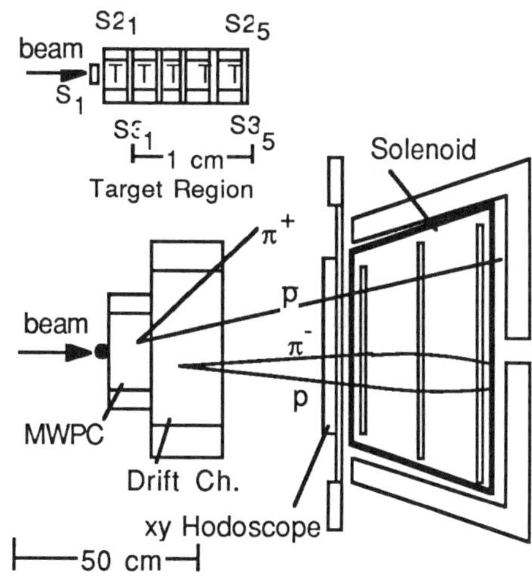

Fig. 1. Experiment PS185 Setup.

The segmentation of the target is used to identify the cell in which the interaction takes place, facilitating corrections for the dE/dx losses of the incident $\bar{p}$ which is used to break the $\bar{p}$ momentum into bins of $\Delta p < 2$ MeV/c.

The two "Vees" from the $\bar{\Lambda}\Lambda$ decays are tracked in a 22-plane stack of MWPC's and drift chambers. The $\bar{\Lambda}$'s are distinguished the Λ's by identifying the sign of their decay products in three drift chambers located inside the 0.1 T magnetic field of a solenoidal magnet .

The differential cross section and polarization data are shown in Fig. 2. The lowest $\overline{p}$ momentum is less than 1 MeV/c above the threshold p_{min}=1435.1 MeV/c and yields Λ kinetic energies less than 0.5 MeV in their center of mass. It can be seen that the $\overline{\Lambda}$ distribution tends to be slightly forward peaked even at this low momentum, indicated non-negligible contributions from L=1 and higher partial waves. This forward peaking rises rapidly with increasing $\overline{p}$ momentum.

The polarization data is consistent with zero at the lowest $\overline{p}$ momentum but becomes positive for forward going $\overline{\Lambda}$'s as the momentum is increased. A zero crossing appears as the $\overline{p}$ momentum is raised giving negative polarizations for four momentum transfers of $-t' = t_{min} - t < 0.2$ $(GeV/c)^2$.

$\overline{\Lambda}\Lambda$ CALCULATIONS

A more detailed look at the $\overline{p} + p \rightarrow \overline{\Lambda} + \Lambda$ calculations reveals the limits of our abilities to deal with close range effects implied by momentum transfers of $q \geq 3$ fm^{-1}. This large momentum transfer has ramifications in modeling the transition amplitudes, in the initial and final state distortion potentials, and forces the calculations to introduce empirical parameters.

For example, there is disagreement on the relative contributions of $K^*(890)$ and $K^{**}(1430)$ exchanges and their inclusion brings in additional uncertainties due to their unknown coupling constants and the need for parameterization of cutoff functions. An alternative approach is to use potentials motivated by quark flow diagrams in which the $\overline{s}s$ pairs are created either with the quantum numbers of a gluon, $^3S_1$, or of the vacuum, $^3P_0$. This approach has been supported by the works of Genz and Tutor[2] and of Rubenstein and Snellman[3] which indicates that perturbative QCD may be more appropriate for $\overline{s}s$ production than for $\overline{u}u$ and $\overline{d}d$ production due to the mass of the s quark.

The initial state $\overline{N}N$ interaction is constrained by other $\overline{N}N$ reaction studies. However, since little is known of the $\overline{\Lambda}\Lambda$ interaction, it is taken as "almost" like the $\overline{N}N$ interaction, allowing the introduction of additional parameters to fit the data. The calculations find that the distortions are critical to obtaining the correct total $\overline{\Lambda}\Lambda$ cross sections.

A coupled channels treatment of the initial state $\overline{N}N$ interaction also requires the transition potentials for coupling to other baryon-antibaryon states. These can be obtained using g-parity transformations on OBE models of NN interactions for the long range part, but normally use effective potentials for the short range contributions.

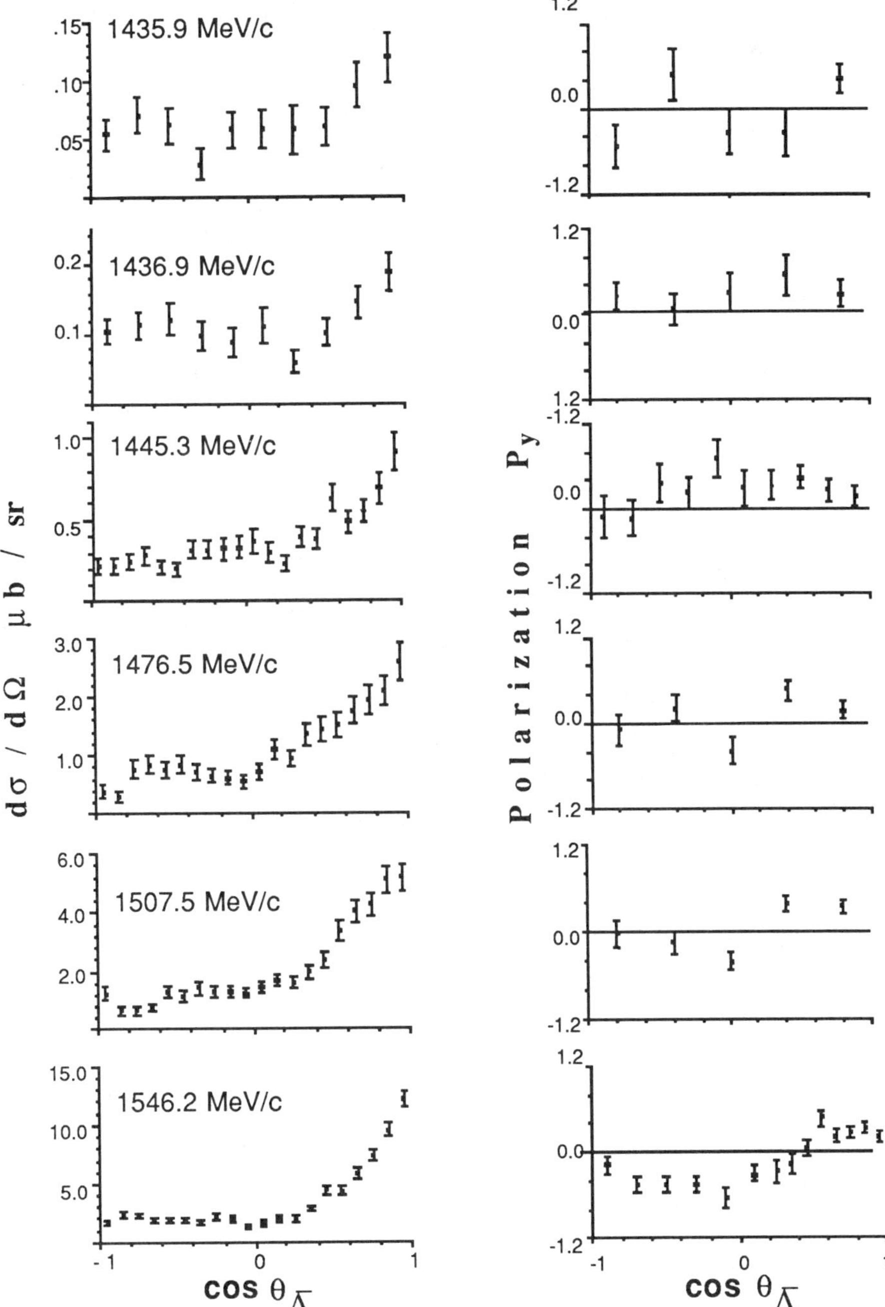

Fig. 2. Differential cross sections and polarization data for the reaction $\overline{p} + p \rightarrow \overline{\Lambda} + \Lambda$ at 6 incident $\overline{p}$ momentum from the PS185 collaboration.

Transitions to the mesonic channels can be included with an annihilation potential fit to the $\bar{N}N$ annihilation data.

A number of DWBA calculations have been performed.[4,5,6] Tabakin and Eisenstein's review of their DWBA calculation[4] provides a detailed look at the inputs to their model. A one-boson exchanges are combined with an eikonal-type treatment of the initial and final state distortions. The initial state distortions are fixed by $\bar{p}p$ annihilation data and their final state $\bar{\Lambda}\Lambda$ parameterization is adjusted to give the measured $\bar{p} + p \rightarrow \bar{\Lambda} + \Lambda$ total cross section. They find a total reduction from their plane wave result of a factor of 10^4 is required to fit the data with a factor of 10^3 coming from the final state. Both K^* and K^{**} exchanges are required and their parameters are adjusted to give the fit to the forward peaking of the $\bar{\Lambda}$. The model fails to give the correct polarization as seen in Fig 3. While it is possible that the model breaks down due to its oversimplification of initial and final state interactions, this has not been verified.

Figure 3 also shows the results of the calculations of Kohno and Weise.[6] They have performed DWBA calculations using one-kaon exchange and one-gluon exchange models for the transition operator. A modified Dover-Richard $\bar{N}N$ potential is used for the initial state distortion and the $\bar{\Lambda}\Lambda$ potential has the same form but its absorption strength parameter is modified to give the correct $\bar{p} + p \rightarrow \bar{\Lambda} + \Lambda$ total cross section. They also find large distortion effects; at 1507 MeV/c incident momentum, the s-wave to p-wave ratio is reduced from 5.9 to 0.1 by the distortions. The fit to the data is obtained without the need of K^* and K^{**} exchanges. Their comparisons of K-exchange with gluon exchange gives only small differences at these energies but the K-exchange is preferred for the higher energy data.

The importance of the distortions in the DWBA calculations has led many to believe a proper treatment requires a coupled channels approach.[7,8,9] This allows the coupling between $\bar{p}p$, $\bar{\Lambda}\Lambda$, and other baryon-antibaryon states to be treated explicitly. A number of detailed coupled channels calculations have been performed and currently being applied to the recent $\bar{\Lambda}\Lambda$ data. The recent results of Timmermans, Rijken, and JJ De Swart will be presented in detail later in this session.

SUMMARY

Although the most recent experimental and theoretical results are just now appearing, some general observations can be made. Most models are forced to include parameters which must be fit to the data, thus the "best fit" does not always imply the best or most complete theory. A comparison of the models must include a careful examination of the extracted parameters used by the models. However, the quality of the data has allowed comparisons of even heavily parameterized calculations. In particular, while the differential cross sections can often be fitted, the polarization data has provided more stringent tests.

The analysis of $\bar{\Lambda}\Lambda$ spin correlation data and the production of $\bar{\Lambda}\Lambda$ near the $\bar{\Sigma}\Sigma$ threshold will soon be available. Future LEAR studies of $\Sigma^o\bar{\Lambda}$ and $\Sigma^+\Sigma^-$ production are planned. These data should provide additional insight into short range effects and the interplay between conventional meson theory and quark-gluon models.

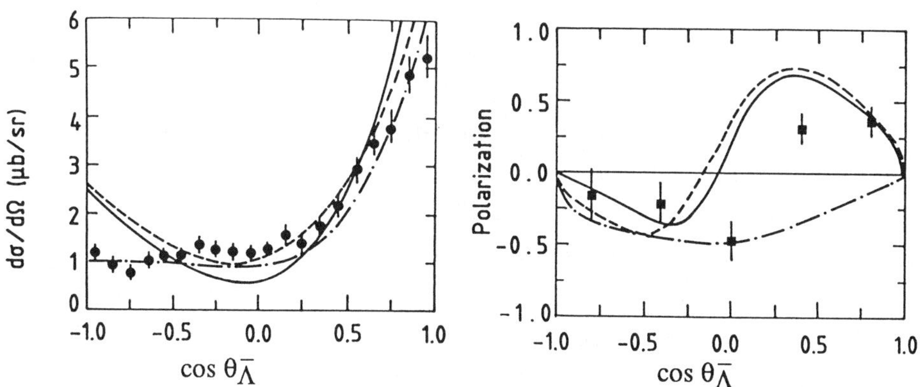

Fig. 3. Differential and total cross section data of PS185 at 1507.5 MeV/c compared to the 1-kaon exch. (solid line) and 1-gluon exch. (dashed line) calculations of ref. 6 and to the K,K*, and K** exch. (dot-dashed line) calculation of ref. 4.

REFERENCES

1. P.D. Barnes et al., Phys. Letts. 189B, 249 (1987).
2. H. Genz and S. Tatur, Phys. Rev D 30, 63 (1984).
3. J. Niskanen and H. Snellman, Phys. Lett. 165B, 187 (1985).
4. F. Tabakin and R. Eisenstein, Phys. Rev. C 31, 1857 (1985).
5. M. Dillig and R. Frankenberg, Antinucleon- and Nucleon-Nucleus Interactions (Plenum Press, N.Y. 1985), p. 399.
6. M. Kohno and W. Weise, Nucl. Phys. A 454 (1986) 429; Phys. Lett. 152 B (1985) 303.
7. J. Niskanen, Helsinki preprint (1985).
8. S. Furui, A. Faessler, and S.B. Khadkikar, Nucl Phys. A 424 (1984) 495; S, Furui, Proceedings of 1986 INS Intern. Symp. on Hypernuclear Physics (Toseisha Printing, Tokyo 1986), p. 371.
9. R.G.E. Timmermans, T.A. Rijken, and J.J. de Swart, Nijmegen preprint (1987); J.J. de Swart, these proceedings.

ANTINEUTRON PHYSICS AT LEAR

Tullio Bressani
Dipartimento di Fisica Sperimentale dell'Università, Torino, Italy
I.N.F.N., Sezione di Torino, Italy

ABSTRACT

The experimental results up to now reported by the antineutron experiment at LEAR are summarized. Plans for future approved experiments are furthermore discussed.

INTRODUCTION

The aims of the antineutron ($\bar{n}$) experiment PS 178 (Cagliari-Padova-Torino Collaboration) at LEAR were that of exploring the experimental aspects of the production and detection of $\bar{n}$'s, and of performing a first series of measurements of $\bar{n}$ cross sections. Two new experimental problems were studied for the first time: the realization of a detector able to localize and measure the energy of $\bar{n}$'s, and the setting-up of a beam of tagged $\bar{n}$. Both solutions we found were successfull, as indicated by the fact that, with minor modifications, they were afterwards adopted by two approved experiments at LEAR now in the setting-up stage (PS 199 and PS 201) and one at the resumed ADONE accumulator ring (FENICE: $e^+e^- \rightarrow \bar{n}n$) at Frascati.

THE ANTIN EXPERIMENT (PS 178)

The first goal of PS 178 was that of designing and operating a detector having a good precision on the measurement of the energy and impact point of $\bar{n}$'s. It consist[1] of ten equal modules:

each one is made of an Iron slab (the converter), 100x100x1 (or 0.5) cm^3, a scintillator wall, 100x100x1 cm^3, and a plane of x-y limited streamer tubes, 96x100 cm^2 with 1 cm pitch. The $\bar{n}$ energy was measured by time-of-flight, with a time resolution of 700 ps fwhm, and the spatial precision on the reconstruction of the annihilation vertex was about ±5 cm fwhm. The $\bar{n}$ detection efficiency, function of the trigger conditions, was 40% in the most favorable case.

With such a detector we studied first of all[2] the inclusive spectra from $(\bar{p},\bar{n})$ reactions at 550 MeV/c on Nuclei (C,Al,Cu and Pb). The data cannot be reproduced by simple models, and need further effort in theoretical investigation. Dover[3] first pointed out possible ways of explaining the results.

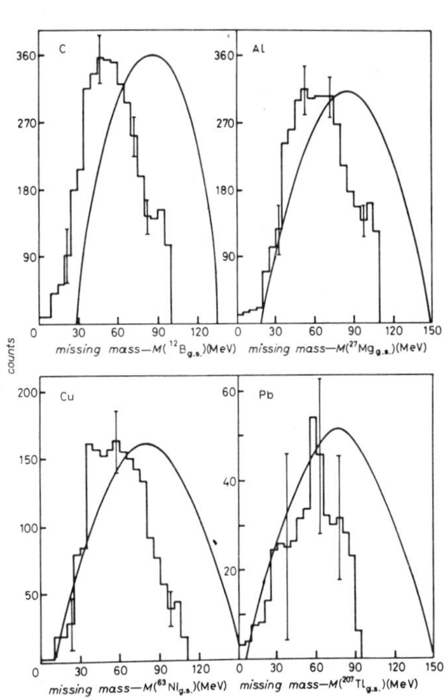

Fig.1. Experimental inclusive $\bar{n}$ spectra for the $\bar{p}+A \rightarrow \bar{n}+X$ reaction, from Ref.2

A second measurement was given by profiting of the particular structure of the $\bar{n}$ detector, that may be considered as a layered absorption set-up. From the study of the attenuation of the $\bar{n}$ beams in the detector[4], we could deduce the $\bar{n}$-Fe annihilation cross section from 100 to 531 MeV/c, shown by Fig.2. The ratio of the ($\bar{n}$ Fe) to the ($\bar{n}\mathcal{N}$)annihilation cross section is consistent with a simple $A^{2/3}$ scaling law, showing that, as for $\bar{p}$, the $\bar{n}$ annihilation is essentially a surface process.

The second goal of PS 178 was that of studying the performance of a tagged $\bar{n}$ beam. The tagging was obtained by detecting in coincidence the neutron associated to the $\bar{n}$ in the

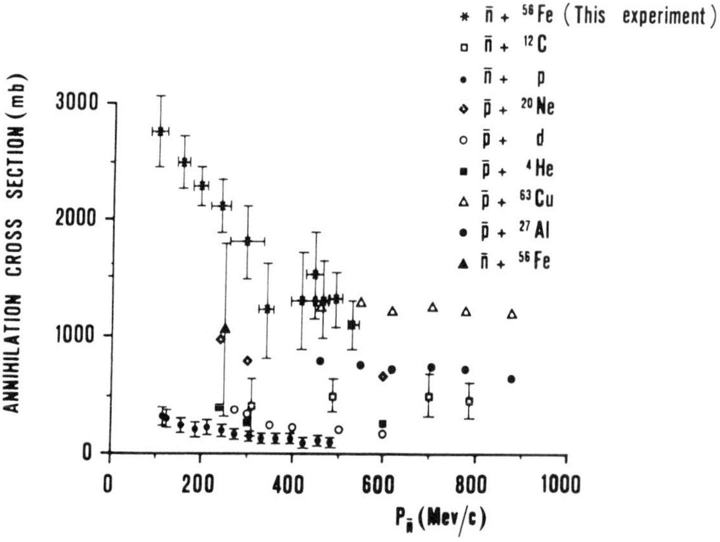

Fig.2 Annihilation cross sections for $\bar{n}$ on Fe, compared with other $\bar{n}$ and $\bar{p}$ measurements. From Ref.4

two-body charge exchange (CEX) reaction $\bar{p}p \rightarrow \bar{n}n$. With a 15 cm long LH_2 target and incident $\bar{p}$'s of 300 MeV/c we obtained[5] a beam of $\bar{n}$ of known momentum, in the range 120-270 MeV/c, as shown by Fig.3. The total rate of tagged $\bar{n}$'s was $\sim 10^{-4}$ $\bar{n}/\bar{p}$, i.e. an order of magnitude lower than a typical non-tagged beam. Indeed the cleanliness of the tagged beam, coupled to the improved $\bar{p}$ beam intensity available now at LEAR, makes it preferable for future experiments.

The analysis of the data from other measurements, such that of the total cross section $\bar{n}$-p and of the angular distribution for CEX at low (< 300 MeV/c) momenta is in progress. We hope that the last measurement could shed light on the present disagreement between the recent data from KEK [6] and LEAR [7].

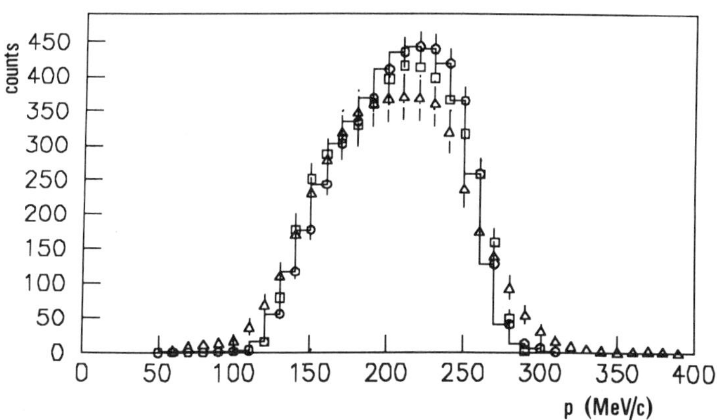

Fig.3. Momentum distribution of the tagged $\bar{n}$'s, from Ref.5

THE APPROVED FUTURE EXPERIMENTS

The experimental layout (Fig.4) of experiment PS 199 looks closely as an enlarged and upgraded version of PS 178. It consists of three neutron hodoscopes and two large $\bar{n}$ detectors (having each an useful area of 4 m^2), again structured as a sequence of Iron converters, scintillators and limited streamer tube planes, with a configuration that looks better for the $\bar{n}$ annihilation vertex reconstruction. The main purpose[8] of the experiment is that of studying spin effects in CEX, with a polarized proton

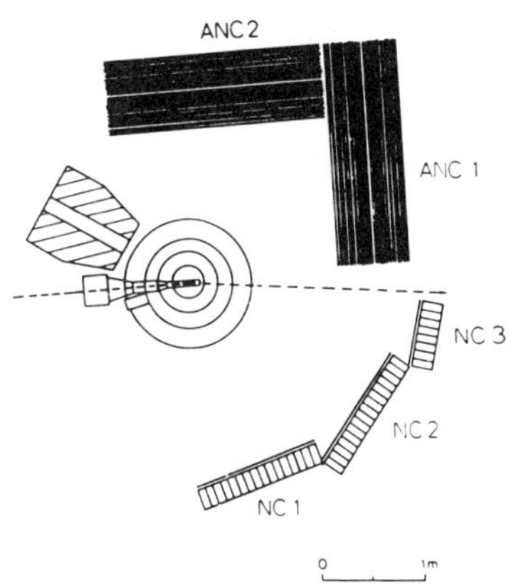

Fig.4. A layout of experiment PS 199, from Ref.8. ANC are $\bar{n}$ detectors,NC neutron hodoscopes

target. The differential cross sections, the polarization parameter P and the polarization transfer parameter D will be measured from 500 to 1500 MeV/c.

Experiment PS 201 (OBELIX) is designed to study[9] exclusive final states of $\bar{p}$ and $\bar{n}$ annihilations at low energies with protons and nuclei. It is based on a multi-purpose apparatus consisting in the magnet and the jet drift chamber of the Open Axial Field Magnet (OAFM) Spectrometer, already used at the CERN ISR, with a Spiral Projection Chamber as vertex detector, two concentric arrays of plastic scintillators providing the first level trigger and four gas sampling electromagnetic calorimeters for detecting the gamma showers. The solid angle covered by the detectors in ~ 2.5 π sr, with the targed located in the middle. Fig.5 shows the scheme[10] of the set-up in the forward inner part of OBELIX for $\bar{n}$ physics. The LH_2 production target of n's is located inside the upstream coil of the magnet. $\bar{n}$'s produced at 0° by CEX will be used, since the mechanical arrangement of the facility prevents the use of a tagged $\bar{n}$ beam. The reaction target (LH_2 or other nuclei) is placed in the center of the magnet, and the surrounding detectors will allow the simultaneous detection of all the final products. With the described set-up and the

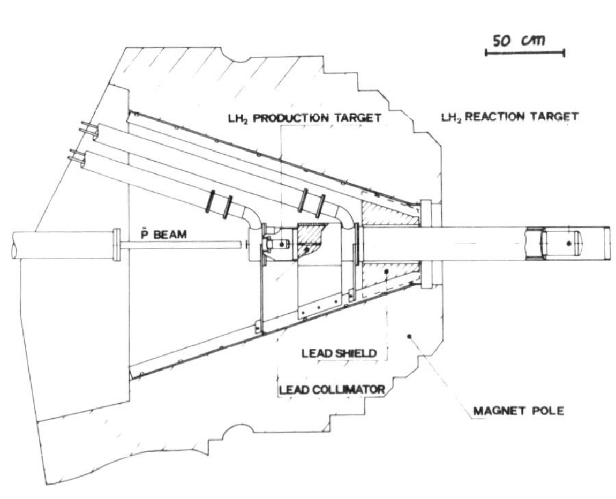

Fig.5. Targets for $\bar{n}$ physics in experiment PS 201, from Ref.10

improved $\bar{p}$ beam intensity we expect to have about 30 $\bar{n}$ interactions per second . The annihilation of $\bar{n}$ on protons will be then studied with a precision and detail comparable to these of the existing $\bar{p}$ experiments. It is planned to use also LD_2 and other light nuclear targets, with the aim of exploring signatures of possible quark degrees of freedom, peculiar to the annihilation process[11].

In conclusions, both experiments will allow in the forthcoming years a precise study of $\bar{n}$ interactions, not investigated up to now and that could deserve some suprise.

REFERENCES

1. T. Bressani et al., Physics at LEAR with Low Energy Antiprotons (Harwood, Chur, 1988), quoted in the following as Villars 1987, p.267.
2. T. Bressani et al., Europhys. Lett.2(2), 587 (1986).
3. C. Dover, Villars 1987, p.649.
4. M. Agnello et al., Villars 1987, p.673.
5. L. Cugusi et al., to be published in Nucl. Instr. Meth. in Phys. Res. A.
6. K. Nakamura et al., Phys. Rev. Lett.53, 885 (1984).
7. W. Brückner et al., Phys. Lett.169B, 302 (1986).
8. M. P. Macciotta et al., Villars 1987, p.283.
9. R. Armenteros et al., Villars 1987, p.507.
10. M. Agnello, private communication.
11. T. Bressani and F. Cannata, in Hadronic Physics at Intermediate Energy II (North Holland, Amsterdam, 1987), 53.

A STUDY OF THE REACTION $\overline{p}p \rightarrow \overline{\Lambda}\Lambda$

J.J. DE SWART, R. TIMMERMANS AND T.A. RIJKEN

University of Nijmegen, Nijmegen, the Netherlands

The experiment PS185 at the LEAR facility of CERN has provided many accurate data on the reaction $\overline{p}p \rightarrow \overline{\Lambda}\Lambda$ close to threshold [1,2,3]. These are the first results of the ongoing investigation of the antihyperon-hyperon channels in $\overline{p}p$ scattering. The weak decay of the hyperons into a nucleon and a pion makes it possible to reconstruct the whole kinematics of these two-body reactions. Also, because of the large asymmetry of the decays with respect to the polarization axis of the hyperon, all spin-observables can be measured without the need for secondary scattering.

To study these reactions we have constructed a coupled channels boundary condition model. All baryonic channels can be taken into account. Consider the relativistic Schrödinger equation:

$$\left(\frac{d^2}{dr^2} - \frac{L^2}{r^2} - U + k^2\right)\Phi(r) = 0 \tag{1}$$

Since some of the $\overline{Y}Y$ channels have only I=0 ($\overline{\Lambda}\Lambda$) or I=1 ($\overline{\Lambda}\Sigma$ and $\overline{\Sigma}\Lambda$), it is advantageous to work on the isospin basis. $U = \sqrt{2m}V\sqrt{2m}$, where V is the OBE antibaryon-baryon potential. This coupled channels potential is constructed via $SU(3)$ from the Nijmegen soft-core OBE potential [4]. The strange mesons included are the $K(495)$, the $K^*(892)$, the $K^{**}(1430)$ and the $\kappa(1000)$. For each partial wave equation (1) is solved numerically with a boundary condition at $r = b = 1$ fm, specified by the P-matrix, which is defined by:

$$P = \left(\frac{d\Phi}{dr}\Phi^{-1}\right)_{r=b} \tag{2}$$

The P-matrix contains the parameters of the model. With the help of this boundary condition the, largely unknown, annihilation effects in initial and final state can be treated properly. We simply assume a complex step potential in the inner region. Thus the diagonal P-matrix elements are of the form:

$$P = \frac{J'(k'b)}{J(k'b)} \tag{3}$$

Here $k'^2 = k^2 - U$, where U is the complex step potential. So far the model can be tested only in the reaction $\overline{p}p \rightarrow \overline{\Lambda}\Lambda$, where we have the three channels { $\overline{N}N$(I=1), $\overline{N}N$(I=0), $\overline{\Lambda}\Lambda$}. For the off-diagonal element of the P-matrix between $\overline{N}N$(I=0) and $\overline{\Lambda}\Lambda$ we take, purely phenomenologically, the first few terms of a power series in $\varepsilon = \sqrt{s} - 2m_\Lambda$, where $\sqrt{s}$ is the total centre of mass energy. The initial state parameters are fixed by fitting the available data on $\overline{p}p \rightarrow \overline{p}p$. The other parameters are then fitted to the data on $\overline{p}p \rightarrow \overline{\Lambda}\Lambda$. To fit all the energies we need 6 parameters, two for the final state. We then achieve $\chi^2 = 1.4$/data, including all the 6 energies available. An example of our fit is shown in figure 1. By adjusting the absorption in

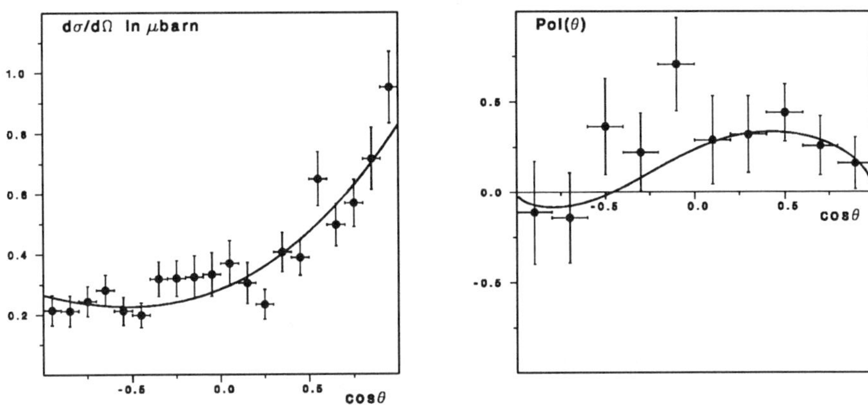

Figure 1: Differential cross-section in μbarn and polarization for $\overline{p}p \rightarrow \overline{\Lambda}\Lambda$ close to threshold at $p_{lab} = 1445.35\ MeV/c$ or $\varepsilon = 3.62\ MeV$.

the final state the total cross-section $\overline{p}p \rightarrow \overline{\Lambda}\Lambda$ can be scaled. If we set the imaginary part of the final state P-matrix element equal to zero this cross-section rises by a factor of about 70, most of which is due to the S-waves. This shows clearly that the S-waves are suppressed.

$\overline{p}p \rightarrow \overline{\Lambda}\Lambda$ partial wave	$p_{lab} =$ 1436.95 MeV/c	$p_{lab} =$ 1445.35 MeV/c
$^1S_0$	0.13	0.54
$^3S_1$	0.11	0.42
$^3D_1 \rightarrow {}^3S_1$	0.94	2.38
$^1P_1$	0.02	0.06
$^3P_0$	0.01	0.03
$^3P_1$	0.04	0.24
$^3P_2$	0.09	0.39
$^3F_2 \rightarrow {}^3P_2$	0.02	0.57
Rest	0.01	0.04
singlet $s = 0$	0.16	0.60
triplet $s = 1$	1.21	4.07
total	1.37 μbarn	4.67 μbarn

Even at the lowest energies measured at LEAR the cross-section is highly anisotropic and there is a sizable polarization. The typical form of the $\overline{p}p \rightarrow \overline{\Lambda}\Lambda$ cross-section is produced in our model by strong off-diagonal tensor transitions $\ell_{\overline{\Lambda}\Lambda} = \ell_{\overline{p}p} - 2$. They are the result of the coherent tensor forces of the tail of the K and K^* potentials. In the table we give the partial wave cross-sections at two energies close to the threshold at $p_{lab} = 1435.07$ MeV/c. The singlet fraction S is quite small especially because the dominant S-wave at threshold is $^3D_1 \rightarrow {}^3S_1$. Note the presence of P-waves even at these values of ε. The momenta given in the table correspond to $\varepsilon = 0.66$ MeV and $\varepsilon = 3.62$ MeV respectively. Most notably the $^3F_2 \rightarrow {}^3P_2$ contribution to the cross-section rises very fast. At higher energies this partial wave is the dominant one. This importance of the tensor force transitions $\ell_{\overline{\Lambda}\Lambda} = \ell_{\overline{p}p} - 2$ can be crudely understood as follows [6]. The transition probability is in the first Born approximation maximal when the wave function overlap is maximal. In the case of the free spherical Bessel functions $prj_\ell(pr)$ this maximum lies at $r \sim \sqrt{\ell(\ell+1)}/p$. In our case $p_{\overline{\Lambda}\Lambda}$ is smaller than $p_{\overline{p}p}$, so the overlap is maximal when $\ell_{\overline{\Lambda}\Lambda}$ is smaller than $\ell_{\overline{p}p}$.

Part of this work was included in the research program of the Stichting voor Fundamenteel Onderzoek der Materie (F.O.M.) with financial support from the Nederlandse Organisatie voor Zuiver- Wetenschappelijk Onderzoek (Z.W.O.).

References

[1] P.D. Barnes et al., Phys. Letts. 189B (1987) 249.

[2] P.D. Barnes et al. Phys. Letts. 199B (1987).

[3] R. von Frankenberg, Thesis University of Erlangen Nürnberg (1987).

[4] M.M. Nagels, T.A. Rijken, and J.J. de Swart, Phys. Rev. D 17 (1978) 768.

[5] E. Eisenhandler et al., Nucl. Phys. B113 (1976) 1.

[6] P.H. Timmers, Thesis University of Nijmegen, unpublished (1984).

NUCLEON–ANTINUCLEON ANNIHILATION DYNAMICS*

Carl B. Dover
Brookhaven National Laboratory, Upton, NY 11973

ABSTRACT

Recent progress towards a microscopic understanding of nucleon–antinucleon $N\overline{N}$ annihilation is reviewed. We consider statistical models, the constraints imposed by SU(2) or SU(3) flavor symmetry and the spin–flavor–color dependence of effective operators for the creation/destruction of quark–antiquark ($Q\overline{Q}$) pairs in $N\overline{N}$ annihilation processes. The importance of dynamical selection rules is emphasized as a means of revealing the underlying reaction mechanism.

INTRODUCTION

Nucleon–antinucleon $N\overline{N}$ annihilation provides a testing ground for models of non–perturbative quantum chromodynamics (QCD). An *ab initio* QCD calculation of $N\overline{N}$ annihilation is beyond our reach at present, so a more phenomenological approach has been adopted, in which one tries various hypotheses for the dependence of the effective $Q\overline{Q}$ creation/destruction operator $\mathcal{O}$ on spin, flavor and color. Even at threshold, the $N\overline{N}$ system offers a rich ensemble of annihilation channels, whose relative branching ratios B offer significant constraints on dynamical models. Recently, the technique of γ–ray tagging at the LEAR facility at CERN has enabled one to prepare the $N\overline{N}$ atomic system in a state of definite orbital angular momentum L. The other quantum numbers such as spin S, isospin I, and charge parity C can also be filtered by comparing particular mesonic final states. The observed ratios B display a non–trivial dependence on L, suggesting dynamical content beyond simple statistical or flavor symmetry models. The key signatures of the underlying annihilation mechanism appear in the guise of **dynamical selection rules** (DSR), i.e., the suppression of transitions $N\overline{N} \rightarrow M_1 M_2$ which are allowed by the conservation of $\{J, \pi, C, G\}$. The emergence of DSR depends on the interplay between the color–spin–flavor structure of the effective $Q\overline{Q}$ operators $\mathcal{O}$ and the topology of the annihilation process. We focus on these features in the present note.

STATISTICAL MODELS

Given the sizable pion multiplicity $< N_\pi > \approx 5$ in $N\overline{N}$ annihilation at threshold, it is natural to try to understand the data in terms of statistical and phase space factors, without any dynamical input. A recent analysis along these lines by Vandermeulen[1] was successful in understanding most features of the data, particularly at higher momenta where numerous partial waves contribute.

* This work has been supported by the U. S. Department of Energy under contract no. DE-AC02-76CH00016.

In this approach, the branching ratios B (or cross sections) for the processes $p\bar{p} \to a + b$ are written as

$$B\left(p\bar{p} \to a+b\right) = C_{s\bar{s}}\left(2J_a+1\right)\left(2J_b+1\right)W_{ab}2^{-\delta_{ab}}\; q\; e^{-A\left(s-s_{ab}\right)^{1/2}} \tag{1}$$

where $J_{a,b}$ are the meson spins, W_{ab} is an isospin Clebsch–Gordon coefficient, $2^{-\delta_{ab}}$ is a Bose–Einstein factor, q is the c.m. meson momentum, $\sqrt{s}$ is the total c.m. energy, and $s_{ab} = \left(m_a + m_b\right)^2$. It is assumed that quasi–two–body (QTB) channels $a + b$ act as "doorway states." The subsequent decay of $\{a, b\}$ into pions builds up the observed value of $< N_\pi >$. The exponential factor in Eq. (1) ensures that each channel $a+b$ peaks close to its threshold $\sqrt{s_{ab}}$; this is related to the constraints of multi–channel unitarity. The constant $A = 1.2$ GeV^{-1} is chosen to reproduce the energy dependence of the $p\bar{p} \to \pi^+\pi^-$ cross section. The constant $C_{s\bar{s}}$ is taken[1] to be 0.15 for modes containing $K\overline{K}$ pairs and unity for non–strange modes. The **strangeness suppression factor** $C_{s\bar{s}}$ expresses the fact that $K\overline{K}+n\pi$ modes represent only about 5% of the low energy annihilation cross section, whereas in a purely statistical picture they should constitute about 30%. This $s\bar{s}$ suppression is also observed[2] in other hadronic reactions. In the context of strong coupling QCD, $C_{s\bar{s}}$ can be understood[3] in terms of the quantum tunneling of massive $s\bar{s}$ quark pairs, in analogy to the Schwinger formula for e^+e^- pair creation in an electric field. The need for a factor $C_{s\bar{s}}$ in meson decay has been detailed by Dosch and Gromes[4].

The exponential factor in Eq. (1) may be viewed as an effective penetrability $P \approx \exp\left(-\widetilde{R}q\right)$, where $\widetilde{R} = 2A \approx 0.5$ fm. In quark bag models for $N\overline{N} \to a+b$, Gaussian wave function overlaps[5] yield

$$P \approx x^{2\ell_f}\; e^{-\beta x^2}\;, \tag{2}$$

where $x = qR$ and ℓ_f = meson orbital angular momentum. Although the functional form of (2) differs from that used in Eq. (1), the numerical values for typical $\ell_f = 0$ channels are rather similar if the bag radius $R \approx 0.5$ fm. Note that $x > 1$ for all channels of interest, so the naive approximation $P \approx x^{2\ell_f}$, as often used in the past, is inadequate.

The statistical approach[1] describes most global features of $N\overline{N}$ annihilation, even at threshold, where one might expect the selectivity associated with particular $L \to \ell_f$ transitions to be important. Nevertheless, there are interesting discrepancies with the data, which allow some dynamical aspects to shine through the dull background of statistics and phase space. For instance, Eq. (1) contains no dependence on the quantum numbers of the $N\overline{N}$ initial state, so it does not allow for the possibility of dynamical selection rules which depend on the intrinsic spin S and isospin I. For example, experiment[6] yields $B\left(\pi^\pm\rho^\mp\right) \approx 2B\left(\pi^0\rho^0\right)$ for both $L = 0, 1$, implying that $\pi\rho$ production is dominated by the ${}^{2I+1,2S+1}L_J = {}^{13}\mathrm{S}_1$ and ${}^{11}\mathrm{P}_1$ channels (this is the "$\pi\rho$ puzzle"). Equation (1) predicts $B\left(\pi^\pm\rho^\mp\right) = 5B\left(\pi^0\rho^0\right)$, which disagrees with experiment.

For the $\rho\rho$ channel, Eq. (1) imples $B(\rho^+\rho^-) = B(\rho^0\rho^0)$, while dynamical models, for instance Ref. 5, lead to $B(\rho^+\rho^-) \gg B(\rho^0\rho^0)$, because of the dominance of the ${}^{33}S_1 \to \rho^+\rho^-$ transition over ${}^{11}S_0 \to \rho^0\rho^0$. A similar situation occurs for the πA_2 channel: data[9] imply $B({}^{11}S_0 \to \pi A_2) \approx (3.6 \pm 0.9)\ B({}^{33}S_1 \to \pi A_2)$, in disagreement with the statistical model[1].

Although the global fit requires $C_{s\bar{s}} \ll 1$, some ratios do not display the expected $(2J+1)$ factors. For instance, for $N\overline{N}$ at rest, we have

$$\frac{B(K^{\pm}K^{*\mp})}{B(K^+K^-)} \approx \begin{cases} 3.6 & \text{(Eq. (1))} \\ 0.8 - 1.4 & \text{(Exper, Ref. 7)} \end{cases} \tag{3}$$

This discrepancy may reflect the delicate isospin interference effects which occur in strange particle channels (no G–parity restriction). These are sensitive to the state dependent admixtures[8] of $p\bar{p}$ and $n\bar{n}$ (and possible $\overline{\Delta}N \pm \Delta\overline{N}$, $\overline{\Delta}\Delta$) which occur at short distances in the atomic state.

In sum, although the statistical–phase space approach is consistent with many features of $N\overline{N}$ annihilation, it fails to account for the observation of **dynamical selection rules**, as seen for example in $N\overline{N} \to \pi\rho$, πA_2. These are then the key to extracting non–trivial information from the data on $N\overline{N} \to$ mesons.

SU(2)/SU(3) FLAVOR SYMMETRY

The implications of flavor symmetry for $N\overline{N}$ annihilation have been examined by Genz[10] and Klempt *et al*[11]. These authors consider the quark line diagrams $A2$ and $R2$ of Fig. 1 corresponding to transitions to two meson final states. By considering non–strange modes only (SU(2)), Genz[10] concludes that the planar topology $A2$ dominates. Klempt *et al*[11] include $K\overline{K}$, $\overline{K}K^*$ and $\overline{K}^*K^*$ modes in the analysis (SU(3)) and conclude that the rearrangement graph $R2$ is dominant, in apparent contradiction to Genz[10]. The argument of Genz is based largely on triangle inequalities for $R2$, such as

$$|B^{\frac{1}{2}}(\eta\eta') - \frac{1}{2}B^{\frac{1}{2}}(\pi^0\pi^0)|^2 \leq 2B(\pi^0\eta') \leq |B^{\frac{1}{2}}(\eta\eta') + \frac{1}{2}B^{\frac{1}{2}}(\pi^0\pi^0)|^2 \tag{4a}$$

$$|B^{\frac{1}{2}}(\rho^0\rho^0) - 2B^{\frac{1}{2}}(\rho^0\omega)|^2 \leq B(\omega\omega) \leq |B^{\frac{1}{2}}(\rho^0\rho^0) + 2B^{\frac{1}{2}}(\rho^0\omega)|^2 \tag{4b}$$

Using recent values $B(\pi^0\eta') \approx 0.5 \times 10^{-3}$ (Ref. 12) and $B(\pi^0\pi^0) \approx 0.2 - 0.25 \times 10^{-3}$ (Refs. 12, 13), Eq. (4a) could only be satisfied if $B(\eta\eta') \approx 10^{-3}$, but Tauscher[14] gives a limit $B(\eta\eta') < 10^{-3}$. Similarly, since $B(\rho^0\rho^0)$ is very small (a value $(0.12 \pm 0.12) \times 10^{-2}$ is given by Diaz *et al*[9]), Eq. (4b) implies $B(\omega\omega) \approx 4B(\rho^0\omega)$. However, the data[15] indicate $B(\omega\omega) \leq B(\rho^0\omega)$. Thus, the constraints (4a,b) from graph $R2$ appear to be violated. The situation is not so clear, however, since the $A2$ topology implies $B(\omega\omega) = B(\rho^0\rho^0)$, which is also not well satisfied by the data, since[16] $B(\omega\omega) \approx (1.4 \pm 0.6) \times 10^{-2}$.

If one assumes that SU(3) symmetry is unbroken, we obtain for the planar $A2$ graph $B({}^{13}S_1 \to \overline{K}K^* + \text{c.c.}) = B({}^{13}S_1 \to \pi\rho)$. The observed $\overline{K}K^*/\pi\rho$

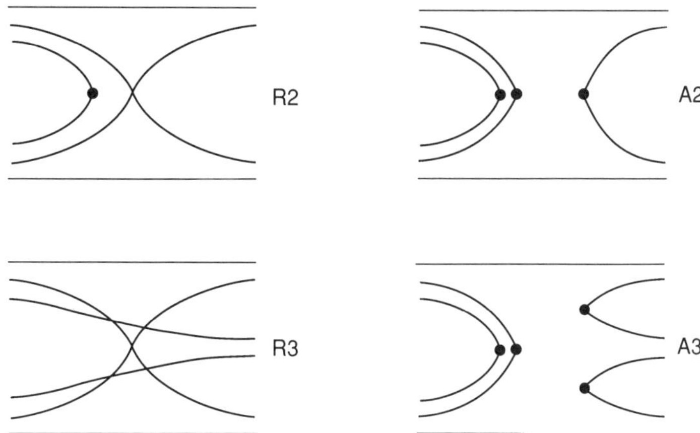

Fig. 1. Quark line diagrams for $N\overline{N}$ annihilation into two ($R2$, $A2$) or three ($R3$, $A3$) mesons. The dot stands for an effective operator for $Q\overline{Q}$ pair creation/destruction.

ratio is small. Since graph $R2$ produces $\pi\rho$ but not $\overline{K}K^*$, Klempt *et al*[11] conclude that $R2$ dominates. A problem with this argument is that it ignores a known mechanism of SU(3) breaking, namely the $s\bar{s}$ suppression factor $C_{s\bar{s}}$ of Eq. (1). In addition, isospin interference is crucial for strange meson modes, i.e., the atomic system is not pure $p\bar{p}$ at short distances where annihilation occurs, as assumed in Ref. 11. The $\bar{n}n$ admixture is important in $S = 1$ channels, where the tensor potential plays an important role[17] in providing a coherent $^{13}\text{S}_1 - {}^{13}\text{D}_1$ mixture[17] at short distances which diagonalizes the tensor operator S_{12}. Thus, the $\overline{K}K^*$ mode, for instance, is suppressed by $C_{s\bar{s}} \approx 1/6$ and also possibly by interference effects. And so it seems inappropriate to attribute the suppression of strange modes to the dominance of $R2$ over $A2$ processes.

For SU(2), certain relations hold for any relative weight of $R2$ and $A2$. These are

$$B\left({}^{13}\text{S}_1 \to \eta\omega\right) = B\left({}^{11}\text{S}_0 \to \omega\omega\right) = B\left({}^{11}\text{S}_0 \to \eta f\right) \tag{5a}$$

$$B\left({}^{13}\text{S}_1 \to \pi\rho\right) = B\left({}^{11}\text{S}_0 \to \pi A_2\right) \tag{5b}$$

The values $B(\eta\omega) \approx 1.3 \times 10^{-2}$ given by Tauscher[14] and $B(\omega\omega) \approx 1.4 \times 10^{-2}$ of Bloch *et al*[16] are consistent with Eq. (5a), while Eq. (5b) is also approximately satisfied. If $R2$ is absent, the above five B values become equal, and proportional to $|A_1^+ + A_2^+|^2$, using the SU(3) invariant amplitudes of Klempt[11]. If the experimental B values in Eq. (5) are corrected for a phase space factor, assumed[11] to be $q^{2\ell_f+1}\exp\left(-R^2q^2/2\right)$, they are consistent with this prediction. Thus, if one restricts attention to non–strange modes (SU(2)) then the topology $A2$ is more or less consistent with the data.

DYNAMICAL MODELS

Thus far we have not assigned any operator structure to the dots in Fig. 1, and relied on general statistical or symmetry considerations. Although one may perform a fit to the data in order to obtain SU(3) amplitudes, this procedure gives no insight into the origin of dynamical selection rules, or the strong L dependence observed for certain modes (examples are πf, which is more prominent for $L = 1$ than for $L = 0$, and $K^0\overline{K}^0$, which is strongly suppressed for $L = 1$). If the effective operator $\mathcal{O}$ corresponding to $Q\overline{Q}$ pair formation depends on spin and orbital angular momentum, the $N\overline{N} \to a + b$ amplitude will not factorize in general into orbital, spin, and flavor dependent terms (à la $0(3) \otimes SU(2)_\sigma \otimes SU(3)_f$).

Recently, there has been a great deal of phenomenological work, in which one combines various prescriptions for the dots in Fig. 1, and tries to determine which topologies among $\{R2, A2, R3, A3\}$ are dominant. Major efforts have been devoted to this problem by the Helsinki[18], Osaka[19] and Tübingen[20] groups, as well as others[5,21–24].

The simplest dynamical approximation consists of using $R2$ only, with the dot replaced by an effective operator $\mathcal{O}$ with the quantum numbers of a single gluon[22–24] $(^3S_1,\ I = 0,\ \text{color } \{8\})$. For meson decay, this approximation fails to account for either relative widths[25] or polarization properties[26] in the decays $B \to \omega\pi$ and $A_1 \to \rho\pi$. Also for $N\overline{N} \to a + b$, the one gluon approximation fails qualitatively: for example[22], in $p\bar{p}$ annihilation at very low momentum, we predict the ratios $\eta\rho^0/\eta\omega \approx 0.1$ and $\eta\rho^0/\pi^0\rho^0 \approx 0.0075$, while the experimental values at rest are much larger, i.e., 0.6 and 0.55, respectively. The transitions $N\overline{N}\,(L = 0) \to a + b\ (\ell_f = 2)$ vanish in one–gluon approximation, while $\pi A_2\ (\ell_f = 2)$ is observed to be one of the largest single two–body modes (3–5%).

One can try to repair the deficiences of the effective one–gluon approximation by adding graph $A2$ of Fig. 1 with two–gluon exchanges[19,23]. This can be made to give reasonable results for some ratios such as $K^+K^-/\pi^+\pi^-$, K_LK_S/K^+K^-, and $\pi^0\omega/\pi^+\pi^-$, but it fails for others such as $K^{*+}K^- + \text{c.c.}/K^+K^-$ (prediction 10 versus exper. $\sim 1 - 1.5$). The above–mentioned ratios $\eta\rho^0/\eta\omega$ and $\eta\rho^0/\pi^0\rho^0$ increase to about $1/4$ and $1/8$, respectively, but remain too small. In a global sense, the use of effective $^3S_1$ $Q\overline{Q}$ operators does not seem promising, even when graph $A2$ is included.

Meson and baryon resonance decay are well described[27,28] phenomenologically using an effective $Q\overline{Q}$ operator with the quantum numbers of the vacuum $(^3P_0,\ I = 0,\ \text{color } \{1\})$. This approximation has been given a basis in the context of QCD in the strong coupling limit[3,29], where $Q\overline{Q}$ pairs are created by breaking a flux tube (formed as a normal (collective) mode of the gluon field). Since the $^3P_0$ model works so well for meson decays, it is reasonable to apply the same technique to $N\overline{N}$ annihilation, although its success is by no means guaranteed. Indeed, the $N\overline{N}$ system is a "heavy meson", which shares the quantum numbers of the ordinary $Q\overline{Q}$ mesons, but differs in its internal quark composition and number of energetically allowed decay modes. Because of competition between different decay channels and topologies (Fig. 1), $N\overline{N}$ annihilation offers a richer dynamics than ordinary meson decay. As an example, $A_1 \to \rho\pi$ is the dominant decay, while $A_1 \to f\pi$ is a closed channel. In contrast, $N\overline{N}\left({}^{33}\mathrm{P}_1\right)$, which has the same quantum numbers as the A_1, can decay into both $\rho\pi$ and $f\pi$ (the latter dominates). This channel competition gives rise to dynamical selection rules, which are the most novel aspect of the $N\overline{N}$ annihilation process.

In Ref. 5, the ${}^3\mathrm{P}_0$ model was applied to two–body decays, and it was concluded that the planar graph $A2$ was more important than the rearrangement graph $R2$. Further support for this conclusion is provided by the recent ASTERIX results[30]

$$\frac{B\left(L=1\to\pi^+\pi^-\right)}{B\left(L=0\to\pi^+\pi^-\right)}\approx 1.44\pm 0.16 \tag{6a}$$

$$\frac{B\left(L=1\to K_sK_s\right)}{B\left(L=1\to K^+K^-\right)}\approx (1.46\pm 0.3)\times 10^{-1} \tag{6b}$$

For $A2$, the ${}^3\mathrm{P}_0$ model predicts[31]

$$\begin{aligned}\frac{B\left(L=1\to\pi^+\pi^-\left(\ell_f=0\right)\right)}{B\left(L=0\to\pi^+\pi^-\left(\ell_f=1\right)\right)}&\approx\left(\frac{45}{13}\right)^2\frac{1}{(qR)^2}\approx\frac{3}{2}\\ \frac{B\left(L=1\to K_sK_s\right)}{B\left(L=1\to K^+K^-\right)}&\approx\frac{1}{8}\end{aligned} \tag{7}$$

in excellent agreement with Eq. (6a,b). For $R2$, on the other hand, the transition $L=0\to\pi^+\pi^-\left(\ell_f=1\right)$ is forbidden and $K\overline{K}$ modes also vanish.

The Tübingen group[20] has done the most extensive analysis of data on $N\overline{N}\to M_1M_2$, $M_1M_2M_3$, comparing the topologies $R2$, $A2$, $R3$, $A3$ of Fig. 1. They conclude that the use of ${}^3\mathrm{P}_0$ $Q\overline{Q}$ operators together with the graphs $A2+A3$ offer the most successful phenomenological description. The rearrangement model $R3$ yields too small a summed cross section and also leads to the selection rule $N\overline{N}\,(L=0)\not\to s\,sp$ for which there seems to be little evidence. In these calculations, initial state distortions are included and are important.

OUTLOOK

On the experimental side, one needs to obtain spin–separated rates for $L=0$, 1 $N\overline{N}$ states and measurement of modes with more than one neutral meson, in order to search for further dynamical selection rules (ex. ${}^{33}\mathrm{S}_1\to\rho^+\rho^-\to\pi^+\pi^-\pi^0\pi^0$). It would also be valuable to measure the polarization of mesons produced in $N\overline{N}$ annihilation; as for meson decays $(B\to\pi\omega)$, this is a delicate test of the reaction mechanism through the interference of amplitudes with different ℓ_f.

On the theoretical side, one needs a "guiding principle" in order to sort out the many somewhat confusing and *ad hoc* phenomenological prescriptions. The "flux tube" picture is successful for meson decays, but needs to be developed into a calculational scheme for $N\overline{N}$ annihilation. One should try to understand the relation of the $^3P_0$ vertex used in models of meson and $N\overline{N}$ decay to the $Q\overline{Q}$ and/or gluon condensates obtained from QCD sum rules. The $N\overline{N}$ system is also potentially rich as a means of producing new "hybrid" mesons containing dynamical gluon excitations, a possibility which merits theoretical and experimental study.

REFERENCES

1. J. Vandermeulen, Z. Phys. C37, 563 (1988).
2. J. L. Bailly *et al*, Phys. Lett. B195, 609 (1987);
   A. Breakstone *et al*, Phys. Lett. B135, 510 (1984).
3. H. G. Dosch and D. Gromes, Phys. Rev. D33, 1378 (1986).
4. H. G. Dosch and D. Gromes, Z. Phys. C34, 139 (1987).
5. C. B. Dover, P. M. Fishbane, and S. Furui, Phys. Rev. Lett. 57, 1538 (1986).
6. M. Foster *et al*, Nucl. Phys. B6, 107 (1968);
   S. Ahmad *et al*, in Proc. VII European Symposium on Antiproton Interactions; Durham, England, 1984, Ed. M. R. Pennington, IOP Conf. Series No. 73 (Hilger, London, 1985), p. 283.
7. C. Baltay *et al*, Phys. Rev. Lett. 15, 532 (1965);
   B. Conforto *et al*, Nucl. Phys. B3. 469 (1967);
   newer results for KK^* are available from LEAR experiment PS183 (B. Smith *et al*, private communication).
8. J. M. Richard and M. Sainio, Phys. Lett. 110B, 349 (1982);
   B. Moussallam, Thesis, University of Paris at Orsay (1985).
9. J. Diaz *et al*, Nucl. Phys. B16, 239 (1970).
10. H. Genz, Phys. Rev. D28, 1094 (1983)and Phys. Rev. D31, 1136 (1985).
11. U. Hartmann, E. Klempt, and J. Körner, Phys. Lett. B155, 163 (1985);
    E. Klempt, contribution to these Proceedings.
12. M. Chiba *et al*, in Physics at LEAR with Low Energy Antiprotons, Proc. of IV LEAR Workshop, Villars–sur–Ollon, 1987, Nuclear Science Research Conference Series, Vol. 14, Eds. C. Amsler *et al*, Harwood Academic Publishers, Chur, Switzerland (1988), p. 401.
13. L. Adiels *et al*, Z. Phys. C35, 15 (1987).
14. L. Tauscher, ref. (12), p. 397.
15. A value $B(\rho^0\omega) \approx 4.1 \times 10^{-2}$ is given by G. Smith *et al* (PS183 at LEAR). This is somewhat larger than older values from BNL and CERN, which were in the range $0.8 - 2.3 \times 10^{-2}$.

16. M. Bloch *et al*, Nucl. Phys. B23, 221 (1970).
17. C. B. Dover and J. M. Richard, Phys. Rev. D17, 1770 (1978).
18. A. M. Green and J. A. Niskanen, Nucl. Phys. A412, 448 (1984), A430, 605 (1984), A446, 543 (1985) and Prog. Part. Nucl. Phys. 18, 93 (1987); A. M. Green and G. Q. Liu, preprints HU–TFT–88–4 and HU–TFT–88–13 (1988).
19. M. Maruyama and T. Ueda, Nucl. Phys. A364, 297 (1981), and Prog. Theor. Phys. 73, 1211 (1985), 74, 526 (1985), 78, (1987).
20. S. Furui, A. Faessler, and S. B. Khadkikar, Nucl. Phys. A424, 495 (1984); S. Furui, Z. Phys. A325, 375 (1986); S. Furui and A. Faessler, Nucl. Phys. A468, 669 (1987); M. Maruyama, S. Furui, and A. Faessler, Nucl. Phys. A472, 643 (1987); M. Maruyama, S. Furui, A. Faessler, and R. Vinh Mau, Nucl. Phys. A473, 649 (1987).
21. C. B. Dover and P. M. Fishbane, Nucl. Phys. B244, 345 (1984).
22. E. M. Henley, T. Oka, and J. Vergados, Phys. Lett. 166B, 274 (1986).
23. M. Kohno and W. Weise, Nucl. Phys. A454, 429 (1986).
24. J. A. Niskanen and F. Myhrer, Phys. Lett. 157B, 247 (1985).
25. E. M. Henley, T. Oka, and J. Vergados, Nucl. Phys. A476, 589 (1988).
26. C. B. Dover, in Proc. Bad Honnef Conf. (1985), Lecture Notes in Physics, Vol. 243, Ed. H. V. von Geramb, Springer Verlag, Berlin (1985), p. 80.
27. A. Le Yaouanc *et al*, Phys. Rev. D8, 2223 (1973); D9, 1415 (1974); D11, 1272 (1975).
28. R. Kokoski and N. Isgur, Phys. Rev. D35, 907 (1987).
29. N. Isgur and J. Paton, Phys. Rev. D31, 2910 (1985).
30. M. Doser *et al*, preprint CERN/EP 88–42 (1988); S. Ahmad *et al*, Proc. III LEAR Workshop, Tignes, France, Editions Frontiéres (1985), p. 353.
31. C. B. Dover and P. M. Fishbane, in preparation.

ANTIPROTON ANNIHILATION ON NUCLEI

J. Cugnon
University of Liège, Institute of Physics, B.5, Sart Tilman,
B-4000 LIEGE 1 (Belgium)

ABSTRACT

A dynamical model is described for the $\bar{p}$ annihilation on nuclei. The energy transfer from the multipion system to the baryonic system is studied, as well as the response of the latter. Annihilation on two nucleons are studied and evidence for these events is searched for in experimental data.

INTRODUCTION

The bulk of the experimental results on $\bar{p}$-annihilation on nuclei obtained so far, especially at LEAR, is consistent[1,2,3] with the following dynamical picture :

(1) the $\bar{p}$ annihilates on a single nucleon at the surface of the nucleus, creating a few pions, with the same properties as those observed in the free space $\bar{p}p$ or $\bar{p}n$ annihilations ;

(2) some of the pions cascade through the nucleus, interacting with the nucleons, ejecting some of them, in a process by which they can also be absorbed ;

(3) after a fast ejection process, the nucleus possibly dissipates its remaining excitation energy in much slower processes, like evaporation and/or fission.

Several questions are however still open. What is the amount of energy transferred to the nucleus ? How is this energy transformed ? Is the annihilation process (point (1)) so simple ? We examine some of them below.

TRANSFER OF ENERGY. NUCLEUS RESPONSE

The 2 GeV or so carried by the primordial pions (those issued from the annihilation) is not entirely used to excite the nuclear system. Some of the pions can escape from the nucleus right away, because of the peripheral nature of annihilation. Furthermore, only a fraction of the pions are eventually absorbed. These effects reduce fairly well the energy transfer to the baryonic system (the nucleus) from the available 2 GeV. For annihilation at rest, the energy pumped from the pion system lies around 500 MeV for annihilation at rest and 800 MeV for annihilation in flight in the LEAR regime. The amount of energy transferred to the baryonic system is given in Fig. 1, as a function of time, for typical systems. Roughly speaking, the energy acquired to the nucleus is dissipated on two time scales. First, energy is removed by the ejection of rapid nucleons. Afterwards, energy is dissipated on a moderate time scale, in a way which is very close to ordinary evaporation (see Fig. 2). For a system like Mo, the amount of energy after the fast

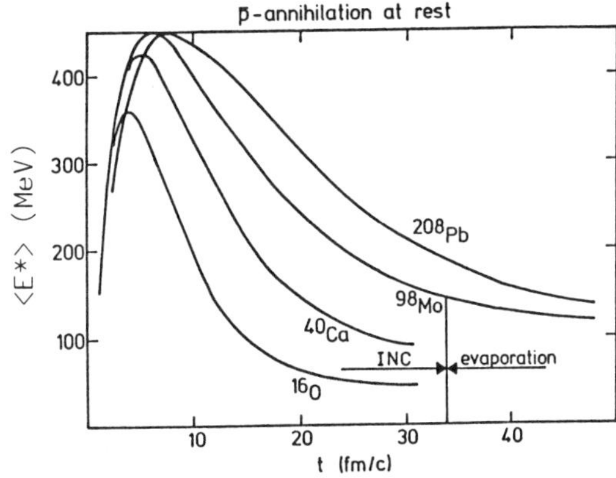

Fig. 1. Time evolution of the target excitation energy after $\bar{p}$-annihilation at rest.

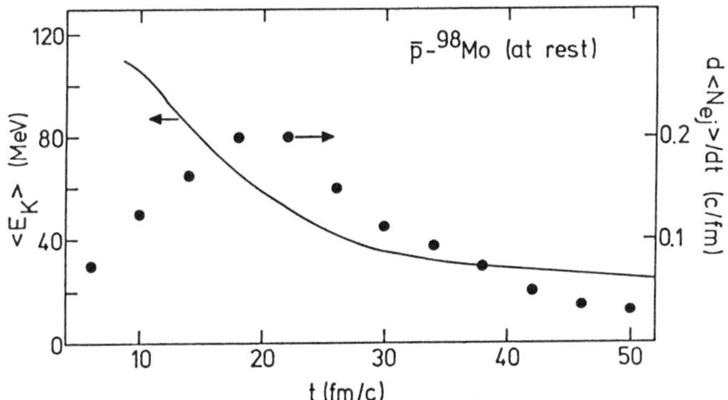

Fig. 2. Time evolution of the mean kinetic energy of the ejected nucleons (full curve, left scale) and of the number of ejected nucleons (dots, right scale) after $\bar{p}$-annihilation at rest on ^{98}Mo.

emission process reaches about 150 MeV (up to about 250 MeV in annihilation in flight), i.e. about 1.5 MeV per baryon. According to current ideas, this energy is removed by evaporation of, mainly, neutrons. A rather elaborate calculation based on the intranuclear cascade for the first step and on evaporation models for the second one reproduces rather nicely the observed residual mass spectrum[3,4] for several nuclei. On the average, a nucleus like ^{98}Mo looses about 15 nucleons, but fluctuations are large and it can loose up to 30 nucleons.

A strongly debated question is to know whether a nucleus can fragment in many pieces of intermediate size (the so-called multifragmentation). Related questions are : what is the minimal excitation energy per baryon E_c^* for the onset of this process ? Is the transition from evaporation to multifragmentation rapid ? What is the effect of "pre-equilibrium" emission ? Those problems are studied with heavy ion beams and are starting to receive partial answers. For instance, it seems that $E_c^* \approx 3$ MeV per particle. Antiproton beams present several advantages : (i) it is a way to deposit energy without large momentum ; (ii) the dynamics is better understood ; (iii) there is no strong impact parameter dependence. Of course, beams of larger energy than at LEAR will be required. The best energy range corresponds to 1-5 GeV incident energy.

UNUSUAL ANNIHILATIONS

A possible deviation from the scenario sketched in section 1 is that the annihilation is not a point-like event in space-time. In other words, one may consider that the annihilating system lasts for some time and has particular properties before the appearance of "asymptotic" pions. On these simple grounds one may wonder whether the annihilation involves two nucleons, both of which participating to the annihilating system. This idea has been exploited in Ref. [5]. Therein, it is assumed that the decay properties of ($\bar{N}NN$), as well as of $\bar{N}N$ are governed by phase space. The approach is a one parameter free model (roughly, the volume of the annihilating system). The parameter is fixed on the $\bar{N}N$ properties, which are well reproduced. The model is then used to make predictions on the $\bar{N}NN$ system. The most striking one is the enhancement of the strange particle yield.

The existence of annihilation on two nucleons is directly demonstrated by the observation of the $\bar{p}d \rightarrow p\pi^-$ process[6]. These ideas have also been used to try to extract from experimental data, the probability of having an annihilation on two nucleons. In all generality, one can write the yield per annihilation for the production of a strange species S as :

$$Y(S) = P_1Q(1,S) + P_2Q(2,S) + P_3Q(3,S) + \dots \qquad (1)$$

where P_1, P_2, P_3 are the probability of having an annihilation on one, two,... nucleons, respectively and where $Q(i,S)$ is the probability of producing the species S after an annihilation on i nucleons. The knowledge of these quantities is rather fragmentary, since they deal with complicated rescattering process following annihilation. What has been done up to now is to compare the observed yield with $Q(1,S)$, which is usually known with good confidence, in some cases at least. We just list a few results in Table I. As can be seen the annihilation on one nucleon cannot explain the experimental yields. The comparison clearly points out the necessity to call for two nucleon annihilations. The case of K_0^S in the last reaction is nevertheless very puzzling since there the production yield seems to be inhibited. This case certainly desserves further investigation. This is also true for the measurements of Ref. [7]. They refer to 4 GeV/c antiproton reactions, for which the evolution of the strange particle abundances is more complicated than in the LEAR regime.

Table I. Comparison of observed strange particle yields and quantity Q(1,S)

Systems	Observable	Experimental yield	Q(1,S)
$\bar{p}$ + C (ref. 8)	K^+/π^-	0.030	0.015
$\bar{p}$+C,Ta,Pb (ref. 9)	Λ	$(1.9\pm0.4)\times10^{-2}$	< 0.01
$\bar{p}$+Bi	hypernucleus (fission)	3×10^{-4}	$\} \lesssim 10^{-4}$
U (ref. 10)	hypernucleus (fission)	9×10^{-4}	
$\bar{p}$ + ^{20}Ne (ref. 11)	Λ K^S_0	$(1.95\pm0.43)\times10^{-2}$ $(0.85\pm0.17)\times10^{-2}$	$< 10^{-2}$ $\{ < 2.5\times10^{-2}$, $> 1.2\times10^{-2}$

CONCLUSION

We have underlined the interest of $\bar{p}$-annihilation in nuclei, in two respects : (i) it is a useful way to study the multifragmentation of the nucleus ; (ii) it is the seat for unusual annihilations. We have here shown a simple example, but one may imagine more complex processes. More generally, $\bar{p}$-annihilation in nuclei is a tool to study hadronic systems in special environments.

REFERENCES

1. P.L. McGaughey et al., Phys.Rev.Lett. 56, 2156 (1986).
2. J. Cugnon, P. Jasselette and J. Vandermeulen, Nucl.Phys. A470, 558 (1987).
3. A.S. Iljinov, V.I. Nazaruk and S.E. Chigrinov, Nucl.Phys. A382, 378 (1982).
4. P. Jasselette, J. Cugnon and J. Vandermeulen, to appear in Nucl. Phys..
5. J. Cugnon and J. Vandermeulen, Phys.Lett. 146B, 16 (1984).
6. R. Bizarri et al., Lett. Nuovo Cim. 2, 431 (1969).
7. K. Miyano et al., Phys.Rev.Lett. 53, 1725 (1984).
8. G.A. Smith, "The Elementary Structure of Matter", ed. by J.M. Richard et al., (Springer-Verlag, Berlin, 1988), p. 197.
9. G.T. Condo et al., Phys.Rev. C29, 1531 (1984).
10. J.P. Bocquet et al., Phys.Lett. 192B, 312 (1987).
11. F. Balestra et al., Nucl.Phys. A452, 573 (1986).

DYNAMICAL SELECTION RULES IN $\bar{p}p$ ANNIHILATION

E. Klempt
Institut für Physik, Johannes Gutenberg-Universität
Staudinger Weg 7, D-6500 Mainz, West Germany

ABSTRACT

The importance of dynamical selection rules observed in $\bar{p}p$ annihilation at rest into two mesons is emphasized. Since these selection rules are independent of spin and angular momentum excitations of the two mesons in the symmetry leading to these selection rules must be SU(3). A SU(3) analysis based on a simple kinematical model for $\bar{p}p$ annihilation shows that the selection rules allow to reduce the number of SU(3) invariant couplings from eight to two. This points to the possibility that quark dynamics might eventually be described in simple terms even in the confinement region.

The annihilation of antiprotons and protons is a fascinating process: annihilation takes place when antiprotons and protons approach each other; the annihilation range is often disputed and estimates vary from 0.1 fm[1] (if baryon exchange is responsible for annihilation) to 1.0 fm[2]. So the energy density exceeds 2 GeV/fm[3], and quark deconfinement may be expected. In this contribution I want to show that dynamical selection rules govern $\bar{p}p$ annihilation at rest into two mesons, and that these selection rules require an interpretation of $\bar{p}p$ annihilation at the level of quark dynamics.

The first dynamical selection rule was discovered as early as 1967 in $\bar{p}p$ annihilation into $K^*\bar{K}$, but to my knowledge it was not yet discussed in the context of annihilation dynamics. This quasi two-body annihilation mode can proceed via all four spin-isospin states with zero angular momentum; their quantum numbers are $I^G(J^{PC}) = 1^+(1^{--})$; $0^-(1^{--})$; $1^-(0^{-+})$; and $0^+(0^{-+})$. The experiment[3] shows that only the first and the last state contribute significantly to $K^*\bar{K}$ + c.c. The difference between the first and last state and the two other states is the sign of the interference between charged and neutral kaons: states which couple to $(K^{*+}K^- + K^{*0}\bar{K}^0)$ + c.c. have small branching ratios (these are the $0^-(1^{--})$ state and the $1^-(0^{+-})$ state), and states which couple to $(K^{*+}K^- - K^{*0}\bar{K}^0)$ + c.c. have large branching ratios.

A similar phenomenon can be observed in $\bar{p}p$ annihilation into K^+K^- and $K^0_SK^0_\ell$. The branching ratios for these processes are approximately the same[4]. This is only possible if there is no interference between the two $0^-(1^{--})$ and $1^+(1^{--})$ initial states from which annihilation into $K\bar{K}$ is possible. Since $\bar{p}n \to K^0_SK^-$ is not negligible[5] the I=0 component of the $\bar{p}p$ system cannot be responsible for $N\bar{N} \to K\bar{K}$ from S-waves. Hence $\bar{p}p$ annihilation into $K^+K^- + K^0_SK^0_\ell$ (from the $0^-(1^{--})$ initial state) is weak and annihilation into $K^+K^- - K^0_SK^0_\ell$ (from the $1^+(1^{--})$ initial state) is strong.

The selection rule observed in $\bar{p}p$ annihilation into two strange mesons does not depend on spin excitations of the kaon or the $\bar{p}p$ system. Hence the symmetry leading to the selection rule is spin independent, and it is natural to assume that the symmetry is SU(3). Within SU(3) there are eight invariant coupling constants for $\bar{p}p$ annihilation into two mesons[6] (OZI rule violating decays are not taken into account). Four of them describe "planar" diagrams and four of them partial rearrangement of quarks. The SU(3) invariant amplitudes are defined by the decomposition of proton and antiproton into S = I = 0 diquarks and "valence" quarks.

Table 1 shows the SU(3) invariant amplitudes for annihilation into vector and pseudoscalar mesons. The $\bar{p}p$ initial state is characterized by its $I^G(J^{PC})$ quantum numbers. The SU(3) invariant amplitudes responsible for annihilation into a given two-meson state are listed in the third column.

The branching ratios into $K^*\bar{K}$ + c.c. from the $J^{PC} = 1^{--}$ and 0^{-+} $\bar{p}p$ initial state are 25.4 (2.5) and $10.6\ (1.8)\cdot 10^{-4}$, resp.; their ratio is inconsistent with their relative number of helicity amplitudes which are 2, respectively 1. This fact and the dynamical selection rule allows to reduce the four planar amplitudes to one amplitude:

$$A_1^{\pm} + A_2^{\pm} \approx 0 \qquad \text{and} \qquad A_2^{+} \approx A_2^{-} = A$$

The result $A_1^{\pm} + A_2^{\pm} = 0$ implies that the annihilation amplitude for annihilating a diquark and an antiquark does not differ (but by the sign) the corresponding amplitude where a diquark is broken up and only one quark of the diquark and the valence quark annihilate[7].

The experimental branching ratios for $\bar{p}p \to M_1M_2$ can be used to define experimental "amplitudes"[8]. These are also listed in Table I. The amplitudes are derived from experimental data by use of the relation

$$\text{branching ratio} = 9/2M_p \cdot 0.094 \cdot n_h \cdot (\text{ampl.})^2.$$

Here $q/2M_p$ is the two-body phase space, n_h the number of helicity amplitudes ($n_h = 2$ for the $J^{PC} = 1^{--}$ state and $n_h = 1$ for the 0^{-+} state). The experimental amplitudes for $\bar{p}p \to \eta\omega$ or $\rho\pi$ are compatible suggesting

$$|R_1{}^+| = |R_1{}^+ + R_2{}^+| \ ;$$

we choose $R_2{}^+ \approx 0$. The amplitude for $\bar{p}p$ annihilation into $\eta\rho$, $\eta'\rho$ is then given by $2A^+ + R_3{}^+$ the experimental result is compatible with $R_3{}^+ = R_1{}^+$.

The second dynamical selection rule is observed in $\bar{p}p$ annihilations from I=0 and I=1 initial states into two isovector mesons. For normalization we compare annihilation into two isovector mesons with annihilations into two strange mesons.

$$\frac{\bar{p}p[0^-(1^{--})] \to \rho\pi}{\bar{p}p[1^{--}] \to K^*\bar{K}+\text{c.c.}} = 23.1 \pm 2.6 \qquad (9)$$

$$\frac{\bar{p}p[0^+(0^{++},2^{++})] \to \pi\pi}{\bar{p}p[0^{++},2^{++}] \to \bar{K}K} = 19.5 \pm 3.4 \qquad (10)$$

$$\frac{\bar{p}p[1^-(0^{-+})] \to \rho\pi}{\bar{p}p[0^{-+}] \to K^*\bar{K}+c.c.} \lesssim 1.3 \qquad (9)$$

$$\frac{\bar{p}p[1^+(1^{--})] \to \pi\pi}{\bar{p}p[1^{--}] \to K\bar{K}} = 1.56 \pm 0.11 \qquad (4)$$

Annihilations into two isovector mesons proceed preferentially from I=0 initial states. This result is confirmed in $\bar{p}p$ annihilations into πa_2 (1320) for which, however, a comparison with annihilations into strange mesons is not possible. But the ratio

$$\frac{\bar{p}p[0^+(1^{--})] \to \pi a_2}{\bar{p}p[1^+(0^{-+})] \to \pi a_2} = 3.6 \pm 1.0/[10.8 \pm 3.0] \quad (11)$$

indicates clearly the preference of I=0 intitial states for annihilation into two isovector mesons. The value of 10.8 for the ratio I=0/I=1 is obtained by correcting for the different number of helicity amplitudes for annihilation into πa_2 from spin triplet and singlet state.

This dynamical selection is hence seen in $\bar{p}p$ annihilation into $\pi\pi$, $\pi\rho$, and πa_2 (1320); hence it does not depend on spin or orbital angular momentum excitations of the mesons or spin excitations of the $\bar{p}p$ atom. This fact strongly supports the assumption that $\bar{p}p$ annihilation needs to be described within SU(3), that annihilation is quark dynamics.

Within SU(3) the annihilation into two isovector mesons from I=0 initial states is described by the amplitude $2A_1^+ + 2A_2^+ - 2R_1^+$; from I=1 initial states by $2A_2^- - R_1^-$. Since $2A_1^+ + 2A_2^+ = 0$, because of the first dynamical selection rule, the simplest description of the second dynamical selection rule is $R_1^- = 0$. Further $\bar{p}p$ annihilation modes into a vector and a pseudoscalar meson are shown in Table I, very simple pattern emerges:

$$A_1^+ = A_1^- = -A_2^+ = -A_2^- = A \approx 0.095$$

$$R_1^+ = R_3^+ = R = 0.250 \ ; \ R_1^- = R_2^+ = 0.$$

It can be shown that also annihilations into two pseudoscalar or into a pseudoscalar and a tensor meson can be described by this choice of parameters. An important exception are annihilations into two isovector mesons: $\bar{p}p \to \rho^0\rho^0$ from the $I^G(J^{PC}) = 0^+(0^{-+})$ initial state has a small branching ratio. If for these annihilations $R_1^+ = R_3^+ = 0$; $R_1^- = R_2^+ = 0.5$ would hold - this is compatible with the data - then $I^G(J^{PC}) = 1^+(1^{--})$ annihilations into $\rho^+\rho^-$ should have a branching ratio of (6.0 ± 1.5)%. This prediction will be tested with the CB detector at LEAR.

In any case the symmetry in the planar and rearrangement amplitudes suggests strongly that quark dynamics is governed by a "simple" quark-gluon interaction. The relation between these symmetries and the quark-gluon dynamics needs, however, further exploration.

REFERENCES

1) I.S. Shapiro, Phys.Rep. C35(1978)129
2) B. Povh and Th. Walcher, Comm.Nucl.Phys.Part.Phys. 16(1985)86
3) A. Bettini et al., Nuovo Cimento 63A(1969)1199
4) R. Armenteros and B. French, High Energy Physics, Academic Press (ed. FHS Burkop), New York 1969; see also CERN proposal P28(1980)
5) A. Bettini et al., Nuovo Cimento 62A(1969)1038
6) U. Hartmann, E. Klempt and J. Körner, Phys.Lett. B155(1985)163
7) E. Klempt: "Strangeness production in p̄p annihilation ...", submitted to Z. Physik A
8) U. Hartmann, E. Klempt and J. Körner, "Antiproton-proton annihilation ...", submitted to Z. Physik A
9) M. Foster et al., Nucl.Phys. B6(1968)107
10) M. Doser et al., submitted to Nucl.Phys. B
11) J. Diaz et al., Nucl.Phys. B16(1970)239

Table 1: SU(3) amplitudes for $\bar{p}p \to V + PS$ mesons

$\bar{p}p$ state $I^G(J^{PC})$	two-meson state	SU(3) amplitude	two-parameter amplitude	amplitude from exp.
$1^+(1^{--})$	$K^{+*}K^- - K^{0*}\overline{K^0}$	$2A_2^+$	2A	0.200 (10)
$0^-(1^{--})$	$K^{+*}K^- + K^{0*}\overline{K^0}$	$2A_1^+ + 2A_2^+$	0	0.035 (10)
$0^-(1^{--})$	$\rho^+\pi^-$, $\rho^0\pi^0$, $\rho^-\pi^+$	$2A_1^+ + 2A_2^+ - 2R_1$	2R	0.506 (14)
$0^-(1^{--})$	$\eta\omega$	$2A_1^+ + 2A_2^+ + 2R_1^+ - 2R_2^+$	2R	0.544 (32)
$1^+(1^{--})$	$\pi^0\omega$, $\rho\eta$, $\rho\pi'$	$2A_2^+ + (R_2^+ + R_3^+)$	2A+R	0.451 (22)
$0^+(0^{+-})$	$K^{+*}K^- - K^{0*}\overline{K^0}$	$2A_2^-$	2A	0.169 (16)
$1^-(0^{-+})$	$K^{+*}K^- + K^{0*}\overline{K^0}$	$2A_1^- + 2A_2^-$	0	0.083 (8)
$1^-(0^{-+})$	$\rho^+\pi^-$, $\rho^-\pi^+$	$-2A_2^- - R_1^-$	-2A	< 0.140

ANTIPROTON-PROTON INTERACTIONS AT THRESHOLD AND ANTIPROTONIC HYDROGEN ATOMS

E. Klempt
Institut für Physik, Johannes Gutenberg-Universität
Staudinger Weg 7, D-6500 Mainz, West Germany

ABSTRACT

Strong interaction effects in $\bar{p}p$ atoms have been observed in recent LEAR experiments: energy shift and broadening of the 1S ground state were deduced from the observed Lyman series; the 2P level width from the intensity balance of the Blamer series feeding the 2P level and the Lyman α line depopulating it. A survey of the experimental results is given.

Antiprotonic hydrogen atoms are formed when antiprotons stop in H_2. The slowing down antiprotons ionize H_2 molecules until their energy is in the few eV range. Then they are captured into high Rydberg states of the $\bar{p}p$ atoms. The atomic electron is ejected by Auger emission in this process. The principal quantum number of the $\bar{p}p$ atom is preferentially large ($n \propto 30$); here the classical trajectory of the antiproton has a large overlap with the electron wave function of the hydrogen atom. The highly excited $\bar{p}p$ atom deexcites into lower energy levels by Coulomb deexcitations (leading to rotational and vibrational excitations of a H_2 molecule), by external Auger effect (leading to ionization of a H_2 molecule), and by radiative transitions. The intense electric fields which the $\bar{p}p$ atom experiences in collisions with H_2 molecules mix the angular momentum states, and S and P wave capture may occur in high-n levels. The intensity of X-rays is therefore strongly quenched, in particular at high H_2 densities.

But even $\bar{p}p$ atoms reaching low-n P states will mostly not emit Lyman X-rays: annihilation is calculated to prevail in comparison to radiative transitions by a factor of about one hundred. Hence the yield of Lyman X-rays is expected to be very small. The strong broadening of the 1S ground state of the $\bar{p}p$ atom makes it even more difficult to detect the Lyman series if the background from various sources is not suppressed sufficiently. Pre-LEAR experiments[1-3] therefore failed to observe the Lyman series from which the strong interaction between antiproton and proton can be deduced. Only the experiment working at the lowest H_2 density (4 atom) and using a 4π gaseous detector was capable to observe at least the Balmer series and to establish the formation of the $\bar{p}p$ atom.

At LEAR three experiments (PS171, PS174, PS175) searched for X-rays of the $\bar{p}p$ atom[4-6]. They use different techniques which I would like to discuss in connection with the data.

PS171 uses a large H_2 gas target at atmospheric pressure. Essentially all antiprotons entering the target stop in H_2. The target is surrounded by a cylindrical drift chamber filled with 50% Argon

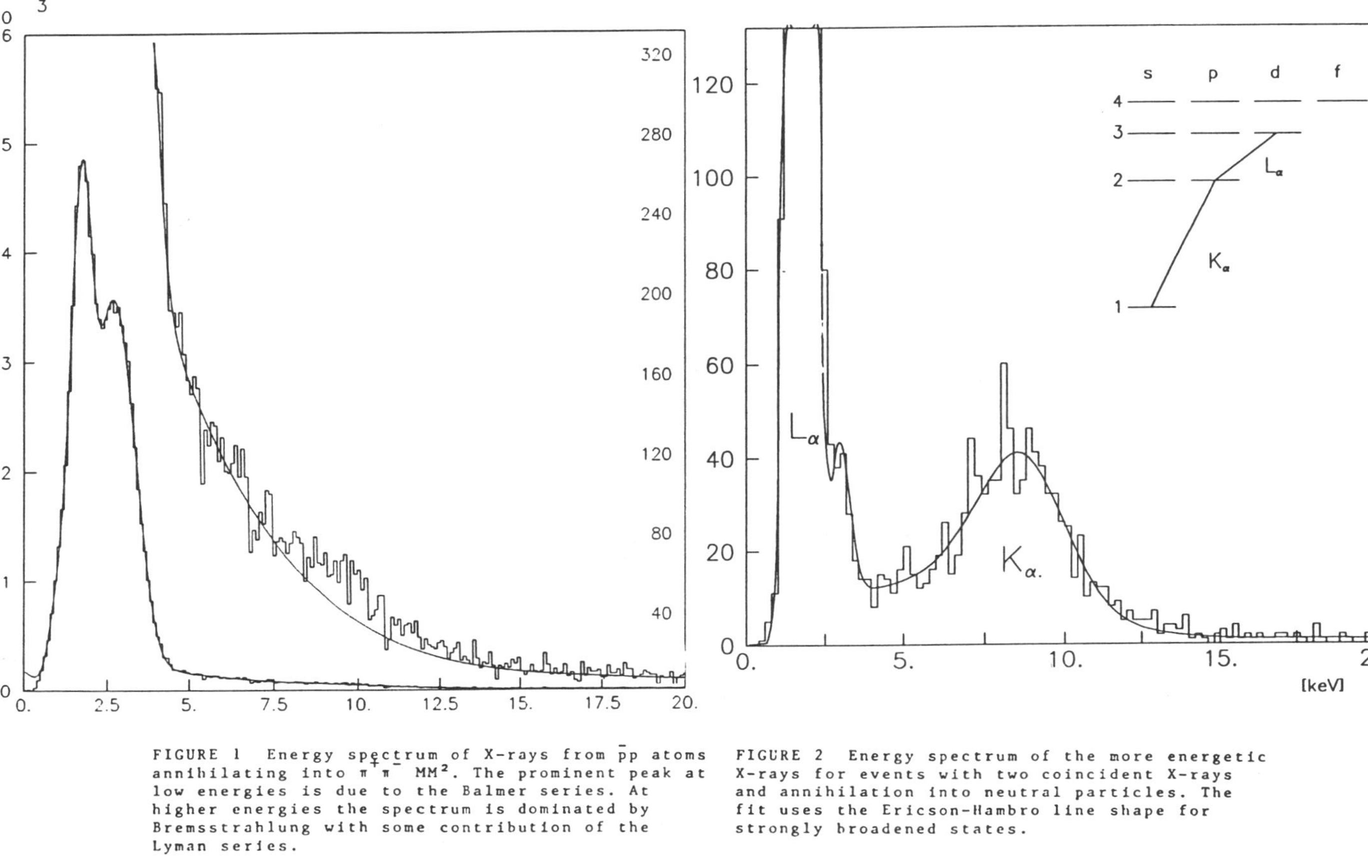

FIGURE 1 Energy spectrum of X-rays from $\bar{p}p$ atoms annihilating into $\pi^+\pi^-$ MM². The prominent peak at low energies is due to the Balmer series. At higher energies the spectrum is dominated by Bremsstrahlung with some contribution of the Lyman series.

FIGURE 2 Energy spectrum of the more energetic X-rays for events with two coincident X-rays and annihilation into neutral particles. The fit uses the Ericson-Hambro line shape for strongly broadened states.

and 50% Ethane. The chamber images tracks of charged particles and serves to identify low-energy X-rays. A solenoidal magnetic spectrometer allows to reconstruct the final state in $\bar{p}p$ annihilation.

The X-ray detector has only a moderate energy resolution (FWHM = 1.9 KeV at 8.7 KeV), but it covers nearly 4π solid angle and has a very good background rejection. X-rays from $\bar{p}p$ atoms are identified by three-dimensional reconstruction of the X-ray conversion point and by mutual coincidences between X-rays belonging to the Lyman-, Balmer-, and Paschen-series.

Fig.1 shows the energy spectrum of single X-rays emitted in events in which the $\bar{p}p$ atom annihilates into two charged pions and some (undetected) neutral pions. The peak at low energy is due to detection of the Balmer series. The smooth line at higher energies represents the spectrum of Bremsstrahlungsquanta emitted during the sudden acceleration of the two charged pions. Energy spectrum and angular distribution of the Bremsstrahlungs X-rays (relative to the emission of the charged pions) agree reasonably well with the data[7]. Since the calculation predicts the energy spectrum and the absolute intensity[8], the good agreement between data and calculation allows only a very small residual background. The additional intensity above the calculated Bremsstrahlungsspectrum may indicate the presence of the Lyman series.

The background due to Bremsstrahlung can be avoided by selection (at the trigger level) of events in which $\bar{p}p$ atoms annihilate into neutral particles only; the residual background can be further reduced by choosing events in which two X-rays are detected in coincidence[9]. Fig.2 shows the energy spectrum of those events in which two X-rays were detected in coincidence and in which annihilation due to neutral particles only took place. The energy distribution of the more energetic X-ray is shown. There is clear evidence for the Balmer-α line at 1.74 KeV, these X-rays are coincident with an X-ray of the Pascher series. (The full peak height is not shown). A second peak is observed which coincides with X-rays of the Balmer series. The distribution of the X-ray conversion points proves that the source of the X-rays are antiprotons stopping in the target; the (measured) gas purity excludes any other sources than antiprotonic hydrogen atoms. The observed line is fitted with a line shape theory develloped by Ericson and Hambro[10]. The theory takes into account the variation of the electric dipole operator over the width of the state. It can be shown that this is a necessary precaution and that the result of a fit does depend on these subtleties[11].

PS174 uses a conventional H_2 gas target; the pressure can be reduced and/or the target can be cooled. The energy of X-rays is measured in Si(Li) detectors and gas scintillation proportional counters. The background in Si(Li) detectors originates mostly from Compton scattering of γ's (from π^0 decays). A Compton shield helps to reduce the overall background. The small solid angle excludes observation of X-ray coincidences; therefore lines of $\bar{p}p$ atoms are identified by their intensity variations with target density and absence of lines when H_2 is replaced by D_2. Cascade calculations indeed predict that the K_α X-ray intensity decreases with increasing pressure. Fig.3 shows X-ray spectra[12] at four different target densities.

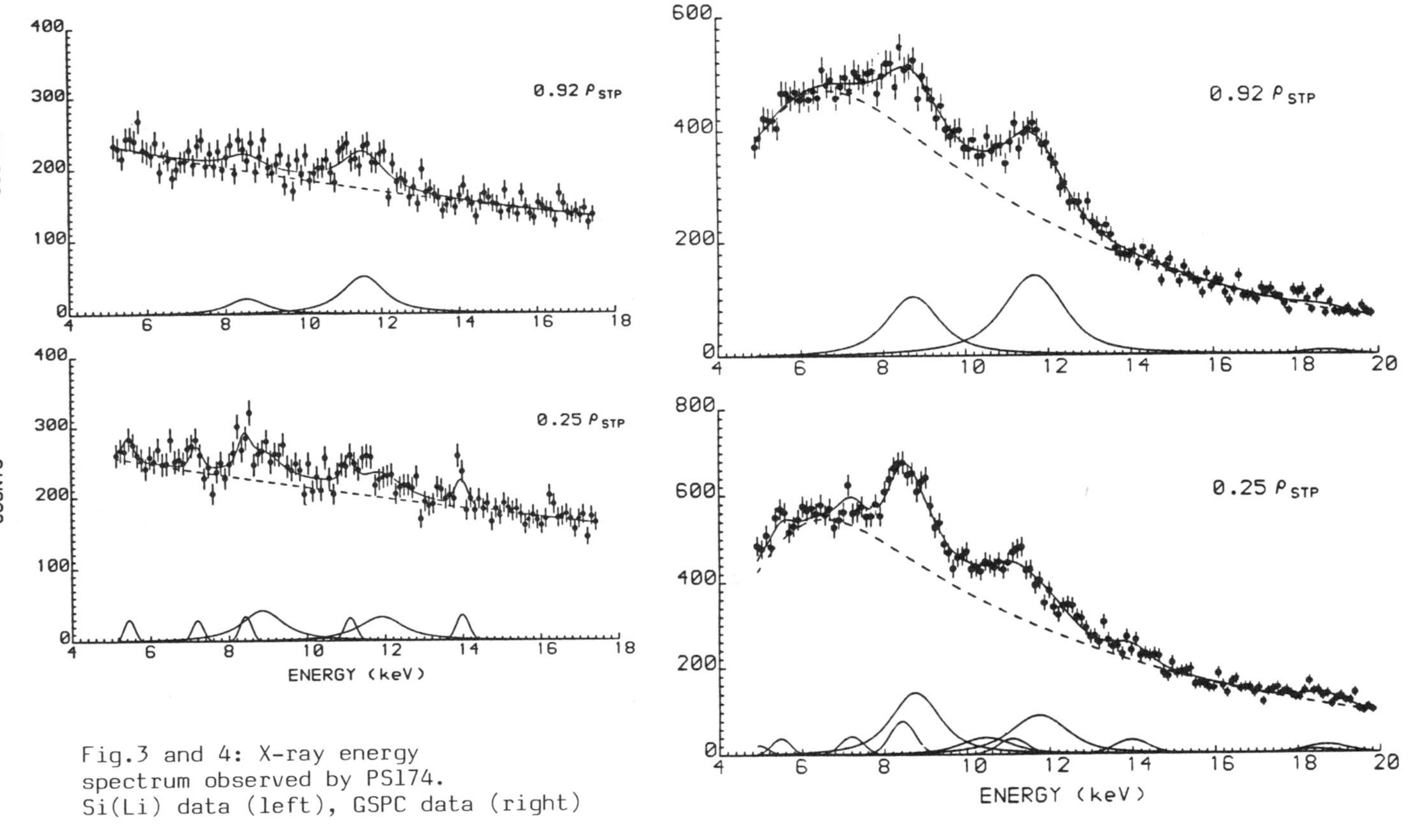

Fig.3 and 4: X-ray energy spectrum observed by PS174. Si(Li) data (left), GSPC data (right)

Two lines are atributed to the $\bar{p}p$ atom, a line at ~ 9 keV to the K_α line and a line at ~ 12 KeV to the series limit. Unfortunately, several background lines disturb the spectrum at the lowest density where the K_α line has a detectable contribution. But nevertheless, the pressure dependent K_α/K_∞ pattern is a very good indication that the Lyman series of the $\bar{p}p$ atom is seen.

The same group has also used two gas scintillation proportional chambers. These detectors offer the advantage of gas counters (low mass, potentially large sizes) but have improved detector resolution: the use of a photon multiplication stage between X-ray conversion point and avalanche region allows to exploit that the Fano factor $F \ll 1$. The data taken with this chamber are shown[13] in Fig.4. The intensity variation within the Lyman series is clearly seen, the lines are absent in D_2. (The annihilation probability of the 2p levels of $\bar{p}d$ is to high to allow sizeable Lyman intensities.)

PS175 uses an inverse cyclotron to stop antiprotons at very low hydrogen gas pressure. Since Stark mixing is reduced at low pressures, a very good signal - to background ratio might be expected. Two different Si(Li) detectors were used, one with a large area to collect data with a larger solid angle ($\sim 10^{-3}$), one with a small area for best resolution. Fig.4 shows the X-ray energy spectrum[14] detected with the large area detector and at a pressure of 30 mbar. The background level is rather high, and some background lines can be identified. The fit finds a χ^2 minimum when K_α and K_∞ lines are introduced above a polynomial background. Energy, width and absolute intensity (extrapolated to this low density in cascade model calculations) agree with those values found in the other data sets. Hence it seems likely that the Lyman series has been identified in all three experiments. Note, however, that experiments PS174, 175 use a Breit-Wigner shape for the K line.

Table I summarizes the experimental results and shows a comparison with recent calculations. There is reasonable agreement within the experimental results and different theoretical approaches. The common feature of all calculations is the pion-exchange potential and an annihilation potential adjusted to describe antiproton-proton scattering data. Presumably this is also the physical effect leading to a repulsive energy shift in the presence of an attractive nuclear potential: the wave function is suppressed at the origin where otherwise strong electromagnetic binding would lead to a more deeply bound state.

The intensity per antiproton stop of the Balmer series is about a factor of 100 higher than that of Lyman α X-rays. Hence their detection should be much easier. On the other hand their energies vary from 1.74 KeV of the Balmer α to 3.1 keV at the series' limit. At this low energy X-rays are easily absorbed in windows, and the determination of absolute yields requires a very good understanding of the detector. The Balmer series of the $\bar{p}p$ atom as observed in PS174 is shown in Fig.5. It should be compared to the Balmer series shown in Fig.1. After taken into account the detection efficiency both experiments agree that the Balmer α line at 1.74 keV gives the strongest contribution! The comparison of the yield of the Balmer series with the yield of the Lyman α line gives the 2P level width, also

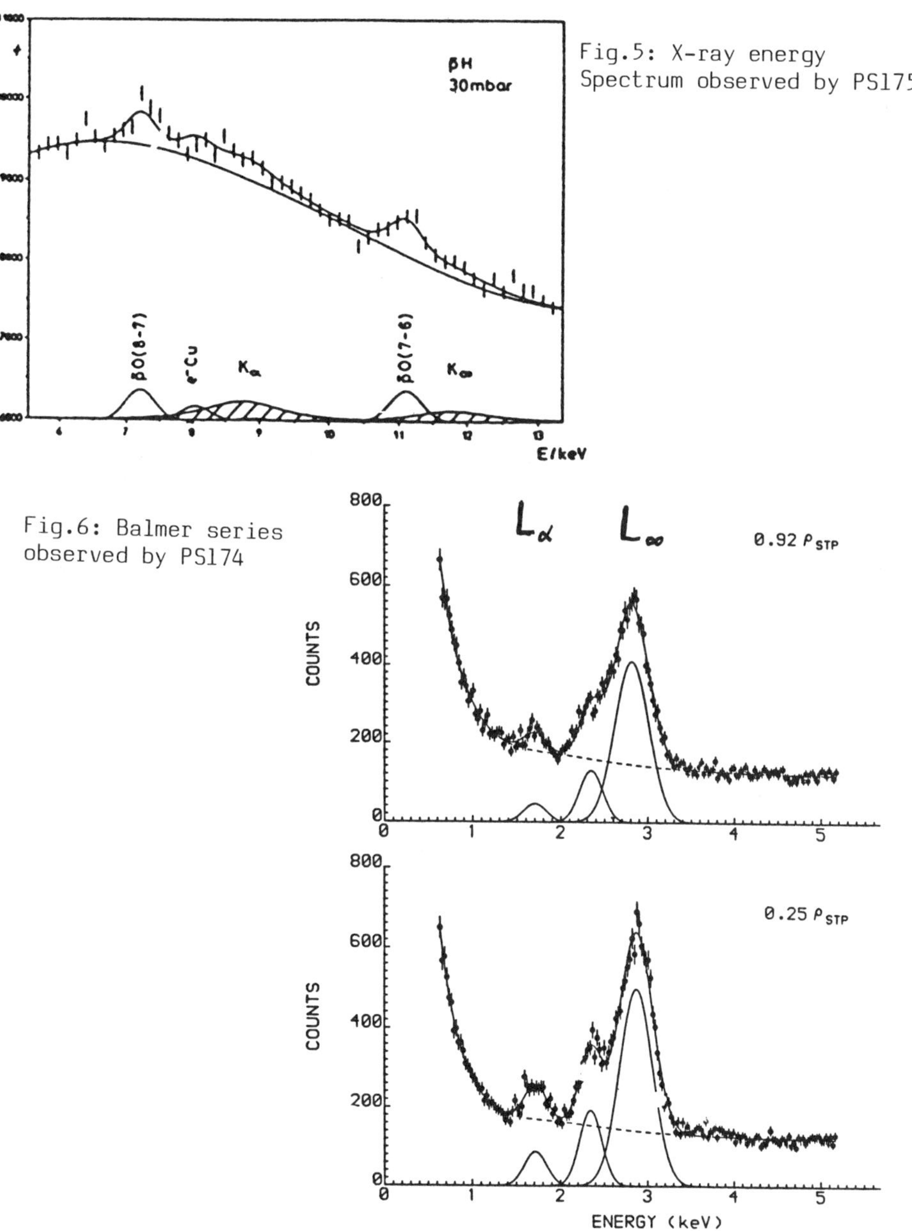

Fig.5: X-ray energy Spectrum observed by PS175

Fig.6: Balmer series observed by PS174

shown in Table I. The agreement with theoretical values is again reasonable. References to theoretical predictions can be found in[12,13].

It should be pointed out that the strong interaction effects are spin-isospin averaged quantities. The four 2P states have different hadronic widths, their fractional population is not known. The 1S

Table I: Strong interaction parameters

Experiment	Energy shift ΔE_{1S}(keV)	Width Γ_{1S}(keV)	Width Γ_{2P}(meV)
PS 171	-0.70 ± 0.15	1.60 ± 0.40	
PS 174	-0.75 ± 0.06	0.90 ± 0.18	45 ± 10
	-0.73 ± 0.05	1.13 ± 0.09	
PS 175	-0.66 ± 0.13	1.13 ± 0.23	40 ± 11

ground state has two hyperfine states; spin dependent nuclear forces will lead to different strong interaction effects. The isospin decomposition of the atomic states is calculable but no experimental information exists. The spin dependence of the 1S strong interaction parameter may be accesible by reconstruction of specific final states; disentangling the 2P states will be even harder. So the overall features of the $\bar{p}p$ atom have been determined experimentally and are well reproduced by theory, but many details wait for new ingenious experiments.

I would like to thank Ch. Batty and L. Simons for sending their data prior to publication, and M. zur Hausen for typing patiently.

REFERENCES

1) E.G. Auld et al., Phys.Lett. B(1978)454
2) M. Izycki et al., Z. Phys. 297(1980)1
3) J.R. Lindemuth et al., Phys.Rev. C30(1984)1740
4) S. Ahmad et al., "Protonium Spectroscopy ..." in Physics at LEAR I (V. Gastaldi and R. Klapisch, eds.) Plenum Press, New York (1984)109
5) J.D. Davies, "Precision Survey of X-Rays from $\bar{p}p$ ($\bar{p}d$) Atoms ...", op.cit., p.143
6) L.M. Simons, "The Cyclotron Trap ..." and D. Gotta, "X-Ray Spectroscopy in $\bar{p}$H ...", ibid, p.155
7) U. Schäfer et al., "X-Rays From $\bar{p}p$ Annihilation into Charged Final States", in preparation
8) R. Rückl and C. Zupancic, Phys.Lett. B150(1985)225
9) M. Ziegler et al., Phys.Lett. B206(1988)151
10) T.E.O. Ericson and L. Hambro, Annals of Physics 107(1977)44
11) S. Ahmad et al., "X-Ray Transitions in $\bar{p}p$ Atoms and their Line Shape", in: Physics at LEAR (C. Amsler et al., eds.) Harwood Academic Publishers, Chur(1988)717
12) C.A. Baker et al., Nucl.Phys. A(1988)
13) C.W.E. van Eijk et al., Nucl.Phys. A(1988)
14) L.M. Simons, "Results from Antiprotonic Atoms at LEAR", in Physics at LEAR IV, op.cit., p.703

QUARK GLUON PLASMA IN ANTIPROTON ANNIHILATIONS ON NUCLEI

Johann Rafelski
Department of Physics, U of Arizona, Tucson, AZ 85721

ABSTRACT

Annihilation of antiprotons on nuclei is capable to create highly excited states of nuclear matter. Conventional interpretation of recent data on strange particle production in 4 GeV/c antiproton annihilations on Ta has not been able to account for the high observed production of strangeness, and in particular the lambda particles. However, assuming quark gluon plasma formation at rather modest temperature T < 60 MeV along with a simple reaction model in which antiprotons deposit energy in the forward cone of nuclear matter within the target nucleus, the observed spectra and total abundances of lambdas and kaons are found to agree with theory.

INTRODUCTION

In the annihilation reaction of antiprotons on nuclei it is possible to deposit in the target nucleus most of the annihilation energy. In such a possibly rare reaction very excited forms of nuclear matter may be created[1]. We will here present an interpretation[2] of the strange particle production experiment[3] in which 4 GeV/c annihilation on heavy nuclei (Ta) was studied. It will be shown that we can understand this experiment taking as reaction mechanism the excitation of the quark matter (quark gluon plasma: QGP) state of nuclear matter[4]. Our work follows on a previous attempt to describe these data in terms of individual hadronic reactions in a hadronic cascade approach, which, however, has not been successful[5]. It is important to appreciate that many features of antiproton-nucleus reactions are appropriately described by such calculations[6]. Hence the failure to account for the strangeness production and in particular for the abundant production of lambdas demonstrates that these degrees of freedom are much more characteristic for the underlying reaction mechanisms[7,8].

In the QGP - phase the identity of individual nucleons is lost, with quarks and gluons being the physical degrees of freedom[4]. An indication that such a state of matter has been formed is of additional fundamental interest at this time, since no other experiment as yet confirmed the existence of QGP. Simple arguments indicate that there is an optimum antiproton energy for excitation of high energy density and presumably QGP in annihilations on heavy nuclei. This is of course a consequence of the annihilation energy being available, aside from the kinetic energy of the projectile. We would like the annihilation energy to be shared in a rather small volume by several of the target nucleus' nucleons. The antiproton momentum must be so large, that the annihilation occurs well below the nuclear surface[1].

KINETICS OF THE 4 GeV/c ANNIHILATION

The experiment performed by Miyano et al [3] at 4 GeV seems to be quite ideally suited for the purpose of deposition of maximum excitation within a heavy target nucleus (Ta). The tube of nucleons in the path through the nuclear center contains 2 $A^{1/3}$ nucleons. Thus it is possible that a fireball of nuclear matter containing the reaction energy will be formed, with baryon number being O(10). The formation of a (thermalised) fireball facilitates abundant formation of strange particles[8]. Observing the rapidity distribution of strange particles one effectively determines the mass of the participating nuclear matter region from a simple kinematic considerations which essentially imply finding the frame of reference, if such exists, in which the spectrum of particles shows spherical symmetry. The rapidity of the fireball with respect to the laboratory is derived from the kinematics of the colliding objects and it depends in particular sensitively on the effective total mass of the target, i.e. the participating nuclear matter cone. In Fig. 1 the dependance of the rapidity of the fireball and the excitation energy per nucleon in the fireball on the number of participants in the target, A_t, is shown, computed for P = 4 GeV/c, ie. E_p = 4.11 GeV.

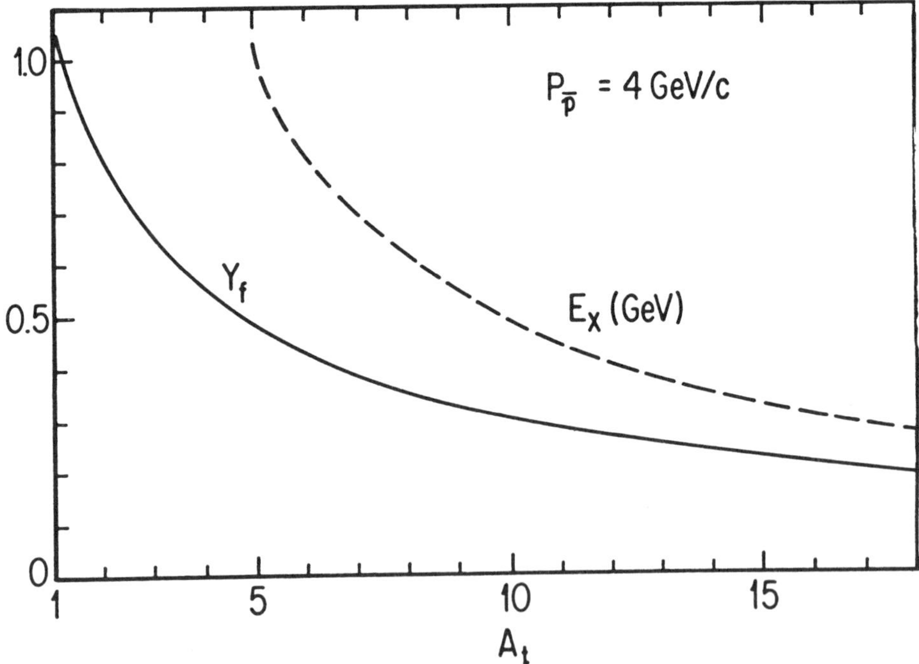

Fig. 1 Y_f, the rapidity of the central fireball as function of A_f, the number of target baryons. Dashed: E_x, the excitation energy in the fireball, per baryon number (b = A_f - 1) in GeV.

The rapidity distribution of lambda particles is observed[3] to peak at $y = 0.25$ ($v/c = 0.245$) indicating that there were about 13 nucleons in the target. This implies further that the excitation energy per baryon number b (= 12) in the fireball is 0.38 GeV (that is, energy per baryon $E/b = 1.32$ GeV), as is also shown in Fig. 1. Further note that at the observed velocity of the fireball the radius of the Tantalum nucleus is transversed in time $t = R/v = 10^{-22}$ sec. This hot, excited volume of nuclear matter will seek during this time a point of relative stability, at the minimum of the energy per baryon curve for a given entropy per baryon. We will return further below to discuss this condition within the context of perturbative QCD-QGP equations of state.

We can compute the energy contained in unobserved pions by utilizing the energy conservation. When balancing the energy, we must remember that each lambda is accompanied by a ($\bar{s}q$)-kaon. Of the total internal excitation of the fireball consisting of 12·380 MeV = 4.6 GeV we find in baryons and strange particles 3.25 GeV at T = 130 MeV (which turns out to be the temperature of the hadronic gas), with the balance of ~1.35 GeV being presumably in the pionic degrees of freedom. Indeed, in the most recent data[3], there is mention of, on average, 1.11 negative tracks in strangeness triggered annihilation events. On the hypothesis that these are mostly π^-, I estimate considering the isospin asymmetry of the fireball that there are 0.71 π^+ and 0.92 π^0, hence in all 2.8 pions, which accounts for the missing energy. We also can estimate the entropy content of the fireball[9]. With $\pi/b = 0.233$ we find about 5 units of entropy per baryon. Any interpretation of the fireball in terms of some fundamental intrinsic structure should have smaller entropy per baryon in order to allow for some subsequent entropy production. This will be in particular the case if quark matter fireballs are formed.

STRANGENESS ABUNDANCE AND SPECTRA

The overall yield of strange baryons (here lambda) originating mainly from the nuclear fireball region y = 0.25 is found not only to be unusually large, but also to be 2.4 times greater than that of kaons (here K_s). In usual $p\bar{p}$ annihilations the opposite is true with the K_s yield being 3.6 times greater. In usual nucleon-antinucleon annihilations occurring primarily on nuclear corona we would expect to create a much greater kaon background (factor eight greater) than is seen in Fig. 2. However, in the kinematic region of $p\bar{p}$ annihilations at y = 1.1 there is a greater yield of kaons than lambdas. The tilt of s-quarks to lambdas in the central fireball kinematic region shows chemical equilibration of the strange particles in the baryon rich fireball[9].

The particle rapidity distribution for a Boltzmann emitter is:

$$N^{-1}dN/dy = \frac{e^{-m/T\cosh y}}{2\,K_2(m/T)}\left(1+2/(m/T\cosh y)+2/(m/T\cosh y)^2\right). \qquad (1)$$

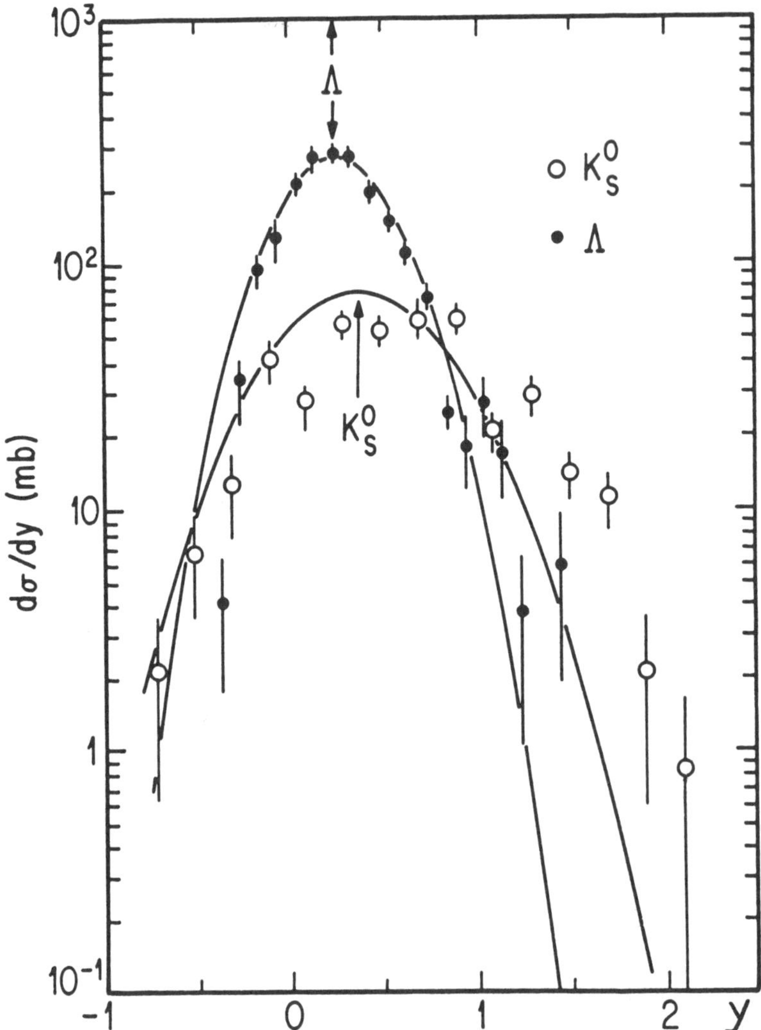

Fig. 2 Data of Miyano et al[3]: strange particle cross sections as function of rapidity are compared with thermal spectra.

It is self evident how one can improve the above expression to allow for presence of several unresolved components. In Fig. 2 the solid line for the lambda spectrum shows such a distribution computed for T = 97 MeV and y_{CM} = 0.25. Fig. 2 confirms the internal consistency of the hypothesis of a central fireball disintegrating into various particles. Even though T=97 MeV fits the rapidity lambda- spectra, it is just a 'mean' temperature. The transverse momentum distributions show two temperature components, one at 60 MeV and one at 130 MeV, with relative strength 2.7, and I find for such a two component system nearly the same rapidity distribution as shown in Fig. 2 .

QUARK GLUON PLASMA

We seek the temperature of quark matter at which as function of baryon density a minimum appears in the energy per baryon expected for the fireball, 1.32 GeV. Taking the value of the QCD coupling constant to be $\alpha_s = 0.6$, and the B(ag) constant at B = (171 MeV)^4, that is 0.112 GeV/fm^3, this turns out to be the case for T = 53 MeV. This minimum is located at a value of baryon density 2.75 times the normal nuclear density (and quark chemical potential of 395 MeV). The entropy per baryon is 4 units per baryon. For strange quark mass of 170 MeV the equilibrium ratio of strangeness to baryon number is 0.0044. For T=60 MeV the minimum of E/b is at 1.35 Gev. At this point the entropy per baryon has increased by 5%, while the strangeness abundance nearly doubled to 0.008 per baryon. Variation of α_s in the range 0.5 - 0.7 and of B within a factor two leave all these qualitative results unchanged, in particular so when an increased value of α_s is associated with a reduced value of B, as required for consistency with other QCD and hadronic spectra studies.

It is thus possible to postulate the formation of (perhaps super cooled) quark matter at T=60 10 MeV, μ_q= 400±100 MeV, S/b = 4.5±0.5 in the $\bar{p}$-Ta annihilations, indeed perhaps to suggest that in most annihilations this is the reaction mechanism. In support of this hypothesis is:

a) The observation that alone the annihilation mechanism generates about 25 units of entropy, half of what we require.

b) The observed large lambda - production cross section, 12 times greater than expected.

c) The total strangeness cross section being twice the scaled $p\bar{p}$ result. I compute σ_s as the sum of the inclusive lambda cross section with twice the inclusive K_s cross section, less four times semi inclusive (K_s/\X)-cross section. For $\bar{p}$-Ta reactions I find 257 ± 20 mb, while for $p\bar{p}$ reactions 4.19 ± 0.15 mb results. Scaling the latter result with $A^{2/3}$ I find 136 ± 10 mb.

d) The equilibrated quark matter abundance of strangeness, $\langle s \rangle$ = (0.004 - 0.02)·b (in dependance on the Temperature of quark matter), where b~2A$^{1/3}$, leads to strangeness production cross section: A·(1.2 fm)^2·(0.04 - .016) = 84 - 210 mb.

e) The yield of antilambdas is at 1-2% of lambdas and this is consistent with the quark matter[9] reaction picture developed here. This yield is at least 1000 times larger than expected from an baryon rich equilibrated hadronic gas. Similarly, there is an unusually high yield of multistrange baryons[3].

It is the natural fate of the annihilation process in which quark antiquark annihilations occur, to proceed via a fireball of quarks and gluons. While the postulation of quark matter formation at such low temperatures may appear bizarre, the antiproton mechanism provides a natural entrance channel into this phase. It is possible that this quark matter state is created in its super cooled form, i.e. it exists in the range of thermodynamic parameters where we would normally find hadronic gas. However, as it is the state of lower entropy, it is rather stable with respect to the global transition to the hadronic gas phase, which requires strong reheating to 130 MeV. The small QGP temperature of 60 MeV assures that the entropy is at the level of 80% of the hadronic gas value found in the hottest reaction component.

An anomalous abundance of strangeness could arise in the hadronic gas fireball if it 'cooks' for an unusually long time. Against such an interpretation speaks the fact that there is no sign of the anomaly in the C-Ta collisions at similar energy[10]. Thus we have here an effect clearly associated with the annihilation process. The essential point in support of the quark matter hypothesis is that strangeness is produced more efficiently and hence more abundantly in the quark gluon plasma. This is particularly true at low temperatures considered here. In the quark matter phase the strangeness production time constant[8] is 3 10^{-22} sec at T = 60 MeV. This time is of similar magnitude as the estimate for the fireball to transverse the nucleus and hence there is chemical equilibrium of strange quark abundance in the quark matter fireball should such be formed. It is worthwhile to reemphasize that in this interpretation the dominant 'cold' sector of the lambda and K_S spectrum originates in the quark matter phase, while the 'hot' component has its origin either in the nuclear gas phase, created either in non - quark matter reactions or in the reheating in the final stages of fireball evolution.

REFERENCES

1. J. Rafelski, Phys. Lett. B91 (1980) 281; J. Rafelski, 'Deep Annihilation on Nuclei', Villars, 1987.
2. J. Rafelski, Phys. Lett. B__ (1988) in press.
3. K. Miyano, Y. Noguchi, M. Fukawa, E. Kohriki, F. Ochiai, T. Sato, A. Suzuki, K. Takahashi, Y. Yoshimura, N. Fujiwara, S. Noguchi, S. Yamashita and A. Ono, Phys. Rev. Lett. 53 (1984) 1725 and 'Neutral Strange Particle Productions and Inelastic Cross Section in p+Ta Reaction at 4 GeV/c', KEK Preprint 87-160 February 1988).
4. B. Muller, 'Physics of Quark Gluon Plasma' Springer Lecture Notes in Physics 225 (1985); J. Rafelski and M. Danos in Springer Lecture Notes in Physics 231 (1985) pp. 361-455.
5. C.M. Ko and R.Yuan, Phys. Lett. B192 (1987) 31.
6. J. Cugnon, these proceedings, and references therein.
7. J. Rafelski, 'Quark-Gluon Plasma in Antiproton Annihilation on Nuclei', In: Physics at LEAR with Low Energy Cooled Antiprotons, U. Gastaldi and R. Klapisch, eds. Plenum Pub. Corp. (1984).
8. J. Rafelski, Phys. Rep. 88 (1982) 331; J. Rafelski and B. Muller, Phys. Rev. Lett. 48 (1982) 1066 and 56 (1986) 2334(E).
9. J. Rafelski, Nucl. Phys. A428 (1984) 215; N.K.Glendenning and J. Rafelski, Phys. Rev. C31 (1985) 823; P. Koch, B. M ller and J. Rafelski, Phys. Rep. 142 (1986) 167.
10. V.D. Toneev, H. Schulz, K.K. Gudima and G. R pke, Sov. J. Part. Nucl. 17 (1986) 485 [Fiz. Elem. Chastits At. Yadra 17 (1986) 1093] and references therein. See in particular page 511.

AN OBSERVATION OF A LEADING MESON IN $\bar{p}$+Ne REACTION AT 607 MeV/c INCIDENT MOMENTUM.

C. Guaraldo

Laboratori Nazionali di Frascati dell'INFN , I-00044 Frascati, Italy

for the

Bergen-Brescia-Dubna-Frascati-Oslo-Pavia-Torino Collaboration

PS 179 Experiment at LEAR (CERN)

ABSTRACT

A pion or a kaon may behave as a leading particle in low-energy antiproton-nucleus reactions, in agreement with a quark model picture describing beam fragmentation.

Antiproton-nucleus interaction is clearly one of the new and exciting tools available for the exploration of the nucleus and hadronic physics in general. Among the effects resulting from the deposition of a large amount of energy in a well defined region of space, strange particle production has stimulated a number of recent experimental works[1÷9] and has been theoretically studied from various points of view[10÷24].

The specific reasons of interest are the dynamics of multinucleon absorption of antiprotons in nuclear matter and the possibility that, under appropriate conditions, a collection of nucleons could be transformed in a quark-gluon plasma. Λ° hyperon production, due to its intrinsic two-body nature in the $\bar{p}$ absorption process, is the ideal "elementary" reaction for a study of collective effect; while one of the typical signatures of the dehadronization of a region of nuclear matter could be an enhancement in strangeness production[12,15]: in fact, according to Rafelski[12], the growing volume of large quark bags due to B>0 annihilations, can favour an increasing production of $s\bar{s}$ pairs. The same conclusion was reached by Cugnon and Vandermeulen[15] with a model based on kinematical constraints operating in a pure hadronic phase. Moreover, if multinucleon annihilations occur, some unusual reactions may also have the possibility of happening. For instance, reactions with doubly strange final state should become feasible. Because of the destruction of a unit of baryon number of the capturing cluster, these reactions possess a larger Q-value than would otherwise be the case.

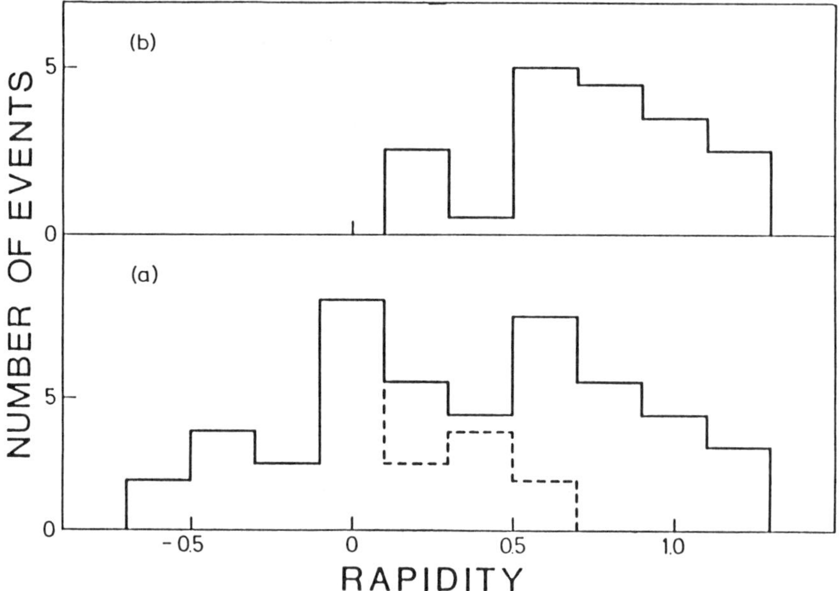

Fig. 1 - For events with a neutral kaon: a) the distribution of rapidity of kaons as sum of two distributions; b) the distribution of the excess.

Therefore, also two groups of pions seem to exist, one emitted from a system with an average rapidity about zero, the other one due to some other emission in the forward direction with an average rapidity about 1.

The fraction of kaons or pions produced, respectively, in the slow system or in the forward direction can be estimated from their excess of positive rapidities. While about 50% of the kaons are produced in each system, about 30% of the lambdas are produced with a negative pion which is emitted in the forward direction.

Pions or kaons as leading particles carry some fraction of the momentum of the incident antiproton in the forward direction in the final state. This fraction of the incident momentum is not absorbed by any target system.

The observation of leading mesons is well known from $\bar{p}p$ reactions[25], where, in the annihilation c.m. system,the negative pions tend to follow the direction of the antiproton, and the positive ones the direction of the proton.

However, despite experimental and theoretical investigations, details of the strangeness production mechanisms are still far from being completely understood.

In a previous work[8] we reported $\Lambda°$ and $K°_S$ production cross sections and rapidity distributions for the reaction $\overline{p}$+Ne→V°+ X (V°=$\Lambda°$,$K°_S$) at 607 MeV/c. In the same paper, rapidity distributions of π^-, both associated and not to strangeness production, were shown. Some conclusions on the effective targets involved were drawn, but without taking into account the role of the associated π's.

In this work we examine in more details our results on rapidity and present them in a different way, trying, within the limits of the present statistics, to give an explanation to the puzzling feature inherent in the previous analysis: the fact that both the π^- rapidity distributions - a parameter not measured in the Japanese experiment[4] - present just quite different average values compared to those of the respective associated strange particles: as if pions were produced from totally different effective targets.

In our data[8] the emission of the $\Lambda°$'s is nearly isotropic in the laboratory system, with an average rapiditiy $<y>=0.07\pm0.03$.

The average longitudinal rapidity of the kaons is significantly larger, namely $<y>=0.33\pm0.07$.

As it is shown in Fig. 1, the rapidity distribution of the kaons can be considered, within the limits of our statistics, as the sum of two distributions: a first one, about symmetrical around zero (Fig. 1a) and a second one given by the difference between the full distribution and the symmetrical one. This difference is characterized by an excess of large rapidity kaons (Fig. 1b). The distribution of the excess extends from $y\simeq 0.1$ to $y\simeq 1.3$, with a average value $<y>\simeq0.7$, which corresponds to a longitudinal kaon momentum of about 400 MeV/c.

Thus, two groups of kaons seem to exist, one emitted from a system with an average rapidity about zero, the other one emitted in the forward direction with an average rapidity about 0.7.

The negative pions have an average longitudinal rapidity, in the laboratory system which, for the events with a lambda, is 0.31 ± 0.08 but, for the events with a neutral kaon, is about $0(0.01\pm0.06)$.

In the same way as for the kaons, also the rapidity of the π^- with a $\Lambda°$ can be considered, within the limits of our statistics, as the sum of two distributions: one with a mean value about zero and the other one with an excess of large rapidity pions (see Fig. 2). The distribution of the excess extends from $y\simeq 0.2$ to $y\simeq1.8$. The average value is $<y>\simeq 1$, corresponding to a pion longitudinal momentum of about 200 MeV/c.

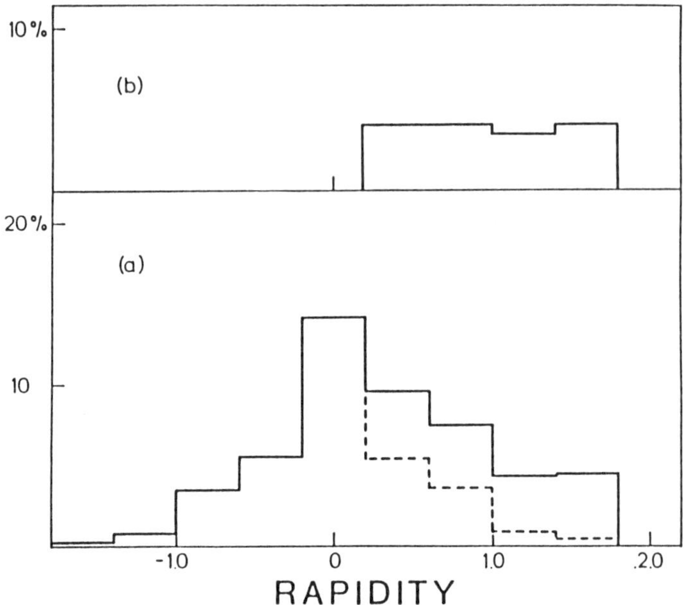

Fig. 2 - For events with a lambda: a) the distributions of rapidity of pions as sum of two distributions; b) the distributions of the excess.

If a leading particle is seen in the final state, the emission of a pion or a kaon in the forward direction may be depicted by diagrams like that shown in Fig. 3. From these kinds of diagrams it can be seen that ΛK production can occur both in the beam vertex through double baryon exchange and in the target vertex through antinucleon exchange and a leading pion emitted from the beam vertex. A pair of $K\bar{K}$ may, in principle, be produced on any vertex, although the production of $K\bar{K}$ in the target vertex with a leading pion from the beam vertex has not been observed in this experiment ($<y> \simeq 0$ for π^- associated to K°_S events).

In terms of a quark model picture describing beam fragmentation, Fig. 4 represent the corresponding quark flow diagram. The leading pion and the leading kaon are produced when one antiquark of the impinging antiproton picks up a light quark, as shown in Fig. 3, or an s quark from the sea-quark of a nucleon of the target system.

Discussions with Profs. H. Olsen, L. Kondratyuk and J. Rafelski are gratefully acknowledged.

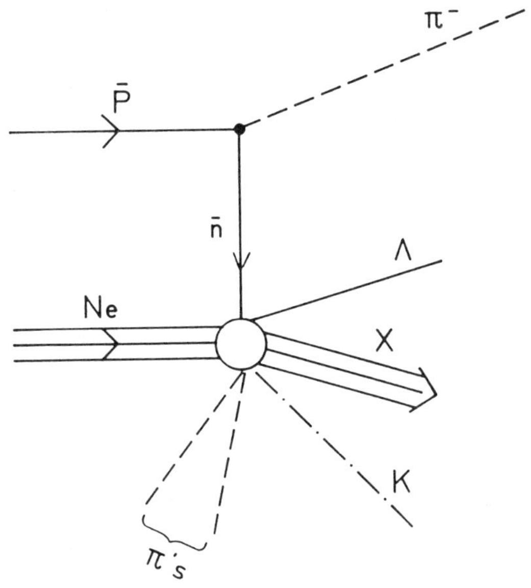

Fig. 3 - Leading meson diagram for Λ° production.

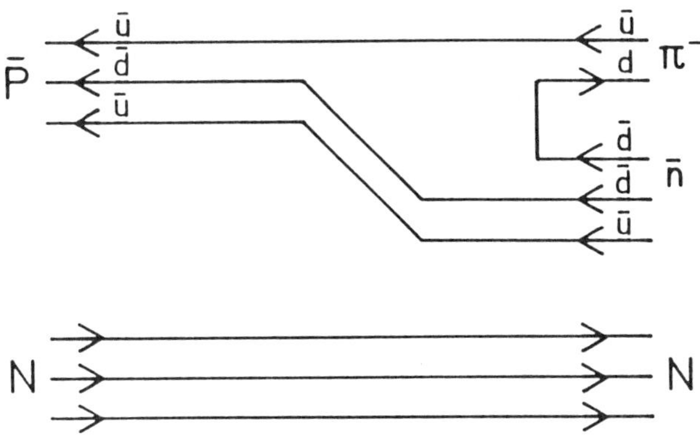

Fig. 4 - Quark flow diagram describing beam fragmentation with a leading pion.

REFERENCES

(1) M.A. Mandelkern et al., Phys. Rev. D27 ,19 (1983) .

(2) G.T. Condo et al., Phys. Rev. C29, 1531 (1984) .

(3) G.T. Condo et al., Phys. Lett. 144B , 27(1984).

(4) K.Miyano et al., Phys. Rev. Lett. 53 , 1725 (1984).

(5) S.J.H. Parkin et al., Nucl. Phys. B277 ,634 (1986) .

(6) G.A. Smith, Preprint PSU/HEP 87-03 (1987).

(7) K. Miyano et al., Proceed. PANIC Conference 1987, abstr. c-86 (1987).

(8) F. Balestra et al., Phys. Lett. 194 B , 192 (1987).

(9) R.A. Lewis and G.A. Smith, Preprint PSU/HEP 87-09 (1987).

(10) J. Rafelski, Phys. Lett. 91 , 281 (1980).

(11) J. Rafelski and M. Danos, Phys. Lett. 97B , 279 (1980).

(12) J. Rafelski and B. Müller, Phys. Rev. Lett. 48 ,1066 (1982) .

(13) J. Rafelski, Nucl. Phys. A418 , 215c (1984) .

(14) N.J. Di Giacomo and M.R. Clover, J. Phys. G: Nucl. Phys. 10 , L 119 (1984).

(15) J. Cugnon and J. Vandermeulen, Phys. Lett. 146 B ,16 (1984) .

(16) D. Strottman and W. R. Gibbs, Phys. Lett. 149 B , 288 (1984).

(17) G.T. Condo, Lett. Nuovo Cimento 42 , 248 (1985).

(18) C. Derreth et al., Phys. Rev. C 31 , 1360 (1985).

(19) S.C. Phatak and M. Sarma, Phys. Rev. C 31 , 2113 (1985).

(20) W.R. Gibbs, in Intersections between Particle and Nuclear Physics, Lake Louise, Canada 1986, editor F. Geesaman, AIP, New York 1986, pag. 505.

(21) C.M. Ko and R. Yuan, Phys. Lett. 192 B ,31 (1987) .

(22) K. Nakai, KEK Preprint 87-53 (1987).

(23) S.C. Phatak and M. Sarma, Phys. Rev. C 36 , 864(1987).

(24) J. Rafelski, Proceed. IV LEAR Workshop, Villars (1987).

(25) R. Armenteros and B. French, in High Energy Physics, edited by E.H.S. Burhop, Academic Press, London 1969, pag. 237.

THE QUARK LINE RULE IN NUCLEON ANTINUCLEON ANNIHILATION

Henning Genz
Institut für theoretische Kernphysik der Universität Karlsruhe, D7500 Karlsruhe, Germany

ABSTRACT

Consequences of the quark line rule are derived from dominance of certain quark line diagrams for nucleon-antinucleon annihilation into two and three mesons.

INTRODUCTION

Any quark line diagram (QLD) such as for example fig. 1 stands for the infinetely many diagrams that are obtained from it by adding gluon lines in an arbitrary way. Since at low energies no reliable estimate of the sum of these diagrams is possible, consequences of QLDs that do not depend on gluon contributions are of particular interest. It is the main purpose of this talk to draw attention to the fact that indeed individual QLDs have this kind of experimental consequences such that their dominance can be tested. Whereas gluons transfer spin, angular momentum and momentum, they do not change the quark flavor flux of any given QLD. Thus predictions following from quark flavor flux alone are the most reliable consequences of assuming that a particular QLD dominates. These consequences should therefore be tested independently of any other assumed properties of such diagrams (such as e.g. the $^3P_0$ or the $^3S_1$ annihilation vertices).

This program has been outlined in references [1, 2, 3]. References [1, 2] concentrate on consequences of only the quark line rule (QLR) whereas reference [3] combines flavor SU(3) with the QLR to obtain the complete set of consequences of quark flavor flux within a given QLD for baryon-antibaryon annihilation into two mesons. Except for technical details, references [3, 4] essentially agree. A more detailed comparison will be contained in reference [5]. We concentrate on the consequences of the QLR in $p\bar{p}$ annihilation.

Two main effective mechanisms of quark-antiquark interactions can be distinguished: annihilation and rearrangement (e.g. figs. 3 and 4). We will obtain consequences of the assumption that either of the two diagrams dominates over the other. If this is the case and, if so, which mechanism dominates depends on the unknown quark-gluon dynamics. From the outset, it is of course unlikely that two different mechanisms are approximately equally important. Before deriving some of the consequences of the assumed dominance of particular diagrams, we write the η and the η' as

$$\eta = x((u\bar{u})_{PS}+(d\bar{d})_{PS})/\sqrt{2}+y(s\bar{s})_{PS}$$
$$\eta' = -y((u\bar{u})_{PS}+(d\bar{d})_{PS})/\sqrt{2}+x(s\bar{s})_{PS} \qquad (1)$$

with $y^2=(1-x^2)$ and define $K=(y/x)^2$. If, contrary to reference [6], there are further contributions to the η and/or η', that however play no part in the η and η' production from $p\bar{p}$, only the numerical factors in the relations to be derived below are changed in an obvious way. We will not discuss meson production processes via the small $\omega-\varphi$-mixing since these may well be smaller than or as large as production processes from the $s\bar{s}$-sea of the nucleon or by other mechanisms.

ANNIHILATION INTO TWO MESONS

Turning to $p\bar{p}\rightarrow M_1M_2$ with M mesons we note first that obviously dominance of fig. 2 implies that production of strange and almost ideally mixed $s\bar{s}$ mesons such as the φ is suppressed. We next consider production of completely neutral mesons via the QLR. These are mixtures of $q\bar{q}$-states with different q, for example $(u\bar{u}+d\bar{d})/\sqrt{2}$ in case of the ω. According to the QLDs, the interaction yields certain coherent mixtures of $q\bar{q}$ that know nothing about the mixing of the different $q\bar{q}$ within the physical mesons. The production cross section of these is computed by taking the scalar product of the $q\bar{q}$ linear combination that has been produced with the linear combination of the physical meson, i.e. via mixing. In order to obtain the corresponding consequences of the QLR it is often easiest and always possible to express those $q\bar{q}$-combinations that cannot be produced via the QLD in question in terms of linear combinations of physical mesons. Linear combinations that cannot be produced according to certain QLDs are for example

$$2xy(s\bar{s})_{PS}=y\,\eta+x\,\eta',\quad \sqrt{2}(d\bar{d})_V=(\omega-\rho^0)\ \text{and}$$
$$x(d\bar{d})_{PS}\sqrt{2}+y(s\bar{s})_V=\eta-x\pi^0. \qquad (2)$$

Obviously, no $s\bar{s}$-component can be produced from any non-strange initial state via the QLDs of either fig. 1 or 2. Within the context of the present paper, this has interesting consequences for the production of η and η' together with (another) η or η' or a π^0, ρ^0 or ω. With $T(M_1M_2)$ the production amplitude for mesons M_1,M_2 and M any one of the mesons listed above, we have $yT(\eta M)=-xT(\eta' M)$ for any linear combination of $p\bar{p}$, $n\bar{n}$, $p\bar{\Delta}$, $n\bar{\Delta}$ or $\Delta\bar{\Delta}$ as initial state. That is, this result is independent of initial state interactions that only lead to isospin mixing. We thus predict for the cross-sections

$$2\sigma(\eta'\eta')=K\sigma(\eta\eta')=2K^2\sigma(\eta\eta),\quad \sigma(\pi^0\eta')=K\sigma(\pi^0\eta),$$
$$\sigma(\rho^0\eta')=K\sigma(\rho^0\eta) \text{ and } \sigma(\omega\eta')=K\sigma(\omega\eta). \tag{3}$$

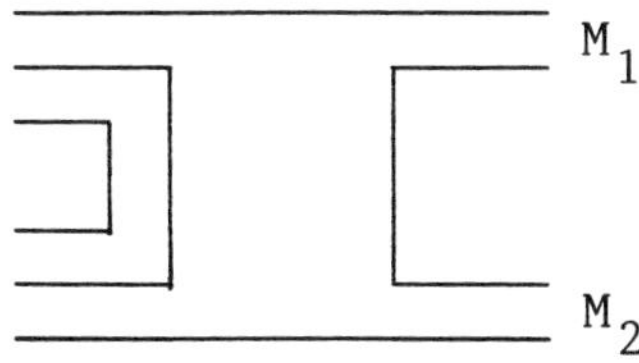

Fig. 1. Annihilation diagram for annihilation into two mesons.

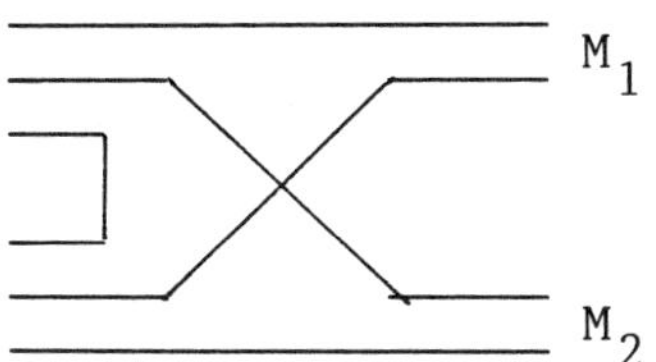

Fig. 2. Rearrangement diagram for annihilation into two mesons.

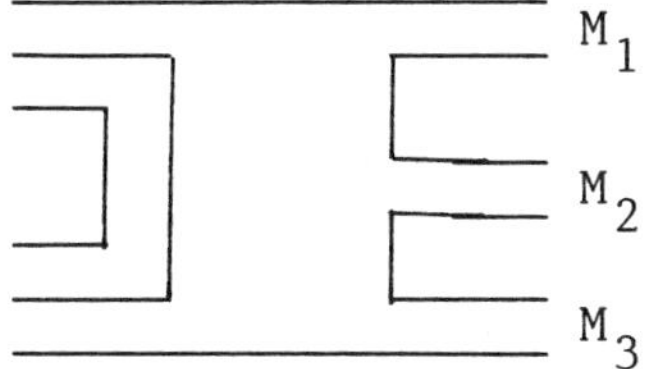

Fig. 3. Annihilation diagram for annihilation into three mesons.

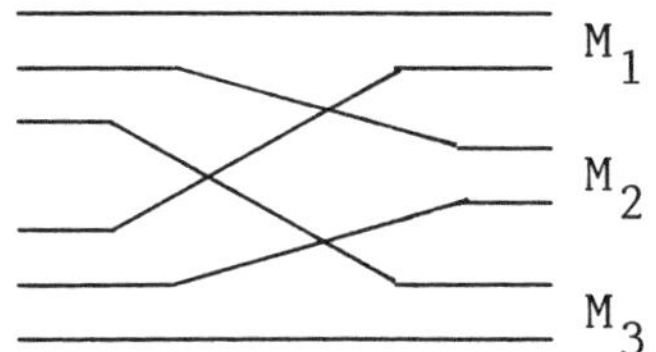

Fig. 4. A rearrangement diagram for annihilation into three mesons.

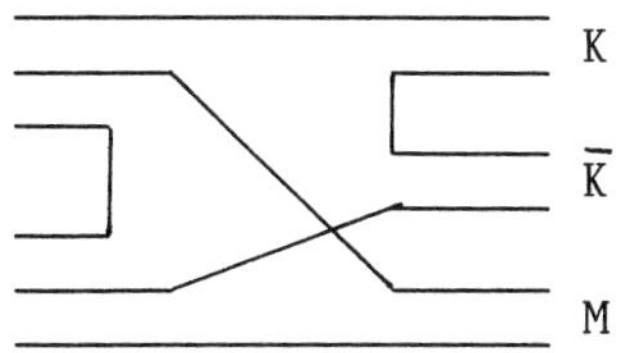

Fig. 5. A mixed diagram for annihilation into a neutral meson and two K.

Here, as everywhere in the paper, <u>phase space correction factors are suppressed</u>. These results are independent of initial state interactions since annihilations that these formulae relate to each other originate from the same initial state. If any one of these is violated, the η and/or η' either are/is more complicated than presently thought or else annihilation into two mesons <u>cannot</u> be described by any linear combination of the amplitudes of figs. 1 and 2. Thus testing these relations is important.

Since via these formulas any <u>two meson</u> η' cross section can be expressed by the corresponding η cross section it suffices to discuss η production. We first turn to production of two mesons via the rearrangement QLD in fig. 2. Since only one d is contained in the proton it <u>is</u> impossible to produce the $d\bar{d}$-component of M_1 together with the $d\bar{d}$-component of M_2 from a $p\bar{p}$ initial state, i.e.

$T((d\bar{d})_1(d\bar{d})_2)=0$ with 1 and 2 independently PS, V or any other space-time meson quantum numbers. This is of particular interest for amplitudes that are even under the interchange of the two mesons. For pseudoscalars, having this property, we may write

$$x^2T(\pi^0\pi^0)+T(\eta\eta)=x(T(\pi^0\eta)+T(\eta\pi^0))=2xT(\pi^0\eta) \qquad (4)$$

We will not assume that these amplitudes are real relatively to each other. Thus we obtain [2] only the triangle inequalities

$$\{\sqrt{(2x^4\sigma(\pi^0\pi^0))}-\sqrt{(2\sigma(\eta\eta))}\}^2 \leqslant 4x^2\sigma(\pi^0\eta) \leqslant \{\sqrt{(2x^4\sigma(\pi^0\pi^0))}+\sqrt{(2\sigma(\eta\eta))}\}^2 \qquad (5)$$

as experimental consequences. This result does not hold up in case of initial state interactions that mix $p\bar{p}$ with $n\bar{n}$, for example, since it depends on the presence of only one d in the initial state of the annihilation. An analogous result can easily be derived for $n\bar{n}$ annihilation. For the weaker results that follow for linear combinations of $p\bar{p}$ and $n\bar{n}$ see reference [5]. In complete analogy to the above it follows that

$$xT(\rho^0\pi^0)+T(\omega\eta)=xT(\omega\pi^0)+T(\rho^0\eta)$$
$$T(\rho^0\rho^0)+T(\omega\omega)=T(\rho^0\omega)+T(\omega\rho^0). \qquad (6)$$

The experimental consequences to be derived [5] from the former relation obviously do not amount to triangle inequalities for three cross sections. From the latter, triangle inequalities for the three cross sections involved follow only under the further assumption that the even ρ^0-ω-amplitude dominates. Other than in reference [2], this assumption is not made here. If the odd part of the amplitude dominates, the same result as from dominance of fig. 1 follows, i.e. $\sigma(\rho^0\rho^0)=\sigma(\omega\omega)$.

Dominance of annihilation (fig. 1) in $p\bar{p}\rightarrow M_1M_2$ has much stronger consequences [1]. Evidently, for any linear combination of baryon-antibaryon (including strange and exited baryons) initial states of the annihilation (i.e. for any initial state interaction that only leads to mixing of these states) and 1 and 2 independently PS, V or any other space-time meson quantum numbers it follows from fig. 1 that $T((q\bar{q})_1(Q\bar{Q})_2)=0$ for $q\neq Q$. Using this in computing meson amplitudes via linear combinations of $q\bar{q}$ amplitudes one finds for example

$$2T(\rho^0\rho^0)=T(((u\bar{u})_V-(d\bar{d})_V)((u\bar{u})_V-(d\bar{d})_V))=$$
$$=T((u\bar{u})_V(u\bar{u})_V)+T((d\bar{d})_V(d\bar{d})_V)=2T(\omega\omega)$$

such that (7)

$$\sigma(\rho^0\rho^0)=\sigma(\omega\omega),\quad x^4\sigma(\pi^0\pi^0)=\sigma(\eta\eta),$$
$$x^2\sigma(\pi^0\rho^0)=\sigma(\eta\omega) \text{ and } x^2\sigma(\pi^0\omega)=\sigma(\eta\rho^0)$$

from dominance of the annihilation diagram. No attempt will be made here to compare these relations to experiment (reference [5]). We note however that the prediction $\sigma(\rho^0\rho^0)=\sigma(\omega\omega)$ only agrees with present experiment within two standard deviations.

ANNIHILATION INTO THREE MESONS

The methods [1, 2] employed here are easily extented to annihilation into three mesons. If [7] rearrangement (fig. 4) dominates, we obtain from the QLR for the $p\bar{p}$ initial state $T((u\bar{u})(u\bar{u})(u\bar{u}))=T((u\bar{u})(d\bar{d})(d\bar{d}))=T((d\bar{d}),(d\bar{d}),(d\bar{d}))=0$ for any space-time quantum numbers and any permutation of the individual $q\bar{q}$. Thus

$$\sigma(\rho^0\rho^0\rho^0)=\sigma(\omega\omega\omega),\ x^2\sigma(\rho^0\rho^0\pi^0)=\sigma(\omega\omega\eta),\ x^4\sigma(\rho^0\pi^0\pi^0)=\sigma(\omega\eta\eta)\ \text{and}\ x^6\sigma(\pi^0\pi^0\pi^0)=\sigma(\eta\eta\eta). \quad (8)$$

Since, according to fig. 4 both η and η' can be produced together with η, η', π^0, ρ^0 and ω only via their $(u\bar{u}+d\bar{d})$-content, we furthermore have (compare eq. (3))

$$\sigma(M_1M_2\eta')=K\sigma(M_1M_2\eta),\ 2\sigma(M\eta'\eta')=K\sigma(M\eta\eta')=2K^2\sigma(M\eta\eta),\ 3\sigma(\eta'\eta'\eta')=K\sigma(\eta\eta'\eta')=K^2\sigma(\eta\eta\eta')=3K^3\sigma(\eta\eta\eta), \quad (9)$$

with M, M_1 and M_2 any meson π^0, ρ^0 or ω.

If, on the other hand annihilation (fig. 3) dominates, the amplitude vanishes if (for any space-time meson quantum numbers) any two of the three meson states M_1, M_2, M_3 are $q\bar{q}$ and $Q\bar{Q}$ with $Q\neq q$. As always in case of annihilation, this remains true for any initial state interaction that only results in isospin mixing. Thus

$$\begin{gathered}3\sigma(\omega\omega\omega)=\sigma(\rho^0\rho^0\omega),\ \sigma(\omega\omega\eta)=\sigma(\rho^0\rho^0\eta),\\ 2\sigma(\omega\eta\omega)=x^2\sigma(\rho^0\pi^0\omega),\ 2\sigma(\omega\eta\eta)=x^2\sigma(\rho^0\pi^0\eta),\\ \sigma(\eta\eta\omega)=x^4\sigma(\pi^0\pi^0\omega),\ 3\sigma(\eta\eta\eta)=x^4\sigma(\pi^0\pi^0\eta),\\ 3\sigma(\rho^0\rho^0\rho^0)=\sigma(\omega\omega\rho^0),\ \sigma(\rho^0\rho^0\pi^0)=\sigma(\omega\omega\pi^0),\\ 2x^2\sigma(\rho^0\pi^0\rho^0)=\sigma(\omega\eta\rho^0),\ 2x^2\sigma(\rho^0\pi^0\pi^0)=\sigma(\omega\eta\pi^0),\\ x^4\sigma(\pi^0\pi^0\rho^0)=\sigma(\eta\eta\rho^0)\ \text{and}\ 3x^4\sigma(\pi^0\pi^0\pi^0)=\sigma(\eta\eta\pi^0).\end{gathered} \quad (10)$$

Eqns. (9) also follow from fig. 3. They therefore are valid in any model based on a linear combination of annihilation (fig. 3) and rearrangemant (fig. 4) amplitudes.

If rearrangement (fig. 4) dominates, K meson production is suppressed. If annihilation (fig. 3) dominates, $T(K^+K^-(d\bar{d}))=T(K^0\bar{K}^0(u\bar{u}))=0$, where K and $q\bar{q}$ can have any space-time quantum numbers. Then

$$\sigma(K^+K^-\rho^0)=\sigma(K^+K^-\omega)\ \text{and}\ \sigma(K^0\bar{K}^0\rho^0)=\sigma(K^0\bar{K}^0\omega) \quad (11)$$

again for any initial state interaction that only leads to isospin mixing. The reader should notice that with M_1

and M_3 K-mesons, the M_2 in fig. 3 may be an $s\bar{s}$ state. Thus the η may also be produced together witk $K\bar{K}$ via its $s\bar{s}$ component. For these processes it only follows that

$$\begin{aligned} &xT(K^+K^-\eta)+yT(K^+K^-\eta')=T(K^+K^-\pi^o) \\ &xT(K^o\bar{K}^o\eta)+yT(K^o\bar{K}^o\eta')=-T(K^o\bar{K}^o\pi^o) \end{aligned} \quad (12)$$

As experimental consequences triangle inequalities for cross sections follow, i.e.

$$\begin{aligned} &\{\sqrt{(x^2\sigma(K\bar{K}\eta))}-\sqrt{(y^2\sigma(K\bar{K}\eta'))}\}^2 \leq \sigma(K\bar{K}\pi^o) \leq \\ &\leq \{\sqrt{(x^2\sigma(K\bar{K}\eta))}+\sqrt{(y^2\sigma(K\bar{K}\eta'))}\}^2 \end{aligned} \quad (13)$$

with $K\bar{K}$ standing for either K^+K^- or $K^o\bar{K}^o$.

If fig. 4 dominates, K meson production is suppressed. K mesons in this case may well be predominantly produced via the mixed diagram in fig. 5 since that introduces only the minimal number of annihilations. From this $T(K^+K^-(s\bar{s}))=T(K^o\bar{K}^o(s\bar{s}))= =T(K^o\bar{K}^o(d\bar{d}))=0$ with the consequences

$$y^2\sigma(K^o\bar{K}^o\pi^o)=\sigma(K^o\bar{K}^o\eta')=K\sigma(K^o\bar{K}^o\eta),\ \sigma(K^+K^-\eta')=K\sigma(K^+K^-\eta) \\ \text{and } \sigma(K^o\bar{K}^o\rho^o)=\sigma(K^o\bar{K}^o\omega) \quad (14)$$

for annihilation from the $p\bar{p}$ initial state only. It also follows that for any initial state that can be reached via isospin mixing, $\sigma(K^+K^-\varphi)$ and $\sigma(K^o\bar{K}^o\varphi)$ should be small compared to the corresponding ω cross section. The reader should notice that $\sigma(K^o\bar{K}^o\rho^o)=\sigma(K^o\bar{K}^o\omega)$ follows from both diagrams for KKM production. Without change, K can be replaced by K* everywhere in the above.

ACKNOWLEDGMENTS

It is a pleasure to acknowledge useful discussions with and contributions from M. Martinis and S. Tatur. This work has been supported by the Bundesministerium für Forschung und Technologie under grant number MEP02-34KAG.

REFERENCES

1. H. Genz, Phys. Rev. D28, 1094 (1983)
2. H. Genz, Phys. Rev. D31, 1136 (1985)
3. M. Malvetti, $SU_f(3)$-Analyse von Baryon-Antibaryon Reaktionen, Diplomarbeit, Karlsruhe (1985)
4. U. Hartmann et.al., Phys. Lett. 155B, 163 (1985)
5. H. Genz, M. Martinis and S. Tatur: Something, but not much follows from the Quark Line Rule in Baryon-Antibaryon interactions, Karlsruhe Preprint, in preparation
6. F. Gilman and R. Kauffman, Phys. Rev. D36, 2761 (1987)
7. H. Rubinstein and H. Stern, Phys. Lett. 21, 447 (1986)

SEARCH FOR CP VIOLATION IN THE $K\bar{K}$-SYSTEM AT LEAR

CP-LEAR COLLABORATION
Basel, Boston, CEN-Saclay, CERN, Coimbra, DAP-Athens, Delft, ETH-Zürich, Fribourg, Ioannina, Liverpool, Marseille, PSI (SIN), Stockholm, Thessaloniki

presented by G. Backenstoss, Basel

ABSTRACT

A novel method is described to measure several CP violating effects with pure K^o- and $\overline{K^o}$-beams proposed at CERN-LEAR and based on measurements of asymmetries due to the interference between CP-violating and CP-conserving amplitudes.

METHODS

CP violation has been established in 1964 in the K_L^o-decay by observing decay into 2π [1]. The admixture of the CP-impurity described by a parameter ε amounted to $\sim 2\cdot 10^{-3}$. Almost 25 years later the origin of the CP violation is not yet understood though the refinements of the mesurements and the knowledge of details have made great progress. It is a pleasure to refer to the review of D. Marlow given at this conference [2]. It is obvious that in view of the high precision required for ε'/ε to distinguish between various models and to trace back the origin of CP violation a new measurement based on a completely different concept should be highly appreciated. The here presented method was proposed at CERN [3] to be performed at its LEAR (low energy antiproton ring) after upgrading by installing the accumulator ring (ACOL). Meanwhile the proposal has been accepted, the apparatus is with some modifications under construction and data taking should start in the coming year.

In contrary to previous and recent experiments which measure decays of K_L and interferences with (regenerated) K_S the present experiment starts with pure $|K^o\rangle$ and $|\overline{K^o}\rangle$ states. These are amply delivered by $\bar{p}$'s from LEAR stopping in hydrogen by the reactions

$$
\begin{aligned}
p\bar{p} &\to K^o K^- \pi^+ \\
&\to \bar{K}^o K^+ \pi^-
\end{aligned}
$$

in equal numbers (BR $\sim 2\cdot 10^{-3}$ each) and with equal momentum distributions. They can be tagged by detecting $K^-\pi^+$ and $K^+\pi^-$ respectively. It becomes evident that resorting to these initial states most of the systematic errors dominating present experimental errors (e.g. acceptance calculations) are avoided. The method is based on the direct observation of the interference term between CP conserving and CP violating amplitudes which allows not only to determine the complex CP violation parameters but is also capable to measure these in all three K^o decay modes

$$K_S \to \pi^+\pi^- \text{ and } \pi^o\pi^o \quad \text{BR } 100\%$$
$$K_L \to \pi^+\pi^-\pi^o \text{ and } \pi^o\pi^o\pi^o \quad \text{BR } 64\%$$
$$K_L \to \pi^-\ell^+r \text{ and } \pi^+\ell^-r \quad \text{BR } 66\%$$

OBSERVABLES

The time dependent decay rate of $K^o(\bar{K}^o)$ into a given final state can be described by

$$R_{K^o\to f}(t) = |a_S|^2 e^{-\gamma_S t} + |a_L|^2 e^{-\gamma_L t} + |a_S|\,|a_L|\; e^{-(\gamma_S+\gamma_L)\frac{t}{2}} \cos(\Delta mt+\phi)$$

for equal amounts of K_S and K_L as realized by pure $K^o(\bar{K}^o)$ states. The third term is the interference term of interest. It can only be observed because $\Delta m \approx \gamma_S/2$, a kind gift of nature; Δm being the mass difference between K_S and K_L and γ the respective widths. With the well known definition of η for the ratio between CP violating and CP conserving amplitudes

$$\eta_f = \frac{A(K_L \to f)}{A(K_S \to f)} = |\eta| e^{i\Phi}$$

We can measure the time dependent asymmetry for any of the above mentioned final states f

$$A_f(t) = \frac{R(\bar{K}^o \to f)(t) - R(K^o \to f)(t)}{R(\bar{K}^o \to f)(t) + R(K^o \to f)(t)}$$

In particular one obtains for the $\pi^+\pi^-$ or $\pi^o\pi^o$ final states keeping in mind the well known definitions for η, the ratio between CP violating and CP conserving amplitudes

$$\eta_{\substack{+-\\oo}} = \frac{\langle \pi^{\substack{+\\o}}\pi^{\substack{-\\o}}|H|K_L\rangle}{\langle \pi^{\substack{+\\o}}\pi^{\substack{-\\o}}|H|K_S\rangle} = |\eta_{\substack{+-\\oo}}|\; e^{i\phi_{\substack{+-\\oo}}}$$

$$\text{and} \quad \eta_{+-} = \varepsilon + \varepsilon'$$
$$\eta_{oo} = \varepsilon - 2\,\varepsilon'$$

$$A_{\pi\pi}(t) = 2\left[\frac{\eta_{|\pi\pi|}\ e^{\gamma_S t/2}\cos(\Delta m\ t - \phi_{\pi\pi})}{1 + |\eta_{\pi\pi}|^2\ e^{\gamma_S t}} - \mathrm{Re}\,\varepsilon\right]$$

A Monte Carlo simulation of $A_{\pi\pi}(t)$ as a function of the eigentime t_S is shown in fig.1).

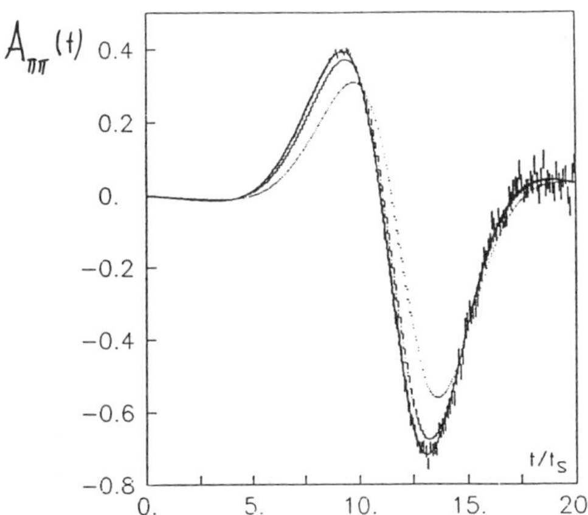

Fig.1) The asymmetry A(t) with perfect time resolution (points with error bars) and with finite time resolutions of $0.1\tau_S$, $0.5\tau_S$ and $1.0\tau_S$. The curve with the resolution $0.1\tau_S$ practically coincides with the curve with perfect resolution.

For both the $\pi^+\pi^-$ and the $\pi^o\pi^o$ channels the parameter $|\eta_{+-}|$ and ϕ_{+-} as well as ϕ_{oo} can be extracted directly being related to ε and ε'. It should be noted that the maximum interference effect occurs for $5\tau_S < t < 15\tau_S$. As can be seen from fig.1) a good resolution in terms of τ_S is required which depends on the accuracy with which the decay vertex can be reconstructed. However, also an asymmetry integrated over time to a point above the interference structure ($\sim 20\tau_S$) leads to a determination of ε and ε', but not the phase ϕ

$$I(t_o) = \frac{\int_o^{t_o} [R(\overline{K^o} \to f) - R(K^o \to f]dt}{\int_o^{t_o} [R(\overline{K^o} \to f) + R(K^o \to f]dt}$$

yielding $I_{+-} = 2\mathrm{Re}\varepsilon + 4\ \mathrm{Re}\varepsilon'$
$I_{oo} = 2\mathrm{Re}\varepsilon - 8\ \mathrm{Re}\varepsilon'$

$$\varepsilon'/\varepsilon = \frac{1}{6}\,(I_{+-}/I_{oo} - 1)$$

A very similar treatment leads to corresponding expressions for 3π- and the semileptonic final states.
For example

$$A_{3\pi}(t) = 2[\mathrm{Re}\varepsilon - |\eta_{3\pi}|\, e^{-(\gamma_S-\gamma_L)t} \cos(\Delta mt + \phi_{3\pi})]$$

and
$$I_{3\pi} = \frac{4}{\gamma_S t_o} \mathrm{Re}\,\eta_{3\pi} - 2\,\mathrm{Re}\,\varepsilon$$

It should be noted that obviously the interference occurs at shorter times, typically $t < 8\tau_S$.

For semileptonic decays the observables are

$$X = \frac{a(\Delta S = -\Delta Q)}{a(\Delta S = \Delta Q)} = \frac{<\pi^-\ell^+\nu|H|\overline{K^0}>}{<\pi^-\ell^+\nu|H|K^0>}$$

Here ReX provides a test for the $\Delta S=\Delta Q$ rule, ImX > 0 would show CP violation, whereas the asymmetry
$A_T = \frac{R_{\bar{K}\to K}(t) - R_{K\to\bar{K}}(t)}{R_{\bar{K}\to K}(t) + R_{K\to\bar{K}}(t)}$ measures T violation provided $\Delta S=\Delta Q$.

EXPERIMENT

In order to measure the quantities described above with sufficient accuracy the detecting system must meet certain partially stringent requirements:

- Tagging K^0 and $\overline{K^0}$ by identifying $K^-\pi^+$and $K^+\pi^-$
- Measuring the momenta of $K^-\pi^+$ and $K^+\pi^-$ in order to determine the momenta of K^0 and $\overline{K}^0$
- Identifying the $K^0(\overline{K}^0)$-decay channels
- Determine decay vertex of $K^0(\overline{K^0})$ in order to know decay time
- High detecting efficiency at high rates
- Fast,efficient trigger possibilities
- Large volume with low density to allow decay lengths up to $20\tau_S$ and prevent regeneration

A cross section of the detector perpendicular to the incoming $\bar{p}$-beam is shown in fig.2) with a typical $\overline{K}^0\to\pi^0\pi^0$ event. The entire detecting system is contained in a solenoid providing a magnetic field of 0.45 T over a length of 3.3m and a diameter of 2m. In the center a gazeous hydrogen target is located pressurized to $\sim$15 bar to stop $\bar{p}$'s entering with $\leq$ 200 MeV/c. The tracking chambers extended to r=.6m corresponding to about $20\tau_S$ for a K^0-momentum of typically 500 MeV/c. The two innermost chambers are cylindrical MWPC followed by 6 drift chambers to measure the momenta of the charged $p\bar{p}$ annihilation products and the charged K^0-decay products. The trackchambers are sur-

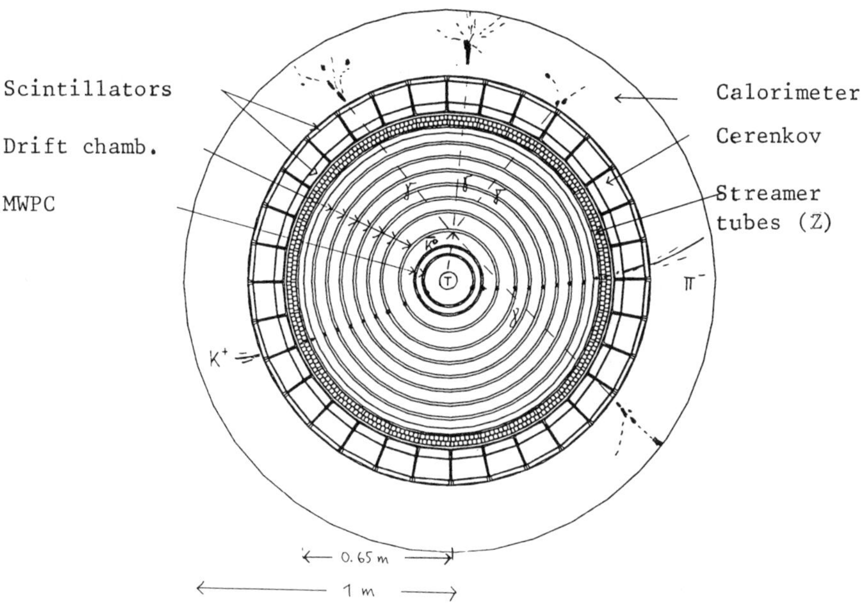

Fig.2) Cross section of the detecting system perpendicular to the solenoid and the $\bar{p}$-beam with a simulation of a $K^+\pi^-\overline{K^0} \rightarrow 2\pi^0$ event. T : H_2 target.

---

rounded by two layers of streamer tubes to add to the track chamber readings a fast readout of the Z-coordinates. The next leyer consists of 32 trapezoidally shaped plastic boxes sandwiched between plastic scintillators for triggering and TOF measurements of a length of 3.1m and a depth of 8cm containing freon (FC 72 and PPO as wave length shifter) acting as threshold Cerenkov counters. These counters have been developped to optimize the rejection of the kaons and the efficiency for detecting pions for the momenta dominant in the events being selected. Extensive tests with pions of varying momenta, positions and incident angles have been performed at SIN. The outmost cylinder is formed by the electromagentic calorimeter to detect γ's originating from decay π^0's. The calorimeter is of the gas sampling type consisting of 19 layers of Pb sandwiching the gas tubes accompagnied by two layers of strips above and beneath, running at $\pm 60^0$ which provide the Z coordinate. The tubes of quadratic cross section (4x4 mm^2) have two brass walls to define the electric field and two high resistivity walls on top and bottom to allow pickup of the pulse by the strips. The total thickness of the detector amounts to 6 radiation lengths. The energy of the shower is determined by counting the number of hits which yields an energy resolution which is more important for the reconstruction of the decay vertex is, however, of the order of a few mm. A prototype detector has been built and was successfully tested at SIN with pions, electrons and γ's with respect to spacial and energy resolution as well as e-π identification.

The experiment needs high statistics, it will be capable to accept $\sim 2\cdot 10^6$ $\bar{p}$/s. Efficient fast triggers are vital to select the good events, about a few 10^2, for further off-line processing. This will be done in several steps. A "strangeness trigger" (2 or 4 hits in the track chambers, 1 or 3 hits in the Cerenkov) reduces the accepted events within 40ns by a factor 5, after 200ns a p_T-cut is available providing a reduction by a factor 10, after 1μs a cut on K^0 mass is possible and after 3μs calorimeter information is available. Thus the dead time should never exceed 10%.

EXPECTED RESULTS

We show in table 1 the quantities accessible in the present experiment and give some indication to which accuracy they are expected to be determined with 10^{13} $\bar{p}$'s delivered by LEAR.

Table 1

Parameter	Current precision	Expected precision	Comments
ε'/ε	$\pm 1.4\cdot 10^{-3}$	$< 2\cdot 10^{-3}$	Small systematic error
$\phi_{+-}-\phi_{00}$	5^0	$< 1^0$	Direct measurement
η_{+-0}	<0.12	$7\cdot 10^{-4}$	Direct measurement
ReX	$<2\cdot 10^{-2}$	$6\cdot 10^{-4}$	$\Delta S=\Delta Q$ rule
ImX	$2.6\cdot 10^{-2}$	$7\cdot 10^{-4}$	CP-violation
$\sigma(A_T)/A_T$	–	$6\cdot 4\cdot 10^{-2}$	T-violation

It is obvious that in view of the importance of ε'/ε and the difficulties of its measurement a completely new method containing less systematic errors should be appreciated even when the accuracy is not improved by a big factor. All other data yield qualitatively new results important for our understanding of the origin of CP-violation.

REFERENCES

1. J.H. Christenson, J.W. Cronin, V.L. Fitch and R.Turlay Phys. Rev. Lett. 13 (1964) 138
2. D. Marlow, These proceedings
3. L. Adiels et al., proposal for experiment PS 195 CERN/PSCC/85-6 and PSCC/P-82 (1985)

SEARCH FOR CP VIOLATION IN HYPERON – ANTIHYPERON DECAYS WITH PROTON – ANTIPROTON EXPERIMENTS

Nikolaus H. Hamann
University of Freiburg, 7800 Freiburg, Fed. Rep. Germany

ABSTRACT

The motivation to search for CP violation in hyperon decays, and the CP violation observables experimentally accessible, are reviewed. Results from a comparison of $\Lambda \to p\pi^-$ and $\bar{\Lambda} \to \bar{p}\pi^+$ decay asymmetries are presented. Possibilities and limitations with $\bar{p}p \to \bar{\Lambda}\Lambda$ experiments at LEAR, and with a $\bar{p}p \to \bar{\Xi}\Xi$ measurement at Super-LEAR, are outlined.

INTRODUCTION

The origin of CP non-conservation is one of the fundamental questions to be addressed in weak interactions [1-2]. In particular, no system other than the $K^0 - \bar{K}^0$ system has exhibited CP violation so far.

The violation of CP invariance was discovered in 1964 by Christenson et al. [3], who measured a non-zero branching ratio for the decay $K_L \to \pi^+\pi^-$. In this experiment and others performed since then [4], the only non-zero measure of CP violation has been the parameter $|\varepsilon| = (2.27 \pm 0.02) \times 10^{-3}$. Since this parameter defines the mass eigenstates K_S and K_L, and the amount of CP impurity in them, the observations have been attributed to a $\Delta S = 2$ CP violating current in the $K^0 - \bar{K}^0$ transitions.

Recently, the decays of K_S and K_L both into $\pi^0\pi^0$ and $\pi^+\pi^-$ final states have been measured [5] by experiment NA31 at CERN – SPS. For the first time, this yielded another non-zero measure of CP violation, $|\varepsilon'/\varepsilon| = (3.3 \pm 1.1) \times 10^{-3}$. Since ε' is related to the decay of the pure CP $= -1$ component of K_L, the result constitutes first experimental evidence for $\Delta S = 1$ direct CP violation in the actual two-pion decays of neutral kaons. A similar experiment with different techniques, E731 at FNAL, has yielded [6] the result $|\varepsilon'/\varepsilon| = (3.2 \pm 3.0) \times 10^{-3}$. More data with improved accuracy can be expected from these important experiments.

In the Standard Model of electroweak interactions with at least three quark generations (u,d′), (c,s′), and (t,b′), the Kobayashi – Maskawa (KM) mixing matrix connects the weak eigenstates d′, s′, and b′ to the mass eigenstates d, s, and b. In addition to mixing angles, some of the KM matrix elements naturally contain a phase which appears to be the one and only source of CP violation. Both types of CP violation discussed above can be explained [1-2]. The $\Delta S = 2$ "indirect" CP violation from $K^0 - \bar{K}^0$ mixing is created by "box diagrams", whereas the $\Delta S = 1$ "direct" CP violation in neutral kaon decays is due to "penguin diagrams".

The mechanism suggested by the Standard Model, however, does not answer the question of the physical origin of CP non-conservation. This will have to be answered mainly by present and future experiments searching for new phenomena of CP violation. If successful, such experiments can be considered a key for new physics.

Among the proposed searches for new signals of CP violation the most prominent are the investigation of rare decays of kaons, searches for CP violation in the $B^0 - \bar{B}^0$ system, and the investigation of decay asymmetries in hyperon – antihyperon systems such as $\bar{\Lambda}\Lambda$ or $\bar{\Xi}\Xi$. The latter is the subject of this paper.

Reactions $\bar{p}p \to \bar{Y}Y$ (where Y denotes Λ or Ξ) can provide a particularly clean laboratory for CP violation studies. To the extent that the hadronic production process is charge-conjugation invariant, Y and $\bar{Y}$ are produced with equal polarizations. Parity conservation requires the polarizations to be transverse to the production plane. Since the $\bar{p}p$ initial state, and thus the $\bar{Y}Y$ final state too, have a definite CP property, final state interactions can not generate a misleading signal. Owing to baryon-number conservation, there is no $Y-\bar{Y}$ mixing, and therefore any signal constitutes a measure of pure $\Delta S = 1$ CP violation.

HYPERON NON-LEPTONIC DECAYS

The non-leptonic decays of hyperons, such as $\Lambda \to p\pi^-$ and $\Xi^- \to \Lambda\pi^-$, are weak and parity-violating decays. They are characterized by mixtures of S and P waves. Following Donoghue et al. [7-8], the partial waves of a given decay mode can be parametrized in terms of S- and P-wave amplitudes, hadronic pion−baryon final state interaction phases δ, and weak CP violation phases ϕ, each summed over isospins $I = 1/2$ and $3/2$.

The observables to be formed from the decay amplitudes are the transition rate

$$\Gamma \sim |S|^2 + |P|^2 ,$$

and the correlation parameters

$$\alpha = 2\,\mathrm{Re}\,(S^*P)/(|S|^2 + |P|^2) \text{ and}$$

$$\beta = 2\,\mathrm{Im}\,(S^*P)/(|S|^2 + |P|^2) ,$$

which govern the asymmetry of the decay angular distribution and the final baryon polarization, respectively. The two decay parameters are related by a phase ψ according to

$$\beta = [\sqrt{(1-\alpha^2)}]\,\sin\psi .$$

Several experimentally observable quantities have been considered [7-8], which, if measured to be non-zero, would signal CP violation. These are

$$\Delta = (\Gamma - \bar{\Gamma})/(\Gamma + \bar{\Gamma}) ,$$

$$A = (\alpha + \bar{\alpha})/(\alpha - \bar{\alpha}) ,$$

$$B = (\beta + \bar{\beta})/(\beta - \bar{\beta}) \text{ , and}$$

$$B' = (\beta + \bar{\beta})/(\alpha - \bar{\alpha}) .$$

To leading order in the $I = 3/2$ contributions and in the hadronic phases, the size of the signals is [7-8]

$$\Delta \approx -2\,(S_3/S_1)\,\sin(\delta^S{}_3 - \delta^S{}_1)\,\sin(\phi^S{}_3 - \phi^S{}_1) ,$$

$$A \approx \sin(\phi^S{}_1 - \phi^P{}_1)\,\sin(\delta^P{}_1 - \delta^S{}_1) ,$$

$$B \approx \sin(\phi^S{}_1 - \phi^P{}_1)\,/\,\tan(\delta^P{}_1 - \delta^S{}_1) \text{ , and}$$

$$B' \approx \sin(\phi^S{}_1 - \phi^P{}_1) .$$

Therefore the orders of magnitude are

$$0.1B \approx B' \approx 10A \approx 100\Delta.$$

The absolute size of the signals as predicted in the framework of the Standard Model is small. The values of A given by Donoghue [9], based on the first non-zero measurement of $|\epsilon'/\epsilon|$, are roughly $A = (0.3 \text{ to } 3.0) \times 10^{-4}$ for $\Lambda \to p\pi^-$, and $A = (0.4 \text{ to } 3.6) \times 10^{-4}$ for $\Xi^- \to \Lambda\pi^-$.

The ratio A is relatively easy to measure, as shown below. It is directly accessible with an accuracy determined by the number of analysed events. The ratios B and B′ have a higher sensitivity to CP violation, but their determination requires the measurement of the final baryon polarization (in the case of Λ hyperons, this would need a secondary scattering). The ratio Δ can also be measured directly, but it has a very low sensitivity to CP violation.

EXPERIMENTAL RESULTS

It is important to note that a dedicated CP violation experiment with hyperons has yet to be performed. The experimental results in this context have been obtained so far more or less as byproducts, and they should therefore be regarded as feasibility tests rather than particularly meaningful numbers.

In all experiments, only the products $(\alpha P)_\Lambda$ and $(\alpha P)_\Lambda$ of the decay asymmetry parameter $\alpha = 0.642 \pm 0.013$ and the polarization P have been evaluated. In the Λ rest frame, the angular distribution of the outgoing protons from $\Lambda \to p\pi^-$ decays is given by

$$W(\theta_p) = W_0[1 + (\alpha P)_\Lambda \cos\theta_p] ,$$

where θ_p is measured between the normal of the production plane and the proton momentum. Parity violation in these decays thus manifests itself in an up−down asymmetry of the decay angular distribution. The degree of asymmetry in these "self-analysing" decays is determined by the product αP. In the absence of CP violation, however, $\alpha = -\bar{\alpha}$ and therefore $A = 0$ are expected.

Below we briefly describe the two initial experimental approaches to the question of CP violation in hyperon decays. The first one [10] refers to a study of inclusive production of Λ and $\bar{\Lambda}$ particles done at CERN−ISR. The second one [11] refers to an experiment at CERN−LEAR in which the exclusive associated production $\bar{p}p \to \bar{\Lambda}\Lambda$ is being investigated [12].

Inclusive production of Λ and $\bar{\Lambda}$

With experiment R608 [10] at CERN−ISR, the forward inclusive production of Λ and $\bar{\Lambda}$ hyperons was measured in the reactions $pp \to \Lambda + X$ and $\bar{p}p \to \bar{\Lambda} + X$, respectively, at $\sqrt{s} = 30.8$ GeV. Charged particles from $\Lambda \to p\pi^-$ and $\bar{\Lambda} \to \bar{p}\pi^+$ decays were detected in a double-septum magnetic forward spectrometer installed above and below the ISR beam pipe.

In this experiment, Λ and $\bar{\Lambda}$ particles were produced in different hadronic reactions. Earlier studies of inclusive hyperon production had shown, however, that the polarization is a consequence of the beam particle fragmentation rather than the nature of the target particle. It was therefore concluded that the Λ's and $\bar{\Lambda}$'s produced in pp and $\bar{p}p$ reactions, respectively, should have essentially the same polarization. This was also supported by the similarity of the measured longitudinal momentum distributions both for charged and neutral reaction products.

The data sample used for the final results consisted of 17028 Λ's and 9553 $\bar{\Lambda}$'s. The polarization of the particles, $(\alpha P)/|\alpha|$, evaluated in bins of transverse momentum, is displayed in figure 1. From this, the ratio $(\alpha P)_\Lambda/(\alpha P)_\Lambda$ was calculated, which reduces to $\bar{\alpha}/\alpha$ under the assumption of equal Λ and $\bar{\Lambda}$ polarizations. The weighted average obtained [10] was $\bar{\alpha}/\alpha = -1.04 \pm 0.29$, which can be converted into $A = -0.02 \pm 0.14$.

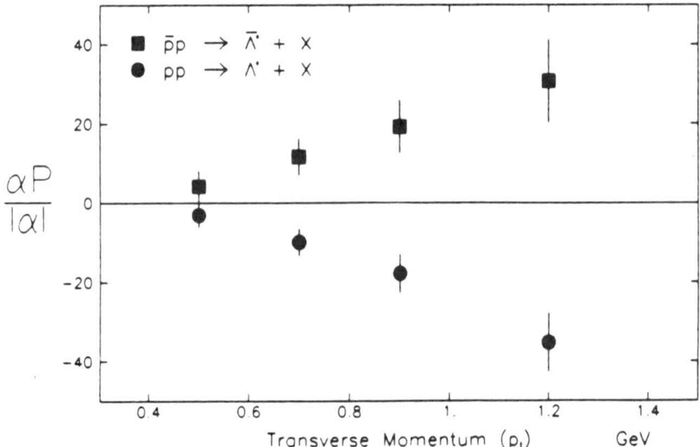

Figure 1: Polarization $(\alpha P)/|\alpha|$ of Λ and $\bar{\Lambda}$ versus transverse momentum at 30.8 GeV centre-of-mass energy [from ref. 10].

Exclusive associated production $\bar{p}p \rightarrow \bar{\Lambda}\Lambda$

A more direct way to compare Λ and $\bar{\Lambda}$ decay properties is the reaction $\bar{p}p \rightarrow \bar{\Lambda}\Lambda$ which is being investigated by experiment PS185 [11-12] at CERN−LEAR from the reaction threshold (1.435 GeV/c) up to 2.0 GeV/c incident $\bar{p}$ momentum. Here the equality of Λ and $\bar{\Lambda}$ polarizations follows directly from charge-conjugation invariance of the hadronic production process. The simultaneous detection of Λ and $\bar{\Lambda}$ decays furthermore reduces systematic errors.

The PS185 apparatus is a non-magnetic forward decay spectrometer. The delayed decays $\bar{\Lambda}\Lambda \rightarrow \bar{p}\pi^+p\pi^-$ are recorded in a large-volume stack of proportional chambers and drift chambers. This chamber stack is sandwiched between devices entering the "charged−neutral−charged" trigger scheme: a CH_2 target system triggering on the $\bar{p}$ beam and on the production of neutral particles, and a scintillator hodoscope triggering on the detection of charged particles from delayed decays. A weak-field solenoid with three drift chambers, well separated from the $\bar{\Lambda}\Lambda$ production and decay regions, distinguishes antiprotons from protons. The apparatus has a 4π centre-of-mass acceptance for the secondary protons and antiprotons.

For 1.546 GeV/c incident $\bar{p}$ momentum, 4063 events $\bar{p}p \rightarrow \bar{\Lambda}\Lambda \rightarrow \bar{p}\pi^+p\pi^-$ were reconstructed and kinematically fitted [13]. The method used for the evaluation of the product αP requires only a symmetric detector acceptance function, $\eta(\theta_p) = \eta(180° - \theta_p)$, without the need to know the acceptance in absolute numbers. Monte Carlo simulations showed, however, that this method did not bias the polarizations extracted from the real data.

Distributions of αP, in bins of the $\bar{\Lambda}$ centre-of-mass angle, are shown in figure 2. Charge-conjugation invariance allows the ratio $A = (\alpha + \bar{\alpha})/(\alpha - \bar{\alpha})$ to be extracted. The average value calculated from the angular distribution [11] is $A = -0.07 \pm 0.09$. Meanwhile the event statistics has been improved by the (preliminary) analysis [14] of data taken at 1.695 GeV/c incident $\bar{p}$ momentum. The combined result from the two measurements, based on a total number of about 16000 reconstructed events, is

$$\langle A \rangle = -0.023 \pm 0.057 \ .$$

The errors quoted here and above are standard deviations from the event statistics only. Systematic errors are estimated to be not significant at the present level of statistical accuracy.

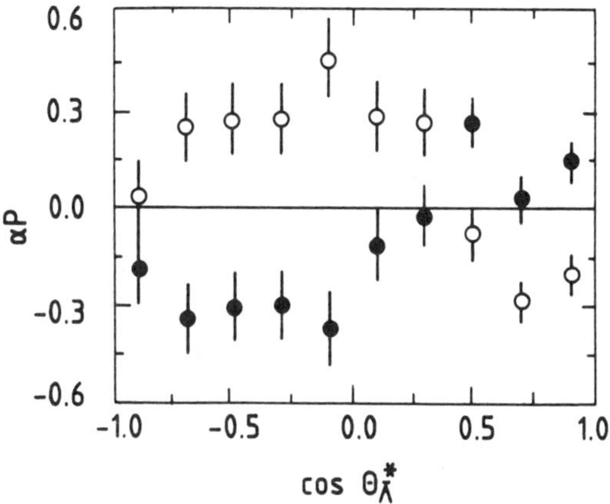

Figure 2: The product αP for Λ (full circles) and $\bar{\Lambda}$ (open circles) versus $\bar{\Lambda}$ centre-of-mass angle at 1.546 GeV/c incident $\bar{p}$ momentum [from ref. 11].

TODAY AND TOMORROW AT "LOW-ENERGY" $\bar{p}p$ MACHINES

In the context discussed here, the key issue is the determination of the production plane and the determination of the decay baryon directions relative to that plane. Furtheron we will consider only the exclusive associated production of hyperon – antihyperon pairs, because of the obvious advantages over inclusive production measurements.

For any dedicated experiment there are three basic choices to be made [15].

- The two processes which seem to be particularly well suited for CP violation studies [7-8] are $\bar{p}p \to \bar{\Lambda}\Lambda$ and $\bar{p}p \to \bar{\Xi}\Xi$.
- The total centre-of-mass energy chosen, and the angular region covered, determine the cross section as well as the polarization. Thus they are of great influence on the sensitivity of the experiment.
- The experiment can be carried out in fixed target mode, including extracted beam and internal gas-jet options, or in a mode using colliding $\bar{p}p$ beams.

Depending on these choices, new experiments can be considered for an existing machine, LEAR at CERN or possibly the $\bar{p}$ accumulator at FNAL, or they may require a machine yet to be built, Super-LEAR [16] designed at CERN.

One limitation of such experiments is due to statistics.

- If $A \approx 10^{-4}$ were to be measured with 3σ accuracy, about 10^9 events would be needed. A corresponding measurement of $B' \approx 10^{-3}$ is technically more difficult, but it would require only about 10^7 events.

Other limitations may be due to systematics or experimental asymmetries, in particular if they affect particles and antiparticles in different ways. It should be noted, however, that there are no normalization problems.

- In $\bar{p}p \to \bar{Y}Y$ the particles are boosted forward. The momentum and decay-vertex distributions are different for Y and $\bar{Y}$. The $\bar{Y}$ angular distribution in the centre-of-mass system is peaked around the direction of the incoming $\bar{p}$.
- If $\bar{Y}Y$ are produced in a magnetic field, the precession of their spin vectors is of the order of 0.5° in a 1 T solenoid, depending on the kinematics.
- Interactions of primary or secondary reaction products can occur in the target material, the beam pipe, or in the detectors. They can disturb the determination of the primary production plane or the determination of the directions of the decay baryons. These interactions may be different for particles and antiparticles. In the worst case they lead to an asymmetric error in the assignment of "above/below the reaction plane".
- Hyperon − antihyperon misidentification can lower the measured signal, but it can not create a misleading one. The same is true for Λ radiative decays which occur with $BR(\Lambda \to p\pi^- \gamma) \approx 10^{-3}$.

Measurements of $\bar{p}p \to \bar{\Lambda}\Lambda$ at LEAR

In an extracted beam approach, with a $\bar{p}$ intensity of $5 \times 10^6\ s^{-1}$ and a 2 cm long liquid hydrogen target, the luminosity is $L = 4 \times 10^{29}\ cm^{-2}\ s^{-1}$. The $\bar{\Lambda}\Lambda$ cross-section at 1.65 GeV/c (just below the $\bar{\Lambda}\Sigma^0$ threshold) is 80 μb, of which a fraction 0.41 corresponds to four charged particles in the detector. With 100 % geometrical acceptance for the decay baryons, in a PS185 type detector, the event rate then is $13\ s^{-1}$. In a clean and realistic experiment with small systematic errors, $A \approx 10^{-3}$ could be measured within a few weeks of full operation.

An internal gas-jet approach has also been considered [17]. The PS202 (JETSET) experiment aims at a peak luminosity of $L = 10^{31}\ cm^{-2}\ s^{-1}$. However, the interaction volume is not so small in this case, the large LEAR beam pipe ($14 \times 6\ cm^2$) is without tracking devices, and, moreover, it introduces matter in front of the nearest detectors. Precision, and control of systematic errors, would be very difficult to achieve. An experiment sensitive to $A \approx 10^{-4}$ needs at least a machine upgrade. If the $\bar{p}$ beam bypasses the experiment during injection, cooling, and acceleration, and is then sent to the experiment down a narrow ($\leq$ 2 cm dia.) beam pipe, the rate of reconstructable events could be high enough.

A measurement of $\bar{p}p \to \bar{\Xi}\Xi$ at Super-LEAR

The cleanest measure of CP violation is given [7-8] by the ratio $B' = (\beta + \bar{\beta})/(\alpha - \bar{\alpha})$, for which both the decay asymmetry and the final state polarization have to be detected. The unique reaction to do this is $\bar{p}p \to \bar{\Xi}^- \Xi^-$ (we consider here only the charged case). It is not accessible at LEAR. However, it provides, besides CP physics, an exciting field of $\bar{q}q$ dynamics and spin physics to be opened by Super-LEAR [15,18].

The produced $\bar{\Xi}^- \Xi^-$ exhibit a sequence of two "self-analysing" delayed decays, which together allow B′ to be determined. The first step, $\bar{\Xi}^- \Xi^- \to \bar{\Lambda}\pi^+ \Lambda\pi^-$ (with $c\tau$ = 4.9 cm), measures α and $\bar{\alpha}$ of the Ξ^- by the up − down asymmetry of the $\bar{\Xi}^- \Xi^-$ decay angular distributions ($\alpha = -0.455 \pm 0.015$ for the Ξ^-). In the second step, $\bar{\Lambda}\Lambda \to \bar{p}\pi^+ p\pi^-$ (with $c\tau$ = 7.9 cm), β and $\bar{\beta}$ are determined by measuring the Λ and $\bar{\Lambda}$ polarizations, i.e. the asymmetry of the $\bar{\Lambda}\Lambda$ decay angular distributions.

Monte Carlo events $\bar{p}p \to \bar{\Xi}^- \Xi^-$ at 3.5 GeV/c incident $\bar{p}$ momentum have been generated isotropically in the centre-of-mass system [19]. The resulting laboratory polar angle distributions are shown in figure 3 for the Ξ^-, its decay products $\Lambda\pi^-$, and the Λ decay products $p\pi^-$. All baryons are confined within $\theta \leq 40°$, and almost all pions within the forward hemisphere.

The $\Xi^-\Xi^-$ cross section at $\sqrt{s} = 3$ GeV is around 2 μb, of which 41 % correspond to (2 + 4) charged particles in the final state. The highest rate can be achieved with internal gas-jet operation at Super-LEAR, for which the peak luminosity is $L = 10^{32}\ cm^{-2}s^{-1}$. The on-line trigger can make use of many kinematical constraints (e.g., charged-multiplicity step "2 to 6", charged and neutral "V" recognition). With about 40 events per second in the detector, $B' \approx 10^{-3}$ could be measured within a few weeks of full operation.

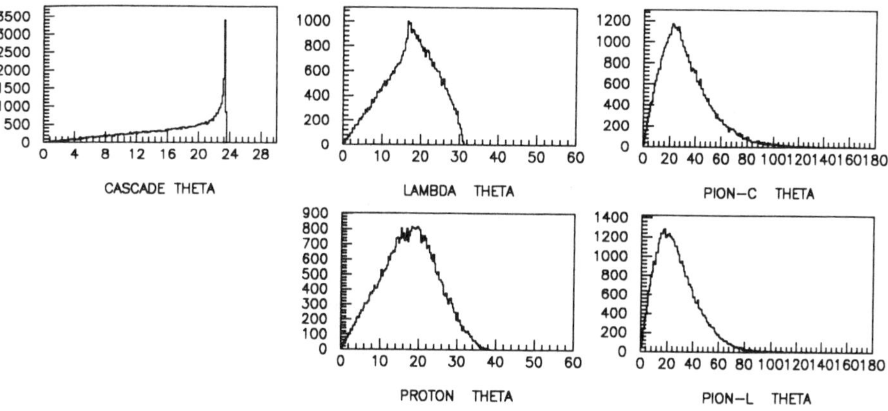

Figure 3: Laboratory polar angle distributions of $\bar{p}p \rightarrow \bar{\Xi}^-\Xi^-$, and of the decays $\bar{\Xi}^-\Xi^- \rightarrow \bar{\Lambda}\pi^+\Lambda\pi^-$ and $\bar{\Lambda}\Lambda \rightarrow \bar{p}\pi^+p\pi^-$, for events produced isotropically in the centre-of-mass system at 3.5 GeV/c incident $\bar{p}$ momentum.

REFERENCES

1. L.L. Chau, Phys. Rep. **95,** 1 (1983).
2. L. Wolfenstein, Annu. Rev. Nucl. Part. Sci. **36,** 137 (1986).
3. J.H. Christenson et al., Phys. Rev. Lett. **13,** 138 (1964).
4. Particle Data Group, Phys. Lett. **170B,** 1 (1986).
5. H. Burkhardt et al., preprint CERN–EP/88–47 (1988).
6. M. Woods et al., Phys. Rev. Lett. **60,** 1695 (1988).
7. J.F. Donoghue, X.G. He and S. Pakvasa, Phys. Rev. **D34,** 833 (1986).
8. J.F. Donoghue, B.R. Holstein and G. Valencia, Phys. Lett. **178B,** 319 (1986).
9. J.F. Donoghue, invited talk at this conference.
10. P. Chauvat et al., Phys. Lett. **163B,** 273 (1985).
11. P.D. Barnes et al., Phys. Lett. **199B,** 147 (1987).
12. G. Franklin, invited talk at this conference.
13. W. Dutty, Ph.D. Thesis, Univ. of Freiburg, in preparation.
14. G. Sehl, private communication.
15. N. Hamann et al., preprint CERN–EP/87–208 (1987), and Proc. IV. LEAR Workshop, Villars-sur-Ollon, 1987, eds. C. Amsler et al. (Harwood, Chur, 1988), p. 619.
16. R. Giannini, P. Lefèvre and D. Möhl, Proc. IV. LEAR Workshop, Villars-sur-Ollon, 1987, eds. C. Amsler et al. (Harwood, Chur, 1988), p. 57.
17. G. Bassompierre, preprint CERN–EP/88–04 (1988).
18. P. Dalpiaz et al., preprint CERN–EP/87–27 (1987).
19. S. Ohlsson, private communication.

THE LEAR CRYSTAL BARREL DETECTOR

K. Braune,S. Keh, L. Montanet, J. Zoll
CERN, Switzerland

R. Beckmann, J. Friedrich, H. Heinsius, T. Kiel, B. Lewendel,
C. Pegel, U. Strohbusch
Univ. Hamburg, W.-Germany

P. Blum, S. Cierjacks*, D. Engelhardt, H. Koch, M. Kunze, H. Matthaey,
W. Rohrbach, W. Schott, D. Walther, Ch. Weddigen*, N. Winter
Univ./ KfK* Karlsruhe, W.-Germany

J. Bistirlich, R. Bossingham, H. Bossy, K.M. Crowe, A. Miller
UC/Berkeley, USA

D.V. Bugg, A.J. Sanjari
Queen Mary College, London, UK

R.P. Haddock
UC /Los Angeles, USA

J. Beden, M. Guckes, H. Kalinowsky, E. Klempt, M. Merkel,
K. Peters, G. Reifenröther, J. Reifenröther, P. Weidenauer, U. Wiedner
Univ. Mainz, W.-Germany

P. Birien, H. Emerich, M. Faessler, G. Folger, D. Jamnik, K. Koenigsmann,
U. Meyer-Berkhout, A. Staude, C. Zupancic
Univ. Munich, W.-Germany

C.A. Baker, C.J. Batty, N. Hessey
Rutherford Appleton Laboratory, UK

M. Suffert
CRN, Strasbourg, France

E. Aker, C. Amsler, C.A. Meyer, B. Schmid
Univ. Zurich, Switzerland

Speaker : H. Koch, Univ. Karlsruhe, Institute for Exper. Nuclear Physics

ABSTRACT

The features of the Crystal Barrel Detector, which is in preparation for LEAR at CERN, are discussed. The physics aims include $\bar{q}q$- and exotics-spectroscopy and a detailed investigation of yet unknown $\bar{p}p$-annihilation channels. An eventual later use on the PSI-B-Meson-Factory is discussed. The paper finishes with a description of the present status of the project.

THE CRYSTAL BARREL DETECTOR

The Crystal-Barrel (CB) detector is a compact detector with equally good performance for the measurement of charged and neutral particles in nearly the full solid angle[1]. Its components are shown in Fig. 1.

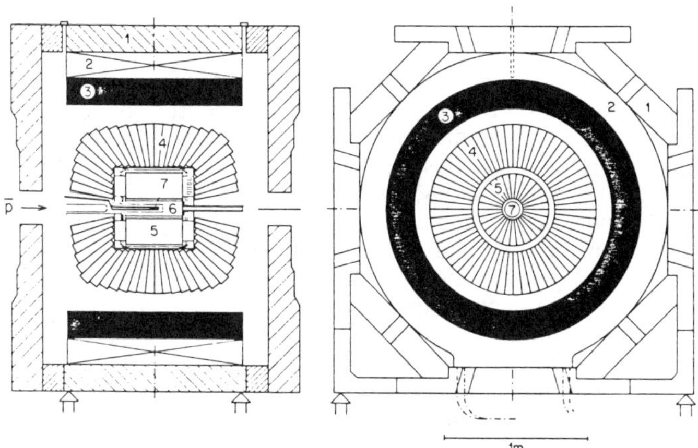

Fig. 1. The Crystal Barrel Detector : (1) Iron Yoke; (2),(3) Coil; (4) CsI-Barrel; (5) Jet-Drift-Chamber; (6) Multiwire-Proportional Chamber; (7) LH_2-Target.

For the measurements at LEAR it comes in two versions : For experiments with antiprotons in flight up to 2 GeV/c a LH_2-target will be used which is surrounded by a two-layer multiwire proportional chamber (PWC), which determines the number of charged annihilation products and is used in the charged-multiplicity trigger and in the $K_s^0(\rightarrow \pi^+\pi^-)$-trigger. For experiments with stopped antiprotons a gaseous target can be used, which is surrounded by an X-ray drift chamber (XDC), which again gives the charged multiplicity, but detects also low energy X-rays (> 1 keV), thus allowing to trigger on pure p-wave initial states[2]. This arrangement is surrounded in both cases by a Jet-Drift-Chamber (JDC), which operates with slow gas (CO_2-Isobutane (90:10)) and thus yields a good spatial resolution ($\leq 100\ \mu$m) and a small Lorentzian-angle. Nearly all neutral (Gammas, up to ≈ 2 GeV) and charged particles are detected in a modular (1380 elements) CsI-detector covering $\approx 97\%\times 4\pi$ solid angle. All this set-up is placed inside a coil producing a field of 1.5 T in the center of the detector.

The detector is tailored for the measurement of multiparticle final states. It will allow a selection of decay channels with minimum background and minimum combinatorial background. Two examples : (1) Instead of analysing $\pi^+\pi^-$ pairs one uses $\pi^0\pi^0$-pairs, where no strong ρ-peak appears; (2) Instead of searching for ω's via the $\pi^+\pi^-\pi^0$-decay channel, one would analyse the $\pi^0\gamma$-channel, thus avoiding $\pi^+\pi^+$, $\pi^-\pi^-$-combinatorial background in a complicated event. In addition, on-line triggers on γ, K_s^0, π^0, η's, . . . will be possible. All these features make the CB-detector an ideal instrument for attacking today's problems in meson/exotics-spectroscopy.

PHYSICS AIMS

$\bar{q}q$ Spectroscopy : It is well known that $\bar{p}p$-annihilation reactions at rest and in flight are a rich source of meson-resonances (X). They are produced in reactions of the

type $\bar{p}p \rightarrow \pi(\pi\pi)+X$. With antiprotons at rest m_x is limited to masses below ≈ 1.6 GeV, with 2 GeV/c antiprotons the limit is $m_x \lesssim 2.4$ GeV.
There are many open problems in light meson spectroscopy, where accurate measurements with CB could contribute :

- Established particles : Look for their quark contents via measurements of yet unknown decay modes (Example : $f_2(1270) \rightarrow \eta\eta$), particularly radiative decays; Compare decay modes involving SU(3) oktett/singlet members and redetermine the corresponding mixing angles; Measure particle decays, and compare with model predictions, e.g. the Flux-Tube-Model[3] (Example $a_2^{\pm}(1320) \rightarrow \eta'\pi^{\pm}, \ldots$)
- Missing particles in Nonetts : Confirm recently found states (Example : $h_1'(1380) \rightarrow \omega\eta$; $\rho_2 \rightarrow \omega\pi^0$) in different channels and look for not yet seen $\bar{q}q$-states.

Exotics-Spectroscopy : Exotics are meson-like states, but with different quark and glue (g)-configurations, e.g. $\overline{qq}qq$-, gg(ggg)- (Glue-Balls) and $\bar{q}qg$-States (Hybrids). The number of candidates for such states is rapidly growing during the last years, but the distinction from ordinary $\bar{q}q$-states is problematic. It seems that many of the candidates like to decay in η's and η'-particles, so that the CB-detector is very suitable for their measurement. A decisive test for the existence of exotic states is the search for states with exotic quantum numbers ($J^{PC} = 0^{--}, 0^{+-}, 1^{-+}, 2^{+-}, \ldots$), which are forbidden in $\bar{q}q$ systems. A first candidate (X(1405), $J^{PC} = 1^{-+}$) has been recently reported[4], decaying into $\eta\pi^0$. This state could be easily detected in the $(\bar{p}p) \rightarrow \eta\pi^0\pi^0$ channel, if it appears with a reasonable branching ratio ($>10^{-4}$) in the $\bar{p}p$-annihilation. Using a liquid hydrogen target, the combination of initial- and final-state quantum numbers operates as a filter and gives rise to a Dalitz-Plot with few and eventually weak background states[5]. In a similar way, but with antiprotons in flight, the Hybrid states predicted around 2 GeV could be searched for[6] (Example : $\bar{p}(2\ \text{GeV/c})p \rightarrow \pi^+X^-(1900)$; $X^- \rightarrow \pi^-b_1(1235)$; $b_1(1235) \rightarrow \omega\pi^0$; $\omega \rightarrow \pi^0\gamma$).

Annihilation-reactions : Because of the lack of a suitable detector 60% of the annihilation reactions, containing more than one $\pi^0(\eta)$, are unknown yet. The knowledge of their branching ratios or cross-sections allows to distinguish between the various quark-models describing annihilation processes[7]. Channels of interest are : $\pi^0\pi^0$, $\eta\eta$, $\eta\eta'$, $\pi^0\eta$, $\rho\rho$, $\omega\omega$, ... Their measurement should help in determining the relative importance of annihilation- and rearrangement diagrams. Of particular interest is the L-dependence of exclusive channels and the search for more dynamic selection rules (e.g. $\rho^+\rho^-, \ldots$) which constrain the model predictions.

Radiative Ψ-decays and $\gamma\gamma$-reactions : The CB-detector is designed such that it could be used later in a collider-machine, like a high intensity B-Meson-Factory, as is planned for PSI in Switzerland. The modularity of the detector is adequate to handle the multiplicities of Ψ-decays and $\gamma\gamma$-collisions. A luminosity of $\mathcal{L} = 8 \times 10^{31} \text{cm}^{-2}\text{cm}^{-1}$ would allow the production of 400 Ψ/s and 210 Ψ'/s and thus the continuation of meson-spectroscopy, which was so successful during the last years at SPEAR, with an improved luminosity and detector.

STATUS OF THE PROJECT

The most critical components of the detector are the CsI-Barrel, the Jet-Drift-Chamber (JDC) and the trigger.
CsI Barrel : A prototype of 39 crystals of final dimensions was tested in a e^-,π^--beam at SIN. With the electronics then available ($\approx$ 1 MeV noise), energy resolutions comparable to the Cleo II-detector were achieved. The new electronics ($\approx$ 400 keV noise) should further improve the resolution, particularly at lower energies. Of particular interest is the light readout, which uses a fluorescent light concentrator, which shifts the wavelength and directs the light to only one photodiode.

JDC : A prototype, consisting of three sectors of full size was tested at SIN. Spatial resolutions of $\sigma_\perp \approx 100$ μm and $\sigma_z \approx 5$ μm were reached, allowing for the anticipated momentum resolution of $\sigma_\perp/p \approx 2\% \times p$. The measurement of pions and protons showed a very efficient dE/dx-separation, thus allowing π/K-discrimination up to 500 MeV/c.
Trigger : The hardware part of the trigger allows a fast (<10 μs) decision on the total multiplicity of the event in the CsI-barrel and on the appearance of $K_S^0(\rightarrow \pi^+\pi^-)$. In two subsequent software levels, cluster energies, invariant masses, etc. are determined in less than 1 ms. Thus, an efficient trigger on η's, e.g., will be possible, yielding about 50-100 η's/sec on tape.

All components are designed and under construction. It is expected to have the detector in full operation in the middle of 1989.

REFERENCES

1. Proposal to the CERN-PSCC; CERN/PSCC/85-56; PSCC/P90.
2. M. Doser et al., CERN-EP/88-42, Submitted to Nucl. Phys. A.
3. R. Kokoski, N. Isgur, Phys. Rev. D35, 907 (1987).
4. D. Alde et al., CERN-EP/88-15, Submitted to Phys. Lett. B.
5. R.L. Jaffe in Proc. of the First Workshop on Antimatter Physics at Low Energies, 10-12. April 1986, Fermilab.
6. N. Isgur et al., Phys. Rev. Lett. 54, 869 (1985).
7. Contributions of C. Dover, H. Genz, E. Klempt to this Conference.

WHY LOOK FOR ANTIPROTONS IN COSMIC RAYS ?

S. A. Stephens*
Particle Astrophysics Laboratory, BOX 3-PAL
New Mexico State University, Las Cruces, NM 88003

ABSTRACT

A brief description of the experiments carried out to measure the energy spectrum of antiprotons is made and the excitement in this field of research is elucidated. Various physical processes by which antiprotons could be created are summarised. Propagation models required to derive the equilibrium spectrum of antiprotons in the Galaxy and the spectral modification which takes place while entering the solar system are discussed. A critical assessment is made of the theoretical models by comparing the predicted spectra with the observations.

INTRODUCTION

Soon after the discovery of antiprotons ($\bar{P}$s) during the fifties in the laboratory[1], the existence of $\bar{P}$s in cosmic rays was speculated on the basis of their production through the interaction of high energy cosmic ray nuclei with interstellar gas. Many attempts were made during the next two decades to search for $\bar{P}$s in cosmic rays, but none gave any positive result. However, theoretical predictions of the spectral shape of $\bar{P}$s continued to keep this search alive. It was during the end of the seventies, that $\bar{P}$s were detected in cosmic rays at relativistic energies using balloon borne magnet spectrometers[2,3]. After this discovery, many theoretical speculations and ideas were put forward and new experiments were proposed. This field of research is so exciting that the total number of papers published far exceeds the number of $\bar{P}$s detected. This paper describes the progress in this field and examines the current status to evaluate its potentiality.

EXPERIMENTAL RESULTS

Many different techniques have been employed to detect $\bar{P}$s, which constitute only a few parts in 10^4 protons in cosmic rays. Techniques based on annihilation of $\bar{P}$s were employed at very low energies, using passive detectors like nuclear emulsions and active detectors using calorimeters with visual detectors. In the case of

* On leave from Tata Institute of Fundamental Research, Homi Bhabha Road, Bombay 400005, India.

experiments using nuclear emulsions, systemetic scanning is carried out for annihilation interactions.These interactions are characterized by the slowing down of one of the particles towards the interaction vertex, which annihilates, and the release of large amount of energy. No $\bar{P}$ event was detected amoung the 25,000 interactions analysed and only upper limits could be set from these experiments. These are shown in Table I. In the calorimeter experiment, event selection is made by measuring ionization using scintillators and by restricting the velocity using a Cerenkov detector. Interactions in the calorimeter are then examined to select $\bar{P}$ events[4] and 14 events have been attributed to $\bar{P}$s. In such experiments one has to carefully eliminate the background due to helium interactions. This result has been recently contradicted by experiments using magnet spectrometer[5,6].

Table I
Measured and estimated fraction of $\bar{P}$ in cosmic rays

Energy in GeV	(P/P)x10^4	Ref.	Energy in GeV	(P/P)x10^4	Ref.
.05-.15	<4.3	7	6.8	6.10+2.72	2
.13-.32	2.2+0.6	4	8.2	9.50+3.17	2
.08-.36	<0.32*	5	10.2	12.23+3.80	2
.205-.64	<0.55*	6	13.4	3.36+2.87	2
.20-2.0	.6(+1.4;-.5)	3	>16	<820	8
2.0-5.0	3.(+4.;-2.)	3	>10^3	<500	9
4.7-11.3	6.8+1.7	2	100-200	<700	10
4.4	2.73+5.46	2	(1-1.5)x10^3	<1700	10
5.0	<.82	2	(1-1.5)x10^4	<1000	10
5.8	5.61+2.99	2	>3.x10^4	<1400	10

* 90% confidence level; others 67% confidence level.

Magnet spectrometer is the most versatile detector to identify $\bar{P}$s as it has the unique capability to determine the sign of the charge. Different detectors have been combined with the spectrometer to identify $\bar{P}$s over a wide range of energies. At non-relativistic energies, mass analysis is carried out for singly charged particles with negative curvature by determining the velocity, using a time of flight device, and the momentum[5,6]. At relativistic energies, velocity measurements using Cerenkov signal has been utilised to eliminate the vast background of secondary particles in the atmosphere like muons[2,3]. When cosmic rays propagate in the atmosphere, the effect of the existence of $\bar{P}$s would persist to greater depths due to the conservation of charge. Based on this, indirect methods have been

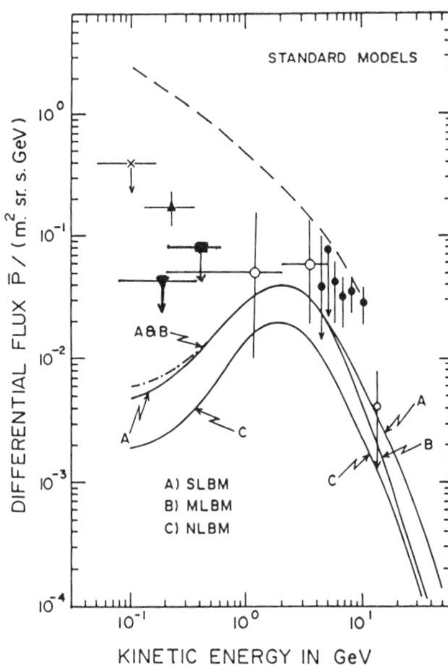

Fig.1 Differential energy Spectrum of $\bar{P}$s. The solid curves are the calculated spectra in interstellar space using models.

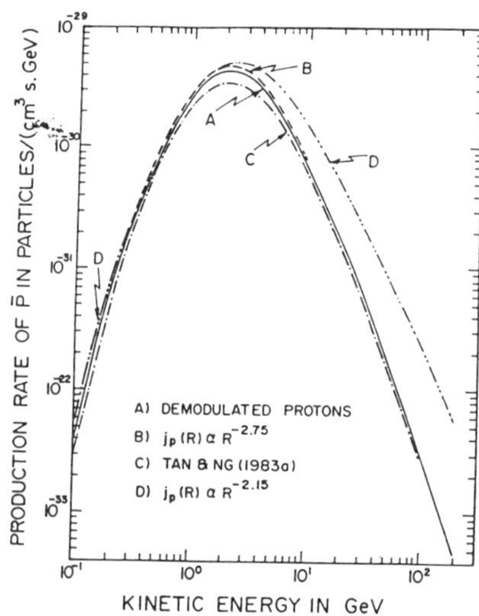

Fig.2 Production spectrum of $\bar{P}$s due to interaction of cosmic ray nuclei with interstellar gas.

employed to derive the fraction of $\bar{P}$s in cosmic rays by examining the measured charge ratio of the secondary particles in the atmosphere[8-10]. Only upper limits could be set by this method, but one could cover an energy region well beyond a TeV. We have plotted in Figure 1 the differential spectrum[11] of cosmic ray $\bar{P}$s from the direct measuements. This spectrum is very flat and does not resemble any of the observed spectrum of cosmic rays. The dashed line is the derived interstellar spectrum by removing the effect due to solar modulation on cosmic rays.

SOURCES OF ANTIPROTONS

The discovery of $\bar{P}$s in cosmic rays has led to many speculations, and almost every conceivable means of producing $\bar{P}$s have been considered. We briefly bring out these in this section.

1. High energy interactions

When high energy nuclei interact with gas, nucleon-antinucleon ($n\bar{n}$) pairs are created. In the case of P-P collision, the threshold kinetic energy is about 5.6 GeV, but the cross-section is insignificant below about 10 GeV. Detailed estimates of the production spectrum of $\bar{P}$s arising from the interaction of cosmic ray nuclei with interstellar gas have been made by a large number of authors by making use of the measured cross-sections at accelerator energies. We have shown in

Figure 2 the calculated production spectrum of $\overline{P}$ in the interstellar space[11]. One can notice the characteristic peak in the spectrum around 2 GeV. The spectral steepening in the high energy side is due to the steep cosmic ray spectrum. This effect is demonstated using two different kinds of proton spectrum as shown by curves B and D. The decrease of the $\overline{P}$ flux at low energies is due to kinematic effect. Note the difference between this spectum and the observed $\overline{P}$s.

2. Antiprotons from thermal plasma

Matter accreting into collapsed objects like neutron star and black holes, can attain temperature in excess of 100 MeV due to the conversion of gravitational energy. As some collisions in the plasma are of head on type, the production spectrum of $\overline{P}$s extends to low energies without depletion and this spectrum should have a cut off at high enrgies, depending upon the temperature of the plasma[12].

3. Neutron oscillations

As a consequence of Grand Unified Theory, neutron may undergo oscillation and transform into $\overline{n}$, which then decay to $\overline{P}$. This oscillation is slightly suppressed by the presence of magnetic field in the Galaxy. Neutrons arise from the collisions of cosmic rays, either through stripping of heavy nuclei or through charge exchange of protons. Neutrons are emitted during supernova explosions. Calculations show that $\overline{P}$s from this process is very small[11].

4. Evaporation of mini black holes

Theory of Quantum Gravity predicts that black holes emit radiation and lose their mass. Further, as the mass decreases the rate of emission increases, leading to an explosive stage at the end of its life[13]. At this stage, quarks and gluons are emitted and the $\overline{P}$ resulting from the evaporation products has a power law spectrum with a spectral index of -3.0 at relativistic energies[14].

5. Annihilation of photinos

A large class of broken Super Symmetry theories imply the existence of fermions with zero baryon and lepton numbers. They decay to photinos and gluinos, and may account for the dark matter in the Universe. Their annihilation would produce $\overline{P}$s and the $\overline{P}$ spectrum would have a cut off at an energy depending upon the mass of the dark matter[15].

6. Antimatter galaxies

If the Universe is symmetric, it is expected that cosmic rays in the intergalactic space should have equal amount of matter and antimatter. One requires about one part in 10^4 cosmic rays to come from outside the Galaxy[15] ,but this depends upon the intensity of intergalactic cosmic rays and their ability to penetrate into the galactic space.

COMPARISON WITH OBSERVATIONS

In order to compare with the observed $\bar{P}$s, one needs to determine the equilibrium spectrum in the Galaxy and to modulate it for the effect of solar wind. It is essential to take into account the energy loss processes , interaction and annihilation processes, acceleration, and leakage from the Galaxy during their propagation in interstellar space. Many propagation models are in use to describe the observed components of cosmic rays. A primary requirement of these models is to describe the observed decrease in the matter traversal of cosmic rays with increasing energy. A detailed description of these models and their consequences on $\bar{P}$s, is given elsewhere[11] and here we summarize them below.

In the standard model it is assumed that cosmic rays leak out of the Galaxy in an energy dependent manner above a certain rigidity, below which a constant amount of about 8 $g.cm^{-2}$ is traversed by them. The expected spectrum of $\bar{P}$s from this model is shown by Curve B in Fig.1. One can also allow the energy dependent matter traversal to be in the source, keeping the escape from the Galaxy at a contant rate. The effect of this model on $\bar{P}$s is shown by Curve C in Fig. 1. One notices that these models predict only a small fraction of the observed $\bar{P}$s. Therefore, a few sources with large amount of matter being traversed by cosmic rays have been included in the standard models. These sources could be either cosmic ray sources surrounded by dense matter or supernova explosions occuring in dense clouds. In the former case, the secondary $\bar{P}$s would have a less conspicuous spectral peak around 2 GeV due to their interactions, while in the latter case, a similar spectrum would be shifted to lower energies due to adiabatic deceleration associated with the expansion of supernova envelope.

We have shown in Figure 3 the expected spectra of $\bar{P}$s at earth from supernova(SN) exploding in dense clouds by Curves A and B[16]. Curves marked as A correspond to SN explosions in an uniformly dense cloud, while Curves

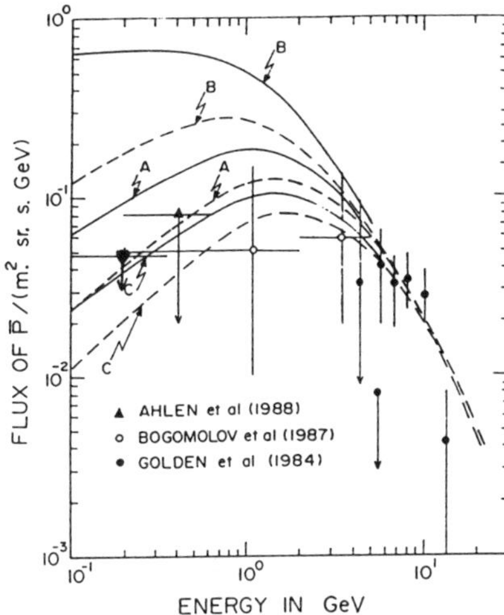

Fig 3. Spectra of Ps from SN are shown during solar minimum & maximum periods by solid & dashed curves respectively. Curve C is for closed galaxy model.

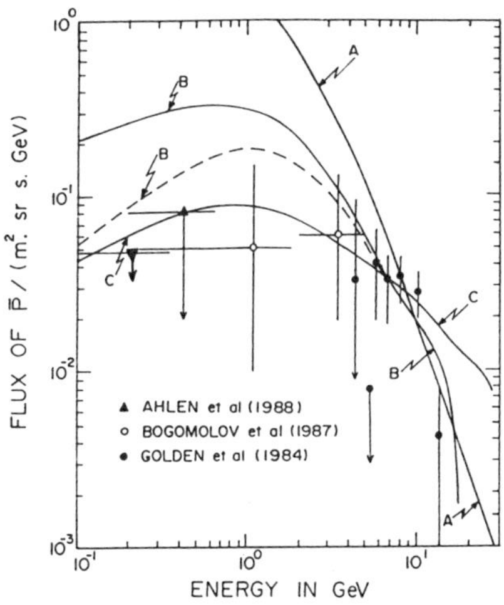

Fig.4 Curves A and B are spectra from mini black holes and dark matter. Curve C is from reacceleration of cosmic rays from dense clouds.

B relates to explosions in a cloud with a distribution of density $n \propto r^{-2}$. The solid and dashed curves relate to the expected spectra during the periods of minimum and maximum solar activity. It can be noticed that Curves marked as B do not fit the data and hence one may conclude that $\bar{P}$s do not undergo large amount of adiabatic deceleration.

If cosmic rays produced in the Galaxy is totally contained in a larger volume they are destroyed only by interactions. As a result large amount of secondary $\bar{P}$s would be created. If the observed nucleons consists of 50% of this old component the expected spectra of $\bar{P}$s are shown by Curves C. This model appears to fit the data very well. We have the expected spectrum of $\bar{P}$s from the evaporation of mini black holes in Figure 4 by Curve A. This spectrum does not fit the data at all. $\bar{P}$s spectrum arising from the annihilation of dark matter in the Galaxy is shown by Curves B in the same figure. It appears from the present data that the contribution from this source of production would also be small. From this brief analysis, one may conclude that the modified closed galaxy model[17] seems to explain the observations very well. Though $\bar{P}$s from SN explosions in dense clouds cannot be excluded, it is inferred that secondary particles created in SN cannot suffer considerable adiabatic deceleration.

While considering the idea of invoking cosmic rays from Symmetric Universe to account for the observed $\overline{P}$s, it is essential to examine the role of galactic wind on these cosmic rays. It is found that[16] a considerable amount of modulation takes place while entering the galactic space. If the diffusion mechanism in the Galaxy is energy dependent, galactic modulation provides an excellent fit to the observed $\overline{P}$ data. However, the predictions of the model are found to be in disagreement with the observed muon charge ratio and the energy spectrum of protons up to 10^5 GeV. In spite of this, the determination of $\overline{P}$ energy spectrum up to a few tens of GeV would be valuable to examine the role of galactic modulation on entering the Galaxy from outside.

From this brief summary of the progress in this field of research it is clear that the meagre data on cosmic ray $\overline{P}$s could not deter any from speculating the origin of the observed $\overline{P}$s. It is very essential to measure the spectrum of $\overline{P}$s over at least 3 decades in energy from about a few hundred MeV. Effort in this direction is being undertaken and one hopes that in the near future many of the puzzles will be resolved.

Acknowledgement: This work is supported by NASA grant NAGW 110.

REFERENCES

1. O. Chamberlain et al., Phys. Rev. 100, 947(1955).
2. R.L. Golden et al., Phys. Rev. Lett., 43, 1196(1979); Ap. Lett., 24, 75(1984)
3. E.A. Bogomolov et al., 16th ICRC, 1, 330(1979); 20th ICRC, 2, 72(1987).
4. A. Buffington et al., Ap. J., 248, 1179(1981).
5. R.E. Streitmatter et al., Bul. Am. Phys. Soc., 33, 1067(1988).
6. S.P. Ahlen et al., to appear in Phys. Rev. Lett.
7. K.M.V. Apparao et al., 19th ICRC, , 326(1985).
8. N. Durgaprasad and P.K. Kunte, Nature, 234, 74(1971).
9. G. Brooke and A.W. Wolfendale, Nature, 202,480(1964).
10. S.A. Stephens, Astr. Ap., 149, 1(1985).
11. S.A. Stephens and R.L. Golden, to appear in Sp. Sci. Rev.(1988).
12. C.D. Dermer and R. Ramaty, Nature, 319, 205(1986).
13. S.W. Hawkins, Nature, 248, 30(1971).
14. P. Kiraly et al., Nature, 293, 120(1981).
15. F.W. Stecker et al., Ap. Sp. Sci., 96, 96(1983).
16. S.A. Stephens and R.L. Golden, to appear in Astron. Ap.(1988).
17. S.A. Stephens, Nature, 289, 267(1981).

MEASUREMENT OF THE IMAGINARY PART OF THE I=1 $\bar{N}N$ S-WAVE SCATTERING LENGTH*

G. S. Mutchler, J. Clement, J. Kruk, R. Moss
Rice University, Houston, TX 77251-1892

E. V. Hungerford, T. Kishimoto, B. Mayes, L. Pinsky, L. Tang, Y. Xue
University of Houston, Houston, TX 77004

T. Armstrong, K. Hartman, A. Hicks,
R. Lewis, W. Lochstet, G. A. Smith
Pennsylvania State University, University Park, PA 16802

B. Bassalleck
University of New Mexico, Albuquerque, NM 87131

D. Lowenstein
Brookhaven National Laboratory, Upton, NY 11973

W. von Witsch
Institut für Strahlen und Kernphysik
University of Bonn, Fed. Rep. of Germany

H. Poth
KFK, University of Karlsruhe, Fed. Rep. of Germany

M. Furic
University of Zagreb, Yugoslavia

ABSTRACT

The survival time spectrum of slow antineutrons produced in a LH_2 target has been measured. From these data the imaginary part of the I=1 spin averaged S-wave antineutron proton scattering length has been deduced to be Im $a_1 = -0.83 \pm 0.07$ fm. The result lies within the range of values calculated from current potential models.

We have applied a method originally proposed by J. Rosner[1] to investigate the antineutron-proton interaction down to very low energies. The basic concept of the experiment is illustrated in Figure 1. A low energy $\bar{n}$ is produced in a large LH_2 target by the $\bar{p}p \rightarrow \bar{n}n$ charge exchange reaction. A fast forward going neutron signals the creation of a low energy $\bar{n}$. If the angle and momentum of the neutron are measured, the energy of the incident $\bar{p}$ and the created $\bar{n}$ can be calculated. The low energy $\bar{n}$ then scatters in the LH_2 target until it annihilates or escapes. The antineutron

*Work Supported by the U. S. Department of Energy

survival time is proportional to the low energy $\bar{n}p$ annihilation cross section. In turn the annihilation cross section, σ_A, can be related to the spin averaged complex scattering length a, as $k \to 0$, by

$$\sigma_A(k) = \frac{-4\pi \mathrm{Im} a/|a|^2}{k|1/a + ik|^2} \ \mathrm{mb.} \qquad (1)$$

Figure 2 shows the experimental set-up used at the AGS LESBII beam line. A total of 4×10^9 antiprotons were incident on a 40 cm diameter x 50 cm length LH_2 target. The 400 Mev/c $\bar{p}$ beam stopped near the center of the target. The neutron angle and T-0-F was measured by a segmented plastic scintillator, 180 cm downstream of the target. The $\bar{n}p$ annhilation was signaled by two or more charged particles orginating from the LH_2 target as determined by the vertex detector, an array of drift chambers and 12 scintillators (M-counters) surronding the LH_2 target. The vertex detector subtended approximately 3π steradians. A valid event consisted of an incident $\bar{p}$ (correct S1-S2 T-0-F and S2 pulse height), a fast neutron in the neutron detector, no hits in the veto counters, and a delayed (>6 ns) annihilation in the LH_2 target, determined by the vertex detector.

The measured quantity, $\Delta n_A(t)$, the number of antineutrons annihilating between t and $t + \delta t$, is shown in Figure 3. The experiment was simulated by a Monte Carlo calculation. The low energy $\bar{n}p$ cross sections were then varied to reproduce the data. The data were binned in two $\bar{n}$ momentum bins, giving values of

$$\beta\sigma_A(22 \pm 12) = 40 \pm 3 \ \mathrm{mb}$$

$$\beta\sigma_A(43 \pm 15) = 32 \pm 5 \ \mathrm{mb}$$

The extrapolated value of $\beta\sigma_A(0) = 44 \pm 3.5$ mb.

The values of $\beta\sigma_A$ are plotted in Figure 4 along with the higher energy results from previous measurements[2,3]. The present results and the previous measurements by this group, Armstrong et al., are in reasonable agreement. The solid curve was calculated from the Dover-Richard[4] potential model and the dashed curve from the Paris[5] potential model. The calculations bracket our result. The dip structure in both models is explained by referring to Figure 5, where the annhilation cross section is decomposed into its partial waves. The S and P wave partial cross sections and the total annhilation cross section (solid curve) is shown for the Dover-Richard potential. The dashed curve is the annhilation cross section calculated from the scattering length approximation using the Dover-Richard scattering length. Thus the total annhilation cross section is already deviating from the S-waves below 100 MeV/c and the dip in $\beta\sigma_A$ is caused by the rapid increase of the P-wave cross section at low energy. The scattering length approximation reproduces the S-wave partial cross section because the potential model does not have I = 1 S-wave resonances near threshold.

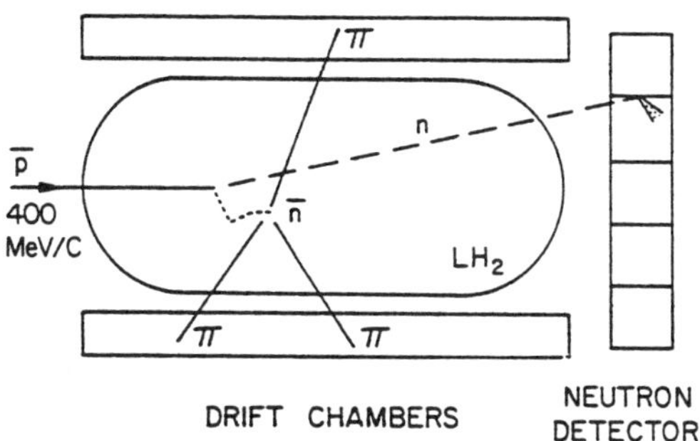

Figure 1

The value of $\beta\sigma_A(0)$ determined in this experiment corresponds to an imaginary part of the I = 1 spin averaged S-wave scattering length of Im a_1 = -0.83 ± 0.07 fm. The only previous determination of the I = 1 scattering length is from the $\bar{p}$-^{16}O,^{18}O atomic data by Poth et al.,[6] who found a_1 = 0.3 ± 1.4 -i(1.0 ± 1.7) fm. The two potential models predict

$$a_1 = 0.94 - i\ 0.63 \text{ Dover-Richard}$$
$$a_1 = 1.04 - i\ 0.91 \text{ Paris}$$

The theories bracket the measured value. The scattering length is sensitive to the tail of the real part of the potential outside the strong absorption region, about 1.1 fm.

REFERENCES

1. J. Rosner, Proc 1975 ANL Symposium on New Directions in Hadron Spectroscopy, Argonne National Laboratory, 1975, p.165.
2. B. Gunderson et al. Phys. Rev. D23 (1981) 587.
3. T. Armstrong et al. Phys. Rev. D36 (1987) 659.
4. C. B. Dover and J.-M. Richard, Phys Rev. C21 (1980) 1466.
5. J. Cote et al. Phys. Rev. Lett. 48 (1982) 1319.
6. H. Poth et al. Nucl. Phys. A294 (9178) 435.

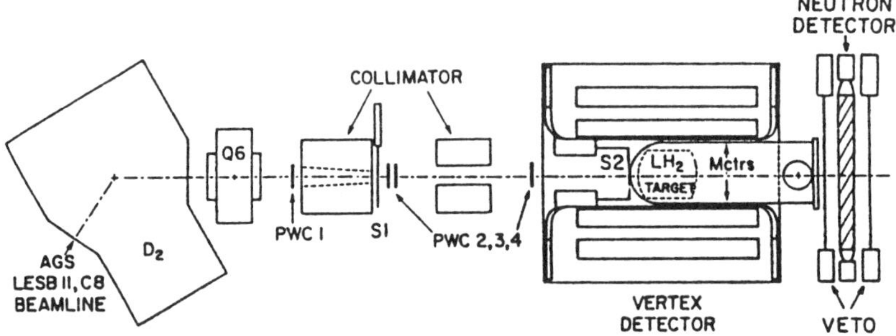

Figure 2: Experimental Set-Up

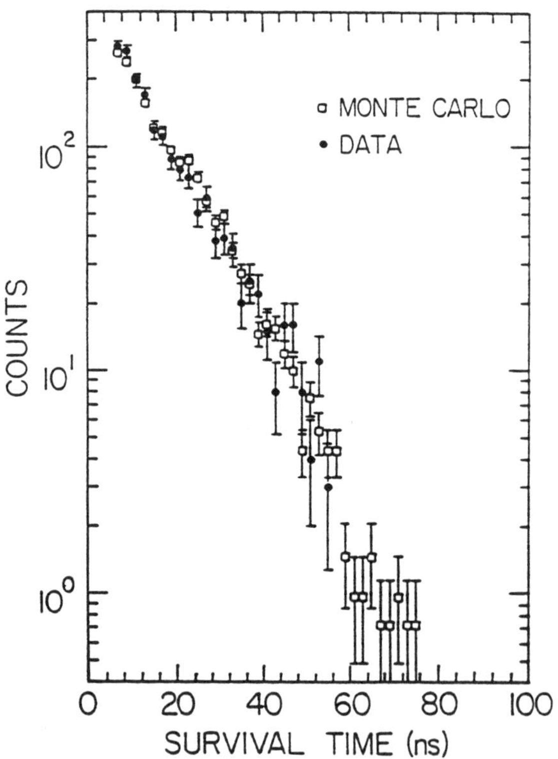

Figure 3

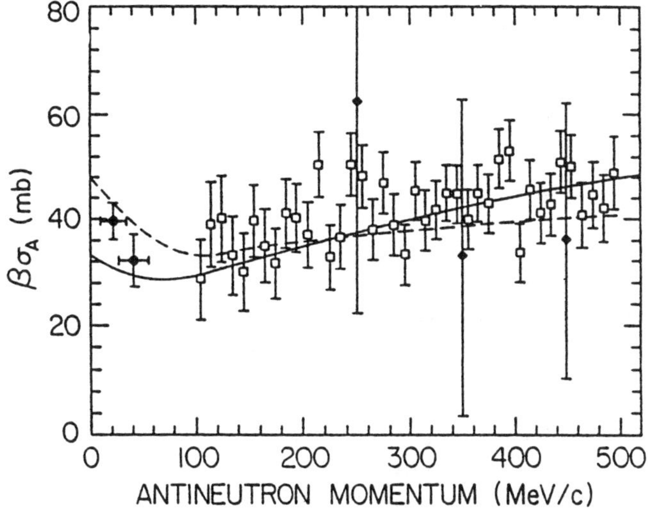

Figure 4: The solid circles are the present result. The open squares are the data of Armstrong et al.[3] and the diamonds are the data of Gunderson et al.,[2] both from n̄p transmission experiments.

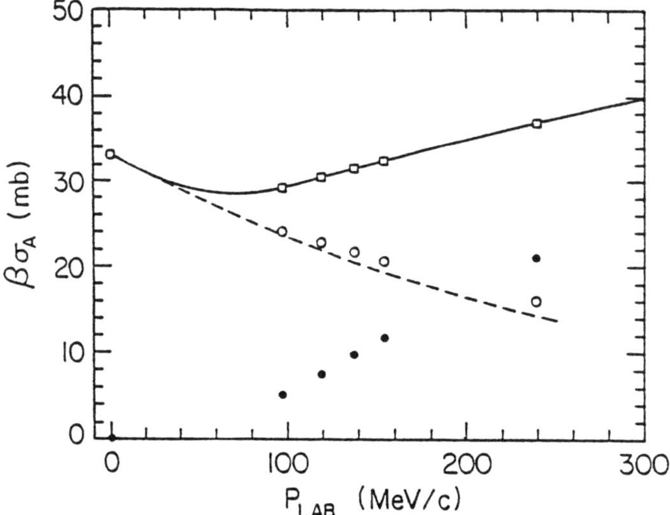

Figure 5: The open squares are the total annihilation cross section, the open circles are the S-wave partial cross section and the solid circles are the P-wave partial cross section.

PHYSICS WITH THERMAL ANTIPROTONS

M. V. Hynes and L. J. Campbell
Physics and Theory Divisions, Los Alamos National Laboratory,
Los Alamos, New Mexico USA 87545

ABSTRACT

The same beam cooling techniques that have allowed for high luminosity antiproton experiments at high energy also provide the opportunity for experiments at ultra-low energy. Through a series of deceleration stages, antiprotons collected and cooled at the peak momentum for production can be made available at thermal or sub-thermal energies. In particular, the CERN, PS-200 collaboration is developing an RFQ-pulsed ion trap beam line for the antiproton gravitational mass experiment at LEAR that will provide beams of antiprotons in the energy range 0.001-1000.0 eV. Antiprotons at these energies make these fundamental particles available for experiments in condensed matter and atomic physics. The recent speculation that antiprotons may form metastable states in some forms of normal matter could open many new avenues of basic and applied research.

THE ANTIPROTON GRAVITY EXPERIMENT AT LEAR

Theoretical approaches to gravitation abound whereas experiments in gravity, until just recently, could be counted on one hand. A fundamental measurement in experimental gravity that has not yet been done is the measurement of the gravitational force on antimatter. In certain extended supergravity models, specific particles have different gravitational masses from their associated antiparticles.[1] However, as yet there has been no direct test of this prediction. The gravity experiment at LEAR (PS200)[2] will test directly the equality of particles and antiparticle gravitational masses in the baryon sector with protons and antiprotons.

Experimentally we plan to use the time-of-flight (TOF) technique pioneered by Witteborn and Fairbank in their measurement of the gravitational force on the electron.[3] In this approach the particles are launched vertically up a drift tube. The TOF of the particle up the tube together with the initial velocity gives a measure of the gravitational force acting on the particle. For such a vertical launch and a fixed drift length, ℓ, where the detector is mounted, there will exist a critical initial velocity such that the particle arrives at the detector with zero velocity. For all initial velocities lower than the critical one, the particle simply does not have enough initial kinetic energy to reach the detector. Thus, there will exist a cutoff time in the TOF spectrum ($t_c = \sqrt{2\ell/g}$). In our experiment we will be dealing with a one-dimensional Maxwell-Boltzmann distribution of velocities. To measure the gravitational effect on antiprotons, the initial velocity and hence the kinetic energy distribution has to be low enough in average value so that the resulting TOF spectrum has sufficient particles near the cutoff time for a significant measurement. The energy required is more conveniently measured in degrees Kelvin rather than the more familiar MeV or keV. The scale for these energies and how they relate to the physics and experimental equipment used at other energies is displayed in Fig. 1. The thermometer shown in the figure is calibrated in electron volts down the left hand side with the corresponding temperature

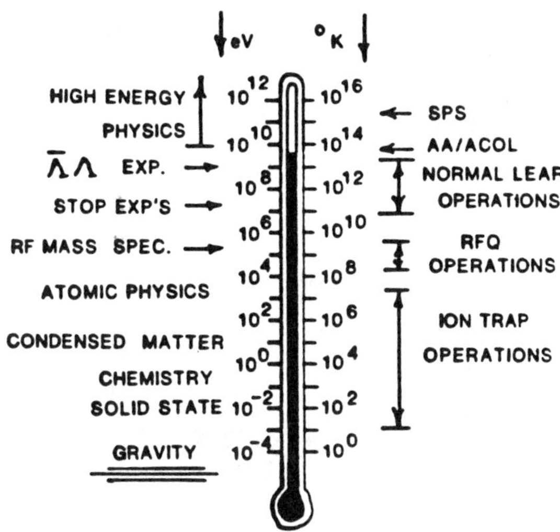

Figure 1: The antiproton thermometer shown is calibrated in electron volts down the left hand side and in degrees Kelvin down the right hand side.

in degrees Kelvin on the right. The physics of interest is schematically indicated on the right hand side whereas on the left, the experimental equipment used is listed. The realm of high energy physics is somewhat arbitrarily indicated to start at 10^{14} K (10^{10} eV) and extend upward without limit thru 10^{16} K (10^{12} eV). The SPS for instance operates at $\sim 2.7 \times 10^{15}$ K (2.5×10^{11} eV). Normal LEAR operations span the temperatures between $1.5 \times 10^{13} - 5.8 \times 10^{10}$ K (1.3 GeV - 5 MeV). On the upper end of this range the $\Lambda\bar{\Lambda}$ experiment runs with all the stop experiments at the lower end. As shown in the figure, antiproton temperatures in the range of 1 - 10 K are required for the gravity experiment. In order to achieve this antiproton temperature we are designing an RFQ decelerator and ion trap system. We are also considering using a foil degrader and plan tests of this concept soon. Such a system will extend the normal LEAR operating temperatures by ten orders of magnitude. Such an extension of antiproton temperatures opens very exciting possibilities in atomic physics, condensed matter, chemistry and solid state.

The overall system we are planning is displayed schematically in Fig. 2. The experimental sequence starts with the extraction of a 2 MeV bunched beam from LEAR. At present, beam at LEAR is available at 5 MeV with poor emittance. However, during the ACOL shutdown, LEAR will be upgraded to provided the 2 MeV fast extracted beam with good emittance that is required by our experiment. The bunched beam in LEAR has a macrotime structure set by the 8th harmonic buncher already in the LEAR ring. A single phase bucket of this bunched beam is fast extracted from LEAR and delivered to the entrance of our experimental beam line. A microtime structure matched to the operating frequency of the RFQ decelerator will also be imposed on the macroburst. This microstructure can be imposed either in, or external to the LEAR ring. Bunching in LEAR makes the most efficient use of antiprotons, whereas bunching

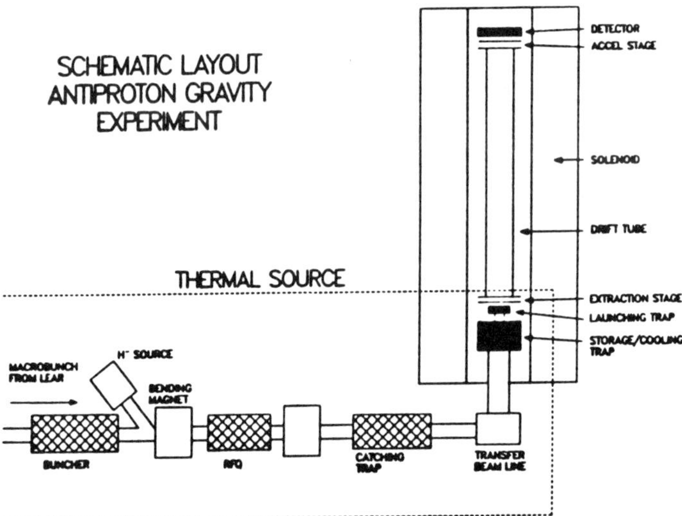

Figure 2: Overall system for the low energy beam line required by the antiproton gravity experiment.

externally will have higher loss rates. In the figure an external bunching section is shown. In any event, this bunched beam at 2 MeV is decelerated in an RFQ currently under design at Los Alamos[4] to 20 keV. With a suitable bending magnet in the beam line the 20 keV beam can be made available to our experiment or to others. Immediately following the RFQ is the first stage of pulsed ion traps that will be used to capture, store, and cool the antiprotons. The first stage catching trap is shown in schematic cross section in Fig. 3. This device is cylindrically symmetric about the horizontal axis and is designed along the lines of a Penning trap.[5] The multiring configuration provides a harmonic potential along the horizontal axis centered at the electrode with the circular cross section when the electrodes symmetric about this one are at specially

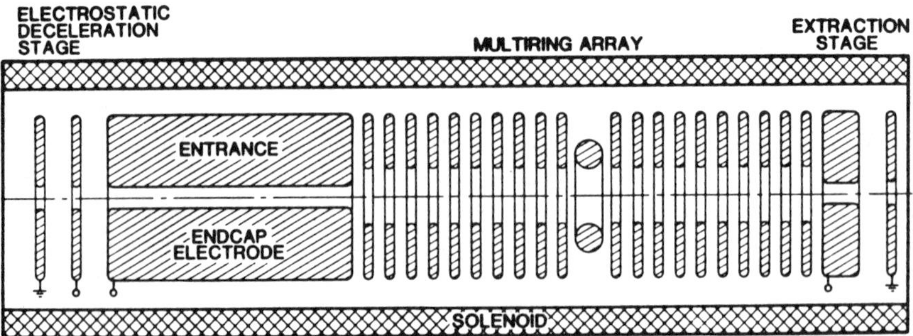

Figure 3: Schematic cross section of the first stage catching trap. The device is azimuthally symmetric about the horizontal axis.

matched potentials. The endcap electrodes at the far ends of the multiring array have small ($\sim$ 0.6 cm diameter) apertures in them to allow the beam burst to enter or exit. A magnetic field along the horizontal axis generated by the solenoid shown in the figure provides the radial confinement of the burst. Initially the beam burst emerging from the RFQ is 50 cm in length and enters the catching trap from the left in the figure. At this time the entrance endcap electrode is at 15 kV so that the burst is decelerated to 5 kV as it enters this elongated electrode. This results in a 25 cm burst length at 5 keV propagating inside this long electrode. When the entire burst is inside, the potential on this electrode is quickly ($\sim$ 25 ns) brought to ground before the burst starts to propagate inside the multiring array. The down-stream endcap electrode is maintained at $\sim$ 8 kV so that the burst decelerates inside the trap and eventually turns around. Before the burst can propagate more than half way back, the potential on the entrance endcap electrode is brought up to $\sim$ 8 kV. Thus the burst cannot leave the way it entered and is therefore caught. Initial testing of a half scale version of the multiring trap has demonstrated the promise of this approach.[6] While in this trap the antiprotons will be cooled to 10-100 eV using resistive, stochastic, or electron cooling. The success of the Genoa and Pisa groups with stochastic cooling in traps suggests that this option will be used.[7] When this is accomplished, the particles will transferred to a highly compensated harmonic trap located near the base of the vertical drift tube for further resistive cooling. In this intermediate trap the particles will be cooled to $\sim$ 5-10 K and stored in preparation for transfer, 100 at a time, into the final launching trap. Once the launching trap is loaded, the potential will be suddenly dropped, thus releasing half of the particles upward in the drift tube. The TOF is measured using a microchannel plate at the top of the drift tube. The vertical ion-trap-drift-tube system must operate at extremely low pressures ($\sim 10^{-14}$ Torr) to avoid loss of antiprotons through annihilation with residual gas atoms. Consequently this system will be cryogenic, operating at liquid Helium temperatures($\sim$ 4 K).

The calibration standard for our measurement of the gravitational acceleration of the antiproton will be the H^- ion. This stable ion has the same charge, magnetic moment, and almost the same inertial mass (within 0.1%) as the antiproton. H^- ions will be injected into our beam line at 20 keV using an ion source upstream of the RFQ (Fig. 2). The RFQ will not be energized during H^- injection. This configuration allows us to exercise the system independent of LEAR beam operations. To measure the gravitational acceleration of antiprotons relative to the H^- ion will require the launching of a total of 10^6 to 10^7 particles for a 1% measurement. There will be $\sim 10^8$ particles in each 8th harmonic macrobunch from LEAR. This is sufficient for one TOF spectrum with some safety margin. Many TOF spectra will be required to assure reproducibility. Each antiproton run will be followed by a run with H^- ions.

The measurement of the gravitational mass of the antiproton is a fundamental measurement that has not yet been performed. No convincing theoretical argument has been made to anticipate the result. This important measurement will test the Equivalence Principle for matter and antimatter and provide a powerful constraint on modern attempts to unify gravity with the other forces of nature.

ANTIPROTONS IN ATOMIC AND CONDENSED MATTER PHYSICS

Originally our interest in antiprotons was solely directed at executing the gravity experiment. However, to accomplish this we will develop in essence a thermal source of antiprotons making these particles available at energies of interest to atomic and condensed matter physics for the first time. This is illustrated in Fig. 1 where the energy regime and applicable physics are listed along with the experimental equipment required. Charged particle probes have provided important information on matter in the condensed state. The arsenal of probes employed up to now includes charged ions, electrons, positrons, and muons. Antiprotons offer a unique new probe of matter in the condensed state.

Antiprotons are, in essence, intrinsically stable, negatively charged particles with a hadronic mass. Unlike electrons, their dynamics in normal matter is not driven by the Pauli exclusion principle. For this and other reasons the photon emissions associated with the downward cascade of thermal antiprotons from high Rydberg states contain new information about the electronic environment of the nucleus, compared to analogous Rydberg transitions involving a leptonic mass.

The intrinsic stability of the antiproton makes it ideal for the study of new stability mechanisms in condensed matter. Dynamic stability, wherein particles are held in a steady state (an excited state) for long times by virtue of compensating instabilities, is particularly interesting. The principle of dynamic stability is illustrated by the familiar alternating gradient synchrotron, but it is not limited to such macroscopic application. The existence of localized states of dynamic stability for antiprotons is currently an open question. Simple arguments based on electrostatics and variational principles that imply the absence of stable antiproton ground states in condensed matter are not valid when excited states are involved. A possible example of nonlocalized dynamic stability is channeling, whereby an antiproton in a certain energy range travels along a particular crystallographic axis without annihilation.

Unlike potential applications of antiprotons in the fields of atomic and nuclear physics, the perspectives in condensed matter physics are still quite open. This is due to the novelty of conceivable applications: an extensive previous literature from which to draw is not available. Nevertheless, four specific condensed matter environments can be identified as possessing unique features that might lead to significant payoffs.

The superfluid state of liquid ^{4}He is remarkable in its display of a coherent quantum state over macroscopic distances. When disturbed by boundaries and impurities this quantum state recovers over a very short distance, approximately the size of the interatomic spacing. Much is known about the behavior of a variety of impurity ions in superfluid helium. Examination of the behavior of $\bar{p}$ impurities offers the possibility of new insights. Unlike free electrons, which create small bubbles in the liquid due to their zero-point motion and the Pauli exclusion principle, $\bar{p}$'s should act more like free protons and create high density clusters of ^{4}He atoms around themselves by electrostrictive attraction. The subsequent annihilation of the $\bar{p}$ should be preceeded by photon emission containing information about its immediate neighborhood. Superfluid helium is essentially transparent to a very wide spectrum of radiation and can be cooled to a temperature where its own excitations are effectively absent. In fact, approximately 11% of the superfluid helium atoms are in a Bose condensed state of zero momentum. These features ensure the greatest possible advantage in observing, with high precision, the $\bar{p}$ transition and decay products in condensed matter because of an extremely small energy spread in these products due to center of mass doppler broadening. It is also possible that the $\bar{p}$ would never see the condensed state of ^{4}He, especially if the electrostrictive cluster formation time is rapid in comparison to $\bar{p}$ annihilation time. In any case, the thermal center of mass motion would probably be the smallest achievable in any context.

It is tempting to speculate that the macroscopic quantum coherence of the superfluid state, including the delocalization of the fraction of the ^{4}He in the Bose condensed state, may give rise to truly dramatic effects in the behavior of the $\bar{p}$, such as coordinated annihilation, stable states, etc. Whether such effects exist is not known, but they seem unlikely in view of the large discrepancies in the size of the antiproton impurity compared with the coherence length of the superfluid state and in view of the coulomb energy available in the $\bar{p}-{}^4$He atom compared to the energy associated with the loss of superfluidity over a comparable volume. Nevertheless, prudence would recommend an open mind – surprises have occurred before in condensed matter physics.

The liquid surface of superfluid ^{4}He also offers a unique environment for $\bar{p}$ studies because it is microscopically smooth, in equilibrium with a vapor of effectively zero density (at low temperatures), and can be charged with ions either above or below the surface. In particular, electrons on the vapor side can be held against the helium surface by applying an electric field; they do not penetrate the surface because of the relatively high energy required to make the bubble state mentioned earlier. Because this electronic surface charge density can be substantial and can be excited in various plasma modes, the possibility exists of finding electron-$\bar{p}$ states that are bound to the surface but have negligible $\bar{p}$ density at the surface. In effect, the $\bar{p}$ would be trapped between the external electric field and the electronic surface charge, which in turn is repelled from the surface by the Pauli principle. As mentioned in the general remarks above, such trapping would have to occur in an excited state.

Many of the features of quantum coherence apply to both superconductors and superfluids: a superconductor is essentially a charged superfluid in a solid, neutralizing background. Electric and magnetic fields are shielded quite effectively in superconductors over distances comparable to the penetration depth: a length scale present only in charged superfluids and typically having a magnitude of many lattice spacings. Thus, one cannot expect known superconductors to shield and stabilize a $\bar{p}$ in any obvious way on an atomic scale, but what *will* happen is not clear either. The origin of the effective electron attraction, which gives rise to superconductivity, is a subtle and delicate interplay between electronic and lattice properties, both of which are disturbed by a $\bar{p}$. A best guess now is that the influence of a $\bar{p}$ impurity may be too localized to probe superconductivity, although it could give information on other electronic structure.

A $\bar{p}$ is attracted to positive charge. In the presence of an effective source of positive charge, other than protons, one could expect $\bar{p}$ trapping. In many respects, especially involving dynamics and transport, the absence of electrons in a bulk medium is equivalent to positive charge. This "positive" charge can be either delocalized as holes in a conduction band or localized as ionic lattice vacancies and certain crystal imperfections. Of course, such pseudo-positive charge cannot violate the laws of electrostatics, and the earlier remarks on the absence of ground state stability still hold. Nevertheless, the existence of localized excited states of the $\bar{p}$-hole system are possible in principle, and the model could serve as a fruitful paradigm.

A charged particle at an interstitial position or a vacancy position in an ionic crystal can be localized at a point where the local electric fields cancel, i.e. a position that is force free but nevertheless unstable. The possibility of stabilizing this position by imposing a time varying external electric field which effectively creates an attractive local potential well, relative to the unstable directions, then exists. The strength and frequency of this field may be tunable to favor particular sites and to avoid combinations that would unduly disrupt the underlying crystal structure. A wide variety of crystal types and laser sources are available to explore this possibility.

Finally, the possibility of forming localized, stable orbits which encompass one or more lattice sites should not be overlooked. This could be envisioned as a microscopic, quantum analog of a storage ring. Again, extensive knowledge from the science of crystal structure is available to aid in the search for the optimum environment.

SUMMARY

A thermal source of antiprotons is currently being developed for the antiproton gravity experiment at LEAR. This source will make antiprotons available at energies of interest to atomic and condensed matter physics. Charged particles have been used as a probe in these fields for many years. Antiprotons offer many unique properties as a probe of normal matter. Moreover, there may exist dynamically stable states of antiprotons in a matrix of normal matter under special conditions. Physics with bottled antiprotons, whether in the bottle or out has many new avenues of research that are as yet unexplored.

REFERENCES

1. J. Scherk in "Unification of the Fundamental Particle Interactions," S. Ferrara, J. Ellis, and P. van Nieuwenhuizen, eds., Plenum, New York, 1981.

2. N. Beverini, J. H. Billen, B. E. Bonner, L. Bracci, R. E. Brown, L. J. Campbell, et al., PS200 proposal, Los Alamos National Laboratory report, LAUR-86-260.

3. F. C. Witteborn and W. M. Fairbank, Rev. Sci. Inst. 48, 1(1977); Nature 220, 436(1968); Phys. Rev. Lett. 19, 1049(1967).

4. J. H. Billen, K. R. Crandall, and T. P. Wangler, Proceedings of the third LEAR Workshop, Tignes-Savoie-France, January 19-26, 1985. "Physics with Antiprotons at LEAR in the ACOL Era," U. Gastaldi, R. Klapisch, J. M. Richard, and J. Tran Thanh Van, eds., Editions Frontieres, Gif sur Yvette, France, 1985.

5. D. J. Wineland, W. M. Itano, and R. S. van Dyck, jr., Adv. At. Mol. Phys. 19, 135(1983).

6. M. H. Holzscheiter, B. S. Birmingham, R. E. Brown, M. V. Hynes, N. Jarmie, S. Graessle, Bull. Am. Phys. Soc., 33, 932 (1988).

7. N. Beverini, V. Lagomarsino, G. Manuzio, F. Scuri, G. Testera, and G. Torelli, "Proceedings of the Workshop and Symposium on the Physics of Low-Energy Stored and Trapped Particles," June 14-18, 1987, Stockholm, Sweden, Physica Scripta, to be published.

ANTIHYDROGEN PRODUCTION SCHEMES

J. Brian A. Mitchell
Department of Physics
The University of Western Ontario
London, Ontario, Canada. N6A 3K7

The formation of neutral antihydrogen requires the blending of high energy physics technology and atomic physics principles. One approach to the problem is to employ laser stimulated radiative recombination of positrons and antiprotons to form $\bar{H}$ atoms in excited states. This can be implemented using the merged beam approach which allows very low energy collisions to occur between fast moving collidants. Methods have been devised for overcoming the problems associated with the survival of the $\bar{H}$ atoms and these will be discussed. Plans for a simulation experiment employing normal matter analogues will be presented.

INTRODUCTION

The formation of neutral antihydrogen is a fascinating prospect and will have important implications in such diverse areas as Quantum Electrodynamics, Gravitational Physics and energy storage technology. Unfortunately it is a difficult goal to achieve not only due to the short supply of the feedstock materials (antiprotons and positrons) but also due to the limitations and constraints concerning the atomic processes involved. A number of methods for producing antihydrogen have been proposed. These include, (a) charge exchange between antiprotons and a positronium target[(1)], i.e. $\bar{p} + e^+e^- \rightarrow \bar{H} + e^-$, (b) three body recombination involving positrons and antiprotons stored at very low temperatures in a trap[(2)], i.e.

$$e^+ + e^+ + \bar{p} \rightarrow \bar{H} + e^+ + h\nu$$

and (c) the radiative recombination of positrons and antiprotons at very low collision energies in a merging beam configuration[(3,4,5)], i.e.

$$e^+ + \bar{p} \rightarrow \bar{H} + h\nu.$$

The general problem of antihydrogen formation involving these three approaches has been recently reviewed[(6)]. The charge exchange approach seems unlikely to be promising due to the short lifetime of positronium, (142ns for the ortho state) leading to inevitable low target densities. The trap approach offers the possibilities of very fast $\bar{H}$ formation rates although considerable research must yet be performed to determine the maximum antiproton and positron densities which can be maintained in such traps. Work must also be done to investigate atomic physics processes at extremely low temperatures. Current estimates indicate that $n(e^+) \sim n(\bar{p}) \sim 10^{10}$ cm^{-3} seems resonable. The present paper will concentrate on the merged beam approach to $\bar{H}$ formation and a number of recent developments in this area will be discussed.

RADIATIVE RECOMBINATION

In order to ionize an atom or molecule it is necessary to supply energy to it either in the forms of the kinetic energy of a bombarding particle or electromagnetic energy of a photon. The neutralization of an ion on the other hand requires the removal of energy and the rate of an electron-ion (positron-anti-ion) recombination is dictated by the expedience of the energy removal process.

During collisions between thermal positrons and antiprotons, close interaction between the collidants takes place during a time of about 10^{-15} secs. Photon emission, however, occurs in a typical time of $\sim 10^{-9}$ seconds. Radiative recombination therefore exhibits a very small rate coefficient (typically $\sim 10^{-11}$ cm^3 sec^{-1} at room temperature.

The cross section for $e^+ + \bar{p}$ recombination may be expressed in the form(7) (assuming matter-antimatter symmetry):

$$\sigma = A \frac{\nu_o}{\nu} \cdot \frac{h\nu_o}{\frac{1}{2}mv^2} \cdot \frac{g}{n^3}$$

where $A = 2.11 \times 10^{-22}$ cm^2, $g \sim 1$ and $h\nu = 13.6$ eV. n is the principal quantum number, ν is the frequency of the emitted radiation and m and v are the positron mass and velocity respectively.

$h\nu = \frac{1}{2} mv^2 + h\nu_0/n^2$ and so it is seen that $\sigma \sim 1/(nE_e)$ where E_e is the kinetic energy of the incoming positron assuming the antiproton to be stationary.

Neumann et al [3] have proposed that the neutralization rate of antiprotons can be enhanced through the process of stimulated radiative recombination, namely: -

$$h\nu + e^+ + \bar{p} \rightarrow \bar{H}\ (n) + 2h\nu$$

where the stimulating photon, supplied by a laser, has the same frequency as the emitted photon. They showed that one can define a gain factor g in the form: -

$$g = \frac{\alpha(\text{STIM})}{\alpha(\text{SPON})} = \frac{Pc^2}{F\Delta\nu 8\pi h\nu^3}$$

where α(STIM) and α (SPON) are the rate coefficients for the stimulated and spontaneous radiative recombination processes, P is the laser power in watts, C the speed of light, F the cross sectional area of the laser beam, h being Planck's constant, ν the laser frequency and $\Delta\nu$ is determined primarily by the velocity spread Δv of the positrons.

$$\Delta\nu = \Delta v \cdot \frac{mv}{h}$$

For the enhancement to be successful, the velocity spread of the collisions should be a small as possible and in order for α(SPON) to have a reasonable value, v should also be small. The merged beam technique has the ability to satisfy both these conditions.

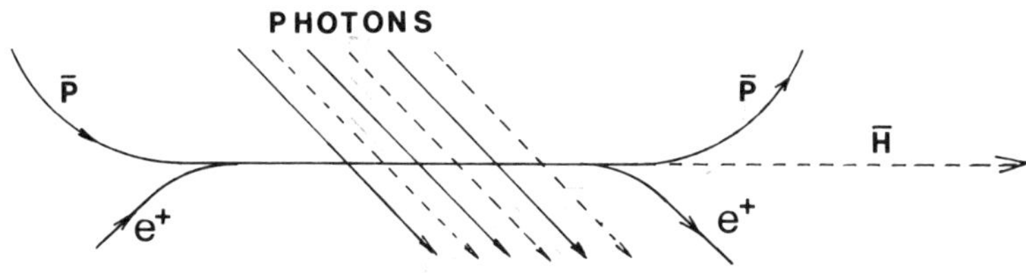

Figure 1 Schematic of antihydrogen experiment. Photons intersect interaction region at angle to minimize illumination time. Pump down photons indicated by dotted lines.

An antihydrogen production experiment is illustrated schematically in fig. 1. Antiprotons stored in a ring such as LEAR at CERN are merged with a beam of positrons and the overlap region is illuminated with a laser of suitable frequency to provide enhancement of the recombination process. As mentioned earlier the enhancement factor scales as ν^{-3} and so it is attractive to consider pumping down from the ionization continuum to a high n state (see fig. 2). Indeed if one considers a positron-antiproton collision energy of 0.01 ev, then pumping to the n = 10 or n = 11 level of $\bar{H}$ can be accomplished using a carbon dioxide laser. This is obviously an asset since high powered dc CO_2 lasers are readily available commercially.

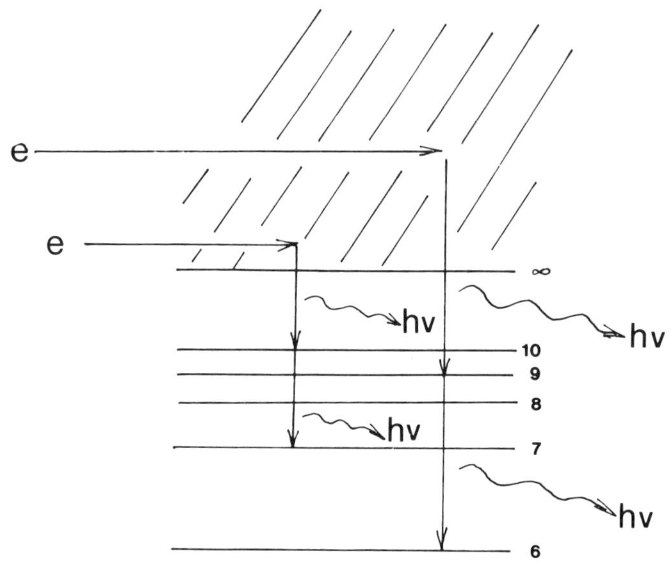

Figure 2 Stimulated recombination of $e^+ + \bar{H}$ followed by de-excitation to a lower level.

Two problems do however arise. Once formed, the excited $\bar{H}$ atoms are very susceptible to reionization by photoionization, if too high a laser power is used ,and to field ionization as the $\bar{H}$ pass through the magnetic fields of the storage ring.

It can be shown[3] that the maximum laser power to avoid subsequent photoionization is given by: -

$$P = \frac{Fh\nu}{\tau\sigma_{ph}}$$

where τ is the time during which the excited state is illuminated and σ_{ph} is the photoionization cross section. For $\bar{H}$ (n = 10) the natural lifetime is ~ 1.9 μs. τ can however be reduced by illuminating the ion beam over a restricted area as illustrated. Since the ions are moving rapidly in the laboratory frame the exposure time can be made very short.

The field ionization is a more severe problem. The electric field required to ionize a hydrogenic atomic in a given state n may be expressed as[8]: -

$$E = \frac{3.2 \times 10^{10}}{n^4}$$

In the context of the antiproton experiment this is the Lorentz field $q\,\bar{v} \times \bar{B}$ encountered by the fast particles passing through the bending magnets of the storage ring. One solution is to reduce the operating energy of the antiproton storage ring although this might lead to a deterioration of stored beam quality. An alternative solution has been proposed by Conti[9] and involves using a second laser beam to pump the excited state of $\bar{H}$ down to a lower state which would be immune to field ionization. It is intriguing to envisage using the same laser to both enhance the recombination and to pump to a lower state (see fig. 2). Since the pump down only occurs when the photon energy corresponds to the energy separation of the two levels, the same laser can only be used for fixed or 'magic' energies. Thus for example stimulation down to the n = 10 level followed by pump down to the n = 7 level can occur for $E_e \sim 6$ meV. If a CO laser was used, stimulation to n = 9 followed by pump down to n = 6 occurs for E_e = 40 meV.

NORMAL MATTER SIMULATION

Clearly there are many uncertainties involved in the antihydrogen production project. Not the least of these concerns the efficiency of the stimulated recombination process. It is desirable therefore to investigate this in a small scale experiment before committing resources to a full scale project at an antiproton storage ring.

A merged electron ion beam experiment has been operating at The University of Western Ontario for the past 12 years[10]. This apparatus has been used primarily for the study of electron-molecular ion recombination processes. Recent experiments[11] performed on vibrationally cold H_2^+ ions displayed narrow resonances, less than 20 meV wide and this has given a clear illustration of the high resolution, low energy capabilities of the technique, alluded to earlier.

An experiment is planned employing electrons and protons as the reactants and with a modest power (10 watt) CO_2 laser to test the stimulated recombination process.

The problem with such an experiment lies in the fact that a large background signal of H atoms will be formed via the charge exchange of protons on the residual gas in the collision chamber. A novel technique however can be used to overcome this problem.

The purpose of the experiment is to demonstrate that the CO_2 laser will lead to an enhancement of the production of H(n = 10) atoms. It is sensible therefore to detect only atoms with n ≥ 10 and this may be accomplished by using a field ionizer to reionize the atoms formed via recombination, the resulting protons being detected fig. 3. The cross section for charge exchange scales as n^{-3} so the H(n = 10) population formed in this way will be 1/1000 of the total H background formation rate and so can be neglected.

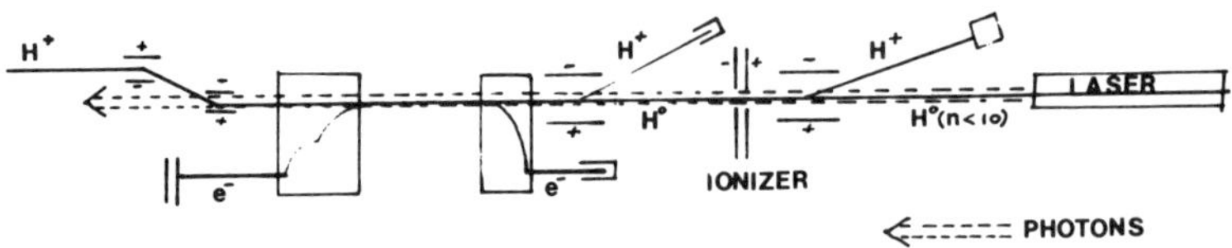

Figure 3 Schematic of merged beam, normal matter simulation experiment at U.W.O.

The objectives of the experiment will be: -

(a) To demonstrate the laser enhancement of the radiative recombination.

(b) To investigate how this scales with laser power.

(c) To demonstrate pump down. This should be apparent as drop in the H(n = 10) signal as the center of mass energy is scanned through the 'magic' energy.

Most of the apparatus for this experiment is currently operational and the project is expected to begin in the summer of 1989. It will be performed in conjunction with Dr. T.J. Morgan of Wesleyan University.

Estimates of $\bar{H}$ formation rates in an actual merged beam type experiment at LEAR have been published elsewhere[4,5,6]. It seems reasonable to expect that if stimulated recombination is successful and if a recirculating positron beam is used, $\bar{H}$ formation rates of ~ 10^6 - 10^8 s^{-1} are achievable.

The atoms of course will exit the storage ring in the form of a high energy beam. This is convenient for a wide variety of experiments including measurements of the Lamb Shift of $\bar{H}$ and studies of collisions between $\bar{H}$ and normal matter species, although obviously the trap approach in which essentially stationary $\bar{H}$'s will be formed, will offer distinct advantages for other applications.

In order to proceed from neutral antihydrogen atoms to larger molecule or cluster formation the most efficient approach is likely to be via the associative ionization process

$$\bar{H}(1s) + \bar{H}(n\ell) \rightarrow \bar{H}_2 + e^+.$$

It may transpire that a hybrid approach involving laser stimulated recombination of antiprotons and positrons inside an ion trap offers the most efficient way of forming $\bar{H}$'s in the correct excited states to maximize the efficiency of this process.

ACKNOWLEDGEMENTS

Many of the ideas presented in this paper have arisen from discussions with Drs. A. Rich, R. Conti, D. Gidley, G. Gabrielse and T.J. Morgan. I would also like to acknowledge the encouragement that I have received from the U.S. AFOSR, U.S. AFAL and the RAND Corporation.

REFERENCES

1. Humberston, J.W., M. Charlton, F.M. Jacobsen and B.I. Deutch, J. Phys. B. 20, L25, 1987.
2. Gabrielse, G., L. Haarsma, S.L. Ralston and W. Kells, Laser Spectroscopy 8 (ed. W. Persson and S. Svandberg, Springer Verlag, N.Y. 1987, p. 26.
3. Neumann, R., H. Poth, A. Winnacker and A. Wolf, Z. Phys. A., 313, 253, 1983.
4. Rich, A., R. Conti, W. Frieze, D.W. Gidley, M. Skalsey, T. Steiger, J. Van House, H. Griffin, W. Zheng and P.W. Zitzewitz To be published in Proc. NATO Adv. Research Workshop, "Atomic Physics with Positrons", University College, London. 1987.
5. Poth, H., Appl. Phys. A43, 287, 1987.
6. Mitchell, J.B.A., in Proceedings of RAND Workshop on Antiproton Science and Technology, Santa Monica, CA, 1987, ed. B. Augenstein.
7. Menzel, D.H., App. J. 85, 330, 1937.
8. Brouillard, F., in Atomic & Molecular Processes in Controlled Thermonuclear Fusion, ed. C.J. Joachain and D.E. Post, (Plenum Press, NY 1983).
9. Conti, R.S. (Private Communication) 1988.
10. Auerbach, D., R. Cacak, R. Caudano, T.D. Gaily, C.J. Keyser, J. Wm. McGowan, J.B.A. Mitchell and S.F.J. Wilk, J. Phys. B. 10, 3797, 1977.
11. Hus, H., F. Yousif, A. Sen and J.B.A. Mitchell, Phys. Rev. Lett. 60, 1006, 1988.

LOW ENERGY NUCLEON-ANTINUCLEON SCATTERING

J.J. DE SWART, R. TIMMERMANS AND T.A. RIJKEN

University of Nijmegen, Nijmegen, the Netherlands

With the advent of LEAR experimental research on the interaction of antiprotons with protons has been boosted. Existing $\overline{N}N$ models do not describe the many new data accurately. We will here present some results of a new model which has some advantages over previous ones. The two main reaction mechanisms are the elastic $\overline{p}p \rightarrow \overline{p}p$ and charge-exchange $\overline{p}p \rightarrow \overline{n}n$ scattering. These two channels are treated as coupled channels. For each partial wave the relativistic Schrödinger equation is solved on the particle basis with a boundary condition prescribed by the P-matrix. This is the logarithmic derivative of the wave function $\Phi(r)$ in channel space:

$$P = \left(\frac{d\Phi}{dr} \Phi^{-1} \right)_{r=b} \tag{1}$$

The P-matrix, specified at a distance $r = b$ =1 fm, represents the physics of the inner region which in the case of $\overline{N}N$ scattering is very complicated. We will describe it for each partial wave by a very simple potential, a coupled channels step potential of the form:

$$V(r < b) = \begin{pmatrix} V_{el} + iW & V_{ce} \\ V_{ce} & V_{el} + iW \end{pmatrix} \tag{2}$$

This form is valid for the partial waves with $\ell = J$. For the partial waves coupled by the off-diagonal tensor force we simply assume that there is no mixing in the inner region, in which case we get two uncoupled potentials of the form given in (2). iW is the complex potential which treats, as an average, the many mesonic annihilation channels. The P-matrix corresponding to this potential can easily be calculated analytically.

The potential matrix of the outer region, used in the Schrödinger equation is given by:

$$V = V_C + V_{HBE} + V_{OPE} \tag{3}$$

where the Coulomb potential $V_C = \alpha'/r$ acts only in the elastic $\overline{p}p \rightarrow \overline{p}p$ channel. V_{HBE} is the heavy boson exchange part of the Nijmegen soft-core potential [1], G-parity transformed. Finally V_{OPE} is the π^0 potential in the elastic channels, with coupling constant $g^2/4\pi = 13.1$, and the π^+ potential in the charge exchange channel with $g^2/4\pi = 14.3$. The experimental masses of the pions are used.

The formalism described above is used for partial waves with total angular momentum $J \leq 5$. Higher partial waves are included in a Coulomb-distorted-wave Born approximation. In figure 1 we show some results at

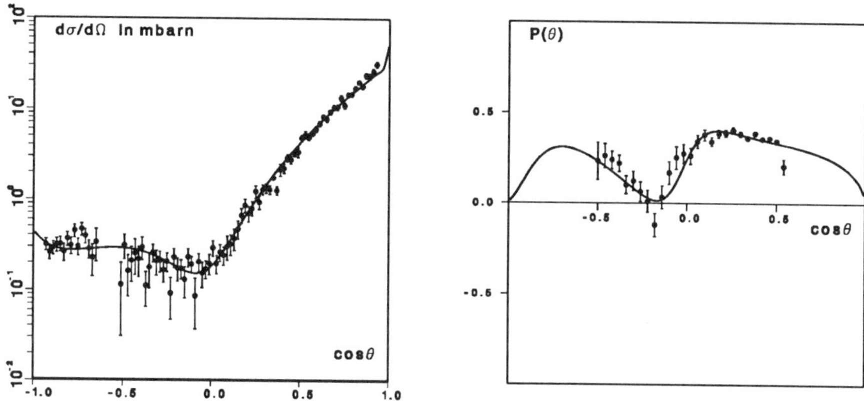

Figure 1: Differential elastic cross-section (data from [2]) and polarization (data from [3]) at $p_{lab} = 690.0\ MeV/c$.

laboratory momentum $p_{lab} = 690.0\ MeV/c$. This data set consists of 135 observables which are fitted with $\chi^2 = 1.4$/data. It includes the recent polarization data of the LEAR experiment PS172 [3]. In order to limit the number of parameters most partial waves are given the same P-matrix. In order to fit the polarization however we have to differentiate between the P-waves. Also the annihilation potential is taken the same for all partial waves. The parameters V_{el}, V_{ce} and iW vary slowly with energy. We are now looking for a (larger) parameter set valid for all momenta below 1.0 GeV/c. A cautious estimate as to the total number of parameters needed to achieve this is 25.

An interesting problem is formed by the ρ-amplitude in $\overline{p}p \rightarrow \overline{p}p$ scattering. LEAR data at low energy suggest values of ρ close to zero, a behavior which present potential models do not reproduce. Taking the full spin dependence of the interaction into account it can be written in terms of the amplitudes $M^s_{mm'}(\theta)$ by:

$$\rho = \frac{\Re(M^0_{00}(0) + M^1_{00}(0) + 2M^1_{11}(0))}{\Im(M^0_{00}(0) + M^1_{00}(0) + 2M^1_{11}(0))} \tag{4}$$

It is extracted from experiment by fitting the Locher [4] formulas to the data. This procedure involves at the same time the input of the spin parameter called η [5]. Two recent LEAR experiments [6,7] give conflicting values for ρ. We take here the data set of [6]. The result of our model at

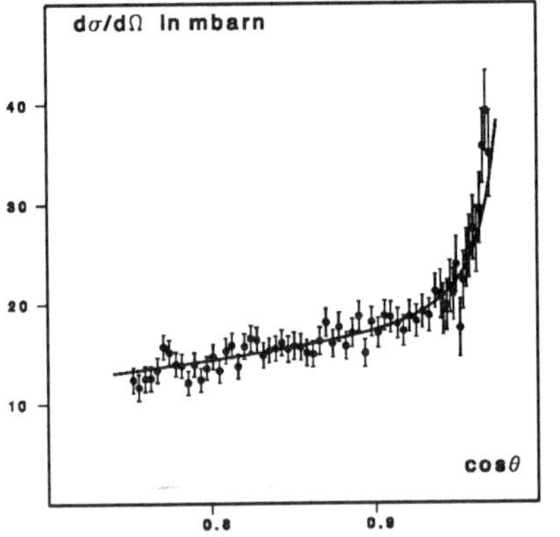

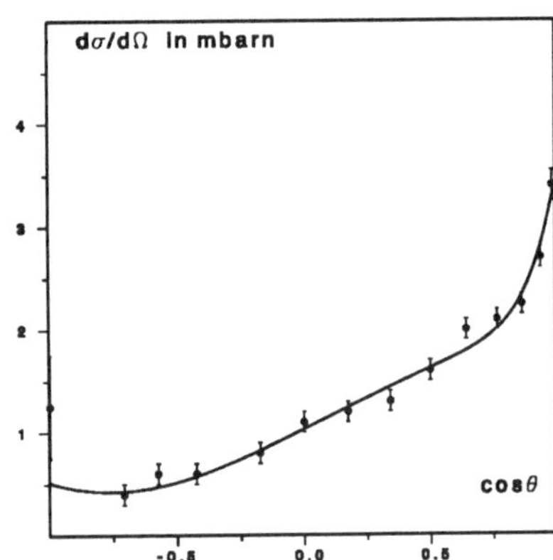

Figure 2: Differential elastic (data from [6]) and charge exchange (data from [8]) cross-sections at $p_{lab} = 272.0\ MeV/c$.

$p_{lab} = 272.0\ MeV/c$ is shown in figure 2. It has $\chi^2 = 0.9$/data. In the table we collect the ρ values of [6] and the exact model values.

	$p_{lab} = 233\ MeV/c$ model	experiment	$p_{lab} = 272\ MeV/c$ model	experiment
σ_{tot}	262 mb	296 mb (input)	247 mb	263 mb (input)
η	0.31	0.0 (input)	0.29	0.0 (input)
ρ	0.016	0.041±0.027	-0.006	-0.014±0.027
χ^2	43.0	45.3/63 d.o.f.	82.9	70.0/52 d.o.f.

We can also test the validity of the Locher method at these low energies by fitting the model cross-sections with the Locher formulas, using the total cross-section and η from the model as input. We find $\rho = 0.076$ at $p_{lab} = 233.0\ MeV/c$ and $\rho = 0.011$ at $p_{lab} = 272.0\ MeV/c$.

To analyse ρ we can make a partial wave decomposition. We find that the smallness of ρ is due to cancellations of the contributions of different P-waves.

Part of this work was included in the research program of the Stichting voor Fundamenteel Onderzoek der Materie (F.O.M.) with financial support from the Nederlandse Organisatie voor Zuiver- Wetenschappelijk Onderzoek (Z.W.O.).

References

[1] M.M. Nagels, T.A. Rijken, and J.J. de Swart, Phys. Rev. D 17 (1978) 768.

[2] E. Eisenhandler et al., Nucl. Phys. B113 (1976) 1.

[3] R.A. Kunne et al., CERN-EP/88-22 February 1988.

[4] M.P. Locher, Nucl. Phys. B2 (1967) 525.

[5] M. Lacombe et al., Phys. Lett. 124B (1983) 443.

[6] L. Linssen et al., Nucl. Phys. A469 (1987) 726.

[7] W. Brückner et al., Phys. Lett. 158B (1985) 180.

[8] W. Brückner et al., Phys. Lett. 169B (1986) 302.

RECENT EMC RESULTS ON THE SPIN-DEPENDENT STRUCTURE FUNCTION g_1 OF THE PROTON

Vassili Papavassiliou
Physics Department, Yale University, New Haven, CT 06520

ABSTRACT

The NA2′ experiment at CERN has measured the spin asymmetry in deep-inelastic polarized muon-proton scattering and extracted the spin-dependent structure function g_1 of the proton. The integral of g_1 over x was found to be $0.114 \pm 0.012 \pm 0.026$, in disagreement with the Ellis-Jaffe sum rule. An implication is that the quarks account for only a small fraction of the proton spin.

INTRODUCTION

The spin structure of the nucleon is described[1] in terms of two spin-dependent structure functions, g_1 and g_2. Of these, g_1 has a transparent interpretation in the context of the quark-parton model, where it can be written in terms of quark densities f_i with helicities the same $(+)$ as or opposite $(-)$ to the helicity of the parent nucleon:

$$g_1(x) = \frac{1}{2}\sum e_i^2[f_i^+(x) - f_i^-(x)] \equiv \frac{1}{2}\sum e_i^2 \Delta f_i(x), \tag{1}$$

where the sums run over the quark flavours and x is the usual Bjorken scaling variable. Although the form of the structure functions cannot be calculated in perturbative QCD, light-cone current algebra predicts relations for the integrals of the structure functions. The Bjorken sum rule[2] relates the difference of the proton and neutron spin-dependent structure functions to the ratio of axial-vector-to-vector weak coupling constants g_A/g_V, measured in nucleon β-decay:

$$\int_0^1 [g_1^p(x) - g_1^n(x)]dx = \frac{1}{6}\frac{g_A}{g_V}(1 - \frac{\alpha_s(Q^2)}{\pi}) \simeq 0.209(1 - \frac{\alpha_s(Q^2)}{\pi}), \tag{2}$$

where the lowest-order QCD correction[3] was included. For the proton and neutron separately, the Ellis-Jaffe sum rules[4] predict, assuming exact flavor-SU(3) symmetry in the decays of the members of the baryon octet,

$$\int_0^1 g_1^{p(n)}(x)dx = \frac{1}{12}\frac{g_A}{g_V}[1 + (-)\frac{1 + 4\lambda(Q^2)}{3}(\frac{3F/D - 1}{F/D + 1})](1 - \frac{\alpha_s(Q^2)}{\pi}), \tag{3}$$

again including the QCD correction. Here F and D are the SU(3) couplings and λ is the ratio of singlet and octet matrix elements, which is Q^2-dependent[5] because of the axial anomaly.[6] The assumption that the strange sea is unpolarized[4] leads to $\lambda = 1$ and the sum rules predict the values 0.189 ± 0.005 (-0.002 ± 0.005) for the proton (neutron) at $Q^2 = 10$ GeV$^2/c^2$, using the most recent value of F/D ($= 0.632 \pm 0.024$).[7]

Experimentally one measures the asymmetry $A = (d\sigma^{\uparrow\downarrow} - d\sigma^{\uparrow\uparrow})/(d\sigma^{\uparrow\downarrow} + d\sigma^{\uparrow\uparrow})$ where $d\sigma^{\uparrow\downarrow}(d\sigma^{\uparrow\uparrow})$ is the differential cross section for beam and target spins antiparallel (parallel), which is related to the virtual-photon–proton asymmetries A_1 and A_2 by $A = D(A_1 + \eta A_2)$. Here D and η are kinematic factors and $A_1 = (\sigma_{1/2} - \sigma_{3/2})/(\sigma_{1/2} + \sigma_{3/2})$, where σ is the total virtual-photoabsorption cross section and the subscript refers to the projection of the total angular momentum of the photon-nucleon spin. A_2 arises from the interference of longitudinal and transverse amplitudes and is kinematically suppressed in this experiment.

The asymmetry A is related to the structure functions g_1 and F_2 by

$$g_1(x) = \frac{F_2(x)}{2x(1+R)}[\frac{A}{D} + (\frac{\sqrt{Q^2}}{\nu} - \eta)A_2], \tag{4}$$

where R is the ratio of longitudinal-to-transverse cross sections. The uncertainty due to the unknown value of A_2 is small because $|A_2| \leq \sqrt{R}$ and because of the kinematic suppression.

Another quantity of interest is the spin asymmetry for semi-inclusive hadron production ($\mu p \rightarrow \mu' h^{\pm} X$, with X being any additional undetected final state). Hadron production can be described, in the frame of the quark-parton model, in terms of quark-fragmentation functions $D_q^h(z)$,[8] which give the probability for a struck quark q to produce a hadron h carrying a fraction z of the virtual-photon energy ν. Neglecting the strange sea, the asymmetries for positive- and negative-pion production are given by[9]

$$A^+ = \frac{4\Delta u(x) + \Delta d(x)\zeta}{4u_v(x) + d_v(x)\zeta + 5(1+\zeta)\bar{q}(x)}, \tag{5}$$

$$A^- = \frac{4\Delta u(x)\zeta + \Delta d(x)}{4u_v(x)\zeta + d_v(x) + 5(1+\zeta)\bar{q}(x)}, \tag{6}$$

where q_v and $\bar{q}$ are the valence- and sea-quark distributions and the suppression factor ζ is the ratio of unfavored-to-favored fragmentation functions integrated from some minimum z_0 to 1: $\zeta = \int_{z_0}^1 D^-(z)dz / \int_{z_0}^1 D^+(z)dz$.[10] Measurement of the semi-inclusive asymmetries can provide information on the spin structure of the up and down quarks separetely, that could otherwise be obtained only by using a polarized-neutron target.

THE EXPERIMENT

The experiment[11] measured the spin asymmetry in deep-inelastic muon-proton scattering at CERN, using polarized beams of energies 100–200 GeV and a polarized target[12] of irradiated ammonia. Data were taken with the EMC Forward Spectrometer.[13]

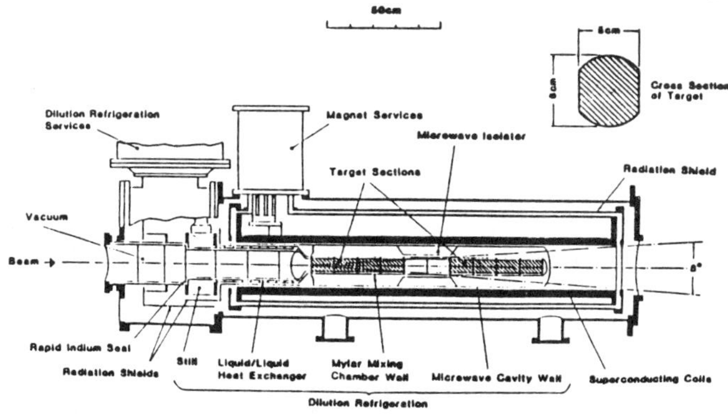

FIG. 1: The EMC polarized target

The positive-muon beam, produced by decay in-flight of a pion beam, was polarized by momentum selection to (negative) 75–85%, depending on the energy. The target consisted of two sections in tandem (see Fig. 1), simultaneously polarized in opposite directions, which made precise monitoring of the beam flux unnecessary. The polarization was periodically reversed, helping to keep false asymmetries under control. Target polarizations reached values of 75–80%.

RESULTS

Results were obtained over the kinematic range $0.01 \leq x \leq 0.7$ and $2.5\ \mathrm{GeV}^2/c^2 \leq Q^2 \leq 70\ \mathrm{GeV}^2/c^2$. This significantly extended the range covered by previous experiments at SLAC[14,15] that used polarized electron beams of lower energies and allowed a precise determination of the integral of g_1 over x, which is very sensitive to the values of the asymmetry at low x. Fig. 2 shows A_1 as a function of x. Inner error bars are the statistical errors and outer the total (statistical plus systematic added in quadrature). Data from the Yale-SLAC experiments[14,15] are also shown; they are in good agreement in the overlap region.

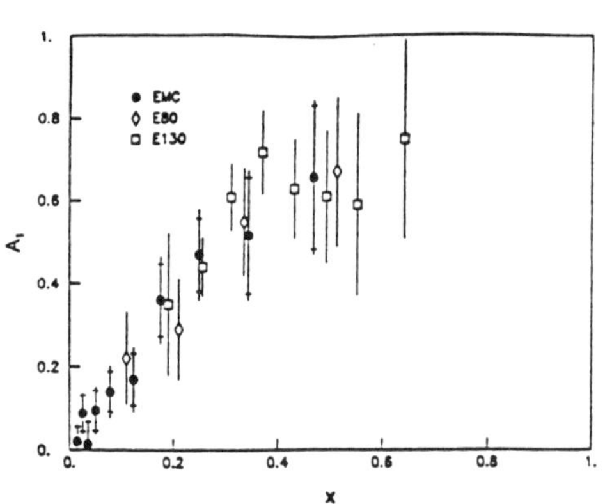

FIG. 2: A_1 vs. x (black dots)
Results from SLAC are also shown

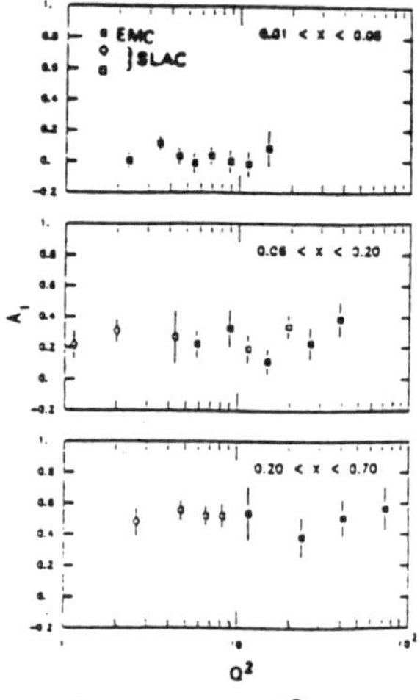

FIG. 3: A_1 vs. Q^2

In order to test the predicted scaling behaviour of A_1 with sufficient statistical accuracy, A_1 is plotted as a function of Q^2 in three wide x bins in Fig. 3. The errors are statistical. No Q^2 dependence can be seen. However, QCD predictions of scaling violations[16] are too small to be tested with the present statistics.

The structure function g_1 was computed from A_1 and the published parameterization of F_2 from a previous EMC experiment,[17] using a QCD calculation for R.[18] In Fig. 4, xg_1 is plotted on a logarithmic scale (right-hand axis and black dots). The solid curve was obtained from a fit to A_1. The area under the points is the integral of g_1. The integral from the low edge of a bin to 1 is also shown in Fig. 4 (left-hand axis and x's). The integral converges towards the value

$$\int_0^1 g_1(x)dx = 0.114 \pm 0.012\ (\text{stat.}) \pm 0.026\ (\text{syst.}), \tag{7}$$

at an average $Q^2 \simeq 11\mathrm{GeV}^2/c^2$ and with the extrapolations to low and high x contributing about 3% of the value. Statistical errors from the different points were combined *in quadrature.* Systematic errors on the other hand were added *linearly* to allow for the possibility of uncertainties affecting all the x bins in the same direction; they include an estimate for possible false asymmetries, uncertainties in the values of R, F_2, A_2 and the beam and target polarizations. Systematic errors due to *different* sources are added *in quadrature,* since they are clearly uncorrelated.

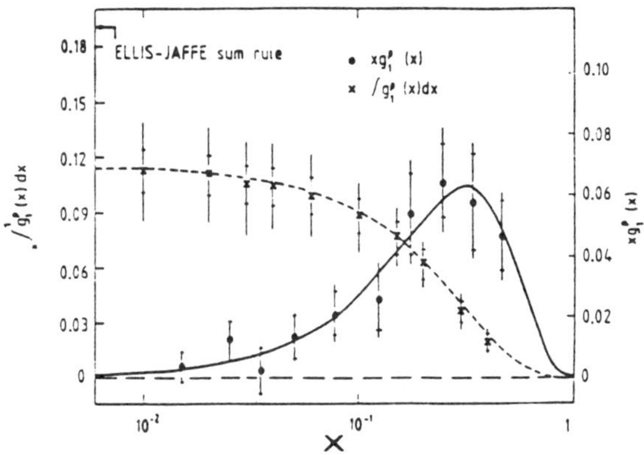

FIG. 4: g_1 vs. x (black dots and left-hand axis) and $\int_x^1 g_1(x)dx$ (x's and right-hand axis)

This result is in disagreement with the prediction of the Ellis-Jaffe sum rule. Since the assumption of flavor-SU(3) symmetry has been verified in a study of hyperon decays,[7] the violation of the sum rule may indicate a net polarization of the strange-quark sea opposite to the proton spin, equal to

$$Ds \equiv \int_0^1 [\Delta s(x) + \Delta \bar{s}(x)]dx = -0.228 \pm 0.036 \pm 0.078. \tag{8}$$

The result however is consistent with that quoted in Ref. 15, which gave $\int_0^1 g_1 dx = 0.17 \pm 0.05$. The error there was dominated by the uncertainty in the extrapolation to the unmeasured region $x < 0.1$.

The disagreement with the sum rule, which is at the level of two standard deviations, if all the errors (including the one due to the uncertainty on the F/D ratio) are added in quadrature, becomes more significant if all the available data are combined. Using the A_1 values from Refs. 14,15 one can obtain $\int_{0.1}^{0.7} g_1^p(x)dx = 0.091 \pm 0.008 \pm 0.013$, in excellent agreement with the EMC value of $0.087 \pm 0.010 \pm 0.015$. Combining the two results and adding the EMC result for the region 0.01–0.1 and the extrapolations to 0 and 1, one obtains

$$\int_0^1 g_1^p(x)dx = 0.116 \pm 0.009 \pm 0.019 \text{ (World average)}, \tag{9}$$

where the correlation between the low- and high-x EMC systematic errors was taken into account, giving

$$\int_0^1 g_1^p(x)dx|_{\text{theory}} - \int_0^1 g_1^p(x)dx|_{\text{experiment}} = 0.073 \pm 0.022, \tag{10}$$

with all errors combined in quadrature.

The semi-inclusive asymmetries A^+ and A^- were also measured, for $z \geq 0.1$;[19] they are shown in Fig. 5. All hadrons were assumed to be pions. A^+ appears to be higher than A^-. Although the statistical significance is not great, the effect is consistent with the expectation that the down quark is polarized anti-parallel to the proton spin.

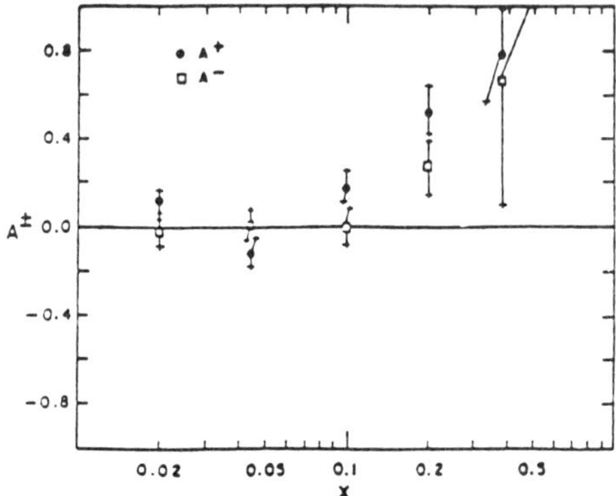

FIG. 5: Semi-inclusive asymmetries vs. x

DISCUSSION AND CONCLUSIONS

The result in Eq. 9 can be used to calculate the contribution of the total quark helicities to the nucleon spin. Assuming the validity of the Bjorken sum rule (Eq. 2), the corresponding integral for the neutron can be calculated. It is

$$\int_0^1 g_1^p(x)dx = -0.075 \pm 0.021 \text{ (predicted)}, \tag{11}$$

at the Q^2 scale of this experiment. Then the definition of g_1 (Eq. 1) can be used to solve for $S_{q_i} \equiv \frac{1}{2}Dq_i$, the net spin of the i-th quark, with S_s given by Eq. 8. The result is $S_u = 0.37 \pm 0.03$, $S_d = -0.26 \pm 0.03$ and $S_s = -0.11 \pm 0.03$, for a total quark spin of

$$S_q \equiv \sum S_{q_i} = 0.00 \pm 0.09. \tag{12}$$

These numbers have been corrected for the lowest-order QCD effects since the quark distributions are only meaningful in the scaling limit.

The implication is that orbital angular momentum and/or gluon spin must account for most if not all the nucleon spin. The conclusion that the quark helicities contribute only a small fraction of the nucleon helicity is not very sensitive to the assumption that the strange sea polarization is responsible for the failure of the Ellis-Jaffe sum rule; setting $S_s = 0$ would still give a net quark spin of only 12% of the nucleon spin, with similar errors. It is interesting to mention here that in the skyrmion model of the nucleon[20] the singlet matrix element (λ in Eq. 3) vanishes[21] in the chiral limit, corresponding to the vanishing of the net quark helicity.

REFERENCES

1. A review of the subject and a summary of all the relevant quantities can be found in V.W. Hughes and J. Kuti, Ann. Rev. Nucl. Part. Sci. **33**, 611 (1983).
2. J.D. Bjorken, Phys. Rev. **148**, 1467 (1966); Phys. Rev. **D1**, 465 (1970).
3. J. Kodaira *et al.*, Phys. Rev. **D20**, 627 (1979); Nucl. Phys. **B159**, 99 (1979).
4. J. Ellis and R.L. Jaffe, Phys. Rev. **D9**, 1444 (1974).
5. J. Kodaira, Nucl. Phys. **B165**, 129 (1980).
6. S.L. Adler, Phys. Rev. **177**, 2426 (1969); J.S. Bell and R. Jakiw, Nuovo Cim. **60A**, 47 (1969).
7. M. Bourquin *et al.*, Z. Phys. **C21**, 27 (1983).
8. R.D. Field and R.P. Feynman, Nucl. Phys. **B136**, 1 (1978).
9. R.L. Heimann, J. Phys. G: Nucl. Phys. **4**, 173 (1978).
10. Here D^+ is the fragmentation function of the up (down) quark to positive (negative) pions while D^- is that of the up (down) quark to negative (positive) pions. An experimental determination of these functions can be found in J.J. Aubert *et al.* (EMC Collaboration), Phys. Lett. **160B**, 417 (1985).
11. J. Ashman *et al.* (EMC Collaboration), Phys. Lett. **206B**, 146 (1988).
12. S.C. Brown *et al.*, in *Proceedings of the 4th International Workshop on Polarized-Target Materials and Techniques,* ed. by W. Meyer, p. 102 (Bonn, 1984).
13. The original EMC apparatus is described in O.C. Allkofer *et al.* (EMC Collaboration), Nucl. Instr. and Meth. **179**, 445 (1981). The apparatus was slightly modified for this experiment.
14. M.J. Alguard *et al.*, Phys. Rev. Lett. **37**, 1261 (1976); Phys. Rev. Lett. **41**, 70 (1978).
15. G. Baum *et al.*, Phys. Rev. Lett. **51**, 1135 (1983).
16. O. Darrigol and F. Hayot, Nucl. Phys. **B141**, 391 (1978).
17. J.J. Aubert *et al.* (EMC Collaboration), Nucl. Phys. **B259**, 189 (1985).
18. G. Altarelli and G. Martinelli, Phys. Lett. **76B**, 89 (1978); M. Glück and E. Reya, Nucl. Phys. **B145**, 24 (1978).
19. V. Papavassiliou, Ph.D. thesis, Yale University, 1988 (unpublished).
20. E. Witten, Nucl. Phys. **B223**, 422 (1983); *ibid* 433; G. Adkins, C. Nappi and E. Witten, Nucl. Phys. **B228**, 552 (1983).
21. S.J. Brodsky, J. Ellis and M. Karliner, "Chiral Symmetry and the Spin of the Proton," SLAC Preprint No. SLAC-PUB-4519 (1988) (to be published).

PRECISION INVESTIGATIONS OF NUCLEI AND NUCLEONS WITH THE $(e, e'\gamma)$ REACTION

C. N. Papanicolas, E. A. Ammons, L. S. Cardman, J. R. Deininger, S. M. Dolfini, J. B. Mandeville, P. E. Mueller and S. E. Williamson
Nuclear Physics Laboratory and Department of Physics
University of Illinois, Urbana, IL 61801 USA

Abstract

Recent theoretical and experimental investigations of the $(e, e'\gamma)$ reaction show that it provides a probe of unparalleled precision and selectivity. Experiments aimed towards the isolation of multipole form factors in mixed transitions, the study of continuum excitations in nuclei, and the measurement of the response of the proton are underway at several laboratories.

1. Introduction

Because it involves only leptons, the $(e, e'\gamma)$ coincident channel is unique among the coincident probes. For this channel the precision inherent in electron scattering is not reduced by the presence of hadronic interactions in the entrance or exit channels. As is the case in single-arm electron scattering, experimental results can be interpreted with an accuracy of few percent, limited primarily by theoretical uncertainties in the dispersive (second-order) corrections[1]. Over the last few years there has been significant progress toward realizing the potential of this probe. The first nuclear $(e, e'\gamma)$ studies[2] demonstrated that the considerable technical obstacles against the detection of nuclear decay γ-rays in the environment of an electron scattering hall could be overcome. In addition, renewed theoretical interest[3,4,5] in this reaction has improved our understanding of the probe and its potential. The $(e, e'\gamma)$ reaction is now understood and treated with the same level of accuracy and sophistication as single arm electron scattering, using the Distorted Wave Born Approximation (DWBA)[3].

2. Features of the $(e, e'\gamma)$ Probe

The main features of $(e, e'\gamma)$ can be understood with the help of Figures 1 and 2. A scattered electron deposits momentum, $\vec{q}$, and energy, ω, in the nucleus, exciting it from its ground state of spin-parity J_i^π to a state J_m^π which subsequently decays to a state J_f^π by emitting a photon of energy E_γ; the final state need not be the ground state. The excited state can be narrow and of well-defined spin and parity, a cluster of broad overlapping resonances, such as the giant resonances, or even nucleon resonances. Depending on the choice of electron kinematics, different m substates of the excited nucleus are populated[3].

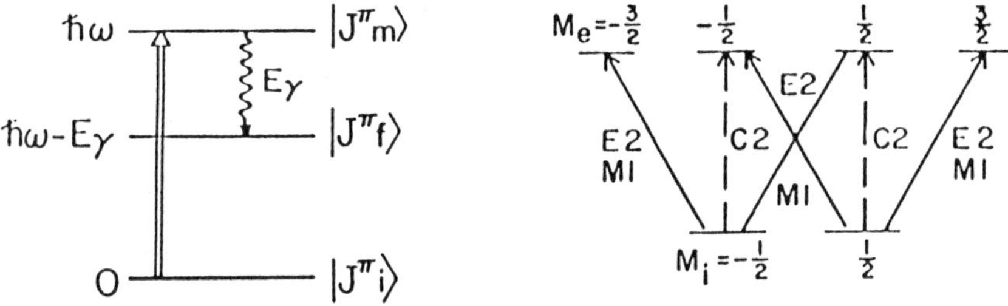

Figure 1: The $(e, e'\gamma)$ process. (a)Energy level diagram. The scattering process determines the population of the m substates of $|J^\pi_m>$. (b)For a spin-$\frac{1}{2}$ system excited to a $J^\pi = \frac{3}{2}$ state (e.g. the 6.32 MeV state of ^{15}N or the N→ Δ(1232) excitations), Coulomb(C2), Electric(E2) and Magnetic(M1) multipoles are allowed.

The emitted γ-ray serves as an analyzer, permitting us to reconstruct the density matrix characterizing the excited state. The use of polarized beams allows us to further manipulate the m-state population. The complementarity and close relationship between $(e, e'\gamma)$ and inclusive electron scattering from polarized targets, where the initial m-state population is prearranged, is apparent. It has been investigated recently by Donnelly *et al.*[5]

The lowest-order Feynman diagrams contributing to the $(e, e'\gamma)$ process are shown in Figure 2. It is evident that bremsstrahlung is of the same order as the nuclear de-excitation process. It is always possible to find bremsstrahlung diagrams where the initial and final nuclear states and the energy of the emitted photon, E_γ, coincide with those of the $(e, e'\gamma)$ process of interest. The two processes are then coherent and interfere.

The technical difficulties associated with $(e, e'\gamma)$ experiments made them impractical until quite recently. The development of cw electron beams lead to the development of $(e, e'\gamma)$ facilities at Illinois and Mainz. Similar programs are

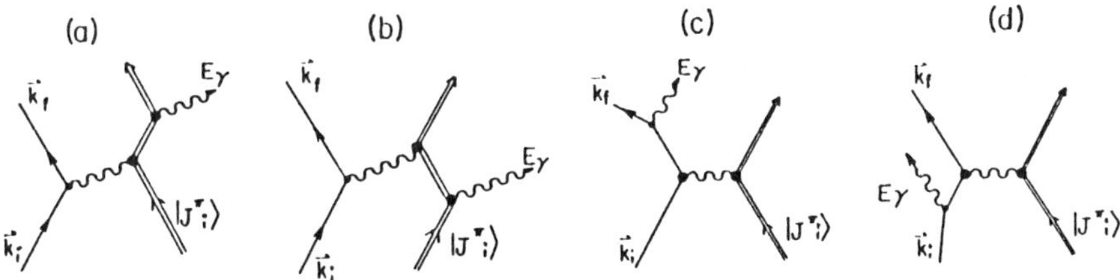

Figure 2: Feynman diagrams contributing to the $(e, e'\gamma)$ process. Note that bremsstrahlung (c,d) is coherent with nuclear de-excitation (a,b)

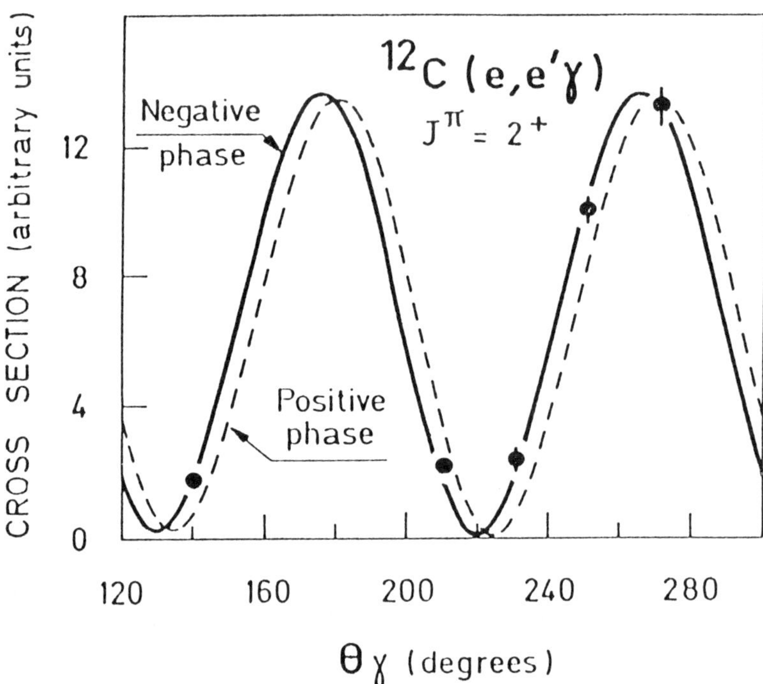

Figure 3: DWBA predictions of the $(e, e'\gamma)$ cross section for the two possible choices of relative phase between F_L and F_T.

planned for other high duty factor accelerators now under construction. The quality and precision of these experiments is nicely exhibited by the angular dependence of the cross section for the $(e, e'\gamma)$ process to the first excited state of ^{12}C shown in Figure 3. Transition charge and current densities reconstructed from single arm electron scattering data were used to predict[3] the angular distributions shown; the two distributions correspond to the two possible choices for the relative phase between the transverse and longitudinal form factors. The excellent agreement between the coincident data and the distribution corresponding to a negative phase (in accord with Siegert's theorem) demonstrates both the compatability of (e, e') and $(e, e'\gamma)$ and the ability of $(e, e'\gamma)$ to access new structure information.

The use of $(e, e'\gamma)$ has been suggested[6] for the study of a variety of outstanding problems in nuclear and nucleon structure including:

1. the isolation and study of multipole form factors in mixed transitions;
2. the study of vorticity in nuclei[7];
3. the achievement of ultra-high resolution in electron scattering by the use of germanium γ-ray detectors in $(e, e'\gamma)$ reactions;
4. the measurement of form factors of giant resonances; and

5. the exploration of the structure of the nucleon.

The first two of these are now rather well understood, and experimental programs at Illinois and Mainz have been developed for their exploitation[6]. The last two involve the study of excitations in the continuum, a region where the presence of large radiative tails and the broad, structureless nature of the excitations present difficult problems.

3. $(e, e'\gamma)$ Studies in the Continuum

An obvious domain for the application of the $(e, e'\gamma)$ reaction is the study of isolated narrow states below the particle emission threshold. As indicated above, the feasibility and precision of such measurements is now proven[6]. In the remainder of this paper we explore the use of $(e, e'\gamma)$ for the study of nuclear and nucleon resonances.

3.1 Studies of Giant Resonances

The ability of the $(e, e'\gamma)$ reaction to separate multipoles will be extremely useful for the study of nuclear and nucleon resonances, where (e, e') studies are hampered by the overlapping nature of the states. These difficulties arise largely from the substantial natural width characterizing states lying above particle threshold. $(e, e'\gamma)$ studies in the continuum are further complicated by the relatively small radiative decay width (typically $\Gamma_\gamma/\Gamma_{Tot} < 0.01$). We are therefore confronted with two very serious issues:

1. In the resonance region the detector resolution is substantially better than the natural width of the states. As a result the resonance decay interferes with the bremsstrahlung process, and it is not possible to subtract the bremsstrahlung contribution by fitting a sharp peak over a smooth background, as can be done for bound state excitations. A far more complicated procedure, which is susceptible to model error, must be employed.
2. Given the small radiative width of continuum excitations, it is not evident that an appropriate kinematic window can always be found to allow the unambiguous detection of all processes of interest.

Fein *et al.*[4] have examined carefully the feasibility of giant resonance studies using the $(e, e'\gamma)$ probe. Working within the PWBA they have considered the most straightforward experiment of this type, in which the giant dipole resonance decays by photon emission back to the ground state. They found for ^{12}C that the effects of bremsstrahlung and bremsstrahlung-nuclear interference are formidable in this case, and suggested that the experiments should be performed at high incident electron energies at relatively small momentum transfers, and for non co-planar kinematics in order to enhance the desired signal. Their conclusions are discouraging. If one is faced with such formidable problems for ^{12}C

(Z=6), there is not much hope for $(e,e'\gamma)$ giant resonance studies in heavy nuclei. However, the authors point out that even a single $(e,e'\gamma)$ giant resonance study will be most valuable as a test of other techniques, which are unfortunately not as "clean" due to the presence of strong interaction effects.

A different approach to this problem has been developed at Illinois. Recognizing that measurement of the decay to the ground state is problematic, we have begun a program of $(e,e'\gamma)$ studies of continuum excitations which focuses on the γ-decay to an excited state. In such cases the bremsstrahlung is suppressed (compared to the case of γ-decay to the ground state) by a factor equal to the ratio of $F_{el}(q)/F_{inel}(q)$, where $F_{el}(q)$ and $F_{inel}(q)$ are the form factors characterizing the electron scattering to the ground state and the final state to which the continuum excitation is decaying. This dramatic background suppression can be seen in Figure 4, where the angular dependence of the $^{12}C(e,e'\gamma)$ cross section is shown for the two cases. This "modified" technique is now being employed for the study of the GDR in ^{12}C and the IVQR in ^{208}Pb . It provides an additional advantage in the high selectivity that the decay gamma provides. The preferential emission of dipole radiation minimizes the contribution to the $(e,e'\gamma)$ cross section of all but the lowest multipoles, and these multipoles can be separated with the help of the angular and momentum transfer dependences of the cross-section.

3.2 Nucleon Studies

The use of $(e,e'\gamma)$ in experiments involving excitations of the nucleon itself is particularly appealing. While the potential of the $(e,e'\gamma)$ probe has not been explored theoretically or experimentally to a degree sufficient for us to outline a coherent program of subnucleonic investigations, two experiments have been suggested[8,9] involving the Δ resonance.

A fundamental question in nuclear physics concerns the intrinsic shape of the nucleon and its excited states, and how these shapes are affected when the nucleon is imbedded in the nuclear medium. Glashow suggested[10] that the persistent difficulties of spherical bag models to account for the axial-vector coupling constant G_A and the $G_{\pi N\Delta}/G_{\pi NN}$ ratio can be resolved by foregoing spherical symmetry. Subsequent investigations[11,12] have demonstrated that a sizable (up to 40%!) D-state admixture is required. One implication of such a deformation is that a quadrupole excitation of the Δ has become possible (such a process is strictly forbidden in the exact SU(6) limit). As a result the E2/M1 and the C2/M1 ratios are observables that can severly constrain various models of the nucleon, and simultaneously provide us with a measure of the quark-quark tensor interaction. While theoretical predictions abound, the experimental situation is unsatisfactory. The E2/M1 ratio is poorly known from photopion data[13], and no information is available for the C2/M1 ratio. The

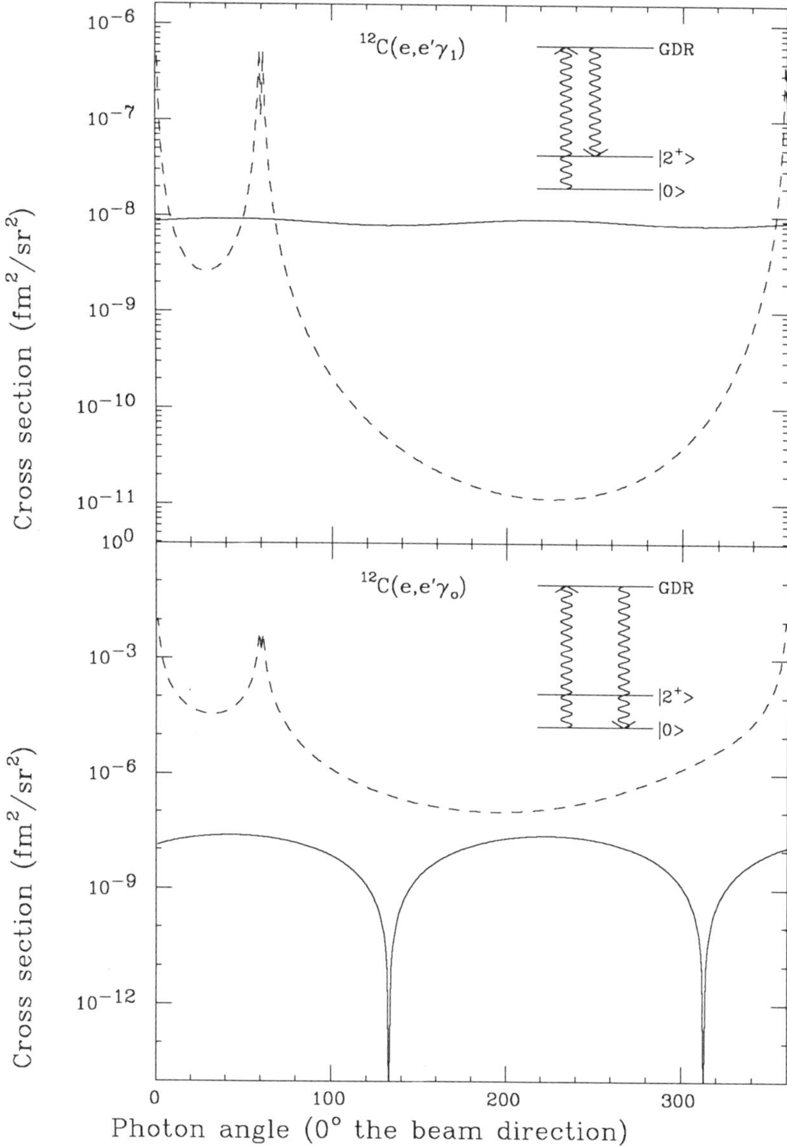

Figure 4: The angular dependence of the GDR ^{12}C $(e, e'\gamma)$ cross section. solid curves show the nuclear decay contribution; dashed curves show the bremsstrahlung contribution. Decay to the 2^+(4.439 MeV) state (top) instead of the ground state (bottom) enhances the signal/background by a factor of 1000.

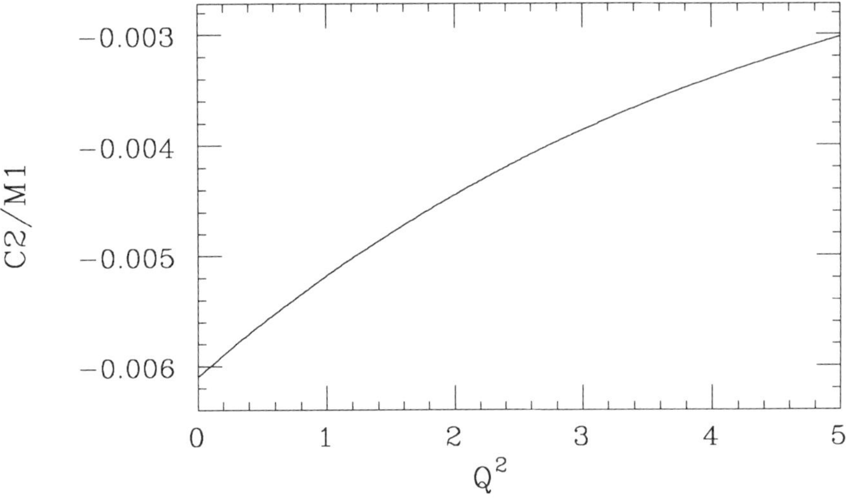

Figure 5: The evolution of the C2/M1 ratio as function of Q^2 in the Constituent Quark Model with Color Magnetic Interaction[12].

C2/M1 ratio, which is predicted by various models of the nucleon to be small (a few percent), is expected to reach unity[14] at asymptotic values of Q^2. Its evolution as a function of momentum transfer and the manner in which the asymptotic value is approached are particularly important for our understanding of the nucleon. Figure 5 shows the evolution of this ratio predicted by a constituent quark model in which the hyperfine color magnetic interaction has been introduced[12].

The similarity of the N→ Δ to the ^{15}N(6.32 MeV) transition, which we are currently investigating, is obvious. Both involve a $\frac{1}{2} \rightarrow \frac{3}{2}$ transition which allows M1, C2, and E2 multipoles. Both are described by the schematic diagram in Figure 1b. However, the experimental difficulties associated with the N→ Δ measurements are formidable. The difficulty of detecting 300 MeV γ-rays amidst the π_0 decay background cannot be underestimated. Moreover because of the enormous width of the Δ resonance, the bremsstrahlung interference, the crossed graphs, and the tails of higher resonances as well as the Born terms (Compton scattering) need to be treated carefully. The difficulty of extracting the desired signal can be appreciated with the help of Figure 6, which shows some of the processes contributing to an $\mathrm{H}(\mathrm{e},\mathrm{e}'\gamma)$ measurement. It is impossible to separate these contributions in a model-independent way, although severe restrictions can be imposed by measuring the ω and q dependence of the cross section. We should observe that one of "background" processes depicted in this figure is Compton scattering with virtual photons. This $(\mathrm{e},\mathrm{e}'\gamma)$ process is in itself intriguing, and has the potential for mapping the nucleon response to any virtual photon mass. Such a program can be realized at one of the planned multi-GeV cw electron accelerators.

The measurement[8] of the C2/M1 ratio of the nucleon through $(e, e'\gamma)$ will be carried out at MIT/Bates. The measurement is feasible at a low duty factor accelerator if we exploit the complete kinematics of the measurement and require a redundant coincidence with the recoiling proton. An experimental arrangement has been proposed which allows the simultaneous measurement of the complementary $(e, e'\pi^0)$ reaction. In Figure 7 the bremsstrahlung contribution is shown to be insignificant over a wide angular range compared to the decay of the delta. The experiment will be performed with polarized electrons, allowing a measurement of the fifth structure function as well as the transverse-longitudinal interference. This experiment should yield the first information on the tantalizing question of Δ "deformation".

Drechsel and Rothhaas have suggested[9] that the nucleon deformation in the nuclear medium can be studied through the ^{12}C $(e, e'\gamma)$ reaction. Count rate estimates indicate that such a measurement will be feasible using the MAMI accelerator provided a γ-ray detector capable of resolving the π^0 decay from the direct γ decay (15 MeV gap) becomes available. Given the recent developments in large volume NaI crystals, one can be optimistic.

4. Conclusion

Recent progress in experimental techniques and in the theoretical understanding of the $(e, e'\gamma)$ reaction have resulted in the development of a versatile and high precision probe for nuclear and nucleon structure.

5. Acknowledgements

It is a pleasure to acknowledge fruitful discussions with many colleagues, in particular W. Donnelly, S. Kumano, D. G. Ravenhall and J. Wambach. We are grateful to the Mainz group (H. Rothhaas *et al.*) for communicating to us their progress in the $(e, e'\gamma)$ program. Support for this work was provided in part by the National Science Foundation under grant NSF PHY 86 10493.

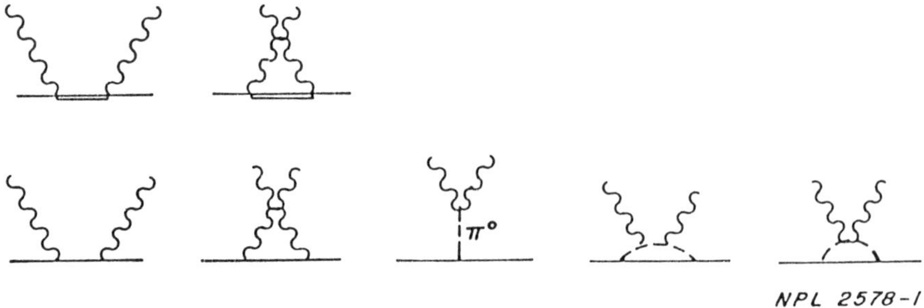

Figure 6: Diagrams contributing to H(e, e′γ) measurements (see text).

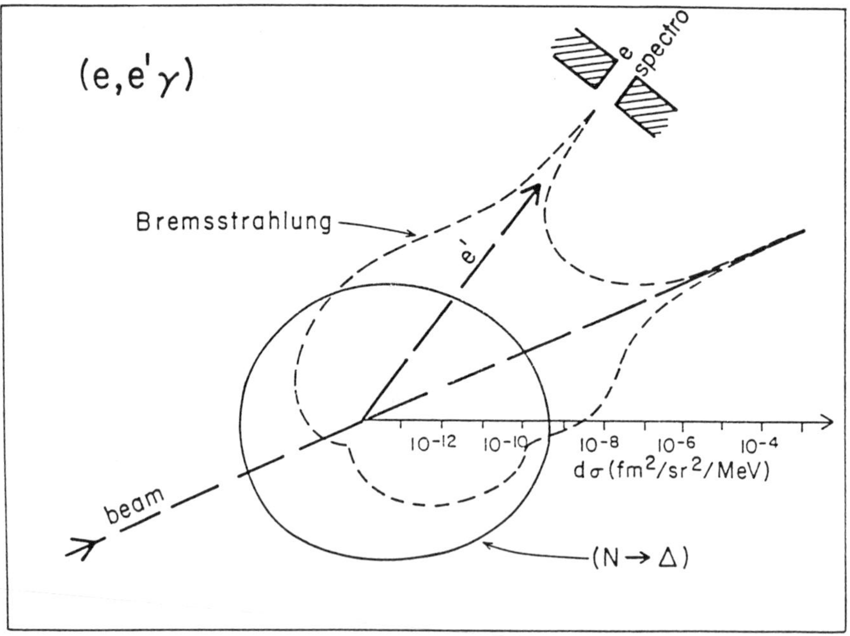

Figure 7: Over a wide angular range the nucleon de-excitation process (solid curve) dominates over bremsstrahlung (dashed curve) in the (e, e′γ) cross section.

REFERENCES

1. D. G. Ravenhall, AIP Conf. Proceedings **161**, 153 (1987) C. N. Papanicolas, L. S. Cardman and R. A. Eisenstein eds.
2. C. N. Papanicolas, S. E. Williamson, H. Rothhaas, G. O. Bolme, L. J. Koester, Jr., B. L. Miller, R. A. Miskimen, P. E. Mueller and L. S. Cardman, Phys. Rev. Lett. **54**, 26 (1985).
3. D. G. Ravenhall, R. L. Schult, J. Wambach, C. N. Papanicolas, and S. E. Williamson, Ann. Phys. **178**, 187 (1987).
4. E. Fein, D. Drechsel, M. Cavinato, M. Marangoni, A. M. Saruis, Nucl. Phys. **A468**, 301 (1987).
5. T. W. Donnelly, A. S. Raskin and J. F. Dubach, Nucl. Phys. **A474**, 307 (1987).
6. C. N. Papanicolas, Proc. Perspectives in Nuclear Physics at intermediate Energies. World Scientific, Singapore (1988), S. Boffi, C. Cioffi degli Atti, and M. M. Giannini eds.
7. D. G. Ravenhall and J. Wambach, Nucl. Phys. **A475**, 468 (1987).
8. C. N. Papanicolas *et al.*, Research Proposal to Bates Linear Accelerator #87-09, "Measurement of the Quadrupole Contribution to the N$\rightarrow$ Δ Excitation".
9. D. Drechsel, Prog. Part. Nucl. Phys. **13**, 63 (1984); H. Rothhaas private communication.
10. S. L. Glashow, Physica **96A**, 27 (1979).
11. V. Vento, G. Baym, A. D. Jackson, Phys. Lett. **102B**, 97 (1981); V. Vento, Phys. Lett. **121B**, 370 (1983).
12. S. A. Gogilidze *et al.*, Yad. Fiz. **45**, 1085 (1987).
13. R. Davidson, N. C. Mukhopadhyay and R. Wittman, Phys. Rev. Lett. **56**, 804 (1986), and references therein.
14. C. E. Carlson, Phys. Rev. **D34**, 2704 (1986).

TRANSVERSE INELASTIC ELECTRON SCATTERING FROM THE DEUTERON AT HIGH MOMENTUM TRANSFERS

P. E. Bosted*, R. G. Arnold, D. Benton,[a] L. Clogher
G. DeChambrier, A. T. Katramatou, J. Lambert,[b] A. Lung
G. G. Petratos, A. Rahbar, S. E. Rock, Z. M. Szalata
The American University, Washington, D.C. 20016

B. Debebe, M. Frodyma, R. S. Hicks, A. Hotta,[c] G. A. Peterson
University of Massachusetts, Amherst, Massachusetts 01003

R. A. Gearhart
SLAC, Stanford University, Stanford, California 94309

J. Alster, J. Lichtenstadt
Tel Aviv University, Tel Aviv, Israel 69978

F. Dietrich, K. van Bibber
Lawrence Livermore National Laboratory, Livermore, CA 94550

(*) Presented by P. Bosted
(a) Present address: Department of Physics, Princeton University, Princeton, NJ 08544
(b) Permanent address: Department of Physics, Georgetown University, Washington, D.C. 20057
(c) Permanent address: School of Physics, Shizuoka University, Shizuoka, 422 Japan.

ABSTRACT

Cross sections for 180° inelastic electron scattering from the deuteron were measured from the break-up threshold to beyond the quasielastic peak for incident beam energies between 0.7 and 1.3 GeV, corresponding to $0.75 \leq Q^2 \leq 2.75$ $(\mathrm{GeV/c})^2$. The data are in reasonable agreement with nonrelativistic models that include final-state interactions and meson exchange currents.

INTRODUCTION

Inelastic electron scattering from the deuteron is of great importance in understanding the nucleon-nucleon interaction. It offers rich grounds for testing detailed calculations that go beyond the Plane-Wave Impulse Approximation (PWIA) by including Final State Interactions (FSI), Meson Exchange Currents (MEC), and Isobar Configurations (IC). Measurements at the top of the quasielastic peak provide good tests of the impulse approximation, since interaction effects are calculated to be small, and thus have been extensively used in extracting the neutron form factors. The high and low momentum sides of the quasielastic peak are the regions where the nuclear structure functions should be most sensitive to the interaction effects, including the role of constituent quarks in the short-range nucleon-nucleon interaction. Of particular value are data at high momentum transfers, where fully relativistic models can be tested.

THE EXPERIMENT

This paper presents new data for transverse inelastic electron scattering from the deuteron. The data extend to higher momentum transfers Q^2 or cover a larger range of scattered electron energy E' than previous experiments.[1,2,3,4]

The measurements were made using the Nuclear Physics Injector and the Stanford Linear Accelerator to deliver electron beams of energy between 0.7 and 1.3 GeV in 1.6 μsec long pulses at average currents of 1 to 5 μA. The beams were transported into End Station A (ESA) and through the chicane magnets of a 180° spectrometer system.[5] After passing through 10 or 20 cm long liquid deuterium cells, the beams were directed to a beam dump in ESA. Electrons scattered at 180° were momentum analyzed by the electron arm spectrometer and detected in a set of six multi-wire proportional chambers. A threshold gas Čerenkov and a lead glass shower counter array were used to reject the large flux of pions. Quasi-elastic sweeps were taken at four beam energies. Typically, ten spectrometer settings were used to cover the E' range from near break-up threshold to the quasielastic peak and into the region where pion production dominates. In addition, data were taken right at break-up threshold for nine beam energies. The data were corrected for detector inefficiencies, trigger and electronic inefficiencies, and the measured contributions from the target end-caps. The variation of the solid angle with relative momentum was calculated using a Monte Carlo program,[6] and verified experimentally to $\pm 2\%$ by a series of measurements in which the central momentum was stepped in small increments. Checks of the absolute solid angle calculations were made by measuring elastic scattering from hydrogen. The results agree within 2% with previous backward angle measurements.[7] The spectra were radiatively corrected using the procedures described in Ref. 8. The total systematic errors ranged from 3.9 to 12.0%.

QUASIELASTIC SPECTRA

The radiatively corrected cross sections at each beam energy E for the four quasielastic sweeps are shown as a function of scattered electron energy E' in Fig. 1. These spectra do not include the momentum bites right at break-up threshold, which are being analyzed separately (see below). The data have been corrected for finite resolution effects. The data are in reasonable agreement with the nonrelativistic PWIA calculations of Laget[9] near the quasielastic peaks, but larger than the calculations by up to a factor of 2 at high E' and up to a factor of 1.5 in the dip region between the quasielastic peak and the onset of pion production at low E'. The Laget calculations use the Paris[10] potential and a nonrelativistic expansion of the interaction operator to terms of order M^{-3}, where M is the nucleon mass. The matrix elements are evaluated in the lab system. Except in the region where pion production dominates, calculations using the PWIA formulas of McGee as modified by Durand[11] are almost indistinguishable from the Laget PWIA curves when the same potential is used. At high E', the McGee–Durand calculations diverge from the data much more when the Bonn[12] potential is used than when the Paris potential is used, showing the great sensitivity to the short-range part of the potentials in this region.

Considerable improvement is provided by the calculations which include FSI and MEC. The model by Laget, which includes real pion production, describes the low E' side of the peaks well, including the dip region. It reproduces the width of the peaks better than the PWIA calculations, but still underestimates the cross sections close to break-up threshold by up to a factor of 1.5. The calculations of Arenhövel[13] also use the Paris potential, and include IC in addition to MEC and FSI. These calculations use a complete nonrelativistic framework (except for the kinematics which are done relativistically), and the matrix elements are evaluated in the final np center-of-mass frame. This model predicts peaks considerably narrower and taller than seen in the data, with the disagreement increasing at large Q^2. The model does not include real pion production, and so should not be compared with the data on the low E' side of the peaks.

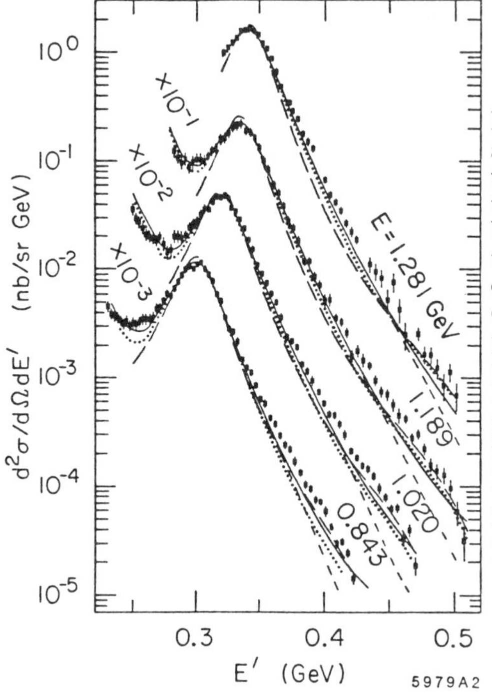

Fig. 1. Cross section for $d(e,e')$ as a function of scattered electron energy E' for four values of the incident energy E. The error bars include statistical and systematic uncertainties. The dotted curves are the PWIA calculations of Laget (Ref. 9) using the Paris potential. The short dashed curves use the PWIA formula of McGee–Durand (Ref. 11) with the Bonn potential, and are indistinguishable from the dotted curves except at high E'. The solid (long dashed) curves are the full calculations of Laget (Arenhövel, Ref. 13) using the Paris potential.

In order to examine the validity of the simple quasi-elastic scattering mechanism, we have plotted in Fig. 2 the ratio of experimental cross sections to the McGee–Durand PWIA model (using the Paris wave function) in five different y regions. The variable $|y|$ represents the smallest initial momentum for the struck nucleon allowed by energy and momentum conservation, and is defined by

$$E + M_d = E' + \sqrt{M^2 + y^2} + \sqrt{M^2 + (y + |\vec{q}|)^2} \quad ,$$

where M_d is the deuteron mass and $\vec{q}$ is the three-momentum transfer. This variable has been used[14] to define functions $F(y)$ which are approximately independent of Q^2 and θ at fixed y. The quasi-elastic peak corresponds to $y = 0$, while the break-up region corresponds to large negative y.

It can be seen that the ratios for forward and backward angle are generally in agreement, except for the largest $|y|$, where the forward angle ratios are significantly larger than the backward angle ones. The trend of the ratios is to decrease with increasing Q^2 and, for the three plots with the smallest $|y|$, to flatten out at a value of Q^2 which increases with $|y|$. This trend can also be seen to some extent in the calculations of Laget and Arenhövel, and is most likely due to the decreasing importance of FSI and MEC with increasing Q^2. In contrast to the $y = 0$ and $y = -0.1$ bins, the ratio for the $y = -0.2$ bin is considerably larger than unity in the region where it is independent of Q^2. This could be ascribed to a lack of high momentum components in the model for the deuteron wave function, or on the influence of effects not included in the model, such as the influence of six-quark states and relativistic corrections.

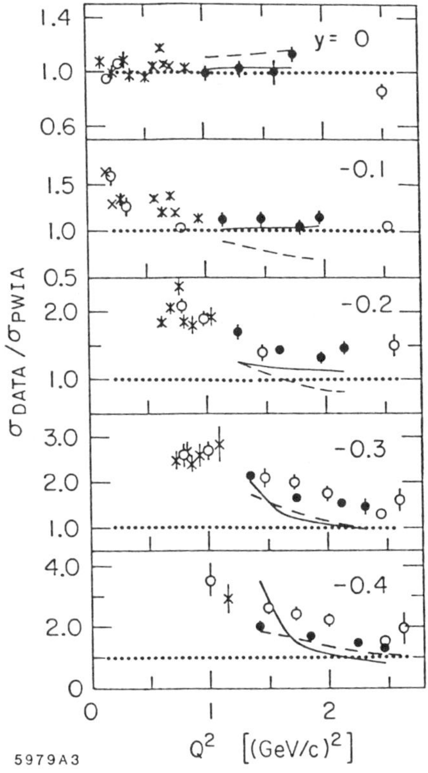

Fig. 2. Ratios of experimental cross sections to the PWIA model of McGee–Durand (Ref. 11) for five values of the scaling variable y. Shown are data from this 180° experiment (solid circles), backward angle data from Refs. 1 and 3 (crosses), and forward angle data from Refs. 1, 15, and 16 (open circles). The solid curves are the full 180° calculations of Laget, while the dashed curves represent those of Arenhövel.

THRESHOLD REGION

Data in the threshold break-up region were taken at nine beam energies in conjunction with measurements of the deuteron elastic magnetic form factor,[17] for which recoil deuterons were detected in coincidence with electrons. The threshold region has been of particular interest because the effects of MEC are large and are required by calculations to fit existing data.[4] New data from this experiment are shown at three representative beam energies in Fig. 3 as a function of the kinetic energy of the final-state np pair. The data are still preliminary, as only approximate radiative corrections have been performed. Due to the long targets used, the energy resolution was relatively poor (between 12 and 20 MeV in E_{np}). We have therefore convoluted the full calculations of Arenhövel shown in Fig. 3 with the experimental resolution at each beam energy. It can be seen that the data lie between the curves that use F_1 for the coupling in the MEC and those that use G_E. It is clear that the choice of coupling, which is of relativistic order, makes large differences in the predictions, implying that fully relativistic calculations should be made. The choice of nucleon form factors can also make substantial differences, but this ambiguity should be resolved when better measurements of the neutron electric form factor (planned at SLAC) become available. Calculations of threshold electrodisintegration have also been performed by many other theorists,[18] but generally only at $E_{np} = 1.5$ MeV. When these calculations are extended to other values of E_{np} so that they can be convoluted with the experimental resolutions and compared with the data, it will be interesting to see if models that include substantial 6-quark clusters ("Hybrid models") can describe the data as well, or better, than the models that do not include this contribution to the short-range nucleon-nucleon interaction.

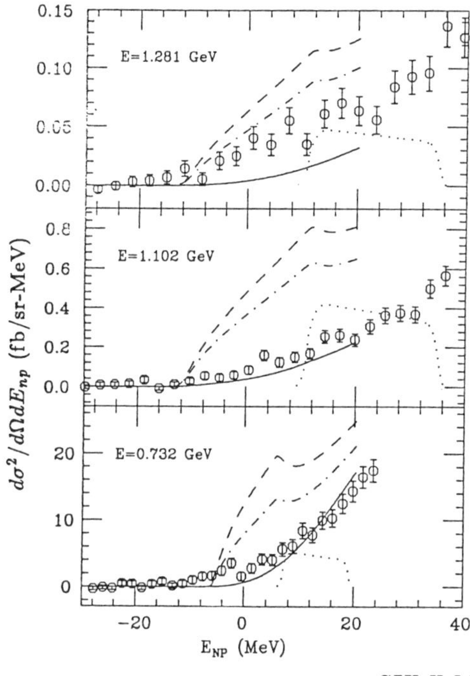

Fig. 3. Preliminary cross sections in the threshold break-up region as a function of E_{np} at three representative beam energies. The errors are statistical only. The dotted curves indicate the shape of the experimental resolutions in E_{np}. The full calculations of Arenhövel have been folded with the experimental resolution and are shown as the dot-dashed curves (G_E coupling, $G_{En} = 0$), the dashed curves (G_E coupling, $G_{En} \neq 0$), and the solid curves (F_1 coupling, results independent of choice for G_{En}).

SUMMARY

We have measured 180° cross sections for $d(e, e')$ which fall by over three orders of magnitude between the quasielastic peak and break-up threshold. The results are generally in agreement with nonrelativistic models that includes MEC, IC and FSI. Detailed agreement has yet to be achieved near break-up threshold, where the uncertainties in the MEC calculations, IC strengths, relativistic corrections, choice of deuteron potential, and percentage of 6-quark states are found to be of increasing importance. Further work will be needed to fully describe the data in this region.

We would like to acknowledge the support of J. Davis, R. Eisele, C. Hudspeth, J. Mark, J. Nicol, R. Miller, L. Otts, and the rest of the SLAC staff. This work was supported in part by the Department of Energy, contracts DOE–AC03–76SF00515 (SLAC), W–7405–ENG–48 (LLNL), DE–AC02-76ER–02853 (U. Mass.) and National Science Foundation Grant PHY85–10549 (A.U.).

REFERENCES

1. B. Parker *et al.*, Phys. Rev. C **34**, 2354 (1986); B. Quinn *et al.*, Phys. Rev. C **37**, 1609 (1988).
2. W. Bartel *et al.*, Nucl. Phys. B **58**, 429 (1973); K. M. Hanson *et al.*, Phys. Rev. D **8**, 753 (1973).
3. A. S. Esaulov *et al.*, Yad. Fiz 45, 410 (1987).
4. S. Auffret *et al.*, Phys. Rev. Lett. **55**, 1362 (1985).
5. A. T. Katramatou *et al.*, Nucl. Instrum. Meth. **A267**, 448 (1988).
6. A. T. Katramatou, SLAC Report No. SLAC–NPAS–TN–86–8, 1986.
7. L. E. Price *et al.*, Phys. Rev. D **4**, 45 (1971).
8. Y. S. Tsai, SLAC Report No. SLAC–PUB–848, 1971; L. W. Mo and Y. S. Tsai, Rev. Mod. Phys. **41**, 205 (1969).
9. J. M. Laget, Can. J. Phys. **62**, 1046 (1984); J. M. Laget, Phys. Lett. **199B**, 493 (1987), and private communication.
10. M. Lacombe *et al.*, Phys. Lett. **101B**, 139 (1981).
11. I. J. McGee, Phys. Rev. **161**, 1640 (1967) and L. Durand, Phys. Rev. **123**, 1393 (1961) as presented in W. Bartel *et al.*, Nucl. Phys. B **58**, 429 (1973).
12. R. Machleidt, K. Holinde, and C. Elster, Phys. Rep. **149**, 1 (1987).
13. H. Arenhövel, Nucl. Phys. A **384**, 287 (1982), W. Leidemann and H. Arenhövel, Nucl. Phys. A **393**, 385 (1983), and private communication.
14. P. E. Bosted *et al.*, Phys. Rev. Lett. **49**, 1380 (1982).
15. W. P. Schütz *et al.*, Phys. Rev. Lett **38**, 259 (1977) and R. G. Arnold, private communication for revised results.
16. S. Rock *et al.*, Phys. Rev. Lett. **49**, 1139 (1982).
17. R. G. Arnold *et al.*, Phys. Rev. Lett. **58**, 1723 (1987).
18. Tan-Sheng Cheng, L. S. Kisslinger, Nucl. Phys. A **457**, 602 (1986): J. F. Mathiot, Nucl. Phys. A **412**, 201 (1984); Y. Yamauchi *et al.*, Nucl. Phys. A **443**, 628 (1985).

MEASUREMENT OF TWO-BODY DEUTERON PHOTODISINTEGRATION AT HIGH PHOTON ENERGIES AND θ_{CM}=90°

J. Napolitano, S. J. Freedman, D. F. Geesaman, R. Gilman, M. C. Green,[a] R. J. Holt, H. E. Jackson, R. Kowalczyk, C. Marchand, J. Nelson, B. Zeidman
Argonne National Laboratory, Argonne, IL 60439-4843

D. Beck, G. Boyd, D. Collins, B. W. Filippone, J. Jourdan, R. D. McKeown, R. Milner, D. Potterveld,[b] R. Walker, C. Woodward
California Institute of Technology, Pasadena, CA 91125

R. E. Segel, T.-Y. Tung
Northwestern University, Evanston, IL 60201

P. Bosted
American University, Washington DC 20016

E. Kinney,[b]
Massachusetts Institute of Technology, Cambridge MA, 02139

Z.-E. Meziani
Stanford University, Stanford, CA 94305

R. Minehart
University of Virginia, Charlottesville, VA 22901

ABSTRACT

We report preliminary results from a measurement of the process $^2H(\gamma,p)n$ at θ_{CM}=90° and for photon energies between 0.8 and 1.6 GeV. The results are compared to a conventional calculation based on meson-exchange and to models based on sub-nucleonic degrees of freedom. We find strong disagreement with the meson-exchange calculation, and best agreement with simple constituent counting at the highest energies.

A clear intersection between nuclear and particle physics is the attempt to experimentally verify the existence of sub-nucleonic degrees of freedom in nuclei. Considerable effort has been expended in order to describe nuclei in terms of the fundamental quark and gluon fields, rather than as collections of nucleons and mesons.[1] However, there is little experimental data to support these descriptions.

In this talk, we present preliminary results of a measurement of the differential cross section for the reaction $^2H(\gamma,p)n$ at θ_{CM}=90° and photon energies between 0.8 and 1.6 GeV. This reaction has several features which suggest that it may serve as a testing ground for nucleon-meson versus quark/parton descriptions of the deuteron. First, it is an exclusive process which according to simple constituent counting rules[2] should be described by $d\sigma/dt \alpha 1/s^{11}$ for large enough values of s. Second, a well-developed picture of this reaction in terms of nucleon and meson degrees of freedom exists and has been tested at lower energies.[3,4] Third, Brodsky and Hiller[5] have formulated a QCD-based description of this reaction in terms of "Reduced Nuclear Amplitudes" which has been demonstrated to be successful in describing the elastic form factor of the deuteron.[6] Our results strongly disagree with an existing meson-exchange calculation and suggest that, at the highest energies of the measurement, the cross section behaves according to the simple constituent counting rule.

The experiment is performed at the Nuclear Physics Injector at SLAC (NPAS) using facilities in End Station A. The experimental arrangement is shown in Fig. 1. Electrons are delivered in ≃1.5 μsec long pulses with peak current up to ≃20 mA at a rate of ≃90 sec^{-1}. The energy spread of the electron beam is defined using collimators which restrict the full width to less than 0.25%. The integrated electron current is monitored by two toroids whose accuracy is better than 1%. The electron beam passes through removable 4% or 6% Cu radiators producing bremsstrahlung photons. Both electron and photon beams pass through the target on the way to a water cooled beam dump.

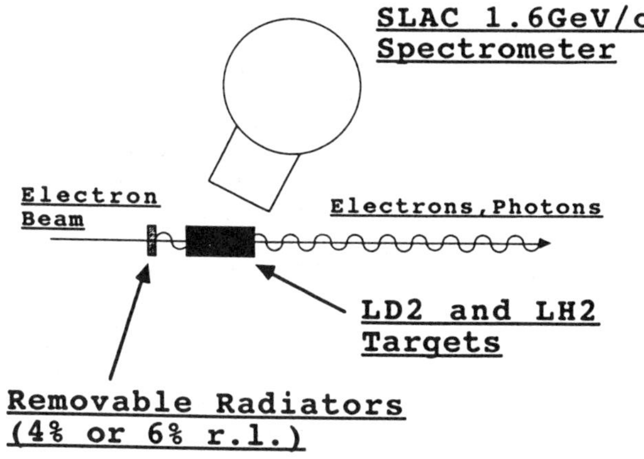

Fig. 1. The experimental setup in End Station A at SLAC. See the text for details.

Protons from the reaction $^2H(\gamma,p)n$ with photon energies near the bremsstrahlung endpoint are momentum analyzed in the SLAC 1.6-GeV/c spectrometer[7] and detected in a multilayer system of plastic scintillator hodoscopes and drift chambers. In addition, an aerogel Cerenkov counter is used to identify pions. The five hodoscope layers, two in the direction of momentum dispersion, are used to trigger the apparatus and to measure time-of-flight (TOF) over a ≃3m flight path. There are three drift chambers, each of which has two planes in both the X and Y directions with ≃1 cm wire spacing. Particles are tracked through the drift chambers allowing full reconstruction of momentum and trajectory. The overall rms momentum resolution obtained, including beam energy spread, is ≃0.3%. For the energies and angles studied, this is sufficient to separate protons due to $^2H(\gamma,p)n$ from those due to reactions with additional particles in the final state. For each beam energy, one momentum setting is sufficient to cover the highest ≃100 MeV of the bremsstrahlung spectrum as well as ≃50 MeV beyond the endpoint.

Data are acquired through CAMAC and read into a VAX-11/780 host computer through an LSI-11 buffering computer. This data acquisition system is limited to one trigger per beam pulse so all triggers are scaled and the extracted yields are corrected. This correction is typically a few percent and at most 10%. Dead time corrections due to triggering electronics are negligible.

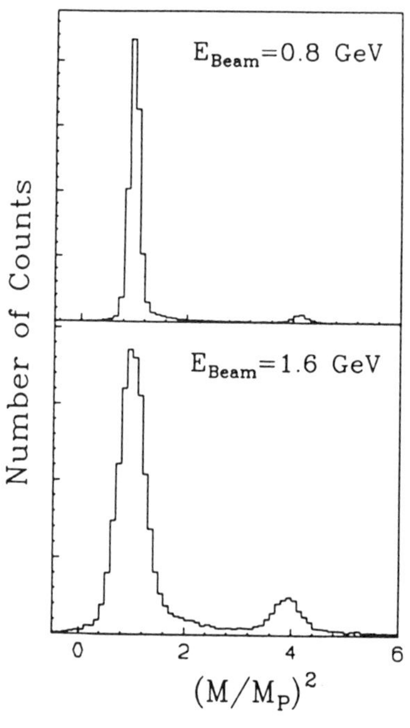

Fig. 2. Histograms of particle mass as determined by the spectrometer and TOF system, at the two extreme beam energies reported here. Particle separation between protons and deuterons is clear, and there is no contamination from pions or positrons.

We use a 15 cm long liquid deuterium (LD_2) target with 0.003" aluminum entrance and exit windows. A liquid hydrogen (LH_2) target of identical dimensions is used for background subtraction. The transverse dimension of the targets fully intercept the electron and photon beams. The dominant background is from the reactions $^{27}Al(\gamma,p)X$ and $^{27}Al(\gamma,d)X$ on the target windows. The mass of particles triggering the detector at the lowest and highest beam energies is histogrammed in Fig. 2. The mass is calculated from the momentum measured in the spectrometer and the particle velocity determined by the TOF system. There is clear separation of protons from deuterons, and we confirm that the deuteron background subtracts to zero using the LH_2 target. The pion and positron rates through the spectrometer are found to be negligible for the data presented here.

We detect protons from elastic electron-proton scattering to determine the product of detection efficiency and spectrometer acceptance. Care needs to be taken to subtract the proton yield from the reaction $H(\gamma,p)\pi^0$ due to real and virtual photons in the target.

We calculate the photon energy E_γ using the reconstructed momentum and scattering angle and assuming the reaction $^2H(\gamma,p)n$. The resulting photon energy spectra are reduced by subtracting the LH_2 target yields from LD_2 target yields separately for radiator-in and radiator-out. Then the remaining radiator-out yield is subtracted from that for radiator-in. The resulting spectrum is then due only to real photons from the external radiator. Sample yield spectra are shown in Fig. 3, again for our lowest and highest beam energies. It is clear from Fig. 3 that the yield beyong the endpoint is consistent with zero and this is an excellent indication that the background is well determined.

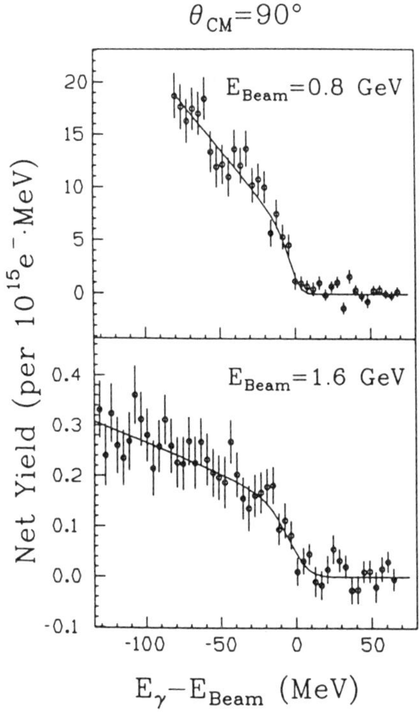

Fig. 3. Background subtracted bremsstrahlung yields from the reaction ^{2}H(γ,p)n at two different beam energies, binned in 1 MeV steps. The solid curve is the product of the calculated bremsstrahlung yield and the measured cross section folded with a gaussian response function. The cross section is assumed to be linear in energy over the range of each spectrum.

Two cross section values are determined from each subtracted yield spectrum of the type shown in Fig. 3. The region below the endpoint is divided into two parts, excluding the highest 25 MeV and also excluding effective photon energies which allow yield from the reaction ^{2}H(γ,p)nπ^0. For each of these two regions, we average the yield and determine the cross section from the target thickness and density, our measured spectrometer acceptance, and the calculated bremsstrahlung yield[8] (corrected for energy loss in the radiator[9]). The photon energy for each of the two regions is calculated using an average weighted by the bremsstrahlung yield. The curves in Fig. 3. are determined by a linear interpolation for the cross section, multiplying by the bremsstrahlung shape, and convoluting the product with a gaussian response function with the expected photon energy resolution. The agreement is quite good over the entire range of photon energies.

Our results are presented in Fig. 4 where we plot the cross section after taking out certain "scaling" factors corresponding to simple constituent counting[2] (Fig. 4a) and for the "reduced nuclear amplitudes" approach[5] (Fig. 4b). Evidence for either approach then takes the form of the data becoming a constant above some photon energy. In each plot, we include the result of a recent calculation based on meson exchange,[4] scaled in the same fashion. Data from previous experiments at lower energies are included.[10,11,12]

According to constituent counting rules, the differential cross section for a particular exclusive process at fixed center of mass angle should approach the form $d\sigma/dt \alpha 1/s^{n-2}$ where n is the total number of elementary fields. (A true QCD calculation modifies this expression through logarithmic and "higher twist" contributions, although the magnitude of these corrections are not known.) Consequently, for the reaction ^{2}H(γ,p)n we might expect the quantity $s^{11} d\sigma/dt$, plotted in Fig. 4a, to approach a constant above some energy.

The approach taken by Brodsky and Hiller[5] using "reduce nuclear amplitudes" implies that the differential cross section should be given

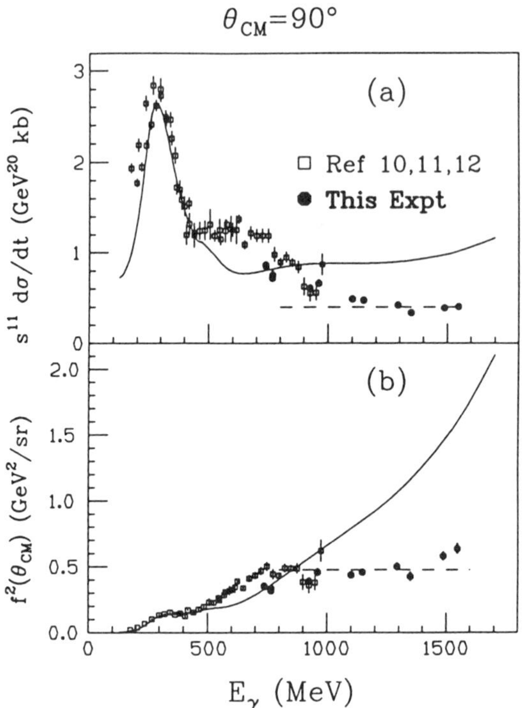

Fig. 4. Results of our experiment along with results of previous experiments at lower energies. The data are plotted so as to elucidate "scaling" as determined by (a) simple constituent counting (Ref. 2), and by (b) a formalism based on "reduced nuclear amplitudes (Ref. 5). The solid line is result of a recent calculation based on meson exchange (Ref. 4). The dashed lines represent constants that approximate the data at high energy but whose magnitude are not predicted by any model.

by the expression

$$\frac{d\sigma}{d\Omega_{CM}} = \frac{1}{\sqrt{s\left(s-M_d^2\right)}} F_p^2(\hat{t}_p)F_n^2(\hat{t}_n) \frac{1}{P_T^2} f^2(\theta_{CM})$$

where

$$\hat{t}_i = (P_i - \frac{1}{2} P_d)^2$$

and $F_N(t)$ is the elastic nucleon form factor. Accordingly, we plot the quantity f^2 in Fig. 4b as a function of photon energy.

It is immediately clear from Fig. 4 that the meson-exchange calculation of Ref. 4 does not describe our data. It is not clear if adjusting the various parameters in the calculation can give agreement with our data while maintaining agreement with data at lower energies and in other reactions.

Despite the very low energy ($s \simeq 2M_d^2$) our data seems to be described by the simple constituent counting relation,[2] although the data does not extend to high enough energy to identify logarithmic or otherwise slowly varying deviations as suggested from QCD. The data do not seem to be well described by the formalism of Brodsky and Hiller,[5] although this conclusion is based largely on the strength of the two highest energy points. We are currently finalizing our determinations of various energy-dependent systematic errors which appear to be small at the present time.

We wish to thank Dr. B. Mecking for his work in identifying possible sources of background prior to the experiment. We are grateful to the SLAC cryogenics target group, especially J. Mark and J. Nichols, and the End Station A support staff, in particular R. Eisele.

This work supported by the U. S. Department of Energy, Nuclear Physics Division, under contract W-31-109-ENG-38.

REFERENCES

a) Present Address: LeCroy Research Systems, Spring Valley, NY 10977
b) Present Address: Physics Division, Argonne National Laboratory, Argonne, IL 60439-4843.
1. See for example S. J. Brodsky and G. P. LePage, Nucl. Phys. A353, 247 (1981); C. E. Carlson and F. Gross, Phys. Rev. Lett. 53, 127 (1984); S. J. Brodsky, Comm. Nucl. Part. Phys. 12, 213 (1984).
2. D. Sivers, Ann. Rev. Part. Sci. 32, 149 (1982); S. J. Brodsky and G. R. Farrar, Phys. Rev. D 11, 1309 (1975).
3. J. M. Laget, Nucl. Phys. A312, 265 (1978).
4. T.-S. H. Lee, to be published.
5. S. J. Brodsky and J. R. Hiller, Phys. Rev. C 28, 475 (1983).
6. S. J. Brodsky and B. T. Chertok, Phys. Rev. Lett. 37, 269 (1976).
7. R. Anderson, et. al., Nucl. Inst. Meth. 66, 328 (1968).
8. J. L. Matthews and R. O. Owens, Nucl. Inst. Meth. 111, 157 (1973).
9. J. L. Matthews, D. J. S. Findlay, and R. O. Owens, Nucl. Inst. Meth 180, 573 (1981).
10. J. Arends, et al., Nucl. Phys. A412, 509 (1984).
11. P. Dougan, et. al., Z. Physik A276, 55 (1976).
12. R. Ching and C. Schaerf, Phys. Rev. 141, 1320 (1966)

THE e⁻d SCATTERING WITH Δ-ISOBARS IN DEUTERON*

R. Dymarz and F.C. Khanna. Theoretical Physics Institute, Physics Department, University of Alberta, Edmonton, Alberta T6G 2J1 and TRIUMF, Vancouver, B.C. V6T 2A3, Canada

In this note we present the results of calculations[1] for e^-d elastic and inelastic scattering with deuteron and the scattering wave functions obtained within a meson exchange model, but with nucleons and isobars as the basic degrees of freedom. We solve a set of coupled-channel Schroedinger equations for nucleons and Δ-isobars using one-boson exchange (OBE) potentials as diagonal and as transition potentials. The potentials in the NN channel are similar to those developed by the Bonn group[2] while the transition potentials and the diagonal potentials in channels involving Δ-isobars are obtained from those in the NN channel through quark model relations. The exchange of six mesons ($\pi,\rho,\sigma,\omega,\eta$, and δ) is considered in diagonal channels. In each meson-baryon-baryon vertex monopole form factors are used with the same cut-off masses for vertices involving nucleons and Δ-isobars. The parameters of five physical mesons can be reasonably chosen empirically while parameters of the σ-meson are fixed by the requirement that deuteron binding energy (for T=0) and low energy scattering parameters (for T=1) are reproduced. Although no fitting to experimental phase shifts was performed it was checked that they are reproduced reasonably well. Also deuteron static properties – the magnetic and quadrupole moments, radius and the D/S states ratio were calculated and found to be in reasonably good agreement with experiment.

A short-range structure of NN interaction in our model is determined by form factors in meson-NN (NΔ, ΔΔ) vertices and by the coupling constants in vertices involving Δ-isobars. By varying cut-off masses and/or relevant coupling constants, different amounts of NΔ and ΔΔ components in the appropriate partial waves can be obtained. Because of isospin conservation only ΔΔ states can be excited in the T=0 channel and it gives for deuteron, in addition to two nucleon components, ${}^3S_1^{NN} - {}^3D_1^{NN}$, four ΔΔ components: ${}^3S_1^{\Delta\Delta}$, ${}^3D_1^{\Delta\Delta}$, ${}^7D_1^{\Delta\Delta}$ and ${}^7G_1^{\Delta\Delta}$. In the T=1 channel both NΔ and ΔΔ states are allowed and it results in coupling of three states: ${}^5D_0^{N\Delta}$, ${}^1S_0^{\Delta\Delta}$ and ${}^5D_0^{\Delta\Delta}$ to the nucleon ${}^1S_0^{NN}$ state.

In the two following paragraphs we discuss a role of the two short-range phenomena: Δ-isobar and meson exchange currents (MEC) in elastic and inclusive inelastic e^-d scattering at large momentum transfers.

(i) Elastic e^-d scattering. In elastic e^-d scattering two form factors, A(q) and B(q), are measured at relatively large momentum transfer. The first one is known for $q \leq 10\ fm^{-1}$ and the second one for $q \leq 8.3 fm^{-1}$. In Fig. 1 we show these two form factors

*Work supported in part by NSERC (Canada).

calculated with the deuteron wave function containing 0.62% of ΔΔ states (mainly $^3S_1^{\Delta\Delta}$ state). The four curves in this figure illustrate contributions to the form factors from nucleon, Δ-isobar and MEC (see caption to Fig. 1 for details). Contributions of two MEC, π-pair and ρπγ-current, are considered. To calculate contributions from nucleon current (NC) and isobar current (IC) the isoscalar

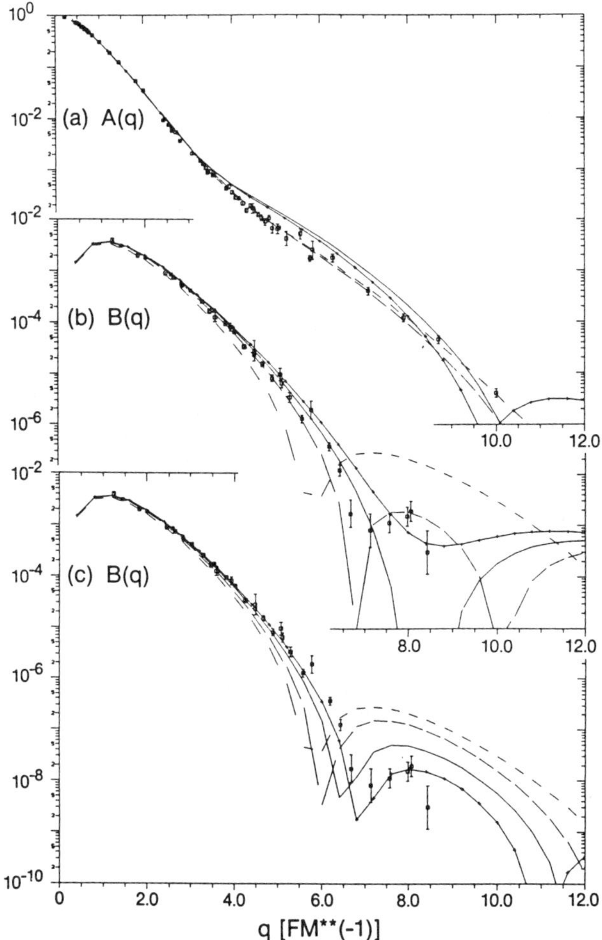

Fig. 1. The elastic scattering deuteron form factors. These in parts (a) and (b) were obtained with the same input parameters while in (c) the strength of IC has been suppressed. The curves represent the following contributions: – –NC,— — NC+IC, ——NC+IC+MEC (π-pair), —×—×–NC++IC+MEC (π-pair and ρπγ). The experimental data are from Ref. 5.

nucleon, $G_N^{IS}(q)$, and Δ-isobar, $G_\Delta^{IS}(q)$, form factors have to be specified. The evaluation of MEC contributions require, in addition, specification of the meson-NN and ρπγ vertices. The results presented in Fig. 1a and 1b were obtained with $G_N^{IS}(q)$ calculated according to a hybrid (meson and quark dynamics) model of Ref. 3 and with $G_\Delta^{IS}(q)$ taken in a dipole form: $G_\Delta^{IS}(q)=\mu_\Delta^{IS}/(1+q^2/\Lambda_\Delta^2)^2$, where $\mu_\Delta^{IS}=\mu_p$ as given by the quark model and $\Lambda_\Delta^2=18.2\text{fm}^{-2}$ as found for the proton. In the meson-NN vertices the same form factors and coupling constants as in the potential model were used. In the $\rho\pi\gamma$ vertex a coupling constant 0.38 and a monopole form factor $(1+q^2/7.0\ \text{fm}^{-2})^{-1}$ were assumed.

As can be seen in Figs. 1a and 1b, with the above specified input, both the A(q) and B(q) are well reproduced already in the IA, with both the nuclear and Δ-isobar currents taken into account, and contributions from MEC destroy fits to both the A(q) and B(q). Does it mean that there is little room for MEC in elastic e^-d scattering? To answer this question we have made calculations with the IC contribution severely suppressed. In Fig. 1c results of calculations with Λ_Δ^2 equal to half of its proton value are shown (the remaining parameters were unchanged). One can see that with such suppressed Δ-isobar strength (the similar effect can be obtained with μ_Δ^{IS} reduced or with smaller ΔΔ state probability in the deuteron wave function) the good fit to B(q) is obtained with contributions from all currents taken into account. As can be expected such manipulation has little effect on A(q). From Fig. 1a one can learn that the IC contribution to A(q) is substantially smaller than that to B(q) and then its reduction will change fits to A(q) very little.

(ii) Backward d(e,e′) scattering at threshold. The experimental cross section for backward electrodisintegration at threshold has been measured for up to 30 fm^{-2} of four momentum transfer squared. It is well established[4] that this reaction is dominated by the M1 transition from the deuteron ${}^3S_1^{NN}$ - ${}^3D_1^{NN}$ states to the scattering ${}^1S_0$ state. The interaction of the virtual photon with NC only is far too weak to reproduce experimental data and substantial contributions from MEC are usually invoked to reconcile theory and experiment. The contributions from seagull and mesonic (involving both π and ρ mesons) currents are necessary to quantitatively reproduce experimental data. Of various contributions from IC the transition to the ${}^5D_0^{N\Delta}$ scattering state is usually evaluated perturbatively and it is found to be small.

Using our coupled channel model wave functions we have calculated all significant M1 matrix elements involving nucleons and Δ-isobars and also MEC mentioned above. The results are shown in Fig. 2. The curves in Fig. 2 illustrate the contributions of four currents to cross section: nucleon, transitional N-Δ, Δ-isobar and meson exchange currents. Inadequacy of IA calculations (even with IC included) and a huge effect of MEC is obvious from inspection of Fig. 2. The two parts of Fig. 2, (a) and (b), represent calculations with different strength of Δ-isobar current. Curves in Fig. 2a and 2b were obtained, as in the case of elastic scattering, with Δ-isobar form factor in dipole form with parameter $\Lambda_\Delta^2=18.2\text{fm}^{-2}$ and half of this value, respectively. The isovector magnetic moment of Δ-isobar was assumed to have its quark model value ($=\mu_p$). Nuclear form factors were taken, as before, from

Ref. 3. In evaluation of the MEC contributions the same characteristics of meson-NN vertices as in the potential model were used. The mesonic currents were calculated with nucleon magnetic form factor and seagull currents with F_1^V nucleon form factor also taken from Ref. 3.

The results presented in Fig. 2a correspond to those shown in Fig. 1b and those presented in Fig. 2b to those shown in Fig. 1c. The strength of IC is identical for each pair of figures. As can be seen, in both the elastic and the inelastic scattering, the weaker IC require stronger MEC to reproduce experimental data. However, unlike in elastic scattering, the IC in inelastic scattering are insufficient to qualitatively describe the experiment. The MEC themselves are quite large and a more careful treatment of these currents is needed. We have reduced some uncertainties in evaluation of MEC by applying the same meson-NN characteristics as in the potential model. However, it is obvious that a closer look at large momentum transfer behaviour of MEC is urgently desirable.

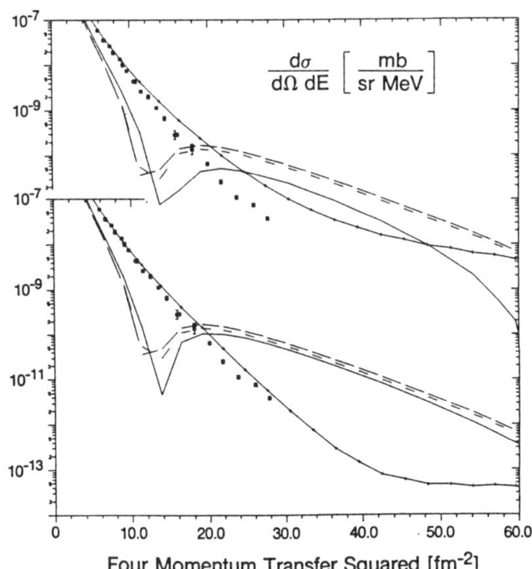

Fig. 2. The backward d(e,e′) scattering. Parts (a) and (b) show results of calculations with different strength of IC (see text for details). The curves represent the following contributions: - -NC, --NC+(N-Δ) current, ——NC+(N-Δ)+IC, —✶—✶—NC+(N-Δ)+IC +MEC (seagull and mesonic for π and ρ). The experimental data are from Ref. 6.

REFERENCES

1. R. Dymarz and F.C. Khanna, to be published.
2. R. Machleidt et al. Phys. Rep. 149, 1 (1987).
3. M. Gari and W. Krümpelmann, Z. Phys. **322**, 689 (1985).
4. J.F. Mathiot, Nucl. Phys. **A412**, 201 (1984).
5. R.G. Arnold et al., Phys. Rev. Lett. **35**, 776 (1975), ibid **58** ,1723 (1987) and references therein.
6. S. Auffret et al., Phys. Rev. Lett. **55**, 1362 (1985); M. Bernheim et al., ibid **46**, 402 (1981).

SCALING IN QUASIELASTIC ELECTRON SCATTERING

R. D. McKeown
California Institute of Technology, Pasadena, CA 91125

ABSTRACT

The inclusive (e, e') cross section on various nuclei in the quasielastic region ($Q^2/2M\nu \sim 1$) has recently been measured in the range $0.25 \leq Q^2 \leq 3.1$. Analysis of these data in the plane-wave impulse approximation allows extraction of a scaling function related to the nucleon momentum distribution. Variation of the scaling function in Q^2 represents a breakdown of these assumptions in the analysis. Recent theoretical work indicates that the observed scaling violations can be understood as corrections to the impulse approximation and the experimental scaling functions at the highest measured Q^2 can in fact be interpreted in terms of nucleon momentum distributions.

The concept of scaling in quasielastic electron scattering from nuclei was introduced by West in 1975.[1] The basic idea is that the quasielastic cross section can be factorized in the form:

$$\frac{d\sigma}{d\Omega\, d\nu} = K \cdot \bar{\sigma}_{eN} \cdot F(y) \tag{1}$$

where ν is the electron energy loss, $\bar{\sigma}_{eN}$ is the electron-nucleon elementary cross section, K is a kinematic factor, and $F(y)$ is the projected nucleon momentum distribution along the $\vec{q}$ direction:

$$F(y) = \int_{|y|}^{\infty} n(p) p\, dp \ . \tag{2}$$

The variable y is the component of nucleon momentum along the $\vec{q}$ direction and is given by energy conservation in the electron scattering process. The essential assumptions in this analysis are: (*i*) "impulse approximation" (single nucleon knockout), (*ii*) no final state interactions, and (*iii*) non-relativistic nucleon kinematics.

This "y-scaling" analysis is quite useful in studying the quasielastic scattering process. The measurement of $d\sigma/d\Omega\, d\nu$ allows extraction of the function $F(y)$ from experimental data. This "scaling function" can be determined for a variety of kinematics (i.e., different Q^2) and one should obtain a "universal" $F(y)$ that is independent of the Q^2 used for the measurement. In principle, this would be an indication that the assumptions in the analysis were valid and lend confidence to the interpretation of $F(y)$ as a measure of the nucleon momentum distribution.

Such an analysis was performed by Sick, Day, and McCarthy[2] for the nucleus ^{3}He. They improved West's treatment by generalizing to relativistic nucleon kinematics, which was necessary due to the large nucleon momenta in the initial and final states. Their results showed a rather impressive scaling behavior, and they concluded that the $F(y)$ they extracted was a reliable determination of the nucleon momentum distribution. The momentum distribution they obtained was compared with that computed using Fadeev techniques, and a discrepancy was noted at high nucleon momentum ($p \gtrsim 300$ MeV/c) where the theory showed much less strength than the experimental data.

The formalism for y scaling has since been developed more carefully in the literature, notably by Pace and Salmé.[3] Again, one assumes the impulse approximation with no final state interactions but one now takes a nucleon spectral function $S(p, E)$ as a description of the initial state. (E is the separation energy of the nucleon and is measure of the degree of excitation of the residual A-1 system). One then obtains

$$\frac{d\sigma}{d\Omega\, d\nu} = \frac{m_n}{|\vec{q}|}\tilde{\sigma}_{eN} \int_{E_{\rm min}}^{E_{\rm max}} dE \int_{p_{\rm min}}^{p_{\rm max}} S(p, E) p\, dp \tag{3}$$

where $\tilde{\sigma}_{eN} = \frac{E'}{m_n}\bar{\sigma}_{eN}$ and $\bar{\sigma}_{eN}$ is the e-nucleon cross section given by deForest.[4] The use of $\tilde{\sigma}_{eN}$ rather than $\bar{\sigma}_{eN}$ allows clearer separation of the kinematic factors.[5] $\bar{\sigma}_{eN}$ (and thus $\tilde{\sigma}_{eN}$) do actually depend on E and p, but only very slowly,[6] so we may factor it out of the integral as in Eq. 3 and use an average $\bar{E}$ and $p_{\rm min}$ for p (since $S(p, E)$ generally falls rapidly with increasing p).

The limits on the integrals in Eq. 3 are very important as pointed out in Ref. 3. The upper limits generally grow with Q^2, and the integrals saturate as Q^2 increases since $S(p, E) = 0$ for large p or E. The lower limit $E_{\rm min}$ is determined by the nuclear ground state masses, and $p_{\rm min}$ is identified as $|y|$. Strictly speaking, $p_{\rm min} = |y-(E-E_{\rm min})|$ but since the E dependence of $S(p, E)$ is not directly measured and we expect $\langle(E - E_{\rm min})\rangle < 100$ MeV one usually simply assumes $p_{\rm min} \simeq |y|$. Nevertheless, one should be cautious in making detailed comparisons with theoretical momentum distribution.

In a recent experiment performed at SLAC, we have measured the quasielastic cross section as a variety of nuclei in the region $0.25({\rm GeV/c})^2 \lesssim Q^2 \lesssim 3.1({\rm GeV/c})^2$. The experiment was a collaboration between the University of Basel, University of Virginia, and Caltech. We obtained data on ^{4}He, C, Al, Fe, and Au and preliminary results were reported previously.[7] As an example, Fig. 1 shows the data for He and Fig. 2 for Al. One can see the quasielastic peak quite clearly in ^{4}He, but the situation for Al (as with the other targets) is less clear. Presumably, as the nucleon momentum distribution is broader in heavier nuclei, the quasielastic peak is a less pronounced feature. In addition, the more inelastic processes (resonances and deep inelastic scattering) are "smeared" into

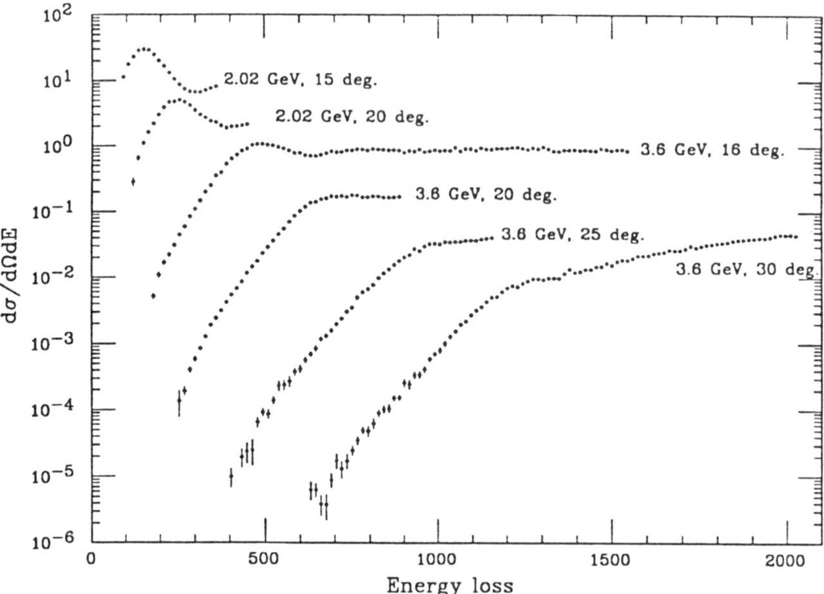

Fig. 1. ^4He cross sections in nb/sr/MeV vs. ν in MeV.

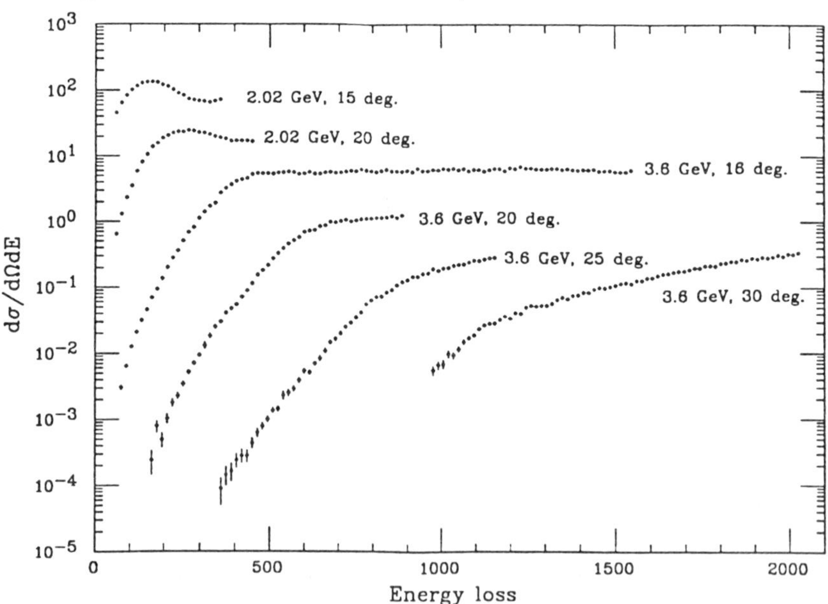

Fig. 2. Al cross sections in nb/sr/MeV vs. ν in MeV.

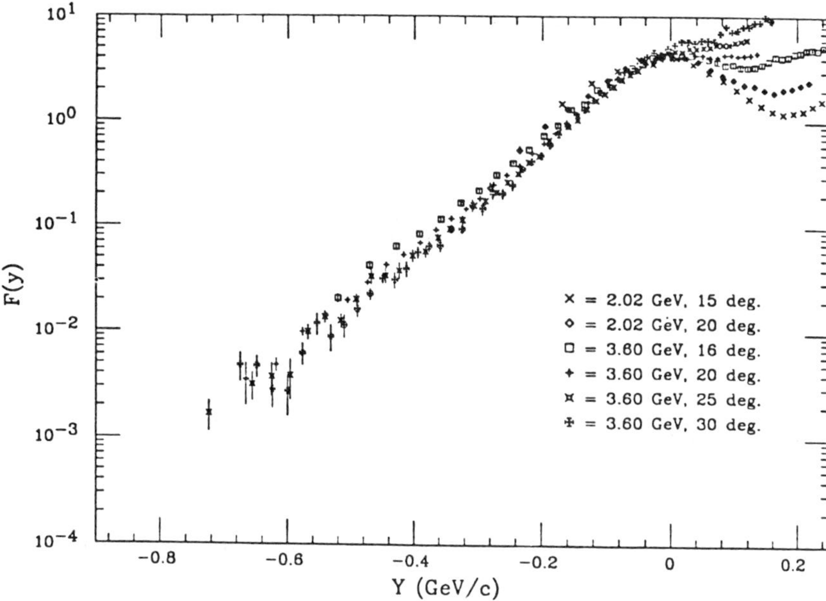

Fig. 3. $F(y)$ for ^{4}He in $(\mathrm{GeV/c})^{-1}$.

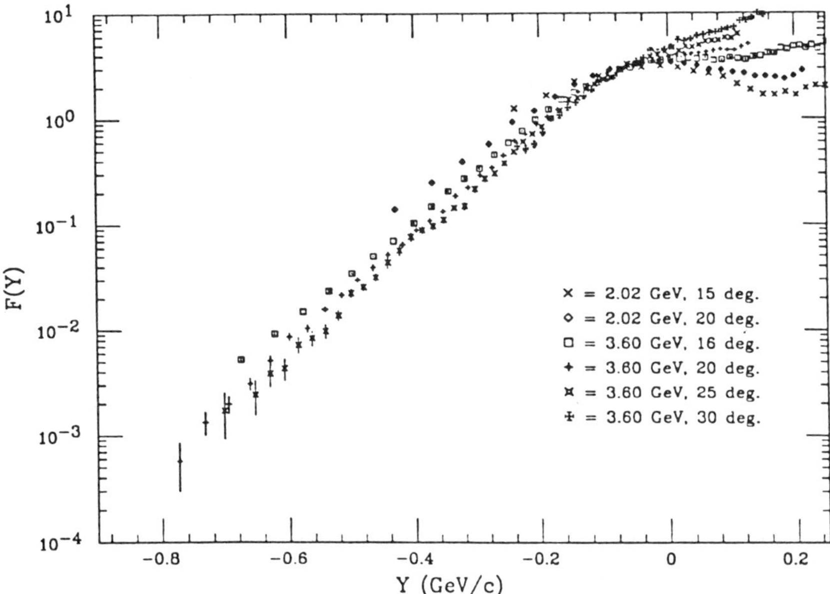

Fig. 4. $F(y)$ for C in $(\mathrm{GeV/c})^{-1}$

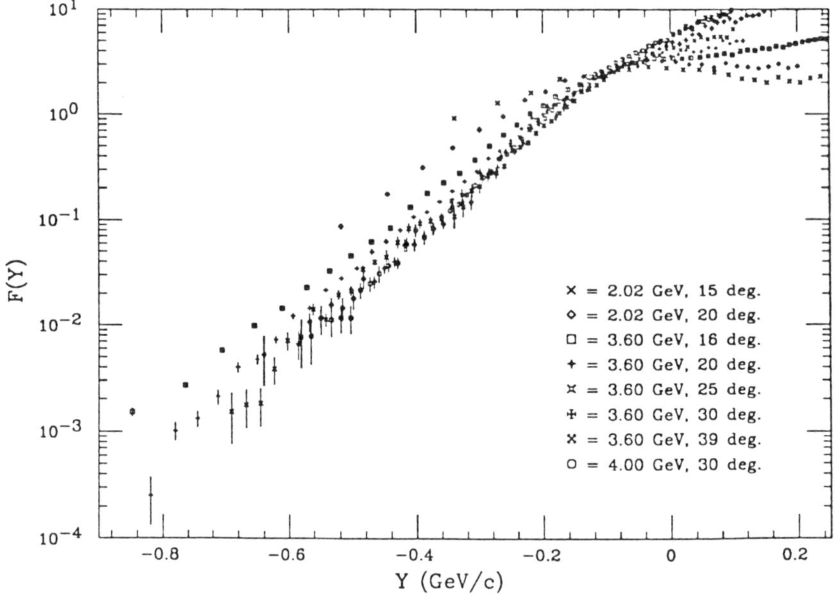

Fig. 5. $F(y)$ for Fe in $(\text{GeV/c})^{-1}$.

the quasielastic region causing the quasielastic peak to be somewhat obscured. Nevertheless, one might expect the low ν side of the quasielastic peak ($y < 0$) to be fairly pure quasielastic scattering, and thus amenable to y scaling. Figure 3,4, and 5 show the extracted $F(y)$ for ^{4}He, C, and Fe. One can see the following features. The region $y \gtrsim 0$ is generally contaminated by non-quasielastic processes as the scaling is severely broken there. The region $y < 0$ shows a scaling function that *decreases* with increasing Q^2 and seems to saturate at the highest Q. Note that the impulse approximation predicts $F(y)$ should *increase* with increasing Q^2 as the integration limits in Eq. 3 expand. Thus, we seem to see a breakdown of the impulse approximation. However, it appears that the situation is understandable using nuclear matter theory.

Before applying such ideas, one should examine the A-dependence of the data to determine whether an infinite nuclear matter approach seems appropriate. Fig. 6 shows the A-dependence of $F(y)$ for $y = -400$ MeV/c and $y = -500$ MeV/c at $Q^2 = 1.8$ GeV/c^2. One can see that the scaling function shows no significant trend in A for $A \geq 12$. This observation is quite valid for all $Q^2 > 1$ GeV/c^2 and $y < -300$ MeV/c. As there appear to be no systematic finite nucleus effects in the data in this kinematic region, one expects that a comparison with infinite nuclear matter calculations is justified.

Butler and Koonin[6] have computed the response function of nuclear matter using Brueckner-Goldstone perturbation theory and a realistic N-N interaction

(Paris). This calculation goes beyond the normal impulse approximation (it includes, for example, final-state interaction and initial state correlation effects) and is valid in the regime where nucleon momenta are larger than the fermi momentum. A shortcoming of the calculation is that it is non-relativistic. However, it is useful to note that more physical processes are accounted for than the $A = 3$ Fadeev spectral response functions computed in the impulse approximation without final state interactions.[2] (Those calculations are also non-relativistic.)

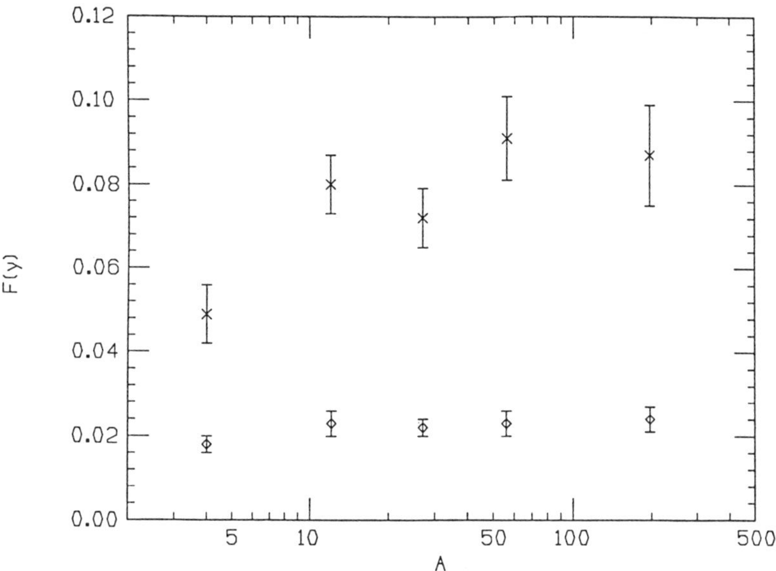

Fig. 6. The A dependence of $F(y)$ [in $(\mathrm{GeV/c})^{-1}$] at $Q^2 = 1.8\,(\mathrm{GeV/c})^2$. The crosses are for $y = -0.4$ GeV/c and the diamonds $y = -0.5$ GeV/c.

Two major conclusions of crucial importance emerged from this work. The first was that the $F(y)$ values *decreased* with increasing Q^2, contrary to the impulse approximation expectation and in agreement with experimental observation. Secondly, the scaling function converged to the impulse approximation result at the larger Q^2 studied in the experiment. (This conclusion is quite non-trivial, as a similar calculation for a Bose gas with a hard-core interaction showed convergence to a function that was *not* the impulse approximation response.[7]) Therefore, this calculation indicates that the momentum distribution inferred from the highest Q^2 studied in the experiment should indeed be interpreted as that of infinite nuclear matter.

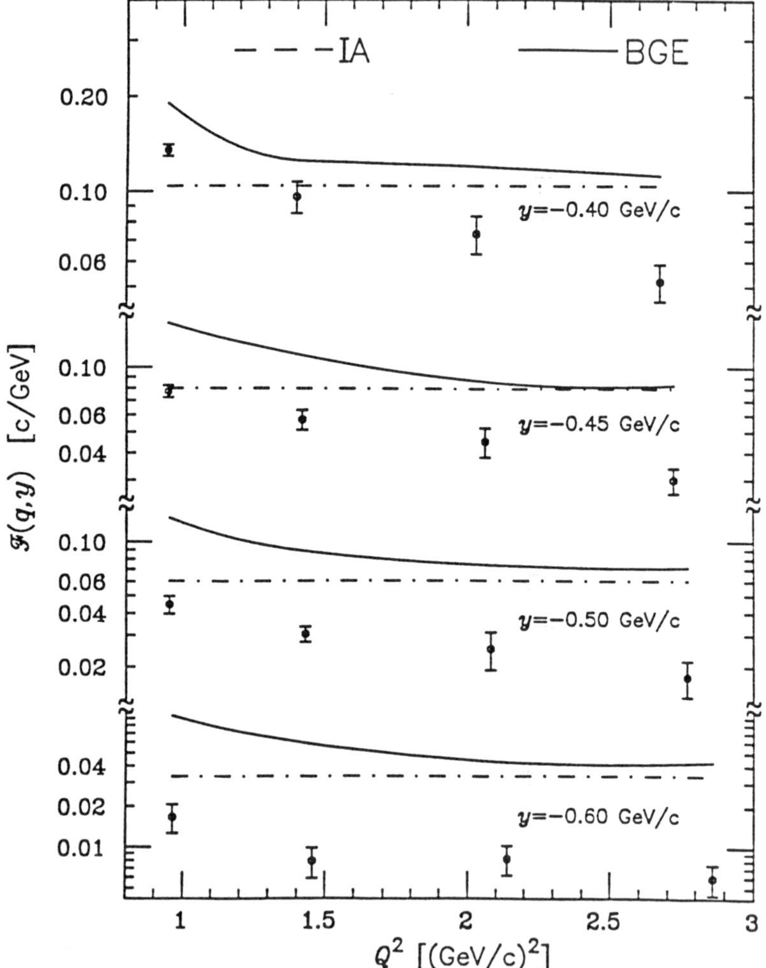

Fig. 7. $F(y)$ vs. Q^2 for a variety of y. The data points are for Fe, the solid curve is the full Brueckner-Goldstone Expansion calculations and the dot-dash curve is the impulse approximation.

I would like to stress that although one should perhaps be skeptical of treating the results of this calculation as numerically very precise (because of its non-relativistic nature), one would expect that as a schematic calculation the gross features should be taken as the first order expectation beyond the impulse approximation. Certainly the calculation is giving us substantial insight into the observed scaling violations. The comparison of the results of the calculation of Butler and Koonin and the experimental $F(y)$ for Fe are shown in Fig. 7. There one can see that the Q^2 dependence of the data is well-reproduced by the calculation, but that the calculated response falls below the experimental values by typically a factor of 3. This seems to be a rather large discrepancy, and if taken seriously indicates that there is something wrong with the short-range correlations in the nuclear matter calculation. The response in this region is dominated by the tensor force, and so this result may indicate that the tensor force in nuclear matter is weaker than predicted by standard theory with the Paris interaction. A similar calculation with a new interaction such as the Bonn potential[8] would be quite interesting, and work along these lines is in progress.[9]

I acknowledge many useful conversations with M. Butler and S. Koonin. This work was supported by NSF grant PHY85-05682.

REFERENCES

[1]G. West, Phys. Rep. **18**, 263 (1975).
[2]I. Sick, D. Day, and J. S. McCarthy, Phys. Rev. Lett. **45**, 871 (1980).
[3]E. Pace and G. Salmé, Phys. Lett. **110B**, 411 (1982).
[4]T. De Forest, Jr., Nucl. Phys. **A392**, 232 (1983).
[5]M. N. Butler and R. D. McKeown, to be published.
[6]D. Potterveld, Ph.D. thesis, California Institute of Technology.
[7]D. Day *et al.*, Phys. Rev. Lett. **59**, 427 (1987).
[8]M. Butler and S. E. Koonin, to be published.
[9]J. J. Weinstein and J. W. Negele, Phys. Rev. Lett. **49**, 1016 (1982).
[10]R. Machleidt, K. Hoinde, and Ch. Elster, Phys. Rep. **149**, 1 (1987).
[11]M. Butler, private communcation.

y-Scaling in a Simple Quark Model.

S. Kumano[*,†,#] and E. J. Moniz[*]

[*]Center for Theoretical Physics, Laboratory for Nuclear Science, and Department of Physics
Massachusetts Institute of Technology, Cambridge, MA 02139

[†]Department of Physics, University of Illinois at Urbana-Champaign, Urbana, IL 61801

[#] Department of Physics and Astronomy, University of Tennessee, Knoxville, TN 37996

and [#] Physics Division, Oak Ridge National Laboratory, Oak Ridge, TN 37831

Abstract

A simple quark model is used to define a nuclear pair model, that is, two composite hadrons interacting only through quark interchange and bound in an overall potential. An "equivalent" hadron model is developed, displaying an effective hadron-hadron interaction which is strongly repulsive. We compare the effective hadron model results with the exact quark model observables in the kinematic region of large momentum transfer, small energy transfer. The nucleon response function in this y-scaling region is, within the traditional framework sensitive to the nucleon momentum distribution at large momentum. We find a surprizingly small effect of hadron substructure. Futhermore, we find in our model that a simple parametrization of modified hadron size in the bound state, motivated by the bound quark momentum distribution, is not a useful way to correlate different observables.

The traditional description of the nucleus, in terms of nucleons interacting through static two-body potentials constrained by free nucleon-nucleon scattering data and deuteron properties, is quite successful in explaining many nuclear properties. This fact is somewhat surprizing because the nucleons are composite objects with the r.m.s. radius 0.8 fm and nucleons overlap strongly at the separation $<$ 1.6 fm, which is slightly smaller than the average nucleon separation 2.2 fm in nuclei. For medium and long range, the potential can be described well by the exchange of mesons. The short range interaction, in most parametrization, is strongly repulsive for $r <$ 0.5 fm. Because of this repulsive short range interactions, the nucleon momentum distribution is thought to be enhanced significantly at large momentum transfer. The nuclear electromagnetic response function in the y-scaling region has been interpreted as providing the first reliable measure of the momentum distribution beyond the Fermi level[(2)].

A major open question in assessing the reliability of this interpretation is that concerned with the role of nucleon substructure. Because it is thought that the short range repulsion has its origin in the quark substructure, one might expect that the internal degrees of freedom will play an important role in the nuclear response function. We study this question within a "minimal" model, that is, a model containing nothing but the most essential features of the problem. We introduce a pair model with two composite hadrons bound in an overall external "shell-model" potential as shown in Fig.(1a). The interaction between the hadrons is generated by the quark model of Lenz et al.[(3)]. The observables corresponding to large-momentum transfer, small energy transfer electron scattering are evaluated. The essential comparison is then to the results generated by an equivalent hadron model; i.e., an effective hadron-hadron potential is introduced through an adiabatic approximation to the quark problem and is then used to compute the same obervables within the "traditional framework", as shown in Fig.(1b).

---

present addresses.

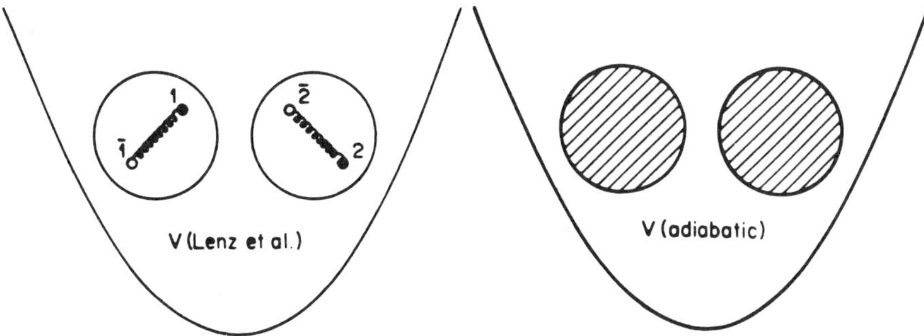

Fig.(1a) Nuclear pair model in the quark description, Fig.(1b) in the hadron description.

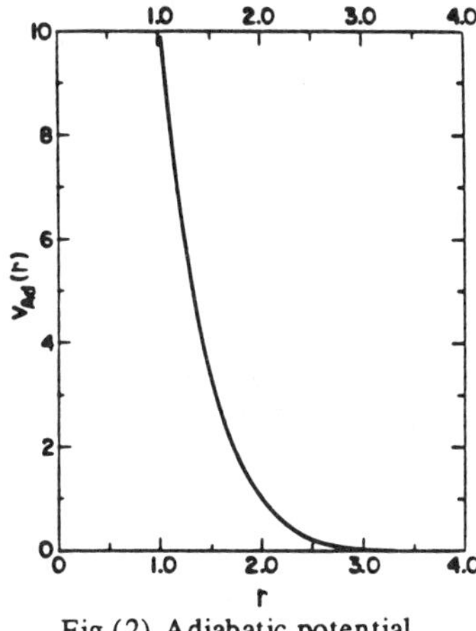

Fig.(2) Adiabatic potential.

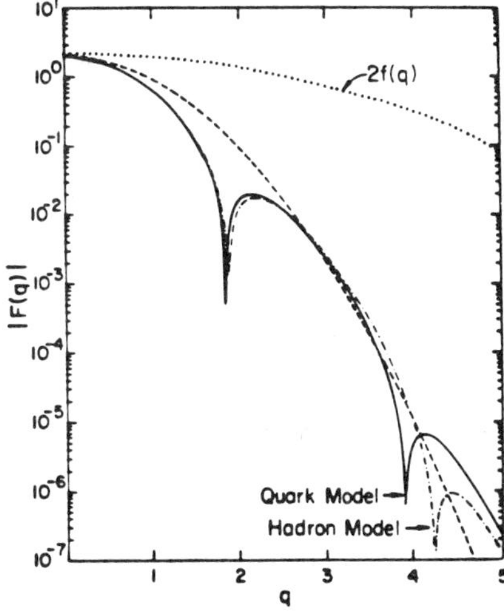

Fig.(3) Elastic form factor.

As a simple model "nucleus", a two hadron system bound in an overall harmonic oscillator potential is studied. The hadron-hadron problem is the antisymmetric $q^2\bar{q}^2$ model of Lenz et al.[(3)]. This is a quantum mechanical model in which they have incoporated confinement, cluster separability, and exchange symmetry and which avoids van der Waals foreces by the requirement of saturation of the confining force. For comparison, an effective hadron model is defined in which the hadron-hadron interaction is given by the adiabatic potential derived from the quark model. We have a purely repulsive potential for the case of antisymmetric spatial exchange as shown in Fig.(2), where the length is shown in the unit of $2r_H/\sqrt{3}(r_H =$ hadron r.m.s. radius) and the energy is in the unit of $\omega/2$.

To get a reasonable overlap of the hadrons in the ground state, the ratio of the nuclear oscillator potential constant(Ω) to the hadron one(ω) is given by $\omega/\Omega = 5$. We calculate both elastic electron scattering and the inclusive response function in the quark model and in the effective hadron model. The Fourier transform of the ground state charge distribution is shown in Fig.(3), where the momentum is shown in the unit of $\sqrt{3}/(2r_H)$ and $f(q)$ is the hadron form factor. To see the overall scale, the form factor in the hadron model without the hadron interaction is shown in the dashed line. It indicates that there are some differences at q> 4, but the form factor is in the order of 10^{-6} in this range and the charge distribution $\rho(r)$ differs very little. Thus, the elastic form factor generated by the quark dynamics can presumably be reproduced with minor changes in the hadron model.

The EMC effect[4] has served to focus much attention on the quark momentum distribution. In the model discussed above, the short range quark exchange dynamics obviously lead to a substantial change in the momentum distribution compared with the free hadron case. In this research, "anti-EMC effect" is expected since the model has a purely repulsive effective interaction. Following the idea of invoking hadron size change to explain the momentum distribution, we need a reduction of the hadron r.m.s. radius by 8.3%. This radius reduction changes the hadron form factor $f(q)$ to $e^{\varepsilon q^2/4} f(q)(\varepsilon = 0.083)$ in the hadron description. The value of such simple prescription would lie in its ability to yield other observables within the hadron framework. However the discrepancy in Fig.(4) can not be removed by the simple rescaling assumption that the hadron changes its size in the nucleus: $f(q) \rightarrow e^{\varepsilon q^2/4} f(q)$.

Next, the longitudinal response functions are calculated in both models. Because of the discrete spectrum of the harmonic oscillator potential, the response function is a sum of delta functions. In order to see how the response function looks, we integrate it over the finite energy range. The results show the response functions differ by a few percent in the region of the quasielastic peak, with the quark model response function larger than that of the hadron approximation for modest momentum transfer. A particularly interesting region is that of large momentum transfer with energy transfer much less than that corresponding to the quasifree peak. A remarkable result is that, when the response function has fallen by four orders of magnitude from the quasi-free peak value, the effective hadron theory prediction is only a factor of two different from the exact quark model value (despite the fact that the free hadron form factor squared is only about 10^{-3} at q= 5). The calculated response functions are compared in Fig.(4) with the y-scaling function evaluated in the impulse approximation in the effective hadron theory. The response functions are above the scaling curve at the momentum transfer q= 1.0, and they reach to the scaling curve at q= 1.8 and small $|y|$ (< 1). At large $|y|$ (> 1), as the momentum transfer becomes larger, the response function approaches to the scaling curve. Since the hadron interaction is strongly repulsive($\sim 1/r^2$), the response function reaches the scaling curve only at very large q>> 5[5].

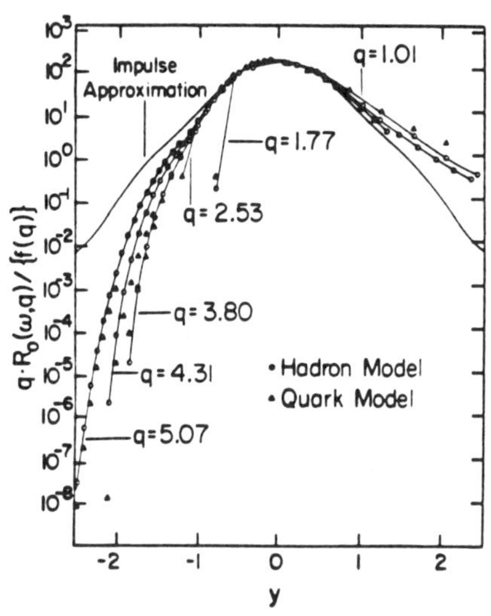

Fig.(4) y-scaling.

We find a surprizingly small effect of hadron substructure, and the discrepancies generated by the quark dynamics can presumably be reproduced with minor changes in the hadron model. Futhermore, we find in our model that a simple parametrization of modified hadron size in the bound state, motivated by the bound quark momentum distribution, is not a useful way to correlate different observables.

Acknowledgement

E.J.M. and S.K. were supported in part through funds provided by the U.S. Department of Energy under contract DE-AC02-76ER03069. S.K. was supported in part by the U.S. National Science Foundation under grant PHY84-15064, by the Division of Nuclear Physics, U.S. Department of Energy under Contract No. DE-AC05-840R21400 with Martin Marietta Energy Systems Inc., and by the State of Tennessee Science Alliance Center of Excellence Program under Contract No. R01-1061-68.

References

(1) S. Kumano and E. J. Moniz, XI Int. Conf. on Part. and Nucl., Kyoto, 128(1987).
S. Kumano and E. J. Moniz, to be published in Phys. Rev. C.

(2) I. Sick, Electron Scattering in Nuclear and Particle Science, ed. C.N. Papanicolas, L.S. Cardman, and R.A. Eisenstein, 1987.

(3) F. Lenz et al, Ann. Phys., **170** ,65(1986).

(4) F. E. Close, Rutherford Laboratory preprint RAL-87-039.
E. L. Berger and F. Coester, Ann. Rev. Nucl. Part. Sci., **37** ,463(1987).

(5) J. J. Weinstein and J. W. Negele, Phys. Lett., **49** ,1016(1982).

NEW RESULTS FROM THE EMC ON STRUCTURE FUNCTION RATIOS AND HADRON PRODUCTION IN NUCLEAR MATTER

Peter B. Renton
Nuclear Physics Laboratory, Keble Road, Oxford OX1 3RH, U.K.

1. STRUCTURE FUNCTION RATIOS

Deep inelastic scattering is a unique tool for studying the partonic nature of both the nucleon and the nucleus. Measurements have recently been carried out by the EMC on the structure function ratios F_2^A/F_2^D for A =C, Cu and Sn[1]. Only one target was in the beam at any one time and they were exchanged every few hours so that any time dependent effects in the efficiency of the apparatus cancelled in the ratios. The results are shown in fig.1 and labelled as expt.1. A second, and complimentary, target set up, consisting of a series of Cu–D–Cu–D–Cu targets, was also used. This is referred to as expt.2 in fig.1 and it can be seen that the results are in good agreement. The data are averaged over Q^2 and are for $Q^2 > 2$ GeV^2. The quantity directly determined in these experiments is the ratio of the cross-sections. However, this ratio is equal to the ratio of the structure functions F_2 on the appropriate targets, provided that the value of R (the ratio of the longitudinal to transverse absorption cross-sections of the virtual photon) is the same for each target. The experimental data[2] is compatible with this hypothesis.

A compilation of the most accurate measurements of the structure function ratios is shown in fig.2. This includes data from SLAC with revised radiative corrections[3]. The different experiments are in quite good agreement. At small x ($x \lesssim 0.1$) the ratio falls below unity. This *shadowing* phenomenum (ie the observation that $\sigma_A \propto A^\alpha$ with $\alpha < 1$) is discussed further in section 3. In the region roughly given by $0.1 \lesssim x \lesssim 0.2$ the ratio rises above unity by a few percent. At larger x, the ratio falls below unity with a cross-over at $x \sim 0.25$. The rise at large x is compatible with the expectations of the Fermi motion of nucleons.

Many models (see e.g. ref.4) have been advanced to explain these observations (the so-called EMC effect). Two main classes of model have emerged. In the first, which is based on conventional nuclear physics ideas, a convolution is made over the possible constituents of the nucleus (N, π, Δ, multi-quark bags etc.). In one variant of the model the effects are ascribed to Fermi-motion alone and parameterised by the mean separation energy $\bar{\epsilon}$. However, no rise above unity is predicted at small x, contrary to what is seen for the data. A model based on nucleons and pions gives a reasonable account of the x and A dependences for $0.1 \lesssim x \lesssim 0.6$. Such agreement with the data is also obtained from the second main model, namely the QCD rescaling model. In this model the effective Q^2 applicable in a nuclear environment is 'rescaled' with respect to that for a free nucleon such that $F_2^A(x,Q^2) = F_2^D(x,\xi(Q^2)Q^2)$, where $\xi \sim 2$ for Fe or Cu.

2. Q^2 DEPENDENCE OF STRUCTURE FUNCTION RATIOS

There is some indication (see e.g. ref.5) that, whereas the observed pattern of scaling violations on light (H, D, C) targets is reproduced by next-to-leading order QCD with $\Lambda_{\overline{MS}} \simeq 100$ MeV, the scaling violations on Fe targets are more rapid and not in good agreement with QCD. If this is the case, then the Q^2 dependence of $b_N = d[F_2^A(x,Q^2)/F_2^D(x,Q^2)]/d\ln Q^2$ would also be expected to reflect this behaviour. Fits have been made to the values of b_N extracted from the SLAC (Fe/D), BCDMS (Fe/D),

EMC (Fe/D) and EMC (Cu/D) data. The combined slopes from these fits (where the systematic errors, if available, have been added to the statistical errors in quadrature) are shown in fig.3 as a function of x. Also shown are the expectations of the pion and QCD rescaling model[6]. The data have a different trend to the predictions. However, a thorough evaluation of the possible systematic errors on all of the experiments must be carried out before the conclusion that a significantly different Q^2 dependence is observed for heavy targets.

3. THE LOW X REGION AND SHADOWING

The region of low x ($0.003 < x < 0.1$) and low Q^2 ($0.3 < Q^2 < 3.2$ GeV^2) has been investigated using a special low angle trigger (EMC NA28) with a view to investigating shadowing behaviour. Fig.4 shows the results of these measurements for the ratios $F_2^{\mathrm{C}}/F_2^{\mathrm{D}}$ and $F_2^{\mathrm{Ca}}/F_2^{\mathrm{D}}$ compared to similar data at higher x and Q^2. There is an additional $\pm 7\%$ overall normalisation uncertainty on the ratios. It can be seen that the ratios are less than unity at small x, and that the size of this shadowing effect increases with increasing A. The ratio at the smallest x values is compatible with the real photon data, plotted arbitrarily at $x = 0.001$. No significant Q^2 dependence is seen in the data, contrary to the expectations of the Vector Dopminance Model (VDM). In the VDM, the virtual photon fluctates to a virtual vector meson ($\rho, \omega, \phi \dots$), which then behaves as a hadron and interacts primarily with the surface nucleons of the nucleus, hence giving shadowing. The cross-section is predicted to fall rapidly with Q^2 because of a propagator type term $1/(Q^2 + M_v^2)$, where M_v is a vector meson mass. From the observed Q^2 dependence, the VDM can be ruled out as the only source of shadowing.

A somewhat different approach is taken in the parton recombination models[7]. At small x the distance over which the parton is localised can become large (eg larger than the nucleus) and partons can no longer be considered to belong to a specific nucleon. These partons can recombine, thus reducing their density and give rise to shadowing. The data are consistent with the slow Q^2 dependence of shadowing expected in this approach and with the prediction (Berger and Qiu[7]) that the cross-over point x_c between shadowing and anti-shadowing increases with A.

4. HADRON PRODUCTION FROM NUCLEAR TARGETS

One of the main areas of motivation in the study of the hadron spectra is to try and get some insight into the space-time evolution of the hadronisation process. Hadrons produced inside the nucleus can interact and will thus, at high z ($z = E_h^{lab}/\nu$), give a depletion in the ratio of the fragmentation functions $R_A(z) = D_A(z)/D_D(z)$, where $D(z) = (1/N_{ev})dN^h/dz$. The secondary products from these intereactions will give an excess ($R_A > 1$) at low z. For the EMC energy range it is estimated that for $z \gtrsim 0.2$ there is little contamination from reinteractions at higher z. If the hadron formation process takes place inside the nucleus at low ν, then, due to the Lorentz boost, it will occur outside for large enough ν. The important parameters are the formation length ℓ_F and the hadron interaction length (ℓ_π), compared to the nuclear radius R_{nuc}. The distribution of ℓ_F expected in the string-model approach to hadronisation can be calculated[8]. However, it is unclear in the model as to whether the constituent quark (formed at length ℓ_c) has hadron-like properties or if these start only at the point where a colour-neutral particle emerges at the so-called yo-yo point (at length ℓ_y).

An analysis has been performed on the hadrons produced in the data described in section 1. These hadrons are restricted to an acceptance region such that they are forward-going in the hadronic centre of mass system, so that target fragments are not observed. In fig.5 the ratios $R_A(z)$, for A =C, Cu and Sn, are shown. The data are consistent with no attenuation for C/D. However, significant attenuation is observed for Cu/D for $z \gtrsim 0.2$, and a rise in R_A is seen at lower z. The Sn/D data has a trend compatible with the Cu/D data, but with larger errors. The curves shown are from a simple model[8], assuming the constituent quark length is relevant for hadronic reinteraction and taking $\ell_\pi = 3$fm. The prediction for the yo-yo length is essentially unity. In fig.6 a comparison of the EMC Cu/D data (in two intervals of ν) and Sn/D data is made with a SLAC experiment[9] performed at low ν. The solid lines correspond to the constituent length and reproduce the data reasonably well. However the yo-yo length (dot-dashed line for the SLAC data) gives a poor description. The ratio integrated over z ($z > 0.2$ for EMC, $z > 0.5$ for SLAC) is plotted in fig.7 versus ν. A rapid rise at low ν is seen and this is again compatible with the string model, provided the hadronic properties begin at about the constituent length.

REFERENCES

1. EMC, J. Ashman *et al*, *Phys. Lett.* **B202** (1988) 603.
2. SLAC E140, S. Dasu *et al*, University of Rochester preprint, UR–1045 (1988).
3. SLAC E140, S. Dasu *et al*, University of Rochester preprint, UR–997 (1988).
4. E.L. Berger and F. Coester, *Ann. Rev. Nucl. Part. Sci.* **37** (1987) 463.
5. R. Voss, Proceedings of Int. Symp. on Lepton and Photon Interactions at High Energies, Hamburg (1987); CERN/EP 87–223.
6. R.P. Bickerstaff and G.A. Miller, *Phys. Rev.* **D34** (1986) 2890.
7. N.N. Nicolaev and V.I. Zakharov, *Phys. Lett.* **B55** (1975) 397; A.H. Müller and J. Qiu, *Nucl. Phys.* **B268** (1986) 427; J. Qiu, *Nucl. Phys.* **B291** (1987) 746; E.L. Berger and J. Qiu, ANL–HEP–PR–87–06 (1988).
8. P. Renton; the calculations are from a model based on the formalism of T. Chmaj, Cracow preprint TPJU 86–9 (1986); A. Bialas and M. Gyulassy, *Nucl. Phys.* **B291** (1987) 793; and M. Bowler, private communication.
9. L.S. Osborne *et al*, *Phys. Rev. Lett.* **40** (1978) 1624.

FIGURE CAPTIONS

Fig.1. Ratios of the structure functions F_2^A/F_2^D for A =C, Cu and Sn, as a function of x, as measured by the EMC.

Fig.2. Measurements of the structure function ratios F_2^A/F_2^D (A =Fe, Cu), as a function of x, from several experiments.

Fig.3. Combined slopes b_N versus x from several experiments, together with some model expectations.

Fig.4. Structure function ratios at low x F_2^A/F_2^D (A =C, Ca), versus x, for two ranges of Q^2.

Fig.5. Ratios $R_A(z)$ of the fragmentation functions, versus z, from the EMC experiment.

Fig.6. Ratios $R_A(z)$ versus z from the EMC and SLAC experiments.

Fig.7. Integrated ratio $\bar{R}_A$ versus ν, from the EMC and SLAC experiments.

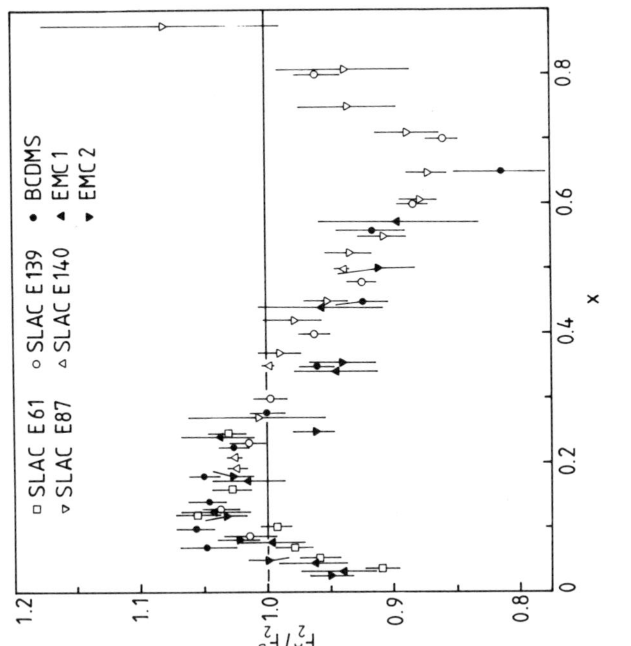
SLAC E61
SLAC E139
BCDMS
SLAC E87
SLAC E140
EMC 1
EMC 2
F_2^A/F_2^D
x

Fig. 2.

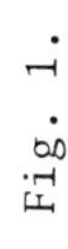

Fig. 1.

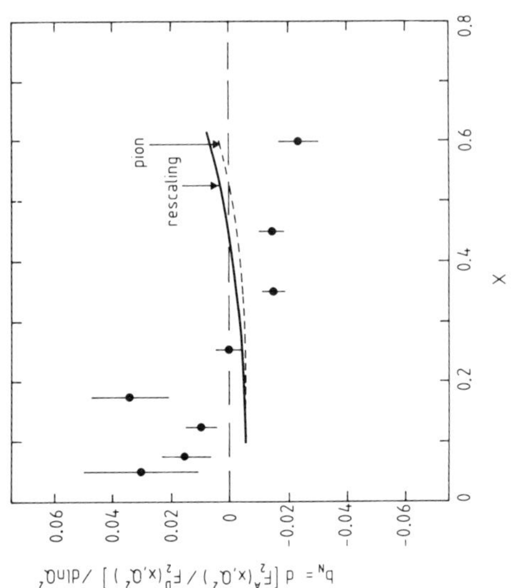

$b_N = d\,[F_2^A(x,Q^2) / F_2^D(x,Q^2)]\,/\,d\ln Q^2$
0.06
0.04
0.02
0
-0.02
-0.04
-0.06
pion
rescaling
0
0.2
0.4
0.6
0.8
x

Fig. 3.

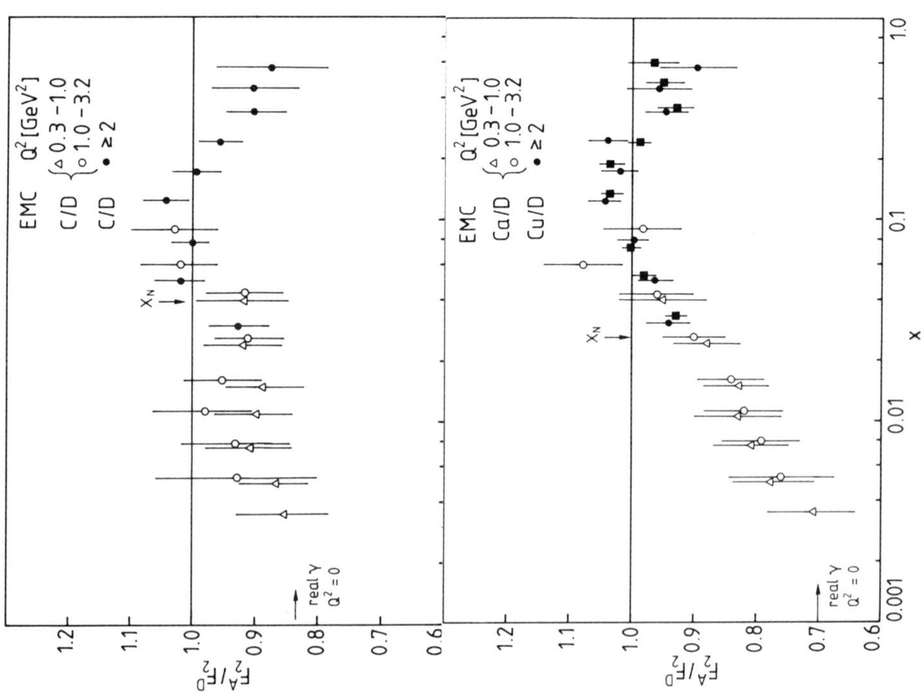

EMC Q^2 [GeV^2]
C/D { △ 0.3 - 1.0
○ 1.0 - 3.2
C/D • ≥ 2
EMC Q^2 [GeV^2]
Ca/D { △ 0.3 - 1.0
○ 1.0 - 3.2
Cu/D • ≥ 2
x_N
real γ
$Q^2 = 0$
F_2^A / F_2^D
1.2
1.1
1.0
0.9
0.8
0.7
0.6
0.001
0.01
0.1
1.0
x

Fig. 4.

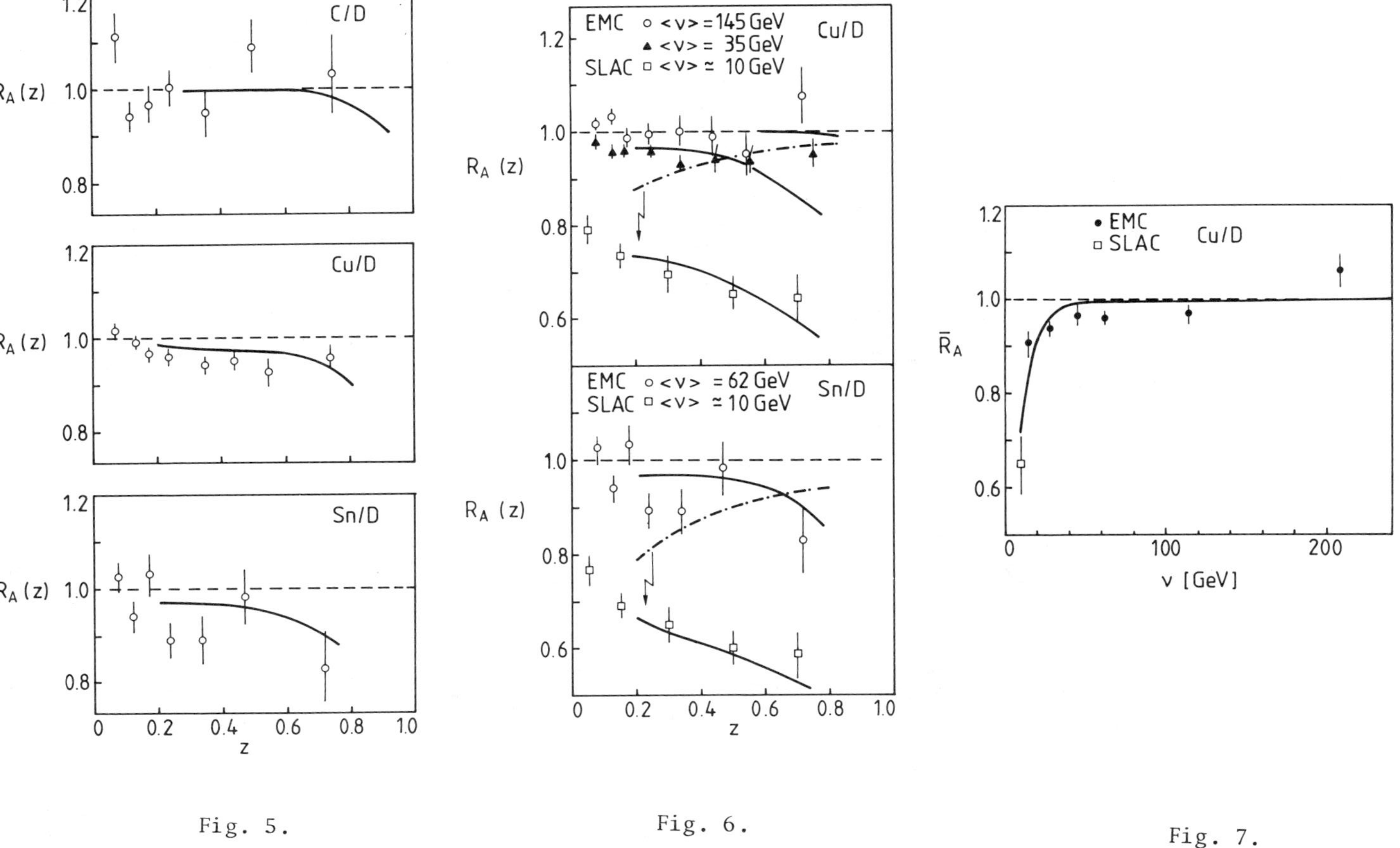

Fig. 5.

Fig. 6.

Fig. 7.

PRECISION MEASUREMENT OF $R = \sigma_L/\sigma_T$ AND F_2 IN DEEP INELASTIC ELECTRON SCATTERING ON DEUTERIUM AND IRON TARGETS

P. Bosted[2*], S. Dasu[1], R. C. Walker[3], L. W. Whitlow[9], J. Alster[6], R. Arnold[2], P. de Barbaro[1], D. Benton[2a], A. Bodek[1], J. Button-Shafer[5], G. deChambrier[2], L. Clogher[2], B. Debebe[5], F. Dietrich[4], B. Filippone[3], R. Gearhart[8], H. Harada[1], R. Hicks[5], J. Jourdan[3], M. W. Krasny[1b], K. Lang[1c], A. Lung[2], R. Milner[3], R. McKeown[3], A. Para[7], D. Potterveld[3], E. M. Riordan[1d], S. Rock[2], Z. M. Szalata[2], K. Van Bibber[4]

1) University of Rochester, Rochester, NY 14627;
2) The American University, Washington, DC 20016;
3) California Institute of Technology, Pasadena, CA 91125;
4) Lawrence Livermore National Laboratory, Livermore, CA 94550;
5) University of Massachusetts, Amherst, MA 01003;
6) University of Tel-Aviv, Ramat Aviv, Tel-Aviv 69978, Israel;
7) Fermi National Accelerator Laboratory, Batavia, IL 60510;
8) Stanford Linear Accelerator Center, Stanford, CA 94305;
9) Stanford University, Stanford, CA 94305.

* Presented by P. Bosted, The American University
a Present address: Department of Physics, Princeton University, Princeton, NJ 08544.
b Present address: Institute of Nuclear Physics, Kawiory 26A, PL-30055, Krakow, Poland.
c Present address: Department of Physics, Stanford University, Stanford, CA 94305.
d Present address: Stanford Linear Accelerator Center, Stanford, CA 94305

ABSTRACT

We report new results on a precision measurement of the ratio $R = \sigma_L/\sigma_T$ from deuterium, iron and gold targets, on the ratio σ^{Fe}/σ^D and on the structure function F_2, for deep inelastic electron-nucleus scattering in the range $0.2 \leq x \leq 0.5$ and $1 \leq Q^2 \leq 10(GeV/c)^2$. We find no difference between R^{Fe} and R^d. Our results show, for first time, a clear falloff of R with increasing Q^2 at SLAC energies. Our R and F_2 results are in good agreement with QCD predictions only when corrections for target mass effects are included.

INTRODUCTION

The ratio $R = \sigma_L/\sigma_T$ measured in deep inelastic lepton-nucleon scattering is a sensitive measure of the spin of the nucleon constituents. Within the naive parton model R is expected to be small and decrease with increasing Q^2 for spin-1/2 partons, but to be large and increase with Q^2 for spin-0 partons. Previous measurements[1] at the Stanford Linear Accelerator Center (SLAC) indicated that scattering from spin-1/2 constituents (e.g. quarks) dominates. However, the results for R were larger than expected (see Fig. 2a), and were consistent with a constant value of 0.2. The errors on these results left room for speculations of additional spin-0 constituents in nucleons[2] (e.g. tightly bound diquarks), and of unexpectedly large primordial transverse momentum for quarks. Experiments[1] in the SLAC Q^2 range $(1 \leq Q^2 \leq 20\ (\mathrm{GeV/c})^2)$ have also indicated deviations from the scaling of the structure functions F_1 and F_2 in the variable $x = Q^2/2M\nu$. In the theory of Quantum Chromodynamics (QCD), scaling violations[3] occur due to quark-gluon interactions $(F_{1,2} \propto 1/\ln(Q^2/\Lambda^2)$ for $x \geq 0.2)$. The effects of target mass (or higher twist) influence[4] scaling, yielding contributions to $F_{1,2}$ which are proportional to powers of M^2/Q^2. These effects lead to non-zero contributions to R which decrease with increasing Q^2.

The discovery of the difference in the structure function $F_2(x)$ between iron and deuterium targets,[5] [6] [7] [8] known as the EMC effect, has sparked considerable activity in the theoretical study of deep inelastic lepton scattering from nuclear targets. There are numerous models[9] for the EMC effect which are built on a variety of ideas: Q^2-rescaling, x-rescaling, convolution of structure functions of clusters of nuclear matter (nucleons and pions in nuclei, Δ-isobars, multiquark clusters ...) and others. Some models[10] predict a large difference in the quantity R for deuterium and iron ($R^{Fe} - R^D \sim 0.1$ to 0.15). Almost all others[11] including those based on QCD, predict a negligible difference ($R^{Fe} - R^D \sim 0.002$). Some authors[12] have conjectured that higher twist effects might be different for different nuclei, and yield an atomic mass (A) dependence of R.

The differential cross section for scattering of an unpolarized charged lepton with an incident energy E, scattering angle θ and final energy E' can be written in terms of the structure functions $F_1(\nu, Q^2)$ and $F_2(\nu, Q^2)$ as:

$$\frac{d^2\sigma}{d\Omega dE'}(E, E', \theta) = \sigma_M \Big[F_2/\nu + \frac{2F_1}{M}\tan^2(\theta/2)\Big] = \Gamma\Big[\sigma_T + \epsilon\,\sigma_L\Big] \tag{1}$$

where $\sigma_M = 4\alpha^2 E'^2 \cos^2(\theta/2)/Q^4$ is the Mott cross section, M is the nucleon mass, $\nu = E - E'$, $Q^2 = 4EE'\sin^2(\theta/2)$, and $x = Q^2/2M\nu$. The quantity $R = \sigma_L/\sigma_T$, with $\Gamma = \frac{\alpha}{4\pi^2}\frac{(2M\nu - Q^2)E'}{Q^2 ME}\left(\frac{1}{1-\epsilon}\right)$, and $\epsilon = \left[1 + 2(1 + \frac{\nu^2}{Q^2})\tan^2(\theta/2)\right]^{-1}$ representing the virtual photon flux and polarization respectively. In this paper the differential cross sections per average nucleon (including the correction for neutron excess in Fe and Au) are represented as σ^A for Fe and Au, and σ^D for deuterium.

EXPERIMENTAL METHOD

The SLAC electron beams and the 8 GeV spectrometer facility were used to measure cross sections accurate to better than 1% in the kinematic range $0.2 \leq x \leq 0.5$ and $1 \leq Q^2 \leq 10 (GeV/c)^2$. Measurements were made on targets of liquid D_2 (2.6% radiation lengths), Fe (2.6% and 6% r.l.) and Au (6% r.l.) at up to five different values of ϵ for each (x, Q^2) point (with a typical ϵ-range of 0.35). To reduce systematic errors the targets were alternated every few minutes.

Extensive measures were taken to minimize uncertainties (summarized in Table I) due to beam steering, charge measurement, target density fluctuation and background contamination. The spectrometer acceptance was studied using floating wire technique, and a careful survey was made of the spectrometer angle and magnetic elements. A considerable effort was made to improve the radiative corrections procedures. We used the "exact" prescription of Bardin et al.,[13] with additional "external" corrections (due to the straggling of electrons in the target material) calculated in the complete formalism of Mo and Tsai.[14] The "internal" corrections obtained using the Bardin formalism agreed to better than 1% at each (x, Q^2, ϵ) point with an improved version of the "exact" formalism of Mo and Tsai. Corrections calculated with different parameterization of structure functions, agreed to better than ±0.2%. In addition, R and F_2 measured with 2.6% and 6% r.l. Fe targets were in agreement. The same radiative correction procedure has also been used in recent neutrino[15] and muon[16] experiments.

		ERROR (±) IN		ERROR (±)	
SOURCE	UNCERTAINITY	σ	R	σ^A/σ^D	$R^A - R^D$
Beam steering	0.003°	0.1%	0.005	0.1%	0.004
Incident energy	0.1%	0.3%	0.014	0.3%	0.014
Charge measurement	0.3%	0.3%	0.014	0.1%	0.004
D Target density	0.2%	0.2%	0.009	0.2%	0.010
D Acceptance vs. θ	0.1%	0.1%	0.005	0.1%	0.004
Acceptance vs. p	0.1%	0.1%	0.005	-	-
e^+/e^- background	0.1%	0.1%	0.005	0.1%	0.004
Scattered energy	0.05%	0.1%	0.005	-	-
Spectrometer angle	0.002o	0.1%	0.005	-	-
Detector efficiency	0.1%	0.1%	0.005	-	-
A Neutron excess	0.2%	-	-	0.2%	-
TOTAL POINT-TO-POINT		0.5%	0.025	0.4%	0.019
Target length error	0.5-1.5%	1%	-	1.0%	-
Radiative corrections	1%	1%	0.030	0.5%	0.015

Table I. Typical systematic errors on R and σ

EMC EFFECT RESULTS: R DIFFERENCE

The difference $R^A - R^D$ was determined by making linear fits, weighted by the statistical and point-to-point (ϵ-uncorrelated) systematic errors, to the ratio of cross sections σ^A/σ^D versus $\epsilon' = \epsilon/(1 + \epsilon R^D)$. The $R^A - R^D$ results are thus independent of absolute normalizations of target length, spectrometer acceptance, beam intensity and energy scale. They are also insensitive to changes in acceptance with ϵ, offsets in beam energy, spectrometer angle, survey errors, long term charge monitor drifts, and "internal" radiative corrections. The average χ^2 per degree of freedom for the ϵ' fits was 0.7, indicating that the estimate of systematic uncertainty is conservative. The results are plotted in Fig. 1a, and are consistent with zero within the errors. The average $R^A - R^D$ is $0.001 \pm 0.018(\text{stat}) \pm 0.016(\text{syst})$. The single measurement for Au is consistent with Fe results.

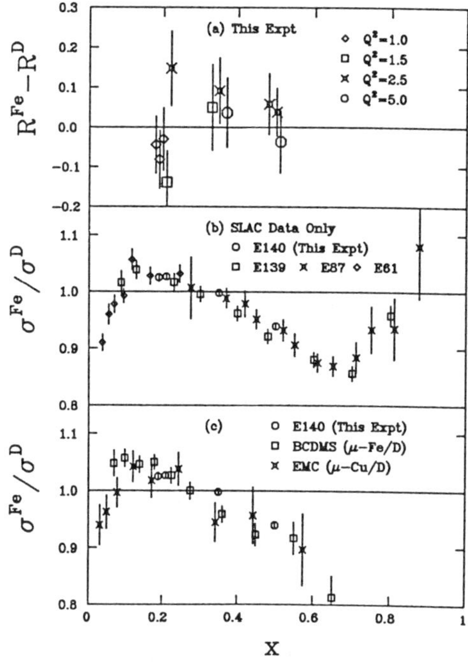

Fig. 1. (a) The results for $R^A - R^D$. Statistical and systematic errors are added in quadrature. (b,c) The results for σ^A/σ^D compared to other (b) electron and (c) muon experiments. Our data from Fe and Au(x=0.2) targets are each averaged over ϵ and Q^2. Statistical and point-to-point systematic errors are added in quadrature for all experiments. The overall normalization errors (Δ) are discussed in text.

The results for the ratio σ^A/σ^D averaged over ϵ are shown in Fig. 1b. The overall normalization error (Δ) in σ^A/σ^D of $\Delta = \pm 1.1\%$ is dominated by the errors in target length measurement and radiative corrections. Our results are compared with data from SLAC-E139 (with improved radiative corrections; $\Delta = \pm 1.3\%$),[7] SLAC-E87 ($\Delta = \pm 1.1\%$)[6] and SLAC-E61 ($\Delta = \pm 4.2\%$)[8] in Fig. 1b. There is excellent agreement between all the SLAC data. In Fig. 1c our data is compared with high Q^2 data from CERN muon experiments BCDMS ($\Delta = \pm 1.5\%$),[16] and EMC (preliminary results $\Delta = \pm 0.8\%$).[17] The lower Q^2 SLAC results are in agreement with these high Q^2 muon scattering results.

We conclude that $R^A - R^D$ is consistent with zero, in agreement with most models. Our data rule out models[10] predicting a large difference $R^A - R^D$ and indicate that possible contributions to R from nuclear higher twist effects and spin-0 constituents in nuclei are not significantly different from those in nucleons. The σ^A/σ^D measurements can now be identified with the structure function ratios F_2^A/F_2^D and F_1^A/F_1^D. The EMC effect is confirmed with very small errors and all data are now in agreement. The ratio σ^A/σ^D is larger than unity at low x, and is therefore inconsistent with models using nuclear binding corrections only, implying a non-trivial difference in the quark distributions between heavy nuclei and the deuteron.[9]

ABSOLUTE VALUE OF R

The values of R, F_1 and F_2 were extracted from cross sections measured at various values of ϵ at fixed (x, Q^2) by making linear fits to σ/Γ verses ϵ. The average χ^2/df for these fits is 0.7. Since the differences $R^A - R^D$ are consistent with zero, the results plotted in Fig. 2a-c and Fig. 3 are averaged over data for different targets at the same x and Q^2. Our results have considerably smaller errors then previous SLAC experiments.[1]. Our results for R at x of 0.2, 0.35 and 0.5 (see Fig. 2a-c) show a clear falloff with increasing Q^2. The agreement with a constant value of $R = 0.2$ is poor (χ^2/df=3.4). The high Q^2 results from CDHS[15] and BCDMS[16] collaborations for ν-Fe and μ-C/H scattering respectively, are also plotted in Fig. 2 and reinforce the conclusion that R decreases with increasing Q^2. Out results at all Q^2 show only a weak x-dependence in the range $0.2 \leq x \leq 0.5$.

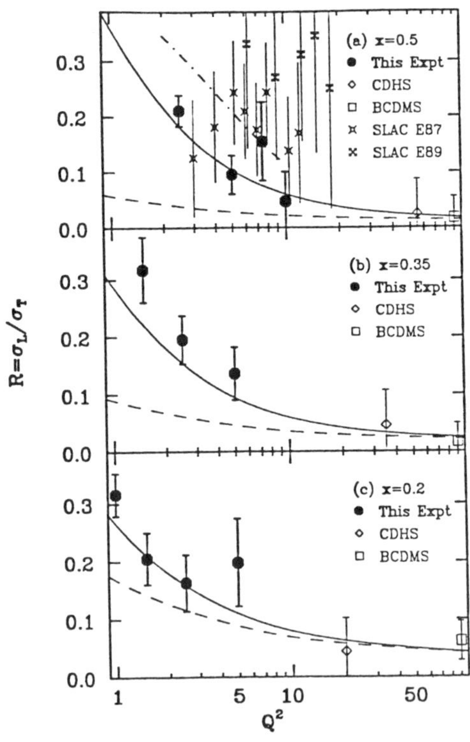

Fig. 2. The values of R at (a) $x = 0.5$, (b) $x = 0.35$, and (c) $x = 0.2$ are plotted versus Q^2. with statistical and systematic errors added in quadrature. Predictions from naive parton model (dot-dash curve), perturbative QCD (quark-gluon interaction effects; the dash curve), QCD with target mass effects (solid curve) are also shown. The data from high Q^2 CDHS (ν-Fe) and BCDMS (μ-C/H) experiments are also plotted.

Our F_2 results, plotted in Fig. 3, are still preliminary since the analysis of absolute normalizations (presently known to ±3%) is not complete. A fall of $F_2^D(x, Q^2)$ with increasing Q^2 due to scaling violations or other effects is evident. Detailed studies of F_2 from all SLAC experiments, including ours, with our improved radiative corrections and new values for R are in progress.

It can be seen in Fig. 2 that the naive parton model $R = 4M^2x^2/Q^2$, gives a poor description of the data (χ^2/df=7.1). Perturbative QCD (to the order α_s) including hard gluon bremsstrahlung from quarks and effective photon-gluon interaction effects[3] also is not in agreement with our data for $x \geq 0.35$

(χ^2/df=9). The scaling violations in F_2 data are also not described completely by these QCD interaction effects alone. The inclusion of target mass effects[4] introduce M^2/Q^2 terms and provide a good description of both the R data (χ^2/df=1.1) and with the Q^2 dependence of the values for F_2. For a complete description of theoretical predictions see ref. 18.

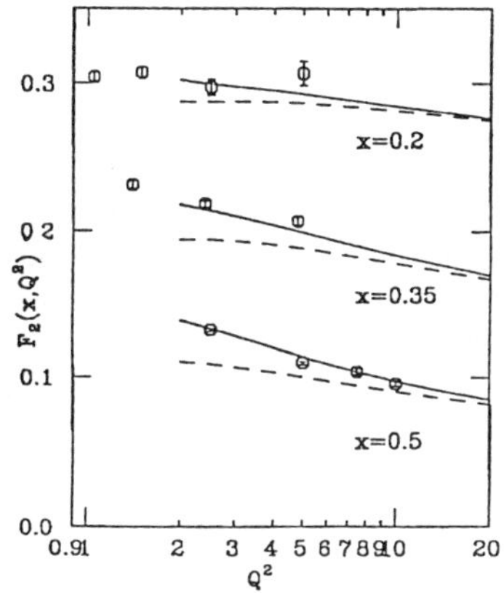

Fig.3. Preliminary values of F_2^D at (a) $x = 0.5$, (b) $x = 0.35$, and (c) $x = 0.5$. Values from Fe and Au are averaged with those from deuterium after correction for the EMC effect. Only statistical and point-to-point systematic errors are shown for our data. There is presently an additional normalization error of 3%. acceptance. QCD calculations of the structure function, F_2^{QCD} (dashed curve), and the target mass effect corrected F_2^{QTM} (solid curve) are also plotted.

In conclusion, our results show for the first time a clear falloff of R with increasing Q^2 in the range $1 \leq Q^2 \leq 10(GeV/c)^2$ for $x = 0.2, 0.35$ and 0.5. Our R and F_2 results are in good agreement with QCD predictions of scaling violations only when corrections for target mass effects are included. Our results do not substantiate any claims of large contributions from diquarks, non-perturbative and higher twist effects within our kinematic range.

We acknowledge the support of Dr. B. Richter and the SLAC staff, which was crucial for the success of this experiment. This research was supported by Department of Energy Contracts Nos. DE-AC02-76ER13065, DE-AC02-76ER02853, and DE-AC03-96SF00515; and National Science Foundation Contracts Nos. PHY84-10549 and PHY85-05682.

REFERENCES

1. M. Breidenbach et al., Phys. Rev. Lett. 23, 930 (1969); G. Miller et al., Phys. Rev. D 5, 528 (1972); A. Bodek et al, Phys. Rev. D 20, 1471 (1979); M. D. Mestayer et al., Phys. Rev. D 27, 285 (1983).

2. L. F. Abbott et al., Phys. Lett. 88B, 157 (1979); S. Ekelin and S. Fredriksson, Phys. Lett. 162B, 373 (1985).

3. G. Altarelli and G. Martinelli, Phys. Lett. 76B ,89 (1978); H. Abramowicz et al. (CDHS), Z. Phys. C 12, 289 (1982); B. Vallage, Ph. D. thesis, University de Paris-Sud, (July 1984).

4. R. Barbieri et al., Nucl. Phys. B117, 50 (1976); A. de Rujula et al., Ann. Phys. 103, 315 (1977); A. Buras et al., Nucl. Phys. B131, 308 (1977); H.Georgi and D. Politzer, Phys. Rev. D14, 1829 (1976)

5. G. Bari et al., Phys. Lett. B 163, 282 (1985).

6. A. Bodek et al., Phys. Rev. Lett. 50, 1431 (1983); 51, 534 (1983).

7. R. G. Arnold et al., Phys. Rev. Lett. 52, 724 (1984).

8. S. Stein et al., Phys. Rev. D 12, 1884 (1975).

9. E. Berger and F. Coester, Ann. Rev. Nucl. Part. Sci. 37, 463 (1987).

10. Bo-Qiang Ma and Ji Sun, Print-86-1217, Beijing University, China (1986).

11. F. E. Close et al., Phys. Lett. B 129, 346 (1983) - QCD model; R. D. Smith, Phys. Lett. B 182, 283 (1986) - convolution model.

12. E. V. Shuryak, Nucl. Phys. A446, 259C (1985).

13. D. Y. Bardin et al, Nucl. Phys. B197, 1 (1982).

14. L. W. Mo and Y. S. Tsai, Rev. Mod. Phys. 41, 205 (1969); Y. S. Tsai, SLAC-PUB-848 (1971). Our procedure included (a) the addition of effects of μ, τ and quark contribution to vacuum polarization, b) correction for multiple soft photon emission (this was used in peaking approximation formulae but not in the exact formula A.24 of SLAC-PUB-848), and c) the bremsstrahlung photon angle integral, for photon energy less than cut off (Δ).

15. P. Buchholz, Report of CDHS collaboration in the Proceedings of the EPS International Conference on High Energy Physics, Bari, Italy, 1985.

16. A. C. Benvenuti et al. (BCDMS), Phys. Lett. 195 B, 91 (1987). A. Milsztajn, the proceedings of the International Europhysics Conference on High Energy Physics, Uppsala (1987); A. C. Benvenuti et al., Phys. Lett. B 189, 483 (1987).

17. M. Dueren and T. Sloan, Proceedings of the International Europhysics Conference on High energy physics (Uppsala, 1987).

18. S. Dasu, Ph.D. Thesis, University of Rochester (1988)

THE PRIMAKOFF EFFECT IN ELECTRON SCATTERING

E. Hadjimichael
Physics Department, Fairfield University
Fairfield, Connecticut, 06430

S. Fallieros
Physics Department, Brown University
Providence, R.I., 02912

ABSTRACT

The lifetime of π° and other pseudoscalar mesons, has been successfully investigated in the past by means of the Primakoff effect, i.e., the production of neutral mesons by incident photons interacting with the Coulomb field of a high Z nucleus. In the present work, the same phenomenon is explored, initiated by virtual, rather than real, photons from electron scattering. Experimental measurements, if feasible, will provide valuable information about the structure of the $\gamma\pi^\circ$ vertex.

INTRODUCTION

Low-energy hadronic physics is dominated by π mesons. They are a basic ingredient in the exchange mechanism which gives rise to the NN interaction phenomenon and also helps determine the nuclear electromagnetic current. Also, pions provide the means for securing chiral invariance and the SU(2)xSU(2) symmetry in effective Lagrangians for strongly interacting systems. The small pion mass testifies to the almost exact validity of this symmetry. Finally, pionic fields are essential elements in models of nucleon structure e.g. chiral quark bags and the Skyrme model.

In view of its multifaceted role in hadronic physics, the structure and other characteristics of the pion itself are subject of intense interest. In the present work, we investigate the structure of the $\gamma\pi$ vertex for neutral pions. These are produced via a $\gamma\gamma\to\pi^\circ$ process, with virtual photons from electron scattering by the Coulomb field of a high Z nucleus. The corresponding process with real incident photons, is known as the Primakoff effect(1). It has been used successfully in the past (1-3) to determine the lifetime of π° and other pseudoscalar mesons.

It should be noted that the kinematics of the process under investigation is such as to allow its differentiation from photoproduction from the nucleons of the nucleus via a hadronic meson-nucleon vertex.

Some information concerning the $\gamma\pi^\circ$ vertex is already available (4) at a kinematically fixed time-like momentum, through the study of the decay

$$\pi^0 \rightarrow \gamma^* \gamma \rightarrow e^+ e^- \gamma \tag{1}$$

The mechanism considered in the present work however, involves variable space-like momenta and yields a kinematically more complete picture.

The main decay of π° is $\pi^\circ \rightarrow \gamma\gamma$ which yields a measured lifetime of $\tau = 0.83 \times 10^{-16}$ sec. The decay, Equ(1), occurs with a branching ratio (5).

$$\frac{\Gamma(\pi^0 \rightarrow e^+ e^- \gamma)}{\Gamma(\pi^0 \rightarrow \gamma\gamma)} = 0.01185 \tag{2}$$

subject to modifications due to radiative corrections. The experimentally measured differential rate for this decay is expressed in terms of a vertex form factor $f(k1^2/m^2,\ k2^2/m^2,\ Q^2/m^2) = f(x,0,1)$, where k1, k2 and Q are the four-momenta of the virtual photon, the real photon and the pion respectively and m is the pion mass. This quantity is normalized by $f(0,0,1) = 1$, and for small x is written as

$$f(x,0,1) = F_\pi(x) = 1 + ax \tag{3}$$

Analysis of data, without regard to radiative corrections yields for the form factor slope, $a=0.05\pm0.03$, while when radiative corrections are taken into account(4), the data yield (6) $a=0.1 + 0.03$. We recall also that the vector meson dominance model(VMD) for the photon-hadron vertex gives $a=0.03$ (7). As the Fourier transform of a density, the form factor can be also written as

$$F\pi(k_1^2) = 1 - \langle r^2 \rangle k_1^2/6 \sim 1/(1 + \langle r^2 \rangle k_1^2/6) \tag{4a}$$

which leads to

$$a = \langle r^2 \rangle m_\pi^2/6 \tag{4b}$$

THE PRIMAKOFF PROCESS WITH VIRTUAL PHOTONS

We investigate electroproduction of π° from the Coulomb field of Pb (Z=82). The microscopic process we are concerned with is shown in Fig(1a), and the corresponding kinematics is displayed in Fig(1b). We calculate the coincidence cross section differential in the scattered electron energy ε_2 and angle Ω_2 and the pion angle Ω_π. All angles are measured with respect to the direction of the three-momentum transfer q, i.e., the direction of the incident virtual photon.

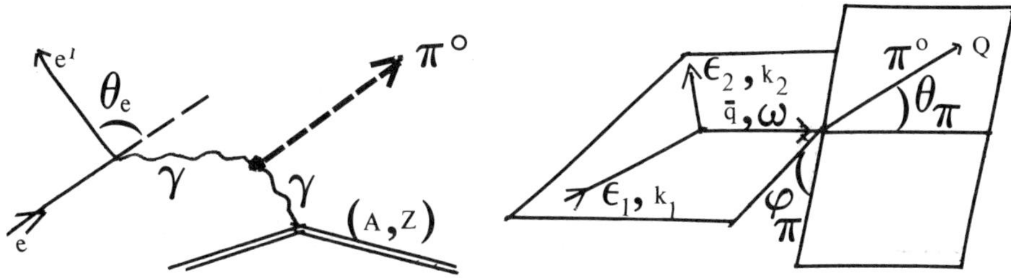

Fig(1a): The Primakoff process with virtual photons

Fig(1b): Kinematics for process in Fig(1a)

We adopt a phenomenological Hamiltonian density for the process in Fig(1a) of the form

$$H_I = i(\eta/4)\,\Sigma\Sigma\,\epsilon_{\mu\nu\sigma\tau}\,F_e^{\mu\nu}\,F_N^{\sigma\tau}\,\Phi_\pi \qquad (5)$$

η=phenomenological coupling constant related to the π° lifetime

F_e and F_N are the electromagnetic tensors for the virtual photon emitted by the scattered electron, and for the nuclear fields. In this latter, we take the electric field E_N to be static, and the magnetic field B_N=0. Φ_π is the field of the outgoing pion which, for our purposes takes the form of an outgoing plane wave. Using Equ(5), and following lengthy manipulations we develop the desired cross section.

$$d^3\sigma/d\epsilon_2 d\Omega_2 d\Omega_\pi = (\eta^2 Z^2/\pi)\,\sigma_M(\theta_e)\sin^2(\theta_e/2)\,|F_N(K^2)|^2/K^4$$

$$\sin^2\theta_\pi\,|F_\pi(q^2)|^2\,Q^4\,\beta_\pi^{-1}/\omega_\pi\,[4\epsilon_1\epsilon_2\sin^2\phi_\pi$$

$$+\,q^2\sec^2(\theta_e/2)] \qquad (6)$$

The quantities appearing in this expression are as follows: Θ_e=the electron scattering angle, q=momentum transfer, K=momentum of recoiling nucleus, Q=momentum of outgoing pion, $\beta_\pi=Q/\omega_\pi$, and ω_π=pion total energy= energy transfer $\omega=\varepsilon_1-\varepsilon_2$. $F_\pi(q^2)$ is the $\gamma\pi^\circ$ form factor taken as in Equ(4), and the nuclear form factor

$$F_N(K^2) = (1/Ze).\int dx\ e^{iK.x}\ \rho(x)$$

is taken equal to 1 since K is quite small and the nucleus remains essentially static. σ_M is the Mott cross section.

The cross section, Equ(6), should be compared to the equivalent expression for the Primakoff process with real incident photons, which is

$$d\sigma / d\Omega_\pi = 4\pi Z^2 \alpha\ \ \eta^2 \sin^2\theta_\pi \sin^2\phi_\pi\ |F_N(K^2)|^2\ |F(q^2)|^2$$
$$Q^4 \beta_\pi^{-1} / K^4 \qquad (7)$$

The angular dependence of the two cross sections is quite similar, but we find a difference in magnitude of about three orders due to the extra power of the fine structure constant in Equ(6) and other factors.

We display typical results of our calculations in Fig(2-5). For selected values of energy transfer $\omega=\omega_\pi$,we show the coincidence cross section as a function of electron angle, Fig(2), and outgoing pion angle, Fig(3).

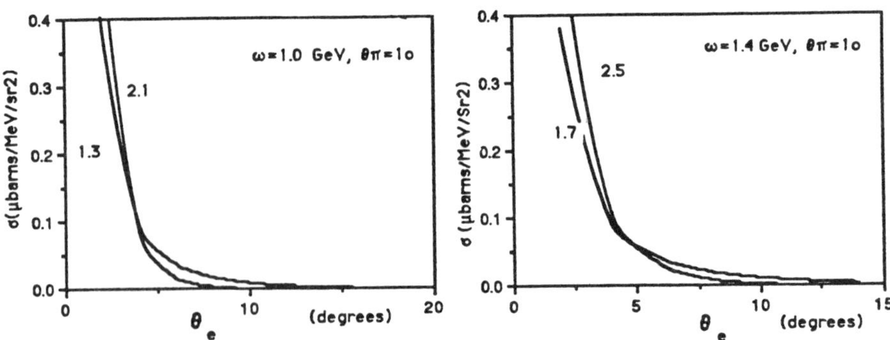

Fig(2): The cross section vs. the electron angle for two values of the energy transfer ω and for $\Theta_\pi=1°$.Curves in each panel are marked by the incident electron energy ε_1.

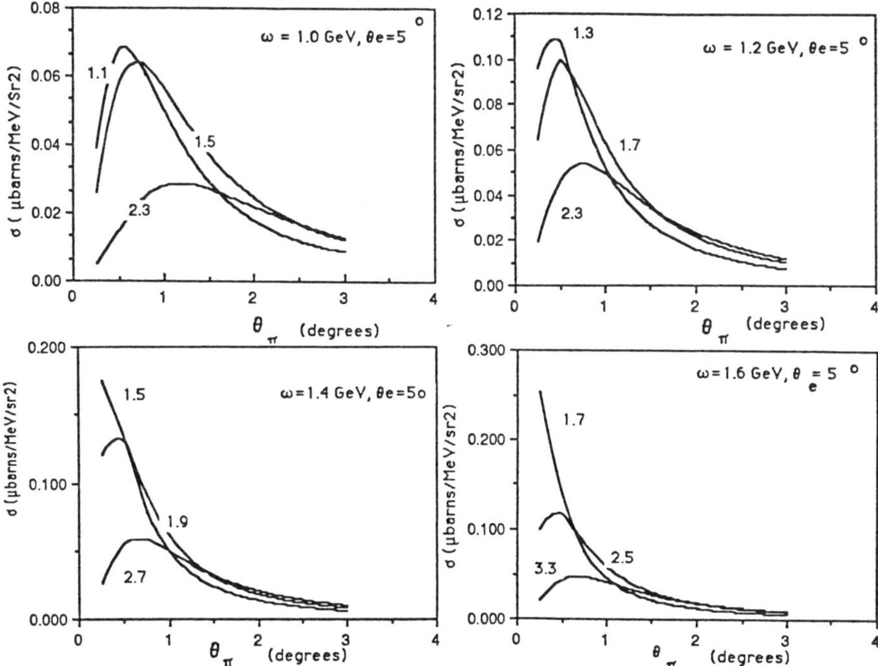

Fig(3): the cross section vs. pion angle for selected values of energy transfer ω. The three curves in each panel are marked by the incident electron energy ε_1.

As expected from inspecting Equ(6), the cross section peaks at extreme forward electron and pion angles. We have taken the pion azimuthal angle $\varphi_\pi=90°$. This 0° peaking of the cross section allows it to be separated from the nuclear photoproduction cross section which starts at zero value and peaks at a pion angle of about 5°(2)

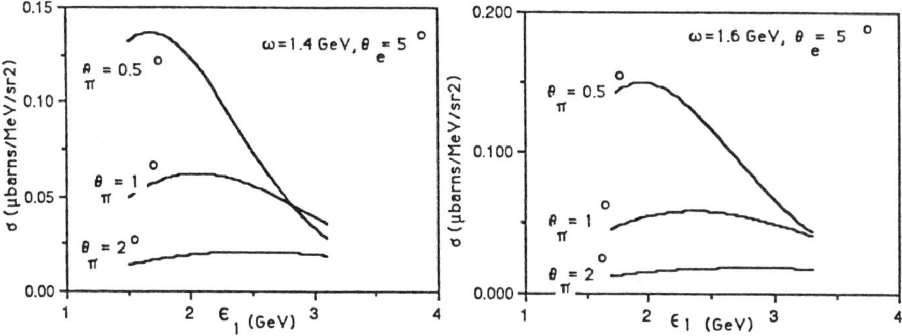

Fig(4): The cross section vs. incident electron energy ε_1.

Fig(4) shows the cross section as a function of electron energy, for $\Theta_e=5°$ and three pion angles. The major observation is that the reaction favors large values of energy transfer but moderate values of four-momentum transfer. The optimum values for ε_1 and $\omega=\omega_\pi$ are above the 1 GeV range.

Finally, Fig(5) shows the sensitivity of this virtual-photon Primakoff effect to the slope of the $\gamma\pi$ form factor, or the radius of the $\gamma\pi$ interaction volume. We show results for two values of a, i.e., a=0.03 derived from VMD considerations, and a=0.1 obtained from the analysis of data on the decay(1) with radiative corrections. It is seen that the sensitivity increases with pion energy. For ω_π in the GeV range and for $\varepsilon_1 \sim \omega_\pi$+600 MeV we encounter optimum conditions for experimental measurements.

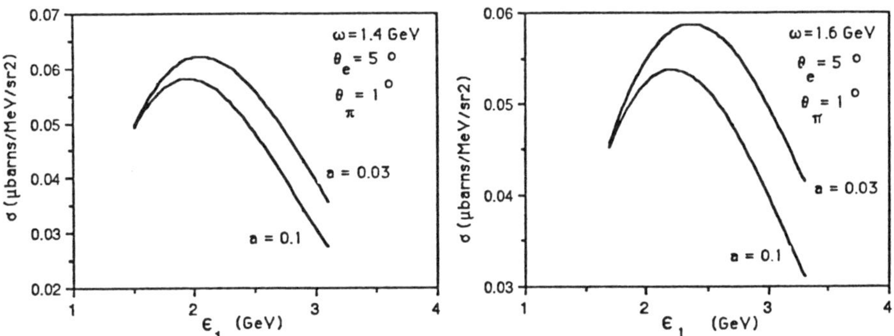

Fig(5): The cross section vs. incident electron energy for two values of the form factor slope, Equ(4).

REFERENCES

1. H. Primakoff, Phys. Rev. 81, 899 (1951).
2. G. Bellettini et al, Nuovo Cimento 40, 1139 (1965) and 66A, 243 (1970); V.I. Kryshkin et al, Soviet Physics JETP 30, 1037 (1970); A. Browman et al, Phys. Rev. Lett. 33, 1400 (1974).
3. V. Glaser and R.A. Ferrell, Phys. Rev. 121, 886(1961); G. Morpurgo Nuovo Cimento 21, 569 (1964); G. Faldt, Nucl. Phys. B43, 591 (1972).
4. K.O. Mikaelian and J. Smith, Phys. Rev. D5, 1763 (1972); G.E. Tupper, T.R. Grosse and M.A. Samuel, Phys. Rev. D28, 2905 (1983).
5. R.W. Dalitz, Proc. Phys. Soc.(London)A64, 667 (1951)
6. J. Fischer et al, Phys. Lett. 73B, 359 (1978)
7. C. Berger et al., Phys. Lett. 142B, 125 (1984)

Quasi-elastic Electron Scattering from ^{3}H and ^{3}He

K.A. Dow
Bates Linear Accelerator Center
Massachusetts Institute of Technology, Middleton, MA 01949
and The Bates Tritium Collaboration

ABSTRACT

Inelastic electron scattering cross sections of ^{3}H and ^{3}He have been measured at the MIT-Bates Linear Accelerator Center, at incident energies from 65 to 790 MeV, and two scattering angles. Longitudinal and transverse response functions have been extracted for 3-momentum transfers between 200 and 550 MeV/c. These response functions are compared for the two nuclei, and the single-particle nature of the quasi-elastic process can be seen. The response functions also are compared with theoretical calculations. The Coulomb sum rule is formed for both nuclei; the results agree with an exact theoretical calculation. Indications of saturation and two-proton correlations in the initial state are seen.

INTRODUCTION

In inclusive inelastic electron scattering, there is a broad "quasi-elastic" peak whose centroid is located at approximately the same electron energy loss as that for electrons scattered elastically from a nucleon at rest.[1] The peak is shifted by the nucleon binding energy, and broadened by the Fermi motion. Thus, quasi-elastic scattering samples the momentum distribution of the nucleons in the nucleus.

The cross section may be decomposed into longitudinal (scattering from charges) and transverse (scattering from spins and currents) contributions. In general, theoretical calculations predict more longitudinal strength than is seen experimentally, and less strength in the transverse response in the region between the quasi-elastic and delta excitation peaks (the dip).[2] These calculations are made in the Plane Wave Impulse Approximation. Attempts to treat additional effects, such as nucleon final state interactions (FSI) and meson exchange currents (MEC), have not produced agreement.

The three-body systems are interesting because there are exact solutions of the non-relativistic wave equation for the ground state. These wave functions describe the elastic form factors reasonably well.[3] Also, ^{3}H and ^{3}He are members of an isospin doublet, so any theory should describe both, only changing the third component of isospin. Work is beginning on solving the wave equation for the continuum of final states. One might also hope that MEC and FSI are more amenable to treatment in the three-body systems than in heavier nuclei.

The Coulomb sum rule (CSR) may be formed by integrating the longitudinal response function over energy loss. The CSR is a property of the ground state, and may be calculated exactly for the three-body nuclei, where bound state wave functions are available.[4,5] Previous experiments on heavier nuclei have not shown saturation of the CSR (at the target charge Z), even at high momentum transfer ($q \geq k_F$, where k_F is the Fermi momentum) where correlation effects should be small.[6,7] One problem may be that there is unmeasured strength for heavier nuclei, due to the width of the quasi-elastic peak.

This is not as severe a problem for light nuclei, where the peak is relatively narrow.

This paper presents a portion of the results of an inclusive electron scattering experiment performed at the MIT-Bates Linear Accelerator Center, using the energy-loss spectometer system.[8] Elastic[9] and inelastic[10,11,12] cross sections for ^{3}H and ^{3}He were measured at two angles (54° and 134.5°) and a range of incident and final energies. In the region of the quasi-elastic peak, Rosenbluth separations were made for $200 \leq q \leq 550$ MeV/c. The ^{3}He response functions agree well with previous measurements made at Saclay.[13] These response functions are compared with theoretical calculations. The Coulomb sum rule has been formed and is compared to a calculation.

RESULTS

Response functions at q =500 MeV/c are shown in Figure 1. Random uncertainties are shown; the systematic uncertainties are approximately the same. Also shown are calculations by the Hannover[14], Rome[15] and Illinois[16] groups. The Hannover and Rome calculations assume single-nucleon ejection, and use a spectral function to describe the initial momentum and energy of the struck nucleon. The ground state wave functions are solutions of the non-relativistic three-body Faddeev wave equation. The inclusive response functions are formed by integrating over all hadronic final states. MEC, FSI and effects of relativistic order are not included in these calculations. As is typical of calculations for heavier nuclei, the longitudinal response is overestimated. The transverse response is described fairly well at the peak, and underestimated in the dip region (which is not surprising, since MEC and Δ production are not considered). Both calculations fall below the experimental responses on the low-ω side of the peak, at kinematics suggestive of knockout of a proton-neutron pair. Evidence for this has been seen in a ^{3}He(e,e′d) experiment.[17]

The Illinois calculation includes the elastic channel and orthogonalizes the final state channels. Thus, some FSI effects are included, although not those due to rescattering of the struck nucleon. This calculation describes the ^{3}He longitudinal response quite well; the ignored FSI effects are expected to be small at this relatively high momentum transfer. However, the calculated longitudinal response for ^{3}H is not as successful. This calculation does not treat the 3-body final state exactly. These approximations affect ^{3}H more than ^{3}He, due to the relative importance of the 2-body final states. The Illinois group has not yet performed a calculation of the transverse response functions.

Figure 2 shows the ratio of ^{3}H to ^{3}He response functions at 400 MeV/c. The simple "single-particle" ratio (the ratio expected in the single-nucleon ejection model) is shown, as well as the ratios of the theoretical calculations. The ratios of the Rome and Hannover calculations are nearly identical, and are shown as one line. The data agree with the single-particle ratio at the quasi-elastic peak (indicated by the arrow). In the transverse dip, the experimental ratio may be tending towards 1. This might indicate scattering from quasi-deuterons, of which there are equal numbers in the two nuclei. A ^{3}He(e,e′p) experiment in the dip region also shows evidence of this.[18]

The Coulomb sum rules are shown in Figure 3. To form the experimental sum rule, the longitudinal response is divided by the square of the proton

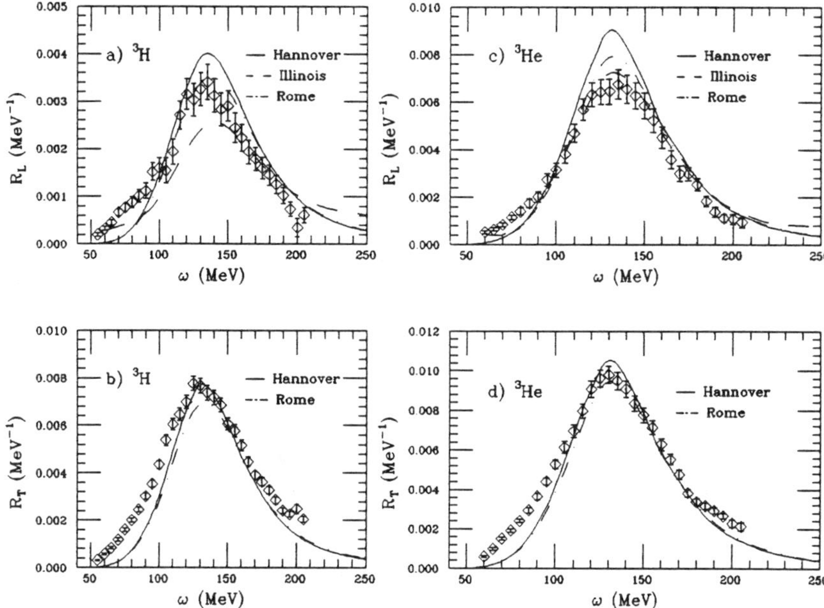

Figure 1. Longitudinal and transverse response functions for ^{3}H and ^{3}He at q =500 MeV/c.

charge form factor (Höhler's parameterization 8.2[19]), to obtain the 'point nucleon' response, and then an integration is performed over all ω:

$$\Sigma_L(q)_{exp} = \sum_{\omega_{min}}^{\omega_{max}} \frac{R_L(q,\omega)\Delta\omega}{[G_E^p(Q^2)]^2\left(\frac{1+Q^2/4M_N^2}{1+Q^2/2M_N^2}\right)},$$

where $Q^2 = q^2 - \omega^2$. The correction to the proton form factor is suggested by deForest in order to account for relativistic effects due to the motion of the proton in the nucleus.[20] This decreases the denominator by as much as 7%. The lower limit of integration was taken to be the energy loss for the two-body breakup threshold, as this is an inelastic sum rule. ω_{max} was chosen such that $R_L(q,\omega_{max}) = .01(R_L(q,\omega_{peak}))$; this meant that extrapolation was required at some momentum transfers. In these cases, an exponential tail was fit to a section of the high-ω side of the quasi-elastic peak. The uncertainty on $\Sigma_L(q)$ is the average of the fractional systematic uncertainties on the response function points.

The Saclay data were summed using the same technique; the results are shown in Figure 3. The curves labeled 'Illinois' are the results of exact calculations performed by the Illinois group using only ground state wave functions.[5]

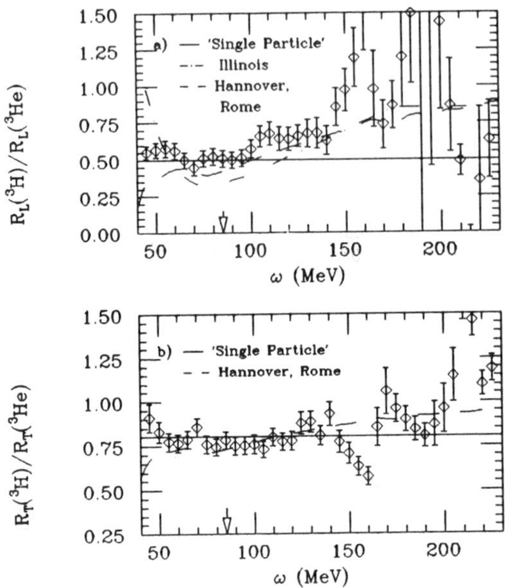

Figure 2. Response function ratios at $q = 400$ MeV/c.

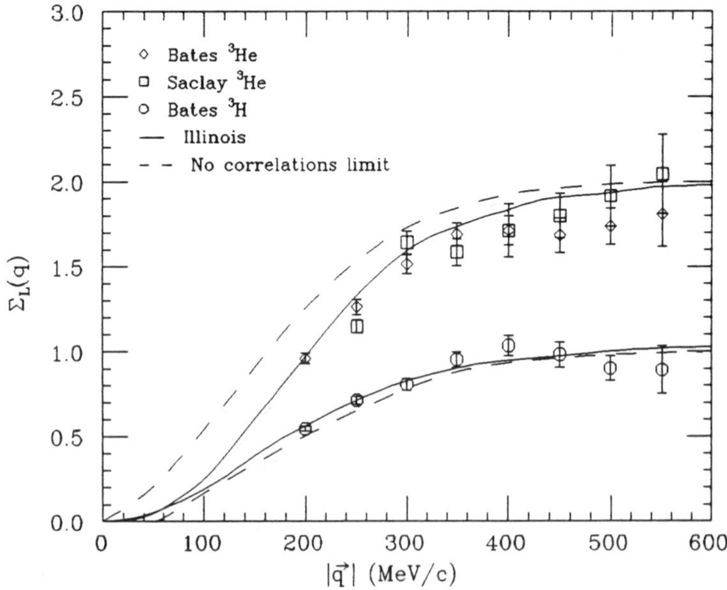

Figure 3. Coulomb sum rules for ^{3}H and ^{3}He.

The curves labeled 'No Correlations' were calculated using the elastic charge form factors measured in this experiment:[9]

$$\Sigma_L^{nc}(q) = Z\big(1 - |F_{ch}(q)|^2/G_E^p(Q^2)^2\big),$$

where G_E^p is again taken from Höhler, and $Q^2 = q^2 - \omega_{qe}^2$. This is the inelastic sum rule in the absence of correlations.

The data approximately saturate the sum rule at high momentum transfer. They agree with the exact calculation of the Illinois group; the ^{3}H experimental sum rule and the Illinois calculation are close to the no-correlation limit. This, taken with the large difference between the no-correlation limit and the ^{3}He data, is evidence for two-proton correlations.

CONCLUSION

Quasi-elastic response functions have been obtained for ^{3}H and ^{3}He. The ^{3}H data are new; the ^{3}He data are in good agreement with previous measurements from Saclay. Simple 'quasi-elastic' models agree with the transverse response functions at the quasi-elastic peak, but fall far below the data in the dip region, where MEC, Δ production and FSI will make up at least some of the difference. However, these models predict too much strength in the longitudinal response functions. A more complete calculation from the Illinois group agrees well with the ^{3}He longitudinal response, but not with the ^{3}H response. This improved agreement with ^{3}He and the failure to reproduce the ^{3}H experimental response are the result of their treatment of the final state wave functions.

The ratio of the data agrees with the single-nucleon ejection model in the region of the quasi-elastic peak; the theories do not. In the dip region, the experimental transverse ratio is approximately 1, suggesting scattering from a p-n pair.

Both nuclei satisfy the Coulomb sum rule, and are in agreement with the Illinois calculation of the sum rule. This, and the agreement of the Illinois calculation with the ^{3}He longitudinal response, suggests that the disagreement between theory and experiment in the quasi-elastic response lies in the treatment of continuum wave functions. MEC will also be important in the transverse response; treatments of FSI should improve agreement in the tails of the quasi-elastic peak.

This work has been supported in part by the U.S. Department of Energy and the National Science Foundation, and the Canadian Natural Sciences and Engineering Research Council.

REFERENCES

1. E. J. Moniz, Phys. Rev. **184**, 1154 (1969), E. J. Moniz, et al. Phys. Rev. Lett. **26**, 445 (1971), R. R. Whitney, et al. Phys. Rev. **C9**, 2230 (1974).
2. P. Barreau, et al. Nucl. Phys. **A402**, 515 (1983), Z. E. Meziani, et al. Phys. Rev. Lett. **52**, 2130 (1984), M. Deady, et al. Phys. Rev. **C33**, 1897 (1986), C. C. Blatchley, et al. Phys. Rev. **C34**, 1243 (1986).
3. See, for example, W. Strueve, et al. Nucl. Phys. **A405**, 620 (1983).
4. T. deForest and J. D. Walecka, Adv. in Phys. **15**, 1 (1966).
5. R. Schiavilla, et al. Univ. of Illinois preprint ILL-(NU)-86-60 (1986), submitted for publication.

6. K. W. McVoy and L. van Hove, Phys. Rev. **125**, 1034 (1962).
7. Z. E. Meziani, Nucl. Phys. **A446**, 113c (1985).
8. W. Bertozzi, et al. Nucl. Inst. Meth. **162**,211 (1979), and references therein.
9. D. Beck, et al. Phys. Rev. Lett. **59**, 1537 (1987).
10. K. Dow, Ph. D. thesis, MIT, unpublished (1987), and K. Dow, et al., to be published (1988).
11. T. S. Ueng, Ph. D. thesis, University of Virginia, unpublished (1988).
12. G. Retzlaff, Ph. D. thesis, University of Saskatchewan, in progress.
13. C. Marchand, et al. Phys. Lett. **153B**, 29 (1985).
14. H. Meier-Hajduk, et al. Nucl. Phys. **A395**, 332 (1983).
15. C. Ciofi degli Atti, E. Pace and G. Salmé, private communication (1987).
16. R. Schiavilla and V. R. Pandharipande, Univ. of Illinois preprint ILL-(NU)-87-43 (1987).
17. P. H. M. Keizer, et al. Phys. Lett. **157B** (1985) 255.
18. C. Marchand, et al. Phys. Rev. Lett. **60** (1988) 1703.
19. G. Höhler, et al. Nucl. Phys. **B114**, 505 (1976).
20. T. deForest, Nucl. Phys. **A414**, 347 (1984).

NUCLEAR PHOTOPRODUCTION OF PIONS

Justus H. Koch
NIKHEF-K, Amsterdam
and
Institute of Theoretical Physics
University of Amsterdam

The interpretation of photon- (or electron-) induced reactions at intermediate energies is closely tied to pion induced reactions. The dominant feature in both cases is the excitation of the Δ-resonance. Consequently, photonuclear and pion-nucleus reactions can be used to study the behaviour of the Δ-resonance inside the nucleus. Photon and pion are in several respects complementary probes for this purpose. First, due to the much stronger π-N interaction, pionic reactions tend to take place in the nuclear surface. The photon, on the other hand, has a much weaker interaction and reaches the entire nuclear volume. Another important difference is the Δ-excitation mechanism: the elementary $\pi N \rightarrow \Delta$ step is a longitudinal spin transition, while $\gamma N \rightarrow \Delta$ is a transverse magnetic excitation. This difference in the elementary excitation mechanism leads to different sensitivity to the Δ-nucleus dynamics. It is, for example, mainly responsible for the different shapes (energy dependence) of the total pion- and photon-nucleus cross sections [1].

Many experiments planned for the new high duty factor electron accelerators will concern the behaviour of the Δ-resonance in nuclei. There exist, however, already several measurements concerning one particular subset of reactions at intermediate energies: the nuclear photoproduction of pions to discrete nuclear states in the resonance region, which allows us to test our present understanding of the Δ-nucleus dynamics.

We will here focus on the resonant pion photoproduction mechanism. In the production to discrete final states, the character of the nuclear transition can be used to emphasize this aspect and therefore to increase the sensitivity to medium modifications of resonant production. Examples will be given for light nuclei for which recent data are available.

The production mechanism: In addition to the dominant resonant production, there is also a sizeable non-resonant photoproduction mechanism on a nucleon. It is mainly important in charged pion production. The major part of this non-resonant amplitude is spin-dependent, while the resonant production through Δ-excitation procceeds mainly through a spin-independent term. These features of the single nucleon amplitude provide a handle to emphasize the resonant production on nuclei: one chooses π^0 photoproduction, where the non-resonant terms are small. Or, in charged pion photoproduction, one picks a nuclear transition or kinematical situation where the spin-transition is suppressed.

In free space, the properties of the Δ are well understood in terms of its coupling (decay) to the πN channel. When produced inside a nucleus, the nuclear medium modifies its properties and propagation significantly. A convenient formalism, which allows to include the Δ-nucleus dynamics and apply it consistently to a variety of intermediate energy reactions, is the Δ-hole formalism. In deciding which of the many dynamical aspects to include, this formalism takes it's cue from the most important decay channels of the Δ-residual nucleus system in pion-nucleus reactions. As known from experiment, the largest contributions to the total π-nucleus cross section are due to quasi-free scattering, elastic scattering and pion absorption.

In the coupling to the quasi-free decay channel of the Δ-resonance, the strong energy dependence makes it necessary to take, for example, binding effects or nucleon Fermi motion into account. Another medium effect influencing the quasi-free decay is Pauli blocking, which reduces the phase-space for the $\Delta \rightarrow \pi N$ decay inside the nucleus, quenching the free width and pushing the resonance energy up. Furtheremore, the Δ decay can occur such that the nuclear ground state is reached again. In pion induced reaction, this is the elastic scattering channel, while in photonuclear reactions it is the coherent π^0 photoproduction channel. It couples very well to excitations initially created by a pion and only poorly to the photon induced transverse Δ-hole excitations. There is therefore a lot of elastic scattering - and a large elastic decay width - for π-nucleus reactions. Photon-nucleus reactions do not have this elastic broadening, leading to a much narrower resonance shape of the total photon-nucleus cross section. Finally, another important mechanism is the coupling to the pion absorption channel through many body mechanisms such as $\Delta N \rightarrow NN$. This coupling is not well understood and one hopes to learn more about it by studying both π- and γ-nucleus reactions with their different density dependence. In the Δ-hole approach, all channel-couplings are included in the propagator of the intermediate Δ-(A-1) nucleus system. It should be stressed that the many body effects enter in the production amplitude on two levels: they modify the resonant production step $\gamma N \to \Delta \to \pi N$ and the subsequent distortion of the outgoing pion through $\pi N \to \Delta \to \pi N$. The medium modification of the production step is absent in DWIA calculations. A consistent treatment of the dynamics of both production and distortion is crucial for a meaningful interpretation of nuclear (γ,π) data.

Coherent π^0 photoproduction: One way to emphasize the resonant production is to look at π^0 photoproduction. Differential cross sections have been measured for ^{12}C and ^{4}He. Fig. 1 shows data for coherent π^0 production on ^{12}C at $k_\gamma = 240$ MeV. The solid curve shows the result of a calculation [2] in the Δ-hole approach, which also provides a good description of π-^{12}C scattering. The agreement between theory and experiment [3] is quite good for small production angles. The dashed line shows what happens if one arbitrarily omits the coupling to the absorption channel in the production mechanism. Intermediate coupling to this channel clearly yields a significant damping of the production step. Since part of the data (dots) were taken with an energy resolution of only 10 MeV, the cross section at larger production angles can be explained by incoherent production, where ^{12}C is left in one of its

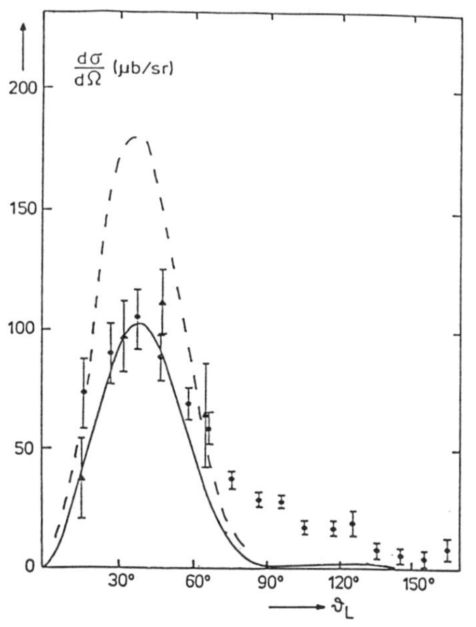

Fig. 1

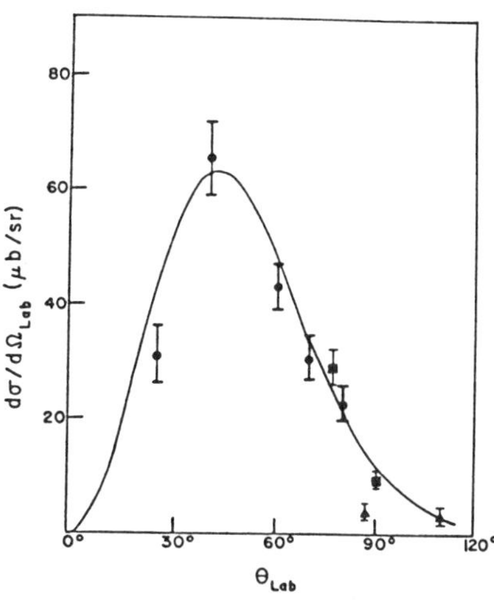

Fig. 2

lowest excited states. The fact that incoherent production can be this large compared to coherent state production is due to the sinθ-dependence of the coherent γ,π amplitude, which suppresses forward production, where the elastic formfactor is large, and favors large angle production, where the inelastic formfactors are sizeable. Good agreement with the same model was also found for ^{4}He [4], shown in Fig. 2. In general, the available π^0 data are therefore well understood and seem to indicate that the large medium effects on nuclear Δ-propagation are adequately described.

Charged pion photoproduction: Charged pion production receives a sizeable non-resonant contribution, which is mainly spin-dependent. For studying resonant nuclear production, one therefore has to try to suppress the spin contribution in order to emphasize the spin-independent Δ-contribution. In photoproduction of a π^+ on ^{12}C, for example, the $J^\pi T = 1^+1$ groundstate of ^{12}N corresponds to an M1 spin transition and is only weakly sensitive to the Δ-nucleus dynamics. The transition to the 2^+1 state, 1 MeV above the ground state, has a large longitudinal spin-independent component. Neglecting the intermediate coupling to the absorption channel in the production step, for example, increases the cross section by a factor five at resonance. At present, measurements studying $(\gamma,\pi^\pm)$ to states other than the residual nuclear groundstate have not been carried out. In the future the study of (γ,π) reactions to different final states in the same target where the pion distortion is almost the same will provide a good test of theoretical models for the production process.

A suitable (γ,π) reaction to the residual ground state is ^{14}N$(\gamma,\pi^+)^{14}$C. It is well known that the Gamov-Teller matrix element for this nuclear transition is strongly suppressed. This implies at low energies a suppression of the usually dominant non-resonant s-wave production through a $\sigma\cdot\varepsilon$-term, making the production amplitude sensitive to the other (momentum-dependent) parts of the amplitude. Similarly, at higher energies it will enhance the role of the Δ. Measurements at low energies were in good agreement with calculations [5,6] which used nuclear wavefunctions constrained by electron scattering. Several measurements [7] have been carried out in the resonance region and here the agreement with the calculations [6,8] is rather poor. The most detailed calculation is shown in Fig. 3. It is difficult to conclude which aspect of the theoretical model is most likely responsible. There is a large cancellation in the non-resonant amplitude, leading to the dip predicted near $\theta = 30^o$, which makes details of the non-resonant production important. In the resonant

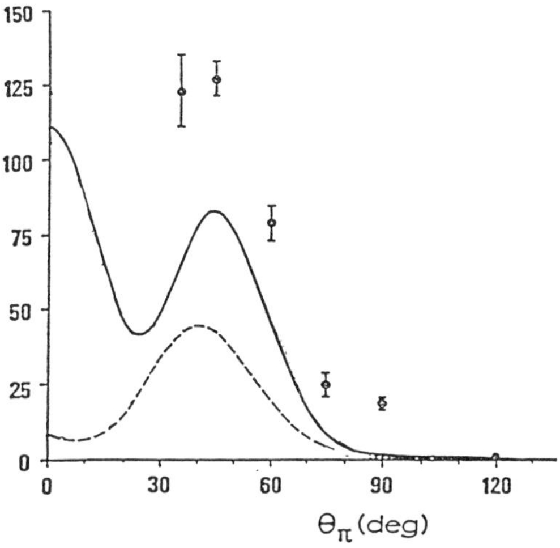

Fig. 3

production, indicated by the dashed curve, there are several large medium effects. These large modifications, however, also tend to cancel [6]. Also $\gamma N \rightarrow \Delta$ vertex corrections or, for example, two-step mechanisms were found [6,7] to be too small to explain the discrepancy between theory and experiment. To further study the Δ-dynamics in this reaction, it will be helpful to measure the pion charge exchange reaction $^{14}C(\pi^+,\pi^0)^{14}N$, which connects the same nuclear states, in the resonance region.

Another example for enhancing resonant production by suppressing the spin component is provided by the $^{13}C(\gamma,\pi^-)^{13}N$ (g.s.) reaction [9]. The nuclear transition has a C0 and M1 part. By carrying out measurements at a momentum transfer of $Q = |\mathbf{k}_\gamma - \mathbf{q}_\pi| = 1\ fm^{-1}$, where the spin dependent M1 formfactor has a deep minimum, the C0 production becomes enhanced and resonant production should dominate. Fig. 4 shows the measurements compared to a calculation [6] in the Δ-hole approach (dash-dotted: CO production; solid: total). The agreement is very poor. Another measurement at $Q = 1.25\ fm^{-1}$, where non-resonant production dominates, can also not be explained. Just as in the case of ^{14}N, there is at present no indication which part of the theoretical assumptions is responsible for this failure. Certainly, the explanation of the (γ,π) data for ^{13}C will be closely coupled to explaining the pion single charge exchange data on the same target, $^{13}C(\pi^+,\pi^0)^{13}N$(g.s.), which have posed a longstanding theoretical puzzle.

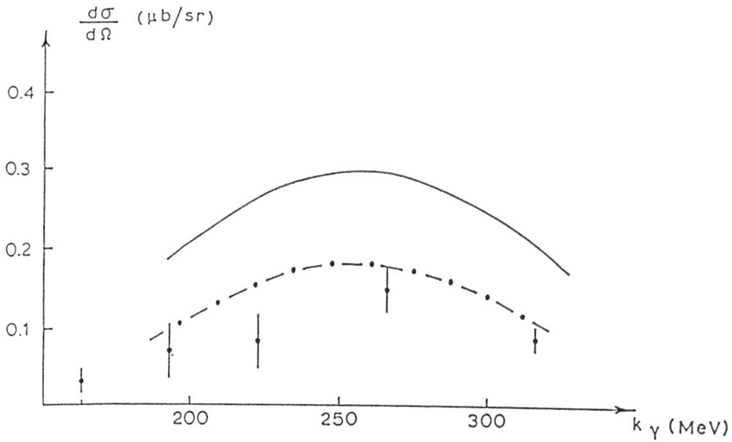

Fig. 4

Summary: The picture emerging from the presently available nuclear photoproduction data in the resonance region is very puzzling. On one hand, coherent π^0 photoproduction data can be explained very well by an approach which takes into account large medium modifications of Δ-propagation inside the nucleus. This reaction is rather closely related to elastic pion scattering, which the same theory describes very well and has been used to constrain some of the model parameters. Furthermore, non-resonant production plays only a minor role in this reaction. On the other hand, charged pion photoproduction at energies around the Δ-resonance, where also non- resonant mechanisms are important, are not well described by the same theory. For the presently available data - even under conditions where the resonant production was enhanced by choosing an appropriate transition or kinematics - non-resonant production still yields a sizeable contribution. This makes it difficult to clearly trace the failure of the calculations for charged pions to a shortcoming in the description of the Δ-dynamics, the non-resonant mechanism or possibly the nuclear wavefunction. Besides the information from electron scattering, a further analysis of nuclear photoproduction has to also include pion reactions (elastic scattering, charge exchange) on the same target to separate the different aspects.

This work is part of the research program of the National Institute for Nuclear Physics and High-Energy Physics (NIKHEF, section K), made possible by financial support from the Foundation for Fundamental Research on Matter (FOM) and the Netherlands Organization of Scientific Research (NWO).

References

1. J.H. Koch, E.J. Moniz and N. Ohtsuka, Ann. Phys. **154**, 99 (1984)
2. J.H. Koch and E.J. Moniz, Phys. Rev. **C20**, 235 (1979); **C27**, 751 (1982)
   T. Takaki, T. Suzuki and J.H. Koch, Nucl. Phys. **A443**, 570 (1985)
3. J. Arends et al., Z. Phys. **A311**, 367 (1983)
4. D.R. Tieger et al., Phys. Rev. Lett. **53**, 755 (1984)
5. B.H. Cottman et al., Phys. Rev. Lett. **55**, 684 (1985)
   K. Röhrich et al., Phys. Lett. **153B**, 203 (1985)
6. T. Suzuki, T. Takaki and J.H. Koch, Nucl. Phys. **A460**, 607 (1986)
7. P.K. Teng et al., Phys. Lett. **177B**, 25 (1986)
8. L. Tiator et al., Mainz preprint 1988
9. P.C. Dunn et al., Phys. Lett. **B196**, 434 (1987)

MEASUREMENT OF $^{16}O(\gamma, n_0)$ AT MEDIUM ENERGIES

E.J. Beise, G.W. Dodson, M. Garçon‡, S. Høibråten, C. Maher,
L.D. Pham, R.P. Redwine, W. Sapp, K. Wilson, and S.A. Wood†
Massachusetts Institute of Technology Cambridge, MA 02139
M. Deady
Department of Physics, Bard College, Annandale-on-Hudson, NY 12504

ABSTRACT

Exclusive differential (γ, n) cross-sections on ^{16}O have been measured at $E_\gamma = 150$, 200 and 250 MeV for $\theta_L = 50°$, 60° and 90°. This measurement is the first of its type to be reported for the energy region above the pion production threshold. Calculations of the (γ, n) reaction indicate that large effects are expected from meson exchange currents. The (γ, n) data will be compared with the predictions of such calculations and with available (γ, p) data.

INTRODUCTION

Exclusive (γ, N) reactions, where the residual nucleus is left in its ground state or a low lying excited state, are of interest primarily because of the large difference in momentum between the incoming probe and the outgoing nucleon. Within the framework of a simple quasi-free knockout picture, the photon is probing the high momentum components of the nucleon wave function. The existing (γ, p) data[1-3] below 100 MeV are described reasonably well by quasi-free knockout[4], but the (γ, n) data[5-7] and the (γ, p) data[8-10] above 100 MeV are underestimated by at least an order of magnitude, even when recoil effects and charge exchange are included in the calculations[11].

The similarities in both magnitude and shape of the existing (γ, p) and (γ, n) data suggest that a two-body mechanism may be a better description. Several different approaches have been taken, from a simple quasideuteron calculation[6,12], to self-consistent RPA calculations[13,14] with long-range N-N correlations, to calculations stressing the dominance of either non-resonant[15] (meson exchange currents) or resonant[16] (Δ production) pion exchange. All can qualitatively account for the similarities between (γ, p) and (γ, n) below 100 MeV. Only ref. 15 has made quantitative predictions for (γ, n) above the pion production threshold; until now the only (γ, n) data[17] available in this region have been on ^{4}He. Presented here are $^{16}O(\gamma, n_0)$ data between 150 and 250 MeV.

EXPERIMENTAL APPARATUS

This experiment has not only provided the first set of (γ, n) data in this energy region, but has demonstrated the feasibility of a rather novel technique for measuring neutrons at an electron facility, namely use of a recoil proton spectrometer (see fig. 1).

Neutrons enter the recoil proton spectrometer through a Pb filter and pass through 10 plastic scintillator conversion planes, where protons are produced from $p(n,p)n$ backward scattering. The protons are then momentum analyzed using two MWPCs, the dipole magnet, and a set of focal plane detectors consisting of

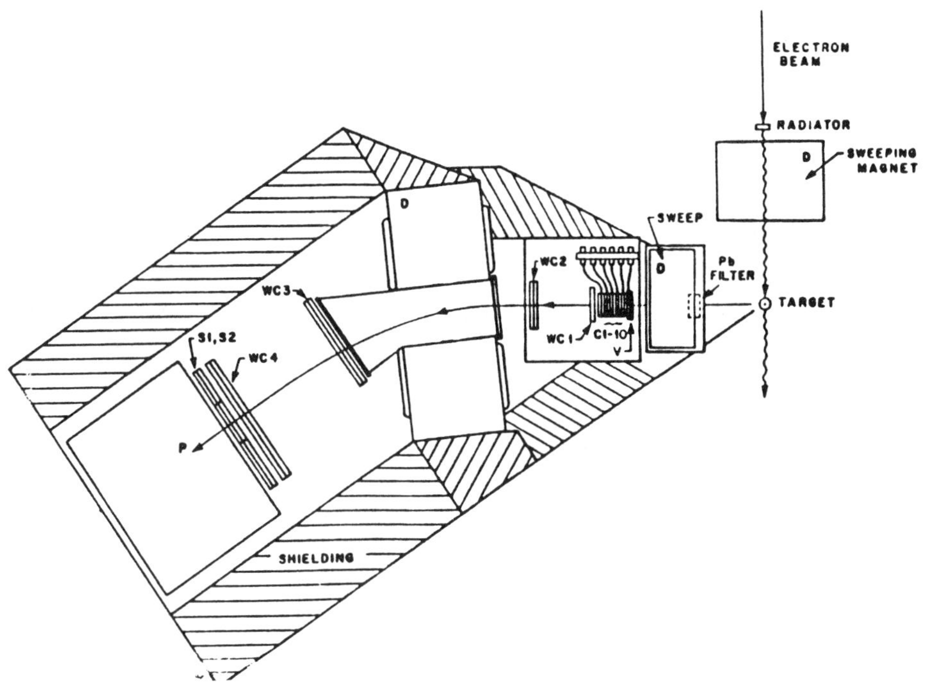

Figure 1. The Bates Recoil Proton Spectrometer

two more MWPCs and two sets of trigger scintillators. By tracing the proton trajectory back to the conversion point, the (n, p) scattering angle is reconstructed, and thus the neutron energy and angle are known. The exclusive (γ, n) cross section to the residual ^{15}O ground state is then deduced from the number of neutrons in the top 5 MeV of the neutron energy spectrum.

The resolution of the spectrometer was estimated from a detailed Monte Carlo calculation to be between 3 and 5 MeV (FWHM). The momentum resolution was checked by measuring $^{12}C(e, e')$ and was consistent with the Monte Carlo result. The efficiency-solid angle product, also determined from the Monte Carlo calculation, was energy and angle dependent but was typically $2{\times}10^{-3}$ msr. The efficiency is mainly determined by the backward angle (n, p) differential cross section and the proton angular acceptance. The maximum proton angle accepted by the dipole was 12°, which guaranteed that the highest energy protons had to come from the H in the scintillator and not from ^{12}C.

The experiment was performed at the Bates linear accelerator in April 1987, with nine data points at $E_\gamma = 150, 200$ and 250 MeV and $\theta_{LAB} = 49°, 59°$ and 88°. This was the first experiment to use the High Flux Pure Photon Facility. A 4% Tungsten radiator was placed in the electron beam 1 m upstream from the experimental target, a plastic cell 10 by 10 cm in area and 5 cm thick filled with H_2O. In between the target and radiator, the electron beam was dumped vertically into a dump approximately 7 m below beam height, shielded by 3 m of concrete. The photon beam continued uncollimated to the target, and was, because of the short target to radiator distance, well contained within the target area. The typical photon flux was $2{\times}10^{10}$ photons/sec in the top 5 MeV, about a factor of 1000 improvement over the collimated low-flux facility previously available at Bates.

RESULTS AND DISCUSSION

The extracted cross sections are still preliminary, but the general trend of the data is that above 100 MeV (γ, n) and (γ, p) are still comparable in magnitude, with (γ, n) actually increasing over (γ, p) for increasing photon energies. Gari and Hebach suggest in their calculation that the dominant contribution to the cross section is meson exchange currents, and that at higher photon energies and forward angles one might expect the (γ, n) cross sections to be higher because of interferences in the (γ, p) process between the quasi-free knockout and both meson exchange currents and final state correlations. The correlation term contributes more to the (γ, n) process because of charge exchange. Qualitatively the data agree with the calculation of Gari and Hebach at 150 MeV, but the calculation fails to account for the extra strength at 200 MeV and above.

In conclusion, the $^{16}O(\gamma, n_0)$ process has been measured in the energy region above the pion production threshold using the technique of a recoil proton spectrometer. Preliminary results show that the (γ, n) cross sections continue to have a shape similar to the (γ, p) cross section, with larger (γ, n) cross sections in the Δ resonance region.

This work was supported under DOE contract #DE-AC02-76ER03069

†Present Address: University of Illinois, Urbana, IL 61801

‡On leave from DPHN/ME, CEN-Saclay, 91191 Gif-Sur-Yvette Cedex, FRANCE

REFERENCES

1 D.J.S. Findlay and R.O. Owens, Nucl. Phys. **A279** 385 (1977).
2 D.J.S. Findlay and R.O. Owens, Nucl. Phys. **A292** 53 (1977).
3 A.C. Shotter et al., Phys. Rev. **C37** 1354 (1988).
4 S. Boffi, C. Guisti and F. Pacati, Nucl. Phys. **A359** 91 (1981).
5 H. Göringer, B. Schoch and G. Lührs, Nucl. Phys. **A384** 414 (1982).
6 M.R. Sené et al., Nucl. Phys. **A442** 215 (1985).
7 P.D. Harty et al., Phys. Rev. **C37** 13 (1988).
8 M. Leitch et al., Phys. Rev. **C31** 1633 (1985).
9 S. Turley et al., Phys. Lett. **157B** 19 (1985).
10 J.L. Matthews, private communication, and G. Adams et al., 1986 Bates Annual Report, 86 (unpublished).
11 S. Boffi, F. Capuzzi, C. Giusti and F.D. Pacati, Nucl. Phys. **A436** 438 (1985).
12 B. Schoch, Phys. Rev. Lett. **41** 80 (1978).
13 M. Cavinato et al., Nucl. Phys. **A373** 445 (1982).
14 J. Ryckebusch et al., Phys. Lett. **194B** 453 (1987).
15 M. Gari and H. Hebach, Phys. Rep. **72** 1 (1981).
16 J.T. Londergan and G. D. Nixon, Phys. Rev. **C19** 19 (1979).
17 R. Schumacher et al., Phys. Rev. **C33** 50 (1986).

The Longitudinal-Transverse Character of the Reaction $^{12}C(e,e'p)$

G. van der Steenhoven[1,2,*], H. P. Blok[1,2], E. Jans[1], L. Lapikás[1]
and P. K. A. de Witt Huberts[1]
[1] *Nationaal Instituut voor Kernfysica en Hoge-Energie Fysica (NIKHEF-K), Amsterdam, The Netherlands*
[2] *Natuurkundig Laboratorium, Vrije Universiteit, Amsterdam, The Netherlands*
* *Department of Physics and Laboratory for Nuclear Science, Massachusetts Institute of Technology, Cambridge, Massachusetts 02139, U.S.A.*

A longstanding problem in intermediate energy nuclear physics is the quenching of the longitudinal (L) response function with respect to the transverse (T) response function in quasi-elastic electron scattering. This problem was first observed in *inclusive* (e,e') experiments on various nuclei[1,2]. More recently several *exclusive* (e,e'p) experiments have been carried out that also show an anomalous TL ratio in the quasi-elastic region[3,4,5]. In this contribution the results of these coincidence experiments are presented and compared to each other. Moreover, the relation between these (e,e'p) data and some older (γ,p) data[6] is discussed. Since the (γ, p) reaction is purely transverse, whereas the kinematics of most (e,e'p) experiments are dominated by the longitudinal response, the comparison of these two reactions may shed light on the nature of the anomalous TL ratio.

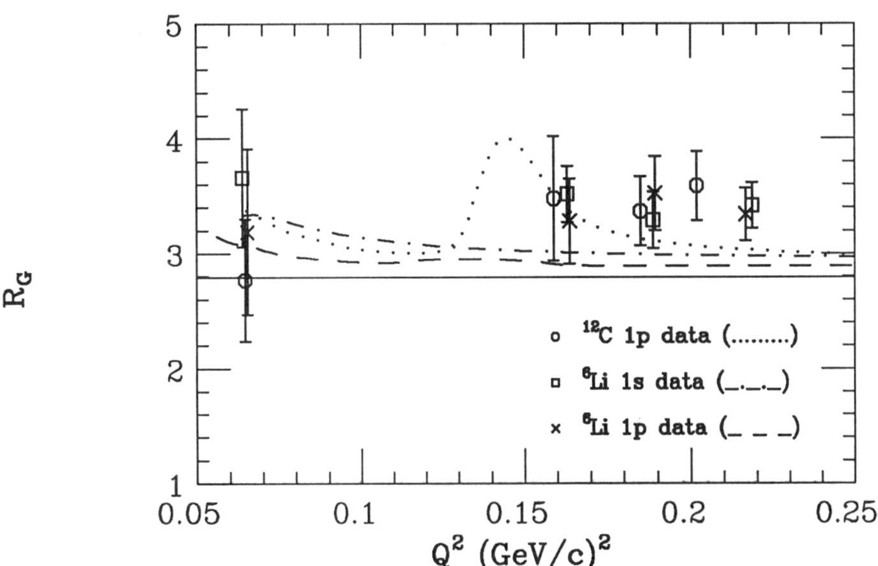

Figure 1. *Measured values of R_G as a function of the four-momentum transfer squared. For a free proton R_G = 2.79 at Q^2 = 0. The curves represent DWIA calculations of R_G.*

In order to study the Transverse and Longitudinal components of the (e,e'p) cross section usually a "Rosenbluth Separation" is performed[4]. This can be done by carrying out an (e,e'p) measurement at two values of the incident electron energy. While the measurement at the low energy has a dominantly transverse character, the one at the high energy is mainly longitudinal. From the data the quantity R_G, which (in PWIA) is the square root of the ratio of transverse and longitudinal response functions, is determined. The results are displayed in fig. 1 together with DWIA curves that represent the differential effect of the final-state interaction (FSI) on the two response functions. The raw data show a similar deviation from $R_G\rfloor_{free}$ in all three cases studied[3,4]: 1p knockout from ^{6}Li and ^{12}C, and 1s knockout from ^{6}Li. On average $R_G\rfloor_{bound}$ deviates from $R_G\rfloor_{free}$ by as much as 14 ± 3 % (after correction for FSI effects).

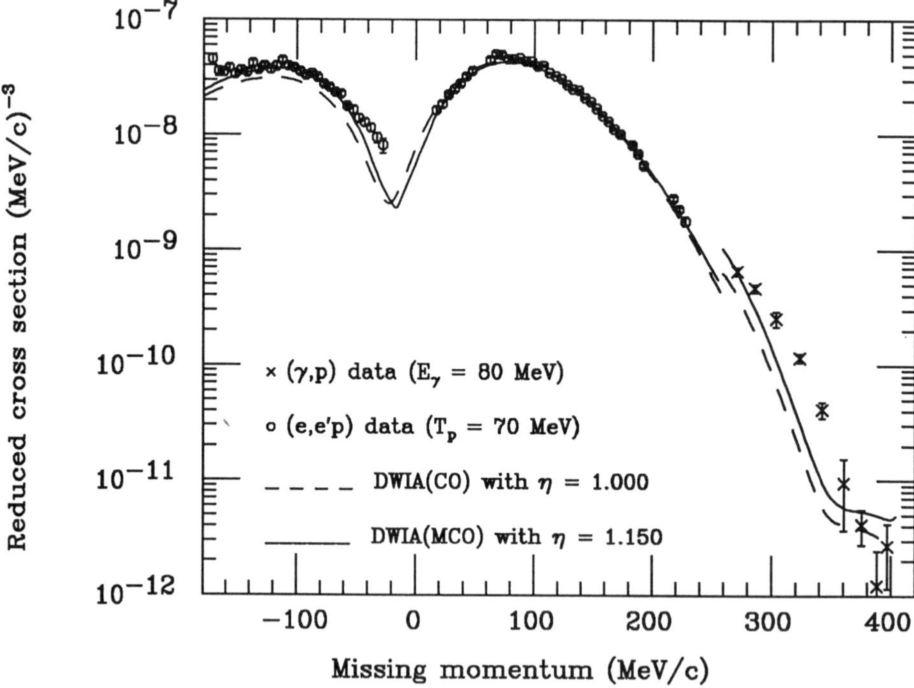

Figure 2. *Reduced cross sections for 1p knockout from ^{12}C leading to $^{11}B_{gs}$. The data below 230 MeV/c are from a ^{12}C(e,e'p) experiment[5], above 230 MeV/c they are from a $^{12}C(\gamma$,p) experiment[6].*

Additional information on η ($\equiv R_G\rfloor_{bound} / R_G\rfloor_{free}$) has been derived from the momentum distributions for 1p knockout from ^{12}C that have been extracted from an (e,e'p) experiment in parallel kinematics[5]. By comparing the data to

DWIA calculations it is possible to obtain precise information on the spectroscopic factor and root-mean-square radius of the orbitals involved and on the anomalous TL ratio[5]. A good description of the 1p knockout data could only be obtained by assuming a 15 $\pm$ 5 % enhancement of η, which is consistent with aforementioned result (see fig. 2).

In both this experiment and the explicit TL study it remains unclear whether the longitudinal response is quenched or the transverse response is enhanced.

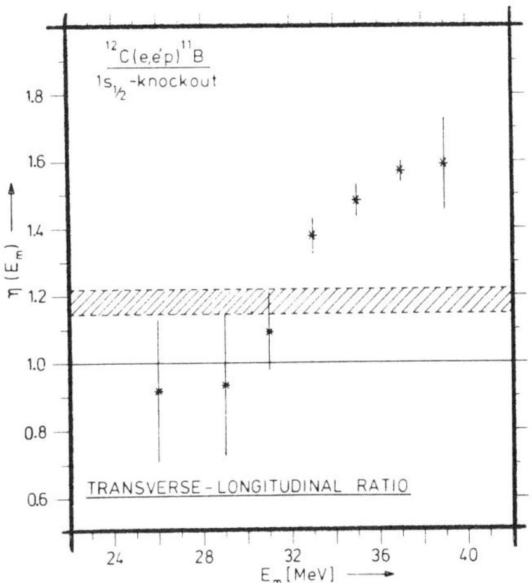

Figure 3. *Anomalous transverse-longitudinal ratio η for 1s knockout from ^{12}C as a function of the mising energy. The two-particle emission threshold is located at about 28 MeV. The hatched area corresponds to the average value of η derived from the momentum distributions for 1p knockout from ^{12}C.*

This same method has also been applied[7] to the momentum distribution for 1s knockout from ^{12}C yielding on average η = 1.40 in the missing energy range from 25 to 40 MeV. The value of η has also been determined for each 2 MeV missing-energy bin, the result of which is shown fig. 3. From this figure it is concluded that a likely mechanism for the relatively large value of η beyond 32 MeV is the onset of (transverse) two-particle emission processes. A similar conclusion was reached by Ulmer *et al.*[8] who performed a Rosenbluth Separation of the two response functions in the 1s-knockout region of the ^{12}C(e,e'p) reaction. Moreover, McGeorge *et al.* [9] observed in a ^{12}C(γ,p) experiment an enhancement relative to a quasi-free calculation that was much larger for 1s knockout than for 1p knockout.

Given this first indication of a similar phenomenon in the (γ,p) reaction, one may now attempt to carry out a more detailed comparison of the two reactions. To that end we applied the accurate description we obtained for the

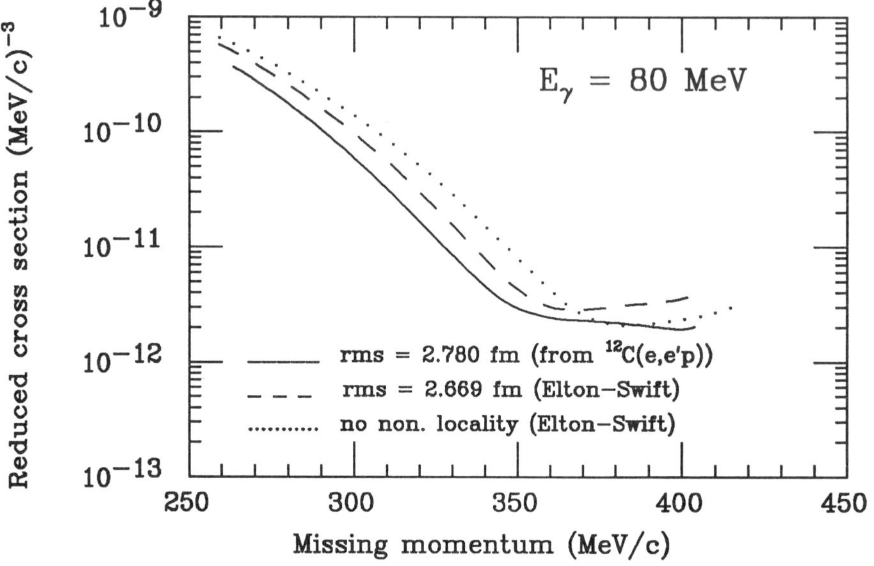

Figure 4. *Differences between previous $^{12}C(\gamma,p)$ calculations[10] and the present ones. The Elton-Swift bound-state wave function is replaced by one with a larger rms and a non-locality correction has been included.*

^{12}C(e,e'p) reduced cross sections[5] to the ^{12}C(γ,p) reaction in the kinematic region ($250 < p_m < 450$ MeV/c) that is closest to the p_m range investigated in the (e,e'p) experiment. Compared to previous direct-knockout calculations of the ^{12}C(γ,p) reaction[10] there are several differences: (i) the parameters of the bound-state wave function are fixed by the description of the (e,e'p) ground-state momentum distribution, while the optical potential is taken from an analysis of (p,p) data[5]; and (ii) a non-locality correction has been included in the calculation of the distorted proton wave. The effect of these differences is a reduction of the (γ,p) cross section (see fig. 4). The new calculations are compared to ^{12}C(γ,p)^{11}B$_{gs}$ data taken at $E_\gamma = 80$ MeV in fig. 2. Both the (γ,p) and the (e,e'p) results are represented in terms of a reduced cross section, which in PWIA corresponds to the square of the overlap function in momentum space. The employed optical potentials are due to Comfort and Karp[11] (CO). We also used a modified version of the same potential (MCO) that incorporates channel couplings to some extent (see ref. 5). It is seen that the use of the MCO potential together with an enhancement of η leads to an improved description of both data sets. In order to be able to discriminate between the effects of the MCO potential and the enhancement of η we also performed a calculations assuming only an η enhancement[5]. The results of these calculations are shown in fig. 5 for several values of E_γ. We observe that an enhancement of η is clearly favored by the data for all shown energies. On the other hand, there is still a sizeable discrepancy between the data and the calculations that increases at lower energies. This might be caused by additional high momentum components in the wave function, or meson-exchange currents, or resonance effects at low energy. However, before drawing such conclusions from the comparison between the (γ,p) and the (e,e'p) reaction it must first be investigated whether the discrepancy is not due to either the incomplete treatment of channel couplings or the neglect of recoil contributions, i. e. processes where the photon couples to the A-1 system and the recoiling proton is detected[12].

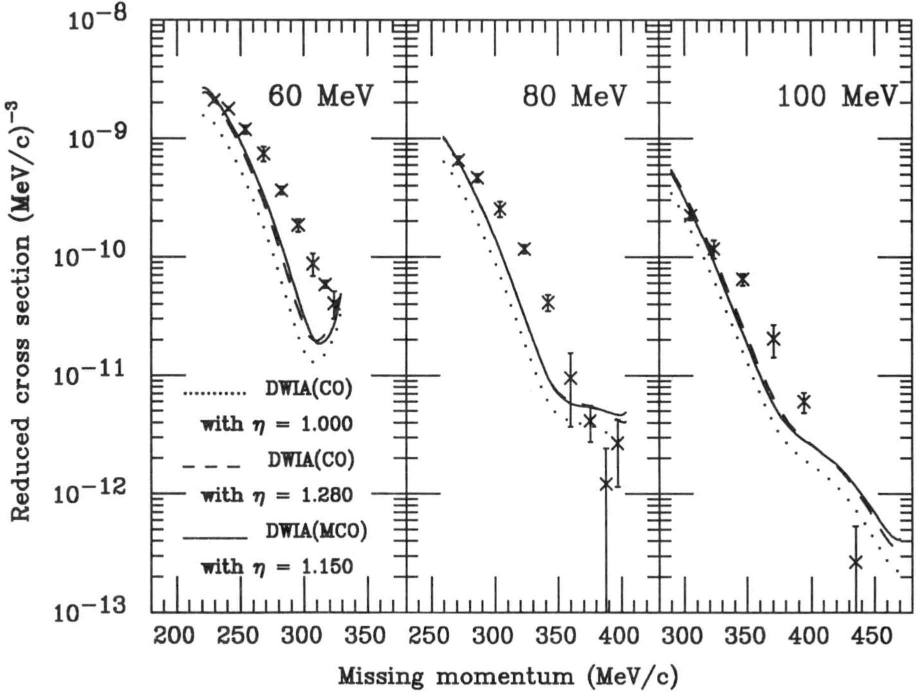

Figure 5. *Quasi-free calculations for the reaction* $^{12}C(\gamma,p)^{11}B_{gs}$ *at three different photon energies. The data are from Matthews et al.*[6]*. The parameters of the calculations are taken from ref. 5*

In summary we have found consistent evidence for an anomalous TL character of the ^{12}C(e,e'p) reaction from a Rosenbluth separation and from an analysis of the momentum distribution. It has also been shown how this anomaly is possibly related to discrepancies between $^{12}C(\gamma,p)$ data and quasi-free DWIA calculations.

References

1. Z.-E. Meziani *et al.*, Phys. Rev. Lett. **52**, 2130 (1982).
2. R. Barreau *et al.*, Nucl. Phys. **A402**, 515 (1983).
3. G. van der Steenhoven *et al.*, Phys. Rev. Lett. **57**, 182 (1986).
4. G. van der Steenhoven *et al.*, Phys. Rev. Lett. **58**, 1727 (1987).
5. G. van der Steenhoven *et al.*, Nucl. Phys. **A480**, 547 (1988).
6. J. L. Matthews *et al.*, Nucl. Phys. **A267**, 51 (1976).
7. G. van der Steenhoven *et al.*, accepted for publication in Nucl. Phys. (1988).
8. P. E. Ulmer *et al.*, Phys. Rev. Lett. **59**, 2259 (1987).
9. J. C. McGeorge *et al.*, Phys. lett. **179B**, 212 (1986).
10. S. Boffi, C. Giusti and F. D. Pacati, Nucl. Phys. **A359**, 91 (1981).
11. J. R. Comfort and B. C. Karp, Phys. Rev. **C21**, 2162 (1980).
12. S. Boffi, R. Cenni, C. Giusti and F. D. Pacati, Nucl. Phys. **A420**, 38 (1984).

PROGRESS REPORT ON THE BATES PARITY EXPERIMENT

P. A. Souder, D.-H. Kim, K. S. Kumar, M. E. Schulze
Syracuse University, Syracuse, NY 13210

M. S. Lubell
City University of New York, New York, NY 11011

J. S. Patch, R. Wilson
Harvard University, Cambridge, MA 02138

G. W. Dodson, K. A. Dow, M. Farkhondeh, J. Flanz,
K. Isakovich, S. Kowalski, C. P. Sargent, W. Turchinetz
Massachusetts Institute of Technology, Cambridge, MA 02139

G. D. Cates, V. W. Hughes, R. Michaels, H. R. Schaefer
Yale University, New Haven, CT 06520

ABSTRACT

We are performing an experiment at the MIT Bates Linear Accelerator Center to measure the parity-violating asymmetry in the elastic scattering of polarized electrons from carbon. The asymmetry, which is due to the isoscalar piece of the neutral weak current, is predicted to be about one part per million for our experimental conditions. We have recently obtained preliminary data with polarized electrons. This paper discusses the status of the experiment, with emphasis on our understanding of the most important systematic errors.

INTRODUCTION

The study of parity violation in neutral weak currents has played a vital role in our understanding of the electroweak interactions. Historically, the observation of parity violation in the deep inelastic scattering of polarized electrons from deuterium[1] as well as in the study of heavy atoms[2] led to the acceptance of the SU(2)×U(1) structure of the standard model[3]. Recently, precise data from the Cs atom[4] has led to important limits on physics beyond the standard model.[5]

At the MIT Bates Linear Accelerator Center, we are measuring parity violation in a new domain; namely, the elastic scattering of polarized electrons from carbon nuclei. The parity-violating asymmetry is defined as

$$A = (\sigma_R - \sigma_L)/(\sigma_R + \sigma_L),$$

where $\sigma_L(\sigma_R)$ is the differential cross section for the scattering of electrons with left(right) helicity. An attractive feature of this experiment is the lack of ambiguity in the theoretical interpretation of its result. Since ^{12}C is spinless and isoscalar, the relevant nuclear physics may be described by a single form factor which precisely cancels in the asymmetry.

At energies where a phenomenological four-fermion interaction is a good approximation[6,7]

$$A = \tilde{\gamma}\frac{3}{2}G_F Q^2(\sqrt{2}\pi\alpha)^{-1},$$

where G_F is the Fermi coupling constant, α is the fine structure constant, Q is the momentum transfer, and $\tilde{\gamma}$ is the parity-violating coupling constant for an axial vector coupling to the electron and an isoscalar coupling to the hadronic matter. In the standard model, $\tilde{\gamma} = \frac{2}{3}sin^2\theta_w$, where θ_w is the weak mixing angle. Since $\tilde{\gamma}$ is relatively small, it is sensitive to possible extensions of the standard model, such as the existence of extra Z bosons, that have been suggested in the literature.

The most precise measurements of $\tilde{\gamma}$ to date come from the study of heavy atoms, especially Cs, which require the computation of complex wavefunctions for their quantitative interpretation. Because of the importance of the value of $\tilde{\gamma}$, we are motivated to measure it by a completely different method, especially one that relies less on elaborate theoretical calculations.

DESCRIPTION OF THE EXPERIMENT

We have chosen to run the experiment with a beam of energy 250 MeV, a scattering angle of about 35°, and a Q of 150 MeV/c. With a beam polarization of about 35%, we expect an experimental asymmetry $A_{exp} = 0.63 \times 10^{-6}$. Since this number is quite small, it places severe demands on the apparatus, both in terms of acquiring sufficient statistical precision as well as providing adequate control of possible systematic errors.

A schematic diagram of the apparatus is given in Fig. 1. The source, which provides an intense beam of polarized electrons, is based on photoemission of polarized light from a GaAs crystal.[8] Light is provided by a DC Kr ion laser chopped to the 1% duty factor of the accelerator. The high average current, ideally on the order of 60 μA at the target, combined with the moderate average laser power available after chopping, requires that the quantum efficiency of the crystal be maintained at a high level (>3%). Unfortunately, operation at such currents tends to degrade the quantum efficiency, and maintenance of the beam intensity over extended periods of time has proven to be difficult. Beam currents on the order of 30 μA were used for the data described in this paper, although frequent interruptions were required to refresh the crystal.

The polarization of the electron beam is proportional to the circular polarization of the laser light. Hence we can control the helicity by applying a voltage to a Pockels cell in the laser beam. The Pockels cell acts as a quarter wave plate or a three-quarter wave plate, depending upon the applied voltage. The helicity may be switched rapidly to eliminate the effects of slow drifts in the apparatus.

A set of monitors in the beam line measures the unwanted effects of helicity reversal: several toroids measure the beam intensity; four position monitors in front of the target determine the position and angle of the beam; and the energy is analyzed by a position monitor at a place where the beam is dispersed in momentum.

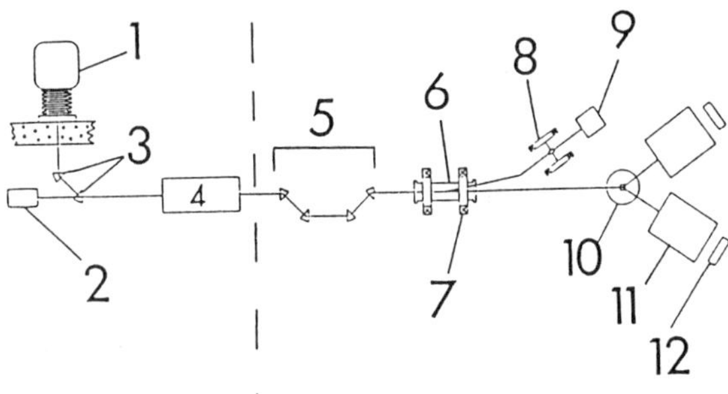

Fig. 1. Schematic of the apparatus, showing the polarized electron source(1), the thermionic source(2), the injection line(3), the linac(4), the instrumented beam line(5), the Møller apparatus(6-9), the target(10), the quadrupole spectrometers(11), and the detectors(12).

The beam impinges on a 5 g/cm^2 carbon target, and the elastically scattered electrons are focused by a pair of single-quadrupole spectrometers onto lucite Cerenkov detectors. Since about 10^5 electrons are detected during each 20 μs burst, individual events are not counted but rather integrated and recorded by 16-bit ADC's.

SYSTEMATIC ERRORS

We have observed two important potential sources of systematic error.

1. The intensity of the laser light reaching the photocathode depends slightly on helicity, causing the energy of the electron beam to depend on the helicity through the effect of accelerator beam loading. Since electromagnetic cross sections depend strongly on energy, the result is a spurious asymmetry.

2. The position of the laser beam depends on helicity, causing the trajectories of the electron beam to be correlated with helicity. Since the number of detected events depends on the position and angle of the beam incident upon the target, spurious asymmetries result. A more insidious problem is the possibility that beam losses due to scraping at various places in the accelerator and beam line will be helicity dependent. The result is a combination of systematic errors with a complex structure that is especially difficult to analyze.

The importance of these effects is illustrated in Table I, which presents some of the first data we have obtained. Here the raw asymmetry is unreasonably large. On the other hand, we note that the differences in the beam monitors are nonzero by many standard deviations. This suggests that it is possible to correct the data. Ideally, if we can eliminate the systematics of the beam at the level of sensitivity of our monitors, these systematic corrections will be negligible.

We correct the raw asymmetries with the use of the equation $A_{exp} = A_{raw} - \Sigma a_i \delta M_i$, where A_{raw} is the uncorrected asymmetry, δM_i are the differences in the beam monitors correlated with helicity, and the a_i are correction coefficients. Data are obtained while the steering coils in the beam line are ramped. This information is used to compute the correction coefficients involving the position and angle of the beam. Since there are large, real fluctuations in the beam current and hence the energy, we are able to use a correlation analysis to extract the coefficient involving energy. (We plan to introduce an energy vernier to independently determine the sensitivity of our apparatus to energy.)

A different approach for detecting and eliminating systematic errors relies on the reversal of the helicity of the beam by an independent method. For our experiment, we can change the direction of the linear polarization of the laser light incident on the pockels cell. The effect is to change the sign of the parity-violating asymmetry without altering the contribution of most of the unwanted effects. Comparing the result for each linear polarization checks for the absence of these problems, and averaging the results cancels them out.

PRESENT STATUS

We have made considerable progress in isolating and reducing the systematic errors apparent in the early data shown in Table I. One important error is related to the deflection of the laser beam induced by the voltage applied to the Pockels cell. We have found two techniques that reduce the effect. In one approach, we alter the position of the Pockels cell laterally in the laser beam until the deflection is minimized. In a second approach, we use a lens to image the Pockels cell onto the photocathode so that the angular changes do not alter the position of the emitted electrons. Each method by itself reduces the problem significantly. In fact, we employ both methods simultaneously.

Another important systematic error is the correlation of intensity with helicity. Such a correlation is easy to produce by the PITA (polarization induced transport asymmetry) effect, in which the transmission efficiency of the light from the Pockels cell to the photocathode depends upon helicity. An error in the voltage on the Pockels cell results in a phase change different from 90° which in turn produces light that is slightly elliptical instead of circular in polarization. The transmission of elliptically polarized light through an optical system generally depends on the direction of the principal axis of the ellipse, giving rise in our case to a helicity-dependent light intensity on the GaAs crystal.

A convenient feature of the PITA effect is that it can be controlled. By making small changes on the voltages applied to the Pockels cell, we can change the appropriate phase and in turn control the asymmetry as shown in Figure 2. Here we see a straight line that passes through zero asymmetry. Such a response is ideal for a slow feedback loop. Moreover, it can be used to eliminate intensity correlations of unknown origin. The PITA effect however, is very sensitive to the Pockels cell voltage, the temperature of various optical elements, and the alignment of these elements. Great care must be taken to achieve acceptable stability.

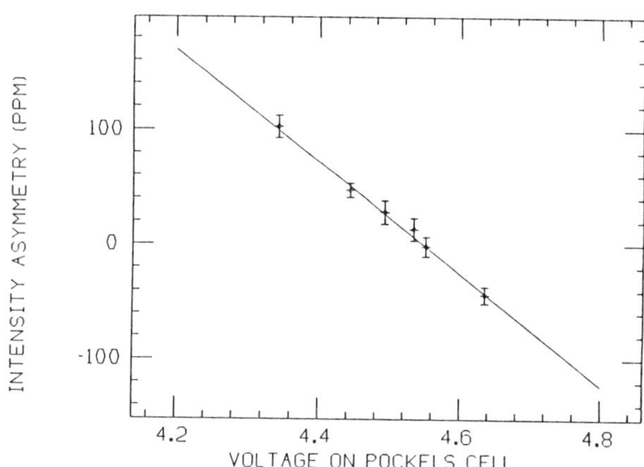

Fig. 2. The PITA effect measured by a diode in front of the vacuum window.

Before we reduced the effect of the helicity-angle correlation, we found that the size of the helicity-intensity correlation depended upon where the intensity was measured. This effect, which presumably is due to beam scraping (as suggested above), would make the simple feedback based on the PITA effect ineffective because only one intensity monitor could be nulled at a given time.

Recently we have obtained data under more favorable conditions. The helicity-angle correlation of the laser beam was reduced by the procedures already outlined. The helicity-intensity correlation was computed on-line and used to adjust the Pockels cell voltage after each half-hour run, thereby nulling the average helicity-intensity correlation. The correction coefficients a_i were measured simultaneously with the data taking.

The preliminary results for about 15 hours of data recorded on tape are given in Table II. These data demonstrate the effectiveness of our methods for reducing systematic errors. The beam monitor differences are consistent with zero, indicating ample control of the beam under helicity reversal. As a result, the corrections to the raw asymmetry are smaller than the statistical error. The experimental asymmetry is consistent with the prediction of the standard model with a statistical error about equal to the expected effect.

As a further check, we obtained some additional data for which we reversed the asymmetry by rotating the polarization of the light incident on the Pockels cell. The results are consistent with those of Table II although they are of lower sensitivity. Our confidence in our measurement is based on this test as well as the insignificance of the corrections we had to apply to the raw data. Combining all of the runs we have analyzed, we obtain a preliminary number of $A_{exp} = 0.6 \pm 0.7$. We are planning to obtain significantly more data in the near future.

This work is supported in part by the U. S. Department of Energy.

Table I. Early data exhibiting large systematic errors.

MONITOR DIFFERENCES (Standard deviations)	
—Position—	
Upstream x	-38
y	2
Target x	68
y	-40
—Energy—	
	65
—Intensity—	
Accelerated	8
On Target	10
PARITY ASYMMETRY (ppm)	
Raw	20± 5

Table II. Recent data. (preliminary results)

MONITOR DIFFERENCES (Standard deviations)	
—Position—	
Upstream x	0.5
y	-0.5
Target x	-0.8
y	-1.5
—Energy—	
	-0.2
—Intensity—	
Accelerated	0.9
On Target	1.0
PARITY ASYMMETRY (ppm)	
Raw	0.9± 0.8
Corrected	0.8± 0.8

References

1. C. Y. Prescott et al., Phys. Lett. **84B**, 524 (1979).
2. M. A. Bouchiat and L. Pottier, Science **234**, 1203 (1986).
3. S. L. Glashow, Nucl. Phys. **22**, 579 (1961); S. Weinberg, Phys. Rev. Lett. **19**, 1264 (1967): A. Salam in **Elementary Particle Theory: Relativistic Groups and Analyticity** (Nobel Symposium No. 8), ed. by N. Svartholm (Almquist and Wicksell, Stockholm, 1978) p. 367.
4. M. A. Bouchiat et al., Phys. Lett. **134B**, 463 (1984); S. L. Gilbert et al., Phys. Rev. Lett. **55**, 2680 (1985); Phys. Rev. A **34**,792 (1986).
5. U. Amaldi et al., Phys Rev. D **36**, 1385 (1987).
6. G. Feinberg, Phys. Rev. D **12**, 3575 (1975).
7. J. D. Walecka, Nucl. Phys. **A285**, 345 (1977).
8. D. T. Pierce and F. Meier, Phys. Rev. B **13**, 5484 (1976).

PROBING PERTURBATIVE QCD AND NUCLEI WITH LARGE ANGLE ELASTIC SCATTERING*

D.S. Barton, G.M. Bunce, A.S. Carroll, S. Gushue, Y.I. Makdisi
AGS Department, Brookhaven National Laboratory, Upton, NY 11973

S. Heppelmann
Pennsylvania State University, University Park, PA 16802

H.C. Courant, G.Y. Fang, K.J. Heller, M.L. Marshak, M.A. Shupe
University of Minnesota, Minneapolis, MN 55455

J.J. Russell
Southeastern Massachusetts University, North Dartmouth, MA 02747

INTRODUCTION

The experiment reported here (AGS Exp. 834) seems particularly appropriate for a conference on the intersections of particles and nuclear physics. We have performed a new probe of aspects of perturbative QCD (pQCD) and the time evolution of hadrons inside nuclei.[1] Simultaneously, we determined the momentum dependence of nuclear spectral functions.[2] That is, we measured the momentum distribution of nucleons inside nuclei with a sensitivity to components well beyond the edge of the normal Fermi distribution. References 1 and 2 describe this experiment with more care and precision.

MOTIVATION

Since the background of the physicists in the group was in particle physics, our original motivation was from that direction. We have been engaged in a program of studying exclusive two-body reactions at large momentum transfer, generally at $\theta_{cm} = 90°$, corresponding to $p_\perp \approx 2$ GeV/c. There is a remarkable relationship known as dimensional scaling which appears to hold for all exclusive processes at large momentum transfer.[3] First, we consider the differential cross section of pp elastic scattering, shown in Fig. 1a. After decreasing by several orders of magnitude, from the forward direction, $d\sigma/dt$ becomes relatively flat in the central region. This is an indication of short ranged, hard processes, as opposed to the peripheral, soft processes of the forward direction. In Fig. 1b, we look at $d\sigma/dt$ at a fixed angle in the center-of-mass (for example, 90°), and then plot $d\sigma/dt$ as a function of the laboratory momentum, p. To good approximation, we see that the cross section falls as a power of s since the center-of-mass energy squared, $s \approx 2p$. For pp scattering the power law is s^{-10} and for πp scattering it is s^{-8}. Brodsky and Farrar showed in the mid-1970's that this behavior is to be expected of any large angle scattering involving a number of constituents obeying a renormalizable vector field theory such as QED and QCD.[3]

---

* Work performed under the auspices of the U.S. Department of Energy and the National Science Foundation. Presented by A.S. Carroll.

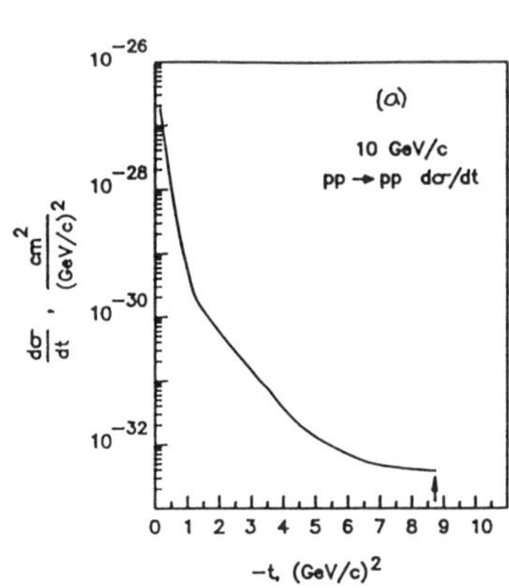

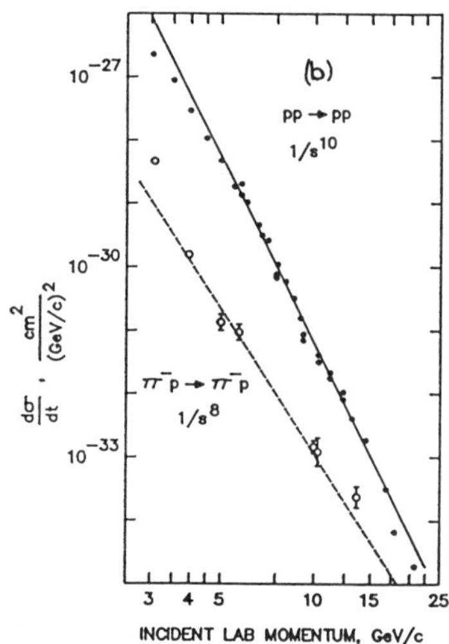

Fig. 1(a) The differential cross section for pp elastic scattering is shown for an incident momentum of 10 GeV/c.

(b) The differential cross sections for pp and $\pi^- p$ elastic scattering at 90° are plotted vs the incident lab momentum to demonstrate the dimensional scaling relationship.

Professor Mueller of Columbia, has presented a nice heuristic picture of what is happening. The elastic scattering of two-point particles such as e^+e^- goes as $1/t^2$. If we study the energy dependence at a fixed angle (a constant ratio of t/s), then $d\sigma/dt \propto 1/s^2$. Mesons and protons are not point particles, however, but are made up of at least two or three constituents, and generally more. A hadron will appear to be a point particle if all of its constituents are contained within a region small compared to a region probed by the interaction. The dimension of that region is $\Delta x \propto 1/\sqrt{t}$. Since Lorentz contraction makes the hadrons appear to be discs, we need only consider areas. So the probability that quarks in both the initial and final state meson will appear to be point-like is (1/t x 1/t). For protons, the probability that the three quarks will be the same area is $(1/t^2 \times 1/t^2)$. Substituting s for t, the power law dependence for πp processes is s^{-8} and pp process is s^{-10}. The general rule is $s^{-(n_a+n_b+n_c+n_d-2)}$. Starting at $p_{lab} \sim 5$ GeV/c, this rule works!

Mueller, and independently, Brodsky, pointed out how this picture could be tested.[4] If the elastic scattering took place on a proton inside a nucleus, then the attenuation of the hadrons leaving the nucleus provides a measure of their size. In more QCD-like language, the quarks with their color charges are very close together at the time of interaction and remain close together for some distance. This distance is somewhat larger than a nuclear radius. In this form, the quark pairs are nearly colorless, so that the hadrons have anomalously small cross sections.

Let me note that there are other pictures of how this exclusive scattering might take place. For example, Landsoff pointed out that there is a singularity when all the quarks stay on shell and are scattered independently to the same angle.[5] In this picture and others, there is no need for the hadrons to be small at the moment of impact. Learning whether or not the interaction selects the size of the hadrons interacting, clearly would be interesting, independent of the particular model.

THE EXPERIMENT

The apparatus used in this experiment was developed to study a variety of exclusive reactions at large momentum transfer.[6] Exclusive reactions are those in which all final state particles are determined. We developed a trigger which required that the momentum of one of the particles be sufficiently close to the elastic limit so as to have a reasonable trigger rate. Since we were studying the elastic scattering of protons on the moving protons inside nuclei, (frequently known as quasielastic scattering), there was only a weak correlation in the angular relationship between the two outgoing protons. Hence, a large aperture system was needed to measure the angle of the recoil proton.

The setup which accomplished these tasks was a magnetic spectrometer for one proton, and a large aperture array of wire chambers to determine the angles of the other, as shown in Fig. 2. There were Čerenkov counters to determine the incident and outgoing particle types.

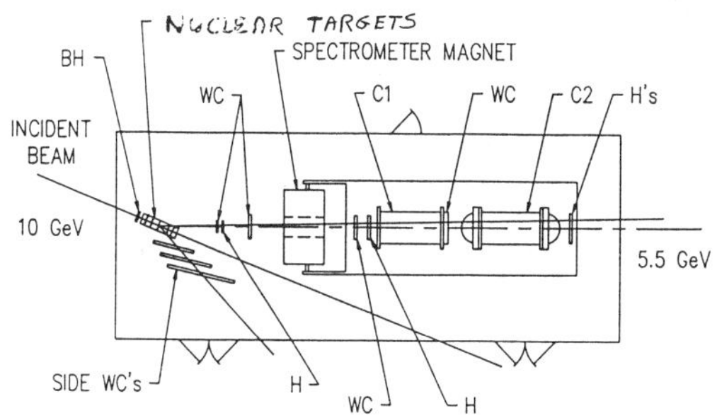

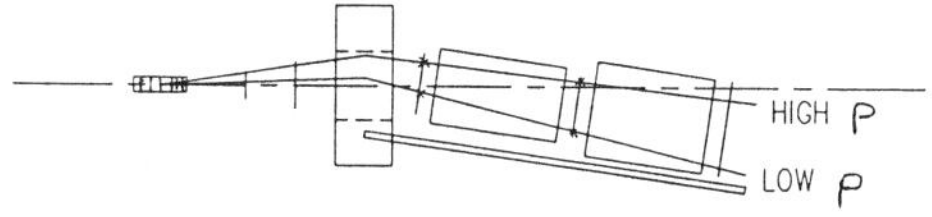

Fig. 2. Top and elevation views of the apparatus used in AGS Exp. 834.

We automatically normalized our experiment by passing the beam through two CH_2 targets which bracket the nuclear targets. Reconstruction of the interaction vertex allowed us to determine which target was struck. The five nuclear targets were Li, C, Al, Cu and Pb.

Two, three layer scintillation counters interspersed with 1.5 rad. length of lead above and below the targets tagged charged particles and gamma rays from inelastic collisions over 2/3 of the solid angle.

NUCLEAR SPECTRAL FUNCTIONS

When we proposed this experiment, the AGS Program Committee had serious doubts as to whether or not we would be able to see the elastic signal of scattering from protons moving inside the nucleus. This happens to be one of those rare occasions when the experiment is easier than it was originally thought to be. In fact, the signal is sufficiently strong and clear as to provide a new technique for studying momentum dependence of nuclear spectral functions, $S(e,\vec{p})$. Because it is extremely difficult to get sufficient energy resolution at these high energies, we simply integrate over all values of the binding energies, e.

Previously this kind of study has been done by scattering electrons of 1 GeV or less from nuclei.[7] The virtues of the simple electron probe have been noted many times before, but in these experiments the proton recoil momenta are comparable to the internal momenta within the nucleus. For incident momenta of 6 and 10 GeV/c, the outgoing momenta are 3.5 and 5.5 GeV/c at $\theta_{cm} = 90°$. A q^2 of 9 $(GeV/c)^2$, probes nuclear distances of ~ 0.2 Fermi, so that the impulse approximation should be justified.

The four kinematic equations that we must solve are

$$E_1 + E_2 = E_3 + E_4, \quad \vec{p}_1 + \vec{p}_2 = \vec{p}_3 + \vec{p}_4$$

where 1 and 2 refer to the beam and target particles and 3 and 4 refer to the outgoing particles.

The energy of the bound target proton, E_2, is set to m_p since the binding energy is small compared to the kinetic energies of the problem. With one unknown outgoing momentum, and the three components of the target momenta unknown, we have a zero constraint equation. A study of the scattering from hydrogen allows us to determine the resolution in the components of the momenta of the struck proton. For the z component, the resolution is 5 MeV/c and for the y component, it is 30 MeV/c.

Restricting the z component to be within ± 200 MeV/c of the origin, clear peak stands out above the background in the p_x-p_y plane. Most of our analysis is done by looking at the y component which is normal to the scattering plane (see Fig. 3). To a good approximation this component is not affected by the kinematics of the interaction and the acceptance varies by < 10% over the range of $|p_y| < 300$ MeV/c. The veto counters above and below the target provide a good method for estimating the background. If we look for events with a penetrating particle (two layers in the veto counter), then the elastic peak disappears. The shape of the background signal is in agreement with the shape of the total distribution for $|p_y| > 500$ MeV/c. After normalizing and subtracting this background, the y component of the target proton momentum is shown in Fig. 4 for five elements at 6 GeV/c and aluminum at 10 GeV/c.

Superimposed on these experimental distributions are the momentum distributions deduced from Fermi gas distribution fits to lower energy electron scattering experiments. In the central region ($|p_y|$ < 200 MeV/c), our distributions are in good agreement with the electron data. However, at momenta greater than Fermi momentum cutoff, we see a significant excess. A first, one might simply ascribe this excess to rescattering of the protons as they leave the nucleus. But the fraction of events within this tail is a constant 20% ± 3% from lithium to copper.

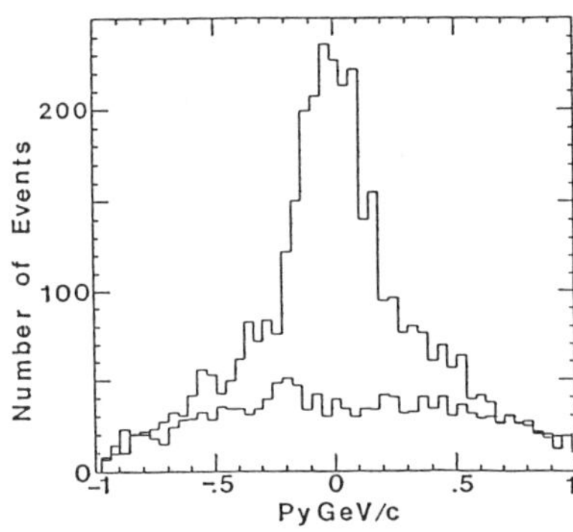

Fig. 3. Distribution of the projection of the y component of the target momenta for aluminum at 6 GeV/c. The upper curve shows the events with 0 or 1 hit in the veto counter out the scattering plane. The lower curve is the background with 2 or more hits.

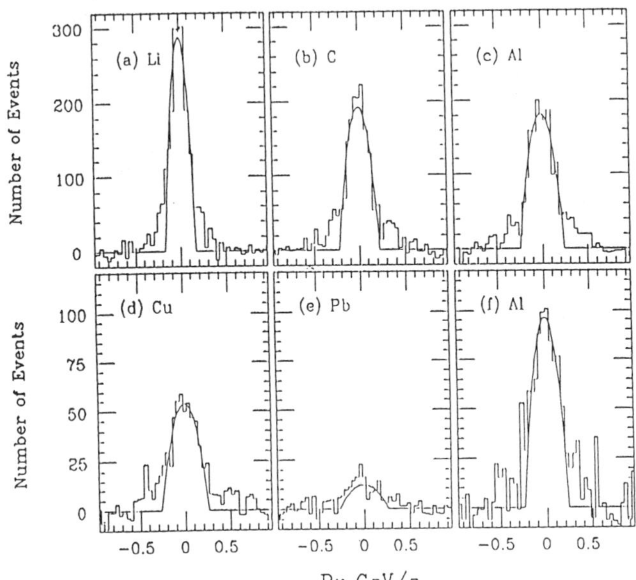

Fig. 4. Distributions of the y components of the target momenta for five elements at 6 GeV/c and aluminum at 10 GeV/c.

TRANSPARENCY

Having determined the momentum dependence of nuclear spectral function, we can turn to the question of determining the transparency of nuclei to large angle pp elastic scattering. In the absence of binding energy and Fermi momentum, the transparency is simply,

$$T = \left[d\sigma/dt(\text{pp elas. in nucleus})\right]/\left[d\sigma/dt(\text{pp elas. in Hyd.})\right]$$

Since the target protons have momenta, we must apply a number of corrections to the data. The measured transparency T which in general is a function of s and hence p_z, is an average over a range in p_z. We must correct for the fact that the acceptance changes over the range of target momenta, and that a particular band in p_z contains only a portion of the spectral function. So the measured T is the ratio of the number of observed elastic scattering events in the nuclei per proton to that from hydrogen, corrected by an integral calculated by the Monte Carlo Method.

$$T = (N_n^{obs}/N_H^{obs}) / \{\int\int\int\int de\,dp_x dp_y dp_z S(e,\vec{p}) A(\vec{p}) [\tfrac{d\sigma}{dt}(s,t)/\tfrac{d\sigma}{dt}(s_o,t_o)]\}$$

where the number of observed particles per target proton is given by N^{obs}, $S(e,\vec{p})$ is the nuclear spectral function and $A(\vec{p})$ is the experimental acceptance. We use measured values of $d\sigma/dt(s,t)$ in evaluating the variation of the differential cross section about the central values of s_o and t_o. The four-fold integral over target momentum and bending energy would be unity if the differential cross sections, $d\sigma/dt(s,t)$ were a constant.

What sort of transparencies do we expect in a conventional picture? The protons scatter elastically and the number observed is reduced by the attenuation length in nuclei of ~ 2.0 Fermi, corresponding to a constant absorption cross section of ~ 36 mb. Wood- Saxon distributions are used to specify the nuclear densities. These calculations result in transparencies of .12 for aluminum to 0.025 for lead. See Ref. 8 for details. The transparencies at 6 and 10 GeV/c as shown in Fig. 5a are both well above the calculated values. Caution should be used in interpreting this result. There is a large uncertainty in the absolute overall normalization (± 25%) due to the uncertainties in treating off-shell scattering, in specifying the nuclear density distribution and in determining the interaction energies.

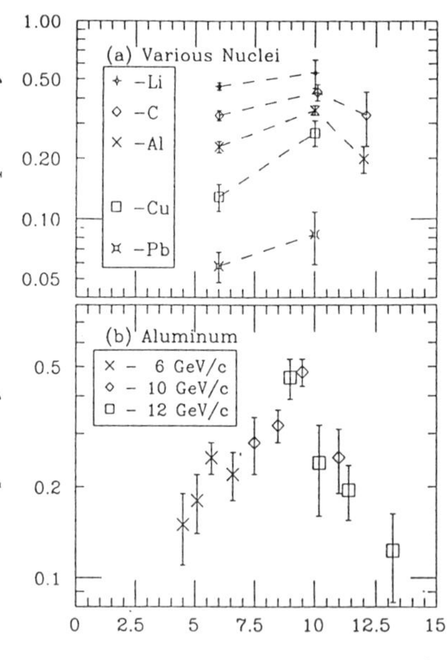

Fig. 5.(a) Transparencies for the five elements measured at 6 and 10 GeV/c, and for C and Al at 12 GeV/c for -0.2 GeV/c $< p_z < 0.1$ GeV/c.

(b) The transparency for aluminum subdivided into four effective incident momenta for each of the three nominal momenta.

Also, the values for 10 GeV/c are significantly higher than for 6 GeV/c. If we plot these transparencies as a function of lab momentum, then we see a consistent rise in all of the nuclei. If we had stopped there, we would have had results which were consistent with the color transparency idea derived from dimensional scaling. However, measurement at 12 GeV/c shows a sudden drop in the transparency indicating that the protons scattered have an increased absorption cross section again. This energy dependence can be seen in more detail for aluminum (see Fig. 5b) where we have used our large number of events to subdivide each incident momentum measurement into four p_z bins. Each bin corresponds to a different s, given by the approximate formula,

$$s \approx s_o \ (1 - p_z/m_p) \quad .$$

Here we seen the transparency rising to a peak at 9 GeV/c and then falling again.

What to make of this result! First, the transparency is energy dependent, a result not expected from normal Glauber theory. Brodsky and de Teramond have suggested a possible model.[9] They say that superimposed on the increasing transparency derived from the hard scattering of the pQCD picture is a soft resonance amplitude possibly associated with the charge threshold which has the normal nuclear attentuation. A soft resonance amplitude unlike the pQCD amplitude implies no decrease from the normal proton size. Such a resonance could also explain the spin dependence in pp elastic scattering.[9] A determination of possible increased charmed production is contained in the data of AGS Exp. 766.[10]

Clearly, it would be useful to go to higher momenta and see if the transparency starts to increase again. Scattering at angles other than 90° would be informative since the cross section there shows much less energy dependence. Finally, interesting comparisons can be made with high energy, large momentum transfer (e,e'p) measurements when these are done at SLAC or CEBAF.

REFERENCES

1. A.S. Carroll et al., "Nuclear Transparency to Large Angle pp Elastic Scattering." Submitted to Phys. Rev. Lett.

2. S. Heppelmann et al., "Momentum Dependence of the Nuclear Quasi-elastic Scattering at Large q^2." Submitted to Phys. Rev. Lett.

3. S.J. Brodsky and G.R. Farrar, Phys. Rev. Lett. 31, 1153 (1973).

4. A. Mueller, Proceedings of the XVII Recontre de Moriond, ed. J. Tran Thank Van, 1982, p. 13; S.J. Brodsky Proceedings of the XIII International Symposium on Multiparticle Dynamics. Eds. W. Kittel, W. Metger, and A. Stergion, 1983, p. 963.

5. P. Landsoff, Phys. Rev. D10, 1024 (1974).

6. B.R. Baller et al. Phys. Rev. Lett. 60, 1118 (1988).
G. Blazey et al. Phys. Rev. Lett. 55, 1820 (1985).

7. S. Frullani and J. Mougey, Advances in Nuclear Physics, eds. J.W. Nagele and E. Vogt, Vol. 14 (1984).

8. G.R. Farrar et al. "Transparency in Nuclear Quasi-Exclusive Processes with Large Momentum Transfer," Rutgers Univ. preprint RU-88-03, to be published.

9. S.J. Brodsky and G.F. de Teramond, Phys. Rev. Lett. 60, 1924 (1988).

10. AGS Exp. 766, Columbia, U. of Massachusetts, Mexico Collaboration, B. Knapp and M. Kreisler, spokesmen (unpublished) (1981).

SPIN-SPIN CORRELATIONS IN PROTON-PROTON COLLISIONS AT HIGH ENERGY AND THRESHOLD ENHANCEMENTS*

GUY F. DE TÉRAMOND
Stanford Linear Accelerator Center,
Stanford University, Stanford, CA 94309

ABSTRACT

The striking effects in the spin structure observed in elastic proton collisions and the Nuclear Transparency phenomenon recently discovered at BNL are described in terms of heavy quark threshold enhancements. The deviations from scaling laws and the broadening of the angular distributions at resonance are also consistent with the introduction of new degrees of freedom in the pp system. This implies new s-channel physics. Predictions are given for the spin effects in pp collisions near 18.5 GeV/c at large p_T^2 where new measurements are planned.

INTRODUCTION

The spin-spin correlation parameter A_{NN} measured in large angle proton-proton collisions at high energy,[1] exhibit structure as a function of large $\sqrt{s}$ and p_T^2. This defies a simple explanation based on the scale invariant nature of the underlying fundamental constituent interactions. As a consequence, a prosperous theoretical industry has developed in an attempt to explain this striking behavior. I have chosen to classify the various models into three categories, according to their predictions for the large-angle behavior of A_{NN} above 12 GeV/c (where the spin effects have not been measured yet),

I) Models where A_{NN} grows to large values[2]

II) Oscillatory models[3]

III) Models where A_{NN} relaxes to the PQCD prediction[4] of $A_{NN} = \frac{1}{3}$.

In the first type I have included models with strong quark correlations, massive quark models, diquark models or models which incorporate other large distance effects.[2] The oscillation of the pp elastic scattering data about the s^{-10} power law has been inspirational for type II models, where an oscillatory behavior of A_{NN} is also expected.[3] As an example, in the model of Pire and Ralston,

*Work supported by the Department of Energy, contract DE–AC03–76SF00515.

one would expect coherent effects in the pp system to arise from the imaginary part of the double log Sudakov corrections. In the third category, I include models such as the present one, based on perturbative QCD (PQCD) which incorporate the onset of new effects due to the opening of a heavy quark threshold at 12 GeV/c. Available data suggest a strong correlation of the spin behavior with other anomalous pp data at 12 GeV/c. The new effect, characteristic of threshold phenomena, would die away at higher energies, the system relaxing to its PQCD behavior.[4,5]

If we closely examine the pp elastic data at 90°, factoring out the PQCD dependence $F^4(t)/sp^2$ (which behaves as s^{-10} at very large s), we find a strict correlation with the A_{NN} data as shown in Fig. 1: a very rapid increase in the total cross section in the 8–12 GeV/c region which tails off at higher energies. This behavior is reminiscent of a threshold effect due to the introduction of new degrees of freedom. The pp elastic angular distribution exhibits the same trend: a rapid anomalous broadening which is maximum at 12 GeV/c [Fig. 2(c)]. This effect also dies away at higher energies. If we take seriously the correlation of anomalous A_{NN} with pp elastic data, this is not encouraging news for the type I models. Finally, the "color transparency" effect,[6] recently discovered at BNL by measuring the energy dependence of absorptive corrections to quasi-elastic pp scattering in various nuclear targets, exhibits an anomalous behavior at 12 GeV/c with respect to the QCD predictions. The QCD transparency disappears and normal attentuation is observed. The above considerations led Stanley Brodsky and myself to propose a simple model based on the opening of new degrees of freedom in the s-channel, which accounts for the observed structure in the pp system.[4]

The PQCD component of the model is the Constituent Interchange Model (CIM),[7] which is the dominant mechanism at large momentum transfer.[8] We do not include diffractive contributions and our model is only valid at large angles where hard scattering is important. To describe the observed structure we need to introduce two broad ($J = L = S = 1$) resonant amplitudes centered at $\sqrt{s} = 2.55$ GeV and 5.08 GeV and a sharp ${}^3F_3$ resonance at $\sqrt{s} = 2.17$ GeV, with widths of 1.6, 1.0 and 0.04 GeV respectively. These amplitudes intefere with the CIM amplitude to reproduce the experimental data. The required $\sqrt{s}$ values correspond to the $pp \to K^+\Lambda p$, $pp \to D^0\Lambda_c p$, and $pp \to \Delta p$ thresholds respectively. Other details of the model are described in Ref. 4.

MODEL PREDICTIONS

The predictions of the model and comparison with experiment are shown in Figs. 2 and 3. As shown in Figs. 2(a) and 2(b), the deviations from the simple scaling predicted by the PQCD amplitudes are readily accounted for by the

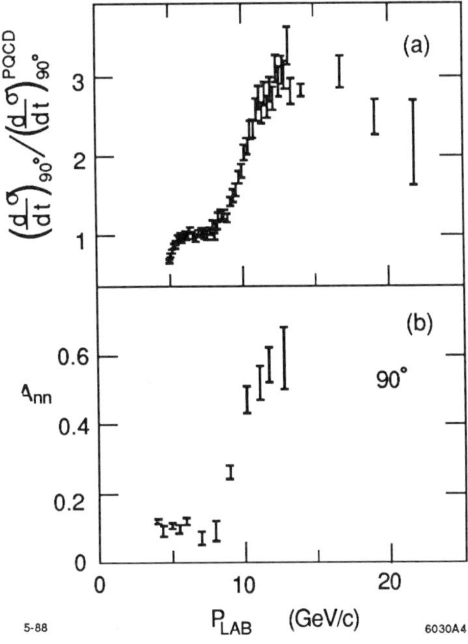

Fig. 1. Onset of threshold effects in $d\sigma/dt$ and A_{NN} at 90° near 12 GeV/c.

resonance structures. In Fig. 2(a) the solid curve is the prediction of the model and the dotted line is the background PQCD prediction. In Fig. 2(b) we show the ratio of $d\sigma/dt$ at 90° to the PQCD prediction. The angular distribution (normalized to the data at $\theta_{\rm cm} = \pi/2$) is predicted to broaden relative to the steeper perturbative QCD form, when the resonance dominates. As shown in Fig. 2(c) this is consistent with experiment, comparing data at $p_{lab} = 7.1$ and 12.1 GeV/c. The solid and dotted lines are predictions for $p_{lab} = 12.1$ and 7.1 GeV/c, respectively. The narrow peak which appears near 1.3 GeV/c in Fig. 3(a) corresponds to the onset of the $pp \to p\Delta(1232)$ channel. The model is also consistent with the recent high energy data point for A_{NN} at $p_{lab} = 18.5$ GeV/c and $p_T^2 = 4.7$ GeV^2 [see Fig. 3(b)]. The data show a dramatic decrease of A_{NN} to zero or negative values. This is explained in our model by the destructive interference effects about the resonance region. The same effect accounts for the depression of A_{NN} for $p_{lab} \approx 6$ GeV/c shown in Fig. 3(a). The comparison of the angular dependence of A_{NN} with data at $p_{lab} = 11.75$ GeV/c is shown in Fig. 2(c). References for the data are given in Ref. 4.

The most striking test of the model is its prediction for the spin correlation A_{NN} shown in Fig. 3(a). The rise of A_{NN} to $\simeq 60\%$ at $p_{lab} = 11.75$ GeV/c is correctly reproduced by the high energy $J = 1$ resonance interfering with a

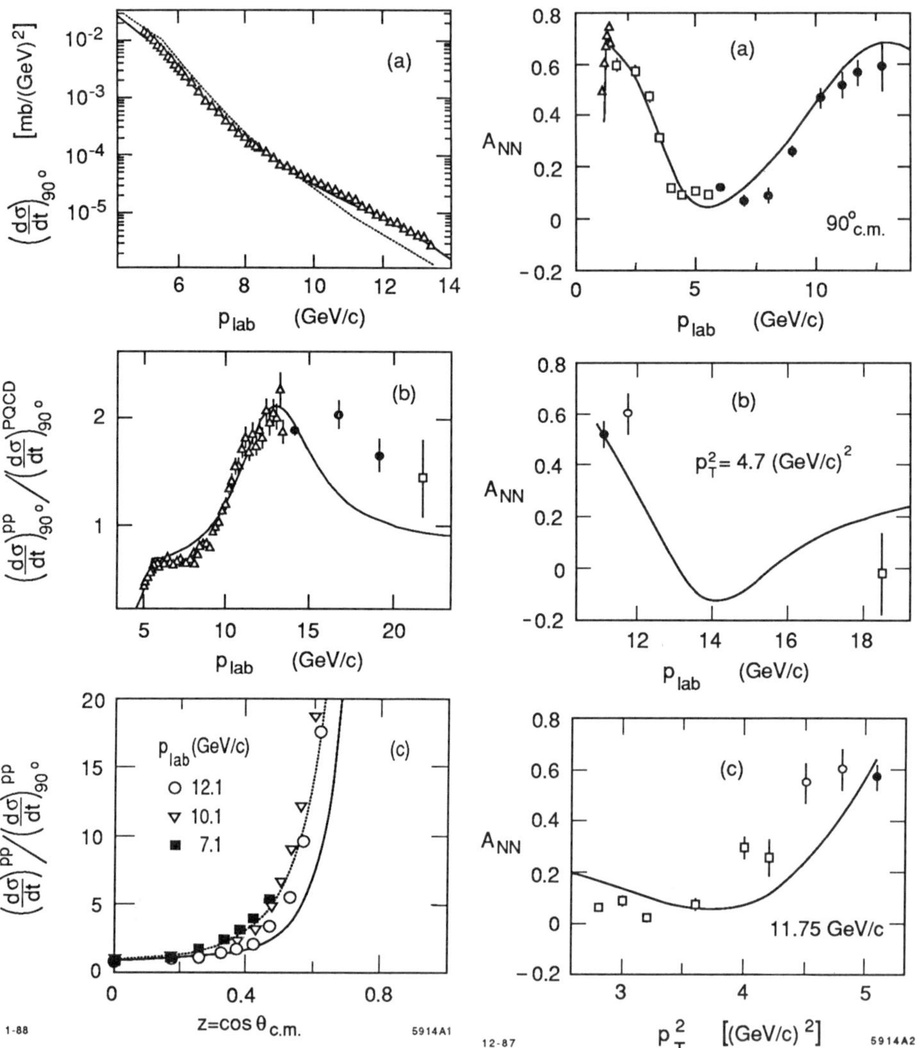

Fig. 2. *pp* elastic cross sections.

Fig. 3. A_{NN}.

PQCD background. In addition using unitarity we predict total charm production cross section of 1 μb near threshold, just below the preliminary limits set by BNL E766.[9] The model prediction for strangeness production near threshold is about 1mb, consistent with experiment (see Ref. 4).

CONCLUSIONS

I have given a brief account of various possible explanations of the anomalous spin-spin correlation data and how different models can be classified according to their prediction of A_{NN} at large angles above 12 GeV/c. Measurements of A_{NN} in this energy region are crucial. Our model predictions above 12 GeV/c at 90° are shown in Fig. 4. We also predict an increase in A_{NN} from 10% at $P_T^2 = 7$ (GeV/c)^2 to 40% at $P_T^2 = 8.2$ (GeV/c)^2 at $P_{lab} = 18.5$ GeV/c.

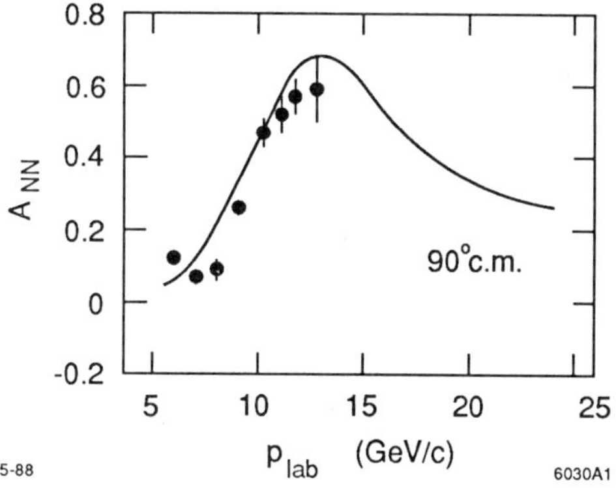

Fig. 4. Model prediction of A_{NN} above 12 GeV/c.

The anomalous pp elastic data is consistent with the onset at 12 GeV/c of a threshold effect due to the opening of new degrees of freedom. The data are well reproduced by the interference of two broad and highly inelastic resonance structures at $\sqrt{s} = 2.55$ GeV and 5.08 GeV. These energies correspond to the threshold value for open strangeness ($pp \to \Lambda K^+ p$) and open charm ($pp \to \Lambda_c D^\circ p$) respectively. The model may be interpreted in terms of actual "hidden-flavor" resonances near the respective thresholds. On the other hand, the successful description of a wide range of data may simply reflect an adequate parametrization of the threshold effects in the NN system arising from opening of new (heavy flavor) inelastic channels.

ACKNOWLEDGEMENTS

Most of this work was done in collaboration with S. J. Brodsky. I wish to thank G. Bounce, W. Dunwoodie and J. Martoff for helpful conversations. I also wish to thank the Guggenheim Foundation for support.

REFERENCES

1. For a review, see A. D. Krisch, UM–HE–86–39 (1987).
2. C. Avilez, G. Cocho and M. Moreno, Phys. Rev. **D24**, 634 (1981); C. Bourrely and J. Soffer, Phys. Rev. Lett. **54**, 760 (1985); M. Anselmino, P. Kroll and B. Pire, Z. Phys. **36**, 89 (1987); S. V. Goloskokov, S. P. Kuleshov and O. V. Seljugin, Proceedings of the VII International Symposium on High Energy Spin Physics, Protuino (1986).
3. A. W. Hendry, Phys. Rev. **D23**, 2075 (1981); B. Pire and J. P. Ralston, in *High Energy Spin Physics*, AIP Conf. Proc. No. 95, p. 347 (1983); S. M. Troshin and N. E. Tyurin, J. Phys., Colloq. **46**, 235 (1985).
4. S. J. Brodsky and G. F. de Téramond, Phys. Rev. Lett. **60**, 1924 (1988).
5. G. R. Farrar, S. Gottlieb, D. Sivers and G. Thomas, Phys. Rev. **D20**, 202 (1979); S. J. Brodsky, C. E. Carlson and H. J. Lipkin, Phys. Rev. **D20**, 2278 (1979).
6. See the contributions of A. S. Carrol and G. R. Farrar to this conference. See also, A. S. Carrol et al., PSU HEP/88–02.
7. R. Blankenbecler, S. J. Brodsky and J. F. Gunion, Phys. Lett. **39B**, 649 (1972).
8. B. R. Baller et al., LK 3848.
9. M. N. Kreisler and C. Avilez, private communications.

Nuclear Quasi-Exclusive Scattering at Large Momentum Transfer

Glennys R. Farrar*
Department of Physics and Astronomy
Rutgers University
New Brunswick, NJ 08903

In collaboration with Leonid Frankfurt, Huan Liu, and Mark Strikman

ABSTRACT

The A dependence of nuclear quasi-exclusive scattering at large momentum transfer is modeled. If "color transparency" applies, these processes should provide an excellent means for experimentally determing the spectrum of 4-momenta of the bound nucleons, giving access to regions of large 4-momenta which are otherwise inaccessible to observation.

*Supported in part by NSF PHY-84-15535-05

I. Introduction

Nuclear quasi-exclusive scattering is defined to be a large momentum transfer lepton-nucleon or hadron-nucleon exclusive scattering in which the target nucleon is in a nucleus and the rest of the nucleus acts as a spectator (Figure 1). In particular the residual $A-1$ nucleus must have small excitation energy ($\ll m_\pi$) and no soft particles should be produced in the process. An example of such a process is $pA \rightarrow p'N(A-1)$, with the momentum transfer, t, between p and p' greater than a few GeV^2.

Perturbative QCD implies that hadrons participating in a large momentum transfer exclusive scattering are in a state of small transverse size, $O(1/\sqrt{-t})$, when they scatter. Moreover, time dilation slows the fluctuations by which the small-transverse-size quark systems return to "normal", so that at large enough energies hadrons participating in an exclusive scattering should be able to penetrate nuclear material without additional scattering. Thus, Mueller[1] argued that the cross section should asymptotically $\sim A^1$. Liu, Frankfurt, Strickman and I[2] have modeled the mechanism by which the small-transverse-size quark systems become fully interacting hadrons, in order to estimate the actual extent of the "color transparency" (the term coined by Mueller to describe the reduced cross sections of hadrons participating in a large transverse momentum exclusive scattering) to be expected at presently accessible energies. Our results are presented here in summary form. In a second paper[3] we argued that if color transparency applies, the components of the nuclear wavefunction having large four-momentum can be experimentally measured by using the rapidly falling elementary scattering cross section to amplify them.

II. Color Transparency Results

We calculated[2] the nuclear transparency, $\frac{A_{eff}}{A}$, as a function of A for: $p + A \rightarrow p + N + (A-1)$ (Figure 2), $\pi + A \rightarrow \pi + N + (A-1)$ (Figure 3), and $e + A \rightarrow e + N + (A-1)$(Figure 4), for three models of the expansion or contraction of the relevant hadrons. Details are given in reference 2. The $\tau = 0$ model corresponds to no transparency effect – the incoming and outgoing hadrons have their standard cross-sections in spite of participating in the large-t exclusive scattering. The $\tau = 1$ and 2 models represent, respectively, "quantum diffusion" and "naive parton" pictures of the expansion process. The beam momenta considered are 6GeV/c, and

20 GeV/c; throughout, we take the scattering to be at 90 degrees in the center of mass. For each energy we plot two curves for $\tau = 1$ and $\tau = 2$ respectively. The case where $\tau = 0$, which is energy independent, is also plotted in each of the three figures. In figures 5 and 6 we plot the nuclear transparency, A_{eff}/A, for 90 degree pA and eA scattering as a function of the incoming momentum, in order to show the energy dependence more directly. Note that the $p_{inc} = 0$ intercept is also the value of the $\tau = 0$ (no transparency) prediction.

As can be seen from the figures, the transparency predictions are model dependent even for 20 GeV/c incident momentum. The important point is that both models predict a significant departure from what would be expected in the absence of the transparency effect, even at low energies. We conclude that experimental investigation of nuclear transparency is worthwhile even at present energies ($t \geq$ a few GeV^2 and $p_{lab} \sim 10$ GeV/c). In spite of the model dependence, perturbative QCD predicts rather unambiguously that A_{eff}/A should be greater than ~ 0.3 for hadron-induced reactions, and that it should increase with increasing p_{lab}.

A BNL experiment[4] has recently reported a level of transparency which is significantly larger than would be expected in the absence of a reduction in σ_{eff}, thus agreeing with the qualitative pqcd expectation. However the momentum dependence they find is in apparent disagreement with the perturbative QCD prediction: although A_{eff}/A increases with increasing momentum up to about 10 GeV/c, it then decreases. The meaning of this result is unclear. It has been interpreted as a charm threshold effect[5] and as evidence for interference effects[6]. The experiment does not measure the three-momenta of both final particles, so that the target particle four-momentum cannot be determined without some additional assumption about the mass or energy of the bound target nucleon. Thus s of the elementary scattering process is not known with certainty, even if the three-momentum of the target nucleon is small. Since the elementary scattering cross-section is an extremely sensitive function of s, altering their assumption that the energy of the target nucleon is equal to the proton mass could significantly alter the A_{eff}/A which they extract from the data.[†] We are currently studying the expected energy spectrum of the target nucleons in harmonic oscillator and mean field models, to estimate the extent of the effect. Another issue which deserves closer scrutiny is the excitation of the "$A - 1$" nucleus, and its implications for the experimental acceptance of the

[†]Monte Carlo studies suggest that their conclusion is not very sensitive to this problem, S. Heppelmann, private communication.

event. Thus far, the theoretical treatment of this aspect of the problem has been very superficial. Clearly, independent verification of the BNL results is desirable, ideally in an experiment capable of measuring both final three momenta.

III. Measuring the nuclear 4-momentum distribution

If color transparency were exact, quasi-exclusive scattering would provide an excellent method to study the nuclear wave function, especially the large momentum component and properties of bound nucleons.[3] The main reason that nuclear quasi-exclusive scattering suits this purpose is that the elementary exclusive reaction at fixed t is an extremely rapidly varying function of s. Therefore, a majority of the scattering events occur in kinematic configurations of the target nucleon fermi momentum such that the elementary reaction is at significantly smaller s than $2m_pE_{beam}$. The important regime of fermi momentum is for the target nucleon traveling in the same direction as the beam with, for example, about 1 GeV/c for quasi-exclusive pp scattering.

Since the elementary scattering reactions are measured, if there were no absorption of the incoming and outgoing particles in the nucleus and if binding effects could be ignored, the distribution of fermi momentum could be measured, with very much enhanced sensitivity to the large momentum tail of the fermi momentum distribution. This simple scenario is disturbed in several ways. First, the target nucleon is not on-mass-shell. We assume that this does not affect the elementary cross-section as a function of s and t, which is plausible when $s, t >> m^2$ and the dimensional scaling laws hold. Moreover determination of s and t is not in general possible due to final and initial state interactions of the scattering hadrons. If the experiment were carried out in the regime of perfect "color transparency", however, these interactions could be neglected and the measured momenta of the projectile and scattered particles would be the same as at the point of interaction.

In order to infer the fermi motion distribution from nuclear quasi-exclusive scattering experiments, it is likely to be necessary to correct for absorption. Since the degree of absorption depends on s and t of the elementary reaction, and the laboratory momenta of the incoming and outgoing hadrons, an extensive set of measurements is necessary, at least in principle, to do this completely. Fortunately, by selecting the kinematics of the outgoing particles, the target 3-momentum can be experimentally fixed to be small, and in principle the target nucleon can be arranged to be nearly on-shell. Then, by comparing model predictions for absorption to experimental results, the correct description of the absorption can be determined and accounted for when interpreting experimental results for large fermi momenta.

To illustrate the nature of the effect, we took[3] three model fermi-motion distributions and calculated the ratio between the differential cross section $\frac{d\sigma^{hA}}{dt}$ and the elementary cross section $\frac{d\sigma^{hN}}{dt}$ as a function of s' at fixed t, assuming the target nucleon is on-mass-shell. The elementary cross section $\frac{d\sigma^{pp}}{dt}$ and $\frac{d\sigma^{p\pi}}{dt}$ were taken from experimental measurments [7,8,9]. The three model fermi-motion distributions which were used[3] are:

A. Simple fermi gas model with $n(p)$ = constant for $p < k_F$ and zero otherwise.

B. Few nucleon correlation model developed by Frankfurt and Strickman [10].

C. Moniz's parameterization based on calculations of nucleon-nucleon correlations in nuclear matter as given by Bodek and Ritchie [11].

The results of our calculations are presented in the figures, where s' is the actual invariant energy of the elementary reaction. Fig. 6 shows the sensitivity of the predictions to the chosen fermi momentum model. Evidently there is a dramatic difference in the predictions of the three models; we take the Moniz model for illustration in subsequent figures since its predictions fall between those of the other models. The s' distribution bears little resemblence to the narrow gaussian around $2P_{inc}$ one would naively expect, except for the fermi-gas model which has no high momentum tail in it. There is, in fact, a large peak at low s', coming from the tail of the fermi-motion distribution. Figures 7 and 8 show the sensitivity of pA and πA to beam energy and to A. As expected, proton beams give greatest sensitivity to the large momentum components of the nuclear wavefunction, because for them the elementary cross sections are the most rapidly falling in s. Not shown, are the analogous quantities for electron induced reactions $eA \rightarrow e + N + (A-1)$. They are less sensitive to large fermi momenta, but nevertheless of crucial importance to study because they are more immune to the problem that determination of the reaction kinematics requires the assumption of non-interaction of the scattering particles except for the exclusive hard scatter itself. They are also simpler to treat from the standpoint of absorption.

We also see that larger beam energy gives greater sensitivity to the tail. For the pA reactions at $p_{inc} = 9$ GeV/c, the most probable value of the momentum of the struck nucleon is about $1 GeV/c$. We can also estimate the characteristic excitation energy of the final nuclear system, as follows. In the pair correlation approximation, the nucleon which balances the momentum of the target nucleon will be ejected with roughly the same light-cone fraction it had in the target nucleus. The excitation energy $M_{Res} - M_{A-1}$ in this case is determined from the equation for the laboratory four-momentum p_R of the recoil nucleon: $2 - \alpha = \frac{\sqrt{m^2+p^2}}{m_N}$ Since the average α for the low s peak is about 0.5-0.7, we have $M_{Res} - M_{A-1} \sim \sqrt{m^2 + p_R^2} - m \sim 100MeV$ in this peak.

In this calculation, perfect nuclear transparency has been assumed for both incoming and outgoing particles. See Ref. 3 for effects of absorption. We have also ignored the differences between the properties of bound and free nucleons. However models accounting for the EMC effect generally have important consequences for the s' dependence of nuclear quasi-exclusive scattering. We intend to discuss this in greater detail in another publication, but note the conclusions briefly here:

(1) Models in which bound nucleons swell (Q^2-rescaling models,etc): Here the quasi-exclusive cross section is expected to be suppressed, due to the reduction of the normalization of the short-distance part of the target nucleon wavefunction since $r_N^*/r_N - 1 \sim 0.1 - 0.2$. This effect imitates absorption for light nuclei, but has a considerablely weaker A-dependence than nuclear transparency, since in these models $r_N^*/r_N - 1 \propto < \rho_A >$, the average nuclear density.

(2) Pion model (x-rescaling) [12]: In this model the EMC effect is interpreted to be due to the depletion of nucleon light-cone fractions, which is assumed to be approximately homogeneous: $\alpha \rightarrow \alpha(1-\eta)$ where $\eta_{Fe} \sim 0.05$. The main effect here would be to shift the maximum in the lower s' peak to larger s'.

(3) "Minidelocalization" model of two of the present authors [13]: Here the the probability of point-like configurations in a bound nucleon with momentum p is suppressed by the factor $\delta(p) = (1 + \frac{x}{2})^{-2}$, where $x = 4\frac{\epsilon_A + p^2/2m_\pi}{\Delta E_A}$, ϵ_A is the binding energy per nucleon, ($\sim 8MeV$ for $A >> 1$). Fits to the EMC effect lead to $\Delta E_A \sim 0.6 - 0.8GeV$, the characteristic energy excitation for point-like configurations. This implies that the height of the low s' peak should decrease by at least a factor of 2 while the peak at $s \sim s_0$ is suppressed very little. To distinguish this model from a Fermi motion model with a more steeply decreasing wave function, one should study $h + D \rightarrow h + p + n$ for nucleon spectator momenta $0.2 < p < 0.5GeV/c$: for these momenta, the deuteron wave function is known reasonably well ($\sim 10\%$ for $p \sim 0.2GeV/c, \sim 30\%$ for $p \sim 0.5GeV/c$) and $\delta(p)$ changes the cross section by a factor of ~ 3 for $p \sim 0.5GeV/c$. Note that for $p \geq 0.3GeV/c$, the average internucleon distances in the deuteron are $\leq$1.2 fm, so the nuclear transparency

condition is easy to satisfy. Thus according to the minidelocalization model, this process is an alternative probe of the dominance of small-transverse-size hadrons in large momentum transfer exclusive scattering.

(4) Multiquark (6q,9q ......) models [14]: In these models it is assumed that at small internucleon distances, i.e., large fermi momenta, nucleons form 6q bags with probability $P \sim 20\%(30\%)$ for $A \sim 60(200)$. Consequently, these models lead to the suppression of the small s' peak by a factor $1 - P$, since the cross section of the reaction $h + (6q) \rightarrow h + N + N$ is strongly suppressed due to the small overlap of the 6q and two nucleon configurations. There is no effect on the $s' \sim s_0$ peak.

In conclusion, we have shown that study of quasi-exclusive scattering is very sensitive to the large momentum component of the nuclear wavefunction, and in some models to the effects of binding on the properties of nucleons. Interpreting the results will be non-trivial, requiring an extensive set of data taken at low fermi-momenta in order to unfold the effects of absorption. For these purposes electron- as well as hadron-induced reactions will be useful. In general it is also difficult to separate effects of binding from effects of fermi momentum. Thus, while experiments such as these will permit study of an important new domain of nuclear physics, a great deal of theoretical as well as experimental work will be necessary to fully understand the results.

REFERENCES

[1]A.Mueller, Proceedings of the XVII Recontre de Moriond, Les Arcs, France(1982).

[2]G.R.Farrar, H.Liu, L.L.Frankfurt and M.I.Strikman, Rutgers preprint RU88-03, to be published in Phys. Rev. Lett.

[3]G.R.Farrar, H.Liu, L.L.Frankfurt and M.I.Strikman, Rutgers preprint RU88-04, submitted to Phys. Rev. Lett.

[4]A.S.Carroll et al., PSU HEP/88-02 submitted to Phys. Rev. Lett. and S. Heppelmann in these Proceedings.

[5]S.J. Brodsky and G.F. deTeramond, SLAC-PUB-4504.

[6]B. Pire and J. Ralston, in preparation. We thank J.R. for communicating their work to us.

[7]M.K.Carter,P.D.B.Collins and M.R.Whalley, Rutherford Appleton Laboratory Report RAL-86-002

[8]Jacques Chauveau, These de Doctorat D'etat et Sciences Physiques, L'Universite Pierre et Marie Curie Paris VI. (1981)

[9]V.Chabaud et al., Phys.Lett. *38B*,441(1972); C.Baglin et al., Nucl.Phys. *B98* 365(1975); D.P.Owen et al., Phys.Rev.*181* 1794(1969).

[10]L.L.Frankfurt and M.I.Strikman, Phys.Rep.*76*, 215(1981).

[11]A. Bodek and J.L.Ritchie, Phys.Rev.*D23*,1070(1981).

[12]C.H.Llewellyn Smith, Phys.Lett.*B128*,127(1983); M.Ericson and A.W.Thomas, ibid *B128* 112(1983).

[13]L.L.Frankfurt and M.I.Strikman, Nucl. Phys. *B250* (1985) 146.

[14]R.L.Jaffe, Phys.Rev.Lett. *50* 228(1983).

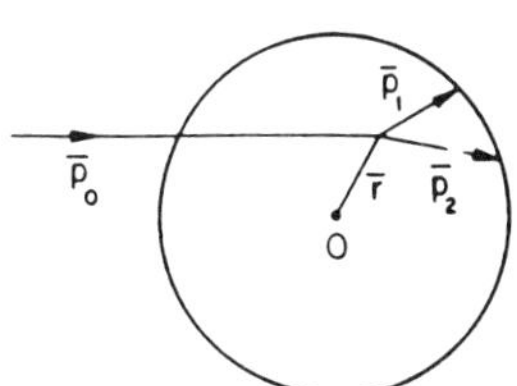

FIG. 1. Nuclear quasi-exclusive scattering.

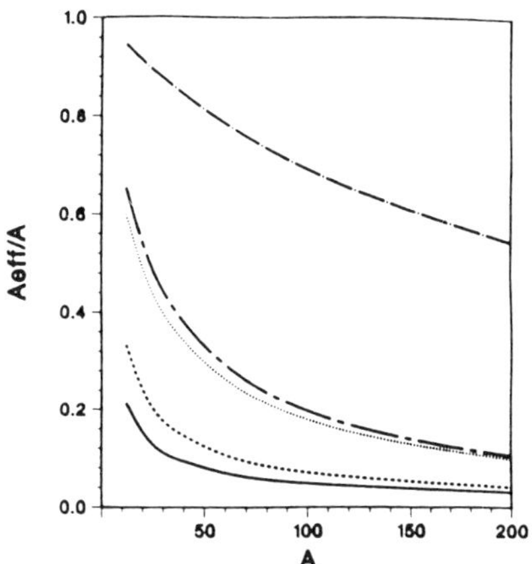

FIG. 2. A_{eff}/A for pA scattering (in this and the following figures all reactions are 90 degrees in the cm system) as a function of A for $\tau = 0$ (solid line), $\tau = 1$ and $p_{lab} = 6$ GeV/c (dashed line), $\tau = 2$ and $p_{lab} = 6$ GeV/c (dash-chain line), $\tau = 1$ and $p_{lab} = 20$ GeV/c (dotted line), and $\tau = 2$ and $p_{lab} = 20$ GeV/c (dot-chain line).

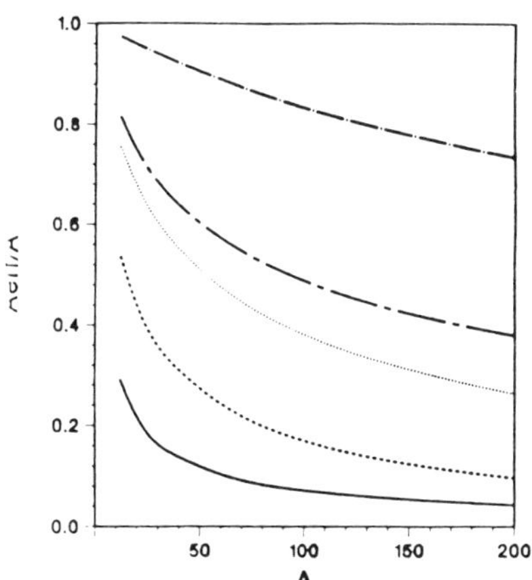

FIG. 3. As in Fig. 2, but for πA scattering.

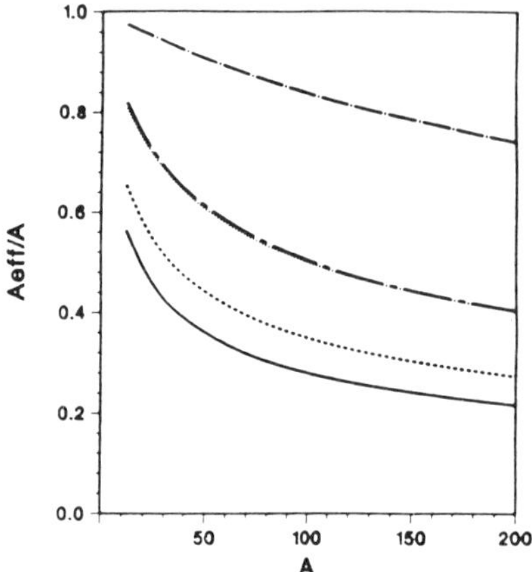

FIG. 4. As in Fig. 2, but for eA scattering.

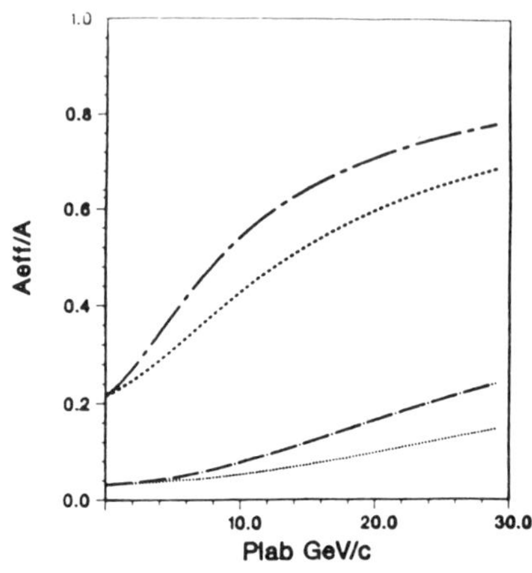

FIG. 5. A_{eff}/A for pA scattering as a function of beam energy for the $\tau=1$ model. For $A=12$, $\langle \Delta M^2 \rangle = 0.7$ GeV^2 (dashed line), and 0.5 GeV^2 (dash-chain line); for $A=200$, $\langle \Delta M^2 \rangle = 0.7$ GeV^2 (dotted line), and 0.5 GeV^2 (dot-chain line).

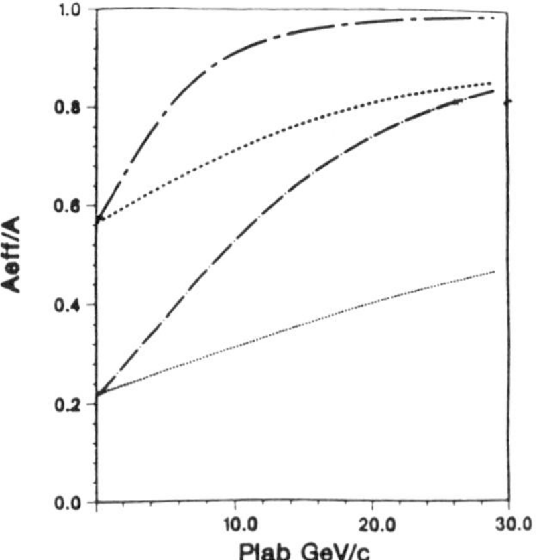

FIG. 6. A_{eff}/A for eA scattering as a function of beam energy. For $A=12$, $\tau=1$ (dashed line) and $\tau=2$ (dash-chain line); for $A=200$, $\tau=1$ (dotted line) and $\tau=2$ (dot-chain line).

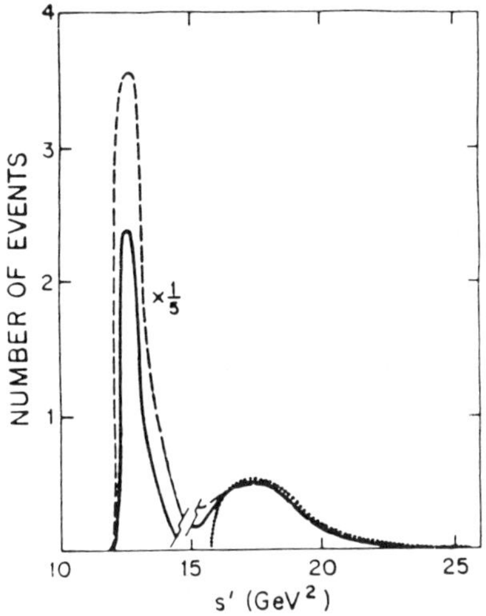

Fig. 7. Number of events as a function of s' calculated using three different models: fermi-gas model (dotted line), Moniz's model (solid line) and the few-nucleon correlation model (dashed line).

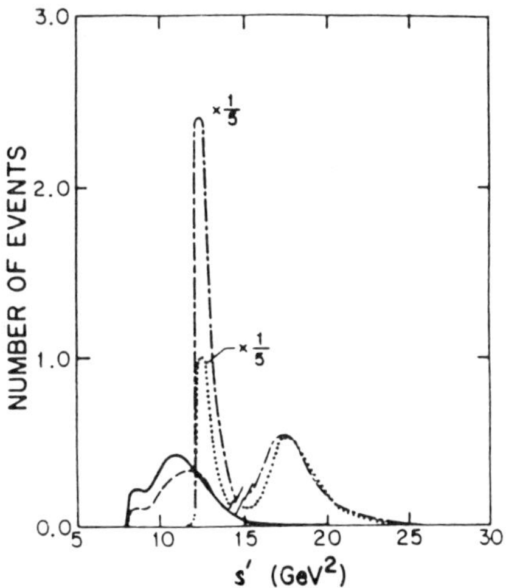

Fig. 8. Number of events as a function of s' using Moniz's model of fermi motion for different beam energies of proton and target masses: $P_{inc} = 6$ GeV, $t = -5$ GeV^2 and $A = 200$ (solid line), $P_{inc} = 9$ GeV, $t = -8$ GeV^2 and $A = 200$ (dash-dot-dash line), $P_{inc} = 6$ GeV, $t = -5$ GeV^2 and $A = 12$ (dashed line), $P_{inc} = 9$ GeV, $t = -8$ GeV^2 and $A = 12$ (dotted line).

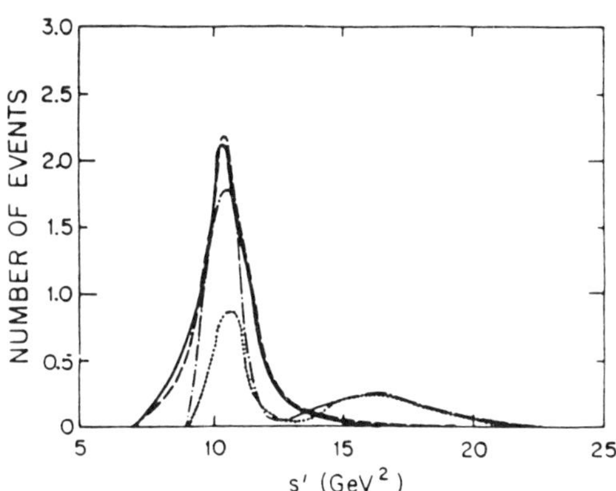

Fig. 9. As for Fig. 7, but for pion beam.

THRESHOLD BEHAVIOR OF DIRECT ELECTRON PRODUCTION IN PROTON-BERYLLIUM COLLISIONS

C. Naudet, G. F. Krebs, E. Lallier(a), A. Letessier-Selvon, H.S. Matis, G. Roche(b), L. Schroeder, P. Seidl, A. Yegneswaren, Z.F. Wang(c)
Nuclear Science Division, Lawrence Berkeley Laboratory, University of California, Berkeley, California 94720

J. Carroll, J. Gordon, G. Igo
Department of Physics, University of California at Los Angeles, Los Angeles, California 90024

T. Hallman, L. Madansky, R. Welsh
Department of Physics, The Johns Hopkins University, Baltimore, Maryland 21218

P. Kirk
Department of Physics and Astronomy, Louisiana State University, Baton Rouge, Louisiana 70803-4001

D. Miller
Department of Physics, Northwestern University, Evanston, Illinois 60201

and

G. Landaud
Universite de Clermont II-Institut National de Physique Nucleaire et de Physique des Particules, 63170 Aubiere, France

ABSTRACT

We report measurements of direct electron pair production in p-Be interactions at 1.0, 2.1 and 4.9 GeV kinetic beam energies. The invariant mass and p_t spectra along with the total cross sections are presented. A rapid decrease in the total cross section is observed at the lowest energy.

INTRODUCTION

The study of low-mass electron pairs (m $<$ 1 GeV/c^2) in proton-nucleus collisions has been a subject of considerable theoretical and experimental interest. Incisive measurements in the early seventies at Fermilab[1] proved that the low-mass dilepton and low-p_t single lepton yields were not fully understood. To address these questions a series of additional experiments ranging from ISR energies to laboratory beam energies of 12 GeV were pursued [2–7]. It has been determined that the low mass continuum could neither be explained by meson and vector meson decay nor by the Drell-Yan model (naive extrapolation to lower masses yields a underestimate by an order of magnitude). While many mechanisms have been suggested to explain the low mass region such as modified Drell-Yan, hadronic Bremsstralung and soft parton annihilations, this region is still not fully understood. Within the available statistics, the low p_t single lepton yield was found to be consistent with its origin predominantly from low mass pairs[2]. The result of a single-electron experiment[8] which did not observe direct electron production in proton-proton collisions at a kinetic beam energy of 800 MeV suggests a *threshold* between 1 and 10 GeV. We present here the study of e^+e^- production in p+Beryllium interactions at 1.0, 2.1, and 4.9 GeV kinetic beam energies.

EXPERIMENTAL APPARATUS

The data were obtained with the Di-Lepton Spectrometer (DLS) at the Bevalac. Physics considerations guided us to employ a large acceptance spectrometer to compensate for the expected low production rate. Adequate detector segmentation were implemented to discriminate between the hadrons and electrons(e^+ and e^-) in the heaviest system we planned to investigate (ie. Ca+Ca at 2.1 GeV). The experimental apparatus consisted of a segmented Beryllium targets at the vertex of two symmetrical arms, each arm covering an angular range with respect to the beam axis of $17^0 < \Theta < 63^0$ and 14^0 in the transverse plane. This translates into a electron pair kinematic range of approximately .1 to 1 GeV/c^2 in mass, 0 to 800 MeV/c transverse momentum (p_t) and 0.5 to 1.9 units of laboratory rapidity (y). The electrons were identified by using large segmented gas Cerenkov counter's having approximately 95 percent electron detection efficiency and 10^{-5} hadron rejection per arm. The momentum reconstruction in each arm was obtained by use of one wire chamber before and two behind a large dipole magnet. The momentum resolution was balanced in favor of maximizing the acceptance of the low mass pairs by setting the magnetic field at 1.5 kilogauss. This yielded a momentum resolution of 15 percent. A sixteen segmented scintillator hodoscope in front and behind the dipole magnets were used in combination with the cerenkov counters to both trigger the apparatus and to give a first order tracking.

The direct lepton signal (true pairs), which originates from a single elementary process is found by subtracting the same sign pairs from opposite sign pairs. This is because the electron pairs are independently produced thus creating charge symmetry of false pairs. Such electrons may be produced through Dalitz decay of a π^0 or gamma-ray conversion. This removes the main background (false pairs). The segmentation of the target and the designed minimal detector thicknesses reduces the rate of external conversions. In p+Be collisions the false pairs are dominated by Dalitz produced electrons. One wide-angle Dalitz decay however may trigger our system at a very low rate and cannot be removed by this subtraction process. This contribution to the invariant mass cross sections is estimated by using the known meson production cross sections in pp interactions, branching ratios and Dalitz decay mass structure[9]. In this analysis we have included the Dalitz decays of π^0, η, ω, and K_0 and the $\Delta(1232) \to Ne^+e^-$.

EXPERIMENTAL RESULTS

In an earlier publication[10] this group reported, the existence of a di-electron signal was clearly established at 4.9 GeV kinetic beam energy. The pair statistics of these results along with 2.1 and 1.0 GeV data is shown in Table 1 (the results of the 1.0 GeV data are still considered preliminary at this stage). The existence of the di-electron signal is clearly established at all three energies. Contributions from empty target interactions were significant only in the 2.1 GeV data where we experienced a beam halo problem.

Interaction	Energy [GeV]	Opposite Sign Pairs	Same Sign Pairs	Direct Pairs $D \pm \sigma_D$	R=D/F	D/σ_D
p + Be	4.90	732	201	531±31	2.6	17.4
p + Be	2.10	567	148	419±27	2.8	15.7
p + Be	1.04	204	58	146±16	2.5	9.1

Table 1. Pair statistics for each kinetic beam energy

The absolute cross sections were obtained by applying the acceptance, efficiency corrections and beam normalization factors to the raw data. The acceptance array, the inverse of which weights each event, was generated via Monte Carlo techniques spanning all the accessible phase-space volume in increments of Δp_t, Δy and Δm. The efficiencies (ie. tracking, vertex cuts...) were all established using analysis techniques on both the raw data and Monte Carlo generated events. Two sets of beam hodoscope counters and a calibrated ion chamber monitored the beam flux. The errors in the table reflect only the experimental statistical uncertainty; the overall systematical error is estimated to be approximately ±30 percent. Additional information concerning the detailed acceptance studies and the data analysis may be found in ref. 10.

Figure 1. shows the cross section per nucleon (assuming an $A_t^{2/3}$ dependance) for p+Be as a function of the invariant mass for all three kinetic beam energies. There exist several interesting features to be compared at each bombarding energy. Noticeable at each energy is a sharp increase in the cross section at masses less than 100 MeV. This is seen to be primarily composed of π_0 Dalitz decay with perhaps some additional hadronic bremsstralung. The total contribution of all higher mass meson Dalitz decays is estimated to be 2 orders of magnitude smaller than the observed data[11]. The rho-omega resonance is seen clearly in the 4.9 GeV data around m = .75 GeV/c^2. It is not observed at either 2.1 or 1.0 GeV. However, this is expected since the available center of mass energy at 2.1 GeV kinetic beam energy is .854 GeV barely

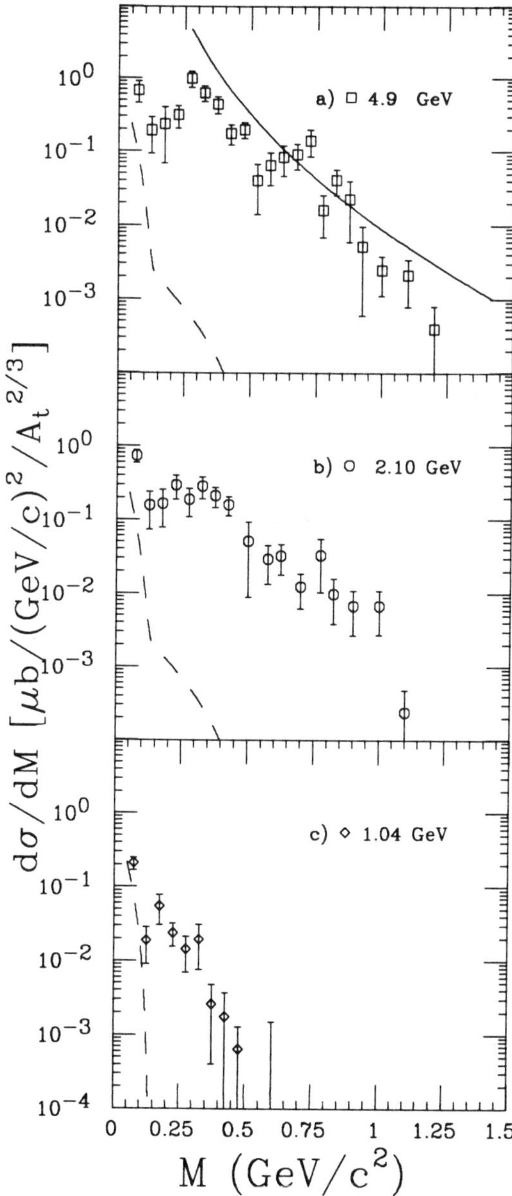

Figure 1. The invariant mass spectra for p+Be at all three kinetic beam energies. (a) The 4.9 GeV DLS data. The solid line is a fit to the KEK 12.1 GeV data.(ref 2) (b) The 2.1 GeV DLS data. (c) The 1.0 GeV DLS data. The dotted line with each figure is the estimated contribution due to the wide angle Dalitz decay at each energy.

above the absolute kinetic threshold for rho-omega production. Comparison with a parameterization of the KEK 12.1 GeV data[2] (the solid curve in Figure 1.(a)) shows that the general shape of the distribution is similar to that at higher energies. For the first time a measurement of the mass structure in regions less than 300 MeV is performed[12]. A structure is observed in both the 4.9 and 2.1 GeV invariant mass distribution near $2m_\pi$. This strongly suggests that a $\pi^+\pi^-$ annihilation mechanisms dominates the production in this energy range. Theoretical studies[13] indicate a thorough understanding of this annihilation structure in heavy systems (ie. Ca+Ca) could yield information concerning the pion dispersion relation and thus information on the density of hot hadronic matter. A rapidly falling cross section is observed in the 1.0 GeV spectra. This is what would be expected for a combination of Dalitz decay, $\pi^+\pi^-$ annihilation and dominating hadronic bremsstralung.

The p_t dependence of our cross section per nucleon, integrated over mass (m>200 MeV/c^2) and y as a function of p_t for all three beam energies is shown in Figure 2. The slopes of all three data sets agree well with one another. The data is adequately fit by a form $d\sigma/dp_t^2 = \exp(-\alpha p_t)$ with $\alpha = 6.$. It is interesting to note that this is similar to the higher energy e^+e^- data and the well known low-p_t hadronic dependence.

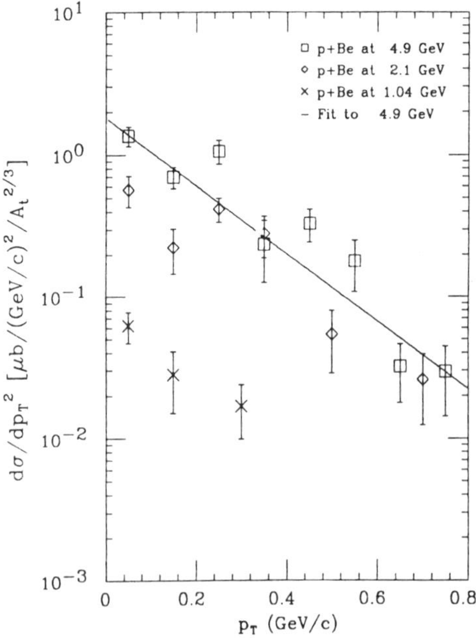

Figure 2. The cross section per nucleon, $d\sigma/dp_t^2$, as a function of p_t for all three kinetic beam energies. The solid line is a fit of the 4.9 GeV data.

The total integrated e^+e^- cross sections per nucleon (200 MeV >m>700 MeV) as a function of the available nucleon-nucleon center-of-mass energy (Q $=\sqrt{s}$-2m$_p$) is shown in Figure 3. A rapid decrease in the cross section is seen at low energy. Naively one would expect that the Q dependence would be similar to that of the total pp inelastic cross section or perhaps the π^0 production cross section, as the e/π ratio is constant at higher energies. However, both cross sections are approximately constant in the energy range where our experimental data shows a threshold. A comparison with the energy dependence of heavier mesons is not feasible since the data is too meagre at these low energies [14]. From bubble chamber data[15] we have computed the inclusive $\pi^+\pi^-$ production cross section at these low energies. Shown as the solid curve in Figure 3. is the Q dependence of $\pi^+\pi^-$ production, scaled down by 75,000 ($= \alpha^2/4$). Both the e^+e^- and $\pi^+\pi^-$ production cross sections have similar threshold behavior. This evidence also supports the interpretation of $\pi^+\pi^-$ annihilation dominating the dielectron production at these energies.

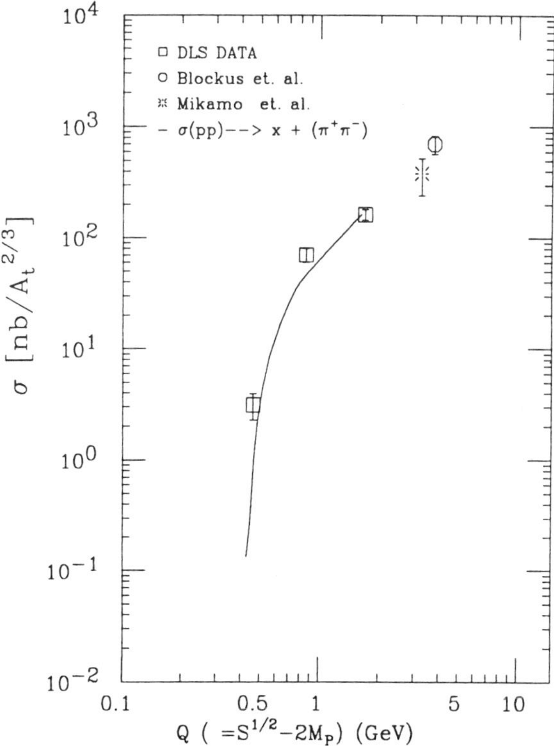

Figure 3. The total integrated e^+e^- production cross section as a function of the available nucleon-nucleon cm. energy. $\sqrt{s}$-2m$_p$: circle, Blockus et al. (ref. 3); triangle, Mikamo et al. (ref 2); squares this experiment. The solid curve shows the $\pi^+\pi^-$ production cross section (scaled down by 75,000). respectively.

SUMMARY

In summary, dielectron production at 4.9, 2.1 and 1.0 GeV has been observed. The production cannot be explained by meson decays. A unique structure in the invariant mass distribution is observed at twice the pion mass at both 4.9 and 2.1 GeV. The total integrated cross sections are found to rapidly decrease at low energies. Both the mass structure and energy dependence is consistent with the interpretation that the dominant production mechanism is the $\pi^+\pi^-$ annihilation.

ACKNOWLEDGMENTS

This work was supported by the Director, Office of Energy Research, Division of Nuclear Physics of the Office of High Energy and Nuclear Physics of the U.S. Department of Energy under the contracts DE-AC03-76SF00098, DE-FG03-88ER40424,DE-FG02-88ER40413 and DE-FG05-88ER40445. We would like to thank the staffs of the Bevalac and of the participating institutions whose contributions made this experiment possible.

REFERENCES

[a] Now at Thomson-lcr, 91401 Orsay, France
[b] On leave from Universite de Clermont II, Aubiere, France
[c] Now at Department of Physics, LSU, Baton Rouge, LA. 70803
[1] L. Lederman Physics Reports **C26**, 4, 149, (1976)
[2] A. Mikamo *et al* Phys. Lett. **106B**, 428, (1981)
[3] D. Blockus *et al* Nucl. Phys. **B201**, 205, (1982)
[4] M. R. Adams *et al* Phys. Rev. **D27**, 1977, (1983)
[5] K. Bunnel *et al* Phys. Rev. Lett. **40**, 136, (1978)
[6] J. Ballam *et al* Phys. Rev. Lett. **41**, 1207, (1978)
[7] B. Haber *et al* Phys. Rev. **D22**, 2107, (19780
[8] A. Browman *et al* Phys. Rev. Lett. **37**, 246, (1976)
[9] The particle data book supplied the branching ratios and inclusive meson cross sections may be obtained in F. Winkelmann LBL-3045. N.M. Knoll and W. Wada Phys. Rev. 98, 1325, (1955).
[10] G. Roche Proceedings of the 8th High Energy Heavy Ion Study nov. 16-20, 1987 LBL-24449 [11] At 4.9 GeV the heavier meson cross sections are uncertain to within a factor of two.
[12] The CCRS experiment did observe low mass pairs down to 50 MeV/c^2 however this was a 90 degree experiment limiting the $p_t >$ 1.3 GeV/c, see ref. 1.
[13] T. Goldman *et al* Phys. Rev. **D20**, 619, (1979) and C. Gale and J. Kapusta Phys. Rev. **C35**, 2107, (1987)
[14] J. Bartke *et al* Nuclear Physics **B107**, 93, (1976)
[15] G. Alexander *et al* Phys. Rev. **154**, 1284, (1967)

SOME IMPLICATIONS OF A POSSIBLE $NN\pi$ BOUND STATE*

HARRY J. LIPKIN†
High Energy Physics Division
Argonne National Laboratory
Argonne, IL 60439
and
School of Physics and Astronomy
Raymond and Beverly Sackler Faculty of Exact Sciences
Tel Aviv University
Tel Aviv, Israel

ABSTRACT

The possible properties of a πNN bound state are considered, in view of recent reports of possible experimental evidence for such a state. A six-quark T=2 state is shown not to be bound. Strong, electromagnetic and weak decays are examined for a πNN bound state with a mass below the πNN threshold and the cases of isospin T=2,1 or 0. States in these multiplets with charge 0,1 and 2 have interesting properties and might be observed as narrow resonances decaying electromagnetically, in particular states with $J^P = 1^+$ with charge 0 and 2 which are Pauli-forbidden for two nucleons. The lifetimes of the charge -1 and +3 states should be much longer than the pion lifetime because of drastically reduced phase space.

A state with the quantum numbers ($T = A = 2$) has been suggested as a possible[1] $NN\pi$ bound state. Recently there have been reports of possible experimental evidence for the charge +3 and charge −1 states in such an isospin multiplet, [2] denoted by T^{+++} and T^-, with binding energies of 12.5 and 18 MeV respectively.

WHY THERE CANNOT BE A T=2 SIX QUARK BOUND STATE

A state with these quantum numbers might arise in a quark model as a state with 6 nonstrange quarks in a relative S wave coupled to isospin 2. However, calculations of the hyperfine interaction in this configuration show that the state ($T = A = 2$) is unbound by $(5/2)[M_\Delta - M_N] - M_\pi$ relative to the relevant πNN threshold. Thus if a bound state with these quantum numbers exists, it cannot be a simple six-quark state, but may be the suggested πNN bound state[1].

The basic physics in the argument against a six-quark state is that any state containing three identical u-quarks or d-quarks in a relative s wave is forced by the Pauli principle to put these three quarks in the Δ color-spin configuration which costs

*Work supported by the U.S. Department of Energy, Division of High Energy Physics, Contract W-31-109-ENG-38.

†Permanent address: Department of Nuclear Physics, Weizmann Institute of Science, Rehovot 76100, Israel.

the $N - \Delta$ mass difference of 300 MeV. This is so much larger than the pion mass of 140 MeV that it is essentially impossible to compensate for this energy by additional attractive or binding forces. Even if an $N - \Delta$ six-quark state is bound by 100 MeV relative to the $N - \Delta$ threshold, which seems extremely unlikely, it will still be 60 MeV above the $\pi - N - N$ threshold. A detailed discussion of the hyperfine interaction is given in Appendix A

POSSIBLE DECAYS OF A $T = A = 2$ ISOSPIN MULTIPLET

It is interesting to examine the selection rules for the decay of such a bound state. The states in a $T = A = 2$ multiplet with $T_z = \pm 2$ have charge +3 and -1 and cannot decay into two nucleons nor into two nucleons plus photons. There are no open channels for strong or electromagnetic decays and only weak decays are allowed. However, the states in the multiplet with charge 2 , 1 and 0 are forbidden to decay into two nucleons only by isospin symmetry which is broken by the mass differences and the electromagnetic interaction. Their dominant decay mode is expected to be two nucleons and a photon, since this decay is first order in α, while a decay without photons would have to be at least second order. Thus in general one would expect such states to appear as narrow resonances in nucleon-nucleon scattering just below the pion production threshold, with the dominant final state being two nucleons and a photon.

$$N + N \rightarrow T \rightarrow N + N + \gamma \tag{1a}$$

The charge +1 state T^+ could also appear in phototdisintegration of the deuteron just below the pion production threshold, with an additional photon in the final state,

$$\gamma + d \rightarrow T^+ \rightarrow p + n + \gamma \tag{1b}$$

Note that in contrast to the case of isobaric analog states with $A > 2$ there is no isospin mixing in the wave functions of the two-nucleon final state. All two-nucleon states are automatically good pure isospin eigenstates with their isospin eigenvalues determined by angular momentum and parity. There are no states available for mixing even if isospin is not a good symmetry. Thus isospin-violating decays of $NN\pi$ states into two nucleons must go via isospin mixtures in the $NN\pi$ state and not in the two-nucleon final state.

ARE THERE ALSO $NN\pi$ BOUND STATES WITH T=1,0 ?

The question now arises whether there is only a bound T=2 multiplet or whether the T=1 and T=0 states are also bound. This clearly depends upon the model for the binding. One might guess that binding results from a strong attractive $\pi - N$ interaction in the T=3/2 p-wave channel which contains the Δ. If this is the case, then the T=0 state would not be bound since it does not have any T=3/2 $\pi - N$ components, while the T=2 state is most strongly bound because both $\pi - N$ pairs are completely in T=3/2. The T=1 state would be in between.

Even if the T=1 and T=0 $NN\pi$ states are also bound their coupling to the two nucleon system is so strong that they would only be observed as very broad resonances in nucleon-nucleon scattering and might not be observed as resonances at all.

THE EXCEPTIONAL CASE OF $J^P = 1^+$

An interesting exceptional case is the $J^P = 1^+$ state which could be formed from a p-wave pion bound to two nucleons in the ^1S_o, $J = 0$ state. This $J^P = 1^+$ state cannot decay into two identical nucleons. The relevant partial waves ($^3S_1$ and $^3D_1$) are forbidden by the Pauli principle for two protons or two neutrons. Thus if a 1^+ state is observed in charge +3 and/or charge -1, the charge 2 and charge 0 states in the multiplet would not show up in nucleon-nucleon scattering even if there were no isospin symmetry because the relevant partial wave does not exist in this channel.

There are two possible charge combinations of the $NN\pi$ system with a total charge +2; namely $pp\pi^o$ and $np\pi^+$. The isospin eigenstates with T=1 and T=2 are linear combinations of these two states. Both charge or isospin states with $J^P = 1^+$ are forbidden to decay into pp or to appear as resonances in pp scattering because there is no available pp state with these quantum numbers, irrespective of whether or not isospin is a good symmetry.

If both states are bound, there will be considerable isospin mixing, since the $\pi^+ - \pi^o$ mass difference of 4.6 MeV and the $n - p$ mass difference of 1.3 MeV gives a mass difference of 5.9 MeV between the $np\pi^+$ and $pp\pi^o$ states. This is of the same order as the binding energy. This effect will be smaller in the charge zero system where the mass difference $np\pi^-$ and $nn\pi^o$ is only 3.3 MeV.

For any 1^+ $NN\pi$ isospin multiplet which cannot decay into two identical nucleons, only the charge +1 state can couple to the two nucleon system. A T=0 1^+ $NN\pi$ state will simply decay strongly into the np channel where there is an available 1^+ state with T=0 and will presumably be too wide to be detected. The np 1^+ state with T=0 can also couple to the T=1 and T=2 states of the 1^+ $NN\pi$ system via isospin breaking.

The charge +1 states with T=1 and T=2 are forbidden by isospin to decay into two nucleons, but allowed to decay into $np\gamma$. However, because of isospin breaking they can also couple to np. These states could then be present as narrow resonances in np scattering just below the pion production threshold. The cross section will depend upon the ratio of the partial width in the nucleon-nucleon channel to the total width, including the radiative channels. One would expect the radiative decay to be dominant, since as noted above it is first order in α, while the Coulomb interaction which produces mixing, as well as any decay mechanisms without photons would have to be at least second order. In this case a T=1 state might show up more strongly in photodisintegration of the deuteron just below the pion production threshold.

WEAK DECAYS OF THE STATES WITH CHARGE +3 AND -1

The states with charge +3 and -1 are stable against strong and electromagnetic decays and would have to decay weakly. Estimates of weak decay lifetimes and branching ratios are difficult because of large unknown model-dependent factors.

The obvious decay mode is the $NN\mu\nu$ decay, which would correspond to the decay of an off-shell or bound pion with two nucleon spectators. But since the available energy for this decay mode of a free pion is only $m_{\pi^\pm} - m_{\mu^\pm} = 34MeV$ and this is reduced for a bound pion by at least the binding energy of about 15 MeV, there is a significant reduction in the phase space factor for the decay. Since the momenta of the nucleon spectators is not changed during the decay, the kinetic energy of the relative motion in the initial $NN\pi$ state remains in the final state and further reduces the available energy and the phase space factor.

The exact calculation of this reduction is highly model dependent as it depends upon the momentum distribution in the bound state, folded into the phase space factor which will vary widely over the this momentum distribution. However, the simple spectator model with a reasonable binding potential gives an estimate which suggests that the additional reduction due to spectator kinetic energies can be serious. The virial theorem relates the expectation values of the kinetic and potential energies,

$$\langle T\rangle = (1/2)\langle rdV/dr\rangle. \tag{2a}$$

For a Coulomb potential this gives

$$\langle T\rangle = -(1/2)\langle V\rangle \tag{2b}$$

Then for a binding energy $B = \langle V - T\rangle$

$$\langle T\rangle = B \tag{2c}$$

In a spectator model for the decay, this kinetic energy of adds to the binding energy in determining the available energy and phase space and the mean available energy E_{avail} is reduced by twice the binding energy,

$$\langle E_{avail}(\pi \to \mu\nu)\rangle = m_\pi - m_\mu - 2B \tag{3}$$

This suggests that a binding energy of 17 MeV results from a potential energy of 34 MeV, a kinetic energy of 17 MeV and an average available energy of zero for the decay. For other potentials the results may be somewhat different, but the same argument is qualitatively correct. Of course there are parts of the wave function in which the kinetic energy is less, and there is more available energy, but the net result would be a fairly drastic reduction in the average available phase space and a much longer lifetime. The high momentum tail of the wave function with total relative kinetic energy greater than 17 MeV cannot contribute to this decay mode at all.

One can speculate on other possible decays involving the nucleons as well as the disappearance of the pion. The $NNe\nu$ decay which has a much larger phase space is strongly suppressed by a helicity factor if the pion decays as a free pion. This decay mode might be enhanced if angular momentum can be transferred to the two nucleons during the decay process. Rough calculations suggest that this enhancement is not sufficient to give a decay rate comparable to the normal pion decay rate. If the suppression of the $NN\mu\nu$ decay below the normal pion decay rate is not very great the $NNe\nu$ decay mode may be difficult to observe. If however, the model-dependent phase space factors suppress the $NN\mu\nu$ decay by more than an order of magnitude, which may be possible, the $NNe\nu$ decay mode may be appreciable and even dominant. Without a detailed model wave function it will be very difficult to obtain reliable estimates of lifetimes and branching ratios.

Some simple calculations which neglect the model-dependent phase space factors are summarized below in Appendix B.

ACKNOWLEGEMENT

Discussions with D. Ashery, E. Piasetzky, A. Rinat and J. P. Schiffer are gratefully acknowledged.

APPENDIX A The Hyperfine Interaction in Six-Quark and other Systems

We now consider the hyperfine interaction in the standard colored quark model, using the simplified form of the general color-spin hyperfine interaction [3] which applies for systems containing only quarks and no active antiquarks:

$$V = -(v/2)[C_6 - C_3 - (8/3)S(S+1) - 16N] \qquad (A1a)$$

where v is a parameter defining the strength of the interaction, C_6 and C_3 denote the eigenvalues of the Casimir operators of the $SU(6)$ color-spin and $SU(3)$ color groups respectively, S is the total spin of the system and N is the number of quarks in the system.

This expression can be further simplified for the case where there are only non-strange quarks, by noting that there is a one-to-one correspondence imposed by the Pauli principle between the color-spin and isospin couplings of the quark.[4] For this case C_6 can be replaced by a function of N and the total isospin T to give

$$V_{ns} = -(v/2)[16N - C_3 - (8/3)N^2 - 8T(T+1) - (8/3)S(S+1)] \qquad (A1b)$$

We now calculate this interaction using a trial wave function which has a two-body density matrix the same for all pairs as the two-body density matrix in a baryon. This allows us to use the experimental $N-\Delta$ mass splitting to determine the strength parameter v.[8,9]

The hyperfine interaction (A1) is easily evaluated for states of interest by substituting the well known and tabulated eigenvalues of the Casimir operators[3,10]:

$$V(N) = \langle 70,1,1/2,3|V|70,1,1/2,3\rangle = -8v \qquad (A2a)$$
$$V(\Delta) = \langle 20,1,3/2,3|V|20,1,3/2,3\rangle = +8v \qquad (A2b)$$
$$V(\Lambda) = \langle 70,1,1/2,3|V|70,1,1/2,3\rangle = -8v \qquad (A2c)$$
$$V(H) = \langle 490,1,0,6|V|490,1,0,6\rangle = -24v \qquad (A2d)$$
$$V(d) = \langle 175,1,1,6|V|175,1,1,6\rangle = +(8/3)v \qquad (A2e)$$

$$V(\alpha) = \langle 1,1,0,12|V|1,1,0,12\rangle = +96v \quad (A2f)$$
$$V(diquark) = \langle 21,3^*,0,2|V|21,3^*,0,2\rangle = -8v \quad (A2g)$$
$$V(T=A=2) = \langle 35,1,0,6|V|35,1,0,6\rangle = +24v \quad (A2h)$$

where we have labeled the states by the quantum numbers $|d_6, d_3, S, N\rangle$ in a conventional notation[10], with d_6 and d_3 denoting the dimensions of the color-spin $SU(6)$ and color SU(3) representations in which the multiquark states are classified and asterisk denoting the conjugate representation. The states d and α denote a "quark-matter" deuteron and α particle with 6 and 12 quarks respectively in the lowest shell. We have included the Λ and the H dibaryon in the flavor $SU(3)$ symmetry limit with no correction for the smaller hyperfine interaction of the strange quark. We have also included the "diquark" in the most attractive symmetric state, which is a color antitriplet (3^*) with spin zero. The state ($T = A = 2$) denotes a state with 6 nonstrange quarks coupled to isospin 2. A state with these quantum numbers has been suggested as a possible πNN bound state[1]

We then obtain

$$M_\Delta - M_N = V(\Delta) - V(N) = 16v \quad (A3a)$$
$$V(d) - 2V(N) = (56/3)v = (7/6)[M_\Delta - M_N] \quad (A3b)$$
$$V(\alpha) - 4V(N) = 128v = 8[M_\Delta - M_N] \quad (A3c)$$
$$V(H) - 2V(\Lambda) = -8v = -(1/2)(M_\Delta - M_N) \quad (A3d)$$
$$V(diquark) - V(N) = 0 \quad (A3e)$$
$$3V(diquark) - 2V(N) = -8v = -(1/2)(M_\Delta - M_N) \quad (A3f)$$
$$V(T=A=2) - 2V(N) = 40v = (5/2)[M_\Delta - M_N] \quad (A3g)$$

where we assume that $M_\Delta - M_N$ is entirely due to the hyperfine energy difference. The quark model α particle is seen to be unbound by $8(M_\Delta - M_N)$; i.e. it costs this enormous hyperfine energy to squeeze four nucleon together so that all 12 quarks are in the same s-wave shell model orbit. The quark model deuteron is seen to be unbound by a much smaller amount, $(7/6)(M_\Delta - M_N)$, but it is still unbound. The H is seen to have a significant attractive hyperfine energy. The state ($T = A = 2$) is seen to be unbound by $(5/2)[M_\Delta - M_N] - M_\pi$ relative to the relevant πNN threshold. Thus if a bound state with these quantum numbers exists, it cannot be a simple six-quark state, but may be the suggested πNN bound state.[1]

APPENDIX B Weak Decays of the State

The $\pi \rightarrow e\nu$ decay is suppressed by a helicity suppression factor because the pion has spin zero, and the two leptons must have the same helicity in the center of mass system in order to have a projection of zero angular momentum on the axis of the relative momentum. If the decay of the bound state allows for a momentum transfer to the two nucleons, the two leptons no longer are required to be in a state of $J = 0$ in their center-of-mass system, and there may be a possibility to avoid the helicity suppression factor.

We consider two possible cases where a large allowed weak decay matrix element might arise:

1. 1. A $\Delta T = 1$ (Gamow-Teller) baryon decay into $Ne\nu$ either preceded by or followed by a pion absorption, possibly producing a Δ.
2. 2. A meson annihilation decay involving an admixed configuration of a ρ or other meson with nonzero spin which can decay into a lepton pair with spin 1.

The baryon decay rate can be estimated from a back-of-the-envelope calculation of an allowed Gamow-Teller transition using an experimentally known lifetime and an E^5 phase space correction. Mutually consistent results are obtained by using the neutron decay and the pion beta decay, which have much lower energy and the $\Sigma^{\pm} \rightarrow \Lambda$ leptonic decays which have a comparable energy. The results show enhancement relative to the $\pi \rightarrow e\nu$ decay, but rates an order of magnitude below the $\pi \rightarrow \mu\nu$ decay.

If we assume a binding energy of about 15 MeV for the $NN\pi$ bound state, to give an available energy of $m_{\pi^{\pm}} - 15 = 125 MeV$ for the decay, and assume an E^5 phase space correction, we obtain the following three estimates for the partial lifetime into the $NNe\nu$ decay mode

$$\tau_{NN\pi \rightarrow NNe\nu} = \frac{\tau_{\Sigma^-}}{BR(\Sigma^- \rightarrow \Lambda e^- \nu)} \cdot [(m_{\Sigma^-} - m_{\Lambda})/125]^5 =$$

$$= \frac{1.5 \times 10^{-10}}{5.8 \times 10^{-5}} \cdot (81/125)^5 = 3 \times 10^{-7} \qquad (B1a)$$

$$\tau_{NN\pi \rightarrow NNe\nu} = \frac{\tau_{\pi^+}}{BR(\pi^+ \rightarrow e^+ \nu \pi^o)} \cdot [(m_{\pi^+} - m_{\pi^o})/125]^5 =$$

$$= \frac{2.6 \times 10^{-8}}{1 \times 10^{-8}} \cdot (4.6/125)^5 = 1.75 \times 10^{-7} = 6.3 \tau_{\pi^+} \qquad (B1b)$$

$$\tau_{NN\pi \rightarrow NNe\nu} = \tau_n \cdot [(m_n - m_p)/125]^5 = 0.9 \times 10^3 \cdot [1.3/125]^5 = 1.1 \times 10^{-7} \qquad (B1c)$$

Additional suppression factors will arise from the pion absorption process which must accompany the beta decay. Even if these factors give an additional suppression of only a factor of two, the $NNe\nu$ decay is already suppressed by an order of magnitude relative to the $\pi \rightarrow \mu\nu$ decay. However we have not taken into account the additional reduction in phase space due to the kinetic energies in the wave function discussed above. These will have a much greater effect on the otherwise dominant $NN\mu\nu$ decay which has an available energy of only 34 - 15 = 19 MeV after the assumed binding energy of 15 is subtracted, in comparison with 125 MeV for the $NNe\nu$ decay. Which factors will dominate remains an open model-dependent question.

An upper limit for the meson annihilation decay amplitude is obtained by neglecting the helicity suppression factor and putting in a phase space enhancement of E^3 for a two-body decay with only one relative momentum. The enhancement factor obtained for a 15 MeV bound state giving an available energy of 125 MeV in comparison with the $\pi - \mu$ mass difference of 34 MeV is $(125/34)^3 = 50$. This may be important but other suppression factors which depend upon the extent of the admixture in the wave function are difficult to estimate and are probably at least an order of magnitude more than enough to compensate for this factor of 50. Here again, the $NNe\nu$ decay is expected to be suppressed by an order of magnitude relative to the $\pi \rightarrow \mu\nu$ decay, but the same uncertainty regarding the relative suppression by phase space factors renders all estimates unreliable.

REFERENCES

1. Humberto Garcilazo Phys. Rev. **C26**, 2685 (1982).
2. D. Ashery et al, Los Alamos preprint, submitted to Physics Letters; C. Morris, these proceedings.
3. R. L. Jaffe, Phys. Rev. Lett. **38**, 195 (1977).
4. K. Johnson, in Fundamentals of Quark Models, Proc. Seventeenth Scottish University Summer School in Physics, 1976, Edited by I. M. Barbour and A. T. Davies, Scottish University Summer School in Physics (1977) p. 245.
5. Harry J. Lipkin, Phys. Lett. **70B**, 113 (1977).
6. Harry J. Lipkin, In Intersections Between Particle and Nuclear Physics, Proc. Conf. on The Intersections Between Particle and Nuclear Physics, Lake Louise, Canada, 1986 Edited by Donald F. Geesaman AIP Conference Proceedings No. 150, p. 657.
7. H. J. Lipkin, Proceedings of the Workshop on Nuclear Chromodynamics, Santa Barbara, August 12-23, 1985, ed. by S. Brodsky and E. Moniz, p. 328.
8. N. Isgur and H. J. Lipkin, Phys. Lett. **99B**, 151 (1981).
9. Harry J. Lipkin, Phys. Lett. **195B**, 484 (1987) and **198B**, 131 (1987).
10. H. Högaasen and P. Sorba, Nucl. Phys. **B145**, 119 (1978).

PROBING TWO-NUCLEON CORRELATIONS WITH PION DOUBLE-CHARGE EXCHANGE

Helmut W. Baer
Los Alamos National Laboratory, Los Alamos, NM 87545, USA

ABSTRACT

A review is given of pion double-charge reactions at incident energies of 30 to 65 MeV and leading to isobaric-analog states. The large effects on cross section magnitudes and angular-distribution shapes produced by two-nucleon correlations are discussed. New results on 1 $f_{7/2}$ shell nuclei are compared to predictions of a shell-model-based two-amplitude model.

INTRODUCTION

There has been some fairly nice progress in the past two years in understanding the role played by two-nucleon correlations in low-energy double-charge-exchange scattering (DCX). The basic reason for this success stems from the relatively weak πN coupling at low energies, which makes it possible to analyze quantitatively DCX using the double-impulse approximation. By studying transitions leading to the double-isobaric-analog states (DIAS), it has been possible to minimize nuclear structure uncertainties and, hence, to bring forth the salient features of the double scattering. Analyses of the considerable body of data at pion energies 30 to 65 MeV shows rather convincingly that the spatial correlations implicit in shell-model wave functions play a major role in shaping DCX cross-sections, and indeed are the cause for the observed "anomalous" cross-section ratios and angular-distribution shapes. What is new in the low-energy studies is the clarity with which the two-body correlations are probed and the ability to isolate the short-range component of the double scattering.

2. ROLE OF TWO-NUCLEON CORRELATIONS IN $^{14}C(\pi^+,\pi^-)^{14}O$

In Fig. 1 are shown the data[1,2] on ^{14}C and the two theoretical curves,[3] which lent considerable impetus to the study of low-energy DCX. The large-forward-angle peak in the data is puzzling by itself in view of the fact that the single-charge-exchange (SCX) scattering[4] at 50 MeV and leading to the isobaric-analog-state (IAS) in ^{14}N has a forward-angle-minimum (Fig. 2). The provocative labelings on the theoretical curves in Fig. 1 as "6-quark" and "nucleons-only" promoted further study. Now after four years of investigations we think we know better what labelings to use. In Fig. 3 the ^{14}C data are compared with two recent calculations of Bleszynski and Glauber[5] and labeled to reflect their distant relationship to the curves in Fig. 1. The "nucleons only" curve, is replaced by an "independent nucleons" curve, and the "6-quark" curve by a "correlated nucleons" curve. The latter describes the data well. From this analysis and several others[6-8] we know that proper treatment of the two-nucleon correlations is paramount in obtaining a quantitative description of DCX. When

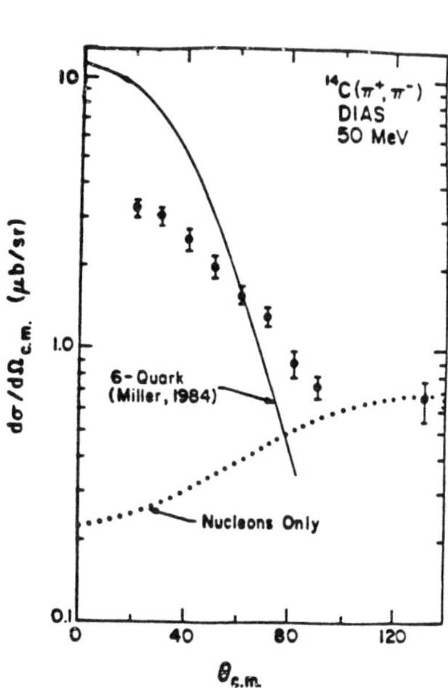

Fig. 1 The ^{14}C data (Ref. 1, 2) and theoretical curves (Ref. 3) for the DIAS transition.

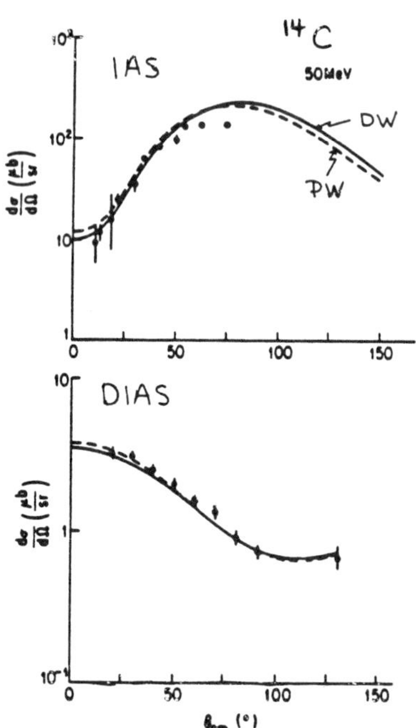

Fig. 2 Analysis of the IAS and DIAS transitions on ^{14}C at 50 MeV recently presented in Ref. 5.

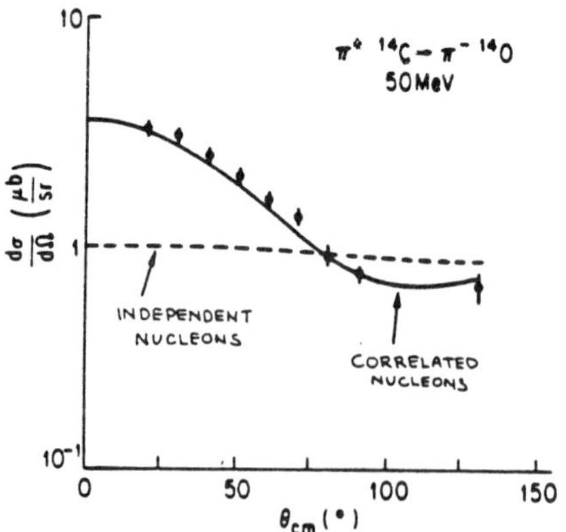

Fig. 3 Calculations of Ref. 5 showing the importance of including the correlated motion of valence nucleons in calculating DIAS cross sections.

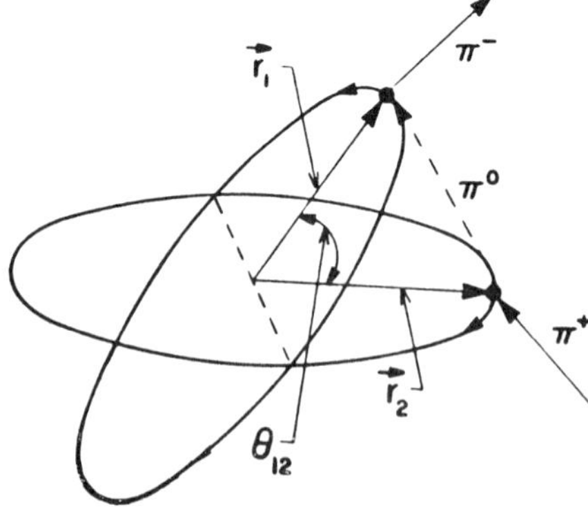

Fig. 4 The coordinates involved in specifying the two-nucleon correlation function.

these are neglected, and nuclear ground states are represented by simple product valance-particle density functions, one gets the curves shown in Figs. 1 and 3 labeled "nucleons only" and "independent nucleons."

Two key ingredients to understanding these results are the two-nucleon correlation function and the range of propagation of the intermediate π°.

2.1 TWO-NUCLEON CORRELATION FUNCTIONS

The nuclear structure involved in the calculation of DIAS cross sections is the pairwise two-body correlation function for an arbitrary pair of valence neutrons. Following Ref. 5, this may be written as

$$\rho^{(2)}(\vec{r}_1,\vec{r}_2) = \rho(r_1)\rho(r_2) \ - \ C_{N-Z}(\vec{r}_1,\vec{r}_2) \ , \tag{1}$$

where the first term on the right-hand side represents independent particle motion and the second term represents the two-body correlations. The coordinates $\vec{r}_1, \vec{r}_2$, and the angle θ_{12} between the planes of single particle orbits are illustrated in Fig. 4. For two nucleons in $\ell = 1$ orbits, it is possible to write the two-body correlation function in the form[6]

$$\rho^{(2)}(\vec{r}_1,\vec{r}_2) \ = \ N\rho(r_1)\ \rho(r_2)\{a \ + \ b\ cos^2\ \theta_{12}\} \ , \tag{2}$$

where N is a normalization constant. The values assumed for a and b depend on the ground-state wave function. The Cohen-Kurath wave functions for ^{14}C yield values[9] (a, b) = (0.393, 1.818). The curve in Fig. 3 labeled "correlated nucleons" follows from these values; the curve labeled "independent nucleons" follows from (a, b) = (1, 0).

The extreme sensitivity to values of a and b in Eq. (2) is illustrated in Fig. 5, taken from Ref. 6. The calculation with (a, b) = (0, 1) gives a very forward-peaked, diffraction-like, angular-distribution shape, whereas the calculation with (a, b) = (1, 0) gives a forward-depressed, featureless, angular distribution.

A more realistic two-body correlation function must take into account somehow, the hard-core repulsion felt by two nucleons as they attempt to occupy the same space. Some calculations have been performed to investigate this.[6] One approach is to introduce a phenomenological factor in the two-body correlation function, such as used in Ref. 10, and illustrated in Fig. 6.

$$\rho^{(2)}(\vec{r}_1,\vec{r}_2) \ = \ N\rho(r_1)\rho(r_2)C(r) \tag{3}$$

$$C(r) \ = \ 1 \ - \ (1 - 8r^2)exp(-2.83r^2) \ . \tag{4}$$

This function simulates short-range repulsion and intermediate-range attraction. Calculations performed with this C(r) for the Ca isotopes[11] have the effect of increasing the 0° DIAS cross sections by a factor of two. Further study of the consequences on DCX of this type of correlation might be informative.

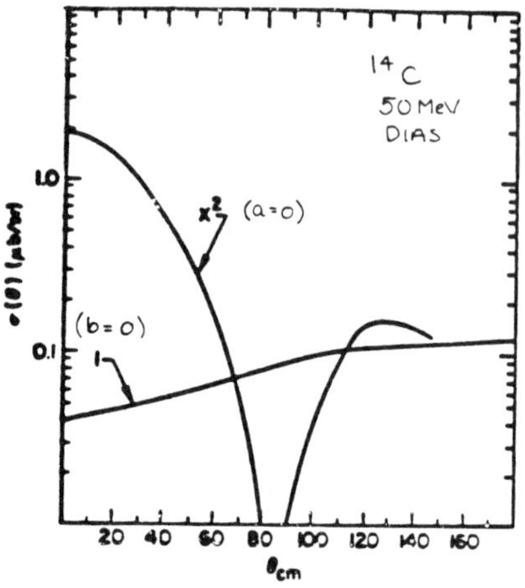

Fig. 5 Calculations of Ref. 6 illustrating the large effects that two-body correlations have on DIAS cross sections. The coefficients a and b are defined in the text.

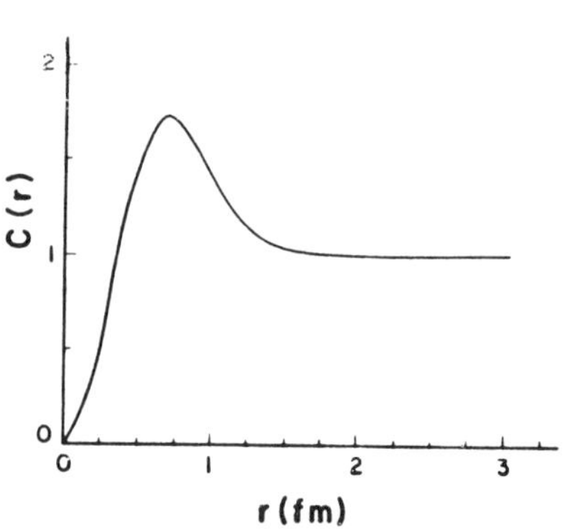

Fig. 6 The two-nucleon correlation function used in Refs. 10 and 11.

2.2 TWO-NUCLEON SEPARATION DENSITY

An instructive way to examine two-nucleon correlations is to examine the separation density function ρ_{SEP}. This is defined in Ref. 12 as follows:

$$\rho_{SEP}(r) = \int \rho^{(2)}_{N-Z} (\vec{R} + \frac{1}{2}\vec{r}, \vec{R} + \frac{1}{2}\vec{r}) d\vec{R} \ , \tag{5}$$

where $\vec{R} = \frac{1}{2}(\vec{r}_1 + \vec{r}_2)$ is the di-nucleon center-of-mass coordinate and $\vec{r} = \frac{1}{2}(\vec{r}_1 - \vec{r}_2)$ is the relative coordinate. Two calculations of ρ_{SEP} are shown[13] in Fig. 7, one for $\rho^{(2)} = \rho(r_1)\rho(r_2)$ and one for $\rho^{(2)}$ corresponding to the Cohen-Kurath wave functions. The latter has approximately twice the value at r = 0 than the former, which has the effect of giving a factor of four increase in the 0° DIAS cross section (Fig. 3).

2.3 RANGE OF DCX

The reason for the heavy weighting of the r = 0 region may be understood by examining the propagation of the π° between the first and second charge-exchange scattering. Following Ref. 5, this propagation may be represented by an effective propagator $\widetilde{G}$, which in coordinate space takes the form

$$\widetilde{G}(r) = -\frac{e^{ikr}}{4\pi r} + \frac{e^{-\Lambda r}}{4\pi r}[1 + \frac{k^2 + \Lambda^2}{2\Lambda} r] \ , \tag{6}$$

where Λ is the range associated with the off-shell πN amplitude. For point πN scattering, $\Lambda = \infty$, and $\widetilde{G}(r)$ reduces to the first term on the right-hand side, which is sharply peaked at r = 0. A typical value used in πN off-shell amplitudes is $\Lambda = 3.5$ fm^{-1}, which gives a somewhat larger range to $\widetilde{G}(r)$. The presence of $\widetilde{G}(r)$ in the DCX scattering matrix integrals gives an inherent weighting to small-r scatterings. In the full calculation of the ^{14}C DIAS transition in Ref. 5, it is found that 1.2 fm is the most probable distance traveled by the intermediate π°. Similar results are found in Ref. 6.

The short-range nature of DCX is further demonstrated in recent calculations by Kaufmann and Gibbs[9] for the $^{14}C(\pi^+,\pi^-)$ ^{14}O(DIAS) cross section at 50 MeV and 0°. This cross section calculated as a function of cut-off radius r is shown in Fig. 8. One sees that excluding the region 0 to 1.25 fm in the radial integral reduces the 0° cross section by 71%.

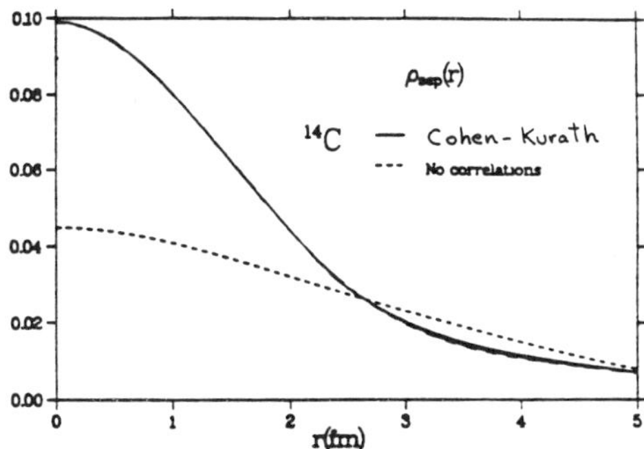

Fig. 7 The separation density for the two valence neutrons of ^{14}C as given in Ref. 13.

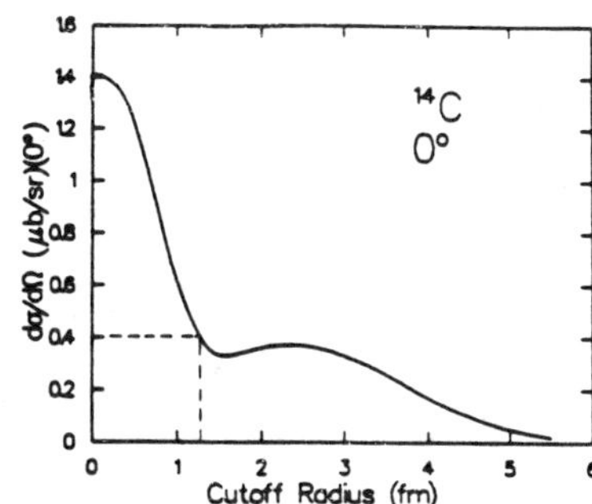

Fig. 8 Calculated[9] DIAS cross for ^{14}C as function of cutoff on the two-nucleon separation distance.

3. RESULTS ON ^{18}O AND ^{26}Mg AT 50 MeV

Angular distributions were measured[14,15] at TRIUMF at 50 MeV for ^{18}O and ^{26}Mg. These data are nearly identical to those of ^{14}C, as seen on Fig. 9. Quite similar angular distributions were measured recently on ^{42}Ca and ^{54}Fe (discussed below). The tentative conclusion that we draw from these results is that the two-particle correlations, as represented by the functions $C_{N-Z}(\vec{r}_1, \vec{r}_2)$ or $\rho_{SEP}(r)$, are probably quite similar for these T = 1 nuclei. The nuclei ^{14}C, ^{18}O, ^{42}Ca, and ^{54}Fe have ground- state wave functions generated by two particles outside a doubly closed-shell core. A second feature to be noted is that the net effect of distortions is similar over this broad range in nuclear mass. This is plausible in view of the distorted wave patterns displayed for ^{14}C and ^{48}Ca in Ref. 7 and the discussion given in Ref. 5.

4. RESULTS ON ^{42,44,48}Ca

Although the data on SCX and DCX are most complete on ^{14}C, it was the puzzles raised by the cross sections for the Ca isotopes that led to the most decisive confirmation on the role of two-nucleon correlations in DCX. For these isotopes, we may study DIAS transitions for 2, 4, 5, 6, and 8 nucleons in the 1 $f_{7/2}$ shell (only cases 2, 4, and 8 have been measured so far). The structure of the DIAS cross section on J = 0 targets with n valence nucleons takes the form[5,11]

$$\frac{d\sigma}{d\sigma_{DIAS}}(\theta) = \frac{n(n-1)}{2} \mid F_{ij}^{(n)}(\theta) \mid^2 , \qquad (7)$$

where $F_{ij}^{(n)}$ is the double-scattering amplitude for any one pair. Its strong dependence on n was discovered on the Ca isotopes.

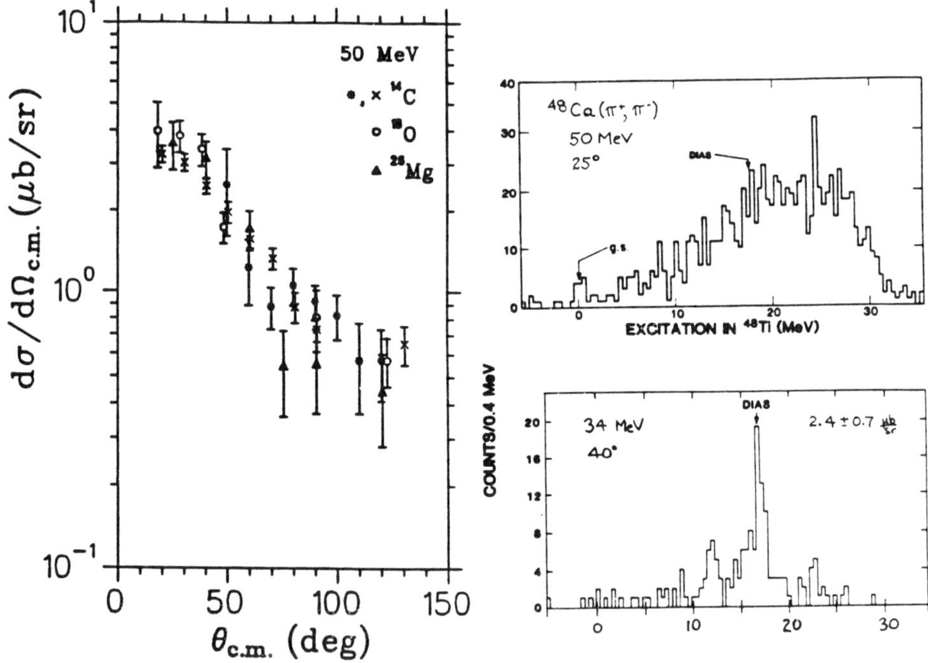

Fig. 9 DIAS cross sections for ^{14}C (Ref. 1,2) ^{18}O (Ref. 14), and ^{26}Mg (Ref. 15).

Fig. 10 DCX spectra for ^{48}Ca at 50 and 35 MeV, as given in Ref. 16.

We started the Ca experiments with ^{48}Ca on the expectation that $F_{ij}^{(n)}$ depends weakly on n. Thus, we expected a DIAS cross section of order 30 to 100 μb/sr at 50 MeV and 25°. Figure 10 shows the outcome of this measurement.[16] We could not detect the DIAS and set an upper limit of 3 μb/sr on the cross section. We then lowered the beam energy and, quite unexpectedly, found[16] that the nonanalog background dropped sharply and the DIAS showed up at the expected energy. Its cross section at 40° is 2.4±0.7 μb/sr. The strong energy dependence in the nonanalog background is still not understood.

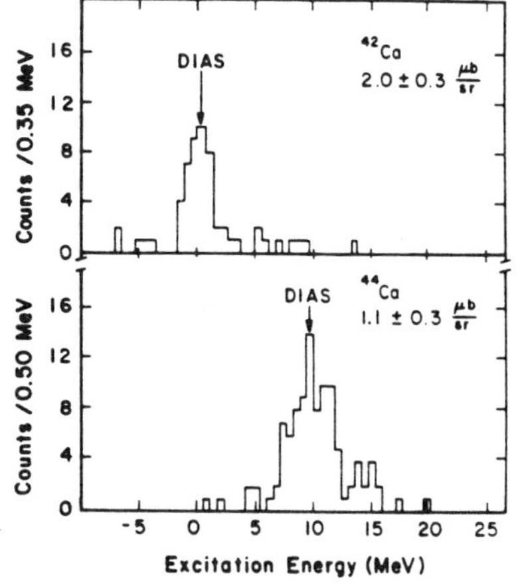

Fig. 11 The data[17] on the DIAS transitions on ^{42,44}Ca at an incident energy of 35 Mev.

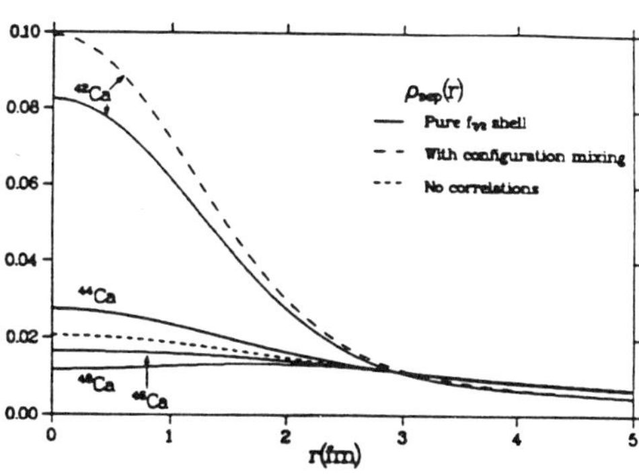

Fig. 12 Separation densities for a pair of valence nucleons in Ca isotopes, as given in Ref. 12.

At this point we were most curious on what the cross sections for ^{42,44}Ca would be and we performed measurements[17] at 35 MeV. Spectra at 40° are shown in Fig. 11. We found that ^{42}Ca and ^{48}Ca had about the same cross section and that ^{44}Ca had about one half this value. These results seemed quite peculiar, in view of our expectations, as stated above. At that time we wondered whether the rather different excitation energies in the residual nuclei (Ti) where the cause of these "anomalous" cross section ratios. These are 0.0, 9.4, 17.1 MeV for Ca -42, -44, and -48, respectively. The reaction Q-values for the DIAS are $\simeq$ 12.3 MeV for all three isotopes.

We now know that the explanation for these results is to be found in the two-body correlations, which are very different for the different isotopes. In Fig. 12 are shown the separation densities as given in Refs. 12,13. This function for the uncorrelated product wave function $\rho(r_1)\ \rho(r_2)$ is nearly flat between 0 and 5 fm. For ^{44,46,48}Ca the functions ρ_{SEP} are only slightly modified from the uncorrelated function, whereas for ^{42}Ca it peaks strongly at r = 0. A second calculation of ρ_{SEP} is shown for ^{42}Ca, which follows from the ground-state wave function

$$\psi(^{42}Ca) \;=\; \sqrt{.91}\ f^2_{7/2} \;+\; \sqrt{.09}p^2_{3/2} \quad . \tag{8}$$

We see that even this modest amount of configuration mixing enhances the short-distance probability significantly. In general, configuration mixing will enhance the short-distance probability.

From the curves shown in Fig. 12, and the short-range nature of DCX as discussed in Sec. 2.3, cross-section ratios for the Ca isotopes can be qualitatively understood.[12] The plane-wave double-impulse-approximation calculations, which follow from the ρ_{SEP} curves in Fig. 12 are shown in Fig. 13. The calculated cross sections are in good agreement with all aspects of the data. In addition to reproducing the measured cross-section ratios at forward angles (30 to 40°), the calculation gives the result that the ^{48}Ca cross section at 80° should be much larger than those of ^{42,44}Ca, as is found in the data.

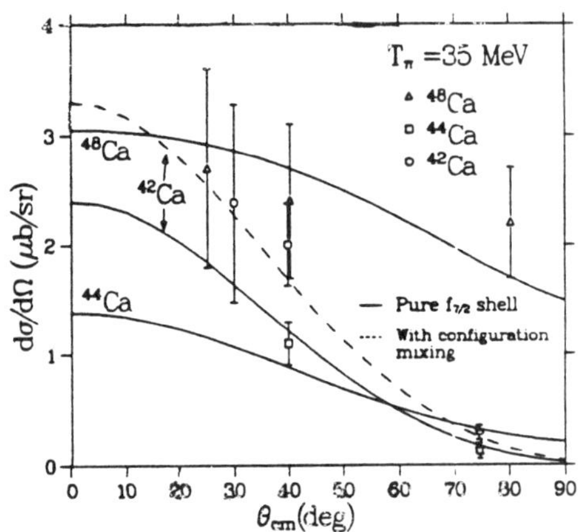

Fig. 13 Comparison of the calculations of Ref. 12 with the data[16,17] on the Ca isotopes.

5. THE TWO-AMPLITUDE SHELL MODEL AND ITS APPLICATION TO 1 $f_{7/2}$ SHELL NUCLEI

The data on the Ca isotopes spurred important theoretical work,[11,18] which relates the double-scattering problem to theorems previously applied to two-body residual interactions in nuclei. In a first article,[18] it was pointed out that a rather general form of the two-body-scattering operator F_{ij} has matrix elements in a space of $(j^n)_J$ configurations with n-even and J = 0 of the form

$$< DIAS \mid F_{ij}|G.S. > = \frac{1}{2}n(n-1)\alpha + \frac{1}{2}n\beta \ , \tag{9}$$

where $\propto$ and β are complex amplitudes independent of n. This leads to DIAS cross sections of form

$$\frac{d\sigma}{d\Omega_{DIAS}}(\theta) = \frac{n(n-1)}{2} \mid \alpha(\theta) + (n-1)^{-1}\beta(\theta)|^2 \ . \tag{10}$$

This expression shows that the β-amplitude, which arises from two-body correlations, has reduced weight for larger n. The values of $\propto$ and β depend on the double-scattering approximations made for DCX. When F_{ij} is written in terms of multipole components, the L = 0 component separates naturally from the L > 0 components. The former is identified as the A(θ) amplitude and the sum of the latter ($\sum_{L>0} F_L$) as the B amplitude. This gives

$$\frac{d\sigma}{d\sigma_{DIAS}}(\theta) = \frac{n(n-1)}{2} \mid A(\theta) + w(n)\, B(\theta) \mid^2 , \qquad (11)$$

where

$$w(n) = (2j + 3 - 2n)(n - 1)^{-1}(2j - 1)^{-1}$$

$$A = \alpha + 2\beta(2j + 1)^{-1}$$

$$B = \beta(2j - 1)(2j + 1)^{-1} .$$

This form of the DIAS cross section holds for spin-dependent F_{ij}.

Equation (11) has been generalized using particle-hole symmetry to include configurations $[j^{n_1}\, j^{n_2}]_{J,v}$ with n_1 neutrons and n_2 protons coupled to $J = 0$ and seniority v. The weighting factor w now depends on n_1 and n_2, or equivalently $n = n_1 + n_2$ and $T = (n_1 - n_2)/2$. For the $j = 7/2$ shell, and $v = 0$ states it is

$$w(n,T) = \frac{9(n + 3)(5 - n) + 7(n - 2T)(n + 2 + 2T)}{27(2T + 3)(2T - 1)} . \qquad (12)$$

This gives the values for $f_{7/2}$ nuclei shown in Table I.

TABLE I

T	Target Nuclei	$d\sigma/d\Omega$(DIAS)
1	Ca-42, Fe-54	$1\|A+B\|^2$
2	Ca-44, Cr-52	$6\|A-0.11B\|^2$
3	Ca-46, Ti-50	$15\|A-0.07B\|^2$
4	Ca-48	$28\|A-0.14B\|^2$
1	Ti-46, Cr-50	$1\|A+1.47B\|^2$
2	Ti-48	$6\|A+0.17B\|^2$

We see that there are two classes of T = 1 nuclei, depending on whether only neutrons participate (^{42}Ca,^{54}Fe), or both neutron and protons participate (^{46}Ti,^{50}Cr). The latter have $w > 1$, thereby giving enhanced weighting to short-range correlations.

To test the generalized two-amplitude model, we performed measurements,[19] in July 1987 on ^{44}Ca, ^{46,50}Ti, and ^{54}Fe. Representative spectra are shown in Fig. 14 and preliminary values of cross sections are shown in Fig. 15. It is interesting to compare model predictions with the preliminary results. At 25°, the Ca(42, 44, 48) cross sections are $(2.4 \pm 0.3, 1.3 \pm 0.2, 2.4 \pm 0.3)$ μb/sr. These values give $|A| = 0.370\sqrt{\mu b/sr}, |B| = 1.317\sqrt{\mu b/sr}$, and $\phi = 45°$, (the relative phase). Inserting these into the expressions given in Table I give the predicted values shown in Table II.

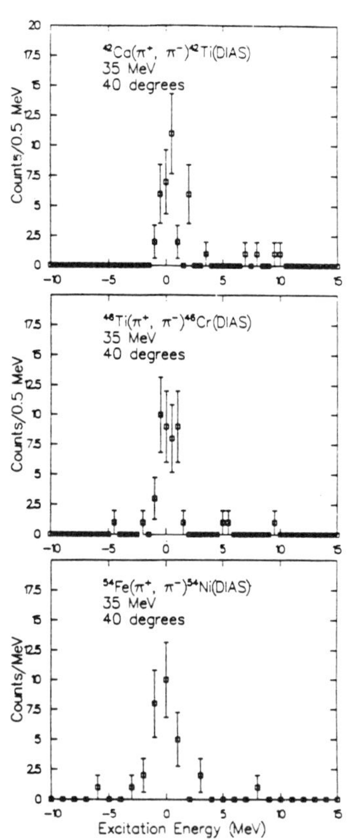

Fig. 14 Spectra showing DIAS transitions on $f_{7/2}$ shell nuclei (Ref. 19).

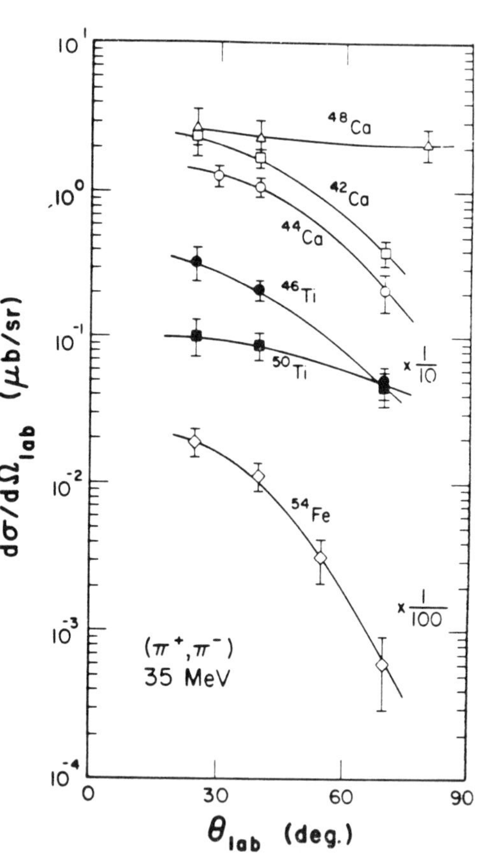

Fig. 15 Preliminary DIAS cross sections $f_{7/2}$ shell nuclei (Ref. 16, 17, and 19).

TABLE II

Cross sections at 25° in μb/sr.

T	Target	Model	Experiment[a]
1	Ca-42	2.4	2.4 ± 0.3
1	Fe-54	2.4	1.9 ± 0.4
1	Ti-46	4.7	3.3 ± 0.8
1	Ca-42	2.4	2.4 ± 0.3
2	Ca-44	1.3	1.3 ± 0.2
3	Ti-50	1.6	1.0 ± 0.3
4	Ca-48	2.7	2.4 ± 0.3

[a]Preliminary values from LAMPF experiment #1045 (Ref. 19)

We see that the preliminary experimental cross sections agree in all cases with the model predictions. ^{46}Ti is expected to have the largest cross section and this was found. The measurements show that the T = 1 nuclei ^{42}Ca, ^{46}Ti, and ^{54}Fe have forward-peaked angular-distribution shapes. For these nuclei $\omega \geq 1$. For ^{50}Ti and ^{48}Ca, $\omega = 0.07$ and 0.14, respectively, and for these the measured angular distributions are relatively flat. ^{44}Ca has $\omega = 0.11$, and thus the model prediction is that the angular-distribution shape should be similar to ^{50}Ti and ^{48}Ca. The preliminary data disagree with this prediction.

In general, the new data on 1 $f_{7/2}$ shell nuclei give good confirmation of the model predictions.

6. CONCLUSION

The following results would seem to hold for DIAS transitions at low pion energies:

(1) The double-impulse approximation works well.
(2) For nuclei with a single-active valence shell j, the DIAS cross section can be written in terms of two amplitudes, as given by Eq. 12.
(3) Two-nucleon correlations produce large effects. These can be understood in terms of interference in $A + \omega B$. Amplitude B results predominantly from short-range correlations.

I believe that the evidence justifies taking the two-amplitude model seriously, and that efforts should be made to understand the empirically determined A and B amplitudes from a fundamental meson-nucleon theory. The measurement of B and its theoretical analysis present a new possibility for understanding short-range two-nucleon dynamics in nuclei.

ACKNOWLEDGEMENTS

I wish to express appreciation to my experimental colleagues M. J. Leitch and E. Piasetzky for many stimulating discussions. Also, numerous enlightening discussions with W. R. Gibbs, W. Kaufmann, M. Bleszynski, and R. Glauber are gratefully acknowledged. I also wish to express my appreciation to Denise Dalmas for her expert assistance in the preparation of this manuscript.

REFERENCES

1. I. Navon et al., *Phys. Rev. Lett.* **52**, 105 (1984).
2. M. J. Leitch et al., *Phys. Rev. Lett.* **54**, 1482 (1985).
3. G. A. Miller, *Phys. Rev. Lett.* **53**, 2008 (1984).
4. J. L. Ullmann et al., *Phys. Rev.* **C33**, 2092 (1986).
5. M. Bleszynski and R. J. Glauber, *Phys. Rev.* **C36**, 681 (1987).
6. W. R. Gibbs et al., LAMPF Workshop on Pion Double Charge Exchange, 1985, Los Alamos National Laboratory report LA-10550-C (H. W. Baer and M. J. Leitch, Eds.) p. 90ff.
7. E. R. Siciliano et al., *Phys. Rev.* **C34**, 267 (1986).
8. T. Karrapiperes and M. Kobayaski, *Phys Rev. Lett.* **54**, 1230 (1985).
9. W. B. Kaufmann and W. R. Gibbs, private communication (1988).
10. P. B. Siegel, Ph.D. dissertation, 1986, Arizona State Univ. (unpublished).
11. N. Auerbach et al., Los Alamos National Laboratory document LA-UR-88-781 (1988).
12. E. Bleszynski, M. Bleszynski, and R. J. Glauber, *Phys. Rev. Lett.* **60**, 1483 (1988).
13. E. Bleszynski et al., in *Proceedings, on Pion-Nucleus Physics: Future Directions and New Facilities at LAMPF*, August 1988.
14. A. Altman et al., *Phys. Rev. Lett.* **55**, 1273 (1985).
15. T. Anderl, Ph. D. dissertation, 1987, Universität at Bonn (unpublished).
16. H. W. Baer et al., *Phys. Rev.* **C35**, 1425 (1987).
17. Z. Weinfeld et al., *Phys. Rev.* **C37**, 902 (1988).
18. N. Auerbach et al., *Phys. Rev. Lett.* **59**, 1076 (1987).
19. LAMPF Proposal #1045 (Spokesman H. W. Baer, J. Leitch, E. Piasetzky).

EXPERIMENTAL DETERMINATION OF THE σ-TERM

H. Matthäy
Institut für Experimentelle Kernphysik, Universität Karlsruhe

ABSTRACT

The situation of experimental low energy pion nucleon data and its impact on the problem of the σ-term is reviewed

INTRODUCTION

Low-energy pion-nucleon scattering has been one of the most important testing grounds of Current algebra and PCAC in the past. In recent years a renewed interest has arisen from the systematic investigation of the low-energy structure of Greens functions in QCD[1]. In this so-called chiral perturbation theory a simultaneous expansion of Green functions in powers of the external momenta and of quark masses is performed. Within the frame of chiral perturbation theory Gasser and Leutwyler calculated the σ-term of pion-nucleon scattering from the bayron mass spectrum[2]. This term is a basic quantity related to the scalar up- and down quark currents of the nucleon and measures the explicit chiral symmetry-breaking of QCD:

$$\sigma = \frac{\hat{m}}{2m_N} < p|\bar{u}u + \bar{d}d|p > , \quad \hat{m} = \frac{m_u + m_d}{2} \tag{1}$$

The authors obtained

$$\sigma = \frac{(35 \pm 5)\text{MeV}}{1-y} \tag{2}$$

where a non zero value of y would indicate a strange sea-quark contribution to the nucleon:

$$y = \frac{2 < p|\bar{s}s|p >}{< p|\bar{u}u + \bar{d}d|p >} \tag{3}$$

Furthermore they derived a relation between the pion-nucleon amplitude Σ at the Cheng-Dashen point $t = 2m_\pi^2$, $\nu = (s-u)/4m_N = 0$ and the σ-term:

$$\Sigma = \sigma + 5\ \text{MeV} \tag{4}$$

The πN-amplitude Σ is, according to a low-energy theorem, related to the isospin-even pion-nucleon amplitude D+ at the Cheng-Dashen point:

$$\Sigma = \frac{f_\pi^2}{2} \left[D^+ (t=2m_\pi^2, \nu = 0) - \frac{g^2}{m_N} \right] \tag{5}$$

where f_π = 0.132 GeV is the pion decay constant and g the pion-nucleon coupling constant. Höhler and Koch[3,4] obtained the amplitude Σ from experimental pion-nucleon scattering data by means of forward dispersion relations extrapolating from the physical region of the ν-t-plane to the Cheng-Dashen point. Their value is $\Sigma = (65 \pm 5)$ MeV and consequently $\sigma = (60 \pm 8$ MeV). The experimental value is almost a factor of two bigger than the value obtained by Gasser and Leutwyler for y=0, and y would have to be of the order of 0.4, if the strange sea quarks in the nucleon are blamed for the discrepancy. Gasser and Leutwyler argue that then the nucleon mass in the chiral limit of vanishing quark masses would be completely different from what it is in the real world. It is this problem of a possible large strange sea quark current in the nucleon which has attracted great attention during the last years[5].

The discrepancy between values for the σ-term on one hand calculated within the frame of chiral perturbation theory and obtained on the other hand from the dispersion analysis of low-energy pion-nucleon scattering is not understood. It is regarded as a serious problem which hints either at theoretical deficiencies (the calculations have been performed at the one-loop-level) or at an insufficient πN data base.

SITUATION OF LOW-ENERGY PION-NUCLEON DATA

The amplitude Σ was determined from scattering phases (referred to as the Karlsruhe-Helsinki (KH) phases in the following) of a phase shift analysis which was carried out almost ten years ago[6]. Due to the fact that only a few low-energy data existed at that time the phase shift analysis relied basically on data above 100 MeV[7,8].

Low energy data ($T_\pi \leq 100$ MeV), however, are of particular significance as input of a dispersion analysis for the determination of Σ. Therefore in the following the results of new experiments will be compared with the predictions of the KH analysis which apart from the CMU-LBL analysis[9] is the only phase shift analysis of πN scattering below 300 MeV to satisfy Mandelstam analyticity and unitarity.

Fig. 1a comprises the π^+p data of Bussey et al.[7] and of Bertin et al.[8] which are part of the KH data base. In addition it shows a recent measurement of π^+p differential cross-section[10] at forward angles to be discussed below which agrees well with the KH prediction.

In Fig. 1b the KH predictions are compared with the results of recent experiments of Frank et al.[11] and Brack et al.[12]. With exception of the 127 MeV/c data of Frank all the data exhibit a tendency to lie below the KH curves. In contrast to this situation the data of Ritchie et al.[13] taken at backward angles are found to be in good agreement with the KH prediction. In the meanwhile the results of Brack have been confirmed by a new experiment carried out by the same group at a pion energy of 66.8 MeV[14]. From the discrepancy between their own data and those of Bertin and Ritchie the authors conclude that the errors of the older experiments have

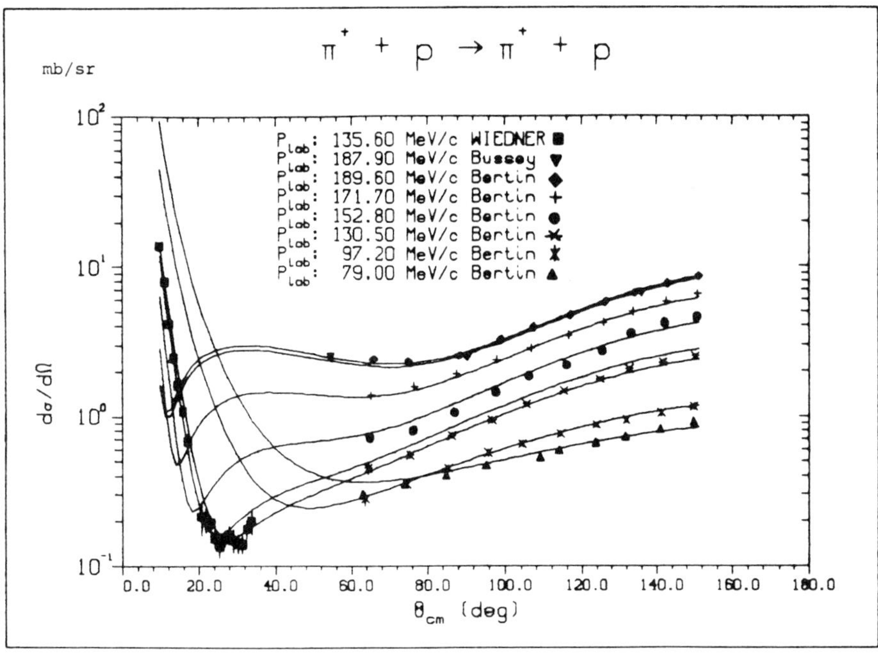

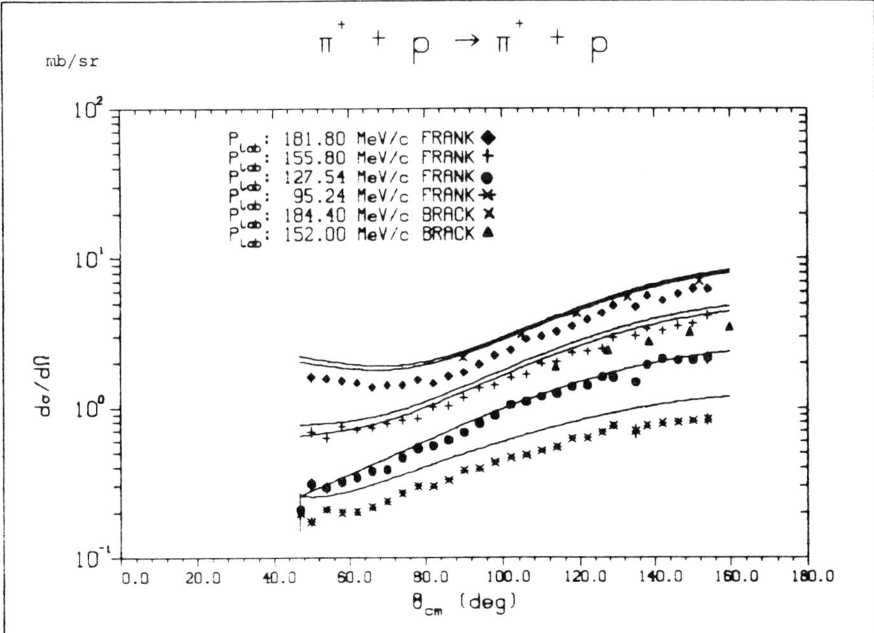

Fig. 1a,b: π^+p data of Bussey, Bertin, Wiedner, Frank, Brack and the corresponding KH predictions

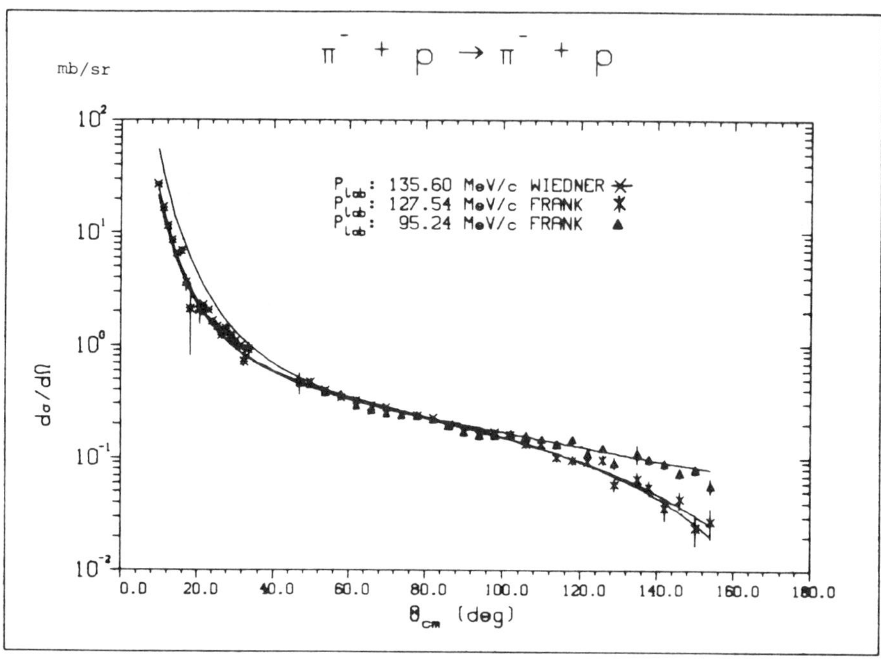

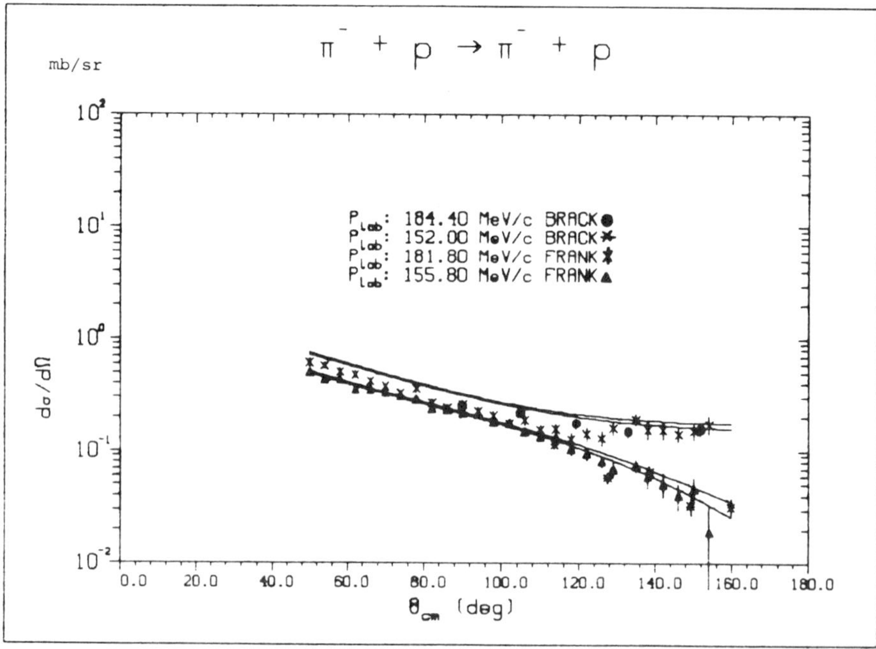

Fig. 2a,b: π^-p data of Frank, Wiedner, Brack and the corresponding KH predictions

been underestimated. It is noteworthy that at the time when the KH analysis has been completed no low energy $\pi^- p$ data were available. It turns out, however, that the $\pi^- p$ data of Frank, Brack and Wiedner shown in Figs. 2a and b with exception of the 182 MeV/c data of Frank agree well with the result of the KH analysis. In good agreement are also the KH isospin odd phases and the data from the charge exchange reaction: $\pi^- p \rightarrow \pi^0 n$ [15].
This, however, has no direct impact on the problem of the σ-term.

In the following two experiments will be discussed in more detail which both have in common that they provide experimental information in a kinematical region which is particularly sensitive to the determination of Σ namely at t=0. The first one is a measurement of $\pi^{\pm} p$ differential cross-sections at 55 MeV and forward angles between 8° and 28° covering the Coulomb-nuclear interference region [10]. The basic idea behind the experiment is that by extrapolation of the expression

$$\Delta(t) = t \cdot \left[\frac{d\sigma^+}{d\Omega} - \frac{d\sigma^-}{d\Omega}\right] \tag{6}$$

to t = o the real part of the isospin even forward scattering amplitude can directly be obtained

$$\lim_{t \to o} \mathrm{Re} D^+(t) = \frac{\pi s}{2\alpha m_N^2 \omega} \lim_{t \to o} \Delta(t) \tag{7}$$

with $\omega = (p_\pi^2 + m_\pi^2)$ and α being the Sommerfeld constant. The value obtained experimentally for Re $D^+(t=o)$ agrees with the KH value. Presently measurements of $\pi^{\pm} p$ angular distributions are continued with the magnetic spectrometer LEPS [16]. The main reason for using the spectrometer is that its response to pions of opposite charge is identical in contrast to the range telescope used before where depending on the charge different πμ discrimination methods had to be applied. The absolute normalization of the data being always a crucial problem with spectrometers is achieved by normalizing the πp cross-sections to the well known ep and μp cross-sections. The second experiment to be mentioned in this context is the measurement of the strong interaction shift of X-rays from the 2P-1S transition in pionic hydrogen by means of crystal diffraction resulting in the direct determination of $a_{o+}^{\pi^- p}$ [17]. The value $a_{o+}^{\pi^- p} = (0.059 \pm 0.006) m_\pi^{-1}$ from this experiment is in disagreement with the KH prediction of $a_{o+}^{\pi^- p} = (0.082 \pm 0.005) m_\pi^{-1}$. As Koch [18] has shown the experimental value would on one hand result in a smaller Σ amplitude compatible with the QCD value of about 35 MeV for the σ-term. On the other hand this value would contradict seriously the scattering data of Frank. At present a new experiment of the same group is underway which allows the simultaneous measurement of the strong interaction shift and the line broadening with an improved technique in order to determine the real and the imaginary part of $a_{o+}^{\pi^- p}$.

Concluding, there is substantial hope that with the forthcoming experiments at PSI and TRIUMF the present problem of contradictory πN data and hopefully the problem of the σ-term will be solved in the next few years.

ACKNOWLEDGEMENT

I thank Dr. J. Gasser and Prof. G. Höhler for many stimulating discussions.

REFERENCES

1. J. Gasser, and H. Leutwyler, Nucl. Phys. B250(1985)465,517,539
2. J. Gasser, Nucl. Phys. B279(1987)65
3. G. Höhler, Landolt-Börnstein, Vol. 9 b2, ed. by H. Schopper, Springer-Verlag, Berlin 1983
4. R. Koch, Z.Physik C15(1982)161
5. J. F. Donoghue, and C. R. Nappi, Phys. Letters 168B(1986)105
6. R. Koch, and E. Pietarinen, Nucl. Phys. A336(1980)331
7. P. F. Bussey et al., Nucl. Phys. B58(1973)363
8. P. Y. Bertin et al., Nucl. Phys. B106(1976)341
9. R. E. Cutkovsky et al., Phys. Rev. D20(1979)2782,2804,2839
10. U. Wiedner et al., Phys. Rev. Letters 58(1987)648
11. J. S. Frank et al., Phys. Rev. D28(1983)1569
12. J. T. Brack et al., Phys. Rev. C34(1986)1771
13. B. G. Ritchie et al., Phys. Letters 125B(1983)128
14. J. T. Brack et al., Preprint University of Colorado, Boulder NPL-1035
15. D. H. Fitzgerald et al., Phys. Rev. C34(1986)619
16. H. Matthäy et al., The LEPS Spectrometer at SIN, Proceedings of the Conference on Dynamics of Collective Phenomena in Nuclear and Subnuclear Long Range Interactions in Nuclei, ed. by P. David, Bad Honnef, 1987, World Scientific Publ. Company, Singapore, in press
17. E. Bovet et al., Phys. Letters 153B(1985)231.
18. R. Koch: Inconsistencies in Low-Energy Pion-Nucleon Scattering, Report TKP 85-5 (unpublished)

SURVEY OF PION ABSORPTION IN LIGHT NUCLEI

H. J. Weyer
Physics Institute, Basel University, CH-4056 Basel, Switzerland

ABSTRACT

The status of the knowledge of pion absorption in nuclei is reviewed. Emphasis is put on light nuclei and on the different reaction modes. The evidence for reaction modes involving more than two nucleons is discussed.

INTRODUCTION

In view of the time allotted this can be only a limited survey of the interesting field of pion absorption. Rather than to try a detailed coverage of the subject, it seemed to be more promising to concentrate on one highly controversial issue, the relative importance of the different reaction mechanisms for the absorption of a pion. The discussion of other important topics e.g. the 2N-absorption component in the deuteron and heavier nuclei, has been kept very short, as there are quite excellent reviews available, and the reader is kindly referred to the literature [1–5] for detailed information.

Whereas in the other main reaction modes of a pion with nuclei our knowledge has reached a remarkable degree of completeness, the microscopic understanding of absorption, with the exception of the deuteron case, is still in its infancy. It is characteristic of the current situation, that in the Δ-hole model [6], which has been so successful in describing pion-nucleus scattering, the whole complexity of the absorption part is taken into account by a phenomenological fit parameter, the imaginary part of the spreading potential V_{sp}.

The investigation of pion absorption is complicated both on the experimental (many-fold coincidence measurements) and on the theoretical side (multi-body problem). But, as one is dealing with typically one third of the total cross section, [3] one needs to understand clearly the absorption part before one achieves a comprehensive picture of the interaction of a pion with a nucleus. Another strong interest in pion absorption stems from the special kinematics of this process. The incoming pion brings a large amount of energy but only small momentum. Therefore large momentum transfers are involved. We will see in addition, that two or more nucleons are involved in the absorption process, which relates absorption to the issue of short-range correlations. Pion absorption is, therefore, regarded [7] as one of the most promising tools in the search for sub-nucleonic degrees of freedom.

REACTION MECHANISMS

The investigation of the reaction mechanisms in pion absorption is strongly related to the question, how many and in which way nucleons in a nucleus are involved in the absorption process. Having to deal with particles subject to the strong interaction both in the initial and in the final state, it is obvious, that one has to worry about a reasonable definition of "participation". In the simplest approach one would require, that a "participating" nucleon has to have acquired a minimum energy, i.e. to be beyond some experimental threshold. For a microscopic understanding this is certainly not sufficient.

As we will see below, one of the most important current topics in the understanding of absorption mechanisms is to distinguish genuine multi-nucleon absorption from 2-step processes consisting of already-known building block reactions.

The most detailed investigations have been performed in the exclusive measurements in ^{3}He and ^{4}He by identification of the different mechanisms with specific distributions in the Dalitz-plot, momentum [8-10] and exchange-mass [9,11] distributions. Because of the nuclear influence (binding, Fermi-momentum) the kinematical signatures are certainly smeared out. It has even been argued[12], that e.g. signatures of 2-step processes could be washed out totally. This, however, seems to be highly improbable; from quasi-elastic reactions we have a conception about the amount of smearing and the expected signatures are quite strong.

1N-ABSORPTION

Absorption of a pion on a free nucleon is exactly forbidden because of momentum conservation. Absorption on a nucleon within a nucleus is strongly suppressed, as the nucleon has to be far off shell in order to satisfy momentum conservation. Although in principle allowed, 1N-absorption, however, is an example of the above mentioned difficulties to identify uniquely a specific reaction mechanism: the kinematical signature is quite clear, e.g. πA$\rightarrow$(A-1)N, but a 1N-mechanism relies on the very high momentum tails of a nuclear wave function with correspondingly low probabilities. Therefore it has to be verified, that those two-particle final states can be reached only by a 1N-mechanism and not by other competing mechanisms, e.g. by a 2N-absorption followed by ("soft") final state interaction (see below). Indeed, in a calculation for the reaction π^-^{4}He$\rightarrow$tn it has been found,[13] that even for this process, which has a signature quite different from 2N-absorption, half of the reaction strength has to be attributed to 2N-absorption followed by final state interactions. 1N-absorption is, however, only a small contribution to the total absorption ($\sim$1% [14]) and can therefore in the following be neglected.

2N-ABSORPTION

In the next simple absorption mode, which has been proposed as the dominant one a long time ago,[15] the pion is absorbed by two nucleons and the rest of the nucleus behaves as a spectator. Consequently the absorbing nucleon pair shows a strong angular correlation, the width of which is governed by the relative momentum distribution of this pair and the spectating system. In the ideal case this distribution is not connected to the absorption process but it is a nuclear property determined by the nuclear wave function. The validity of this assumption could be demonstrated quite clearly in the case of ^{3}He. Here the momentum distribution of the spectating nucleon has been determined from the ^{3}He(π,pp) data [9]. Quite good agreement has been found [9] with the nucleon momentum distribution, which is known very well, e.g. from (e,e′p) measurements.[16]

The 2N-absorption can be subdivided into two components according to the isospin of the absorbing nucleon pair:

- In the isoscalar component the initial nucleon pairs have the quantum numbers of the deuteron. This part could be studied therefore in detail in the πd$\rightarrow$pp reaction. [3,17] It has been found that the dominant amplitude describes a reaction proceeding through an intermediate $^5S_2$ ΔN state ending in a $^1D_2$ pp state. The πd system is supposed to be theoretically understood, including differential cross section and polarisation data, on the level of 10%. In the next simple system, ^{3}He, detailed measurements [9,10,18–22] have been performed, which showed clearly the dominance of the 2N-mechanism in the measured angular correlations and in the momentum distributions. Also the increase in the integrated cross section has been found to be near the factor 1.5, which is to be expected from the ratio of numbers of isoscalar nucleon pairs in ^{2}H and ^{3}He.

For A >3 there have been measurements of the 2N-absorption in various nuclei.[23,24] The measured angular correlations have been fitted by a sum of a "narrow" and a "broad" Gaussian. Then only the component described by the "narrow" Gaussian has been identified[23] with the 2N-mechanism. This component has been integrated and by comparison found to be only 1/10 of the total[25] absorption cross section. Even after correction for the estimated loss due to "hard FSI" the contribution of 2N-absorption was found to be only 1/3. In the ^{58}Ni-data the 2N-contribution is estimated[24] to 1/2. This procedure has been heavily disputed in literature. It has been pointed out,[26] that the two-Gaussian approach is much too simplified, e.g. in the case of non-zero angular momentum of the absorbing pair the measured angular correlation can be much more complicated. This has been known from measurements with stopped pions[27] and has been found also now for pions in flight.[28] The issue is not yet settled and there is still appreciable uncertainty in the interpretation of the experimental data, but a reasonable estimate may be that the 2N-absorption may explain 1/2 to 4/5 of the total absorption strength.

- In contrast to the isoscalar case the absorption on isovector nucleon pairs cannot be studied in the deuteron system. The lightest nucleus providing isovector nucleon pairs is ^{3}He. Here detailed investigations have been performed. [9,18,19,29,30] Because of quantum numbers the $^5S_2/^1D_2$ partial wave, which dominated the isoscalar case, is now forbidden and one has access to amplitudes, which are usually overwhelmed by Δ-mechanisms. Significant asymmetries in the angular distributions [9,18,19,29] indicate, that not only weak NΔ-amplitudes contribute but even amplitudes with no intermediate Δ-state at all. This explains, why Δ-dominance models[31] predict an isospin ratio, which is too high and that the inclusion of NN′ intermediate states[32] improves the agreement[33] with the experimental data. An energy-independent partial wave analysis[34] of the TRIUMF data at 65MeV yielded 2 solutions with different strengths for the $^3D_1$ and $^3S_1$ waves, both of which describe non-Δ intermediate states. It has been argued by referring to the time-reversed reaction,[35] that the right solution is the one with a dominant $^3D_1$ amplitude. This is certainly an important issue and it should be possible to find an unique solution by a polarisation measurement[34] or by an energy-dependent analysis, when more data for the isovector component are available.

MULTI-STEP PROCESSES

We have seen, that the dominant absorption mechanism for pions is by two nucleons. It is, however, conceivable, especially when the direct 2N-mechanism is hindered, that in addition the pion of the initial state or one of the nucleons of the final state interact with the nuclear medium.

- In an initial state interaction (ISI) the pion may interact first with another nucleon by quasi-elastic scattering or charge exchange and is then absorbed by two nucleons.

- Similarly, the outgoing nucleons may interact with other nucleons in final state interactions (FSI) after a 2N-absorption.

  - After a 2N-absorption there may be billiard-type collisions between one of the nucleons and a third nucleon where the colliding partners are near the mass shell. In this kinematics the laboratory angle between these nucleons should be always near 90° in the case of on-shell nucleons, which will within a nucleus be smeared out because of Fermi momentum. This case with a high relative momentum between the FSI-nucleons is defined as 'hard' FSI.

  - In another type of FSI, the 'soft' FSI, there is an enhancement in the cross section in those kinematical regions, where the relative momentum of two final nucleons is very small. This effect has been observed in numerous hadronic reactions e.g. break-up, pion-production. The enhancement is determined by the low energy behaviour of the nucleon-nucleon interaction and is described conveniently in terms of the effective range approximation[36] parameters.

In the on-shell limit ISI-processes preceding or FSI-processes following a 2N-absorption cannot be seen as new mechanisms. They must be regarded as combination of two known sub-components in a factored way. E.g. a 2N-absorption, preceded by an initial state interaction, is not a new reaction mechanism, but rather a modification of an ideal 2N-absorption by the inevitable influence of the surrounding nuclear medium. Such multi-step processes should therefore in principle be able to be described by known building blocks (quasi-elastic π-nucleus scattering, charge exchange, 2N-absorption and NN-scattering).[37]

It has been argued[38] and demonstrated by cascade code calculations,[23,24] that in a 2-step process with an ISI preceding a 2N-absorption there should be an asymmetry in the NN angular correlation in comparison with unpertubed reactions, as in a 2-step process the π's have some energy lost. The calculations have been compared[23,24] with the experimental data and no 2-step ISI-2N processes have been found with an upper limit of 30%.

Another method of identification exploits the fact, that the differential cross section for quasi-elastic scattering is forward/backward peaked. In a 2-step process the distributions should reflect the $\pi N \rightarrow \pi N$ reaction of the first step by exhibiting a peak at the corresponding kinematics. An analogous argument has been used in the search for multi-nucleon photon-induced reactions.[39]

In two three-proton coincidence experiments[40,41] on ^{12}C the authors have searched for such structures at small angles. Whereas in one experiment[40] no clear indication for such a 2-step mechanism has been identified, the authors of the other experiment[41] report a peak at small angles with the signature for a ISI/2N-process with, however, a small part of the yield.

In another experiment the measured angular correlation[42] for $^6Li(\pi^- pp)$ and $^{12}C(\pi^- pp)$ is interpreted by the authors as showing double peaks originating both from ISI and FSI. Recent preliminary results from TRIUMF[43] suggest, however, a more conservative interpretation of the data as one broad peak, which would be also much more consistent with the results from measurements with stopped pions.[44,45] In the ^3He data also "hard-FSI" signatures have been searched for. Such a mechanism should be indicated by bands of events parallel to the coordinate axes in the Dalitz-plot. No indication has been found.[9]

MULTI-NUCLEON-ABSORPTION

What do we expect a genuine 3N- or 4N-mechanism to look like? As there is not yet a generally accepted model for those reaction modes, the only way of definition is by exclusion of known mechanisms. For the above discussed ISI and FSI processes we expect, as also for 2N-absorption, a significant peaking of the cross section in clearly defined regions of phase space. Therefore good candidates for multi-N-absorption are "phase-space-flat" distributions.

It has been discussed in the previous sections, that the comparison of the total absorption cross section and the FSI-corrected 2N-components indicate the presence of an additional mechanism with an estimated strength of 30-50%. In

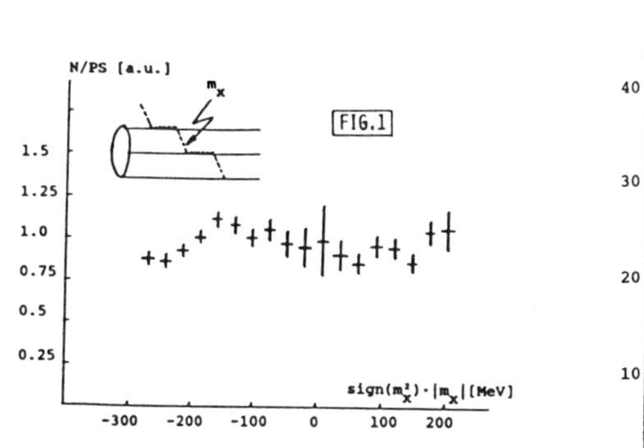

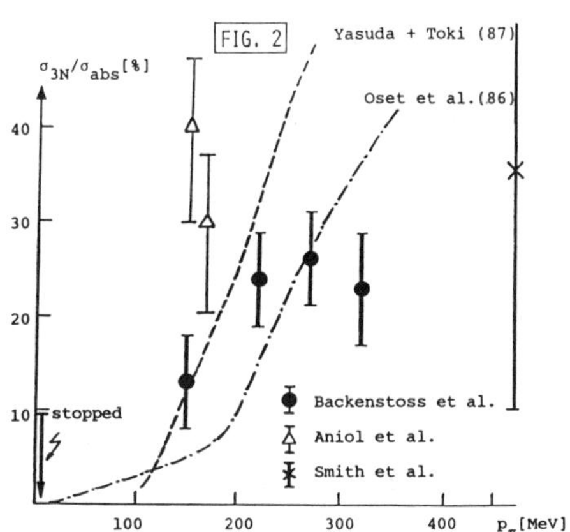

Figure 1: Distribution of the mass of the exchanged object in a sequential 3N-absorption graph determined from 3N-absorption data[30] on ^{3}He(π^+,pp). The distribution is normalised to phase space.

Figure 2: Relative contribution of 3N-absorption to the total absorption in ^{3}He as function of the incoming pion momentum. The data are from Aniol et al.[18], Backenstoss et al.[30], Smith et al.[19]. Theoretical calculations are from Oset et al.[46] and Yasuda et al.[47].

heavier nuclei it is hard to disentangle a conceptually new reaction mechanism from 2-step processes consisting of known building blocks. In a light nucleus like ^{3}He, however, the situation is much more favorable. Here the reaction can be measured fully exclusively and there is only a limited combinatorial ambiguity to identify a measured final configuration with a specific reaction mechanism. From the angular correlation and recoil distribution data measured in conjugate counter configurations one sees clearly,[30] that there is an obvious deviation from the exponential behaviour of a 2N-mechanism in the region of high relative momenta between the absorbing pair and the spectator.[18,30] This component is even more clearly present in off-conjugate positions and is found[30] to be, within experimental errors, proportional to the phase space factor("phase-space-flat").

Nevertheless, it may be conceivable, that this cross section may be generated by multi-step mechanisms and that this "phase-space-flat" behaviour would be only a result of an unfortunate projection.

The most probable reaction mechanism is[11] a sequential 3-body mechanism as shown in the inset of Fig. 1. This graph is able to describe both a genuine 3N-mechanism, where the object exchanged between the first and the second nucleon is far off-shell, and a 2-step process, where the object will be a pion near to its mass shell. From an exclusive measurement on ^{3}He [9] this mass of the exchanged object has been determined in a frozen nucleus approximation. This is shown in Fig. 2, normalized to phase space and corrected for acceptance. The theoretical prediction[11] for a genuine 3N-process exhibits a rather flat distribution, whereas in the case of an on-shell 2-step process a strong enhancement is predicted at $m_x = m_\pi$. Within the experimental resolution the data are totally constant over the whole accessible mass region. This lends strong support for a genuine 3-body mechanism in ^{3}He. The theoretical calculation[11] does not include the effect of Fermi motion, which will certainly smear out the distributions. It is, however, improbable, that the whole signature will disappear.

The contribution of 3N-absorption to the total absorption in ^{3}He is shown in Fig. 2. There seems to be some normalisation difference between the TRIUMF and the PSI data, but all experiments agree, that 3N-absorption is an important contribution. Furthermore, the energy-dependence indicates a mechanism involving intermediate Δ-states. First theoretical calculations[46,47] are in reasonable agreement with the data. Going from ^{3}He to ^{4}He there is in addition to the known mechanisms now the possibility for a mechanism involving 4 nucleons (e.g. double-Δ[48]). The absorption cross section[25,49] shows a strong increase for A $\geq$4 in comparison with the lighter nuclei, which has been conjectured[38] to be generated by a 4N-mechanism.[48]

In a recent exclusive measurement ^{4}He(π^+,3N)N the different multi-nucleon contributions could be identified. [8] It has been found, that the cross section for the 3N-component is about two times the one in ^{3}He, a 4N-contribution, however, is an order of magnitude lower.

CONCLUSION

The new series of experiments on pion-absorption have provided us with exiting results, which some years nobody had expected. With exclusive measurements it was possible to perform detailed investigations of the reaction mechanism. Strong evidence has been found, that in addition to the long-known 2N-mechanism there is also a significant 3N-mechanism. A 4N-mechanism, about which there has been much speculation about, seems to be, however, much weaker.

The third generation of absorption experiments with large solid angle detectors, which have been recently built or are under construction (BGO-ball at LAMPF, LADS at PSI and CLASS at TRIUMF), will certainly provide us with more exciting information about this physics.

ACKNOWLWEDGEMENTS

I thank my colleagues, especially the collaborators in the Basel-Karlsruhe-Zagreb and the LADS collaborations for invaluable discussions. Various authors are grateful acknowledged providing me with information prior to publication. I thank R.J. Powers for carefully reading the manuscript.

REFERENCES

1. C.H.Q. Ingram, Nucl. Phys. A374, 319c (1982).
2. R. P. Redwine Nucl. Phys. A434, 239c (1985).
3. D. Ashery, J. P. Schiffer, Ann. Rev. Nucl. Part. Sci. 36, 207 (1986).
4. R. Mach, M.G. Sapozhnikov, I.V. Falkomin, Sov. J. Part. Nucl. 17, 549 (1986).
5. W. R. Gibbs, B. F. Gibson, Ann. Rev. of Nucl. Part. Phys. 37, 411 (1987).
6. L. S. Kisslinger, and W. L. Wang, Ann. Phys. 99, 374 (1976);
   M. Hirata, F. Lenz and K. Yazaki, Ann. Phys. 108, 116 (1977);
   M. Hirata, J.H. Koch, F. Lenz and E.J. Monitz, Ann. Phys. 120, 205 (1979);
   for a recent review of the theory see F. Lenz, Nucl. Phys. A478, 581c (1988).
7. E. Vogt, AIP Conference Proc. 163, 551 (1988).
8. G. Backenstoss et al., PSI-Preprint PR-88-04 and to be publ. in Phys. Rev. Lett.
9. G. Backenstoss et al., to be published.
10. H.J. Weyer, Helv. Phys. Acta 60, 667 (1987).
11. L.L. Salcedo et al., preprint and to be published.
12. J. Silk, Phys. Rev. C37, 891 (1988).
13. K. Sakamoto et al. Phys. Rev. C31, 1987 (1985).
14. B. Bassalleck et al. Nucl. Phys. A319, 397 (1079).
15. M. Rudermann, Phys. Rev. 81, 383 (1952).
16. E. Jans et al., Phys. Rev. Lett. 49, 974 (1982).
17. B. Blankleider, Nucl. Phys. A463, 77c (1987).
18. K. Aniol et al., Phys. Rev. C33, 1714 (1986).
19. C. Smith et al., Contr. C 26, Panic Conf., Kyoto (1987).
20. D. Gotta et al., Phys. Lett. 112B, 129 (1982).
21. G. Backenstoss et al., Phys. Lett. 137B, 329 (1984).
22. M. Moinester et al., Phys. Rev. Lett. 52, 1033 (1984).
23. A. Altman et al., Phys. Rev. C34, 1757 (1986).
24. W. J. Burger et al., Phys. Rev. Lett. 57 ,58 (1986).
25. D. Ashery et al., Phys. Rev. C23, 2173 (1981).
26. B. G. Ritchie et al., Phys. Rev. C30, 969 (1984).
27. e.g. B. Bassalleck et al., Nucl. Phys. A343, 365 (1980); Phys. Rev. C19, 1893 (1979).
28. R. Schumacher et al., Contr. C 66, Panic Conf., Kyoto (1987) and to be published.
29. H.J. Weyer et al., Czech. J. Phys. B36, 243 (1986).
30. G. Backenstoss et al., Phys. Rev. Lett. 55, 2782 (1985).
31. T.-S. H. Lee, K. Ohta, Phys. Rev. Lett. 49, 1079 (1982).
32. R. Silbar, E. Piasetzky, Phys. Rev. C29, 1116 (1984); Phys. Rev. C30, 1365 (1984).
33. O.V. Maxwell, C.Y. Cheung, Nucl. Phys. A454, 606 (1986).
34. E. Piasetzky et al., Phys. Rev. Lett. 57, 2135 (1986).
35. S.E. Vigdor et al., Phys. Rev. Lett. 58, 840 (1987).
36. M. L. Goldberger, K. M. Watson, Collision Theory, Wiley & Sons (1964);
    K. M. Watson, Phys. Rev. 88, 1163 (1952);
    A. B. Migdal, JETP (Sov. Phys.) 1, 2 (1955).
37. For a most recent example of a cascade calculation see e.g. W. R. Gibbs and W. B. Kaufmann, AIP Conference Proc. 163, 279 (1988).
38. J. P. Schiffer, Comm. Nucl. Part. Phys. 14, 15 (1985).
39. J.M. Laget et al., Phys. Lett. 151B, 325 (1985);
    G. Tamas, Nucl. Phys. A463, 295c (1987)
40. R. Tacik et al., Phys. Rev. C32, 1335 (1985).
41. W. Brückner et al., Nucl. Phys. A469, 617 (1987).
42. H. Yokota et al., Phys. Rev. Lett. 58, 191 (1987).
43. G. Lolos et al., private information
44. M. Doerr et al., Nucl. Phys. A445, 557 (1985).
45. U. Sennhauser et al., Nucl. Phys. A386, 429 (1982).
46. E. Oset et al., Nucl. Phys. A448, 597 (1986).
47. T. Yasuda et al., Progr. Theor. Phys. 77, 905 (1987).
48. G. E. Brown et al., Phys. Lett. 118B, 39 (1982);
    B. Schwesinger et al., Phys. Lett. 132B, 269 (1983).
49. M. Baumgartner et al., Nucl. Phys. A399, 451 (1983).

COINCIDENCE MEASUREMENTS OF PION SINGLE CHARGE EXCHANGE IN THE Δ–RESONANCE REGION

S. Høibråten, S. Gilad[a], W. J. Burger[b], G. W. Dodson, G. Gatoff[c], L. D. Phạm, R. P. Redwine
Massachusetts Institute of Technology, Cambridge, Massachusetts 02139

E. Piasetzky[d], H. W. Baer, J. D. Bowman, F. H. Cverna, F. Irom[e], M. J. Leitch, U. Sennhauser[b]
Los Alamos National Laboratory, Los Alamos, New Mexico 87545

J. N. Knudson[f]
Arizona State University, Tempe, Arizona 85287

Th. S. Bauer[g], C. H. Q. Ingram, G. S. Kyle[h]
Paul Scherrer Institute, formerly SIN, CH–5234 Villigen, Switzerland

D. Ashery, S. A. Wood[i]
School of Physics and Astronomy, Raymond and Beverly Sackler Faculty, Tel Aviv University, Tel Aviv, Israel

S. H. Rokni[j]
Utah State University, Logan, Utah 84322

Z. Fraenkel
Weizmann Institute, Rehovot, Israel

ABSTRACT

We discuss the first coincidence measurements of the reaction $(\pi^+, \pi^0 p)$. These were performed primarily on ^{16}O at $T_{\pi^+} = 165$ and 245 MeV. The reaction is shown to have a clear quasi-free signature. The π^0 energy spectra for events corresponding to removal of a p-shell nucleon are compared to predictions of Δ–hole model calculations;

---

[a] Present address: Cubital Ltd., Herzlia B, Israel.
[b] Present address: Paul Scherrer Institute, CH–5234 Villigen, Switzerland.
[c] Present address: Division de Physique Théorique, Institut de Physique Nucléaire, F–91406 Orsay, France.
[d] Present address: School of Physics and Astronomy, Tel Aviv University, Tel Aviv, Israel.
[e] Present address: VG Systems, Woodland Hills, CA 91367.
[f] Present address: Los Alamos National Laboratory, Los Alamos, NM 87545.
[g] Present address: NIKHEF–K, NL–1009DB Amsterdam, The Netherlands.
[h] Present address: Department of Physics, NMSU, Las Cruces, NM 88003.
[i] Present address: Department of Physics, University of Illinois, Urbana, IL 61801.
[j] Present address: SLAC, Stanford University, Stanford, CA 94305.

the calculations generally underestimate the $(\pi^+, \pi^0 p)$ cross sections. At forward pion angles experimental ratios between the $(\pi^+, \pi^+ p)$ and $(\pi^+, \pi^0 p)$ reactions are found to differ significantly from estimates using free π–N cross sections. A study of the A-dependence of the $(\pi^+, \pi^0 p)$ process indicates that the cross section is roughly constant over the examined range of nuclear masses.

MOTIVATION

Coincidence measurements of pion induced nucleon knockout reactions have made a significant contribution to our understanding of medium energy nuclear interactions. These interactions appear to be dominated by the creation of a $\Delta(1232)$ in the nucleus. The charged pion channels $(\pi^\pm, \pi^\pm p)$ have been studied through inclusive[1,2] and exclusive[3] coincidence experiments. In these studies the quasi-free process was identified, and the quasi-free $(\pi^+, \pi^+ p)/(\pi^-, \pi^- p)$ cross section ratio was found to be larger than the free $\pi^+ p/\pi^- p$ cross section ratio. This implies that the quasi-free scattering process is substantially modified in the nuclear medium. Similar modifications are expected for the quasi-free single charge exchange (SCX) process. Measurement of the $^{16}O(\pi^+, \pi^0 p)$ reaction should provide information about the isospin structure of interactions which modify the quasi-free process. This paper discusses the first coincidence measurements of this process. Initial results for $T_{\pi^+} = 245$ MeV have already been published[4].

The most promising theoretical approach to the problem of medium effects appears to be the Δ–hole model[5,6]. Figure 1 shows some of the $(\pi^+, \pi^+ p)/(\pi^-, \pi^- p)$ cross section ratios along with the calculated Δ–hole model values. For the $(\pi^+, \pi^0 p)$ process this model predicts a significant enhancement (about 30%) of the cross section in nuclei relative to the free process.

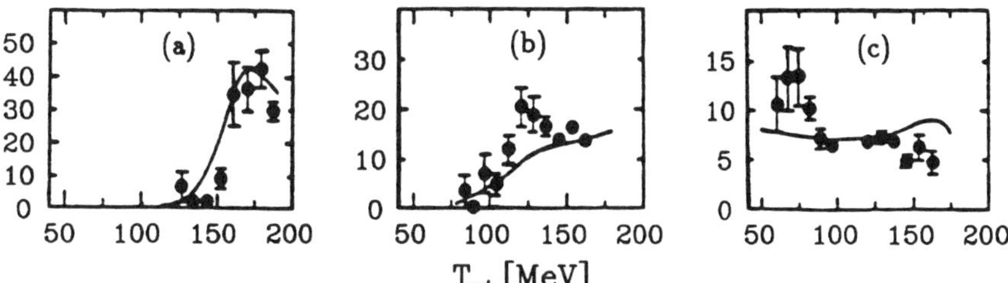

Figure 1. Ratios between $d^3\sigma/dE_{\pi'}d\Omega_{\pi'}d\Omega_p$ for the $^{16}O(\pi^+, \pi^+ p)$ and $^{16}O(\pi^-, \pi^- p)$ reactions as a function of the kinetic energy of the scattered pion. The incident beam energy was 240 MeV. The pions were detected at (a) 35°, (b) 60°, and (c) 130°, and the protons were detected at the corresponding conjugate quasi-free angles. Only events where the proton had been removed from the p-shell were included in the analysis. The solid curves are Δ–hole model calculations of the ratios. (From Reference 3.)

EXPERIMENT

The measurements were performed at the Clinton P. Anderson Meson Physics Facility (LAMPF) using the Low Energy Pion channel (LEP). The $^{16}O(\pi^+, \pi^0 p)$

reaction was studied at a number of π^0 angles from 50° to 130° at incident π^+ energies of 165 MeV and 245 MeV. Water was used as target material. The $(\pi^+, \pi^0 p)$ process requires a neutron, so the presence of hydrogen nuclei did not complicate the data analysis. At the lower energy, measurements at $\theta_{\pi^0} = 110°$ were also taken using ^{56}Fe, ^{120}Sn, and ^{208}Pb targets.

Based on Monte Carlo simulations[7] of the experiment, the target thickness and the target angle were chosen for each measurement to maximize the counting rate while maintaining energy resolution at about 8 MeV for the determination of the excitation energy of the residual nucleus.

The two photons from the decay of the ejected pion were detected by the LAMPF π^0 spectrometer[8]. The protons were detected by a vertical array of seven telescopes covering an angular range of ±51° around the nominal free proton emission angle (the "quasi-free" angle). This configuration allowed measurements over the entire quasi-free angular correlation region. Each telescope covered a solid angle of 8.5 msr and consisted of a ΔE–E pair of plastic scintillators. This allowed excellent particle identification over a proton energy range of 25–175 MeV. The FWHM energy resolution of the proton detectors was about 5 MeV at $T_p = 100$ MeV.

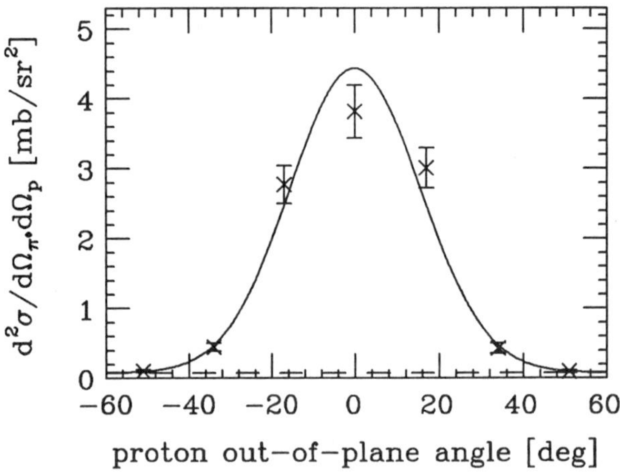

Figure 2. $d^2\sigma/d\Omega_{\pi^0}d\Omega_p$ for $^{16}O(\pi^+, \pi^0 p)$ as a function of the proton out-of-plane angle for $T_{\pi^+} = 165$ MeV and $\theta_{\pi^0} = 110°$. Only events where the nucleon was removed from the p-shell are included. The solid curve is a fit to the data assuming a Gaussian distribution on top of a flat background (dashed line).

RESULTS

In the analysis of the $^{16}O(\pi^+, \pi^0 p)$ measurements, events corresponding to removal of a p-shell nucleon were identified by the π^0 and p energies involved. A typical spectrum of the doubly differential cross section $d^2\sigma/d\Omega_{\pi^0}d\Omega_p$ as a function of the proton out-of-plane angle is shown in Figure 2. The solid curve is a Gaussian fit to the data assuming a flat backgroud. This clearly shows the quasi-free nature of the

Table 1. Measured differential cross sections for $T_{\pi^+} = 165$ MeV. The table lists $d\sigma/d\Omega_{\pi^0}$ in mb per sr.

(a) Angular dependence, oxygen target.

Process	$\theta_{\pi^0} = 70°$	$\theta_{\pi^0} = 80°$	$\theta_{\pi^0} = 110°$	$\theta_{\pi^0} = 130°$
$(\pi^+, \pi^0 p_{\text{p-shell}})$	0.72 ± 0.07	0.88 ± 0.06	2.01 ± 0.12	1.98 ± 0.10
$(\pi^+, \pi^0 p)$	1.68 ± 0.33	2.33 ± 0.32	4.43 ± 0.33	4.89 ± 0.27
(π^+, π^0)	6.12 ± 0.57	6.94 ± 0.57	11.56 ± 1.04	11.96 ± 1.17

(b) A-dependence at $\theta_{\pi^0} = 110°$.

Process	^{16}O	^{56}Fe	^{120}Sn	^{208}Pb
$(\pi^+, \pi^0 p)$	4.4 ± 0.3	4.4 ± 0.4	4.9 ± 0.6	5.8 ± 0.7
(π^+, π^0)	11.6 ± 1.0	16.2 ± 1.5	25.6 ± 2.4	39.1 ± 3.7

process studied. The rather flat shape of the cross section near the central (quasi-free) angle is observed only when we limit the analysis to p-shell events. It can be explained by the general momentum distribution of p-shell nucleons which has a low probability for low-momentum particles, and hence will lead to very few nucleons being ejected into the near vicinity of the nominal quasi-free angle.

Earlier studies have shown that the shape of the quasi-free angular correlation is roughly symmetric in all directions about its center[2]. We used this to calculate $d\sigma/d\Omega_{\pi^0}$ by integrating $d^2\sigma/d\Omega_{\pi^0}d\Omega_p$ over all proton angles. We also measured the inclusive, single-arm process $^{16}O(\pi^+, \pi^0)$ directly. Table 1a compares these cross sections to the exclusive measurements discussed above. We find that 30–40% of the inclusive cross sections can be accounted for by the coincidence process, and that about 40% of these events correspond to removal of a single p-shell nucleon.

Figure 3 shows a π^0 energy spectrum of $d^3\sigma/dE_{\pi^0}d\Omega_{\pi^0}d\Omega_p$ for events where the proton was detected at the conjugate quasi-free angle. The solid curve is a Δ–hole model calculation of the process[9]. The theory exhibits the p-shell characteristics already discussed. Two features of the experimental process contribute to the smoothing out of this effect in the data: (1) the calculation assumes a point-like proton detector, and (2) due to the fairly modest resolution, the data may have some contributions from s-shell removal and other processes favoring low-momentum nucleons. These features do not, however, account for the significant difference between the size of the measured and the predicted cross sections. A similar discrepancy between experiment and theory is found in all our measurements at both 165 MeV and 245 MeV.

We have also calculated ratios between the triply differential cross sections measured for $^{16}O(\pi^+, \pi^+ p)$ and $^{16}O(\pi^+, \pi^0 p)$. This was done for $T_{\pi^+} = 245$ MeV at π^0 angles of 60° and 130°. For the forward angle this ratio was found to be significantly less (by about a factor 2) than the ratio of free cross sections. This is in qualitative agreement with the enhancement of the $(\pi^+, \pi^0 p)$ process in nuclei predicted by the Δ–hole model[5]. As expected we did not observe a very pronounced difference at the backward angle.

To study the dependence of the $(\pi^+, \pi^0 p)$ process on the nuclear mass we made

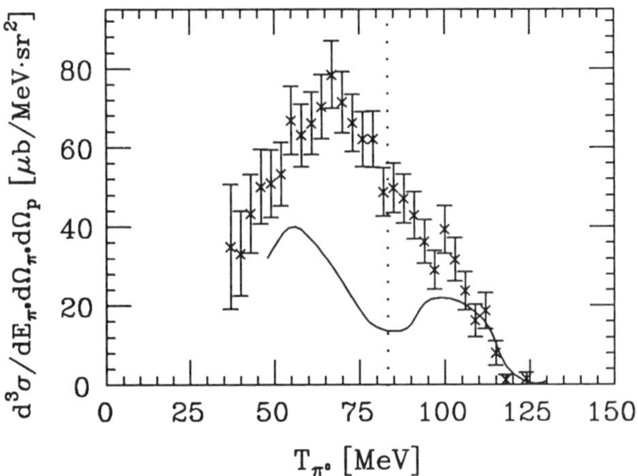

Figure 3. π^0 energy spectrum of $d^3\sigma/dE_{\pi^0}d\Omega_{\pi^0}d\Omega_p$ for the $^{16}O(\pi^+, \pi^0 p)$ reaction with $T_{\pi^+} = 165$ MeV and $\theta_{\pi^0} = 110°$. The protons were detected at the conjugate quasi-free angle. The data are restricted to events where the nucleon was removed from the p-shell. A normalization uncertainty of approximately 9% is not included in the figure. The dotted line indicates the quasi-free π^0 energy, and the solid curve is a Δ–hole model calculation of the process assuming a point-like proton detector.

some measurements using heavier targets (see Figure 4). For this comparison we did not limit the oxygen data to p-shell removal only. The quasi-free structure is clearly observed in all cases. Comparing the cross sections in Table 1b we see that as the nuclear mass increases, the fraction of the inclusive cross section that appears as $(\pi^+, \pi^0 p)$ decreases. This is to be expected since final state interactions will be more important in complex nuclei. As a net result of this, the cross section shows little variation over the whole range of nuclei studied (A = 16–208).

FINAL REMARKS

Our measurements have shown that the $(\pi^+, \pi^0 p)$ process is predominantly quasi-free, in agreement with earlier observations of the $(\pi^\pm, \pi^\pm p)$ process. The pion SCX data should provide a stringent test for present and future intermediate energy interaction models. In particular, we have noted that the Δ–hole model does not seem able to account for the full cross section. It may be necessary to include more reaction channels such as $T = 2$ coupling of the Δ–N interaction.

More information to assist the theoretical attempts should come from a recently completed $^3He(\pi^\pm, \pi^0 p)$ experiment[10]. Initial and final state interactions are small effects for this nucleus. Also, having only one neutron and two protons should lead to an interesting enhancement of important second-order processes. In addition, an experiment to measure $^{16}O(\gamma, \pi^- p)$ is expected to run soon at the Bates Linear Accelerator Center.

This work was supported in part by the U.S. Department of Energy, the Paul Scherrer Institute, and the U.S.–Israeli Binational Science Foundation.

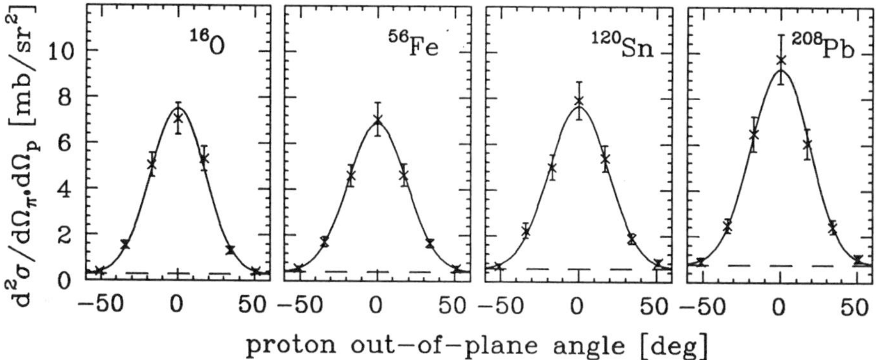

Figure 4. The A-dependence of $d^2\sigma/d\Omega_{\pi^0}d\Omega_p$ for $(\pi^+, \pi^0 p)$ as a function of the out-of-plane proton angle for $T_{\pi^+} = 165$ MeV and $\theta_{\pi^0} = 110°$. The solid curves are fits to the data assuming Gaussian distributions on top of flat backgrounds (dashed lines). The FWHM's are $42° \pm 2°$, $44° \pm 3°$, $45° \pm 3°$, and $44° \pm 3°$, respectively.

REFERENCES

1. E. Piasetzky et al., Phys. Rev. Lett. 46, 1271 (1981).
2. E. Piasetzky et al., Phys. Rev. C 25, 2687 (1982).
3. G. S. Kyle et al., Phys. Rev. Lett. 52, 974 (1984).
4. S. Gilad et al., Phys. Rev. Lett. 57, 2637 (1986).
5. M. Hirata, F. Lenz, and M. Thies, Phys. Rev. C 28, 785 (1983).
6. T. Takaki and M. Thies, to be published.
7. S. Gilad, Ph. D. thesis, Tel-Aviv University, 1979 (unpublished).
8. H. W. Baer et al., Nucl. Instrum. Methods 180, 445 (1981).
9. T. Takaki, private communication.
10. S. Gilad and E. Piasetzky, spokesmen.

CHARGE INDEPENDENCE BREAKING IN THE NN-INTERACTION

J.J. DE SWART AND V.G.J. STOKS

University of Nijmegen, Nijmegen, the Netherlands

The charge independence of the NN-interaction has been an important assumption in nuclear physics. It has been clear for a long time that this isospin symmetry is broken due to the electromagnetic force between the protons and due to mass differences between protons and neutrons and between neutral and charged pions. However, the charge independence breaking (CIB) due to these effects is rather small. A much larger breaking will occur if the neutral and charged pion-nucleon coupling constants are substantially different.

The coupling constant g_c of the charged pions to the nucleons has been determined from $\pi^{\pm}p$ scattering data. The recommended value is [1]

$$g_c^2 = 14.3 \pm 0.2 \ . \tag{1}$$

The coupling constant g_0 of the neutral pion to the nucleons, on the other hand, has always been hard to determine accurately. The values as found in the literature range from $g_0^2 = 13$ to $g_0^2 = 15$ (see e.g. Table I in Ref. 2).

The effect of the CIB in the pion-nucleon coupling constants can best be understood when we neglect for a moment all effects due to mass differences. We can then write the one-pion-exchange (OPE) potential for pp scattering as

$$V_{OPE}(pp) = g_0^2 V_\pi \ .$$

The OPE-potential for np scattering in the $I = 1$ states then reads

$$V_{OPE}(np) = -g_0^2 V_\pi + 2g_c^2 V_\pi = (g_0^2 + 2\,\Delta g^2)V_\pi \ ,$$

where $\Delta g^2 \equiv g_c^2 - g_0^2$. Thus, $V_{OPE}(np)$ in the $I = 1$ states is stronger than $V_{OPE}(pp)$ by $2\,\Delta g^2/g_0^2$.

We have performed a phase shift analysis of all pp scattering data below $T_{\rm lab} = 350$ MeV and an analysis of all np scattering data below $T_{\rm lab} = 30$ MeV. Detailed descriptions of these analyses can be found elsewhere [3,4] The lower partial wave ($J \leq 4$) pp phase shifts in our 0–350 MeV analysis are parametrized with an energy dependent P-matrix at $r = b$ and for

$r > b$ a potential $V_{EM} + V_{OPE} + V_{HBE}$ is used. Here V_{EM} is the electromagnetic potential, consisting of the relativistic Coulomb potential [5] and the vacuum polarization potential [6], and V_{HBE} represents the shorter-range non-OPE boson exchanges. For V_{HBE} we use the non-OPE part of the Nijmegen soft core potential [7]. When we use the parametrized Paris potential [8], then our results do not change significantly. The highest partial waves ($J \geq 10$) are computed in Coulomb distorted wave Born approximation and for $5 \leq J \leq 9$ the phase shifts are determined by optimal mapping techniques. Including the neutral pion-nucleon coupling constant, 28 parameters are used and we arrive at a χ^2 per degree of freedom of $\chi^2/N_{DF} = 1.07$. We then find [2]

$$g_0^2 = 13.1 \pm 0.1 \ . \tag{2}$$

Comparing this result with the value for the charged coupling constant g_c, we observe a CIB in the pion-nucleon coupling constants of $\Delta g^2 \equiv g_c^2 - g_0^2 = 1.2 \pm 0.2$.

For the np phase shift analysis below $T_{\rm lab} = 30$ MeV the isospin $I = 0$ phase shifts with $J \leq 2$ and the $^1S_0$ phase shift are parametrized with the P-matrix at $r = b$, and for $r > b$ the OPE potential is used. The $I = 1$ lower partial waves (^3P_J and $^1D_2$) are taken from our pp analysis after correcting them not only for electromagnetic and mass difference effects, but also for the difference between g_0^2 and g_c^2. All higher partial waves ($J \geq 3$) are taken to be pure OPE. Including Δg^2, we have 7 parameters and arrive at $\chi^2/N_{DF} = 1.08$. We find [9]

$$\Delta g^2 = 1.28 \pm 0.15 \ ,$$

in excellent agreement with Eqs. 1 and 2. No CIB in the pion-nucleon coupling constants (i.e. $\Delta g^2 = 0$) yields $\chi^2/N_{DF} = 1.18$, a rise in χ^2 of 50.

We have also performed a combined pp and np phase shift analysis below $T_{\rm lab} = 30$ MeV. Now the $I = 1$ phase shifts are not only determined by the pp scattering data, but also by the np scattering data. We now obtain $\chi^2/N_{DF} = 1.01$ (a drop of 31.1 in χ^2) and $\Delta g^2 = 1.62 \pm 0.27$. Again we have agreement with the expected value of $\Delta g^2 = 1.2 \pm 0.2$.

The drop in χ^2 is due to the following. The CIB-corrections can not change the spin-orbit combination of the np ^3P-waves, since the OPE potential does not contain a spin-orbit part and the spin-orbit part of the electromagnetic potential is negligible in this energy range. In the combined analysis the spin-orbit combination of the *pp* ^3P-waves is changed and the central and tensor combinations remain the same, causing a rise of $\Delta\chi^2 = 0.9$ on the *pp* scattering data. This change then enables a change in the spin-orbit combination of the *np* ^3P-waves, causing an enormous drop of $\Delta\chi^2 = -32.0$ on the *np* scattering data. In Fig. 1 we have plotted the

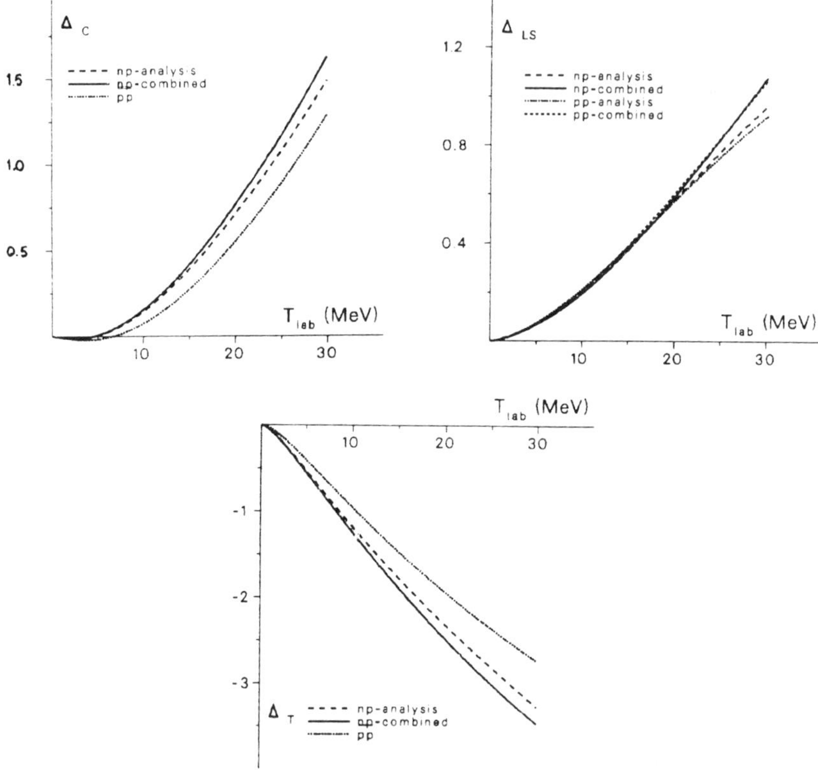

Figure 1: Central, tensor and spin-orbit combinations of the ^3P-phase shifts.

central, tensor, and spin-orbit combinations of the ^3P-waves of the separate *pp* and *np* analyses and of the combined *pp* and *np* analysis. The central and tensor combinations of the *pp* phase shifts are practically the same for both separate and combined analyses, so only the latter are shown. Next to these combinations, the *np* ^3P-phase shifts themselves are very much different from other analyses [10] (up to 35%).

References

[1] O. Dumbrajs *et al.*, Nucl. Phys. B216 (1983) 277.

[2] J.R. Bergervoet, P.C. van Campen, T.A. Rijken, and J.J. de Swart, Phys. Rev. Lett. 59 (1987) 2255.

[3] J.R. Bergervoet, P.C. van Campen, W.A. van der Sanden, and J.J. de Swart, Nijmegen Report THEF-NYM-87.02 (1987), to be published in Phys. Rev. C.

[4] V.G.J. Stoks, P.C. van Campen, and J.J. de Swart, in progress; J.R. Bergervoet, Ph.D. thesis, University of Nijmegen, 1987 (unpublished)

[5] G.J.M. Austen and J.J. de Swart, Phys. Rev. Lett. 50 (1983) 2039.

[6] L. Durand, III, Phys. Rev. 108 (1957) 1597.

[7] M.M. Nagels, T.A. Rijken, and J.J. de Swart, Phys. Rev. D 17 (1978) 768.

[8] M. Lacombe *et al.*, Phys. Rev. C 21 (1980) 861.

[9] V.G.J. Stoks, P.C. van Campen, T.A. Rijken, and J.J. de Swart, Nijmegen Report THEF-NYM-88.04 (1988), submitted for publication.

[10] see e.g. R.A. Arndt, J.S. Hyslop III, and L.D. Roper, Phys. Rev. C 35 (1987) 128.

THE A-DEPENDENCE OF ψ PRODUCTION IN π^- NUCLEUS COLLISIONS AT 530 GeV/c

T. Marshall, T. Bowyer, R. Crittenden, A. Dzierba, S. Kartik,
J. Martin, C. Neyman, P. Smith, T. Sulanke, K. Welch, A. Zieminski
Indiana University, Bloomington, IN 47405
R. Gomez, G. Ludlam
California Institute of Technology, Pasadena, CA 91125
J. Krider
Fermilab, Batavia, IL 60510
H. Goldberg, S. Margulies, J. Solomon
University of Illinois at Chicago, Chicago, IL 60680
C. Davis, H. Xiao
University of Louisville, Louisville, KY 40292
L. Dauwe
University of Michigan at Flint, Flint, MI 48503
V. Abramov, Yu. Antipov, B. Baldin, S. Denisov, V. Glebov,
Yu. Gorin, V. Koreshev, V. Kryshkin, S. Petrukhin, S. Polovnikov,
V. Sirotenko, R. Sulyaev
Institute for High Energy Physics, Serpukhov, USSR

ABSTRACT

The E672/E706 Spectrometer, located in the MW beam at Fermilab, was used to collect data on events containing a pair of muons in the final state with large effective mass. The momentum of incident pions and protons was 530 GeV/c. Nuclear targets included Be, C, Al, Cu and Pb. We report on a preliminary measurement of the A - dependence of the per nucleus cross section for forward J/ψ production. The apparatus also detected charged particles and γ's produced in association with the muon pair. The expected physics results on the hadroproduction of χ states and beauty particles are discussed.

1. Introduction

This paper summarizes recent results from Fermilab experiment E672, a fixed target experiment in the MW beam, the aim of which is to study hadronic processes yielding high mass dimuons (the trigger) and associated particles. The specific goals are:

— the characteristics of ψ production and Drell-Yan processes including A-dependence, multiplicities and momenta of associated particles

— the hadronic production of $\chi \to \psi + \gamma$ states and associated particles

— BB production where one of the Bs decays into ψ +X , or where the semileptonic decays of both B and B yield a $\mu\mu$ pair with large invariant mass.

In this paper we concentrate on the effectiveness per nucleon of heavy targets in producing ψ's. The effect is usually parameterized by the expression: $\sigma(hA \to \psi + X) = \sigma_o \times A^\alpha$, with $\alpha < 1$ implying a decreasing effectiveness of ψ production with increasing A of the nuclear target. This A - dependence has been studied by several experiments spanning the c.m. energy, $\sqrt{s}$, of 15 to 24 GeV (for this experiment $\sqrt{s} = 31.4$ GeV). For the inclusive ψ production $\alpha \simeq 1.$. However, for ψ 's produced at large x_f and with $p_T < 1.2$ GeV/c the values of α were found to be less than one [1–6].

A number of mechanisms for the A-dependence of ψ production and nuclear suppresion at large x_f have been proposed. At large $\sqrt{s}$ ψ's x_f can be approximated as: $x_f = x_1 - x_2$, where x_1 and x_2 are fractions of the beam and target momenta, respectively, carried by interacting partons. Therefore, any mechanism which hardens the momentum distribution (suppression of the low x component) of target parton will decrease the probability of producing a ψ at large x_f. This observation rules out some mechanisms, as for example the EMC effect, which is described as a softening of the momentum distribution of target partons and would predict $\alpha > 1$ at large x_f.

Since a large contribution of ψ production is expected to be due to gluon fusion[7] , any mechanism which affects the momentum distribution of target gluons would also influence the ψ production off nuclei. In particular the nucleons surrounding the target nucleon of a heavy nucleus could produce a "shadowing" effect by which the soft gluon component of the target is absorbed by the other nucleons, leaving a harder component intact[8].

Another class of models explains the observed A-dependence of ψ production as due to the scattering of the beam particle before the ψ is produced or the scattering of the ψ itself before it escapes the nuclear target[9]. These effects decrease the number of ψ s at large x_f and increase the average p_T of the ψ.

The E672 experiment took data during the Fermilab fixed target 1987/88 run. In this paper we present the experimental setup, trigger rates and preliminary results on the A - dependence of ψ production in π^- nucleus interaction at 530 GeV/c. We also discuss expected physics results on the hadroproduction of χ states and beauty particles.

2. Apparatus

Experiment E672 studies dimuon production in an open-geometry configuration. The experiment uses a dimuon detector consisting of a toroid magnet, proportional chambers, scintillator hodoscopes and a trigger processor. The E672 detector shares the MW beam line with E706 (see Fig. 1). The physics goal of E706 is a study of direct photon production. The E706 detector consists of a liquid argon calorimeter (LAC) and a spectrometer including a silicon strip detector (SSD; 14 x-y planes), 16 planes of PWC's and a forward calorimeter .[10] The two experiments take data simultaneously with the information from the entire apparatus being available to both groups.

The the E672 dimuon detector, located 21 m to 33 m downstream of the target, consists of:

— a PWC station (D) consisting of 3 planes

— a tungsten and steel beam dump

— a toroid providing an average p_T kick of 1.3 GeV/c

— four PWC chambers (μ1 through μ4), each with three planes

— two planes of scintillator,each with sixteen wedge-shaped counters

— associated electronics and pretrigger and trigger processors.

Signals from the scintillator planes are used to form the dimuon pretrigger (2 or more hits in each of the two planes). Data from the muon chambers are read via the LeCroy PCOS III system into a trigger processor [11] which: (a) finds space points at two of the chambers; (b) finds tracks; (c) computes momenta assuming the muons orginated in the target and (d) calculates effective masses. The calculations of the momenta and masses take into account correction for the magnetic field in the dipole magnet. The mass resolution of the processor is set by multiple scattering to about 700 MeV. The average trigger formation time has been measured to be less than 10 μs.

The pretrigger rate in the final configuration for πC interactions at 530 GeV/c is 2×10^{-4} interactions and is approximately four times higher than a genuine two muon tracks signal. The minimum energy required for muons to pass through the two pretrigger scintillator planes is 15 GeV. The trigger processor dimuon mass cutoff at 0.5 GeV reduces the trigger rate to 4×10^{-5} interactions.

3. The A - dependence for the forward ψ production

During September 1987 data were taken with the E672 part of the apparatus only (no SSD or PWC data were available). The incident beam was 530 GeV/c π^-. We ran with a number of thick targets: C (4" and 8"), Al(5" and 6.5"), Cu (4") and Pb (2" and 3"). Events were selected using the pretrigger only, however the trigger processor results were also recorded. The off-line track and mass reconstruction was based entirely on muon chamber and D-station. The momentum and mass resolution are therefore determined by the multiple scattering of the muons. We have checked that the mass resolution function can be roughly approximated by a gaussian with a width of 550 MeV. It improves considerably once the upstream tracking information is available, when it is expected to better than 150 MeV.

The measured, uncorrected x_f distribution for $\mu^+\mu^-$ with mass > 2.5 GeV and the acceptance of our apparatus are shown in Fig. 2. For the further analysis we have restricted our data to the x_f region $0.2 < x_f < 0.8$ where our integrated acceptance is approximately 40%.

The $\mu^+\mu^-$ mass distribution for all targets combined is shown in Fig. 3. The fit of an exponential background and a Monte Carlo predicted mass resolution function yielded a total sample of 900 ψ 's. The results of the fit are shown in Fig. 3.

The p_T acceptance of the apparatus is essentially flat up to momentum of 5 GeV/c. The average observed p_T is 1.4 GeV/c, which is consistent with results of previous experiments and effects of the smearing due to the measurement resolution.

Mass fits were repeated for different data samples to establish contributions from individual targets and to check effects of target thickness, backscatter of target debris toward the muon veto wall, halo muons contamination etc. The resulting cross sections are presented in Fig. 4. They were corrected for the acceptance estimated to be 0.41 ± 0.03 for $0.2 < x_f < 0.8$. Other corrections: for the pretrigger and chambers efficiencies, software losses, Drell-Yan pairs, target thickness, non-vertex interactions, x_f smearing, etc. have not yet been applied. We estimate that their net effect could increase the ψ production rate by at most 15% and should be almost target independent. The observed rates are consistent with the extrapolated πp data [12] . A fit to an expression $\sigma \sim A^\alpha$ yields $\alpha = 0.87 \pm 0.06$.

4. Conclusions

Preliminary results of our experiment indicate that the forward production of ψ exhibit $A^{0.9}$ rather than $A^{1.0}$ dependence in π^- - A collisions at 530 GeV/c.

The data presented here were collected in the Fall of 1987. During January of 1988 the entire E672/E706 apparatus was operating. Almost 2 million dimuon triggers were recorded with pion and proton beams at 530 GeV/c incident on Be and Cu targets. The projected total data sample is: 5000 ψ events; 600 $\chi \rightarrow \psi + \gamma$ events with $E_\gamma > 8$ GeV; 8 beauty events with either one $B \rightarrow \psi + X$ decay or two B semileptonic decays with $\mu\mu$ mass > 2.5 GeV.

During the 1989/1990 Fermilab fixed target run we expect to collect an order of magnitude more data.

REFERENCES

1) Yu. Antipov et al., *Phys. Lett.* **72B**, 278 (1977) , *Phys. Lett.* **76B**, 225 (1978) .
2) K. J. Anderson et al., *Phys. Rev. Lett.* **42**, 944 (1979) .
3) M. Corden et al., *Phys. Lett.* **110B**, 415 (1982) .
4) J. Badier et al., *Z.Phys.* **C20**, 101 (1983) .
5) S. Katsanevas et al., *Phys.Rev.Lett.* **60**, 2121 (1988) .
6) K. McDonald, Proceedings of the QCD Workshop, St. Croix, V.I., Oct. 1987; to be published
7) L. Clavelli et al., *Phys. Rev.* **D32**, 612 (1985) .
8) J. Qui, *Nucl.Phys.* **B291**, 746 (1987) .
9) B. Kopeliovich and F. Niedermayer, JINR-E2-84-834, Dubna (1984)
10) G. Fanourakis, Proceedings of the QCD Workshop, St. Croix, V.I., Oct. 1987; to be published
11) R. Crittenden et al., to be published in Nucl. Instr. and Methods, 1988; also R. Crittenden and A. Dzierba, Proceedings from the Fermilab b-physics Workshop, Nov. 1987
12) Y. Lemoigne et al., *Phys. Lett.* **113B**, 509 (1982) .

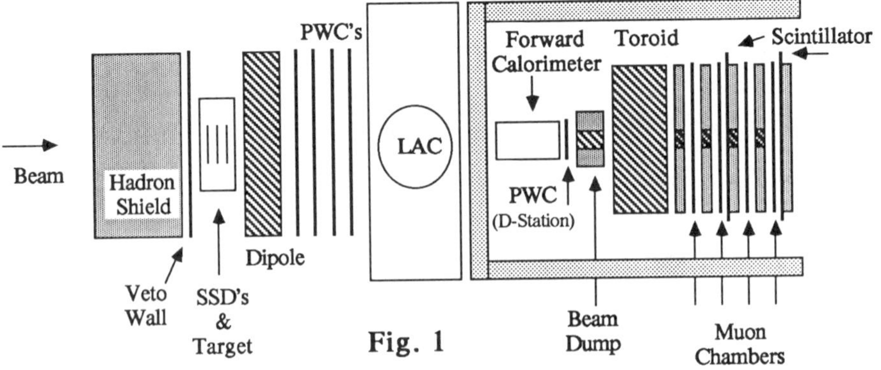

Fig. 1

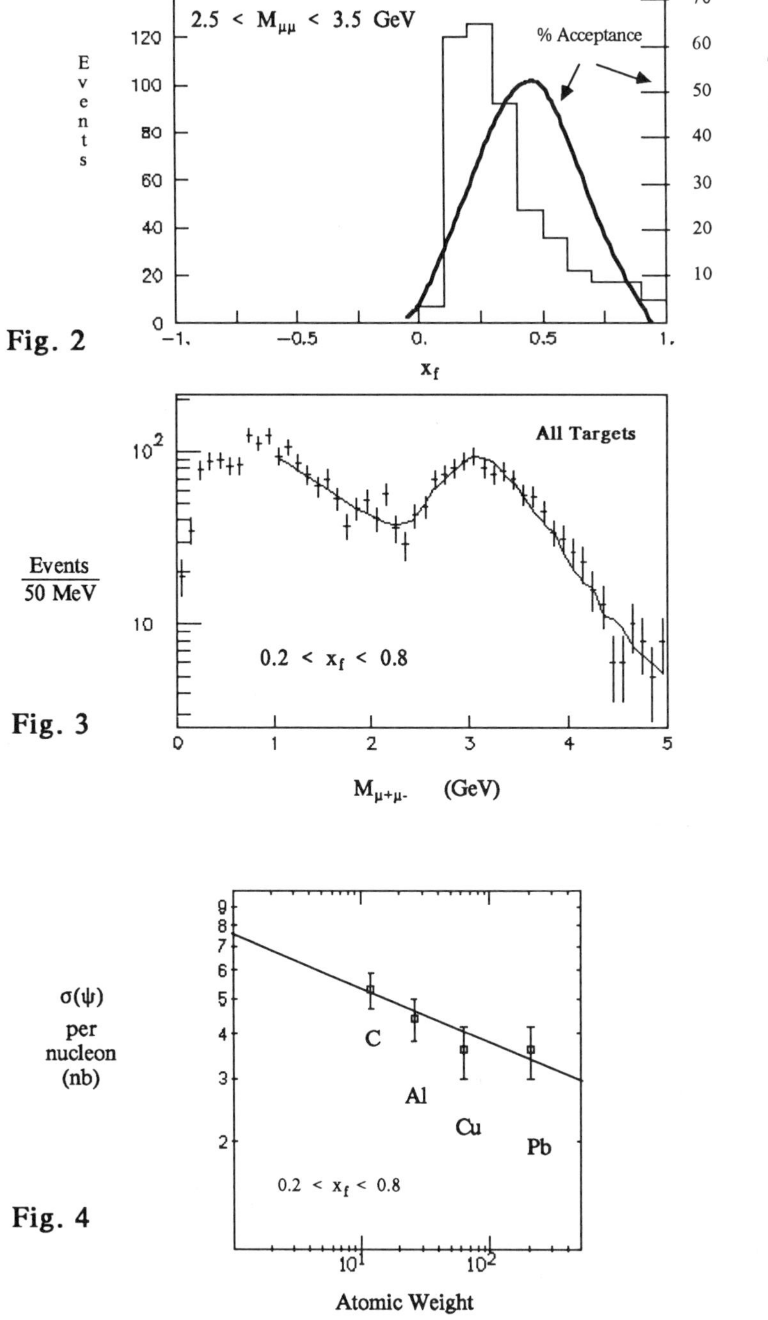

Fig. 2

Fig. 3

Fig. 4

PRODUCTION OF HIGH p_T JETS IN PROTON-NUCLEUS COLLISIONS

C. Stewart and A. Zieminski
Indiana University, Bloomington, IN 47405

ABSTRACT

Production of "jet-like" clusters in 800 GeV/c proton-nucleus (pA) collisions is discussed. Events with high values of transverse energy in the central region were selected for six nuclear targets: H, Be, C, Al, Cu and Pb. A standard jet finding algorithm was applied to the data. The A-dependence of the jet cross sections at a given transverse momentum, p_T, was parameterized as A^α. The value of α reaches a plateau at p_T>8 GeV/c of approximately 1.6. The dependence of α on the assumed size of jets in azimuthal angle and psuedorapidity is also presented. The collimation of observed "jet-like" clusters decreases with A, and the fragmentation is softer for heavier nuclei.

1. Introduction

This paper summarizes recent results from Fermilab experiment E557 concerning production of the "jet-like" clusters in pA collisions at the c.m. energy, $\sqrt{s}$= 39 GeV.[1] In our previous publication[2] we reported a strong A-dependence of cross sections for producing events with a given E_T. The A-dependence of the cross section at a given E_T was parametrized as A^α. The values of α were close to 1.6 for high-E_T global events and somewhat less (1.4) for small-aperture triggers. Moreover, we noticed that "jetty" events, selected by requiring large value of the planarity variable in the transverse plane exhibit an A-dependence consistent with $\alpha = 1.1 - 1.2$, while for more isotropic events α is much larger. The planarity variable[3] was commonly used by previous experiments to extract the jet signal.[4] Therefore, values of α at high planarity could reflect the A-dependence for the "jet" production rates. However, we noticed that values of α determined by this analysis are sensitive to both the A-dependence for the "jet" production rates and possible changes in the event structure with A as described by the planarity variable. In this paper we attempt to separate the two effects.

2. Apparatus

The experimental procedure as well as details of the apparatus have already been described previously.[2] The E557 apparatus consisted of a magnetic spectrometer followed by series of highly segmented calorimeters. Each of these calorimeters consisted of an electromagnetic section followed immediately downstream by a fully absorptive hadronic section. Segmentation of the calorimeters in azimuth-psuedorapidity ($\phi - \eta$) space is shown in Fig. 1. For pp data, the beam was incident on a 45-cm liquid hydrogen target; for pA data, nuclear targets of Be,C,Al,Cu and Pb replaced the hydrogen target and were constructed of three successive foils thin enough to avoid significant rescattering.

We triggered on events which deposited a large amount of transverse energy, E_T, within the aperture of a calorimeter located at 90° in the c.m.s. Only two classes of triggers are discussed here: (a) a sum of signals from two apertures back to back in azimuth (denoted BB trigger), with $\Delta\phi = \pi/2$ azimuthal and 45° - 135° c.m. polar angle coverage for each component, and (b) small aperture trigger (SA) corresponding to a c.m. coverage $\Delta\phi = 90°$ and $60° < \theta^* < 110°$.

In the off-line analysis the events were processed through the standard E557 cluster-finding procedure for the calorimeters to form "calorimeter tracks".[5] The cluster was designated to be electromagnetic or hadronic, depending on its energy deposition pattern in the electromagnetic and hadronic calorimeters. Our Monte Carlo studies imply that the calorimeter tracks reproduce the actual track multiplicity at the interaction vertex to within 10%. We have also tested that the track distributions are fairly symmetric around c.m. 90° polar angle for the pp data.

3. Event Structure for Di-jet E_T Triggers

We shall concentrate on discussing events selected with the BB trigger for which two jets were found and the sum of their transverse momenta exceeds 15 GeV/c. We have checked that our conclusions are independent of the trigger, i.e. di-jet (BB) or single-jet (SA) trigger. The transverse-momentum flows as a function of the azimuthal angle are shown in Fig. 2. The azimuthal angle is calculated with respect to the "jet" axis approximated by calculating a vector sum of particle momenta in a cone with a radius R = 1.0 in the ϕ-η space around jet axis. The transverse-momentum in Fig. 2 is normalized by the p_T of the "jet". The data for hydrogen exhibit clear maxima attributed to the trigger (higher p_T) and away jets. The observed transverse-momentum flow for pp data is in good agreement with published results from the AFS experiment at the ISR.[6] The carbon and lead data indicate significant widening of jets with A, which results in smearing of the di-jet structure for heavier nuclei. Similar widening of jets with A is also observed in η (not shown). The width of the distribution implies the size of the jet to be contained within R of approximately 1.0 ± 0.15.

The events were processed through a jet-finding algorithm.[7] We shall refer to the clusters of energy found by the algorithm as jets, (the name of "jet-like" cluster would be more appropriate). The analysis was done for three values of

R: 0.85, 1.0 and 1.15. The search for jets covered the c.m. psuedorapidity (η^*) range from -1.0 to 1.5.

4. Jet Production Cross Sections

The acceptance corrected cross sections for single jets obtained with R=1.0 are shown in Fig. 3. Only jets within -0.4 $< \eta^* <$ 0.4 were included in the cross section calculations. There is an overall 7% uncertainty in the p_T scale. In view of this, the pp data agree well with the jet cross sections found in other experiments.[4] The particular jet size of R=1.0 is approximately equal to that assumed by other experiments. Therefore we conclude that our results are consistent with previous analyses and "jet-like" clusters observed for pp interactions are produced with rates attributable to jets. The measured rates change by a factor of approximately four when R increases from R=0.85 to R=1.15.

The rates for the central "jet-like" cluster production off various nuclei are also shown in Fig. 3. We have checked that the standard A^α parametrization for the A-dependence is consistent with the data. The pp data were included in the fit.

Fitted values of α for p_T's greater than 5 GeV/c are plotted vs p_T in Fig. 3 b. For different values of R, α increases with p_T from values below 1.0 to plateaus with an approximate value of 1.6 for $p_T > 8$ GeV/c. There is a rather small variation of α (less than 0.25) with R increasing from 0.85 to 1.15. Therefore we observe a strong nuclear enhancement for all reasonable dimensions of "jet-like" clusters. The di-jet cross sections are shown in Fig. 4. The observed nuclear enhancement again corresponds to values of α equal 1.6. There is an indication for α decreasing for large E_T's, where E_T is the sum ot the two jet transverse-momenta.

The interpretation of the observed enhancement requires some caution. The values of α were not corrected for two compensating effects which could influence real jet cross sections. The possible widening of jets with A should lead to an even stronger nuclear enhancement. The subtraction of the underlying background would tend to reduce the enhancement. At this stage of the analysis we have not yet attempted a rigorous subtraction of the underlying event, which seems to be clearly A-dependent (see Fig. 2). However, we have estimated that the last effect should reduce α by less than 0.20.

The relatively large value of α found in this analysis does not contradict our previous report on value of α close to 1.0 for highly planar events. The old observation combined two compensating effects: the nuclear enhancement for "jet-like" cluster production reported here and broadening of those clusters with A, to be discussed quantitatively in next section.

5. Properties of "Jet-like" Clusters

In order to study the properties of "jet-like" clusters described in the previous section, we have defined variables using calorimeter tracks.

(a) "Jet-like" cluster collimation

We formed a new variable which we called the collimation of the jet, C. The collimation is defined as the ratio of jet transverse momenta contained within cones of radii R/2 and R around the jet axis (determined using R), respectively. The collimation distributions for $R = 1.0$ are shown in Fig. 5. The distributions for nuclear targets are significantly shifted towards smaller values, reflecting a broadening of "jet-like" clusters with A. The broadening effect is present for different values of R.

(b) "Jet-like" cluster z - distributions

The longitudinal structure within the jet can be described by the variable $z = p_{\|}/p(\text{jet})$, where $p_{\|}$ is the projection of the track momentum onto the jet direction. The z distributions for hadronic tracks are shown in Fig. 6 for jets with $p_T > 8$ GeV/c. The jets produced off nuclear targets are significantly softer than those observed for pp data. A dramatic disappearance of the most energetic particles is clearly observed for heavy nuclei.

6. Conclusions

We have demonstrated that production of "jet-like" clusters is strongly enhanced for nuclear targets with values of parameter α close to 1.6. The clusters observed in pA collisions are significantly broader than jets known from pp interactions.

There are several possible interpretatons for the observed enhancement and the broadening of the "jet-like" clusters. It is feasible that at $\sqrt{s}$ and jet p_T's available in this experiment the hard scattering mechanism already dominates the pp data, whereas the soft scattering is still a dominant mechanism for the heavy nuclei data.

Another less likely interpretation explains the enhancement as a result of jet rescattering within the nucleus. Such rescattering could also be responsible for drastic changes in the properties of our jets as observed for the heavy target data, for which the "core" of the jet completely disappears. The values of α close to 1 for high planarity events would then reflect cases where there was no jet rescattering and nuclear enhancement was proportional to the atomic weight.

REFERENCES

1) The members of the E557 Collaboration are: R. Gomez (Caltech), L. Dauwe, H. Haggerty, E. Malamud, M. Nikolic (Fermilab), S. Hagopian, B. Pifer (Florida State), R. Abrams, J. Ares, H. Goldberg, C. Halliwell S. Margulies, D. McLeod, A. Salminen, J. Solomon, G. Wu (UIC), S. Blessing, R. Crittenden, P. Draper, A. Dzierba, R. Heinz, J. Krider, T. Marshall, J. Martin, A. Sambamurti, P. Smith, A. Snyder, C. Stewart, T. Sulanke, A. Zieminski (Indiana), R. Ellsworth (George Mason), J. Goodman, S. Gupta, G. Yodh (Maryland), S. Ahn, T. Watts (Rutgers) V. Abramov, Yu. Antipov, B. Baldin, S. Denisov, V. Glebov, Yu. Gorin, V. Kryshkin, S. Petrukhin, S. Polovnikov, R. Sulyaev (Serpukhov)

2) R. Gomez et al., *Phys. Rev.* **D35**, 2736 (1987)

3) C. De Marzo et al., *Nucl.Phys.* **B211**, 375 (1983)

4) L. R. Cormell et al., *Phys. Lett.* **150B**, 322 (1985)

5) B. C. Brown et al., *Phys. Rev.* **D29**, 1895 (1984)

6) T. Akesson et al., *Z. Phys.* **C30**, 27 (1986)

7) C. Stewart et al., paper submitted to the XXIV Inter. Conference on High Energy Physics, Munich, August 1988

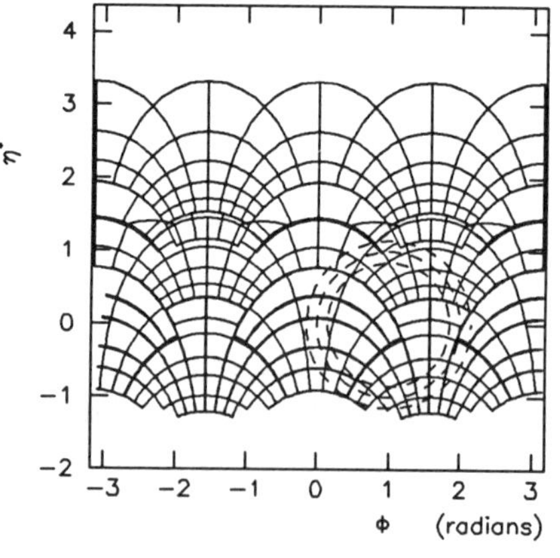

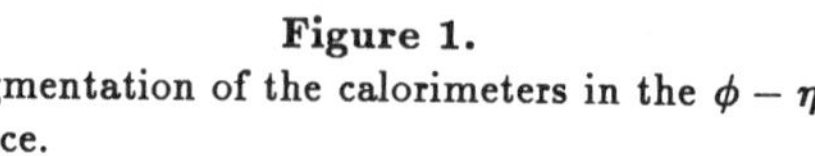

Figure 1.
Segmentation of the calorimeters in the $\phi - \eta$ space.

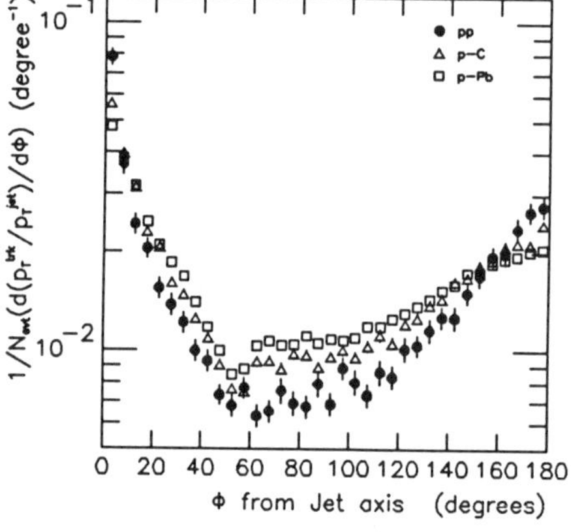

Figure 2.
Azimuthal p_T flow *vs* ϕ from the jet axis, normalized to the p_T of the larger "jet". BB trigger data with $E_T > 15$ GeV are shown.

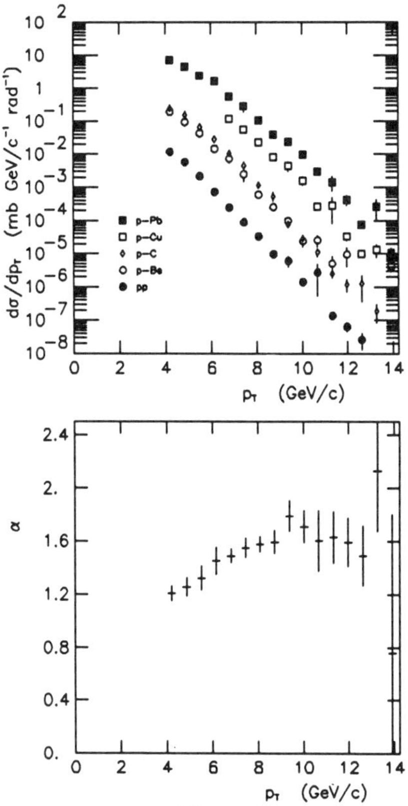

Figure 3.
(*a.*) The $d\sigma/dp_T$ *vs.* inclusive single jet p_T for *pA* interactions at 800 GeV/c. R=1.0
(*b.*) α *vs.* p_T.

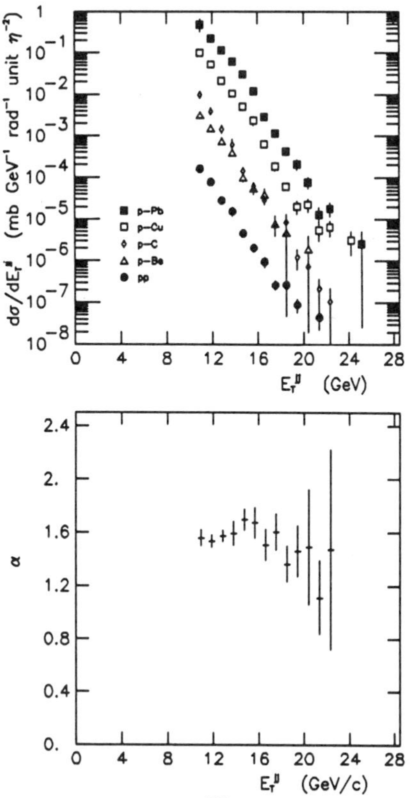

Figure 4.
(*a.*) The $d\sigma/dE_T$ *vs.* di-jet E_T for *pA* interactions at 800 GeV/c. R=1.0
(*b.*) α *vs.* E_T.

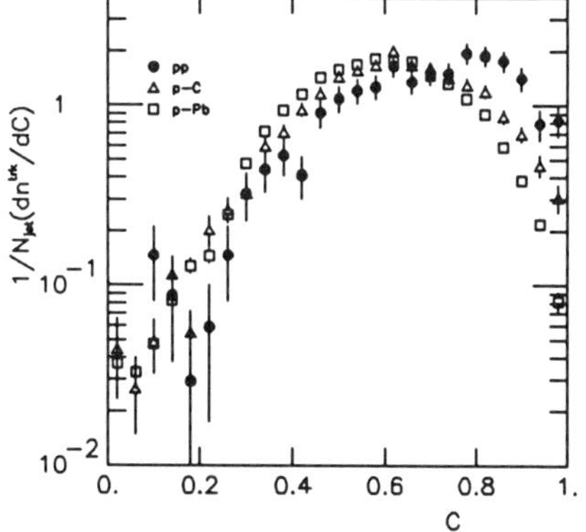

Figure 5.
Distribution of jet collimation (see text) for *pp* and *pA* data.

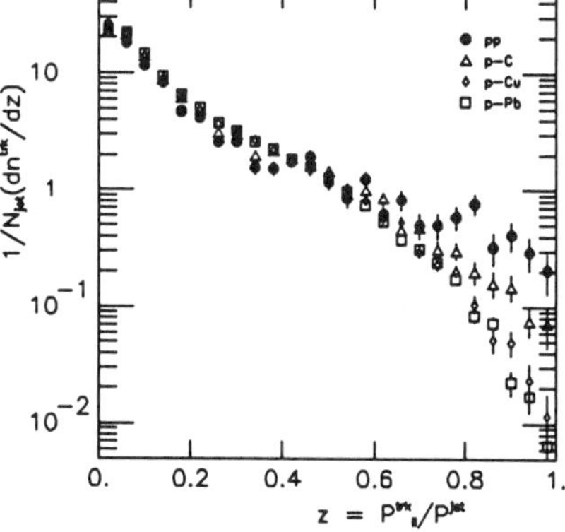

Figure 6.
z distributions (see text) of hadronic tracks for jets with $p_T > 8$ GeV/c for *pp* and *pA* data.

RELATIVISTIC HARTREE-FOCK CALCULATIONS

Peter G. Blunden
TRIUMF, 4004 Wesbrook Mall, Vancouver, B.C., Canada V6T 2A3

ABSTRACT

Within the framework of the relativistic Dirac-Hartree-Fock approach, the properties of nuclear matter and finite nuclei are studied based on a 'realistic' one-boson exchange interaction. With a minimal number of parameters fit to the saturation properties of nuclear matter one is able to give a reasonably good account of the binding energies, single particle energies and density distributions of finite nuclei.

INTRODUCTION

In this contribution I will discuss some recent work which attempts to describe the properties of nuclear matter and of finite nuclei in a relativistic framework with a 'realistic' interaction.

'Realistic' in principle means that one should start with an interaction which fits the free NN scattering data. For example, the Bonn potential[1] may be parametrized in the one-boson exchange form

$$V = \sum_i V_i(|\, r_1 - r_2 \,|)\, \Gamma^i(1) \cdot \Gamma^i(2), \tag{1}$$

where the Γ^i's are the Lorentz invariants $1, \gamma^\mu, \sigma^{\mu\nu}, \gamma_5\gamma^\mu$ and γ_5 associated with various meson fields, and the functions V_i have the Yukawa form of the meson propagators.

Such an approach requires a Brückner-Hartree-Fock calculation, which has been performed for nuclear matter by Machleidt[1] and also by ter Haar and Malfliet[2]. These authors both claim to approximately fit the empirical saturation density and binding energy of 1.30 fm.$^{-1}$ and -15.75 MeV respectively, with a bare interaction fit to the NN phase shifts. Brückner calculations for finite nuclei are exceedingly difficult and have not yet been attempted. I will therefore use an interaction which qualitatively contains the features expected of more realistic interactions, with a minimal number of parameters (three) fit to the properties of nuclear matter.

RELATIVISTIC MODELS

The prototype for these relativistic approaches is the sigma-omega model, studied most extensively by Walecka and collaborators[3,4]. The equations of motion in this model take a particularly simple form in the mean field or Hartree approximation (Fig. 1a). Saturation in nuclear matter is achieved by the competition between an attractive scalar interaction and a repulsive vector interaction, interpreted as the sigma and omega meson contributions of one-boson exchange models.

The scalar and time-like vector interactions 1 and γ^0 are the only terms of Γ in Eq. 1 which contribute to the self-energy diagram of Fig. 1a — tensor, pseudo-vector and pseudo-scalar terms vanish by symmetry.

The Hartree model works remarkably well for the bulk properties of finite nuclei. It contains a spin-orbit force of about the right magnitude to give the

Figure 1. Baryon self-energy contributions of (a) Hartree, and (b) Fock terms.

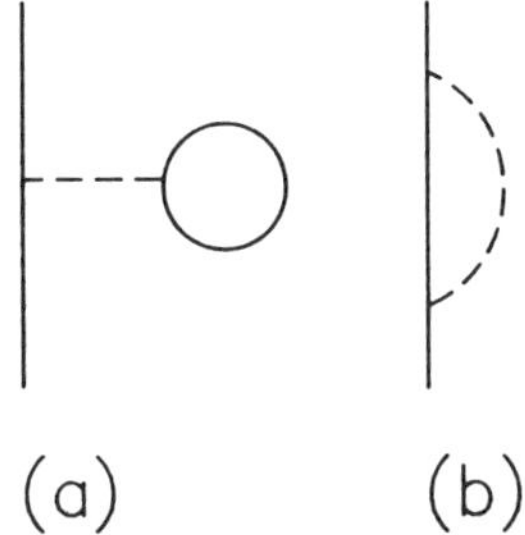

nuclear shell model, and it more or less fits the charges densities of closed-shell nuclei deduced from electron scattering.

On the negative side, it doesn't describe the nuclear surface particularly well. This is related in part to the large value of 550 MeV for the nuclear compressibility (empirically 210±30 MeV), implying a very stiff equation of state. Hartree calculations also significantly underbind finite nuclei — a feature which is very likely also related to the large compressibility. A final drawback is the philosophical objection that there is no role for the pion in this theory.

Incorporating new terms into the interaction requires going beyond the simple Hartree approximation to Hartree-Fock (Fig. 1b). Consider the following Lagrangian, which contains the main meson-nucleon interaction terms of more sophisticated one-boson exchange models:

$$\begin{aligned} \mathcal{L}_{int} &= g_\sigma \bar{\psi}\psi\, \sigma \quad - \quad g_\omega \bar{\psi}\gamma_\mu\psi\, \omega^\mu \\ &- \frac{g_\pi}{2M}\bar{\psi}\gamma_\mu\gamma_5\partial^\mu\psi\, \vec{\pi}\cdot\vec{\tau} \quad - \quad g_\rho\bar{\psi}[\gamma_\mu - \frac{K_\rho}{2M}\sigma_{\mu\nu}\partial^\nu]\psi\, \vec{\rho}^\mu\cdot\vec{\tau}. \end{aligned} \tag{2}$$

The new ingredients here are a pseudo-vector pion interaction and a rho meson interaction containing both vector and tensor couplings. In the non-relativistic limit these interactions give rise to the following spin-isospin dependent potentials:

$$V_\pi(q) = -(\frac{g_\pi}{2M})^2 \frac{\sigma_1\cdot\vec{q}\,\sigma_2\cdot\vec{q}}{q^2+m_\pi^2}\vec{\tau}_1\cdot\vec{\tau}_2 \tag{3}$$

$$V_\rho(q) = -(\frac{g_\rho}{2M})^2(1+K_\rho)^2 \frac{\sigma_1\times\vec{q}\,\cdot\,\sigma_2\times\vec{q}}{q^2+m_\rho^2}\vec{\tau}_1\cdot\vec{\tau}_2 \tag{4}$$

The tensor coupling of the rho is clearly an important ingredient. Vector dominance would suggest a value of $K_\rho = 3.7$, but experimentally it is closer to $K_\rho = 6.6$, and this is the value adopted here. The relative importance of the rho versus the pion may be seen by comparing the coupling strengths in the non-relativistic expressions: $g_\rho^2(1+K_\rho)^2/g_\pi^2 \approx 2$. This is compensated somewhat by the shorter range of the rho, but they are clearly comparable.

The interaction parameters are given in Table I. There are three free parameters: the sigma coupling constant g_σ and mass m_σ, and the omega coupling constant g_ω. The other parameters are fixed at their experimental values. As in the Hartree model[4], the sigma and omega coupling constants are chosen to fit the nuclear matter saturation point, while the sigma mass is fixed by requiring a finite nucleus calculation to reproduce the r.m.s. charge radius of ^{40}Ca.

Table I. Parameters of the various mesons in the interaction, chosen to saturate nuclear matter at 1.30 fm.$^{-1}$. The ρ tensor coupling is $K_\rho = 6.6$.

	σ	ω	ρ	π
g	9.58	10.16	2.63	13.45
m (MeV)	530	783	770	138

The saturation curve for nuclear matter is shown in Fig. 2. The solid curve is the Hartree result and the dashed curve is the Hartree-Fock. The Hartree-Fock shows a softer equation of state, with a compressibility of 400 MeV. The symmetry energy a_4 (recall $E \sim a_4(N-Z)^2/A$) is 35.2 MeV, which compares well with the empirical result of 33.2 MeV. In the Hartree model the symmetry energy comes out too low, and can only be fixed by making the rho meson coupling artificially large.

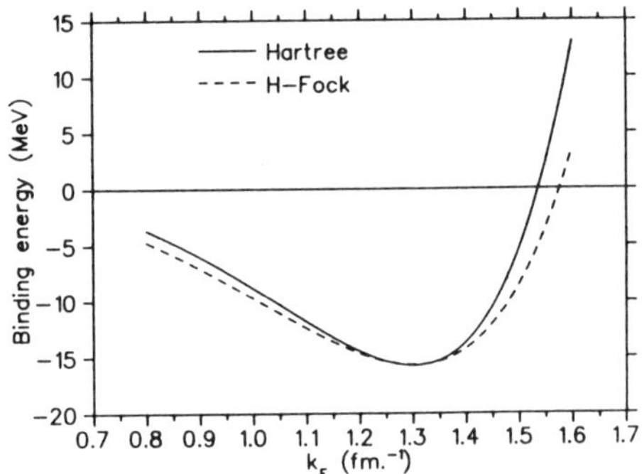

Figure 2. Nuclear matter saturation curve for Hartree model of Ref. 3 and Hartree-Fock model with parameters of Table I.

The Hartree-Fock equations for finite nuclei are summarized schematically below. First is an equation for the various meson fields,

$$(\nabla^2 - m^2)\ \phi_{ij} \ = \ -g\bar{\psi}_j \Gamma \psi_i, \tag{5}$$

where i and j refer to nucleon single-particle states, and Γ are the various Lorentz invariants of Eq. 1. For the baryon (nucleon) fields we have

$$(-i\vec{\alpha}\cdot\vec{\nabla} + \beta M)\psi_i \ + \ g\sum_j \phi_{jj}\Gamma\psi_i \ - \ g\sum_j \phi_{ij}\Gamma\psi_j \ = \ \epsilon_i\psi_i. \tag{6}$$

The first term is the usual Dirac equation, the second term is the Hartree contribution, and the third term is the non-local Fock or exchange term.

These form a coupled set of integro-differential equations, which are solved iteratively on a grid in coordinate space. The non-locality of the Dirac equation may be handled either by converting it to an approximately equivalent local form, or by Green function methods.

The charge densities of closed shell nuclei are shown in Fig. 3, where they are compared with the densities deduced from electron scattering. The agreement is quite good, although there is too much shell structure in the nuclear interior.

Fig. 4 shows a comparison of the point neutron and proton densities of Hartree and Hartree-Fock. The Hartree-Fock predicts a more diffuse surface, in

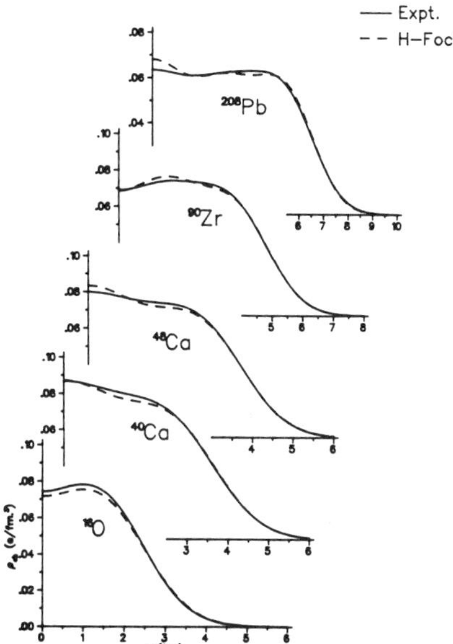

Figure 3. Hartree-Fock predictions for charge densities of closed shell nuclei, compared with results from electron scattering.

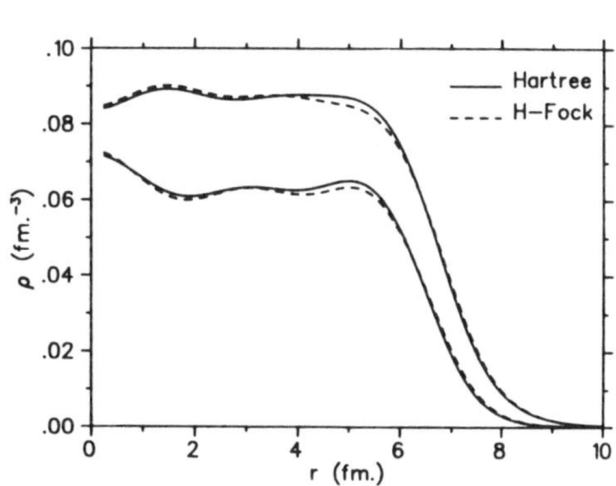

Figure 4. Comparison of Hartree (solid curve) and Hartree-Fock (dashed curve) predictions for point neutron and proton densities in ^{208}Pb.

agreement with experimental observations. The theory is able to describe the spatial distribution of both heavy and light nuclei. This is an improvement over Hartree calculations[4], which give too large an r.m.s. radius for light nuclei and too small an r.m.s. radius for heavy nuclei. For example, in ^{208}Pb the Hartree-Fock gives the correct r.m.s. charge radius of 5.50 fm., whereas Hartree gives 5.45 fm.

A comparison of the binding energies for these nuclei is shown in Fig. 5. There is a clear improvement for the Hartree-Fock over Hartree. Following Horowitz and Serot[4], we can fit the discrepancy between the experimental and theoretical results with a volume and a surface term of the form

$$\delta E \equiv E_{expt} - E_{th} = aA + bA^{2/3}. \tag{7}$$

For the Hartree one finds $a = 0.4$ MeV and $b = 5.0$ MeV, indicating that the discrepancy is associated with the surface properties of the theory. The Hartree-Fock gives $a = -0.1$ MeV and $b = 3.0$ MeV, reflecting an improvement in the description of the nuclear surface due, primarily, to the smaller compressibility.

The better surface properties of the Hartree-Fock are also reflected in spin-orbit splittings, shown in Table II. The Hartree-Fock gives somewhat smaller results than Hartree, and is able to reproduce the reduction seen in going from ^{40}Ca to ^{48}Ca.

As a final example consider the excited state spectra of ^{16}O computed in the random phase approximation (RPA). In the RPA the various multipole excitations explore different aspects of the nuclear interaction. Since there are no additional parameters introduced, this constitutes a severe test of the interactions as well as the underlying theory.

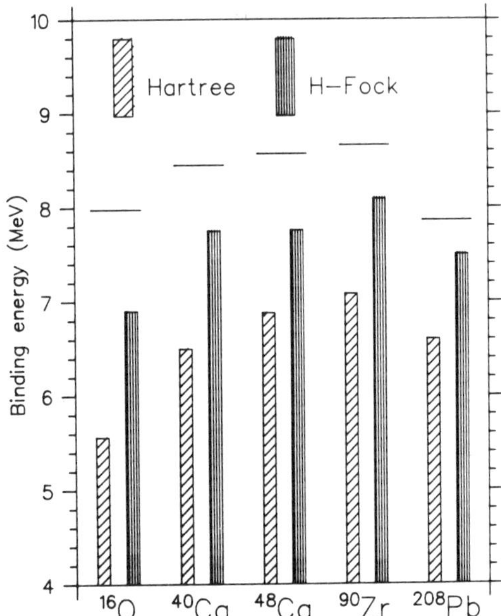

Figure 5. Binding energies of closed shell nuclei in Hartree and Hartree-Fock. Horizontal bar is the experimental value.

The results are shown in Fig. 6 for the isoscalar and isovector modes. The Hartree-Fock column is based on the Hartree-Fock basis states and includes exchange terms in the residual interaction, whereas the Hartree column includes only direct terms. Differences in the $T = 0$ spectra are not significant, but the $T = 1$ states show significant effects, particularly for the unnatural parity states $(0^-, 2^-, 4^- ...; T = 1)$ which carry the quantum numbers of the pion. In the Hartree case, the attractive pion interaction pulls these states below their unperturbed configurations and their isoscalar counterparts, in strong disagreement with the data. It is very likely that the Hartree leads to pion condensation at nuclear matter densities.

However, the exchange interactions in the Hartree-Fock case are strongly repulsive, and are sufficient to counteract the pion interaction and push these states above their unperturbed configurations by several MeV. This results in a very reasonable spectrum, which would indicate that the Hartree-Fock interaction is to be preferred in this case.

Table II. Spin-orbit splittings near the Fermi surface for Hartree and Hartree-Fock, compared with experiment.

		Hartree	H-Fock	Expt.
^{16}O	$p_{3/2} - p_{1/2}$	8.0	6.9	6.3
^{40}Ca	$d_{5/2} - d_{3/2}$	8.4	7.7	7.2
^{48}Ca	$d_{5/2} - d_{3/2}$	8.2	5.5	4.3

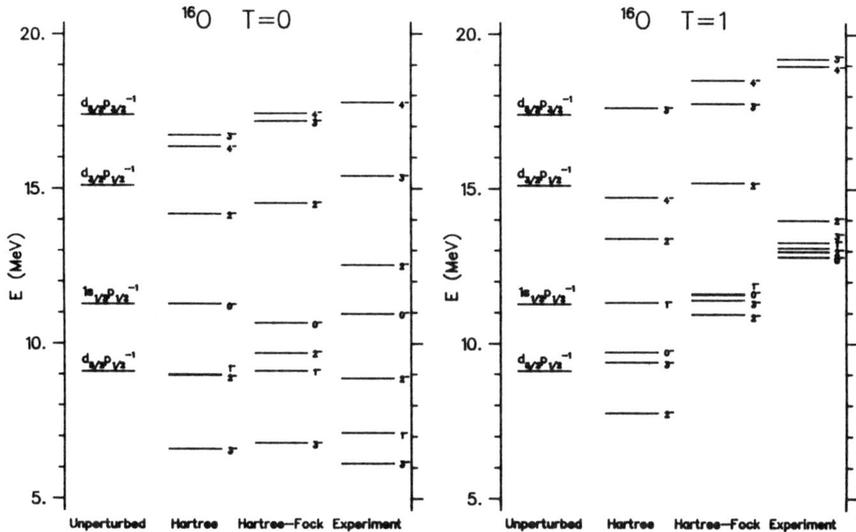

Figure 6. RPA spectra for isoscalar and isovector negative parity states in ^{16}O. Shown are the unperturbed basis states in a Hartree model, the Hartree-based RPA, Hartree-Fock RPA, and the experimental levels.

SUMMARY AND OUTLOOK

With a 'realistic' interaction one is able to obtain a reasonably good description of the properties of finite nuclei. The model has the same essential features as the simpler sigma-omega model, but provides a better description of the nuclear surface. Indeed, comparison of a variety of Hartree-Fock interactions shows that those models with a small compressibility tend to do better for surface properties and binding energy than those models with a large compressibility.

The Brückner-Hartree-Fock calculations of Machleidt[1] and ter Haar and Malfliet[2] saturate nuclear matter at the empirical point with a realistic interaction, and give a compressibility of 250 MeV. Clearly any improvement for finite nuclei will require a Brückner calculation of this sort.

The above discussion suggests that a model which gets the empirical compressibility right would be able to simultaneously fit the binding energies and densities of finite nuclei. Calculations in this direction are in progress.

REFERENCES

1. R. Machleidt, Lectures presented at the Workshop on Relativistic Dynamics and Quark-Nuclear Physics, Los Alamos (1985).

2. B. ter Haar and R. Malfliet, Phys. Rev. Lett. **56**, 1237 (1986).

3. J.D. Walecka, Ann. Phys. **83**, 491 (1974).
   B.D. Serot and J.D. Walecka, Adv. in Phys. **16**, 1 (1986).

4. C.J. Horowitz and B.D. Serot, Nucl. Phys. **A368**, 503 (1981).

CHARACTERISTICS OF HADRON-NUCLEUS INTERACTIONS AT 100 GeV/c

W.S. Toothacker
The Pennsylvania State University, Mont Alto, Pennsylvania 17237

J. Whitmore
Laboratory for Elementary Particle Science, Department of Physics
The Pennsylvania State University, University Park, Pennsylvania 16802

P.A. Elcombe, J.C. Hill, W.W. Neale
University of Cambridge, Cambridge CB3 OHE, England

W. Kowald, W.D. Walker
Duke University, Durham, North Carolina 27706

P. Lucas, L. Voyvodic
Fermilab, P.O. Box 500, Batavia, Illinois 60510

R. Ammar, D. Coppage, R. Davis, J. Gress, S. Kanekal, N. Kwak
University of Kansas, Lawrence, Kansas 66045

J.M. Bishop, N.N. Biswas, N.M. Cason, V.P. Kenney,
M.C.K. Mattingly, R.C. Ruchti, W.D. Shephard and S.J.Y. Ting
University of Notre Dame, Notre Dame, Indiana 46556

ABSTRACT

We report on 100 GeV/c interactions of p and $\overline{p}$ with Ag and Au targets. This is a subset of the data from Fermilab experiment E597 and was performed with the 30-inch bubble chamber and Downstream Particle Identifier. Final state protons with laboratory momentum less than 1.4 GeV/c have been identified by their ionization in the bubble chamber. Final state protons/antiprotons with laboratory momentum greater than 10 GeV/c have been identified using CRISIS, an ionization sampling drift chamber. The cross section and mean transverse momentum squared of the leading baryon from the reactions $p + (Ag, Au) \rightarrow p + X$ and $\overline{p} + (Ag, Au) \rightarrow \overline{p} + X$ are presented as a function of the rapidity loss of the leading baryon. The laboratory rapidity and transverse momentum squared of the associated pions are also presented.

INTRODUCTION

Fermilab experiment E597, a study of hadron-heavy nucleus interactions and antiproton-proton annihilations, was performed in the 30-inch bubble chamber spectrometer with the Downstream Particle Identifier which is shown in Figure 1. Beam trajectories were measured with three proportional wire chamber triplets (A,B,C), and mass identification for the beam particles was provided by three Cerenkov counters in the beam line. Downstream of the bubble chamber, which was in a 2T magnetic field, were seven planes of proportional wire chambers (D,E), and three drift chamber triplets (F,G,H). Charged particle identification was obtained from two devices, CRISIS and OSIRIS. CRISIS was a 1m by 1m by 3m ionization sampling drift chamber containing an 80% Argon/ 20% Carbon Dioxide mixture at atmospheric pressure. Particle identification was possible because of the logarithmic rise in ionization for relativistic particles. OSIRIS was an eight-cell, one-atmosphere He/N_2 threshold Cerenkov detector which could distinguish between pions and protons above 25 GeV/c. In addition, there was a segmented iron/scintillator calorimeter to provide neutron and antineutron identification.

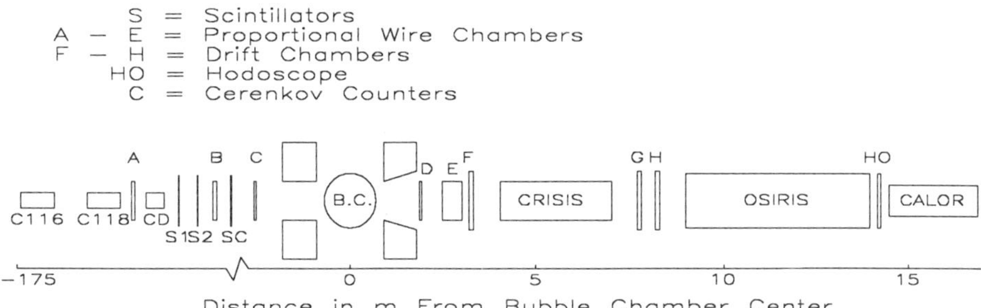

Figure 1. A schematic of the 30-inch bubble chamber spectrometer and Downstream Particle Identifier.

Interactions occurred in thin nuclear targets of Mg, Ag, and Au which were placed in the bubble chamber with incident 100 GeV/c beams of p, $\overline{p}$, π^+, π^-, and K^+ particles as well as 320 GeV/c π^- particles. We report here on the p and $\overline{p}$ beams interacting with Ag and Au targets. Table I shows the number of events. For each event all charged tracks were measured, and if possible the momentum of the fast tracks was improved by using the wire chambers. Details of the particle identification methods used in CRISIS will appear elsewhere[1]. Multiplicity[2] and stopping power[3] results from this experiment have been previously published.

Table I. Numbers of Events by Beam and Target Type

Beam Particle	Target Nucleus Ag	Target Nucleus Au	Total
p	360	181	541
$\overline{p}$	586	467	1053

BARYON SPECTRA

We are able to identify protons by their ionization in the bubble chamber for momenta up to 1.4 GeV/c. In addition, by using CRISIS we are able to identify protons with momenta between 10 GeV/c and 100 GeV/c. Figure

2 shows the laboratory momentum spectra of all identified protons from the reactions $p + Ag \to p + X$ and $\overline{p} + Ag \to p + X$. The low momentum portion of the spectra consists of the isotropic, "evaporation" protons emitted by the nucleus as a whole, and the forward "recoil" protons which have been knocked out of the nucleus. It is apparent that proton and antiproton beam particles are equally efficient at dislodging these nuclear protons.

It can be seen in this figure that the protons knocked out of the nucleus by the antiproton beam extend up to at least 15 GeV/c in momenta. For the proton beam data the points above 10 GeV/c represent a combination of this high momentum tail of the nuclear proton spectra (presumed to be the same as that from the antiproton beam data) and the energy degraded incident proton.

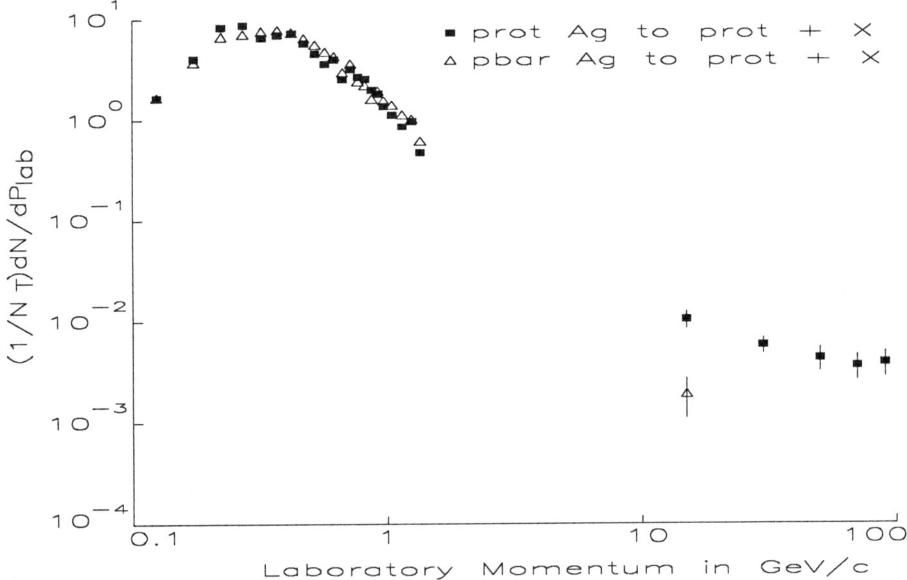

Figure 2. Laboratory momentum spectra from the reactions $p + Ag \to p + X$ and $\overline{p} + Ag \to p + X$. The protons with momenta less than 1.4 GeV/c were identified by ionization in the bubble chamber while those with momenta greater 10 GeV/c were identified in CRISIS. N_T is the total number of events.

The nature of pA events where the leading proton ($P_{lab} \geq 10$ GeV/c) has been identified are very similar to $\overline{p}A$ events (not shown) where the leading antiproton has been identified (no annihilation). We are thus able to improve our statistics by combining results for leading baryons and associated pions from the reactions $p + A \to p + X$ and $\overline{p} + A \to \overline{p} + X$. In Figure 3 we show the mean transverse momentum squared ($< P_t^2 >$) of the leading baryon as a function of the rapidity lost by the baryon in traversing the nucleus ($\Delta y = 5.36 - y_{baryon}$). It can be seen that the $< P_t^2 >$ of the leading baryon increases with the rapidity which it has lost, which is consistent with having suffered an increased number of nuclear collisions.

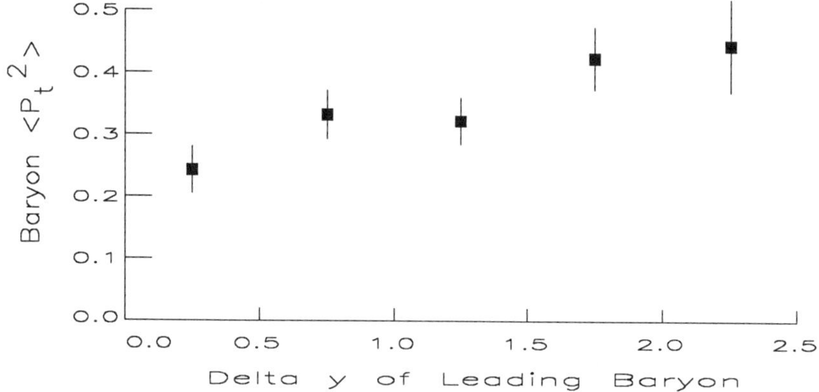

Figure 3. Mean transverse momentum squared of the leading baryon from the reactions $p + (Ag, Au) \rightarrow p + X$ and $\overline{p} + (Ag, Au) \rightarrow \overline{p} + X$ as a function of the rapidity loss of the leading baryon.

ASSOCIATED PION SPECTRA

We have previously shown[3] that the number of pions produced increases with increasing energy loss of the leading baryon. These pions are produced with the same mean laboratory momentum, independent of the rapidity loss of the leading baryon. In Figure 4 we plot the mean transverse momentum squared and in Figure 5 a) the mean laboratory rapidity of these associated pions as a function of the rapidity loss of the leading baryon. We see almost no correlation between these variables, indicating that the pions produced in interactions having a large amount of energy deposited in the nucleus are produced with the same $< P_t^2 >$ and $< y >$ as when little energy is deposited, only the number of pions produced increases.[3]

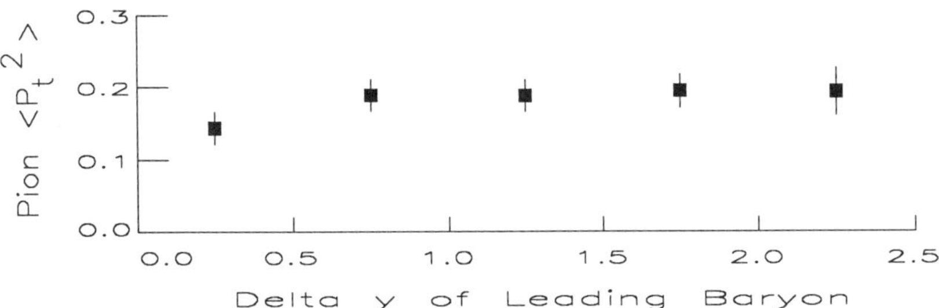

Figure 4. Mean transverse momentum squared of the associated pions from the reactions $p + (Ag, Au) \rightarrow p + \pi + X$ and $\overline{p} + (Ag, Au) \rightarrow \overline{p} + \pi + X$ as a function of the rapidity loss of the leading baryon.

Even though the mean laboratory momentum, laboratory rapidity, and transverse momentum squared of the associated pions is seen to be independent of the rapidity loss of the leading baryon, it would be possible for the shape (width) of the pion rapidity distribution to change as a function of Δy of the leading baryon. Specifically, it would be possible for the rapidity of the fastest few pions in an event to increase with the Δy of the leading baryon. In Figure 5 b) we plot the mean laboratory rapidity of the fastest charged particle (other than the identified leading baryon), the second fastest, and the third fastest. We see, except for a kinematic effect at low values of Δy (not enough energy has been deposited to produce "fast" fastest, second fastest, and third fastest pions), that these distributions are also not correlated with Δy.

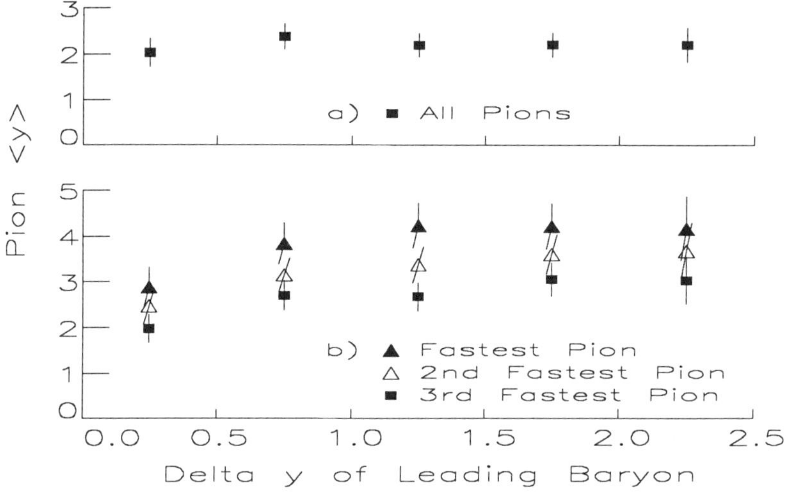

Figure 5. Mean laboratory rapidity of the associated pions from the reactions $p + (Ag, Au) \rightarrow p + \pi + X$ and $\overline{p} + (Ag, Au) \rightarrow \overline{p} + \pi + X$ as a function of the rapidity loss of the leading baryon for (a) all pions and (b) separately for the fastest, second fastest, and third fastest pion.

To investigate further the partition of energy deposited by the incident fast baryon, in Figure 6 a) we show the mean fraction of energy deposited which is carried away by the one or two fastest pions. We see that at large values of Δy (where substantial energy has been deposited) on average 27% of the deposited energy is carried by the fastest pion and 34% of the deposited energy is carried by the two fastest pions. In Figure 6 b) we show the fraction of events in which the one or two fastest pions carry more than 50% of the deposited energy. For large Δy, only about 10% of the time is more than half the energy carried by one pion. Thus we conclude that when large amounts of energy are deposited by an incident baryon, that energy is most likely to appear in the form of many relatively slow pions rather than a few fast pions.

CONCLUSIONS

We have studied p and $\overline{p}$ interactions on Ag and Au targets at 100 GeV/c. We have previously shown[3] that the number of slow protons knocked out of the nucleus increases with the energy loss of the baryon. We have shown here that the mean transverse momentum also increases with increasing energy loss.

These observations are consistent with a picture in which the projectile baryon suffers a series of nuclear collisions and deposits energy in the nucleus. This energy is used to produce pions and to eject nucleons. The greater the amount of energy deposited, the greater is the number of pions produced and the greater is the number of nucleons liberated. Most of these liberated nucleons are given only small amounts of momentum; however the spectrum does continue up to at least 15 GeV/c. The pions that are produced have the same character (transverse momentum, laboratory momentum and rapidity) independent of the amount of energy lost by the incident baryon.

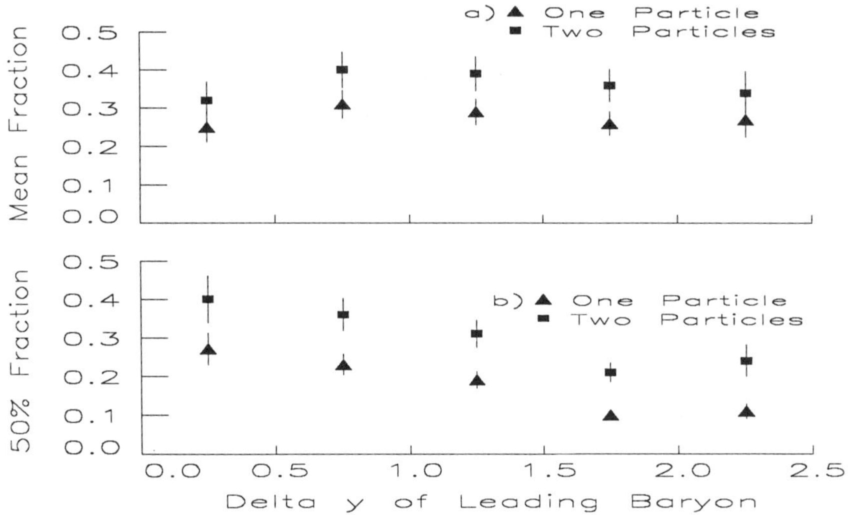

Figure 6 a) The mean fraction of energy deposited by the incident baryon which is carried away by the one or two fastest pions as a function of the rapidity loss of the baryon. b) The fraction of events in which the one or two fastest pions carry more than 50% of the deposited energy as a function of the rapidity loss of the baryon.

ACKNOWLEDGEMENTS

This work was supported in part by a US National Science Foundation Research Opportunity Award (Penn State/Mont Alto), the US National Science Foundation (Kansas, Notre Dame, Pennsylvania State), US Department of Energy (Duke, Fermilab), and the UK Science and Engineering Research Council (Cambridge).

REFERENCES

1. W.S. Toothacker et al, *Nuclear Instruments and Methods,* in press (1988).
2. N.N. Biswas et al, *Phys Rev* D33, 3167 (1986).
3. W.S. Toothacker et al, *Physics Letters* 197B, 295 (1987).

Nuclear Response in N-Nucleus Scattering at Intermediate Energy

M.C. Vetterli*
Simon Fraser University, Burnaby, B.C., Canada, V5A 1S6

ABSTRACT

The nuclear response is examined by considering quasi-elastic spin observables. These are compared to both non-relativistic (Schrödinger) and relativistic (Dirac) models and the effects of medium modifications are emphasized. Furthermore, the decomposition of the nuclear response into its spin, isospin, and orbital angular momentum components using (p,p'), and (n,p) reactions is presented.

INTRODUCTION

There is much current interest in inelastic scattering of nucleons to the continuum. The region of excitation energy $E_x = \frac{q^2}{2m}$, where q is the momentum transfer and m is the nucleon mass, corresponds to quasi-elastic scattering of the incoming proton. This implies that the projectile scatters from only one of the nucleons which is knocked out leaving (A-1) nucleons in the nucleus unchanged. Quasi-elastic scattering is analogous to free nucleon-nucleon (NN) scattering except that the interaction occurs in the nuclear medium. Hence this process is a suitable tool to study how the NN interaction is modified at nuclear densities. Results for a measurement of the full set of spin observables in the quasi-elastic region are presented. These data show marked deviations from the free NN observables indicating that medium effects are significant. At lower momentum transfers, the nuclear response is different with collective giant resonances appearing. In the second part of the paper, new techniques for the decomposition of the response into its spin, isospin, and angular momentum components are discussed. The experiments described are possible at TRIUMF through the recent development of a focal plane polarimeter and of a longitudinally polarized beam, as well as of the CHARGEX facility for (n,p) and (p,n) reactions.

QUASI-ELASTIC SCATTERING

Data presented for the full set of spin observables in the quasi-elastic region are compared to both a non-relativistic RPA surface response model [1,2] and a relativistic model based on the Dirac equation [3,4]. The cross-section for

* Mailing address: TRIUMF, 4004 Wesbrook Mall, Vancouver, BC, Canada, V6T 2A3

quasi-elastic scattering can be written:

$$\frac{d\sigma}{d\Omega} = N_{eff} \sum S_{T,S}(q,\omega) \left(\frac{d\sigma}{d\Omega}\right)_{NN}$$

where N_{eff} is the effective number of target nucleons, $S_{T,S}$ is the nuclear response, and $(\frac{d\sigma}{d\Omega})_{NN}$ is the free NN cross-section. N_{eff} is calculated in a Glauber model [5] in both the relativistic and non-relativistic cases. Since the reaction is surface peaked, $N_{eff} \ll A$ and depends on energy. For example at 400 MeV, N_{eff}= 4.5 for ^{12}C and 12.8 for ^{208}Pb [4]. In the non-relativistic case, $S_{T,S}$ is calculated in a surface RPA model while a Fermi gas response is used in the relativistic calculation. Pauli blocking and Fermi momentum averaging are included in both models although not in the same way. The non-relativistic curves include double scattering, which is found to be a small effect [6]. Finally, the free

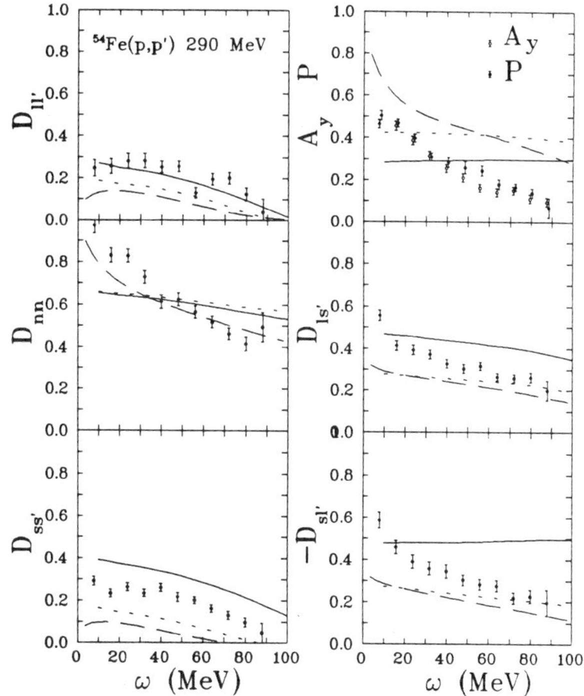

Fig.1. Spin observables for ^{54}Fe(p,p') at 290 MeV and 20°. The dotted curve is the free response while the solid and dashed curves are the result of relativistic and non-relativistic calculations respectively.

NN cross-section is modified in the relativistic case by using Dirac spinors with an effective mass m*. This enhances the lower components of the spinors. Quasi-elastic spin observables were previously obtained for ^{2}H and Pb at 500 MeV [7].

The spin observables shown in Fig. 1 are recent TRIUMF data for the (p,p') reaction on ^{54}Fe at 290 MeV and 20° [8]. This corresponds to a momentum transfer of 1.36 fm^{-1} which was chosen small to introduce some sensitivity to the nuclear response function resulting from RPA correlations [1]. The coefficients D_{ij} are defined as the probability that the spin of the proton is transfered from direction i to direction j, where i and j are either normal(n), longitudinal(l), or sideways(s) with respect to the proton momentum, and i(j) is in the frame of the incident (outgoing) proton (see K coefficients in ref. [9]). Although the RPA model does not reproduce all the data, the slopes (ω dependence) are well described. This, along with the failure of the relativistic model to reproduce the ω dependence, leads to the conclusion that a good description of the nuclear response is essential. The RIA uses a Fermi gas response which is reliable only near the quasi-elastic peak (E_x= 40 MeV). The relativistic model does well for A_y, D_{nn}, and D_{ll} for quasi-elastic kinematics. Furthermore, the RIA always corrects the free observables in the correct direction (i.e. always towards the data) although it does overshoot in some cases. It is hoped that the inclusion of a better nuclear response will improve the ω dependence of the RIA. Perhaps it will also remedy the overcorrection of D_{ss}, D_{sl}, and D_{ls}. Such relativistic RPA calculations are being done [10]. The corrections of the non-relativistic surface RPA model are sometimes in the wrong direction. The results of a calculation using a density dependent interaction (of the von Geramb type) would be interesting; such a calculation should be done. In conclusion, although quasi-elastic scattering is not quantitatively described by either the surface RPA or the modified RIA, it would appear that both models include many of the features observed in the data.

EFFECT OF THE NUCLEAR RESPONSE ON S_{nn}

Fig.2a shows data for the spin-flip probability S_{nn} on ^{54}Fe at 290 MeV [11]. The dashed curve is the S_{nn} value for free pp scattering. It is clear that there are large deviations from the free value. This effect has been seen in many nuclei [12]. The full curve which gives a much better fit, is an RPA surface response calculation which includes the residual nuclear interaction. It is the strong attractive isoscalar non spin-flip part of this interaction which is responsible for the slope of S_{nn}. It concentrates ΔS=0 strength at low excitation energy therefore reducing the value of S_{nn} which is 0 for ΔS=0 states. The ΔS=1 strength is pushed to higher excitation and S_{nn} is therefore enhanced over the free value in this region. The enhancement over the RPA calculation is discussed in other contributions to these proceedings [12]. A further indication of the role of the isoscalar interaction is given by the spin-flip probability in (p,n) reactions which are purely isovector. Fig.2b shows S_{nn} data for the ^{208}Pb(p,n) reaction at 135 MeV [13]. The curve is the free response and fits the data reasonably well verifying that the isoscalar interaction is responsible for the deviations from the free value in (p,p') reactions.

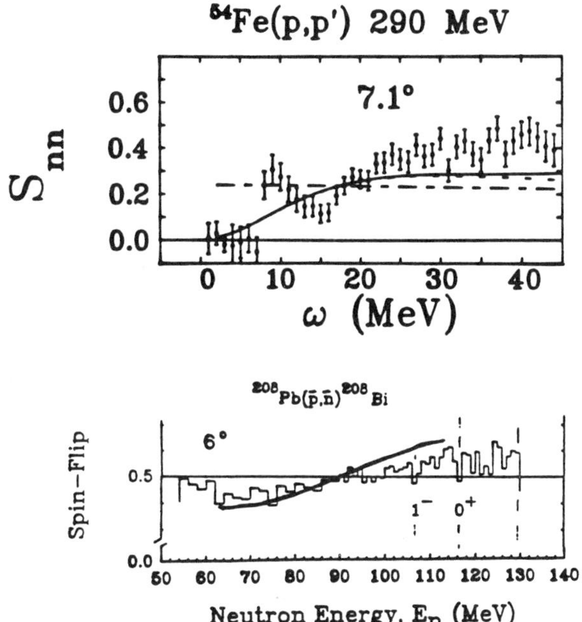

Fig.2. a) Spin-flip probability for ^{54}Fe(p,p') at 290 MeV. The dot-dashed line is the free response and the solid line is a surface RPA calculation. The dotted line includes two-step processes. b) S_{nn} for ^{208}Pb at 135 MeV [13]. The solid line is the free response.

DECOMPOSITION OF THE NUCLEAR RESPONSE INTO ITS SPIN, ISOSPIN, AND ANGULAR MOMENTUM COMPONENTS

The determination of the background underlying giant resonances has been controversial. Methods range from using shapes generated from a gaussian plus an exponential tail [14], to using the surface response model to predict the background. By separating the nuclear response into its spin-isospin components, the contribution of the background in each channel is reduced permitting a cleaner determination of the strength of the different resonances. The availability of data on the spin-flip probability and of charge exchange data at intermediate energy has made such a separation possible.

a) Isospin Decomposition.

The (n,p) reaction is the most selective of the (N,N) reactions since it only populates $T=T_0+1$ states where T_0 is the isospin of the target. The isovector strength in (p,p') is split between $T=T_0$ and $T=T_0+1$ states. The proportions are determined by Clebsch-Gordan coefficients. Therefore, we can determine the T_0+1 component of the (p,p') cross-section by multiplying the (n,p) cross-section by an appropriate factor. For ^{54}Fe, $T_0=1$ and this ratio is 1/2. We must also include a factor of 1/2 because of the isospin flip of the projectile. Taking into account the Coulomb shift in the final nuclei, the prescription for obtaining the T_0+1 component of the ^{54}Fe(p,p') cross-section is to divide the (n,p) cross-section by 4 and shift the spectrum by 8.6 MeV. The ^{54}Fe(n,p)^{54}Mn data were taken from reference [15]. The results are plotted in fig.3a. The T_0+1 component is seen to be small since the (p,p') reaction is dominated by the isoscalar piece of the interaction. Fig.3b shows the T_0+1 component of the spin-flip cross-section (σS_{nn}) which is a measure of the $\Delta S=1$ strength as we shall see below. The T_0+1 piece is a larger component of σS_{nn} than of the cross-section because the large isoscalar $\Delta S=0$ response is removed. The T_0+1 component of σS_{nn} was obtained in a similar way to the cross-section. The value of S_{nn} multipying the (n,p) cross-section was calculated from the surface response model.

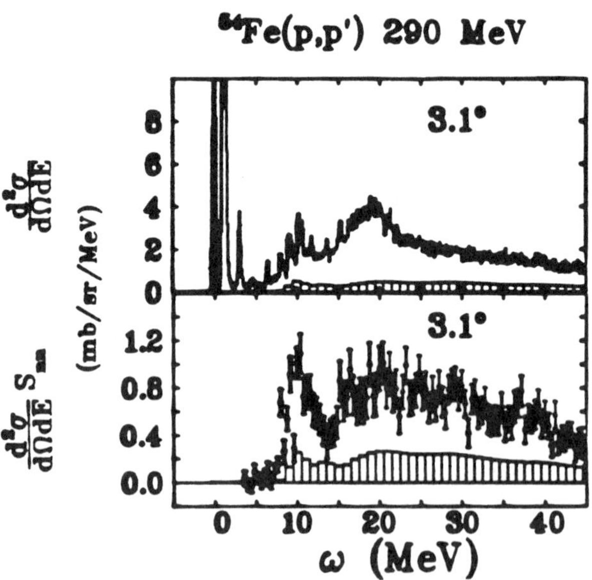

Fig.3. Cross-section and spin-flip cross-section for ^{54}Fe(p,p') at 290 MeV. The vertically hatched sections are the T_0+1 components deduced from the ^{54}Fe(n,p) reaction.

b) Spin Decomposition.

The spin-flip probability, defined as:

$$S_{nn} = \frac{1}{2}(1 - D_{nn})$$

where D_{nn} is the spin transfer coefficient for spin in the normal direction, is a measure of the $\Delta S=1$ strength. Since S_{NN} is not 1 for pure $\Delta S=1$ transitions, separating the response into its spin components must be model dependent. The $\Delta S=0$ response is obtained as follows [12]:

$$\sigma(\mathrm{S}=0) = \sigma_{\mathrm{tot}} - \sigma \mathrm{S}_{\mathrm{nn}}/\alpha \quad \text{where} \quad \alpha = \mathrm{S}_{\mathrm{nn}}(\mathrm{S}=1)$$

$\alpha(E_x)$ cannot be measured for the reason stated above and was obtained from a linear interpolation between a microscopic M1 calculation at $\omega = 10$ MeV and the surface response model value at $\omega = 40$ MeV. A multipole decomposition was done for the resulting $\Delta S=0$ spectra. The results show the isovector giant dipole resonance (IVGD) at 19 MeV and the isoscalar giant quadrupole (ISGQ) at 16 MeV. The position of the IVGD agrees with previous photonuclear data [16]. The summed strength in the peak of the IVGD is 0.84 of the energy weighted sum rule, while the total integral up to 30 MeV is 1.24. The corresponding numbers for the ISGQ are 0.70 and 0.94. A similar analysis has been done for ^{40}Ca(p,p') [17] and the results are essentially the same as for ^{54}Fe [11].

A multipole decomposition was also done for the $\Delta S=1$ response. The M1 resonance is found to be quenched by approximately 50% [11].

CONCLUSIONS

Quasi-elastic scattering provides evidence of modifications of the NN interaction at nuclear density. A full set of spin observables shows that although quantitative fits are not yet available, the theories seem to include many of the correct features. A better nuclear response must be included in the relativistic models; relativistic RPA calculations are in progress [10]. A non-relativistic calculation which uses a density dependent interaction such as that of von Geramb would be useful. A new technique for the decomposition of the response into its spin-isospin components has been developed. This permits a less model dependent determination of the strength of the giant resonances. It is hoped that the availability of these complete data in the continuum will help constrain the theories and separate the different effects included in the models.

The results described here are the work of many collaborators. I would like to thank O. Häusser, C. Glashausser and the MRS group at TRIUMF, as well as J. Watson for the use of his (p,n) S_{nn} data.

REFERENCES

[1] H. Esbensen and G.F. Bertsch, Phys. Rev. C34, 1419 (1985).
[2] R.D. Smith and S.J. Wallace, Phys. Rev. C32, 1654 (1985).
[3] C.J. Horowitz and M.J. Iqbal, Phys. Rev. C33, 2059 (1986).
[4] C.J. Horowitz and D.P. Murdock, Phys. Rev. C35, 1442 (1987).
[5] R.J. Glauber and G. Matthiae, Nucl. Phys. B21, 135 (1970).
[6] R.D. Smith and J. Wambach, Phys. Rev. C36, 2704 (1987).
[7] T.A. Carey et al, Phys. Rev. Lett. 53, 144 (1984).
[8] O. Häusser et al, submitted to Phys. Rev. Lett.
[9] G.G. Ohlsen, Rep. Prog. Phys. 35, 717 (1972)
[10] C.J. Horowitz, private communication.
[11] O. Hausser et al, to be published.
[12] C. Glashausser and K. Jones, contributions to this conference.
[13] J.W. Watson et al, Conference on Spin and Isospin Excitations of the Nucleus, Telluride 1988.
[14] D.K. McDaniels et al, Phys. Rev. C33, 1943 (1986), and L.W. Swenson et al, to be published.
[15] M.C. Vetterli et al, Phys. Rev. Lett. 59, 439 (1987).
[16] B.S. Ratner et al, Nucl. Phys. A285, 71 (1977).
[17] F.T. Baker et al, Phys. Rev. C37, 1350 (1988).

QUASIFREE HADRON PRODUCTIONS IN 3.9-GeV/c NUCLEAR REACTIONS

K. Nakai
KEK, Oho, Tsukuba-shi, Ibaraki 305, Japan

ABSTRACT

Quasifree productions of Δ in the target-rapidity region, and of ρ^0 and f_2^0 mesons in the projectile region in hadron-nucleus reactions at 3.9 GeV/c have been studied with a large acceptance spectrometer 'FANCY' at KEK. Exclusive observation of the Δ produced in nuclei via the (p,p'Δ) reaction was consistent with a quasifree picture. The position and width of Δ-peak in missing mass spectra were explained by the binging energy and Fermi motion. Cross sections for pion-induced productions of ρ^0 and f_2^0 mesons on C, Al and Cu targets were analyzed to deduce the $\rho^0 - N$ and $f_2^0 - N$ elementary cross sections with an intranuclear-cascade calculation as, $\sigma_{\rho^0-N} = 31^{+10}_{-7}$ mb and $\sigma_{f_2^0-N} = 60^{+20}_{-10}$ mb, respectively. While σ_{ρ^0-N}, $\sigma_{\omega-N}$ and $\sigma_{\pi-N}$ are all nearly equal as expected from a naive quark model, $\sigma_{f_2^0-N}$ is significantly larger probably because of the orbital excitation.

INTRODUCTION

Studies of hadron-nucleus reactions in GeV region provide interesting information on both particle and nuclear physics. In a series of experiments at KEK-PS using a large acceptance spectrometer, "FANCY", we have been studying two types of processes in hadron-induced reactions:

- Multiparticle productions in highly inelastic processes.
- Quasifree hadron productions.

Previously, we have reported on the multiparticle productions in which we studied space-time structures of nuclear reactions and energy deposition of GeV particles in nuclei [1]. In this paper we shall discuss the following two subjects on the quasifree hadron production studies with the FANCY spectrometer [2].

- The Δ production in the target-rapidity region [3].
- The ρ^0 and f_2^0 productions in the projectile region.

While the former subject is an example of using high-energy particles for nuclear studies, the latter is an example of using nuclei to study properties of particles.

We have built a large acceptance spectrometer, FANCY, for high-energy nuclear physics experiments at the KEK 12-GeV proton synchrotron [2] (Fig. 1). The name, FANCY, is an abbreviation of 'Forward ANd CYlindrical'. Particle productions

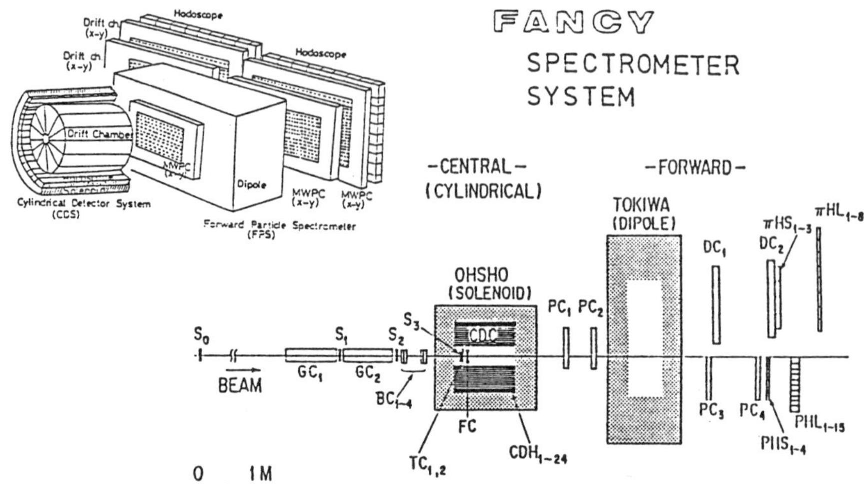

Figure 1: Schematic view and layout of the FANCY spectrometer.

in hadron-nucleus reactions in GeV region are characterized by a clear separation between particles emitted in the forward projectile-rapidity region and those in the target-rapidity region. Accordingly, the spectrometer was designed to cover the projectile-rapidity region by a forward partiticle spectrometer (FPS) and the target region by a cylindrical spectrometer (CDS) [2]. This system provides a unique tool for the study of quasifree production of hadrons both in projectile- and target-rapidity regions.

QUASIFREE Δ PRODUCTION

The production of Δ in a target-rapidity region via quasifree processes provides unique information on behaviors of the Δ in nclei. We were interested in the quasifree Δ production with protons at several GeV which would provide information on real Δ's in nuclei. Especially we were stimulated by the interesting data with the $(^3He, t)$ reaction [4] which indicated a shift of the Δ mass.

We have studied the (p,p'Δ) process with data taken by the FANCY spectrometer [3]. Proton beam at 3.88 GeV/c were used on targets of CD_2, C, Al and Cu. Data were taken with a trigger condition requesting at least one charged particle in each of the forward(FPS) and the cylindrical(CDS) spectrometers, and events were selected in the off-line analysis requesting;

(1) one proton in the forward spectrometer,

(2) one proton and one pion in the cylindrical spectrometer and

(3) no other charged particle.

Invariant-mass spectra of proton-and-pion pairs detected in the cylindrical spectrometer showed peaks of Δ and N* depending upon event-selection windows set on the forward proton spectrum (Fig. 2). Through the (p,p'Δ) processes only Δ^0 and Δ^+ are produced: p +'n'→ p + Δ^0, p +'p'→ p + Δ^+.

Indeed, the invariant-mass spectra with a deuteron (D) target showed the peak corresponds to Δ^0 in the (pπ^-) spectra, but no peak in the (pπ^+). The data from the deuteron target were deduced by subtracting C data from CD_2 data. The clean result in Fig. 2 indicates that the subtraction method worked very well. In contrast to the D-target data, the (p,π^+) spectra with the C target in Fig. 2 showed a peak corresponding to Δ^{++}. These Δ^{++}must be produced in secondary processes such as charge-exchange in the nucleus.

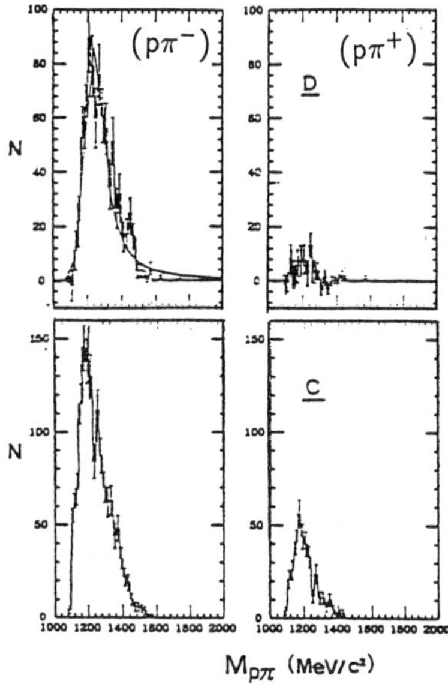

Figure 2: Invariant-mass spectra of (pπ^-) and (pπ^+) following the (p,p'Δ) process on D and C targets.

Shown in Fig. 3(a) are momentum spectra of inelastically scattered protons obtained with a cut on the pπinvariant-mass spectra ($M(p\pi) < 1400$ MeV/c^2). Once again it is seen that the peak corresponding to Δ^0 appears only with a (pπ^-) gate in the case of the D target. In the momentum spectra with heavier targets, however, continuum backgrounds are seen in both (pπ^-)- and (pπ^+)-gated spectra. These are probably due to secondary processes in which additional neutral particles, or charged particles to the blind angles of the detector, are emitted.

In order to reduce the continuum background and to observe quasifree processes exclusively, we made a further selection of events imposing a kinematical constraint. We calculated the mass (M_2) of the target nucleon (2) from the four momenta of (1) incident proton, (3) inelastically scattered proton and decay products of Δ; (4) $\pi^-$and (5) proton: $(M_2)^2 = (E_1\text{-}E_3\text{-}E_4\text{-}E_5)^2$ - $(\mathbf{p}_1\text{-}\mathbf{p}_3\text{-}\mathbf{p}_4\text{-}\mathbf{p}_5)^2$.
The distribution of $(M_2)^2$ thus obtained showed a sharp peak at the nucleon mass squared. When we set a gate on this peak for selction of events, the continuum background was strongly suppressed as shown in Fig. 3(b), and the 'genuine' quasifree events were selected.

We, then, compared the momentum spectra of forward protons in the 'genuine' quasifree Δ production with those calculated with simple assumptions taking into account the Fermi motion and the binding energy of nucleons in nuclei. In order to determine the Fermi motion and the binding energy we analyzed data on the quasielastic scattering of protons taken simultaneously on C and CH_2 targets.

Figure 3(c) shows the momentum spectra of forward protons from p-p and p-C scatterings. The solid curve drawn along the C data is the best-fit curve with the binding energy of 25 MeV and the Fermi momentum distribution with a width of 120 MeV/c assuming a Gaussian-type function. Using these parameters determined in the analysis of quasielastic process, we calculated the momentum spectra of protons for quasifree Δ production. Solid curves in Fig, 3(b) are the results of the calculation. The width of Δ used in the calculation was 115 MeV. The calculation reproduced experimental data quite well. In contrast to other inclusive experiments, the exclusive data were well reproduced with the quasifree picture.

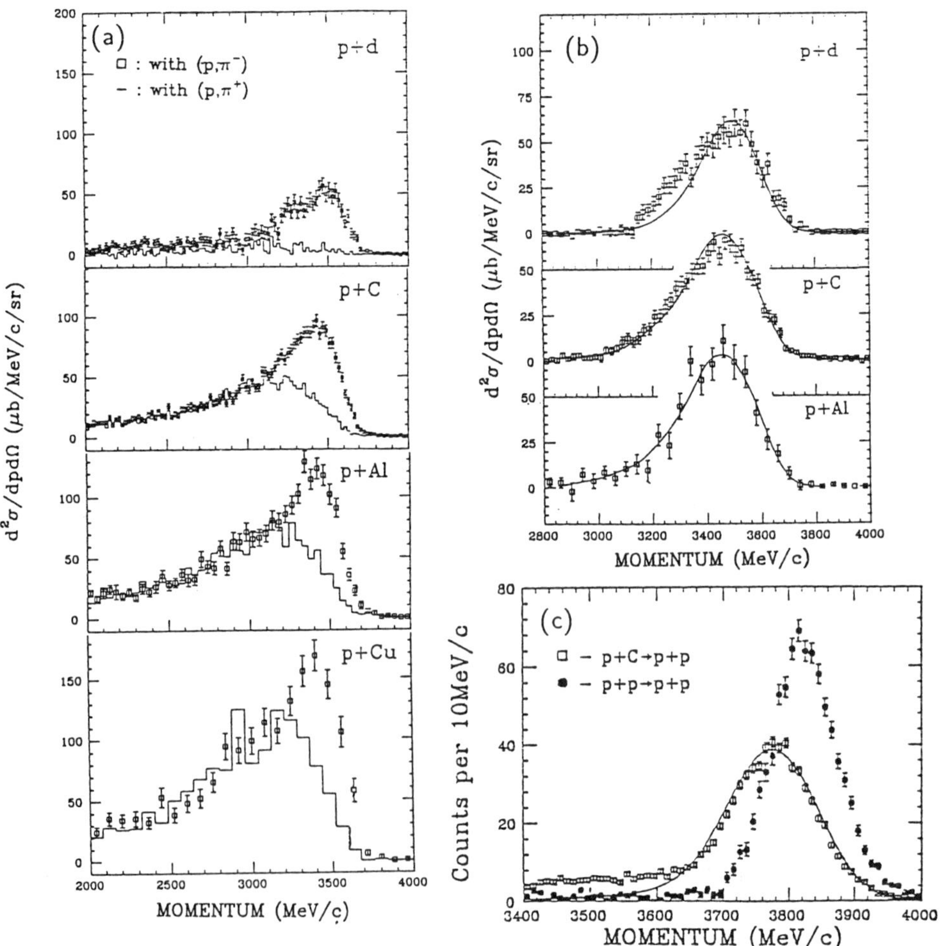

Figure 3: Momentum spectra of scattered protons from (a) the (p,p'Δ) processes, (b) the 'genuine' quasifree (p,p'Δ) processes, and (c) the quasielastic scattering. Solid lines in (b) show results of a calculation taking into account the width of the Δ, the binding energy, and the Fermi motion of nucleons.

CROSS SECTIONS FOR ρ^0 and f_2^0 PRODUCTIONS

Production cross sections of mesons on nuclei are determined by the elementary production cross section and the interactions of hadrons with nucleons in nuclei. Thus, from the target-mass dependence of the production cross sections, the elementary cross sections for interactions of unstable mesons with nucleons can be deduced through analyses using multiple-scattering theories [5].

In the past, there have been many measurements of the meson-production cross sections on nuclei to deduce the cross sections for interactions of the mesons, $\eta^0, \rho^0, \omega^0, \phi^0$ and f_2^0, with nucleons (Fig. 4). Experimental data were either from coherent photoproduction or from incoherent production in pion-nucleus reactions. The deduction of the meson-nucleon cross sections from production cross sections was made by use of the Glauber's multiple-scattering theory.

In one of those previous bubble-chamber experiments [6], it has been reported that $\sigma_{f_2^0-N}$ may be larger than σ_{ρ^0-N} though a naive quark model predicted $\sigma_{f_2^0-N} = \sigma_{\rho^0-N} = \sigma_{\pi-N}$. However, this interesting result had to be further investigated because of two reasons. First the cross sections were determined from data with only one nuclear target, and the authors pointed out importance of additional data. In addition, the validity of using the Glauber theory for the analysis of the f_2^0 production cross section was questionable because the mean-decay-length $(\beta\gamma c\tau)$ was comparable to the nuclear radius.

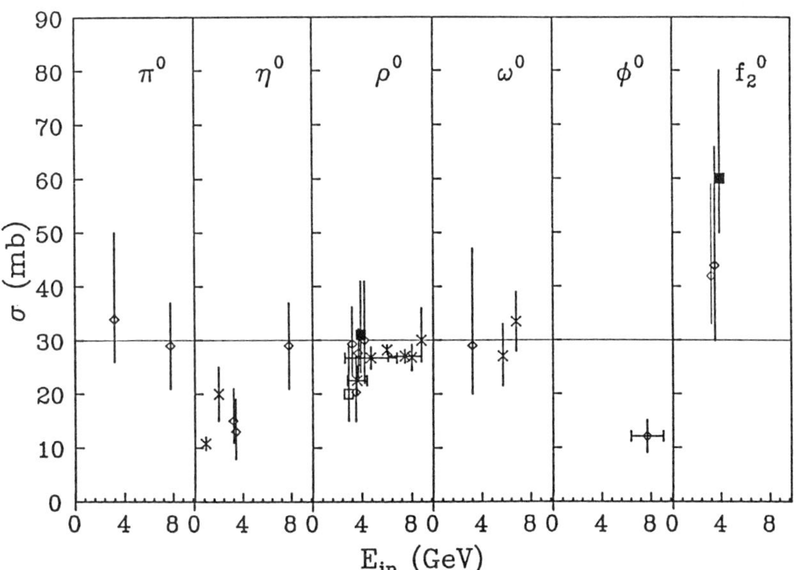

Figure 4: Summary of meson-nucleon cross sections deduced from data on coherent photoproductions (crosses), or incoherent production in pion-nucleus reactions (diamonds). Present data are shown with squares.

In the present work we have measured the cross sections using D, Li, C, Al and Cu, and analyzed data with an intranuclear-cascade calculation. We made analysis also with the Glauber's multiple-scattering theory to test the calculation by comparing results for σ_{ρ^0-N}. Production cross sections of ρ^0 and f_2^0 mesons on nuclei were determined by analyses of data taken with the FANCY spectrometer. Unseparated beams of protons and pions of 3.88 GeV/c were incident on a nuclear target (CD_2, Li, C, Al or Cu).

Data were taken with a trigger condition requiring at least one particle in each of the CDS and FPS. In off-line analyses, we selected events with at least one pion in each region of the CDS and FPS. Then, we could identify the ρ^0 and f_2^0 production events not only inclusively but also exclusively by identifying an additional recoil proton in the CDS. Figure 5 shows the $\pi^+\pi^-$ invariant-mass spectra of the exclusively selected events with different nuclear targets. The deuterium data were obtained by a subtraction of the C-target data from the CD_2 data. Two peaks corresponding to the ρ^0 and f_2^0 mesons are clearly seen. One of the solid lines in the spectra are results of least-square fitting assuming two Breit-Wigner functions for ρ^0, f_2^0 and a phase-space background component.

It was noted that the production ratio of the two mesons changes with the target mass. The decrease of the f_2^0 production relative to ρ^0 with the target mass indicates the larger cross section for the f_2^0-N than the ρ^0-N, ($\sigma_{f_2^0-N} > \sigma_{\rho^0-N}$), in accord with the suggestion from the previous bubble-chamber experiment [6].

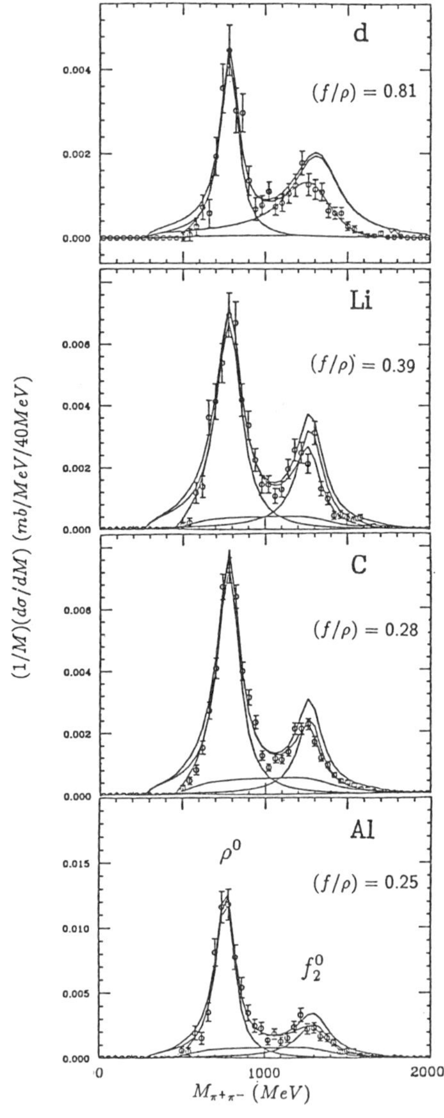

Figure 5: Invariant $\pi^+\pi^-$ mass spectra of exclusive events for D, Li, C and Al targets. The yield ratio of f_2^0 to ρ^0, (f/ρ), for each target is given for comparison. Solid curves drawn along the data are the best-fit curves, the curves corrected for the mass acceptance, and their decompositions to ρ^0, f_2^0 and background components.

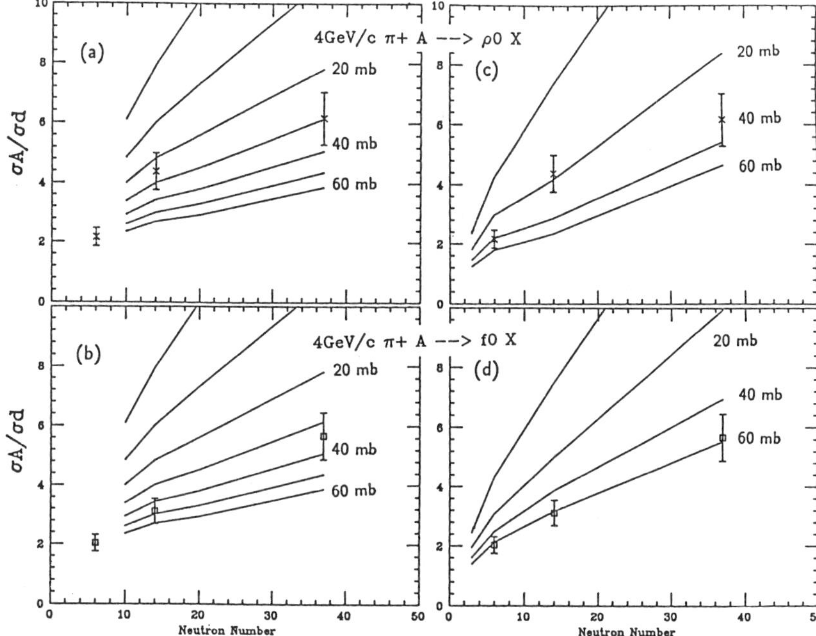

Figure 6: Comparisons of experimental production cross sections of (a)(c) ρ^0 and (b)(d) f_2^0 with calculated ones, by the Glauber theory (a)(b) and by the cascade code (c)(d).

In the analyses to deduce the $f_2^0 - N$ and $\rho^0 - N$ cross sections ($\sigma_{f_2^0-N}$, σ_{ρ^0-N}), we used, however, only the inclusive data because the Glauber method was not applicable to the exclusive cross sections. In this case the ρ^0 production cross sections suffered a relatively large uncertainty due to corrections for a mass-acceptance cut.

In Fig. 6(a)(b), experimental production cross sections for C, Al and Cu targets are compared with calculated ones from the Glauber-type multiple-scattering theory assuming several different meson-nucleon cross sections. The calculated cross sections were taken from reference [5], and an interpolation was made to deduce the best-fitted cross sections as,

$$\sigma_{\rho^0-N} = 27^{+9}_{-5} \text{ mb}, \quad \sigma_{f_2^0-N} = 41^{+11}_{-8} \text{ mb} \qquad \text{(by the Glauber method).}$$

The cross sections are in good agreement with the previous experiment [6].

As discussed earlier, the Glauber-type analysis is hardly justified especially for the deduction of the $f_2^0 - N$ cross section, because the recoil velocity of the f_2^0 meson was small and consequently the mean-decay-length ($\beta\gamma c\tau$) was comparable to the nuclear radius. We, therfore, made analyses with an intranuclear cascade code written by ourselves [7].

The production cross sections by the cascade calculation are also shown in Fig. 6(c)(d), and compared with experimental data. The $\rho^0 - N$ and $f_2^0 - N$

cross sections deduced by this method were,

$$\sigma_{\rho^0-N} = 31^{+10}_{-7} \text{ mb}, \quad \sigma_{f_2^0-N} = 60^{+20}_{-10} \text{ mb} \quad \text{(by the cascade method)}.$$

The good agreement between the ρ^0-N cross sections deduced by the Glauber-type analysis and the cascade method convinced ourselves the validity of our calculations.

The result indicates that the f_2^0-N cross section is significantly larger than the ρ^0-N. The uncertainties in the deduced cross sections includes not only experimental errors but also uncertainties in the analyses, However, we wish to stress that this conclusion was supported more directly by the stronger attenuation of f_2^0 mesons shown in the exclusive mass spectra (Fig. 5) as discussed previously.

Present ρ^0-N and f_2^0-N cross sections are also plotted in Fig. 4. In the figure we noted that while the σ_{ρ^0-N}, $\sigma_{\omega-N}$ and $\sigma_{\pi-N}$ are all nearly equal as expected from a naive quark model, the cross section $\sigma_{f_2^0-N}$ is larger. The small ϕ^0-N cross section may be because of the narrower $s-\bar{s}$ potential than $u-\bar{u}$ or $d-\bar{d}$ due to the heavier mass, and thus the wave function of ϕ would have a narrower distribution than ρ, ω or π. In contrast to ρ, ω or π, the f_2^0 meson has an orbital angular momentum L=1, so that its wave function is more spread out. We may, therefore, conclude that the $\sigma_{f_2^0-N}$ is significantly larger because of the orbital excitation.

Further systematic studies of the meson-nucleon cross sections should be very interesting. For instance, the $\eta-N$ cross section seems to be smaller than the $\rho-N$ or $\omega-N$, possibly reflecting a unique structure of the η meson. Deductions of cross sections for higher-mass, or higher-L mesons are also extremely interesting.

This paper is based on a result of a KEK-PS experiment E-133. The author is indebted to all members of the experimental group. The present work was partly supported by the Grant-in-Aid for Scientific Research by the Japanese Ministry of Education, Science and Culture.

References

[1] H. En'yo et al., *Phys. Lett.* **159B** 1 (1985): K. Tokushuku et al., *submitted to Phys. Lett.* (KEK-Preprint): K. Nakai *Proc. of Changchun Summer School* (KEK-Preprint).

[2] T. Nagae et al., *in preparation*, KEK-Preprint.

[3] T. Nagae et al., *Phys. Lett.* **191B** 31 (1987): J. Chiba, *Nucl. Phys.* **A478B** 491c (1988).

[4] D. Contardo et al., *Phys. Lett.* **B168** 331 (1986).

[5] B. Margolis, *Phys. Lett.* **26B**, (1968) 524: K.S. Kolbig and B. Margolis, *Nucl. Phys.* **B6**, (1968) 85.

[6] B.S. Chaudhary et al., *Nucl. Phys.* **B67**, (1973) 333.

[7] H. Sano, PhD thesis, University of Tokyo, in preparation.

Light ions induced charge exchange reactions at Saturne

T. HENNINO*

I.P.N. B.P. N° 1 91406 Orsay Cedex FRANCE

Abstract : Spin isospin excitations are studied at Laboratoire National Saturne using charge exchange reactions. New data, completing the (^{3}He,t) one, have been recently obtained. We present measurements done on the (d,2p) reaction with a special emphasis on the quasi free and Δ regions. Spin observables are also discussed. We also present the results of a systematic study of the characteristics of the Δ excitation by ^{12}C, ^{14}N, ^{16}O and ^{20}Ne scattering. The cross section dependence with respect to projectile and ejectile is qualitatively understood. As observed in all other charge exchange reactions, the Δ peak on nuclei is shifted by ~70 MeV with respect to the free Δ.

Introduction

Charge exchange reactions have been studied at Saturne for now 6 years using various projectiles with an energy ranging from 200 MeV to 1100 MeV per nucleon. Above 500 MeV aside from the nuclear excitations already seen in the Indiana (p,n) experiments [1], a well developed bump shows up with a strength comparable to the one corresponding to the particle hole states sector[2]. This corresponds to the spin isospin flip of one quark in a target nucleon leading to the Δ. These reactions appear to be direct reactions selectively exciting the spin isospin modes of the nucleus. With the good quality heavy ions beam delivered by Saturne, we were able to study both (p,n) like and (n,p) like reactions induced by ^{12}C, ^{14}N, ^{16}O and ^{20}Ne projectiles. In parallel, a program was developed to study the ($\vec{d}$,2p[$^1S_0$]) reaction with a polarised beam. All these experiments were performed using the SPES4 magnetic spectrometer[3]. The properties of the spectrometer and of the associated detection system is quite suitable to realize measurements at small angle (including 0°) up to a maximum rigidity of 4.2 GeV/c. The ion identification is done by the association of two thick scintillators which determine the charge and a 17 m long time of flight basis which gives the mass. Ray tracing is carried out by two sets of drift chambers. Their resolution (300 μm FWHM) allows us to measure angular distributions with a resolution of ~ 2 mrd. I will in this paper mainly focus on quasi free charge exchange and Δ production.

The (d,2p) reaction

The 2 protons were detected simultaneously in the SPES4 spectrometer. Its angle and momentum acceptance selects the two protons in the L=0 state. One then has a pure $\Delta S=1$ $\Delta T=1$ probe. Statistics are not very good in the low energy loss region for two reasons. On one hand, as the cross section for charge exchange is forward peaked, we have to face the intense background due to the deuteron breakup. This leads us to use a rather low

* work done in collaboration with:
D. Bachelier†, M. Bedjidian††, J. L. Boyard†, A. Brockstedt†††,
D. Contardo††, C. Ellegaard‡, C. Gaarde‡, C. Goodman‡‡,
T. Jorgensen‡, J. C. Jourdain†, J. S. Larsen‡, B. Million‡,
M. Osterlund†††, P. Radvanyi‡‡‡, M. Roy-Stephan†, P. Zupranski‡‡‡
† IPN Orsay BP N°1 F 91406 Orsay FRANCE
†† IPN Lyon F 69622 Villeurbanne FRANCE
††† Inst. of Phys. University of Lund S-223 62 Lund SWEDEN
‡ NBI DK 2100 Copenhagen DENMARK
‡‡ Indiana University Bloomington IN 47405 USA
‡‡‡ LNS F 91191 Gif sur Yvette FRANCE

beam intensity, especially in the region of the discrete states and at small angles ($\leq 2''$), in order to be able to separate the (d,2p) process from the very large amount of accidentals induced by the (d,pn) reactions. On the other hand the detection efficiency for two protons, independent of angle, is of the order of 15% at 2 GeV and 3% at 400 MeV. The complete excitation spectrum can be divided in three parts : the discrete states,the quasi free and the Δ excitations.

The discrete states

On hydrogen, the neutron peak is easily separated. In an impulse approximation calculation[4] with the NN amplitudes of H. Esbensen[5], one follows the cross section over 3 decades. The agreement with the polarisation observable M, shown by the curve on fig 1, is good. The deviation at q$\geq$ 2 fm^{-1} might be explained either by distortion effects or by our poor knowledge of the nucleon nucleon charge exchange amplitude in this momentum transfer region. On nuclei, the statistics is too poor to allow the extraction of a valuable nuclear structure information. With the future availability of a polarised ^{6}Li beam at Saturne, one then should be able to extract nuclear informations of good quality analogous to those obtained by the ($\vec{p}$,$\vec{n}$) reaction.

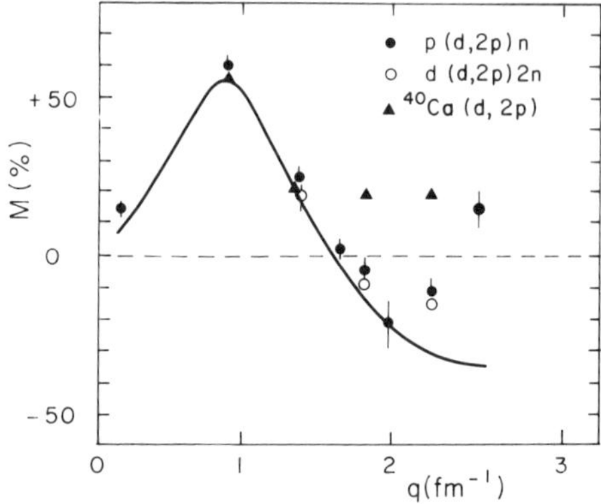

Fig 1. Polarisation signal of the free and quasi free peaks at 2 GeV. M is defined by $M = \frac{1}{2}\rho_{20}(T_{20} + \sqrt{6}T_{22} < \cos 2\varphi >)$ where ρ_{20} is the beam polarisation and the mean value takes into account the finite size angular aperture effects.

The quasi free bump

Fig 2 displays in a ($\|\vec{q}\|$,ω) plot the position of the maximum of the quasi free peak. One notes that (^{3}He,t) and (d,2p) data are consistent. However, the peak position seen in (p,p') [6] and (e,e') [7] is different. For these two reactions, the experimental slope corresponds to the free nucleon one with a shift in the electron case. For the two charge exchange reactions the departure from the electron line increases with tranfer to reach 40 MeV at 2.5 fm^{-1}. Such

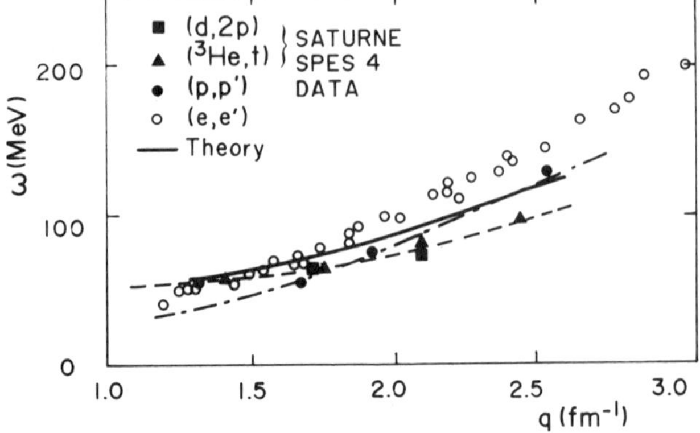

Fig 2. Comparison of the quasi free peak position. The dot dashed line is the free position. The dashed one is drawn as a guide line for our data. The full curve is the theoretical calculation [9].

an effect can be attributed neither to distortions nor to the (d,2p) form factor variation across the peak[8]. One can interpret it as due to NN^{-1} attractive correlations in the longitudinal channel. Such an interpretation has been tested by a recent calculation[9]. Nevertheless, although this calculation gives an effect of the correct sign, it is still unable to explain the size of the effect. The polarization response in the quasi free was also measured(fig 1). Up to 1.2 fm^{-1}, the response on ^{1}H, ^{2}H and ^{12}C coïncide. Above, the nuclear response deviates appreciably indicating that the (d,2p) reaction is mainly a transverse probe. This fact is difficult to reconcile with the shift attributed to nuclear correlations in the longitudinal channel and should be further studied on both experimental and theoretical sides.

The Δ bump

The unpolarized response shows similar effects to those already observed in the (^{3}He,t) reaction, namely that the position of the Δ peak on nuclei is shifted with respect to the free Δ^o created from a free proton by ~ 60±10 MeV. The cross section on the deuterium is twice the proton one. This is consistent with the 1/2 absorption factor, deduced from the comparison of the reactions d(d,2p)2n and p(d,2p)n, if one takes into account the Clebsh-Gordan coefficients for the production of the Δ^0 on a proton and of the Δ^- on a neutron. Spin observables were also measured, namely iT_{11} and M(see fig 1 for definition of M), at 2 GeV and 1.6 GeV from 0° to 7.2° over the Δ (Fig 3). With a pure longitudinal probe (π exchange)one would expect a negative signal (~ -40%) whereas a pure transverse one (ρ exchange) would give a positive signal (~ +20%). The ($\vec{d}$,2p) reaction in the Δ region appears to be mostly transverse. Moreover, the contrast between hydrogen, deuterium and carbon target is rather weak, indicative of the smallness of medium effects.

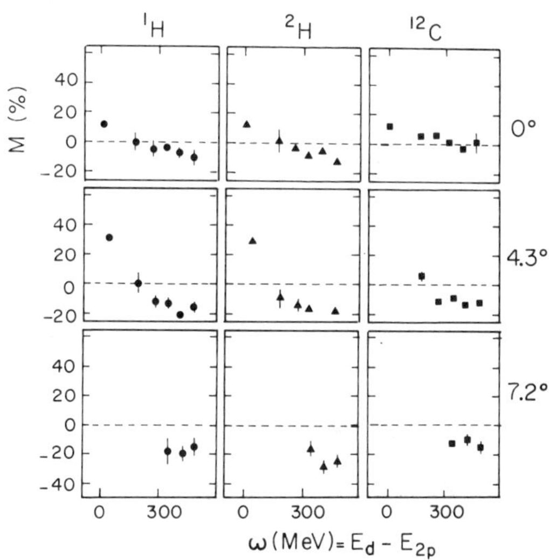

Fig 3. Polarisation signal M at 2 GeV and 0° in the Δ region. See fig 1 for definition of M.

Heavy ion reactions

With such reactions,we were able to measure the charge exchange at 0° for both (p,n) like and (n,p) like reactions and to investigate the Δ production in different experimental conditions. Of importance is of course the important question whether medium effects in the longitudinal channel, advocated by several groups[10] to explain the shift observed in (^{3}He,t) , subsist in so peripheral reactions. Because of the high incident energy, one covers, with SPES4[3], the response of the target nucleus over 1 GeV within only one magnetic setting. Moreover, with the large entrance square collimator (± 0.8° in both directions), one integrates the angular distribution almost entirely. One very important feature to have in mind is that we measure a yield which corresponds in fact to a summation over all transitions to states of the ejectile which are bound against particle emission. The resolution of the spectrometer in the GeV per nucleon energy domain (20 to 30 MeV) does not allow to separate these different contributions. This effect of mixing will be discussed in greater details later. We have measured the (^{12}C,^{12}N) , (^{12}C,^{12}B) , (^{14}N,^{14}O) , (^{14}N,^{14}C) , (^{16}O,^{16}N) , (^{20}Ne,^{20}F) and (^{20}Ne,^{20}Na) at energies between 400 MeV and 1100 MeV per nucleon on ^{1}H, ^{2}H, ^{12}C, ^{89}Y and ^{208}Pb. Only qualitative data had been obtained before 1987. With Mimas, a small synchrotron inserted between the ebis type source Dioné and the Saturne ring, the gain in intensity was important ($\sim$ 20), then allowing us to measure precisely position, width and intensity of the different structures appearing in energy loss spectrum. On fig 4, one sees a typical spectrum. It reveals essentially two structures : one at low energy loss value ($\omega \leq m_\pi$) which corresponds to nuclear spin isospin transitions and an another one around 300 MeV, the Δ bump.

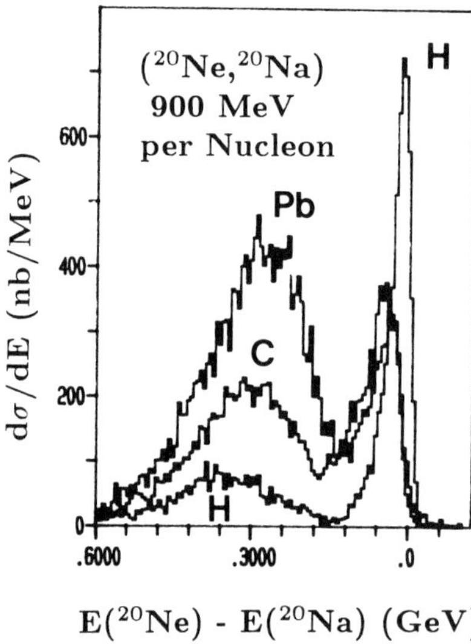

Fig 4. Angle integrated energy loss spectrum at 900 MeV per nucleon for the (^{20}Ne,^{20}Na) reaction.

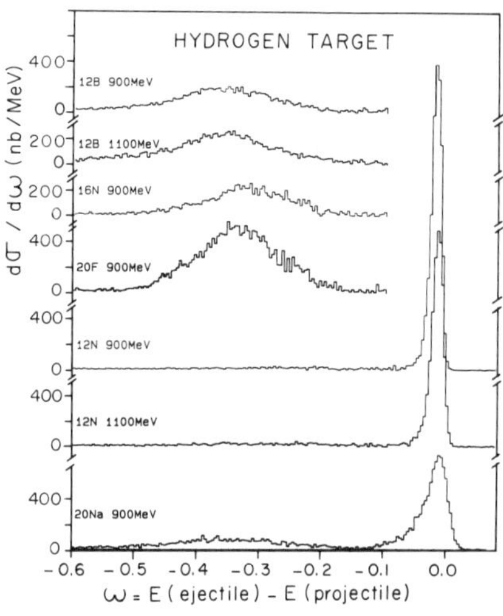

Fig 5. Angle integrated energy loss spectra on hydrogen.

The region below the π threshold

On hydrogen in (n,p) like reactions one observes the neutron peak(fig 5). Its asymmetrical shape reflects the combined effects of the strong kinematics and of the angular distribution. The fact that the width is larger in (^{20}Ne,^{20}Na) indicates that the transition form factor $^{20}Ne \rightarrow {}^{20}Na$ is less steep than the corresponding one for the (^{12}C,^{12}N) reaction. The comparison of 900 MeV and 1100 MeV per nucleon data indicates that the $\sigma\tau$ term in the NN amplitude is still decreasing with energy. This is consistent with a recent analysis[11]. On nuclei it includes the Gamow-Teller as well as the other multipole spin resonances. Since the angular distributions of all these different contributions is weighted by sinθ, the fraction of the cross section corrresponding to the L=0 transition (very forward peaked) is low and the spectrum is dominated by higher multipoles. An estimate based on the (^{3}He,t) data indicates that the relative contribution of the Gamow Teller in the (^{12}C,^{12}N) on ^{12}C is of the order of 8%. The quasi free charge exchange cannot be isolated on such an integrated spectrum though it is likely to contribute to the high energy loss tail of the low energy peak. From the (^{3}He,t) one knows that this process is important at q=1.4 fm^{-1} (4$''$ in (^{3}He,t)). This corresponds, for an incident ^{12}C, to the angular acceptance limit of our collimator.

The Δ bump

Fig 5 displays on the same scale free Δ production spectra for seven different experimental conditions. First one notices the weakness of the Δ^0 excitation by (^{12}C,^{12}N) reaction. This can be easily understood if one realizes that the only state involved in the ejectile is the ground state which corresponds to a L=0 transition. This transition form factor can be deduced from the inverse reaction $^{12}C(p,n)^{12}B$ measured at 800 MeV [12]. It falls off by a factor 100 between 0 and 1 fm^{-1} which is the transfer at the Δ peak position. Detailed calculations by M. Soyeur et al[13] reproduce quantitatively both the absolute cross section and the energy dependence on ^1H and ^{12}C. On the (^{12}C,^{12}B) side the Δ^{++} cross section is large for two reasons : first, the Clebsch-Gordan coefficient is three times bigger and secondly the contribution of the 2^+ state at 0.96 MeV in ^{12}B, which cannot be separated from the ground state contribution, is large. This state corresponds to a L=2 transfer which presents a maximum at q ~ 1 fm^{-1}. This interpretation explains also the variation of the cross section with the energy. As the energy increases the minimum momentum transfer decreases and the variation of the cross section then just reflects the variation of the form factor in the 1 fm^{-1} transfer region. A steep form factor as in the (^{12}C,^{12}N) reaction leads to a rapid increase with energy[13] as observed experimentally. On the contrary, for the (^{12}C,^{12}B) reaction, a smoother form factor gives an moderate increase. The cross sections for the Δ^{++} are also large in the (^{16}O,^{16}N) and (^{20}Ne,^{20}F) reactions as several states corresponding to L=1 transfer in the ejectile contribute. The transition form factor has also important effects on the position and width of the free Δ. The steeper it is, the more shifted and the narrower the Δ peak is. One then should be very careful when comparing raw positions and widths.

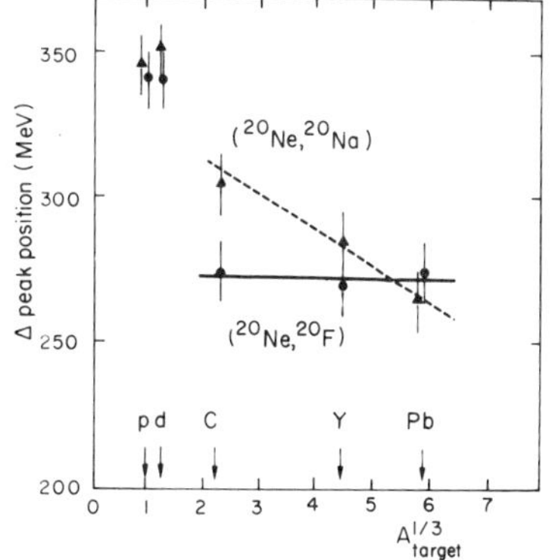

Fig 6. Δ bump position versus target mass.

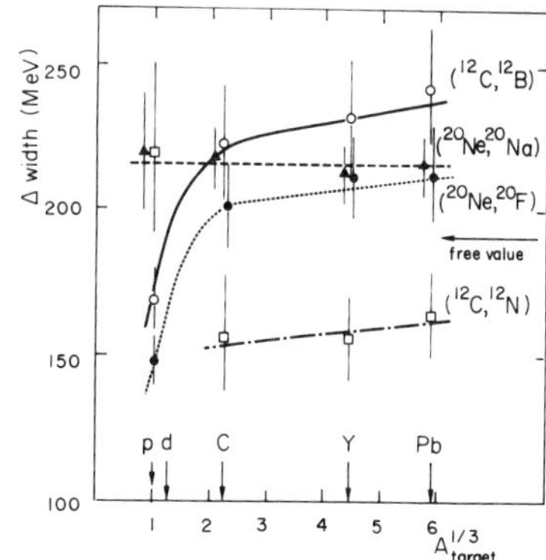

Fig 7. Δ bump width versus target mass.

On nuclei the Δ excitation is comparable or even bigger than to the excitation of the low energy sector. The important difference between the Δ^0 and the Δ^{++} excitations is washed out by the presence of neutrons which can be turned respectively into Δ^- and Δ^+. As a consequence, the ratio of the Δ cross section for (p,n) like and (n,p) like reactions on a T=0 target nucleus is droven by the number of possible transitions between the incident ion and the outgoing ion and the associated form factors. The effects of the distortions is very important : without distortions one would expect a cross section proportional to N+3Z in (n,p) like and Z+3N in (p,n) like reactions since there is no Pauli blocking for Δ's in nuclei. Taking out these factors one has a measure of the distortion. For the (^{20}Ne,^{20}F) reaction on ^{208}Pb one has only 4 effective target nucleons. A comparison with theory can be made for the ^{12}C target. The reaction appears to be very peripheral : a skin of only 2 fm gives 71% of the cross section[13]. Fig 6 shows the position of the Δ peak for different reactions. Besides the absolute position with is sensitive to the projectile ejectile form factor, as for the hydrogen case, one notices different target dependences. In (p,n) like reaction the shift with respect to the free Δ^{++} is stable from ^{12}C to ^{208}Pb. On the contrary, in (n,p) like reactions the Δ peak moves continuously from ^{1}H to ^{208}Pb. This is the first indication that medium effects can be different in mirror reactions. If one now plots the width of the Δ versus the target mass one observes also differences(fig 7). For (p,n) like reactions, the variation of the Δ width with the target mass seems to be explainable by the onset of the Fermi momentum. For the (^{20}Ne,^{20}Na) reaction the target dependence is small. For the (^{12}C,^{12}N) case the width of the free Δ^{o} is very large. In that case, because of the weakness of the excitation and the subtraction method used, one cannot rule out completely a possible background which would enhance artificially the width.However, from this systematics on widths and positions one may think that we have at hand a way to determine the isospin dependence of the Δ interaction with the nucleus. In heavy target nuclei proton and neutron density differences might suppress certain decay channels (For example, in such a model for neutron rich nucleus the decay process through the absorption of a Δ^{-} with a neutron is forbidden). Future experiments are planned to look for the decay channels of the Δ bump.

conclusion

A complete set of data on the Δ production have been obtained at Laboratoire National Saturne using various charge exchange reactions. The position, width and intensity, as well as polarisation observables were measured. A common observation is made : the Δ peak in nuclei is shifted with respect to the free one by $\sim$ 70 MeV. Further experiments are still on the way at Saturne. They should lead to a better understanding of the interaction of a Δ with the nuclear field.

References

1) G. E. Brown et al NP A372 (81) 397.
   C. Gaarde et al NP A369 (81) 258.
2) D. Bachelier et al PL B172 (86) 23.
   D. Contardo et al PL B168 (86) 331.
3) E. Grorud et al NIM 188 (81) 549.
   M. Bedjidian et al NIM A 257 (85) 132.
4) C. Ellegaard et al PRL 59 (87) 974.
5) H. Esbensen Private communication.
6) R. E. Chrien et al PR C21 (80) 1014.
7) U. B. Glawe DESY Report F23-80/1.
   J. S. O'Connell et al PRL 53 (84) 1627.
   P. Barreau et al NP A402 (83) 515.
8) I. Bergqvist et al NP A469 (87) 648.
9) W. M.Alberico et al Preprint CERN-TH.4862/87.
10) G. Chanfray et al PL B141 (84) 163.
    V. F. Dmitriev et al NP A438 (85) 697.
    J. Delorme et al to be published.
11) M. Soyeur private communication.
12) N. S. P. King et al PL B175 (86) 279.
13) M. Soyeur et al to be published.

Measurement of 2-nd and 3-rd Order Spin Observables in Elastic Scattering of 1.6 GeV Tensor and Vector Polarized Deuteron Beam from a Polarized Hydrogen Target

V. Ghazikhanian[1], B. Aas[1,5], D. Adams, E. Bleszynski[1,6]
M. Bleszynski, J. Bystricky[1,2], G.J. Igo[1], F. Sperisen[1,7]
C.A. Whitten[1], P. Chaumette[2], J. Deregel[2], J. Fabre[2]
F. Lehar[2], A. de Lesquen[2], L. van Rossum[2], J. Arvieux[3]
J. Ball[3], A. Boudard[3,4], F. Perrot[4]

ABSTRACT

Seven previously unmeasured spin asymmetries and ten previously unmeasured spin transfer observable have been determined in the momentum transfer range, $0.12 < -t < 0.85$ $(\mathrm{GeV/c})^2$. We used a 1600 MeV polarized deuteron beam incident on a polarized hydrogen target in these measurements. This is equivalent to using an 800 MeV proton beam on a deuteron target. These observables are poorly predicted by the relativistic impulse approximation, which has been used successfully to reproduce spin observables for protons on spin-zero nuclei.

I. Introduction

Due to the phenomenal success of the Relativistic Impulse Approximation (RIA) in reproducing the spin observables in elastic proton scattering from medium and heavy weight nuclei, it has become of interest to test the RIA in other situations. It is well known that elastic proton-nucleus (spin zero) scattering, being a spin 0-$\frac{1}{2}$ system, is only sensitive to two of the five relativistically invariant amplitudes of the N-N interaction, i.e., the spin- independent and pseudoscalar (alternatively, pseudovector) amplitudes. The remaining amplitudes, of tensor character, will enter when the RIA is used to predict proton- deuteron elastic scattering, a spin 1-$\frac{1}{2}$ system. Thus it is possible, in proton-deuteron elastic scattering, to test the RIA in such a way that all five invariant NN amplitudes enter sensitively.

The measurements reported here were made in two 10-day running periods in 1985 and 1986 at the Saturne II accelerator at Saclay.

(1) Physics Department, UCLA, Los Angeles, CA 90024
(2) DPhPE, CEN-Saclay, France
(3) LNS, CEN-Saclay, France
(4) DPhN/ME, CEN-Saclay, France
(5) Present Address: Rogland Research Institute, Ullandhung, N4001 Stavanger, Norway
(6) Present Address: Rockwell International Corporation, Los Angeles, CA
(7) Present Address: IUCF, Indiana University, Bloomington, IN 47405

II. Formalism

Spins are expressed in two right-handed coordinate systems $(\hat{S}, \hat{N}, \hat{L})$, one for the incident deuteron and target proton and another for the recoil proton. The polarization of the scattered deuteron was not measured. The longitudinal vector $\hat{L}$ is parallel to k_i and the normal vector $\hat{N}$ is perpendicular to the scattering plane along $\vec{k}_i \times \vec{k}_f$. The sideways direction is defined by $\hat{S} = \hat{N} \times \hat{L}$. Note that $\hat{S}$ and $\hat{L}$ change sign under a parity transformation while $\hat{N}$ is unaffected.

The observables are defined in terms of the scattering matrix F:

$$C(a, \alpha; b, \beta) = \frac{\mathrm{Tr}(F\sigma_a J_\alpha F^\dagger \sigma_b J_\beta)}{\mathrm{Tr}\,(FF^\dagger)} \quad \text{(Vector type)} \tag{1}$$

and

$$C(a, \alpha\eta; b, \beta) = \frac{\mathrm{Tr}(F\sigma_a J_{\alpha\eta} F^\dagger \sigma_b J_\beta)}{\mathrm{Tr}\,(FF^\dagger)} \quad \text{(Tensor type)} \tag{2}$$

In Eqs. (1) and (2), $\sigma_i (i = L, N, S)$, $J_k (k = L, N, S)$, and $J_{kj} (k, j = L, N, S)$ are the Pauli spin operators and vector and tensor spin operators for the deuteron respectively. The quantities we measure are the relative yield Y $\propto$ Tr (ρ_{out}) and components of polarization of the recoil proton,

$$p'_i = \frac{\mathrm{Tr}\,(\rho_{out}\sigma'_i)}{\mathrm{Tr}\,(\rho_{out})} \tag{3}$$

where the prime refers to the fact that the polarization is measured in the coordinate frame of the recoil proton. Here ρ_{out} can, as usual, be expressed in terms of the initial density matrix ρ_{in} for the p-d system as $\rho_{out} = F\rho_{in}F^\dagger$. Because of parity restrictions mentioned above, only those coefficients $C(a, \alpha; b, 0)$ and $C(a, \alpha\eta; b, 0)$ involving *even* numbers of spin components, which are either L or S, i.e., a, α, b, in the first and $a, \alpha\eta, b$ in the second, will be non-zero. This is illustrated in Fig. 1 where $C(0, S, 0, 0)$ which can be extracted from the data, is plotted. In this case the sum is equal to unity and the data is consistent with a null value for this observable.

Figure 2 is a schematic design of the N-N area at Saturne where the experiment was done. In the beam line, which is not shown, the asymmetry for quasi-elastic p-p scattering in the collision between the deuterons and a CH_2 target was measured in a polarimeter to determine the vector polarization of the beam. The tensor (and vector) polarization of the polarized beam was measured with another polarimeter before acceleration in Saturne II. In Fig. 2, the position of the frozen spin target is indicated. The polarized material was Pentanol and the typical polarization was 80-95%.

On the exit side of the cryostat a CH_2 target (unpolarized) was attached. Figure 3 shows vertex reconstruction of elastic deuteron events. Events from the polarized target and the CH_2 target are readily distinguishable. Elastic events originating in the CH_2 target were used in the analysis to monitor the relative tensor polarization of the beam.

The four momentum transfer interval of interest $0.12 < -t < 0.85$ $(GeV/c)^2$ could be covered in one angular setting. Elastically scattered deuterons, in the corresponding angular interval, 8.5° to 19.5°, could be readily distinguished from the large breakup cross section in the off-line analysis by the imposition of a cut on the bend angle in the magnet (Gouppillon). Breakup events result in forward- going protons with velocities roughly equal to those of the deuteron and thus bend approximately twice as much. This is illustrated in Fig. 4 where the sharp group (deuterons) is readily distinguishable from the broad group (protons). The N and S components of the recoil proton's spin are determined by measuring the asymmetry produced from scattering in the carbon target shown in Fig. 2. Note that the carbon target was actually wedge-shaped to account for the broad range of energy of the recoil protons corresponding to a range of $-t$ extending from 0.2 to 0.85 $(\mathrm{GeV/c})^2$.

Since the p-d cross section is more than one order of magnitude smaller than the p-p cross section at 800 MeV, it was necessary to provide a tighter trigger than was necessary in the pp measurements where the OR of the hodoscopes WH and H3 is used to initialize a trigger to fire the MWPC detectors. This was simply done by correlating each of the WH hodoscope elements with two or three of the H3 hodoscope elements which will be struck in a p-d elastic event when account is taken of the kinematics and of the geometry involved. The vector asymmetries C(0,N,0,0), C(N,N,0,0) and C(L,S,0,0) have been measured. The tensor asymmetries C(0,NN,0,0), C(N,NN,0,0), C(0,SS,0,0), C(N,SS,0,0), C(0,SL,0,0), C(N,SL,0,0) and C(L,NS,0,0) have also been measured. Of these, C(L,S,0,0), C(N,N,0,0), C(N,NN,0,0), C(N,SS,0,0), C(0,SL,0,0), C(N,SL,0,0) and C(L,NS,0,0) have not been measured previously in the momentum transfer range of this experiment. As for the triple-spin or spin transfer observables C(0,N,N,0), C(N,N,N,0), C(0,S,S,0), C(N,S,S,0), C(0,NN,N,0), C(0,SS,N,0), C(0,NS,S,0), C(N,NN,N,0), C(N,SS,N,0) and C(N,NS,S,0) have been measured. None of the above spin-transfer observables have been measured in this momentum transfer range before. A publication is being prepared of this work.

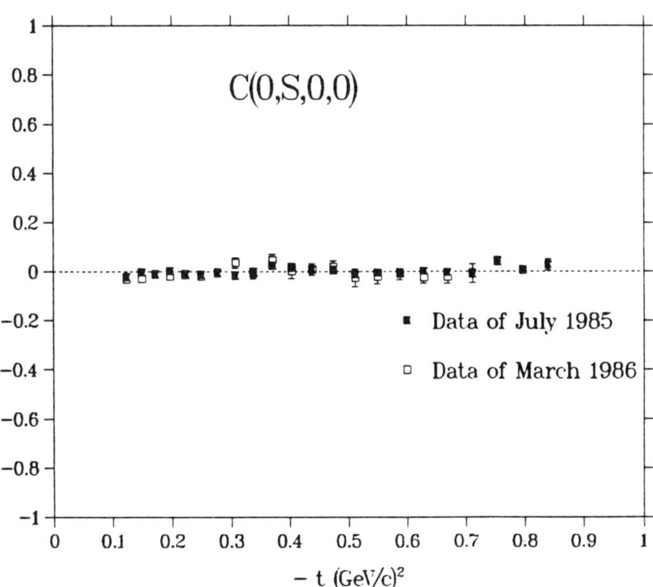

Fig. 1 The null observable C(0,S,0,0).

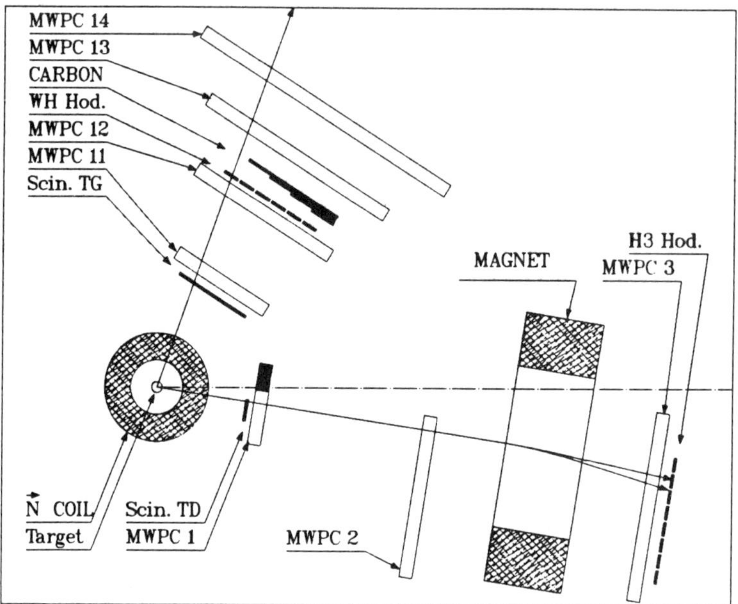

Fig. 2 Schematic diagram of the N-N experimental area at Saturne II. The locations of the spin frozen target, the bending magnet Goupillon, the plastic scintillator hodoscopes and the wedge shape polarimeter target are shown.

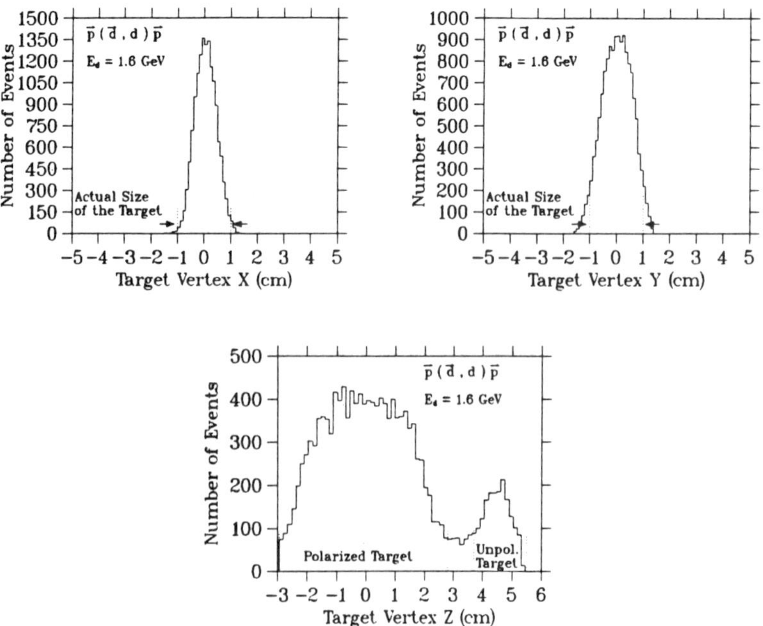

Fig. 3 Vertex reconstruction of the target region.

RADIATIVE KAON CAPTURE

E.K. McIntyre[a], E.C. Booth[a], W. Fickinger,[b], K.P. Gall[a], M.D. Hasinoff[c], N.P. Hessey[d], D. Horváth[e], J. Lowe[d], D.F. Measday[c], J.P. Miller[a], A.J. Noble[c], B.L. Roberts[a], D.K. Robinson[b], M. Sakitt[f], M. Salomon[g], J. Skelly[f] and D.A. Whitehouse[a].

[a] *Boston University, Boston, MA, USA 02215*
[b] *Case Western Reserve University, Cleveland, OH, USA 44106*
[c] *University of British Columbia, Vancouver, BC, Canada V6T 2A3*
[d] *University of Birmingham, Birmingham, U.K. B15 2TT*
[e] *KFKI, Budapest, Hungary and TRIUMF, Vancouver, BC. Canada*
[f] *Brookhaven National Laboratory, Upton, NY, USA 11973*
[g] *TRIUMF, Vancouver, BC, Canada V6T 2A3*

Abstract

Negative kaons were stopped in liquid hydrogen and liquid deuterium targets to study radiative capture rates. Branching ratios for the radiative kaon capture reactions $K^-p \to \Lambda\gamma$ and $K^-p \to \Sigma^0\gamma$ provide information about the quark structure of the $\Lambda(1405)$. In the reaction $K^-d \to \Lambda n\gamma$, the shape of the photon spectrum in the endpoint region provides information on the final state Λ-n interaction. Superior photon resolution is needed to distinguish the signal photons from the background of π^0 decay photons. Such resolution is provided by the B.U. NaI detector, which has a resolution of 1.3% (FWHM) near 300 MeV. A description of the experimental technique and data reduction is presented, along with preliminary branching ratio results.

The $\Lambda(1405)$ is the lowest mass, $J^\pi = (\frac{1}{2})^-$, strangeness −1 baryon resonance, with a width of 40 MeV. Because of its proximity to the mass of the K^-p system (1432 MeV), it has been interpreted as either a $\bar{K}N$ bound state[1], or a true three quark state[2]. The radiative capture reactions $K^-p \to \Lambda\gamma$ and $K^-p \to \Sigma^0\gamma$ provide information on the quark structure of the $\Lambda(1405)$ since the $\Lambda(1405)$ is an important states through which the low energy K^-p system interacts. Within the non-relativistic quark model (NRQM) of Isgur and Karl[2], the $\Lambda(1405)$ is considered to be strictly a three quark state. A calculation[3] has been made of the radiative capture branching ratios, using Isgur-Karl wavefunctions, and the $\Lambda(1405)$ is found to be the dominant intermediate state resonance. Within the cloudy bag model (CBM), analysis of low energy K^-p scattering data leads to a description of the $\Lambda(1405)$ as primarily a $\bar{K}N$ bound state.[1] Branching ratios are predicted[4] and the relative intensities of the two radiative capture reactions are found to be independent of the parameters of the cloudy bag model. The published branching ratio predictions, as well as previous experimental results and the preliminary results of this experiment are shown in Table 1. An alternate theoretical approach is to use measured branching ratios as input to a pole model calculation of the radiative widths of the $\Lambda(1405)$. Here[5] the Born diagrams are found to be important, in contrast to the assumptions of reference 2.

Branching ratios for the radiative capture reactions $B.R. = \frac{K^-p \to Y\gamma}{K^-p \to anything}$ have been measured. The experiment was performed in the Low Energy Separated Beam II at the AGS. The beam of 655 MeV/c negative kaons had a $\pi : K$ ratio of 7:1. Beam particles were defined by four in-beam plastic scintillators. Kaons were distinguished from pions by means of a velocity selecting Cerenkov counter and dE/dx information from three plastic scintillators downstream of a copper degrader.

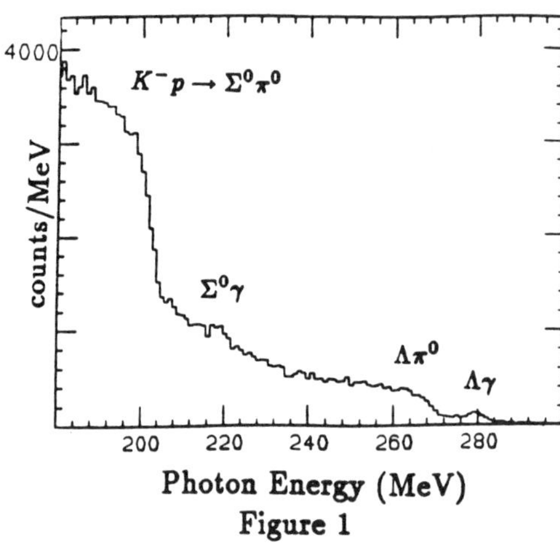

Figure 1

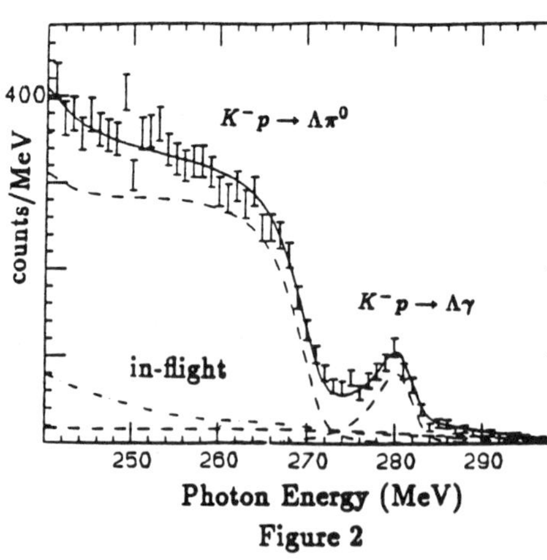

Figure 2

The reaction $K^-p \to \Lambda\gamma$ produces a monoenergetic photon of 282 MeV. This is the most energetic photon produced in a stopping reaction on the proton. The reaction $K^-p \to \Lambda\pi^0$ has a 6.7% branching ratio[6], with π^0 decay photons ranging in energy up to 271 MeV. In the 240 to 270 MeV range, this channel is the primary contributor to the photon spectrum, and serves as a normalization for the measurement of the radiative capture branching ratio. Because the 282 MeV signal photon is kinematically separated by only about 10 MeV from the copious π^0 decay spectrum, superior energy resolution is needed to distinguish the signal photons from the background (normalization) photons. Attaining and maintaining high photon energy resolution was the chief experimental challenge of the experiment. This resolution is provided by a large NaI detector designed at Boston University and built by the Bicron Corporation.[7]

The B.U. NaI detector is composed of five optically isolated pieces. These are a 10″ diameter by 22″ long cylindrical core, surrounded by a 5″ thick annulus, segmented into four quadrants. The NaI is surrounded by a 5″ thick annulus of plastic scintillator. The resolution of the detector was measured using 330 MeV electrons from the MIT-Bates Linear Accelerator and with 129.4 MeV photons at the AGS. These measurements yielded an intrinsic resolution of approximately 1.3% (FWHM) in the 280 MeV region.

The upper end of the photon spectrum (from 180 to 300 MeV) resulting from stopping kaons in hydrogen is shown in figure 1. The major features of the spectrum are the sharply falling edge near 200 MeV corresponding to π^0 decay photons from the $\Sigma^0\pi^0$ channel, and a second edge near 260 MeV corresponding to the $\Lambda\pi^0$ channel. The radiative capture peaks are clearly visible near 220 and 282 MeV. A Monte Carlo spectrum derived from known reaction channels[6] is fit to the data. The simulated response function of the NaI detector, based on EGS[8] calculations, and the measured pile-up shape are included in the simulation. Pile-up, plus backgrounds from energetic neutrons and kaon interactions both with aluminum and in-flight account for the observed yield above the $\Lambda\gamma$ peak. The branching ratio is determined by varying the amplitude of the $\Lambda\gamma$ (or $\Sigma^0\gamma$) channel, with all other branching ratios fixed. Figure 2 shows the data and fits in the region of the $\Lambda\gamma$ channel. The results of these fits are presented in Table 1. These results are smaller than the previous measurements, and smaller (by about a factor of two) than the theoretical predictions. The $\Sigma^0\gamma$ branching ratio is as large or larger than the $\Lambda\gamma$ ratio. In contrast to previous measurements of the photon spectrum, the peaks measured here are unambiguously observed.[9,10] Systematic errors in the present work are indicated in the table by ranges of possible branching ratio values. These errors come from several sources. Since the peaks are small, the estimates of their area are sensitive to the details of the fit. This is especially true for the $\Sigma^0\gamma$ channel. Additionally, resolution broadening due to small errors in gain shifting the data have been modeled. Finally, fits using other features of the spectrum (such as the $\Sigma^0\pi^0$ edge) yield branching ratio values similar to those in Table 1. Since only about 30% of the data have been examined, both systematic and statistical errors will be reduced in the final results. An improved calculations within the non-relativistic quark model is underway and will include a contact diagram and a better estimate of the total K^-p decay width.[11]

Table 1

Theoretical predictions and experimental results in units of 10^{-3}. Results for the current experiment are preliminary, based on 30% of the data. Statistical errors for the two channels are $\pm.08$ and $\pm.15$ respectively. Systematic errors are indicated by ranges of branching ratios.

	B.R.($\Lambda\gamma$)	B.R.($\Sigma^0\gamma$)	Ratio ($\frac{\Lambda\gamma}{\Sigma^0\gamma}$)
NRQM	3.4	2.6	1.31
CBM	1.9	2.3	.82
Ref. 9	$2.8\pm.8$		
Ref. 10	3 ± 1.5		
This exp.	.8 – 1.0	1.0 – 1.5	.5 – 1.0

Kaons were also stopped in a liquid deuterium target, allowing the study of the reaction $K^-d\rightarrow\Lambda n\gamma$. The experimental technique was the same as that described above. This three body final state gives a photon spectrum with a maximum photon energy at 293 MeV. The maximum photon energy of the $\Lambda n\pi^0$ final state is 280 MeV. This allows a measure of the branching ratio analogous to the reaction in hydrogen, where the $\Lambda n\pi^0$ channel serves as the normalization. The $\Lambda n\gamma$ reaction provides information on the interaction of the Λ and n in the final state. The Λ - n interaction leads to an enhancement of the associated photon spectrum near its end point. The effect of the Λ-n interaction on the shape of the photon spectrum has been calculated by Workman[12]. The final state interaction is characterized by the singlet and triplet scattering lengths The results for singlet and triplet scattering lengths $a_s = -2.40fm$ and $a_t = -1.84fm$ (suggested by Nagels, et. al.[13]) are shown in figure 3. The components of the fitted synthetic spectrum are shown, including the $\Lambda n\pi^0$ channel and in-flight backgrounds. The result is a branching ratio of

$1.75 \pm .17 \times 10^{-3}$. This is independent of any assumptions about the scattering lengths, but with a less satisfactory fit for a 'no interaction' spectral shape. One systematic uncertainty in the branching ratio value is in the effect of the ΣN threshold cusp in the spectra for the $\Lambda n \pi^0$ channel.[14]

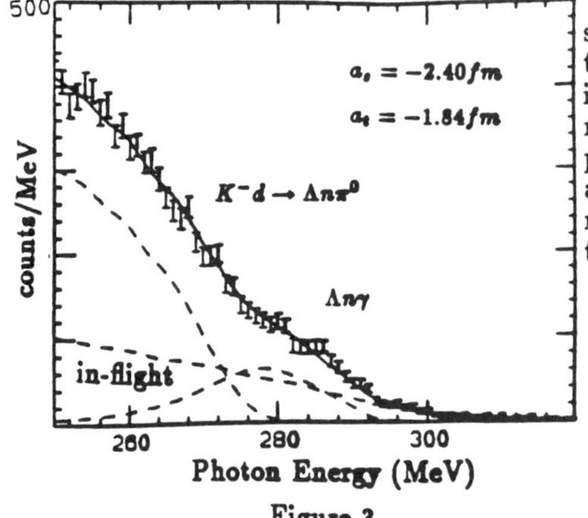

Figure 3

A lack of knowledge of the π^0 spectrum makes the yield of photons in the region used for normalization uncertain. When the data is normalized to the number of stopping kaons instead of the $\Lambda n \pi^0$ yield, a branching ratio $1.94 \pm .12 \pm .20 \times 10^{-3}$ results, where the errors are statistical and systematic, respectively.

In summary, branching ratios for radiative capture reactions have been measured for stopped kaons in both hydrogen and deuterium. The reactions on the proton provide insight into the nature of the $\Lambda(1405)$, which can be viewed as either a $\bar{K}N$ bound state or a three quark state. Due to the high energy resolution of a large NaI detector used for photon detection, the monoenergetic peaks in the hydrogen spectra have been observed for the first time. The branching ratios are below those predicted by either the non-relativistic quark model or the cloudy bag model. In addition, the branching ratio for the reaction $K^- d \to \Lambda n \gamma$ has been measured. The enhancement of the photon spectrum at the endpoint is indicative of non-zero $\Lambda - n$ final state interactions.

Footnotes

1) E. Veit, et. al. Physical Review D31 (1985) p 1033
2) N. Isgur, G. Karl Physical Review D18, 11, (1978) p. 4187
3) J. Darewych, R. Koniuk, N. Isgur Physical Review D32, 7, (1985)
4) Y. Zhong et. al. Physics Letters B171, 4, p. 471
and subsequent revision, to be published in Summer, 88
5) R. Workman and H. Fearing TRIUMF preprint, and this conference.
6) D. J. Miller et.al. , Low and Intermediate Energy Kaon-Nucleon Physics, p. 251-257, Editors, E. Ferrari, G. Violini, 1981
7) J. P. Miller et. al. to be published, Nuclear Instruments and Methods
8) W. Nelson et. al. SLAC Report 265, Dec. 1985
9) J. Lowe et al., Nuclear Physics B209 (1982) 16-28
10) W. Humphreys and R. Ross, Physical Review 127,1305, 1962
11) Nathan Isgur, private communication.
12) R. L. Workman, Ph. D. Thesis, University of British Columbia, 1988
13) M. Nagels, T. Rijken, J. de Swart, Physical Review D20, (1979), 1633
14) M. Torres, R. Dalitz, A. Deloff Physics Letters B174, 2, 1986

FEW BODY KAON REACTIONS

Harold W. Fearing
TRIUMF, 4004 Wesbrook Mall, Vancouver, B. C. Canada V6T 2A3

R. L. Workman
TRIUMF, 4004 Wesbrook Mall, Vancouver, B. C. Canada V6T 2A3
and
Department of Physics, Virginia Polytechnic Institute, Blacksburg, Va. 24061

ABSTRACT

We have calculated the reactions $K^- p \to \Lambda\gamma$ and $K^- p \to \Sigma^0\gamma$ for at-rest kaons in a simple pole model using both pseudoscalar and pseudovector couplings. Our results suggest that the contribution of the $\Lambda(1405)$ is important, particularly for the Σ case. Unlike some other models, they allow in a natural way for the branching ratio for $\Sigma\gamma$ to be greater than that for $\Lambda\gamma$. This model has been extended to in-flight capture, and some preliminary results are given. Finally the interaction generated in this model is applied to the reaction $K^- d \to \Lambda n\gamma$ and the sensitivity of this reaction to the Λ-n scattering length is investigated.

INTRODUCTION

We summarize results from a series of calculations of few body kaon reactions which are of particular interest in light of some new experiments at Brookhaven, some reported at this conference,[1] and in view of the possibilities for much improved kaon beams at the kaon factories now under discussion.

We first consider the radiative capture reactions $K^- p \to \Lambda\gamma$ and $K^- p \to \Sigma^0\gamma$ for stopped kaons. Such processes are expected to be sensitive to the rather poorly understood $\Lambda(1405)$ and thus are interesting for the prospect they present of learning more about this resonance.

These calculations have then been extended to the in-flight capture of kaons. This introduces a new degree of freedom, but also additional complications in terms of many resonances which may or may not be important depending on the energy. As we shall see however, it provides one way of distinguishing between competing theories of the basic process.

Finally the interaction derived for the simple radiative capture is applied to the process $K^- d \to \Lambda n\gamma$ where the aim is to explore the feasibility of gaining information on the Λ-n scattering length.

CALCULATION OF $K^- p \to Y\gamma$

We consider first the at-rest process $K^- p \to Y\gamma$ where Y is either Λ or Σ and where the stopped kaon is captured from an S-state. The rate is given by

$$\Gamma_{K^- p \to Y\gamma} = |\phi_K(0)|^2 \frac{m_Y k}{8\pi m_K(m_K + m_p)} \sum |M|^2 \qquad (1)$$

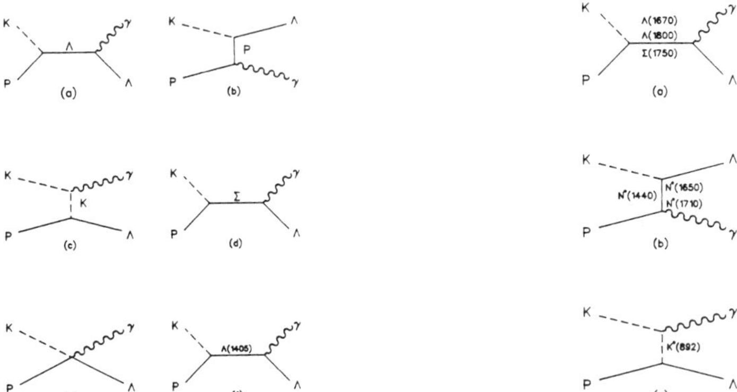

Fig. 1. Dominant diagrams included: a), b), and c) are the Born terms; d) is the $\Sigma \leftrightarrow \Lambda$ transition term; e) is a contact term required for PV coupling; f) is the $\Lambda(1405)$ contribution.

Fig. 2. Other s-, t-, and u-channel diagrams which were considered.

where ϕ_K is the kaon wave function, k is the photon momentum and the m's are the particle masses. To get the branching ratio, which is what is measured, we must divide by the rate for $K^- p \rightarrow$anything which is given by

$$\Gamma_{K^- p \rightarrow \text{anything}} = 2W_p |\phi_K(0)|^2 \tag{2}$$

where W_p is the imaginary part of the kaon-nucleon pseudopotential and is taken to be[2] 560 ± 135 MeV-fm^3.

The physics of the problem is contained in the square of the matrix element $|M|^2$. We evaluate this in a simple pole model based on a set of diagrams with effective couplings. Our aim is to provide a definitive calculation in this model. This is a standard approach which has been used successfully in the past for a variety of medium energy processes. It is also the approach used in Ref. 2, which however contains some technical errors. It is gauge and relativistically invariant and allows easy contact with other calculations and with evaluations of the coupling constants. It is, however, only an effective interaction model. Rescattering and distortion effects are ignored as is anything resulting from the composite nature of the hadrons.

The primary diagrams are those given in Fig. 1. These are the ones for which the effective couplings are best known, and fortunately they, together with the $\Lambda(1405)$ in the s-channel, give the largest contribution. We have also considered the s-, t-, and u-channel resonances of Fig. 2 but they are generally less important. In each case we have used both pseudoscalar (PS) and pseudovector (PV) couplings at the KYN vertex. Couplings have been taken from experimental information on KN scattering or from measured decay widths or, when no experimental information is available, estimated from the nonrelativistic quark model (NRQM). Specific values of the couplings used, as well as the explicit expression for M, can be found in Refs. 3,4.

A number of interesting results emerge from this calculation. First consider the situation without the $\Lambda(1405)$. In that case for the $\Lambda\gamma$ final state the Born graphs together with the $\Sigma \to \Lambda$ dominate. The K^* contribution is of the order of 10% and the N^* and Y^* pieces are negligible. PS and PV couplings give roughly the same result. A typical branching ratio is $\sim 3 \times 10^{-3}$. For the $\Sigma\gamma$ final state the $\Lambda \to \Sigma$ term is very important. The K^* and N^* terms are negligible but the Y^* contributions can be significant compared to the Born terms and the PS and PV cases are somewhat different. The most significant feature however is the size of the branching ratio, $\sim$ 3-5 $\times 10^{-4}$, depending on the diagrams chosen. Thus in the absence of the $\Lambda(1405)$ the $\Sigma\gamma$ final state could be nearly an order of magnitude lower than the $\Lambda\gamma$ final state.

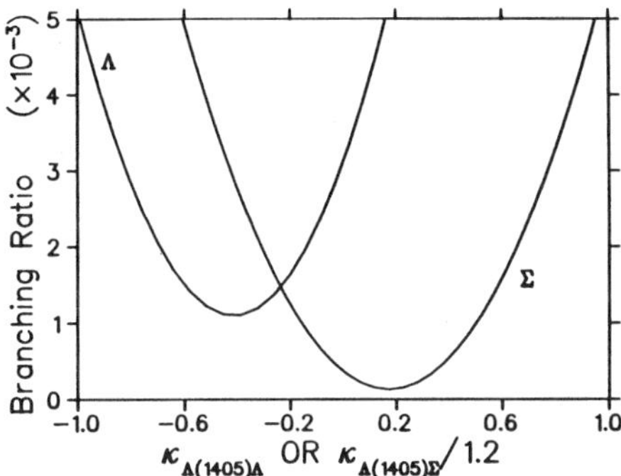

Fig. 3. Branching ratio for $K^-p \to \Lambda\gamma$ and $K^-p \to \Sigma^0\gamma$ as a function of $\kappa_{\Lambda(1405)\Lambda}$ and $\kappa_{\Lambda(1405)\Sigma}/1.2$.

Consider now the effects of a contribution from the $\Lambda(1405)$ according to Fig. 1(f). There are two unknown couplings involved, $g_{Kp\Lambda(1405)}$ at the strong vertex and $\kappa_{\Lambda(1405)\Lambda}$ or $\kappa_{\Lambda(1405)\Sigma}$ at the photon vertex. For definiteness we take[2] $g_{Kp\Lambda(1405)} = 3.2$ and give results in terms of the κ's. Without further input the two κ's are in principle unrelated. However if the $\Lambda(1405)$ is a pure singlet three quark state then $\kappa_\Sigma = \sqrt{3}\kappa_\Lambda$. Alternatively in the more realistic NRQM,[5] where the $\Lambda(1405)$ is a mixture of singlet and octet components, $\kappa_\Sigma = 1.2\kappa_\Lambda$.

In Fig. 3 we show the branching ratio for the two final states plotted against κ_Λ or against $\kappa_\Sigma/1.2$. We observe that if we require agreement with the old published value for the $\Lambda\gamma$ branching ratio,[6] 2.8×10^{-3}, then the branching ratio for the $\Sigma\gamma$ case is either very small or too large to be consistent with the old published upper limit[7] of 4×10^{-3}. However a branching ratio for the $\Lambda\gamma$ case of about 1×10^{-3}, which is consistent with the new data from Brookhaven,[1] is obtained with $\kappa_\Lambda \approx -0.4$. This, with the ratio 1.2 for $\kappa_\Sigma/\kappa_\Lambda$, implies that the branching ratio for the $\Sigma\gamma$ case is a few times 10^{-3}, which is also consistent with the new experiment.

It is hard to tell what the uncertainty in these numbers is. The branching ratio at the minimum point depends directly on the width of the $\Lambda(1405)$ and the position of the curve for Σ relative to that for Λ depends on the ratio $\kappa_\Sigma/\kappa_\Lambda$, so that variations in these would change the relative positions of the two curves. Furthermore the 25% uncertainty in the value of W_p introduces a corresponding uncertainty in the absolute values of the branching ratios.

This solution implies that for the $\Lambda\gamma$ case the $\Lambda(1405)$ is moderately important, comparable to the other contributions, and is in destructive interference with them. For the $\Sigma\gamma$ case, however, the $\Lambda(1405)$ is the dominant contribution.

What is most significant however is that in this model, unlike some of the others, one can find values of the parameters consistent with the NRQM which lead in a natural way to the branching ratio for $\Sigma\gamma$ being greater than that for $\Lambda\gamma$, which seems to be the result of the new experiment.

It is of some interest to compare these results with those of some of the other models of these processes. The pole model of Burkhardt, et al.[2] is very similar to this one, but unfortunately contained some errors which invalidated their numerical results. A calculation by Zhong, et al.[8] in the cloudy bag model uses a completely different approach, which is somewhat hard to compare with the one here. They include some contributions not included here, notably diagrams involving rescattering of the K and p in the initial state, but they leave out some other the diagrams which we have included which are necessary for full gauge invariance or which we find (in a specific gauge) non-negligible. The $\Lambda(1405)$ is for them a KN bound state rather than a simple three quark state. Originally they found $\Lambda\gamma > \Sigma\gamma$ but have since found some errors which reverse this result. The other main calculation is one in the NRQM by Darewych, et al..[9] They keep just the $\Lambda(1405)$ diagram, which is the dominant contribution in their calculation, plus some higher mass s-channel resonances and some quark exchange corrections, all of which are small. The advantage of this approach is that the vertex functions can be calculated from the model and there is not the freedom to adjust couplings. However the result implies $\Lambda\gamma > \Sigma\gamma$ which is apparently in conflict with the new experiment.[1]

$K^-p \rightarrow Y\gamma$ IN FLIGHT

One possibility of distinguishing among the various models is to examine these reactions in flight. This introduces new degrees of freedom, namely angular and energy distributions, but also some additional complications, notably contributions from a large number of resonances. In the capture at-rest process we could show that most of these gave minimal contributions. However for the in-flight process there will always be an energy which will put s-channel resonances nearly on shell, where they might be expected to dominate, or at least be important in the rate. This leads to sensitivity to poorly known couplings and, at least at the present stage, to probably too much freedom in the calculation.

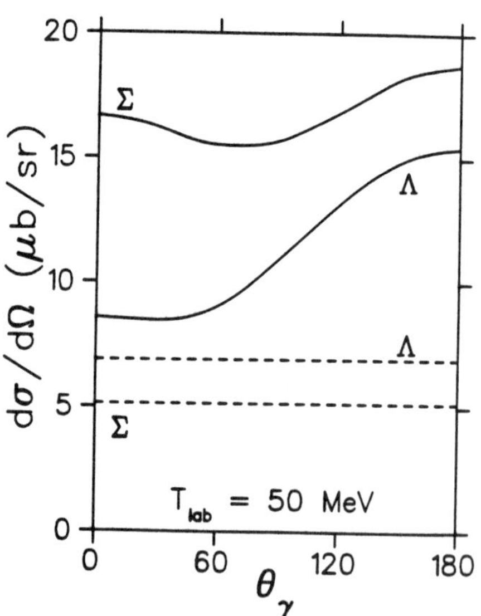

Fig. 4. Angular distributions as a function of the center of mass scattering angle for the in-flight processes $K^-p \rightarrow \Lambda\gamma$ and $K^-p \rightarrow \Sigma\gamma$ for a laboratory kaon kinetic energy of 50 MeV. The dashed curves are for the $\Lambda(1405)$ contribution only and the solid curves are for the full calculation including all the other diagrams as well.

However one can stay near threshold and thus suppress the higher resonance contributions. We have thus extended our calculations to the in-flight case, in the near-threshold kinematics, and have asked if one can find any features which distinguish the various models. In particular can one distinguish models, such as the NRQM, which include primarily only the $\Lambda(1405)$ from others, such as the present one, which include a number of other diagrams?

The answer to this question seems to be yes. As one increases the energy above threshold and thus gets farther from the peak of the resonance, the relative importance of the $\Lambda(1405)$ decreases. Thus by choosing the energy one can dial the relative importance of the $\Lambda(1405)$. Furthermore the angular distribution of the $\Lambda(1405)$ contribution is completely flat, as expected from an s-channel resonance, while the other terms tend to give a backward peaked angular distribution.

An example of these features is shown in Fig. 4. The dashed curves show the $\Lambda(1405)$ contribution alone, while the solid curves are the full calculation. For the Λ case there is a rather dramatic difference in these two curves. The effect is somewhat less significant for the Σ case because at this energy the $\Lambda(1405)$ still dominates the other diagrams.

These results are clearly suggestive. However for a definitive result a number of refinements must be included. In particular in the pole model calculation some additional resonances, such as the $\Lambda(1520)$, should be included to be sure that the tails from these resonances do not affect the angular distributions even at this low energy. Nevertheless it is clear that the additional degrees of freedom allowed by the in-flight process lead to some features which may allow one to distinguish between various models.

$Kd \to \Lambda n\gamma$ AT REST

We now extend these considerations to the process $Kd \to \Lambda n\gamma$. The motivation here is to learn about the low energy Λ-n interaction.[10,11] Thus we look at this process in the geometry corresponding to the highest energy photons where the Λ and the neutron have low relative momentum. We show that the shape of the photon spectrum in this region is sensitive to the Λ-n scattering length and, like the analogous process $\pi d \to nn\gamma$, relatively insensitive to most other ingredients in the calculation.

To evaluate this photon spectrum we first obtain an effective operator for the process $K^-p \to \Lambda\gamma$ by taking a nonrelativistic reduction of the matrix element described above. This operator is then taken between the kaon-deuteron initial wave function and a Λ-n scattering wave function parameterized in terms of scattering length and effective range.

There is some new physics involved here. First the initial proton is no longer at rest as a consequence of the Fermi motion in the deuteron. Hence the nonrelativistic operator needs in principle to be generalized to include terms containing p_p/m which were not present for the capture on a proton at rest considered above. Second there are some additional channels which may be relevant here because of the possibility of the strong rescattering in the final state. In particular it is possible to get the Λ-n final state via the process $K^-p \to \Sigma^0\gamma$ followed by the rescattering $\Sigma^0 n \to \Lambda n$ or by $K^-n \to \Sigma^-\gamma$ followed by the rescattering $\Sigma^-p \to \Lambda n$. Thus one must also obtain nonrelativistic effective interactions for these processes, as was done for $K^-p \to \Lambda\gamma$, and include them in the calculation.

We find, just as was the case for $\pi d \to nn\gamma$, that changes in essentially any of the ingredients of the calculation change the absolute value of the rate. Thus calculating this absolute rate will be very difficult. However the shape of the spectrum very near the end point seems to be fairly insensitive to most ingredients of the calculation except the scattering length. In particular the p_p/m terms are unimportant, as are the uncertainties in the contributions of the $\Lambda(1405)$ and of the various N^*'s, K^*'s and Y^*'s. Likewise the choice of the deuteron S-state wave function, Bonn or Hulthen, is relatively unimportant for the shape, as is the presence or absence of the D-state. The other channels involving $K^-p \to \Sigma^0\gamma$ and $K^-p \to \Sigma^-\gamma$ contribute only far below the endpoint

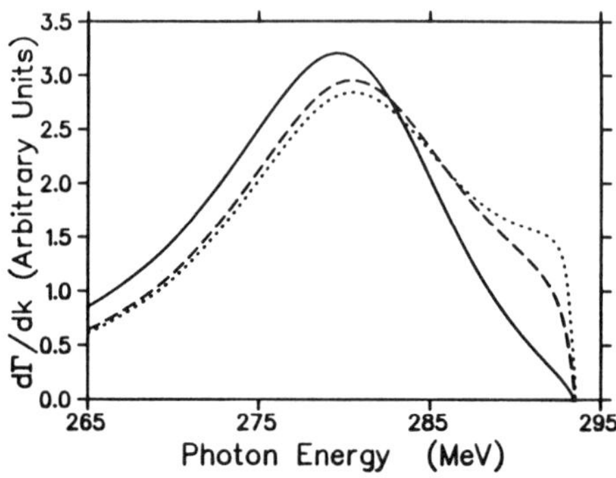

Fig. 5. Photon spectrum near the upper end, normalized to constant area, for several sets of triplet and singlet Λ-n scattering lengths and ranges in fm. The curves have $a_s = 0$, $a_t = 0$, $r_s = 0$, $r_t = 0$ (solid); $a_s = -2.40$, $a_t = -1.84$, $r_s = 3.15$, $r_t = 3.37$ (dashed); $a_s = -5.0$, $a_t = -3.0$, $r_s = 3.5$, $r_t = 3.5$ (dotted).

of the photon spectrum and so for the present purposes can be neglected. Finally the shape in this endpoint region is not very sensitive to the Λ-n effective range. Thus as long as we consider just the shape of the photon spectrum near the endpoint, the results are sensitive primarily to the scattering length.

Some of these results are shown in Fig. 5 which shows the photon spectrum near the endpoint normalized to equal area so as to isolate the differences in the shape. The solid curve corresponds to no rescattering in the Λ-n system, while the dashed curve corresponds to a Λ-n scattering length obtained from the potential of Nagels et al.[12] and the dotted curve to a somewhat larger scattering length. The shape is obviously sensitive to the Λ-n rescattering. It is encouraging that the Brookhaven experiment[1] indicates, as a consequence of fitting preliminary data to these curves, a preference for a non zero scattering length. Thus this may be a feasible method to obtain the scattering length, though clearly a precision experiment, such as may be possible with the improved beams at a one of the proposed kaon factories, will be necessary to obtain reliable values of the scattering length or to distinguish among similar interactions.

This work was supported in part by a grant from the Natural Sciences and Engineering Research Council of Canada.

REFERENCES

[1]E. K. McIntyre, et al., this proceedings.

[2]H. Burkhardt, J. Lowe, and A. S. Rosenthal, Nucl. Phys. **A440**, 653 (1985).

[3]R. L. Workman, Ph.D. Thesis, University of British Columbia, 1987.

[4]R. L. Workman and Harold W. Fearing, Phys. Rev. D **37**, (1988), in press.

[5]J. Darewych, M. Horbatsch, and R. Koniuk, Phys. Rev. D **28**, 1125 (1983).

[6]J. Lowe et al., Nucl. Phys. **B209**, 16 (1982).

[7]J. D. Davies et al., Nucl. Phys. **B160**, 492 (1979).

[8]Y. S. Zhong, et al., Phys. Lett. **B171**, 471 (1986). A corrected version of this paper can be found in an Adelaide preprint #ADP87-72-T/52.

[9]J. W. Darewych, R. Koniuk and N. Isgur, Phys. Rev. D **32**, 1765 (1985).

[10]B. F. Gibson, et al., *Proc. of the Summer Study on Nuclear and Hypernuclear Physics*, Brookhaven National Lab, 1973, ed. by H. Palevsky, p 296.

[11]A. I. Akhiezer et al., Sov. Jour. Nucl. Phys. **27**, 115 (1978).

[12]M. M. Nagels, T. A. Rijken and J. J. de Swart, Phys. Rev. D**20**, 1633 (1979) and references cited therein.

Perturbative QCD Predictions for Meson-Baryon Scattering

H. Zhang*and G.R. Farrar†
Department of Physics and Astronomy
Rutgers University
New Brunswick, NJ 08903

Presented by H. Zhang in the Third Conference on
the Intersection between Particle and Nuclear Physics

30 June 1988

ABSTRACT

We have completed the leading order (order α_s^4) perturbative QCD calculation of all the quark scattering amplitudes involved in the meson-baryon scattering reactions that have been measured at Brookhaven. Ratios of differential cross sections for these reactions have been computed by convoluting the quark scattering amplitudes with hadronic wave functions. When the QCD sum-rule-derived wave functions are used, the calculation yields rough agreement with data. We have also investigated the sensitivity of the predictions to the treatment of singularities and the numerical reliability of the integration method that we used.

*Supported by Rutgers Supercomputer Postdoctoral Fellowship and DOE DE-AS05-76ER04928 at Virginia Polytechnic Institute & State University
†Supported in part by NSF PHY-84-15535-05

I. Introduction

Exclusive meson-baryon scattering at wide angle is an intereting testing ground for perturbative QCD since there is a rich set of reactions which depend on a relatively simple set of fundamental quark scattering amplitudes[1,2]. For example, in the scattering subclass

$$\left\{\pi^+ \text{or } K^{\pm,0}\right\} + p \rightarrow \{\pi, K, \eta, \eta' \text{or } \phi\} + \{B_8 \text{ or } B_{10}\} \tag{1.1}$$

where mesons have spin 0 and and baryons have spin $\frac{1}{2}$, there are 45 reactions which can be completely determined by the convolution of 12 fundamental quark scattering amplitudes shown in Figure 1 and the wave functions for the relevant hadrons. Moreover, meson-baryon scattering involves contributions from diagrams containing single, multiple or combinations of three- and four-gluon vertices, which are characteristic of non-Abelian field theories. We present here preliminary predictions for the meson-baryon scattering reactions that have been measured at Brookhaven[3].

A major obstacle to theoretical progress on this problem has been the enormous number of Feynman diagrams involved in these processes, about 30,000 order α_s^4 diagrams for the complete set of meson baryon scatterings. We have eliminated this problem by developing a computer program[4] to perform this task for any exclusive hadronic processes up to baryon-baryon scattering. Details of this aspect of the calculation are given in[5].

The analytic expressions for the quark scattering amplitudes, when coded in FORTRAN, are typically 10,000 lines long and occupy 1 megabyte of disk space. These quark scattering amplitudes must be integrated over the hadronic wavefunctions, a six-dimensional integration. With such an integrand, an 8-point per dimension Gauss quadrature requires more than 8 CPU hours on a VAX 8600 or 14 CPU minutes on the JvNC (John von Neuman National Computing Center) Cyber 205. Furthermore, the integrands contain singularities caused by gluons and quarks going on-shell. Theoretically, this integration region is suppressed by the so-called Sudakov factor which is still not well understood. Leaving a better regulating scheme to the next step, we simply cut-off the singularities, as follows. After simplification, including cancellation between numerators and denominators, let the remaining factors in the denominators be denoted $d_1, d_2, \ldots$. Each diagram is then multiplied by a product of step functions

$$H(d_1 - d_{cut})H(d_2 - d_{cut})\ldots$$

with the cut-off parameter d_{cut}. While absolute predictions are clearly very sensitive to the cut-off, the hope is that cross-section ratios will be insensitive, so that this procedure can give useful information.

Another uncertain element is the appropriate choice of wavefunction for the hadrons. Asymptotically, the wave function for the baryons is SU(6) symmetric and for the proton can be written as

$$\phi_N(x) \sim \frac{\phi_{as}(x)}{\sqrt{6}} [2|u_+u_+d_- > -|u_+d_+u_- > -|d_+u_+u_- >] .$$

More generally, the coefficients of the three terms are different functions we shall call T, B, and C below. Due to isospin and fermi statistics, there is actually only one independent function, called $\phi(x,y)$, which determines T, B, and C. The most popular model for ϕ is that proposed by Chernyak and Zhitnitsky, where ϕ is taken to be a quadratic in x, y whose moments are fixed using QCD sum-rules[6,7] See Ref. 5,8,9 for details.

Table I presents scattering amplitudes for various reactions in terms of convolutions of quark scattering amplitudes $A_0 - A_7$ and components of wave functions. An expression of $TT + BB + CC$ in the A_0 column for the reaction $\pi^+p \to \pi^+p$ represents the integral:

$$\int d_3x d_3y \phi_\pi(x)\phi_\pi(y) [T(x)T(y) + B(x)B(y) + C(x)C(y)] A_0(x,y)$$

The scattering amplitude for each reaction is obtained by adding the columns of the corresponding line of Table I. SU(6) symmetry results from taking $\frac{1}{2}T = B = C$.

In order to compare with the Brookhaven data, we did our calculations for 90 degree center-of-mass scattering. We first performed the calculations with a series of cut-off values and investigated the sensitivity of the integral to the cut-off. Table II presents the dependence of $d\sigma(\pi^-p \to \pi^-p)/d\sigma(\pi^+p \to \pi^+p)$ on the value of the cut-off d_{cut}. As it shows, the ratio becomes independent of the cut-off when the cut-off goes below 10^{-7} and keeps independent until the cut-off reaches 10^{-38}. In the following, we use $d_{cut} = 10^{-7}$. Table II also compares the results of 4- and 8-point gaussian quadrature, indicating that our results have a numerical error of about 10While gaussian quadrature gives much better accuracy than Monte Carlo integration, the apparent insensitivity of the ratios to the cut-off for d_{cut} may be an artifact of the integration method. We are investigating this question now. Until this analysis is completed, the present results are inconclusive.

After normalizing the differential cross sections to that of $\pi^+p \to \pi^+p$, we plot the results against the data in Figure 2, with sum-rule wave functions, SU(6) sym-

Reaction	A_0	A_1	A_2	A_3	A_4	A_5	A_6	A_7
$\pi^+ p \to \pi^+ p$	TT+BB+CC	TT+BB+CC	BB	TT	TT+CC	BB+CC	CC	2BT
$\pi^- p \to \pi^- p$	TT+BB+CC	TT+BB+CC	TT+CC	BB+CC	BB	TT	BB	2TB
$K^+ p \to K^+ p$	TT+BB+CC	TT+BB+CC			TT+CC	BB+CC		
$K^- p \to K^- p$	TT+BB+CC	TT+BB+CC	TT+CC	BB+CC				
$\pi^+ p \to \rho_0^+ p$	TT+BB+CC	-TT-BB-CC	BB	-TT	TT+CC	-BB-CC	CC	
$\pi^- p \to \rho_0^- p$	TT+BB+CC	-TT-BB-CC	TT+CC	-BB-CC	BB	-TT	BB	
$\pi^+ p \to \pi^+ \Delta_{\frac{1}{2}}^+$			BT_Δ	$-TT_\Delta$	$(C-T)T_\Delta$	$(B+C)T_\Delta$	CT_Δ	$(T-B)T_\Delta$
$\pi^- p \to \pi^+ \Delta_{\frac{1}{2}}^-$							$-TT_\Delta$	$-(C+T)T_\Delta$
$\pi^+ p \to K^+ \Sigma^+$			BB_Σ	TT_Σ			CC_Σ	$2BT_\Sigma$
$\pi^- p \to K^+ \Sigma^-$							$-TC_\Sigma$	$CT_\Sigma - BC_\Sigma$
$\pi^- p \to K^0 \Lambda^0$			$-(C+T)B_\Lambda$	$(C-B)T_\Lambda$			$-BC_\Lambda$	$-TT_\Lambda \cdot BB_\Lambda$
$\pi^- p \to K^0 \Sigma^0$			$(C-T)B_\Sigma$	$-(B+C)T_\Sigma$			BC_Σ	$-TT_\Sigma - BB_\Sigma$

TABLE I. Scattering amplitudes for $MB \to MB$ in terms of the fundamental quark scattering amplitudes integrated over sum-rule wave functions

d_{cut}	$d\sigma(\pi^- p)/d\sigma(\pi^+ p)_{4pt}$	$d\sigma(\pi^- p)/d\sigma(\pi^+ p)_{8pt}$
10^{-1}	1.2968	2.0998
10^{-2}	0.2873	0.9425
10^{-3}	1.6195	0.9010
10^{-4}	1.0313	1.2874
10^{-5}	0.0877	0.3783
10^{-6}	0.0877	0.1353
10^{-7}	0.0877	0.1419
10^{-8}	0.0877	0.1082
10^{-9}	0.0877	0.1082

TABLE II. Cut-off dependence and numerical reliability of the ratio $d\sigma(\pi^- p)/d\sigma(\pi^+ p)$ for both 4-point and 8-point Gauss quadrature.

metric wave functions, and sum-rule wave functions for the baryon octet and pseudoscalar meson nonet neglecting SU(3) breaking. The predictions with SU(6) wave functions do not bear much resemblance to the data, whereas when sum-rule wave functions are used the predictions resemble the data fairly well. It is also encouraging that accounting for SU(3) breaking yields better agreement than ignoring it. The sensitivity of the above conclusions to cutoff is addressed in Ref. 5.

II. Conclusion

We have made QCD predictions for meson-baryon scattering which require both perturbative and non-perturbative methods. The perturbation theory is used to evaluate the quark scattering amplitudes, and the sum-rule-derived wave functions contain non-perturbative effects. As the results show, the non-perturbative effect, i.e., the choice of the wave functions, is important. The results from asymptotic and QCD sum-rule wavefunctions are substantially different. Although still preliminary, the predictions with sum-rule wave functions agree with data reasonably well. SU(3) symmetry breaking effects, while not very significant, are still noticeable.

Another interesting discovery of our calculation is that quark scattering amplitudes which differ only in the quark helicity in the meson, and are identical otherwise, are considerably different in their integrated values. One might hope that this helicity dependence of the amplitudes may make the polarized pp data comprehensible.

Before this calculation, effort[3] was devoted to the assessment of which types of quark scattering amplitudes dominate the process. As is clear from this calculation, the question of domination can not be asked and answered in this simple way. Table III lists the integrals of each quark scattering amplitude with both asymptotic wave functions and sum-rule wave functions. For the purpose of comparison, all the sum-rule integrals have been evaluated with $[T(x)T(y) + B(x)B(y) + C(x)C(y)]$ in their integrands. If only the asymptotic wave functions are used, it is obvious that A_4 and A_5 dominate and followed by A_0 and A_1. However, with sum-rule wave functions, the scattering amplitude of a reaction is no longer a linear combination of the same integrals of the quark scattering amplitude as are listed in Table III. As is shown in Table I, each A_i has to be sandwiched by proper flavor-spin states for a given reaction before the sum can be found. Furthermore, amplitudes A_0 and A_1 can no longer be treated as a single class as gluon exchange, nor can A_4 and A_5 to be the same as quark interchange. Therefore, the magnitude of the scattering amplitude

wfn	$< A_0 >$	$< A_1 >$	$< A_2 >$	$< A_3 >$	$< A_4 >$	$< A_5 >$	$< A_6 >$	$< A_7 >$
asy	1	0.90	0.044	0.078	-2.2	-2.4	0.061	0.56
s.r.	1	0.45	-0.016	0.022	0.45	-1.59	0.0069	0.13

TABLE III. Relative magnitude of integrated fundamental quark scattering amplitudes with asymptotic and sum-rule wave functions.

for a given reaction depends not only on which quark scattering amplitudes are involved but also on the detailed combination of the flavor-spin states and the quark scattering amplitudes.

At this time, the most serious obstacle to obtaining reliable perturbative QCD predictions for meson-baryon scattering is in the theoretical understanding of the properties of the singularities and their correct treatment. Much theoretical effort is needed in this aspect and only then can we make sensible and reliable absolute predictions for meson-baryon scatterings as well as other exclusive processes. Hopefully ratios of cross-sections will prove to be insensitive to these problems, as these preliminary indications suggest.

REFERENCES

[1]G.R. Farrar, Phys. Rev. Lett. **53**, 28 (1984).

[2]G.P. Lepage and S.J. Brodsky, Phys. Rev. D **22**, 2157 (1980).

[3]B.R. Baller et al., Phys. Rev. Lett **60**, 1118 (1988).

[4]G.R. Farrar and H. Zhang, Rutgers Preprint in preparation ().

[5]H. Zhang and G.R. Farrar, Rutgers preprint, RU-88-32 (1988).

[6]V.L. Chernyak and A.R. Zhitnitsky, Nucl. Phys. B **201**, 492 (1982); V.L. Chernyak, A.R. Zhitnitsky, and I.R. Zhitnitsky, Nucl. Phys. B **204**, 477 (1982); V.L. Chernyak and I.R. Zhitnitsky, Nucl. Phys. B **246**, 52 (1984).

[7]I.D. King and C.T. Sachrajda, Nucl. Phys. B **279**, 785 (1987).

[8]V.L. Chernyak, A.A. Ogloblin, and I.R. Zhitnitsky, Preprint INP-136-1987, Novosibirsk (1987).

[9]G.R. Farrar, H. Zhang, A.A. Ogloblin, and I.R. Zhitnitsky, to be published in Nucl. Phys. B ().

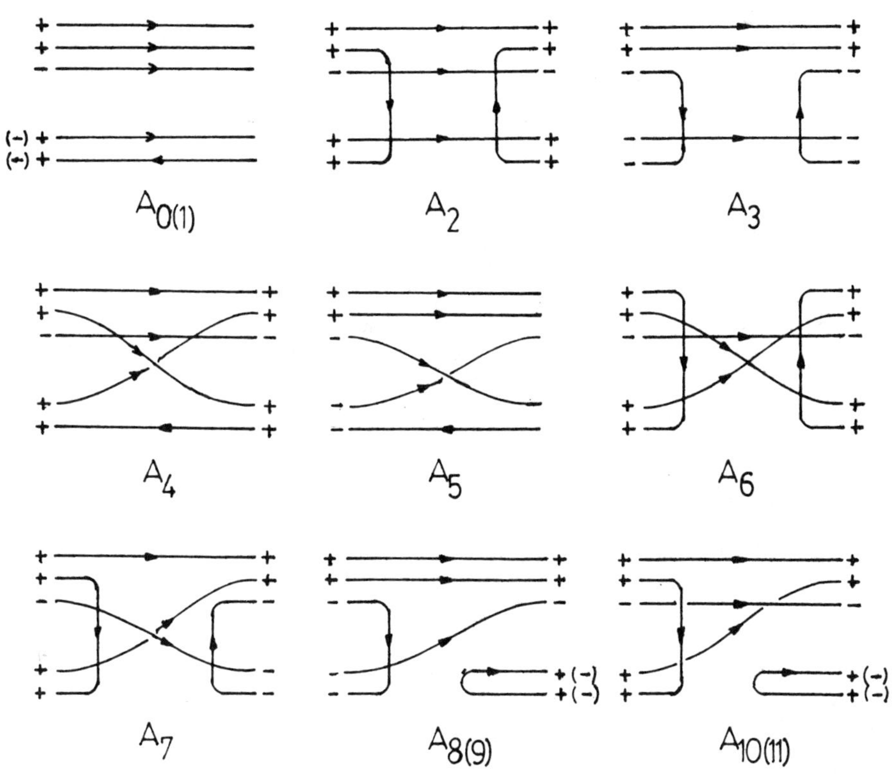

FIG. 1. The fundamental quark scattering amplitudes for $MB \to MB$

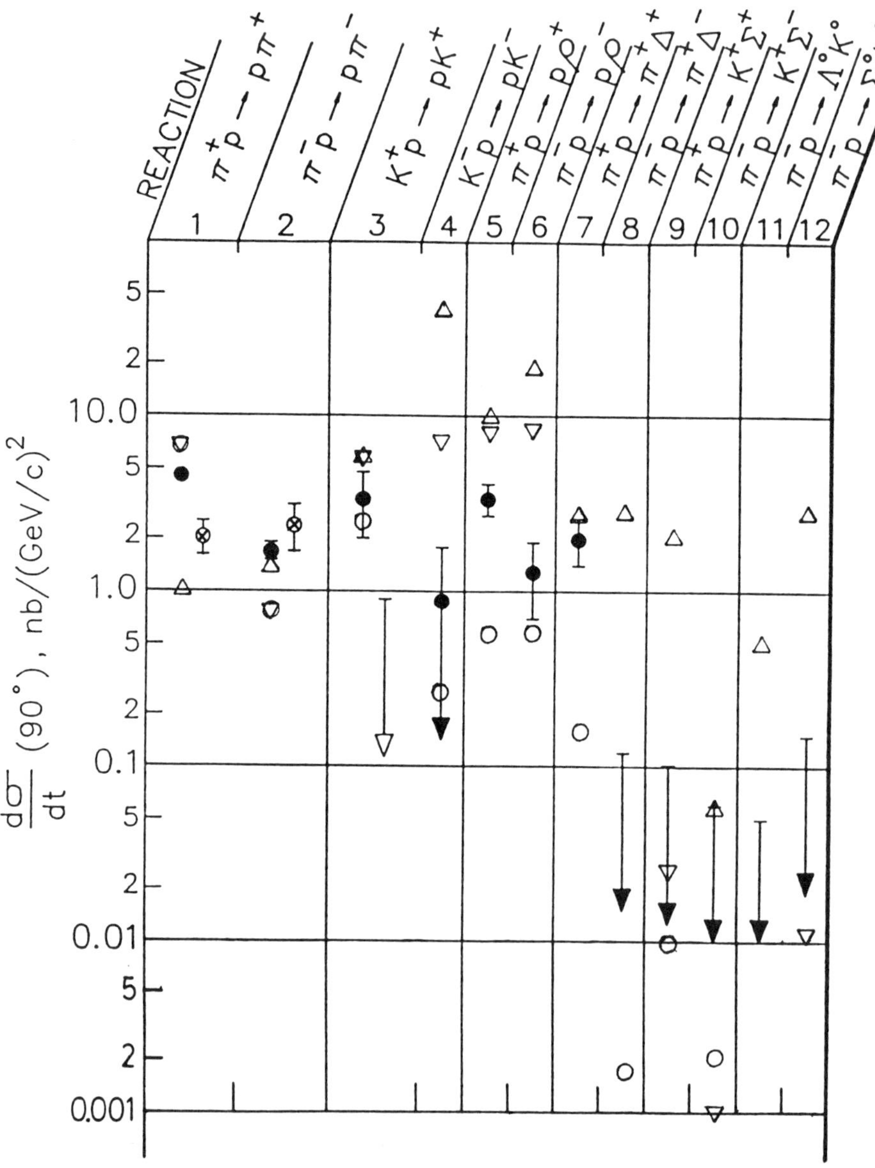

FIG. 2. Comparison of perturbative QCD predictions with data. △ are predictions with SU(6) wavefunction, ○ are that with sum-rule wavefunctions and ▽ are that with sum-rule wavefunction and assuming SU(3). Each set of predections is normalized independently to fit the data best. The data graph is taken from Reference 5.

INTRANUCLEAR CASCADE STUDY OF THE HADRON–NUCLEUS INTERACTIONS AT 1 GeV/c

M. Divadeenam and T.E. Ward
Neutral Beam Division,* Brookhaven National Laboratory, Upton, NY 11973

ABSTRACT

The proton inclusive spectra measured with 800 MeV/c π^+ beam (BNL) on light nuclei, and the 600 MeV/c $\bar{p}$ beam (LEAR) on ^{6}Li and ^{63}Cu have been successfully interpreted in terms of the Intranuclear Cascade (INC) Model. According to the INC model predictions the $\bar{p}$ induced reaction results in high pion yield for the light ^{6}Li target case. This is attributed to the nonavailability of nucleons to be knocked out by the pions produced following the $\bar{p}$ annihilation. While in the case of the heavier ^{63}Cu target some of the internally produced pions are subsequently absorbed leading to thermal equilibration of the nuclear medium, resulting in a softer proton spectrum at the high energy end. Based on the INC model predictions the K^+ induced reactions could be studied to search for the "true resonances" beyond 1 GeV K^+ energy.

I. INTRODUCTION

The hope that a relativistic quantum mechanical many–body problem could reliably calculate exotic processes that take place with high and medium energy projectile induced reactions is very remote. A number of models[1] that identify the conventional and the exotic processes with suitable approximations are able to predict most of the experimentally observable quantities. Among these the Intranuclear Cascade Model (INC) is the most convenient model at intermediate energies for conceptual simplicity, with fewer approximations, and including multiple scattering processes. The model is suitable for different kinds of projectiles, projectile energy and in principle for almost all target nuclei except the very light ones.

Recently, a π^+ 800 MeV/c investigation at BNL[2] was undertaken to search for the η–meson nuclear bound states superimposed on the (π^+, p) continuum spectrum. The experimental search for sharp η–meson nuclear bound states failed to observe such states.

Analogous to the BNL π^+–nucleus study, the search for $\bar{p}$–nucleus states was carried out[3] at the LEAR $\bar{p}$ beam facility with the expectation of finding sharp well–defined $\bar{p}$–nucleus states, superimposed on a continuous proton spectrum arising from the annihilation of the antiprotons, and the subsequent proton knockout by the annihilation of pions. For the initial study the 600 MeV/c LEAR beam was used on several targets to measure the 0° proton spectra. No evidence for the narrow bound or resonant $\bar{p}$–nucleus states could be found. The small angle proton spectra measured in both the reactions $\bar{p}$–nucleus at 600 MeV/c and π^+–nucleus at 800 MeV/c are similar in shape.

In the present study a theoretical interpretation of the ($\bar{p}$, p) and (π^+, p) spectra was undertaken within the framework of the INC model. The INC model calculations were extended to include the (K^+, p) reaction even though data are not available. Details

*Research performed under the auspices of the U.S. Department of Energy under contract no. DE–AC–02–76CH00016.

In Figure 1 a comparison is shown between the INC model calculated 15° double–differential proton inclusive spectrum for the $A\ell$ 800 MeV/c (π^+, p) reaction. The agreement between theory and experiment is very good both in shape and magnitude. Unfortunately, the experimental data covers a very limited energy range; a similar situation holds for ^{6}Li, carbon and oxygen. Figures 2a and 2b show the 5° proton and neutron double–differential inclusive spectra. The energy integrated angular distribution for (π^+, p) and (π^+, π^+) inclusive reactions are displayed in Fig. 3. The (π^+, n), (π^+, π^o) and (π^+, π^-) angular distributions are also similar in trend and magnitude. The angular distributions are forward peaked. There are no corresponding experimental data at this medium energy to compare with theory.

IIIb. 600 MeV/c $\bar{\text{P}}$ Induced Reactions on ^{6}Li and ^{63}C

The ISABEL results which predict the relative inclusive production of various outgoing particles is displayed in Figs. 4a and 4b for ^{6}Li + $\bar{p}$ and ^{63}Cu + $\bar{p}$ reactions respectively. In the case of the light target ^{6}Li, since there are not enough nucleons that could be knocked out by pions produced after the initial $\bar{p}$–nucleon annihilation takes place, we notice a copious production of pions. The situation is reversed in the case of the ^{63}Cu target. Lighter targets like ^{9}Be + $\bar{p}$ reaction would produce abundant π^- which would decay into μ mesons suitable for μ meson catalyzed fusion process and subsequent energy release.

Figures 4c and 4d display a comparison of the INC model predictions for the inclusive proton production at 0° and the corresponding LEAR experimental data. The energy dependence and magnitude observed in the LEAR experiment are well reproduced. Increasing the number of cascades from 15,000 to a higher figure might improve the statistics for the predicted double differential cross sections.

The angular distributions for ($\bar{p}$, n), ($\bar{p}$, p), ($\bar{p}$, π^+) and ($\bar{p}$, K$^+$) are shown in Fig. 5a. The angular distributions, are rather isotropic implying equilibration of the nucleus after the initial annihilation of the incident antiproton. This feature is also evident in fig. 5b, which displays the inclusive proton energy spectrum. The histogram could be fitted with two Maxwellian distributions with different temperatures. The spectrum up to about 40 MeV (or less) corresponds to the evaporation spectrum and the rest of the spectrum does not correspond to a typical fast–particle spectrum as observed in nucleon induced reactions or the K$^+$ induced reaction to be discussed in the next paragraph. After the incident $\bar{p}$ is annihilated, the subsequent pion production and some absorption leads to equilibration of the nuclear medium. The ($\bar{p}$, p) spectrum beyond 40 MeV demonstrates this point.

Due to space limitations a detailed discussion of $P_\perp$ distribution vs. rapidity, energy deposition description, particle multiplicities etc. is deferred to a later publication.

IIIc. 1 GeV K$^+$ + $A\ell$ Reaction

Figure 6 displays the K$^+$ induced reaction total cross sections[10] for the (K$^+$, p), (K$^+$, d) and (K$^+$, n) reactions. The variation of the total cross section with the K$^+$ incident energy increases slowly with energy beyond 1 GeV and the difference σ_t–$\sigma_{o\ell}$ for the (K$^+$, p) case increases with K$^+$ energy. This indicates that several reaction channels are opening beyond 1 GeV K$^+$ incident energy. The ISABEL calculations at 1 GeV were performed to check the relative inclusive production of the several outgoing

of the INC model calculations and a comparison of the INC model predictions with the corresponding experimental data are presented.

II. INTERMEDIATE–ENERGY REACTION MECHANISM

IIa. Intranuclear Cascade and Evaporation Model

Serber[4] in 1947 proposed the idea that the high energy nucleon induced reactions may be described in terms of essentially a two–step model. At high energies ($E_{proj} \geq$ 100 MeV/nucleon) the reaction is assumed to proceed in two steps; 1) a fast process, and 2) a slow process based on the interaction time scale. In the first step, the incident nuclear interaction develops a series of binary nucleon–nucleon collision cascades with allowance for some particles to escape. At the end of the first step, the residual nucleus deexcites (after immediate equilibration) through the second step or statistical evaporation of other nucleons/light ions. In this second step, it is tacitly assumed that the residual nucleus is in statistical equilibrium prior to the commencement of the evaporation process.

We used two different INC model codes; 1) MECC7 developed by Bertini[5] at ORNL in the sixties and early seventies, and 2) ISABEL,[6] the generalized version of the original BNL code VEGAS eveloped by Chen et al.[7] The modification and improvements were made by Yariv and Frankel[5] and later by Clover et al.[8] Joseph Cugnon[9] has developed his own version of the INC model at the University of Liege. Monte Carlo techniques are used to simulate the intranuclear interactions.

In the ISABEL code the nuclear density distribution is approximated by a series of sixteen concentric shells of constant density which approximate the measured nuclear charge distribution functions. Whereas in MECC7 the nuclear density distribution is approximated by 3 concentric shells. The details of the models may be found in references 6 and 8 on ISABEL and 5 on MECC7.

At the end of the cascade process described above, the intermediate nucleus after ejecting a few nucleons is in an excited mode which can deexcite via evaporation process. The statistical evaporation theory was used in conjunction with MECC7 and ISABEL to calculate the process of deexcitation via evaporation by using Monte Carlo method. Relative probability of emission of the particles is based on the statistical theory of a fermi gas model. The step–wise Monte Carlo method has been adopted for following the fate of a given excited nucleus and the average behavior is deduced from analysis of a large number of cascades. Six particles n, p, d, t, h and α are allowed to be emitted from the excited nucleus.

III. MODEL CALCULATIONS AND COMPARISON TO EXPERIMENT

IIIa. 800 MeV/c π^+ Induced Reaction on Light Nuclei

As pointed out in the introduction π^+ induced proton inclusive spectra were measured at BNL on ^{6}Li, C, O and $A\ell$ at 15° with the π^+ beam momentum at 800 MeV/c. The MECC7 code was used to predict the proton inclusive spectra besides other quantities. The INC calculations were performed with 30,000 to 50,000 cascades. Due to space limitation the discussion is restricted to the $A\ell + \pi^+$ reaction.

particles. Figure 7a presents the ISABEL predicted relative production cross sections for p, n, π, K^+, K° etc. It is noticeable that in addition to the production of K^+ and the nucleons, the K° production cross section is not negligible as is the situation for the other particles. It would be interesting to look into the K^+ induced reaction "true resonances" in the K° channel, i.e. delta and N–stars resonances.

Figure 7b displays a comparison of the (K^+, p) and (K^+, K^+) production cross section angular distributions, which are forward peaked–typical of hadron induced reactions other than the $\bar{p}$ projectile.

Experimental data is lacking for a comparison with ISABEL predictions.

IV. SUMMARY AND CONCLUSIONS

Theoretical interpretation of the 15° BNL 800 MeV/c (π^+, p) inclusive spectra from light target nuclei and the 0° LEAR 600 MeV/c ($\bar{p}$, p) inclusive spectra from ^{6}Li and ^{63}Cu targets was undertaken within the frame–work of the Intranuclear Cascade (INC) Model. The INC model codes MECC7 and ISABEL were used in predicting the relative production cross section and the doubly–differential cross section besides several other related quantities. The INC model predictions compare very well with the BNL and LEAR measured low angle proton spectra. The ^{6}Li + $\bar{p}$ and ^{63}Cu + $\bar{p}$ reaction study reveals interesting features regarding the relative pion production vs. nucleon production. The phenomena of phase transition in terms of equilibration temperatures in the case of ^{63}Cu + $\bar{p}$ reaction reveals that the pion production and absorption and subsequent thermalization of the nuclear system leads to a higher energy softer spectrum compared to the spectra observed in non–antiparticle induced reactions. The K^+ + $A\ell$ reaction was considered as a test case to explore the possibility of measuring the reaction products beyond 1.4 GeV/c. Beside the nucleon and K^+ production cross sections, the K° production cross section is non–negligible compared to the rest of the particle production cross sections.

In conclusion, it is demonstrated that quantitative reproduction of proton inclusive spectra measured both in 800 MeV/c pion and 600 MeV/c $\bar{p}$ induced reactions on light to medium mass target is possible within the framework of the INC model. It would be desirable to compare the ISABEL and MECC7 predictions for the π^+ induced reactions.

The ISABEL predicted (K^+, p), (K^+, n), (K^+, K^+) and (K^+, K°) inclusive spectra for the 1 GeV K^+ induced reaction on $A\ell$ suggest the use of one of these channels to measure the "true resonance structure" with K^+ beam momenta above 1.4 GeV/c.

Experimental data are in general sparse for the π^+, $\bar{p}$ and K^+ induced reactions for a critical test of the INC model predictability in the intermediate to high energy region.

One of the authors (M.D.) acknowledges useful discussions with M. R. Clover. We thank Bob Chrien (BNL) for allowing us to use the (π^+, p) spectral data before publication.

REFERENCES

1. D. H. Boal, Ann. Rev. of Nucl. Science, 37, 1, (1987).
2. R. E. Chrien et al., Phys. Rev. Lett. 60, 2595 (1988), and private communication (April, 1988).
3. E. Aslanides et al., Nucl. Phys. A470, 445 (1987); D. Garreta et al., Phys. Lett. 150B, 95 (1985).

4. R. Serber, Phys. Rev. 72, 1116 (1967).
5. H. Bertini, Phys. Rev. 6C, 631 (1972), ibid 188, 1711 (1969).
6. Y. Yariv and Z. Frankel, Phys. Rev. C20, 2227 (1979), ibid, C24, 488 (1981).
7. Chen, et al., Phys. Rev. 166, 949 (1968).
8. M. R. Clover et al., Phys. Rev. 26C, 2138 (1982); Y. Yariv and Z. Frankel, Phys. Rev. 24, 488 (1981).
9. J. Cugnon, Phys. Rev. C22, 1885 (1980).
10. "Review of Particle Properties," Phys. Lett. 170B, 87 (1986).

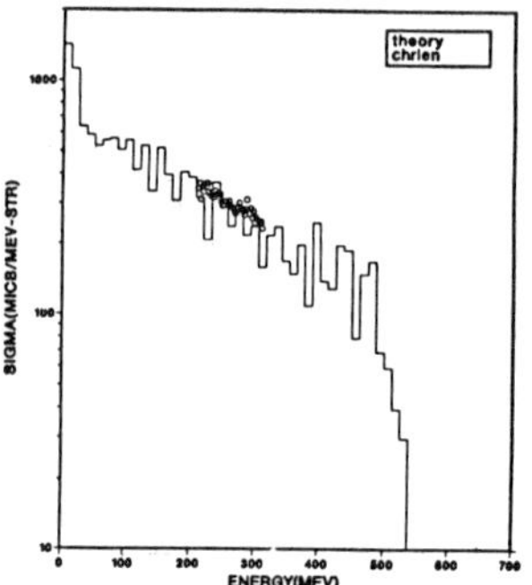

Fig. 1a. 15° inclusive proton spectrum; INC model calculations with 30000 cascades.

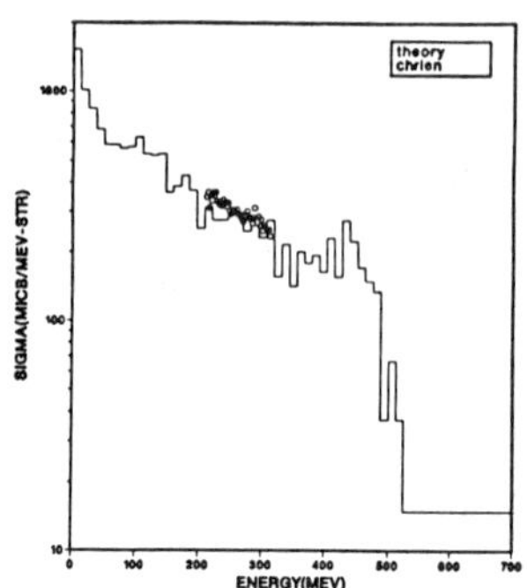

Fig. 1b. Inclusive proton spectrum; INC model calculations with 40000 cascades.

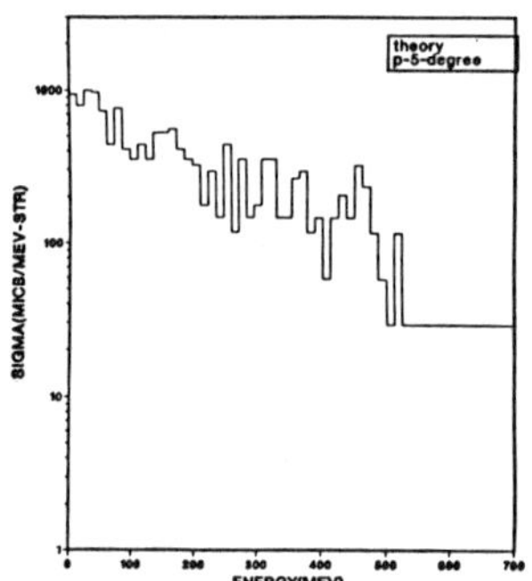

Fig. 2a. 0–5° inclusive neutron spectrum.

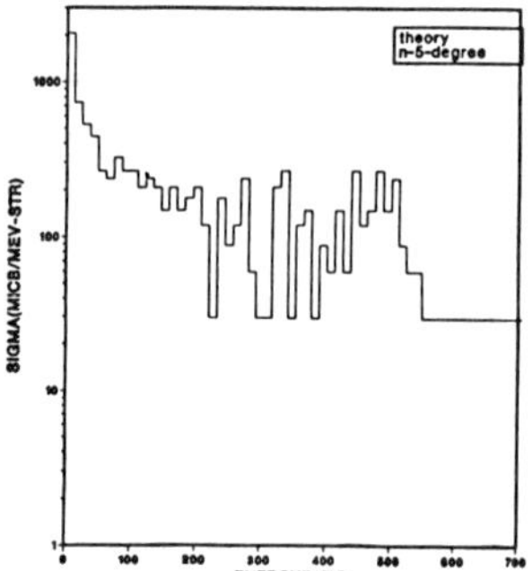

Fig. 2b. 0–5° inclusive proton spectrum.

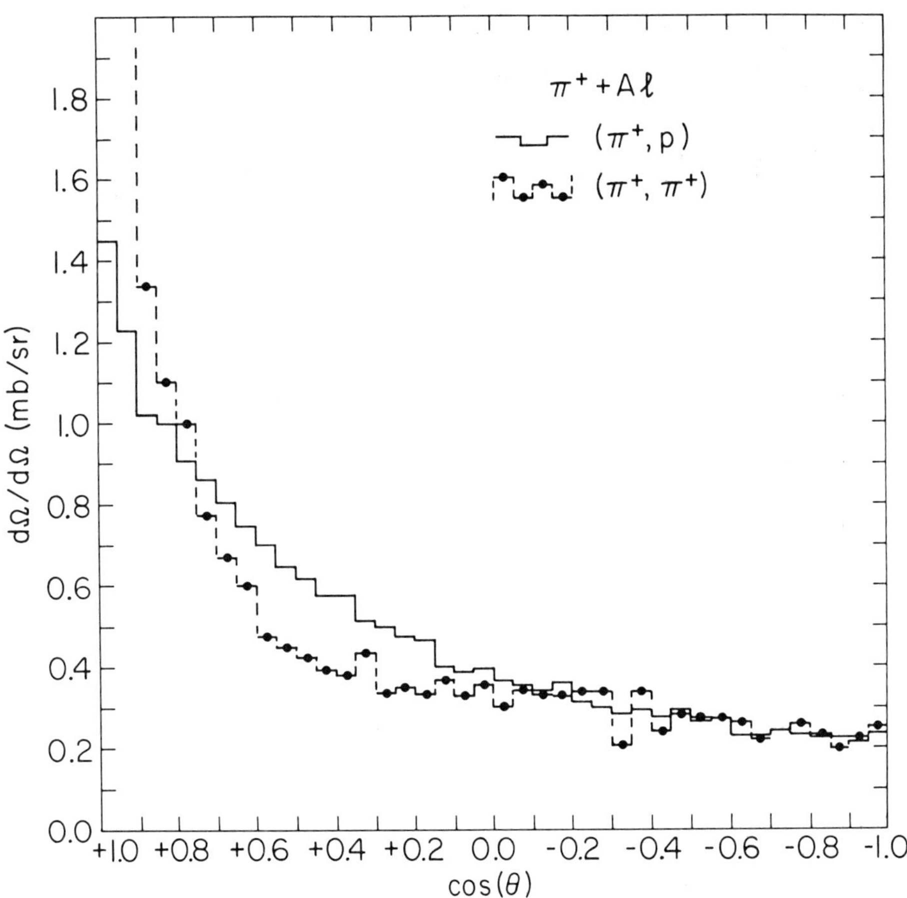

Fig. 3. $A\ell(\pi^+, p)$ and $A\ell(\pi^+, \pi^+)$ INC model predicted angular distributions.

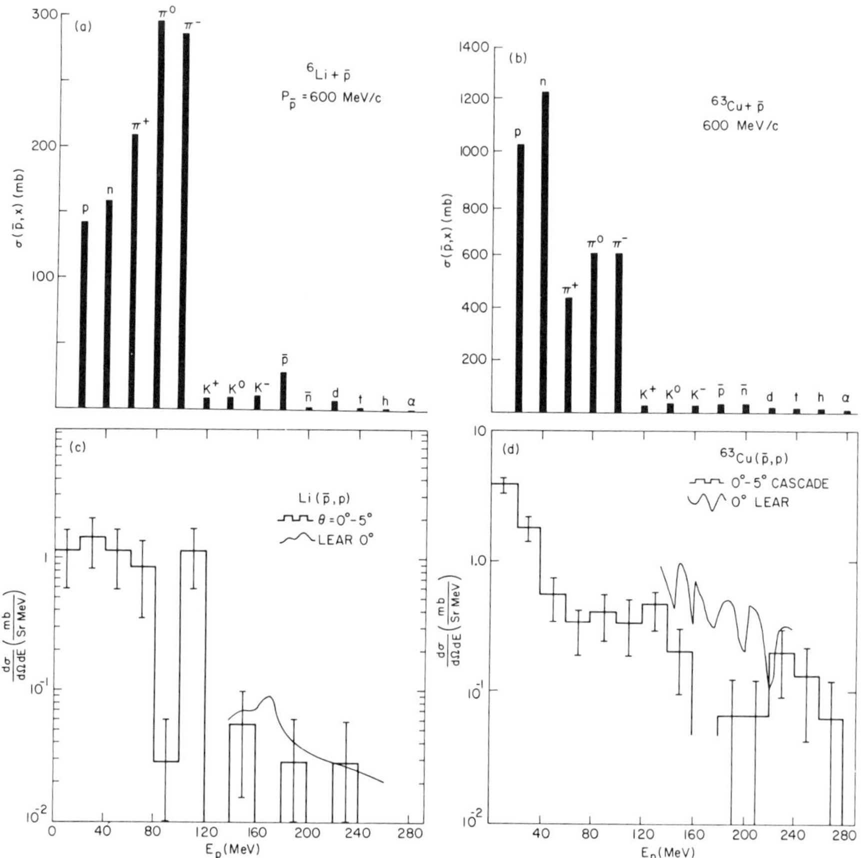

Fig. 4a. ^{6}Li + p̄ 600 MeV/c reaction–product cross sections: INC model predictions.
Fig. 4b. ^{63}Cu + p̄ 600 MeV/c reaction–product cross sections: INC model predictions.
Fig. 4c. ^{6}Li (p̄, p) 0° inclusive proton spectrum: comparison of theory with experiment.
Fig. 4d. ^{63}Cu (p̄, p) 0° inclusive proton spectrum: comparison of theory with experiment.

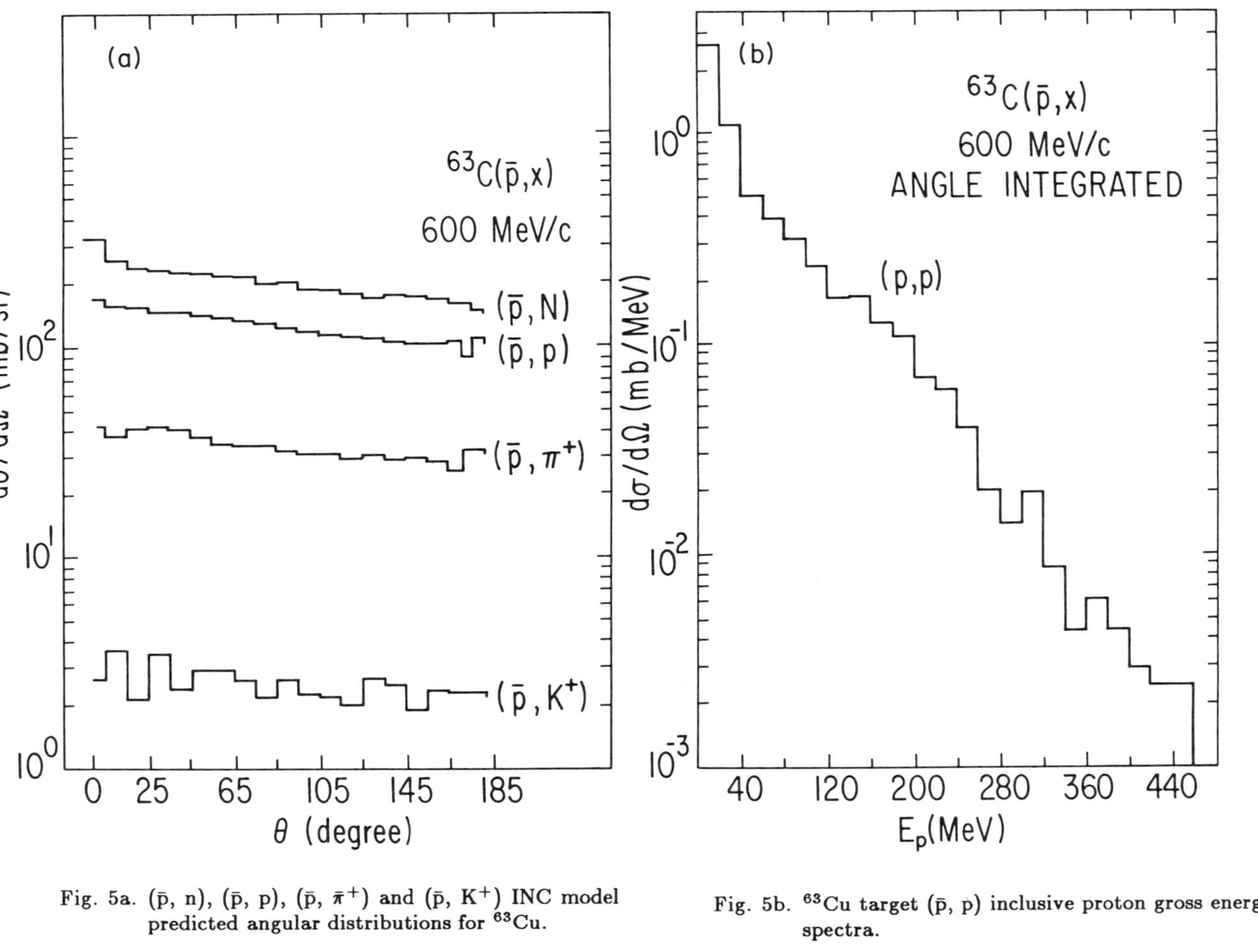

Fig. 5a. $(\bar{p}, n)$, $(\bar{p}, p)$, $(\bar{p}, \bar{\pi}^+)$ and $(\bar{p}, K^+)$ INC model predicted angular distributions for ^{63}Cu.

Fig. 5b. ^{63}Cu target $(\bar{p}, p)$ inclusive proton gross energy spectra.

Fig. 6. K^+p, K^+d, K^+n cross sections. Plots reproduced from reference 10.

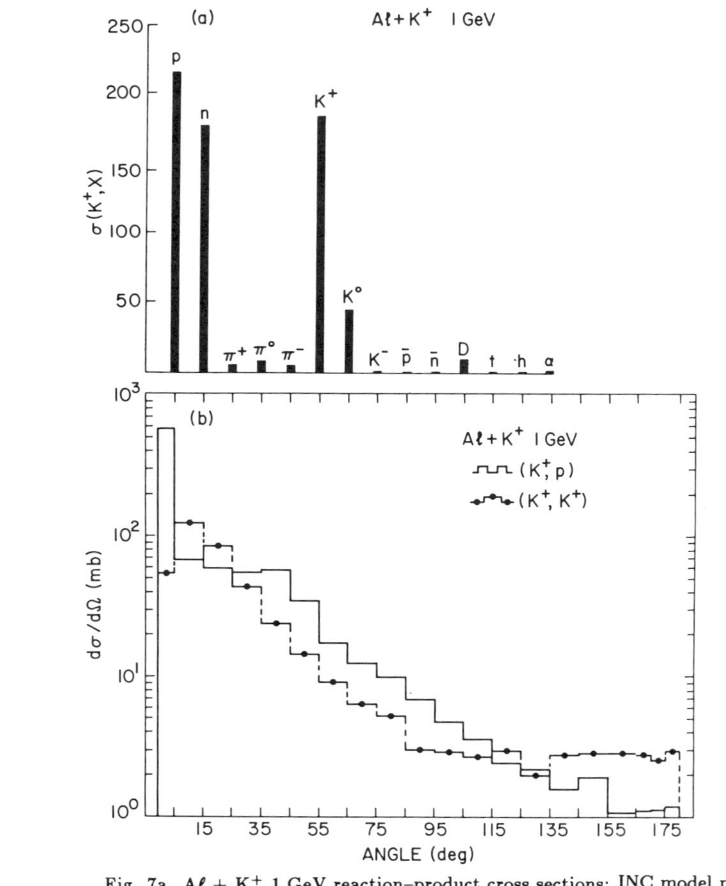

Fig. 7a. $A\ell + K^+$ 1 GeV reaction–product cross sections: INC model predictions
Fig. 7b. $A\ell(K^+, p)$ and $A\ell(K^\pm, K^+)$ angualr distributions: INC model predictions.

QCD AND PHENOMENOLOGY OF NUCLEAR AND HADRONIC PROCESSES AT INTERMEDIATE ENERGIES

Chueng-Ryong Ji
North Carolina State University, Raleigh, NC 27695-8202

QCD can be connected with an effective nuclear theory by Fadeev-Popov trick[1] in the path integral quantization method[2]. In the path integral quantization method, the pertinent vacuum-to-vacuum amplitude can be given basically by

$$e^{iW} = \int[d\bar{\psi}][d\psi][dA_\mu]e^{i\int d^4x L_{QCD}(\bar{\psi},\ \psi,\ A_\mu)}, \qquad (1)$$

where QCD Lagrangian density L_{QCD} is a function of quark $\psi(x)$, antiquark $\bar{\psi}(x)$, gluon $A_\mu(x)$ field and their derivatives. The transformation of quark and gluon degrees of freedom to effective fields $u(x)$ and $v(x)$ can be observed in the following equation:

$$\begin{aligned} 1 &= \int[du]\delta(u - f(\bar{\psi},\ \psi,\ A_\mu)) \\ &= \int[du][dv]e^{-i\int d^4x v\cdot(u-f(\bar{\psi},\ \psi,\ A_\mu))}, \end{aligned} \qquad (2)$$

where the effective field $u(x)$ is related to quark and gluon degrees of freedom by a functional f and the $v(x)$ field is introduced because δ-function is a Fourier transform of plane waves. If we now insert the factor given by Eq. (2) into Eq. (1) and integrate over $\bar{\psi}$, ψ, and A_μ, then one can obtain an effective Lagrangian density L_{eff} (u, v) in terms of effective fields u and v without changing any physics[3]. As J. Donoghue mentioned[4], higher derivative terms in L_{eff} are neglected and thus the physics described by L_{eff} without higher derivative terms is basically low energy physics. In order to describe high energy physics, one should either include higher order derivative terms in L_{eff} or use L_{QCD}. Therefore phenomenology of nuclear and hadronic processes at intermediate energies may be one of the most fertile ground of strong interaction phenomenology.

If one is interested in analyzing various strong interaction phenomena with L_{QCD} based on quark and gluon degrees of freedom, one might consider QCD convolution formalism[5,6] as a unified approach to strong interaction phenomenology. The various phenomena include form factors of hadrons and nuclei at large momentum transfer Q and large angle scattering reactions such as photoproduction $\gamma p \to \pi^+ n$, nucleon-nucleon scattering, photodisintegration $\gamma d \to np$ at large angles and energies, etc. As we can see in Fig. 1, the formalism is a convolution of quark and gluon scattering part with nonperturbative bound state part. The formalism has perturbative aspect because quark and gluon scattering part can be expanded in terms of order $\alpha_s(Q^2)$. However, I think that the terminology of "PQCD" to this formalism is somewhat misleading because it certainly includes nonperturbative bound state part.

A key result is that such amplitudes factorize at large momentum transfer in the form of a convolution of a hard scattering amplitude T_H which can be computed perturbatively from quark-gluon subprocesses multiplied by process-independent "distribution amplitudes" $\phi(x, Q)$ which contain all of bound-state non-perturbative dynamics of each of the interacting hadrons. To leading order in 1/Q the scattering amplitude has the form[5]

$$M = \int_0^1 T_H(x_j, Q) \prod_{H_i} \phi_{H_i}(x_j, Q)[dx]. \tag{3}$$

Here T_H is the probability amplitude to scatter quarks with fractional momentum $0 < x_j < 1$ from the incident to final hadron directions, and ϕ_{H_i} is the probability amplitude to find quarks in wavefunction of hadron H_i collinear up to scale Q, and

$$[dx] = \prod_{j=1}^{n_i} dx_j \delta(1 - \sum_k^{n_i} x_k). \tag{4}$$

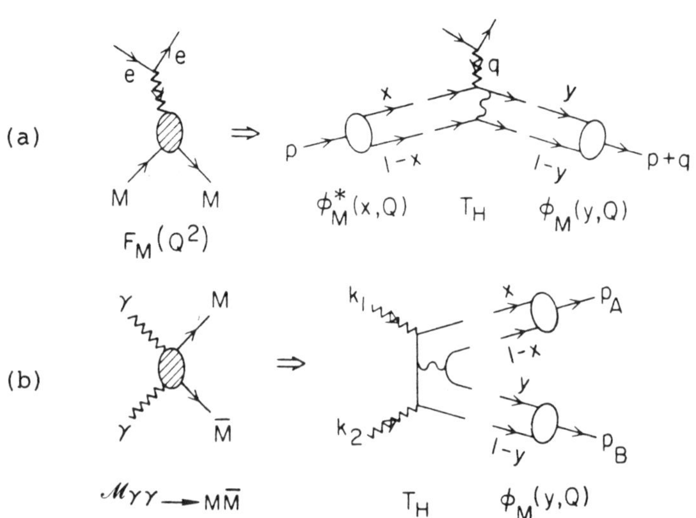

Fig. 1. QCD subprocesses and factorization for high momentum transfer exclusive processes. (a) The meson form factor, (b) Contributions to the $\gamma\gamma \rightarrow M\bar{M}$ amplitude.

However, as N. Isgur pointed out[7], this formalism is potentially dangerous for certain class of processes because the first order kernel of hard scattering part is singular at end points of longitudinal momentum fraction x and y (see Fig. 1(a)). For example, the hard scattering part of pion form factor from Fig. 1(a) is given by

$$T_H \sim e_1 \frac{\alpha_s((1-x)(1-y)Q^2)}{(1-x)(1-y)Q^2} + e_2 \frac{\alpha_s(xyQ^2)}{xyQ^2}, \tag{5}$$

where e_1 and e_2 are the charges carried by particles 1 and 2 whose longitudinal momentum fractions are x(y) and 1-x (1-y) respectively in the initial (final) quark distribution amplitude. Therefore, even though the bound state quark distribution amplitude suppresses the end points the dominant contribution of the form factor comes from small momentum transfer of the exchanged gluon. If one uses naive running coupling constant $\alpha_s(Q^2)$, the calculation cannot be a good approximation.

Furthermore, one can think of another potential danger on this formalism if one assumes Drell-Yan-West connection[8] between deep inelastic lepton hadron structure function and exclusive form factor. For the n-body system, both theoretically and experimentally consistent structure function near x=1 (Bjorken variable $x = -q^2/2p\cdot q$) is given by

$$\nu W_2(x \to 1) \sim (1 - x)^n. \tag{6}$$

From Drell-Yan-West connection, one can calculate the contribution of end point region to the form factor of n-body system:

$$F_n(Q^2)|_{\text{end point region}} \sim \int_{1-\lambda/Q}^{1} (1-x)^n dx \cdot F_{\text{Sudakov}}(Q^2)$$

$$\sim \left(\frac{\lambda^2}{Q^2}\right)^{\sqrt{n+1}} \cdot \left(\frac{\Delta^2}{Q^2}\right)^{c \ln\ln\left(\frac{Q^2}{\Delta^2}\right)}, \tag{7}$$

where λ is some typical hadronic scale and a Sudakov form factor[5,9] $F_{\text{Sudakov}}(Q^2)$ appears since the struck quark is off-shell by only $\Delta^2 \sim \lambda^2/1-x \sim \lambda Q \ll Q^2$. On the other hand, the quark counting rule predicted by QCD convolution formalism under the assumption that the contribution from the end point region is negligible gives (modulo small logarithmic corrections)

$$F_n(Q^2) \sim \left(\frac{1}{Q^2}\right)^{n-1}. \tag{8}$$

There is no a priori reason to believe that $1 - \lambda/Q$ (lower bound of the integration in Eq. (7)) should be replaced by $1 - (\lambda/Q)^2$. Thus, if n > 3, there may be a potential danger in using convolution formalism at finite Q^2. More careful study on the end point region and the validity of Drell-Yan-West connection in QCD should be carried out.

Nevertheless, there are several reasons why it is worthwhile to pursue QCD convolution formalism for the phenomenology of strong interactions.

First, so far, we didn't yet see any better phenomenology.

Second, using convolution formalism, one can relate different two exclusive processes so that the ambiguity from nonperturbative bound state part could be very much suppressed. An example can be shown in the analysis of $\gamma\gamma \to \pi^+\pi^-$ or K^+K^-. With the convolution formalism, $\gamma\gamma \to \pi^+\pi^-(K^+K^-)$ process (see Fig. 1(b)) is connected with pion (kaon) form factor (see Fig. 1(a)) and the QCD prediction[10] is consistent with experimental data as shown in Fig. 2.

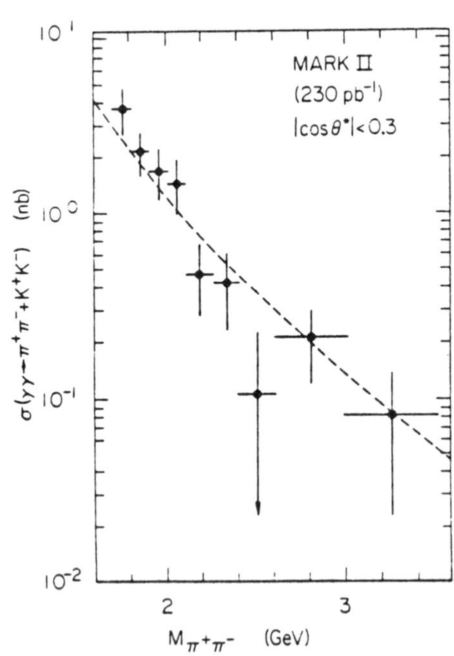

Fig 2. Measured cross section for $\gamma\gamma \to \pi^+\pi^-$ plus $\gamma\gamma \to K^+K^-$ integrated over the angular region $|\cos\theta^*| < 0.3$. The errors contain systematic as well as statistical contributions. The curve is a perturbative QCD prediction.

Third, exclusive pair production of heavy hadron $|Q_1Q_2\rangle, |Q_1Q_2Q_3\rangle$ consisting of higher generation quarks (Q_i = t, b, c, and possibly s) can be reliably predicted[11] within the framework of QCD convolution formalism, since the required wavefunction input is essentially determined from nonrelativistic considerations. The results can be applied to e^+e^-

annihilation, $\gamma\gamma$ annihilation, and $W^{\pm}$ and Z^{o} decay into higher generation pairs. Recent prediction[12] for heavy quark fragmentation function from the same formalism is consistent with experimental data as shown in Fig. 3.

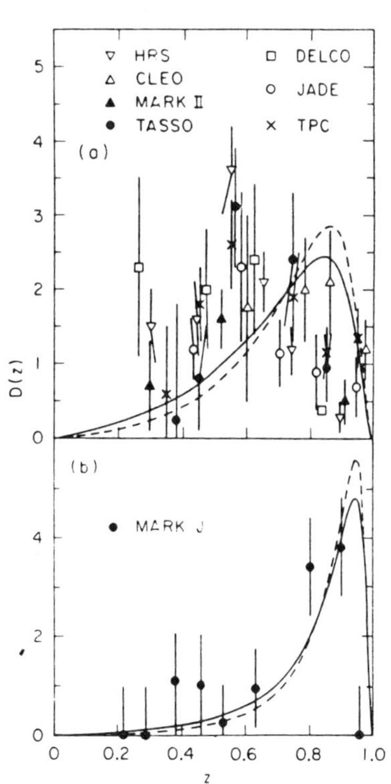

Fig. 3. Charm (a) and bottom (b) fragmentation function predictions compared to various experimental data. The solid curves in (a) and (b) are pseudoscalar mesons and the dashed curves are vector mesons. For the detailed analysis, see Ref. 12.

Last but not least, the notion of color confinement is not the divergence (or blowing up) of QCD running coupling constant at small momentum transfer ($Q^2 \leq \Lambda^2 \sim 0.01$ GeV^2). But it could be a frozen coupling constant.[13] QCD vacuum may be an analogue of a superconductor for the color. A relation between the nontrivial vacuum $|vac\rangle$ which has vacuum condensate and the trivial vacuum $|0\rangle$ may be given by

$$|vac\rangle \sim e^{\sqrt{n}a^{+}}|0\rangle, \qquad (9)$$

where a^{+} is a creaction operator of condensed field and $a|0\rangle = 0$. Then one can show that $a|vac\rangle \sim \sqrt{n}|vac\rangle$ and $\langle vac|a^{+}a|vac\rangle \sim n$. Thus n is the number of vacuum condensate and plays the role of an order parameter. An implication from this observation is that the normal ordering of field operators in the nontrivial vacuum is

annihilation, $\gamma\gamma$ annihilation, and $W^{\pm}$ and Z^{o} decay into higher generation pairs. Recent prediction[12] for heavy quark fragmentation function from the same formalism is consistent with experimental data as shown in Fig. 3.

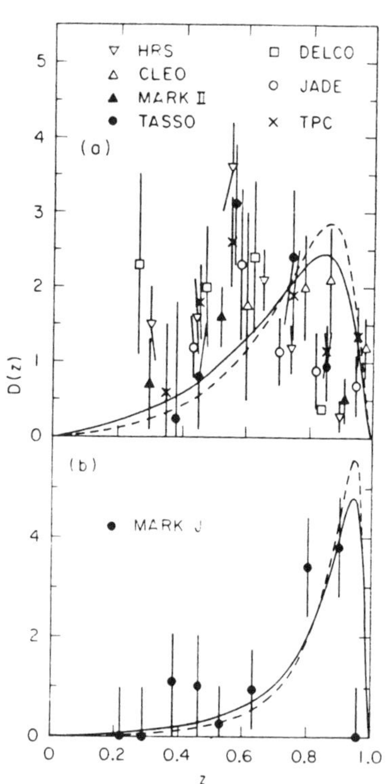

Fig. 3. Charm (a) and bottom (b) fragmentation function predictions compared to various experimental data. The solid curves in (a) and (b) are pseudoscalar mesons and the dashed curves are vector mesons. For the detailed analysis, see Ref. 12.

Last but not least, the notion of color confinement is not the divergence (or blowing up) of QCD running coupling constant at small momentum transfer ($Q^2 \leq \Lambda^2 \sim 0.01\ \mathrm{GeV}^2$). But it could be a frozen coupling constant.[13] QCD vacuum may be an analogue of a superconductor for the color. A relation between the nontrivial vacuum $|vac\rangle$ which has vacuum condensate and the trivial vacuum $|0\rangle$ may be given by

$$|vac\rangle \sim e^{\sqrt{n}a^{+}}|0\rangle, \tag{9}$$

where a^{+} is a creaction operator of condensed field and $a|0\rangle = 0$. Then one can show that $a|vac\rangle \sim \sqrt{n}|vac\rangle$ and $\langle vac|a^{+}a|vac\rangle \sim n$. Thus n is the number of vacuum condensate and plays the role of an order parameter. An implication from this observation is that the normal ordering of field operators in the nontrivial vacuum is not zero while it is zero in the trivial vacuum. The QCD vacuum condensate has been obtained[14] by QCD sum rule from PCAC and instanton solution as follows:

$$\langle vac|:\bar{u}u:|vac\rangle = \langle vac|:\bar{d}d:|vac\rangle \simeq -(250\ \text{MeV})^3$$

and

$$\langle vac|:\frac{\alpha_s}{\pi}F^a_{\mu\nu}F^{a,\mu\nu}:|vac\rangle \simeq 0.012\ (\text{GeV})^4. \qquad (10)$$

Using a special set of Schwinger-Dyson equations, the formation of dimensionful parameters, for example given by Eq. (10), has been studied[15]. The numerical solution of the Schwinger-Dyson equation was consistent with the idea of frozen coupling constant:

$$\alpha_s(Q^2) = \frac{4\pi}{(11 - \frac{2}{3}n_f)\ln(\frac{Q^2+4m_g^2(Q^2)}{\Lambda^2})}, \qquad (11)$$

where n_f is the number of flavor and the effective gluon mass is obtained by

$$m_g^2(Q^2) = m^2[\ln\frac{Q^2 + 4m^2}{\Lambda^2}/\ln\frac{4m^2}{\Lambda^2}]^{-12/11}, \qquad (12)$$

and $m = 500 \pm 200$ MeV. If one uses $\alpha_s(Q^2)$ given by Eq. (11), then $\alpha_s(Q^2) < 0.3$ in the entire Q^2 range ($\alpha_s(0) \sim 0.3$).

For form-factor calculations in QCD, it has been shown[16] that the argument of the running coupling constant should be taken as the square of the momentum transfer of the exchanged gluon in order to make the fast convergence of the perturbation series. Careful analysis based on this idea and frozen coupling constant has been undertaken.[17] This analysis shows that it is possible to fit the data (see Fig. 4) for proton Dirac form factor F_1^p in the range $10 \leq Q^2 \leq 30\ (\text{GeV}/c)^2$ when one uses the distribution amplitudes proposed from QCD sum-rule calculations[18] and an effective gluon mass consistent with values given by Eq. (12) and with the results of a lattice QCD calculation[19] and a recent discussion of dynamical mass generation in QCD.[20] However, it should be observed that the sensitivity of the above results (see Fig. 5) to the choice of cutoff parameter (effective gluon mass) indicates the potential importance of checking the same procedure on other processes and dealing gluon propagators more carefully.

In conclusion, both points of support and criticism on the PQCD phenomenology based on convolution formalism are well taken. The convergence of both points may be obtained from the idea of a nontrivial QCD vacuum.

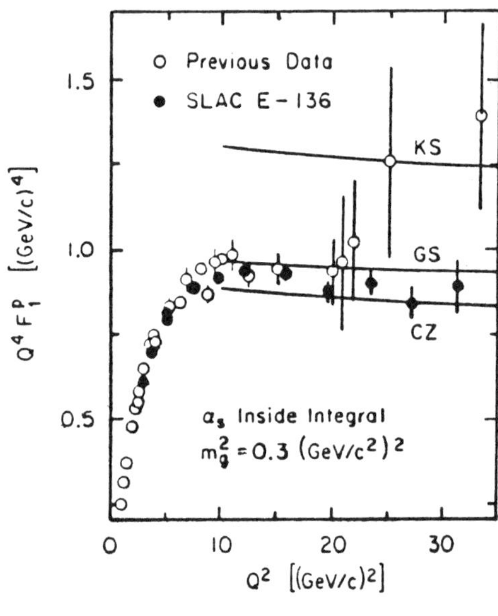

Fig. 4. Leading-twist calculation of the proton Dirac from factor with three different models[18] for the distribution amplitude and with the arguments of the running coupling constant illustrated in the text. For the detailed analysis, see Ref. 17.

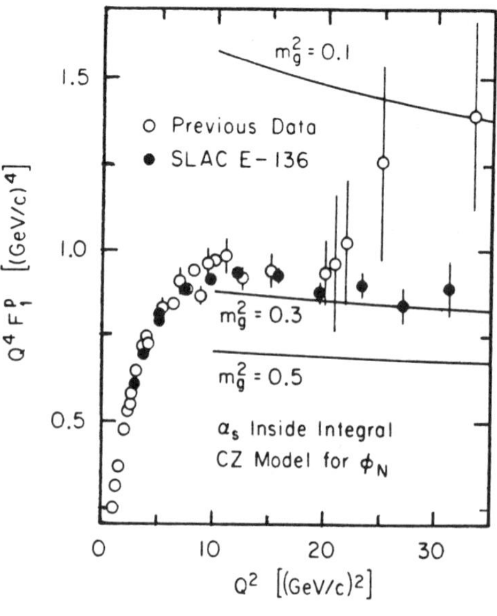

Fig. 5. Leading-twist calculation of the proton Dirac form factor with the distribution amplitude of Chernyak and Zhitnitsky (Ref. 18) and with three choices for the effective gluon mass.

This work was supported by the U. S. Department of Energy under contract No. DE-AS05-79ER10407.

REFERENCES

1. L. S. Faddeev and U. N. Popov, Phys. Lett. 25B, 29 (1967).
2. For a review, see E. S. Abers and B. W. Lee, Phys. Reports 9, 1 (1973).
3. L. S. Celenza, C.-R. Ji and C. M. Shakin, Phys, Rev. C37, 265 (1988); Phys. Rev. D36, 2144 (1987).
4. J. Donoghue, Structure of the Proton-Chiral Perturbation Theory, in this conference.
5. G. P. Lepage and S. J. Brodsky, Phys. Rev. D22, 2157 (1980).
6. S. J. Brodsky, SLAC-PUB-3034 (invited talk presented at the 1982 meeting of the Division of Particles and Fields); S. J. Brodsky and C.-R. Ji, SLAC-PUB-3747 (invited lectures presented at the Stellenbosch Advanced Course in Theoretical Physics, 1985).
7. N. Isgur, Critical Assessment of Models for Strong Interactions, in this conference.
8. S. D. Drell and T. M. Yan, Phys. Rev. Lett. 24, 181 (1970); G. West, Phys. Rev. Lett. 24, 1206 (1970); It was mentioned in Ref. 5 that Drell-Yan-West connection might not be valid in QCD.
9. A. Duncan and A. H. Mueller, Phys. Lett. 90B, 159 (1980); Phys. Rev. D21, 1636 (1980).
10. S. J. Brodsky and G. P. Lepage, Phys. Rev. D24, 1808 (1981); B. Nizic, Phys. Rev. D35, 80 (1987).
11. S. J. Brodsky and C.-R. Ji, Phys. Rev. Lett 55, 2257 (1986); C.-R. Ji and A. F. Sill, Phys. Rev. D34, 3350 (1986);
12. F. Amiri and C.-R. Ji, Phys. Lett. 195B, 593 (1987); C.-R. Ji and F. Amiri, Phys. Rev. D35, 3318 (1987).
13. G. Parisi and R. Petronzio, Phys. Lett. 94B, 51 (1980).
14. M. A. Shifman, A. I. Vainshtein and V. I. Zakharov, Nucl. Phys. B147, 385 (1979).
15. J. M. Cornwall, Phys. Rev. D26, 1453 (1982).
16. S. J. Brodsky, G. P. Lepage and P. B. Mackenzie, Phys. Rev. D22, 228 (1983).
17. C.-R. Ji, A. F. Sill, and R. M. Lombard-Nelson, Phys. Rev. D36, 165 (1987).
18. V. L. Chernyak and I. R. Zhitnitsky, Nucl. Phys. B246, 52 (1984), referred to as CZ; M. Gari and N. G. Stefanis, Phys. Lett. 175B, 462 (1986), referred to as GS; I. D. King and C. T. Sachrajda, Nucl. Phys. B279, 785 (1987), referred to as KS.
19. C. Bernard, Phys. Lett. 108B, 436 (1982).
20. L. S. Celenza, C.-R. Ji and C. M. Shakin, Phys. Rev. D36, 895 (1987).

HYPERNUCLEAR INTERACTIONS AND THE BINDING ENERGIES OF Λ AND ΛΛ HYPERNUCLEI

A. R. Bodmer

Department of Physics, University of Illinois at Chicago, Chicago, IL 60680 and Physics Division, Argonne National Laboratory, Argonne, IL 60439-4843 USA

and

Q. N. Usmani

Department of Physics, Jamia Millia, New Delhi, India and Department of Physics, University of Illinois at Urbana-Champaign, Urbana, IL 61801 USA

ABSTRACT

By use of variational calculations a reasonable hadronic description is obtained of the s-shell hypernuclei, of ${}^{9}_{\Lambda}$Be, and of the well depth, with ΛN forces which are consistent with Λp scattering and which are quite strongly spin-dependent, with reasonable TPE ΛNN forces and with strongly repulsive dispersive-type ΛNN forces. For the latter we also consider a spin-dependent version which is somewhat favored by our analysis. ${}^{9}_{\Lambda}$Be is treated as a $2\alpha+\Lambda$ system and is significantly overbound, ≃1 MeV, if only $\alpha\alpha$ and $\alpha\Lambda$ potentials are used. An $\alpha\alpha\Lambda$ potential obtained from the ΛNN forces nicely accounts for this overbinding. The ΛΛ hypernuclei ${}^{6}_{\Lambda\Lambda}$He and ${}^{10}_{\Lambda\Lambda}$Be are treated as $\alpha+2\Lambda$ and $2\alpha+2\Lambda$ systems. Use of the ${}^{10}_{\Lambda\Lambda}$Be event gives ≃1.5 MeV too little binding for ${}^{6}_{\Lambda\Lambda}$He. The ${}^{1}S_{0}$ ΛΛ potential obtained from ${}^{10}_{\Lambda\Lambda}$Be is quite strongly attractive, comparable to the ΛN and also to the NN potential without OPE.

INTRODUCTION

We review our recent work on the calculation of binding energies of Λ and ΛΛ hypernuclei using reasonable ΛN, ΛNN and ΛΛ interactions. Reasonable here means to be generally consistent with meson-exchange models. Effects of baryon quark structure are assumed to be short-range and capable of parameterization in the conventional way through repulsive cores and cutoffs. The aim is to learn about these interactions and in particular whether such a hadronic approach is adequate or whether more explicit quark effects must be invoked. An essential element in our approach is the use of 3-body ΛNN forces considered to be the result of eliminating Σ, Δ, ... degrees of freedom (from a coupled channel approach which includes these and which represents a more sophisticated level of phenomenology) to obtain a reduced description in terms of only Λ and nucleon degrees of freedom. We consider phenomenological ΛNN forces both of "dispersive" and two-pion-exchange (TPE) type. A new feature of our dispersive ΛNN forces is that we not only consider spin independent forces but also spin dependent forces, depending on the nucleon spins, which are suggested by the suppression mechanism due to ΛN-ΣN coupling. The ΛN and ΛNN forces we use are central and so mostly are our NN forces. We have studied the following systems.

1. Λp scattering, the s-shell hypernuclei ($A \lesssim 5$) including the excited state of the A=4 hypernuclei, and also the Λ binding in nuclear matter ($A=\infty$), i.e. the Λ well depth $D \simeq 30$ MeV.[1]
2. The α-cluster hypernuclei ${}^{9}_{\Lambda}Be(2\alpha+\Lambda)$, ${}^{6}_{\Lambda\Lambda}He(\alpha+2\Lambda)$ and ${}^{10}_{\Lambda\Lambda}Be(2\alpha+2\Lambda)$ together with the α-Λ system ${}^{5}_{\Lambda}He$ which is used to determine the αΛ potential.[2] The α is considered as an "elementary" constituents for these systems.

For the few-body systems (s-shell hypernuclei and α-cluster hypernuclei) we use well developed variational (MC) methods;[3] D is calculated variationally with the Fermi hypernetted chain method.[4] The needed extensions are discussed in Refs. 1 and 2.

ΛN POTENTIAL AND Λp SCATTERING

We use a central Urbana-type ΛN potential[5] $V_{\Lambda N}=V_c-V_{2\pi}$ where V_c is a repulsive core close to that for the NN potential and where $V_{2\pi}$ has a TPE (square of the OPE tensor potential) shape involving the (attractive) singlet and triplet strengths as adjustable parameters, or equivalently the spin-average strength $\bar{V}=1/4V_s+3/4V_t$ and the spin-dependent strength $V_\sigma=V_s-V_t>0$. The low-energy Λp scattering data (average over the singlet and triplet states) determines $\bar{V}^{\Lambda p}$ for Λp scattering quite well but V_σ only very poorly. The charge symmetric strength $\bar{V}$ (average over Λn and Λp) can then be determined from $\bar{V}^{\Lambda p}$ using the CSB interaction determined from the A=4 hypernuclei.[6] $\bar{V}$ is essentially the only ΛN strength relevant for hypernuclei with core nuclei having zero spin, i.e. ${}^{5}_{\Lambda}He$, ${}^{9}_{\Lambda}Be$ and D. With only ΛN forces one then has the well known problem of overbinding: thus for B_Λ

(experimental) vs. B_Λ (calculated) one has in MeV: 3.1 vs. ≃ 6($^5_\Lambda$He), 6.7 vs. ≃ 12($^9_\Lambda$Be), 11.7 vs. ≃ 22($^{13}_\Lambda$C) and 30 vs. ≃ 60(D).

ΛNN POTENTIALS

For a reasonable phenomenology which accounts for this overbinding we include ΛNN forces, assumed to arise mostly from coupling of the ΛN to the ΣN channel, which when eliminated will give many-body ΛNN, etc. forces. An alternate approach is to explicitly include the ΣN channel. However, except for lowest order G matrix calculations of D, this has not been adequately implemented within a variational framework either for A≲5 or for D.

We consider both dispersive ΛNN forces, of the form used in Ref. 7 for NNN forces, and two-pion exchange (TPE) ΛNN forces.[8] Both arise from elimination of the ΣN channel, with the former, which are repulsive, assumed to arise mostly from medium suppression of the ΣN channel in the TPE ΛN interaction. For the dispersive-type forces we consider two types: $V^D_{\Lambda NN}$ which is spin-independent, and $V^{DS}_{\Lambda NN}$ which has a novel spin-dependence suggested by a spin-dependence of the ΛN-ΣN suppression mechanism where this operates in the triplet but not in the singlet ΛN channel.[9] (Thus, $V^{DS}_{\Lambda NN} = V^D_{\Lambda NN}\,[1 + 1/6\, \vec{\sigma}_\Lambda \cdot (\vec{\sigma}_1 + \vec{\sigma}_2)]$, where 1 and 2 label the two nucleons.) $V^D_{\Lambda NN}$ and $V^{DS}_{\Lambda NN}$ are completely equivalent when the core nuclei have zero spin, in particular for $^5_\Lambda$He, $^9_\Lambda$Be and D, but differ for A≲4. The TPE ΛNN potential $V^{2\pi}_{\Lambda NN}$ is spin-independent but angle-dependent and its effect is quite sensitive to appropriate ΛNN correlations for A≲5 which can result in an overall attractive contribution of $V^{2\pi}_{\Lambda NN}$ for the s-shell hypernuclei.[1]

INTERACTIONS FROM THE s-SHELL HYPERNUCLEI AND D

We very briefly summarize, together with appropriate caveats, the main conclusions obtained from Λp scattering, the s-shell hypernuclei and D.[1] Overall, we obtain a satisfactory description of all these systems.

1. Our ΛN potential is consistent with Λp scattering and the CSB interaction as determined from A=4.
2. The s-shell hypernuclei require TPE ΛNN forces whose strength is consistent with theoretical expectation.[8] In particular, these forces cannot account for the overbinding of $^5_\Lambda$He and give only a relatively small reduction in D.

3. Strongly repulsive dispersive ΛNN forces are required, essentially to account for the overbinding obtained with only ΛN forces. For ${}^{5}_{\Lambda}$He ΛN tensor forces could possibly significantly reduce the overbinding;[10] however, for A>5, in particular for D, this seems quite unlikely.[11]
4. The spin-dependent ΛNN potential $V^{DS}_{\Lambda NN}$ is somewhat preferred by our analysis when all systems (A≲5 and D) are considered.
5. We obtain an appreciable spin-dependence V_σ for the ΛN potential, the magnitude of which depends on whether or not a spin-dependence is included in the dispersive ΛNN potential.

The ΛN spin-dependence is rather directly related to the observed splitting ≃1 MeV between the ground 0^+ and the excited (spin flip) 1^+ state of ${}^{4}_{\Lambda}$H, ${}^{4}_{\Lambda}$He. With $V^{D}_{\Lambda NN}$, i.e. no ΛNN spin-dependence, this splitting is essentially determined entirely by the ΛN spin-dependence, whereas with $V^{DS}_{\Lambda NN}$, i.e. with a ΛNN spin-dependence, ≃1/3 of this splitting is due to $V^{DS}_{\Lambda NN}$, leading to a reduced ΛN spin-dependence. The effect of the ΛNN spin-dependence is consistent with a larger ΛN-ΣN suppression (more repulsion) in the triplet than in the singlet state since the former is more heavily weighted in the 1^+ than in the 0^+ state.[9] Our ΛNN spin dependence should be considered as a phenomenological representation of this differential suppression effect which maximizes this. However, our results show that even with ΛNN spin-dependence there is still a very appreciable ΛN spin-dependence. Thus, the scattering lengths we obtain are:

With $V^{D}_{\Lambda NN}$: $a_s \simeq -3.0$ fm, $a_t \simeq -1.5$ fm

With $V^{DS}_{\Lambda NN}$: $a_s \simeq -2.5$ fm, $a_t \simeq -1.6$ fm

Our analysis thus shows that the 0^+-1^+ splitting for A=4 makes it very difficult to avoid the conclusion that there is an appreciable ΛN spin-dependence even if a significant part of the splitting is due to the spin-dependence of a ΛNN force. This strong ΛN spin-dependence is not consistent with the Nijmegen potential D and F,[12] but is in better agreement with the more recent potential of the Nijmegen group.[13]

THE α-CLUSTER HYPERNUCLEUS ${}^{9}_{\Lambda}$Be

Calculations for ${}^{9}_{\Lambda}$Be are made with a 2α+Λ cluster model.[2] The potential between the constituents are local and accurately describe the component subsystems ^{8}Be and ${}^{5}_{\Lambda}$He. A variety of (s-state) αα potentials $V_{\alpha\alpha}$ (which include the Coulomb interaction) are used, all fitted to the $\alpha\alpha$ scattering data. Several $\alpha\Lambda$ potentials $V_{\alpha\Lambda}$ were used, all determined from ${}^{5}_{\Lambda}$He for some $V_{\Lambda N}$ and $V_{\Lambda NN}$ and all giving the experimental $\alpha\Lambda$ separation energy of $B_{\Lambda}({}^{5}_{\Lambda}\mathrm{He})=3.12$ MeV.

In particular, some of our $V_{\alpha\Lambda}$ were obtained from our 5-body (ΛNNNN) variational Monte-Carlo (MC) calculations of ${}^{5}_{\Lambda}$He.[1] Thus from the 5-body wave function the Λ density $\rho_{\Lambda}(r_{\alpha\Lambda})$ relative to the four nucleons (i.e. relative to the α particle) in ${}^{5}_{\Lambda}$He is obtained numerically. The "trivially equivalent" local $\alpha\Lambda$ potential is then obtained from the $\alpha\Lambda$ "wave function" $\phi_{\Lambda}=\rho_{\Lambda}^{1/2}(r_{\alpha\Lambda})$ by solving the corresponding Schrödinger equation for $V_{\alpha\Lambda}$. These (MC) $V_{\alpha\Lambda}$ include many-body effects not present in $\alpha\Lambda$ potentials which are the variational equivalent of Brueckner-Hartree calculations. The latter are obtained with effective ΛN and ΛNN potentials which include just the effect of 2-body ΛN correlations obtained from well depth calculations for a Λ in nuclear matter. In particular, these many-body effects give a central repulsion in the MC $V_{\alpha\Lambda}$ even with only ΛN potentials (adjusted in strength to fit $B_{\Lambda}({}^{5}_{\Lambda}\mathrm{He})$ but then no longer fitting the Λp data). This effect was first pointed out in the context of reaction matrix calculations.[14] Repulsive ΛNN potentials give an additional large central repulsion in $V_{\alpha\Lambda}$.

Experimentally one has $B_{\Lambda}=6.71\pm0.04$ MeV for the ground state of ${}^{9}_{\Lambda}$Be. Our variational calculations give $B_{\Lambda}\simeq8.2\pm0.2$ MeV where the error is mostly due to uncertainties in $V_{\alpha\alpha}$ and $V_{\alpha\Lambda}$. For ${}^{9}_{\Lambda}$Be the ΛN p-state potential V_p can made a significant contribution. The Λp scattering data is consistent with $V_p\simeq1/2V_{\Lambda N}$ ($V_{\Lambda N}$ is the s-state potential) which gives a reduction of $\simeq$0.4 MeV from a number of independent estimates[1,15] (see Ref. 1 for a discussion). With this reduced p-state interaction we then obtain $B_{\Lambda}\simeq7.8$ MeV.

Thus our calculations predict ${}^{9}_{\Lambda}$Be to be overbound by about 1 MeV for $\alpha\alpha$ and $\alpha\Lambda$ potentials which accurately describe the component two-body systems. Because we have made a variational calculation any improvement in the trial wave function can only given even more overbinding. Furthermore, the α-cluster model is internally consistent in that the calculated rms $\alpha\alpha$ separation of $\simeq$3.7 fm is appreciably larger than twice the rms α-particle radius of 1.47 fm. We thus believe that this overbinding is a significant result whatever its detailed explanation may be.

In fact it is quite striking that this overbinding is quite naturally and quantitatively accounted for by the repulsive ΛNN dispersive forces. Thus, the ΛNN potential not only gives a repulsive contribution to $V_{\alpha\Lambda}$ (due to the Λ interacting with a pair of nucleons in the same α, and required to avoid overbinding $^{5}_{\Lambda}$He), but will also give a contribution due to the interaction of the Λ with a pair of nucleons each in a different α. By a suitable folding procedure this gives an effective $\alpha\alpha\Lambda$ potential $V_{\alpha\alpha\Lambda}$ which is repulsive and completely determined. With inclusion of this $V_{\alpha\alpha\Lambda}$ our variational calculations then give $B_{\Lambda}\simeq 6.9$ MeV, corresponding to $\langle V_{\alpha\alpha\Lambda}\rangle\simeq 1$ MeV, i.e. just about the necessary decrease in B_{Λ} to compensate for the overbinding obtained with only $V_{\alpha\alpha}$ and $V_{\alpha\Lambda}$!

THE ΛΛ HYPERNUCLEI $^{6}_{\Lambda\Lambda}$He AND $^{10}_{\Lambda\Lambda}$Be

These provide the only presently available information about the ΛΛ potential $V_{\Lambda\Lambda}$ in the $^{1}S_{0}$ state, appropriate to the ground state of ΛΛ hypernuclei. Our variational calculations treat $^{6}_{\Lambda\Lambda}$He as an $\alpha+2\Lambda$ system and $^{10}_{\Lambda\Lambda}$Be as a $2\alpha+2\Lambda$ system. The potentials $V_{\alpha\alpha}$, $V_{\alpha\Lambda}$ and $V_{\alpha\alpha\Lambda}$ are those discussed for $^{9}_{\Lambda}$Be. For $V_{\Lambda\Lambda}=V_{c}-V_{A}$ we used a variety of shapes, with a range of repulsive cores V_{c} (including the same V_{c} as used for $V_{\Lambda N}$, and also ω-exchange potentials), and attractive potentials V_{A} (including a TPE potential $V_{2\pi}$ and a σ-meson exchange potential V_{σ}). The experimental separation energy $B_{\Lambda\Lambda}$ of both Λs then determines, through the variational calculations, just one strength parameter, that of V_{A}, for each ΛΛ hypernucleus. We give more weight to the $^{10}_{\Lambda\Lambda}$Be event[16] since this has been thoroughly checked and since also the quoted error for $B_{\Lambda\Lambda}$ is smaller than for the $^{6}_{\Lambda\Lambda}$He event.[17]

For <u>any</u> $V_{\Lambda\Lambda}$ (including some strength for V_{A}) our variational calculations of both $^{6}_{\Lambda\Lambda}$He and $^{10}_{\Lambda\Lambda}$Be give the $B_{\Lambda\Lambda}$ of each. From calculations for a large variety of $V_{\Lambda\Lambda}$ one then obtains essentially a linear relation between $^{6}B_{\Lambda\Lambda}\equiv B_{\Lambda\Lambda}(^{6}_{\Lambda\Lambda}\mathrm{He})$ and $^{10}B_{\Lambda\Lambda}\equiv(^{10}_{\Lambda\Lambda}\mathrm{Be})$: $^{6}B_{\Lambda\Lambda}\simeq -4.17+0.76\ ^{10}B_{\Lambda\Lambda}$ (MeV). The experimental value $^{10}B_{\Lambda\Lambda}=17.7\pm0.1$ MeV then predicts, effectively independently of any details of $V_{\Lambda\Lambda}$, that $^{6}B_{\Lambda\Lambda}=9.25\pm0.08$ MeV[16] compared to the quoted value of 10.92 ± 0.6 MeV.[17] Thus, $^{6}_{\Lambda\Lambda}$He is underbound by $\simeq 1.5$ MeV. In view of the uncertainties associated with the $^{6}_{\Lambda\Lambda}$He event, the significance of this underbinding is unclear. However, this comparison of the $B_{\Lambda\Lambda}$ of two

$\Lambda\Lambda$ hypernuclei does demonstrate very clearly that more $\Lambda\Lambda$ hypernuclear events could lead to valuable information about both the $\Lambda\Lambda$ and Λ-nuclear interactions.

THE $\Lambda\Lambda$ INTERACTION FROM $^{10}_{\Lambda\Lambda}$Be

We use the $^{10}_{\Lambda\Lambda}$Be event to obtain the $\Lambda\Lambda$ interaction since the experimental errors are less than for $^{6}_{\Lambda\Lambda}$He and also since this event has been thoroughly checked. For each $V_{\Lambda\Lambda}$ the experimental $^{10}B_{\Lambda\Lambda}$, through the use of the variational calculations, determines the strength of the attractive part V_A of $V_{\Lambda\Lambda}$ and hence the corresponding scattering length $a_{\Lambda\Lambda}$ (and also the effective range). For a given intrinsic range b (the effective range when $V_{\Lambda\Lambda}$ just gives a bound state) $a_{\Lambda\Lambda}$ is found to be approximately independent of the detailed shape of $V_{\Lambda\Lambda}$. A $V_{2\pi}$ potential together with the same V_c as for $V_{\Lambda N}$ has $b \simeq 2$ fm and gives $a_{\Lambda\Lambda} \simeq -4$ fm, whereas a σ exchange potential with this same V_c has $b \simeq 2.5$ fm and gives $a_{\Lambda\Lambda} \simeq -6$ fm. As the range b increases $-a_{\Lambda\Lambda}$ increases and becomes ∞ at $b \simeq 3$ fm corresponding to a bound state. Since such large repulsive cores seem implausible, it seems quite unlikely that the $^1S_0$ $\Lambda\Lambda$ system has a bound state. (In fact, quark exchange models of the repulsive core suggest this could be appreciably softer for the $\Lambda\Lambda$ than for the NN system.) In any case, the $\Lambda\Lambda$ interaction obtained from $^{10}_{\Lambda\Lambda}$Be, is quite strongly attractive with $2 \lesssim -a_{\Lambda\Lambda} \lesssim 5$ fm.

The phenomenological $V_{\Lambda\Lambda}$ obtained from $^{10}_{\Lambda\Lambda}$Be is more or less consistent with at least one published meson-exchange potential, namely the Nijmegen potential D.[11] Thus, Bando et al.[18] have shown that this potential gives a reasonable value for $^6B_{\Lambda\Lambda}$ and hence will also give a more or less reasonable value for $^{10}B_{\Lambda\Lambda}$. Thus, the large value of $-a_{\Lambda\Lambda}$ cannot be taken as a possible indication of an H dibaryon which is just unbound.

The existence of $\Lambda\Lambda$ hypernuclei would seem to exclude an H strongly bound with respect to $2M_\Lambda$. Clearly, further $\Lambda\Lambda$ hypernuclear events would very much strengthen this argument. However, the binding of $\Lambda\Lambda$ hypernuclei would permit a weakly bound H with $2M_\Lambda - M_H \lesssim 18$ MeV, in which case the strong decay $^{10}_{\Lambda\Lambda}\mathrm{Be} \rightarrow {}^8\mathrm{Be} + \mathrm{H}$ would be energetically forbidden.

Finally, it is interesting to compare the $^1S_0$ NN, ΛN and $\Lambda\Lambda$ potentials. For this comparison to be significant the same shape V_c-

$V_{2\pi}$ is used for all three potentials, and for the NN potential we consider V_{NN}-V_{π}, i.e. with the OPE part subtracted out. Then a_{NN}= -3.5 fm, instead of -17.5 fm as for V_{NN}, which is to be compared with $a_{\Lambda N}\simeq$ -3 fm, and $a_{\Lambda\Lambda}\simeq$ -4 fm. It seems remarkable that these (and the corresponding strengths of $V_{2\pi}$) are so close to each other, indicating that, excluding OPE for V_{NN}, the three $^1S_0$ interactions are effectively very similar.

Work supported in part by the U.S. Department of Energy, Nuclear Physics Division, under contract W-31-109-ENG-38, and by grants NSF-PHY-84-15064 and NSF-INT-8319827.

REFERENCES

1. A. R. Bodmer and Q. N. Usmani, Nucl. Phys. A477, 621 (1988); Nucl. Phys. A450, 257c (1986).
2. A. R. Bodmer and Q. N. Usmani, Nucl. Phys. A463, 221c (1987); A468, 653 (1987); A. R. Bodmer, Q. N. Usmani and J. Carlson, Nucl. Phys. A422, 510 (1984).
3. J. Lomnitz-Adler, V. R. Pandharipande and R. A. Smith, Nucl. Phys. A361, 399 (1981).
4. Q. N. Usmani, Nucl. Phys. A340, 397 (1980); J. Dabrowski and W. Piechocki, Ann. Phys. 126, 317 (1980); W. Piechocki and J. Dabrowski, Acta Physica Polorica B12, 475 (1981).
5. I. E. Lagaris and V. R. Pandharipande, Nucl. Phys. A359, 331 (1981).
6. A. R. Bodmer and Q. N. Usmani, Phys. Rev. C 31, 1400 (1985).
7. J. Carlson and V. R. Pandharipande, Nucl. Phys. A371, 301 (1981); J. Carlson, V. R. Pandharipande and R. B. Wiringa, Nucl. Phys. A401, 59 (1983).
8. R. K. Bhaduri, B. A. Loiseau and Y. Nogami, Ann. Phys. 44, 57 (1967); A. Gal, Adv. in Nucl. Phys., ed. M. Baranger and E. Vogt, 8, 1 (1975); R. H. Dalitz, R. C. Herndon and Y. C. Tang, Nucl. Phys. B47, 109 (1972).
9. J. Dabrowski and E. Fedorynska, Nucl. Phys. A210, 509 (1973); B. F. Gibson and D. R. Lehman, in Proc. of Int. Conf. on Hypernuclear and Kaon Physics, Heidelberg, Germany 1982, ed. B. Povh (Max-Planck Institute für Kernphysik, Heidelberg, 1982); Phys. Rev. C 23, 573 (1981).
10. H. Bando and T. Shimodaya, Prog. Theor. Phys. 63, 1812 (1980); S. Shinmura, Y. Akaishi and H. Tanaka, Prog. of Theor. Phys. 71, 546 (1984).
11. A. R. Bodmer, D. M. Rote and A. L. Mazza, Phys. Rev. C 2, 1623 (1970).
12. M. M. Nagels, T. A. Rijken and J. J. deSwart, Phys. Rev. D 15, 2545 (1977); 20, 1633 (1979).
13. J. J. deSwart and T. A. Rijken, Proc. of 1986 INS Int. Sym. on Hypernuclear Physics, August 1986, Tokyo, eds. H. Bando, O. Hashimoto and K. Ogawa (Institute of Nuclear Study, Univ. of Tokyo, 1988), p. 303.
14. Y. Kurihara, Y. Akaishi and H. Tanaka, Prog. Theor. Phys. 71, 561 (1984); Phys. Rev. C 31, 971 (1985).

15. Wang Xi-chang, H. Takaki and H. Bando, Prog. Theor. Phys. 76, 865 (1986).
16. M. Danysz et al., Nucl. Phys. 49, 121 (1963); Phys. Rev. Lett. 11, 29 (1963); J. Pniewski and D. Zieminska, Proc. of Conf. on Kaon-Nuclear Interaction and Hypernuclei (Zvenigorod, Moscow, 1977) p. 33.
17. J. Prowse, Phys. Rev. Lett. 17, 782 (1966).
18. K. Ikeda, H. Bando and T. Motba, Prog. Theor. Phys. Sup. 81, 147 (1985).

A SEARCH FOR BOUND STATES OF THE η-MESON IN LIGHT NUCLEI

Philip H. Pile
Brookhaven National Laboratory, Upton, New York, 11973

ABSTRACT

This paper describes an experiment designed to search for a new form of nuclear matter--a bound η-nucleus system. The (π^+,p) reaction was used to study the possible formation of an η-mesic nucleus. No narrow η-nuclear bound states were observed using ^7Li, ^{12}C, ^{16}O and ^{27}Al targets. The results of this experiment have now been published elsewhere by Chrien et al.[1]

INTRODUCTION

No meson has been observed to bind with an atomic nucleus to form a nuclear bound state. Both π^- and K^- mesons are observed to be captured in atomic orbits via the Coulomb attraction but have not been observed to bind to the nucleus through the nuclear force.

In a recent theoretical analysis of η-nuclear systems using a potential model approach, Haider and Liu[2,3] have shown that for nuclei with $A>10$ the η meson should be bound to the nucleus. They predict the binding should be a few MeV with a decay width of about 10 MeV. They predict the bound η states should be observed using the (π^+,p) reaction with a π^+ momentum around 800 MeV/c. Li, Chen and Kuo[4] verified the result of Haider and Liu using a standard many-body Green's function method.

NATURE OF THE η-NUCLEUS BINDING

Bhalerao and Liu[5] have shown that the low energy η-nucleon interaction is attractive. This is a result of the threshold for the $\pi N \to \eta N$ reaction lying below the lowest s-wave resonance channel of the ηN system--the $N^*(1535)$. Later, Haider and Liu's[2,3] theoretical studies showed that the η should be bound in nuclei with $A>10$ with decay widths on the order of 10-13 MeV for $A<30$. Their study was based on an optical model approach using a momentum-space relativistic three-dimensional integral equation. They studied six spin zero nuclei. Although the results of their calculation depended upon the particular πN phase shift set used (Arndt or CERN), the disagreement between the sets is no greater than 2 MeV. The average of the two results are shown in Table I.

The η can be produced inside the nucleus using the reaction

$$\pi^+ + n \to \eta + p$$

which allows the η to be produced with a low momentum in the nucleus under selected kinematical conditions. In fact, the η laboratory momentum is identically zero with about a 900 MeV/c incident pion momentum and a zero degree recoil proton. The proper choice of incident pion momentum depends on the elementary η

Table I Binding energy and decay widths of selected η-mesic nuclei

Core Nucleus	η-binding energy (MeV)	FWHM decay width (MeV)
^{207}Pb	17.3	20.0
^{89}Zr	13.7	17.4
^{39}Ca	7.9	13.0
^{25}Mg	5.5	12.5
^{15}O	2.7	9.9
^{11}C	0.6	6.5

production cross section as well as the probability that an η of a given momentum will bind in a particular nuclear state.

The calculated[3] total cross section for η production using the (π^+,p) reaction on a nucleus peaks at about a 740 MeV/c pion momentum, reflecting the effects of the $N^*(1535)$ production. The differential cross section for s-state η-mesic nucleus production is predicted to peak near a 15 degree laboratory angle for light nuclei at this incident pion momentum. The ^{16}O target case gives the largest predicted differential cross section and is therefore an ideal case for the initial η-mesic nucleus search.

Figure 1 shows the Liu and Haider[3] prediction for the (π^+,p) singles proton spectrum for the ^{16}O case and a 740 MeV/c incident π^+. Only the η-nucleus ground state is shown. The solid line under the η peak is due to the $^{16}O(\pi^+,p)X$ process exclusive of processes involving η's in the final state. The dot dashed line

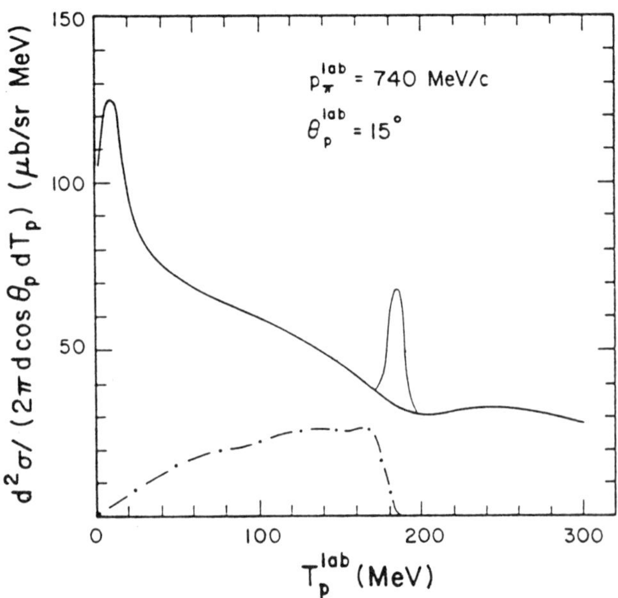

Fig. 1. The proton energy spectrum for the $^{16}O(\pi^+,P)$ case.

represents quasi-free η production. As can be seen in the figure, if the theory is correct, the η-bound state will be easily seen. The signal to background is expected to slowly worsen as the incident pion momentum is increased above 740 MeV/c.

THE EXPERIMENT

The experiment was designed to look for η-mesic nuclei using the Brookhaven National Laboratory AGS hypernuclesr spectrometer system. The system uses two spectrometers. The first is used to determine the incident particle momentum and a second rotatable spectrometer is used to determine the scattered particle momentum. Drift chambers and time-of-flight scintillators are used in both arms for trajectory definition and particle identification respectively. The incident pion beam was provided by the LESB-I beam line whose end magnets served as the beam spectrometer. An incident momentum of 800 MeV/c was chosen in order to allow proton elastic scattering to be used as a calibration reaction and to maximize the missing mass acceptance of the proton spectrometer. The choice of 800 MeV/c rather than the optimum 740 MeV/c was not expected to appreciably change either the η production cross section or the signal to noise ratio.

A system resolution of 1 MeV/c^2 rms was obtained as is shown in Figure 2. The figure shows a CH(p,p) spectra taken at a 20 degree laboratory angle. The three sharp peaks are a result of proton scattering from the ^{12}C ground state, the 4.4 MeV 2^+ state, and the 9.6 MeV 3^- state. The broad bump is due to proton scattering from hydrogen in the scintillator target with a ^{12}C(p,p) kinematical correction applied. The cross section scale for the experiment was confirmed using the p(π^+,p)π^+ reaction at a 525 MeV/c π^+ momentum.

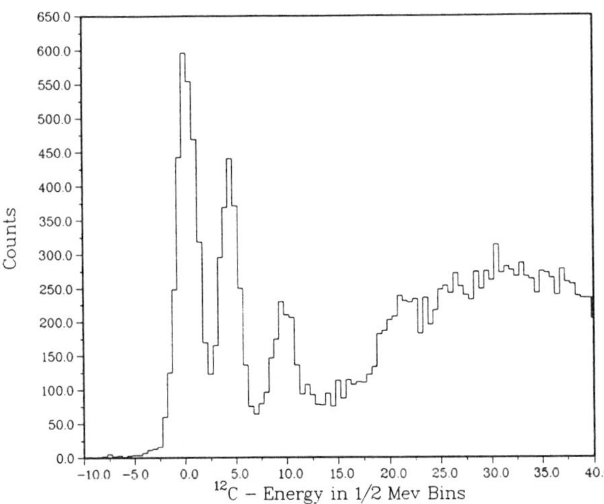

Fig. 2. The CH(p,p) calibration spectrum.

Four targets were chosen--^{7}Li, ^{12}C, ^{16}O and ^{27}Al. For the ^{7}Li case no bound states were expected. The spectrum was expected to show only η-quasielastic events riding on a $^{7}Li(\pi^{+},p)X$ background. The ^{12}C case could produce a η-bound state; however, the binding is expected to be small. Both the ^{16}O and ^{27}Al targets were expected to yield clear bound state η production. The ^{16}O target was considered the optimum case in the Liu and Haider[3] theoretical work and thus received the most beam time.

THE RESULTS

Figure 3 shows the results of the experiment as published in Ref. 1. The data represent a convolution of three different proton spectrometer momentum settings--657, 700 and 740 MeV/c. The data were taken at a 800 MeV/c π^{+} momentum and with the proton

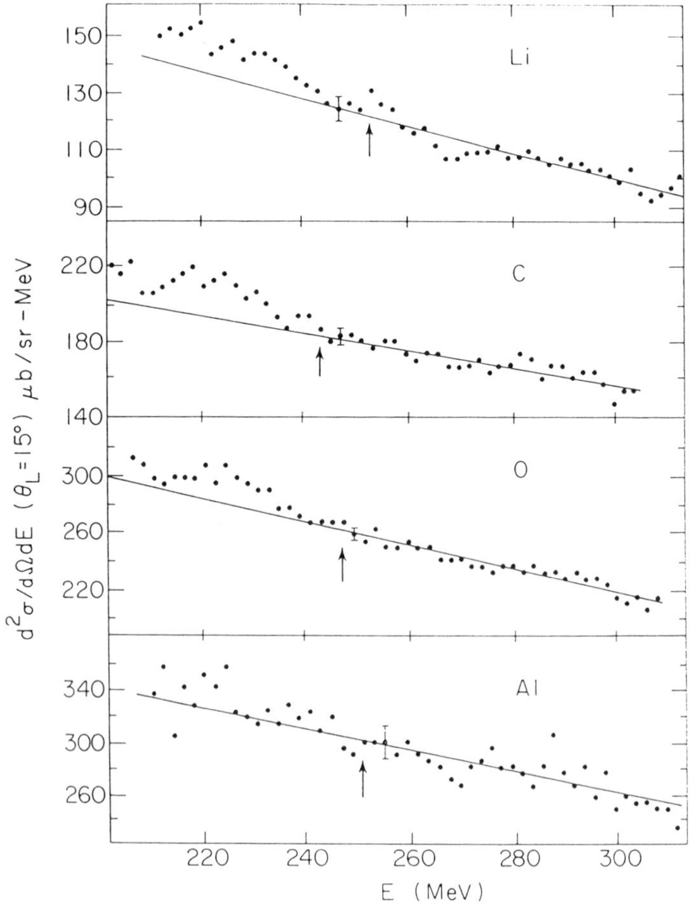

Fig. 3. The (π^{+},p) spectra for the 4 different targets. E represents the kinetic energy of the outgoing proton.

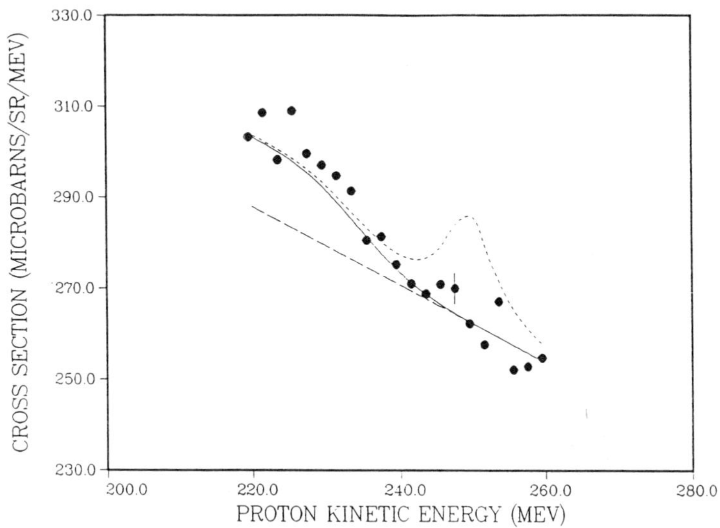

Fig. 4. The $^{16}O(\pi^+,p)$ data with Haider and Liu's[3] predicted η-mesic nuclear bound state drawn in.

spectrometer at 15 degrees in the laboratory. The solid line represents a Maxwellian fit to the data below the η threshold (indicated by the arrows) and then extrapolated into the η quasielastic region. The data clearly show the onset of η production for the ^7Li, ^{12}C and ^{16}O targets and give some indication for quasielastic η production for the ^{27}Al target. Bound states of the η would be expected to appear to the right of the arrow in Fig. 3. The data, therefore, do not give support for η bound states for any of the targets.

The data of Fig. 3 show a general similarity to the $(\bar{p},p)$ data of Garreta et al. The $(\bar{p},p)$ reaction is expected to proceed through the $\bar{p}$ annihilation into mostly pions which then interact with the residual nucleus and eject protons in a manner similar to the (π^+,p) reaction of this experiment. Their data were characterized by smooth Maxwellian curves similar to those drawn in Fig. 3.

Figure 4 shows a detailed blow-up of the $^{16}O(\pi^+,p)$ spectrum around the η threshold. The long dashed curve shows the Maxwellian fit of Figure 3 and the (solid) curve the assumed (π^+,p) background plus η-quasielastic background. The short dashed curve shows the Haider and Liu[3] predicted η bound state peak for the ground state η-mesic nucleus $_\eta^{15}O$. The assumed binding is 2.7 MeV with a 10 MeV Lorentzian width. As can be seen, a bound state peak would clearly have been seen if the width and production cross section were as predicted by Haider and Liu. In fact, a peak with the predicted 10 MeV width would have been seen at the 3 σ level if the production cross section were as low as 1/3 of the predicted size.

The measured (π^+,p) background cross section of Fig. 4 is about a factor of 7 higher than the predicted background of Fig. 1. The choice of a 800 MeV/c pion momentum was expected to increase the background by only about 10%. Divadeenam and Ward[7] have calculated the (π^+,p) spectra for all the targets of this experiment within the framework of an Internuclear Cascade Code and find their predicted cross sections agree rather well with our data and are much larger than the calculated (π^+,p) backgrounds of Ref. 3. The sensitivity of this experiment was seriously compromised by the (π^+,p) background. Nevertheless, this experiment has yielded a significant negative result.

THE CONCLUSION

This search for bound η-nuclear states has yielded a negative result. Either the width of η-nuclear states are much larger than predicted and/or the production cross sections are much smaller than predicted.

ACKNOWLEDGEMENT

Research has been performed under contract DE-AC02-76CH00016 with the United States Department of Energy.

REFERENCES

1. R. E. Chrien, S. Bart, P. Pile, R. Sutter, N. Tsoupas, H. O. Funsten, J. M. Finn, C. Lyndon, V. Punjab, C. F. Perdrisat, B. J. Lieb, T. Kishimoto, L. C. Liu, R. Estep, B. Dropesky, C. E. Stronach, and R. L. Stearns, Phys. Rev. Lett. 60, 2595 (1988).
2. Q. Haider and L. C. Liu, Phys. Lett. 172B, 257 (1986).
3. L. C. Liu and Q. Haider, Phys. Rev. C 34, 1845 (1986).
4. G. L. Li, W. K. Cheng, and T. T. S. Kuo, Phys. Lett. 195B, 515 (1987).
5. R. S. Bhalerao and L. C. Liu, Phys. Rev. Lett. 54, 865 (1985).
6. D. Garetta, P. Birien, G. Bruge, A. Chaumeaux, D. M. Drake, S. Janouin, D. Legrand, M. C. Lemaire, B. Mayer, J. Pain, J. C. Peng, M. Berrada, J. P. Bocquet, E. Monnand, J. Mougey, P. Perrin, E. Aslanides, O. Bing, J. Lichtenstadt, and A. I. Yavin, Phys. Lett. 150B, 95 (1985).
7. M. Divadeenam and T. E. Ward, these proceedings.

THRESHOLD ETA PHOTOPRODUCTION AND THE ROLE OF THE S_{11} RESONANCE[1]

M. Benmerrouche and Nimai C. Mukhopadhyay
Department of Physics
Rensselaer Polytechnic Institute
Troy, New York 12180–3590

ABSTRACT

We study the reaction $\gamma p \to \eta p$ with a view to examining the extraction of the $\gamma N \to S_{11}$ resonant amplitude. A difficulty to be overcome is the role of the background. We investigate it in the effective Lagrangian theory (ELT) and show that there is a large difference between pseudoscalar (PS) and pseudovector (PV) theories, as in the neutral pion case. We also study the importance of the vector meson exchanges in the t–channel and show that ρ and ω contributions can be substantial. Finally, we make contact between $\gamma p \to S_{11}$ amplitude determinations from the extant analyses of the pion and eta photoproduction experiments.

INTRODUCTION

With the advent of quantum chromodynamics as the theory of strong interaction among the constituents of hadrons, the hadron spectroscopy, somewhat dormant in the resonance region for a decade, has got a shot in the arm, soon to be bolstered by the new experimental facilities now under construction, such as the Continuous Electron Beam Accelerator Facility (CEBAF). In this paper, we shall concentrate on the reaction $\gamma p \to \eta p$, for CM energies from threshold (1.487 GeV) up to 1.6 GeV. In contrast to photoproduction of pi–zero meson at threshold, where the role of $\Delta(1232)$ is relatively small[1], we expect a rather large contribution of the $S_{11}(1535)$ resonance even near eta photoproduction threshold. Our main interest in this paper is somewhat technical, to explore the background contributions in the ELT. To our knowledge, the background has been so far determined from smooth fits to various resonance contributions, without any serious attempt to explore its physics content from an underlying theory. One of our eventual goals is to construct an impulse operator for the nuclear photoproduction of eta meson, for which there is already some theoretical interest[2], parallel in spirit to our work of photoproduction of pions in the delta region[3].

First, let us make some remarks on the $S_{11}(1535)$ resonance. In the SU(6) quark model, it belongs to the $\underline{70}\,(1^-)$ representation with the SU(3) content $\underline{8}$.

[1]Supported by the U.S. Department of Energy Contract #DE–FG02–88ER40448

In the $SU(6)_W$ theory, originally due to Melosh[4], the helicity couplings for the excitation of nucleons to the 70 (1^-) family of resonances can be expressed in terms of the three parameters A,B,C. Thus, the $N \rightarrow S_{11}(1535)$ helicity amplitudes for p and n targets are, in pure SU(6) limit,

$$A^p_{1/2} = \frac{1}{6}(A + B + C), \text{ and } A^n_{1/2} = -\frac{1}{6}(A + \frac{B}{3} - \frac{C}{3}). \qquad (1)$$

However, the physical mixing of the S_{11} resonance in terms of two pure SU(6) states is important[5] and must be taken into account in the actual analysis. Thus, the photoproduction amplitude provides a crucial test of the SU(6) mixings, which, in turn, is connected to the fundamentals of the interactions among the hadronic constituents.

It is useful to recall that the analyses of single–pion photoproduction up to a cm energy of 2.5 GeV have yielded what appears to be a fairly consistent set of values for the helicity amplitudes for the proton target. Thus, these are

$$A^p_{1/2} = 77 \pm 21,\ 83 \pm 7,\ 65 \pm 16\ (53 \pm 15),$$

$$A^n_{1/2} = -35 \pm 16,\ -75 \pm 9,\ -112 \pm 34\ (-98 \pm 26)$$

in units of 10^{-3} $GeV^{-1/2}$, from Nagoya, Tokyo and Glasgow groups respectively[6], the last group reporting two set of values. These should be compared with the quark model values[7] ranging from 68 to 156 in the first case and –83 to –119 in the second. The Isgur–Konyuk theoretical estimate[7] for proton is about a factor of two too large compared to the experimental estimates. Given the arbitrary nature of the background contribution in the extraction[6] of resonant amplitudes, we are keen on examining this aspect more theoretically. Considering the fact that single pion decay channel is less selective on the nature of the resonance, we hope that the more selective η meson decay channel would put a better constraint on the radiative amplitude. Given the paucity of data base on the η channel, this hope, of course, has to be tested in the future.

For the $\gamma p \rightarrow \eta p$ process, we shall compare our conclusion with the isobar analysis by Hicks et al. of $\gamma p \rightarrow p\eta$ of all available data below 2.24 GeV collected in the sixties and early seventies.

EFFECTIVE LAGRANGIAN THEORY

In the effective Lagrangian theory, the nucleon Born terms for the photoproduction of eta meson involves s–channel and u–channel nucleon exchange. Since η is a neutral particle, there is no t–channel η exchange and no sea–gull contribution for the PV coupling. Thus, the near threshold value of the E_{0+} multipole is given for the proton target, in terms of the nucleon Born term contributions, as follows:

$$E_{0+}^{PV} = -\chi_1 \chi_2 (1 - \tfrac{1}{2} \beta k_p), \quad (3a)$$

$$E_{0+}^{PS} = -\chi_1 \chi_2 (1 + k_p), \quad (3b)$$

for PV and PS theories respectively. Here

$$\chi_1 = \frac{e f_\eta}{4\pi \; m_{\pi+}}, \; \chi_2 = \frac{\beta}{(1+\beta)^{3/2}}, \quad (4)$$

where $\beta = m_\eta/M \simeq 0.6$, $f_\eta = (m_{\pi+}/2M)g_\eta$; $m_{\pi+}$, m_η, M are pion, eta and nucleon masses, g_η is the ηNN coupling constant, k_p is the anomalous magnetic moment of the proton. Expressions (3a–b) demonstrate the "equivalence theorem" explicity in that PV and PS theories give identical result in the absence of the anomalous magnetic moment interaction. In reality, $\beta\, k_p > 1$, hence $|E_{0+}^{PS}| > |E_{0+}^{PV}|$ for a given value of χ_1, χ_2. This fact is illustrated in Fig. 1.

In practice, the coupling constant g_η is highly uncertain. Information on it comes from sources such as NN scattering, hyperon decay, ratio of η to π^0 cross section and so on. A conservative range of values for g_η is: $0.6 \leq g^2_{\eta NN}/4\pi \leq 6.3$. A much better estimate of this coupling constant is obviously needed.

VECTOR MESON EXCHANGES

Even for the neutral pion photoproduction, the role of t–channel vector meson exchange is unmistakable. It is, therefore, interesting to look at the t–channel vector meson contributions to the eta photoproduction. The strong and electromagnetic effective Lagrangians are:

$$L_I^S = -g_V \, \overline{\psi} \, \gamma_\mu \, \psi \, V^\mu + g_T \, \overline{\psi} \, \sigma_{\mu\nu} \, \psi \, V^{\mu\nu},$$

$$L_I^{em} = \frac{e \lambda_V}{4m_{\pi+}} \, \epsilon_{\mu\nu\lambda\sigma} \, F^{\mu\nu} V^{\lambda\sigma} \eta, \quad (5)$$

where the notations have obvious significance. We consider ρ for the isovector

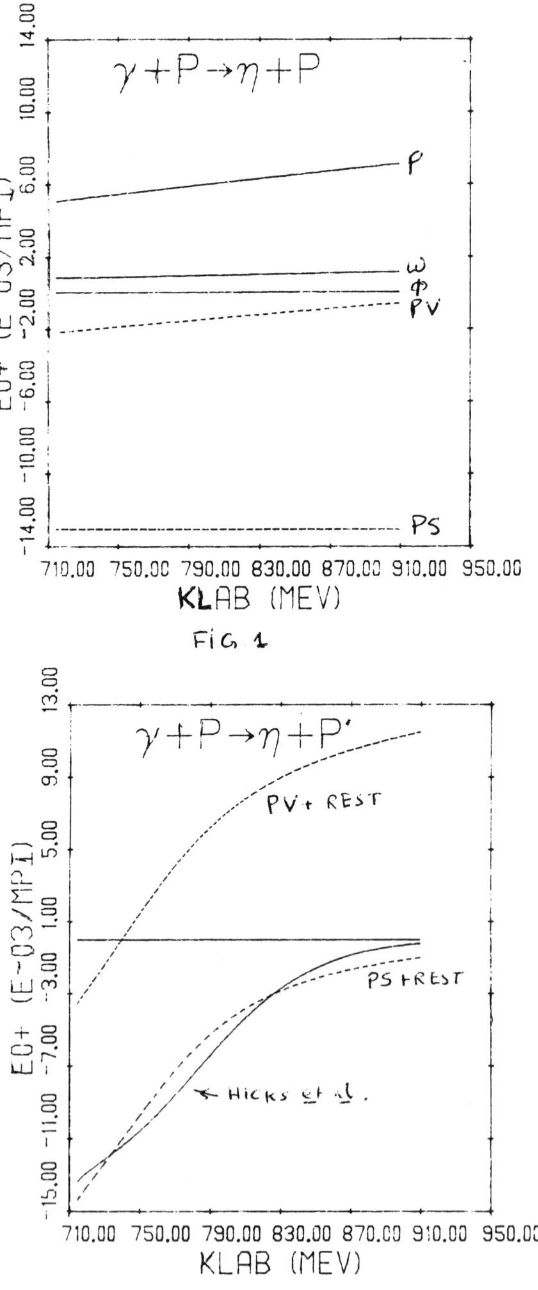
γ+P→η+P
ρ
ω
φ
PV
PS
EO+ (E-03/MPI)
14.00
10.00
6.00
2.00
-2.00
-6.00
-10.00
-14.00
710.00 750.00 790.00 830.00 870.00 910.00 950.00
KLAB (MEV)
γ+P→η+P'
PV+ REST
PS+REST
← Hicks et al.
13.00
9.00
5.00
1.00
-3.00
-7.00
-11.00
-15.00

FIG. 1

FIG. 2

and ω, ϕ for the isoscalar exchanges. From the $\eta\gamma$ decay[10] of ρ^{o}, ω and ϕ mesons, the couplings λ_{ρ} and λ_{ω} cannot be uniquely determined and two sets of values are possible. These are:

$$\lambda_{\rho} \approx 0.24,\ \lambda_{\omega} \approx 0.055, \quad \text{(6a)}$$

and

$$\lambda_{\rho} \approx 0.31 \text{ and } \lambda_{\omega} \approx 0.174, \quad \text{(6b)}$$

the former set being close to the quark model estimates. λ_{ϕ} is $\approx$ 0.096. For g_V and g_T, a variety of values are possible.

Our conclusion concerning vector mesons is that the contribution from ρ and ω is quite important, while ϕ contribution is negligible (Fig. 1).

RESONANT S_{11} CONTRIBUTION

The s– and u– channel contributions of the S_{11} excitation are given by the effective Lagrangian

$$L_I = g_{RN\eta}\, \overline{\psi}\, \gamma_{\mu}\, R\partial^{\mu}\eta + \frac{e}{4M}\, \overline{R}\, (G^S + G^V\tau_3)\gamma_5\, \sigma_{\mu\nu}\psi F^{\mu\nu} + \text{H.c.} \tag{7}$$

Here $g_{RN\eta}$, G^S and G^V are parameters of the Theory to be determined. We have investigated the role of the u–channel and found it to be negligible for reasonable values of couplings in (7). We have fixed parameters in (7) using the helicity amplitudes for the resonance excitation combined from the single pion experiments. Resultant E_{0+} amplitude has a real part as displayed in Fig. 2. Here the $g_{\eta NN}$ coupling is maximum allowed in the literature, as in Fig. 1. Surprisingly, the PS background plus the resonance describes the Hick et al. amplitude well. The PV background throws this agreement off.

To check the consistency between the available single pion and eta data, we have compared the $A_{1/2}$ resonant amplitude from the analysis of Hicks et al. We obtain this to be 80.3 x 10^{-3} $\text{GeV}^{-1/2}$, compared with (73±14)x10^{-3} $\text{GeV}^{-1/2}$, from the Particle Data Group, extracted from the single–pion data. Barring a happy coincidence, this is a remarkable agreement.

CONCLUSIONS

The effective Lagrangian Theory seems to give a good account of the E_{0+} amplitude extracted by Hicks et al., given the uncertainties of the parameters, if we use the known $\gamma N \rightarrow S_{11}$ resonant amplitude from the single–pion data. Surprisingly, the pseudoscalar Theory for the nuclear Born terms, along with vector meson and the s–channel resonance contributions, reproduces the Hicks et al. amplitude. the vector meson contributions are surprisingly large. The resonant amplitudes from the single pion and eta channels mutually agree.

ACKNOWLEDGEMENTS

We thank R. Davidson for valuable discussions.

[1]R. Davidson and N.C. Mukhopadhyay, Phys. Rev. Lett. 60, 748 (1988).

[2]F. Tabakin et al., private comm. (1987).

[3]R. Wittman and N.C. Mukhopadhyay, Phys. Rev. Lett. 57, 1113 (1986).

[4]H.J. Melosh, Phys. Rev. D9, 1095 (1974).

[5]F. Foster and G. Hughes, Rep. Progr. Phys. 46, 1445 (1983).

[6]R. Crawford and W.T. Morton, Nucl. Phys. B 211, 1 (1983) and refs. therein.

[7]See Feynman et al., Kubota–Ohta, Barbour–Ponting, Koniuk–Isgur and Forsyth, cited in Ref. 6.

[8]H.R. Hicks et al., Phys. Rev. D7, 2614 (1973).

[9]M.G. Olsson, Nucl. Phys. B78, 55 (1974).

[10]D.E. Andrews et al., Phys. Rev. Lett. 38, 198 (1979).

FLAVOR-OCTET DIBARYONS IN THE QUARK MODEL

Makoto Oka

Department of Physics, University of Pennsylvania
Philadelphia, PA 19104, U.S.A.

ABSTRACT

The possibile existence of the flavor $SU(3)$ octet dibaryon resonances is studied in the quark model.

INTRODUCTION

The role of quark dynamics in hadron-hadron interactions has been studied by several different approaches. Among them the quark cluster model is successful in providing a mechanism of the nucleon-nucleon short range repulsion and a realsitic baryon-baryon interaction potential.[1] There the color-magnetic gluon exchange interaction and the quark antisymmetrization play essential roles. The color-magnetic interaction (CMI) with the quark exchange between nucleons is responsible for the short-range NN repulsion. However, contribution of CMI to the baryon-baryon interaction is not universally repulsive for all the two-baryon systems. Some two-baryon interactions are attractive at a short distance due to CMI.[2] The quark cluster model predicts, for instance, an S-wave bound state in $\Delta\Delta$ with spin $J = 3$ and isospin $I = 0$. Because of the absence of short-range repulsion, such a bound state is a compact "6-quark" like object, where the baryons, Δ, lose their identity. Another possible dibaryon state has been proposed by Jaffe in the $SU(3)_f$ singlet six-quark system, where CMI is strongly attractive.[3] The state couples the two-baryon systems, $\Lambda\Lambda - N\Xi - \Sigma\Sigma$ with $I = 0$ and $J = 0$. CMI is also attractive in the $SU(3)_f$ octet six-quark systems. Here we study the possible existence of the $SU(3)_f$ octet dibaryon resonances in the quark model.[4] The importance of two-baryon channel couplings and quark antisymmetrization is stressed for low-lying bound/resonance states. We argue that, in general, $J = 2^+$ resonances are more likely than $J = 1^+$ ones.

FLAVOR-OCTET 6-QUARK STATES

First we study qualitative features of the 6-quark system using a general quark model with the color-magnetic interaction (CMI). Suppose that all the six (valence) quarks are in the ground state in a bag or a potential well. Then the orbital wave function is totally symmetric and therefore the color-spin symmetry and the flavor symmetry are conjugate with each other in order to make a totally antisymmetric state. For such a state, the expectation value of CMI

$\mathcal{S}$	I	BB
-3	$1/2$	$(1/\sqrt{5})\left(\sqrt{2}\lvert N\Omega\rangle - \lvert\Lambda\Xi^*\rangle - \lvert\Sigma\Xi^*\rangle + \lvert\Sigma^*\Xi\rangle\right)$
-2	0	$(1/\sqrt{5})\left(\sqrt{2}\lvert N\Xi^*\rangle - \sqrt{3}\lvert\Sigma\Sigma^*\rangle\right)$
-2	1	$(1/\sqrt{15})\left(\sqrt{2}\lvert N\Xi^*\rangle - \sqrt{3}\lvert\Lambda\Sigma^*\rangle - \sqrt{2}\lvert\Sigma\Sigma^*\rangle + \sqrt{8}\lvert\Delta\Xi\rangle\right)$
-1	$1/2$	$(1/\sqrt{5})\left(\lvert N\Sigma^*\rangle + \sqrt{4}\lvert\Delta\Sigma\rangle\right)$

Table 2 Flavor-octet dibaryon states with $J = 2^+$.

is proportional to $\Gamma_{cm} \equiv -\sum_{ij}(\lambda_i \cdot \lambda_j)(\sigma_i \cdot \sigma_j)$, which are determined by the color-spin $SU(6)$ symmetry and the total spin of the state:

$$\langle\Gamma_{cm}\rangle^J_{[f]} = 48 - \frac{1}{2}C_6([f]) + \frac{4}{3}J(J+1) \tag{1}$$

where $C_6([f])$ is the $SU(6)$ Casimir operator and J is the total spin. It is easy to see that CMI is most attractive in the flavor singlet, [222] symmetric, state. The corresponding spin-color symmetry is [33], where only $J = 0$ is allowed for the color singlet state, and $\langle\Gamma_{cm}\rangle^{J=0}_{[33]} = -24$. This state is known as the H dihyperon with strangeness $\mathcal{S} = -2$ and isospin 0.[3] In the two-baryon (BB) configuration, H corresponds to a combination of $\Lambda\Lambda - N\Xi - \Sigma\Sigma$:

$$|H\rangle = 1/\sqrt{8}\left(|\Lambda\Lambda\rangle + \sqrt{4}|N\Xi\rangle - \sqrt{3}|\Sigma\Sigma\rangle\right). \tag{2}$$

Several different calculations[5,6] suggest either a bound state or a low-lying resonance in the $\Lambda - \Lambda$ system and experimental searches are going on.[7]

The next low lying state is labeled by the flavor symmetry [321], the octet. Spins $J = 1$ and $J = 2$ are allowed. The flavor octet consists of ($\mathcal{S} = -1$, $I = 1/2$); ($\mathcal{S} = -2$, $I = 0$); ($\mathcal{S} = -2$, $I = 1$) and ($\mathcal{S} = -3$, $I = 1/2$). Tables 1 and 2 summarize the two-baryon content of these flavor-octet states with $J = 1^+$ and $J = 2^+$, respectively. Note that the $J = 2$ states are given only by combinations of an octet and a decuplet baryon, while the $J = 1$ states are also made of two octet baryons. The eigenvalue of the color-magnetic operator Γ_{cm} is $-12 + (4/3)J(J+1) = -28/3$ for $J = 1$ and -4 for $J = 2$.

$\mathcal{S}$	I	BB
-3	$1/2$	$(1/3)\left(\sqrt{2}\lvert\Lambda\Xi\rangle - \sqrt{2}\lvert\Sigma\Xi\rangle\right.$ $\left.-\sqrt{2}\lvert N\Omega\rangle + \lvert\Lambda\Xi^*\rangle + \lvert\Sigma\Xi^*\rangle + \lvert\Sigma^*\Xi\rangle\right)$
-2	0	$(1/3)\left(\sqrt{4}\lvert N\Xi\rangle + \sqrt{2}\lvert N\Xi^*\rangle - \sqrt{3}\lvert\Sigma\Sigma^*\rangle\right)$
-2	1	$(1/\sqrt{27})\left(-\sqrt{4}\lvert N\Xi\rangle + \sqrt{8}\lvert\Sigma\Sigma\rangle\right.$ $\left.+\sqrt{2}\lvert N\Xi^*\rangle - \sqrt{3}\lvert\Lambda\Sigma^*\rangle - \sqrt{2}\lvert\Sigma\Sigma^*\rangle - \sqrt{8}\lvert\Delta\Xi\rangle\right)$
-1	$1/2$	$(1/3)\left(-\sqrt{2}\lvert N\Lambda\rangle + \sqrt{2}\lvert N\Sigma\rangle + \lvert N\Sigma^*\rangle - \sqrt{4}\lvert\Delta\Sigma\rangle\right)$

Table 1 Flavor-octet dibaryon states with $J = 1^+$.

The 2-baryon systems which participate in the flavor-octet states may exhibit a bound or a resonance state. For instance, $N\Omega$ with $S = -3$, $I = 1/2$, and $J = 2$ has been discussed in this context.[8] Because the flavor-octet states are given as particular combinations (given in Tables 1 and 2) of the BB systems, the BB channel coupling should be substantial in the resonances. The threshold energies of those channels, however, are different due to the flavor symmetry breaking and also the octet-decuplet mass difference. If their thresholds are distributed in a wide energy range, a bound or a low-lying resonance state around the threshold of the lowest BB state is unlikely. Thus we expect that $J = 1^+$ dibaryon resonances are less likely than $J = 2^+$ ones because of the large octet-decuplet splitting.

QUARK CLUSTER MODEL

We apply the quark cluster model[1] to a quantitative study of the dibaryon spectrum, which takes into account the full quark antisymmetrization, coupling of various two-baryon channels and the $SU(3)_f$ breaking effect.[5] Baryon-baryon scattering problem is solved for the flavor-octet dibaryon channels by using the resonating group method.[4] In figures, the calculated S-wave phase shift, δ, and the elasticity, η, for the lowest lying BB channel are shown for each ($\mathcal{S}$, I, $J = 2^+$) state. The full and the dashed (or dotted) curves are obtained by the full and a partial channel-coupling calculations, respectively. One sees, in general, the importance of channel couplings within the BB configurations which appear in the flavor-octet state. Coupling of the decuplet-decuplet channels, which do not participate in the flavor-octet states, is not significant. There is no bound state obtained in any of the calculated channels. The $J = 2^+$ channels show rich structure, a quasibound state for ($\mathcal{S} = -2$, $I = 0$, $J = 2^+$) and ($\mathcal{S} = -3$, $I = 1/2$, $J = 2^+$), and low energy resonances ($E < 100$ MeV) for ($\mathcal{S} = -1$, $I = 1/2$, $J = 2^+$) and ($\mathcal{S} = -2$, $I = 1$, $J = 2^+$). At the resonance energies, the scattering wave function shows a superposition of the BB channels similar to the flavor-octet combination given in Table 2.

By contrast, the $J = 1^+$ channels do not show low energy resonances, although effect of the channel coupling is sometimes significant. The difference between the $J = 1^+$ and 2^+ states comes from the fact that the $J = 1^+$ flavor-octet states are distributed over many BB channels, the octet-octet as well as the octet-decuplet. They have a wide range of the threshold energies due to the octet-decuplet mass difference and the flavor $SU(3)$ breaking.

CONCLUSION

The flavor-octet 6–quark systems seem to have rich structure. We have studied the roles of the color-magnetic interaction (CMI) and the quark exchange processes by using the quark cluster model approach. We showed that all the $J = 2^+$ channels have a resonance and/or a quasi-bound state. Couplings among the octet-decuplet two-baryon channels are important because

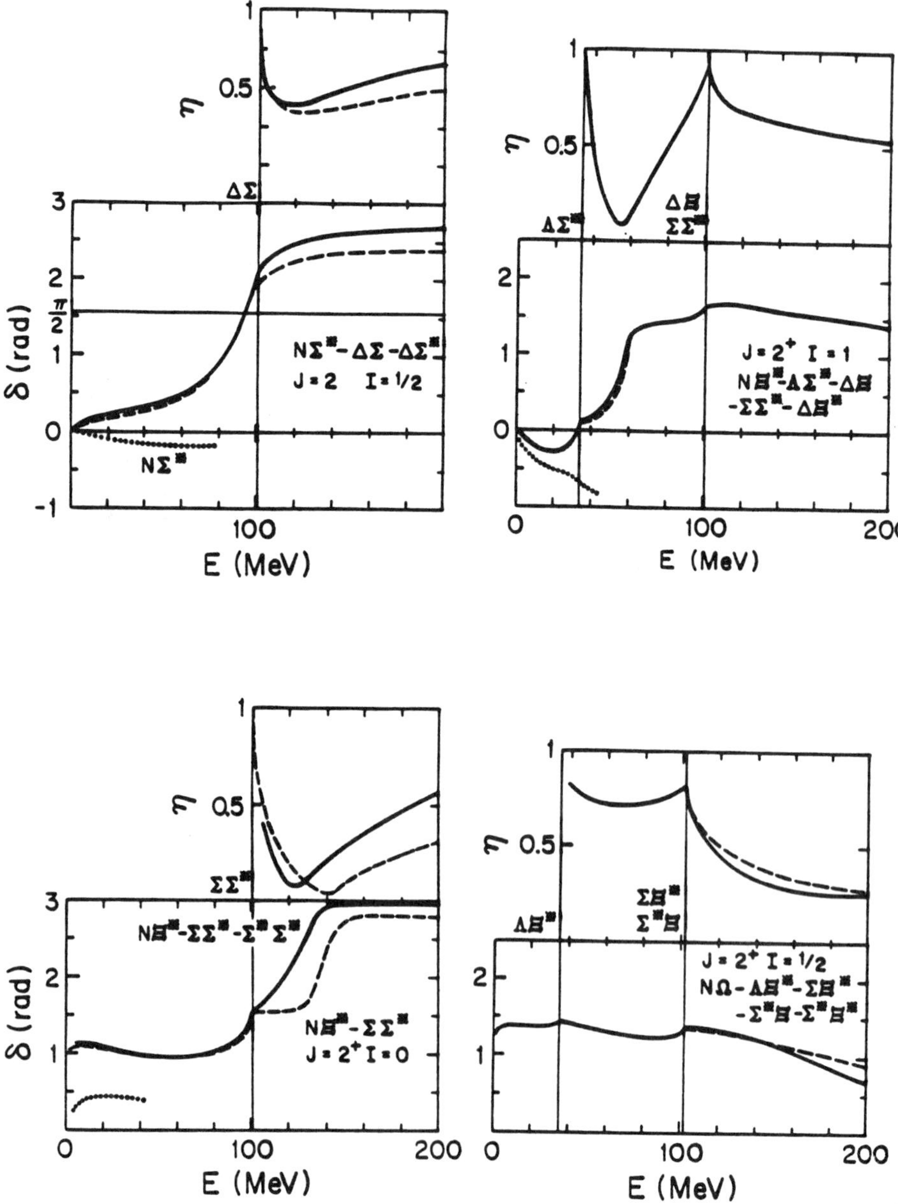
η
0.5
1
ΔΣ
3
2
π/2
1
0
-1
δ (rad)
NΣ* - ΔΣ - ΔΣ*
J = 2 I = 1/2
NΣ*
100
E (MeV)
ΔΣ*
ΔΞ
ΣΣ*
J = 2+ I = 1
NΞ* - ΔΣ* - ΔΞ
- ΣΣ* - ΔΞ*
0
100
200
E (MeV)
ΣΣ*
NΞ* - ΣΣ* - Σ* Σ*
NΞ* - ΣΣ*
J = 2+ I = 0
ΔΞ*
ΣΞ*
Σ*Ξ
J = 2+ I = 1/2
NΩ - ΔΞ* - ΣΞ*
- Σ*Ξ - Σ*Ξ*

the flavor-octet combinations are favored by CMI. It should be stressed that CMI plays a major role in the baryon-baryon interactions. The magnitude of CMI is determined by the $N - \Delta$ mass difference and therefore is model independent. The quark exchange effect, *i.e.*, the antisymmetrization among the six quarks, is also crucial. In fact, without the antisymmetrization, CMI does not induce any interbaryonic interaction. For example, the $N - \Omega$ system does not have a quark exchange contribution unless it is coupled with other channels like $\Lambda - \Xi^*$, $\Sigma - \Xi^*$ or $\Sigma^* - \Xi$. The quark cluster model approach shows no $N - \Omega$ interaction in the single channel calculation. Coupling of other channels is crucial.

No $J = 1^+$ resonance is predicted up to 200 MeV above the lowest two-baryon threshould. There the flavor-octet state is distributed to many two-baryon channels, where thresholds are split by the octet-decuplet mass difference as well as the flavor $SU(3)$ breaking. Resonances do not appear in low energy region.

This work was supported in part by a grant from the National Science Foundation.

REFERENCES

1. M. Oka and K. Yazaki, in *Quarks and Nuclei*, ed. by W. Weise (World Scientific, 1985), and references therein.
2. M. Oka and K. Yazaki, Phys. Lett. **B 90**, 41 (1980); Prog. Theor. Phys. **66**, 556 (1981); *ibid.* **66**, 572 (1981); M. Cvetic, B. Golli, N. Mankoc-Borstnik and M. Rosina, Phys. Lett. **B 93**, 489 (1980); *ibid.* **99**, 486 (1981).
3. R.L. Jaffe, Phys. Rev. Lett. **38**, 195 (1977)
4. M. Oka, Phys. Rev. D in press.
5. M. Oka, K. Shimizu and K. Yazaki, Phys. Lett. **B 130**, 365 (1983); Nucl. Phys. **A464**, 700 (1987); K. Yazaki, Nucl. Phys. **A479**, 217c (1988).
6. P.B. Mackenzie and H.B. Thacker, Phys. Rev. Lett. **55**, 2539 (1985); Y. Iwasaki, T. Yoshié, and Y. Tsuboi, Phys. Rev. Lett. **60**, 1371 (1988).
7. P.D. Barnes, in *Proceedings of the 2nd Conference on the Intersection between Particle and Nuclear Physics*, (Lake Louise, 1986), ed. by D.F. Geesman, AIP Conference Proceedings 150, (AIP, New York 1986), and references therein.
8. T. Goldman, *et al.*, Phys. Rev. Lett. **59**, 627 (1987).

Testing the Standard Model with Rare and Forbidden Decays of the η and Other Light Mesons

B.M.K. Nefkens
Department of Physics, University of California, Los Angeles, CA 90024

Abstract

The traditional tests of the Standard Model involving charged and neutral weak currents should be augmented with sensitive searches for new interactions and new particles such as are possible with the rare and forbidden decays of mesons, particularly the eta meson. Some examples which will be discussed are $\eta \to e^+e^-$ (leptoquark), $\eta \to \pi^\circ H$ (light Higgs), $\eta \to \mu e$ (lepton family violation), $\eta \to \pi^+\pi^-$ (CP violation), and $\eta \to 3\gamma$ (C violation).

The Standard Model[1,2,] (SM) has been *the* standard by which new theories and fundamental particle physics are evaluated. The SM enjoys all the successes of the Electroweak Theory and the Quark Model. Tests involving weak neutral currents and the W and Z masses are summarized in Table I; they cover ten decades in q^2 and range from atomic parity tests to neutrino interactions, and from polarized lepton-nucleus scattering asymmetries to the masses of the intermediate vector boson.[3] We have included four tests of the forward-backward asymmetry and of the total cross sections for $e^+e^- \to l^+l^-$ interactions. Table II gives a smorgasbord of tests based on the universality of the charged weak current in which often the predictions[4] are more precise than the experiments.[5]

There are few physicists who believe that the SM is our ultimate particle physics theory for reasons of principle and practice. The SM in its present three-family format has no room for unification of the strong, electromagnetic, and weak interactions with gravity. There is a glaring inconsistency between quarks and leptons since lepton family number is conserved while the quark family number is not. Finally, CP violation is not understood at a fundamental level. Among the practical problems with the SM is the fact that it requires too many input parameters and no argument is provided for the number of families nor any reason for the mass scale of the 16 fundamental constituents.

To test the SM there are basically two types of experiments:

1. Searches for new particles in order to a) discover the predicted new particles, the t-quark, and the Higgs bosons or b) find a new particle that does not fit into the present model such as a monopole, sparticle, leptoquark, and so forth.
2. Searches for new interactions, this may be accomplished via measurements of forbidden and rare decay modes. For instance, one can search for a new lepton-family-violating interaction in the "forbidden decay"

$\eta \rightarrow \mu e$. One may also search for a new, non-K-M, CP-violating interaction via an enhancement of the rare decay $\eta \rightarrow \pi^+\pi^-$ which in the SM can have a branching ratio (BR)$< 10^{-16}$ when there is a strong CP violation that is governed by the θ parameter.

Various conservation laws and new phenomena that can be used to test the SM are listed in Table III. Many experiments at different laboratories are already under way in an attempt to provide new limits on the validity of the SM.

In this paper we consider some possibilities for testing the SM offered by the rare and forbidden decays of the eta and other light mesons. These decays have been neglected so far, presumably for lack of a suitable source of tagged etas; we shall return to this aspect later. The advantage of the eta meson is the fact that it has all the quantum numbers of the vacuum

$$\mathrm{J} = \mathrm{S} = \mathrm{Q} = \mathrm{I} = \mathrm{B} = \mathrm{C} = 0$$

except for the intrinsic parity. Thus, certain transitions can be investigated that are not accessible in decays of the K-meson or muon. The disadvantage of the eta is its short half-life which can make it less competitive for certain interactions by a factor $\sim 10^{10}$ compared to K-decay. Table IV shows various interesting decay modes of the η-meson, the physics that can be probed or tested, the expected branching ratio in the SM, and the current experimental limit. Below, we elaborate on some decay modes.[6]

A1. $\eta \rightarrow e^+ + e^-$

This decay is suppressed by α^2 compared to $\eta \rightarrow 2\gamma$ and by a helicity factor. The standard estimate consistent with the SM is about 1.3 times the unitary limit or BR($\eta \rightarrow e^+e^-$) $\simeq 6 \times 10^{-9}$. This rare decay mode may be enhanced if leptoquarks or certain other new particles exist. Because the mass of the η is four times the π° mass, the η is a much better candidate, when searching for new particles, than the decay $\pi^\circ \rightarrow e^+e^-$ (for which the BR $\simeq 10^{-7}$). The experimental upper limit is BR($\eta \rightarrow e^+e^-$) $< 3 \times 10^{-4}$.

A2. $\eta \rightarrow \mu^+ + \mu^-$

This decay is similar to the previous one, but the branching ratio is larger because the muon is much heavier than the electron. The world supply of this decay mode is only 27±8 events corresponding to a BR = $(6.5\pm2.1)\times10^{-6}$.

A3.

$$\eta \rightarrow \pi^\circ + H \rightarrow \pi^\circ + \mu^+ + \mu^-$$
$$\eta' \rightarrow \pi^\circ + H \rightarrow \pi^\circ + \mu^+ + \mu^-$$
$$\eta' \rightarrow \eta + H \rightarrow \eta + \mu^+ + \mu^-$$

One of the embarrassing failures of the SM is its inability to predict the mass of the Higgs particle, H. The coupling of the Higgs to fermions is proportional to the fermion mass which favors the decay into muons over electrons. There is theoretical prejudice for a Higgs doublet with a mass in excess of several GeV, but a light Higgs is not excluded. The decay of the η and η' offer excellent opportunities to search for a possible light Higgs of mass$>$ 218 MeV. Estimates for BR($\eta \rightarrow \pi$H) yield O(10^{-8}). More attractive is $\eta' \rightarrow \eta$H for which the BR is O(10^{-4}).

A4. $\eta \rightarrow \pi^\circ + a \rightarrow \pi^\circ + e^+ + e^-$

The axion, a, is a novel pseudoscalar particle that has been postulated in connection with the strong CP puzzle. Estimates are that BR($\eta \rightarrow \pi^\circ a \rightarrow \pi^\circ e^+ e^-) \simeq 10^{-10}$.

B1. $\eta \rightarrow \mu + e$

Another glaring shortcoming of the SM is a reason for conservation of the lepton family number. The three lepton families are each characterized by a conserved lepton family quantum number (L_l). The five known quarks all can decay outside their family so there is no quark family quantum number. Therefore, there is a great theoretical prejudice for the lepton family number to be only approximately conserved. The reaction $\eta \rightarrow \mu + e$ is unique among tests of lepton family number conservation in that it concerns a $\Delta I = \Delta S = 0$ transition in contrast to $\Delta I = 1$ transitions such as $\pi^\circ \rightarrow \mu + e$, BR $< 7 \times 10^{-8}$, or $\pi^+ \rightarrow \mu^+ + \nu_e$, BR$< 8 \times 10^{-3}$. The $\Delta S = 1$ transition studies, such as in $K_L \rightarrow \mu + e$, are very sensitive due to the fact that the half-life of K is $\sim$ 10 orders of magnitude larger than the half-life of the eta. There is no known experimental limit for $\eta \rightarrow \mu + e$. It would be very interesting to look for this at a level of 10^{-10} where it could put limits on grand unification theories, supersymmetry, and superstrings.

B2. $\eta \rightarrow \pi^+ + \pi^-$

The decay $\eta \rightarrow \pi^+ + \pi^-$ is forbidden by P and CP invariance. It is allowed in the SM and could proceed for instance, through $K^\circ - \eta$ mixing but at a discouragingly small rate, BR$\leq 10^{-16}$. If it were observed much more frequently it would imply a new $\Delta S = \Delta I = 0$ CP-violating interaction and a breakdown of the SM. As there are very few tests of CP outside the K° system, it is of interest to push a search for $\eta \rightarrow \pi\pi$ to the limit.

B3. Test of C-invariance in η decay

The decay $\eta \rightarrow 3\gamma$ is a test of C-invariance of the electromagnetic interaction of hadrons. Again, the SM allows a very small branching ratio whose estimates are of order 10^{-13}. Other tests of C-invariance include the decays $\eta \rightarrow \pi^\circ l^+ l^-$, which are forbidden in first order only, and a charge asymmetry in the decays $\eta \rightarrow \pi^+\pi^-\pi^\circ$ and $\eta \rightarrow \pi^+\pi^-\gamma$.

C. Exotica

There are very few opportunities for testing the strength of so-called second class charged weak currents which involve a $\Delta I = 1$ transition. The decay mode $\eta \rightarrow \pi + e + \nu$ does provide such an opportunity; it requires a sensitivity $> 10^{-10}$ which is the order of magnitude of the first class currents. The decay mode $\eta \rightarrow \pi + \mu + e$ is forbidden by lepton family number conservation also.

The decay mode $\eta \rightarrow \pi + \pi + e + \nu$ is suppressed because it involves two pseudoscalar bosons in the final state, but can take place through the Wess-Zumino term in the effective Lagrangian. The decay $\eta \rightarrow K + e + \nu_e$ is an ordinary weak interaction and is thus expected at the level 10^{-10}, any large deviation would spell trouble for the SM. The rate for the "massless" decay $\eta \rightarrow \gamma + \nu + \bar{\nu}$ is sensitive to the number of neutrino families. The estimated rate is very small which makes an experimental search extremely difficult.

D. Electron-muon Universality

The SM implies that the electrons and muons have the same coupling strength to quarks. This can be tested via the equalities:

$$\Gamma(\rho \rightarrow \mu^+ + \mu^-) = \Gamma'(\rho \rightarrow e^+ + e^-)$$

$$\Gamma(\omega \rightarrow \mu^+ + \mu^-) = \Gamma'(\omega \rightarrow e^+ + e^-)$$

$$\Gamma(\phi \rightarrow \mu^+ + \mu^-) = \Gamma'(\phi \rightarrow e^+ + e^-).$$

The prime on the Γ indicates a small adjustment for the difference in phase space of $\mu\mu$ compared to ee. The experimental situation is given below:

$$BR(\rho \rightarrow \mu^+ + \mu^-) = 0.0067 \pm 0.0012$$

$$BR(\rho \rightarrow e^+ + e^-) = 0.0045 \pm 0.0002$$

$$BR(\omega \rightarrow e^+ + e^-) = 0.0067 \pm 0.0004$$

$$BR(\phi \rightarrow \mu^+ + \mu^-) = 0.025 \pm 0.003$$

$$BR(\phi \rightarrow e^+ + e^-) = 0.031 \pm 0.001.$$

Until recently there were no suitable sources of tagged eta mesons available. The obvious production reaction is $\pi^- + p \rightarrow \eta + n$ which has a maximum cross section of about 200 μb/sr for $p_\pi \simeq 776$ MeV/c. This falls just outside the available energies at LAMPF, while the AGS and KEK accelerators do not have the necessary quality pion beam available. A breakthrough occurred recently at Saclay. Measurements of $d + p \rightarrow ^3\text{He}+\eta$ and $p + d \rightarrow ^3\text{He}+\eta$ have found a cm cross section of about 40 nb/sr very near threshold where the Jacobian is $\sim 10^4$. By tagging this reaction through detection of the recoil ^3He with a good spectrometer at 0°, a clean, tagged monochromatic η beam of 10^3 η/sec was found with every expectation that this may be boosted to 10^4 η/sec later. The spectacular Saturne II result is illustrated in Fig. 1 which shows on a log plot the ^{3}He yield at 0° measured in the reaction $p+d \rightarrow ^3\mathrm{He} + X$ for maximum allowed invariant mass of X as a function of the incident proton energy. The spike at $T_p = 0.8$ GeV consists of several points. It is remarkable that the η production is comparable to the π° production in this reaction. The etas move in the lab with a momentum of ~ 250 MeV/c. The chief difficulty with using this method in searches for low branching ratios is the unavoidable flood of charged particles that are produced besides the etas. It is interesting that it is even possible for LAMPF to become an active participant in eta research through a production reaction analogous to the one used at Saclay namely, $p+^3\mathrm{H} \rightarrow ^4\mathrm{He} + \eta$.

Needless to say, the decay modes of the η can also be applied to the η'. There is no easy production mechanism known, and with a half-life of 10^{-21} sec, the η' decay modes are often not competitive.

In summary, the rare and forbidden decays of light ordinary mesons, especially the η, provide a rich variety of tests of the Standard Model and should be pursued vigorously.

Work supported in part by the USDOE.

References

[1] S. Weinberg, Phys. Rev. Lett. **19**, 1264 (1967); A. Salam, In *Elementray Particle Theory*, ed. N. Svartholm, p. 367. Stockholm, Almquist & Wiksells (1968).

[2] H. Fritzsch, M. Gell-Mann, and H. Leutwyler, Phys. Rev. Lett. **47B**, 365 (1973); S. Weinberg, Phys. Rev. Lett. **31**, 494 (1973); D. Gross, and F. Wilczek, Phys. Rev. D **8**, 3633 (1973).

[3] U. Amaldi, et al., Phys. Rev. D **36**, 1385 (1985).

[4] W.J. Marciano and Z. Parsa, Ann. Rev. Nucl. Part. Sci. **36**, 171 (1986).

[5] Branching ratios quoted are from the 1986 Edition of the review of particle Properties.

[6] An extensive discussion of the physics of light, ordinary mesons is given by A. Soni in *Proceedings of Physics with Light Mesons*, ed. W. Gibbs and B.M.K. Nefkens, LANL Conference Proceedings LA–11184–C, p. 55, (1987).

TABLE I Tests of the Standard Model involving neutral weak currents. Tests 1–7 are expressed as a determination of the Weinberg angle as quoted in Ref. 3.

Reaction	$\sin^2\theta_w$	exp error	theor. error	q^2 GeV^2
ν A scatt (NC/CC)	0.233	0.003	0.005	10
$\bar{\nu}_\mu e \rightarrow \bar{\nu}_\mu e$	0.232	0.018	0.002	10^{-1}
$\bar{\nu}p \rightarrow \bar{\nu}_\mu p$	0.210	0.033		1
atomic parity violation	0.209	0.018	0.014	10^{-6}
asymm $\vec{e}$d scatt.	0.221	0.015	0.013	10^2
asymm $\vec{\mu}$C scatt.	0.25	0.08		10^2
mass W,Z	0.228	0.007	0.002	10^4
f-b asymm $e^+e^- \rightarrow \mu^+\mu^-$				
f-b asymm $e^+e^- \rightarrow \tau^+\tau^-$				
$\sigma(e^+e^- \rightarrow \mu^+\mu^-)$/QED				
$\sigma(e^+e^- \rightarrow \tau^+\tau^-)$/QED				

TABLE II Tests of the Standard Model involving charged weak currents. The experimental data are from Ref. 5, and the theoretical predictions are discussed in Ref. 6.

Process	Experiment	Prediction
$\Gamma(\pi \rightarrow e\nu)/\Gamma(\pi \rightarrow \mu\nu)$	$(1.228\pm0.022) \times 10^{-4}$	1.233×10^{-4}
$BR(\pi^+ \rightarrow \pi^\circ e^+\nu)$	$(1.025\pm0.034) \times 10^{-8}$	$(1.038\pm0.003) \times 10^{-8}$
$\Gamma(\tau \rightarrow e\nu\bar{\nu})$	$(4.06\pm0.25) \times 10^{-13}$	4.103×10^{-13}GeV
ft $(^{10}C \rightarrow ^{10} B^* e^+\nu)$	3065 ± 31	(3099 ± 7)sec
$\tau_{1/2}$(neutron)	635 ± 11	(620.9 ± 2.6)sec
$BR(K^+ \rightarrow \pi^+\nu\bar{\nu})$	$< 1.7 \times 10^{-7}$	$< 10^{-10}$
$\Gamma(B \rightarrow e\nu x)$		$< (0.6 \pm 0.2)10^{14}$sec^{-1}

TABLE III Some general tests of the Standard Model

1. Lepton family conservation.
2. Baryon conservation (e.g. proton decay).
3. C and P invariance of strong and em interactions.
4. CP and/or T violation outside the K-M matrix.
5. No first-order flavor-changing neutral currents.
6. No first-order $\Delta S=2$, $\Delta C=2$, and $\Delta S=\Delta Q$ transitions.
7. Universality of charged weak currents.
8. e, μ, τ universality.
9. Weak-decay parameters of $\mu, \Lambda, \Sigma \cdots$.
10. Neutrino mass, half life, and oscillations.

TABLE IV Rare and forbidden decays of the eta meson to test the Standard Model.

		Physics motivation	branching ratio experiment	SM
A.	Search for new particles			
1	$\eta \to e^+e^-$	leptoquark	$< 3 \times 10^{-4}$	$\sim 6 \times 10^{-9}$
2	$\eta \to \mu^+\mu^-$	leptoquark	$(6.5 \pm 2.1) \times 10^{-6}$	$\sim 1.3 \times 10^{-5}$
3	$\eta \to \pi^\circ H$	Light Higgs	$< 5 \times 10^{-6}$	
4	$\eta \to \pi\pi a$	axion		0
B.	Search for new interactions			
1	$\eta \to \mu e$	lept. fam. viol.		0
2	$\eta \to \pi^+\pi^-$	CP viol.	$< 1.5 \times 10^{-3}$	$0(10^{-14})$
3	$\eta \to 3\gamma$	C viol.		$0(10^{-13})$
	$\eta \to \pi^\circ \mu + \mu -$	C viol(1st order)	$< 5 \times 10^{-6}$	
	$\eta \to \pi^\circ e + e -$	C viol(1st order)	$< 5 \times 10^{-5}$	1
	asym $\eta \to \pi^+\pi^-\pi^\circ$	C viol.		
	asym $\eta \to \pi^+\pi^-\gamma$	C viol.		
C.	Exotica			
1	$\eta \to \pi e \nu$	2nd class current		$0(10^{-14})$
2	$\eta \to \pi \mu e$	2nd class + l.f.v.		0
3	$\eta \to \pi\pi e \nu$	chiral anomally		$0(10^{-10})$
4	$\eta \to K e \nu$	$\Delta S=1$ weak curr.		$0(10^{-16})$
5	$\eta \to \gamma \nu \nu$	$\#\nu$ families		$0(10^{-15})$
6	$\eta \to \pi\pi\nu\nu$	$\#\nu$ families		$0(10^{-10})$

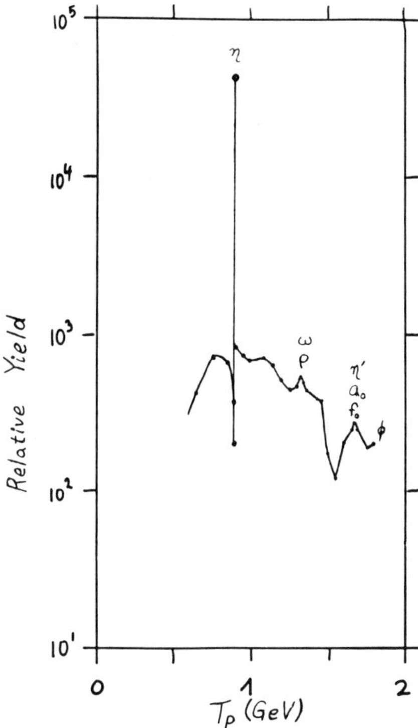

Fig. 1. Relative yield of ^{3}He produced in the reaction $p+d \rightarrow {}^3$He$+X$ at 0° as a function of the incident proton energy with the ^3He detector tuned for maximum possible missing mass of X.

WHY THE PROTON SPIN IS NOT DUE TO QUARKS

Marek Karliner

Stanford Linear Accelerator Center
Stanford University, Stanford, California 94309

ABSTRACT

Recent EMC data on the spin-dependent proton structure function suggest that very little of the proton spin is due to the helicity of the quarks inside it. We argue that, at leading order in the $1/N_c$ expansion, none of the proton spin would be carried by quarks in the chiral limit where $m_q = 0$. This model-independent result is based on a physical picture of the nucleon as a soliton solution of the effective chiral Lagrangian of large-N_c QCD. The Skyrme model is then used to estimate quark contribution to the proton spin when chiral symmetry and flavor $SU(3)$ are broken: this contribution turns out to be small, as suggested by the EMC. Next, we discuss the other possible contributions to the proton helicity in the infinite-momentum frame – polarized gluons (ΔG), and orbital angular momentum (L_z). We argue on general grounds and by explicit example that $\Delta G = 0$ and that if the parameters of the chiral Lagrangian are adjusted so that gluons carry $\sim$ 50% of the proton momentum, most of the orbital angular momentum L_z is carried by quarks. We mention several experiments to test the EMC results and their interpretation.

The EMC data[1] on polarized structure functions of the proton signals the need to re-examine our understanding of the various contributions to the proton spin. In the non-relativistic quark model (NQM) the proton is constructed as a bound state of three heavy quarks ($m_q \sim$ 300 MeV) and its spin results from combining the spins of these objects. The structure of the proton as suggested by QCD and the deep inelastic scattering (DIS) experiments is very different. The proton contains an infinite number of partons, i.e. quarks and gluons, and the quarks are light. Both the quarks and the gluons can contribute to the proton angular momentum, either by combining their intrinsic spins or through their orbital angular momentum. This is reflected in the sum rule

$$\frac{1}{2}\sum_q \Delta q + \Delta G + \langle L_z \rangle = \frac{1}{2}. \tag{1}$$

where

$$\Delta q = \int_0^1 \Delta q(x) = \int_0^1 dx \left[q_\uparrow(x) + \bar{q}_\uparrow(x) - q_\downarrow(x) - \bar{q}_\downarrow(x)\right]$$

$$\Delta G = \int_0^1 \Delta G(x) = \int_0^1 dx\, [G_\uparrow(x) - G_\downarrow(x)]$$

The net quark helicities Δq are related to matrix elements of the various axial currents between proton states, *e.g.*

$$\langle p|\, A^0_\mu\, |p\rangle = \sqrt{2/3}(\Delta u + \Delta d + \Delta s)\cdot\Sigma_\mu(p) \tag{2}$$

where $\Sigma_\mu(p)$ is the proton spin.

What are the experimental sources of information about the axial form factors? Historically, the first piece of information comes from charged-current weak interactions. Because these currents are almost conserved, i.e. have soft divergence $\propto m_q$, they have no anomalous dimension. This allows us to relate, through the operator product expansion, their low-energy matrix elements to parton distributions observed in DIS. Thus from neutron decay we obtain

- $$\Delta u - \Delta d = g_A = 1.25 \tag{3}$$

Hyperon β–decay, combined with $SU(3)$ flavor symmetry yields[2]

- $$(\Delta u + \Delta d - 2\Delta s)\,/\sqrt{3} = 0.39 \tag{4}$$

So far we have two equations in three unknowns. The third equation can be obtained from DIS involving the electromagnetic current. Because of the vector nature of the electromagnetic interaction, information about axial form factors can only be obtained if both the proton and the photon are polarized and their spins are either parallel or antiparallel. The difference A_1 between the antiparallel ($\equiv \sigma_{1/2}$) and the parallel cross section ($\equiv \sigma_{3/2}$) is expressed in the parton model as

$$A_1 \equiv \frac{\sigma_{1/2} - \sigma_{3/2}}{\sigma_{1/2} + \sigma_{3/2}} \xrightarrow[\text{Bjorken limit}]{} \frac{\sum_q e_q^2 \Big[q_\uparrow(x) + \bar{q}_\uparrow(x) - q_\downarrow(x) - \bar{q}_\downarrow(x)\Big]}{\sum_q e_q^2 \Big[q_\uparrow(x) + \bar{q}_\uparrow(x) + q_\downarrow(x) + \bar{q}_\downarrow(x)\Big]} \tag{5}$$

Using the measured values of the unpolarized structure function

$$F_2 = \sum_q e_q^2 x \Big[q_\uparrow(x) + \bar{q}_\uparrow(x) + q_\downarrow(x) + \bar{q}_\downarrow(x)\Big] \tag{6}$$

one can extract from A_1 the structure function

$$g_1(x) = \frac{1}{2}\sum_q e_q^2 \Big[q_\uparrow(x) + \bar{q}_\uparrow(x) - q_\downarrow(x) - \bar{q}_\downarrow(x)\Big]$$
$$= \frac{1}{2}\sum_q e_q^2 \Delta q(x). \tag{7}$$

$g_1(x)$ was obtained in this way by the SLAC-Yale collaboration in the 1970's[3] and more recently, for a wider range of x, by the EMC collaboration.[1] Their combined[4] result is

$$\int_0^1 dx g_1^p(x) = \frac{1}{2}\Big(\frac{4}{9}\Delta u + \frac{1}{9}\Delta d + \frac{1}{9}\Delta s\Big) =$$

$$\bullet \qquad = 0.112 \pm 0.009 \pm 0.019 \tag{8}$$

In 1974, using the experimental values of the charged current matrix elements, taken together with $SU(3)$ flavor symmetry and the assumption $\Delta s = 0$, Ellis and Jaffe[5] wrote down a sum rule

$$\int_0^1 dx g_1^p(x) = 0.19 \tag{9}$$

which is violated by the EMC result. Now that we have three equations denoted by $\bullet$, its failure can be traced back[6,7,8] to the assumption $\Delta s = 0$, as the solution of the equations is

$$\left.\begin{aligned} \Delta u &= 0.73 \pm 0.07 \\ \Delta d &= -0.52 \pm 0.07 \\ \Delta s &= -0.24 \pm 0.07 \end{aligned}\right\} \quad \begin{aligned} \Delta u + \Delta d + \Delta s &= \\ -0.01 \pm 0.21 & \end{aligned} \tag{10}$$

The first surprise is the large value of Δs. But perhaps we shouldn't have been so surprised. The large value of the σ-term in πN scattering has for some years been known to indicate[9] rather large strange sea in the proton, $\langle p|\,\bar{s}s\,|p\rangle$.

A more striking conclusion is that the total contribution of quark helicities to proton helicity is zero. Loosely speaking, the contribution of valence quarks is cancelled out by the sea quarks. As noted, a crucial ingredient is the relatively large and negative Δs. An independent corroboration of the above estimate of Δs can be obtained[10,11] from weak neutral current, elastic $\nu p \to \nu p$ and $\bar{\nu} p \to \bar{\nu} p$ scattering:[12] since Z^0 couples to $(\bar{u}\gamma_\mu\gamma_5 u - \bar{d}\gamma_\mu\gamma_5 d - \bar{s}\gamma_\mu\gamma_5 s)$ the deviation of the axial form factor $G_1(q^2 = 0)$ from $g_A = \Delta u - \Delta d$ provides an estimate $\Delta s = -0.15 \pm 0.09$.[11] The neutral current result in itself would not be sufficient to establish that $\Delta s < 0$, but is very important as independent verification of the EMC result.

It has recently been observed[13,14] that the Δu, Δd, Δs appearing in the parton model expression for $\int_0^1 dx g_1^p(x)$ and elsewhere acquire QCD radiative corrections and should be replaced[14] by $\widetilde{\Delta u} = \Delta u - (\alpha_s/2\pi)\Delta G$, etc... It has been suggested[14] that perhaps $\Delta s = 0$ and the discrepancy between the EMC result for $\int_0^1 dx g_1^p(x)$ and the previously expected value of 0.19[5] might be entirely due to ΔG. This would require $\Delta G \simeq 8 \pm 2$ at $Q^2 \simeq 10\,\mathrm{GeV}^2$, where $\alpha_s \simeq 0.2$, and $L_z \simeq -8$, surprisingly large values. We will in fact argue in the following that $\Delta G \simeq 0$.

The result $\Delta u + \Delta d + \Delta s \simeq 0$ can be rephrased as a statement about the matrix element of the ninth axial current,

$$J_{\mu 5}^0 = \bar{u}\gamma_\mu\gamma_5 u + \bar{d}\gamma_\mu\gamma_5 d + \bar{s}\gamma_\mu\gamma_5 s;$$

$$\langle p|\,J_{\mu 5}^0\,|p\rangle = -0.01 \pm 0.21 \tag{11}$$

It should be stressed here that u, d and s in $\Delta u + \Delta d + \Delta s \simeq 0$ are current, not constituent, quarks. Because of the many successes of NQM we sometimes forget the difference between the two and tend to apply our NQM intuition to DIS phenomena and that is part of the reason why the result (10) is so surprisng. In fact, it turns out that $\langle p| J^0_{\mu 5} |p\rangle = 0$ occurs naturally in large-N_c QCD in the chiral limit, i.e. with current masses of quarks taken to be zero.[8] Given that we are interested in the matrix element of an axial current at zero momentum transfer, it is natural to calculate it in an effective Lagrangian. Since the early sixties it has been known that chiral Lagrangians provide a very successful description of soft pion physics.[15] One approximates the QCD Lagrangian with an effective Lagrangian describing low energy dynamics of a chiral field U:

$$\mathcal{L}_{QCD}(q,g) \to \mathcal{L}_{eff}(U);\ U = \exp\left(2i\pi_a\tau_a/f_\pi\right) \tag{12}$$

More recently it has been realized that in large-N_c QCD the chiral Lagrangians describe baryon, as well as pion physics, provided only that the momentum transfer is small compared to the QCD scale.[16,17] Baryons appear as solitons of the chiral Lagrangian - "Skyrmions". Baryon number is identified with topologically conserved winding number. The solitons, when quantized, have precisely the same spin and flavor quantum numbers as lowest lying baryons - J=1/2 isodoublet for $SU(2)$ flavor and J=1/2 octet together with J=3/2 decuplet for $SU(3)$ flavor.[18] All the qualitative counting rules of large-N_c QCD are correctly reproduced, including N_c dependence of baryon masses, radii and hadronic cross sections.[18] On a more quantitative level, N_c independent ratios of experimental quantities as well as pion-nucleon partial-wave amplitudes are reproduced rather well.[19,20] Thus, to the extent that the real world with $N_c = 3$ is well described by large-N_c QCD, that description is also present in the chiral Lagrangian language. A useful analogy is the Thomas-Fermi model of the atom:[21] where one replaces the electron wave function by the average of its bilinear. Similarly, in this framework one replaces the quark field operator by the average of its chiral bilinear:

$$\text{atom: } \rho(r) \sim \psi^*\psi\,;\ \Leftrightarrow\ \text{proton: } U \sim \bar{q}_L q_R.$$

The full effective Lagrangian contains a very large number of couplings and fields. The Skyrme model is only a rough approximation to the full $\mathcal{L}_{eff}$. It does, however, have all the right symmetries and can be used to illustrate model independent results which are valid in any chiral Lagrangian in which the nucleon corresponds to a hedgehog soliton (see below). The result (11) is precisely of this kind.[8] To see this, consider a "generic" Lagrangian of the form

$$\mathcal{L} = \frac{f_\phi^2}{16}\mathrm{Tr}\left(\partial_\mu U \partial^\mu U^\dagger\right) + \dots \quad ;$$

$$U(x) = \exp\left[\left(\frac{2i}{f_\phi}\right)\sum_{i=0}^{8}\lambda_i\phi_i(x)\right] \tag{13}$$

where $\{\phi_i \equiv \eta_0,\ \underline{\pi},\ K_a,\ \eta_8\}$.

$\mathcal{L}$ is invariant under $SU(3)_L \times SU(3)_R$: $U \to VUW^\dagger$. The corresponding Noether currents can be written explicitly in terms of U. Since U has non-zero expectation value, $SU(3)_L \times SU(3)_R$ is spontaneously broken down to vector $SU(3)$ and the remaining axial $SU(3)$ is realized in Goldstone mode. The vacuum corresponds to $\langle U\rangle = 1$, while in the sector with baryon number = 1 the classical ground state is given by a "hedgehog" soliton $U_0 = \exp[iF(r)\hat{\boldsymbol{r}}\cdot\boldsymbol{\tau}]$. This ground state has a large degeneracy, $U_0 \to VU_0V^\dagger$ where V is any constant $SU(3)$ matrix. This degeneracy is removed when V-s are treated as collective coordinates and the corresponding Hamiltonian is diagonalized. Baryon wavefunctions $B(V)$ are the eigenstates of the collective coordinate hamiltonian. Matrix elements of the currents can now be evaluated explicitly.[18] For example, for the axial isovector current

$$\langle B| J^a_{i5} |B\rangle \propto \langle B(V)|\,\mathrm{Tr}(\lambda_i V\lambda_a V)\,|B(V)\rangle \tag{14}$$

where a is the isospin index and i is a space-like Lorentz index. For the isoscalar current, $\lambda_a \to \lambda_0 = \sqrt{\frac{2}{3}}\mathbf{1}$ and therefore[8]

$$\langle B| J^0_{i5} |B\rangle \propto \langle B(V)|\,\mathrm{Tr}(\lambda_i V\lambda_0 V)\,|B(V)\rangle = 0 \tag{15}$$

which is equivalent to $\Delta u + \Delta d + \Delta s = 0$, (*cf.* (10) and (11).) Let me stress again that this is a model independent result, relying only on the particular symmetry of the

U_0. The vanishing of $\langle p| J_5^0 |p\rangle$ can also be understood by considering the soliton topology. The soliton exists because the mappings from the real space to internal $SU(2)$ or $SU(3)$ flavor group space fall into distinct classes which cannot be continuously deformed into each other: $\Pi_3(SU(2)) = \mathbb{Z}$. But the same is not true for when the internal target space is $U(1)$, for $\Pi_3(U(1)) = 0$. That means that the soliton has no tail in the isoscalar direction and that the corresponding current decouples.

We have just obtained the matrix element of J_5^0 at $Q^2 = 0$. Unlike the flavor non-singlet currents however, J_5^0 has a "hard" divergence, due to the triangle anomaly[22]. Because of that, it also has non-zero anomalous dimension[23] and its matrix elements have some Q^2 dependence. We should therefore proceed with caution[24] when attempting to relate the $Q^2 = 0$ result to DIS data. Fortunately, the renormalization in this case is multiplicative, so if $\langle p| J_5^0 |p\rangle = 0$ at some Q^2, it will remain so at all Q^2 so that (15) which is derived in a low Q^2 effective Lagrangian remains valid in the kinematic region explored by the EMC[8].

We thus see that in the double limit $N_c \to \infty$ and $m_q = 0$ the result (10) occurs naturally. We do not know at present how to compute the $1/N_c$ corrections, but we can estimate corrections of $\mathcal{O}(m_q/\Lambda)$. This is done by adding to $\mathcal{L}$ a mass term $\sim$ $\mathrm{Tr}[m_q(U + U^\dagger - 2)]$ and an extra kinetic term,

$$\Delta\mathcal{L}_K=\epsilon\frac{f_\phi^2}{16}\mathrm{Tr}[\frac{\lambda_8}{2}(U^\dagger U_{\mu L}U^{\mu L}+U^\dagger U_{\mu R}U^{\mu R})] \tag{16}$$

which have the effect of introducing $\eta - \eta'$ mixing and $f_\pi \neq f_K$. When these effects are taken into account, we obtain

$$\frac{\langle p| J_5^0 |p\rangle}{\langle p| J_5^8 |p\rangle} = -0.38\,(cf.\ \sqrt{2}\text{ in NQM}), \tag{17}$$

leading to a corrected estimate[8] $\Delta u + \Delta d + \Delta s = -0.18$ (*vs.* exp. value -0.01 ± 0.21). Please keep in mind, though, that this does not take into account possible $1/N_c$ or higher order m_s/Λ_{QCD} corrections.

Given the sum rule (1) and the result $\sum_q \Delta q \simeq 0$, we would like to find out where the proton spin does come from. In the chiral soliton approach the proton angular momentum is purely orbital. To see this explicitly and to make sure that the glue does not contribute to L_z, we will make $\mathcal{L}(U)$ scale invariant, as expected[25] of $\mathcal{L}_{eff}$ for QCD. To that effect we introduce a scalar gluonium field χ.[26,27] The modified kinetic term in $\mathcal{L}$ reads

$$\mathcal{L} = \frac{f_\phi^2}{16}\mathrm{Tr}\left(\partial_\mu U \partial^\mu U^\dagger\right)\chi^2 + \dots \tag{18}$$

The classical solution of (18) is given in terms of U_0 and the glue "profile" $\chi(r)$. U and χ indicate the relative contribution of quarks and gluons, respectively, to the energy-momentum tensor $\theta_{\mu\nu}$, and through it to the various observables. For example, soliton mass M_0 is given by

$$M_0 = \int d^3\boldsymbol{r}\theta_{00}(\boldsymbol{r}) \tag{19}$$

With this in mind, we first compute $\theta_{++} = \theta_{00} + \theta_{33}$ and require that half of proton's linear momentum in the infinite momentum frame be carried by gluons. The adjustable parameters in $\mathcal{L}$ are thereby fixed. Next, we consider the angular momentum. In the collective coordinate approach the spin of the proton is due to rotation of the soliton as a whole, i.e. $\langle\Delta G\rangle = 0$, $J_z = L_z = \frac{1}{2}$. In other words, $J_z = \omega I$, where $I \sim N_c$ is the moment of inertia and $\omega \sim 1/N_c$ is the (slow) angular frequency of rotation. The slow rotation justifies the semi-classical treatment of the problem. The moment of inertia I is given by

$$I = \int d^3\boldsymbol{r}\theta_{00}(\boldsymbol{r})r^2 \tag{20}$$

The relative contribution of quarks and gluons to the spin is determined by their relative contributions to I. In the chiral limit these turn out to be 36% and 64%, respectively.[11] The glue contributes less, because its energy density is concentrated in a small region of space $\sim$ 1 fm, as suggested by the bag model, while the chiral field extends farther away.

The physical picture of the proton spin as suggested by this work has several rather interesting experimental consequences.

Clearly it is of great importance to confirm the EMC result (8) and to measure also

$\int_0^1 dx g_1^n(x)$ using polarized neutrons, so as to check the Bjorken sum rule.[28] The theoretical interest in new experiments to measure these quantities is enhanced by the fundamental information about chiral symmetry and its breaking that they provide. We also remind the reader of the relevance of $\langle p|\, A_\mu^0\, |p\rangle$ to dark matter searches[6,7] and to axion couplings.[29] Assuming that the EMC measurement (8) is essentially correct, the next priority is to determine the origin of the bulk of the proton spin, which must be carried by gluons and/or orbital angular momentum: $\frac{1}{2}(\Delta u + \Delta d + \Delta s) + \Delta G + \langle L_z\rangle = \frac{1}{2}$. There are various possibilities for measuring ΔG, including the following.*

(*a*) Measurement of J/ψ production and decay properties in deep inelastic muon scattering off polarized targets;[31]

(*b*) Measurements of charm distributions in deep inelastic scattering off a polarized target using dimuon events from $c(\bar{c}) \to \mu^+(\mu^-) + X$ decays;

(*c*) Hadronic jet asymmetries in polarized *pp* collisions;[32]

(*d*) Direct photon production at large p_T by polarized protons;[32]

(*e*) Hyperon production at large p_T in polarized *pp* collisions;[33]†

(*f*) Higher order effects in polarized *ep* collisions;[34]

(*g*) Drell-Yan l^+l^- production with polarized beams;[35]

(*h*) Large p_T hadron production in photoproduction off polarized targets.[36]

Acknowledgements: The work described in this talk[8,11] was done in collaboration with Stan Brodsky and John Ellis. It was supported by the Department of Energy, contract DE–AC03–76SF00515.

⋆ For a review and other references on spin physics at short distances, see Ref. 30.

† The fact that $\Delta s < 0$ suggests that there may be significant spin anticorrelation for hyperons produced by polarized protons, even at low p_T.

REFERENCES

1. EMC collaboration, J. Ashman *et al.*, *Phys. Lett.* **206**(1988), 364.
2. M. Bourquin *et al.*, *Z. Phys.* **C21**(1983), 27; for a review see: J.-M. Gaillard and G. Sauvage, *Ann. Rev. Nucl. Part. Sci.* **34**(1984), 351.
3. M. J. Alguard *et al.*, *Phys. Rev. Lett.* **37** (1976), 1261, *ibid* **41**(1978)70; G. Baum *et al.*, *Phys. Rev. Lett.* **51**(1983), 1135.
4. G. Baum, V.W. Hughes, K.P. Schüler, V. Papavassiliou and R. Piegaia, Yale preprint, May 1988.
5. J. Ellis and R. Jaffe, *Phys. Rev.* **D9**(1974), 1444.
6. J. Ellis, R. Flores and S. Ritz, *Phys. Lett.* **198B**(1987), 393; M. Glück and E. Reya, Dortmund preprint, DO-TH-87/14, Aug. 1987.
7. J. Ellis and R. Flores, CERN preprint CERN-TH-4911/87, Dec. 1987.
8. S. Brodsky, J. Ellis and M. Karliner, *Phys. Lett.* **206**(1988), 309.
9. T.P. Cheng, *Phys. Rev.* **D13**(1976), 2161; see also J. Donoghue and C. Nappi, *Phys. Lett.* **B168**(1986), 105; J. Donoghue, in Proc. of II-nd Int. Conf. on πN Physics; V.M. Khatymovsky, I.B. Khriplovich and A.R. Zhitnitsky, *Z. Phys.* **C36**(1987), 455; Riazudin and Fayyazuddin, Dharan Univ. preprint, 1988.
10. D.B. Kaplan and A. Manohar, Harvard preprint HUTP-88/A024, May 1988.
11. J. Ellis and M. Karliner, SLAC preprint SLAC-PUB-4592.
12. L. A. Ahrens *et al.*, *Phys. Rev.* **D35**(1987), 785.
13. A.V. Efremov and O.V. Teryaev, Dubna preprint E2-88-287(1988).
14. G. Altarelli and G. Ross, CERN preprint TH.5082/88(1988).
15. M. Gell-Mann, R. J. Oakes and B. Renner, *Phys. Rev.* **175**(1968), 2195, and references therein.
16. E. Witten, *Nucl. Phys.* **B160**(1979), 57
17. E. Witten in Lewes Workshop Proc.; A. Chodos *et al.*, Eds; Singapore, World Scientific, 1984.
18. E. Witten, *Nucl. Phys.* **B223**(1983), 422,*ibid* 433; G. Adkins, C. Nappi and E. Witten, *Nucl. Phys.* **B228**(1983), 433; for the 3 flavor extension of the model see: E. Guadagnini, *Nucl. Phys.* **236**(1984), 35; P. O. Mazur, M. A. Nowak and M. Praszałowicz, *Phys. Lett.* **147B**(1984)137.
19. M. P. Mattis and M. Karliner, *Phys. Rev.* **D31**(1985), 2833; *ibid*, **D34**(1986), 1991;, *Phys. Rev. Lett.* **56**(1986), 428; M. Karliner, *Phys. Rev. Lett.* **57**(1986), 523; M. P. Mattis and M. Peskin, *Phys. Rev.* **D32**(1985), 58.
20. A. Hayashi, G. Eckart, G. Holzwarth and H. Walliser, *Phys. Lett.* **147B**(1984),

5; H. Walliser and G. Eckart, *Nucl. Phys.* **A429**(1984), 514.
21. This analogy is due to V. Gribov, private communication.
22. S. L. Adler, *Phys. Rev.* **177**(1969), 2426); J. S. Bell and R. Jackiw, *Nuov. Cim.* **A51**(1967), 47.
23. J. Kodaira, S. Matsuda, T. Muta, T. Uematsu and K. Sasaki, *Phys. Rev.* **D20**(1979), 627; J. Kodaira, S. Matsuda, K. Sasaki and T. Uematsu, *Nucl. Phys.* **B159**(1979), 99; J. Kodaira, *Nucl. Phys.* **B165**(1979), 129.
24. R. Jaffe, *Phys. Lett.* **193B**(1987), 101.
25. N. K. Nielsen, *Nucl. Phys.* **B210**(1977), 212; J. C. Collins, A. Duncan and S. D. Joglekar, *Phys. Rev.* **D16**(1977), 438); see also S. L. Adler, J. C. Collins and A. Duncan, *ibid.* **15**, 1712 (1977); M. S. Chanowitz and J. Ellis, *ibid.* **7**(1973)2490; R. J. Crewther, *Phys. Rev. Lett.* **28**(1972), 1421
26. J. Schechter, *Phys. Rev.* **D21**(1980), 3393; J. Schechter and T. Tudron, *ibid.* **23**(1981)1143; A. A. Migdal and M. A. Schifman, *Phys. Lett.* **114**(1982), 445; J. M. Cornwall and A. Soni, *Phys. Rev.* **D29**(1984), 1424; J. Ellis and J. Lánik, *Phys. Lett.* **150B**(1985), 289; H. Gomm and J. Schechter, *ibid.* **158B**(1985)449.
27. R. Gomm, P. Jain, R. Johnson and J. Schechter, *Phys. Rev.* **D33**(1986), 3476.
28. J. Bjorken, *Phys. Rev.* **148**(1966), 1467.
29. R. Mayle, J. R. Wilson, J. Ellis, K. Olive, D. N. Schramm and G. Steigman, preprint EFI-87-104;UMN-TH-637/87;CERN-TH.4887/87, Dec. 1987.
30. N. S. Craigie, K. Hidaka, M. Jacob and F. M. Renard, *Phys. Rep.* **99C**(1983), 69.
31. C. Papavassiliou, N. Mobed and M. Svec, *Phys. Rev.* **D26**(1980), 3284; A. D. Watson, *Nuov. Cim.* **81A**(1984), 661; J. P. Guillet, Marseille preprint, CPT-87/P.2037, Sep. 1987.
32. M. B. Einhorn and J. Soffer, *Nucl. Phys.* **274**(1986), 714; C. Bourrely, J. Soffer and P. Taxil, Marseille preprint CPT-87/P-1992, March 1987.
33. N.S. Craigie, V. Roberto and D. Whould, *Zeit. Phys.* **C12**(1982), 173.
34. K. Hidaka, *Phys. Rev.* **D21**(1980), 1316; P. Chiappetta, J.P. Guillet and J. Soffer, *Phys. Lett.* **183B**(1987), 215.
35. E. Richter-Wąs and J. Szwed, *Phys. Rev.* **D31**(1985), 633; E. Richter-Wąs, *Acta Phys. Pol.* **B16**(1985), 739.
36. M. Fontannaz, B. Pire and D. Schiff, *Zeit. Phys.* **C8**(1981), 349.

STRANGE QUARK SPECTROSCOPY FROM THE LASS SPECTROMETER*

D. Aston,[1] N. Awaji,[2] T. Bienz,[1] F. Bird,[1] J. D'Amore,[3]
W. Dunwoodie,[1] R. Endorf,[3] K. Fujii,[2] H. Hayashii,[2] S. Iwata,[2]
W.B. Johnson,[1] R. Kajikawa,[2] P. Kunz,[1] D.W.G.S. Leith,[1] L. Levinson,[1]
T. Matsui,[2] B.T. Meadows,[3] A. Miyamoto,[2] M. Nussbaum,[3] H. Ozaki,[2]
C.O. Pak,[2] B.N. Ratcliff,[1] D. Schultz,[1] S. Shapiro,[1] T. Shimomura,[2]
P. K. Sinervo,[1] A. Sugiyama,[2] S. Suzuki,[2] G. Tarnopolsky,[1]
T. Tauchi,[2] N. Toge,[1] K. Ukai,[4] A. Waite,[1] S. Williams[1]

[1] *Stanford Linear Accelerator Center, Stanford University,*
[2] *Department of Physics, Nagoya University,*
[3] *University of Cincinnati,*
[4] *Institute for Nuclear Study, University of Tokyo*

ABSTRACT

A brief summary is presented of results pertinent to strange quark spectroscopy derived from high statistics data on K^-p interactions obtained with the LASS spectrometer at SLAC.

INTRODUCTION

This paper summarizes recent results on strange quark spectroscopy obtained from an exposure of the Large Aperture Superconducting Solenoid (LASS) spectrometer at SLAC to a K^- beam of 11 GeV/c. The spectrometer and relevant experimental details are described elsewhere.[1,2] The raw data sample contains $\sim$ 113 million triggers, and the resulting useful beam flux corresponds to a sensitivity of 4.1 events/nb. The acceptance is approximately uniform over almost the full 4π solid angle.

---

* Work supported in part by the Department of Energy under contract No. DE-AC03-76SF00515; the National Science Foundation under grant Nos. PHY82-09144, PHY85-13808, and the Japan U.S. Cooperative Research Project on High Energy Physics.

Invited talk presented by W.Dunwoodie at the Third Conference on the Intersections between Particle and Nuclear Physics, Rockport, Maine, May 14-19, 1988

Ω^{*-} SPECTROSCOPY

Although the Ω^- was discovered in 1962, claims for the observation of Ω^{*-} resonances have been made only recently.[3,4,5] Preliminary evidence for an Ω^{*-} of mass ~ 2.26 GeV/c^2 was reported in the present experiment.[3] Subsequently, evidence for states with masses of 2251 and 2384 MeV/c^2 produced in Ξ^- interactions in beryllium was published.[4] More recently,[5] the full data sample from the present experiment has yielded clear evidence for the production of an Ω^{*-} resonance of mass 2253 ± 13 and width 85 ± 40 MeV/c^2, thus confirming the preliminary result.[3] The state is observed in the $\Xi^*(1530)\overline{K}$ decay mode, as shown in fig.1a, and the corresponding inclusive cross section is estimated to be 630 ± 180 nb. It is reasonable to consider this result as confirmation of the 2251 MeV/c^2 state of ref. 4, however there is no evidence for the production of the state of mass 2384 MeV/c^2 in the K^-p data.

An ongoing study of inclusive Ω^- production in the present experiment has resulted in the $\Omega^-\pi^+\pi^-$ mass distribution of fig.1b.[6] A clear peak is observed at ~ 2.47 GeV/c^2. Interpreting this as being due to the production of a new Ω^{*-}, the fit denoted by the solid curve yields mass and width estimates of 2466 ± 13 and 61 ± 35 MeV/c^2, respectively, and the signal has $\sim 5\sigma$ significance.

In summary, the existence of an Ω^{*-} of mass $\sim$ 2250 MeV/c^2 decaying primarily to $\Xi^*(1530)\overline{K}$ seems confirmed; moreover, the mass and principal decay mode are quite consistent with theoretical expectations.[7,8] The status of the state at 2384 MeV/c^2 is uncertain, since it is not observed in the present experiment. Finally, the new Ω^{*-} at $\sim$ 2470 MeV/c^2 needs confirmation from other experiments; theoretical predictions concerning possible states in this mass range would also be of interest.

K^* SPECTROSCOPY

A principal objective in studying meson spectroscopy is the precise definition of the level structure of the anticipated $q\overline{q}$ meson states. In this regard, the strange sector is particularly favored since it is free, not only from the need for isoscalar-isovector separation, but also from confusion resulting from the possible production of e.g. $K\overline{K}$ molecule and/or glueball states. The quark model level diagram is expected to take the form[9] illustrated qualitatively in fig.2 for the charmonium states.[10] Here S and L denote total quark spin and orbital angular momentum, respectively, and C indicates the charge conjugation parity of the resulting meson state. The para-charmonium levels are singlets, whereas the ortho-charmonium levels, other than ^{3}S, are triplets separated in mass due to spin-orbit interaction. Within each column, the lowest-lying state is the ground state, and the higher states correspond to radial excitations of this state. The leading orbital excitations are the ground states with largest total angular momentum, J, and, for the triplet levels, the remaining states are termed the underlying states. In $e^+ e^-$ collisions, the ^3S tower of states is readily accessible, while information on the leading orbital excitations with J $\geq$ 3 is lacking, and that on the corresponding underlying states is sparse. Results of a complementary nature are obtained from hadronic interaction experiments, where production of the leading orbital states predominates, and information on the underlying and radially- excited states requires careful amplitude analyses of high statistics data.

The present experiment provides such data on the natural parity strange meson (i.e.K^*) states. The resulting light quark level structure is related to the nature of the long-range (i.e.confining) part of the $q\overline{q}$ interaction, and thus, in principle, provides information on non-perturbative QCD. In practice, it enables us to learn about QCD potential models incorporating relativistic corrections.[11]

The results relevant to K^* spectroscopy in the present experiment are obtained from amplitude analyses of the $\overline{K}^0\pi^+\pi^-$ system in the reaction

$$K^-p \rightarrow \overline{K}^0\pi^+\pi^-n \tag{1}$$

and of the $K^-\pi^+$ and $K^-\eta$ systems produced in the reactions

$$K^-p \rightarrow K^-\pi^+n \tag{2}$$

and

$$K^-p \rightarrow K^-\eta\ p\ \ ; \tag{3}$$

these analyses are discussed in detail in refs. 12-14 respectively, and so only the main features will be summarized here.

The analysis of reaction (1)[12] yields the intensity distributions of fig.3 for the natural spin-parity (J^P) states of the $\overline{K}^0\pi^+\pi^-$ system. Signals corresponding to the $\overline{K^*}_2(1430)$, the $\overline{K^*}_3(1780)$ and the $\overline{K^*}_4(2060)$ are observed in the 2^+, 3^- and 4^+ distributions, respectively. The 2^+ distribution exhibits a second peak at $\sim$ 2 GeV/c^2, while the 1^- distribution has a shoulder at $\sim$ 1.4 GeV/c^2 followed by a peak at $\sim$ 1.8 GeV/c^2. The $\overline{K^*}(892)\pi$ and $\overline{K}\rho(770)$ contributions to the 1^- spectrum are shown in figs.4a,b; the curves result from a description in terms of two Breit-Wigner(BW) resonances. The lower mass state (M=1420$\pm$17, Γ=240$\pm$30 MeV/c^2) approximately decouples from the $K\rho$ channel, and its production characteristics indicate weak coupling to $K\pi$ also; it is most readily interpreted as the first radial excitation of the $\overline{K^*}(892)$. The second state (M=1735±30, Γ=423±48 MeV/c^2) may be the underlying member of the ^{3}D ground state; however its width is significantly larger than that obtained from reaction (2)[13] so that it could be a mixture of this state and the second radial excitation of the $\overline{K^*}(892)$.

A similar fit to the 2^+ data yields mass and width estimates of 1973 ± 33 and 373 ± 93 MeV/c^2 respectively for the higher mass state, which corresponds most probably to the first radial excitation of the $\overline{K^*}(1430)$.[12]

The raw $K^-\pi^+$ mass spectrum for reaction (2) is shown in fig.5; the shaded region corresponds to events with M(n π^+)≥ 1.7 GeV/c^2. In addition to an elastic BW amplitude describing the $\overline{K^*}(892)$ (cf.fig.9), the amplitude analysis[13] yields the resonant D-H wave signals of fig.6, thereby extending the observed leading orbital states to $J^P= 5^-$. The corresponding behavior of the underlying S wave amplitude up to 1.6 GeV/c^2 is shown in fig.7. The curves correspond to an effective range parametrization plus a BW amplitude (M=1412 ± 4 ± 5, Γ=294 ±10 ±21 MeV/c^2), and the resultant S wave is approximately elastic up to ~ 1.5 GeV/c^2. The state described by the BW is the $^3P_0$ partner of the $\overline{K^*}(1430)$. Above 1.8 GeV/c^2, there are two S wave solutions, as shown in fig.8. Both resonate in the 1.9-1.95 GeV/c^2 region, have width ~ 0.2 GeV/c^2 and elasticity ~ 0.5. In the quark model such a state can only be the first radial excitation of the $^3P_0$ ground state.

Finally, the behavior of the P wave $\overline{K}\pi$ scattering amplitude from reaction (2) is shown in fig.9 for the region up to 1.8 GeV/c^2. The solid curve provides a satisfactory description in terms of three BW resonances:[13] (i) the $\overline{K^*}(892)$; (ii) a resonance at ~ 1.4 GeV/c^2 of elasticity ~ 0.07 which is consistent with the radial excitation of the $\overline{K^*}(892)$ observed in reaction (1); and (iii) a resonance (M=1677 ±10 ±32, Γ=205 ±16 ±34 MeV/c^2) of elasticity ~ 0.39 which is readily interpreted as the $^3D_1$ ground state.

In general, theoretical predictions[11] are in quite good agreement with these measurements. However, the $^3P_0$ ground state is predicted ~ 170 MeV/c^2 low, and the $^3S_1$ first radial excitation is predicted ~ 160 MeV/c^2 high with respect to the measured mass value. Also, for the $^3S_1$ and $^3P_0$ levels the expected splitting between the ground state and first radial excitation is ~650-680 MeV/c^2, whereas it is observed to be ~500-550 MeV/c^2. It should be noted also that for the $^3S_1$ states this splitting is predicted to increase from the charmonium to strange meson sector; it appears actually to decrease by ~70 MeV/c^2.

Reaction (3) has been measured with large statistics for the first time in the present experiment, and the resulting $K^-\eta$ mass spectrum is shown in fig.10. An amplitude analysis of the $K^-\eta$ system[14] has established that the peak at ~1.8 GeV/c^2 results from production of the $\overline{K^*}_3(1780)$ (cf.fig.11), and that the branching fraction to $\overline{K}\eta$ is 7.9±2.4%, in accord with SU(3). SU(3) predicts also that K^* states of even spin couple only weakly to $K\eta$. This is confirmed by the D wave intensity distribution of fig.11. No $\overline{K^*}_2(1430)$ signal is observed, and a 95% confidence level upper limit on the branching fraction is established at 0.45%. This value is an order of magnitude smaller than that quoted in ref.10.

$s\bar{s}$ SPECTROSCOPY

The processes most relevant to the study of $s\bar{s}$ (i.e.strangeonium) spectroscopy are those in which a $K\overline{K}$ or $K\overline{K}\pi$ system is produced against a recoil hyperon. The objectives in studying such processes are basically the same as for the strange sector. However, in recent years candidate glueball states which also couple to these meson systems have emerged from the study of J/ψ decay. Consequently, it has become increasingly important to define the $s\bar{s}$ level structure in order to distinguish those states which may be of a different dynamical origin. In the present experiment, the relevant reactions which have been studied are

$$K^- p \to K^0_S K^0_S \Lambda_{seen} \tag{4}$$

$$K^- p \to K^- K^+ \Lambda_{seen} \tag{5}$$

and

$$K^- p \to K^0_S K^\pm \pi^\mp \Lambda_{seen} \tag{6}$$

They are characterized by peripheral (i.e.quark exchange) production of the meson system, and it follows that the creation of glueball states should be disfavored in such processes. Although the data samples are considerably smaller than for reactions (1) and (2), amplitude analyses have been performed,[2,15,16,17] the results of which are summarized here.

The raw $K\overline{K}$ mass distributions for reactions (4) and (5) (fig.12) differ, since for the former only even spin states are produced. In addition, for reaction (5), the mass region above ~1.7 GeV/c^2 is dominated by the reflection of diffractive production of low mass ΛK^+ systems.[18] In fig.13, the $K^0_S K^0_S$ mass spectrum from reaction (4) is compared to the MARK III data on radiative J/ψ decay[19] in the mass region below 1.9 GeV/c^2 (the LASS data have been scaled to the MARK III data at the peak of the $f_2'(1525)$). Both spectra show a small, but intriguing, threshold rise,[2,20] followed by activity in the $f_2(1270)/a_2(1320)$ region and then the large $f_2'(1525)$ peak. At higher mass, the MARK III spectrum is dominated by the $f_2(1720)$. There is no evidence for any such signal in the LASS distribution, and the upper limit on the production cross section is 94 nb. at the 95% confidence level; it should be noted also that there is no evidence for $f_2(1720)$ production in reaction (6) in the present experiment[15] (cf.fig.19f). These results indicate that the $f_2(1720)$ is indeed a strong candidate glueball state.

The amplitude analysis of reaction (4)[2] yields the S wave intensity distribution of fig.14a. Although the uncertainties are large, the data seem to peak in the range ~ 1.5-1.6 GeV/c^2. A similar distribution (fig.14b) has been obtained in an analysis of reaction (5) at 8.25 GeV/c,[21] and there are indications of S wave structure at this mass from the data on reaction (5) in the present experiment. These results suggest the existence of a 0^+ state in this mass region which is naturally interpreted as a partner of the $f_2'(1525)$ in the ^{3}P ground state. It is not clear whether this state might be identified with the $f_0(1590)$.[22] An immediate consequence of this observation is that the $f_0(975)$, which is usually assigned to this multiplet, may well be a weakly bound $K\overline{K}$ system,[23] or be of some other non-$q\overline{q}$ origin. This is discussed at greater length in ref.2.

For the $K\overline{K}$ system in reaction (5), the acceptance corrected spherical harmonic moments ($t_L^M = \sqrt{4\pi}N\langle Y_{LM}\rangle; L \leq 8, M{=}0$) in the t-channel helicity frame are shown in fig.15 for the mass region 1.68-2.44 GeV/c^2, and $t' \leq 0.2$ (GeV/c)^2. It should be noted that amplitudes with spin J can contribute to moments with L$\leq$2J. There is a peak in the mass spectrum, t_0^0, at ~ 1.86 GeV/c^2, and similar structure is present for all moments with $L \leq 6$, but is absent for $L \geq 7$. This, together with the absence of such a signal for reaction (4),[2] indicates the presence of a $J^{PC} = 3^{--}$, mostly $s\overline{s}$ state. A detailed analysis of this region[16] yields the total F wave intensity distribution of fig.16a for $t' \leq 1.0$(GeV/c)^2; the fitted curve gives BW mass and width 1855 ± 22 and 74 ± 67 MeV/c^2, respectively. A similar fit to the mass spectrum (fig.16b) gives parameter values 1851 ± 9 and 66 ± 29 MeV/c^2, in agreement with previous measurements from the mass distribution only.[24,25] It should be noted that the spin of this state, which is listed as the $\phi_J(1850)$,[10] is established by the LASS data, and that the mass and width have been estimated for the first time on the basis of an amplitude analysis.

A significant $J^P{=}3^-$ signal (fig.19g) has been obtained in the $\phi_3(1850)$ region from the partial wave analysis of reaction (6).[15] This signal, after all corrections, is shown in fig.16c (open dots) in comparison with that from reaction (5) (solid dots); the branching ratio obtained is $BR[(\phi_3(1850) \rightarrow (K^*\overline{K} + c.c.))/(\phi_3(1850) \rightarrow K\overline{K})] = 0.55^{+0.85}_{-0.45}$, in agreement with theory.[11]

The $\phi_3(1850)$ is interpreted as belonging to the $^3D_3$ quark model nonet which also includes the $\rho_3(1690)$, the $\omega_3(1670)$, and the $K_3^*(1780)$. Using current mass values,[10] the mass formula[26] yields an octet-singlet mixing angle of ~ 30°. This indicates that the multiplet is almost ideally mixed, and that the $\phi_3(1850)$

is an almost pure $s\bar{s}$ state, in accord with its production characteristics.

The analysis of the ϕ_3 region[16] has shown that the peaks observed in fig.15 are due to interference between the resonant F wave and the approximately imaginary amplitude describing the diffractive production of the low mass ΛK^+ system. In fig.15, there is also a small signal in every moment with $L \geq 1$ at $\sim$ 2.2 GeV/c^2; no such signal is observed for $L \geq 9$. This indicates the existence at this mass of a small, resonant 4^+ amplitude which interferes with the large, imaginary, diffractive background, just as for the ϕ_3. An analysis based on this interpretation[17] yields preliminary mass and width values of $2209 {}^{+17}_{-15}$ and $60 {}^{+107}_{-57}$ MeV/c^2 for this $J^{PC} = 4^{++}$ state. These values are consistent with those obtained by MARK III for the X(2220),[19] and with a signal in the $\eta\eta'$ mass spectrum observed by the GAMS collaboration.[27] In addition, the $K\overline{K}$ mass distribution from reaction (4) in this region has been shown[2] to be very similar to that observed for radiative J/ψ decay (fig.17). This suggests that the X(2220), which has been conjectured to be a glueball state, may instead be a member of the quark model $^3F_4$ ground state nonet.

The raw $K\overline{K}\pi$ mass spectrum for reactions (6) (fig.18) exhibits a small signal at the $f_1(1285)$, followed by a rapid rise at $K\overline{K^*}$ threshold to a peak at $\sim$ 1.5 GeV/c^2, and a second peak at $\sim$ 1.9 GeV/c^2. The low mass structure is similar to that observed at 4.2 GeV/c.[28] An amplitude analysis[15] yields the partial wave intensity distributions of fig.19. The low mass peak is associated primarily with 1^+ waves, while the 1.9 GeV/c^2 bump is due mainly to 2^- and the 3^- contribution discussed previously. Only K^* or $\overline{K^*}$ isobar production is important; amplitudes involving the $a_0(980)$ are negligibly small. The 1^+ intensity at low mass shows a pronounced asymmetry in favor of the $\overline{K^*}$ isobar, and also exhibits K^*-$\overline{K^*}$ interference, which is destructive near threshold, and constructive at $\sim$ 1.5 GeV/c^2. This suggests the existence of two 1^+ states of opposite G-parity in this region. The corresponding J^{PG} amplitude combinations yield 1^{++} and 1^{+-} production intensity distributions (fig.20) which are well described as BW resonances of mass 1.53 ± 0.01 and 1.38 ± 0.02 GeV/c^2 respectively, with corresponding widths 0.10 ± 0.04 and 0.08 ± 0.03 GeV/c^2. The 1^{++} state confirms an earlier observation,[28] while the 1^{+-} state is new. If these are isoscalars,[28] the lower mass state, the $h_1(1380)$, completes the J^{PC}=1^{+-} ground state nonet, while the higher mass state, the $f_1(1530)$, is a strong candidate to replace the E/$f_1(1420)$ in the corresponding 1^{++} nonet. Use of the $f_1(1530)$ in this nonet gives a mixing angle $\sim 55°$; this implies that the $f_1(1530)$ is mainly $s\bar{s}$ and that the $f_1(1285)$ has little $s\bar{s}$ content, in agreement with their production characteristics. This is not the case when the E is used, since its production properties are not consistent with the predicted large $s\bar{s}$ content. Finally, a previous analysis[29] of the 1^{++} nonet made use of the

E mass and width to predict an a_1 mass of ~ 1.47 GeV/c^2; when the $f_1(1530)$ mass and width are used instead, the predicted mass is ~ 1.28 GeV/c^2, in much better agreement with the accepted value.[10]

The 1^+ states observed in the present experiment satisfactorily complete the ground state $^1P_1$ and $^3P_1$ quark model nonets. Furthermore, the small spin-orbit interaction implied when the new information on the 0^+ is also taken into account contrasts markedly with the corresponding behavior in the charmonium sector. The status of the E meson as a $q\overline{q}$ state is dubious. However, it may be that its proximity to $K\overline{K^*}$ threshold is significant in that it could be a $K\overline{K^*}$ molecule or four-quark state of some kind.

CONCLUSION

The present experiment has made significant contributions to the various areas of strange quark spectroscopy discussed above. Data analysis is continuing, and new results will be forthcoming on inclusive Ξ^- and Ξ^* production, the $\overline{K}^0\pi^-p$ final state, and the production of meson systems such as $K^*\overline{K^*}$ against a recoil Λ.

REFERENCES

1. D. Aston et al., The LASS Spectrometer, SLAC-REP-298 (1986).
2. D. Aston et al., Nucl. Phys. B301, 525 (1988).
3. D. Aston et al., contributed talk given by B. Ratcliff in session P09 at the International Europhysics Conference on High Energy Physics, Bari, Italy, July 1985; see S. Cooper, Rapporteur Talk in the Proceedings of this Conference, p. 947 and SLAC-PUB-3819.
4. S.F. Biagi et al., Z. Phys. C31, 33 (1986).
5. D. Aston et al., Phys. Lett. 194B, 579 (1987).
6. D. Aston et al., SLAC-PUB-4657, in preparation.
7. N. Isgur and G. Karl, Phys. Rev. D19, 2653 (1979).
8. K.T. Chao, N. Isgur and G. Karl, Phys. Rev. D23, 155 (1981).
9. W. Grotrian, Graphische Darstellung der Spektren von Atomen und Ionen mit ein, zwei und drei Valenzelektronen (Springer, Berlin, 1928).
10. Particle Data Group, Phys. Lett. 170B (1986).
11. S. Godfrey and N. Isgur, Phys. Rev. D32, 189 (1985).
12. D. Aston et al., Nucl. Phys. B292, 693 (1987).

13. D. Aston et al., Phys. Lett. 180B, 308 (1986); Nucl. Phys. B296, 493 (1988).
14. D. Aston et al., Phys. Lett. 201B, 169 (1988).
15. D. Aston et al., Phys. Lett. 201B, 573 (1988).
16. D. Aston et al., DPNU-88-15 / SLAC-PUB-4202 (1988), to be published in Phys. Lett. B.
17. D. Aston et al., SLAC-PUB-4661 (1988), in preparation.
18. D. Aston et al., SLAC-PUB-4202 / DPNU-87-08 (1987).
19. R.M. Baltrusaitis et al., Phys. Rev. Lett. 56, 107 (1986).
20. K.L. Au et al., Phys. Rev. D35, 1633 (1987).
21. M. Baubillier et al., Z. Phys. C17, 309 (1983).
22. D. Alde et al., Phys. Lett. 201B, 160 (1988), and references therein.
23. J. Weinstein and N. Isgur, Phys. Rev. Lett. 48, 659 (1982); Phys. Rev. D27, 588 (1983), and references therein.
24. S. Al-Harran et al., Phys. Lett. 101B, 357 (1981).
25. T. Armstrong et al., Phys. Lett. 110B, 77 (1982).
26. M. Gell-Mann, Caltech report TSL-20 (1961) unpublished;
S. Okubo, Prog. Theor. Phys. (Kyoto) 27, 949 (1967).
27. D. Alde et al., Phys. Lett. 177B, 120 (1986).
28. Ph. Gavillet et al., Z. Phys. C16, 119 (1982).
29. R.K. Carnegie et al., Phys. Lett. 68B, 287 (1977).

PARTIAL WAVE ANALYSIS OF THE $K^+\overline{K}^\circ\pi^-$ SYSTEM PRODUCED IN THE REACTION $\pi^- p \rightarrow K^+\overline{K}^\circ\pi^- n$ AT 8 GeV/c*

S. Blessing, R. Crittenden, A. Dzierba, T. Marshall, S. Teige,[†] D. Zieminska
Physics Department, Indiana University, Bloomington, Indiana 47405

A. Birman,[*] S.U. Chung, D. Peaslee,[‡] R. Fernow,
H. Kirk, S.D. Protopopescu, D. Weygand, H.J. Willutzki
Physics Dept., Brookhaven National Laboratory, Upton, New York 11973

A. Boehnlein, D. Boehnlein, J.H. Goldman, V. Hagopian, D. Reeves[♮]
Physics Department, Florida State University, Tallahassee, Florida 32306

Z. Bar-Yam, J. Dowd, W. Kern, E. King, H. Rudnicka, P. Rulon 3
Physics Department, Southeastern Massachusetts University,
North Dartmouth, Massachusetts 02747

ABSTRACT

We have performed a partial wave analysis of the $K^+K^0_S\pi^-$ system produced in the reaction $\pi^- p \rightarrow K^+\overline{K}^\circ\pi^- n$ at 8 GeV/c. We present the results of the analysis of 30,740 events in the mass range 1.24 - 1.60 GeV/c^2, with $0.0 \leq -t < 1.0$ GeV^2/c^2. In the 1.28 GeV/c^2 mass region, we see evidence for a J^{PG}=0^{-+} resonance in addition to the resonant 1^{++} wave. We observe a 0^{-+} resonance and possibly a small 1^{++} resonance in the 1.42 GeV/c^2 region. We also observe a 1^{++} state near 1.5 GeV/c^2.

The 1.4 - 1.5 GeV/c^2 region of nonstrange mesons is a subject of great controversy. Various experiments studying the $K\overline{K}\pi$ and $\eta\pi\pi$ final states in hadronic reactions, hadronic J/ψ decay, and two-photon interactions have reported a state near 1.42 GeV/c^2 with J^{PC} determined alternatively as 0^{-+}[1,2] or 1^{++}.[3] The $K\overline{K}\pi$ spectrum observed in radiative J/ψ decay[4] and in one experiment studying hadronic interactions[5] is dominated by a

★ Work supported by the Department of Energy under contract numbers DE-AC02-84ER40125, DE-AC02-76CH00016, and DE-FG05-87ER40319 and NSF grant PHY87-01401.

† Present address: Department of Physics and Astronomy, Rutgers-The State University of New Jersey, Piscataway, NJ 08854

∗ Permanent address: Physics Department, Technion, Haifa 32000 Israel

‡ Permanent address: Physics Department, University of Maryland, College Park, MD 20742

♮ Present address: Fermilab, P.O. Box 500, Batavia, IL 60510

broad structure near 1.45 GeV/c^2 with $J^{PC}= 0^{-+}$ sometimes interpreted as a glueball with a small admixture of quarks.[6] Our present experiment reveals the complicated structure of the $K\overline{K}\pi$ system in the mass range from 1.24 to 1.6 GeV/c^2.

We present results of a partial wave analysis of the $K^+K^0_S\pi^-$ system produced in the reaction $\pi^-p \to K^+\overline{K}^\circ\pi^-n$ at 8 GeV/c. A total of 30,740 events in the mass range 1.24 to 1.60 GeV/c^2 with $0.0 \leq -t < 1.0$ GeV^2/c^2 was collected using the BNL Multiparticle Spectrometer facility. The trigger required an increase in the number of charged particles from two to four, a forward going charged particle with momentum greater than 1.5 GeV/c^2 which was not a pion, and no signal from a lead-scintillator box surrounding the target to veto slow charged particles. Further experimental details can be found in Reference 7.

The observed $K^+K^0_S\pi^-$ mass spectrum is shown in Figure 1. The distribution is dominated by two peaks at 1.28 GeV/c^2 and 1.42 GeV/c^2. A smaller peak is also seen near 1.51 GeV/c^2. A fit of three Breit-Wigner functions and a polynomial background gives the following mass, width and number of events for each peak:

Mass (MeV/c^2)	Width (MeV/c^2)	Number of Events
1285±1	22±2	4750±100
1419±1	66±2	8800±200
1512±4	35±15	600±200

The $K^+\overline{K}^\circ\pi^-$ system was described as a superposition of states characterized by spin, parity and G-parity, J^{PG}, absolute value of the projection of spin on the quantization axis, M, and reflectivity (naturality of the exchanged particle), η. We applied the parametrization of the spin density matrix of Chung and Trueman[8] which ensures that the conditions of positivity, parity conservation, and rank two are automatically satisfied with the minimum number of parameters.

We assumed that the final three meson system is produced in a two step process involving two isobars: $K^*(890)$ and $a_0(980)$ [was $\delta(980)$]. For a given $K^+\overline{K}^\circ\pi^-$ mass, the state is fully specified by five variables: the isobar mass and the polar and azimuthal angles for the primary and secondary decays $\{\mathrm{m}, \theta, \phi, \alpha, \beta\}$.

The intensity distribution is described by

$$I(\tau) = \sum_a \rho_{aa'} A_a(\tau) A^*_{a'}(\tau)$$

where $a = \{J^{PG}M^{\eta}(\text{isobar})\}$ and $\tau = \{\text{m}, \theta, \phi, \alpha, \beta\}$. The A_a are decay amplitudes calculated from the data which contain Breit-Wigner functions for the secondary decays. The spin-density matrix is given by

$$\rho_{aa'} = \sum_{k=1}^{2} V_{ak} V^*_{a'k}$$

where the V_{ak} are complex parameters obtained in the fit and are related to production amplitudes; k allows for spin flip and spin non-flip of the nucleon.

The following waves were needed to fit the data: $J^{PG}= 0^{-+}$, 1^{++}, 1^{+-}, and 1^{--} with all possible values of M and $J^{PG}M^{\eta}=1^{-+}0^{-}$. For $J^{PG}=0^{-+}$ and 1^{++} both decay channels, $a_0\pi$ and K^*K, were allowed. However, we found that the waves $1^{++}(a_0)$ and $1^{++}(K^*)$ were too much alike to distinguish between them below the K^*K threshold and only one of them could be included at a time. We also found that in the region $1.40< M(K^+\overline{K}^\circ\pi^-)<1.60$ GeV/c^2, where both $1^{++}(a_0)$ and $1^{++}(K^*)$ could be included in the fit, $1^{++}(a_0)$ was small. The best choice was to allow $1^{++}(a_0)$ for $M(K^+\overline{K}^\circ\pi^-)< 1.34$ GeV/c^2 and $1^{++}(K^*)$ for $M(K^+\overline{K}^\circ\pi^-)> 1.34$ GeV/c^2. The waves $0^{-+}(a_0)$ and $0^{-+}(K^*)$ were varied simultaneously with an arbitrary degree of coherence between them.

The analysis was done for $1.24 \leq M(K^+\overline{K}^\circ\pi^-) < 1.60$ GeV/c^2 in 20 MeV/c^2 mass bins and three t bins: $0.0 \leq -t < 0.2$ GeV^2/c^2, $0.2 \leq -t < 0.45$ GeV^2/c^2, and $0.45 \leq -t < 1.0$ GeV^2/c^2.

Figure 2 shows the comparison between the data and the PWA fit for $1.40 \leq M(K^+\overline{K}^\circ\pi^-) < 1.42$ GeV/c^2 and $0.2 \leq -t < 0.45$ GeV^2/c^2. Our PWA fit reproduces the data very well.

Figure 3 shows the major spin-parity components of the $K^+\overline{K}^\circ\pi^-$ system in terms of the effective number of events attributed to each spin-parity state. The observed peak in the $K^+\overline{K}^\circ\pi^-$ mass spectrum at ~ 1.42 GeV/c^2 has contributions from the 0^{-+}, 1^{++}, and 1^{+-} waves. Figure 4 shows the contribution of the $0^{-+}(a_0)$ and $0^{-+}(K^*)$ waves. There is a small peak in $0^{-+}(a_0)$ at ~ 1280 MeV/c^2 along with the larger peak at ~ 1.40 GeV/c^2 in both $0^{-+}(a_0)$ and $0^{-+}(K^*)$. The main peaks in $0^{-+}(a_0)$ and $0^{-+}(K^*)$ are offset by ~ 50 MeV/c^2.

The phases of $0^{-+}(a_0)$ and $0^{-+}(K^*)$ with respect to $1^{++}0^{+}(K^*)$, for the three t regions discussed, are shown in Figure 5. For all three t regions, the waves $0^{-+}(a_0)$ and $1^{++}0^{+}(K^*)$ are highly coherent and their relative phase is well measured. For $0.2 \leq -t < 0.45$ GeV^2/c^2 and $0.45 \leq -t < 1.0$ GeV^2/c^2, the phase of $0^{-+}(a_0)$ is rising through about 300 degrees in the mass range 1.35 to 1.50 GeV/c^2, and then begins to fall. This phase variation is evidence of a resonant $0^{-+}(a_0)$ wave and predominantly nonresonant $1^{++}0^{+}(K^*)$ wave at $-t$ above 0.2 GeV^2/c^2. Both the phase motion and the intensity distribution of the $0^{-+}(a_0)$ wave are consistent with the presence of one resonance near 1.40 GeV/c^2 or two resonances between 1.40 and 1.50 GeV/c^2. Contrary to the two larger $-t$ regions, for the low $-t$ region ($0.0 \leq -t < 0.2$ GeV^2/c^2) the phase of $0^{-+}(a_0)$ relative to $1^{++}0^{+}(K^*)$ is almost constant. Therefore, the possibility of a resonance in 1^{++} at $-t$ below 0.2 GeV^2/c^2 cannot be ruled out. The phase of $0^{-+}(K^*)$ for $0.2 \leq -t < 0.45$ GeV^2/c^2 and $0.45 \leq -t < 1.0$ GeV^2/c^2 seems to be similiar to the phase of $0^{-+}(a_0)$, although not as well measured.

In the 1200 - 1350 MeV/c^2 $K^+\overline{K}^\circ\pi^-$ mass region, we see a 1^{++} resonance at $\sim$ 1285 MeV/c^2 which is identified as the $f_1(1285)$ [was $D(1285)$] along with a 0^{-+} resonance at $\sim$ 1280 MeV/c^2, the $\eta(1275)$.

In the 1350 - 1500 MeV/c^2 region, the 0^{-+} resonance composes $\sim$ 50 % of the observed peak. (The possibility of two resonances is not excluded. It is unclear whether $a_0\pi$ and K^*K are different decay modes of the same resonance(s) or different resonances.) A 1^{+-} resonance composes $\sim$ 20% of the observed peak and a 1^{++} resonance is seen only at $0.0 \leq -t < 0.2$ GeV^2/c^2 and composes $\sim$ 10 % of the observed peak. The balance of the observed peak is non-resonant 1^{++} K^*K threshold effect and small amounts of 1^{-+} and 1^{--}.

Above 1500 MeV/c^2, we see a 1^{++} resonance at $\sim$ 1560 MeV/c^2, the $f_1(1530)$ [was $D'(1530)$].

REFERENCES

1) P. Baillon, et al., *Nuovo Cimento* **50A**, 393 (1967)
2) A. Ando, et al., *Phys. Rev. Lett.* **57**, 1296 (1986)
3) C. Dionisi, et al., *Nucl. Phys.* **B169**, 1 (1980)
4) C. Edwards, et al., *Phys. Rev. Lett.* **49**, 259 (1982)
   J.D. Richman, *Q.C.D. and Beyond*, J. Tran Thanh Van, ed. (Editions Frontiers, Gif Sur Yvette, France, 1985), p. 471; PhD thesis (Caltech), CALT-68-1231 (1985)
5) M.G. Rath, et al., submitted to *Phys. Rev. Lett.*
   M.G. Rath, PhD thesis (University of Notre Dame), 1988
6) For a review of glueball candidates see F.E. Close, to be published in *Rep. Prog. Phys.* (1988)
7) S.U. Chung, et al., *Phys. Rev. Lett.* **55**, 779 (1985)
8) S. U. Chung and T. L. Trueman, *Phys. Rev.* **D11**, 633 (1975)

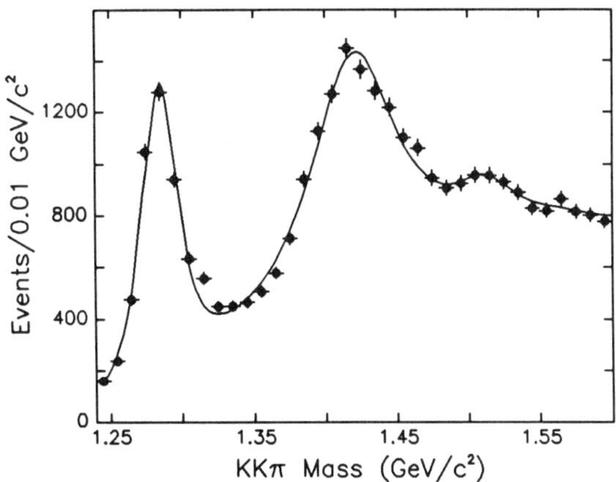

Figure 1. Observed $K^+K^0_S\pi^-$ mass spectrum, $0.0 \leq -t < 1.0$ GeV^2/c^2.

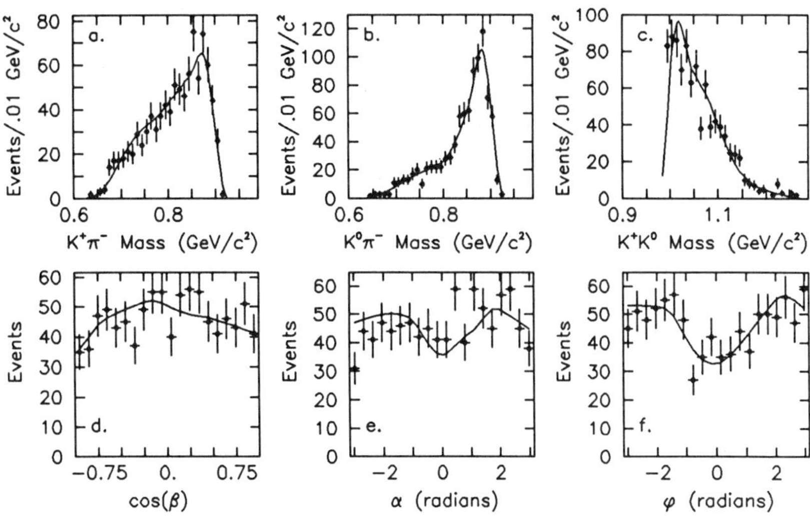

Figure 2. Comparison between data and the PWA fit of two-body masses and angular distributions of $K^+\pi^-$ for $1.40 \leq M(K\overline{K}\pi) < 1.42$ GeV/c^2, $0.2 \leq -t < 0.45$ GeV^2/c^2.

a. $K^+\pi^-$ mass b. $\overline{K}^\circ\pi^-$ mass

c. $K^+\overline{K}^\circ$ mass d. Polar angle of primary decay

e. Azimuthal angle of primary decay f. Azimuthal angle of secondary decay

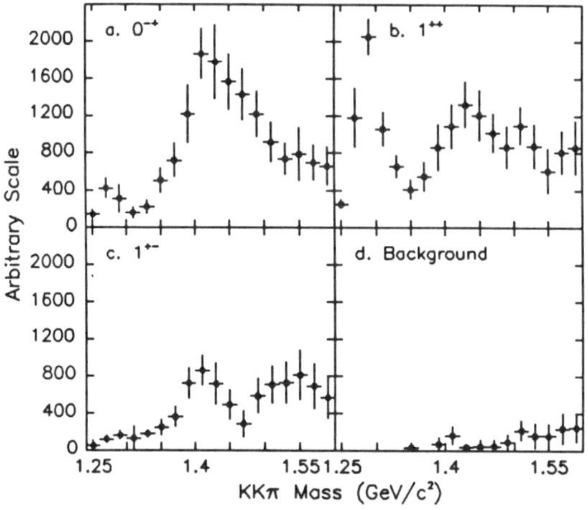

Figure 3. Major spin-parity components of the $K^+\overline{K}^0\pi^-$ system, $0.0 \leq -t < 1.0$ GeV^2/c^2.
a. 0^{-+} (a_0) and (K^*)
b. 1^{++} (a_0) and (K^*)
c. $1^{+-}(K^*)$
d. Background

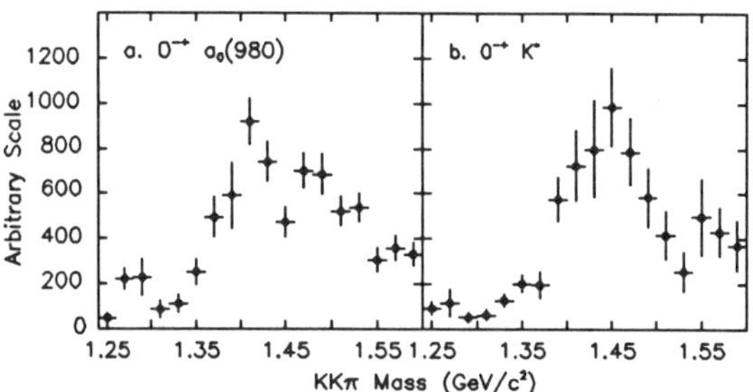

Figure 4. Contribution of 0^{-+} waves, $0.0 \leq -t < 1.0$ GeV^2/c^2.
a. $0^{-+}(a_0)$
b. $0^{-+}(K^*)$

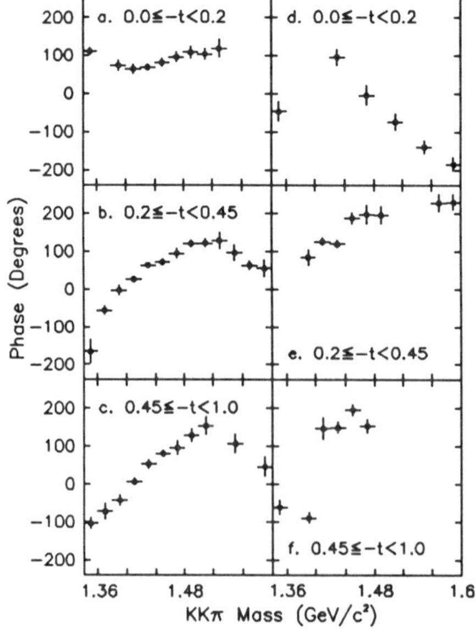

Figure 5. Phases with respect to $1^{++}0^+(K^*)$ for the three t regions discussed in the text. If the density matrix element is within two standard deviations of zero, the corresponding phase is not shown.
a-c. $0^{-+}(a_0)$
d-f. $0^{-+}(K^*)$

THE NUCLEAR DEPENDENCE OF PARTON DISTRIBUTIONS

F. E. Close
Physics Department, University of Tennessee
Knoxville, TN 37996 U.S.A.

and

Oak Ridge National Laboratory*
Oak Ridge, TN 37831-6373 U.S.A.

3rd Conference on the Intersections between
Particle and Nuclear Physics
Rockport, Maine
May 14-19, 1988

*Operated by Martin Marietta Energy Systems, Inc. under contract DE-AC05-84OR21400 with the U.S. Department of Energy.

THE NUCLEAR DEPENDENCE OF PARTON DISTRIBUTIONS

F. E. Close
Physics Department, University of Tennessee
Knoxville, TN 37996 U.S.A.

and

Oak Ridge National Laboratory*
Oak Ridge, TN 37831-6373 U.S.A.

ABSTRACT

I review the emerging information on the way quark, antiquark, and gluon distributions are modified in nuclei relative to free nucleons. I place particular emphasis on Drell-Yan and ψ production on nuclei and caution against premature use of these as signals for quagma in heavy-ion collisions.

INTRODUCTION

If we are to identify the formation of quark-gluon plasma in heavy-ion collisions by changes in the production rates for ψ relative to Drell-Yan lepton pairs, then it is important that we first understand the "intrinsic" changes in parton distributions in nuclei relative to free nucleons. So, I will review our emerging knowledge on how quark, antiquark, and gluon distributions are modified in nuclei relative to free nucleons, and briefly summarize the emerging theoretical concensus.

PARTONS IN NUCLEI

The best known nuclear distortion is that of the EMC effect which reveals a modification of the valence quark distributions in nuclei relative to those in free nucleons.

All experiments now show broad agreement.[1,2] The rise in F_A/F_N at $x > 0.7$ is due to Fermi motion causing the structure function F_A to leak out to $x > 1$; dramatic as this appears, it occurs where $F_{A,N} \simeq 0$, and is, in fact, a very minor contributor to the overall phenomenon. Indeed, overall, the effect is a subtle 10% affair, and

*Operated by Martin Marietta Energy Systems, Inc. under contract DE-AC05-84OR21400 with the U.S. Department of Energy.

we don't need to rewrite the nuclear physics textbooks. As $x \to 0$, we are beginning to see evidence for shadowing, a subject on which theory is now also starting to develop.[3]

QUARKS IN NUCLEI

In the "intermediate" region $0.2 \lesssim x \lesssim 0.6$ the ratio falls below unity as (valence) quarks lose momentum due to nuclear binding. The A dependence was successfully predicted in advance of data[4] and is rather well understood. It was first predicted in the context of the rescaling analysis and subsequently verified by experiments at SLAC.[1] Qualitatively, the effect is driven by the chance that there is a nearby nucleon correlated with the target nucleon. Reference 4 made a geometrical model and predicted δF_2^A for all A. (As the volume of a nucleon is only about 40% of the available volume, the two-nucleon contribution is $\lesssim$ 10% effect, and three-body and higher contributions can be ignored.) The fit is excellent and shows that the EMC effect is sensitive to details of nuclear structure, reflected in $\rho^A(r)$ as a result of which significant fluctuations are predicted at small A which have yet to be studied. At large A the behavior is smooth and it is safe to interpolate. Thus, one can infer the $F_A(x)$ when A = tungsten, say, and use this as input to $\pi W \to \mu\bar{\mu}$ ... analyses for example.

A common feature of models is that the degradation of the valence quarks transfers energy momentum to some other component (gluons and $q\bar{q}$ in rescaled QCD[5] or the partons in the π that are responsible for nuclear binding).[6,7] Thus, they generate an increased sea in nuclei relative to that measured in free nucleons. In turn, this implies that $F_A/F_N > 1$ as $x \to 0$. However, this predicted enhancement will probably be blacked out by shadowing (which has not been incorporated in these models so far). Mueller and Qiu[4] have begun to illuminate us about the x and Q^2 dependence of nuclear shadowing; the first attempts at making a quantitative combination of their work with "soft π"[6,7] or rescaled QCD[5] are now appearing.[8,9] I will discuss these ideas later; first I will concentrate on the empirical situation concerning parton distributions in nuclei with specific reference to the Drell-Yan process.

If $x_{1,2}$ refers to the beam and target partons, $x_F = x_1 - x_2$, and $x_1 x_2 = Q^2/s$, then the ratio of cross sections for some fixed Q^2/s is

$$\frac{\sigma^{bA}}{\sigma^{bN}} \sim \frac{\bar{q}^b(x_1) q^A(x_2) + q^b(x_1)\bar{q}^A(x_2)}{\bar{q}^b(x_1) q^N(x_2) + q^b(x_1)\bar{q}^N(x_2)}$$

where sum over flavors weighted by their squared charge is understood. In the case of π^- beams, if $x_2 > 0.2$ so that $\bar{q}^A \ll q^A$, the DY process is dominantly due to $\bar{q}^\pi$ annihilating with $q^{A,N}$. Thus, in this kinematic regime

$$\frac{\sigma^{\pi-A}}{\sigma^{\pi-N}} \sim \frac{u^A(x_2)}{u^N(x_2)} \sim \frac{F^A(x_2)}{F^N(x_2)}$$

This is the same ratio as measured in inelastic lepton scattering ("EMC effect") and must be obtained here too if factorization is valid. Thus, we should not be surprised by the results from NA10[10] who study

$$\frac{\sigma(\pi^- W \to \mu^+ \mu^- \ldots)}{\sigma(\pi^- D \to \mu^+ \mu^- \ldots)}$$

and by varying Q^2 and x_F can separate both the pion and target structure functions. In Ref. 10 they exhibit the resulting ratio of $q^A/q^N(x_2)$ and $\bar{q}^{\pi(A)}/\bar{q}^{\pi(N)}(x_1)$.

The latter should be unity and is within errors when one combines data from two energies, 140 GeV and 286 GeV incident π beams. (However, if one restricts attention to the lower energy sample, the situation is more messy and the pion distributions do not seem to factorize. Why this should be is unclear to me, but bear it in mind as an empirical observation for later reference.)

So the message is: do not be misled by $\sigma = A^\alpha$, $\alpha \simeq 1$ for Drell-Yan pair production on nuclei. While this may be approximately true for the total rate, there can be (and are) non-trivial effects in x and p_t^2. A depletion at large x may be compensated by an enhancement as $x \to 0$ for example. The kinematic conditions of experiments may emphasize different regions of x for the beam and target. The extent to which A-dependent effects will arise depends on $\bar{q}^A(x)$ and $g^A(x)$, about which we know almost nothing. Rescaling and pion models imply that both $\bar{q}(x,Q^2)$ and $g(x,Q^2)$ have non-trivial A dependence; moreover, shadowing effects will modify them as $x \to 0$. Empirical information is only now beginning to emerge.

ANTIQUARKS IN NUCLEI

The Drell-Yan process with incident nucleons can probe $\bar{q}$ in the target if suitable kinematics are chosen, e.g., $x_1 \simeq 0.7$ and $x_2 \ll x_1$. An investigation of this in various models has been made by Bickerstaffe et al.[11] and by Berger et al.[7] As an example, in Fig. 1, I show the predictions for $\bar{q}^A/\bar{q}^N(x)$ in iron in three models compared with information gleaned from CDHS.[13] The dramatic rise in the Berger-Coester model at $x > 0.3$ is due to their prediction that $\bar{q}$ leak out to moderate x values in nuclei. However, it is illusory to some degree as both $\bar{q}^A$ and $\bar{q}^N$ are vanishingly small; even so, experiment E772 may be able[12] to test this. Independent of specific models, it is an interesting question whether $\bar{q}$ leak to "large" x in nuclei as this will have a bearing on the Q^2 shape of Drell-Yan pairs in nuclei which may differ from the ψ production (produced by gluons) and potentially provide a background to the plasma signal sought in heavy-ion collisions.

Recently WA25 and WA59 in collaboration have studied the EMC effect using ν and $\bar{\nu}$ interactions in neon and deuterium. The x,y distributions allow separation of quark and antiquark distributions, and there is some indication that the sea decreases in going from deuterium to neon; however, these data are probably being dominated by nuclear shadowing, like the electromagnetic data for $x \lesssim 0.1$, again highlighting the need to see how shadowing modifies the curves in Fig. 1 at small x.

GLUONS IN NUCLEI

Insofar as inelastic $\gamma A \to \psi + \ldots$ proceeds via photon-gluon fusion and $NA \to \psi + \ldots$ involves gluon-gluon fusion, these processes probe $g^A(x)$. There have been early claims that $g^A > g^N$ $(x \sim 0.05)$, this based on the EMC data[14] for $(\gamma Fe \to \psi)/(\gamma D \to \psi \ldots)$. Expressed as

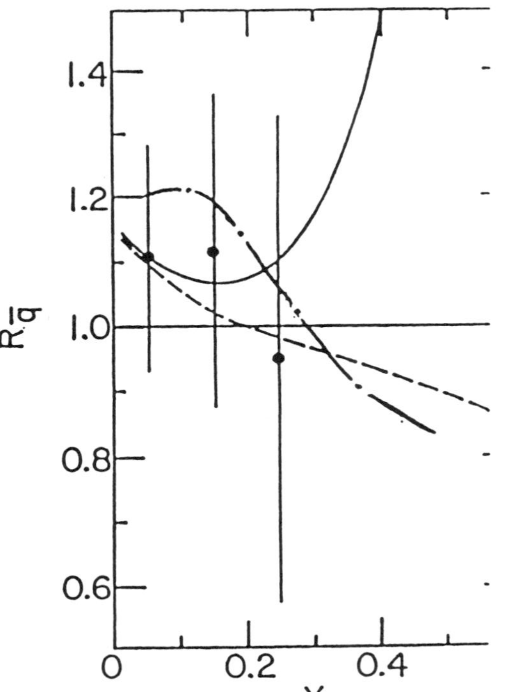

Fig. 1. Data from Ref. 13 on the ratio $\bar{q}_-^{Fe}/\bar{q}^D(x)$ from neutrino scattering. Here $\bar{q} = (\bar{u}=\bar{d}=2\bar{s})$. The solid curve illustrates predictions of pion exchange model (Refs. 7 and 11), the dot-dash is the pion model of Ref. 6, and the dashed curve is the rescaling model, Ref. 5.

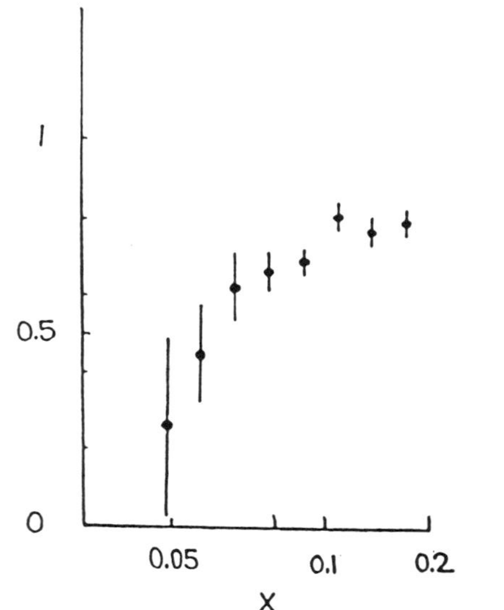

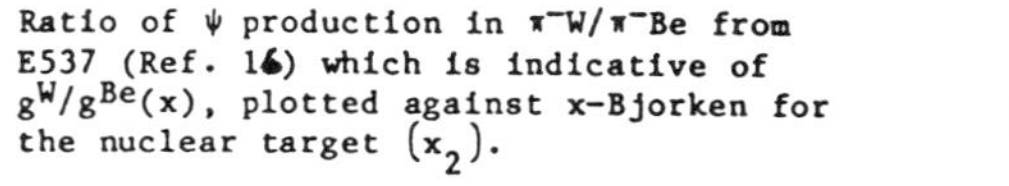

Fig. 2. Ratio of ψ production in π^-W/π^-Be from E537 (Ref. 16) which is indicative of $g^W/g^{Be}(x)$, plotted against x-Bjorken for the nuclear target (x_2).

A^α this gave $\alpha = (1.10 \pm 0.03 \pm 0.04)$. However, it is now less clear whether coherent ψ production (for which $\alpha \simeq 4/3$) has been entirely removed from the data. Indeed E691 (Sokoloff *et al.*) report[15] that (at $Q^2 = 0$)

$$\alpha_{\text{coherent}} = 1.40 \pm 0.06 \pm 0.04$$

$$\alpha_{\text{incoherent}} = 0.94 \pm 0.02 \pm 0.03$$

This suggests that gluons are shadowed in nuclei.

There are also confusing signals[16] coming from E537 who probe the gluon distribution with $\pi^- W/Be \to J/\psi \ldots$ at 125 GeV, measuring the x_F distributions and thereby enabling $g^{W/Be}$ to be measured for x_{Bj} 0.2. If $x_F = x_1 - x_2$, then for J/ψ production at 125 GeV

$$x_F \simeq \frac{1}{25x_2} - x_2$$

Thus, we can replot the data from E537 against X_2 (Fig. 2). This is equivalent to $g^W(x)/g^{Be}(x)$ only if $g^{\pi(W)} \neq g^{\pi(Be)}$ cancels out. However, we have no immediate way of knowing if this is true empirically as $Q^2(=m_\psi^2)$ is fixed. *Prima facie*, one may justifiably be worried. First, which is theoretical prejudice, if Fig. (2) is interpreted as $g^{W/Be}(x)$, it implies that gluons are significantly shadowed for x as large as 0.2. Our understanding of shadowing is still rather primitive, but such behavior would be against all current models. However, the Na(10) data at 140 GeV were confusing, as I mentioned earlier, and things only became clear when the 286-GeV data were considered. Therefore, in the E537 case I will remain cautious until higher energy ψ production data are available.

Experiment E672 at Fermilab is measuring[17] $\pi^- A \to \psi$ on four nuclei at 530 GeV. I await their "high energy" extraction of $g^{A/N}(x)$. Until the conundrum of energy dependence (i.e. the non-factorization of the partons in the incident beam) is settled, I conclude that $g^{A/N}(x)$ probably falls as $x \to 0$, in qualitative agreement with the shadowing phenomenon, but the quantitative measure is unclear.

Thus, with the exception of valence quarks for $x > 0.2$, there is little or no evidence for non-trivial behavior for $\bar{q}^{A/N}$ and $g^{A/N}$. It is imperative to know these quantities much better, and to understand the anomalous energy dependence manifested by NA(10), and implicitly hinted at by E537. Until we do, then we cannot, with any confidence, use J/ψ relative to Drell-Yan production in AA collisions as a signal for quark-gluon plasma formation. Note that if the E537 experiment's dramatic suppression of nuclear glue is true, then ψ production per nucleon in heavy-ion collisions will be markedly suppressed relative to that in pp or even pA interactions. There is no reason to anticipate such drama for Drell-Yan.

SHADOWING

With the recent work of Mueller and Qiu[3] we have a model for shadowing that is rooted in QCD. As $x \to 0$, the evolution equations imply that the number densities of gluons become very large. Gluons (and quarks) from different nucleons should interact with each other and reduce the number densities through annihilation. These annihilations are then responsible for nuclear shadowing.

For our present purposes, we need only be concerned with the x dependence of the input shadowed distributions. Arguments based on the longitudinal length scales probed suggest that shadowing vanishes if $x > x_n \simeq 1/2\ rm$ (m = nucleon mass, r its radius) and becomes total if $x < x_A \simeq 1/2\ Rm$ (R = nuclear radius). Qiu then interpolates; his formula for the shadowing ratio $F^A/F^N(x)$ may be written (Fig. 3)

$$F^A/F^N(x) = R(x) = \begin{cases} 1 - K\left(\frac{x_n}{x} - 1\right) & (x_A < x < x_n) \\ 1 - K\left(A^{1/3} - 1\right) & (0 < x < x_A) \end{cases}$$

where K is a constant to be fitted by $x \to 0$ data. We thus see that R is a universal curve whose only A dependence is the saturation point.

We can now see the effect of shadowing combined with EMC-inspired models. For heavy nucleus B, the enhancement is generally expected to be greater than for a light nucleus A. On shadowing these expectations with Qiu's universal curve, one finds that the crossover from enhancement to shadow at small x will be $x(B) < x(A)$ (Fig. 4). This will be a subtle effect and more analysis is needed to see precisely what the values of the crossover should be. The empirical extraction

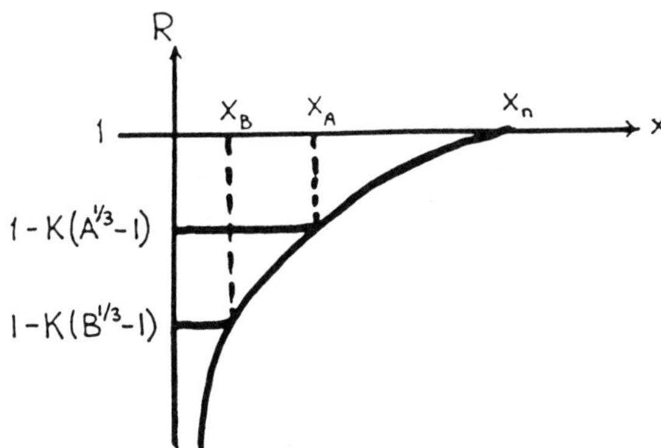

Fig. 3. A dependence of shadowing in Mueller-Qiu model.

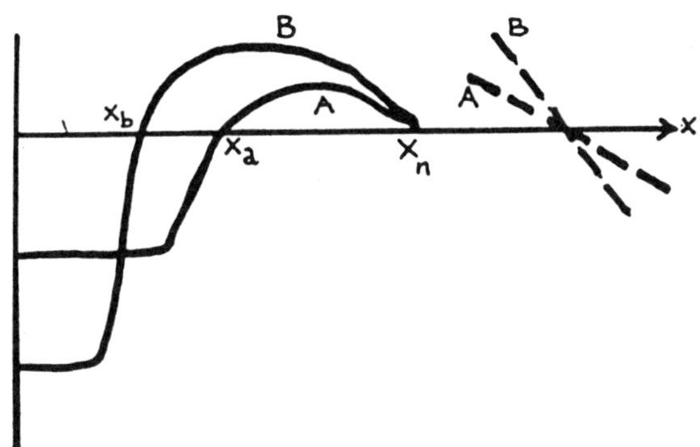

Fig. 4. Effect of Mueller-Qiu shadowing combined with enhancements in naive EMC models. Note the crossover $x_b < x_a$.

of the crossover $x_c(A)$ will be an interesting exercise in the future.

The emerging data suggest that $x(B) > x(A)$, contrary to the above. This may imply that x_n is A-dependent. Qiu assumed, for simplicity in a first orientation, that $x_n \sim 1/R$ where R is the nucleon size — this being a measure of the distance scale for quarks from neighboring nucleons to "interfere". However, the A-dependence of the large x data ("EMC effect") suggests that quarks in neighboring nucleons have a greater chance to interfere as A increases; therefore the interference distance scale decreases and x_n increases with A. This qualitatively mimics the trend of the shadowing data but, of itself, does not seem to be sufficient. Berger and Qiu[9] in a recent ANL preprint have proposed a similar idea but have refrained from making a quantitative comparison. I will now look at the A dependence of small x data in more detail.

We take the original rescaled QCD, or the similar curves of the pion models of the EMC effect, and shadow them using Qiu's shadowing ansatz. Note that, for a given A, a single scale controls shadowing since he assumes that saturation $(x = x_A)$ and onset $(x = x_n)$ are related as follows

$$\frac{x_A}{x_n} \simeq \frac{R_n}{R_A} \simeq A^{-1/3}.$$

When we compare the resulting curve with the data for carbon (smallest A, largest x_A), we find that the crossover point (x_a) is predicted at larger x than in the data. One must find some way of translating the curve to smaller x values.

If we are to preserve the $Q^2 = 0$ $(x < x_A)$ saturation value, one way to achieve an effective shift is to reduce x_A. However, this is probably unphysical: $x_A \sim R_A^{-1}$ and so one achieves saturation at dimensions greater than those of the entire nucleus. Our present taste is that shadowing is complete at distances no greater, and maybe even less, than nuclear dimensions. Therefore to attain the desired translation, we choose to shadow only the sea (a rationale being that, in Müller and Qiu's model, this component, and not valence quarks, is responsible for shadowing).

One gets an acceptable fit to carbon data in this way: shadow the sea in the rescaled QCD or pion models of EMC.

Now proceed to the A dependence. As pointed out earlier, the naive prediction that x_a (Fig. 4) decreases as A increases is inconsistent with data. This bad result is probably generally true in any single scale model. Berger and Qiu have recently suggested that two length scales are involved (surface to volume effects provide the new degree of freedom in their work) and can thereby reverse the dependence, at least qualitatively.

Inspired by their idea, we have begun to develop a unified picture of the A dependence for all x. This model has similar empirical features, but differences in detail, to that of Ref. 8. It is rooted in our previous successful prediction[4] of the A dependence at large x. With increasing A, there is an increasing chance for a parton in one nucleon to be "aware" of constituents of neighboring nucleons; the mean distance between the members of such "overlapping" nucleons decreases and hence x_n increases with A.

This is similar to Ref. 8. While it may agree with the trend of the data for $x \gtrsim 0.1$, it does not, alone, describe the most dramatic A dependence in the new EMC data, namely $x_a(A)$. Reference 8 and the present ideas have very little to say about the actual data; the data at $x \lesssim 0.1$ seem to require a radical re-evaluation of $x_a(A)$.

So we return to carbon, with which there exists a reasonable fit, and look for guidance: we were able to fit carbon with its canonical $x_A \sim R_A^{-1} \sim x_n A^{-1/3}$. One easily verifies that the most dramatic effects arise from changing $x_A(A)$. Qiu's assumption is that $x_A \sim A^{-1/3}$; we illustrate the other physical extreme — namely that x_A = constant ($\equiv x_{carbon}$, motivated by our successful "fit" to that data). Physically this corresponds to, at large Q^2, saturation of shadowing occurring once a few nucleon dimensions only are exceeded. Such a behavior is consistent with data, but it is far too soon to claim that data imply this. At this stage, we note only that it is $x_A(A)$ which seems to control most radically the data, and that if $x_A/x_n \sim A^{-n}$, then n is probably nearer to zero than 1/3. A dynamical model for this is awaited.

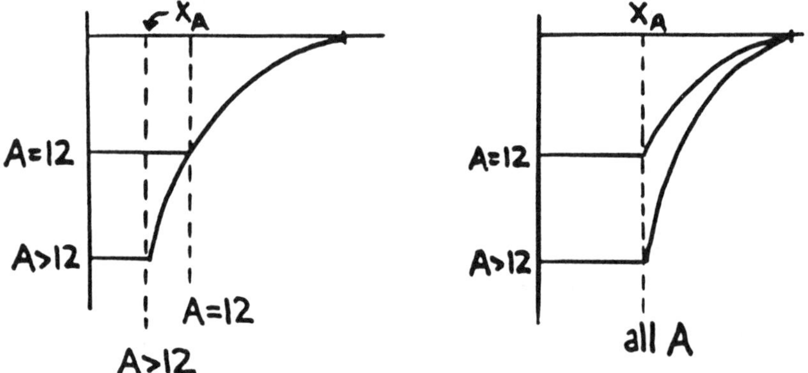

Fig. 5. Effect of preserving the depth ($x \to 0$) but modifying x_A.

REFERENCES

1. EMC collaboration: Phys. Lett. 123B, 275 (1983); BCDMS: Phys. Lett. 163B, 282 (1985); A. Bodek *et al*., Phys. Rev. Lett. 50, 1431 and 51, 534 (1983); R. Arnold *et al*., Phys. Rev. Lett. 52, 727 (1984).

2. T. Sloan, Proceedings of Uppsala International Conference on HEP (1987).

3. A. Mueller and J. Qiu, Nucl. Phys. B268, 427 (1986); J. Qiu, Nucl. Phys. B291, 746 (1987). Early qualitative ideas are in N. N. Nikolaev and V. Zakharov, Phys. Lett. 55B, 397 (1975).

4. F. E. Close, R. L. Jaffe, R. G. Roberts, and G. G. Ross, Phys. Letts. 134B, 449 (1984).

5. F. E. Close, R. G. Roberts, and G. G. Ross, Phys. Letts. 129B, 346 (1983).

6. C. H. Llewellyn Smith, Phys. Lett. 128B, 107 (1983); M. Ericson and A. Thomas, Phys. Lett. 128B, 112 (1983).

7. E. Berger and F. Coester, Phys. Rev. D32, 1071 (1985); E. Berger, F. Coester, and R. Wiringa, Phys. Rev. D29, 398 (1984); E. Berger, Nucl. Phys. B267, 231 (1986).

8. E. Berger and J. Qiu, ANL-HEP-PR88-06.

9. F. E. Close and R. G. Roberts (in preparation).

10. NA(10): Phys. Letts. 193B, 368 (1987).

11. R. Bickerstaffe, M. Birse, and G. Miller, Phys. Rev. Lett. 52, 2532 (1984).

12. J. Moss *et al*., E772 collaboration.

13. CDHS: H. Abramowicz *et al*., Z. Phys. C25, 29 (1987); E. Berger and F. Coester, Argonne report ANL-HEP-PR-87-13 (to appear in Ann. Rev. of Nuclear and Particle Science).

14. EMC: J. Aubert *et al*., Phys. Letts. 152B, 433 (1985); Nucl. Phys. B213, 1 (1983); F. E. Close, R. G. Roberts, and G. G. Ross, Z. Phys. C26, 515 (1985).

15. E691, M. D. Sokoloff *et al*., Flab 86/120E.

16. E537, S. Katsanevas *et al*.,

17. E672, reported by A. Zieminski in Proceedings of St. Croix Workshop on QCD, October 1987.

SEARCH FOR QGP SIGNALS AT AGS WITH A TPC SPECTROMETER, AND COMPARISON OF OUR EVENT GENERATOR PREDICTIONS FOR PLASMA MODEL AND CASCADE INTERACTIONS†

S.J. Lindenbaum
Brookhaven National Laboratory and City College of New York

K.J. Foley, S.E. Eiseman, A. Etkin, R.W. Hackenburg,
R.S.Longacre, W.A. Love, T.W. Morris, E.D. Platner,
A.C. Saulys
Brookhaven National Laboratory, Upton, New York 11973

C. Chan, M.A. Kramer
City College of New York, New York, New York 10031

T.J. Hallman, L. Madansky
Johns Hopkins University, Baltimore, Maryland 21218

B.E. Bonner, J.A. Buchanan, J.M. Clement, M.D. Corcoran,
J.W. Kruk, H.E. Miettinen, G.S. Mutchler, F. Nessi-Tedaldi,
M. Nessi, G.C. Phillips, J.B. Roberts
Rice University, Houston, Texas 77251

ABSTRACT

We have developed and successfully tested a TPC Magnetic Spectrometer to search for QGP signals produced by ion beams at AGS. We also developed a cascade and plasma event generator the predictions of which are used to illustrate how our technique can detect possible plasma signals.

INTRODUCTION

In this talk we presented the following topics under investigation by the BNL/CCNY/Johns Hopkins/Rice Collaboration: 1) Quark Gluon Plasma and ordinary cascade event simulation in ion-nucleus collisions; 2) A TPC Tracking Magnetic Spectrometer system for plasma signal detection; 3) Plasma signals detectable at AGS.

In order to simulate what non-plasma events would do we modified ISAJET to conserve baryon number, strangeness, energy and momentum. We then improved the remainder of HIJET to include cascading of slow produced particles inside the nucleus leading to N-N, π-N and K-N scattering. Then we interfaced a version of the fragment program

---

† This research was supported by the U.S. Department of Energy under Contract Nos. DE-AS05-81ER40032, DE-AC02-76ER03274, DE-AC02-76CH00016, DE-AC02-83ER40107 and the City University of New York PSC-BHE Research Award Program.

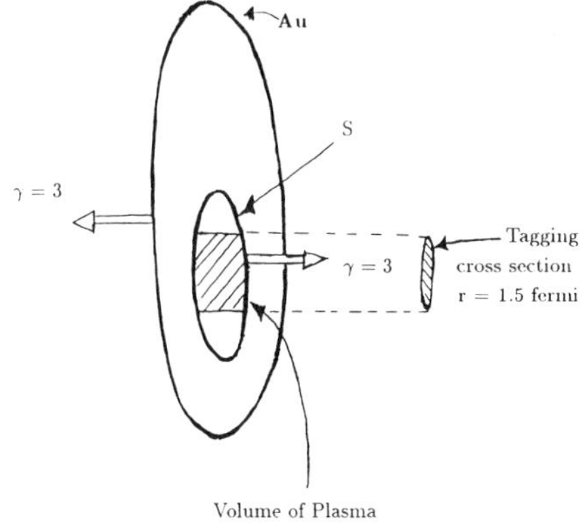

Figure 1a: Tagged Plasma Volume

FAIHI with HIJET. This gave us a more realistic cascade (non-plasma) generator for ordinary events.

Central ordinary cascade events led to essentially complete stopping (in the nucleon-nucleon cms) of incident sulphur ions by a gold target. This is consistent with experimental results[1] for AGS energies. As an illustration of what our technique can be expected to do we added a simple plasma model. We assumed that in some of the 15 GeV/c × A sulphur on Au events the nucleon-nucleon central collisions in a tagged region of radius 1.5 fermi were stopped in a compressed volume in the nucleon-nucleon cm system corresponding to the $\gamma = 3$ (Fig. 1a). This determined the energy and chemical potential. These tagged nucleons made a spherical plasma bubble of the Van Hove type of critical temperature ≈ 170 MeV, with chemical potential and volume which conserves baryon number and energy.

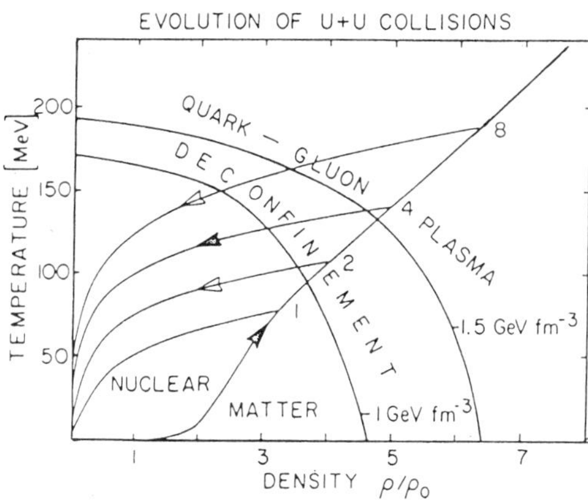

Figure 1b: T vs. ρ (Stöcker, Nucl. Phys. A418, 587c (1984)

If one looks at an example of the canonical temperature vs. pressure diagram (Fig. 1b) in conjunction with the dynamics (Fig. 1b) it appears that approximately complete stopping at AGS energies could have a reasonable chance of producing plasma in some of the central events at AGS by providing a combination of high enough baryon density and temperature. The plasma breakup was based on the work of Koch, Muller and Rafelski.[2] In their model the particle production probabilities depends on the critical temperature and the gluon fragmentation function.

We conserve all relevant quantum numbers, energy and momentum. The plasma bubble typically contains ≈ 30 GeV (i.e. < 20% of the available energy) and spreads over ≈ 1 unit of rapidity. A given chemical potential μ and critical temperature T_c for the plasma bubble determines N_q, $N_{\bar{q}}$, N_s, $N_{\bar{s}}$, N_g where N stands for the number. The subscripts q, $\bar{q}$, s, $\bar{s}$ and g stand for (u,d) quarks, ($\bar{u}$,$\bar{d}$) quarks, strange quarks, anti-strange quarks and gluons respectively. The effective number of quarks when they hadronize are: $\tilde{N}_q = N_q + f_q N_g$; $\tilde{N}_{\bar{q}} = N_{\bar{q}} + f_q N_g$; $\tilde{N}_s = N_s + f_s N_g$; $\tilde{N}_s = \tilde{N}_{\bar{s}}$. where f_q = gluon fragmentation function for u and d. f_s = gluon fragmentation functions for s. The average number of particles produced each with a Boltzman distribution corresponding to the critical temperature is: $N_\pi = \alpha\tilde{N}_q\tilde{N}_{\bar{q}}$, $N_K = \alpha\tilde{N}_q\tilde{N}_{\bar{s}}$, $N_{\bar{K}} = \alpha\tilde{N}_{\bar{q}}\tilde{N}_s$, $N_\phi = \alpha\tilde{N}_s N_{\bar{s}}$, $N_n = 1/3!\ \beta\tilde{N}_q{}^3$, $N_{\bar{n}} = 1/3!\ \beta\tilde{N}_{\bar{q}}{}^3$, $N_\Lambda = 1/2!\ \beta\tilde{N}_q{}^2\tilde{N}_s$, $N_{\bar{\Lambda}} = 1/2!\ \beta\tilde{N}_{\bar{q}}{}^2\tilde{N}_{\bar{s}}$, $N_\Xi = 1/2!\ \beta\tilde{N}_q\tilde{N}_s{}^2$, $N_{\bar{\Xi}} = 1/2!\ \beta\tilde{N}_{\bar{q}}\tilde{N}_{\bar{s}}{}^2$, $N_\Omega = 1/3!\ \beta\tilde{N}_s{}^3$, $N_{\bar{\Omega}} = 1/3!\ \tilde{N}_{\bar{s}}{}^3$. The plasma production and breakup models we have used may be far from correct. However that may also be true of all others proposed to date.

REQUIREMENTS ON THE DETECTOR SYSTEM

From the foregoing investigation it is clear that a tracking detector system capable of working in this high particle density environment with sufficient space, angular and momentum resolution and a means of identifying $\bar{p}$ is essential. A TPC is a three dimensional point detector and thus can track in much higher multiplicity enviroments than projective geometry detectors. We have built a specially designed TPC to maximize two-track separability and also minimize costs. As a result dE/dx information for the TPC is not available. However it is supplemented with a time-of-flight hodoscope (TOF) which will identify ≈ half of the $\bar{p}$ particles of interest. Secondly negative particles are expected to be > 90% pions so that pion identification can be made to a good approximation. Figure 2 shows the experimental arrangement for the first phase of AGS E-810. The TOF $\bar{p}$ identifier is 50' downstream of target and is not shown in Fig. 2.

Four modules comprise the TPC system. They are placed along the beam with the S (or Si) and O ion beams passing through the TPC to provide large solid angle coverage. From ion beam tests we conclude that this will be satisfactory for these runs. However it should be noted that one can deaden the beam area if necessary. When the booster becomes available at AGS to accelerate Au ions; this will likely become necessary. The anode readout wires are 20 μ gold-plated tungsten 1 cm long rows parallel to the beam direction. There are 10 wires to the inch between cathode structures. A gate which opens only when events of interest occur is included for operation at high ion beam rates (~ 1/2 10^4/pulse).

A TPC module was tested in > 25K Si ions per pulse beam without magnetic field. The hits can clearly be associated with several tracks consistent with beam particles dispersed in y because of late arrival. When we reduced the anode structure gain to correspond to detecting minimum ionizing particles the silicon ions were detected with high efficiency > 90%. The next tests were in an 18 GeV/c proton beam. Figure 3a shows accidental tracks from several million per pulse incident 18 GeV/c protons. Figure 3b shows detected events from a target. Our pattern recognition program led to fits with efficiencies of > 90%. Figure 3c shows the TPC module inside the MPS magnet with a 5 KG field. A 3 GeV π^- beam is incident and low momentum tracks from a target are clearly reconstructed with high efficiency (> 90%). The resolutions in x (transverse to the beam and magnetic field) and y (along the drift and magnetic field direction but perpendicular to the beam) were both measured to be < 1 mm. The two track separability was measured to be ≈ 1-3 mm. The gas is a stable high gain, low difusion mixture of Argon, Isobutane and Methylal which is a slightly modified mixture of the standard MPS II drift chamber gas. Minimum ionizing particles gave clear well-shaped pulses averaging 8 μamps with peaks of > 20 μamps. Setting a threshold of 2-3 μamps ensured clean pulse operation with very high efficiency. The readout electronics was vintage 86 LeCroy hybrids we designed for TPC readout. An improved version of prototypes has in tests shown that even lower thresholds (i.e. a fraction of a microamp) will be useable with the new readout system.

Thin targets will be used to minimize secondary interactions. The triggering system includes beam halo counters in veto, counters to detect interactions in the target and a $\sum Z^2$ counter to select candidates for events varying from central to peripheral depending on the setting (see Fig. 2). We prefer minimal triggering in the early stages since according to our Monte Carlo's and plasma model one can get surprises if one triggers tightly. For example our plasma model calculations predict that plasma events have a smaller multiplicity distribution than cascade events at AGS energies. The pattern recognition program previously developed by BNL/CCNY was described in Ref. 3. It has been modified somewhat. A local pattern recognition forms a single hit from adjacent and nearby readout wires. A subroutine corrects $\vec{E} \times \vec{B}$ effects which varies from less than 2 mm to a fraction of a mm. Track recognition and reconstruction starts downstream. The PR forms and fits > 3 consecutive hit chains. Finally the chains are joined to form tracks. The existing MPS vertex finding and fitting program has been adapted to the TPC analysis. High efficiency track reconstruction > 95% has been attained in Monte Carlo's and actual MPS and beam tests which were monitored on-line.

The acceptance in the forward half of the nucleon-nucleon cm rapidity region (i.e. y_{lab} = 1.75 to 3.5 correspooonding to y_{cm} = 0 - 1.75 in the cm system which extends from central to maximum rapidity). We get excellent (< 0.1 unit) pseudorapidity and rapidity

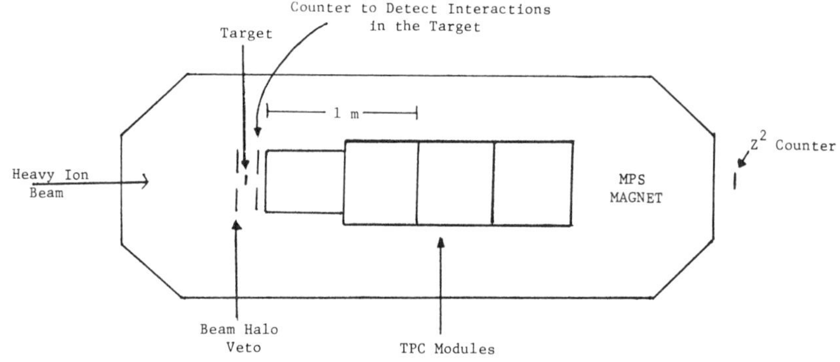

NOTE: Counters and target are not to scale

Figure 2: AGS E-810 Setup

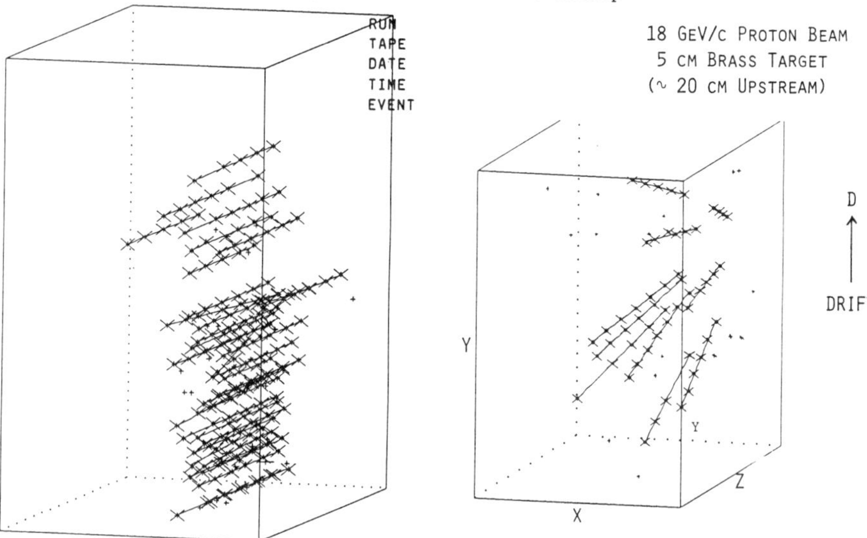

Fig. 3a: High Rate 18 GeV/c Protons

Fig. 3b: Interaction Detected

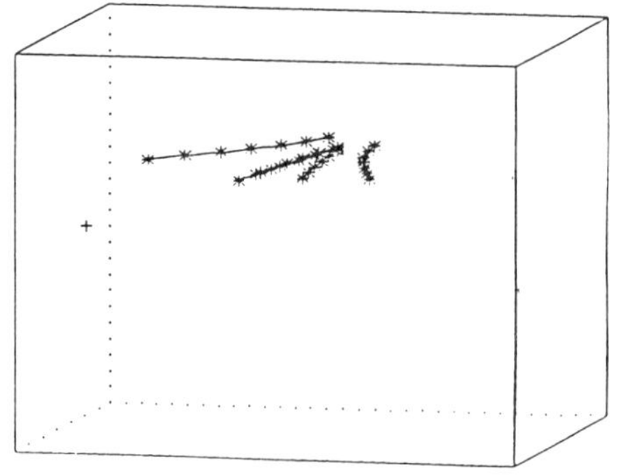

Fig. 3c: TPC in MPS Magnet

resolution (when particles are identified). Fig. 4 shows that central cascade and plasma events have similar enough rapidity distributions that we cannot use those as a plasma signal.

Figure 5 shows that the multiplicity distributions for cascade and plasma events are similar except that plasma events are shifted toward lower multiplicity by about 20% because they have enhanced production of heavier objects such as $\bar{p}p$ pairs, $\bar{\Lambda}$, and K_S, etc. Thus multiplicity cannot be used as a plasma detector. Comparisons between plasma and central cascade transverse and longitudinal momentum distribution of π^- show no essential difference. Thus they cannot be used as a plasma detection signal.

PLASMA SIGNALS

However we did find that $\bar{p}\bar{\Lambda}$ and $\bar{p}\bar{\Lambda}$ + Vees from the same event provide an excellent plasma signal which is detectable.

The $\bar{\Lambda}$ can be identified by cuts which ensure the $\bar{\Lambda}$ vertex or particles came from the interaction point in the target and that the $\bar{\Lambda}$ momentum points to the interaction point. K_S^0 and Λ can be identified in a similar manner. The results of this type of kinematic identification obtained by our Monte Carlo simulation for plasma events and for central events shows that we can identify Λ and K_S about an order of magnitude above background. In plasma events this is also true for $\bar{\Lambda}$. About half the $\bar{p}$ are identified by the time-of-flight hodoscope referred to previously. This will further improve the signal-to-background ratio.

In Table I we compare various plasma singals (according to the model described) in events with plasma and events with no plasma per 10,000 S-Au central events. Since we plan to gather ~ 50,000 effectively central collision events. If ≈ 5% of the central collisions produce plasma in an effectively central region covering a radius of $r \leq 1.5$ fermi, we should expect to observe the following: 1) 37 events containing a $\bar{p}\bar{\Lambda}$ with a background of ≈ 10; 2) 12 events containing $\bar{p}\bar{\Lambda}$ and 1 V with background ~ 1; 3) 2.5 events

TABLE I – PLASMA SIGNALS

Detected Event Rates and Background Rates per 10,000 $S - Au$ Central Events

Reaction Selector	Signal Required	Events With Plasma	No Plasma
(1)	$\bar{p}\bar{\Lambda}$	150	1.0
(2)	$\bar{p}\bar{\Lambda}$ and 1V	50	0.1
(3)	$\bar{p}\bar{\Lambda}$ and 2V	10	0.01
(4)	$\bar{p}\bar{\Lambda}$ and 3V	2	0.001

The background due to fake $\bar{\Lambda}$ is expected to be comparable with the events with no plasma.

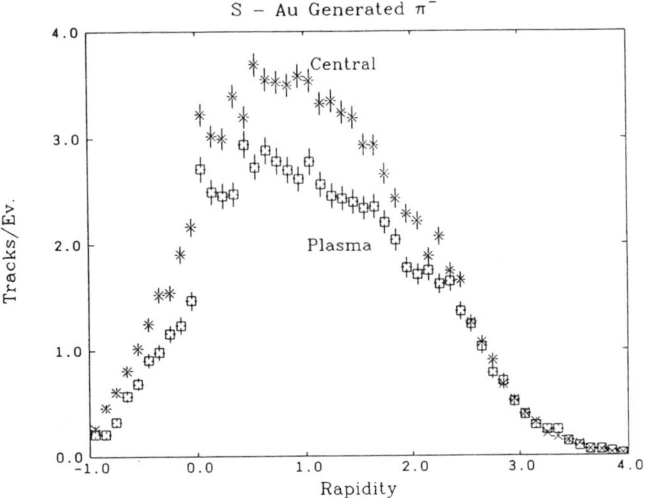

Figure 4: Rapidity Distribution

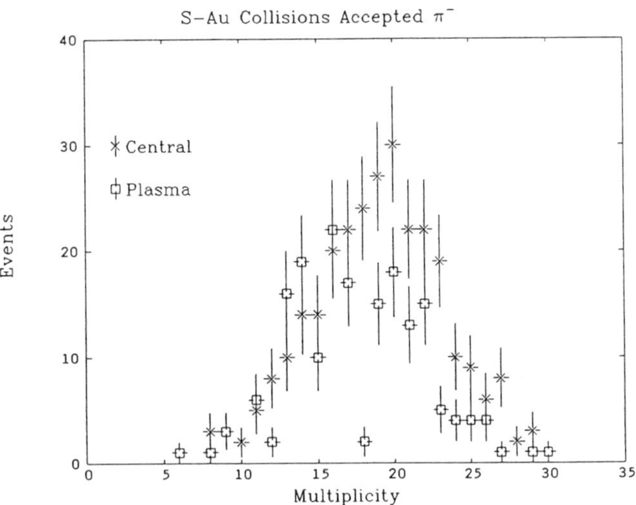

Figure 5: Multiplicity

containing $\bar{p}\bar{\Lambda}$ and 2 Vees with a background of ≈ 0.1; 4) ≈ 1/2 event $\bar{p}\bar{\Lambda}$ and 3V with a background ~ 0.01.

We plan to run with several targets including p, O, and Si beams. We obviously can increase our running time especially to detect reactions (3) and (4). We have also from our Monte Carlo's shown that we detect Hanbury-Brown twiss effects. We have used a particular plasma model to show the power of our technique. Nature may well reveal a different scenario. However we have a technique which is very well suited to detect any hadronic anomolies ad in fact we plan to use similar techniques for RHIC detectors.[4]

It should be noted that while some of the TPC module tests previously reported were being made in the heavy ion beam some preliminary data on Λ production was gathered in Si + Al collisions using MPS II with its drift chambers. Figure 6 shows the clear Λ peak.

Since the conference we have just now in progress tests of 15 GeV/c × A ion beams incident on Al etc. targets. The preliminary results for TPC with MPS magnetic field detection of the events are satisfactory.

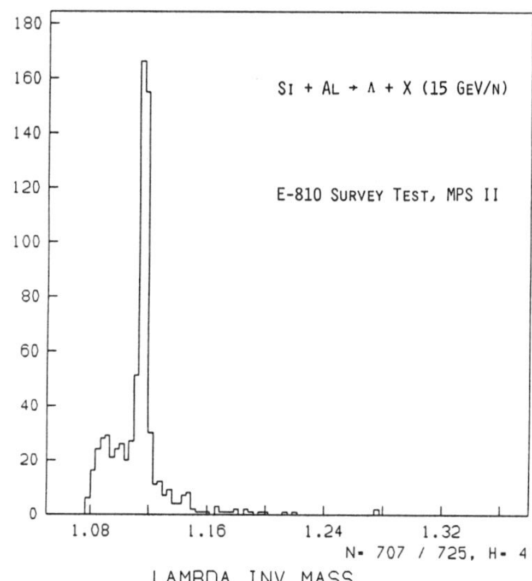

Figure 6:

Λ Peak with Drift Chambers

SUMMARY

1) We have improved HIJET in the following respects: a) We have modified ISAJET to conserve B, S, E, P, etc; b) Included cascading of N-N, π-N and K-N; c) Interfaced, the fragmentation program FAIHI; d) Used our model for plasma formation and treated plasma breakup according to Koch, Muller and Rafelski.

2. a) We have found with these models that at AGS energies the best plasma signals that are clearly detectable are $\bar{p}\bar{\Lambda}$ and $\bar{p}\Lambda$ plus Vees. We can detect these effects at a level of $<$ 5% plasma events with $r \leq 1.5$ fermi for effectively central collisions using the TPC spectrometer system of AGS E-810. Hanbury-Brown Twiss effects can also be detected.

3. Tests show the TPC works well with reconstruction efficiency $>$ 90%.

4. AGS E-810 data run is planned for fall of 88.

5. A Phase II of this program has been planned to fill a much larger part of the MPS magnet with a TPC system to increase the acceptance.

6. Our plasma and cascade event generators are being used for RHIC calculations.

7. We have proposed a large TPC spectrometer for RHIC.[4]

REFERENCES

1. Abbot *et al*., Phys. Lett. B 197, 285 (1987).

2. Koch, Müller and Rafelski, Phys. Reports, February 1986.

3. Lindenbaum *et al*., Nucl. Phys. A461, 431 (1987).

4. Lindenbaum, S.J. An Approximately 4π Tracking Spectrometer for RHIC, Proc. of Second Workshop on Experiments and Detectors for RHIC, LBL, Berkeley, California, May 25-29, 1987).

GLUEBALL CANDIDATES IN RADIATIVE J/ψ DECAYS

William J. Wisniewski ‡
California Institute of Technology, Pasadena, CA 91125

representing the Mark III Collaboration

Introduction

The J/ψ decays via several mechanisms: three gluon decay ($\sim 62\%$), electromagnetic decay ($\sim 30\%$), radiative decay via η_c ($\sim 1\%$), and radiative decay to two gluons ($\sim 7\%$). The last mode is of great interest for the study of unusual states predicted by QCD, such as glueballs (resonant states of two gluons) and hybrids (resonant states of two quarks and a gluon). Complications arise in interpreting the nature of states seen in this mode because of the presence of $q\bar{q}$ states. The experimenter has two rules-of-thumb for glueball candidates: they have a large branching ratio in J/ψ radiative decays and they do not fit into $q\bar{q}$ multiplets. Observed states which satisfy these criteria include $\eta(1440)(\iota(1440))$ and $f_2(1720)(\theta(1720))$. Results on these two states are presented, as well as new results on the radiative decay to *vector-vector* final states.

The Mark III Collaboration has acquired a data set of $5.8 \times 10^6 J/\psi$. The data were taken using the Mark III detector[1], a general purpose solenoidal detector installed at the e^+e^- storage ring SPEAR. The detector features relevant to the analyses presented are: a central drift chamber which provides momentum resolution $\delta p/p = 1.5\%(1+p^2)^{1/2}$ with 2σ $\pi - K$ separation by dE/dx below 0.6 GeV/c; time-of-flight(TOF) counters which provide at least 2σ $\pi - K$ separation up to 1.1 GeV/c; a calorimeter which covers 0.95 of 4π sr, has resolution $\delta E/E = 17\%/E^{1/2}$, and has photon detection efficiency which is 50% at 0.05 GeV and rises to $\sim 100\%$ at 0.10 GeV; and a muon detector which is efficient above 0.6 GeV/c and covers 0.65 of 4π sr.

$\eta(1440)$

The $\eta(1440)$ was first observed in radiative J/ψ decays by the Mark II experiment.[2] Its spin-parity was found to be 0^- by the Crystal Ball collaboration.[3] This state has a large branching ratio in J/ψ radiative decays ($\sim 5 \times 10^{-3}$, comparable to the η_c rate) and does not readily fit into the established 0^- $q\bar{q}$ multiplets.

The Mark III has studied the $\eta(1440)$ in three decay modes: $K^0_S K^{\pm}\pi^{\mp}$, $K^0_S K^0_S \pi^0$, and $K^+K^-\pi^0$. Clear $\eta(1440)$ signals are observed in the 1.3-1.6 GeV/c^2 region in all three modes. Since the efficiencies for these modes are nearly identical and are approximately constant in this mass region, the spectra may be summed. The resulting spectrum is fit with a single Breit-Wigner line shape (Fig 1a). The fit yields:

$$m_{\eta(1440)} = 1451 \pm 2 \text{ MeV/c}^2,\ \Gamma_{\eta(1440)} = 104 \pm 5 \text{ MeV/c}^2$$
$$B(J/\psi \to \gamma\eta(1440)) \cdot B(\eta(1440) \to K\overline{K}\pi) = (5.1 \pm 1.2) \times 10^{-3}.$$

However, the fit probability, $P(\chi^2)$, is 1.4×10^{-13}.

The $\eta(1440)$ decay may proceed via two intermediate states, $\delta\pi$ and $K^*\overline{K}$. The $\eta(1440)$ is ~ 60 MeV/c^2 above $K^*(892)\overline{K}$ threshold. If this mode is present, threshold effects may be responsible for the distorted line shape. Dalitz plots are studied for the mass regions 1.36-1.46 GeV/c^2 and 1.46-1.58 GeV/c^2. In the high mass region, there is clear evidence of $K^*(892)$.

A coupled channel analysis is performed to understand the effects of the $K^*(892)\overline{K}$ threshold on the $K\overline{K}\pi$ spectrum. Two $K^*(892)\overline{K}$ amplitudes are used, which interfere constructively, along with a $\delta\pi$ amplitude whose magnitude and phase are allowed to vary in the fit. The $K^*(892)$ is parameterized by a P-wave relativistic Breit-Wigner. The Flatte parameterization[4] of the δ amplitude is used. A Breit-Wigner line shape for the $f_1(1285)$ is also included in the fit. The best fit is shown superposed on the $K\overline{K}\pi$ spectrum in Fig. 1b, as are the $K^*(892)\overline{K}$ (largest component), $\delta\pi$, and interference term (smallest component). The resonance mass and width resulting from the fit are

$$m_{\eta(1440)} = 1454 \pm 3 \text{ MeV/c}^2,\ \Gamma_{\eta(1440)} = 160 \pm 11 \text{ MeV/c}^2,$$

with product branching ratio of

$$B(J/\psi \to \gamma\eta(1440)) \cdot B(\eta(1440) \to K\overline{K}\pi) = (6.3 \pm 1.4) \times 10^{-3}.$$

The probability of the fit is 47%.

It is also possible to explain the $K\overline{K}\pi$ line shape by the presence of two resonances. For the case where the resonances do not have the same-spin parity, the spectrum is fit to two incoherent Breit-Wigner line shapes. This yields,

$$m_1 = 1422 \pm 2 \text{ Mev/c}^2,\ \Gamma_1 = 43 \pm 7 \text{ MeV/c}^2,$$
$$m_2 = 1490 \pm 5 \text{ MeV/c}^2,\ \Gamma_2 = 95 \pm 10 \text{ MeV/c}^2$$

with product branching ratios of $(1.8 \pm 0.5) \times 10^{-3}$ and $(3.2 \pm 0.8) \times 10^{-3}$ respectively. The fit probability is 15%. If two coherent Breit-Wigner line shapes are fit, as is required for the like spin-parity case, the results are:

$$m_1 = 1409 \pm 5 \text{ Mev/c}^2,\ \Gamma_1 = 69 \pm 11 \text{ MeV/c}^2,$$
$$m_2 = 1499 \pm 9 \text{ MeV/c}^2,\ \Gamma_2 = 138 \pm 25 \text{ MeV/c}^2$$

with product branching ratios of $(1.5 \pm 0.4) \times 10^{-3}$ and $(2.7 \pm 0.8) \times 10^{-3}$ respectively. The fit probability is 58%.

Vector Vector

The Mark III Collaboration has observed structure in both $\gamma\rho\rho$ [5] and $\gamma\omega\omega$[6]

near threshold. In both cases, multichannel spin-parity analyses indicate that the pseudoscalar component dominates in this region. In the $\omega\omega$ final state, $85 \pm 19\%$ of the signal below 2 GeV/c^2 is 0^-. The product branching ratio for J/ψ radiative decay to $\omega\omega$ in this mass range is $(1.22 \pm 0.07 \pm 0.31) \times 10^{-3}$. In the $\gamma\pi^+\pi^-\pi+\pi-$ and $\gamma\pi^+\pi^-\pi^0\pi^0$ final states, the pseudoscalar component is found to account for $51 \pm 9\%$ of the low mass region (<2 GeV/c^2). The product branching ratio for J/ψ radiative decay to pseudoscalar $\rho\rho$ in this region is $(4.7 \pm 0.3 \pm 0.9) \times 10^{-3}$.

The $\phi\phi$ final state has been studied in a continuation of the exploration of the radiatively produced *vector-vector* final states. Additional impetus was given to this study by the observation[7,8] of structure near $\phi\phi$ threshold in the reaction $\pi^-p \rightarrow n\phi\phi$. One group[7], after performing a partial wave analysis, has resolved three resonances with $J^P = 2^+$. These states have been interpreted[9] as glueballs and are expected to be produced in radiative J/ψ decays.

The $\gamma\phi\phi$ final state is studied in both $\gamma K^+K^-K^+K^-$ and $\gamma K^+K^-K^0_SK^0_L$. Both modes are relatively background free since J/ψ decay to $\phi\phi$ or $\phi\phi\pi^0$ is forbidden by C-parity. In both cases, significant acceptance problems exist. In the $K^+K^-K^+K^-$ mode, K decays affect 60% of events in the low mass region ($m < 2.4$ GeV/c^2). In the $K^+K^-K^0_SK^0_L$ mode, K^0_L detection efficiency is less than 50%. In order to overcome these problems, constrained fits to the hypotheses $J/\psi \rightarrow \gamma K^+K^-K^\pm K^\mp_{miss}$ and $J/\psi \rightarrow \gamma K^+K^-K^0_SK^0_{L_{miss}}$ are performed. Particle identification by TOF and dE/dx is applied. Clear evidence for a $\phi\phi$ signal is seen in both modes when the $m_{K^+K^-}$ vs. $m_{K\overline{K}_{miss}}$ scatter plots are studied. The $\phi\phi$ signal events are selected by requiring $\delta \leq 3$, where

$$\delta^2 \equiv ((m_{K^+K^-} - m_\phi)/\sigma_{K^+K^-})^2 + ((m_{K\overline{K}_{miss}} - m_\phi)/\sigma_{miss})^2$$

The resolutions for the good K pair are ~ 4 MeV/c^2 while the resolutions for the pair with the omitted track are ~ 6 MeV/c^2 in both modes. Acceptance corrections are applied. The resulting spectra for the $K^+K^-K^+K^-$ and $K^+K^-K^0_SK^0_L$ modes are displayed in Fig. 2a and b respectively. Clear structure is seen in the low mass region in both modes. The background, determined by requiring $5 \leq \delta \leq 6$, is negligible. The spectra are fit to two Breit-Wigner line shapes, one of which is the η_c, convoluted with a phase space background. The parameters which result for the low mass state in the $K^+K^-K^+K^-$ mode are:

$$m_X = 2220 \pm 15 \pm 20 \text{ MeV/c}^2,\ \Gamma_X = 114 \pm 45 \pm 35 \text{ MeV/c}^2$$
$$B(J/\psi \rightarrow \gamma X) \cdot B(X \rightarrow \phi\phi) = (3.45 \pm 0.8 \pm 0.7) \times 10^{-4}.$$

In the $K^+K^-K^0_SK^0_L$ mode the results are:

$$m_X = 2206 \pm 20 \pm 25 \text{ MeV/c}^2,\ \Gamma_X = 150 \pm 46 \pm 35 \text{ MeV/c}^2$$
$$B(J/\psi \rightarrow \gamma X) \cdot B(X \rightarrow \phi\phi) = (3.0 \pm 0.6 \pm 0.6) \times 10^{-4}.$$

An indication of the spin-parity of the low mass $\phi\phi$ structure can be obtained from a study of χ, the angle between the decay planes of the ϕ's in the $\phi\phi$ center of mass, and θ_K, the polar angle of the K in the ϕ rest frame. Pseudoscalar assignment is clearly favored in both $\phi\phi$ decay modes, as can be seen in Fig.3a-b, for the region $2.14 \leq m_{\phi\phi} \leq 2.3$ GeV/c^2 in $K^+K^-K^+K^-$, and in Fig.3c-d, for $K^+K^-K^0_SK^0_L$. Fig. 3e-f displays these angles for the η_c region. In all cases, the acceptance for these angles is flat.

The $\gamma K^{*^0}(892)\overline{K^{*0}(892)}$ final state has been studied in the $\gamma K^+\pi^-K^-\pi^+$ decay mode. TOF and dE/dx are used to select events with at least one well identified K. These events are kinematically fit to the hypothesis $J/\psi \rightarrow \gamma K^+\pi^-K^-\pi^+$. The fit probability is required to be greater than 2%. Background events are identified by kinematic fits: events satisfying $J/\psi \rightarrow K^+K^-\pi^+\pi^-$ and $K^+K^-\pi^+\pi^-\pi^0$ are rejected. Clear evidence for $K^{*^0}\overline{K^{*0}}$ is seen when the $m_{K^+\pi^-}$ vs. $m_{K^-\pi^+}$ scatter plot is studied. In addition, K^{*^0} bands, which are not associated with the $K^{*^0}\overline{K^{*0}}$ signal, are present. The quantity δ^2 is defined as:

$$\delta^2 \equiv ((m_{K^+\pi^-} - m_{K^{*0}})/\sigma_{K^{*0}})^2 + ((m_{K^-\pi^+} - m_{K^{*0}})/\sigma_{K^{*0}})^2$$

where $\sigma_{K^{*0}}$ is measured to be 26 MeV/c^2. Signal events are required to have $\delta^2 \leq 6.25$. The resulting signal is seen as the histogram in Fig.4. The background is determined using events from the K^{*^0} bands which do not lie in the signal region along with events which lie in neither the K^{*^0} bands nor the $K^{*^0}\overline{K^{*0}}$ signal region. The background regions' areas are carefully weighted and normalized to the signal region area. The background events are shown shaded in Fig.4. Structure, which can not be attributed to background, is seen for $m_{K^{*0}\overline{K^{*0}}} \leq 2.5$ GeV/c^2, as well as at the η_c. Using the acceptance shown overplotted in Fig. 4, the branching ratios are found to be:

$$B(J/\psi \rightarrow \gamma K^{*^0}\overline{K^{*0}}) = (2.8 \pm 0.1 \pm 0.4) \times 10^{-3}$$
$$B(J/\psi \rightarrow \gamma\eta_c) \cdot B(\eta_c \rightarrow K^{*^0}\overline{K^{*0}}) = (0.7 \pm 0.2 \pm 0.2) \times 10^{-4}.$$

An indication of the spin-parity of the structure is obtained from the χ and θ_K angles. Fig. 5a-b show the angles for the region $1.8 \leq m_{K^{*0}\overline{K^{*0}}} \leq 2.2$ GeV/c^2; Fig. 5c-d for the region $2.2 \leq m_{K^{*0}\overline{K^{*0}}} \leq 2.55$ GeV/c^2; and Fig. 5c-d for the η_c. These figures suggest substantial pseudoscalar content.

$f_2(1720)$

The $f_2(1720)$ was discovered by the Crystal Ball Collaboration[10] in $\eta\eta$. The Mark III Collaboration[11] has observed $f_2(1720)$ in K^+K^- and $K^0\overline{K^0}$ and measured

$$B(J/\psi \rightarrow \gamma f_2(1720)) \cdot B(f_2(1720) \rightarrow K^+K^-) = (4.8 \pm 0.6 \pm 0.9) \times 10^{-4} \text{ and}$$
$$B(J/\psi \rightarrow \gamma f_2(1720)) \cdot B(f_2(1720) \rightarrow K^0\overline{K^0}) = (4.5 \pm 1.2 \pm 1.1) \times 10^{-4}.$$

The spin parity was measured to be 2^+, with ratios of helicity amplitudes unlike those of the $q\overline{q}$ mesons $f_2(1270)$ and $f_2'(1525)$. Limits were set, at 90% CL, on the product branching ratio from J/ψ of $< 5.5 \times 10^{-4}$ for $\rho\rho$[5] and $< 2.4 \times 10^{-4}$ for $\omega\omega$[6].

The radiatively produced $\pi^+\pi^-$ final state has been studied using the full data sample. Events are kinematically fit to the four constraint hypothesis $J/\psi \to \gamma\pi^+\pi^-$. The fit probability is required to be greater than 1%. Background events from QED (γe^+e^- and $\gamma\mu^+\mu^-$) are removed by identifying electrons and muons using the shower counter and muon system. Events from $J/\psi \to \rho\pi \to \pi^+\pi^-\pi^0$ are suppressed by rejecting those which have satisfactory fits to the $J/\psi \to \pi^+\pi^-\gamma\gamma$ hypothesis and have $m_{\gamma\gamma}$ within 0.05 GeV/c^2 of m_{π^0}. The $\pi^+\pi^-$ mass spectrum obtained is seen in Fig. 6. An $f_2(1720)$ signal appears, as well as $f_2(1270)$, a peak near 2.1 GeV/c^2, a clear shoulder on the $f_2(1270)$ that is attributed to the $f_2'(1525)$, and substantial $\rho\pi$ background. The spectrum is fit in the region from 0.9 to 2.7 GeV/c^2. Three interfering Breit-Wigner line shapes, with masses and widths fixed, are included for the $f_2(1270), f_2'(1525)$, and $f_2(1720)$, as well as an incoherent Breit-Wigner line shape for the structure near 2.1 GeV/c^2. The shape of the $\rho\pi$ background, which has been parameterized by Monte Carlo studies, is also included. The fit result, which models the spectrum well, is shown overplotted in Fig. 6. The product branching ratios, $B(J/\psi \to X) \cdot B(X \to \pi^+\pi^-)$, are found to be: $(6.7 \pm 0.5 \pm 0.5) \times 10^{-4}$ for the $f_2(1270)$, $(2.2 \pm 1.0 \pm 0.6) \times 10^{-5}$ for the $f_2'(1525)$, $(1.1 \pm 0.2 \pm 0.5) \times 10^{-4}$ for $f_2(1720)$, and $(1.5 \pm 0.2 \pm 0.7) \times 10^{-4}$ for the structure near 2.1 GeV/c^2. The mass and width of this state are found to be $2089 \pm 10 \pm 15$ MeV/c^2 and $127 \pm 34 \pm^{150}_{50}$ MeV/c^2, respectively.

The $\gamma\eta\eta$ final state has been studied in the $\gamma(\pi^+\pi^-\pi^0)(\gamma\gamma)$ decay mode. Events are kinematically fit to the five constraint hypothesis $J/\psi \to \gamma\gamma\gamma\pi^+\pi^-\pi^0$. Events satisfactorily fitting hypotheses with two π^0's are eliminated. The quantity δ^2 is defined to be:

$$\delta^2 \equiv ((m_{\gamma\gamma} - m_\eta)/\sigma_{\gamma\gamma})^2 + ((m_{3\pi} - m_\eta)/\sigma_{3\pi})^2$$

Signal events are required to have $\delta^2 < 4$, while background is determined using events with $16 < \delta^2 < 20$. The background is found to be flat. Two additional kinematic constraints, $\eta \to \gamma\gamma$ and $\eta \to \pi^+\pi^-\pi^0$, are imposed in order to improve $\eta\eta$ mass resolution from 40 MeV/c^2 to 25 MeV/c^2. The spectrum displayed in Fig. 7 results. The branching ratio for $J/\psi \to \gamma\eta\eta$ is $(7.3 \pm 1.2 \pm 2.0) \times 10^{-4}$. The spectrum is fit to a single Breit-Wigner line shape with a phase space background. The result is overplotted in Fig. 7. The Breit-Wigner is found to have mass $1655 \pm 30 \pm 20$ MeV/c^2, with width $189 \pm 130 \pm 50$ GeV/c^2 and product branching ratio for the J/ψ of $(3.1 \pm 1.2 \pm 1.5) \times 10^{-4}$. The spectrum may also be satisfactorily fit with two resonances, $f_2'(1525)$ and $f_2(1720)$.

‡ This work was supported in part by the Department of Energy under contract DE-AC03-85ER40050.

References

1. D. Bernstein *et al.*, Nucl. Instr.Meth.**226**,301(1984).
2. D. L. Scharre *et al.*, Phys.Lett. **97B**,329(1980).
3. C. Edwards *et al.*, Phys.Rev.Lett. **49**,259(1982).
4. S. M. Flatte, Phys.Lett. **63B**, 224 (1976).
5. R.M.Baltrusaitis *et al.*, Phys.Rev.**D33**,1222(1986).
6. R.M.Baltrusaitis *et al.*, Phys.Rev.Lett.**55**,1723(1985).
7. A. Etkin *et al.*, Phys.Lett.**B165**,127(1985).
8. C. Daum *et al.*, Phys.Lett.**B104**,246(1981)
   T.F.Davenport *et al.*, Phys.Rev.**D33**,2519(1986)
   P.S.L.Booth *et al.*, Nucl.Phys.**B273**,677(1986).
9. S.J.Lindenbaum & R.S.Longacre, Phys.Lett.**B165**,202(1985)
10. C. Edwards *et al.*, Phys.Rev.Lett. **48**,458(1982).
11. R.M.Baltrusaitis *et al.*, Phys.Rev.**D35**,2077(1987).

Figure Captions

1. $K\overline{K}\pi$ mass spectrum with superposed fit results. a.) Single Breit-Wigner line shape. b.) The coupled-channel spectrum fit including $K^*(892)\overline{K}$, $\delta\pi$, interference term and $f_1(1285)$ component. (The $K^*(892)\overline{K}$ component is the largest; the $\delta\pi$ component is the next largest.) c.) Two incoherent Breit-Wigner line shapes. d.) Two coherent Breit-Wigner line shapes.
2. Acceptance corrected $\phi\phi$ mass spectrum with superposed fit to two Breit-Wigner line shapes with a phase space term. a.) $K^+K^-K^+K^-$ decay mode. b.) $K^+K^-K^0_SK^0_L$ decay mode.
3. Angular distributions for $\gamma\phi\phi$. χ is the angle between the ϕ decay planes in the $\phi\phi$ rest frame. θ_K is the polar angle of the K in the ϕ rest frame. a.-b.) low mass $K^+K^-K^+K^-$. c.-d.) low mass $K^+K^-K^0_SK^0_L$. e.-f.) η_c in $K^+K^-K^+K^-$.
4. The $K^{*^0}\overline{K^{*^0}}$ mass spectrum is shown with background events shaded. The acceptance is superposed.
5. Angular distributions for $\gamma K^{*^0}\overline{K^{*^0}}$. χ is the angle between the K^{*^0} decay planes in the $K^{*^0}\overline{K^{*^0}}$ rest frame. θ_K is the polar angle of the K in the K^{*^0} rest frame. a.-b.) $1.8 \leq m_{K^{*0}\overline{K^{*0}}} \leq 2.2$ GeV/c^2. c.-d.) $2.2 \leq m_{K^{*0}\overline{K^{*0}}} \leq 2.55$ GeV/c^2. e.-f.) η_c.
6. $\pi^+\pi^-$ mass spectrum.
7. $\eta\eta$ mass spectrum.

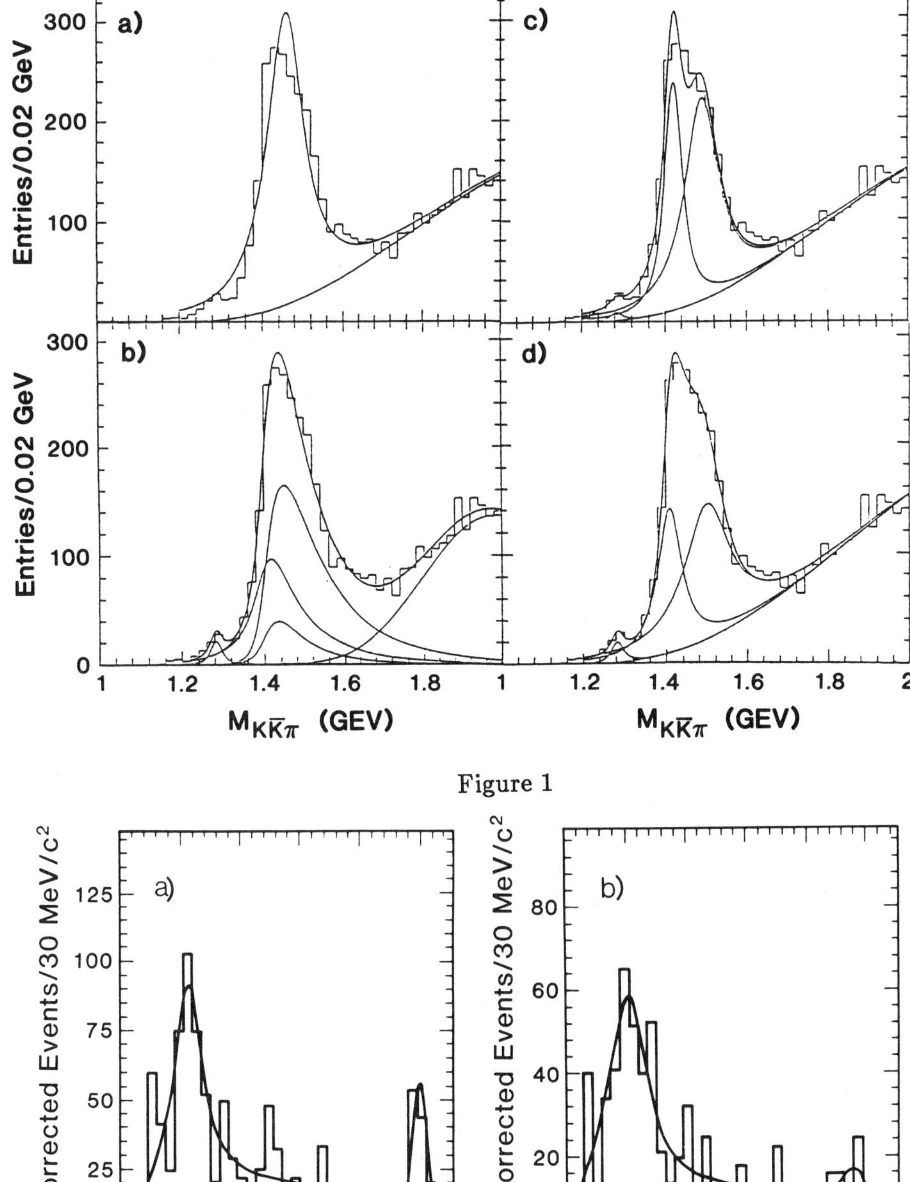

Figure 1

Figure 2

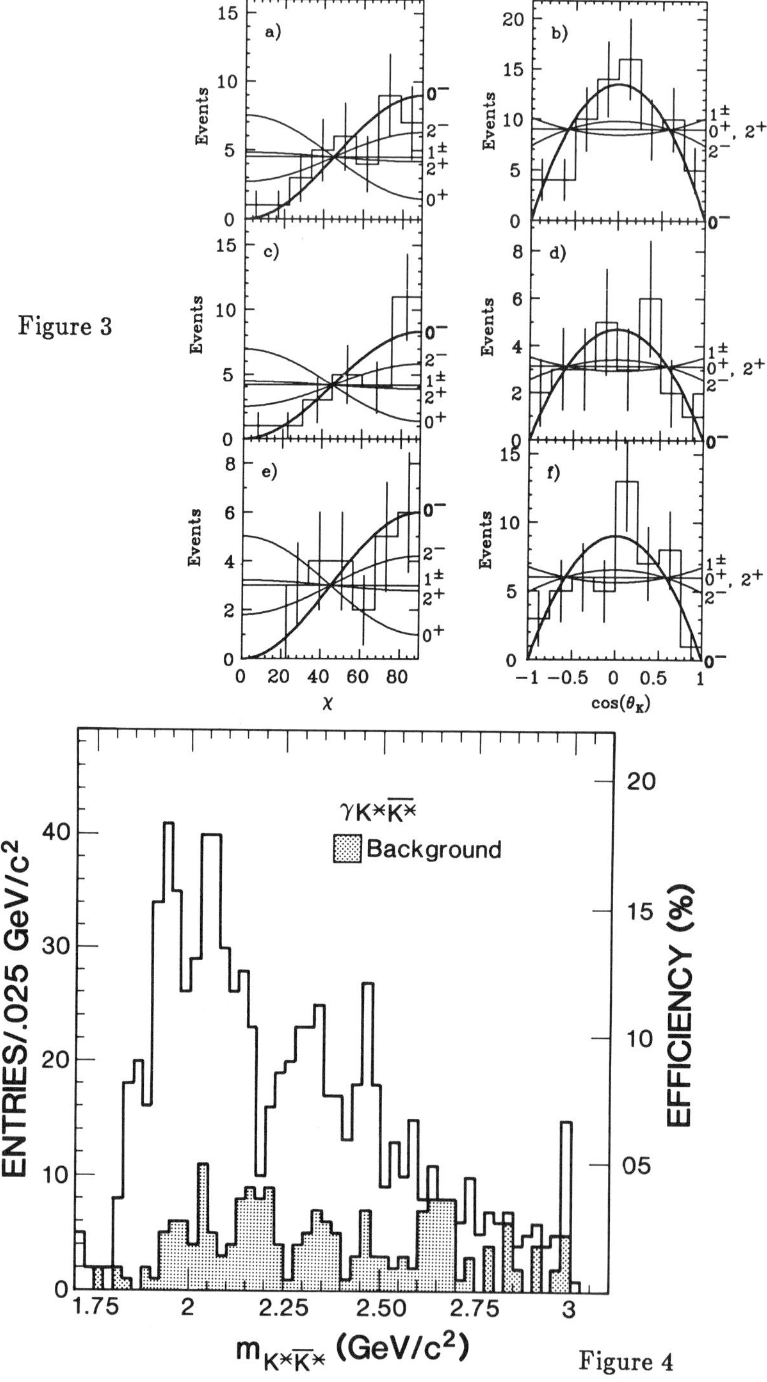

Figure 3

Figure 4

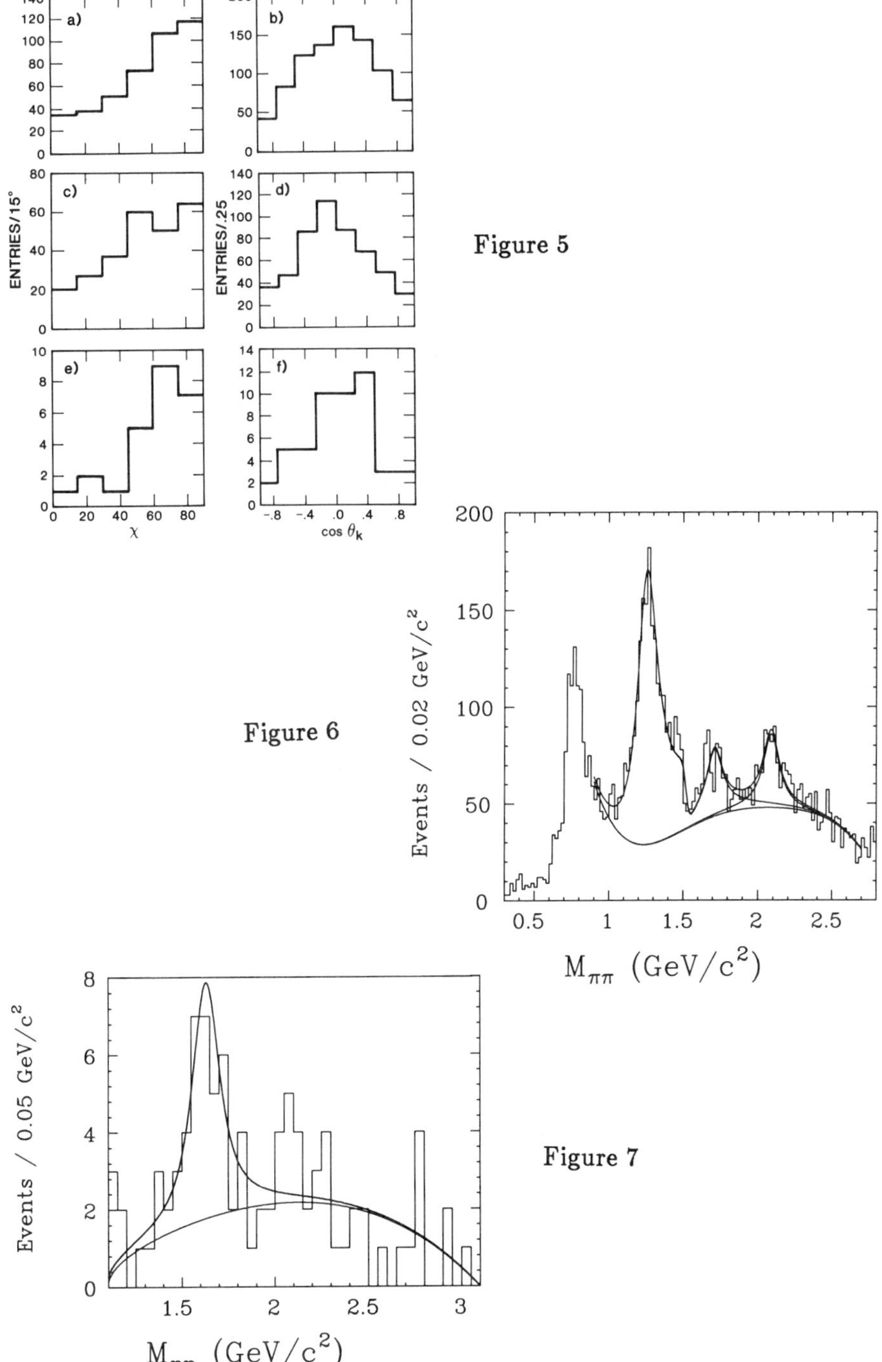

Figure 5

Figure 6

Figure 7

GLUEBALLS AND HADRON SPECTROSCOPY

Sydney Meshkov

California Institute of Technology, Pasadena, CA 91125
and
National Bureau of Standards, Gaithersburg, MD 20899

ABSTRACT

Identifying glueballs is intimately intertwined with the study of meson spectroscopy. In this brief review, recent experimental results are compared with QCD inspired predictions for the masses of the 0^{-+}, 1^{+-}, 1^{--}, 1^{++} mesons. They agree remarkably well. For the 0^{-+} and 1^{++} systems, these agreements lead to the possibility of identifying additional states as candidates for exotica. A major mystery is why the 2^{++} g_T states are not seen in $J/\psi \rightarrow \gamma\phi\phi$, in view of the fact that the OZI-violating mechanism should be a good way to produce glueballs.

INTRODUCTION

The existence of glueballs is one of the most striking predictions of QCD.[1] Identifying these bound states of gauge bosons is an important current experimental question. By now it is clear that the study of glueball spectroscopy is intertwined with the study of meson spectroscopy. In this brief review, I discuss the impact of recent experimental results, some presented for the first time at this conference, which confirm earlier QCD predictions for the masses of the J^{PC} (J = spin, P = parity, C = charge conjugation) = 0^{-+}, 1^{+-}, 1^{--}, 1^{++} mesons, and of radial excitations for the 0^{-+} and 1^{--} systems. Firmly established extra states with these quantum numbers must be interpreted as exotica, i.e., as glueballs, hybrids or multiquark states. Candidates for such extra states exist in the 0^{-+}, 1^{++}, and 2^{++} systems. The confirmation of the $\eta(1275)$ and the exhaustion of the first few radial excitations of the η and η' resonances by the Mark III and DM2 experiments highlight the importance of unscrambling the 1400–1500 MeV region experimentally to see whether there are extra pseudoscalar states that can be interpreted as glueballs, hybrids or multiquark states. Analogously, there is an extra axial state, the $E(1420)$.

GLUEBALL SPECTROSCOPY

Level order. This subject has been reviewed often[2] so we only present the lowest levels for glueballs made from two gluons moving with a relative orbital

⋆ This work supported in part by the U.S. Department of Energy under Contract No. DE-AC0381-ER40050

angular momentum L. Note the preponderance of low lying 0^{++}, 2^{++}, and 0^{-+} states in Figure 1.

$$\begin{array}{ll} L=2 & 2^{++},0^{++},\ldots,4^{++} \\ L=1 & 0^{-+},|1^{-+}|,2^{-+} \\ L=0 & 0^{++},2^{++} \end{array}$$

Figure 1. $2g$ glueball levels. | | denotes an oddball, a glueball with quantum numbers that do not occur in $q\bar{q}$, L mesons.

Masses. The two main questions are: 1) Where is the ground state?; 2) What sets the scale for the excited states? Lattice Gauge Theories are our best current hope for answering these questions. After years of confusion, we are down to the only firm result — a Mass Gap $m(0^{++})$ exists. The three most recent calculations for $m(0^{++})$ are in essential agreement,[3–5] all finding $m(0^{++})$ to be in the range 1200–1400 MeV. Au *et al.*[6] recently found an extra 0^{++} state at 991 MeV in a coupled channel K-matrix $\pi\pi$ and $K\bar{K}$ analysis that they put forth as a glueball candidate. Patel *et al.*[4] had no evidence for a 2^{++} glueball, disagreeing with earlier calculations by de Forcrand *et al.*,[7] who obtained $m(2^{++})/m(0^{++}) \simeq 2.9$, and with Berg *et. al.*[8] who found $m(0^{++}) \simeq m(2^{++})$.

A recent paper by Chin *et al.*[9] has shown that the ratio $m(2^{++})/m(0^{++})$ can be expressed in terms of two ground state correlation coefficients; a simple variational estimate of these coefficient for $SU(3)$ with no dynamical fermions shows that $m(2^{++})/m(0^{++}) \geq 1$, and never < 1. Gocksch and Heller[10] have shown that fermion loops in LGT calculations for glueball masses are unimportant.

HADRONS

Meson spectroscopy is intimately related to glueball spectroscopy because mesons and glueballs with the same quantum numbers can mix. Clearly, an ideal way to determine the existence of glueballs is to find an oddball, i.e., a glueball with quantum numbers not allowed in the $q\bar{q}$; L model. Although there are several oddball candidates, most of the experimental data and glueball candidates lie in the realm where mixing can occur. The strategy here is to look for extra states. If we know meson spectroscopy so well that we can assign all of the mesons for a given J^{PC} to $SU(3)$ multiplets, including radial excitations, then any extra states must be glueballs, hybrids or multiquark states.

Do we understand meson spectroscopy well enough at present to do this? We certainly know it better than we did before, and probably well enough to make detailed comparisons between experiment and theory. Consequently, we take seriously older QCD calculations of meson spectroscopy by Stanley and Robson (SR)[11] and by Godfrey and Isgur (GI)[12] and compare their predictions with many new results in the past year for 1^{--}, 0^{-+}, 1^{++}, 1^{+-} states.

A. 1^{--}

The predicted lowest excitation of the $\rho(770)$ are listed in Table I. Until recently only a single state $\rho'(1600)$ was observed. A recent experiment at KEK in $\pi^- p \to \eta\pi^+\pi^- n$ at 8.95 GeV/c (Expt. 1 in Table I)[13] examines the spin-parity structure of the $\eta\pi^+\pi^-$ system. A single Breit-Wigner fit gives a ρ' at 1635 ± 26 MeV, $\Gamma = 317 \pm 35$ MeV. A two Breit-Wigner resonance fit gives masses for the 2^3S_1 and 1^3D_1 states in clear agreement with theory. Also listed are similar results of meson photoproduction experiments (Expt. 2).[14]

Table I. Predicted and experimental $I = 1$ $J^{PC} = 1^{--}$ Masses (MeV)

State	SI	GI	Expt. 1[13]	Expt. 2[14]
2^3S_1	1521	1450	1446 ± 10	1465 ± 25
1^3D_1	1712	1660	1701 ± 15	1700 ± 25

B. 0^{-+}

This oft-examined system has some important new results. The $\eta(1275)$ has now been confirmed by a new BNL experiment $\pi^- p \to K^+ \bar{K}^0 \pi^-$ at 8 GeV/c as reported by Susan Blessing[15] at this conference. I stress this because it tells us to take the $\eta(1275)$ very seriously as the first radial excitation of $\eta(548)$. Inasmuch as both $\eta(548)$ and $\eta'(960)$ are not ideally mixed ($\theta_p \simeq -20°$) there is no pressing need to regard their radial excitations as ideally mixed states.

The confirmation of the $\eta(1275)$ means that we can take seriously the predictions (SR and GI) for the 0^{-+} $I = 0$ excitations (Table II). [The GI masses (Solution P2) are tailored to accommodate a first radial excitation at 1270 MeV by the inclusion of an annihilation term.]

Let us consider all of the 0^{-+} states listed in Table II in turn.

$\eta(548)$ and $\eta'(960)$ are finally understood phenomenologically, if not theoretically. It is generally accepted that the 8–1 mixing angle $\theta_p \simeq -20°$.[16] Such a mixing angle indicates that *there is no need to mix anything into* η'! η and η' seem to be decoupled from the rest of the 0^{-+} system.

A promising approach to describe the pseudoscalar system theoretically was discussed by Akhoury.[17] W. A. Bardeen, J. Frére, and Akhoury, using an effective Lagrangian approach and the Gluon Anomaly, obtain a prediction of the η, η' mixing angle of $\sim -18°$, describe gluonic production of $\eta'\pi\pi$, $K\bar{K}\pi$, predict Γ $(\eta' \to 3\pi^0)$ and obtain other relevant properties of the pseudoscalars.

Table II. Predicted and Experimental $I = 0$ $J^{PC} = 0^{-+}$ Masses (MeV).

State	Stanley–Robson (1980)	Godfrey–Isgur (P2) (1985)	Expt.[18,19] (MK III, DM 2)
1η	565	530	548
$1\eta'$	937	960	960
2η	1250	1270	1275
$2\eta'$	1550	1540	$\sim$ 1550 $(\rho\rho)(\iota?)$
3η	1850	1870	$\sim$1850 $(\rho\rho, \omega\omega, K^*\bar{K}^*)$
$3\eta'$	2060	2120	$\sim$2100–2200 $(\rho\rho, K^*\bar{K}^*, \phi\phi)$

B1. HIGHER 0^{-+} EXCITATIONS (1350–1500 MeV)

As we know, the mass region (1350–1500) MeV is very complicated. The BNL analysis[15] gives evidence for one or two peaks with 0^{-+}, as well as 1^{+-} and 1^{++} resonating structures. The $(K\bar{K}\pi)$ peaks with $a_0(980)$ and K^* final states are offset by $\sim$ 50 MeV. The Mark III preliminary analysis[18,19] of the $K\bar{K}\pi$ final states allows various interpretations ranging from a one resonance 0^{-+} fit at $\sim$ 1460 MeV, to either two coherent (both 0^{-+}) or two incoherent (low 1^{++}, high 0^{-+}) BW's. The low mass state is $\sim$ 1410–1420 MeV and the high mass state is $\sim$ 1490–1500 MeV.[18,19] In these fits there may be 1 or 2 extra 0^{-+} states or even none.

B2. TWO PHOTON ANNIHILATION

A. Untagged $\gamma\gamma$ (real on-shell photons) yields nothing in the $K\bar{K}\pi$ final state above η'. 0^{-+} states, such as the $\eta(1275)$ should show up as peaks in this experiment if they are made of quarks.

B. $\gamma\gamma^* \rightarrow K\bar{K}\pi$ yields $f_1(1420)$, $J^C \equiv 1^+$. It is probably 1^{++}, but 1^{-+} cannot be ruled out. $M(f_1) = 1426 \pm 9$ MeV.[20]

A and B together are usually taken as "proof" that a 0^{-+} state in the mass region 1400 – 1500 MeV is a glueball, since glueballs don't couple to photons.

C. In this conference, Barker[21] has presented new data for $\gamma\gamma^* \rightarrow \eta\pi\pi$. *TPC* /$2\gamma$ finds the masses and J^{PC} of $\eta'(958)$ and $f_1(1285)$ to be

$$M(\eta') = 961 \pm 3 \text{ MeV} \quad 0^{-+}, \; M(f_1) = 1273 \pm 11 \text{ MeV} \quad 1^{++}$$

D. Caveat to "Proof"

In a recent published paper A. Lundin and H. Snellman[22] have shown that chiral symmetry in an effective Lagrangian suppresses 2γ couplings to radial excitations of 0^{-+} mesons made of quarks. This "explains" why $\eta(1275)$ and higher

0^{-+} quark states are not seen in $TPC/2\gamma$ experiments. The lack of ι production in the $\gamma\gamma$ production of $K\bar{K}\pi$ is thus not an indication that ι is a glueball.

B3. 0^{-+} STATES IN $J/\psi \to \gamma VV$

0^{-+} states have been obtained at Mark III and DM2 for $VV = \rho\rho$, $\omega\omega$, $K^*\bar{K}^*$, and $\phi\phi$ as reported by Wisniewski[18] and by Eigen[19]. The thresholds are: $\rho\rho = 1540$ MeV, $\omega\omega = 1566$ MeV, $K^*\bar{K}^* = 1794$ MeV, and $\phi\phi = 2040$ MeV. The second and third radial excitations of η and η' (Table II) at 1550, 1850, and 2100 MeV correspond to the experimentally observed 0^{-+} peaks as follows:

1550 — $\rho\rho$
1850 — $\rho\rho$, $\omega\omega$, $K^*\bar{K}^*$
2100 — $\rho\rho$, $K^*\bar{K}^*$, $\phi\phi$

The $\phi\phi$ peak is actually slightly higher, at 2200 MeV. Eigen[19] has also suggested that the $\rho\rho$, $\omega\omega$ and $\phi\phi$ bumps may be radially excited η and η' states. Hitlin,[23] in his 1986 Aspen review, has discussed the importance of doing $SU(3)$ checks for the different decay modes at each mass, i.e., 1850 and 2100 MeV.

The net result of this discussion is that these VV states together with $\eta(1275)$ exhaust the $n = 2, 3$ states of η and η'. This successful fit to the theoretical predictions for excited 0^{-+} states makes the question of how many 0^{-+} states there are in the 1400–1500 MeV mass interval even more crucial.

A heavy $\iota(1490\text{–}1500)$ as obtained in the two Breit–Wigner fits discussed by Wisniewski and Eigen, might be regarded as strongly coupled to the tail of the $\rho\rho(1550)$ 0^{-+} resonance. In fact, the lowest DM2 $\rho\rho$ state at 1497 ± 3 MeV corresponds almost exactly to the upper Mark III 0^{-+} $K\bar{K}\pi$ state. The Mark III fits give either a low, 1420 MeV 1^+ state (no additional 0^{-+} states) or a low 1420 0^{-+} state (one additional 0^{-+} state).

Independent of the Mark III and DM2 experiments, there are three experiments[24,25,15] that give an extra 0^{-+} state at 1420 MeV (extra compared to the calculations of SR and GI).*

B4. 0^{-+} CONCLUSIONS

If we take the spectroscopic calculations of Table II seriously, then there may be one or even two extra states in the ι region. What could they be? They could

---

⋆ In an effort to find a glueball selection criterion that might resolve the problem of experiments that seem contradictory, but which may all be correct, D. Hitlin, W. Wisniewski and I, aided by discussions with S. J. Lindenbaum and H. J. Lipkin, have tried to use an analog of the idea that glueballs like to be produced centrally in non charge exchange reactions, i.e., for $x_F = 0$. Consider the production of glueballs with quantum numbers that forbid coupling to two pions (0^{-+} for example), and hence cannot be produced in a pion beam by π-exchange. For such production, the analog of small x_F is large t. The production of such a glueball should be manifested by a relatively large differential cross section at large t.

be glueballs, hybrids or multiquark states. It is crucial that this region of 1400–1500 MeV be analyzed with more data, preferably with a mass dependent partial wave analysis.

We have not discussed the familiar problems like the lack of $\delta \to \eta\pi$, or the smallness of $B(\iota \to \eta\pi\pi)/B(\iota \to K\bar{K}\pi)$. Maybe calculations like those of Akhoury, Bardeen and Frére will help solve them.

C. 1^{++} AND 1^{+}

Five recent experiments and analyses give us new information: 1. BNL partial wave analysis in which a $D'(1530)$ is seen;[15] 2. $\gamma\gamma^* \to f_1(1285)$ and $f_1(1420)$;[21] 3. Lass Data: $D'(1530)$ 1^{++}, $H'(1380)$ 1^{+-};[26,27] 4. $f_1(1285) \to \phi\gamma$ implies $s\bar{s}$ content of $f_1(1285)$;[28] so does $J/\psi \to \phi K\bar{K}\pi$);[19] 5. Iizuka, Masuda and Miura[20] analyze the shape of the $K\bar{K}$ spectrum in the E/ι region. They want an $E = 1^{+-}$ and a 0^{-+} state, not an $E = 1^{++}$ and a 0^{-+} state.

Taking all experiments in the 1^{++} system at face value seems to require that the $f_1(1285)$ and $D'(1530)$ are the $I = 0$ members of a non-ideally mixed nonet, with the exact mixing dependent on the A_1 mass. The old A_1 mass was $\sim$ 1100 MeV; PDG lists $\sim$ 1275 MeV.[30] If we use 1275 MeV, $m(A_1) \simeq m(f_1)$, implying ideal mixing for a 1^{++} nonet; then $f_1 = (u\bar{u}+d\bar{d})/\sqrt{2}$ and $D' = s\bar{s}$. This is clearly inconsistent with item 4 above. However, if we use the recently reported Argus value of $m(A_1) = 1046 + 11$ MeV[31] the 1^{++} nonet may be analogous to the 0^{-+} nonet; f_1 and D', with a low-lying $I = 1$ A_1, may be non-ideal mixtures of $(u\bar{u}+d\bar{d})/\sqrt{2}$ and $s\bar{s}$ (or of $SU(3)$ octet and singlet) as are the $\eta(548)$ and $\eta'(960)$ with the low-lying $I = 1$ pion.

Table III. Predicted and Experimental 1^{++} and 1^{+-} Masses.

	SR	GI	Expt.
1^{+-}			
h_1	1106	1220	1190
b	1106	1200	1235
K_{1a}	1289	1340	$K_1 = 1400$, $K_2 = 1280$
H'	1424	1470	1380
1^{++}			
D	1273	1240	1285
A_1	1273	1240	1046-1275
K_{1b}	1381	1380	$K_1 = 1400$, $K_2 = 1280$
D'	1487	1480	1530

Table III contains a comparison between theory and experiment for the lowest 1^{++} and 1^{+-} $SU(3)$ multiplets which yields fair agreement. (A similar analysis has been carried out by Chung.[2d])

As may be seen from Table III, a $D'(1530)$ mass corresponds well to the SR and GI predictions of $\sim$ 1485 MeV.

Such an assignment gives us an extra 1^{++} state, $E(1420)$. This low mass state seems unlikely to be a glueball candidate, but may possibly be a $q\bar{q}g$ or a $q\bar{q}q\bar{q}$ state.

D. 2^{++}

This system has been reviewed often for it has prominent glueball candidates.

The $\theta(1720)$[19] has $J^{PC} = 2^{++}$ but $J^{PC} = 0^{++}$ is not entirely ruled out. It is a glueball candidate, being produced in $J/\psi \to \gamma\theta$, and not produced in Lass. It apparently doesn't like to be produced hadronically, but DM2 has observed it in the hadronic decay $J/\psi \to \phi K_S^0 K_S^0$.[32] A persistent problem for the θ's glueball assignment is that as a glueball it should decay into $\pi\pi/K\bar{K}/\eta\eta$ as 3/4/1. Experimentally, the ratio is 1.7/27/14. Various excuses for this have been made, but it is still a problem.[2a]

The g_T states,[33,34] $g_T(2011)$, $g_{T'}(2297)$, and $g_{T''}(2339)$, the BNL glueball candidates produced in the OZI forbidden process $\pi^- p \to \phi\phi n$, have an increased number of events. Their masses and widths have changed slightly, but the three 2^{++} resonances are clearly evident.

We are faced with a major theoretical challenge. These 2^{++} states are not seen in $J/\psi \to \gamma\phi\phi$.[18,19] The experimental indications from Mark III and DM2 are that a resonance in the $\phi\phi$ system is produced with $J^P = 0^{-+}$, with $M \simeq 2200$ MeV that corresponds to a radial excitation of η'. Yet there is nothing wrong with the OZI-forbidden mechanism for producing glueballs.

How do we reconcile this?

E. 0^{++}

The $G(1590)$, discussed by Alde,[35,36] has been produced at GAMS, centrally, in the process $\pi^- N \to M^0 \pi^- N$ where $M^0 \to \eta\eta \to 4\gamma$. It has been proposed as a glueball candidate. As such it should pass the same flavor 1 test as the $\theta(1720)$, $G \to \pi\pi/K\bar{K}/\eta\eta$ as 3/4/1. It is possible that this state has been seen by LASS.[26]

F. EXOTICS

Two exotics are currently under consideration.

1. $M(1405) \to \eta\pi^0$, $I = 1, 1^{-+}$. This state has been proposed by GAMS and was discussed by Alde.[36] The partial wave analysis leading to the assignment has been questioned by Dunwoodie.[26]
2. $M(1480 \pm 40) \to \phi\pi$ $I = 1, 1^{--}$ (Bityukov *et al.*).[37] The properties of this resonance have been discussed in detail by Close and Lipkin.[38]

G. SUMMARY

Advances

1. The $\eta(1275)$ has been confirmed.

2. The QCD spectroscopy predictions have been confirmed for $0^{-+}, 1^{+-}, 1^{--}, 1^{++}$.

3. We have a possible "explanation" for $\gamma\gamma \not\to 0^{-+}$ quark radial excitations.

4. Akhoury, Bardeen, and Frére may have the beginnings of a theory for the 0^{-+} resonances, both quark and glue.

Problems or Opportunities

1. There may be extra 0^{-+} states (1 or 2).

2. There is an extra 1^{++} state ($E(1420)$),

3. There are exotic candidates.

Major Mystery

Why are the $2^{++} g_T$ states not seen in $J/\psi \to \gamma\phi\phi$, in view of the fact that the OZI-violating mechanism should be a good way to produce glueballs?

H. ACKNOWLEDGEMENTS

The author is indebted to G. Eigen, D. Hitlin, and W. Wisniewski for countless discussions, and to S. Lindenbaum and H. J. Lipkin for very helpful conversations.

REFERENCES

1. H. Fritzsch and M. Gell-Mann, *XVI Int. Conf. on High Energy Physics, Chicago-Batavia*, Vol. **2**, pp. 135.
2. Some recent reviews with extensive references to earlier work are:
   a. S. Meshkov, Proc. Second Aspen Winter Particle Physics Conf., Aspen, CO, Jan. 1986. New York Academy of Sciences, **490**, 87-101, (1987).
   b. S. Meshkov, W. F. Palmer, and S. S. Pinsky, Proc. XXII Rencontre de Moriond, Les Arcs, France, pp. 683-689 (1987).
   c. F. E. Close, RAL-87-072 (preprint, Sept. 1987).
   d. S. U. Chung, BNL 40599 (preprint, Sept. 1987).
   e. G. Eigen, CALT-68-1483 (preprint, 1988).
3. T. deGrand, Phys. Rev. D **36**, 176 (1987).
4. A. Patel *et al.*, Phys. Rev. Lett. **57**, 1288 (1986).
5. S. A. Chin and M. Karliner, Phys. Rev. Lett. **58**, 1803 (1987).
6. K. Au, D. Morgan and M. Pennington, Phys. Rev. Lett. **167B**, 229 (1986).
7. Ph. de Forcrand *et al.*, CERN TH 4167/85 (preprint 1985).

8. B. Berg *et al.* Phys. Rev. Lett. **57**, 400 (1986).
9. S. A. Chin *et al.*, Phys. Rev. Lett. **60**, 1467 (1988).
10. A. Gocksch and U. M. Heller, Phys. Rev. Lett. **60**, 1809 (1988).
11. D. P. Stanley and D. Robson, Phys. Rev. **D21**, 3180 (1988).
12. S. Godfrey and N. Isgar, Phys. Rev. **D32**, 189 (1985).
13. S. Fukui *et al.*, Phys. Lett. **202B**, 441 (1988).
14. B. Diekmann, Proc. Second Int. Conf on Hadron Spectroscopy, April 1987, KEK, Tsukuba, Japan, pp. 7-18.
15. S. Blessing *et al.* (these proceedings).
16. F. J. Gilman and R. Kauffman, Phys. Rev. **D36**, 2761 (1987).
17. R. Akhoury (these proceedings).
18. W. Wisniewski (these proceedings).
19. G. Eigen, CALT-68-1483 (preprint, 1988).
20. H. Aihara *et al.*, Phys. Rev. Lett. **57**, 51 and 2500 (1986).
21. T. Barker (these proceedings).
22. A. Lundin and H. Snellman, Phys. Lett. **202B**, 251 (1988).
23. D. Hitlin, Proc. Second Aspen Winter Particle Phys. Conf. Aspen, CO, Jan. 1986, New York Academy of Sciences, **490**, 47-60 (1987).
24. A. Ando *et al.*, Phys. Rev. Lett. **57**, 1296 (1986).
25. D. F. Reeves *et al.*, Phys. Rev. **D34**, 1960 (1986).
26. W. Dunwoodie (these proceedings).
27. D. Aston *et al.*, Phys. Lett. **B201**, 573 (1988).
28. S. I. Bityukov *et al.*, Phys. Lett. **203B**, 327 (1988).
29. J. Iizuka, F. Masuda and T. Miura, Prog. Theo. Phys. **79**, 141 (1988).
30. Particle Data Group, Phys. Lett. **170B**, 1 (1986).
31. Soding, P., Aspen Winter Particle Phys. Conf., Aspen, Co, Jan. 1987.
32. J. E. Augustin, *et al.*, LAL 85/27, 1985 (Orsay preprint).
33. A. Etkin *et al.*, Phys. Lett. **201B**, 568 (1988) and references contained therein.
34. S. J. Lindenbaum (these proceedings).
35. D. Alde *et al.*, Phys. Lett. **201B**, 160 (1988).
36. D. Alde (these proceedings).
37. S. I. Bityukov, *et al.*, Phys. Lett. **188B**, 383 (1987).
38. F. E. Close and H. J. Lipkin, Phys. Lett. **196B**, 245 (1987).

MAGNETIC MOMENTS NEAR Pb CLOSED SHELL

I.S. TOWNER

Atomic Energy of Canada Research Company,
Chalk River Nuclear Laboratories,
Chalk River, Ontario, Canada K0J 1J0

ABSTRACT

Magnetic moments in the Pb region can be quantitatively understood in an impulse-approximation calculation augmented by meson-exchange current corrections.

1. INTRODUCTION

Magnetic moments in nuclei whose configurations are predominantly closed-shells-plus-(or minus) one nucleon deviate from the single-particle shell-model estimates (Schmidt values) by amounts typically of order 0.2 μ_N in light nuclei and of order 1.0 μ_N in heavy nuclei. These deviations arise from the inadequacy of the single-particle description of the nuclear state or the inadequacy of the one-body magnetic moment operator whose coupling constants have been determined from the anomalous magnetic moments of the free proton and the free neutron. Corrections for these two inadequacies have been computed to second order in perturbation theory[1,2,3] and it is generally found that the two corrections are large and of opposite sign.

In this note we present the results of a new calculation[3]. Our goal is to provide as complete a calculation as possible in the Hilbert space of nucleons and mesons. It is very popular these days to discuss nuclear properties in the quark model looking for evidence of quark degrees of freedom in nuclei. One manifestation of these ideas has been the notion that nucleons swell in size in the nuclear medium. The relativistic σ-ω models also lead to such notions. Magnetic moments can provide a testbench for these ideas. But before conclusions can be drawn it is necessary to have in place a comprehensive calculation of the traditional type such as the one presented here. Only if the traditional calculation is found wanting can specific statements be made on alternatives. What is more likely, however, is that notions such as nucleons swelling in nuclei are just different parameterisations of much the same physics as that being addressed in the traditional framework.

2. CORE POLARISATION

In restricting the discussion of magnetic moments to nuclei described as closed-shell-plus (or minus) one nucleon, we are hoping to minimise the uncertainty in the nuclear structure calculation. Nevertheless excitations from the closed shell core cannot be ignored. There will be small pieces in the wavefunction of say the ground-state of ^{209}Bi of 2p-1h and 3p-2h configuration. These admixtures can be estimated in perturbation theory. The first-order calculation is trivial, the only uncertainty concerns the choice of residual interaction. We have chosen the one-boson-exchange potential (OBEP) for reasons outlined in ref. 3. The calculation can be algebraically extended to higher orders in the Random Phase Approximation and the computation is straightforward. However in second order there are additional terms involving unrestricted intermediate-state summations which pose a formidable calculation. We take our results from the work of Shimizu[4], where simplifying closure approximations have been made.

3. MESON-EXCHANGE CURRENTS

The magnetic moment operator formally comes from the curl of the electromagnetic current operator. In nuclei, the electromagnetic current is taken to be the sum of the individual currents of all the individual nucleons. If the coupling constants representing the orbital and spin contributions to these currents are fixed to reproduce the magnetic moments of the free proton and free neutron and are used unchanged in finite nuclei, then this procedure is known as the impulse approximation. It leads to the familiar one-body magnetic moment operator.

Corrections to the impulse approximation arise because nucleons in nuclei interact through the exchange of mesons and this exchange can be perturbed by the action of the electromagnetic current. Since this perturbation requires at least two nucleons to be involved, corrections from this origin lead to two-body operators. Consider first the pion-exchange graphs which can be classified into two types: Born graphs, involving only nucleons and pions; and non-Born terms in which the electromagnetic current either excites a nucleon to an excited state, such as the isobar Δ, or converts the pion into a heavy meson such as the ρ. The important point is that the Born terms are largely model independent and can be calculated unambiguously. The non-Born terms require the injection of model-dependent assumptions. The Born terms are also the largest terms, but what spoils this nice argument is that there are two contributions to the Born terms: one where the pion "in flight" interacts with the electromagnetic current (the pion-current term), and one where the current interacting with a nucleon prompts a pion to be emitted and reabsorbed on a second nucleon (the pion-pair term). These two terms give large corrections to the magnetic moment in a heavy nucleus (see table 1), but are of opposite sign. The net effect is comparable to the model-dependent non-Born terms.

There are two further points to consider: The core polarisation calculation corrects the matrix element for 2p-1h and 3p-2h admixtures in the single-particle state. The perturbation is carried out to second order in the residual interaction (or to the fourth power in the meson-nucleon coupling constants). To be consistent the matrix elements of two-body MEC operators should also be corrected for 2p-1h and 3p-2h admixtures. Since the MEC operator already

Table 1: The correction to the magnetic moment of a proton in an $h_{9/2}$ orbit and a neutron in an $i_{13/2}$ orbit as a percentage of the single-particle Schmidt value coming from meson-exchange Born graphs (first two lines) and non-Born graphs (second two lines).

	$\Delta\mu/\mu$ (%)	
MEC	proton $h_{9/2}$	neutron $i_{13/2}$
π-current	-36%	-53%
π-pair	+44%	+62%
heavy-meson	+25%	+15%
isobar	+2%	-5%
Total	+35%	+18%

involves the meson-nucleon coupling constants to the second power, it is sufficient to estimate this correction to first order in the residual interaction.

The second point concerns the one-body magnetic moment operator which is obtained as a leading term in the nonrelativistic reduction of a relativistic current. There is a few per cent correction coming from the next order terms in this reduction. We call this a relativistic correction. In the following sections we incorporate both these additional points under the generic heading of "MEC".

4. RESULTS

In tables 2 and 3 we give the calculated corrections to eight magnetic moment and three B(M1) transition matrix elements in the Pb region for which there are experimental data. It is useful to characterize the results of the calculation in terms of an equivalent effective magnetic moment operator that contains three independent rank-one tensors

$$\underset{\sim}{\mu}_{eff} = g_{L,eff}\underset{\sim}{L} + g_{S,eff}\underset{\sim}{S} + g_{P,eff}\,[Y_2,\underset{\sim}{S}] \tag{1}$$

where $g_{L,eff} = g_L + \delta g_L$ with g_L the bare impulse-approximation value and δg_L the correction to it. Note that in the bare operator $g_P = 0$. The agreement between theory and experiment in table 2 is within 0.2 μ_N in all cases except one, the neutron $i_{13/2}$ state in ^{207}Pb, where the strong cancellation between core polarisation and meson-exchange currents seems to be too severe.

Table 2: Correction to magnetic moments and B(M1) in the Pb region from core polarisation, CP(RPA) and CP(2nd), and meson-exchange currents compared to experiment (from ref. 3).

	$\delta\mu$ or $\delta[B(M1;i\to f)]^{1/2}$				
	CP(RPA)	CP(2nd)	MEC	Sum	Expt
$\pi h_{9/2}$	0.45	-0.33	1.34	1.45	1.49 ± 0.01
$\pi i_{13/2}$	-0.44	-1.69	1.30	-0.83	-0.78 ± 0.10
$\pi s^{-1}_{1/2}$	-0.45	-0.60	0.13	-0.92	-0.92 ± 0.01
$\pi d^{-1}_{3/2}$	0.21	0.04	0.49	0.74	0.64 ± 0.19
$\pi f_{5/2}\to f_{7/2}$	-0.37	-0.22	-0.09	-0.67	-0.82 ± 0.09
$\nu g_{9/2}$	0.38	0.49	-0.44	0.44	0.58 ± 0.06
$\nu p^{-1}_{1/2}$	-0.09	-0.02	-0.14	-0.24	-0.05 ± 0.00
$\nu f^{-1}_{5/2}$	-0.30	0.15	-0.60	-0.74	-0.58 ± 0.03
$\nu i^{-1}_{13/2}$	0.50	0.78	-0.76	0.52	0.90 ± 0.03
$\nu p^{-1}_{3/2}\to p^{-1}_{1/2}$	-0.23	-0.08	-0.13	-0.44	-0.44 ± 0.05
$\nu f^{-1}_{7/2}\to f^{-1}_{5/2}$	-0.30	-0.04	-0.13	-0.47	-0.52 ± 0.11

Table 3: Effective magnetic moment operator, eq. (1), calculated for the single-particle orbitals in the Pb region.

Orbit	δg_L	δg_S	δg_P
Proton			
0h	0.11	-2.4	0.3
0i	0.07	-2.5	0.1
2s		-1.8	
1d	0.17	-1.9	1.1
1f	0.16	-1.7	-0.1
Average	0.13±0.03	-2.0±0.2	-
Neutron			
1g	-0.06	1.5	-0.6
2p	-0.05	1.5	-0.3
1f	-0.10	1.4	-0.1
0i	-0.07	1.7	0.5
Average	-0.07±0.01	1.5±0.1	-

The effective g-factors are listed in table 3, together with an average value. We see that MEC enhances the proton g_L value by roughly 13%, while the core-polarisation calculation quenches the spin matrix element by roughly 40%, ie. $g_{S,eff}/g_S = 0.6$.

5. DIRECT MEASUREMENT OF δg_L

A direct measurement of $g_{L,eff}$ can be obtained from two-nucleon configurations in which a nucleon in an orbital $j_1 = \ell_1+1/2$ is coupled to a nucleon in an orbital with $j_2 = \ell_2-1/2$ to a resultant high spin $I = j_1+j_2 = \ell_1+\ell_2$. The spin contribution to the magnetic moment roughly cancels. A case in point[5] is the 11^- isomer in ^{210}Po whose wavefunction is

$$|11^-\rangle = |\pi i_{13/2} \times \pi h_{9/2}\rangle + \ldots .$$

where additional small terms in the wavefunction are not written down but are included in the analysis. The g-factor for the isomer, $g = \mu/I$, assuming a state-independent effective operator, eq. (1), is

$$g = g_{L,eff} + \frac{1}{I(2\ell_2+1)} \left[g_{S,eff} + g_{L,eff}\right] + \ldots .$$

Note that $g_{S,eff}$ occurs only in a small term. The measurement yields $g(^{210}Po;11^-)_{expt} = 1.18 \pm 0.01$ which on correcting for small terms produces for a proton

$$g_{L,eff} = 1.16 \pm 0.01$$

This compares well with the calculated average value of 1.13 ± 0.03 from table 3, confirming the importance of MEC.

Recently a two-neutron isomer has been found in ^{190}Os and its magnetic moment measured[6]. This is a deformed nucleus and the analysis involves using Nilsson orbitals but the essence is the same as the ^{210}Po case. The spin contributes only in small terms so the measurement yields essentially the neutron $g_{L,eff}$ value. The result is

$$-0.12 \le g_{L,eff} \le -0.07$$

again in good agreement with our average neutron value of -0.07 ± 0.01.

6. SUMMARY

In summary we note the following points:

(i) Calculations based on the one-boson-exchange model for core-polarisation and meson-exchange current corrections taken to second order are very successful in reproducing experimental data in Pb region. (The only failure in eight moments and three transitions is the neutron $i_{13/2}$ orbital).

(ii) The proton and neutron $g_{L,eff}$ calculated values are corroborated by experimental measurements in ^{210}Po and ^{190}Os.

(iii) The core polarisation calculation leads to a strong quenching of the spin matrix element $g_{S,eff}/g_S \simeq 0.6$ in heavy nuclei.

(iv) While there are uncertainties in the calculation due to approximations in the second-order core-polarisation evaluation and the model dependence of the non-Born MEC contributions there is no evidence for a failure of impulse approximation when augmented by MEC. In confronting data there is little room in the theory for explicit quark effects in nuclei or the swelling of nucleons in nuclei other than that automatically implied by the presence of two-body meson-exchange currents.

The study of magnetic moments in Pb region spans 35 years, since the pion contribution to δg_L was discussed by Miyazawa[7] in 1951 and first-order core-polarisation calculations were initiated in 1954 by Arima & Horie[8]. Since then the Tokyo group[2] have been mainly responsible for the increasing precision of these calculations. Our contribution has been a careful study of all MEC operators of heavy-meson range involving π, ρ, ω, σ and A_1 mesons, and the construction of the two-body magnetic moment operator from a chosen set of Lagrangians that respects chiral invariance and builds in vector-meson dominance[3]. We also examine the question of gauge invariance that leads to a continuity equation connecting the exchange current with the potential. In the one-boson-exchange model this consistency condition can be met, but it is necessary to repeat the core-polarisation calculations using a OBEP residual interaction.

References

1. I.S. Towner and F.C. Khanna, Nucl. Phys. A399(1983)334
2. A. Arima, K. Shimizu, W. Bentz and H. Hyuga, Adv. in Nucl. Phys. 17(1987)
3. I.S. Towner, Phys. Reports 155(1987)263
4. K. Shimizu, Z. Physik A278(1976)201
5. T. Yamazaki et al., Phys. Rev. Lett. 25(1970)547;
   O. Häusser et al., Nucl. Phys. A273(1976)253
   T. Yamazaki, in Mesons in Nuclei, eds. D.H. Wilkinson and M. Rho (North-Holland, Amsterdam, 1979) p.651
6. R. Beck et al., Phys. Rev. Lett. 59(1987)2923
7. H. Miyazawa, Prog. Theor. Phys. 6(1951)801
8. A. Arima and H. Horie, Prog. Theor. Phys. 11(1954)509

UNDERSTANDING THE NUCLEAR SPIN RESPONSE TO INTERMEDIATE ENERGY PROTONS

C. Glashausser
Physics Department, Rutgers University, Piscataway, NJ 08854

ABSTRACT

Measurements of the spin-flip probability S_{nn} for inclusive inelastic proton scattering around 300 MeV determine the relative $\Delta S=1/(\Delta S=0 + \Delta S=1)$ response of the nucleus provided S_{nn} is a "robust" observable. Recent data then give a general picture of the nuclear spin response up to about 40 MeV of excitation for momentum transfers up to about 200 MeV/c. Slab model calculations give qualitative agreement with much of the data; they explain the small values of the relative response at low excitation, and the larger values at high excitation. They do not explain the values approaching 90% for intermediate momentum transfers. RPA calculations which localize the S=0 strength indicate that the enhanced values of the relative response are caused primarily by the exhaustion of S = 0 strength at lower excitation energies, coupled with a shift in S=1 strength to higher excitations.

INTRODUCTION

Experiments carried out over the last several years at LAMPF and TRIUMF have measured the spin-flip probability S_{nn} for inclusive proton scattering around 300 MeV.[1,2,3] Targets include ^{12}C, $^{40,44,48}Ca$, ^{54}Fe, and ^{90}Zr. The typical range of angles and energy loss ω covered are shown by the data for ^{40}Ca in Fig. 1. The solid lines are theoretical curves which will be discussed below; the dashed lines correspond to the values of S_{nn} for isospin-averaged free nucleon-nucleon scattering. The measured values of S_{nn} are small at low ω; they tend to increase with increasing ω with a slope and maximum value which depend on the scattering angle. The maximum value in the 7° region (three-momentum transfer q about 115 MeV/c) is about 0.4, more than 50% larger than the free value. The qualitative features of the S_{nn} data for a given laboratory scattering angle are similar for all targets studied.

Such data, together with the standard differential cross sections, can be analyzed to determine the approximate position and shape of spin-dipole and quadrupole resonances in nuclei. They also determine the relative nuclear spin response, R_s, provided S_{nn} is a "robust" observable in the sense described below. Preliminary

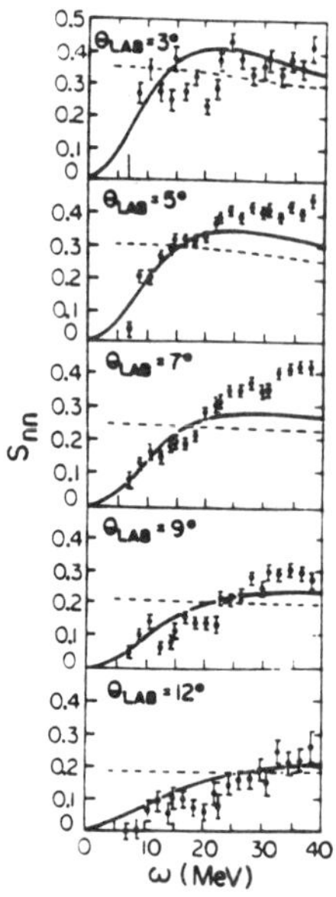

Fig. 1. Spin-flip probabilities S_{nn} for inelastic scattering from ^{40}Ca at 319 MeV as a function of excitation energy ω. The solid line is a prediction of the slab-response model described in the text. The dashed lines correspond to the free NN values of S_{nn}.

values of R_s for ^{48}Ca at 7° are shown in Fig. 2; at high ω, this figure shows that the percentage of the total nuclear response which is due to the transfer of spin to the nucleus is about 80%. These features of the data are discussed in detail by Kevin Jones in his contribution to this conference. Here we will concentrate on the attempts by several theorists to explain the data.

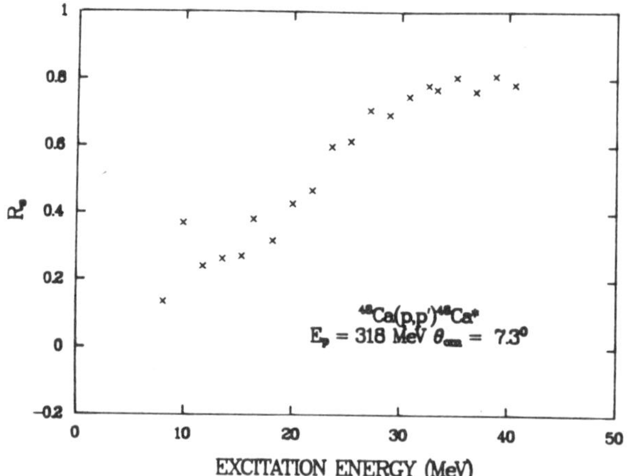

Fig. 2. Preliminary values of the relative nuclear spin-response R_s for ^{48}Ca at 7° (lab). This is the percentage of the total S=0 and S=1 response which is due to S=1 excitations.

IS S_{nn} ROBUST?

Many of the features of the S_{nn} data can be described by the surface response model developed by Bertsch, Esbensen, and Scholten,[4,5] in which the nucleus is considered to be a semi-infinite slab with an absorbing surface. Pauli blocking and RPA correlations are included in a nuclear Fermi gas. There is no nuclear structure input, so the standard S=0 giant resonance strength, for example, is not predicted to be localized in the 15 to 20 MeV region. The scattering is assumed to take place via a one-step process with the free NN amplitude. The solid curves in Fig. 1 show the results of calculations in this model by Esbensen.[1] They describe the S_{nn} data at low ω very well; at high ω, the basic shape is qualitatively described, but the strongly enhanced values around 7° are not predicted. Straightforward interpretation of these calculations suggests that the crucial ingredient in the qualitative success of this model is the RPA. The S=0 correlations in the nucleus are strongly attractive, so that S=0 strength is pulled down to low excitation energy; the S=1 correlations are repulsive and this strength is pushed up to higher excitation energies relative to the uncorrelated positions. But it should be noted that the good agreement between theoretical and experimental values of S_{nn} at low ω is somewhat artificial, since the slab response concentrates S=0 strength unrealistically close to zero excitation.

The differences between the measured values of S_{nn} and the free scattering values as illustrated in Fig. 1 are an important indication of the specifically nuclear response only if the effects of distortions, multistep processes, Fermi averaging, relativistic modifications in the nuclear medium, and the like are negligible. We have called S_{nn} a "robust" observable because initial calculations suggested that such processes did indeed change the expected values of S_{nn} only slightly; they indicated that if the nucleus were a non-interacting Fermi gas, then S_{nn} should be very close to its free value. For example, calculations by Smith using the slab response model showed that two-step processes make only a small contribution to the cross section in this region, and that, even if they were important, the value of S_{nn} would hardly be changed.[1] He was able to make a preliminary estimate of the effect of spin-orbit distortions in this model as well. Fermi averaging and relativistic effects were shown to be small by Horowitz and Murdock[6] in a plane wave model which should be valid close to the quasielastic peak. While there are reasons to suspect that S_{nn} should be insensitive to some of these effects, it should be emphasized that it is in many ways fortuitous that the spin transfer observable which is intrinsically the most informative about spin excitations in nuclei should also be robust. Many of these effects <u>are</u> important for other spin observables, and it may be that better calculations will reveal some sensitivity also for S_{nn}. In this regard, it is useful to look at the very recent calculations by Smith[7] for ^{40}Ca shown in Fig. 3. These include a careful treatment of Fermi motion and the full optical potential with central and spin-orbit distortions. The basic ingredients of the slab model are included also, but the details of this calculation are significantly different from that previously published. The dotted curves show the free response with the effect of Fermi motion included in the optimum frame of Gurvitz.[8] The dashed curves include the effects of distortion on the free response, and the solid curves show the complete calculation with the RPA response. Comparison of the curves in Fig. 3 with those of Fig. 1

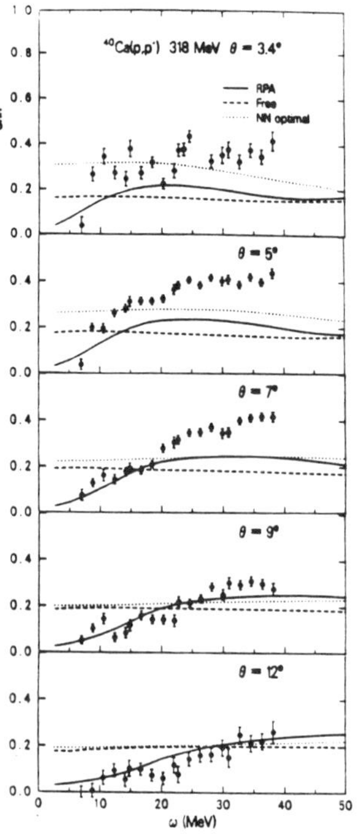

Fig. 3. New theoretical calculations of S_{nn} for ^{40}Ca by R. D. Smith (Ref. 7). The dotted curves show the free response with Fermi motion included. The dashed curves include the effects of distortion on the free response, and the solid curves show the complete calculation with the RPA response.

indicates little change at the larger angles, but at 5° and especially at 3° the new predictions are significantly below the old (and thus in worse agreement with the data). The difference is said to be due mostly to the momentum transfer in the distortion which causes the NN amplitudes to be evaluated on the average at a higher momentum transfer where the S_{nn} is smaller. The results of such calculations for the complete set of spin observables for proton scattering at 290 MeV from ^{54}Fe at 20° are also available.[9]

THE ENHANCEMENT OF S_{nn}

Perhaps the first possibility to consider in attempting to explain the enhanced values of S_{nn} is a spin resonance at high excitation. Indeed a mutipole decomposition of the spin-flip cross section data suggests a spin-quadrupole resonance at about 35 MeV in ^{40}Ca, with a width of about 10 MeV. This is part of the answer, but it is not complete, as shown by the data for ^{44}Ca at 7° in Fig. 4.

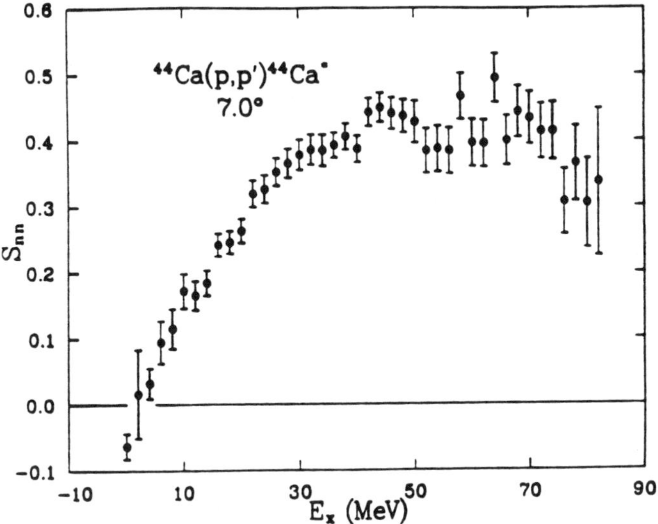

Fig. 4. The spin-flip probability for the $^{44}Ca(p,p')$ reaction at an incident energy of 290 MeV as a function of excitation energy at 7°.

The data for ^{44}Ca are similar to those for ^{40}Ca up to 40 MeV excitation at all angles studied; these data at 7° extend up to 80 MeV of excitation with little deviation from the value of about 0.4 for S_{nn} seen around 40 MeV. The ratio of S=1 to S=0 transfer thus is about constant over a wide region of excitation; this suggests a more general feature of the response than a particular relatively narrow resonance.

The enhancement might be a peculiarity of scattering around 300 MeV, rather than a characteristic of the nuclear response. To test this possibility we have measured S_{nn} at 800 MeV, where the free scattering value is about 0.05, rather than the 0.25 found in free scattering at 300 MeV; free scattering at 800 MeV is much more dominated by the central spin-independent term of the scattering amplitude. If the relative nuclear spin response is the same at 800 MeV as it is at 300 MeV, then the spin-flip probability for scattering from the ^{40}Ca target should be roughly 0.20 at high excitation at the same laboratory momentum transfer where the enhancement was seen at 300 MeV. Thus 7° scattering at 300 MeV should correspond to 4° at 800 MeV, and, in fact, our preliminary values are 0.23 ± 0.03 for the high excitation energy region at this angle. In 7° scattering at 800 MeV (which corresponds to about 13° at 300 MeV), our preliminary value corresponds closely to the free value; this is consistent with the high excitation behavior at large angles for 300 MeV scattering. The relative spin response thus appears to be the same at 300 MeV and 800 MeV as expected if it is really a nuclear structure phenomenon that we measure. (Actually, the results could in principle be different at the two energies if the spin dependent part of the amplitude changed noticeably between the two energies in the ratio of tensor to spin-spin components, for example, and the nucleus strongly preferred one to the other.)

In our discussion in Ref. 1 of the enhancement in S_{nn}, we noted that a depletion of the S=0 strength might be as important as an enhancement of the S=1 strength in achieving large values of S_{nn}. At 7°, the S=0, T=0 quadrupole is close to its maximum, but its strength is mostly exhausted in the 15 to 20 MeV excitation region.[10] Only the tails of other S=0 giant resonances and S=0, T=1 strength are left in the 40 MeV region. There were several difficulties with this explanation. Primary among them was the fact that, if this qualitative idea was true, then it should occur naturally in an RPA calculation; yet the one RPA calculation available, in the slab model as described above, failed to describe S_{nn} in just the interesting region. New RPA calculations suggest that S=0 depletion is indeed crucial in the explanation of the enhancement; what was missing in the previous calculation seems to be the localization of the S=0 strength that comes from the nuclear structure absent in the slab response. The calculations of Boucher et al. are shown in Fig. 5; these were made in the schematic model and include 2p-2h damping.[11] The positions of the S=0 resonances are adjusted to fit the positions of the resonances in the cross section data. Preliminary microscopic RPA calculations by Wambach[12] show similar features at high excitation. The latter calculations, in fact, have more difficulty with very low excitations where the predicted values of S_{nn} are too large. Preliminary schematic model calculations by Castel et al. also agree quite well with our S_{nn} data for ^{12}C at high excitation.[13]

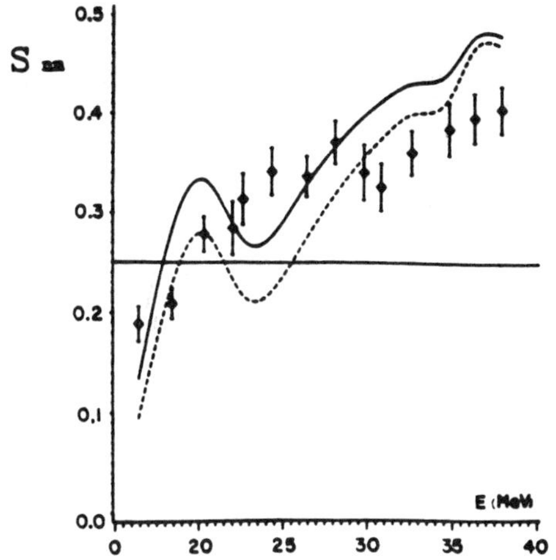

Fig. 5. Calculated and experimental values of S_{nn} for ^{40}Ca at 7° at high excitation energies. The horizontal line represents S_{nn} for free scattering without q dependence. The curves have been calculated by Boucher et al. (Ref. 11) in the schematic model with (solid) and without (dashed) meson exchange currents.

THE FUTURE

It appears, then, that the S_{nn} data provide a general picture of nuclear collectivity in the S=0 and S=1 channels which should ultimately be described by detailed calculations in the RPA framework. An important test of such calculations will be the corresponding differential cross sections which we have not discussed here, partly because they are not robust observables. They demand that distortions, in particular, be carefully treated, and that Fermi motion, relativity, and the like be included. New experiments will soon measure all the spin observables for ^{40}Ca over the same (q,ω) region as the S_{nn} data shown here; they should allow a longitudinal/transverse separation of the spin response. If the explanation of the enhancement depends essentially on the S=0 depletion, then both the longitudinal and transverse responses should be about equally enhanced; if it is primarily an S=1 effect, then one or the other response might be dominant. The A dependence of the enhancement also poses an interesting test. The S_{nn} data now available seem (roughly) independent of A. The slab response, for example, contains no A dependence in S_{nn} and it does describe the qualitative features; even the enhancement seems to occur at similar angles in ^{12}C and ^{90}Zr. Data for ^{4}He and ^{208}Pb will test this further. Yet an A dependence occurs naturally in the finite nucleus RPA; to explain a 7° enhancement requires different multipoles in ^{208}Pb than in ^{12}C, and the same resonance occurs at a different excitation energy in the two nuclei. Finally, it will be useful to try to separate the relative spin response into T=0 and T=1 components; the proton probe sees essentially only T=1 in the spin channel at the smallest angles but the larger angles in the present data contain significant contributions from T=0. The (p,n) probe sees only the T=1 channel, of course, but it is essentially spin-saturated at intermediate energies and so it is insensitive to the enhancement seen here. The T=0 component can be isolated in deuteron scattering. Such experiments will be attempted at Saturne II, but both measurement and interpretation are complicated by the spin-one nature of the deuteron. Understanding the nuclear spin response has only begun.

REFERENCES

1. C. Glashausser et al., Phys. Rev. Lett. 58, 2404 (1987).
2. O. Hausser et al., Can J. Phys. 65, 691 (1987).
3. K. Jones, International Conference on Spin Observables of Nuclear Probes, Telluride, Colorado, 1988.
4. G. Bertsch and O. Scholten, Phys. Rev. C 25, 804 (1982).
5. H. Esbensen and G. Bertsch, Ann. Phys. 157, 255 (1984), and Phys. Rev. C 34, 1419 (1986).
6. C. Horowitz and D. Murdock, Indiana University preprint IUNTC-87-8.
7. R. D. Smith, International Conference on Spin Observables of Nuclear Probes, Los Alamos National Laboratory preprint LA-UR88-1604.
8. S. A. Gurvitz, Phys. Rev. C 33, 422 (1986).
9. O. Hausser et al., preprint, 1988.
10. F. T. Baker et al., Phys. Rev. Rap. Comm. C 37, 1350 (1988).
11. P. M. Boucher et al., Phys. Rev. Rap. Comm. C 37, 906 (1988).
12. J. Wambach, private communication.
13. B. Castel, private communication.

ACKNOWLEDGEMENTS

Major contributors to the work at LAMPF are F. T. Baker, L. Bimbot, R. W. Fergerson, A. Green, K. W. Jones, and S. K. Nanda. At TRIUMF, the major participants, led by O. Hausser, include R. Abegg, R. Henderson, K. Hicks, K. P. Jackson, R. Jeppesen, J. Lisantti, C. A. Miller, R. Sawafta, M. Vetterli, and S. Yen. I have learned much from conversations with B. Castel, H. Esbensen, C. Horowitz, J. Iqbal, R. D. Smith, and J. Wambach and I am grateful to them for permission to use their calculations prior to publication. This work was supported in part by the National Science Foundation and the Department of Energy of the United States.

SYSTEMATICS OF THE Σ^-, Σ^0 and Σ^+ HYPERNUCLEAR SINGLE PARTICLE STATES

M. Divadeenam and T.E. Ward
Neutral Beam Division,* Brookhaven National Laboratory, Upton, NY 11973

D.J. Millener
Physics Department,* Brookhaven National Laboratory, Upton, NY 11973

ABSTRACT

Local density dependent Hartree-Fock potentials in analogy with the Λ^0 hypernuclear single particle state description by Millener, Dover and Gal were employed to investigate the systematics of the Σ hypernuclear single particle states. Both charge and isospin dependent (Lane potential) one–body potentials were used to describe the Σ^-, Σ^0 and Σ^+ single particle (s.p.) states for Σ hypernuclei in the mass range 6–40 with core–nucleus isospin $T_c = 1/2$. Due to the smaller magnitude of the Σ–nucleus potential the coulomb effects are dominant, especially for nuclei towards A = 40, while the isospin dependent potential has strong effect for the lighter nuclei due to the A^{-1} dependence of the Lane potential.

INTRODUCTION

Experimentally, narrow structures ($\Gamma < 5$-10 MeV) have been found in (K^-, π^-) and (K^-, π^+) reactions on light nuclei (A = 6-16).[1] The interpretation of these structures, which occur a few MeV into the Σ continuum, remains controversial. The fact that the observed peaks are a few MeV less bound than well established Λ substitutional states, which are formed by the transformation of a p–shell neutron into a Λ in the same orbit, suggests a similar interpretation in terms of Σ^- hypernuclear states. However, Σ single–particle p states 3 MeV or more in the continuum are very broad and the bound–state shell–model approach[2] used for Λ hypernuclei should not be applicable. The analysis of the experimental data is also complicated by a quasi–free Σ production background. Thus, questions of Σ–hypernuclear structure are best addressed in terms of the continuum shell model.[3] Even then, there is as yet no satisfactory way of including the effects of the $\Sigma N \rightarrow \Lambda N$ conversion process which is expected to give rise to large widths for Σ–hypernuclear states. For example, there is no evidence for the s–wave Σ states which are believed to be located in the bound region.

The ΣN interaction is believed to give rise to an average central well which is somewhat shallower than the well for the Λ particle.[4] This is consistent with experimental observations described above and with the Σ^- atom data. The central ΣN interaction based on two–body data has a strong spin and isospin dependence[1] so that the ΣN effective interaction is not necessarily weak even if the spin–isospin average force is weak.

*Work performed under the auspices of the U.S. Department of Energy under contract no. DE–AC–02–76CH00016.

We investigate the behaviour of Σ single–particle states for the three charge states Σ^-, Σ^0 and Σ^+ where the Coulomb force plays an important role. A Lane potential is used to roughly estimate the effect of the isospin dependence of the Σ–nucleus interaction.

THE POTENTIAL MODEL

For Λ hypernuclei, recent data[5] from the (π^+, K^+) reaction have determined the binding energies of many Λ single particle states up to A = 90. The Λ single–particle energies can be accounted for[6] by a density dependent potential of the Skyrme type.[7] The equivalent local potential, $V_L(r,E)$, is of the approximate form

$$V_L(r,E) = \frac{m*(r)}{m}\ U(r) + \left(1 - \frac{m*(r)}{m}\right)E \qquad (1)$$

$$U(r) = \tilde{t}_0\ \rho(r) + \frac{3}{8}\ t_3\ \rho^2(r) + \frac{1}{4}\ (t_1+t_2)\ T(r)$$

$$T(r) = \frac{3}{5}\ \left(\frac{3}{2}\ \pi^2\right)^{2/3}\ \rho^{5/3}(r)$$

$$\frac{\hbar^2}{2m*(r)} = \frac{\hbar^2}{2m} + \frac{1}{4}\ (t_1+t_2)\ \rho(r)$$

With a proper choice of $\tilde{t}_0$, t_3 and $(t_1 + t_2)$ Millener, Dover and Gal[6] were able to describe Λ hypernuclear single–particle states for mass ranging from 6–90. Experimentally determined charge densities were used for $\rho(r)$. The effect of the repulsive $\rho^2(r)$ is to give a potential well with half–value radius greater than that of the underlying density. This, and the non–locality, are important in obtaining the correct level spacings over a wide range of A.

HYPERNUCLEAR SINGLE PARTICLE STATES

We use a potential of the form given in Eq. (1) for Σ hypernuclear single–particle states. There is little information on Σ single particle states and we simplify the potential by setting $t_1 + t_2 = 0$, i.e. $\frac{m*}{m} = 1$, with some justification since space exchange terms in the ΣN interaction are small.[1] For the nuclear densities we use a two parameter Fermi distribution with diffuseness 0.54 fm and

$$r_0 = 1.15 - 1.12\ A^{-2/3} - 0.5\ A^{-4/3}$$

The Coulomb potential is taken to be that due to a uniformly charged sphere of radius $r_{0c}\ A^{1/3}$, where[8]

$$r_{0c} = 1.614 - 0.1067\ell nA + 0.005456(\ell nA)^2 + 6.112/(A-1.76)^2$$

We introduce a Lane potential in the form

$$\tilde{t}_0 = t_0 + t_{0\ell}\frac{< t \cdot T_c >}{A} \qquad (2)$$

where T_c and A refer to the isospin and mass number of the nuclear core and t refers to the isospin of the Σ particle. To maintain the shape of the potential we scale t_3 in the same way, i.e. $\tilde{t}_3 = t_3(\tilde{t}_0/t_0)$.

We choose the potential parameters with reference to the data[9] obtained from (K^-, $\pi^\pm$) reactions on a ^{12}C target. Crudely, speaking, if the three peaks seen are identified with $^{11}Bgs \times p_{3/2}\ \Sigma^-$, $^{11}Cgs \times p_{3/2}\ \Sigma^0$ and $^{11}Bgs \times p_{3/2}\ \Sigma^+$ configurations, the Σ p states are 3–4 MeV unbound for all three charge states. For the charge states with a $T^c = \frac{1}{2}$ core, Eq. (2) becomes

$$\tilde{t}_0 = t_0 - \frac{t_{0\ell}}{2A} \times Z_\Sigma \tag{2'}$$

Thus $t_{0\ell} > 0$ compensates for the Coulomb interaction to some degree. In the shell–model,[10,11] the contributions from the ΣN interaction to the binding energies of the three states are $B_{3/2}$, $\frac{1}{3}B_{1/2} + \frac{2}{3}B_{3/2}$, $\frac{2}{3}B_{1/2} + \frac{1}{3}B_{3/2}$ where B_T is the contribution for total hypernuclear isospin T. The relative energies $-\frac{1}{3}(B_{1/2} - B_{3/2})$, 0, $\frac{1}{3}(B_{1/2} - B_{3/2})$ show a behavior similar to that for the potential well in Eq. (2′). We expect[11] $B_{1/2} > B_{3/2}$ corresponding to $t_{0\ell} > 0$.

The parameters chosen give rise, for the mass 12 hypernuclei, to the potentials listed in Table I, where a typical Λ potential[6] is given for comparison. The ratio $t_3/\tilde{t}_0$, which controls the radial size of the well for a given density distribution, is arbitrarily chosen since there is no information on single particle level spacings for Σ hypernuclei. The parameters chosen correspond to a large, shallow well with a central depth of 12.1 MeV for Σ^0. To get the same s–state binding energy for Σ^0 with $t_3 = 0$ requires a central depth of 22.9 MeV.

The Σ^-, Σ^0 and Σ^+ single–particle binding energies for A = 9–40 are shown in Fig. 1(a)–(c). The solid curves are for no Lane potential, which we can interpret as energies for an isospin zero core. The strong influence of the Coulomb potential is clearly evident. The parameters were chosen so that, when the Lane potential is switched on, the mass 12 s–state binding energies are similar for all three charge states and the mass 12 p states are unbound.

The curves for unbound levels are terminated when the width of the resonant states becomes comparable to the resonance energy. In the Σ^- case, the attractive Coulomb potential cancels out the centrifugal barrier and there are no narrow resonance states.

The binding energies shown in Fig. 1 give some guidance as to what might be expected for the basic Σ single–particle states. Then the effects of the ΣN residual interaction, isospin mixing and $\Sigma N \to \Lambda N$ conversion must be considered. However, until the experimental Σ hypernuclear spectroscopic information and the theoretical treatments are both improved it will be difficult to make comparisons between the predicted Σ hypernuclear states and experiment.

Table 1. Density Dependent Potential Parameters[a]

	Λ	Σ^-	Σ^0	Σ^+	Units
$\tilde{t}_0$	-403	-160	-220	-280	MeV fm^3
t_3	3393	1600	2200	2800	MeV fm^6
$t_1 + t_2$	103	0	0	0	MeV fm^5
V(r=0)	27	8.8	12.1	15.4	MeV

[a]The Λ potential parameters come from ref. (5). The Σ^- and Σ^+ potential parameters are given for mass 12 [$t_{0\ell} = 1320$ MeV fm^3, A = 11 in Eq. (2′)].

REFERENCES

1. C. Dover, A. Gal and D.J. Millener, in preparation, to be submitted to Physics Reports.
2. E. H. Auerbach et al., Ann. Phys. 148, 381 (1983).
3. D. Halderson and R.J. Philpott, Phys. Rev. C37, 1104 (1988).
4. C. Dover and A. Gal, Progress in Particles and Nuclear Physics, 12, 171 (1984).
5. R. Chrien, Nucl. Phys. A478, 705c (1988).
6. D. J. Millener, C. Dover, and A. Gal, preprint, BNL–41295 (May 1988).
7. M. Rayet, Nucl. Phys. A367, 381 (1981).
8. Nuclear Physics with Heavy Ions and Mesons, Les Honches 1977. Eds., R. Balian, D. H. Wilkinson, M. Rho and G. Ripka. Vol. 2, p. 877 (North–Holland Publishing Co.).
9. R. Bertini et al., Phys. Lett. 136B, 29 (1984).
10. D. J. Millener, Nuclear Physics 450, 199C, (1986).
11. C. Dover, A. Gal and D. J. Millener, Phys. Lett. 138B, 337 (1984).

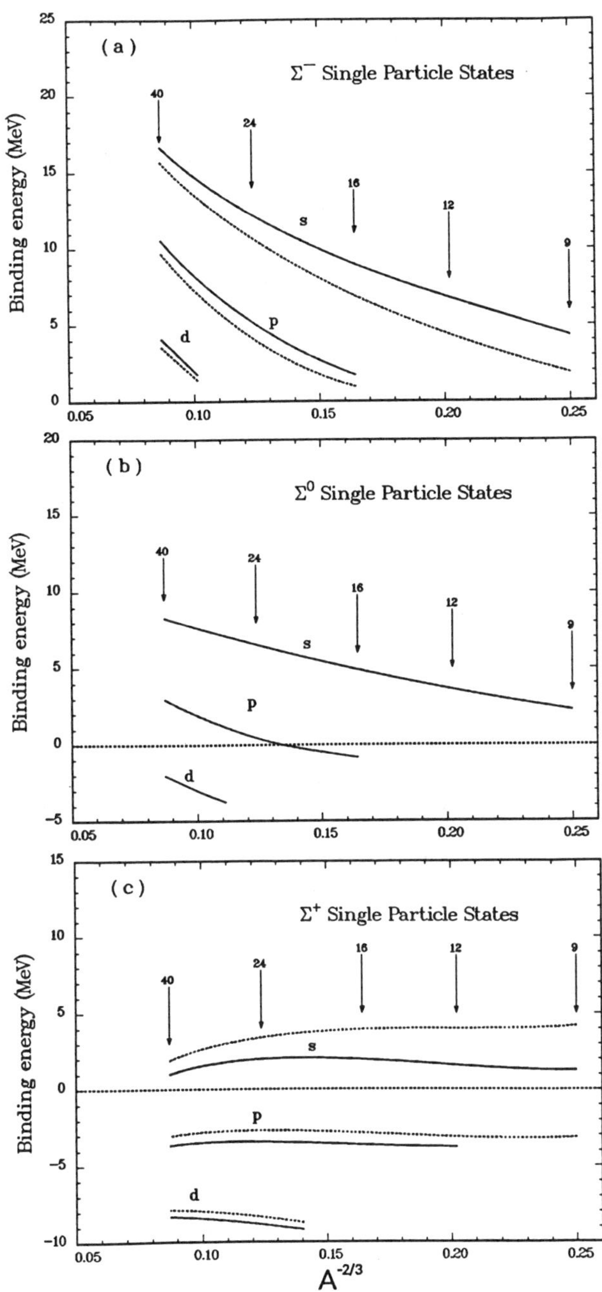

Figure 1. Binding energies of Σ-hypernuclear single–particle states obtained with the density dependent potentials of Table 1. Solid lines correspond to $t_{\ell 0} = 0$ (no lane potential) and dashed lines to $t_{\ell 0} = 1320$ MeV tm^3.

The widths of the dibaryon with I = 1 and $J^P = 0^-, 1^-, 2^-$

N. Konno and H. Nakamura
Department of Physics, College of Science and Engineering,
Aoyama Gakuin University, Setagaya, Tokyo, 157 Japan
H. Noya
Institute of Physics, Faculty of Economics,
Hosei University at Tama, Machida, Tokyo, 194-02 Japan

The diquark cluster model[1-3] predicts that some narrow dibaryon resonances with negative parity are degenerate at three energy levels of approximately 2.13, 2.19 and 2.25 GeV. This prediction may be quite consistent with the experimental analysis of the missing mass spectrum in the $p+{}^3He \longrightarrow d+X$ reaction studied by Tatischeff et al., in which they found three narrow peaks at 2.121, 2.192 and 2.240 GeV.[4] In this short article, we report our predictions for the widths of these narrow resonances calculated using the diquark cluster model as well as their precise resonating energies.

In the diquark cluster model, the matrix elements for NN and πd scattering involve an energy denominator

$$\underline{\mathbf{D}} = (E - M)\underline{\mathbf{I}} - \underline{\mathbf{M}}(E)$$

where E and M represent total energy and the dibaryon rest mass without fine structure, respectively, $\underline{\mathbf{I}}$ is the unit $n \times n$ matrix and $\underline{\mathbf{M}}(E)$ is a complex $n \times n$ matrix of which definition is given in ref.2. (n : the number of the q^6-states with the same quantum number). The eigen values $\lambda^j (j = 1, 2, \cdots, n)$ of the matrix $\underline{\mathbf{D}}$ are the function of E, which have a zero-point $E = E^j - i\Gamma^j/2$. The resonating energy and the width of the j-th dibaryon resonance are given by E^j and Γ^j, respectively. To find approximate values for E^j and Γ^j, we examined all the real E-values $\overline{E}^j$ where λ^j takes an imaginary eigen value $i\overline{\Gamma}^j/2$. When the value of $\overline{\Gamma}^j$ is sufficiently small, $\overline{E}^j$ and $\overline{\Gamma}^j$ approximate closely the values of E^j and Γ^j, respectively. The $\overline{E}^j$ and $\overline{\Gamma}^j$ calculated with the standard parameters which were used in ref.2 are illustrated in Fig.1 together with the experimental evidence by Tatischeff et al. (shaded region). Clearly, our prediction agrees well with their experiment. A uniform shift lower of the mass spectra by $\sim$ 10 MeV improves the agreement. This is realized by a slight reduction of the effective quark mass m (= 300 MeV) by $\sim$ 1 MeV. We remark that, in this case, the mass spectrum of the I = 0 dibaryons with $J^p = 1^-$ and 3^- is also shifted lower by $\sim$ 10 MeV and then the resonating energy of the lowest 1^- state becomes 2.12 GeV.

The existence of a group of dibaryon resonances with the same isospin, spin and parity as illustrated in Fig.1 has some implications for the structure of the NN and πd T-matrices. Namely, the T-matrices for I = 1 and $J^p = 0^-, 1^-$ and 2^- must have fine structures and may vary rapidly with total energy E. The investigation of such fine structures may be one of the more interesting future experimental problems. Finally, we note that the NN T-matrices may be reduced considerably by the possible strong repulsive forces in the entrance and exit channels, the existence of which in the $^3P_0$, $^3P_1$ (and $^1P_1$) states is suggested by phase shift analysis.[5] Therefore, so far as the widths of the 0^- and 1^- dibaryons are concerned, the actual values might be considerablly smaller than the present results.

References

[1] N. Konno and H. Nakamura, Lett. Nuovo Cimento **34**, 313 (1982)

[2] N. Konno, H. Nakamura and H. Noya, Phys. Rev. **D35**, 239 (1987)

[3] N. Konno, H. Nakamura and H. Noya, Phys. Rev. **D37**, 154 (1988)

[4] B. Tatischeff et al., Phys. Rev. **C36**, 1955 (1987)

[5] See, for example, R. A. Arndt et al., Phys. Rev. **D35**, 128 (1987)

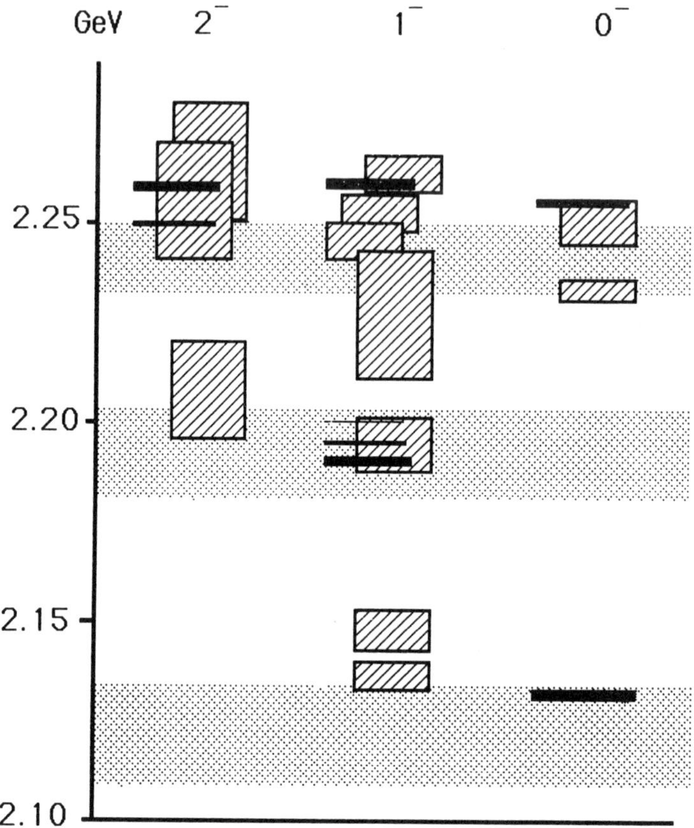

Fig.1 The predicted mass spectra of dibaryons with I = 1 and $J^p = 0^-, 1^-, 2^-$. Parameters R_π, Δ_{01} and Δ_{10} are taken to be 2.15 fm, 10 MeV and − 60 MeV, respectively. Each rectangle and its length represent a possible dibaryon state and its full width, respectively.(Some broad states with $\Gamma \gtrsim 100$ MeV are omitted.) The shaded regions represent the corresponding experimental values by Tatischeff et al.[4]

THEORY OF RARE K AND μ DECAYS

Robert E. Shrock
Institute for Theoretical Physics
State University of New York
Stony Brook, NY 11794

ABSTRACT

A review is given of theoretical issues of interest for rare K and μ decays.

I. INTRODUCTION

At present there is intensive experimental effort at several laboratories to search for certain possible K and μ decays which are forbidden in the successful standard $SU(3) \times SU(2) \times U(1)$ model of QCD and electroweak interactions. Any signal which is seen constitutes a far-reaching discovery of new physics beyond this standard model. There are also efforts to push down the limits for rare but allowed decays; the point here is that if a signal is seen at a level above that predicted by conventional physics, then again, one has observed something new. These search experiments are complementary to high-energy collider experiments in their ability to probe for physics beyond the standard model; indeed, as will be seen, they already are yielding limits on the mass scale of lepton flavor-changing interactions which are in the 10's of TeV. These are impressive limits indeed and will be improved in the near future. In this review we shall first recall the theoretical possibilities beyond the standard model which might give rise to such decays, and then describe the interest in various specific decay modes.[1] We shall not cover CP-violating decays, since these will be discussed by Sanda at this conference.

Historically, studies of K and μ decays yielded a number of important discoveries and many valuable limits on possible new physics. In the case of K decays, besides the discoveries of parity and CP violation, the extreme suppression of $K^0 - \bar{K}^0$ mixing, as measured by $\Delta m_K/m_K = [m(K_L) - m(K_S)]/m_K \simeq 0.7 \times 10^{-14}$, indicated the absence of tree-level neutral strangeness-changing currents and thus led to the GIM mechanism[2] and the inference of the c quark. Indeed, this also showed that there must be further suppression of such currents at the one-loop level.[3] Regular μ decay helped to confirm the $V - A$ nature of charged weak currents. The so-far successful empirical rule of lepton family number conservation was established in part from the null search for such decays as $\mu \to e\gamma$.

II. SOME POSSIBILITIES FOR NEW PHYSICS BEYOND THE STANDARD MODEL

Why does one expect new physics beyond the standard model? Although it is very successful, this model is clearly incomplete. It does not give any explanation of the different fermion generations. Why are there three such generations (so far), why do the fermion mass ratios have the values that they do, and why are the observed fermion masses small compared to m_W and m_Z? Although the standard model unifies strong, electromagnetic and weak interactions, there are three different, independent gauge couplings corresponding to the three factor groups, and consequently, in particular, there is no understanding of why the weak mixing angle in $\sin^2\theta_W$ has the value that it does. Appealing efforts have been made to extend the model further by embedding it in a single grand unified group, such as[4] SU(5), with only one gauge coupling. Quarks and leptons are unified in the same representations, electric charge quantization is explained, and the value of $\sin^2\theta_W$ is predicted. However, the simplest and most predictive theory, minimal SU(5), predicts proton decay at too rapid a rate to agree with experimental limits. The standard model assumes the Higgs mechanism and an associated physical Higgs particle. Especially in the context of grand unification, but more generally, this gives rise to the hierarchy problem: if there is a scale at which new physics sets in (e.g. $\Lambda_{GUT} \sim 10^{14}$ GeV), then it is unnatural for the mass scale characterizing the Higgs sector (and hence also the electroweak symmetry breaking) to be much smaller than Λ since there is a quadratic shift in the mass squared parameter in the Higgs potential $\delta\mu^2 \sim \alpha\Lambda^2$ at the one-loop level. This problem is addressed by two extensions of the standard model (among others), viz., technicolor and supersymmetry. The puzzle of multiple generations has led to models involving generational gauge groups and compositeness. We briefly describe the implications of these models below. Another reason for expecting new physics is that studies of the full path integral which defines the scalar sector of the standard model (subsuming, but going beyond, perturbation theory) indicate that it actually describes a free theory.[5] These studies include lattice formulations, in which the coupling is found to vanish as the lattice cutoff is removed. Similar behavior is found even when gauge fields are added, suggesting that the theory should only be used as an effective model, with a cutoff in place which represents the onset of new physics.[6]

1. Technicolor

In technicolor (TC) models[7] there is no Higgs. It is assumed that there is a new gauge interaction with a running coupling which gets large at a scale Λ_{TC} of order TeV. This is supposed to give rise to a technifermion condensate, in a manner reminiscent of the usual chiral symmetry breaking in QCD. This condensate couples to the weak gauge bosons and has the effect of giving them masses. However, in minimal technicolor models, there is no obvious way to give the fermions masses, since the Yukawa couplings are absent. For this reason, extended technicolor (ETC) models were developed; these contain a new set of massive vector bosons which mediate transitions between ordinary fermions and technifermions, and can give the usual fermions their masses. These models have their own problems: for example, what gives masses to the ETC gauge bosons? Of greatest interest here is the fact that they naturally lead to flavor-changing neutral interactions (FCNI) at rather large levels. In order to suppress $K^0 - \bar{K}^0$ mixing sufficiently in order not to conflict with experiment, one has to increase the ETC mass scale, but this in turn renders it difficult to obtain sufficiently large fermion masses, which have the form $m_f \sim (4\pi^2)^{-1}(\Lambda_{TC}/m_{ETC})^2\Lambda_{TC}$. Indeed, there can even be tree-level FCNI in such theories, mediated by ETC gauge bosons. This can give rise to both $K^0 - \bar{K}^0$ mixing from the quark process $d+\bar{s} \to \bar{d}+s$, and to the leptonic flavor-number violating decays $K_L^0 \to \mu^\pm e^\mp$ and $K^+ \to \pi^+\mu^\pm e^\mp$ from the quark process $d+\bar{s} \to \mu^\pm + e^\mp$. Typical values of the ETC gauge boson mass of about 30 TeV can give rise to flavor-changing decays at levels which are accessible to current K decay experiments. It should be noted that the problems with large NFCI's in ETC models have led theorists to modify these models. Subsequent developments have concentrated on suppressing the NFCI's without rendering the fermion masses too small. This can be achieved in models with slowly running technicolor couplings.[8]. Models have also been proposed which combine features of compositeness and technicolor, and have a GIM mechanism.[9] These models suppress NFCI's sufficiently that the latter would not be likely to be observable in the current K-decay experiments searching for lepton family number violation.

2. Generational Group

In response to the continuing mystery of the multiple generations of fermions, one proposal is that they be gauged. This would mean that a particle of one generation can make a transition to a particle of another generation via emission or absorption of a gauge boson. For example, a simple model might assign the three generations to triplets of an SU(2) gauge group[10], as $(d,s,b)^T, (e,\mu,\tau)^T$, etc. Such a "horizontal" group would give rise to a wealth of processes including the $|\Delta G| = 2$ process $K^0 \leftrightarrow \bar{K}^0$ via $A^{(+)} - A^{(-)}$ mixing, where $A^{(+)}$ is the [electrically neutral] gauge boson increasing generation number G by one unit, as indicated by the superscript $(+)$. In this model one would also have the $\Delta G = 0$ quark process $d + \bar{s} \to \mu^+ + e^-$ which yields $K^+ \to \pi^+\mu^+e^-$, and, together with the conjugate quark process, contributes to $K_L \to$

$\mu^{\pm}e^{\mp}$. Assuming that $A^{(+)} - A^{(-)}$ mixing involves some suppression, one would infer that the $|\Delta G| = 2$ decay $K^+ \to \pi^+\mu^-e^+$ would be commensurately suppressed in this model, relative to $K^+ \to \pi^+\mu^+e^-$. Among leptonic processes, one has the $|\Delta G| = 1$ decay $\mu^+ \to e^+e^+e^-$. Since all of these are tree-level processes, it is easy to calculate their rates, as functions of the masses and couplings in the model. From the non-observation of the various decays, one then obtains correlated lower bounds on the masses of the generation-changing vector boson mass eigenstates, as follows:

$$m_A \geq (13 \text{ TeV})\left|\frac{g_A}{g}\right|\left(\frac{|a|^2 + |b|^2}{2}\right)^{1/2}|c_V|^{1/2}\left[\frac{10^{-9}}{B(K^+ \to \pi^+\mu^+e^-)}\right]^{1/4} \tag{2.1}$$

where g_A is the generational gauge boson-fermion coupling, and the leptonic and hadronic currents are given by $\bar{e}\gamma_\lambda(a + b\gamma_5)\mu$ and $\bar{s}\gamma_\lambda(c_V + c_A\gamma_5)d$. Note that (2.1) would vanish if the hadronic generational current were purely axial-vector. For typical values of these quantities, the current upper bound on this decay ((3.1) below) yields a lower limit on m_A of order 10 TeV.

Similarly, from a consideration of the $K_L \to \mu^+e^-$ decay one calculates

$$m_A \geq (40 \text{ TeV})\left|\frac{g_A}{g}\right|\left(\frac{|a|^2 + |b|^2}{2}\right)^{1/2}|c_A|^{1/2}\left[\frac{10^{-9}}{B(K_L \to \mu^{\pm}e^{\mp})}\right]^{1/4} \tag{2.2}$$

Taking the present upper limit ((3.2) below) on this decay and making the same assumptions about the couplings as before, one obtains the bound $m_A > 25$ TeV. We have chosen the generational group to calculate rates for these decays and to obtain consequent mass and coupling limits because of the simplicity of the tree-level graphs which contribute; however, similar estimates are obtained from (extended) technicolor models, and from the other possible sources of new physics discussed here. This is quite interesting, especially in the case of (conventional) ETC models, since for them one has a definite theoretical expectation of the mass scale of the ETC vector bosons (in contrast, for example, to the generational group) and it is accessible to current K-decay experiments.

3. Flavor-Changing Higgs Bosons

In a theory with multiple Higgs fields, one can get neutral flavor- changing interactions. These again lead to interesting decays like those mentioned above. The value of $\Delta m_K/m_K$ puts a very severe limit on the coupling strengths of the Higgs-mediated NFCI. If the Yukawa coupling $h_{\mu e} \sim h_{sd}$, then the constraint on $K^0 - \bar{K}^0$ mixing would limit the contibutions to such decays as $K^+ \to \pi^+\mu^{\pm}e^{\mp}$ and $K_L \to \mu^{\pm}e^{\mp}$ to negligibly small levels. But it is, of course, possible that $h_{\mu e} >> h_{sd}$; in this case, this mechanism might be significant.

4. Compositeness

The proliferation of fermions and the belief that particle physics must involve a few simple building blocks has led some to doubt whether the known particles really are fundamental. If one views the fermions, in particular, as bound states of more fundamental constituents, then it is natural to regard the particles of the lowest generation as the ground states and those of higher generations as excited states. In this theoretical framework, decays of higher-generation fermions to lower-generation fermions are quite natural. These would give rise to decays such as those which are currently being sought. For example, among other modes, one would expect $\mu \to e\gamma$ and $\mu \to ee\bar{e}$ to occur at possibly observable levels. There are, however, problems with this direction of theoretical work. There is no direct experimental evidence for compositeness. It is not known how to get masses for the bound states which are nonzero but many orders of magnitude smaller than the mass scale of the compositeness, which latter scale is known to be greater than about 1 TeV (depending on the model).[11] This contrasts sharply with the one theory where we have a reasonable understanding of the bound states, namely, QCD: there the masses are comparable to the QCD scale, measured by quantities such as the scale for the running coupling, Λ_{QCD}, or the value of the chiral symmetry breaking constant f_π. Related to this problem, there is no known realistic dynamical composite model. This is unfortunate, since if such a model existed, it would help to solve longstanding mysteries such as the value of m_e/m_μ and the other fermion mass ratios.

5. Supersymmetry

Supersymmetric extensions of the standard model have been intensively studied over the past years.[12] They have the potential of keeping Higgs scalars light compared with mass scales 10^{14} GeV or higher, since the scalars would be degenerate with their fermionic supersymmetry (SS) partners, which in turn can be kept massless by chiral symmetry. The supersymmetry must, of course, be broken, since no supersymmetric partners have been detected; the nature of this breaking is not well understood. Recent interest in such models has been spurred partly because they occur as the low-energy sector of superstring theories. Since supersymmetric theories contain a plethora of new particles, it is not surprising that they feature new mechanisms for getting flavor-changing processes For example, there is a new contribution to $K^0 - \bar{K}^0$ mixing from a box diagram involving virtual $\tilde{w}$ and $\tilde{q}(2/3)$ lines, where these are, respectively, the SS partners of the W and of charge 2/3 quarks. From mixing in the $\tilde{q}$ sector, one can get tree-level contributions to the decay[13] $K^+ \to \pi^+\tilde{\gamma}\tilde{\gamma}$, where $\tilde{\gamma}$ is the SS partner to the photon, and to the decay[14] $\mu \to e\tilde{\gamma}\tilde{\gamma}$. The first of these might produce a signal in the BNL experiment searching for $K^+ \to \pi^+ +$ missing neutrals which is substantially above the conventional branching ratio due to $K^+ \to \pi^+\nu\bar{\nu}$. If the tree level decay is sufficiently suppressed, or is absent, then the decay still receives SS contributions at the one-loop level; however, these are estimated to be smaller than the standard model contributions.[15,16] (Actually, the photino would, in general, mix with the higgsino and zino to form mass eigenstates.)

6. Massive Neutrinos

The extension of the standard model to include massive neutrinos is well-known. By adding right-handed singlet ν_R fields, one can get Dirac masses in the same way as in the quark sector. By adding an $I = 1$, $Y = 2$ Higgs field, one can get Majorana neutrino masses. A natural concomitant of massive nondegenerate neutrinos is leptonic mixing (and hence lepton family number violation), similar to the mixing in the quark sector. Thus, the weak eigenstates ν_e, ν_{mu}, and ν_{tau} would be linear combinations of the neutrino mass eigenstates, with $\nu_e = \sum_{i=1}^N U_{1i}\nu_i$, $\nu_{mu} = \sum_{i=1}^N U_{2i}\nu_i$, etc. This mixing would give rise to the various leptonic-family number-violating decays, such as $\mu \to e\gamma$, $\mu \to ee\bar{e}$, $K^+ \to \pi^+\mu^\pm e^\mp$ and $K_L \to \mu^\pm e^\mp$. Calculations of the rates for $\mu \to e\gamma$ [17,18] and the latter three decays[18] have been carried out in the extended standard model. Furthermore, in the presence of neutrino mixing, a decay such as $K^+ \to \mu^+\nu_\mu$ really comprises the set of decays into all of the kinematically allowed mass eigenstates ν_i contained in the weak eigenstate ν_μ. This leads to very sensitive tests for heavy neutrino mixing in two-body leptonic K^+ and π^+ decays.[19] and a test using the spectral distribution in regular μ decay.[20] Both analysis of old data and a number of experiments at KEK, SIN, and TRIUMF (see below) have now yielded excellent upper limits on such mixing. Using these limits in conjunction with the previous calculations[17,18], one finds that the contributions of massive neutrinos to the muon and K decays listed above are negligibly small, well below experimental sensitivities.

III. K Decays

Here we list some specific K decays, their importance, and their current experimental status.

1. $K^+ \to \pi^+\mu^\pm e^\mp$, $\quad K_L \to \mu^\pm e^\mp$

These decays are of interest because they are lepton family-number violating. Since any signal constitutes new physics, it is clearly valuable to push the upper limits on these decays down as far as possible. One can estimate the relative sensitivites of the two decays as follows: for a typical elementary amplitude, the observed branching ratio $B(K_L \to \mu^\pm e^\mp) \simeq 40B(K^+ \to \pi^+\mu^\pm e^\mp)$ because of (a) the fact that $\Gamma(K^+)_{tot} \simeq 4\Gamma(K_L)_{tot}$ and (b) the relative suppression due to the 3-body phase space factor in the K^+ mode compared to the K_L mode. However, the situation may be reversed by special characteristics of the new physics responsible for these decays; for example, if the effective hadronic current is purely vectorial, then $c_A = 0$ in eq. (2.2), and the matrix element $< 0|(J_{sd})_\lambda|K_L> = 0$. Even if the effective hadronic current is not exactly vectorial, it could have an axial-vector part which is suppressed sufficiently relative to the vector part that the signals in both modes could be comparable. More generally, the $K_L \to \mu^\pm e^\mp$ decay can proceed via a pseudoscalar as well as axial-vector coupling, while $K^+ \to \pi^+\mu^\pm e^\mp$ can proceed via a scalar as well as vector coupling.

There are currently several experiments searching for these decays:

(i) E777 at BNL, a BNL-Seattle-SIN-Yale collaboration (M. Zeller, spokesman) is searching for $K^+ \to \pi^+\mu^+e^-$ with an ultimate goal in sensitivity of $\sim 10^{-10}$ in branching ratio. It has currently set the limit[21] (here and elsewhere, 90 % CL)

$$B(K^+ \to \pi^+\mu^+e^-) < 1.1 \times 10^{-9} \tag{3.1}$$

and expects to reduce this by summer, 1988.

(ii) E780 at BNL, a BNL-Yale collaboration (W. Morse and M. Schmidt, spokesmen) is searching for $K_L \to \mu^\pm e^\mp$ with an ultimate goal in sensitivity of $B \sim 10^{-10}$. It has currently set the limit[22]

$$B(K_L \to \mu^+e^-) < 6.7 \times 10^{-9} \tag{3.2}$$

and expects to reduce this by summer, 1988.

(iii) E791 at BNL, a UCLA-LANL-Penn-Princeton-Stanford-Temple collaboration (Cousins, Molzon, Wojcicki et al.) is searching for $K_L \to \mu^\pm e^\mp$ and other rare K_L decays with an ultimate goal in sensitivity stated in the proposal to be 2×10^{-12}.

(iv) E137 at KEK, a KEK-Tokyo-Kyoto collaboration (Inagaki, spokesman) is searching for $K_L \to \mu^\pm e^\mp$.

$K_L \to \mu^+\mu^-$, $K_L \to e^+e^-$

The $K_L \to \mu^+\mu^-$ decay played an important role historically in showing the suppression of neutral strangeness changing currents, both at the tree level, and at the one-loop level.[3,23] It has also been used to obtain constraints on quark mixing angles.[24] Experiments E780 and E791 at BNL are accumulating events from this decay. E791 plans to measure the μ polarization, which could give information on possible exotic decay mechanisms.[25] The ee mode is helicity-suppressed relative to the $\mu\bar{\mu}$ mode and, from the absorptive part, one gets an estimate $B(K_L \to e^+e^-) \sim$ several $\times 10^{-12}$, probably below current experimental sensitivities. However, precisely because the conventional branching ratio is so small, this decay mode provides an excellent probe for new physics which might produce a rate significantly above the conventional level and perhaps observable. The current limit is set by E780 [22]:

$$B(K_L \to e^+e^-) < 1.6 \times 10^{-9} \tag{3.3}$$

an improvement over their published limit of 4.5×10^{-9}.

2. $K^+ \to \pi^+ + X^0$

If X^0 is the standard model Higgs, which then decays as $H \to e^+e^-$, the resulting rate is significantly above the data[26,27], thereby ruling out the Higgs in much of the accessible mass region. (One should note that although the standard Higgs with a mass $m_H < 2m_\mu$ would normally decay dominantly to e^+e^-, it might also have invisible decays[28], into aa, MM, and/or $\nu\bar{\nu}$, where a is the axion[29] and M is the Majoron.) A KEK experiment has searched over a wide mass region for $K^+ \to \pi^+ + X^0$.[30] This experiment has established a correlated upper bound on the branching ratio for this decay which varies from $\sim 10^{-6}$ for $m(X^0) < 75$ MeV to $\sim 10^{-5}$ for $170 < m(X^0) < 260$ MeV. Recently, there has been renewed interest in this decay mode because of the suggestion from heavy ion experiments[31] that there might be an X^0 of mass 1.8 MeV, decaying to e^+e^-. E777 at BNL has carried out a search for a peak in the e^+e^- invariant mass distribution in $K^+ \to \pi^+e^+e^-$ as would result from $K^+ \to \pi^+X^0$; $x^0 \to e^+e^-$ and has set bounds on such a mode as a function of the X^0 lifetime[32] (see also the analysis of previous searches in the second paper of this reference). A previous experiment on $\eta' \to \eta\mu^+\mu^-$ has set the limit $m_H > 409$ MeV.[33]) It should be noted that under certain assumptions, there is a lower bound on the standard model Higgs of about 7 GeV[34]; however, these assumptions can be evaded.

The $K^+ \to \pi^+X^0$ decay mode also constitutes a test for the conventional axion and yields evidence against it.[35]

3. $K^+ \to \pi^+ +$ missing neutrals

This decay has a conventional source, viz., $K^+ \to \pi^+\nu_\ell\bar{\nu}_\ell$, $\ell = e, \mu, \tau$. The decay is well short-distance dominated, and the GIM mechanism is operative. An estimate for the rate in the standard model requires an input assumption for the t-quark mass and values for the quark mixing angles; one finds a branching ratio of order 10^{-10} with a theoretical uncertainty of a factor of 3-4 up or down.[3,36–38] If a signal is observed above this level, it would suggest new physics. The current published limit, from a KEK experiment, is[27] $B(K^+ \to \pi^+ + \text{missing neutrals}) < 1.4 \times 10^{-7}$, so that there is a wide range in which new physics might be discovered. E787 at BNL, a BNL-Princeton-TRIUMF collaboration (Littenberg, Smith, Bryman et al.) is currently searching for this decay, with an ultimate goal in sensitivity of 2×10^{-10} in branching ratio. The possible new physics which might contribute here includes $N > 3$ generations of fermions. The contributions of a possible fourth generation are constrained by such other processes as $K^0 - \bar{K}^0$ mixing and $K_L \to \mu^+\mu^-$. As discussed in II.5, possible decays into supersymmetric particles such as photinos could also contribute.[13,15,16].

4. Some K Decays Involving Photons

These include $K^+ \to \ell^+\nu_\ell\gamma$, $\ell = \mu, e$ and $K^+ \to \mu^+\nu_\ell e^+e^-$. These decays yield information about structure-dependent contributions, where the real or virtual photon is emitted from the initial K. The latter modes, where the virtual γ converts to an $e\bar{e}$ pair, are particularly useful since one can vary the q^2 of the probe. E777 at BNL plans to study these latter decays. It would also be useful to measure $K^+ \to \pi^+\gamma\gamma$ for which the branching ratio is estimated[3] to be $10^{-5} - 10^{-6}$.

5. Limits on Heavy Neutrino Mixing from K Decays

Since the theoretical proposal and analysis of existing data[19], there have been a number of experiments to search for a heavy neutrino signal in two-body leptonic decays of K and π. These include

(i) $K^+ \to \mu^+\nu_i$, in two KEK experiments[39,40]
(ii) $\pi^+ \to \mu^+\nu_i$, in three SIN experiments[41,42,43] and a TRIUMF experiment[44]
(iii) $\pi^+ \to e^+\nu_i$ in experiments at TRIUMF[45] and SIN[46].

These have yielded quite stringent upper limits on heavy neutrino mixing; for example, with K decays, $|U_{2i}|^2 < 10^{-5}$ for $m(\nu_i) = 100$ MeV down to $|U_{2i}|^2 < 2\times10^{-6}$ for $m(\nu_i) = 200$ MeV from the second KEK experiment[40], and $|U_{1i}|^2 < 4 \times 10^{-6}$ from analysis of K_{e2} data.19. Excellent limits have also been obtained from π decays; see Refs. 1 and 47 for a compilation of limits).

IV. μ Decays

1. Regular μ Decay (including possible lepton family number violating contributions)

New physics could affect regular μ decay in several ways. For example, models such as $SU(2)_L\times SU(2)_R\times U(1)$ feature right-handed currents, which would contribute to μ decay and change the observed spectrum. A LBL-TRIUMF experiment has measured the combination of spectral parameters $\xi P_\mu \delta/\rho$ and thereby set upper limits on the effects of such right-handed currents.[48] For no $W_L - W_R$ mixing, they find $m_{W_2} > 480$ GeV. Recalling that the observed decay is really $\mu \to e+$ missing neutrals, one sees that decays such as $\mu \to e\tilde{\gamma}\tilde{\gamma}$ could also contribute and could, again, affect the observed spectral distribution.[14] Finally, one can use this distribution to set a good limit on the emission of massive neutrinos.[20]

2. Muon Decays Which Violate Lepton Family Number

These include $\mu \to e\gamma$, $\mu \to ee\bar{e}$, and $\mu \to e\gamma\gamma$. These decays can arise from the various types of possible physics beyond the standard model discussed above. Indeed, $\mu \to e\gamma$ was searched for long ago when it was thought that the muon might be an excited electron. The upper limits on the branching ratios for these decays have been improved greatly since the late 1970's. The current limits are $B(\mu \to e\gamma) < 4.9\times 10^{-11}$ [49] and $B(\mu \to e\gamma\gamma) < 7.2\times 10^{-11}$ [50], from the LASL Crystal Box experiment, and $B(\mu^+ \to e^+e^+e^-) < 1.0\times 10^{-12}$ from the SINDRUM experiment[51]. As an illustration of the kind of limits on new physics which these results imply, in a generational group model, one obtains

$$m_A > (80\text{TeV})\left|\frac{\epsilon g_{\mu e} g_{ee}}{g^2}\right|^{1/2}\left[\frac{10^{-12}}{B(\mu \to ee\bar{e})}\right]^{1/4} \tag{4.1}$$

Using the current upper limit[51] and assuming the plausible values $\epsilon \sim 0.1$, $|g_{\mu e}g_{ee}/g^2| \sim 1$, one gets $m_A > 25$ TeV, which is an excellent bound.

There have also been analyses of data and experimental searches to set limits on the decays[48,52–53] $\mu^+ \to e^+X^0$ and[54] $\mu^+ \to e^+\gamma X^0$, where X^0 is a light scalar boson. Such a boson might, for example occur in a model in which there is spontaneous breaking of the (global) lepton family number (the majoron).[55]

Another process which violates lepton family number is muon conversion in the field of a nucleus, subsuming both $\mu^- + (A,Z) \to e^- + (A,Z)$, and $\mu^- + (A,Z) \to e^+ + (A,Z-2)$, which also violates total lepton number. The upper limits on these processes have been greatly improved over the past several years by experiments at SIN and TRIUMF. The current upper bounds on these respective processes, normalized by the correponding muon capture rates, are $< 4.6\times 10^{-12}$ and $< 1.7\times 10^{-10}$ from TRIUMF experiments.[56].

V. Conclusions

In conclusion, searches for decays which are forbidden in the standard model, or for signals above the level at which allowed decays are predicted to occur, provide very powerful probes for new physics beyond the standard model. Because of the incompleteness of the standard model, there is good reason to expect such new physics. In at least some models, such as extended technicolor, the mass scale is reasonably well determined, and this model is already severely constrained by experiment. From eqs. (2.1), (2.2), and (4.1), it is clear that recent and ongoing experiments already yield bounds on new physics which are competitive with what colliders can achieve. These experiments are continuing to push down the limits, and their forthcoming results are eagerly awaited.

REFERENCES

1. Some details not covered here can be found in the reviews R. E. Shrock, "Rare K Decays as Probes of New Physics", in *Proceedings of the Third LAMPF II Workshop*, 1983, vol.

I, p. 316, and G. Kane and R. E. Shrock, "Rare K and μ Decays: Theoretical and Experimental Status and Prospects", in Theoretical Symposium on Medium-Energy Sources of Strangeness, A.I.P. Conf. Proc. **102**, 123 (1983). For other recent reviews, see, e.g., M. Zeller, Nucl. Phys. **A450**, 491C (1986) and L. Littenberg, Nucl. Phys. **B279**, 171 (1987).
2. S. L. Glashow, J. Iliopoulos, and L. Maiani, Phys. Rev. **D2**, 1285 (1970).
3. M. K. Gaillard and B. W. Lee, Phys. Rev. **D10**, 897 (1974).
4. H. M. Georgi and S. L. Glashow, Phys. Rev. Lett. **32**, 438 (1974).
5. For a review, see, e.g., D. Callaway, Phys. Repts, in press.
6. A. Hasenfratz and P. Hasenfratz, Phys. Rev. **D34**, 3160 (1986) and references therein.
7. For a review, see, e.g., E. Farhi and L. Susskind, Phys. Rept. **74**, 277 (1981).
8. B. Holdom, Phys. Lett. **150B**, 301 (1985); T. Appelquist et al., Phys. Rev. Lett. **57**, 957 (1986); **60**, 1114 (1988); T. Appelquist and L. Wijewardhana, Phys. Rev. **D35**, 774 (1987); **36**, 568 (1987).
9. R. S. Chivukula and H. M. Georgi, Phys. Lett. **188**, 99 (1987), and to be published.
10. See, e.g., R. Cahn and H. Harari, Nucl. Phys. **B176**, 135 (1980); G. Kane and R. Thun, Phys. Lett. **94B**, 513 (1980).
11. E. Eichen, K. Lane, and M. Peskin, Phys. Rev. Lett. **50**, 811 (1983).
12. For reviews of supersymmetry with phenomenological discussions, see, e.g., P. Fayet and S. Ferrara, Phys. Rept. **32C**, 249 (1977); H. Haber and G. Kane, ibid., **117**, 75 (1985).
13. R. E. Shrock, in *Proceedings of the 1982 DPF Summer Study on Elementary Particle Physics and Future Facilities*, Snowmass, p. 291.
14. J. S. Barber and R. E. Shrock, Phys. Lett. **139B**, 427 (1984).
15. M. K. Gaillard, Y.-C. Kao, I-H. Lee, and M. Suzuki, Phys. Lett. **123B**, 241 (1983).
16. J. Ellis and J. Hagelin, Nucl. Phys. **B217**, 189 (1983).
17. W. J. Marciano and A. I. Sanda, Phys. Lett. **67B**, 303 (1977).
18. B. W. Lee and R. E. Shrock, Phys. Rev. **D16**, 1444 (1977).
19. R. E. Shrock, Phys. Lett. **96B**, 159 (1980); Phys. Rev. **D24** 1232 (1981).
20. R. E. Shrock, Phys. Rev. **D24**, 1275 (1981); Phys. Lett. **112B**, 382 (1982).
21. M. Zeller et al, to be published.
22. H. Greenlee et al., Phys. Rev. Lett. **60**, 893 (1987); S. Schaffner, talk at AGS Users Group Annual Meeting, BNL, May, 1988; to be published.
23. M. K. Gaillard, B. W. Lee, and R. E. Shrock, Phys. Rev. **D13**, 2674 (1976).
24. R. E. Shrock and M. B. Voloshin, Phys. Lett. **87B**, 375 (1979).
25. P. Herczeg, Phys. Rev. **D27**, 1512 (1983).
26. R. S. Willey and H. L. Yu, Phys. Rev. **D26**, 3287 (1982); R. S. Willey, Phys. Lett. **173B**, 480 (1986); A. Vainshtein, V. Zakharov, and M. Shifman, Sov. Phys. Usp. **23**, 429 (1980).
27. Y. Asano et al., Phys. Lett. **107B**, 159 (1981).
28. R. E. Shrock and M. Suzuki, Phys. Lett. **110B**, 150 (1982).
29. R. Peccei and H. Quinn, Phys. Rev. Lett. **38**, 1440 (1977); S. Weinberg, Phys. Rev. Lett. **40**, 223 (1978); F. Wilczek, Phys. Rev. Lett. **40**, 223 (1978).
30. T. Yamazaki et al., Phys. Rev. Lett. **52**, 1089 (1984).
31. T. Cowan et al., Phys. Rev. Lett. **56**, 444 (1986).
32. N. J. Baker et al., Phys. Rev. Lett. **59**, 2832 (1987); see also L. M. Krauss and M. Zeller, Phys. Rev. **D34**, 3385.
33. R. Dzhelyadin et al., Phys. Lett. **105B**, 231 (1981).
34. A. D. Linde, JETP Lett. **23**, 73 (1976); S. Weinberg, Phys. Rev. Lett. **36**, 294 (1976).
35. See, e.g., T. Goldman and C. Hoffman, Phys. Rev. Lett. **40**, 220 (1978); W. Bardeen and S.-H. H. Tye, Phys. Lett. **74B**, 229 (1978).
36. T. Inami and C. S. Lim, Prog. Theor. Phys. **65**, 297 (1981).
37. E. Ma and J. Okada, Phys. Rev. **D18**, 4219 (1978).
38. F. Gilman and J. Hagelin, Phys. Lett. **133B**, 443 (1983).
39. Y. Asano et al., Phys. Lett. **104B**, 84 (1981).

40. R. Hayano et al., Phys. Rev. Lett. **49**, 1305 (1982).
41. R. Abela et al., Phys. Lett. **105B**, 263 (1981).
42. R. Minehart et al., Phys. Rev. Lett. **52**, 804 (1984).
43. M. Daum et al., Phys. Rev. **D36**, 2624 (1987).
44. D. Bryman et al., Phys. Rev. Lett. **50**, 1546 (1983).
45. G. Azuelos et al., Phys. Rev. Lett. **56**, 2241 (1986).
46. N. DeLeener-Rosier et al., Phys. Lett. **177B**, 228 (1986).
47. Particle Data Group, Review of Particle Properties, Phys. Lett. **170B**, 1 (1986) and 1988 edition, in press.
48. A. Jodidio et al., Phys. Rev. **D34**, 1967 (1986); erratum, Phys. Rev. **D37**, 237 (1988).
49. R. Bolton Phys. Rev. Lett. **56**, 2461 (1986).
50. D. Grosnick et al., Phys. Rev. Lett. **57**, 3241 (1986).
51. U. Bellgardt et al., Nucl. Phys. **B299**, 1 (1988).
52. R. Eichler et al., Phys. Lett. **175B**, 101 (1986).
53. D. Bryman and E. Clifford, Phys. Rev. Lett. **57**, 2787 (1986).
54. T. Goldman et al., Phys. Rev. **D36**, 1543 (1987).
55. Y. Chikashige, R. Mohapatra, and R. Peccei, Phys. Lett. **98B**, 265 (1981); Gelmini and M. Roncadelli, Phys. Lett. **99B**, 411 (1981).
56. D. Bryman et al., Phys. Rev. Lett. **55**, 465 (1985); **59**, 970 (1987).

SEARCH FOR MUON-ELECTRON AND MUON-POSITRON CONVERSION

S. Ahmad,[(1)] G. Azuelos,[(2)] M. Blecher,[(3)] D.A. Bryman,[(1)] R.A. Burnham,[(4)] E.T.H Clifford,[(1)] P. Depommier,[(5)] M.S. Dixit,[(6)] K. Gotow,[(3)] C.K. Hargrove,[(6)] M. Hasinoff,[(4)], M. Leitch,[(1)] J.A. Macdonald,[(2)], H. Mes,[(6)] I. Navon,[(1)] T. Numao,[2] J-M. Poutissou,[(2)] R. Poutissou,[(5)] P. Schlatter,[(1)] J. Spuller,[(2)] and J. Summhammer[(1)]

[(1)] *TRIUMF and University of Victoria, Victoria, B.C., Canada V8W 2Y2*

[(2)] *TRIUMF, 4004 Wesbrook Mall, Vancouver, B.C., Canada V6T 2A3*

[(3)] *VPI and State University, Blacksburg, VA 24061, USA*

[(4)] *University of British Columbia, Vancouver, B.C., Canada V6T 2A6*

[(5)] *Université de Montréal, Montréal, Québec, Canada H3C 3J7*

[(6)] *National Research Council, Ottawa, Canada K1A 0R6*

presented by D.A. Bryman

ABSTRACT

Limits on the lepton flavor violating reactions $\mu^- + Z \to e^- + Z$ and $\mu^- + Z \to e^+ + (Z-2)$, muon-electron and muon-positron conversion, have been obtained. Upper limits (90% C.L.) for the branching ratios compared to ordinary muon capture are: $R_-(\mathrm{Ti}) = \frac{\Gamma(\mu^-\mathrm{Ti}\to e^-\mathrm{Ti})}{\Gamma(\mu^-\mathrm{Ti\ capture})} < 4.6 \times 10^{-12}$, $R_+(\mathrm{Ti}) = \frac{\Gamma(\mu^-\mathrm{Ti}\to e^+\mathrm{Ca}^*)}{\Gamma(\mu^-\mathrm{Ti\ capture})} < 1.7 \times 10^{-10}$ and $R_-(\mathrm{Pb}) < 4.9 \times 10^{-10}$.

INTRODUCTION

Muon-electron conversion, $\mu^- + Z \to e^- + Z$, where Z is a nucleus of atomic number Z, is a lepton flavor violating reaction which may be enhanced by the coherent action of the nuclear quarks. In muon-positron conversion, $\mu^- + Z \to e^+(Z-2)$, a double charge changing current is required and neither lepton flavor nor lepton number are conserved. In this paper, searches for muon-electron and muon-positron conversion in titanium and lead targets performed using the TRIUMF time projection chamber (TPC) are described.[1]

The experimental signature of $\mu^- \to e^-$ conversion is a monoenergetic electron with kinetic energy determined by the muon mass, electron mass, and the muonic atom binding energy. However, for muon-positron conversion, where nuclear excitation and breakup are likely, the expected positron spectrum covers a range of kinetic energies depending also on the mass difference between initial and final nuclear states.

EXPERIMENT

The main features of the detector system used are shown schematically in Fig. 1. Scintillation counters B, and vetoes BV and V, were used to define incoming beam particles and a muon stop in the target. Charged particles emitted from the target were detected and analyzed in the TPC located in a uniform axial magnetic field of 0.9 T. The trigger was derived from a system of scintillators and multiwire proportional chambers I, IWC, E, EWC, W which detected charged particles entering and exiting the TPC. To identify cosmic rays, layers of scintillators and drift chambers covered the top and sides of the magnet. A 73 MeV/c negative muon beam from the TRIUMF M9

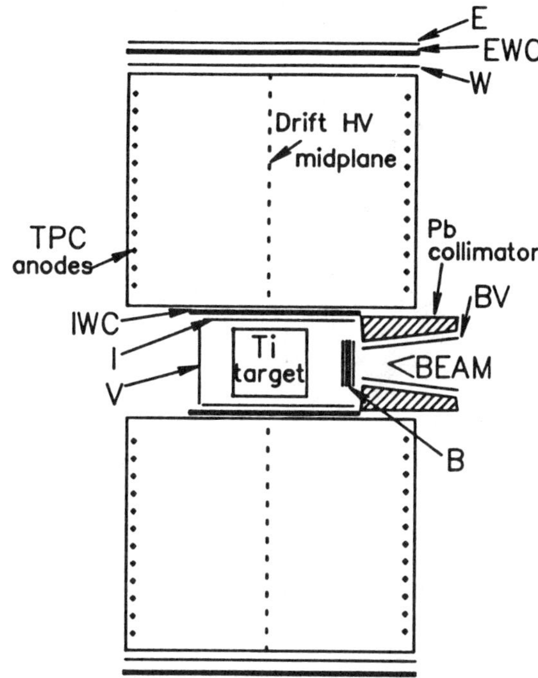

Fig. 1. Schematic cross-section view of the time projection chamber showing the target, beam counters B, veto counters BV, V, and trigger counters: scintillators I, E, W; wire chambers IWC and EWC. The assembly was contained within the magnet.

channel was stopped in a 2 g/cm^2 shredded natural Ti target at an average rate of 10^6 s^{-1}. Pion and electron contaminants in the beam were suppressed to the levels $\pi/\mu \simeq 10^{-4}$ and $e/\mu \simeq 10^{-2}$, respectively, by an rf separator. Single electrons or positrons with momenta $p > 70$ MeV/c emitted after a muon stop were accepted by the trigger system. In total 1.1×10^{13} μ^- were stopped in the Ti target. For the lead target, 5.3×10^{11} μ^- stops were recorded.

The system resolution and efficiency were periodically monitored with the 70 MeV/c positrons from the decay $\pi^+ \to e^+\nu_e$ by stopping a 89 MeV/c pion beam in the target and lowering the magnetic field to 0.6 T.

ANALYSIS

The initial data analysis applied very loose cuts on the number of points in the track, the momentum, and the track origin. In the second analysis, separate sets of cuts were used for the $\mu^- \to e^-$ conversion and $\mu^- \to e^+$ conversion data. Efficiencies for these cuts were determined from the $\pi^+ \to e^+\nu$ calibration data where possible. Exceptions were cuts to reject pions and cuts related to the resolution of the TPC, since the position resolution for positrons was poorer than that for electrons. Therefore, less restrictive values of several cut variables were allowed for the $\pi \to e\nu$ calibration and positron conversion data.

To eliminate pion-induced reactions, candidate events in prompt coincidence with any incoming particle signal in the beam scintillation counters were rejected. This procedure resulted in a 10% loss of events due mostly to accidental coincidences with incident beam particles.

Cosmic ray-induced backgrounds were studied using data taken with the beam off and identical trigger conditions except for the beam requirement. These events were identified by several characteristics including multiple hits in the trigger counters, multiple tracks in the TPC, and hits in the cosmic ray-veto counters. The effectiveness of the latter is illustrated in Fig. 2 which shows events obtained at zero magnetic field with the beam off, using a trigger optimized for cosmic rays. The events are distributed on the upper hemisphere; ϕ is the azimuthal angle, and θ is the polar angle with respect to a vertical polar axis. Figure 2(a) shows all such events, while Fig. 2(b) is an unnormalized distribution of those events ($\approx 3\%$) in which no cosmic ray scintillators or drift chambers fired. The remaining peaks reveal the gaps in the veto counters.

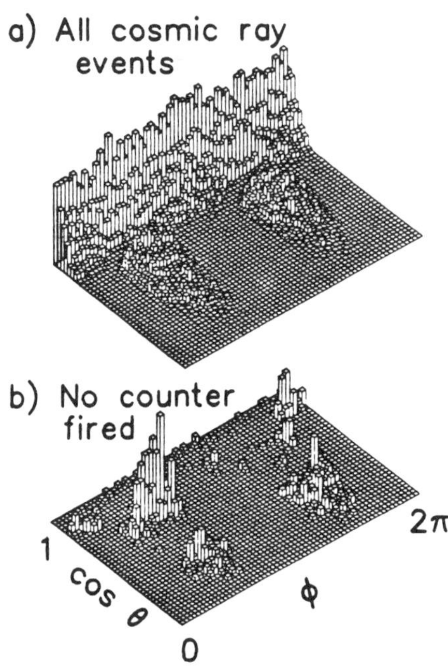

Fig. 2. Distribution of cosmic-ray events on the upper hemisphere as a $(\phi, \cos\theta)$ plane, to show the coverage of the cosmic-ray veto counters. (a) All events, (b) those events [$\sim 3\%$ of those in (a)] missed by the veto counters.

RESULTS

The total electron spectrum obtained from the titanium target is shown in Fig. 3(a). No events were found in the expected region of the μ-e conversion peak from 96.5–106 MeV/c. From the Monte Carlo simulation of the process shown normalized to the level $R_- = 7 \times 10^{-11}$ in Fig. 3(c), 85% of μ-e events would be expected in this momentum range. The e^- spectrum below 96.5 MeV/c is consistent in shape and magnitude with the spectrum shown in Fig. 3(b) from a Monte Carlo calculation for bound muon decay in orbit. The nine events observed above 106 MeV/c could have been eliminated by tightening cosmic-ray and pion cuts at the expense of reducing the overall acceptance by a factor of 2. The 90% confidence level upper limit to the branching ratio is

$$R_-(\mathrm{Ti}) = \frac{\Gamma(\mu^- \mathrm{Ti} \to e^- \mathrm{Ti})}{\Gamma(\mu^- \mathrm{Ti\ capture})} < 4.6 \times 10^{-12} \,. \tag{1}$$

The experimental positron spectrum is shown in Fig. 4(a). The signature for μ^--e^+ conversion is an excess of positrons over those expected from background processes, primarily radiative muon capture (RMC) with asymmetric γ conversion. The Monte Carlo calculation of the RMC background is shown in Fig. 4(b) and is consistent with the observed spectrum. In order to obtain the total branching ratio limit a model calculation of the positron momentum spectrum is needed. A Lorentzian giant resonance excitation was assumed for the final state. Figure 4(c) shows the positron energy spec-

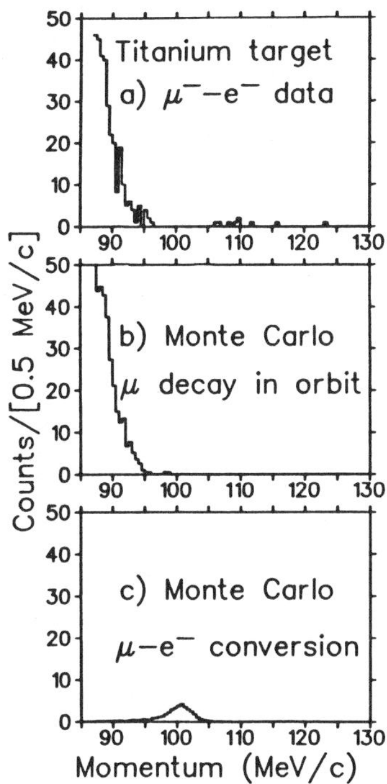

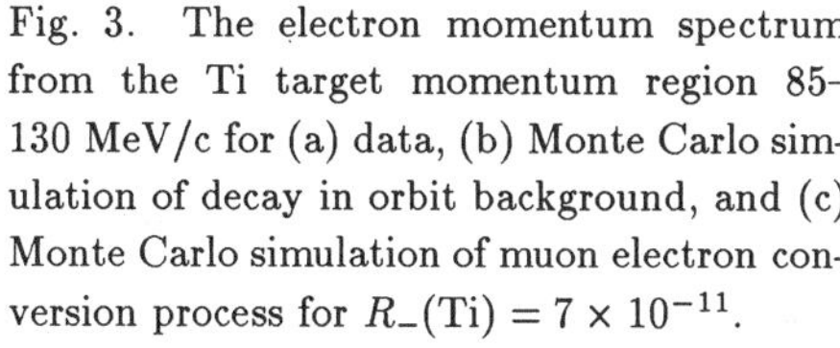
Fig. 3. The electron momentum spectrum from the Ti target momentum region 85–130 MeV/c for (a) data, (b) Monte Carlo simulation of decay in orbit background, and (c) Monte Carlo simulation of muon electron conversion process for $R_-(\text{Ti}) = 7 \times 10^{-11}$.

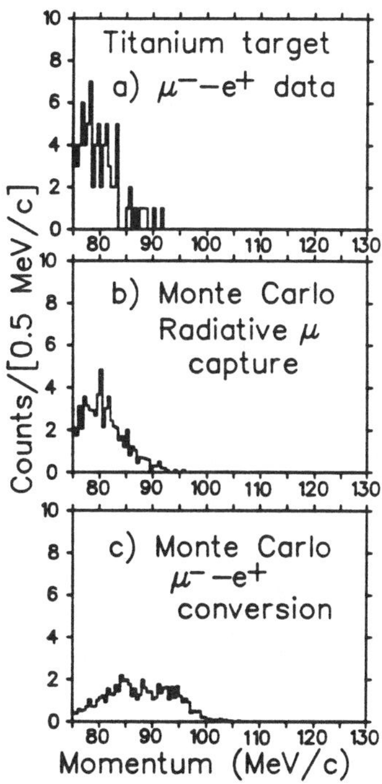

Fig. 4. The positron momentum spectrum from the Ti target in the momentum region 75–130 MeV/c for (a) data, (b) Monte Carlo simulation of the radiative muon capture with asymmetric photon conversion background, and (c) Monte Carlo simulation of the muon-positron conversion process assuming average nuclear excitation $\langle E_x \rangle = 20$ MeV with Lorentzian width $\Gamma = 20$ MeV and $R_+(\text{Ti}) = 10^{-9}$.

trum generated by the Monte Carlo method for the $\mu^- \to e^+$ process in Ti assuming an average nuclear excitation $\langle E_x \rangle = 20$ MeV, a width $\Gamma = 20$ MeV and a branching ratio of 10^{-9}. This leads to a model-dependent limit (90% C.L.) on the overall branching ratio

$$R_+(\text{Ti}) = \frac{\Gamma(\mu^-\text{Ti} \to e^+\text{Ca}^*)}{\Gamma(\mu^-\text{Ti capture})} < 1.7 \times 10^{-10} \,, \tag{2}$$

which would vary from $R^+(\text{Ti}) < 6 \times 10^{-11}$ to $R^+(\text{Ti}) < 2.7 \times 10^{-10}$ for $\langle E_x \rangle = 10$ MeV to $\langle E_x \rangle = 30$ MeV, respectively.

A smaller data set was obtained for the search for $\mu \to e$ conversion in lead using an experimental setup similar to that for the Ti measurement. The resultant e^-

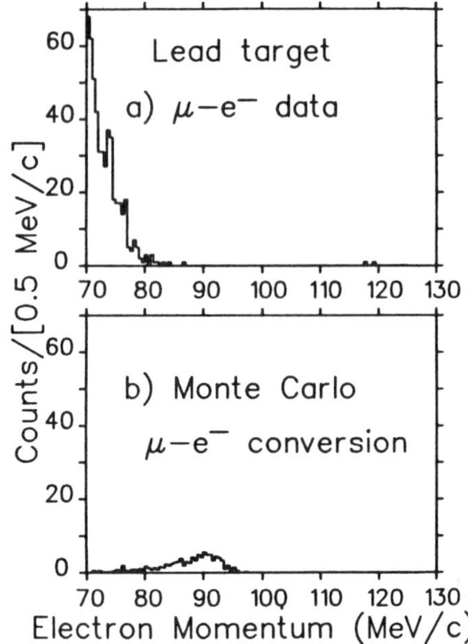

Fig. 5. Electron momentum spectrum from the Pb target. (a) Data and (b) Monte Carlo simulation of muon-electron conversion for $R_-(\text{Pb}) = 10^{-8}$.

momentum spectrum is shown in Fig. 5(a). There is one candidate event in the region 84.5–98.0 MeV/c, expected to contain 83% of conversion events. Based on a Monte Carlo generated e^- spectrum for bound μ^- decay in Pb, a few events are to be expected in this region. In addition, backgrounds from radiative μ and π capture processes can be expected to be enhanced considerably for a heavy target such as Pb compared to Ti. Figure 5(b) shows the expected $\mu \to e$ conversion peak in Pb assuming a branching ratio of 10^{-8}.

Using Poisson statistics for one observed event, a 90% confidence level upper limit on $\mu^- \to e^-$ conversion in Pb is

$$R_-(\text{Pb}) = \frac{\Gamma(\mu^-\text{Pb} \to e^-\text{Pb})}{\Gamma(\mu^-\text{Pb capture})} < 4.9 \times 10^{-10} \,. \tag{3}$$

According to Shanker,[2] the branching ratio for muon-electron conversion $R_-(Z)$ can be expressed phenomenologically in a model-independent way as

$$R_-(Z) \approx \left\{ g_v^0 + g_v^1 \left(\frac{1}{3} \frac{Z-N}{Z+N} \right) \right\}^2 \tag{4}$$

for the vector dominant case of the ground state to ground state transition. In Eq. (4), Z and N are atomic and neutron numbers, respectively, and g_v^0 and g_v^1 are isoscalar and isovector coupling constants. Results from at least two nuclei are required to place bounds on both g_v^0 and g_v^1. The allowable phase space for these coupling constants is shown in Fig. 6 by the shaded area bounded by the three pairs of parallel lines obtained from the limits for R_- in ^{32}S from Ref. 3 and in ^{208}Pb and ^{48}Ti from the present work.

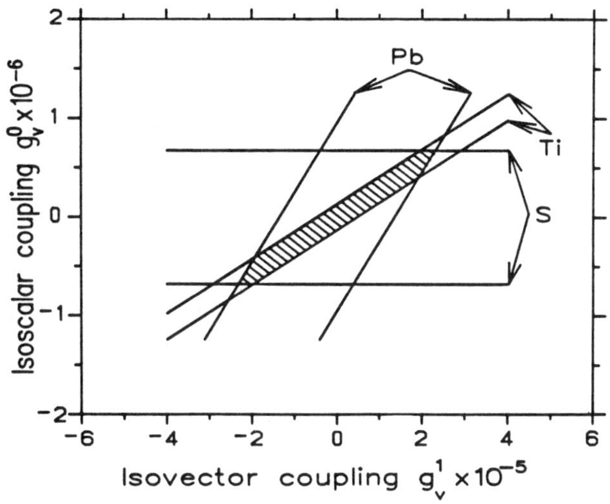

Fig. 6. Phase space of hypothetical isoscalar and isovector coupling constants constrained to the shaded area by the limits in Ti, Pb and S for muon-electron conversion.

REFERENCES

1. S. Ahmad *et al.*, submitted to Phys. Rev. D;
   S. Ahmad *et al.*, Phys. Rev. Lett. **59**, 970 (1987).
2. O. Shanker, Phys. Rev. D **20**, 1608 (1979).
3. A. Badertscher *et al.*, Nucl. Phys. **A377**, 406 (1982).

Status of E811: Hyperon Weak Radiative Decay

A.J. NOBLE, M.D. HASINOFF, D.F. MEASDAY, C.E. WALTHAM, University of British Columbia, Vancouver, B.C., Canada V6T 2A6; N.P. HESSEY, J. LOWE,University of Birmingham, U.K. B15 2TT; E.C. BOOTH, K.P. GALL, E.K. MCINTYRE, J.P. MILLER, B.L. ROBERTS, D. WHITEHOUSE, Boston University, Boston, MA, USA 02215; M. SAKITT, J. SKELLY, Brookhaven National Laboratory, Upton, NY, USA 11973; W. FICKINGER, D.K. ROBINSON, Case Western Reserve University, Cleveland, OH, USA 44106; D. HORVÁTH, KFKI, Budapest, Hungary and TRIUMF, Vancouver, B.C., Canada V6T 2A3; B. BASSALLECK, J.R. HALL, K.D. LARSON, D. WOLFE, University of New Mexico, Albuquerque, NM 87131; A. HALLIN, Princeton University, Princeton, NJ 08544; M. SALOMON, TRIUMF, Vancouver, B.C., Canada V6T 2A3.

Abstract

We report the status of branching ratio measurements of $\Sigma^+ \rightarrow p + \gamma$ and $\Lambda \rightarrow n + \gamma$. For the Σ^+ decay a preliminary result is presented.

Introduction.

Weak radiative decays of hyperons are appealing as they probe the weak part of the non-leptonic decay without the encumbrance of a final state interaction, containing instead the well-known and calculable radiative vertex. Such decays are rare and hence there are few experimental tests of the present theories. In non-leptonic decays there are still many unresolved problems. Two leading examples are the origin of the empirical $|\Delta I| = 1/2$ rule, and the relative magnitude and phase of the parity conserving and parity violating amplitudes.

The $|\Delta I| = 1/2$ rule is only partially explained by the pure $|\Delta I| = 1/2$ penguin diagram and by enhancements due to hard radiative gluon corrections. (The contributions from soft gluons, which cannot be treated perturbatively, are unknown but could be very important). For the process $B(p_i) \rightarrow B(p_f) + \gamma(k)$ the asymmetry and the rate,

$$\alpha = \frac{2Re(a^*b)}{|a|^2 + |b|^2} \qquad \Gamma = \frac{1}{\pi}\left(\frac{m_i^2 - m_f^2}{2m_i}\right)^3 \left(|a|^2 + |b|^2\right)$$

are derived directly from the most general Lorentz invariant transition matrix;

$$\mathcal{M} = \varepsilon_\nu \overline{U}_f[a + b\gamma_5]\sigma_{\nu\mu}k_\mu U_i,$$

where ε and k are the photon polarization and 4-momenta and a and b are the parity conserving and parity violating amplitudes.

The theories so far have found the asymmetry term most difficult to calculate. In the decay $\Sigma^+ \rightarrow p+\gamma$ the experimental asymmetry has been determined to be large and negative whereas the theories tend to predict zero or a positive value. However if one constrains the free parameters to fit the $\Sigma^+ \rightarrow p+\gamma$ branching ratio then measurements of other decays will provide sensitive tests of the theory. Of these, the decay $\Lambda \rightarrow n+\gamma$ has been singled out as being particularly sensitive.[1]

Theoretical calculations fall into two general categories. The first of these, pole models, assume the decay is factorable and treat the weak and radiative transitions independently with a variety of intermediaries. In the second category, at quark level, one calculates via Feynman rules the transition amplitudes for the applicable graphs.

Experimental status.

Here we report on the present status of the BNL AGS experiment E811 conducted in the LESB-II kaon beam line. For both measurements, $\Sigma^+ \to p + \gamma$ and $\Lambda \to n + \gamma$ we stopped negative kaons in a liquid hydrogen target. Detectors in the beam included particle identifying Čerenkov and dE/dx counters as well as a position determining hodoscope.

In the Σ^+ measurement we tagged production via $K^- p \to \Sigma^+ + \pi^-$. The π^- was tracked through a MWPC above or below the target and then followed through a range stack. The γ was detected in a 49 element NaI array[2] adjacent to the target. The principal background in this measurement came from the much more prolific decay $\Sigma^+ \to p + \pi^0$ with only one of the 2 γ's from the $\pi^0 \to \gamma\gamma$ being detected. The endpoint of the π^0 box and the weak radiative decay gamma (WRD) are separated by only 14 MeV. Hence the resolution of the array of 49 (6% FWHM at 129 MeV) was the primary contributor to the background under the Σ^+ WRD peak.

We collected data for this measurement in two running periods during 1987. The analysis is almost complete and a fit to our data with Monte Carlo shapes yields the <u>*preliminary*</u> value; $\Gamma(\Sigma^+ \to p + \gamma)/\Gamma(\Sigma^+ \to \textit{anything}) = (1.45 \pm 0.25) \times 10^{-3}$, in good agreement with previous measurements.[3]

We have just completed an engineering run in preparation for measuring the $\Lambda \to n + \gamma$ decay. Using a combination of fast FERA adc's, and LeCroy CAB preprocessors we were able to take data at about 110 Hz with an event length of 950 words. The general philosophy was to run with a loose trigger which would then be optimized between the engineering run and the main production run. This combined with enhanced preprocessing will allow us to run at a manageable rate with a more selective sample.

In $\Lambda \to n + \gamma$ we produce Λ's via the reaction $K^- + p \to \Lambda + \pi^0$. Here the π^0 is monoenergetic at 288 MeV and is used to tag the Λ production. We measure the energy and position of the γ's using the Crystal Box[2] NaI array, and if the neutron is detected, we also have its position and a lower limit on the energy.

The principal backgrounds here come from the decay $\Lambda \to n + \pi^0$ with only one decay γ detected, and from charge exchange in flight $K^- + p \to \overline{K}^0 + n$ with subsequent decay $K_S \to \pi^0 + \pi^0$. The $\overline{K}^0$ contribution has been eliminated in Monte Carlos via strong angular cuts. The separation between the π^0 edge and the WRD γ is 25 MeV.

For 3γ events we see the π^0 production peak clearly and also evidence for the π^0 box from $\Lambda \to n + \pi^0$. In 4γ events we can identify both background channels previously mentioned. If we assume the only published data for the $\Lambda \to n + \gamma$ branching ratio ($(1.02 \pm 0.33) \times 10^{-3}$ based on 23.7 events by Biagi et al.[4]), we would expect at least 150 events in our current data sample (even with our rather severe software cuts). With the optimized hardware trigger and enhanced preprocessing planned for our 1989 production run we would anticipate a further significant improvement in this $\Lambda \to n+\gamma$ measurement.

References.

1. F.E. Close and H.R. Rubinstein, Nucl. Phys. **B173**, 477 (1980).
2. S.L. Wilson et al., Nucl. Instrum. Methods **A264**, 263 (1988).
3. Particle Data Group. Rev. Mod. Phys. **56**, S1 (1984).
4. S.F. Biagi et al., Z. Phys. **C30**, 201 (1986).

Limits on $K^0_L \to \mu e$, ee, and $\pi^0 ee$

W.M. Morse, E. Jastrzembski, R.C. Larsen, and L.B. Leipuner
Department of Physics
Brookhaven National Laboratory, Upton, NY 11973

H.B. Greenlee, H. Kasha, E.B. Mannelli, M. Mannelli,
K.E. Ohl, S.F. Schaffner, and M.P. Schmidt
Department of Physics
Yale University, New Haven, CT 06511

I present here some recent results from AGS experiment 780. This is a dedicated search for the decays $K^0_L \to \mu^{\pm}e^{\mp}$ and e^+e^-. The decay $K^0_L \to \mu e$ violates lepton number and is not expected to occur in the standard SU(3) × SU(2) × U(1) model. If it occurs it will be suppressed by the mass of the carrier of the new interaction even as the weak interaction is suppressed by the mass of the W. The decay $K^0_L \to e^+e^-$ is expected in the standard model, but is suppressed by the ratio m_e/m_K. The unitary limit is $BR(K^0_L \to e^+e^-) > 3 \times 10^{-12}$. This decay is then an excellent window to search for new pseudoscalar interactions which are not suppressed by helicity.

We have now analyzed the 1987 data. Details of the analysis can be found in Ref. 1. No evidence of $K^0_L \to \mu e$ or $K^0_L \to e^+e^-$ was found. 90% confidence limits were obtained $BR(K^0_L \to e^+e^-) < 4.5 \times 10^{-9}$ and $BR(K^0_L \to \mu e) < 6.7 \times 10^{-9}$. This result implies the mass of the carrier of a lepton number violating current with standard couplings must be greater than 25 TeV.

We have finished analysis of the 1988 data for the decays $K^0_L \to e^+e^-$ and $K^0_L \to \pi^0e^+e^-$. A report on that analysis has been submitted to Physical Review Letters. The 90% confidence level limit $BR(K^0_L \to e^+e^-) < 1.2 \times 10^{-9}$ was obtained. The decay $K^0_L \to \pi^0e^+e^-$ via the one photon diagram violates CP and is expected at around 10^{-11} in the standard model. There is then a large window between the present limit of 2.3×10^{-6} and the standard model prediction to search for new physics. The 90% confidence level limit $BR(K^0_L \to \pi^0e^+e^-) < 3.2 \times 10^{-7}$ was obtained.

REFERENCE

1. H. Greenlee, et al. Phys. Rev. Lett. 60, 893 (1988).

The MEGA Experiment: A Search for $\mu \rightarrow e\gamma$*

M. D. Cooper
Los Alamos National Laboratory, Los Alamos, NM 87545, USA

ABSTRACT

The MEGA experiment is designed to search for the $\mu \rightarrow e\gamma$ process with a branching ratio sensitivity of 10^{-13}. This decay violates the empirically established rule of lepton family number conservation and lies outside the Standard Model of electroweak interactions. In order for the experiment to make a factor of 500 improvement over the existing limit, a new design was adopted that employs highly modular, fast detectors and state-of-the-art electronic readout. The detectors are contained in a 15 kG solenoidal field produced by a superconducting magnet. The central region is a positron spectrometer, and the outer region is four layers of pair spectrometers. Data taking is expected to commence in 1989.

The search for $\mu \rightarrow e\gamma$ at LAMPF is called MEGA, an acronym standing for Muon decays into an Electron and a GAmma ray, and is an experiment with a branching ratio sensitivity of roughly 10^{-13}. The motivation for pressing the search for this neutrinoless muon decay is sufficiently strong to attract over 60 scientists and engineers from 13 institutions to carry out the measurement.[1] Such enthusiasm is engendered because the observation of the process would herald physics not explained by the minimal Standard Model of electroweak interactions that currently accounts for all observations of relevant processes. In particular, observing a radiative connection between the muon and electron would be a strong clue toward understanding the family structure of the Standard Model.

Candidates for new mechanisms to allow $\mu \rightarrow e\gamma$ abound. Theoreticians have invented many extensions to the Standard Model to explain its possible shortcomings. Most of the extensions allow $\mu \rightarrow e\gamma$ by design or accident. Some models with reasonable parameters allow the rate to be observable by the new generation of experiments. In addition to $\mu \rightarrow e\gamma$, there are many other candidate processes such as $\mu \rightarrow eee$, $\mu \rightarrow e\gamma\gamma$, $\mu A \rightarrow eA$, $K_L \rightarrow \mu e$, and $K^+ \rightarrow \pi^+ \mu e$ that may be triggered. Depending on the model and its parameters, a different one of these processes may have the biggest branching ratio. An important point is that each of these decays is complementary and must be pursued experimentally because the theories lack detailed predictive power. The MEGA experiment searches for one of these which is prominent in models that embody more generations, right-handed neutrinos, a complicated Higgs sector, composite particles, and supersymmetry. Models containing horizontal symmetries and technicolor do not favor $\mu \rightarrow e\gamma$.

The design of the MEGA experiment calls for a factor of 500 improvement in the branching ratio for $\mu \rightarrow e\gamma$. The signature for $\mu \rightarrow e\gamma$ is particularly simple and purely kinematic; it is the observation of coincident photons and positrons arising from a common vertex that are back-to-back and each of 52.8 MeV. There are two major backgrounds for $\mu \rightarrow e\gamma$, the muon radiative decay $\mu \rightarrow e\gamma\nu\bar{\nu}$ and accidental coincidences. The latter comes from two separate muon decays that mock up the kinematic conditions and is the more severe. Both backgrounds are suppressed with good resolution. A comparison of the MEGA design to the best previous experiment[2] shows that MEGA has 70 times the rate capability, one-quarter the solid-angle-efficiency product, and improved resolutions that eliminate background 50,000 times more effectively.

The experimental layout is shown in Figs. 1 and 2. The apparatus is divided into two concentric spectrometers that are both contained in a large solenoidal magnet. The muon target is centered in the magnet, is a planar ellipse canted at a steep angle to the beam, and is just thick enough to stop the 3 x 10^7 muons taken from the channel.

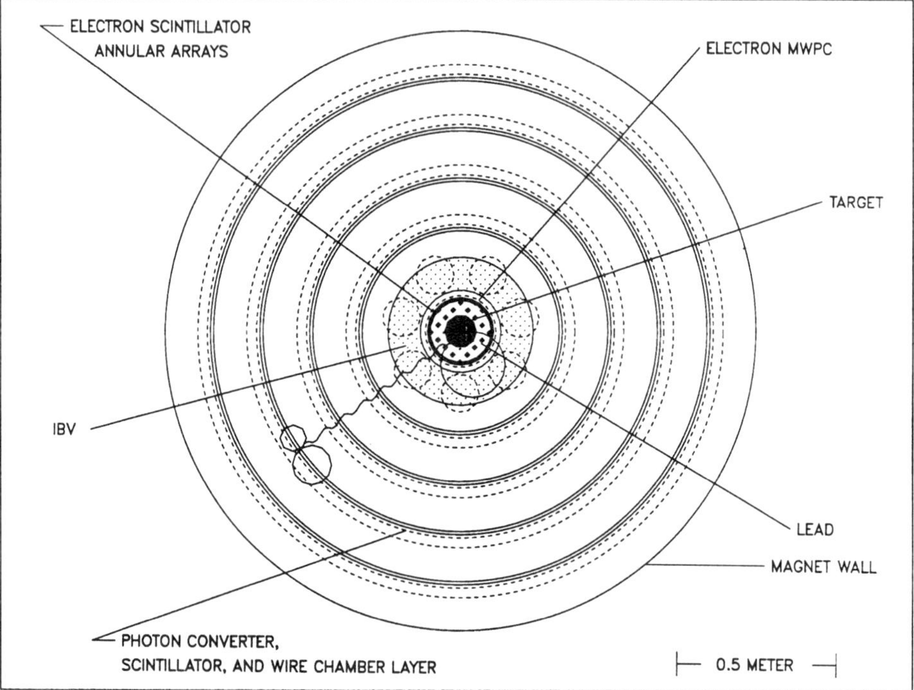

Fig. 1. The MEGA detector configuration viewed along the direction of the beam and the magnetic field.

The 15 kG magnetic field is sufficient to contain all decay positrons inside the 30 cm radius spectrometer. The momentum of the decay positron is found by measuring the curvature of the helix in the spectrometer with multi-wire proportional chambers (MWPC) and its time is ascertained with plastic scintillators. The positron chambers present only a total of 10^{-3} radiation lengths of material to the particles in the central region. The chamber arrangement involves one chamber that is surrounded by seven smaller chambers. This design keeps the MWPC out of the highest rate region about the target, but gives sufficient tracking for pattern recognition. After less than eight revolutions, the positron hits one of the 200 scintillators before stopping in the lead shielding. The spectrometer resolution should be 0.6% (FWHM) in energy, 0.2 cm in position, 0.6° in angle, and 0.5 ns in time.

The electron chambers are cylindrical, with 0.1 cm pitch wires running 132 cm parallel to the symmetry axis. At a half-gap of 0.175 cm, there are cathode foils of 0.0025 cm copperized Kapton. With overpressure inside, the foils keep a cylindrical

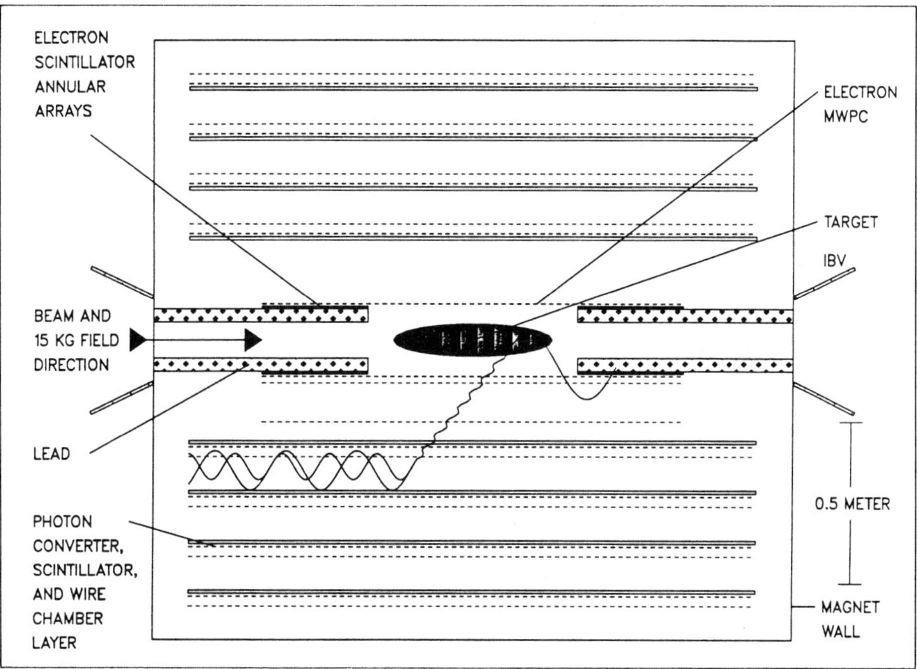

Fig. 2. A sectioned plan view of the MEGA detector.

shape to 0.01 cm. Spiral cathode readout is possible by scratching away the copper to make stripes. In a planar prototype, the geometry has been checked, with both the anode wires and cathode foils coming on plateau with a 79.5%-CF_4:20%-isobutane:0.5%-isopropyl alcohol gas mixture. Electro-mechanical instabilities need to be prevented by attaching the wires to garlands, and these structures have been fabricated from 0.01 cm nylon filaments attached to the wires with epoxy. Data taken at two angles of incidence compare well to a Monte Carlo simulation. A wire winder for cylindrical chambers has been developed that strings two quadrants of a chamber in a few hours. A cylindrical chamber has yet to reach plateau.

The decay photon propagates out to a series of four pair spectrometers that measure its kinematic properties. The four spectrometers consist of two layers of 0.025 cm lead converters separated by a MWPC to identify the layer of conversion. The converters surround a barrel of 1 x 5 x 180 cm^3 scintillators for timing. Outside the lead each layer has three drift chambers to measure the momenta of the electron-positron pair. The resolutions of the spectrometer are expected to be 3% in energy, 10° in angle, and 0.5 ns in time. Some background suppression of high energy photons is obtained by looking at low energy positrons that accompany high energy photons from radiative muon decay. These low energy positrons follow the field lines and are detected at the magnet ends by the internal bremsstrahlung veto (IBV) counters.

Many aspects of the pair spectrometers have been simulated by computer or prototype. Pattern recognition of the spiraling pairs has yielded an algorithm for finding the vertex and the perpendicular momenta. The parallel momenta are still under study. A 180 cm prototype with 0.25 cm pitch planar chamber has come on plateau. The scintillators have been connected to photomultiplier tubes at each end with 300 cm fiber optics light guides to yield 0.6 ns resolution.

MEGA has a staged trigger. The trigger depends on the rapidly falling photon spectrum that implies a 38 MeV energy cut will eliminate all but one in 10^4 decays. Such a cut is imposed by demanding a transverse spread of the pair of 16 cm in the scintillators and MWPC's. Data that passes the level one hardware is encoded and stored in FASTBUS latches, time-to-digital converters, and analog-to-digital converters with on-board memories that store a macro-pulse worth of events.[3] A schematic of the data acquisition system is given in Fig. 3. The memories are read out as

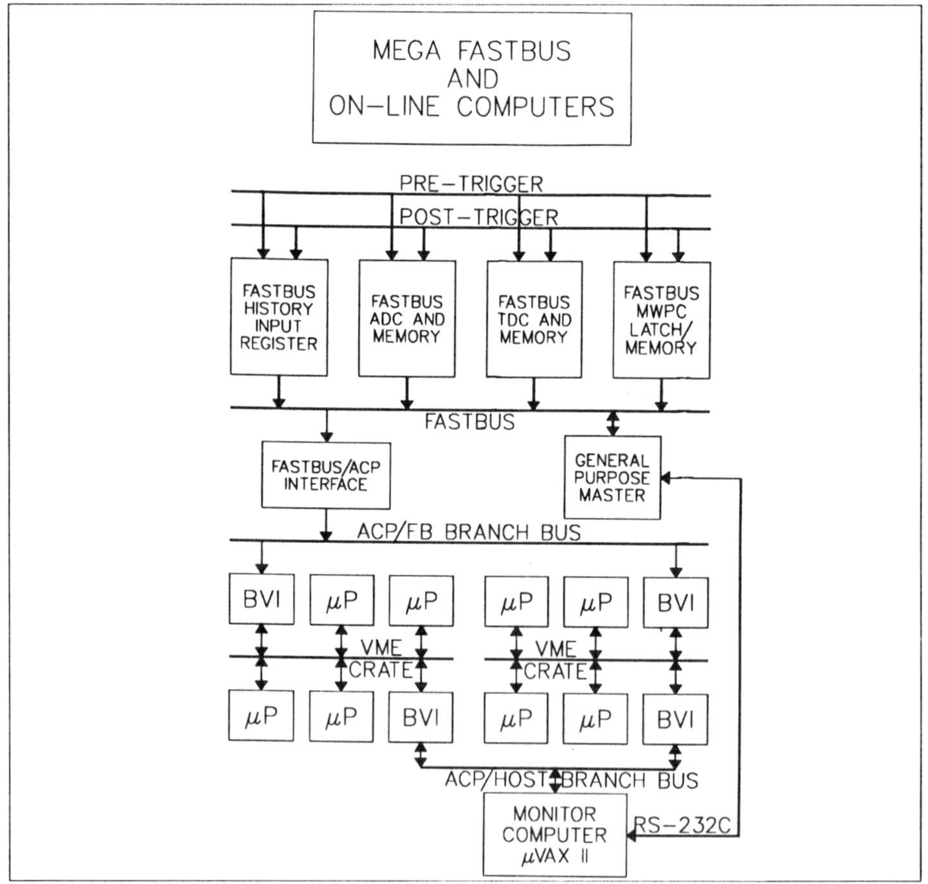

Fig. 3. A schematic of the MEGA data acquisition hardware.

a single block by the General Purpose Master[4] and transferred into a VME microprocessor via the FASTBUS to Branch Bus Connector.[5] The microprocessor farm is the Advanced Computer Project from Fermilab.[6] There is no level two trigger, but the microprocessors provide the level three cuts by analyzing the positron spectrometer data to find high-energy, in-time candidates that point at the photon of the level one trigger. Eventually there will be nearly 200 FASTBUS modules and 32 microprocessors.

MEGA is in the middle of development and construction. First data taking is expected in 1989, with the main data acquisition commencing in 1990.

REFERENCES

*Supported by the U. S. Department of Energy and the National Science Foundation.

1. The MEGA collaborators and institutions include: D. Barlow, R. Kessler, G. Kim, J. Labrentz, B. M. K. Nefkens, C. Pillai, J. Wightman, UCLA, S. C. Wright, U. of Chicago, P. S Cooper, Fermilab, M. Dzemidzio, J. Flick, E. V. Hungerford III, B. W. Mayes II, L. Pinsky, W. von Witsch, W. Wong, Y. Xu, Y. Xue, U. of Houston, A. Hallin, Princeton, J. F. Amann, K. Black, R. D. Bolton, S. Carius, M. D. Cooper (Spokesman), W. Foreman, C. Hansen, R. Harrison, V. Hart, C. M. Hoffman, N. Hoffman, T. Hunter, G. E. Hogan, N. June, D. Kercher, J. Little, T. Kozlowski, R. E. Mischke, F. J. Naivar, J. Novak, M. A. Oothoudt, W. Smith, J. Sturrock, J. Szymanski, J. Van Dyke, R. D. Werbeck, C. Wilkinson, Los Alamos Nat. Lab., E. B. Hughes, C. Jui, J. N. Otis, M. W. Ritter, Stanford U., C. Gagliardi, F. Liu, R. E. Tribble, L. Van Ausdeln, Texas A & M U., R. J. Fisk, D. D. Koetke, R. Manweiler, Valparaiso U., R. Marshall, W. Stephens, B. Tippens, K. O. H. Ziock, U. of Virginia, L. E. Piilonen, VPI, A. R. Kunselman, U. of Wyoming, K. Hahn, and J. K. Markey, Yale U.
2. R. D. Bolton, M. D. Cooper, J. S. Frank, A. L. Hallin, P. A. Heusi, C. M. Hoffman, "Search for Rare Muon Decays with The Crystal Box," et al., submitted to *Physical Review D* (1988).
3. Phillips Scientific, 305 Island Road, Mahwah, NJ 07430.
4. Creative Electronics Systems, 70, Route du Pont-Butin, Case Postale 107, CH-1213 Petit-Lancy 1, Geneve, Suisse.
5. M. Larwell, Fermilab CDF453 (1986).
6. I. Gaines, H. Areti, J. Biel, G. Case, A. Cook, M. Fischler, et al., *IEEE Trans. on Nucl. Sci.* NS-32 1397 (1985).

A STUDY OF THE DECAY $\pi \longrightarrow e\nu$

D.I. Britton*, S. Ahmad*, D.A. Bryman*, R.A. Burnham*, E.T.H. Clifford*, M.S. Dixit†, P. Kitching*, Y. Kuno*, J.A. Macdonald*, T. Numao*, A. Olin*, C.E. Picciotto‡, J-M. Poutissou*, and J. Summhammer*.

*TRIUMF, 4004 Wesbrook Mall, Vancouver, B.C. Canada, V6T 2A3
†National Research Council of Canada, Ottawa, Ont., Canada, K1A 0R6
‡University of Victoria, Victoria, B.C., Canada V8W 2Y2

presented by D.I. Britton

ABSTRACT

A new measurement of the $\pi \rightarrow e\nu$ branching ratio is described. The $\pi \rightarrow e\nu$ positron spectrum, in which the background from $\mu \rightarrow e\nu\bar{\nu}$ has been highly suppressed, has been examined for evidence of massive neutrinos and other exotic processes postulated in extensions to the standard model of electroweak interactions.

INTRODUCTION

Detailed study of the decay $\pi \rightarrow e\nu$ continues to be an important testing ground for weak interaction theory. The branching ratio, calculated with the standard model incorporating the assumptions of zero-mass neutrinos, lepton universality, and the V-A form of the weak interaction, is[1]

$$R_{e\nu}^{Th} = \frac{\Gamma(\pi \rightarrow e\nu + \pi \rightarrow e\nu\gamma)}{\Gamma(\pi \rightarrow \mu\nu + \pi \rightarrow \mu\nu\gamma)} = 1.233 \times 10^{-4} \ . \tag{1}$$

Pion structure effects could lead to variation of this value up to 0.3%, but significant deviation of the experimental value beyond that would signal inconsistency with the standard theory. The measured value is $R=1.218 \pm 0.014 \times 10^{-4}$, from previous work at TRIUMF[2].

A variety of extensions to the standard model lead to exotic decay modes observable in $\pi \rightarrow e\nu$ decay. For example, the decay $\pi \rightarrow e\nu M$ where M could be a new scalar, pseudoscalar, vector or pseudovector particle with a mass ranging from zero to the kinematic limit is of considerable interest. In addition, massive neutrinos arise in many standard model extensions. Sensitivity is enhanced in $\pi \rightarrow e\nu$ decay since admixtures of massive neutrinos would be much less suppressed by helicity than are the zero-mass neutrino components.

This is a progress report on a study of $\pi \rightarrow e\nu$ decay at TRIUMF. In order to extract a precise branching ratio result, a detailed understanding of a variety of systematic effects is crucial and the main effort ongoing in the new work is to reduce and quantify the major sources of systematic uncertainty. Meanwhile, the high statistics data for pion decay obtained in this experiment have been used to search for evidence of exotic decay phenomena.

EXPERIMENT

The experiment was carried out in a π^+ beam of momentum 83 MeV/c using the setup shown schematically in Fig. 1. The incoming beam was detected in scintillator B1 and stopped near the downstream side of B3. Veto scintillators defined the stopping region and rejected penetrating particles. Decay positrons were detected passing through two wire chambers for position monitoring, and scintillators T1-4 before being energy analysed in a large NaI(Tl) crystal "TINA". The acceptance and decay time were determined by T4. The positron spectrum consists of two components: a peak at 69 MeV from $\pi \rightarrow e\nu$ and a distribution from 0-53 MeV from the decay chain $\pi \rightarrow \mu\nu$ followed by $\mu \rightarrow e\nu\overline{\nu}$

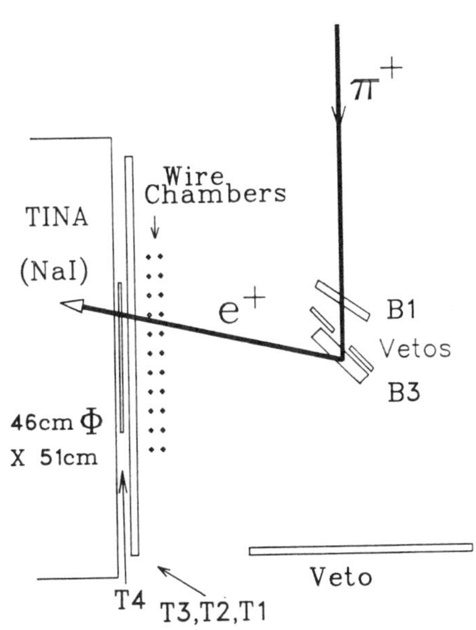

Fig. 1. Schematic layout of the experimental set up.

The measurement requires determination of the ratio of positrons in the $\pi \rightarrow e\nu$ peak to those from the $\pi \rightarrow \mu \rightarrow e$ chain, applying suitable timing, and geometrical corrections. The largest source of systematic uncertainty arises from the long low energy tail in the $\pi \rightarrow e\nu$ peak due mostly to the NaI response function. (The component due to radiative losses is small since NaI is also sensitive to the forward peaked gamma rays in $\pi \rightarrow e\nu\gamma$ decay). The tail extends under the 10^4 times more copious $\pi \rightarrow \mu \rightarrow e$ distribution. In order to determine the tail precisely it is a considerable advantage to be able to suppress the dominant $\pi \rightarrow \mu \rightarrow e$ component.

Two techniques were applied to suppress the $\pi \rightarrow \mu \rightarrow e$ component. First, by accepting decays only within about one pion lifetime ($\sim$ 30 ns), muon decay positrons ($\tau_\mu \sim 2\mu s$) are discriminated against by a factor of $\sim$ 100. The second technique is more subtle and utilizes the energy deposited as well as pulse shape information from the stopping counter B3. For $\pi \rightarrow e\nu$ decay the energy consists of the kinetic energy of the stopping pion plus a small contribution from the exiting decay electron on the pulse tail distributed with the 26 ns pion lifetime. For $\pi \rightarrow \mu \rightarrow e$ decay there is, in addition, the 4 MeV kinetic energy of the decay muon which stops in the counter. The essence of the technique is to reject events with the additional 4 MeV signalling decay to a muon. The stopping counter pulse is digitized in two ADC's: One with a short early gate which integrates the part of the pulse containing primarily the pion kinetic energy; the second ADC has a long gate to integrate the full energy including decays. By suitable cuts on the ratio of long gate/short gate ADC the copious $\pi \rightarrow \mu \rightarrow e$ decays are suppressed.

RESULTS

Figure 2 shows a positron spectrum resulting from both the early timing cut plus an additional suppression factor of 470 from the dual ADC cut. The efficiency of the latter cut is about 35%. A small component of $\pi \to \mu \to e$ remains on the tail and the continuing work is to refine the subtraction of this component to reveal the tail of the $\pi \to e\nu$ distribution (which is about 1.7% of the peak).

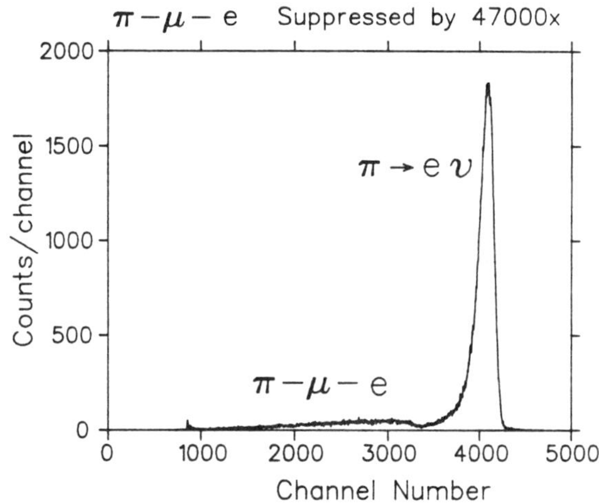

Fig. 2. Positron spectrum after timing and $\pi \to \mu$ decay suppression showing the peak from $\pi \to e\nu$ decay at 69 MeV. The tail region below ch 3400 (0-53 MeV) shows a small residual component due to the $\pi \to \mu \to e$ decay chain.

However, without knowing the exact magnitude of the tail, and with minimal assumptions about its shape it has been possible to set limits for massive neutrinos and on various exotic decays. To do this, the suppressed spectrum is fit with terms to describe the known $\pi \to e\nu$ and $\pi \to \mu \to e$ components and a third term attributable to the hypothetical process.

One such decay involves the Majoron, m, a massless Nambu- Goldstone boson arising in gauge models in which the breaking of the symmetry responsible for neutrino masses also spontaneously breaks global B-L symmetry, where B is baryon number and L lepton number. In the model proposed by Gelmini and Roncadelli[3] neutrino masses arise from the vacuum expectation value of a weak isotriplet scalar Higgs boson χ. The expected positron spectrum from $\pi \to e\nu m(\chi)$ decay is shown in Fig. 3. A fit to the data using this as the third term yielded an upper limit for the branching ratio[4]

$$R_{m(\chi)} = \frac{\Gamma(\pi \to e\nu m(\chi))}{\Gamma(\pi \to \mu\nu)} < 4 \times 10^{-6} \ . \tag{2}$$

from which an upper limit for the Majoron-neutrino coupling is

$$g^2 < 2 \times 10^{-4} \ . \tag{3}$$

A somewhat better limit can be derived from the ratio of the standard model prediction of eq. (1) to the experimental result[3]:

$$g^2 < 7.6 \times 10^{-5} \ . \tag{4}$$

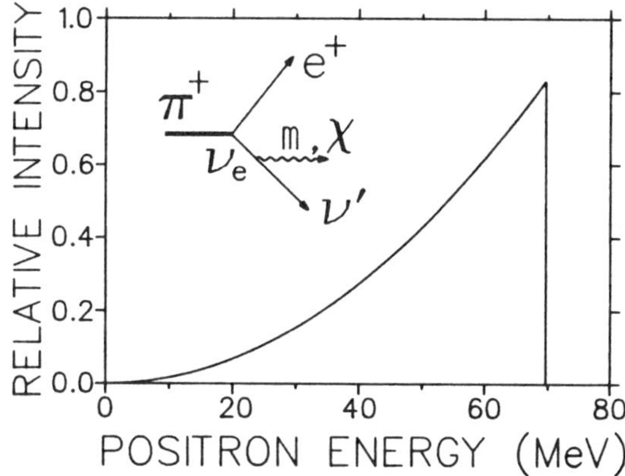

Fig. 3. Positron spectrum expected from $\pi \to e\nu m(\chi)$ decay The inset shows the diagram for the process emitting a Majoron, m, and Higgs boson, χ .

The spectrum in Fig. 2 can also be used to estimate limits on the branching ratio

$$R_M = \frac{\Gamma(\pi \to e\nu M)}{\Gamma(\pi \to \mu\nu)} \tag{5}$$

for any unobserved particle M with mass in the range 0-139 MeV/c^2. The inclusive limits for R_M where M has a rest-frame lifetime greater than 2 ns, are in the range from 4×10^{-6} at m_M=0 to 4×10^{-7} at m_M=125 MeV, as shown in Fig. 4. For lifetimes less than 2 ns, these limits increase by up to an order of magnitude, depending on the value of m_M.

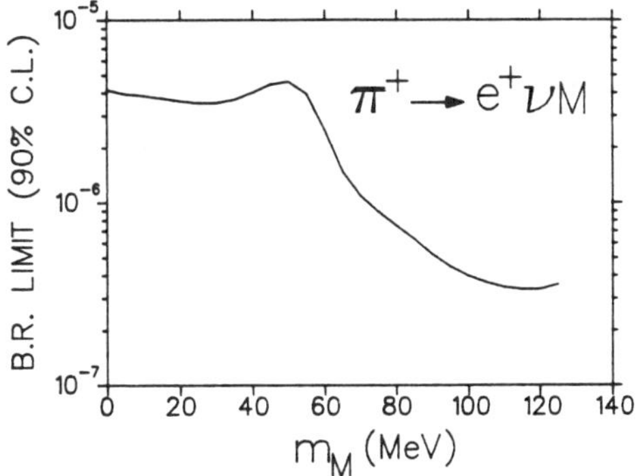

Fig. 4. Branching ratio for $\pi \to e\nu M$ for any massive particle M with lifetime >2 ns as a function of its mass m_M.

A limit on the branching ratio for $\pi^+ \to e^+\nu\nu\overline{\nu}$ can also be determined following the same procedure used to deduce the $\pi \to e\nu m$ branching-ratio limit. The shape of the positron energy spectrum for the $\pi^+ \to e^+\nu\nu\overline{\nu}$ decay was obtained from that used by Pang et al.[5] for $K^+ \to \mu^+\nu\nu\overline{\nu}$ with appropriate masses and the approximation m_e=0. The limit (90% C.L.) obtained for the $\pi^+ \to e^+\nu\nu\overline{\nu}$ branching ratio is

$$R_{\nu\overline{\nu}} = \frac{\Gamma(\pi^+ \to e^+\nu\nu\overline{\nu})}{\Gamma(\pi \to \mu\nu)} < 5 \times 10^{-6} \, . \tag{6}$$

For neutrinos of nonzero mass, the mass eigenstates, ν_i, could be related to the weak eigenstates, ν_ℓ, by a unitary matrix $U_{\ell i}$,

$$\nu_\ell = \sum_i U_{\ell i}\nu_i \ , \tag{7}$$

where $\ell = e, \mu, \tau, \ldots$. According to Shrock[6], the contribution of a mixed state of a massive neutrino to $\pi \to e\nu$ decay can be written as

$$R_{ei} = \frac{\Gamma(\pi \to e\nu_i)}{\Gamma(\pi \to e\nu_1)} = | U_{ei} |^2 \rho_e(\delta_i) \ . \tag{8}$$

where the branch involving ν_1 refers to the conventional massless neutrino and $\rho_e(\delta_i)$ is a kinematic factor with $\delta_i = (m_{\nu_i}/m_\pi)^2$. The data of Fig. 2 were examined for evidence of peaks attributable to the presence of mass eigenstates ν_i .

In the search, the fitting function included an additional peak whose shape was a Gaussian with an exponential tail scaled to approximate the response function of TINA at any given energy. The peak was stepped through the energy range 7 to 60 MeV to find the most probable peak area at each energy. The distribution of the sample peak areas was consistent with random statistical fluctuations about a mean value of zero. Upper limits deduced from this analysis are shown by the solid curve in Fig. 5, in terms of the mixing coeffi cients $| U_{ei} |^2$ for heavy neutrinos coupling to electrons. In Fig. 5 curves 1,2,3 show limits on $| U_{ei} |^2$ derived recently from the CERN PS191 experiment[7] searching for heavy neutrino decay in $\pi \to e\nu_h$ and $K \to e\nu_h$.

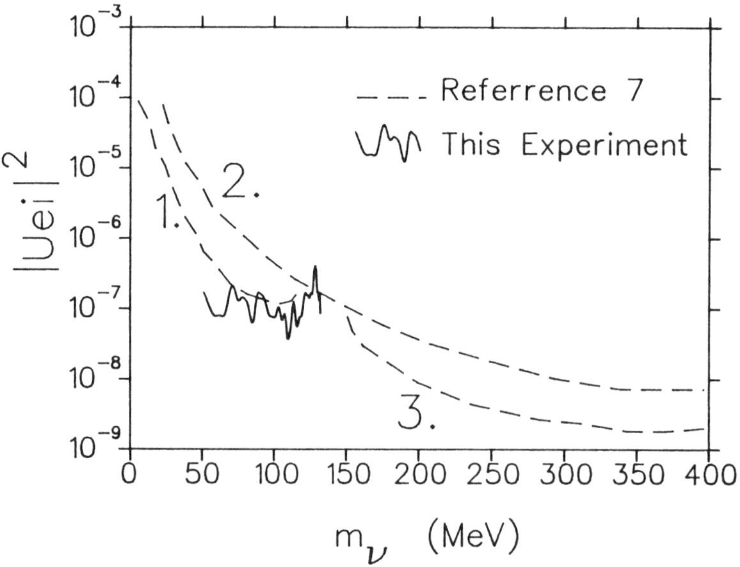

Fig. 5. Limits on the neutrino-electron mixing coefficient $| U_{ei} |^2$ as a function of neutrino mass m_ν. The solid curve shows the result of this work. Curves 1,2,3 are recent results from neutrino decay experiments, (Ref. 7): curve 1 from $\pi \to e\nu_h, \nu_h \to e^+e^-\nu_e$; curve 2 from $K \to e\nu_h, \nu_h \to e^+e^-\nu_e$; curve 3 from $K \to e\nu_h, \nu_h \to \pi^+e^-$.

REFERENCES

[1]W. Marciano and A. Sirlin, Phys Rev. Lett. 36, 1425 (1976).
[2]D.A. Bryman et al., Phys. Rev. D33, 1211 (1986).
[3]G.B. Gelmini and M. Roncadelli, Phys. Lett. 99B, 411 (1981).
[4]C.E. Picciotto et al., Phys. Rev. D37, 1131 (1988).
[5]C.Y. Pang et al., Phys. Rev. D8, 1989 (1973).
[6]R.E. Shrock, Phys. Rev. D24, 1232, (1981).
[7]G. Bernardi et al., Phys. Lett. B203, 332, (1988).

TAU DECAYS IN A FUTURE B-FACTORY AS A TEST OF THE STANDARD MODEL: A PROPOSITION FOR NOVEL EXPERIMENTS

W. Fetscher
Institute for Intermediate Energy Physics, ETHZ,
CH-5234 Villigen, Switzerland

ABSTRACT

The leptonic τ decay interaction, contrary to widespread belief, cannot be determined by measuring the Michel parameter ρ precisely. From muon decay, however, it is known that the interaction can be fixed by measuring the life time and four additional parameters. The measurement of three of these seems feasible with the advent of a B meson factory. They are: The longitudinal polarization of the μ^+ from τ^+ decay and the decay asymmetry described by the parameters ξ and δ. In addition, the helicity of the τ neutrino, $|h_{\nu_\tau}|$, can be determined. The implications of these measurements on the decay interaction will be discussed.

INTRODUCTION

Leptonic decays of heavy leptons are especially important for the study of the charged weak interaction. Thus the Fermi coupling constant G_F is deduced from muon decay assuming a $V - A$ type interaction. This assumption is not necessary any more, since it had been shown just recently that $V - A$ follows for muon decay from existing measurements [1]. The method used in that analysis can readily be applied to the two decays $\tau \rightarrow \mu\overline{\nu}_\mu\nu_\tau$ and $\tau \rightarrow e\overline{\nu}_e\nu_\tau$ as well.

MUON DECAY INTERACTION

All of these decays can be described by the most general, local, derivative-free, lepton-number conserving four-fermion point interaction hamiltonian [2]. It contains ten complex coupling constants corresponding to 19 independent parameters to be determined by experiment. The experimental quantities are described most conveniently in terms of a hamiltonian in charge changing and "helicity projection" form [3, 4] which is characterized by fields of definite handedness.

The matrix element for μ decay is given by [1]

$$M \sim \sum_{\substack{\gamma=S,V,T \\ \varepsilon,\mu=R,L}} g^{\gamma}_{\varepsilon\mu} <\overline{e}_\varepsilon|\Gamma^\nu|(\nu_e)_n> < (\overline{\nu}_\mu)_m|\Gamma_\gamma|\mu_\mu> .$$

γ labels the type of interaction: Γ^S, Γ^V, Γ^T (scalar, vector, tensor). The indices ε and μ indicate the chiral projections (left-handed, right-handed) of the spinors of the experimentally observed particles, $\varepsilon \hat{=}$electron, $\mu \hat{=}$muon. The helicities n and m for the ν_e and the ν_μ, respectively, are uniquely determined for given γ, ε, μ. In this picture, the standard model corresponds to $g^V_{LL} = 1$, all other couplings being zero.

MINIMAL SET OF MEASUREMENTS

Fortunately it is not necessary to perform all of the 19 possible measurements. In fact it has been shown that the Lorentz structure can be determined completely by measuring the following five observables [5]:

1) Lifetime τ
2) polarization ξ' of the decay electron
3) and 4) decay asymmetry relative to the spin of the muon, described by ξ and δ, and
5) total cross section for the inverse decay $\nu_\mu e^- \to \mu^- \nu_e$ with ν_μ of known helicity.

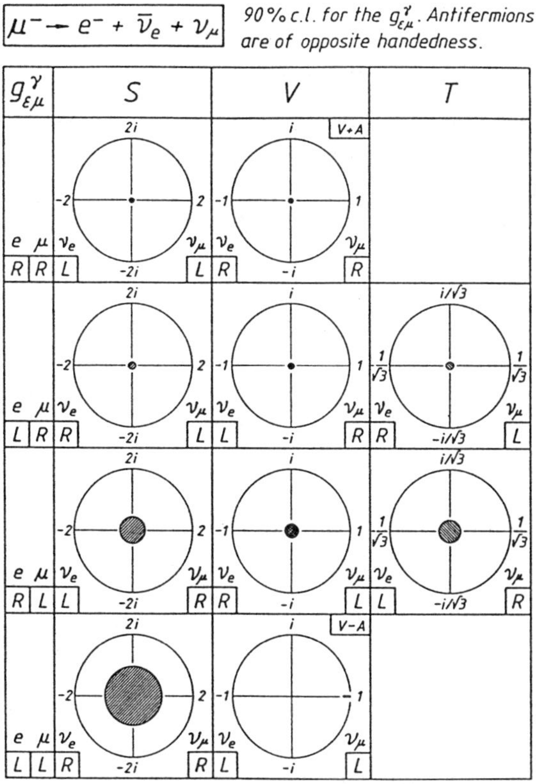

$\mu^- \to e^- + \bar{\nu}_e + \nu_\mu$

90% c.l. for the $g^\gamma_{\varepsilon\mu}$. Antifermions are of opposite handedness.

Fig. 1. 90% c.l. limits for the coupling constants $g^\gamma_{\varepsilon\mu}$. Each coupling is uniquely determined by the handedness ε and μ of the electron and the muon, respectively, and the type of interaction $\gamma = S, V$ or T.

$m_\mu/m_\tau \approx 1/17$. With $|\eta| \leqq 1$ this gives a relative error $\Delta G_F/G_F \leqq 12\%$! η can be determined by a) direct measurement and b) by deriving upper limits for the coupling constants by measuring ξ'_μ, ξ_μ and δ_μ.

τ-LEPTON DECAYS

The two parameters ξ and δ describing the decay asymmetry in leptonic decays and the τ neutrino helicity h_{ν_τ} from semileptonic decays can be determined by making use of the spin correlation of the τ pair. As an example we consider the following decays of τ pairs:

$$\tau^+ \to \mu^+\nu_\mu\bar{\nu}_\tau \quad \text{and} \quad \tau^- \to \pi^-\nu_\tau .$$

The distribution of the μ^+ and π^+ has been calculated in terms of the c.m. variables [9]. A suitable choice of laboratory variables are p_μ and p_π, the momenta of the μ and the π, and the angle $\vartheta_{\mu\pi}$ between $\vec{p}_\mu$ and $\vec{p}_\pi$. The decay distributions have been calculated by integrating over unobserved variables [6] and can be expressed by

$$\frac{d^3\Gamma}{dp_\mu dp_\pi dcos\vartheta_{\mu\pi}} = f(p_\mu, p_\pi, cos\vartheta_{\mu\pi}, \rho, \eta)$$

$$\times \{1 + [h_{\nu_\tau}\xi_\mu] \cdot A(p_\mu, p_\pi, cos\vartheta_{\mu\pi}, \rho, \eta, \delta)\}$$

ρ and η describe the energy dependence of the isotropic part of the muon spectrum in the c.m. system of the τ^+, while δ describes the energy dependence of the anisotropic part. The analyzing power A is sufficiently large (say $> 10\%$) in a large part of phase space so that the product $[h_{\nu_\tau} \cdot \xi_\mu]$ can be measured with good precision.

Similar distributions can be derived for all correlations between leptons, scalar and vector mesons from τ decays. Fig. 2 shows the correlation between K^+ and π^- from the decays $\tau^+ \to K^+\bar{\nu}_\tau$ and $\tau^- \to \pi^-\nu_\tau$ for $cos\vartheta_{K\pi} = -0.92$. The boundary is determined by kinematics and geometry, the lines inside are for constant A. A minimal set of measurements needed to determine ξ_μ, ξ_e and h_{ν_τ} separately is given in Table I.

Table I Minimal set of measurements needed to determine ξ_μ, ξ_e and h_{ν_τ}.

$\tau^+ \to$	$\tau^- \to$	Quantity measured
$\mu^+\nu_\mu\bar{\nu}_\tau$	$e^-\bar{\nu}_e\nu_\tau$	$\xi_\mu \cdot \xi_e$
$\mu^+\nu_\mu\bar{\nu}_\tau$	$\pi^-\nu_\tau$	$h_{\nu_\tau} \cdot \xi_\mu$
$e^+\nu_e\bar{\nu}_\tau$	$\pi^-\nu_\tau$	$h_{\nu_\tau} \cdot \xi_e$

The experimental limits for the coupling constants are shown in Fig. 1 from ref. [5]. It is seen that $V-A$ is clearly established while there is still room for other, especially scalar couplings.

Leptonic τ decays can be treated in the same way, but up to now only the lifetime and the Michel parameter ρ have been determined. The observables ξ'_{μ}, ξ_{μ}, ξ_e, δ_{μ} and δ_e can be measured at a B meson factory in the region of the Υ resonance [6]. Their influence on the coupling constants can be seen readily:

$$\left(\frac{1}{4}\left|g^S_{RR}\right|^2 + \left|g^V_{RR}\right|^2\right) + \left(\frac{1}{4}\left|g^S_{RL}\right|^2 + \left|g^V_{RL}\right|^2 + 3\left|g^T_{RL}\right|^2\right) = \frac{1}{2}(1-\xi')$$

and

$$\left(\frac{1}{4}\left|g^S_{RR}\right|^2 + \left|g^V_{RR}\right|^2\right) + \left(\frac{1}{4}\left|g^S_{LR}\right|^2 + \left|g^V_{LR}\right|^2 + 3\left|g^T_{LR}\right|^2\right) = \frac{1}{2}\left\{1 - \frac{1}{9}\xi\,(16\delta - 3)\right\}$$

$V-A$ predicts $\xi' = 1$, $\xi = 1$ and $\delta = 3/4$.

For the decay $\tau \to \mu\bar{\nu}_{\mu}\nu_{\tau}$ one can therefore, by the measurement of ξ'_{μ}, ξ_{μ} and δ_{μ}, get upper limits for eight of the ten complex coupling constants. For the decay $\tau \to e\bar{\nu}_e\nu_{\tau}$, where a measurement of the electron polarization ξ'_e seems hardly feasible, one can get upper limits for five coupling constants. The remaining ambiguity for the muonic decay cannot be resolved without a neutrino measurement; the interaction could still be an arbitrary mixture of $V-A$ ($g^V_{LL} \neq 0$) and a scalar, charged Higgs coupling ($g^S_{LL} \neq 0$).

We note that the measurement of the Michel parameter δ, in contrast to widespread belief, does not yield much information, if it turns out to have the $V-A$ value 3/4: Any combination of the couplings g^S_{LL}, g^S_{LR}, g^S_{RR}, g^V_{LL} and g^V_{RR} will yield the same value! In particular right-handed vector bosons W_R ($g^V_{RR} \neq 0$) cannot be excluded by the measurement of ρ.

The implications of ρ being different from 3/4 have been worked out in ref. [7]; we note especially that for $\rho > 3/4$ both scalar and tensor currents have to be present.

UNIVERSALITY

The universality of the charged weak interaction can be checked by testing if a) the couplings are equal and b) the strength is the same. The Fermi coupling constant is given by [8]

$$G_F^2 = \frac{1}{\tau_{\ell}} \cdot \frac{192\pi^3}{m_{\ell}^5} \cdot \frac{1}{1 + 4\eta \cdot \frac{m_{\ell'}}{m_{\ell}}} ,$$

where τ_{ℓ} is the lifetime of the lepton ℓ, m_{ℓ} its mass and $m_{\ell'}$, the mass of the daughter lepton. Higher order corrections have been neglected here. η is the low energy parameter and is important in the muonic decay of the τ, since

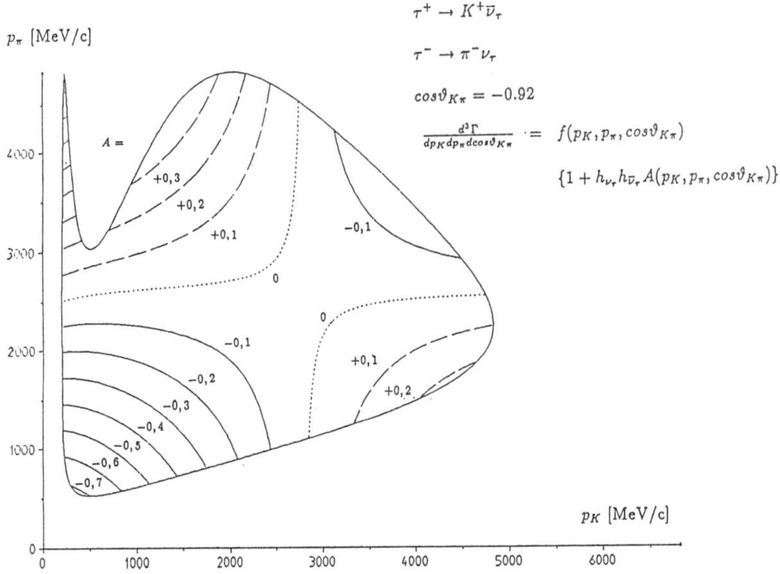

Fig. 2. Asymmetry A of the decay distribution $d^3\Gamma/(dp_K dp_\pi dcos\vartheta_{K\pi})$ of the K^+ and π^- from the correlated decays $\tau^+ \to K^+\bar{\nu}_\tau$ and $\tau^- \to \pi^-\nu_\tau$ for a total beam energy 10 GeV and for $cos\vartheta_{K\pi} = -0.92$; $\vartheta_{K\pi}$ is the angle between the momenta $\vec{p}_K$ and $\vec{p}_\pi$. The boundary is determined by kinematics and geometry, the inside lines are for constant A.

The combination of the three results yields $h^2_{\nu_\tau}$, ξ^2_μ, ξ^2_e and relative signs. For 10^7 produced $\tau^+\tau^-$ pairs and a detection threshold of 1 GeV/c one gets the following statistical errors:

$$\Delta|h_{\nu_\tau}| \;=\; \Delta|\xi_\mu| \;=\; \Delta|\xi_e| \;=\; 3\%.$$

The parameters δ_e and δ_μ can be derived from the same distributions with comparable precision. The function $f(p_\mu, p_\pi, cos\vartheta_{\mu\pi}, \rho, \eta)$ allows the determination of ρ and η.

The sign of the parity violation can be determined in a separate experiment as proposed by Kühn and Wagner [10]: In the decay of unpolarized $\tau^- \to a_1^- \nu_\tau$ the a_1 decay through the ρ-resonance into $\pi^+\pi^-\pi^-$ is used as an analyzer of the a_1 polarization which in turn yields the helicity of the ν_τ. The analyzing power for this process is in the order of a few percent and thus sufficiently large to determine the sign of the helicity, which can be combined with the results from the correlation experiments to give h_{ν_τ}, ξ_μ and ξ_e.

Finally the polarization ξ'_μ of μ^+ from τ^+ decay can be determined in a third experiment. The μ^+ are stopped in a weak magnetic field in which the μ^+ polarization processes. The μ^+ decay distribution is modulated by the

μSR signal. Its analyzing power is 1/3 and has been measured with a precision of $< 1\%$ [11]. A similar experiment has been performed for the reaction $\overline{\nu}_\mu Fe \to X\mu^+$ [12], where the helicity transfer onto the μ^+ had been measured. For 10^7 $\tau^+\tau^-$ pairs produced one can determine ξ'_μ with an error of (10-20)%.

CONCLUSIONS

With the proposed B-Meson-Factory at PSI (formerly SIN) the first measurements of the parity violating parameters ξ_e, δ_e, ξ_μ, δ_μ and ξ'_μ in leptonic τ decays and of the τ neutrino helicity in semileptonic decays become possible. These measurements in turn allow to derive model-independent upper limits for right-handed gauge bosons W_R (characterized by the couplings g^V_{RR}, g^V_{RL} and g^V_{LR}), for charged Higgs bosons (g^S_{RR}, g^S_{RL} and g^S_{LR}) and for tensor bosons (g^T_{RL}, g^T_{LR}) and thus to test the universality of the charged weak interaction.

REFERENCES

1. W. Fetscher, H.-J. Gerber and K. F. Johnson, Phys. Lett. 173B, 102 (1986).
2. L. Michel, Proc. Phys. Soc. A63, 514 (1950).
3. F. Scheck, Leptons, Hadrons and Nuclei (North Holland, Amsterdam, 1983).
4. K. Mursula and F. Scheck, Nucl. Phys. B253, 189 (1985).
5. W. Fetscher, in "12th International Conference on Neutrino Physics and Astrophysics", Sendai, Japan, June 3-8, 1986.
6. W. Fetscher, to be published;
   W. Fetscher, Leptonic τ Lepton Decays, in "Proposal for a B-Meson-Factory, 1988.
7. H.-J. Gerber, Lepton Properties, in "International Europhysics Conference on High Energy Physics, Uppsala (Sweden), 25 June–1 July, 1987.
8. F. Scheck, Muon Physics, Phys. Rep. 44 (1978)
9. So-Young Pi and A. I. Sanda, Ann. Phys. 106, 171 (1977).
10. H. Kühn and F. Wagner, Nucl. Phys. B236, 16 (1984).
11. I. Beltrami et al., Phys. Lett. 194B, 326 (1987).
12. M. Jonker et al., Phys. Lett. 86B, 229 (1979).

MESON AND GLUEBALL DECAYS

R. Akhoury
Randall Lab. of Physics, Univ. of Michigan, Ann Arbor, MI 48109

ABSTRACT

The role played by the gluon anomaly in certain Pseudoscalar meson and glueball decays is discussed. A chiral Lagrangian framework is used to calculate the hadronic decays of the η and η' in agreement with experiments. The gluonic production of pseudoscalar mesons is also discussed.

The hadronic decays of the η and η' mesons and the gluonic production of pseudoscalar mesons are studied using mostly an effective chiral Lagrangian approach. As we discuss below, these are good systems for studying the role of the gluon anomaly and the non perturbative QCD vacuum structure.

In order to motivate the form of the effective Lagrangian let us first consider the 3π decay modes of the η and η', using current algebra and certain orthogonality relations.[1] The physical η and η' fields are expressed in terms of the octet and singlet fields η_8 and η_0 by :

$$\begin{aligned} |\eta >&= \cos\theta|\eta_8 > - \sin\theta|\eta_0 > \\ |\eta' >&= \sin\theta|\eta_8 > + \cos\theta|\eta_0 > \end{aligned} \tag{1}$$

where, θ is the $\eta - \eta'$ mixing angle. A^0_μ and A^8_μ are the axial currents corresponding to the singlet and the light component of the octet. In the presence of quark mass terms, these currents are not conserved and further A^0_μ has a gluon anomaly (ignoring electromagnetism)

$$\begin{aligned} \partial_\mu A^8_\mu &= \frac{1}{2\sqrt{3}}\left(2im_u\bar{u}\gamma_5 u + 2im_d\bar{d}\gamma_5 d - 4im_s\bar{s}\gamma_5 s\right) \\ \partial_\mu A^0_\mu &= \frac{1}{\sqrt{6}}\left(2im_n\bar{u}\gamma_5 u + 2im_d\bar{d}\gamma_5 d + 2im_5\bar{s}\gamma_5 s\right) \\ &\quad + \frac{1}{\sqrt{6}}\frac{3\alpha_s}{4\pi}G\widetilde{G} \end{aligned} \tag{2}$$

The non zero couplings of these divergences to the physical pseudoscalar nonet is defined through the relations:

$$\begin{aligned} &< 0|\frac{A^8_\mu}{f_8}\cos\theta - \frac{A^0_\mu}{f_0}\sin\theta|\eta >= \frac{ip_\mu}{(2\pi)^{3/2}} \qquad (p^2 = m^2_\eta) \\ &< 0|\frac{A^8_\mu}{f_8}\sin\theta + \frac{A^0_\mu}{f_0}\cos\theta|\eta' >= \frac{ip_\mu}{(2\pi)^{3/2}} \qquad (p^2 = m^2_{\eta'}) \end{aligned} \tag{3}$$

with the orthogonal combinations vanishing between the respective states. f_8 and f_0 are respectively the octet and singlet decay constants which together with f_π may be determined from the two photon decay modes of the η, η' and the pion. Using the above relations one may determine $< 0|\frac{3\alpha_s}{4\pi}G\widetilde{G}|\eta(\eta') >$, $< 0|2im_s\bar{s}\gamma_5 s|\eta(\eta') >$ in terms of f_8, f_0 and θ and then using soft pion theorems and current algebra the amplitudes for $\eta \rightarrow 3\pi$ and $\eta' \rightarrow 3\pi$. These amplitudes are:

$$\text{amp}(\eta \rightarrow 3\pi) \propto \frac{\sqrt{3}}{4}i\left(\frac{m_d - m_u}{m_s}\right)\cos\theta\frac{f_8}{f_\pi^3}\left(\frac{\cos\theta - \sqrt{2}\sin\theta}{\cos\theta + \dfrac{\sin\theta}{\sqrt{2}}}\right)m_\eta^2$$

$$\text{amp}(\eta' \rightarrow 3\pi) \propto \frac{\sqrt{3}}{4}i\left(\frac{m_d - m_u}{m_s}\right)\sin\theta\frac{f_8}{f_\pi^3}\left(\frac{\cos\theta + \dfrac{\sin\theta}{\sqrt{2}}}{\cos\theta - \sqrt{2}\sin\theta}\right)m_{\eta'}^2 \qquad (4)$$

The rates for these as well as the branching ratio $\frac{\psi\rightarrow\eta'\gamma}{\psi\rightarrow\eta\gamma}$ are consistent with experiment for $-20° \leq \theta \leq -16°$, in agreement with the value of the mixing angle obtained from other phenomenological analyses.[2] The effect of η–η' mixing for $\eta \rightarrow 3\pi$ is quite significant: It increases the rate by about a factor of 3 from the naive current algebra predictions.

The amplitudes (4) can be reproduced (to lowest order in chiral perturbation theory) using an effective Lagrangian:[3]

$$\begin{aligned}\mathcal{L} =& \frac{1}{4}f_\pi^2 tr\left(\partial_\mu U^+\partial^\mu U\right) + \frac{1}{2}f_\pi^2 r_0 tr\left(MU^+ + U^+M\right)\\ &-\frac{1}{2}\mu^2\frac{f_\pi^2}{6}\left(Im\,Tr\,\ln U\right)^2\end{aligned} \qquad (5)$$

where, $U = e^{2i\tilde{\pi}/f}$ and $\tilde{\pi}$ is the nonet of pseudoscalar mesons, M is the diagonal quark mass matrix $M =$ diag. (m_u, m_d, m_s). The parameter r_0 is related to the meson masses through (ignoring electromagnetism) $m_{\pi^+}^2 = r_0(m_u + m_d)$, etc. In the above Lagrangian, the effect of the gluon anomaly and the nonperturbative QCD vacuum structure is incorporated by including a mass μ for the singlet field. Its justification lies in the large N_C studies of QCD.

An analysis[4] of the neutral mass matrix obtained from (5) shows that the predicted $\eta - \eta'$ mixing angle is $\theta \cong -20°$. The dominant decay mode of the η' is $\eta' \rightarrow \eta\pi\pi$ and an explanation of this decay has hitherto been lacking, at least, within the chiral Lagrangian framework. As is now shown,[4] this decay can be obtained in reasonable agreement with experiment by supplementing the Lagrangian (5) with certain four derivative terms.

The contribution from (5) to the decay $\eta' \to \eta\pi\pi$ can be easily seen to be

$$\mathcal{L}_{\rm int} = \frac{m_\pi^2\sqrt{2}}{3f_\pi^2}\left(\cos 2\theta - \sqrt{2}\sin 2\theta\right)\eta\eta'\left(\frac{1}{2}\pi^{02} + \pi^+\pi^-\right) \tag{6}$$

and the decay rate computed from this is around 3 keV. The experimental, total decay rate is[5] 156 ± 20 keV and so the prediction from 91) is small by about a factor of 60. We claim that (6) predicts such a small rate for $\eta' \to \eta\pi\pi$ because its contribution vanishes in the chiral limit. Thus, to properly account for this decay one has to supplement (5) by additional interactions that are of leading order in $1/N_C$ and which do not vanish in the chiral limit. For these interacations we use the results[6] of recent QCD inspired attempts at obtaining the effective Lagrangians for pseudoscalar mesons. In fact, from ref [6] we have:

$$\mathcal{L}^{(4)} = \frac{N_c}{192\pi^2}Tr\left\{\frac{1}{2}\left[U^+\partial_\mu U,\, U^+\partial_\mu U\right]^2 + \left(\partial_\mu U^+\partial_\mu U\right)^2 - 2\left[\partial_\mu(U^+\partial_\mu U)\right]^2\right\}. \tag{7}$$

where $N_C = 3$ is the number of colors. The contribution of (7) to $\eta' \to \eta\pi\pi$ may be expressed in terms of the following effective Lagrangian:

$$\begin{aligned}\mathcal{L}_{\rm int} = \frac{N_C}{192\pi^2}\left(\frac{2}{f\pi}\right)^4\frac{\sqrt{2}}{6}(\cos 2\theta - \sqrt{2}\sin\theta)\Big\{&\partial_\mu\eta'\partial_\mu\eta\partial_\nu\vec{\pi}\cdot\partial_\nu\vec{\pi}\\ &+ 2\partial_\mu\eta'\partial_\nu\eta\partial_\mu\vec{\pi}\cdot\partial_\nu\vec{\pi}\Big\}\end{aligned} \tag{8}$$

which clearly does not vanish in the chiral limit. In fact the total decay rate for $\eta' \to \eta\pi\pi$ evaluated using (8) is $\Gamma \sim 300$ keV, which is in reasonable agreement with experiment. The slope parameter α[7] obtained is also of the correct order of magnitude. This contribution (8), to $\eta' \to \eta\pi\pi$ is the only surviving one in the chiral limit. For non zero quark masses, there are additional contributions[8] including one meson loop contributions which must be taken into account, and whose contribution may be aabout 40%.

Finally, we will now discuss the gluonic production of Pseudoscalar mesons[9] which exemplifies the role of the gluon anomaly in pseudoscalar glueball decays. It has been established that the predominant radiative decay mode of the ψ is $\psi \to \gamma gg$ which then implies that this is a good channel to look for glueball states. In fact, a possible glueball candidate is the $i(1440)$ seen in this channel with the peculiarity that it decays more often to $K\bar{K}\pi$ states than to the $\eta\pi\pi$ states.[10] Using the Lagrangian (5) supplemented by the anomalous interaction

$$\mathcal{L}_{\rm anomaly} = \frac{\alpha_s}{8\pi}TrG\widetilde{G}\cdot Tr\tilde{\pi}, \tag{9}$$

the amplitude for two gluons in the 0^- state decaying to final states of $\eta\pi\pi$ or $K\bar{K}\pi$ via intermediate η', η and π^0 states can be determined. In the limit $m_\pi = 0$ we find (for example):

$$\text{Amp}(gg \to K\bar{K}\pi) = \frac{\alpha_s}{8\pi}\left(TrG\widetilde{G}\right)\left(\sqrt{2}f_\pi^3\right)^{-1}\left(P^2 - m_{\eta'}^2\right)^{-1}\left(P^2 - m_\eta^2\right)^{-1}$$
$$\left\{4m_K^2 P \cdot P_{\pi^-}\right\}$$
$$\text{Amp}(gg \to \eta\pi\pi) = 0 \tag{10}$$

where P is the total momentum of the final state and P_π that of the pion. Thus the prediction of Lagrangian (5) and (9) for the amplitude ratio of $gg \to K\bar{K}\pi$ to $gg \to \eta\pi\pi$ in approximately m_K^2/m_π^2 when $m_\pi \neq 0$, providing the needed enhancement. However, since the Lagrangian (5) is insufficient to explain the $\eta' \to \eta\pi\pi$ state by itself and needs to be supplemented by (7), we have to also include the contributions of (7) in determining the ratio. When this is done, the enhancement of the $K\bar{K}\pi$ decay over the $\eta\pi\pi$ one is smaller than m_K^2/m_π^2.

In conclusion, we have shown that the Lagrangian (5) based on large N_C consideration and supplemented by (7) and (9) is remarkably successful in explaining the hadronic decay modes of the η and η' and the peculiarities of the gluonic production of pseudoscalar mesons. In the conjunction with the Wess-Zumino term, a large number of the radiative decay modes of η and η' are also explained.

ACKNOWLEDGEMENTS

I am grateful to W.A. Bardeen, J.M. Frère, M. Leurer and F.J. Ynduráin for discussions and collaboration. This work was supported by the U.S. Department of Energy.

REFERENCES

1. R. Akhoury and J.M. Frère, Michigan preprint UM–TH–87–19 (to appear in Phys. Lett. B).
2. F.J. Gilman and R. Kaufman, Phys. Rev. D36 , 2761 (1987).
3. W.A. Bardeen, Private communication.
4. R. Akhoury and M. Leurer, Univ. of Michigan preprint, UM–TH–88–10.
5. Particle Data Group, Phys. Letts. 170B 1 (1986).
6. J. Balog, Phys. Letts. 149B , 197 (1984).
7. IHEP, IISN, LANL, LAPP Collab., Phys. Letts. 177B , 115 (1986).
8. J. Gasser and H. Leutwyler, Nucl. Phys. B250 , 465 (1985).
9. R. Akhoury, W.A. Bardeen and F.J. Ynduráin, in preparation.
10. W. Wisniewski, these proceedings.

Status of a Study of the Decay $K^+ \rightarrow \pi^+\nu\bar{\nu}$

Brookhaven, Princeton, TRIUMF
(Presented by Richard C. Strand, BNL Physics Department)

ABSTRACT

AGS-787, BNL-Princeton-TRIUMF, began taking data in February 1988 on rare K^+ decays into π^+ and non-interacting neutrals. Recent observations of B-$\bar{B}$ mixing have sharpened the branching ratio prediction of the three generation standard model, (SM), to ~ 1-5 × 10^{-10} for $\pi^+\nu\bar{\nu}$. The detector has 2π acceptance for this decay and 4π acceptance for background gammas. A very large window for viewing new physics exists between the SM value and the previous experimental limit of 1.4 × 10^{-7}. Several other predicted rare K^+ decays can be detected. The performance of the detector is described.

AGS Experiment 787 can yield an important constraint on the parameters of the SM, probe for new generations in SM, and explore for new physics beyond SM by seeking K^+ decays into π^+ and weak neutral(s).

The K^+ has a large kinematic window for π^+ emission and a large branching fraction window for $\pi^+\nu\bar{\nu}$ decays between 1.4 × 10^{-7}, the current limit[1,2], and ~ 10^{-10}, SM's prediction[3]. The appearance of more frequent decays into π^+ and non-interacting neutral particle(s) would be a clear signature for new physics beyond SM. AGS 787 is sensitive to decays involving a variety of neutral, light, and weakly interacting particles such as axions[4], familons[5], hyperphotons, Higgs, and the supersymmetric photino, goldstino, and scalar Higgs (Shiggs)[6], predicted by current theoretical models.

The observation of $\pi^+\nu\bar{\nu}$ at the SM value would constrain the mass of the top quark and the element linking the top and down quarks in the Kobayashi-Maskawa matrix. The decay's interpretation is free of uncertainties from light hadronic matrix elements and final state interactions. An extra generation of quarks and leptons in SM could modify the $\pi^+\nu\bar{\nu}$ decay rate considerably.

The experimental challenge of studying $\pi\nu\bar{\nu}$ decays to the 10^{-10} level requires high rates and rejection of many backgrounds from ordinary K^+ decays. The challenge is confronted with adequate segmentation in fully active tracking systems with redundant momentum, kinetic energy, range, and time measurements to select π^+ and reject μ^+ decays of K^+. The trackers are surrounded by "seamless" EM shower units to reject decays to photons.

At the AGS, Low Energy Separated Beam, LESB1 delivers an intense beam of 1 K^+/3 π^+ at 775 MeV/c. The K^+ are identified by a differential Cerenkov counter, tagged by fast wire chambers, marked by counter elements, degraded in beryllium oxide, and stopped at the center of the spectrometer in a segmented scintillating fiber degrader, that is called the "target". The isometric drawing of the magnetic spectrometer in Fig. 1 reveals the architecture of a collider experiment appropriate for K^+ decays at rest. More detail appears on the cross-section drawing in Fig. 2.

The cylindrical drift chamber and the 24 scintillator ranging sectors, the "range stack(s)", subtend 2π for π^+ decays from the target. A "seamless" 4π coverage for γ detection is provided by all-active tracking units surrounded by one barrel and the two endcap segmented EM shower units. The long active target segments provide γ veto coverage for the hole in the downstream endcap.

The decay π^+ gets three independent energy-momentum measurements: drift chamber arc in the one Tesla field, energy deposited in the target and range stack scintillator, and range in the segmented target and range stack. The preliminary rms resolution of momentum is ± 2.5 at 236 MeV/c, the "Kμ2" peak for μ^+ in $K^+ \rightarrow \mu^+\nu$ and ± 3% at 205 MeV/c, the "Kπ2" peak for π^+ in $K^+ \rightarrow \pi^+\pi^0$. The resolution of the π^+ mass calculated from a fit to momentum, energy and range is about ±8%.

Figure 3 shows the three sectors of a kaon's-eye view of the target. There are 9 × 14 clusters in each sector for a total of 378 segments. Every segment of the target is a triangular cluster of six state-of-the-art scintillating optical fibers 2 mm in diameter and 3.1 meters long. (The full length can register γ conversions.) For 2 meters the fibers are epoxied together to provide mechanical stability needed to locate each cluster and to help support the cantilevered stopping region. The six free ends are cemented to an optical mixing block. The blocks are fanned out to the back of a spherical surface supporting the 378 photomultiplier tubes. Six "I-counters" surrounding the 20 cm stopping region and the target elements, participate in the trigger logic. The target is instrumented with both ADC's and TDC's. Figure 4 is an xy drawing of the target, with K^+ and π^+ tracks from the run.

The drift chamber is designed to provide momentum resolution matched to scintillator range and energy resolution for background rejection of "Kμ2" ($K^+ \rightarrow \mu^+\nu$) and Kπ2 and to provide tracking to the segmented target. This is achieved in an active cylinder 50 cm long between radii of 12 and 43 cm containing five layers of six axial sense wire cells used to measure r-ϕ coordinates. The wires of layers 2 and 4 are at 3° and 4.5° stero angles for z position measurement. Graphite-epoxy cylindrical walls minimize scattering and inert material. High rate capability is achieved in short drifts (1 cm to 1.7 cm) with short memories (200 to 350 nsec). The inner layer sees one MHz rate spread over 36 cells. Reliability is enhanced by conservative selection of wires. The sense wires are 20 micron diameter gold-plated tungsten tensioned to 50 gm and the field wires are 170 micron diameter Be-Cu tensioned to 150 gm. There are 300 micron diameter field wires at the ends of the cells. The fields at the surface of the sensewires are in the range 270 kV/cm and the field wire surface fields are all less than 20 kV/cm to prevent whisker growth.

The range stack scintillator bars are read out at both ends to provide signal localization at high rates. Twenty four azimuthal sectors are each 40 cm thick. Each sector has 15 layers that are 2 cm thick in the stopping region. Two PWC's are imbedded after layers 4 and 8 in each range stack for tracking. The inner portion of the range stack below the $K\pi 2$ peak is instrumented with TDC's and ADC's. The balance of the range stack is instrumented with ADC's, TDC's, and new 500 MHz transient digitizers designed and built for AGS 787. The transient digitizer captures the $\pi^+ \rightarrow \mu^+ \rightarrow e^+$ sequence in a charge-time plane whereby a π^+ is positively identified.

The barrel photon veto covers 360° in azimuth ϕ between 45 and 135° in polar θ. Photons from $K\pi 2$ decays have between 20 and 225 MeV. The acceptance of the π^+ in the range stack forces most of the high energy photons into the barrel veto. For the $K^+ \rightarrow \mu^+ \nu\gamma$ decay, 85% of the photons above 20 MeV enter the barrel. Seventy-five layers of 1 mm lead to 5 mm scintillator providing fourteen radiation length coverage is segmented into 4 radial layers in 48 azimuthal segments. The segmentation allows separation of the rare $K^+ \rightarrow \pi^+\gamma\gamma$ decay from the $K\pi 2$. To achieve seamless coverage none of the cracks between the segments project back to the target.

The two endcap γ veto units also present 14 radiation lengths with the same 1 mm lead and 5 mm scintillator layers. The scintillator light is collected with BBOT wave length shifter bars. Mostly low energy photons fall on the end caps because of the restricted acceptance of the π^+ within the range stacks.

Triggering $\pi^+\nu\bar{\nu}$ occurs in three levels (0,1,2). Level 0 signals the K^+'s stopping within the target and the decay π^+'s entering the first layer of a range sector, with appropriate π^+ range limit and γ veto condition met. The deadtime is about 40 ns. Level 1 involves memory lookup units that permit muon rejection by range and dip angle. The deadtime is seven microseconds. Level 2 software processing of transient digitizer data for the decay μ^+ from the π^+ and of target and range stack energy for a 135 MeV maximum occur in a SLAC Scanner Processor (SSP) over 500 microseconds. Parallel triggers selected "$\pi^+\mu^+\mu^-$" and $\pi^+\gamma\gamma$ decays.

Data is acquired in a microVAXII via FASTBUS and CAMAC bus networks. SSP's are distributed on FASTBUS networks servicingTDC's and the Transient Digitizers. ADC's and scalers are on CAMAC buses.

During the first engineering/data run from February to May, 1988, about twenty times the world sample of K^+ stops for $\pi\nu\bar{\nu}$ were written to tape. Most of the data were recorded at 130 thousand K^+ stops per AGS spill. Some running took place at 300 thousand K^+ stops per spill. The resolutions of all of the systems are approaching design values. The tails of distributions must be thoroughly understood. Data analysis is proceeding for decays into weak neutrals and π^+ above the $K\pi 2$ peak.

Preliminary accounting of the data obtained by AGS 787 indicates that new limits can be obtained for the following decays for which present limits exist:

Decay	Sensitivity[7]	Present Limit
$K^+ \rightarrow \pi^+\nu\bar{\nu}$	$< 10^{-8}$	$(< 1.7 \times 10^{-7})^2$
$\rightarrow \pi^+X^0$	$< 3 \times 10^{-9}$	$(< 4 \times 10^{-8})^2$
$\rightarrow \pi^+\mu^+\mu^-$	$\lesssim 10^{-7}$	$(< 2.4 \times 10^{-6})^8$
$\rightarrow \pi^+\gamma\gamma$	$\lesssim 5 \times 10^{-7}$	$(< 8 \times 10^{-6})^8$

The goal of the present program is to press the study of $K^+ \rightarrow \pi^+\nu\bar{\nu}$ to a level of 2×10^{-10}.

This work is supported in part by the U.S. Department of Energy under Contract Nos. DE-AC02-76CH00016, DE-AC02-76ER03072 and the National Science and Engineering Research Council of Canada.

REFERENCES

1. G.D. Cable, et al., Phys. Rev. D8, 3807 (1973).
2. Y. Asano, et al., Phys. Lett. 107B, 159 (1981).
3. J. Ellis, J. Hagelin, S. Rudaz, Phys. Lett. 192B, 201 (1987).
4. R. Peccei and H. Quinn, Phys. Rev. Lett. 38, 1440 (1977) and Phys. Rev. D16, 1791 (1977); S. Weinberg, Phys. Rev. Lett. 40, 223 (1978); F. Wilczek, Phys. Rev. Lett. 40, 220 (1978).
5. F. Wilczek, Phys. Rev. Lett. 49, 1549 (1982).
6. R. Shrock. "The value of experiments on decays of the type $K \rightarrow \pi$ + missing neutral(s)", Snowmass Proceedings, 291 (1983).
7. L. Littenberg, private communication.
8. Particle Data Group, Phys. Lett. 170B April (1986).

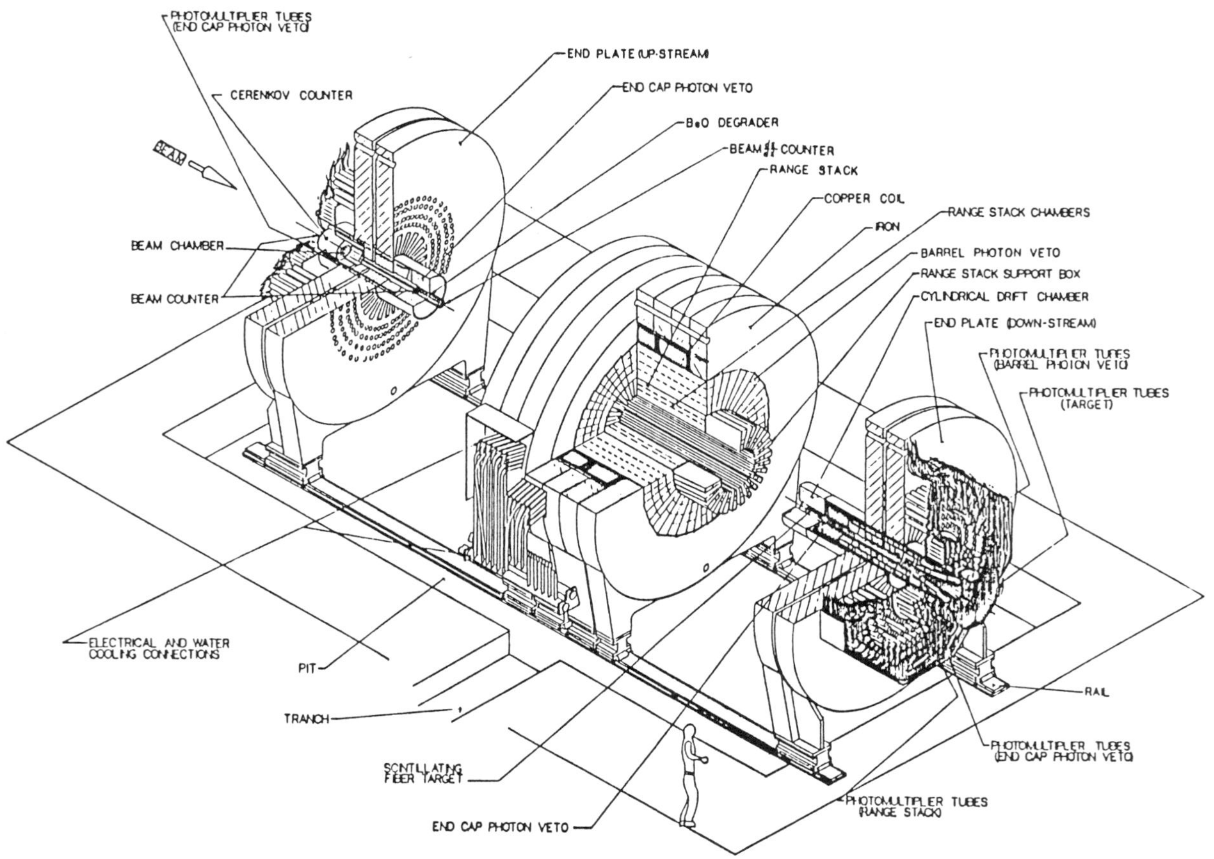

Fig. 1. Isometric Drawing of AGS 787 Spectrometer.

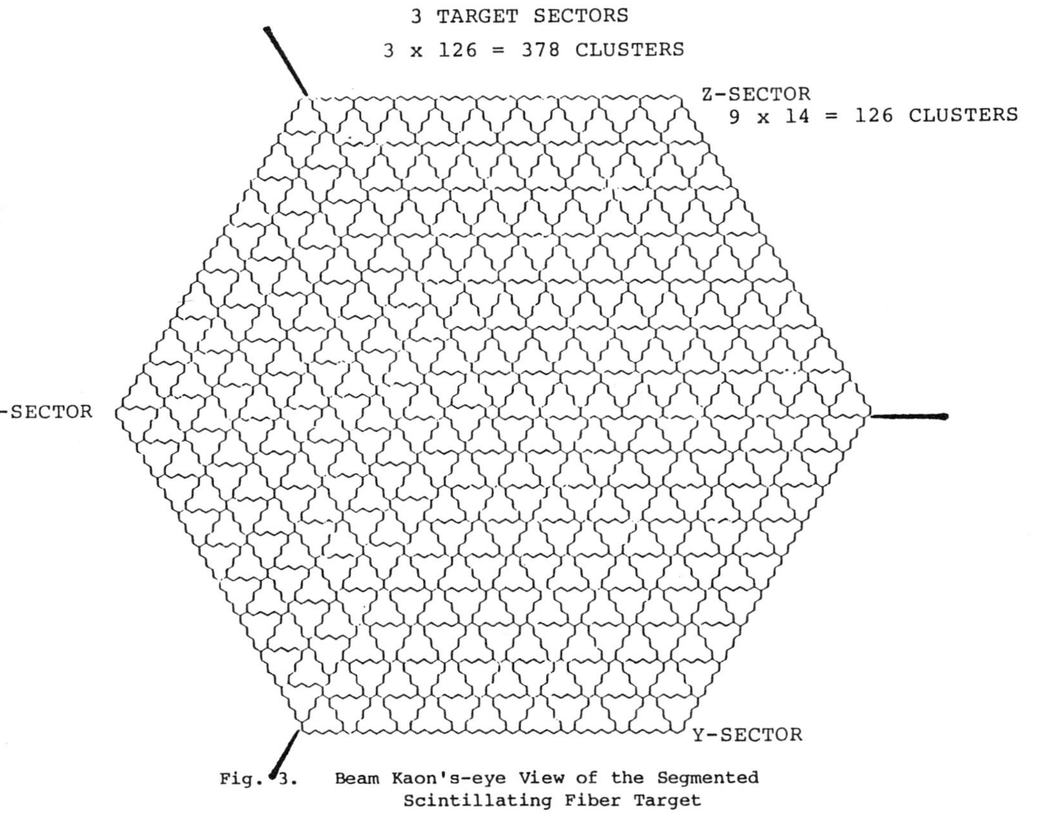

Fig. 3. Beam Kaon's-eye View of the Segmented Scintillating Fiber Target

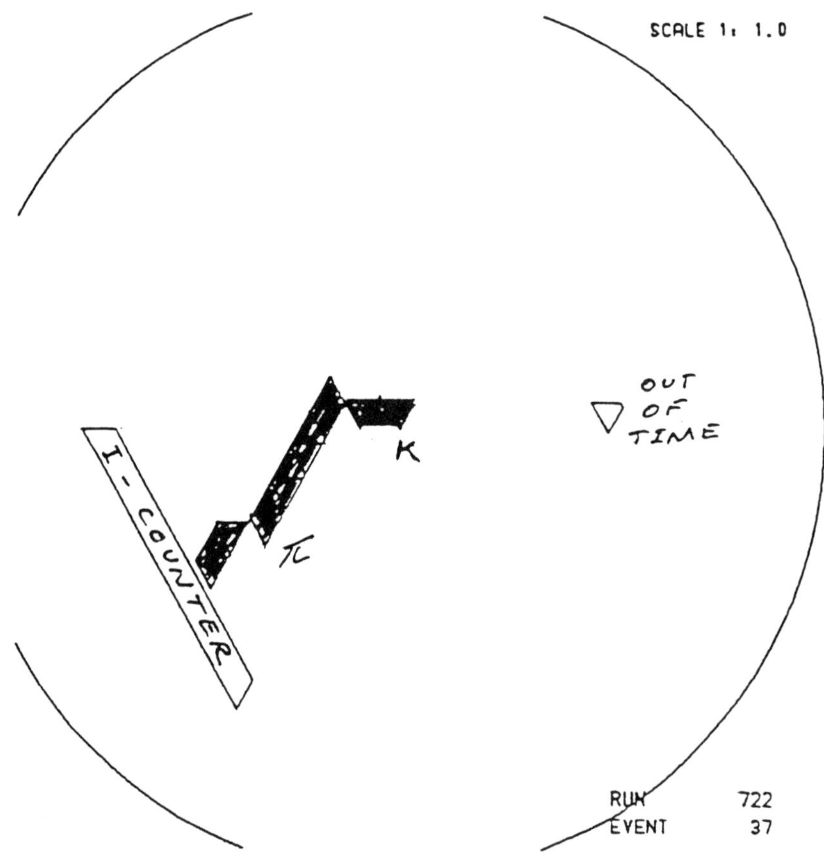

Fig. 4. $K^+ \to \pi^+ + \pi^0$ in the Target: K^+ visits 3 triangles coming to rest; π^+ visits 11 triangles and an I-counter; one triangle has out of time energy not associated with this decay. Each triangle of six scintillating fibers has time and energy recorded. Photons from π^0 decay convert in the barrel veto system. See text.

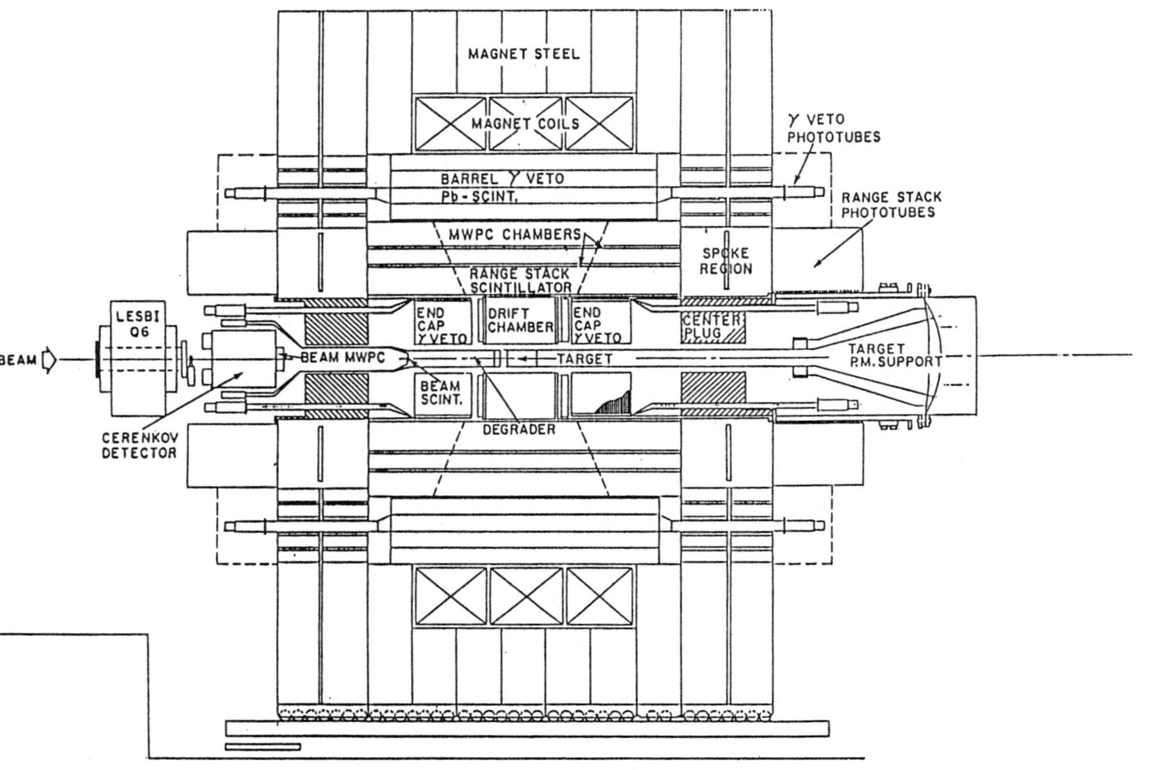

Fig. 2. AGS 787 Spectrometer Longitudinal Section.

IN SEARCH OF $K^+ \to \pi^+\mu^+e^-$ AND OTHER RARITIES SELDOM OR NEVER BEFORE SEEN *

D.M. Lazarus, H.A. Gordon and P. Rehak
Brookhaven National Laboratory

P.S. Cooper
Fermi National Accelerator Laboratory

J. Egger, W.D. Herold and H. Kaspar
Paul Scherrer Institute (SIN)

V. Chaloupka, E.A. Jagel, H.J. Lubatti and T. Zhou
University of Washington

C. Alliegro, C. Campagnari, N.J. Hadley
A.M. Lee and M.E. Zeller
Yale University

ABSTRACT

The progress of AGS Experiment 777, a search for $K^+ \to \pi^+\mu^+e^-$ is described. Results on the decay $K^+ \to \pi^+A^0$, where A^0 is a weakly interacting neutral particle that decays into e^+e^- and preliminary results on the direct decay $K^+ \to \pi^+e^+e^-$ are presented as well as new limits on $\pi^0 \to \mu^+e^-$.

In recent years considerable interest in rare decays has developed in order to probe the validity of the Standard Model. Decays such as $K_L^0 \to \mu e$ and $K^+ \to \pi^+\mu^+e^-$ violate conservation of lepton family number (LFN) and are strictly forbidden in the Standard Model and hence any observation of these decays would unambiguously imply the discovery of new physics. Although very stringent limits on LFN have been obtained for other rare decays such as $\mu \to eee$, $\mu^+A \to e^+A$ and $\mu \to e\gamma$, these decays also violate generation number[1] conservation which may lead to an additional suppression. Searches for $K_L^0 \to \mu e$ and $K^+ \to \pi^+\mu^+e^-$ are complementary since the former can only occur if the form of the interaction is tensor or axial vector whereas the latter can proceed if the interaction transforms as a scaler or vector.

Advantages of searching for $K^+ \to \pi^+\mu^+e^-$ include the momentum definition of the kaons and the possibility of tracking in a charged particle beam, the three charged particles in the final state yield good vertex reconstruction, better constraint to the decay hypothesis, better rejection of accidentals and finally low background resulting from good particle identification and the requirement of an electron in the final state. The most prolific source of K+ decay electrons is

$$K^+ \to \pi^+\pi^0 \qquad \text{(Branching Ratio=0.2117)}$$
$$\hookrightarrow e^+e^-\gamma \qquad \text{(Branching Ratio=1.198 x } 10^{-2}\text{)}$$

* Work supported in part by the U.S. Dept. of Energy under Contract No. DE-AC02-76CH00016 and by the National Science Foundation Grant No. PHY-861 3003.

whereas the most troublesome is

$$K^+ \rightarrow \pi^+\pi^+\pi^- \qquad \text{(Branching ratio = 5.6 x } 10^{-2}\text{)}$$

$$\qquad\qquad\quad \hookrightarrow e^-\nu \qquad \text{(Branching ratio = 1.2 x } 10^{-4}\text{)}$$

$$\qquad\quad \hookrightarrow \mu^+\nu \qquad \text{(Branching ratio = .635)}$$

where the π^- beta decays as indicated or is misidentified as an e-.

The experimental apparatus is shown in Figure 1. The 5.8 GeV/c secondary beam is produced by the interaction of .5 x 10^{12} protons per pulse in a one interaction length platinum target. The unseparated beam consists of three dipole magnets providing momentum dispersion and definition followed by a quadrupole doublet which focusses the beam on the first proportional wire chamber, P1 in order to minimize the beam size in the acceptance region of the detectors. Collimators in the magnets limit the acceptance to .4 msr-GeV/c and were designed to minimize beam scraping and secondary sources. There are some 2 x 10^8 π^+ and protons in the beam passing through the apparatus per pulse which requires a dead region in the center of the experiment. Ten percent of the 10^7 K^+ per pulse in the beam decay in a five meter-long vacuum box which extends through the aperture of the large dipole magnet M1 whose function is to separate decay products from the beam with positive particles deflected to the right and negatives to the left permitting specialization of particle identification on the two sides of the experiment. The first MWPC, P1 is at the end of the decay vacuum and with P2 upstream of the spectrometer magnet M2 and with P3 and P4 downstream, momentum analysis of the decay products is achieved. Two atmospheric Cerenkov counters are located between the MWPCs on either side of M2 to give redundent electron identification. The Cerenkov counters are segmented by thin membranes separating hydrogen gas on the left from carbon dioxide on the right. For electron detection, an average of 2.2 photoelectrons were detected in hydrogen and 6.6 in carbon dioxide. The threshold for muons in hydrogen is 6.3 GeV/c.

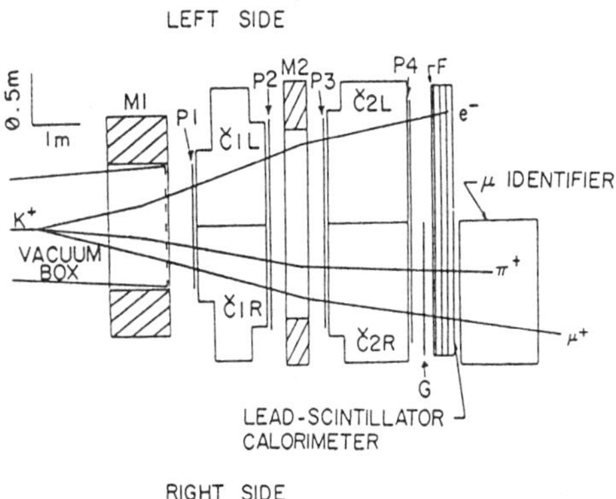

Fig. 1. Plan view of the apparatus. M1 and M2 are spectrometer magnets, P1-P4 are proportional wire chambers, C1L-C2R are Cerenkov counters. F and G are scintillation counter hodoscopes used in triggering.

A lead-scintillator shower counter containing 11.4 radiation lengths of material provides further redundancy for electron identification. The shower counter is segmented into twenty-four horizontal sections and vertically in two to allow association of individual hits with tracks as determined by the proportional wire chamber spectrometer. It was further segmented into three longitudinal sections to provide information on shower development. A muon identifier was located on the right side at the rear of the system and consisted of eight packages of proportional tubes sandwiched between 9 cm thick steel plates. Each package consisted of one horizontally and one vertically oriented plane of tubes with an effective wire spacing of 6.5 mm. Muons of 0.6 GeV/c, the minimum possible for $K^+ \to \pi^+\mu^+e^-$ at this momentum, would penetrate the first four layers while the entire array would be traversed by a 1.1 GeV/c muon. Finally, two scintillation counter hodoscopes F and G were used for triggering. The F hodoscope has the same segmentation as the shower counter.

In order to deal with the backgrounds noted above, excellent particle identification is essential. The performance of the particle identification system allowed a background limited sensitivity below 10^{-11} in branching ratio. Table I gives the measured efficiencies and rejections for the two sides of the appararatus.

Table I Measured Particle Identification Probabilities

		Identified as		
		e^-	μ^-	π^--
Particle on Left	e-	0.80	$< 2 \times 10^{-2}$	2×10^{-4}
	μ^-	$< 2 \times 10^{-7}$		
	π^-	6×10^{-7}		

		Identified as		
		e^+	μ^+	π^+
Particle on Right	e+	0.94	$< 8 \times 10^{-7}$	8×10^{-5}
	μ+	3×10^{-3}	.95	.05
	π+	$(0.05\text{-}5.0) \times 10^{-3}$	0.04	.94

Data was taken with three simultaneous triggers:

1. τ decays. Three charged particles ($K^+ \to \pi^+\pi^+\pi^-$), one on the left and two on the right. These were prescaled by 8192.

2. $K^+ \to \pi^+e^+e^-$. Cerenkov counters all fire with two charged particles on the right. These were prescaled by 8. A subset of these for high mass e+e- pairs requires one particle above and one below the vertical midplane of the apparatus taking advantage of the large opening angle for high mass pairs.

3. $K^+ \to \pi^+\mu^+e^-$. Both Čerenkov counters on the left fire with one particle on the left and two on the right, one of which penetrates four layers of the muon identifier.

Data analysis involved track reconstruction and association of tracks with a common vertex with one and only one track on the left side of the apparatus and at least two on the right. A parameter S, the square root of the sum of the squares of the distance of closest approach of each track to the calculated best vertex was computed for each event. The momentum and direction of the parent particle were calculated and loose cuts applied corresponding to the momentum spread in the beam and the location of the production target. These cuts allowed a significant number of events with missing neutral energy such as Dalitz decays to pass.

Light in Cerenkov counter sections, shower counter elements and range in the muon identifier were required to give unambiguous particle identification for each event. The effect of cuts on various distributions were compared with their effect on τ and Dalitz decays as well as with Monte Carlo simulations in order to understand efficiencies and possible biases. The mass resolution for taus is 3.7 MeV/c² (standard deviation). The distribution in S and calculated mass is displayed in Figure 2a. Using this as a template for $K^+ \to \pi^+\mu^+e^-$ decays and allowing for the increased energy released in the latter process and its consequent degradation of the experimental resolution as indicated by the larger signal region, Fig. 2b shows our results for $K^+ \to \pi^+\mu^+e^-$. Events to the left of the signal region mainly arise from τ's where decay pions were misidentified or decayed prior to particle identification. Those at higher mass are due to association of accidental hits with tracks in an event.

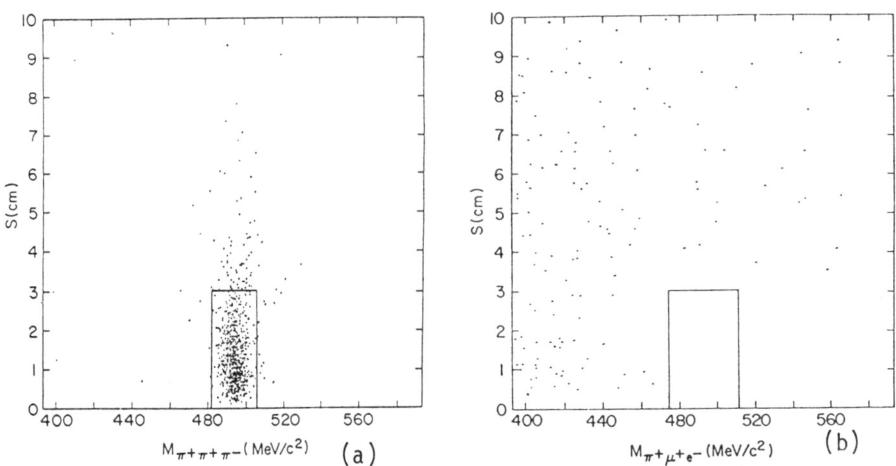

Fig. 2(a) S vs. $M_{\pi^+\pi^+\pi^-}$ scatter plot for $K^+ \to \pi^+\pi^+\pi^-$. The signal region is designated by the rectangle for 483 MeV/c² < Mass < 505 MeV/c² and S < 3 cm, where S is the square root of the sum of the squares of the distances of the tracks to a common vertex. There are 873 events in the plot. (b) Scatter plot of S vs. Mass for πμe candidates. There are 133 events in the plot.

A limit on the branching ratio is then obtained by normalizing to Dalitz decays which were chosen because they had the same Cerenkov counter requirements on the left side in the trigger. The upper limit for the branching ratio is given by:

$$BR(K^+ \rightarrow \pi^+\mu^+e^-) = BR(\text{Dalitz}) \times \frac{N(\pi^+\mu^+e^-)}{N(\text{Dalitz})} \times \frac{G(\text{Dalitz})}{G(\pi^+\mu^+e^-)} \times C$$

where $BR(\text{Dalitz}) = BR(K^+ \rightarrow \pi^+\pi^0) \times BR(\pi^0 \rightarrow e^+e^-\gamma) = 2.54 \times 10^{-3}$, $N(\pi^+\mu^+e^-) = 2.3$, the 90% confidence limit on the number of $\pi^+\mu^+e^-$ events obtained, N(Dalitz) is the number of observed Dalitz decays multiplied by the prescaling factor, G(Dalitz) and $G(\pi^+\mu^+e^-)$ are the geometric acceptances for the respective decays, C is the product of correction factors accounting for differences in the effects of cuts, contamination from other decay modes, and particle identification and trigger efficiencies. The result is:

$$BR(K^+ \rightarrow \pi^+\mu^+e^-) < 1.1 \times 10^{-9} \text{ (90\% C.L.)} \quad ,$$

an improvement of a factor of 4.4 on the previous limit.[2] This implies a lower limit of 27 TeV on the mass of a gauge boson propagator for a new interaction by which this decay would proceed assuming standard model couplings.[1]

The limit on $K^+ \rightarrow \pi^+\mu^+e^-$ implies a limit on $\pi^0 \rightarrow \mu^+e^-$ since the decay of the K^+ into $\pi^+\pi^0$ accounts for 21 percent of all K^+ decays. The limit obtained is:

$$BR(\pi^0 \rightarrow \mu^+e^-) < 7.8 \times 10^{-8} \text{ (90\% C.L.)}$$

which is comparable to the previously determined value which was inferred by authors not involved in the measurements.[3,4] This data was obtained in FY1987. More than 8 times as much data has been accumulated in the present run which just ended.

The observation of low mass correlated e^+e^- peaks produced in heavy ion collisions by the EPOS experiment[5] at GSI, Darmstadt, and the spate of theoretical papers on short lived axions and light Higgs bosons that followed stimulated us to search for 1.8 MeV/c^2 e^+e^- pairs emitted in $K^+ \rightarrow \pi^+e^+e^-$. The results of this search are summarized in Fig. 3 where the data is compared to a Monte Carlo distribution for a neutral particle A^0, produced in $K^+ \rightarrow \pi^+A^0$ with a mass of 1.8 MeV/c^2 that subsequently decays into e^+e^- with a branching ratio of 10^{-6}. Resolution smearing, brehmsstrahlung and multiple Coulomb scattering were included in the Monte Carlo calculation. The branching ratio upper limit obtained is plotted vs lifetime for A^0 masses of 1.8 Mev/c^2 and 100 MeV/c^2 in Fig. 3(b). Normalization to Dalitz decays which have the same observed final state particles was again employed with appropriate acceptance corrections as determined by Monte Carlo simulation. Limits obtained by Asano et al.[6,7,8] are also plotted. This experiment set more stringent limits for lifetimes shorter than 10^{-12} second. For longer lifetimes the probability of the decay occurring before the detector is traversed is reduced whereas in the KEK experiment the decay occurred at rest enhancing their sensitivity to longer lifetimes. This work is complete and has been published in the Physical Review Letters.[9]

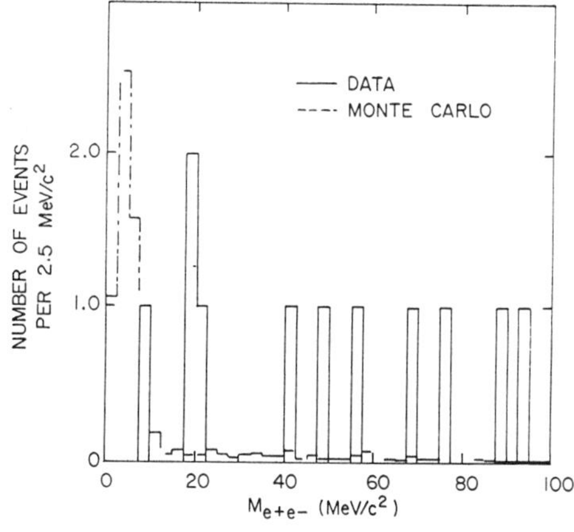

Fig. 3(a) Distribution of M_{ee} for 492.5 MeV/c^2 $\lesssim M_{\pi ee} \lesssim$ 520 MeV/c^2. The dashed line indicates the signal expected from the decay of a 1.8 MeV/c^2 A^0 with BR(A^0) equal to 10^{-6}. The high mass tail on this Monte Carlo distribution is the result of electron bremsstrahlung and multiple Coulomb scattering.

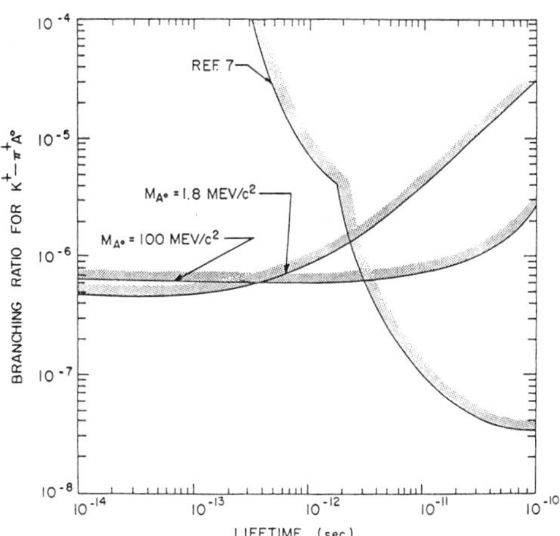

(b) BR(A^0) for A^0 mass equal to 1.8 MeV/c^2 and 100 MeV/c^2 (this experiment), and for A^0 mass equal to 1.8 MeV/c^2 from Ref. (6) for data from Refs. (7) and (8). Areas above the respective lines are excluded with greater than 90% confidence level.

The direct decay $K^+ \to \pi^+e^+e^-$ has been previously detected at CERN[10] where 41 events were obtained yielding a branching ratio of $(2.6 \pm 0.5) \times 10^{-7}$ assuming a uniform phase space distribution. A scatter plot of our data is given in Fig. 4 where the high mass e^+e^- events can be seen above the Dalitz decays which reconstruct to a lower kaon mass because of the unobserved energy carried away by the gamma. In Fig. 5 the e^+e^- invariant mass distribution is plotted with Monte Carlo distributions generated from scalar (a) and vector (b) matrix elements. The vector interaction provides a better fit to the data which consists of 160 events obtained in the FY1987 run. In the FY1988 data more than 800 events are anticipated which will allow for a detailed analysis of the Dalitz plot.

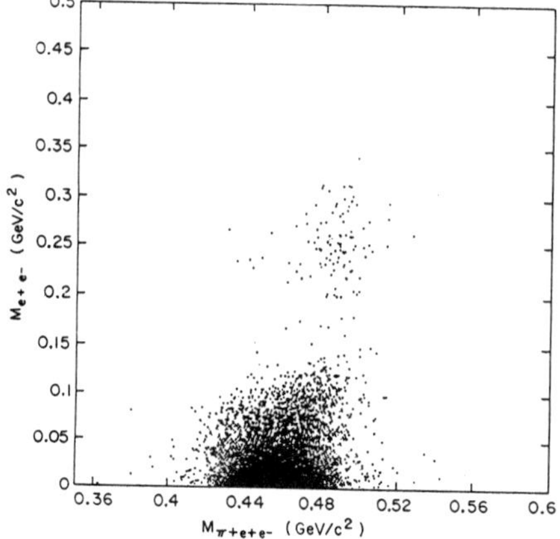

Fig. 4 Scatter plot of M_{ee} vs. $M_{\pi ee}$ for $K^+ \to \pi^+e^+e^-$ decays with the requirement that the reconstructed K^+ momentum vector originate at the production target.

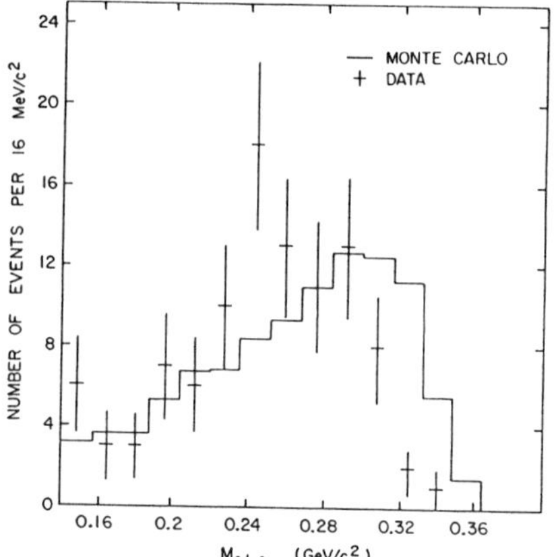

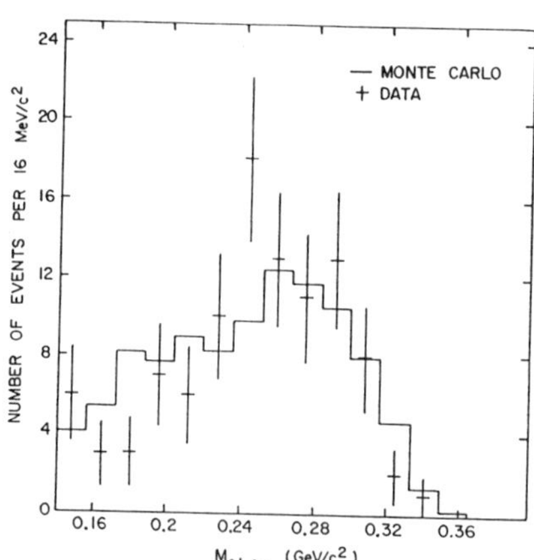

Fig. 5(a) Distribution of e^+e^- invariant mass for $K^+ \to \pi^+e^+e^-$ with .475 GeV/c^2 < $M_{\pi ee}$ < .520 MeV/c^2. A phase space Monte Carlo generated mass distribution is indicated by the histogram. (b) The histogram was generated by a Monte Carlo calculation based on a vector interaction.

SUMMARY AND FUTURE PLANS

In FY87 we achieved a new limit for $K^+ \to \pi^+\mu^+e^-$ of 1.1×10^{-9} (90% C.L.) and more than eight times as much data was obtained in FY1988 implying a sensitivity approaching 10^{-10}, a factor of fifty improvement on previously published results. A limit was also set on $\pi^0 \to \mu^+e^-$ of 7.8×10^{-8} in FY 87 and the FY1988 data should provide a sensitivity below 10^{-8}. The experiment should be sensitive to $\pi^0 \to e^+e^-$ which has been observed in two previous experiments,[11 12] which have reported branching ratios of $(22.3\ ^{+24}_{-11})\times10^{-8}$ and $(17\pm9\pm3)\times10^{-8}$ based on about 6 and 59±21 events respectively. Model dependent calculations limit this branching ratio to twice the unitarity lower bound of 4.75×10^{-8}.

We anticipate about 80 events in our present data sample of which approximately 30 were recorded with a high rate beam chamber operating. This chamber was installed late in the FY1988 run and has operated successfully with beam intensities in excess of 10^8 particles per second. The 0.7 mm resolution of this device should allow additional kinematic constraint to suppress the Dalitz decay background as well as improve the mass resolution and accidental suppression for all decays. In FY 1989, several hundred $\pi^0 \to e^+e^-$ events could be easily obtained.

We have set branching ratio limits of 4.5×10^{-7} on the decay $K^+ \to \pi^+A^0$, for A^0 lifetimes shorter than 10^{-13} second where A^0 is a weakly interacting neutral particle that decays into e^+e^- as discussed in several axion variant models proposed to explain the e^+e^- peaks observed in heavy ion collisions at GSI.

Data from the FY 1987 run yielded 160 fully analyzed $K^+ \to \pi^+e^+e^-$ events whose e^+e^- mass spectrum favors a vector interaction. With the FY1988 data in hand, it should be possible to carry out a Dalitz plot analysis with more than 800 events.

In the next two years we anticipate additional running on K^+ $\pi^+e^+e^-$ and $\pi^0 \to e^+e^-$, as well as new limits on $K^+ \to \pi^+\mu^+e^-$ which should be close to an order magnitude improvement on our present results. Additional physics to be explored includes searches for boson decays resulting in peaks in the e^+e^- mass spectrum of $\pi^+e^+e^-$ decays and more sensitive searches for other forbidden decays.

REFERENCES

1. R.N. Cahn and H. Harari, Nuc. Phys. B176, 135 (1980).
2. A.M. Diamant-Berger et al., Phys. Lett. 62B, 485 (1976).
3. D. Bryman, Phys. Rev. D26, 2538 (1982).
4. P. Herczeg and C.M. Hoffman, Phys. Rev. D29, 1954 (1984).
5. T. Cowan et al., Phys, Rev. Lett. 56, 444 (1986).
6. L.M. Krause and M.E. Zeller, Phys. Rev. D34, 3385 (1986).
7. Y. Asano et al., Phys. Lett. 107B, 159 (1981).
8. Y. Asano et al., Phys. Lett. 113B, 195 (1982).
9. N.J. Baker et al., Phys. Rev. Lett. 59, 2832 (1987).
10. P. Bloch et al., Phys. Lett. 56B, 201 (1975).
11. J. Fischer et al., Phys. Lett. 73B, 364 (1978).
12. J.S.Frank et al., Phys. Rev. D28, 423 (1983).

NEW RESULTS ON WEAK DECAYS OF CHARMED MESONS

Krzysztof Sliwa
Fermilab
(representing The Fermilab Tagged Photon Spectrometer Collaboration[1])

ABSTRACT

Results of a study of the exclusive semileptonic decay modes $D^0 \to K^- e^+ \nu_e$ and $D^+ \to K^- \pi^+ e^+ \nu_e$ are presented. We have measured their branching fractions relative to those of dominant hadronic modes, $BF(D^0 \to K^- e^+ \nu_e)/BF(D^0 \to K^- \pi^+) = 0.91 \pm 0.07 \pm 0.11$ and $BF(D^+ \to K^* e^+ \nu_e)/BF(D^+ \to K^- \pi^+ \pi^+) = 0.53 \pm 0.05 \pm 0.06$. The non-resonant $K\pi$ component in the $D^+ \to K\pi e\nu$ decay mode was found to be small. Combining our results with measurements with of the D^0 lifetime and a branching fraction for the mode $D^0 \to K^- \pi^+$ we find $|V_{cs}|^2 |f_+(0)|^2 = 0.50 \pm 0.07 \pm 0.08$. We have observed signals for decays D_s^+ and $D^+ \to \pi^- \pi^+ \pi^+$. We have analysed these modes for the fraction due to $\rho^0 \pi^+$ and $f_0(975)\pi^+$ and measured their relative branching fractions. Also, we present limits on flavour-changing neutral current decays $D^0 \to e^+ e^-$ and $D^0 \to \mu e$.

1. Introduction

The results presented in this paper are based on an analysis of the full data sample from E691, a high energy photoproduction experiment performed at the Fermilab Tagged Photon Spectrometer. The experiment recorded 10^8 triggers. The detector, a two-magnet spectrometer of large acceptance, very good mass resolution, particle identification (Cerenkov counters, electromagnetic and hadronic calorimetry, muon filter) and equipped with a high resolution silicon microstrip detector, has been described elsewhere[2]. Based on experience with the spectrometer from previous experiment, the E516, we have increased the magnetic field in the analysing dipoles (momentum resolution), added six new drift chamber planes (tracking efficiency) and installed new mirrors in Cerenkov counters (particle identification). Also, very importantly, we have taken advantage of advances in technology. We have added a high resolution silicon microstrip vertex detector (transverse resolution of $20\mu m$, longitudinal of $300\mu m$). This allows to identify charm particles decay vertices and further reduce backgrounds. Finally, we made use of the Advanced Computer Program[3] system of parallel 32-bit microprocessors at Fermilab, offering very large computing power.

2. Semileptonic decays of D mesons

The Cabibbo-favoured semileptonic decays can proceed only through flavour decay (spectator) processes and, unlike the situation in hadronic decays, there is no uncertainty due to the possible presence of other diagrams. Also, there is no interference or final state interactions between leptons and hadrons in the final state. Since the leptonic part of the matrix element is well understood, the study of semileptonic decays probes the structure of the hadronic part.

The electron identification, essential for a study of the semileptonic modes, used (a)the ratio of energy seen in the electromagnetic calorimetry to track momentum, (b)the sizes of the signals in the electromagnetic and hadronic calorimeters, (c)the transverse shower shapes, and (d)the comparison of the positions of showers in calorimeters with projected charged tracks positions. The electron efficiency and the pion misidentification probability, while being position and energy dependent, had the typical values of 61% and 0.3% respectively.

2.1 $D^0 \to K^- e^+ \nu_e$

In the $D^0 \to K^- e^+ \nu_e$ decay, because of the $V - A$ nature of the weak current and D, K being pseudoscalars, the $D - K$ interaction is purely vector. (Throughout the paper the charge conjugate states are implicitly included.) There are two formfactors involved, $f_+(t)$ and $f_-(t)$, the latter always appears in a final result with m_e, the electron mass and its contribution to the decay rate is negligible. The decay rate can then be shown (in the D^0 center of momentum) to be proportional to

$$\Gamma \propto G^2 |V_{cs}|^2 |f_+(t)|^2 [(E_K)^2 - (M_K)^2 - (M_D - E_K - 2 \times E_e)^2] \qquad (1)$$

Analysis of the t distribution makes it possible to extract the vector form factor $f_+(t)$. This, combined with branching fraction and lifetime measurements (plus theoretical input[4,5] about $f_+(0)$), allows a measurement of the $|V_{cs}|$ element of K-M matrix.

We have selected the candidate events through the cascade decay $D^* \to D^0 \pi^+$ followed by $D^0 \to K^- e^+ \nu_e$. The technique used is based on the fact that it is possible to reconstruct the missing neutrino momentum providing that the D^0 direction is measured with sufficient precision in the vertex detector. The algebra is easiest in the Lorentz frame with z-axis chosen along the D^0 path, and such that p^Z_{Ke} is equal to zero. Setting $M_{Ke\nu} = M_D$ and $M_\nu = 0$, one can easily obtain the longitudinal component of the neutrino momentum, p^Z_ν :

$$(p^Z)^2 = (p^Z_\nu)^2 = (p^Z_D)^2 = \frac{F^2}{4 \times (E_{Ke})^2} - (p^T_{Ke})^2; \qquad (2)$$

$$\begin{aligned} F &= (M_D)^2 - (p^T_{Ke})^2 - (E_{Ke})^2 \\ &= 2 \times E_{Ke} E_\nu \geq 0. \end{aligned} \qquad (3)$$

Because of the finite vertex position resolution F and $(p^Z)^2$ can acquire non-physical, negative values. We have required $F > 0$, which reduces background considerably while keeping about 60% of signal. In the case of $(p^Z)^2 < 0$ (about 40% of events), we have set[6] $(p^Z)^2 = 0$. In the remaining events, because equation (2) is quadratic there exist two solutions for the $E_{Ke\nu}$. In some cases, one of them is non-physical and can be discarded (*e.g.* $E_{Ke\nu} > 260\ GeV$). In the remaining events, for every π^+ we will obtain two D^{*+} solutions, corresponding to the two p^Z_ν solutions. We choose the one which gives the *lower* D^* mass[6].

The experimental procedure consists of selecting $K^- e^+$ pairs originating from a common vertex significantly separated from a primary one, solving for the ν_e, and then combining the $K^- e^+ \nu_e$ four-momentum (constrained to M_D) with that of a π^+ candidate. The $K^- e^+$ vertex was required to be well separated from the primary vertex ($\Delta\vec{x} \geq 7\sigma_{\vec{x}}$) and both vertices were required to be of good quality. The primary vertex should have at least two tracks associated with it, with a slow pion from the D^* decay being one of them. Background distributions were obtained using the same approach, but using the *wrong charge* $K^+e^+\nu_e\pi^+$, $K^+e^+\nu_e\pi^-$ and $K^+e^-\nu_e\pi^+$ combinations. These were added together, and subtracted from the final $M_{K^-e^+\nu_e\pi^+}$ distribution after being normalized to the integral over the mass interval $2.03 - 2.40$ GeV.

We required the kaon and electron candidates to be good quality, well identified tracks. A cut on electron momentum, $p_e \geq 12$ GeV was applied to improve the signal to noise ratio in the electron identification. Due primarily to this momentum cut-off the event detection efficiency is sensitive to the electron radiation in the material dowstream of the decay vertex and the radiative corrections, including real and virtual photons[7]. The combined effect of the two is estimated to reduce the reconstruction efficiency by a factor of 0.84 ± 0.04.

The largest physics background comes from another semileptonic decay mode, namely $D^0 \rightarrow K^-e^+\pi^0\nu_e$. What we actually measure is the sum of contributions from the $K^-e^+\nu_e$ and $K^-e^+\pi^0\nu_e$ modes, with obviously different efficiencies. To determine the amount of $K^-e^+\pi^0\nu_e$ in our signal we assume $\Gamma(D^0 \rightarrow K^*e\nu) = \Gamma(D^+ \rightarrow K^*e\nu)$ (isospin symmetry) and make use of our measurement (see below) of the ratio $\Gamma(D^+ \rightarrow K^*e^+\nu_e)/\Gamma(D^0 \rightarrow K^-e^+\nu_e)$. Taking the ratio $\Gamma(K^-e^+\nu_e)/\Gamma(K^*e^+\nu_e) = 2$, and assuming K^* dominance, we find that the correction for the $D^0 \rightarrow K^-e^+\nu_e$ is small, 7% of the total rate.

In Figures 1, 2 and 3 we present $M_{Ke\nu\pi}$ distributions for the signal, normalized background and background subtracted signal respectively. We find in the signal region ($2.000 - 2.025$ GeV) 347 events, out of which 250 are identified (after background subtraction) as signal. The reconstruction efficiency for this set of cuts was 1.45%.

Comparing the number of events *produced* in the mode $D^{*+} \rightarrow D^0\pi^+$, $D^0 \rightarrow K^-e^+\nu_e$ (corrected for the reconstruction efficiencies and with 7% of signal subtracted to account for contribution from $K^-e^+\pi^0\nu_e$) with the number of events *produced* in the mode $D^{*+} \rightarrow D^0\pi^+$, $D^0 \rightarrow K^-\pi^+$ we have measured the ratio of decay rates $\Gamma(D^0 \rightarrow K^-e^+\nu_e)/\Gamma(D^0 \rightarrow K^-\pi^+) = 0.91 \pm 0.07(stat) \pm 0.11(syst)$. Assuming the Mark III[8] branching fraction for $D^0 \rightarrow K^-\pi^+ = 4.2 \pm 0.4 \pm 0.4\%$ we have obtained the result $BF(D^0 \rightarrow K^-e^+\nu_e) = 3.8 \pm 0.5 \pm 0.6\%$. Combining this result with the D^0 lifetime[2] measurement, $\tau_{D^0} = (4.22 \pm 0.08 \pm 0.10) \times 10^{-13}$ s, we find the semileptonic partial rate to be $\Gamma(D^0 \rightarrow K^-e^+\nu_e) = (9.1 \pm 1.1 \pm 1.2) \times 10^{10}$ s^{-1}.

Figure 4 presents the distribution of the four-momentum transfer t (or $M_{e\nu}^2$). Fitting to this distribution, assuming a single pole form of the vector form factor, we obtain for the mass of the exchanged particle, $M_{F^*} = 2.1 \pm^{0.3}_{0.2}$ GeV, consistent with the value of $M_{F^*} = 2.11$ GeV, as measured *directly* by Mark III[9]. Assuming this latter value for M_{F^*}, and after integrating the expression (1), one finds $\Gamma(K^-e^+\nu_e) = |V_{cs}|^2|f_+(0)|^2 1.818 \times 10^{11} s^{-1}$. Combining the predicted and measured values of the semileptonic partial rates we find $|V_{cs}|^2|f_+(0)|^2 = 0.50 \pm 0.07 \pm 0.08$.

With a knowledge of $|f_+(0)|$, this measurement would translate directly into a measurement of $|V_{cs}|$. At present, form factor models do not offer enough precision to regard a result of such an exercise too seriously. (Assuming[4] $|f_+(0)| = 0.76$, and a single pole form of the vector form factor, yields $|V_{cs}| = 0.93 \pm 0.06 \pm 0.08$.) Reversing the argument, we can adopt a value of

$|V_{cs}| = 0.975$ (assuming three families and imposing unitarity condition on K-M matrix) and obtain a measurement of $|f_+(0)| = 0.73 \pm 0.05 \pm 0.07$.

2.2 $D^+ \to K^-\pi^+e^-\nu_e$

The theoretical description of this mode in more complicated than in $D^0 \to K^-e^+\nu_e$ mode. Here, five formfactors are involved, instead of two. There exist several calculations[4,5,10] of their behaviour in the case of $D^+ \to K^*e\nu_e$. Models predicting the relative branching fractions for $D \to Ke\nu_e$ and $D \to K\pi e\nu_e$ are relying on assumption that the $K\pi$ system is composed entirely of the $K^*(890)$ resonance. Mark III reported recently[11] found $BF(D^+ \to K^*e\nu_e) \approx BF(D^+ \to K\pi e\nu_e)$. Finding a large contribution from the non-resonant mode raises doubts in validity of such an assumption, which is routinely made not not only in case of D mesons, but also in analogous case of semileptonic decays of B mesons.

The technique used in this study exploits a long D^+ lifetime. We have simply required a large separation between the $K\pi e$ and primary vertices, $\Delta\vec{x} \geq 12\sigma_{\vec{x}}$. The primary vertex should have at least two tracks associated with it. If any other track passed within $65\mu m$ from the secondary vertex, the event was rejected. In Figure 5, we present $K\pi$ mass distributions for events in which $K\pi e$ were originating from a secondary vertex. We find 66 events with wrong sign (dashed line) and 318 events with correct sign (solid line). A $K^*(890)$ peak is clearly visible in the latter sample. Comparing the numbers of *produced* (corrected for reconstruction efficiencies) events in $K^*e^+\nu_e$ and $K^-\pi^+\pi^+$ modes we find $BF(D^+ \to K^*e^+\nu_e)/BF(D^+ \to K^-\pi^+\pi^+) = 0.53 \pm 0.05 \pm 0.06$. (The electron radiation in the spectrometer and the radiative corrections reduce the reconstruction efficiency by factors[7] of 0.867 ± 0.044 and $0.972 \pm .009$ respectively.) Taking[8] $9.1 \pm 1.3 \pm 0.4\%$ for $BF(D^+ \to K^-\pi^+\pi^+)$ we obtain $BF(D^+ \to \bar{K}^{*0}e^+\nu_e) = 4.90 \pm 0.45 \pm 0.92\%$. The non-resonant component in the $D^+ \to K\pi e\nu$ decay mode is found to be small, $BF_{NR}(D^+ \to K^-\pi^+e^+\nu_e) = 0.38 \pm .27 \pm .20\%$.

Taking[2] D^+ lifetime, $\tau_{D^+} = (10.90 \pm 0.30 \pm 0.25) \times 10^{-13}$ s, we find a decay rate for $D^+ \to \bar{K}^{*0}e^+\nu_e$ equal to $4.48 \pm 0.41 \pm 0.84 \times 10^{10} s^{-1}$. It is an interesting result, this rate has been expected to be larger by about a factor of two. (Because of isospin invariance $\Gamma(D^0 \to K^{*-}e^+\nu_e)$ should be equal to $\Gamma(D^+ \to \bar{K}^{*0}e^+\nu_e)$, and most of the models predict $\Gamma(D \to K^*e\nu) \approx \Gamma(D \to Ke\nu)$. The ratio of our two results, $\Gamma(D^0 \to K^-e^+\nu_e)/\Gamma(D^+ \to \bar{K}^{*0}e^+\nu_e) = 2.03 \pm 0.34 \pm 0.47$.)

3. Non-strange decays of D^+ and D_s^+ mesons

Among the proposed explanations for the difference in the lifetimes of charmed mesons the most important are (a) the significant presence of the non-spectator diagrams (not available in D^+ decay) and (b) destructive interference of spectator diagrams (a process which can occur for the D^+ but not for the D^0 and D_s^+). A large contribution from $D_s^+ \to \pi^+\pi^-\pi^+$ decay mode, which can occur via the Cabibbo allowed annihilation process (but is unlikely to occur via the spectator diagram), would indicate the importance of annihilation diagrams in charmed meson decays.

We required the separation between the primary and the secondary vertex, $\Delta z > 15\ \sigma_z$, where σ_z is the error in Δz. The secondary vertex should have three and only three tracks pointing to it. Also, the secondary vertex was required not to share tracks with any vertex candidate upstream of it. A $\pi^+\pi^-\pi^+$ mass plot is shown in Figure 6. There are clear peaks for both D^+ and D_s^+. A maximum likelihood fit, shown in the figure, gives 82.6 ± 15.3 D^+ events and 68.1 ± 12.4 D_s^+ events. To determine the contribution due to $\rho\pi^+$ and $f_0\pi^+$ a maximum likelihood fit was performed. Projections of Dalitz plots, for D^+ and D_s^+ events, are shown in Figure 7. There is evidence for $D^+ \to \rho\pi^+$ and for $D_s^+ \to f_0(975)\pi^+$. Results for relative branching fractions are shown in Table 1. We find $D_s^+ \to \rho\pi^+$ and $D_s^+ \to (\pi^+\pi^-\pi^+)_{NR}$, which can decay via the annihilation process, small. The ratio of $BF_{NR}(D_s^+ \to \pi^+\pi^-\pi^+)/BF(D_s^+ \to K^-K^+\pi^+)$[12], which compares the rate due to annihilation to that due to the spectator diagrams, is found to be 0.14 ± 0.04. The suppression of the mode $D^+ \to \pi^+\pi^-\pi^+$, compared to the decay $D^+ \to K^+K^-\pi^+$, can be interpreted as evidence for the destructive interference of spectator diagrams.

4. Rare decays of D mesons

Neutral current interactions are known from studies of K meson decays to conserve flavour to a high precision. Although flavour changing neutral currents (FCNC) are forbidden in the Weinberg-Salam model, many extensions of the Standard Model allow such processes. In general, limits for the K decays are $\sim 10^4$ times more stringent than for the corresponding decays of charmed particles. However, there exist models which avoid the K limits by introducing new FCNC interactions. By intruducing different couplings to the "up-type" quarks (u,c,t) and to the "down-like" (d,s,b) quarks, the flavour non-conserving decays of charmed particles can be enhanced with respect to the corresponding decays of strange particles[13].

We present upper limits on FCNC-mediated decays of charmed particles, namely $D^0 \to e^+e^-$ and $D^0 \to \mu e$. In the $D^0 \to e^+e^-$ mode we find, with a 3 σ vertex separation cut, 0.0 ± 2.6 events. After all the corrections we obtain a $(90\% C.L)$ limit $BF(D^0 \to e^+e^-) < 5 \times 10^{-5}$. The fit to the μ^+e^- mass distribution (a vertex separation cut of 6 σ was used here) gives 0.0 ± 2.8 events. After correcting for reconstruction efficiencies, we set a limit $BF(D^0 \to \mu^+e^-) < 5 \times 10^{-5}(90\% C.L.)$.

4. Future plans

A study of $D^0 \to \pi^-e^+\nu_e$ is under way. Combined with analysis of $K^-e^+\nu_e$ mode, presented above, it would yield a measurement of the $|V_{cd}|/|V_{cs}|$. Also, we are trying to measure the semileptonic rate for decay $D^0 \to K^*e\nu$, $K^* \to K^0\pi^-$, allowing us to measure $\Gamma(D^0 \to Ke\nu)/\Gamma(D^0 \to K^*e\nu)$.

REFERENCES

1. The members of The Tagged Photon Collaboration are : J.C. Anjos, J.A. Appel, S.B. Bracker, T.E. Browder, L.M. Cremaldi, J.R. Elliott, C.O. Escobar, P. Estabrooks, M.C. Gibney, G.F. Hartner, P.E. Karchin, B.R. Kumar, M.J. Losty, G.J. Luste, P.M. Mantsch, J.F. Martin, S. McHugh, S.R. Menary, R.J. Morrison, T. Nash, U. Nauenberg, P. Ong, J. Pinfold, G. Punkar,

M.V. Purohit,J.R. Raab, A.F.S. Santoro, J.S. Sidhu, K. Sliwa, M.D. Sokoloff, M.H.G. Souza, W.J. Spalding, M.E. Streetman, A.B. Stundžia, M.S. Witherell;

2. J. Raab *et al.*, Tagged Photon Collaboration, Phys. Rev. **D37** (1988) 2391.
3. Nash, T., *et al.*, Proc. of "Computing in High Energy Physics", Amsterdam 1985.
4. M. Wirbel, B. Stech and M. Bauer, Z. Phys. **C29**, 637 (1985);
5. B. Grinstein, M.B. Wise, Phys. Lett **B197**, 249 (1987); and references therein.
6. It has been checked with Monte Carlo events that this procedure does not introduce any serious biases. It removes the ambiguity problem in an experimentally clean way, making the subsequent interpretation of results straightforward.
7. The radiative corrections have been evaluated using the modified calculation of K_{e3} decay, see: M. Dremin, JETP 35 (1959) 355; E. S. Ginsberg, Phys. Rev. 142 (1966) 1035; E. S. Ginsberg, Phys. Rev. 162 (1967) 1570.
8. J. Adler *et al*, Mark III Collaboration, Phys. Rev. Lett. **60** (1988) 89.
9. R. Schindler, Mark III Collaboration, SLAC-PUB-4055, September 1986.
10. Suzuki, M., Phys. Lett. **155B**, 112 (1985), Ali, A. and T.C. Yang, Phys. Lett. **65B**, 275 (1976).
11. Coffman, D., Ph. D. Thesis, California Institute of Technology, CALT-68-1415 (1987).
12. J. Anjos *et al.*, Tagged Photon Collaboration, Phys. Rev. Lett. **60** (1988) 10.
13. A. Hadeed and B. Holdom, Phys. Lett. **159B** (1985) 379; W. Buchmuller, CERN-TH-4499/86, July 1986.

TABLE I. Relative branching ratios (B) for D_s^+ and $D^+ \to$ non-strange states. Errors are given in the form: ±statistical ±systematic.

Decay mode 1	Decay mode 2	B(decay mode1)/B(decay mode2)
$D_s^+ \to \pi^-\pi^+\pi^+$	$D_s^+ \to \phi\pi^+$	.44±.10±.04
$D_s^+ \to (\pi^-\pi^+\pi^+)_{NR}$	$D_s^+ \to \phi\pi^+$	.29±.09±.03
$D_s^+ \to \rho^0\pi^+$	$D_s^+ \to \phi\pi^+$	<. 08 (90% c.l.)
$D_s^+ \to f_0\pi^+$	$D_s^+ \to \phi\pi^+$	.28±.10±.03
$D^+ \to \pi^-\pi^+\pi^+$	$D^+ \to K^-\pi^+\pi^+$	.035±.007±.003
$D^+ \to (\pi^-\pi^+\pi^+)_{NR}$	$D^+ \to K^-\pi^+\pi^+$	.027±.007±.002
$D^+ \to \rho^0\pi^+$	$D^+ \to K^-\pi^+\pi^+$	.008± .005± .001
$D_s^+ \to \pi^-\pi^+\pi^-\pi^+\pi^+$	$D_s^+ \to \phi\pi^+$	< 0.29 (90% c.l.)
$D^+ \to \pi^-\pi^+\pi^-\pi^+\pi^+$	$D^+ \to K^-\pi^+\pi^+$	< 0.019 (90% c.l.)

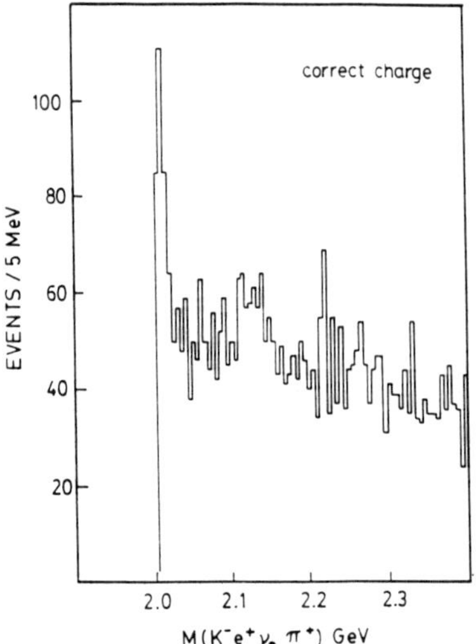

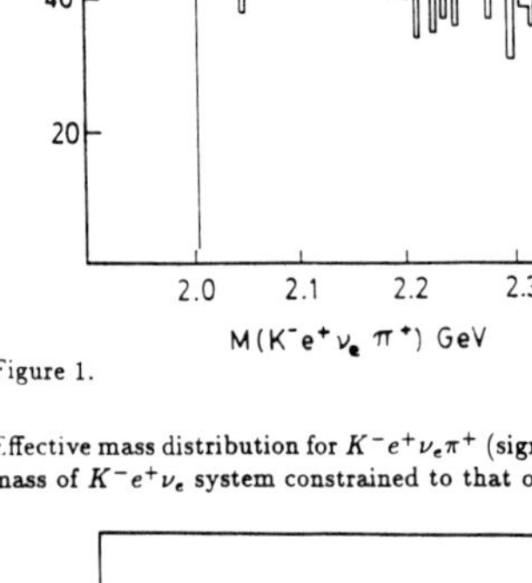

Figure 1.

Effective mass distribution for $K^-e^+\nu_e\pi^+$ (signal) combinations, mass of $K^-e^+\nu_e$ system constrained to that of a D^0.

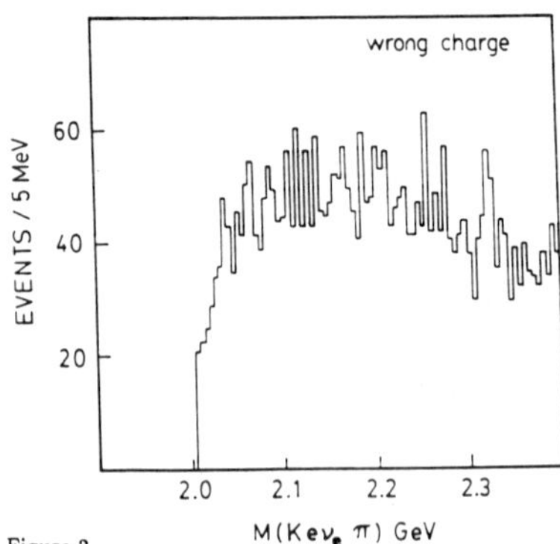

Figure 2.

Effective mass distribution for $K^+e^+\nu_e\pi^+$, $K^-e^-\nu_e\pi^+$ and $K^+e^-\nu_e\pi^+$ combinations (background), normalized to the integral over the mass interval 2.03 – 2.40GeV of the correct sign (signal) distribution.

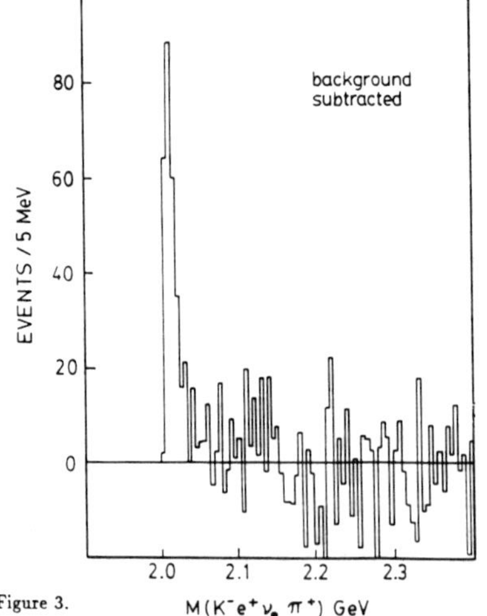

Figure 3.

Background subtracted (see above) effective mass distribution for $K^-e^+\nu_e\pi^+$ (signal) combinations.

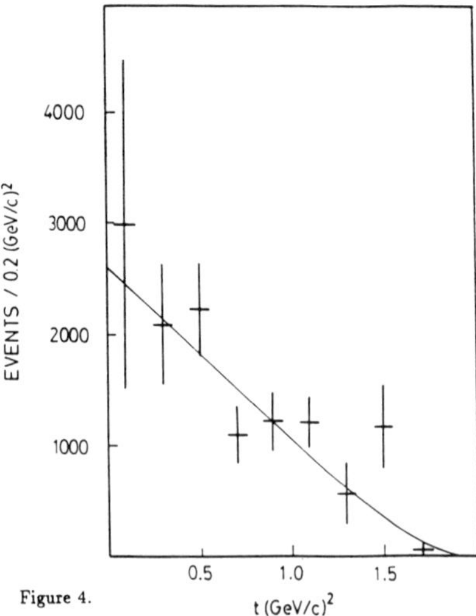

Figure 4.

Four-momentum transfer from D to K (or $M^2_{e\nu}$) distribution. The superimposed curve is the result of a fit to a t-distribution expected, after integration over phase space, from the assumed single pole form for the vector formfactor. The fit yields a value of $M_{F_*} = 2.1 \pm^{0.3}_{0.2}$ GeV.

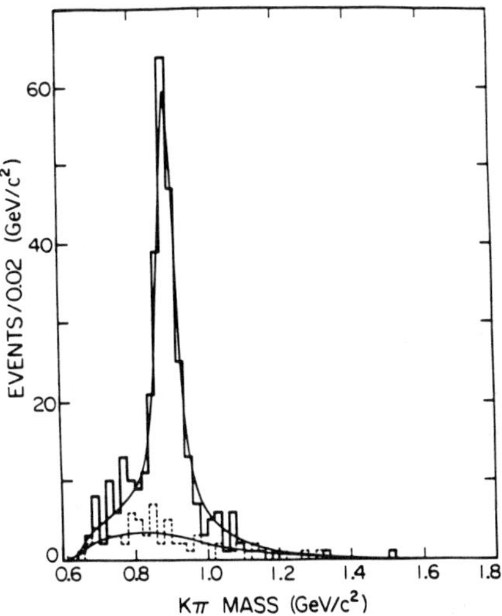

Figure 5.

Effective mass distributions of $K\pi$ system for events with wrong $K\pi e$ (dashed line) and correct $K\pi e$ charge combinations (solid line).

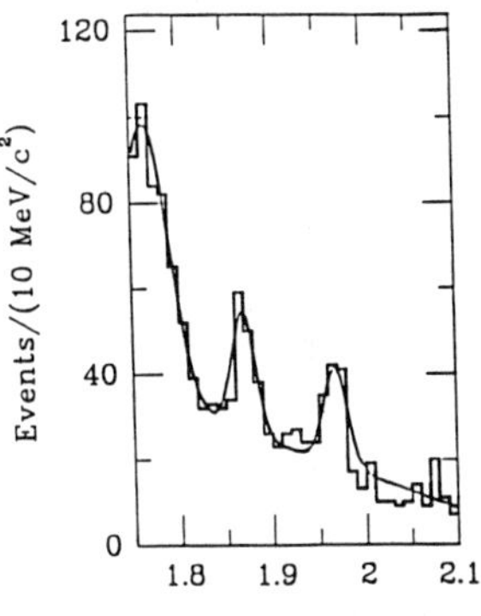

Figure 6.

Effective mass distribution of $\pi^+\pi^-\pi^+$ system.

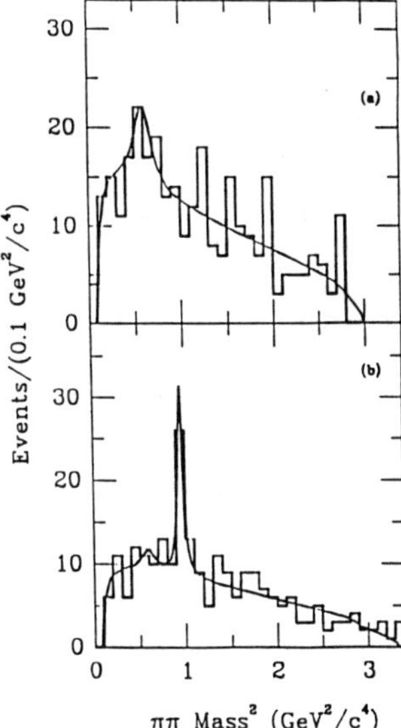

Figure 7.

Sum of Dalitz plot projections onto $\pi^-\pi_1^+$ mass squared axis and $\pi^-\pi_2^+$ mass squared axis for a). D^+ candidates ($1.854\ GeV < M_{3\pi} < 1.884\ GeV$) and b). D_s^+ candidates ($1.953\ GeV < M_{3\pi} < 1.983\ GeV$).

Experimental studies of CP violation in K meson decays through measurements of $\varepsilon' / \varepsilon$

George D. Gollin

Joseph Henry Laboratories, Princeton University
Princeton, New Jersey 08544

(presented at the 3rd Conference on the Intersections between Particle and Nuclear Physics, Rockport, ME, May, 1988)

ABSTRACT

Experimental evidence today is consistent with the dominant source of CP violation arising from a $\Delta S=2$ interaction which mixes K^0 and $\bar{K}^0$ in a time-asymmetric fashion. Fermilab experiment 731 and CERN experiment NA31 are performing precision measurements of the ratio $|\eta_{00} / \eta_{+-}|$ in hopes of shedding new light on the origins of CP violation. These two experiments are discussed.

I. INTRODUCTION

Experiments performed over the twenty-five years since the discovery of CP violation have failed to find evidence for any CP noninvariance outside of the K^0- $\bar{K}^0$ system. CP nonconservation in $K_L \to \pi\pi$ decays and in the charge asymmetry for $K_L \to \pi\mu\upsilon$ and $K_L \to \pi e\upsilon$ can be accounted for largely by a T-violating imbalance in the $\Delta S=2$ transitions $K^0 \leftrightarrow \bar{K}^0$. Gauge theories of the weak interactions suggest that a $\Delta S=1$ contribution to CP violation also should exist. In $K \to \pi\pi$ decay, this contribution arises from interference between the $\Delta I=1/2$ and $\Delta I=3/2$ K^0, $\bar{K}^0 \to \pi\pi$ decay amplitudes and is expected to produce a measurable difference between the ratios $\Gamma(K_L \to \pi^0\pi^0) / \Gamma(K_S \to \pi^0\pi^0)$ and $\Gamma(K_L \to \pi^+\pi^-) / \Gamma(K_S \to \pi^+\pi^-)$.[1] Defining $\eta \equiv$ Amplitude (CP violating K→pions) / Amplitude (CP conserving K→pions), one has $\eta_{00} = \varepsilon - 2\varepsilon'$ and $\eta_{+-} = \varepsilon + 2\varepsilon'$, or $\eta_{00} / \eta_{+-} \approx 1 - 3\varepsilon' / \varepsilon$. The ratio $\varepsilon' / \varepsilon$ is indicative of the relative amounts of $\Delta S=1$ and $\Delta S=2$ CP violation in K_L decay. Neither $\Delta S=1$ nor $\Delta S=2$ amplitudes are expected to be capable of splitting the phases of η_{00} and η_{+-} in the complex plane by more than a fraction of a degree. Generally, a non-negligible phase difference can only be caused by a CPT-violating interference between the K^0, $\bar{K}^0 \to \pi\pi$ decay amplitudes.[2]

Calculations using the Standard Model tend to predict $\varepsilon' / \varepsilon \approx .005$ while those based on models with CP violation in the Higgs sector yield larger values. There is still a fair amount of uncertainty in predictions of a lower bound on $|\varepsilon'/\varepsilon|$.[1] The goal of Fermilab experiment 731 and CERN experiment NA31 is to measure ε'/ε to an accuracy of $\pm .001$. E731 is currently analyzing a data set which contains approximately 250 000 $K_L \to \pi^0\pi^0$ decays and has just published results from an earlier test run;[3] NA31 has recently submitted for publication an $\varepsilon' / \varepsilon$ result based on approximately 100 000 $K_L \to \pi^0\pi^0$ events.[4]

II. ε'/ε EXPERIMENTS: FERMILAB E731 and CERN NA31

A. Experimental considerations

Since $\Gamma(K_L\to\pi\pi)\,/\,\Gamma(K_S\to\pi\pi) = |\eta|^2$ and since the phases of the complex numbers ε' and ε are approximately equal, one may write $\dfrac{\Gamma(K_L\to\pi^0\pi^0)}{\Gamma(K_S\to\pi^0\pi^0)} \Big/ \dfrac{\Gamma(K_L\to\pi^+\pi^-)}{\Gamma(K_S\to\pi^+\pi^-)} \approx 1 - 6\,\dfrac{\varepsilon'}{\varepsilon}$.

A determination of the four $K\to\pi\pi$ rates yields a value for ε'/ε. Because experiments generally record substantially more $K_S\to\pi\pi$ and $K_L\to\pi^+\pi^-$ than $K_L\to\pi^0\pi^0$, the statistical error on a measurement of ε'/ε is dominated by the number of $K_L\to\pi^0\pi^0$ decays obtained. The level of precision needed to test current ideas about CP violation requires experiments to measure ratios of kaon decay rates to an accuracy of better than one percent. As a result, detectors must be designed to minimize systematic uncertainties as well as to record large numbers of $K\to\pi\pi$ events. Some possible sources of systematic error are the following:

1. Imperfect knowledge of K_L and K_S fluxes.
2. Different acceptances for K_L and K_S decays.
3. Different reconstruction efficiencies for $\pi^0\pi^0$ and $\pi^+\pi^-$ final states.
4. Contamination from K_S produced by inelastic interactions in a regenerator or in beam-defining collimators.
5. Changes in apparatus resolution and calibration during data taking.
6. Background subtractions.
7. Accidentals influencing detection of the four final states $\{K_L, K_S\} \times \{\pi^0\pi^0, \pi^+\pi^-\}$ differently.

These effects can be quite subtle, and must be controlled or understood to great accuracy.

The E731 group produced a pair of vertically-separated K_L beams of approximately 10^7 K_L per twenty second spill by striking a single 30 cm beryllium target with 10^{13} 800 GeV protons per pulse. Fermilab delivered a spill to the experimental areas once per minute. After passage through absorbers, the 4.5 mrad production angle beam contained about as many neutrons as kaons. K_S were produced through coherent regeneration by passing one of the neutral beams through additional absorber near the target and then through a two-interaction-length boron-carbide regenerator at the start of the spectrometer's fiducial decay volume. The regenerator and extra absorber switched beams after each spill. Decays of K_L and K_S to one final state, either $\pi^0\pi^0$ or $\pi^+\pi^-$, were recorded simultaneously. The experiment switched between $\pi^0\pi^0$ mode and $\pi^+\pi^-$ mode several times during data taking.[5]

The NA31 group generated a single beam, either K_L or K_S, by passing a 450 GeV primary proton beam through one of two 40 cm beryllium targets. The upstream (K_L) target was struck by 10^{11} protons per 2.8 second machine pulse, yielding 1.9×10^6 kaons and 1.5×10^7 neutrons after collimation in a 3.6 mrad production angle beam. CERN delivered a proton spill every 14.4 seconds. The downstream (K_S) target was struck by 2×10^7 protons at the same targeting angle, generating about 3800 kaons and 30 000 neutrons per spill. The K_S target was mounted on rails and could be moved through a distance of 48 meters, allowing the region of kaon proper time sampled by the NA31 decay volume to be changed. NA31 recorded both $\pi^0\pi^0$ and $\pi^+\pi^-$decays simultaneously, switching between K_L and K_S running periodically.[6]

B. Fermilab E731

The Chicago-Elmhurst-Fermilab-Princeton-Saclay[7] detector is shown below in Figure 1. A pair of neutral beams enters from the left in the figure. K_L decays from the unobstructed beam

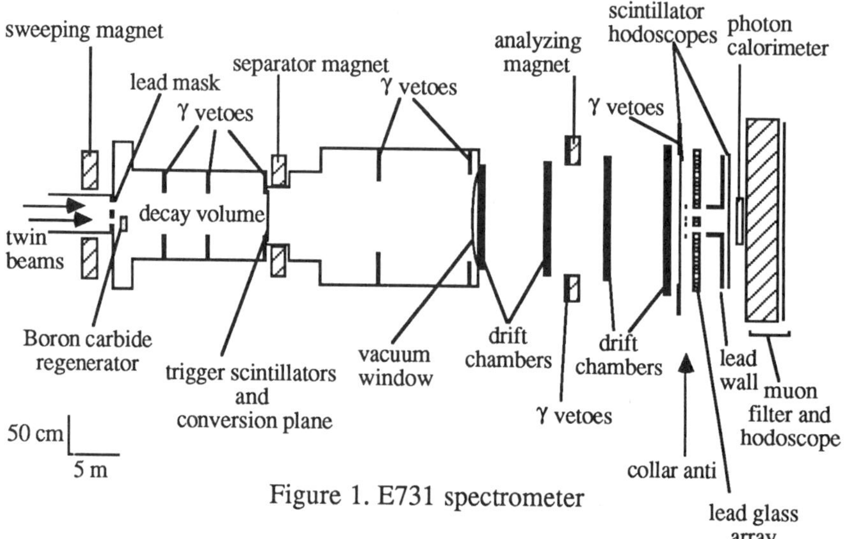

Figure 1. E731 spectrometer

were detected by the drift chamber spectrometer and 804-block lead glass array. Most decays in the beam passing through the boron-carbide regenerator were $K_S \rightarrow \pi\pi$; the regenerator was instrumented with four planes of scintillator to permit rejection of inelastic interactions which might yield K_S. The regenerator switched beams after every machine pulse. The detector simultaneously recorded K_L and K_S decays to one of the two relevant final states. Changing from $\pi^0\pi^0$ to $\pi^+\pi^-$ mode required the removal or insertion of a thin lead sheet between the scintillators of the conversion plane.

The charged ($\pi^+\pi^-$) mode trigger required a signal at the conversion plane and signals corresponding to two charged pions in downstream scintillator hodoscopes. Events are reconstructed by calculating the invariant mass, total energy, and momentum for pairs of tracks. The charged mode mass resolution is 3.5 MeV/c^2. Coherently regenerated K_S are selected by requiring the apparent change in p_T^2, the square of the kaon's transverse momentum after passing through the regenerator, to be less than 250 (MeV/c)^2. The major source of background to K_L decays in the charged mode comes from $K_L \rightarrow \pi\mu\nu$ decays where the muon stopped in the iron muon filter before reaching the downstream scintillator hodoscope.

The neutral mode ($\pi^0\pi^0$) trigger used during the 1985 test run required one photon to convert in a 0.1 radiation length lead converter at the end of the decay volume. The currents in the separator and analyzing magnets were chosen to split the electron-positron pair apart, then reconverge them at the lead glass. The trigger also required at least 30 GeV to be deposited in the lead glass array. A processor (the "mass box") performed a coarse invariant mass calculation based on the pattern of energy deposition in the lead glass, vetoing about half the events satisfying the rest of the trigger. The trigger for the 1987-88 data run allowed decays with or without a conversion in the lead sheet to be recorded. The minimum energy required in the glass was lowered to 25 GeV for this run. In addition, the mass box was replaced by a hardware cluster finder which counted the number of photons striking the lead glass. Information from the glass array is used offline to determine the event's mass and the distance from decay vertex to lead glass. Extrapolation of the e^+/e^- tracks back to this point yields the transverse position of the vertex for events in which a gamma converted. For these events, p_T is determined by measuring the angle between the line from the target to the regenerator and the line from the regenerator through the decay point to the center-of-energy in the lead glass. Events in the test run's data were required to have p_T^2 less than 4000 $(MeV/c)^2$. Events without a conversion in the 1987-88 data set will be selected or rejected based on the location of the lead glass center-of-energy relative to the beam profile. Mass resolution in the neutral mode is 6.1 MeV/c^2 and is dominated by the lead glass' resolution $\sigma/E \approx (1.5 + 5E^{-1/2})\%$. The dominant source of background to $K_L \to \pi^0\pi^0$ events is the copious $K_L \to 3\pi^0$ decay mode. Ten planes of anticoincidence counters with sensitivity above about 100 MeV were employed to recognize photons from $3\pi^0$ decays which would miss the lead glass.

The ratio of the numbers of reconstructed $\pi\pi$ decays in the K_L and regenerated beams, as a function of proper time and kaon energy, is

$$\frac{N(K_L \to \pi\pi)}{N(K_S \to \pi\pi)} \approx \frac{|\eta\, e^{-\Gamma_L \tau/2}|^2 \times K_L \text{ flux} \times \text{acceptance}}{A\,|\rho\, e^{-\Gamma_S \tau/2} + \eta\, e^{-\Gamma_L \tau/2}| \times K_L \text{ flux} \times \text{acceptance}} . \qquad (1)$$

η is $\varepsilon - 2\varepsilon'$ for $\pi^0\pi^0$ decays and $\varepsilon + \varepsilon'$ for $\pi^+\pi^-$ decays. Γ_L and Γ_S are K_L and K_S decay rates, respectively, and τ is the proper time for the kaon, with τ=0 corresponding to the downstream end of the regenerator. ρ is the regeneration amplitude, typically ten times greater than η, and A is an attenuation factor due to scattering in the regenerator and upstream absorber. The twin beam technique makes the ratio insensitive to the K_L flux; terms involving A and ρ are eliminated by comparing the ratios for charged and neutral mode data. Hardware failures and inefficiencies caused by noise, halo muons, and accidentals interfere nearly equally with K_L and K_S because decays from both beams to one particular final state are registered simultaneously.

Since the detector's fiducial decay volume is about 3 τ_S but only .005 τ_L long, the spatial distributions of K_S and K_L decays differ appreciably. The E731 group uses a Monte Carlo calculation to correct for minor acceptance variations between the up- and downstream regions of the decay volume, permitting all K_L decays (including those upstream of the regenerator in the unobstructed beam) to contribute to the numerator of the above ratio. The copious $K_L \to 3\pi^0$ signal provides a check on neutral mode Monte Carlo assumptions while semileptonic decays and $K_L \to \pi^+\pi^-\pi^0$ can be used to check the charged mode Monte Carlo. Since the mean decay position for K_S differs from that for K_L by 2-3 meters, and since the detector's acceptance changes by about a per cent per meter, the Monte Carlo acceptance correction is not large. Figure 2 shows the mass distribution for $K_L \to \pi^0\pi^0$ decays with a superimposed background distribution from a $K_L \to 3\pi^0$ Monte Carlo calculation for neutral mode data from the 1985 test run; the background is approximately 1.5% and is well understood. Figure 3 shows reconstructed and simulated decay vertex distributions; signal and background levels from all modes in the test run are listed below:

Mode	Events after subtractions	Background (%)	Improvement factor over Ref. 10	Systematic uncertainty (%)
$K_L \to \pi^0\pi^0$	6747	1.56	5.4	0.30
$K_S \to \pi^0\pi^0$	21788	2.90	5.0	0.20
$K_L \to \pi^+\pi^-$	35838	1.23	2.5	0.18
$K_S \to \pi^+\pi^-$	130025	0.30	5.7	0.03

If η_{+-} and η_{00} were known, equation (1) would allow the extraction of $\rho(p_K)$, the regeneration amplitude as a function of kaon momentum, from either neutral or charged mode data samples. The E731 analysis extracts ε'/ε by determining the value of ε' which gives the best fit when data from both modes are used to measure $\rho(p_K)$. Important sources of systematic error on ε'/ε include uncertainties in the Monte Carlo correction for the K_L-K_S acceptance difference and background subtractions. Other effects, such as uncertainties in the relative energy scales for neutral and charged final states and uncertainties in the effects of accidentals are thought to be unimportant.

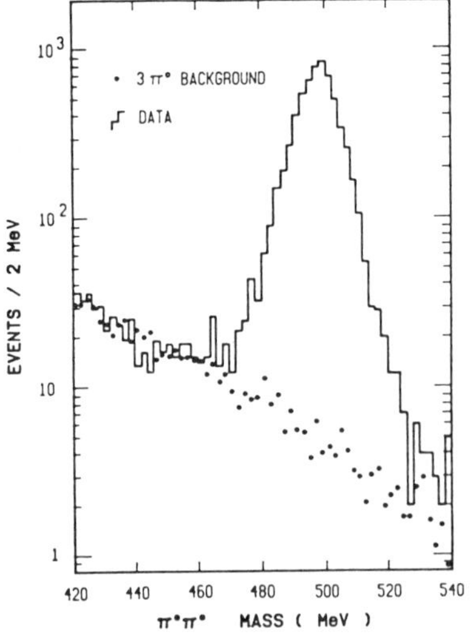

FIG. 2. Reconstructed $\pi^0\pi^0$ invariant mass for K_L. The $3\pi^0$ background shape determined from Monte Carlo is superimposed.

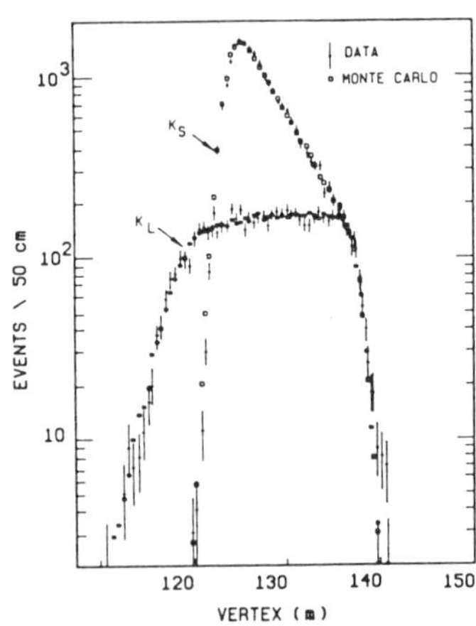

FIG. 3. Reconstructed decay-vertex distribution for $K_{L,S} \to 2\pi^0$, data and Monte Carlo simulation.

The result from E731's test run is that $\varepsilon'/\varepsilon = 0.0032 \pm 0.0028$ (statistical) $\pm\, 0.0012$ (systematic). Approximately 40 times more data were recorded in the 1987-88 run than in the test run;

sensitivity from this much larger data set is expected to be better than ± 0.001 including both statistical and systematic errors.

C. CERN NA31

The CERN-Dortmund-Edinburgh-Mainz-Orsay-Pisa-Siegen[8] detector is shown below in Figure 4.

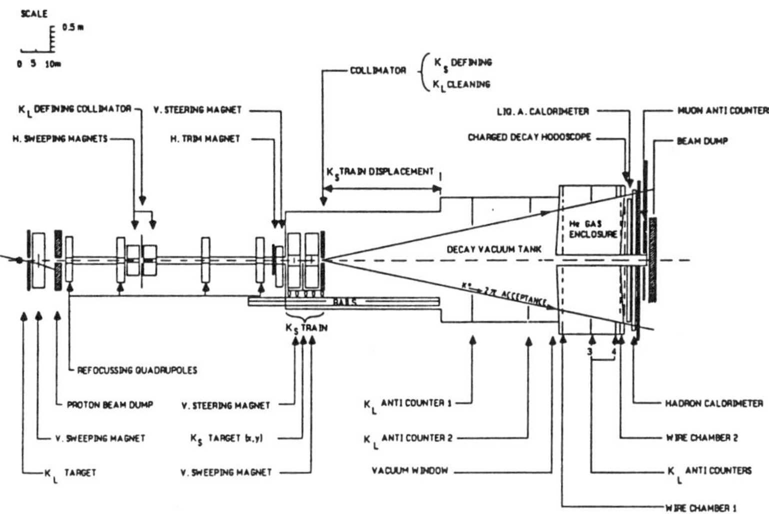

Figure 4. NA31 detector

A single primary proton beam enters from the left, striking either a K_L target far from the spectrometer or a moveable K_S target in vacuum. Kaon decays were detected in a pair of drift chambers (each with four sense wire planes), a liquid argon electromagnetic calorimeter, and an iron-scintillator hadron calorimeter. No magnet is used in the measurement of charged pion momenta. The detector recorded simultaneously decays to charged and neutral pions from either K_L or K_S; the NA31 group frequently switched between K_L and K_S running. By moving the K_S target along the beam direction, K_L and K_S data samples with similar decay distributions along the beam direction are obtained.

The charged mode trigger required hits in the first drift chamber and in a scintillator hodoscope between the second drift chamber and the liquid argon calorimeter. Analog sums of signals from both the liquid argon and hadron calorimeters were required to be above thresholds to satisfy the charged mode trigger. To reject K_{e3} decays, no track was allowed to deposit more than 80% of its energy in the liquid argon calorimeter. About 2% of the good $\pi\pi$ final states were rejected by this K_{e3} veto. Since the spectrometer has no analyzing magnet, the total energy of the $\pi\pi$ system is calculated using the opening angle between pion tracks, the ratio of the pions' calorimeter signals, and the known kaon mass. To avoid contamination from $\Lambda \rightarrow p\pi$ decays, the ratio of pion signals is required to be less than 2.5. This cut eliminates most Λ decays, but loses about 50% of the good $\pi\pi$ decays. The hadron calorimeter's energy resolution is $(65 \times E^{-1/2})\%$; mass resolution in the charged mode is between 20% and 25%, depending on kaon energy. A distribution of $\pi^+\pi^-$ masses is shown for K_L decays in Figure 5. The peak centered at 360 MeV/c^2 is from $\pi^+\pi^-\pi^0$ decays where photons from the π^0 were not registered

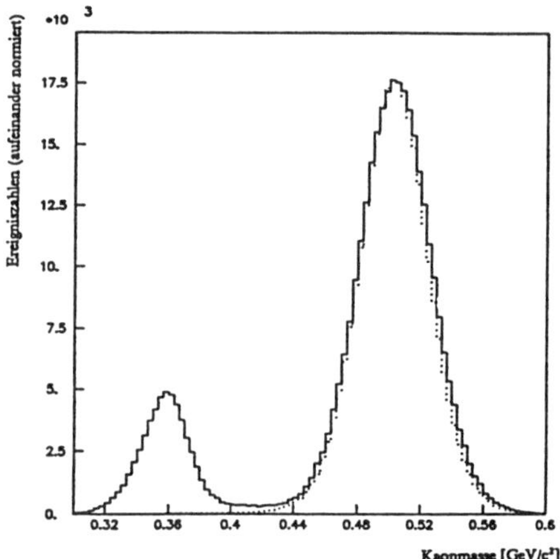

Figure 5. NA31 $\pi^+\pi^-$ mass distribution

by the NA31 veto counters. The background consists mainly of $K_L \rightarrow \pi e \upsilon$ decays.

The neutral mode trigger demanded signals from a scintillator hodoscope submerged in the liquid argon shower calorimeter. A minimum energy threshold for liquid argon signals and a maximum energy limit for hadron calorimeter signals was imposed at the trigger level to select electromagnetic showers. A neutron accompanying a $\pi^0\pi^0$ event could veto a good K decay. $K_L \rightarrow 3\pi^0$ events are rejected at the trigger level by requiring that neither x nor y projections of the liquid argon signals contain more than four isolated depositions of energy, or "clusters". Events are reconstructed by chosing the best of three possible pairings for the four photons. The distance from the kaon decay point to the liquid argon is calculated by assuming that a kaon decayed to the four observed photons. The invariant masses of the two "pions" are calculated and plotted in a scatter plot of "pion" mass vs. "pion" mass. Good $\pi\pi$ decay events appear in a well defined peak; the background, primarily from $3\pi^0$ decays, is flat and must be subtracted from the volume under the peak. The mass resolution for a π^0 is about 2 MeV/c^2. The π-π mass scatter plot for K_L decays is shown in Figure 6. Anticoincidence planes, sensitive above about 500 MeV, were used to veto $3\pi^0$ events with wide angle photons.

Since NA31 recorded decays to two different final states simultaneously, with different vetoes acting on the two triggers, accidentals in veto counters could influence the detected rates of neutral mode decays differently from charged mode decays. In addition, changing calibrations and efficiencies could influence $\pi^0\pi^0$ events differently from $\pi^+\pi^-$ events in a fashion that would not cancel when comparing K_L and K_S running. Consequently, apparatus performance and running conditions had to be carefully monitored and controlled by the NA31 group. Running with different K_S target positions produced similar, but not identical decay distributions for K_L and K_S. As a result, a Monte Carlo correction of 1% was made by the NA31 group to compensate for effects related to the differences in the two distributions.

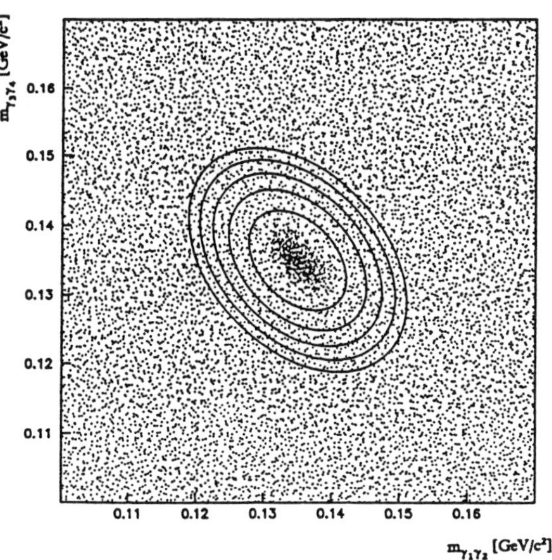

Figure 6. NA31 π^0vs π^0 mass distribution

Signal and background levels for the result presented in reference 4 follow:

Mode	Events	Background (%)	Systematic uncertainty (%)
$K_L \to \pi^0\pi^0$	109 000	4.0	0.2
$K_S \to \pi^0\pi^0$	932 000	< 0.1	
$K_L \to \pi^+\pi^-$	295 000	0.6	0.2
$K_S \to \pi^+\pi^-$	2 300 000	< 0.1	

Important sources of systematic error include a possible energy scale difference between charged and neutral decays, the background subtractions in the K_L beam, and the net loss of events due to accidentals. The final result is that $\varepsilon'/\varepsilon = 0.0032 \pm 0.0007$ (statistical) ± 0.0008 (systematic). The NA31 group plans to take more data with comparable statistical power in hopes of improving systematic errors.

III. CONCLUSION

An accurate determination of ε'/ε is one of the most important high energy measurements being performed today. The required level of precision is greater than that usually obtained by a high energy physics experiment; many subtle systematic errors must be understood before results can be considered reliable. E731 and NA31 are performing the same measurement with rather different methods and consequently must cope with vastly different systematic uncertainties. Confidence in the acceptance correction used by E731 to compare the K_S and K_L samples ultimately may be that experiment's dominant source of systematic uncertainty. The ability to understand (and compensate for) changing efficiencies and accidental rates may be NA31's dominant systematic error. It is fortunate that these two very different experiments are pursuing the same physics goal. Preliminary indications are that ε'/ε is different from zero, and in rough agreement with the predictions of the Standard Model, but the issue is far from settled.

REFERENCES

1. See L. Wolfenstein, Ann. Rev. Nucl. Part. Sci. **36**, 137 (1986) for a discussion of the predictions.

2. V.V. Barmin *et al.*, Nucl Phys. **B247**, 293 (1984).

3. M. Woods *et al.*, Phys. Rev. Lett. **60**, 1695 (1988) and M. Woods, Ph.D. thesis, University of Chicago, 1988 (unpublished).

4. H. Burkhardt *et al.*, CERN-EP/88-47 (1988) (submitted to Physics Letters B).

5. See reference 3 and G.L. Grazer, Ph.D. thesis, Princeton University DOE/ER/3072-44, 1988 (unpublished).

6. H. Burkhardt *et al.*, CERN-EP/87-166 (1987) (submitted to Nucl. Instrum. Methods in Phys. Research).

7. The E731 group consists of L. Gibbons, K. Nishikawa, V. Papadimitriou, R. Patterson, Y. Wah, B. Winstein, R. Winston, M. Woods, and H. Yamamoto from the University of Chicago; E. Swallow from Elmhurst College; G. Bock, R. Coleman, Y. Hsiung, K. Stanfield, R. Stefanski, and T. Yamanaka from Fermilab; G. Blair, G. Gollin, G. Grazer, M. Karlsson, and J. Okamitsu from Princeton University; J. Brisson, P. Cabeza, R. Daudin, P. Debu, P. Jarry, B. Peyaud, R. Turlay, and B. Vallage from CEN-Saclay.

8. The NA31 group consists of H. Burkhardt, P. Clarke, D. Coward, D. Cundy, N. Doble, L Gatignon, V. Gibson, R. Hagelberg, G. Kessler, J. van der Lans, I. Mannelli, T. Miczaika, A.C. Schaffer, P. Steffen, J. Steinberger, H. Taureg, H. Wahl, and C. Youngman from CERN; G. Dietrich and W. Heinen from Univ. Dortmund; R. Black, D.J. Candlin, J. Muir, K.J. Peach, B. Pijlgroms, I.P. Shipsey, and W. Stephenson from Univ. Edinburgh; H. Blümer, M. Kasemann, K. Kleinknecht, B. Panzer, and B. Renk from J. Gutenberg Univ., Mainz; E. Auge, R.L. Chase, M. Corti, D. Fournier, P. Heusse, and A.M. Lutz from Univ. Paris Sud, Orsay; A. Bigi, M. Calvetti, R. Carosi, R. Casali, C. Cerri, R. Fantechi, S. Galeotti, E. Massa, A. Nappi, D. Passuello, and G. Pierazzini from INFN, Pisa; C. Becker, D. Heyland, M. Holder, G. Quast, M. Rost, W. Weihs, and G. Zech from Siegen.

9. M. Kasemann, Ph. D. thesis, Johannes Gutenberg-Universität in Mainz, 1987 (unpublished) and M. Kasemann, private communication.

10. R.H. Bernstein *et al.*, Phys. Rev. Lett. **54**, 1631 (1985).

SEARCH FOR MUONIUM TO ANTIMUONIUM CONVERSION: OVERVIEW AND RECENT EXPERIMENTAL RESULTS

H. Reiner Schaefer
Yale University, New Haven, CT 06511

ABSTRACT

The motivation for an experimental search for muonium to antimuonium conversion is briefly discussed. Results of measurements on limits for the strength of the coupling constant involved in this interaction are summarized with emphasis on the recently completed search, LAMPF E985. Plans are presented for an upcoming next generation experiment, LAMPF E1073, which aims at an improvement in the sensitivity by more than two orders of magnitude.

INTRODUCTION

The possibility of the spontaneous conversion of a muonium atom, ($\mu^+ e^-$ or M) to its antiatom, antimuonium ($\mu^- e^+$ or $\overline{M}$) is very much analogous to the (K^0, $\overline{K^0}$) system, as was first pointed out by Pontecorvo[1].Two different neutral degenerate particles are coupled by the weak interaction and hence become mixed.The latter system has been the subject of discourse by many speakers at this conference, therefore I will discuss the conversion process in a more general context of muon physics, namely the framework of conservation laws. Since its discovery the muon played a vital role in efforts to illuminate principles of fundamental interactions among elementary particles. At the center of interest stood its decay and attempts to formulate rules which explain the observed decay mode or respectively the absence of others which are energetically and in any other known respect possible yet unobserved. From the variety of laws proposed since the early 1950's the one surviving the discovery of the muon-neutrino, the introduction of the electro-weak unified theory[2] and all experimental evidence to date is the postulate of an additively conserved quantum number for each type of lepton involved in the interaction, see table I.

Table I. Lepton Number Assignment[3]

Particle	L_e	L_μ	L	P_L
e^-,ν_e	+1	0	+1	+1
μ^-,ν_μ	0	+1	+1	-1
$e^+,\bar{\nu}_e$	-1	0	-1	+1
$\mu^+,\bar{\nu}_\mu$	0	-1	-1	-1

$\Sigma L_e, \Sigma L_\mu$ = const.; Additive Lepton # Conservation

$\Pi P_L, \Sigma L$ = const.; Multiplicative Lepton # Conservation

As all prominent candidates for rare muon decay, e. g. $\mu \to e\gamma$, $e\gamma\gamma$ and $\mu \to eee$, the muonium conversion is also forbidden by this rule. However, the current standard model represents an incomplete theory and the lepton number convention summarizes empirical evidence rather than describing a fundamental conservation law, so it is important to improve the experimental limits on a possible violation of this symmetry. Furthermore, alternative schemes[4] suggesting a multiplicative character for lepton number conservation and extensions of the standard model with left-right symmetry[5] are consistent with the muonium conversion.

THEORETICAL ASPECTS OF THE (M, $\overline{M}$) SYSTEM

In a phenomenological approach one chooses a Hamiltonian term for a four-fermion interaction of the V-A type to represent the M -> $\overline{M}$ conversion:

$$H_{M\overline{M}} = \frac{1}{\sqrt{2}} G_{M\overline{M}} \overline{\psi}_\mu \gamma_\lambda(1+\gamma_5)\psi_e \overline{\psi}_\mu \gamma^\lambda(1+\gamma_5)\psi_e + \text{h.c.} \qquad (1)$$

in which $G_{M\overline{M}}$ is a coupling constant characterizing the strength of the interaction in a model independent way. This form leads to an expression[6] for the relevant matrix elements coupling M and $\overline{M}$ states in terms of the ratio of $G_{M\overline{M}}$ to the Fermi coupling of weak interaction G_F:

$$\left\langle \overline{M} \middle| H_{M\overline{M}} \middle| M \right\rangle = \frac{\delta}{2} \approx (1 * 10^{-12}\ \text{eV}) \frac{G_{M\overline{M}}}{G_F} \qquad (2)$$

Since $H_{M\overline{M}}$, if it exists at all, is very small, the Schrödinger equation can be solved by perturbation theory and one gets the time-dependent state wave function:

$$\psi(t) = a(t)\,|M> + b(t)\,|\overline{M}> \qquad (3)$$

with the initial conditions a(0)=1 and b(0)=0 if M is formed in vacuum at time t=0. Assuming the antimuonium amplitude is allowed to develop over an observation time T, the probability per muonium atom for a transition is given by:

$$P(\overline{M},T) = \int_0^T |b(t)|^2 \exp(-t/\tau)\, \frac{dt}{\tau} \qquad (4)$$

where τ is the muon lifetime. This quantity can then be compared to the experimentally determined ratio of antimuonium to muonium atoms in order to extract the measured value of $G_{M\overline{M}}$. This can easily be converted into a value for the branching ratio,R, describing the probability of the system to decay in vacuum as $\overline{M}$ rather than M.

$$R = \frac{\delta^2}{2(\delta^2 + \Delta^2 + \lambda^2)} \qquad (5)$$

where δ is the matrix element from equation (2), Δ represents any energy splitting between M, $\overline{M}$ caused by external fields, usually neglible, and λ is the muon decay constant ($3*10^{-10}$ eV).

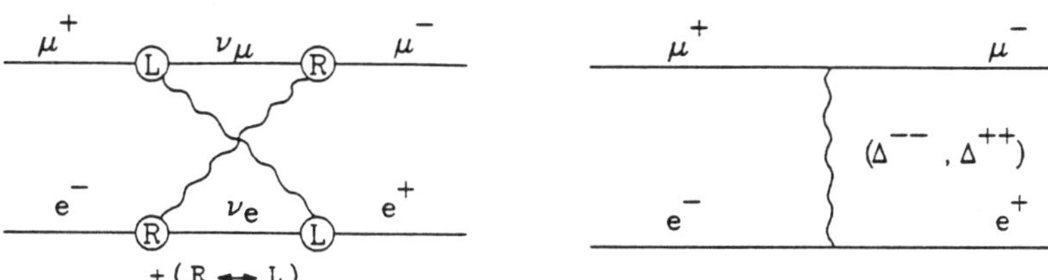

Fig. 1 Feynman diagrams for suggested conversion mechanisms.

Theoretical predictions for the size of the coupling arise outside the standard model from processes related to neutrinoless double beta-decay[7] and muon to 3 electron decay[8]. Halprin discusses the Feynman diagrams shown in Fig. 1 and estimates values for $G_{M\overline{M}}$, based on limits from double beta decay data, ranging from 10^{-6} - 10^{-5} G_F for the process involving a massive Majorana neutrino as mediator and 1 - 10 G_F for the exchange of a doubly charged Higgs particle. These less than rigid predictions reflect both the freedom of adjustment in the choices for the parameters involved and the uncertainty in the nuclear decay data. In pointing out that more stringent bounds come from μ->3e data and cosmology constraints for $\nu_{\mu,\tau}$ decay, Mohapatra gives an upper limit of about 10^{-2} G_F. However, he also emphasizes the speculative character of the theory and underlines the importance for sensitive searches of the M -> $\overline{M}$ conversion.

EXPERIMENTAL PROGRESS

Fig.2 gives a glimpse of the published experimental history of the conversion process up to now[9]. Most of the experiments establishing upper limits on $G_{M\overline{M}}$ used a common approach in defining the signature of a $\overline{M}$ event. The presence of antimuonium was to be confirmed by the detection of muonic X-rays which would be emitted upon the atomic capture of the μ^- following the collision of $\overline{M}$ with a target atom. No true signal was observed in either case and the results from the different experiments depicted as data points in the graph stem from background limited analyses.The first experimental attempt reported by Amato et al.[10] utilized a μ^+ beam from the Columbia U. NEVIS cyclotron. The beam was stopped in a 1 atm purified argon target cell for muonium production. Subsequent conversion was to be indicated through the detection of the 643 keV K_α muonic Ar X-ray with NaI(Tl) scintillation counters.

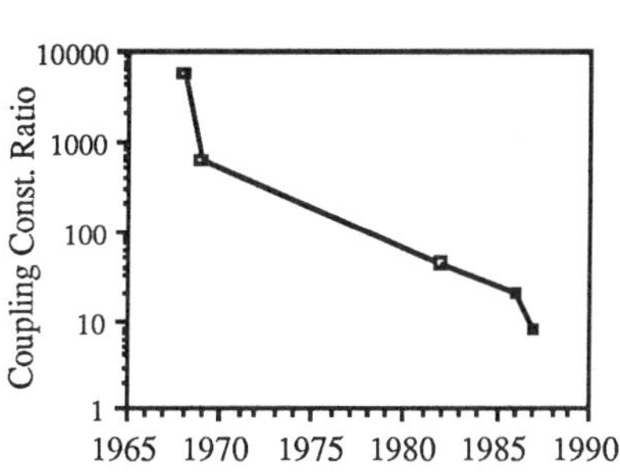

Fig. 2 Experimental history for M -> $\overline{M}$

Although the background limited result of $G_{M\overline{M}} < 5800\ G_F$ presented no challenge to theoretical arguments, it indicated potential focal points for future more sensitive pursuits. Early experiments were at a considerable disadvantage, since muonium production in vacuum was not developed until 1981[11] and collisions of muonium with gas atoms reduce the conversion probability significantly as a consequence of the removal of the degeneracy of M and $\overline{M}$ states due to the interaction. Well aware of this deficiency Glashow[12] proposed early on to search instead for the cross-channel reaction $e^- e^- \to \mu^- \mu^-$ using colliding electron beams. This was experimentally realized by Barber et al.[13] at the Stanford storage rings with an energy of 525 MeV per beam. An arrangement of spark chambers, absorbers and scintillation counters served to detect the collision products under transverse directions. They tried to distinguish electrons from the more penetrating muons in the analysis but found only events compatible with Möller scattering, $e^- e^- \to e^- e^-$. From the measured cross-section for this process they deduced a limit of $G_{M\overline{M}} < 610\ G_F$ on the muonium-antimuonium coupling. With the advent of surface muon beams and sources providing higher muonium formation fraction in vacuum, experiments were designed to get closer proximity to the theoretical threshold. The next step in this direction was completed by a group at TRIUMF, where a novel technique was employed to produce muonium by thermalizing an incoming surface μ^+ beam in a layered SiO_2 powder target supported by a collodion film which was itself coated with CaO. The presence of antimuonium, which could develop inhibited from muonium diffused into the voids, was searched for by again using a muonic X-ray (Ca 784 keV K_α) as a signature. The result published by Marshall et al.[14] of $G_{M\overline{M}} < 42\ G_F$ was later revised to $G_{M\overline{M}} < 20\ G_F$ after a reassesment of the muonium fraction used in the previous analysis[15].

For the remainder of this paper I will concentrate on the work of our group and want to refer for new results from TRIUMF to the paper presented by G. Marshall. The last data point in the graph of Fig.2 represents the achievement of our previous collaboration[16] at LAMPF (E 985) which measured an upper limit of $G_{M\overline{M}} < 7.5\ G_F$, a result beginning to probe theoretical predictions with a doubly charged Higgs triplet. A schematic diagram of the apparatus is shown in Fig. 3.

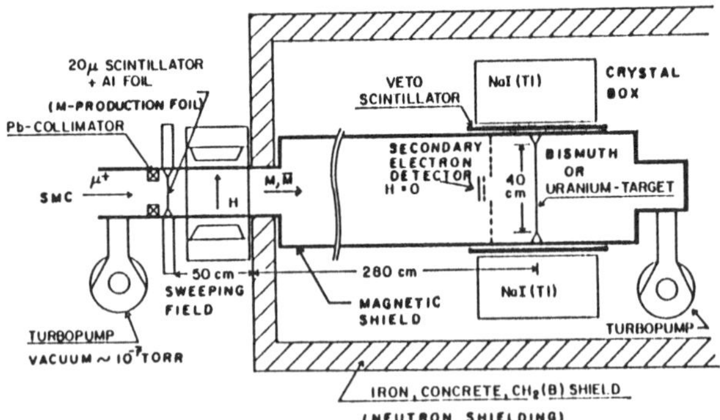

Fig. 3 Experimental arrangement for LAMPF E 985

Following the tradition of earlier approaches, the principle of our experimental method comprised basically three steps. First, M in vacuum was produced via the beam-foil technique from an incident separated subsurface μ^+ beam with a momentum of 10 MeV/c. After traversing a section where charged components were magnetically swept out, the M beam traveled through a field suppressed drift region, shielded to < 30 mG, allowing the $\overline{M}$ conversion to occur. Finally the neutral beam was stopped on a MgO coated Bi target and a pair of muonic X-rays was looked for in conjunction with a signal indicating a correlated impact on the target. For the detection of the K_α (6.05 MeV) and L_α (2.55 MeV) X-rays from Bi we employed the LAMPF NaI(Tl) Crystal Box detector[17], modified to extend its low energy threshold to below 2 MeV. The emitted secondary electrons from the Bi target, indicating the target impact of M or $\overline{M}$, were focused and accelerated onto a microchannel plate detector (μCP). The $\overline{M}$ event signature was defined as the coincidence of the X-ray pair and, as a requirement only functional in the off-line analysis, a correlated count in the μCP. Extensive neutron shielding was applied to reduce background counts in the Crystal Box arising from neutron capture γ rays. While it was straightforward to simulate experimentally a real event for the NaI(Tl) detector with an incident μ^- beam, the association of a μCP count to a particular event turned out to be more intricate. The correct relationship between X-ray and μCP count could be deduced from the timing information gained from a TDC spectrum displaying the difference in arrival time of the two signals in the respective detectors. Fig.4 shows the equivalent spectrum with the box operating in the 1 charged particle trigger mode. Following a list of experimentally verified arguments, omitted here because of space constraints, one can be convinced that the X-ray correlated μCP counts would have to appear in the region of the large peak. Also added in software was the requirement of a larger than 30 degree opening angle between the two X-rays, which reduced the background considerably without severely affecting the signal rate. Our quoted result for the coupling strength derived from a maximum likelihood analysis is presumably close to the experimentally achievable limit for the traditional muonic X-ray signature method.

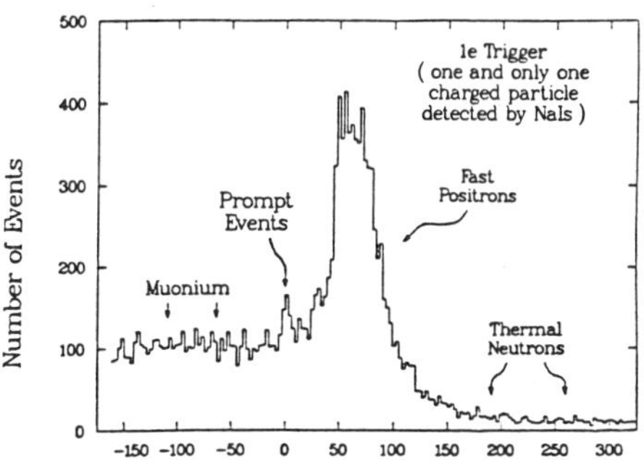

Fig. 4 Time difference, $\Delta T = T_{\mu CP} - T_{NAI}$ [ns].

FUTURE PROSPECTS

In closing I want to comment briefly on our plans for a significantly more sensitive search of the M -> $\overline{M}$ conversion, LAMPF E1073[18]. Despite its radical departure from earlier discussed attempts, the experimental approach, schematically outlined in Fig. 5, is directly oriented towards the detection of the components of the atomic system $\overline{M}$. With the decay of the μ^-, the atom is broken up into a Michel electron and the atomic positron. Our apparatus is designed to detect both products in coincidence, which in principle should provide for a rather clean signature of a $\overline{M}$ event. This undertaking became possible only after the discovery of abundant formation of thermal muonium[19]. The principal advantage offered by a M_{th} source, besides the increased number of M atoms created, e.g. with a SiO_2 target partly stopping the incoming muon beam, lies in the relatively tight localization of the emerging M_{th} in space. This allows for observation of the (M, $\overline{M}$) system throughout several lifetimes, as will be fully exploited by our scenario. The fast electron from $\overline{M}$ decay is looked for with a magnetic spectrometer incorporating MWPC detector planes for track reconstruction. The slow positron is first collected, then accelerated and focused onto a μCP detector located downstream of a bending magnet and a solenoid. The main background source will presumably originate from knock-off electron processes induced by Michel positrons from μ^+ decay. Taking into account our stringent event trigger requirements, we project, under rather conservative assumptions about primary intensities, detector efficiencies etc., a $\overline{M}$ signal rate of several mHz $*(G_{M\overline{M}}/G_F)^2$ compared to a background rate of < 0.1 μHz. This would allow us to set a new upper limit on the coupling strength of $< 10^{-2}\, G_F$. Our first beam time, scheduled for this summer, will hopefully enable us to challenge current theoretical beliefs on this interesting process.

I wish to thank B. Matthias for fruitful discussions on this subject.

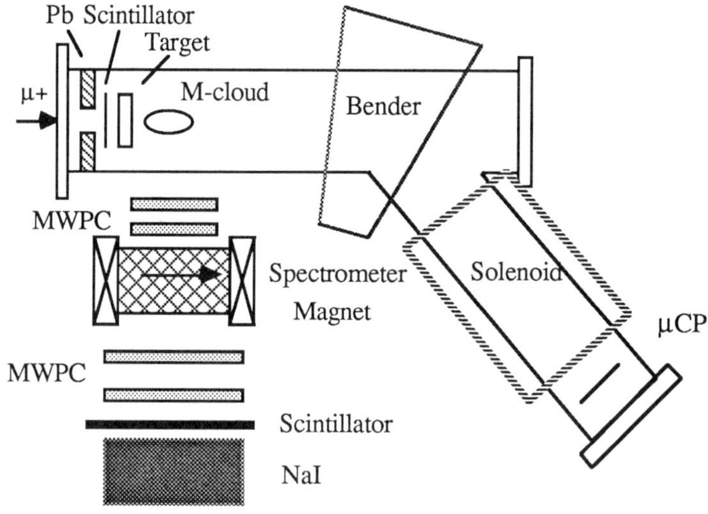

Fig. 5 Schematic layout for LAMPF E1073 apparatus.

1. B. Pontecorvo, Zh. Eksp. Teor. Fiz. 33, 549 (1957) [Sov. Phys.JETP 6, 429 (1958)].
2. S.L. Glashow, Nucl.Phys. 22, 579 (1961); S. Weinberg, Phys. Rev. Lett. 19, 1264 (1967); A. Salam, Elementary Particle Theory: Relativistic Groups and Analyticity (Nobel Symposium No. 8),ed. by N. Svartholm (Almquist and Wicksell, Stockholm, 1978) p.367.
3. tau generation assignment omitted here
4. G. Feinberg and S. Weinberg, Phys. Rev. Lett. 6,381 (1961); C.N. Yang, J. Tiomno, Phys. Rev.79, 495 (1950); G. Feinberg, P. Kabir and S. Weinberg,

Phys. Rev. Lett. 3, 527 (1959); N. Cabibbo and R. Gatto, Phys. Rev. Lett. 5, 114 (1960)
5. R.N. Mohapatra, J.D. Vergados, Phys. Rev. Lett.47, 1713 (1981); R.N. Mohapatra, Quarks, Leptons, and Beyond, ed. by H. Fritzsch (Plenum Publishing, 1985), p. 219
6. G. Feinberg and S. Weinberg, Phys. Rev.123, 1439 (1961)
7. A. Halprin, Phys. Rev. Lett.48,1313 (1982)
8. R.N. Mohapatra, Eigth Workshop on Grand Unification, ed. by K. Wali (World Scientific, Singapore, 1988), p. 200
9. new results from TRIUMF (T.M. Huber et al.,TRI-PP-88-30) reported elsewhere in the proceedings are not included, see G. Marshall et al.
10. J.J. Amato, P. Crane, V.W. Hughes, J.E. Rothberg and P.A. Thompson, Phys. Rev. Lett.21 1709 (1968)
11. P.R. Bolton et al., Phys. Rev. Lett.47,1441 (1981); C.J. Oram et al., J. Phys. B 14, L789 (1981)
12. S.L. Glashow, Phys. Rev. Lett. 6, 196 (1961)
13. W.C. Barber, B. Gittelman, D.C. Cheng and G.K. O'Neill, Phys. Rev. Lett.22, 902 (1968)
14. G.M. Marshall, J.B. Warren, C.J. Oram and R.F. Kiefl, Phys. Rev. D 25, 1174 (1982)
15. G.A. Beer et al. Phys. Rev. Lett. 57, 671 (1986)
16. B. Ni, K.P. Arnold, F. Chmely, V.W. Hughes, S.H.Kettell, Y. Kuang, J. Markey, B.E. Matthias, H. Orth, H.R. Schaefer, K. Woodle, M.D. Cooper, C.M. Hoffman, G.E. Hogan, R.E. Mischke, L.E. Piilonen, R.A. Williams, M. Eckhause, P. Guss, J. Kane, J. Reidy, and G.zu Putlitz, Phys. Rev. Lett.59, 2716 (1987).
17. R.D. Bolton et al., Nucl. Instrum. Methods 188, 275 (1981); G.H. Sanders et al., Nucl. Instrum. Methods 180, 603 (1981); R.D. Bolton et al., Phys. Rev. Lett. 53,1415 (1984)
18. H.R. Schaefer et al., LAMPF proposal E1073, (1987)
19. A.P. Mills et al., Phys. Rev. Lett.56, 1463 (1986); G. A. Beer et al., Phys. Rev. Lett.57, 671 (1986), K.A. Woodle et al., forthcoming, Zeit. f. Phys. (1988)

AN IMPROVED LIMIT FOR CONVERSION OF (μ^+e^-) TO (μ^-e^+)

G.M. Marshall,[1],[a] G.A. Beer,[1] T. Bowen,[2] C.A. Fry,[3],[b] Z. Gelbart,[4] P.G. Halverson,[2] B. Heinrich,[5] T.M. Huber,[6],[a] A.C. Janissen,[1] K.R. Kendall,[4] A.R. Kunselman,[6] G.R. Mason,[1] K. Myrtle,[5] A. Olin,[1],[a] J.B. Warren[7]

[1] University of Victoria, Victoria, British Columbia V8W 2Y2, [2] University of Arizona, Tucson, Arizona 85721, [3] University of Rochester, Rochester, New York 14627, [4] TRIUMF, Vancouver, British Columbia V6T 2A3, [5] Simon Fraser University, Burnaby, British Columbia V5A 1S6, [6] University of Wyoming, Laramie, Wyoming 82071, [7] University of British Columbia, Vancouver, British Columbia V6T 1W5

ABSTRACT

An improved upper limit is reported for muonium (μ^+e^-) conversion to antimuonium (μ^-e^+) which corresponds to a coupling strength approximately equal to that of the weak interaction. Muonium was produced in vacuum from a thin silica powder layer. The conversion signal was radioactivity induced by negative muon capture in tungsten in an adjacent surface. No events were seen, establishing an upper limit on the effective four fermion weak coupling constant for conversion of $1.1G_F$ (90% confidence level). The corresponding lower limit on the mass of a possible doubly charged scalar boson in a left-right symmetric model is approximately the W boson mass.

INTRODUCTION

The possible conversion of muonium (μ^+e^- or Mu) into antimuonium (μ^-e^+ or $\overline{\text{Mu}}$) was first discussed by Pontecorvo [1] in the context of lepton number conservation. This reaction is allowed in many extensions of the standard model. In particular, observable rates for this process are predicted in certain left-right symmetric models [2,3] incorporating an additional Higgs triplet which violates lepton number. Mu mixes with $\overline{\text{Mu}}$ through the exchange of the doubly charged member of this triplet, while neutrino masses are generated by the neutral partner.

Feinberg and Weinberg [4] postulated a four fermion effective interaction

$$\mathcal{H} = \frac{G}{\sqrt{2}}\overline{\psi}_\mu\gamma_\lambda(1-\gamma_5)\psi_e\overline{\psi}_\mu\gamma^\lambda(1-\gamma_5)\psi_e + \text{h.c.} \tag{1}$$

which mixes Mu and $\overline{\text{Mu}}$ with strength G. The branching ratio relative to normal muon decay is found by integration to be

$$\frac{\Gamma(\mu^+e^- \rightarrow \mu^-e^+)}{\Gamma(\mu^+ \rightarrow e^+\overline{\nu_\mu}\nu_e)} \approx C\left(\frac{G}{G_F}\right)^2 \tag{2}$$

where G_F is the Fermi weak coupling constant, λ is the muon decay rate, and $C = 2.5\times10^{-5}$. Several previous experiments have searched for conversion [5,6] with the most recent limit [7] being $G < 7.5G_F$ (90% confidence). The present experiment probes for the first time interactions having $G \sim G_F$, i.e. the Fermi coupling strength.

TECHNIQUE

Positive muons are stopped in the silica powder target (Fig. 1(a)) where they form thermal muonium which diffuses into vacuum with a yield Y per incident μ^+. In

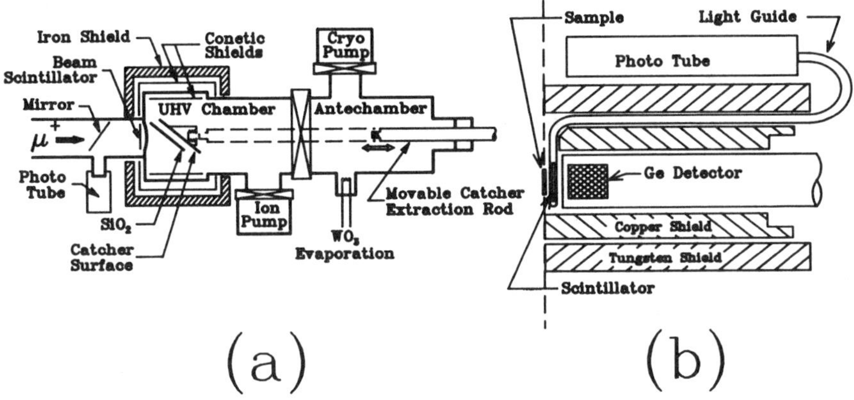

Figure 1: (a)Schematic diagram of the vacuum system and muonium production target, showing the catcher foil insertion system. (b)One half of the low level counting (LLC) apparatus, showing one beta scintillator and one Ge gamma detector. The other half is a mirror image about the dashed line through the sample.

a large field-free vacuum region, Mu will convert to $\overline{\text{Mu}}$ with probability $C \cdot (G/G_F)^2$. In the vacuum drift space adjacent to the silica, however, conversion must occur before the Mu atom strikes the catcher surface coated with WO_3 , on which the μ^- can be captured to form a radioisotope of Ta. The fraction of $\overline{\text{Mu}}$ atoms reaching the surface is **e(catcher)**, which includes the time dependence of the evolution of Mu to $\overline{\text{Mu}}$ as well as the catcher geometry. Further reductions **e(field)** and **e(surface)** are due to small residual magnetic fields and to surface contamination of WO_3 , respectively.

When an antimuonium atom strikes the WO_3 surface, the μ^- will undergo atomic capture and nuclear capture on W with probabilities **f(capture)** and **f(nuclear)** respectively. A fraction **f(184)** of the nuclear captures produces a recoiling ^{184}Ta ion, and a further fraction **e(recoil)** remains in the surface layer. This layer is chemically removed and concentrated after each μ^+ exposure. A fraction **f(chemistry)** of the ^{184}Ta extracted by the chemical process is counted in the shielded low level counting (LLC) apparatus (Fig. 1(b)) in a separate location.

The ^{184}Ta nuclei decay by β emission to excited states of ^{184}W. A dominant mode includes a prompt 414 keV gamma, having a branching ratio of **g(decay)** to a metastable state. The LLC apparatus detects both radiations with efficiencies **g(β)** and **g(γ)** respectively, including the contributions from incomplete solid angle. The metastable nuclear state emits typically two or more delayed gammas to reach the ground state. One or more of these delayed gammas is detected, with efficiency $\mathbf{g}(\gamma_{del})$, in a delayed time interval after the prompt signal. A delayed gamma detected in the same Ge detector as the prompt gamma may shift the measured prompt photopeak energy, reducing the efficiency by **g(pileup)**. Due to finite exposure and counting times, only a fraction **h** of any ^{184}Ta extracted from the catcher foil will decay in the LLC apparatus. The values of all of these factors are summarized in Table 1.

The number, N, of detected antimuonium events is then given by the product of all factors times the total number, N_{μ^+}, of μ^+ entering the muonium production target, and can be written

$$N = N_{\mu^+} \cdot C \cdot (G/G_F)^2 \cdot Y \cdot \mathbf{e} \cdot \mathbf{f} \cdot \mathbf{g} \cdot \mathbf{h} \tag{3}$$

where **e**, **f**, and **g** represent products of the similarly labelled quantities previously defined. N_{μ^+} is 4.0×10^{11}, based on about 100 hours of exposure to μ^+. The yield Y was measured with a procedure described elsewhere [5].

As a consistency check, we were able to verify the product **f·g** by a 12 hour μ^- irradiation of a 700 nm WO_3 layer on a Mo substrate. The residue from chemical removal of the layer was counted, and the LLC spectra obtained are shown in Fig. 2. After allowing for exposure and counting times, the number of detected β-γ-γ_{del}

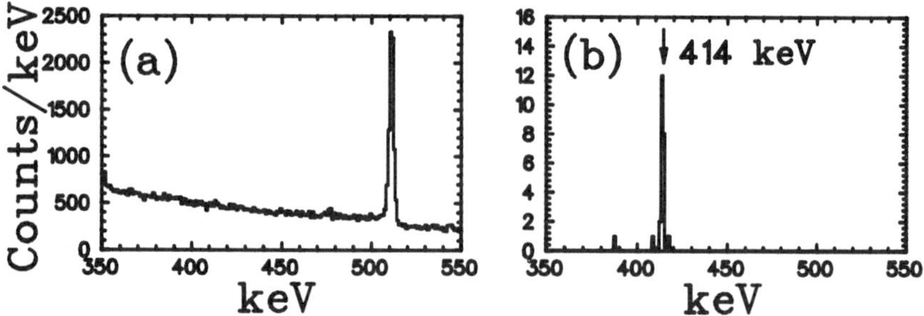

Figure 2: Gamma energy spectra from μ^- exposure of an evaporated WO_3 layer: (a) all events, (b) events in prompt coincidence with a count in a beta scintillator and followed by a delayed gamma.

events per stopped μ^- is $(3.3 \pm 1.1)\times10^{-4}$, which is consistent with the product of the individual factors, $\mathbf{f}\cdot\mathbf{g} = (3.4 \pm 0.5)\times10^{-4}$.

RESULTS AND CONCLUSION

The combined LLC spectra from all μ^+ exposures are shown in Fig. 4. No events within 4 keV of 414 keV are observed in the β-γ-γ_{del} histogram. With the quoted sensitivities, Eq. (3) implies a limit [11] on the coupling constant of $G < 1.1G_F$ (90% confidence). The corresponding upper limit on the branching ratio is 3×10^{-5}.

Table 1: Factors which determine the sensitivity of the experiment.

Symbol	Description	Factor	Method
Y	yield into vacuum	0.013 ± 0.003	positron imaging [5]
$C \cdot (\frac{G}{G_F})^2$	conversion in vacuum	$2.5 \cdot 10^{-5}(\frac{G}{G_F})^2$	calculation [4]
e(field)	magnetic fields	> 0.98	measured field [4]
e(catcher)	conversion correction	0.16 ± 0.01	Monte Carlo
e(surface)	surface contamination	> 0.90	Auger analysis
e(recoil)	recoil of ^{184}W	0.49 ± 0.01	Geometry and range
Product of e factors		0.070	
f(capture)	atomic capture on W	0.67 ± 0.03	X-ray intensities [8]
f(nuclear)	nuclear capture	0.97 ± 0.02	μ^- lifetime [9]
f(184)	^{184}Ta production	0.096 ± 0.014	μ^- in W
f(chemistry)	^{184}Ta recovery	> 0.90	^{187}W source
Product of f factors		0.056	
g(decay)	^{184}Ta branching ratio	0.74	decay scheme [10]
g(β)	β efficiency	0.55 ± 0.01	^{198}Au source
g(γ)	γ efficiency	0.052 ± 0.001	^{198}Au source
g(γ_{del})	delayed γ efficiency	0.34 ± 0.02	^{184}Ta source
g(pileup)	loss from pileup	0.85 ± 0.01	^{184}Ta source
Product of g factors		0.0061	
h	^{184}Ta decay correction	0.68 ± 0.01	Calculation
OVERALL PRODUCT		$5.2 \times 10^{-12} \cdot (\frac{G}{G_F})^2$	

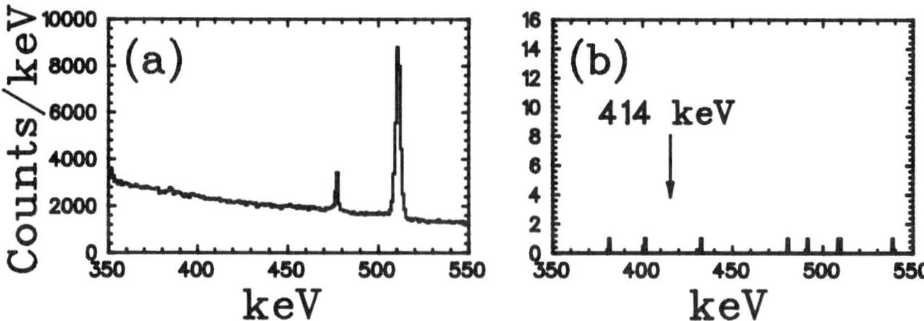

Figure 3: Gamma energy spectra from all μ^+ exposures (antimuonium runs): (a) all events, (b) events in prompt coincidence with a count in a beta scintillator and followed by a delayed gamma.

This places a constraint [3] on some left-right symmetric models of

$$\frac{g_{\mu\mu}g_{ee}}{g^2} \cdot \left(\frac{M_W}{M}\right)^2 \sim \frac{G}{G_F} < 1.1 \tag{4}$$

where $g_{\mu\mu}$ and g_{ee} are coupling constants of muons and electrons respectively to a doubly charged scalar boson of mass M. M_W is the W mass and g is the dimensionless coupling constant of the standard model. Under the assumption that the couplings $g_{\mu\mu}$ and g_{ee} are of the same order as g, the mass of the doubly charged boson is at least M_W.

ACKNOWLEDGMENTS

The authors wish to acknowledge the capable assistance of Dr. D. Harrington, Dr. T. J. Ruth, Dr. A. E. Pifer, Mr. T. Wu, and the professional and technical staff of TRIUMF. An essential contribution to the understanding of the surface cleanliness of WO_3 was made by the Surface Physics Laboratory at Simon Fraser University. This work was supported by the Natural Science and Engineering Research Council of Canada and the National Science Foundation.

[a]Present address: TRIUMF, 4004 Wesbrook Mall, Vancouver, B.C. V6T 2A3, Canada.
[b]Present address: National Laboratory for High Energy Physics (KEK), Oho-machi, Tsukuba-gun, Ibaraki-ken 305, Japan.

REFERENCES

1. B. Pontecorvo, Zh. Eksp. Teor. Fiz. 33, 549 (1957); translation, Sov. Phys. JETP 6(33), 429 (1958).

2. G. B. Gelmini and M. Roncadelli, Phys. Lett. 99B, 411 (1981); R. N. Mohapatra and J. D. Vergados, Phys. Rev. Lett. 47, 1713 (1981); A. Halprin, Phys. Rev. Lett. 48, 1313 (1982).

3. G. K. Leontaris, K. Tamvakis, and J. D. Vergados, Phys. Lett. 162B, 153 (1985); J. D. Vergados, Phys. Rep. 133, 1 (1986).

4. G. Feinberg and S. Weinberg, Phys. Rev. 123, 1439 (1961), and Phys. Rev. Lett. 6, 381 (1961).

5. G.A. Beer et al., Phys. Rev. Lett. 57, 671 (1986).

6. G. M. Marshall et al., Phys. Rev. D 25, 1174 (1982); J.J. Amato et al., Phys. Rev. Lett. 21, 1709 (1968).

7. B. Ni et al., Phys. Rev. Lett. 59, 2716 (1987).

8. T. von Egidy et al., Phys. Rev A23, 427 (1981).

9. T. Suzuki, D. F. Measday, and J. P. Roalsvig, Phys. Rev. C 35, 2212 (1987).

10. Table of Isotopes (7th Edition), edited by C. M. Lederer and V. S. Shirley (John Wiley and Sons, Inc., New York, 1978), and references therein.

11. See, for example, W. T. Eadie et al., Statistical Methods in Experimental Physics (North-Holland, Amsterdam, 1971).

ANALYSIS OF THE EXCLUSIVE INTERACTION $\nu_\mu{}^{-12}C \rightarrow \mu^{-12}N$ NEAR THRESHOLD

T. Bowles, A. Brown, R. E. Brown, R. Burman, D. Clark, S. Clearwater, T. Dombeck, H. Kruse and D. Lee
Los Alamos National Laboratory, Los Alamos, NM

L. Auerbach, S. Datta, and Y. Huang
Physics Department, Temple University, Philadelphia, PA

B. Aas and G. Igo
Physics Department, University of Calif., Los Alamos, NM

D. Beavis, S. Y. Fung, B. Gorn, R.Poe, and G. Van Dalen
Physics Department, University of Calif., Riverside, CA

C. Newsom
Physics Department, University of Iowa, Iowa City, IA

B. Bassalleck, B. Dieterle, R. Hill, J. Kang, and C. Leavitt
Physics Department, University of New Mexico, Albuquerque, NM

R. Fisk, D. D. Koetke, D. S. Koetke, and R. W. Manweiler
Physics Department, Valparaiso University, Valparaiso, IN

Presented by R. W. Manweiler

ABSTRACT

We report on the first observation of ν_μ-induced nuclear transitions $\nu_\mu{}^{12}C \rightarrow \mu^{-12}N$. From a sample of ν_μ interactions on carbon near the muon production threshold, events with a ^{12}N nucleus in the final state were isolated by observing the subsequent beta decay of the ^{12}N ground state. This exclusive process represents (31+12)% of the inclusive cross section for $\nu_\mu{}^{12}C \rightarrow \mu^- X$. (Theoretical estimates of this fraction suggest a maximum of only 3%.) The muon kinetic energy for both the inclusive and exclusive channels was also measured. Using the muon kinetic energy distribution, we identify the contributions to the cross sections from the $^{12}N^*$ and compare our results to theory.

INTRODUCTION AND RESULTS

This experiment performed at the Los Alamos Meson Physics Facility, reports the first observation of ν_μ-induced transitions between specific nuclear states. The neutrino source and detector are described in detail elsewhere[1,2]. Our neutrinos came from the decays of positive pions produced by 800 MeV protons incident on a water target. Since a pion focussing system was not employed, negative pions were also present and gave a muon anti-neutrino contamination of less than 6% of the total neutrino flux and with a softer characteristic energy spectrum than the muon-neutrino flux.

Time-of-flight (TOF) techniques were used to discriminate against beam-associated neutrons generated in the target. The TOF region for neutrinos was well separated from the low energy neutron arrivals in the detector.

The experimental signature for this nuclear transition is a three-step process, which is exceptionally clear and free from backgrounds: (1) the primary charged current ν_μ interaction leaves the ionization trail of the μ^- which stops in less than 10 cm, usually within the detector volume; (2) the μ^- decays with its lifetime in scintillator of 2.026 μsec[3] leaving an electron track with the Michel energy distribution; and (3) the ^{12}N beta decays with a lifetime of 15.87 msec and an endpoint energy of 16.3 MeV[3] producing an electron track in the scintillator. Our data analysis was designed to select these three ionizating events with the correct time sequence, correlated in space, and having correct energies. A detailed analysis of the exclusive events with ^{12}N beta decays can be found in Ref. 4.

RESULTS

Of the 844,667 triggers with a muon decay signature, 68% occurred within the neutrino TOF region. Almost all of these triggers result from a spurious combination of a low-energy, activation induced, prompt trigger with a neutron arriving anywhere in the detector during the 20 μs muon decay gate interval. No neutron induced events had sufficient energy to produce pions in the detector, which might have produced a background source of muons from pion decay in the detector. We rejected spurious triggers in offline analysis by requiring spatial and timing correlations between prompt and muon decay signals, and by imposing energy cuts.

The final inclusive sample had 104 muon event candidates whose decay time was less than 10 μsec. Because of higher backgrounds in the sub-sample of very low energy events (T_μ<12 MeV), we found that analysis was only possible for a limited neutrino TOF interval.[2] Therefore 21 of the 104 events enter our analysis with a weight of two. The final sample had 125 weighted events. Elsewhere we discuss the properties of the inclusive ν_μ interaction in detail.[2]

Figure 1 shows the muon decay time histogram for the weighted muon events surviving all of our cuts. The events with short decay time indicates the neutrino-induced signal, but there are also events at large muon decay time (>10 μsec) where backgrounds are expected to dominate. The background level was estimated from obvious background events where the prompt and muon decay signals were spatially separated within the detector, but otherwise passed all other cuts. The parameter free background curve is shown as the dashed curve. In the signal region below 10 μsec, the area under the background curve is 48.6 events, leaving an excess signal of 76.4 ν_μ-induced events. The signal is well fit by an exponential with a lifetime of 2.026 μsec. A beam-off data sample shows that cosmic-ray-induced backgrounds were negligible.

The final ^{12}N data sample was derived from the 104 muon events by requiring a beta decay candidate with 35 msec after the beam

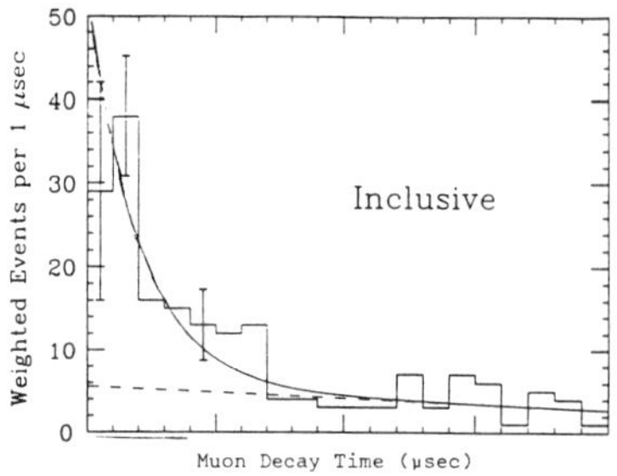

Fig. 1 Muon decay time for 165 weighted inclus. events.

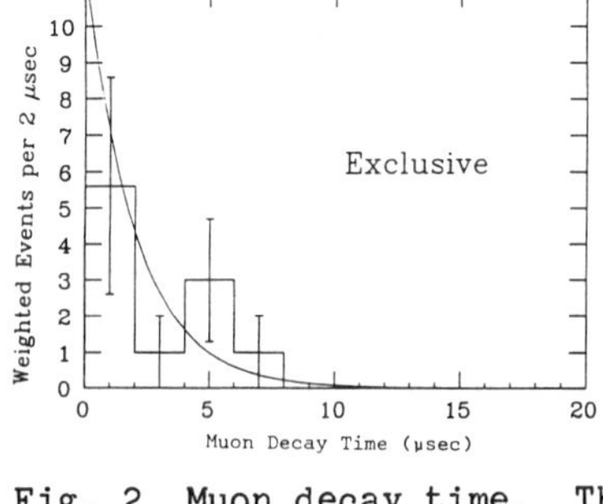

Fig. 2 Muon decay time. The solid curve is an EGS4 calc.

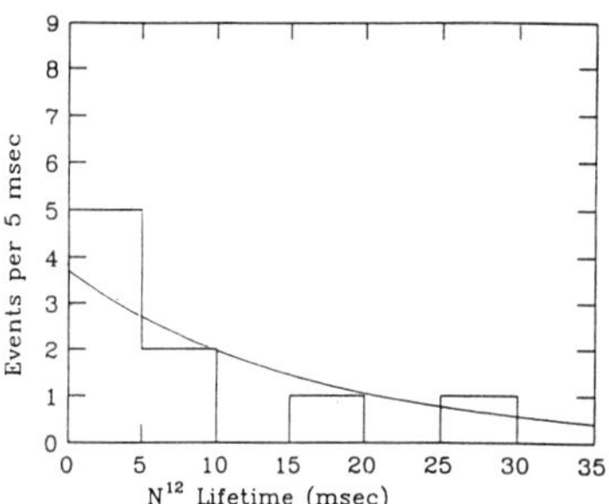

Fig. 3 The ^{12}N decay time distribution

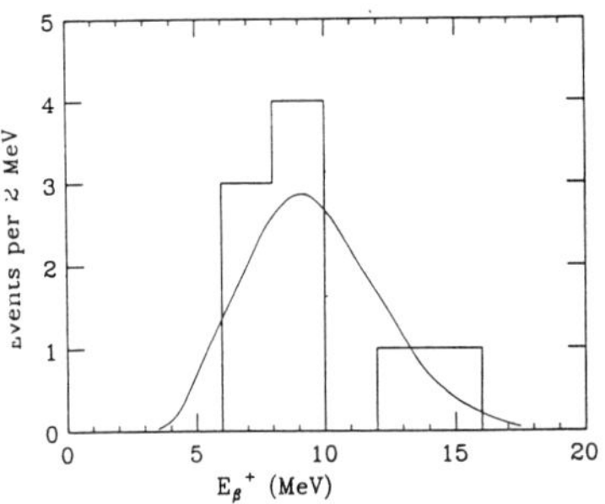

Fig 4. The ^{12}N energy distribution.

macropulse. Nine of the events had a ^{12}N decay electron candidate spatially associated with the prompt trigger; each electron energy was less than 20 MeV. We estimate the efficiency for an ^{12}N trigger in the detector at 58% using our detector simulation based on EGS4[2]. The spatial cuts had an efficiency of 85%, and the ^{12}N decay live-time internal efficiency was 86%. The combined efficiency for the additional requirement of a ^{12}N signature is then 42%.

Figure 2 shows the muon decay time histogram for the ^{12}N candidate events. There are no events with a decay time greater than 7 μsec, indicating that the background level is low. A high statistics study of background events with spatially dissociated prompt, muon decay, and ^{12}N decay signals has shown that the random background level is 0.8 event, giving a signal of 8.2 events above background. Superimposed on the histogram in Fig. 2 is a fit of our 9 events to an exponential with a lifetime fixed at 2.026 μsec.

Figure 3 shows the observed ^{12}N decay times for the nine events. The smooth curve is an exponential with the accepted lifetime of 15.87 msec. The observed ^{12}N beta energy distribution is shown in Fig. 4. We have used the EGS4 shower Monte Carlo together with detector acceptance and resolution models to calculate the expected beta energy distribution in our detector which is shown as the smooth curve on Fig. 4.

Figure 5 shows the muon kinetic energy histogram for inclusive events[5]. The nine events with ^{12}N beta decays are shown with cross-hatching. The dashed curve is the predicted muon energy spectrum obtained from the Fermi Gas Model[6] (FGM) and our energy spectrum distribution[7]. The FGM is corrected for the Coulomb interaction[8] between the negative muon and the nuclear charge[9].

Several significant observations can now be made. From Fig. 5, we find that the total number of events predicted by the FGM accounts for only 61% of all inclusive events. Moreover, it fails to include those events with T_μ > 60 MeV. In addition, the ratio of σ N(g.s.)/ σ Inclusive, including all efficiency corrections, is (31±12)% independent of the overall normalization. This is to be compared with 1% for the same ratio of theoretical energy averaged σ's $\{ <\sigma_{N(g.s.)}>/<\sigma_{Inclusive}> \}$ defined below. Our experiment therefore shows an anomalously large cross section for the ν_μ-induced reactions to the ^{12}N(g.s.).

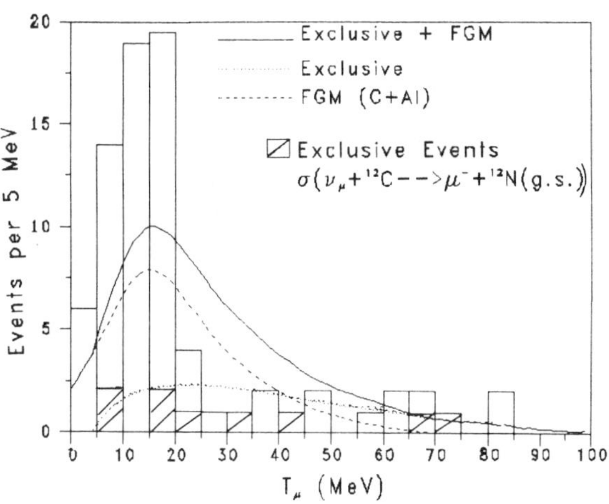

Fig 5. Muon kinetic energy distribution for the 9 ^{12}N events. The curves are described above.

We take these observations to be indicative of the fact that the FGM does not well describe the ^{12}N exclusive reactions, including those in which the $^{12}N^*$ is left in an excited state. If we assume that all events not included in the FGM cross section are ^{12}N events[10], we can easily obtain an estimate of the cross section for the exclusive channel $\nu_\mu\ ^{12}C \dashrightarrow \mu^-\ ^{12}N^*$ since we have directly measured the cross section for ν_μ-induced ^{12}N(g.s.). The results are shown in Table I. Shown also on Table I are the cross section predictions from the particle-hole model[11,12], averaged over our flux distribution $\phi(E_{\nu_\mu})$ and our detector efficiency $\varepsilon(E_\nu)$ as

$$<\sigma> = \int \phi_\nu(E_\nu)\varepsilon(E_\nu)\sigma(E_\nu)dE_\nu \Big/ \int \phi_\nu(E_\nu)\varepsilon(E_\nu)\, dE_\nu \qquad (1)$$

The various functions entering Eq. (1) are shown in Fig. 6 below. The mean interaction energy for the FGM cross section is 217 MeV.

As seen in Table I, our measured cross section for the inclusive channel is in good agreement with the theoretical prediction [6,11]. This is likewise true the $^{12}N^*$ final states cross section obtained as above described.

Our measurement of the ^{12}N(g.s.) is substantially larger than the theoretical[12] prediction. Furthermore, the difference between the measured and predicted inclusive cross sections can be attributed to the difference between the measured and predicted ^{12}N(g.s.) contributions to the inclusive channel.

TABLE I. Cross Section Summary using preliminiary normalization and σ in 10^{-39} cm^2.

Channel	%	Evt	σ_{EXP}	σ_{THEORY}
Inclusive	100%	63	10.9±2.8	8.2[6,11]
Fermi Gas	61%	38	6.6	6.6[6]
^{12}N(g.s.)	31%	19	3.4±1.6	0.08[11,12]
$^{12}N^*$	8%	5	0.9± .5	1.6[11,12]

The T_μ distribution further supports the appropriateness of adding the particle-hole[11,12] and FGM[6] cross sections. We approximate T_μ for the ^{12}N events as

$$T_\mu(E_\nu) = E_\nu - M_\mu - \Delta M \qquad (2)$$

where ΔM is the ^{12}C - ^{12}N mass difference. Eq. (2) is good for our q^2 region.[13] Using our neutrino flux, Eq. (2) and the ^{12}N particle-hole cross sections[11,12] normalized to the total ^{12}N events, we obtain the dotted curve in Fig. 5. We normalize to our ^{12}N event total since the measured ^{12}N(g.s.) cross section is much larger than that predicted[12]. We note that this curve is a good description of the region T_μ > 60 MeV which is not well described by the FGM (dashed curve). The solid curve is then our predicted T_μ distribution for all inclusive channels and is in reasonable agreement with our measured event distribution.

The only major background which might give a false ^{12}N(g.s.) signal is the reaction $\bar{\nu}_\mu$ ^{12}C, μ^+ ^{12}B. The ^{12}B may be formed in its ground state or low level excited state which then gamma decays to the ground state. ^{12}B ground state beta decays to ^{12}C with a lifetime and end point energy not measurably different from that of ^{12}N for so few events. We estimate only about 6% $\bar{\nu}_\mu$ contamination in our beam. Furthermore the flux will

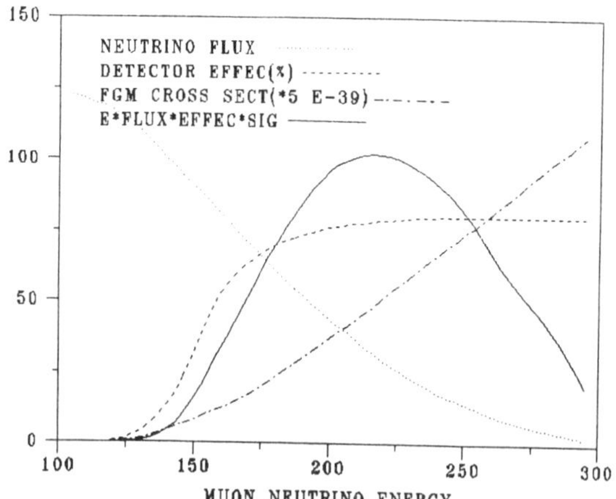

Fig. 6 The functions for calculating averaged cross sections: relative energy spectra, detect. effec. (in %), & σ_{FGM} (to be mult. by $5\cdot10^{-39}cm^2$).

be lower in energy than that for neutrinos. We estimate that with a ^{12}B cross section enhanced by a factor of ten over ^{12}N(g.s.) we can account for no more than 1.5 events. Consequently the N(g.s.) cross section is still anomalously large.

Our anomalous ^{12}N(g.s.) result coupled with the agreement between experiment and theory in ν_e processes[14] has raised the question of possible contributions from induced pseudoscalar or second class currents. It is then very desirable to repeat this experiment in this important q^2-region just above the muon threshold with significantly higher statistics. The use of a π focusing device would enhance the ν_μ flux while reducing the component, and diminish $\bar{\nu}_\mu$ ^{12}B events.

REFERENCES

1. T. Dombeck et al., Phys. Lett. B194, 591 (1987).
2. D. D. Koetke et al., "A Study of Inclusive Muon-Neutrino-Carbon Charged-Current Interaction in the Energy Range Just Above the Muon Threshold", to be submitted to Phys. Rev. C.; see also, S. Datta, Ph.D. thesis, Temple University (1987) and J. Kang, Ph.D. thesis, University of New Mexico (1987).
3. F. Ajzenberg-Selove, Nuc. Phys. A433, 1 (1985).
4. G. J. Van Dalen, et al., "Observation of Anomalous ν_μ12C --> μ^- ^{12}N Exclusive Interactions Near Threshold", submitted to Phys. Rev. Lett.
5. Approximately 88% of these muons arise from neutrino charged current interactions on ^{12}C and approximately 12% from the aluminum in the detector.
6. J. S. O'Connell,T.W. Donnelly, H.H. Chen,R.L. Burman & B. Cortez, Proceedings of the Los Alamos Neutrino Workshop, June 8-12, 1981, eds. F. Boehm and G. J. Stephenson, Jr., Report No. LA-9358-C, p. 46. See also T. K. Gaisser and J. S. O'Connell, Phys. Rev. D34, 822 (1986).
7. T. Dombeck, Los Alamos report LA-UR-1589 (1982).
8. See e.g., G. Audit et al., Phys. Rev. C15, 1415 (1977).
9. Since the inclusive events shown are for reactions from both ^{12}C (88%) and other nuclei (12%), we have plotted the ^{12}C FGM enhanced to include these other target nuclei. However in calculations we only use the ^{12}C FGM σ. We correct the number of inclusive events to correspond solely to those from ^{12}C.
10. We recognize that some small fraction of the ^{12}N final states also may be accounted for in the FGM, but we adopt our assumption as reasonable for this analysis.
11. J. S. O'Connell, T. W. Donnelly and J. D. Walecka, Phys. Rev. C6, 719 (1972). See also Proceedings of the Los Alamos Neutrino Workshop, June 8-12, 1981, Rept No. LA-9358-C, p. 13.
12 T. W. Donnelly, Progress in Particle and Nuclear Physics 13, 183 (1985). See also, J. D. Walecka, "Electroweak Interact. with Nuclei, in "Intersec. Nuclear and Part. Phys.," AIP No. 123, ed. R. Mischke, Steamboat Springs, CO (1984).
13. E.g., a 200 MeV incident neutrino leaving ^{12}N in the ground state emits muons with energies between 77.4 and 72.3 MeV.
14. R. C. Allen et. al., this proceedings, as reported by X. Lu.

ANALYTIC TREATMENTS OF NEUTRINO OSCILLATIONS IN MATTER

A.B. Balantekin

Department of Physics, University of Wisconsin, Madison, Wisconsin 53705, U.S.A.

ABSTRACT

Matter-enhanced neutrino oscillations for the case of two neutrino flavors are studied and three different approximation methods for calculating the unaveraged oscillation probability are presented. Two of these methods provide analytical expressions, valid for small or large vacuum mixing angles, which are very simple to evaluate numerically. The third method provides an expression uniformly valid over the entire range of mixing angles. In their respective regions of validity, the approximate calculations using these methods are found to be in very good agreement with the exact calculations.

After it was experimentally determined that the measured solar electron neutrino capture rate on earth is about one third of the standard solar model prediction, neutrino oscillations were proposed as one of the possible mechanisms, perhaps the most appealing one, to resolve the discrepancy between experiment and theory[1]. More than a decade ago, it was shown that[2] even a small mixing angle can result in complete conversion of electron neutrinos into muon neutrinos due to the modifications caused by the coherent forward scattering in electronic matter. However, at that time, such a resonant conversion was assumed to be too restrictive to be applicable to the solar neutrino problem. More recently Mikheyev and Smirnov demonstrated that the resonance condition can be satisfied in a medium with varying electron density[3]. Bethe rederived this result in the adiabatic approximation and showed that Mikheyev-Smirnov-Wolfenstein (MSW) mechanism has an adiabatic solution for the solar electron density distribution[4]. MSW mechanism was subsequently discussed in much detail by many authors. A partial list of references is given in[5-11].

As was stressed by Bethe[4] and Messiah[11], the main contribution of Mikheyev and Smirnov is to recognize the importance of adiabatic level crossing. Haxton emphasized that level crossing and other salient features of the underlying physics of matter-enhanced neutrino oscillations can be understood analytically and discussed an extension of the Landau-Zener level-crossing approximation for neutrino propagation in matter[6]. In this talk I will elaborate further on analytical aspects of matter-enhanced neutrino oscillations to gain a better insight into the MSW mechanism.

I consider two flavors of neutrinos. The Dirac equation, describing propagation of neutrinos in matter, can be viewed as two coupled, first-order differential equations. Differentiating once with respect to time, however, one can decouple the equations for the electron and the muon neutrinos. For the electron neutrino we get

$$-\hbar^2 \frac{d^2\psi}{dt^2} - \left[\phi^2 - i\hbar\frac{d\phi}{dt}\right]\psi = \left[\frac{(m_2^2 - m_1^2)\sin 2\theta}{4E}\right]^2 \psi \qquad (1),$$

where

$$\phi(t) = \frac{m_2^2 - m_1^2}{4E}\left[\frac{2\sqrt{2}G_F N_e(t) E}{(m_2^2 - m_1^2)} - \cos 2\theta\right].$$

In the above equation m_1 and m_2 are the mass eigenvalues, θ is the vacuum mixing angle, E is the energy of the neutrinos, G_F is the Fermi constant and N_e is the number of electrons per unit volume, which, in the ultrarelativistic limit ($R \sim ct$), can be taken as a function of time. Because of the simultaneous existence of a macroscopic (solar radius) and a microscopic (quantum mixing) scale, numerical integration of the Eq. (1) can be very time-consuming.

The equation (1) is formally equivalent to a one dimensional, time independent Schrödinger equation with a complex "potential". (Indeed, it very much resembles the equation studied in supersymmetric quantum mechanics with ϕ being the superpotential[12]. Since we can solve the Schrödinger equation exactly for a number of such "potentials", it is possible to write down analytic expressions for the neutrino conversion probability in matter for a number of specific electron densities. Such expressions were written down for the exponential density[13] (which approximates the solar density rather accurately), and for other density profiles[14]. In all these cases, the conversion probability was obtained in terms of hypergeometric functions. Unfortunately, those exact solutions are of limited value since numerical evaluation of hypergeometric functions of complicated arguments is not much easier than exact numerical integration of the Dirac equation.

Accurate approximate expressions for the unaveraged conversion probability not only would be very useful for calculating expected neutrino capture rates at the detectors, but also could facilitate our understanding of various aspects of the MSW mechanism. Exploiting its analogy to the Schrödinger equation, we applied three different approximations to Eq. (1). Here I only quote the results, since the detailed analysis will be published elsewhere[12].

Throughout this talk I assume that initially ($t = 0$) there are only electron neutrinos present. In the primitive WKB approximation the wavefunction for the electron neutrino takes the form

$$\Psi_e(t) = T_-(0)T_-(t)\exp\left\{+\frac{i}{\hbar}\int_0^t \left[\phi^2(t) + \left(\frac{\Delta\sin 2\theta}{2}\right)^2\right] dt\right\}$$

$$+T_+(0)T_+(t)\exp\left\{-\frac{i}{\hbar}\int_0^t \left[\phi^2(t) + \left(\frac{\Delta\sin 2\theta}{2}\right)^2\right] dt\right\}, \tag{2}$$

where we have defined

$$T_\pm(t) \equiv \left[1 \pm \frac{\phi}{\sqrt{\phi^2 + (\frac{\Delta\sin 2\theta}{2})^2}}\right], \tag{3}$$

with

$$\Delta = \frac{m_2^2 - m_1^2}{2E}.$$

The condition of validity of this semiclassical approximation is the same as the adiabatic condition given by Bethe, hence this approximation is applicable for relatively large vacuum mixing angles.

The second approximation we utilized is a logarithmic perturbation[15] approximation, where $g \equiv (1 - \cos 2\theta)$ is taken as the expansion parameter. Neglecting contributions which are second order in g, we get

$$\Psi_e(T) = \exp\left[-\frac{iQ(T)}{2} - g\frac{i\Delta T}{2\hbar} - g\frac{\Delta^2}{2\hbar^2}\int_0^T dt\; e^{+iQ(t)}\int_0^t dt' e^{-iQ(t')}\right] + \mathcal{O}(g^2), \tag{4}$$

where

$$Q(t) = \frac{2}{\hbar}\int_0^t \phi(t')dt', \tag{5}$$

calculated at $\theta = 0$. Rosen and Gelb pointed out that, for very small mixing angles, the MSW enhancement region is very small[5]. In this limit our approximation further simplifies and our results agree with theirs.

The third approximation we studied is a uniform WKB approximation, valid for all mixing angles, which generalizes Haxton's Landau-Zener approximation[6]. In this approximation we get

$$\Psi_e(t) = \frac{K(t)}{K(0)} \frac{\Gamma(-\nu)}{\sqrt{2\pi}} \Bigg[- D_\nu(y(t)) D'_\nu(-y(0)) - D_\nu(-y(t)) D'_\nu(y(0))$$

$$+ \left[-i\phi(0) - \frac{K'(0)}{K(0)} \right] \Big(D_\nu(-y(t)) D_\nu(y(0)) - D_\nu(y(t)) D_\nu(-y(0)) \Big) \Bigg] \tag{6}$$

where $y(t) = (1+i)[S_o(t) + \hbar S_1(t)]$, $\nu = -(1 - \eta + i\Omega/2)$,

$$K(t) = \left[\frac{\Omega + S_o^2(t)}{\Lambda + \phi^2(t)} \right]^{\frac{1}{4}}. \tag{7}$$

and

$$S_1(t) = \frac{i}{2\sqrt{\Omega + S_o^2(t)}} \Bigg[\log \frac{\phi(t) + \sqrt{\phi^2(t) + \Lambda}}{i\sqrt{\Lambda}}$$

$$- \eta \log \frac{i\sqrt{\Omega}}{S_o(t) + \sqrt{S_o^2(t) + \Omega}} \Bigg]. \tag{8}$$

In the equations above, Ω is determined by

$$\int_{t_o}^{t_o^*} \sqrt{\Lambda + \phi^2(y)}\, dy = \frac{-i\Omega\pi}{2} \tag{9}$$

and $S_o(t)$ by

$$\int_{t_o}^{t} \sqrt{\Lambda + \phi^2(y)}\, dy = \int_{i\sqrt{\Omega}}^{S_o(t)} \sqrt{\Omega + \sigma^2}\, d\sigma. \tag{10}$$

where

$$\Lambda \equiv \left(\frac{\Delta \sin 2\theta}{2} \right)^2.$$

and the turning points t_o and t_o^* are chosen such that $\phi(t_o) = +i\sqrt{\Lambda}$ corresponds to $S_o(t_o) = +i\sqrt{\Omega}$ and $\phi(t_o^*) = -i\sqrt{\Lambda}$ corresponds to $S_o(t_o^*) = +i\sqrt{\Omega}$ where t_o has a positive imaginary part.

In their respective regions of validity, we found our approximate calculations to be in very good agreement with the exact calculations. Since our approximations yield the unaveraged neutrino oscillation probability, they are especially suitable to use when the oscillation length is comparable to the distance traveled in matter. For neutrinos traveling through the earth that might be the case and we plan to apply our results to this problem. We can also investigate analytically the interference of two resonances very close to each other. Such a situation could be realized when there are macroscopic density fluctuations, as in the convective zone of the sun extending from about seventy-five percent of the solar radius outwards. The influence of macroscopic density fluctuations on solar density conversion was numerically studied by Koonin and Schäfer[7]. An inquiry, using analytical expressions, could systematically rule out certain neutrino mass difference and mixing angle values that might invalidate the MSW mechanism. Furthermore, for neutrinos originating on the far side of the sun or traversing earth, there could be two resonance locations. Especially for the earth those two resonance locations can be very close for certain parameter values. In consequence, our results would have a bearing on the interpretation of the experiments designed to measure regeneration of solar neutrinos in the interior of the earth.

ACKNOWLEDGEMENTS

This research was done in collaboration with S.H. Fricke and P.J. Hatchell. It was supported in part by the University of Wisconsin Research Committee with funds granted by the Wisconsin Alumni Research Foundation, in part by a Presidential Young Investigator Award administered by the National Science Foundation, and in part by the National Science Foundation Grant No. PHY-8610851.

REFERENCES

1. J.N. Bahcall, *et. al.*, Rev. Mod. Phys. **54**, 767 (1982); J.K. Rowley, B.T. Cleveland, and R. Davis, *Solar Neutrinos and Neutrino Astronomy*, AIP Conference Proceedings No. 126 (AIP, New York, 1985), p.1; W.C. Haxton, Comments Nucl. Part. Phys. **16**, 95 (1986).
2. L. Wolfenstein, Phys. Rev. **D17**, 2369 (1978); **D20**, 2634 (1979).
3. S.P. Mikheyev and A. Yu. Smirnov, Nuovo Cim. **9C**, 17 (1986).
4. H.A. Bethe, Phys. Rev. Lett. **56**, 1305 (1986).
5. S.P. Rosen and J.M. Gelb, Phys. Rev. **D34**, 969 (1986).
6. W.C. Haxton, Phys. Rev. Lett. **57**, 1271 (1986); Phys. Rev. **D35**, 2352 (1987).
7. A. Schäfer and S.E. Koonin, Phys. Lett. **B185**, 417 (1987).
8. V. Barger, R.J.N. Phillips, and K. Whisnant, Phys. Rev. **D34**, 980 (1986).
9. E.W. Kolb, M.S. Turner, and T.P. Walker, Phys. Lett. **B175**, 478 (1986).
10. S.J. Parke, Phys. Rev. Lett. **57**, 1275 (1986).
11. A. Messiah, in *'86 Massive Neutrinos in Astrophysics and Particle Physics* (Editions Frontieres, Gif-sur-Yvette, 1986), p.373.
12. A.B. Balantekin, S.H. Fricke, and P.J. Hatchell, Phys. Rev. **D38**, in press (1988).
13. P. Pizzochero, Phys. Rev. **D36**, 2293 (1987); S. Toshev, Phys. Lett. **B196**, 170 (1986); S.T. Petcov, Phys. Lett. **B200**, 373 (1988).
14. D. Nötzold, Phys. Rev. **D36**, 1625 (1987).
15. Y. Aharanov and C. Au, Phys. Rev. Lett. **42**, 1582 (1979).

CONTRIBUTION OF THE AXIAL CHARGE OPERATOR TO $2\nu\ \beta\beta$-Decay

A. G. Williams and W. C. Haxton
Institute for Nuclear Theory, Dept. of Physics, FM-15
University of Washington, Seattle, Washington 98195

ABSTRACT

We examine the possibility that the double axial charge operator could have a significant effect on the rates for $2\nu\ \beta\beta$-decay. Although the new $\beta\beta$-decay operator is nominally suppressed by $(v/c)^2$, it is not subject to the Pauli effects that weaken allowed matrix elements. In standard shell model treatments of the double beta decay nuclei ^{76}Ge, ^{82}Se, and ^{130}Te, this operator contributes as a first-order correction. By using the schematic Nilsson pairing model, we argue that core-polarization contributions generated by the quadrupole-quadrupole interaction will systematically cancel against the dominant Gamow-Teller amplitude. The contribution from the tensor force may also be important, but has not yet been evaluated.

INTRODUCTION

Double beta decay is an important low-energy probe of the standard model, the nature of the neutrino, and grand unified theories. It also tests our understanding of conventional nuclear physics issues. Because of the pairing force between nucleons, single β-decay is energetically forbidden for many even-even nuclei, leaving $\beta\beta$-decay as the only open decay made. As the ground states of such nuclei have $J^\pi = 0^+$, $0^+ \to 0^+$ transitions are expected to dominate total $\beta\beta$-decay rates.

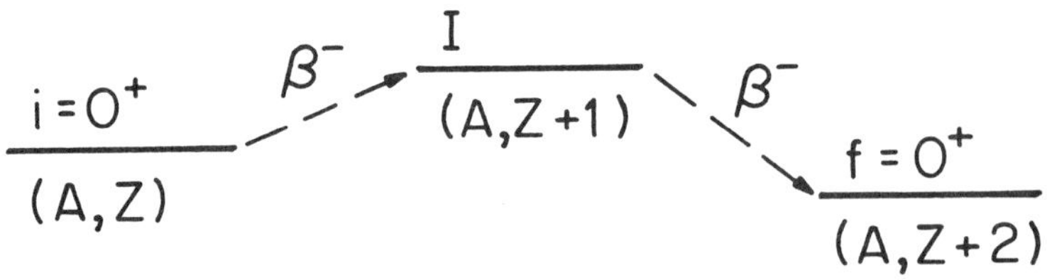

Fig. 1. Level scheme for $2\nu\ \beta\beta$-decay of an even-even nucleus where i, I, and f are the initial, intermediate, and final nuclear states respectively.

Although it has not yet been observed, neutrinoless $\beta\beta$-decay provides a significant test of physics beyond the standard model. Upper limits on the $0\nu\ \beta\beta$-decay rates are used to set limits on the masses and right-hand couplings of possible Majorana neutrinos. The validity of such limits depends on the accuracy with which we can calculate the associated nuclear matrix elements. $2\nu\ \beta\beta$-decay is a second order weak process in the standard model, occurring through two sequential β-decays

$$n \to p + e^- + \bar{\nu}_e \ . \qquad (1)$$

with excitation of virtual states in the intermediate nucleus. Thus a comparison of predicted and observed 2ν $\beta\beta$-decay rates provides a meaningful test of our ability to calculate two-body nuclear matrix elements. Success in reproducing experiment rates would enhance our confidence in related calculations of 0ν matrix elements.

A comparison is made in Table I between experimental results for 2ν $\beta\beta$-decay rates and predictions from a weak-coupling shell model, with the β-decay operators evaluated in the allowed approximation (only Gamow-Teller and Fermi operators). The only significant discrepancy is for ^{130}Te, where the theoretical half-life is significantly shorter than the geochemically determined total half-life. The theoretical predictions are those of Haxton and Stephenson[1], when $g_A^{eff} = 1.0$ is used for the Gamow-Teller operator[2], as indicated by studies of β-decay (recall $g_A^{free} \simeq 1.25$). The predicted ^{130}Te matrix element is similar in size to those found in ^{76}Ge and ^{82}Se, while that extracted from the geochemical measurements is comparable to or smaller than the limit on the celebrated "suppressed" matrix element in ^{48}Ca (where an approximate rotational selection rule operates). This anomaly becomes even more stark when one considers that the total double Gamow-Teller strength for a given nucleus will scale crudely as the number of pairs of unpaired neutrons. Thus the relative fraction of this strength exhausted by the transition to the ^{130}Xe ground state is unusually small.

Table I. A comparison of experimental 2ν $\beta\beta$-decay half lives with those predicted from a weak-coupling shell model and the allowed approximation.

	Expt. $\tau_{1/2}(10^{20}$ yr)	Theory $\tau_{1/2}(10^{20}$ yr)
^{48}Ca $\rightarrow$ ^{48}Ti	(lab.) ≥ 0.36 Ref. 3	0.72
^{82}Se $\rightarrow$ ^{82}Kr	(lab.) $1.1^{+0.8}_{-0.3}$ Ref. 4 (geochem.) 1.45 Ref. 5 (geochem.) 1.0 Ref. 6	0.64
^{130}Te $\rightarrow$ ^{130}Xe	(geochem.) 24.9 Ref. 7 (geochem.) 7.0 Ref. 6	0.42

The allowed approximation for β-decay consists of taking the nonrelativistic limit for the hadronic current, and of using the long-wavelength approximation for the outgoing e^- and $\bar{\nu}_e$. This approximation is universally used in nuclear physics calculations of 2ν double beta decay rates. The result is an amplitude for $0^+ \rightarrow 0^+$ double beta decay

$$\sum_I \langle f|H_{\beta^-}|I\rangle \frac{1}{E_i - E_e - E_{\bar{\nu}} - E_I} \langle I|H_{\beta^-}|i\rangle \tag{2}$$

where H_{β^-} is either the Gamow-Teller $(\vec{\sigma}\tau_+)$ or Fermi (τ_+) operator. In the large-basis shell model calculations of Ref. 1 this expression is simplified by replacing the energy denominator by an average value $< E >$, where $< E >$ was chosen to match measured β^--decay strength functions. (There is a rather elegant means of avoiding this approximation that can be implemented even in very large shell model calculations. The technique depends on an expression for the nondiagonal matrix elements of the resolvent operator $(E - H)^{-1}$ in terms of continued fractions. The terms in the continued fraction can be evaluated efficiently with the Lanczos algorithm.[8]) With these approximations only two nuclear operators arise

$$M_{GT} \equiv g_A^{eff\;2} \langle 0_f^+ \| \frac{1}{2} \sum_{ij} \vec{\sigma}(i) \cdot \vec{\sigma}(j) \tau_+(i) \tau_+(j) \| 0_i^+ \rangle$$

$$M_F \equiv g_V^2 \langle 0_f^+ \| \frac{1}{2} \sum_{ij} \tau_+(i) \tau_+(j) \| 0_i^+ \rangle = g_V^2 \; \frac{1}{2} \; \langle 0_f^+ \| (T_+)^2 \| 0_i^+ \rangle \; . \qquad (3)$$

where $g_V = 1$ is the vector coupling constant. M_F contributes only if isospin symmetry is broken. First-order perturbation estimates of M_F (which arise through the tensor piece of the Coulomb interaction) yield $M_F \ll M_{GT}$. The 2ν $\beta\beta$-decay rate can then be written[1]

$$\omega_{2\nu} = A |M_{GT} - M_F|^2 \; , \qquad (4)$$

where

$$A \simeq \frac{(G_F \cos\theta_c)^4 2 m_e^{11}}{7!\pi^7} \left[\frac{\mathcal{F}^{PR}(Z_f)}{E_i - < E > - \frac{1}{2} W_0} \right]^2 \left[1 + \frac{\tilde{T}_0}{2} + \frac{\tilde{T}_0^2}{9} + \frac{\tilde{T}_0^3}{90} + \frac{\tilde{T}_0^4}{1980} \right] \tilde{T}_0^7 \; , \qquad (5)$$

where $m_e \equiv$ electron mass, $\mathcal{F}^{PR}(Z_f) \equiv$ Coulomb correction factor for outgoing electrons, $W_0 \equiv M_i - M_f =$ total energy available in the decay, $T_0 \equiv W_0 - 2m_e$ $=$ total kinetic energy of leptons, and $\tilde{T}_0 \equiv T_0/m_e$. In actual calculations $< E >_{GT} \neq < E >_F$ and the exact phase space integrals are carried out rather than using the approximation of Eq. (5) (see Ref. 1 for details).

AXIAL CHARGE OPERATOR

The axial charge operator is a first forbidden correction to β-decay and is given by $g_A^{eff} \vec{\sigma} \cdot \vec{P}/M$, where $\vec{P}$ is the nucleon momentum operator and M is the nucleon mass. It has been fairly well established[9] from studies of axial charge β-decay (e.g. ^{16}O) that $g_A^{eff} \simeq 1.7 \; g_A^{free} \simeq 2$; this enhancement arises from the strong one-pion exchange current contributions, which are of the same order (v/c) as the one-body operator. In the context of double beta decay, this operator is intriguing. In the limit $q_\mu \to 0$ it is the only nonvanishing correction of order $\frac{v}{c}$ $(\sim \frac{1}{5})$ to the Gamow-Teller and Fermi operators. Because of the parity change, it is not Pauli-blocked appreciably: the axial-charge operating on either the initial or final state in double beta decay will be strong. This is in contrast to the Gamow-Teller operator, where the β^+ strength for the final nucleus can be sharply reduced by the neutron excess in heavier nuclei. Indeed, it was our expectation that the favorable nuclear structure physics could compensate in part for the v/c suppression that led us to investigate this operator.

The corresponding double axial charge operator in double beta decay couples the initial and final states through 0^- excitations of the intermediate nucleus. In the closure approximation the results analogous to Eqs. (3) and (4) are

$$M_{AC} \equiv (g_A^{eff})^2 \langle 0_f^+ \| \frac{1}{2} \sum_{ij} \vec{\sigma}(i) \cdot \frac{\vec{\nabla}(i)}{iM} \vec{\sigma}(j) \cdot \frac{\vec{\nabla}(j)}{iM} \tau_+(i)\tau_+(j) \| 0_i^+ \rangle \tag{6}$$

$$\omega_{2\nu} = A|M_{GT} - M_F - M_{AC}|^2 \; . \tag{7}$$

However, in the schematic calculations we describe in the next section, we avoid the closure approximation and do the 1/E-weighted sum explicitly.

In the canonical shell model spaces for ^{76}Ge/^{82}Se $(2p - 1f_{5/2} - 1g_{9/2})$ and ^{130}Te $(3s - 2d - 1g_{7/2} - 1h_{11/2})$ there are no single-particle orbitals with the same spin but opposite parity. Thus in such models nonzero axial charge matrix elements arise only through core polarization. One important aspect of the effective NN interaction in nuclei is the strong quadrupole-quadrupole interaction responsible for nuclear deformation. In Fig. 2 we illustrate schematically how this can act in first order to produce a significant matrix element of M_{AC}. In this illustration M_{GT} would also vanish in the absence of core polarization: the β^+ strength is completely blocked.

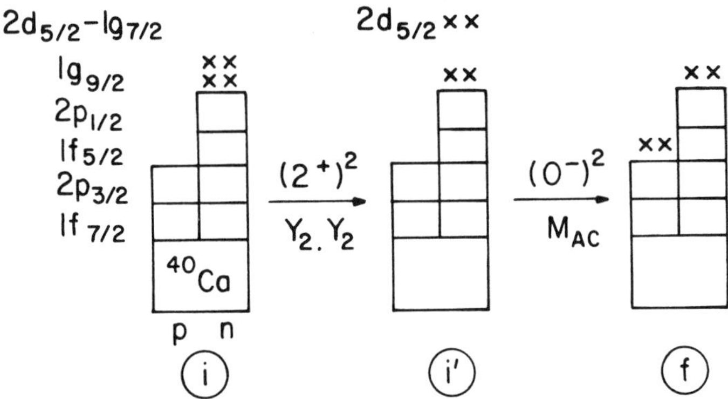

Fig. 2. Schematic illustration of how the quadrupole-quadrupole interaction $(Y_2 \cdot Y_2)$ can enhance the contribution of the double axial charge operator (M_{AC}) to 2ν $\beta\beta$-decay. The example shown is for $^{76}\mathrm{Ge} \rightarrow {}^{76}\mathrm{Se}$. Similar arguments can be made for other nuclei of interest. There is strong mixing of i and i' in the initial state due to $(Y_2 \cdot Y_2)$.

RESULTS AND CONCLUSIONS

The matrix elements of M_{GT} and M_F were taken from Ref. 1, which diagonalized in a weak coupling shell model an interaction based on the Kuo g-matrix with small effective multipole-multipole terms and the Coulomb force added. The closure approximation was used.

For the new calculations of M_{AC} presented here we used the Nilsson pairing model, which provides a simple prescription for including the quadrupole-quadrupole interaction. We employed harmonic oscillator wavefunctions, which have the attractive property that the underlying sum rules governing the axial charge operator can be satisfied exactly by including all excitations with principal quantum numbers $N' = N \pm 1$. The pairing gaps used for ^{76}Ge/^{82}Se and for the Te isotopes were 1.35 MeV and 1.05 MeV, respectively. We described the 0_i^+ and 0_f^+ ground states as zero quasiparticle excitations; we then could avoid the closure approximation, completing the sum over a complete set of two-quasiparticle excitations in the intermediate nucleus. The calculated excitation energies were adjusted so that the lowest two quasiparticle states would correspond to the ground state of the intermediate nucleus.

The results of these calculations were to increase the predicted 2ν $\beta\beta$-decay half-lives of ^{76}Ge, ^{82}Se, ^{128}Te, and ^{130}Te by 15%, 25%, 15%, and 20% respectively. The results for ^{82}Se and ^{130}Te are shown in Table II. We see that M_{AC} cancels against M_{GT} and so increases the half-lives. Although not proved here, we believe this is an exact result in the asymptotic limit of the Nilsson pairing model. Thus the magnitude of M_{AC} in a more realistic model (which would include the tensor interaction, for instance) is clearly of great interest. Note that the discrepancy between the theoretical and laboratory rates for ^{82}Se is reduced by about a factor of two when M_{AC} is included. The net result is agreement at about the level of the 1σ uncertainty in the laboratory number.

Table II. Effect on predicted 2ν $\beta\beta$-decay half lives from the inclusion of the axial charge operator (indicated by [...]).

	Expt. $\tau_{1/2}(10^{20}\text{yr})$	Theory $\tau_{1/2}(10^{20}\text{ yr})$
$^{82}\text{Se} \rightarrow {}^{82}\text{Kr}$	$1.1^{+0.8}_{-0.3}$ 1.45 1.0	$0.64 + [0.17] = 0.81$
$^{130}\text{Te} \rightarrow {}^{130}\text{Xe}$	24.9 7.0	$0.42 + [0.09] = 0.51$

While the size of the M_{AC} correction in the case of ^{130}Te is small compared to the existing discrepancy, M_{AC} is still significant compared to the measured geochemical rate. In fact, we see that $|M_{AC}|^2$ is roughly comparable to $|M_{exp}|^2$

needed to explain the ^{130}Te half life. This is another illustration of the suppression implied by that experimental number. Also, this suggests that dialing nuclear physics matrix elements to reproduce the ^{130}Te half life involves a new uncertainty: the magnitude of M_{GT}^{exp} is only known up to $\pm M_{AC}$.

In closing, we stress the preliminary nature of these results. If we interpret our Nilsson-pairing results in terms of a core-polarization shell model calculation, we see that the quadrupole-quadrupole contribution corresponds to those two-nucleon diagrams where the parity of each nucleon is changed. There are also substantial shell model two-nucleon density matrix elements where neither parity changes. The tensor force would contribute through these diagrams, and nothing we could do in the Nilsson-pairing model would tell us about this magnitude or sign. Clearly a careful calculation with a realistic G-matrix is needed. Furthermore, there are other forbidden terms that should be studied. For instance, we are in the process of examining the 3-current vector operator, which has some interesting properties because of Siegert's theorem. These studies will be published shortly.[10]

REFERENCES

1. W.C. Haxton and G.J. Stephenson, Jr., Prog. Part. Nucl. Phys. 12, 409 (1984).
2. C.D. Goodman and S.D. Bloom, *Spin Excitations in Nuclei*, edited by F. Petrovich *et al.*, (Plenum, New York, 1982), p. 143.
3. R.K. Bardin, P.J. Gollan, J.D. Ullman, and C.S. Wu, Phys. Lett. 26B, 112 (1967); Nucl. Phys. A158, 337 (1970).
4. S.R. Elliott, A.A. Hahn, and M.K. Moe, Phys. Rev. Lett. 59, 2020 (1987).
5. T. Kirsten, in *Science Underground*, AIP Conference Proc. 96, edited by M.M. Nieto *et al.* (1983).
6. O.K. Manuel, in *Proceedings of the International Symposium on Nuclear Beta Decay and Neutrinos, Osaka, Japan, 1986*, edited by T. Kotani, H. Ejiri, and E. Takasugi (World Scientific, Singapore, 1986), p. 71.
7. T. Kirsten, H. Richter, and E. Jessburger, Phys. Rev. Lett. 50, 474 (1983); Z. Physik C16, 189 (1983).
8. W.C. Haxton, in preparation.
9. E.G. Adelberger and W.C. Haxton, Ann. Rev. Nucl. Part. Sci. 35, 501 (1985); I.S. Towner, *ibid.* 36, 115 (1986).
10. W.C. Haxton and A.G. Williams, in preparation.

EXCHANGE EFFECTS IN DOUBLE BETA DECAY†∮

G. J. Stephenson Jr.*
University of New Mexico, Albuquerque, NM 87106

T. Goldman
Los Alamos National Laboratory, Los Alamos, NM 87545

K. R. Maltman
York University, North York, ON M3J 1P3, CANADA

K. E. Schmidt
New York University, New York, NY 10003

Over the past decade there has been very impressive progress in the laboratory study of double beta decay with very precise limits on 0-neutrino decay in ^{76}Ge[1], the imminent prospect of the observation of 2-neutrino decay in ^{100}Mo[2] and the first laboratory observation of 2-neutrino decay in ^{82}Se[3]. For the last case, the laboratory rate is in essential agreement with geochemical results and in reasonable agreement with theoretical predictions based on a full shell model calculation[4].

The motivation underlying the resurgence of interest in double beta decay is the hope that the observation of or limits on the 0-neutrino mode will provide information about the nature of the neutrino[5]. This clearly requires confidence in the nuclear matrix elements involved in the transition. The shell model calculations[6] do not agree well with the geochemical values for ^{130}Te, which has led to a spate of papers offering specific fixes for the problem. In this contribution we shall not comment on any of the specific nuclear calculations, rather we make some remarks which should be relevant to any model calculation.

These remarks, elaborated below are:

a) Due to quark interchange (viewed as tunneling, as pion exchange or as multiquark bag components) there are components of the transition depending on the short range behavior of the two neutron wave function for 2-neutrino decay as well as for 0-neutrino decay.

b) Although the specific matrix elements differ, there is a similar dependence on the structure of the wave function for pion double charge exchange.

c) Due to improved experimental techniques, it should be possible to measure a few double beta decay rates and couple this with more extensive pion double charge exchange measurements to constrain the nuclear matrix elements.

∮ Presented by G. J. Stephenson, Jr.
† Work supported in part by the USDOE
* Permanent address: Los Alamos National Laboratory

Were it not for the existence of pairing in nuclear ground states, double beta decay would be prohibitively hindered by the mismatch, for neutron rich nuclei, between the last filled neutron orbitals and the first empty proton orbitals. Within a single parity shell, the relative wave function for paired nucleons peaks at 0 and 180 degrees. If both parites are active, the backward peaking is suppressed. Therefore we expect the dominant pair of nucleons in the double beta decay to be physically close, subject to the usual Brueckner anti-correlation. In this configuration the requirements of quark antisymmetry generate exchange effects which can give rise to distinct pieces of the matrix element. In addition, important effects may occur as a result of pion exchange currents [7] and the distortion of nucleon quark wave functions in the medium (by which we also include effects induced by possible multiquark bag components of the nuclear wave function.[8]) Using a relativistic potential model for hadrons which we have applied to ^4He[9] we have calculated the overlap between quark wave functions centered in different wells as a function of the separation between the well centers, which we display in Figure 1.

Taking the full space-color-isospin-spin symmetry into account, including all 9 pairs, the exchange term leads to a small enhancement of the matrix element, 19/75 X $(\text{overlap})^2$ X normal matrix element. As Vergados noted[7] in the context of pion exchange, such short ranged terms can be dominant for 0-neutrino decay. It thus becomes important to find a method for sorting out the contributions to the normal, 2-neutrino decay from correlated and uncorrelated components in the nuclear wave function.

This leads us to consider pion charge exchange. Many people have suggested for some time that the nuclear matrix for forward double charge exchange should be the same as for double beta decay. Given the very different momentum dependencies involved, especially in the intermediate state, this claim seems to be overly optimistic. However, the discrete operations are the same and recent theoretical studies at Los Alamos, which we shall now summarize, indicate that the same aspects of the nuclear wave functions come into play.

Auerbach, Gibbs and Piasetzky[10] have shown that, for isotopes with only active neutrons in a single j-shell (the Calcium isotopes), the pion double charge exchange (non-analog) cross-section can be parameterized as

$$\sigma(\Theta) = \frac{n(n-1)}{2} \left| \alpha + \frac{\beta}{(n-1)} \right|^2 \tag{1}$$

where the first term refers to sequential scattering from uncorrelated neutrons and the second term to (short-ranged) scattering from correlated neutrons. These formulas have been extended[11], using generalized seniority wave functions, to allow both neutrons and protons in the active space and including more than one j-shell. The basic structure of the amplitude remains the same. This allows one to study the relative importance of the short-range correlations systematically, using both isotopes and isotones.

Note that such a program does not directly lead to double beta decay matrix elements, but does give a strong handle on the relevant description of the nuclear wave functions. To have a serious impact on double beta decay, eventually allowing one to convert 0-neutrino limits into statements about neutrino properties, it will be necessary to have measurements of 2-neutrino rates for a (small) number of nuclei in the same region of the periodic table. In Table 1, which is a partial reproduction of Table 1A in Haxton and Stephenson[5], we note some candidates with large energy release that cluster near ^{82}Se, ^{100}Mo, or ^{130}Te.

Given the existing efforts in ^{82}Se and ^{76}Ge, as well as the relative success of shell model calculations there, this may well be the most fruitful region to study, provided that it is possible to carry out the necessary systematic pion double charge exchange measurements throughout the region.

Some caveats are in order. First, the seniority scheme is presented as an approximation to the full interacting shell model. Therefore the proper blend of short range correlation and "uncorrelated" nucleons should be present, awaiting a proper treatment of exchange effects. The approach suggested here will not replace complete calculations, but may allow one to focus on the dominant features of the wave function and study their systematics over several cases.

Second, the quark exchange described above is, in some sense, already present in nuclear wavefunctions, albeit masquerading as effective operators. For example, using a quenched value of G_A and enumerating quark exchanges may double count. This requires careful study. Third, while the double beta decay and pion double charge exchange transitions may probe the same pieces of the initial and final wave functions, the range of intermediate states may be different. Nevertheless, the possibility of including a few distinct double beta decays with systematic pion studies should provide a handle on the difference.

To summarize, the very great strides in experimental techniques now open the possibility of a systematic study of double beta decay, coupled with pion double charge exchange, in a limited region of the periodic table to minimize the uncertainties due to models for nuclear structure. The elucidation of the effect of short range correlations will permit us to extract reliable information about neutrino properties from the results on the 0-neutrino mode.

Table 1. Selected possible $\beta^-\beta^-$ decay parent isotopes and the corresponding energy release.

Transition	Energy Release (keV)
$^{76}Ge \rightarrow ^{76}Se$	2045
$^{82}Se \rightarrow ^{82}Kr$	3005
$^{86}Kr \rightarrow ^{86}Sr$	1249
$^{94}Zr \rightarrow ^{94}Mo$	1148
$^{96}Zr \rightarrow ^{96}Mo$	3350
$^{100}Mo \rightarrow ^{100}Ru$	3033
$^{104}Ru \rightarrow ^{104}Pd$	1301
$^{110}Pd \rightarrow ^{110}Cd$	2014
$^{124}Sn \rightarrow ^{124}Te$	2278
$^{130}Te \rightarrow ^{130}Xe$	2533
$^{136}Xe \rightarrow ^{136}Ba$	2481

REFERENCES

1. D. O. Caldwell, R. M. Eisberg, D. M. Grumm, M. S. Witherell, F. S. Goulding, and A. R. Smith, Phys. Rev. Lett. 59, 419 (1987); P. Fisher, F. Boehm, E. Bovet, J. P. Egger, K. Gabathuler, H. Henrikson, and J. L. Vuilleumier, Phys. Lett. B192, 460 (1987); F. T. Avignone III, R. L. Brodzinski, H. S. Miley, and J. H. Reeves, Phys. Lett. B198, 253 (1982).
2. M. Alston-Garnjost, B. Dougherty, R. Kenney, J. Krivicich, R. Tripp, H. Nicholson, S. Sutton, B. Dieterle, J. Kang, and C. Leavitt, Phys. Rev. Lett. 60, 1928 (1988).
3. S. R. Elliot, A. A. Hahn and M. K. Moe, Phys. Rev. Lett. 59, 1649 (1987).
4. W. C. Haxton, G. J. Stephenson, Jr., and D. Strottman, Phys. Rev. Lett. 47, 153 (1981).
5. see e.g., W. C. Haxton and G. J. Stephenson, Jr., Progress in Particle and Nuclear Physics 12, 409 (1984).
6. W. C. Haxton, G. J. Stephenson, Jr., and D. Strottman, Phys. Rev. D25, 2360 (1982).
7. J. D. Vergados, Phys. Rev., D25, 914 (1982).
8. M. Lusignoli and M. G. Schepkin, I.N.F.N. preprint n. 449 (1985).
9. T. Goldman, K. R. Maltman, G. J. Stephenson, Jr., and K. E. Schmidt, Nucl. Phys. A, A481, 621 (1986).
10. N. Auerbach, W. R. Gibbs, and E. Piasetzky, Phys. Rev. Lett. 59, 1076 (1987).
11. N. Auerbach, W. R. Gibbs, J. N. Ginocchio, and W. B. Kaufman, Los Alamos National Laboratory document preprint LA-UR-88-781 (1988).

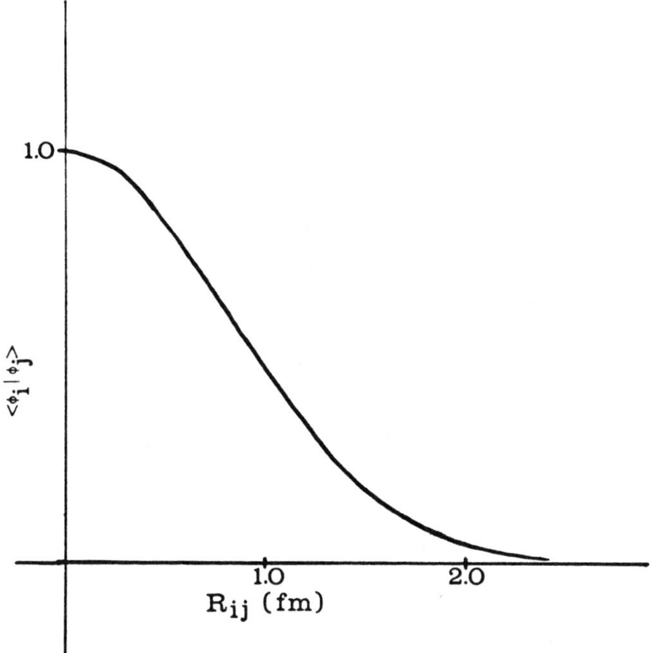

Fig. 1. Overlap integral as a function of well separation[9].

NEUTRINO MASS*

Boris Kayser
Lawrence Berkeley Laboratory, University of California
Berkeley, California 94720
and
Division of Physics, National Science Foundation, Washington, D.C. 20550†

ABSTRACT

An introduction is given to the search for evidence of neutrino mass.

From the standpoint of the grand–unified theories, it is natural for neutrinos to have non–zero masses.[1] However, we know experimentally that, at most, the neutrino in each generation is much lighter than the associated charged lepton and quarks. The most popular explanation of this relative lightness is the "see–saw mechanism." This predicts that each neutrino mass M_ν obeys a "see–saw relation" of the form $M_\nu M \approx$ [Typical quark or charged lepton mass]^2, where M is a very large mass scale. Very importantly, the see–saw mechanism also predicts that neutrinos are their own antiparticles. A recent variation of the see–saw picture is a model which is constructed to account for the observed lightness of the electron relative to the W boson, but which then automatically predicts the see–saw relation $M_{\nu_e} M \approx M_e^2$, where M is once again a heavy mass scale.[2]

Unfortunately, even if one accepts the see–saw mechanism, gauge theories do not predict neutrino masses in detail. Indeed, with reasonable variations of the parameters in the see–saw models, one can obtain neutrino masses from 10^{-11}eV to 10^{+11}eV. Considering how small the neutrino masses may be, it is important to try to find ways of observing them even if they are much less than 1eV.

The evidence concerning M_{ν_e} coming from studies of tritium beta decay is discussed in these Proceedings by Wilkerson. Interestingly, the upper limit on M_{ν_e} derived from studies of the arrival times of the neutrinos from supernova 1987A[3] is of the same order as the upper limits reported by several tritium experiments.

Evidence of neutrino masses smaller than 1 eV can be sought by searching for the flavor oscillation of neutrinos for which $L(km)/E(\text{GeV}) > 1$. Here L is the distance travelled by the neutrinos between production and detection, and E is their energy. Among the neutrinos with $L/E > 1$ are some of those produced in the atmosphere by cosmic ray protons, and detected in underground detectors.

---

*This is a very–highly abbreviated version of the talk given at the Conference, meant primarily to call attention to the new developments mentioned in the talk, and to give references to some of the neutrino mass literature.

†Permanent address

These cosmic–ray neutrinos are generated largely through the reaction [cosmic ray p] + [Air] $\rightarrow \pi + \ldots$, followed by $\pi \rightarrow \mu\nu_\mu$, and then by $\mu \rightarrow e\nu_e\nu_\mu$. (We shall not distinguish between a ν and a $\bar{\nu}$.) Obviously, this chain of reactions produces neutrinos in the ratio $\nu_\mu : \nu_e = 2 : 1$. Of course, if one does not detect all cosmic ray neutrinos, but only those in a certain energy range, then the ratio of detected muon neutrinos to detected election neutrinos need not be $2 : 1$, but one should be able to correct for this effect. Now, the Kamiokande–II collaboration has reported that the observed $\nu_\mu : \nu_e$ ratio does not agree with theoretical expectations.[4] For charged leptons produced by the cosmic ray neutrinos in the detector with momenta in the range (200 – 700) MeV/c, the observed anomaly is $R \equiv (\nu_\mu : \nu_e)_{observed}/(\nu_\mu : \nu_e)_{expected} = 0.5 \pm 0.1$.[5] *Conceivably*, this anomaly is due to neutrino oscillation. For example, it could be due to oscillation between ν_μ and ν_τ, with a large mixing angle, and with a neutrino mass (carried by a neutrino mass eigenstate) between 0.03 eV and 0.6 eV.[6] This mass is large enough to allow the neutrinos to oscillate between their production and detection, but not so large as to violate bounds from other experiments. It should be noted, however, that the theory of $(\nu_\mu : \nu_e)_{expected}$ is undergoing further analysis, and the value of R may change.[7]

With the discovery of the Mikheyev–Smirnov–Wolfenstein effect, we have learned that future observations of solar neutrinos may provide evidence for neutrino masses in the range between 10^{-4} eV and 10^{-2} eV. Such masses are, of course, even smaller than those probed by studies of cosmic ray neutrinos.

The most sensitive test of whether neutrinos are their own antiparticles remains the search for neutrinoless double beta decay, the nuclear process $(A, Z) \rightarrow (A, Z + 2) + 2e^-$.[8] Barring exotic mechanisms, this process can occur only if neutrinos are their own antiparticles. Needless to say, it has not as yet been seen, although the somewhat related rare process $(A, Z) \rightarrow (A, Z+2)+2e^- +2\bar{\nu}$, which can occur whether or not neutrinos are their own antiparticles, has been seen in the laboratory.[9] The observation of neutrinoless double beta decay would imply not only that neutrinos are their own antiparticles, but also that at least one of them has a mass large enough to be sought in other types of present–generation experiments.[10]

In summary, there are good theoretical reasons to look for neutrino mass, but the neutrino masses may be much smaller than 1 eV. Thus, experiments which are sensitive to tiny masses are of great interest. The see–saw mechanism suggests that neutrinos are their own antiparticles, a possibility which is best explored, so far as we now know, by searching for neutrinoless double beta decay.

REFERENCES

1. For recent general treatments of the physics of neutrino mass, see S. Bilenky and S. Petcov, *Rev. Mod. Phys.* **59**, 671 (1987), F. Boehm and

P. Vogel, *Physics of MassiveNeutrinos* (Cambridge University Press, Cambridge, 1987), and B. Kayser, F. Gibrat–Debu, and F. Perrier, *The Physics of Massive Neutrinos*, to be published by World Scientific.

2. A. Davidson and K.C. Wali, *Phys. Rev. Lett.* **59**, 393 (1987).

3. See, for example, E. Kolb, A. Stebbins, and M. Turner, *Phys. Rev.* **D35**, 3598 (1987), and references therein.

4. K. Hirata et al., University of Tokyo preprint UT–ICEPP–88–02.

5. This statement is an expression of the data of Ref. 4 due to J. Learned, S. Pakwasa, and T. Weiler, University of Hawaii preprint UH–511–643–88.

6. J. Learned et al., Ref. 5, and V. Barger and K. Whisnant, University of Wisconsin preprint MAD/PH/414.

7. E. Beier, private communication.

8. For a recent discussion of the question of whether neutrinos are their own antiparticles, and of the special advantages of neutrinoless double beta decay as a probe of this question, see B. Kayser, to appear in the proceedings of the Eighth Moriond Workshop on Neutrinos and Exotic Phenomena, held in Les Arcs, France, January, 1988.

9. S. Elliott, A. Hahn, and M. Moe, *Phys. Rev. Lett.* **59**, 2020 (1987); A. Hahn, these Proceedings.

10. B. Kayser, in *Proceedings of the Seventh Moriond Workshop on New and Exotic Phenomena*, edited by O. Fackler and J. Tran Thanh Van (Editions Frontieres, Gif–sur–Yvette, France, 1987), p. 349; B. Kayser, S. Petcov, and S. P. Rosen, in preparation.

DOUBLE BETA DECAY IN ^{82}Se

S.R. Elliott, A.A. Hahn, and M.K. Moe
Physics Department, University of California, Irvine 92717

ABSTRACT

A Time Projection Chamber (TPC) has been constructed to search for double beta decay in ^{82}Se. The lifetime for the $2\nu2\beta$ is $\tau_{1/2} = 1.4(^{+6}_{-3})\ 10^{20}$ years (68%C.L.). Limits for $0\nu2\beta$ decay with and without majoron emission are also given.

INTRODUCTION

The detection of neutrinoless double beta ($0\nu2\beta$) decay would provide unique information on the nature of the neutrino. $0\nu2\beta$ decay requires the neutrino to be a majorana particle. By definition a majorana particle is one which is indistinguishable from (or self-conjugate to) its own antiparticle. Prior to the discovery of parity violation, the absence of a $0\nu2\beta$ signal was taken as evidence that the neutrino was a Dirac particle, distinct from the antineutrino. However since parity violation is incorporated in the standard model by allowing only left-handed weak currents, the absence of $0\nu2\beta$ decay might just signify the mismatch of helicities, i.e. the virtual neutrino is emitted right-handed while, to be absorbed, it must be left-handed. This suppression can be broken if the neutrino has mass as the helicity is exactly ±1 only in the massless limit. In addition, a small right-handed component in the standard weak interaction current would also drive $0\nu2\beta$ decay. Kayser[1] has pointed out that a right-handed current would actually generate a neutrino mass. Therefore, the observation of $0\nu2\beta$ decay requires the neutrino to have mass and be a majorana particle.

Some models[2,3,4] predict another form of $0\nu2\beta$ decay involving the emission of a massless neutral particle known as the majoron (B). The majoron is the Goldstone Boson associated with the spontaneous breaking of a global (Baryon-Lepton) number symmetry. Experimentally $0\nu2\beta$ decay and $0\nu(B)2\beta$ are distinguishable by the energy spectra of the two electrons. In $0\nu2\beta$ decay, the entire kinetic energy release is given to the electrons and this appears as a spike in the sum energy spectrum. The sum energy spectrum for the $0\nu(B)2\beta$ decay is continuous since the majoron, which is undetectable, carries off a portion of the energy. Figure 1 depicts the sum energy spectrum of the $0\nu2\beta$, $0\nu(B)2\beta$, and $2\nu2\beta$. The last decay is just the expected second order weak interaction decay.

An important factor in all 2β decay modes is the nuclear matrix element between the initial and final nucleus. The interpretation of a $0\nu2\beta$ decay result in terms of the particle properties of the neutrino depends upon the value of this matrix element. Currently there are several intensive efforts being made to improve the calculation of the matrix elements for both the $0\nu2\beta$ and $2\nu2\beta$ decay modes. Even though the $0\nu2\beta$ decay matrix element differs from the $2\nu2\beta$ decay one, it is hoped that an experimental measurement of the $2\nu2\beta$ decay rate will provide a calibration or "figure of merit" for the various calculations.

The ^{82}Se TPC experiment is sensitive to all modes of 2β decay. The TPC is described in detail elsewhere.[5] Essentially we have a foil containing 14 grams of 97% enriched ^{82}Se (9.8 10^{22} atoms). The TPC allows us to identify two electrons leaving the same source point. The energies are determined by fitting helices to the ionization tracks left by the electrons as they spiral about a 700 gauss magnetic field. The tracking capability enables this experiment to be very sensitive to the continuous distributions of the $2\nu2\beta$ and $0\nu(B)2\beta$ electron sum spectrum in spite of the low mass of the 2β decay isotope in comparison with other experiments. Only in the $0\nu2\beta$ decay where the sum energy spike itself is a strong background suppression does the TCP lose ground to other experiments with greater mass and/or better energy resolution.

$2\nu2\beta$ DECAY

Even though (or perhaps because) it is theoretically expected, $2\nu2\beta$ decay is experimentally the most daunting mode of 2β decay. The electron sum ($T_1 + T_2$) energy spectrum is soft, peaking, in the case of ^{82}Se, at 1.1 MeV. The opening angle distribution between the two electrons is $1-\beta_1\beta_2\cos(\Theta_{12})$ ($\beta = v/c$). The background for 2β decay in the TPC comes primarily from two sources: gamma ray activity external to the source and natural beta decay activity on or within the source itself. The external gamma activity creates Compton electrons in the source foil, a small fraction of which, through Møller scattering, produce two electrons emerging from the same point. Although Møller scattering has a unique opening angle, multiple scattering within the foil broadens this angle (especially since one of the two emerging electrons usually has low energy). The second source of background, beta decay, can produce two electrons from either Møller scattering of the original β or from an internal conversion electron from the excited daughter nucleus. In principle, the internal conversion electron from a particular transition has a unique energy, but the energy resolution of the TPC and the many possible transition energies largely prevent us from using this technique to eliminate these events.

The Møller produced background spectrum can be calculated from the measured single electron spectrum from the foil. The internal conversion (I C.) background spectrum is taken from the measured $^{208}Tl \xrightarrow{\beta} {}^{208}Pb$ + (I.C.) spectrum (the only known natural radioactive source which we have not eliminated by our background reduction techniques). The opening angle spectrum from both cases have also been taken from actual measured distributions in the TPC.

In order to obtain the $2\nu 2\beta$ halflife, we have applied several kinematic cuts to the data. The first required both electrons to have kinetic energies greater than 150 keV. This greatly reduces the number of Møller background events, but decreases the efficiency for $2\nu 2\beta$ by only 20%. Furthermore, to eliminate low and high energy backgrounds, we have restricted the sum energy to be between 1.3 and 2.0 MeV. Finally we required the topology of the events to be those in which the electrons are emitted on opposite sides of the source foil. This last requirement was imposed to simplify the understanding of the TPC opening angle bias. Events which are on the same side of the source foil intertwine with each other, complicating the opening angle bias of our automated event finder. The bias for opposite side events has been calculated both by Monte Carlo and by observing the acceptance for actual single electron events. Both methods agree quite well. The total $2\nu 2\beta$ decay efficiency for this analysis is 6.2(5)%.

We have applied a maximum likelihood analysis for the data from TPC1. This run of 7960 hours had 46 events between 1.3 and 2.0 MeV. The likelihood function was calculated using the Sum ($T_1 + T_2$), the Singles (T_1 and T_2), and cos (Θ_{12}) spectra. The experimental spectra are shown in figure 2 along with the expected spectra for the $2\nu 2\beta$, Møller and I.C. components, each normalized to the integral rate. The confidence levels were calculated by projecting the likelihood values onto the $2\nu 2\beta$, Møller, and internal conversion (I.C.) axes. The results of this process are given in the following table.

	Number of Events		
	68% C.L.	90% C.L.	Best
$2\nu 2\beta$	21-39	15-44	30
Møller	4-14	1-20	9
I.C.	0-6	0-15	0

From the 68% confidence level values we calculate the $2\nu 2\beta$ halflife,

$$\tau_{1/2} = 1.3(5)\ 10^{20} \text{ years (68\% C.L.).}$$

In addition from independent observations, we are able to calculate the absolute values of the background within these regions. We expect 9.8(2.3) events due to Møller scattering and 2.4(1.4) I.C. events. These numbers compare quite favorably with the above best fit values. We also did a maximum likelihood analysis with the background constrained to the expected values and uncertainties. Following a similar maximum likelihood procedure as above we find a $2\nu 2\beta$ halflife of

$$\tau_{1/2} = 1.2\ (^{+3}_{-2}) \text{ (68\% C.L.) (constrained background).}$$

This value is basically identical to a halflife calculated from simply subtracting the expected backgrounds from the total number of events.[6]

We also have 6766 hours of data from TPC2. TPC2 uses the same source plane as TPC1, but has a reduced background rate due to its construction from lower activity materials. We have not yet completed a maximum likelihood analysis for TPC2. However between 1.3 and 2.0 MeV the single electron spectrum shape (from which we calculate our Møller spectrum) remains nearly identical (but a factor 1.9 lower) to that in TPC1. Therefore we can simply subtract from the total observed events (22) the expected number of events from the Møller (4.7) and I.C. (1.2) backgrounds. The halflife value is $\tau_{1/2} = 2.0\ (^{+10}_{-5})\ 10^{20}$ years (68% C.L.). The halflife values from TPC1 and TPC2 just overlap at the 68% C.L.

Although TPC1 and TPC2 share the same systematic error, the primary contribution to their uncertainties is statistical. Therefore we combine the results from TPC1 and TP2 in a weighted mean giving a $2\nu 2\beta$ decay halflife of

$$\tau_{1/2} = 1.4\ (^{+6}_{-3})\ \text{(68\% C.L.) (TPC1 + TPC2).}$$

The value of χ^2, for this averaging process is 2.6 for 1 degree of freedom. This has a confidence level of only 12%. Following the example of the Particle Data Group[7] the 68% C.L. range given above has already been increased by $(\chi^2)^{1/2}$. The data from TPC1 and TPC2 are shown together with the 68% C.L. halflife curves from the combined average in figure 3.

0ν(B)2β DECAY

We use only the data from TPC2 since the uncut number of events above 2.0 MeV is much lower than in TPC1 (9 vs 25). For this mode we take a sum energy cut from 2.0 - 3.0 MeV. The majoron mode has 55% of its total strength above 2.0 MeV. We also require that all events have a ratio of T_{lo}/T_{hi} greater than 0.2. This discriminates against the Møller background which is the dominant background at this energy. This Møller cut removes 60% of all Møller events in this energy range which remain after the 150 keV single electron energy cut. Of the 9 events in this sum energy range, three events are removed by the T_{lo}/T_{hi} cut. If we attribute all the remaining events to 0ν(B)2β, the halflife is

$$\tau_{1/2} \geq 6.8\ 10^{20}\ \text{yrs (90\% C.L.).}$$

Since we expect the $2\nu 2\beta$ tail to have (at our halflife) about 2 events, and the Møller feedthrough another 2 events, the actual halflife limit is probably a factor of 2-3 longer. It should be noted that the phase space in ^{82}Se is a factor 8.2 larger than in ^{76}Ge. The square of the ratio of the nuclear physics matrix elements ranges between 0.53-0.78 depending upon the particular author.[8,9,10,11] Thus our result corresponds to a ^{76}Ge halflife limit of 30-43 10^{20} yrs.

$0\nu 2\beta$ DECAY

The endpoint of ^{82}Se is 3.0 MeV. In all the running of this experiment, we have never had a count within a 300 keV window about 3.0 MeV. The total time includes an early 1680 hour run of TPC1 (dubbed TPC0). This corresponds to a $0\nu 2\beta$ halflife of

$$\tau_{1/2} \geq 2.1\ 10^{22} \text{ years (68\% C.L.)}.$$

Using the current range of matrix element calculations,[8,9,10,11] this would correspond to (assuming the neutrino to be majorana particle and ignoring any right-handed current)

$$m_\nu \leq (1.9 - 18) \text{ eV}.$$

This may be compared to the ^{76}Ge values[12] of

$$m_\nu \leq (0.55 - 4.2) \text{ eV}.$$

Both these limits correspond to the (so-called) 68% C.L. As one can see, the our limits are less stringent than the ^{76}Ge values.

CONCLUSION

The TPC has proven to be a powerful device for double beta decay. The pattern recognition enables us to search for the $2\nu 2\beta$ and the $0\nu(B)2\beta$ modes of double beta decay with minimum background subtraction. The major experimental limitation is the amount of mass we have available (38 g or approximately three times what we are currently measuring). We are searching for an underground site to put the TPC. This should lower the Møller background contribution until it is negligible and allow us to insert the remaining 24 g of ^{82}Se. We are also investigating other isotopes for the source, but this is hampered by the lack of available isotopes from Oak Ridge.

The future of double beta decay rests equally upon the reliability of the calculations of the nuclear matrix elements. The $2\nu 2\beta$ decay mode places severe constraints upon these calculations. However, it is also obvious that if we are ever to be serious about the limits we place upon the particle properties of the neutrino, these calculations must be believable and not simply "tuned" to reproduce the $2\nu 2\beta$ rate. Therefore it is important that these calculations have some means of independent confirmation other than the $2\nu 2\beta$ rates.

REFERENCES

1. B. Kayser, these proceedings.
2. H. M. Georgi, S. L. Glashow, and S. Nussinov, Nucl. Phys. B193, 297 (1981).
3. G. B. Gelmini and M. Roncadelli, Phys. Lett. 99B, 411 (1981).
4. Y. Chikashigi, R. N. Mohapatra, and R. D. Peccei, Phys. Lett. 98B, 265 (1981).
5. S. R. Elliott, A. A. Hahn, and M. K. Moe, submitted to

NIM.
6. S. R. Elliott, A. A. Hahn, and M. K. Moe, Phys. Rev. Lett. 59, 2020 (1987). The difference between the $2\nu 2\beta$ decay $\tau_{1/2}$ value reported here and in this earlier paper is due only to our projection of the likelihood function onto the axes. This results in a tighter limit than before.
7. Review of Particle Properties, Phys. Lett. 170b, pp. 6-9, (1986).
8. W. C. Haxton and G. H. Stephenson, Jr., Prog. Part. Nucl. Phys. 12, 409 (1984).
9. K. Grotz and H. V. Klapdor, Phys. Lett. 153, 1 (1985).
10. T. Tomoda and A. Faessler, Phys. Lett. 199b, 475 (1987).
11. J. Engel, P. Vogel, and M. R. Zirnbauer, Phys. Rev. C37, 731 (1988); and P. Vogel, Proceedings of the XIII Rencontres de Moriond, Jan 23-30, 1980.
12. D. O. Caldwell, et al., Phys. Rev. Lett. 59, 419 (1987).

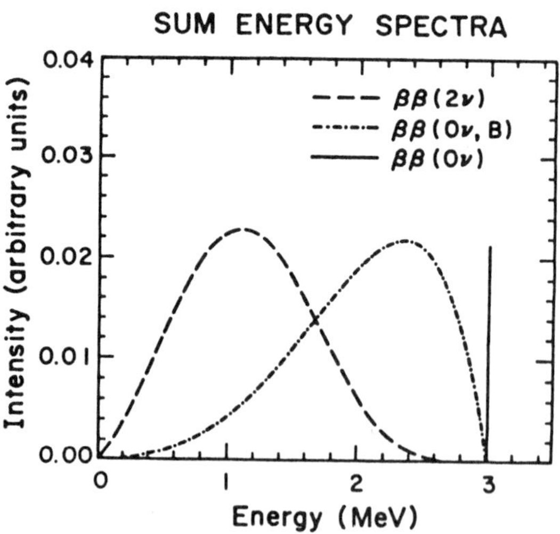

Fig. 1: The theoretical sum energy spectra of the two electrons for three modes of double beta decay in ^{82}Se. The normalizations are arbitrary.

Fig. 2: The histograms are the observed spectra for two-electron events with a sum energy between 1.3 and 2.0 MeV and which have a single energy threshold of 150 keV. Also these events were required to have the electrons emitted on opposite sides of the source plane. The curves show the shapes (normalized to the total intensity) of the three assumed contributions to the data: $2\nu\beta\beta$ (solid curve), Møller scattering (dashed curve), and $^{208}Tl(\beta + I.C.)$ (dot-dashed curve).

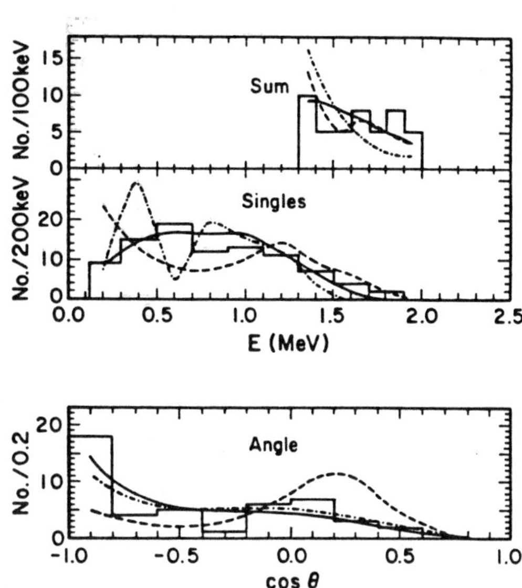

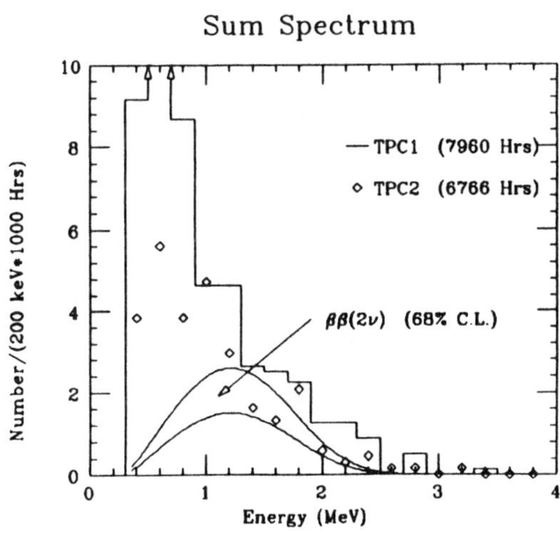

Fig. 3: Sum energy spectrum of all 2β decay candidates with $T_1, T_2 > 150$ keV. The 68% C.L. $2\nu\beta\beta$ decay rates as determined from both TPC1 and TPC2 data are superimposed over the data.

STATUS OF SELECTED NEUTRINOLESS $\beta\beta$-DECAY EXPERIMENTS

F.T. Avignone, III
Department of Physics and Astronomy, University of South Carolina, Columbia, South Carolina 29208

R.L. Brodzinski, J.H. Reeves and H.S. Miley
Pacific Northwest Laboratory[(a)], Richland, Washington 99352

ABSTRACT

A brief update is given on a number of ongoing 0ν $\beta\beta$-decay experiments. Progress is reported in background reduction, limits are given on the various modes of decay, and projections of potential sensitivity are made in some cases. In particular, a new result from the PNL-USC 0ν $\beta\beta$-decay experiment yields a half-life of $(3.2^{+2.2}_{-1.2}) \times 10^{21}$ y (1σ) for decay with the emission of a majoron. While a positive result is still observed, it is now statistically less significant. A new experiment is in progress utilizing ~200-cm^3 detectors fabricated from recently deep-mined Ge to decrease background cosmogenic radioactivity.

INTRODUCTION

One of the best examples of the intersection of nuclear and particle physics is neutrinoless double-beta decay (0ν $\beta\beta$-decay). Observation of this process would unambiguously require that neutrinos must be Majorana particles and, according to the theorem by Kayser, Petkov, and Rosen[1] (KPR), that one flavor must have a mass at least as large as the mass parameter, $\langle m_\nu \rangle$, determined in the experiment. However, we should keep in mind that 0ν $\beta\beta$-decay might also be driven by exotic exchange diagrams involving, for example, Higgs or supersymmetric particles[2]. In addition, the existence of 6-quark states in nuclei could enhance 0ν $\beta\beta$-decay for all such processes significantly, particularly for heavy exchange particles.

The status of the nuclear structure calculations is still not clear, although several Quasiparticle Random Phase Approximation (QRPA) calculations of 2ν $\beta\beta$-decay rates for ^{82}Se and ^{76}Ge agree reasonably well with the shell model[3]. The Caltech group still predicts a nuclear matrix element about one half as large[4] as that of the Shell Model. A fundamental question still exists concerning how much of the particle-particle contribution to short-range correlations is already included in the renormalized two-body interaction matrix elements used in the shell model. This question cannot be declared settled at this time, although calculations now appear to be in better general agreement than before.

A new review of $\beta\beta$-decay through the fall of 1987 should appear in print by the time these proceedings are published[5]. This brief note is

---

[(a)]Operated for the U.S. Department of Energy by Battelle Memorial Institute under contract DE-AC06-76RLO 1830.

intended to present an update of the results on a selection of the experiments which have made, or have the potential to make, a major impact.

GERMANIUM-76 $\beta\beta$-DECAY EXPERIMENTS

The current best limit on the half-life of ^{76}Ge for 0ν $\beta\beta$-decay to the ground state of ^{76}Se is from the 11.2 kg.y of data from the large multi-crystal, Compton-suppressed, University of California, Santa Barbara-Lawrence Berkeley (UCSB-LBL) experiment[6]. The system is comprised of 8 detectors each approximately 160 cm^3. It has an average background of about 1.4 counts/keV/kg/y at the decay energy of 2041 keV. The limit on the half-life derived from the data is 7×10^{23} y if a statistical fluctuation analysis is used, or 1.3×10^{24} y according to a maximum likelihood analysis.

The authors of ref. 6 prefer to present the conservative limit. Assuming 7×10^{23} y, the corresponding limit on the Majorana mass parameter, $\langle m_\nu \rangle$, depends on the theoretical matrix element used. For the Tübingen and the Heidelberg nuclear structure calculations $\langle m_\nu \rangle \lesssim 2$ eV; while the Caltech calculations yield $\langle m_\nu \rangle \lesssim 5.8$ eV. These values, of course, depend on the choice of the values of the coupling constant, g_{pp}, and are not very different from the value $\langle m_\nu \rangle \lesssim 2.9$ eV obtained when the value $g_A/g_\nu = 1.0$, used in all of the current QRPA calculations, is also used with the shell model results.

The UCSB-LBL group also obtain the limits $T^{0\nu}(0^+ \to 2^+) > 2 \times 10^{23}$ y when a simple fluctuation analysis is used and $> 3 \times 10^{23}$ y in a maximum likelihood analysis. This effort continues, and the group is applying semiconductor detector technology to search for dark matter candidates.

Several other large-volume detectors are in operation or are soon to come on line. The Caltech-Neuchatel-SIN 8-crystal detector has been operated in the Gotthard Tunnel since the fall of 1987[7]. The present limits are $T^{0\nu}(^{76}\text{Ge}) \geq 3 \times 10^{23}$ y for the $0^+ \to 0^+$ transition and $\gtrsim 3.4 \times 10^{22}$ y for the $0^+ \to 2^+$ transition. The background is approximately 3.9 counts/keV/kg/y at 2041 keV. The volume is comparable to that of the UCSB-LBL detector, but the background is higher. Some of this background is due to cosmogenic radioactivity in the Ge and will decay with time. In this work, the background spectrum was corrected for a number of gamma-ray peaks and their continua calculated by Monte Carlo modeling techniques. In addition, the continuum from cosmogenically generated ^{68}Ge was determined from the gallium x-ray and was subtracted. The residual spectrum was used to set the following limits: $T^{2\nu} > 5 \times 10^{20}$ y and $T^{0\nu,M} > 1 \times 10^{21}$ y, where 0ν,M indicates 0ν $\beta\beta$-decay with the emission of a majoron.

The Guelph-Queens detector has been upgraded and now has a background of about 1 count/keV/kg/y at 2041 keV. Approximately 2.7 kg of Ge, in 3 crystals, have been operated in the Windsor Salt Mine for several months. The limit is $T^{0\nu}(0^+ \to 0^+) > 1.6 \times 10^{23}$ y. This group has enjoyed impressive success with radioactive background reduction. They have, for example, reduced their background gamma-ray peaks at 239, 583, 911, 1461, and 2615 keV by factors of 72, 139, 47, 1.7, and 47, respectively. The background at 2 MeV continues to decay, and it is not yet clear what the ultimate background will be.

The University of Milano ^{76}Ge experiment has been terminated, with a final result of $T^{0\nu}(^{76}Ge) \gtrsim 3 \times 10^{23}$ y[8]. The detector is now being used to evaluate materials for the group's efforts in ^{136}Xe $\beta\beta$-decay, discussed below, and an exciting new bolometric experiment. Alpha particles have been detected with a 1-cm^3 Ge crystal bolometer. This is being developed for a future ^{128,130}Te $\beta\beta$-decay experiment.

The Pacific Northwest Laboratory-University of South Carolina (PNL-USC) detector has achieved an improvement in background since last year by a factor of 20 for energies above 3 MeV. The background at 2 MeV can be almost completely accounted for by the spectrum of ^{68}Ga, following the electron capture decay of ^{68}Ge. This activity is formed in the crystal via the fast cosmogenic neutron reaction ^{70}Ge(n,3n)^{68}Ge and has a half-life of 271 days. The first two large crystals ($\sim$ 200 cm^3) have been grown from Ge that was deep mined last year and stored underground to avoid the formation of ^{68}Ge, ^{65}Zn, ^{56}Co, and ^{58}Co. The avoidance of ^{68}Ge should result in a background of $\lesssim$ 0.05 counts/keV/kg/y at 2041 keV, or a factor of about 28 lower than the average background in the data from the UCSB-LBL detector after Compton suppression. The planned large array will be in individual modules similar to our current prototype. The $\sim$ 120-cm^3 fiducial volume prototype detector has been used to set the limit $T^{0\nu} \gtrsim 3.0 \times 10^{23}$ y. The search for the continuum $\beta\beta$-decay spectra followed corrections for the ^{68}Ga spectrum and the γ-ray lines from the decay of ^{40}K and ^{60}Co. Small corrections were also made for the Compton continua from these γ-ray lines using Monte Carlo modeling techniques. A maximum likelihood analysis on the residual spectrum results in the limit $T^{2\nu} \gtrsim 1 \times 10^{21}$ y (1σ) and $T^{0\nu,M} = (3.2^{+2.2}_{-1.2}) \times 10^{21}$ y (1σ). While this is still a positive result, it is not statistically very significant, and we prefer to quote a limit $T^{0\nu,M} \gtrsim 2.0 \times 10^{21}$ y.

The Frejus Tunnel ^{76}Ge experiment was a collaboration between the universities of Bordeaux and Zaragoza[9]. It consisted of four Ge detectors with a combined fiducial volume of 417 cm^3. It was housed in a large Compton suppression shield which was also operated in coincidence with the Ge detectors to search for the $0^+ \rightarrow 2^+$ 0ν $\beta\beta$-decay transition at a summed electron energy of 1482 keV. A small peak did appear in the data at 1483.7 keV at a confidence level between 2 and 3σ which would correspond to a half-life of $\sim 2 \times 10^{22}$ y. The search for the peak at the anticipated energy of 1482 keV yields a limit of $T^{0\nu}(0^+ \rightarrow 2^+) \gtrsim 6 \times 10^{22}$ y. The detector has been disassembled, and the Compton suppression detectors have been moved to the University of Zaragoza underground laboratory in the Canfranc Tunnel. A 200-cm^3 detector, designed and fabricated in collaboration with the PNL-USC group, will be installed in the summer of

1988, and several other detectors will follow. The group has not made any claims concerning the peak at 1483.7 keV; however, the resolution of this issue is important. A University of Zaragoza-PNL-USC collaboration is organizing the International Germanium Experiment (IGEX), an effort to fabricate a number of 1-kg Ge detectors enriched to 80-85% in ^{76}Ge.

^{100}Mo EXPERIMENTS

The Osaka experiment, called ELEGANTS IV, is described in the 1988 doctoral dissertation of T. Watanabe of Osaka University[10]. Ten foils, enriched to 94.5% in ^{100}Mo, are located between Si(Li) detectors and placed inside the live NaI(Tl) shield of the old ELEGANTS ^{76}Ge experiment. The array was operated for 1866 hours, and the following limits were deduced from the data: $T^{0\nu}(0^+ \rightarrow 0^+) > 1.6 \times 10^{20}$ y and $T^{0\nu,M}(0^+ \rightarrow 0^+) > 1.9 \times 10^{19}$ y, to a 68% level of confidence. The background in the foils has been significantly reduced, and the experiment continues. It is anticipated that significantly more sensitive results will be forthcoming.

The LBL-Mt. Holyoke-UNM ^{100}Mo experiment consists of 40 Si(Li) detectors with slots for ^{100}Mo foils between them. In their recent report[11], analysis of data from 1284 hours operating with 20 foils and 884 hours with no foils in the array resulted in the following limits: $T^{0\nu,M}(0^+ \rightarrow 0^+) > 3.3 \times 10^{20}$ y, and $T^{2\nu}(0^+ \rightarrow 0^+) > 3.8 \times 10^{18}$ y, to 90% levels of confidence. The limit on $T^{0\nu,M}$ is equivalent to 4.6×10^{21} y for ^{76}Ge, considering phase space alone and assuming that the matrix elements of the two decays are equal. The expectation is that the background in the foils can be reduced by a factor of 5, and the 2ν $\beta\beta$-decay should be observable. This hypothesis will be tested in the near future.

The Baksan ^{100}Mo experiment consists of 4 layers of 50-cm x 50-cm x 5-cm plastic scintillators with an energy resolution of 17.5% at 1 MeV. Two similar scintillators located on the outside serve as anticoincidence shields to reduce background. A powder sample of Mo_2O_3 with a density of 70 mg/cm^2 was placed between two inner layers of the scintillators. The sample consisted of 4.2×10^{23} atoms of ^{100}Mo and was counted for 300 h. A sample-out background spectrum was accumulated for the same amount of time. The subtraction of these spectra leads to the following: $T^{0\nu}(0^+ \rightarrow 0^+) > 4.4 \times 10^{20}$ y, $T^{0\nu}(0^+ \rightarrow 2^+) > 2.0 \times 10^{20}$ y, $T^{0\nu,M}(0^+ \rightarrow 0^+) > 2.8 \times 10^{19}$ y, and $T^{2\nu}(0^+ \rightarrow 0^+) > 6.2 \times 10^{18}$ y, at confidence levels of 68%.

One ^{100}Mo experiment that is frequently overlooked is the trial UCI TPC experiment[12]. The same TPC used in the UCI ^{82}Se experiment, described by Alan Hahn in this volume, was operated in a trial run of only $\sim$253 hours with a foil containing 26.3 g of natural Mo. They were able to place limits of $T^{0\nu,M}(0^+ \rightarrow 0^+) > 7.5 \times 10^{18}$ y and $T^{2\nu}(0^+ \rightarrow 0^+) > 6.8 \times 10^{17}$ y. The foil had significant radioactive contamination that can be reduced, and enrichment would lead to a factor of 10 more sensitivity, which could well make this experiment very competitive.

^{136}Xe EXPERIMENTS

A number of experiments involve searches for the $\beta\beta$-decay modes of ^{136}Xe. The University of Milano experiment has been operating since late 1986 in the Gran Sasso Laboratory[13]. It is a multi-element wire proportional chamber made up of 61 cells 80 cm long and 2.5 cm on a side formed by axial wires arranged in hexagonal arrays. The fiducial volume of 46 liters corresponds to 1.13 x 10^{24} nuclei of ^{136}Xe when natural Xe is used. The chamber has operated for several months and has set the following limits: $T^{0\nu}$ $(0^+\to0^+) > 2 \times 10^{22}$ y and $T^{0\nu,M}(0^+\to0^+) > 6 \times 10^{20}$ y, to 68% levels of confidence. The experiment will be upgraded with enriched xenon in the near future.

The Caltech xenon TPC has been tested on campus and will be shipped to the Gotthard Tunnel in the summer of 1988. It has a design sensitivity of 8 x 10^{22} y, in one year of counting, for 0ν $\beta\beta$-decay. It will also employ xenon enriched in ^{136}Xe.

High pressure Xe chambers are being developed at Yale University and at the INR in Moscow[14]. These show some promise of yielding high resolution for the observation of 0ν $\beta\beta$-decay; however, to our knowledge this has not been demonstrated.

The INR ^{136}Xe high-pressure proportional counter in the Baksan laboratory was filled alternately with natural Xe and with 94% enriched ^{136}Xe at a pressure of 25 atm. The fiducial volume is 3.14 liters. The gases were alternated 4 times for counting times of 710 hours with the natural Xe and 685.4 hours with the enriched gas. The subtracted spectra result in the following limits: $T^{0\nu}$ $(0^+\to0^+) > 3.3 \times 10^{21}$ y, $T^{0\nu}$ $(0^+\to2^+) > 1.5 \times 10^{21}$ y, $T^{0\nu,M}(0^+\to0^+) > 1.9 \times 10^{20}$ y, and $T^{2\nu}$ $(0^+\to0^+) > 7.8 \times 10^{19}$ y, to 68% confidence levels. The energy resolution is 3.8% FWHM at 2.5 MeV.

THE BAKSAN ^{150}Nd EXPERIMENT

The plastic scintillator assembly used for the ^{100}Mo experiment discussed above was also used with a $Nd_2 0_3$ powder source[14]. Seventy grams (2.8 x 10^{23} atoms of ^{150}Nd) were counted for 600 h, and the background was counted for the same amount of time. Subtraction of these data sets yields a positive result. If all of the net counts are attributed to 2ν $\beta\beta$-decay, then $T^{2\nu}$ (^{150}Nd) = $(1.8 \pm 1.0) \times 10^{19}$ y with a 2σ level of confidence. No claim has been made, and a series of tests were undertaken to determine if the source is contaminated with ^{232}Th, ^{238}U, and/or ^{40}K. In any case, the sensitivity of this experiment is reaching interesting levels.

CONCLUSIONS

A number of very interesting experiments are under way to search for all modes of $\beta\beta$-decay in various isotopes. In many cases, the sensitivities continue to improve to levels at which 2ν $\beta\beta$-decay will be observed and the half-lives measured. Table I summarizes the results discussed in this paper.

In considering the best possibility to search for 0ν $\beta\beta$-decay, the idea of operating tens of kilograms of germanium enriched to ~80% in ^{76}Ge is very appealing. This is not only because of excellent energy resolution (~ 0.12% at 2 MeV) but also because all of the necessary technology is already available. Nevertheless, it is important to continue on all fronts to understand the nuclear physics of $\beta\beta$-decay and to develop new technologies for 0ν $\beta\beta$-decay.

ACKNOWLEDGEMENTS

This work was supported by the U.S. Department of Energy under contract DE-AC06-76RLO 1830 and the National Science Foundation under Grant PHY-8604581.

REFERENCES

1) B. Kayser in Proc. XXIII Int. Conf. on High Energy Physics ed. S.C. Loken 945 (1987), World Scientific, Singapore.
2) J. Schechter and J.W.F. Valle, Phys. Rev. D25, 2951 (1982); L. Wolfenstein, Phys. Rev. D26, 2507 (1982), see also the work of Vergados and Mohapatra referenced therein; W.C. Haxton, S.P. Rosen, and G.J. Stephenson, Jr., Phys. Rev. D26, 1805 (1982); C.O. Escobar and V. Pleitez, Phys. Rev. D28, 1166 (1983); R.N. Mohapatra, Phys. Rev. D34, 3457 (1986); J.D. Vergados, Phys. Lett. 184, 55 (1987).
3) K. Muto and H.V. Klapdor, Phys. Lett. B201, 420 (1988); O. Civitarese, A. Faessler, and T. Tomoda, Phys. Lett. B194, 11 (1987), W.C. Haxton and G.J. Stephenson, Jr., Progress in Particle and Nuclear Physics 12, 409 (1984).
4) J. Engel, P. Vogel, and M.R. Zirnbauer, Phys. Rev. C37, 731 (1988).
5) F.T. Avignone, III and R.L. Brodzinski, PNL-SA-15651, to appear in Progress in Particle and Nuclear Physics, ed. A. Faessler (1988).
6) D.O. Caldwell, R.M. Eisberg, D.M. Grumm, M.S. Witherell, B. Sadoulet, F.S. Goulding, and A.R. Smith, UCSB-HEP-88-2 (1988).
7) P. Fisher, F. Boehm, E. Bovet, J.-P. Egger, H. Henrikson, K. Gabathuler, D. Reusser, and J.-L. Vuilleumier, CALTECH V-88-CO47, June 1988.
8) E. Bellotti, O. Cremonesi, E. Fiorini, C. Liguori, A. Pullia, P. Sverzallati, and L. Zanotti, Phys. Lett. 146B, 450 (1984).
9) A. Morales, J. Morales, R. Nuñez-Lagos, J. Puimedón, J.A. Viullar, and A. Larrea, University of Zaragoza Preprint FANZ-9-87.
10) Takayuki Watanabe, Ph.D. Dissertation, Osaka University, Toyonaka, Osaka 560, Japan (Jan. 1988).
11) M. Alston-Garnjost, B. Dougherty, R. Kenney, J, Krivicich, R. Tripp, H. Nicholson, S. Sutton, B. Dieterle, J. Kang, and C. Leavitt, Phys. Rev. Lett. 60, 1928 (188); C.S. Sutton, Mount Holyoke College preprint for Proc. Neutrino-88, June 1988.
12) S.R. Elliot, A.A. Hahn, and M.K. Moe, Phys. Rev. C36, 2129 (1987).
13) C. Liguori, Proc. TELEMARK IV ed. V. Barger et al., World Scientific 248 (1987). (An update was given by E. Fiorini at

NEUTRINO-88 from which the present limits were obtained).
14) A.A. Klimenko et al., INR Preprint 3·I4–526T·10 March 1988.

TABLE I. Double-beta decay half-lives (years) from selected experiments.

Experiment	Isotope	$T^{0\nu}_{1/2}(0^+\to0^+)$	$T^{0\nu,M}_{1/2}(0^+\to0^+)$	$T^{0\nu}_{1/2}(0^+\to2^+)$	$T^{2\nu}_{1/2}$
UCSB/LBL	^{76}Ge	$>7 \times 10^{23}$*	$>1.4 \times 10^{21}$		
CALTECH/ NEUCHATEL/ SIN	^{76}Ge	$>3 \times 10^{23}$	$>1 \times 10^{21}$	$>3.4 \times 10^{22}$	$>5 \times 10^{20}$
GUELPH/ QUEENS	^{76}Ge	$>1.6 \times 10^{23}$	---	---	---
PNL/USC	^{76}Ge	$>3 \times 10^{23}$	$>2 \times 10^{21}$		1.0×10^{21}
OSAKA	^{100}Mo	$>1.6 \times 10^{20}$	$>6 \times 10^{18}$	---	2.6×10^{17}
LBL/ MT. HOLYOKE/ UNM	^{100}Mo	---	$>3.3 \times 10^{20}$	---	$>3.8 \times 10^{18}$
BAKSAN	^{136}Xe	$>3.3 \times 10^{21}$	$>1.9 \times 10^{20}$	$>1.5 \times 10^{21}$	$>7.8 \times 10^{19}$
BAKSAN	^{100}Mo	$>4.4 \times 10^{20}$	$>2.8 \times 10^{19}$	$>2.0 \times 10^{20}$	$>6.2 \times 10^{18}$
MILANO	^{136}Xe	$>2.0 \times 10^{21}$	$>6 \times 10^{20}$	---	---
BAKSAN	^{150}Nd	$>1.7 \times 10^{21}$	$>1 \times 10^{20}$	$>1.1 \times 10^{21}$	$>1.8 \times 10^{19}$

*This result is based on statistical fluctuations alone. A maximum likelihood analysis yields 1.3×10^{24} y.

A SEARCH FOR NEUTRINO OSCILLATIONS AT LAMPF

L.S. Durkin, R. Harper, T.Y. Ling, J. Mitchell,
T.A. Romanowski, E. Smith, M. Timko
The Ohio State University, Columbus, Ohio.

S. Freedman, J. Napolitano
Argonne National Laboratory, Argonne, Il. 60439

C. Choi, A. Fazeley, R. Imlay, W.J. Metcalf
Louisiana State University, Baton Rouge, La. 70803

B. Fujikawa, R.D. McKeown
California Institute of Technology, Pasadena, Ca. 91125

R. Carlini, J. Donahue, G. Garvey, V. Sandberg,
Las Alamos National Laboratory, Los Alamos, Nm. 87545

Presented by: L.S. Durkin

If neutrinos are massive and lepton number is not conserved then the phenomenon of neutrino oscillations may occur. This experiment searches for the appearance of $\bar{\nu_e}$ resulting from the oscillation mode $\bar{\nu_\mu} \to \bar{\nu_e}$. Assuming just these two types of neutrinos, the probability of an initial neutrino $\bar{\nu_\mu}$ of energy E (in MeV) oscillating into a neutrino $\bar{\nu_e}$ after a path length L (in meters) is given by

$$P(\bar{\nu_\mu} \to \bar{\nu_e}) = \sin^2(2\theta)\sin^2(1.27\delta m^2 L/E), \quad (1)$$

where δm^2 (in eV^2) is the difference of the squares of the masses of neutrino mass eigenstates and θ is the mixing angle between them.

The experiment was carried out at the Los Alamos Meson Physics Facility (LAMPF). Neutrinos incident on our detector result from decays of stopping π^+ and μ^+ produced by 780 MeV protons in a beam stop. The resulting beam is nearly isotropic with equal admixtures of ν_μ, $\overline{\nu}_\mu$, and ν_e in the energy range 0 to 52.8 MeV. Since π^- are absorbed in the beam stop no significant contamination of $\overline{\nu_e}$ is expected. $\overline{\nu}_e$ from neutrino oscillation would be detected by the inverse beta reaction

$$\overline{\nu}_e + p \to e^+ + n \quad (2)$$

The modular detector, located 26.8 m from the beam stop, is constructed of vertical planes of liquid scintillator followed with proportional drift tubes (PDTs). The transverse dimension of the detector is 3.66x3.66 m^2 and it is 7m long. The liquid-scintillator (15 tons of CH_2) provide free proton targets for the inverse beta decay reactions and measure the ionization energy loss of charged particles. The PDT's provide track coordinates for charged particles and also measure their ionization energy loss.

Since the duty cycle of the LAMPF accelerator is relatively long (6.4%) the detector needs to be well shielded against cosmic rays, both charged and neutral ones. An active 4π cosmic-ray shield in the shape of a horizontal cylinder with closed ends completely covered the detector. The cylinder is 7.6m in diameter and consists of two layers: a 15 cm thick outer layer of liquid scintillator and an inner layer made of 13 cm of lead and 5 cm of iron. During data taking the detector and the active shield are placed in a tunnel covered with a minimum of $2000g/cm^2$ overburden of steel to further reduce the cosmic ray flux incident on the detector.

The experimental trigger requires the coincidence of cells in three out of four contiguous scintillation planes, and the absence of hits in the active shield. Triggers are accepted during 780 μs long beam spills and during a cosmic-ray gate about three times longer in duration. Cosmic ray muons are used to monitor the performance the detector. Electrons from the decays of muons stopped in the detector are used to calibrate the detector, and provide a control sample to backcheck the data analysis.

Data for this analysis were collected during LAMPF running cyles 48,49 and 50 from July to December of 1987. A total of 5,100 coulombs of protons incident on the beam stop resulted in 9.1×10^5 triggers written onto tapes.

The raw data tapes were first processed to filter out cosmic ray induced events which constitute the majority of all the triggers : electrons from stopping muon decays, through going muons and low energy non-trackable debris induced by neutrons. The events which remained after the software filter consisted mainly of neutron-induced knock-on protons and electrons arising from neutron induced photons or from the bremsstrahlung of stopping muons in the lead shield surrounding the detector. The likelihood of a track being an electron was calculated based on the differential energy depositions in the liquid scintillator cells and the PDT cells traversed. Protons are removed from the data sample with a rejection factor better than 10^{-3} by requiring that the track must pass as an electron in the likelihood analysis and that the energy deposition in the scintillator cells at either end of the track must be less than 1.5 times minimum-ionizing. Events from neutron induced photons were found to have large in-time energy deposition in cells separated from the track (associated energy). These background are eliminated by requiring that the associated energy be lower than 12 MeV. Finally the events in the remaining data sample are required to have no hits in the shield from -30μs to the trigger time . The efficiency for keeping signal electrons in this analysis was measured to be 0.42$\pm$0.03.

The events which pass these cuts are contained electrons or positrons. In the energy range 0-52.8 MeV we found 20 events in the beam-on gate (total live-time of 70.3 LD's) and 27 events in the beam-off gate (total live-time of 249.8

LD's), where a LD(live-day) is equal to 5529.6s. After subtracting the beam-off(cosmic-ray) background of 0.10±0.02 events/LD, we obtain a beam-excess of 12.4±4.9 events.

There are four possible sources of the beam excess events: i) electrons produced by beam associated neutrons; ii) electrons or positrons resulting from charged-current (CC) interactions of background ν_μ or $\overline{\nu}_\mu$ due to the decay-in-flight(DIF) of pions produced upstream of the beam stop, iii) electrons from CC reactions of ν_e on nuclei and elastic νe^- reactions and iv) neutrino oscillation, i.e., positrons from reaction (2).

We have empirically rule out the first two possibilities. Incident neutron flux was monitored by the detected rate of contained knock-on proton events. The resulting background from neutron induced events was calculated to be negligible. The DIF ν_μ (or $\overline{\nu}_\mu$) were monitored by looking for events with visible muons which stop and decay visibly in our dectector. A simple extrapolation from these visible muons to the case where the muon would go undetected ruled out these events as a source of background.

Neutrino backgrounds have been studied using Monte Carlo events combined with known fluxes and cross sections [1,2]. Fig.1 shows the energy distribution for the beam-excess electron events. The data is found to be consistent with, both in shape and in normalization, the predicted spectrum due to ν_e CC and νe^- elastic reactions. Also shown in Fig.1 is the energy spectrum for the inverse beta reaction (2), assuming maximum oscillation. The signal for neutrino oscillation peaks at substantially higher energy than the data.

We conclude that the beam-excess events are well explained by ν_e interactions and there are no evidence for neutrino oscillations through the mode $\overline{\nu}_\mu \rightarrow \overline{\nu}_e$.

A binned maximum likelihood analysis was performed on the data shown in Fig.1. Limits on δm^2 and $\sin^2(2\theta)$ are established by using equation (1) and the calculated rate and energy spectrum of the inverse beta reaction [3] and those of the neutrino background reactions. Expected systematic errors are also included in this analysis. Fig.2 gives the 90% confidence contour for the region excluded by this experiment. Other published limits from previous searches are also shown for comparison.

References

[1] T.W.Donnelly *et al.*, Phys. Lett. 34B, 93 (1973)

[2] B.Kaiser *et al.*, Phys. Rev. D20, 87 (1979)

[3] J.S.O'Connell, LAMPF report LA-9858-C, p.43, (1981);
S.A.Fayans, Soviet Journ. Nucl. Phys. 42, 590 (1986)

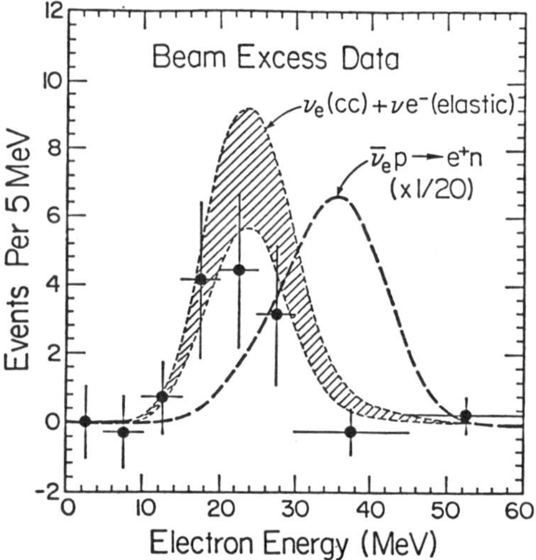

Fig. 1 Visible energy distribution for the Beam-Excess electrons. The solid curve is the expected electron spectrum for ν_e CC reactions and νe elastic reactions. The dashed curve shows the maximum oscillation signal, i.e., the positron spectrum from inverse beta reaction (2), scaled down by a factor of 20.

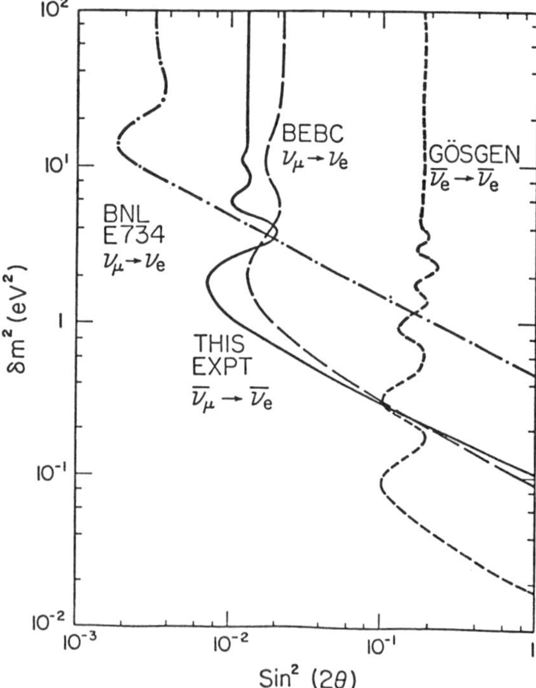

Fig. 2 The 90% confidence level contour for the region of δm^2 and $\sin^2(2\theta)$ excluded by this experiment. Other recently published limits are shown for comparison.

BNL E776: A SEARCH FOR NEUTRINO OSCILLATIONS

R. Seto
Columbia University, New York, NY 10027

ABSTRACT

We report on the results of BNL E776, carried out at the Brookhaven National Laboratory in the summer and fall of 1986. Its purpose was to search for neutrino oscillations, through the appearance of ν_e in a ν_μ beam. The experiment used a low energy narrow band beam of ν_μ with a mean energy of about 1.4 GeV, and a detector at a distance of 1 km from the target. With 3×10^{19} protons on target, no significant excess of electron neutrinos was detected. The 90% confidence limits obtained are $\Delta m^2 \leq 0.11$ eV^2 in the limit of large mixing angle, and $\sin^2 2\theta \leq 0.018$ in the limit of large Δm^2.

INTRODUCTION

If neutrinos are massive, and lepton numbers are not independently conserved, then the weak eigenstates are not necessarily identical to the mass eigenstates but can be mixtures of the mass eigenstates. For simplicity, let us assume two flavor dominance, then the probability for muon neutrinos to oscillate to electron neutrinos over a distance L is the following:

$$P(\nu_\mu \to \nu_e) = \sin^2 2\theta \, \sin^2(1.27 \frac{L}{E} \Delta m^2) \qquad (1)$$

where θ is the lepton flavor mixing angle, L is the distance between the source and detector in kilometers, E is the neutrino energy in GeV, and Δm^2 is the difference in the masses squared of the eigenstates in eV^2.

E776 was a search for the appearance of electron neutrinos in a muon neutrino beam. The two primary backgrounds are first, ν_e contamination of the ν_μ beam, and second, π^0's from ordinary muon neutrino interactions which fake electrons in our detector. The mean energy of the beam was 1.4 GeV. The detector was located in the middle of the RHIC ring at the Brookhaven National Laboratory at a distance of L=1km from the proton target.

THE BEAM

A high intensity narrow band beam was used for this experiment. A Monte Carlo program was written to calculate the fluxes produced by this beam. Target production was simulated using NUCEVT[1] in conjunction with a model by Grote, Hagedorn, and Ranft.[2] Dominant contributions to the ν_e background came from $K^+ \to \pi^0 e^+ \nu_e$ (45%), $\pi^+ \to \mu^+ \nu_\mu \to e^+ \overline{\nu_\mu} \nu_e$ (34%), and $K_L \to \pi^- e^+ \nu_e$ (17%). Figure 1 shows the energy spectra for the various neutrino flavors. The integrated ν_e/ν_μ ratio is about 6.5×10^{-3}. The production model dependency of the calculation was estimated by repeating the calculation using the Grote, Hagedorn and Ranft model alone, as well as a semiempirical model by Sanford and Wang.[3] From the study of the various production models the uncertainty in the ν_e/ν_μ flux is about 20%.

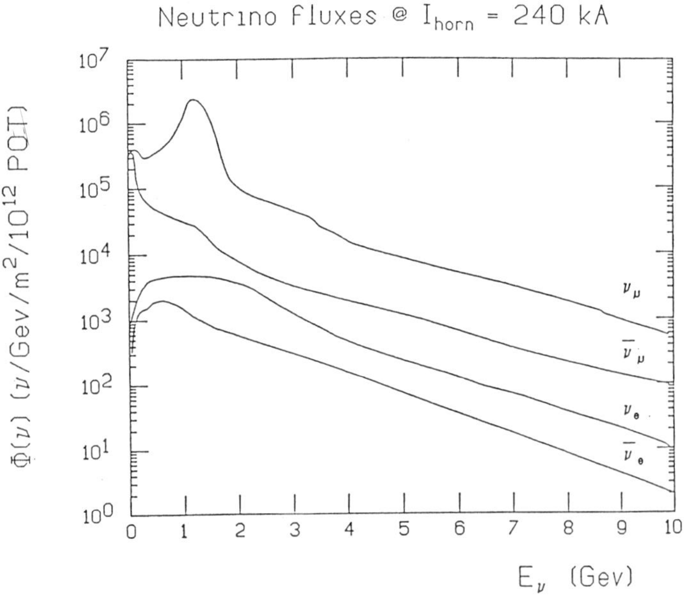

Fig. 1. Calculated spectra for ν_μ, ν_e, $\overline{\nu_\mu}$, and $\overline{\nu_e}$.

THE DETECTOR AND DATA TAKING

A schematic of the E776 detector is shown in figure 2. It was composed of two sections. The front "electron detector" was a finely segmented electromagnetic calorimeter. This was followed by the "muon spectrometer",

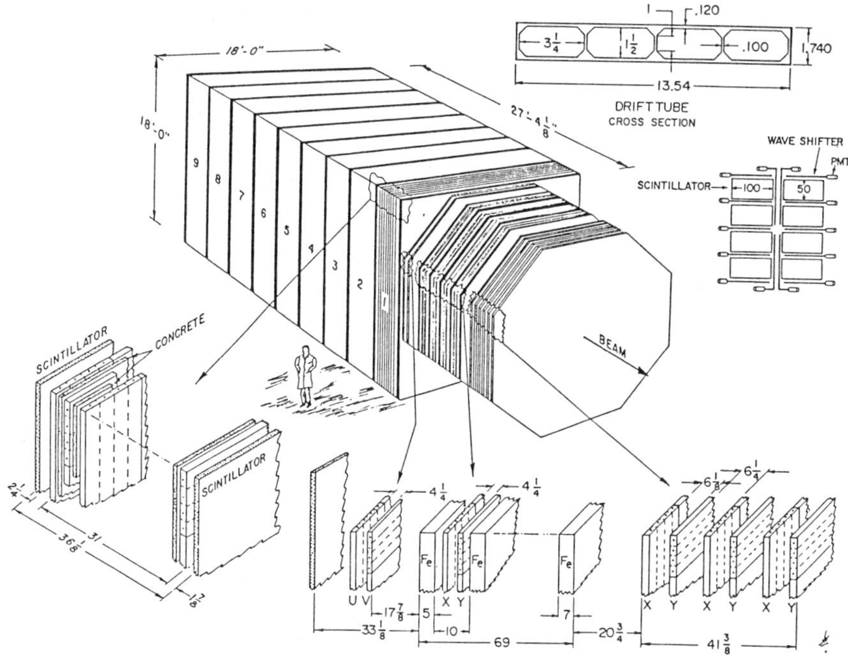

Fig. 2. The E776 detector.

a toroidal magnet with an 18 kG field for measuring the momentum of muons.

The electron detector was composed of 90 planes of proportional drift tubes (PDT's) interleaved with 1-in. thick (1/4 radiation length) concrete absorbers, for a total of 240 tons. Every tenth plane of concrete was replaced by a plane of scintillator which was used for timing. Sequential PDT planes were placed at 90 degrees with respect to each other. Each PDT plane contained sixty-four 18 ft. x 3.25 in. x 1.5 in. drift tubes made of extruded aluminum. The PDT signals were amplified and read out by a 6-bit flash ADC running with a 22.4 ns clock in conjunction with a 256 bin memory, providing us with a recorded time span of 5.7 μs.

The flash ADC's allowed us to examine the details of the pulses within the PDT's. Electromagnetic showers were characterized by multiple peaks and large pulses which resulted from the fact that several tracks cross each PDT as the shower developed. In contrast, muons were characterized by a single minimum ionizing peak in each PDT. The energy resolution of the detector for electromagnetic showers was measured to be $20\%/\sqrt{E}$.

Data were taken during the summer and fall of 1986 at two neutrino energies. Two-thirds of the data were taken with a neutrino energy of 1.3 GeV (horn current= 240 kA). Later the neutrino energy was increased to 1.5 GeV (horn current= 280 kA). The beam trigger came from the AGS clock. A "free trigger" was set up between beam pulses to monitor the cosmic ray background. Straight through cosmic rays were recorded to monitor the detector's performance. A total of 2.6×10^{6} beam triggers were taken corresponding to 3×10^{19} protons on target. The events were passed through three filter programs which selected events with an event vertex in the detector. This left us with about 12.8×10^{3} events.

Although programs were available to separate electrons from π^{0}'s, it became clear that the best e/π^{0} separation was done by the human eye. In order to prevent bias in scanning and to monitor the contamination of electrons from π^{0}'s, a sample was prepared which was a mixture of Monte Carlo events and real events. This sample was scanned and analyzed. The sample consisted of the 12,800 data events, and about 10,000 ν_{μ} and ν_{e} Monte Carlo events. In addition, a sample of Monte Carlo single π^{0}'s were put into the sample to monitor the rate at which π^{0}'s were called electrons. In the first scan, PDT hits were grouped into tracks and clusters which were called showers. The muon analysis which followed, utilized the computer with no further scanning.

THE MUON ANALYSIS

The muon analysis provided us with a normalization sample for the electrons. Events with tracks found in the scan were subjected to the following cuts: 1) The track vertex had to be within the fiducial volume defined as being within 5 planes from the front of the detector, 15 planes from the back, and 4 wires from the side. 2) The longest track was required to be either contained in the detector or pass though the muon spectrometer, allowing us to measure the muon energy. 3) The track had to be longer than 20 planes, that is 1.6 interaction lengths. 4) There could be no kinks in the track. (Cuts 3 and 4 insure that the track is a muon.) 5) The event time, as measured by the scintillator planes, was required to be within the beam bucket. 6) The polar angle had to be less than 40°. 7) The number of hits not associated with the primary muon had to be less than or equal to 10. The data gave 1021 events for the muon normalization sample, with an acceptance of 11.4%.

The Monte Carlo predicted 1029 events, in good agreement with the data.

THE ELECTRON ANALYSIS

The initial scan picked up 1721 events with showers. At this point the majority of "showers" were small clusters of hits with low total pulse area, or overlapping tracks and showers which could not be separated in the detector. A second scan identified clear electromagnetic showers. The following cuts were used: 1) in an electromagnetic shower at least 30% of the PDT's had to have multiple peaks <u>or</u> 2) there had to be skipped PDT's where the shower energy was completely carried by photons; 3) the shower angle was required to be greater than 45°; and 4) the shower length was required to be greater than twelve planes. After these cuts we were left with 318 events.

A third scan was done, whose primary purpose was to separate π^{0}'s from electrons. More stringent cuts on the shower were also made at this stage as follows: 1) the shower length was now required to be greater than or equal to fifteen planes; 2) the shower had to be contained so that an energy measurement was possible. 3) the shower had to have a well defined vertex within the fiducial volume; and 4) events containing three or more showers were rejected.

The following categories of showers were set up:

"e": "e" events were essentially showers which were well collimated. The requirements were: 1) 90% of the shower energy had to be contained within ±1 wire of the shower axis; 2) the shower had to be symmetric; and 3) no overlapping tracks were allowed.

"π^{0}": "π^{0}" events in contrast, were showers which were 1) clearly two overlapping showers or 2) an asymmetric shower.

"γ": γ's were showers separated from a vertex defined by another track.

"X": X events were events which were rejected because they did not pass the cuts mentioned above. For example, if a shower was exiting the detector it was rejected as an X because an energy measurement was not possible.

In addition to the shower, events were allowed to contain a track. Ninety-three events were kept as e, π^{0}, or γ, however only the "e" category will concern us here.

The results of the scan were then run through the final histogramming programs. The final cuts were the same as those made on the muons, namely: 1) the polar angle was required to be less than 40°; 2) events with greater than 10 hits other than the primary shower were thrown out; and 3) the timing from the scintillators was required to be within the beam bucket. This left us with a final sample of 17 ν_e events in the data and 1 ν_e event in the "free trigger" sample. The acceptance for electron neutrino events was about 10.6%.

The π^0 background obtained from the Monte Carlo is 4.6 events. We assigned a 50% systematic error because of the lack of knowledge of the cross sections. The ν_e background from the beam is 7.9 events with a systematic error of about 20% coming from the model dependence of the target production. The background from $\overline{\nu}_e$ which was not included in the scan is calculated to be 1.1 events with a 50% systematic error. The 1 "free" trigger event gives us a cosmic ray background of 0.2 events after multiplication by the ratio of the gate lengths. Systematic errors coming from scanning efficiencies, acceptance, and normalization add up to about 10%. Table I summarizes the backgrounds. The total background is 13.9 ± 3.7 (stat) ± 3.0 (sys).

Table I. A Summary of the Backgrounds to ν_e

π^0	4.6 ± 2.3 (sys)
ν_e (Beam)	7.9 ± 1.6 (sys)
Cosmics 1*.2	0.2 ± 1.0 (sys)
$\overline{\nu}_e$	1.1 ± 0.5 (sys)
Total Background	13.9 ± 3.7 (stat) ± 3.0 (sys)

RESULTS

Figure 3 shows the neutrino energy distribution of the electron neutrino events. The data is well fit by the expected background distributions. Note that in these plots the contribution of $\overline{\nu}_e$ is not included.

We are left with a data sample of 17 electron events consistent with the background prediction of 13.9±4.7 events. Hence, at the 90% confidence level, $P(\nu_\mu \to \nu_e) < 1.14\%$, or $\Delta m^2 < 0.11$ eV^2 for maximal mixing and $\sin^2 2\theta < .018$ at large Δm^2. Figure 4 shows the region excluded at the 90% confidence level by this experiment.

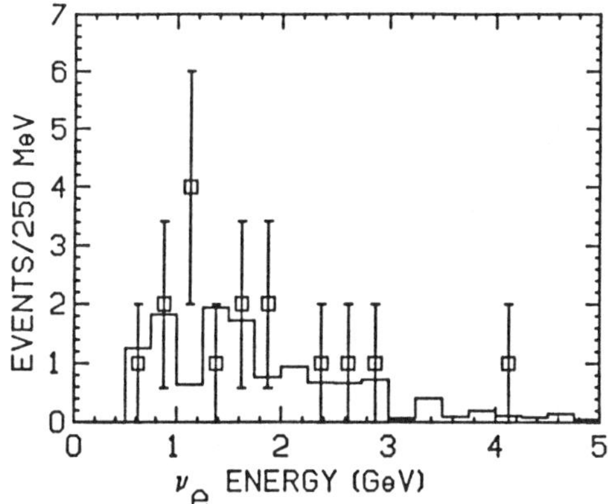

Fig. 3. The measured ν_e energy spectrum. The solid line corresponds to the expected background.

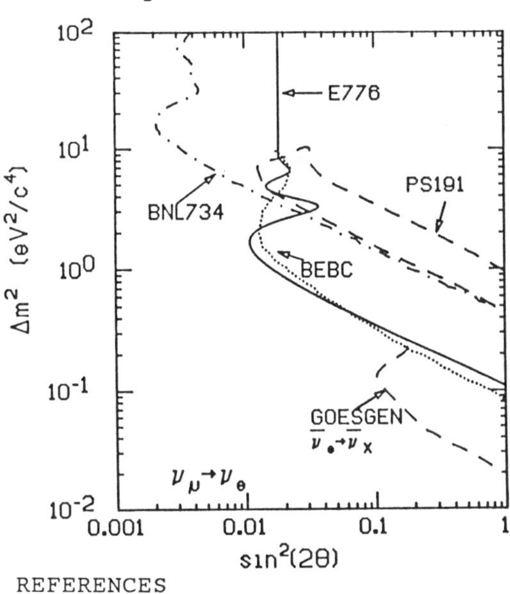

Fig. 4. Limits on ν_e appearance from this experiment. Also plotted are the E734, BEBC and Goesgen limits. The PS191 result, which reports an excess of ν_e, is also shown.

REFERENCES

1. J. Ranft and S. Ritter, The Monte Carlo Codes NUCEVT and HADEVT to Simulate Hadron Production in Hadron-Nucleus and Hadron-Hadron Collisions, CERN Internal Report TIS-RP/IR/83-23 (1983).
2. H. Grote, R. Hagedorn and J. Ranft, Atlas of Particle Production Spectra, CERN report, 1970 (unpublished).
3. J. R. Sanford and C. L. Wang, "Empirical Formulas for Particle Production in p-Be Collisions Between 10 and 35 BeV/c". AGS internal report, BNL # JRS/CLW-1 and 2, March 1, 1967. BNL report No. 11479, 1967, (unpublished).

THE NEUTRON ELECTRIC DIPOLE MOMENT

Bruce H. J. McKellar and Xiao-gang He
University of Melbourne, Parkville, Victoria, Australia 3052

Sandip Pakvasa
University of Hawaii at Manoa, Honolulu HI 96822, USA

INTRODUCTION

There are many different theories of CP violation, all of which can fit the experimental information available from K decays. There have been many attempts to compute the neutron electric dipole moment in these theories, which have all concentrated on one particular mechanism for the generation of the neutron edm. Here we consider both quark level and hadronic level processors, and argue that there is no double counting in simply adding these terms. We then survey the results obtained in the standard model, in the Weinberg Model and in the Left Right symmetric model. The present experimental upper bound is 2.6×10^{-25} e cm, We show that the Weinberg model would give a dipole moment greater than 1×10^{-25} e cm, and the L-R symmetric model gives a value in the range 1.9×10^{-26} e cm to 1.9×10^{-27} e cm - the range being fixed by the recently observed value of $\varepsilon'/\varepsilon = (3.3 \pm 1.1) \times 10^{-3}$. The standard model gives much smaller values - of order 10^{-32} e cm.

GENERATION OF THE ELECTRIC DIPOLE MOMENT

The simplest way to generate the electric dipole moment of the neutron is to simply generate the electric dipole moments of the quarks and to simply add them together. In principle, contribution of order are possible but in the standard model, these contributions are proportional to $|U_{qi}|^2$ and thus cannot introduce a CP violation phase. It is a remarable result that in the standard model the simple order G_F^2 diagrams sum to zero, and the lowest order contribution in the standard model is of order $G_F^2\,\alpha_s$.

The neutron edm is then given in terms of the quark edm's by

$$D_n = \frac{1}{3}\left(4D_d - D_u\right)$$

A second mechanism which can generate an edm in the neutron is the quark colour electric dipole moment (cedm). The replacement of the photon in any diagram contributing to the quark edm by a gluon will transform a T and P violating interaction of the quark with the electric field, $- d_q \underset{\sim}{\sigma}.\underset{\sim}{E}$, into a T and P violating interaction of the quark with the colour electric field, $- g_s f_q \frac{\lambda^a}{2} \underset{\sim}{\sigma}.\underset{\sim}{E}_a$, defining the quark cedm, f_q. This induces T and P violating components in the neutron wave function, and hence a neutron edm which is given by

$$D_n = \frac{1}{3} e\left(\frac{4}{3} f_d + \frac{2}{3} f_u\right)$$

There are two further mechanisms which need to be considered. It is possible for exchange currents at the quark level to induce a neutron edm. In practice the exchange current effect is reduced compared to the single quark edm because it requires the two quarks to be close together. This effect thus only becomes potentially important in situations such as those of the standard model when the quark edm contribution is already reduced.

One must also consider long range contribution to the neutron edm. In a model in which these long range effects are represented through the coupling of mesons to baryons, meson loops (some of which are illustrated in figure 2) may contribute to the neutron edm, as emphasised many years ago by Barton and White. Once again the qualitative difference between the standard model and other models is apparent - to obtain a non zero neutron edm in the standard model one must go to contirbutions of second order in the weak interaction, i.e. of order G_F^2. Crewther et al. [2] emphasised the logarithmic singularity in these meson loop diagrams in the chiral limit as the meson mass goes to zero. In practice these terms are singular but not divergent in this limit, behaving as $m_M^2 \ln m_M^2$.

With the information that one expects the neutron edm in the standard model to be of order G_F^2, whereas in other models it is expected to be of order G_F, one can make dimensional estimates of D_n:

$$D_n^{(SM)} \sim \frac{e}{2m_n}\left(G_F m_\pi^2\right)^2 \eta_{+-} \sim 10^{-23} \text{ e cm}$$

$$D_n^{(other)} \sim \frac{e}{2m_n}\left(G_F m_\pi^2\right)\eta_{+1} \sim 10^{-30} \text{ e cm}$$

So we see that the present experimental limits are just on the threshold of an interesting phase in which they will be able to distinguish standard and non standard models of CP violation. It is also clear that it is now important for theoretical calculations of the neutron edm to be both reliable and believable if we are to achieve these goals.

We believe that the calculations reported in the literature do not meet these criteria because they have emphasised one of the possible mechanisms outlined above which contribute to the neutron edm, without considering other possible contributions. It is important to ask whether these contributions are genuinely independent and additive. We believe they are, in that the quark edm, quark cedm and quark exchange current effects are short distance effects, while the meson baryon loop are long distance effects. The separation between short distance and long distance effects is seen most clearly in the cloudy bag model, in which it is clear that both the proton baryon loops and the quark level terms will contribute in an additive way.

We present the results of a comprehensive calculation of the neutron edm expected in the standard model, the L-R symmetric model and the Weinberg multi Higgs model of CP violation in the report[3),

THE STANDARD MODEL

We have already seen that the neutron edm arises in the standard model at the second order in the weak interaction. Indeed the quark edm contributions are of order $G_F^2 \alpha_s$, as emphasised by Shabalin[1)] and Khriplovich[4)], and are further suppressed because the quark edm would vanish in a chirally symmetric world with massless quarks. The final result is that the quark dipole moments are of the order 10^{-34} e cm in the standard model. Quark level exchange currents also give contributions of this order.

Meson-baryon loops give significantly larger contributions. As we emphasised elsewhere[5)], when calculating CP violating effects at order G_F^2 in a hadronic basis it is essential that the CP violating phase associated with each of the two interactions is included. If this is not done the results are not invariant under changes in the phase of the s quark wave function.

Subsequent to the publication of our letter we realised that there is also a contribution from neutral meson loops with the photon coupled to the baryon through a Pauli type of coupling. If we classify our results according to the type of photon coupling we obtain the contributions to D_n listed in table I.

TABLE I. Contributions of meson-baryon loops to the neutron edm in the standard model

Photon coupling	Contribution to D_n in e cm
meson and Dirac coupling to baryon	$4.95 \times 10^{-29}\ s_2 s_3 s_\delta$
Pauli coupling to charged baryon	$1.95 \times 10^{-29}\ s_2 s_3 s_\delta$
Pauli coupling to neutral baryons	$-3.13 \times 10^{-29}\ s_2 s_3 s_\delta$

Note that the listed contribution from Pauli type coupling is that obtained on the assumption that the anamolous moment retains its full strength off mass shell. If this assumption is relaxed the contribution will reudce in magnitude. Noting this fact, and noting that $s_2 s_3 s_\delta$ is limited by

$$2 \times 10^{-4} \le s_2 s_3 s_\delta \le 2 \times 10^{-3}$$

we obtain our prediction for the magnitude f D_n in the standard model

$$7.5 \times 10^{-33} \text{ e cm} \le |D_n| \le 7.5 \times 10^{-32} \text{ e cm.}$$

THE WEINBERG MODEL

In the Weinberg Model CP violation enters through the phase between the many Higgs vacuum expectation values. In this case there are contirbutions to the neutron edm of order G_F. After detailed calculations we find contributions to D_n from:

quark electric dipole moment: -1.2×10^{-25} e cm

quark colour electric dipole moment: -0.2×10^{-25} e cm

meson baryon loops: $0 \ge D_n \ge -8.3 \times 10^{-25}$ e cm.

Thus we obtain

$$-1.4 \times 10^{-25} \text{ e cm} \ge D_n \ge -9.7 \times 10^{-25} \text{ e cm,}$$

i.e. D_n is negative, and larger in magnitude than 10^{-25} e cm in the Weinberg model.

We remark in passing that confirmation of the recent CERN value of ε'/ε, which is positive, would rule out the Weinberg model, which predicts a negative value for ε'/ε.

L-R SYMMETRIC MODELS

It is characteristic of the $SU(2)_L \times SU(2)_R$ models that many additional parameters enter calculations of CP violation effects: these are the mixing angle, ξ, of W_L and W_R, and the phases γ, δ_1 and δ_2 which appear in the 2x2 mixing matrix for the right handed currents[6]. Explicity the W mass eigenstates are

$$\begin{pmatrix} W_1 \\ W_2 \end{pmatrix} = \begin{pmatrix} \cos\xi & \sin\xi \\ -\sin\xi & \cos\xi \end{pmatrix} \begin{pmatrix} W_L \\ W_R \end{pmatrix}$$

and the right handed KM matrix is

$$U_R = e^{i\gamma} \begin{pmatrix} \cos\theta e^{-i\delta_2} & \sin\theta e^{-i\delta_1} \\ -\sin\theta e^{i\delta_1} & \cos\theta e^{i\delta_2} \end{pmatrix}.$$

The phases in U_R give the principal contribution to $K_1^0 - K_2^0$ mixing, and ε'/ε and D_n both receive their leading contributions from W_L-W_R mixing coupled with the phases in R_R.

It has been calculated that[7)]

$$\left|\frac{\varepsilon'}{\varepsilon}\right| = 276 \tan\xi \left|\sin(\gamma-\delta_2) + \sin(\gamma-\delta_1) - 0.23\left[\sin(\gamma+\delta_2) + \sin(\gamma+\delta_1)\right]\right|$$

and

$$|D_n| = \sin 2\xi \left\{4.5\sin(\gamma-\delta_2) + 74\sin(\gamma+\delta_1) - 1.1\sin(\gamma-\delta_1) + 15.8\sin(\gamma+\delta_2)\right\} \times 10^{-23} \text{ e cm.}$$

Unfortunately different combinations of phases enter the two CP violating observables, so it is not possible to use the CERN value for ε'/ε to constrain the dipole moment prediction without making further assumptions.

The simplest model of the phases is due to Chang[8)] using this model and using the 95% confidence limits on ε'/ε, viz

$$5.5 \times 10^{-3} \geq \frac{\varepsilon'}{\varepsilon} \geq 1.1 \times 10^{-3}$$

gives the range of values

$$1.9 \times 10^{-26} \text{ e cm} \gtrsim |D_n| \gtrsim 1.9 \times 10^{-27} \text{ e cm}$$

as our prediction for D_n in the LR models.

CONCLUSION

It is clear that the experimental limits on $|D_n|$ are on the threshold of interesting developments. A modest imprevement in sensitivity could rule out or see the Weinberg model, whereas an imprevement by two orders of magnitude would rule out or see the LR model. The predictions for the neutron edm in the standard model remain well out of reach of experiment.

REFERENCES

1. E.P. Shablin, Sov. Phys. Usp. 26, 296 (1988)

2. R.J. Crewther, P. Di Vecchia, G. Venezianno and E. Witten, Phys. Lett. B88, 123 (1979).

3. X-G. He, B.H.J. McKellar and S. Pakvasa, in preparation.

4. I.B. Khriplovich, Phys. Lett. B173, 193 (1986).

5. B.H.J. McKellar, S.R. Choudhury, X-G He and S. Pakvasa, Phys. Lett B197, 556 (1987).

6. See R.N. Mohapatra, Maryland Preprint UMPP #88-003 (1988), for a review.

7. X-G He, B.H.J. McKellar and S. Pakvasa, University of Melbourne preprint UM<-P-88-35 (1988).

8. D. Chang, Nucl Phys. B214, 435 (1983).

IS THERE ANYPLACE LEFT FOR A MASSIVE MAGNETIC MONOPOLE TO HIDE?

D. E. Groom
SSC Central Design Group, LBL90-4040, Berkeley CA 94720

ABSTRACT

The search for the magnetic monopoles predicted by grand unified theories continues, with smaller number of experiments and larger number of experimenters than was the case a few years ago. The search is complicated by a variety of attributes the monopole might possess in addition to its magnetic charge. Our preconceptions about its mass and velocity are not to be trusted. It may or may not catalyze proton decay. It may have intrinsic electric charge, and if not it probably attached a proton at an early epoch. It is difficult to probe interesting flux levels with induction detectors, and the monopole's other possible attributes complicate the analysis of all other experiments and astronomical observations. In this discussion we attempt to classify the available limits according to the input assumptions.

INTRODUCTION

Valentines Day, 1982, marked the start of a very exciting adventure for many of us. Anyone with a few counters or other devices quickly made modifications to join the search for the magnetic monopole. We had been assured by our theorist friends that this supermassive primordial remnant *had* to be there, and Cabrera's result, however cautiously stated, was sufficient to trigger an enormous effort. The learning curve was steep as we argued about ionization at low velocities, atomic induction, and how squids work, and delved into all kinds of physics which had never before caught our attention. The only problem with the field was the absence of monopoles.

Six years later, the field belongs to a few hardy professionals who are willing to spend a decade or more building and operating very large and sophisticated detectors, undaunted by the likelihood of just setting improved upper limits on the monopole flux. The chance of discovery is slim, but the stakes are enormous: Few discoveries would have a comparable impact on physics. The observation of a monopole would at once place experimental physics at mass scales many orders of magnitude beyond the standard model and at epochs even more remote than the element synthesis era.[1]*

What is to be learned from more than 500 papers and the results obtained by at least 33 experimental groups? The monopole carries at least one magnetic charge and is massive, but it might also have other properties and do other things than just those implied by its magnetic charge. The experiments have all set flux limits, but in many cases the analysis depends upon conjectured properties. This is particularly true of those setting the best limits. In this short paper we try to

* The review papers and conference proceedings given in Ref. 1 contain several hundred references to work published prior to 1986. Specific references to this material are therefore avoided here, although several recent papers are cited.

classify the experimental limits from this point of view. Under what rocks should we still bother looking for the most enduring nonexistent particle in physics?

ATTRIBUTES AND INTERACTIONS

1. *Magnetic charge.* To qualify as a monopole, it must certainly have this property, and Dirac long ago pointed out that if electron wave functions are to be single-valued the magnetic charge must be $n\hbar c/2e$, where $|n| = 1, 2, \ldots$ In designing detectors it is sufficient to assume $n = 1$, because for $n > 1$ even larger effects will be observed. Only induction experiments make no further assumptions, except perhaps that the location of the detector is not special. (Perhaps monopoles are repelled by the earth?)

2. *Mass.* The monopole required by grand unified theories has a mass of about 10^{15} GeV, but the prediction is soft. Monopoles lighter than a few times 10^{18} GeV cannot still be gravitationally bound to the galaxy, and if they are lighter than about 10^{14} they cannot penetrate a neutron star's magnetic field.

3. *Velocity.* Primordial monopoles would have cooled to extremely low velocities by the present epoch, but would have been accelerated by various magnetic fields. Relativistic monopoles would have been detected by earlier experiments. A velocity near $v_g = 10^{-3}c$ is characteristic of the galaxy's or supercluster's gravitational potential, and the earth's velocity is $10^{-4}c$. It is hard to imagine a monopole velocity smaller than $10^{-3}c$, but searches must include the possibility.

   At one time there were very lively discussions as to whether a slow monopole could be detected by organic scintillators. This problem has been more or less put to rest by a recent elegant experiment by Ficenec *et al.*,[2] who demonstrated that protons with velocities as low as $2.5 \times 10^{-4}c$ produce light in plastic scintillator. The light output falls below the linear relationship between velocity and electronic stopping power given by Lindhard-Scharff theory, but substantially above that predicted by both Ritson and Ahlen *et al.*. The substitution of magnetic charge for electric charge is reasonably well understood, and there is now little doubt that scintillators will work at low velocities.

4. *Nucleon decay catalysis.* Monopoles might catalyze nucleon decay with cross sections not much less than those characteristic of the strong interactions. Even if this is not the case, catalysis might occur (via the "weak anomaly") at rates characteristic of the weak interactions.

5. *Electric charge.* It is possible that monopoles have an electric charge in addition to magnetic charge. This would make their detection by scintillation or proportional wire chambers (PWCs) much easier, but would prevent substantial energy transfer via the atomic induction mechanism described by Drell *et al.*.

6. *Bound proton.* A monopole can bind a proton, even in the absence of an electric charge. As a result, most monopoles would have acquired a proton

at a very early epoch. Exactly what "most" means remains the subject of some debate. A bound proton is not easily removed, remaining firmly attached even when a monopole passes through granite.

7. *Expected abundance.* In a simple big bang–grand unified theory scenario, about one monopole per nucleon should exist, leading to a baryonic contribution to closure density which is at least 10–14 times lower than is observed. With inflation occurring between monopole production and baryon production, there might be one monopole within the present event horizon. The range of predicted abundances is 10^{80}.

1988 AND THE NEAR FUTURE

The present state of the art is summarized by the diagram shown in Fig. 1.

Since there is little doubt about the veracity of induction experiments, we take the combined flux limit set by these experiments ($1.1 \times 10^{-12}\,\mathrm{cm}^{-2}\mathrm{sr}^{-1}\mathrm{s}^{-1}$) as a starting point.[3]* It should be doubled if one disputes whether monopoles can traverse the earth. This result is dominated by data from the two 1 m^2 superconducting detectors presently in operation. One is Cabrera's detector at Stanford, and the other is the IBM detector operating at Brookhaven. Work at IBM on the 4 m^2 "prototype" continues, but has detoured into exploring the use of the new high-T_c superconductors. The Chicago group has now achieved a useful signal/noise ratio using normal conductors, and is pursuing this approach. No dramatic improvement of the above limit is to be expected in the next few years.

Parker and his co-workers have argued that the survival of magnetic fields in the galaxy means that the monopole flux implies an (optimistic) upper flux limit at about $10^{-15}\,\mathrm{cm}^{-2}\mathrm{sr}^{-1}\mathrm{s}^{-1}$. The limit can be somewhat higher if monopoles are gravitationally bound to the galaxy, but they would have acquired escape velocity over the lifetime of the galaxy if their masses were less than several times 10^{18} GeV. Parker's scenario has been challenged with various models in which the galactic field is actually maintained by monopoles, but in a recent analysis he rather effectively refutes these models.[4] If his arguments can be extended to clusters of galaxies the limits are reduced by about four orders of magnitude. Parker's ideas are convincing, and flux limits below $10^{-15}\,\mathrm{cm}^{-2}\mathrm{sr}^{-1}\mathrm{s}^{-1}$ may be taken as a target for useful experiments.

The next question concerns whether or not catalysis exists. If the catalysis cross section is even 10^{-5} of the strong interaction cross section, then limits based on the luminosity of neutron stars may be set at $10^{-8\pm1}$ of the Parker bound. These limits are very much lower than those achievable by looking for catalysis in a terrestrial proton decay detector. However, if catalysis occurs with only a weak interaction cross section, it would be observable as the result of a two-step process (formation of a bound state followed by catalysis after a

---

* A 90% confidence level is quoted, as with all experimental limits quoted in this report.

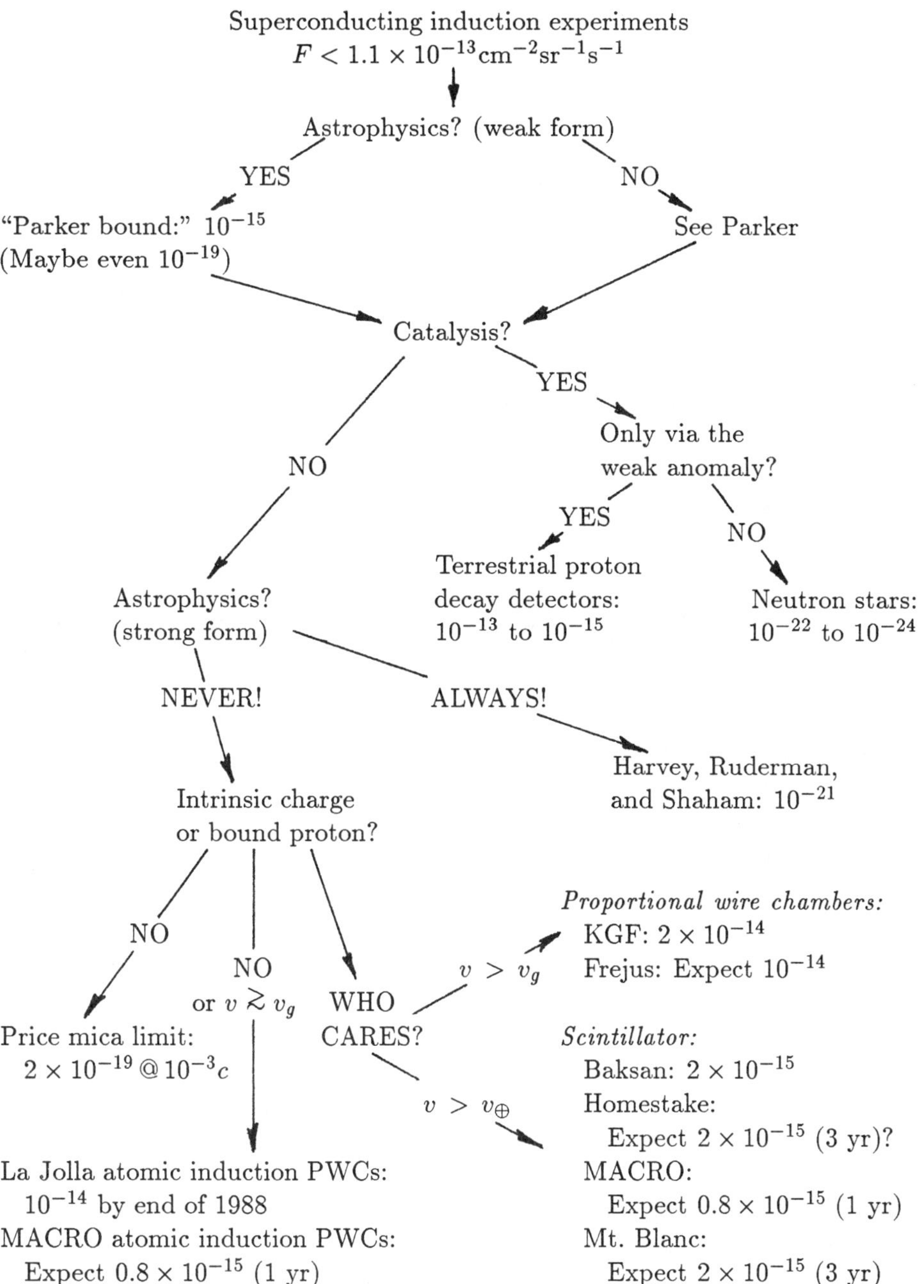

Fig. 1. A highly abbreviated summary of the status and prospects for monopole detection. All the figures quoted are flux upper limits in units of $cm^{-2}sr^{-1}s^{-1}$. The galactic escape velocity is approximately $v_g = 10^{-3}c$, and the earth's orbital velocity is $v_\oplus = 10^{-4}c$.

nanosecond or so) in a terrestrial detector, while not contributing to neutron star luminosity.

In the absence of catalysis more conventional physics obtains. This is a highly relative statement, given the arguments of Harvey, Ruderman, and Shaham concerning field survival in neutron stars.[5] If one accepts their analysis of such matters as the viscous drag experienced by a monopole with two flux quanta attached as it is pulled through the superconducting core of a neutron star, the limit is near $10^{-21}\,\mathrm{cm^{-2}sr^{-1}s^{-1}}$.

The final watershed concerns whether the monopole carries an electric charge, either intrinsically or as a bound proton. In the absence of intrinsic charge it seems likely that at least a large fraction of the monopoles attached protons at an early epoch. The main question is *what* fraction. Fractions between 15% and 98% seem to be reasonable, but such results are exponentially sensitive to a variety of parameters such as the poorly known monopole-proton binding energy.

Price's search for tracks in ancient mica provides by far the best laboratory limit, but the analysis assumes uncharged monopoles. The experiment makes use of an enormous integration time (0.5×10^9 years) and a relatively small detector area ($1000\ \mathrm{cm^2}$). The monopole is assumed to capture a nucleus with a large magnetic moment (such as ^{27}Al) while passing through rock, and the bound highly-charged system then leaves an etchable track in mica. The velocity-sensitive flux limit has a minimum at nearly $10^{-3}c$, the most likely monopole velocity. The limit at $2\times10^{-19}\,\mathrm{cm^{-2}sr^{-1}s^{-1}}$ should be corrected upward by a factor of 50 by those who expect 98% of the monopoles to arrive with bound protons.

According to the atomic induction mechanism proposed by Drell *et al.*, many ^{4}He atoms are left in their metastable 2^3S state after the passage of a slow monopole. Energy transfer to an organic quencher such as CH_4 produces ionization detectable in a proportional wire device. This is the basis for several operating detectors, notably the one at La Jolla, which will have achieved a flux limit of $10^{-14}\,\mathrm{cm^{-2}sr^{-1}s^{-1}}$ when it shuts down in December.

Although a fascinating exploration in its own right, atomic induction originally seemed to provide the only way to investigate the monopole flux for velocities in the region $1\times10^{-4} \lesssim v/c \lesssim 6\times10^{-4}$. Two developments have undercut this advantage: the observation of scintillation light from very low-velocity protons which was discussed above, and the growing conviction that most monopoles either carry a bound proton or have intrinsic charge. This charge very much reduces the cross section for helium excitation by atomic induction. A He-CH_4 PWC should still detect charged monopoles by conventional means, but with a sensitivity which drops dramatically below $10^{-3}c$.

The final branch in our summary diagram is labeled "Who Cares?," because the experiments listed under this heading work with bare monopoles but work even better if the monopoles carry electric charge. Because of the low-velocity proton scintillation studies, we can be relatively confident that scintillators are sensitive to monopoles down to perhaps twice the earth's velocity ($v_\oplus = 10^{-4}c$). Conventional PWCs are probably sensitive to just below the galactic velocity ($v_g \approx 10^{-3}c$).

An interesting University of Tokyo experiment is not shown in the figure because it is difficult to classify. Japanese CR39 exhibits a lower velocity threshold for track-etch experiments than does CR39 manufactured in the US, and when properly etched can reach $10^{-3}c$ for even an uncharged monopole. 4000 m^2 were deployed underground in 1986, and after an average three-year exposure will be etched to set a limit at about $2 \times 10^{-16}\,\mathrm{cm}^{-2}\mathrm{sr}^{-1}\mathrm{s}^{-1}$.

The MACRO detector is listed in three of the final scenarios shown in the figure, and in fact contains CR39 sheets in addition to scintillator and He-CH_4 proportional wire detectors. It is now being assembled in a hall of the Gran Sasso tunnel, in Italy, and represents our best hope for monopole detection in the coming decade. The design has evolved somewhat in the pass two years. Each 6.2 m $\times$ 12 m "supermodule" is now 9 m (almost two stacked 4.7 m modules) rather than 4.7 m high. There will probably be 12 such supermodules (rather than 18 modules). The aperture is reduced by a factor of somewhat less than 2/3 by the change, so that the flux limit in 1 year will be about $0.8 \times 10^{-15}\,\mathrm{cm}^{-2}\mathrm{sr}^{-1}\mathrm{s}^{-1}$. As of May 1988, one supermodule has been built and is in the tunnel. External wiring on the proportional chambers is progressing, and the scintillator tanks will be filled this month. It will be turned on during the summer.[6]

CONCLUSIONS

The flux of massive magnetic monopoles is severely constrained by a combination of induction experiments, astrophysical arguments, track etch experiments, and measurements with conventional particle detectors. The flux limit is highly dependent upon assumed monopole properties. The new MACRO detector will set important new limits for monopoles which do not catalyze nucleon decay but might carry electric charge.

REFERENCES

1. R. A. Carrigan and W. P. Trower, eds., *Magnetic Monopoles*, Plenum, Ny (1983); J. L. Stone, ed., *Monopole '83*, Plenum, NY (1984); J. Preskill, Ann. Rev. Nucl. Part. Sci. **34**, 461 (1984); G. Giacomelli, La Rivista del Nuovo Cimento **7**, No. 12 (1984); D. E. Groom, Phys. Reports **140**, 324-373 (1986).
2. D. J. Ficenec, S. P. Ahlen, A. A. Marin, J. A. Musser, and G. Tarlé, Phys. Rev. D**36**, 311 (1987).
3. B. Cabrera, private communication (April 1988).
4. E. N. Parker, Ap. J. **321**, 349-354 (1987).
5. J. A. Harvey, M. A. Ruderman, and J. Shaham, Phys. Rev. D **33**, 2084 (1986).
6. B. C. Barish and C. W. Peck, private communications (May 1988).
7. S. Orito, private communication (May 1988).

TYPE II SUPERNOVAE: HOW DO THEY EXPLODE?†

E. Baron
Dept. of Physics, SUNY, Stony Brook, NY 11794-3800

ABSTRACT

I discuss what has been learned from the neutrino observations of Supernova 1987A. The neutrino detections confirmed our basic theoretical scenario that Type II supernovae involve the gravitational collapse of a massive star. The small number of events makes it difficult to infer details about the actual mechanism of collapse. I discuss the current theoretical situation on the mechanism of explosion.

INTRODUCTION

The detection of neutrinos from SN 1987A[1,2] confirmed our basic scenario that Type II supernovae involve the gravitational collapse of a massive star. The one surprise was that the progenitor star was not the expected red supergiant, but was rather a blue supergiant. That the star collapsed while it was blue rather than red seems to be understandable due to the low metallicity of the LMC.[3,4] Observations with IUE seem to show that the star lost mass in a red giant phase prior to returning to the blue.[5] Stellar evolution seems to be able to accommodate such a history if the assumptions about mixing are slightly altered.[6,7]

The physics that can be gleaned from the neutrinos detections has been discussed by very many authors and I shall simply summarize the results from our own analysis[8] which is in accord with that of most other authors. From the events observed in IMB and Kamiokande one can determine the approximate mass and binding energy of the collapsed remnant. The binding energy (BE) of the neutron star is in the range $1.5 < BE < 2.5$ in units of 10^{53} ergs. This corresponds to a mass in the range $1.1M_{\odot} < M < 1.7M_{\odot}$. These values give us no reason to assume that this remnant is not a neutron star, whose presence as a pulsar we should detect in the next $\sim 1-5$ years, if our ideas about the pulsar mechanism are correct. The initial cooling temperature at the neutrinosphere is in the range $4 < T < 6$ in units of MeV. Simple assumptions about the duration of the neutrino emission at the star lead to an upper limit of the neutrino mass of about 20 eV. Calculations to fit the optical light curve give estimates of the explosion energy in the range $1.3 - 1.8 \times 10^{51}$ ergs.[9,10]

†Supported in part by the Department of Energy grant no. DE-FG02-88ER40388.

THE MECHANISM

While these results are extremely exciting, the paucity of events makes it difficult to determine more details about the mechanism of the explosion, for that we will have to wait for a closer supernova. Theoretically, two mechanisms for turning the gravitational collapse of a massive star into an explosion have received the most scrutiny: the prompt mechanism, where the shock wave generated by the bounce of the iron core has enough energy to traverse through that core and expel the mantle and envelope; and the delayed mechanism, where the shock wave is not sufficiently energetic, so it stalls becoming an accretion shock. In this model, studied by Wilson and collaborators[11–13] the shock is resuscitated on the deleptonization timescale of about 1 second.

Calculating stellar collapse is quite difficult, since the explosion energy of about 10^{51} ergs is only about 1% of the energy released in the collapse. In addition, there is an enormous amount of physics involved, from weak interactions to the nuclear physics of warm dense matter. Also, the computation is large and complicated involving hydrodynamics, general relativity, and non-LTE radiation transport.

The delayed mechanism tends to give explosions of rather low energy, but convection may help here. It also seems to be sensitive to the exact details of the neutrino transport. In any case, the model needs to be confirmed by other workers in the field. We hope to present our own calculations of this mechanism in the very near future.[14]

The prompt mechanism, upon which my calculations have been focused seems to hinge on essentially three factors: 1) the mass of the iron core, 2) the final value of the trapped lepton fraction Y_l, and 3) the stiffness of the high density equation of state. The previous statement is rather simplistic, since for example the equation of state at low densities is also extremely important and it assumes that e.g. general relativity is used in the calculations. Nevertheless, it is convenient to focus on these three effects. In so doing, I will attempt to identify likely ranges for these parameters.

The mass of the iron core is an extremely important parameter. It is an input to stellar collapse calculations, being the result of millions of years of stellar evolution. The mass of the iron core is important since it basically sets the energy budget of the shock wave. The less mass of iron the shock wave must traverse, the less dissociation energy the shock must pay. Now the dissociation energy of iron is enormous ~ 8 MeV/nucleon or $\sim 2 \times 10^{51}$ ergs/$0.1 M_\odot$. This number isn't exactly correct because if the shock is successful the temperature is reduced and dissociation proceeds only to alpha particles, reducing the cost by roughly a factor of 4. Now the stellar evolution calculations have been slowly reducing the mass of the iron core over the last ten years or so. Refinements in electron capture rates, nuclear burning rates, the equation of state, and assumptions about convection have only served to bring the core mass down. Recent stellar cores have masses 1.18 $M_\odot$ for a 13 $M_\odot$ star,[15] 1.28 $M_\odot$ for a 15 $M_\odot$ star,[6] and 1.33 $M_\odot$ for a 18 $M_\odot$ star.[16] In the 18 $M_\odot$ model although the beginning of the silicon shell is a 1.33 $M_\odot$, there is a sharp drop in density at 1.17 $M_\odot$ which is close to the Chandrasekhar mass for this core. Since it is the density that the shock wave measures, it may be somewhat misleading to simply quote the value of the iron core. Due to different assumptions about convection, the nominally larger 15 $M_\odot$ core is as easy, or easier to explode than the 13 $M_\odot$ core.

All of these calculations used a value for the $^{12}C(\alpha,\gamma)^{16}O$ rate[17] of roughly three times the rate of Fowler, Caughlan, and Zimmerman.[18] In a difficult experiment, the Caltech group has done a coincidence measurement of this rate.[19] A three level R–matrix analysis of their results leads to an allowed range of 0.0–0.14 Mev barn for $S_{E1}(300)$, and a hybrid R–matrix analysis gives a value 0.08 MeV barn. This leads to a rate roughly in accord with the old result.[18] This lower rate will tend to further decrease the iron core mass, since it lowers the total amount of carbon depleted at the end of helium burning, and allows a stage of convective core and shell carbon burning. This alters the entropy structure which prevents silicon from burning further out in the core. Weaver and Woosley[16] have recently calculated the evolution of an 18 $M_\odot$ model with the new $^{12}C(\alpha,\gamma)^{16}O$ rate. This core looks like it will be extremely easy to explode. Although the iron core mass is still about 1.3 $M_\odot$, it has a break in density at about 1 $M_\odot$. This lower density will help the shock to get through the important region 1.0 to 1.3 $M_\odot$. This development is extremely helpful to the prompt mechanism, since iron cores larger than about 1.4 $M_\odot$ do not seem to be able to explode promptly.

The final trapped lepton fraction Y_l also determines how much matter the shock wave must traverse, since it determines where in mass the shock wave will begin. The size of the homologously collapsing core is proportional to Y_l^2 and hence small changes in Y_l result in large changes in the mass where the shock forms M_{form}. In calculations, done with Jerry Cooperstein, using a completely general relativistic hydrodynamic non-LTE radiative transport code we find that Y_l is about 0.39 and the 13 $M_\odot$ and 15 $M_\odot$models explode with energies greater than 10^{51} ergs. These calculations use our two fluid model[20] and Levermore-Pomraning[21] flux-limited diffusion. They ignore the effects of neutrino electron scattering and electron capture on heavy nuclei. Electron scattering will reduce Y_l since it lowers the density at which the neutrinos thermalize, lowering their average energy, and increasing their mean free paths, allowing more of them to escape the homologous core.[22,23] Preliminary calculations, that we have performed with electron scattering seem to show that Y_l is not reduced that much, but these still need to be checked and compared with the results of other workers.[22,23] In addition, we have neglected capture on nuclei which will also reduce Y_l in the early stages of collapse. Again, calculations using the best nuclear physics[24] have not been done. From parameter studies that we have performed, it is my feeling that if Y_l drops below 0.365-0.37 then the prompt mechanism probably cannot succeed.

All of the above calculations have used the so-called BCK parameterization of the equation of state at high density.[25,26] This parameterization is good for the somewhat asymmetric matter found in the first few milliseconds after bounce. We assume for the cold part of the equation of state that the pressure can be parameterized as

$$P = \frac{K_0(x)\rho_0(x)}{9\gamma}[(\rho/\rho_0(x))^\gamma - 1],$$

where K_0 and ρ_0 are the nuclear incompressibility and density at saturation, both of which depend on the asymmetry $x = Z/A$. This equation of state is used only for $\rho > \rho_0(x)$. In the above calculations, $K_0(0.5) = 180$ MeV and $\gamma = 2.5$. We have shown[25,26] that the use of such an equation of state combined with general relativity allows the collapse to proceed to much higher density and thus to dig deeper into the gravitational well, providing the shock wave with much greater initial energy. Now this parameterization cannot be used for neutron matter since neutron matter doesn't saturate. Nevertheless the equation of state cannot be too soft or else the maximum mass of neutron stars falls below the value of measured neutron stars of about 1.45 $M_\odot$. Now the stiffness of the equation of state is not measured by the value of K_0 or of γ alone, and for comparison with microphysical calculations of the equation of state, I think it is best to compare the pressure at a given density. It is my feeling that the pressure of asymmetric matter ($x \sim 0.33$) *including leptons* should be less than about 75 MeV/nucleon at a density of about $3\rho_0$. Pressures much greater than this will cause the core to bounce at a density less than about $3\rho_0$ which is what seems to be needed for the prompt mechanism. Again, the decreasing core mass may lessen this requirement.

CONCLUSIONS

Since the whole process of the supernova is a delicate one, it is possible with known physics to make the prompt mechanism fail. At the same time, and particularly with the low values the iron core masses seem to be taking, it is possible to make it work and to give explosions with the required energy. Also, one must remember that the calculations depend on two complicated calculations, one the of collapse itself, and the other of stellar evolution to provide the initial models. If the stellar evolution calculations, which are uncertain in the treatment of convection, were to further alter the density structure in the outer part of the iron core, then the shock wave may have no problems getting out. Along these lines Haxton[27] has suggested that the inclusion of inelastic scattering of neutrinos on nuclei may pre-heat the matter in front of the shock wave and help to keep the density low. I don't think the effect will be large enough to be decisive, but every little bit helps.

ACKNOWLEDGEMENT

I would like to thank my collaborators J. Cooperstein and M. Aufderheide for allowing me to discuss our joint work prior to publication.

References

[1]K. Hirata et al., Phys. Rev. Lett. **58**, 1490 (1987).

[2]R. Bionta et al., Phys. Rev. Lett. **58**, 1494 (1987).

[3]W. Brunish and J. Truran, Astrophys. J. **49**, 447 (1982).

[4]W. D. Arnett, Astrophys. J. **319**, 136 (1987).

[5]R. P. Kirshner, in *George Mason University Workshop on SN 1987A*, edited by M. Kafatos (Cambridge University Press, Cambridge, 1988) [in press].

[6]S. E. Woosley and T. A. Weaver, Phys. Repts. (1988) [in press].

[7]H. Saio, M. Kato, and K. Nomoto, Astrophys. J. (1988) [in press].

[8]S. Kahana, J. Cooperstein, and E. Baron, Phys. Lett. **B196**, 259 (1988).

[9]S. E. Woosley, Astrophys. J. (1988) [in press].

[10]K. Nomoto, T. Shigeyama, and M. Hashimoto, in *Proceedings of ESO Workshop on SN 1987A*, edited by J. Danziger (1988).

[11]J. R. Wilson, in *Relativistic Astrophysics*, edited by J. Centrella, J. LeBlanc, and R. Bowers (Jones and Bartlett, Boston, 1985).

[12]H. A. Bethe and J. R. Wilson, Astrophys. J. **295**, 14 (1985).

[13]R. Mayle, PhD thesis, University of California, Berkeley, 1985, issued as Livermore Report UCRL-53713.

[14]E. Baron and J. Cooperstein, (1988) [in preparation].

[15]K. Nomoto and M. Hashimoto, Phys. Repts. (1988) [in press].

[16]T. A. Weaver and S. E. Woosley, (1988) [private communication].

[17]G. R. Caughlan, W. A. Fowler, M. J. Harris, and B. A. Zimmerman, Atomic Data and Nuclear Data Tables **32**, 197 (1985).

[18]W. A. Fowler, G. R. Caughlan, and B. A. Zimmerman, Ann. Rev. Astron. Ap. **13**, 69 (1975).

[19]R. M. Kremer et al., Phys. Rev. Lett. **60**, 1475 (1988).

[20]J. Cooperstein, L. J. van den Horn, and E. Baron, Astrophys. J. **309**, 653 (1986).

[21]C. D. Levermore and G. C. Pomraning, Astrophys. J. **248**, 321 (1981).

[22]S. Bruenn, Astrophys. J. Suppl. **58**, 771 (1985).

[23]E. S. Myra, S. A. Bludman, Y. Hoffman, I. Lichtenstadt, N. Sack, and K. A. Van Riper, Astrophys. J. **318**, 744 (1987).

[24]J. Cooperstein and J. Wambach, Nucl. Phys. **A420**, 591 (1984).

[25]E. Baron, J. Cooperstein, and S. Kahana, Nucl. Phys. **A440**, 744 (1985).

[26]E. Baron, J. Cooperstein, and S. Kahana, Phys. Rev. Lett. **55**, 126 (1985).

[27]W. Haxton, (1988) [preprint].

SEARCH FOR HIGH ENERGY GAMMA-RAYS FROM SN 1987A

Akira MASAIKE*

Department of Physics, Kyoto University, Sakyo-ku, Kyoto, Japan

ABSTRACT

High energy γ-rays from SN 1987A were searched for at the Black Birch Range in New Zealand, using a scintillation counter array and mirrors for Cerenkov light. An upper limit on the γ-ray flux greater than 100 TeV was about 3.6×10^{-12} $cm^{-2}\cdot s^{-1}$ from 13 October to 3 December, 1987. An upper limit on the γ-ray greated than 3 TeV was $6\times10^{-12} cm^{-2}\cdot s^{-1}$ during December 1987 and January 1988. Data obtained on January 14 and 15 have excess counts corresponding to a flux of $(1.9\pm0.5)\times10^{-11} cm^{-2}\cdot s^{-1}$.

The observation of high energy γ-rays from supernova 1987A in Large Magellanic Cloud (LMC) is interesting from various viewpoints. Especially it provides information on the formation of a neutron star and the dynamic behavior of the explosion of supernova. Furthermore, it may suggest the origin of the cosmic ray. More than 10 years ago, H. Sato pointed out that high energy γ-rays and neutrinos may be produced from pions generated by collisions of protons, which are accelerated around the neutron

* JANZOS Collaboration: I.A. Bond, E. Budding, M.J. Conway, K.B. Fenton, H. Fujii, Z. Fujii, M. Fujimoto, H. Hasegawa, N.Hayashida, M. Honda, N. Hotta, J.E. Humble, S. Kabe, K. Kasahara, T.Kifune, A. Masaike, Y. Matsubara, K. Mitsui, Y. Miura, M. Mori, K. Murakami, Y. Muraki, M. Nagano, K. Nakamura, T. Nakamura, P.M. Norris, Y. Ohashi, A. Okada, To. Saito, M. Sakata, H. Sato, S. Shibata, M. Shima, J.R. Storey, T. Tanimori, M. Teshima, S. Torii, K. Uchino, Y. Watase, M.D. Woodhams, Y. Yamamoto, P.C.M. Yock, and T. Yuda

star, with supernova ejecta.[1] The flux of γ-rays increases with time, as the column density of the expanding ejecta decreases. The γ-ray intensity should be high when the column density is thin enough for γ-rays to pass through the ejecta without interaction and thick enough for protons to interact with ejecta. Detailed calculations suggested that the maximum intensity of γ-rays occur about a half year to 2 years after the explosion.[2,3] Collisions with background radiation of 2.7K reduce the flux at energies higher than about 200 TeV.

We installed two types of detectors at the Black Birch Range in New Zealand (1640m a.s.l., 41°45'S, 173°47'E).[4,5] An air shower array for the measurement of γ-rays in the 100 TeV region consists of seventy-six scintillation counters (Fig.1). Forty-five scintillators are of 0.5 m^2 in area and 5 cm in thickness with lead sheet of 5 mm in thickness on the top, viewed by fast photomultipliers (Hamamatsu H1949) from the bottom, whereas 31 scintillators are of 1 m^2 in area and 10 cm in thickness, viewed by rather slow photomultipliers, which are used for particle density measurements. The timing resolution of scintillators of 0.5 m^2 is 1.2 ns. The system is triggered by any four-fold coincidence of the 0.5 m^2 scintillators within 150 ns.

The mirror system was used for the detection of γ-rays greater than 1 TeV. It consists of three spherical aluminum mirrors of 2 m aperture and 2 m focal length, situated with the distance of 80 m from each other (Fig. 2). Ten fast photomultipliers (Hamamatsu H1531) were installed in the focal plane of each mirror along the path of SN 1987A. The triggering of the system is made when any one of the 10 sets of 3 mirrors which direction in the sky is the same. The accuracy of the timing signal is measured with a pulsed laser system and found to be better than 1 ns. After the correction of curvature and the thickness of the Cerenkov light disk, the angular accuracy is about 0.5°∿1.0°.

The observation with the scintillation counter array started on October 13, 1987. The data until 3 December was analyzed. Fig. 3 represents a right ascension scan of events near the declination of the supernova. There is no clear excess of events. An upper limit on the flux of ultra high energy (UHE) γ-rays is obtained using the maximum likelihood method. The effective exposure of the γ-rays from the supernova was 2.54×10^{13} $cm^2\cdot sec$ and the median energy was 180 TeV. Assuming a spectral index $\alpha=2.0$ and complete absorption by the microwave background radiation, the γ-ray luminosity of the supernova is deduced to be

$$L_\gamma \lesssim 3.3\times10^{38}\ (d/50\ \text{kpc})^2\ \text{erg}\cdot\text{s}^{-1}$$

where d is the distance to SN 1987A. The intensity of UHE γ-rays observed at earth is related to proton flux at the source, using various models. According to the models of I. Yamada et al.[3)] assuming 10^{17} eV monoenergetic proton injection and complete absorption by the background radiation, the total energy of protons (L_p) is calculated from our data to be less than 1.4×10^{40} $erg\cdot s^{-1}$, when the UHE γ-ray flux reaches its maximum value at the column thickness of the shell of 76.3 $g\cdot cm^{-2}$. Gaisser et al.[6)] assumed a proton spectrum of $E^{-2}dE$ with a cut-off at 10^{17} eV and regeneration of γ-rays by electron inverse Compton scattering on the background radiation in the average field of $\ll10^{-10}$ gauss. Using the Gaisser's model, the upper limit of L_p was found to be 2.6×10^{39} $erg\cdot s^{-1}$ from our data.

The observation using the mirror detectors for Cerenkov light were performed during December 1987 and January 1988. The data obtained for 42 hours during the period were analyzed. Fig. 4 shows the distribution of observed events in the declination strip -69.3°±0.5° as a function of right ascension. The smooth curve indicates the average level of cosmic ray background. There is no significant excess in the direction of SN 1987A. From those results an upper bound on the γ-ray flux, at the 95%

confidence level, of 6.1×10^{-12} $cm^{-2}\cdot s^{-1}$ for energies greater than 3 TeV was found. A corresponding limit of the γ-ray luminosity above 3 TeV is 8.4×10^{37} $erg\cdot s^{-1}$ with the same assumption for (UHE) ray detection. The ratios of observed events to the average level of cosmic ray background events on January 14 and 15 are plotted against right ascension in Fig. 5. The data of January 14 and 15 are found to have 2.2σ and 3.3σ excesses, respectively. The combined data of the two nights show an excess of 4.0σ. The excess in two nights corresponds to a flux $(1.9\pm0.5)\times10^{-11}$ $cm^{-2}\cdot s^{-1}$ for $E\gamma\geq$ 3 TeV. The total energy emitted in TeV γ-rays is estimated to be $(1.9\pm1.1)\times10^{43}$ erg for 2 days. From these data we conclude that it is likely that an outburst of TeV γ-rays were emitted from SN 1987A. The observed time of the excess apparently coincides with the time of an X-ray flux increase detected by the Ginga satellite,[7)] whose total energy is reported to be $\sim10^{43}$ erg.

REFERENCES

1. H. Sato, Prog. Theor. Phys. 58, 549 (1977).
2. T. Nakamura et al., Prog. Theor. Phys. 78, 1065 (1987).
3. Y. Yamada et al, Prog. Theor. Phys. 79, No.2 (1987).
4. I.A. Bond et al., Phys. Rev. Lett. 60, 1110 (1988).
5. I.A. Bond et al., ICR-Report-161-88-7 (1988).
6. T.K. Gaisser et al., Nature 329, 314 (1987).
7. Y. Tanaka, a talk presented at 20th Yamada Conference, March 28-April 1, 1988 in Tokyo.

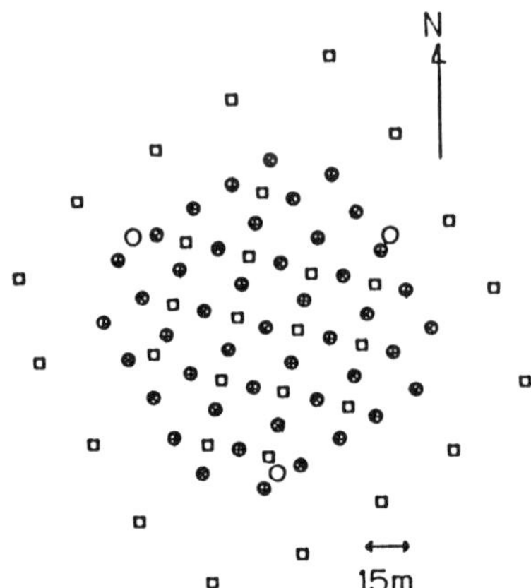

Fig. 1 The arrangement of detectors on the Black Birch Range.
⊗ , ⊕ : 0.5 m^2 scintillators, □ : 1 m^2 scintillators,
○ : mirrors.

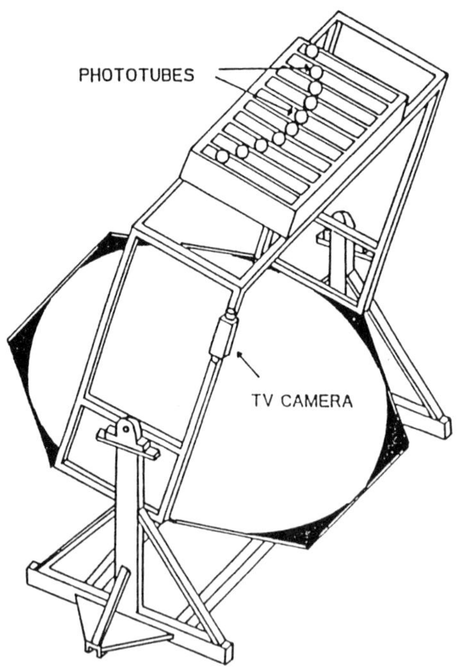

Fig. 2 A schematic view of the mirror telescope.

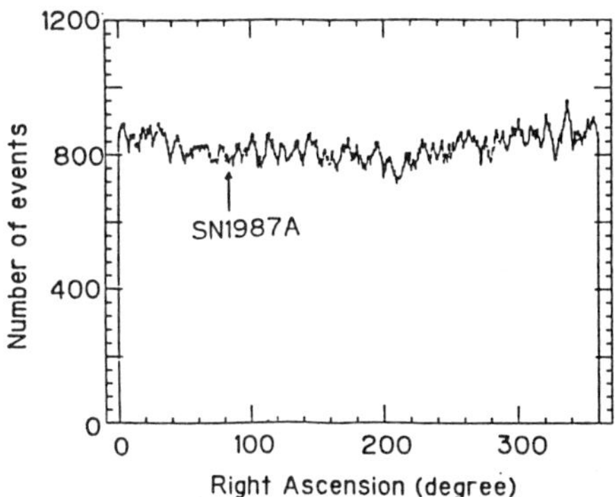

Fig. 3 Right ascension scan of events on the declination of the supernova with the scintillation counter array.

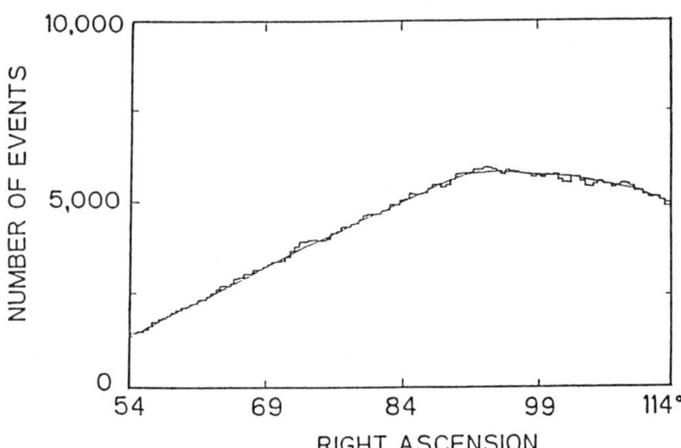

Fig. 4 The distribution of observed events with the mirror telescope in the declination strip -69.3°±0.5° as a function of right ascension.

Fig. 5 The ratios of observed events to the average level of cosmic ray background events on January 14, 15, 1988 protted against right ascension.

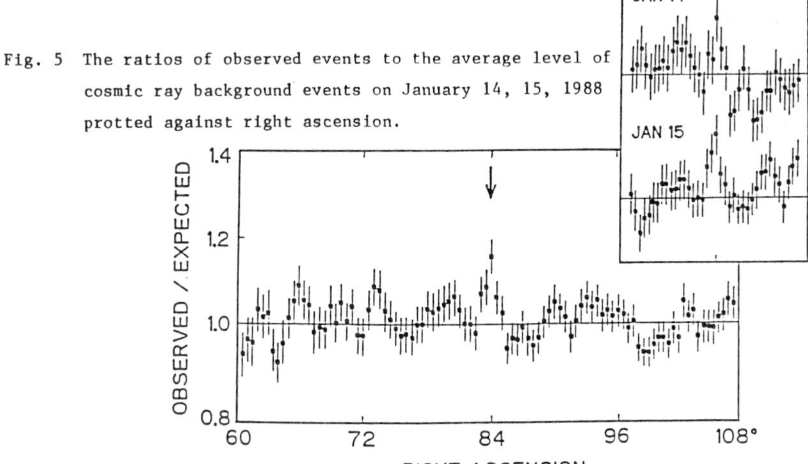

PRIMORDIAL NUCLEOSYNTHESIS AS A PROBE OF COSMOLOGICAL QCD

James H. Applegate*
Department of Astronomy, Columbia University
New York, New York 10027

ABSTRACT

The cosmological consequences of a first order phase transition in quantum chromodynamics are discussed. The most likely observable relic of the cosmological QCD transition is the production of heavy elements in an r-process during primordial nucleosynthesis.

The development of modern gauge theories in particle physics has brought about the realization that the universe may have undergone several phase transitions during its early evolution. The richness of the cosmology associated with a first order phase transition in quantum chromodynamics (QCD) was first described by Witten[1], who argued that large amplitude baryon density inhomogeneities could be produced in the transition, and by Applegate and Hogan[2], who realized that these density inhomogeneities would profoundly affect primordial nucleosynthesis. Density inhomogeneities produced in the QCD transition will have a characteristic length scale which is smaller than the horizon by a factor of order $4 \ln M_p/T_c$, where $M_p \simeq 10^{19}$GeV is the Planck mass and $T_c \simeq 0.1$GeV is the critical temperature for transition[3]. This length scale is in an interesting regime – small enough to allow neutrons to diffuse to uniformity by the onset of primordial nucleosynthesis, but large enough to prevent protons from doing so. Neutron diffusion will convert regions of low baryon density into environments in which neutrons locally outnumber protons. Nucleosynthesis in these neutron rich regions is sufficiently different from the homogeneous standard model to produce an observable signature. The most promising relic of neutron rich nucleosynthesis is the production of a heavy elements in a primordial r-process.

Witten's[1] original scenario envisioned concentrating $\simeq 90\%$ of the baryon number of the universe into nuggets of stable[4] strange quark matter. This would allow a closure density of baryons because only the $\simeq 10\%$ of the baryons not in nuggets would participate in primordial nucleosynthesis; the $\simeq$

---

* Alfred P. Sloan Foundation Fellow

90% in nuggets would be inert. The nuggets would account for the dark matter in the universe. This scenario was criticized on the grounds that strange quark matter was unlikely to be stable at zero pressure[5], that the nuggets, if formed, would evaporate into nucleons before nucleosynthesis[6], that the strength of the baryon number concentration has been overestimated[2], and that forming the nuggets would leave the baryons not in nuggets so inhomogeneously distributed that primordial nucleosynthesis constraints would probably still be violated[2]. These arguments removed strange quark nuggets from the list of viable dark matter candidates. However, the nugget scenario was important because it led to the realization that primordial nucleosynthesis is potentially a probe of cosmological QCD.

The production of density inhomogeneities in the QCD transition depends on the degree of supercooling of the high temperature phase, the expansion rate of the universe, and the diffusion rate of baryon number while the universe is evolving through the two-phase region. If the expansion of the universe is slow enough to allow baryon number to diffuse to equilibrium while the universe is in the two-phase region, then baryon number will be concentrated in the quark matter phase[1]. Adding unit baryon number to the hadron matter phase requires an energy of at least the proton rest energy $m_p c^2$. Unit baryon number can be added to the quark matter phase with very little energy cost because the quarks are nearly massless. Therefore, in equilibrium the baryon density in the hadron phase is suppressed by a Boltzmann factor $e^{-m_p c^2/T}$ relative to the baryon density in the quark matter phase. However, estimates of the diffusion coefficient for baryon number in the quark phase[1,7] indicate that baryon number cannot diffuse fast enough to maintain chemical equilibrium. Equilibrium concentration of baryon number will not occur unless some mechanism hydrodynamically mixes the quark matter phase.

If the hadron matter phase nucleates rapidly once the universe cools to slightly below T_c the universe will rapidly enter a near-equilibrium two-phase configuration. The rate of cooling by expansion is balanced by the rate of release of latent heat, and the universe remains near T_c throughout the transition. The flow of matter through the phase boundary has the character of a condensation discontinuity[8]. The density perturbations produced by this type of transition have been calculated[2]. Their amplitudes are too small to affect nucelosynthesis.

If the universe supercools by a significant amount ($\Delta T/T_c \gtrsim 0.1$ is probably required) the transition will nucleate with some violence because of the large departure from equilibrium. This will result in the uneven release of latent heat. Regions that have been heated more will have a higher entropy per baryon, and hence a lower baryon density once pressure equilibrium is reestablished, then regions that have received little latent heat. The nucleation and hydrodynamics of the QCD transition have been discussed by several authors[9,10,11], however the amplitude of density inhomogeneities produced by hydrodynamics has not been calculated.

The most important physical process affecting the density inhomogeneities is neutron diffusion[2,12,13]. Because they are charged protons are efficiently scattered by electrons and positrons, and the proton mean free path is very short. Neutrons scatter much less strongly because they are neutral. The most important neutron scattering mechanisms are hadronic collisions with protons and scattering of electrons or positrons by the neutron magnetic dipole moment. These mechanisms give the neutron a mean free path about 10^6 times longer than the proton.

Despite the large difference in neutron and proton mean free paths the diffusion lengths of neutrons and protons are equal as long as the weak interactions are in equilibrium. This is because a proton diffuses by changing into a neutron, diffusing, and changing back. When the weak interactions fall out of equilibrium at $T \sim 1$ MeV nucleons retain their identity as neutrons or protons, and diffusive separation occurs. Neutron diffusion before nucleosynthesis has been treated in detail[12]. The comoving neutron diffusion length can be as much as 10^3 to $10^{3.5}$ times larger than the comoving proton diffusion length.

The character of the neutron diffusion changes once nucleosynthesis begins[13]. Once nucleosynthesis begins ^{4}He is produced rapidly until one of the primary reactants, neutrons or protons, is exhausted. Neutrons are exhausted in a proton rich environment. Protons are exhausted in the neutron rich regions, giving these regions a late-time supply of neutrons. This has the effect of producing a neutron density gradient which drives neutron diffusion out of the neutron rich regions. Neutrons which diffuse back into proton rich environments are captured rapidly by $n(p,\gamma)^2$H, and ultimately wind up in ^{4}He. Neutron diffusion during nucleosynthesis was treated using a plane symmetric model and neglecting the expansion of the universe[13]. The diffusion gives the neutron an effective decay rate κ larger than the free decay rate of $\kappa_{\text{free}} = 1.09 \times 10^{-3}\text{s}^{-1}$. The requirement that the neutron rich regions hold on to their neutrons long enough to do much nucleosynthesis gives a constraint $\kappa \lesssim 10^{-2}\text{s}^{-1}$, which produces a limit on the comoving size of a neutron rich region. This limit is slightly larger than the comoving horizon scale at $T = 100$ GeV, so the electroweak transition will probably not affect nucleosynthesis. The larger neutron rich regions that can be filled to uniformity will hold their neutrons long enough to do interesting nucleosynthesis (larger regions will also affect nucelosynthesis, but they cannot be filled to uniformity). In particular, the neutron exposure in large regions is large enough to produce a runaway primordial r-process.

Light element nucleosynthesis in neutron-proton segregated universes has been calculated by several authors[11–16]. The proton rich regions in inhomogeneous universes nucleosynthesize like homogeneous standard models, but with somewhat fewer neutrons than a pure standard model because some neutrons have diffused away. This lowers the ^{4}He production relative to a homogeneous model of the same global Ω_b. In a proton rich environment two neutrons are required to make a ^{4}He nucleus. Most of the baryons in a neutron rich region end up in ^{4}He, but four neutrons are needed to make ^{4}He because two neutron beta decays are needed to produce the protons. The same global number of neutrons is used less efficiently, so the ^{4}He production is decreased. The reduction is by a factor $\Delta Y/Y = -\frac{1}{2}f_n$, where Y is the ^{4}He mass fraction and the f_n is the fraction of neutrons which undergo neutron rich nucleosynthesis.

The products of incomplete burning, particularly ^{2}H, are boosted in inhomogeneous models by neutron rich nucleosynthesis. The boost comes from the fact that neutron beta decay followed by $p(n,\gamma)^2$H can produce deuterium at temperatures that are low enough so that charged particle reactions are strongly suppressed by Coulomb barriers. This, and the reduction of ^{4}He production, raised the possibility that an $\Omega_b = 1$ inhomogeneous universe could be consistent with the observed light element abundances. The numerical calculations showed that this was possible, but only if a high primordial ^{7}Li production (a few times the pop I star value) could be tolerated. This raised the interesting possibility of using ^{7}Li to constrain cosmological QCD, but this line of investigation is hampered by our lack of understanding of the galactic nucleosynthesis[17] of ^{7}Li. Fortunately, the inhomogeneous models have a much more dramatic prediction.

A major surprise in the results of the numerical calculations was that the abundance of $A \geq 12$ elements, most of which was ^{14}C, was orders of magnitude higher in the neutron rich regions in inhomogeneous universes than in homogeneous proton rich models with the same global Ω_b. This result was completely unexpected because the density is orders of magnitude too low for the triple alpha or $\alpha - \alpha - n$ reactions to play a role. The reaction sequence that produces ^{14}C is ^{4}He$(^3$H$,\gamma)^7$Li$(n,\gamma)^8$Li$(\alpha,n)^{11}$B$(n,\gamma)^{12}$B$(\beta^-)^{12}$C(n,γ) ^{13}C$(n,\gamma)^{14}$C; an alternate chain, which replaces ^{7}Li$(n,\gamma)^8$Li$(\alpha,n)^{11}$B with ^{7}Li$(\alpha,\gamma)^{11}$B, contributes at the 1% to 10% level, depending on the density. This sequence is strongly suppressed in proton rich environments because ^{11}B$(p,2\alpha)^4$He dominates the neutron capture by many orders of magnitude.

Once produced, ^{14}C is difficult to get rid of. Several strong (n,α) and (p,α) reactions serve to cycle material that has nucleosynthesized past carbon back to the carbon isotopes. The strong cycling reactions are: $^{14}\mathrm{N}(n,p)^{14}\mathrm{C}$, $^{15}\mathrm{N}(p,\alpha)^{12}\mathrm{C}$, $^{17}O(n,\alpha)^{14}\mathrm{C}$, and $^{18}\mathrm{F}(n,\alpha)^{15}\mathrm{N}$. The existence of these reacitons explains why most of the $A \geq 12$ material found in the numerical integrations was ^{14}C.

The reaction sequence that produces heavy elements has three distinct stages: the match, the fuse, and the bomb. The match is the reaction sequence that avoids the cycling reactions. It is $^{14}\mathrm{C}(\alpha,\gamma)^{18}\mathrm{O}(n,\gamma)^{19}\mathrm{O}(\beta^-)$ $^{19}\mathrm{F}(n,\gamma)^{20}\mathrm{Ne}$. This sequence produces a very small amoung of ^{20}Ne. Detailed calculation[13] showed that $X(^{20}\mathrm{Ne})/X(^{14}\mathrm{C})$, where X denotes mass fraction, was 10^{-5} for a global $\Omega_b = 1$ universe and 5×10^{-9} for $\Omega_b = 0.06$. The fuse is a series of (n,γ) reactions and beta decays that converts neon into $A \simeq 60$ seeds for a classical r-process. The fuse is characterized by rather small (n,γ) cross sections because the nuclei are light, but no strong (p,α) or (n,α) reactions. Both high and low Ω_b universes have enough neutron exposure for the $A \simeq 60$ r-process seeds to be produced.

The bomb is a classical r-process seeded with the $A \simeq 60$ nuclei produced by the fuse. The neutron density and temperature are not very different from their values in a standard stellar r-process site. The difference between a stellar r-process site and the big bang is that the neutron exposure in a stellar site lasts a few seconds, whereas the exposure in the big bang lasts a few minutes. The long duration of the neutron exposure in the big bang means that the r-process will be dominated by fission cycling[18]. The amount of time for a $A \simeq 125$ nucleus to capture 125 neutrons and undergo the requisite β-decays to reach mass number 250, and then fission into two mass 125 nuclei is called the cycle time. The cycle time is estimated to lie in the range of 3–10 seconds. The number of r-process nuclei doubles each fission cycle. Both the global $\Omega_b = 1$ and $\Omega_b = 0.06$ universes will have fission cycling dominated r-processes, and observable, and possibly excessive, levels of r-process enrichment can occur[13].

The primordial r-process should be detectable as a cosmological floor to the r-process abundance in very metal poor stars. No stars with r-process abundances before the cosmological production should exist. In addition, the abundance pattern should shift to the abundance pattern produced by fission cycling. This is strikingly different from the solar pattern because $A \lesssim 120$ r-process elements are not produced by fission cycling.

Current r-process observations are already a significang constraint on the global $\Omega_b = 1$ model[13]. In order to produce an acceptable light element nucleosynthesis at least 10% of the neutrons must undergo nucelosynthesis in neutron rich environments. These regions must be large enough to have an effective neutron decay rate smaller than $\simeq 10^{-2}s^{-1}$. However, in order to avoid r-process overproduction less than 10^{-7} of the neutrons can be in regions whose effective decay rate is less than $3 \times 10^{-3}s^{-1}$. Such a special distribution of sizes of neutron rich regions argues against the global $\Omega_b = 1$ model. This model is on the verge of being ruled out by r-process overproduction.

REFERENCES

1. E. Witten *Phys. Rev.*, **D30**, 272 (1984).
2. J.H. Applegate and C.J. Hogan, C.J., *Phys. Rev.*, **D31**, 3037 (1985).
3. C.J. Hogan, *Phys. Lett.*, **133B**, 172 (1983).
4. E. Farhi and R.L. Jaffe, *Phys. Rev*, **D30**, 2379 (1984).
5. H.A. Bethe, G.E. Brown, and J. Cooperstein, *Nuc. Phys.*, **A462**, 791 (1987).
6. C. Alcock and E. Farhi, *Phys. Rev.*, **D32**, 1273 (1985).
7. H. Kurki-Suonio, *Phys. Rev.*, **D37**, 2104.
8. L.D. Landau and E.M. Lifshitz, *Fluid Mechanics* (Pergamon-Oxford, 1959).
9. T. DeGrand and K. Kajantie, *Phys. Lett.*, **147B**, 273 (1983).
10. K. Kajantie and H. Kurki-Sonio, *Phys. Rev.*, **D34**, 1719 (1986).
11. G.M. Fuller, G.J. Mathews, and C. Alcock, *Phys Rev.*, **D37**, 1380 (1988).
12. J.H. Applegate, C.J. Hogan, and R.J. Scherrer, *Phys. Rev.*, **D35**, 1151 (1987).
13. J.H. Applegate, C.J. Hogan, and R.J. Scherrer, *Ap.J.*, to appear June 15, 1988.
14. C. Alcock, G.M. Fuller, and G.J. Mathews, *Ap.J.*, **320**, 439 (1987).
15. R.A. Malaney and W.A. Fowler to appear in *Origin and Distribution of the Elements*, ed. G.S. Mathews (Word Scientific – Singapore 1988).
16. R.A. Malaney and W.A. Fowler, submitted to *Ap.J.* (1988).
17. A.M. Boesgaard and G. Steigman, *Ann. Revs. Astr. Ap.*, **23**, 319 (1985).
18. P.A. Seeger, W.A. Fowler, and D.D. Clayton, *Ap.J. Supp.*, **11**, 121.

PROBES OF COLLECTIVE BEHAVIOR OF DENSE HADRONIC MATTER

P.V. Ruuskanen
Department of Physics, University of Jyväskylä
Jyväskylä, Finland

ABSTRACT

Collective transverse flow in the central rapidity region of ultrarelativistic heavy ion collisions is described briefly. The effects of flow on transverse momentum spectra are discussed with special emphasis on the correlation between the average transverse momentum and the scaled multiplicity, dN/Ady. In the case of unequal nuclei or non-central collisions the mass number A is related to the transverse dimension of the overlapping collision region. Transverse momentum dependence of the ρ peak in the dilepton spectrum is directly related to the flow of pions whereas for the kaons and nucleons nonequilibrium phenomena may change the $\langle p_T \rangle$ correlations predicted from the flow.

INTRODUCTION

When hot, dense matter can freely expand, thermal energy is transformed into a collective flow motion through the work done by pressure. If the high energy collisions of heavy nuclei produce strongly interacting matter at so high energy (or number) densities that a QCD plasma of quarks and gluons (QGP) is formed, we should expect collective motion to build up. The reason for this is that the mean free paths of the quanta would be considerably smaller than the size of the system[1,2,3]. At T=200 MeV the mean free path, averaged over quarks and gluons, is estimated to be 0.2–0.5 fm and could become smaller than 0.1 fm as the temperature increases. If the latter possibility is true, even a system of hadronic dimensions might exhibit collective behavior.

There are two energy regimes in the heavy ion collisions where hydrodynamics may be a simple way to describe the evolution of the hadronic matter. A simple situation is believed to occur in the Central rapidity Region (CR) at ultrarelativistically high collision energies. The colliding nuclei appear as thin plates which will pass through one another in a time interval $2R/\gamma_{cm}$ (<.15 fm at 100+100 AGeV, the planned energy at RHIC). It is then reasonable to assume that the final–state quanta are formed and thermalized in a time interval <1 fm/c which is much smaller than the expected total expansion time when matter is dense enough for the particles to interact. Production and expansion then separate, the former providing the initial conditions for the latter.

In the low energy regime of E_{lab} < few GeV/nucleon, the nuclei exhibit strong stopping. A hydrodynamic description of the whole collision process seems then reasonable. The colliding nuclei are considered as two droplets of nuclear matter fusing into a single fluid, which is first compressed and heated up and which then expands, finally decoupling into nucleons, light nuclei and few pions.

The "intermediate" energy region of the present CERN experiments, E_{lab}= 200 AGeV, appears more involved for the application of hydrodynamics. One reason for this is that particle production overlaps with the expansion, especially in the fragmentation regions, for a considerable fraction of the total expansion time making the description of the production in terms of initial conditions useless. On the other hand the one–fluid picture, adequate at low energies, breaks down because of the increasing transparency of the colliding nuclei.

In this talk I will mainly deal with the ultrarelativistic case. Since the hydrodynamics of transverse expansion in the central rapidity region, in the case of head–on collisions of equal nuclei, has been discussed in detail elsewhere[4–6], I will not repeat it here. I will discuss in some detail the collision of unequal nuclei and the effects arising from finite collision energies. The outlook for experimentally observing the flow effects from the transverse momentum spectra of hadrons is considered and I will also comment on the recent results from the FNAL p̄p collider[7] where the average transverse momenta of charged particles are measured at different multiplicities. Finally the effects of flow on the dilepton spectra are discussed.

PRODUCTION, EXPANSION AND FREEZE–OUT

When a high energy collision is viewed in the center–of–mass frame, the Lorentz–contracted nuclei pass through one another in a short time interval of order $2R_A/\gamma_{cm}$ exchanging color charges. As the nuclei recede from each other, a strong color field is created. In the beginning the energy in the central region between the nuclei is in the color field but pair creation of color–charged quanta soon transfers the energy into particles which are likely to thermalize on a timescale similar to that of the production. One important prediction from the production models based on pair creation in color field[8] is that the timescale of production is <1 fm/c and gets shorter with increasing initial field strength.

The short production time in the central region gives us the possibility to treat the production and the expansion separately, the production providing the initial state for the expansion. In this situation we can hide our ignorance about the production dynamics into the parametrization of initial conditions. To start the hydrodynamic calculation one has to specify the initial time τ_i when the production is over and the matter has reached (approximate) thermal equilibrium. Lacking a reliable quantitative production model to specify τ_i and the amount of produced matter, one can take the approach of estimating what is a reasonable range of values for these initial conditions and then vary them over this range. Essentially this allows us to study how the collective flow, which builds up in the course of expansion, leads to correlations between observable quantities.

In order to be more quantitative let me first define the appropriate variables, the longitudinal proper time τ and the space-time rapidity η as

$$\tau = \sqrt{t^2 - z^2}, \qquad \eta = \frac{1}{2}\ln\frac{t+z}{t-z} = \frac{1}{2}\ln\frac{1+z/t}{1-t/z}. \tag{1}$$

At very high energies the invariance under longitudinal Lorentz transformations fixes the longitudinal flow velocity to $v_z = z/t$ and reduces the problem to finding the transverse flow at $z = 0$ (η=0) where $\tau = t$. We always assume that the initial transverse velocity of the matter vanishes, $v_r(t_i, r) = 0$.

Assuming that the expansion is approximately isentropic one can relate τ_i and T_i directly to the measured multiplicity by

$$\frac{dN}{dy} \simeq \frac{1}{3.7}(\frac{dS}{dy})_i = \frac{1}{3.7}(\tau_i \pi R_A^2) 4 a_q T_i^3 \tag{2}$$

Solving the equation for $\tau_i T_i^3$ and inserting the appropriate numbers, we get

$$\frac{\tau_i}{\mathrm{fm/c}}\left[\frac{T_i}{200\mathrm{MeV}}\right]^3 \simeq \frac{1}{17A^{2/3}}\frac{dN}{dy}. \tag{3}$$

The highest multiplicities observed at CERN for the oxygen beams are $dN/dy \simeq$ 150 which gives $\simeq$ 1.5 for the r.h.s., indicating that temperatures up to T=200 MeV are possible with the quark–gluon degrees of freedom. One should notice the limitations of this estimate since in deriving it a connection $dz = \cosh y dy$ between the rapidity and the length of the comoving fluid element is assumed. This is only true at times τ which are clearly larger than the passage time of the nuclei and the size of the source which in the CERN case are of order 1-2 fm. This is illustrated in fig. 1 which shows the $O - Pb$ collision in the frame where the matter at the maximum of the rapidity distribution is at rest ($y_{lab} \simeq 2.5$). It is seen that the quanta originating from the first and last primary interactions, the origin and the point P, and having $y \simeq 0$ in this frame, cannot be at the same space-time point before $\tau \simeq 2$ fm/c. Thus going below $\tau \simeq 2$ fm/c means going beyond the applicability of the above estimate as well as that of Bjorken[9]. From the space–time geometry of fig. 1 it seems reasonable that the qualitative time dependence of e.g. the entropy density, s, looks as indicated in fig. 2 with the late time τ^{-1} behavior starting only after $\simeq$ 2 fm/c.

On the other hand, if $a_q \simeq 4$ in eq. (2) is replaced with a_π=0.33, assuming pion gas throughout the collision, one would get temperatures and number densities which are not reasonable in terms of the hadron degrees of freedom. These estimates are almost certainly too high for the average values because the effective radius of the oxygen is, due to the 1 fm range of interactions, larger than the geometrical radius and also because secondary interactions with the spectator nucleons contribute to the observed multiplicity.

As another example let me mention the recent $\bar{\mathrm{p}}\mathrm{p}$ result from FNAL[7] where charged particle multiplicities up to $(dN/dy)_{ch} = 30$ are observed. At time $\tau_i = 1$ fm/c this indicates a temperature $T_i \simeq 300$ MeV for the quark-gluon degrees of freedom. For a transverse radius of 1 fm the data indicates up to 15 quanta/fm^3 which would be an extremely high density for ordinary hadrons.

From the above considerations one has a reasonably good notion on how to vary the quantity $\tau_i T_i^3$. For the initial conditions τ_i and T_i must be specified separately. This we have done by using a relation (see refs. 1 and 2) $\tau_i T_i = \kappa = 250$ MeVfm. With $v_r = 0$ initially, the initial state for a given nuclear size is fixed with one parameter, e.g. T_i, which is directly related to the multiplicity of the event.

The expansion of matter is governed by the hydrodynamic equations which for an azimuthally symmetric head-on collision can be written into a simple form[10]

$$\frac{\partial}{\partial t}(rts\cosh\eta) + \frac{\partial}{\partial r}(rts\sinh\eta) = 0,$$

$$\frac{\partial}{\partial t}(T\sinh\eta) + \frac{\partial}{\partial r}(T\cosh\eta) = 0, \tag{4}$$

which is useful in discussing the dependence of the solutions on initial conditions t_i, T_i and the nuclear radius R_A. Here s is the comoving entropy density.

The bag model equation of state is used as a parametrization for the first order phase transition, which seems to emerge from the lattice simulations[11]. A special feature of the solutions of the hydrodynamic equations which results from a sharp rise of entropy density in a very small temperature interval, is a slowly propagating rarefaction discontinuity, the deflagration wave, between the pure hadron gas and the mixed phase or the QGP close to critical temperature[4]. This is connected with the large ratio between the number of degrees of freedom in the QGP and the hadron gas, $r = g_{pl}/g_{\pi} \simeq 12 - 14$.

In order to check the reliability of the assumption concerning the existence of the homogeneous mixed phase during the expansion, we have modeled[12] the mixed phase in terms of plasma droplets embedded in pion gas. The droplets have a radius of 1 fm and an entropy content of 70-80 for n_F=2-2.5 at T_c= 200 MeV that we have used in the hydrodynamic calculations[1,4-6]. Each pion which hits a dropled is assumed to be absorbed. The emission rate of pions is then fixed from the requirement of thermal equilibrium between the pion gas and plasma droplets at the transition temperature to be the thermal emission rate at T_c. To account for the finite pion size and interaction range, a skin thickness of 1 fm is assumed for the droplets. The results of this cascade simulation support the hydrodynamic calculations and will not be discussed further in this connection[5].

Since the properties of the transverse flow as well as the folding of the thermal motion of hadrons and the thermal emission rates of dileptons are discussed in detail elsewhere[4-6], I will not repeat it but instead go directly to the results.

4. RESULTS ON TRANSVERSE MOMENTA

(i) p_T distributions of hadrons. The flow affects both the shapes of the p_T distributions and the relative size of the average values of the transverse momenta of particles with different mass. The difference between the thermal spectra and the spectra with the flow effects becomes apparent if they are plotted in terms of the transverse mass. This is done in figs. 3a and b for uranium-uranium collision with T_i = 350 MeV at τ_i = 0.7 fm/c corresponding to dN/Ady=13.5. For the thermal case the spectra are independent of mass (except for a small difference between fermions and bosons), but with the flow they clearly separate.

As the initial temperature of the collision increases, the transverse flow grows stronger and the flow effects become more visible. This is depicted in fig. 4 in terms of the average transverse momenta as a function of multiplicity. The dashed curves, marked 'thermal', show the $\langle p_T \rangle$ of nucleons and kaons at the temperature which would reproduce the $\langle p_T \rangle$ of pions. For large multiplicities the differences in $\langle p_T \rangle$ of different particles are much larger with flow than for thermal spectra or what is observed in hadron-hadron collisions. Similarly the difference in $\langle p_T \rangle$ with or without phase transition grows fast with increasing multiplicity. This is also shown in fig. 4, where the dashed-dotted line is the $\langle p_T \rangle$ of pions in the case of no phase transition. The possibility of observing these effects experimentally depends crucially on the range of multiplicities which will occur in the collisions. For this it is important to consider the case of unequal nuclei and noncentral collisions which I will do below.

*(ii) Scaling of $\langle p_T \rangle$ as a function of dN/Ady.** With $v_r(t_i, r) = 0$, the initial conditions are fixed by three parameters, t_i, T_i, and R_A. If the variables t and r are rescaled with t_i and R_A respectively, the only initial parameter will be T_i. It is easily seen from eqs. (4) that only the ratio t_i/R_A will appear in the scaled hydrodynamic

* This treatment is based on discussions with with T. Matsui and M. Kataja.

equations. Thus, independent of R_A, or equivalently A, the flow will be similar for a given T_i provited t_i/R_A is kept fixed. From eq. (2) it is seen that for a fixed t_i/R_A the value of dN/Ady is determined by T_i alone and is independent of A. This means that if e.g. $\langle p_T \rangle$ is plotted as a function of dN/Ady, a universal curve, independent of A, will result. Note that in the case of unequal nuclei the same argument holds for central collisions if the mass number A is that of the smaller nucleus. The reason for this is that in eq. (2) the factor A arises from $R_A^3 \sim A$ where R_A is the transverse initial radius of the expanding disk.

For the initial conditions which we have used in the calculations, t_i/R_A is not in general independent of A. Instead, as mentioned earlier, t_i is related to the initial temperature through $t_i T_i = \kappa$. However, t_i and R_A usually satisfy the relation $t_i/R_A << 1$ which is enough for approximate scaling of $\langle p_T \rangle$ versus dN/Ady as is shown in fig. 5 for mass numbers 16, 56 and 238. The scaling is especially good in the region $dN/Ady \simeq$ 1-10, which is the range of multiplicities most readily accessible for experiment. Unfortunately in this range the final flow is still rather weak and e.g. the mass effect of fig. 4 is not yet prominent. On the other hand the differences between the cases with or without phase transition are already quite clear.

(iii) The range of dN/Ady. The present experiments at CERN[13,14] show that for the heaviest targets and for the central events the maximum value of dN/Ady is of order 10 and that the fluctuations towards larger multiplicities are not big. Smaller multiplicities occur in several ways; away from the maximum of dN/Ady, for noncentral collisions and for lighter targets. Of these the last alternative is the best since in the other cases the geometry soon becomes complicated. E.g. with increasing impact parameter the azimuthal symmetry is lost and the ϕ dependence differs at different rapidities. For the light targets the value of dN/Ady might decrease by a factor of 3 to 4 as compared to the case of the heaviest targets. It is seen that the total range for the scaled multiplicity which is available in these experiments is quite small from the point of view of the situation in fig. 4.

Heavier beams and larger collision energies like those at RHIC will help to extend the range. The dependence of dN/Ady on A will be crucial. If it turns out that $\frac{dN}{dy}$ scales as A^α with $\alpha = 1$, increasing A will not help (exept that the system will become more macroscopic). On the other hand if $\alpha \sim 1\frac{1}{3}$, we could observe a variation of 5 or more in dN/Ady. The increase in energy when moving from CERN to RHIC energies will add at least a factor of two to this variation. In summary, the overall range of dN/Ady which will be available with nuclear beams and targets may not exceed 20, which could be too low for observing a strong increase in $\langle p_T \rangle$ due to the increase of initial temperature and the resulting flow.

(iv) $\bar{p}p$ collisions. In the above discussion I have implicitly excluded nucleon-(anti)nucleon collisions. In a recent FNAL $\bar{p}p$ experiment[4] a wide range of multiplicities, N_{ch}, was observed and the average transverse momentum of the secondaries was measured as a function of N_{ch}. The data for negative particles is shown in fig. 6 together with the hydrodynamic results with phase transition. It is seen that the hydrodynamic calculation with T_c of order 170 MeV reproduces the data reasonably well. At low multiplicities the thermalization can not be complete[15] and hydrodynamics probably leads to too high values of $\langle p_T \rangle$.

Even though encouraging, one must be cautious in interpreting these results. The basic reason to this is that the transverse size of the $\bar{p}p$ system is the same as the scale of granularity of the mixed phase which must be of the order 2 fm. This may prohibit the formation of mixed phase, confine the transverse expansion and lead to an evolution of the system which is different from the one presented here. One possibility is that the initial plasma expands transversely against the vacuum pressure[16] until the longitudinal expansion breaks it into separate droplets which evaporate through thermal emission as assumed in the cascade calculation[12]. In this case one would not expect flow effects. At very high initial temperatures the system could grow large enough during the plasma phase for the breaking in the transverse direction to become possible. It will be extremely interesting to see whether the increase of p_T, which is suggested by the last two data points, will be confirmed by data with better statistics and whether the transverse momenta of kaons and protons show the mass effect of the flow.

(v) Dilepton spectra. Thermal dileptons will be emitted, first from the plasma mainly through quark-antiquark annihilations and then from hadron gas through $\pi^+\pi^-$ annihilations. To calculate the dilepton spectra, the thermal emission rates must be folded over the whole expansion time. If the initial temperature is clearly above the transition temperature the high mass dileptons will be emitted mainly at the earliest times from the plasma phase. Since the transverse flow has not yet had time to evolve, the high mass region of the dilepton spectrum is independent of the hydrodynamic flow. Another simple feature of the contribution from the plasma phase is that both the mass and the p_T dependence enter always through the transverse mass $M_T^2 = M^2 + p_T^2$. If the mass spectrum is plotted at a fixed transverse mass, the contribution from plasma is flat.

In the mass region below 1 GeV, the dominant source of dileptons is the $\pi^+\pi^-$ annihilation through the ρ pole. Now the mass dependence enters both through the form factor of the pion and the transverse mass M_T from the thermal distributions and the flow. At a fixed M_T the shape of the mass spectrum is given by the form factor and the magnitude is fixed by the thermal distributions and the flow. Since the temperature in the hadron gas is lower than in the plasma, the contribution from the hadron gas would have without flow a steeper dependence on M_T than that from the plasma. This case is shown in fig. 7 as dashed lines and is seen as a diminishing ρ peak in the mass spectrum of dileptons as M_T is increased[17]. Like in the case of kaons and nucleons, the flow can contribute also to the transverse mass of the dileptons and since the flow is stronger at later times during the hadron gas phase, it has a much larger effect on the ρ peak than on the contribution from the plasma. Depending on the initial state, the ρ contribution relative to the flat plasma contribution may even grow with M_T as is shown in fig. 7 by the solid lines[18].

In a way the behavior of the ρ peak might be a better signal of the collective flow than the relative behavior of $\langle p_T \rangle$ of pions, kaons and nucleons. The reason for this is that at least in the central region the pions dominate the flow and since the ρ's arise from the interactions among the pions, they directly measure the flow of the pions. E.g. in the case of kaons the flavour kinetics is coupled with the space-time behavior of the event and kaons may be out of chemical equilibrium by different amounts in regions of different final flow velocity[19].

5. SUMMARY AND OUTLOOK

Transverse momentum spectra of hadrons and dileptons can provide a means of deducing whether collective flow phenomena occur or not in the central rapidity region of ultrarelativistic collisions. At the moment the multiplicities available at the CERN experiments are too low to show the predicted upward bend in the average transverse momentum with increasing multiplicity or to show the mass effect on $\langle p_T \rangle$. In analyzing the data it is not enough to measure the average transverse momenta in high multiplicity events; they must be measured in the region of phase space where dN/Ady is large. At the CERN experiments the largest values of this scaled multiplicity are $\simeq 10$ which is well below the values where p_T is predicted to grow. On the other hand the hadronic equation of state without phase transition leads to a clear increase of $\langle p_T \rangle$ already in this multiplicity range.

The recent results from FNAL on $\langle p_T \rangle$ of negative particles in $\bar{p}p$ collisions show a mild increase of $\langle p_T \rangle$ consistent with the hydrodynamic picture. The small transverse dimension of the system could constrain the system in such a way that the mixed phase scenario which may work for the transverse expansion of nuclei is not valid in this case.

From the point of view of collective flow the behavior of the ρ peak is of special interest due to its direct connection to pions which dominate the flow.

REFERENCES

1. H. von Gersdorff, L. McLerran, M. Kataja, and P.V. Ruuskanen, Phys. Rev. D34, 794 (1986).
2. J.P. Blaizot, Acta Phys. Pol. B18, no 7 (1987).
3. A. Hosoya and K. Kajantie, Nucl. Phys. B250, 666 (1985).
4. M. Kataja, P.V. Ruuskanen, L. McLerran, and H. von Gersdorff, Phys. Rev. D34, 2755 (1986).
5. P.V. Ruuskanen, Z. Phys. C - Particles and Fields 38, 219 (1988).
6. P. V. Ruuskanen, Acta Phys. Pol. B18, no 7 (1987).
7. T. Alexopoulos et al., Phys. Rev. Lett. 60, 1622 (1988).
8. T. Matsui, Nucl. Phys. A461, 27c (1987).
9. J.D.Bjorken, Phys.Rev.D27,140(1983).
10. Gordon Baym, B.Friman, J.-P.Blaizot, M.Soyeur and W.Czyz, Nucl. Phys. A407, 541 (1983).
11. F. Karsch, Z. Phys. C - Particles and Fields 38, 147 (1988).
12. G. Bertsch, M. Gong, L. McLerran, P.V. Ruuskanen, and E. Sarkkinen, Phys. Rev. D37, 1202 (1988).
13. R. Albrecht et al., Phys. Lett. 202B, 596 (1988).
14. A. Bamberger et al., Phys. Lett. 203B, 320 (1988).
15. S. Gavin, 'Dynamics of Pions in Ultrarelativistic Heavy Ion Collisions', Thesis, University of Illinois at Urbana–Champagne, 1987, unpublished.
16. Bengt L. Friman, Phys. Lett. 159B, 369 (1985).
17. P. Siemens and S.A.Chin, Phys.Rev.Lett. 55, 1266 (1985).
18. K. Kajantie, M. Kataja, L. McLerran and P.V. Ruuskanen, Phys. Rev. D34, 811 (1986).
19. K. Kajantie, M. Kataja and P.V. Ruuskanen, Phys. Lett. 179B, 153 (1986).

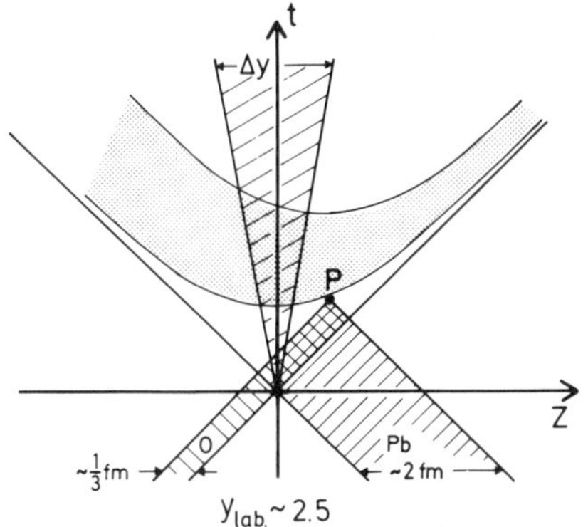

Fig. 1. At finite collision energies the primary interactions take place over a finite time interval and in a finite volume. The figure is drawn to scale for $O - Pb$ collision at $E_{lab} = 200$ GeV/nucleon.

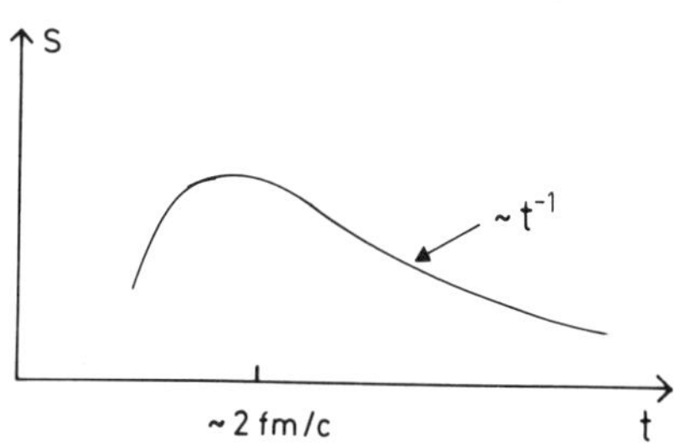

Fig. 2. Qualitative time dependence of the entropy density s in a collision like the one in fig. 1.

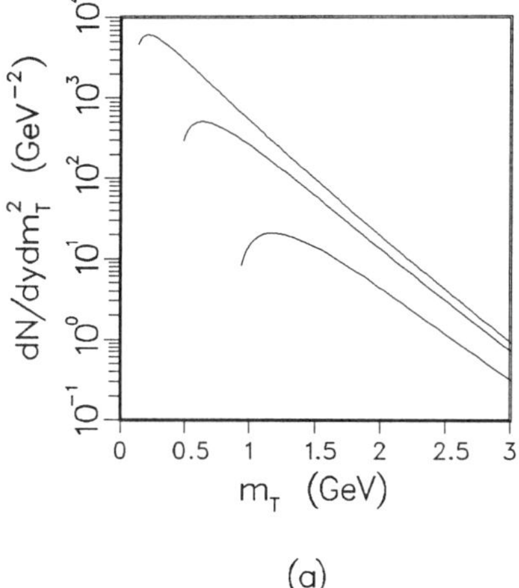

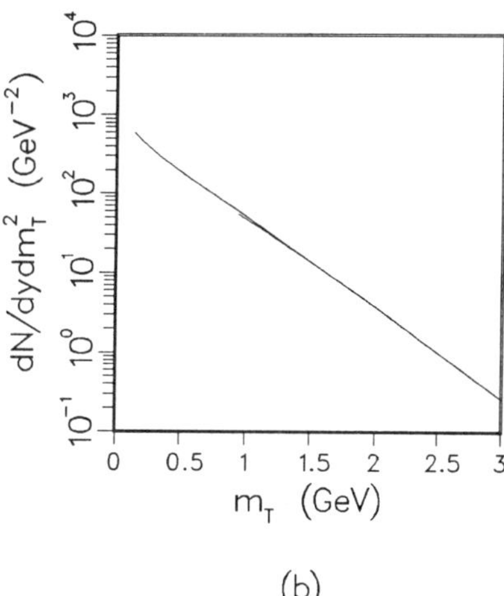

Fig. 3. Transverse mass spectra of pions, kaons and nucleons: (a) The hydrodynamic calculation with $T_i = 350$ MeV at $\tau_i = 0.7$ fm, $R_A = 6.8$ fm, $T_c = 200$ MeV and freeze-out temperature $T_{dec} = 140$ MeV. (b) Thermal distributions for $T = 350$ MeV.

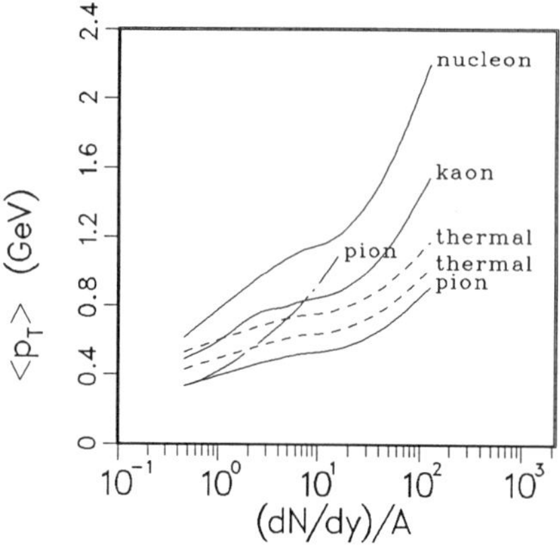

Fig. 4. Average transverse momentum of pions, kaons and nucleons as a function of scaled multiplicity. See text.

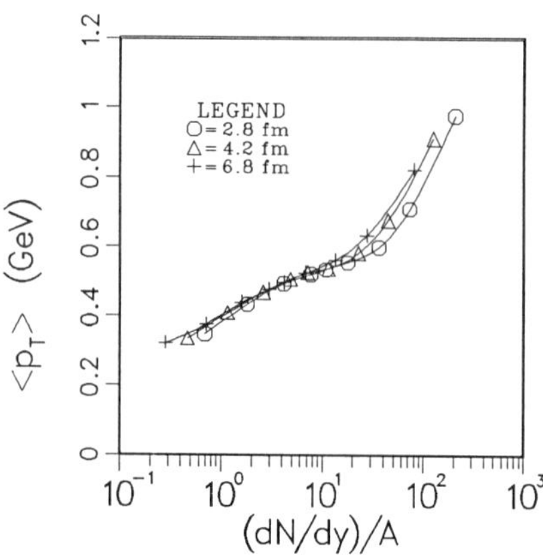

Fig. 5. Average transverse momentum of pions as a function of scaled multiplicity for different nuclear radii R_A.

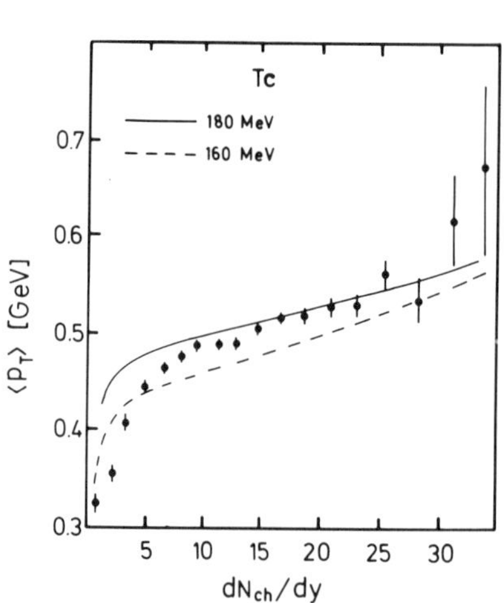

Fig. 6. Transverse momenta of negative particles (*ref.* 4) in $\bar{p}p$ collisions as a function of multiplicity. The curves are the results of hydrodynamic calculations with T_c = 160 and 180 MeV.

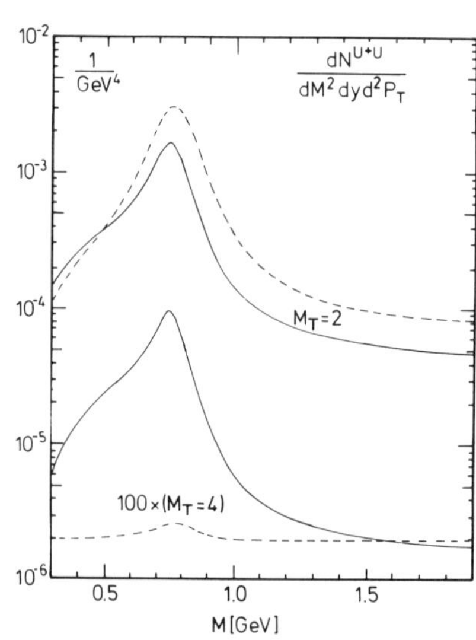

Fig. 7. Mass distributions of dileptons at two different fixed values of transverse mass. The dashed and solid curves, respectively, give the results when the transverse flow is ignored or included.

TWO-PION INTERFEROMETRY AS A SIGNAL FOR THE QUARK-GLUON PLASMA

Scott Pratt
Department of Physics, Oregon State University, Corvallis, OR 97331

ABSTRACT

Long lived drops of plasma are discussed as a signal of the existence for the quark-gluon plasma. The correlation measurements required to infer such a state are discussed.

INTRODUCTION

If one were able to build a suitable container, thermometer and barometer which could withstand temperatures of several hundred MeV, the detection of the deconfinement phase transition would be quite straightforward. During the mixed phase the pressure P and temperature T would remain constant while the energy density ρ changed by an order of magnitude. This would dramatically differ from the hadronic equation of state, $P = \rho/3$.

This lower pressure does not manifest itself quite so clearly in the context of ultrarelativistic collisions. In fact the behavior of the mixed phase is very poorly understood. One scenario is that the region of mixed phase hadronizes only through a shock wave originating from the outer boundary of the collision[1]. The speed of this shock is quite slow compared to the rate at which a hadronic gas might expand. The lengthening of the reaction's duration could be erased however by the initial collective expansion due to incomplete stopping. A second scenario is that globs of higher energy density plasma would separate from the less dense hadrons[2]. Each individual glob would then have no internal expansion and hadronize quite slowly.

The principles of interferometry are reviewed in the first section with attention focussed on the ability to extract lifetimes from the correlation functions. Three theoretical scenarios of the collision and their implications are explained. A model of a spherically symmetric exploding ball[3] is reviewed to demonstrate the clarity of the signal of a low pressure phase if the matter completely stops to form a single drop. The most simplistic model of a Bjorken expansion where the pions are all emitted at the same local proper time[3,4] is discussed along with its predictions. Finally, the recent interferometric predictions[5] of the cascade model[2] with plasma globs are discussed. Rather tantalizing data[6] from the NA35 collaboration at CERN are also reviewed where there are hints of a signal for a long lived state.

PRINCIPLES OF TWO-PARTICLE INTERFEROMETRY

Correlation measurements measure the relative probability of seeing two particles (usually π^-) in the same event to the probability of detecting them in separate events. There is an enhancement due largely to the identical particle interference of the two pions. This enhancement is felt only over small relative momenta with a width of the inverse source size. The correlation function $C(\mathbf{K},\mathbf{k})$ for noninteracting identical bosons can be extracted from the final single-body phase-space distribution $f(\mathbf{p},\mathbf{r})$ in the following way:

$$C(\mathbf{K},\mathbf{k}) = \frac{P(\mathbf{p},\mathbf{q})}{P(\mathbf{p})\ P(\mathbf{q})} \tag{1}$$

$$\mathbf{K} \equiv (\mathbf{p}+\mathbf{q})/2 \qquad \mathbf{k} \equiv (\mathbf{p} - \mathbf{q})/2$$

$$C(\mathbf{K},\mathbf{k}) = 1 + \int d^3r\, F(\mathbf{K},\mathbf{r}) \exp(2i\mathbf{k}\cdot\mathbf{r})$$

$$F(\mathbf{K},\mathbf{r}) \equiv \frac{\int d^3(\mathbf{r}_1+\mathbf{r}_2)\, f(\mathbf{K},\mathbf{r}_1)\, f(\mathbf{K},\mathbf{r}_2)}{\int d^3\mathbf{x}\, f(\mathbf{K},\mathbf{x}) \int d^3\mathbf{x}\, f(\mathbf{K},\mathbf{x})}$$

The function $F(\mathbf{K},\mathbf{r})$ can be thought of as the probability that two uncorrelated particles with the same momentum, $\mathbf{K}$ are separated by a distance, $\mathbf{r} = \mathbf{r}_1 - \mathbf{r}_2$, where the probability is normalized such that $\int d^3r\, F(\mathbf{K},\mathbf{r}) = 1$. There are several derivations of the same formula with somewhat different looking results, however all become identical when the source size is considered to be much larger than the thermal wavelength. This semiclassical approximation is inherent in all the formalisms.

Extracting the lifetime from the experimentally determined correlation function can only be accomplished by looking at the spatial shape of $F(\mathbf{K},\mathbf{r})$. A long lived source would emit particles whose final phase-space distribution would be elongated along the direction of $\mathbf{K}$. For a static source with an emission function $g(\mathbf{K},\mathbf{x},t)$ characterized by $\mathbf{R}_x,\mathbf{R}_y,\mathbf{R}_z$ and a lifetime τ:

$$g(\mathbf{K}_x,\mathbf{r},t) = \exp - \left(\frac{x^2}{R_x^2} + \frac{y^2}{R_y^2} + \frac{z^2}{R_z^2} + \frac{t^2}{\tau_0^2} \right), \qquad (2)$$

the phase-space distribution becomes:

$$f(\mathbf{K}_x,\mathbf{r}) = \exp - \left(\frac{x^2}{R_x^2 + v_K^2\tau^2} + \frac{y^2}{R_y^2} + \frac{z^2}{R_z^2} \right)$$

and the resulting correlation function is:

$$C(\mathbf{K},\mathbf{k}) = 1 + \exp - \frac{1}{2}\left\{ k_x^2\,(R_x^2 + v_K^2\tau^2) + k_y^2 R_y^2 + k_z^2 R_z^2 \right\}$$

The lifetime is not distinguishable from R_x unless some assumption is made about the spatial dimensions being equivalent. This is due largely to the fact that since the correlation function is plotted for only three dimensions of the relative momenta, only three spatial dimensions can be extracted. One would be tempted to assume that by measuring the increase of the elongation in $F(\mathbf{K},\mathbf{R})$ for growing values of the total momentum $\mathbf{K}$ and therefore v_K, the lifetime could be inferred. But the lifetime is a function of the total momentum, particularly in the case of a cooling system as argued in the case of proton interferometry in lower energy collisions[7,8].

Other correlations besides those due to Bose-Einstein statistics influence the correlation function. There is a clear discussion of the effects of Coulomb and strong interaction effects in reference 5. A treatment of the joint symmetrization of the many-particle system and its influence on two-body correlations has been made by Zajc[9].

SINGLE-DROP MODEL

As mentioned in the introduction the bag model equation of state yields a large range of energy densities wherein matter would be in the mixed phase and therefore have little pressure and no pressure gradients. Monte Carlo analyses reaffirm these hypotheses. This latent heat sucks up energy in such a way that a drop of plasma in a mixed phase has little impetus to expand, and with no pressure gradients can only hadronize through a shock wave approaching from the outside. The implications of this are demonstrated in reference 3. The lifetime can be extended from under 10 fm/c to greater than 50 fm/c when the bag model equation of state is introduced. Also shown in reference 3, are the resulting correlation functions. Not only does the phase transition extend the lifetime and narrow the width of the correlation function when the relative momentum is along the direction of the total momenta, but the behavior of the width as a function of the magnitude of the relative momentum is reversed. It has been shown that a rapidly expanding system tends to lead to a source size that falls with increasing momentum[10]. However, a long lived system where the temperature of the emitting surface remains fairly constant has the opposite behavior. This can be seen by inspecting equation 2 when τ does not depend strongly on the momentum of the pions.

Reference 3 includes an example where the spherical drop is given initial collective expansion. This is shown to erase the signal for sufficiently large initial velocity gradients. Physically, this is due to the fact that a short lifetime and rapid dissolution of the matter is no longer dependent on internal pressure. Thus a concrete signal will only occur if the drops of plasma are in some sense internally static. This is discussed in the context of the droplet cascade model later in this paper. At lower energies, it is perhaps even reasonable to talk about stopped matter, although it is not clear that the mixed phase will then be reached and the equation of state in reference 3 does not include baryonic charge, which becomes very important in this regime.

PURE BJORKEN SCALING

Bjorken[11] proposed a simple analytic one-dimensional solution to hydrodynamic equations, which remains invariant to boosts along the beam axis. All thermodynamic variables such as pressure P, energy density ρ, temperature T and entropy density s are functions of a proper time τ, while the local collective velocity and τ depend on the space-time coordinates in the following fashion:

$$\gamma v = \frac{z}{\tau} \qquad \tau = \sqrt{t^2 - z^2} \tag{3}$$

The solutions are for isentropic or isoergic solutions respectively:

$$s(\tau) = s(\tau_0)\frac{\tau_0}{\tau} \quad , \quad \rho(\tau) = \rho(\tau_0)\frac{\tau_0}{\tau} \tag{4}$$

Calculations[3,4] have been performed to predict the correlation functions assuming the matter breaks up at a given proper time. The correlation functions then only depend on the final proper time τ_f, the temperature T and the transverse radius R. It has been shown[3] that the apparent longitudinal size from the correlation functions should decrease when looking at pions of higher transverse momentum. Improving on the calculation of reference 3, reference 4 includes an analytic solution for the correlation function for arbitrary directions of the relative momentum. This can then be compared to data[4].

DROPLET CASCADE MODEL

Recent work[2,5] has demonstrated that even if a single drop does not exist as in a scenario with complete stopping, long lived globs of plasma distributed in a Bjorken-like fashion still lead to a similar signature. In reference 2 a cascade model was worked out where after using Bjorken's initial conditions, the matter phase separates into globs of plasma surrounded by pions in the hadronic phase. The motion of the pions and globs are then simulated with the pions being absorbed any time they come within one fermi of a blob. The authors arbitrarily choose the blobs to be of one fermi radius. Pions are emitted from the blob with a probability derived from standard time reversal arguments such as those used in models of the compound nucleus.

The authors state that the lifetime of the drops is about 9 fm/c, which is enough to significantly affect the correlation function, reducing the width when the relative momentum is parallel to the direction of the total momentum of the two-pion pair. To be more exact, this direction is chosen when in the rest frame of the blob, so in the laboratory frame both the longitudinal and outward (in-plane) correlation functions are narrowed. The out-of-plane or sideward dimension to the final distribution is however still of the order of 5 fm. Thus a lifetime of 9 fm/c is on the border of being detectable. One should note that the author's estimate of the lifetime is somewhat conservative, since the drops were chosen to be quite small which allows them to decay quickly, and the extra capture radius of one fermi shortens the lifetime by a factor of four due to the time-reversal arguments.

Long lifetimes (>20 fm/c) would lead to an unambiguous signature in the correlation function of a long lived state. Long lifetimes seem to occur regardless of the model for hadronization assuming the drops or globs are sufficiently large. This was demonstrated with a hydrodynamic picture from the previous section. In a microscopic model Matsui and Glendenning[12] concluded that the hadronization at a surface boundary between plasma and a hadronic vacuum is quite difficult. It is this author's opinion that the time-reversal arguments give an upper bound to the lifetime, since it seems reasonable that all pions entering such a drop would disassociate into partons. The only other assumption involved in a time-reversal argument is that the lifetime of the blob is much greater than the equilibration time for a glob to sample all available states after accepting a pion. If hadronization occurs easily, and mean free paths are short compared to the glob size, the hydrodynamic and shock wave solutions become valid, and even in this picture lifetimes become quite long. Basically, the phase transition lowers the pressure and some extension of the reaction's duration is inevitable.

NA35 DATA

Analysis[6] of 105 "central impact" streamer chamber events from a 200 GeV/nucleon Oxygen beam incident on a gold target show a large transverse size for midrapidity particles. Unfortunately, no cuts were made with regard to the two transverse directions of the relative momenta when constructing the correlation functions. Thus the large transverse size of of 8.1 fm could be due to either either transverse expansion of the original 4 fm oxygen radius, or a lifetime of the order of 20 fm/c. The data are reviewed below with midrapidity corresponding to $2 < y < 3$.

Transverse and Longitudinal Source Sizes

Rapidity	R_T(fm)	R_L(fm)
$1 < y < 4$	4.1 ± .4	3.1 + .7 / -.4
$1 < y < 2$	4.3 ± .6	2.6 ± .6
$2 < y < 3$	8.1 ± 1.6	5.6 + 1.2 / -.8
$3 < y < 4.5$	4.3 + 1.2 / -.8	5.8 ± 2.2

Another sign of a long lived state emitting particles at a steady temperature would be a transverse in-plane ("outward in the notation of reference 5) source size that increased with the transverse momenta of the two particle pair. This is due to the fact that faster particles would separate further from each other than would the slow ones if their emission was characterized by a common lifetime.

It is possible that the two transverse dimensions will turn out to be identical and large as the data would suggest. This would necessitate a transverse expansion that was rapid enough to overcome any escaping particles. A consequence of this would be transverse dimensions falling with increasing transverse momentum as discussed previously. A hadron-like equation of state with no large latent heat would be signaled.

A most peculiar possibility would be the transverse sizes remaining equal and large while the size increases with the transverse momentum of the two-pion pair. This does not seem physically reasonable under any scenario currently under consideration. The last few paragraphs are hoped to motivate the experimentalists towards deriving both a directional and transverse momentum dependence to their inferred source sizes.

The most sobering aspect of this technique is the "coherence parameter" which fits the height of the experimental correlation function's bump as the relative momentum vanishes. This should go to unity but it is usually measured to be somewhat less. In the CERN data it is found to be .3 for the low rapidity particles and .8 for midrapidity. This could be due to a number of factors such as particle misidentification or pions emitted far outside the reaction zone from long lived resonances. Past experiments in both pp collisions and heavy-ion collisions from the Bevalac have derived coherence parameters ranging from .3 to 1.0. In ultrarelativistic heavy-ion collisions the reaction zone is large enough and the number of uncorrelated sources is large enough that one really expects the coherence parameter to approach unity. No quantitative analysis of the streamer chamber data has been performed which could rule out the possibility of it being an experimental artifact due to particle misidentification for example. Bertsch[13] has argued that the traditional explanation of a "coherent phenomenon" is unlikely. It is of major importance that this facet of the analysis is understood before great credence is given to the interpretation of interferometric results.

CONCLUSIONS

In principle one can obtain a wealth of information from two-particle correlation measurements. This can answer the most important questions about the space-time behavior of the emission of pions and in particular whether there exist long lived sources. This gets to the heart of the question of whether or not there is a tremendous latent heat as in the predictions of the bag model and lattice calculations. Aside from the specter of the coherence phenomenon not being well understood, the statistics will only improve as the multiplicity is increased with more energetic and heavier projectiles. All the cuts proposed here then become well within the bounds of possibility.

---

[1]G. Baym et al., Nucl. Phys. A407, 541 (1983)
[2]G. Bertsch, M. Gong, L. McLerran, V. Ruuskanen, and E. Sarkkinen, Phys Rev. D37, 1202 (1988)
[3]S. Pratt, Phys. Rev. D33, 1314 (1986)
[4]K. Kolehmainen and M. Gyulassy, Phys. Lett. B180, 203 (1986)
[5]G. Bertsch, M. Gong, and M. Tohyama, Phys. Rev. C37, 1896 (1988)
[6]T.J. Humanic, Proc. of the Quark Matter 1987 Conf. at Nordkirchen, W. Germany
[7]D.H. Boal and H. DeGuise, Phys. Rev. Lett. 57, 2901 (1986)
[8]S. Pratt and M.B. Tsang, Physical Review C36, 2390 (1987)
[9]W. Zajc, Phys. Rev. D35, 1798 (1987)
[10]S. Pratt, Phys. Rev. Lett. 53, 1219 (1984)
[11]J.D. Bjorken, Phys. Rev. D27, 140 (1983)
[12]N.K. Glendenning and T. Matsui Phys. Lett. B141, 419 (1984)
B. Banerjee, N.K. Glendenning and T. Matsui, Phys. Lett. B127, 453 (1983)
[13]G.F. Bertsch, "Coherent Source Interferometry", Mich. State Univ. preprint, May 1988

PARTICLE PRODUCTION RESULTS FROM HELIOS

Barbara Jacak
Los Alamos National Laboratory, Los Alamos NM, 87545

THE HELIOS COLLABORATION

The availability of ultrarelativistic heavy ion beams at CERN and Brookhaven allows investigation of the physical properties of a state of high energy density in nuclear matter. The interest is driven by the expectation that confinement of the partons in a hadron cannot survive when the density of hadrons is large compared to the density inside ordinary hadrons. This prospect is supported by QCD calculations[1], and collisions between heavy nuclei at ultrarelativistic energies are predicted to provide the energy densities required for deconfinement (greater than 2-3 GeV/fm^3). Among the goals of the first experiments are investigation of the energy density and spatial extent of the highly excited central rapidity region to address whether the necessary conditions have been satisfied, and a closer examination of events with large amounts of energy deposited in the target. The HELIOS (High Energy Lepton and Ion Spectrometer) experiment therefore combines 4 π calorimetric coverage with measurement of charged particle multiplicity, inclusive particle spectra, two particle correlations, lepton pairs and photons.

The target region is surrounded by a box of calorimeter modules that cover the pseudorapidity interval $-0.1 < \eta < 2.9$. The forward region is covered by a beam calorimeter placed further downstream. The granularity of this calorimeter allows extension of the E_t measurement to $\eta = 4.9$. The transverse energy cross-sections $d\sigma/dE_t$ are presented in Figure 1 for ^{16}O beams at 60 and 200 GeV per nucleon incident on W, Ag and Al targets[2]. The main feature of the distributions is a long plateau followed by a steeply falling tail, reflecting the shape of the overlap integral of nuclear densities in collisions with random impact parameters[3]. The large value of E_t attained is striking. At 200 GeV/A, values of 200 GeV are reached; adding the E_t measured for $\eta > 2.9$ increases this to 280 GeV. This is ~0.73 of the value of the kinematic limit, and shows that nuclei are not transparent to one another at these beam energies. An important result of the first experiments is that the calculated energy density is still increasing at 200 GeV/A.

The curves in fig. 1 are fits to the geometrical overlap integral, folded with a Gaussian distribution in E_t per N-N collision[3]. This procedure yields an excellent description of the data, with the fitted mean of approximately 1 GeV, which may be compared with the mean E_t of ~ 1.4 GeV in p-p collisions at $\sqrt{s} = 20$ GeV.

Figure 2 shows the transverse energy distributions measured for ^{16}O and ^{32}S beams at 200 GeV/A, incident on a W target. The maximum transverse energy achieved with the S beam is approximately twice that of the O beam, suggesting that the energy density stays approximately constant while the "thermalized" volume increases. The bands depict predictions made with IRIS, an event generator incorporating the collision geometry with the dual parton model[4]. The width of the

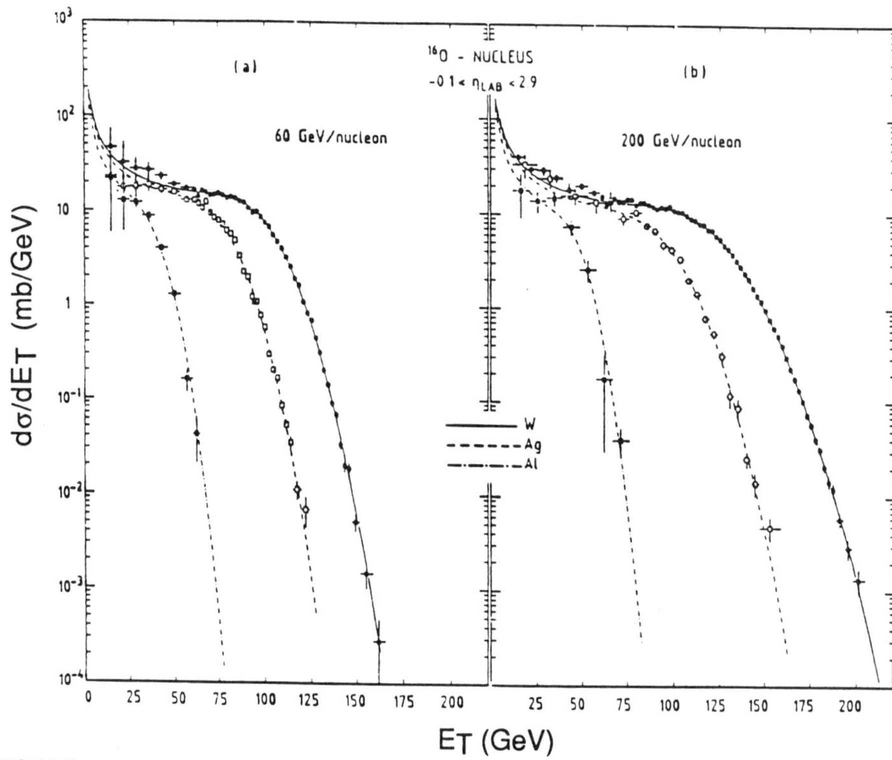

FIGURE 1
The transverse energy distributions measured at a) 60 GeV per nucleon, b) 200 GeV per nucleon. The curves are fits to the geometrical overlap integral, folded with a Gaussian distribution in E_t per N-N collision [2].

bands represents the effect of the absolute E_t scale uncertainty. IRIS is seen to underpredict the transverse energy for both beams. Such a result is not surprising as the calculation does not account for cascading of secondary particles in the target.

In parallel with the E_t measurements, the charged particle multi- plicities have been measured with silicon detectors. These detectors have approximately 400 segments each, and are positioned 3 and 9 cm downstream from the target. The multiplicity distribution exhibits a similar plateau and tail to the E_t distributions, and the multiplicity is seen to increase linearly with Et. One can infer from p-p collisions and HIJET, together with the known calorimeter response, that 55% of the E_t is produced by charged particles. This allows one to estimate an average pt per particle of ~ 340 MeV/c. Though this may increase very slightly with E_t, we conclude that high transverse energies arise via formation of many charged particles, rather than by a large change in the momentum distributions of the particles.

Figure 3 shows the pseudorapidity distributions of both E_t and charged particle multiplicity for central collisions of 200 GeV/A O + W. Though the comparison is limited by the calorimeter granularity, the two distributions agree rather well. They show a broad peak with maximum density between $\eta = 2$ and 2.5, clearly shifted backward from the p-p center of mass, $\eta \sim 3$. The values are close to $\eta = 2.4$, the center of mass rapidity characteristic of a system of 16 projectile nucleons stopped by 50 nucleons from the target. Also in figure 3 are the $dE_t/d\eta$ predictions from IRIS. The calculation is clearly more forward peaked than the data, and the rapidity densities in the backward region are underpredicted. These effects are both consistent with expectations from target cascading.

From the E_t and multiplicity measurements we have learned that there are no drastic changes in the nature of the charged particles emitted; larger E_t is the result of more charged particles. Nevertheless, a direct measurement is necessary to study the details of the charged particle momentum spectra. These may indicate the degree of thermalization reached in the collision, and allow a search for evidence of radial flow in the boost of heavy particle spectra. To aid in understanding the charged particle spectra, the transverse energy flow can be used to sort events according to impact parameter and energy deposition.

HELIOS measures charged particles with a magnetic spectrometer which views the target through a narrow slit in the calorimeter wall, covering the pseudorapidity interval $0.9 < \eta < 2.0$. A magnet with a p_t kick of ~ 80 MeV/c and two high resolution (175 μm) drift chambers provide the momentum measurement of charged particles. Particles are tracked back through the slit to the target to eliminate bacground from the calorimeters around the slit. Particle identification is achieved via time-of-flight and Aerogel Cerenkov counters.

Three data samples are presented: p + W, ^{16}O + W and ^{32}S + W, all at 200 GeV per nucleon. The data are triggered on the presence of a valid interaction and transverse energy observed in the calorimeters. Varying scale-down factors were used at the trigger level to ensure uniform statistical coverage over the entire E_t range. The measured pt spectra were corrected for various detector effects, namely geometrical acceptance, decays in flight, photon conversion in upstream material, reconstruction efficiency, and finite momentum resolution.

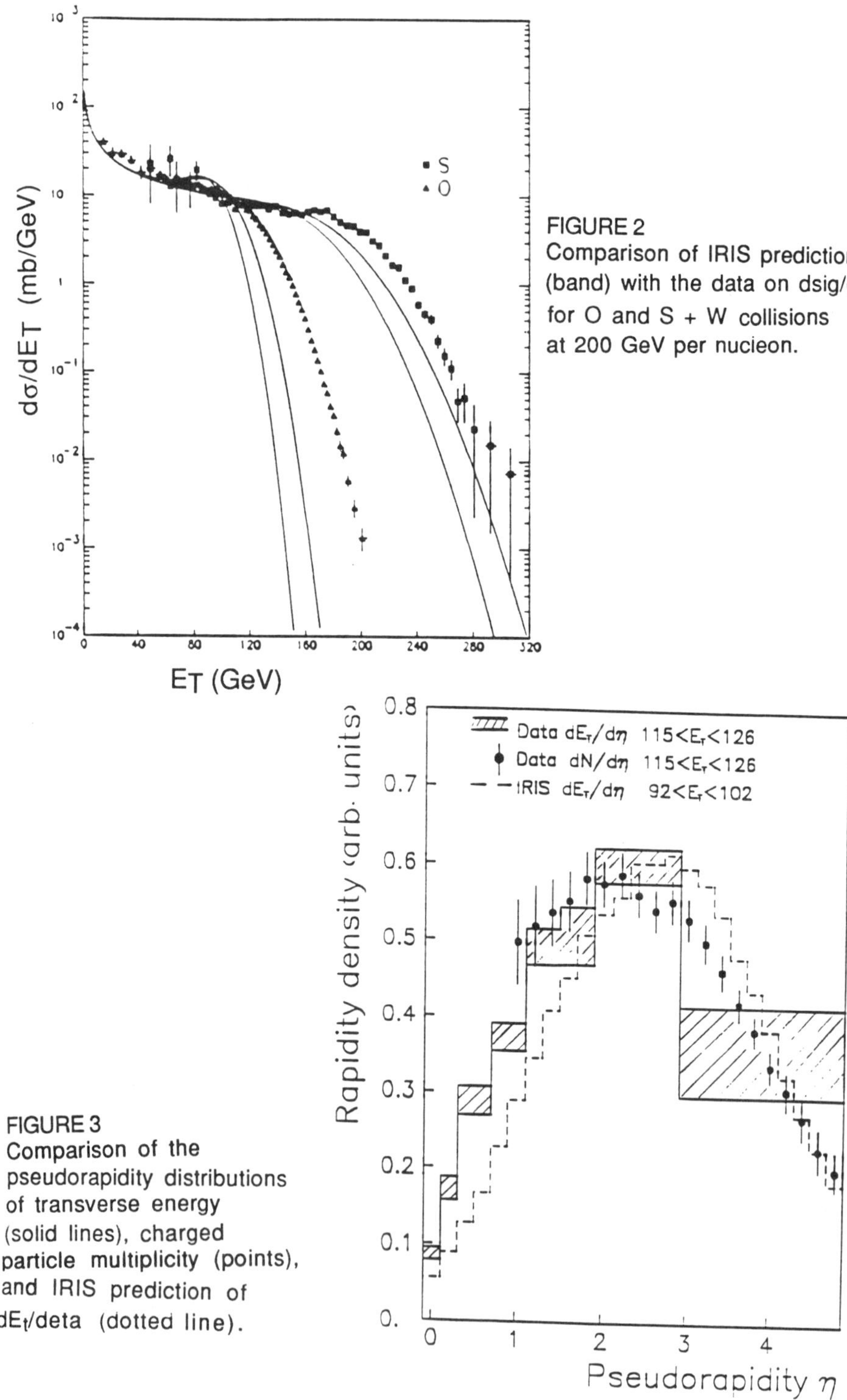

FIGURE 2
Comparison of IRIS predictions (band) with the data on dsig/dEt for O and S + W collisions at 200 GeV per nucleon.

FIGURE 3
Comparison of the pseudorapidity distributions of transverse energy (solid lines), charged particle multiplicity (points), and IRIS prediction of dE$_t$/deta (dotted line).

p_t spectra were simulated including all detector properties, and reconstructed using the full analysis chain; the ratio of input and reconstructed spectra is used to correct the data presented below. To generate the spectra we used a parameterization combining FNAL and CERN ISR measurements of charged particles, taking into account the c.m. energy dependence, the rapidity, and the well-known hardening of the p_t spectra in p+A collisions[5,6]. Systematic errors were estimated by comparing the corrected p_t spectra for nearby regions in pseudorapidity; in addition small variations of the detector response between the data sets were investigated. The systematic error is approximately 20%, but it should be noted that this yields an error in the slope of less than 3%.

Figure 4 shows the transverse momentum spectra for negative particles entering the spectrometer in p + W, O + W and S + W collisions at 200 GeV/A. The data are in the rapidity range $0.9 < y < 1.9$ (the transformation from pseudorapidity to rapidity neglects the small admixture of K^- in the negative particle spectrum) and the pt range $0.075 < p_t < 2.0$ GeV/c. The dashed line is the result of the parameterization described above. It is important to note that these data are taken from E_t triggered events, and are therefore biased toward central collisions.

All three data sets are well described by the parameterization, indicating that charged particles emitted in central nucleus-nucleus collisions show the same behavior as those from proton-nucleus collisions. S + W collisions yield a charged particle distribution identical to that from O + W collisions. The p_t spectra clearly cannot be described by a single exponential, and we have derived the inverse slopes in the regions $0.5 < p_t < 2.0$ GeV/c and $0.1 < p_t < 0.2$ GeV/c. The results are ~ 190 MeV/c and 85 MeV/c, respectively. The rise in cross section at low p_t has also been reported in heavy ion reactions in the streamer chamber at CERN[7] and in p - nucleus collisions at Fermilab 8. In all three experiments this could be a target fragmentation effect.

Figure 5 presents ratios of negative particle spectra for events in different regions of E_t, as a function of p_t. This plot allows careful comparison of the spectral shape over the entire p_t range. It is important to note that the ratios shown are normalized by the number of events in each spectrum; absolute normalizations may change these ratios from unity. All four curves are flat within statistical errors, and should be relatively insensitive to systematic errors. They indicate that the transverse momentum spectrum of charged particles does not change as the E_t of the collision increases. Note that this analysis probes the character of the charged particles from events in the tail of the E_t distribution. These have $E_t \geq 100$ MeV, and represent only about 10% of the total cross section, i.e. the most central collisions. The reference spectrum is taken from events with $55 < E_t < 100$ GeV, which lie in the plateau of the E_t distribution. Figure 1 and the geometrical description of the collision indicate that collisions included in figure 5 range from near overlap of the projectile and target to the smallest impact parameters. Thus we do not see the hardening of the p_t spectra at very high multiplicity (or E_t) that was predicted as a signature of deconfinement[9].

In order to study the charged particle spectra over the entire range of impact parameters, we calculate the mean p_t of the distribution over a limited range. This allows a detailed study of the variation of $\langle p_t \rangle$ with E_t. The $\langle p_t \rangle$ vs. E_t

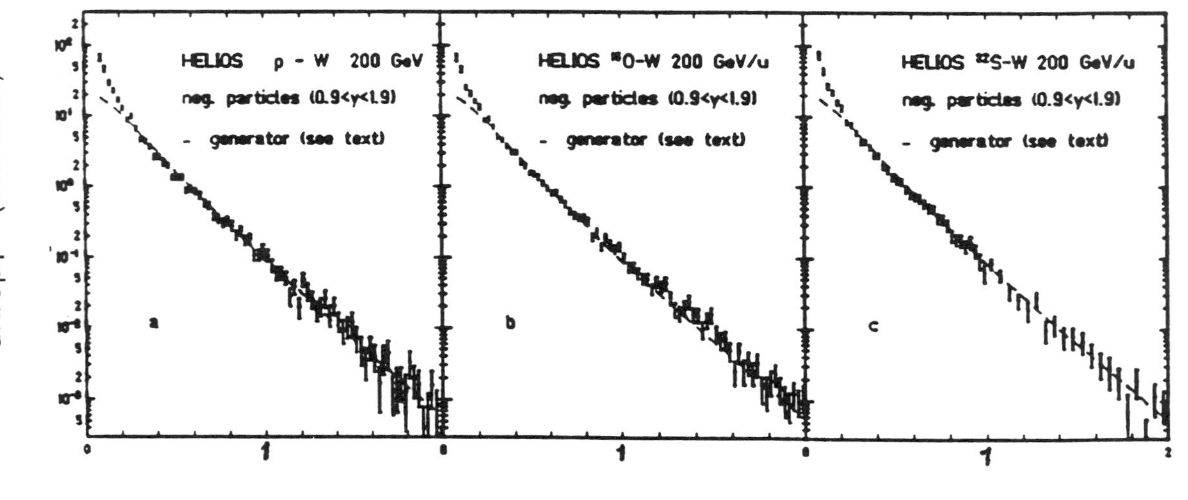

FIGURE 4
P_t spectra of negative particles detected in the external spectrometer for a) p + W, b) O + W, c) S + W at 200 GeV/A. The dashed line shows the results of a parameterization of p-p and p-A data described in the text.

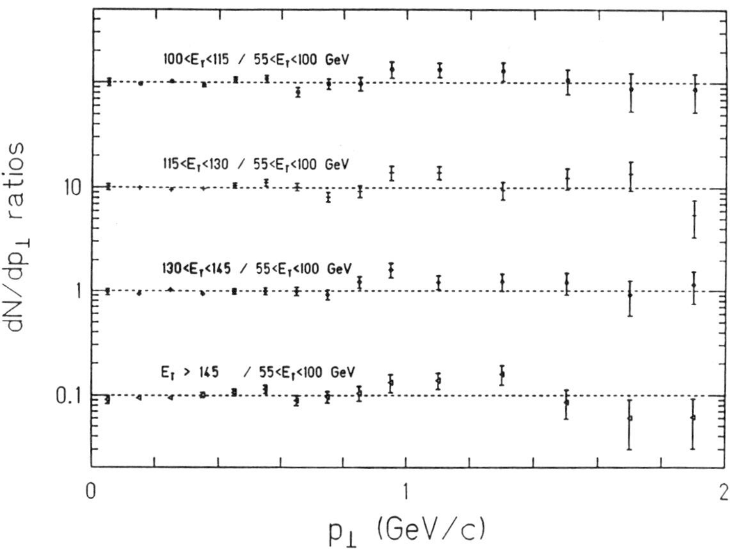

FIGURE 5
Ratios of negative particle spectra from O + W collisions at 200 GeV/A, plotted as a function of p_t. The ratios are normalized by the number of events in each spectrum. The four curves correspond to four regions of E_t, all normalized to the negative particle spectrum in events with 55 GeV < E_t < 100 GeV. The curves are offset by factors of 10 for clarity.

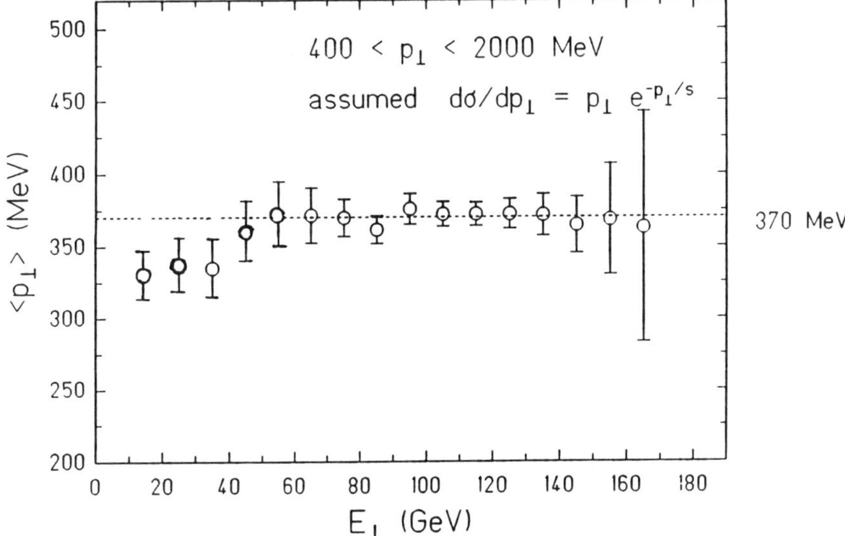

FIGURE 6
$<p_t>$ as a function of E_t for 200 GeV/A O + W. The $<p_t>$ is derived from the measured spectra by correcting the mean value in the range $0.4 < p_t < 2.0$ GeV/c under the assumption of an exponential spectrum.

for negative particles from 200 GeV/A O + W is plotted in Figure 6. The values are calculated by determining mean pt in the range $0.4 < p_t < 2.0$ GeV/c and extrapolating to the mean of the entire spectrum under the assumption of a purely exponential shape. This method yields $<p_t>$ which is not sensitive to the low p_t rise; the resulting values should be free of target effects and may be compared with p-p collisions. The $<p_t>$ thus obtained can still, however, be affected by any deviations in the spectral shape away from exponential. Figure 4 shows that the spectra begin to follow a power law behavior above p_t of $\sim$ 1.5 GeV/c, where one would expect to see effects of multiple hard scattering processes[6]. Thus the calculated $<p_t>$ will be somewhat higher than from a slope fitted to the purely exponential part of the spectrum. Figure 6 shows that central collions, with $E_t >$ ~60 GeV, yield a constant $<p_t>$ of 370 MeV/c. This should be compared with the IRIS prediction of 360 MeV/c for all charged particles and with 350 MeV/c for pions from p-p collisions[5]. For $E_t \leq 40$ GeV, corresponding to relatively peripheral collisions, $<p_t>$ is lower, falling by ~ 40 MeV/c for the events with the lowest E_t. An increase in $<p_t>$ with multiplicity or E_t of the collisions is predicted in models incoporating a hydrodynamical expansion of the highly excited central region[10]. However, the pt spectrum should also be strongly influenced by the hadronization process, making interpretation of the observe change more difficult.

In summary, we have seen that global variables such as transverse energy and multiplicity can be used to characterize events according to impact parameter. E_t and multiplicity show a linear relationship, indicating that higher E_t arises from production of more particles rather than a large change in the particle spectra. The shapes of the distributions can be explained by superposition of nucleon-nucleon scattering. There is no large difference among the observed charged particle spectra from p + W, O + W, and S + W. We observe no sharp rise in $<p_t>$ vs.E_t, but it must be noted that the detected p_t spectrum is affected by the hadronization process and reflects conditions at the time of freezeout of hadrons. Also, an increase in the cross section for large pt is expected from hard scattering of partons and does not necessarily indicate more exotic processes.

REFERENCES

1. See for example, Proc. of the Quark Matter Conferences in Brookhaven (1984), Helsinki (1984) and Asilomar (1986)
2. T. Akesson, et al., Z. Phys. C38, 383 (1988).
3. G. Baym, P. Braun-Munzinger and V. Ruuskanen, Phys.Lett. 190B, 29 (1987); A.D. Jackson and H. Boggild, Nucl. Phys. A470, 669 (1987).
4. A. Capella and J. Tran Thanh Van, Phys. Lett. 93B, 146 (1980).
5. B. Alper, et al., Nucl. Phys. B100, 237 (1975);
6. D. Antreasyan, et al., Phys. Rev. D19, 764 (1979).
7. H. Stroebele, et al., Z. Phys. C38, (1988).
8. D.A. Garbutt, et al., Phys. Lett. 67B, 355 (1977).
9. L. Van Hove, Phys. Lett. 118B, 138 (1982).
10. M. Kataja, P.V. Ruuskanen, L. McLerran and H. von Gersdorff, Phys. Rev. D34, 2755 (1986); X. Wang and R. Hwa, Phys. Rev. D35, 3409 (1987).

J/ψ SUPPRESSION IN ULTRA-RELATIVISTIC HEAVY-ION COLLISIONS *

M. -C. Chu

Center for Theoretical Physics
Laboratory for Nuclear Science
Massachusetts Institute of Technology
Cambridge, Massachusetts 02139, U.S.A.

ABSTRACT

We present a simple phenomenological model of J/ψ suppression in ultra-relativistic heavy-ion collisions, which gives good agreement with the recent CERN NA38 data on 200 GeV/A Oxygen-Uranium collisions. The essential ingredients of our model are the existence of a characteristic energy density ϵ_s above which no J/ψ could be formed, and Bjorken's longitudinal scaling hydrodynamics. With these assumptions, we deduce a lower bound on ϵ_s from the NA38 data: $\epsilon_s \geq 1$ GeV/fm^3. Fixing parameters from the Oxygen data, we also make predictions for 200 GeV/A Sulphur-Uranium interactions.

INTRODUCTION

With the high energy heavy-ion beams at CERN coming on-line and producing interesting data, it becomes somewhat urgent for the theorists to sort out possible signatures of the formation of a quark-gluon plasma (QGP). One particular proposal has gathered a great deal of attention recently, namely, the observation of Matsui and Satz[1] that a J/ψ bound state would not be formed in a deconfined medium due to the screening of the Coulomb potential. The rate of J/ψ production in heavy-ion collisions is expected to be suppressed if a quark-gluon plasma is formed. Karsch and Petronzio[2] in addition argued that because there is a finite formation time for the J/ψ bound state to develope, charm quark pairs ($c\bar{c}$) with high transverse momenta (p_T) would have escaped the plasma region before the formation of J/ψ's. It therefore follows that the suppression of J/ψ should be more severe at low p_T. The recent NA38 J/ψ data[3] indeed show qualitative agreement with these expectations.

Here, we will present a simple model analysis of the transverse momentum (p_T) dependence of J/ψ suppression in ultra-relativistic heavy-ion collisions.[4] We assume that in heavy-ion collisions at high enough energies, a region of high energy density ($\epsilon \geq \epsilon_s$) is formed, and J/ψ formation is forbidden for all charm quark pairs inside this region at the J/ψ formation time, $\tau_{J/\psi}$.[1] Our model is a modification of an earlier one by Karsch and Petronzio[2], the difference being that we shall incorporate the plasma expansion dynamics in our calculations. Our analysis differs from that of Blaizot and Ollitrault[5] in that we follow the energy density of the system in the hydrodynamics. This allows us to use the experimental data of neutral transverse energy together with estimates of $\tau_{J/\psi}$ to extract a lower bound on ϵ_s.

---

* This work is supported in part by funds provided by the U.S. Department of Energy (D.O.E.) under contract #DE-AC02-76ER03069.

MODEL

We consider central collisions of heavy ions at high enough energies such that Bjorken's scaling hydrodynamics[6] is a good description in the central rapidity region. We can then concentrate on the central plane ($z = 0$) in the center-of-mass frame along the collision axis (z), and the solution in any other $z \neq 0$ plane within the central rapidity region can be obtained by simple scaling. Furthermore we shall ignore transverse expansion, which is considerably slower than the longitudinal one. Bjorken's scaling solution then gives the proper-time dependence of the local energy density, $\epsilon(\tau) = \epsilon(\tau = \tau_0)(\tau_o/\tau)^{1+c_s^2}$, where τ_o is the plasma formation time, and c_s is the speed of sound, which is assumed to be constant throughout. For our calculations we take a simple parametrization,

$$\epsilon(\tau_o, r) = \epsilon_o \left[1 - \left(\frac{r}{R}\right)^2\right]^\beta \theta(R - r) \quad , \tag{1}$$

for the transverse coordinate (r) dependence of the initial proper energy density, where $\epsilon_o \equiv \epsilon(\tau = \tau_o, r = 0)$, and $\theta(x)$ is the ordinary step function. $R = 1.2A^{1/3}$ fm is the radius of the smaller one of the colliding ions (with atomic number A). When the system is cooled down to some characteristic energy density (ϵ_s) at time $\tau_s(r)$, J/ψ formation is no longer suppressed. We can calculate $\tau_s(r)$ simply from scaling and Eq. (1):

$$\tau_s(r) = \tau_o \left(\frac{\epsilon(\tau_o, r)}{\epsilon_s}\right)^{\frac{1}{1+c_s^2}} = \tau_o \left(\frac{\epsilon_o}{\epsilon_s}\right)^{\frac{1}{1+c_s^2}} \left[1 - \left(\frac{r}{R}\right)^2\right]^{\frac{\beta}{1+c_s^2}} \quad . \tag{2}$$

Equating the screening duration $\tau_s(r)$ given in Eq. (2) with the J/ψ formation time $t_\psi = \tau_{J/\psi}\gamma$ gives the critical radius r_s, which marks the boundary of the J/ψ suppression region:

$$r_s = R \left[1 - \left(\frac{\gamma\tau_{J/\psi}}{\tau_{so}}\right)^{\frac{1+c_s^2}{\beta}}\right]^{\frac{1}{2}} \theta\left(1 - \frac{\gamma\tau_{J/\psi}}{\tau_{so}}\right) \quad , \tag{3}$$

where $\tau_{so} \equiv \tau_s(r = 0)$, and $\gamma = E_T/M_\psi$ is the Lorentz factor associated with the transverse motion of the pair (see Fig. 1). A J/ψ bound state can be formed if the charm quark pair is outside of the screening region at the J/ψ formation time. Given an initial distribution of charm quark pairs (eg. $f(r) \propto \left[1 - (r/R)^2\right]^\alpha \theta(R - r)$) and assuming that they move in classical straight line trajectories, we can then calculate the acceptance function, $A(p_T)$, which is the ratio of charm quark pairs that escape the screening region to the total number of pairs created.

In general we have to do the calculation of $A(p_T)$ numerically. There are two simple limiting behaviors of $A(p_T)$, which we can find analytically. If p_T is large enough so that the dilated J/ψ formation time becomes longer than the screening duration, there would be no suppression of J/ψ formation. The

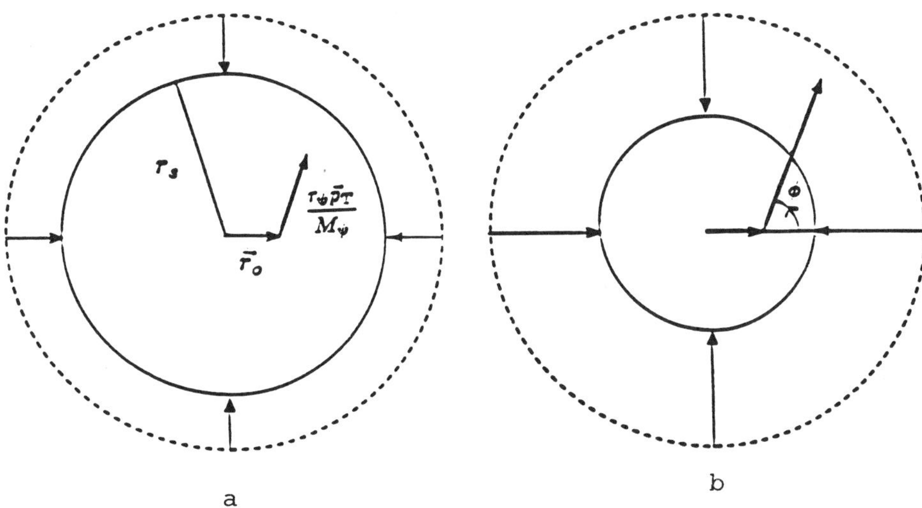

Fig. 1: Creation of a charm quark pair at $\vec{r}_o$ and its subsequent motion in the central plane in ultra-relativsitic heavy-ion collisions. In a), the charm quark pair remains inside the screening region at the J/ψ formation time $\tau_{J/\psi}$, and J/ψ formation is suppressed. In b), the charm quark pair escapes and forms J/ψ.

critical value of p_T for this condition is given by $\gamma\tau_{J/\psi}/\tau_{so} = 1$, as can be seen from Eq. (3). We therefore find

$$A(p_T) = 1 \quad , \quad \text{for } p_T \geq p_T^c = M_\psi \sqrt{\left(\frac{\tau_{so}}{\tau_{J/\psi}}\right)^2 - 1} \ .$$

Another limit is when $p_T = 0$; then

$$A(p_T = 0) = \left(\frac{\tau_{J/\psi}}{\tau_{so}}\right)^{\frac{(1+c_s^2)(1+\alpha)}{\beta}} \ . \tag{4}$$

In between these two limits, the acceptance function increases monotonically as p_T increases. We show a plot of $A(p_T)$ in Fig. 2 for a fixed ratio of $\tau_{J/\psi}$ to τ_{so} and assuming $\alpha = 1/2$, $\beta = 1/3$, $c_s^2 = 1/3$ to illustrate the above points.

COMPARISION WITH NA38 DATA

In the CERN NA38 experiment, the dilepton spectrum is measured together with the total "neutral"(neutral + about 30% of charged) transverse energy, E_T^o, which is approximately 50% of total transverse energy E_T.[3] We can extract the freezeout time (τ_f) from E_T using scaling hydrodynamics, once the freezeout energy density (ϵ_f) is specified. Knowing ϵ_f and τ_f, we can then determine the energy density at any proper time using Bjorken's scaling.

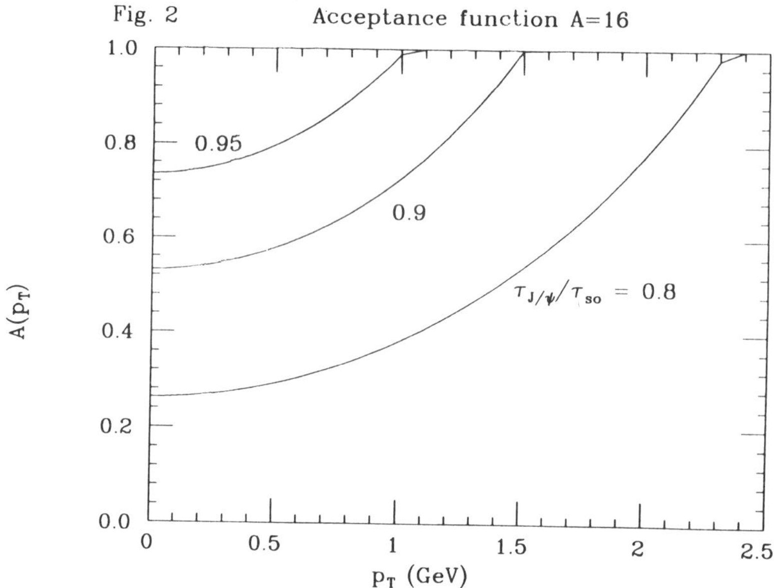

Fig. 2: The acceptance function vs. p_T for fixed ratio of J/ψ formation time to plasma lifetime, $\tau_{J/\psi}/\tau_{so}$, which is taken to be 0.8, 0.9 and 0.95 respectively for the three solid curves shown. Central collisions of Oxygen ions are assumed.

In Fig. 3 we show a plot of $\epsilon(\tau, r = 0)$ as a function of τ for collisions of Oxygen ions with $\beta = 1/3$, $\epsilon_f = 0.1$ GeV/fm^3, and $E_T^o = 73$ GeV. The two solid curves correspond to two extreme equations of states: $c_s^2 = 1/3$ and 0. The dotdashed curve represents a bag model equation of state with a first order phase transition ($c_s^2 = 0$ for $\epsilon_h \leq \epsilon \leq \epsilon_q$, and $c_s^2 = 1/3$ otherwise; we choose $\epsilon_h = 0.2$ GeV/fm^3 and $\epsilon_q = 2.0$ GeV/fm^3). We expect the true equation of state to lie between the two solid curves. For each value of E_T^o and a given equation of state then, we have a one-to-one correspondence between τ_{so} and ϵ_s. On the other hand, the observed p_T dependence of the acceptance function gives a narrow constraint on the possible values of $\tau_{J/\psi}/\tau_{so}$ (see Fig. 2): 0.8 $< \tau_{J/\psi}/\tau_{so} < 0.95$ (these bounds are not very sensitive to c_s^2). We can therefore infer the value of ϵ_s from an estimate of $\tau_{J/\psi}$. In particular, if we consider the range 0.7 fm/c $< \tau_{J/\psi} <$ 2.5 fm/c to be reasonable[2,5,7], we obtain an upper bound on τ_{so} ($\tau_{so} \leq 3.1$ fm/c, indicated by the dashed line in Fig. 3), and hence a lower bound on ϵ_s: (see Fig. 3)

$$\epsilon_s \geq \begin{cases} 1.0\ \text{GeV/fm}^3, & c_s^2 = 0 \text{ (free expansion);} \\ 1.5\ \text{GeV/fm}^3, & \text{Bag model E.O.S.;} \\ 2.2\ \text{GeV/fm}^3, & c_s^2 = 1/3. \end{cases}$$

This is our primary result: the observed p_T dependence of J/ψ suppression in ultra-relativistic heavy-ion collisions can be understood in our model if the

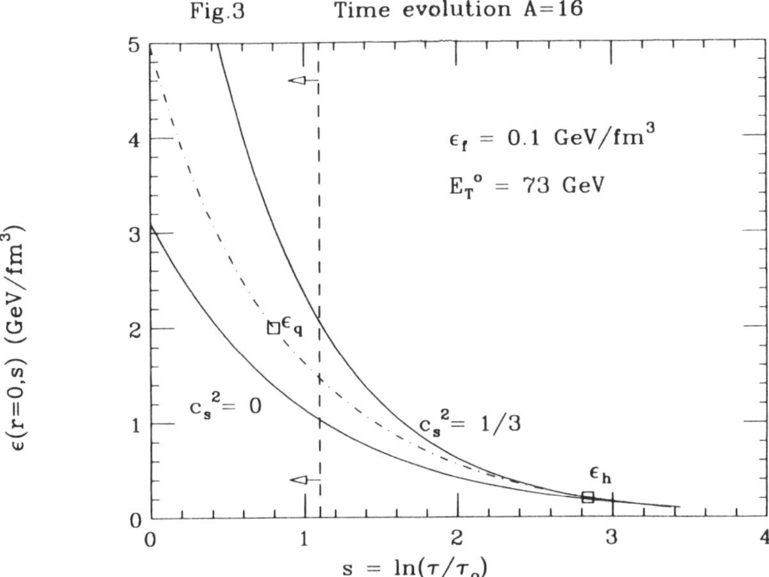

Fig. 3: Proper energy density at $r = 0$ in the central rapidity region as a function of $s \equiv \ln(\tau/\tau_o)$ for central collisions of Oxygen ions with total transverse "neutral" energy $E_T^o = 73$ GeV. The two solid curves correspond to two limiting equation of state: $c_s^2 = 0$ and 1/3 respectively, while the dotdashed line depicts expansion with a bag model equation of state. The allowed range of the screening duration τ_{so} extracted from the p_T-dependence of the acceptance function (see Fig. 2) is indicated by the dashed line and the arrows. The points where the dashed line cross with the energy density curves give the lower bounds on the screening energy density ϵ_s.

suppression of J/ψ formation is associated with high energy density ($\epsilon > 1$ GeV/fm^3).

Once ϵ_s is determined, we can compare results from our model calculations to experimental data, after folding in the experimental E_T^o spectrum:

$$\mathcal{A}(p_T, E_{\min}, E_{\max}) = \int_{E_{\min}}^{E_{\max}} dE_T^o\, W(E_T^o) A(p_T; E_T^o) \quad , \tag{5}$$

where $W(E_T^o)$ is the normalized observed E_T^o distribution in the J/ψ mass range (2.7 - 3.5 GeV). One more modification is in order: so far we have ignored the possibility that J/ψ is formed from the cascade of χ. In fact, in pp collisions, it was observed that roughly 40% of the J/ψ is formed this way[8]. Assuming the same branching ratio in heavy-ion collisions, we can include this effect by repeating the calculation for χ with a longer formation time τ_χ and adding up the acceptance functions due to J/ψ and χ weighted with the appropriate branching ratios.

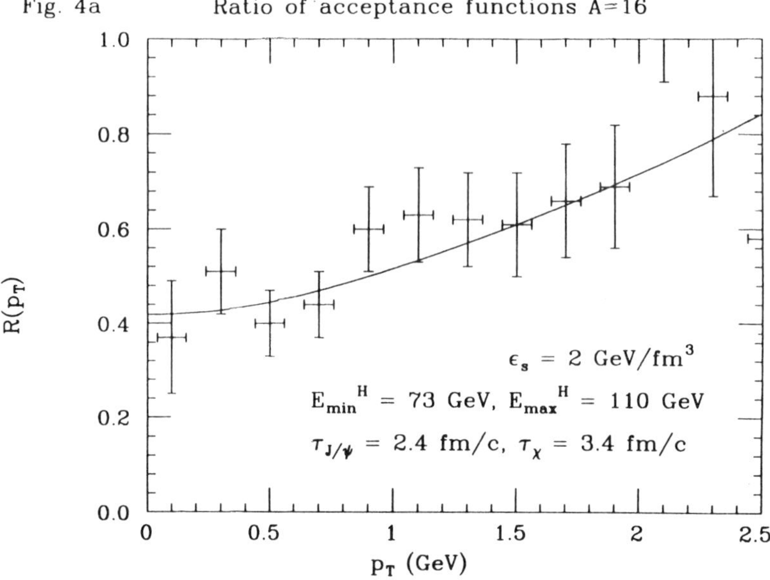

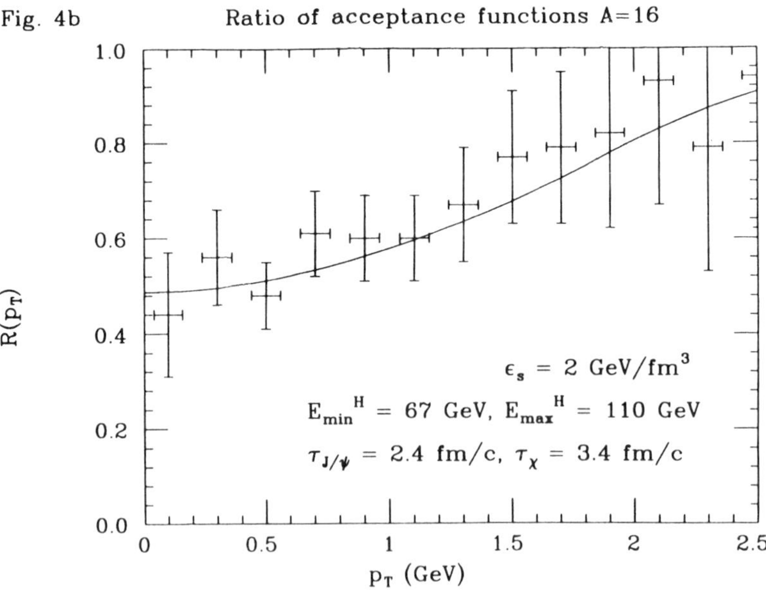

Fig. 4: Continued next page.

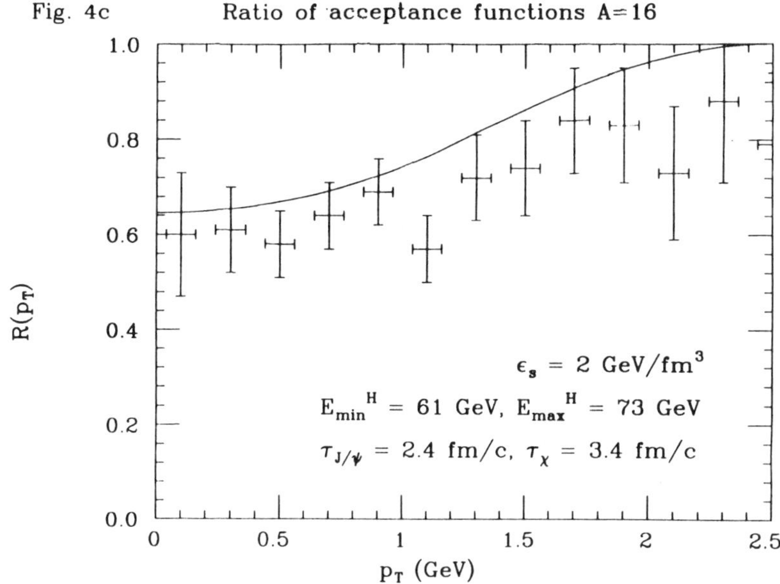

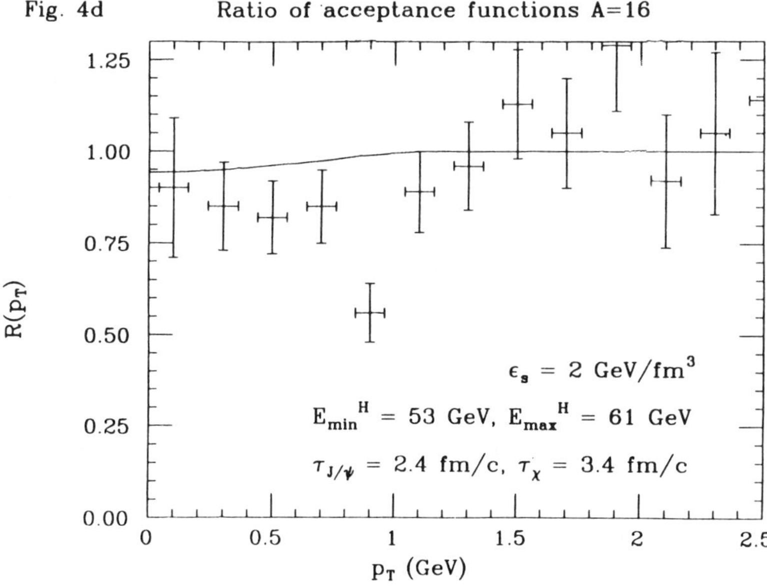

Fig. 4: Transverse momentum (p_T) dependence of the ratio of the acceptance functions for Oxygen beam at 200 GeV/A on Uranium target compared to NA38 data points[3] for different energy bins: $(E^H_{\rm min}, E^H_{\rm max})$ = a) (73, 110), b) (67,110), c) (61,73), d) (53, 61) GeV.

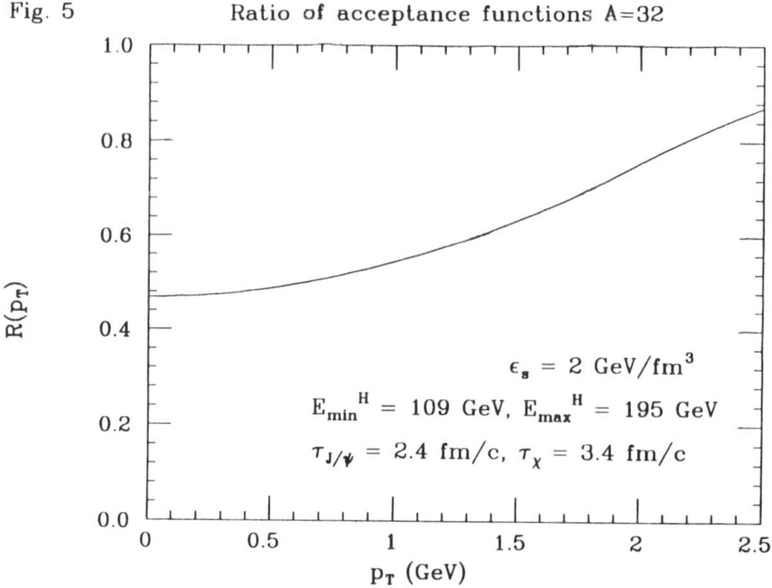

Fig. 5: Same as Fig. 4 except that Sulphur beam at 200 GeV/A is assumed. $(E^H_{\rm min}, E^H_{\rm max}) = (109, 195)$ GeV, and $(E_{\rm min}, E_{\rm max}) = (0, 58)$ GeV.

In Fig. 4 we show the calculated ratio of J/ψ produced in collisions with E^o_T between $E^H_{\rm min}$ and $E^H_{\rm max}$, to those produced in $E^o_T \leq 32$ GeV as a function of p_T,

$$R(p_T) \equiv \frac{\mathcal{A}(p_T,\ E^H_{\rm min},\ E^H_{\rm max})}{\mathcal{A}(p_T,\ E_{\rm min} = 0 \text{ GeV},\ E_{\rm max} = 32 \text{ GeV})} \ . \tag{6}$$

In these figures we also plotted the NA38 data for the ratio,

$$Y(p_T) \equiv \frac{r_{\psi/c}(p_T,\ E^H_{\rm min},\ E^H_{\rm max})}{r_{\psi/c}(p_T,\ E_{\rm min} = 0 \text{ GeV},\ E_{\rm max} = 32 \text{ GeV})} \ ,$$

where $r_{\psi/c}(p_T,\ E_{\rm min},\ E_{\rm max})$ is the ratio of J/ψ to continuum yields with transverse momentum p_T in events associated with total transverse "neutral" energy $E_{\rm min} < E^o_T < E_{\rm max}$. We have assumed that $\epsilon_s = 2$ GeV/fm^3. In Fig. 4a, we compare the calculated results to data for the highest energy bin: $E^H_{\rm min} = 73$ GeV, $E^H_{\rm max} = 110$ GeV. We obtain a good fit to the data if we use $\tau_{J/\psi} = 2.4$ fm/c and $\tau_\chi = 3.4$ fm/c. Fixing these two parameters, we can repeat the calculations for other energy bins. These are shown in Fig.s 4b-d for $(E^H_{\rm min}, E^H_{\rm max}) = (67, 110)$ GeV, (61, 73) GeV, and (53, 61) GeV respectively. Clearly if E^o_T is higher, the energy density reached is also higher, and the screening duration is longer, leading to more suppression of J/ψ formation. The overall trend of the data is well reproduced with our model. Using the parameters deduced from the Oxygen data, we also make predictions for the acceptance function in 200 GeV/A $S^{32} + U$ collisions (Fig. 5).

Another concrete prediction of our model is that τ_{so} depends on the atomic number, A, only through the factor $R^{-2} = A^{-2/3}$. We therefore expect that the suppression at $p_T = 0$ would depend only on $E_T^o/A^{2/3}$ (see Eq. (4)). Since low p_T events dominate in abundances, this implies that the p_T-integrated suppression should fall on an approximate universal curve for all ions as a function of $E_T^o/A^{2/3}$.

To summarize, we find that the dependence of J/ψ cross section on p_T can be understood in a simple model where the formation of J/ψ is forbidden when the energy density is above some characteristic value. Our results strongly suggest that high energy density matter ($\epsilon \geq 1$ GeV/fm^3) is responsible for the observed J/ψ suppression.

ACKNOWLEDGEMENT

I thank T. Matsui for his helpful advices and collaboration. I also thank Jean-Paul Blaizot, B. Chaurand, C. Gerschel, L. Kluberg and Helmut Satz for useful discussions.

REFERENCES

1. T. Matsui and H. Satz, *Phys. Lett.* **178B**, 416 (1986).
2. F. Karsch and R. Petronzio, *Phys. Lett.* **193B**, 105 (1987).
3. M. C. Abreu *et al.*, NA38 collaboration, "The Production of J/ψ's in 200 GeV/N Oxygen-Uranium Interactions", in *Quark Matter '87* and in the *International Conference on Physics and Astrophysics of Quark-Gluon Plasma*, Bombay, India, 1988. B. Chaurand and L. Kluberg, private communication.
4. M.-C. Chu and T. Matsui, *Phys. Rev.* **D37**, 1851 (1988).
5. J. P. Blaizot and J. Y. Ollitrault, *Phys. Lett.* **B199**, 499 (1987).
6. J. D. Bjorken, *Phys. Rev.* **D27**, 140 (1983).
7. F. Karsch, M. T. Mehr and H. Satz, *Z. Phys.* **C38**, in press (1988); J. Kapusta, *Phys. Rev.* **D36**, 2857 (1987).
8. J. H. Cobb *et al.*, *Phys. Lett.* **72B**, 497 (1978).

Study of High Energy Nucleus Nucleus Collisions

Achim Franz, LBL, for the WA80 - Collaboration

GSI - LBL - LUND - ORNL - MÜNSTER

Introduction

The CERN experiment WA80 studies ultrarelativistic heavy ion collisions by calorimetry and charged particle measurements over a large fraction of 4π. The setup and its components are shown in Fig.1 and described elsewhere in detail [1]. Global observables like the transverse

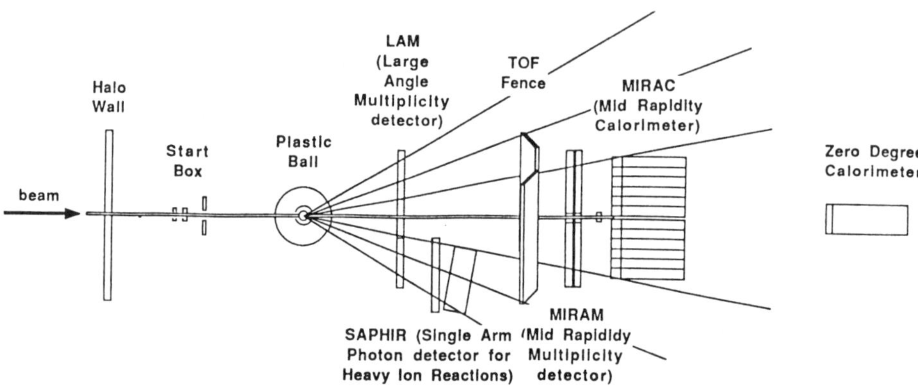

Figure 1: WA80 setup

energy distribution, charged particle multiplicities and p_T distributions for photons and neutral pions are published in [2,3,4]. Here I want to concentrate on a closer study of nucleus nucleus collisions and recent results from the lead glass spectrometer Saphir.

Target Spectators and Participants

The Plastic Ball can identify target fragments in the angular range of $30° \leq \vartheta \leq 160°(-1.7 \leq \eta \leq 1.3, \eta = -ln(tan(\vartheta/2.)))$, together with the Zero Degree Calorimeter (ZDC), which measures the energy flow in the beam direction ($\vartheta \leq 0.3°$) one can get an insight into what happens to the target nucleus in a central collision with the $3.2TeV^{16}O$ projectile. Fig.2 shows the number of baryons measured by the Plastic Ball as a function of the energy in the ZDC, which is a measure of the centrality of the event. The bands for the four different targets, ^{197}Au, ^{108}Ag, ^{64}Cu and ^{12}C, indicate the error mainly due to double hit corrections. For low energy in the ZDC, i.e. central collisions, the Plastic Ball measures about 120 baryons from the ^{197}Au target, including a N/Z corrections for the neutrons, about 60 from ^{108}Ag and 30 from ^{64}Cu. These baryons emerging from the target are called spectators, because they are not directly involved in the collision, i.e. they don't lie in the pathway of the projectile. Even though the mean transverse energy, for the protons as measured by the PB is with $175MeV$ higher than for target participants in $800AMeV^{197}Au +^{197} Au$ collisions measured by the PB at the Bevalac [5]. The number of participants in the collisions studied at CERN can be related to the energy in the ZDC by:

- Comparing it with FRITIOF, the LUND Monte Carlo for nucleus nucleus collisions

- Assuming a monotonic dependence of the number of participants on the ZDC energy.

These procedures are described in detail in [2] and one example, $200 GeV/A\, ^{16}O + ^{197} Au$, is shown in Fig.3. If one assumes that the participant nucleons are boosted into the angular range between the PB and the ZDC $\Delta\vartheta = 29.7°(\Delta\eta = 4.7)$ the sum of the spectators measured by the PB and the number of participants from the ZDC gives an insight into what happens to the target nucleus. These numbers are:

Target	nucleons	spectators + participants
Au	197	174
Ag	108	104
Cu	64	64

This indicates that the whole target nucleus is disintegrated into small fragments.

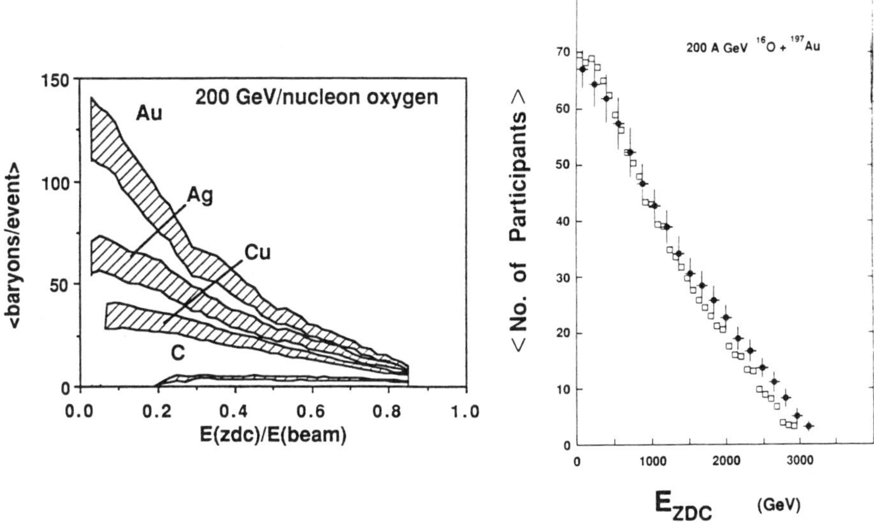

Figure 2: Average number of baryons measured in the Plastic Ball (left)

Figure 3: Number of participants for 200AGeV O+Au collisions (right)

Average E_T per particle

Two other detector components are used to analyse the produced particle flow:

- Streamer tubes with capacitively coupled pad readout to measure charged particles, covering $1.7° \leq \vartheta \leq 32°(1.2 \leq \eta \leq 4.8)$ set up as three individual planes (LAM, MIRAM1 and MIRAM2 in Fig.1)

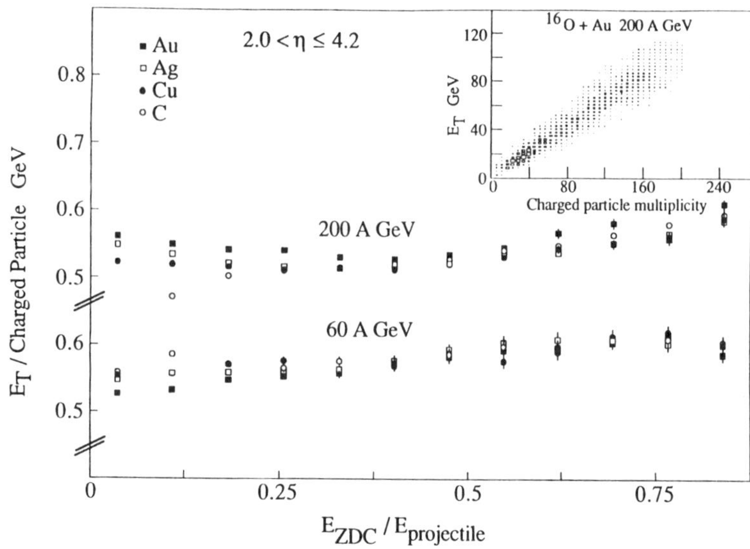

Figure 4: Average transverse energy per charged particle, see text for details

- Pb-Scintillator - Fe-Scintillator Mid-Rapidity Calorimeter, MIRAC see [6] for details, covering $0.4° \leq \vartheta \leq 12.6°(2.4 \leq \eta \leq 5.5)$ with full polar angle coverage.

The insert in Fig.4 shows the number of charged particles detected in a pseudorapidity interval of $2.4 \leq \eta \leq 4.0$ plotted against the transverse energy, E_T, measured by MIRAC in the same region. The correlation between these observables, with an average of $550 MeV/charged\ particle$, is striking. This number corresponds to $370 MeV/particle$ if one assumes equal production for all three charges. In the main part of Fig.4 this correlation is shown for the two beam energies, $200AGeV/c$ and $60AGeV/c$, the four targets used, ^{197}Au, ^{108}Ag, ^{64}Cu and ^{12}C, and plotted against the energy in the ZDC. Under all those different conditions the average E_T per charged particle stays constant.

Saphir

The invariant cross section for neutral pions as a function of p_T was published in [4] and is shown in Fig.5. The plot is a combination of oxygen and proton induced interactions of ^{197}Au and ^{12}C targets at $200AGeV/c$ and $60AGeV/c$. The inverse slope (T_0) of these exponential distribution $(1/p_T)dN/dp_T \sim exp(-p_T/T_0)$ is also included and shows only slow variation going from $200AGeV/c^{16}O+^{197}Au$ collisions to $60GeV/cp+Au$. The data in this plot is a summary of all "minimum bias" events, which in our case corresponds to less than 88% of the beam-energy measured in the ZDC plus at least one charged particle measured in the multiplicity arrays.
To investigate changes of the p_T distribution on the centrality the cross section as a function of p_T for $200AGeV/c\,^{16}O +^{197} Au$ is plotted in Fig.6 for central and medium plus peripheral collisions separately. The centrality of the collision is again defined by the energy observed in the ZDC. Here events selected as central collisions have less than 30% of the beam energy left in the forward direction. The change in inverse slope between the peripheral and central collisions are compatible with going from $200AGeV/c\,^{16}O +^{197} Au$ to $60AGeV/c\,p +^{197} Au$ collisions.
More details about the source of the π^0 and a possible single photon signal are currently under

investigation. A preliminary study of the correlation length for π^0 pairs done with HBT correlation indicate a rather small size of $\sim 1fm$. For the single photon signal more investigations have to be done to understand the signal coming from misidentification of photon pairs coming from resonances like π^0's and η's. The current analysis indicate a π^0 to η ratio within the expected limits.

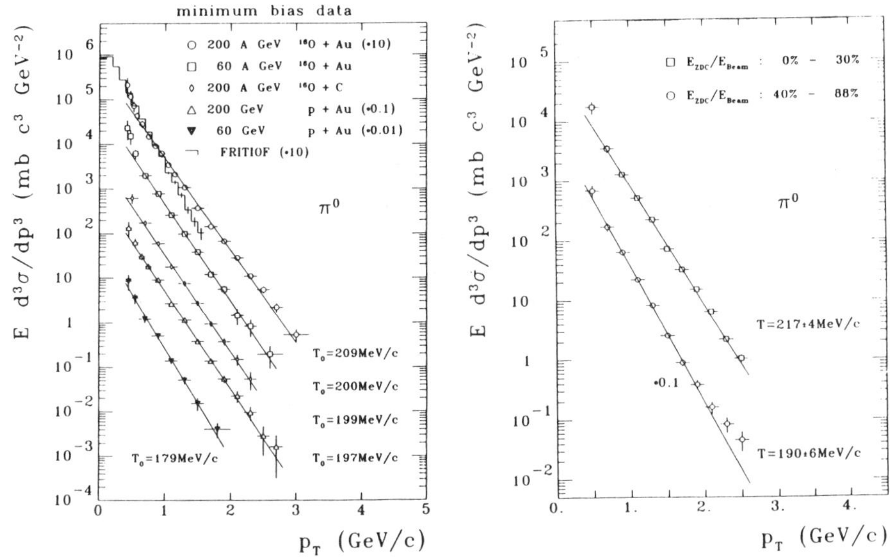

Figure 5: π^0 invariant cross section as a function of p_T (left)

Figure 6: π^0 invariant cross section for $200AGeV/c\ {}^{16}O + {}^{197}Au$ separated for peripheral and central collisions (right)

References

[1] "Study of Relativistic Nucleus-Nucleus Collisions at the CERN SPS", R.Albrecht et al., GSI preprint, GSI-85-32

[2] Physics Letters B, 199, Dec. 1987, 297-303

[3] Physics Letters B, 202, Mar. 1988, 596-602

[4] Physics Letters B, 201, Feb. 1988, 390-396

[5] K.G.R. Doss et al., "Transverse Energy Production in High Energy Nuclear Collisions and the Equation of State of Nuclear Matter," to be published in Modern Phys. Letters.

[6] T.C.Awes et al. Nucl.Instr. Methods. to be submitted

QUASIPARTICLES AND SPIN-STATISTICS IN THE DECONFINED PHASE*

Janos Polonyi[†‡], Center for Theoretical Physics, Laboratory for Nuclear Science and Department of Physics, Massachusetts Institute of Technology, Cambridge, Massachusetts 02139 U.S.A.

ABSTRACT

Due to the multi-connected nature of the group of global gauge transformations of pure Yang-Mills systems the triality is permanently confined. The deconfined phase of QCD is characterized by the appearance of unusual gluon states which screen the center charge of the quarks. It is shown that chromomagnetic monopoles changes from boson to fermion as they develop triality charge in the deconfined phase. A distinguished signature of the high temperature phase of QCD is that the isolated quarks screened by the monopoles appear as bosons.

I. INTRODUCTION

The studies of the high temperature properties of QCD were initiated by the naive hope that the momentum scale set by the thermal motions renders QCD perturbative and the basic constituents become visible. The absence of perturbative magnetic mass gap spoils the perturbation expansion at arbitrary high temperature[1] but numerical computations in lattice gauge theory support the existence of the high temperature deconfined phase[2]. Another motivation is the hope that the dynamics of the confinement-deconfinement phase transition which is formed in front of us during the computer simulations hold the key to the understanding of the confinement mechanism.

The phenomenon of confinement is the absence of colored states among the observed asymptotic states of QCD. The conventional scenario for quark confinement supposes the infinitely rising potential between static color charges when quark-antiquark vacuum polarization effects are ignored. The confinement-deconfinement "order parameter" used in the determination of the finite temperature phase structure is the thermal and color average of the heavy quark

* This work is supported in part by funds provided by the U. S. Department of Energy (D.O.E.) under contract #DE-AC02-76ER03069.

† Alfred P. Sloan Research Fellow.

‡ On leave of absence from CRIP, Budapest, HUNGARY.

propagator in imaginary time. The nonvanishing of this propagator signals the dynamical breakdown of the symmetry with respect global gauge transformations which are periodic in Euclidean time up to a center element of the gauge group.

A mechanism is presented in this talk which gives an account of the vanishing of the quark propagator in an arbitrary gauge of the operator formalism of Yang-Mills theories. We take the point of view that the vanishing of the quark propagator in each gauge signals the impossibility of isolating quarks. The vanishing of the propagator is obvious as long as manifestly gauge invariant state vectors are used. Rather than trying to prove that the gauge invariance is preserved in the vacuum state we concentrate on the consequences of the dynamical breakdown of the center symmetry.

This dynamical symmetry breaking occurs in the functional integral representation of the projection operator which is supposed to impose Gauss' law. Thus the constraint what we intend to implement is not realized by the functional integral formalism. One might wonder whether such a symmetry breakdown *i.e.* the deconfinement phase transition is characteristic of the path integral formalism only or reflects real physical effects. We hope that the computation of the deconfinement phase transition in the operator formalism in terms of manifestly gauge invariant states which is beyond our present analytic power will ultimately yield the same result as the path integral method. The difficult point in such a computation is to break a symmetry which is obeyed explicitly by the physical states.

It will be shown that an isolated quark can propagate in the symmetry broken phase only after its center charge *i.e.* triality is completely screened by gluons. In this sense the confinement of triality is permanent. The deconfined phase is characterized by the appearance of gluon states in the partition function which carry triality. The existence of the isolated quarks is the result of an unusual screening rather than the weakening of the nonperturbative confining interactions[3].

This picture of the deconfined phase leads us to investigate the nature of the gluon states which carry triality. The apparent difficulty of this question *i.e.* of the construction of charged states by neutral operators will be resolved by exploiting the multiply connected topology of the space of global gauge transformations for Yang-Mills systems. Gluonic quasiparticles with center color charges will be introduced in such a manner. These states contribute to some matrix elements since the Gauss' law is not properly implemented in the deconfined phase.

The ultimate screening of the chromomagnetic sector of the deconfined phase suggests the use of quasiparticles with chromomagnetic charges[4]. It will be shown that chromomagnetic monopoles which are the prototypes of gauge-hedgehogs may carry triality too. Thus the known nonperturbative features of the deconfined phase[4] may be reproduced by the help of chromomagnetic monopoles only.

The magnetic monopoles lead not only to an economic way to saturate the nonperturbative effects at high temperature but to rather exceptional effects of the deconfined phase. It will be shown that the spin-statistics of gauge-hedgehogs which screen quarks changes at the deconfinement temperature. These quasi-particles which appear as bosons in the confined phase acquire Fermi statistics in the deconfined-screened environment. The "deconfined" quarks which are screened by such quasiparticles behave as bosons. This mechanism may have important effects on the stability of the deconfined plasma due to the reduction of the overlap of the quark wave functions.

This situation can be best illustrated by the hypothetical world where only charmed quarks are present in the electro-weak currents. The charmed mesons appear as "deconfined", bosonic Cooper-pair quarks for the observers of this world. The role of the screening antiquarks which are invisible for other interactions is played by the gauge-hedgehogs in the real deconfined plasma.

Chromomagnetic monopoles with triality can be used to describe some bound states and resonances of the screened-deconfined phase. These states, constructed analogously to the ordinary nonstrange hadrons of the confined phase are color neutral.

II. THE PROJECTION OPERATOR OF YANG-MILLS THEORIES

The $SU(3)$ Yang-Mills Hamilton operator is of the form

$$H = -\operatorname{tr}\int\left(g^2\vec{E}^2(\vec{x},t) + \frac{1}{g^2}\vec{B}^2(\vec{x},t)\right)d^3\vec{x} \tag{1}$$

in temporal gauge $A_0 = 0$ where $\vec{A}(\vec{x}) = \vec{A}^a(\vec{x})\frac{\lambda^a}{2i}$, $\vec{E}(\vec{x}) = \vec{E}^a(\vec{x})\frac{\lambda^a}{2i}$, $E_i^a(\vec{x}) = -i\frac{\delta}{\delta A_i^a(\vec{x})}$ and $B_i = \partial_j A_k - \partial_k A_j + [A_j, A_k]$. The canonical commutators give

$$\left[\left(\vec{D}\vec{E}(\vec{x})\right)^a, \left(\vec{D}\vec{E}(\vec{y})\right)^b\right] = 2if^{abc}\left(\vec{D}\vec{E}(\vec{x})\right)^c\delta(\vec{x}-\vec{y}) \tag{2}$$

and

$$[H, \vec{D}\vec{E}(\vec{x})] = 0 \tag{3}$$

One has to introduce counterterms in (1). All what we need to know is that they preserve gauge invariance and (3) remains valid.

Gauge transformations act as $\vec{A}^g = g(\vec{\partial} + A)g^\dagger$. The important point is that the space of global gauge transformations $SU(3)_{\text{glob}}$ is multiply connected. In fact, the center $C = Z_3$, *i.e.* the subgroup consisting of elements which commute with $SU(3)_{\text{glob}}$ leaves the gauge field invariant, $SU(3)_{\text{glob}} = SU(3)/Z_3$. The situation is particularly simple for $SU(2)$ Yang-Mills theory. The gauge field is characterized by a three vector in color space $\vec{A}(\vec{x}) = \vec{A}^a(\vec{x})\frac{\sigma^a}{2i}$ which is rotated in the color space uniformly by global gauge transformations. The field theory can be viewed as the interactive system of infinitely many particles moving in the three dimensional color space. Global gauge symmetry is the rotational symmetry for this system with symmetry group $SU(2)/Z_2 = SO(3)$.

One may diagonalize the A and a maximal commuting set of generators $(\vec{D}\vec{E}(\vec{x}))^\alpha$ simultaneously. A complete set of states can be labeled by the eigenvalues of these operators and the Casimireans of the local gauge rotations. The eigenvalues of $(\vec{D}\vec{E}(\vec{x}))^\alpha$ and the Casimireans specify the transformation properties of the states under local gauge transformations which give the static external charge imposed on the system according to Gauss' law.

The reduction of the representation $T(g)$ of the group G can be achieved by means of the operator

$$P^{(\mu)} = \int dg T(g) \chi^{\mu\dagger}(g) \tag{4}$$

which project into the μ-th irreducible representation of G. $\chi^\mu(g)$ is the character of the μ-th irreducible representation and dg is the invariant integral over G. The operator projecting into the physical space with vanishing external charge *i.e.* into the gauge invariant subspace is given by

$$\mathcal{P}_0 = \mathcal{N}^{-1} \int \mathcal{D}_\mathcal{G}\left[tA_0(\vec{x})\right] e^{it\int \operatorname{tr}\left(A_0 \vec{D}\vec{E}\right) d^3\vec{x}} \tag{5}$$

where the integration is over the group $\mathcal{G}$ which is generated by $\vec{D}\vec{E}(\vec{x})$.

It is important to note that *any* group with the given set of generators may be used in the projection operator. If the group $\mathcal{G}$ has a discrete subgroup K than integration over $\mathcal{G}' = \mathcal{G}/K$ is sufficient and K can be represented non-trivially even in the vacuum sector of the theory. Now we chose $\mathcal{G}$ to be the direct product of the local gauge groups. In case of continuum regularization some smoothness conditions are to impose as well. The boundary conditions will be periodic or free in three space[5].

The center charge is defined by the transformation properties under global gauge transformations from the center. The corresponding projection operators are $\mathcal{P}_n = \mathcal{P} P_n$, $n = 0, 1, 2$ where

$$\mathcal{P} = \mathcal{N}^{-1} \int \mathcal{D}_{\mathcal{G}/Z_3}\left[tA_0(\vec{x})\right] e^{it\int \operatorname{tr}\left(A_0 \vec{D}\vec{E}\right) d^3\vec{x}} \tag{6}$$

and

$$P_n = \frac{1}{3}\sum_{\ell=1}^{3} e^{-i\frac{2\pi}{3}n\ell} z^\ell \tag{7}$$

Here z is the global gauge transformation by the center element $e^{\frac{2i\pi}{3}}$. The states prepared by $\mathcal{P}_n$ give

$$z\mathcal{P}_n|\rangle = e^{\frac{2i\pi n}{3}} \mathcal{P}_n|\rangle \tag{8}$$

States with nonvanishing center charge are described by wave functionals which are multi-valued in the space of global gauge transformations. In fact, this is precisely the set of wave functionals which gives a nontrivial realization of the center of global gauge transformations. In the case of $SU(2)$ Yang-Mills theory these are the states whose wave functional involves half-integer spin spherical functions in their dependence on global gauge transformations. States with multi-valued wave functional may carry fundamental charges and thus screen quarks.

The projection operator which prepares the physical states with an external charge in the multiplet r at location $\vec{x}_0$ is given as

$$\mathcal{P}[r,\vec{x}_0] = \mathcal{N}^{-1}\int \mathcal{D}_{\mathcal{G}}\left[tA_0(\vec{x})\right] e^{i\int \operatorname{tr}\left(tA_0\vec{D}\vec{E}\right)d^3\vec{x}} \operatorname{tr} e^{-tA_0^a(\vec{x}_0)\frac{\lambda_r^a}{2i}} \tag{9}$$

where the generators λ_r^a act in the representation r of the gauge group. It is important to note that the surface term is to be neglected when integration by part is carried out in the exponent[5].

The propagator in the zero external charge sector is given by the matrix element

$$\left\langle \vec{A}^{(f)}\left|\mathcal{P}_0\, e^{-itH}\right|\vec{A}^{(i)}\right\rangle \tag{10}$$

which is written as

$$\mathcal{N}^{-1}\int \mathcal{D}_{\mathcal{G}}\left[tA_0(\vec{x})\right]\left\langle \vec{A}^{(f)}\left|e^{-itH_{A_0}}\right|\vec{A}^{(i)}\right\rangle \tag{11}$$

where

$$H_{A_0} = -\operatorname{tr}\int\left\{g^2\vec{E}^2(\vec{x}) + \frac{1}{g^2}\vec{B}^2(\vec{x}) + A_0(\vec{x})\vec{D}\vec{E}(\vec{x})\right\}d^3\vec{x} \tag{12}$$

These expression can be interpreted by saying that the time evolution is generated by the gauge non-invariant Hamiltonian H_{A_0} in the presence of the random external field $A_0(\vec{x})$. This formalism allows us to gain some insight into the dynamics of the matrix element of $e^{-itH_{A_0}}$ between unconstrained states. Gauss' law is recovered after the integration over the random external field only.

III. THE CONFINED PHASE

We are interested in the quenched quark propagator

$$\begin{aligned}\left\langle \vec{A}^{(f)}|T\psi(y)\bar{\psi}(x)|\vec{A}^{(i)}\right\rangle_{\text{phys}} &= \left\langle \vec{A}^{(f)}|\mathcal{P}_0 T K^{-1}[x,y;\vec{A},A_0]e^{-itH}|\vec{A}^{(i)}\right\rangle \\ &= \mathcal{N}^{-1}\int \mathcal{D}_{\mathcal{G}}\left[tA_0(\vec{x})\right]\left\langle \vec{A}^{(f)}\left|TK^{-1}[x,y;\vec{A},A_0]e^{-itH_{A_0}}\right|\vec{A}^{(i)}\right\rangle\end{aligned} \tag{13}$$

where the matrix $K[x,y;\vec{A},A_0]$ appears in the action for the quarks

$$S_q = \int d^4x d^4y \bar{\psi}(x) K[x,y;\vec{A},A_0]\psi(y) \tag{14}$$

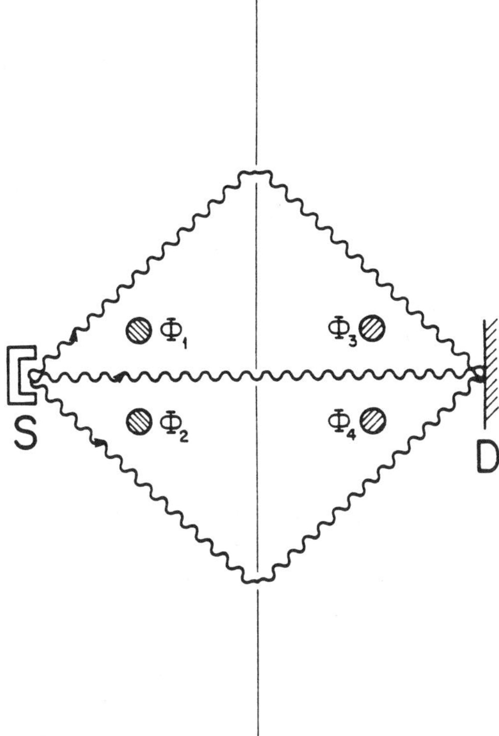

Fig. 1: Thought experiment illustrating the interference of the propagation in the space of global gauge transformations.

and depends on the operators $\vec{A}(\vec{x}, x^0)$ and the c-number field $A_0(\vec{x})$.

Consider first the integrand of (11) for a given configuration $A_0(\vec{x})$. This amplitude is represented pictorially by one of the wavy lines of Fig. 1. This Figure displays a thought experiment of measuring the propagation of particles with charge e from the source S to the detector D via one of the three small slits on the surface represented by the vertical line in the middle. The amplitudes corresponding to the configurations $A_0^z(\vec{x})$ are the degenerate if

$$e^{tA_0^z(\vec{x})} = z\, e^{tA_0(\vec{x})} \tag{15}$$

and z is an element of the center of the global gauge group. In fact, global gauge transformations corresponding to the center Z_3 leave the gauge field invariant and the states $|\vec{A}^{(i)}\rangle$ and $|\vec{A}^{(f)}\rangle$ are represented by well defined single valued wave functionals. The three type of propagations depicted in Fig. 1 belong to three different homotopy classes in the multiply connected space of global gauge transformations $SU(3)_{\text{glob}} = SU(3)/Z_3$. Note that the invariance of the amplitude with respect of the change of $A_0(\vec{x})$ given by (15) is precisely the one used in the conventional finite temperature path integral computations to detect the deconfinement phase.

We are now ready to extend our thought experiment to the quenched quark propagator (13). It is easy to verify that $K^{-1}[x,y;\vec{A},A_0^z] = e^{\frac{2\pi i}{3}} K^{-1}[x,y;\vec{A},A_0]$ if z is the global gauge transformation by the center element $e^{\frac{2\pi i}{3}}$. Thus the integrand of (13) is multiplied by a cubic root of one as $A_0(\vec{x})$ is changed according to (15). This extra phase is realized in our thought experiment by the magnetic flux $\Phi_3 = \Phi_4 = \frac{-2\pi}{3e}$. The additive interference of the propagation of the gauge field (11) becomes a completely destructive one when the quark propagator is considered in (13). This destructive interference between the homotopically different trajectories in the space of global gauge transformations leads to the non-localizability of the quarks.

The vanishing of a propagator is usually explained by the diverging energy of the corresponding state. Note that one can have nonpropagating states without using such a singular objects by realizing that the configuration space is multiply connected . In fact, our construction gives completely destructive interference between homotopically different trajectories. The application of algebraic and topological methods allows us to generalize this construction for arbitrary multiply connected group space as well[6].

The presented argument is valid for periodic boundary conditions for $A_0(\vec{x})$ in the three space which is used in lattice computations. The boundary conditions $A_0(\vec{x}) \rightarrow A_0^{\infty}$ as $|\vec{x}| \rightarrow \infty$ used in the semiclassical analysis formally excludes the contributions of different Z_3 sectors in the functional integral (14). The matrix element (14) called confining when it includes the summation over different Z_3 sectors, *i.e.* when the symmetry breaking boundary conditions are irrelevant in the thermodynamical limit. The dynamical content of quark confinement is that the vacuum expectation values belong to this class.

IV. DECONFINED-SCREENED PHASE

The numerical evidence for the high temperature deconfined phase is that the symmetry with respect the change of variables (15) in the heavy quark propagator (13) is broken dynamically. Since the actual critical temperature of the lattice regulated gauge theory changes appropriately as the cutoff is removed it is believed that this symmetry breaking also occurs in the renormalized theory. The absence of the summation over the center elements in (7) can be taken into account by using the projection operator $\mathcal{P}_{\rm dec} = \mathcal{P}$. It is advantageous to use charge diagonal basis which gives

$$\mathcal{P}_{\rm dec} = \sum_{k=1}^{N} P_k \mathcal{P} \tag{16}$$

In order to keep the contribution of the identity transformation only in the projection operator P_k one has to sum up the contributions of different charge sectors by equal weight. This is the way how the symmetry breaking will be made manifest in the operator formalism.

We are lacking the precise mathematical construction of the Hilbert space of the Yang-Mills Hamiltonian (1) for infinite volume. Our discussion of the properties of the wave functionals is subject of the usual assumptions: It is assumed that the dynamical breakdown of the symmetry with respect the transformation (15) of the infinite volume system can be approximated by placing the system into a sufficiently large but finite quantization box and by constraining the functional integration into an appropriate smaller space characterized by a fixed value of the order parameter.

The appearance of P_k with $k = 1, 2$ generates fundamental changes in some matrix elements. To see this consider the nonvanishing quark propagator (13) with $\mathcal{P}$ replaced by $\mathcal{P}_{\text{dec}}$ in (16). The action of P_1 on the right creates a states described by a multi-valued wave functional. The change of variable (15) generates the phase factor $e^{i\frac{2\pi}{3}}$ due to this multi-valuedness. This phase factor is incorporated in our simple thought experiment by installing two additional magnetic fields with flux $\Phi_1 = \Phi_2 = -\Phi_3 = -\Phi_4$. The interference becomes constructive by canceling out the phase factor originating from the propagator. The action of P_k with $k = 0, 2$ leaves uncancelled phase factors and the propagator receives no contributions from these sectors. Thus the quark propagator selects the single and multi valued states from the left and the right sides of (13), respectively. Since the antiquark propagator acts similarly in the confined phase we may say that a confining force is generated between the apparently "free" quark and the gluon state described by multi-valued wave functional.

It is easy to construct such a gluon state. Choose a gauge field configuration $\vec{A}(\vec{x})$ and form the linear combination

$$\int dg \psi(g) | g \vec{A}(\vec{x}) g^\dagger \rangle \tag{17}$$

Here the invariant integration over global gauge transformations involves the wave function $\psi(g)$ which is a matrix element of the representation of $SU(3)$ where the center is realized non-trivially. Localized configuration in space describes localized gluon state carrying triality. The Hamiltonian (1) has single valued matrix elements thus states with different triality charge are not mixed. Though the state (17) is not stable in general. In fact, the magnitude of the gauge field may approach zero when the energy is lowered in a variational scheme. But the nonvanishing of the quark propagator requires the contributions of such states and provides the dynamical origin of their stability in the presence of an isolated quark. Quasiparticles with multi-valued wave functional described in the next Section are subject to stabilizing forces independently of the quarks due to their magnetic charges.

V. THE GAUGE-HEDGEHOG

Any nonzero gauge field configuration $\vec{A}(\vec{x})$ may be used in (17). One finds additional effects when the directions in the three space and the color space are correlated. More precisely a gauge configuration is called gauge-hedgehog if global gauge transformations can be compensated by three rotations. Consequently such configurations can be characterized by specifying a map from the space of three rotations $SO(3)$ into the gauge group. These maps fall into homotopy classes and the corresponding topological invariant, the "winding number" provides us with a conserved quantum number. The space $SO(3)$ may "wind around" different $SU(2)$ subspaces of the gauge group. The group $SU(3)$ has eight parameter so at most two $SU(2)$ subgroup may enter for each map. Thus gauge-hedgehogs of QCD carry two integer quantum numbers.

Nontrivial gauge-hedgehog configurations posses magnetic charges. The magnetic flux is defined by projecting the non-Abelian flux into an appropriately chosen color direction $\phi(\vec{x})$

$$\Phi = \text{tr} \int d\vec{x} \phi(\vec{x}) \vec{B}(\vec{x}) \tag{18}$$

where the field $\phi(\vec{x})$ is characterized by the same map and transforms homogeneously under local gauge transformations. The parameter of the projection operator $A_0(\vec{x})$ transforms in the same manner under local gauge transformations. In fact, performing a local gauge transformation on the initial and final state of the matrix element (11) the integrand remains invariant as long as the field $A_0(\vec{x})$ transforms homogeneously.

The construction of simple gauge-hedgehog configurations proceeds by forming covariant quantities in terms of the coordinate vector $\vec{x}$ and the Pauli matrices $\vec{\sigma}$. A typical gauge-hedgehog configuration is the 't Hooft-Polyakov magnetic monopole[7]. Replacing the adjoint Higgs field by $A_0(\vec{x})$ one obtains

$$\begin{aligned} A_0(\vec{x}) &= F(|\vec{x}|)\vec{x}\vec{\sigma} \\ A_i^a(\vec{x}) &= G(|\vec{x}|)\epsilon_{iaj}x^j \end{aligned} \tag{19}$$

The choice of $A_0(\vec{x})$ is obvious. The relation

$$gA_i(\vec{x})g^\dagger = \mathcal{R}(g^\dagger)_{ij}A_j(\mathcal{R}^{-1}(g^\dagger)\vec{x}) \tag{20}$$

which characterizes the gauge-hedgehogs with unit winding can be verified by the help of the identity

$$\epsilon_{ijk} = \epsilon_{\ell mn}\mathcal{R}_{i\ell}\mathcal{R}_{jm}\mathcal{R}_{kn} \tag{21}$$

where the $SO(3)$ matrices $\mathcal{R}(g)$ give the adjoint representation for the group $SU(2)$. The generalization for the gauge group $SU(3)$ is discussed in Ref. [8] with the explicit construction of the monopole configurations.

Magnetic monopoles are stable in the semiclassical approximation for an appropriate Higgs-potential. The strong coupling environment can be studied systematically in lattice regularization. The investigations of magnetic monopoles on the lattice leads to the conserved magnetic current[9]

$$J_m^\mu = \epsilon^{\mu\nu\rho\lambda}\nabla_\nu n_{\rho\lambda} \tag{22}$$

in terms of an integer valued tensor field $n_{\rho\lambda}$ indicating the algebraic stability of the monopoles.

The relevance of the gauge-hedgehogs becomes clear when they are used in constructing multi-valued state in (17). Consider first the case of gauge-hedgehogs with unit winding in $SU(2)$ gauge theory. These states are described by double-valued wave functional. Rotation by 2π in the color space leaves the gauge field invariant so it must correspond to 2π rotation in the three space as well. The wave functional changes sign under rotation by 2π either in the color or in three space. Thus the angular momentum is half-integer. According to the spin-statistics theorem for kinks[10] the phase acquired by the wave function when the particle is rotated by 2π agree with the phase which multiplies the wave function when the particles are exchanged. Thus states with half-integer angular momentum must have odd exchange parity and obey the Pauli exclusion principle. Gauge-hedgehogs with unit winding behave as fermions if their wave functional is multi-valued.

This mechanism generalizes to $SU(3)$ gauge theories. Since the magnetic monopoles of the $SU(3)$ gauge models correspond to some $SU(2)$ subspace of the gauge group (20) remains valid by choosing g from an appropriate $SU(2)$ subgroup. Integrating over this subgroup in (17) we can construct a state transforming according to the fundamental representation of the global subgroup. But this state then belongs to the fundamental representation of the complete gauge group as well and has nonvanishing triality. In such a manner these states obey ordinary Fermi statistics rather than fractional one. The theory contains three versions of each monopole in the deconfined-screened phase. One of these which is the result of the action of P_0 remains boson and neutral. The other two species become dyon and acquire Fermi statistics in agreement with Ref. [10].

VI. QUASI PARTICLES OF THE DECONFINED PHASE

The arguments presented above establish the relevance of some localized states when deconfined matrix elements are considered. These states have multi-valued wave functional and are bound to the "free" quarks by confining forces. In addition they screen the triality of the quark. When these states have the gauge-hedgehog properties then they behave as fermions.

This picture of the deconfined phase suggests the relevance of some multi-hedgehog states which are constructed analogously to the ordinary hadrons. $SU(3)$ gauge theory has two kinds of monopoles which can be used like an isodublet. Due to their vanishing triality charge these states may be observed though all of them are resonances only except those with nonvanishing magnetic charges.

The "deconfined" quark screened by a gauge-hedgehog borrows an unusual spin-statistics form the hedgehog and appears as composite Bose particle similarly to the Cooper-pairs. Quarks screened by ordinary states preserve their fermionic appearance. Thus the high temperature plasma contains bound states with quark quantum numbers in fermionic and in bosonic versions. The fraction of this latter should be finite due to finite dependence of the magnetic monopole density on the temperature in the continuum limit of $SU(3)$ lattice gauge theory[4].

Finally we consider the case of dynamical quarks. After integrating over the Grassmann variables we are left with an effective action containing closed Wilson loops and open Polyakov line type contributions. The latter arises from the integration over nonzero baryon number sectors and describe the propagation of the "background" quarks. We shall restrict ourselves to the case of isolated quarks which assumes vanishing "background" triality[12]. The effective action contains those combination of the Polyakov line type contributions which remain invariant under the change of variable (15). Thus our arguments remain valid in this case.

REFERENCES

1. A. Linde, *Phys. Lett.* **39b** (1980) 327; S. Nadkarni, *Phys. Rev.* **D33** (1986) 3738.
2. L. McLerran and B. Svetitsky, *Phys. Lett.* **98B**, 196 (1980); J. Kuti, J. Polonyi and K. Szlachanyi, *Phys. Lett.* **98B**, 199 (1980); J. Engels, F. Karsch, H. Satz and I. Montvay, *Phys. Lett.* **101B** (1981) 89.
3. S. Manousakis and J. Polonyi, *Phys. Rev. Lett.* **58** (1987) 847.
4. J. Polonyi, *Nucl. Phys.* **A461** (1978), 279c. (1985).
5. J. Polonyi, "Excited States of Gauge Theories", in *Proceedings of the Workshop on Higgs Theories on the Lattice*, SRI (Florida State University, Tallahassee, FL, May 1988).
6. J. Polonyi, "Quantum Mechanics on Multiply Connected Spaces and Quark Confinement", MIT Preprint, CTP# 1507.
7. 't Hooft, *Nucl. Phys.* **B79** (1974) 276; S. Polyakov, *JETP Lett.* **20** (1974) 194.
8. A. Sinha, *Phys. Rev.* **D14** (1976) 2016.
9. T. DeGrand and D. Toussaint, *Phys. Rev.* **D22** (1980) 2478; J. Polonyi, in *Lattice Gauge Theory 86*, H. Satz, I. Harrity and J. Potvin, eds. (Plenum Press, 1986).
10. D. Finkelstein and J. Rubinstein, *J. Math. Phys.* **9**, 1762 (1967); R. Feynman, in *The 1986 Dirac Memorial Lectures* (University of Cambridge, 1987).

11. R. Jackiw and C. Rebbi, *Phys. Rev. Lett.* **36** (1976) 1116, P. Hasenfratz and G. 'tHooft, *Phys. Rev. Lett.* **36** (1976) 1119, A. S. Goldhaber, *Phys. Rev. Lett.* **36** (1976) 1122.

12. A. Roberge and N. Weiss, *Nucl. Phys.* **B275** (1986) 734.

THE MYSTERIOUS e^+e^- PEAKS - A THEORETICAL PERSPECTIVE

A. Chodos
Yale University, Sloane Physics Laboratory, New Haven, Ct. 06511

In a brief note like this, it will not be possible to survey all the many interesting and ingenious theoretical ideas[1] that have been put forward to explain the e^+e^- peaks observed at GSI.[1-7] Rather, I should like to take this opportunity to make a few simple points.

In so doing, I shall invoke the theorist's license of assuming that, at least in their major features, the experimental data are to be believed. It is not my purpose here to attempt a critical analysis of the data.

Among the points I wish to make is this: that the data as they now exist cannot be explained either by conventional nuclear physics or by conventional particle physics. It is a nice illustration of the theme of this volume, since both nuclear and particle ideas can be brought to bear on the problem, and both fail equally to solve it.

By conventional nuclear physics I mean, for example, a process in which the two heavy ions collide, and form a compound nucleus of charge $Z_1 + Z_2$ in an excited state. This decays to the ground state via the emission of an e^+e^- pair, and then the compound nucleus breaks up into its constituents which are detected in coincidence with the pair. One can obviously invent many variations on this theme (for example, it could be one of the fragments that decays, and not the compound nucleus itself - this actually results in further discrepancies with the data), but all such scenarios, it seems to me, face qualitative difficulties in confronting the data. The fact that the e^+ and e^- have essentially equal energies seems to indicate that the object which produces them does not decay in the presence of the large nuclear charge; a fortiori it is not the compound nucleus, or one of its fragments. Also, it is known that the e^+ and e^- have very nearly equal and opposite momenta, leading to the conclusion that there is no recoiling third body present when the decay takes place. Furthermore, any purely nuclear effect would be expected to exhibit a strong dependence on the total charge $Z_1 + Z_2$; no such dependence is seen. Finally, experience with the decays of lower-Z nuclei leads one to expect that the e^+e^- energy distributions should be much broader; a conventional nuclear picture offers no explanation of one of the most striking features of the data, the extreme narrowness of the peaks.

By conventional particle physics I mean that there exists a new neutral elementary particle, ϕ, of mass about 1.8 MeV, which can decay into e^+e^-. One supposes that the ϕ is produced in the complicated environment generated by the heavy-ion collision; for example, either through strong interaction effects[8] or via the strong background electromagnetic fields produced by the ions.[9] Attempts to study this possibility, however, have generally led to the following conclusion: (i) A realistic estimate of the strength of the ϕ's couplings either to photons or to hadronic matter fails to account for the size of the production cross-section needed to explain the peaks; (ii) In order to give rise to the observed narrow peaks in the kinetic energy spectra of the e^+ and e^-, the ϕ's momentum distribution must lead to a non-negligible probability of producing the ϕ at rest in the center of mass of the heavy ions. No model has been able to do this.

A further difficulty is that there is no evidence from other experiments of the existence of such a particle. These other experiments include sensitive QED tests,[10] as well as direct particle searches,[11] either using beam dumps or via nuclear decays. Finally, the evidence for more than one peak (by now it seems that three or more peaks have been firmly established) leads to the conclusion that several new elementary particles must exist in the mass range between e^+e^- threshold and about 1.8 MeV. I regard this as the coup de grace for the elementary particle idea.

What is left? Resisting the temptation to throw up one's hands and to hope that the data goes away (I for one believe that the time is long past when such a hope or fear might have been realistic), one is led fairly naturally to the idea that the peaks may be due to a new species of composite object. The multiple peak structure is obviously accounted for by this hypothesis. The models along these lines have generally fallen into two categories: Those that invoke a new type of constituent for the composite,[12] and those that retain the e^+ and e^- as the constituents, but postulate new dynamics to give rise to the new structure.[13-17]

I have been thinking mainly about the latter possibility. My collaborators and I[14] have found it attractive to suppose that, within QED itself there are as yet unexplored non-perturbative effects that can lead to these e^+e^- states. In particular, we have been looking at the idea that there may be a new phase of QED that is triggered by the strong background fields of the heavy ions. The e^+e^- peaks are then manifestations of the analogues of positronium in the new phase.

I do not have the space here to go into any detail. Let me just remark that, in the first place, there is a well-known phase transition in QED at a critical value α_c of the coupling constant. The new, strong coupling phase is confining and breaks chiral symmetry spontaneously. From lattice studies[18] and from an investigation of the Schwinger-Dyson equation in the ladder approximation[19] one finds $\alpha_c \sim 1$. A major difficulty is understanding how the background fields produced in the heavy ion collisions can mimic strong coupling. The problem can be seen, for example, in Schwinger's formula[20] for the effective Lagrangian for photons in a constant background field, taking into account to all orders the effects of fermion loops in vacuum polarization:

$$\mathscr{L}_{eff} = -\frac{1}{4} F_{\mu\nu}F^{\mu\nu} + \frac{\alpha}{2\pi} \mathscr{L}_1 (\mathscr{F},\bar{\mathscr{G}})$$

where $\mathscr{F} = \frac{1}{2}(E^2-B^2)$; $\mathscr{G} = E\cdot B$, and $\mathscr{L}_1$ is a non-linear function that can be expressed as a proper-time integral. The trouble is that, for large $\mathscr{F}$ and $\mathscr{G}$, $\mathscr{L}_1$ grows only logarithmically, so that one needs field strengths of order $e^{1/\alpha} m^2$ in order to compete with the Maxwell term.

At the moment, I would say that the situation regarding a new phase of QED can be summarized as follows: if one simply assumes its existence, then many of the features of the data can be understood, at least qualitatively. The more challenging question appears to be to justify this assumption within the context of QED. If one insists on identifying the new phase with the known strong-coupling phase, then one faces the problem mentioned above. If one imagines that the phase invoked to explain the heavy ion data is totally new, then one awaits the discovery of exactly what characterizes it, and what peculiar attributes of the background field are needed to trigger it. Either way, it seems that some fresh theoretical input would be very welcome.

Let me close with a few remarks about the Bhabha scattering experiments that are currently underway.[21] I have one phenomenological (or perhaps psychological) comment, and one concerning the input of these experiments on the idea of a new phase.

(a) One should bear in mind that the heavy ion experiments themselves place only very weak bounds on the lifetime τ of the object that decays into e^+e^-, roughly 10^{-19} sec. $< \tau < 10^{-9}$ sec. One can cut this interval in half by invoking (g-2) measurements to restrict τ to greater than about 10^{-14} sec. (although a purely QED effect, such as the new phase I have been discussing, is already included in the g-2 calculations and therefore places no restrictions on τ).[22] Now the Bhabha scattering experiments are generally sensitive to lifetimes fairly close to this limit; they require, optimistically, $\tau < 10^{-13}$ sec. (and maybe $\tau < 10^{-14}$ sec.). Thus there is at worst a considerable window, 10^{-13} sec. $< \tau < 10^{-9}$ sec., in which the Bhabha scattering experiments will see nothing, and yet no inconsistency will arise with the heavy ion data. If, when all the results are in, the Bhabha experiments do not see anything, there will still be a mystery in want of an explanation.

(b) It is sometimes taken for granted that if the Bhabha scattering experiments show a positive result, then the idea of a new phase of QED is ruled out. The point is simple: the new phase requires intense background fields, which are present in the heavy-ion scattering experiments but not in Bhabha scattering.

A moment's reflection, though, leads to a counter-argument of the following kind:[23] We know that the data require that whatever is producing the e^+e^- pair must propagate some distance away from the heavy-ions before it decays. By time reversal, then, we can deduce that it must be possible to create this object in e^+e^- scattering. The picture that emerges is that of a metastable object, which perhaps one can think of as a blob of false vacuum (or, in another language, as a soliton-like extended object), that has an appreciable amplitude to decay into e^+e^- and nothing else.

If one's picture of the heavy-ion scattering experiments is that a new phase of QED is formed in a finite region of space, which then decays into e^+e^-, then I believe it must be possible to reverse this process and produce the blob of new phase in an e^+e^- collision. On the other hand, if the phase survives only in the presence of background fields (which seems to contradict the data) or if it decays into e^+e^- plus a lot of other stuff, presumably soft photons (ditto) then the time reversal argument will not work and a positive signal in Bhabha scattering will be inconsistent with the new phase idea. Both theoretically and experimentally, it will be interesting to see what develops.

REFERENCES

1. For an extensive list of references, see for example T. Cowan and J. Greenberg in "Physics of Strong Fields", W. Greiner, ed. (Plenum, 1987), p. 111.
2. J. Schweppe et al., Phys. Rev. Lett. 51, 2261 (1983).
3. M. Clemente et al., Phys. Lett. 137B, 41 (1984).
4. H. Tsertos et al., Phys. Lett. 162B, 273 (1985).
5. T. Cowan et al., Phys. Rev. Lett. 54, 1761 (1985).
6. T. Cowan et al., Phys. Rev. Lett. 56, 444 (1986).
7. See also K. Danzmann et al., Phys. Rev. Lett. 59, 1885 (1987).
8. A. B. Balantekin, C. Bottcher, M.R. Strayer, and S.J. Lee, Phys. Rev. Lett. 55, 461 (1985); N. Mukhopadhyay and A. Zehnder, Phys. Rev. Lett. 56, 206 (1986).
9. A. Chodos and L.C.R. Wijewardhana, Phys. Rev. Lett. 56, 302 (1986); D. Carrier, A. Chodos, and L.C.R. Wijewardhana, Phys. Rev. D34, 1332 (1986); K. Lane, Phys. Rev. Lett. 169B, 97 (1986); B. Müller and J. Reinhardt, Phys. Rev. Lett. 56, 2108 (1986); B. Müller and J. Rafelski, Phys. Rev. D34, 2896 (1986).
10. J. Reinhardt et al., Phys. Rev. C33, 194 (1988); A. Zee, Phys. Lett. B172, 377 (1986).
11. E.M. Riordan et al., Phys. Rev. Lett. 59, 755 (1987); F.P. Calaprice et al., Phys. Rev. Lett. 56, 302 (1986); M.J. Savage et al., Phys. Rev. Lett. 57, 178 (1986); for a more extensive list, see Refs. 102-117 in Ref. 1 above.
12. G.L. Shaw, Physics Letters B199, 560 (1987); S. Schramm, B. Müller, J. Reinhardt and W. Greiner, Frankfurt preprint (1987).
13. L.S. Celenza, V.K. Mishra, C.M. Shakin, and K.F. Liu, Phys. Rev. Lett. 57, 55 (1986); L.S. Celenza, C.R. Ji and C.M. Shakin, Phys. Rev. D36, 2144 (1987).
14. D.G. Caldi and A. Chodos, Phys. Rev. D36, 2876 (1987); D.G. Caldi, A. Chodos, K. Everding, D. Owen and S. Vafaeisefat, manuscript in preparation.
15. Y.J. Ng and Y. Kikuchi, Phys. Rev. D36, 2880 (1987).
16. R.D. Peccei, J. Solà, and C. Wetterich, Phys. Rev. D37, 2492 (1988).
17. C.-Y. Wong and R.L. Becker, Phys. Lett. B182, 251 (1986); H.W. Crater, C.-Y. Wong, R.L. Becker and P. Van Alstine, Oak Ridge Preprint (1987); A.O. Barut in "Physics of Strong Fields", W. Greiner, ed. (Plenum, 1987).

18. A. Guth, Phys. Rev. D21, 2291 (1980); J. Fröhlich and T. Spencer, Commun. Math. Phys. 83, 411 (1982); M. Creutz, L. Jacobs, and C. Rebbi, Phys. Rev. D20, 1915 (1979); B. Lautrup and M. Nauenberg, Phys. Lett. 95B, 63 (1980): G. Bhanot, Phys. Rev. D24, 461 (1981); T.A. DeGrand and D. Toussaint, Phys. Rev. D24, 466 (1981); D.G. Caldi, Nucl. Phys. B220, [FS8], 48 (1983).
19. T. Maskawa and H. Nakajima, Progr. Theor. Phys. 52, 1326 (1974), ibid. 54, 860 (1975); R. Fukuda and T. Kugo, Nucl. Phys. B117, 250 (1976).
20. J. Schwinger, Phys. Rev. 82, 664 (1951).
21. K. Maier et al., "Evidence for a Resonance in Positron-Electron Scattering at 810 keV Centre-of-Mass Energy", Stuttgart preprint (1988), (submitted to Z. Physik A), and numerous other works in progress. See also the contribution by Koenig to this volume.
22. I thank R. Robinett for this remark.
23. I am grateful to Jack Sandweiss for making this argument to me in forceful terms.

Pseudorapidity Distributions and Correlations in Central ^{16}O Interactions at 200 A GeV

Presented by R. J. Wilkes, University of Washington

The EMU-01 Collaboration :

M. I. Adamovich[12], Y. A. Alexandrov[12], S. A. Asimov[13], S. K. Badyal[3], K. B. Bhalla[2], A. Bharti[2], A. Bhasin[3], T. H. Burnett[11], X. Cai[7], L. P. Chernova[13], M. M. Chernyavsky[12], B. Dressel[9], E. M. Friedlander[10], E. R. Ganssauge[9], S.Garpman[8], S. G. Gerassimov[12], A. Gill[2], J. Grote[11], K. G. Gulamov[13], V. K. Gupta[3], S. Hackel[9], H. H. Heckman[10], B. Jakobsson[8], B. Judek[4], H. Kallies[9], Y. J. Karant[10], S. P. Kharlamov[12], S. Kitroo[3], J. Kohli[3], G. L. Koul[3], V. Kumar[2], V. G. Larivnova[12], P. J. Lindstrom[10], L. S. Liu[7], S. Lokanathan[2], J. J. Lord[11], E. Malinina[12], L. K. Mangotra[3], N. Maslenikova[12], E. Monnand[1], S. Mookerjee[2], C. Muller[9], V. S. Navotny[13], G. I. Orlova[12], I. Otterlund[8], N. G. Peresadko[12], S. Persson[8], R. Raniwala[2], S. Raniwala[2], N. K. Rao[3], J. T. Rhee[9], N. G. Salmanova[12], W. Schulz[9], F. Schussler[1], V. S. Shukla[2], D. Skelding[11], K. Soderstrom[8], E. Stenlund[8], R. Storey[4], S. Strausz[11], J. F. Sun[5], M. I. Tretyakova[12], Z. Q. Weng[5], R. J. Wilkes[11], G. F. Xu[6], P. Y. Zheng[6]

1) Univerity of Grenoble, 2) University of Jaipur, 3) University of Jammu, 4) NRC, Canada, 5) Shanxi Normal University, 6) IHEP, Beijing, 7) Hua-Zhong Normal University, 8) University of Lund, 9) Phillips University, Marburg 10) LBL, 11) University of Washington, 12) Lebedev Institute of Physics, 13) Physical-Technical Institute, Tashkent.

1. Introduction

Whether or not adequate energy densities for a transition to the Quark-Gluon Plasma (QGP) phase are possible in ^{16}O and ^{32}S induced reactions is still an open question [1]. Before unambiguous signals can be identified it is necessary to understand the background upon which the expected QGP signatures can be sought. An important aim of early experiments in this new energy density regime is thus to understand the process of hadronization in a nuclear environment. In any discussion of high energy nucleus-nucleus collisions formation time is an important feature. If, for example, the final state particles are produced instantaneously in a bound hadron-nucleon collision, all these particles will have a chance to reinteract inside the nuclear medium. If, on the other hand, the formation time is long, i.e. longer than the nuclear diameter, the energy flux will stay together as a unit throughout its passage through the target nucleus. While these extreme cases are clearly unrealistic, one expects interaction parameters such as the multiplicity and pseudorapidity distributions to be sensitive functions of the formation time.

Here we present results on charged hadron production in an experiment using tracking detectors with the highest possible spatial resolution: emulsion chambers. A sample of the most central collisions will be compared to the predictions of the Lund Model [2], which does an accurate job of representing conventional (non-QGP) physics in nuclear collisions. We can use the high resolution of our detectors to examine spatial correlations in several ways. In particular, the radial size of the particle formation region can be estimated using pion interferometry techniques [3].

2. Experimental Techniques

In this experiment, emulsion chambers (arrays of thin nuclear emulsion plates exposed perpendicular to the beam direction) were exposed to ^{16}O at 60 and 200 A GeV, and ^{32}S at 200 A GeV at the CERN SPS. Similar chambers were exposed to 14.6 A GeV oxygen at the Brookhaven AGS. For this report we concentrate on the 200 A GeV oxygen data. The structure of the chambers and the data reduction techniques have been reported in detail elsewhere [4]. Pseudorapidity ($\eta = -ln(tan\theta/2)$) is determined for all singly charged particles with $v \geq 0.7c$. As determined from repeated measurements, the error in η is $\sim$0.01 rapidity units for $1 \leq \eta \leq 7$, and is still only $\sim$0.2 at η=10.

The experimental data can be compared to the predictions of the Lund Model for high-energy nucleus-nucleus interactions [2], as embodied in a Monte Carlo program called FRITIOF [5]. A basic feature of the model is long formation time, where all cascading is neglected. A significant fraction of the incident energy is carried from one collision to the next, and all collisions in a long chain are important. Thus, in the most central collisions where long chains often develop, the results will differ substantially, especially in the target region, from the particle density predicted by models with a short formation time, such as HIJET [6]. We have shown elsewhere [7] that FRITIOF provides a significantly better approximation to the data than HIJET.

3. Psuedorapidity distributions

We measured pseudorapidity (η) distributions of singly charged particles with $v \geq 0.7c$ in events occuring in the emulsion target layers of the chambers. The number of such particles, n_s, serves as a measure of the centrality. In particular, high multiplicity interactions with no spectator fragments can be assumed to be central collisions. Due to the experimental procedure only particles emitted at angles less than $\sim$30 degrees ($\eta \geq 1.3$) with respect to the beam axis are measured.

Data presented here come from a minimum-bias sample of 352 events. A subsample of 84 high multiplicity (n_s >150) events is also defined. No scanning losses are expected for events with such high multiplicity; n_s in each event has, however, an estimated loss of about 5%. Events with $n_s \geq 150$

were found to have $< n_s > = 196 \pm 1$ for the Lund model and 180 ± 4 for the experimental data. The model indicates that only interactions with the heaviest target components in nuclear emulsion, i.e. Ag and Br, can produce such high multiplicities.

In Fig 1a we show the average pseudo-rapidity distributions $\rho = \frac{1}{N}(dn/d\eta)$ for events having $n_s > 200$, in Fig 1b for events having $150 < n_s < 200$, and in Fig 1c for events with $10 < n_s < 50$. In each case, the histograms correspond to the experimental data and the curves to the monte carlo events. The experimental data show rather smooth distributions with some fluctuations presumably due to the limited statistics. The FRITIOF sample slightly overestimates the densities in the central region, but the shape of the distribution is in good agreement with the data, with its peak position close to the experimental one.

In Fig. 1c the low-multiplicity events represent non-central interactions. Here it is expected that projectile fragments which do not interact with the target nucleus will continue along the beam axis and give rise to spectator particles at large values of pseudorapidity. With transverse momentum limited by their Fermi motion prior to breakup, spectator protons are expected to appear around $\eta = ln(p_{beam}) + 3 \simeq 8.3$ [8]. Fig. 1c shows the expected distributions for spectators to be in good agreement with the experimental data. These features are almost totally absent in the high multiplicity sample, indicating that very high multiplicity alone is a good criterion for centrality.

Plots of pseudorapidity density ρ versus multiplicity n_s for four different rapidity ranges are shown in Fig 2, with curves from the Lund model. As expected, the correlation between $\rho(\eta)$ and n_s is strongest in the central region. An estimate of the range of energy densities, ϵ, reached in ^{16}O + Em interactions is also shown on the right-hand scale in Fig 2. We use the "naive" Bjorken formula [9] $\epsilon = \frac{3}{2} < m_t > \rho(\eta)/\pi R^2 \tau_0$ (GeV/fm^3), with the conventional assumptions $< p_t >$=0.35 GeV/c, τ_0=1 fm/c, and R=3 fm. One should note that some authorities suggest this procedure overestimates ϵ by as much as a factor of 4 [1]. In any case, only the highest multiplicity events begin to approach energy densities predicted for the QGP phase transition, so the absence of anomalies is not surprising. The general trend as well as the fluctuations around the average densities (Fig 2) are consistent with predictions from the Lund model. The sulfur data, currently under analysis, will probe a region of higher ρ.

4. Correlations

The radial size of the hadron emission region can be estimated using the Bose-Einstein correlation between produced pairs of tracks. Independent sources uniformly distributed over the surface of sphere of radius R, with lifetime τ, produce a ratio of the number of pairs of like particles N_l to the number of pairs of unlike particles N_u, given approximately by [10] $N_l/N_u = 1 + \lambda\ exp(-q_t^2\ R^2/4)\ /\ [1 + (q_0\ \tau)^2]$ where q_t is the component of momentum difference $\mathbf{p_1} - \mathbf{p_2}$ transverse to $\mathbf{p_1} + \mathbf{p_2}$, q_0 is the component of momentum difference $\mathbf{p_1} - \mathbf{p_2}$ parallel to $\mathbf{p_1} + \mathbf{p_2}$, and the Gaussian is an approximation to the first order Bessel function [11]. The pure Bose Einstein correlation is distorted by the electromagnetic and strong interactions of hadrons which accounts for the incoherence parameter λ.

The emulsion chamber technique permits the measurement of large numbers of closely spaced tracks which would be difficult for a counter experiment to resolve. The lifetime τ cannot be determined without a measurement of the momenta, but the radius R can be determined by angular measurements alone, given the average transverse momentum. The ratio of observed density of particle pairs with given angular separation to that of uncorrelated particle pairs is $N_{pairs}/N_{pairs}^{random} = (N_l + N_u)/N_{pairs}^{random} = 1 + \alpha\ exp(-q_t^2\ R^2\ /4)$. The parameter α includes the lifetime τ and incoherence λ variables.

The background distribution for pair separation is obtained by randomizing the azimuthal angles while retaining the pseudorapidity values $(\eta_n, \phi_n) \rightarrow (\eta_n, \phi_n^{random})$ in the experimental data. In this way the background displays the same rapidity distribution as the data, but with uncorrelated space angles. To increase the statistical significance of the two particle separation distribution, the background calculation is repeated 10 times for the smaller data samples. The momentum difference q_t is approximated by assuming a conventional distribution of transverse momentum. For each pair we take $q_t = < |p_1 - p_2| > = < p_t >\ \Delta\omega$ where $\Delta\omega$ is the opening angle of the pair. An average transverse momentum of 0.4 GeV/c is used for all data samples. The results for the high multiplicity sample of oxygen-emulsion interactions are shown in Fig 3a. To examine the effect of correlations between the decay products of short-lived particles on this analysis, the same distribution was plotted for interactions generated by FRITIOF, which produces the hadronic decay products but not an overall Bose Einstein correlation. Figure 3b shows that hadronic decays do not have a significant effect on the procedure described.

Results of this analysis indicate a transverse formation zone radius of $3.1 \pm 0.6\ fm$ for the oxygen-emulsion high multiplicity sample (central collisions) and $3.0 \pm 0.7\ fm$ for the low multiplicity sample (peripheral collisions), suggesting that no new processes are being excited in the central collisions. These results are consistent with previous experiments which indicate that the radius of the hadronic emission region is approximately equal to that of the smaller of the two interacting nuclei. For example, the NA35 experiment [12] finds a radius of 3.6 $\pm$0.3 for central O+Au interactions.

This work supported in part by US Dept. of Energy contract DEAT06-88ER40423.

References

1. Contributions to this conference by L. McLerran and W. Zajc.
2. B. Andersson, *et al*, Nucl. Phys. **A447** 165c (1986).
3. R. Hanbury-Brown and R. Q. Twiss, Nature **120** 1046 (1960); G. Goldhaber,

et al, Phys. Rev. **120** 300 (1960).
4. M. Adamovich, *et al,* Phys. Lett. **B201** 397 (1988).
5. B. Nilsson-Almqvist and E. Stenlund, *Computer Phys Com* **43** 373 (1987).
6. T. W. Ludlam, BNL Report 51921 (1985).
7. M. Adamovich, *et al,* U. Washington preprint VTL-PUB-118 (1987).
8. I. Otterlund, *Nucl Phys* **A461** 113c (1987).
9. J. D. Bjorken, *Phys Rev* **27D** 140 (1983).
10. G. I. Kopylov & M. I. Podgoretskii, Sov. J. Nucl. Phys. **18**, 336 (1974).
11. A. Giovannini *et al.,* Nuovo Cim. **2** 1, (1979).
12. A. Bamberger, *et al,* Phys. Lett. **B203** 320 (1988).

Figure Captions

Fig 1: Pseudorapidity distribution of minimum-ionizing particles (a) for events with $n_s \geq 200$; (b) for events with $150 \leq n_s \leq 200$; (c) for peripheral interactions (events with $n_s < 50$ and projectile fragments).

Fig 2: Pseudorapidity density versus n_s for four different η bins. Energy density estimated by the Bjorken formula is shown on the right-hand scale.

Fig 3: Ratio of particle pair densities for data and an uncorrelated background sample (see text) for a) experimental data (central collisions) with curve showing fit to R = 3.1 fm; b) monte carlo events (no correlations).

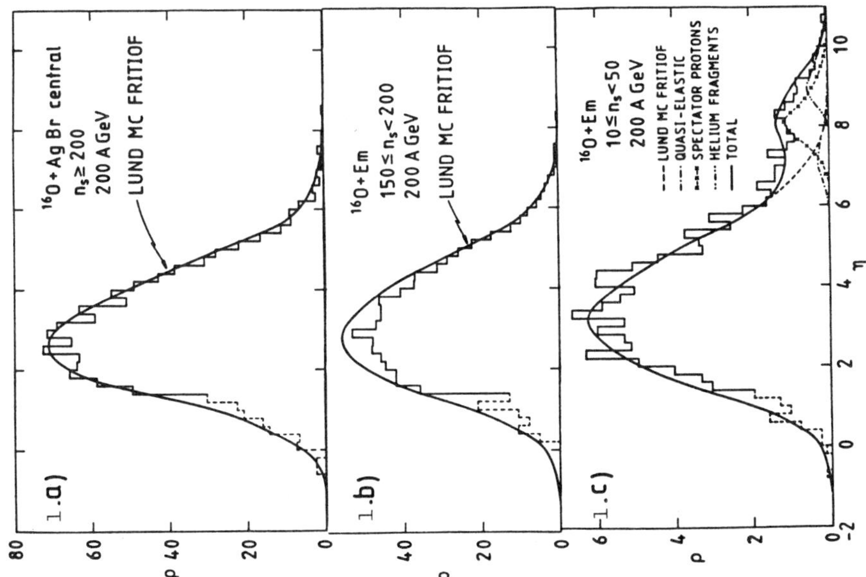

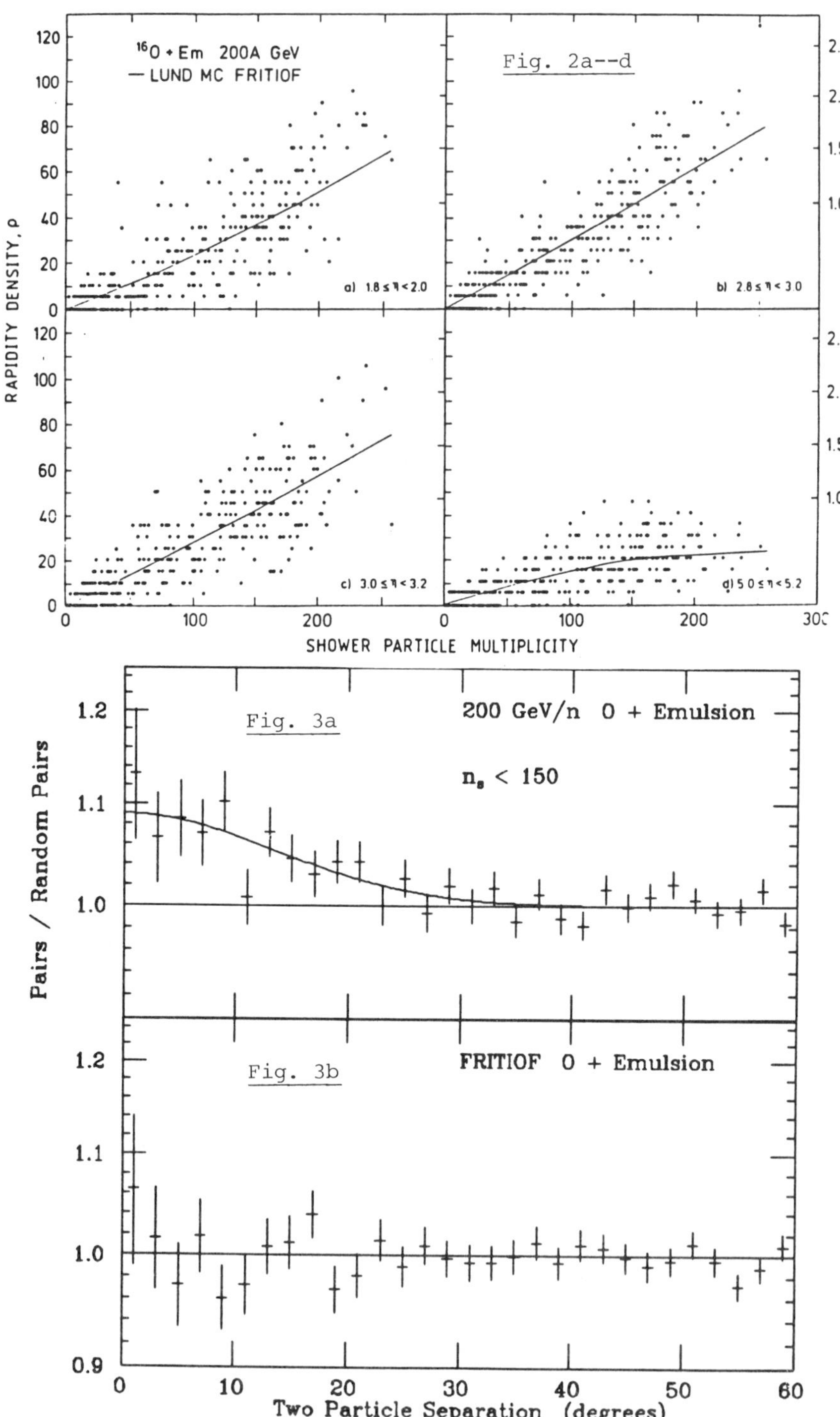

^{16}O + Em 200A GeV
— LUND MC FRITIOF
Fig. 2a--d
RAPIDITY DENSITY, ρ
ε GeV/fm^3
a) 1.8 ≤ η < 2.0
b) 2.8 ≤ η < 3.0
c) 3.0 ≤ η < 3.2
d) 5.0 ≤ η < 5.2
SHOWER PARTICLE MULTIPLICITY
Fig. 3a
200 GeV/n O + Emulsion
n$_s$ < 150
Pairs / Random Pairs
Fig. 3b
FRITIOF O + Emulsion
Two Particle Separation (degrees)

STRING MODELS FOR ULTRARELATIVISTIC HEAVY ION COLLISIONS: LUND MODEL VERSUS COLOR EXCHANGE MODELS

K. WERNER
Brookhaven National Laboratory, Upton, NY 11973

ABSTRACT

We discuss similarities and differences between several string models for ultrarelativistic nucleus nucleus scattering, being proposed to account for "normal" collisions (with no quark gluon plasma involved).

INTRODUCTION

Although the aim of ultrarelativistic heavy ion collisions is to find new phenomena like matter of free quarks and gluons (quark gluon plasma (QGP)), the very first task should be to try to understand nucleus nucleus (AA) collisions in terms of "conventional" physics. This has certainly no unique meaning, but throughout this paper we are referring to "conventional" descriptions as models, being extrapolations of nucleon nucleon scattering models.

The simplest realization of such an extrapolation is a superposition of nucleon nucleon (NN) collisions: whenever an incident nucleon comes closer than $\sqrt{\sigma_{NN}/\pi}$ to a target nucleon, a collision is performed and the produced leading baryon plays the role of the projectile for the next collision. HIJET[1] is constructed in this way. It has to be criticized, that timescales are not taken into account properly: the hadronization time $\tau_h = m^{-1} \cosh y$ is for an ultrarelatvistic collision (which is defined as $\gamma = \cosh y \gg 1$) much larger than the mean free path λ in nuclear matter ($\lambda \approx 2$). The time scale problem can be cured by considering an individual NN collision only as a mechanism to form color strings, the fragmentation (particle production) of the strings occurring at an appropriate time after the collision.

In the following we are going to investigate AA extrapolations of string models for NN scattering. String fragmentation should be consistent with e^+e^- annihilation and deep inelastic lepton scattering, which reduces the number of free parameters considerably. We further expect the models to be available in Monte Carlo formulation (i.e., event generator) which allows sensitive tests by comparing with all available pp, pA and AA data. Models matching these requirements are the "color exchange models" IRIS (J. P. Pansart[2]), the multichain fragmentation model (MCFM, J. Ranft[3]), and VENUS (K. Werner[4]) as well as the Lund model FRITJOF (B. Anderson et al[5]) and the FRITJOF modification ATTILA (M. Gyulassi[6]). The "color exchange models" (IRIS, MCFM, VENUS) are motivated by Regge theory and based on the work of Capella et al[7] in the sense that a nucleon nucleon interaction is realized by color exchange between the two nucleons. In the Lund model, a nucleon nucleon collision results in two excited objects which are assumed to be color strings. All the models show a three-step structure:

1. Nuclear geometry: Nuclear geometry and the mean free path λ of a nucleon in nuclear matter determine individual NN collisions.
2. String formation: Each NN collision results in the formation of color strings.
3. String fragmentation: After all strings are formed they fragment independently of each other into observable hadrons.

In the following we are going to investigate similarities and differences between the models separately for these three aspects. This is meant to be a theoretical investigation rather than a competition to see who is doing best compared with data!

NUCLEAR GEOMETRY

All the models treat the geometrical aspect in the same way, as described briefly in the following. Since particle production in pp collisions is known to be very forward peaked ($<p_t> \lesssim 0.4$ GeV for 200 GeV incident energy), nucleons are assumed to move on straight lines parallel to the beam through the other nucleus. A constant nucleon nucleon cross section determines collisions, although after a collision the nucleon may be far from being a ground state nucleon. Yet, the fact that the nucleon may be colored and carries a string "behind" should not matter for a nucleon moving with practically the velocity of light. A simple way to actually perform the calculations goes as follows: distribute the nucleons in the nuclei according a standard Wood Saxon density distribution. Project the nucleon coordinates (including a shift due to the impact parameter) on a plane perpendicular to the beam axis. Whenever two nucleons are closer than $\sqrt{\sigma_{NN}/\pi}$ a collision occurs.

STRING FORMATION

In this chapter we are going to discuss how individual nucleon nucleon collisions are realized, first for VENUS, then for MCFM and IRIS compared to VENUS and finally for Lund (FRITJOF, ATTILA) compared to the three color exchange models VENUS, MCFM and IRIS.

Let us first discuss the interaction of two nucleons — not necessarily "fresh" nucleons — in VENUS[4]. The whole contribution is assumed to be a superposition of contributions σ_i with i color exchanges: $\sigma = \sum_i w_i \sigma_i$ with w_i being the probability for such a contribution. We first describe the one color exchange contribution σ_1. As shown in fig. 1 four graphs (labeled a,b,c,d) contribute to σ_1. In the dominant contribution (a), color exchange (arrow) between a projectile and a target nucleon rearranges the color structure such that instead of two nucleons in singlet states we find two singlets each consisting of a diquark and a quark of the other nucleon. In contributions (b) and (c) one of the

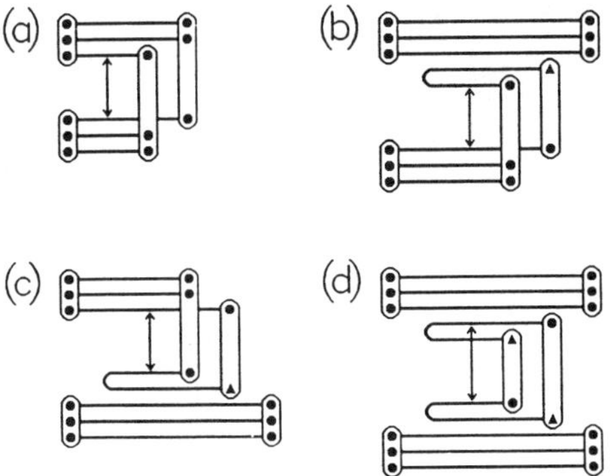

Fig. 1
One–color–exchange diagrams

quarks participating in the color exchange is part of a color neutral $q\bar{q}$ pair so that the diquark-quark ($qq-q$) string from (a) is replaced by a quark-antiquark ($q-\bar{q}$) string and a baryon. In contribution (d) both quarks are part of white $q\bar{q}$ pairs, resulting in two $q-\bar{q}$ strings and two baryons. Higher order contributions are obtained by applying contributions (a), (b), (c), (d) several times (→ 16 contributions for two–color exchange etc.). For reasons discussed in Ref. 4 contributions (a)-(d) are referred to as (a) nondiffractive, (b) diffractive target excitation, (c) diffractive projectile excitation, and (d) double Pomeron exchange.

IRIS and MCFM are closer to the Capella's Dual Parton Model[7] and differ from VENUS in the following way: the first collision in a pA collision is always of type (a) in fig. 1 (no diffractive scattering!), subsequent collisions are of type (b), higher order contributions are achieved by just adding contribution (d) several times. From the point of view of a projectile nucleon colliding with several target nucleons, this means that in the first collision the nucleon breaks up into a fast diquark and a slow (held back) quark, and in subsequent collisions the diquark survives. In VENUS always (a),(b),(c),(d) are possible. As a consequence VENUS produces more baryons in the central region than IRIS and MCFM. On the other hand the number of $qq-q$ strings are the same in VENUS, IRIS and MCFM ($qq-q$ strings are more energetic and thus more important for multiparticle production than $q-\bar{q}$ strings).

In all color exchange models (IRIS, MCFM, VENUS) momentum fractions of particpating quarks are calculated according to either measured (VENUS) or Regge motivated (IRIS, MCFM) quark structure functions. In the ultra-relativistic limit rapidity and mass of a $qq-q$ string from fig. 1a is given as

$$
\begin{aligned}
y &= \frac{1}{2}\log\frac{1-x_1}{x_2} \\
m^2 &= s x_2 (1-x_1)
\end{aligned}
\tag{1}
$$

with x_2 being the momentum fraction of the quark and $1-x_1$ being the diquark momentum fraction.

In the Lund model (FRITJOF, ATTILA) a string is characterized by its light cone momenta $(p^+, p^-) = (p_0 + p_{\parallel}, p_0 - p_{\parallel})$ and its transverse momentum $\vec{p}_t$. It is not specified how the string momenta are shared among the constituents of the string (like in the color exchange models). A collision between two strings with momenta (p_1^+, p_1^-) and (p_2^+, p_2^-) is defined as

$$\left.\begin{array}{l}(p_1^+ \,,\, p_1^-)\\ (p_2^+ \,,\, p_2^-)\end{array}\right\} \rightarrow \left\{\begin{array}{l}\left(\left(1-x^+\right)p^+ \,,\, x^-p^-\right)\\ \left(x^+p^+ \,,\, \left(1-x^-\right)p^-\right)\end{array}\right. \tag{2}$$

with $p^{\pm} = p_1^{\pm} + p_2^{\pm}$ and with $0 \leq x^{\pm} \leq 1$ generated according a phenomenological distribution $g(x)$. In the ultrarelativistic limit and for collisions of two "fresh" nucleons, Eq. (2) can be written as (in the NN cm system):

$$\left.\begin{array}{l}(2P \,,\, 0)\\ (0 \,,\, 2P)\end{array}\right\} \rightarrow \left\{\begin{array}{l}\left(\left(1-x^+\right)2P \,,\, x^-2P\right)\\ \left(x^+2P \,,\, \left(1-x^-\right)2P\right)\end{array}\right. \tag{3}$$

which is identical to a picture where the two strings consist of two components each with momenta $\left(1-x^+\right)P$ and $x^-(-P)$ and correspondingly with $\left(1-x^-\right)(-P)$ and x^+P. This is completely equivalent to the dominant one-color-exchange contribution (fig. 1a) in the color exchange models (MCFM, IRIS, VENUS) where the two partons defining a string are a diquark and a quark. This equivalence can also be seen by evaluating (in the ultrarelativistic limit) expressions for string rapidity and mass for the forward moving string from Eq. (2):

$$\begin{aligned} y &= \frac{1}{2}\log\frac{1-x^+}{x^-} + y_{\text{cm}} \\ m &= s\,x^-\left(1-x^+\right) \,. \end{aligned} \tag{4}$$

This is exactly what we obtained for the color exchange models (see Eq. (1))!

Although, as discussed, for NN scattering the Lund model (in its simplest version — without gluon kinks) is practically identical to the dominant one–color exchange diagram in the color exchange models (MCFM, IRIS, VENUS), the extrapolation towards pA and AA collisions is different. As shown in Ref. 6, in the Lund model the mass of a projectile increases considerably with the number of collisions, whereas in the color exchange models due to the slowing down of the diquark the mass decreases slightly. As also discussed in Ref. 6, this increase of string mass in the Lund model is needed to reproduce data, whereas the color exchange models reproduce AA data equally well without such an effect.

STRING FRAGMENTATION

In all models string fragmentation is done consistently with e^+e^- annihilation and deep inelastic lepton nucleon scattering data. Therefore, one could consider the fragmentation procedures being used as black boxes, and it shouldn't be terribly important what happens inside the box. Nevertheless, we are going to discuss the fragmentation prescriptions applied in the models. There are two string fragmentation procedures being used: MCFM and VENUS use the Field Feynman fragmentation model[8], being the first widely used recursive method for string fragmentation, whereas IRIS and the Lund models FRITJOF and ATTILA use the Lund fragmentation model[9].

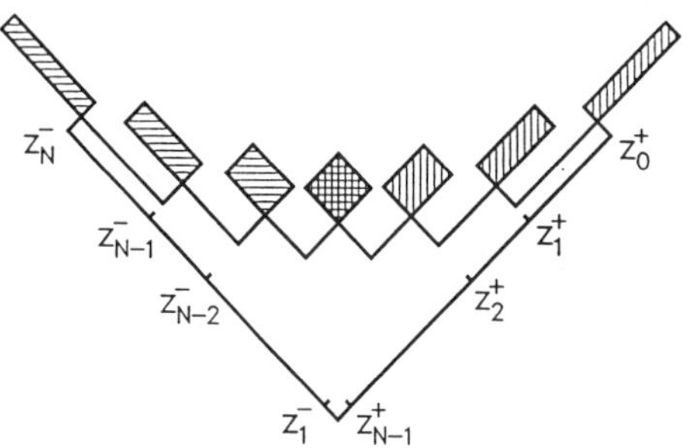

Fig. 2.
Space–time evolution of string fragmentation

Although the Field Feynman model is usually described in momentum space[8,4], we are going to use a space–time picture (see fig. 2) in order to discuss similarities with the Lund fragmentation model (we use the language of the Lund model, introduced in Ref. 9, to describe the Field Feynman model). The link between momentum and space–time picture is the linear relation between the string energy E and the elongation z

$$E = \kappa\, z \tag{5}$$

with κ being the string tension. We introduce light cone variables $z^+ = t+z$ and $z^- = t-z$. The two massless partons, which define the string, are moving along the light cone axes, initially with momenta p_0^+ and p_0^-. The forward moving parton has to stop at $z_0^+ = p_0^+/\kappa$ because all the momentum went into string energy. Attracted from the color field, the parton moves back (parallel to the z^--axis, forming a particle with one of the partons created at $(z_1^+,\ z_1^-)$. Due to energy conservation the transverse mass μ_1 of the particle is proportional to the rightmost shaded area in fig. 2, defined by the two parton lines:

$$\mu_1^2 = p_1^+ p_1^- = \kappa^2 \Delta z_1^+ \Delta z_1^- \equiv \kappa^2 \left(z_0^+ - z_1^+\right)\left(z_1^- - z_0^-\right)$$

The second parton created at $(z_1^+,\ z_1^-)$ and moving parallel to the z^--axis assumes the role of the original parton, yet with reduced initial momentum

$p_0^+ - p_1^+ = \left(z_0^+ - \Delta z_1^+\right)/\kappa = z_1^+/\kappa$, and combines correspondingly with a parton created at $\left(z_2^+, z_2^-\right)$, etc. In this way we obtain a recursive outside–inside fragmentation cascade for the forward parton which stops (per def) in the middle (we require $z > 0$ or $z^+ > z^-$). In the same way the backward moving parton fragments iteratively from the "outside" towards the inside. Finally the two remainder partons in the middle are combined to form a hadron.

After having discussed the Field Feynman model by using "Lund language" it is trivial to explain the basic version of the Lund model and how it differs from the Field Feynman model. The Lund fragmentation starts in the same way as Field Feynman, namely fragmenting the forward–moving parton from "outside" towards "inside". Yet, the fragmentation is not terminated at $z = 0$ $\left(z^+ = z^-\right)$; the cascade continues over the central region until the other end of the string is reached (the backward–moving parton). Since the Lund fragmentation is not right–left symmetric by definition (as Field Feynman), a specific form for the distribution function $f(x)$ — which determines $x_i = p_i^+ / \left(p_0^+ - \sum_{j=1}^{i-1} p_j^+\right)$ — has to be chosen to achieve symmetry. Higher order effects like gluon kinks, which might be important at very high energies, are not incorporated in the Field Feynman model.

CONCLUSION

All the models we investigated — the color exchange models MCFM, IRIS, and VENUS, as well as the Lund model (FRITJOF and ATTILA) — are constructed according to a three–step structure: (i) geometry; (ii) string formation; (iii) string fragmentation. The geometrical part is essentially the same in all models. String formation in VENUS is different compared to MCFM and IRIS in the sense that the selection of diagrams is different. The calculation of the diagrams is done according to the same principle. Whereas all the color exchange models (MCFM, IRIS, VENUS) calculate string properties from the two string constituents (diquark and quark, for example), the Lund model just gives a phenomenological excitation law for the whole string. This is done such that for a forward–moving string (or nucleon) a collision results in a reduction of the "forward" light cone variable and a gain in the "backward" one. For *NN* collisions this Lund prescription is identical to the dominant diagram (fig. 1a) in the color exchange models! Yet, the extrapolation towards *AA* collisions is different. The two fragmentation procedures, Field Feynman model (used in MCFM, VENUS) and Lund fragmentation model (used in IRIS, FRITJOF, ATTILA) are essentially identical.

ACKNOWLEDGEMENTS

This work has been supported by the U. S. Department of Energy under contract no. DE-AC02-76CH00016.

Die Arbeit wurde mit Unterstützung eines Stipendiums des Wissenschaftsausschusses der NATO über den DAAD ermöglicht.

REFERENCES

1. T. W. Ludlam, BNL Report 51921, 373 (1985).
2. T. P. Pansart, Nucl. Phys. A461, 521c (1987).
3. J. Ranft, Phys. Rev. D37, 1842 (1988).
4. K. Werner, preprint BNL 40981, to appear in Phys. Lett. B..
5. B. Anderson, G. Gustafson, B. Nielsson-Almquist, Nucl. Phys. B281, 289 (1987).
6. M. Gyulassy preprint CERN-Th 4794/87.
7. A. Capella, T. A. Casado, C. Pajares, A. V. Ramello and J. Tran Thanh Van, Z. Phys. C33, 541 (1987).
8. R. D. Field and R. P. Feynman, Nucl. Phys. B136, 1 (1978).
9. B. Anderson, G. Gustafson, G. Ingelman and T. Sjöstrand, Phys. Rep. 97, 31 (1983).

SPECTROMETER RESULTS FROM BNL E802

E-802 Collaboration, Brookhaven National Laboratory, Upton, NY 11973 USA
ANL-BNL-Columbia-Hiroshima-INS-Kyushu-MIT-UCBerkeley-UCRiverside-Tokyo

T. ABBOTT[j], Y. AKIBA[e], D. ALBURGER[b], D. BEAVIS[b], R.R. BETTS[a], L. BIRSTEIN[b], M.A. BLOOMER[h], P.D. BOND[b], C. CHASMAN[b], Y.Y. CHU[b], B.A. COLE[h], J.B. COSTALES[h], H.J. CRAWFORD[i], J.B. CUMMING[b], R. DEBBE[b], E. DUEK[b], J. ENGELAGE[g], S.Y. FUNG[j], L. GRODZINS[h], S. GUSHUE[b], H. HAMAGAKI[e], O. HANSEN[b], R.S. HAYANO[k], S. HAYASHI[e,*], S. HOMMA[e], H.Z. HUANG[h], Y. IKEDA[f], S. KATCOFF[b], S. KAUFMAN[a], K. KIMURA[f], K. KITAMURA[d], K. KURITA[c], R.J. LEDOUX[h], M. J. LEVINE[b], Y. MIAKE[b,***], R.J. MORSE[h], S. NAGAMIYA[c], J. OLNESS[b], C.G. PARSONS[h], L.P. REMSBERG[b], H. SAKURAI[k], M. SARABURA[h], A. SHOR[b], P. STANKUS[c], S.G. STEADMAN[h], G.S.F. STEPHANS[h], T. SUGITATE[d], M. TANAKA[b], M. J. TANNENBAUM[b], M. TORIKOSHI[e], J.H. VAN DIJK[b], F. VIDEBAEK[a], M. VIENT[j], P. VINCENT[b], V. VUTSADAKIS[h], W.A. WATSON III[b], H.E. WEGNER[b], D.S. WOODRUFF[h], Y. WU[c], AND W. ZAJC[c]

Presented on behalf of the E-802 Collaboration by: Robert J. Ledoux

[a] Argonne National Laboratory, Argonne, IL 60439
[b] Brookhaven National Laboratory, Upton, NY 11973
[c] Columbia University, New York, NY 10027 and Nevis Laboratories, Irvington, NY 10533
[d] Hiroshima University, Hiroshima 730, JAPAN
[e] Institute for Nuclear Study, University of Tokyo, Tokyo 188, JAPAN
[f] Kyushu University, Fukuoka 812, JAPAN
[g] Lawrence Livermore National Laboratory, Livermore, CA 94550
[h] Massachusetts Institute of Technology, Cambridge, MA 02139
[i] University of California, Space Sciences Laboratory, Berkeley, CA 94720
[j] University of California, Riverside, CA 92507
[k] University of Tokyo, Department of Physics, Tokyo 113, JAPAN
* JSPS Fellowship for Japanese Junior Scientist
*** On leave of absence from University of Tokyo

ABSTRACT

Measurements of collisions of 14.5 GeV/c per nucleon ^{28}Si ions with nuclear targets using the BNL E802 spectrometer are presented. A description of the experiment and preliminary particle production ratios are discussed.

INTRODUCTION

In this talk I would like to report the latest results from E802 at Brookhaven National Lab. The main objective of E802 is to measure semi-inclusive spectra from nucleus-nucleus collisions with full particle identification (PID) and large kinematic coverage. In these respects it is unique among existing relativistic heavy-ion experiments at BNL and CERN. The experiment is centered around a magnetic spectrometer, augmented with other detectors, which provide global event characterization. The first run using the spectrometer was performed in April, 1987, using 14.5 GeV/c per nucleon ^{28}Si ions obtained from the AGS. Preliminary results from this run were reported on at Quark Matter 87[1,2]. A subsequent run in the Fall of 1987 used ^{28}Si beams to bombard Al, Cu, Ag, and Au targets. The data from this run were of greater statistics and covered a larger kinematic range. The data presented here will be limited to a discussion of particle production ratios from Si+Au collisions.

EXPERIMENTAL SETUP

The experimental setup for E802 is shown in Figure 1. The main spectrometer has an acceptance of approximately 25 msr. The magnet can produce maximum integrated field strengths of 1.5 Tm. The spectrometer is very compact (approx. 6.5 m) which allows a large fraction of the kaons to traverse the entire spectrometer before they decay. There is a second arm of the spectrometer, the Čerenkov Complex (Č-C), which is used to measure high momentum particles (p > 4 GeV/c). The spectrometer can measure particles emitted with lab angles between 5° and 55°. The acceptance of the spectrometer for pions, kaons and protons, as a function of rapidity in the nucleon-nucleon center-of-mass frame and p_T, is displayed in Figure 2. Lines of constant lab angle and momentum are plotted. The 5° and 55° degree lines show the limits of the angular acceptance, while the 14° and 28° lines illustrate the angular range subtended by one setting of the spectrometer. The lower momentum cutoff of the spectrometer is approximately 0.5 GeV/c. The upper momentum lines will be discussed in connection with the limits of the PID component of the spectrometer.

The spectrometer data can be correlated with other detectors which provide both triggers and global event characterization. A minimum bias trigger is provided by beam defining scintillators just upstream of the target and a downstream scintillator (not shown in Figure 1) which measures the charge (Z) of the projectile fragment. For ^{28}Si beams, an event with a projectile fragment of $Z \leq 12$ is deemed an interaction. A hadronic calorimeter positioned at 0° measures the energy contained in a 1.5° cone. The target multiplicity array (TMA) measures the charged particle pseudorapidity distribution ($-1.2 < \eta_{lab} < 3.0$) and provides an online hardware trigger for central events, these being the upper 7% of the total multiplicity distribution. A 245 element Pb-glass array is used to measure neutral energy flow in the central rapidity region[2,3]

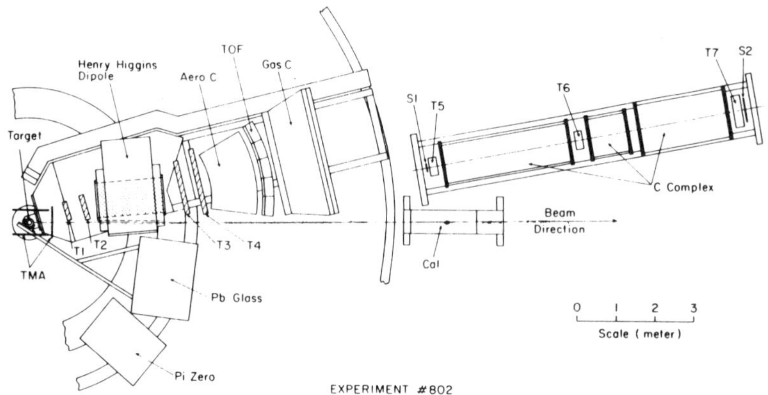

Fig. 1 Schematic view of BNL E802.

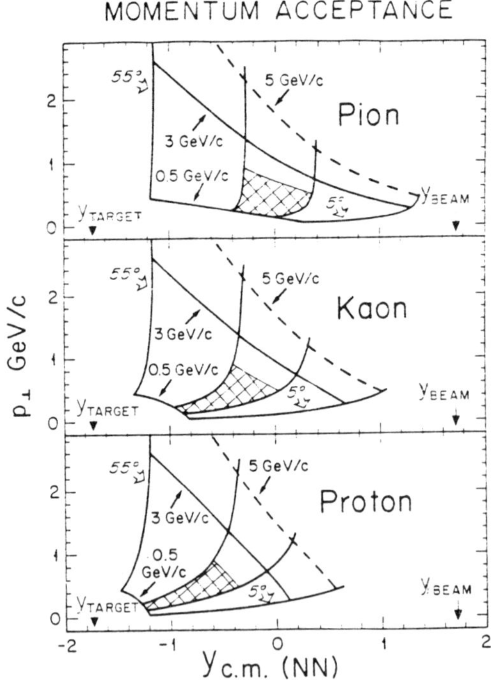

Fig. 2. Acceptance of the E802 spectrometer as a function of rapidity in the nucleon-nucleon center of mass frame and p_T.

(-0.5 < η_{cm} < 0.7).

The large multiplicity expected in central Si+Au collisions poses a number of technical challenges to tracking and PID. The main spectrometer was designed to handle charged particle multiplicities of up to 15 behind the magnet. This design criterion is approximately twice the predicted mean multiplicity for central Si+Au collisions with the spectrometer at its most forward position. The analysis of the spectrometer data to date has concentrated on measurements in the central rapidity region. Thus, there has been no verification yet whether or not such high multiplicities can be properly reconstructed in the spectrometer. The measured average multiplicity behind the magnet for central Si+Au collisions with the spectrometer spanning 14°-28° is 1-2. Less than one percent of these events have multiplicities greater than 6 behind the magnet.

The tracking and PID portions of the spectrometer were designed by exploiting some existing detector technologies, but with high segmentation and performance. The tracking of particles is accomplished with two sets of projective drift chambers before (T1 and T2) and after (T3 and T4) the magnet. Each set of drift chambers contains 10 wire planes with 5 views in T1 and 4 views in T2-T4. All wires have multi-hit capability which is provided by the use of 1ns sampling pipeline TDCs. The measured single track resolution of the drift chambers is $\sigma \approx 150$ μm and the double track resolution is estimated to be $\sigma \approx 2$ mm. Good single track resolution is important, since vectors formed in one chamber are used to point to another chamber. Thus, the number of vector pairs which must be considered in track reconstruction is significantly reduced as the single track resolution is improved.

PID is accomplished behind the magnet by a segmented Aerogel Čerenkov detector (AEROČ) and a time-of-flight scintillator array (TOF). The segmented gas Čerenkov counter (GASČ) was not in place for the runs reported on here. The AEROČ consists of 96 elements of Aerogel with an index of refraction of 1.02. The TOF wall consists of 160 plastic scintillator slats (1.6 cm square $\times$ 78 cm) with a photomultiplier at both ends. The TOF resolution with software slewing correction is $\sigma \approx 75$ ps. Without slewing correction, the resolution is approximately 100 ps. The TOF wall alone allows $\pi - K$ separation at the $5-\sigma$ level for p < 2.2 GeV/c and $K - p$ separation for p < 3 GeV/c. The AEROČ provides redundancy and extends the $\pi - K$ separation to p < 2.8 GeV/c. For the results shown here, only the PID obtained from TOF will be discussed.

PARTICLE SPECTRA

Figure 3 is a scatter plot of 1/p for reconstructed tracks versus their flight time to the TOF wall. The data are from central ^{28}Si+Au collisions from the Spring 1987 run, with the spectrometer spanning mid-rapidity ($\theta_{lab} = 14^o$-28°). Comparable PID spectra were obtained in the Fall 1987 run. There are distinct particle bands for $\pi^{\pm}$, $K^{\pm}$, p and deuterons. The tracks included in the plot are those which were fully reconstructed, i.e., tracks which project back

TABLE 1.

Preliminary particle ratios for 14.5 GeV/c per nucleon ^{28}Si+Au collisions at θ_{lab} = 14° - 28°.

	K^+/π^+	K^-/π^-	π^+/p	d/p
Min. bias trigger	19 ± 5%	5 ± 4%	50 ± 10%	11 ± 3%
Central Trigger	23 ± 5%	5 ± 4%	41 ± 10%	9 ± 3%

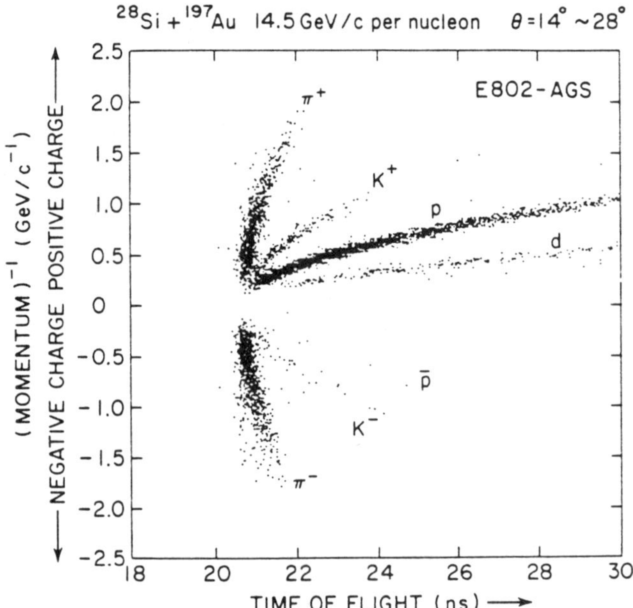

Fig. 3: Scatter plot of the inverse momentum for reconstructed tracks versus TOF for negative and positive particles from Si+Au collisions. The spectrometer spanned θ_{lab} = 14°-28°.

through the magnet from T3-T4 and are consistent with hits on T1-T2 and the target location. Also, only tracks which are associated with a hit on TOF and whose energy loss is consistent with Z=1 particles were included. Preliminary results from the April 1987 run[1] for the K^+/π^+ and K^-/π^- ratios found in the region $14^o < \theta_{lab} < 28^o$ for ^{28}Si+Au collisions were 19 ± 5% and 6 ± 5% for minimum bias events, respectively, and 24 ± 5% and 4 ± 4% for central events, respectively. The data are from the region indicated by the cross-hatched area in Fig.2 and $p_{lab} < 2$ GeV/c. These results contain decay corrections but do not contain tracking efficiency corrections. The data were obtained using one magnet polarity. Preliminary results for Si+Au collisions for the same angle setting for the Fall 1987 run are tabulated in Table 1. Also included in Table 1 are the π^+/p and d/p ratios from the Fall run. The agreement between the two runs is excellent.

Examination of Figure 2 reveals that with the spectrometer at the 14°-28° setting, the π and K come from somewhat different regions of phase space. However, by selecting particles with rapidities in the nucleon-nucleon center-of-mass frame from -0.44 to -0.14, and combining runs of both magnet polarities, it is possible to obtain preliminary invariant cross sections as a function of p_T which have small acceptance corrections. The p_T integrated K^+/π^+ ratio in this restricted region is consistent with that in Table 1.

Before discussing the significance of these particle ratios, I will briefly discuss the question of systematic errors in these measurements. It is important to note that the errors quoted in Table 1 are not indicative of the statistical error in the measurements but rather reflect an estimate of the systematic errors. Monte Carlo simulations are now under way to place the discussion of systematic errors on a quantitative footing. So far, Monte Carlo simulations which include the correct geometry and magnet map and the measured resolutions of the tracking chambers and TOF wall have been made. The Monte Carlo event generator "Fritiof"[4] has been employed to simulate Si+Au events. The simulated tracks are propagated through the spectrometer and hits in the tracking chambers and TOF are "digitized". The events are then reconstructed using different parameters and cuts and compared to the input event. The K^+/π^+ and K^-/π^- ratios for input tracks with $p < 2$ GeV/c for Si+Au minimum bias events in the angular range 14° - 28° are 7% and 2%, respectively. The reconstructed K^+/π^+ and K^-/π^- ratios are 8 ± 1% and 2 ± 0.3%, respectively. Thus, the systematic error in particle ratios from this simple Monte Carlo is on the order of 10%.

DISCUSSION

The most striking feature of the particle ratios presented in Table 1 is the large K^+/π^+ ratio. There is an increase of this ratio for central collision over that in minimum bias events, although both agree within the quoted errors. The K^-/π^- ratio is approximately a factor of 4 smaller for both triggers.

Since these preliminary measurements are the first of their kind in nucleus-nucleus collisions, they can only be compared to results from p+p and p+A collisions. The K^+/π^+ and K^-/π^- ratios from 19.2 GeV/c p+p collisions[5] are approximately 6% and 4%, respectively. The available data from p+A collisions are consistent[1] with $K^+/\pi^+ \approx 10\%$ and $K^-/\pi^- \approx 4\%$. It should be emphasized that these ratios depend on both the particle momentum and p_T. The K^+/π^+ ratio as a function of p_T for particles with p = 6 GeV/c from p+Be, Al and Pb collisions[6] at 19.2 GeV/c are shown in Figure 4. The K^+/p^+ ratio systematically increases with target mass and p_T. The data for these measurements were from $\theta_{lab} \approx 1^o$ - 4° and hence correspond to particles with larger center of mass rapidities than those in Table 1.

There is, of course, no substitute for measuring particle ratios for p+A collisions in our apparatus. These measurements will be performed in the near future. Until then, it would appear that the K^+/π^+ ratio in Si+Au collision is higher than that extrapolated from p+A data, whereas the K^-/π^- data ratio is consistent with such an extrapolation. I would like to emphasize once again that the data in Table 1 have not been corrected for tracking efficiencies.

Mechanisms have been proposed[7] which qualitatively predict an increased K^+/π^+ ratio in relativistic nucleus-nucleus collisions. For example, in a high baryon density gas or plasma, K^+ "distillation" may take place, whereby, $s\bar{s}$ pairs produced in the gas are preferentially found in the final state most often as K^+ and Λ due to the large relative abundances of u and d quarks as compared to $\bar{u}$ and $\bar{d}$ quarks. However, it is also possible that more mundane mechanisms, such as associated production from collisions of the primary nucleons or rescattering of secondary particles in the surrounding nuclear medium could increase the production of K^+'s. The reason for this is essentially the same, that as in a high baryon density gas, there are many available u and d valence quarks.

Which, if any, of the above scenarios is the most probable explanation of the measured particle ratios has yet to be determined. The data presented here are only a part of a large body of data from a variety of targets and covering a large kinematic range. We are in the process of analyzing these data and obtaining absolute cross sections. I would challenge theorists to develop models which are useful at predicting cross sections at AGS energies. There will shortly be plenty of data to explain!

SUMMARY

A fairly comprehensive set of data from collisions of 14.5 GeV/c per nucleon ^{28}Si beams with nuclear targets have been obtained using the BNL E802 spectrometer. Preliminary results from central Si+Au collisions from two runs yield K^+/π^+ and K^-/π^- ratios for $\theta_{lab} = 14^o$ - 28° (mid-rapidity) of 23 $\pm$ 5% and 5 $\pm$ 4%, respectively. The K^+/π^+ ratio is larger than that extrapolated from existing $p + A$ data.

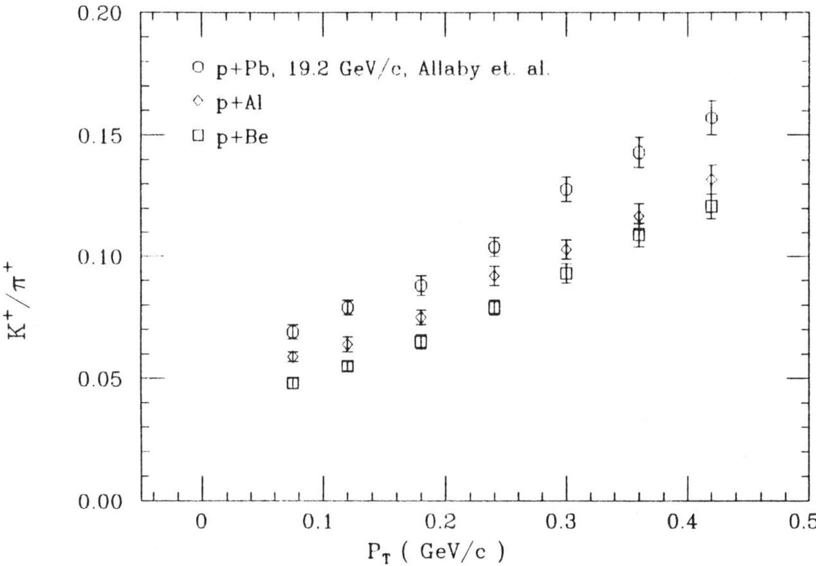

Fig. 4: The K^+/π^+ ratio for p = 6 GeV/c particles for $\theta_{lab} \approx 1^o - 4^o$ as a function of p_T from 19.2 GeV/c p+Be, Al and Pb collisions.[6]

ACKNOWLEDGEMENTS

E802 experiments are supported in part by the U.S. Department of Energy under contracts with ANL, BNL, Columbia, LBL, MIT, and UC Riverside, in part by the NASA through Grant NGR05-003-513 under contract with the University of California and by the US-Japan High Energy Physics Collaboration Treaty. I am indebted to all my collaborators on E802 as well as the AGS and Tandem Operation Technical Staffs.

REFERENCES

1. Y. Miake et al., Z. Phys. C38, 135 (1988).
2. L. Remsberg et al., Z. Phys. C38, 35 (1988).
3. T. Abbott et al. (E802 Collaboration), Phys. Lett. 197B, 287 (1987).
4. B. Anderson et al. Nucl. Phys. B281, 289 (1987).
5. A. N. Diddens et al., Nuovo Cimento 31, 961 (1964).
6. J. V. Allaby et al., CERN 70-12 (1970).
7. M. Gyulassy, Z. Phys. C38, 361 (1988).

NUCLEAR COLLISIONS AND STRANGE PARTICLES IN QGP

Johann Rafelski
Department of Physics, U of Arizona, Tucson, AZ 85721

and

Allard Schnabel
Department of Physics, U of Cape Town, Rondebosch 7700, RSA

ABSTRACT

The properties of quark gluon plasma fireballs created in central inelastic interactions of high energy Oxygen and Sulphur nuclei are determined within a simple reaction model, and the expected rapidity distributions of strange baryons and antibaryons are given.

OVERVIEW

Recent experimental results involving collisions of heavy nuclei indicate that we are creating the desired conditions permitting the study of hot, dense nuclear matter in its quark gluon form. There is a high transverse energy flow, signifiying a surprisingly high inelasticity. Furthermore, the transverse energy generated in central collisions increases[1] roughly as $A_p{}^{2/3}$. These developments confirm the hypothesis[2] that a dense and hot fireball with a baryon number originating from both the projectile and target nuclei is formed. We will interpret the fireball in terms of quark gluon matter. Subsequently, we discuss what this implies for the expected abundance of strange baryons and antibaryons, concentrating here in particular on the rapidity distributions. Our expectations appear to be in agreement with informally reported preliminary results from the 200 GeV/A O and S beam CERN experiments NA35, NA36 and WA85, in that enhancement of strange antibaryons in the 'hot' central fireball is expected, while in the 'cold' target fragmentation region an enhancement of strange baryon abundances results. First results also reported informally from ANL experiments further indicate an anomalous presence of strange baryons and antibaryons in (Si-Al) collisions at 15 GeV/A.

FIREBALL KINEMATICS

Upon the formation of a single fireball between projectile and target nuclei, relativistic kinematics leads to the relationship between the projectile and fireball rapidities $y_f=\frac{1}{2}y_p-\frac{1}{2}\ln(A_t/A_p)$. Assuming that the projectile collides with all matter in the central tube it hits on impact in the heavy target nucleus A_T, the mean number of participants from the target nucleus is $A_t=1.5A_T^{1/3}A_p^{2/3}$, which gives $A_t=55$ for the oxygen projectile, $A_p=16$ and for a gold target, $A_T=197$. Hence, $y_f=2.4$. For sulphur projectiles, $A_p=32$ and hence $A_t=88$ with y_f at about 0.1 units higher, since the colliding matter approaches the symmetric limit $A_p=A_t$. Counting the projectile participants we thus expect 1.7 times more participating baryons in the central fireball created by Sulphur as compared to Oxygen, but somewhat less slowing down, since the colliding systems are more symmetric. It is worthwhile to remember that events with higher rapidity have less energetic fireballs, as the lost rapidity of the projectile particles becomes the internal excitation of the nuclear fireball. We thus anticipate similar energy per baryon in the fireball, and correspondingly only 1.7 times higher transverse energy flow as one compares O and S induced central collisions.

As the above discussion indicates, in order to determine the mean energy of the fireballs created one should look at the shape of the multiplicity distribution in rapidity, with the width and position of the distribution being the key parameters. Assume first that the rapidity shape is described by a gaussian: $f(y) = N^{-1}dN/dy = (\pi\Gamma)^{-1/2} \exp((y-y_f)^2/\Gamma)$. Clearly, the greater the transparency, the greater will be the width parameter Γ; ultimately for total transparency the Gaussian approximation becomes inadequate, as the target and projectile rapidity regions separate and are peaked in projectile and target fragmentation regions.

The energy contents of all particles is now obtained by recalling that for each particle i we have $E_i=E_{Li}\cdot\cosh y_i$. Since the particles are observed in the laboratory frame, their energy must be equal to the energy E_p+m_t, i.e. the energy of the projectile plus target as measured in the same frame. Hence:

$$\begin{aligned} E_p+m_t \equiv E_f &= \Sigma E_i = \int dE_I/dy \cosh y \, dy \\ &= E_I \int N^{-1}dN/dy \cosh y \, dy = E_I \cosh y_f \, e^{\Gamma/4} \end{aligned} \qquad (1)$$

where in the penultimate equality we assumed that the rapidity distribution of E_I, the total transverse energy, is (nearly) equal to the multiplicity distribution, and in the last equality we have taken the gaussian distribution. Note that assuming the rapidity density of a massless (quark-gluon) gas, i.e. $N^{-1}dN/dy = 0.5/\cosh^2(y-y_f)$ (the temperature T enters only in the normalization in the absence of a scale such as quark mass), the factor $\pi/2$ takes place of the last exponential in eq. 1 , as well as in other considerations given below.

A further simplification arises if we recall the physical meaning of the factor $\cosh y_f$; it is the Lorentz factor of the fireball with reference to the laboratory, $\cosh y_f = \gamma_f = E_f/M_f$. Inserting into eq. 1 we obtain in the case of a single (gaussian) fireball model of the reaction:

$$M_f = E_I \, e^{\Gamma/4} \ . \qquad (2)$$

This expression relates two measurable quantities to the intrinsic mass of the disintegrating fireball object. Indeed, it clearly relates the experimentally observed energy flows in longitudinal and transverse directions to the intrinsic energy (mass) of the fireball. The observed[3] FWHM of negative particles (pions) originating in most inelastic collisions has a distribution in rapidity peaked; as expected, at $y_f=2.4$ with a FWHM=2.8, corresponding to Γ=2.8. Here the expectation for transverse energy according to eq. 2 is 210 GeV, if one takes for M_f the invariant mass of the colliding matter as given by $\sqrt{s}=m_N(\tau_p A_p A_t)^{\frac{1}{2}}$. This is consistent with the maximum transverse energy flow as reported in Ref. 4, in which, y>2.4 up to 110-120 GeV transverse energy in the forward "hemisphere" is observed. This kinematic consistency supports the hypothesis of one central fireball formed in these reactions.

QUARK GLUON PLASMA FIREBALL

At this point we turn to consider the maximum transverse energy and the energy per baryon which could be reached if the reaction were to evolve according to the geometric picture described here, and were to involve a gas of quarks and gluons; for $E_I = \sqrt{s}\cdot 2/\pi$, one finds 250 GeV for 200 GeV/A O collisions with A_T = 200 target nuclei, e.g. gold. This value is somewhat larger than has been observed, but we have to remember that the experiment observes the hadronic matter following its hadronization and (spherical?) expansion. It has been shown[2] that the adiabatic expansion of the quark fireball alone consumes 20% of the thermal energy, which is transferred to the kinetic flow energy.

We can further use the observed pion abundance in order to determine the entropy content of the fireball[5]. Together with the estimates of the energy per baryon given above, this determines the statistical parameters T (temperature) and $\mu_b = 3\mu_q$ when equations of state are employed - e.g. those of perturbative QCD matter. Specifically, each pion at $T > m_\pi$ carries away just above 4 units of entropy, and each baryon $2.5 + m_N/T$. The π^- and all charged particle yield has been given[6] for the O-Au collisions with central trigger: $\langle n^-\rangle=124\pm2$ and $\langle n_c\rangle=286\pm4$. Only 83% ± 5% of the negative particle multiplicity is in the T = 153 MeV component[3], which we associate with central fireball π^-. Thus the central fireball $\langle\pi^-\rangle$ multiplicity is $\langle n_{\pi^-}\rangle=103\pm4$. Accounting for the expected fireball neutron to proton ratio of 1.37, we find $\langle n_{\pi^+}\rangle=92\pm4$, and the average $\langle n_\pi\rangle/3=\langle n_{\pi^0}\rangle=97$. Taking the baryon multiplicity 71 for the central hot fireball $\langle n_\pi\rangle/b=4.1$ at the hadronization point leads to S/b = 25.5 . This is the entropy of the hadronic gas. Since the hadronization process of the hot and entropy rich QGP is not strongly entropy generating, the entropy of the quark gluon plasma should have been just a little less.

Given the initial conditions, viz E/b=O(5.5 GeV) and S/b=O(25) we can find the initial values of T and μ_b in the QGP phase. Using the perturbative QCD equations of state with parameters $\alpha_s = 0.6$, $B^{1/4}$=170 MeV[4], the result is T=220 MeV, μ_b=450 MeV, ϵ=2.5 GeV/fm^3 and n/n_0=2.5 . The isentropic evolution of the QGP to the phase transition temperature[2] leaves the ratio $\mu_b/T \approx 0.7$ constant, as shown in Fig. 1. We note in the Figure the evolution for S/b = 25 and 20 . The energy density of 1 GeV/fm^3 is reached at a temperature of 180 MeV, and the range of T considered is just that recorded by the WA80 collaboration for the neutral pions[7]. Varying α_s and B by ±30% and ±40% respectively, leads to ±15% variations in the initial values of μ and T. This sensitivity indicates that we can measure the key QCD parameters in nuclear reactions, once their entropy and energy contents are understood in detail. We notice in Fig. 1 that the evolution of the fireball with $\epsilon > 1 GeV/fm^3$ proceeds completely in the regime of parameters T, μ_b, which favors the retention of antistrange quarks in the plasma and emission of strange quarks. Thus heavy ion collisions at SPS energies create conditions favorable to the enrichment of antistrange quarks in the hot, baryon rich plasma phase, while emitting strange quarks into the hadronic gas envelope[8,9]. Further studies of the entropy and energy contents of the fireballs, and how these values relate to T, μ_b show that there is a sensitive relation between the collision kinematics and T. This is illustrated in Fig. 2, where a remarkable and here very useful feature[2] of the quark gluon plasma equation of state is shown; there is almost a unique correspondence between the temperature of the fireball, and the energy density for large E/b values (Fig. 2a) or large S/b (Fig.2b). The baryochemical potential μ_b varies strongly, but the ϵ-T relation is rather unique.

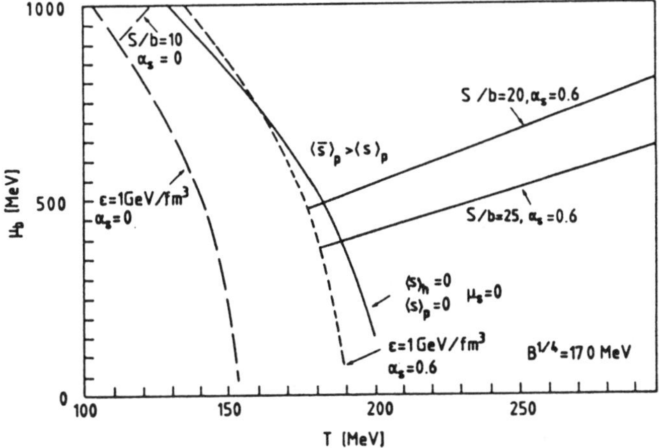

Fig. 1: Full lines in T,μ_b - plane: entropy per baryon 20 and 25 Dashed curves - energy density 1 GeV/fm^3 for α_s = 0 and 0.6;

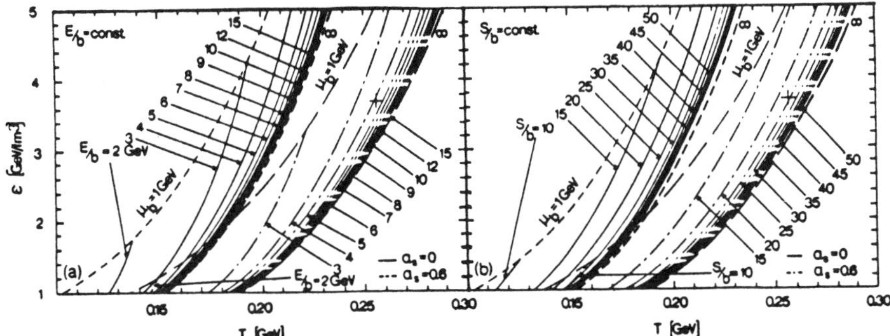

Fig. 2 : Energy density as function of temperature at (a) fixed E/b and (b) fixed S/b for α_s = 0, 0.6

STRANGENESS AS AN OBSERVABLE OF QUARK PLASMA

At the hadronization of the quark gluon plasma, the numerous $\bar{s}$-quarks which have been separated from the s-quarks, and enriched in the hot quark gluon plasma, can form strange antibaryons, which are thus comparatively enriched in comparison to strange baryons (this occurs only in the rapidity region of the central fireball). This is in striking contrast to circumstances arising in 'cold' quark matter, $S/b < 10$, possibly excited in the target fragmentation region, $y<1$, here s-quarks remain in quark matter, while $\bar{s}$-quarks are radiated before hadronization mostly as Kaons. The remaining strange and baryon rich quark matter then hadronizes, forming abundant lambdas and other (multi) strange baryons. This circumstance associated with cold quark matter has also been observed in $\bar{p}$-annihilations[10].

As this discussion shows, it is worthwhile to measure, as function of rapidity, the abundance of strange and antistrange baryons; the target rapidity region should contain an enhancement of 'cold' strange baryons, while the fireball rapidity region should contain an enhancement of 'hot' strange antibaryons. The problem such a measurement poses, is that we have to include the abundances of strange particles very likely formed at varying trigger conditions, hence including a different folding of impact parameters. For such a study to be of relevance to nuclear collisions, it is therefore essential to consider the ratio of strangeness to the transverse energy in the considered rapidity region. The situation is of course much simpler in the pp and $p\bar{p}$-collider experiments, which are usually taken as a reference, since we do not have the unknown baryon number and baryon density to deal with. At the ISR the observed anti-cascade to anti-hyperon ratio[11] was about 1/15 (in QGP at central rapidity we expect about 7 times greater value). A significantly enhanced value may have been observed[11] by the UA 5 collaboration in central $p\bar{p}$ interactions[12].

It is easy to see that the likelihood of making a strange or antistrange particle in a fireball model scales linearly with the fireball energy content, which in turn scales linearly with fireball transverse energy (see eq. 2). By thus weighing for each rapidity window the observed strangeness, the transverse energy permits us to recognize that the reaction mechanisms which produce transverse energy, are also likely to produce strange baryons or antibaryons at a rate which differs as function of projectile energy and projectile target size. Such an approach may also permit us to group the results of different experiments.

In order to account for the partial suppression of the abundance of anti-lambdas for high chemical potentials (very baryon rich area), it may be interesting to further consider the product of lambda with anti-lambda abundance as function of rapidity. This quantity is to a first approximation independent of the baryochemical potential and is eliminating so potential distortions arising from reaction mechanisms. Whenever this quantity is large, we expect that quark gluon plasma is formed as function of rapidity. Numerous other such products of particle abundances can be considered, also on the event basis in order to amplify on these remarks. Laborious calculations carried out previously[13] indicate that one anti-hyperon may be expected per each 100 fm^3 volume of the central fireball plasma state. With the energy density in the final stages of the plasma phase at 1GeV/fm^3 this in turn implies one anti-hyperon per 50 GeV transverse energy. Since according to our prior discussion, in Sulphur collisions with heavy nuclei fireballs with more than 300 GeV transverse energy are expected. We can thus anticipate several antihyperons (and fewer anticascades) per central interaction, should QGP be formed.

We have argued that in view of recent experiments it is consistent and indeed rather compelling to consider the formation of quark matter in relativistic nuclear collisions. Having discussed the general physical properties of quark gluon fireball formed at 200 GeV/A, we have addressed the expected abundances of strange baryons and antibaryons in 'cold' and 'hot' quark matter, and proposed to measure the production rate of antihyperons and anticascades per unit of transverse energy.

REFERENCES

1. W. Heck et al, GSI-88-08, January 1988, 'Study of the Energy Flow in O-Nucleus Collisions at 60 and 200 GeV/Nucleon', see in particular Fig. 12 .
2. J. Rafelski and A. Schnabel, Phys. Lett. B in press;
3. H. Ströbele et al, GSI-88-07, January 1988, 'Negative Particle Production in Nuclear Collisions at 60 and 200 GeV/nucleon'.
4. R. Albrecht et al, WA80 collaboration, Phys. Lett. B199 (1987) 297.
5. N.K. Glendenning and J. Rafelski, Phys. Rev. C31 (1985) 823.
6. NA35 Collaboration, A. Bamberger et al, GSI-88-09, January 1988, 'Charged Particle Multiplicities and Interaction Cross Sections in High Energy Nuclear Collisions'.
7. WA80 Collaboration, R. Albrecht et al, GSI-87-82, December 1987, 'Photon and Neutral Pion Distributions in 60 and 200 A GeV O + Nucleus and p+Nucleus Reactions'.
8. C. Greiner, Diploma Thesis, Frankfurt/M 1988; Phys. Rev. Lett. 58 (1987) 1825 .
9. J. Rafelski, Phys. Lett. B190 (1987) 167.
10. J. Rafelski, 'Quark Gluon Plasma in Antiproton Annihilations on Nuclei', these proceedings.
11. T. Akesson et al, The Axial Field Spectrometer Collaboration, Nucl. Phys. B246 (1984) 1 .
12. G.J. Alner et al, UA5 Collaboration, Phys. Lett B151 (1985) 309.
13. P. Koch, B. Muller and J. Rafelski, Physics Reports 142 (1986) 167 .

SOME THEORETICAL ASPECTS OF J/ψ SUPPRESSION

R. L. Thews
Department of Physics, University of Arizona, Tucson, AZ 85721

ABSTRACT

All current explanations of the variation of J/ψ suppression with transverse and longitudinal momentum depend on the existence of a classical resonance formation time, before which time the quark–antiquark system is assumed to be unbound. We point out that under the conditions of quark–antiquark formation in heavy–ion collisions via some elementary hard process, this time delay does not exist. The energy and length scales involved require a nonrelativistic quantum mechanical tratment of the quark–antiquark system. Therefore the probabilities for observation of bound states are determined uniquely by the initial wave functions at t=0, supplemented by time evolution determined by the eigenvalue spectrum. A simple one–dimensional model has been worked out explicitly to illustrate these properties, using an infinite square well for the quark–antiquark interaction and square initial wave packets with gaussian average momentum distributions. Suppression due to plasma formation is simulated by expanding the walls of the square well. In the sudden expansion limit, the suppression factor increases with total momentum in a similar manner to the experimental data, but the details depend strongly on the interplay between initial correlated or uncorrelated momentum distributions and the bound state spectrum. It is argued that a calculation of this type using realistic three–dimensional interactions and time–varying plasma boundaries and densities must be performed in order to have any confidence that the kinematic properties of J/ψ suppression may be a signature for plasma formation.

INTRODUCTION

Experimental evidence has been presented recently[1] showing that the production of the J/ψ is considerably suppressed in events in relativistic ion collision having a large transverse neutral energy. Such a suppression has been suggested a few years ago[2,3] on the basis of the relatively low temperature (T $\approx$ 200 MeV) which is expected to prevail in a quark gluon plasma. Such a low temperature makes it very difficult to create J/ψ's via a mechanism like gluon–gluon fusion. The argument for J/ψ suppression was made stronger by Matsui and Satz[4], who pointed out that J/ψ's produced in the early stages in relativistic ion collisions dissociate due to the Debye screening in the quark–gluon plasma. More recently, a number of estimates[5] have been made to quantify this effect and relate it to the kinematic properties of the J/ψ. All of these make use of a classical formation time for the resonance, typically based on the revolution period in a Bohr model. The kinematic dependence of the suppression then enters when the quark–antiquark pair either does not form the resonance before the plasma density falls below some critical value, or when the total momentum of the pair carries it outside the spatial boundary of the plasma before the resonance formation time elaspses. All of this assumes, however, that the quarks can be described by classical trajectories[6] during this time. It is simple to show that this is not the case. The hard process which produces the quark–antiquark pair has a distance scale of order $1/m_q$, which is of order a few tenths of a fermi for the charm quarks. This is much smaller than the range of interquark potentials, so that the production process is unaffected by the subsequent interaction. In order for the pair to form the nonrelativistic bound state J/ψ, the quark momenta must certainly have major components much less than the quark mass. Hence one arrives at a situation where $\Delta p/p$ is of order unity, i.e. in the nonrelativistic quantum mechanical regime. The probability of the system to be in any of the bound states of the potential is completely determined at t=0 by projection of the initial wave function onto the appropriate state. The influence of the external medium comes in as a space–time variation in the parameters of the potential, due to formation of a quark–gluon plasma and charge screening effects. One follows the time evolution of the bound state probabilities through the period of plasma formation and decay. Given these final probabilities for an initial wave function localized as above and with some average initial momentum, one must then convolute over a distribution of initial momenta from the hard production process. The final result is the probability to find the system in a particular bound state with a given total momentum. The ratio of this probability to that without the time variation of interaction potential due to plasma formation is the appropriate quantity to compare with experiment.

A SIMPLE MODEL[7]

For simplicity, we consider initially a one–dimensional situation in the direction transverse to the beam axis. We use an infinite square–well confining potential whose width is typically of the order of fermis (much larger than the width of the initial quark wave functions), and simulate the deconfining effect of a plasma by allowing this width to increase to larger values. The wave function of each heavy quark is taken as a square wave packet of width 2a, centered at the origin, with average momentum p_j,

$$\varphi_j(x) = e^{ip_jx}/\sqrt{2a}\ , \quad |x| < a \tag{1}$$

with $a \approx 1/m_q$. The quark–antiquark total initial wave function can then be written in terms of the relative $r = (x_1 - x_2)$ and center of mass $R = \frac{1}{2}(x_1 + x_2)$ coordinates as

$$\Psi(r,R) = e^{i(p_1+ p_2)R}\, e^{\frac{1}{2}i(p_1- p_2)r}/2a, \quad |R \pm \tfrac{1}{2}r| < a \tag{2}$$

The effect of the finite spatial boundary of the plasma is quite difficult to accomodate in this approach, since the correlation of quark–antiquark potential with center of mass position does not allow the usual separation of center of mass and relative motion for the two body system. We concentrate here on the time variation and leave the spatial problem to a later study, keeping in mind that one must then limit the duration of the time evolution of our system to prohibit the center of mass position of the pair wave function from crossing the plasma boundary. We start by expanding the pair wave function in products of $\Gamma_p(R) = \frac{1}{\sqrt{2\pi}} e^{ipR}$, the factor for center of mass motion with total momentum p, and $\chi_n(r,t) = \frac{1}{\sqrt{L}} \cos\left[(n+\frac{1}{2})\frac{\pi r}{L(t)}\right]$, the bound state eigenfunctions in the confining potential (for purposes of this example, we restrict to even solutions only):

$$\psi(r,R,t) = \int dp \sum_n c_{pn}(t)\, \Gamma_p(R)\, \chi_n(r,t). \tag{3}$$

The probability amplitudes at the initial time (t=0) can be evaluated by inverting equation (3), using the initial wave function (2).

$$c_{pn}(t{=}0) = \frac{2}{a\sqrt{2\pi L}} \left[\frac{\cos[(p_1-p_2+2k_n)a] - \cos(qa)}{q^2-(p_1-p_2+2k_n)^2} + p_1 \leftrightarrows p_2\right] \tag{4}$$

where $k_n = (n+\frac{1}{2})\frac{\pi}{L}$ are the proper momenta inside the potential and $q = p_1 + p_2 - p$ is the deviation of the wave packet average total momentum from that of the center of mass. This formula clearly shows the expected maximum in probability amplitude when the quark pair relative and total momenta match the values appropriate for the bound state. It also can be maximized when q, the mismatch in center of mass momentum, is compensated by an equal deviation of the average quark relative momentum from one of the potential well eigenvalues. To follow the time evolution of the probability amplitudes, one must solve a set of coupled equations generated by the time–dependent Hamiltonian

$$\frac{dc_{pn}}{dt} = \sum_m c_{pm}(t)\left[-iE_{pm}(t)\delta_{mn} + \tau\frac{(m+1/2)(n+1/2)(-1)^{m+n}}{(m+n+1)\ (m-n)}\right] \tag{5}$$

where $\tau = 1/L(t)\ dL(t)/dt$ is a measure of the plasma formation time. Estimates[4] of this time are of order 1 fm/c, which is of the same order of magnitude as the wave function oscillation times for nonrelativistic charmonium. For an initial estimate of what might occur, we calculate the relative bound state probabilities in the "sudden" approximation, which just means using (4) with a confinement length scale L appropriate for the plasma phase. To compare with experiment, one must average these probabilities over the initial distribution $f(p_1,p_2)$ of the quark–antiquark pair. This average will then be a function of the total momentum p and the bound state label n,

$$g_n(p) = \int dp_1 \int dp_2 \,|c_{pn}(p_1,p_2,p,n,a,L)|^2 f(p_1,p_2) \tag{6}$$

where the distribution f is assumed to be normalized. One would ordinarily calculate this distribution from some model of quark pair production, e.g. gluon–gluon fusion (with another gluon in the final state to allow for formation of vector quarkonium). To get some idea of how this simple one–dimensional model works, we choose instead to guess some reasonable distribution functions. We parameterize such a distribution with a gaussian form $f(p_1,p_2) = \frac{2\sqrt{\gamma_1\gamma_2}}{\pi} e^{-\gamma_1(p_1+p_2)^2} e^{-\gamma_2(p_1-p_2)^2}$. We show in Figure 1 the ratio of L=1.0 fm to L= 0.5 fm (simulating plasma/normal) for the ground state (n=1). The line labeled "uncorrelated" uses $\gamma_1 = \gamma_2 = 0.5$ GeV/c^{-2} and $a = 1/m_c = .13$ fm. The suppression at low p is almost entirely due to the normalization factor for increasing the potential width L. The variation with p comes from matching relative and total momentum to the ground state energies, and shows a small enhancement with increasing p. The line in Figure 1 labeled "correlated" shows the same suppression factor for $\gamma_1 = 2.0$ GeV/c^{-2} and $\gamma_2 = 0.5$ GeV/c^{-2}. The factor starts out idential with the uncorrelated case, but then shows a much more pronounced rise with p, which is quite close to the trend of the NA38 data. The magnitude of the effect is obviously a complicated interplay between the energy scales of the bound state model and the initial momentum distributions.

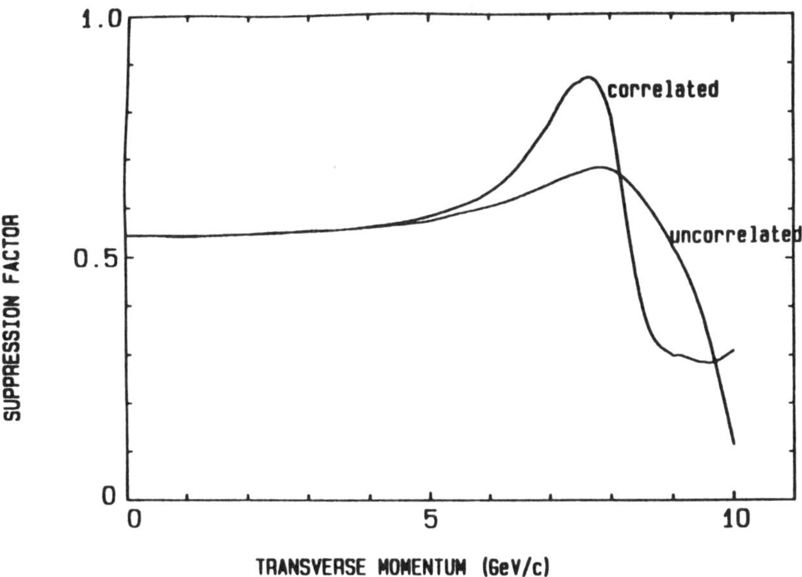

Figure 1. Ground State Probability Ratio

SUMMARY

The physical conditions of J/ψ production in hadronic collisions is shown to require a quantum–mechanical calculation for time evolution of suppression factors due to possible quark–gluon plasma formation. A classical resonance formation time, which is central to other treatments of kinematic features of this phenomenon, plays no part in this picture. An approximate solution for a simple one–dimensional model shows the possibility of producing momentum–dependent suppression factors which are similar to those found in the data. A quantitative comparison with data is in progress, using a gluon–gluon fusion calculation of initial quark momentum distributions along with a 3–dimensional potential model of quarkonium. This treatment will also follow the time evolution of probabilities to check the validity of the impulse approximation. The problem of incorporating spatial plasma boundaries remains to be addressed.

REFERENCES

1. NA38 collaboration, "The Production of J/PSI's in 200 GeV/N Oxygen–Uranium Interactions". Presented at the "Quark Matter" Conference, August 1987, Schloβ Nordkirchen.
2. J. Cleymans, Proceedings of the "Reunion sin la Physique des Ions Ultra–Relativistes", June 1984, Centre de Recherches Nucleaires de Strasbourg.
3. J. Cleymans, and C. Vanderzande, Physics Letters **147B**, 186 (1984).
4. T. Matsui and H. Satz, Physics Letters **178B**, 416 (1986).
5. J. Kapusta, Phys. Rev. **D36**, 2857 (1987).
F. Karsch and R. Petronzio, Physics Letters **193B**, 105 (1987).
P. V. Ruuskanen and H. Satz, Brookhaven National Laboratory preprint BNL–40438, 1987.
M. –C. Chu and T. Matsui, M.I.T. preprint, 1988.
6. Even if a classical decription of quark motion were valid, there seems to be no justification for the concept of resonance formation time. The initial position and velocity, or equivalently energy and angular momentum of the system determines the trajectories of the particles in a given potential. Hence the bound state nature of the system is completely determined at t=0. What should be done in the classical case is to follow the trajectory evolution in the presence of the time–varying interaction produced by formation and decay of a background plasma, and determine what the state of the two–body system is after the plasma phase has disappeared. This is precisely what is advocated for the actual physical situation, but for which a quantum–mechanical time evolution must be followed.
7. This report is based on work initiated in collaboration with J. Cleymans. See J. Cleymans and R. L. Thews, University of Arizona preprint, January 1988.

Thermalization and Stopping in Relativistic Heavy Ion Collisions

J.Stachel*
Physics Department, SUNY Stony Brook, NY 11794

INTRODUCTION

Over the last year first experimental results from the Relativistic Heavy Ion facilities both at Brookhaven National Laboratory and CERN have become available. One of the prime motivations for such experiments is the search for a quark-gluon plasma. Since both machines operate with fixed targets and at still relatively moderate beam momenta of 10-14.5 GeV/c (BNL) and of 60 and 200 GeV/c (CERN) the conjectured plasma is expected to form in a baryon-rich region. A necessary condition besides high baryon density is that a sufficient fraction of the incident projectile kinetic energy is thermalized. It therefore seems worthwhile to try and extract the degree of thermalization from the present experimental data and investigate whether sufficiently high temperatures and energy densities can be achieved. Here this will be analyzed in terms of a simple model involving formation of a hadronic fireball and its decay after thermalization. Two decay scenarios have been investigated: i) Isotropic emission from a thermal source (see also reference 1). ii) Decay after hydrodynamic expansion as originally proposed by Landau[2]. A more detailed discussion of this latter scenario including references can be found in reference 3.

FIREBALL FORMATION AND DECAY

It is assumed that in a collision between two relativistic heavy ions a (hadronic) fireball is formed with the number of participant nucleons from target and projectile, $N_t(b)$ and $N_p(b)$, given by the geometric overlap of the two nuclei for a given impact parameter b. In the case of full stopping the incident kinetic energy is transformed into fireball cm motion with

$$\beta_f(b) = \beta_p \gamma_p N_p(b)/(\gamma_p N_p(b)+N_t(b))$$

and excitation energy

$$E_{cm}{}^{f} = m_n \sqrt{[N_p(b)^2+N_t(b)^2+2\gamma N_p(b)N_t(b)]}$$

including the restmass energies of the constituent nucleons. To accomodate the situation of incomplete stopping it is assumed[3] that in this case the fireball is formed of $p_s N_t(b)$ target nucleons and $p_s N_p(b)$ projectile nucleons in which case its excitation energy is

$$E_{cm}{}^{fs} = p_s E_{cm}{}^{f}.$$

The stopping fraction p_s is a parameter to be obtained from comparison to experimental data.

Both decay scenarios assume that a thermalized fireball decays statistically by emitting the constituent nucleons and as many pions produced in the reaction as allowed by imposing energy conservation.

To compare to experimental data, the acceptance of the various experiments is sampled in a Monte-Carlo procedure and the resulting spectral distributions are integrated over impact parameter. If experimental data have been obtained under special trigger conditions the same constraints are applied to the simulation. The only difference between the two decay scenarios lies in the (vector) momentum distributions of the emitted particles.

In the case of an <u>isotropically decaying fireball</u> only the freeze-out temperature T_f determines the momentum spectra of the particles emitted:

$$d\sigma/dp \propto p^2 \exp(-E/T_f).$$

In all calculations presented T_f=0.15 GeV was chosen to sample this momentum distribution. The equivalent distributions in transverse momentum p_T and rapidity y, integrated over the respective other variable, are

$$d\sigma/dp_T \propto \sqrt{m_T}\, \exp(-m_T/T_f)$$

and

$$d\sigma/dy \propto 1/\cosh^2(y).$$

In the case of the <u>Landau model</u> the hydrodynamical expansion alters the momentum distributions and an additional scale enters through the initial condition. Therefore, more interestingly, the observable quantities carry some information about the initial compressed stage of the fireball. Although the exact solutions of classical relativistic hydrodynamics are complicated, it has been shown[4], that in an approximate solution the invariant cross section factorizes into transverse and longitudinal motion:

$$\sigma_{inv} = [\sqrt{m_T}\, \exp(-m_T/T_f)]\ [1/(\sqrt{2\pi\sigma_y{}^2})\, \exp(-y^2/2\sigma_y{}^2)]$$

where the first term is governed by the thermal motion and the gaussian rapidity distribution has a variance σ_y determined by the change in entropy at the initial instant of the collision.

In the Landau model the expansion is at first one-dimensional along the beam axis followed by a three-dimensional stage. At moderate beam energies when the compression is not too large one may assume the quasi-one-dimensional approximation[4] to be valid: In this case $\sigma_y{}^2=3\ln(T_i/T_f)$ with an initial temperature T_i. In the high energy limit the three dimensional stage of the expansion is important and σ_y then scales logarithmically with the ratio of radius to thickness of the initial compressed fireball (γ_f for a symmetric system). This quantity should vary for different target-projectile combinations. On the other hand none of the experimental data[5,6,7,8] show such a variation at a given beam energy confirming that, for the present experiments, the three-dimensional stage is probably not yet important. For the calculations shown here the quasi-one-dimensional solution is adopted and σ_y was taken from experiment for each beam energy.

COMPARISON TO EXPERIMENTAL DATA

The transverse energy $E_T = \Sigma \sin\theta_i E_i(\theta_i)$ is a good measure for thermalization since it reflects the energy removed from the beam direction and reemitted into larger angles. Therefore, the discussion here will be concentrated on this variable, although of course, one has to test whether other variables support the assumption of energy emitted from a thermalized, possibly expanding, fireball. This has been done in reference 3.

In the top part of Fig. 1 is displayed the experimental E_T spectrum for 10 GeV/c ^{28}Si + Pb together with the results from the isotropic fireball assuming full stopping. The solid line displays the impact parameter integrated calculation, while the dotted lines show contributions from the various impact parameter (1: 0-1.5 fm etc.). It is clear that the shoulder at large E_T arises from the fact that, below a certain impact parameter, the thickness of the Pb nucleus changes very little. It can be concluded that this feature of the experimental data is in agreement with the assumption of complete stopping of the projectile in the fireball center of mass frame. For the 10 GeV/c ^{28}Si reactions calculations using the Landau model give practically identical E_T distributions as can be seen from the bottom part of Fig. 1. The calculations for E_T shown are for σ_y=0.6 and were found to be little sensitive to this variable. A comparison[3] to $dE_T/d\eta$ (E814, ref.5) and, at a little higher beam momentum of 14.5 GeV/c, $dN_c/d\eta$ (E802, ref.6) shows good agreement. For the latter case the distribution from the Landau model gives better agreement with experiment; the isotropic fireball yields a distribution that is somewhat too narrow.

The approach discussed here has also been tried at higher beam energies and, as an example, in Fig. 2 a comparison to the transverse energy distributions of one experiment, WA80, is shown. All theoretical curves are from the Landau model. The dashed lines are for the Au target and full stopping or thermalization. It can be seen that they are not very far from the data. The solid lines are obtaines by adjusting the stopping fraction p_S to the data and give p_S=0.92-0.93 at 60 GeV/c and p_S=0.76-0.84 at 200 GeV/c (lower values for lightest target). While at the beam energies of 10-15 GeV/c the difference between isotropic and hydrodynamically expanding fireball was small, at higher beam energies evidently the compression and therefore also expansion plays a larger role and the predictions from an isotropic fireball are quite different. This is indicated in Fig. 2 by the arrows marking the transverse energy for the average central collision in this model (to be compared to the shoulder in the experimental distributions). Along the same line, also the multiplicity distribution from the isotropic fireball is significantly narrower than the data while the hydrodynamic model gives good agreement[3]. This is not too surprising since in the Landau model more particles are emitted into large angles but the particles along the beam axis carry more energy.

Summarizing, the comparison of transverse energy distributions from the Landau model to experimental data indicates a very large amount of stopping/thermalization up to the highest beam energies explored so far experimentally in heavy ion collisions. At the largest beam energies a model that does not include the influence of expansion on the energyflow, like the isotropic fireball, gives quite different predictions and is probably not applicable any more. Initial fireball temperatures as obtained from the rapidity distributions in the hydrodynamic model range from 0.17 to 0.24 GeV for the reactions discussed here.

* A.P.Sloan research fellow. Supported in part by the NSF.

REFERENCES

1. J.Stachel and P.Braun-Munzinger, Proceedings of the Texas A & M Symposium on Hot Nuclei, Dec.1987, in print
2. L.D.Landau, collected papers No.88, Izv.Akad.Nauk SSSR 17, 51 (1953)
3. J.Stachel and P.Braun-Munzinger, submitted to Phys.Lett.B
4. G.A.Milekhin, Soviet Physics JETP 35, 829 (1959)
5. B.Bassaleck et al. (E814 collaboration), Z.Physik C38, 45(1988)
6. T.Abbot et al. (E802 collaboration) Proceedings IXth Autumn School on the Physics of the Quark-Gluon Plasma, Lisbon, Dec. 1987, in print
7. R.Albrecht et al. (WA80 collaboration) Phys.Lett. 199B, 297(1987)
8. M.Murray (Helios collaboration), Proceedings IXth Autumn School on the Physics of the Quark-Gluon Plasma, Lisbon, Dec.1987, in print

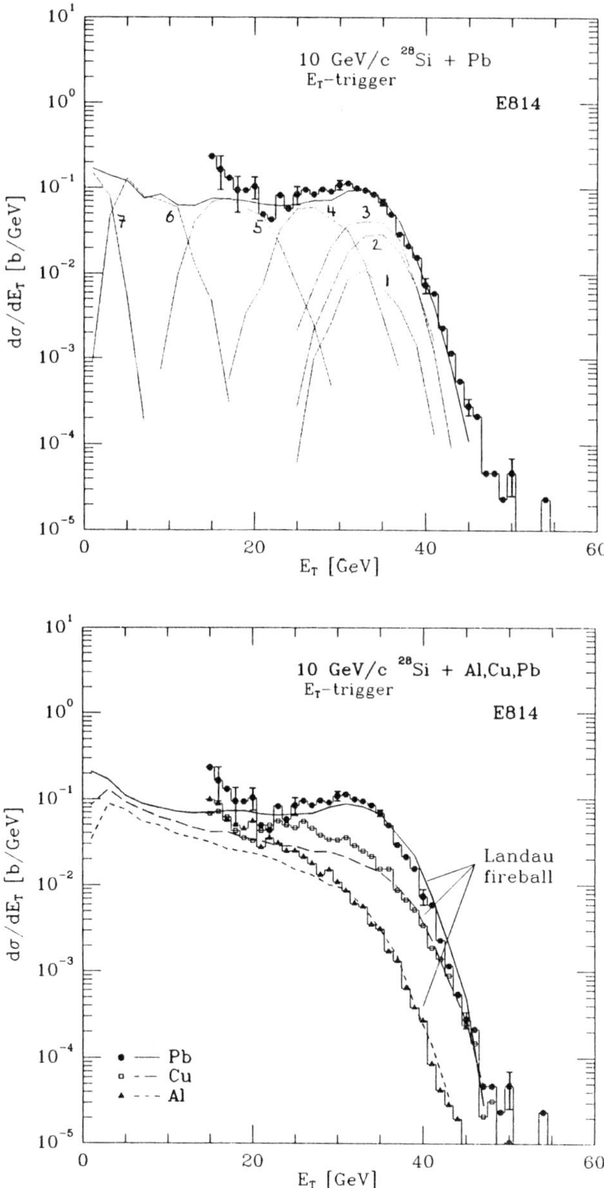

Fig. 1: Comparison of predictions from the isotropic (top) and the Landau (bottom) fireball models to experimental data[5]. The curves labelled 1-7 correspond to 1.5 fm wide impact parameter bins.

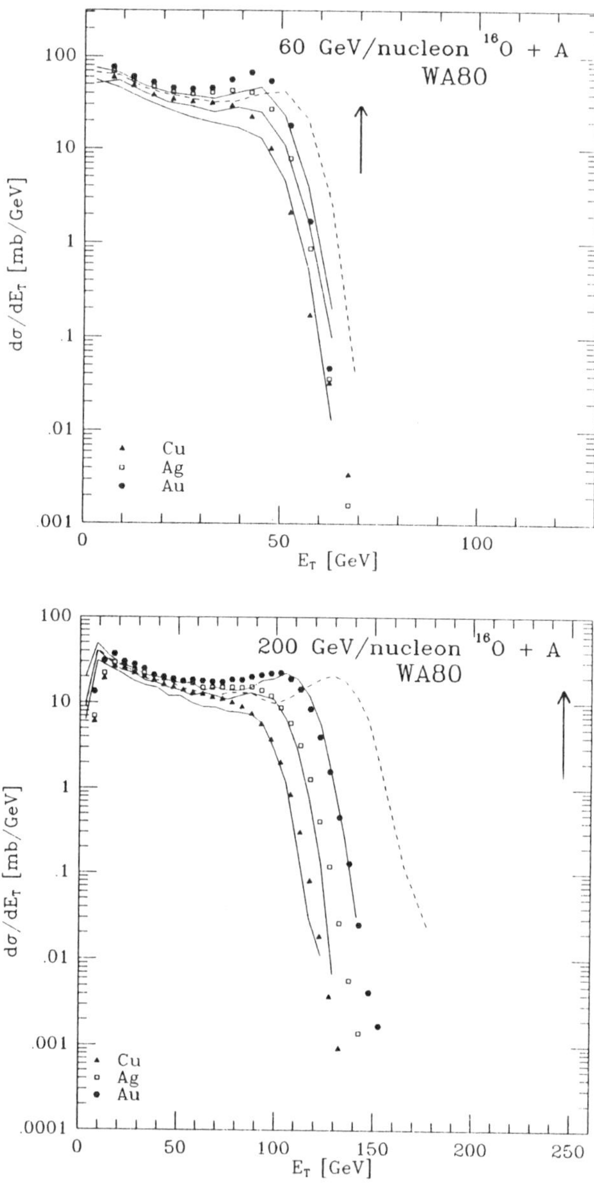

Fig. 2: Transverse energy distributions from the Landau model to experimental data[7]. The dashed lines are for O + Au and full stopping. Solid lines are for Cu, Ag and Au corresponding to partial stopping (see text) and σ_y=1.1 and 1.2 at 60 GeV/c and 200 GeV/c, respectively. The arrows mark the transverse energy for the average central collision from the isotropic fireball model.

LATEST RESULTS ON SUBTHRESHOLD KAONS AND ANTIPROTONS*

George Igo
University of California, Los Angeles, CA 90024

ABSTRACT

The bombarding energy dependence up to 1.2 GeV/A and the dependence on target-projectile of K^- production have been studied. In one case, K^+ production has been measured at the same Q-value as in K^- production. Preliminary results on antiproton production are reported.

By 'subthreshold production' we refer to those nuclear reactions yielding particles whose production threshold (in free nucleon-nucleon collisions) is significantly higher than the energy available in the average N+N collision. Some sort of collectivity is thus required by this definition. It is useful to distinguish between two types of 'collectivity' – that which is intrinsic to the initial state nuclei (such as Fermi motion, clusters of nucleons or quarks . . .); and that which is characteristic of the collision itself (possible equilibration of the various kinetic and chemical degrees of freedom, for example). One expects to be able to measure the intrinsic nuclear effects by studying $q + A \rightarrow q + X$ reactions, where q represents a 'non-composite' probe, thus it is primarily the collisional collectivity that is the object of study in nucleus-nucleus subthreshold production. By choosing to detect particles of various masses (e.g., pions, kaons, antiprotons), one may study these collisional collective effects in a range of excitation energies where the dynamics of the collision process may be expected to change significantly. Much work has been done in the past few years in studying subthreshold pion production.[1] Our effort has focused on studies at higher excitation energies with the aim to understand how, and how far, the collision dynamics proceed toward equilibration when the excitation energies are far removed from the realm of conventional nuclear physics. In the following, due simply to considerations of time, I will present only our recent results[2] and must leave until another occasion interesting comparisons with relevant work of others, both experimental and theoretical.

Our initial measurements showed that in Si+Si collisions at 2.1 GeV K^- are produced at a level more than 10 times higher than that obtained from a somewhat careful calculation which included the effects of internal nuclear motion, and that the invariant cross section shows the usual exponential behavior with a slope of about 95 MeV.[3] Our program since then has been to outline, within the capabilities of available facilities, the systematics governing this process. We thus set out to measure the variation of the K^- yield

*Supported by USDOE under contract DE-AT03-81ER40027, PA DE AM03-765SF00034.

with incident energy and with projectile/target mass combinations; to measure the K^- yield at a cm angle of 90°; and to measure sub-threshold K^+ production. We have accomplished these goals, although we did not achieve as large a range in mass and energy as we had originally hoped for. The data presented in this report are still preliminary in the sense that not all systematic corrections have been identified and carried out, nor have systematic errors been calculated – all errors shown represent counting statistics only.

Figure 1 shows center of mass cross sections at 0° for Si + Si $\rightarrow K^- + X$. Points belonging to the same incident energy have been connected by straight lines to indicate the groupings, with the exception of the data at 2.1 GeV/A, where the dots indicate a fit to the combined data from several past measurements. Note that the measurements cover more than two orders of magnitude in cross section, but that no large changes are apparent in the slope (at least

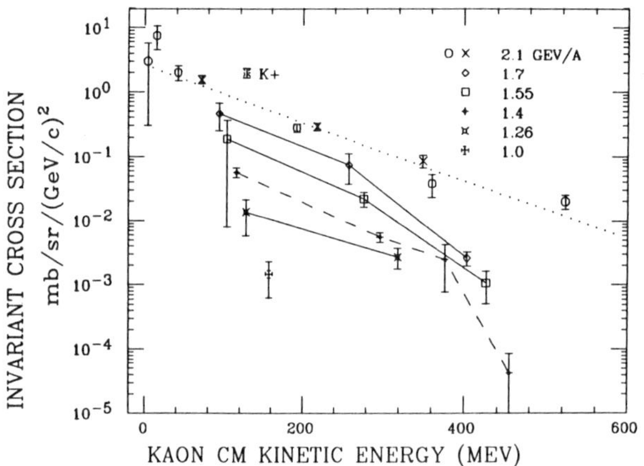

Figure 1. CM Spectra for Si + Si $\rightarrow K^- + X$

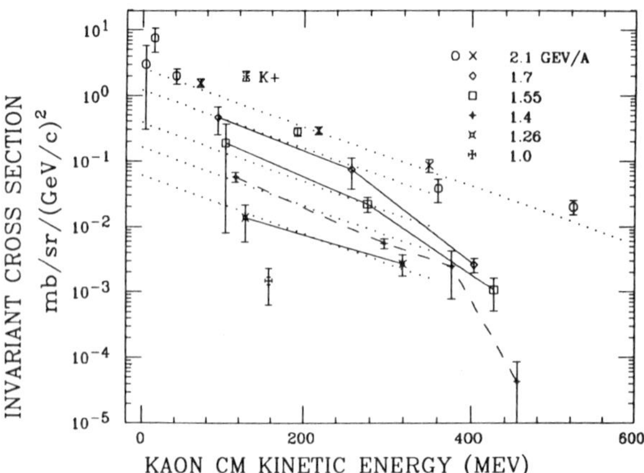

Figure 2. CM Spectra for Si + Si $\rightarrow K^- + X$

for low values of kaon c.m. kinetic energy). There are increases in the slope in the region where incident energies are low and kaon c.m. energies are high. This is the behavior which would be expected from the phase space limitations. In Figure 2, the similarity of the initial slopes has been emphasized by passing a line parallel to the 2.1 GeV/A data through the first two data points of each of the other data sets. From these lines we have made a crude estimate of the total cross section (assuming isotropy) that is shown in Figure 3, plotted against the available excitation energy in the center of mass. No error bars are given at this stage of analysis. The slope of the exponentially rising part of the curve is 54 MeV. Speculation is invited. When I look at this curve I see only the effect of the K^+K^- threshold and no influence of the Λ threshold, suggesting that the contribution of the strangeness exchange channel is small.

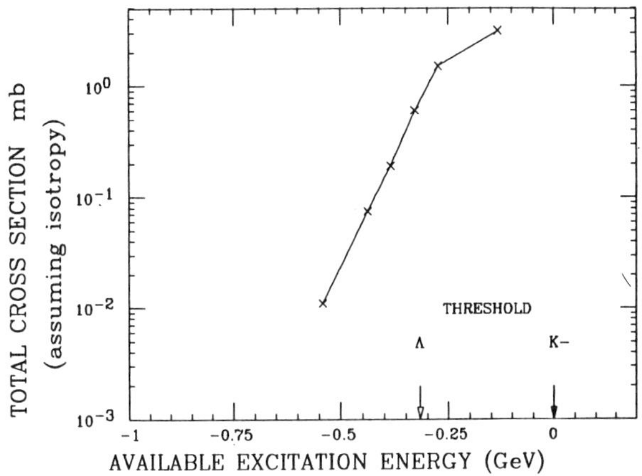

Figure 3. Total Cross Section Versus Q

Figure 1 also contains the first result of our measurement of the cross section for subthreshold production of K^+, taken at an incident energy such that the Q (available c.m. energy) for the NN $\rightarrow$ $N\Lambda K^+$ channel was the same as that for the NN $\rightarrow$ NNK^+K^- channel at 2.1 GeV/A. Note that the K^+ cross section is higher than that for the K^-, even though no strangeness exchange channel is available to contribute to this yield. It may prove to be possible to account for the larger K^+ cross section by noting that the absorption of K^- in nuclear matter is larger than for K^+, while the 3-body K^+ phase space is larger than the 4-body K^- phase space at the same value of Q. This approximate equality of cross sections suggests that the production mechanism is directly related to the Q of the NN collisions rather than to secondary (or tertiary) production mechanisms such as strangeness exchange.

Figure 4 shows the cm spectra for three equal-mass projectile-target combinations C+C, Si+Si, and Ca+Ca, all at 2.1 GeV/A incident energy. (One of the yet unresolved systematics problems also shows itself.) Again we note that there is no appreciable change in slope when going from carbon to silicon to calcium. (The lines simply connect the points and are not fits.) These same data are shown in another form in Figure 5, where the cross section at fixed kaon c.m. energy is plotted versus the mass number of the target. When interpreted as an A^k dependence the lower two points give a value of k = 3.6, while the upper points give a value k = 1.2. Again, at this stage of analysis we prefer not to assign errors to these derived quantities. Attempts were made to measure the yields from the Nb+Nb and La+La systems, but the combined effects of lower beam intensity and lower beam energy, and the need for thinner targets to maintain a fixed fragmentation loss reduced the yield below that measurable in the allotted time. This apparent saturation, or roll-over, of the cross section is one of the tantalizing features that will remain for future workers in this field to explore.

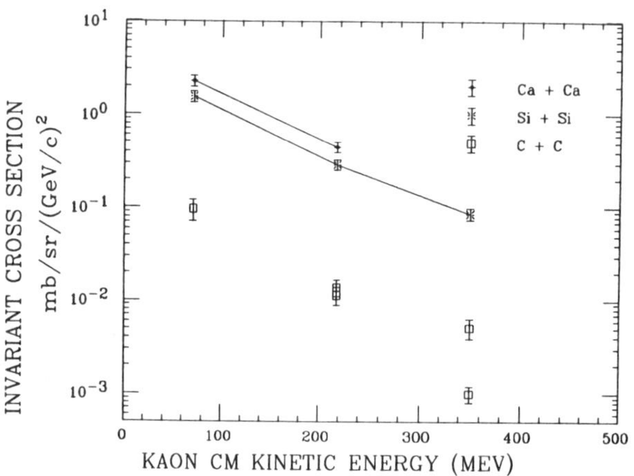

Figure 4. Mass Dependence at 2.1 GeV/A

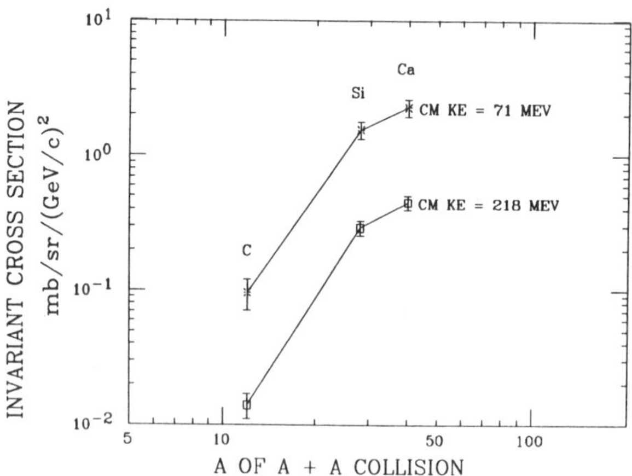

Figure 5. Mass Dependence at 2.1 GeV/A

This experimental program began as a search for anti-proton production in relativistic heavy ion collisions, and during each of the experiments we have usually produced a single event with the characteristics of an anti-proton, but have never had sufficient integrated flux to collect a number of such events, nor particle ID quite sufficient to completely persuade ourselves that the identification was correct. During our last data acquisition period, however, we have accumulated 5 such events, all other events being clearly identified as either pions or kaons; that is to say, no background. These 5 events have the timing characteristics of anti-protons; the mean time-of-flight of the group is within 50ps of the measured time-of-flight of protons under the same experimental conditions, and the width of this distribution is also consistent with the 50ps sigma we measure for pions under these conditions. (The pion-proton time-of-flight difference is 2.9 ns.) The time-of-flight measured by a third high-resolution detector, placed roughly mid-way between the other two, is consistent, for edach event, with that of protons, and inconsistent with that of pions or kaons. A lead glass calorimeter shows that 4 out of the 5 events give pulse heights much larger than those produced by pions and kaons. (Since the calorimeter does not absorb all of the incident energy, this ratio of 4/5 is what is expected.) Auxillary apparatus shows that the events in question are 'clean', with no pile up in crucial detectors, or multiple particles traversing the detector assemblies. The collaboration feels that the present level of evidence that these events are in fact anti-protons is at least as good as that on which the discovery of the anti-proton was based, but we are still working on ways of making this more quantitative. The observed yield (p-bar/π^-) is about 4×10^{-7} (at 1.89 GeV/c), corresponding to a laboratory cross section of about 80 nb/sr/(GeV/c).

REFERENCES

[1] See Stachel, et al., Phys. Rev. C 57 (198?) and references therein.

[2] The collaboration which has done the work presented here:
UCLA — S. Carlson, J. Carroll, J. Gordon, G. Igo, S. Trentalange, J.-F. Wang, K. Ganezer, S. Abachi.
LBL — B. Keay, G. Krebs, P. Lindstrom, V. Perez-Mendez, T. Mulera.
BNL — A. Shor.
LSU — P. Kirk.
Johns Hopkins U. — T. Hallman.
UC Davis — E. Barasch.

[3] Shor, et al., Phys. Rev. Lett. 48 (1982); Barasch, et al., Phys. Lett. 161B (1985).

INVESTIGATION OF RESONANCE STATES IN THE GLUEBALL-ENHANCED CHANNEL $\pi^- p \rightarrow \phi\phi n$ †

S.J. Lindenbaum
Brookhaven National Laboratory and City College of New York

K.J. Foley, S.E. Eiseman, A. Etkin, R.W. Hackenburg,
R.S.Longacre, W.A. Love, T.W. Morris, E.D. Platner,
A.C. Saulys
Brookhaven National Laboratory, Upton, New York 11973

C. Chan, M.A. Kramer
City College of New York, New York, New York 10031

ABSTRACT

In this paper we discuss the striking evidence for the $g_T(2010)$, $g_{T'}(2300)$ and $g_{T''}(2340)$ $I^G J^{PC} = 0^+2^{++}$ resonances which comprise virtually all of the $\pi^- p \rightarrow \phi\phi n$. The complete breakdown of the expected OZI suppression, and the striking differences of these states from conventional states and background in other channels has so far only been successfully explained by assuming they are produced by 1-3 2^{++} glueballs. The comparison with J/ψ radiative decay results is made. The forthcoming search for an exotic J^{PC} glueball is discussed.

INTRODUCTION AND GENERAL INTERPRETATION

QCD has glueballs embedded in its guts since locally gauge invariant $SU(3)_{color}$ (without quarks) would have glueballs as the only hadrons in the world. The addition of quarks can in no way remove the glueballs. However the Particle Data Group Table has lots of $q\bar{q}$ states including several nonets of (uds) quarks but no glueball section. This indicates that glueballs are suppressed, or not recognizeable in the usual OZI-allowed (quark dominated) production. If at least one glueball is not established in our opinion QCD will become just another failed theory in the strong coupling non-perturbative region.

† Research carried out under the auspices of the U.S. Department of Energy under Contract Nos. DE-AC02-76CH00016 (BNL), DE-AC02-83ER40107 (CCNY), and the City University of New York PSC-BHE Research Award Program.

Since 1978[1-2] we have used the OZI-forbidden reaction $\pi^- p \to \phi\phi n$ hoping the pure glue intermediate state would resonate for glueballs (see Fig. 1) while suppressing $q\bar{q}$ states etc. Thus this reaction would act as a very selective filter which readily passes glueballs (if they exist) while strongly rejecting $q\bar{q}$ states and thus provide an excellent vehicle for detecting and establishing glueball states. In strong interactions where there are so many possibilities of explanation due to the strong coupling, one has traditionally needed very striking and unusual characteristics of the data which forces you to accept a new phenomenon in order to establish it. This was true for Δ, J/ψ, Υ, etc.

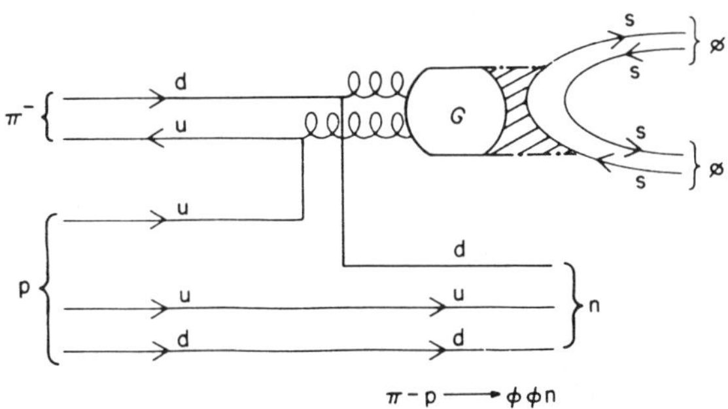

Figure 1: $\pi^- p \to Gn \to \phi\phi n$ Quark Line Diagram

In a partial wave analysis and unitary K-matrix fit to these results we have found three [g_T, $g_{T'}$, $g_{T''}$][2-3] relatively very high cross section Breit-Wigner resonances in the $\phi\phi$ with $I^G J^{PC} = 0^+ 2^{++}$ which completely break down the OZI suppression and no continuium or other states (within errors). They tower (by a factor ~ 50) over the $\phi K^+ K^-$ OZI-allowed physical reaction which has mostly a featureless background and no evidence for any particular resonances. They also tower over the $K^+K^-K^+K^-$ reaction by a factor ~ 1,000 which again appears to be featureless. The production mechanism is π-exchange ($\pi^+\pi^-$ annihilation). There is no indication whatsoever of the h(2030) ($I^G J^{PC} = 0^+ 4^{++}$) which should be strongly produced in π-exchange thus showing how good the filtering action of this channel is against $q\bar{q}$ states. This is consistent with what one would expect from the reaction shown in Fig. 1. In contrast in $\pi^- p \to \phi K^+ K^- n$ where the K^+K^- pair is just above the ϕ in mass so that the kinematics etc. are very similar we find ≈ 67% is structureless (flat background), ≈ 30% is 1^{--} (a kinematic effect for an s-wave $\phi K^+ K^-$ system produced by π-exchange), and only 3% is $J^{PC} = 2^{++}$ which appears different and non-resonant. Also in $\pi^- p \to K_S^o K_S^o n$ in the region of the g_T resonances[4] we find a structureless behavior with no indication of resonances in the $J^{PC} = 2^{++}$ in the 2.0 - 2.5 GeV mass region where our $\phi\phi$ data lie, but we clearly find the h(2030) in $J^{PC} = 4^{++}$ in striking contrast to its absence in $\phi\phi$.

These very striking phenomena are very naturally explained in the context of QCD by production of 1-3 $I^G J^{PC} = 0^+2^{++}$ glueballs.[2-3] At least one broad primary $J^{PC} = 2^{++}$ glueball is required to explain the complete breaking of the Zweig suppression and selection of only $J^{PC} = 2^{++}$ resonant states, and the unique observation of these states only in the $\phi\phi$ system. One could in principle then mix with one or two other conventional $q\bar{q}$ states. However the simplest and natural explanation of our data within the context of QCD is that we have found a triplet of $J^{PC} = 2^{++}$ glueballs, which is the expected number of low-lying $J^{PC} = 2^{++}$ glueballs. Although there have been several attempts to explain our data by various other assumptions they have been shown to be incorrect, do not fit the data or both.[3] Another arguement advanced is that perhaps we have seen radially excited mixed states such as the η' which are formed via ordinary u,d quarks and decay via the $s\bar{s}$ quark content.*

The 2^{++} nonet is well-known to be ideally mixed. In QCD there are only two basic flavor mixing mechanisms (i.e. $q\bar{q} \rightarrow s\bar{s}$ where q = u,d): 1) Vacuum mixing which mixes the η and η' about as far as possible from ideal mixing. Vacuum mixing is expected to be most important for J = 0. It clearly does not mix the $J^{PC} = 2^{++}$ nonet which is well-established and ≈ ideally mixed nor does it affect the well-established 1^{--} nonet, and the reasonably established 3^{--} nonet. 2) Glueball mixing. In a coupled channel partial wave analysis of the 2^{++} world sample of data[4] we have shown that even the $f_r(1810)$ the radially excited f (the singlet of the octet) is composed of ud quark and anti-quark pairs, and does not have any appreciable $s\bar{s}$ content. Thus it clearly follows that the radially excited singlet should be ≈ pure $s\bar{s}$. Hence its production would be Zweig suppressed. Furthermore there is no explanation why in $\phi\phi$ we see three closely spaced isosinglet $J^{PC} = 2^{++}$ resonances, and nothing else whereas in ϕK^+K^- and $K^0_S K^0_S$ we see virtually all non-resonant background and no evidence of resonances. Thus our data can naturally be explained within the context of QCD by production of 1-3 $J^{PC} = 2^{++}$ glueballs and there is no viable alternate explanation to date[3] which fits its unusual characteristics.[2-3]

THE PARTIAL WAVE ANALYSIS AND K-MATRIX FITS

We used the standard LBL/SLAC isobar model program to analyze our 6658 number of events. Due to the narrowness of the ϕ the analysis becomes independent of the isobar model characteristics. Of course we replaced the spectator particle with the second ϕ which also decays giving us six angles the Gottfried-Jackson polar and azimuthal angles (β,γ) in the $\phi\phi$ system and the polar and azimuthal angles ($\theta_{1,2}$ and $\phi_{1,2}$) in each ϕ rest system. The spin J = 1 of the ϕ makes each of these six angles and their correlations incredibly powerful wave selectors. We used all waves which satisfy Bose symmetry and have L = 0-4 and J = 0-6 for a total of 114 waves. We simultaneously analyzed the small ϕK^+K^- physical background which

---

* Comment by F. Close during discussion.

occurred at the level of a few percent but used wide cuts (14 MeV) on the ϕ to ensure lack of biases and allow a search for a phase reference. This raised the level of this background to $\approx 13\%$. We allowed up to J = 2 (27 waves) in its analysis, so that interference effects could be most accurately taken into account and result in maximum precision. The only other physical effect involved was the four kaon background which occurred at a level ~ 0.1% and thus was of negligible significance.

Figure 2 shows the result of the partial wave analysis of the $\phi\phi$. The three $J^{PC} = 2^{++}$ amplitudes and the phase behavior of the

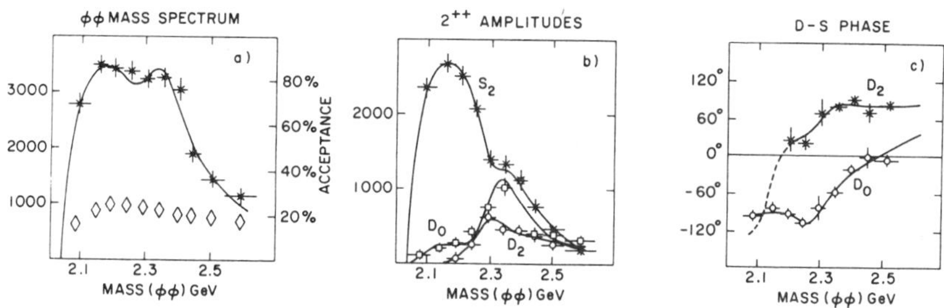

Figure 2: (a) The acceptance corrected $\phi\phi$ mass spectrum, (b) intensity and (c) phase difference for the three $J^{PC} = 2^{++}$ waves. The curves show the fit by three Breit-Wigner resonances (i.e. K-matrix poles).

two D-waves relative to the S-wave as a phase reference clearly show resonant phase behavior. The significance of this particular set of waves was 13σ and they gave a good fit to the data. $M^{\eta} = 0^{-}$ for all three waves which together with the $d\sigma/dt' \approx 9.5$ for $|t'| < 0.3$ (i.e. $\approx 95\%$ of the data) demonstrates that pion exchange is the production mechanism (see Fig. 3). We fit the partial wave analysis results with K-matrix poles which preserve unitarity and correspond on a one-to-one basis with Breit-Wigner resonances. The curves shown in the various figures correspond to our best fit $\approx 2\sigma$. The unitary effects turned out to be small so that even if we had fit with simple complex Breit-Wigners we would have obtained consistent values. A partial wave analysis of the physical background reaction ϕK^+K^-n (Fig. 4) shows $\approx 67\%$ flat (in all angular variables) background, only $\approx 3\%$, $J^{PC} = 2^{++}$ in the $\phi\phi$ and $\approx 30\%$ $J^{PC} = 1^{--}$ (the expected result for all particles in relative S-wave produced by pion exchange).

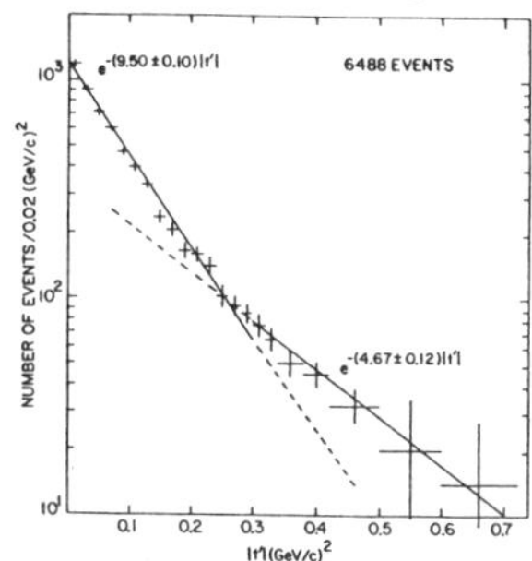

Fig. 3: log(const. × $d\sigma/dt'$) plotted versus $|t'|$.

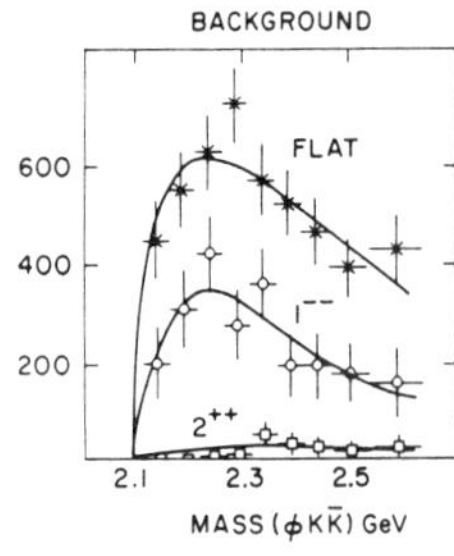

Fig. 4: Intensity of the partial waves in the background reaction $\pi^- p \rightarrow \phi K^+ K^- n$.

In order to calculate absolute phase motion we used the 1^{--} ϕK^+K^- wave (which is coherent with the $\phi\phi$ waves as shown by interference) as a phase reference. As shown in Fig. 5 we then converted our $\phi\phi$ phases to absolute phase for the two extreme cases. a) If one assumes ϕK^+K^- is caused by a peripheral reggeized Deck type mechanism one expects to obtain the minimum phase taversal of the 1^{--} wave over the mass region of the $\phi\phi$ data. b) Assuming that the 1^{--} wave corresponds to a resonance (K-matrix pole) gives the maximum phase traversal and the resultant absolute phase, gave the Argand diagrams shown in Fig. 5a and 5c. Both extreme cases gave similar classic Breit-Wigner three resonance behavior. Thus the actual case which lies in-between these extremes obviously has classic Breit-Wigner resonance behavior for all three resonances.

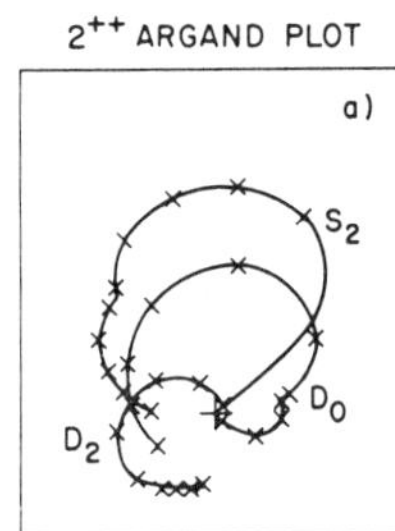

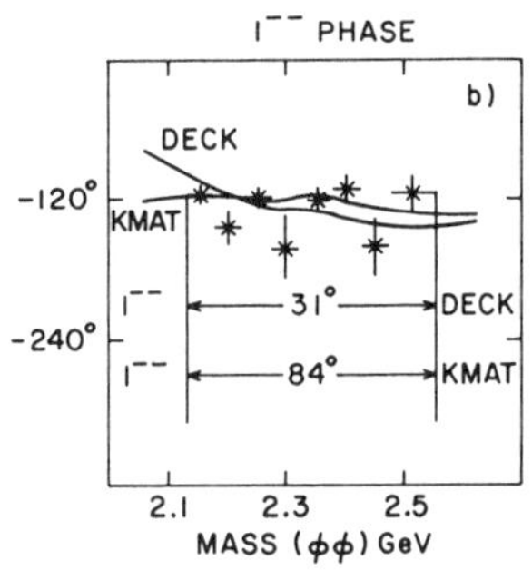

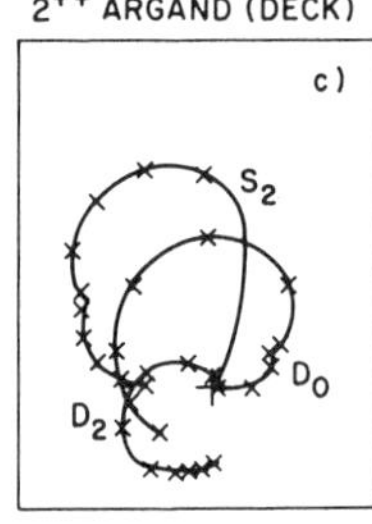

Figure 5: (a) Argand plots for the three 2^{++} waves with the absolute phase based on the 1^{--} $\phi K\bar{K}$ wave being a Breit-Wigner resonance (K-matrix), while (c) is based on the 1^{--} $\phi K\bar{K}$ resulting from a deck mechanism. (b) shows the 1^{--} phase (relative to the S-wave in $\phi\phi$). The curves come from the two models stated for the 1^{--} absolute phase.

The parameters of the Breit-Wigner resonances etc. are given in Table I.

COMPARISON WITH SLAC MK III RESULTS

Another filter studied for enrichment of glueballs compared to conventional $q\bar{q}$ states is the radiative decay of the $J/\psi \rightarrow \gamma gg$ or hopefully preferentially $J/\psi \rightarrow \gamma G$ where G is a glueball. The question hs been raised at this conference and elsewhere as to why our $\phi\phi$ states have not been seen in this channel. In order to answer this we must consider the characteristics of this channel and compare to our channel. In the radiative decay of the J/ψ conventional $q\bar{q}$ states such as η, η' in $J^{PC} = 0^{-+}$ and f, f' in $J^{PC} = 2^{++}$ are seen

TABLE I

Parameters of the Breit Wigner Resonances (corresponding to the K-matrix poles) and percentage of the resonances going into 2^{++} S_2, D_2, and D_0 channels. The errors come from a complete study of the χ^2 surface.

State	% of $\phi\phi$ Data	Mass (GeV)	Width (GeV)	S_2(%)	D_2(%)	D_0(%)
g_T	45	$2.011^{+0.062}_{-0.076}$	$0.202^{+0.067}_{-0.062}$	98^{+1}_{-3}	0^{+1}	2^{+2}_{-1}
$g_{T'}$	20	2.297±0.028	0.149±0.041	6^{+15}_{-05}	25^{+18}_{-14}	69^{+16}_{-27}
$g_{T''}$	35	2.339±0.055	$0.319^{+0.081}_{-0.069}$	37±19	4^{+12}_{-4}	59^{+21}_{-19}

with relatively sizeable cross sections compared to the glueball candidates iota and θ respectively. Furthermore there are relatively large cross sections of the general continium (at least unresolved into resonances) type with both quantum numbers. Therefore the J/ψ radiative decay channel clearly cannot be a tight glueball filter or these conventional objects would not appear in it. This is in contrast to the $\pi^-p \to \phi\phi n$ where no conventional objects such as the flat featureless background that appears in $\pi^-p \to K_S^0K_S^0n$ does not appear in $\phi\phi$. The h(2030) with $J^{PC} = 4^{++}$ does not show even a trace in $\pi^-p \to \phi\phi n$ even though we can very easily detect even small amounts of it with the enormous analysis power of the $\phi\phi$ system. In fact nothing conventional or anything else is found except the three $J^{PC} = 2^{++}$ g_T, $g_{T'}$ and $g_{T''}$ resonances. Thus there are huge factors in the filter action for glueballs in favor of the $\pi^-p \to \phi\phi n$ channel. Secondly the width of the $J/\psi \to ggg$ agrees with perturbative calculations. The branching ratio $J/\psi \to \gamma gg$ also agrees with perturbative calculations. Therefore if J/ψ radiative decay were strongly coupled to glueballs one might expect that the process $J/\psi \to \gamma G$ would enhance the radiative decay branching ratio and in the process increase the width of the J/ψ. As far as we know no theorist has properly addressed this question although we have raised it before.

The MK III collaboration[5] reported that in J/ψ radiative decay they observe several hundred $\phi\phi$ which they conclude have $J^{PC} = 0^{-+}$ and see no evidence for $J^{PC} = 2^{++}$ $\phi\phi$. In J/ψ radiative decay one would expect the 0^{-+} channel to be strongly enhanced, since the process $J/\psi \to \gamma\eta_c$(virtual) $\to \gamma$ hadrons (with $J^{PC} = 0^{-+}$) would be expected to be strongly enhanced because it proceeds via spin flip of the $c\bar{c}$ pair and is also enhanced at high masses (2.0 - 2.5 GeV) by the influence of the η_c pole at 3.1 GeV. Therefore with limited statistics (a few hundred) a 0^{-+} $\phi\phi$ continuum (or other resonances) could well be enhanced and accompanied by relatively few $\phi\phi$ in a 2^{++} state. The expected relative strength of the 0^{-+} channel in J/ψ

radiative decay is consistent with MK III data. It should be noted that in $\pi^-p \rightarrow \phi\phi n$, the production process for glueballs would be as previously noted virtually all $\pi^-\pi^+ \rightarrow G$ and thus only 0^{++}, 4^{++}, 6^{++} .... etc. would be possible. Thus we would not see the 0^{-+} channel in our experiments. Secondly as previously discussed since conventional $q\bar{q}$ objects or continuum are obviously from our results filtered out, we would not expect to see anything that does not contain resonating gluons. Therefore we do not feel there is any basis for believing there is an inconsistency between the two experiments. However in comparing our investigation on another reaction $\pi^-p \rightarrow K_S^0K_S^0n$ and our coupled channel K-matrix unitary analysis using the world data[4] including MK III, we cannot reconcile our experiment, the LASS experiment on $\pi^-p \rightarrow K_S^0K_S^0\Lambda$, and the MK III data on the 2^{++} channel which contains the f and f' and θ. We find in that analysis that LASS should have seen the θ at a level about half the f' peak. If we leave SLAC MK III data out we can fit all hadronic experiments easily. If we leave LASS data out we can also fit the remaining data but we have not succeeded in fitting both. Meshkov[7] pointed out in his talk although $J^{PC} = 2^{++}$ is assigned to the θ, 0^{++} is not ruled out. If $J^{PC} = 0^{++}$ then our coupled channel analysis which assumed θ has 2^{++} would not be relevant. We leave this as an unresolved problem at present.

What about Meshkov oddballs[6] (Exotic Glueballs)? One calls a particle exotic if its J^{PC} cannot be made from a $q\bar{q}$ pair. This obviously does not apply to glueballs which one could expect could easily have exotic quantum numbers. However our $\phi\phi$ production process obviously involves π-exchange (i.e. $\pi^+\pi^-$ annihilation into gluons) therefore we can only have $J^{PC} = 0^{++}$, 2^{++}, 4^{++} ... and thus cannot make exotic glueballs. However as you can see from Fig. 3 for $|t'| \geq 0.3$ there is a break in slope toward smaller slopes implying that $\bar{A}$-exchange etc. which can make exotic glueballs can occur. At present only about 5% of our data is in this region which are clearly too few events to allow a serious analysis. We plan in our next runs to enhance observation of A-exchange by a factor ~ 8 by going to lower energies and increasing our apparatus acceptance. This will also improve our statistics per unit time in π-exchange. Our goal will be to search for possible Meshkov oddballs (exotic J^{PC}). The question of the mass of such glueballs is of course quite uncertain. However an experimental search in the mass region we have available is clearly the next step in this glueball program.

SUMMARY

We have gathered and partial-wave analyzed 6658 events of $\pi^-p \rightarrow \phi\phi n$, and for maximum precision simultaneously analyzed the small physical background reaction $\pi^-p \rightarrow \phi K^+K^-n$ accompanying the $\phi\phi$. We find the $\phi\phi$ which is OZI-forbidden towers above the ϕK^+K^- which is OZI-allowed by a factor ~ 50 when corrected for resolution. The $\phi\phi$ events are composed entirely within errors of three $I^GJ^{PC} = 0^+2^{++}$ resonances the $g_T(2010)$, $g_{T'}(2300)$ and $g_{T''}(2340)$ whereas in contrast the ϕK^+K^- is mostly structureless background, shows no evidence for resonances and has only 3% $J^{PC} = 2^{++}$ compared to the

$\phi\phi$ which has 100% resonant 2^{++}. We were able to use the coherent 1^{--} wave of the ϕK^+K^- to calculate absolute phase ranges for all three $\phi\phi$ resonances and show that they exhibit classic Breit-Wigner resonance Argand plot behavior. They have been shown to be produced by the mechanism of pion exchange. In contrast all other $J^{PC} = 2^{++}$ channels in the 2.0 - 2.5 GeV region (the mass range of the g_T's) do not show any of these resonances.

Thus we have found a set of very striking phenomena which have not been explainable by conventional means. The striking characteristics of our data can be naturally explained by assuming that 1-3 primary glueballs with $J^{PC} = 2^{++}$ produce these states. At least one primary $J^{PC} = 2^{++}$ glueball is necessary to explain the selective breakdown of the OZI suppression and the strong filtering which led to the absence of other states such as the h(2030) and background found in other experiments. Other attempts to explain our data have been shown to be incorrect, do not fit the data or both.[3]

A comparison with the SLAC MK III J/ψ radiative decay was made and it was concluded that due to the fact that they must have a much weaker glueball filter as evidenced by their seeing $q\bar{q}$ states such as η, η', f, f', etc. and background there is no inconsistency in their not seeing our $\phi\phi$ states. Furthermore their radiative decay process should favor 0^{-+} especially at the high masses. We cannot observe 0^{-+} due to the π-exchange production mechanism. We also found from a unitary coupled channel analysis of the world's data that we cannot understand the absence of the θ in the LASS experiment $K^-p \to K_S^0 K_S^0 \Lambda$.

REFERENCES

1. A. Etkin *et al.*, Phys. Rev. Lett. 40 (1978) 422; 41 (1978) 784; 49 (1982) 1620.

2. A. Etkin *et al.*, Phys. Lett. B 201 (1988) 568.

3. S.J. Lindenbaum, Comm. Nucl. Part. Physics 13 (1984) 285; in: *Superstrings, Supergravity and Unified Theories*, the ITCP Series in Theoretical Physics, Vol. 2 (World Scientific, Singapore, 1986) pp. 548-593.

4. R.S. Longacre *et al.*, Phys. Lett. B177 (1986) 223; and, paper to be published.

5. W. Wasniewski, "Glueball Candidates in Radiative J/ψ Decay," this conference.

6. S. Meshkov, "Theoretical Overview of Glueballs and Hadron Spectroscopy," these proceedings.

7. S. Meshkov, *Experimental Meson Spectroscopy - 1983 (Seventh International Conference, Brookhaven)*, S.J. Lindenbaum, Editor, AIP Conf. Proc. No. 113, pp. 125-152; P.M. Fishbane and S. Meshkov, Comm. Nucl. Part. Physics 13 (1984) 325.

PARITY MIXING OF THE 0^+–0^-;T=1 DOUBLET IN ^{14}N†

V.J. Zeps, E.G. Adelberger, A. Garcia, C.A. Gossett, H.E. Swanson
University of Washington, Seattle Wa. 98195

W. Haeberli, P.A. Quin, J. Sromicki*
University of Wisconsin, Madison, Wi. 53706

ABSTRACT

We have examined parity mixing of the 0^+,0^-;T=1 doublet in ^{14}N ($E_x \sim 8.6$MeV) to study the isoscalar part of the weak nucleon-nucleon interaction. The weak matrix element connecting the two levels is extracted by measuring the longitudinal analyzing power (A_L) in the elastic scattering cross-section across the narrow $^{13}C + \vec{p}$ $J^\pi = 0^+$–resonance at $E_p = 1.16$ MeV.

INTRODUCTION

Parity violation (PV) in nucleonic systems is the only available probe for studying strangeness conserving non-leptonic weak interaction processes. (See eg.– [1] for a detailed review of the subject). The objective of low-energy parity violation studies (E< 300 MeV), is to determine the strengths of the weak nucleon-nucleon-meson (NNM) coupling constants that characterize the weak N–N interaction. Ideally, the weak N–N force would be determined by six independent N–N experiments that are sensitive to different combinations of the π, ρ, ω–meson exchanges. However, the strong interaction dwarfs weak interaction effects so that pseudoscalar observables in the N–N system are predicted to be $\sim 10^{-7} - 10^{-8}$ rendering these measurements difficult at best. To date, parity violation in the N–N system has been observed only in the longitudinal analyzing power (A_L) of the elastic proton-proton scattering cross-section [2,3]. The quantity A_L is defined as

$$A_L = \frac{1}{|P_z|}\left(\frac{\sigma^+ - \sigma^-}{\sigma^+ + \sigma^-}\right), \tag{1}$$

where P_z is the beams longitudinal polarization, and $\sigma^+(\sigma^-)$ represents the cross-section for the incident beam's polarization being parallel(anti-parallel) to it's momentum. By far the most precise result is the elegant SIN measurement of $A_L = (-1.50 \pm 0.22) \times 10^{-7}$ [2]. Most impressive was the treatment of systematic effects, allowing elimination or correction of all systematic errors to a level of $A_L^{false} < 1 \times 10^{-8}$. To be able to use this result, however, additional experiments that can be interpreted in terms of a weak N–N interaction must achieve sufficiently high precision. Much of the recent emphasis has been on measuring parity violation through the mixing of isolated opposite parity doublets in light nuclei [4,5,6]. There are several reasons for the utility of these measurements.

1. In light nuclei, the true many-body wavefunctions should be a fairly reliably reproduced using the shell model.

2. Isolated doublets offer the best hope for extracting weak N–N interaction parameters from shell model calculations, since the parity impurity is due primarily to one near-by opposite parity level (simple two-level mixing). The closeness of the levels also helps to enhance any PV effect due to the small energy denominator in the perturbation expansion.

3. For good candidates, one of the two levels is a strongly hindered decay, while the other is allowed, again serving to amplify any PV observable.

†supported in part by DOE and NSF

*present address Eidgenossische Technische Hochschule, Zurich, Sw.

4. One can take advantage of the nuclear structure to isolate the different isospin dependences of the weak N–N interaction.

Shown in figure 1, is a two parameter analysis of existing parity violation measurements interpreted in terms of the two dominant weak coupling constants, F_π (isovector weak π–exchange) and F_o ($\approx$isoscalar weak ρ–exchange). The SIN measurement is not analyzed here due to a strong isotensor contribution to A_L. The data is sparse yet already some discrepancies exist. The interpretation of the ^{21}Ne result is suspect, however, since there is no check of the nuclear structure calculation, as was available for the ^{18}F and ^{19}F measurements [7]. The prospect for improving these measurements is slim, thus new information must result from new experiments.

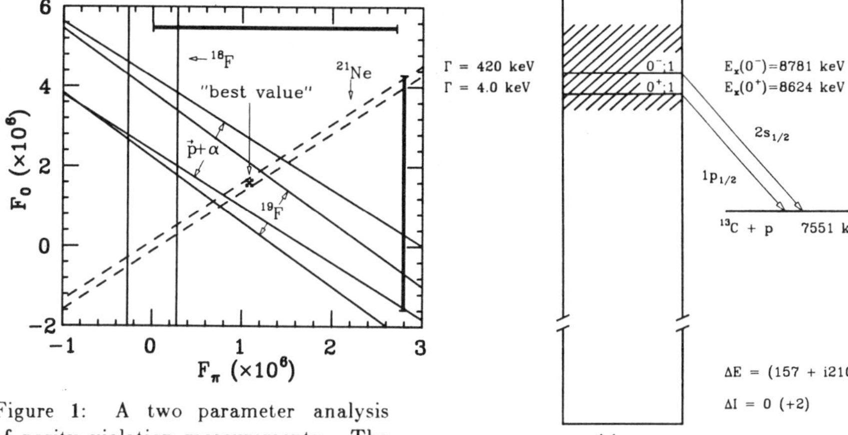

Figure 1: A two parameter analysis of parity violation measurements. The bands are 1σ error bars. The "best value" is shown for the best available bag-model prediction [8], with the double bars repesenting the models reasonable range.

Figure 2: $0^+,0^-;T = 1$ doublet in ^{14}N energy level parameters and proton decay channels.

The $0^+,0^-$;T=1 doublet at $E_x \sim 8.6$ MeV in ^{14}N is another system that meets the above stated criteria for complex nucleus PV experiments (see fig. 2). The parity mixing is predominantly isoscalar, thus the PV observable in ^{14}N is sensitive to F_o, complimenting previous measurements.. Since the levels are unbound to proton decay, the obvious PV observable is the circular polarization of the decay protons, or equivalently, the helicity dependence (i.e.–longitudinal analyzing power, A_L) of the scattering cross-section. A_L, which is a measure of the weak matrix element connecting the two levels, can be described as a first order perturbation in the standard S-matrix formalism of elastic scattering [9]. The fact that the levels are unbound would normally introduce large theoretical uncertainties due to complicated strong interaction dynamics. The simplicity of the ^{14}N system resides in the unique allowed L–S channels of the resonant scattering, and the fact that elastic scattering is the only open hadronic channel, thus circumventing the usual uncertainties associated with multi-channel coupling effects. Based on the S-matrix theory, and a weak matrix element calculation which used the DDH "best value" PV N–N interaction and a full $2\hbar\omega$ and $1\hbar\omega$ model space for the 0^+ and 0^- levels, respectively, our expected PV observable is $\approx -2.8 \times 10^{-5}$ (see ref. [9] for further theoretical details).

EXPERIMENTAL APPARATUS

The experimental apparatus was developed with symmetry and stability as major design considerations. The detection scheme is shown in figure 3. The maximum count rate attainable for this experiment is governed by the 4.0 keV width of the 0^+–resonance, since the optimum

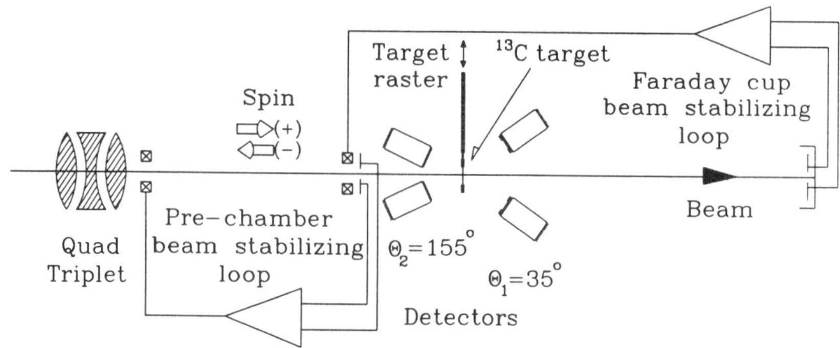

Figure 3: Schematic of the experimental apparatus. The apparatus is essentially identical in both x and y planes, where z defines the beam axis.

target thickness is of that order, and the maximum possible current on target, I$\leq$ 1μA. Thus, the experiment could be done by counting the scattered protons individually (expected rates $<$ 1MHz). Since elastic scattering is the only open particle decay channel, detector resolution could be sacrificed for speed. We elected to use thin plastic scintillators mounted on 5.1 cm PMT's. The relevant features of the detectors are; $\sim$ 10nsec pulsewidth, $\sim$ 25 – 30% resolution, a stiff base current for gain stability, robustness toward radiation damage, and large solid angle. We used a four-fold (left-right, up-down) detection geometry, with eight detectors aximuthally symmetric about the beam axis, four at $\Theta_1 = 35^o$ and four at $\Theta_2 = 155^o$. The parity violating signal is the difference in longitudinal analyzing powers, $A_L(\Theta_2) - A_L(\Theta_1)$ (which we define as A_L(back-front)). The symmetry of this detection scheme is apparent. The four-fold symmetry desensitizes the detector to position and angle modulations that may be correlated with beam helicity reversal. The back-front difference signal gives us a self-normalizing system, and desensitizes the detector to small target thickness variations. To further desensitize the apparatus to the target irregularities, a target rastering device is used, allowing us to precisely vary the position of the target w.r.t. the beam in both lateral directions. This averaging smears out carbon build-up over a larger area, as well. The rastering feature is also used in beam diagnostic tests with strip targets described below.

The beam position and angle are stabilized on target by a pair of fast-ferrite core based feedback magnets (correction up to 2kHz), controlled by the difference signals of the pre-chamber slits and split faraday cup, allowing stability at the spin-flip frequency of better than 1μm and 10μrad in position and angle, respectively. The long-term constancy of the beam centroid is held to $<$ .2mm($<$ .3mrad); dominated by drifts in beam's emittance, since the slits that drive the feedback loops intercept only the halo of the beam. The spin direction is stabilized by taking advantage of the small transverse analyzing power observed in the front (Θ_1) detectors. The computer controlled feedback loop calculates the spin direction, based on the observed transverse asymmetries ($P_x A_T$, $P_y A_T$) of each run, and known transverse analyzing power (A_T). The necessary correction to the spin precessor is automatically applied after each run if statistically non-zero asymmetries are observed. The long-term correction capabilities are remakable. For a typical two week run, transverse asymmetries are kept to a level of $\overline{P}_T A_T < 2 \times 10^{-5}$.

SYSTEMATIC ASYMMETRIES

The feedback loops mentioned above allow us to reduce the amount of position/angle modulations and net transverse polarizations correlated with position/angle misalignments, so that their resultant systematic asymmetry is satisfactorily small: $A_L^{false}(b - f) < 2 \times 10^{-7}$. The remaining sources of systematic error result from effects we were unable to control effectively,

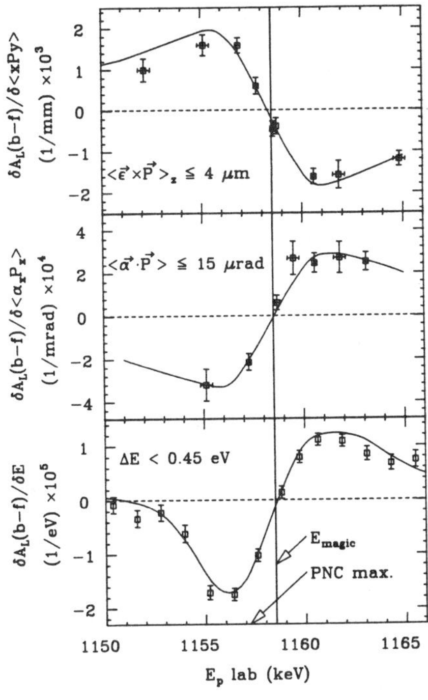

Figure 4: The sensitivities to the three dominant sources of systematic error are plotted vs. energy, shown with the "worst case" limits for the given beam attributes.

Panels 1 & 2–Sensitivity to a transverse polarization gradient in position ($< \vec{\epsilon} \times \vec{P} >_z$) and angle ($< \vec{\alpha} \times \vec{P} >$). The finite beam size (angle spread) coupled with a position (angle) dependent transverse polarization and a non-zero transverse analyzing power (A_T) in the reaction produces a false asymmetry. We determine these sensitivities by measuring the position (angle) dependence of the count rates with a transversely polarized beam. The solid lines are predicted sensitivities obtained by numerically integrating the cross-section over beam emittance, detector solid angle, and target thickness.

Panel 3–Energy modulation sensivity (Δ E). The data points were measured by modulating the target bias by ± 50V. The curve is simply $1/2\delta \ln(B/F)/\delta E$, where B/F is just the ratio of count rates in the back (Θ_2) vs. front (Θ_1) detectors.

namely, energy modulation and second order transverse spin-phase space correlations.

Energy modulations—Suppose that when the helicity of the proton beam was reversed, the mean beam energy changed by an amount ΔE. This would lead to an $A_L^{\Delta E}(b-f)$, due to the difference in the energy dependence of the count rates for front vs. back detectors (panel 3, figure 4). At the maximum of the PV signal, the sensitivity to energy modulation is $\delta A_L(b-f)/\delta E = 15 \times 10^{-6}/eV$, thus energy modulation in the beam would need to be $\Delta E < 0.2eV$ to eliminate it as a significant source of systematic uncertainty. To measure the amount of energy modulation, we used two techniques. The first technique took advantage of the narrow (140 eV) 2^+-resonance in the ^{13}C(p,γ) reaction at $E_p = 1.75$ MeV. The extremely sharp energy dependence of the cross-section leads to a maximum fractional γ-yield change of $\Delta R/2R \sim .003$ per electron-Volt of energy modulation. With this technique, after two days of counting with a nearly 2π detection efficiency, an upper limit of $\Delta E = 0.3 \pm 0.6eV$ was set [10]. To improve upon this, we considered the elastic scattering cross-section from the narrow ($\Gamma \sim 800eV$) s-wave resonances in the ^{27}Al(p,p) reaction at $E_p = 1.35$ MeV. We could now take advantage of the much larger elastic cross-section with our PV detector. The measured sensitivity with this technique was $\delta A_L(b-f)/\delta E = 9 \times 10^{-5}/eV$; statistical accuracy of $\Delta E = 0.2eV$ was achieved with 1.5 days of running time. Several measurements were taken using this technique, giving no clear indication for any energy modulation. We combine the data from both methods to produce limits on the amount of energy modulation, distinguishing between a possible energy modulation correlated with the R.F. transition unit, helicity on target, sign of the ionization solenoid, or intensity modulation. For this purpose, we adopt the "worst" value from this set for a strict upper limit of ΔE < 0.45 eV.

Transverse Polarization Gradients—The most important systematic effect arises from residual transverse polarization gradients inherent in the nominally longitudinally polarized beam (see panels 1,2 of fig. 4). The feedback loops null the average transverse polarization of the beam on target; inhomogeneities in the transverse polarization must be either eliminated or corrected for. It was not possible to use the technique of continuously measuring and correcting for the transverse polarization distibution developed for the SIN PV proton-proton scattering experiment [11]. At our low energy, producing such scanners is complicated by the lack of good polarization analyzers and the necessity for using thin (fragile) targets. The best we could do was to determine how bad the gradients could be, and select conditions (see below) where these inhomgeneities gave an acceptably small systematic error. To measure the transverse polarization distribution in the beam, we took advantage of the large analyzing power ($A_T(\Theta_2)=0.83$) in the elastic scattering from the $^{12}C + \vec{p}$ anomaly at $E_p = 1.8$ MeV. Strip targets were prepared by evaporating $\sim 100\mu gm/cm^2, 0.6mm$ wide carbon strips onto commercially available 3–5 $\mu gm/cm^2$ carbon foil backings. Scans showed clear evidence for a linear transverse distribution in position corresponding to as much as 2% at the beam edge. Similar correlations were also observed in the angular spread of the beam. At the beam energy of the maximum PV effect, these gradients would result in an error in the A_L (b–f) measurement which is ~50% of the expected effect. Consequently, a meaningful A_L measurement at this energy was not possible.

Fortunately, there exists a "magic" energy (E_{magic}) where the sensitivity to both energy modulation and the transverse polarization gradients roughly vanishes, while the PV signal is reduced by only 20% (see fig. 4). This allowed us to perform the experiment without the need to continuously monitor and correct the data for these effects.

RESULTS

The data were taken during three 2-week runs using the 1.16 MeV polarized proton beam from the crossed-beam polarized ion source and EN tandem accelerator at the Univ. of Wisconsin. We took data not only at the "magic" energy, but also at energies on the wings of the resonance where the sensitivity to these systematic effects is maximal. The optimum energy is stabilized to better than ±150eV by periodically precessing the spin 30° away from longitudinal and using the measured transverse analyzing power, $A_T(\Theta_2)$, to correct the energy. After acounting for the ~ 0.86 beam polarization and ~ 5.5% ^{12}C contaminant in the target, we observe A_L (b–f)= $(8.6 \pm 5.9) \times 10^{-6}$ with a systematic error estimate of 2.4×10^{-6} (fig.). The systematic uncertainty is derived from a "worst case" set of measured beam modulations and known detector sensitivities at $E_{magic} \pm 150$eV. The A_L results, when interpreted as a weak matrix element, yield $\langle H_{weak} \rangle = +i3.6 \pm 2.7$eV which is inconsistent with Haxton's prediction of $-i1.04$eV [9]. Here we define

$$\langle H_{weak} \rangle \equiv \langle 0^- | H_{pnc} | 0^+ \rangle \times \text{sign} \left(\frac{\langle {}^{13}C + p | H_{strong} | 0^- \rangle}{\langle {}^{13}C + p | H_{strong} | 0^+ \rangle} \right) \tag{2}$$

to relate the shell model phase to the scattering phase.

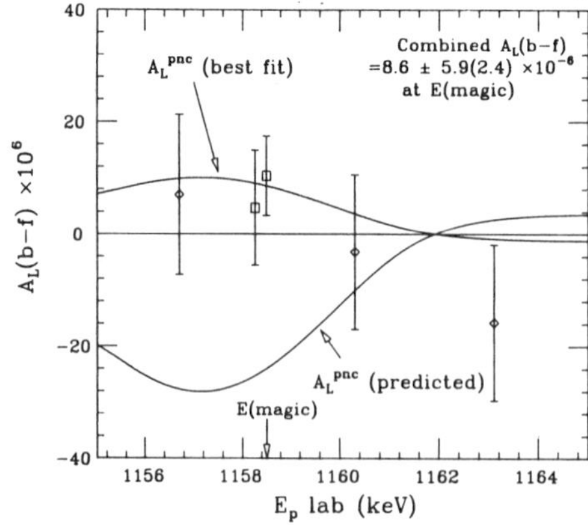

Figure 5: A_L (b–f) measurements (with 1σ error bars) are plotted vs. energy, along with the predicted and "best fit" A_L curves. We expect systematic effects as large as $\sim 15 \times 10^{-6}$ for the three data points (diamond) away from E_{magic}. At E_{magic}, "worst case" systematic effects are $\leq 2 \times 10^{-6}$. A_L^{pnc}(best fit) was determined from the two points near E_{magic}. Inclusion of the other data in the analysis, however, does not significantly alter our result.

INTERPRETATION

The discrepancy between our experimental results and the theoretical expectation is disconcerting to the extent that our results are inconsistent with the weak N–N interaction inferred from previous experiments. Before we can extract a reliable value for the F_o NNM coupling constant, we must have confidence in the nuclear structure considerations (shell model, scattering theory) used to predict the PV effect.

Several tests and refinements are being examined to help solidify the theoretical foundation for the expected A_L signal.

Shell Model— We tested the ^{14}N shell model wavefunctions, by comparing spectroscopic factors and electromagnetic transition rates for all the levels below 10.5 MeV excitation energy with shell model predictions. While there is general agreement with experimental values for these quantities, some interesting discrepancies exist, particularly for transitions for the predominantly $2\hbar\omega$ states. The trend of these discrepancies is symptomatic of too much configuration mixing. We hope to remedy this by doing the calculation with a better residual interaction.

A more likely problem is the question of the basis state representation used in the shell model. The unbound character of the 0^+ and 0^- levels suggest that a harmonic oscillator potential may not be suited for an accurate description. To demonstrate this effect, we consider a simple model

$$|0^+\rangle = \alpha|1p_{1/2}^{-2}\rangle + \beta|1p_{1/2}^{-4}2s_{1/2}^2\rangle + \gamma|1p_{1/2}^{-4}1d_{5/2}^2\rangle \quad (3)$$

$$|0^-\rangle = |1p_{1/2}^{-3}2s_{1/2}\rangle \quad (4)$$

with respect to an ^{16}O core. where shell model calculations and spectroscopic factor measurements suggest that $\alpha, \gamma \ll \beta$. The one-body reduction of the two-body parity violating operator is $\propto \vec{\sigma}\cdot\vec{p}$. In this simplification, the parity mixing matrix element is

$$\langle 0^-|H_{weak}|0^+\rangle \propto \alpha\langle 2s_{1/2}|\vec{\sigma}\cdot\vec{p}|1p_{1/2}\rangle + \beta\langle 1p_{1/2}|\vec{\sigma}\cdot\vec{p}|2s_{1/2}\rangle \quad (5)$$

If we use more realistic Woods–Saxon (WS) rather than the usual HO radial wavefunctions, this weak matrix element is reduced by ~40%. We have therefore written code to calculate the two-body weak matrix element using WS wavefunctions.

We are also investigating other questions reguarding the shell model. We are studying the effect of the neglected $3\hbar\omega$ component of the 0^- level, which connects to the dominant $2\hbar\omega$ component of the 0^+ level and should, in principle, be important. We also believe, that in order to describe properly the $2\hbar\omega$ interaction in ^{14}N, $4\hbar\omega$ states must be included in the shell model space. Preliminary results of a first-principles $4\hbar\omega$ calculation in ^{16}O support this contention.

S-matrix—It is well known that the S-matrix formalism is appropriate when applied to resonances whose widths are narrow compared to their excitation energies. The 0^- level, however, is fairly broad, (E_p = 1.35 MeV, Γ = 0.42 MeV), thus we are investigating possible inaccuracies in the S-matrix as applied to this level. We will compare the predicted to empirically determined scattering phase shifts, thus removing ambiguity in the scattering theory prediction of the PNC effect.

When these tests and refinements have been completed, we should be able to use the A_L measurement in ^{14}N to provide reliable constraints for the F_o weak NNM coupling constant.

References

[1] E.G. Adelberger and W.C. Haxton. *Annual Reviews of Nuclear and Particle Science* **35**, 501 (1985).

[2] S. Kistryn *et al.*, *Physical Review Letters* **58**, 1616 (1987).

[3] D.E. Nagle *et al.*, In *AIP Conf. Proc. 51*, page 218 (1978).

[4] H.G. Evans *et al.*, *Physical Review Letters* **55**, 791 (1985); M. Bini *et al.*, *Physical Review Letters* **55**, 795 (1985).

[5] E.G. Adelberger *et al.*, In AIP Conf. Proc. 69, page 1367 (1981); K. Elsener *et al.*, *Nuclear Physics* **A461**, 579 (1987).

[6] K.A. Snover *et al.*, *Physical Review Letters* **41**, 145 (1978); E.D. Earle *et al.*, *Nuclear Physics* **A396**, 211c (1983).

[7] E.G. Adelberger *et al.*, *Physical Review C* **27**, 2833 (1983)

[8] B. Desplanques, J.F. Donoghue, and B.R. Holstein. *Annals of Physics* **124**, 449 (1980).

[9] E.G. Adelberger, P. Hoodboy, and B.A. Brown. *Physical Review C* **30**, 456 (1984) and **33**, 1840 (1986).

[10] J. Sromicki *et al.*, *Nuclear Instruments and methods in Physics Research* **A255**, 611 (1987).

[11] W. Haeberli *et al.*, *Nuclear Instruments and methods in Physics Research* **A163**, 403 (1979).

PARITY VIOLATION IN PROTON-PROTON SCATTERING AT 230 MEV

J. Birchall[(1)], J.D. Bowman[(2)], C.A. Davis[(1)], N.E. Davison[(1)], P.W. Green[(3)], S.A. Page[(1)], W.D. Ramsay[(1)], N.L. Rodning[(3)], G. Roy[(3)], P.W. Schmor[(4)], A.V. Sekulovich[(1)], J. Soukup[(3)], G.M. Stinson[(3)], W.T.H. van Oers[(1)]
[(1)]Department of Physics, University of Manitoba, Winnipeg, Manitoba R3T 2N2
[(2)]Los Alamos National Laboratory, Los Alamos, New Mexico 87545
[(3)]Department of Physics, University of Alberta, Edmonton, Alberta T6G 2N5
[(4)]TRIUMF, Vancouver, B.C. V6T 2A3

ABSTRACT

Below ~300 MeV six meson-nucleon coupling strengths are required to describe the weak interaction of nucleons. Many experiments have found parity-violating effects in nuclear systems, but only four significant, independent constraints exist. A new measurement is proposed where measurable effects are predicted with minimal dependence on nuclear interaction models, yielding information complementary to previous results. This is a measurement of the parity-violating analyzing power A_z in p-p scattering at 230 MeV, which is shown to be sensitive to the weak ρ-nucleon coupling. This measurement, at a precision of $\pm 2 \times 10^{-8}$, together with a proposed measurement by the University of Washington group at I.L.L. of the parity-violating neutron spin rotation, will provide the fifth and sixth independent constraints needed to determine the weak meson-nucleon coupling constants.

INTRODUCTION

The parity-violating nucleon-nucleon interaction at low and intermediate energies is described in a meson exchange model in terms of one weak and one strong interaction vertex. Conventional meson exchange interactions describe the strong vertex, while the weak meson-nucleon coupling can be predicted from the standard Weinberg-Salam model with W and Z exchanges between quark constituents of the nucleon and meson. There is a factor of ~3 uncertainty in the calculation of the weak meson-nucleon couplings, however, due to the renormalization of weak vertices by the strong interaction, which cannot be treated in perturbation theory. This problem persists in all calculations of weak hadronic matrix elements at low energy. It constitutes a major uncertainty in predictions of CP violation in the kaon system, for example[1]. Progress can be made through the measurement of parity violation in nucleonic systems for which the standard model provides a complete description.

Below the pion production threshold the weak N-N interaction is described in terms of π, ρ and ω exchanges, effects due to the exchange of neutral scalar mesons being excluded by CP conservation[2]. The six couplings are termed $f_\pi^1, h_\rho^{0,1,2}, h_\omega^{0,1}$, the superscript denoting isospin change. The most comprehensive theoretical predictions of the weak meson-nucleon couplings are due to Desplanques et al.[3]. Large uncertainties are typical.

To determine the weak meson-nucleon couplings at least six independent experiments must be carried out. The observables measured should be readily interpretable in terms of weak coupling constants, which implies experiments in two-nucleon and light nuclear systems, and should be technically feasible at a level to test the coupling constants derived from earlier data. Only four independent constraints exist at present on the six weak couplings. They are obtained from electromagnetic decays of parity-mixed doublets in nuclei of the mass-20 region and from the helicity dependence of elastic proton scattering from protons and light nuclei[4]. The experiments and constraints are summarized in table I.

Table I Experimental Constraints on Weak Meson-Nucleon Couplings

Case	Coupling	Expt. Performed	Additional Requirements	Proposed Expt.
1	f_π^1	^{18}F: P_γ (1081 keV)	None	
2	$h_\rho^0 + 0.6h_\omega^0$	^{19}F: A_γ(110 keV) $\vec{p} + \alpha$: A_z(46 MeV) ^{21}Ne: P_γ(2789 keV) ^{14}N: $A_z(0^+$; 8624 keV)	Case 1	
3	$h_\rho^0 + h_\rho^1 + h_\rho^2/\sqrt{6}$	None	None	$\vec{p} + p$: A_z, 230 MeV TRIUMF
4	$h_\omega^0 + h_\omega^1$	$\vec{p} + p$: A_z (15 MeV) $\vec{p} + p$: A_z (45 MeV)	Case 3	
5	$h_\rho^1 + h_\omega^1$	$\vec{p} + d$: A_z (15 MeV)	Cases 1,2,3,4	
6	$h_\omega^0 + 2.8h_\rho^2$	None	Cases 1,2	$\vec{n} + p$: ϕ_{PNC}/λ (Wash/ILL)

As an example, the low energy measurements of A_z in p-p scattering determine $h_\omega^0 + h_\omega^1$ when input is also got from experiments of case 3. A measurement of A_z in p-p scattering at 230 MeV at TRIUMF and of the parity nonconserving neutron spin rotation, which is proposed by the University of Washington group, will provide the two further constraints needed to determine the six coupling constants.

P-P SCATTERING AT 230 MEV

A measurement of A_z near 230 MeV to a precision of 2×10^{-8} is planned at TRIUMF. At low energy, mixing occurs mostly between the $^1S_0$ and $^3P_0$ partial waves, and a mixture of ρ and ω weak couplings is determined. Near 230 MeV, however, the contribution to A_z from S-P mixing integrates over angle to zero (purely a strong interaction effect). Mixing remains between the $^3P_2$ and $^1D_2$ waves, which is dominated by the h_ρ weak couplings, as illustrated in table I.

A_z will be measured simultaneously from the helicity dependence of the transmission of 230 MeV protons through a 20 cm LH_2 target, and from the helicity dependence of scattering from the target. The transmission asymmetry will be measured from the ratio of beam currents downstream and upstream of the target with the aid of transverse field ionization chambers (TRICs). The design of the TRICs is based on that of the low-noise ionization chambers used in the parity-violation measurement at 800 MeV at LAMPF. The helicity dependence of scattering is measured by a planar ionization chamber covering laboratory scattering angles from 7° to 43°, in conjunction with the upstream TRIC.

Measurements of A_z are susceptible to changes of beam properties correlated with helicity flip. The approach is to minimize these changes, measure the sensitivity of the apparatus to them and to constantly monitor them. The following will be monitored throughout the experiment: beam position, beam size and shape, net transverse polarization, transverse polarization moments. Each of the beam properties will be monitored at two points. Because of its expected stability of beam properties on spin flip, a laser-pumped polarized source will be used in this experiment.

Beam position, size and shape will be measured by two secondary electron monitoring devices (SEMs) which contain both split foils and a multiwire arrangement. The split foils are sensitive to the median of the beam current. Signals from the foils will serve as input to a double loop feedback system (x and y at two positions along the beamline) to control beam position to 3 μ in 1 ms. Beam size and shape will be measured on a 1 s timescale by means of the multiwire SEMs.

Transverse polarization components of the beam are expected to be the prime cause of systematic error in this experiment. It is therefore very important to measure them. This will be achieved with two "scanning polarimeters". They will each contain a thin carbon blade that can be scanned across the beam, either up-down, or left-right. Detectors downstream from the blade will measure the left-right and up-down asymmetry of protons scattered from the blade as a function of position in the beam, thus providing a picture of the polarization profile of the beam. Systematic errors in these polarimeters are reduced by nearly two orders of magnitude by moving the detectors with the blade, thereby keeping the detection geometry fixed.

The first phase of the experiment consists of a two year program of building and testing. The ability to control beam position to the required accuracy has already been demonstrated using split plate ionization chambers. Prototype SEM beam intensity profile monitors are now being built and will be tested this year. The scanning polarimeter is at the advanced design stage and should be available for testing within a year. A detailed design of the large planar scattering detector is ready, construction will follow. The new laser-pumped polarized source at TRIUMF is at the stage of on-line testing.

This work was supported in part by the Natural Sciences and Engineering Research Council of Canada.

REFERENCES

1. J.F. Donoghue, E. Golowich, B.R. Holstein, Phys. Rep. 131, 319 (1986)
2. G. Barton, Nuovo Cimento 19, 512 (1961)
3. B. Desplanques, J.F. Donoghue, B.R. Holstein, Ann. Phys. (N.Y.) 124, 449 (1980)
4. E.G. Adelberger and W.C. Haxton, Ann. Rev. Nucl. Part. Sci. 35, 506 (1985)

A DETAILED STUDY OF THE T=0, NN INTERACTION VIA $\vec{N}$P-SCATTERING AT 68 MEV

R. Henneck, J. Campbell, C. Gysin, M. Hammans, W. Lorenzon, M.A. Pickar [*], I.Sick
Institut für Physik der Universität Basel, CH-4056 Basel, Switzerland

J.A. Konter, S. Mango, B. van den Brandt
Paul Scherrer Institut, CH-5234 Villigen, Switzerland

S. Burzynski
Institute for Nuclear Studies, Hoza 69, PL-00681 Warsaw, Poland Poland

ABSTRACT

We report on first results for a measurement of the spin correlation parameter A_{zz} in $\vec{n}\vec{p}$ scattering at 68 MeV, which is highly sensitive to the mixing parameter ϵ_1. A phase shift analysis of these data, together with forthcoming results of our measurements of $\frac{d\sigma}{d\Omega}$ and A_y at the same energy is expected to allow a determination of ϵ_1 to within $\sim \pm 0.3^o$.

The precise determination of the T=0, NN phase shifts via np- scattering has been hampered in the past by the low intensity of the neutron beams available, especially when polarized neutrons were required. With the advent of high intensity polarized ion sources, which are capable of delivering up to 5μA of polarized protons onto a neutron production target, beams of sufficient intensity ($\sim 10^5 - 10^6$ neutrons$\cdot s^{-1} \cdot cm^{-2}$) and polarization ($\sim 35 - 50\%$), combined with good energy resolution (~ 2 MeV) have become available [1]. This will make it possible to solve a longstanding problem in the study of the basic NN interaction, namely the precise extraction of the mixing parameters ϵ_i, which can only be extracted from spin correlation experiments [2,3]. Of particular interest is the $^3S_1 -^3 D_1$ mixing parameter ϵ_1, which is essential to our understanding of the tensor component of the nucleon-nucleon potential. The tensor component is related to the D-state probabilities of light nuclei, (A=2,3) and its accurate determination is crucial in order to assess the

[*]present address: Department of Physics and Astronomy, University of Kentucky, Lexington,Kentucky 40506-0055, U.S.A.

influence of non-nucleonic degrees of freedom and possible 3 - body forces. Binstock and Bryan [2] and Bugg [4] have demonstrated that in the 50 MeV region, the spin correlation parameter A_{zz} and the spin-dependent total cross section $\Delta\sigma_L$ are the most sensitive observables which can be measured with present techniques.
In order to reduce the uncertainty in the determination of ϵ_1 from a phase shift analysis, which arises from the correlation of ϵ_1 with other phase shifts - notably with $^1P_1$ [2] - additional observables have to be measured. Specifically, a precise measurement of $\frac{d\sigma}{d\Omega}$ near 180° would largely constrain $^1P_1$, while A_y is particularly sensitive to D- and F- waves. Since $\frac{d\sigma}{d\Omega}$ at 90° is almost independent of the value of $^1P_1$, the problem of the absolute normalization can be circumvented by observing the ratio $\frac{d\sigma}{d\Omega}(\theta)/\frac{d\sigma}{d\Omega}(90^o)$ for angles near 180°.

We have measured A_{zz}, $\frac{d\sigma}{d\Omega}$, and A_y at $E_n = 68MeV$ at the injector cyclotron polarized neutron facility [1] of the Paul Scherrer Institut (PSI, formerly Schweizerisches Institut fur Nuklearforschung, SIN). Here we shall concentrate on the measurement of A_{zz} and present first results.
Figure 1 shows an outline of the experimental setup. Longitudinally polarized neutrons bombarded a longitudinally polarized hydrogen target. Two plastic scintillator telescopes allowed forward angle measurements of recoil protons in the range $5^o \leq \theta^p_{lab} \leq 35^o$, corresponding to $110^o - 170^o$ in the neutron center of mass. The spatial resolution of the wire chambers (2mm for chamber 1, 4mm for chambers 2 and 3) permitted a determination of the proton scattering angles to approximately 0.5° (lab). The neutron flux monitor consisted of a thick CH_2 target and two symmetric scintillator telescopes.

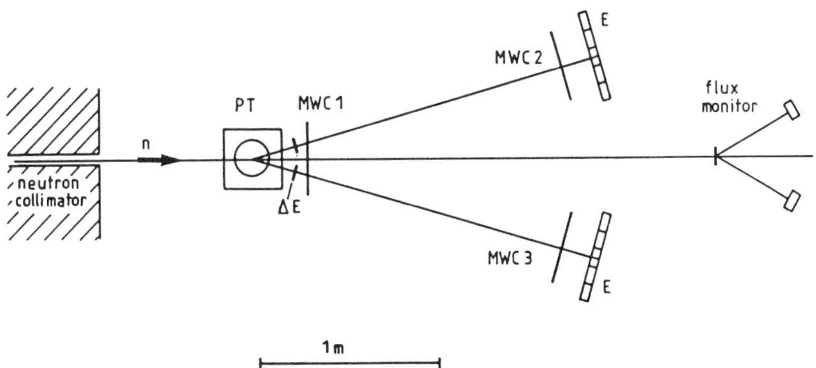

Figure 1: Experimental Setup. PT: polarized target, MWC: multi-wire chamber, E: scintillator array.

The polarized neutron facility [1] delivered a collimated neutron flux of approximately $2 \cdot 10^5$ neutrons $s^{-1} \cdot cm^{-2}$ with 2.3 MeV FWHM and $\sim 34\%$ polarization. The polarization of the target (3mm butanol) was typically 0.4 to 0.6 and was known with an accuracy of $\pm$ 0.025. The target and neutron spins were flipped in approximately 24 hour (target) and 10 second (neutron) intervals.

Figure 2 displays a preliminary result of our 1987 run. A_{zz} was computed as the weighted average of the formula

$$A_{zz} = \frac{1}{P_T \cdot P_n} \cdot \frac{N^+ - N^-}{N^+ + N^- - 2N_b} \tag{1}$$

taken over 66 half hour runs. $N^+(N^-)$ is the yield corresponding to an incident neutron beam with helicity + (-). N_b is the corresponding yield of the background under the hydrogen peak, obtained from measurements with a graphite target, whose thickness was taken so as to match the proton energy loss in the polarized target. A representative value of 34% was used for the neutron polarization, based on a previous calibration using the reaction $^4He(n,n)^4He$ [1]. The target polarization was determined from NMR measurements for each run.

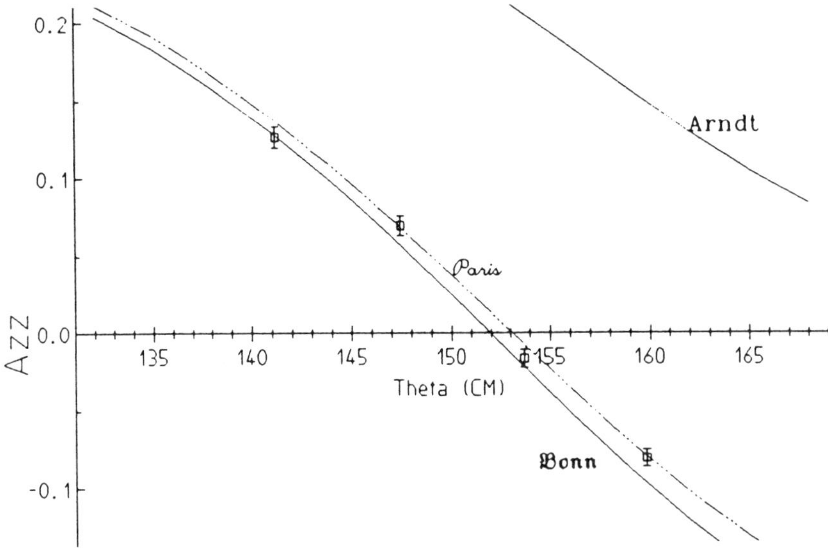

Figure 2: Preliminary results for A_{zz} compared with the predictions of the Bonn ($\epsilon_1 = 2.2^o$,[1] $P_1 = -12.4^o$) and Paris ($\epsilon_1 = 2.0^o$,[1] $P_1 = -12.9^o$) potentials at E_n = 68 MeV. Also shown is the prediction of the phase shift solution (C 50,WI87) obtained by Arndt ($\epsilon_1 = 1.3^o$,[1] $P_1 = -6.0^o$).

The final extraction of A_{zz} will require (i) a detailed Monte Carlo simulation of the experiment in order to precisely determine the incident neutron polarization, and (ii) consideration of the proton deflection in the solenoidal magnetic field of the polarized target. As the corrections are small, the final results are not expected to be significantly different from those shown.

The error bars shown in Fig. 2 are purely statistical. The combined systematic error due to uncertainties in neutron beam polarization and in proton target polarization will be approximately 5%. In the region of the observed zero crossing of A_{zz} this uncertainty will play a minimal role. All corrections applied, we expect to be able to determine the zero crossing to $\pm 1^o$ in the center of mass.

Our preliminary results are in excellent agreement with the predictions of both the Paris and Bonn potentials, and in clear contradiction to the prediction based on the phase shift solution of Arndt [5]. The discrepancy is primarily due to the value of the $^1P_1$ phase shift obtained in the phase shift analysis and the sensitivity of A_{zz} to $^1P_1$ at back angles.

Apart from the results shown, we have measured A_{zz} over the whole angular region $110^o \leq \theta_{cm} \leq 170^o$ with similar accuracy. Given the projected errors of our measurements we expect the final uncertainty in the determination of ϵ_1 from a phase shift analysis of A_{zz}, $\frac{d\sigma}{d\Omega}$, and A_y to be $\sim 0.3^o$.

REFERENCES

1. R. Henneck, C. Gysin, M. Hammans, J. Jourdan, W. Lorenzon, M.A. Pickar, I. Sick, S. Burzynski, T. Stammbach , Nucl. Instr. and Meth. A259, 329 (1987).
2. J. Binstock, R. Bryan, Phys. Rev. D9, 2528 (1974).
3. G.S. Chulick, C. Elster, R. Machleidt, A. Picklesimer, R.M. Thaler, Phys. Rev C 37, 1549 (1988) .
4. D.V. Bugg, J. Phys. G: Nucl. Phys. 6, 1329 (1980).
5. R.A. Arndt, Programme SAID, Wi 87.

TESTS OF TIME REVERSAL IN NEUTRON-NUCLEUS SCATTERING

J. David Bowman
Los Alamos National Laboratory, Los Alamos, NM 87544, USA

ABSTRACT

Experiments to test time-reversal invariance are discussed. The experiments are based on observables constructed from the momentum and spin vectors of epithermal neutrons and from the spin of an aligned or polarized target. It is shown that the proposed tests are detailed balance tests of time-reversal invariance. The status of the experiments is briefly reviewed.

I will report on tests of time-reversal invariance (TRI), which are in the beginning stages of experimental work at the Proton Storage Ring (PSR) at Los Alamos. I discussed plans for this work two years ago at the Lake Louise conference.[1] A workshop on Tests of Time-Reversal Invariance in Neutron Physics was held in Chapel Hill, North Carolina in April 1987, and the reader is referred to the proceedings[2] for a more complete discussion of many points than will be presented here. The initial work that will be discussed has been carried out by a collaboration between Los Alamos, Harvard, and Princeton. During 1988 colleagues from TUNL, TRIUMF, Delft, and University of Virginia joined in the work.

Figure 1 illustrates the classical meaning of TRI. A system evolves from an initial state i to a final state f. The system is described by momenta p and

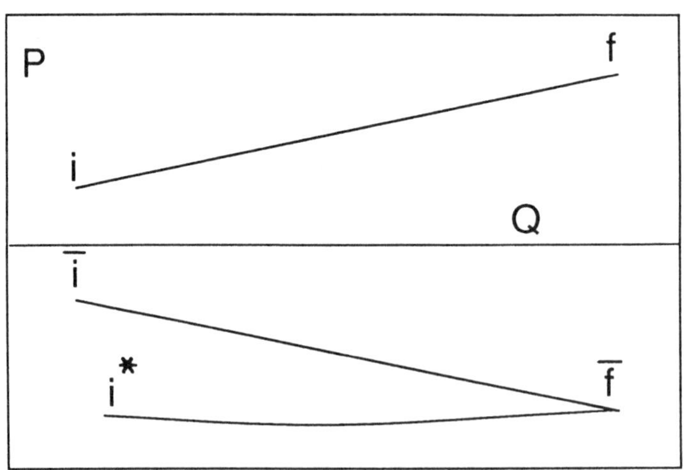

Fig. 1. *Ilustrates classical trajectories that obey $\bar{f} \rightarrow \bar{i}$, and violate $\bar{f} \rightarrow i^*$ time-reversal invariance (see text).*

coordinates q. A new initial state $\bar{f}$, is formed by time-reversing the state f. The signs of the p's are changed and the q's are kept. The system evolves from the state $\bar{f}$ to some state i^*. TRI holds if the state i^* is the same as the time-reversed initial state $\bar{i}$. For a quantum system TRI holds if the amplitude for the system to make a transition from i to f has the property

$$S(i \rightarrow f) = S^*(\bar{f} \rightarrow \bar{i}) \ . \tag{1}$$

Since only the squares of amplitudes are measurable, Eq. (1) implies that tests of TRI must be based on measurements that test whether or not

$$\sigma(i \rightarrow f) = \sigma(\bar{f} \rightarrow \bar{i}) \ . \tag{2}$$

In order to see a TRI-violating phase change, the amplitude S must be a sum of two or more amplitudes whose phases change differently under time reversal (TR).

Interest in the use of neutron-spin observables as probes of symmetry violations in nuclear states was sparked by the observation by V. P. Alfimenkov, et al.,[3] of a large helicity ($\sigma \cdot K$) dependence of the total scattering cross section at the energy of a p-wave resonance in ^{139}La. Table I summarizes the properties of the neutron spin (σ), the neutron momentum (K), and the target spin (J) under parity (P) and TR transformations. The idea of the present work is to use the spin of epithermal, 0.1 to 10 keV, neutrons to construct observables with which to test TRI. Using the fluxes available at the PSR, 10^{14} neutron-scattering events from a p-wave resonance can be detected in 10^7 seconds. Thus, transmission asymmetries as small as 10^{-7} can be studied. Two types of measurements have been suggested. P. K. Kabir,[4] L. Stodolsky,[5] and V. E. Bunakov and V. P. Gudkov [6] suggested the parity-odd observable $\sigma \cdot J \times K$. P. K. Kabir[7] proposed the parity-even observable $(J \cdot K)(\sigma \cdot J \times K)$.

TABLE I

Table I shows the parity- and time-reversal-transformation properties of various quantities discussed in the text.

Observable	Symbol	Parity	Time Reversal
Neutron Spin	σ	+	−
Neutron Momentum	K	−	−
Target Spin	J	+	−
Neutron Helicity	$\sigma \cdot K$	−	+
Triple Product	$\sigma \cdot J \times K$	−	−
Five-Fold Product	$(J \cdot K)(\sigma \cdot J \times K)$	+	−

V. E. Bunakov and V. P. Gudkov[8] showed that the large helicity dependence of the total cross section could be understood as resulting from the mixing of nearby s-wave resonance into the p-wave resonance by the parity-violating weak nucleon-nucleon force. Two enhancement factors increase the size of the observed effect: 1) the small energy spacing, 10 eV, in the compound nucleus and 2) the large ratio of s-wave to p-wave neutron decay amplitudes. They argued that the enhancement of the dependence of the cross section on the TR-odd P-odd scalar triple product $\sigma \cdot J \times K$ should be the same as that of the helicity. Since the helicity dependence of the cross section has been measured to be as large as 10%, a 10^{-7} measurement of the dependence of the cross section on the scalar triple product would test for a TR-odd P-odd interaction at the level of 10^{-5} of the weak force between nucleons, a sensitivity competitive with searches for a non-zero electric dipole moment of the neutron.

V. E. Bunakov[9] has recently estimated the nuclear enhancement of the cross-section dependence of the TR-odd P-even five-fold product, which arises from the mixing of p-wave resonances, to be between 10^3 and 10^5. Thus, a 10^{-7} measurement of the dependence of the cross section on the five-fold product would test for a TR-odd P-even force with a sensitivity of 10^{-10} of the strong force between nucleons. P. Herczeg[10] has studied the relationships between theories of CP violation in the kaon system and TRI violation in $\Delta S = 0$ nuclear systems.

Figure 2 shows an experiment designed to look for a dependence of the total cross section on the scalar triple product. This approach is flawed. The

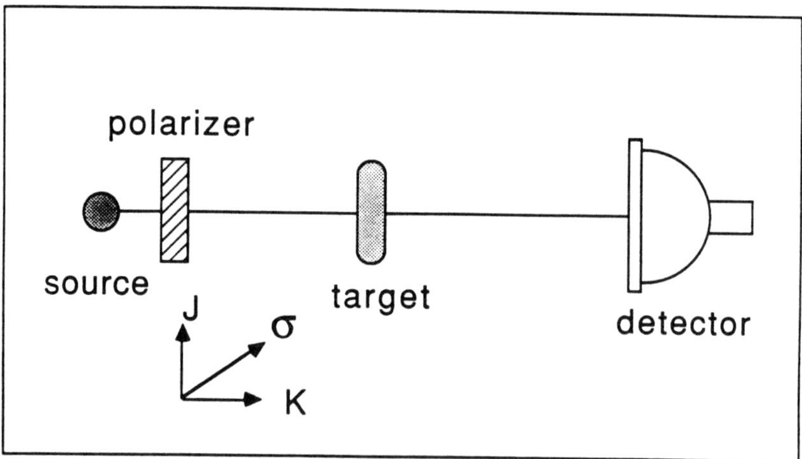

Fig. 2 *A neutron beam with momentum K passes through a polarizer and energies with a spin σ. The target spin J is perpendicular to K. The neutron spin is switched so that the triple product $\sigma \cdot J \times K$ changes sign. A change in the neutron transmission upon switching σ is not an ambiguous test of TRI (see text.)*

real part of the spin-orbit force ($\sigma \cdot J$) will cause the neutron spin, which is initially perpendicular to both J and K, to precess around J. The vector σ will then develop a component along K and the $\sigma \cdot K$ interaction will produce an asymmetry even in the absence of TRI violation. The appearance of a fake TRI-violating asymmetry in the above gedanken experiment is closely related to the falsification of TRI violation by final-state interactions in scattering experiments. L. Stodolsky[11] proposed that by adding an analyzer after the target, as shown in Fig. 3, the apparatus could be made TR invariant and that detailed-balance TRI experiments in the sense of Eq. 2 above would result. In order to test for

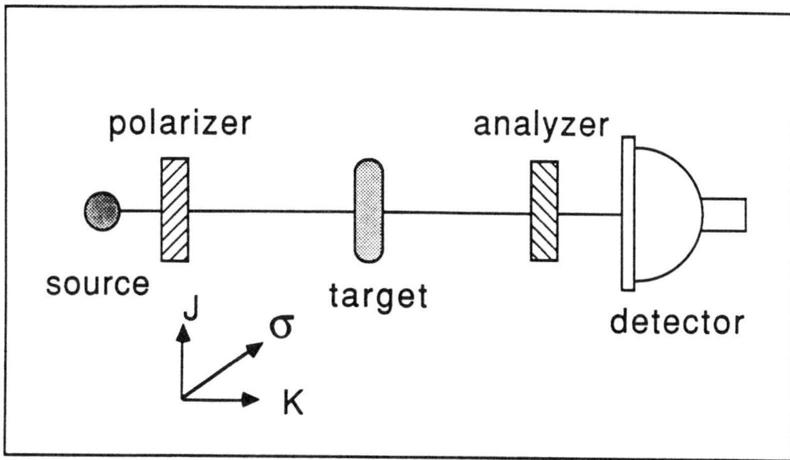

Fig. 3 *Adding an analyzer, which is switched simultaneously with the polarizer leads to an unambiguous test of TRI (see text).*

TRI in the scalar triple product the polarizer-target-analyzer configuration is necessary. In the Chapel Hill workshop I[12] showed that since the five-fold-product experiment could be carried out with an aligned target, where the average value of J is zero, the analyzer is not necessary.

We began work on parity and time-reversal experiments at Los Alamos in 1986. In our first experiment we measured the degree of helicity dependence in the total cross section of the 0.734 eV resonance in ^{139}La. The Alfimenkov, et al.,[3] experiment, which measured an asymmetry of 7.2 $\pm$ 0.4%, and the A. Masaike, et al.,[13] experiment, which measured an asymmetry of 10.4 $\pm$ 0.3%, both used polarized proton filters to prepare a polarized neutron beam. In 1986 no polarized neutron beam was available at Los Alamos. We carried out an asymmetry measurement as shown in Fig. 4. The helicity dependence in the total cross section was used both to polarize and to analyze the neutron beam. We obtained an asymmetry of 8.2 $\pm$ 1.7%. To my knowledge, this is the first time the weak interaction has been used both to polarize and analyze a beam.

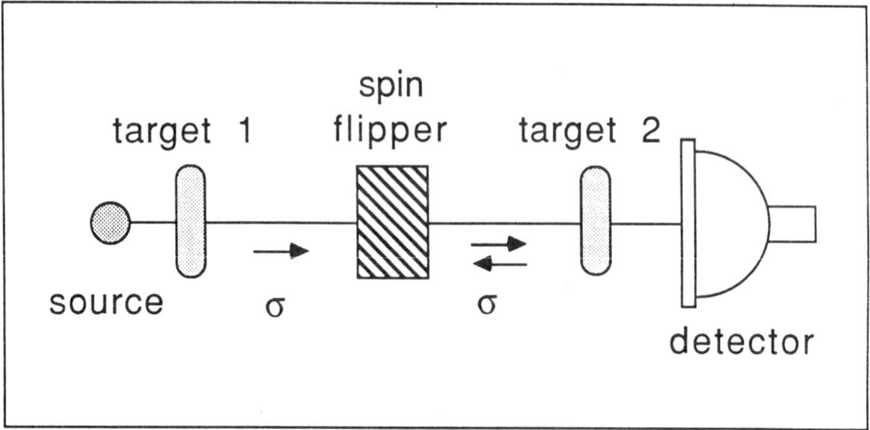

Fig. 4. *Los Alamos experiment to measure the helicity dependence of the total scattering cross section. Target 1 prepares a polarized neutron beam, which is analyzed by target 2.*

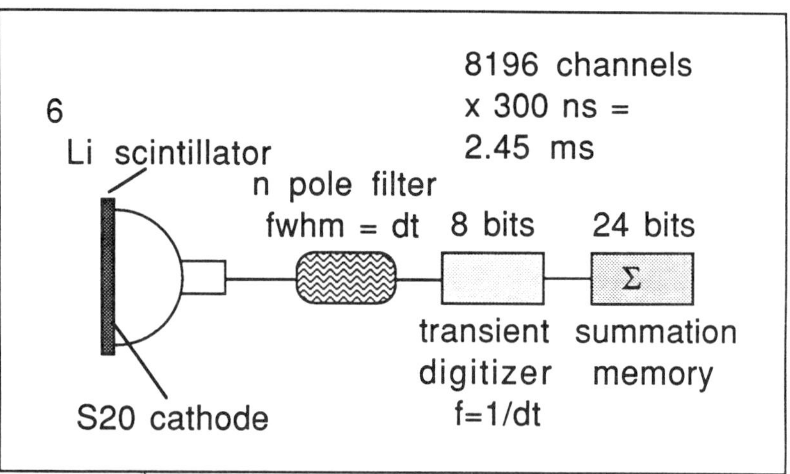

Fig. 5 *The system which detects neutrons at rates up to 10^{12}Hz in a current mode.*

The 1986 work was done by counting individual neutrons using a ^{6}Li loaded glass scintillator. Pulse-counting techniques limited us to instantaneous rates of less than 10^6 Hz. The PSR delivers neutron beams into a few micro steradians of 10^{12} Hz. In order to utilize fully the available neutron fluxes we developed and tested neutron-current measuring techniques in 1987. Our detection apparatus is shown in Fig. 5. Using these techniques, we will be able to take advantage of the full intensity available from the PSR without loss of neutron-energy resolution or degradation on the statistical accuracy of the number of neutrons detected. Using this detector we began a survey of p-wave resonances in atoms that allow nuclear polarization or alignment.

In 1987, our colleagues from Princeton and Harvard set up a polarized ^{3}He filter, which prepared a polarized neutron beam. About 10-atmosphere-cm^3 of ^{3}He were polarized using optical pumping techniques. This apparatus has been described by K. Coulter[14] in the Chapel Hill workshop. Using this polarized neutron beam we repeated the ^{139}La experiment, and searched for new examples of p-wave resonances with helicity-dependent cross sections. One such resonance was identified in ^{151}Gd. This techniques has the attractive feature that it is modest in scale and is noncryogenic. Furthermore, the ^{3}He spin, and, hence, the neutron spin, can be reversed relative to a weak holding field using an adiabatic fast passage. Reversing the neutron spin in this way is attractive for symmetry-test experiments because the neutron spin is reversed without changing the magnetic fields used for spin transport. The weakness of this technique is that since the n$-^3$He cross section decreases as the inverse of the neutron velocity, neutrons can only be polarized up to a few eV, and over a fraction of a square centimeter using a 10-atmosphere-cm^3 cell. The ^{3}He technique is attractive as a polarization-sensitive detector or as a polarizer/analyzer if thicker cells can be developed.

For 1988, we are developing a dynamically polarized proton filter with which to polarize the neutron beam up to 10 keV. This project is well under way, with our collaborators from TRIUMF and Delft playing a leading role. We expect to have an area of 10 cm^2 with a polarization of 70%. We have developed a seven-detector array, which is suitable for measurements over a 60-m flight path. We will have adequate energy resolution up to energies of 1 keV. We have designed and built a spin flipper that can reverse the neutron's spin between energies of 0.01 and 10 keV. A new data acquisition and analysis code has been written by the TUNL team. With this apparatus we plan to continue our search for p-wave resonances suitable for TRI tests to higher energies and additional targets. The data obtained in this survey will determine the distribution and strength of parity violation as a function of atomic mass. It will be interesting to see whether or not meson-exchange models of the weak nucleon-nucleon force with experimentally determined couplings combined with theories of the compound nucleus can correctly explain the observed distribution of parity-violating mixing amplitudes.

REFERENCES

1. J. D. Bowman in *Proceedings of the Conference on the Intersections Between Particle and Nuclear Physics*, Lake Louise, Canada, 1986. Donald F Geesaman Editor. American Institute of Physics, New York 1986.
2. *Proceedings of the Workshop on Tests of Time-Reversal Invariance in Neutron Physics*, Chapel Hill, North Carolina, April 1987, N. R. Roberson, C.

FIRST OPERATION OF THE FERMILAB POLARIZED BEAM[†]

David G. Underwood[§]
High Energy Physics Division
Argonne National Laboratory, Argonne, Illinois 60439

ABSTRACT

We describe the first operation of a 200 GeV/c $p\bar{p}$ beam at Fermilab, including beam tuning and momentum and polarization tagging. The present status of analysis of the data from two polarimeter experiments is described here. A third experiment is described in a separate paper.

The fundamental aspects of the beamline including optics were described in the proceedings of the 1986 Intersections Conference.[1] During this initial operation, we tuned the beam optics and set up the tagging electronics so that the polarized p or $\bar{p}$ from Λ or $\bar{\Lambda}^0$ decay would be transported without depolarization and the polarization could be tagged. We ran three physics experiments which were planned to have the potential to give either absolute or relative beam polarization measurements. The results of one of these, a π^0 inclusive measurement at large x, are described in another paper in these proceedings.[2] The experiment to measure the polarization by means of Coulomb-nuclear interference in elastic scattering has isolated a clean elastic signal at the low t required but does not yet have an asymmetry result. An absolute measurement of the beam polarization by means of the inverse Primakoff effect (Coulomb dissociation) has been made and confirms the tagged values of polarization within the error bars. It also checks the operation of the spin precession snake.

The first few months of operation of the beamline involved tuning a parasitic positron beam through both the section that would normally transport primary accelerated beam to our production target

† Work supported by the U.S. Department of Energy, Division of High Energy Physics, Contract W-31-109-ENG-38.

§D. Grosnick, D. Hill, D. Lopiano, T. Shima, H. Spinka, R. Stanek, D. Underwood, A. Yokosawa Argonne National Lab.; A. deLesquen, L. van Rossum, F. Lehar CEN-Saclay, France; D. Carey, R. Coleman, D. Cossairt, L. Read Fermilab; A. Derevschikev, A. Meschanin, V. Solovyanov, Y. Matulenko, S. Nurushev, A. Vasiliev IHEP, Serpukhov, USSR; K. Imai, A. Konaka, S. Makino, A. Masaike, T. Nakano, K. Miyake, T. Nagamine, N. Tamura, T. Yoshida Kyoto, U. Japan; F. Takeuchi Kyoto Sangyo U., Japan.; A. Maki, R. Takashima Kyoto U. of Education, Japan; K. Kuroda, A. Michalowitz LAPP, Annecy, France; F. Luehring, D. Miller Northwestern U.; M. Corcoran, F. Nessi-Tedaldi, M. Nessi, G. Phillips, J. Roberts Rice U.; G. Pauletta U. of Texas; A. Penzo, G. Salvato, P. Schiavon, A. Villari, A. Zanetti INFN, Trieste, Italy

and through the secondary line that would normally transport polarized protons. These tuning activities, which were necessary for the use of the calibration beam for the lead glass and other polarimeter elements, also gave invaluable experience for the tuning of actual primary and polarized beams. During the polarized beam running, we typically got 10^{12} primary during a 20 second spill each minute. From this we obtained 10^7 protons/spill from Λ decay with about 13% background pions tagged by Cherenkov. With a negative secondary beam we got about 10^6 $\bar{p}$ (from neutrals decaying) and roughly a factor of five more pions than $\bar{p}$.

The hodoscopes at the central focus, midway down the secondary beamline, are shown in Fig. 1. The primary purpose of these is to tag the momentum (vertically) and tag the x position which correlates with polarization (horizontally). In operation we use pipelined ECL memory lookup to encode simultaneously to find polarization. The polarization and momentum signals are available for triggers or to be latched in experiments. A plot of tagged polarization is shown in Fig. 2.

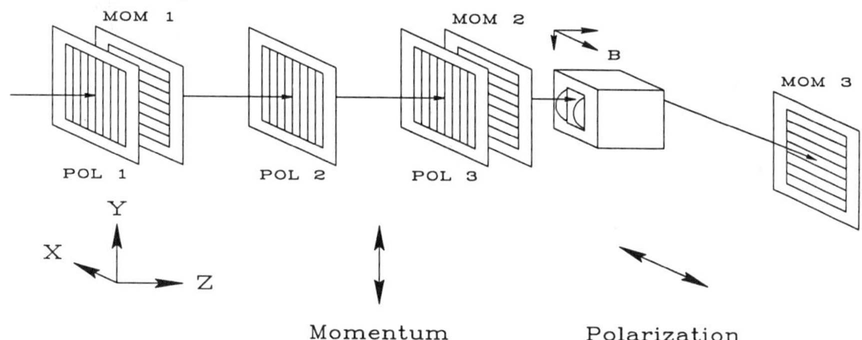

Fig. 1 Momentum and polarization tagging hodoscope arrangement. The central focus of the beam is at the hodoscope labeled POL 2.

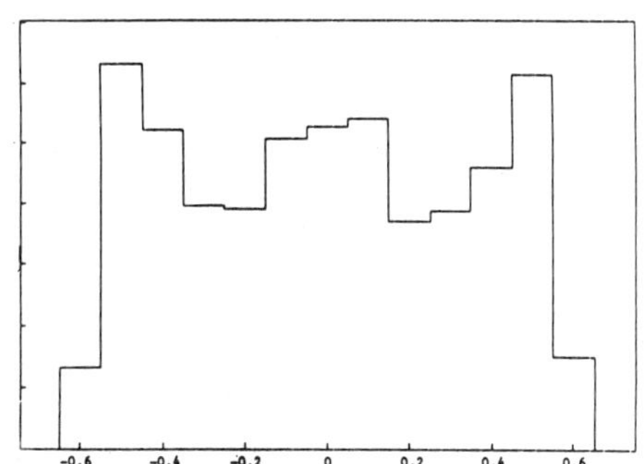

Fig. 2 Tagged polarization distribution.

These hodoscopes, in conjunction with similar ones near the final focus at the experiment, provide very powerful beam diagnostics. We can get online phasespace plots of x vs. x' or y vs. y' at 8 z positions near the central focus and at 8 positions near the final focus. These are used to determine the locus of the focus in beam tuning. We can also get plots of space magnification and angular magnification between the two foci. A plot of angular magnification is shown in Fig. 3.

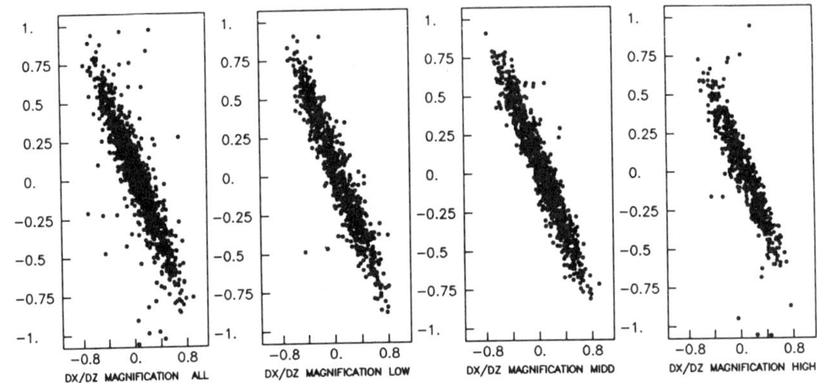

Fig. 3 Slope dx/dz at the final focus vs. dx/dz at the central focus (in mr) shows angular magnification for all momenta, and low, central and high momenta.

In the following analysis we have available tagged polarization in 13 bins from -.65 to +.65. We reduce this to three categories, -,0,+, (-.65 to -.3, -.3 to +.3, +.3 to +.65). We also have two snake states which transform left to up and right to down or vice-versa. We can then calculate asymmetries in five ways: 1) tagging + with reversing snake, 2) tagging - with reversing snake, 3) snake + with reversing tagging, 4) snake- with reversing tagging, 5) tagging 0 with reversing snake.

The asymmetry in photoproduction of π^0 by 620 MeV/c photons on polarized protons is about 90% at 620 MeV/c. This is the s channel of the reaction which we employ in the t channel, with a beam of 185 GeV/c protons and low energy photons from the Coulomb field of a lead nucleus as shown in Fig. 4. We must be careful to average over the recoil angle about the beam in order to have unpolarized photons. The asymmetry we observe is in the ϕ of the π^0 about the effective p/π^0 mass direction. This electromagnetic process excites both delta states and n* states. Strong diffraction dissociation excites n* states and provides our main background. Preliminary fitting indicates the interference term between these processes is small as expected. A mass and t spectrum for this experiment are shown in Fig. 5.

We have analyzed the Primakoff data for beam tracks which are tagged as having polarizations between 30% and 55% of either sign. The polarization direction has been rotated from horizontal to vertical by the snake. At the present state of analysis we measure a polarization of 40% ± 12% statistical +14 -11% from the scale factor (the signal to background ratio). We expect to reduce the scale factor errors significantly by fitting both the mass and t spectra for production on both carbon and lead. The asymmetry in the vertical direction is consistent with zero, indicating that the snake is working. The asymmetry for the delta mass region is consistent with zero, as it would be if amplitudes other than delta production are small there.

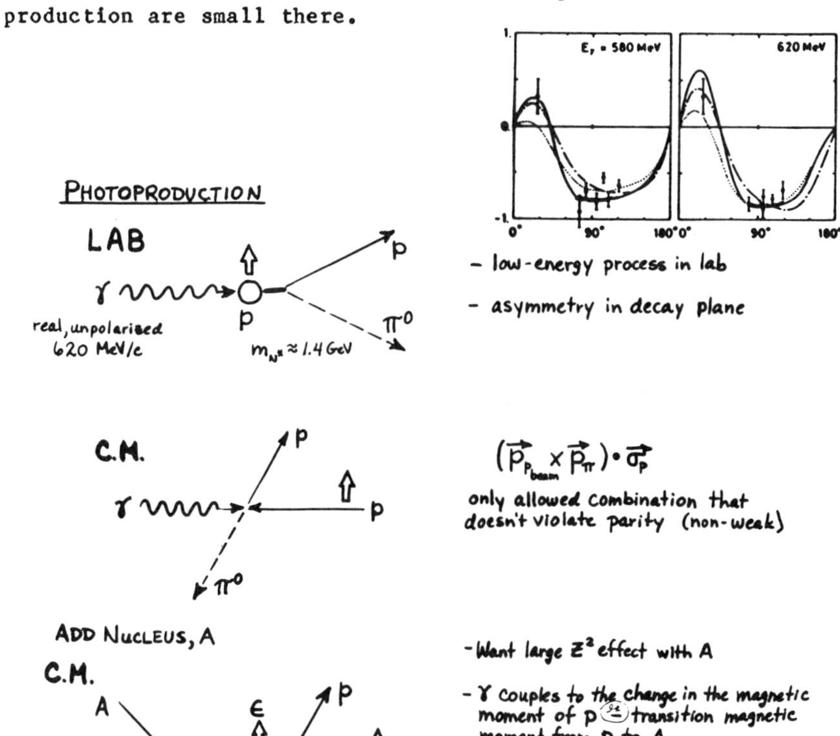

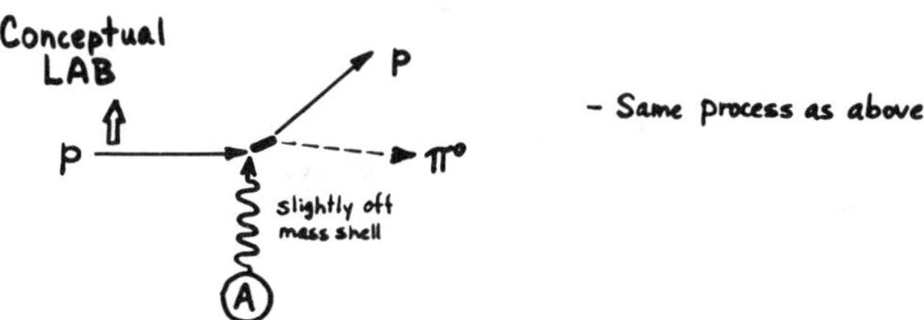

Fig. 4 Photoproduction s channel polarized target asymmetry is related to t channel polarized beam Coulomb dissociation.

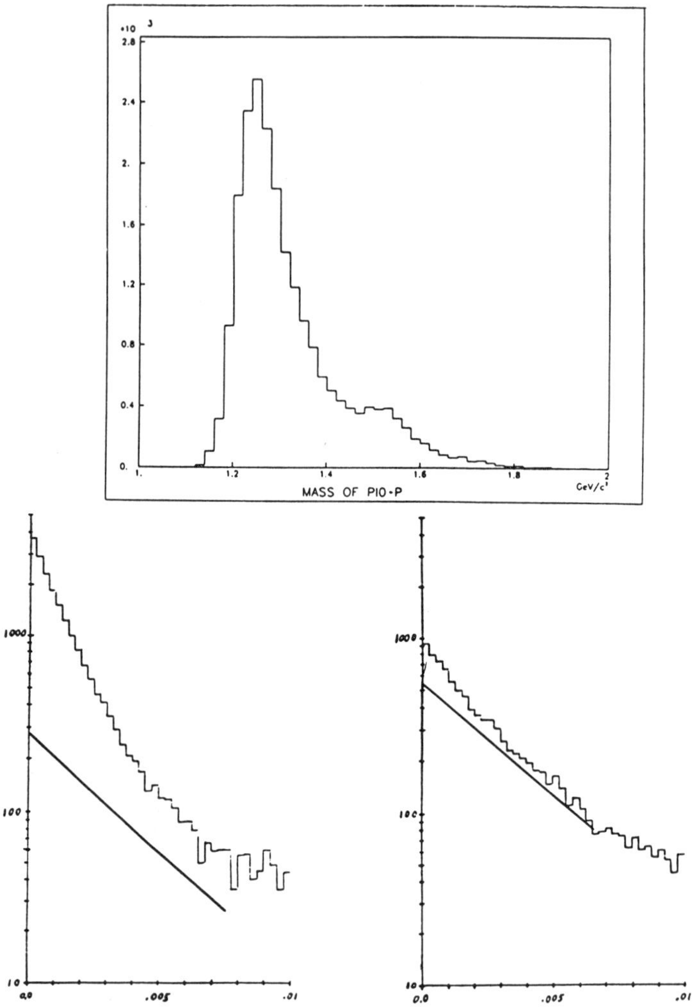

Fig. 5 Coherent production mass spectrum for t' < .001, and t spectra for delta mass (p π^0) region (1.1 to 1.3 GeV/c^2) and polarization analysis mass region (1.35 to 1.5 GeV/c^2). Fitted values of the strong interaction dissociation part are indicated.

REFERENCES

1) D. Underwood, High Energy Polarized Beams from Hyperon Decays, AIP Conference Proceedings 150, Intersections Between Particle and Nuclear Physics, Lake Louise Canada, 1986, Editor: D. F. Geesaman.

2) M. Nessi, following talk in these proceedings, Intersections 1988.

ANALYZING POWER MEASUREMENT IN INCLUSIVE π^0 PRODUCTION AT HIGH x_F

Marzio Nessi*
Rice University, Houston, TX 77251

ABSTRACT

A measurement of inclusive π^0 production has been performed with the new 185 GeV/c Fermilab polarized proton beam. First results for A_N in the x_F range between 0.2 and 0.8 and p_T~0.8 GeV/c are presented.

During the last weeks of the most recent fixed target run at Fermilab, our collaboration (E581/E704) has succeded in commissioning tertiary tagged polarized proton and anti-proton beams of momentum around 185 GeV/c.[1] Along with the measurement of the beam polarization, the availability of a polarized beam enabled us to make simple parasitic measurements. Based on preceding inclusive π^-, π^+ and K_S production asymmetry measurements at lower energies[2] and the predictions that follow from the quark fragmentation model presented in Ref.3, a non-vanishing analyzing power is expected in inclusive π^0 production in the beam fragmentation region and at moderate values of transverse momentum, of about $((A_N(\pi^+)+A_N(\pi^-))/2 \simeq 0{,}08$. We were also strongly motivated to provide E704 with a simpler polarization monitor than the more sophisticated absolute polarimeters based on Coulomb-Nuclear interference and the Primakoff effect.[1] The layout of the measurement consisted essentially of an electromagnetic calorimeter placed on one side of the beam axis, 50.5 m downstream of the target, with a surface of ~0.5 m². The calorimeter had two sequential sections; upstream 129 leadglass modules (2.5" x 2.5", $13X_0$), downstream was a lead-scintillator sandwich (~$20X_0$). The leadglass provided the necessary position resolution, whereas the Pb-scintillator part allowed the total absorption of the γ-energy from the observed $\pi^0 \rightarrow \gamma\gamma$ decay. The high-x_F π^0 trigger consisted of energy deposition in the leadglass of >30 GeV in anticoincidence with a charged particle veto counter upstream. We required the interaction to be originated by a

*B. Bonner, J. Buchanan, M. Corcoran, J. Kruk, H. Miettinen, G. Mutchler, F. Nessi-Tedaldi, M. Nessi, C. Nguyen, G. Phillips, J. Roberts, P. Stevenson, J. White, A. Zadeh, Q. Zhu Rice U.; D. Grosnick, D. Hill, D. Lopiano, Y. Ohashi, T. Shima, H. Spinka, R. Stanek, D. Underwood, A. Yokosawa, ANL- Argonne; A. de Lesquen, F. Lehar, L. van Rossum, CEN-Saclay; D. Carey, R. Coleman, C. Cossairt, L. Read, Fermilab; A. Derevschikev, Y. Matulenko, A. Meschanin, S. Nurushev, V. Solovyanov, A. Vasiliev, IHEP-Serpukhov; K. Imai, A. Maki, A. Masaike, K. Miyaka, T. Nagamine, R. Takashima, F. Takeuchi, T. Yoshida, Kyoto; K. Kuroda, A. Michalowitz, LAPP; F. Luehring, D. Miller, Northwestern U.; G. Pauletta, U.Texas; A. Penzo, G. Salvato, P. Schiavon, A. Villari, A. Zanetti, Trieste.

beam proton, i.e. a particle with momentum and polarization tagged successfully, giving a coincidence in a beam telescope placed immediately upstream of the target, and giving no signal in two threshold beam Cherenkhov counters. A vertical polarization was obtained through rotation of the tagged x-polarization by means of a Siberian snake, consisting of 12 dipoles. The polarity of the snake, and therefore the correlation between each beam ray and its polarization sign, was reversed every 10 Tevatron spills, in order to suppress systematic errors. Each beam particle track was reconstructed by means of four segmented hodoscopes placed at each end of the snake magnet system. The data acquisition was done with a PDP11/45. The calorimeter stability was monitored continuously with an LED-light source coupled to the modules via quartz-fiberoptics; before each accelerator spill LED-data as well as one pedestal measurement were acquired. A total of 285000 triggers were collected, for a beam flux of $1.16 \cdot 10^{10}$ protons. Half the data were obtained with a 4"x4"x4" plastic target, where for an average rate of 10^7 beam particles/Tevatron spill the trigger rate was typically 300, 7% of which reconstructed to a π^0. The other half of the data were collected parasitically with a 7 mm scintillator target. The reconstructed mass resolution was typically ±17 MeV/c^2. We have binned our data into 4 regions of x_F (see Fig.1). We have determined the analyzing power A_N for π^0 production from data acquired with beam polarization values between 30% and 55%, taking the average of the results obtained with the two snake polarities evaluated separately. The results are presented in Fig. 2. The average analyzing power we observe for $\langle x_F \rangle=0.52$ and $\langle p_T \rangle=0.79$ GeV/c is 0.10±0.03 (0.13±0.04 for the plastic target), for an average polarization of 44%, in excellent agreement with the predictions. Systematic effects on the result are negligible: their magnitude, which is determined as the contribution to the measured analyzing power that does not reverse sign under snake polarity reversal is 0.02±0.03. The analyzing power of the events constituting the background was also determined to be consistent with zero. The backround underneath the π^0 peak was fitted and the measured asymmetry scaled accordingly. We were also able to perform a very short measurement with the polarized

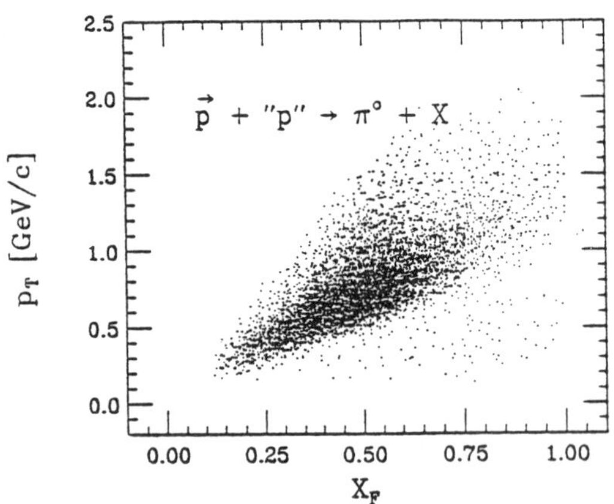

Fig.1: x_F-p_T distribution

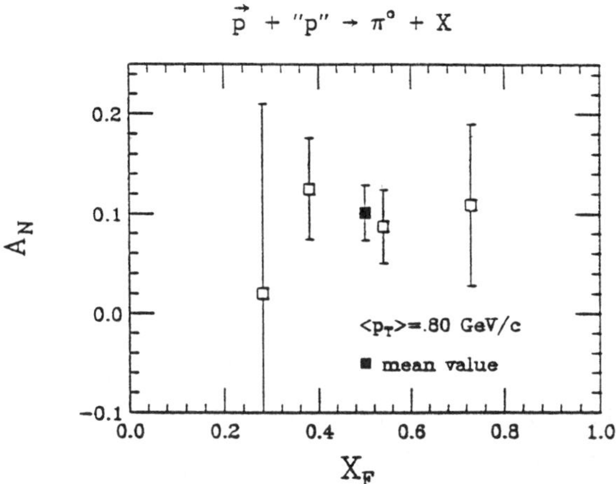

Fig.2: Analyzing power results vs. Feynman x

antiproton beam. We collected 4300 π^0 triggers for a beamflux of $8.5 \cdot 10^7$ antiprotons; We obtain A_N=0.26±0.19. This result encourages us to extend these measurements to other mesons and hyperons in order to provide a complete picture of the mechanisms involved in these processes.

This work was supported by U. S. Department of Energy Contract DE-AS05-76ER05096.

REFERENCES

1. D. G. Underwood, this conference
2. R. D. Klem et al., Phys. Rev. Lett. 36, 929 (1976)
   W. H. Dragoset et al., Phys. Rev. D18, 3939 (1978)
3. T. DeGrand and H. Miettinen, Phys. Rev. D24, 2419 (1981)

RECENT EXPERIENCE IN ACCELERATING POLARIZED BEAM AT THE AGS*

L.G. Ratner
Brookhaven National Laboratory, Upton, NY 11973

INTRODUCTION

The most recent operation of the AGS for polarized protons occurred in December, 1987 and January, 1988. The primary purpose during this period was to tune up the accelerator as soon as possible and to provide a usable polarized beam for high energy physics. We succeeded in providing 1-2 x 10^{10} polarized protons per pulse at 18.5 GeV/c with an average polarization of 43 ± 3% and a peak of 52%. The conditions for this run differed in some respects from the previous run done in 1986.[1] Due to problems with the main ring power supply, we were forced to use a back-up MG set which was only capable of 60% of the normal field rate of rise. This, of course, enhanced the effect of the depolarizing resonances. A second difference was the fact that a complete horizontal and vertical realignment of the ring magnets was done during the 1987 summer shutdown. In addition, the fast pulsed quadrupole positions were readjusted with respect to the equilibrium orbit. It had been suspected that misalignment of these quads was responsible for large transverse emittance growth in both planes. We will look at the effects of these differences, but the bottom line is that the "standard correction techniques" worked as expected.

SYSTEMS AND "STANDARD CORRECTION TECHNIQUES"

Compared to a standard accelerator, there are several additional systems needed for polarized proton acceleration:

(1) Fast pulsed quadrupoles to tune jump the "intrinsic" depolarizing resonances. The resonances are characterized by $G\gamma = nP \pm \nu_y$ where G is the proton anomolous magnetic moment, P the periodicity of the accelerator (12 for the AGS), n an integer, and ν_y the vertical tune (number of vertical betatron oscillations per revolution). These are intrinsic resonances because they arise from the accelerator focusing field which are necessary to keep the beam in the accelerator during the acceleration process. The fast quads change ν_y before a resonance is reached and keep ν_y displaced from the resonance until this point in the acceleration cycle is passed.

(2) Pulsed dipole magnets to compensate for vertical closed orbit distortions which cause "imperfection" resonance depolarization. These are imperfection resonances because the closed orbit distortions are due to imperfections in construction or alignment of the ring magnets and are not necessarily inherent in an accelerator. These are characterized by $G\gamma = n$, where n is an integer.

---

*Work performed under the auspices of the U.S. Department of Energy.

(3) A source to produce polarized protons, which for the AGS is an atomic beam source with a Cs charge exchange stage to produce polarized H^-. The source gave a 500 μsec pulse of 25 μAmps into an RFQ which accelerated the beam to 750 keV with 75% transmission efficiency. The beam was then accelerated to 200 MeV with 10 μAmps surviving and with a polarization of 78% ± 2%. The beam is then transported to the AGS where the H^- is stripped into polarized protons during the injection process.

(4) A polarimeter to measure the beam polarization during the acceleration cycle so that one can tune the correction elements and maximize the surviving beam polarization. The AGS uses an internal polarimeter which has a nylon fiber (0.006" diameter) target that is spooled at a rate of about 1 M per second and flipped into the beam on magnetic field flattops. These flattops are set above the particular resonance that one is correcting and was set at 10 different values during this run. This polarimeter measures relative polarization and is essentially only calibrated at those momenta where beam is extracted and an absolute measurement can be made. In this run we have a calibration point for the injected beam and for the 18.5 GeV/c full energy beam.

EXPERIENCE WITH INTRINSIC RESONANCE CORRECTIONS

As in previous runs, the correction amplitude was preset and only the jump turn-on time was varied. These times agreed with calculated values within 50 GCC (Gauss clock counts which are proportional to the accelerator's magnetic field). However, the faster rate of rise of field caused the tim plateau over which a correction can be made to become shorter. This became especially worrisome on $G\gamma = 24 - \nu$ where the plateau was reduced to 55 GCC instead of the normal 88 (Figure 1). To prevent loss of polarization requires that the Gauss clock be stable to the order of 1 part in 10^4 for both long and short intervals.

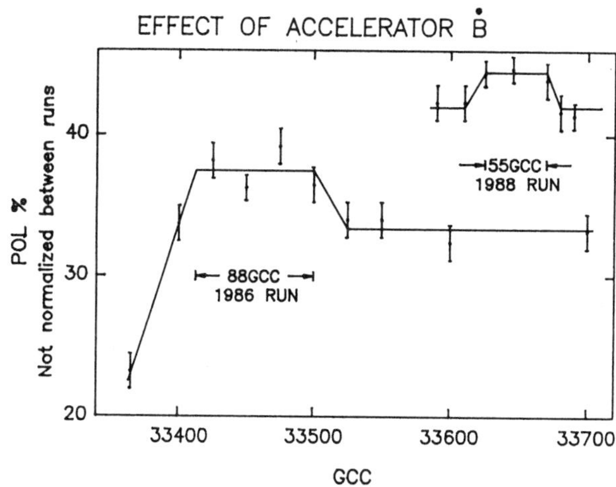

Figure 1.

We also again used the technique of slowly adjusting the machine tune before the fast jump so that a larger jump can be made without driving the tune into an integer or one-half integer stopband. With the AGS nominal tune of 8.75 and the need to make a fast jump of +0.3 in tune, we would normally go to 9.05 and therefore lose beam at the 9.0 integer stopband. The fix is to slowly lower the tune to 8.55 before the jump, and then end up at 8.85. In the earlier runs, this was used at several intrinsic resonances since it appeared to reduce the emittance growth which apparently was produced by the fast non-adiabatic quadrupole pulse. This was not necessary during this last run and this tune shift was used only as described above on one resonance.

The realignment of the AGS was partially motivated by our desire to reduce closed orbit errors for polarized protons and because we thought that the large emittance growth observed in the 1986 run might be due to misalignment between the equilibrium orbit and the fast pulsed quads. The misalignment of the dipoles could produce a non-adiabatic kick to the beam leading to the observed growth. After the main ring realignment, we pulsed the fast quads and the effect on the equilibrium orbit was measured. From this we could deduce the quad offset and physically reposition the quads. During the run we maintained the radial beam position at the quad centers. Figure 2 compares the emittance growth between these two runs. The greatly reduced emittance reduced the depolarization at the intrinsic resonances and improved the beam efficiency.

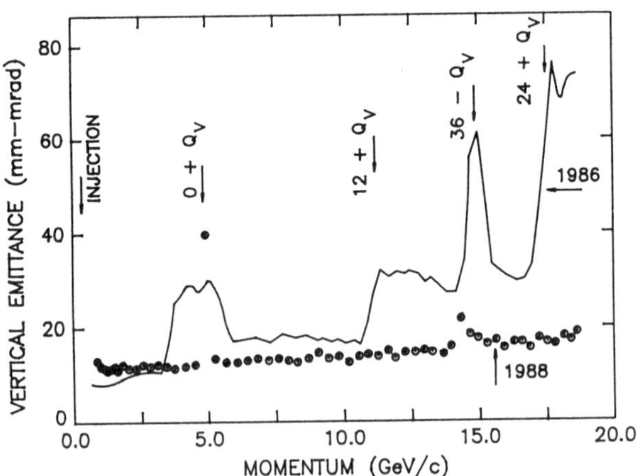

Figure 2. Emittance comparison

EXPERIENCE WITH THE IMPERFECTION RESONANCE CORRECTIONS

In contrast to the intrinsic resonances, here the timing is preset and the amplitude of the closed orbit vertical correction is varied. The fields of the 96 correction dipoles are energized with the appropriate harmonic correction at the energy where the resonance occurs $\left(E = \frac{nM_p}{G}\right)$. Since the vertical orbit distortions of harmonic number n is driven by imperfection fields of harmonic n, we must correct by cancelling the nth harmonic of the horizontal imperfection field. Since there are two parameters, namely, amplitude and phase, we apply orthogonal corrections of sin $n\theta$ and cos $n\theta$ in an iterative process. This then assures us that our correction vector has the correct amplitude and phase to minimize the alignment error. It was hoped that the survey and realignment would reduce the necessary corrections even though we know that survey errors of the order of mils could cause significant problems. The results were a mixed bag and we infer that survey techniques must be significantly improved to be able to address this problem. Figure 3 shows the ratio of the correction amplitudes for the 1988/1986 run and shows that less correction was needed on fifteen and more on twelve resonances--a sort of random change. One significant occurrence was the twelve-fold enhancement of the 29th harmonic. The correction was large enough so that we decided to correct $G\gamma = 29$ with $n = 7$ harmonic, the so-called "beat harmonic" technique described in References 1 and 2 and one of our "standard techniques". This was also used on $G\gamma = 27, 28$. Even though the amount of correction, i.e., whether we needed more or less appears to be random, there is a strong correlation between the two runs as far as the change in amplitude as a function of energy (Figure 4).

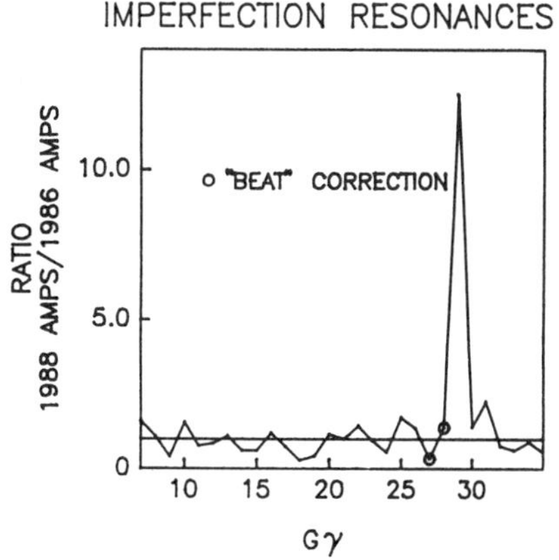

Figure 3.

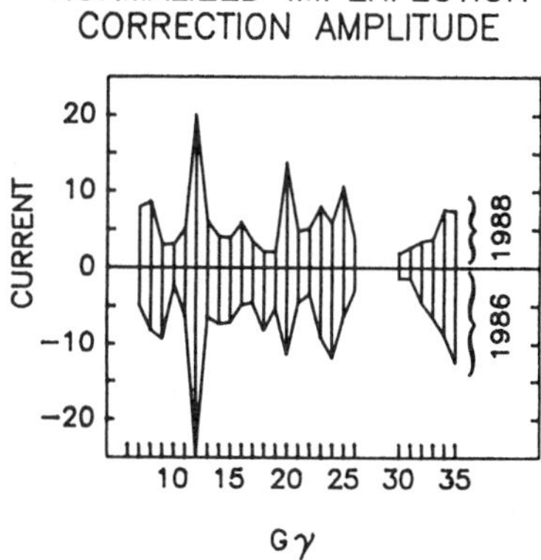

Figure 4.

On the other hand, the phase correction shows no correlation (Figure 5). It would appear that the phase and the amplitude required is changed by the realignment in a random manner and without essentially improving the closed orbit distortions. However, the shape of the amplitude versus energy curve seems to be unaffected by the realignment and shows a strong correlation between runs.

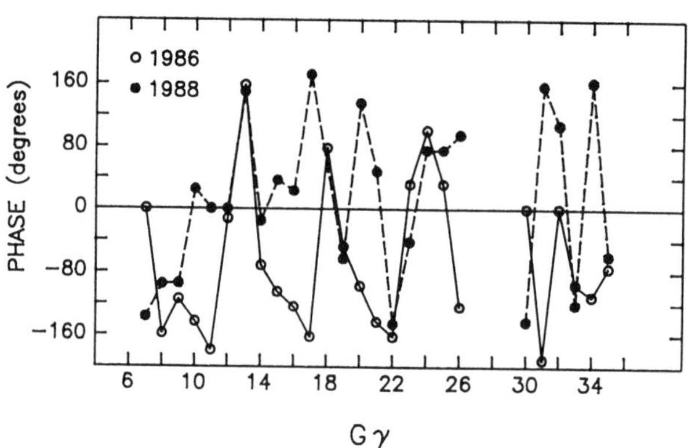

Figure 5.

SUMMARY

The 1988 run led to no surprises and to the continued successful use of our standard techniques. We saw no great benefits from realigning the ring magnets and guess that this might require precisions of the order of 0.001" to show improvements. We had sufficient corrections to accommodate the slower rate of rise, but we were getting close to the limit for $G\gamma = 24 - \nu$. We did see an improvement in the emittance growth on the fast quad pulse and conclude that it is desirable to have these quads as close to the beam axes as possible. The beam polarization as determined by the internal polarimeter is shown in Figure 6 and in contrast to the 1986 run we see no strong evidence for a large loss at 13 GeV/c. We do see a change in slope after a few days shutdown which is unexplainable. We hope to be able to re-establish that level as we eventually bring the polarized beam energy up to the design goal of 26 GeV/C.

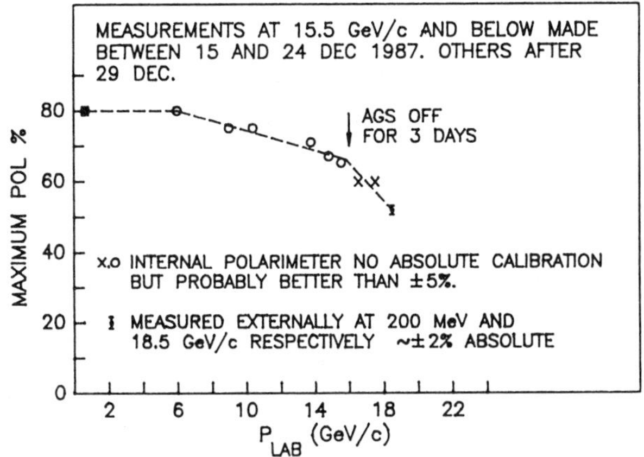

Figure 6. Maximum polarization as a function of energy.

ACKNOWLEDMENTS

We wish to acknowledge the many dedicated people at BNL and to the continued strong support from Alan Krisch's group from the University of Michigan, who all made this run go so well. We also acknowledge the past contributions from Argonne National Laboratory and the Universities of Rice and Yale. I would like to especially thank Leif Ahrens for his strong leadership in the BNL effort and especially for Figures 2, 4, and 5.

REFERENCES

1. L.G. Ratner, Proceedings of the International Workshop on Hadron Facility Technology, Los Alamos, 302 (1987).
2. K.M. Terwilliger, et al., IEEE Trans. on Nucl Sci., NS-32, 2635 (1985).

An Ultra-Cold Jet of Polarized Atomic Hydrogen

R.S. Raymond, D.G. Crabb, T. Roser
University of Michigan, Ann Arbor, MI 48109

G.R. Court
Massachussets Institute of Technology, Cambridge, MA 02139
and
University of Liverpool

ABSTRACT

Equipment is being tested to determine the applicability of cryogenic atomic physics techniques to the production of dense polarized atomic hydrogen beams. Such beams, with expected densities of $10^{14}/\mathrm{cm}^3$, would be useful for internal targets.

Our group is working to determine the applicability of cryogenic atomic physics techniques to the production of polarized atomic hydrogen beams. At low temperatures (0.5 K) and high magnetic fields (5 T) the magnetic energy of atomic hydrogen, approximately $\mu_e B$, is considerably greater than the thermal energy, k T. If, then, cold atomic hydrogen approaches a high field region, atoms in the two upper hyperfine states are repelled. Atoms in the lower two states, with electron spins antiparallel to the field, are attracted into the field, where they thermalize by hitting cold walls and are trapped. All cold walls seen by the atoms are coated with ^{4}He to suppress recombination. We shine microwaves onto these trapped atoms in order to drive transitions to the upper hyperfine states. The flipped atoms are then on a potential hill and will be accelerated out of the field forming, we hope, a dense intense beam.

Shown in the figure is our apparatus to test this scheme. The vertical section contains parts of a dilution refrigerator and a liquid nitrogen-temperature dissociator. The horizontal section contains the mixing chamber of the refrigerator and around it, a high-uniformity solenoid. Atoms from the dissociator flow down through the teflon transfer tube and then emerge into the accomodator region, which has ^{4}He coated walls cooled to 0.5 K. The atoms then are guided into the fringe field of the magnet, where electron state selection and trapping occur. Microwaves (140 GHz for 5 T) enter from the left and atoms expelled to the right form the beam.

Note: Since the CIPAP meeting we have seen signals of atomic hydrogen expelled from the trap when the magnetic field and microwave frequency are matched at either of the allowed atomic transitions. Typically, 10 mW of microwave power were used. The detector consists of a light copper cup at about 4 K. Atomic hydrogen recombines on the copper with high efficiency and the energy of recombination is measured by the increase in temperature of the detector. Work is continuing to understand loading, trapping, and extraction.

We would like to thank A.D. Krisch, D. Kleppner, and T.J. Greytak for support and advice and T.O. Niinikoski and P.R. Cameron for assitance in the early stages. Brookhaven National Laboratory has provided lab space and cryogenic support. This work is supported by the U.S. Department of Energy.

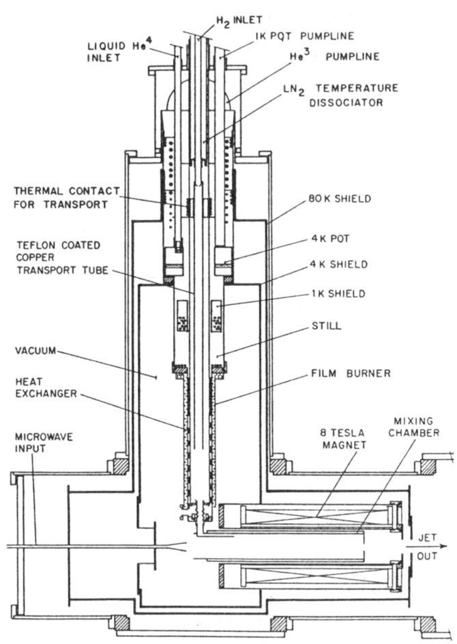

Fig. 1. Polarized jet test apparatus

Measurement of The Deep-inelastic Spin-dependent Structure Functions of The Proton and Neutron at HERA

D.H. Beck, B.W. Filippone, J. Jourdan, R.D. McKeown, R.G. Milner, C.E. Woodward
W.K. Kellogg Radiation Laboratory, California Institute of Technology, Pasadena, CA 91125, USA

S.J. Freedman, D.F. Geesaman, R.J. Holt, H.E. Jackson, R. Kowalczyk, J. Napolitano, L. Young, B. Zeidman
Argonne National Laboratory, Argonne, IL 60439, USA

ABSTRACT

It is possible to measure the deep-inelastic spin-dependent structure functions $g_1^p(x)$ and $g_1^n(x)$ for the proton and neutron using internal polarized hydrogen, deuterium, and ^{3}He targets of polarization 50% and thickness 10^{14} to 10^{15}cm^{-2} and the 60 mA longitudinally polarized 30 GeV electron beam in the HERA electron storage ring at DESY, Hamburg, W. Germany. The measurement of the deep-inelastic spin-structure of both isospin states of the nucleon at the same kinematics and using the same apparatus allows the Bjorken sum rule to be experimentally checked. In addition, it uniquely constrains the spin distribution of the u and d quarks as a function of x in any model of the nucleon. Possible target and detector configurations are described and an estimate of the accuracy of such a measurement is presented.

Recently precise measurements have been presented of the asymmetry $A_1^p(x)$ in deep inelastic scattering of polarized muons from polarized protons.[1] The spin-dependent structure function $g_1^p(x)$ was deduced from these measurements. The data indicate that the Ellis-Jaffe sum rule[2] for the proton is violated. Assuming the Bjorken sum rule[3] is valid, the data imply that the neutron must have a much larger contribution to the sum rule than hitherto believed[4], with the asymmetry for the neutron being largely negative over at least part of the x range. No data for the neutron exist. Also, the new data can be interpreted to indicate that the fraction of the proton spin carried by the quarks is small[1].

To measure the asymmetries in deep inelastic scattering of polarized electrons from polarized protons and neutrons we propose a new technique[5]. This is to use internal polarized atomic gas targets of thickness 10^{14} to 10^{15}cm^{-2} and polarization 50% placed in the 60 mA circulating polarized electron beam of the HERA storage ring. To study the proton, a polarized atomic hydrogen target will be used. For the neutron, polarized deuterium and ^{3}He targets will be used[6]. Since the deuteron is an isoscalar target we will measure the sum of proton and neutron asymmetries. As pointed out by Close and Roberts, these measurements allow a precise determination of the fraction of proton spin carried by the quarks[7]. The neutron asymmetry can be deduced by subtracting the proton asymmetry from the deuteron asymmetry. ^{3}He is of interest because to a good approximation the two protons in this nucleus have opposite spins, and so the asymmetry is due to the neutron alone. Thus, the deep-inelastic neutron asymmetry can be measured in two independent ways. These measurements become feassible because of both the development of a new generation of polarized hydrogen, deuterium, and ^{3}He targets based on the method of optical pumping and because of the expected availability of longitudinally polarized high energy electrons in the HERA electron storage ring. In the case of the hydrogen and deuterium targets, a holding field of order several kgauss is necessary to prevent depolarization by the intense magnetic field of the circulating beam[8]. The targets will provide a luminosity of about 10^{32}cm^{-2}s^{-1}. The proposed technique does not suffer the disadvantages of conventional polarized target technology. In the conventional approach, polarized deuterons in the form of deuterated ammonia would be used. Thus, the asymmetry is diluted by scattering from large amounts of unpolarized material in the ammonia target. In the internal target method, the polarized atoms are pure atomic species. In addition, since the proposed target thickness is of order 10^{-10} radiation lengths, the contribution from external radiative corrections in the target is negligible.

Development of few-body polarized internal targets by optical pumping has been in progress for some years. The common approach is to polarize atoms via optical pumping with circularly polarized light from a laser. These atoms are then fed into a storage bottle[9] with vacuum pumping constrictions at each end. The pumping constrictions serve to generate the desired target density while allowing the circulating beam to pass through the target. At Argonne National Laboratory polarized hydrogen and deuterium gas targets have been under development using the technique of optical pumping spin-exchange. Dissociated hydrogen (deuterium) is introduced into a cell containing a small amount ($\approx 1\%$) of potassium. The potassium is polarized by optical pumping with a dye laser, and the polarization is transferred to the hydrogen (deuterium) by spin-exchange collisions. The Argonne group has achieved a polarization of 30% at a flux rate of 8×10^{16} polarized atoms per second. Work is in progress to achieve a polarization of at least 50% at a flux rate of 4×10^{17} polarized atoms per second[10]. At Caltech a polarized ^{3}He target has been developed[11] for an experiment[12] to study the neutron form-factors by spin-dependent quasielastic scattering of polarized electrons from polarized ^{3}He. The technique used is polarization of the 2^3S metastable state via optical pumping with a laser[13], followed by polarization of the ground state through metastability exchange. Feed rates of 5×10^{17} per sec and polarizations of 52% have been achieved. Work is in progress to achieve higher laser powers, which will allow a higher flux rate and polarization.

A collaboration between Argonne and Novosibirsk has been formed to study storage cells for polarized deuterium atoms in the VEPP-3 electron ring and to measure the analyzing power in electron-deuteron elastic scattering. A cell coated with drifilm has been fabricated at Argonne and has been installed in the VEPP-3 electron storage ring[14]. Tests of this cell are expected to be performed during the Summer of 1988. Design of a high density storage cell is in progress at Argonne and is to be used with a state-of-the-art conventional polarized atomic beam source[15] in the VEPP-3 ring. The expected target thickness that can be achieved with the conventional source and a 30 cm long elliptical storage tube is 10^{13} cm^{-2}. In order to achieve the 10^{14} cm^{-2} target thickness required for the HERA experiment, the high flux optically pumped polarized source of the type being developed at Argonne will be necessary. In addition, an atomic beam source and storage bottle for polarized hydrogen and deuterium atoms has been developed at the University of Wisconsin-Madison by the group of W. Haeberli[16]. In the case of polarized ^{3}He, wall depolarization and beam ionization[17] problems are negligible, since it is a noble gas. Storage times of hundreds of seconds have been obtained at Caltech. The presence of an internal target in an electron ring will give rise to multiple scattering and bremsstrahlung losses. The beam lifetime due to these effects has been estimated to be of order 100 hours[18].

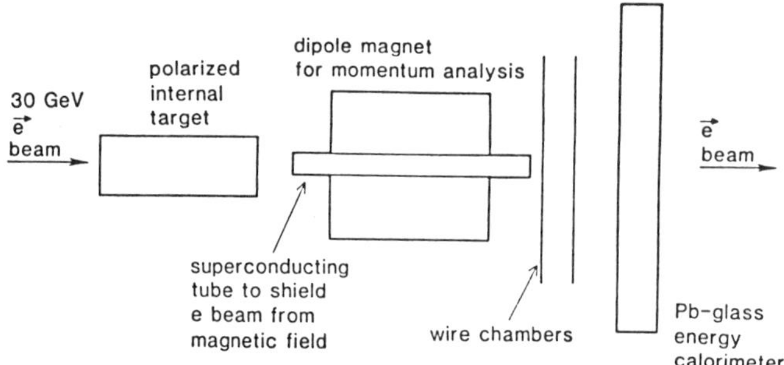

Figure 1. This figure shows a possible detector configuration for the proposed experiment. It is based on the detector used in the experiment described in reference 19.

Fig. 1 shows a possible detector configuration for this experiment. This is based on a detector used at SLAC in the 1970's to measure exclusive states in deep inelastic scattering[19]. Using the measured rates in the SLAC experiment, we predict a trigger rate of several hundred Hz for the proposed HERA experiment, which is quite acceptable. Clearly we require that the proton beam does not pass through the target. This implies that either the experiment runs in parasitic mode with the proton beam displaced from the electron beam or that the experiment is run in dedicated mode with the proton beam turned off. The detector consists of a 1 Tm × 1 m gap dipole magnet through which both the scattered particles and the stored electron beam pass. A superconducting tube shields the electron beam from the dipole field. Planes of MWPCs allow the tracking of charged particles and a wall of lead-glass serves as a calorimeter to measure the energy. The detector should have large acceptance for electrons with scattering angles between 5° and 25°.

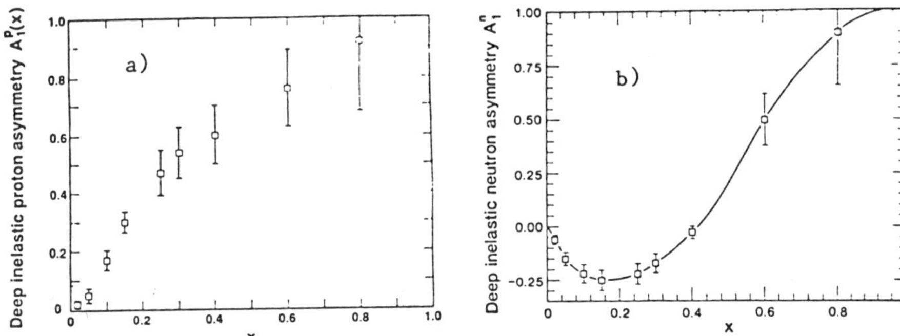

Figure 2. This figure shows the expected precision in a 240 hour run at HERA on (a) a polarized hydrogen internal target of thickness 10^{14} cm^{-2} and (b) on a polarized ^3He internal target of thickness 10^{15} cm^{-2}. The assumptions used are described in the text.

Figures 2a and 2b show the precision as a function of x attainable in a 240 hour run at HERA in a measurement of A_1^p(x) and A_1^n(x) on polarized proton and ^{3}He targets. It is assumed that the 30 GeV electron beam and target polarizations are each 0.5; the electron beam current is 60 mA; the incident electron energy is taken to be 30 GeV. The polarized proton and ^{3}He target thicknesses are 10^{14} cm^{-2} and 10^{15} cm^{-2} respectively. A dilution factor of 0.33 is included in the ^{3}He estimate becaused the electron can also scatter from the two protons in the ^{3}He nucleus. The region in x extends from 0.02 to 0.8 and in Q^2 from 1 to 20 $(GeV/c)^2$. The low x limit is determined by elastic radiative tails and backgrounds at high $y = \frac{\nu}{E}$. A total systematic error of ±15% has been included. The value of $A_1^p(x)$ used was the measured value from the recent EMC data. The solid line in figure 2b is the prediction of a modified Carlitz-Kaur model which is in good agreement with the new EMC data and is constrained to obey the Bjorken sum rule[20].

We see that the precision attainable at low x, where the main contribution to the sum rule arises, is very good. It should be possible to perform a precise test of the fundamental Bjorken sum rule, deduce the fraction of nucleon spin carried by the quarks for both isospin states of the nucleon, and contribute greatly to our understanding of the structure of the nucleon. It is important that these measurements be carried out, especially in the light of recent data, which indicate that our understanding of the spin-structure of the nucleon is very incomplete. At present we are actively pursuing the possibility of performing these measurements at DESY in collaboration with groups from Heidelberg, Wisconsin and Saclay.

We wish to thank J. Goral, G.E. Thomas, and J. Gregar (polarized hydrogen and deuterium targets) and W. Schick and R. Carr (polarized ^{3}He target) for technical assistance. This research was supported in part by the U.S. National Science Foundation under grant PHY85-05682 (Caltech) and in part by the U.S. Department of Energy, Nuclear Physics Division, under contract W-31-109-ENG-38 (Argonne).

References

1 J. Ashman *et al.* [The EMC Collaboration], submitted to Phys. Lett.
2 J.Ellis and R.Jaffe, Phys. Rev. **D9**, 1444 (1974)
3 J.D.Bjorken, Phys. Rev. **148**, 1467 (1966)
4 R.Carlitz and J.Kaur, Phys. Rev. Lett. **38** 673 (1976)
5 Memorandum to DESY Research Director, R. Milner and T. Sloan, September 1987; Letter to HERA Physics Research Committee, Caltech-Argonne collaboration, (PRC88/1); Letter to HERA Physics Research Committee, Heidelberg-NMC collaboration, (PRC88/2)
6 R.G. Milner, *Proceedings of the Workshop on Polarized ^{3}He Beams and Targets*, Princeton, NJ October 1984, p.186 (AIP Conference Proceedings No. 131); R.G. Milner, *Proceedings of the Workshop on Electronuclear Physics with Internal Targets*, SLAC January 1987, p.195
7 F.E. Close and R.G. Roberts, Phys. Rev. Lett. **60**, 1471 (1988)
8 R.D. McKeown, *Proceedings of the Workshop on Electronuclear Physics with Internal Targets*, SLAC January 1987, p.99
9 M.D. Barker *et al.*, *Polarization Phenomena in Nuclear Physics* edited by G.G. Ohlsen *et al.* American Institute of Physics, 1981
10 L. Young *et al.*, Nucl. Instr. and Meth., **B24**, 963 (1987)
11 R.G. Milner, R.D. McKeown, and C.E. Woodward, *to be published*
12 Bates experiment 88-02, Caltech-MIT-Harvard-NBS
13 C.L. Bohler, L.D. Schearer, M. Leduc, P.J. Nacher, L. Zachorowski, R.G. Milner, R.D. McKeown, and C.E. Woodward, J. Appl. Phys. **63** 2497, (1988)
14 R. Holt, *Proceedings of the International Conference on Spin Observables in Nuclear Physics*, Telluride (1988)
15 A.V. Evstigneev, S.G. Popov and D.K. Toporkov, Nuclear Instr. and Meth. **A238**, 12 (1985)
16 W. Haeberli, Helv. Phys. Acta., **59** 597 (1986)
17 R.G. Milner, R.D. McKeown, and C.E. Woodward, Nucl. Instr. and Meth. **A257**, 286 (1987)

[18] T. Weiland, DESY, private communication
[19] J.F. Martin *et al.*, Phys. Rev. D **20**, 5 (1979)
[20] Andreas Schäfer, Caltech preprint MAP-104, *to be published* Phys. Lett. B

Relativistic Effects on Quasielastic Spin Observables

C. J. Horowitz
Indiana University, Bloomington, IN 47405

Abstract

Spin observables in proton quasielastic scattering are calculated in a relativistic plane wave impulse approximation using relativistic mean field calculations for nuclear matter in a local density approximation. Results are compared to (p, p') data at 420 MeV on O and C and 290 MeV on Fe. The analyzing power and polarization are substantially reduced from free NN values by relativistic dynamics in good agreement with experiment. Polarization transfer coefficients $D_{s's}$ and $D_{l'l}$ are predicted to be enhanced. The need for further theoretical calculations and experimental measurements of (p,n) spin observables is discussed. Finally, the effect of vacuum polarization on the longitudinal response in quasielastic electron scattering is discussed.

INTRODUCTION

There have been several relativistic calculations[1,2] for elastic proton scattering. These find large relativistic effects on spin observables. Perhaps the largest effect is on the spin rotation function Q in forward angle diffraction minima at low energy. The relativistic calculations provide a much better description of elastic spin observables (for a variety of nuclei and energies) than do similar nonrelativistic calculations.

These (possible) relativistic effects are interesting for a variety of reasons. Any small effect in normal nuclei could have dramatic consequences when one extrapolates to the extreme conditions of density or temperature reached in neutron stars or in relativistic heavy ion collisions. Several relativistic nuclear matter calculations[3] find that nuclear matter saturates at a lower density and binding energy with these effects. Furthermore, the relativistic equation of state of nuclear or neutron matter is predicted to be quite stiff at high densities[4].

Therefore, it is important to document relativistic effects. Unfortunately, there are limitations to elastic scattering results. There is a complicated relationship between Q at forward angles and relativistic changes in the NN interaction. This relationship involves many theoretical assumptions and empirical inputs such as detailed density distributions for the target. In contrast, the spin observables in quasielastic scattering may be more directly related to relativistic effects.

Elastic scattering is primarily sensitive to the *isoscalar* Lorentz scaler and vector parts of the NN interaction. One would also like to investigate relativistic effects in the *isovector* NN interaction or in other parts of the isoscalar interaction. This can be done with relativistic inelastic (p, p') calculations both to discrete final states[5] and to the continuum. Unfortunately, uncertainties in the relativistic final state wave function complicate the analysis of discrete states. [Note, there is now much work to improve these wave functions[6].]

Here we consider quasielastic scattering to the continuum. If the excitation energy is high enough ($\approx 50+$ MeV or well above giant resonances), a statistical or Fermi gas wave function may already provide a first approximation.

Furthermore, by considering spin observables rather than cross sections one minimizes the sensitivity to distortion effects. This is because most of the effects of spin independent distortions cancel in spin observable ratios. [The effects of spin orbit distortions have been estimated in ref. 7 to be small.]

Finally, there is a clear benchmark to compare to relativistic predictions. At high energies most nonrelativistic calculations, such as ref. 8, predict the spin observables should be within 10 percent of the free NN spin observables. For example, the analyzing power should be close to the incoherent isospin average of the free pp and pn analyzing powers. Relativistic effects may show up by simply comparing the quasielastic results to free NN observables.

In section II we briefly review the relativistic plane wave impulse approximation formalism of ref. 9. Next, results are compared in section III to experimental data. Section IV discusses the decrease in the analyzing power and the difference between (p,p') and (p,n) spin observables. Relativistic effects on quasielastic *electron* scattering is discussed in section V and section VI concludes.

RELATIVISTIC PLANE WAVE IMPULSE APPROXIMATION FORMALISM

We wish to calculate the response of nuclear matter, treated in a relativistic mean field approximation, to a proton probe. This was done in ref. 9 (here after I) in a plane wave impulse approximation formalism. This work should be consulted for any details. Here we provide only a brief review. We emphasize the use of a plane wave approximation in I. This should be a good first approximation for spin observables but needs to be checked with full distorted wave (DWIA) calculations. [Note, the first such relativistic DWIA calculation has now been done for (p,2p)[10].] Also, ref. 1 uses nuclear matter results in a local density approximation. This provides a qualitative introduction but more sophisticated finite nucleus calculations are clearly needed.

The NN interaction in I is assumed to depend on four component Dirac spinors for the nucleons. These spinors change in nuclear matter because of strong scaler potential. The self-energy of a nucleon in the medium Σ,

$$\Sigma = \Sigma_s + \gamma_0 \Sigma_0, \tag{1}$$

has lorentz scalar Σ_s and vector Σ_0 pieces. The scalar self-energy is large and changes the effective nucleon mass M^*.

$$M^* = M + \Sigma_s \tag{2}$$

Here M is the free nucleon mass. The reduced effective mass $M^* < M$ enhances the lower components of the nucleon spinors U.

$$U(p) = \sqrt{\frac{E^* + M^*}{2M^*}} \begin{pmatrix} 1 \\ \frac{\vec{\sigma}\cdot\vec{p}}{E^*+M^*} \end{pmatrix} \chi_\lambda \tag{3}$$

Here χ_λ is a Pauli spinor and $E^* = \sqrt{\vec{p}^2 + M^{*2}}$. These spinors are completely characterized by M^* and have lower components of order p/M^*.

The NN amplitude is assumed to be

$$A(M^*) = \overline{U}_1\overline{U}_2\hat{F}U_1U_2, \tag{4}$$

where $\hat{F}$ is an operator in the spinor space of the two nucleons. The form of $\hat{F}$ used is shown in table I. With eq. (4) the amplitude changes as the spinors change. This introduces model dependence because only free spinor matrix elements of $\hat{F}$ are constrained by the experimental NN phase shifts. However, the form used in table I is known to work well in relativistic impulse approximation calculations of elastic scattering.

Using $\hat{F}$ it is straight forward to calculate the spin observables in quasielastic proton scattering in a relativistic plane wave impulse approximation. This was done in I assuming the target was described by a mean field model of nuclear matter at a density of about 1/3 of saturation density. This is a simple average for the surface peaked (p, p') reaction.

In the results which follow, we refer to nonrelativistic (NR) calculations as *relativistic* calculations using *free* Dirac spinors of mass M. Thus NR calculations include "relativistic kinematics". The NR results are very close to the isospin average of free pp and pn spin observables. The only difference is a small correction from Fermi motion averaging. Relativistic (R) calculations use Dirac spinors of mass M^*. Therefore the (primary) difference between R and NR calculations is the change in NN amplitudes $A(M^*)$ as the spinors change.

We close this section by listing the approximations used in I. The calculations assume a *plane* wave impulse approximation (PWIA) using nuclear matter results in a local density approximation. The simple local form of $\hat{F}$, table I, is used. Finally, there are no corrections for multiple nucleon knockout.

RESULTS

The number of published (p, p') quasielastic polarization transfer spin observables is still very small. The T. Carey et. al.[11] experiment at LAMPF was compared to theory in I. Since then, two TRIUMF experiments have been performed. The first by C. Chan et al[12] was on ^{12}C and ^{16}O at 420 MeV and a laboratory scattering angle of 23.8 degrees. The results for O and C are very similar.

The polarization (which is equal to the analyzing power in the approximations used in I) is seen to be substantially below the NR prediction. Estimates of corrections to the NR result from spin orbit distortions or multiple scattering are all much to small.[8] This reduction in the polarization or analyzing power has now been seen in a variety of nuclei from C to Pb at energies of 300 to 800 MeV. It is explained naturally by the relativistic calculation. As discussed below, this reduction in the analyzing power may be the cleanest relativistic signature found to date. Furthermore, there have been no alternative nonrelativistic explanations (so far).

The NR and R results for the depolarization D_{nn} are very similar and either is in rough agreement with the data. Thus for D_{nn} there is no relativistic signature. However, the agreement with data of either calculation supports the single step impulse approximation reaction mechanism. It is difficult to get a strong reduction in the polarization from multiple scattering with out also changing D_{nn}.

The R calculation for $D_{s's}$ is substantially above the NR result. The data is clearly enhanced compared to the NR result but not by as much as predicted. Some of the relativistic enhancement may be canceled by the effects of spin orbit distortions. Furthermore, the simple nuclear matter calculation of I could substantially overestimate the relativistic effects in this surface peaked reaction. A full relativistic distorted wave impulse approximation calculation using a finite nucleus RPA response function could test both effects.

Finally, the non-diagonal $D_{l's}$ is predicted to be unchanged by relativity (the R and NR are very close). This agrees with the data at low but not at high excitation energy. Perhaps the rise at high energy is do to nuclear structure effects or spin-orbit distortions. We note that this problem is common to both the NR and R calculations and that the experimental errors are large.

The other TRIUMF experiment by Otto Hauser and collaborators,[13] $^{54}Fe(p,p')$ at 290 MeV and 20 degrees in the lab., is shown in figure 1. In addition to NR and R curves the nonrelativistic RPA calculations of R. Smith are also shown (see also ref. 14). These are based on the semi-infinite slab responses of Esbensen and Bertsch[15] and include effects of nuclear structure not in our nuclear matter response. We will refer to these calculations as NR RPA.

The analyzing power of the NR RPA has a larger slope vrs excitation energy then the NR curve. This is because the isoscalar response, which has

$\hat{\mathcal{F}} = \sum_i F^i(q,E)\lambda^i_{(1)}\lambda^i_{(2)}$	
i	λ^i
S (Scalar)	1
V (Vector)	γ_μ
P (Pseudoscalar)	γ_5
A (Axial-Vector)	$\gamma_5\gamma_\mu$
T (Tensor)	$\sigma_{\mu\nu}$

TABLE I

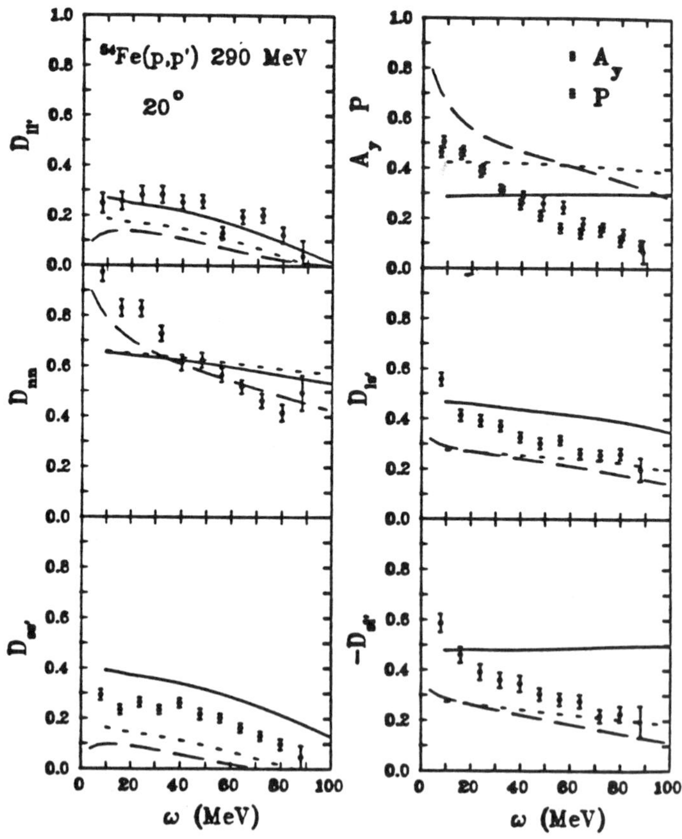

Fig. 1. Quasielastic spin observables at 290 MeV for ^{54}Fe from reference 13. The solid curves are relativistic (R), the short dashed nonrelativistic (NR) and the long dashed nonrelativistic RPA (NR RPA) calculations (see text).

a larger analyzing power then the isovector, is shifted by the RPA to lower energy. The slope of the NR RPA is in good agreement with data. However, its magnitude is substantially above the data. Indeed the R calculation predicts just this reduction. Presumably a relativistic RPA calculation could explain both the slope and magnitude of the analyzing power and polarization.

The $D_{ll'}$ values are above both the NR and NR RPA in agreement with the R results. For D_{nn} there is no difference between the NR and R curves. Indeed, the data is in good agreement with the NR RPA calculation. We see that RPA corrections to nuclear structure might also explain the slight slope seen in the C and O data. For $D_{ss'}$ R predicts an enhancement compared to NR which is clearly seen in the data. The fact that the data is still slightly below the R curve could be do to nuclear structure. Indeed comparing the NR and NR RPA curves we see such a reduction.

Finally, the off diagonal $D_{ls'}$ and $-D_{sl'}$ are both predicted by R to be enhanced compared to NR results. The data, while showing some enhancement, lies in between the curves. Clearly, finite nucleus relativistic RPA calculations would be useful to see if the remaining disagreement is serious.

In summary, the change in Dirac wave function with $U(M)$ going to $U(M^*)$ reduces the analyzing power, enhances $D_{ll'}$, $D_{ss'}$ and $D_{ls'}$ and has no effect on D_{nn}. Qualitatively all these effects are seen in the data. To examine these effects further, relativistic calculations more sophisticated then the simple local density approximation of I would be very useful.

DISCUSSION

The reduction in analyzing power is directly related to the change in NN interaction in the medium. At small angles the analyzing power is approximately (note this is only qualitative!)

$$A_y \approx Im F_s^* F_v / |F_s + \lambda F_v|^2. \tag{5}$$

Here λ is a kinematic factor which describes the importance of the vector interaction compared to the scalar (see table I). This factor depends on the four component spinors and changes in the medium.

$$\lambda = 1 + T_{lab}/M^* \tag{6}$$

As the effective mass M^* decreases, λ grows (T_{lab} is the laboratory kinetic energy) and the repulsive vector becomes more important then the attractive scalar interaction. This greatly increases the denominator in eq. (5) and reduces the analyzing power. Small changes in M^* can be important because there are sensitive cancellations between vector and scalar invariants.

This decrease in the analyzing power is very closely related to a predicted relativistic effect on nuclear saturation. As the wave functions change, the intermediate range scalar attraction in the NN interaction is reduced compared to the short range vector repulsion. Therefore, nuclear matter saturates at a lower density and binding energy then in a similar nonrelativistic calculation. This theoretical prediction should carry more weight now that the reduction in A_y has been observed.

The (p,n) reaction allows one to examine relativistic effects in the isovector NN interaction. The isovector interaction is very different from the isoscalar one. The lorentz scalar F_s and vector F_v amplitudes are no longer dominant. Instead, the pseudoscalar invariant F_p from one pion exchange is very important. As a result D_{nn} is now very sensitive to relativistic effects. It would be extremely useful to have measurements of (p,n) quasielastic spin observables. [Relativistic plane wave impulse approximation predictions for specific kinematics are available from the author upon request.]

ELECTRON SCATTERING

The longitudinal response function in quasielastic electron scattering is interesting because of the nonrelativistic Coulomb sum rule. This sum rule was derived by summing over a complete set of states. However, in relativistic models the positive energy states alone are not complete. Therefore, there is no rigorous relativistic coulomb sum rule.[16]

Indeed, quasielastic scattering measures the sum of the *squares* of the charges. When one includes the vacuum this is not conserved. A virtual nucleon antinucleon pair can contribute to the scattering if there is enough momentum transfer to resolve the individual charges in the pair.

In ref. 17 we have calculated the Coulomb sum in infinite nuclear matter in a relativistic RPA approximation which includes both particle-hole and nucleon-antinucleon excitations (see also ref. 18). As the effective mass decreases, it becomes easier to excite virtual pairs and these pairs do a better job of screening the original charges. Therefore, less of the charge is visible to the electron probe (even at relatively modest momentum transfers). The Coulomb sum is reduced by about 30 percent *in nuclear matter* compared to relativistic Hartree calculations. This reduction may be in good agreement with Saclay experiments on ^{40}Ca. Finite nucleus calculations including vacuum polarization are underway.

CONCLUSIONS

Spin observables in quasielastic proton scattering have been calculated in a relativistic plane wave impulse approximation for nuclear matter. The spin dependence of the NN interaction changes in nuclear matter as the four component Dirac spinors change. This leads to a reduction in the analyzing power and an enhancement in some polarization transfer coefficients. All of these relativistic effects have now been observed in recent TRIUMF experiments. To examine these effects further, finite nucleus relativistic DWIA calculations would be vary useful as would experimental measurements of (p,n) quasielastic spin observables. Finally, relativistic RPA calculations for electron scattering have a reduced longitudinal response because of an enhanced screening from vacuum polarization.

REFERENCES

1. J. R. Shepard *et al.*, Phys. Rev. Lett. **50** 1443 (1983); L. Ray and G. Hoffmann, Phys. Rev. **C31** 538 (1985).
2. D. P. Murdock and C. J. Horowitz, Phys. Rev. **C35** 1442 (1987).
3. C. J. Horowitz and B. D. Serot, Nucl. Phys. **464A** 613 (1987).
4. B. D. Serot and J. D. Walecka, Advances in Nucl. Phys. **16**, J. W. Negele and E. Vogt, eds. (Plenum, NY 1986).
5. J. R. Shepard *et al.*, Phys. Rev. **C30** 1604 (1986).
6. R. J. Furnstahl, "International Conference on Spin Observables of Nuclear Probes", C. J. Horowitz *et al.*, eds. (Plenum, in press).
7. C. J. Horowitz and M. J. Iqbal, Phys. Rev. **C33** 2059 (1986).
8. R. D. Smith and S. J. Wallace, Phys. Rev. **C32** 1654 (1985).
9. C. J. Horowitz and D. P. Murdock, Phys. Rev. **C37** 2032 (1988).
10. E. D. Cooper, "International Conference on Spin Observables of Nuclear Probes", C. J. Horowitz *et al.*, eds. (Plenum, in press).
11. T. Carey *et al.*, Phys. Rev. Lett. **53** 144 (1984); private communication.
12. C. Chan *et al.*, "Spin Observables For Quasielastic Proton Scattering From ^{12}C and ^{16}O", submitted to Phys. Rev. Lett.
13. O. Hausser *et al.*, "Complete Spin Observables for Quasielastic Proton Scattering From ^{54}Fe at 290 MeV", submitted to Phys. Rev. Lett.
14. R. D. Smith, Phys. Rev. Lett. **58** 2404 (1987).
15. H. Esbensen and G. Bertsch, Ann. of Phys. **157** 255 (1984) and Phys. Rev. **C34** 1419 (1986).
16. G. Do Dang *et al.*, Phys. Rev. **C35** 1637 (1987).
17. C. J. Horowitz, "Vacuum Polarization and the Coulomb Sum Rule", Preprint IU/NTC 88-4, Phys. Lett. B, in press.
18. H. Kurasawa and T. Suzuki "Effects of the Dirac Sea on the Coulomb Sum Value for Electron Scattering", RIFP-739, submitted to Nucl. Phys. A; Xiangdong Ji, "Effects of Vacuum Polarization on Quasielastic Electron Scattering", W. K. Kellogg Lab. Preprint 1988.

ISOVECTOR SPIN RESONANCES IN ^{90}Y FROM THE ^{90}Zr(n,p)^{90}Y REACTION AT 198 MeV

K.J. Raywood, S.A. Long, B.M. Spicer
School of Physics University of Melbourne
Parkville, Victoria, 3052, AUSTRALIA

S. Yen, K.P. Jackson
TRIUMF 4004 Wesbrook Mall
Vancouver, B.C. CANADA V6T 2A3

M.A. Moinester
University of Tel Aviv
Tel Aviv, ISRAEL

W.P. Alford
Department of Physics, University of Western Ontario
London, Ontario, CANADA

ABSTRACT

Proton spectra from the ^{90}Zr(n,p)^{90}Y reaction have been measured at laboratory angles of 0°, 3°, 6°, 8°, 13°, 17° and 21°, for 198 MeV incident neutrons. Spectra are presented, and the separation of the spin-flip giant resonances from the quasi-free scattering is discussed. Results of the analysis are presented.

This experiment was performed with the TRIUMF nuclear charge exchange facility[1]. A proton beam from the TRIUMF cyclotron, of energy 200 MeV, struck a ^{7}Li target of 220 mg cm^{-2} thickness. The neutrons produced in this way, of energy 198 MeV, struck the secondary target assembly which consisted of 5 ^{90}Zr foils, each 250 mg cm^{-2} thick, followed by a foil of CH_2, 46 mg cm^{-2} thick. These 6 foils were separated by multi-wire proportional chambers (MWPC). The target assembly is fully described in ref.[2]. The protons produced by the (n,p) reactions in the secondary target foils were momentum analyzed by the medium resolution magnetic spectrometer (MRS). The main contributions to the overall energy resolution, which was about 1.3 MeV, came from the energy spread in the incident beam and the proton energy loss in the primary and secondary targets.

The proton spectra measured by the MRS were collected in event mode. The tapes were replayed off-line for the purpose of identifying the Zr-foil in which the reaction occurred, and shifting the spectra to take account of the energy loss of the outgoing protons in succeeding target foils. They were further corrected for efficiency of the MWPC's in the target box before summing the contributions from all five Zr foils to give the total spectrum at each angle. These spectra were then corrected for spectrometer acceptance. The appropriate "target empty" spectrum, which had been treated in exactly the same way, was then subtracted, and the ^{7}Li(p,n) neutron spectrum lineshape was unfolded to give the final spectra.

The $^{90}Zr(n,p)$ reaction cross section is small enough to make it necessary to consider seriously the contribution of quasi-free charge exchange scattering. This has been done using a method which closely resembles that which Erell et al.[3)] used in their analysis of pion charge exchange reactions at 165 MeV. In our analysis, the magnitude of the quasi-free scattering cross section was allowed to vary freely in the peak-fitting process; the end result, however, was a quadratic variation such as specified by Erell et al.[3)].

The peak fitting process also revealed four peaks, whose differential cross sections were compared with sample angular distributions calculated for various ℓ-transfers using distorted wave theory. Figure 1 shows two spectra, at 5.8° and 12.9°, fitted with the 4 peaks and the quasi-free scattering contribution.

Comparison of the deduced differential cross sections for the 4 peaks with the sample angular distributions calculated indicates that the 3 lowest excitation peaks (at excitations of 2.8, 5.5 and 10.3 MeV in ^{90}Y) are dipole in the nature (ℓ=1). Figure 2 shows the differential cross sections for these three spin dipole peaks, compared with a calculated angular distribution using the program ALLWORLD[5)] to calculate the particle-hole amplitudes, and the DWIA to obtain the angular distributions. The agreement is fair at small angles, but the 5.5 and 10.3 MeV peaks show a shoulder commencing at about 10°. This signals the presence of the 2^- components of the spin dipole resonance, which can be reached by ℓ=3, as well as ℓ=1, charge exchange.

The spin dipole resonance is thus well localized in energy, though not to the same extent as indicated in the calculation of Auerbach and Klein[4)].

REFERENCES

1. R.L. Helmer, Can. J. Phys. 65, 588 (1987).

2. R.S. Henderson, W.P. Alford, D. Frekers, O. Hausser, R.L. Helmer, K.H. Hicksa, K.P. Jackson, C.A. Miller, M.C. Vetterli and S. Yen, Nuclear Instruments and Methods A257, 97 (1987).

3. A. Erell, J. Alster, J. Lichtenstadt, M.A. Moinester, J.D. Bowman, M.D. Cooper, F. Irom, H.S. Matis, E. Piasetzky and U. Sennhauser, Phys. Rev. C34, 1822 (1986).

4. N. Auerbach and A. Klein, Phys. Rev. C30, 1032 (1984).

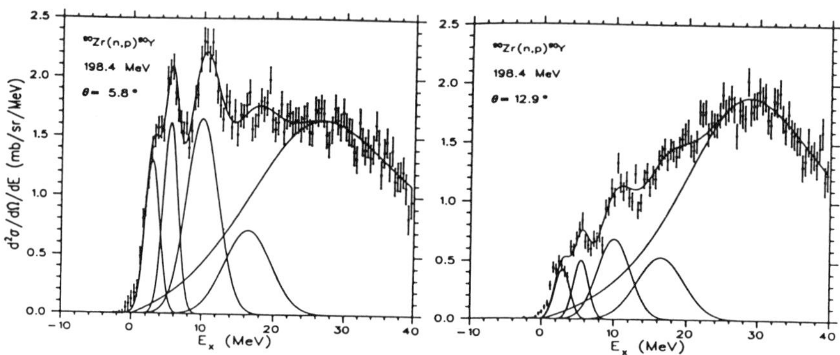

Figure 1. Proton spectra measured at center of mass angles of 5.8° and 12.9°, showing the results of the peak fitting.

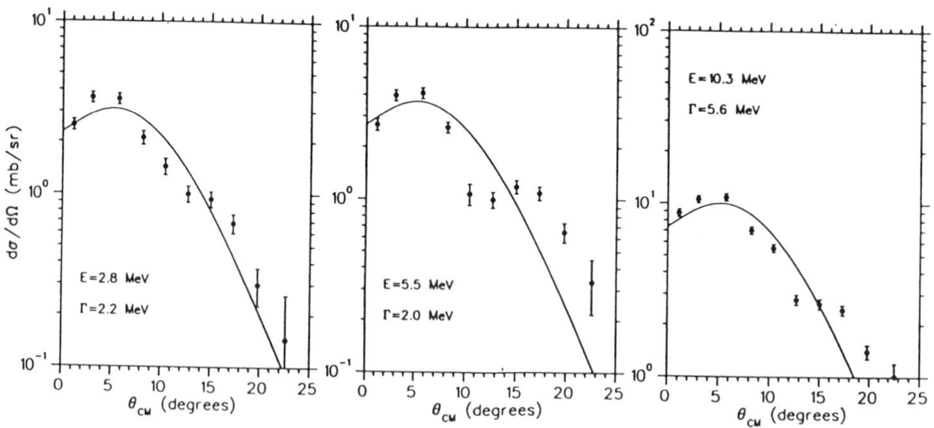

Figure 2. Differential cross sections for the three dipole peaks, shown with the results of RPA-DWIA calculations.

Spin Rotation Parameters A and R for $\pi^{\pm}p$ Elastic Scattering at the Onset of the Roper Resonance

B.M.K. Nefkens, D.B. Barlow, R.S. Kessler, G.J. Kim
C. Pillai, J.W. Price, J.A. Wightman
University of California, Los Angeles, CA 90024

S. Hall, D.W. Lane, S.R. Loe, L.K. Morton, M.E. Sadler
Abilene Christian University, Abilene, TX 79699

S.D. Adrian, W.J. Briscoe, L.H. Kramer, A. Mokhtari, M.F. Taragin
George Washington University, Washington, DC 20052

J.F. Davis
Los Alamos National Laboratory, Los Alamos, NM 87545

I. Supek
Institute Rudjer Boscovic, Yugoslavia

ABSTRACT

The verification of the existing sets of πN scattering phases is sought via the determination of the A and R spin rotation parameters in π^+p and π^-p elastic scattering. The preliminary results of this first measurement of these parameters for incident pion momenta from 427 to 657 MeV/c, show agreement with the gross features of the partial-wave analyses by the groups from VPI, Karlsruhe-Helsinki, and Carnegie-Mellon-LBL. There is room for improvement.

Hadronic processes at energies up to several GeV are rich in QCD-related problems that are approachable via experimentation. Theoretical efforts are severly hampered by a large QCD coupling constant that renders the perturbative approaches unworkable. Among the basic problems are the following:

1. How does confinement work?
2. How does one identify exotic hadronic matter such as $q^4\bar{q}$ states, glueballs, and hybrids?
3. Why does the simple quark model work so well in describing baryonic resonances?
4. How to quantify the gluon contributions in resonances? Estimates are that the quarks make up less than half of the mass and the spin of the nucleon.

To this list we may add a few other interesting questions:

5. What is the strange quark, $s\bar{s}$, content of the nucleon?
6. Why do many of the baryonic resonances appear in clusters?
7. How good are isospin invariance and charge symmetry? Are these symmetries better observed away from threshold when Coulomb energies are not important?

The πN system is particularly suited for shedding light on some of these problems because:

1. it is easily accessible experimentally;
2. it has a simple spin-isospin structure that greatly simplifies the theoretical analysis.

There are two spin amplitudes, the non-spin-flip and spin-flip, and two isospin amplitudes, I = 1/2 and 3/2. The complete description of the πN system requires only 4 amplitudes which can be determined in 8 experiments, except for the overall phase factor, see Table I. Since the measured quantities are given by quadratic relations among the four amplitudes some redundancy in the data is needed.

Our group has undertaken a program of measuring all necessary quantities in the πN system with a few redundancies at five incident pion energies accessible in the P^3 channel at LAMPF. It is important that all experiments be done at precisely the same incident energies to avoid interpolation and extrapolation of the measured data to the PWA-grid energies when performing a full partial-wave analysis, PWA. Here we report on the last phase of the program, which is the most complicated one, namely, the first measurement of the spin rotation parameters A and R in $\pi^+ p$ and $\pi^- p$ elastic scattering. Results on $d\sigma/d\Omega$ and A_N have already appeared, see Refs. 1–3.

The data on the πN resonances come from the three modern πN PWA's by the groups from Karlsruhe-Helsinki, K-H,[4] Carnegie-Mellon University-Lawrence-Berkeley Laboratory, C-L,[5] and Virginia Polytechnic Institute and State University, VPI.[6] These PWA's are done as follows. The published data consisting of $d\sigma/d\Omega$ and A_N measurements are amalgamated by fitting to polynomials which are then extrapolated to the nearest PWA-grid energies using a smooth extrapolation; conflicting data are "massaged" by enhancing the published errors or discarding a set altogether. A fixed-energy PWA is performed for each grid energy resulting in several sets of phases as there are no A and R data available to resolve the ambiguities. K-H and C-L use strong theoretical constraints based on dispersion relations to arrive at a unique set of phases; similarly, VPI obtains one solution by using a strong ansatz. There are three occasions in this procedure where things can go wrong: a) in the amalgamation phase, the wrong data set could be picked from among conflicting sets; b) the smoothing of the data may wipe out small structures; c) the theoretical constraints may lead the analysis astray especially at the opening of a large inelastic channel. These pitfalls are avoided in our program in which all necessary data are measured at precisely the same beam energies avoiding amalgamation, data smoothing, and theoretical constraints, as the program includes measurements of the A and R parameters.

The first complete set of measurements of both A and R for $\pi^+ p$ and $\pi^- p$ elastic scattering was carried out at LAMPF in the P^3 channel with the setup shown in Fig. 1. The apparatus includes the HERA superconducting solenoid and a dilution refrigerator. The hydrogen in the propanediol target was polarized at 0.5 K with HERA at 2.5 T. When the maximum polarization of about 75% was achieved, the spin was "frozen in" by lowering the temperature to $\lesssim$ 50 mK. This allowed the holding field to be reduced to 0.5 T which was necessary for the charged particles to get in and out of the target. The spin relaxation time was several hundred hours. Target spin reversal was done by a small adjustment of the frequency of the microwaves used for polarizing. To allow for different scattering angles, the entire HERA system was mounted on a rotatable platform. No magnetic material was allowed within 2 m of HERA because of the large stray field.

The principal spin orientations are illustrated in Fig. 2. The $\hat{s}_f$ component is measured from the up-down asymmetry in the polarimeter. Instead of measuring the $\hat{s}_f$ component separately for the $\hat{l}_i$ and $\hat{s}_i$ target spin orientation, we reduced the measuring time to half by bending the scattered proton vertically such that the spin precessed by 90° which allows the $\hat{l}_f$ component to be measured from the left-right asymmetry in the polarimeter. The bend angle needed for this is calculated from the precession equation $\theta_R = \gamma\theta_b(1/2g - 1)$. With g = 5.586 and $\theta_R = 90°$, this gives a bend angle $\theta_b \simeq 42°$ for this experiment. The LAS spectrometer was used both as a spin precessor and momentum analyzer. The polarization of the scattered proton was measured with the JANUS polarimeter consisting of 6 drift chambers and a 5–10 cm thick carbon plate scatterer. The scattered pions were detected in a counter hodoscope with 3° resolution.

In terms of the non-spin-flip amplitude, f, and the spin-flip amplitude, g, and pion cm scattering angle θ, we have

$$A = \frac{(|f|^2 - |g|^2)sin\theta + 2\mathrm{Re}(fg^*)cos\theta}{|f|^2 + |g|^2} \tag{1}$$

$$R = \frac{(|f|^2 - |g|^2)cos\theta - 2\mathrm{Re}(fg^*)sin\theta}{|f|^2 + |g|^2} \,. \tag{2}$$

A useful consistency check is provided by the relation

$$A^2 + R^2 + P^2 = 1.$$

The polarization components measured by JANUS are related to the recoil proton polarization by a 3×3 spin precession matrix that depends on the magnetic fields of HERA and LAS; it was evaluated using an extensive Monte Carlo program. The experiment was performed for π^+ and π^- incident momenta of 427, 471, 547, 625, and 657 MeV/c at several angles in the backward hemisphere.

Some preliminary results for A and R are shown in Fig. 3. Comparisons are made with the VPI, K-H, and C-L πN PWA's. All three PWA's *predict* the gross features of the data very well indicating that the major features of the existing πN PWA's are correct. This is the first ever confirmation of this. Upon closer inspection, however, we recognize there are several smaller deviations between our data and the PWA's indicating the need for a new round of πN partial wave analyses in the energy region of this experiment.

This work was supported in part by the USDOE.

References

[1] M.E. Sadler, W.J. Briscoe, D.H. Fitzgerald, B.M.K. Nefkens, and C.J. Seftor, Phys. Rev. D **35**, 2718 (1987).

[2] A. Mokhtari, A.D. Eichon, G.J. Kim, B.M.K. Nefkens, J.A. Wightman, D.A. Fitzgerald, W.J. Briscoe, and M.E. Sadler, Phys. Rev. D **35**, 810 (1987).

[3] J.A. Wightman, UCLA Ph.D. Dissertation; and J.A. Wightman *et al.*, Phys. Rev. D **36**, 3529 (1987).

[4] G. Hohler, Pion-Nucleon Scattering, Landolt-Bornstein **I/9b2** edited by H. Schopper (Springer-Verlag, 1983).

[5] R.E. Cutkosky, *et al.*, Phys. Rev. D **20**, 2804 and 2839 (1979); and in "Proceedings of the IVth Int. Conf. on Baryon Resonances," edited by N. Isgur, Toronto (1980).

[6] R.A. Arndt, J.M. Ford, and L.D. Roper, Phys, Rev. D **32**, 1085 (1985).

[7] N.W. Dean and Ping Lee, Phys. Rev. D **5**, 2741 (1972).

TABLE I Experiments in πN scattering. We use the abbreviations $\pi^+ \equiv \pi^+ p \rightarrow \pi^+ p$, $\pi^- \equiv \pi^- p \rightarrow \pi^- p$, and $\pi^\circ \equiv \pi^- p \rightarrow \pi^\circ n$ for the three accessible channels.

Measurement	Channels	PWA	
		required	optional
$d\sigma/d\Omega$	π^+, π^-, π°	3	
$P = A_N$	π^+, π^-, π°	3	
spin rotation A∗	any	1	2
spin rotation R∗	any	1	2
$\sigma_t \rightarrow I_m A(0^\circ)$	π^+, π^-, π°		3
		tot 8	tot 7

∗ Either $A^+$and A^-, $A^-$and R^+, or $A^+$and R^- are necessary and sufficient to resolve the ambiguities, but $R^+$and R^- are not, see Ref. [7].

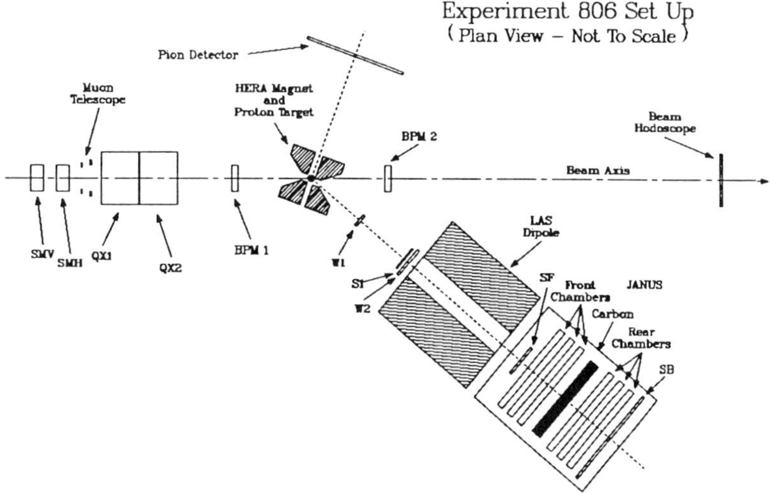

Fig. 1 Setup used to measure the A and R spin rotation parameters in π^+p and π^-p elastic scattering.

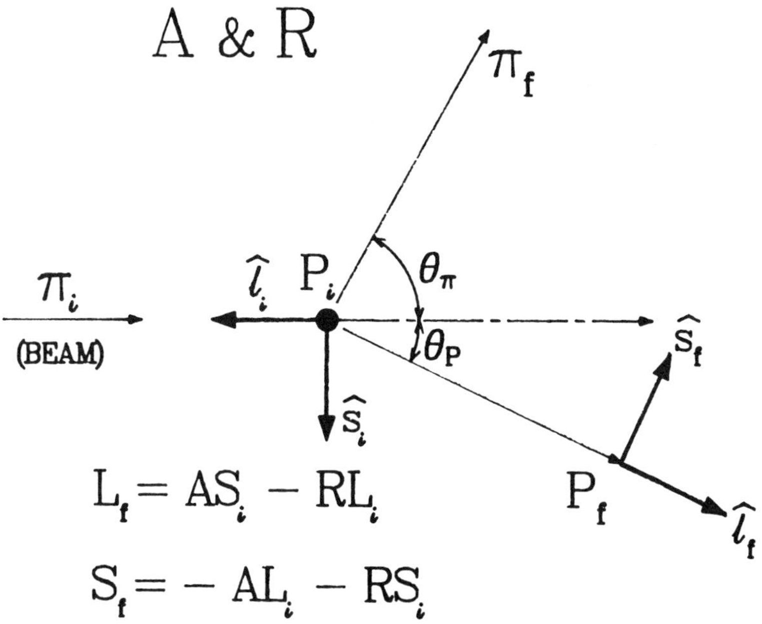

Fig. 2 Definitions of the initial and final-state proton polarization.

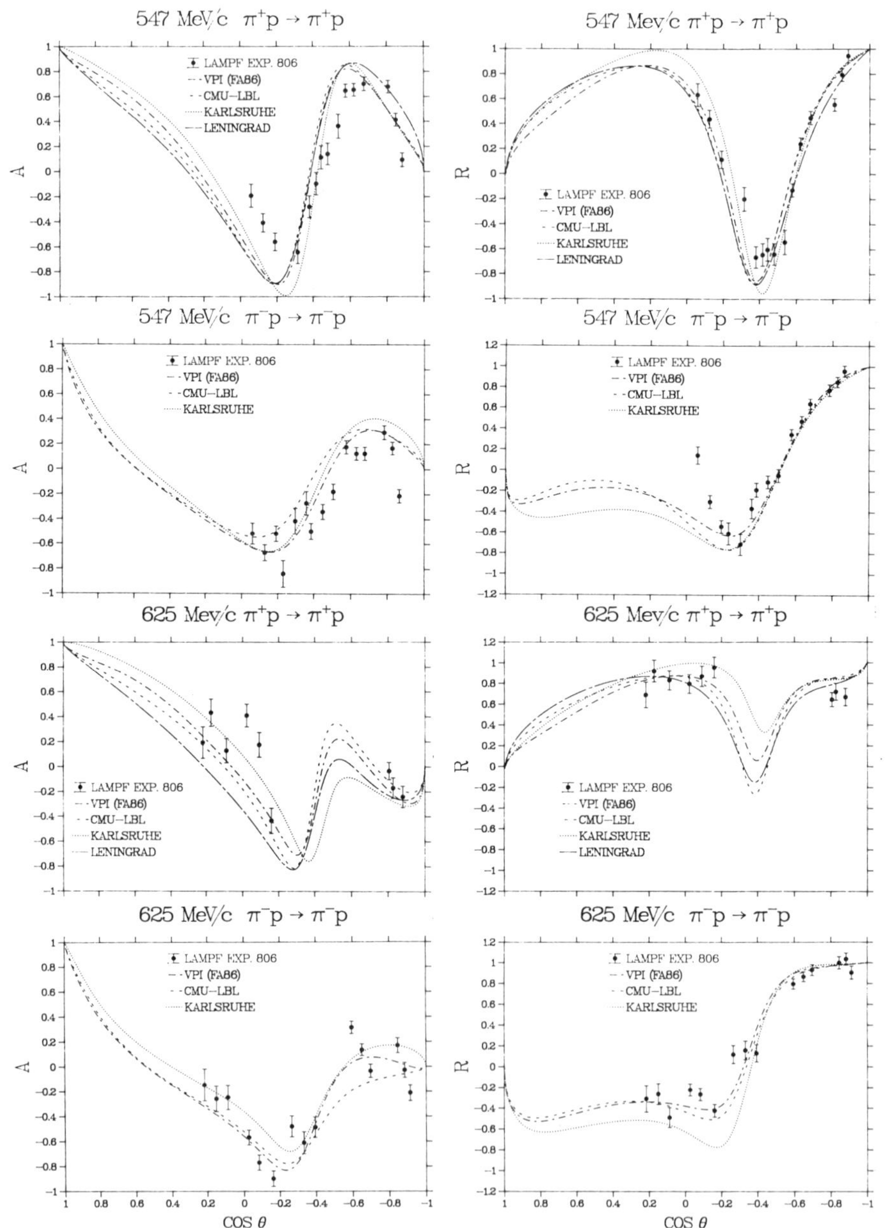

Fig. 3 Preliminary results of the measurements of the A and R spin rotation parameters.

ENERGY DEPENDENCE OF T_{20} AND τ_{21} IN $\pi\vec{d}$ ELASTIC SCATTERING

G.R. Smith, D.R. Gill, D. Healey, D. Ottewell, G.D. Wait, P. Walden
TRIUMF, Vancouver, B.C., V6T 2A3

R.R. Johnson, G. Jones, R. Olszewski,
F.M. Rozon, R. Rui, M.E. Sevior, R.P. Trelle
Univ. of British Columbia, Vancouver, B.C., V6T 1W5

E.L. Mathie, G. Lolos, S.I.H. Naqvi, V.Pafilis
Univ. of Regina, Regina, Sa., S4S 0A2

N.R. Stevenson, R.B. Schubank
Univ. of Saskatchewan, Saskatoon, Sa., S7N 0W0

W. Gyles, C.R. Ottermann
KFK and Univ. Karlsruhe, 7500 Karlsruhe, W. Germany

G.S. Kyle
New Mexico State Univ., Las Cruces, N.M., 88003

ABSTRACT

Angular distributions of the tensor analyzing powers T_{20} and τ_{21} have been measured for $\pi\vec{d}$ elastic scattering, in a single scattering experiment employing a tensor polarized deuteron target. Measurements of T_{20} were obtained for pion bombarding energies of 134, 151, 180, 220, and 256 MeV. Measurements of τ_{21} were obtained for pion bombarding energies of 134, 180, and 220 MeV. The results are compared with three-body calculations where effects relating to pion absorption are seen to play an important role.

INTRODUCTION

The study of the πd elastic scattering reaction is of fundamental importance in intermediate energy pion physics. The deuteron is the simplest composite nucleus which can be studied with pions, and forms our basis for understanding more complicated systems on a microscopic level. It is the only system which can in principle be calculated exactly, using three-body theories with measured πN and NN phases as input. The more advanced of these theories are unified in the sense that they address the πNN system as a whole. The simultaneous prediction of many observables for all πNN channels provides a severe constraint on the theory. It is therefore a goal of the experimentalist to provide as broad and diverse a data base as possible with which to confront the theory.

FORMALISM

Before describing the experiment or the results it is necessary to define the observables of interest, T_{20} and τ_{21}, in terms of the quantities actually measured during the experiment. In accordance with the usual Madison convention we adopt a reference frame defined in terms of the incident and scattered pion trajectories, in which the z-axis lies along the incident pion momentum, and the y-axis is perpendicular to the scattering plane defined by the incident and scattered pion momenta. The orientation of the spin alignment axis of the polarized deuterons is described using Euler angles α and β which refer respectively to the polar angle between the incident pion beam and the target magnetic field, and the angle between the y-axis and the projection of the target magnetic field on the x-y plane.

Measurement of a given spin observable T_{kq} is accomplished by choosing appropriate values of α and β such that the contribution of other spin observables is eliminated or minimized. Measurements of T_{20} are performed in an experimental configuration such that α is 0°. This implies the target magnetic field axis is aligned along the incident beam direction (longitudinally). The result is a clean, simple expression for T_{20} which involves only the polarized and unpolarized πd elastic cross sections and the target tensor polarization according to:

$$T_{20} = \frac{\sqrt{2}}{p_{zz}} \left(\frac{\sigma(pol)}{\sigma(unp)} - 1 \right). \qquad (1)$$

In order to emphasize the spin observable T_{21}, the appropriate choices of α and β are $\alpha = 57.3°$ (to eliminate the T_{20} term) and $\beta = 90°$ (to eliminate the iT_{11} term). For purely experimental reasons, α was chosen to be 45° for the T_{21} measurements, and β 90°. In terms of the experimental configuration, this implies that the target magnetic field is oriented in the horizontal plane, at an angle of 45° to the incident beam. With these choices for α and β, we construct the quantity

$$\tau_{21} \equiv T_{21} + \frac{1}{2} \left(\frac{T_{20}}{\sqrt{6}} + T_{22} \right) \qquad (2)$$

$$= \frac{2}{\sqrt{3}\, p_{zz}} \left(\frac{\sigma(pol)}{\sigma(unp)} - 1 \right). \qquad (3)$$

The experimentally measured quantities are the target tensor polarization, p_{zz}, and the relative πd elastic scattering differential cross sections σ(pol) and σ(unp). In a T_{21} experiment the quantity actually measured is a mixture of T_{21}, T_{20}, and T_{22} according to Eq. 2, which we refer to as τ_{21}. The dominant contribution to τ_{21} comes from T_{21}.

EXPERIMENTAL TECHNIQUE

The main characteristics of the detection system are as follows: A solid angle of 27 msr for each of six independent arms was defined by a pion scintillator located 1 m from the polarized target. Together with another scintillator at 0.5 m radius, this constituted one of six pion telescopes, each of which was placed in coincidence with an associated recoil deuteron arm consisting of 3 scintillators. The first scintillator, at a radius of 1.3 m from the target, provided TOF as well as energy loss (ΔE) information. Following this scintillator was an aluminum absorber, whose thickness was adjusted so that deuterons stopped in the following thick scintillator. The third was a veto scintillator. The flux of the incident beam was counted directly with in–beam scintillators. Protons in the incident beam were reduced by using a differential degrader (2 mm) near the midplane of the TRIUMF M11 channel. The incident flux was typically $2x10^6$ π^+/sec.

The final analysis of the data was performed by constructing polygons around the πd elastic events identified in two dimensional histograms of the deuteron TOF vs the deuteron total energy E+ΔE. The resulting scatterplot provides good particle identification and clearly separates deuterons from protons. Explicit measurements of the background arising from quasi-elastic πd scattering from the contaminant carbon and oxygen nuclei in the polarized target material were also made by replacing the deuterated butanol target with an equivalent amount of non-deuterated butanol (C_4H_9OH).

The tensor polarized target consisted of 2.4 cc of frozen 1 mm diameter beads contained in a thin walled teflon basket measuring $22x18x6$ mm^3. The beads were composed of a mixture of 95% fully deuterated N-butyl alchohol (C_4D_9OD) and 5% D_2O into which deuterated EHBA-Cr^V was dissolved to a molecular density of $6x10^{19}$ atoms/ml. The packing density of the butanol beads was approximately 0.7. The teflon basket, which also served as a support for an NMR pickup coil, was immersed in a mixture of $^3He/^4He$ in the mixing chamber of a dilution refrigerator. The polarizing field of 2.5T was provided by two superconducting coils in the "thick" Helmholtz configuration with a horizontal magnetic field axis either 0° (for the T_{20} measurements) or 45° (for the τ_{21} measurements) to the incident beam axis. The target tensor polarizations achieved in this experiment varied from run to run, from 0.10 ± 0.005 to 0.17 ± 0.009. The target polarization was obtained from analysis of the NMR signals using two independent techniques, the thermal equilibrium (TE) technique and the asymmetry technique. The validity of these techniques has been confirmed in an independent experiment [1] which measured the target tensor polarization directly. In practice, the values of target polarization used in the experiment reported here were obtained from the average of the two (consistent) NMR signal techniques.

RESULTS

The results of this experiment consist of new measurements of T_{20} at pion bombarding energies of 180, 220, and 256 MeV. At each of these bombarding energies, approximately six angles were obtained. Furthermore, new measurements of τ_{21} have also been obtained at 134 and 220 MeV. Approximately twelve angles were obtained at each of the bombarding energies studied in the τ_{21} measurements. These new measurements complement previous ones published by this collaboration. The previous results consist of a twelve point angular distribution of τ_{21} at 180 MeV [2], as well as six point angular distributions of T_{20} at 134 and 151 MeV [3]. Taken as a complete set, the old and new data provide a reasonably complete, systematic measurement of the behaviour of the tensor analyzing powers T_{20} and τ_{21} over the range of incident pion energies spanned by the (3,3) resonance.

All of the measured values of T_{20} are negative. The general trend of T_{20} in the backward hemisphere is towards increasingly negative values as the bombarding energy is raised from 134 to 256 MeV. No dramatic changes in the shape of the angular distributions, which in general tend toward more negative values with increasing angle, is observed.

Good agreement is observed between the projected tensor polarizations T_{20} (lab or c.m.) of this experiment and the t_{20} (lab) data published previously from double scattering experiments [4,5]. This agreement is quite significant, given the completely different experimental approaches used for the measurements of t_{20} and T_{20}. The only discrepancy arises at 134 MeV, where the results of the double scattering experiments of Ref. (6) are in complete disagreement with the results of this (single scattering) experiment, as well as those of the other double scattering experiments [4,5]. Clearly, the weight of experimental evidence strongly suggests that the results of Ref. (6) are in error, as pointed out in Ref. (3).

The τ_{21} measurements were collected in two overlapping angular settings of the six detectors. The consistency obtained with the two angular settings is excellent at each bombarding energy. Extensive angular coverage was possible due to the more open geometry associated with the 45° orientation of the polarized target magnet coils. The behaviour of τ_{21} is similar to that of T_{20} in that the data are all negative. The angular distributions become more negative as the bombarding energy is raised from 134 MeV to 220 MeV. The angular distributions appear to be smooth.

THEORETICAL PREDICTIONS

Of the many theoretical predictions available, we have chosen a restricted set [7] with which to make comparisons to the results of this experiment. Most are three body, Faddeev calculations which use the same formalism in principle, but differ significantly in several practical aspects. In particular, the methods used to split the P_{11} πN partial wave input into pole and non–pole parts are quite different among the different Faddeev groups, and have a noticeable impact on the ultimate results of the calculations.

The ambiguity in the exact shape of these potentials is overshadowed by another effect on which the results of the calculations are extremely sensitive. It is now clear that for some partial waves, the pole term contribution will be Pauli blocked in the two–nucleon intermediate state. In such cases the non–pole term acts alone and, with no pole term contribution to cancel it, has a considerable impact on the results of the calculations. The Pauli blocking of the P_{11} pole term is the major source of problems in the theory at the present time.

To summarize, in general the comparison of theory and both the T_{20} and τ_{21} experimental results strongly favors approaches in which the P_{11} contribution is minimized. There is apparently a failure in the theory with respect to this particular aspect. The most sensitive observable to the treatment of the P_{11} appears to be T_{20}. The main features of the overall situation can be seen in Fig. 1. This figure is a matrix whose rows denote the pion bombarding energies covered in this experiment (134, 180, 220 and 256 MeV), and whose columns denote all the πd observables which have been measured to date (the differential cross section, iT_{11}, T_{20}, and τ_{21}). The predictions shown in each of the elements of this matrix represent the calculations of the Flinders group, with and without the P_{11} contribution. The two predictions diverge as the bombarding energy gets farther away from 180 MeV in either direction. Whereas the T_{20} column stands out strongly favoring the predictions without the P_{11} as does the 134 τ_{21} matrix element, the situation is much less clear for the other observables.

A number of interesting suggestions have been proposed for dealing with the P_{11} problem described above. Since the heart of the difficulty has to do with the Pauli blocking of the P_{11} pole term in the two nucleon intermediate state, some workers have raised the intriguing possibility that there may be a limit to the applicability of the Pauli principle. The nucleon which has absorbed the pion is far off–shell, and may no longer be identical to an on–shell nucleon in the sense of the Pauli principle. Were this so, the pole term would not be Pauli blocked and would always largely cancel the non–pole term, eliminating the problem. These speculations are reminiscent of the open question in hypernuclear physics of whether the non–strange quarks in the hyperon are Pauli blocked in a hypernuclear ground state or not. A more mundane way of stating the problem of the P_{11} contribution in πd scattering would be to ascribe it to the unknown off–shell behavior of the nucleons in the absorption process.

Recently, Jennings [8] has suggested what may well provide a much more acceptable solution. He has pointed out that there are two diagrams which are left out of the P_{11} contribution that tend to cancel the diagrams that give rise to the Pauli blocking. Inclusion of the missing diagrams is expected to greatly reduce the problems associated with Pauli blocking in the P_{11} amplitude. The missing diagrams are crossed diagrams which have two pions simultaneously in the intermediate state, and are therefore not explicitly included in the three body calculations. Work is presently being carried out to incorporate these diagrams quantitatively into the theory, and to study their influence on the πd observables.

REFERENCES

1) G.R. Smith *et al.*, NIM **A254**, 263 (1987).
2) G.R. Smith *et al.*, Phys. Rev. **C35**, 2343 (1987).
3) G.R. Smith *et al.*, Phys. Rev. Lett. **57**, 803 (1986).
4) E. Ungricht *et al.*, Phys. Rev. Lett. **52**, 333 (1984);
   E. Ungricht *et al.*, Phys. Rev. **C31**, 934, (1985).
5) Y.M. Shin *et al.*, Phys. Rev. Lett. **55**, 2672 (1985).
6) J. Ulbricht *et al.*, Phys. Rev. Lett. **48**, 311 (1982);

W. Gruebler *et al.*, Phys. Rev. Lett. **49**, 444 (1982);
V. Koenig *et al.*, J. Phys. **G9**, L211 (1983).
7) B. Blankleider and I.R. Afnan, Phys. Rev. **C24**, 1572 (1981);
B. Blankleider, private communication (1986).
8) B.K. Jennings, submitted to Phys. Lett. **B**, (1987);
B.K. Jennings and A.S. Rinat, private communication (1987).
9) K. Gabathuler *et al.*, Nucl. Phys. **A350**, 253 (1980).
10) C. Ottermann *et al.*, Phys. Rev. **C32**, 928 (1985).
11) G.R. Smith *et al.*, Phys. Rev. **C29**, 2206 (1984).

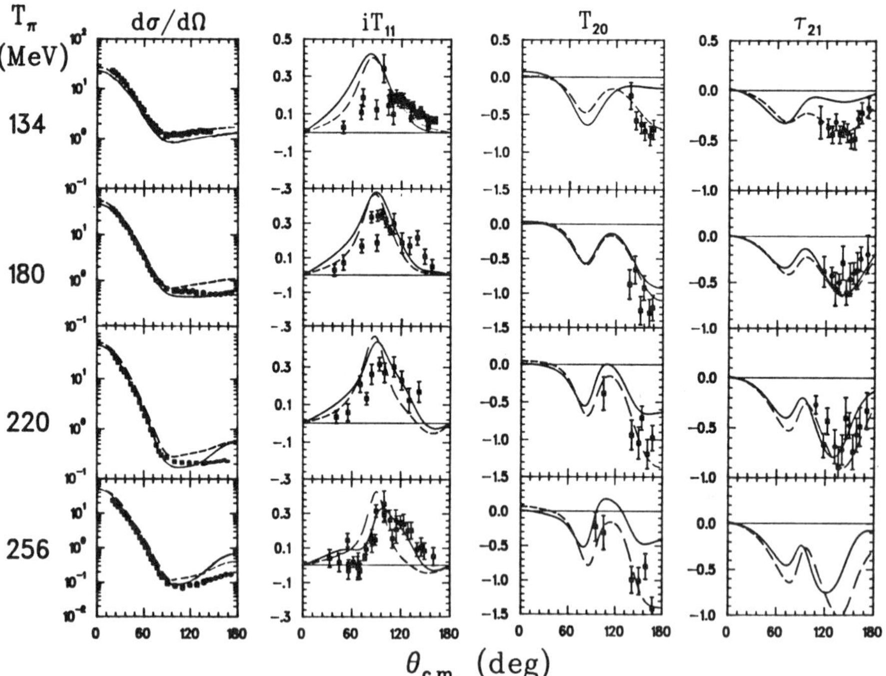

Figure 1: At each of four representative pion bombarding energies for πd elastic scattering, values of the differential cross section [9,10], iT_{11} [11], T_{20} [3], and τ_{21} [2] are shown. The new measurements of T_{20} and τ_{21} described here are also shown. The solid curves depict the full three body predictions of the Flinders group [7], and the dashed lines correspond to the same calculations without the P_{11} contribution.

Measurement of A and A_{NN} in $p_\uparrow$ $p_\uparrow$ Elastic Scattering at 18.5 GeV/c*

D.G. Crabb †
Randall Laboratory of Physics
University of Michigan
Ann Arbor, Michigan 48109

As part of a continuing study of spin effects in pp-interactions, we have measured the analyzing power, A, and the spin correlation parameter, A_{NN}, in $p_\uparrow$ $p_\uparrow$ elastic scattering in the $p_\perp^2$ range 1.4 to 3.35 $(GeV/c)^2$.[1] The experiment was carried out using a polarized proton beam at the AGS at Brookhaven National Laboratory together with a polarized proton target. The intensity, polarization, and reliability of the polarized proton beam have improved significantly since the first commissioning in 1984. These improvements have allowed a systematic study at a single energy with good precision, rather than the more scattered measurements possible in the earlier runs.

The parameters for the accelerated beam and extracted beam to the experiment are shown in Table I.

Table I
Polarized Beam Parameters

18.5 GeV/c	Intensity		Polarization	
	Peak	Average	Peak	Average
Accelerated	2.5 10^{10} pppp	2.10^{10} pppp	52%	43%
Extracted to D Line		1.2 10^{10} pppp		41%

The experimental layout is shown in Fig. 1. The polarized protons enter from the left and first strike the liquid hydrogen target of the high energy polarimeter. The polarimeter measures the left-right asymmetry in proton-proton elastic scattering at $p_\perp^2 = 0.3$ $(GeV/c)^2$ where an analyzing power of 3.9 ± 0.3 % is used to calculate the beam polarization.

The polarized protons continue on to interact in the polarized proton target from which elastic scattering events are detected in the double arm spectrometer. Six magnets and an eight channel hodoscope system ensure that elastic events are selected cleanly with little background.

* Work supported by the U.S. Department of Energy

† I. Gialas, A.D. Krisch, A.M.T. Lin, D.C. Peaslee, R.A. Phelps, R.S. Raymond, T. Roser, J.A. Stewart, K.M. Terwilliger; University of Michigan: K.A. Brown, G.T. Danby, F.Z. Khiari, L.G. Ratner; Brookhaven National Laboratory: J.R. O'Fallon; U.S. Department of Energy: G. Glass; Texas A & M University.

$$A_B = -\frac{1}{P_B}[\frac{N(\uparrow\uparrow) + N(\uparrow\downarrow) - N(\downarrow\uparrow) - N(\downarrow\downarrow)}{N(\uparrow\uparrow) + N(\uparrow\downarrow) + N(\downarrow\uparrow) + N(\downarrow\downarrow)}]$$

$$A_T = -\frac{1}{P_T}[\frac{N(\uparrow\uparrow) - N(\uparrow\downarrow) + N(\downarrow\uparrow) - N(\downarrow\downarrow)}{N(\uparrow\uparrow) + N(\uparrow\downarrow) + N(\downarrow\uparrow) + N(\downarrow\downarrow)}]$$

where P_B and P_T are the beam and target polarizations respectively.

The results are shown in Fig. 2. Both A and A_{NN} show considerable structure. The A data are similar to data at lower energies[4] where the value of A peaks at around $p_\perp^2 = 1.5$ (GeV/c)^2, falls to a minimum then rises again. In the older data there appear to be variations in peak width and value of the minimum. The quality of the data made it difficult to quantify any systematic variations. In the current data it is clear that the value of A is negative at the minimum. However, comparing these data with those at 24 and 28 GeV/c[5] it appears that A in this $p_\perp^2$ region is varying with energy, in that the peak value of A is considerably suppressed, dropping from $\sim$ 15% at 18.5 GeV/c to $\sim$ 5% at 24 and 28 GeV/c. Moreover, the peak width increases, delaying the onset of the minimum to about $p_\perp^2 = 3.5$ (GeV/c)^2.

The A_{NN} data are plotted again on the three dimensional graph shown in Fig. 3. It is not clear how the structure of these data correlates with the structure of the older data from the ZGS. We note that the rise of A_{NN} to 60% around 12 GeV/c has never been satisfactorily explained, although there have been many attempts. The most recent is from Brodsky and de Teramond[6]. However, two general statements have been made on the trend of the data: one is the apparent oscillation in A_{NN} at 90°cm has been linked to the oscillation of the 90°cm pp differential cross-section around the general s^{-10} dependence[7]; second is that the peaking towards 90°cm at fixed momentum is due to particle identify effects[8].

It is clear from Fig. 3 that a least two sets of measurements of A_{NN} are necessary to address the validity of these two statements. One is a continuation of the 90°cm measurement from 12.75 GeV/c and the second is to extend a fixed momentum measurement to large values of θcm. The simultaneous measurement of A for the second case may give some insight into whether perturbative QCD is applicable in this $p_\perp^2$ region. Though PQCD predicts that A = 0, it is clearly not the case at 28 GeV/c and $p_\perp^2 = 6.5$ (GeV/c)^2.

The polarized target consists of a conventional evaporation refrigerator, using a $^3He/^4He$ mixture as the coolant, operating at 0.5K in a 2.5T field. The target materials used during the course of the experiment were radiation doped ammonia (NH_3)[2] and chemically doped ethylamine-borane ammonia (EABA)[3]. The performance of these materials are summarized in Table II.

Table II

Target Material Performance

Material	Polarization	
	Maximum	Average
NH_3	55%	44%
EABA	80%	65%

The ammonia gave rather disappointing values of polarization as past experience with a number of samples had shown a much better performance. However, the beam intensities were such that EABA could be used without serious radiation damage problems arising. The majority of the data was taken with EABA. The measurement of background events from non-hydrogenous material was made by replacing the polarizable material with teflon (CF_2).

The beam position and size at both the hydrogen and polarized target were monitored continuously by segmented wire ion chambers S_1, S_2, S_3, S_4 while the relative beam intensity was measured with an ion chamber, secondary emission chamber, and three counter telescopes M, N, and K. The absolute beam intensity was calculated from an aluminum foil irradiation.

A pp-elastic scattering event in the spectrometer was defined by a sevenfold FB coincidence between the appropriate channels of the $F = F_0F_1F_2F_3$ and $B = B_1B_2B_3$ arms. The normalized event rates N(i,j) in the four possible spin states for beam [$i =\uparrow$ *or* $\downarrow$ (up or down)] and target ($j =\uparrow$ *or* $\downarrow$) were obtained by measuring the quantities:

$$N(i,j) = \frac{Events(i,j)}{I(i,j)}$$

where N(i,j) is the number of elastic events corrected for accidentals and non-hydrogen background and I(i,j) is the relative intensity obtained by averaging the monitors. A and A_{NN} are then obtained from the relations

$$A_{NN} = \frac{1}{P_B P_T}[\frac{N(\uparrow\uparrow) - N(\uparrow\downarrow) - N(\downarrow\uparrow) + N(\downarrow\downarrow)}{N(\uparrow\uparrow) + N(\uparrow\downarrow) + N(\downarrow\uparrow) + N(\downarrow\downarrow)}]$$

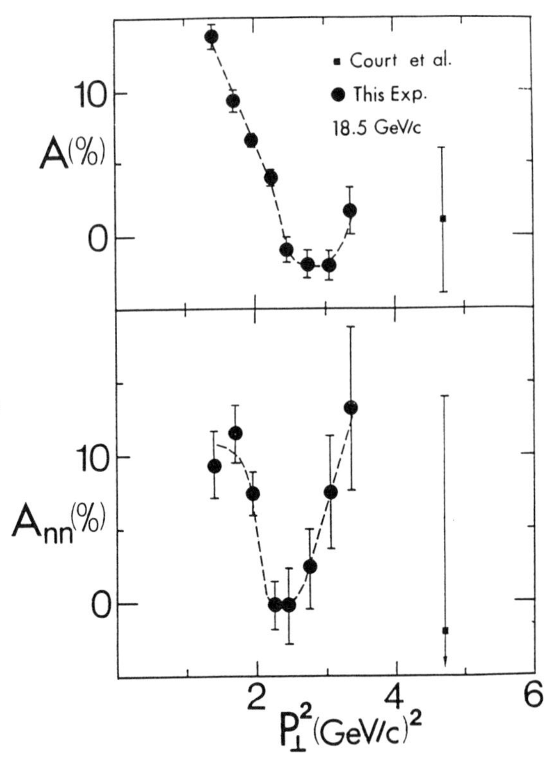

Fig. 2. Plot of the analyzing power, A, and the spin-spin correlation parameter, A_{nn}, as a function of momentum transfer squared for proton-proton elastic scattering at 18.5 GeV/c. The error bars include both statistical and systematic errors. The dashed lines are hand-drawn curves to guide the eye.

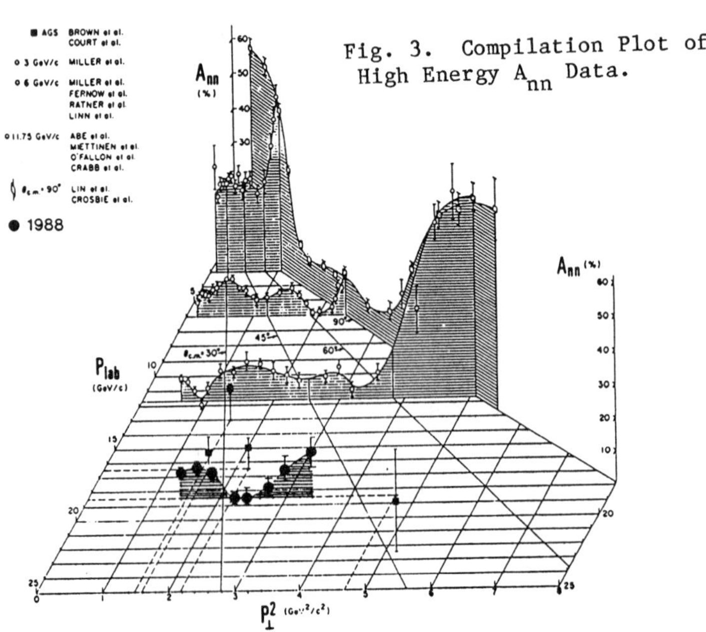

Fig. 3. Compilation Plot of High Energy A_{nn} Data.

References

1. D.G. Crabb et al., Phys. Rev. Letts., 60 , 2351 (1988).
2. R.S. Raymond et al., Helvetica Physica Acta., 59 , 777 (1986).
3. D.G. Crabb and D. Hill, Proc. 4TH International Workshop on Polarized Target Materials and Techniques, Bonn University, ed. W. Meyer, p. 100 (1984).
4. J. Antille et al., Nuc. Phys., B185 , 1 (1981).
5. P.R. Cameron et al., Phys. Rev. Rapid Comm., D32 , 600 (1985).
6. S.J. Brodsky and G.F. de Teramond, Phys. Rev. Letts., 60 , 1924 (1988).
7. A.W. Hendry, Phys. Rev., D23 , 2075 (1981).
8. H.J. Lipkin, Phys. Rev. Letts., B181 , 164 (1987).

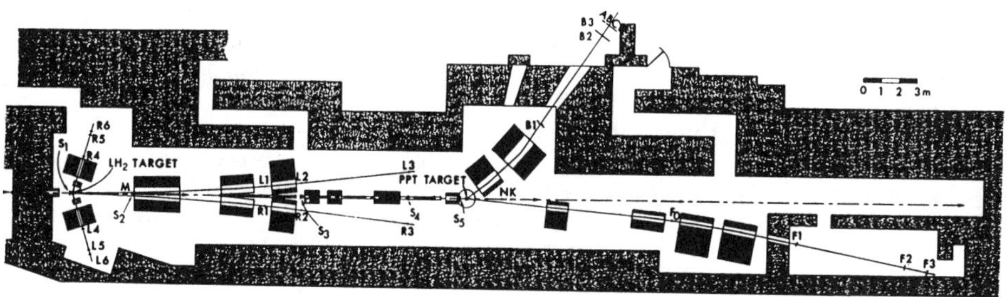

Fig. 1 Layout of the experiment. The high energy polarimeter on the left uses a liquid hydrogen target to measure the left-right asymmetry in p-p elastic scattering. The polarized proton beam then scattered in the vertically polarized proton target (PPT) and the elastic events were detected by the spectrometer which contained magnets for momentum analysis and the F and B scintillation counter hodoscopes. The M, N, and K counters were intensity monitors, while the S_1, S_2, S_4 and S_5 segmented wire ion chambers monitored the beam's position, size and angle.

A NEW PRECISION EXPERIMENT TO MEASURE THE MUON g-2 VALUE

Morgan May
Brookhaven National Laboratory, Upton, New York, 11973

ABSTRACT

A new measurement[1] of the g-2 value of the muon with a sensitivity of .35 parts per million is being prepared at the Brookhaven National Laboratory Alternate Gradient Synchrotron (BNL AGS). The weak interaction contributes significantly to the magnetic moment anomaly at the planned level of sensitivity.

INTRODUCTION

An experiment to measure the magnetic moment anomaly of the muon[1], $a_\mu = (g-2)/2$, with a precision of .35 parts per million has recently been approved at the BNL AGS. The planned level of accuracy is a factor of 20 improvement over the present experimental value[2].

Precision experiments can complement high energy experiments in probing physics at higher mass scales. Within the standard model the muon magnetic moment anomaly is calculable[3] with no adjustable parameters. Any undiscovered particles which couple to muons or undiscovered interactions of the muon will alter a_μ. The new experiment is thus a standard model test and a search for an indication of particles and interactions beyond the standard model.

The planned level of accuracy brings the experiment into a new and interesting region. The weak interaction[4] contribution to a_μ is approximately 1.5 PPM, and thus the new experimental value will be sensitive to it while the present experimental value is not. The magnetic moment anomaly is the classic test for particle substructure. The new experiment will be sensitive to muon substructure at the 4-5 TeV level. A further discussion of contributions to a_μ arising from theories beyond the standard model can be found in ref. 5.

EXPERIMENTAL APPROACH

The experiment measures the spin precession of polarized muons which are trapped in a magnetic storage ring. For a Dirac particle g=2 and the spin precession frequency ω_s is equal to the cyclotron frequency ω_c. The spin precession in excess of the cyclotron frequency is termed the anomalous spin precession ω_a. In a g-2 experiment ω_a is measured directly. Shower counters distributed around the inside of the storage ring (fig. 1) detect electrons from muon decay. The counting rate for high energy electrons (emitted forwards) is modulated at the frequency ω_a.

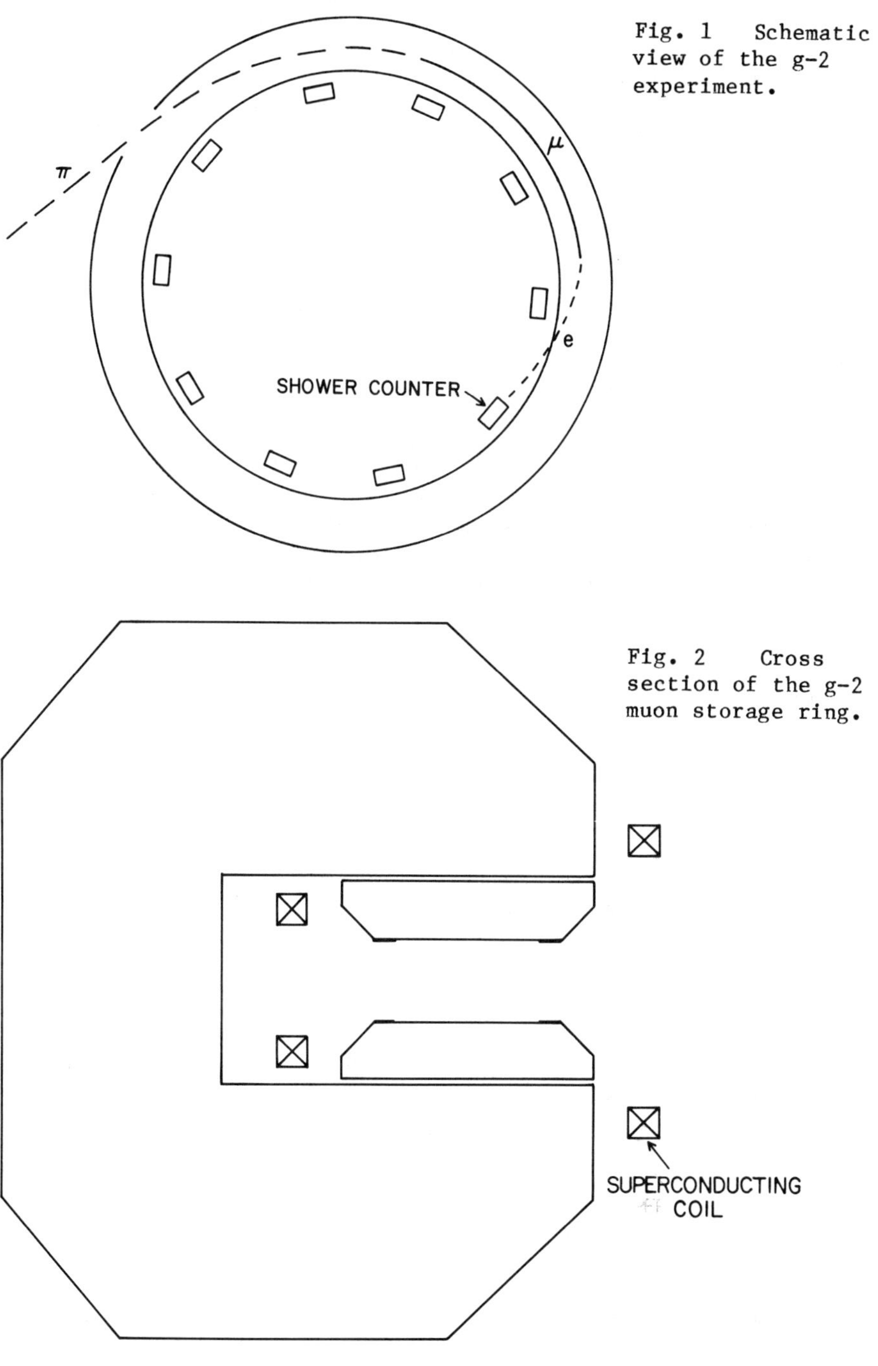

Fig. 1 Schematic view of the g-2 experiment.

Fig. 2 Cross section of the g-2 muon storage ring.

The new experiment is similar in concept to the 1978 CERN experiment[2] in that it uses a constant magnetic field with electrostatic quadrupoles for vertical focusing. In the presence of an electric field E the anomalous spin precision is given by[6]

$$\omega_a = e/mc \left[a_\mu B + (a_\mu - 1/(\gamma^2-1)\left|\vec{\beta} \times \vec{E}\right| \right]$$

For $\gamma = 29.3$ corresponding to muon momenta 3.094 GeV/c, the coefficient of the electric field term vanishes and a_μ is determined solely by ω_a and the constant magnetic field B.

The experiment will utilize a 14-meter diameter 1.5 Tesla storage ring. The ring is powered by superconducting coils. The ring will have azimuthal symmetry interrupted only by the field-free inflector region through which the pion beam enters. Forward decay muons are captured in stable orbits and are polarized due to the angular asymmetry of the pion decay process. The muons make 440 revolutions per lifetime and undergo 15 anomalous spin precessions per lifetime. The ring is a C magnet in cross section open on the inside where the shower counters are located. Shower counters of a lead scintillation sandwich design are planned, but other electron calorimeter designs are under study.

CONTRIBUTIONS TO $(g-2)_\mu$

In contrast to $(g-2)_e$, which even at the level of 4 parts per billion (present experimental[7] precision) involves only electrons and photons, $(g-2)_\mu$ is sensitive to the presence of other particles. For example, vacuum polarization loops involving e, μ, π, and τ are significant for $(g-2)_\mu$ (figs. 3a and b).

Within the standard model, $(g-2)_\mu$ has contributions from (1) QED involving leptons and photons only, (2) QED involving hadrons, (3) the weak interaction. At present the pure QED contribution has been calculated to about .1 PPM, the hadronic contribution to 1 PPM, and the weak contribution to .01 PPM. A factor of 4 improvement in the value for the hadronic contribution is required to match the sensitivity of the new experiment.

The calculation of the hadronic contribution requires experimental input in the form of $\sigma(e^+e^- \to \text{hadrons})$ over the range of energies from $\sqrt{s} = 2m_\pi$ up to about 1 GeV. An experiment[8] at the VEPP-2M ring at Novosibirsk using the CMD-2 detector is planned to take the relevant data. Fixed target experiments with energetic positron beams (>100 Gev) can also measure these cross sections[9].

Diagrams of the weak interaction contributions to $(g-2)_\mu$ are shown in figs. 3c and d. The contribution involving a virtual W boson is particularly interesting since it involves the W interacting directly with the magnetic field and thus is sensitive to the W magnetic moment. In the standard model the magnetic moment of the W is equal to e/M_W. A deviation from this value would imply W substructure[9].

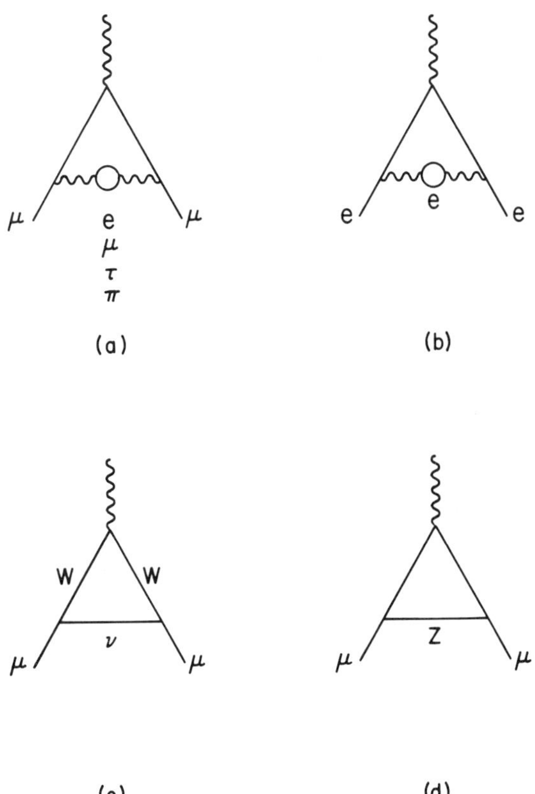

Fig. 3
(a) Vacuum polarization contribution to $(g-2)_\mu$ includes e, μ, τ, and π loops. Lowest-order diagram is shown. (b) Vacuum polarization contribution to $(g-2)_e$ includes only electron loops. (c) and (d) Weak interaction contribution to $(g-2)_\mu$.

EXPERIMENTAL CHALLENGES

In the g-2 experiment the magnetic moment anomaly is determined as a ratio of the anomalous spin precession frequency to the magnetic field.

$$a_\mu = \frac{mc}{e}\frac{\omega_a}{B}$$

To attain a sensitivity of .35 PPM, 3 10^{10} muon decays must be observed with systematic timing errors less than 30 ps. The average magnetic field seen by the muons must be known to .1 PPM. Since the magnetic field is measured by proton NMR, knowledge of the constant μ_μ/μ_p is required[10].

The improvement in statistical precision over the previous $(g-2)_\mu$ experiment is due to the hundredfold greater proton intensity available at the AGS. The freedom from systematic timing error required for the frequency measurement is well within the capability of modern electronic circuits. The challenge will be to maintain timing accuracy in the presence of high counting rates which vary over the course of the measurement.

In addition to the high counting rates due to muon decays, there is an initial burst of radiation in the pion injection approach described above (the pion beam is lost after less than one turn in the ring). Studies in a high-intensity pion beam indicate that this initial flash is manageable. However, the beam line to the experiment is long enough to accommodate a muon beam, and it may be preferable to avoid the flash problem through muon injection.

The magnetic field will be measured by proton NMR. With appropriate diamagnetic shielding corrections an absolute field measure to .1 PPM will be achieved. The challenge in this experiment is the large volume over which the field must be known--a ring 14 m in diameter with a 9 cm diameter storage region. The use of superconducting coils has advantages in current regulation, thermal stability and cost. The magnet will be continuously powered during the course of measurement to avoid problems of hysteresis. An elaborate system of measurement and control of the field is planned with 176 fixed NMR probes positioned around the ring above and below the vacuum chamber and a matrix of 25 probes which can be moved around the inside of the vacuum vessel.

The experiment will begin taking data in four years under the present construction schedule.

Research has been performed under contract DE-AC02-76CH00016 with the United States Department of Energy.

REFERENCES

1. A New Precision Measurement of the Muon g-2 Value at the Level of 0.35 ppm. AGS Proposal 821, Sept. 1985, revised Sept. 1986. V. W. Hughes, spokesman. E. Hazen, C. Heisey, B. Kerosky, F. Krienen, E.K. McIntyre, D. Magaud, J.P. Miller, B.L. Roberts, D. Stassinopoulos, L.R. Sulak, W. Worstell-Boston U.; H.N. Brown, E.D. Courant, G.T. Danby, C.R. Gardner, J.W. Jackson, M. May, A. Prodell, R. Shutt, P.A. Thompson-BNL; J.A. Johnson, M.S. Lubell-City College NY; A.M. Sachs-Columbia; T. Kinoshita-Cornell; D. Winn-Fairfield U.; M. Janousch, H.-J. Mundinger, G. zu Putlitz, J. Rosenkranz, W. Schwarz-U. Heidelberg; W.P. Lysenko-LANL; A. Rich-U. Michigan; J. J. Reidy-U. Mississippi; F. Combley-Sheffield U.; K. Nagamine, K. Nishiyama-U. Tokyo; K. Endo, H. Hirabayashi, S. Kurokawa, T. Sato-KEK; K. Ishida-Riken; L.M. Barkov, B.I. Khazin, E.A. Kuraev, Ya. M. Shatunov-Inst. Nucl. Phys., Novosibirsk, USSR; J.M. Bailey, S.K. Dhawan, A.A. Disco, F.J. M. Farley, V.W. Hughes, Y. Kuang, H. Venkataramania-Yale.

2. J. Bailey et al., Nucl. Phys. B150, 1 (1979).

3. T. Kinoshita, B. Nizic, Y. Okamoto, Phys. Rev. Lett. 52, 717 (1984).

4. R. Jackiw, S. Weinberg, Phys. Rev. D5, 2396 (1972); G. Altarelli, N. Cabibbo, L. Maiani, Phys. Lett. 40B, 415 (1972); I. Bars, M. Yoshimura, Phys. Rev. D6, 374 (1972); K. Fujikawa, B.W. Lee, A.I. Sanda, Phys. Rev. D6, 2923 (1972); W.A. Bardeen, R. Gastmans, B.E. Lautrup, Nucl. Phys. B46, 319 (1972).

5. V. Hughes in Fundamental Symmetries, eds. P. Bloch, P. Pavlopoulos, R. Klapisch (Plenum Publishers, 1987), p. 271-85.

6. V. Bargmann et al., Phys. Rev. Lett. 2, 435 (1959); (F.J.M. Farley, Cargese Lectures in Physics (Gordon and Breach, 1968), Vol. 2, p. 55; J. Bailey, E. Picasso, Progr. in Nucl. Phys. 12, 43 (1970)).

7. R.S. Van Dyck, P.B. Schwinberg, H.G. Dehmelt, Phys. Rev. Lett. 59, 722 (1987).

8. L.M. Barkov et al., Novosibirsk Inst. of Nucl. Phys. Preprint 85-118 CMD-2 Detector for VEPP-2M; L.M. Barkov et al., Nucl. Phys. B256, 365 (1985).

9. S.R. Amendolia et al., Phys. Lett. 138B, 454 (1984); 146B, 116 (1984).

10. W. Marciano, private communication; M. Suzuki, Phys. Lett. B153, 289 (1985).

11. The constant μ_μ/μ_p was measured to .36 PPM by F.G. Mariam et al., Phys. Rev. Lett. 49, 993 (1982). Improvement to better than .1 PPM is expected in the next few years.

RECENT STUDIES ON THE POLARIZATION PHENOMENA AT KEK

Akira MASAIKE

Department of Physics, Kyoto University, Sakyo-ku, Kyoto, Japan

Abstract

The slow neutron from the KEK booster synchrotron has been polarized using a polarized proton filter. It is a useful tool for the test of P and T violations.

The polarized beam has been successfully accelerated up to 5 GeV at KEK proton synchrotron. Experimental projects on nucleon-nucleon interactions using it are in progress.

§1. Slow Neutron Polarization and Its Application for P and T Violation Experiments

Slow neutrons transmitted through the polarized proton filter become polarized, because the cross section for the neutron-proton scattering in the singlet state is substantially larger than that in the triplet state. The method has a unique tool to obtain polarized epithermal neutrons with high intensity. An experiment of the polarization of slow neutron beam using a filter, whose protons were polarized perpendicular to the neutron beam direction, was reported previously.[1)] Recently it was found that the neutron can be polarized in the direction of the beam using a longitudinally polarized filter[2),3)] as shown in Fig. 1. The degree of polarization is nearly the same as that in the case of transversely polarized filter.

The protons in a polycrystalline sample of ethylene glycol containing a few percent of Cr^{V} complexes were dynamically polarized by means of microwave of 70 GHz in the magnetic field of 2.5T at the temperature of 0.4 K in a ^{3}He cryostat. In order to avoid the absorption of neutrons in ^{3}He, thin filter material was imersed in liquid ^{4}He,[4)] which was cooled by liquid ^{3}He from outside through a heat exchanger. The heat exchanger consists of copper fins. The difference of the temperature in liquid ^{3}He and in liquid ^{4}He was about 0.02 K when the micro-wave power of 30 mW was supplied in the cavity, in which liquid ^{4}He was filled. The beam intensity of the polarized neutron at 6 m from the moderator was about 10^{5} eV^{-1} cm^{-2} sec^{-1} at 1 eV, when the intensity of 500 MeV proton is 8×10^{11} ppp with 40 pulse/2.5 sec. The typical neutron polarization was 75%.

Recently the parity violating effects have been tested in resonance absorption reactions of slow neutrons on heavy nuclei. The difference in the total cross section for neutrons of different helicity was measured by the neutron transmission method at Dubna,[5)] while the helicity asymmetry in the neutron radiative capture reaction was observed by means of detection of γ-rays in neutron capture reactions, as well as the transmission method at KEK.[6)] In the measurement at KEK the γ-rays emitted from the target were detected by an annular scintillation counter of BaF_2, which was placed at $\Theta=90°$ with respect to the beam direction. Phototubes, in which photoelectrons were selectively emitted for the fast component of the scintillation light, were used. In Fig. 2 are shown the TOF spectra of the γ-ray around the p-wave resonance of ^{139}La at the neutron energy of 0.734 eV. Crosses and dots in the figure correspond to γ-ray counts with positive and negative neutron helicity states, respectively. The average asymmetry of γ-ray count in the resonance of ^{139}La at 0.734 eV is $A_L=10.2\pm0.5\%$.

The asymmetry in the neutron transmission was also measured using a ^{10}B-loaded scintillator installed in the neutron beam. The measured asymmetry is consistent with the capture γ-ray asymmetry. These values are slightly different from Dubna's result ($A_L=(7.3\pm0.5)\%$). The origin of the difference has not been clearly explained.

Such a large enhancement of the parity non-conserving effect can be understood as a parity mixing between compound states of the nucleus. Recently, theoretical explanations were performed using the S-matrix formalism by Y. Yamaguchi.[7] He discussed the large parity violating effect in the ^{139}La nucleus relating to several possibilities, including CP-conserving and non-conserving nuclear interactions.

The enhancement may exist for time-reversal violation in neutron induced reactions. In the experiment of time reversal violation, the target nucleus as well as the neutron must be polarized. The nucleus of ^{139}La can be polarized by means of the dynamic polarization in a dilution refrigerator. We hope a time reversal violation experiment can be done in a year using polarized neutrons and a polarized target.

§2. Acceleration of Polarized Beam in Proton Synchrotron and Experimental Projects using It

Recently the polarized proton beam has been successfully accelerated up to 5 GeV in the KEK proton synchrotron.[8]

As an ion source, a method of optical pumping of sodium atoms with a dye laser was used. The polarization of electrons in the sodium atoms is transfered to hydrogen atoms by means of a charge exchange reaction between fast hydrogen ions extracted from an ECR ion source and sodium atoms. The polarized H^- beam is accelerated up to 750 KeV by a Cockcroft-Walton pre-accelerator. The polarization of 55±5% was obtained at 750 KeV with 60 μA. The polarized H^- beam is accelerated to 40 MeV by a linac and then up to 500 MeV by a booster synchrotron (Fig. 3).

Two strong depolarizing resonances must be crossed during the acceleration in the booster. One is an imperfection resonance ($\gamma G=2$) at 108 MeV and the other is an intrinsic resonance ($\gamma G=\nu_Z$) at 260 MeV, where ν_Z is the vertical betatron-tune and G is $(g/2-1)$. The imperfection resonance can be corrected by adjusting the vertical COD. Two deflector magnets were placed in straight sections of the booster to control the COD. The spin flip occurs on this resonance without COD correction using the deflector magnet. On the other hand, the beam passes through the resonance without change of spin direction, if the current of the deflector magnet is adjusted carefully. The intrinsic resonance of $\gamma G=\nu_z$ is strong enough to flip the proton spin completely. However, the effect of the synchrotron oscillation cannot be ignored when the rf voltage of acceleration is about 11 KV as in the usual operation. Therefore, the booster must be operated with the rf voltage of 8.9 KV at the expense of beam loss of 50%, in order to minimize the depolarization. The polarization of proton beam at 500 MeV was about 45%, when the polarization of the beam from the linac was 50%.

In the main ring there are 4 intrinsic resonances and 9 imperfection resonances up to 5 GeV. Among them an intrinsic resonance of $\gamma G=\nu_z$ at 2.3 GeV is strong enough to flip the spin direction with small depolarization. Significant depolarization ocurrs in intrinsic resonances of $\gamma G=12-\nu_z$ at 2.1 GeV and $\gamma G=16-\nu_z$ at 4.2 GeV in the case of no correction. In order to overcome these depolarization resonances 4 pulsed quadrupole magnets were used for fast passage of the resonances. The rise time of the magnet is about 40 μsec. In the intrinsic resonance of $\gamma G=4+\nu_z$ at 4.4 GeV the beam passes through the resonance slowly for complete spin flip, using the falling tail of the pulsed quadrupole magnets.

In order to minimize the depolarization at imperfection resonances, COD was corrected by 56 dipole magnets. The polarization of the beam was measured by a polarimeter installed in a long straight section of the main ring. Typical polarizations at 3.5 GeV and 5 GeV was 41% and 26%, respectively.

The polarized proton beam at the energy from 3 GeV to 5 GeV is the unique facility. Several experimental proposals using the polarized beam have been approved at KEK. One of the experiments is the systematic measurement of the energy dependence of $p_\uparrow p \rightarrow pp$, $p_\uparrow n \rightarrow pn$, and $p_\uparrow p \rightarrow \pi d$ reactions with the energy resolution of 20 MeV.[9] The motivation of the experiment is the search for narrow structures, by means of the observation of spin dependent parameters. The measurement will be performed using the polarimeter installed in the main ring.

Measurement of spin parameters of pd scattering was performed at 3.5 GeV recently.[10]

An experiment of the measurement of A_{nn} parameters on $pn \rightarrow pn$ scattering at 12 GeV has been proposed.

We are grateful to Dr. Y. Masuda for discussion and giving us unpublished data.

References

1) S. Hiramatsu et al., J. Phys. Soc. Jpn. 45 (1978) 949.
2) N. Hoshizaki and A. Masaike, Jpn. J. Appl. Physcis 25 (1986)

L244.
3) S. Ishimoto et al., ibid 25 (1986) L246.
4) Y. Masuda et al., Nucl. Inst. Meth. Phys. Research A264 (1988) 169.
5) V. P. Alfimenkov et al., Nucl. Phys. A398 (1983) 93.
6) Y. Masuda et al., Nucl. Phys. A478 (1988) 737c,
Y. Masuda et al., in preparation.
7) Y. Yamaguchi, J. Phys. Soc. Japan 57 1518(L), ibid 1522(L), ibid 1525(L).
8) H. Sato et al., Nucl. Inst. Meth. Phys. Research A, to be published.
9) K. Imai and J. Holt, private communication.
10) N. Horikawa, private communication.

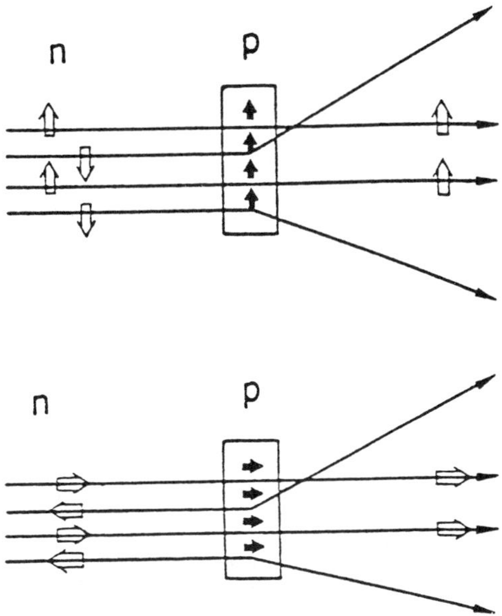

Fig. 1 Schematic view of the slow neutron polarization using the polarized proton filter.

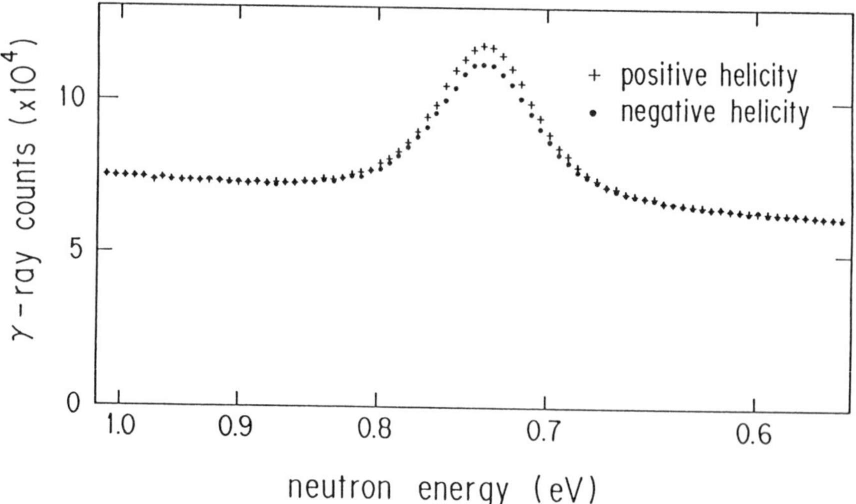

Fig. 2 TOF spectra around the 0.734 eV resonance of ^{139}La.

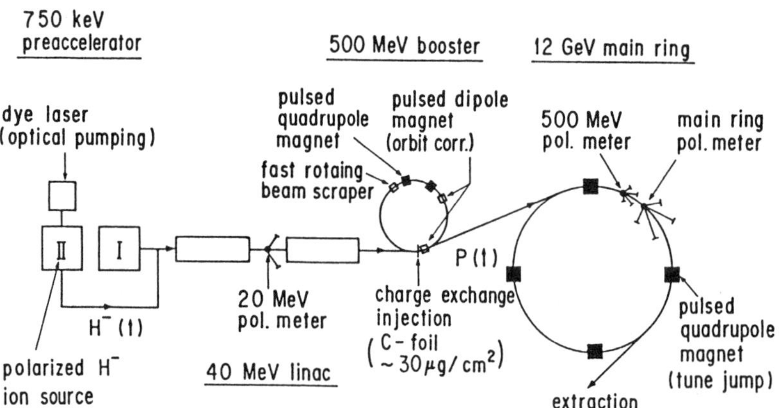

Fig. 3 Schematic view of the KEK proton synchrotron for acceleration of polarized beam

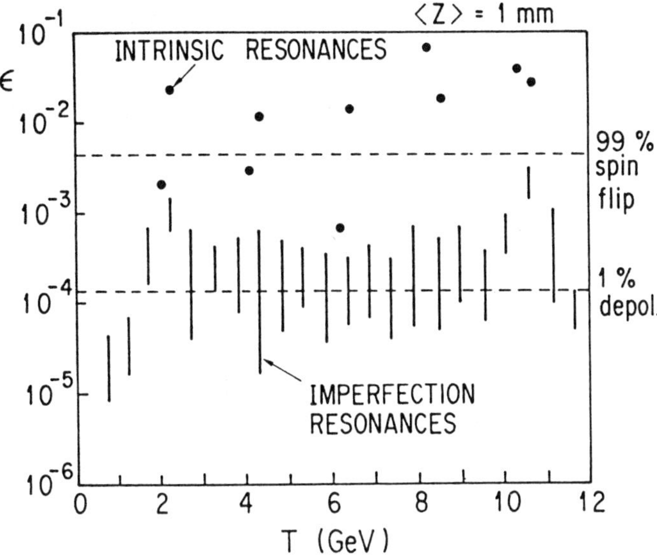

Fig. 4 Calculated resonance strength (ϵ) for ν_Z=6.25 in the main ring.

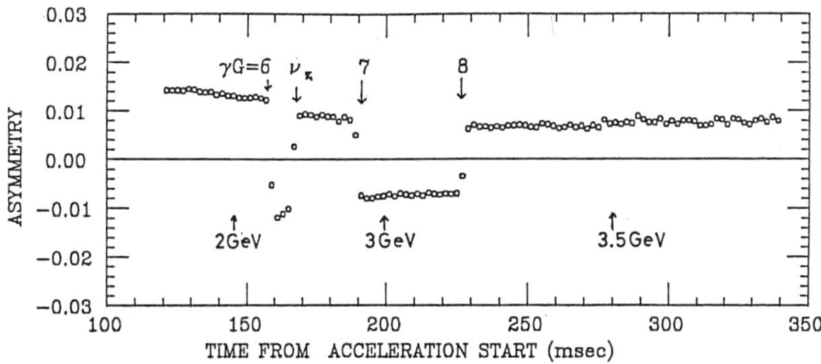

Fig. 5 Asymmetry measured by a polarimeter installed in the main ring.

SPIN PARAMETER MEASUREMENTS IN INCLUSIVE HYPERON PRODUCTION

Francesca Nessi-Tedaldi*
T. W. Bonner Nuclear Laboratories
Rice University, Houston, Texas 77251-1892

ABSTRACT

The present status of polarization parameter measurements is briefly summarized. First results for analyzing power A and polarization transfer D in inclusive Σ^0 production are presented, as well as new results for Λ, K^0_S and $\bar{\Lambda}$ inclusive production.

OVERVIEW

It is well known today that in inclusive reactions hyperons produced with high values of Feynman x have a large polarization.[1] It has also been observed that the values are independent of $\sqrt{s}$ from 5 to 60 GeV. The results indicate that $P_{\Sigma^0} \simeq P_{\Sigma^-} \simeq P_{\Sigma^+} \simeq -P_{\Lambda} \simeq -P_{\Xi^-} \simeq -P_{\Xi^0}$ and $P_{\bar{\Lambda}} = 0$ for proton induced reactions, whereas P_{Λ} is positive and greater in magnitude for production with a K^- beam. Since the polarization appears at low values of transverse momentum (~1 GeV/c) and is therefore a long-distance effect, a perturbative QCD calculation cannot be performed. However, the production processes can be described through the recombination of fragments of the incident particle with quarks that are produced from the sea. This model has been used for a phenomenological parametrization[2] that reproduces the observed behavior rather well. In this framework, a spin-dependent asymmetry is introduced, by means of small parameters, in the quark and diquark production amplitudes. Spin observables are then calculated using static SU(6) wave functions, under the assumption that the spins of the beam particle fragments are preserved in the scattering and recombination processes. The relative signs of the measured polarization values are reproduced very well, as are in general the magnitudes. Possible mechanisms that are proposed to explain the origin of polarization involve string breaking[3] or Thomas precession[2] in the quark-recombination process. However, the predictions are mostly independent from the underlying polarization mechanism. The polarized proton beams now available both at the Brookhaven AGS and at Fermilab make it possible to test the predictions for the analyzing power A and the polarization transfer D. Such measurements test all the features of the model, since they not only involve the existence of a polarization mechanism, but also the spin flavor structure of the wavefunctions and the assumption that the

*E817-collaboration: B. Bonner, J. Buchanan, J. Clement, M. Corcoran, N. Krishna, J. Kruk, D. Lincoln, H. E. Miettinen, G. Mutchler, F. Nessi-Tedaldi, M. Nessi, G. Phillips, J. Roberts, P. Stevenson, S. Tonse, J. White, Rice U.; S. Chung, A. Etkin, R. Fernow, S. Protopopescu, H. Willutzki, Brookhaven Nat. Lab.; T. Hallman, L. Madansky J. Hopkins U., Baltimore; L. Pinsky U. of Houston.

proton spin is transferred undisturbed to fragments emerging from the reaction. Our group has undertaken in recent years a series of measurements to determine polarization transfer and analyzing power in inclusive production using the AGS polarized proton beam as well as the Fermilab 185 GeV/c polarized beam.[4] A first measurement[5] on Λ's gave values of A and D consistent with zero within ± 0.006 for A and ±0.015 for D, while reproducing earlier Λ polarization data. Values of A = -0.094 ± 0.012 for K_s^0 production and A= 0.03 ± 0.10 for $\bar{\Lambda}$ production were also observed. Our newest results confirm these previous ones.

ANALYZING POWER AND POLARIZATION TRANSFER MEASUREMENTS

A measurement of spin parameters in Σ^0 production was highly motivated by the above results. For direct Λ production A and D are expected to vanish, since in this case the (ud) diquark arising from the proton fragmentation is in a singlet state and does not carry any information about the spin of the projectile. In Σ^0 production however, the (ud) diquark is in a triplet state and is expected to carry a polarization correlated to the one of the proton, which is transmitted to the decay Λ. A small contribution to the measured spin parameters in Λ production was therefore expected (about 0.06 for A and -0.06 for D), whereas smaller values were observed.

The measurement of A and D in Σ^0 production has been performed during the most recent polarized beam run in Brookhaven (January '88). Details are given here for this measurement, where for the decay-Λ detection essentially the same setup was adopted as used in the previous experiment[5]. We have measured A and D for Σ^0 produced by a 18.35 GeV/c proton beam hitting a Berillium target. The decay Λ's were observed through their decay mode $p\pi^-$ in the BNL-Multiparticle Spectrometer (Fig.1), placed on the left side of the incoming beam, behind which a leadglass calorimeter was used to detect the γ's arising from the Σ^0 decay. The calorimeter was placed in a slightly low position, in order to favor the acceptance of transverse decays, where the Λ carries the biggest polarization.

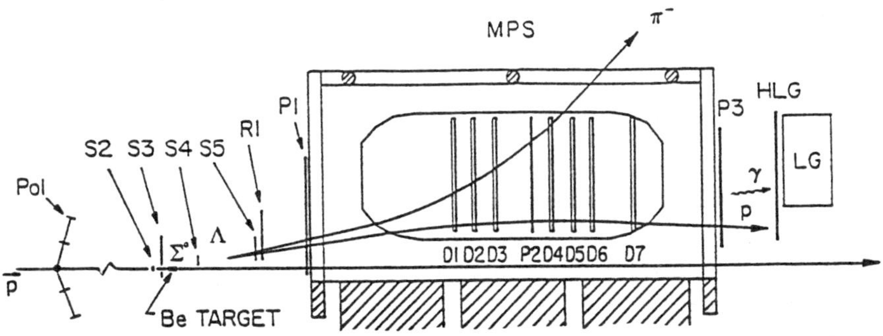

Fig.1 Top view of the experimental setup

A set of seven drift chambers (D1-D7) and four proportional chambers (R1 and P1-P3) served for the charged particle reconstruction. The Λ trigger was mainly constituted by a requirement of a neutral going through scintillator S4 and two minimum ionizing particles going through S5. Multiplicity requirements in the proportional chambers P1-P3 were used in a second stage of the Λ trigger which, together with a minimum energy deposition requirement in the leadglass, defined the basic sigma trigger. A hodoscope covering the whole MPS aperture upstream of the leadglass was used for two purposes: vetoing charged particles hitting the layer in which an energy deposition in the leadglass was observed, it suppressed hadronic showers; and requiring at least one charged particle hit, it favored events in which a fast proton traversed all the chambers. The differences between the time a beam particle hit the target and an energy deposition in each layer of leadglass were fed into TDC's, for accidentals suppression. The transverse components of the beam polarization $\vec{P}_B$ were monitored with a polarimeter consisting of a polyethylene target viewed by four scintillator telescopes and whose analyzing power was periodically checked against the University of Michigan absolute polarimeter located in another beamline. The average vertical beam polarization component was 39%, whereas the horizontal polarization component was -20%. The incident beam was counted as $S2\cdot\overline{S3}$, and with an average intensity of $3\cdot10^6$ per ~500 ms AGS pulse. We collected $1.5\cdot10^6$ Σ^0 triggers for a total flux of $7.5\cdot10^{11}$ protons. 171000 events reconstructed to Λ's and 10000 of them reconstucted to a Σ^0, as shown in Fig.2. We also reconstructed 50000 K^0_S and 1167 $\bar{\Lambda}$ from our data sample. Several cuts were applied in the data analysis: The neutral decay vertex was required to fall inside the region between S4 and S5, the produced particle momentum had to extrapolate back to the target. In the Σ^0 recontruction, we required proton and pion to reconstruct a Λ effective mass within ±6 MeV/c^2 of the nominal Λ mass; in order to suppress showers induced by charged hadrons hitting the leadglass, the particle tracks reconstructed inside the spectrometer were extrapolated to the leadglass position, and showers closer than 15 cm to any track were rejected. In addition, TDC cuts were applied. The analyzing power for each one of the production processes (see Fig.3) was determined from the total number of events reconstructed for spin up versus spin down. For Σ^0 production the measured values of A and D had to be corrected for the background contamination present underneath the mass peak. This background is Λ production events, for which we measured A and D values consistent with zero. The data

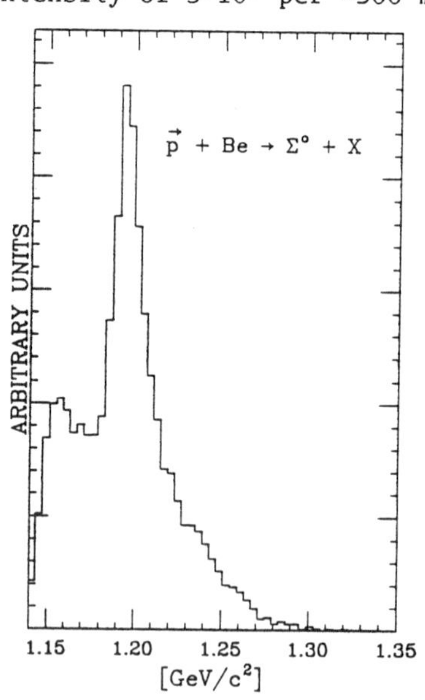

Fig.2 Σ^0 effective mass plot

cover a region of x_F between 0.0 and 0.75 (average 0.30) and p_T between 0.5 and 2.0 GeV/c (average value 1.18 GeV/c), and they show a value for A = 0.013±0.024. A correction of ~10% due to the presence of a horizontal polarization component in the beam combined with the slightly asymmetric detector setup was also included. The Λ's arising from the decay have a polarization $\vec{P}_\Lambda = -(\vec{P}_\Sigma \cdot \hat{p}_\Lambda)\hat{p}_\Lambda$, where $\hat{p}_\Lambda$ is a unit vector in the direction of the Λ-momentum in the Σ^0 restframe. Σ^0 production polarization parameters can therefore be measured through its electromagnetic decay, taking into account that the polarization of the particle has a component P along the direction normal to the production plane and an additional component proportional to the polarization transfer D. The polarization transfer parameter was determined in a fully acceptance independent way: the phase space was subdivided into single elements $d(\cos\delta)\cdot d(\cos\Theta)$, where δ is the angle between Σ^0 polarization and Λ direction in the Σ^0 restframe, and Θ is the angle between Λ polarization and proton direction in the Λ restframe. For each bin of $d(\cos\delta_i)$ and $d(\cos\Theta_i)$ a value D_{ij} can be determined, that is acceptance independent. The results for D is determined from the average over all D_{ij}, weighted with their statistical errors. For all our data we obtain an average value D=0.38±0.18.

DISCUSSION AND RESULTS

The quark fragmentation model of Ref.2 leads to a prediction for A of 20% and for D of 67% in Σ^0 production. For direct Λ production, A and D are expected to be zero. However, in our experiment we do not distinguish direct production from Λ produced in the decay of other hyperons. Making use of the Σ^0/Λ production ratio and the known polarization transfer mechanism in the $\Sigma^0 \to \Lambda\gamma$ decay[5,6], values of A ≃ 0.06 and D ≃ -0.06 were predicted. For Λ production we measure an A value which is slightly negative (mean

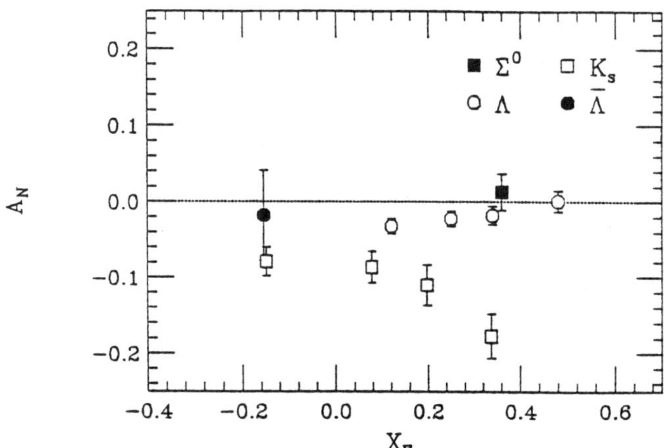

Fig.3 Analyzing power results for different inclusive processes

value -0.019±0.005 for $\langle x_F \rangle=0.25$ and $\langle p_T \rangle=1.10$ GeV/c) and a polarization transfer value of 0.001±0.021, in agreement with the predictions for direct Λ production, not however with the estimate that includes indirect Λ production through Σ^0 decay. The new results for Σ^0 production are, especially for A, considerably smaller than the expectations. In this sense, they can be considered consistent with the previous ones for Λ's. However, the disagreement with the model predictions cannot be easily understood. If Λ's and Σ^0's are produced through intermediate states (like Σ(1385)), the expectations could be altered considerably, but it is difficult to understand how we could obtain the same value, consistent with zero, for both hyperons.

For K_S^0 production we observe that A is large and negative, with an average value of -0.115±0.011 and a trend to increasing magnitude towards high x_F. The values we observe here are, if anything, larger than the theoretical expectations we obtain from the quark fragmentation model. If we include in the estimate also the contribution from K^* to K_S^0 production, a value of -7% is predicted, which is somewhat smaller in magnitude than the observed analyzing power. We also observe an unexpectedly large A value at negative values of Feynman x. In addition, in $\bar{\Lambda}$ production we observe that A is consistent with zero, with a value of -0.018±0.059.

A higher statistics measurement of D in Σ^0 production is compelling, both at the same energy as well as at Fermilab energies. Other measurements (e.g. $p \to \Sigma^+$) are also urgent to provide more independent measurements to better understand the mechanisms underlying this kind of reactions.

This experiment was supported by the U.S. Department of Energy. We are grateful to Dr. L. Ratner and the entire staffs of the AGS Department, the MPS group and the BNL-Applied Mathematics Department.

REFERENCES

1. For a review, see e.g. K. Heller, 7th Int. Symp. on High Energy Spin Physics, Protvino, (1986).
2. T. A. DeGrand and H. I. Miettinen, Phys. Rev. D24, 2419 (1981).
3. B. Anderson, G. Gustafson, A. Ingelman, Phys. Lett. 85B, 417 (1979).
4. See talks by D. G. Underwood and M. Nessi, this conference.
5. B. E. Bonner et al., Phys. Rev. Lett. 58, 447 (1987) and B. E. Bonner et al., to be published in Phys. Rev. D.
6. M. W. Sullivan et al., Phys. Rev. D36, 674 (1987).

ELECTRON–POSITRON COLLISION PHYSICS: 1 MeV TO 2 TeV*

Martin L. Perl
Stanford Linear Accelerator Center
Stanford University, Stanford, California 94309

ABSTRACT

An overview of electron-positron collision physics is presented. It begins at 1 MeV, the energy region of positronium formation, and extends to 2 TeV, the energy region which requires an electron-positron linear collider. In addition, the concept of searching for a lepton-specific force is discussed.

TABLE OF CONTENTS

A. *Introduction*

The initial purpose of this talk was to tell an audience of nuclear physicists and particle physicists about electron-positron collision physics: what we have learned and what we are doing. As I wrote the talk, I found that it provided me with the opportunity to give a broader view of electron-positron collision physics, to point out the areas where future research might be most fruitful. Some of these areas are well-known: electron-positron annihilation physics at the Z° and studying the properties of hadrons containing the b–quark. Less recognized fruitful areas are: understanding the decay modes of the tau lepton and studying hadrons containing the c–quark. There are two areas which I have never seen discussed: electron–positron collision physics below 500 MeV in the barycentric system and the search for what I call a lepton-specific force. In the written version, I have shifted the emphasis to pointing out these less recognized and unrecognized areas of electron-positron collision physics.

---

*Work supported by the Department of Energy, contract DE–AC03–76SF00515.

Figure 1 is the route map for this talk. I start at 1 MeV and go up in energy, dividing the discussion into energy regions according to the physics, the experimental method and the accelerator technology.

I have restricted the references to a few special areas because I am covering a broad subject.

Unless otherwise noted, all energies are total energies in the barycentric system denoted by $\sqrt{s}$ or $E_{c.m.}$.

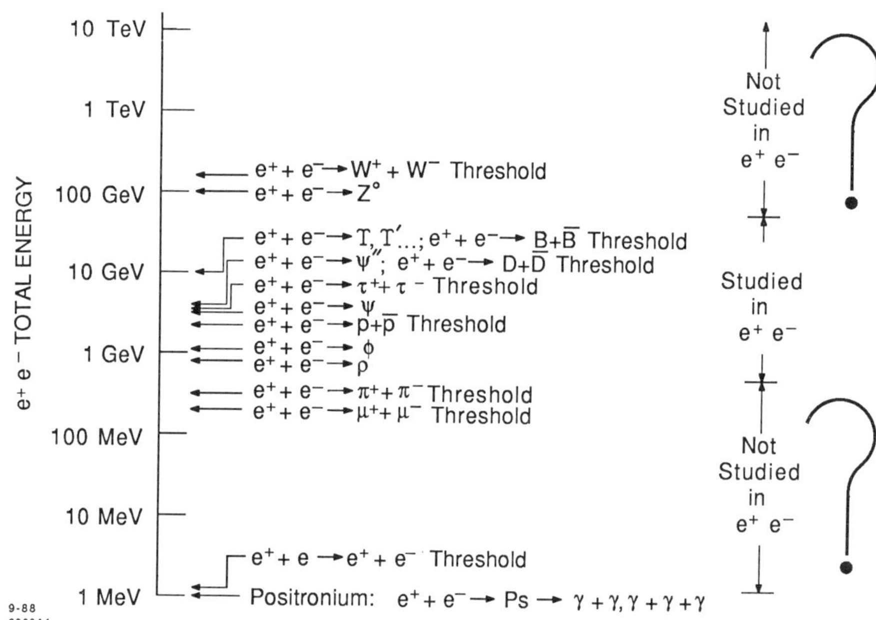

Figure 1

B. 1 MeV to 500 MeV and Positronium

I begin with the regions of 1 MeV to 500 MeV total energy. At the 1 MeV boundary of this region lies a vast area of e^+e^- physics: the formation of positronium

$$e^+ + e^- \to Ps\ , \tag{1a}$$

and its decay

$$Ps \to \gamma + \gamma\ , \quad \gamma + \gamma + \gamma\ . \tag{1b}$$

A minor subject in the region is the study of Bhabha scattering

$$e^+ + e^- \to e^+ + e^-\ . \tag{2}$$

Three decades ago, Bhabha scattering was studied using a positron beam on a fixed target, then work stopped in most of the energy region. Research continued only at the low-energy end, the area of atomic physics. In the last few years, a special interest in Bhabha scattering at an $E_{c.m.}$ of 1.6 to 1.8 MeV has developed, as I discuss later.

There are two reasons for there being no interest in general studies of Bhabha scattering or photon pair production,

$$e^+ + e^- \rightarrow \gamma + \gamma \,, \tag{3}$$

in the 1 MeV to 500 MeV region. First, the consensus is, or at least was, that quantum electrodynamics explains all e^+e^- physics in this region. Second, an e^+e^- circular collider of traditional design has low luminosity in this energy region, and no e^+e^- collider has been built in this energy region. The two reasons reinforce each other. If there is no physics interest, there is no incentive to build a collider. If a collider is very hard to build, why get interested in the physics. One set of perhaps anomalous measurements has revived interest in studying Bhabha scattering in the 1 MeV to several MeV region. Some experiments[1] show e^+e^- pairs produced in collisions of high Z ions, these pairs having masses of 1.6 to 1.8 MeV/c^2. The data on these pairs is confusing; the production mechanism is unknown. Electron beam dump experiments[2] have excluded a production mechanism involving a conventional, unstable elementary particle: $\phi \rightarrow e^+ + e^-$. Bhabha scattering at an $E_{c.m.}$ of 1.6 to 1.8 MeV might show a peak in the cross section if the unknown production mechanism can occur in a pure e^+e^- system; that is, if the presence of high Z ions is not required. Numerous measurements of Bhabha scattering in the energy region have been made in the past few years. An e^+ beam and a fixed target, usually of low Z, are used. At present, the measurements are contradictory and there is no confirmed observation of a peak.[3]

Interest in the several MeV region has also been stimulated by possible problems in precisely understanding the properties of positronium itself; the lifetime of orthopositronium, for example.[4]

I have been thinking about the best strategy for exploring the 1 MeV to 500 MeV region; measuring with precision the cross sections for

$$e^+ + e^- \rightarrow e^+ + e^- \,, \tag{4a}$$

$$e^+ + e^- \rightarrow \mu^+ + \mu^- \,, \tag{4b}$$

$$e^+ + e^- \rightarrow n\gamma \,, \quad n \geq 2 \,; \tag{4c}$$

looking for anomalous differential cross sections and resonances in the total cross section. We do not know how to design an e^+e^- circular collider which can operate from several MeV to 500 MeV, because known design principles limit the dynamic range to three or four. Therefore, I am considering fixed target experiments. The SLAC linear accelerator can produce e^+ beams with a maximum energy of 50 GeV, corresponding to $E_{c.m.} = 220$ MeV. Above about 500 MeV, the VEPP–2M e^+e^- circular (Sec. C) can take up studies of the reactions in Eq. 4; but there would still be a gap in energy coverage.

In Sec. L on lepton-specific forces, I will again take up the 1 MeV to 500 MeV region. There are indirect ways to explore this region, using processes such as

$$e^+ + e^- \to e^+ + e^- + e^+ + e^- \ ,$$

but these indirect methods have less sensitivity.

C. 500 MeV to 3 GeV and Hadron Production

The history of high-energy, e^+e^- collision physics began in this energy region: ADONE in Italy, the CEA collider in the United States, the DCI in France and the VEPP colliders in the Soviet Union. Today, there is only one e^+e^- collider in this region, VEPP–2M.

In this energy region, the production of hadrons becomes important

$$e^+ + e^- \to \text{hadrons} \ ,$$

and it was in this region that two main processes for hadron production were elucidated. Resonances such as the ρ, ω, ϕ, and at higher energies, the ψ and Υ, are produced through the process in Fig. 2a. The continuum production of hadrons,

$$e^+ + e^- \to \text{many hadrons} \ ,$$

occurs through the process in Fig. 2b, with the cross section per quark type and color,

$$\sigma(e^+e^- \to q\bar{q} \to \text{hadrons}) \ = \ \frac{4\pi\alpha^2 Q_q^2}{3s} \ . \tag{5}$$

The recognized research to be done in this region concerns more precise studies of hadron production. I also see valuable research to be done on the reactions in Eq. 4, searching for anomalous behavior. Looking ahead, a high luminosity collider producing

$$e^+ + e^- \to \phi \to K^\circ + \bar{K}^\circ$$

can extend our knowledge of the CP violating mechanism in the $K^\circ\bar{K}^\circ$ system.

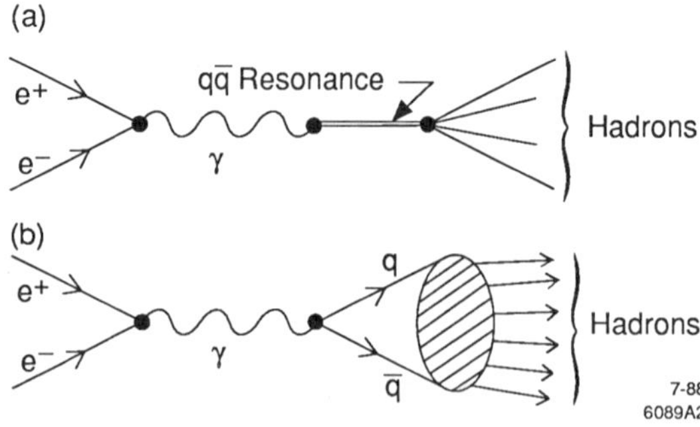

Figure 2

D. 3 GeV to 10 GeV: the ψ, the τ, the c–Quark and Hadron Jets

This is the energy region where four great discoveries were made in the 1970's:

(*i*) The ψ, ψ', ψ'' family of $c\bar{c}$–hadrons was discovered.

(*ii*) The τ heavy lepton was found through

$$e^+ + e^- \to \tau^+ + \tau^- \ . \tag{6}$$

(*iii*) The D–hadrons containing a single c–quark were found.

(*iv*) Hadron jets produced by the process

$$e^+ + e^- \to q + \bar{q} \ , \qquad q \to \text{hadron jet} \ , \quad \bar{q} \to \text{hadron jet} \ ,$$

were identified.

One e^+e^- collider is now operating in this region, SPEAR in the United States. A new, higher luminosity e^+e^- collider, BEPC, is now being built in the Peoples Republic of China to operate in this region.

There are a number of recognized research areas in this region. There is much more to be done in studies of the ψ, D and F particle families. The complicated energy dependence of $\sigma(e^+e^- \to \text{hadrons})$ from $E_{c.m.} = 3.5$ *GeV* to about 5 GeV, Fig. 3, is not understood. The τ decay mode puzzle[5,6] needs to be unraveled. Some properties[7] of the τ, such as the mass of ν_τ, can be measured precisely using data from this region. I add to this list more precise studies of the lepton-photon vertices in Eqs. (4) and (6) and the search for a lepton-specific force.

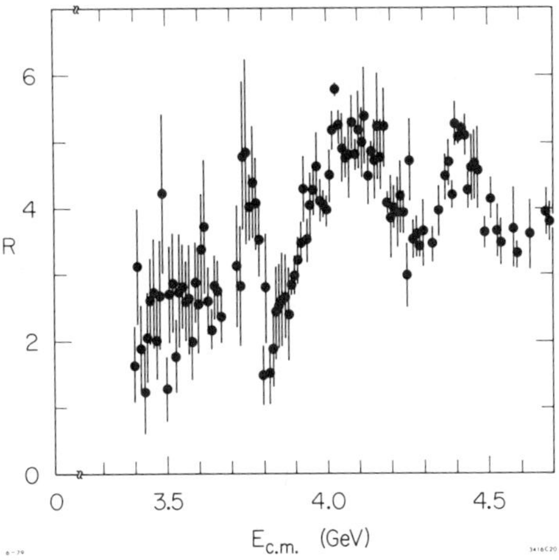

Figure 3

The need for large amounts of data to further the research listed in the previous paragraph has stimulated a proposal for a very high luminosity e^+e^- circular collider in this energy region.[8,9]

E. 10 GeV to 56 GeV: b–Quark Physics and Other Successes

At the lower end of the energy region lies the threshold for the production of the upsilon family of particles, Υ, Υ', Υ'', ..., made up of a $b\bar{b}$–quark pair, and the threshold for the production of hadrons containing a single b–quark. Three e^+e^- circular colliders operate in the 10 GeV region: CESR in the United States, DORIS in Germany and VEPP–4 in the Soviet Union.

At higher energies, there is the PEP collider in the United States, 20 to 30 GeV, and the new TRISTAN collider in Japan. The TRISTAN collider sets the upper end of this region, 56 GeV, the highest e^+e^- collision energy at which there is data. In the next few years, TRISTAN's energy will move into the 60–70 GeV range.

There have been four successes in the 10–56 GeV range. One triumphant success is the comprehensive research on the properties of hadrons containing the b–quark. The most recent research indicates the possible existence of substantial mixing of B° and $\overline{B}^\circ$ mesons. This in turn allows the possibility of searching for CP violation in the B°–$\overline{B}^\circ$ system and, if it exists, studying CP violation in this new system.

Future detailed studies of b–quark physics and searching for CP violation requires e^+e^- collider luminosities in the range of 5×10^{32} to 10^{34} cm^{-2} s^{-1}. Existing e^+e^- colliders have luminosities in the range of 10^{31} to 10^{32} cm^{-2} s^{-1}. Therefore, a great amount of discussion and design work is being devoted to higher luminosities. Plans and proposals[10] include: increasing the luminosities of the existing single-ring, circular colliders; building new double-ring, circular colliders; building a linear collider; and building a mixed linear-circular collider. These e^+e^-–collider proposals have to be compared with proposals to use the large number of B mesons produced in hadron-hadron collisions, in fixed target or collider experiments.

The second great success is the elucidation of the theory of quantum chromodynamics through the study of hadron production in the continuum. An important part of this elucidation came from the discovery and study of events with three hadron jets from the process

$$e^+ + e^- \to q + \bar{q} + g \ ,$$
$$q \to \text{quark} - \text{hadron jet} \ ,$$
$$\bar{q} \to \text{quark} - \text{hadron jet} \ ,$$
$$g \to \text{gluon} - \text{hadron jet} \ .$$

The third success is the study of hadron production and related physics in the two-virtual photon process,

$$e^+ + e^- \to e^+ + e^- + \gamma_v + \gamma_v$$
$$\gamma_v + \gamma_v \to \text{hadrons}$$

in Fig. 4. Here γ_v is a virtual photon which in some kinematic conditions is almost a real photon.

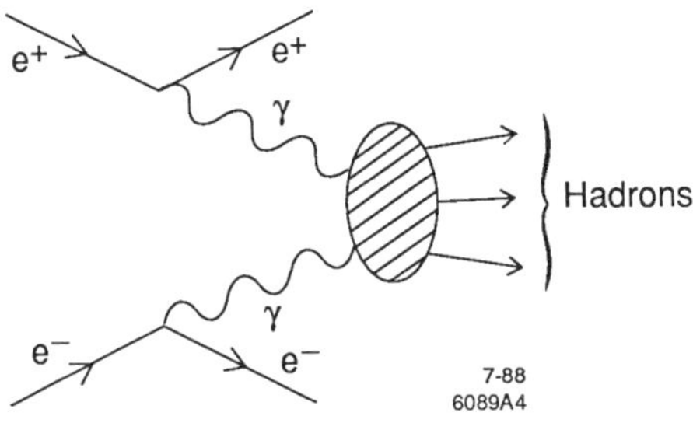

Figure 4

The fourth success is the measurement of the interference of the electromagnetic and weak amplitudes, Fig. 5, as the energy advances up the lower tail of the Z°. A taste of greater things to come.

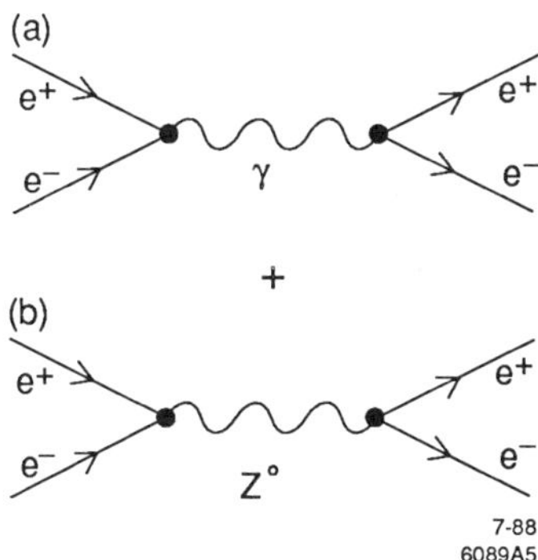

Figure 5

The PETRA e^+e^- collider in Germany contributed data to all of the mass, along with the other colliders listed in this section. PETRA which reached a maximum energy of about 46 GeV is not operated at present.

Out of the large amount of data collected in this 10 to 56 GeV region has also come a strange result. The many measurements of the decay modes of the τ lepton have resulted in a problem in understanding the decay modes with one-charged particle, Table I. There is a discrepancy[5,6] between the inclusive

Rows	Symbol	Decay Mode of τ^-	Branching Fraction (%)
1	B_1	1–charged particle inclusive	86.6 ± 0.3
2	B_e	$\nu_\tau + e^- + \bar{\nu}_e$	17.6 ± 0.4
	B_μ	$\nu_\tau + \mu^- + \bar{\nu}_\mu$	17.7 ± 0.4
	B_π	$\nu_\tau + \pi^-$	10.8 ± 0.6
	B_ρ	$\nu_\tau + \rho^-$	22.5 ± 0.9
		Sum for modes in Rows 2	68.6 ± 1.2
3	$B_{\pi 2\pi^0}$	$\nu_\tau + \pi^- + 2\pi^0$	7.6 ± 0.8
		$\nu_\tau + mK + n\pi^0$ $\longrightarrow$ 1–charged particle $m \geq 1,\ n \geq 0,\ K = K^0$ or K^-	1.8 ± 0.3
		Sum for modes in Rows 3	9.4 ± 0.9
4		$\nu_\tau + \pi^- + n\pi^0$ $n \geq 3$ $\nu_\tau + \pi^- + m\eta + n\pi^0$ $\longrightarrow$ 1–charged particle $m \geq 1,\ n \geq 0,$	
		Sum for modes in Rows 4	≤ 2.7
5		Sum for modes in Rows 2, 3 and 4	80.7 ± 1.5

Table 1. Summary of present knowledge of 1–charged particle branching fraction in percent from Refs. 5 and 6. The numbers in Rows 1, 2, and 3 are the average of measured values and the associated standard deviation. The sum in Row 4 is the 95% upper limit obtained from other data and accepted theory. Note that Refs. 5 and 6 used Gaussian error distributions.

one-charged particle branching fraction and the sum of the known exclusive one-charged particle branching fractions.

The 10 to 56 GeV region has brought great disappointment, as well as great success. The top quark has not been found; a fourth generation quark has not been found; a fourth generation charged lepton has not been found; nor have additional neutral leptons been found—but the neutral leptons searches are not definitive.[11] Further e^+e^- searches for these particles requires higher energy: TRISTAN above 56 GeV, the SLAC Linear Collider (SLC) and LEP in the Z° region of 70 to 110 GeV, LEP above the Z° to about 200 GeV and, ultimately, e^+e^- linear colliders into the TeV region.

Before moving on from what we have measured below 56 GeV to what we hope to measure above 56 GeV, I will summarize our knowledge and expectations about the total cross sections for

$$e^+ + e^- \to \ell^+ + \ell^- \; ; \quad \ell = e, \mu, \tau ,$$

and

$$e^+ + e^- \to \text{hadrons} .$$

These reactions make up the explored land of known e^+e^- collision physics. We have to look beyond that land for new particles or new phenomena.

F. *Overview of Cross Section for $e^+e^- \to \ell^+\ell^-$*

The total cross section for

$$e^+ + e^- \to \ell^+ + \ell^- \; ; \quad \ell = e, \mu, \text{ or } \tau , \tag{7}$$

follows from electroweak interaction theory. In writing down these formulas, I keep in mind that the formulas have only been confirmed directly in the e^+e^- collisions below 56 GeV, and indirectly in the decays of Z°'s produced in $p\bar{p}$ collisions. There may be surprises. Consider three energy regions, Fig. 6: the region centered on the Z° mass of about 93 GeV; the region below the Z°; and the region far above the Z°, above 200 GeV. In these formulas I ignore threshold effects and radiative corrections. The latter may be substantive, changing cross sections by tens of percent.

Below the Z°, the electromagnetic process, Fig. 5a, dominates with

$$\sigma(e^+e^- \to \gamma \to \ell^+\ell^-) = \frac{4\pi\alpha^2}{3s} \approx \frac{87}{s} \text{ nb} , \tag{8a}$$

where s is in GeV^2. The weak process itself gives, Fig. 5b,

$$\sigma(e^+e^- \to Z^\circ \to \ell^+\ell^-) = \frac{G^2 s}{96\pi} = 1.8 \times 10^{-7} s \text{ nb} , \tag{8b}$$

where s is again in GeV^2. Thus, when $\sqrt{s} \lesssim 50$ GeV, the weak process is detected through its interference with the electromagnetic.

In the vicinity of the Z°, the weak process dominates as a real $Z^\circ$is produced. Here

$$\sigma(e^+e^- \to Z^\circ \to \ell^+\ell^-) = \frac{G^2 s}{96\pi} \left[\frac{m_Z^4}{(s - m_Z^2)^2 + \Gamma_Z^2 m_Z^2} \right] , \tag{9}$$

This is about 1.6 nb at $E_{c.m.} = m_Z$. Radiative corrections reduce this to about 1.2 nb.

Above 200 GeV, if there are no surprises, the electromagnetic process contributes according to Eq. (8a), but it is convenient to put s in TeV^2 and use picobarns instead of nanobarns. Then,

$$\sigma(e^+e^- \to \gamma \to \ell^+\ell^-) \approx \frac{0.087}{s} \text{ pb} \approx \frac{0.1}{s} \text{ pb} , \tag{10a}$$

a very small cross section. The weak process, ignoring interference, gives an even smaller cross section:

$$\sigma(e^+e^- \rightarrow Z^\circ \rightarrow \ell^+\ell^-) \approx \frac{G^2 m_Z^4}{96\pi s} = \frac{0.013}{s} \text{ pb} , \qquad (10b)$$

with s in TeV^2.

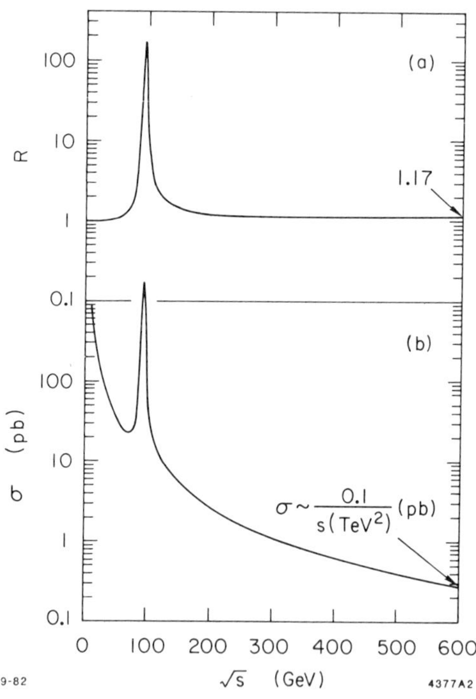

Figure 6

Figure 6 also gives the cross section in terms of R, where

$$R = \frac{\sigma}{\sigma_0} , \qquad \sigma_0 = \frac{4\pi\alpha^2}{3s} . \qquad (11)$$

Thus far above this Z° region, $R(e^+e^- \rightarrow \ell^+\ell^-)$ is again about one.

G. *Overview of Cross Section for $e^+e^- \rightarrow$ hadrons*

Again, I consider three energy regions: below the Z°, in the vicinity of the $Z^\circ$and far above the Z°.

Below the Z°, hadron production is dominated by the electromagnetic process

$$e^+ + e^- \rightarrow \gamma \rightarrow \text{ hadrons} , \qquad (12)$$

operating through the two mechanisms described in Fig. 2: resonance production and continuum production. The cross section for continuum production is obtained from Eq. (5) by summing over the number of quarks multiplied by three, for the three colors. Ignoring quantum chromodynamic corrections

$$\sigma(e^+e^- \to \gamma \to \text{ hadrons, continuum}) \approx \frac{4\pi\alpha^2}{s} \sum_q Q_q^2 \,. \tag{13a}$$

Usually, as in Fig. 7,[12] R is displayed. Measurements confirm the expectation

$$R(e^+e^- \to \gamma \to \text{ hadrons, continuum}) \approx 3 \sum_q Q_q^2 \,. \tag{13b}$$

Once quantum chromodynamics is taken into account, as $E_{c.m.}$ rises to the lower tail of the Z°, the weak process contribution

$$e^+ + e^- \to Z^\circ \to \text{ hadrons} \,, \tag{14}$$

becomes obvious, Fig. 7b.

In the vicinity of the Z°, we expect an enormous increase in $\sigma(e^+ + e^- \to Z^\circ \to \text{ hadrons})$, Fig. 8. At the Z°, $\sigma(e^+ + e^- \to Z^\circ \to \text{ hadrons}) \approx 40$ nb, including the radiative corrections.

Far above the Z°(above, say, 200 GeV), the production of hadrons occurs through the combined amplitudes for the reactions in Eqs. (12) and (14). In analogy to Eq. (10)

$$\sigma(e^+e^- \to q\bar{q} \to \text{ hadrons}) = \frac{0.087\, r_q\, T_q\, (m_q, s)}{s} \text{ pb} \,. \tag{15}$$

Here s is in TeV^2, T_q is a threshold factor with $T_q = 1$ for $s \gg m_q^2$, m_q is the mass of the q–quark and

$$r_q = 1.9 \,, \quad \text{charge } \frac{2}{3} \text{ quarks : } u, c \,,$$

$$r_q = 1.1 \,, \text{ charge } \frac{1}{3} \text{ quarks : } d, s, b \,,$$

These five known quarks will give

$$\sigma\left(e^+e^- \to \text{ hadrons}\right) = \frac{0.6}{s} \text{ pb} \,. \tag{16}$$

With these completed overviews of charged lepton and hadron production, I move above the 56 GeV boundary to energies yet to be explored in e^+e^- collision physics.

H. 56 GeV to 70 GeV

The new TRISTAN collider will soon explore this region. The great interest is to search for new particles, particularly the top quark, and to search for new phenomena. Other interests are the study of the interference between electromagnetic and weak amplitudes and the study of two-virtual photon physics.

I have though of a speculative possibility in this region where the cross sections for lepton and quark production are at a minimum before the rise of

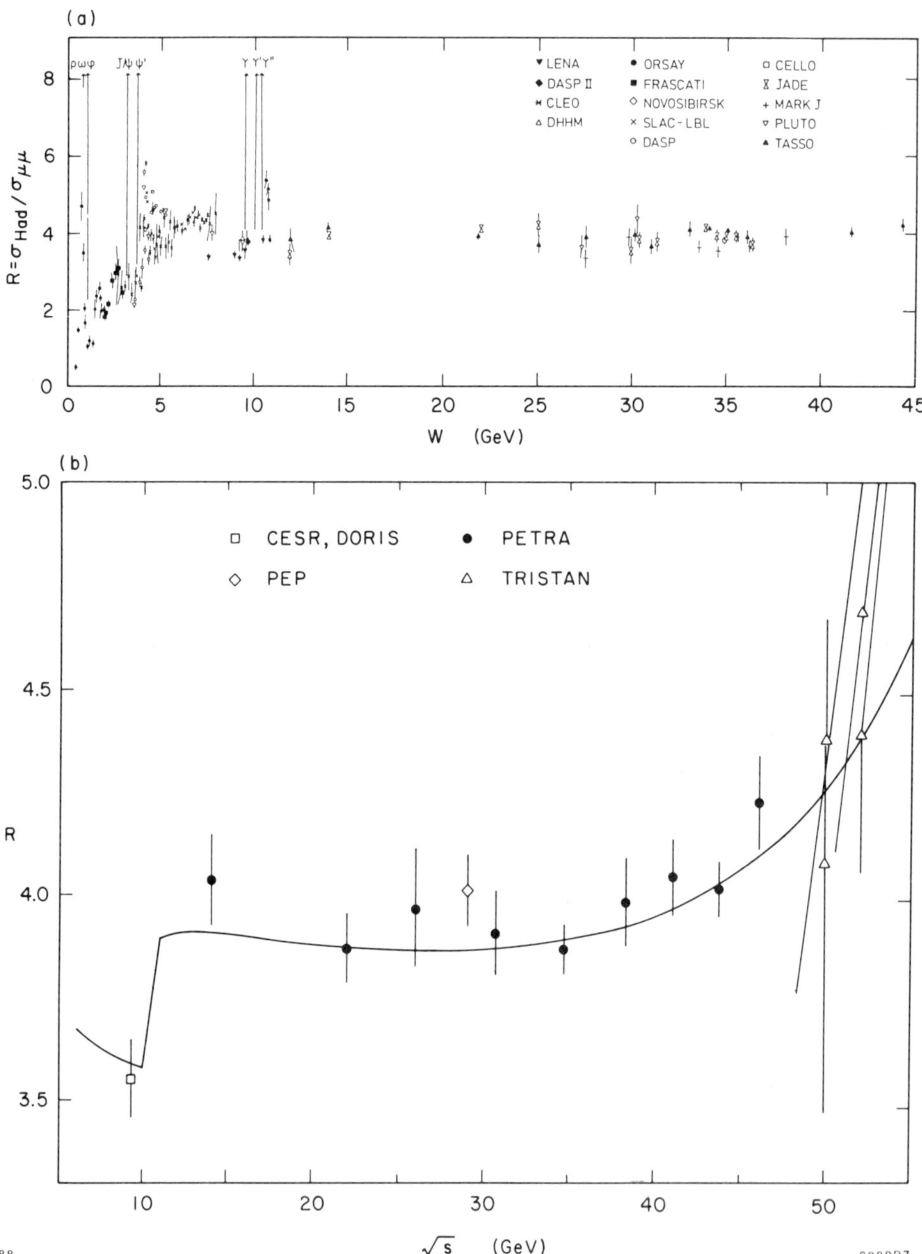

(a)
R = σ Had / σ μμ
W (GeV)
ρωφ
J/ψ ψ'
Υ Υ' Υ''
LENA
DASP II
CLEO
DHHM
ORSAY
FRASCATI
NOVOSIBIRSK
SLAC-LBL
DASP
CELLO
JADE
MARK J
PLUTO
TASSO
(b)
CESR, DORIS
PEP
PETRA
TRISTAN
R
√s (GeV)
7-88
6089B7

Figure 7

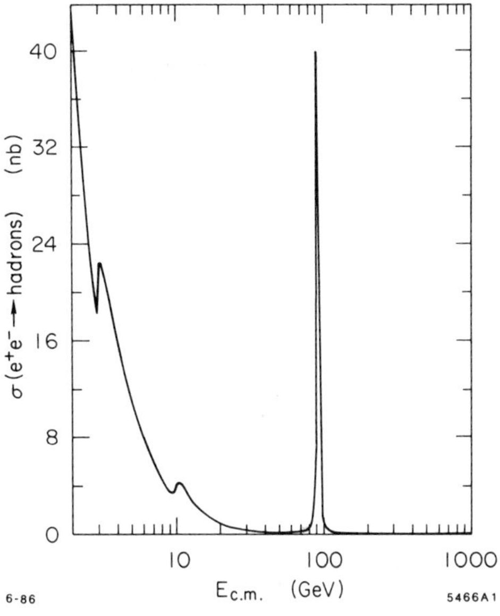

Figure 8

the Z°. Suppose there is a new phenomenon in e^+e^- collision physics which occurs at much higher energy. At lower energies, the cross section for effects due to this phenomenon might be proportional to s in analogy to Eq. (8b). The minimum in the cross section for lepton and quark production is an ideal place to look for such effects.

I. The Z° Region: SLC and LEP

We are entering this energy region with two new e^+e^- colliders. The SLAC Linear Collider (SLC) in the United States, Fig. 9, is starting operation at the Z°. Its energy range is 70 to 110 GeV. It is the first collider to use the linear collider principle,[13] pointing the way for this new collider technology.

In about a year, the LEP e^+e^- circular collider, Fig. 10, in Switzerland will begin operation; the largest diameter accelerator or collider in the world. It will begin in the Z° energy region. As the supply of radio frequency power to the circulating beams is increased, the energy range will be extended to about 180 GeV.

So much has been dreamed and written about e^+e^- collision physics at the Z°; too much to summarize. I will mention a few main points.[14]

The Z° resonance, Figs. 6 and 8, provides an enormous cross section for lepton and quark production at high energy. Neutral elementary fermions, such as neutrinos, are produce as copiously as charged fermions. At $E_{c.m.} = m_z$,

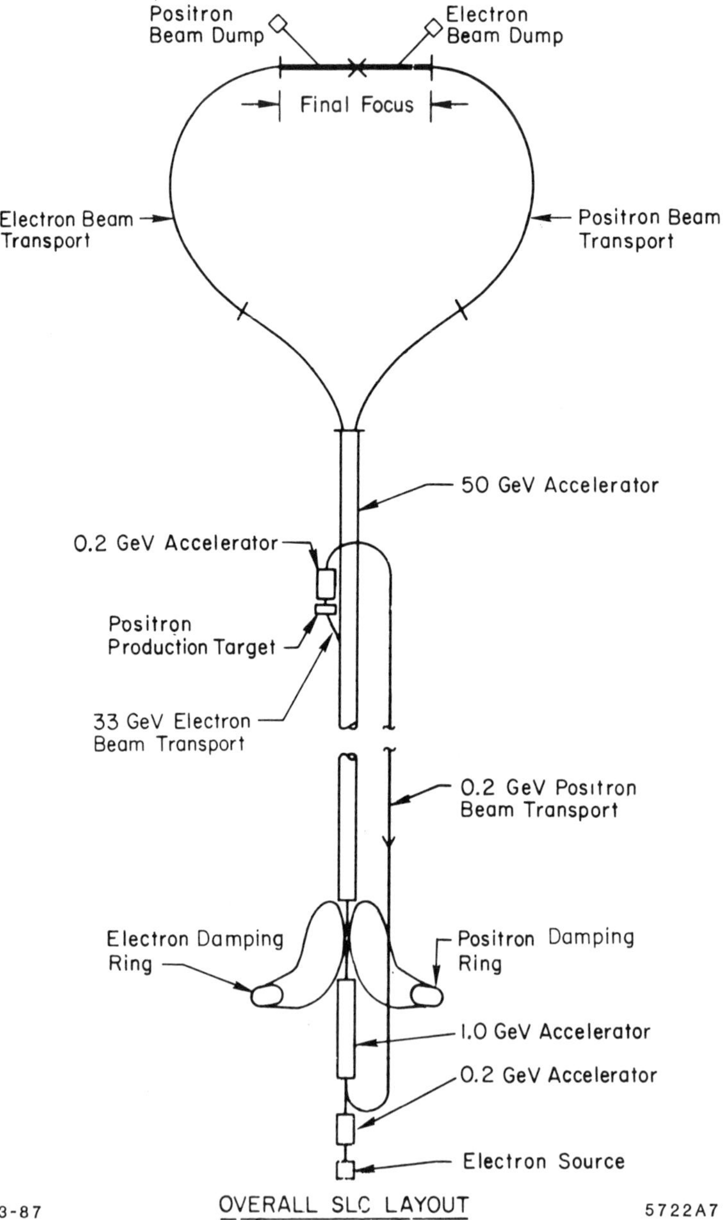
Positron
Beam Dump
Electron
Beam Dump
Final Focus
Electron Beam
Transport
Positron Beam
Transport
50 GeV Accelerator
0.2 GeV Accelerator
Positron
Production Target
33 GeV Electron
Beam Transport
0.2 GeV Positron
Beam Transport
Electron Damping
Ring
Positron Damping
Ring
1.0 GeV Accelerator
0.2 GeV Accelerator
Electron Source
3-87
OVERALL SLC LAYOUT
5722A7

Figure 9

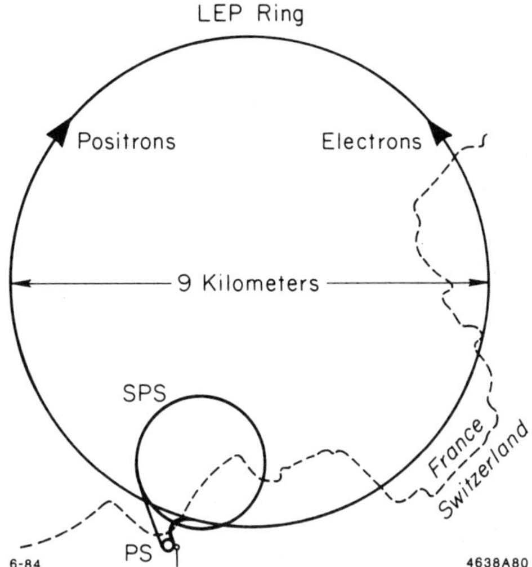

Figure 10

ignoring radiative corrections:

$$\sigma(e^+e^- \rightarrow Z^\circ \rightarrow f\bar{f}) = \frac{G^2\, m_z^4\, r_f\, T_f\,(m_f, m_z)}{96\pi\, \Gamma_Z^2} . \tag{17}$$

Here T_f is a threshold factor and

$$\begin{aligned} r_f &= 1.0 , & f &= \ell^- = e^-, \mu^- \text{ or } \tau^- \\ r_f &= 2.0 , & f &= \nu_\ell , \\ r_f &= 1.2 , & f &= u \text{ or } c \text{ quark} , \\ r_f &= 1.5 , & f &= d, s, \text{or } b \text{ quark} , \end{aligned}$$

Searches for new particles can be direct or indirect. If a new particle is charged, stable or unstable, or if a new particle is neutral and unstable, the search can be direct—unexplained events from the Z° decay.

The production of new stable neutral particles—massive neutrinos, for example—can be detected indirectly through additions to the predicted width of the Z°:

$$\begin{aligned} \Gamma_Z &= \sum_f \Gamma_{Zf} , \\ \Gamma_{Zf} &= \frac{G m_Z^3\, r_f\, T_f}{24\sqrt{2}\,\pi} \end{aligned} \tag{18}$$

This, again, ignores radiative corrections which must be precisely calculated. If the mass of the new, neutral particle is of the order of a GeV/c^2 or less, the cross section for the reaction

$$e^+ + e^- \to Z^\circ \to \gamma + \text{missing energy} , \tag{19a}$$

will be augmented above that contributed by

$$e^+ + e^- \to Z^\circ \to \gamma + \nu_\ell + \bar{\nu}_\ell , \qquad \ell = e, \mu, \tau . \tag{19b}$$

If you believe in the existence of a physical Higgs particle, the elusive H° can be sought through

$$e^+ + e^- \to Z^\circ \to Z^\circ + H^\circ .$$

Returning to known physics, production and decay of the Z° at LEP and the SLC is a source of c–quarks, b–quarks and τ leptons, allowing extension of the knowledge gained in the 3 GeV to 56 GeV regions. In addition, electroweak theory and quantum chromodynamics can be tested in more detail.

J. 100 GeV to 200 GeV: LEP

Moving above the Z°, the energy region up to about 200 GeV will be explored by the LEP collider.[15] New searches will be made for more massive quarks and leptons, for the Higgs particle (if not yet found), and for speculative particles of all sorts.

The reaction

$$e^+ + e^- \to W^+ + W^- , \tag{20}$$

is of great interest because one of its amplitudes, Fig. 11, has a W–Z°–W vertex. Electroweak theory predicts the cross section in Fig. 12. At 200 GeV, $\sigma(e^+e^- \to W^+W^-) = 20$ pb compared to $\sigma(e^+e^- \to \mu^+\mu^-) = 2.5$ pb from Eq. (10). Thus, the cross section for $e^+e^- \to W^+W^-$ is relatively substantial at the upper end of this energy range. Once W's are directly produced, their properties and their decay products can be studied in detail.

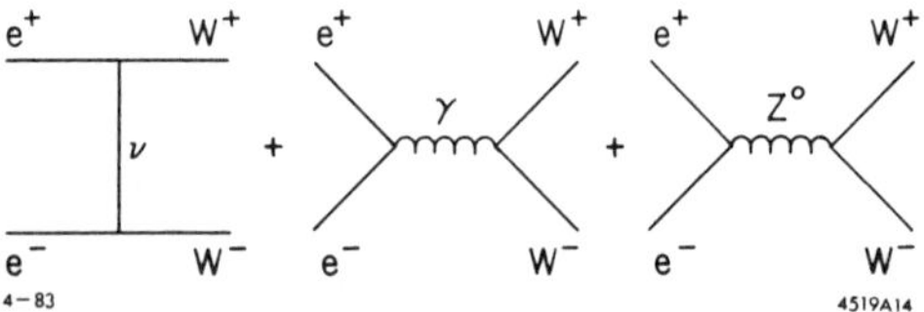

Figure 11

For most of us, the 100 to 200 GeV region is both fascinating and frightening. We expect to understand e^+e^- collision physics thorough the Z° region, although this may be hubris; but the 100 to 200 GeV region is the entrance to *terra incognita*. What if no new particles are found below 200 GeV, not even the top quark? What if $\sigma(e^+e^- \to f\bar{f})$ continues to decrease as $1/s$, beautifully simple, but ever smaller?

It will be eight to ten years before experimenters at LEP have concluded a through exploration of this region. We are too impatient to wait those years

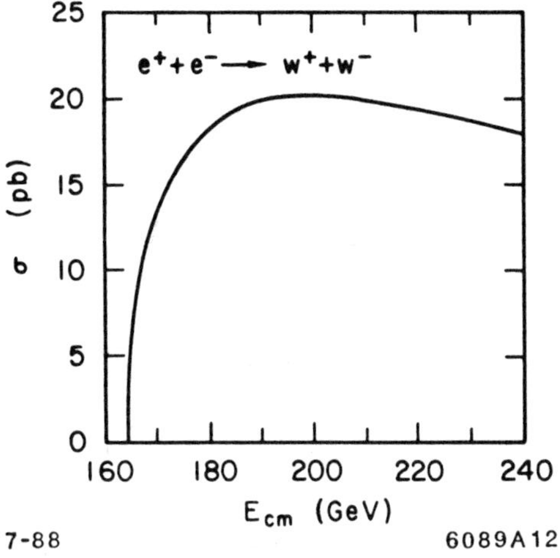

Figure 12

before deciding to push above 200 GeV in e^+e^- collision physics. Hence, our great interest in e^+e^- linear colliders[13] and e^+e^- collision physics from 200 GeV to 2 TeV.

K. 200 GeV to 2 TeV: e^+e^- Linear Colliders

We are excited about with e^+e^- linear colliders, Fig. 13, because the traditional e^+e^- circular collider costs too much to build and operate above 200 or 300 GeV. We don't know yet how to design a 500 GeV or 2 TeV linear collider, but we know the general principles; we know what we have to do. Cross sections of the order of

$$\sigma \sim \frac{.1}{s}\ \text{pb}\ ,$$

with s in TeV^2, require luminosities in the range of 10^{33} to 10^{34} $\text{cm}^{-2}\ \text{s}^{-1}$. This requires the transverse dimensions of the colliding e^+ and e^- bunches to be 10^{-3} to 10^{-1} μm, putting large demands on controlling the size and position of the bunches as they move through the linear accelerator. Large amounts of microwave power must be produced efficiently. Bunch halos and backgrounds from passage of the bunches through the magnets and collimators must be eliminated so as not to overwhelm the detector.

Linear collider research and development work[13] is going on in Europe, Japan, the Soviet Union and the United States. There is a great deal to learn and invent. It may not be wise to try to jump from the 100 GeV, pioneering SLAC Linear Collider to a 1 or 2 TeV linear collider. The best course may be to build an intermediate energy facility of, say, 500 GeV and then use that experience to go on.

As I have already said, the 200 GeV to 2 TeV region is *terra incognita.* Even conventional processes become strange and wonderful.[16,17] Suppose there

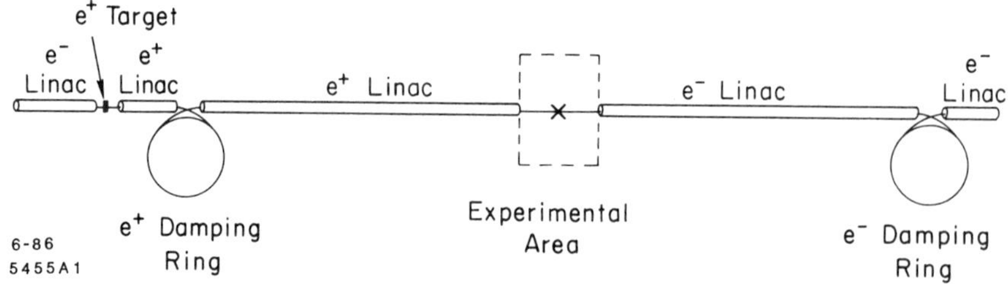

Figure 13

is a massive lepton pair L^-, L° with

$$m_- - m_0 > m_W \ ,$$

where m_-, m_0 and m_W are the masses of the L^-, L° and W. Then the major decay mode is

$$L^- \to L^\circ + W^- \ ,$$

where the W^- is real and itself decays. The W^- here is like the π^- or ρ^- in

$$\tau^- \to \nu_\tau + \pi^- \ , \quad \nu_\tau + \rho^- \ .$$

Similarly, the major decay mode of a massive quark pair q, q' with

$$m_q - m_{q'} \ > \ m_W \ ,$$

is

$$q \to q' + W^- \ .$$

Another wonderful effect of very high energy on conventional processes is that the "two virtual-photon" process of Fig. 14a is joined by the "two virtual–W" process of Fig. 14b. A variation of the process in Fig. 14b provides a neat way to produce a physical Higgs particle,

$$e^+ + e^- \to \bar{\nu}_e + \nu_e + H^\circ \ ,$$

via the process in Fig. 14c.

By now we are far in the future; it is almost time to conclude this journey from 1 MeV to 2 TeV. Before concluding, I want to speculate on something we can search for now—a lepton-specific force.

L. *Speculations on a Lepton–Specific Force*

It is strange to me that it is conventional to hope or expect that (a) lepton generations mix just as quark generations mix, and (b) neutrinos have nonzero mass. It is strange because experiment leads to opposite conclusions. There is no evidence for $e - \mu$, $e - \tau$ or $\mu - \tau$ generation mixing. Indeed, one of the most precise measurements in particle physics is the upper limit

$$\frac{\Gamma(\mu^- \to e^- \gamma)}{\Gamma(\mu^- \to e^- \bar{\nu}_e \nu_\mu)} \ < \ 10^{-10} \ .$$

There is no confirmed evidence for neutrino masses being other than zero.

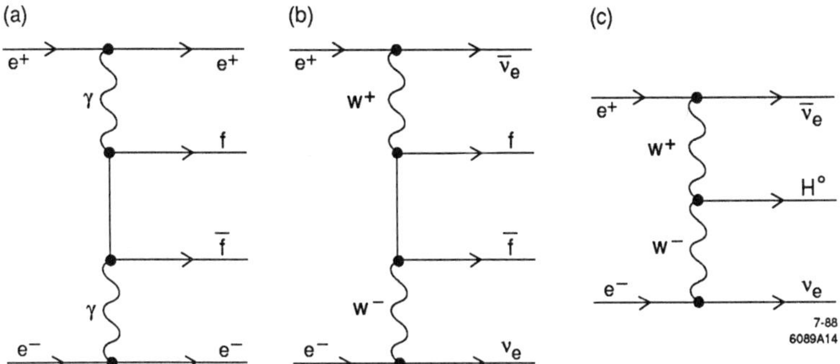

Figure 14

I see the leptons as very different from the quarks, and this has led me to speculate as to the existence of a force only exerted by and on leptons—a lepton-specific force.

I think of the force as being carried by a neutral particle λ, with mass m_λ and an L–λ–L coupling of strength $\sim \sqrt{\alpha_\lambda}$. Here L is a charged lepton or neutrino. This speculation overlaps axion and Higgs particle ideas, but the λ doesn't couple to nonleptons. The precise strength of the coupling may depend on the type of lepton or generation, but I don't think of $\sim \sqrt{\alpha_\lambda}$ as depending, like the Higgs particle, on the lepton mass.

Christopher Hawkins and I[18] have considered what ranges of m_λ and α_λ are ruled out by: atomic measurements such as $g_e - 2$, by previous particle physics measurements and searches, such as axion searches, and by deductions from astrophysics. We find that when

$$m_\lambda \gtrsim 10 \text{ to } 100 \text{ MeV/c}^2 \ .$$

There are few experimental limits on the existence of a particle λ carrying a lepton-specific force.

Three hypothetical processes in e^+e^- collision physics, Fig. 15, allow searches for the λ. The process in Fig. 15a, with a λ in the t–channel, would affect small angle Bhabha scattering—a reaction which has never been tested to better than several percent in e^+e^- collisions physics.

The process in Fig. 15b, annihilation through the λ, would show up as an energy resonance at $E_{c.m.} = m_\lambda$ in

$$e^+ + e^- \rightarrow \ell^+ + \ell^- \ , \quad \ell = e, \mu, \tau \ . \tag{21}$$

In Sec. B, I noted that there are precise experiments searching for a resonance in $e^+e^- \rightarrow e^+e^-$ at $E_{c.m.} \sim 1.8$ MeV; but many energy regions in $E_{c.m.}$collision physics have not been explored for an energy resonance in the reactions in Eq. 21.

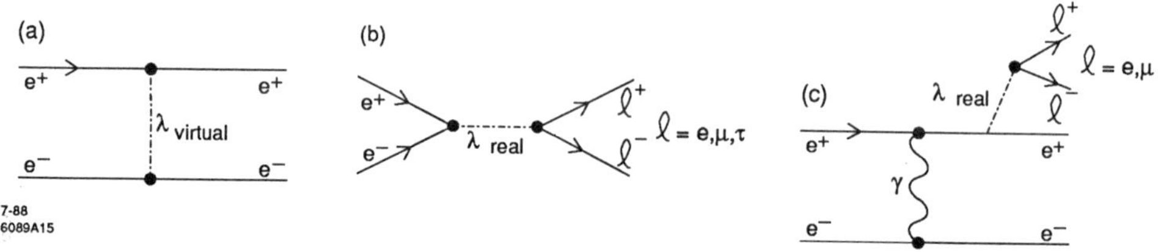

Figure 15

Hawkins and I are looking for the process in Fig. 15c, using 29 GeV data acquired with the Mark II detector at PEP:

$$e^+ + e^- \rightarrow e^+ + e^- + \lambda \ , \tag{22a}$$

$$\lambda \rightarrow e^+ + e^- \quad \text{or} \quad \lambda \rightarrow \mu^+ + \mu^- \ . \tag{22b}$$

One final e^+ or e^- in Eq. (22a) will usually have an angle close to the beam line; then the detected event is

$$e^+ + e^- \rightarrow e^\pm + \ell^\pm + \ell^\mp + \text{missing energy} \ ,$$

the missing energy being carried off approximately along the beam line. The background is "two virtual-photon" processes.

ACKNOWLEDGMENTS

I have had valuable discussions with my colleagues T. L. Barklow, D. L. Burke, J. M. Dorfan, G. J. Feldman, K. K. Gan, C. A. Hawkins, K. G. Hayes, J. A. Jaros, M. E. Peskin and Y. S. Tsai.

REFERENCES

1. An excellent set of papers is in Physics of Strong Fields (Plenum Press, N.Y., 1987), W. Greiner, ed.
2. M. Davier, Proc. XXII Int. Conf. on High Energy Physics (World Scientific, Singapore, 1987), S. C. Loken, ed.; E. M. Riordan *et al.*, Phys. Rev. Lett. **59**, 755 (1987).
3. W. Koenig, paper in this proceeding.
4. C. I. Westbrook *et al.*, Phys. Rev. Lett. **58**, 1328 (1987).
5. M. L. Perl, SLAC–PUB–4632 (1988).
6. K. G. Hayes and M. L. Perl, Phys. Rev. (to be published); also issued as SLAC–PUB–4471.
7. R. Stroynowski, CALT–68–1431 (1988).
8. J. Kirkby, CERN–EP/87–210 (1987).
9. J. M. Jowett, CERN LEP–TH/87–56 (1987).
10. See, for example, E. D. Bloom, Proc. 8th Int. Conf. on High Energy Physics (Nashville, 1987); also issued as SLAC–PUB–4604(1988).

11. K. K. Gan and M. L. Perl, Int. J. Mod. Phys. **A3**, 531 (1988).
12. Figure 7 is adapted from S. L. Wu, Phys. Rep. **107**, 60 (1984); S. L. Wu in Proc. 1987 Int. Sym. Lepton and Photon Interactions at High Energies (Hamburg, 1987).
13. See, for example, papers in Proc. 1987 ICFA Seminar on Future Perspectives in High Energy Physics, BNL52114 (1987).
14. J. M. Dorfan in New Frontiers in Particle Physics, (World Scientific, Singapore, 1986), J. M. Cameron, B. A. Campbell, A. N. Kamal, and F. C. Khanna, eds.
15. G. Barbiellini *et al.*, in Physics at LEP, Vol. 2. (CERN 86–02, 1986), J. Ellis and R. Peccei, eds.
16. C. Ahn *et al.*, SLAC–329 (1988).
17. Proc. Workshop on Physics at Future Accelerators, CERN 87–07 (1987).
18. C. A. Hawkins and M. L. Perl, to be published.

3rd CONFERENCE ON THE INTERSECTIONS BETWEEN PARTICLE AND NUCLEAR PHYSICS

Rockport, Maine • • May 14-19, 1988

Plenary Program

Knox County Ball Room

Saturday, May 14

	Chairman, V. Hughes, Yale
8:15 a.m.	V. Hughes, Yale • Welcome
8:20 a.m.	E. Vogt, TRIUMF • Introduction
8:30 a.m.	S. Errede, University of Illinois Collider Physics at Tevatron
9:15 a.m.	K. Hicks, TRIUMF Spin-Isospin Response Functions
10:00 a.m.	Coffee
	Chairman, A. Krisch, Michigan
10:30 a.m.	J.-C. Peng, Los Alamos Associated Production of Hypernuclei
11:15 a.m.	D. Marlow, Princeton K-decay Physics
12:00 noon	Lunch
	Chairman, E. Vogt, TRIUMF
8:30 p.m.	H. Bethe, Cornell Dynamics of Supernova and Collapse Explosion
9:20 p.m.	G. Brown, Stony Brook Nuclear Equation of State at High Density
10:10 p.m.	End

Sunday, May 15

	Chairman, G. Farrar, Rutgers
8:30 a.m	J. Donoghue, University of Massachusetts Structure of the Proton-Chiral Perturbation Theory
9:15 a.m.	N. Isgur, University of Toronto Critical Assessment of Models for Strong Interactions
10:00 a.m.	Coffee
	Chairman, F. Mills, Fermilab
10:30 a.m.	M. Perl, SLAC Electron-Positron Collider Physics
11:15 a.m.	R. Kirshner, Harvard-Smithsonian All Aspects of Supernova 1987a
12:00 noon	Lunch

Monday, May 16

	Chairman, M. Macfarlane, Indiana
8:30 a.m.	K. Rith, Heidelberg Structure Functions in Nucleons and Nuclei
9:15 a.m.	G. Gabrielse, Harvard Trapped Antiprotons
10:00 a.m.	Coffee
	Chairman, J. McCarthy, Virginia
10:30 a.m.	W. Zajc, Columbia Recent Results from AGS & CERN on R.H.I. Collisions

11:15 a.m.	L. McLerran, Fermilab A Theoretical Overview of Relativistic Heavy-Ion Collisions
12:00 noon	End

••

Tuesday, May 17

	Chairman, E. Moniz, MIT Bates
8:30 a.m.	R. Lourie, MIT Nuclear Response in the Quasi-Elastic Region
9:15 a.m.	K. Schubert, Univerity of Heidelberg B Decay Results from Argus
10:00 a.m.	Coffee
	Chairman, D. Geesaman, Argonne
10:30 a.m	G. Charpak, CERN New Development in Detectors
11:15 a.m.	M. Marshak, University of Minnesota An Overview of Non-Accelerator Physics
12:00 noon	Lunch

••

Wednesday, May 18

	Chairman, T. O'Halloran, Illinois
8:30 a.m.	J. Rutherfoord, University of Washington Aspects of Hard Interactions in Nuclei
9:15 a.m.	G. Fidecaro, CERN High Energy Hadron Scattering
10:00 a.m.	Coffee
	Chairman, R. Mischke, Los Alamos
10:30 a.m.	Accelerator Physics Summary - C. Leemann
10:50 a.m.	Electron and Muon Physics Summary - H. Montgomery
11:10 a.m.	Neutrino Physics Summary - D.H. White
11:30 a.m.	Relativistic Heavy-Ion Physics Summary - J. Harris
11:50 a.m.	Hadron Spectroscopy Summary - D. Hitlin
12:10 p.m.	Lunch
9:00 p.m	R. Serber, Columbia - It Was Easier Then

••

Thursday, May 19

	Chairman, W. van Oers, TRIUMF
8:30 a.m.	Non-Accelerator & Astrophysics Summary - S. Kahana
8:50 a.m.	Spin Physics Summary - T. Roser
9:10 a.m.	Hadron Scattering Summary - S. Heppelmann
9:30 a.m.	Antiproton Physics Summary - R. Cester
9:50 a.m.	Meson & Lepton Decays Summary - M. Zeller
10:10 a.m.	Coffee
	Chairman, L. Schroeder, LBL
10:40 a.m	W. Marciano, Brookhaven Neutrino Masses, Mixing and Matter
11:25 a.m.	L. Teng, Fermilab Spin Dynamics in Accelerators & Storage Rings
12:10 p.m.	R. Jaffe, MIT Summary Remarks
12:50 p.m.	End

Third Conference on the Intersections Between
Particle and Nuclear Physics
Rockport, Maine, U.S.A.
May 14-19, 1988

PARALLEL SESSIONS

Saturday, May 14, 1988

2:30 - 3:40 p.m.	Room	4:10 - 5:50 p.m.	Room
Neutrino	Ebb Tide	Accelerator	Ebb Tide
Non-Accelerator	Rockport	Non-Accelerator	Rockport
Spin	Rockland	Spin	Rockland
Hadron Spectroscopy	Camden	Hadron Spectroscopy	Camden
Relativistic Heavy Ion	Penobscot	Relativistic Heavy Ion	Penobscot

Sunday, May 15

1:30 - 3:10 p.m.	Room	3:40 - 5:20 p.m.	Room
Electron and Muon	Rockland	Electron and Muon	Rockland
Neutrino	Ebb Tide	Accelerator	Ebb Tide
Hadron Scattering	Camden	Hadron Scattering	Camden
Antiprotons	Penobscot	Relativistic Heavy Ion	Penobscot
Meson & Lepton Decays	Rockport	Meson & Lepton Decays	Rockport

Tuesday, May 17

3:00 - 4:40 p.m.	Room	5:10 - 6:50 p.m.	Room
Accelerator	Ebb Tide	Neutrino	Ebb Tide
Electron and Muon	Rockland	Electron and Muon	Rockland
Non-Accelerator	Rockport	Hadron Spectroscopy	Camden
Relativistic Heavy Ion	Penobscot	Relativistic Heavy Ion	Penobscot
Antiprotons	Camden	Antiprotons	Rockport

Wednesday, May 18

2:30 - 4:10 p.m.	Room	4:40 - 6:10 p.m.	Room
Hadron Spectroscopy	Camden	Hadron Scattering	Camden
Spin	Penobscot	Spin	Penobscot
Hadron Scattering	Ebb Tide	Hadron Scattering	Ebb Tide
Antiprotons	Rockland	Antiprotons	Rockland
Meson & Lepton Decays	Rockport	Meson & Lepton Decays	Rockport

ACCELERATOR PHYSICS

Session I Saturday, May 14, 4:10-5:25 p.m. — L. Teng, Chairman

4:10 R. Rossmanith, CEBAF
Polarized Antiprotons with the Spin Splitter
4:35 H. Thiessen, Los Alamos
Planning for an Advanced Hadron Facility at Los Alamos
5:00 A. Ruggiero, Brookhaven
Current Issues in the Beam Dynamics of Heavy Ion and Hadron Colliders

Session II Sunday, May 15, 3:40-5:10 p.m. — R. Rossmanith, Chairman

3:40 J. Flanz, MIT Bates
MIT Bates South Hall Ring
4:10 D. Peaslee, Maryland
Potential Low Energy $\bar{p}$ Source at Brookhaven National Laboratory
4:40 E. Colton, Los Alamos
The Present Status of the Los Alamos 800 MeV Proton Storage Ring

Session III Tuesday, May 17, 3:00-4:00 p.m. — A. Chao, Chairman

3:00 J. Bisognano, CEBAF
Beam Breakup in Superconducting Recirculating Linacs
3:30 D. Rice, Cornell
Issues, Options and Performance for Storage Ring Based $B\bar{B}$ Factories
4:25 R. Pollock, Indiana
Update on the Indiana University Cooler and Siberian Snake Test

Wednesday, May 18

10:30 C. Leemann, CEBAF
Accelerator Physics Summary

ANTIPROTON PHYSICS

Session I Sunday, May 15, 1:30-3:10 p.m. — P. Barnes, Chairman

1:30 J. Peoples, Fermilab
Charmonium Spectroscopy with $\bar{P}$ at FNAL
1:55 B. Quinn, Carnegie-Mellon
Status of the Search for $\xi(2230)$
2:20 D. Hertzog, Illinois
Physics with a Jet Target at LEAR
2:35 J. Donoghue, Univ. of Massachusetts
CP Violation Amplitudes in $K\bar{K}$ and $\Lambda\bar{\Lambda}$ Systems

Session II Tuesday, May 17, 3:00-4:40 p.m. P. Barnes, Chairman

3:00 J. Richard, Grenoble
$\bar{P}$ Nucleon and Nucleus Interactions
3:25 W. Haeberli, Wisconsin
Physics with a Polarized Hydrogen Internal Target at LEAR
3:50 G. Franklin, Carnegie-Mellon
$\Lambda\bar{\Lambda}$ Production: Experiment and Theory
4:15 T. Bressani, Torino
Anti-Neutron Physics
4:35 J. de Swart, Nijmegen
A Study of the Reaction: $P + \bar{P} \rightarrow Y + \bar{Y}$

Session III Tuesday, May 17, 5:10-6:50 p.m. P. Barnes, Chairman

5:10 C. Dover, Brookhaven
Nucleon-Antinucleon Annihilation Dynamics
5:35 J. Cugnon, Liegé
Antiproton Annihilation in Nuclei
5:50 E. Klempt, Mainz
The Origin of Dynamical Selection Rules
6:05 J. Rafelski, Arizona
Super Cooled Quark Gluon Plasma in $\bar{P}$ Annihilation
6:20 C. Guaraldo, Frascati
Leading Meson Effects in Antiproton Neon Interactions
6:36 H. Genz, Karlsruhe
The Quark Line Rule in $P\bar{P}$ Annihilation

Session IV Wednesday, May 18, 2:30-4:10 p.m. R. Cester, Chairman

2:30 G. Backenstoss, Basel
Search for CP Violation in the $K\bar{K}$ System at LEAR
2:55 N. Hamann, Freiberg
Search for CP Violation in Hyperon-Antihyperon Decays with $P\bar{P}$ Experiments
3:20 H. Koch, Karlsruhe
The LEAR Crystal Barrel Detector
3:35 A. Stephens, New Mexico State University
Why Look for Antiprotons in Cosmic Rays?
4:00 G. Mutchler, Rice
Measurement of the Imaginary Part of the I=1 $N\bar{N}$ S-Wave Scattering Length

Session V Wednesday, May 18, 4:40-6:10 p.m. R. Cester, Chairman

4:40 E. Klempt, Mainz
Static Properties of Antiprotons
5:05 M. Hynes, Los Alamos
Experiments with Thermal Antiproton Beams
5:30 B. Mitchel, Western Ontario
Anti-hydrogen Production Schemes
5:55 J. deSwart, Nijmegen
Low Energy Nucleon-Antinucleon Interaction

Thursday, May 19

9:30 R. Cester, Torino
Antiproton Physics Summary

ELECTRON AND MUON PHYSICS

Session I Sunday, May 15, 1:30-3:10 p.m. H. Montgomery, Chairman

1:30 V. Papavassiliou, Yale
EMC Polarized Target Experiment
2:00 C. Papanicolas, Illinois
$(e,e'\gamma)$ Including N-Δ
2:30 P. Bosted, American University
Transverse Electron Deuterium Scattering at High Q
2:50 J. Napolitano, Argonne
$D(\gamma,p)n$ NPAS NE-8 Results

Session II Sunday, May 15, 3:40-5:20 p.m. W. Donnelly, Chairman

3:40 J. McCarthy, Virginia
Overview of Electromagnetic Few Body Studies
4:10 R. Dymarz, Alberta
Deuteron Form Factors
4:30 R. McKeown, Caltech
Y-Scaling
5:00 S. Kumano, Tennessee
Quark Model of a Nuclear Pair

Session III Tuesday, May 17, 3:00-4:40 p.m. H. Montgomery, Chairman

3:00 P. Renton, Oxford
EMC Structure Functions/Hadrons
3:30 P. Bosted, American University
SLAC E140 Results
4:00 P. McGaughey, Los Alamos
E772, Drell-Yan Structure Functions
4:20 E. Hadjimichael, Fairfield
Primakoff Effect in Electron Scattering

Session IV Tuesday, May 17, 5:10-6:50 p.m. W. Donnelly, Chairman

5:10 K. Dow, MIT
^{3}He-^{3}H Quasi-Elastic
5:35 J. Koch, NIKHEF
Pion Production
6:00 E. Biese, MIT
$^{16}O(\gamma,n)^{15}O$ at 250 MeV
6:20 G. van der Steenhoven, MIT/NIKHEF
Sigma L/Sigma T for $^{12}C(e,e'p)$
6:35 P. Souder, Syracuse
Parity Violating Electron Scattering

Wednesday, May 18

10:50 H. Montgomery, Fermilab
Electron and Muon Physics Summary

HADRON SCATTERING

Session I Sunday, May 15, 1:30-3:10 p.m. S. Heppelmann, Chairman

1:30 A. Carroll, Brookhaven
Probing QCD and Probing Nuclei with Large Angle Elastic Scattering

2:00 G. De Terramond, SLAC
Color Transparency, Spin Correlations, Heavy Quark Thresholds in pp Scattering

2:30 G. Farrar, Rutgers
Nuclear Transparency and Fermi Motion

2:55 C. Naudet, LBL
Threshold Behavior of Direct Electron Production in p-Be Collisions at the Bevalac Inelastic p-Nucleus Collisions

Session II Sunday, May 15, 3:40-5:20 p.m. O. Häusser, Chairman

3:40 H. Lipkin, Weizmann
Some Implications of a Possible NNπ Bound State

3:55 H. Baer, Los Alamos
Nucleon-Nucleon Correlations Studied by Pion Double Charge Exchange

3:20 H. Matthaey, Karlsruhe
Experimental Determination of the Sigma Term in Pion-Nucleon Scattering

3:45 H. Weyer, Basel
Pion Absorption Models in Light Nuclei

4:05 S. Hoibraten, MIT
Coincidence Measurements of Pion Single Charge Exchange in the Delta Resonance Region

Session III Wednesday, May 18, 2:30-4:10 p.m. S. Heppelmann, Chairman

2:30 T. Marshall
J/ψ Production in π-Nucleus Collisions

2:50 A. Zieminski, Indiana
Production of High PT Jets in Proton Nucleus Collisions

3:00 A. Sambamurti, Indiana
Nuclear Dependence of Event Structure in Highly Inelastic Proton Nucleus Collisions

3:10 P. Blunden, TRIUMF
Relativistic Hartree-Fock Calculations

3:30 W. Toothaker, Pennsylvania State
Characteristics of Particle Production in Hadron-Nucleus Interactions at 100 GeV/c

3:50 S. Frankel, Pennsylvania
Fact and Fancy in Relativistic Heavy Ion Collisions

Session IV Wednesday, May 18, 4:40-6:10 p.m. O. Häusser, Chairman

4:40 M. Vetterli, Simon Fraser
Nuclear Response in Intermediate Energy N-Nucleus Scattering
5:05 K. Nakai, KEK
Quasifree Hadron Production in 4 GeV/c Nuclear Reactions
5:30 T. Hennino, Orsay
Light-Ion Induced Charge Exchange Reactions at Saturne
5:55 G. Igo, UCLA
Elastic Scattering of Vector and Tensor Polarized Deuterons from Polarized Hydrogen at 1600 MeV

Session V Wednesday, May 18, 4:40-6:10 p.m. S. Heppelmann, Chairman

4:40 E. McIntyre, Boston
Radiative Kaon Capture
5:00 H. Fearing, TRIUMF
Few Body Kaon Reactions
5:15 H. Zhang, Virginia Polytechnic
Meson-Baryon Perturbative QCD Calculations
5:35 M. Divadeenam, Brookhaven
Intranuclear Cascade Study of Hadron-Nucleus Interactions at 1 GeV/c
5:55 C.-R. Ji, North Carolina State
QCD and Phenomenology of Nuclear and Hadronic Processes at Intermediate Energies

Thursday, May 19

9:10 S. Heppelmann, Pennsylvania State
Hadron Scattering Summary

HADRON SPECTROSCOPY

Session I Saturday, May 14, 2:00-3:30 p.m. D. Geesman, Chairman

2:00 A. Bodmer, Illinois
Hypernuclear Interactions and the Binding Energies of λ and λ-λ Hypernuclei
2:30 P. Pile, Brookhaven
Search for ε-Nucleus Bound States
3:00 N.C. Mukhopadhyay, Rensselaer Polytechnic
Threshold ε Photoproduction and the Role of the S11 Resonance
3:15 M. Oka, Pennsylvania
Flavor-Octet Dibaryons in the Quark Model

Session II Saturday, May 14, 4:10-5:50 p.m. D. Hitlin, Chairman

4:10 B. Nefkens, UCLA
Precision Tests of the Standard Model Using Rare Decays of the η, η', etc.

4:25 M. Karliner, SLAC
The EMC Structure Functions: Why the Proton Spin is Not Due to Quarks
4:50 D. Alde, Los Alamos
Results from GAMS
5:00 W. Dunwoodie, SLAC
Strange Quark Spectroscopy from the LASS Spectrometer
5:25 T. Barker, U.C. Santa Barbara
Formation of $f_1(1285)$ and $f_1(1420)$ in Photon-Photon Fusion
5:40 S. Blessing, Indiana
Partial Wave Analysis of the $K^+K^0\pi^-$ System

Session III Tuesday, May 17, 5:10-6:50 p.m. D. Hitlin, Chairman

5:10 F. Close, Tennessee
Parton Distributions in Nuclei: Quagma and Quagmire?
5:40 S. Lindenbaum, Brookhaven
Investigation of Resonance States in the Glueball Enhanced Channel $\pi^-p \to \phi\phi n$
5:55 W. Wisniewski, Caltech
Glueball Candidates in J/ψ Decays
6:20 S. Meshkov, Caltech
Theoretical Overview of Glueballs and Hadron Spectroscopy

Session IV Wednesday, May 18, 2:30-4:05 p.m. E. Vogt, Chairman

2:30 C. Morris, Los Alamos
Search for πNN Bound Systems
3:00 I. Towner, AECL
Magnetic Transitions
3:25 C. Glashausser, Rutgers
Understanding the Nuclear Spin Response to Intermediate Energy Protons
3:50 M. Divadeenam, Brookhaven
Systematics of the $\sigma^\pm$ Hypernuclear Single Particle States

Wednesday, May 18

11:50 D. Hitlin, Caltech
Hadron Spectroscopy Summary

MESON AND LEPTON DECAYS

Session I Sunday, May 15, 1:30-3:10 p.m. M. Zeller, Chairman

1:30 R. Shrock, SUNY
Theory of Rare K and Muon Decays
2:00 R. Engfer, Paul Scherrer Inst. (SIN)
Studies of Pion and Muon Decays at PSI
2:30 C. Waltham, British Columbia
Decays of π^0 into Electron-Positron Pairs
2:40 D. Bryman, TRIUMF
A Review of Searches for Muon to Electron Conversion

3:00 A.J. Noble, University of British Columbia
Radiative Kaon Capture and Σ and Λ Weak Radiative Decays

Session II Sunday, May 15, 3:40-5:20 p.m. N. Reay, Chairman

3:40 W. Morse, Brookhaven
Status of a Search for $K \rightarrow \mu\, e$ (BNL E780)

4:00 R. Whyley, William and Mary
Status of a Search for $K \rightarrow \mu\, e$ (BNL E791)

4:20 M. Cooper, Los Alamos
Searches for $\mu \rightarrow e\gamma$ and an Update on MEGA

4:40 D. Britton, TRIUMF
A Study of the Decay $\pi^+ \rightarrow e^+$ Neutrino

4:50 W. Fetscher, IMP/ETH Zurich
τ Decays in a Future B Factor as a Test of the Standard Model

5:10 R. Akhoury, Michigan
Mesons and Glueball Decays

Session III Wednesday, May 18, 2:30-4:10 p.m. M. Zeller, Chairman

2:30 R. Strand, Brookhaven
Status of a Search for the Decay $K^+ \rightarrow \pi^+$ XX, Where XX Does Not Interact in the Detector (BNL E787)

2:50 D. Lazarus, Brookhaven
Status of a Search for the Decay $K \rightarrow \pi^+\mu^+e^-$ (BNL E777)

3:10 K. Sliwa, Fermilab Results on Weak and Strong Decays of Charmed Particles (FNAL E691)

3:30 W. Marciano, Brookhaven
Charged Current Unitarity Tests

3:50 G. Gollin, Princeton,
The Experimental Status of CP Violations

Session IV Wednesday, May 18, 4:40-6:10 p.m. N. Reay, Chairman

4:40 A. Sanda, Rockefeller
Theoretical Status and Outlook for CP Plysics

5:10 R. Schaefer, Yale
Overview and Recent Results of a Search for Muonium to Anti-muonium Conversion

5:30 G. Marshall, TRIUMF
Recent Results of a Search for Muonium to Anti-muonium Conversion

5:45 D. Macfarlane, McGill
Results of B decay Studies from ARGUS

Thursday, May 19

9:50 M. Zeller, Yale
Meson and Lepton Decays Summary

NEUTRINO PHYSICS

Session I Saturday, May 14, 2:00-3:40 p.m. W. Haxton, Chairman

2:00 X.Q. Lu, U.C. Irvine
$\nu_e + {}^{12}C$ Scattering
2:25 R. Manweiler, Valparaiso
$\nu_\mu + {}^{12}C$ Scattering
2:50 A. Balantekin, Wisconsin-Madison
Analytic Treatments of Matter Oscillations
3:15 A. Williams, Univ. of Washington
The Axial-Charge Operator in Double Beta Decay
3:20 G. Stephenson, Jr., Los Alamos
Exchange Currents and Double Beta Decay

Session II Sunday, May 15, 1:30-3:10 p.m. W. Haxton, Chairman

1:30 B. Kayser, LBL
Massive Neutrinos
1:55 J. Wilkerson, Los Alamos
Tritium Beta Decay
2:20 A. Hahn, U.C. Irvine
The 2ν Double Beta Decay of ${}^{82}Se$
2:45 F. Avignone, South Carolina
Neutrinoless Double Beta Decay

Session III Tuesday, May 17, 5:10-6:50 p.m. D.H. White, Chairman

5:10 S. Durkin, Ohio State
$\nu_\mu \rightarrow \nu_e$ Oscillation Limit (Los Alamos)
5:35 R. Seto, Columbia
Recent Results on Neutrino Oscillations (BNL/E776)
6:00 W. Frati, Pennsylvania
Neutrino Flux Measurements at Kamioka
6:25 M. Diwan, Brown
ν_μ-e Scattering

Wednesday, May 18

11:10 D.H. White, Los Alamos
Neutrino Physics Summary

NON-ACCELERATOR AND ASTROPHYSICS

Session I Saturday, May 14, 2:00-3:30 p.m. E. Beier, Chairman

2:00 P. Boynton, Washington
Experimental Search for a Fifth Fundamental Force
2:30 B. McKellar, Melbourne
The Neutron Electric Dipole Moment
3:00 D. Groom, LBL
Where Could the GUT Monopole be Hiding?

Session II Saturday, May 14, 4:10-5:40 p.m. S. Kahana, Chairman

4:10 E. Baron, SUNY
How Do Supernovae Explode?
4:40 E. Beier, Pennsylvania
Detection of Supernova Neutrinos
5:10 A. Masaike, Kyoto
Ultra-High Energy γ-ray from Supernova 1987a

Session III Tuesday, May 17, 3:00-4:30 p.m. S. Kahana, Chairman

3:00 C. Pennypacker, LBL
Recent Results and Future Prospects for the Berkeley Automated Supernova Search
3:30 E. Loh, Utah
A Study of Ultra High Energy Cosmic Rays
4:00 J. Applegate, Columbia
Phase Transitions in the Early Universe and Primordial Nucleosynthesis

Thursday, May 19

8:30 S. Kahana, Brookhaven
Non-Accelerator and Astrophysics Summary

RELATIVISTIC HEAVY ION PHYSICS

Session I Saturday, May 14, 2:00-3:40 p.m. J. Harris, Chairman

2:00 V. Ruuskanen, Univ. of Jyvaskyla
Probes of Collective Behavior of Dense Hadronic Matter
2:25 T. Humanic, GSI Darmstadt
Hanbury-Brown-Twiss Results from NA35
2:50 S. Pratt, Oregon State
Hanbury-Brown-Twiss Effects in Relativistic Heavy Ion Collisions
3:15 B. Jacak, CERN
Particle Production Results from NA34

Session II Saturday, May 14, 4:10-5:50 p.m. T. Matsui, Chairman

4:10 G. Landaud, Clermont-Ferrand
J/ψ Production Results from NA38 with O and S Beams at 200 GeV/n
4:35 M. Chu, MIT
J/ψ Suppression in Ultra-Relativistic Heavy Ion Collisions
5:00 A. Franz, LBL
γ, π^0 and Transverse Energy Results from WA80
5:25 J. Polonyi, MIT
Quasiparticles in the Deconfined Phase

Session III Sunday, May 15, 3:40-5:20 p.m. T. Matsui, Chairman

3:40 W. Koenig, GSI Darmstadt
Evidence for Monoenergetic e^+e^- Decay in Heavy Ion Collisions and Search for Corresponding Resonances in Bhabha Scattering

4:05 K. Danzmann, Stanford
A New Light Boson from Nuclear Reactions?

4:30 A. Chodos, Yale
The e^+e^- Peaks: Theoretical Perspective

Session IV Tuesday, May 17, 3:30-4:40 p.m. J. Harris, Chairman

3:00 K. Werner, Brookhaven
String Models for Multiparticle Production in Nuclear Collisions

3:20 R. Thews, Arizona
Some Theoretical Aspects of the J/ψ Suppression

3:45 J. Stachel, SUNY
Transverse Energy Results from E814

4:05 R. Ledoux, MIT
Results from E802

4:25 J. Rafelski, Arizona
Strangeness Production and the Quark Gluon Plasma

Session V Tuesday, May 17, 5:10-6:50 p.m. T. Matsui, Chairman

5:10 H. Ritter, LBL
Flow, Transverse Energy and Entrophy at the Bevalac-Nuclear Matter Equation of State

5:30 C. Gale, AECL
Dilepton Production at Bevalac and AGS Energies

5:50 G. Igo, UCLA
Subthreshold Kaon and Antiproton Production at the Bevalac

6:20 R.J. Wilkes, Univ. of Washington
Rapidity Distributions and Correlations

6:35 S. Lindenbaum, Brookhaven
Quark-Gluon Plasma Model Event Simulation

Wednesday, May 18

11:30 J. Harris, LBL
Relativistic Heavy Ion Physics Summary

SPIN PHYSICS

Session I Saturday, May 14, 2:00-3:40 p.m. T. Roser, Chairman

2:00 M. Simonius, ETH Zürich
Parity Violation in Proton-Proton Scattering

2:25 V. Zeps, Univ. of Washington
Parity Mixing of the 0^+, 0^-, T=1 Doublet in ^{14}N

2:50 J. Birchall, Manitoba/TRIUMF
Parity Violation in Proton-Proton Scattering at 230 MeV

3:05 R. Henneck, Basel
Detailed Study of the T=0, NN Interaction via np-Experiments at ~ 70 MeV

3:20 D. Bowman, Los Alamos
P and T Violation in Neutron Resonance Reactions

Session II Saturday, May 14, 4:10-5:55 p.m. T. Roser, Chairman

4:10 D. Underwood, Argonne
Polarized Protons at FNAL

4:35 M. Nessi, Rice
Assymmetry Measurements in Inclusive π° Production at High x_F with a 180 GeV/c Polarized Proton Beam

4:50 L. Ratner, Brookhaven
Recent Experience with the Acceleration of Polarized Beam at the AGS

5:15 T. Chupp, Harvard
A Polarized ^3He Target for Electron Scattering Experiments

5:30 R. Raymond, Michigan
An Ultracold Jet of Polarized Atomic Hydrogen

5:45 R. Milner, Caltech
Measurement of the Deep Inelastic Spin Dependent Structure Function of Proton and Neutron at HERA

Session III Wednesday, May 18, 2:30-4:10 p.m. J. Moss, Chairman

2:30 C. Horowitz, Indiana
Continuum Polarization Observables and Relativistic Models

2:55 K. Jones, Los Alamos
Excitation of Magnetic Resonances in Inelastic Proton Scattering

3:20 B. Spicer, Melbourne
Isovector Spin Resonance in ^{90}Y from the $^{90}Zr(n,p)^{90}$ Reaction at 200 MeV

3:35 B. Nefkens, UCLA
Spin Rotation Parameters A and R in $\pi^\pm p$ Elastic Scattering at the Onset of the Roper Resonance

3:50 G. Smith, TRIUMF
Tensor Analyzing Powers T_{20} and T_{21} of π^+d Elastic Scattering

Session IV Wednesday, May 18, 4:40-6:10 p.m. J. Moss, Chairman

4:40 D. Crabb, Michigan
Spin Effects in Proton-Proton Elastic Scattering Measurements at the AGS

5:05 M. May, BNL
Muon (g-2) measurment at Brookhaven

5:20 A. Masaike, Kyoto
Studies of Particle and Nuclear Interactions Using Spin as a Tool at KEK

5:45 F. Nessi-Tedaldi, Rice
Spin Parameter Measurements in Hyperon Production

6:10 G. Burleson, New Mexico State
Measurement of C_{LL}, C_{SL}, and C_{SS} in np Elastic Scattering at 484, 634 and 784 MeV

Thursday, May 19

8:50 T. Roser, Michigan
Spin Physics Summary

CONFERENCE PARTICIPANTS

Robert Adair	Brookhaven National Laboratory
David L. Adams	University of Texas-Austin
Ralf Adelseck	Ohio University
Michelangelo Agnello	Lab. Nazionali di Frascati
S. Ahmad	TRIUMF
Ratindranath Akhoury	University of Michigan
Douglas M. Alde	Los Alamos National Laboratory
James Applegate	Columbia University
Richard A. Arndt	Virginia Polytechnic Institute
Frank T. Avignone	University of South Carolina
Andrew Bacher	Indiana University
G. K. Backenstoss	Basel University
Helmut Baer	Los Alamos National Laboratory
Baha Balantekin	University of Wisconsin-Madison
Tony Barker	University of California-Santa Barbara
Peter D. Barnes	Carnegie-Mellon University
Edward Baron	State University of New York-Stony Brook
D. Beck	California Institute of Technology
Eugene W. Beier	University of Pennsylvania
Elizabeth Beise	Massachusetts Institute of Technology
Aron M. Bernstein	Massachusetts Institute of Technology
Hans Bethe	Cornell University
James Birchall	University of Manitoba/TRIUMF
Joseph J. Bisognano	CEBAF
Ewart W. Blackmore	TRIUMF
Susan K. Blessing	Indiana University
Matthew A. Bloomer	Massachusetts Institute of Technology
Peter G. Blunden	TRIUMF
Arnold R. Bodmer	University of Illinois at Chicago
Werner Boeglin	Massachusetts Institute of Technology
Peter Bond	Brookhaven National Laboratory
Bernardino Bosco	Universita di Firenze
Peter Bosted	SLAC/American University
J. David Bowman	Los Alamos National Laboratory
Paul E. Boynton	University of Washington
James Bradbury	Los Alamos National Laboratory
Tullio Bressani	Universita di Torino
David I. Britton	University of Victoria/TRIUMF
Gerlad E. Brown	State University of New York-Stony Brook
Stanley G. Brown	Physical Review Letters
Douglas A. Bryman	TRIUMF
Gerry Bunce	Brookhaven National Laboratory
G. Burleson	New Mexico State University
J. Cameron	Indiana University
Alan S. Carroll	Brookhaven National Laboratory
Rosanna Cester	Universita di Torino
Alexander W. Chao	SSC/URA
Georges Charpak	CERN

Donald F. Geesaman	Argonne National Laboratory
Henning Genz	Universitat Karlsruhe
Charles M. Glashausser	Rutgers University
George D. Gollin	Princeton University
Gordon L. Greeniaus	University of Alberta
Donald E. Groom	Lawrence Berkeley Laboratory
Frank L. Gross	CEBAF/College of William and Mary
Carlo Guaraldo	INFN Frascati
E. Hadjimichael	Fairfield University
Willy Haeberli	University of Wisconsin-Madison
Donald Hagerman	Los Alamos National Laboratory
Alan A. Hahn	University of California-Irvine
Nikolaus H. Hamann	CERN
John W. Harris	Lawrence Berkeley Laboratory
Dietrich Hartwig	GSI
Otto Hausser	TRIUMF/SFU
Wick C. Haxton	University of Washington
David L. Hendrie	United States Department of Energy
R. Henneck	Universitat Basel
Thierry Hennino	Institute de Physique Nucléaire
Steven F. Heppelmann	Pennsylvania State University
David W. Hertzog	University of Illinois
Kenneth H. Hicks	TRIUMF
Virgil L. Highland	Temple University
David L. Hill	Valutron N.V.
David Hitlin	California Institute of Technology
Steiner Høbraten	Massachusetts Institute of Technology
Harry D. Holmgren	University of Maryland
Charles J. Horowitz	Indiana University
E. Barrie Hughes	Stanford University
Vernon Hughes	Yale University
Thomas J. Humanic	CERN
Michael V. Hynes	Los Alamos National Laboratory
George J. Igo	University of California-Los Angeles
C. H.Q. Ingram	PSI (SIN)
Donald Isenhower	Abilene Christian University
N. Isgur	CERN/University of Toronto
Milton Isidro-Filho	Massachusetts Institute of Technology
Barbara Jacak	CERN
K. Peter Jackson	TRIUMF
Richard J. Jacob	Arizona State University
Robert Jaffee	Massachusetts Institute of Technology
Chueng-Ryong Ji	North Carolina State University
Kevin J. Jones	Los Alamos National Laboratory
Sidney H. Kahana	Brookhaven National Laboratory
Marek Karliner	Stanford Linear Accelerator Center
Boris Kayser	Lawrence Berkeley Laboratory
J. D. King	University of Toronto
Nicholas S.P. King	Los Alamos National Laboratory

Robert Kirshner	Harvard-Smithsonian Center
Andi Klein	Los Alamos National Laboratory
E. Klempt	Johannes Guttenberg Universitat
Helmut Koch	Universitat Karlsruhe KPK
Justus H. Koch	NIKEHF-K
Wolfgang Koenig	GSI Darmstadt
Stanley Kowalski	Massachusetts Institute of Technology
Alan D. Krisch	University of Michigan
Shunzo Kumano	University of Tennessee
Ray Kunselman	University of Wyoming
Gary Kyle	New Mexico State University
Gilbert A. Landaud	University of Clermont-Ferrand
L. Dean Larson	University of New Mexico
Donald M. Lazarus	Brookhaven National Laboratory
Robert J. Ledoux	Massachusetts Institute of Technology
Professor W.-Y. Lee	Columbia University
Christopher W. Leemann	CEBAF
Michael Leitch	Los Alamos National Laboratory
Catherine LeLuc	University of Geneva
S. J. Lindenbaum	Brookhaven National Laboratory
Harry J. Lipkin	Weizmann Institute of Science
Eugene C. Loh	University of Utah
Robert W. Lourie	Massachusetts Institute of Technology
X-Q. Lu	University of California-Irvine
David B. Macfarlane	McGill University
Malcolm H. Macfarlane	Indiana University
Angelo Maggiora	Universita di Torino
Yousef I. Makdisi	Brookhaven National Laboratory
Kim Maltman	York University
Robert W. Manweiler	Valparaiso University
William J. Marciano	Brookhaven National Laboratory
Daniel Marlow	Princeton University
Marvin Marshak	University of Minnesota
Glen Marshall	University of Victoria/TRIUMF
Tom R. Marshall	Indiana University
Akira Masaike	Kyoto University
Alberto Masoni	INFN Torino
E. Mathie	University of Regina
Tetsuo Matsui	Massachusetts Institute of Technology
Helmut F. Matthaey	Universitat Karlsruhe
Oren V. Maxwell	Florida International University
Morgan May	Brookhaven National Laboratory
James S. McCarthy	University of Virginia
Patrick L. McGaughey	Los Alamos National Laboratory
Edward K. McIntyre	Boston University
Bruce H.J. McKellar	University of Melbourne
Robert D. McKeown	California Institute of Technology
Larry McLerran	Fermilab
Sydney Meshkov	California Institute of Technology
James Miller	Boston University

Frederick E. Mills	Fermilab
Deborah Mills-Errede	University of Wisconsin
Richard G. Milner	California Institute of Technology
R. Mischke	Los Alamos National Laboratory
J. Brian A. Mitchell	University of Western Ontario
E. Moniz	MIT/Bates Linear Accelerator
Hugh E. Montgomery	Fermilab
Christopher Morris	Los Alamos National Laboratory
William Morse	Brookhaven National Laboratory
Joel M. Moss	Los Alamos National Laboratory
Nimai C. Mukhopadhyay	Rensselaer Polytechnical Institute
Gordon S. Mutchler	Rice University
Kozi Nakai	KEK
James Napolitano	Argonne National Laboratory
Charles J. Naudet	Lawrence Berkeley Laboratory
Bernard M. K. Nefkens	University of California-Los Angeles
Marzio Nessi	Rice University
Francesca Nessi-Tedaldi	Rice University
Anthony J. Noble	University of British Columbia
Blaine E. Norum	University of Virginia
Hiroshi Noya	Hosei University
T. O'Halloran	University of Illinois
Makoto Oka	University of Pennsylvania
M. Oleszczuk	Universitat Regensburg
Arthur Olin	TRIUMF/University of Victoria
Catherine Olmer	Indiana University
H. Ottewitte	Idaho National Engineering Laboratory
Shelley A. Page	University of Manitoba
Costas Papanicolas	University of Illinois
Vassili Papavassiliou	Yale University
Zohreh Parsa	Brookhaven National Laboratory
David C. Peaslee	University of Maryland
Jen-Chieh Peng	Los Alamos National Laboratory
Carl Pennypacker	Lawrence Berkeley Laboratory
John Peoples, Jr.	SSC Central Design Group
Martin L. Perl	Stanford Linear Accelerator Center
Philip H. Pile	Brookhaven National Laboratory
Lawrence S. Pinsky	University of Houston
W. Karl Pitts	University of Wisconsin
R. Pollock	Indiana University
Janos Polonyi	Massachusetts Institute of Technology
Scott Pratt	Oregon State University
B. Quinn	Carnegie-Mellon University
Johann Rafelski	University of Arizona
W. Ramsay	University of Manitoba/TRIUMF
Ronald D. Ransome	Rutgers University
Lazarus G. Ratner	Brookhaven National Laboratory
Richard S. Raymond	University of Michigan

N. Reay	Ohio State University
Peter B. Renton	University of Oxford
David H. Rice	Cornell University
J. Richard	ISN Grenoble
Clarence R. Richardson	U.S. Department of Energy
Klaus Rith	Max Planck Institute
Hans George Ritter	Lawrence Berkeley Laboratory
B. Lee Roberts	Boston University
Richard W. Robinett	Pennsylvania State University
Thomas Roser	University of Michigan
Robert Rossmanith	CEBAF
Gerald Roy	University of Alberta
Alessandro G. Ruggiero	Brookhaven National Laboratory
John P. Rutherfoord	University of Washington
Vesa Ruuskanen	University of Jyvaskyla
Aditya K. Sambamurti	Indiana University
Anthony Sanda	Rockefeller University
Reyad Sawafta	Indiana University Cyclotron Facility
H. Reiner Schaefer	Yale University
L. Schroeder	Lawrence Berkeley Laboratory
Klaus K.R. Schubert	Universitat Karlsruhe
Peter Schuler	Yale University
Reinhard Schumacher	Carnegie-Mellon University
Urs J. Sannhauser	PSI (SIN)
Robert Serber	Columbia University
Richard Seto	Columbia University
Robert Shrock	State University of New York-Stony Brook
Markus Simonius	ETH Zurich
Krzysztof Sliwa	Fermilab
Gregory R. Smith	TRIUMF
Paul A. Souder	Syracuse University
Brian M. Spicer	University of Melbourne
Johanna Stachel	State University of New York-Stony Brook
Steven Steadman	Massachusetts Institute of Technology
S. A. Stephens	New Mexico State University
Gerard J. Stephenson	Los Alamos National Laboratory
Morton Sternheim	University of Massachusetts
Richard Strand	Brookhaven National Laboratory
Mitsuyoshi Tanaka	Brookhaven National Laboratory
Lee C. Teng	Fermilab
R. L. Thews	University of Arizona
Henry A. Thiessen	Los Alamos National Laboratory
William Toothacker	Pennsylvania State University
Ian S. Towner	Chalk River Nuclear Laboratory
David G. Underwood	Argonne National Laboratory
Gerard van der Steenhovem	Massachusetts Institute of Technology
W.T.H. van Oers	University of Manitoba/TRIUMF
Michael C. Vetterli	TRIUMF/SFU
Erich Vogt	TRIUMF

William Wallenmayer	SURA
Chris Waltham	TRIUMF
John D. Weinstein	University of Tennessee
Klaus Werner	Brookhaven National Laboratory
Heinz Josef Weyer	Basel University
Christopher G. White	University of Minnesota
D. Hywel White	Los Alamos National Laboratory
David Whitehouse	Boston University
Robert Whyley	College of William and Mary
John Wilkerson	Los Alamos National Laboratory
R. Jeffrey Wilkes	University of Washington
A. G. Williams	University of Washington
William J. Wisniewski	California Institute of Technology
David M. Wolfe	University of New Mexico
Samuel S.M. Wong	University of Toronto
Mieko T. Yamawaki	North Adams State College
William A. Zajc	Nevis Laboratory
Michael E. Zeller	Yale University
Valdis J. Zeps	University of Washington
Huayi Zhang	Virginia Polytechnic Institute
Andrsej Zieminski	Indiana University

CONFERENCE STAFF

Debbie Burton	University of Michigan
Joan Depken	Brookhaven National Laboratory
Sheila Dodson	Massachusetts Institute of Technology
Lorraine King	TRIUMF
Roberta Marinuzzi	Los Alamos National Laboratory
Terry Murphy	TRIUMF
Karen Poleakker	Los Alamos National Laboratory

AUTHOR INDEX

AIP Conference Proceedings

		L.C. Number	ISBN
No. 1	Feedback and Dynamic Control of Plasmas – 1970	70-141596	0-88318-100-2
No. 2	Particles and Fields – 1971 (Rochester)	71-184662	0-88318-101-0
No. 3	Thermal Expansion – 1971 (Corning)	72-76970	0-88318-102-9
No. 4	Superconductivity in *d*- and *f*-Band Metals (Rochester, 1971)	74-18879	0-88318-103-7
No. 5	Magnetism and Magnetic Materials – 1971 (2 parts) (Chicago)	59-2468	0-88318-104-5
No. 6	Particle Physics (Irvine, 1971)	72-81239	0-88318-105-3
No. 7	Exploring the History of Nuclear Physics – 1972	72-81883	0-88318-106-1
No. 8	Experimental Meson Spectroscopy –1972	72-88226	0-88318-107-X
No. 9	Cyclotrons – 1972 (Vancouver)	72-92798	0-88318-108-8
No. 10	Magnetism and Magnetic Materials – 1972	72-623469	0-88318-109-6
No. 11	Transport Phenomena – 1973 (Brown University Conference)	73-80682	0-88318-110-X
No. 12	Experiments on High Energy Particle Collisions – 1973 (Vanderbilt Conference)	73-81705	0-88318-111–8
No. 13	π-π Scattering – 1973 (Tallahassee Conference)	73-81704	0-88318-112-6
No. 14	Particles and Fields – 1973 (APS/DPF Berkeley)	73-91923	0-88318-113-4
No. 15	High Energy Collisions – 1973 (Stony Brook)	73-92324	0-88318-114-2
No. 16	Causality and Physical Theories (Wayne State University, 1973)	73-93420	0-88318-115-0
No. 17	Thermal Expansion – 1973 (Lake of the Ozarks)	73-94415	0-88318-116-9
No. 18	Magnetism and Magnetic Materials – 1973 (2 parts) (Boston)	59-2468	0-88318-117-7
No. 19	Physics and the Energy Problem – 1974 (APS Chicago)	73-94416	0-88318-118-5
No. 20	Tetrahedrally Bonded Amorphous Semiconductors (Yorktown Heights, 1974)	74-80145	0-88318-119-3
No. 21	Experimental Meson Spectroscopy – 1974 (Boston)	74-82628	0-88318-120-7
No. 22	Neutrinos – 1974 (Philadelphia)	74-82413	0-88318-121-5
No. 23	Particles and Fields – 1974 (APS/DPF Williamsburg)	74-27575	0-88318-122-3
No. 24	Magnetism and Magnetic Materials – 1974 (20th Annual Conference, San Francisco)	75-2647	0-88318-123-1
No. 25	Efficient Use of Energy (The APS Studies on the Technical Aspects of the More Efficient Use of Energy)	75-18227	0-88318-124-X

No. 26	High-Energy Physics and Nuclear Structure – 1975 (Santa Fe and Los Alamos)	75-26411	0-88318-125-8
No. 27	Topics in Statistical Mechanics and Biophysics: A Memorial to Julius L. Jackson (Wayne State University, 1975)	75-36309	0-88318-126-6
No. 28	Physics and Our World: A Symposium in Honor of Victor F. Weisskopf (M.I.T., 1974)	76-7207	0-88318-127-4
No. 29	Magnetism and Magnetic Materials – 1975 (21st Annual Conference, Philadelphia)	76-10931	0-88318-128-2
No. 30	Particle Searches and Discoveries – 1976 (Vanderbilt Conference)	76-19949	0-88318-129-0
No. 31	Structure and Excitations of Amorphous Solids (Williamsburg, VA, 1976)	76-22279	0-88318-130-4
No. 32	Materials Technology – 1976 (APS New York Meeting)	76-27967	0-88318-131-2
No. 33	Meson-Nuclear Physics – 1976 (Carnegie-Mellon Conference)	76-26811	0-88318-132-0
No. 34	Magnetism and Magnetic Materials – 1976 (Joint MMM-Intermag Conference, Pittsburgh)	76-47106	0-88318-133-9
No. 35	High Energy Physics with Polarized Beams and Targets (Argonne, 1976)	76-50181	0-88318-134-7
No. 36	Momentum Wave Functions – 1976 (Indiana University)	77-82145	0-88318-135-5
No. 37	Weak Interaction Physics – 1977 (Indiana University)	77-83344	0-88318-136-3
No. 38	Workshop on New Directions in Mossbauer Spectroscopy (Argonne, 1977)	77-90635	0-88318-137-1
No. 39	Physics Careers, Employment and Education (Penn State, 1977)	77-94053	0-88318-138-X
No. 40	Electrical Transport and Optical Properties of Inhomogeneous Media (Ohio State University, 1977)	78-54319	0-88318-139-8
No. 41	Nucleon-Nucleon Interactions – 1977 (Vancouver)	78-54249	0-88318-140-1
No. 42	Higher Energy Polarized Proton Beams (Ann Arbor, 1977)	78-55682	0-88318-141-X
No. 43	Particles and Fields – 1977 (APS/DPF, Argonne)	78-55683	0-88318-142-8
No. 44	Future Trends in Superconductive Electronics (Charlottesville, 1978)	77-9240	0-88318-143-6
No. 45	New Results in High Energy Physics – 1978 (Vanderbilt Conference)	78-67196	0-88318-144-4
No. 46	Topics in Nonlinear Dynamics (La Jolla Institute)	78-57870	0-88318-145-2
No. 47	Clustering Aspects of Nuclear Structure and Nuclear Reactions (Winnepeg, 1978)	78-64942	0-88318-146-0
No. 48	Current Trends in the Theory of Fields (Tallahassee, 1978)	78-72948	0-88318-147-9

No. 49	Cosmic Rays and Particle Physics – 1978 (Bartol Conference)	79-50489	0-88318-148-7
No. 50	Laser-Solid Interactions and Laser Processing – 1978 (Boston)	79-51564	0-88318-149-5
No. 51	High Energy Physics with Polarized Beams and Polarized Targets (Argonne, 1978)	79-64565	0-88318-150-9
No. 52	Long-Distance Neutrino Detection – 1978 (C.L. Cowan Memorial Symposium)	79-52078	0-88318-151-7
No. 53	Modulated Structures – 1979 (Kailua Kona, Hawaii)	79-53846	0-88318-152-5
No. 54	Meson-Nuclear Physics – 1979 (Houston)	79-53978	0-88318-153-3
No. 55	Quantum Chromodynamics (La Jolla, 1978)	79-54969	0-88318-154-1
No. 56	Particle Acceleration Mechanisms in Astrophysics (La Jolla, 1979)	79-55844	0-88318-155-X
No. 57	Nonlinear Dynamics and the Beam-Beam Interaction (Brookhaven, 1979)	79-57341	0-88318-156-8
No. 58	Inhomogeneous Superconductors – 1979 (Berkeley Springs, W.V.)	79-57620	0-88318-157-6
No. 59	Particles and Fields – 1979 (APS/DPF Montreal)	80-66631	0-88318-158-4
No. 60	History of the ZGS (Argonne, 1979)	80-67694	0-88318-159-2
No. 61	Aspects of the Kinetics and Dynamics of Surface Reactions (La Jolla Institute, 1979)	80-68004	0-88318-160-6
No. 62	High Energy e^+e^- Interactions (Vanderbilt, 1980)	80-53377	0-88318-161-4
No. 63	Supernovae Spectra (La Jolla, 1980)	80-70019	0-88318-162-2
No. 64	Laboratory EXAFS Facilities – 1980 (Univ. of Washington)	80-70579	0-88318-163-0
No. 65	Optics in Four Dimensions – 1980 (ICO, Ensenada)	80-70771	0-88318-164-9
No. 66	Physics in the Automotive Industry – 1980 (APS/AAPT Topical Conference)	80-70987	0-88318-165-7
No. 67	Experimental Meson Spectroscopy – 1980 (Sixth International Conference, Brookhaven)	80-71123	0-88318-166-5
No. 68	High Energy Physics – 1980 (XX International Conference, Madison)	81-65032	0-88318-167-3
No. 69	Polarization Phenomena in Nuclear Physics – 1980 (Fifth International Symposium, Santa Fe)	81-65107	0-88318-168-1
No. 70	Chemistry and Physics of Coal Utilization – 1980 (APS, Morgantown)	81-65106	0-88318-169-X
No. 71	Group Theory and its Applications in Physics – 1980 (Latin American School of Physics, Mexico City)	81-66132	0-88318-170-3
No. 72	Weak Interactions as a Probe of Unification (Virginia Polytechnic Institute – 1980)	81-67184	0-88318-171-1
No. 73	Tetrahedrally Bonded Amorphous Semiconductors (Carefree, Arizona, 1981)	81-67419	0-88318-172-X

No. 74	Perturbative Quantum Chromodynamics (Tallahassee, 1981)	81-70372	0-88318-173-8
No. 75	Low Energy X-Ray Diagnostics – 1981 (Monterey)	81-69841	0-88318-174-6
No. 76	Nonlinear Properties of Internal Waves (La Jolla Institute, 1981)	81-71062	0-88318-175-4
No. 77	Gamma Ray Transients and Related Astrophysical Phenomena (La Jolla Institute, 1981)	81-71543	0-88318-176-2
No. 78	Shock Waves in Condensed Matter – 1981 (Menlo Park)	82-70014	0-88318-177-0
No. 79	Pion Production and Absorption in Nuclei – 1981 (Indiana University Cyclotron Facility)	82-70678	0-88318-178-9
No. 80	Polarized Proton Ion Sources (Ann Arbor, 1981)	82-71025	0-88318-179-7
No. 81	Particles and Fields –1981: Testing the Standard Model (APS/DPF, Santa Cruz)	82-71156	0-88318-180-0
No. 82	Interpretation of Climate and Photochemical Models, Ozone and Temperature Measurements (La Jolla Institute, 1981)	82-71345	0-88318-181-9
No. 83	The Galactic Center (Cal. Inst. of Tech., 1982)	82-71635	0-88318-182-7
No. 84	Physics in the Steel Industry (APS/AISI, Lehigh University, 1981)	82-72033	0-88318-183-5
No. 85	Proton-Antiproton Collider Physics –1981 (Madison, Wisconsin)	82-72141	0-88318-184-3
No. 86	Momentum Wave Functions – 1982 (Adelaide, Australia)	82-72375	0-88318-185-1
No. 87	Physics of High Energy Particle Accelerators (Fermilab Summer School, 1981)	82-72421	0-88318-186-X
No. 88	Mathematical Methods in Hydrodynamics and Integrability in Dynamical Systems (La Jolla Institute, 1981)	82-72462	0-88318-187-8
No. 89	Neutron Scattering – 1981 (Argonne National Laboratory)	82-73094	0-88318-188-6
No. 90	Laser Techniques for Extreme Ultraviolt Spectroscopy (Boulder, 1982)	82-73205	0-88318-189-4
No. 91	Laser Acceleration of Particles (Los Alamos, 1982)	82-73361	0-88318-190-8
No. 92	The State of Particle Accelerators and High Energy Physics (Fermilab, 1981)	82-73861	0-88318-191-6
No. 93	Novel Results in Particle Physics (Vanderbilt, 1982)	82-73954	0-88318-192-4
No. 94	X-Ray and Atomic Inner-Shell Physics – 1982 (International Conference, U. of Oregon)	82-7401[illegible]	0-88318-193-2
No. 95	High Energy Spin Physics – 1982 (Brookhaven National Laboratory)	83-70154	0-88318-194-0
No. 96	Science Underground (Los Alamos, 1982)	83-70377	0-88318-195-9

No. 97	The Interaction Between Medium Energy Nucleons in Nuclei – 1982 (Indiana University)	83-70649	0-88318-196-7
No. 98	Particles and Fields – 1982 (APS/DPF University of Maryland)	83-70807	0-88318-197-5
No. 99	Neutrino Mass and Gauge Structure of Weak Interactions (Telemark, 1982)	83-71072	0-88318-198-3
No. 100	Excimer Lasers – 1983 (OSA, Lake Tahoe, Nevada)	83-71437	0-88318-199-1
No. 101	Positron-Electron Pairs in Astrophysics (Goddard Space Flight Center, 1983)	83-71926	0-88318-200-9
No. 102	Intense Medium Energy Sources of Strangeness (UC-Sant Cruz, 1983)	83-72261	0-88318-201-7
No. 103	Quantum Fluids and Solids – 1983 (Sanibel Island, Florida)	83-72440	0-88318-202-5
No. 104	Physics, Technology and the Nuclear Arms Race (APS Baltimore –1983)	83-72533	0-88318-203-3
No. 105	Physics of High Energy Particle Accelerators (SLAC Summer School, 1982)	83-72986	0-88318-304-8
No. 106	Predictability of Fluid Motions (La Jolla Institute, 1983)	83-73641	0-88318-305-6
No. 107	Physics and Chemistry of Porous Media (Schlumberger-Doll Research, 1983)	83-73640	0-88318-306-4
No. 108	The Time Projection Chamber (TRIUMF, Vancouver, 1983)	83-83445	0-88318-307-2
No. 109	Random Walks and Their Applications in the Physical and Biological Sciences (NBS/La Jolla Institute, 1982)	84-70208	0-88318-308-0
No. 110	Hadron Substructure in Nuclear Physics (Indiana University, 1983)	84-70165	0-88318-309-9
No. 111	Production and Neutralization of Negative Ions and Beams (3rd Int'l Symposium, Brookhaven, 1983)	84-70379	0-88318-310-2
No. 112	Particles and Fields – 1983 (APS/DPF, Blacksburg, VA)	84-70378	0-88318-311-0
No. 113	Experimental Meson Spectroscopy – 1983 (Seventh International Conference, Brookhaven)	84-70910	0-88318-312-9
No. 114	Low Energy Tests of Conservation Laws in Particle Physics (Blacksburg, VA, 1983)	84-71157	0-88318-313-7
No. 115	High Energy Transients in Astrophysics (Santa Cruz, CA, 1983)	84-71205	0-88318-314-5
No. 116	Problems in Unification and Supergravity (La Jolla Institute, 1983)	84-71246	0-88318-315-3
No. 117	Polarized Proton Ion Sources (TRIUMF, Vancouver, 1983)	84-71235	0-88318-316-1

No. 118	Free Electron Generation of Extreme Ultraviolet Coherent Radiation (Brookhaven/OSA, 1983)	84-71539	0-88318-317-X
No. 119	Laser Techniques in the Extreme Ultraviolet (OSA, Boulder, Colorado, 1984)	84-72128	0-88318-318-8
No. 120	Optical Effects in Amorphous Semiconductors (Snowbird, Utah, 1984)	84-72419	0-88318-319-6
No. 121	High Energy e^+e^- Interactions (Vanderbilt, 1984)	84-72632	0-88318-320-X
No. 122	The Physics of VLSI (Xerox, Palo Alto, 1984)	84-72729	0-88318-321-8
No. 123	Intersections Between Particle and Nuclear Physics (Steamboat Springs, 1984)	84-72790	0-88318-322-6
No. 124	Neutron-Nucleus Collisions – A Probe of Nuclear Structure (Burr Oak State Park - 1984)	84-73216	0-88318-323-4
No. 125	Capture Gamma-Ray Spectroscopy and Related Topics – 1984 (Internat. Symposium, Knoxville)	84-73303	0-88318-324-2
No. 126	Solar Neutrinos and Neutrino Astronomy (Homestake, 1984)	84-63143	0-88318-325-0
No. 127	Physics of High Energy Particle Accelerators (BNL/SUNY Summer School, 1983)	85-70057	0-88318-326-9
No. 128	Nuclear Physics with Stored, Cooled Beams (McCormick's Creek State Park, Indiana, 1984)	85-71167	0-88318-327-7
No. 129	Radiofrequency Plasma Heating (Sixth Topical Conference, Callaway Gardens, GA, 1985)	85-48027	0-88318-328-5
No. 130	Laser Acceleration of Particles (Malibu, California, 1985)	85-48028	0-88318-329-3
No. 131	Workshop on Polarized ^{3}He Beams and Targets (Princeton, New Jersey, 1984)	85-48026	0-88318-330-7
No. 132	Hadron Spectroscopy–1985 (International Conference, Univ. of Maryland)	85-72537	0-88318-331-5
No. 133	Hadronic Probes and Nuclear Interactions (Arizona State University, 1985)	85-72638	0-88318-332-3
No. 134	The State of High Energy Physics (BNL/SUNY Summer School, 1983)	85-73170	0-88318-333-1
No. 135	Energy Sources: Conservation and Renewables (APS, Washington, DC, 1985)	85-73019	0-88318-334-X
No. 136	Atomic Theory Workshop on Relativistic and QED Effects in Heavy Atoms	85-73790	0-88318-335-8
No. 137	Polymer-Flow Interaction (La Jolla Institute, 1985)	85-73915	0-88318-336-6
No. 138	Frontiers in Electronic Materials and Processing (Houston, TX, 1985)	86-70108	0-88318-337-4
No. 139	High-Current, High-Brightness, and High-Duty Factor Ion Injectors (La Jolla Institute, 1985)	86-70245	0-88318-338-2

No. 140	Boron-Rich Solids (Albuquerque, NM, 1985)	86-70246	0-88318-339-0
No. 141	Gamma-Ray Bursts (Stanford, CA, 1984)	86-70761	0-88318-340-4
No. 142	Nuclear Structure at High Spin, Excitation, and Momentum Transfer (Indiana University, 1985)	86-70837	0-88318-341-2
No. 143	Mexican School of Particles and Fields (Oaxtepec, México, 1984)	86-81187	0-88318-342-0
No. 144	Magnetospheric Phenomena in Astrophysics (Los Alamos, 1984)	86-71149	0-88318-343-9
No. 145	Polarized Beams at SSC & Polarized Antiprotons (Ann Arbor, MI & Bodega Bay, CA, 1985)	86-71343	0-88318-344-7
No. 146	Advances in Laser Science–I (Dallas, TX, 1985)	86-71536	0-88318-345-5
No. 147	Short Wavelength Coherent Radiation: Generation and Applications (Monterey, CA, 1986)	86-71674	0-88318-346-3
No. 148	Space Colonization: Technology and The Liberal Arts (Geneva, NY, 1985)	86-71675	0-88318-347-1
No. 149	Physics and Chemistry of Protective Coatings (Universal City, CA, 1985)	86-72019	0-88318-348-X
No. 150	Intersections Between Particle and Nuclear Physics (Lake Louise, Canada, 1986)	86-72018	0-88318-349-8
No. 151	Neural Networks for Computing (Snowbird, UT, 1986)	86-72481	0-88318-351-X
No. 152	Heavy Ion Inertial Fusion (Washington, DC, 1986)	86-73185	0-88318-352-8
No. 153	Physics of Particle Accelerators (SLAC Summer School, 1985) (Fermilab Summer School, 1984)	87-70103	0-88318-353-6
No. 154	Physics and Chemistry of Porous Media—II (Ridge Field, CT, 1986)	83-73640	0-88318-354-4
No. 155	The Galactic Center: Proceedings of the Symposium Honoring C. H. Townes (Berkeley, CA, 1986)	86-73186	0-88318-355-2
No. 156	Advanced Accelerator Concepts (Madison, WI, 1986)	87-70635	0-88318-358-0
No. 157	Stability of Amorphous Silicon Alloy Materials and Devices (Palo Alto, CA, 1987)	87-70990	0-88318-359-9
No. 158	Production and Neutralization of Negative Ions and Beams (Brookhaven, NY, 1986)	87-71695	0-88318-358-7

No. 159	Applications of Radio-Frequency Power to Plasma: Seventh Topical Conference (Kissimmee, FL, 1987)	87-71812	0-88318-359-5
No. 160	Advances in Laser Science–II (Seattle, WA, 1986)	87-71962	0-88318-360-9
No. 161	Electron Scattering in Nuclear and Particle Science: In Commemoration of the 35th Anniversary of the Lyman-Hanson-Scott Experiment (Urbana, IL, 1986)	87-72403	0-88318-361-7
No. 162	Few-Body Systems and Multiparticle Dynamics (Crystal City, VA, 1987)	87-72594	0-88318-362-5
No. 163	Pion–Nucleus Physics: Future Directions and New Facilities at LAMPF (Los Alamos, NM, 1987)	87-72961	0-88318-363-3
No. 164	Nuclei Far from Stability: Fifth International Conference (Rosseau Lake, ON, 1987)	87-73214	0-88318-364-1
No. 165	Thin Film Processing and Characterization of High-Temperature Superconductors	87-73420	0-88318-365-X
No. 166	Photovoltaic Safety (Denver, CO, 1988)	88-42854	0-88318-366-8
No. 167	Deposition and Growth: Limits for Microelectronics (Anaheim, CA, 1987)	88-71432	0-88318-367-6
No. 168	Atomic Processes in Plasmas (Santa Fe, NM, 1987)	88-71273	0-88318-368-4
No. 169	Modern Physics in America: A Michelson-Morley Centennial Symposium (Cleveland, OH, 1987)	88-71348	0-88318-369-2
No. 170	Nuclear Spectroscopy of Astrophysical Sources (Washington, D.C., 1987)	88-71625	0-88318-370-6
No. 171	Vacuum Design of Advanced and Compact Synchrotron Light Sources (Upton, NY, 1988)	88-71824	0-88318-371-4
No. 172	Advances in Laser Science–III: Proceedings of the International Laser Science Conference (Atlantic City, NJ, 1987)	88-71879	0-88318-372-2
No. 173	Cooperative Networks in Physics Education (Oaxtepec, Mexico 1987)	88-72091	0-88318-373-0
No. 174	Radio Wave Scattering in the Interstellar Medium (San Diego, CA 1988)	88-72092	0-88318-374-9
No. 175	Non-neutral Plasma Physics (Washington, DC 1988)	88-72275	0-88318-375-7